MW01383264

Synthesis of lactones and lactams

THE CHEMISTRY OF FUNCTIONAL GROUPS

A series of advanced treatises under the general editorship of Professor Saul Patai

The chemistry of alkenes (2 volumes)
The chemistry of the carbonyl group (2 volumes)
The chemistry of the ether linkage
The chemistry of the amino group
The chemistry of alkenes (2 volumes)
The chemistry of carboxylic acids and esters
The chemistry of the carbon–nitrogen double bond
The chemistry of amides
The chemistry of the cyano group
The chemistry of the hydroxyl group (2 parts)
The chemistry of the azido group
The chemistry of acyl halides
The chemistry of the carbon–halogen bond (2 parts)
The chemistry of the quinonoid compounds (2 volumes, 4 parts)
The chemistry of the thiol group (2 parts)
The chemistry of the hydrazo, azo and azoxy groups (2 parts)
The chemistry of amidines and imidates (2 volumes)
The chemistry of cyanates and their thio derivatives (2 parts)
The chemistry of diazonium and diazo groups (2 parts)
The chemistry of the carbon–carbon triple bond (2 parts)
The chemistry of ketenes, allenes and related compounds (2 parts)
The chemistry of the sulphonium group (2 parts)
Supplement A: The chemistry of double-bonded functional groups (2 volumes, 4 parts)
Supplement B: The chemistry of acid derivatives (2 volumes, 4 parts)
Supplement C: The chemistry of triple-bonded functional groups (2 parts)
Supplement D: The chemistry of halides, pseudo-halides and azides (2 parts)
Supplement E: The chemistry of ethers, crown ethers, hydroxyl groups and their sulphur analogues (2 parts)
Supplement F: The chemistry of amino, nitroso and nitro compounds and their derivatives (2 parts)
The chemistry of the metal–carbon bond (5 volumes)
The chemistry of peroxides
The chemistry of organic selenium and tellurium compounds (2 volumes)
The chemistry of the cyclopropyl group (2 parts)
The chemistry of sulphones and sulphoxides
The chemistry of organic silicon compounds (2 parts)
The chemistry of enones (2 parts)
The chemistry of sulphinic acids, esters and their derivatives
The chemistry of sulphenic acids and their derivatives
The chemistry of enols
The chemistry of organophosphorus compounds (2 volumes)
The chemistry of sulphonic acids, esters and their derivatives
The chemistry of alkanes and cycloalkanes
The chemistry of sulphur-containing functional groups
Supplement E2: The chemistry of hydroxyl, ether and peroxide groups

UPDATES

The chemistry of α-haloketones, α-haloaldehydes and α-haloimines
Nitrones, nitronates and nitroxides
Crown ethers and analogs
Cyclopropane derived reactive intermediates
Synthesis of carboxylic acids, esters and their derivatives
The silicon–heteroatom bond
Synthesis of lactones and lactams

Patai's 1992 guide to the chemistry of functional groups—*Saul Patai*

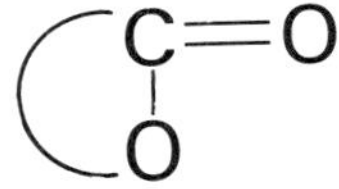

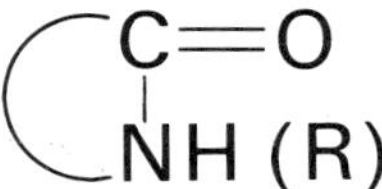

Synthesis of lactones and lactams

by

MICHAEL A. OGLIARUSO

and

JAMES F. WOLFE

Virginia Polytechnic Institute and State University

Edited by

SAUL PATAI and ZVI RAPPOPORT

The Hebrew University, Jerusalem

Updates from the Chemistry of Functional Groups

1993

JOHN WILEY & SONS

CHICHESTER · NEW YORK · BRISBANE · TORONTO · SINGAPORE

An Interscience® Publication

Other Wiley Editorial Offices

John Wiley & Sons, Inc., 605 Third Avenue,
New York, NY 10158-0012, USA

Jacaranda Wiley Ltd, G.P.O. Box 859, Brisbane,
Queensland 4001, Australia

John Wiley & Sons (Canada) Ltd, 22 Worcester Road,
Rexdale, Ontario M9W 1L1, Canada

John Wiley & Sons (SEA) Pte Ltd, 37 Jalan Pemimpin #05-04,
Block B, Union Industrial Building, Singapore 2057

Library of Congress Cataloging-in-Publication Data

Ogliaruso, Michael A.
Synthesis of lactones and lactams / by Michael A. Ogliaruso and James F. Wolfe ; edited by Saul Patai and Zvi Rappoport.
p. cm.—(Updates from the Chemistry of functional groups)
'An Interscience publication.'
Includes bibliographical references and indexes.
ISBN 0 471 93734 7
1. Lactones—Synthesis. 2. Lactams—Synthesis. I. Wolfe, James F. II. Patai, Saul. III. Rappoport, Zvi. IV. Title. V. Series.
QD305.A2033 1993 92-28932
547'.637—dc20 CIP

British Library Cataloguing in Publication Data

A catalogue record for this book is available from the British Library

ISBN 0 471 93734 7

Typeset by Thomson Press (India) Ltd, New Delhi, India
Printed and bound in Great Britain by Biddles Ltd, Guildford, Surrey

List of contributors

M. A. Ogliaruso — Department of Chemistry, College of Arts and Sciences, Virginia Polytechnic Institute and State University, Blacksburg, Virginia 24061-0122, USA

J. F. Wolfe — Department of Chemistry, College of Arts and Sciences, Virginia Polytechnic Institute and State University, Blacksburg, Virginia 24061-0122, USA

Foreword

This new volume contains the original chapter on 'The synthesis of lactones and lactams' by Professors J. F. Wolfe and M. A. Ogliaruso, which appeared in 1979 in *Supplement B: The chemistry of acid derivatives.* To this original chapter, the same two authors have added an appendix of about three times the size, albeit the period discussed in the appendix is less than ten years. This situation is similar in the Update volume on the *Synthesis of carboxylic acids, esters and their derivatives*, also by the same two authors.

We would be very grateful to our readers if they would bring to our attention omissions or mistakes in this or in any of the other volumes of the series.

Jerusalem
July 1992

SAUL PATAI
ZVI RAPPOPORT

Preface

Because of the importance of lactones and especially lactams as starting materials for the preparation of a large range of antibacterial agents, methods of synthesis and interconversions of these functional groups are of great significance to a large group of practicing organic chemists. Because of this, and in order to provide the series, *The Chemistry of Functional Groups*, with a single source of information on general methods for the synthesis of lactones and lactams, a chapter on the synthesis of lactones and lactams was authored and appeared in *Supplement B: The Chemistry of Acid Derivatives*, Chapter 19, pp. 1063–1330. That chapter contained descriptions of the most common methods for the synthesis of lactones and lactams, with emphasis on preparative techniques that appeared in the primary literature during the period 1967 through 1976.

The present monograph volume on the synthesis of lactones and lactams is our response to an invitation from the Editors and publishers of the 'Functional Groups' series of books to combine the material contained in the original chapter with new methodology from the literature for the period 1976 to the present. The format for this combination consists of the original text as published in 1979, along with an up-to-date Appendix containing the newer material in the same format as used earlier.

This monograph is designed for the practicing chemist who seeks a convenient, single source for synthetic methods leading to lactones and lactams. An attempt has been made to include sufficient detail and examples of typical preparations to allow the reader to make a rational choice from among several alternative methods. Obviously there will be synthetic procedures that have not been included or that have been given only cursory attention, but it is sincerely hoped that what has been included will be of help and interest to our colleagues in the international organic chemistry community.

Deep gratitude is expressed to the Department of Chemistry at Virginia Polytechnic Institute, while sincere appreciation is extended to Marion Bradley Via, without whose generosity this project would not have been possible.

Blacksburg,
Virginia
1992

MICHAEL A. OGLIARUSO
JAMES F. WOLFE

Contents

Abbreviations

The following abbreviations, arranged alphabetically below, are used consistently throughout this volume

Ac	acetyl
acac	acetylacetonate
AIBN	azobisisobutyronitrile
An	anisyl
anhy.	anhydrous
Ar	argon, aryl
atm	atmospheres
Bipy	bipyridyl
Bn	benzyl
BOC	*t*-butoxycarbonyl
BSA	bis(trimethylsilyl)acetamide
Bu	butyl
Bz	benzoyl
c	cyclo
CAN	ceric ammonium nitrate
Cp	η^5-cyclopentadienyl
CSI	*N*-chlorosulphonyl isocyanate
CTAB	cetyltrimethylammonium bromide
DBN	1,5-diazabicyclo[4.3.0]non-5-ene
DBU	1,8-diazabicyclo[5.4.0]-undec-7-ene
DCCD	*N*,*N'*-dicyclohexyl-carbodiimide
DCE	1,2-dichloroethane
DDQ	2,3-dichloro-5,6-dicyano-1,4-benzoquinone
DEAD	diethyl azodicarboxylate
DIAD	diisopropyl azodicarboxy-late
DIBAL	di(*t*-butyl)aluminium
DICD	diisopropylcarbodiimide
diglyme	diethylene glycol dimethyl ether
DMAP	4-(*N*,*N*-dimethylamino)-pyridine
DME	1,2-dimethoxyethane
DMF	*N*,*N*-dimethylformamide
DMSO	dimethyl sulphoxide
2,4-DNPH	2,4-dinitrophenylhydrazine
DPPA	diphenylphosphoryl azide
dppb	1,4-bis(diphenylphosphino)-butane
EDTA	ethylenediaminetetraacetic acid
EEDQ	2-ethoxy-1-ethoxycarbonyl-1,2-dihydroquinoline
e.e.	enantiomeric excess
Et	ethyl
eV	electron volts
Fc	ferrocene
fod	tris(6,6,7,7,8,8,8)heptafluoro-2,2-dimethyl-3,5-octanedionate
Fu	furyl
glyme	ethylene glycol dimethyl ether
h	hours
Hex	hexyl
HMPA	hexamethylphosphortri-amide
hv	irradiation with light
HOMO	highest occupied molecular orbital
i	iso
IPNS	isopenicillin N synthase
IR	infrared
kbar	kilobar
LAH	lithium aluminium hydride

LCAO	linear combination of atomic orbitals
LDA	lithium diisopropylamide
LTEA	lithium triethoxyaluminium hydride
M	metal
MCPA	*m*-chloroperbenzoic acid
Me	methyl
Mes	mesyl (methanesulphonyl)
min	minutes
MOM	methoxymethyl
n	normal
Naph	naphthyl
NBA	*N*-bromoacetamide
NBS	*N*-bromosuccinimide
NHPI	*N*-hydroxyphthalimide
NMR	nuclear magnetic resonance
N.R.	no reaction
PDC	pyridinium dichromate
Pent	pentyl
PI—N	phthalimido
Pip	pideridyl
Ph	phenyl
PMR	proton magnetic resonance
PPA	polyphosphoric acid
Pr	propyl
pyr	pyridyl
R	any alkyl group
r.t.	room temperature
Rx	reaction
s	secondary
$(salen)_2$	bis(salicylaldehyde)ethylene diimine
t	tertiary
TBAF	tetra(*n*-butyl)ammonium fluoride
TCNE	tetracyanoethylene
TfO	triflate
THF	tetrahydrofuran
Thi	thienyl
THP	tetrahydropyranyl
TMEDA	tetramethylethylene diamine
TMS	trimethylsilyl
TMSCN	trimethylsilyl cyanide
Tol	tolyl
Tos	tosyl(*p*-toluenesulphonyl)
TPP	triphenylphosphine
Triton B	benzyltrimethylammonium hydroxide
Trityl	triphenylmethyl
UV	irradiation with ultraviolet light
Xyl	xylyl

CHAPTER **1**

The synthesis of lactones and lactams

I.	INTRODUCTION	2
II.	SYNTHESIS OF LACTONES	3
	A. By Intramolecular Cyclization of Hydroxy Acids, Hydroxy Acid Derivatives and Related Compounds	3
	B. By Intramolecular Cyclization of Unsaturated Acids and Esters	16
	1. Acid-catalysed cyclizations	16
	2. Photochemical and electrochemical cyclizations	18
	3. Halolactonization	19
	4. Intramolecular Diels–Alder reactions	30
	C. By Acetoacetic Ester and Cyanoacetic Ester Condensations	34
	D. By Aldol Condensations	36
	E. By Malonic Ester or Malonic Acid Condensation	40
	F. By Perkin and Stobbe Reactions	46
	G. By Grignard and Reformatsky Reactions	48
	H. By Wittig-type Reactions	53
	I. From α-Anions (Dianions) of Carboxylic Acids	54
	J. From Lithio Salts of 2-Alkyl-2-oxazolines	58
	K. By Direct Functionalization of Preformed Lactones	61
	L. From Ketenes	63
	M. By Reduction of Anhydrides, Esters and Acids	76
	N. By Oxidation Reactions	83
	1. Oxidation of diols	83
	2. Oxidation of ketones	106
	3. Oxidation of ethers	109
	4. Oxidation of olefins	110
	O. By Carbonylation Reactions	111
	P. By Cycloaddition of Nitrones to Olefins	118
	Q. By Rearrangement Reactions	120
	1. Claisen rearrangements	120
	2. Carbonium ion rearrangements	122
	3. Photochemical rearrangements	124
	R. Lactone Interconversions	125
	S. Miscellaneous Lactone Syntheses	127
	1. The Barton reaction	127
	2. Photolysis of α-diazo esters and amides	127
	3. Photolysis of 2-alkoxyoxetanes	129
	4. α-Lactones by photolysis of 1,2-dioxolane-3,5-diones	130
	5. Oxidation of mercaptans, disulphides and related compounds	130

6. Addition of diazonium salts to olefins 130
7. Addition of diethyl dibromomalonate to methyl methacrylate . . 130
8. Dehydrohalogenation of 2,2-dimethoxy-3-chlorodihydropyrans . . 131
9. Preparation of homoserine lactone 131

III. SYNTHESIS OF LACTAMS 132
A. By Ring-closure Reactions (Chemical) 133
1. From amino acids and related compounds 133
2. From halo, hydroxy and keto amides 138
B. By Ring-closure Reactions (Photochemical) 144
1. Cyclization of α,β-unsaturated amides 144
2. Cyclization of benzanilides 145
3. Cyclization of enamides 145
4. Cyclization of N-chloroacetyl-β-arylamines 162
5. Cyclization of α-diazocarboxamides 162
6. Miscellaneous cyclizations 162
C. By Cycloaddition Reactions 162
1. Addition of isocyanates to olefins 162
2. From imines 168
a. Reaction of imines with ketenes 168
b. Reformatsky reaction with imines 171
c. Other imine cycloadditions 203
3. From nitrones and nitroso compounds 207
D. By Rearrangements 210
1. Ring contractions 210
a. Wolff rearrangement 210
b. Miscellaneous ring contractions 211
2. Ring expansions 212
a. Beckmann rearrangement 212
b. Schmidt rearrangement 222
c. Miscellaneous ring expansions 223
3. Claisen rearrangement 236
E. By Direct Functionalization of Preformed Lactams 237
F. By Oxidation Reactions 240
1. Using halogen 240
2. Using chromium or osmium oxides 242
3. Using manganese oxides 243
4. Using platinum or ruthenium oxides 244
5. Via sensitized and unsensitized photooxidation 247
6. Via autooxidation 250
7. Using miscellaneous reagents 250
G. Miscellaneous Lactam Syntheses 252

IV. ACKNOWLEDGMENTS 253
V. REFERENCES 253

I. INTRODUCTION

This chapter is devoted to a discussion of recent developments in the synthesis of lactones and lactams, and is meant to supplement our earlier chapter in this volume dealing with the synthesis of carboxylic acids and their acyclic derivatives (Chapter 7).

The primary literature surveyed for this review consists mainly of articles listed in *Chemical Abstracts* from 1966 through mid-1976. In order to treat topics which have not been reviewed before, and to lend continuity and chronological perspective to certain sections, a number of references which appeared prior to 1966 are also included.

Although we have not attempted to make this chapter encyclopaedic, we hope that the numerous lactone and lactam preparations presented in tabular form will be helpful to practitioners of the fine art of organic synthesis in spite of inevitable, but unintentional, omissions.

II. SYNTHESIS OF LACTONES

The first extensive review of lactones covered the synthesis and reactions of β-lactones, and was published in 1954 by Zaugg[1]. A review[2] in 1963, while not concerned with lactones *per se*, discusses many reactions which do give rise to lactones. In 1964 three reviews appeared: the first, by Etienne and Fischer[3], was on the preparation, reactions, etc. of β-lactones; the second, by Rao[4], was on the chemistry of butenolides; and the third, by Ansell and Palmer[5] discussed the cyclization of olefinic acids to ketones and lactones. In 1967 and 1968 three reviews appeared which discussed the synthesis of 2-pyrone[6], the preparation of macrocyclic ketones and lactones from polyacetylenic compounds[7], and the synthesis of substituted lactones, their odour and some transformations[8]. A review in 1972 discussed the preparation, properties and polymerization of β-lactones, ε-caprolactone and lactides[9], and another reported on the preparation, properties and polymerization of hydroxy acids and lactones[10]. The synthesis of α-methylene lactones was reviewed[11] in 1975, while in 1976 Rao reported[12] on recent advances in the chemistry of unsaturated lactones.

Because of the large number and variety of reviews published on all aspects of lactone preparation, this section will mainly be concerned with discussion of newer methods of lactone preparation along with selected recent applications of traditional synthetic methods.

A. By Intramolecular Cyclization of Hydroxy Acids, Hydroxy Acid Derivatives and Related Compounds

Numerous hydroxy acids, hydroxy esters and hydroxylated acid derivatives can be converted to lactones by intramolecular reactions similar to those employed in the synthesis of acyclic esters. Acids containing enolizable carbonyl functions can also serve as useful lactone precursors.

Acid-catalysed cyclization of hydroxy acids comprises a widely used procedure for lactone formation. Examples of intramolecular acid-catalysed condensations yielding γ- and ε-lactones are the reaction of sodium *o*-hydroxymethylbenzoate with concentrated hydrochloric acid, which affords[13] a 67–71% yield of phthalide (equation 1), and cyclization[14] of (*R*)-(+)-6-hydroxy-4-methylhexanoic acid and

N—H —(Zn–Cu / NaOH)→ COO^- Na^+ / CH_2OH —(HCl)→ (1)

(*R*)-(+)-6-hydroxy-3-methylhexanoic acid to (*R*)-(+)-γ-methyl-ε-caprolactone (35%) and (*R*)-(−)-β-methyl-ε-caprolactone (59%), respectively (equation 2). Similarly D-gulonic-γ-lactone has been prepared from gulonic acid[15].

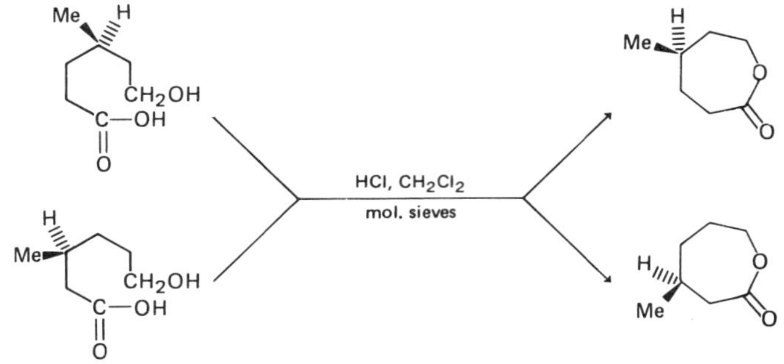

$$\text{HCl, } CH_2Cl_2 \text{, mol. sieves} \qquad (2)$$

The most popular acidic reagent for effecting direct cyclization of hydroxy acids appears to be *p*-toluenesulphonic acid in a variety of solvents[16–22] (Table 1).

The direct cyclization of β-hydroxy acids with benzenesulphonyl chloride in pyridine at 0–5°C (equation 3) has been shown to be a general reaction for the

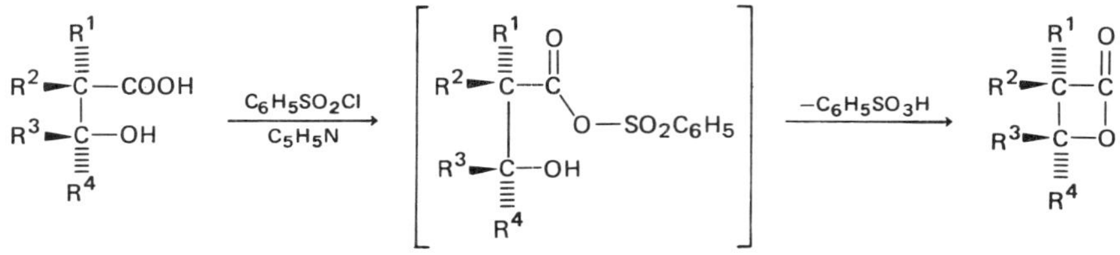

$$\xrightarrow[C_5H_5N]{C_6H_5SO_2Cl} \qquad \xrightarrow{-C_6H_5SO_3H} \qquad (3)$$

formation of tri- and tetra-substituted β-lactones in high yields (Table 2). During these investigations it was observed that hydroxy acids **1**, **2** and **3** afforded olefins rather than lactones upon treatment with benzenesulphonyl chloride. Although no explanation was advanced for the absence of lactone formation from **1** or **2**, the

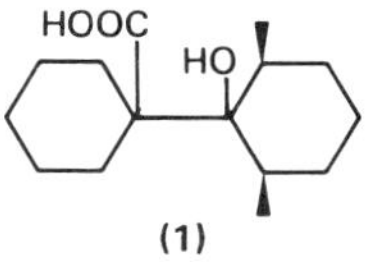

(1)
(Ref. 26)

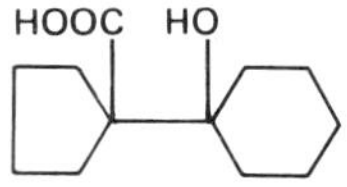

(2)
(Ref. 26)

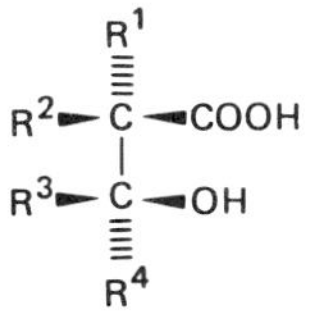

(3)
(Ref. 25)

R^1 = H, H, H, H
R^2 = H, H, H, H
R^3 = Ph, $PhCH_2$, Et, *i*-Pr
R^4 = Ph, Ph, Ph, Ph

preferred linear dehydration of acids **3** was explained[25] in terms of the absence of substituents at the β-carbon of the hydroxy acid, a structural feature which is essential for cyclization.

Stereoselective cyclizations of hydroxy acids to trisubstituted β-lactones have been reported using methanesulphonyl chloride[27]. For example, the diastereomers

TABLE 1. Cyclization of hydroxy acids to lactones using *p*-toluenesulphonic acid

Hydroxy acid	Conditions	Product	Yield (%)	Reference
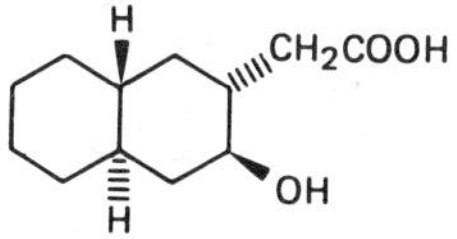 H, H, CH_2COOH, OH	Heat, 10 min.	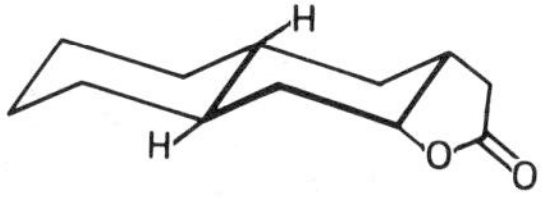 H, H, O, O	97	16
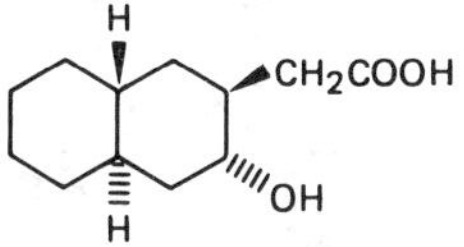 H, H, CH_2COOH, OH	Xylene, heat	 O, O	73	16
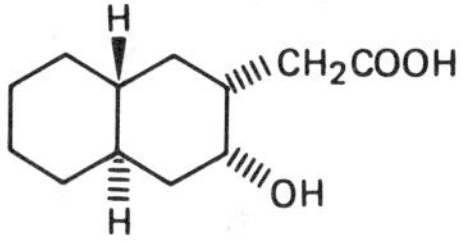 H, H, CH_2COOH, OH	Benzene, heat	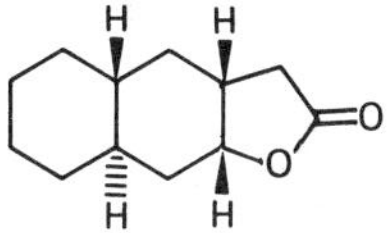 H, H, H, H, O, O	40	16
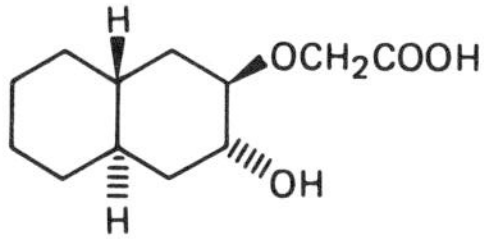 H, H, OCH_2COOH, OH	Benzene, heat	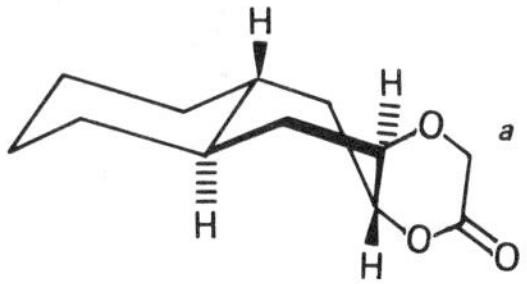 H, H, H, H, O, O, O [a]	51	16
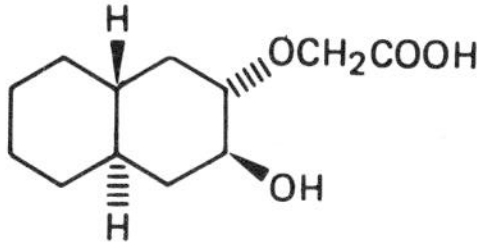 H, H, OCH_2COOH, OH	Heat[a]	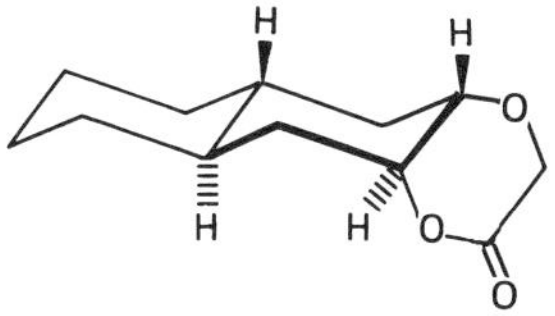 H, H, H, H, O, O, O	40	16

TABLE 1. (Continued)

Hydroxy acid	Conditions	Product	Yield (%)	Reference
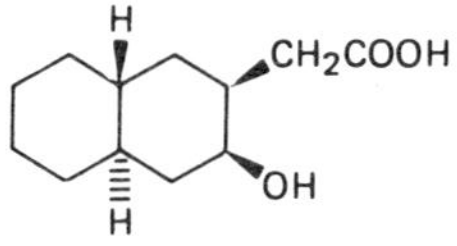	Heat[a]	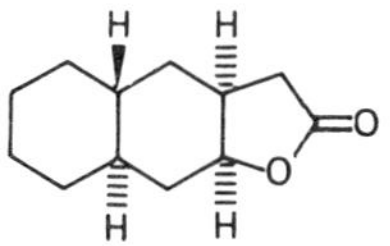	75	16
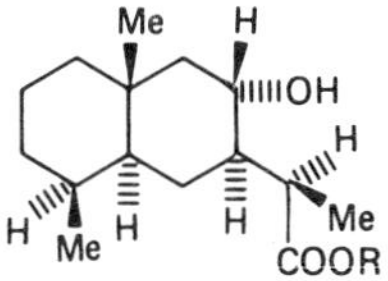	(R = H) Benzene, heat (R = H) Acetic acid, heat (R = Me) Benzene, heat (R = H)[b]	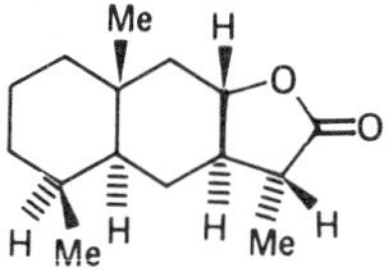	80 88 82 30	17
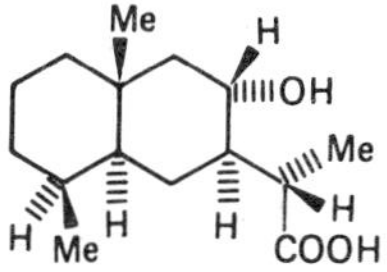	Benzene, acetic acid, heat	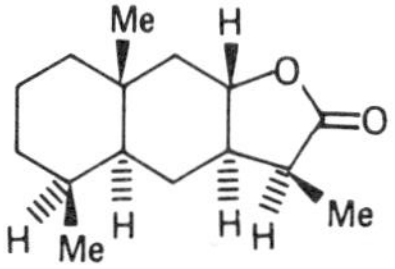	90	17
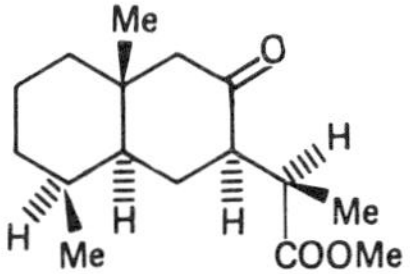	*c*	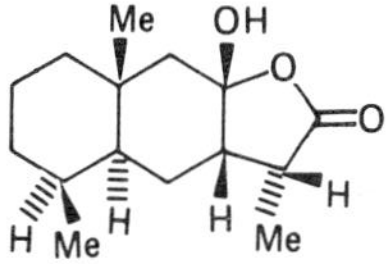	5	17
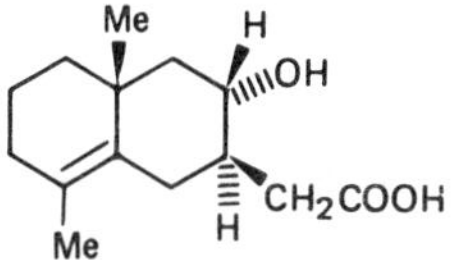	Benzene, heat	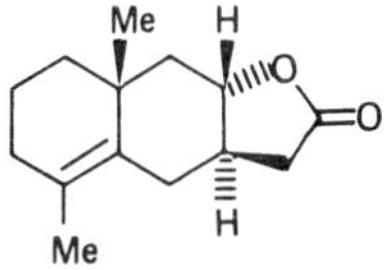	68	19

Substrate	Conditions	Product	Yield (%)	Ref.
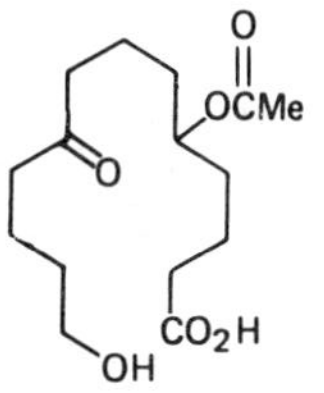 O, OCMe, O, CO_2H, OH	Benzene, heat Et_3N[d]	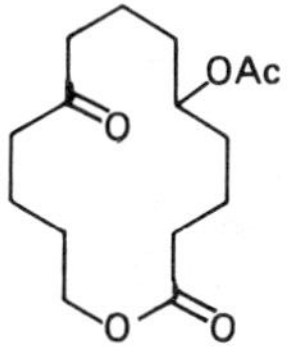 OAc, O, O, O	52	21
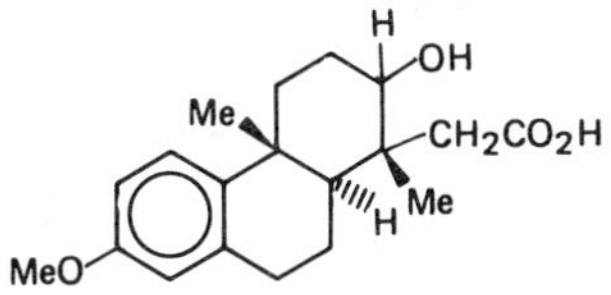 H, OH, Me, CH_2CO_2H, H, Me, MeO	Benzene, heat	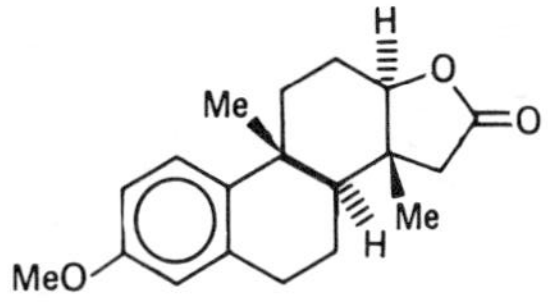 H, O, O, Me, H, Me, MeO	95	22

[a] These products were obtained by heating without *p*-toluenesulphonic acid.
[b] Using boron trifluoride–etherate in ether without *p*-toluenesulphonic acid.
[c] Conversion occurred after hydrolysis of the ester, by allowing the mixture to stand at 0°C for 24 hours.
[d] Also prepared in 40% yield by heating a benzene solution containing 1,1-carbonyldiimidazole followed by treatment with a catalytic amount of sodium *t*-amylate in benzene.

TABLE 2. Preparation of β-lactones from β-hydroxy acids with benzenesulphonyl chloride in pyridine

R^1	R^2	R^3	R^4	Yield (%)	Reference
Me	OMe	$—(CH_2)_5—$		82	23
Me	OMe	$—(CH_2)_4—$		83	23
Me	OMe	n-C_3H_7	n-C_3H_7	77	23
Me	OMe	Me	n-C_6H_{13}	45	23
Me	OMe	n-C_6H_{13}	Me	45	23
$—CH_2CH{=}C(Me)CH_2CH_2—$		Me	Me	82	24
Me	Me	H	$—(CH_2)_2CHMeC_6H_4Me$-p	77	24
H	Ph	Ph	Ph	70	25
H	Me	Ph	Ph	37	25
H	t-Bu	Ph	Ph	100	25
Me	H	Me	Ph	87	25
Me	Me	Ph	Ph	95	25
Me	Me	Me	Ph	92	25
Me	Me	$PhCH_2$	Ph	85	25
Me	$PhCH_2$	$PhCH_2$	Ph	30	25
Me	Me	H	Ph	95	25
Me	Me	$—(CH_2)_5—$		67	25
H	Me	$PhCH_2$	Ph	93	25
H	Me	Ph	$PhCH_2$	90	25
$—(CH_2)_3—$		$—(CH_2)_3—$		92	26
$—(CH_2)_3—$		$—(CH_2)_5—$		90	26
$—(CH_2)_4—$		$—(CH_2)_3—$		65	26
$—(CH_2)_4—$		$—(CH_2)_4—$		86	26
$—(CH_2)_4—$		$—(CH_2)_5—$		80	26
$—(CH_2)_5—$		$—(CH_2)_3—$		88	26
$—(CH_2)_5—$		$—(CH_2)_4—$		88	26
$—(CH_2)_5—$		$—(CH_2)_5—$		94	26
$—(CH_2)_6—$		$—(CH_2)_6—$		88	26
$—(CH_2)_7—$		$—(CH_2)_7—$		88	26
$—(CH_2)_2—CH{=}CH—(CH_2)_2—$		$—(CH_2)_5—$		77	26
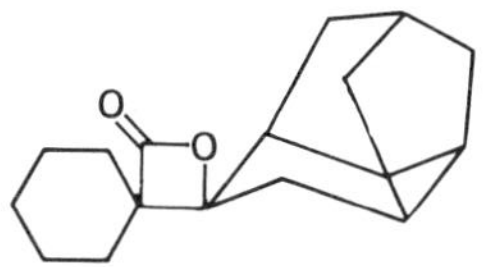				82	26

of α-methyl-α-*n*-butyl-β-hydroxyheptanoic acid afford the corresponding β-lactones. These lactonizations proceed through formation of intermediate mesyl derivatives, which then undergo internal nucleophilic displacement by the carboxylate group (Scheme 1).

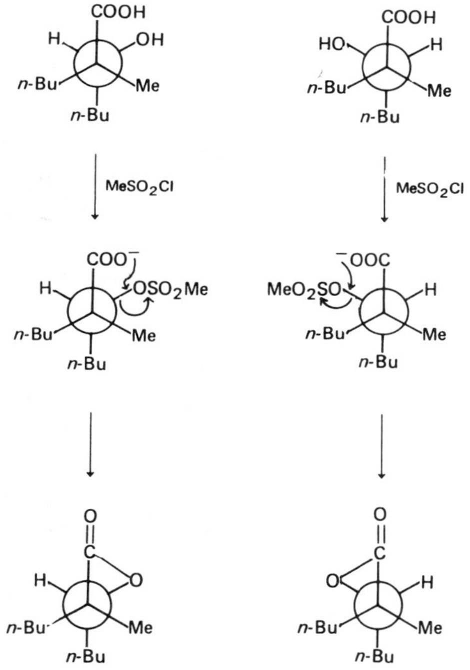

SCHEME 1.

The reaction of hydroxy acids with sodium acetate in acetic anhydride–benzene mixtures is a very effective method of lactonization, which has been used by Woodward and coworkers[28] in the total synthesis of reserpine (equation 4), and by

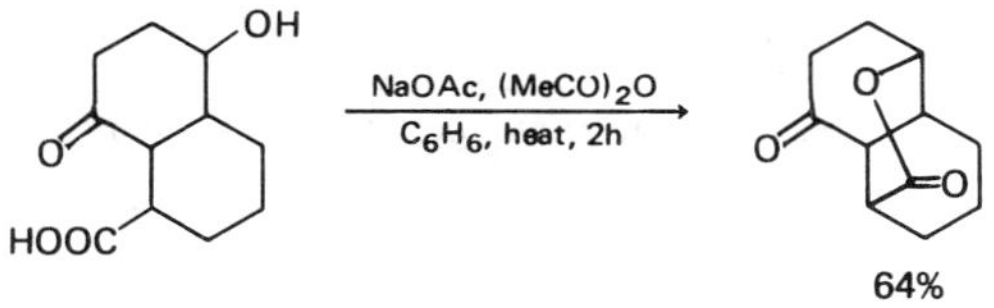

(Ref. 28) (4)

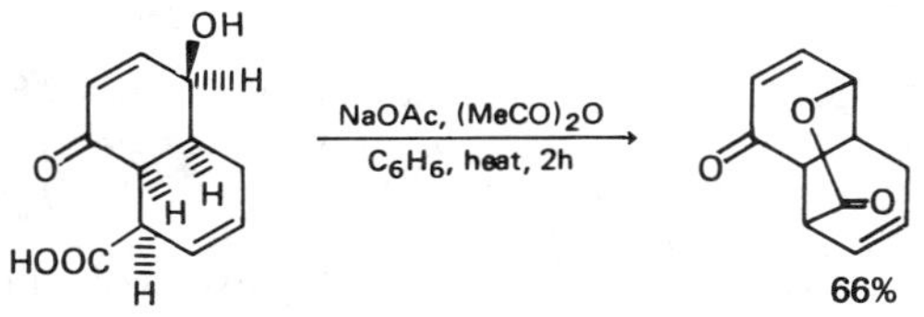

(Ref. 28) (5)

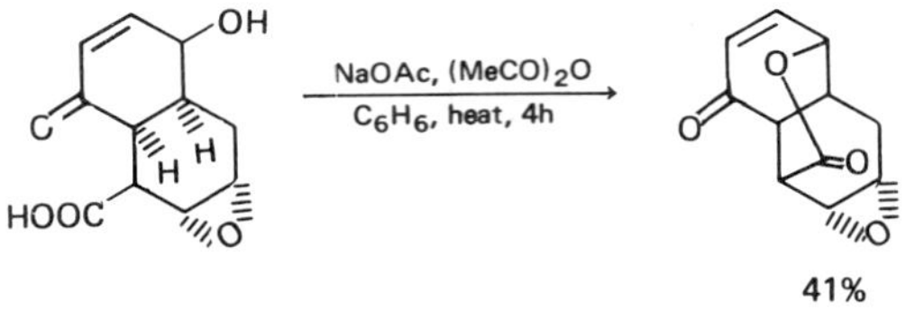
OH
NaOAc, $(MeCO)_2O$
C_6H_6, heat, 4h
HOOC
41%

(Ref. 28) (6)

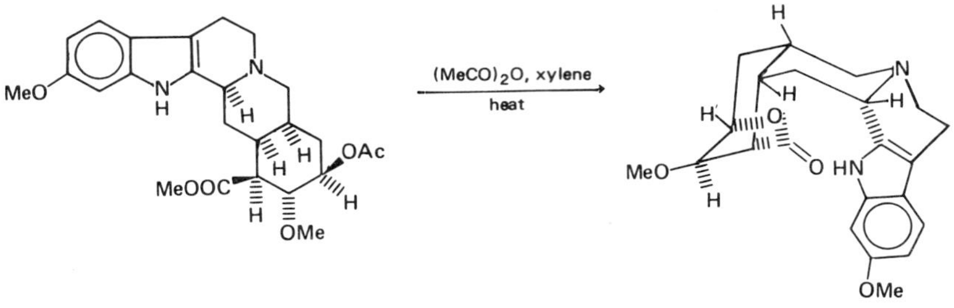
MeO
MeOOC
OAc
OMe
$(MeCO)_2O$, xylene
heat
HN
OMe

(7)

(Ref. 28)

Meinwald and Frauenglass[29] for the synthesis of various bicyclic lactones (equations 8 and 9).

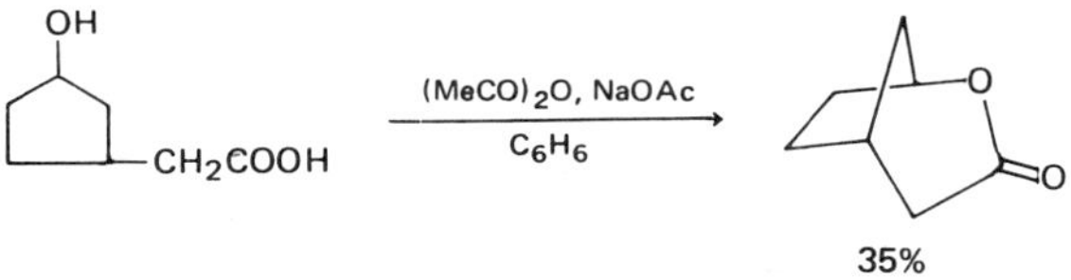
OH
CH_2COOH
$(MeCO)_2O$, NaOAc
C_6H_6
35%

(Ref. 29) (8)

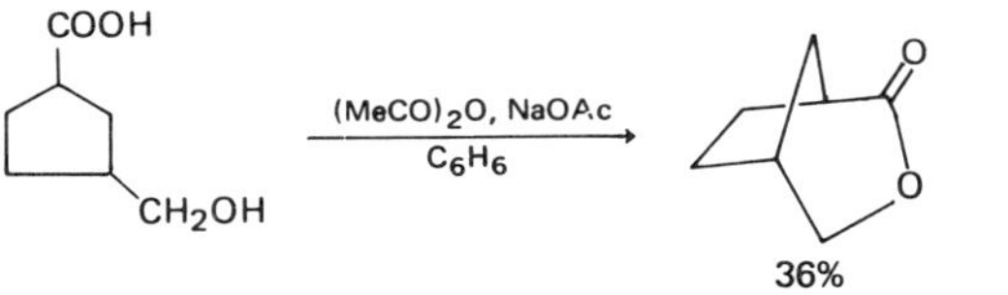
COOH
CH_2OH
$(MeCO)_2O$, NaOAc
C_6H_6
36%

(Ref. 29) (9)

N,N'-Dicyclohexylcarbodiimide (DCCD) is also an effective reagent for lactone formation from hydroxy acids[16,30–32] as illustrated in Table 3.

Reaction of 3,4-dimethoxyphenylacetic acid with formalin in the presence of acetic acid and aqueous hydrochloric acid affords 6,7-dimethoxy-3-isochromanone (equation 10) in a process which may be regarded as an *in situ* formation and cyclization of a hydroxy acid[33,34]

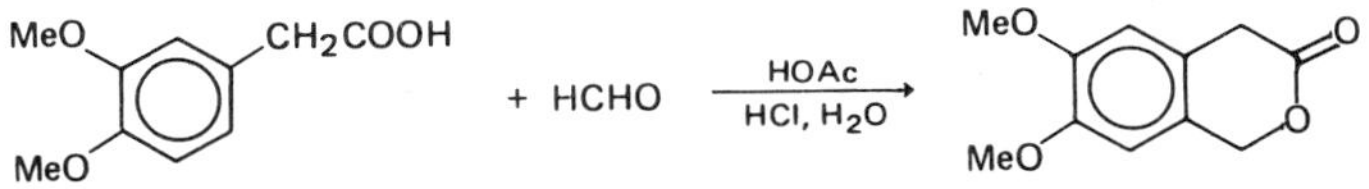
MeO
MeO
CH_2COOH
+ HCHO
HOAc
HCl, H_2O
MeO
MeO

(10)

Lactones can be prepared by acid-catalysed cyclization of hydroxy esters, as shown by the reaction of the methyl or ethyl esters of γ-alkyl-γ-carboethoxy-δ-hydroxyhexanoic acids with metaphosphoric acid to afford[35] the expected δ-lactones in 95–99% yield (equation 11).

Intramolecular acid-catalysed cyclization of γ-hydroxy esters has been found

TABLE 3. Lactonization of hydroxy acids by DCCD

Hydroxy acid	Product	Yield (%)	Reference
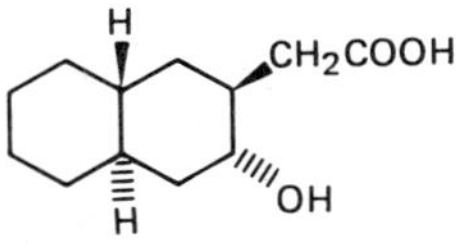 H, H, CH_2COOH, OH	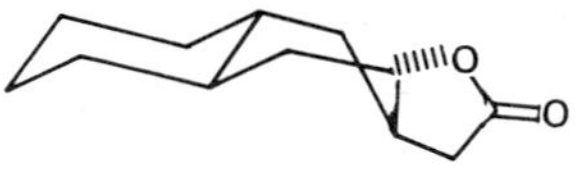 O, O	55–86	16
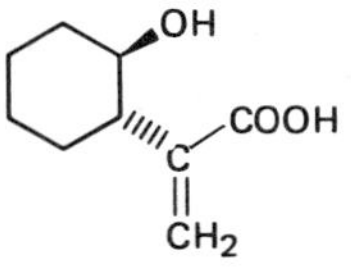 OH, COOH, C, CH_2	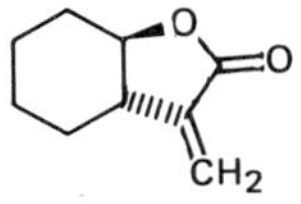 O, O, CH_2	60	30, 31

MeCHCCH$_2$CH$_2$COOR^1 (with OH, R^2 and COOEt substituents) $\xrightarrow{HPO_3}$ lactone (R^2, COOEt, Me) (11)

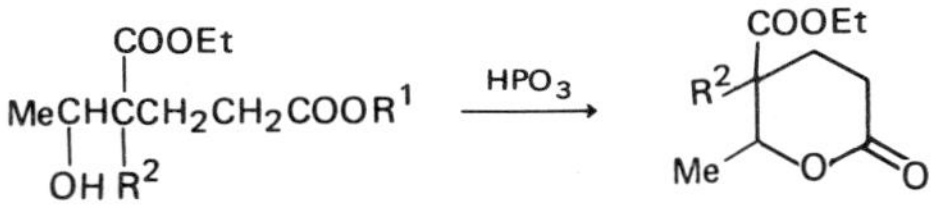

R^1 = Me, Et

R^2 = Et, *n*-Pr, *n*-Bu, $CH_2CH_2CH(Me)_2$

useful in the preparation of bicyclic lactones. Thus, reaction of diethyl Δ^4-cyclohexene-*cis*-1,2-dicarboxylate oxide with dilute aqueous sulphuric acid at 40–50°C gave[36] a 73% yield of diethyl *trans*-4,5-dihydroxycyclohexane-*cis*-1,2-dicarboxylate, which upon partial acid hydrolysis at 80°C afforded the bicyclic lactone shown in equation (12).

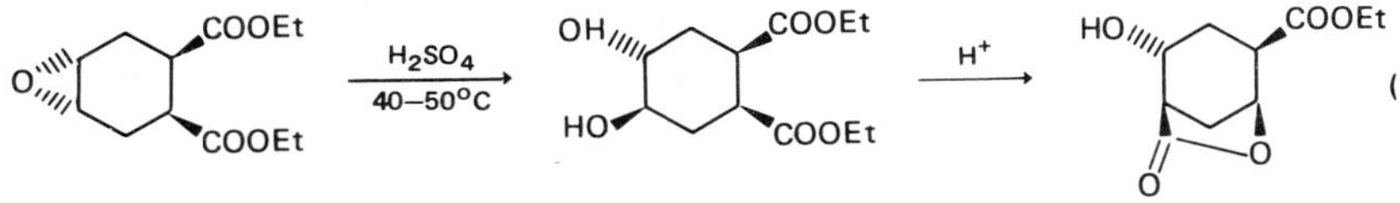

COOEt, COOEt, O $\xrightarrow[40\text{–}50^{\circ}C]{H_2SO_4}$ OH, HO, COOEt, COOEt $\xrightarrow{H^+}$ HO, COOEt, O, O (12)

Sulphuric acid-catalysed lactonization of *cis*- and *trans*-*N*-(carboxymethyl)-4-phenyl-4-ethylpyrrolidin-3-ols, as well as their corresponding methyl and ethyl esters or their 3-acetates, all afforded[37] the bicyclic lactone, 6-phenyl-6-ethyl-1-aza-4-oxabicyclo[3.2.1]octane-3-one (equation 13). The fact that the same lactone was obtained from either the *cis* or *trans* compounds, indicates that the probable mechanism for this transformation involves initial protonation of the $C_{(3)}$-OH or -OR function with subsequent elimination of water or alcohol to create a positive centre at $C_{(3)}$, followed by intramolecular nucleophilic attack by the carbonyl as shown in equation (14).

The use of boron trifluoride–etherate for direct lactonization of hydroxy esters is demonstrated by reaction of methyl 4-hydroxy-6-phenylhex-5-en-1-yne-1-carboxylate to give the lactone of 4-hydroxy-2,2-dimethoxy-6-phenylhex-5-en-1-carboxylic acid, which upon heating at 150°C afforded[38–40] (±)-kawain (**4**), a constituent of the kawa root (equation 15)[41–44]. Hydroxy esters can also be converted to lactones by means of DCCD (equations 16 and 17)[28,45].

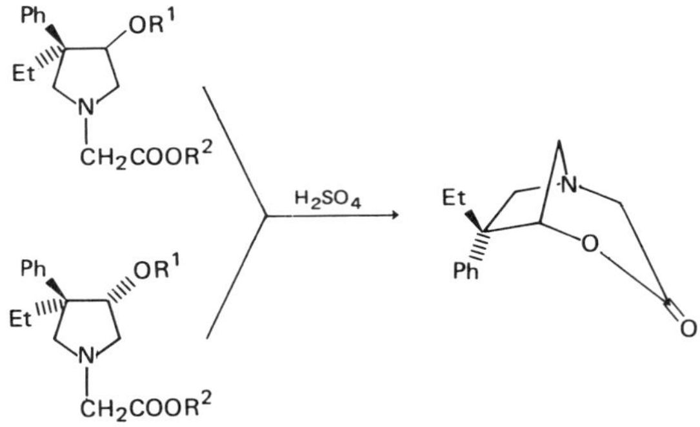

(13)

$R^1 = R^2 = H$

$R^1 = H$; $R^2 = Me$

$R^1 = H$; $R^2 = Et$

$R^1 = COMe$; $R^2 = Me$

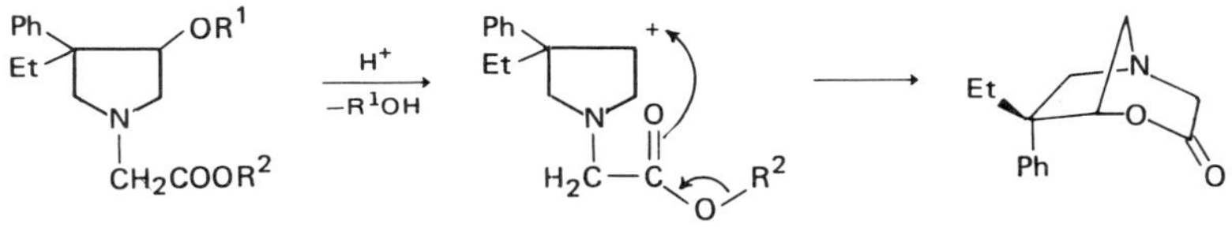

(14)

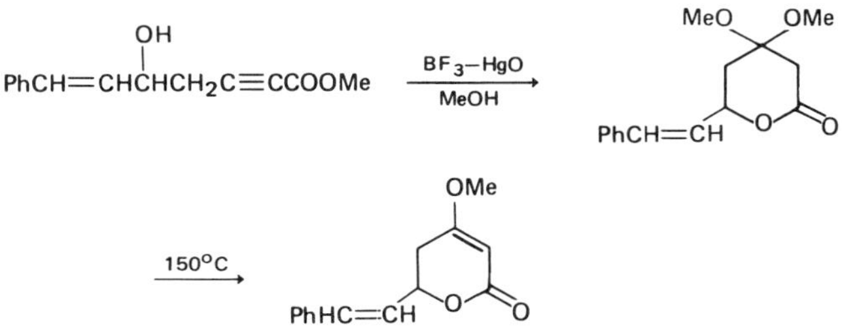

(15)

(4)

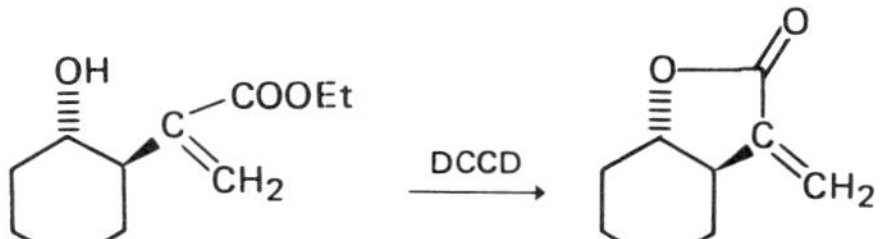

(Ref. 45) (16)

In some instances of lactone synthesis from hydroxy esters, the ester function is first saponified, and subsequent acidification leads to lactone formation. Such a procedure has been employed in an alternative synthesis of (±)-kawain[38,46] as shown in equation (18). The synthesis of β-carboxy-γ-tridecyl-γ-butyrolactone is accomplished in a similar fashion (equation 19)[47].

In general, uncatalysed thermal lactonization of hydroxy esters tends to give significant amounts of polymeric material. For example, distillation of a series of

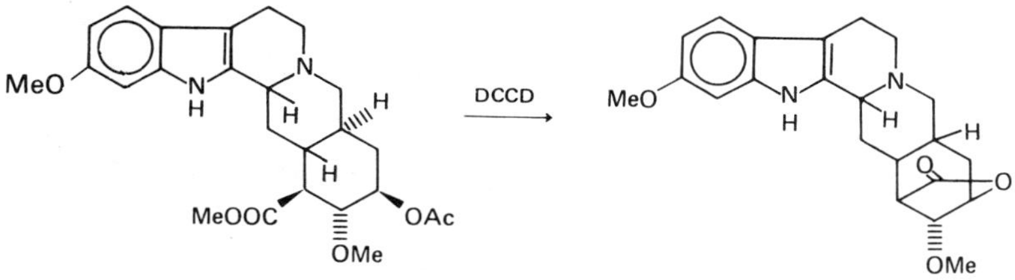

(Ref. 28) (17)

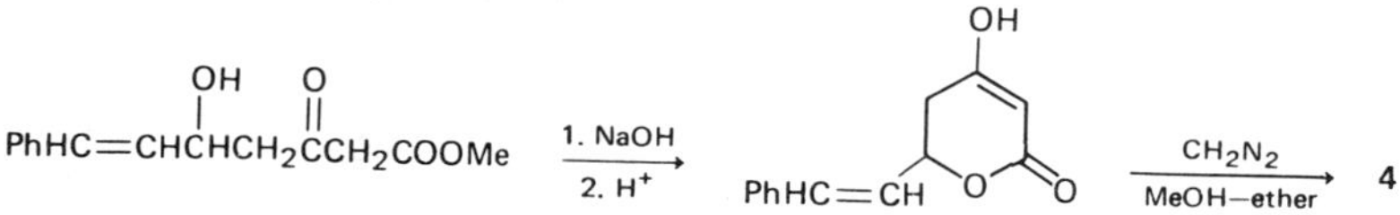

$$\text{PhHC=CHCH(OH)CH}_2\text{C(=O)CH}_2\text{COOMe} \xrightarrow[\text{2. H}^+]{\text{1. NaOH}} \text{PhHC=CH-(4-hydroxy-5,6-dihydro-2-pyranone)} \xrightarrow[\text{MeOH-ether}]{\text{CH}_2\text{N}_2} \mathbf{4} \quad (18)$$

$$\text{Me(CH}_2)_{12}\text{CH(OH)CH(COOMe)CH}_2\text{COOMe} \xrightarrow[\text{2. H}^+]{\text{1. KOH-MeOH}} \text{HOOC-, Me(CH}_2)_{12}\text{-}\gamma\text{-butyrolactone} \quad (19)$$

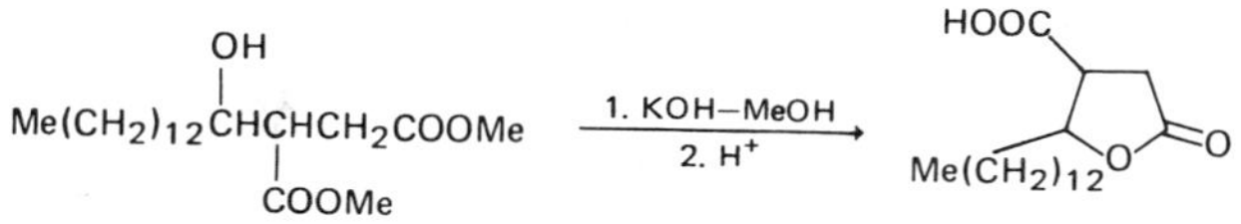

ethyl α-alkyl-δ-hydroxyhexanoates affords a mixture of unidentified polymers. However, depolymerization of this mixture by distillation in the presence of concentrated sulphuric acid or phosphoric acid produces the corresponding α-alkyl-δ-hydroxyhexanoic acid lactones in good yields (equation 20)[48].

$$\text{MeCH(OH)CH}_2\text{CH}_2\text{CH(R)COOEt} \xrightarrow{\text{distillation}} \text{polymers} \xrightarrow[\text{H}_2\text{SO}_4 \text{ or H}_3\text{PO}_4]{\text{distillation}} \text{Me, R-}\delta\text{-valerolactone}$$

R = Et, Pr, *n*-Bu, *i*-Bu, *i*-amyl (20)

Corey and Nicolaou[49] have recently reported an ingenious method for lactone synthesis in which ω-hydroxy carboxylic acids are first converted to ω-hydroxy-2-pyridinethiol esters, which subsequently undergo facile thermal lactonization (equations 21 and 22). This appears to be one of the most general methods for large-ring lactone synthesis currently available.

Conversion of α,α-dialkyl-β-hydroxy acids to 4-oxo-1,3-dioxanes by reaction with methyl orthopropionate followed by heating these compounds at 150–200°C affords β-lactones in good yields via a proposed concerted mechanism (equation 23)[50].

Reaction of ethyl 1-hydroxymethylcyclopropanecarboxylate with zinc bromide in 48% hydrobromic acid results in cyclopropane ring enlargement to afford αmethylene-γ-butyrolactone in 25% yield (equation 24)[51]. Reaction (25) and (26) provided similar results[51]. This rearrangement has also been observed with cyclopropylmethyl methyl ethers and cyclopropylmethyl bromides (equations 27 and 28)[51].

The conversion of γ- or δ-keto acids to enol lactones is a well-known process, illustrated here by the synthesis[52] of the enol lactone of 1,4,4-trimethylcyclohexane-2-oneacetic acid (equation 29).

$HO(CH_2)_nCOOH$ + (2,2'-dipyridyl disulfide) ⟶ $SC(O)(CH_2)_nOH$ (2-pyridyl thioester)

$\xrightarrow{\text{reflux}}$ ⟶ ⟶ $O{=}C{-}O{-}(CH_2)_n$ (21)

n	Solvent	Reflux time (h)	Isolated yield (%)
5	C_6H_6	10	71
7	$Me_2C_6H_4$	30	8
10	$Me_2C_6H_4$	20	47
11	$Me_2C_6H_4$	10	66
12	$Me_2C_6H_4$	10	68
14	$Me_2C_6H_4$	10	80

1. $(C_5H_4N{-}S)_2$
2. C_6H_6, reflux
3. HOAc, H_2O, THF (22)

THP = tetrahydropyranyl

$R_2C(COOH)(CH_2OH)$ + $(MeO)_3C{-}Et$ $\xrightarrow{50-100^{\circ}C}$ $\xrightarrow{150-200^{\circ}C}$ (23)

R = Me, 87%

R = Et, 84%

$\xrightarrow[100^{\circ}C]{ZnBr_2,\ HBr}$ (24)

(25)

Treatment of 2-carboethoxymethyl-2-methylcyclohexane-1,3-dione with polyphosphoric acid (equation 30) results in ring-opening, followed by ring-closure to form lactone **6**, presumably by isomerization of intermediate enol lactone **5**[53]. When the analogous 2-carboethoxymethyl-2-(3-ketopentyl)cyclohexane-1,3-dione is treated under similar conditions, a new mode of cyclization is observed[53], affording fused δ-lactone **7** in 34% yield (equation 31). Formation of unexpected products was also observed when 2,2-di(carboethoxymethyl)cyclohexane-1,3-dione was treated[53] with polyphosphoric acid to afford a 64% yield of dilactone **8** (equation 32). Treatment of 2,2-dimethylcyclohexane-1,3-dione under similar conditions

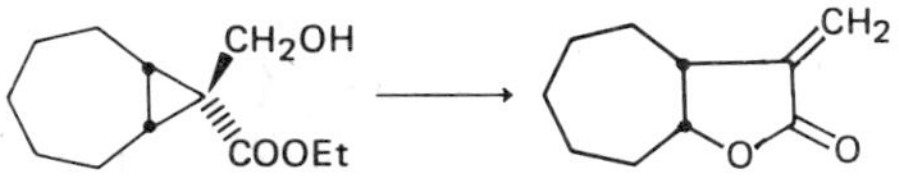

(26)

Reaction	Reaction conditions	Yield (%)
(25)	$ZnBr_2$, 48% HBr, EtOH, 100°C, 6 h	50
(25)	Conc. H_2SO_4, 0°C, 2 h	30
(25)	F_3CSO_4H, C_6H_6	—
(25)	p-$MeC_6H_4SO_3H$	—
(26)	$ZnBr_2$, 48% HBr, EtOH, 100°, 6 h	43

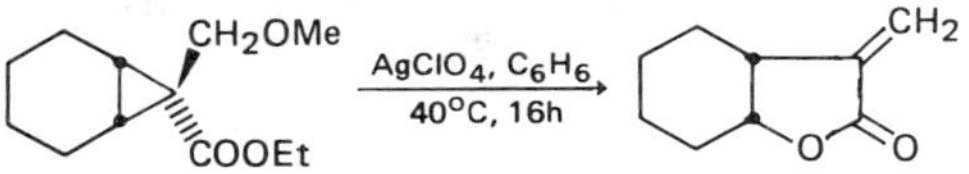

$AgClO_4$, C_6H_6, 40°C, 16h (27)

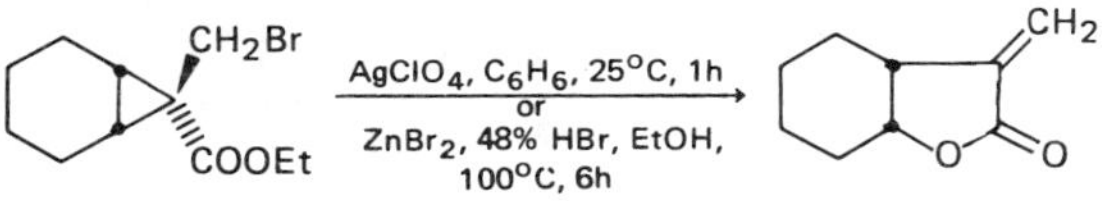

$AgClO_4$, C_6H_6, 25°C, 1h or $ZnBr_2$, 48% HBr, EtOH, 100°C, 6h (28)

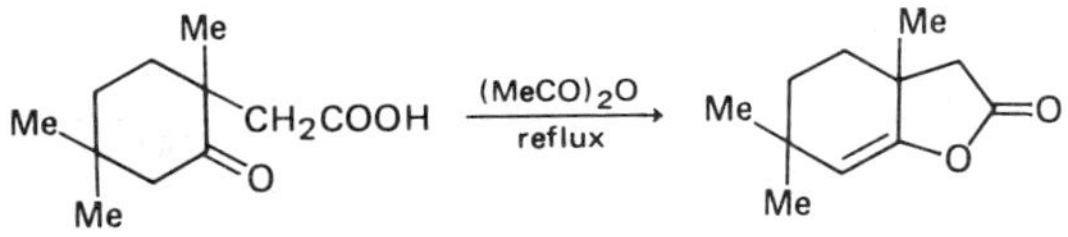

$(MeCO)_2O$ reflux (29)

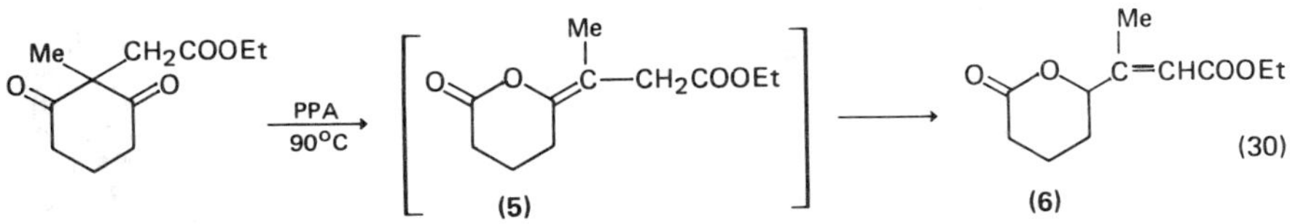

PPA 90°C (5) (6) (30)

PPA

$R^1 = CH_2COOEt$, $R^2 = MeCH_2\overset{O}{\overset{\|}{C}}CH_2CH_2$

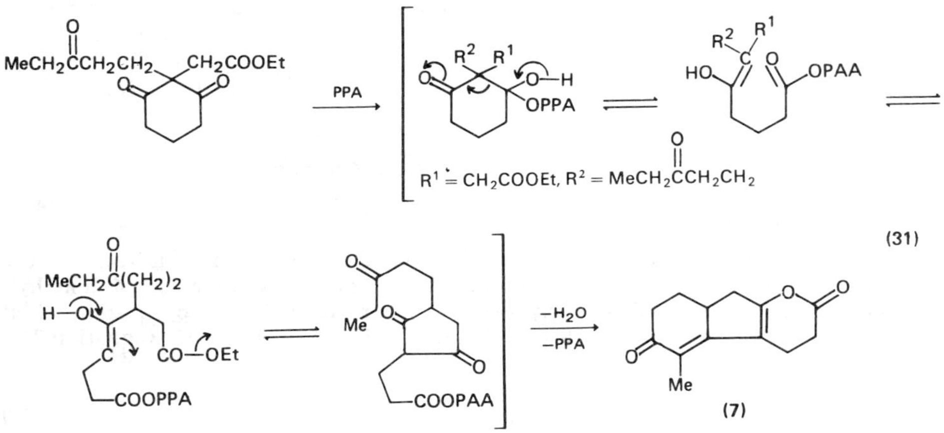

(31)

$-H_2O$ $-PPA$ (7)

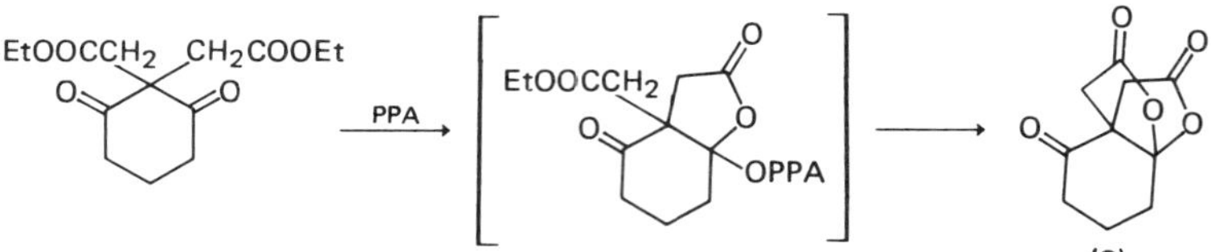

(32)

(8)

afforded no reaction, which was attributed to the deactivating effect of the two methyl groups[53].

Aldehydic acids can be cyclized to enol lactones by treatment with *p*-toluenesulphonic acid in benzene (equations 33–35)[54].

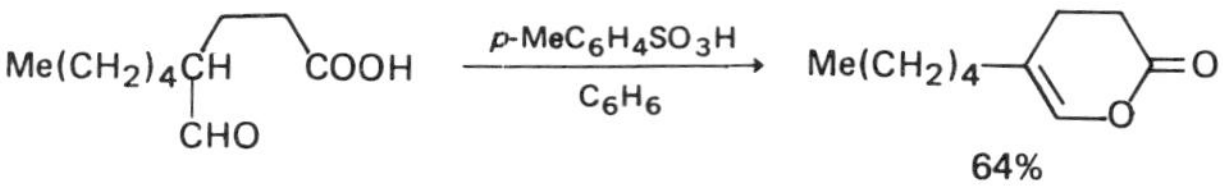

(33)

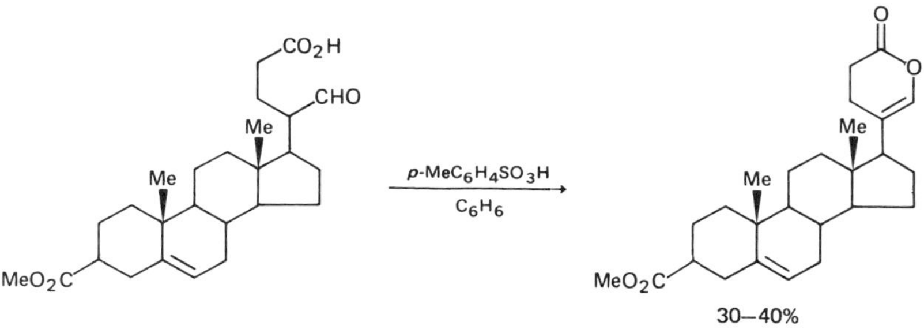

64%

(34)

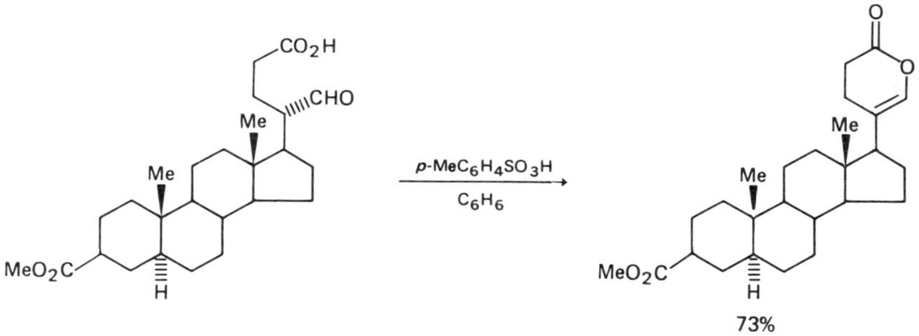

30–40%

(35)

73%

B. By Intramolecular Cyclization of Unsaturated Acids and Esters

1. *Acid-catalysed cyclizations*

Various unsaturated acids and esters have been converted to lactones in the presence of acids[5]. Recent examples include the preparation of 4,4-dimethylbutyrolactone[55,56] and 4-methyl-4-phenylbutyrolactone[55] by cyclization of 4-methyl-3-pentenoic acid and 4-phenyl-3-pentenoic acid, respectively (equation 36).

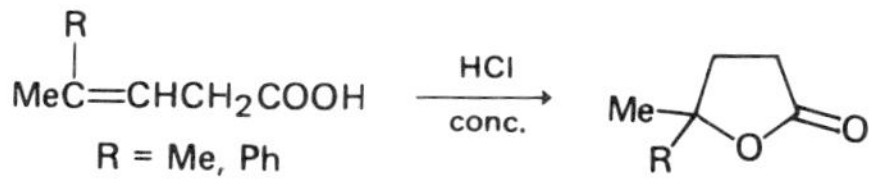

$$\mathrm{MeC(R){=}CHCH_2COOH} \xrightarrow[\text{conc.}]{\text{HCl}} \text{Me(R)-butyrolactone} \qquad R = \text{Me, Ph} \qquad (36)$$

Treatment of alkenyl-substituted malonic esters with aqueous acid affords the expected γ-lactones in good yields, while basic hydrolysis produces the γ,δ-unsaturated acid (equation 37)[57].

$$R^2CH{=}C(R^3){-}CH_2C(R^1)(COOEt)_2 \xrightarrow[-CO_2]{H^+,\ H_2O} \text{γ-lactone } (R^2H_2C,\ R^3,\ R^1) \qquad \xrightarrow[-CO_2]{^-OH,\ H_2O} R^2CH{=}C(R^3)CH_2CH(R^1)COOH \tag{37}$$

R^1 = H, *n*-Bu

R^2 = H, *n*-Bu, *n*-C_5H_{11}

R^3 = H, Me

2-Hydroxy-2,6,6-trimethylcyclohexylideneacetic acid γ-lactone[58–62] has been synthesized[52] by treatment of 2,6,6-trimethylcyclohexene-1-glycolic acid with aqueous sulphuric acid or by simply heating the glycolic acid at 200–220°C (equation 38). Alternatively, this lactone can be prepared[52] by base-catalysed ring closure of **9**. Interestingly, treatment of 2,6,6-trimethylcyclohexene-1-glycolic acid with chromic anhydride–pyridine[52] affords 'hydroxyionolactone', which can also be prepared[52] by permanganate oxidation of β-ionone (equation 39).

$$\xrightarrow[\text{or heat}]{H_2SO_4,\ H_2O} \qquad \xleftarrow[\text{toluene}]{Me_3CO^-\ K^+} \quad \textbf{(9)} \tag{38}$$

$$\xrightarrow[\text{pyridine}]{CrO_3} \qquad \xleftarrow{KMnO_4} \tag{39}$$

2,5-Dienoic acids and esters, such as those shown in Table 4, can be converted into α,β-unsaturated δ-lactones upon treatment with 80% sulphuric acid at

TABLE 4. Cyclization of 2,5-dienoic acids and esters using sulphuric acid[63]

Acid or ester	Lactone	Yield (%)
cis-2,5-Hexadienoic acid	6-methyl-5,6-dihydro-2H-pyran-2-one (Me, O, =O)	84.6
Methyl *cis*-2,5-hexadienoate	6-methyl-5,6-dihydro-2H-pyran-2-one (Me, O, =O)	77.0

0–5°C[63]. Carboxylic acids containing multiple unsaturation can undergo rather complex cyclizations[64] in the presence of sulphuric acid as shown in equation (40).

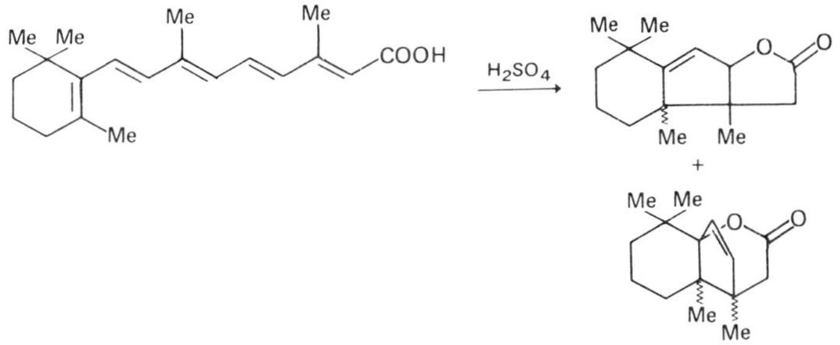

(40)

Cyclization of acids and esters containing acetylenic bonds has seen wide application in the preparation of lactones[65–80]. A number of representative examples are presented in Table 5.

o-Phenylbenzoic acids undergo cyclization upon treatment with hydrogen peroxide or chromic anhydride to form lactones in moderate yields (equation 41)[81].

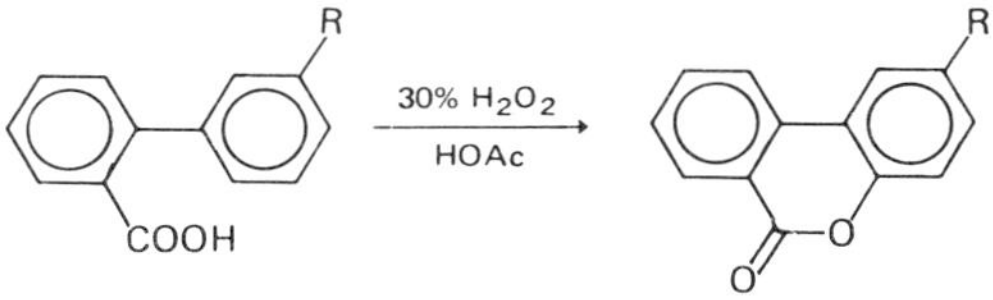

(41)

2. *Photochemical and electrochemical cyclizations*

Preparations of lactones by photochemical cyclization of unsaturated acids or esters have also been reported in the literature[83–89]. Irradiation[82] of a series of α-substituted cinnamic and crotonic acids afforded the corresponding substituted β-lactones (equation 42).

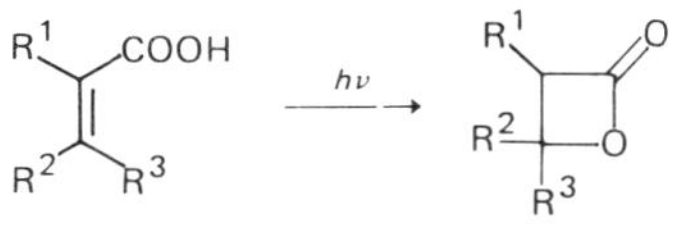

(42)

R^1 = Ph, Ph, Me, Ph, Ph, Ph

R^2 = H, Ph, Ph, Me, Ph, *p*-MeC_6H_4

R^3 = Ph, H, H, H, Ph, H

γ-Lactones have been prepared by the irradiation-induced addition of alcohols to α,β-unsaturated acids or esters in the presence[83–85] or absence[86,87] of a sensitizer, by the use of ^{60}Co γ-rays[88,89], and by reductive electrochemical addition of acetone[89]. Some γ-lactones prepared by these various methods are listed in Table 6. The mechanism suggested[87] for the photolytic addition in the absence of a sensitizer is shown in equation (43) and involves initial hydrogen abstraction by the excited ester carbonyl, which then leads to α-hydroxyalkyl and allylic radicals. Coupling of the former to the β-carbon of the latter and tautomerization affords a γ-hydroxy ester which then cyclizes.

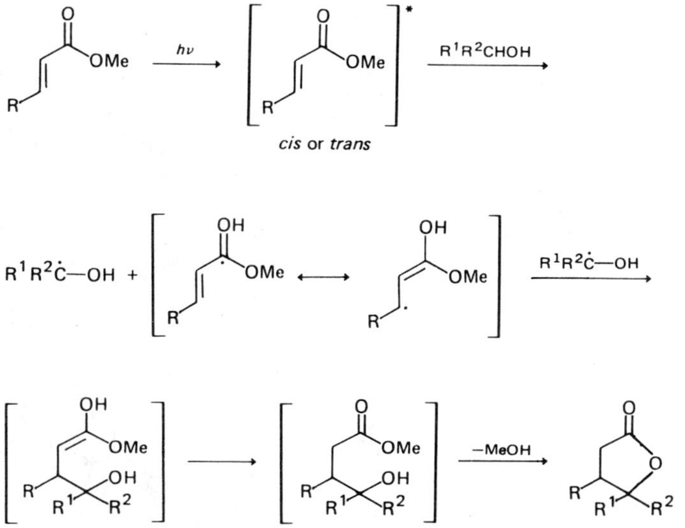

$$\text{RCH=CHCO}_2\text{Me} \xrightarrow{h\nu} [\text{RCH=CHCO}_2\text{Me}]^* \; (\textit{cis} \text{ or } \textit{trans}) \xrightarrow{R^1R^2CHOH}$$

$$R^1R^2\dot{C}\text{—OH} + [\ldots] \xrightarrow{R^1R^2\dot{C}\text{—OH}} \quad (43)$$

$$[\ldots] \longrightarrow [\ldots] \xrightarrow{-\text{MeOH}} \text{lactone}$$

3. *Halolactonization*

The reaction of unsaturated acids with iodine–potassium iodide and bicarbonate in aqueous medium affords iodolactones. This iodolactonization, first reported[90] in 1908, was originally believed to exhibit the following characteristics: (*a*) α,β-unsaturated acids do not give iodolactones (*b*) β,γ- as well as γ,δ-unsaturated acids do afford iodolactones, (*c*) δ,ε acids or acids with the unsaturation further removed from the carboxyl group yield only poorly characterized unsaturated acid iodohydrins and (*d*) α-keto β,γ-alkenoic acids and α,β,γ,δ-alkenoic acids are exceptional in that no iodolactones are obtained from them. Since α,β-unsaturated acids do not give iodolactones but β,γ-unsaturated acids afford β-iodo-γ-lactones via this procedure, a number of workers[91–93] have used this approach to distinguish α,β-unsaturated acids from β,γ isomers. In a more involved study of this reaction, Van Tamelen and Shamma[94] showed that although β,γ-butenoic acid does not afford any iodolactone even upon long standing, β,γ-pentenoic acid (equation 44)

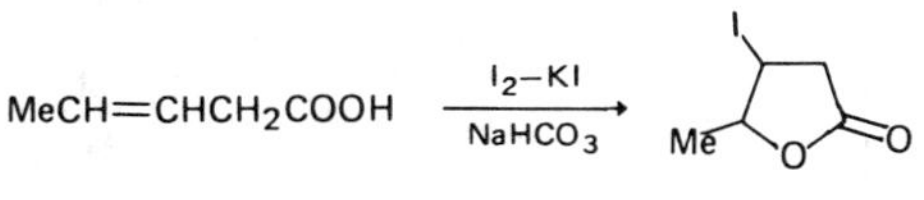

$$\text{MeCH=CHCH}_2\text{COOH} \xrightarrow[\text{NaHCO}_3]{I_2\text{–KI}} \text{iodolactone} \quad (44)$$

$$\text{cyclohexenyl–CH}_2\text{COOH} \xrightarrow[\text{NaHCO}_3]{I_2\text{–KI}} \text{iodolactone} \quad (45)$$

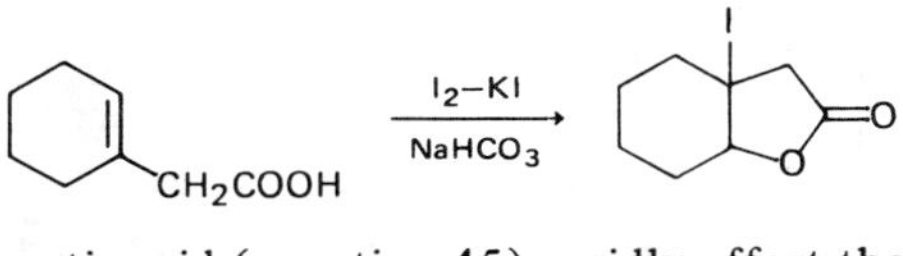

and Δ^1-cyclohexeneacetic acid (equation 45) rapidly affort the corresponding iodolactones[95]. Van Tamelen further established[94] that although there are two structural possibilities, δ-iodo-γ-lactones and γ-iodo-δ-lactones, for lactones derived

TABLE 5. Preparation of lactones via intramolecular cyclization of acetylenic acids

Acetylenic acid	Reaction conditions	Product	Yield (%)	Reference
Me–C≡C–C≡C–C≡C–CH=CH–COOH *cis*	$KHCO_3$, H_2O	Me—C≡C—C≡C—CH= (furanone, O, =O)	55	65–67
PhC≡C–CH=C(COOH)$_2$	190°C, 10–15 min or $AgNO_3$	PhHC= (lactone, O, =O, COOH)	85	68
(tetrahydropyranyl, O) OCH$_2$—C≡C—C(CH(OMe)$_2$)=CH—COOH	MeOH, $AgNO_3$	(tetrahydropyranyl, O) OCH$_2$—CH= (furanone, O, =O, (MeO)$_2$CH)	80	74
	MeOH, $AgNO_3$	PhCH= (furanone, O, =O, Me)	63	76

$PhC{\equiv}C{-}CR^1{=}CR^2{-}COOH$ *cis*	H^+ 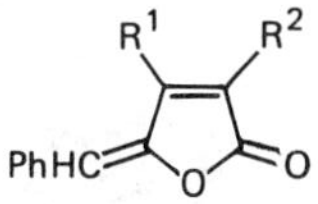	(5-benzylidene-2(5H)-furanone: PhHC=, R^1, R^2, ring O, C=O)	79
		R^1 = H; R^2 = Ph — 80	
		R^1 = H; R^2 = p-$O_2NC_6H_4$ — 85	
		R^1 = Me; R^2 = Ph — 70	
		R^1 = Me; R^2 = p-$O_2NC_6H_4$ — 75	
		R^1 = R^2 = Ph — 40	
$PhC{\equiv}C{-}C(R){=}CH{-}COOH$ *cis*	50% H_2SO_4, 3h or $HgSO_4$, dioxane or acetone	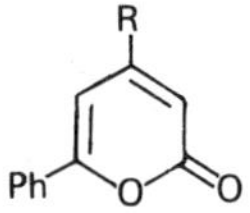 (6-phenyl-4-R-2-pyranone: R, Ph, ring O, C=O)	80
		R = COOH — 54–80	
		R = Ph — 40–60	
		R = β-naphthyl — 40	
		R = p-biphenyl — 20	
		R = p-$MeOC_6H_4$ — 20	
		R = m-$MeOC_6H_4$	

TABLE 6. γ-Lactones prepared via irradiation-induced and electrochemical addition of alcohols to unsaturated acids or esters

Acid or ester	Alcohol	Method[a]	Product	Yield (%)	Reference
HOOCCH=CHCOOH *cis*	MeC(OH)HR		HOOC, Me, R, O, O		
	R = Me	A		60	83
	R = Et	A		57	83
	R = C_6H_{13}-*n*	B		40	84
	R = H	B		20	84
MeCH=CHCOOH *trans*	*i*-PrOH	B	Me, Me, Me, O, O	60	84
HOOCC≡CCOOH	*i*-PrOH	C	Me, Me, O, O, O, O, Me, Me	15	85
HC≡CCOOEt	*i*-PrOH	D	Me, Me, O, O	20	86
MeOOCCH=CHCOOMe *cis* or *trans*	*i*-PrOH	E	HOOC, Me, Me, O, O — from *cis*,	64	87
			from *trans*	70	87

MeCH=CHCOOMe *trans*	*i*-PrOH	E	+	50 + 12	87
MeOOCCH=CHCOOMe *cis* or *trans*	EtOH	E	from *cis* from *trans* *cis* + *trans*	68 71	87 87
MeCH=CHCOOMe *trans*	EtOH	E	*cis* + *trans*	59	87
MeCH=CHCOOEt	MeOH				
		F		1	88
		G		1	88
		H		1	88
	EtOH				
		F	*trans: cis*, 4:5	16	88
		G	*trans: cis*, 4:5	4	88
		H	*trans: cis*, 4:5	1	88

TABLE 6. (Continued)

Acid or ester	Alcohol	Method[a]	Product	Yield (%)	Reference
	i-PrOH		Me, Me, Me		
		F		54, 44	88, 89
		G		12	88
		H		12	88
		I		30	89
	n-PrOH		Me, Et		
		F	*trans: cis,* 2:3	9	88
		G	*trans: cis,* 4:5	4	88
		H	*trans: cis,* 4:5	5	88
	sec-BuOH		Me, Me, Et		
		F	*trans: cis,* not determined	18	88
		H	*trans: cis,* not determined	22	88
	Et_2CHOH		Me, Et, Et		

		F		10	88
		H		9	88
	$PhCH_2OH$	F	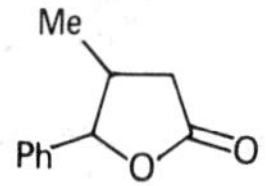 *trans: cis*, 2:1	23	88
$H_2C{=}CHCOOEt$	*i*-PrOH		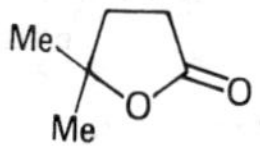		
		F		0	89
		I		95	89
$H_2C{=}CHCOOH$	*i*-PrOH		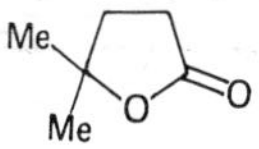		
		F		41	89
		I		36	89
$Me_2C{=}CHCOOEt$	*i*-PrOH		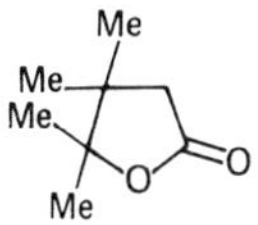		
		F		29	89
		I		trace	89
EtOOCCH=CHCOOEt *cis* or *trans*	*i*-PrOH		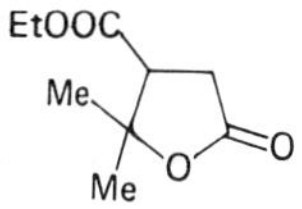		
		F	from *cis*	70	89
		F	from *trans*	22	89
		I	from *cis*	59	89
		I	from *trans*	21	89

TABLE 6. (Continued)

Acid or ester	Alcohol	Method[a]	Product	Yield (%)	Reference
HOOCCH=CHCOOH *trans*	*i*-PrOH	F	HOOC, Me, Me, O, O	69	89
PhCH=CHCOOEt *trans*	*i*-PrOH		Ph, Me, Me, O, O		
		F		0	89
		I		0	89

[a]A = Irradiation with an ultraviolet light source for 18 h at 16°C using benzophenone as sensitizer.
B = As above but irradiated for 25 h in the cold.
C = As above but irradiated for 60 h at 35°C.
D = Irradiated using a quartz-contained mercury arc.
E = Irradiated using a 450 W-Hanovia medium-pressure mercury arc.
F = Irradiated using γ-rays in a ^{60}Co cavity source at room temperature.
G = Irradiated using a quartz tube for 50 h. with a 500 W high-pressure mercury vapour lamp in the pressure of benzophenone as sensitizer.
H = Irradiated using a Pyrex tube for 72 h. with a 500 W high-pressure mercury vapour lamp in the presence of benzophenone as sensitizer.
I = Electrolysis with acetone, 20% sulphuric acid and water for 1 h with a terminal voltage of 75–95 V in a cylindrical vessel using a mercury pool cathode and a platinum plate anode.

from γ,δ-unsaturated acids, both γ,δ-pentenoic acid (equation 46) and Δ^2-cyclohexeneacetic acid (equation 47) give rise to the γ-lactones rather than the

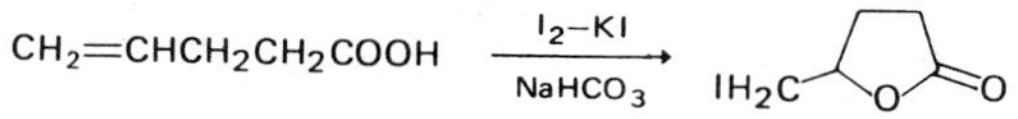

(46)

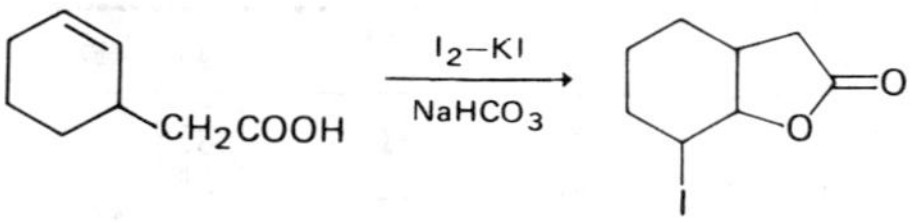

(47)

δ isomers originally proposed by Bougault[90]. It was also established[94] that δ,ε-hexenoic acid affords (probably) δ-iodomethyl-δ-valerolactone, again contrary to Bougault's findings, while ε,ζ-heptenoic acid and ω-undecylenic acid led to unstable, poorly defined products.

Halolactonization reactions have also been used to separate[96,97] mixtures of *endo*- and *exo*-norborn-5-enyl acids and *endo*- and *exo*-methylenenorborn-5-enyl acids[98] In the former case[96,97] the *endo* isomer reacts to produce a γ-lactone while the *exo* isomer remains in the aqueous layer as the carboxylate salt; in the latter, both isomers react with bromine in methylene chloride to give lactone products, whereas reaction with iodine in methylene chloride affords the iodolactone from the *endo* isomer only (equation 48)[98]. With the carboxylate salt of the *exo* acid β-lactones are obtained during the bromolactonization but none have been detected during iodolactonization[98].

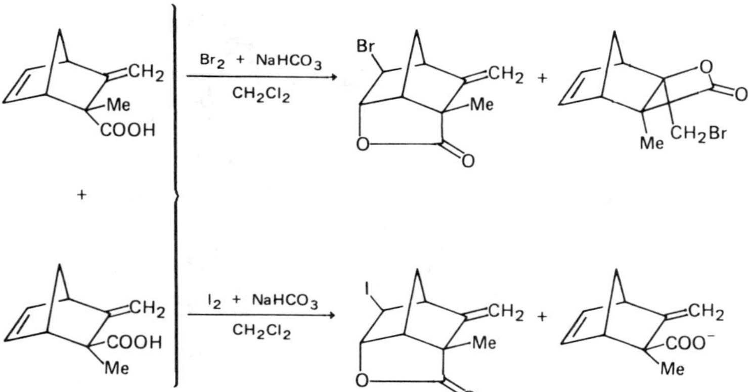

(48)

In order to determine if β-lactone formation was only associated with rigid systems or if conformationally more flexible β,γ-unsaturated acids would also form β-lactones, open-chain β,γ-unsaturated carboxylate salts in aqueous solutions were treated[99] with carbon tetrachloride or methylene chloride solutions containing bromine and were observed to readily cyclize to γ-bromo-β-lactones (equation 49).

In 1972, Barnett and Sohn[100] explained the seeming anomaly that iodolactonization of β,γ-unsaturated carboxylate salts affords β-iodo-γ-lactones, whereas bromolactonization of the same salts affords γ-bromo-β-lactones, and showed that the size of the lactone ring obtained did not depend upon the kind of halogen used, but rather the differences in the experimental procedures used. In iodolactonization conducted under conditions similar to bromolactonization the products obtained

$$H_2C{=}C(R^3)\text{—}C(R^1)(R^2)\text{—}COO^- \, Na^+ \xrightarrow[CCl_4 \text{ or } CH_2Cl_2]{Br_2} \text{β-lactone } (R^3, CH_2Br; R^1(R^2), R^2(R^1)) \tag{49}$$

R^1	R^2	R^3	Yield (%)
Me	Me	H	83^a
COOEt	Me	Me	88^b
H	H	H	50

[a]Pure product.
[b]Mixture of isomers.

are indeed γ-iodo-β-lactones. Thus, if iodolactonization is performed using excess potassium iodide and long reaction times, β,γ-unsaturated acids produce γ-lactones, whereas if short reaction times are used in the absence of potassium iodide, γ-iodo-β-lactones are produced (equation 50). One exception to this rule in stytylacetic acid, which is concerted[100] to the γ-lactone regardless of the procedure employed (equation 51).

$$H_2C{=}CH\text{—}CR_2COOH \xrightarrow{NaHCO_3} H_2C{=}CHCR_2CO^- \, Na^+ \tag{50a}$$

$$H_2C{=}CHCR_2CO^- \, Na^+ \; (R = Me, H) \xrightarrow[ether]{I_2} \text{I—}CH_2\text{-β-lactone} \quad R = Me, 70\text{–}80\%; \; R = H, 20\text{–}30\%$$
$$\xrightarrow{\text{excess } KI\text{—}I_2} \text{iodo-γ-lactone} \quad R = Me \tag{50b}$$

$$PhCH{=}CHCH_2COOH \longrightarrow \text{4-iodo-5-phenyl-γ-lactone} \tag{51}$$

Bromolactonization of Δ^4-cyclohexene-*cis*-1,2-dicarboxylic acid esters has also been reported (equations 52 and 53)[101], while application to this method to linear di- and tetra-carboxylic acid esters affords[102] substituted γ,γ-dilactones (equations 54 and 55).

An interesting application of a bromolactonization-type of reaction in the field of steriod synthesis[6,103] has been reported[104] in the preparation of 5β,14α-bufa-20(22)-enolide using *N*-bromosuccinimide in carbon tetrachloride (equation 56). Quinone-mediated dehydrogenation[105] of the product in refluxing dioxane con-

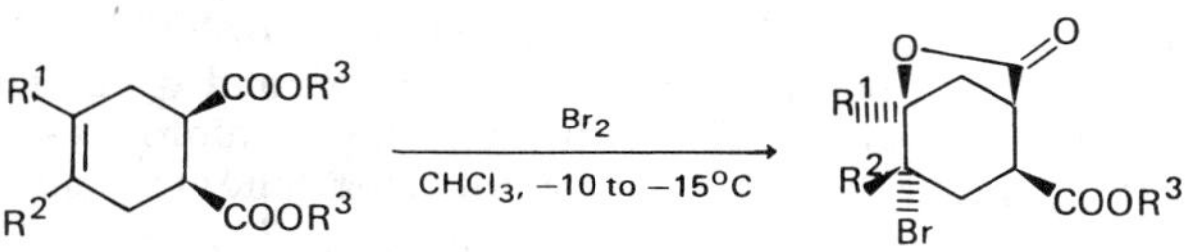

Br_2, $CHCl_3$, -10 to $-15\,^{\circ}C$ (52)

R^1	R^2	R^3	Yield (%)
Me	H	Me	36
Me	Me	Me	72
Me	Me	Et	75

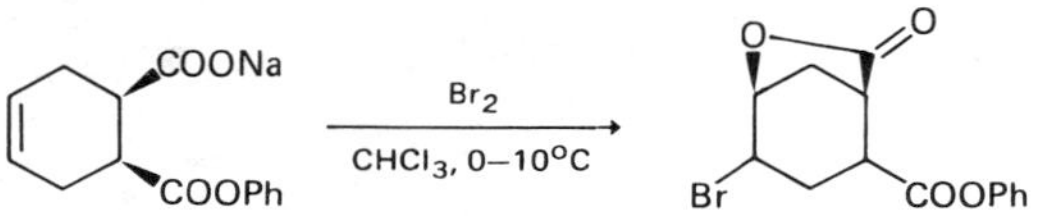

Br_2, $CHCl_3$, 0–10 °C (53)

39%

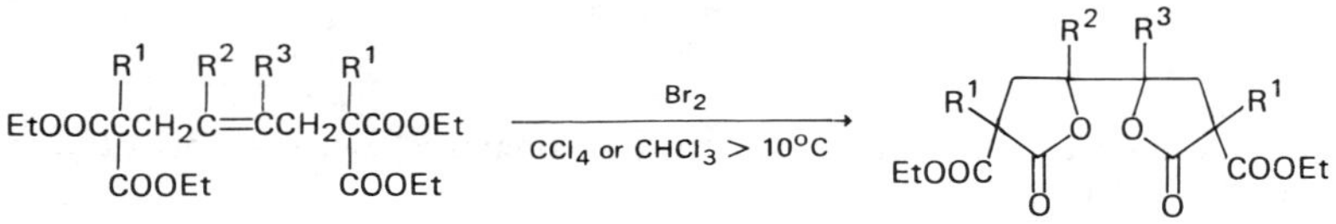

Br_2, CCl_4 or $CHCl_3 > 10\,^{\circ}C$ (54)

R^1 = C_{1-5}, alkyl or benzyl

R^2 = H, H, Me

R^3 = H, Me, Cl

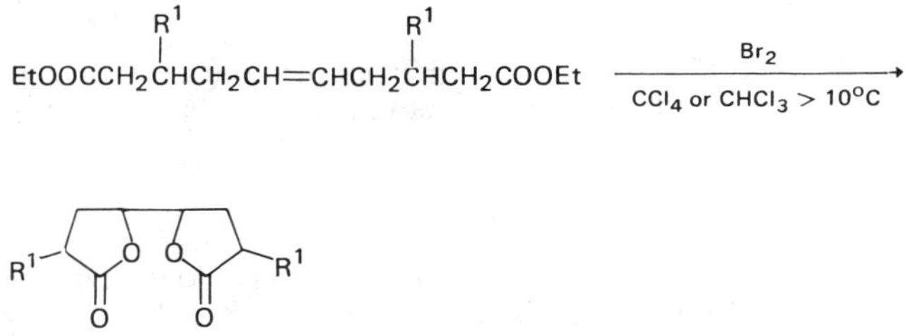

$EtOOCCH_2CH(R^1)CH_2CH{=}CHCH_2CH(R^1)CH_2COOEt$ — Br_2, CCl_4 or $CHCl_3 > 10\,^{\circ}C$ (55)

R^1 = C_{1-5}, alkyl or benzyl

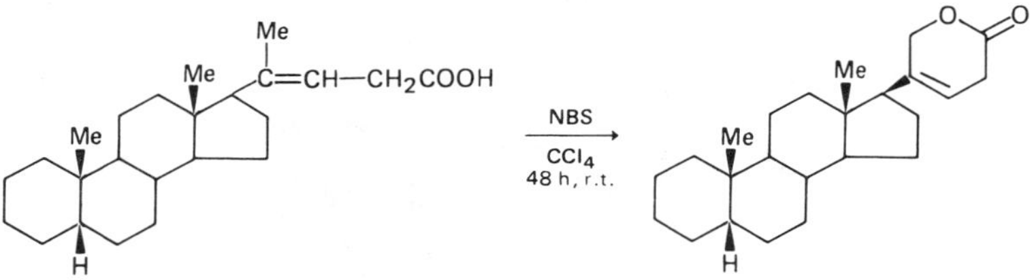

NBS, CCl_4, 48 h, r.t. (56)

57%

taining anhydrous hydrogen chloride or pyridine affords[104,106] the isomeric bufa-17(20),22-dienolides (equation 57). Similar dehydrogenation results are obtained[104,106] using chloranil. However, with 2,3-di-chloro-5,6-dicyanoquinone (DDQ) in refluxing dioxane containing *p*-toluenesulphonic acid, the dehydrogenation is specific at $C_{(21)}$ and $C_{(23)}$ to produce 5β,14α-bufa-20,22-dienolide in quantitative yield.

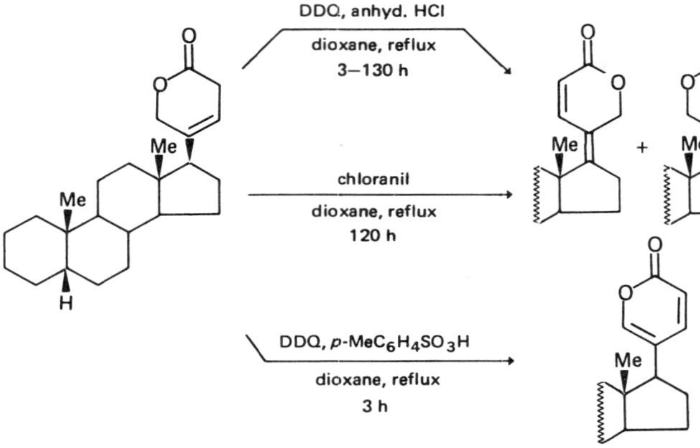

DDQ, anhyd. HCl; dioxane, reflux; 3–130 h

chloranil; dioxane, reflux; 120 h

DDQ, p-$MeC_6H_4SO_3H$; dioxane, reflux; 3 h

(57)

Bromolactonization has also been used to prepare[107] precursors of gibberellic acid (equation 58).

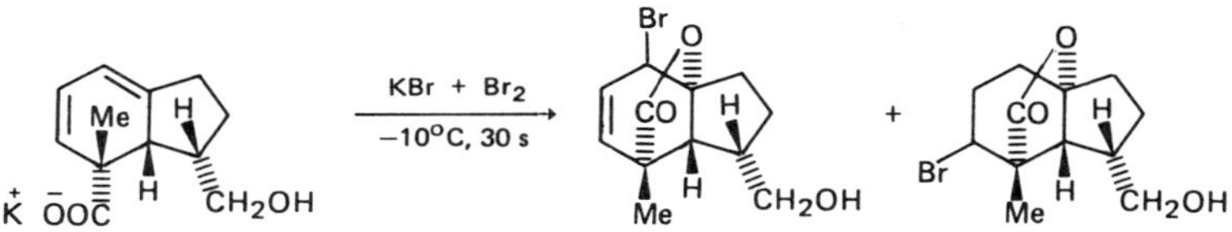

$KBr + Br_2$; $-10^{\circ}C$, 30 s

(58)

Arnold and Lindsay[108] have shown that iodolactones can be obtained by the use of cyanogen iodide in place of iodine–potassium iodide and bicarbonate.

4. *Intramolecular Diels–Alder reactions*

Intramolecular cycloaddition reactions of the Diels–Alder type have been employed in a number of interesting lactone syntheses[109,110]. These reactions may be generalized by viewing them as addition of the dienophilic triple bond of an acetylenic acid ester to a diene function contained in the alkoxy moiety of the ester (equation 59). A number of representative examples of such reactions are given in

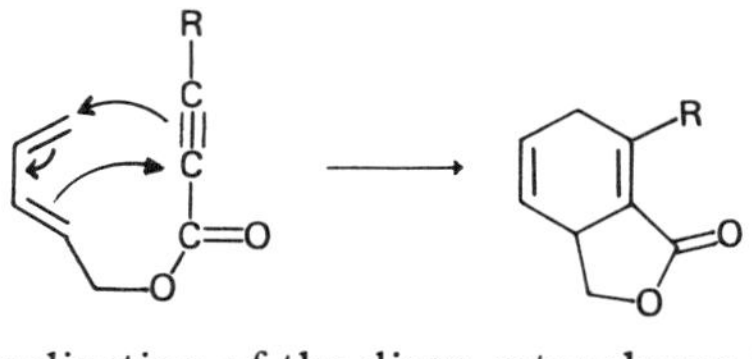

(59)

Table 7. Diels–Alder cyclization of the diene ester shown in equation (60) has been observed to occur thermally[115] via a [1,5] sigmatropic hydrogen shift in which the

TABLE 7. Preparation of lactones by intramolecular Diels–Alder cycloadditions

Acetylenic acid	Reaction conditions	Product	Yield (%)	Reference
2-Thienyl–CH=CHCH$_2$OC(=O)C≡CPh *trans*	$(MeCO)_2O$, reflux, 6 h	[structure: Ph] + [structure: Ph]	24 10	111
PhCH=CR^1–CR$^2_2$–O–C(=O)–C≡CPh	$(MeCO)_2O$, reflux, 6 h	[structure: R^1, R^2, Ph] R^1 = R^2 = H R^1 = H; R^2 = D R^1 = R^2 = D	 46 28 30	112
PhC≡CCH$_2$–O–C(=O)–C≡C–Ph	$(MeCO)_2O$, reflux, 5 h	[structure: Ph]	39	112

TABLE 7. (Continued)

Acetylenic acid	Reaction conditions	Product	Yield (%)	Reference
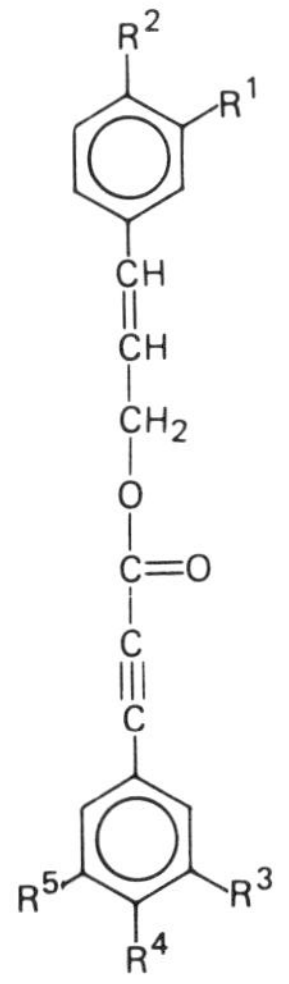 trans	$(MeCO)_2O$, reflux 6 h or 240°C (0.3 mm)	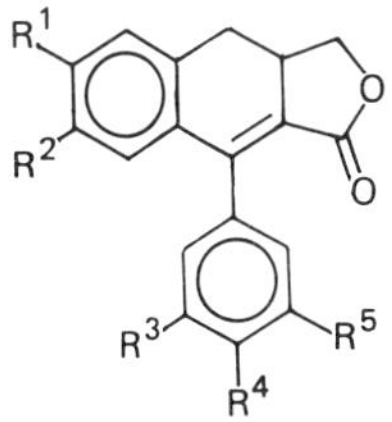 (A) 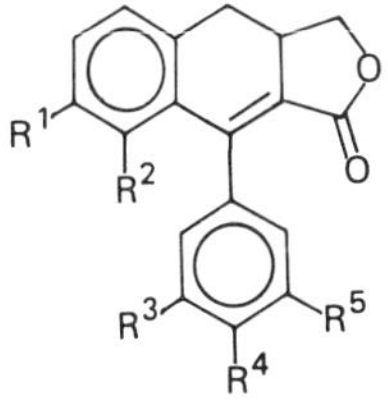 + (B)		113, 114
		$R^1, R^2 = -OCH_2O-$; $R^3 = R^4 = R^5 = OMe$	45(A) + 10(B)	
		$R^1, R^2 = -OCH_2O-$; $R^3 = R^4 = OMe$; $R^5 = H$	43(A + B)	
		$R^1 = R^2 = R^3 = R^4 = OMe$; $R^5 = H$	32	
		$R^1 = R^2 = OMe$; $R^3 = R^4 = R^5 = H$	22	
		$R^1 = R^2 = R^5 = H$; $R^3 = R^4 = OMe$	43	

Reactant	Conditions	Product	Yield (%)	Ref.
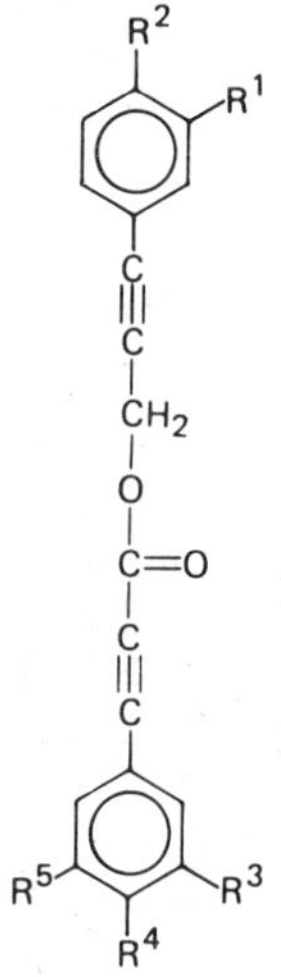	$(MeCO)_2O$, reflux 6 h	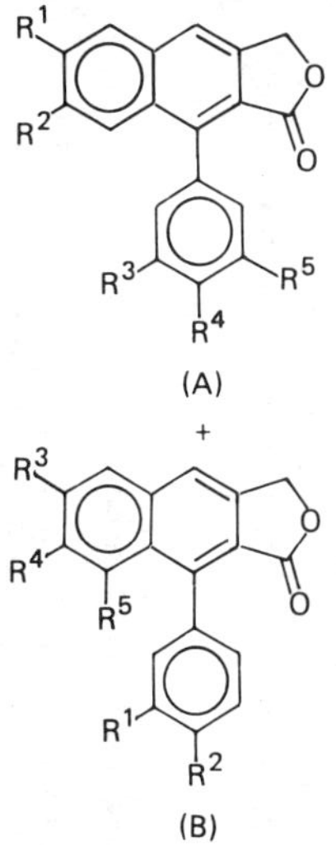 (A) + (B)		114
		$R^1 = R^2 = R^3 = R^4 = OMe$ $R^5 = H$	12(B)[a] + 4.5(A)[b]	
		$R^1, R^2 = -OCH_2O-$ $R^3 = R^4 = R^5 = OMe$	13(B) + 11(A)[c]	
		$R^1 = R^2 = R^5 = H$; $R^3 = R^4 = OMe$	70(A)[d]	
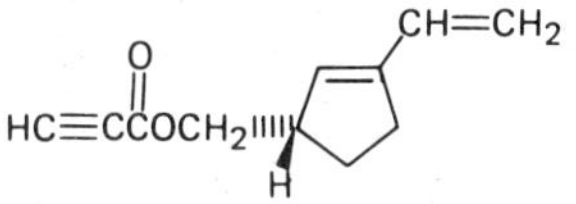 $HC{\equiv}CCOCH_2$ (vinylcyclopentenyl ester)	Heat >135°C or $(MeCO)_2O$, reflux, 20 h	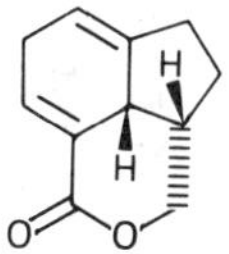	70	107

[a]Also prepared by dehydrogenation of dimethyl-α-conidendrin using *N*-bromosuccinimide.
[b]Also prepared by dehydrogenation of 1-(3,4-dimethoxyphenyl)-3-hydroxymethyl-6,7-dimethoxy-3,4-dihydro-2-naphthoic acid lactone using lead tetraacetate in glacial acetic acid at 80°C.
[c]Also prepared by sublimation of a mixture of podophyllotoxin and 30% Pd–C at 275°C (0.15 mm) for 8 h, followed by cyclization of the sublimate.
[d]Prepared only by hydrogenation of 1-(3,4-dimethoxyphenyl)-3-hydroxymethyl-3,4-dihydro-2-naphthoic acid lactone over 30% Pd–C in *p*-cymene for 44 h.

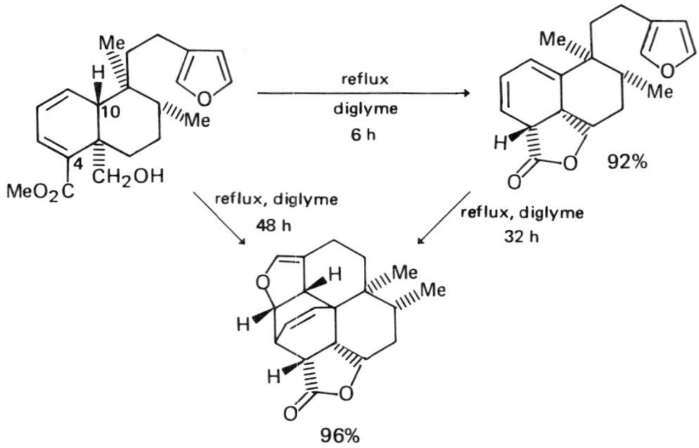

(60)

10β-hydrogen migrates suprafacially to the $C_{(4)}$ position to produce the intermediate lactone. This intermediate then undergoes a Diels–Alder reaction between the disubstituted double bond of the furan (dienophile) and the cyclohexadiene (diene).

C. By Acetoacetic Ester and Cyanoacetic Ester Condensations

Heating ethyl acetoacetate with a trace of sodium bicarbonate affords[116] a 53% yield of dehydroacetic acid (equation 61); however, if ethyl acetoacetate is treated

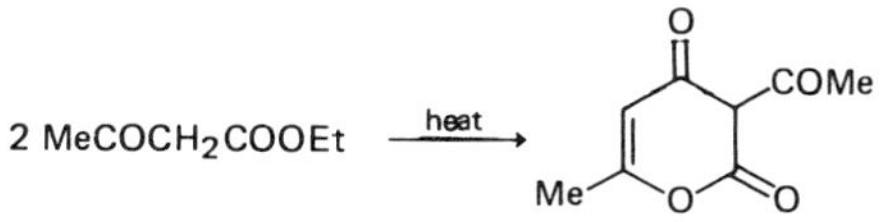

$$2\ MeCOCH_2COOEt \xrightarrow{heat} \tag{61}$$

with concentrated sulphuric acid at room temperature for 5–6 days a mixture of 22–27% of isodehydroacetic acid (4,6-dimethylcoumalic acid) and 27–36% of ethyl isodehydroacetate (ethyl 4,6-dimethylcoumaloate) is obtained (equation 62)[117].

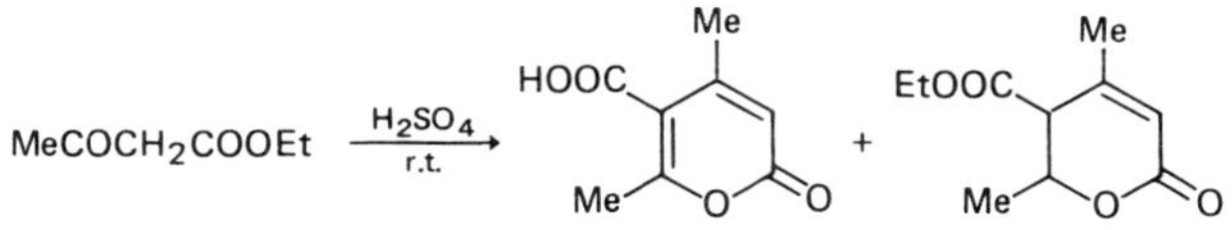

$$MeCOCH_2COOEt \xrightarrow[r.t.]{H_2SO_4} \quad + \tag{62}$$

Condensation of the monocarbanions of acetoacetic ester or ethyl cyanoacetate with a series of α-keto alcohols has been reported[118] to give the corresponding 2-acetyl- and 2-cyano-2-buten-4-olides in 55–96% yield (equations 63 and 64). Alcoholysis[118] of the 2-cyano-3,4,4-trimethyl analogue in the presence of sulphuric acid affords the corresponding ethyl ester (equation 65).

When ethyl acetoacetate is allowed to react[119] with 3-chloro-1,2-epoxypropane (epichlorohydrin) at 45–50°C for 18 hours in the presence of sodium ethoxide, a 61–64% yield of α-acetyl-δ-chloro-γ-valerolactone is obtained (equation 66). A similar report[120] involves the reaction of the carbanion generated from ethyl 2-furoylacetate and propylene oxide, which affords α-furoyl-γ-valerolactone in 54% yield (equation 67).

$MeCOCH_2COOEt + R^2C(R^1)(OH)COMe \xrightarrow[NaOEt]{EtOH}$ Me, COMe, R^1, R^2, O, O (63)

R^1 = Me, Me, $(CH_2)_5$
R^2 = Me, Et,

$NCCH_2COOEt + R^2C(R^1)(OH)COMe \xrightarrow[NaOEt]{EtOH}$ Me, CN, R^1, R^2, O, O (64)

Me, CN, Me, Me, O, O $\xrightarrow[H_2SO_4]{EtOH}$ Me, COOEt, Me, Me, O, O (65)

$MeCOCH_2COOEt + ClCH_2CHCH_2$ (O) $\xrightarrow[2.\,HOAc]{1.\,NaOEt}$ COMe, ClH_2C, O, O (66)

$COCH_2COOEt + H_2C{-}CH{-}Me$ (O) $\xrightarrow[C_6H_6]{Na,\ EtOH}$ O, O, Me, C, O, O, O (67)

$Me(CH_2)_3$, H, C=O, S, S $+ \bar{C}H_2CO\bar{C}HCOOEt \longrightarrow$

O, $Me(CH_2)_3$, O, O, S, S **(10)** $\xrightarrow[2.\ Hg(II),\ H_2O]{1.\ (CH_3)_2SO_4}$ OMe, $Me(CH_2)_3$, O, O, S, S **(11)** $\xrightarrow[H_2O]{Hg(II)}$ (68)

OMe, $Me(CH_2)_3C$, O, O, O **(12)**

Reaction of the dianion of acetoacetic ester with 2,2-(propane-1,3-dithio)-hexanal affords[121] the oxolactone **10**, which in turn gives the enol ether **11**. Hydrolysis of the thioacetal provides (±)-didehydropestalotin (**12**).

The condensation of phenols with β-keto esters, β-keto acids or malic acid in the presence of concentrated sulphuric acid affords coumarins, and is known as the von Pechmann reaction[122]. A series of representative preparations of coumarins[125–131] by the von Pechmann reaction are given in Table 8. It may be noted that treatment of malic acid with fuming sulphuric acid in the absence of a phenol affords coumalic acid[123,124].

$$\mathrm{HOOCCH(OH)CH_2COOH} \xrightarrow[SO_3]{H_2SO_4} \text{[2-pyrone-5-COOH (coumalic acid)]} \quad 65\text{–}70\% \qquad (69)$$

D. By Aldol Condensations

Base-catalysed aldol condensations of substituted malonic and acetoacetic esters with paraformaldehyde afford good yields of substituted γ-butyrolactones (equations 70–72)[132].

$$\mathrm{(EtOOC)_2CH\underset{\displaystyle COOEt}{C}{=}CH_2} + \mathrm{HOCH_2(OCH_2)_{\mathit{n}}OCH_2OH} \xrightarrow[EtOH]{EtOAc,\ KOH}$$

$$\text{[3,3-(EtOOC)}_2\text{-4-methylene-γ-butyrolactone]} \xrightarrow[2\ days]{HCl,\ 40^{o}C} \text{[3-HOOC-4-methylene-γ-butyrolactone]} \qquad (70)$$

$$\mathrm{MeCO\underset{\displaystyle}{\overset{\displaystyle COOEt}{C}}HCH_2COOEt} + \mathrm{HOCH_2(OCH_2)_{\mathit{n}}OCH_2OH} \xrightarrow{MeONa}$$

$$\text{[EtOOC, MeCO-substituted γ-butyrolactone]} \xrightarrow[55\ h]{HCl,\ 55^{o}C} \text{[MeCO-substituted γ-butyrolactone]} \qquad (71)$$

$$\mathrm{MeCO\overset{\displaystyle COOEt}{C}HCHMeCOOEt} + \mathrm{HOCH_2(OCH_2)_{\mathit{n}}OCH_2OH} \xrightarrow{MeONa}$$

$$\text{[EtOOC, MeCO, Me-substituted γ-butyrolactone]} \xrightarrow[55\ h]{HCl,\ 55^{o}C} \text{[MeCO, Me-substituted γ-butyrolactone]} \qquad (72)$$

Although it has been found that steroidal 17-β-hydroxy-16-β-acetic acids[133,134] and 17-β-hydroxy-16-β-propionic acids are easily converted into their respective *cis*-fused γ- and δ-lactones by simple intramolecular acid-catalysed condensation, the formation of the *trans*-fused δ-lactones from 17-β-hydroxy-16-α-propionic acids requires a more complex approach[135,136]. The procedure involves base-catalysed

TABLE 8. Synthesis of coumarins by the von Pechmann reaction

Phenol	β-Keto ester (acid)	Product	Yield (%)	Reference
Phenol	Ethyl acetoacetate	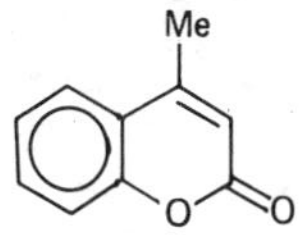	40–55	125
Resorcinol	Ethyl acetoacetate	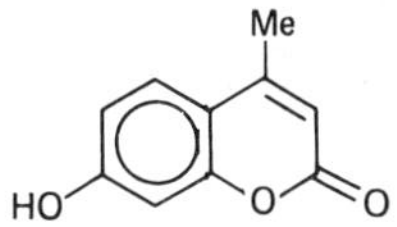	82–90	126
Hydroxyhydroquinone triacetate	Ethyl acetoacetate	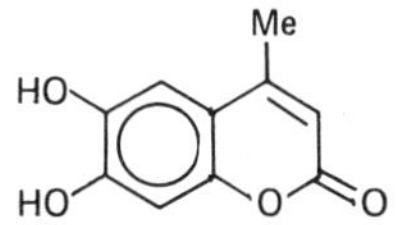	92	127

aldol condensation of 3-β-hydroxy-5-α-androstan-17-one with glyoxylic acid to afford 3-β-hydroxy-17-oxo-5-α-androstan-$\Delta^{16,\alpha}$-acetic acid, the key intermediate in the synthesis. Several additional steps convert this compound into 3-β,17-β-dihydroxy-5-α-androstane-16-α-propionic acid, which upon warming in a solution of acetic anhydride and acetic acid[137,138] afford 3-β-acetoxy-17-β-hydroxy-5-α-androstane-16-α-propionic acid δ-lactone (equation 73). Condensation of 5-α-androstanolone with glyoxylic acid in aqueous methanolic sodium

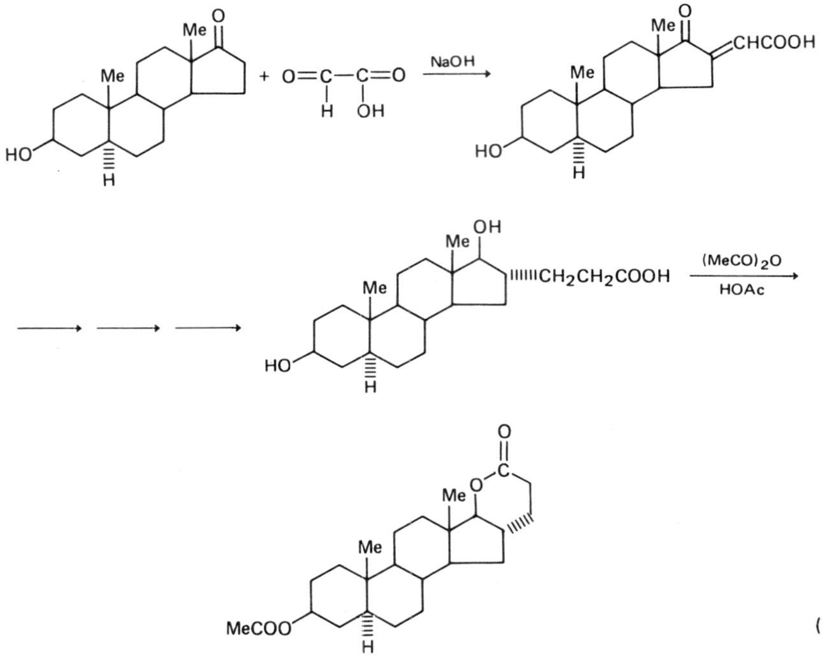

NaOH

(MeCO)$_2$O, HOAc

(73)

NaOH, MeOH, H_2O

MeOH, HCl

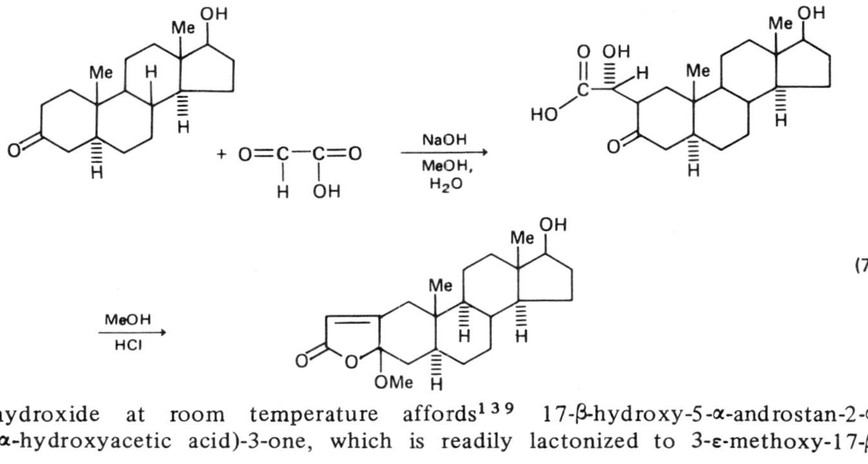

(74)

hydroxide at room temperature affords[139] 17-β-hydroxy-5-α-androstan-2-α-(α-hydroxyacetic acid)-3-one, which is readily lactonized to 3-ε-methoxy-17-β-hydroxy-5-α-androstan-2-α-(α-hydroxyacetic acid)-3-one-lactol upon treatment with methanolic hydrogen chloride (equation 74). Similarly, 17-α-hydroxy-3-oxo-5-α-androstan-$\Delta^{2,\alpha}$-acetic acid was prepared in 85% yield[139] via condensation of glyoxylic acid with 5-α-androstanolone. Several additional steps converted this product into 3-β,17-β-dihydroxy-5-α-androstan-2-β-acetic acid, which was lactonized upon refluxing with *p*-toluenesulphonic acid (equation 75).

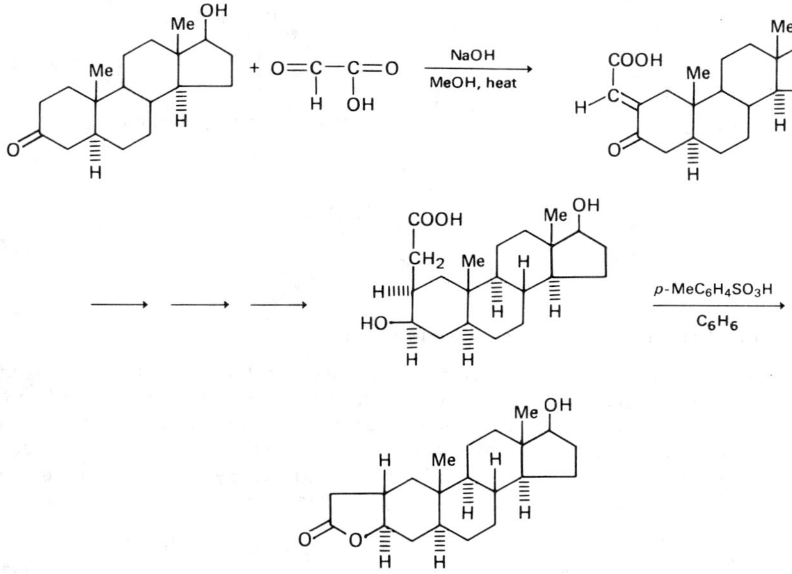

NaOH, MeOH, heat

p-$MeC_6H_4SO_3H$, C_6H_6

(75)

This approach to the preparation of key intermediates in the syntheses of isocardenolides[140], cardenolides[140–142], isobufadienolides[140] and bufadienolides[140] has been investigated.

Intermediates in the total synthesis of fomannosin (**13**), a biologically active metabolite from *Fommes annosus*, have also been prepared[143] via the intramolecular aldol condensation of **14** to form **15**. This product, which contains the formannosane skeleton, appears to be a promising intermediate in the total synthesis of fomannosin.

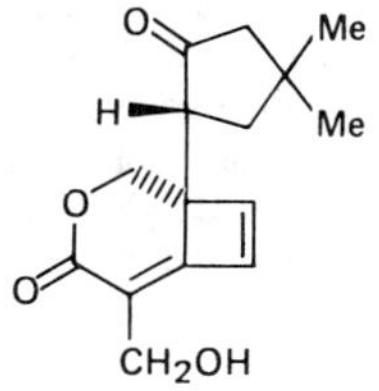

(13)

An interesting example of the use of an intramolecular aldol condensation for construction of α,β-unsaturated butyrolactones may be found in the mercuric sulphate-catalysed hydration of the acetylenic ester of acetoacetic acid (equation 77)[118].

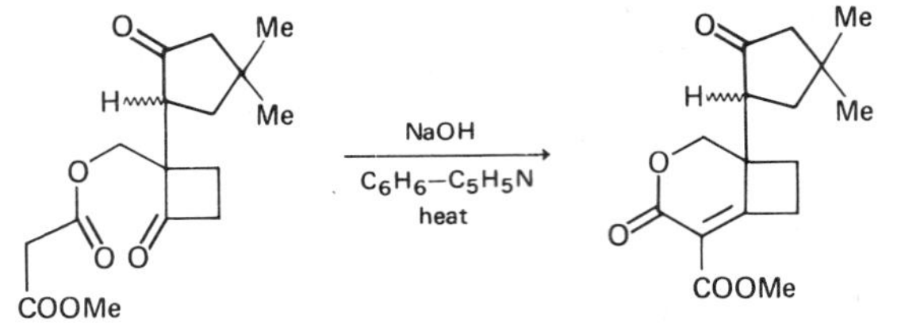

$$\text{(14)} \xrightarrow[\text{heat}]{\text{NaOH, } C_6H_6\text{–}C_5H_5N} \text{(15)} \qquad (76)$$

$$MeCOCH_2COOCMe_2C{\equiv}CH \xrightarrow[HgSO_4]{EtOH,\ H_2O} [MeCOCH_2COOCMe_2COCH_3]$$

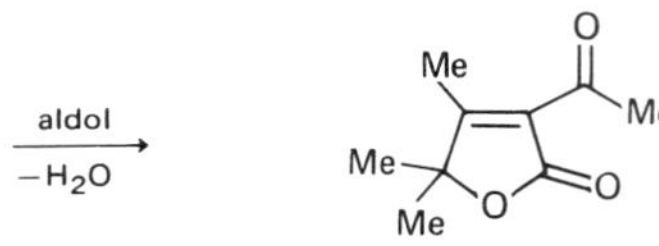

$$\xrightarrow[-H_2O]{\text{aldol}} \qquad (77)$$

E. By Malonic Ester or Malonic Acid Condensation

The condensation of malonic acid or diethyl malonate with *o*-hydroxybenzaldehydes or β-alkoxy-α,β-unsaturated aldehydes in piperidine has proved to be a very convenient route to 5,6-fused and 6-substituted-2-pyrones. Using this approach, salicylaldehyde and ethyl malonate were condensed in piperidine–glacial acetic acid solutions to yield[144] a 78–83% conversion to 3-carboethoxycoumarin (ethyl 2-oxo-2*H*-1-benzopyran-3-carboxylate) (equation 78). The scope of this

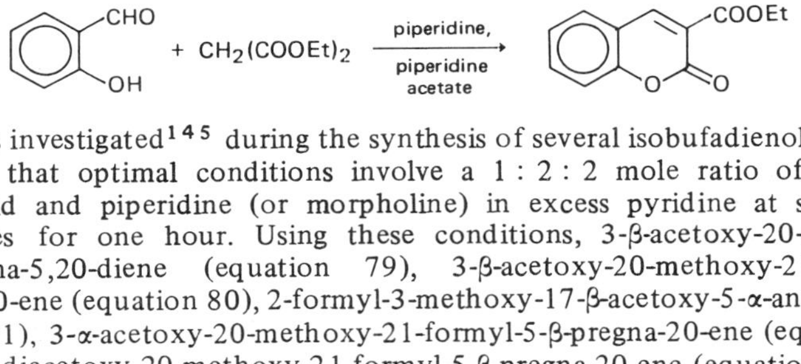

$$+ CH_2(COOEt)_2 \xrightarrow[\text{piperidine acetate}]{\text{piperidine,}} \qquad (78)$$

method was investigated[145] during the synthesis of several isobufadienolides, and it was found that optimal conditions involve a 1 : 2 : 2 mole ratio of aldehyde, malonic acid and piperidine (or morpholine) in excess pyridine at steam bath temperatures for one hour. Using these conditions, 3-β-acetoxy-20-ethoxy-21-formylpregna-5,20-diene (equation 79), 3-β-acetoxy-20-methoxy-21-formyl-5-α-androst-20-ene (equation 80), 2-formyl-3-methoxy-17-β-acetoxy-5-α-androst-2-ene (equation 81), 3-α-acetoxy-20-methoxy-21-formyl-5-β-pregna-20-ene (equation 82) and 3-α,6-α-diacetoxy-20-methoxy-21-formyl-5-β-pregna-20-ene (equation 83) were converted into 3-β-acetoxy-17-β-(6′α-pyronyl)-androst-5-ene (54%), 3-β-acetoxy-17-β-(6′α-pyronyl-5-α-androstane (54%), 17-β-acetoxy-5-α-androstano-[2,3-*c*]-2-pyrone (20%), 3-α-acetoxy-17-β-(6′α-pyronyl)-5-β-androstane (21%) and 3-α,6-α-di-

$$\longrightarrow \qquad (79)$$

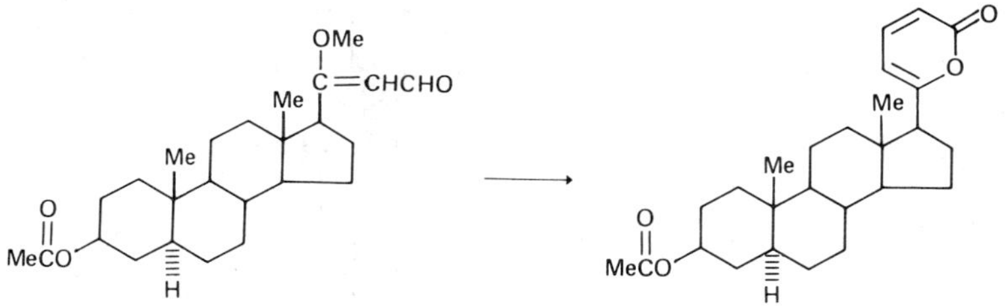

(80)

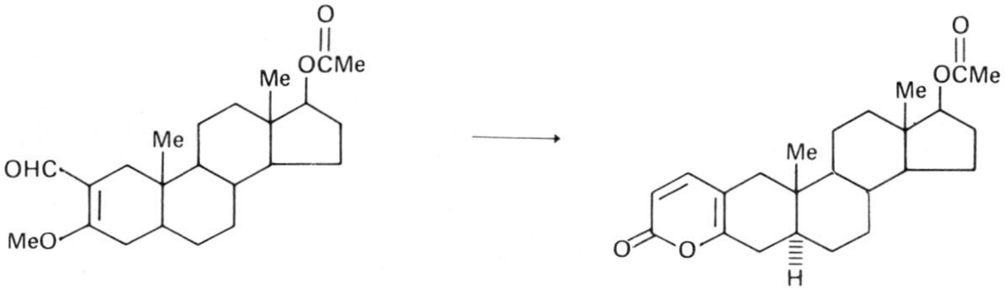

(81)

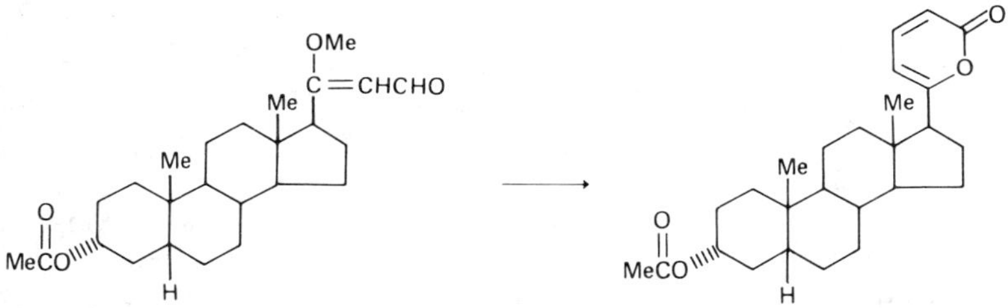

(82)

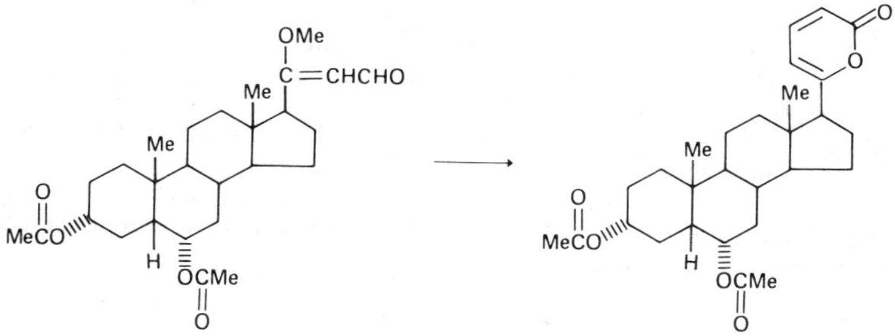

(83)

acetoxy-17-β-(6′α-pyronyl)-5-β-androstane (57%), respectively. The mechanistic pathway proposed[145] for these conversions is shown in equation (84).

Condensation of tertiary α-hydroxy ketones (acyloins) with malonic ester (equation 85) affords unsaturated γ-lactones in good yields (Table 9). The proposal that this reaction occurs via initial transesterification, with subsequent intramolecular condensation of the resulting keto ester, was confirmed by several observations. For example, when weaker bases such as pyridine and triethylamine were used as catalysts it was possible to isolate the intermediate keto esters, which were converted into the unsaturated γ-lactones upon treatment with sodium ethoxide (equation 86).

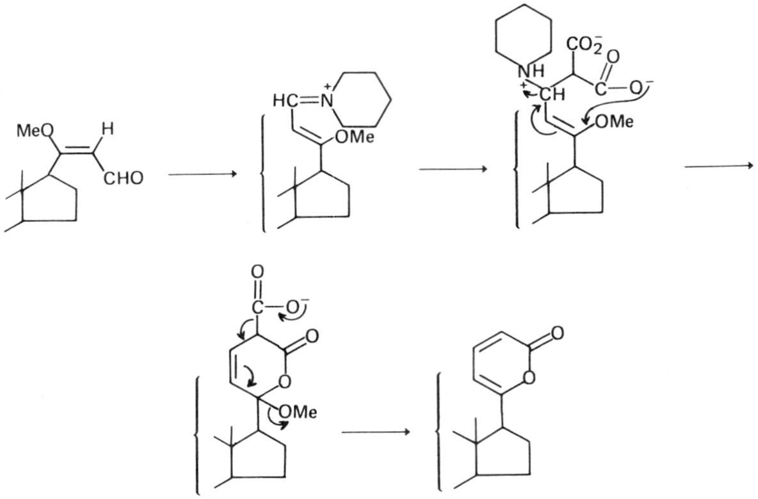

$$\text{(84)}$$

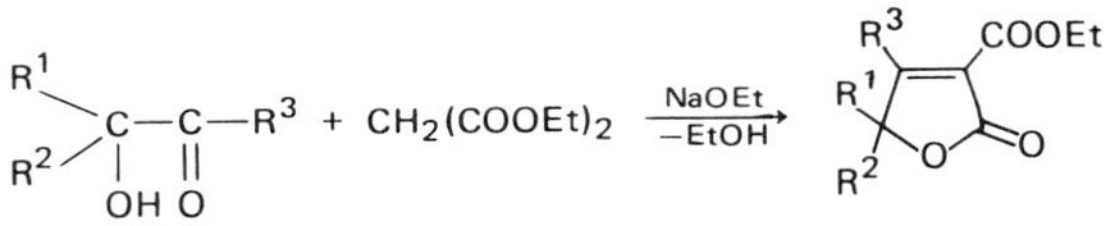

$$R^1R^2C(OH){-}C(=O){-}R^3 + CH_2(COOEt)_2 \xrightarrow[-EtOH]{NaOEt} \text{(butenolide: } R^1, R^2, R^3, \text{COOEt)} \qquad (85)$$

$$R^1R^2C(OH){-}C(=O){-}Me + CH_2(COOEt)_2 \xrightarrow{-EtOH} Me{-}C(=O){-}CR^1R^2{-}O{-}C(=O){-}CH_2COOEt \xrightarrow{-H_2O} \text{(butenolide: } R^1, R^2, \text{Me, COOEt)} \qquad (86)$$

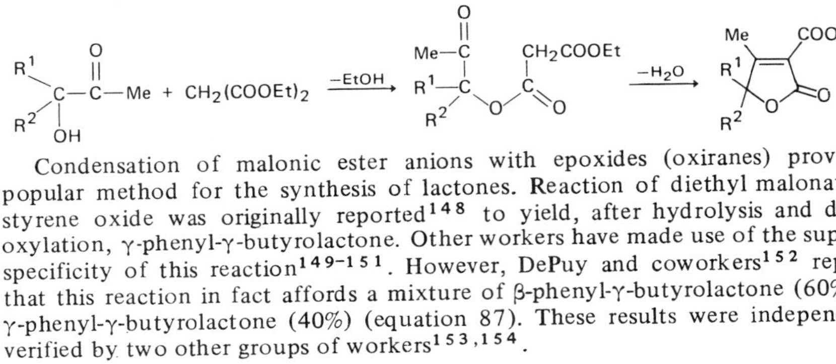

Condensation of malonic ester anions with epoxides (oxiranes) provides a popular method for the synthesis of lactones. Reaction of diethyl malonate and styrene oxide was originally reported[148] to yield, after hydrolysis and decarboxylation, γ-phenyl-γ-butyrolactone. Other workers have made use of the supposed specificity of this reaction[149–151]. However, DePuy and coworkers[152] reported that this reaction in fact affords a mixture of β-phenyl-γ-butyrolactone (60%) and γ-phenyl-γ-butyrolactone (40%) (equation 87). These results were independently verified by two other groups of workers[153,154].

TABLE 9. Condensations of acyloins with malonic ester to form unsaturated γ-lactones

R^1	R^2	R^3	Yield (%)	Reference
H	Et	Et	–	146
H	Pr	Pr	–	146
H	*n*-Bu	*n*-Bu	–	146
H	*n*-C_5H_{11}	*n*-C_5H_{11}	–	146
Me	Me	Me	65.5	147
Me	Et	Me	60	147
$-(CH_2)_5-$		Me	61	147

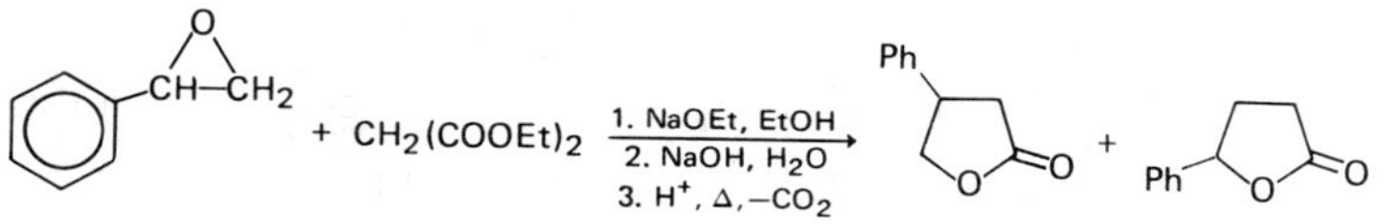

(87)

Van Tamelen and Bach[47] used the reaction of malonic ester anion with methyl tridecyl glycidate to prepare α,β-dicarbomethoxy-γ-tridecyl-γ-butyrolactone, an important intermediate in the synthesis of *d,l*-protolichesterinic acid (equation 88).

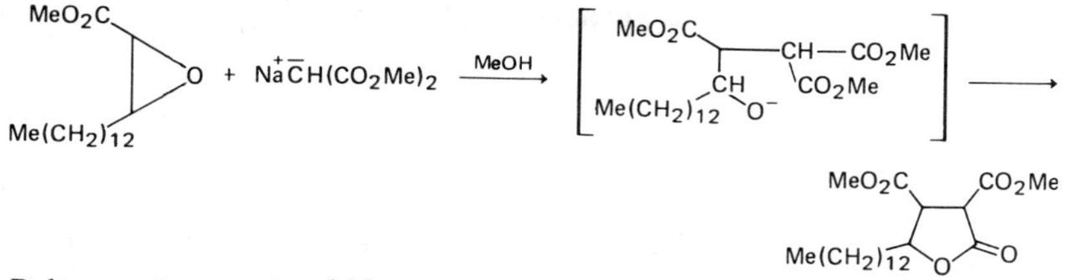

(88)

Dalton and coworkers[155] employed a similar approach to prepare fluorene-9-spiro-4′-(2′-carboxybutyrolactone) (**16**) and fluorene-9-spiro-3′-(2′-ethoxycarbonyl-butyrolactone) (**17**). Thus, condensation of sodium diethylmalonate with fluorene-9-spira-2′-oxiran afforded a 28% yield of **17** and a 20% yield of a diacid, which upon heating under vacuum afforded **16**. Both of these products were also decarboxylated to form spiro butyrolactones **18** and **19**. Similarly these workers[155]

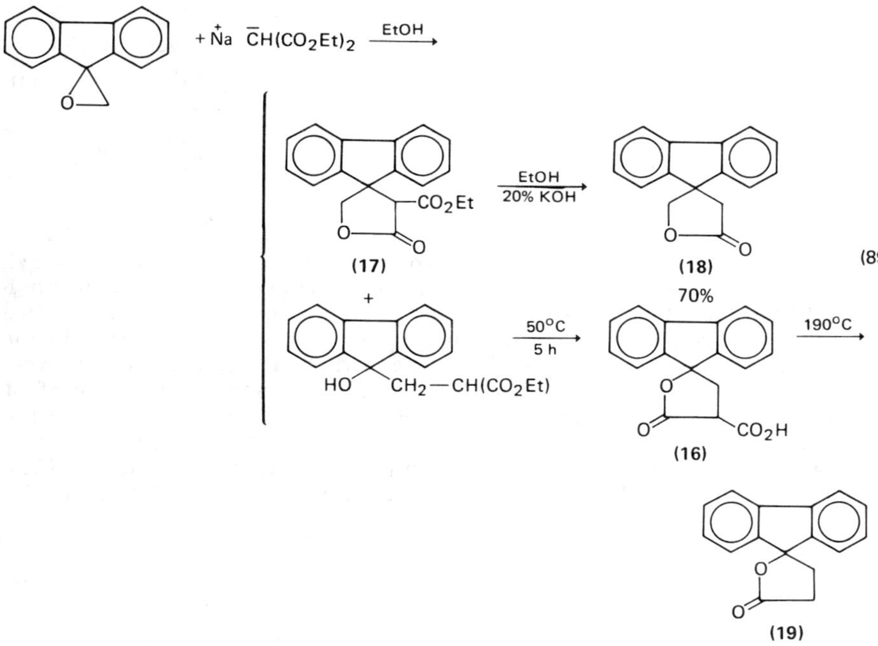

(89)

prepared 2-carboxy-4,4-diphenylbutyrolactone from 2,2-diphenyloxirane, while condensation of sodium diethylmalonate with 2-chloro-1,1-diphenylethanol afforded the same product (equation 90).

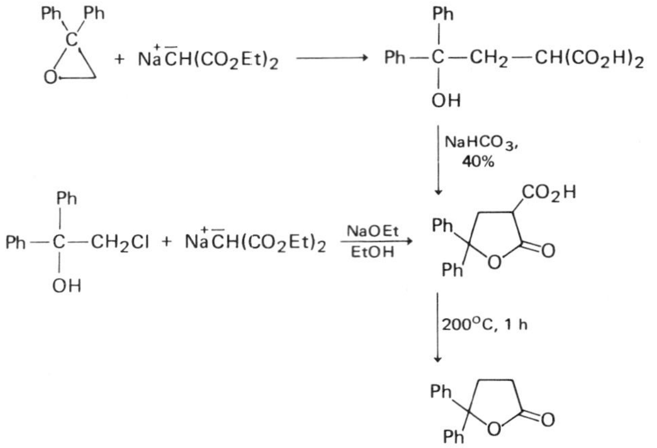

(90)

The condensation of 2-chloro-2-methylpropanal with malonic esters in the presence of potassium carbonate to produce γ-butyrolactones (equation 91) has also been studied[156]. At room temperature, in THF, using one equivalent each of

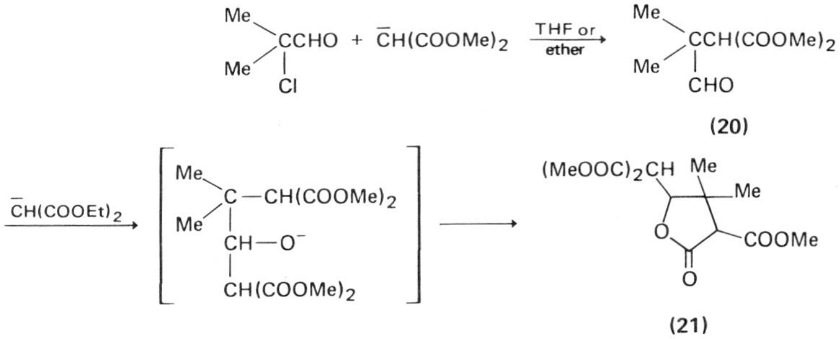

(91)

dimethyl malonate and 2-chloro-2-methylpropanal, two products, methyl-3-formyl-2-methoxycarbonyl-3-methylbutanoate (**20**) and α-methoxycarbonyl-β,β-dimethyl-γ-dimethoxycarbonylmethyl-γ-butyrolactone (**21**), are obtained in 60% and 26% yields, respectively. The mechanistic course of this reaction was established by the observations that in a separate experiment the methyl butanoate **20** and dimethyl malonate condensed to produce the γ-butyrolactone **21**, and that the yield of **21** was significantly increased when two equivalents of malonate in THF were used in the initial experiment. However, when **20** was treated with sodium methoxide, a new lactone, α-methoxycarbonyl-β,β-dimethyl-γ-methoxy-γ-butyrolactone (**22**), was obtained in 65% yield via intramolecular cyclization (equation 92). Similarly,

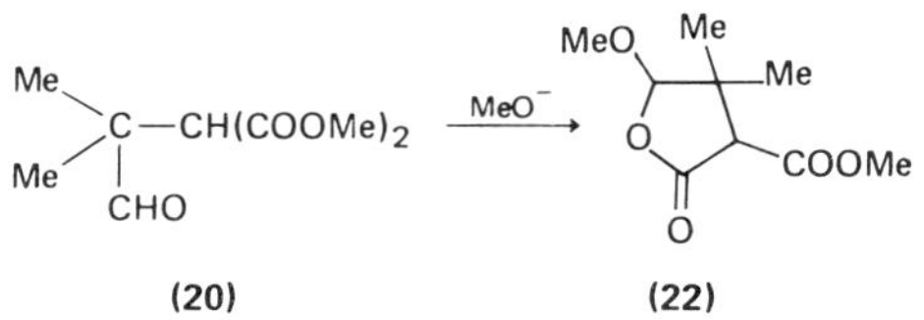

(92)

the reaction of 2-chloro-2-methylpropanal with dimethyl malonate in ether containing sodium methoxide (equation 93) also afforded lactone **22**, albeit in 20% yield.

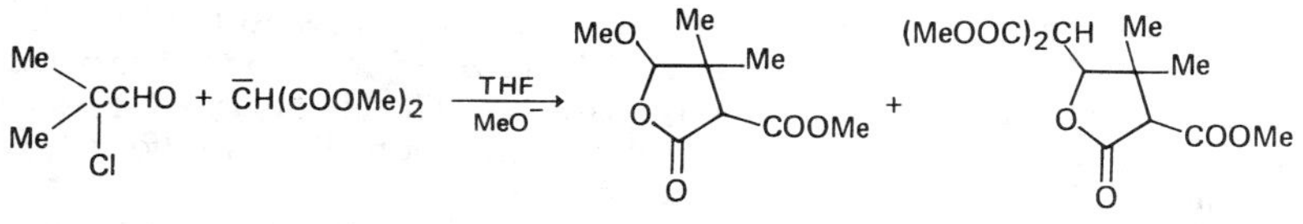

(93)

(22) (21)

The major product from this reaction was still **21** (46%). Ester cleavage of **21** to α-carboxy-β,β-dimethyl-γ-carboxymethyl-γ-butyrolactone[156] was effected in 97% yield upon heating with concentrated hydrochloric acid (equation 94) at 70–80° for 24 hrs. Heating this product at 180–200°C for 30 minuted afforded[156] a 98% conversion to β,β-dimethyl-γ-carboxymethyl-γ-butyrolactone.

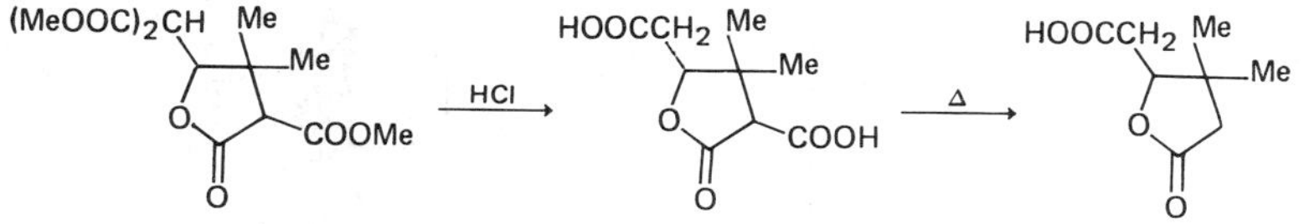

(94)

(21)

When 2-chloro-2-methylpropanal is condensed with the methyl or ethyl ester of malonic acid in *aqueous* potassium carbonate[156], α-alkoxycarbonyl-β-dialkoxycarbonylmethyl-γ,γ-dimethyl-γ-butyrolactones (**23**) are formed in 70–82% yield. This is explained by assuming an epoxide intermediate, which reacts further with malonate as shown in equation (95).

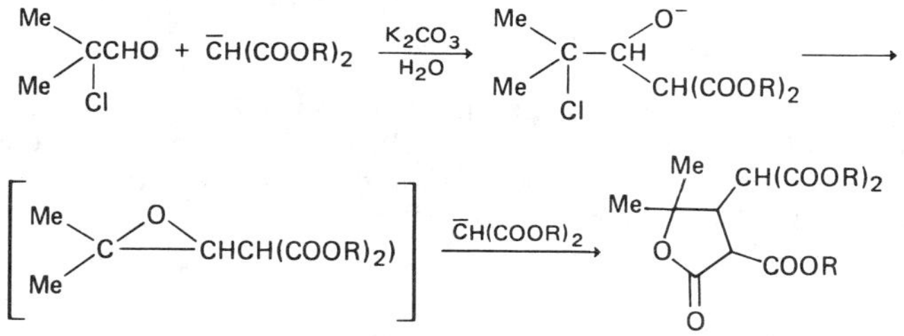

(95)

(23)

Hydrolysis of lactone **23** gave the expected diacid (98%) which upon heating afforded terpenylic acid (**24**) (equation 96). These results contradicted a previous

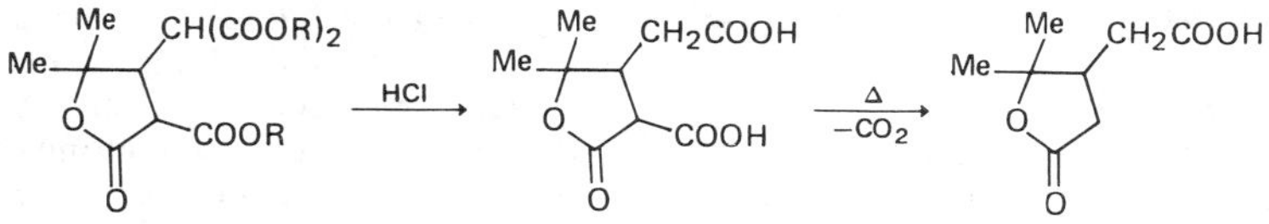

(96)

(23) (24)

report[157] that 2-bromo-2-methylpropanal reacted with diethyl sodiomalonate in ethanol to afford α-ethoxycarbonyl-γ,γ-dimethyl-$\Delta^{\alpha,\beta}$-γ-butenolide. Reinvestigation[156] showed that α-ethoxycarbonyl-β-diethoxycarbonylmethyl-γ,γ-dimethyl-γ-butyrolactone (**23**, R = Et) was indeed formed in 53% yield.

The explanation[156] advanced for the results discussed above maintains that in aprotic solvents such as THF, the carbanion of malonic esters becomes more nucleophilic than in protic solvents and thus attacks the α-carbon of the α-haloaldehyde forming a C—C bond via an S_N2 reaction. This is followed by an intramolecular cyclization to afford **21** and **22**. However, in protic solvents such as water, the carbanion attacks the carbonyl carbon, which is polarized by solvent molecules, forming a C—C bond by nucleophilic addition. This is followed by an intramolecular cyclization to afford the lactones **23**.

Malonic ester anion has also been used to obtain *trans*-fused γ-lactones. For example, reaction[158] of sodium diethylmalonate with 3,4-epoxy-1-cyclooctene, 5,6-epoxy-1-cyclooctene and 1,2: 5,6-diepoxycyclooctane, followed by hydrolysis, affords 10-oxo-9-oxabicyclo[6.3.0] undec-2-en-11-carboxylic acid (65%), 10-oxo-9-oxabicyclo[6.3.0] undec-4-en-11-carboxylic acid (70%) and 4,5-epoxy-10-oxo-9-oxabicyclo[6.3.0]-undecan-11-carboxylic acid (60%), respectively (equation 97).

1. $Na^+ \bar{C}H(COOEt)_2$
2. OH^-, H_2O

(97)

These acids were converted into their methyl esters by reaction with diazomethane and were decarboxylated to 9-oxobicyclo[6.3.0]undec-2-en-10-one (61%), 9-oxabicyclo[6.3.0]undec-4-en-10-one (90%) and endo-4,5-epoxy-9-oxabicyclo[6.3.0] under-10-one (62%) by heating at 160–180°C.

F. By Perkin and Stobbe Reactions

Although the Perkin reaction[159–162] is not widely used for the direct synthesis of lactones, several applications of this condensation have found some utility in lactone preparation.

A Perkin-type reaction of 2,6-dimethoxy-*p*-benzoquinone with propionic anhydride affords[163] the two products shown in equation (98) along with the propyl diester of 2,6-dimethoxy-*p*-hydroquinone. Using isobutyric acid anhydride, the isobutyl diester of 2,6-dimethoxy-*p*-hydroquinone and the fused lactone (exclusive structure not determined) are formed[163]. To establish the mechanistic course[164] the standard Perkin reaction procedure was modified by using shorter reaction times and lower temperatures. Under these conditions it was possible to isolate the β-lactones **25** and **26**, respectively. Transformation of **25** to the mixture of products initially obtained was easily accomplished by heating at 100°C for 48 hours in the presence of sodium propionate and propionic anhydride. β-Lactone **26** could similarly be transformed upon prolonged heating with isobutyric acid anhydride in the presence of sodium isobutyrate, but could not be so transformed upon treatment with acetic acid–sulphuric acid mixtures. Although these and other experiments

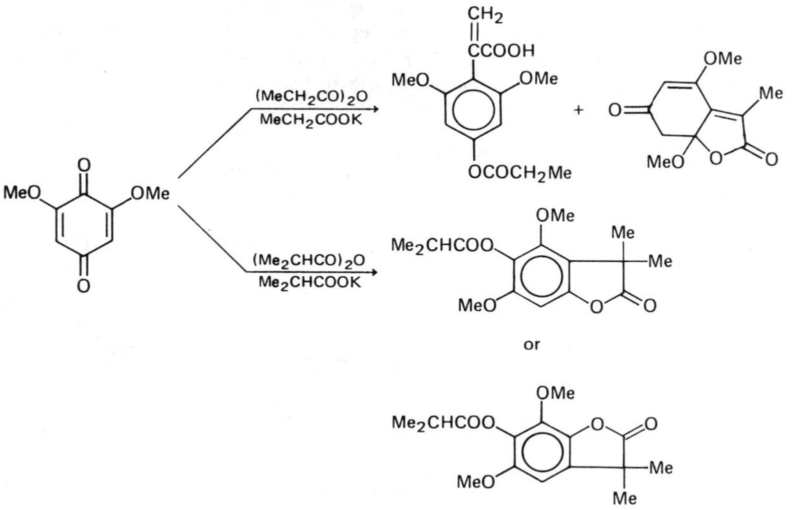

(98)

did not establish with certainty that β-lactones are intermediates in the formation of the observed γ-lactones, they did establish that β-lactones could be formed under Perkin-like reaction conditions.

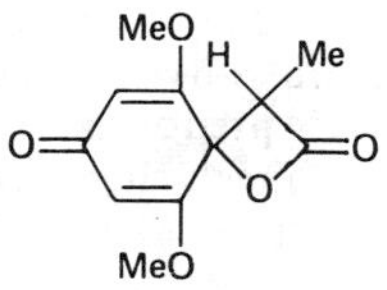

(25) 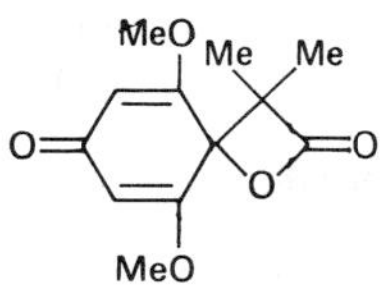 **(26)**

A variety of α-benzylidene-γ-phenyl-$\Delta^{\beta,\gamma}$-butenolides substituted in the aralkyl idene ring with either electron-withdrawing or electron-donating substituents have been prepared[165] by a Perkin-type condensation of 3-benzoylpropionic acid with substituted benzaldehydes in the presence of sodium acetate in acetic anhydride (equation 99).

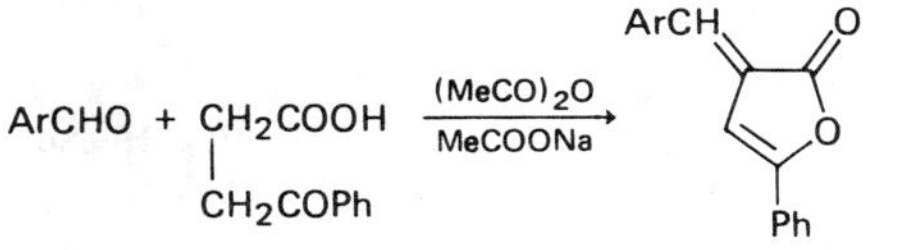

$$\mathrm{ArCHO} + \mathrm{CH_2COOH(CH_2COPh)} \xrightarrow[\mathrm{MeCOONa}]{(\mathrm{MeCO})_2\mathrm{O}} \quad (99)$$

The preparation[166–169] of aralkylidine- and subsequently arylmethylphthalides[168,169], originates with the condensation of phthalic anhydride with arylacetic acids (equation 100).

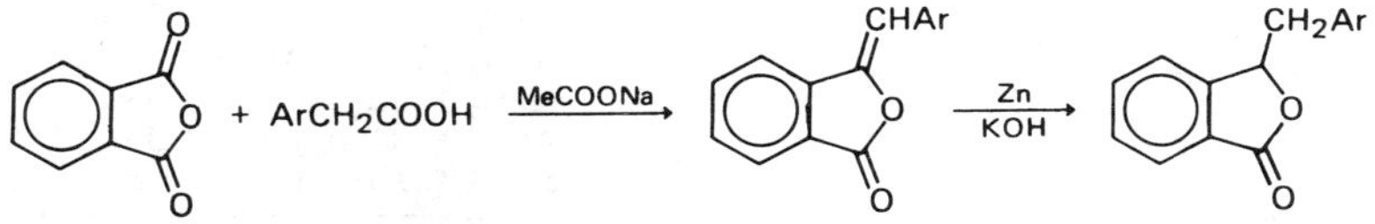

$$+ \ \mathrm{ArCH_2COOH} \xrightarrow{\mathrm{MeCOONa}} \xrightarrow[\mathrm{KOH}]{\mathrm{Zn}} \quad (100)$$

Various applications of the Stobbe condensation to the synthesis of lactones have been reviewed[170]. Recently, it has been reported[171] that β-carboethoxy-$\Delta^{\beta,\delta}$-δ-valerolactones can be prepared by an intramolecular Stobbe reaction preceded by condensation of tertiary α-keto alcohols with diethyl succinate (equation 101).

$$R^3—C(=O)—C(R^1)(R^2)—OH + (CH_2COOEt)_2 \xrightarrow{Me_3C\bar{O}\overset{+}{Na}} R^3—C(=O)—C(R^1)(R^2)—O—C(=O)—(CH_2)_2—COOEt \longrightarrow$$

(101)

R^1	R^2	R^3
Me	Me	Me
Me	Me	Me
—$(CH_2)_5$—		Me

G. By Grignard and Reformatsky Reactions

During a series of studies[172,173] involving the synthesis of steroids, a general synthesis of δ-lactones was developed. This method[174,175] consists of the reaction of Grignard reagents with glutaraldehyde to afford δ-hydroxyaldehydes in good yields (equation 102). These aldehydes, which exist predominately in cyclic hemiacetal (δ-lactol) form, were then oxidized to δ-lactones using a variety of reagents as shown in Table 10.

TABLE 10. Synthesis of δ-lactols and δ-lactones by reaction of Grignard reagents (RMgX) with glutaraldehyde[175]

RMgX	Yield of δ-lactol (%)	Oxidizing agent	Yield of δ-lactone (%)
$MeCH_2MgBr$	68.5	Ag_2O	50
$Me(CH_2)_3CH_2MgBr$	—	Ag_2O	41
$MeCH(OCMe_3)(CH_2)_2CH_2MgCl$	66	Br_2, HOAc	83
2-Me-1,3-dioxolan-2-yl-$(CH_2)_2CH_2MgCl$	52	Br_2, HOAc	77
2,4,5-trimethyl-1,3-dioxolan-2-yl-$(CH_2)_2CH_2MgCl$	64	Br_2, HOAc	88
		MnO_2, C_6H_6	45
2-Me-1,3-benzodioxol-2-yl-$(CH_2)_2CH_2MgCl$	78	Ag_2O	86
		$Na_2Cr_2O_7$, HOAc	60
		MnO_2, C_6H_6	35
		Ag_2CO_3, C_6H_5Me	33
		air, MeCOOEt, Pt	90

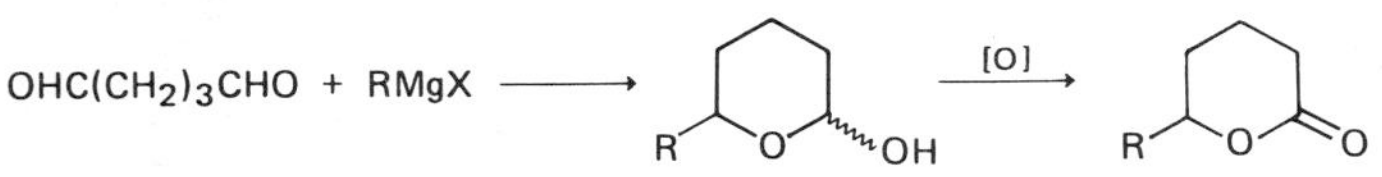

$$OHC(CH_2)_3CHO + RMgX \longrightarrow \text{(2-R-6-hydroxytetrahydropyran)} \xrightarrow{[O]} \text{(6-R-δ-valerolactone)} \quad (102)$$

Similarly, addition of methylmagnesium iodide to diethyl acetoglutarate[176] produces a racemic δ-lactone, ethyl terpenylate, which can be easily hydrolysed to terpenylic acid (equation 103).

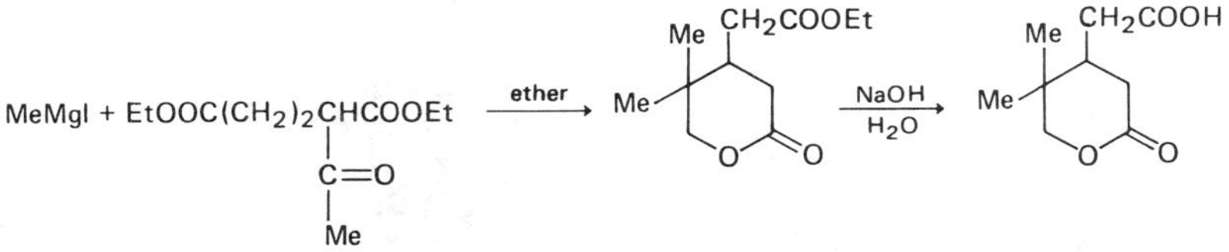

$$MeMgI + EtOOC(CH_2)_2CH(COMe)COOEt \xrightarrow{\text{ether}} \text{(ethyl terpenylate; } CH_2COOEt\text{)} \xrightarrow[H_2O]{NaOH} \text{(terpenylic acid; } CH_2COOH\text{)} \quad (103)$$

Contrary to previous reports[177,178], it has now been found[179] that addition of the Grignard or Reformatsky reagent formed from ethyl α-bromoisobutyrate to α,β-ethylenic ketones occurs via conjugated addition to produce a mixture of δ-keto

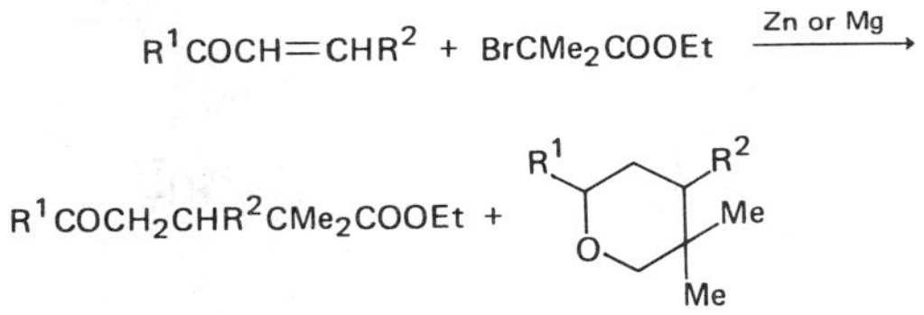

$$R^1COCH{=}CHR^2 + BrCMe_2COOEt \xrightarrow{\text{Zn or Mg}} R^1COCH_2CHR^2CMe_2COOEt + \text{(enol δ-lactone: } R^1, R^2, Me, Me\text{)} \quad (104)$$

R^1	R^2	Method
Et	Ph	Reformatsky
i-Pr	Ph	Reformatsky
Ph	Ph	Grignard
p-$MeOC_6H_4$	*p*-$MeOC_6H_4$	Grignard
Ph	Me	Grignard
n-Pr	Me	Grignard

esters and enolic δ-lactones (equation 104). In addition, several cyclic α,β-unsaturated ketones underwent reaction to afford the δ-lactones shown below:

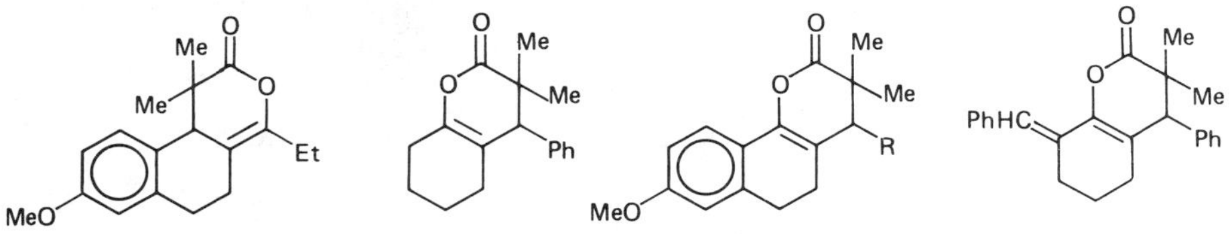

R = Et, Ph

Although these unsaturated ketones underwent smooth conjugate addition, the α,β-unsaturated methyl ketones, 1-acetyl-cyclohex-1-ene, methyl styryl ketone and 3-pentene-2-one, did not undergo conjugated addition with either the Grignard or Reformatsky reagent of ethyl-α-bromoisobutyrate[179].

Using the above approach, the reaction of 16-dehydropregnenolone acetate with

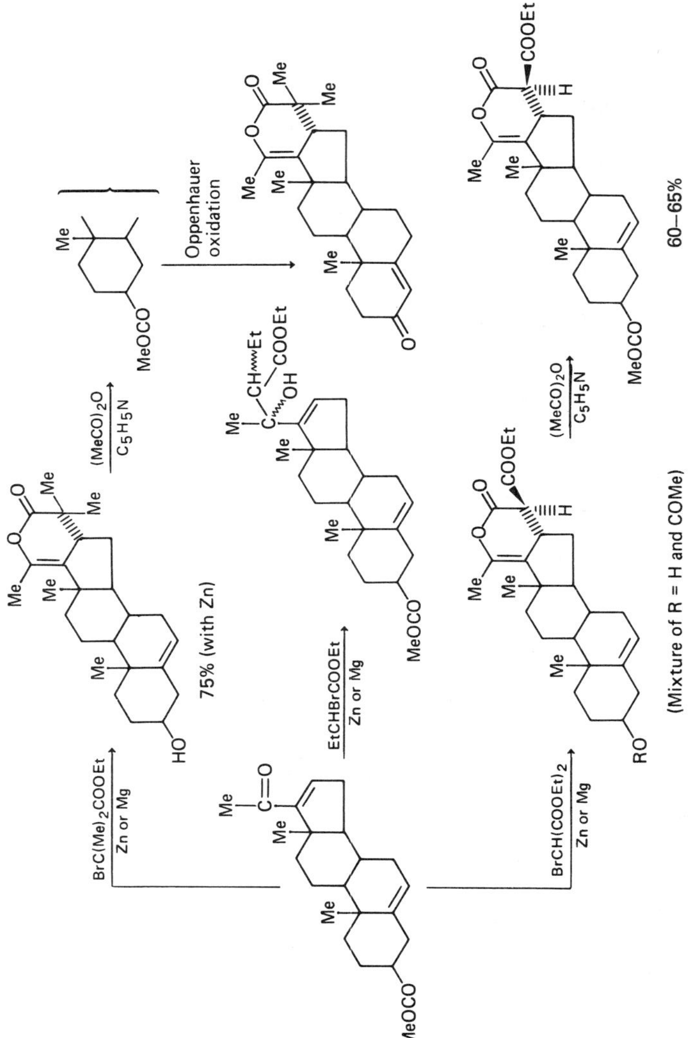
BrC(Me)$_2$COOEt
Zn or Mg
75% (with Zn)
(MeCO)$_2$O
C_5H_5N
Oppenhauer
oxidation
EtCHBrCOOEt
Zn or Mg
BrCH(COOEt)$_2$
Zn or Mg
(Mixture of R = H and COMe)
60–65%

SCHEME 2.

the Grignard or Reformatsky reagents obtained from the ethyl esters of α-bromoisobutyric, α-bromomalonic and α-bromobutyric acids was investigated[180]. It was found that, although the results depended largely upon the type of α-bromo ester used[181,182], the best yields were obtained with 1 : 6 molar ratio of steroid to Reformatsky reagent, with the Grignard reagent giving less reproducible results. A flow chart listing the reactants used and the products obtained is shown in Scheme 2.

The Reformatsky reagent prepared from diethyl α-methyl-α-bromomalonate has been added[183] to β-acetylenic alcohols to effect the synthesis of various δ-valerolactones (equation 105).

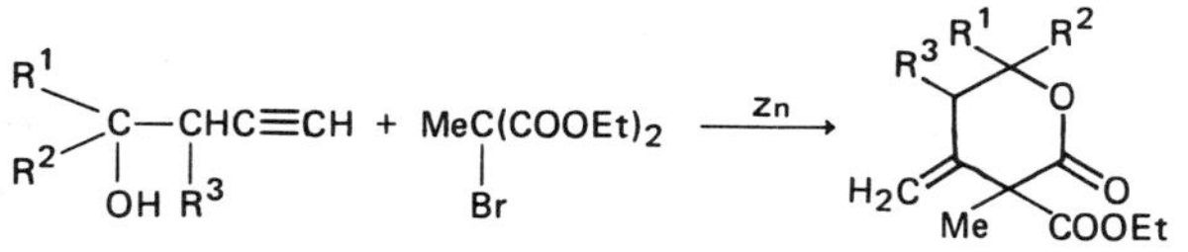

$$R^1R^2C(OH)-CH(R^3)C{\equiv}CH + MeC(COOEt)_2Br \xrightarrow{Zn} \text{product} \qquad (105)$$

R^1 = H, H, H, H, Me, Me, Me, Me

R^2 = H, Me, Ph, $CHMe_2$, H, Me, Ph, $CHMe_2$

R^3 = H, H, H, H, Me, Me, Me, Me

Addition of the organozinc reagents derived from α-(bromomethyl)acrylic esters to a variety of aldehydes and ketones in THF affords[184] a single-step synthesis of α-methylene γ-lactones (equation 106). This technique affords good yields (Table

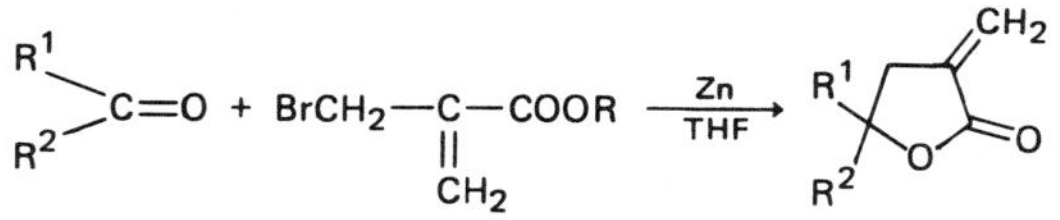

$$R^1R^2C{=}O + BrCH_2-C({=}CH_2)-COOR \xrightarrow[THF]{Zn} \text{product} \qquad (106)$$

11) except in the case of formaldehyde, where a mixture of α-methylenebutyrolactone and γ-hydroxy-α-methylenebutyric ester is formed in low yields. The analogous reaction of methyl β′-bromotiglate with ketones to produce α-methylene-β-methylbutyrolactones has also been reported (equation 107)[185]. This study also

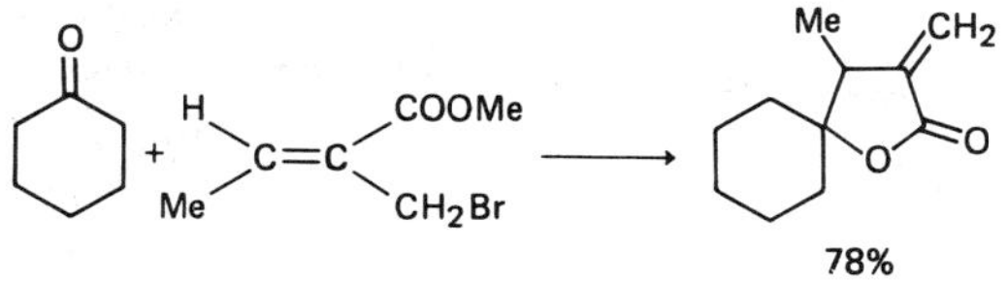

$$\text{cyclohexanone} + (H)(Me)C{=}C(COOMe)(CH_2Br) \longrightarrow \text{spiro lactone (78\%)} \qquad (107)$$

$$MeOOC-CH{=}C(COOMe)CH_2Br + C_{13}H_{27}CHO \xrightarrow{Zn}$$

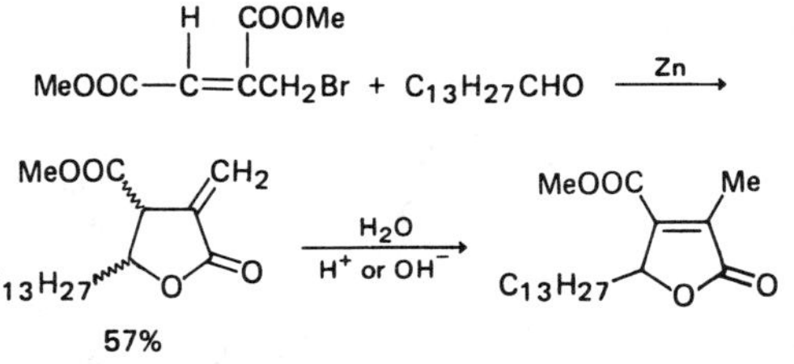

$$\text{lactone (57\%, } cis : trans = 1 : 1) \xrightarrow[H^+ \text{ or } OH^-]{H_2O} \text{butenolide} \qquad (108)$$

included the synthesis of both the *cis* and *trans* isomers of protolichesterinic ester (equation 108) and *trans*-protolichesterinic acid itself (equation 109) by reaction of myristic aldehyde with β′-bromomesaconic acid or β′-bromocitraconic anhydride in the presence of zinc.

TABLE 11. Synthesis of α-methylene γ-butyrolactones by reaction of aldehydes and ketones with the zinc reagent derived from α-(bromomethyl)acrylic esters[184]

R^1	R^2	Yield (%)
$-(CH_2)_4-$		66
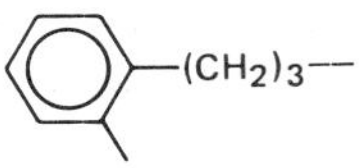 2-methylphenyl$-(CH_2)_3-$		75
Me	Me	42
Ph	Me	78
Ph	Ph	100
Ph	H	100
i-Pr	H	76
PhCH=CH—	Me	92
and		
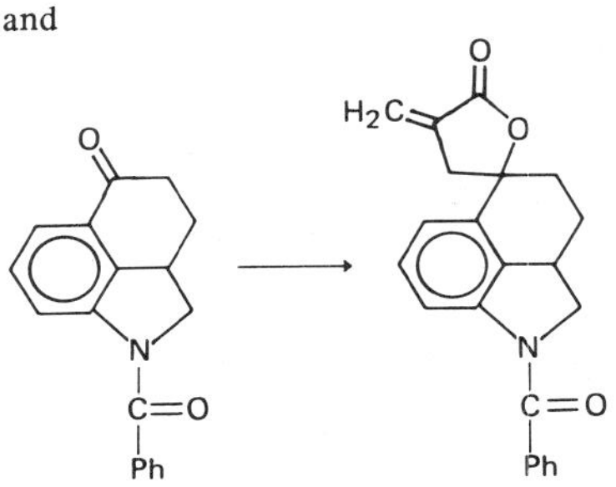 N-benzoyl tricyclic ketone (C=O, N—C(=O)Ph) ⟶ spiro α-methylene γ-butyrolactone (H_2C=, O, N—C(=O)Ph)		100

$$BrH_2C\text{-maleic anhydride} + C_{13}H_{27}CHO \xrightarrow{Zn} \text{HOC(=O), H, =CH}_2\text{, H, } C_{13}H_{27} \text{ lactone} \quad (109)$$

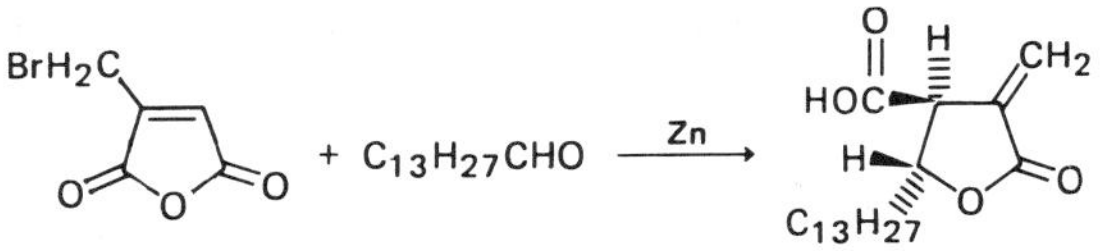

12%
trans

$$\text{3,4-methylenedioxyphenyl-CH=CH—CHO} + BrCH_2C(OMe)=CHCOOMe \xrightarrow[\text{2. } NH_4Cl,\ H_2O]{\text{1. Zn, THF}}$$

$$\text{(27) CH=CH, OMe, O, O} \xrightarrow[Pd/C]{H_2} \text{(28) } CH_2CH_2\text{, OMe, O, O} \quad (110)$$

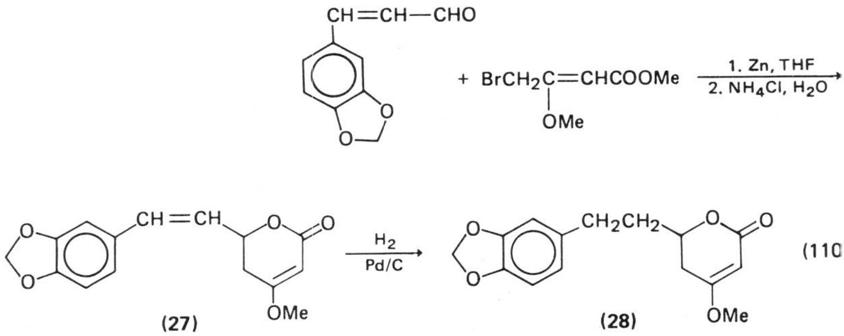

(27) 38% (28)

The synthesis of *d,l*-methysiticin (**27**), another lactone constituent of the kawa root[41–44], has been accomplished[186] by a vinylogous Reformatsky-type condensation of 3,4-methylenedioxycinnamaldehyde and methyl γ-bromo-β-methoxycrotonate in THF (equation 110). Catalytic reduction of *d,l*-methysiticin affords *d,l*-dihydromethysiticin (**28**).

d,l-Mevalonolactone (**29**) has been synthesized by the Reformatsky condensation[187,188] of methyl or ethyl bromoacetate with either 1,1-dimethoxy-3-oxobutane or 1-acetoxy-3-oxobutane, by the use of ethyl lithioacetate in liquid ammonia[189,190], and by the Reformatsky reaction modification using trimethyl borate[191]. A new synthesis of *d,l*-mevalonolactone (**29**), which is superior to the methods mentioned above, and which holds considerable promise of generality in lactone preparation[192], consists of condensation of ethyl lithioacetate, with 1-acetoxy-3-oxobutane (equation 111). Hydrolysis of the resulting diester, followed by acidification, affords **29**.

$$MeC(=O)(CH_2)_2C(=O)Me + LiCH_2COOEt \longrightarrow MeC(=O)OCH_2CH_2C(OH)(Me)CH_2COOEt \xrightarrow[2.\ H^+]{1.\ KOH,\ MeOH} \text{(29)} \quad (111)$$

(29)

H. By Wittig-type Reactions

Preparation of α-pyrones has been effected via reaction of ethoxycarbonylmethylenetriphenylphosphorane with a variety of β-diketones (equation 112)[193]

$$H_2C(COR^1)(COR^2) + Ph_3P{=}CHCOOEt \longrightarrow \left[R^1C(=O)CH{=}C(R^2)C(=O)OEt \rightleftarrows R^1C(OH){=}CHC(R^2){=}... C(=O)OEt \right] \xrightarrow{-EtOH} \text{α-pyrone} \quad (112)$$

R¹	R²	Yield (%)
Ph	Ph	–
p-MeOC$_6$H$_4$	p-MeOC$_6$H$_4$	–
p-ClC$_6$H$_4$	p-ClC$_6$H$_4$	20
2-thienyl	2-thienyl	17

The mechanism apparently involves initial reaction between the ylide and one of the keto groups of the diketone to form an intermediate keto ester, the enol form of which immediately forms the lactone by ring closure. When 2-benzoylcyclohexanone was used (equation 113)[193] only a 5% yield of 4,5-(tetramethylene)-6-phenyl-2*H*-pyran-2-one was obtained.

$$\text{2-benzoylcyclohexanone} + Ph_3P{=}CHCOOEt \longrightarrow \text{4,5-(tetramethylene)-6-phenyl-2}H\text{-pyran-2-one} \qquad (113)$$

γ-Butyrolactone has been prepared[194] in 62% yield as shown in reaction (114), which, although it is not strictly a Wittig reaction, does involve an intermediate phosphonium salt.

$$Ph_3P + ClCH_2CH_2CH_2COOH \longrightarrow Ph_3\overset{+}{P}CH_2CH_2CH_2COOH\ Cl^-$$

$$\xrightarrow{NaOH} Ph_3\overset{+}{P}CH_2CH_2CH_2COO^- \xrightarrow{200-225^\circ C} \gamma\text{-butyrolactone} \qquad (114)$$

Wittig reactions which have been employed for functionization of preformed lactones rather than for ring-closure are discussed in Section II.K

I. From α-Anions (Dianions) of Carboxylic Acids

Because of their potential as aldosterone inhibitors, steroidal spiro γ-lactones have been the subject of considerable interest. In 1972 Creger[195] published a method for the preparation of a wide variety of such compounds. This procedure involved dimetalation of several aliphatic carboxylic acids using lithium diisopropylamide (LDA), followed by reaction of the resulting lithio α-anions (dianions) with (17*S*)-spiro[androst-5-ene-17,2′-oxiran]-3-β-ol (equation 115). Oppenauer oxidation of these spiro lactones afforded the substituted 4′,5′-dihydro-(17*R*)-spiro-[androst-4-ene-17,2′-(3′*H*)-furan]-3,5′-diones, which upon further oxidation with chloroanil in *t*-butyl alcohol followed by treatment with chloroanil in toluene–acetic acid mixtures produced substituted 4′,5′-dihydro-(17*R*)-spiro[androsta-4,6-diene-17,2′-(3′*H*)-furan]-3,5′-diones (equation 116). Treatment of these products with thiolacetic acid afforded the 7α-thioacetyl derivatives. The parent unsubstituted 4′,5′-dihydro-(17*R*)-spiro[androst-4-ene-17,2′(3′H)-furan]3,5′-dione was not prepared via the oxidation technique discussed above but by the condensation shown in equation (117).

Other conversions reported by Creger[195] include the preparation of 4′,5′-dihydro-3β-hydroxy-4′-vinyl-(17*R*)-spiro[androst-5-ene-17,2′-(3′*H*)-furan]-5′-one via the reaction of (17*S*)-spiro[androst-5-ene-17,2 -oxiran]-3β-ol with the crotonic acid anion (equation 118). A large number of steroidal lactones were also synthesized by hydrolysis of various amide or nitrile derivatives as shown in equations (119) and (120).

The generality of the reaction of metalated carboxylic acids with simple and steroidal epoxides is demonstrated by the results summarized in Table 12[195]. A similar approach to the preparation of γ-butyrolactones[196] consists of the reaction

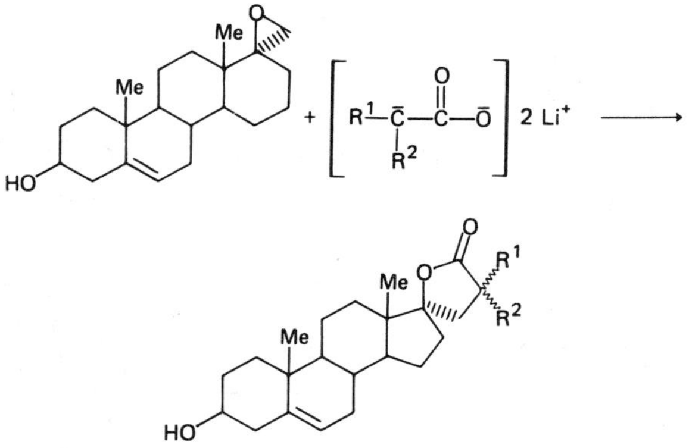

Me
Me
HO
+ $\left[R^1-\bar{C}(R^2)-C(=O)-\bar{O} \right] 2\,Li^+$ ⟶

(115)

Me
Me
HO
O
R1
R2

R¹	R²	Yield (%)
H	H	55
Me	Me	81
H	Me	70
H	Et	87
H	*n*-Bu	76
H	Ph	75
H	OMe	19

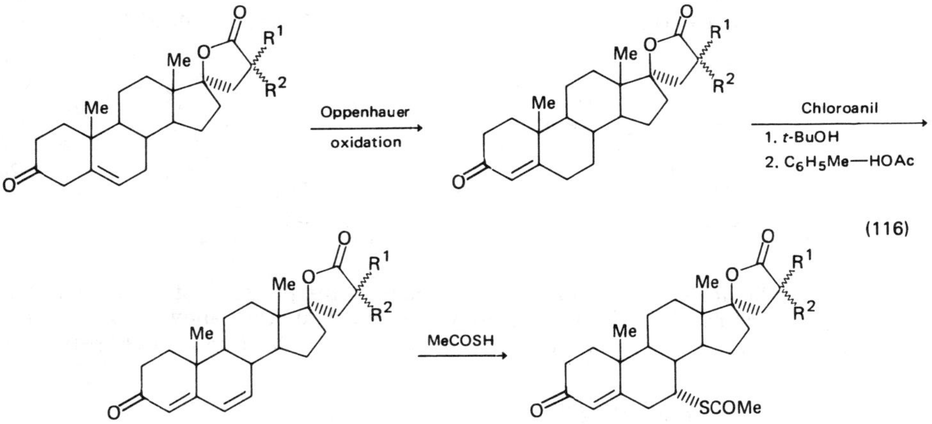

Oppenhauer oxidation ⟶

Chloroanil ⟶
1. *t*-BuOH
2. C_6H_5Me—HOAc

(116)

MeCOSH ⟶

SCOMe

R¹	R²	Yield (%) Monoene	Diene	Thioacetyl
H	Me	90	85	64
H	Et	84	82	64
Me	Me	77	61	61
H	*n*-Bu	35	—	—
H	Ph	71	—	—

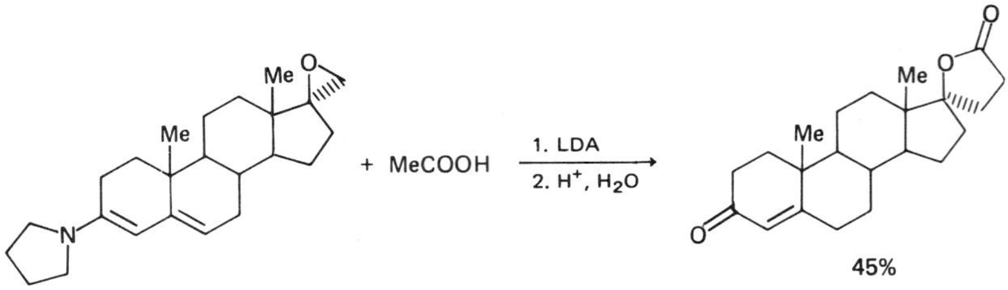

(117)

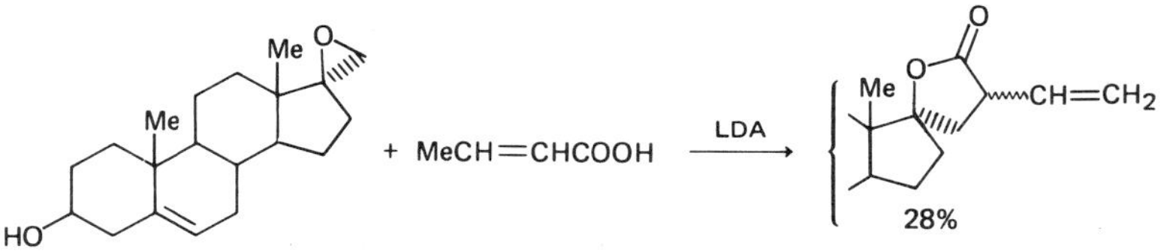

(118)

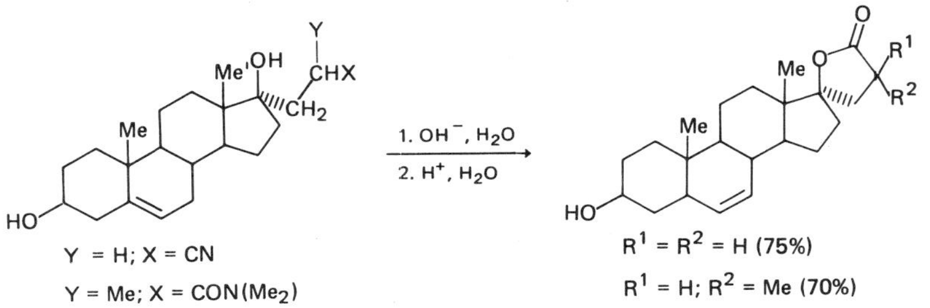

(119)

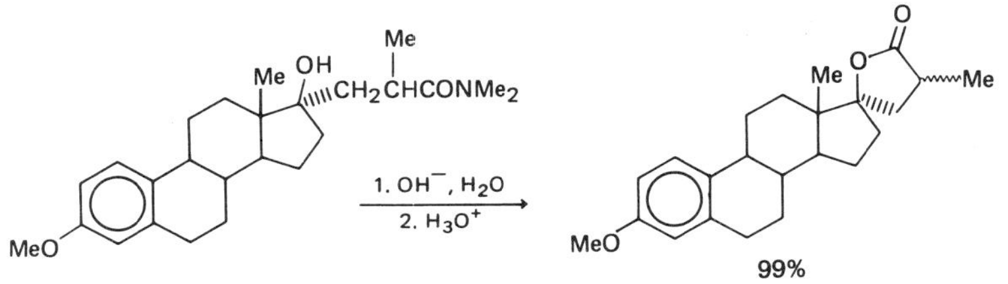

$Y = H; X = CN$

$Y = Me; X = CON(Me_2)$

$R^1 = R^2 = H$ (75%)

$R^1 = H; R^2 = Me$ (70%)

(120)

of carboxylic acids with lithium naphthalenide in the presence of diethylamine to produce α-anions of the lithium carboxylates, which are then allowed to react with epoxides to afford γ-hydroxy acids (equation 121). Cyclization of these γ-hydroxy

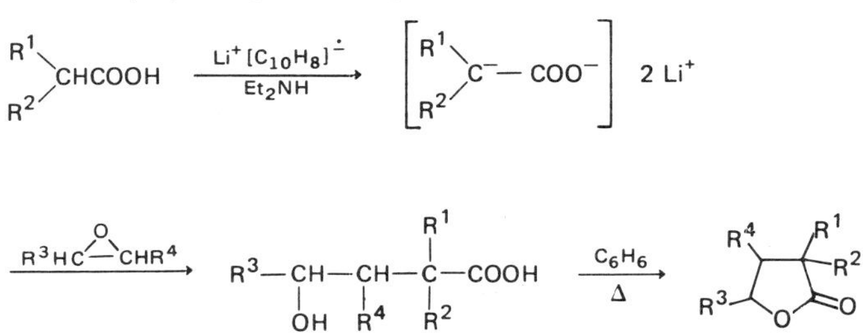

(121)

acids in refluxing benzene provides the lactones shown in Table 13. As may be seen from these results, monosubstituted epoxides react more readily than do disub-

TABLE 12. γ-Lactones by reactions of metalated acids with epoxides

Epoxide	Li^+ Na^+ Anion	Product	Yield (%)
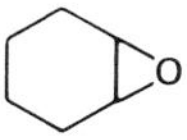 O	$Me_2\bar{C}COO^-$, then heat in C_6H_5 Me	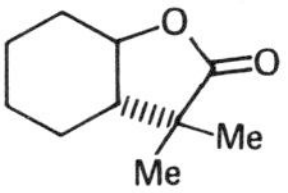 O, O, Me, Me	83
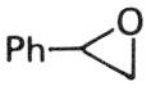 Ph, O	$Me_2\bar{C}COO^-$, then heat in C_6H_6	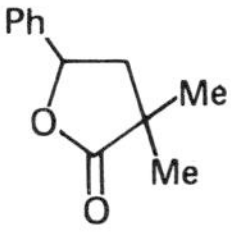 Ph, O, Me, Me, O	84
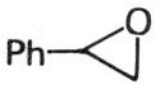 Ph, O	$Me_3C\bar{C}HCOO^-$	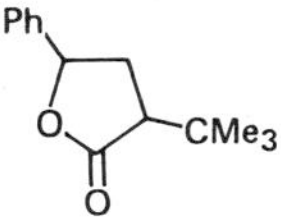 Ph, O, CMe_3, O	100
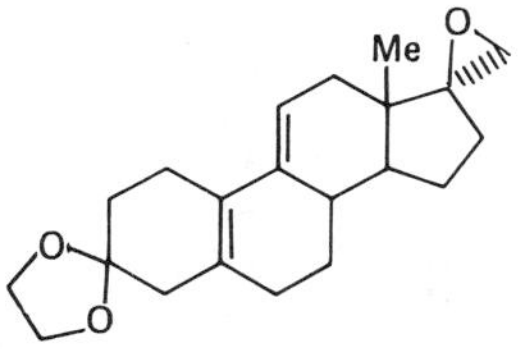 Me, O, O, O	$Me_2\bar{C}COO^-$	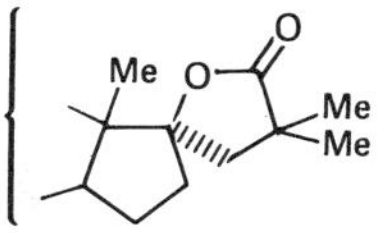 O, Me, O, Me, Me	73
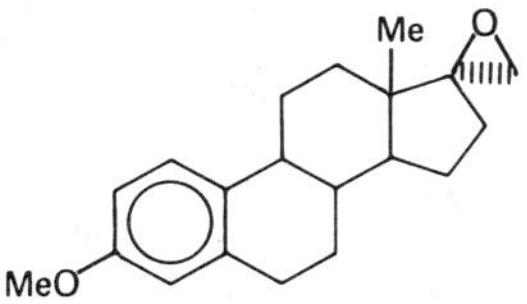 Me, O, MeO	$Me_2\bar{C}COO^-$	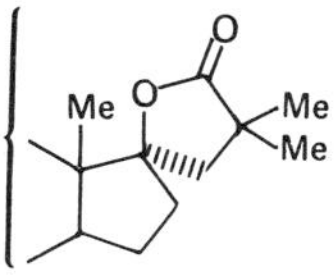 O, Me, O, Me, Me	82
As above	$Ph\bar{C}HCOO^-$	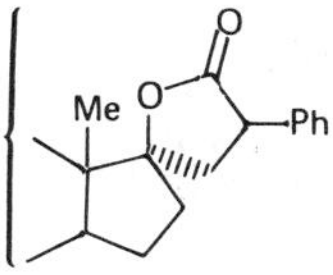 O, Me, O, Ph	85
As above	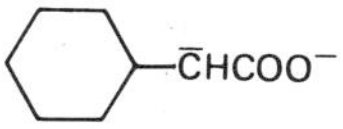 $\bar{C}HCOO^-$	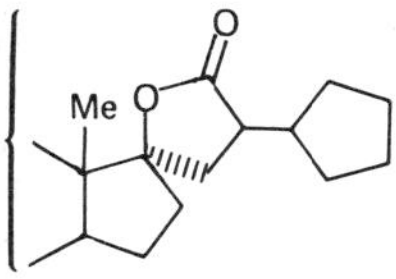 O, Me, O	89

stituted epoxides. This reaction difficulty was found to increase to the point where no product was obtained when the di- and trisubstituted epoxides shown in reaction (122) were used.

TABLE 13. γ-Butyrolactones prepared by lithium naphthalenide-promoted reactions of carboxylic acids with epoxides

R^1	R^2	R^3	R^4	Yield (%)
H	H	Me	H	5
H	H	Et	H	22
H	H	Ph	H	31
H	H	$-(CH_2)_4-$		18
Me	H	Me	H	47
Me	H	Et	H	51
Me	H	Ph	H	57
Me	H	$-(CH_2)_4-$		39
Et	H	Me	H	38
Et	H	Et	H	41
Et	H	Ph	H	53
Et	H	$-(CH_2)_4-$		35
Me	Me	Me	H	48
Me	Me	Et	H	73
Me	Me	Ph	H	69
Me	Me	$-(CH_2)_4-$		55
n-Pr	H	Me	H	58
n-Pr	H	Et	H	64
n-Pr	H	Ph	H	69
n-Pr	H	$-(CH_2)_4-$		52
i-Pr	H	Me	H	44
i-Pr	H	Et	H	71
i-Pr	H	Ph	H	53
i-Pr	H	$-(CH_2)_4-$		14
n-Bu	H	Me	H	49
n-Bu	H	Et	H	53
n-Bu	H	Ph	H	66
n-Bu	H	$-(CH_2)_4-$		31
Ph	H	Me	H	52
Ph	H	Et	H	68
Ph	H	Ph	H	55
Ph	H	$-(CH_2)_4-$		23
$Me_2C{=}CH(CH_2)_2CHMe$	H	Me	H	66
$Me_2C{=}CH(CH_2)_2CHMe$	H	Et	H	71
$Me_2C{=}CH(CH_2)_2CHMe$	H	Ph	H	80
$Me_2C{=}CH(CH_2)_2CHMe$	H	$-(CH_2)_4-$		54
Me	Me	$EtOCH_2$	H	35
Me	Me	*i*-Pr	H	33
Me	Me	$CH_2{=}CHCH_2OCH_2$	H	38
Me	Me	*n*-$BuOCH_2$	H	70
Me	Me	*i*-$BuOCH_2$	H	81
Me	Me	$PhOCH_2$	H	35
Me	Me	$C_6H_{11}OCH_2$	H	52

J. From Lithio Salts of 2-Alkyl-2-oxazolines

The synthetically versatile[197,198] lithio derivatives of 2,4,4-trimethyl-2-oxazoline and its 2-alkyl homologues have recently been employed[199] in the

$$Me_2CHCOOH + R^1R^2C\overset{O}{—}CHR^3 \xrightarrow[Et_2NH]{Li^+[C_{10}H_8]^{\dot{-}}} \text{No product} \qquad (122)$$

R^1	R^2	R^3
H	$-(CH_2)_6-$	
H	$-(CH_2)_{10}-$	
Ph	Me	H
$CH_2=CMe-CH=CH-CH_2-$	Me	Me
Me	Me	$CH_2=CMe(CH_2)_2$
Me	Me	$-CH_2CH(i\text{-Pr})CH_2CH_2-$

preparation of a variety of butyrolactones substituted in the α,β- and/or γ-positions with alkyl groups (equation 123). The procedure involves reaction of lithiated

(123)

oxazolines with an appropriate epoxide. This produces 2-(β-hydroxyalkyl)oxazolines, which upon hydrolysis with aqueous acid, acidified ethanol or *p*-toluenesulphonic acid in benzene afford the butyrolactones shown in Table 14.

TABLE 14. γ-Butyrolactones from epoxides and lithio salts of 2-alkyl-2-oxazolines[199]

R^1	Epoxide	Hydrolysis method[a]	Lactone		Overall yield (%)
H		A		R = H R = Me R = Et	75 72 85
H		C	(94%) + (6%)		89
Me		C	(60%) + (40%)		65

TABLE 14. (Continued)

R^1	Epoxide	Hydrolysis method[a]	Lactone	Overall yield (%)
$Me(CH_2)_4$—	 O, Et	C	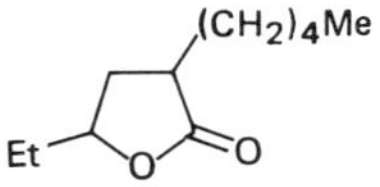 $(CH_2)_4Me$, Et, O, O	76
H	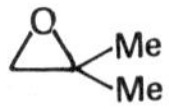 O, Me, Me	A	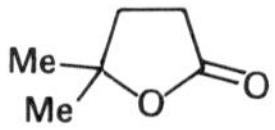 Me, Me, O, O	72
H	 O	B	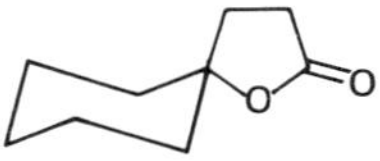 O, O	56
$PhCH_2$	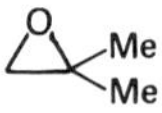 O, Me, Me	C	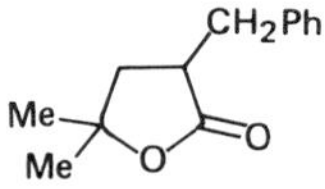 CH_2Ph, Me, Me, O, O	70
H	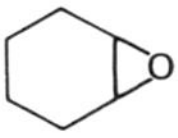 O	C	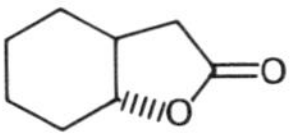 O, O	65
H	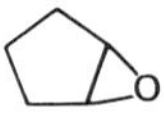 O	C	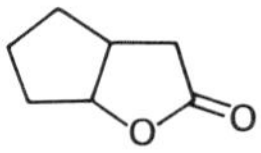 O, O	5–6
H	 Me, O, Me (*cis* or *trans*)	A	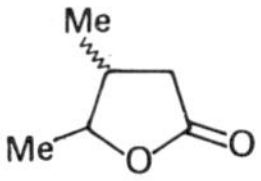 Me, Me, O, O (50:50)	16
Me	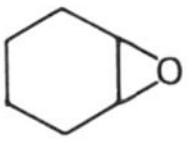 O	B	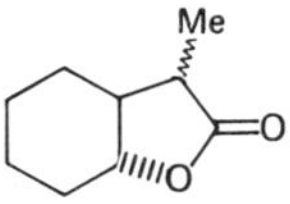 Me, O, O	9
H	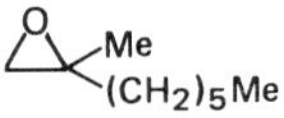 O, Me, $(CH_2)_5Me$	C	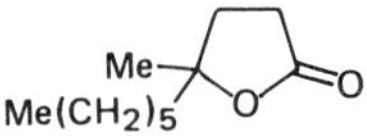 Me, $Me(CH_2)_5$, O, O	70

[a] Hydrolysis performed in: A = acidic EtOH, B = wet benzene–toluenesulphonic acid, C = acidic aqueous methanol.

It was observed that certain 1,2-disubstituted epoxides, especially those with *trans* substituents, gave low yields of lactones, or in some cases, no product at all.

The oxazoline procedure has also been used in the asymmetric synthesis of 2-substituted γ-butyrolactones as shown in equation (124)[200].

$$\text{Ph-oxazoline-CH}_3 \xrightarrow[\text{2. ethylene oxide}]{\text{1. BuLi, DME}} \text{Ph-oxazoline-(CH}_2)_3\text{OLi} \xrightarrow{\text{Me}_3\text{SiCl}}$$

$$\text{Ph-oxazoline-(CH}_2)_3\text{OSiMe}_3 \xrightarrow[\text{2. RX, }-98^\circ\text{C}]{\text{1. LDA, }-78^\circ\text{C}} \text{Ph-oxazoline-CH(R)(CH}_2)_2\text{OSiMe}_3 \xrightarrow{\text{H}_3\text{O}^+} (R)\text{-lactone}$$

(124)

R	Optical purity (%)	Yield (%)
Me	64.2	58
Et	–	68
n-Pr	73.3	75
Allyl	72.0	60
n-Bu	–	71

K. By Direct Functionalization of Preformed Lactones

The acidity of lactone α-hydrogens permits structural elaboration at the α-position of the lactone nucleus via certain carbanion condensations. One of the earliest examples[201–204] of this type of reaction involved dehydrative aldol condensations of aromatic aldehydes at the α-methylene group of 2,3-dihydrofuran-2-ones. A more recent study[205] of analogous aldol condensations of 2(3*H*)-coumaranone with 2-hydroxybenzaldehydes in the presence of triethylamine revealed that the expected 3-(2-hydroxybenzylidene)2-(3*H*)-coumaranones were produced upon dropwise addition of triethylamine to the reaction mixture at 15°C, while an increase in temperature to 25–40°C during the condensation increased the yield of 3-(2-hydroxyphenyl)coumarins at the expense of the benzylidene products (equation 125). If the temperature were raised to 70°C or if the 3-(2-hydroxybenzylidene)2(3*H*)-coumaranones were treated at 80°C with additional triethylamine, 3-(2-hydroxyphenyl)coumarins resulted via an intramolecular, *in situ* cyclization. Analogous results were obtained[206] in condensations of substituted 2-hydroxybenzaldehydes with γ-aryl-$\Delta^{\beta,\gamma}$-butenolides (equation 126).

Several methods for α-alkylation of lactones have appeared in the literature[207–212]. Best results have been obtained by formation of the lactone enolate with a strong base such as LDA, lithium isopropylcyclohexylamide or trityl lithium, followed by treatment of the enolate with an alkyl halide (Table 15). A similar approach[212,213] affords dialkylated products, while attempts to use benzylbromomethyl sulphide as an alkylating agent have failed[214].

Various methods for introducing an α-methylene group into preformed lactones have been discussed in a 1975 review[11] on α-methylene lactones. A procedure[215] which is not discussed in this review involves reactions of an α-phosphono-γ-butyrolactone carbanion with aldehydes, ketones, heterocumulenes and nitrosobenzene to form α-ylidene-γ-butyrolactones (Method A; equation 127). The α-bromo-γ-butyrolactone employed as the starting material for these reactions has also been used in

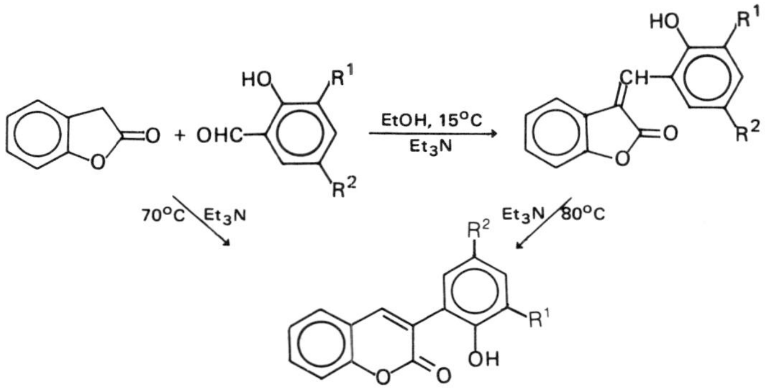

(125)

R^1	R^2	Yield (%) 3-(2-hydroxybenzylidene)-2(3*H*)-coumaranones	3-(2-Hydroxy-phenyl)coumarins
H	H	76	100
H	Cl	62	89
H	Br	82	96
H	NO_2	62	97
Cl	Cl	91	84
Br	Br	93	81

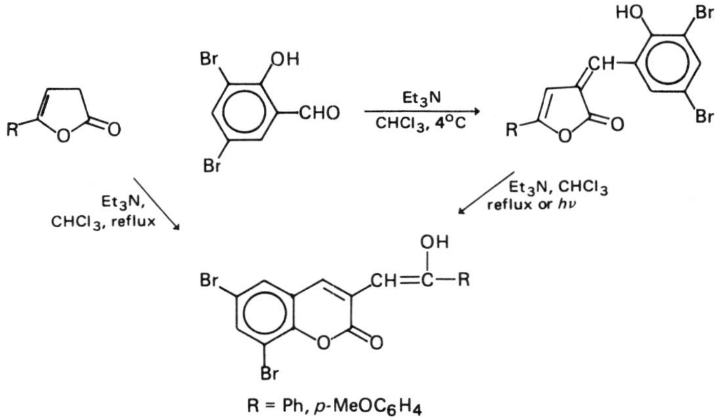

(126)

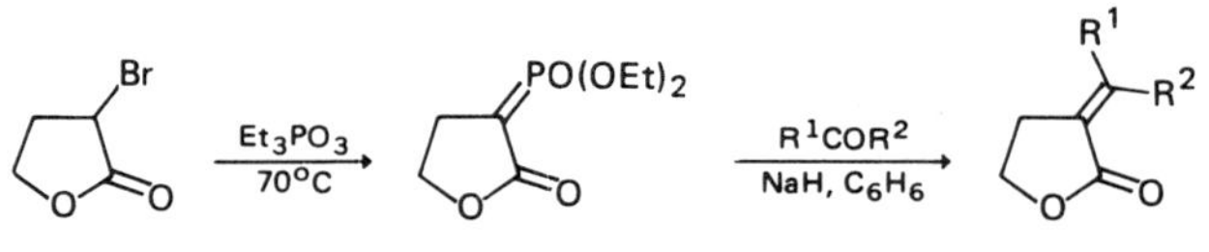

R = Ph, *p*-$MeOC_6H_4$

(127)

Reformatsky-type reactions to afford similar products (Method B; equation 128)[216]. Results from these two procedures are given in Table 16.

$$\text{3-bromo-}\gamma\text{-butyrolactone} + R^1COR^2 \xrightarrow[C_6H_6]{Zn} \text{3-}(C(OH)R^1R^2)\text{-}\gamma\text{-butyrolactone} \xrightarrow{-H_2O} \text{3-}(=CR^1R^2)\text{-}\gamma\text{-butyrolactone} \qquad (128)$$

A recent, facile method for the preparation of β-methoxycarbonyl γ-substituted γ-butyrolactones proceeds via generation of the enolate of succinic anhydride in the presence of carboxyl compounds[217]. Thus, addition of a THF solution of 3-phenylpropanal and succinic anhydride at $-78°C$ under argon to a THF solution of lithium 1,1-bis(trimethylsilyl)-3-methyl-1-butoxide afforded, after hydrolysis and treatment with diazomethane, an 80% yield of β-methoxycarbonyl-γ-phenethyl-γ-butyrolactone. The corresponding substituted γ-butyrolactones can be obtained in moderate yields when ketones are used in place of aldehydes in this reaction (equation 129). Methylsuccinic anhydride produces the enolate on the methylene

$$R^1\text{-succinic anhydride} + R^2\text{—}\overset{O}{\overset{\|}{C}}\text{—}R^3 \xrightarrow[\text{2. HCl; 3. } CH_2N_2]{\text{1. LiOCH(TMS)}_2CH_2CHMe_2} \text{lactone } (R^1, MeO_2C, R^2, R^3) \qquad (129)$$

R^1	R^2	R^3	Yield (%)
H	Ph	H	84
H	$PhCH_2CH_2$	H	80
H	*n*-PrCH=CH	H	82
H	C_5H_{11}	H	78
H	Me_2CH	H	76
H	$-(CH_2)_5-$		51
Me	Ph	H	84
Me	$PhCH_2CH_2$	H	75
Me	*n*-PrCH=CH	H	85
Me	C_5H_{11}	H	72
Me	Me_2CH	H	70
Me	$-(CH_2)_5-$		57

site and affords the corresponding adduct exclusively, while generation of the enolate from glutaric anhydride does not afford the butyrolactones in yields as high as when succinic anhydride is used.

L. From Ketenes

Simple, as well as substituted ketenes react with aldehydes and ketones via a $(2\pi + 2\pi)$ cycloaddition to afford β-lactones (equation 130).

$$R^1R^2C{=}O + R^3R^4C{=}C{=}O \longrightarrow \text{β-lactone } (R^1, R^2, R^3, R^4) \qquad (130)$$

TABLE 15. α-Alkylation of γ-butyrolactone and δ-valerolactone

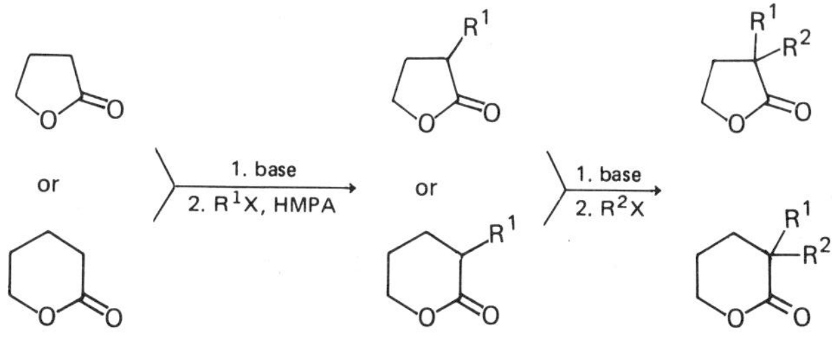

or $\xrightarrow[\text{2. } R^1X\text{, HMPA}]{\text{1. base}}$ or $\xrightarrow[\text{2. } R^2X]{\text{1. base}}$

Lactone	Base[a]	R^1	Yield (%)	R^2	Yield (%)	References
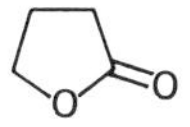	A	Me	56	—	—	211
	A	$H_2C{=}CHCH_2$	74–>90	—	—	211, 212
	A	*n*-Bu	low	—	—	211
	B	Et	>90	—	—	212
	B	$HC{\equiv}CCH_2$	>90	—	—	212
	B	$Br(CH_2)_2CH_2$	80	—	—	212
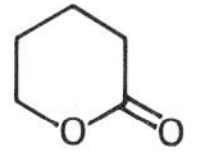	B	Et	>90	—	—	212
	B	$H_2C{=}CHCH_2$	>90	—	—	212
	B	$HC{\equiv}CCH_2$	>90	—	—	212

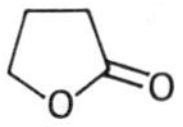	A	Me	80	Me	13	211
	C	Me	—	Me	—	213
	B	Et	—	Et	95	212
	B	Et	—	$H_2C{=}CHCH_2$	95	212
	B	Et	—	$HC{\equiv}CCH_2$	95	212
	B	Et	—	$Br(CH_2)_2CH_2$	95	212
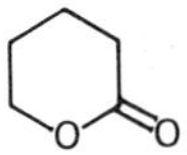	B	Et	—	Et	95	212
	B	Et	—	$H_2C{=}CHCH_2$	95	212
	B	Et	—	$HC{\equiv}CCH_2$	95	212
	B	Et	—	$Br(CH_2)_2CH_2$	95	212

[a] A = Lithium isopropylcyclohexylamide; B = lithium diisopropylamide; C = trityllithium.

TABLE 16. Preparation of α-ylidene-γ-butyrolactones from α-bromo-γ-butyrolactone via Wittig (method A) and Reformatsky (method B) reactions

R^1	R^2	Method	Product		Yield (%)	Reference
H	Ar	A	[structure: α-(arylmethylene)-γ-butyrolactone, Ar, H]			
				Ar = Ph	100	215
				Ar = p-$O_2NC_6H_4$	71	215
				Ar = PhCH=CH	55	215
H	CCl_3	A	[structure: α-(2,2,2-trichloroethylidene)-γ-butyrolactone, CCl_3, H]		100	215
H	Et or *i*-Pr	A	[structure: α-alkylidene-γ-butyrolactone, H, R^2] (*cis* + *trans*)			
				R = *i*-Pr	89	215
				R = Et	100	215
H	[structure: α-(2-methylbenzylidene)-γ-butyrolactone]	A	[structure: 1,2-bis(γ-butyrolacton-α-ylidenemethyl)benzene, H, H] +		47	215

H	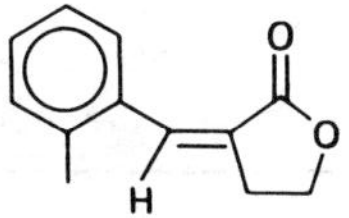	A	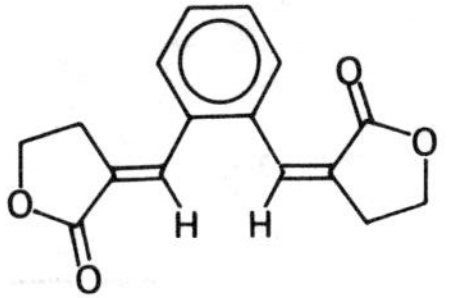	7.3	215
H	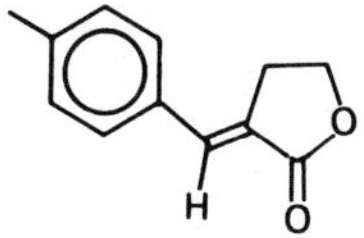	A	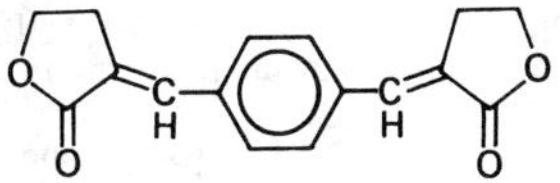	67	215
H	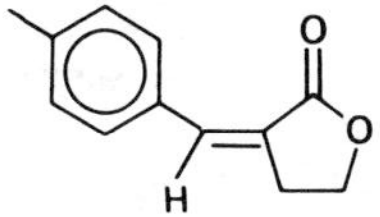	A	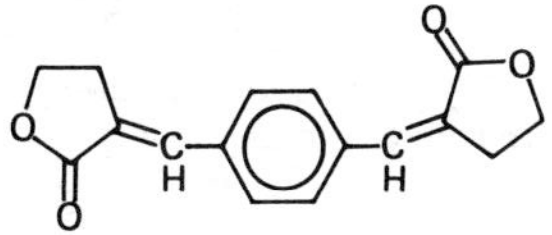	4.4	215
	$—(CH_2)_5—$	A	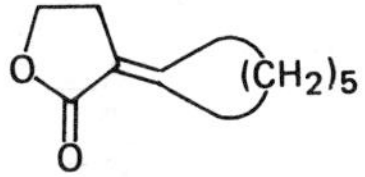	92	215
	=N—Bu-*n*	A	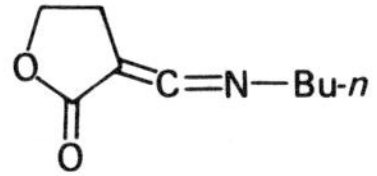	91	215
	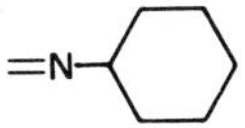	A	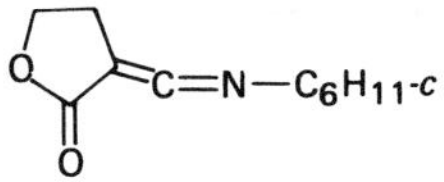	95	215
	$=CPh_2$	A	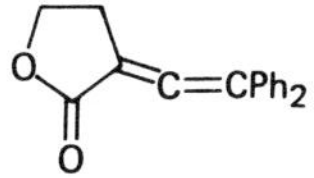	100	215

TABLE 16. (Continued)

R^1	R^2	Method	Product	Yield (%)	Reference
	$=C(Ph)(Et)$	A	O, O (lactone) $=C=C(Et)(Ph)$	100	215
	$=C(Me)(Ph)$	A	O, O (lactone) $=C=C(Me)(Ph)$	81	215
17α-Methyldihydro-testosterone		A	3-(γ-Butyrolacton-α-ylidene)-17α-methylandrostan-17β-ol + 3-(γ-Butyrolacton-α-yl)-17α-methylandrost-2(or 3)-ene-17β-ol	75 25	216
Cortisone acetate		B	3-(γ-Butyrolacton-α-ylidene)-17α-hydroxy-11-dehydrocorticosterone-21-acetate	–	216

In the preparation of β-butyrolactone, β-propiolactone and β-caprolactone by reaction of the appropriate aldehyde with ketene it was found[218] that boron trifluoride or its etherate complex in THF could be used to increase both the yield and purity of the product. The versatility of both catalysed and uncatalysed reactions of ketenes with aldehydes and ketones may be seen by inspection of Table 17,[219–224], where a representative series of lactone preparations are collected. The first four entries in Table 17 involve γ-lactone formation[219]. Generation of these products is explained[219] by a mechanism involving initial formation of the expected β-lactone, followed by ring-opening, carbonium ion rearrangement and recyclization as shown in equation (131) with methyl *t*-butyl ketone.

$$Me_3C{-}\overset{O}{\overset{\|}{C}}{-}Me + H_2C{=}C{=}O \xrightarrow{BF_3} \text{(β-lactone: Me, }Me_3C\text{)} \xrightarrow{BF_3}$$

$$Me_3C\overset{Me}{\overset{|}{\underset{+}{C}}}CH_2COO^- \longrightarrow Me\underset{+}{C}(Me){-}\overset{Me}{\overset{|}{\underset{Me}{\underset{|}{C}}}}CH_2COO^- \longrightarrow \text{(γ-lactone: Me Me, Me, Me)} \qquad (131)$$

Ketenes undergo 1,3-dipolar addition with carbenes derived from diazo ketones to form enol lactones (butenolides)[225] as shown in equation (132) and summarized in Table 18.

$$R\overset{O}{\overset{\|}{C}}CHN_2 \xrightarrow{-N_2} \left[R{-}\overset{O}{\overset{\|}{C}}{-}\ddot{C}H \longleftrightarrow R{-}\overset{\bar{O}}{\overset{|}{C}}{=}\overset{+}{C}H \right] \xrightarrow{H_2C=C=O}$$

$$\text{(butenolide: O—C=O, C(R)=CH—CH}_2\text{)} \qquad (132)$$

An interesting preparation of lactones has been observed during irradiation of several α β-epoxy diazoketones in benzene[226]. The butenolide products obtained are explained in terms of an intermediate epoxy ketene, formed by a Wolff rearrangement, which then undergoes intramolecular cyclization (equation 133).

$$\text{(epoxide } R^1, R^2, R^3{-}\overset{O}{\overset{\|}{C}}{-}CHN_2) \xrightarrow{h\nu} \text{(epoxide } R^1, R^2, R^3{-}CH{=}C{=}O) \longrightarrow \text{(butenolide } R^1, R^2, R^3) \qquad (133)$$

R^1	R^2	R^3	Yield (%)
Ph	Ph	H	90
H	Ph	H	90
H	Ph	Ph	70
Me	Me	H	43

TABLE 17. Preparation of lactones by reaction of aldehydes and ketones with ketene

Aldehyde or ketone	Ketene	Product	Yield (%)	Reference
$Me_2CHCOMe$	$H_2C{=}C{=}O + BF_3$	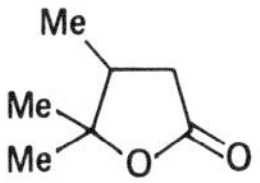 Me; Me; Me; O; O	41	219
Me_3CCOMe	$H_2C{=}C{=}O + BF_3$	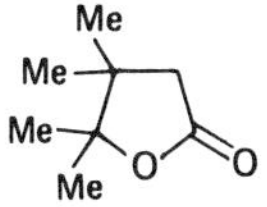 Me; Me; Me; Me; O; O	67	219
$(CH_2)_5CHCOMe$	$H_2C{=}C{=}O + BF_3$	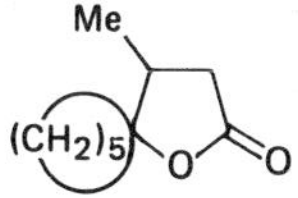 Me; $(CH_2)_5$; O; O	44	219
$(CH_2)_5C(Me)COMe$	$H_2C{=}C{=}O + BF_3$	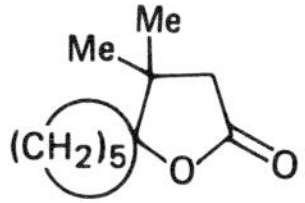 Me; Me; $(CH_2)_5$; O; O	49	219
Cl_3CCHO	$H_2C{=}C{=}O$[a]	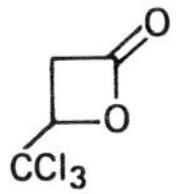 O; O; CCl_3	72.2	220, 221
Cl_3CCHO	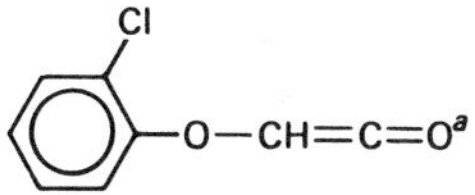 Cl; —O—CH=C=O[a]	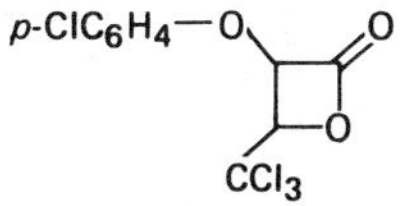 p-ClC_6H_4—O; O; O; CCl_3; (−)[a]	45	221

Cl_3CCHO	PhO—CH=C=O	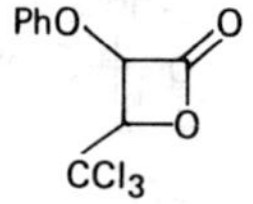	36	221
Cl_3CCHO	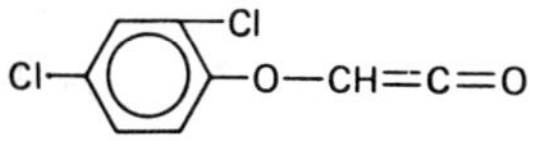	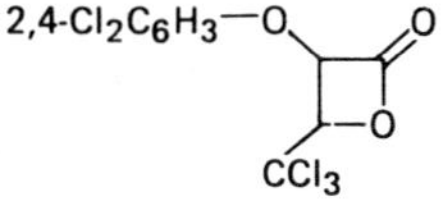	63	221
Br_3CCHO	ClCH=C=O	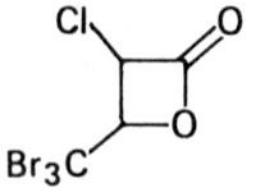	11	221
$Cl_3CCOCCl_3$	$H_2C{=}C{=}O$	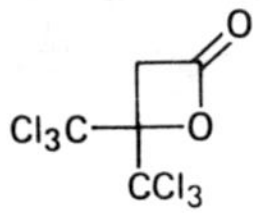	6	221
F_3CCHO	$H_2C{=}C{=}O$	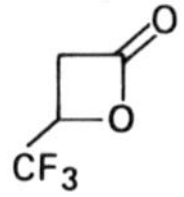	20	221
Cl_3CCHO	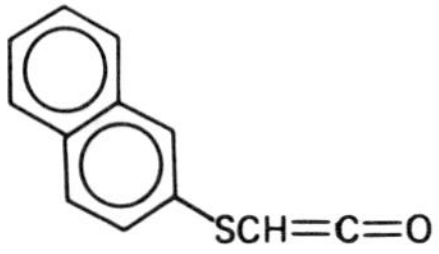 SCH=C=O	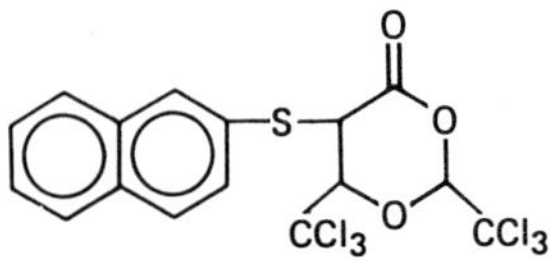	7.8	221
CCl_3CHO	$O{=}C{=}CH(CH_2)_2CH{=}C{=}O$	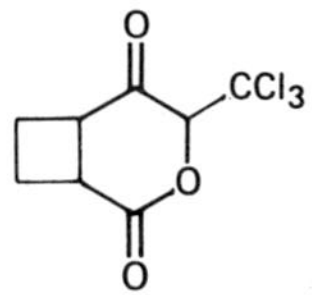	47	221

TABLE 17. (Continued)

Aldehyde or ketone	Ketene	Product		Yield (%)	Reference
Cl_3CCHO	$Cl_2C{=}C{=}O$	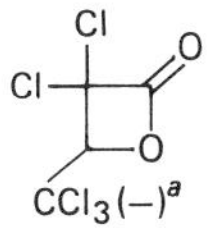 Cl, Cl, O, O, CCl_3 (−)[a]		39	221
RCHO	$Cl_2C{=}C{=}O$	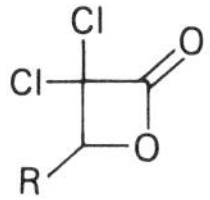 Cl, Cl, O, O, R			
			R = Me	51	222
			R = Me_2Ch	40	222
ArCHO	$Cl_2C{=}C{=}O$	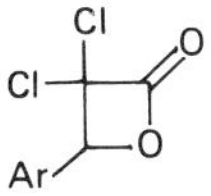 Cl, Cl, O, O, Ar			
			Ar = Ph	30	222
			Ar = p-ClC_6H_4	66	222
MeOC—COOEt	$Cl_2C{=}C{=}O$	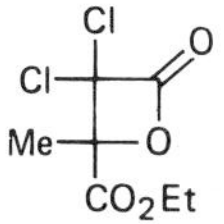 Cl, Cl, O, O, Me, CO_2Et		33	222
EtOOC—CO—COCOEt	$Cl_2C{=}C{=}O$	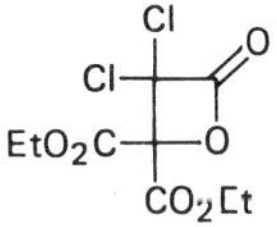 Cl, Cl, O, O, EtO_2C, CO_2Et		76	222

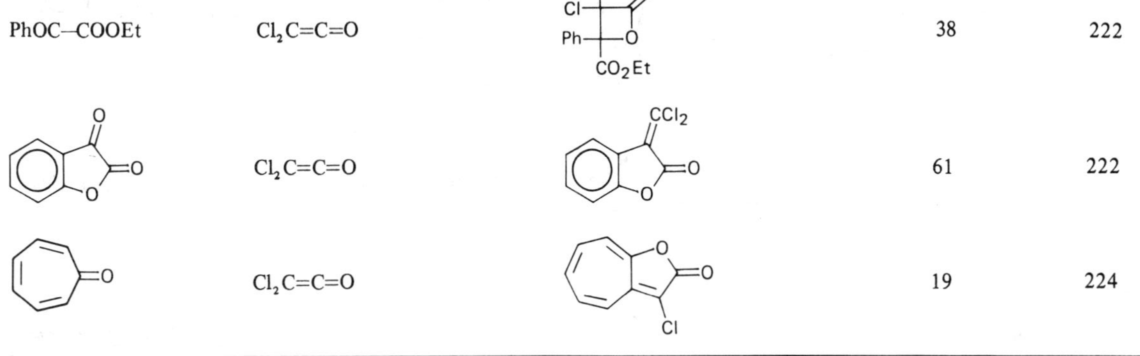

PhOC—COOEt	$Cl_2C=C=O$		38	222
	$Cl_2C=C=O$		61	222
	$Cl_2C=C=O$		19	224

[a] Ketene prepared from acetyl chloride with *N*,*N*-dimethyl-α-phenethylamine afforded (−) product, from acetyl chloride with brucine afforded (+) product.

TABLE 18. Preparation of enol lactones by reaction of ketenes with diazoketones[225]

Diazoketone	Ketene	Product	Yield (%)
$MeCH_2COCHN_2$	$H_2C{=}C{=}O$	Et O O	43
$PhCOCHN_2$	$H_2C{=}C{=}O$	Ph O O	34
$COCHN_2$	$H_2C{=}C{=}O$	O O	11

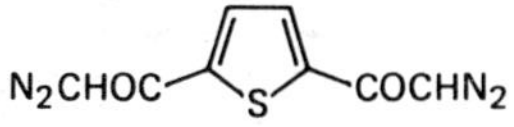 N_2CHOC–(S)–$COCHN_2$	$Ph_2C{=}C{=}O$	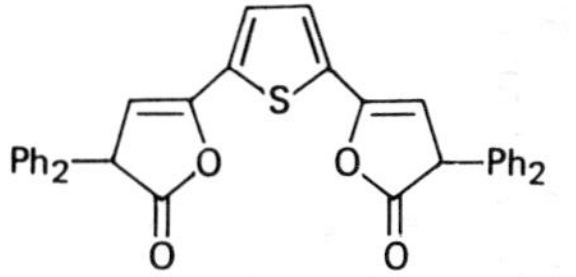 Ph_2, O, O, S, O, O, Ph_2	18
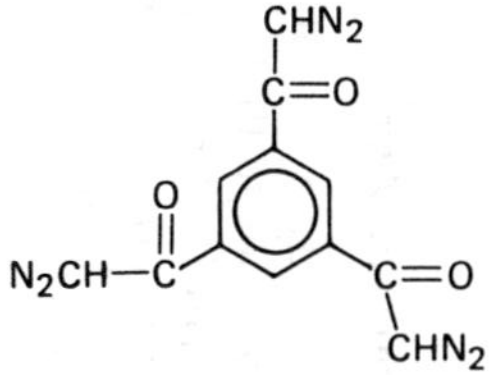 CHN_2, C=O, O, N_2CH—C, C=O, CHN_2	$Ph_2C{=}C{=}O$	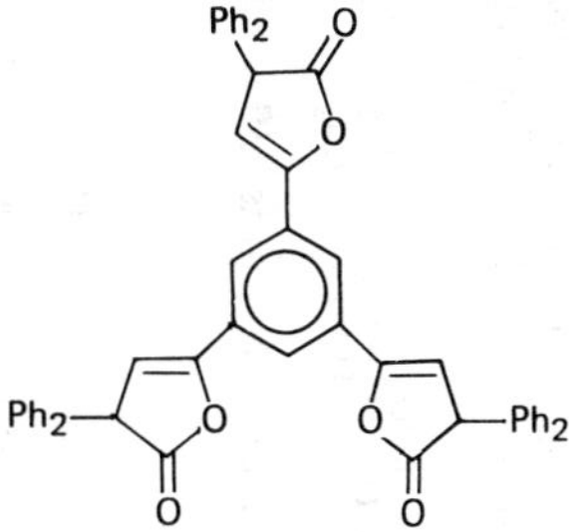 Ph_2, O, O, Ph_2, O, O, O, O, Ph_2	29
$N_2CHOC(CH_2)_nCOCHN_2$	$H_2C{=}C{=}O$	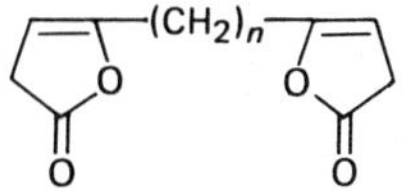 $(CH_2)_n$, O, O, O, O	$n = 3$ 40 $n = 4$ 90 $n = 5$ 47 $n = 6$ 32

Since it is known[227–229] that ozone is an effective epoxidizing agent toward highly hindered alkenes, Wheland and Bartlett[230] treated an emulsion of diphenylketene in ethyl acetate and hexafluoroacetone with ozone at −78°C expecting an α-lactone. Instead they obtained the product shown in equation (134), the structure of which was established by spectroscopy and its alkaline hydrolysis to benzilic

$$\mathrm{Ph_2C{=}C{=}O} + \mathrm{CF_3{-}\overset{O}{\overset{\|}{C}}{-}CF_3} \xrightarrow[\text{EtOAc}]{\mathrm{O_3},\ -78^\circ\mathrm{C}} \text{[2,2-bis(CF}_3\text{)-5,5-diphenyl-1,3-dioxolan-4-one]} \xrightarrow[\text{2. HCl}]{\text{1. KOH, EtOH}} \mathrm{Ph_2\overset{OH}{\overset{|}{C}}{-}COOH} \qquad (134)$$

acid. A similar approach was used to prepare di-*t*-butylacetolactone (equation 135)[231]; however, when hexafluoroacetone was added to the chlorotrifluoromethane (Freon 11) used as the solvent at −78°C and the mixture brought to room

$$\mathrm{(Me_3C)_2C{=}C{=}O} \xrightarrow[\mathrm{FCCl_3}]{\mathrm{O_3},\ -78^\circ\mathrm{C}} \text{[3,3-di-}t\text{-butyl-}\alpha\text{-lactone]} \qquad (135)$$

temperature, the two rearrangement products **30** and **31** were isolated upon distillation[230].

$\mathrm{Me_2C{=}C(Me)CMe_3}$ **(30)** 29%

[2,3-dimethyl-3-*t*-butyl-β-lactone] **(31)** 21%

Although attempts[231] to cause nitrous oxide to react with hydroxyacetylenic compounds in inert solvents have not been very successful, 3-butyn-1-ol did react to afford γ-butyrolactone, presumably via formation and cyclization of the intermediate 2-hydroxyethylketene (equation 136).

$$\mathrm{HOCH_2CH_2C{\equiv}CH} \xrightarrow[\mathrm{C_6H_{12}}]{\mathrm{N_2O}} [\mathrm{HOCH_2CH_2CH{=}C{=}O}] \longrightarrow \text{[γ-butyrolactone]} \qquad (136)$$

M. By Reduction of Anhydrides, Esters and Acids

Although the first report of the sodium borohydride reduction of an acid anhydride appeared in 1949[232], it was not until 1969 that this method of lactone preparation was thoroughly investigated[233,234]. Since that time, a variety of reagents such as sodium borohydride, lithium aluminium hydride, lithium tri-*t*-butoxyaluminohydride and sodium in ethanol have been used to reduce numerous acid anhydrides to lactones (Table 19)[235–248].

One of the most interesting aspects of this preparative method is the controversy that has developed[236,237,244] concerning which carbonyl group of the anhydride is reduced when one carbonyl function is hindered and the other is relatively free. The majority of unsymmetrical anhydrides undergo reduction at the more hindered carbonyl, irrespective of the reducing agent employed (see first entry in Table 19).

TABLE 19. Preparation of lactones by reduction of acid anhydrides

Anhydride	Reducing agent	Product	Yield (%)	References
	Na + EtOH		–, 35	235, 236
	$LiAlH_4$, THF		78–82	237
	$NaBH_4$, THF		80	236
	$NaBH(OMe)_3$, THF		78	236
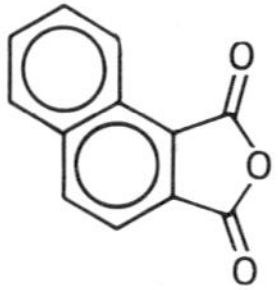	$LiAlH_4$, ether	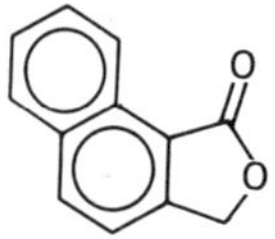	62	238
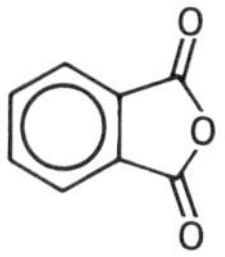	$LiAlH_4$, THF or $NaBH_4$, DMF	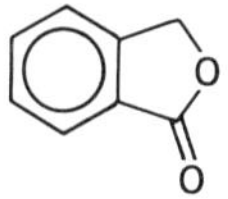	–, 71–73	239, 236
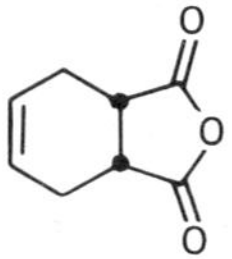	$LiAlH_4$, ether or THF	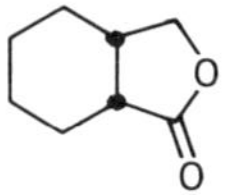	––, 75	239, 237

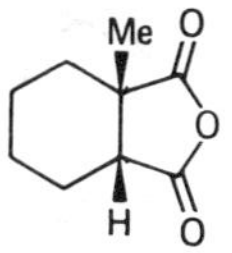

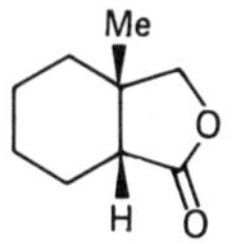

TABLE 19. (Continued)

Anhydride	Reducing agent	Product	Yield (%)	References
	$LiAlH_4$, ether		–	239
	THF		72.8	237
	$NaBH_4$, THF		76	236
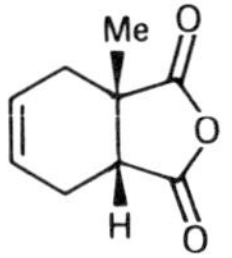 Me, H	$LiAlH_4$, ether or $NaBH_4$, THF	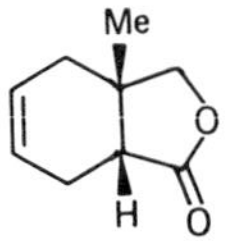 Me, H	– –, 65	240, 236
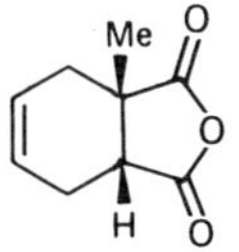 Me, H	H_2, Pt or $LiAlH_4$, THF	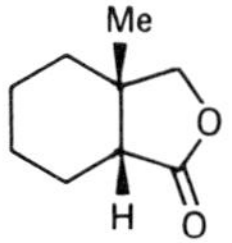 Me, H	–, 89	240, 237
Me, H, CH_2, CHO, Me		Me, H, CH_2, CH_2OH, Me		
	$NaBH_4$ in THF		54	241
	$NaBH_4$ in THF–MeOH		40–86	248

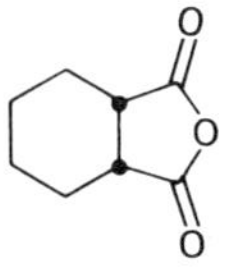

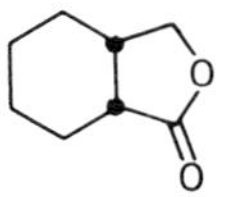

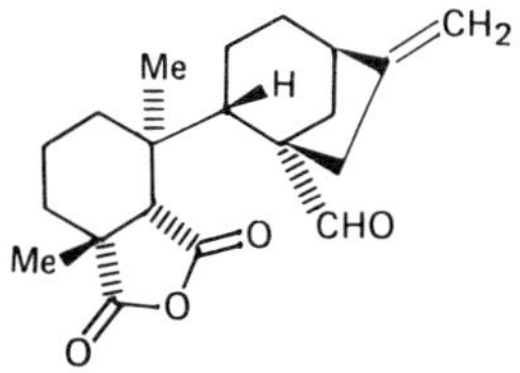

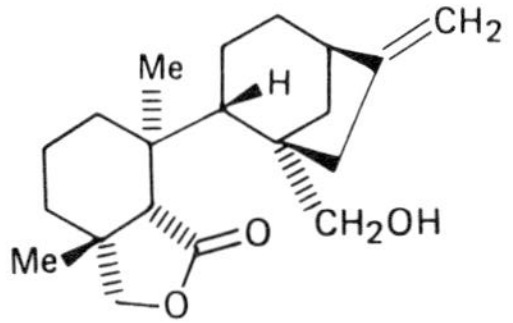

	$NaBH_4$ in MeOH		8	248
	$LiAlH_4$ in ether, dioxane		60	248
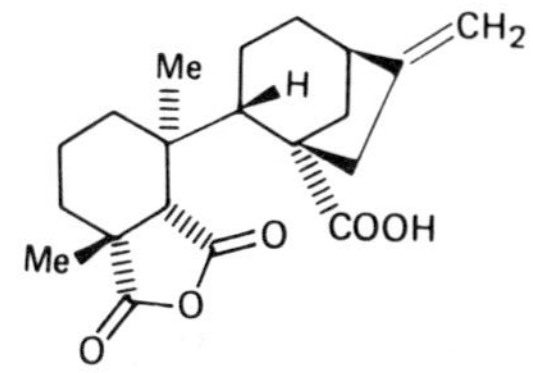	$LiAlH_4$, ether	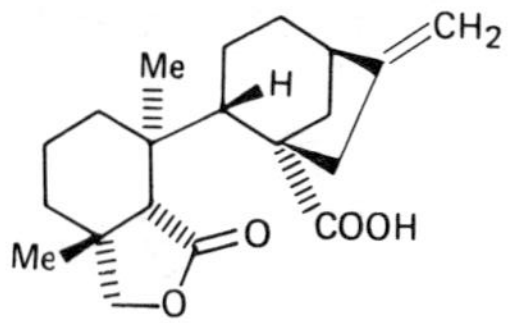	82, 90	241, 248
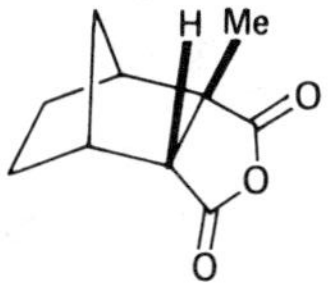	$NaBH_4$, *i*-PrOH	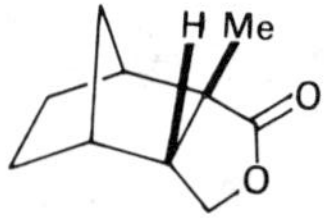	83	242
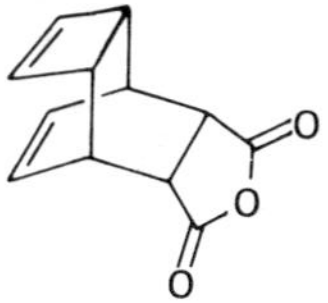	$LiAlH_4$, THF	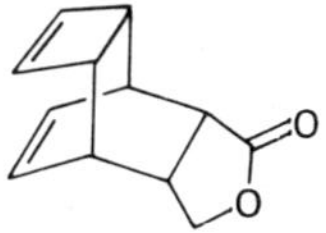	––	243
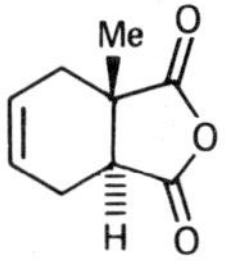	$LiAlH_4$, THF	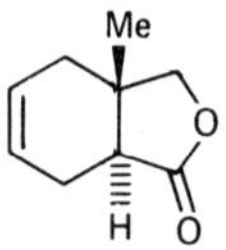	75	237
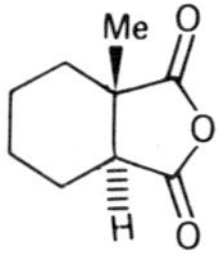	$LiAlH_4$, THF	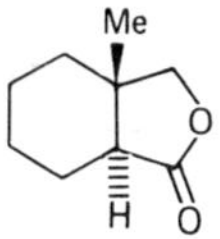	85	237

TABLE 19. (Continued)

Anhydride	Reducing agent	Product	Yield (%)	References
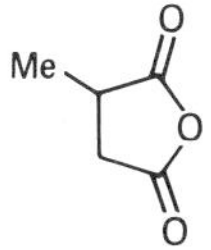	$LiAlH_4$, THF	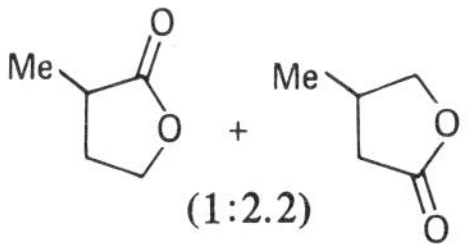 (1:2.2)	69	237
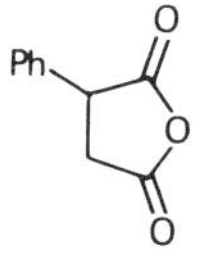	$LiAlH_4$, THF or $NaBH_4$, THF	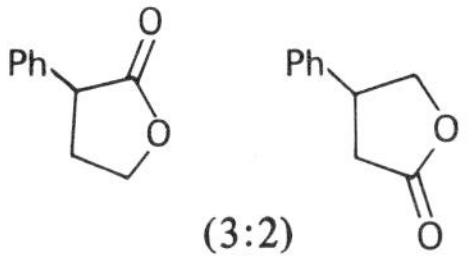 (3:2)	72, 67	237, 236
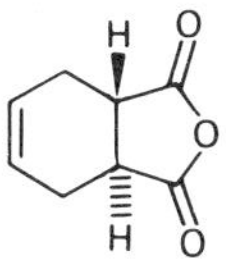	$LiAlH_4$, THF	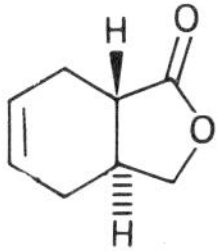	70.4	237
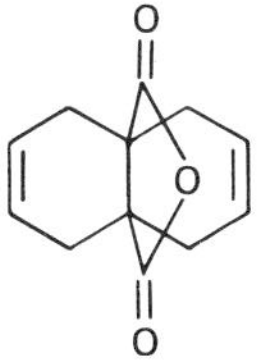	$LiAlH_4$, THF	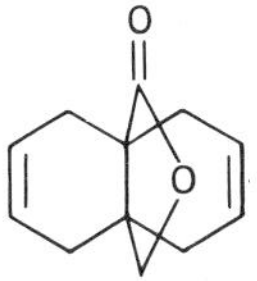	79–83	237
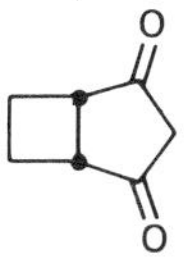	$LiAlH_4$, THF	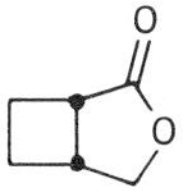	70.5	237

Substrate	Reagents	Product	Yield (%)	Ref.
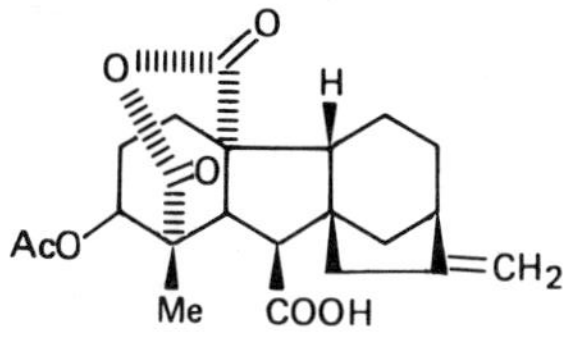	$LiAlH_4$, THF	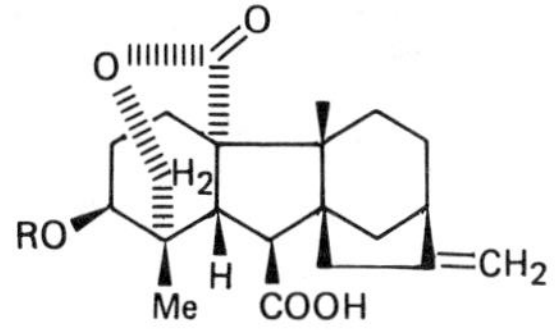 R = H, major product R = Ac, minor product	50	244
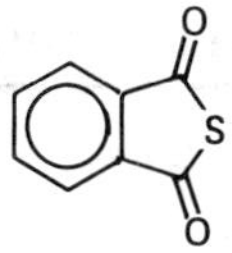	$NaBH_4$, C_6H_6–MeOH	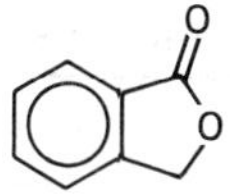	80	245
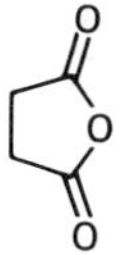	$NaBH_4$, THF, 10*N* HCl in EtOH	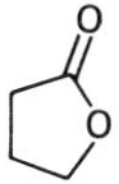	51	236
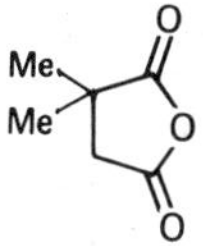	$NaBH_4$, THF, 6*N* HCl in H_2O	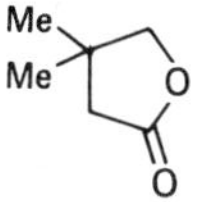	74	236
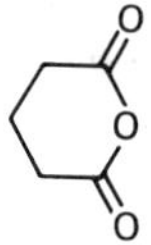	$NaBH_4$, THF, 10*N* HCl in EtOH	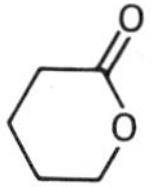	67	236
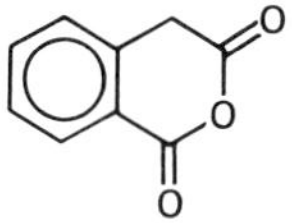	$NaBH_4$, THF, 10*N* HCl in EtOH	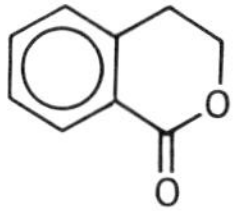	55	236

TABLE 19. (Continued)

Anhydride	Reducing agent	Product	Yield (%)	References
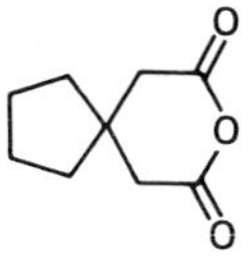	$NaBH_4$, THF, 10N HCl in EtOH	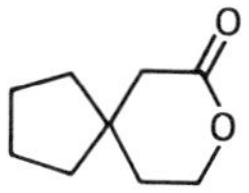	68	236
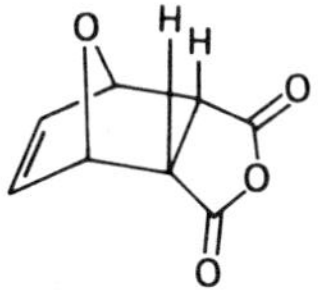	$NaBH_4$, EtOH	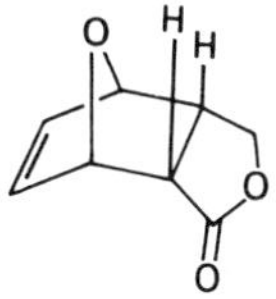	80[a]	247

[a] Heating at 140–150°C (20 Torr) afforded 79% of γ-crotonolactone via a retro Diels–Alder reaction.

Although reactions which have been reported to exhibit the opposite trend are apparently not in question, a uniform explanation for the anomalies is still unavailable.

Lithium aluminium hydride and catalytic[249] reductions of dicarboxylic acids (equation 137)[250,251], and their diesters[249] and monoesters[237,239] have been employed with only modest success for lactone synthesis.

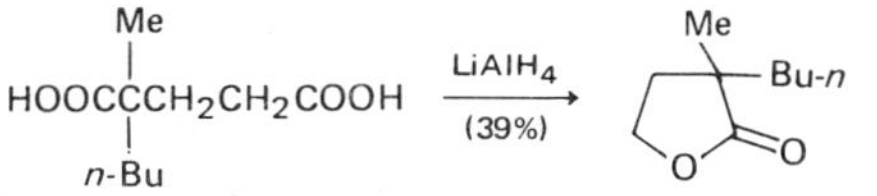

$$\mathrm{HOOCC(Me)(\textit{n}\text{-}Bu)CH_2CH_2COOH} \xrightarrow[(39\%)]{\mathrm{LiAlH_4}} \text{3-methyl-3-}n\text{-butyl-}\gamma\text{-butyrolactone} \quad \text{(Ref. 250)} \qquad (137)$$

Carboxylic acids and esters containing an aldehyde or ketone carbonyl function at the γ- or δ-position often provide good yields of lactones upon treatment with various reducing agents (Table 20).[252–262]. The choice of reduction conditions is often governed by whether the carboxyl group is free or esterified, for in the latter instances the reducing agent should be capable of reducting the ketone or aldehyde carbonyl without affecting the carboalkoxy function.

N. By Oxidation Reactions

Diols, ketones, ethers, olefins and several other miscellaneous types of compounds can be converted to lactones by oxidative reactions employing a variety of reagents. The following discussion is organized in terms of the type of compound used as starting material.

1. *Oxidation of diols*

A wide variety of 1,4- and 1,5- diols have been oxidized to lactones by reagents such as copper chromite, chromic acid, manganese dioxide, potassium permanganate and silver carbonate on celite (equation 138)[263–277]. Table 21 contains a representative series of diols along with their lactone oxidation products.

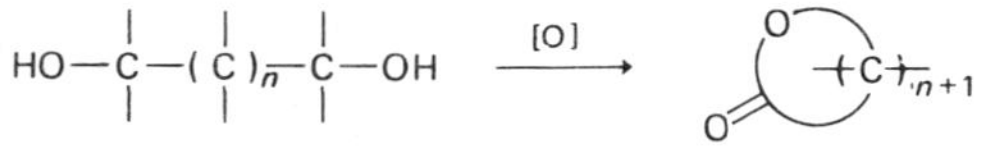

$$\mathrm{HO{-}C{-}(C)_{\mathit{n}}{-}C{-}OH} \xrightarrow{[O]} \text{lactone ring } (\mathrm{C})_{n+1} \qquad (138)$$

Oxidative cleavage of unsaturated keto diols using lead tetraacetate or sodium periodate has been found to be an effective method for the production of steroidal lactones (equations 139–142)[255,256,278].

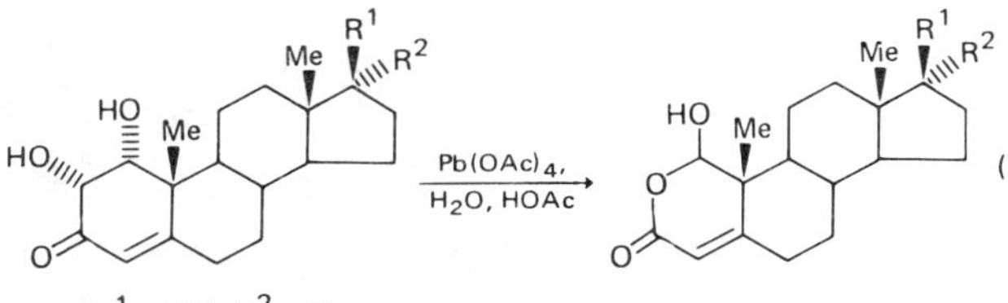

$$\text{steroid diol} \xrightarrow[\mathrm{H_2O,\ HOAc}]{\mathrm{Pb(OAc)_4,}} \text{steroidal lactol lactone} \quad \text{(Ref. 255)} \qquad (139)$$

R^1 = OH, R^2 = Me

R^1, R^2 = O

TABLE 20. Preparation of lactones by reduction of keto and aldehydic acids and esters

Acid or ester	Reducing agent	Product	Yield (%)	References
$ROCCH_2CH_2CH_2COOH$	$Al(i\text{-}PrO)_3$, i-PrOH	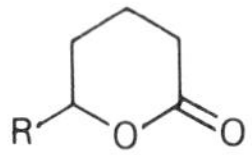		20
		$R = n\text{-}C_4H_9$	69	
		$R = n\text{-}C_6H_{13}$	64	
		$R = n\text{-}C_7H_{15}$	66	
$R^1OCCH_2CH_2CH_2COOR^2$		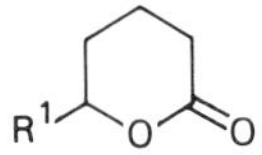		20
	$Al(i\text{-}PrO)_3$, i-PrOH	$R^1 = n\text{-}C_3H_7$	89	
		$R^1 = n\text{-}C_5H_{11}$	69	
		$R^1 = n\text{-}C_8H_{17}$	62	
		$R^1 = n\text{-}C_9H_{19}$	73	
	$NaBH_4$, NaOH	$R^1 = n\text{-}C_4H_9$	87	
		$R^1 = n\text{-}C_6H_{13}$	68	
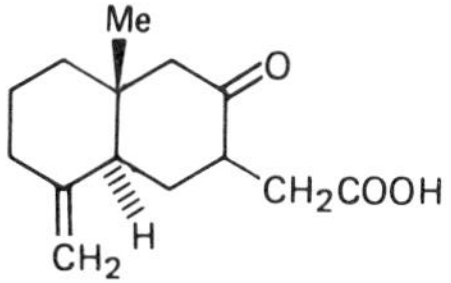	$NaBH_4$, MeOH	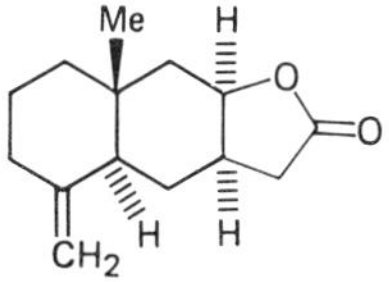	66	253
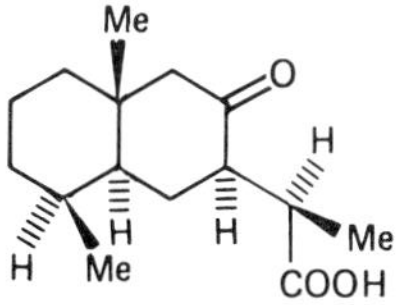	Na, i-PrOH	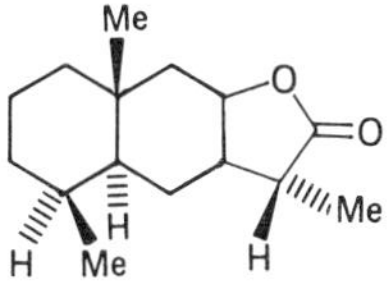	10	17
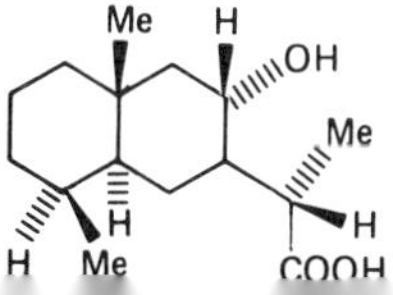	Na, i-PrOH			

Substrate	Reagent	Product	Yield (%)	Ref.
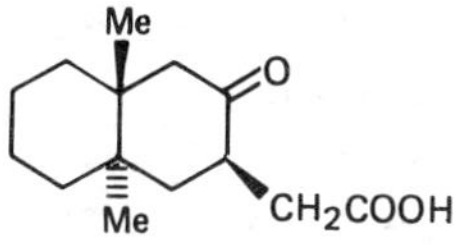 CH2COOH; Me; Me; O	H_2, PtO_2	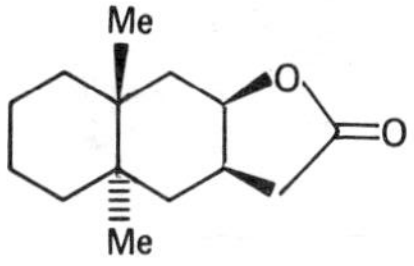 Me; Me; O; O	73	19, 254
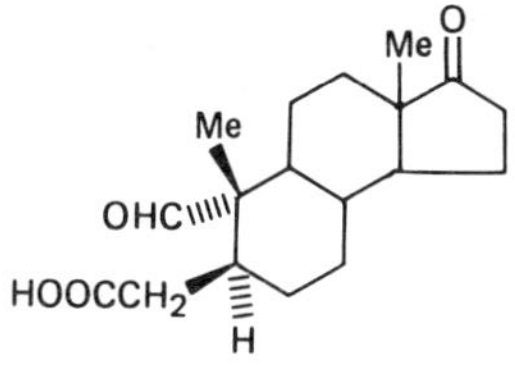 Me; O; Me; OHC; $HOOCCH_2$; H	$NaBH_4$	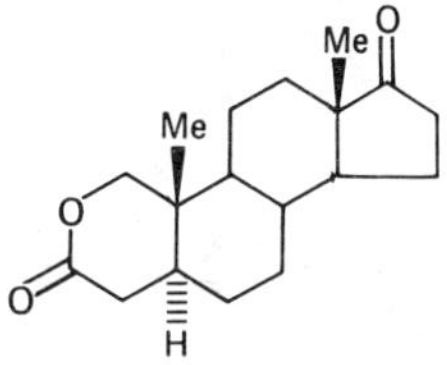 Me; O; Me; O; O; H	50–60	255
Me; C_8H_{17}; Me; HOOC; O	$NaBH_4$[a], NaOH, EtOH Na + EtOH Na + *i*-PrOH	Me; C_8H_{17}; Me; O; O; H	34 75[b] 41	257 258 250
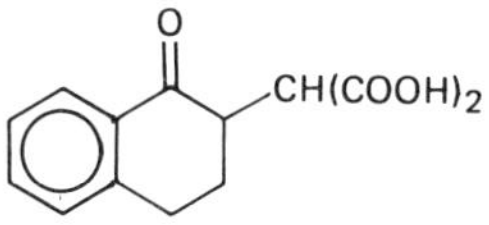 O; $CH(COOH)_2$	$NaBH_4$, NaOH, H_2O	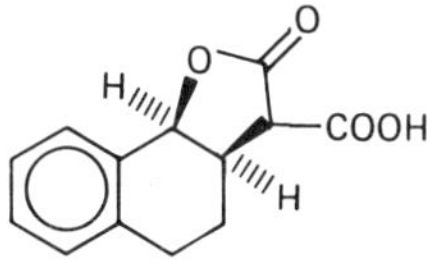 H; O; O; COOH; H	90	154
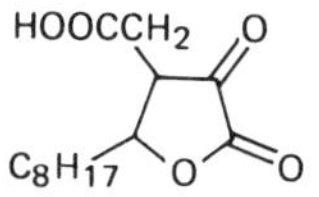 $HOOCCH_2$; O; C_8H_{17}; O; O	H_2, 5% Rh on alumina, then HCl	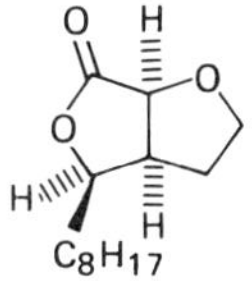 O; H; O; O; H; H; C_8H_{17}	>90	261

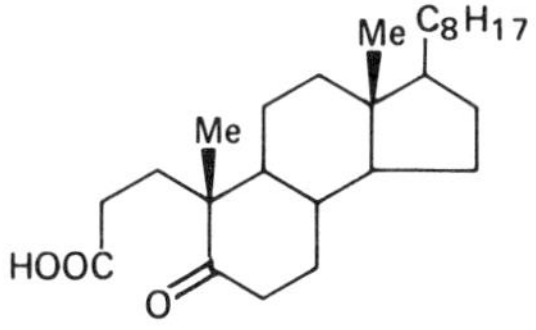

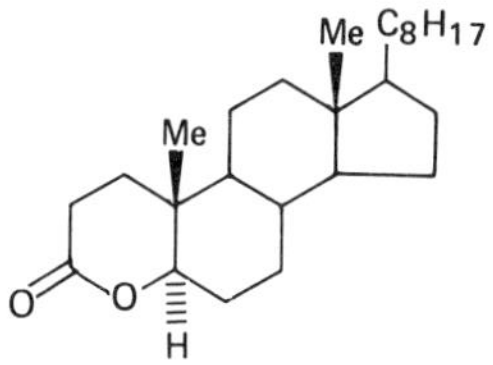

TABLE 20. (Continued)

Acid or ester	Reducing agent	Product	Yield (%)	References
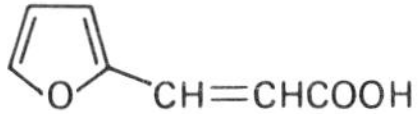 CH=CHCOOH	Ni(Al)[c], NaOH, H_2O	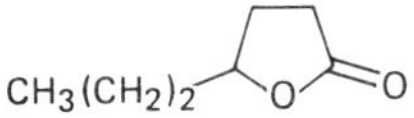 $CH_3(CH_2)_2$	33–37	262
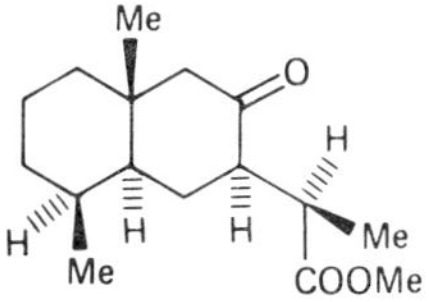 Me, O, H, Me, H, H, COOMe, Me, H	1. 3% NaOMe, MeOH 2. KBH_4, MeOH	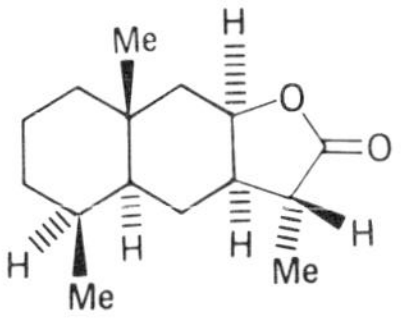 Me, H, O, O, H, Me, H, H, Me	42	17
Me, O, Me, H, COOMe, H, Me, H	KBH_4, MeOH		45	17
Me, O, Me, H, COOR, H, H, Me, H		Me, H, O, O, Me, H, H, H, Me, H		17
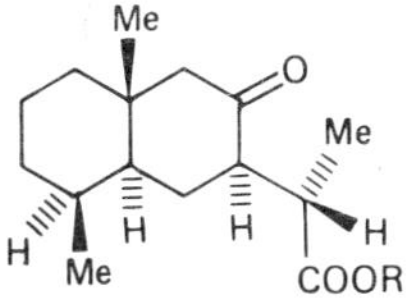	KBH_4, MeOH $NaHCO_3$	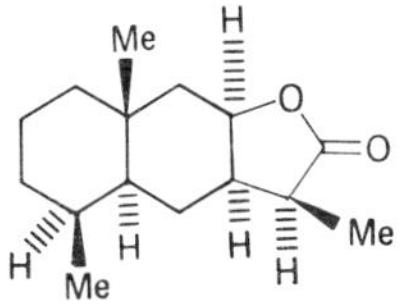	88 (R = H) 45 (R = Me)	
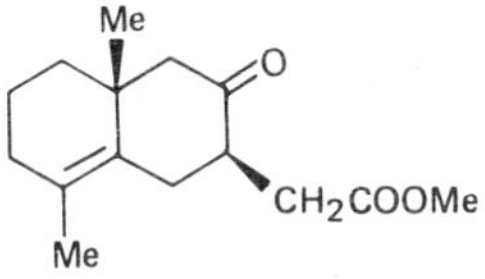 Me, O, CH_2COOMe, Me	KBH_4, MeOH	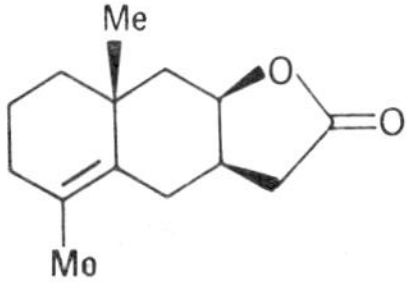 Me, O, O, Mo	74	19

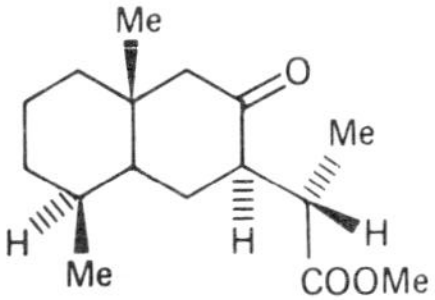

Starting material	Conditions	Product	Yield (%)	Ref.
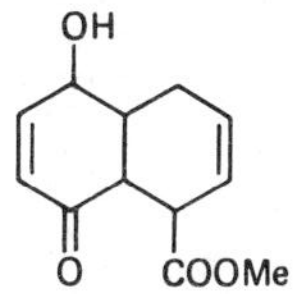 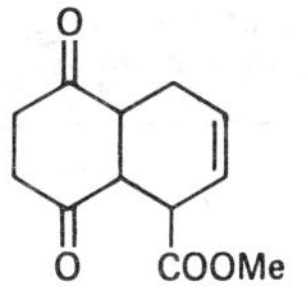	Al(i-PrO)$_3$, i-PrOH	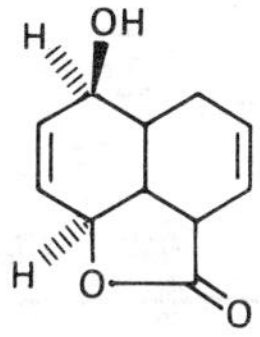	65	28
	Al(i-PrO)$_3$, i-PrOH		45	
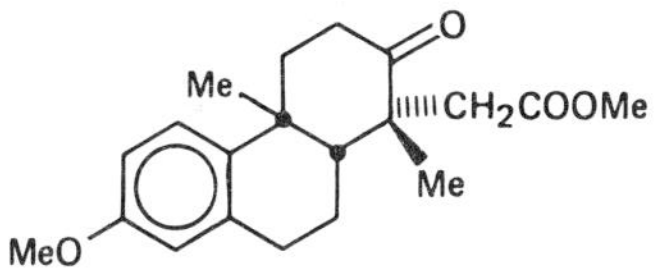	$NaBH_4$, MeOH	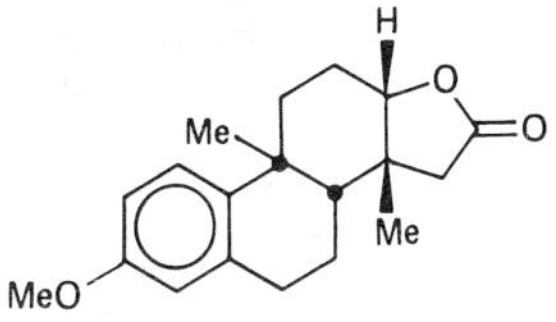	79	22
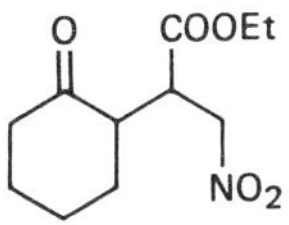	$NaBH_4$, MeOH 1 h, 0°C	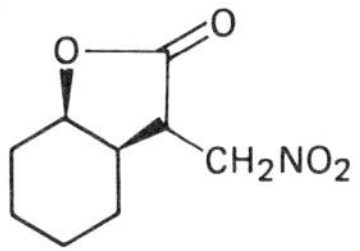	50	45
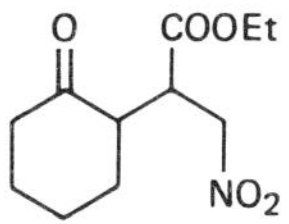	1. $NaBH_4$, MeOH, 0°C 2. Reflux, 10 h	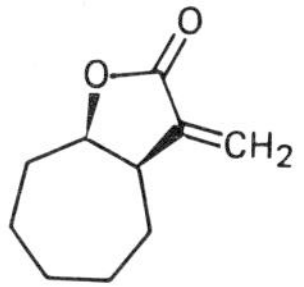	45	45

TABLE 20. (Continued)

Acid or ester	Reducing agent	Product	Yield (%)	References
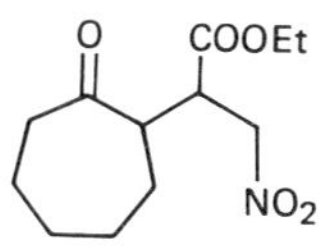 O, COOEt, NO_2	$NaBH_4$, MeOH	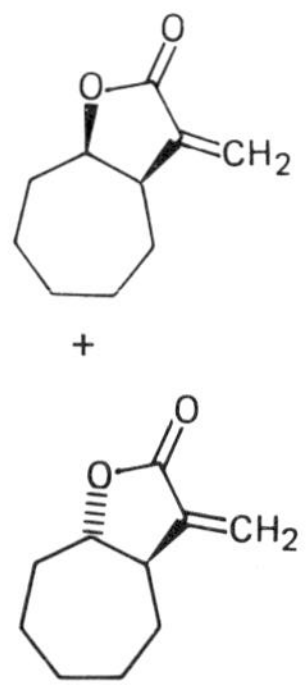 O, O, CH_2 + O, O, CH_2	91 (86:14)	45

[a]Reduction with $Al(i\text{-}PrO)_3$ in *i*-PrOH gave an oily product containing *cis* and *trans* lactones.
[b]Crude product.
[c]Raney nickel–aluminium alloy.

TABLE 21. Preparation of lactones by oxidation of diols

Diol	Oxidizing agent	Product	Yield (%)	References
$HOCH_2CH_2CH(Me)CH_2CH_2OH$	Copper chromite or copper on pumice	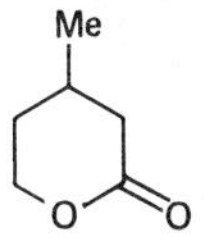 4-Me-δ-valerolactone (structure)	90–95	263, 264
$MeCH(OH)CH_2CH_2CH_2OH$	Copper chromite	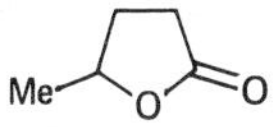 γ-Me-γ-butyrolactone (structure)	87	265
$HOCH_2CH_2CH_2CH_2CH_2OH$	Copper chromite	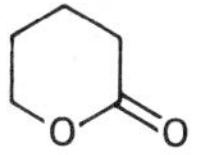 δ-valerolactone (structure)	71	266
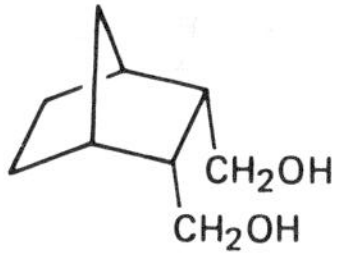 bicyclo[2.2.1]heptane-2,3-dimethanol (structure; CH_2OH, CH_2OH)		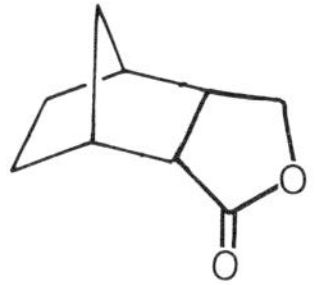 fused bicyclic lactone (structure)		267
	Raney Ni, C_6H_6		80	
	$KMnO_4$, NaOH		10	
$MeCH_2CH_2CH(OH)CH(Me)CH_2CH_2OH$	$K_2Cr_2O_7$, AcOH	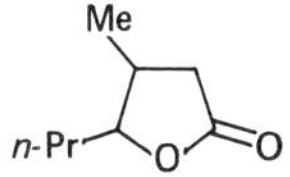 (structure; Me, n-Pr)	75–80	252
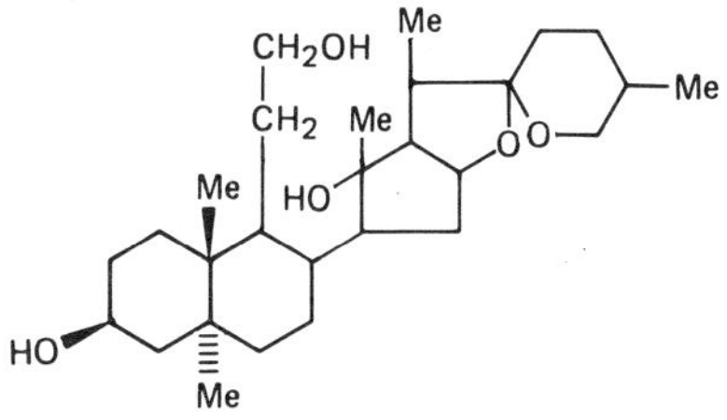 steroidal spiroketal diol (structure; HO, Me, CH_2OH, CH_2, Me, HO, Me, Me)	CrO_3, C_5H_5N[a]	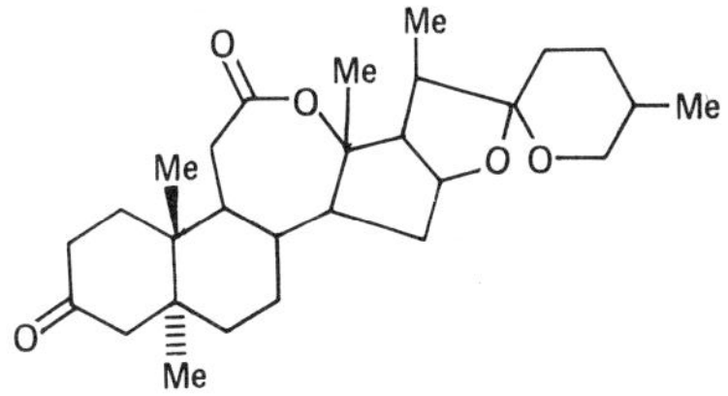 steroidal spiroketal keto-lactone (structure; Me, Me, Me, Me, Me)	86	268

TABLE 21. (Continued)

Diol	Oxidizing agent	Product	Yield (%)	References
$Me(CH_2)_5C(OH)(Me)(CH_2)_2CH_2OH$	CrO_3, H_2SO_4, H_2O	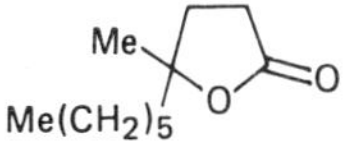 Me, $Me(CH_2)_5$	71	270
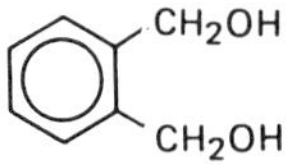 CH_2OH, CH_2OH		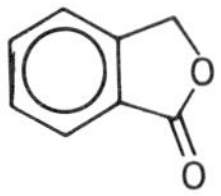		
	$KMnO_4$, H_2O		71	271
	CrO_3, C_5H_5N[a]		60	271
	$Na_2Cr_2O_7$, H_2SO_4, H_2O		95	271
	Ag_2CO_3–celite, C_6H_6		95	272
$HOCH_2CH{=}CHCH_2OH$	CrO_3, C_5H_5N[a]	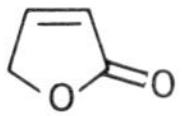	51	271
$HOCH_2CH_2CH_2CH_2OH$	CrO_3, C_5H_5N[a]	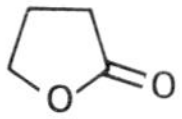	34	271
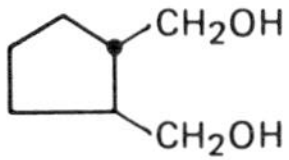 CH_2OH, CH_2OH	CrO_3, C_5H_5N[a]	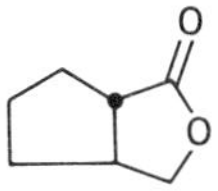	Trace	271
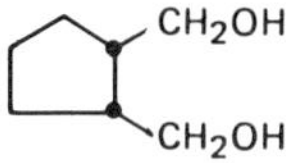 CH_2OH, CH_2OH		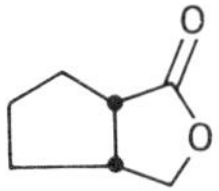		271
	CrO_3, C_5H_5N[a]		Trace	
	$Na_2Cr_2O_7$, H_2SO_4, H_2O		60	

Substrate	Reagents	Product	Yield	Ref.
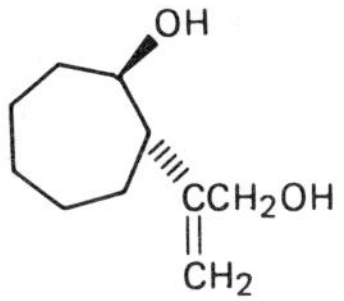	MnO_2, C_6H_6	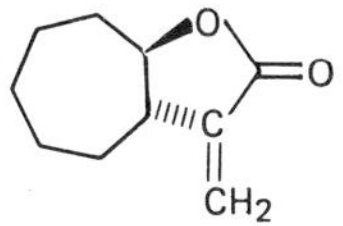	87	30, 31
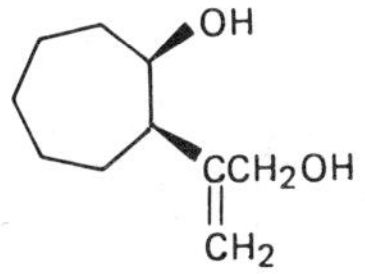	MnO_2, C_6H_6	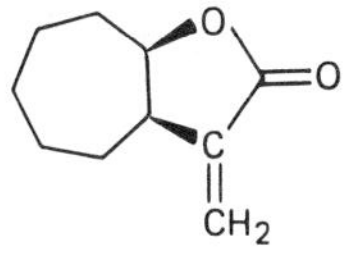	58	31
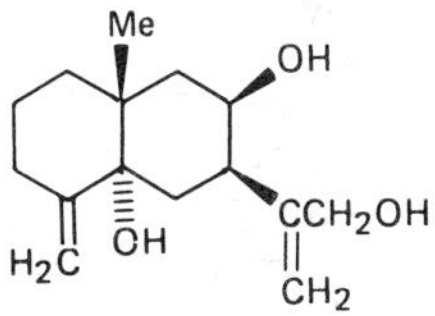	MnO_2, C_6H_6	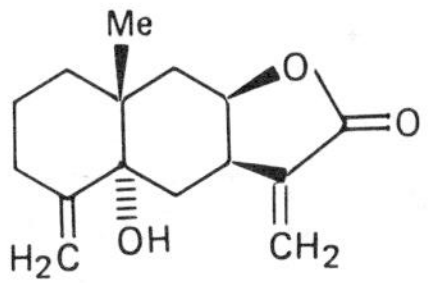	83	19
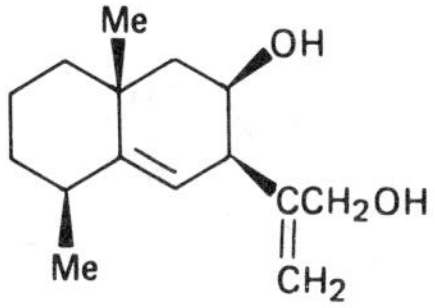	MnO_2, C_6H_6	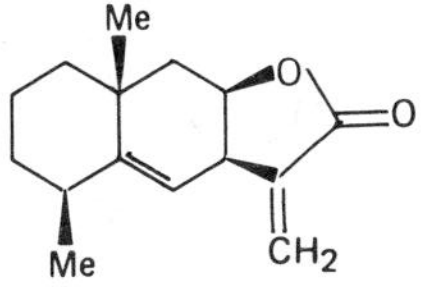	80	19
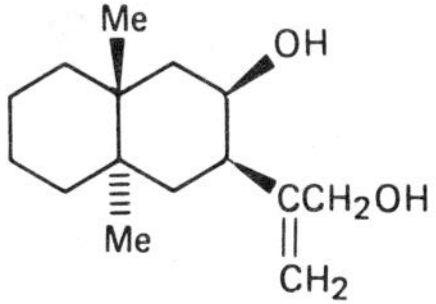	MnO_2, C_6H_6	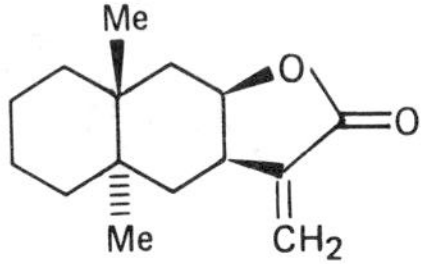	63	254
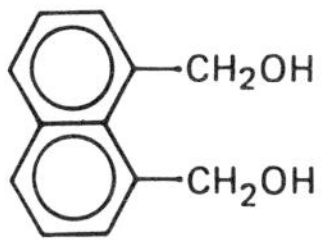	MnO_2, C_6H_6	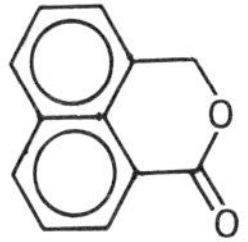	76	273

TABLE 21. (Continued)

Diol	Oxidizing agent	Product	Yield (%)	References
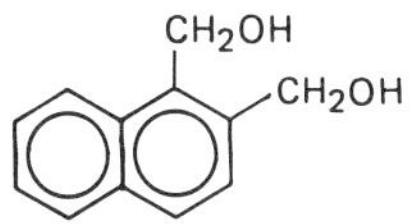 CH2OH, CH2OH	MnO_2, C_6H_6	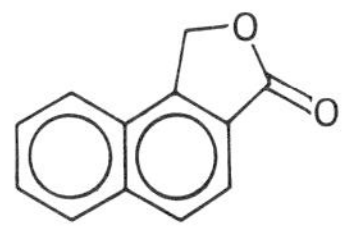 O, O	65	273
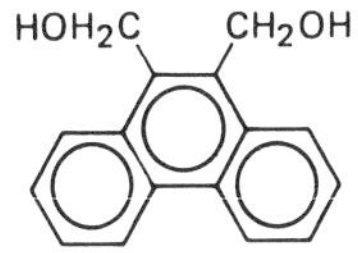 HOH2C, CH2OH	MnO_2, C_6H_6	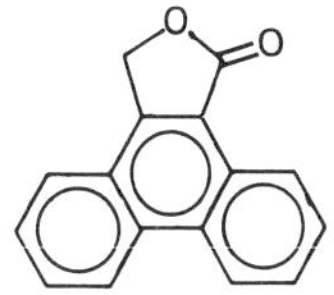 O, O	73	273
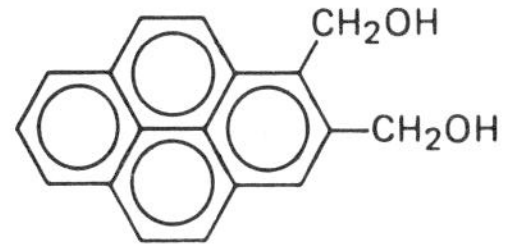 CH2OH, CH2OH	MnO_2, C_6H_6	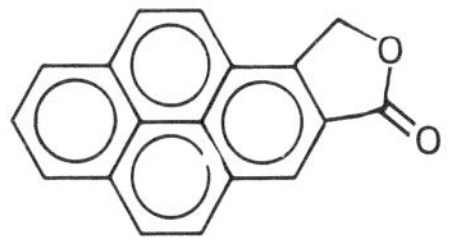 O, O	58	273
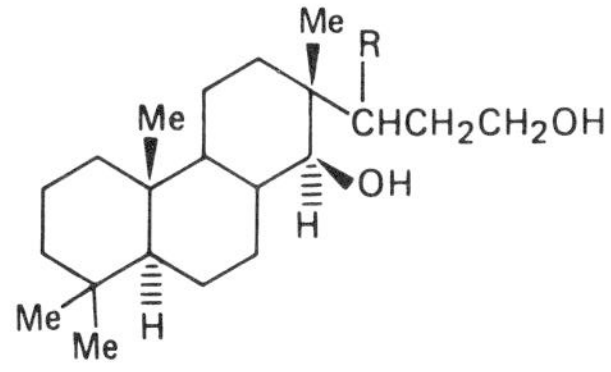 Me, R, CHCH2CH2OH, Me, OH, H, Me, Me, H		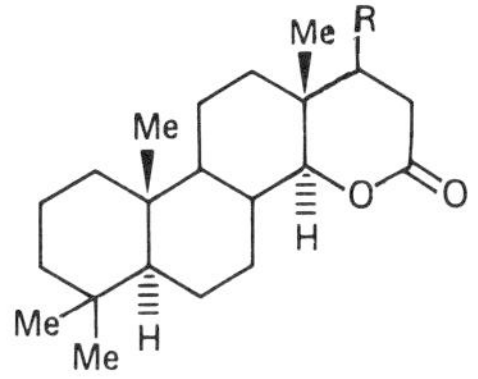 Me, R, Me, O, O, H, Me, Me, H		274
	CrO_3, C_5H_5N[a] (R = Me)		55	
	MnO_2, MeCN (R = Me)		90	
	MnO_2, MeCN (R = OH)		60	

Substrate	Reagent	Product	Yield (%)	Ref.
Me, Me, Me, H, Me, $CH_2CH_2CH_2OH$, H, OH	CrO_3, C_5H_5N[a]	Me, Me, Me, H, Me, H, O, O	50	274
Me, Me, Me, H, H, OH, $CH_2CH_2CH_2OH$, Me	MnO_2, C_6H_6	Me, Me, Me, H, H, O, O, Me	84	276
$HOCH_2(CH_2)_nCH_2OH$	Ag_2CO_3–celite, C_6H_6	O, O, $(CH_2)_n$, CH_2	n = 2 52 n = 3 90–94 n = 4 96–100	272, 277
O, O, CH_2OH, CH_2OH	Ag_2CO_3–celite, C_6H_6	O, O, O, O	60–79	272, 277
Me, Me, O, O, H, H, CR_2OH, CR_2OH R = H, D	Ag_2CO_3–celite, C_6H_6	Me, Me, O, O, H, H, O, O, R, R	50–65	272, 277

TABLE 21. (Continued)

Diol	Oxidizing agent	Product	Yield (%)	References
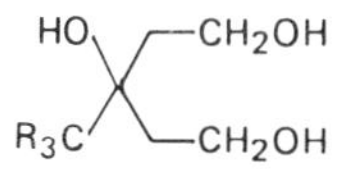 HO, R_3C, CH_2OH, CH_2OH	Ag_2CO_3–celite, C_6H_6	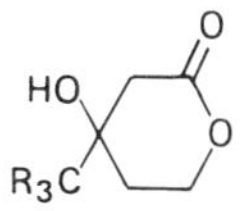 HO, R_3C, O, O		277
		R = H	74	
		R = D	––	
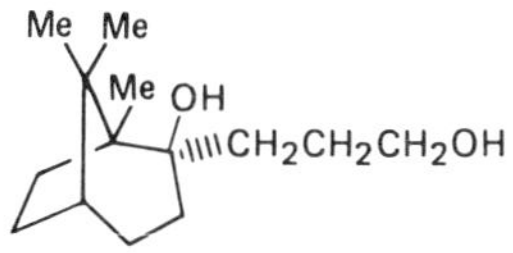 Me, Me, Me, OH, $CH_2CH_2CH_2OH$	Ag_2CO_3–celite, C_6H_6	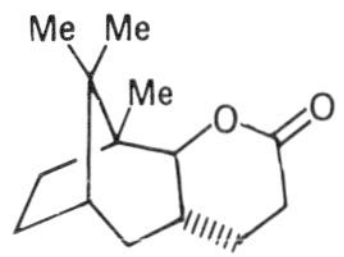 Me, Me, Me, O, O	~100	272
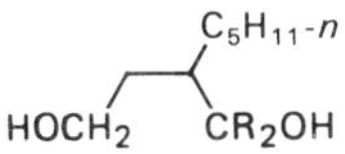 C_5H_{11}-*n*, $HOCH_2$, CR_2OH	Ag_2CO_3–celite, C_6H_6	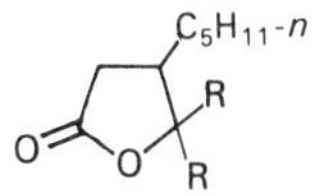 C_5H_{11}-*n*, R, R, O, O		272
		R = H	60	
		R = D	90	
		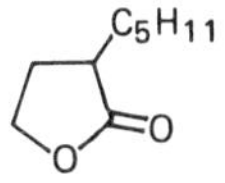 + C_5H_{11}, O, O		
		R = H	40	
		R = D	10	
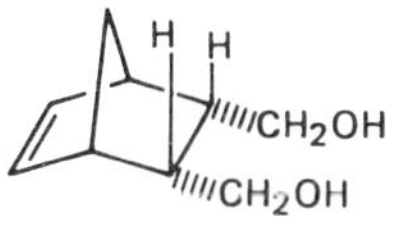 H, H, CH_2OH, CH_2OH	Ag_2CO_3–celite, C_6H_6	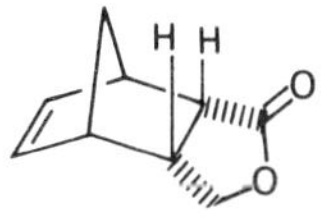 H, H, O, O	96	272

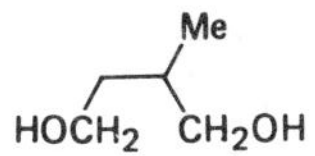 Me; $HOCH_2$ CH_2OH	Ag_2CO_3–celite, C_6H_6	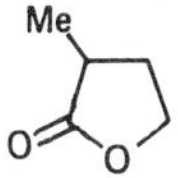 Me; O; O	66	272
		+ Me; O; O	28	
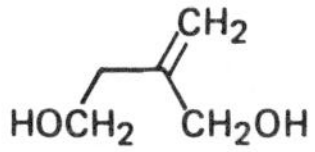 CH_2; $HOCH_2$ CH_2OH	Ag_2CO_3–celite, C_6H_6	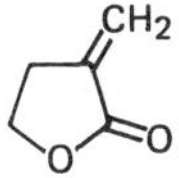 CH_2; O; O	80	272
$X(CH_2CH_2OH)_2$	Ag_2CO_3–celite, C_6H_6	X; O; O	X = O 95 X = S 9	272
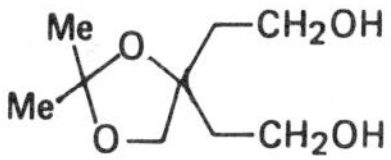 Me; Me; O; O; CH_2OH; CH_2OH	Ag_2CO_3–celite, C_6H_6	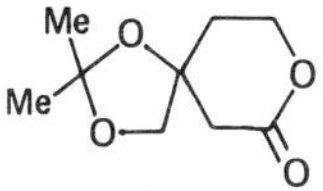 Me; Me; O; O; O; O	77	272
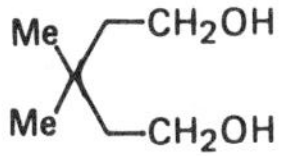 Me; Me; CH_2OH; CH_2OH	Ag_2CO_3–celite, C_6H_6	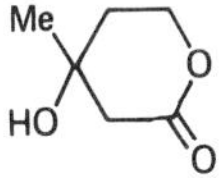 Me; HO; O; O	74	272
Me–$CHCH_2CH_2CH_2OH$ (OH on CH)		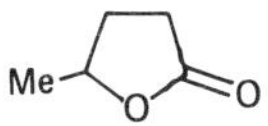 Me; O; O		272
	Ag_2CO_3–celite, C_6H_6 $CHCl_3$		41 72	

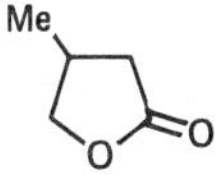

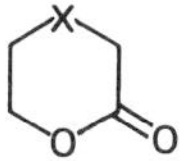

TABLE 21. (Continued)

Diol	Oxidizing agent	Product	Yield (%)	References
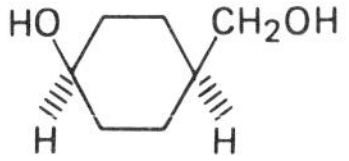 HO, CH_2OH, H, H		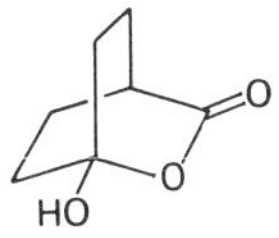 HO, O, O		272
	Ag_2CO_3–celite, C_6H_6		56	
	$CHCl_3$		14	
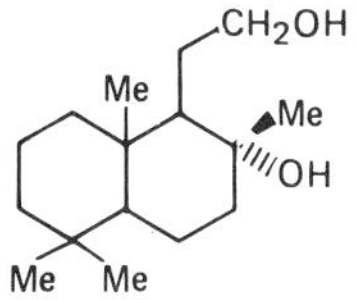 CH_2OH, Me, Me, OH, Me, Me	Ag_2CO_3–celite, C_6H_6	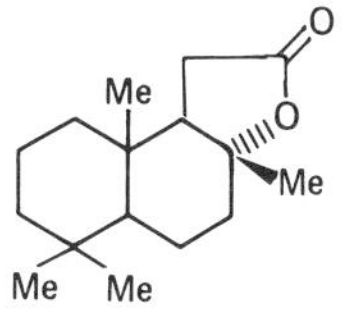 O, Me, O, Me, Me, Me	100	272

[a]Chromic anhydride–pyridine complex; see G. I. Poos, G. E. Arth, R. E. Beyler and L. H. Sarett, *J. Amer. Chem. Soc.*, **75**, 427 (1953).

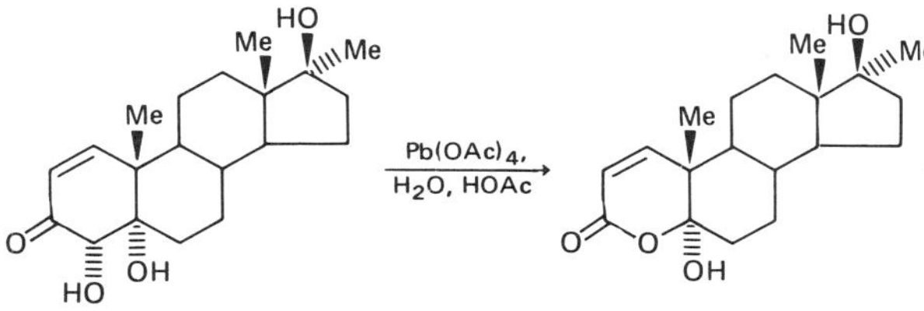

Pb(OAc)$_4$, H$_2$O, HOAc (Ref. 255) (140)

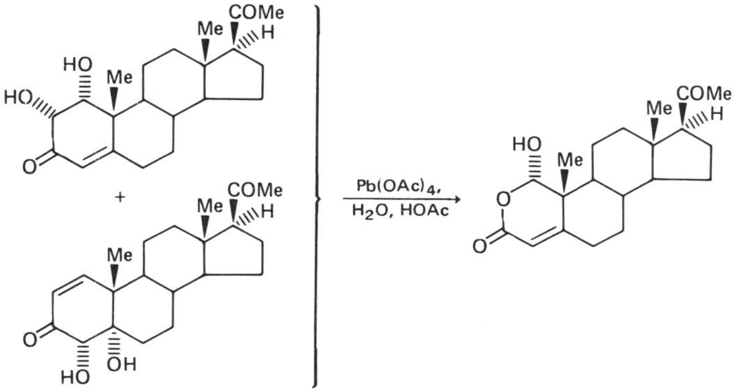

Pb(OAc)$_4$, H$_2$O, HOAc (Ref. 255) (141)

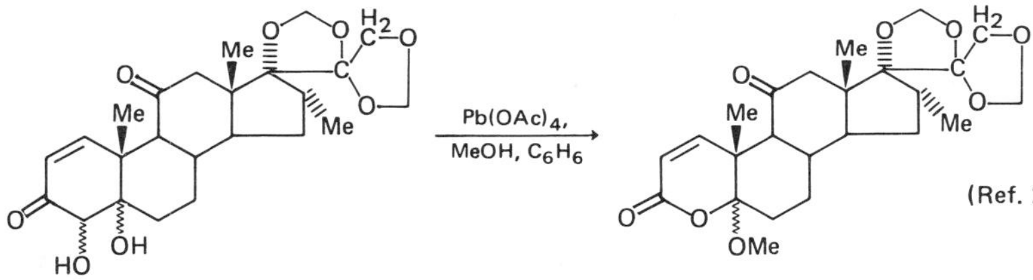

Pb(OAc)$_4$, MeOH, C$_6$H$_6$ (Ref. 278) (142)

Steroidal δ-hydroxy oximes and lactols derived from the free δ-hydroxy aldehydes can be oxidized to lactones with sodium dichromate[256] or chromic anhydride (equation 143)[279].

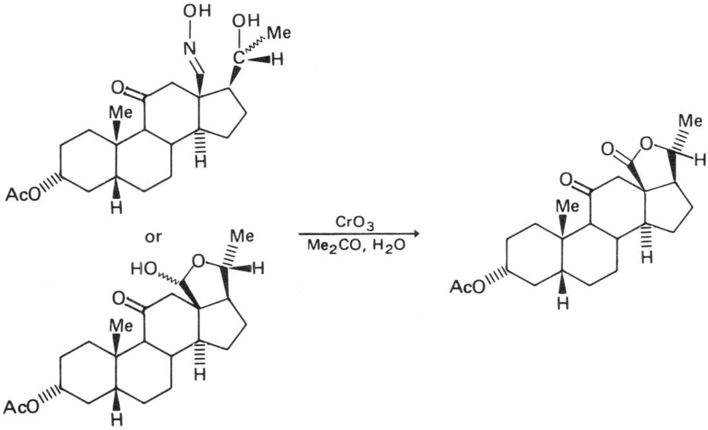

CrO_3, Me_2CO, H_2O (Ref. 279) (143)

TABLE 22. Preparation of lactones via Baeyer–Villiger oxidation of ketones

Starting material	Reagent	Product	Yield (%)	Reference
AcOH$_2$C, H, O, H, Me, Me, H	$MeCO_3H$, 25°C, 80 h.	O, O, H, Me, Me, H, H, CH$_2$OAc	50	282
		+ O, H, O, Me, Me, H, H, CH$_2$OAc	18	
RO, Me, Me, O, Me, Me, O, O, Me	$PhCO_3H$, $CHCl_3$, H_2SO_4, 25°C	RO, Me, Me, O, O, Me, Me, O, O, Me		268
	60 h	R = OAc	80	
	12 days	R = OAc	98	
	64 h.	R = H	90	
	30% H_2O_2, HOAc, 12 days	R = OAc	trace	
EtO—C(=O)—O, Me, Me, O, Me, Me, CHCH$_2$CH$_2$COOMe	$PhCO_3H$, $CHCl_3$ H_2SO_4, HOAc	EtO—C(=O)—O, Me, Me, O, O, Me, Me, CHCH$_2$CH$_2$COOMe	90	268

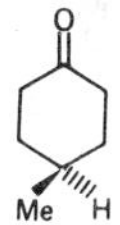	$(CF_3CO)_2O$, 90% H_2O_2	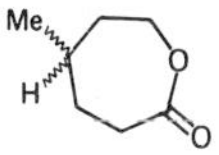	79.2	283
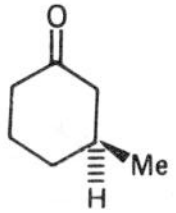	$(CF_3CO)_2O$, 90% H_2O_2	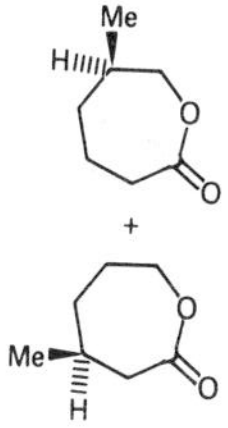 +	64.3	283
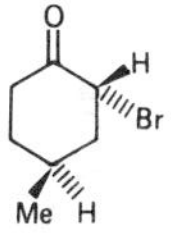	$(CF_3CO)_2O$, 90% H_2O_2, $CHCl_3$	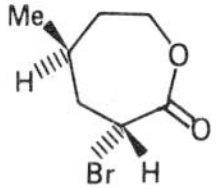	86.7	283
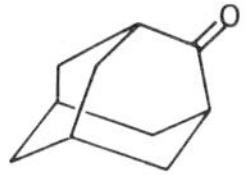	m-$ClC_6H_4CO_3H$ 30% H_2O_2, HOAc	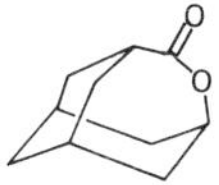	– 90–95	284 285
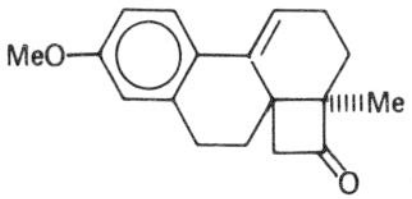	H_2O_2, OH^- or m-$ClC_6H_4CO_3H$, CH_2Cl_2	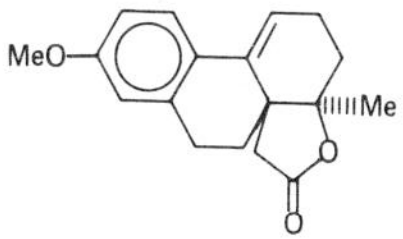	90	286

TABLE 22. (Continued)

Starting material	Reagent	Product	Yield (%)	Reference
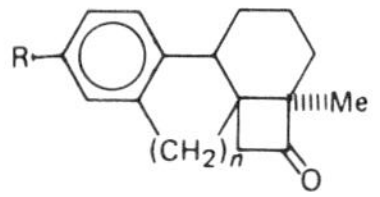 R, $(CH_2)_n$, Me, O	m-$ClC_6H_4CO_3H$, CH_2Cl_2	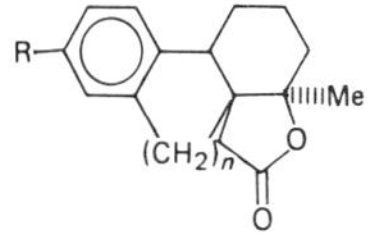 R, $(CH_2)_n$, Me, O		286
		$n = 2$, R = H	68–72	
		$n = 2$, R = OMe	–	
		$n = 1$, R = H	80	
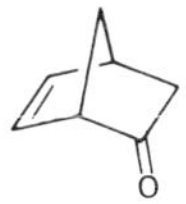 O	40% $MeCO_3H$, NaOAc, $CHCl_3$	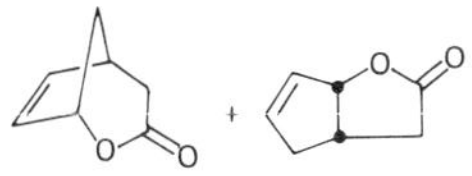 O + O	56	289
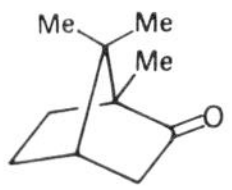 Me, Me, Me, O		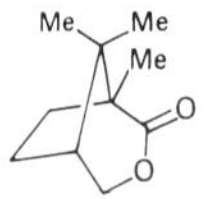 Me, Me, Me, O		290
	40% $MeCO_3H$, 2:3 H_2SO_4:HOAc		30	
	$MeCO_3H$, HOAc, NaOAc		82	
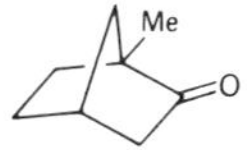 Me, O	40% $MeCO_3H$, HOAc, NaOAc	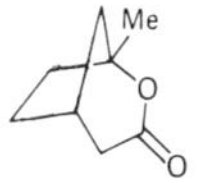 Me, O	42	291
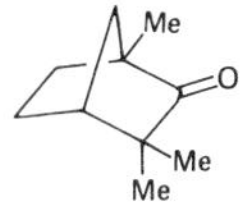 Me, O, Me, Me	40% $MeCO_3H$, HOAc, NaOAc	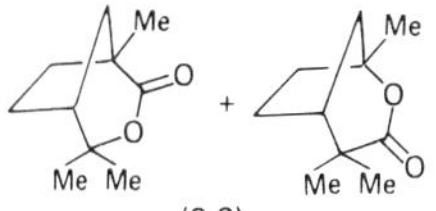 Me, O, Me Me + Me, O, Me Me (3:2)	35	291

Substrate	Conditions	Product	Yield (%)	Ref.
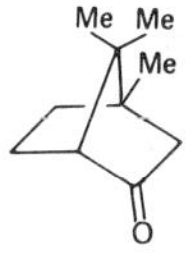	40% $MeCO_3H$, HOAc, NaOAc	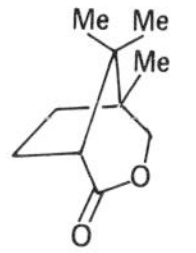	94	291
				29
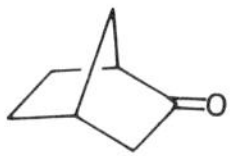	28% $MeCO_3H$, HOAc, NaOAc 40% $MeCO_3H$, HOAc, H_2SO_4	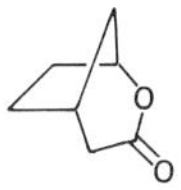	88 97	
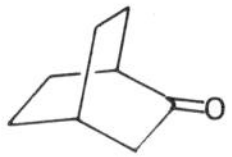	28% $MeCO_3H$, HOAc, NaOAc	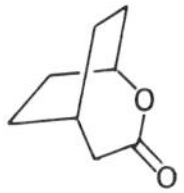	85	29
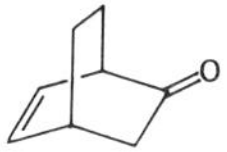	28% $MeCO_3H$, HOAc, NaOAc	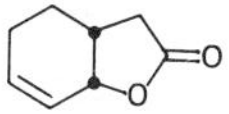	80	29
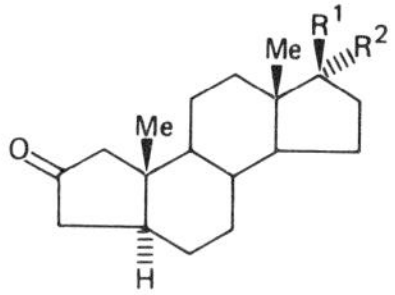	$PhCO_3H$, $CHCl_3$	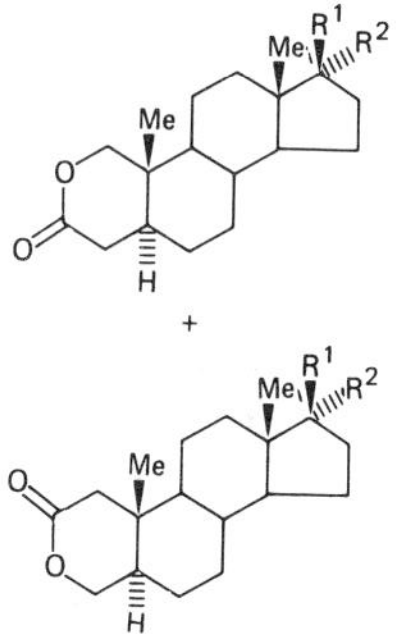 + $R^1 = C_8H_{17}$, H, OH, OAc, OH R^2 = H, H, H, H, Me	70–90	293

TABLE 22. (Continued)

Starting material	Reagent	Product	Yield (%)	Reference
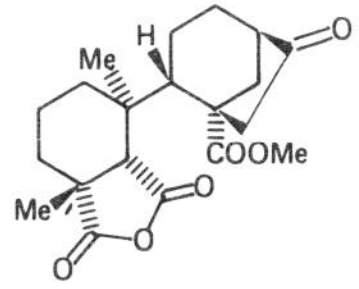	PhCO$_3$H, CHCl$_3$, p-MeC$_6$H$_4$SO$_3$H	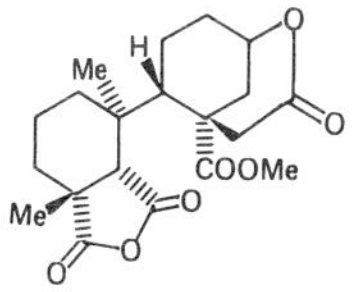	~60	248
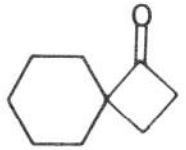	30% H$_2$O$_2$, MeOH, NaOH HOCl	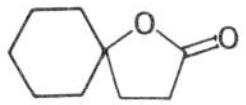	82 90	294, 295 295
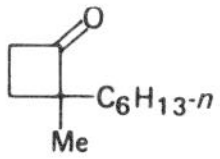	30% H$_2$O$_2$, MeOH, NaOH	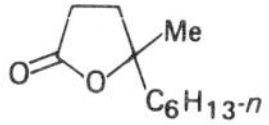	100	294, 295
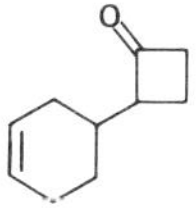	30% H$_2$O$_2$, MeOH, NaOH	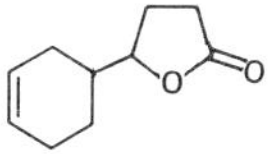	82	294, 295
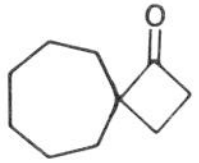	30% H$_2$O$_2$, MeOH, NaOH NaOBr	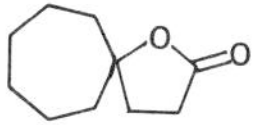	94 94	294 295
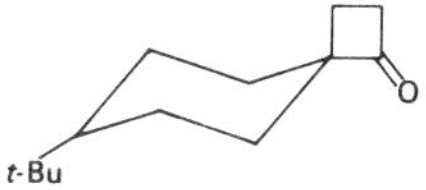	30% H$_2$O$_2$, MeOH, NaOH	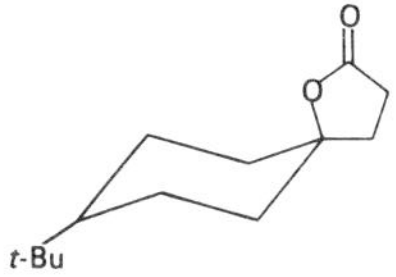	100	294

Substrate	Reagent	Product	Yield (%)	Ref.
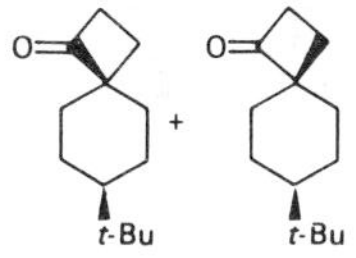 t-Bu, t-Bu; 70:30	30% H_2O_2, MeOH, NaOH	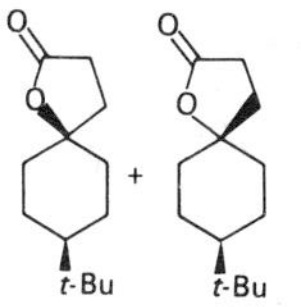 t-Bu, t-Bu	100	295
	HOCl, pH 4[a]		83	296
	Me_3CCO_2H		22	297
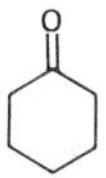	H_2O_2-urea, 85% HCO_2H	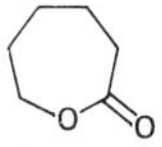	95	298
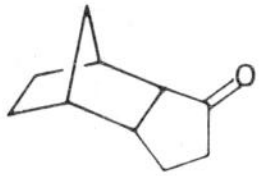	30% H_2O_2, HOAc	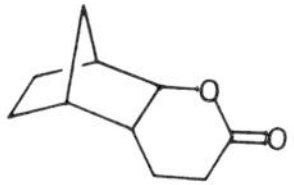	85–90	285
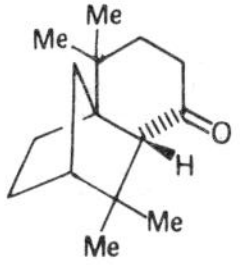 Me, Me, H, Me, Me	30% H_2O_2, HOAc	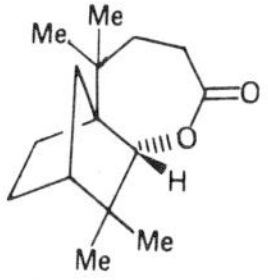 Me, Me, H, Me, Me	90–95	285
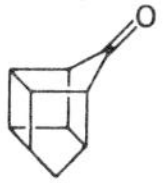	30% H_2O_2, HOAc	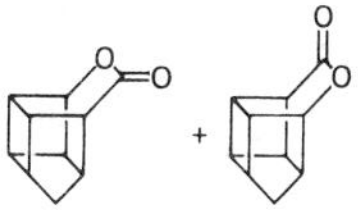	80–85	285

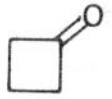

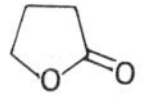

TABLE 22. (Continued)

Starting material	Reagent	Product	Yield (%)	Reference
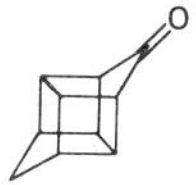	30% H_2O_2, HOAc	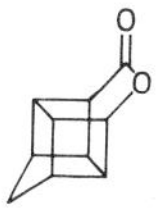	80–85	285
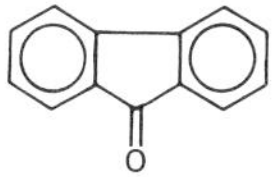	30% H_2O_2, HOAc	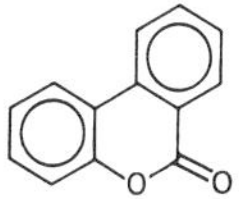	80–85	285
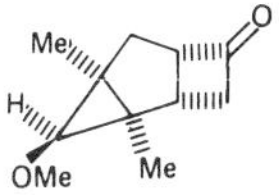 Me, H, OMe, Me	30% H_2O_2, HOAc	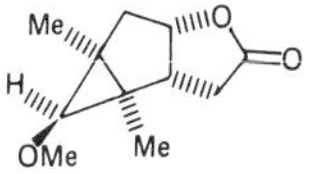 Me, H, OMe, Me	>90	299
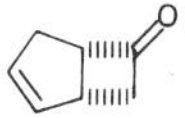	30% H_2O_2, HOAc, H_2O	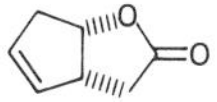	90	299, 300
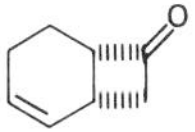	30% H_2O_2, HOAc, H_2O; 30% II_2O_2, MeOII, II_2O, OII^-	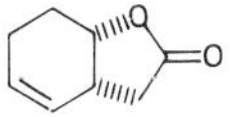	95	303
Me, Me	$PhCO_3H$, C_6H_6, p-$MeC_6H_4SO_3H$	Me, Me	68	307
	or	+ (4:1)		
	$PhCO_3H$, C_6H_6	Me, Me, H, H	78	

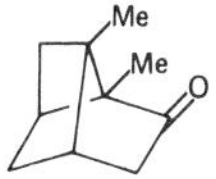

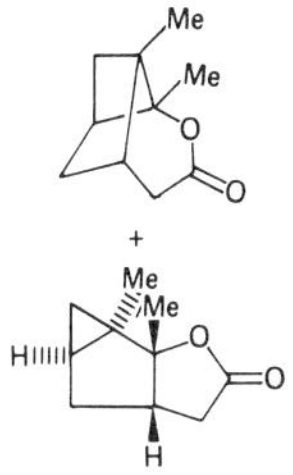

	Reagent	Product	Yield (%)	Ref.
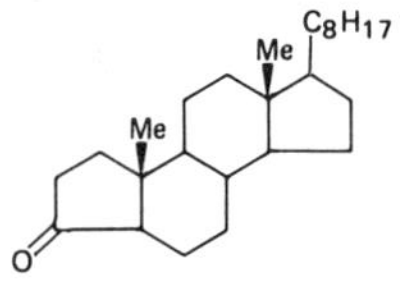 Me, Me, C_8H_{17} (steroid ketone)	$MeCO_3H$, HOAc, *p*-$MeC_6H_4SO_3H$	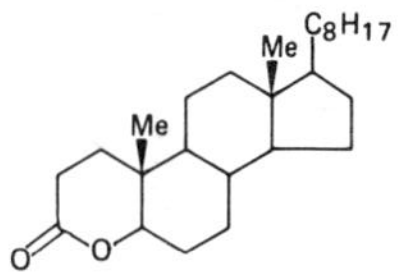 Me, Me, C_8H_{17} (lactone)	90	308
	$MeCO_3H$, HOAc, H_2SO_4		40	

[a] Using this reagent with cyclopentanone and cyclohexanone did not afford any lactone.

2. Oxidation of ketones

The Baeyer–Villiger[280,281] reaction remains the premier oxidative method for the preparation of lactones from cyclic ketones. The mechanism of this reaction has been reviewed in detail[281] and will not be discussed here. Table 22 contains a number of recent examples[282–308].

Oxygen or ozone have been used to convert ketones to lactones. For instance, reaction of cyclopentanone with oxygen in the presence of 1-benzyl-1,4-dihydronicotinamide has been reported[296] to afford a 12% yield of butyrolactone (equation 144). When similar reactions were conducted under nitrogen or in the absence of

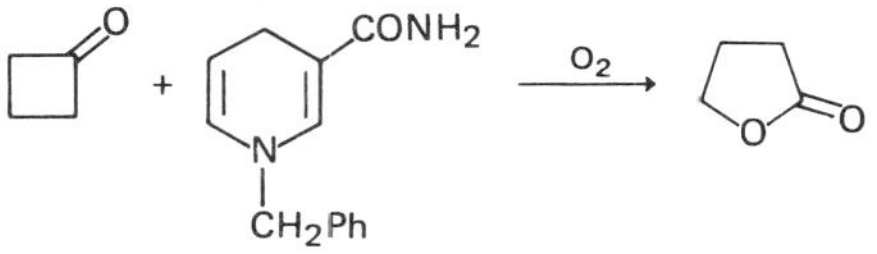

CONH2 + O2 CH2Ph (144)

the nicotinamide, no lactone was produced. These findings led the authors[296] to conclude that the dihydronicotinamide probably functions as an oxygen carrier, and is converted by oxygen into its hydroperoxide, which then produces the lactone via Baeyer–Villiger oxidation of the ketone.

Various ketones can be oxidized to lactones using potassium *t*-butoxide and atmospheric oxygen (equations 145–147)[306], Rose Bengal-sensitized photo-oxidation (equation 148)[307] or potassium *t*-butoxide and oxygen followed by reduction with sodium borohydride (equations 149–151)[308].

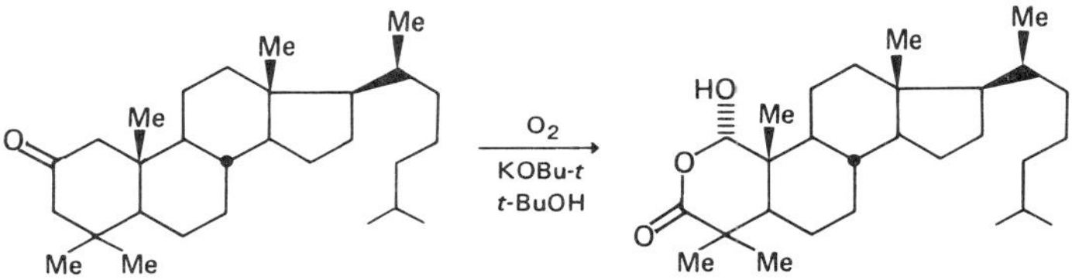

O2, KOBu-*t*, *t*-BuOH (145)

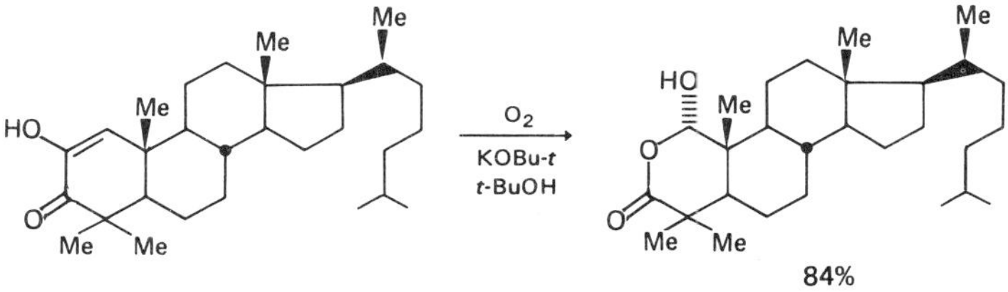

O2, KOBu-*t*, *t*-BuOH 84% (146)

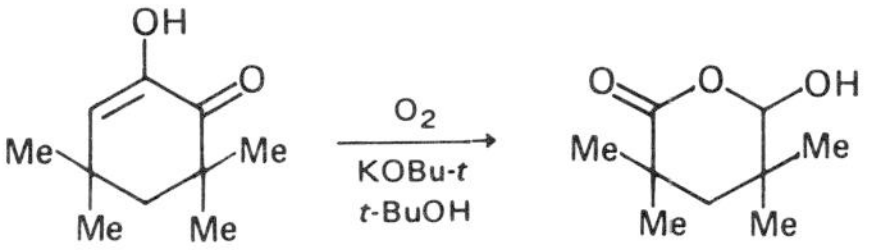

O2, KOBu-*t*, *t*-BuOH (147)

Ozonolysis of silyloxyalkenes followed by treatment with sodium borohydride has also been reported[309] to afford lactones (equation 152).

Anodic oxidation of the sodium bisulphite addition products of cyclopentanone and cyclohexanone[310] afford mixtures of γ- and δ-lactones as shown in equations (153) and (154). Since the relative amounts of the lactones obtained by this

Rose Bengal, $h\nu$ → 37% (148)

1. O_2, KOBu-*t*, *t*-BuOH; 2. $NaBH_4$, MeOH, H_2O → 50%

or

1. O_2, KOBu-*t*, $(Me_2NCO)_3PO_4$; 2. $NaBH_4$, H_2O, MeOH → 20% (149)

1. O_2, KOBu-*t*, *t*-BuOH; 2. $NaBH_4$, MeOH, H_2O → 21% + 9% (150)

1. O_2, KOBu-*t*, *t*-BuOH; 2. $NaBH_4$, MeOH, H_2O → + (151)

1. O_3, MeOH, CH_2Cl_2, $-78°C$
2. $NaBH_4$
3. H^+, H_2O

R = CH=CH$_2$ 93%
R = Me 70%

(152)

$-2e$
Na_2CO_3, H_2O

HO SO_3Na

17–20% trace

(153)

$-2e$
Na_2CO_3, H_2O

HO SO_3Na

Et Me

(3:2)

(154)

method correspond to the relative proportions of the same lactones obtained by acid-catalysed cyclization of 5-hexenoic acid[311,312], the authors find it reasonable to assume that the electrolytic oxidation proceeds via a carbonium ion or oxonium intermediate[310].

Oxidation of 2-adamantanone with ceric ammonium nitrate in aqueous acetonitrile at 60°C has been reported[284] to afford a 73% yield of the corresponding lactone, while similar oxidation of 2-adamantanol gave[284] the same lactone in 50% yield (equation 155).

$Ce(NH_4)_2NO_3$
MeCN

(155)

Addition of aqueous methanolic sodium periodate to a crude sample of the hydroxymethylene ketone shown in equation (156) effected[313] a direct conversion to $R(-)$-mevalonolactone, since the acetal group was hydrolysed during isolation of the product. In a similar manner[313] $S(+)$-mevalonolactone was prepared from the analogous hydroxymethylene ketone precursor.

$NaIO_4$,
MeOH, H_2O

CHOH

Me Me

(156)

3. Oxidation of ethers

The oxidative conversion of cyclic esters to lactones is not a commonly encountered synthetic procedure; however, it has been found to be useful in several cases, and should not be ignored.

Ruthenium tetroxide has been reported[314] to oxidize tetrahydrofuran to γ-butyrolactone, and tetrahydrofurfuryl alcohol to a compound tentatively identified as the corresponding aldehyde lactone. Attempts to convert ethylene oxide to an α-lactone with this reagent were unsuccessful[314].

t-Butyl chromate has been used[315] to obtain spiro lactones from spiroethers. Thus, reaction of 3β-acetoxy-2′,3′α-tetrahydrofuran-2 ,-spiro-17(5-androstene) with *t*-butyl chromate under standard conditions[316] afforded[315] a 23% yield of 3-(3β-acetoxy-17β-hydroxy-7-oxo-5-androsten-17α-yl) propionic acid lactone (equation 157). Similar results[315] were obtained with the spiro ethers shown in equations (158) and (159).

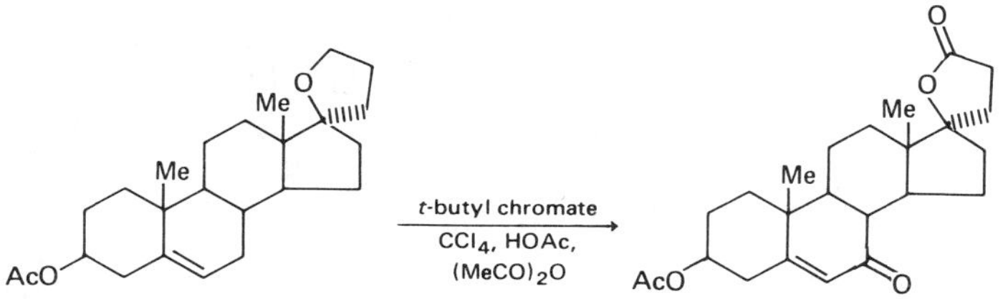

t-butyl chromate; CCl_4, HOAc, $(MeCO)_2O$ (157)

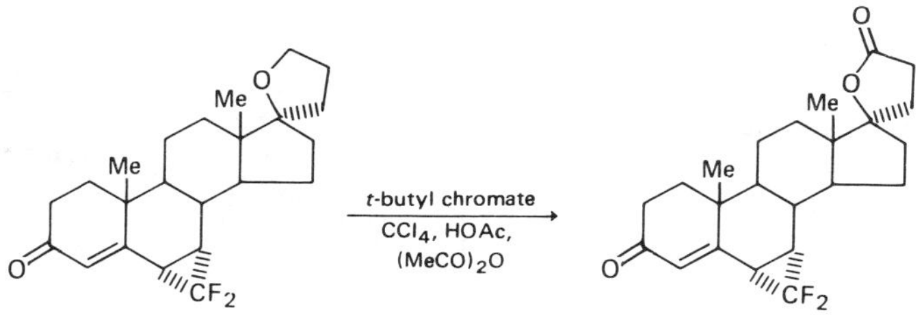

t-butyl chromate; CCl_4, HOAc, $(MeCO)_2O$ (158)

40%

t-butyl chromate; CCl_4, HOAc, $(MeCO)_2O$ (159)

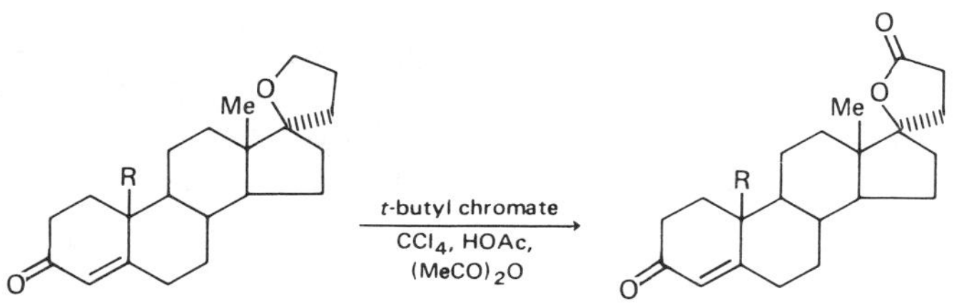

R = Me, 48%

R = H, 30%

Photosensitized oxygenation of furan and furan derivatives in the presence of an appropriate sensitizer such as Rose Bengal can be employed for the synthesis of certain butenolides (equations 160 and 161)[317-321].

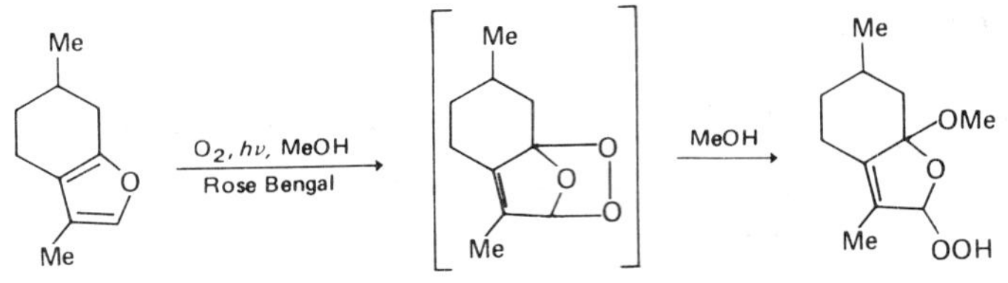

(Ref. 317, 318) (160)

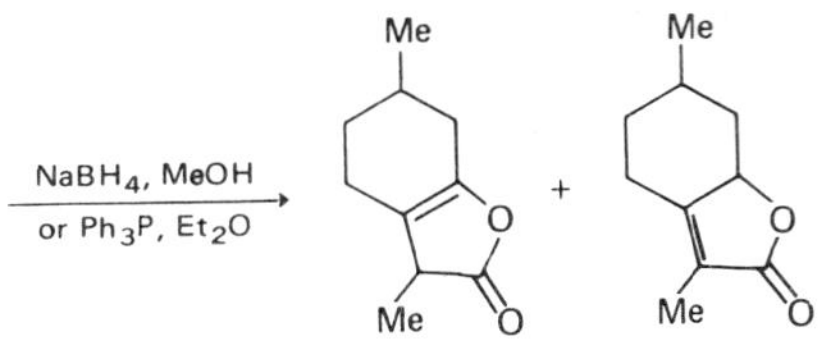

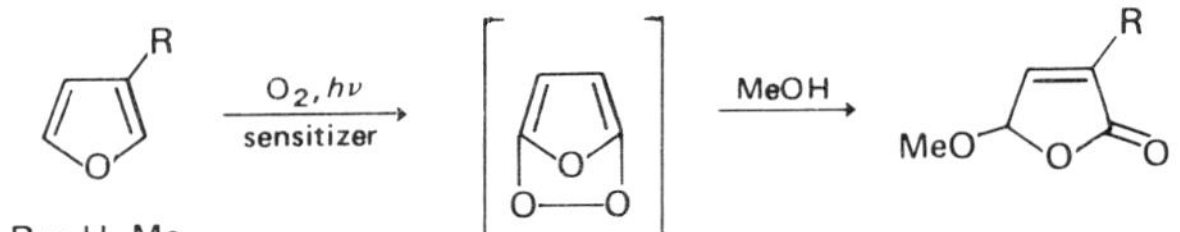

(Ref. 319–321) (161)

4. *Oxidation of olefins*

Oxidation of olefins with excess manganese (III) acetate affords γ-lactones in moderate to good yields (Table 23)[322,323]. The mechanism of this reaction, illustrated in equation (162) with styrene, involves addition of a carboxymethyl

$$PhCH{=}CH_2 + \dot{C}H_2COOH \longrightarrow Ph\dot{C}HCH_2CH_2COOH \xrightarrow{[O]} Ph\overset{+}{C}HCH_2CH_2COOH \longrightarrow \quad (162)$$

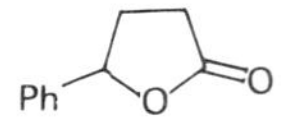

radical to the double bond, oxidation of the resulting radical to a carbonium ion, and then ring-closure to form the lactone. Similar results have been observed with manganese dioxide in the presence of acetic anhydride and acetic acid[324].

Manganese (III) acetate, as well as certain cerium and vanadium salts, have been found effective in catalysing the addition of carboxylic acids, having an α-hydrogen across the double bond of various olefins (equation 163) to produce γ-lactones (Table 24)[325].

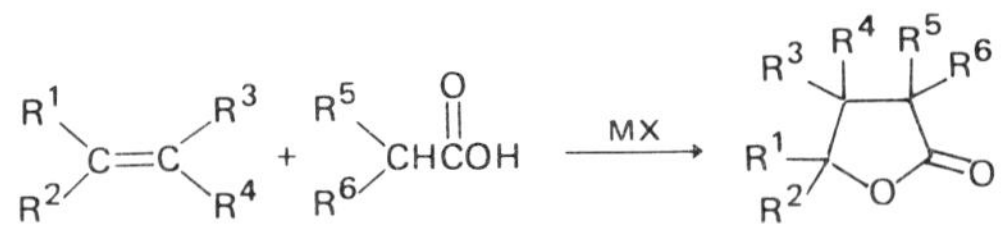

$$R^1R^2C{=}CR^3R^4 + R^5R^6CHCOOH \xrightarrow{MX} \quad (163)$$

MX = $Mn(OAc)_3 \cdot 2H_2O$, $Mn(OAc)_3$, MnO_2, Mn_2O_3, $Ce(OAc)_4$, $Ce(NH_4)_2(NO_3)_6$, NH_4VO_3

TABLE 23. Oxidation of olefins to lactones by manganese (III) acetate

$$R^1R^2C{=}CR^3R^4 \xrightarrow[\text{HOAc, (MeCO)}_2\text{O}]{\text{Mn(OAc)}_3 \cdot 2\,\text{H}_2\text{O}} \gamma\text{-lactone (}R^1,R^2\text{ at C-5; }R^3,R^4\text{ at C-4)}$$

R^1	R^2	R^3	R^4	Yield (%)	References
Ph	H	H	H	75, 60	322, 323
Ph	Me	H	H	83, 74	322, 323
Ph	H	Me	H	21[a], 79	322, 323
$PhCH_2$	H	H	H	16[a]	322
Ph	H	H	Ph	20[a]	322
Ph	H	Ph	H	16[b]	323
Me_3C	H	H	H	12[a]	322
H	$-(CH_2)_4-$		H	10[a]	322
n-C_6H_{13}	H	H	H	74	323
n-Pr	H	n-Pr	H	44[c]	323
H	$-(CH_2)_6-$		H	62	323

[a]Yields were not maximized.
[b]Only one isomer was obtained (presumably *trans*).
[c]Two isomers in the ratio of 5:1 were obtained.

Oxidation of olefins with lead tetracetate has been shown[326] to produce γ-lactones (equation 164), but yields are generally inferior to those obtained with manganese (III) acetate.

In a rather specialized example of olefin oxidation, *p*-nitroperbenzoic acid has been reported[327] to produce β-lactones from allylallenes (equation 165).

$$R^1CH{=}CHR^2 \xrightarrow[\text{KOAc, HOAc}]{\text{Pb(OAc)}_4} \gamma\text{-lactone (}R^1\text{ at C-5, }R^2\text{ at C-4)} \tag{164}$$

$$R^1R^2C{=}C{=}CR^3{-}C(Me)_2CH{=}CHR^4 \xrightarrow{p\text{-NO}_2\text{C}_6\text{H}_4\text{CO}_3\text{H}} \beta\text{-lactone: }R^1R^2C\text{ and }R^3C{-}C(Me)_2{-}CH{=}CHR^4 \tag{165}$$

O. By Carbonylation Reactions

Unsaturated esters undergo carbonylation with carbon monoxide in the presence of hydrogen and dicobalt octacarbonyl to afford lactones (Table 25)[328,329]. These reactions are believed[328] to occur via hydroformylation of the double bond followed by cyclization of the intermediate hydroxy ester under the reaction conditions (equation 166).

Alkenyl and acetylenic alcohols are converted to lactones by carbonylation by nickel tetracarbonyl in the presence of aqueous acid[154,155,330] or dicobalt

TABLE 24. Preparation of γ-lactones by addition of carboxylic acids to olefins[325]

Olefin	Acid	Lactone[a]	Yield (%)
$C_6H_{13}CH{=}CH_2$	$MeCO_2H$	$R^1 = C_6H_{13}$	74
$PhCH{=}CH_2$	$MeCO_2H$	$R^1 = Ph$	60
$PhC(Me){=}CH_2$	$MeCO_2H$	$R^1 = Ph, R^2 = Me$	74
$Me_2C{=}CH_2$	$MeCO_2H$	$R^1, R^2 = Me$	30
$Me_3C{-}CH{=}CH_2$	$MeCO_2H$	$R^1 = Me_3C$	48
PrCH=CHPr (*trans*)	$MeCO_2H$	$R^1, R^4 = Pr$	44
PhCH=CHPh (*trans*)	$MeCO_2H$	$R^1, R^4 = Ph$	16
PhCH=CHMe (*trans*)	$MeCO_2H$	$R^1 = Ph, R^4 = Me$	79
Cyclooctene	$MeCO_2H$	$R^1, R^3 = -(CH_2)_6-$	62
$PhCH{=}CHCO_2Me$	$MeCO_2H$	$R^1 = Ph, R^4 = CO_2Me$	45
1,5-Hexadiene	$MeCO_2H$	$R^1 = CH_2{=}CH(CH_2)_2-$	24
1,7-Octadiene	$MeCO_2H$	$R^1 = CH_2{=}CH(CH_2)_4-$	26
Butadiene	$MeCO_2H$	$R^1 = CH_2{=}CH-$	30
Isoprene	$MeCO_2H$	$R^1 = CH_2{=}C(Me)-$ +	13
		$R^1 = CH_2{=}CH-, R^2 = Me$	37
$Me(CH_2)_4C{\equiv}CCH_2CH{=}CH_2$	$MeCO_2$	$R^1 = Me(CH_2)_4C{\equiv}CCH_2-$	50
$PhCH{=}CH_2$	$MeCH_2CO_2H$	$R^1 = Ph\ R^5 = Me$	50
$PhCH{=}CH_2$	$NCCH_2CO_2H$	$R^1 = Ph, R^5 = CN$	41
$C_6H_{13}CH{=}CH_2$	$NCCH_2CO_2H$	$R^1 = Ph, R^5 = CN$	60
$PhC(Me){=}CH_2$	$NCCH_2CO_2H$	$R^1 = Ph, R^2 = Me, R^5 = CN$	43
4-Octene	$NCCH_2CO_2H$	$R^1, R^4 = Pr, R^5 = CN$	49
PhCH=CHMe	$NCCH_2CO_2H$	$R^1 = Ph, R^4 = Me, R^5 = CN$	51
Isoprene	$NCCH_2CO_2H$	$R^1 = CH_2{=}C(Me)-, R^5 = CN +$	5
		$R^1 = CH_2{=}CH, R^2 = Me, R^5 = CN$	39
$C_6H_{13}CH{=}CH_2$	$(CH_2CO_2H)_2$	$R^2 = C_6H_{13}, R^5 = CH_2CO_2H$	25

[a]Where not specified R = H.

$$R^1R^2C{=}CR^3COOR^4 \xrightarrow[H_2]{CO} OHCCR^1R^2{-}CHR^3COOR^4 \xrightarrow{H_2}$$

$$HOCH_2CR^1R^2{-}CHR^3COOR^4 \xrightarrow{-R^4OH} \text{(γ-lactone with } R^1, R^2, R^3\text{)} \qquad (166)$$

octacarbonyl in the presence of carbon monoxide and hydrogen[331] (Table 26). Lactone formation in these cases may be viewed as proceeding by hydrocarboxylation of the unsaturated function with subsequent cyclization of an intermediate hydroxy acid.

Reaction of certain diols and dienes with carbon monoxide or formic acid and a strong mineral acid in the presence of Group IB metal compounds results in Koch–Haaf[332,333] hydrocarboxylation followed by ring-closure to form lactones[334]. As may be seen from equations (167) and (168), these reactions are accompanied by deep-seated carbonium ion rearrangements.

$$HO(CH_2)_7OH + CO \xrightarrow{H^+} \text{(δ-lactone with Et, Me)} \qquad (167)$$

TABLE 25. Preparation of lactones by carbonylation of unsaturated esters with CO, H_2 and $Co_2(CO)_8$ at 200–350°C

Starting material	Product	Yield (%)	Reference
$H_2C{=}C(R^1)CO_2R^2$	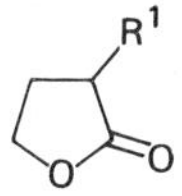		328
$R^1 = H, R^2 = Me$		69	
$R^1 = H, R^2 = Et$		88	
$R^1 = R^2 = Me$		51	
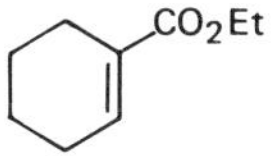 1-(ethoxycarbonyl)cyclohexene (CO_2Et)	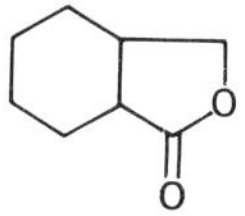	23	328
$MeCH{=}CHCO_2R$	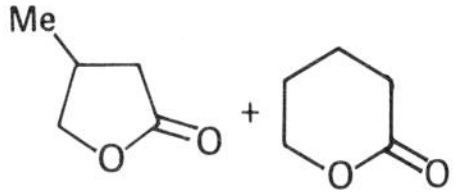		328
R = Me		20 + 72	
R = Et		23 + 67	
$MeCH{=}CMeCO_2Et$	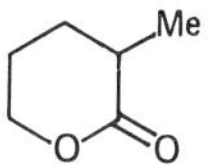	31	328
	+ 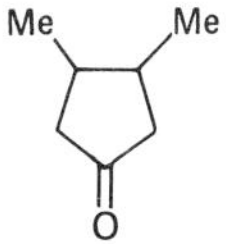	21	
	+ 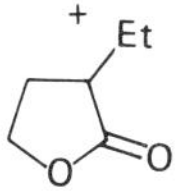	7	

TABLE 25. (Continued)

Starting material	Product	Yield (%)	Reference
CH_2=$CHCH_2CO_2Et$	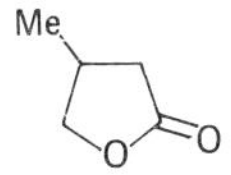	17	328
	+		
	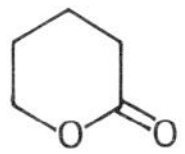	52	
Me_2C=$CHCO_2Et$	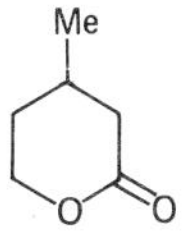	88	328, 329
	+		
	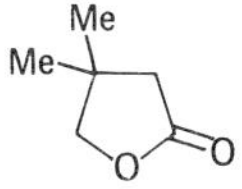	1	
Me_2C=$CMeCO_2Et$	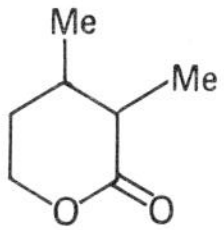	55	328
	+		
	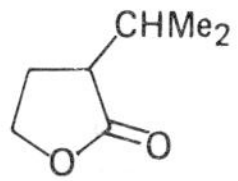	31	

$H_2C{=}CH{-}C(Me)_2{-}CO_2Et$	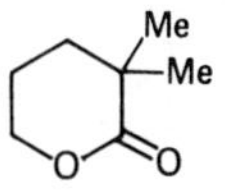	93	328
	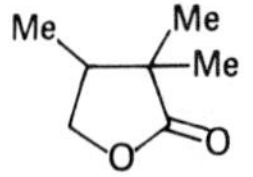 +	1	
$MeCH{=}CHCH{=}CHCO_2Et$ *cis*, *cis*	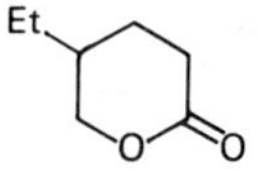	49	328
	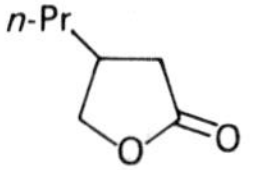 +	33	
$PhCH{=}CHCO_2Et$	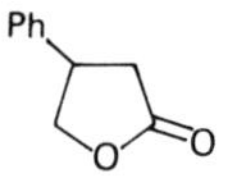	8.5–49	328, 329
$EtO_2CCH{=}CHCO_2Et$	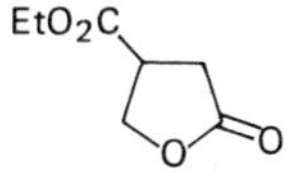		328
cis		47	
trans		49	

TABLE 26. Preparation of lactones by carbonylation of unsaturated alcohols

Starting material	Reagent[a]	Product	Yield (%)	Reference
$R^1CH{=}CR^2CH_2OH$	A	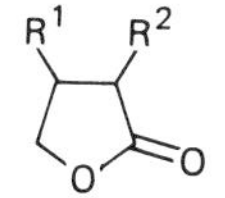		331
		$R^1 = H, R^2 = Me$	2	
		$R^1 = Me, R^2 = H$	2	
		$R^1 = R^2 = H$	61	331a
$H_2C{=}CR^1{-}C(R^2)(R^3){-}CH_2OH$		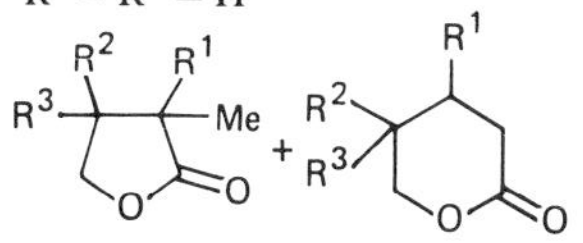		331
	A	$R^1 = H, R^2 = R^3 = Me$	51 + 14	
	A	$R^1 = R^2 = R^3 = Me$	3 + 25	
	A	$R^1 = H, R^2 = Me, R^3 = Et$	40 + 13	
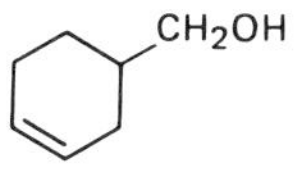	A	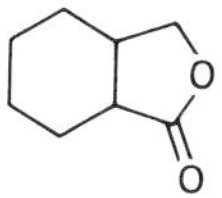	16	331
$H_2C{=}CHCH_2CH(R)OH$		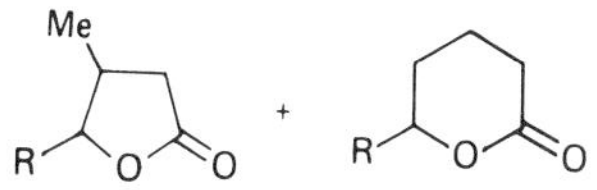		331

	A	R = Et	2 + 73	
	A	R = *n*-Pr	2 + 0	
$H_2C{=}CHCH_2C(R)(Me)OH$		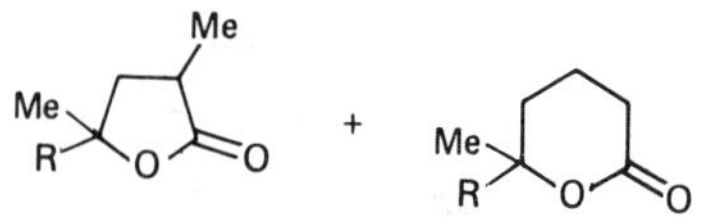 5-Me-5-R-3-methyl-γ-butyrolactone + 6-Me-6-R-δ-valerolactone		331
	A	R = Me	10 + 2	
	A	R = Et	29 + 6	
	A	R = *i*-Bu	10 + 2	
$HOC(R^1)(R^2)CH_2C{\equiv}CH$		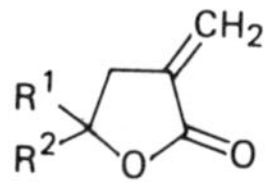 α-methylene-γ-butyrolactone (R^1, R^2 at C-5)		
	B or C	$R^1 = R^2 = H$	23	330
	B or C	$R^1 = H, R^2 = Me$	30–50	330
	B	$R^1 = R^2 = Me$	10	330
	D	$R^1 = H, R^2 = Ph$	44	154
$HOCH_2CH_2CH_2C{\equiv}CH$	B	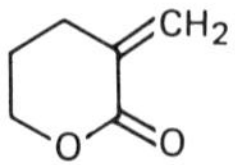 α-methylene-δ-valerolactone	20	330

[a] A = CO, H_2, $Co_2(CO)_8$, 100–350°C; B = $Ni(CO)_4$, HOAc, EtOH, H_2O, 80°C; C = $Ni(CO)_4$, MeOH, HCl; D = $Ni(CO)_4$, HOAc, EtOH, H_2O, hydroquinone.

$$H_2C{=}CHCH_2CH_2CH{=}CH_2 + CO \xrightarrow{H^+}$$ (168)

$$RCOCl + HC{\equiv}CH \xrightarrow[H_2O,\ X^-]{Ni(CO)_4}$$ (169)

A recent publication[335] describes the synthesis of unsaturated butyrolactones by reaction of acetylenes with acyl chlorides in the presence of nickel tetracarbonyl and halide ion (equation 169).

P. By Cycloaddition of Nitrones to Olefins

An interesting general method for the preparation of γ-lactones from olefins involves initial silver ion-induced addition of *N*-cyclohexyl-α-chloroaldonitrones to olefins to produce the $(2\pi + 4\pi)$ cycloadduct, which is then treated with base and hydrolysed (equation 170)[336,337].

1. $AgBF_4$, CH_2Cl_2; 2. KCN, H_2O; *t*-BuOK, *t*-BuOH, H_2O; H_3O^+ (170)

R^1	R^2	R^3	Yield (%)[a]
H	H	H	91
H	H	Me	82[b]
Me	H	H	83
Me	Me	H	70

[a]Yields reported are only for the hydrolysis (last) step.
[b]Diastereomeric mixture, α:β ≈ 4:1.

Use of the diastereomeric 2-butenes in this reaction (equations 171 and 172)[337] showed the addition to be a stereospecific *cis* process. The reaction may also be performed using *N*-cyclohexyl-α-chloroethanaldonitrone (equations 173 and 174)[336], *N*-(*t*-butyl)-α-chloroethanaldonitrone (equation 175)[336] and *N*-cyclohexyl-α,β-dichloropropionaldonitrone (equation 176)[337].

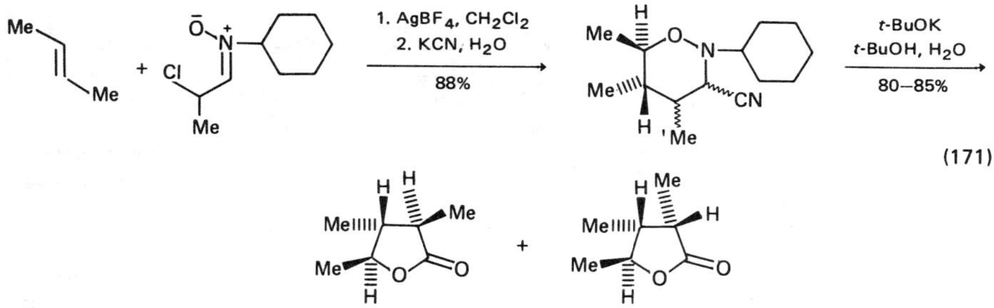

1. $AgBF_4$, CH_2Cl_2
2. KCN, H_2O
88%

t-BuOK
t-BuOH, H_2O
80–85%

(171)

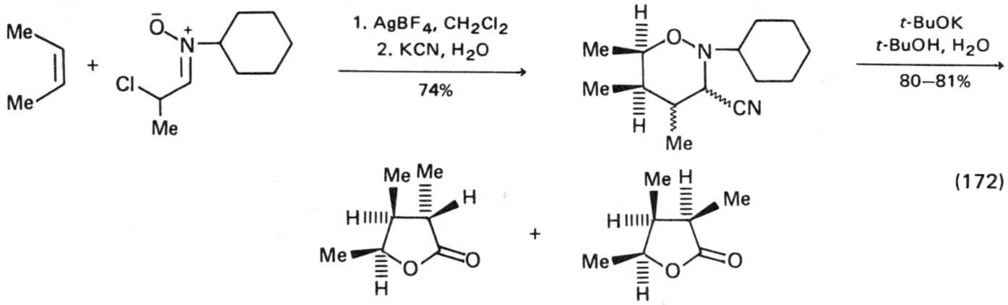

1. $AgBF_4$, CH_2Cl_2
2. KCN, H_2O
74%

t-BuOK
t-BuOH, H_2O
80–81%

(172)

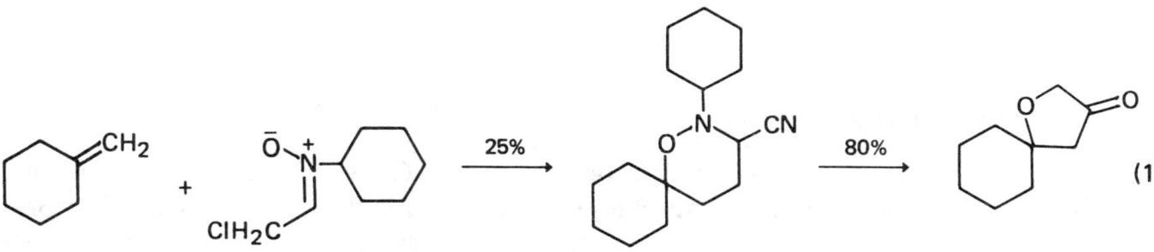

25%

80%

(173)

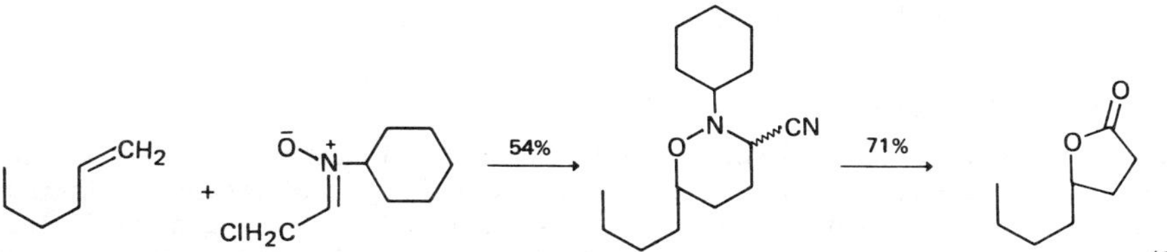

54%

71%

(174)

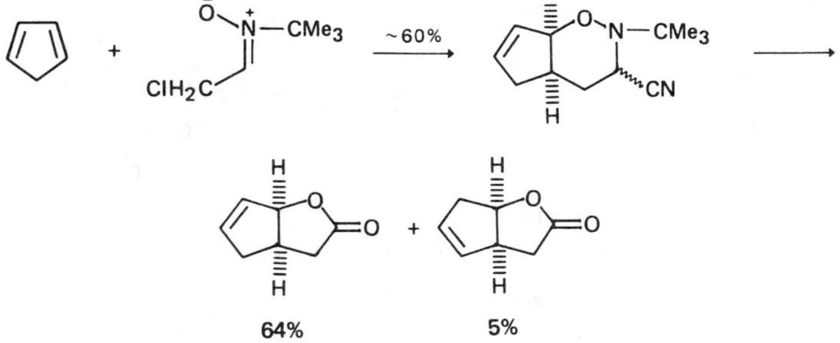

~60%

(175)

64% 5%

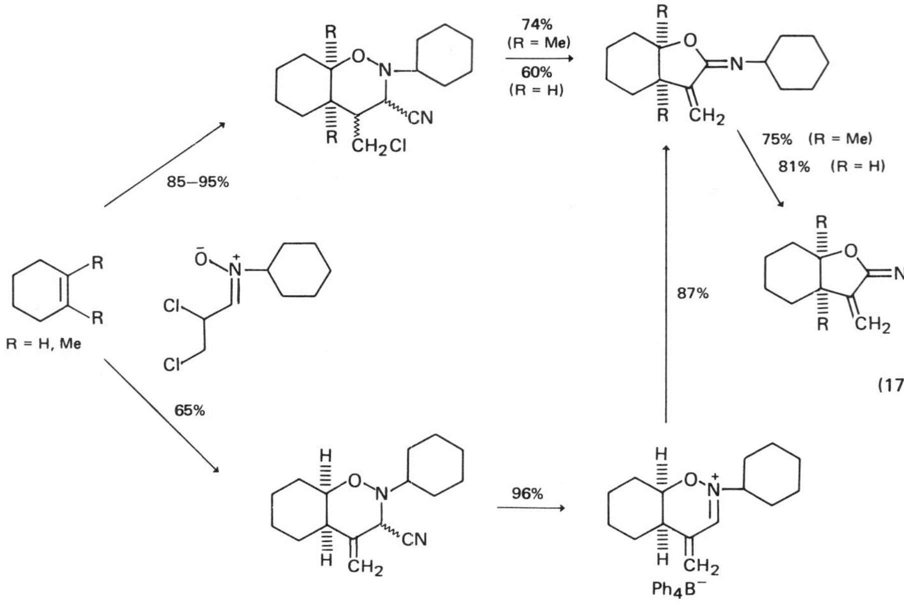

(176)

Q. By Rearrangement Reactions

This section deals with lactone preparations by Claisen, carbonium ion and photochemical rearrangements. The Baeyer–Villiger reaction and certain lactone interconversions, which might also be regarded as rearrangements, are discussed in Sections II. N.2. and II.R, respectively.

1. Claisen rearrangements

Reaction of a series of 2-alkene-1,4-diols with orthocarboxylic esters in the presence of a catalytic amount of hydroquinone or phenol results[338] in the formation of various β-vinyl-γ-butyrolactones via a Claisen rearrangement (Table 27). The proposed mechanism, illustrated in equation (177) involves an exchange of the alkoxy group of the *ortho* ester with the diol, followed by elimination of ethanol to produce a mixed ketene acetal. Rearrangement of this intermediate to a

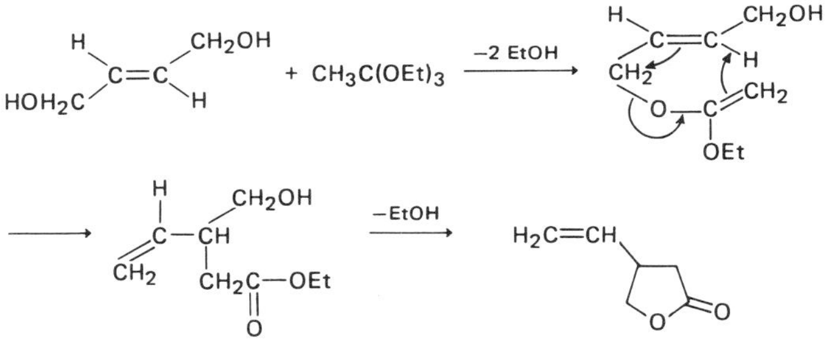

(177)

TABLE 27. γ-Lactones by reaction of *ortho* esters $RCH_2C(OEt)_3$ with unsaturated 1,4-diols[338]

Diol	R	Product	Yield (%)
HOH_2C, H; C=C; H, CH_2OH	H	$H_2C{=}CH$; O; O	89
$HO(Me)_2C$, H; C=C; H, CH_2OH	H	$H_2C{=}CH$; Me; Me; O; O	91
HO(Me)HC, H; C=C; H, CH(Me)OH	H	MeHC=CH; Me; O; O *cis–trans* mixture	52
$HO(Me)_2C$, H; C=C; H, $C(Me)_2OH$	H	$Me_2C{=}CH$; Me; Me; O; O pyrocin	70
HOH_2C, Me; C=C; H, CH_2OH	H	Me; $H_2C{=}C$; O; O + $H_2C{=}CH$; Me; O; O	81 (ratio 6:4)
$HO(Me)_2C$, H; C=C; H, $C(Me)_2OH$	Me	$Me_2C{=}CH$; Me; Me; Me; O; O *cis-trans* mixture	60

β-vinyl-γ-hydroxy carboxylic ester and lactonization under the conditions of the reaction affords the observed lactones. It should be noted that all of the entries in Table 27 are *trans* diols. With substituted *cis*-2-alkene-1,4-diols, γ-lactones were obtained in lower yields. For example, condensation of *cis*-2-butene-1,4-diol with ethyl orthoacetate afforded β-vinyl-γ-butyrolactone in 45% yield, along with 20% of 2-methyl-2-ethoxy-1,3-dioxacyclohept-5-ene. Condensations of allyl alcohols with cyclic orthoesters have also been used to prepare γ- and δ-lactones (equations 178–181)[339].

+ $Me_2C{=}CH{-}CH(OH){-}Me$ (racemic) ⟶ racemic, 65% (178)

racemic + ⟶ racemic, 65% (179)

+ racemic ⟶ racemic, 80% (180)

racemic + ⟶ racemic and its epimer (181)

2. *Carbonium ion rearrangements*

A number of cyclopropane carboxylic acids undergo acid-catalysed and/or thermal rearrangements to form γ-butyrolactones. The former reactions may be envisioned as occurring via concomitant protonation at the cyclopropyl carbon holding the carboxyl group, and ring-opening to form the most highly substituted carbonium ion, which then interacts with the carboxy group to generate the lactone

$\xrightarrow{H^+}$ $\xrightarrow{-H^+}$ (182)

ring (equation 182). The specific examples given in equations (183)–(185) are representative of this scheme for lactone formation.

$\xrightarrow[\text{18 days}]{H_2SO_4,\ \text{r.t.}}$ (Ref. 340) (183)

$\xrightarrow[\text{r.t.}]{\text{HBr, HOAc}}$ (Ref. 341) (184)

$\xrightarrow{150^{\circ}C}$ (Ref. 342) (185)

Certain other monocarboxylic acids containing ring systems which are susceptible to carbonium ion rearrangements can be converted to lactones upon treatment with acid. Thus, both the *endo* and *exo* isomers of (+)-1.5,5-trimethylbicyclo[2.1.1]hexane-6-carboxylic acid produce dihydro-β-campholenolactone in 49% yield (equation 186)[343]. The [4.1.0] bicyclic hydroxy ester shown in equation (187) affords an 88% yield of *trans*-fused cycloheptene butyrolactone[344].

$\xrightarrow{H_2SO_4}$ (Ref. 343) (186)

$\xrightarrow[H_2O,\ MeOH]{HClO_4}$ (Ref. 344) (187)

Cyclopropane-1,1-dicarboxylic acids can serve as useful starting materials for γ-butyrolactones as shown by the reaction of several such acids with deuterated sulphuric acid (equation 188)[345]. The location of the deuterium labels in the final

$\xrightarrow[D_2O,\Delta]{D_2SO_4}$ $\xrightarrow{-CO_2}$ (188)

$R^1 = R^2 = H$

$R^1 = H, R^2 = Me$

$R^1 = R^2 = Me$

products is consistent with operation of a mechanism analogous to that described above for cyclopropanecarboxyclic acids. Thermal decarboxylation of related diacids also affords lactones (equation 189)[342].

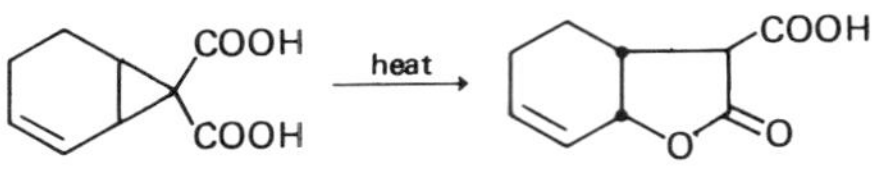

(189)

3. *Photochemical rearrangements*

Irradiations of β,γ-epoxy cyclic ketones and simple substituted epoxides produce lactones in 35%–65% yields (equations 190–193).

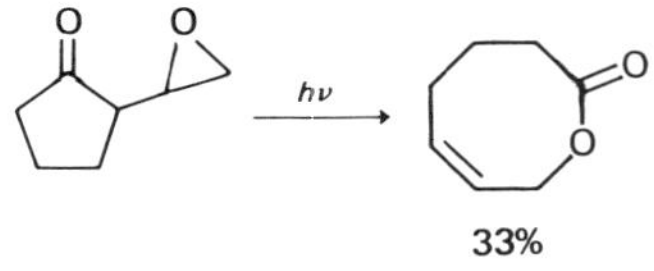

$h\nu$ (Ref. 346) (190)

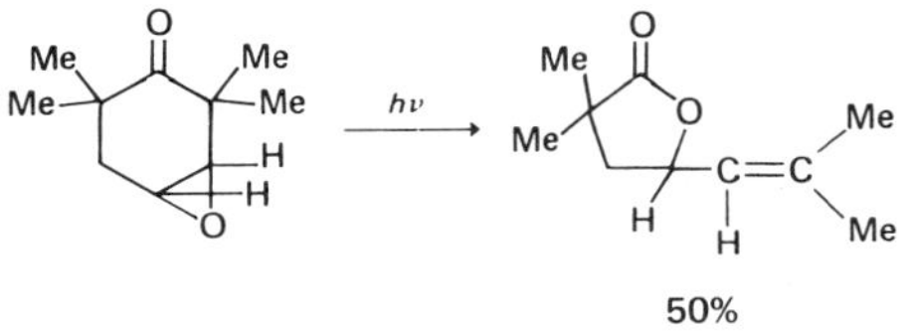

33%

$h\nu$ (Ref. 347) (191)

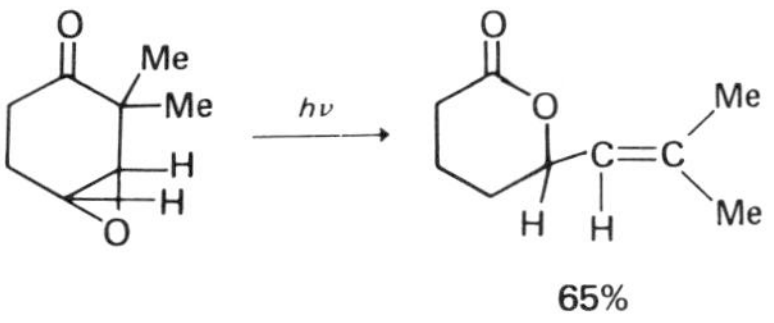

50%

$h\nu$ (Ref. 348) (192)

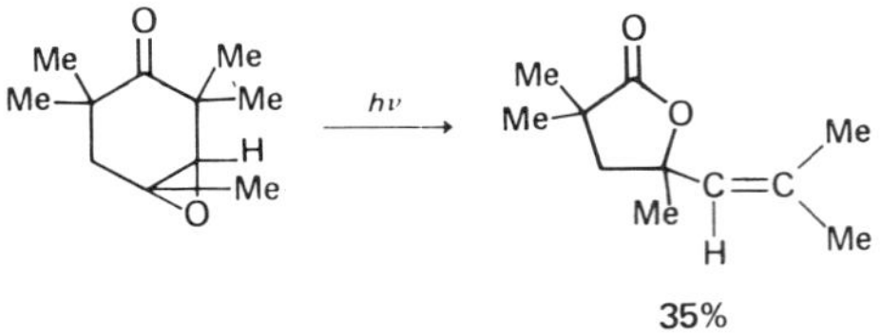

65%

$h\nu$ (Ref. 348) (193)

35%

The photochemical behaviour of the non-enolizable β-diketone, 2,2,5,5-tetramethyl-1,3-cyclohexanedione, has been studied by several groups of workers[349–352] and all are in essential agreement concerning the products obtained in benzene (equation 194). However, in ethanol or cyclohexane, one group of

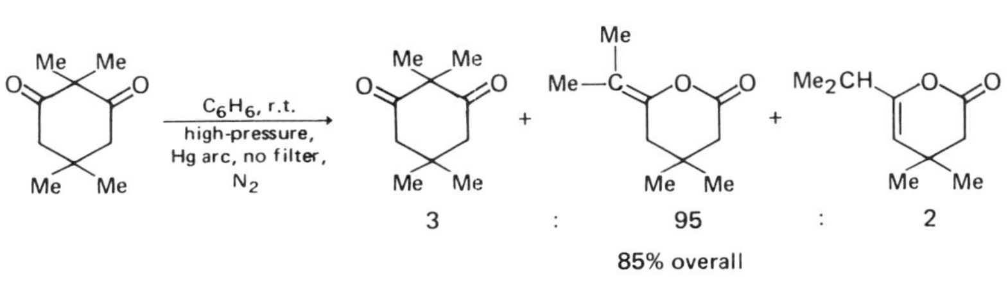

C_6H_6, r.t., high-pressure, Hg arc, no filter, N_2 (194)

3 : 95 : 2

85% overall

workers[349] reported a single product, while a second group[352] obtained all the products shown in equation (194).

Interestingly, irradiation of the exocyclic enol lactone, 5-hydroxy-3,3,6-trimethyl-5-heptenoic acid δ-lactone afforded[352] a pseudo-equilibrium mixture (equation 195). Treatment of 2,2-dimethyl-1,3-cyclohexanedione in a similar

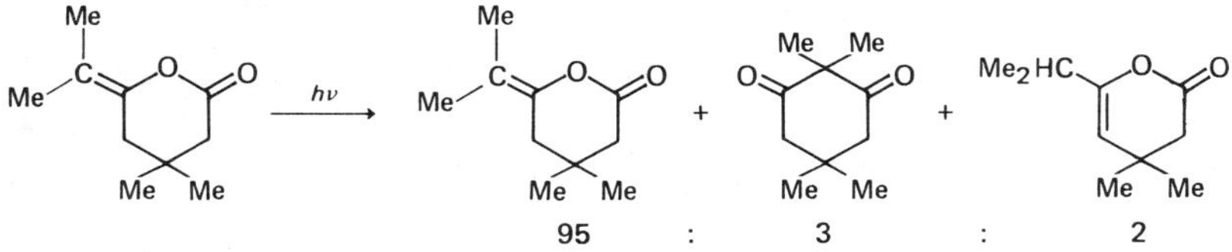

(195)

manner afforded[352] exclusively the corresponding enol lactone in 70% yield (equation 196).

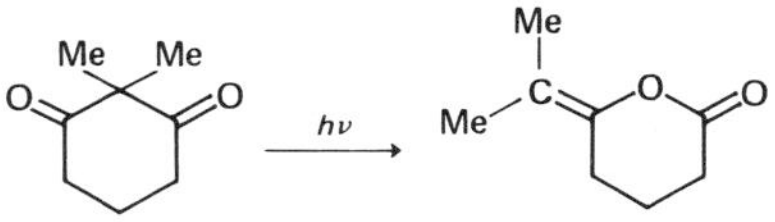

(196)

R. Lactone Interconversions

Although there are not enough literature reports to permit generalization, the following reactions provide some examples of the synthetic potential of lactone interconversions.

Treatment of *d,l*-α-campholenic acid lactone with sulphuric acid has been reported[290] to produce the isomeric dihydro-β-campholenolactone (equation 197);

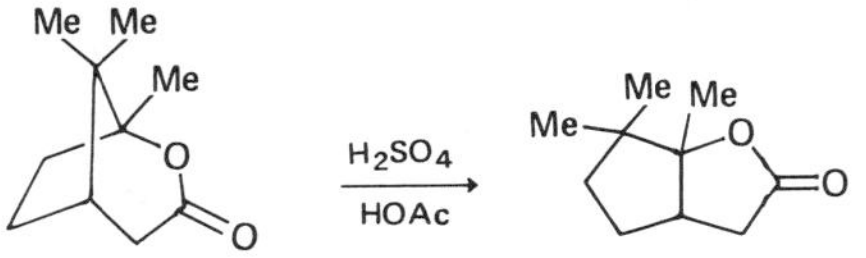

(197)

however, when the isomeric bicyclic lactone was treated in the same manner no interconversion was observed (equation 198)[291]. This difference in reactivity has

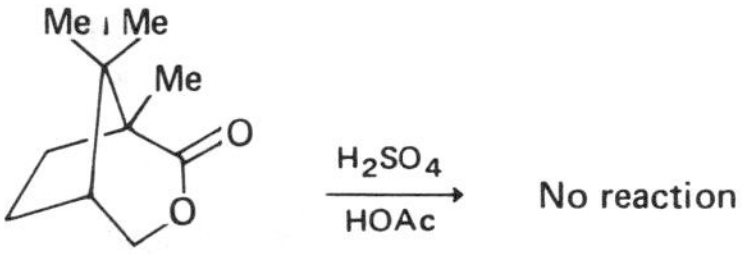

(198)

been used[291] to obtain analysis of the lactone products obtained from peracetic acid oxidation of camphor (equation 199).

During the elegant synthesis of reserpine, Woodward and coworkers[28] have observed a number of lactone interconversions (equations 200 and 201).

The γ- to δ-lactone interconversion shown in equation (202) has recently[307] been observed during the total synthesis of Rhoeadine alkaloids.

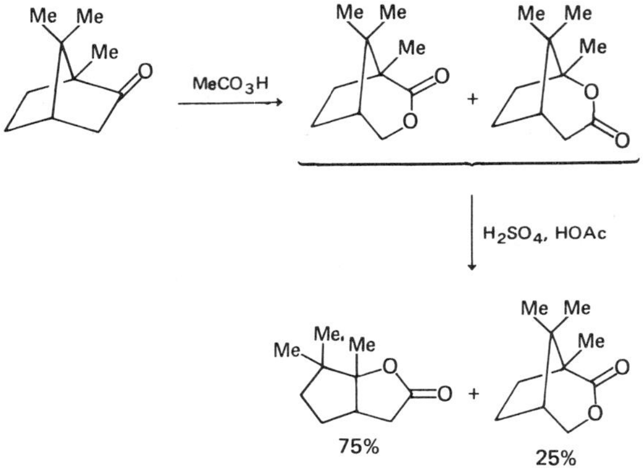

MeCO$_3$H

H_2SO_4, HOAc (199)

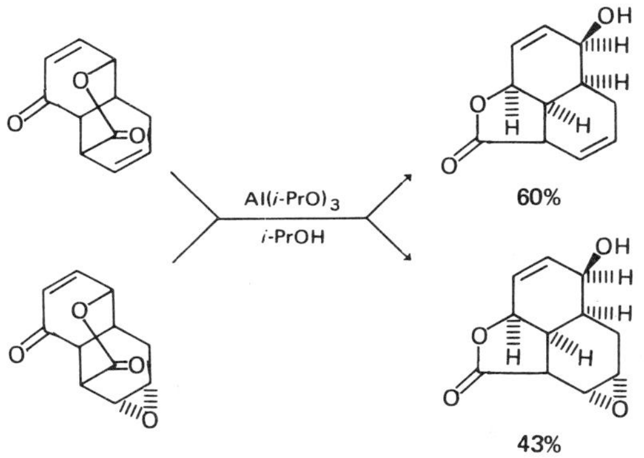

75% 25%

Al(*i*-PrO)$_3$ / *i*-PrOH (200)

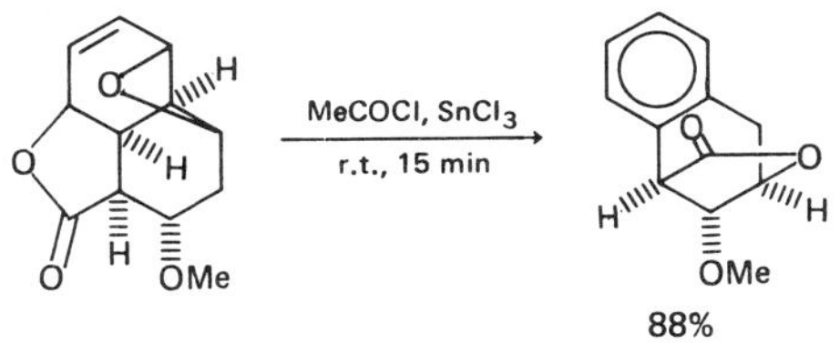

60%

43%

MeCOCl, $SnCl_3$ / r.t., 15 min (201)

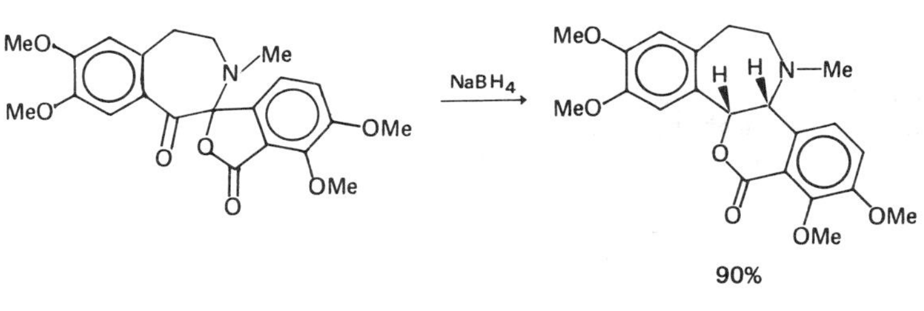

88%

$NaBH_4$ (202)

90%

S. Miscellaneous Lactone Syntheses

The following preparations do not fall conveniently into any of the preceding categories; nevertheless several of them are extremely attractive as general lactone syntheses.

1. The Barton reaction

This useful synthesis of lactones[353] consists of reaction of primary or secondary amides with lead tetraacetate or *t*-butyl hypochlorite in the presence of iodine to form *N*-iodo amides, which then undergo a free radical cyclization to lactones when the reaction mixture is photolysed.

$Pb(OAc)_4$, I_2, C_6H_6, $h\nu$ or *t*-BuOCl, I_2, C_6H_6, $h\nu$ (203)

R = H, Ph

$R^1CH(R^2)CH_2CH_2CONH_2$ —*t*-BuOCl, I_2, C_6H_6, $h\nu$→ (204)

R^1 = H, C_6H_6, Me

R^2 = H, H, Et

$Pb(OAc)_4$, I_2, $CHCl_3$, $h\nu$ or *t*-BuOCl, I_2, $CHCl_3$, $h\nu$ (205)

$Me(CH_2)_{16}CONH_2$ —$Pb(OAc)_4$, I_2, C_6H_6, $h\nu$ or *t*-BuOCl, I_2, C_6H_6, $h\nu$ or *t*-BuOI, C_6H_6, $h\nu$→ $Me(CH_2)_{13}$ (γ-lactone) + $Me(CH_2)_{12}$ (δ-lactone) (206)

In a reaction somewhat related to the Barton reaction, photolysis of *N*-acetyl-3-methyl-3-phenylpropionamide was reported to accord the lactone of 4-phenyl-4-hydroxy-3-methylbutyric acid[354].

2. Photolysis of α-diazo esters and amides

Photolysis of certain esters of α-diazo carboxylic acids gives rise to lactones by insertion of the resulting α-carbene into a carbon–hydrogen bond of the alkoxy residue[355]. These reactions are, however, often characterized by low yields. Thus, photolysis of the *t*-butyl esters of diazoacetic acid in cyclohexane affords only a 4% yield of γ,γ-dimethylbutyrolactone (equation 207)[355]. Performing the same

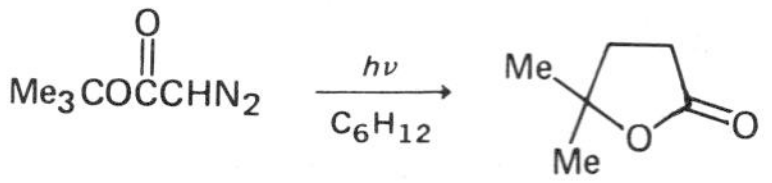

$$\mathrm{Me_3COCCHN_2} \xrightarrow[\mathrm{C_6H_{12}}]{h\nu} \text{(lactone)} \quad (207)$$

reaction on the *t*-amyl ester of diazoacetic acid[355] affords β,γ,γ-trimethyl- and γ-methyl-γ-ethylbutyrolactone, both in low yields (equation 208). Interestingly,

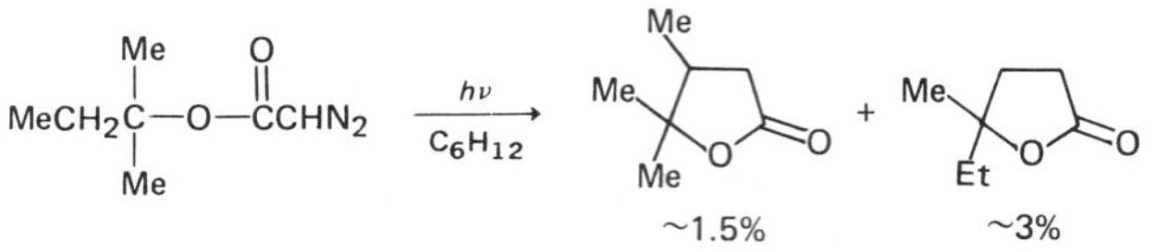

$$\mathrm{MeCH_2C(Me)_2{-}O{-}C(O)CHN_2} \xrightarrow[\mathrm{C_6H_{12}}]{h\nu} \text{(lactone, } \sim 1.5\%) + \text{(lactone, } \sim 3\%) \quad (208)$$

photolysis[356] of *N*-[(*t*-butoxycarbonyl)diazoacetyl] piperidine produced only *cis*-7-*t*-butoxycarbonyl-1-azabicyclo[4.2.0] octan-8-one and its *trans* isomer (equation 209), but no γ-lactone. Using *N*-[(*t*-butoxycarbonyl)diazoacetyl] pyrro-

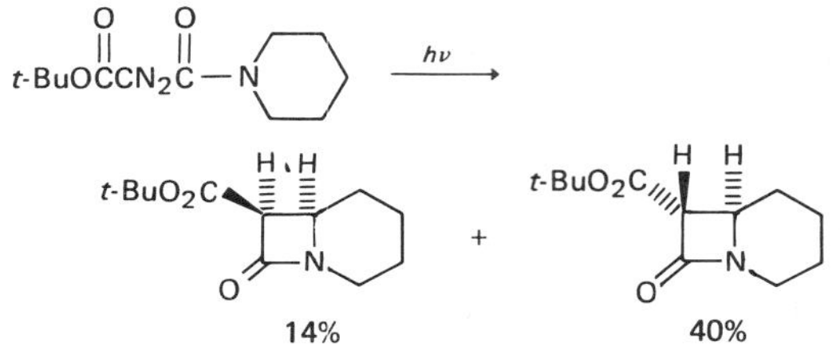

$$t\text{-BuOCOCN}_2\text{CO{-}N(piperidine)} \xrightarrow{h\nu} cis\ (14\%) + trans\ (40\%) \quad (209)$$

$$\mathrm{RO_2CCN_2C(O){-}N(pyrrolidine)} \xrightarrow[R\,=\,t\text{-Bu}]{h\nu} \gamma\text{-lactone}; \quad \xrightarrow[R\,=\,\mathrm{Et}]{h\nu} \beta\text{-lactone} \quad (210)$$

$$t\text{-BuO}_2\text{CCN}_2\text{C(O){-}N(thiazolidine)CO}_2\text{CH}_2\text{Ph} \xrightarrow[1\ \text{h}]{h\nu} \beta\text{-lactam}\ (50\%) + \gamma\text{-lactone} \quad (211)$$

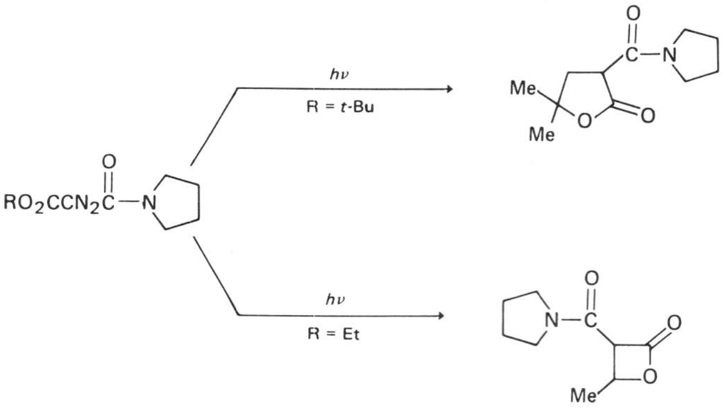

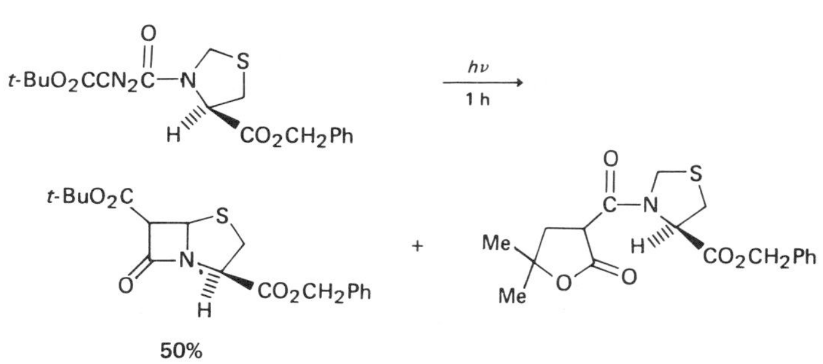

lidine, only the γ-lactone forms, while from *N*-[(ethoxycarbonyl)diazoacetyl]pyrrolidine only the β-lactone is obtained (equation 210)[356]. Application of this reaction[356] to *N*-[(butoxycarbonyl)diazoacetyl]-L-thiazolidine-4-carboxylate substantiated the expectation that the 2-methylene group in the thiazolidine is very susceptible to carbene insertion, since a mixture of β-lactam and its isomeric γ-lactone was obtained (equation 211).

A similar photochemically induced intramolecular insertion has been reported[357] during the photolysis of diethyl diazomalonate with thiobenzophenone in cyclohexane (equation 212).

$$N_2C(CO_2Et)_2 + Ph_2CS \xrightarrow{h\nu} [\text{2-oxotetrahydrofuran-3-}CO_2Et] \longrightarrow \gamma\text{-butyrolactone} \quad (212)$$

3. *Photolysis of 2-alkoxyoxetanes*

A novel synthesis[358] of tetramethyl-β-propiolactone involves irradiation of an acetonitrile solution of any of the 3,3,4,4-tetramethyloxetanes shown in equation (213) with acetone. This lactone may also be prepared[358] via irradiation, of either

$$\text{2-OR-3,3,4,4-tetramethyloxetane} + Me_2CO \xrightarrow{h\nu} \text{tetramethyl-}\beta\text{-propiolactone} \quad (213)$$

R = —OMe, —OEt, —OPr-*n*, —OBu-*n*

$$R{-}O{-}CH{=}C(Me){-}Me + Me_2CO \xrightarrow{h\nu} \underset{37\%}{\text{tetramethyl-}\beta\text{-propiolactone}} + \underset{(1:1)\ 29\%}{\text{2-OR-3,3,4,4-tetramethyloxetane} + \text{3-OR-2,2,3,4-tetramethyloxetane}} \quad (214)$$

R = Me or *n*-Pr

methyl or *n*-propyl β,β-dimethyl vinyl ether with acetone (equation 214). Similar irradiation[358] of acetone with ethyl β,β-diethyl vinyl ether affords α,α-diethyl-β,β-dimethyl-β-propiolactone, which has also been prepared by irradiation of a mixture of isomeric oxetanes with acetone or benzophenone (equation 215). Preparation of

$$EtOCH{=}C(Et)Et \xrightarrow[Me_2CO]{h\nu} \text{α,α-diethyl-β,β-dimethyl-β-propiolactone} \xleftarrow[\substack{Me_2CO\ (37\%) \\ Ph_2CO\ (28\%)}]{h\nu} \text{2-OEt-3,3-diethyl-4,4-dimethyloxetane} + \text{3-OEt-2,2-dimethyl-3-ethyl-4-ethyloxetane} \quad (215)$$

the α,α,β-triethyl-β-propiolactone was accomplished[358] via irradiation of a mixture of the corresponding 2- and 3-methoxyoxetanes with acetone.

4. α-Lactones by photolysis of 1,2-dioxolane-3,5-diones

Methods of preparation of α-lactones are not very common; however, a rather unique, high-yield photochemical synthesis of these elusive compounds via photochemical decarboxylation of 4,4-disubstituted-1,2-dioxolane-3,5-diones has recently been reported[359]. Thus, irradiation of substituted 1,2-dioxalane-3,5-diones as neat liquids at 77 K produces disubstituted α-lactones (equation 216). If the irradiation is performed at room temperature or if the α-lactone is warmed above -100°C a polyester is the only product obtained.

$$\text{1,2-dioxolane-3,5-dione }(R^1,R^2) \xrightarrow[77\text{ K}]{h\nu} CO_2 + \alpha\text{-lactone }(R^1,R^2) \xrightarrow[77\text{ K}]{h\nu} CO + R^1R^2C{=}O$$

$$\alpha\text{-lactone} \xrightarrow{h\nu \text{ or heat}} \left[-C(R^1)(R^2)-C(=O)-O- \right]_n \qquad (216)$$

$R^1 = R^2 = $ Me, *n*-Bu

$R^1 + R^2 = (CH_2)_2, (CH_2)_3, (CH_2)_4$

5. Oxidation of mercaptans, disulphides and related compounds

When mercaptans and disulphides are treated with an oxidizing agent such as dimethyl sulphoxide under basic conditions in a polar solvent, lactones have been reported[360] as the products. Also prepared were the δ-lactones where $R = n\text{-}C_6H_{13}$, Me, Et and Ph.

$$2\ Me(CH_2)_3CH_2SH \xrightarrow[NaOH,\ Me_2SO]{MeOH,\ H_2O} \text{3,5-dialkyl-δ-valerolactone (R, R)} \qquad (217)$$

R = *n*-Pr

The sulphur-donor ligand *ortho*-metalated complexes shown in equation (218) afford lactones upon treatment with 30% hydrogen peroxide or *m*-chloroperbenzoic acid[361].

6. Addition of diazonium salts to olefins

Treatment of olefins with substituted benzenediazonium chlorides in the presence of cuprous chloride and an alkali metal halide affords aryl-substituted butyrolactone esters (equation 219)[362].

7. Addition of diethyl dibromomalonate to methyl methacrylate

Condensation of diethyl dibromomalonate with methyl methacrylate in the presence of iron pentacarbonyl produces the substituted butyrolactone shown in equation (220)[363].

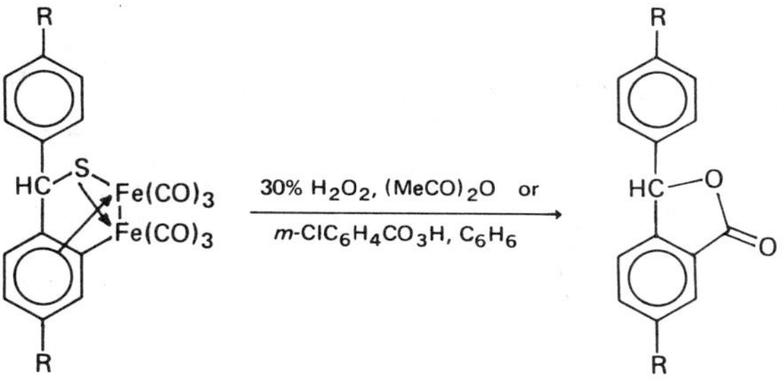

(218)

R	Reagent	Yield (%)
OMe	30% H_2O_2	73
OMe	m-$ClC_6H_4CO_3H$	57
Me	30% H_2O_2	45
H	30% H_2O_2	49

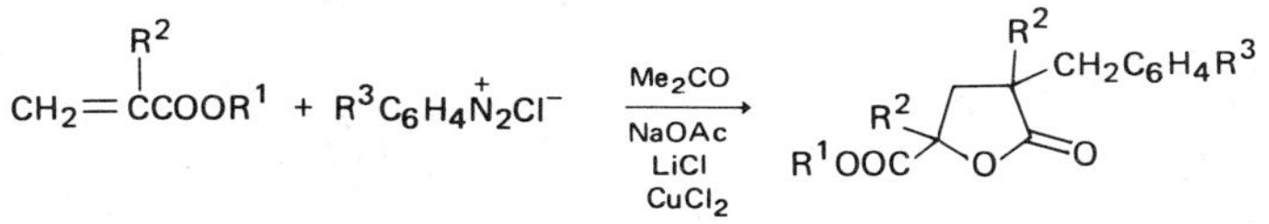

(219)

$R^1 = R^2 = Me$; $R^3 = p$-Me, p-Cl

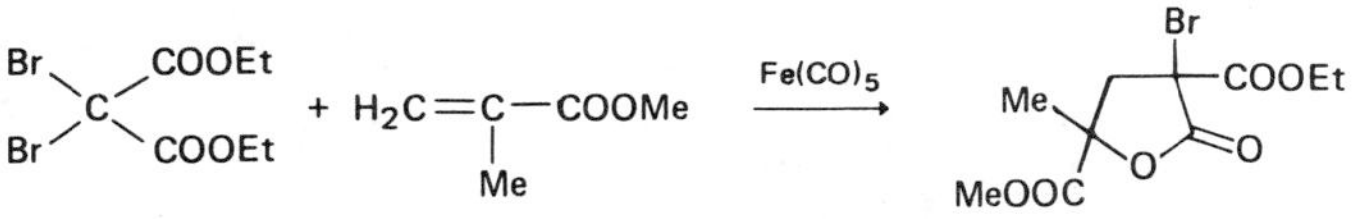

(220)

8. *Dehydrohalogenation of 2,2-dimethoxy-3-chlorodihydropyrans*

Treatment of a series of substituted 3-chlorodihydropyrans with sodium methoxide in dimethyl sulphoxide or dimethylformamide at room temperature affords the corresponding α-pyrones in good yields (equation 221)[364].

9. *Preparation of homoserine lactone*

α-Amino-γ-butyrolactone (homoserine lactone), an important intermediate in the synthesis of various amino acids, has been prepared by a two-step sequence in which N-tosyl- or N-benzoylglutamine is converted into N-tosyl- or N-benzoyl-α,γ-diaminobutyric acid with potassium hypobromite, followed by diazotization[365]. A second route involves the reaction of N-acyl methionines with methyl iodide in a mixture of acetic and formic acids to produce their corresponding sulphonium salts, which are then hydrolysed under reflux at pH 6–7 (equation 222). The resulting N-acyl-α-amino-γ-hydroxybutyric acids are then converted into their corresponding lactones using hydrogen chloride[366].

(221)

R^1	R^2	R^3	Yield (%)
Ph	H	H	85
H	H	Ph	64
Ph	H	Ph	72
Et	$-(CH_2)_4-$		78
H	$C_{10}H_{21}$	H	52

(222)

R	Yield (%)
PhCO	73
p-$MeC_6H_4SO_2$	92
EtOCO	81
$PhCH_2OCO$	80
Me_3COCO	29
O=C–C=O (cyclic)	45

III. SYNTHESIS OF LACTAMS

Information about the synthesis of lactams may be found in numerous review articles, most of which, however, have been limited to the preparation of one particular class of lactam or to the general synthesis of amides.

In 1957 Sheehan and Corey[367] published a review on 'The synthesis of β-lactams'. The synthesis of lactam monomers was reviewed in 1962 by Dachs and Schwartz[368] and by Testa[369]. The synthesis of β-lactams was again reviewed in 1962 by Graf and coworkers[370], while in 1966 a review of the preparation, properties and pharmacology of amides, amino acids and lactams was published by Piovera[371], and in 1967 a discussion of the preparation of β-lactams was published by Muller and Harmer[372].

The first review on 'α-Lactams (aziridinones)' appeared in 1968 from Lengyel and Sheehan[373], while the synthesis of all types of lactams was reviewed first by

Beckwish[374] in 1970 in his chapter on 'Synthesis of amides' for this series, by L'Abbé and Hassner[375] in 1971 in their review of 'New methods for the synthesis of vinyl azides', by Millich and Seshadri[376] in their chapter on lactams in *High Polymers*, by Manhas and Bose[377] in *Chemistry of β-Lactams, Natural and Synthetic*, by Hawkins[378] in his review of 'α-Peroxyamines', and finally, by Mukerjee and Srivastava[379] in a review entitled 'Synthesis of β-lactams'.

A. By Ring-closure Reactions (Chemical)

1. *From amino acids and related compounds*

Intramolecular reaction of a carboxylic acid or ester function with an appropriately positioned amino group is quite often the method of choice for the synthesis of γ- and δ-lactams. Lactams of smaller and larger ring size are somewhat less frequently synthesized by such procedures, although α-, β- and ε-lactams can be prepared by careful choice of reaction conditions and starting materials. Thermal cyclization of a mixture of *cis*- and *trans*-4-aminocyclohexanecarboxylic acid to produce 3-isoquinuclidone[380] is representative of a typical δ-lactam synthesis (equation 223). Preparation of the γ-lactam, 1,5-dimethyl-2-pyrrolidone[381],

COOH / NH_2 (4-aminocyclohexanecarboxylic acid) $\xrightarrow[\text{reflux}]{\text{Dowtherm}}$ bicyclic lactam (O, NH) (223)

involves a related cyclization of the methylammonium salt of γ-(methylamino) valeric acid (equation 224).

$$CH_3\underset{\displaystyle NHCH_3}{CH}CH_2CH_2COO^-\ \overset{+}{N}H_3CH_3 \xrightarrow{\text{heat}} \text{1,5-dimethyl-2-pyrrolidone } (CH_3,\ N\text{-}CH_3,\ O) \qquad (224)$$

An interesting example[382] of α-lactam (aziridinone) formation involves the synthesis of optically active 3-substituted-1-benzyl-oxycarbonylaziridin-2-ones from *N*-benzyloxycarbonyl L-amino acids by use of phosgene, thionyl chloride or phosphorus oxychloride in THF at −20 to 30°C (equation 225). The cyclization appears to involve initial formation of a mixed anhydride between the *N*-protected amino acid and the dehydrating agent.

$$R^1O\overset{O}{\overset{\|}{C}}NH\overset{R^2}{\overset{|}{C}H}COOH \xrightarrow[\text{THF, }-20^\circ C]{COCl_2,\ Et_3N} R^1O\overset{O}{\overset{\|}{C}}N\text{—}CH\text{—}R^2 \ (\text{aziridinone ring, C=O}) \qquad (225)$$

$R^1 = —CH_2C_6H_5, —CH_2C_6H_4Br\text{-}p, —CH_2C_6H_4Cl\text{-}p,$

$R^2 = —CH_2C_6H_5$

Intramolecular cyclization of amino esters has found numerous applications in lactam synthesis. In some cases the desired cyclizations are accomplished thermally as in the preparations of 5,5-dimethyl-2-pyrrolidone (equation 226)[383], α-(equation 227)[384], β-(equation 228)[385] and δ-methylcaprolactam (equation 229)[385].

$$(CH_3)_2C(NH_2)CH_2CH_2COOMe \xrightarrow[88-96\%]{200^\circ C} \text{5,5-dimethyl-2-pyrrolidinone} \quad (226)$$

$$H_2N(CH_2)_4CH(CH_3)COOEt \xrightarrow[162-165^\circ C,\ 44\%]{\text{ethylene glycol}} \text{3-methylcaprolactam} \quad (227)$$

$$H_2N(CH_2)_3CH(CH_3)CH_2COOEt \xrightarrow[56\%]{\text{heat}} \text{4-methylcaprolactam} \quad (228)$$

$$H_2NCH_2CH(CH_3)(CH_2)_3COOEt \xrightarrow[41\%]{\text{heat}} \text{6-methylcaprolactam} \quad (229)$$

Cyclization of dienamino esters, obtained by addition of enamino esters to methyl and ethyl propiolate, has been accomplished at 160–190°C in dipolar aprotic solvents to afford α-pyridones in good yields (equation 230)[386]. Reaction

$$HC(COOEt)=C(R^1)NHR^2 + HC{\equiv}CCOOR^3 \longrightarrow R^3OOCCH=CH\text{–}C(COOEt)=C(R^1)NHR^2 \longrightarrow \text{EtOOC-substituted } 6\text{-}R^1\text{-}1\text{-}R^2\text{-}2\text{-pyridone} \quad (230)$$

R^3 = Me, Et

R^1	R^2
H	$CH_2C_6H_5$
Me	H
Me	Me
Ph	H
Ph	Me
o-$CH_3C_6H_4$	H

of the α,β-unsaturated triester prepared from malonic ester and ethyl pyruvate, with the diethyl acetals of a series of *N,N*-dimethylamides affords the corresponding dienamino triesters, which in turn undergo cyclization with benzyl amine in refluxing ethanol to afford a series of 1-benzyl-3,4-dicarboethoxy-2(1*H*)-pyridones (equation 231)[387].

Cyclization of 2-piperidinylacetates to form β-lactams has been effected by means of ethylmagnesium bromide (equation 232)[388]. Yields increase with increasing substitution at the α-carbon of the ester. Similar cyclization of the methyl ester of 3-(methylamino)butyric acid produces[389] *N*-methyl-β-butyrolactam (equation 233); however, the reaction failed with ethyl 2-pyrrolidinylacetate[388].

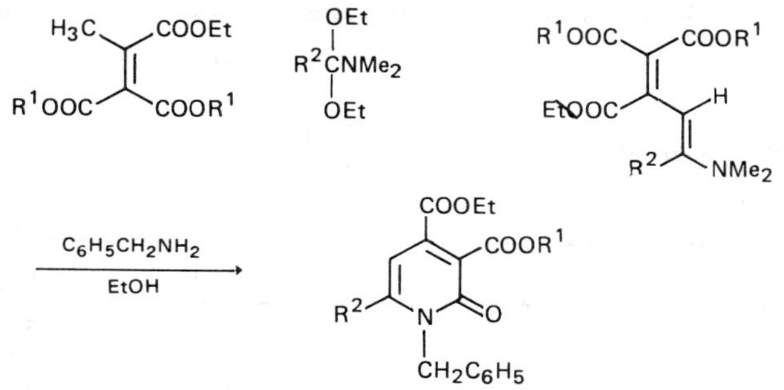

(231)

R^1	R^2
Et	H
Me	H
Me	3-Pyridyl
Me	2-Cyanophenyl
Me	3-Cyano-2-quinolyl

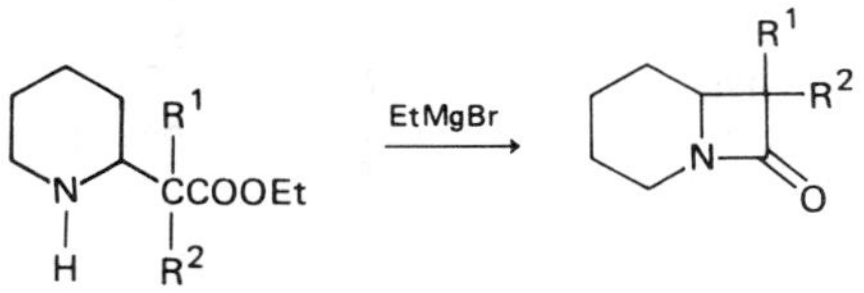

(232)

R^1 = Me, Me, Me, Et
R^2 = H, Me, Et, Et

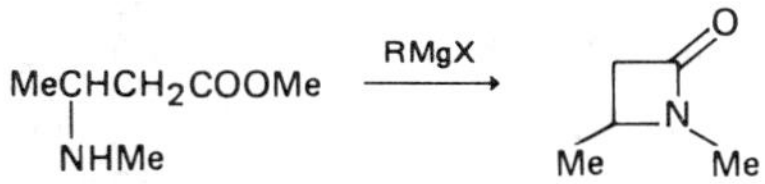

(233)

Listed in Table 28 are various β-aminopropionic acid esters which have been converted to β-lactams by a Grignard reagent. Other examples of β-lactam preparation using this method include the conversion of ethyl 3-phenyl-β-aminopropionate to 4-phenyl-2-azetidinone (equation 234)[401], the conversion of several ethyl

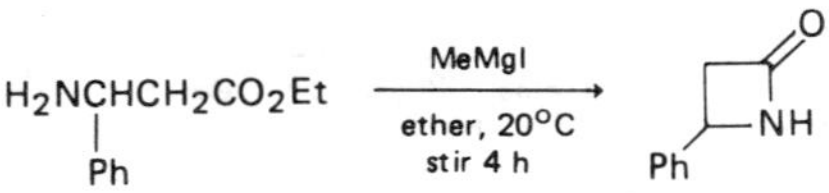

(234)

N-substituted 2-ethyl-2-phenyl-3-aminopropionic acid esters to their corresponding *N*-substituted 3-ethyl-3-phenyl-2-azetidinones (equation 235)[402], the conversion of methyl 2-substituted 3-phenyl-3-(phenylamino) propionates to a mixture of *cis* and *trans* 1-phenyl-3-substituted-4-phenyl-2-azetidinones (equation 236)[403] and the conversion of the methyl, ethyl, isopropyl and benzyl esters of 2-phenyl-3-(benzylamino)propionic acid to 1-benzyl-3-phenyl-2-azetidinone (equation 237)[404].

In connection with a new synthesis of oxindoles[405-407], Gassman and co-

TABLE 28. β-Lactams prepared via the reaction of substituted β-aminopropionic acid esters with a Grignard reagent

$$H_2NCH_2C(R^1)(R^2)COOEt + R^3MgX \xrightarrow{\text{ether}} \text{3-}R^1\text{-3-}R^2\text{-azetidin-2-one (NH)}$$

R^1	R^2	R^3	Yield (%)	References
H	*n*-Pr	Me	22	390
H	*i*-Bu	Me	67	391
H	*c*-C_6H_{11}	Me	54	391
H	Ph	Et	–	392
H	*p*-MeC_6H_4	Me	54	393
H	*p*-$MeOC_6H_4$	Me	20	393
H	$C_6H_5CH_2$	Me	43	391
H	α-naphthyl	Me	49	391
H	*p*-$C_6H_5C_6H_4$	Me	11	391
Me	Me	Et	80	394
Me	Ph	Et	51	395
CH_2OH	Ph	Et	–	396
Et	Et	Me	32	390
Et	Et	Et	92	394
Et	Ph	Me	79	390
Et	Ph	Et	86	394–398
Et	*p*-MeC_6H_4	Et	88	399
Et	$C_6H_5CH_2$	Et	64	395
n-Pr	*n*-Pr	Et	91	394
n-Pr	Ph	Et	56	395
i-Pr	Ph	Et	75–79	394, 395
n-Bu	*n*-Bu	Et	99	394
n-Bu	Ph	Et	92	394
$Me_2N(CH_2)_3$	Ph	Et	32	395
$Et_2N(CH_2)_2$	Ph	Me	16	399
c-C_6H_{11}	Ph	Et	80	394
Ph	$C_6H_5CH_2$	Et	83–87	394
H (as hydrochloride salt)	Ph	Me	52	400

$$RNHCH_2C(Et)(Ph)CO_2Et \xrightarrow[\text{ether, 0°C, stir 2 h, then 4 h at room temp.}]{\text{EtMgBr}} \text{3-Et-3-Ph-1-R-azetidin-2-one} \qquad (235)$$

R	Yield (%)
n-Pr	23
i-Pr	27
n-Bu	60
$C_6H_5CH_2$	74

$$\text{PhNHCH(Ph)—CH(R)COOMe} \xrightarrow{\text{EtMgBr}} \text{(cis and trans)} \qquad (236)$$

R	Yield (%)
Me	88
i-Pr	95

$$PhCH_2NHCH_2CH(Ph)COOR \xrightarrow[\text{or EtMgBr}]{\text{MeMgI}} \sim 40\% \qquad (237)$$

R = Me, Et, *i*-Pr, $PhCH_2$

$$\textbf{(32)} \xrightarrow{H^+} \textbf{(33)} \qquad \text{(Refs. 405, 407)} \qquad (238)$$

R^1	R^2	R^3	R^4	Yield (%)
Me	H	H	H	34
H	H	Me	H	67
H	H	H	Me	46
NO_2	H	H	H	51
H	NO_2	H	H	61
COOEt	H	Me	H	66

workers found that amino esters **32** afford 3-methylthiooxindoles **33** upon treatment with dilute acid.

Cyclization of amino esters with 2-pyridone as catalyst[408,409] is quite effective, as illustrated by a recent example (equation 239)[410].

$$\xrightarrow[\text{dioxane, 82\%}]{\text{2-pyridone}} \qquad (239)$$

Reductive cyclization of nitro esters such as ethyl 3-carboethoxy-4-nitropentanoate[411] can be used for the preparation of γ- and δ-lactams[412,413]. The required nitro esters can often be obtained by Michael addition of a nitroalkane to an appropriate α,β-unsaturated ester (equation 240)[411].

$$RCH_2NO_2 + \text{EtOOC-CH=CH-COOEt} \longrightarrow \text{EtOOC-CH(CHR-NO}_2\text{)-CH}_2\text{-COOEt} \xrightarrow[100^\circ C,\ 500\ psi]{H_2,\ Pd/C} \text{4-ethoxycarbonyl-5-R-2-pyrrolidone (81\%)} \quad (240)$$

R = Me, Et

It may be noted that Michael addition of diethyl acetamidomalonate to ethyl acrylate or ethyl crotonate can be accompanied by cyclization of the intermediate adduct to form 2-pyrrolidones[414,415].

In a study of lactam formation from a series of *o*-aminophenoxyacetamides Cohen and Kirk[416,417] have drawn the conclusion that the mechanism involves simultaneous attack of the aromatic amino function and an external proton donor at the amide carboxyl (equation 241).

$$o\text{-}H_2N\text{-}C_6H_4\text{-}OCH_2C(=O)NHR \cdots H\text{-}B \longrightarrow \longrightarrow \text{2H-1,4-benzoxazin-3(4H)-one} \quad (241)$$

2. *From halo, hydroxy and keto amides*

Treatment of α-, β-, γ- or δ-halo amides with a suitable basic reagent results in ionization of the amide proton to form a nitrogen anion, which then reacts by intramolecular displacement of halide ion to produce the appropriate lactam (equation 242). The scope and limitations of this method as applied to α-lactam

$$X(CH_2)_nCONHR \xrightarrow{\text{base}} (CH_2)_n\text{-}C(=O)\text{-}N\text{-}R \ (\text{ring}) \quad (242)$$

$$n = 1\text{–}4$$

synthesis have been discussed[418–424]. Successful preparations require the presence of one or more alkyl or aryl substituents at the α-carbon as well as a bulky *N*-alkyl group such as *t*-butyl or 1- or 2-adamantyl. Syntheses of β-, γ- and δ-lactams, but not ε-lactams[425], by cyclizations of prerequisite halo amides are much more general, as may be seen from equations (243)–(250). Some of the basic reagents which have been used include sodium in liquid ammonia[425], sodium hydride in DMSO[425], potassium *t*-butoxide in DMSO[425] and sodium ethoxide in ethanol[426].

A lactam synthesis first reported by Sheehan and Bose[432], and later exploited by numerous investigators[433–436] consists of intramolecular *C*-alkylation[435] of *N*-substituted α-haloacetamides and β-halopropionamides. Alkylation is effected through generation of a carbanion centre in the *N*-alkyl substituent, where one or preferable both of the substituents R^2 and R^3 shown in the generalized equation

$$\mathrm{BrCH_2C(R^1)(R^2)CONHR^3} \xrightarrow{\text{base}} \text{1-}R^3\text{-3-}R^1\text{-3-}R^2\text{-azetidin-2-one} \qquad (243)$$

R^1	R^2	R^3	Yield (%)	Reference
H	H	Ph	68–95	425
H	H	o-BrC_6H_4	71	425
H	H	o-FC_6H_4	90	425
H	H	p-BrC_6H_4	58	425
Me	Me	p-BrC_6H_4	55	425
H	H	p-ClC_6H_4	73	425
H	H	p-IC_6H_4	80	425
H	H	p-$MeOC_6H_4$	50	425
n-Pr	n-Pr	Me	52	427
Me	Ph	H	—	427
Me	Ph	Me	61	427
Me	Ph	$C_6H_5CH_2$	—	427
Me	Ph	Ph	54	427
Me	Ph	o-$O_2NC_6H_4$	54	427
Ph	Ph	Me	56	427

$$\mathrm{PhCH(X)CH_2CONHR} \xrightarrow[\mathrm{NaNH_2}]{\mathrm{KNH_2\ or}} \text{1-R-4-Ph-azetidin-2-one} \quad \text{(Ref. 428)} \qquad (244)$$

X	R	Yield (%)
Cl	c-C_6H_{11}	75
Cl	H_2NCOCH_2	76
Br	$Me_2CCH(CO_2Et)_2$	83
Cl	$Me_2C{=}C(CO_2Et)_2$	85

$$\mathrm{R^1C(R^2)(Br)CH_2CONHPh} \xrightarrow{\mathrm{NaNH_2}} \text{1-Ph-4-}R^1\text{-4-}R^2\text{-azetidin-2-one} \quad \text{(Ref. 429)} \qquad (245)$$

R^1	R^2	Yield (%)
H	Me	26
Me	Me	28

(251) are electron-withdrawing functions such as carboalkoxy, phenyl or cyano. In most cases where both R^2 and R^3 are activating groups, triethyl amine[432–438], sodium acetate[439], ethanolic potassium hydroxide[439–441], or basic ion-exchange resins[440] function satisfactorily as the base. With a single activating group sodium hydroxide[443] has proved to be effective. A number of representative examples of this procedure, which have appeared since 1966, are presented in Table 29.

PhCH(Br)CH(Br)CONHPh $\xrightarrow{\text{base}}$ [3-bromo-4-phenyl-1-phenylazetidin-2-one] (Ref. 429) (246)

Base	Yield (%)
NaOH, liq. NH_3	96
$NaNH_2$, liq. NH_3	78
KNH_2, liq. NH_3	86
NH_3 alone	38

$BrCH_2C(Ph)_2$—CONHR $\xrightarrow{\text{base}}$ [3,3-diphenyl-1-R-azetidin-2-one] (Ref. 430) (247)

R	Base	Yield (%)
H	KNH_2 in liq. NH_3	92
	EtONa in EtOH	90
Ph	KNH_2 in liq. NH_3	96
	KOH in MeCOEt	91
p-$MeOC_6H_4$	EtONa in EtOH	98
p-$O_2NC_6H_4$	EtONa in EtOH	96
$C_6H_5CH_2$	NaSH in EtOH	92

TABLE 29. Synthesis of β- and γ-lactams by base-catalysed intramolecular alkylation of N-substituted α-haloacetamides and β-halopropionamides

Starting amide	Base	Lactam	Yield (%)	Reference
$ClCH_2CON(R^1)CH(Ph)COOEt$	KOH	[4-Ph-4-HOOC-1-R^1-azetidin-2-one]		439
		R^1 = Ph	90 (90)[a]	
		R^1 = C_6H_4Cl-p	80	
		R^1 = C_6H_4Br-p	80	
		R^1 = C_6H_4Me-p	89 (90)[a]	
$ClCH_2CON(R^1)CH(COOEt)_2$	DMF[b]	[4,4-$(EtOOC)_2$-1-R^1-azetidin-2-one]		439
		R^1 = Ph	98	
		R^1 = C_6H_4Cl-p	95–98	
		R^1 = C_6H_4Br-p	95–97	
		R^1 = C_6H_4Me-p	95–98	
		R^1 = β-$C_{10}H_7$	95–98	

TABLE 29. (Continued)

Starting amide	Base	Lactam	Yield (%)	Reference
$ClCH_2CON(CH(CH_3)C_6H_5)CH_2CN$	NaH	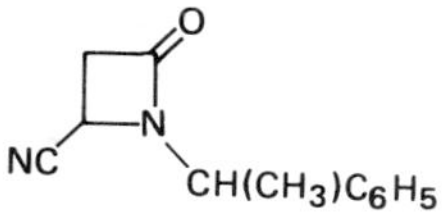 1-[$CH(CH_3)C_6H_5$]-4-cyano-azetidin-2-one (NC, $CH(CH_3)C_6H_5$)	70	443
$Br(CH_2)_2CON(R^1)$–$CH(COOEt)_2$	KOH	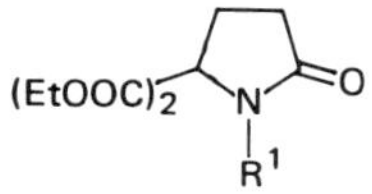 1-R^1-5,5-$(EtOOC)_2$-pyrrolidin-2-one		440
		R^1 = Ph	85	
		R^1 = C_6H_4Cl-*p*	90	
		R^1 = C_6H_4Br-*p*	84	
		R^1 = C_6H_4Me-*p*	80	
$Br(CH_2)_2CON(R^1)CH(Ph)COOEt$	KOH	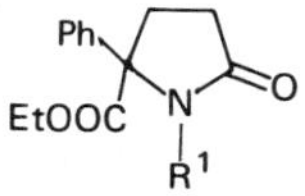 1-R^1-5-Ph-5-EtOOC-pyrrolidin-2-one		440
		R^1 = Ph	85	
		R^1 = C_6H_4Cl-*p*	80	
		R^1 = C_6H_4Br-*p*	80	
		R^1 = C_6H_4Me-*p*	80	
$Br(CH_2)_2CON(R^1)$–$CH(COOH)COOEt$	KOH	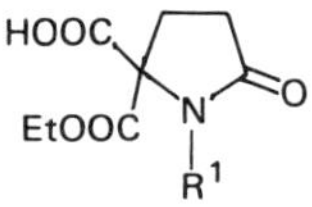 1-R^1-5-HOOC-5-EtOOC-pyrrolidin-2-one		440
		R^1 = Ph	80	
		R^1 = C_6H_4Cl-*p*	86	
		R^1 = C_6H_4Br-*p*	80	
		R^1 = C_6H_4Me-*p*	80	

[a]Yield of ethyl ester obtained by heating sodium acetate and starting amide without solvent at 140–145°C.
[b]Reactions carried out in refluxing DMF without added base.

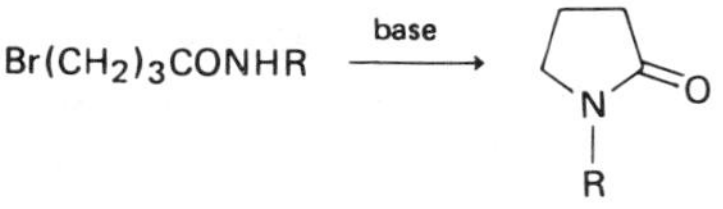

$$Br(CH_2)_3CONHR \xrightarrow{\text{base}} \text{1-R-pyrrolidin-2-one} \quad \text{(Ref. 425)} \qquad (248)$$

R	Yield (%)
Ph	48
o-BrC_6H_4	50
o-FC_6H_4	61
p-BrC_6H_4	67
p-ClC_6H_4	54
p-IC_6H_4	79

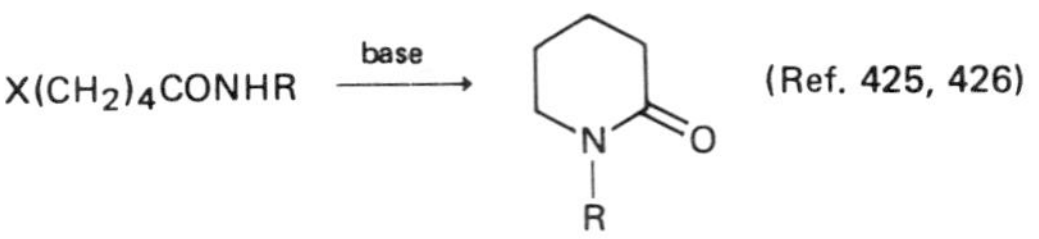

$$X(CH_2)_4CONHR \xrightarrow{\text{base}} \text{(1-R-2-piperidone)} \quad \text{(Ref. 425, 426)} \qquad (249)$$

X	R	Yield (%)
Br	Ph	61
Cl	H	82
Cl	Me	48
Cl	Et	33
Cl	*n*-Pr	28
Cl	*n*-Bu	36
Cl	*n*-C_8H_{17}	37
Cl	*c*-C_6H_{11}	11
Cl	Ph	96

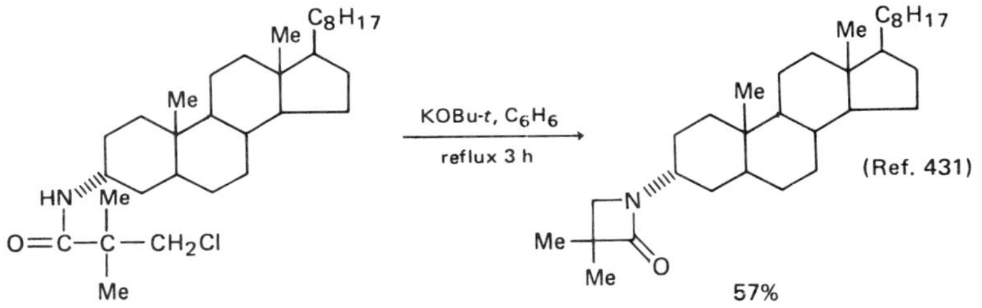

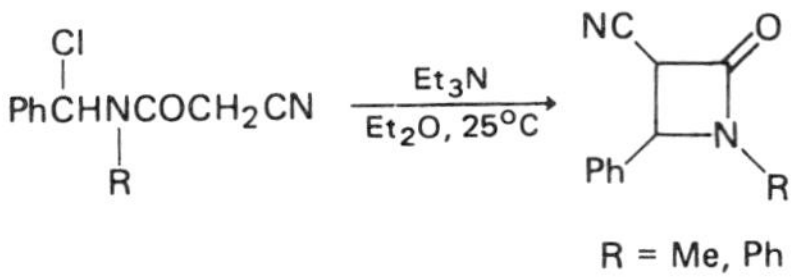

$$\xrightarrow[\text{reflux 3 h}]{\text{KOBu-}t\text{, } C_6H_6} \quad 57\% \quad \text{(Ref. 431)} \qquad (250)$$

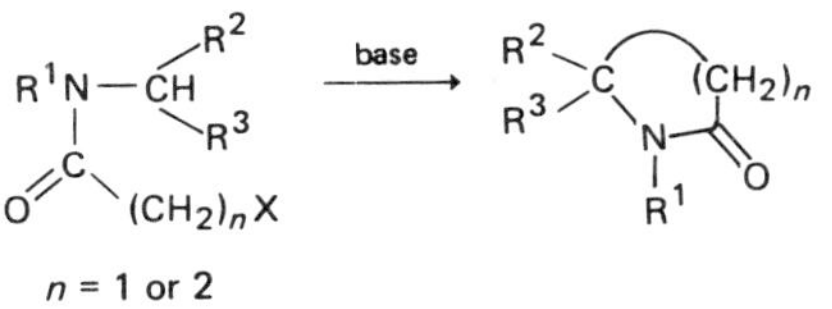

$$R^1N(CHR^2R^3)C(=O)(CH_2)_nX \xrightarrow{\text{base}} \qquad (251)$$

n = 1 or 2

$$PhCH(Cl)N(R)COCH_2CN \xrightarrow[Et_2O,\ 25^\circ C]{Et_3N} \qquad (252)$$

R = Me, Ph

β-Lactams have also been prepared by base-catalysed cyclization of *N*-(α-chlorobenzyl)-β-cyanoamides (equation 252)[444] and by intramolecular Michael addition (equations 253–255)[445].

In addition to the nucleophilic displacements of halide ion shown above, *N*-aryl-α-halo amides can be cyclized via intramolecular Friedel–Crafts alkylation of the *N*-aryl moiety to produce oxindoles, as shown in the synthesis of 3-ethyl-1-methyloxindole from *N*-methyl-α-bromo-*n*-butyranilide (equation 256)[446].

An interesting approach to the cyclization of bromo amides may be seen in the reaction of *N*-(2-bromopropanoyl)aminoacetone with triethyl phosphite to afford an intermediate phosphonate ester, which can then be converted into 2-oxo-3,4-dimethyl-Δ^3-pyrroline via and intramolecular Wittig reaction (equation 257)[447].

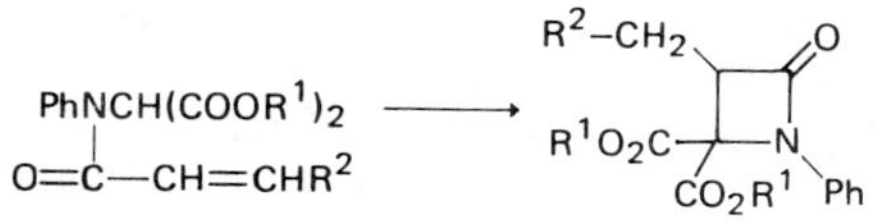

(253)

R^1	R^2	Yield (%)
Et	p-$O_2NC_6H_4$	94
Me	p-$O_2NC_6H_4$	74
Et	o-$O_2NC_6H_4$	70
Et	CO_2Et	–

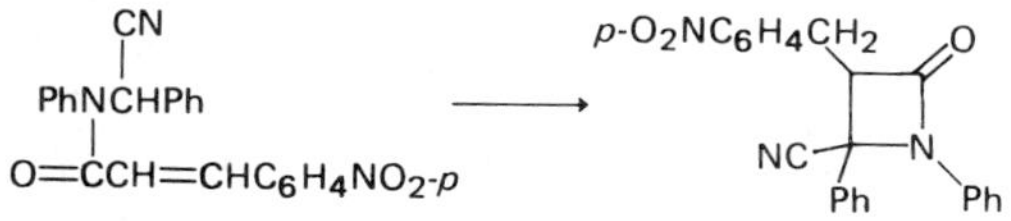

(254)

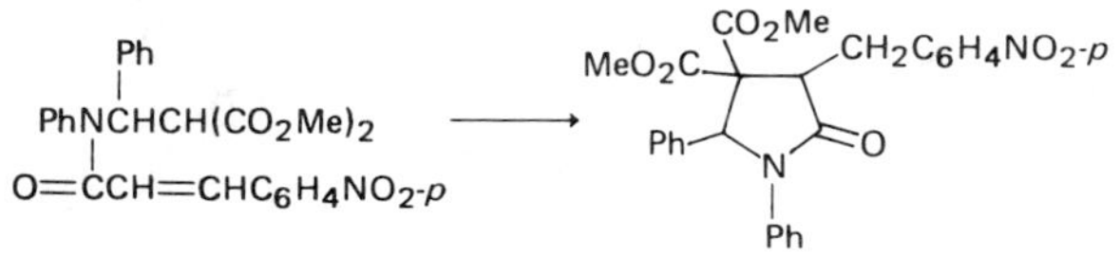

(255)

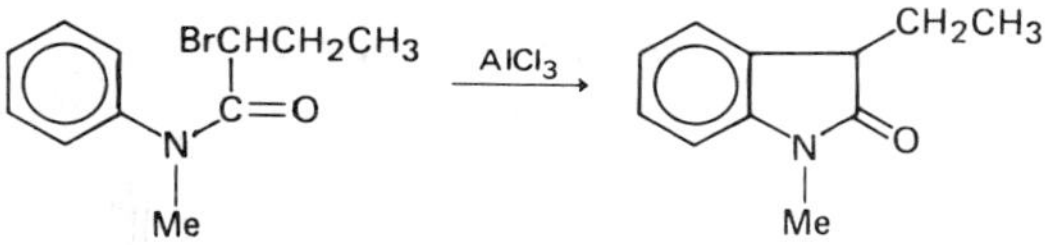

(256)

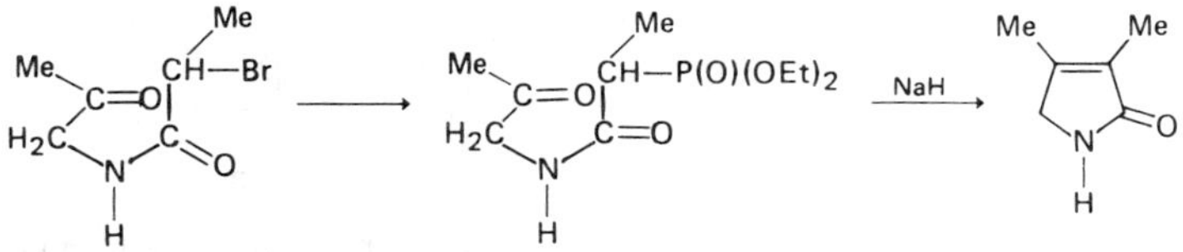

(257)

γ- and δ-Hydroxy amides obtained from reactions of aldehydes and ketones with the dilithio derivatives of *N*-substituted benzamides[448] and *N*-substituted *o*-toluamides[449] have been cyclized in the presence of cold, concentrated sulphuric acid to form γ- and δ-lactams, respectively (equations 258 and 259)[450-452]. A

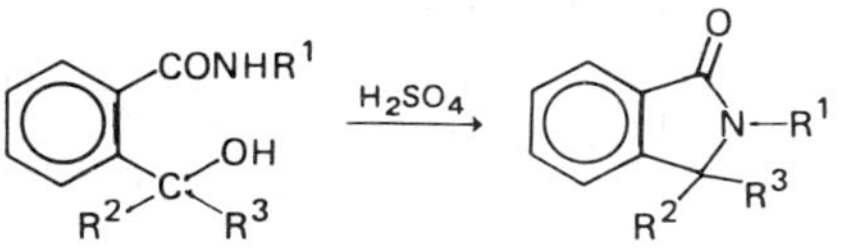

(258)

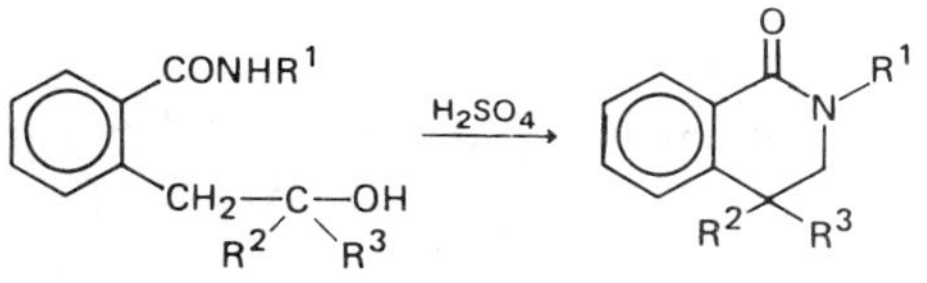

(259)

mechanistic study[453] of reactions of this type revealed that in addition to lactam formation, linear dehydration to form olefin amides and cyclodeamination to form δ-lactones also occurred. The major course of reaction was found to be dependent upon the nature of the acidic medium, the temperature and the structure of the hydroxy amide.

A recent patent[454] claims the preparation of β-lactams by reaction of *N*-methyldiarylglycolamides with concentrated sulphuric acid in acetic acid (equation 260).

$$(Ar)_2C(OH)CONHMe \xrightarrow[HOAc]{H_2SO_4} \text{3,3-diaryl-2-azetidinone (N–H)} \tag{260}$$

$$\xrightarrow[\text{toluene, reflux}]{p\text{-}CH_3C_6H_4SO_3H} \tag{261}$$

$R^1 = H$, $R^2 = Me$; $n = 2$
$R^1 = R^2 = H$; $n = 1$
$R^1 = H$, $R^2 = Me$; $n = 1$
$R^1 = OMe$, $R^2 = Me$; $n = 1$

$$MeCOCOHN\text{–} \ \textbf{(34)} \xrightarrow[THF]{NaH} \textbf{(35)} \tag{262}$$

Acid-catalysed cyclization of a series of δ-keto carboxamides has been found[455] to afford unsaturated lactams in 80–90% yield (equation 261).

An interesting intramolecular aldol cyclization of α-keto amide **34** afforded the tricyclic lactam **35** (equation 262)[456].

B. By Ring-closure Reactions (Photochemical)

A large variety of substituted amides have been found to produce lactams upon exposure to ultraviolet and ultraviolet–visible irradiation[457–492].

The type of lactam obtained is dependent upon the structural features of the starting amide (Table 30).

1. Cyclization of α,β-unsaturated amides

Irradiation of α,β-unsaturated anilides affords 3,4-dihydrocarbostyrils via ring-closure involving the *ortho* position of the *N*-aryl substituent and the β-carbon the acyl moiety (equation 263)[457–461]. Unsaturated amides possessing an *N*-heteroaryl

$$R^1CH{=}C(R^2)CON(R^3){-}C_6H_5 \xrightarrow{h\nu} \text{3,4-dihydroquinolin-2-one } (4\text{-}R^1, 3\text{-}R^2, N\text{-}R^3) \tag{263}$$

substituent react similarly (equation 264)[461]. In certain cases where $R^1 = R^2$ = Ph, β-lactam formation can become the major reaction pathway[457,458] (Table 30).

$$H_2C{=}C(Me){-}CONH{-}(2\text{-pyridyl}) \xrightarrow{h\nu} \text{3-methyl-3,4-dihydro-1,8-naphthyridin-2(1H)-one} \tag{264}$$

2. *Cyclization of benzanilides*

Prolonged irradiation of a benzene solution of benzanilide in the presence of iodine produces phenanthridone in 20% yield (equation 265); however, without

$$C_6H_5{-}C(=O){-}N(H){-}C_6H_5 \xrightarrow{h\nu} \text{phenanthridone} \tag{265}$$

iodine lactam formation drops to less than 1%[462]. The reaction proceeds more satisfactorily of one or the other of the aromatic residues contains an *ortho* halogen or methoxy group (equation 266)[462–465]. Anilides of thiophene-2-carboxylic acid,

$$\left.\begin{array}{l} 2\text{-}X{-}C_6H_4{-}C(=O){-}N(H){-}C_6H_5 \quad X = I, Br \\ C_6H_5{-}C(=O){-}N(H){-}C_6H_4{-}2\text{-}X \quad X = I, Br, OMe \end{array}\right\} \xrightarrow{h\nu} \text{phenanthridone} \tag{266}$$

furan-2-carboxylic acid, indole-2-carboxylic acid and indole-3-carboxylic acid participate in similar photoinduced cyclizations (Table 30).

3. *Cyclization of enamides*

Various enamides of the general type shown in equation (267) have been cyclized in connection with the synthesis of a number of isoquinoline alkaloids[465,466]. Related enamide photocyclizations appear in Table 30.

TABLE 30. Preparation of lactams by intramolecular photocyclization

Amide	Conditions	Product	Yield (%)	Reference
PhCH=C(Ph)CONH$_2$ *cis*	C_6H_6, 70h	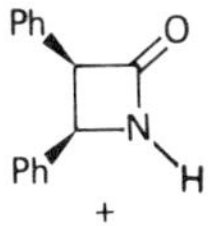 cis-3,4-diphenyl-2-azetidinone (Ph, Ph, O, N–H) +	13	457, 458
		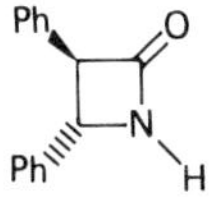 trans-3,4-diphenyl-2-azetidinone (Ph, Ph, O, N–H)	3	
PhCH=C(Ph)CONHPh *cis*	C_6H_6, 23 h	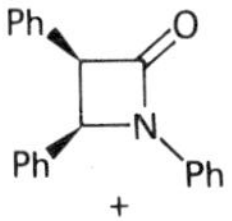 cis-1,3,4-triphenyl-2-azetidinone (Ph, Ph, O, N–Ph) +	37	457, 458
		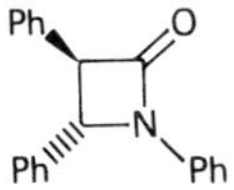 trans-1,3,4-triphenyl-2-azetidinone (Ph, Ph, O, N–Ph) +	2.3	
		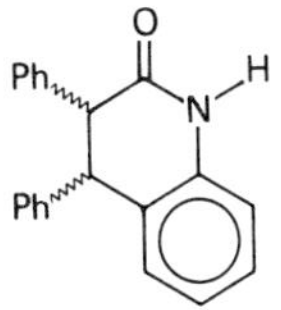 3,4-diphenyl-3,4-dihydro-2(1H)-quinolinone (Ph, Ph, O, N–H)	5	
MeCH=CMeCONHPh *cis*	Ether, HOAc, 6 h	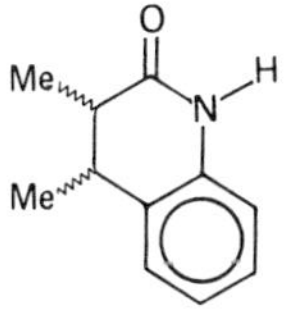 3,4-dimethyl-3,4-dihydro-2(1H)-quinolinone (Me, Me, O, N–H) *cis* and *trans*	58	459

MeCH=CHCONHPh *trans*	Ether, HOAc, 5 days	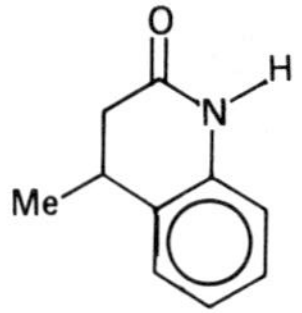	25	459
H_2C=CMeCONHPh	Ether, HOAc, 9 h	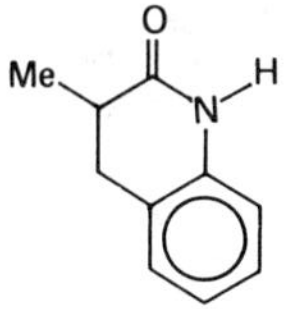	50 (82[a])	459
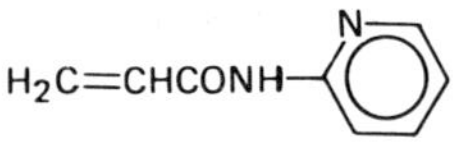 H_2C=CHCONH-2-pyridyl	C_6H_6, HOAc, 3 h	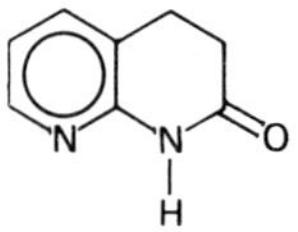	17	460
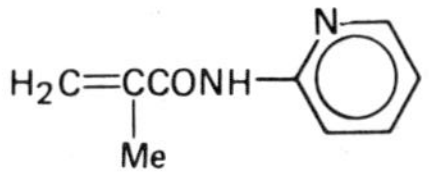 H_2C=C(Me)CONH-2-pyridyl	C_6H_6, HOAc, 3 h	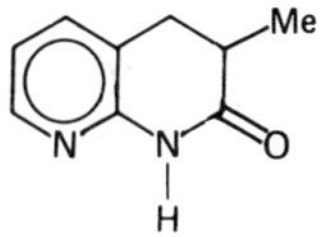	78	460
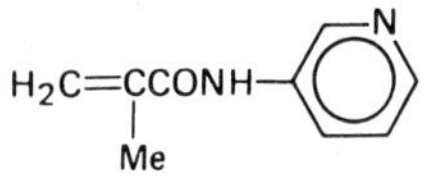 H_2C=C(Me)CONH-3-pyridyl	C_6H_6, HOAc, 3 h	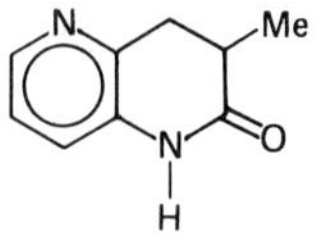	53	460
		+ 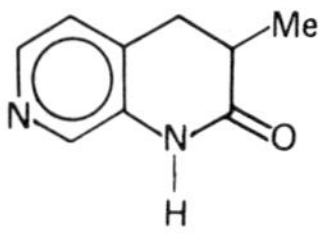	22	

TABLE 30. (Continued)

Amide	Conditions	Product	Yield (%)	Reference
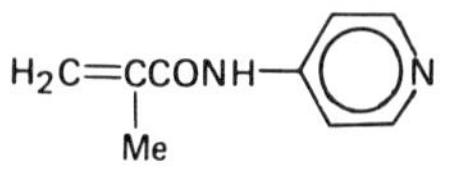 $H_2C{=}C(Me)CONH$–(N)	C_6H_6, HOAc, 3 h	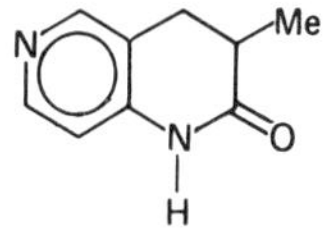 N; Me; N–H; O	72	460
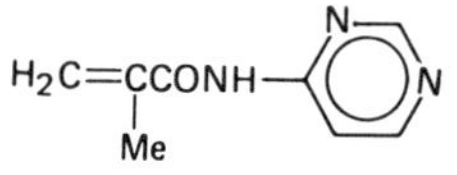 $H_2C{=}C(Me)CONH$–(N, N)	C_6H_6, HOAc, 3 h	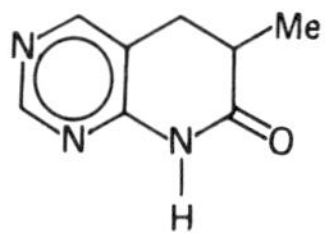 N; N; Me; N–H; O	24	460
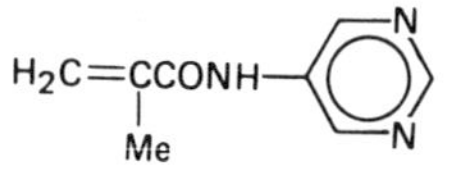 $H_2C{=}C(Me)CONH$–(N, N)	C_6H_6, HOAc, 3 h	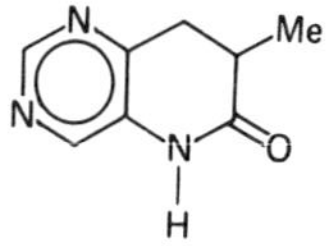 N; N; Me; N–H; O	14	460
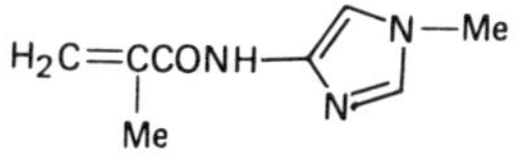 $H_2C{=}C(Me)CONH$–(N—Me, N)	C_6H_6, HOAc, 3 h	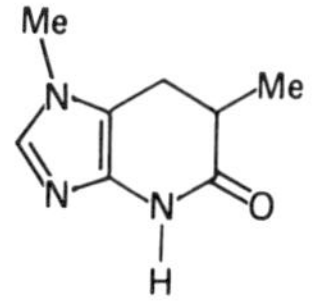 Me–N; N; Me; N–H; O	6	460
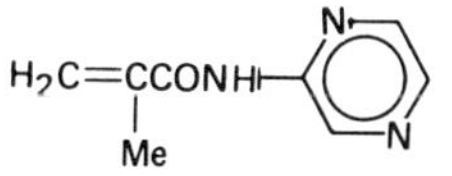 $H_2C{=}C(Me)CONH$–(N, N)	C_6H_6, HOAc, 3 h	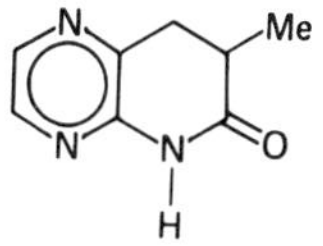 N; N; Me; N–H; O	19	460
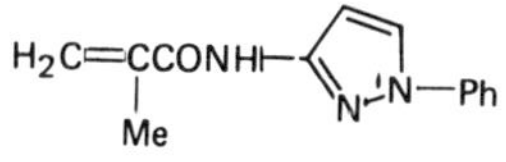 $H_2C{=}C(Me)CONH$–(N, N—Ph)	C_6H_6, HOAc, 3 h	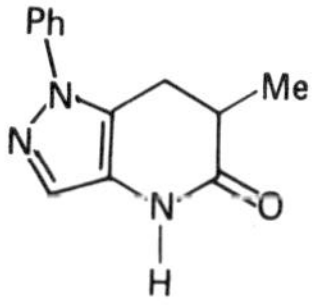 Ph–N; N; Me; N–H; O	3	460

$H_2C{=}C(Me)CONH$–(2-Cl-3-pyridyl)	C_6H_6, HOAc, 3 h		25	460
$R^3CH{=}C(R^2)CONR^1Ph$	C_6H_6, r.t.			461
$R^1 = R^2 = R^3 = H$	150 h		4	
$R^1 = R^2 = Me$, $R^3 = H$	80 h[b]		57	
$R^1 = R^2 = H$, $R^3 = Me$	80 h[b]		61	
PhCONHPh	C_6H_6, I_2, 148h		20	462
2-I-C_6H_4CONHPh				
	C_6H_6, 126 h		9	462
	C_6H_6, r.t. 30 h		18[c]	463
PhCONH–(2-I-C_6H_4)	C_6H_6, 160 h		48	462

TABLE 30. (Continued)

Amide	Conditions	Product	Yield (%)	Reference
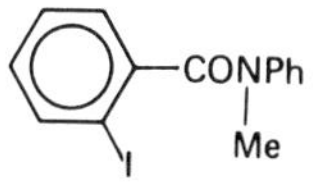 2-IC6H4CONPh(Me)	C_6H_6, 30 h, r.t.	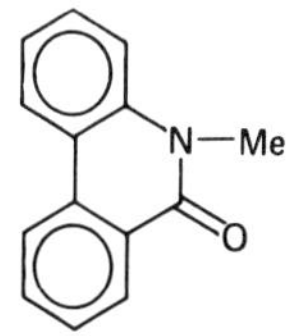 5-methylphenanthridin-6(5H)-one	34–36	463
		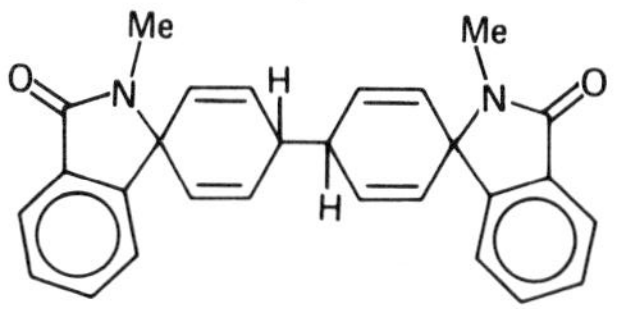 + dimer (spiro-isoindolinone)	30–36[d]	
		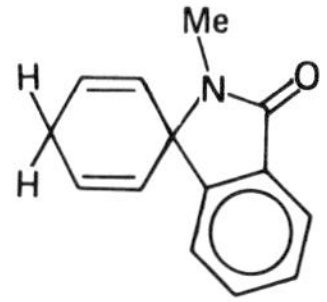 + spiro-isoindolinone	4–5	
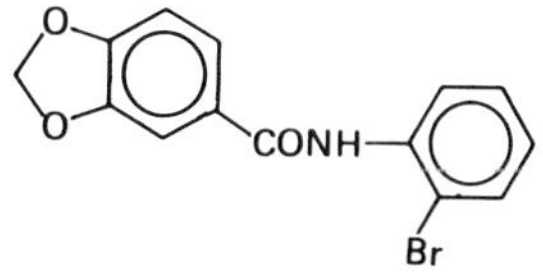 3,4-methylenedioxybenzoyl 2-bromoanilide	C_6H_6, MeOH, 24 h	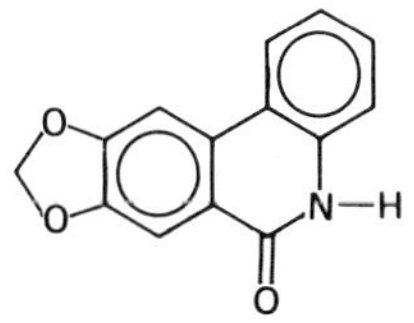 methylenedioxy-phenanthridinone	15–20	464
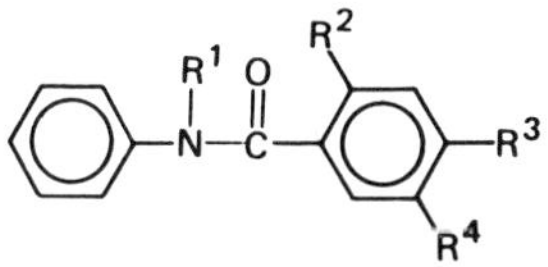 PhN(R¹)C(O)C6H2(R²)(R³)(R⁴)		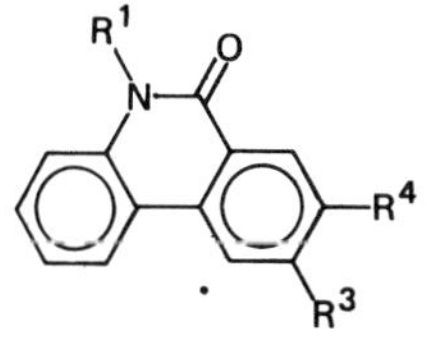 phenanthridinone (R¹, R³, R⁴)		465

$R^1 = R^3 = R^4 = H$, $R^2 = OMe$	MeOH		0	
$R^1 = R^3 = R^4 = H$, $R^2 = O_2CMe$	EtOAc, 21 h		0	
$R^1 = H$, $R^2 = R^3 = R^4 = OMe$	EtOAc, 20 h		55	
$R^1 = Me$, $R^2 = R^3 = R^4 = OMe$	EtOAc, 12 h		41	
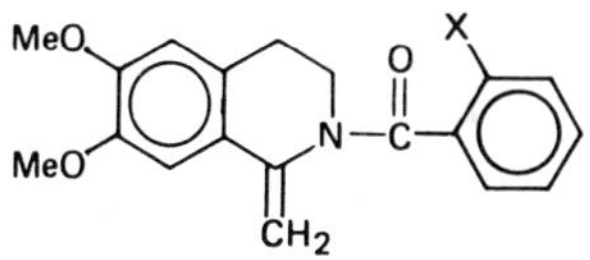	*t*-BuOH, 2 h	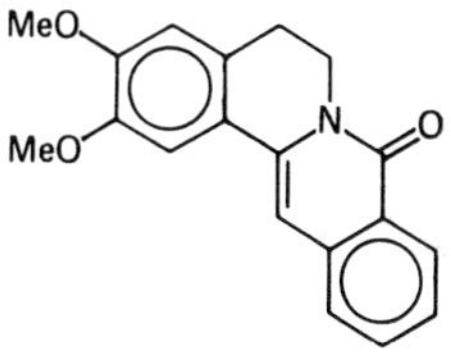		465
X = F			85	
X = Cl			50	
X = Br			50	
$X = O_2CMe$			76	
X = SMe			55	
$X = NO_2$			17	
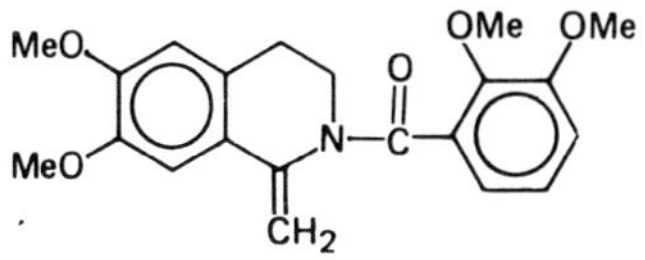	C_6H_6, 4.5 h	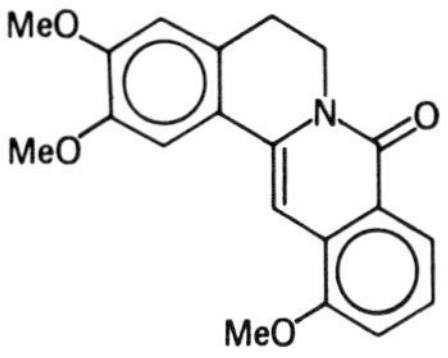	85	465
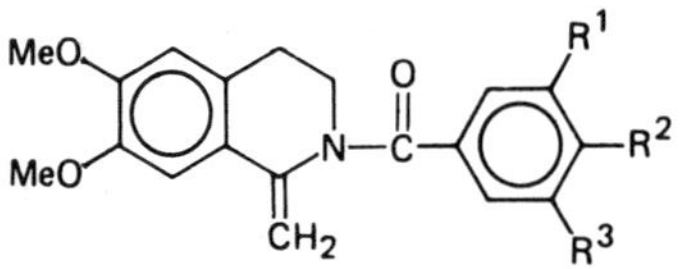	*t*-BuOH	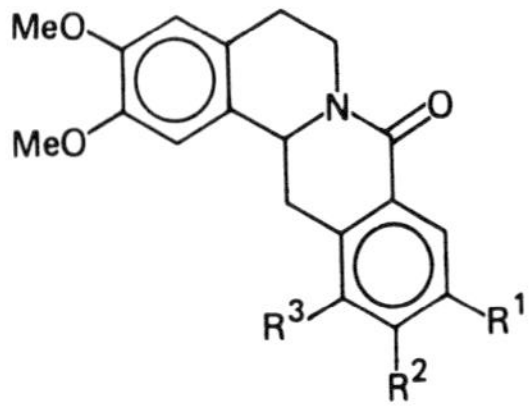		466

TABLE 30. (Continued)

Amide	Conditions	Product	Yield (%)	Reference
$R^1 = R^2 = R^3 = H$	1.5 h		97	
$R^1 = R^2 = OMe, R^3 = H$	2.5 h		94	
$R^1R^2 = —OCH_2O—, R^3 = OMe$	2.5 h		75	
$R^1 = R^2 = R^3 = OMe$	5 h		70	
$R^1 = OMe, R^2 = O_2CMe, R^3 = H$	12 h		45	
$R^1 = R^3 = H, R^2 = Me$	4.5 h		85	
$R^1 = R^3 = H, R^2 = Cl$	16 h		75	
$R^1 = R^3 = H, R^2 = Ph$	12 h		76	
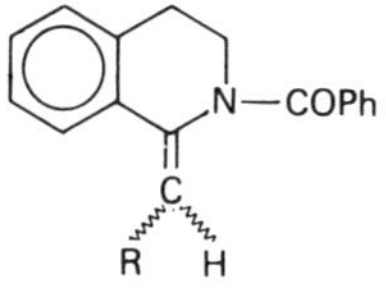 N—COPh; R H; R = H, Me	MeOH, 1–20 h, r.t.	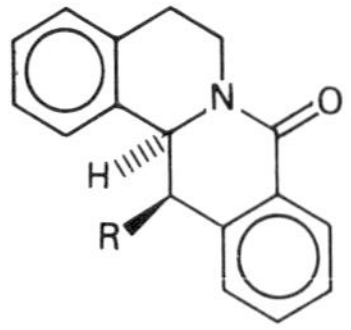 H, R, N, O	~70	471
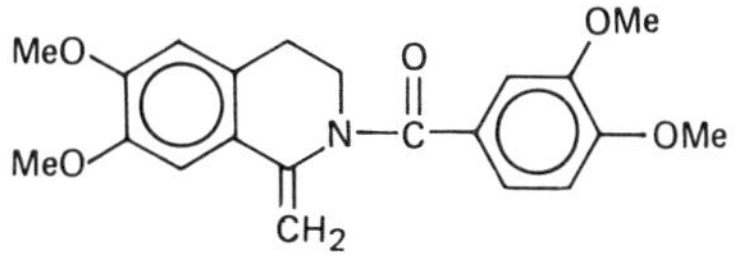 MeO, MeO, N—C(=O)—, OMe, OMe, CH_2	MeOH, 3 h, r.t.	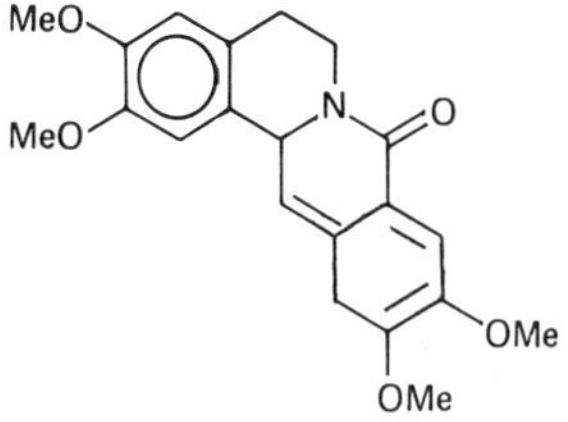 MeO, MeO, N, O, OMe, OMe	5	471
		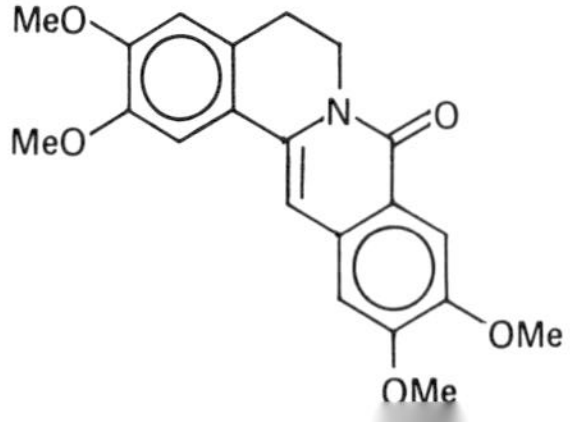 + MeO, MeO, N, O, OMe, OMe	40	

Substrate	Conditions	Product	Yield (%)	Ref.
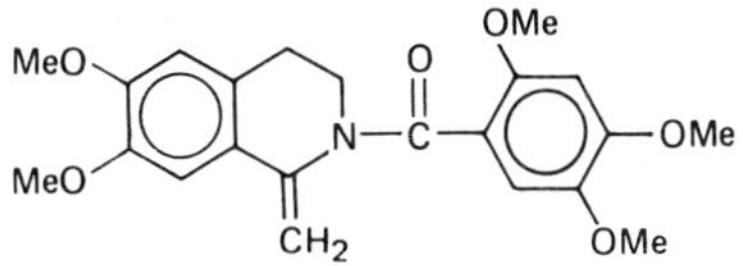	t-BuOH, 2 h	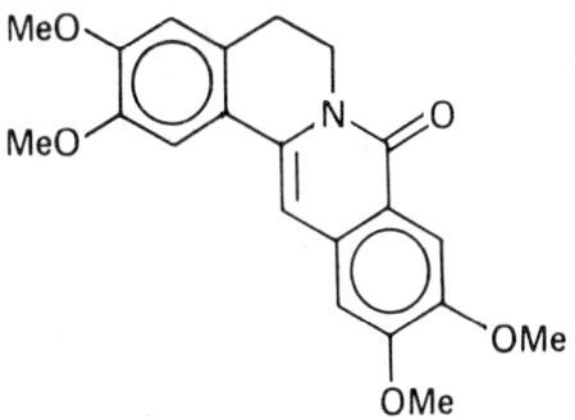	85	465
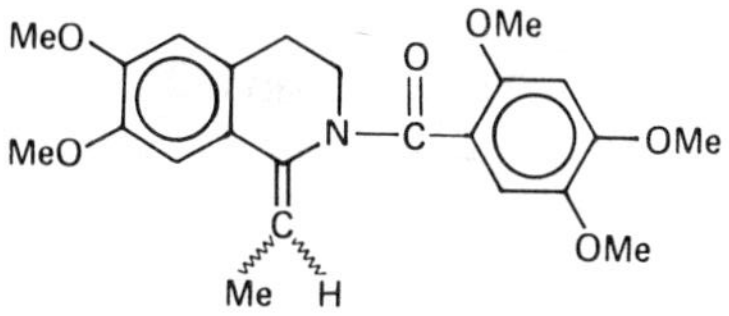	EtOAc, 12 h	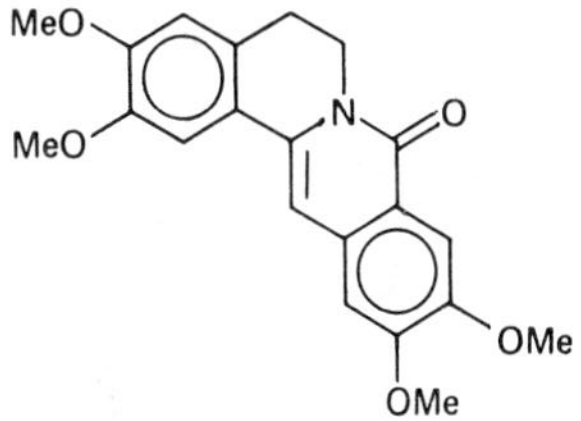	69	465
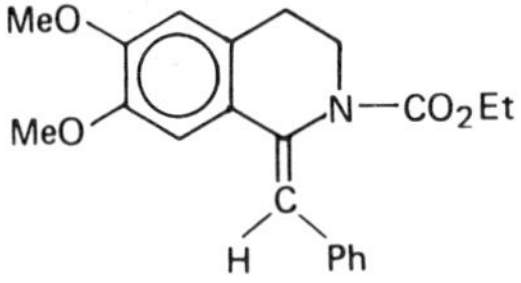 *trans*	EtOH, 120 h	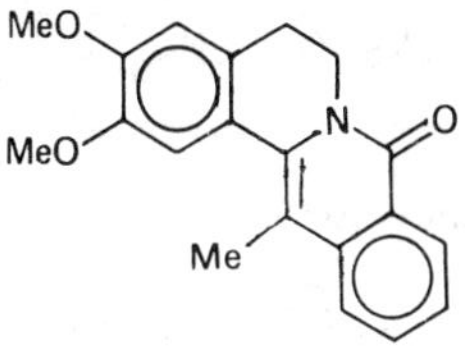	64	465
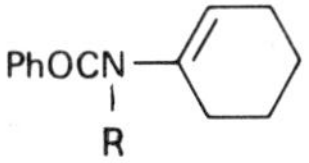	MeOH, 40 h	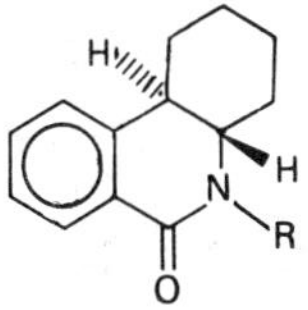 R = $CH_2C_6H_5$ R = Me	35 15	475

TABLE 30. (Continued)

Amide	Conditions	Product	Yield (%)	Reference
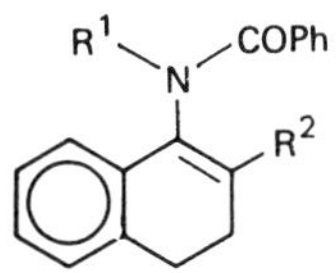 R^1, N, COPh, R^2	MeOH	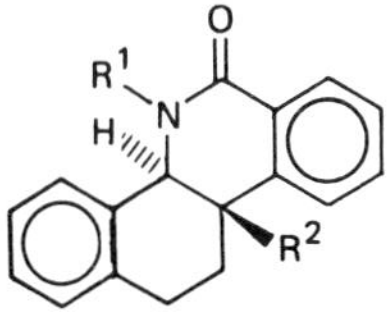 O, R^1, N, H, R^2		474
	40 h	$R^1 = CH_2C_6H_5$, $R^2 = H$	55	
	40 h	$R^1 = Me$, $R^2 = H$	51[e]	
	106 h	$R^1 = CH_2C_6H_5$, $R^2 = Me$	20	
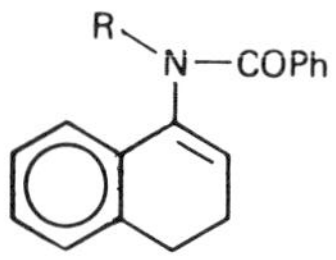 R, N—COPh	MeOH, r.t., 40 h	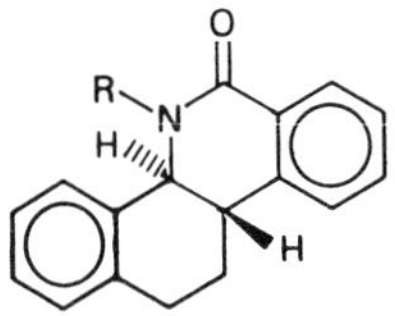 O, R, N, H, H		477
		$R = CH_2C_6H_5$	55	
		$R = Me$	51	
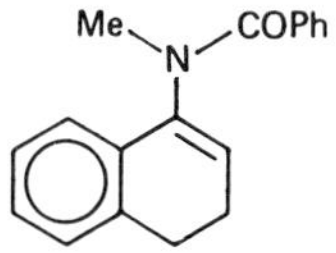 Me, N, COPh	MeOH, I_2, 40 h	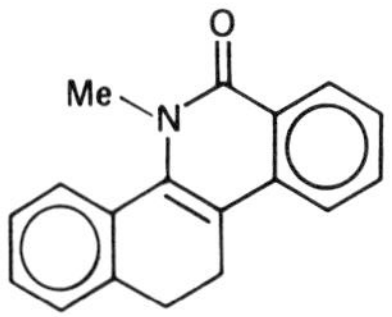 O, Me, N	21	474
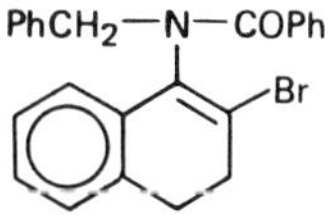 $PhCH_2$—N—COPh, Br	1. MeOH, 15 h 2. KOH, MeOH, reflux 1.5 h	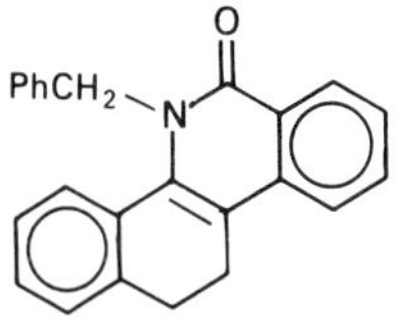 O, $PhCH_2$, N	28	474

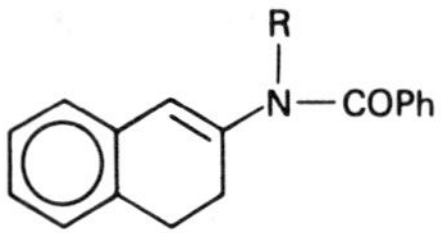 R–N–COPh	Ether, r.t., 24 h	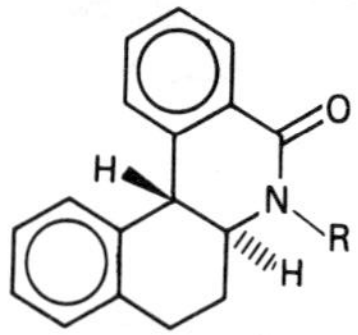		476
		R = $CH_2CH{=}CH_2$	71	
		R = *n*-Bu	63	
		R = Me	40	
		R = $CH_2C_6H_5$	47	
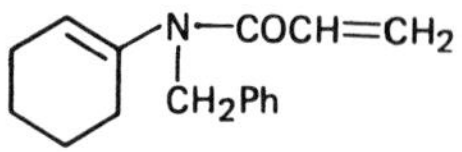 N–COCH=CH_2, CH_2Ph	Ether, r.t.	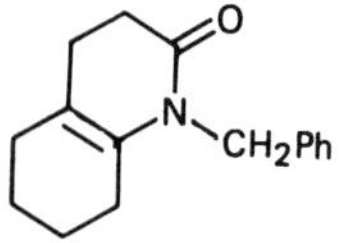 N–CH_2Ph	61[f]	473
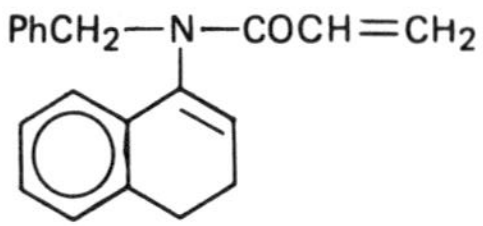 $PhCH_2$–N–COCH=CH_2	Ether, r.t.	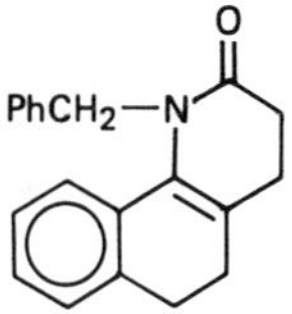 $PhCH_2$–N	42	473
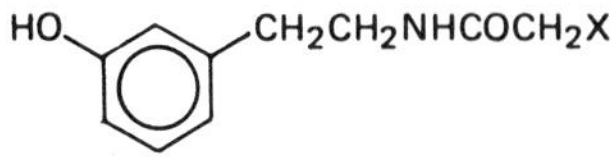 HO, $CH_2CH_2NHCOCH_2X$	EtOH, H_2O	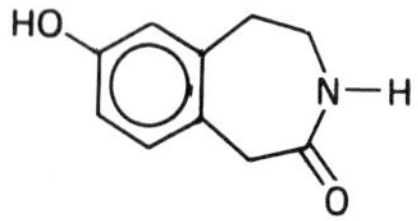 HO, N–H		478
X = Cl	2 h		70	
X = I	20 min		11	
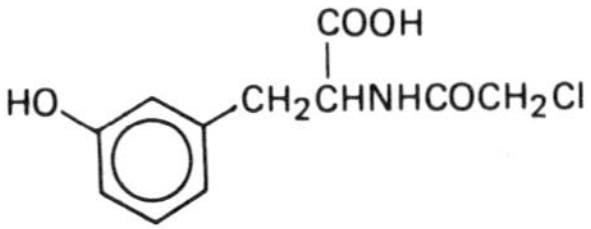 HO, $CH_2CH(COOH)NHCOCH_2Cl$	H_2O, NaOH, (pH 6.5), 45 min	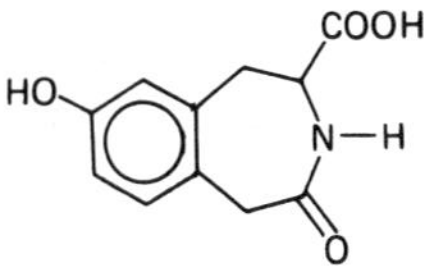 HO, COOH, N–H	25	478

TABLE 30. (Continued)

Amide	Conditions	Product	Yield (%)	Reference
3,4-(HO)$_2C_6H_3CH_2CH_2NHCOCH_2Cl$	MeOH, (pH 6), 1 h	HO, HO, O, N—H	—	478
		+ HO, HO, O, N—H	—	
3,4-(MeO)$_2C_6H_3CH_2CH_2NHCOCH_2Cl$	EtOH, H_2O	MeO, OMe, N—H, O	9 (4)[g]	480
		+ MeO, MeO, N—H, O	33 (6)[g]	
		+ MeO, MeO, N, O	11 (10)[g]	
		+ MeO, O, N—H, O	0 (4)[g]	

Substrate	Conditions	Product	Yield (%)	Ref.
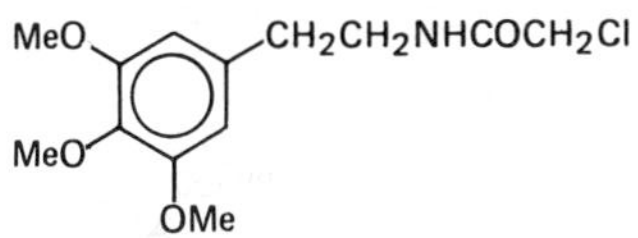 MeO, MeO, OMe; $CH_2CH_2NHCOCH_2Cl$	EtOH, H_2O	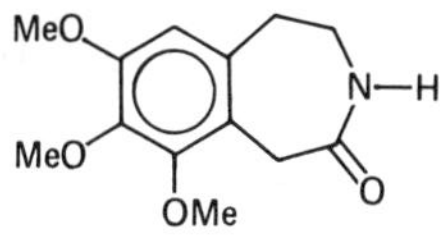 MeO, MeO, OMe; N—H; O	10	481
		+ OMe, OMe, OMe, HO, H N, O	33	
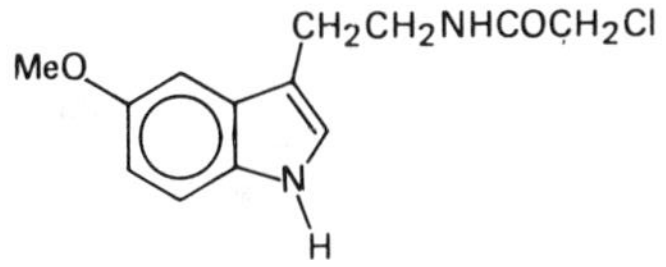 MeO; $CH_2CH_2NHCOCH_2Cl$; N H	H_2O, THF, NaOAc, 5 h	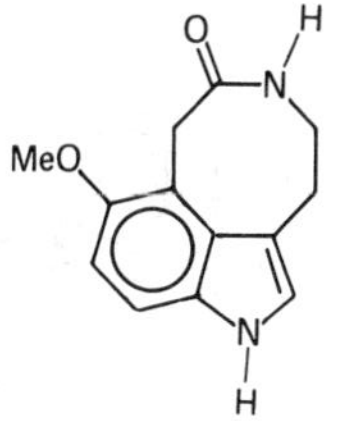 MeO; O; N H; N H	47	482
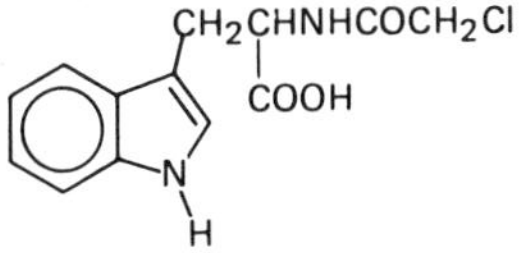 $CH_2CHNHCOCH_2Cl$ COOH; N H	H_2O, $NaBH_4$, (pH 9.5–10), 30 min	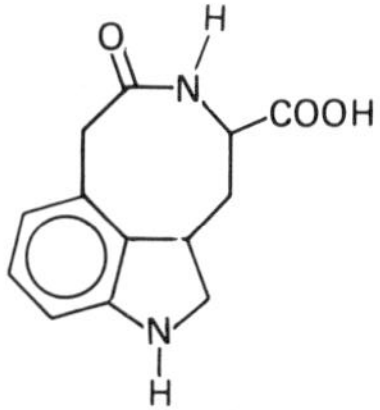 O; N H; COOH; N H	18	483

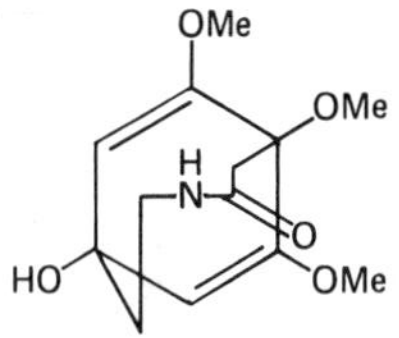

TABLE 30. (Continued)

Amide	Conditions	Product	Yield (%)	Reference
	NaOH, H_2O, 45 min		18	
	H_2O, 45 min		40	
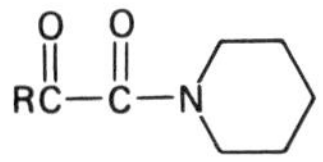	$h\nu$	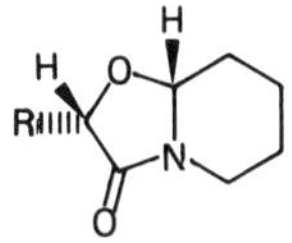	1.2 (R = Me) 2.4 (R = Ph)	484
R = Me, Ph		+ 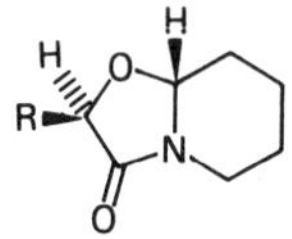	1.6 (R = Me or Ph)	
		+ 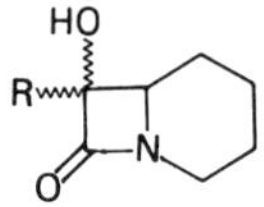	1.2 (R = Me) 1.0 (R = Ph)	
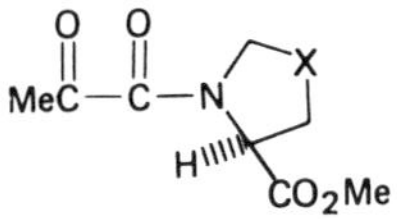	C_6H_6, N_2, $h\nu$, 0–10°C	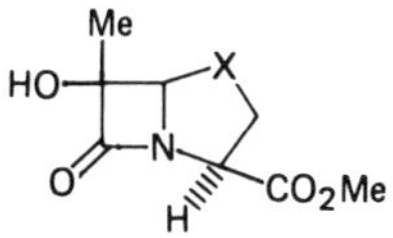		485
		X = S	11	
		X = SO	8–40	
		X = SO_2	70	

Diazo compound	Conditions	Product	Yield (%)	Ref.
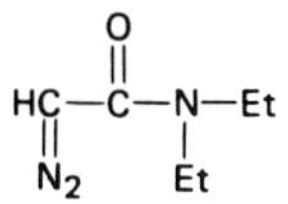	Dioxane	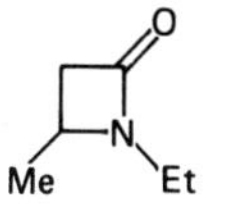	57 (43)[h]	486
		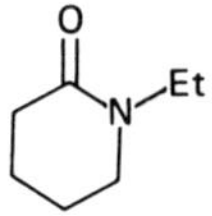	43 (5)[h]	
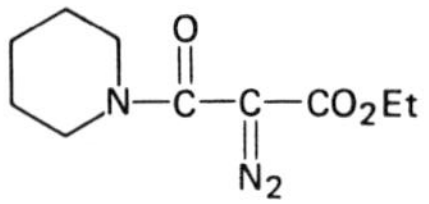	CCl_4, r.t., 2 h	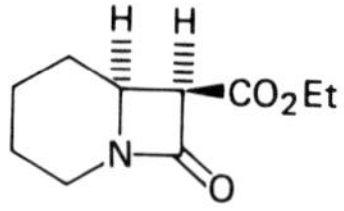	80 (*cis:trans*, 1:2)	488
		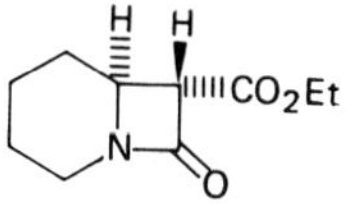		
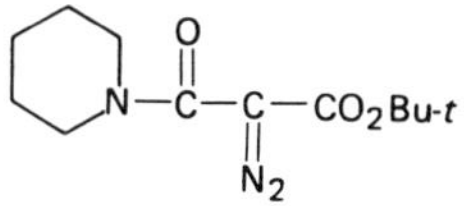	CCl_4, r.t., 1 h	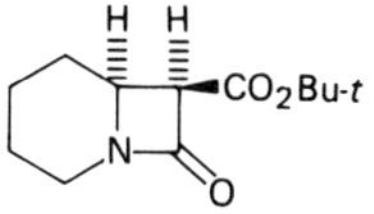	14	488
		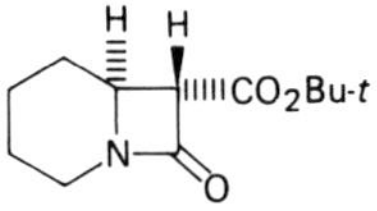	40	
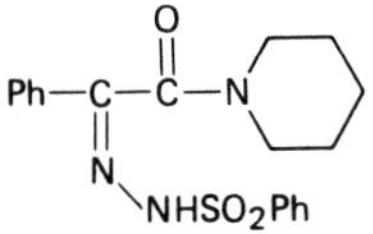	1. NaH, 60°C 2. *hν*, CH_2Cl_2, 10°C	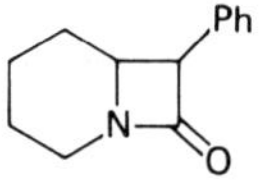	50	489

TABLE 30. (Continued)

Amide	Conditions	Product	Yield (%)	Reference
$m\text{-}RC_6H_4C(=N_2)\text{—}C(=O)\text{—}N(CH_2)(CH_2)_n$ (cyclic)	Heat only	bicyclic β-lactam, $C_6H_4R\text{-}m$		
R = H; n = 4			40	487
R = H; n = 3			—	
R = H; n = 5			--	
R = NO_2; n = 4			32	
$t\text{-}BuO_2C\text{—}C(=N_2)\text{—}C(=O)\text{—}N$ (thiazolidine, H, CO_2CH_2Ph)	CCl_4, r.t., 1 h	S, N bicyclic β-lactam, $CO_2Bu\text{-}t$, H, CO_2CH_2Ph	50	488
$R^1COCH_2CON(CH_2R^3)\text{—}CH_2R^2$	C_6H_6	OH, R^1, R^2, N—CH_2R^3, =O (pyrrolidinone)		467
$R^1 = R^2 = R^3$ = Ph			80	
R^1 = Ph, $R^2 = R^3$ = H			88	
R^1 = Ph, $R^2 = R^3$ = Me			60	
R^1 = Me, $R^2 = R^3$ = H			73	
$R^1 = R^2 = R^3$ = Me			76	
R^1 = Ph, R^2, R^3 = —CH_2OCH_2—			80	

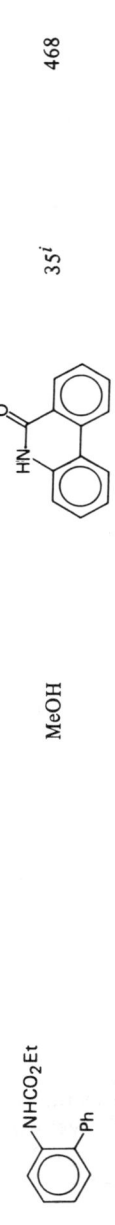

(starting material)	MeOH	(product)	35[i]	468

[a] Yields based upon recovered starting material.
[b] The effect of solvent was studied in the photocyclizations and the results obtained are shown below:

Solvent	= MeCN,	MeOH,	$Me_2C{=}O$,	*i*-PrOH,	*n*-PrBr,	Et_2O,	C_6H_6,	*n*-C_6H_{14}
Yield (%)	= 0,	0,	0,	0,	22,	24,	33,	63
Irrad. time (h)	= 8,	8,	8,	8,	8,	8,	8,	5 min

[c] This product was also obtained in 35–37% yield by the copper-catalysed decomposition[465] of *N*-methylbenzanilide-2-diazonium fluoroborate.
[d] This product was also obtained in 38–40% yield by the copper-catalysed decomposition[465] of *N*-methylbenzanilide-2-diazonium fluoroborate.
[e] This product was also prepared by sodium in liquid ammonia reduction of *trans*-5-benzyl-4*b*, 10*b*, 11, 12-tetrahydrobenzo[*c*]phenanthridin-6-(5*H*)-one followed by methylation of the resulting *trans*-4*b*,10*b*,11,12-tetrahydrobenzo[*c*]phenanthridin-6-(5*H*)-one.
[f] This product was also prepared by the reaction of benzylamine with methyl 3-(2-cyclohexanone-1)propionate.
[g] Yield of product when irradiation was performed in THF.
[h] Yields of product when irradiation was performed in methanol.
[i] By an analagous reaction (irradiation of ethyl *o*-biphenylyl carbonate) the corresponding lactone was prepared also in 85% yield.

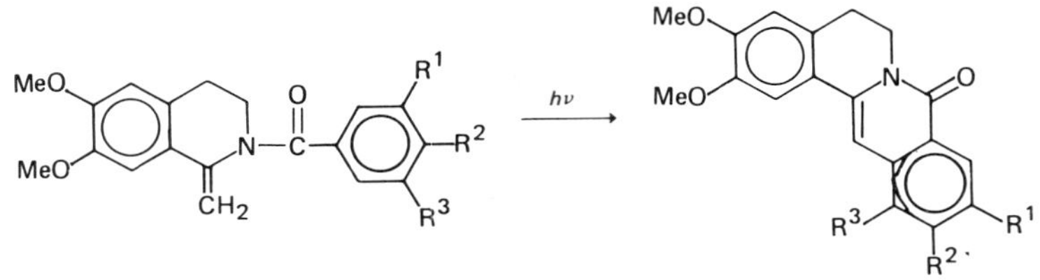

(267)

4. *Cyclization of N-chloroacetyl-β-arylamines*

These photocyclizations may be generalized by equation (268). A majority of such reactions have been carried out with *N*-chloroacetyl-β-phenylethyl amines

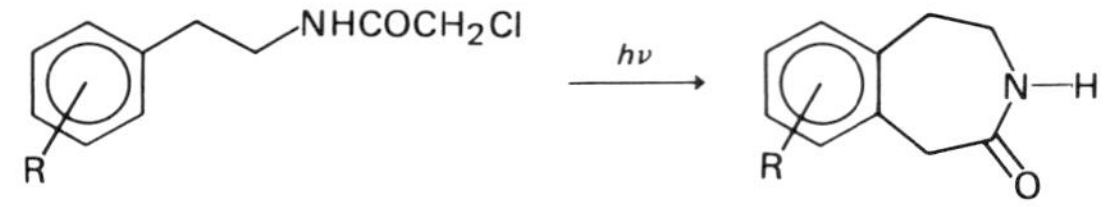

(268)

containing one or more electron-furnishing groups in the aromatic ring (Table 30). It is interesting to note that the *N*-chloroacetyl derivatives of the biologically important amines – tryptamine, tyramine, dopamine and normescaline – participate in these cyclizations

5. *Cyclization of α-diazocarboxamides*

Certain β-lactams have been synthesized by photolysis of α-diazocarboxamides. These reactions proceed by photolytic decomposition of the azo compound to form a carbene intermediate, which then undergoes insertion into a carbon–hydrogen bond of the *N*-alkyl substituent.

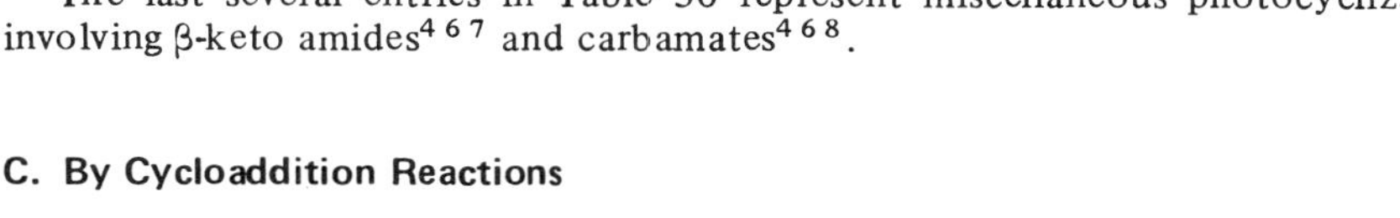

(269)

6. *Miscellaneous cyclizations*

The last several entries in Table 30 represent miscellaneous photocyclizations involving β-keto amides[467] and carbamates[468].

C. By Cycloaddition Reactions

1. *Addition of isocyanates to olefins*

In theory, cycloaddition of isocyanates to olefins should lead directly to β-lactams. In practice, however, simple *N*-alkyl and *N*-aryl isocyanates add only to electron-rich olefins such as enamines[493], while successful cycloadditions with simple olefins require the use of an 'activated' isocyanate possessing a strong electron-withdrawing substituent on nitrogen. Since its discovery in 1956[494] chlorosulphonyl isocynate (CSI)[495] has emerged as one of the most widely used isocynate addends for conversion of olefins into β-lactams. The chemistry of CSI has been reviewed[496–498] along with its applications to β-lactam synthesis[367,370,372,377,379,499]. Reaction of CSI with olefins is presumed[499] to involve equilibrium formation of a π complex, which then rearranges to a 1,4-dipolar intermediate having positive charge on the more highly substituted carbon of the original olefin. Combination of the termini of this intermediate completes the stepwise process to form an *N*-chlorosulphonyl β-lactam (equation 270). In order for the cycloaddition reaction to serve as a viable route to β-lactams,

$$\mathrm{RCH{=}CH_2} + \mathrm{ClSO_2NCO} \rightleftharpoons \pi\ \text{complex} \longrightarrow \mathrm{R\overset{+}{C}H{-}CH_2{-}C({=}O){-}\bar{N}{-}SO_2Cl} \longrightarrow \text{R-CH–CH}_2\text{ / ClO}_2\text{SN–C=O (ring)} \qquad (270)$$

the *N*-chlorosulphonyl group must be removed reductively, preferably by treatment in a suitable organic solvent, with a 25% aqueous sodium sulphite solution[500], or with an aqueous solution of a sulphur oxo acid, or its salt, in the presence of sodium bicarbonate (equation 271)[501,502].

$$\text{N-}SO_2Cl\text{ β-lactam } (R^1, R^2) \xrightarrow[CH_2Cl_2,\ H_2O]{Na_2S_2O_4,\ NaHCO_3} \text{N-H β-lactam } (R^1, R^2) \quad \text{(Ref. 501)} \qquad (271)$$

R^1 = Ph, Ph, H, Me

R^2 = H, Me, Me, Me

Table 31 contains a number of recent examples. Examination of these reactions reveals that addition of CSI is both a regiospecific and stereospecific reaction. Some regiospecificity is lost with olefins of the type $R^1CH{=}CHR^2$, where both R^1 and R^2 are simple alkyl groups. Dienes can easily be converted to monoadducts, but

TABLE 31. Synthesis of β-lactams by addition of chlorosulphonyl isocyanate to olefins followed by reduction

Olefin	β-Lactam	Overall yield (%)	Reference
$Me_2C{=}CH_2$	4,4-dimethylazetidin-2-one (Me, Me, N–H)	51–53	502
MeCH=CHMe *cis*	*cis*-3,4-dimethylazetidin-2-one (Me, Me, H, H, N–H)	85	503

TABLE 31. (Continued)

Olefin	β-Lactam	Overall yield (%)	Reference
MeCH=CHMe *trans*	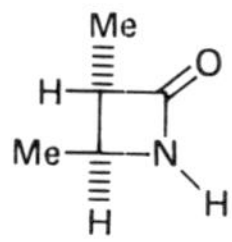 (trans-3,4-dimethyl β-lactam)	85	503
MeCH=CHPr-*n* *cis*	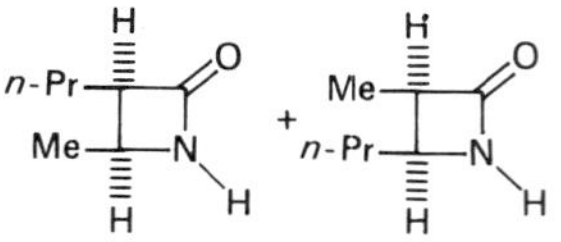 (cis-3-Me-4-n-Pr + cis-3-n-Pr-4-Me β-lactams) 1:3	55	503
MeCH=CHPr-*n* *trans*	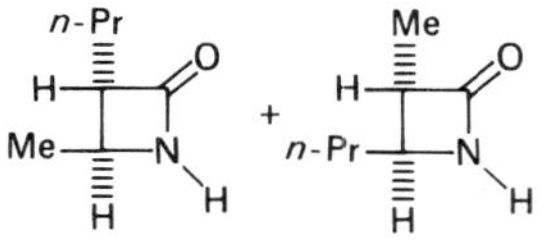 (trans-3-Me-4-n-Pr + trans-3-n-Pr-4-Me β-lactams) 2:3	55	503
Me_2C=CRMe	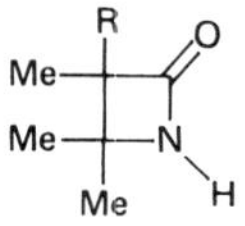 (3-R-3,4,4-trimethyl β-lactam)		
	R = Me	92	498, 500
	R = H	98	
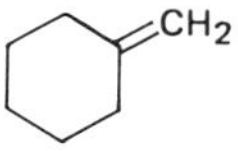 Methylenecyclohexane (CH_2)	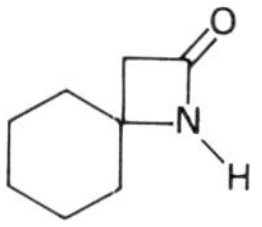 (spiro β-lactam)	94	500
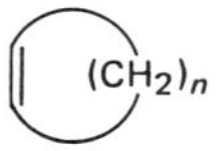 Cycloalkene $(CH_2)_n$	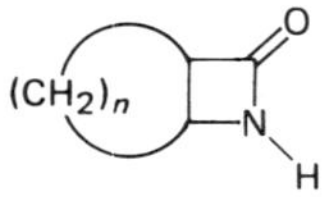 Fused β-lactam $(CH_2)_n$		
	n = 3	63	503
	n = 4	57	
	n = 6	75	
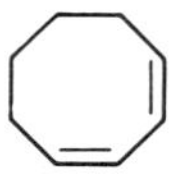 Cycloocta-1,3-diene	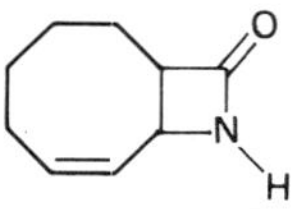 (fused bicyclic β-lactam)	86	500
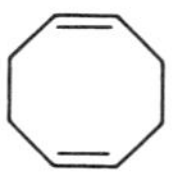 Cycloocta-1,5-diene	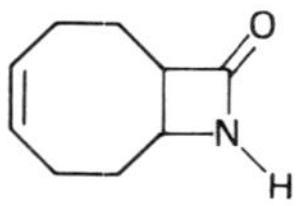 (fused bicyclic β-lactam)	41	503

TABLE 31. (Continued)

Olefin	β-Lactam	Overall yield (%)	Reference
	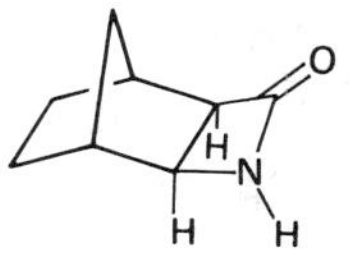	66	499, 500
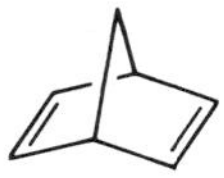	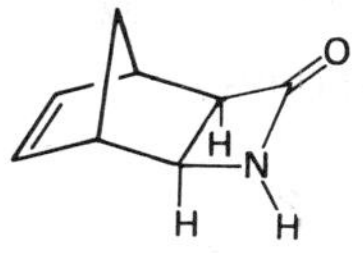	68	499
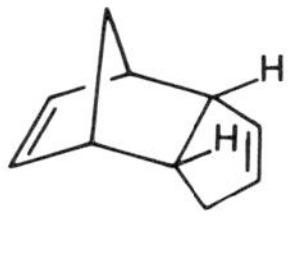	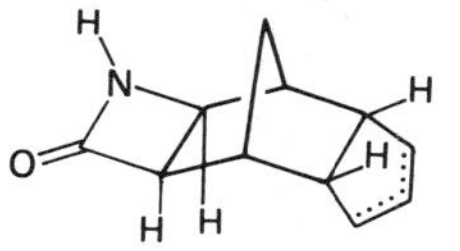	57	499, 503
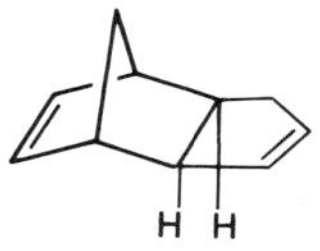	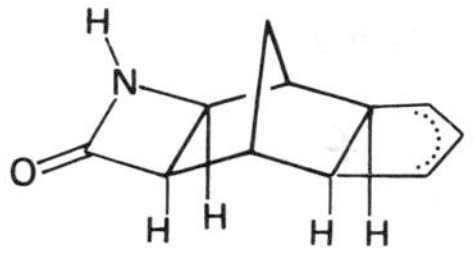	30	499
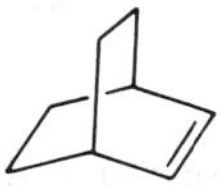	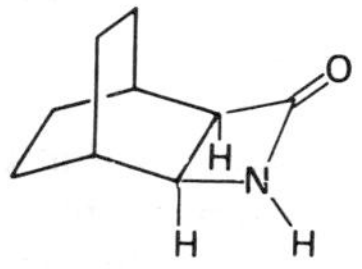	49	499
$H_2C{=}CR{-}CH{=}CH_2$	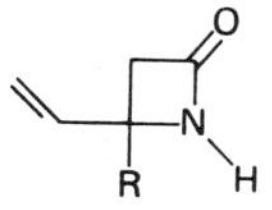	72	500, 504
	R = H	72	
	R = Me	68	
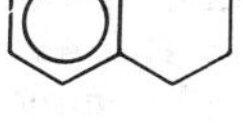	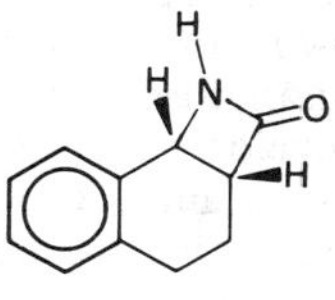	35	505

TABLE 31. (Continued)

Olefin	β-Lactam	Overall yield (%)	Reference
$Me_2C{=}C{=}CMe_2$	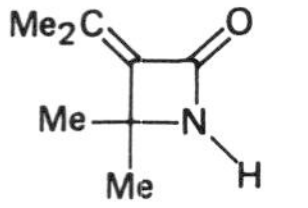 [β-lactam: 3-isopropylidene-4,4-dimethylazetidin-2-one]	52	506
	+ $CH_2{=}C(Me)C(CONH_2){=}CMe_2$	22	
[cyclohexylidene]$=C=CH_2$	[spiro β-lactam with $H_2C{=}$]	26	506
	+ [cyclohexenyl]$C(=CH_2)CONH_2$	32	
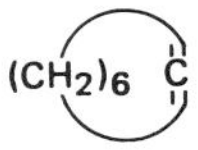 $(CH_2)_6$ [cyclic allene]	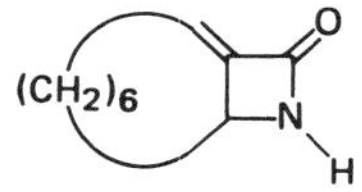 $(CH_2)_6$ [bicyclic β-lactam]	36	506

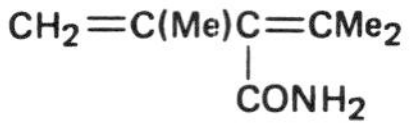

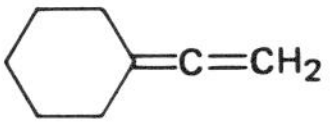

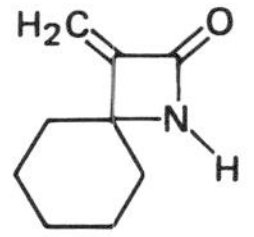

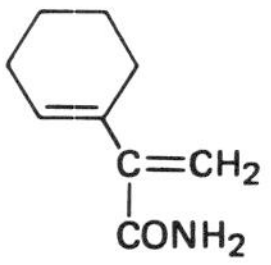

diadducts have not been isolated. Strained double bonds, such as those in a bicyclo [2.2.1] heptene system, tend to react more rapidly than normal unstrained olefins.

As mentioned previously, enamines react with simple isocyanates to afford β-amino-β-lactams (equation 272)[493,507,508].

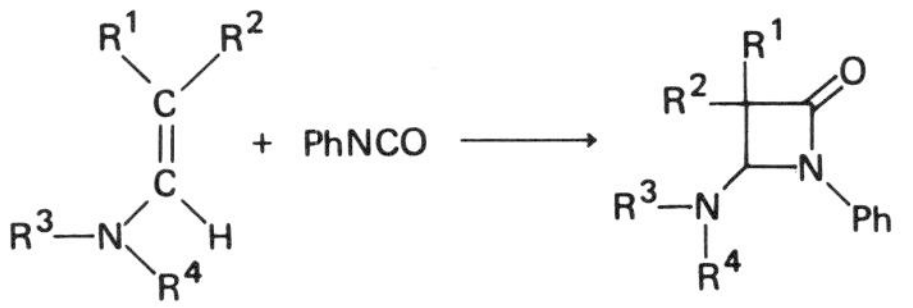

$$R^1R^2C{=}CH{-}NR^3R^4 + PhNCO \longrightarrow \text{[β-lactam: } R^1,R^2 \text{ at C-3; } NR^3R^4 \text{ at C-4; N-Ph]} \qquad (272)$$

Pentahaptocyclopentadienyl dicarbonyl (olefin) iron complexes[509], represented in equation (273) as Fp–olefin complexes **36a,b**, fail to react with either ethyl or phenyl isocyanate, but react with 2.5-dichlorophenyl isocyanate, CSI, *p*-toluenesulphonyl isocyanate and methoxysulphonyl isocyanate in a 1,3-addition process to afford butyrolactams **37a–e**[510]. The cycloalkenyl complexes **38**, **40** and **42** react similarly[510] to give lactams **39**, **41a** or **b** and **43**, while the butynyl complex **44** affords the unsaturated lactam **45** upon reaction with tosyl isocyanate.

Isocyanates of various types undergo [2π + 2π] cycloaddition with ketenimines to afford β-imino-β-lactams in good yields (equation 278)[511].

$$\text{(36)} \xrightarrow{R^2NCO} \text{(37)} \tag{273}$$

(36)
(**a**) R^1 = H
(**b**) R^1 = Me
Fp = h^5-$C_5H_5Fe(CO)_2$

(37)
(**a**) R^1 = H, R^2 = $C_6H_3Cl_2$-2,5
(**b**) R^1 = H, R^2 = SO_2Cl
(**c**) R^1 = Me, R^2 = Ts
(**d**) R^1 = Me, R^2 = Ts
(**e**) R^1 = H, R^2 = $MeOSO_2$

$$\text{(38)} \xrightarrow{TsNCO} \text{(39)} \tag{274}$$

$$\text{(40)} \xrightarrow{RNCO} \text{(41)} \tag{275}$$

(41) (**a**) R = Ts; (**b**) R = $MeOSO_2$

$$\text{(42)} \xrightarrow{TsNCO} \text{(43)} \tag{276}$$

$$FpCH_2C{\equiv}CMe\ \text{(44)} \xrightarrow{TsNCO} \text{(45)} \tag{277}$$

$$R^1_2C{=}C{=}NC_6H_4Me\text{-}p + R^2NCO \longrightarrow \text{azetidinone } (p\text{-}CH_3C_6H_4N{=}) \tag{278}$$

R^1	R^2	Yield (%)
Ph	Ph	83
Ph	p-MeC_6H_4	78
Ph	p-$MeOC_6H_4$	78
Ph	p-$MeC_6H_4SO_2$	72
Me	Ph	76
Me	p-$MeC_6H_4SO_2$	73

Phenyl isocyanate reacts with various acetylenes in the presence of aluminium chloride to afford 3,4-disubstituted carbostyrils (equation 279)[512].

Treatment of *o*-benzoylbenzaldehyde with aryl isocyanates affords 2,3-disubsti-

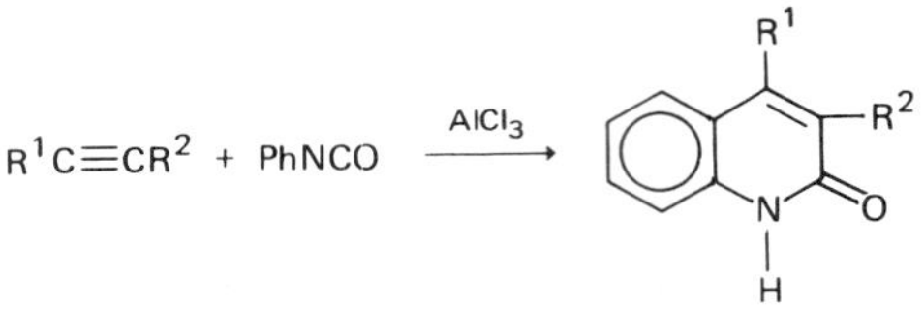

$R^1C{\equiv}CR^2$ + PhNCO $\xrightarrow{AlCl_3}$ (279)

R^1	R^2	Yield (%)
Ph	H	43
Ph	Me_3Si	44
Et	Et	–
n-Bu	Me_3Si	50
Me_3Si	Me_3Si	57
H	Me_3Si	25
Me_3Si	H	89

tuted phthalimidines by a reaction pathway involving intermediate formation of *o*-benzoylbenzylideneanilines followed by phenyl group migration (equation 280)[513]. Similar results have been observed with aromatic isocyanates and phthaldehyde (equation 281)[514].

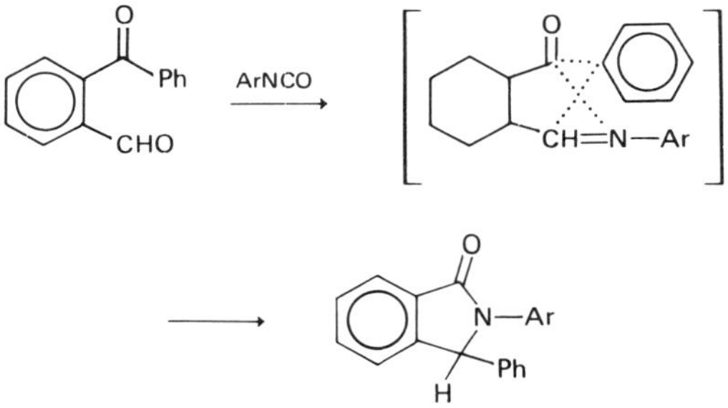

ArNCO

(280)

Ar	Yield (%)
Ph	67
C_6H_4Me-*m*	54
α-Naphthyl	84
β-Naphthyl	81

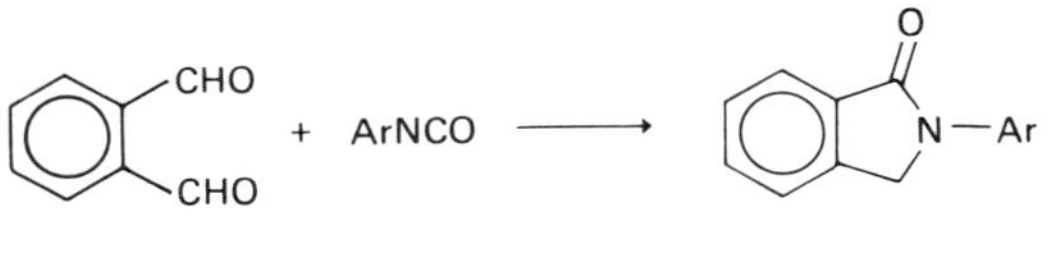

+ ArNCO ⟶ (281)

2. *From imines*

a. Reaction of imines with ketenes. The most frequently used method for the preparation of lactams involves the reaction of a large variety of imines with ketenes, which are prepared prior to or during the reaction.

In one of the earliest reviews[515] on this method, Staudinger pointed out that the reactivity of ketenes towards benzophenone anil exhibited the following order:

$$\text{(fluorenylidene)C=O} > Ph_2C{=}C{=}O > Ph(Me)C{=}C{=}O > Me_2C{=}C{=}O > H_2C{=}C{=}O$$

A similar order of ketene reactivity was observed by Brady[516] in a recent investigation of the cycloaddition of ketene itself and fluoro-, chloro-, dibromo-, methylchloro-, phenylchloro-, diphenyl-, phenylethyl-, butylethyl- and dimethylketenes to dicyclohexyl- and diisopropylcarbodiimide.

The mechanism and stereochemistry of the reaction have both been recently elucidated. In 1967, Gomes and Joullie[517] investigated the cycloaddition of ketene to benzylideneaniline in sulphur dioxide as the solvent and obtained the product shown in equation (282) in 52% yield. They concluded from their results, that

$$PhCH{=}NPh + H_2C{=}C{=}O \xrightarrow{SO_2} \text{(cyclic product: Ph, } SO_2\text{, N–Ph, C=O)} \qquad (282)$$

although the cycloaddition may proceed through a concerted mechanism or through the formation and subsequent reaction of a 1,4-dipolar intermediate (equation 283), the latter mechanism appeared more probable. Extension of this

$$R{-}N{=}C(H){-}R^1 + O{=}C{=}CH_2 \longrightarrow R{-}N{-}\overset{+}{C}(H){-}R^1 \text{ / } O{=}C{-}\overset{-}{C}H_2 \xrightarrow{SO_2} \text{(cyclic product: } R^1\text{, } SO_2\text{, N–R, C=O)} \qquad (283)$$

mechanism to the reaction of a ketene and an imine in an inert solvent would produce a 1,4-dipolar intermediate as shown in equation (284), which would then cyclize to produce the lactam.

$$R{-}N{=}C(H){-}R^1 + O{=}C{=}CH_2 \longrightarrow R{-}N{-}\overset{+}{C}(H){-}R^1 \text{ / } O{=}C{-}\overset{-}{C}H_2 \longrightarrow \text{(β-lactam: N–R, } R^1\text{, C=O)} \qquad (284)$$

In 1968, Luche and Kagan[818] reported that regardless of the method used to generate the ketenes, they added to benzylidene aniline to produce *trans*-β-lactams exclusively (equation 285). This work in conjunction with the study of Sheehan[519] and Bose[520] or the stereochemistry of the β-lactams formed by the reaction of an acid chloride and an imine in the presence of a tertiary amine has produced a controversy in the literature. Based upon the original suggestion of Sheehan[519] that the formation of a ketene from the acid chloride and tertiary amine and subsequent cycloaddition of the ketene to the imine was probably not the pathway

$$\left.\begin{array}{l} MeCOCHN_2 + h\nu \\ MeCOCHN_2 + Ag_2O \\ MeC{\equiv}COEt + \text{heat} \\ MeCH_2COCl + 2\,NEt_3 \\ MeCH_2COCl + 4\,NEt_3 \end{array}\right\} \longrightarrow MeHC{=}C{=}O + PhN{=}CHPh \longrightarrow \text{(285)}$$

to the β-lactams produced, Bose[520] investigated the initial adduct formed from the reaction of a series of acid chlorides and anils in carbon tetrachloride solution using ^{1}H n.m.r. spectroscopy. He found that the adduct could best be represented by the covalent structure shown in equation (286), and that an equilibrium is established

$$R^1{-}CH(R^2){-}C({=}O){-}Cl + R^3{-}C(R^4){=}N{-}R^5 \rightleftharpoons \text{adduct} \qquad (286)$$

between the starting materials and the adduct. He further found that in all cases where a β-lactam was formed both the *cis* and the *trans* isomers were obtained. It was thus concluded, that although the addition of a preformed ketene to an imine produces a *trans*-β-lactam in every case, it appears 'that "the acid chloride reaction" for β-lactam formation by-passes the ketene pathway – at least in those cases where the *cis*-β-lactams are produced'.

Table 32 contains a representative series of lactams produced from the reaction of imines with ketenes[516–594], and although many of the reactions shown do not necessarily involve a ketene intermediate, as can be seen from the discussion presented above, the products obtained are identical with those expected from a formal cycloaddition of a ketene and an imine.

It is interesting to note that the reaction of ketenes with double bonds has also been used to produce diazetidinones when the ketene is allowed to undergo a cycloaddition with an azo compound[515,595–599]. Selected examples of this approach are shown in Table 33. In one instance[597] it has been noted that the diazetidinone obtained by the cycloaddition of diphenylketene and azobenzene dissociates upon heating at 220°C into benzophenone anil, phenyl isocynate and the starting materials, diphenylketene and azobenzene. Recombination of these

$$Ph_2C{=}C{=}O + PhN{=}NPh \underset{}{\overset{220^\circ C}{\rightleftharpoons}} \text{diazetidinone} \xrightarrow{220^\circ C} PhN{=}C{=}O + Ph_2C{=}NPh \xrightarrow{Ph_2C{=}C{=}O} \text{β-lactam} \qquad (287)$$

compounds via a ketene–imine interaction affords 1,3,3,4,4-pentaphenylazetidin-2-one (equation 287)[515].

It has also been reported[600] that irradiation of diphenylacetylene and nitrobenzene for 3 days with a mercury arc lamp affords a 1.8% yield of 1,3,3,4,4-pentaphenylazetidin-2-one. The mechanism proposed involves initial formation of diphenylketene and benzylideneaniline, followed by their subsequent cycloaddition to produce the β-lactam (equations 288 and 289).

$$PhC{\equiv}CPh + PhNO_2 \xrightarrow[\text{ether, } N_2]{h\nu} [PhNO_2]^* + PhC{\equiv}CPh \longrightarrow Ph_2C{=}C{=}O + PhNO \longrightarrow [\text{oxazetidine intermediate}] \longrightarrow Ph_2C{=}NPh \quad (288)$$

$$Ph_2C{=}NPh + Ph_2C{=}C{=}O \longrightarrow \text{1,3,3,4,4-pentaphenylazetidin-2-one} \quad (289)$$

b. Reformatsky reaction with imines. The main interest in the Reformatsky reaction with imines has not been with their preparative potential, but with their stereochemistry, since both *cis* and *trans* isomers may be expected from the addition of a Reformatsky reagent to an anil (equation 290). Studies of this

$$R^1CH(Br)COOR^2 + R^3CH{=}NR^4 \xrightarrow[Zn]{C_6H_5Me} [R^1HC(COOR^2)-CH(R^3)-N(R^4)-ZnBr] + [\text{diastereomer}] \longrightarrow \textit{cis}\text{-}\beta\text{-lactam} + \textit{trans}\text{-}\beta\text{-lactam} \quad (290)$$

reaction using a variety of α-bromo esters have shown[602,603] that as the size of the R^1 group increases the *cis–trans* product ratio decreases, and that the *cis–trans* product ratio is influenced by the solvent (equation 291)[603].

A comparison[519] of the stereochemistry of the Reformatsky reaction with the stereochemistry of the $[2\pi + 2\pi]$ cycloaddition of a ketene and an imine shows the former reaction to yield mixtures of *cis* and *trans* β-lactams, while the latter reactions afford mainly *trans* β-lactam. Also of interest is the observation[605] that a competitive Reformatsky reaction using 1 equivalent of methyl α-bromophenyl acetate and 1 equivalent each of benzylideneaniline and α-deuteriobenzylideneaniline showed a secondary isotope effect of k_H/k_D 0.86 (equation 292), whereas a similar reaction of 1 equivalent of diphenylketene with the same mixture of Schiff bases showed no isotope effect.

In Table 34 are listed β-lactams which have been prepared using a Reformatsky

TABLE 32. Production of lactams by reaction of ketenes with imines

Ketene or ketene precursor	Imine	Conditions	Product	Yield (%)	Reference
$H_2C{=}C{=}O$	i-PrN=C=NPr-i	R.t., 8 h	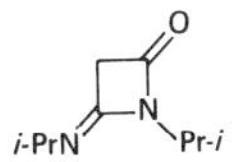 [structure]	5	516
$H_2C{=}C{=}O$	PhCH=NPh	SO_2	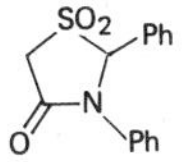 [structure]	52	517
$H_2C{=}C{=}O$	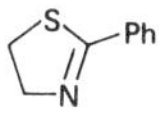 [structure]	SO_2	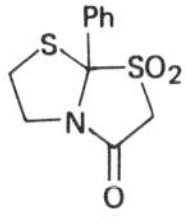 [structure]	80	517
$H_2C{=}C{=}O$	$RC_6H_4CH{=}NPh$	180–200°C, 1 h	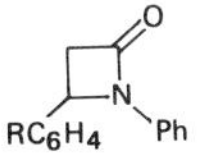 [structure]		523
			R = o-Me	11.5	
			R = m-Me	12	
			R = p-Me	22	
			R = o-MeO	39	
			R = m-MeO	16	
			R = m-Cl	32	
			R = p-NMe_2	62	
$H_2C{=}C{=}O$	$PhCH{=}N{-}C_6H_4R$	180–200°C, 1 h	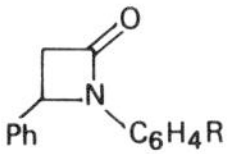 [structure]		523
			R = o-Me	7	
			R = m-Me	18	
			R = p-Me	13	
			R = o-MeO	—	
			R = p-MeO	19	
			R = m-Cl	34	

$H_2C{=}C{=}O$	$PhCH{=}CHCH{=}NPh$	MeOH, ether, 1 h, reflux	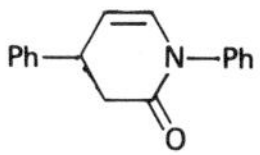	69	523
$MeCH{=}C{=}O$	$PhCH{=}NPh$		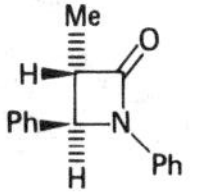		
			$MeCOHCN_2$ + hν	50, 47	518, 525
			$MeCOCHN_2$ + Ag_2O	17	518
			MeC≡COEt, heat	30	518
			$MeCH_2COCl + 2\ NEt_3$	2	518
			$MeCH_2COCl + 4\ NEt_3$	39	518
$MeCH{=}C{=}O$	$Ph_2C{=}NPh$	C_6H_6, N_2, *hν*, 5 h	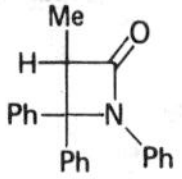	48	525
$RCH{=}C{=}O$	*i*-PrN=C=NPr-*i*	Hexane, reflux, 2 h C_6H_6, reflux	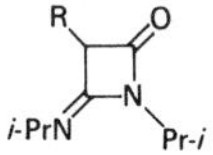		516, 526
			R = F	40	
			R = Cl	20	
$RCH{=}C{=}O$	$PhCH{=}NPh$		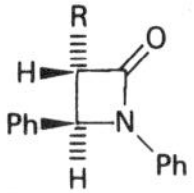		
		EtC≡COEt, heat	R = Et	35, 24	518, 525
		$EtCH_2COCl + 2\ NEt_3$		2	518
		i-PrC≡COEt, heat	R = *i*-Pr,	78	518
		i-$PrCH_2COCl + 2.5\ NEt_3$		32	518
		t-$BuCH_2COCl + 2\ NEt_3$	R = *t*-Bu,	2	518

TABLE 32. (Continued)

Ketene or ketene precursor	Imine	Conditions	Product	Yield (%)	Reference
$Me_2C{=}C{=}O$	PhC(SMe)=NPh	EtOAc, 2 days, r.t.	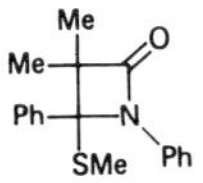	60	528[a]
$Me_2C{=}C{=}O$	*i*-PrN=C=NPr-*i*	Hexane, reflux, 8 h	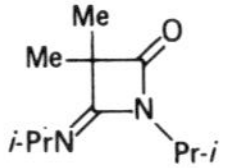	32	516
$R^1R^2C{=}C{=}O$	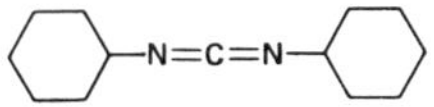 —N=C=N— (cyclohexyl on each N)	Hexane, reflux, 5 h	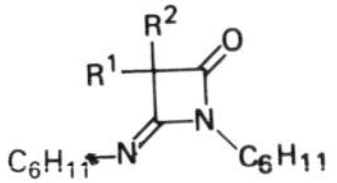		516
			R^1 = Me, R^2 = Cl	25	
			R^1 = R^2 = Br	59	
n-BuC(Et)=C=O	*i*-PrN=C=NPr-*i*	Hexane, reflux, 2 h	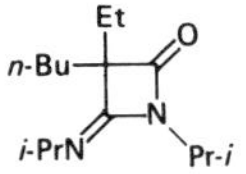	15	516
PhCH=C=O	PhCH=NPh		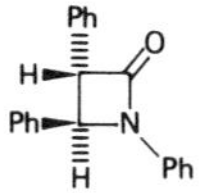		
		C_6H_6, N_2, 4 h, 40–50°C		35	523
		C_6H_6, N_2, $h\nu$, 5 h		74	525
		$PhCOCHN_2$ + Ag_2O		44	518, 532
		PhCOCl + 4 NEt_3		6	518, 532
PhCH=C=O	$RC_6H_4CH{=}NPh$	C_6H_6, N_2, 4 h 40–50°C	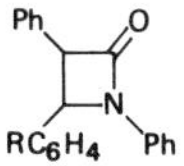		

			R = *o*-Me	21	523
			R = *m*-Me	20	523
			R = *p*-Me	14	523
			R = *o*-MeO	25	523
			R = *m*-MeO	29	523
			R = *m*-Cl	13.5	523
			R = *p*-NO_2	–, 78	523, 525
			R = *p*-NMe_2	90, 79	523, 525
PhCH=C=O	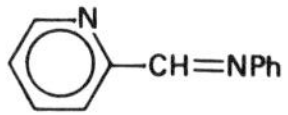 2-pyridyl–CH=NPh	C_6H_6, N_2, *hν*, 5 h	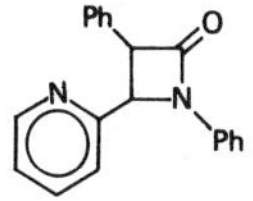	56	525
PhCH=C=O	PhCH=N—C_6H_4R	C_6H_6, N_2, *hν*, 4 h, 40–50°C	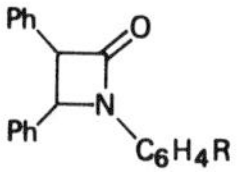		
			R = *o*-Me	12	523
			R = *m*-Me	15	523
			R = *p*-Me	15	523
			R = *o*-MeO	5	523
			R = *p*-MeO	19	523
			R = *m*-Cl	28	523
		C_6H_6, N_2, *hν*, 5h	R = *p*-NMe_2	70	525
PhCH=C=O	Ph_2C=NPh	C_6H_6, N_2, *hν*, 4–5 h, 40–50°C	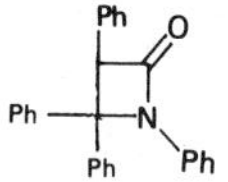	42, 76	523, 525
PhCH=C=O	PhCH=CHCH=NPh	MeOH, reflux, 40–50°C	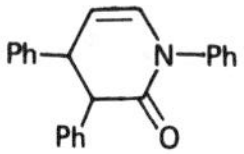	32	523
PhCH=C=O	PhN=CMe–CMe=NPh	C_6H_6, N_2, *hν*, 4–5 h, 40–50°C	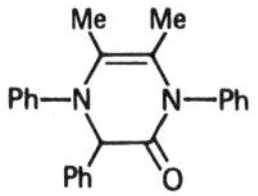	47	523

TABLE 32. (Continued)

Ketene or ketene precursor	Imine	Conditions	Product	Yield (%)	Reference
PhHC=C=O	p-$O_2NC_6H_4$—CH=N—$CH_2C_6H_5$	C_6H_6, N_2, $h\nu$, 5 h	Ph, O, N, p-$O_2NC_6H_4$, $CH_2C_6H_5$	65	525
p-RC_6H_4CH=C=O	PhCH=NPh	C_6H_6, N_2, $h\nu$, 5 h	p-RC_6H_4, O, N, Ph, Ph		525
			R = MeO	65	
			R = Cl	54	
p-$MeOC_6H_4$CH=C=O	PhCH=N—$CH_2C_6H_5$	C_6H_6, N_2, $h\nu$, 5 h	p-$MeOC_6H_4$, O, N, Ph, $CH_2C_6H_5$	36	525
PhRC=C=O	PhCH=NPh		H, O, Ph, Ph, N, R, Ph		532
			R = Me; *cis* : *trans* = 1:4		
			R = Et; *cis* : *trans* = 2:1		
			R = *i*-Pr; *cis* : *trans* = 9:1		
PhC(COOMe)=C=O	PhCH=NPh	C_6H_6, N_2, $h\nu$, 5 h	CO_2Me, O, Ph, N, Ph, Ph	14	521, 525
PhC(COOMe)=C=O	Ph_2C=NPh	C_6H_6, N_2, $h\nu$, 5 h	CO_2Me, O, Ph, Ph, N, Ph, Ph	35	525

PhEtC=C=O	i-PrN=C=NPr-i	Hexane, 48 h, r.t.	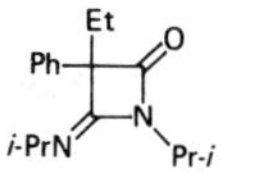	57	516
PhClC=C=O	$H_{11}C_6$–N=C=N–C_6H_{11}	Hexane, reflux, 2 h.	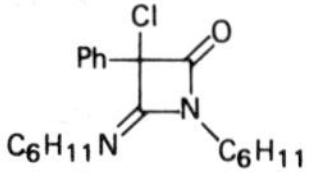	65	516
Ph_2C=C=O	PhCH=NMe		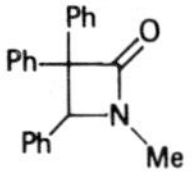		538
		MeCN, r.t. molar ratio 1:1		82	538
		C_6H_6, r.t. molar ratio 1:1		71	538
		C_6H_6, r.t. molar ratio 2:1		95	538
Ph_2C=C=O	PhCH=NMe	MeCN, r.t. molar ratio 2:1	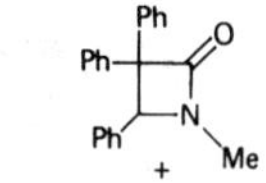	19	538
			+ 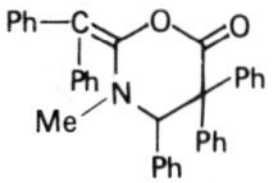	81	
Ph_2C=C=O	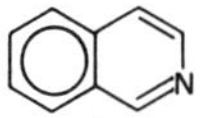 + PhN=C=O	R.t.	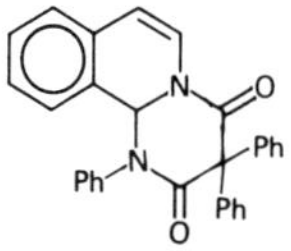 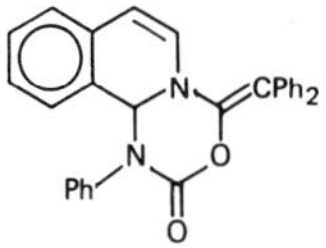	31	538[b]
			or		

TABLE 32. (Continued)

Ketene or ketene precursor	Imine	Conditions	Product	Yield (%)	Reference
$Ph_2C{=}C{=}O$	PhCH=NPh		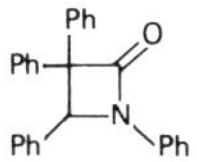 1,3,3,4-tetraphenyl-2-azetidinone (structure)		
		Heat		–	533, 534, 536
		C_6H_6, N_2, $h\nu$, 5 h		71	525
		Ether, stand 1 day		70	535
		C_6H_6, stir 20 min, r.t.		53–65	537
$Ph_2C{=}C{=}O$	$RC_6H_4CH{=}NPh$		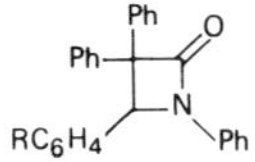 3,3-Ph, 4-RC_6H_4, N-Ph azetidinone (structure)		
		Waterbath, 70–80°C	R = *o*-OMe	52	523
			R = *m*-Me	63	523
		C_6H_6, 70–80°C, 1 h	R = *p*-Me	72	523
		Waterbath, 70–80°C	R = *o*-MeO	30	523
			R = *m*-MeO	21	523
			R = *m*-Cl	71	523
		C_6H_6, stir 20 min, r.t.	R = *p*-MeO	53–65	537
		C_6H_6, 70–80°C, 1 day	R = *p*-NMe_2	67	523
$Ph_2C{=}C{=}O$	$PhCH{=}N{-}C_6H_4R$		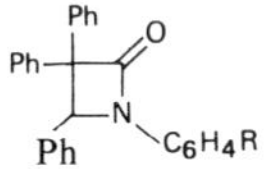 3,3,4-Ph, N-C_6H_4R azetidinone (structure)		
		Waterbath, 70–80°C	R = *o*-Me	72	523
			R = *m*-Me	69	523
		Ether, 2 days, r.t.	R = *p*-Me	21	523
		EtOAc, r.t.	R = *o*-MeO	55	523
			R = *p*-MeO	16	523
		C_6H_6, stir 20 min, r.t.		53–65	537
		Waterbath, 70–80°C	R = *m*-Cl	98	523
		Without solvent, 1 week; in solvent (C_6H_6, ether or EtOAc) 5 h on waterbath; without solvent, melt at 200°C	R = *p*-NMe_2	65	535[c]

$Ph_2C{=}C{=}O$	PhN=CMe–MeC=NPh	EtOAc, reflux, 3 h	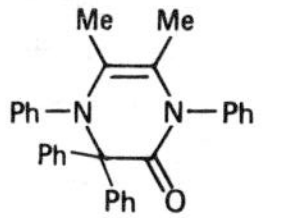 [structure: Me, Me, Ph–N, N–Ph, Ph, Ph, O]	61	523
$Ph_2C{=}C{=}O$	$PhCH{=}NHCH_2C_6H_5$	C_6H_6, N_2, $h\nu$, 5 h	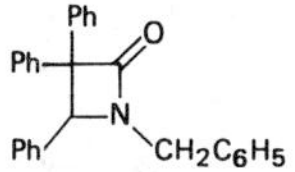 [structure: Ph, Ph, Ph, O, N, $CH_2C_6H_5$]	46	525
$Ph_2C{=}C{=}O$	$Ph_2C{=}NPh$	C_6H_6, N_2, $h\nu$, 5 h	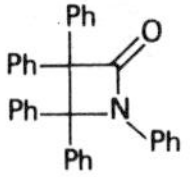 [structure: Ph, Ph, Ph, Ph, O, N, Ph]	72	525[a]
$Ph_2C{=}C{=}O$	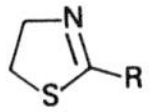 [structure: N, S, R]		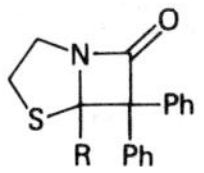 [structure: N, O, S, Ph, R, Ph]		
	R = SH	EtOAc, N_2, stir 20 min	R = $SCOCHPh_2$	68, –	523, 531
	R = $NHCOCHPh_2$	EtOAc, N_2, stir 5 min	R = $NHCOCHPh_2$	54, –	523, 531
	R = NHCOMe	EtOAc, C_6H_6, N_2, stir 20 min	R = NHCOMe	44	523
	R = Ph	C_6H_6, N_2, $h\nu$, 5 h	R = Ph	20 –	525, 531
	R = Me	(molar ratio 2:1); 25°C, 1 week	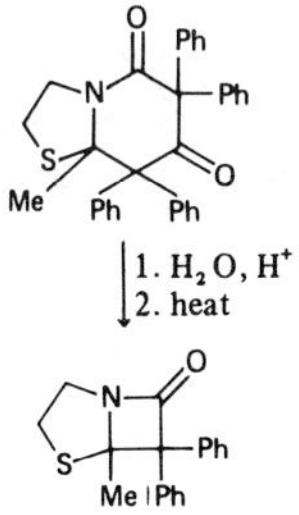 [structure: O, Ph, N, Ph, S, O, Me, Ph, Ph] ↓ 1. H_2O, H^+ 2. heat [structure: N, O, S, Ph, Me, Ph]	–, 73	531, 543

TABLE 32. (Continued)

Ketene or ketene precursor	Imine	Conditions	Product	Yield (%)	Reference
$Ph_2C=C=O$	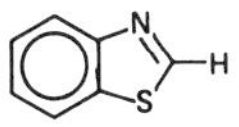 [benzothiazole structure]	25°C, 1 day (molar ratio 2:1)	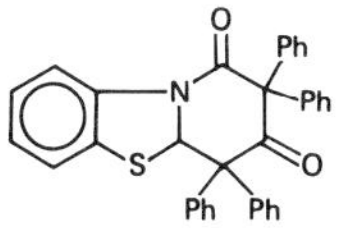 [structure]	–, 86	531, 543
$Ph_2C=C=O$	$PhC(SR^1)=NR^2$		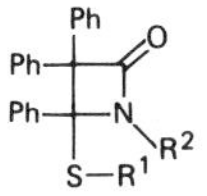 [structure]		
	$R^1 = Me$, $R^2 = -CH_2CO_2Me$	C_6H_5Me, 16 h, r.t.		67–69	539
	$R^1 = Me$, $R^2 = -CH(i\text{-}Pr)CO_2Me$	C_6H_6, reflux, 20 h	(2 diastereomers)	72	539
	$R^1 = Me$, $R^2 = -C(CO_2Me)=CMe_2$	C_6H_6, reflux, 20 h		61–63	539
	$R^1 = -CH_2CH_2CO_2Me$, $R^2 = -C(CO_2Me)=CMe_2$	C_6H_6, reflux, overnight		53–64	539
$Ph_2C=C=O$	$R^1SCH=NCH(CO_2Me)CHMe_2$	C_6H_5Me, reflux 12 h	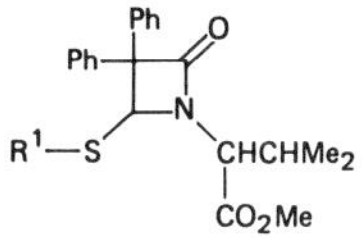 [structure]		540
			$R^1 = Me$	47[d]	
			$R^1 = CH_2C_6H_5$	69[e]	
$Ph_2C=C=O$	$Ph_2C=NNHCOPh$	C_6H_5Me 100°C, 3 h	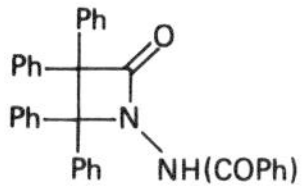 [structure]	75	541

$Ph_2C{=}C{=}O$	(9-fluorenylidene)C=N—NH(COPh)	Ether	[structure] NH(COPh); ⌐ = 9-Flurenyl	66	541
$Ph_2C{=}C{=}O$	$R^1R^2C{=}NC_6H_4R^3$	Ether, $h\nu$, 3.5 h	[structure] $C_6H_4R^3$		542
	R^1 = morpholino (O⟨ ⟩N—), $R^2 = R^3 = H$			44–48	
	R^1 = piperidino (⟨ ⟩N—), $R^2 = R^3 = H$			79	
	R^1 = piperidino (⟨ ⟩N—), $R^2 = H$, $R^3 = p\text{-}NO_2$			75	
	R^1 = morpholino (O⟨ ⟩N—), $R^2 = Ph$, $R^3 = p\text{-}OMe$			61	
$Ph_2C{=}C{=}O$	benzoxazole	R.t. 1 week (molar ratio 2:1)	[structure]	88	543
$Ph_2C{=}C{=}O$	EtOCH=NPh	R.t. 1 week (molar ratio 2:1)	[structure]	60	543

TABLE 32. (Continued)

Ketene or ketene precursor	Imine	Conditions	Product	Yield (%)	Reference
$Ph_2C{=}C{=}O$	[structure: 1-R-benzimidazole]		[structure: N, O, Ph labels]		543
	R = H	100°C, 1 h (molar ratio 3:1)		83	
	R = $-COCHPh_2$	100°C, 1 h (molar ratio 2:1)		86	
	R = Me	Ether, r.t., 1 day (molar ratio 2:1)		85	
$Ph_2C{=}C{=}O$	[structure: 1-R-imidazole]		[structure: N, O, Ph labels]		543
	R = Me	Ether, r.t., 1 day (molar ratio 1:2)		64	
	R = Me	Ether, 100°C, 1 h (molar ratio 3:1)		19	
	R = $-COCHPh_2$	THF, r.t., 1 day (molar ratio 2:1)		56	
$Ph_2C{=}C{=}O$	RN=C=NR	C_6H_6, r.t., 2 h	[structure: Ph, Ph, O, N–R, RN= labels]		516
			R = C_6H_{11}	90	
			R = *i*-Pr	88	
RCH_2COCl	[structure: cyclohexylidene=NPh]		[structure: R, O, N–Ph labels]		

		CH_2Cl_2, NEt_3	R = Cl	Trace	545
			R = OMe	14	545
			R = Ph	–	545, 547
		CH_2Cl_2, $N(Pr\text{-}i)_3$, 0°C, 3 h stir	R = OPh	48	545
		CH_2Cl_2, NEt_3	R = N_3	54	545
		CH_2Cl_2, NEt_3, 0°C, 4 h stir		54	547
		CH_2Cl_2, NEt_3	R 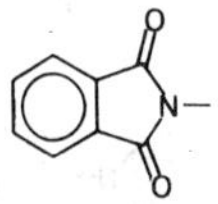 = (O, N—, O)	33	545
RCH_2COCl	Ph NPh		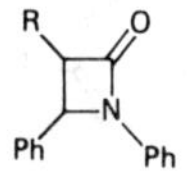 R, O, N, Ph, Ph		
R = Cl		CH_2Cl_2, NEt_3, N_2	*trans*	19, 62, 20	520, 545
		C_6H_6, NEt_3, 70–75°C,	*trans*	4	551
		DMF, 180°C	*cis* : *trans* = 45:55	–	552
		DMF, NEt_3, 25°C	*trans*	–	552
		C_6H_6, NEt_3, reflux	*trans*	60	590
R = Me		CH_2Cl_2, NEt_3, N_2	*trans*	35, 42, 50	520, 545
R = CH=CH_2			*trans*	10, 49	520
R = CMe_3			*trans*	34	520
R = OMe			*cis* : *trans* = 3:1	50	520, 545
			cis : *trans* = 1.7:1	85	
R = CH_2OMe			*trans*	0.5	545
R = Ph			*trans*	20, 59	520, 545
R = OPh				89	520, 545
			cis	(38)	
			trans	(51)	
R = p-$MeOC_6H_4$			*trans*	42	520
R = p-$O_2NC_6H_4$			*trans*	0, 38	520
R = N_3			*cis* : *trans* = 3:1	40	545
			cis : *trans* = 1.3:1	98	545
			cis	35–45	548[k]
			trans	50	548[k]
R = $SCH_2C_6H_5$			*trans*	40	545

TABLE 32. (Continued)

Ketene or ketene precursor	Imine	Conditions	Product	Yield (%)	Reference
R 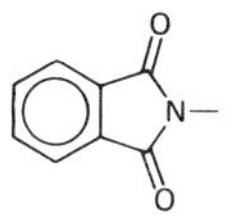 = [phthalimido structure]			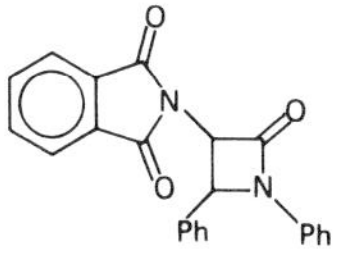 [structure] *trans*		
		CH_2Cl_2, NEt_3, N_2	1. EtOH, N_2H_4 reflux 2 h	30	545
		C_6H_6, NEt_3, r.t.	2. HCl	50	519
		C_6H_6, NEt_3, r.t. stir 1 h	3. KOH, H_2O	–	549
			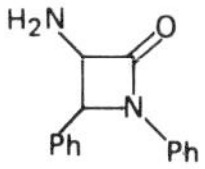 [structure]		371, 401
R 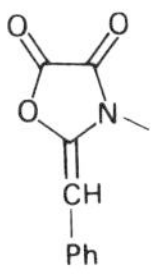 = [structure]		CH_2Cl_2, dioxane, NEt_3, r.t., 0.5 h	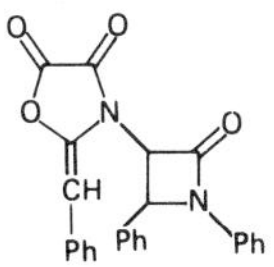 [structure]	16	580
RCH_2COCl	EtOCH=NPh		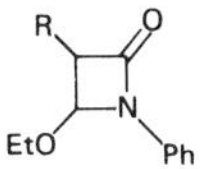 [structure] *trans*		
		CH_2Cl_2, NEt_3	R = Cl	Trace	545
			R = OMe	18	545
			R = OPh	31	545
			R = $SCH_2C_6H_5$	Trace	545
			R = N_3	31	545, 548

		CH_2Cl_2, NEt_3	R 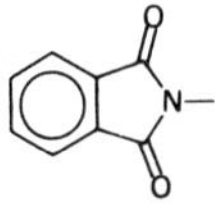 = N— (O, O)	42	545, 548
		C_6H_6, NEt_3, reflux 3 h		31	553
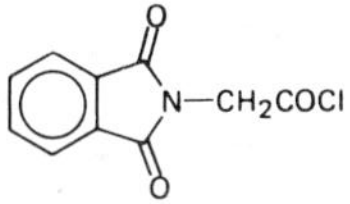 N—CH_2COCl (O, O)	EtSCH=NPh	Ether, NEt_3, r.t., 2 h	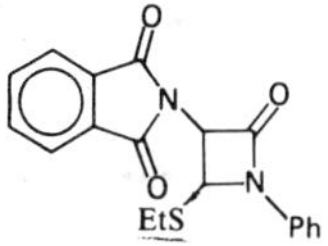 N, O, O, O, N, EtS, Ph	33	553
$ClCHR^1COOH$	$R^2C_6H_4CH{=}NC_6H_4R^3$	$POCl_3$, DMF, reflux	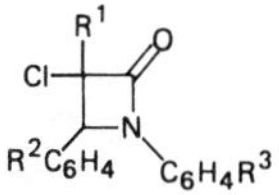 Cl, R^1, O, N, $R^2C_6H_4$, $C_6H_4R^3$		
R^1 = H	$R^2 = R^3$ = H	2 h		78	550[f]
		2 h	(*cis* : *trans* = 53:47)	59	552[f, h]
	R^2 = H, R^3 = *p*-Me	3 h		52	550[f]
	R^2 = H, R^3 = *p*-MeO	4 h		38	550[f]
	R^2 = H, R^3 = *p*-Cl	2.5 h		36	550[f]
	R^2 = *p*-Cl, R^3 = H	5 h		41	550[f]
R^1 = Cl	$R^2 = R^3$ = H	$POCl_3$, C_6H_5Me, reflux 2 h		30	550[f]
		$POCl_3$, DMF, reflux 1.5 h		89	550[f]
	R^2 = H, R^3 = *p*-Cl	1.5 h		66	550[f]
	R^2 = H, R^3 = *p*-MeO	140°C, 4 h		44	550[f]
	R^2 = H, R^3 = *m*-NO_2	140°C, 4 h		80	550[f]
	R^2 = H, R^3 = *p*-NO_2	140°C, 6 h		15	550[f]
	R^2 = H, R^3 = *p*-Me	1.5 h		56	550[f]
	R^2 = H, R^3 = *p*-Cl	4 h		53	550[f]
R^1 = Me	$R^2 = R^3$ = H	1 h		54	550[f]
	R^2 = *p*-Me, R^3 = H	1 h		22	550[f]

TABLE 32. (*Continued*)

Ketene or ketene precursor	Imine	Conditions	Product	Yield (%)	Reference
$BrCH(R^1)COOH$	$R^2C_6H_4CH{=}NC_6H_4R^3$	$POBr_3$, DMF, reflux	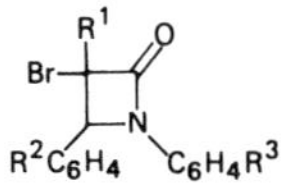		550[f]
$R^1 = H$	$R^2 = R^3 = H$	130–140°C, 2 h		53	
	$R^2 = H$, $R^3 = p$-Me	150°C, 7 h		47	
	$R^2 = p$-Cl, $R^3 = H$	130°C, 5 h		36	
	$R^2 = H$, $R^3 = p$-MeO	140°C, 3 h		27	
$R^1 = Br$	$R^2 = R^3 = H$	130–140°C, 2 h		66	
	$R^2 = p$-Cl, $R^3 = H$	140°C, 4 h		55	
$R^1 = Me$	$R^2 = R^3 = H$	140°C, 3.5 h		35	
CH_3COOH	PhHC=NPh	$POCl_3$, C_6H_5Me, C_5H_5N, reflux 1 h	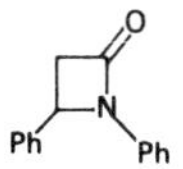	5	550[f, g]
$NC{-}CH(R^1)COOH$	$R^2R^3C{=}NC_6H_4R^4$	$POCl_3$, DMF, C_6H_5Me, reflux 90 min	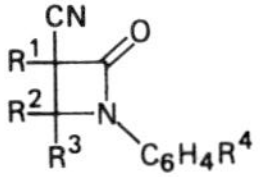		586[f]
$R^1 = Me$	$R^2 = H$, $R^3 = Ph$, $R^4 = H$			57	
	$R^2 = H$, $R^3 = Ph$, $R^4 = o$-Cl			92	
	$R^2 = H$, $R^3 = Ph$, $R^4 = p$-Me			11	
	$R^2 = H$, $R^3 = Ph$, $R^4 =$ 2,4-dimethyl			52	
$R^1 = Et$	$R^2 = H$, $R^3 = Ph$, $R^4 = H$			18	
	$R^2 = H$, $R^3 = Ph$, $R^4 =$ 2,4-dimethyl			25	
	$R^2 = R^3 = Ph$, $R^4 = H$			53	
	$R^2 = R^3 = Ph$, $R^4 = p$-MeO			78	
$R^2 = C_6H_5CH_2$	$R^2 = H$, $R^3 = Ph$, $R^4 = H$			60	
R^1CH_2COCl	$R^2R^3C{=}NR^4$		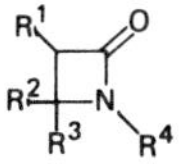		

$R^1 = N_3$	$R^2, R^3 = -(CH_2)_5-$, $R^4 = C_6H_{11}$	CH_2Cl_2, NEt_3, 0°C, stir 3–4 h		20	547
	$R^2, R^3 = -(CH_2)_6-$, $R^4 = Ph$			30	547
	$R^2, R^3 = -(CH_2)_5-$, $R^4 = p\text{-}MeOC_6H_4$			16	547
	$R^2, R^3 = -(CH_2)_5-$, $R^4 = p\text{-}MeC_6H_4$			30	547
	$R^2, R^3 = -(CH_2)_5-$, $R^4 = p\text{-}ClC_6H_4$			14	547
	$R^2, R^3 = -(CH_2)_5-$, $R^4 = o\text{-}MeC_6H_4$			31	547
	$R^2 = H$, $R^3 = p\text{-}O_2NC_6H_4$, $R^4 = Ph$	Method A[k]	*cis*	35	548
	$R^2 = H$, $R^3 = p\text{-}MeOC_6H_4$, $R^4 = Ph$	Method A[k]	*cis*	25–30	548
		Method B[k]	*trans*	53	
	$R^2 = H$, $R^3 = p\text{-}BrC_6H_4$, $R^4 = Ph$	Method A[k]	*cis*	30	548
		Method B[k]	*trans*	65	
	$R^2 = H$, $R^3 =$ 3,4-methylenedioxyphenyl, $R^4 = p\text{-}BrC_6H_4$	Method A[k]	*cis*	30–35	548
		Method B[k]	*trans*	31	
	$R^2 = H$, $R^3 = Ph$, $R^4 = p\text{-}FC_6H_4$	Method A[k]	*cis*	23	548
	$R^2 = H$, $R^3 = p\text{-}BrC_6H_4$, $R^4 = Ph$		*cis*	30	548
	$R^2 = H$, $R^3 = p\text{-}FC_6H_4$, $R^4 = Ph$		*cis*	19	548
	$R^2 = Me$, $R^3 = R^4 = Ph$			30	548
	$R^2 = R^3 = R^4 = Ph$	˙Et_3		60	548
$R^1 = PhCH_2OCONH$	$R^2, R^3 = -(CH_2)_2N(Me)(CH_2)_2-$, $R^4 = p\text{-}MeOC_6H_4$	CH_2Cl_2, NEt_3, 0°C, stir 3–4 h		68	547
$R^1 = OMe$	$R^2, R^3 = -(CH_2)_5-$, $R^4 = Ph$	CH_2Cl_2, NEt_3, 0°C, stir 3–4 h		14	547
	$R^2 = Ph$, $R^3 = SMe$, $R^4 = Ph$	CH_2Cl_2, NEt_3, stir 10 h		90	560, 566
$R^1 = Ph$	$R^2 = Ph$, $R^3 = SMe$, $R^4 = Ph$			76	560, 566
$R^1 = OPh$	$R^2 = Ph$, $R^3 = SMe$, $R^4 = Ph$			80–90, 64	560, 566
	$R^2 = H$, $R^3 = p\text{-}MeOC_6H_4$,		*cis* and *trans*	32 (*cis*)	563

TABLE 32. (Continued)

Ketene or ketene precursor	Imine	Conditions	Product	Yield (%)	Reference
	$R^4 = CH(CO_2Me)(Pr\text{-}i)$				
	$R^2 = Ph$, $R^3 = CO_2Me$, $R^4 = p\text{-}MeOC_6H_4$	CH_2Cl_2, NEt_3, N_2, stir overnight	2 isomers	73	564
	$R^2 = Ph$, $R^3 = SCH_2C_6H_5$, $R^4 = Ph$	CH_2Cl_2, NEt_3, stir 10 h		71	566
	$R^2 = Ph$, $R^3 = p\text{-}NO_2C_6H_4CH_2S$, $R^4 = Ph$			81	566
$R^1 = SCH_2C_6H_5$	$R^2 = H$, $R^3 = p\text{-}MeOC_6H_4$, $R^4 = p\text{-}MeC_6H_4$		*trans*	56	566
R^1 = phthalimido (N-)	$R^2, R^3 = -(CH_2)_5-$, $R^4 = Ph$	CH_2Cl_2, NEt_3, 0°C stir 3–4 h		33	547
	$R^2 = OMe$, $R^3 = R^4 = Ph$	Ether, NEt_3, 35°C, 4–5 h		50	559
	$R^2 = Ph$, $R^3 = SMe$, $R^4 = Ph$	CH_2Cl_2, NEt_3, stir 10 h		69	560, 506
	$R^2 = OEt$, $R^3 = R^4 = Ph$			55	553
	$R^2 = OCHMe_2$, $R^3 = R^4 = Ph$			51	553
	$R^2 = SMe$, $R^3 = R^4 = Ph$			70	553
	$R^2 = H$, $R^3 = SCH_2C_6H_5$, $R^4 = CH(CO_2Me)CHMe_2$	CH_2Cl_2 or C_6H_5Me, NEt_3. reflux 3 h	2 *trans* isomers	39	540, 556
	$R^2 = H$, $R^3 = SMe$, $R^4 = CH(CO_2Me)CHMe_2$	C_6H_5Me, NEt, r. t. 2 h	2 *trans* isomers	40	540
$R^1 = Cl$	$R^2 = H$, $R^3 = SCH_2C_6H_5$, $R^4 = CH(CO_2Me)CHMe_2$	CH_2Cl_2 or C_6H_5Me, NEt_3, r. t. 2.5 h	2 *trans* isomers	45	540, 556
	$R^2 = R^3 = R^4 = Ph$	C_6H_6, NEt_3, 20°C		100	587
	$R^2 = H$, $R^3 = R^4 = Ph$		*trans*	70	587
R^1CH_2COCl	$R^2HC{=}N\text{-}C_6H_4\text{-}CO_2SiMe_3$ (para)	1. CH_2Cl_2, Et_3N, N_2, stir overnight 2. MeOH	β-lactam: 3-R^1, 4-R^2, N-$C_6H_4CO_2H$-*p* azetidin-2-one		557

R^1 = OMe	R^2 = p-$MeOC_6H_4$		*cis*:*trans* = 70:30	78	
R^1 = OPh	R^2 = p-$MeOC_6H_4$		*cis*	76	
	R^2 = o-HOC_6H_4		*cis*:*trans* = 65:35	89	
	R^2 = p-$Me_2NC_6H_4$		*cis*:*trans* = 65:35	82	
R^1 = N_3	R^2 = p-$MeOC_6H_4$		*trans*	80	
R^1CH_2COCl	Me_3SiO_2C–C_6H_4–$CH{=}NR^2$	1. CH_2Cl_2, Et_3N, N_2, stir overnight 2. MeOH	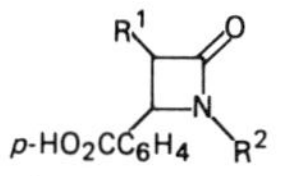 R^1, O, N, p-$HO_2CC_6H_4$, R^2 *cis*		557
R^1 = OMe	R^2 = p-$MeOC_6H_4$			89	
R^1 = OPh	R^2 = p-$MeOC_6H_4$			86	
R^1 = N_3	R^2 = p-$MeOC_6H_4$			95	
R^1 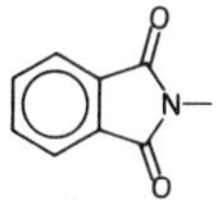 = (phthalimido, N–)	R^2 = p-$MeOC_6H_4$			79	
R^1 = OPh	R^2 = 3,4-di $MeOC_6H_3CH_2$–			91	
	R^2 = p-$MeOC_6H_4CH_2$–			75	
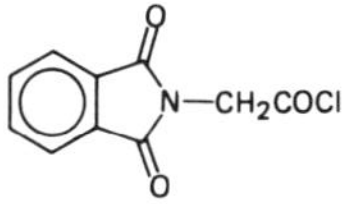 (phthalimido) N–CH_2COCl	$R^1C(NR^2_2){=}NR^3$	Ether, NEt_3, r. t., 1.5 h	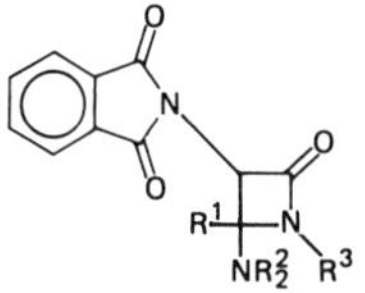 (phthalimido) N, O, O, N, R^1, NR^2_2, R^3	~100	554[j]
R^1 = Ph, R^2 = Me, R^3 = Ph					
R^1 = Ph, R^2 = Et, R^3 = Ph					
R^1 = Ph, R^2 = Et, R^3 = p-MeC_6H_4					
R^1 = p-MeC_6H_4, R^2 = Et, R^3 = Ph					
N_3CH_2COCl	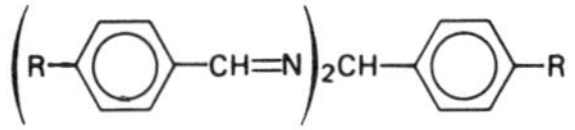 $(R$–C_6H_4–$CH{=}N)_2CH$–C_6H_4–R	1. CH_2Cl_2, NEt_3, r. t. 2. 10% HCl	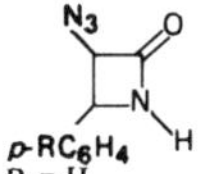 N_3, O, N, H, p-RC_6H_4		558
			R = H	40	
			R = OMe	44	
			R = Me	36	

TABLE 32. (Continued)

Ketene or ketene precursor	Imine	Conditions	Product	Yield (%)	Reference
N_3CH_2COCl	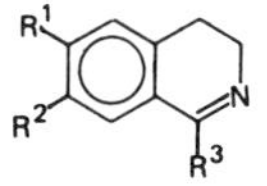	CH_2Cl_2, NEt_3	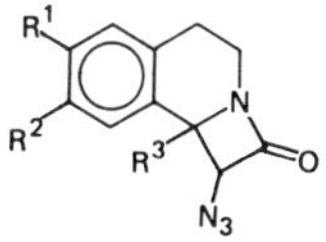		548
$R^1 = R^2 = H$, $R^3 = Ph$				66	
$R^1 = R^2 = H$, $R^3 = p\text{-}O_2NC_6H_4$				77	
$R^1 = R^2 = OMe$, $R^3 = p\text{-}O_2NC_6H_4$				73	
$ClCH_2COCl$	$R^1C_6H_4CH{=}NC_6H_4R^2$	C_6H_6 NEt_3, 70–75°C, 2 h	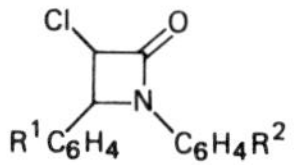		551
			cis:trans		
$R^1 = o\text{-}NO_2$, $R^2 = p\text{-}MeO$			1:1	19	
$R^1 = o\text{-}NO_2$, $R^2 = p\text{-}Cl$			32:68	6	
$R^1 = o\text{-}NO_2$, $R^2 = 2,4\text{-}diMe$			22:78	9	
$R^1 = o\text{-}NO_2$, $R^2 = o\text{-}Br$			1:4	2	
$R^1 = o\text{-}NO_2$, $R^2 = H$			44:56	9	
$R^1 = m\text{-}NO_2$, $R^2 = p\text{-}MeO$			0:100	9	
$R^1 = p\text{-}NO_2$, $R^2 = p\text{-}MeO$			0:100	16	
$R^1 = o\text{-}Cl$, $R^2 = p\text{-}MeO$			18:82	30	
$R^1 = o\text{-}Cl$, $R^2 = H$			13:87	28	
$R^1 = p\text{-}Cl$, $R^2 = p\text{-}MeO$			0:100	7	
$R^1 = o\text{-}MeO$, $R^2 = p\text{-}MeO$			1:9	28	
$R^1 = o\text{-}MeO$, $R^2 = H$			0:100	25	
$R^1 = p\text{-}MeO$, $R^2 = p\text{-}MeO$			0:100	65	
$R^1 = o\text{-}Me$, $R^2 = p\text{-}MeO$			1:9	45	
$R^1 = o\text{-}Me$, $R^2 = H$			0:100	20	
$R^1 = o\text{-}(t\text{-}Bu)$, $R^2 = p\text{-}MeO$			1:3	10	
$ClCH_2COCl$	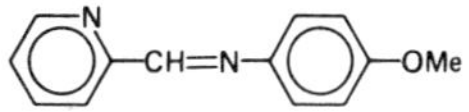 2-pyridyl–CH=N–C₆H₄–OMe	C_6H_6, NEt_3 70–75°C, 2 h	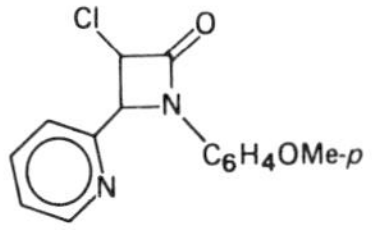 $C_6H_4OMe\text{-}p$; *cis:trans* = 27:73	5	551

$MeCH_2COCl$	o-$O_2NC_6H_4CH{=}NC_6H_4OMe$-$p$	C_6H_6, NEt_3, 70–75°C, 2 h	Me, O, N, o-$O_2NC_6H_4$, C_6H_4OMe-p; *cis:trans* = 36:64	25	551
$ClCH_2COOH$	$R^1C_6H_4CH{=}NC_6H_4R^2$	$POCl_3$, DMF, reflux 2 h	Cl, O, N, $R^1C_6H_4$, $C_6H_4R^2$		552[f]
			cis:trans		
$R^1 = o$-NO_2, $R^2 = H$			1:1	1	
$R^1 = o$-NO_2, $R^2 = p$-MeO			1:1	1	
$R^1 = p$-NO_2, $R^2 = p$-MeO			53:47	53	
$R^1 = p$-NO_2, $R^2 = H$			1:1	40	
$R^1 = o$-Cl, $R^2 = H$			54:46	38	
$R^1 = p$-Cl, $R^2 = p$-MeO			48:52	42	
$R^1 = p$-Cl, $R^2 = H$			1:1	33	
$R^1 = o$-MeO, $R^2 = H$			1:1	46	
$R^1 = p$-MeO, $R^2 = p$-MeO			1:1	53	
$PhCHClNCOCH_2Cl$ (N–Ph)			Cl, O, N, Ph, Ph		552[i]
		DMF, 25°C or C_6H_6, reflux	No reaction	–	
		DMF, reflux	*cis:trans* = 55:45	–	
		DMF, NEt_3, 25°C	*trans*	–	
		C_6H_6, NEt_3, 25°C	*trans*	–	
O, N—CH_2COCl, O	R^2, O←$P(OR^3)_2$, C=N—R^1	Ether, NEt_3, r.t.	O, N, O, O, p-$R^2C_6H_4$, N—R^1, $P(OR^3)_2$, O		555

TABLE 32. (Continued)

Ketene or ketene precursor	Imine	Conditions	Product	Yield (%)	Reference
R^1 = Ph, R^2 = H, R^3 = Me				18	
R^1 = Ph, R^2 = H, R^3 = Et				46	
R^1 = Ph, R^2 = Me, R^3 = Me				29	
R^1 = Ph, R^2 = Me, R^3 = Et				32	
R^1 = Ph, R^2 = Cl, R^3 = Me				17	
R^1 = Ph, R^2 = Cl, R^3 = Et				28	
R^1 = Ph, R^2 = Br, R^3 = Me				22	
R^1 = Ph, R^2 = Br, R^3 = Et				27	
R^1 = Ph, R^2 = OMe, R^3 = Me				36	
R^1 = Ph, R^2 = OMe, R^3 = Et				40	
R^1 = Me, R^2 = H, R^3 = Et				24	
R^1CH_2COCl		CH_2Cl_2, NEt			
R^1 = N_3	R^2 = Ph, R^3 = R^4 = H	Reflux 24–26 h.		70	566
	R^2 = Ph, R^3 = Me, R^4 = H			87	566
	R^2 = Ph, R^3 = Me, R^4 = CO_2Me			20–25	574
	R^2 = CO_2Bu-*t*, R^3 = Me, R^4 = H	Stir overnight		86	568
	R^2 = H, R^3 = Me, R^4 = CO_2Me		*trans*	5–8	572
R^1 = MeO	R^2 = Ph, R^3 = R^4 = H	Reflux 24–26 h	*cis*	90	566
	R^2 = MeS, R^3 = R^4 = H	Stir overnight		73	565
	R^2 = R^3 = H, R^4 = CO_2Et	Reflux 24–26 h		11	566
	R^2 = Ph, R^3 = Me, R^4 = CO_2CHPh_2			90	566
	R^2 = *p*-$PhCH_2CO_2C_6H_4$, R^3 = R^4 = H		*cis*	70	566
R^1 = PhO	R^2 = Ph, R^3 = R^4 = H			70	560, 566
	R^2 = *p*-$PhCH_2CO_2C_6H_4$, R^3 = R^4 = H			63	566
R^1 = Ph	R^2 = 2-furyl, R^3 = R^4 = H		*cis*	70	566
R^1 = $PhCH_2OCONH$	R^2 = *p*-$O_2NC_6H_4$, R^3 = R^4 = H		*trans*	–	575

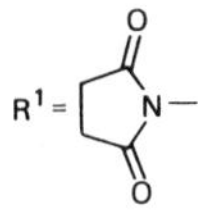 R^1 =	R^2 = Ph, R^3 = Me, R^4 = CO_2Me	Reflux 24–26 h	–	574
		Reflux 6.25 h	13	577. 578
	R^2 = Ph, R^3 = R^4 = H	Reflux 6 h, NEt_3, C_6H_6, reflux 4 h	56	577
			14	577
		Reflux 10 h	16	577
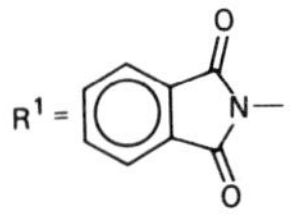 R^1 =	R^2 = $PhCH_2CO_2$–, R^3 = R^4 = H	CH_2Cl_2, NEt_2, reflux 24–26 h	85	569
	R^2 = Ph, R^3 = Me, R^4 = CO_2Me	CH_2Cl_2, NEt_3, reflux		574
		CH_2Cl_2, NEt_3, reflux 6 h	58.3	582
	R^2 = Ph, R^3 = Me, R^4 = H	Ether, NEt_3, reflux 2 h	5	577
	R^2 = Ph, R^3 = R^4 = H		40	581
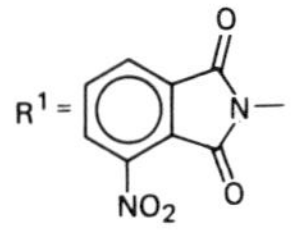 R^1 =	R^2 = Ph, R^3 = R^4 = H		17	581
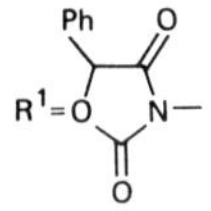 R^1 =	R^2 = Ph, R^3 = R^4 = H	CH_2Cl_2, NEt_3, reflux 7 h	28.4	579
		Ether, NEt_3, reflux	poor	579
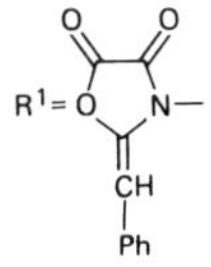 R^1 =	R^2 = Ph, R^3 = R^4 = H	CH_2Cl_2, dioxane, NEt_3, N_2, reflux 6.5 h	45	580

TABLE 32. (Continued)

Ketene or ketene precursor	Imine	Conditions	Product	Yield (%)	Reference
N_3CH_2COCl	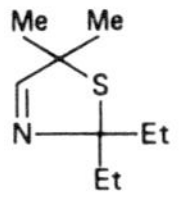	CH_2Cl_2, NEt_3, reflux	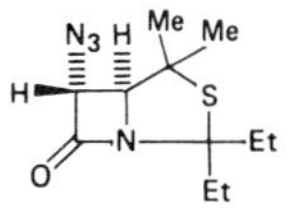	18	570, 573
N_3CH_2COCl	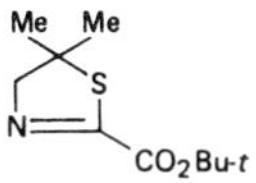	CH_2Cl_2, NEt_3	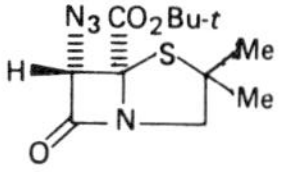	Good	573
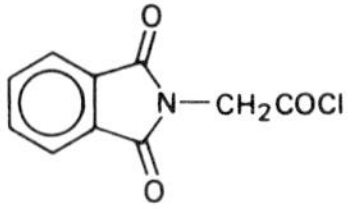	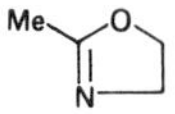	C_6H_6, NEt_3, r.t. 2 h	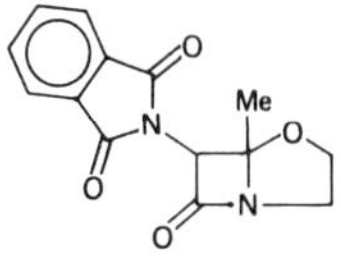	26	583
$ClCH_2COCl$	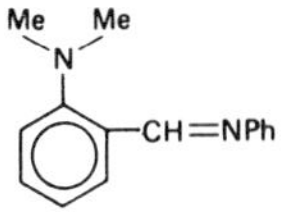	C_6H_6, NEt_3, reflux	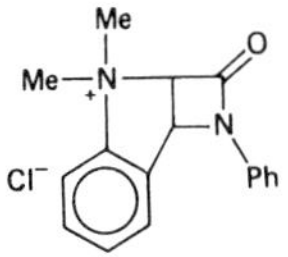	49	591
$ClCH_2COCl$	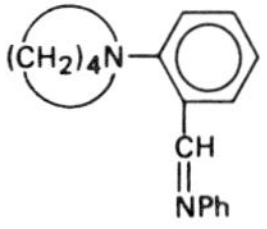	C_6H_6, NEt_3, reflux	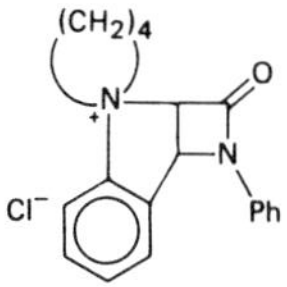	41	591

R^1CH_2COCl	[structure: 2-R^2-5,6-dihydro-4H-1,3-thiazine]		[structure: bicyclic β-lactam with R^1, R^2]		
R^1 = MeO	R^2 = Ph	CH_2Cl_2, NEt_3, reflux, then stir overnight	*cis*	63	560, 566
R^1 = PhO	R^2 = Ph		*cis*	81	560, 566
R^1 = N_3	R^2 = Ph		*cis*	23	571
	R^2 = p-$O_2NC_6H_4$		*cis*	55–60	571, 574
R^1 = $PhCH_2OCONH$	R^3 = Ph		*trans*	50–70	575
R^1 = $PhCH_2OCONH$	R^2 = p-$O_2NC_6H_4$		*trans*	50–70	575
R^1 = [structure: succinimido, N—]	R^2 = Ph, R^3 = R^4 = H	C_6H_6, NEt_3, 80°C, 12 h		53	584
R^1 = [structure: maleimido, N—]	R^2 = Ph, R^3 = R^4 = H	C_6H_6, NEt_3, 54°C, 2.5 h		55	584
R^1 = [structure: phthalimido, N—]	R^2 = Ph, R^3 = R^4 = H	C_6H_6, NEt_3, r.t. 2 h		43	583
		C_6H_6, NEt_3, 80°C, 2.5 h		70.5	584
N_3CH_2COCl	[structure: thiazine with Me, CO_2Me]	CH_2Cl_2, NEt_3	[structure: N_3 β-lactam cephem with Me, CO_2Me]	52	567

TABLE 32. (Continued)

Ketene or ketene precursor	Imine	Conditions	Product	Yield (%)	Reference
N_3CH_2COCl	S, N, CH_2OAc, $CO_2CH_2—C_6H_4OMe$-*p*	CH_2Cl_2, NEt_3, $-78°C$	N_3, H, H, S, N, O, CH_2OAc, $CO_2CH_2—C_6H_4OMe$-*p*	56	567
N_3CH_2COCl	Me, Me, Ph, S, N	CH_2Cl_2, NEt_3	Ph, N_3, Me, Me, H, S, N, O	30	571
N_3CH_2COCl	Me, Me, Me, N, S, Me, Me	CH_2Cl_2, NEt_3	N_3, Me, Me, Me, H, N, S, O, Me, Me	10	571
O, N—CH_2COCl, O	R, O, Me, N, Me, Me	C_6H_6, NEt_3, reflux	O, R, N, O, Me, O, N, O, Me, Me		585
	R = Ph	4 h		57	
	R = $C_6H_5CH_2$	4 h		25	
	R = *p*-$O_2NC_6H_4$	3 h, then stir at r.t.		65	

R^1CH_2COCl	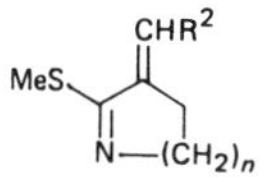	CH_2Cl_2, NEt_3, N_2, r.t.	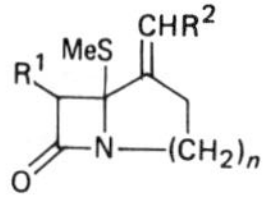		
R^1 = MeO	R^2 = Ph; $n = 1$		*trans*	68	561, 562
	R^2 = Ph; $n = 2$			75	561
	R^2 = 3,4-$(MeO)_2C_6H_3$; $n = 2$			67	561
R^1 = PhO	R^2 = 3,4-$(MeO)_2C_6H_3$; $n = 1$			33	561
	R^2 = 3-$O_2NC_6H_4$; $n = 1$			60	561
	R^2 = Ph; $n = 1$			50	561
RCH_2COCl	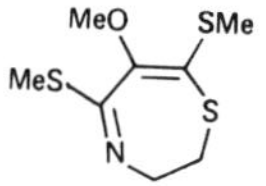	CH_2Cl_2, NEt_3, N_2, stir overnight, r.t.	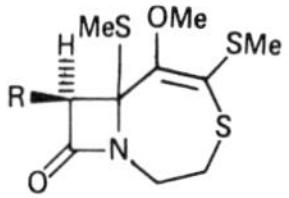		565
			R = OMe	25	
			R = OPh	62	
			R = N_3	31	
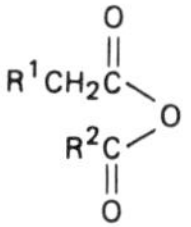	$R^3CH{=}NR^4$	CH_2Cl_2, NEt_3, reflux 1 h, then stir overnight	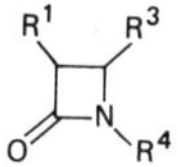	30–70	576
R^1 = CF_3, OEt, OBu-*i*; R^2 = N_3	R^3 = R^4 = Ph		*cis* and *trans*		
	R^3 = *p*-$MeOC_6H_4$, R^4 = *p*-$HO_2CC_6H_4$		*trans*		
R^2 = PhO	R^3 = R^4 = Ph		*cis* and *trans*		
	R^3 = *p*-$MeOC_6H_4$, R^4 = $CHPh_2$		*cis*		
$R^1R^2CHCOCl$	$R^3R^4C{=}NR^5$		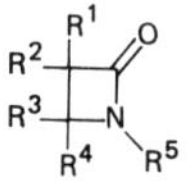		
R^1 = N_3, R^2 = Me	R^3 = H, R^4 = R^5 = Ph	CH_2Cl_2, NEt_3		10	548
R^1 = N_3, R^2 = Et	R^3 = H, R^4 = R^5 = Ph			9	548
R^1 = N_3, R^2 = Ph	R^3 = H, R^4 = R^5 = Ph			5.7	548

TABLE 32. (Continued)

Ketene or ketene precursor	Imine	Conditions	Product	Yield (%)	Reference
R^1 = CN, R^2 = Me	R^3 = H, R^4 = R^5 = Ph	C_6H_6, heat 3 h		53	586
	R^3 = H, R^4 = Ph, R^5 = *p*-MeOPh	C_6H_6, heat 3 h		31	586
	R^3 = R^4 = R^5 = Ph	C_6H_6, heat 2 h		77	586
	R^3 = R^4 = Ph, R^5 = *p*-MeOPh			82	586
R^1 = CN, R^2 = Ph	R^3 = H, R^4 = R^5 = Ph			49	586
	R^3 = H, R^4 = Ph, R^5 = *p*-MeOPh			45	586
R^1 = R^2 = Cl	R^3 = R^4 = R^5 = Ph	C_6H_6, NEt_3, 20°C		100	587
	R^3 = H, R^4 = R^5 = Ph			100	587[l]
	R^3 = H, R^4 = Ph, R^5 = *n*-Bu			70	587
	R^3 = H, R^4 = Ph, R^5 = C_6H_{11}			90	587
R^1 = Ph, R^2 = OAc	R^3 = H, R^4 = R^5 = Ph	NEt_3	*trans*	90–95	588
	R^3 = MeS, R^4 = R^5 = Ph		*cis*	–	588[m]
R^1–CH(R^2)COCl	R^3N=C=NR^3		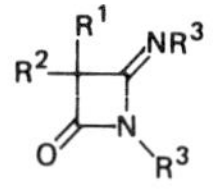		
R^1 = CN, R^2 = Me	R^3 = C_6H_{11}	C_6H_6, 140°, 6 h		88	586
R^1 = R^2 = Cl	R^3 = C_6H_{11}	Cyclohexane, NEt_3, reflux 50 min		88	589
	R^3 = *i*-Pr	Cyclohexane, NEt_3, reflux 100 min		76	589
$Cl_2CHCOCl$	PhCH=CHCH=NR	C_6H_6, NEt_3	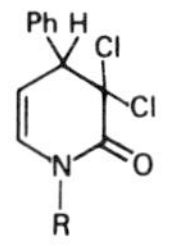		589
			R = Ph	45	
			R = *p*-MeC_6H_4	67	
			R = PhHC=CHCH=N–	75	

$R^1R^2CClCOCl$	Imine (X-ring N-aryl, $CH{=}NR^4$, R^3)	Conditions	Product (β-lactam)	Yield (%)	Ref.
$R^1 = R^2 = Cl$	$R^3 = NO_2$, $R^4 = Ph$, $X = (CH_2)_4$	C_6H_6, NEt_3, reflux		75	591
	$R^3 = NO_2$, $R^4 = Ph$, $X = (CH_2)_5$			85	
	$R^3 = NO_2$, $R^4 = Ph$, $X = (CH_2)_6$			49	
	$R^3 = NO_2$, $R^4 = Ph$, $X = (CH_2)_2O(CH_2)_2$			52	
$R^1 = H$, $R^2 = Cl$	$R^3 = H$, $R^4 = Ph$, $X = (CH_2)_5$		*trans*	55	
	$R^3 = H$, $R^4 = Ph$, $X = (CH_2)_6$		*trans*	52	
	$R^3 = H$, $R^4 = Ph$, $X = (CH_2)_2O(CH_2)_2$		*trans*	25	
$R^1 = Cl$, $R^2 = Me$	$R^3 = H$, $R^4 = Ph$, $X = Me_2$			64	
	$R^3 = H$, $R^4 = Ph$, $X = (CH_2)_4$			61	
	$R^3 = H$, $R^4 = Ph$, $X = (CH_2)_5$			65	
	$R^3 = H$, $R^4 = Ph$, $X = (CH_2)_6$			54	
$R^1 = Cl$, $R^2 = Ph$	$R^3 = H$, $R^4 = Ph$, $X = Me_2$	C_6H_6, NEt_3, r.t.		10	
	$R^3 = H$, $R^4 = Ph$, $X = (CH_2)_4$	C_6H_6, NEt_3, reflux		63	
	$R^3 = H$, $R^4 = Ph$, $X = (CH_2)_5$			75	
	$R^3 = H$, $R^4 = Ph$, $X = (CH_2)_6$			32	
	$R^3 = H$, $R^4 = Ph$, $X = (CH_2)_2O(CH_2)_2$			60	
$R^1 = R^2 = Cl$	$R^3 = H$, $R^4 = Ph$, $X = (CH_2)_5$			83	
	$R^3 = H$, $R^4 = Ph$, $X = Me_2$			65	
$R^1 = H$, $R^2 = Me$	$R^3 = H$, $R^4 = Ph$, $X = (CH_2)_5$		*trans*	66	
$R^1 = H$, $R^2 = OPh$	$R^3 = H$, $R^4 = Ph$, $X = (CH_2)_5$		*trans*	75	
$R^1 = H$, $R^2 = Cl$	$R^3 = H$, $R^4 = C_6H_{11}$, $X = (CH_2)_5$		*trans*	24	
	$R^3 = H$, $R^4 = n$-Bu, $X = (CH_2)_5$		*trans*	44	
	$R^3 = H$, $R^4 = CH_2C_6H_5$, $X = (CH_2)_5$		*trans*	66	
$R^1 = t$-Bu, $R^2 = CN$	$R^3 = H$, $R^4 = Ph$, $X = (CH_2)_5$			67	
	$R^3 = H$, $R^4 = Ph$, $X = Me_2$			68	
$R^1 = R^2 = Cl$	$R^3 = NO_2$, $R^4 = Ph$, $X = (CH_2)_5$	C_6H_6, NEt_3, r.t.		15	
$R^1 = H$, $R^2 = Ph$	$R^3 = H$, $R^4 = Ph$, $X = Me_2$	C_6H_6, NEt_3, reflux 2 h	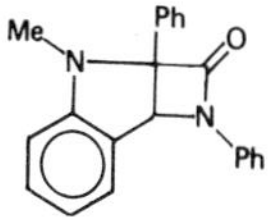 (fused tricyclic: Me–N, Ph, O, N–Ph)	50	

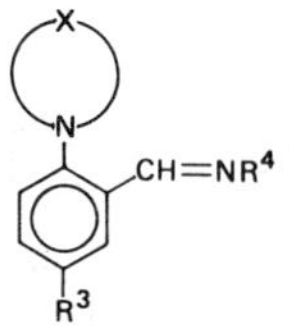

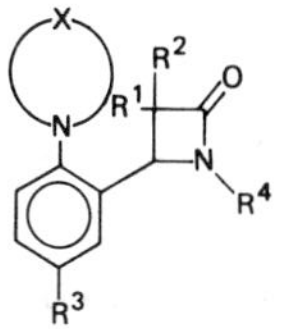

TABLE 32. (Continued)

Ketene or ketene precursor	Imine	Conditions	Product	Yield (%)	Reference
$R^1CH(COR^2)_2$	$R^3R^4C{=}NR^5$		R^2OC–azetidin-2-one (R^1, R^3, R^4, N–R^5)		
$R^1 = Et, R^2 = Cl$	$R^3 = H, R^4 = R^5 = Ph$	C_6H_6, 120°C, 15 min		44.8	592
		No solvent, reflux		44.8	592
$R^1 = Et, R^2 = OEt$	$R^3 = H, R^4 = R^5 = Ph$	C_6H_6, reflux, 4 h		21.7	592
$R^1 = Et, R^2 = Cl$	$R^3 = H, R^4 = R^5 = Ph$	1. C_6H_6, 120°C, 15 min 2. EtOH, reflux 1 h	$R^1 = Et$ $R^2 = OEt$ $R^3 = H$ $R^4 = Ph$	46.5	592
$R^1 = CH_2C_6H_5, R^2 = Cl$	$R^3 = H, R^4 = R^5 = Ph$	1. C_6H_6, reflux 4 h 2. EtOH, reflux 1 h or 1. No solvent, 120°C, 15 min 2. EtOH, reflux 1 h	$R^1 = CH_2C_6H_5$ $R^2 = OEt$ $R^3 = H$ $R^4 = Ph$	70	592
		1. C_6H_6, reflux 4 h 2. MeOH, reflux 1 h or 1. No solvent, 110–120°C, 15 min 2. MeOH, reflux 1 h	$R^1 = CH_2C_6H_5$ $R^3 = H$ $R^4 = Ph$	–	592
$R^1 = Ph, R^2 = Cl$	$R^3 = H, R^4 = R^5 = Ph$	C_6H_6, reflux	*cis*	–	593, 594
	$R^3 = H, R^4 = R^5 = Ph$	1. C_6H_6, reflux 4 h 2. MeOH, reflux 1 h	$R^1 = Ph$ $R^2 = OMe$ $R^3 = H$ $R^4 = Ph$	75.8	592
$R^1 = i\text{-}Pr, R^2 = Cl$	$R^3 = H, R^4 = Ph, R^5 = 2,4\text{-di}MeC_6H_3$	C_6H_6, reflux 6 h		72	592

[a]Treatment of this product with Raney Ni in 95% EtOH for 1 h afforded a 30% conversion to 3,3-dimethyl-1,4-diphenylazetidine-2-one.
[b]Structure uncertain. Probable mechanism:

isoquinoline + $Ph_2C{=}C{=}O$ ⟶ isoquinolinium–N–C(O⁻)=CPh_2 + $PhN{=}C{=}O$ ⟶ Products

[c]This product was also prepared in this paper by the reaction of 2 moles of diphenylketene and 1 mole of *p*-nitroso-*N,N*-dimethylaniline in ether.

$$Ph_2C{=}C{=}O + p\text{-}Me_2NC_6H_4N{=}O \longrightarrow \underset{p\text{-}Me_2NC_6H_4N-O}{Ph_2C-C{=}O} \longrightarrow CO_2 + \underset{p\text{-}Me_2NC_6H_4N}{Ph_2C{=}} \quad Ph_2C{=}C{=}O \longrightarrow \text{Product}$$

[d]Product isolated as a 5:2 mixture of 2 diastereomers.
[e]Product isolated as a 3:1 mixture of 2 diastereomers.
[f]The mechanism for this reaction is believed[550] to be:

$$R^1-CH(R^2)-C(=O)-OH + POCl_3 \longrightarrow \left[HO-C^+(OPOCl_2)-CH(R^1)(R^2)\right]Cl^- + H(R^4)C{=}N-R^3 \longrightarrow \left[HO-C(OPOCl_2)(CH(R^1)R^2)-\overset{+}{N}(R^3){=}CH-R^4\right]Cl^- \xrightarrow[-HOPOCl_2]{-HCl} \text{azetidin-2-one } (R^1, R^2, R^3, R^4)$$

[g]This product is also prepared in this paper by treatment of 1,4-diphenyl-3-bromozaetidin-2-one or 1,4-diphenyl-3,3-dibromazetidin-2-one with zinc in MeOH and liq. NH_3 for 8 h.
[h]Treatment of *cis*-1,4-diphenyl-3-chloroazetidin-2-one with $POCl_3$ and $ClCH_2COOH$ in DMF for 7 h. afforded 1,4-diphenyl-3-chloroazetidin-2-one in a *cis:trans* ratio of 53:47. This same ratio of isomers was also obtained by treatment of *trans*-1,4-diphenyl-3-chloroazetidin-2-one with the same mixture for 22 h. Treatment of the *cis* isomer as above for only 2 h gave 18% *trans*, while treatment of the *trans* isomer for the same length of time afforded no isomerization to the *cis* isomer.
[i]This reaction illustrates that the lactam product is formed by direct acylation rather than *in situ* generation of ketene[519,520].
[j]The products from this reaction were not isolated but were cleaved directly by distillation in vacuum to

$$R^1-C(NR^2_2){=}CH-N(\text{phthalimido})$$

[k]Dropwise addition of a soln. of acid chloride in CH_2Cl_2 to a CH_2Cl_2 soln. of benzylidineaniline and NEt_3 at r.t. was found to produce *cis*-lactam exclusively (Method A), while addition of NEt_3 to a CH_2Cl_2 solution of acid chloride and benzylidineaniline afforded *trans*-lactam exclusively (Method B).
[l]Reaction of dichloroacetyl chloride with benzylideneaniline without any NEt_3 afforded $Cl_2CHCONPhCHClPh$, which upon heating to 150°C or refluxing in benzene afforded a 20–30% yield of 1,4-diphenyl-3,3-dichloroazetidin-2-one.
[m]This product was desulphurized using Raney Ni to afford the previous product with opposite stereochemistry.

TABLE 33. Production of diazetidinones by the reaction of ketenes with azo compounds

Ketene	Azo compound	Conditions	Product	Yield (%)	Reference
$R^1R^2C{=}C{=}O$	$R^3N{=}NR^4$		1-R^3-2-R^4-4-R^1-4-R^2-1,2-diazetidin-3-one		
$R^1 = R^2 = H$	$R^3 = R^4 = Ph$	Hexane, 15°C	–		595
$R^1 = H, R^2 = Ph$	$R^3 = R^4 = Ph$	C_6H_6		–	515,596
$R^1 = R^2 = Ph$	$R^3 = R^4 = Ph$	100°C, or ether		–	515,596
	(*trans*)	125–130°C, 42 h, CO_2		25	597
	(*cis*)	r.t.		92	597
	$R^3 = R^4 = o\text{-}MeC_6H_4$	Ether, $h\nu$, stand overnight		80	597
	$R^3 = R^4 = m\text{-}MeC_6H_4$				
	(*cis*)			35	597
	(*trans*)			90	597
	$R^3 = R^4 = C_6H_5CH_2$	C_6H_6, N_2		96	515, 599

$$PhCH{=}NPh + R^1CH(Br)COOMe \xrightarrow[\text{Solvent}]{Zn} \text{1-Ph-3-}R^1\text{-4-Ph-azetidin-2-one (cis and trans)} \quad (291)$$

Solvent	cis : trans					
	R^1 = Me	Et	*i*-Pr	C_6H_{11}	*t*-Bu	Ph
C_6H_5Me	73:27	64:36	55:45	45:55	25:75	0:100
Et_2O, C_6H_6 (50:50)	–	63:37	80:20	76:24	–	–
THF	80:20	74:26	100:0	100:0	100:0	0:100

$$\begin{array}{c} PhCH{=}NPh \\ + \\ PhCD{=}NPh \end{array} + PhCH(Br)COOMe \xrightarrow[\substack{C_6H_5Me \\ \text{reflux} \\ 1.5\ h}]{Zn,\ HgCl_2} \text{(3-H, 4-H) 56.5 + 0.5\%} + \text{(3-H, 4-D) 43.5 + 0.5\%} \quad (292)$$

reaction with imines. One interesting sidelight to these investigations is the study[606] of the time required to epimerize the *cis* β-lactams prepared into their *trans* counterparts. The results obtained[606] are shown in the table below equation (293). In addition, it was also found that in 50 h at 75 °C 92% of pure *trans* 1,3,4-triphenylazetidin-2-one was converted into its *cis* epimer.

$$\textit{cis}\text{-3-R-1,4-diphenylazetidin-2-one} \xrightarrow[75^\circ C]{NaOH,\ t\text{-BuOH}} \textit{trans}\text{-3-R-1,4-diphenylazetidin-2-one} \quad (293)$$

R	% *trans*				
	t(h) = 25	50	75	100	150
Me	42				
i-Pr	55			70	75
C_6H_{11}			80		95
t-Bu	95		98		98

c. Other imine cycloadditions. Reaction of anils containing a methyl or methylene group in the α-position of the imine double bond with monosubstituted malonyl chloride has been reported[607] to afford acceptable yields of *n*-aryl-

TABLE 34. β-Lactam preparation by the reaction of a Reformatsky reagent with an imine

$R^1CH(Br)COOR^2 + R^3HC{=}NR^4 \xrightarrow[Zn]{Solvent}$ β-lactam (R^1, R^3 on ring carbons; N–R^4; C=O)

R^1	R^2	R^3	R^4	Solvent	Stereo-chemistry (*cis:trans*)	Yield (%)	Reference
H	Et	Ph	Me	C_6H_5Me	–	52	601
H	Et	Ph	Ph	C_6H_5Me	–	56	604
Me	Me	Ph	Ph	C_6H_5Me	73:27	90	518,602,603
				THF	–	94	603
				Et_2O, C_6H_6	–	75	603
Me	Me	Ph	*p*-BrC_6H_4	THF	–	90	603
Me	Et	Ph	Me	C_6H_5Me	–	81	601
Me	Et	Ph	Ph	C_6H_5Me	–	85	604
Me	Et	Ph	$C_6H_5CH_2$	C_6H_5Me	–	76	601
Me	*i*-Pr	Ph	Ph	C_6H_5Me	55:45	–	603
Me	*t*-Bu	Ph	Ph	C_6H_5Me	25:75	–	603
Me	2-(*i*-Pr)-5-MeC_6H_3	Ph	*p*-BrC_6H_4	C_6H_5Me	–	80	603
Et	Me	Ph	Ph	C_6H_5Me	64:36	94	518,602,603
				THF	74:26	96	603
				Et_2O, C_6H_6	63:37	72	603
i-Pr	Me	Ph	Ph	C_6H_5Me	55:45	98	518,602,603
				THF	100:0	92	603
				Et_2O, C_6H_6	80:20	98	603

i-Pr	Me	Ph	p-MeC_6H_4	THF	–	95–97	603,606
i-Pr	Me	Ph	c-C_6H_{11}	THF	–	98	603
i-Pr	Me	Ph	p-BrC_6H_4	C_6H_5Me or THF	–	95–98	603,606
i-Pr	Me	p-$MeOC_6H_4$	Ph	C_6H_5Me	–	70	603
i-Pr	i-Pr	Ph	Ph	C_6H_5Me	34:66	93	603
i-Pr	t-Bu	Ph	Ph	C_6H_5Me	2:98	–	603
i-Pr	2-(i-Pr)-5-MeC_6H_3	Ph	Ph	C_6H_5Me	–	90	603
C_6H_{11}	Me	Ph	Ph	C_6H_5Me	45:55	92	518,602,603
				THF	100:0	96	603
				Et_2O, C_6H_6	76:24	98	603
t-Bu	Me	Ph	Ph	C_6H_5Me	25:75	71	518,602,603
				THF	100:0	96	603
t-Bu	Me	Ph	p-BrC_6H_4	C_6H_5Me or THF	–	98	603
t-Bu	i-Pr	Ph	Ph	C_6H_5Me	–	95	603
Ph	Me	Ph	Ph	C_6H_5Me or THF	0:100	–	518,602,603
Ph	Et	Ph	Ph	C_6H_5Me	–	7	601
H	2-(i-Pr)-5-MeC_6H_3	Ph	Ph	C_6H_6	–	82	603
2-(i-Pr)-5-MeC_6H_3	Me	Ph	Ph	C_6H_5Me	38:62	–	603
2-(i-Pr)-5-MeC_6H_3	i-Pr	Ph	Ph	C_6H_5Me	10:90	–	603
$MeC(Me)(Br)CO_2Et$		Ph	$C_6H_5CH_2$	C_6H_5Me		84	601[a]

[a] Product is: 3,3-dimethyl-4-phenyl-1-($CH_2C_6H_5$)azetidin-2-one (Me, Me, Ph, $CH_2C_6H_5$ on β-lactam ring)

4-hydroxy-2-pyridones. The mechanism proposed for this reaction is shown in equation (294).

Initial attack of one acid chloride function on the anil nitrogen affords an imine salt, which then loses two moles of hydrogen chloride consecutively to afford product.

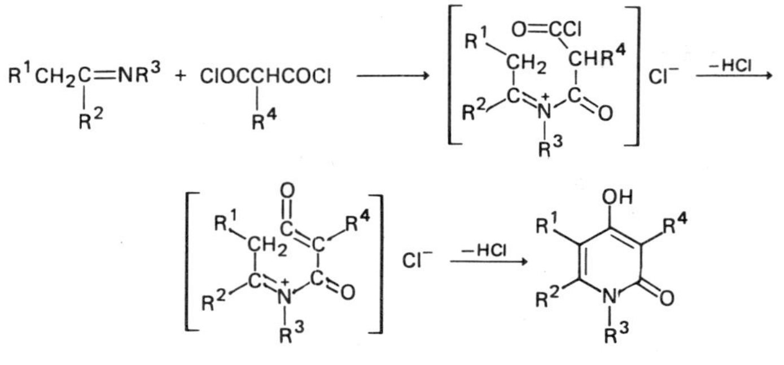

$$R^1CH_2C(R^2){=}NR^3 + ClOCCH(R^4)COCl \longrightarrow [\text{cyclic imine salt}]\,Cl^- \xrightarrow{-HCl} [\text{cyclic intermediate}]\,Cl^- \xrightarrow{-HCl} \text{4-hydroxy-2-pyridone } (R^1, R^2, R^3, R^4) \quad (294)$$

R^1	R^2	R^3	R^4	Yield (%)	Conditions
H	Ph	Ph	$CH_2C_6H_5$	42–44 31.2	C_6H_5Me, reflux 90 min C_6H_5Me, reflux 30 min
H	Ph	Ph	*i*-Pr	55.7	C_6H_6, reflux 1 h
H	Ph	*p*-MeC_6H_4	$CH_2C_6H_5$	81.7	C_6H_6, reflux 45 min
H	Ph	*p*-MeC_6H_4	*n*-Bu	42	C_6H_6, reflux 45 min
H	Ph	*p*-MeC_6H_4	*i*-Pr	65.9	C_6H_6, reflux 80 min
Me	Ph	Ph	$CH_2C_6H_5$	38.2	C_6H_6, reflux 90 min
Et	Ph	Ph	$CH_2C_6H_5$	39.4	C_6H_6, reflux 2 h
H	*s*-Bu	Ph	$CH_2C_6H_5$	21	C_6H_6, reflux 80 min

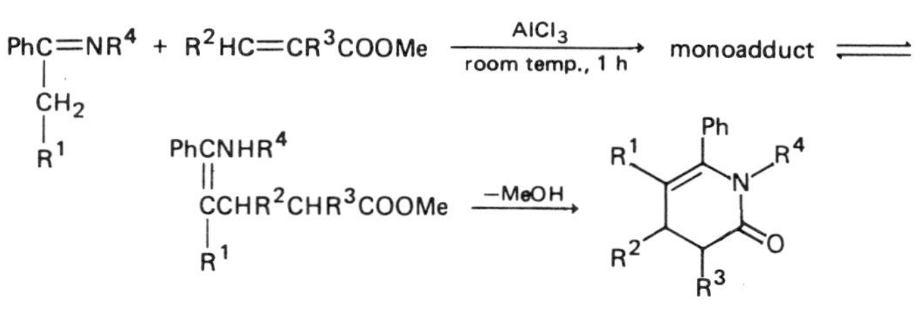

$$PhC(CH_2R^1){=}NR^4 + R^2HC{=}CR^3COOMe \xrightarrow[\text{room temp., 1 h}]{AlCl_3} \text{monoadduct} \rightleftharpoons PhC(NHR^4){=}C(R^1)CHR^2CHR^3COOMe \xrightarrow{-MeOH} \text{dihydropyridone } (Ph, R^1, R^2, R^3, R^4) \quad (295)$$

R^1	R^2	R^3	R^4	Yield (%)
H	H	H	Ph	85
H	H	Me	Ph	55
Me	H	H	Ph	85
Me	H	Me	Ph	60
H	Ph	H	Ph	30[a]
Me	Ph	H	Ph	25

[a]This yield was obtained at 80°C for 24 hr.; using the conditions shown above (r. t., 1 h) afforded no reaction.

Dihydropyridones have been similarly prepared[608] by the reaction of aromatic ketimines and acrylic esters. The proposed mechanism involves initial formation of an unspecified monoadduct which is proposed to be in equilibrium with the α-substituted anil shown in equation (295). Elimination of methanol affords dihydropyridone via intramolecular condensation. 2-Azetidinylidene ammonium salts (**46**) afford upon hydrolysis the corresponding 2-azetidinones (**47**)[609]. The salt **46** is prepared by addition of a *N,N*-dimethyl-1-chloroalkenylamine to a Schiff base. The mechanism proposed (equation 296) involves initial aminoalkenylation of

(296)

(47) (46)

R^1	R^2	R^3	R^4	R^5	Yield (%) 46	Yield (%) 47
Me	Me	Ph	H	Me	74	82
Me	Me	Ph	H	Ph	47	68
Me	Me	$C_6H_5CH_2S$	H	Me_3C	60	42

the Schiff base to give the intermediate shown, which then cyclizes to afford the salt. Since the intermediate is also in principle available from the reaction of α-chloroalkylideneammonium chloride with Schiff bases, followed by elimination of hydrogen chloride, the authors utilized the reaction of tertiary amides with phosgene followed by reaction with Schiff bases and triethylamine to produce β-lactams as shown in Scheme 3.

3. From nitrones and nitroso compounds

In 1919 Staudinger and Miescher[610] first investigated the reaction of diphenylketene and various nitrones (anil *N*-oxides) and proposed the reaction course shown

$$R^1R^2CHC(O)NMe_2 \xrightarrow[\text{2. } CH_2Cl,\text{ 2 days, r.t.}]{\text{1. } COCl_2,\ 0^{\circ}C} R^1R^2CH{-}C(Cl){=}\overset{+}{N}Me_2\ Cl^-$$

$$+ \quad R^3R^4C{=}N{-}R^5$$

$$\longrightarrow \left[R^1R^2CH{-}C(={\overset{+}{N}Me_2}){-}N(R^5){-}C(R^3)(R^4){-}Cl \quad Cl^- \right] \xrightarrow{NEt_3,\ r.t.} \left[R^1R^2C{=}C(NMe_2){-}\overset{+}{N}(R^5){=}CR^3R^4 \quad Cl^- \right]$$

$$\longrightarrow \textbf{(46)}\ Cl^- \longrightarrow \textbf{(47)}$$

R^1	R^2	R^3	R^4	R^5	Yield (%)	
					46	**47**
Me	Me	Ph	Ph	Ph	65	70
H	CMe_3	Ph	H	Ph	80 (*trans*)	0

SCHEME 3.

in equation (297). A similar reaction was investigated in 1938 by Taylor, Owen and Whittaker[611] who proposed the reaction course (298). More recently, however,

$$Ph_2C{=}C{=}O + Ph{-}\overset{O}{\overset{\uparrow}{N}}{=}CPh_2 \longrightarrow \text{[4-membered ring: } C(=O){-}O{-}N(=CPh_2){-}CPh_2\text{; substituents Ph, Ph, Ph]}$$

$$\longrightarrow Ph_2C{=}N(Ph){=}CPh_2 \xrightarrow{H_2} Ph_2CH{-}N(Ph){-}CHPh_2 \tag{297}$$

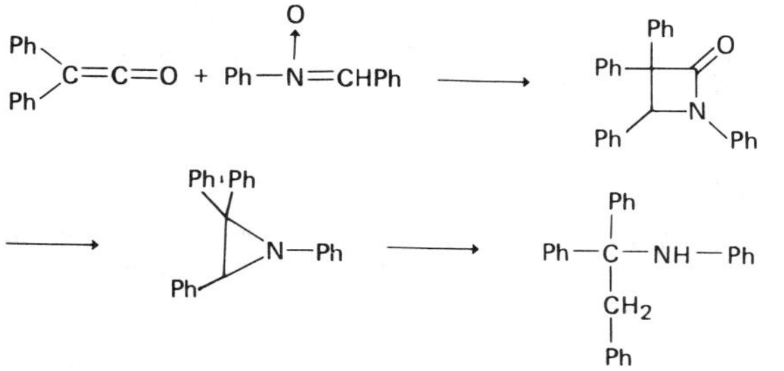

(298)

Hassall and Lippman[612] have found that the reaction of diphenylketene and benzylideneaniline oxide affords *o*-benzylideneaminophenyldiphenylacetic acid, which upon treatment with Adams catalyst in ethyl acetate produces 1-benzyl-3,3-diphenyloxindole and not a β-lactam (equation 299).

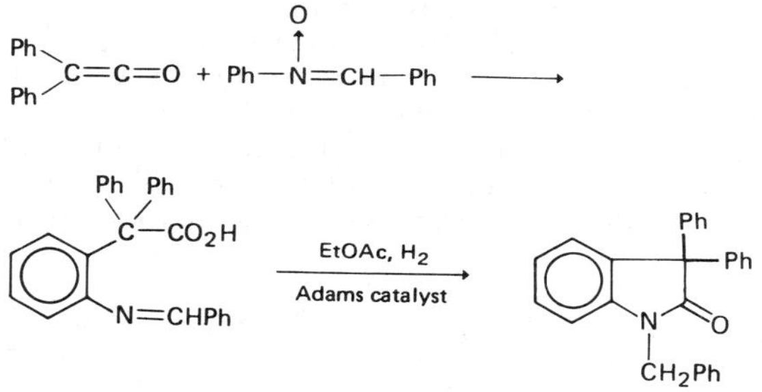

(299)

However, β-lactams have been produced from nitrones by reaction of the nitrones with copper phenylacetylide (equation 300)[613].

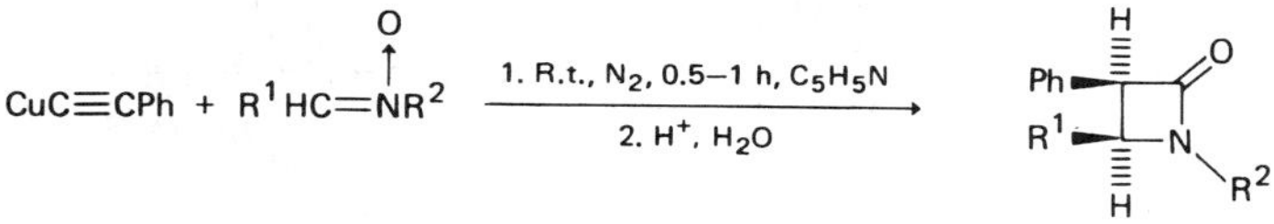

(300)

R^1	R^2	Yield (%)
Ph	Ph	55
Ph	*p*-ClC_6H_4	60
o-MeC_6H_4	Ph	51
o-ClC_6H_4	Ph	51

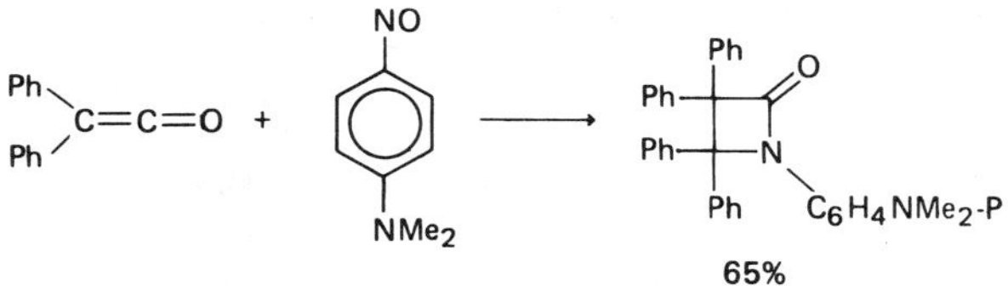

(301)

The reaction of a ketene with a nitroso compound to produce a lactam has been used[614]. in the preparation of 1-(*p*-dimethylaminophenyl)-3,3,4,4-tetraphenylazetidin-2-one (equation 301).

D. By Rearrangements

A number of rearrangements have been used to prepare lactams of varying ring size. In this section, preparations of lactams are presented in terms of the type of rearrangement employed.

1. Ring contractions

a. Wolff rearrangement. By far the most common method for effecting lactam syntheses by ring contraction has been the photolytic Wolff rearrangement. Recently this approach has been studied by Lowe and Ridley[615,616] for the generation of β-lactams from diazopyrrolidinediones. Thus, *N*-(*t*-butoxycarbonylacetyl)-*d,l*-alanine ethyl ester and KOBu-*t* in xylene afforded[615,616] a 60% yield of 5-methylpyrrolidine-2,4-dione, which upon treatment with methane sulphonyl azide in triethylamine produced a 95% yield of 3-diazo-5-methylpyrrolidine-2,4-dione. Photolysis of this product in the presence of *t*-butyl carbazate afforded a 36% yield of the *cis*-β-lactam and 55% of the *trans*-β-lactam shown in equation (302). Addition of dibenzyl acetylenedicarboxylate to 3-diazo-5-methylpyrrolidine-

t-$BuO_2CCH_2C(O)NHCH(Me)CO_2Et$ —(1. KOBu-*t*; 2. Xylene)→ 5-methylpyrrolidine-2,4-dione —($MeSO_2N_3$, Et_3N)→ 3-diazo-5-methylpyrrolidine-2,4-dione —($h\nu$ >300 nm, t-BuO_2CNHNH_2)→ [ketene β-lactam intermediate] → t-$BuO_2CHNHNC(O)$-β-lactam (Me, N—H)

cis 36%

+ t-$BuO_2CHNHNC(O)$-β-lactam (Me, N—H)

trans 55%

(302)

2,4-dione (equation 303) affords[616] both the (*E*)- and (*Z*)-dibenzyl (3-diazo-5-methyl-2,4-dioxopyrrolidin-1-yl)fumarates, and irradiation of the (*Z*) adduct for 0.5 h in the presence of *t*-butyl carbazate affords both the (*E*)- and (*Z*)-*trans*-β-lactams, dimethyl [*cis*-3-(3-*t*-butoxycarbonylcarbazoyl)-2-methyl-4-oxoazetidin-1-yl]maleate. By irradiation for 2 h the (*E*)-*trans*-β-lactam is generated exclusively.

3-diazo-5-methylpyrrolidine-2,4-dione + $C_6H_5O_2CCH_2CC{\equiv}C-CH_2CO_2C_6H_5$ —(Na, C_6H_6 N_2, 24 h, r.t.)→

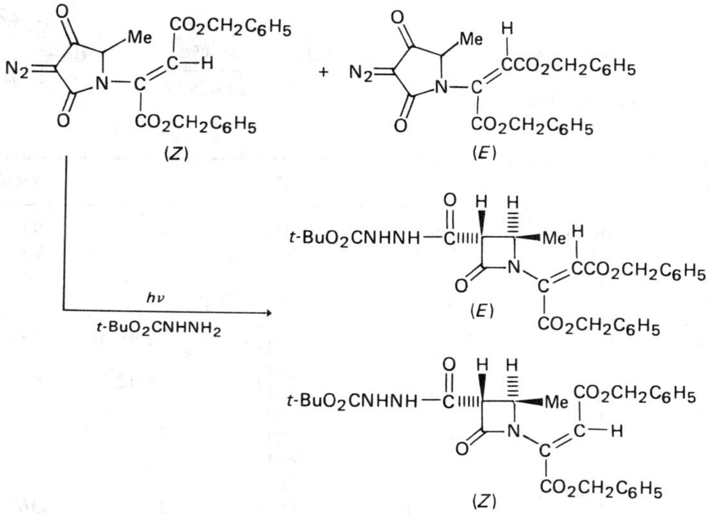

(303)

In a similar manner, irradiation[615,616] of benzyl 6-diazo-5,7-dioxohexahydropyrrolidine-3-carboxylate afforded benzyl 7-oxo-6α-[3-(2-phenyl-2-propyloxycarbonyl)carbazoyl]-5α*H*-1-azabicyclo-[3.2.0]hexane-2α-carboxylate (equation 304).

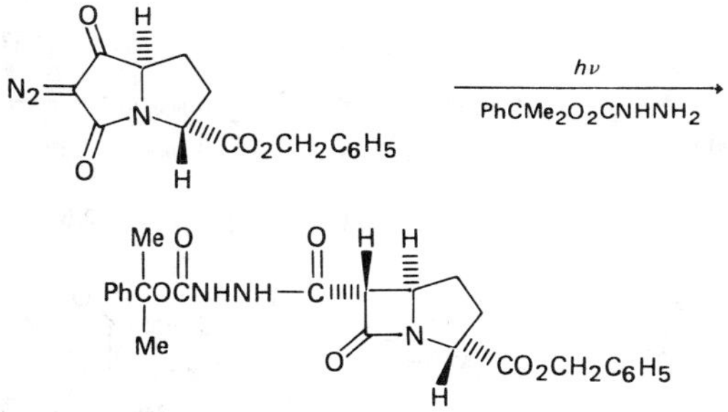

(304)

Using a series of 2,4-pyrrolidinediones ('tetramic acids') as starting materials, Stork and Szajewski[617] demonstrated that the photolytic Wolff rearrangement was a general method for the preparation of carboxy β-lactams (equations 305 and 306).

Ring contraction of 2-acylpyrazolidin-3-ones to afford β-lactams has also been reported[618] to occur upon photolysis (equation 307).

b. Miscellaneous ring contractions. Treatment of substituted α,α-dichlorosuccinimides with sodium methoxide in a variety of solvents has been reported[619] to produce both the corresponding ring-opened α-chloroacrylamide and the β-lactam as products, with the proportion of the two products depending upon the nature of the substituents and the solvent used (Scheme 4). Interestingly, if potassium *t*-butoxide is used as the base, the imide again affords the corresponding α-chloroacrylamide (35%) by the mechanism shown, along with a less substituted β-lactam (47%), which arises through a proposed ketene intermediate (equation 308).

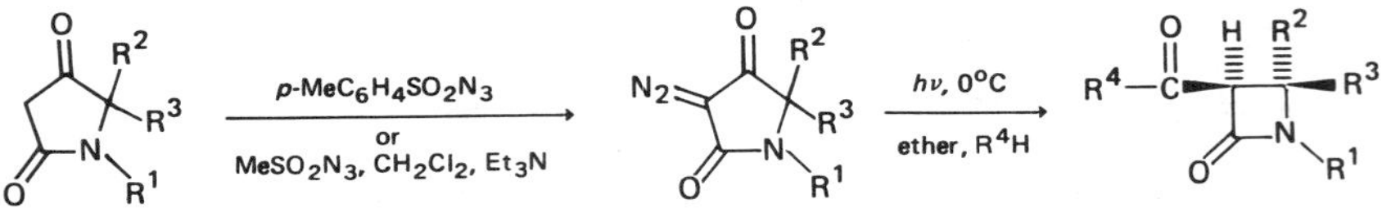

(305)

R^1	R^2	R^3	R^4		Yield (%)
Me	Me	Me	OMe		93
Me	Me	Me	$NHNH_2$		95
Me	Me	Me	OH		98
Me	Me	Me	OCH_2CF_3		–
Me	Me	Et	OMe	56:44	64
Me	Et	Me	OMe		
Me	Me	Et	OH	58:42	80
Me	Et	Me	OH		
Me	Me	n-C_6H_{13}	OMe	3:1	85–93
Me	n-C_6H_{13}	Me	OMe		
Me	Me	n-C_6H_{13}	NH_2	3:1	86
Me	n-C_6H_{13}	Me	NH_2		
Me	Me	n-C_6H_{13}	OH	2:1	96
Me	n-C_6H_{13}	Me	OH	or 5:1	80
Me	Me	n-C_6H_{13}	$NHNH_2$	3.5:1	94
Me	n-C_6H_{13}	Me	$NHNH_2$		
H	i-Bu	H	OMe		47
Me	i-Bu	H	OMe		54

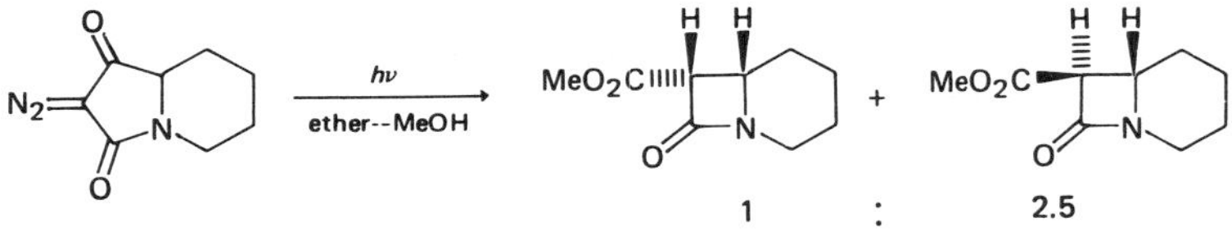

(Ref. 617)

(306)

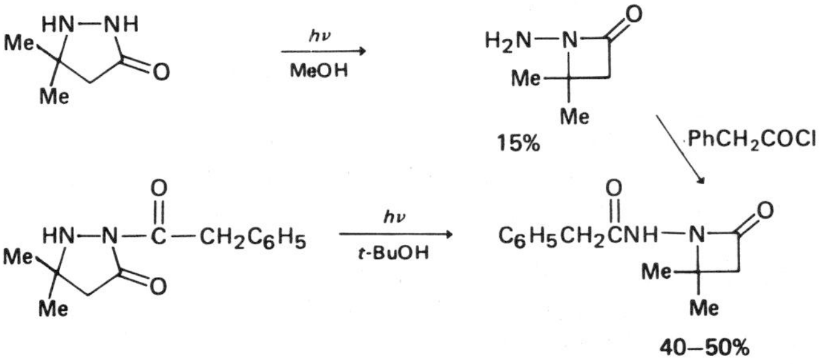

(307)

2. *Ring expansions*

a. Beckmann rearrangement. The Beckman rearrangement[620–622] has found extensive use in the preparation of lactams. This reaction generally involves treatment of the oximes of cyclic ketones with H_2SO_4, PCl_5, HCl–HOAc–Ac_2O or polyphosphoric acid[622] to convert the hydroxyl group of the oxime into a

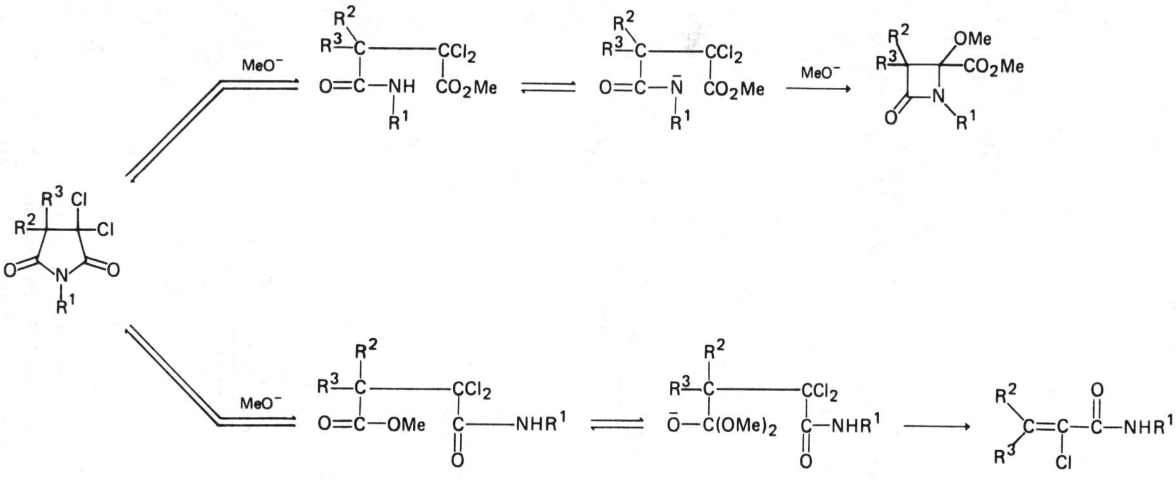

R^1	R^2	R^3	Solvents	Yield (%)	
				Lactam	Amide
Ph	Ph	Me	DMSO	0	100
Ph	Ph	Me	MeOH	0–23	100–77
Ph	Ph	Me	MeOH, DME (50%)	10	90
Ph	Ph	Me	MeOH, dioxane (50%)	25	75
Ph	Ph	Me	MeOH, *t*-BuOH (50%)	25	75
Ph	Ph	Me	HMPA, MeOH (50%)	5	95
Ph	Ph	Ph	MeOH	20	80
Ph	Me	Me	MeOH	100	0
$-(CH_2)_4-$		Ph	MeOH	100	0
$-(CH_2)_4-$		Ph	MeOH, DMSO (50%)	90	–

SCHEME 4.

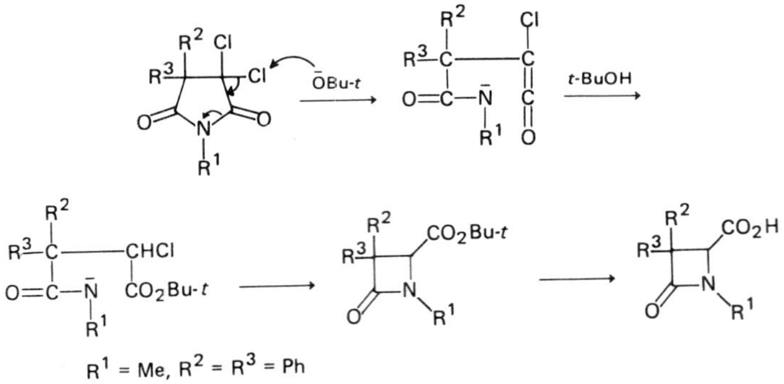

(308)

R^1 = Me, R^2 = R^3 = Ph

better leaving group, followed by rearrangement and tautomerization (equation 309). The group which migrates is normally the one which is *anti* to the hydroxyl

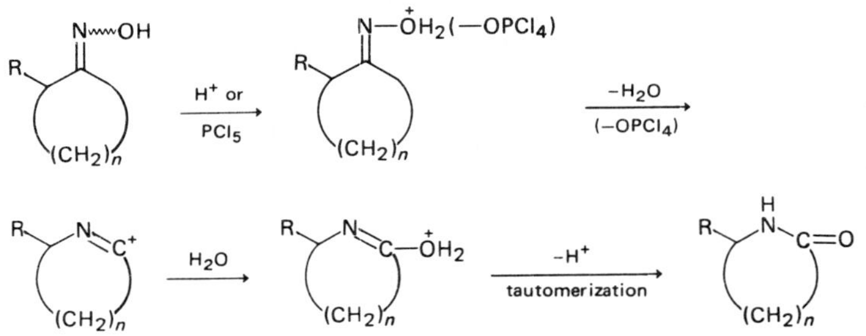

(309)

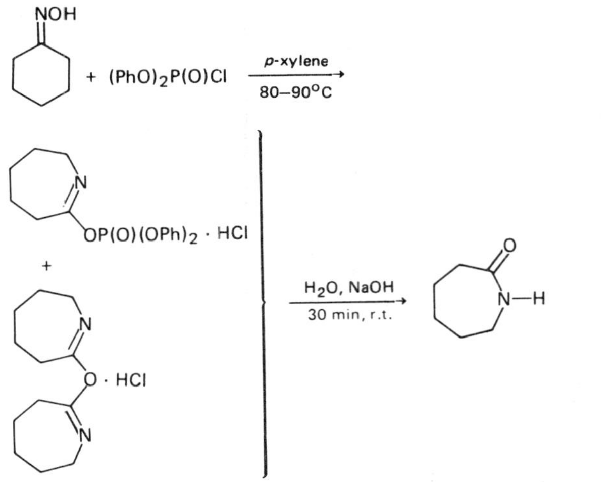

(310)

group in the oxime. Exceptions have been observed; however, these may involve a *syn* to *anti* isomerization prior to rearrangement. A representative series of recent lactam syntheses via the Beckmann rearrangement[623–642] are compiled in Table 35.

TABLE 35. Preparation of lactams via the Beckmann rearrangement

Oxime	Conditions	Product	Yield (%)	Reference
Me =NOSO$_2$ —Me Me	Basic alumina, MeOH	O N—H	81	624
=NOH	190°C, tetralin, 1 h, boric anhydride	O NH	82	627
	1. MeCH, 1 h, 80°C 2. HCl (g)	O NH · HCl	–	628
	AcOH–HCl–MeCN, heat		>90	623
Me =NOSO$_2$ —Me Me	Basic alumina, MeOH	O N—H	77	624
$(CH_2)_{11}$ C=NOH		$(CH_2)_{11}$ N—H O		
	1. MeCN, 1 h, 80°C 2. HCl (g)		–	628
	H_2SO_4 or H_3PO_4, 101–103°C		95	625

TABLE 35. (Continued)

Oxime	Conditions	Product	Yield (%)	Reference
N~OH; *syn* or *anti*		N–H, O		
	PPA, 148°C, 10 min	from *syn* from *anti*	25 0	631
	1. 19% oleum, 140°C, 1 h 2. MeOH–KOH, 0°C	from *syn* from *anti*	59 59	632
N~OH, Me, Me; *syn:anti* mixture	PPA, 130°C	O, Me, N–H, Me	30	633
NOH, Me, Me, Me; *syn*		O, Me, N–H, Me, Me		
	PCl_5, ether PPA, 132–135°C, 10 min PPA		25 21 64	634 636 638

Substrate	Conditions	Product	Yield (%)	Ref.
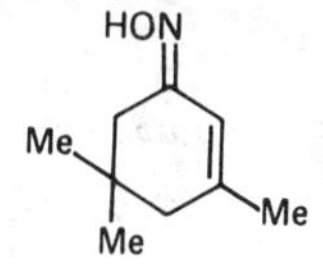 [structure] *anti*	PCl_5, ether PPA, 132–135°C, 10 min	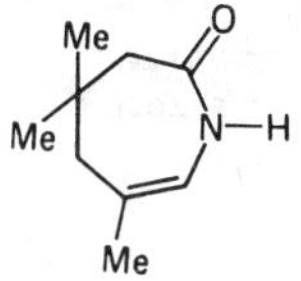 [structure]	20 72	634 636
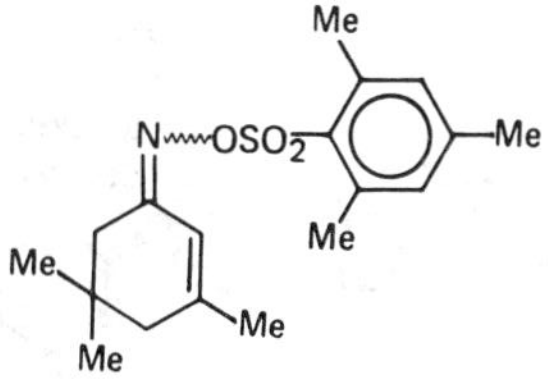 [structure]	Basic alumina, MeOH	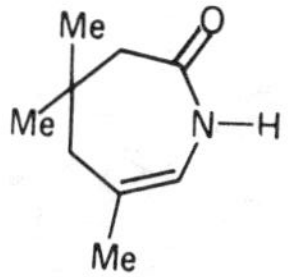 [structure]	50	624
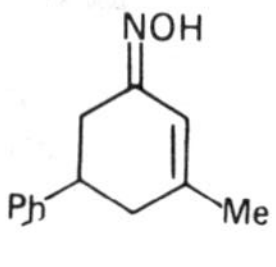 [structure] *syn*	PCl_5, ether–C_6H_6	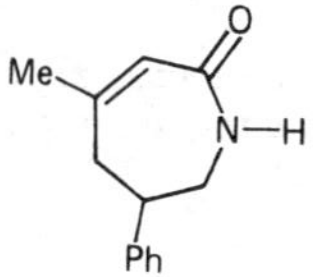 [structure]	25	634
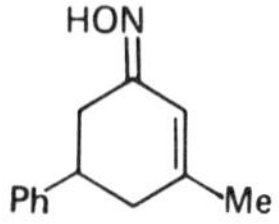 [structure] *anti*	PCl_5, ether–C_6H_6	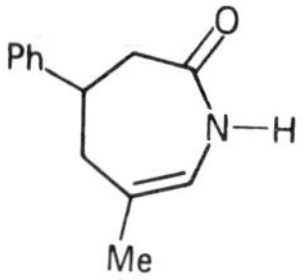 [structure]	15	634
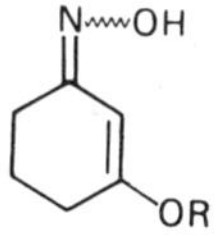 [structure]	PPA, 120°C, 1 h	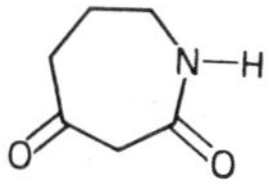 [structure]		
R = Me		from 2:1 *syn:anti* mixture from *syn*	66–75 77	635,637 637
R = Et		from 4:3 *syn:anti* mixture	88–98	635,637

TABLE 35. (Continued)

Oxime	Conditions	Product	Yield (%)	Reference
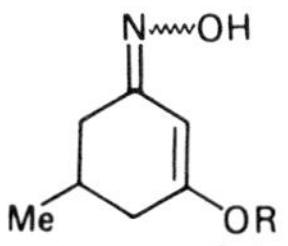 N～OH, Me, OR *syn:anti* = 2:1 R = Me or Et	PPA, 120°C, 1 h	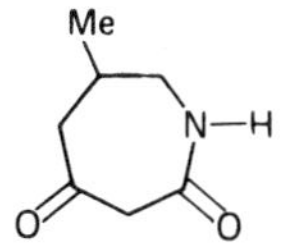 Me, N—H, O, O	78–91	635,637
N～OH, Me, Me, OR R = Me R = Et	PPA, 120°C, 1 h	Me, Me, N—H, O, O from 5:4 *syn:anti* mixture from 3:2 *syn:anti* mixture	 78–79 81–96	635,637
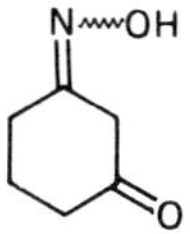 N～OH, O *syn:anti* mixture	PPA, 120°C, 1 h	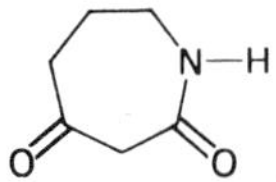 N—H, O, O	30	637
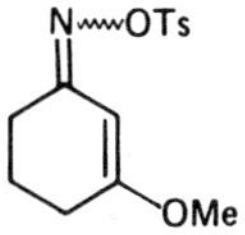 N～OTs, OMe *syn* or *anti*	PPA	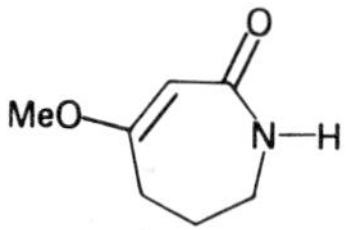 MeO, O, N—H from *syn* from *anti*	 87 0	637

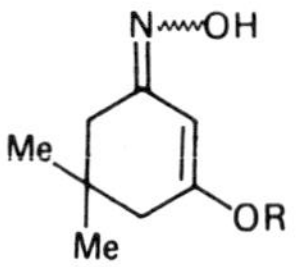

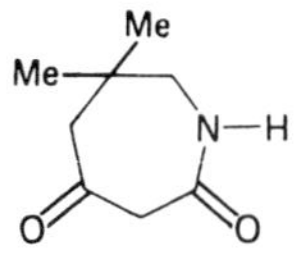

PPA, 120°C

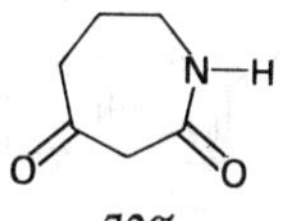

72%

Reagents	Product	Yield (%)	Ref.
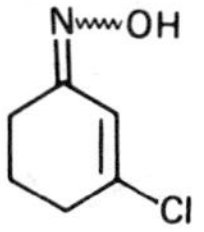 PPA, 120°C	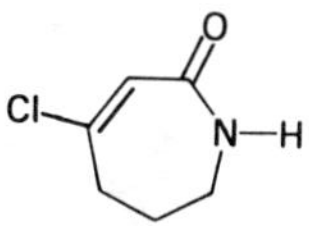		637
	from 4:7 *syn:anti* mixture from *anti*	26 28	
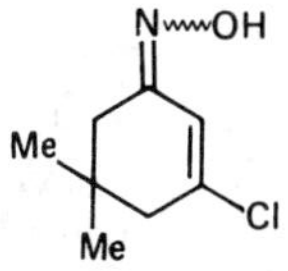 PPA, 120° (*syn:anti* = 4:7)	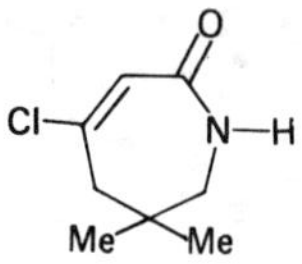	34	637
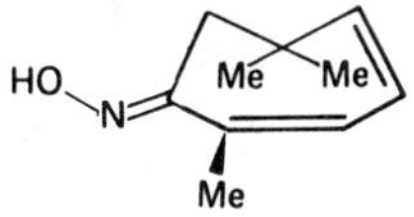 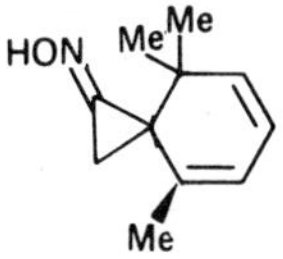 Me_2CO, H_2O, p-$MeC_6H_4SO_2Cl$	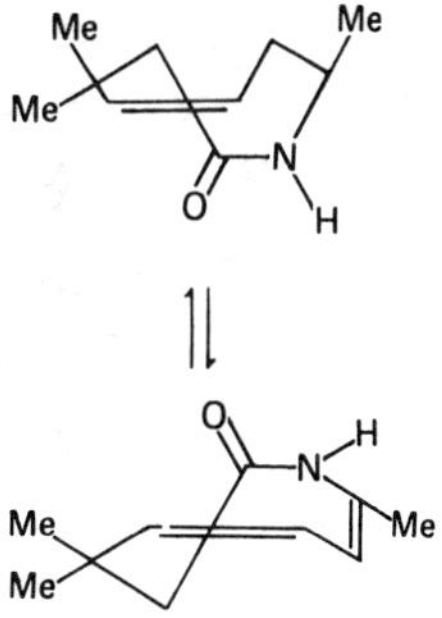	–	639

TABLE 35. (Continued)

Oxime	Conditions	Product	Yield (%)	Reference
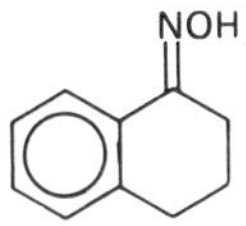 NOH	PPA, 120–130°C, 10 min	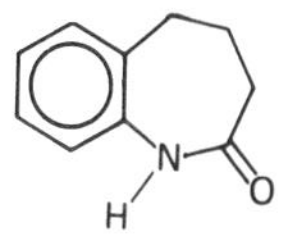 N, H, O	91	633
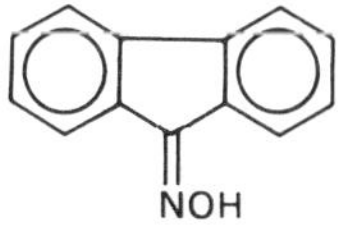 NOH	PPA, 175–180°C, 10 min	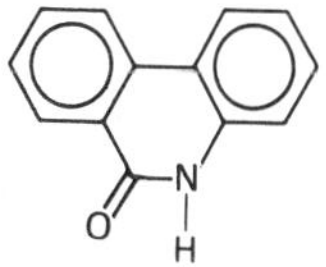 O, N, H	93	633
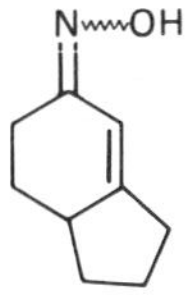 N～OH $Z{:}E = 3{:}2$	PPA, 130–135°C, 20 min	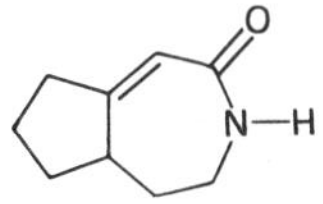 O, N—H	52	640
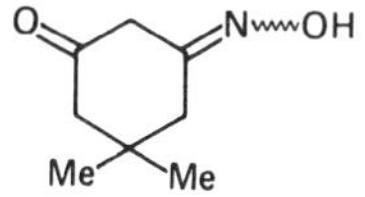 O, N～OH, Me, Me $Z{:}E = 6{:}1$	PPA	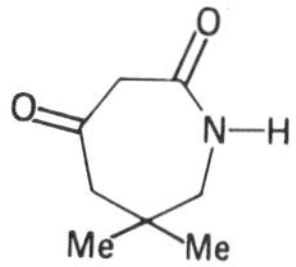 O, O, N—H, Me, Me	75	640
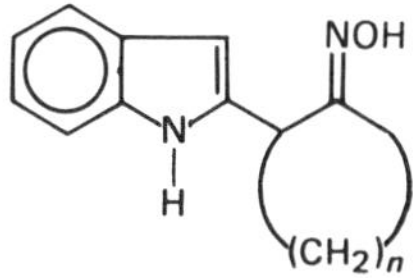 N, H, NOH, $(CH_2)_n$	PPA, H_2SO_2	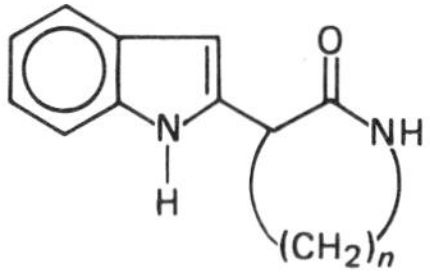 N, H, O, NH, $(CH_2)_n$		641
		$n = 4$	35	
		$n = 5$	40	

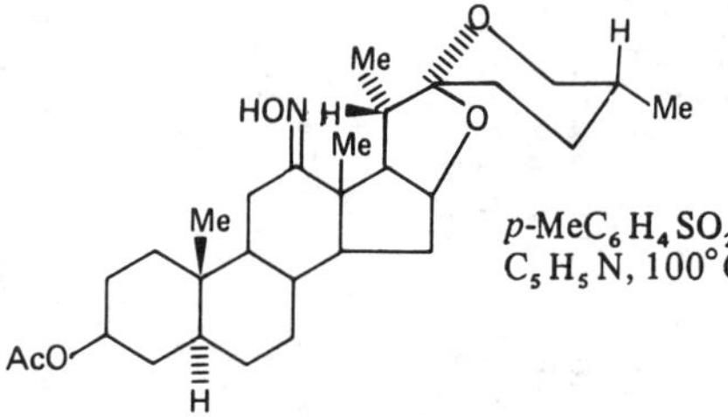	p-$MeC_6H_4SO_2Cl$, C_5H_5N, 100°C, 3 h	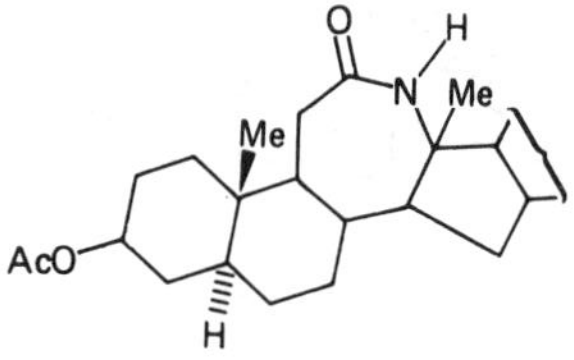	74	642
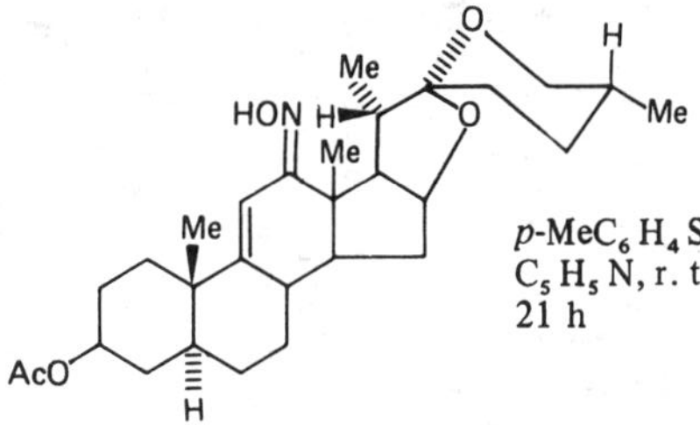	p-$MeC_6H_4SO_2Cl$, C_5H_5N, r. t., 21 h	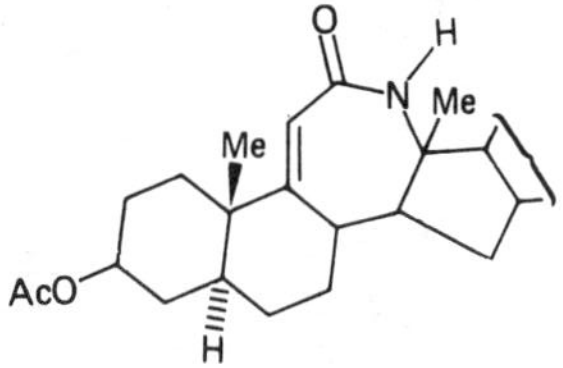	50	642

A lactam synthesis which mechanistically resembles a Beckmann rearrangement involves the reaction of cyclohexanone ketoxime in *p*-xylene with diphenyl chlorophosphite at 80–90°C for 18 hours. This reaction afforded[643] a mixture of the lactim phosphate hydrochloride and bislactim ether hydrochloride shown in equation (310), both of which produced ε-caprolactam upon hydrolysis. Similar preparation of a series of C_4–C_{12} lactams has also been reported[643].

b. Schmidt rearrangement. Among various Schmidt rearrangements[644,645], only the reaction of cyclic ketones with hydrazoic acid gives rise to lactams. The mechanism for this rearrangement is shown in equation (311), and the question of

(311)

which group migrates during the loss of nitrogen in several different systems has produced errors in the literature and many lively published debates. Table 36[640–656] contains a representative sampling of lactams prepared over the last 25 years by means of the Schmidt rearrangement.

These studies indicate that with saturated cyclic ketones possessing an electron-donating substituent (R = Me, Et, *n*-Pr, *i*-Pr, *n*-Bu, CH_2R, $CHMeC_6H_5$) in the 2-position, route (1) is the path observed. This route is also observed with some electron-withdrawing substituents such as CN and CO_2Et. However, when the substituent at position 2 is either Cl or Ph, or if the cyclic ketone contains an α,β-double bond, then route (2) appears to be the preferred reaction path even though mixtures usually result. In the case of cyclic diketones, the azide ion appears to attack preferentially the less hindered, more basic carbonyl function, and this attack is followed by preferential migration of the larger adjacent group[653,654].

Sodium azide in acetone at pH 5.5 (KH_2PO_4–NaOH buffer) has been reported[657] to convert the ethyl hemiketal of cyclopropanone into γ-butyrolactam in 21% yield via the mechanism shown in equation (312). This reaction was

(312)

subsequently extended to the preparation[658] of fused-ring β-lactams from 1,1-disubstituted cyclopropanones in the bicyclo[4.1.0] series (equation 313), and made more general by preparation of the corresponding carbinolamines of the cyclic ketones and then affecting the ring enlargement reaction through the

$\xrightarrow{NaN_3}$ (313)

X = HO, EtO, N(piperidino)

$H_2C{=}C{=}O + CH_2N_2 \xrightarrow[\text{ether}]{-78^\circ C}$ [cyclopropanone] $\xrightarrow{RNH_2}$ HO NHR $\xrightarrow{Me_3COCl}$

HO N(Cl)—R $\xrightarrow[\text{MeCN}]{Ag^+}$ [HO $\overset{+}{N}$—R] $\longrightarrow$ R–N (β-lactam) (314)

R	Yield (%)
Cyclohexyl	61
$Me(CH_2)_3$—	43
$MeCH_2CHMe$	38
Me_2C—	52
MeCHCOOEt	65

cyclopropanone + $PhCOONHCMe_3$ $\longrightarrow$ H—O N(O—CPh(=O))—CMe_3 $\longrightarrow$ N-CMe_3 β-lactam + PhCOOH (315)

nitrenium ion produced from these intermediates (equation 314)[658]. In an effort to extend this method, the same authors[658] investigated leaving groups other than Cl^- and N_2 in the ring enlargement, and found that the *o*-benzoyl derivative of *N*-(*t*-butyl)hydroxylamine reacted directly with cyclopropanone in ether at $-78^\circ C$ to produce *N*-(*t*-butyl)-β-propiolactam in 40% yield (equation 315). It was also found that alkyl hydroxylamines (equation 316)[659] and amino acid esters (equation 317)[660] can be employed in this transformation.

c. Miscellaneous ring expansions. A novel ring expansion reaction for the preparation of γ-lactams involves the carbonylation of cyclopropylamine using rhodium catalysts (equation 318)[661].

An interesting disproportionation rearrangement for the preparation of ε-caprolactam involves heating peroxy amines in the presence of a Group I or II element salt in a non-hydrocarbon organic solvent[662]. Thus, 1,1-peroxydicyclohexylamine afforded a 100% conversion to caprolactam and cyclohexanone (equation 319).

A novel photochemical ring expansion which allows conversion of fused β-lactams to fused bicyclic ring-expanded lactam ethers has also been

TABLE 36. Preparation of lactams via the Schmidt rearrangement

Ketone	Conditions[a]	Product	Yield (%)	Reference
cyclopentanone	A, 3–7°C	2-piperidone (N–H)	80	646
	B, 50°C, 8.5 h		83	647
2-R-cyclopentanone		6-R-2-piperidone (N–H)		
	A, 3–7°C	R = Me, Et, Pr, or *i*-Pr	78–94	646
	B, 50°C, 8.5 h	R = Me or Pr	82–87	647
cyclohexanone	B, 50°C, 8.5 h	caprolactam (N–H)	89	647
2-R-cyclohexanone		7-R-caprolactam (N–H)		
	R = Me	A, 3–7°C	74	646
		B, 50°C, 8.5 h	96	647
		B, 50°C	96	640
	R = Et	A, 3–7°C	84	646
		B, 50°C	95	640
	R = Pr	A, 3–7°C	92	646
		B, 50°C, 8.5 h	95	647
		B, 50°C	95	640
	R = *i*-Pr	B, 50°C	98	640
	R = Bu	B, 50°C	94	640

Substrate		Conditions / Product	Yield (%)	Ref.
	R = CO_2Et	A, 3–7°C	80	646
		B, 50°C, 8.5 h	75	647
	R = $PhCH_2CH_2$	B, 0°, 1.5 h then r.t., 6 h	49	640
	R = p-$MeOC_6H_4CH_2CH_2$		40	640
	R = o-$MeOC_6H_4CH_2CH_2$		42	640
	R = p-$MeC_6H_4CH_2CH_2$		45	640
	R = MeCHPh	B, 0°C, 1.5 h then r.t., 6 h	37	640
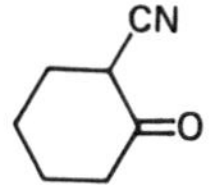 CN, O		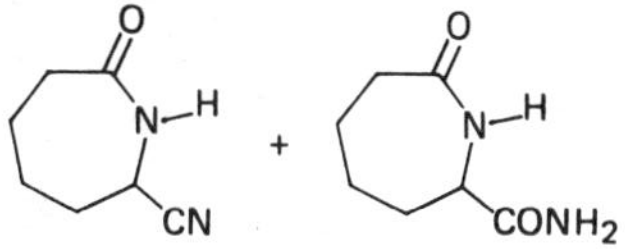 O, N–H, CN + O, N–H, $CONH_2$		
	A, 3–7°C		70 + 0	646
	B, 25°C, 8.5 h		76 + 8	647
	B, 35°C, 8.5 h		54 + 30	647
	B, 45°C, 8.5 h		37 + 45	647
	B, 65°C, 8.5 h		0 + 83	647
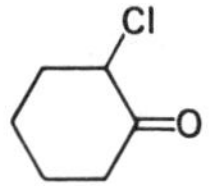 Cl, O		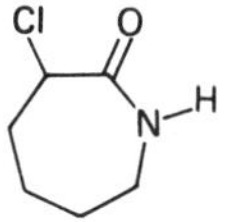 Cl, O, N–H		
	A, 3–7°C		9	646
	HN_3, $CHCl_3$, HCl, 30°C		26	646
	A, EtOH, 40–45°C		31	646
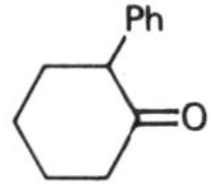 Ph, O		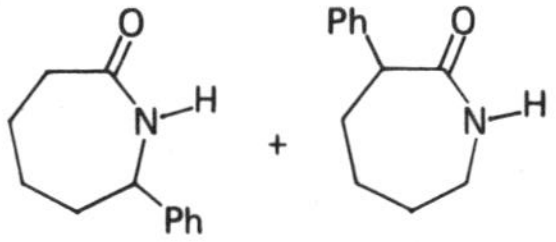 O, N–H, Ph + Ph, O, N–H		
	A, r.t., 90 min		67 + 0	641
	A, r.t., 90 min	(1:2)	60	640
	B, r.t., 18 h	(1:10)	55	640
	B, 55°C	(1:3)	30	640

TABLE 36. (Continued)

Ketone	Conditions[a]	Product	Yield (%)	Reference
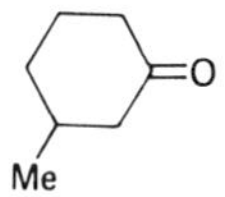	C	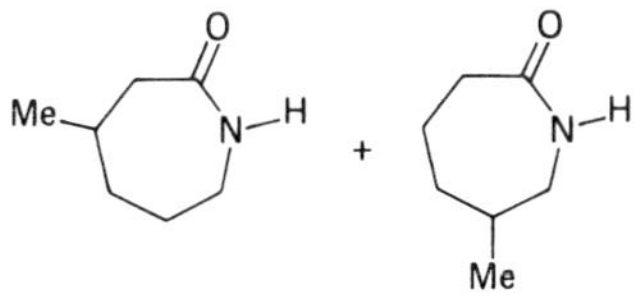 +	–	640
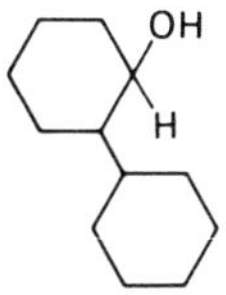	B, r.t.	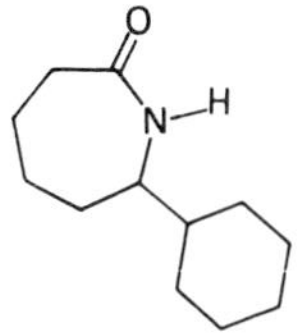	80	640
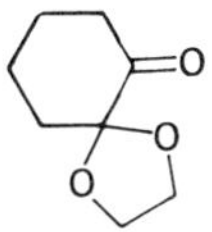	B, 50–55°C, 24 h	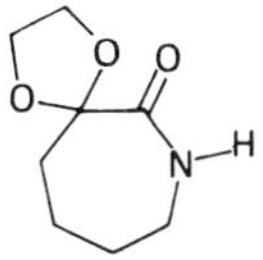	1.6	648
		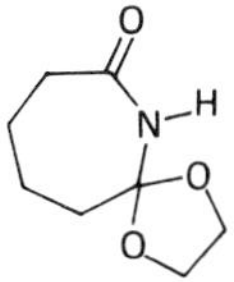 +	1.6	
		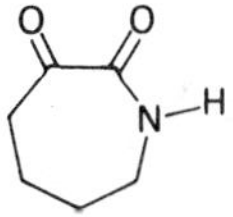 +	2.0	

		+	2.2	
	B, 120°C, 2 h		54	648
	B, 0°C, 1.5 h then 55°C, 3.5 h		30	640
	D, <35°C		39	638
	B, 120°C, 1 h		11 + 15	649
	B, 120°C, 2 h		32 + 36	649

TABLE 36. (Continued)

Ketone	Conditions[a]	Product	Yield (%)	Reference
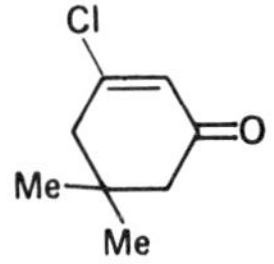 3-Chloro-5,5-dimethylcyclohex-2-enone	NaN_3, MeOH, H_2O reflux, 4 h	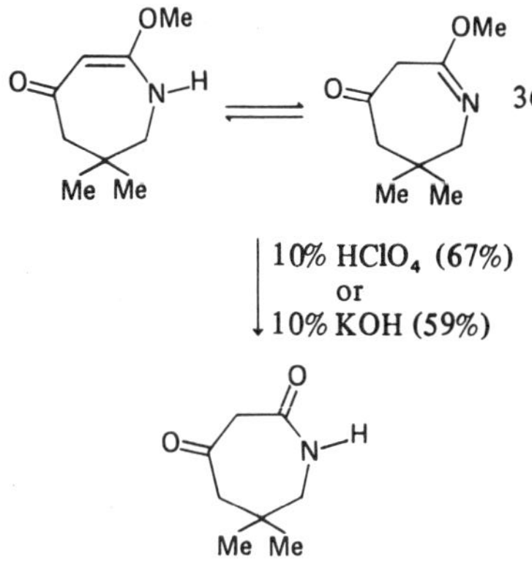 (structures: OMe, N–H, O, Me Me ⇌ OMe, N, O, Me Me); ↓ 10% $HClO_4$ (67%) or 10% KOH (59%)	36	650
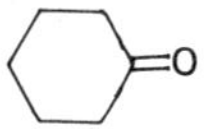 Cyclohexanone	B, 50°C	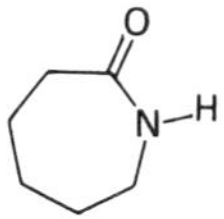 (structure: O, N–H)	83	647
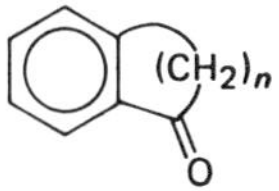 (structure: $(CH_2)_n$, O)	B, 50°C	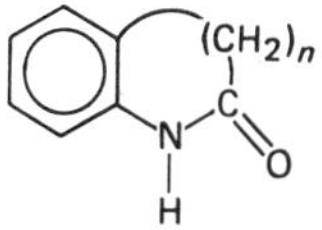 (structure: $(CH_2)_n$, C, O, N–H) $n = 2$	90	647
		$n = 3$	95	647
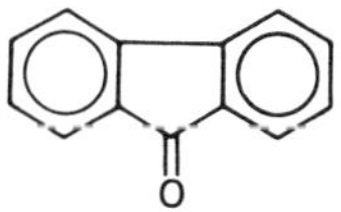 Fluorenone	B, 50°C	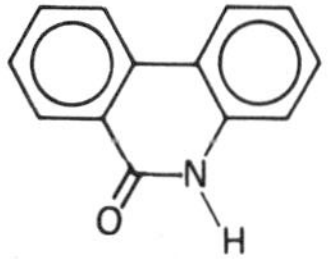 (structure: O, N–H)	92	647

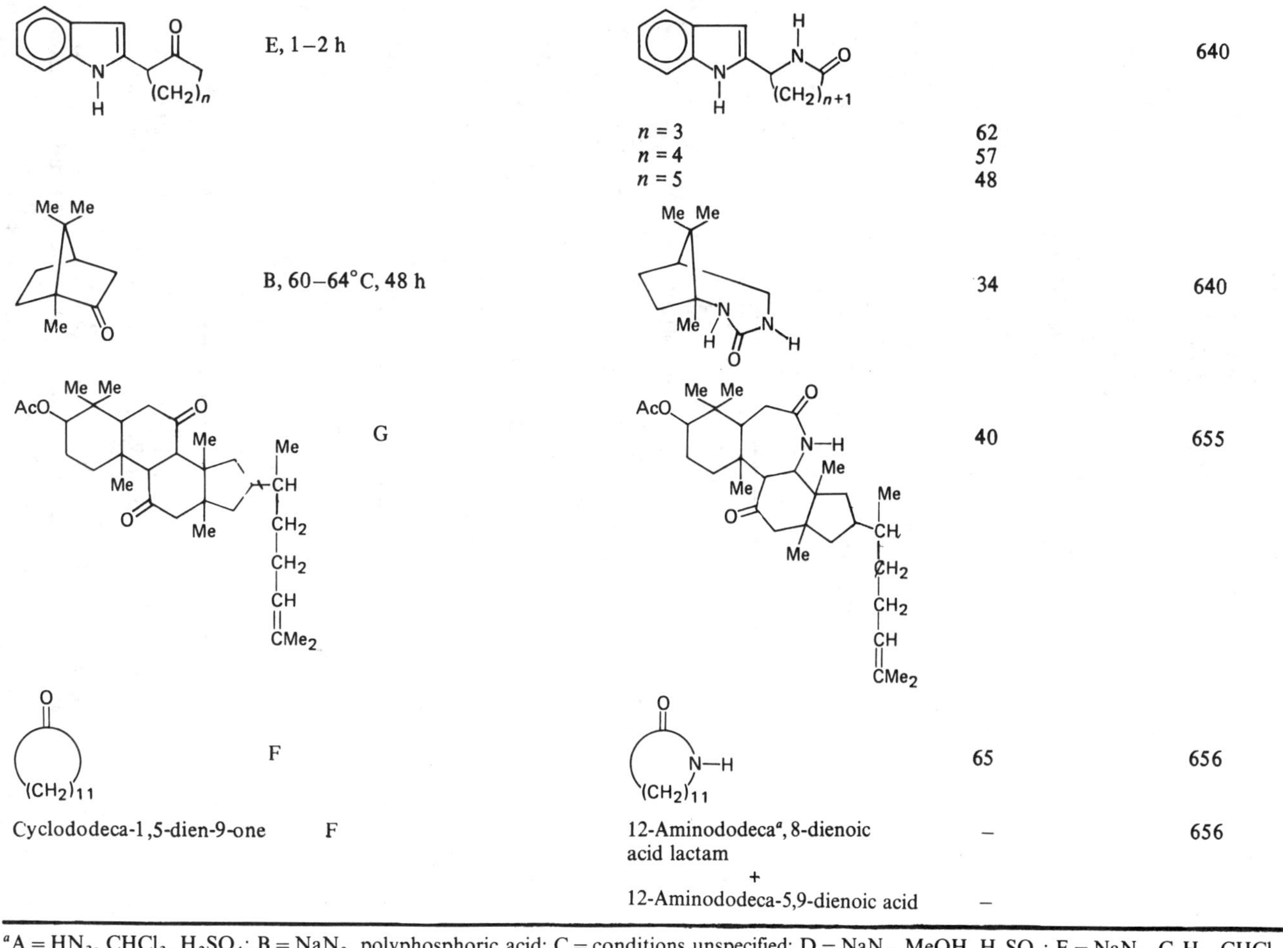

Starting material	Conditions[a]	Product	Yield (%)	Ref.
(indolyl ketone, $(CH_2)_n$)	E, 1–2 h	(lactam, $(CH_2)_{n+1}$)		640
		$n = 3$	62	
		$n = 4$	57	
		$n = 5$	48	
(bicyclic ketone)	B, 60–64°C, 48 h	(bicyclic urea)	34	640
(steroidal diketone)	G	(steroidal lactam)	40	655
(cyclododecanone, $(CH_2)_{11}$)	F	(lactam, $(CH_2)_{11}$)	65	656
Cyclododeca-1,5-dien-9-one	F	12-Aminododeca[a], 8-dienoic acid lactam	–	656
		+ 12-Aminododeca-5,9-dienoic acid	–	

[a] A = HN_3, $CHCl_3$, H_2SO_4; B = NaN_3, polyphosphoric acid; C = conditions unspecified; D = NaN_3, MeOH, H_2SO_4; E = NaN_3, C_6H_6, $CHCl_3$, H_2SO_4; F = NaN_3, H_2SO_4; G = NaN_3, $CHCl_3$, H_2SO_4.

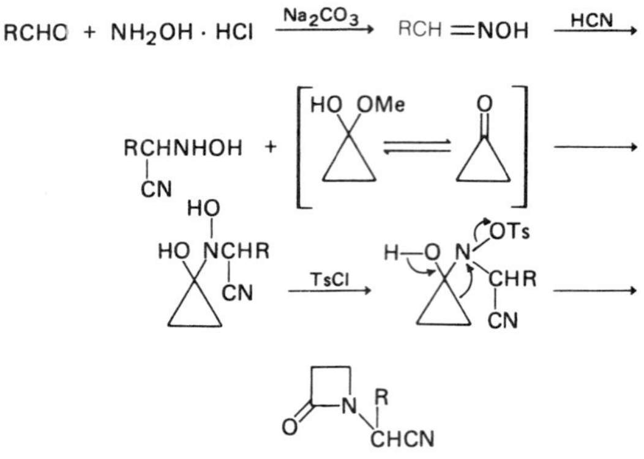

$$RCHO + NH_2OH \cdot HCl \xrightarrow{Na_2CO_3} RCH{=}NOH \xrightarrow{HCN}$$

(316)

R	Yield (%)
$MeCH_2CH_2$—	45
Me_3C—	41
$Me(CH_2)_3$—	40
Me_2CHCH_2—	45

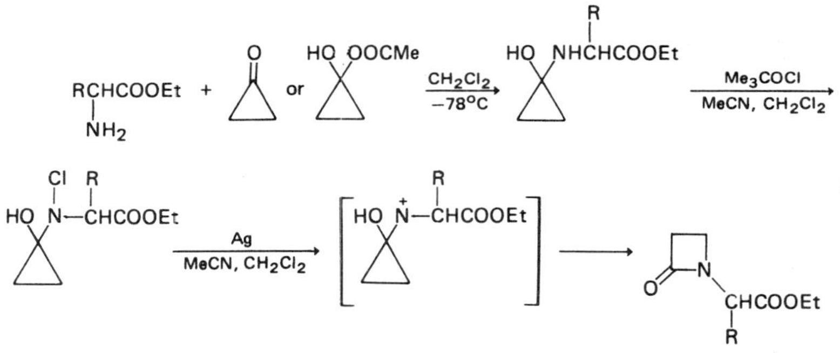

(317)

R	Yield (%)
H	33
Me	47
Me_2CH—	65
Me_2CHCH_2—	65
$PhCH_2$—	70

reported[663,664]. This conversion is limited however, and occurs only when the β-lactam moiety is fused to a bicyclo [2.2.1] system (equations 320–323). A mechanism for this reaction has been proposed[664].

The spirooxiranes, prepared as shown in equation (324), can be anticipated to ring-open in two ways upon irradiation giving rise to two different intermediate diradicals[665,666]. Recombination can be expected to lead to two different products. The regioselectivity and the effect of the solvent upon it, have been

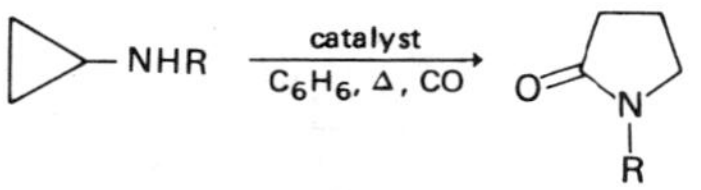

(318)

T (°C)	CO pressure (atm)	Total yield lactam (%)	Composition of mixture (%)			
			R = $(CH_2)_3$	n-Pr	Allyl	H
100[a]	130	10	92	2	5	1
120[a]	150	55	75	19	1	5
140[a]	145	60	62	24	1	12
130[a]	145	22	89	7	2	2
130[a]	150	40	81	16	1	2
130[b]	150	40	28	4	1	67

[a]Catalyst = Rh_6CoO_{16}.
[b]Catalyst = $ClRh(PPh_3)_3$.

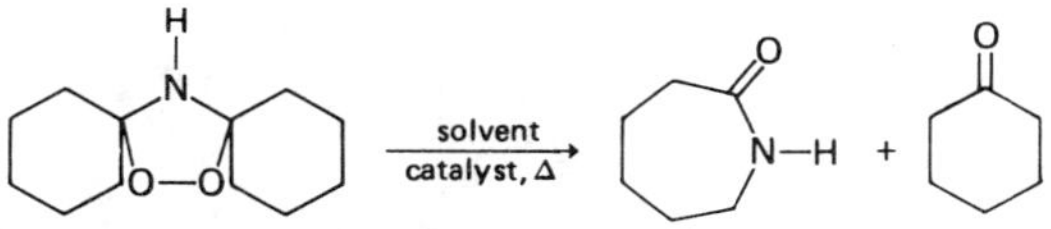

(319)

Solvent = Me_2SO, MeOH, Me_2CO, EtOH, MeCN, or $HO(CH_2)_2OH$

Catalyst = LiBr, $CaCl_2$, NaCl, AgOAc, LiCNS, KF or $SrCl_2$

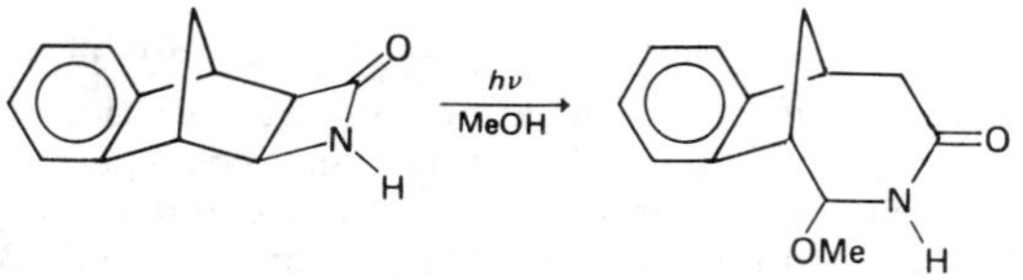

(320)

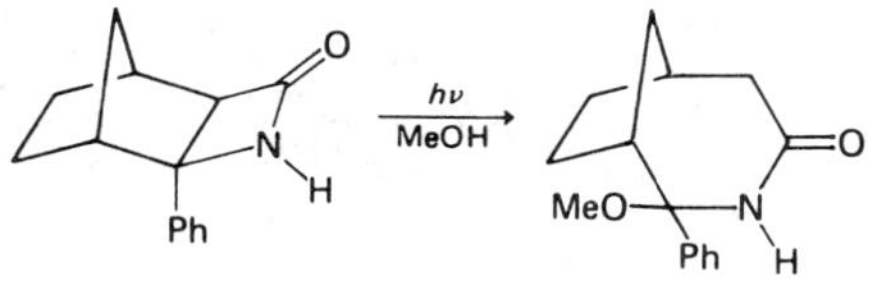

(321)

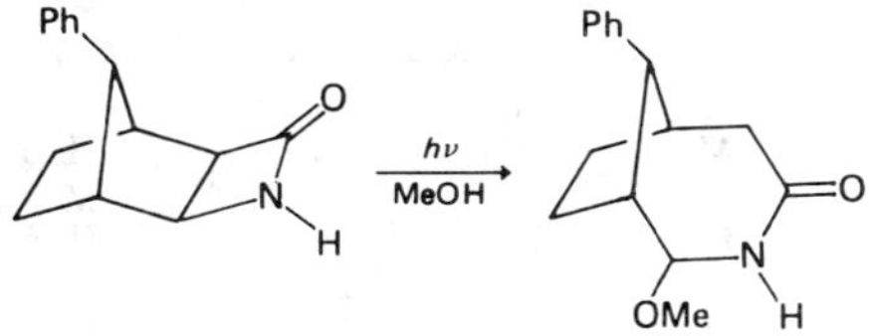

(322)

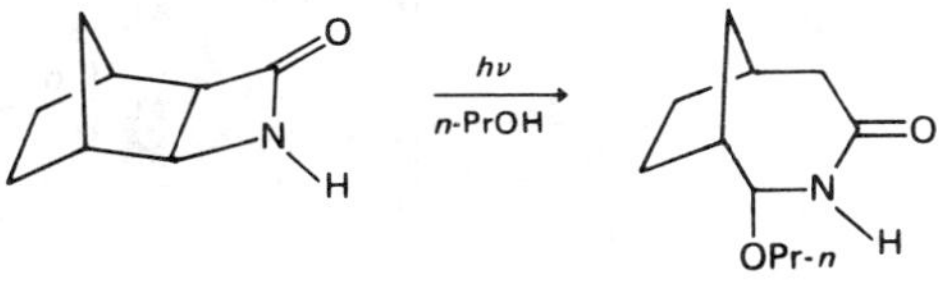

(323)

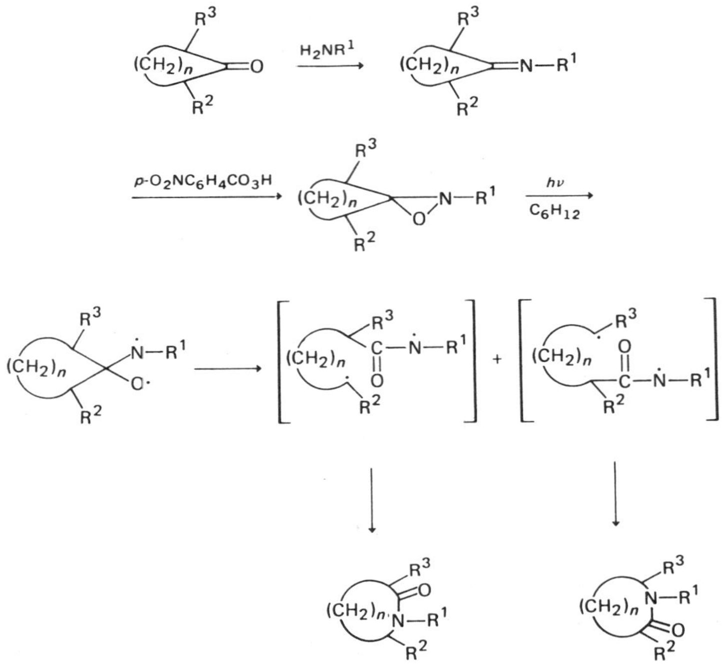

(324)

investigated, and as the results in equations (325)–(329) indicate, a high degree of regioselectivity is observed. Application of this reaction to the synthesis of *N*-phenyl- and *N*-(*p*-chlorophenyl)caprolactam gave poor results, and failed completely in attempts to prepare *N*-(*p*-methoxyphenyl)caprolactam[666]. This reaction can also be performed thermally (equations 330 and 331)[667,668].

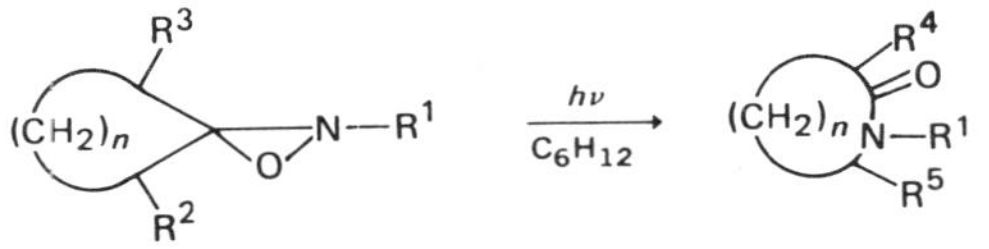

(325)

n	R^1	R^2	R^3	R^4		R^5	Yield (%)
2	$C_6H_5CH_2$	H	H	H		H	75
2	$C_6H_5CH_2$	Me	H	Me H	and	H Me } 95:5	80
3	$C_6H_5CH_2$	H	H	H		H	85
3	C_6H_{13}	H	H	H		H	95
3	Me_2CH	H	H	H		H	85
3	Me	Me	H	Me H	and	H Me } 95:5	80
3	$C_6H_5CH_2$	Me	H	Me H	and	H Me } 95:5	80
3	$C_6H_5CH_2$	Me	Me	Me		Me	50
4	$C_6H_5CH_2$	H	H	H		H	85
5	$C_6H_5CH_2$	H	H	H		H	85
9	$C_6H_5CH_2$	H	H	H		H	85

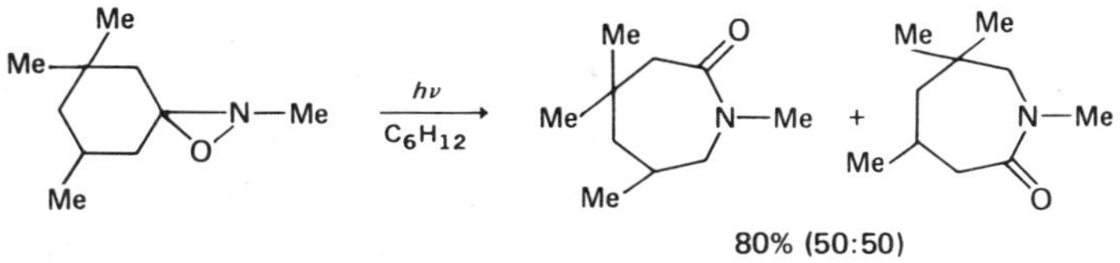

80% (50:50) (Ref. 666) (326)

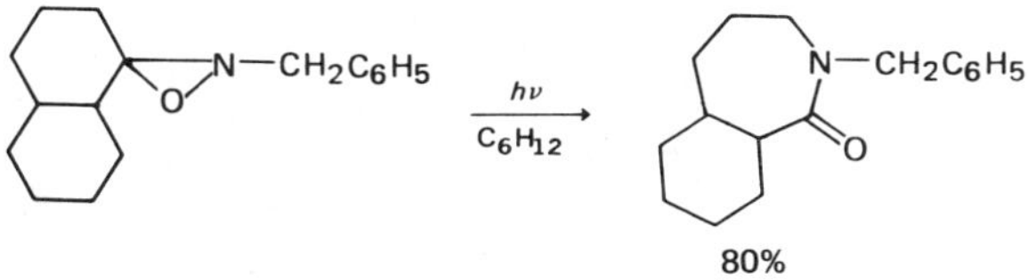

80% (Ref. 665) (327)

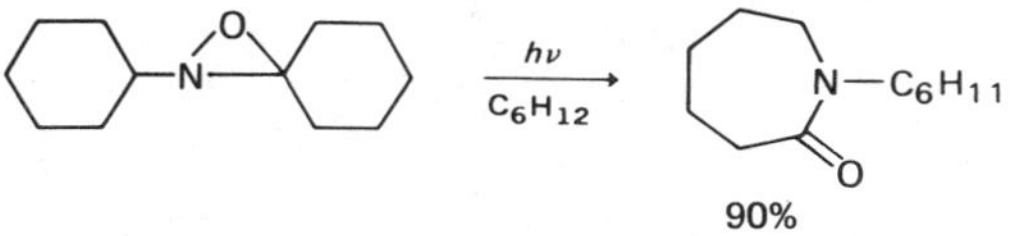

90% (Ref. 648) (328)

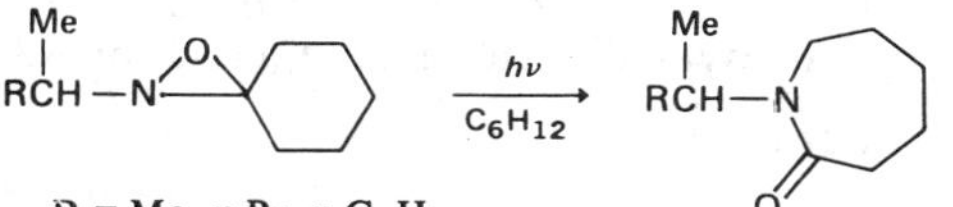

$R = Me, n\text{-}Pr, n\text{-}C_6H_{13}$ (Ref. 648) (329)

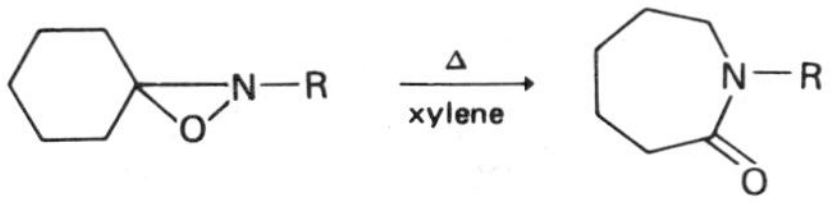

(Ref. 667) (330)

R	Yield (%)
Ph	83
p-ClC_6H_4	90
c-C_6H_{11}	85

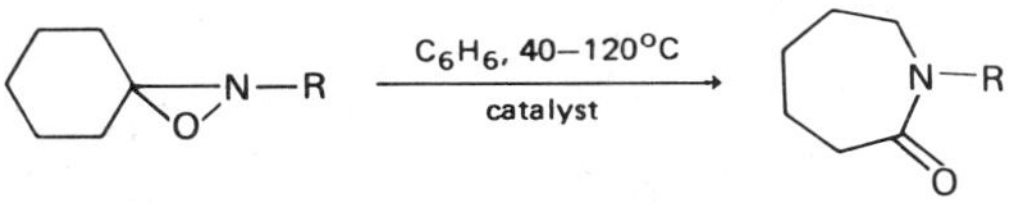

C_6H_6, 40–120°C, catalyst (Ref. 668) (331)

Catalyst = vanadyl bis(acetylacetonate), V_2O_5, P_2O_5, B_2O_3 or MnO_3
R = H, alkyl

The ring expansion of spirooxiranes is also believed to be involved in the production of lactams during irradiation of primary and secondary nitroalkanes in cyclohexane[669]. Thus, irradiation of nitroethane in cyclohexane leads to *N*-ethylcaprolactam presumably via a mechanism which involves intermediate formation of a spirooxirane (equation 332). Other primary and secondary nitroalkanes which have also been found[669] to produce lactams are shown in equation (333).

Treatment of cycloalkanecarboxylic acid with nitrosyl pyrosulphuric acid, prepared[670] as shown in Scheme 5, affords the corresponding ring-expanded

$$EtNO_2 \xrightarrow[C_6H_{12}]{h\nu} Et\dot{N}O_2H + \text{cyclohexyl radical} \longrightarrow$$

$$\text{cyclohexyl}-N^+(O^-)(OH)-Et \xrightarrow{-H_2O} \text{cyclohexylidene}=N^+(O^-)-Et \xrightarrow{h\nu} \qquad (332)$$

$$\text{spiro-oxaziridine (N—Et)} \xrightarrow{h\nu} \text{N-ethyl caprolactam}$$

$$R—NO_2 \xrightarrow[C_6H_{12}]{h\nu} \text{caprolactam}—N—C_6H_{11} + \text{caprolactam}—N—R \qquad (333)$$

R = $C_6H_5CH_2$, $C_6H_5CH_2CH_2$, cyclohexyl, *i*-Pr, *n*-PrCHMe, *n*-C_6H_{13}CHMe

lactam (equation 334)[670]. Similar results were obtained[671] when the same cycloalkanecarboxylic acids were treated with nitrosyl chlorosulphonate (equation 335).

$(NO)_2S_2O_7$ + H_2SO_4 (100%) → $ONOSO_2OSO_2OH$

$NOHSO_4$ + SO_3 → $ONOSO_2OSO_2OH$

$(NO)_2S_2O_7$ + HSO_3Cl + H_2SO_4 (100%) → $ONOSO_2OSO_2OH$

$NOHSO_4$ + HSO_3Cl → $ONOSO_2OSO_2OH$

N_2O_3 + HSO_3Cl + H_2SO_4 (100%) —(SO_3)→ $ONOSO_2OSO_2OH$

$NOHSO_4$ + H_2SO_4 (100%) —(SO_3)→ $ONOSO_2OSO_2OH$

$NOSO_3Cl$ + H_2SO_4 (100%) → $ONOSO_2OSO_2OH$

NOCl + H_2SO_4 (100%) → $ONOSO_2OSO_2OH$

SCHEME 5.

$$(CH_2)_n\,\text{CHCOOH (ring)} + ONOSO_2OSO_2OH \longrightarrow (CH_2)_n \text{ ring with } N—H \text{ and } C{=}O \qquad (334)$$

n	Yield (%)
4	71
6	87
7	81
11	82

Reaction of the aziridine ring with thionyl chloride has also been reported[672] to afford β-lactams via ring expansion. Thus, reaction of the sodium salt of 1-(*t*-butyl)-

$$\mathrm{R(CH_2)_n\,CHCOOH} + \mathrm{ClSO_3NO} \longrightarrow \mathrm{R(CH_2)_n}\,\text{lactam (N—H, C=O)} \qquad (335)$$

R = H; n = 4, 5, 6, 7, 10, 11

R = 4-Me; n = 5

2-aziridincarboxylic acid with either thionyl chloride and sodium hydride or oxalyl chloride affords 1-(*t*-butyl-3-chloro-2-azitidinone (equation 336). Similar reaction of the two isomeric 3-methyl-substituted aziridincarboxylates (equation 337) showed this reaction to be stereospecific and led to the conclusion that the rearrangement involved intermediate formation of a mixed anhydride, which ionized to give a novel bicyclic ion which in turn captured Cl$^-$ to give the final product (equation 338).

(336): sodium 1-*t*-Bu-aziridine-2-carboxylate → ($SOCl_2$, NaH; THF, 25°C) or (ClCOCOCl, C_6H_6 or ClCOCOCl, C_6H_6, Et_3N) → 3-chloro-1-(Bu-*t*)-2-azetidinone

(337): aziridine carboxylate (R^1, R^2, H; N–*t*-Bu; C—O$^-$ Na$^+$) → ($(COCl)_2$, C_6H_6) → 3-Cl, 3-H, 4-R^1, 4-R^2 azetidinone, N–Bu-*t*

R^1	R^2	Yield (%)
Me	H	79
H	Me	63

(338): aziridine carboxylate → (XCl) → mixed anhydride (N:, C(=O)X, H, R^1, R^2, *t*-Bu) → bicyclic ion (N$^+$–*t*-Bu, C=O, H, R^1, R^2) + Cl$^-$ → 3-Cl, H, R^1, R^2 azetidinone, N–Bu-*t*

3. Claisen rearrangement

Thermal treatment of the allyl imidates, 7-allyloxy-3,4,5,6-tetrahydro-2*H*-azepine (**48**) and (2′*E*)-7-(3′-phenylallyloxy)-3,4,5,6-tetrahydro-2*H*-azepine (**49**). prepared by extended heating of the methyl imidate 7-methoxy-3,4,5,6-tetrahydro-2*H*-azepine, with excess allyl and cinnamyl alcohol, respectively (equation 339),

RCH=CH—CH$_2$OH, 140°C, 18 h (339)

R = H (**48**), 49%

R = Ph (**49**), 61%

affords in both cases the *N*-allyl and *C*-allyl lactams via a sigmatropic Claisen rearrangement[673]. Thus, heating **48** afforded two products, the *C*-allyl lactam (3-allylhexahydro-2*H*-azepin-2-one) (**50**) and the *N*-allyl lactam (1-allylhexahydro-2*H*-azepin-2-one (**51**). The *O,N*-ketene acetal shown in equation (340) was postulated as the intermediate, and its formation the rate-determining step, in the

(**48**) heat → [intermediate] → 60% (**50**) (340)

(**51**) 30%

TABLE 37. Effect of temperature on the product distribution for the thermal rearrangement of allyl (**48**) and cinnamyl (**49**) imidates

	Yield (%)			
	Products for **48**		Products for **49**	
T (°C)	**50**	**51**	**52**	**53**
197–199	32	68	–	–
202.5–205	–	–	95	5
211–213	76	24	–	–
212–214	–	–	78	22
222.5–224.5	–	–	36	64
234–236	69	31	–	–

(341)

(49) (52) (53)

preparation of **50**. This view was supported by the observation that the yield of **50** was greatly increased by the presence of the bifunctional catalyst 2-pyridone[673]. Similar thermal treatment of the cinnamyl imidate **49**, afforded the *C*-allyl lactam [3-(1′-phenylallyl)hexahydro-2*H*-azepin-2-one] (**52**) and the *N*-propenyl lactam [(*E*)-1-(1′-phenylpropenyl)hexahydro-2*H*-azepin-2-one (**53**). The effect of temperature on the product distribution for the rearrangement of the allyl and cinnamyl imidates was also investigated and the results are given in Table 37[673].

E. By Direct Functionalization of Preformed Lactams

Generation of a carbanion centre adjacent to the lactam carbonyl provides a convenient method for structure elaboration. Gassman and Fox reported[674] that 1-methyl-2-pyrrolidone could be alkylated to afford a series of 3-substituted-1-methyl-2-pyrrolidones (equation 342). Using two molecular equivalents of sodium amide and of methyl iodide afforded 1,3,3-trimethyl-2-pyrrolidone in 45% yield.

(342)

R = $CH_3O(CH_2)_3$, CH_3CH_2, $(EtO)_2CHCH_2CH_2$

(343)

R^1	R^2	R^3	E	Yield (%)
Me	$H_2C{=}CH$	H	Me_2COH	80
Me	$H_2C{=}CH$	H	Me	59
Me	$H_2C{=}CH$	H	PhCO	28
Me	$H_2C{=}CH$	Me	Me_2COH	77
Me	$H_2C{=}CH$	Me	Ph_2COH	75
Me	Ph	H	Ph_2COH	50
Ph	Ph	H	1-hydroxycyclohexyl	41
Ph	Ph	H	Me_2COH	58
Ph	Ph	H	I	29
Ph	Ph	H	PhCO	61

A similar carbanion approach[675] to the synthesis of 3-substituted β-lactams consists of treatment of *N*-alkyl and *N*-aryl β-lactams with lithium diisopropylamide (LDA) in THF at –78°C to generate the lithio salt, which can then react with various electrophiles (equation 343).

A later study[676] revealed that β-lactams having no substituents at the 1- and 3-position can be converted into 1,3-dilithio salts by means of *n*-butyllithium in THF at 0°C. These salts react regiospecifically with electrophilic reagents to give 3-substituted β-lactams (equation 344).

(344)

R^1	R^2	E	Yield (%)
Ph	H	Me_2COH	55
Ph	H	Me	53
Ph	H	*n*-Bu	66
Ph	H	I	16
$H_2C{=}CH$	H	Ph_2COH	88
$H_2C{=}CH$	H	(1-hydroxycyclohexyl) OH	55
$H_2C{=}CH$	H	*n*-Bu	77
$H_2C{=}CH$	H	*i*-Pr	45
$H_2C{=}CH$	Me	Ph_2COH	57
Et	H	Ph_2COH	65
Et	H	*n*-Bu	65

It has been reported[640,648] that attempts to alkylate caprolactam through the dianion intermediate have given a mixture of 1,3-dialkyl and 1-alkyl derivatives. However, lithiation of caprolactim methyl ether with LDA followed by alkylation and hydrolysis of the resulting 3-alkyllactim ether (equation 345) affords a useful alternative[648,677,678] to the dianion method.

1. LDA; 2. RX; 3. hydrolysis (Ref. 648) (345)

R = Me, Et, *n*-Pr

In view of the pharmaceutical importance of penicillin and cephalosporin antibiotics, it is not surprising that carbanions have been investigated as intermediates for substitution at the position adjacent to the β-lactam carbonyl[679]. Among the more successful approaches to the type of functionalization are those involving generation and reactions of carbanions derived from penicillins and cephalosporins containing a 6- or 7-*N*-arylidene group, which prevents β-elimination of the thiolate ion derived from the fused thiazolidine and dihydrothiazine rings during carbanion formation. The examples given in equations (346)–(349) are typical of this synthetic strategy in the penicillin series.

A new related synthesis of β-lactams[683], involves oxidative decarboxylation of

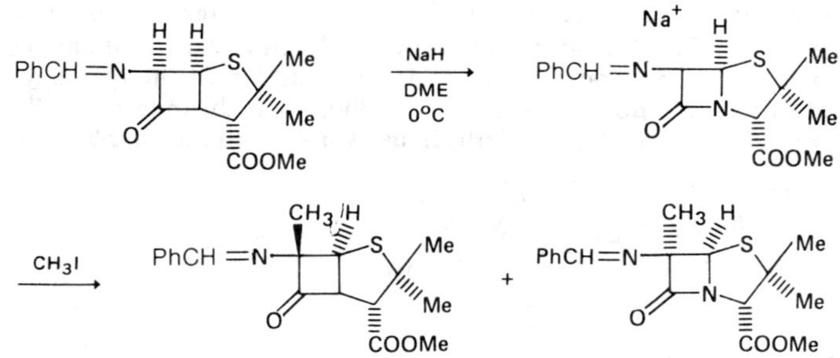

NaH, DME, 0°C; CH_3I (Ref. 680) (346)

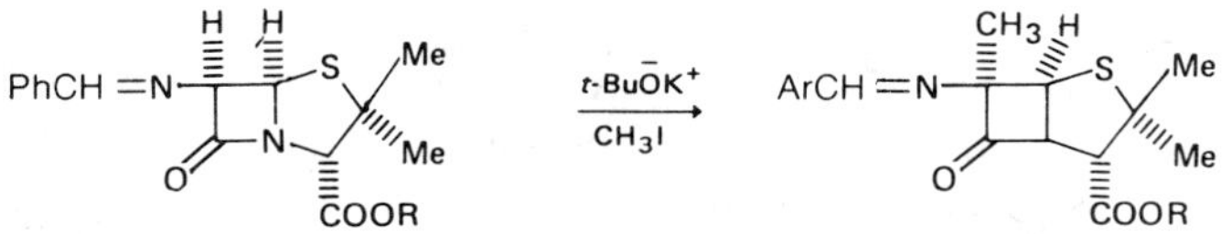

t-BuOK$^+$, CH_3I (Ref. 680) (347)

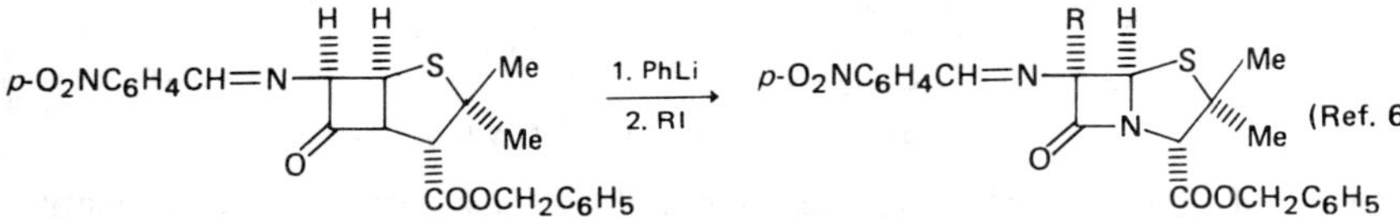

1. PhLi, 2. RI (Ref. 681) (348)

R = Me, Et

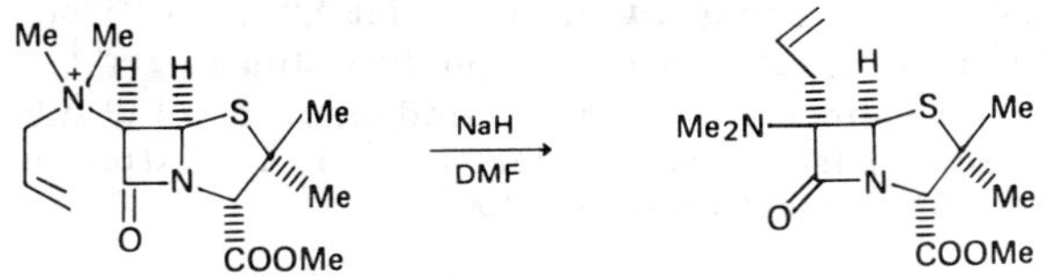

NaH, DMF (Ref. 682) (349)

azetidine 2-carboxylic acids. Oxygenation of the dianion formed from the appropriate acid and two equivalents of LDA in THF and subsequent acidification of the dilithium salt of the resulting hydroperoxy acid leads to decarboxylation and formation of the desired lactam (equation 350).

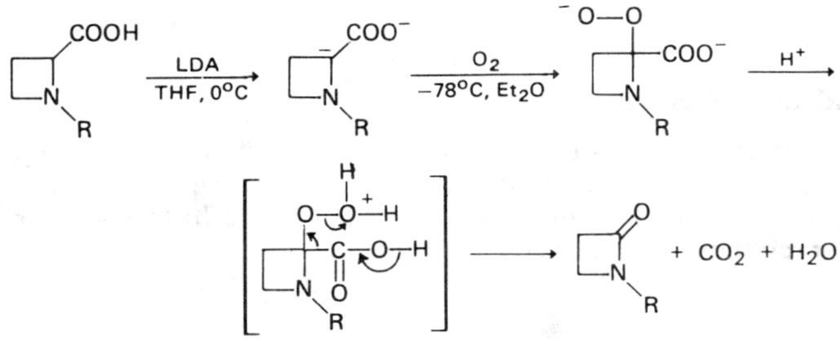

LDA, THF, 0°C; O_2, −78°C, Et_2O; H^+ → + CO_2 + H_2O (350)

R = CMe_3, n-C_5H_{11}, $C_6H_5CH_2CH_2$, cyclohexyl, cyclooctyl, $CH_2CH(OCH_3)_2$

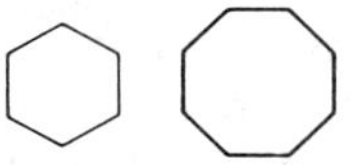

An interesting route[684] to *N*-(2-arylethyl) lactams containing 5-, 6- and 7-membered rings consists of initial reaction of *O*-methyl lactims with a phenacyl halide to form *N*-phenacyl lactams. Sodium borohydride reduction of the phenacyl carbonyl group affords the corresponding benzylic alcohols, which undergo facile hydrogenolysis to give the desired *N*-(2-arylethyl) derivatives (equation 351). It

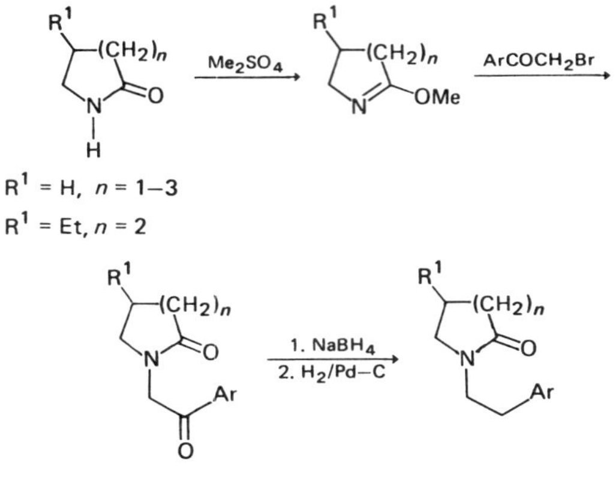

R^1 = H, n = 1–3

R^1 = Et, n = 2

(351)

Ar = Ph and $C_6H_3(OMe)_2$-3, 4

should be noted that this rather elaborate method of *N*-alkylation is not necessary with halides that do not undergo facile β-elimination. More routine procedures include reaction of lactams with alkyl halides and sulphates in the presence of sodium hydride[685], reactions with expoxides[686], acetylenes[686] and aldehydes[687], and by thermal rearrangement of allylic lactim ethers[648,688].

A potentially general method[640] for the introduction of alkyl substituents at the 4-position of caprolactam involves reaction of a mixture of Δ^2- and Δ^3-caprolactam with triethylborane (equation 352).

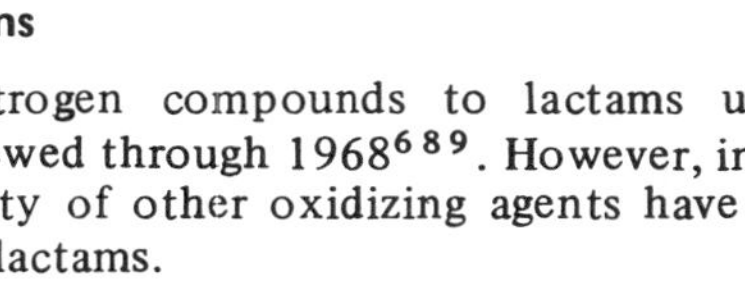

(352)

F. By Oxidation Reactions

The oxidation of nitrogen compounds to lactams using transition metal compounds has been reviewed through 1968[689]. However, in addition to transition metal compounds a variety of other oxidizing agents have been used to convert nitrogen compounds into lactams.

1. *Using halogen*

The use of bromine under acid conditions to effect the oxidation of nicotine has been studied since 1892[690]. In the original work[690–692] it was reported that treatment of nicotine with bromine in the presence of hydrogen bromide (equation 353) resulted in oxidative bromination of nicotine affording two products identified as dibromocotinine (**54**) and dibromoticonine (**55**).

(353)

(54) (55)

Reinvestigation of the structure of **54** using n.m.r.[693] and mass spectra[694] led to the disclosure that it should be represented as compound **56**. A more recent

(56)

study[695] on the structure of **55** using chemical and spectral (including ^{13}C-n.m.r.) techniques has established its correct structure to be 3,4-dibromo-5-hydroxy-1-methyl-5-(3-pyridyl)-Δ^3-pyrolin-2-one (**57**).

aqueous 20% HBr, Br_2, 130°C, 5 min → Dibromoticonine hydrobromide → 1. 10% NaOH 2. HOAc (354)

(57)

Bromine under basic conditions has been used[696] to oxidize cyclic tertiary amines into lactams (equations 355 and 356). This conversion may be done directly using excess bromine, or via the intermediate formation of the iminium salts, which are easily isolated and convertible into the lactam upon further treatment with additional bromine.

21-Nor-5α-conanine —Br_2 (2 moles), CH_2Cl_2, NaOH, r.t., stir 3 h→ (80%) 21-Nor-5α-conanine-20-one

(355)

21-Nor-5α-conanine —Br_2 (1 mole), CH_2Cl_2, Na_2CO_3, r.t., stir 2 h→ 21-Nor-5α-coneninium-20(*N*) bromide (100%) —Br_2 (1 mole), CH_2Cl_2, NaOH, 80°C, 4 h→ 21-Nor-5α-conanine-20-one (80%)

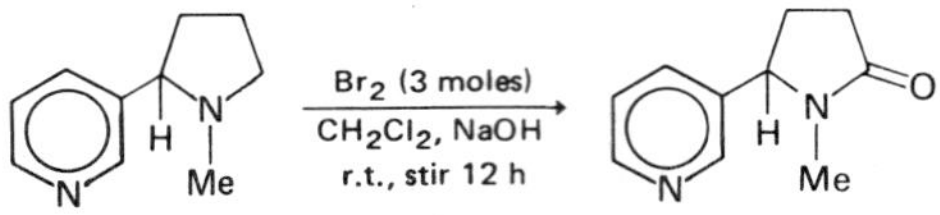

Br_2 (3 moles), CH_2Cl_2, NaOH, r.t., stir 12 h (356)

Cotinene

This reaction sequence may also be performed using *N*-bromosuccinimide[695], but slightly different results are obtained if the intermediate iminium salt is further treated with aqueous sodium hydroxide without bromine (equation 357)[695].

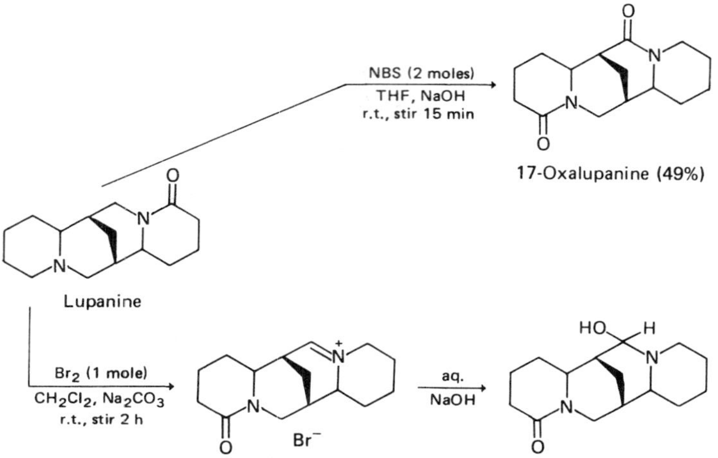

NBS (2 moles), THF, NaOH, r.t., stir 15 min

17-Oxalupanine (49%)

(357)

Lupanine

Br_2 (1 mole), CH_2Cl_2, Na_2CO_3, r.t., stir 2 h

Br^-

aq. NaOH

Δ-16-Dihydrolupaninium bromide

17-Hydroxylupanine

Basic solutions of iodine in tetrahydrofuran have been reported to convert ibogamine[697], ibogaine[697] and voacangine[699] to their respective lactams (equation 358). Voacangine lactam has also been prepared[698] by the basic iodine oxidation of dihydrovoacamine followed by acid cleavage of the resulting product.

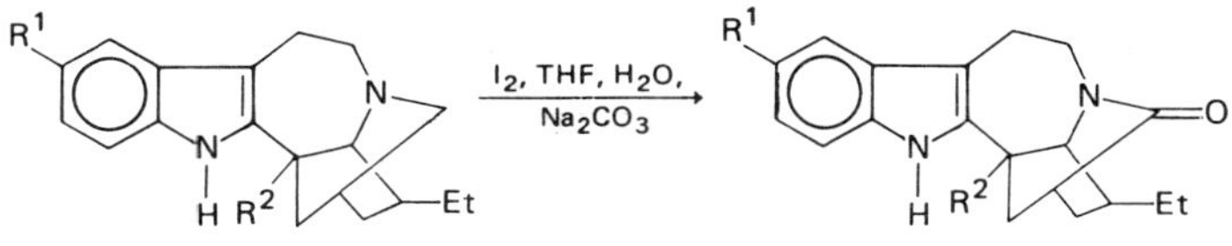

I_2, THF, H_2O, Na_2CO_3 (358)

Ibogamine ($R^1 = R^2 = H$)	Ibogamine lactam
Ibogaine (R^1 = MeO, R^2 = H)	Ibogaine lactam (89%)
Voacangine (R^1 = MeO, R^2 = COOMe)	Voacangine lactam (10%)

2. *Using chromium or osmium oxides*

In addition to the use of basic iodine to convert ibogamine and ibogaine to their respective lactams, chromium trioxide in pyridine has also been used[697]. This reagent has also been used to effect the conversion of iboquine (equation 359)[697], iboluteine[697], conanine[700], 3-oxoconanine[700], kopsine and both epimers of

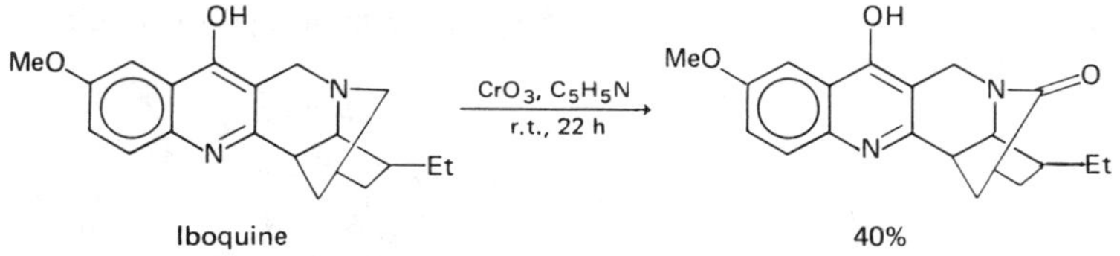

OH
MeO
N
N
Et
CrO_3, C_5H_5N
r.t., 22 h
OH
MeO
N
O
N
Et
Iboquine
40%

(359)

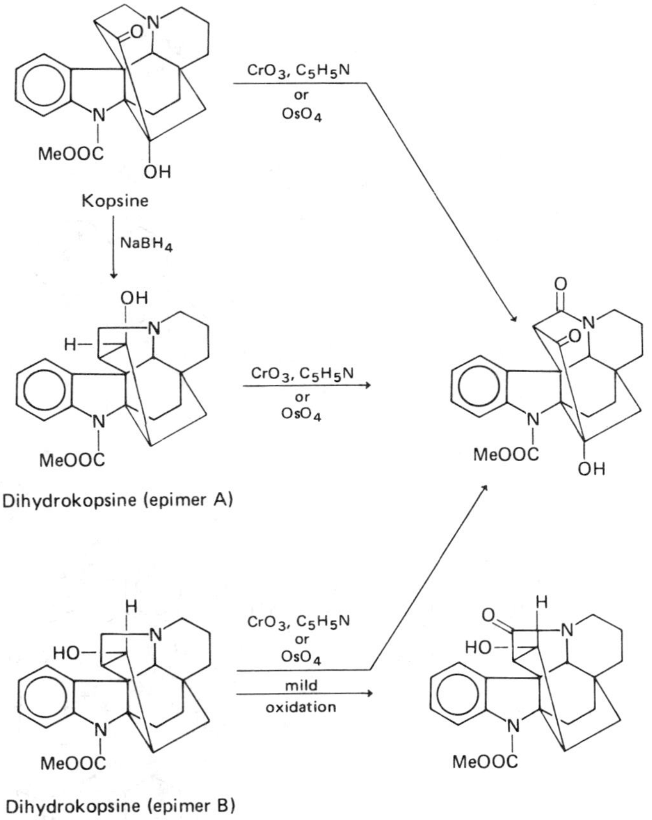

O
N
N
MeOOC
OH
Kopsine
CrO_3, C_5H_5N
or
OsO_4
$NaBH_4$
OH
H
N
N
MeOOC
Dihydrokopsine (epimer A)
CrO_3, C_5H_5N
or
OsO_4
O
N
O
N
MeOOC
OH
H
HO
N
N
MeOOC
Dihydrokopsine (epimer B)
CrO_3, C_5H_5N
or
OsO_4
mild
oxidation
O
H
HO
N
N
MeOOC

(360)

dihydrokopsine (equation 360)[701] into their respective lactams. These latter conversions have also been accomplished using osmium tetroxide[701].

3. *Using manganese oxides*

Manganese dioxide in acetone has been used to oxidize 4-(3,4-dimethoxyphenacetyl)- and 4-benzoyl-2-methyl-1,2-dihydroisoquinoline to 4-(3,4-dimethoxyphenacetyl)-2-methylisocarbostyril and 4-benzoyl-2-methylisocarbostyril, respectively (equation 361)[702], while acetone solutions of potassium permanganate have been used to oxidize *dl*-lupanine to *dl*-oxylupanine[703], *d*-lupanine and 17-hydroxylupanine to *d*-oxylupanine (equation 362)[704] and *N*-formyldihydrovindoline to the two lactams shown in equation (363)[705].

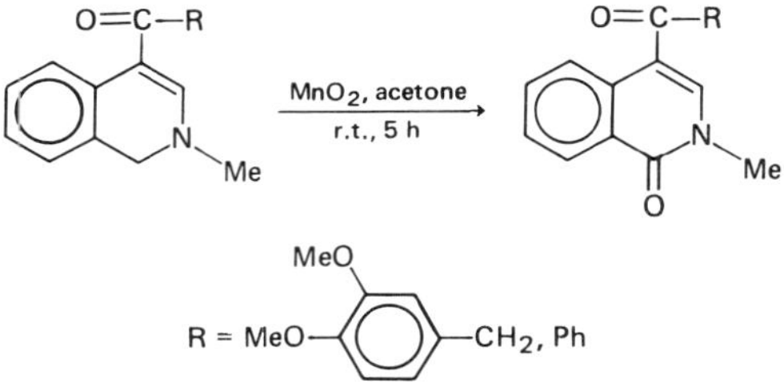

MnO_2, acetone, r.t., 5 h (361)

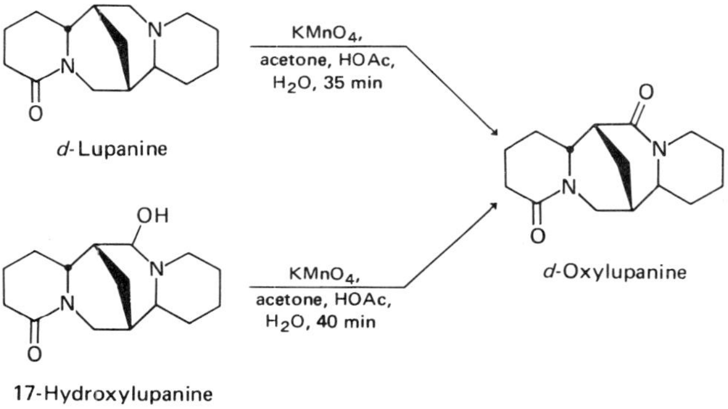

R = 3,4-(MeO)$_2$C$_6$H$_3$CH$_2$, Ph

d-Lupanine $\xrightarrow{KMnO_4,\ acetone,\ HOAc,\ H_2O,\ 35\ min}$ *d*-Oxylupanine (362)

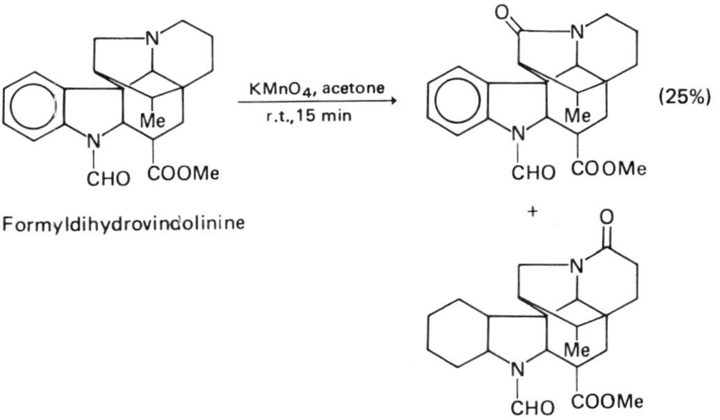

17-Hydroxylupanine $\xrightarrow{KMnO_4,\ acetone,\ HOAc,\ H_2O,\ 40\ min}$ *d*-Oxylupanine

N-Formyldihydrovindolinine $\xrightarrow{KMnO_4,\ acetone,\ r.t.,\ 15\ min}$ (25%) + (363)

4. *Using platinum or ruthenium oxides*

It was originally reported[706] that voacangine, the major alkaloid of *Rejoua aurontiaca* Gaud., was converted into β-hydroxyindolenine by controlled oxidation using platinum and oxygen followed by catalytic reduction. However, a more recent study[699] of this reaction has shown the product to be voacangine lactam

Voacangine → (1. Pt, O_2; 2. H_2, catalyst) → β-Hydroxyindolenine

Voacangine → (1. PtO_2, EtOAc; 2. H_2, catalyst; 3. O_2) → Voacangine lactam (364)

(equation 364), identical to the product obtained[698] from the basic iodine oxidation of dihydrovoacamine followed by acid cleavage of the product.

Although unsubstituted amines[707], aziridine[708] and piperidine[708] react with ruthenium tetroxide to produce imides in good yields without oxidation of the nitrogen atom directly, suitable substitution on nitrogen followed by oxidation with ruthenium tetroxide affords β-lactams (equation 365)[708]. This reaction

$$\text{(CH}_2)_n\text{-cyclic amine } H_2C\text{–N(R)–}CH_2 \xrightarrow[\text{r.t., 2–6 days}]{RuO_2,\ NaIO_4,\ CH_2Cl_2 \text{ or } CHCl_3} H_2C\text{–N(R)–C=O} \quad (365)$$

R	Yield (%)			
	$n = 0$	1	2	3
p-$MeC_6H_4SO_2$	a	a	46	3–5
$MeSO_2$	a	a	90	85
MeO–C(=O)–C(=O)–		22	68	59
HCO		a	34	a
MeCO	a	9–13	45–69	42–60
EtOCO	5	15–33	65	63

^{a}No product could be isolated.

appears more likely to succeed as the ring size increases, and appears to be effected by the electronegativity of the nitrogen substituent. The rate of reaction has also been noted to decrease as the ring size decreases and the electronegativity of the nitrogen substituent increases[708]. By use of the methyloxalyl protecting group it was possible to prepare lactams of varying ring size according to equation (366)[708].

Ruthenium tetroxide has also been reported[709] to oxidize 2-substituted-N-acetyl pyrrolidines and piperidines regiospecifically to their corresponding lactams in about 60% yields with retention of absolute configuration (equations 367 and 368).

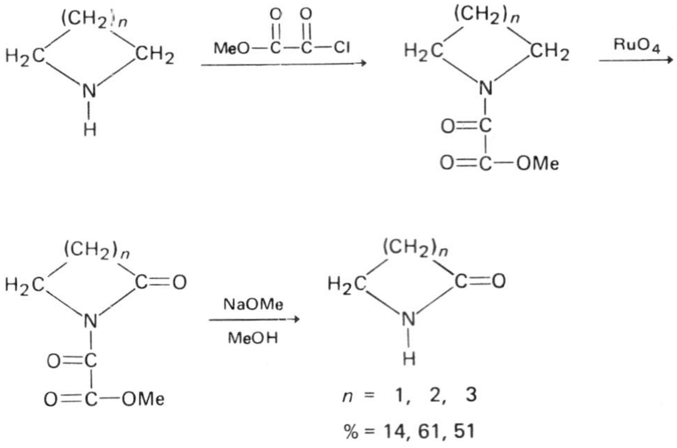

$(CH_2)_n$ ring amine (H_2C, CH_2, N–H) $\xrightarrow{MeO-C(=O)-C(=O)-Cl}$ N–C(=O)–C(=O)–OMe $\xrightarrow{RuO_4}$

(366)

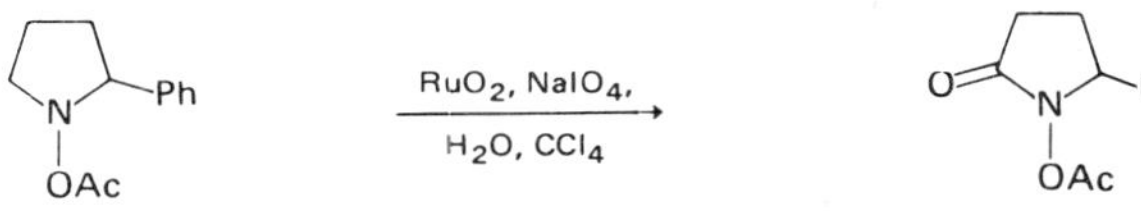

$\xrightarrow[MeOH]{NaOMe}$

n = 1, 2, 3

% = 14, 61, 51

$\xrightarrow[H_2O,\ CCl_4]{RuO_2,\ NaIO_4,}$ (367)

R-(+)-*N*-Acetyl-2-phenylpyrrolidine

R-(+)-*N*-Acetyl-5-phenyl-2-pyrrolidinone

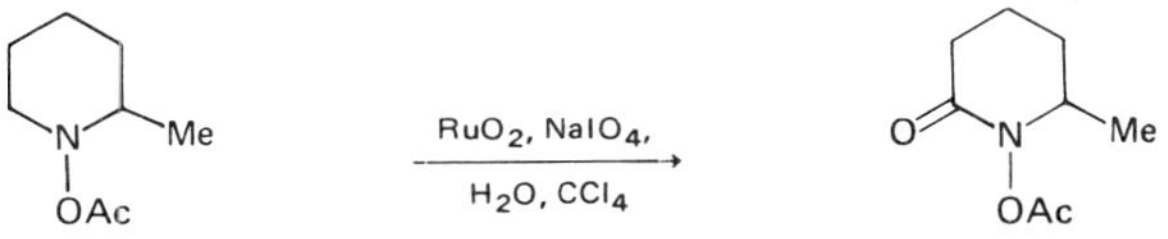

$\xrightarrow[H_2O,\ CCl_4]{RuO_2,\ NaIO_4,}$ (368)

R-(+)-*N*-Acetyl-2-methylpiperidine

R-(+)-*N*-Acetyl-6-methylpiperidone

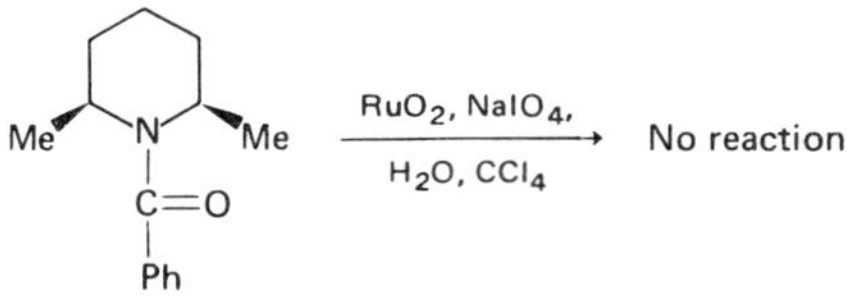

$\xrightarrow[H_2O,\ CCl_4]{RuO_2,\ NaIO_4,}$ No reaction (369)

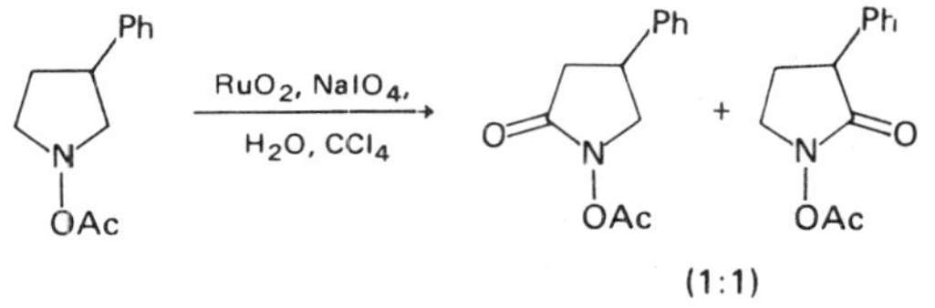

$\xrightarrow[H_2O,\ CCl_4]{RuO_2,\ NaIO_4,}$ + (370)

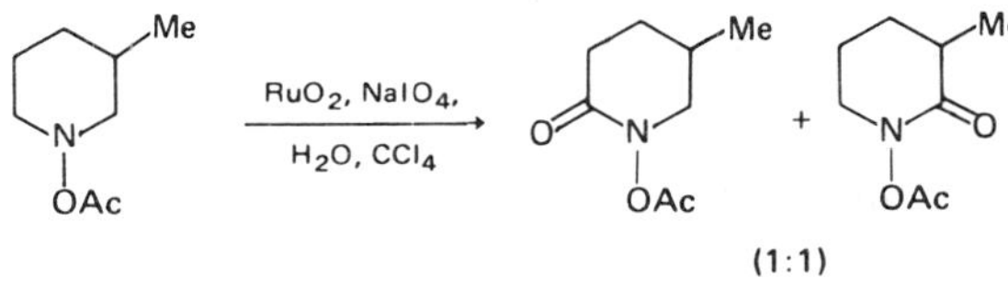

(1:1)

$\xrightarrow[H_2O,\ CCl_4]{RuO_2,\ NaIO_4,}$ + (371)

(1:1)

However, similar oxidation[709] of *N*-benzoyl-*cis*-2,6-dimethylpiperidine afforded only recovered starting material (equation 369), while oxidation of similarly 3-substituted *N*-acylpyrrolidine and piperidine afforded a 1 : 1 mixture of corresponding lactam isomers (equations 370 and 371). Application of this oxidation to *N*-benzoyl- and *n*-acetylpiperidine afforded the expected products in good yields (equation 372)[709].

RuO_2, $NaIO_4$, H_2O, CCl_4 (372)

R = PhCO, MeCO

5. *Via sensitized and unsensitized photooxidation*

Although reaction of ibogaine with ethylmagnesium bromide followed by treatment with oxygen has been reported[699] to produce a 20% yield of iboluteine, benzophenone-sensitized photolysis of this compound affords[699] a 35% yield of ibogaine lactam. Similar treatment[699] of voacangine affords a 5% yield of voacangine lactam, whereas sensitization using Rose Bengal affords a 10% yield of the same product. Rose Bengal-sensitized photooxidation has also been used[710] to effect the conversion shown in equation (373).

Rose Bengal, $h\nu$, O_2

37% (373)

methylene blue, MeOH, $h\nu$, O_2 (374)

Conanine 49%

In addition to benzophenone and Rose Bengal, methylene blue has also been used to sensitize several photooxidations, including the conversion of conanine (equation 374)[711] and sparteine (equation 375)[711] to their corresponding lactams, and lupanine (equation 376)[711] to its corresponding lactam dimer. It has also been employed in the photooxidation of laudanosine (equation 377)[712], a reaction which affords a better yield of product when performed unsensitized[712].

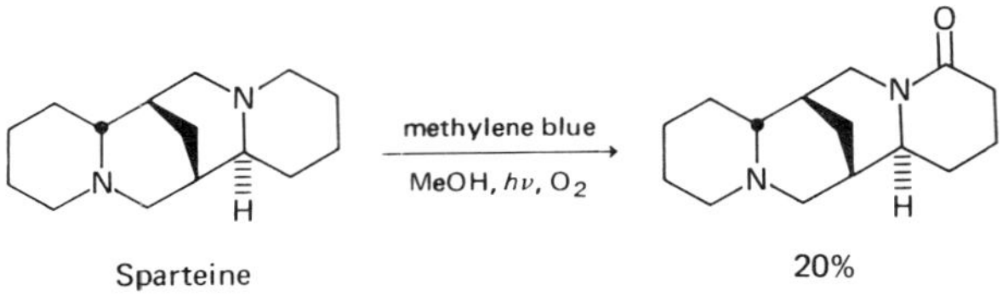

Sparteine —(methylene blue; MeOH, $h\nu$, O_2)→ 20% (375)

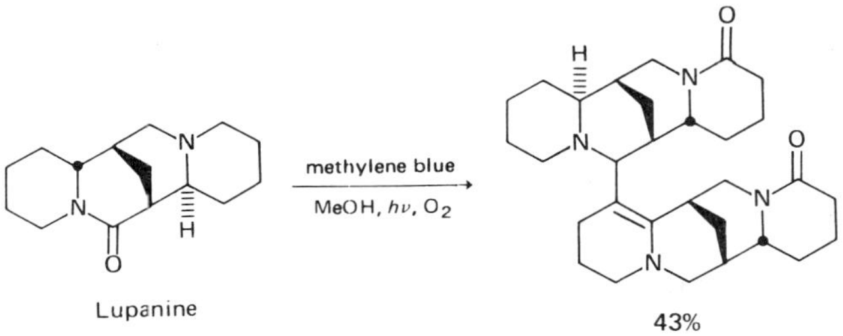

Lupanine —(methylene blue; MeOH, $h\nu$, O_2)→ 43% (376)

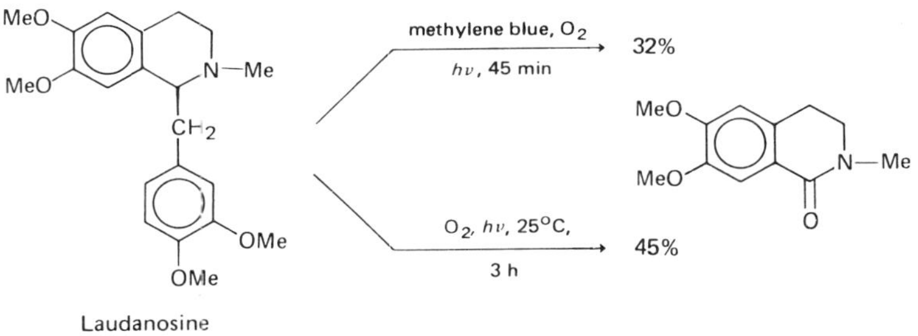

Laudanosine —(methylene blue, O_2; $h\nu$, 45 min)→ 32%; —(O_2, $h\nu$, 25°C, 3 h)→ 45% (377)

Unsensitized photooxidation has also been found to be effective in the production of lactams from a variety[712] of bisbenzylisoquinoline derived alkaloids such as isotetrandrine and berbamine (equation 378), tenuipine and micranthine.

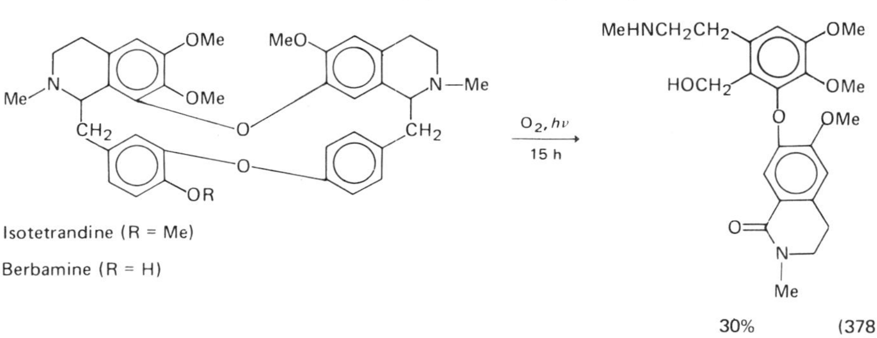

Isotetrandine (R = Me)

Berbamine (R = H)

—(O_2, $h\nu$; 15 h)→ 30% (378)

Photooxidation in the presence of base has been found useful in the conversion of 2′-bromoreticuline to thalifoline (equation 379)[713], and 10-phenyl-9,10-dihydrophenanthridine to *N*-phenylphenanthridone (equation 380)[714]. This latter conversion has also been accomplished[714] without the use of base via the peroxide dimer followed by cleavage under reflux as shown in equation (381).

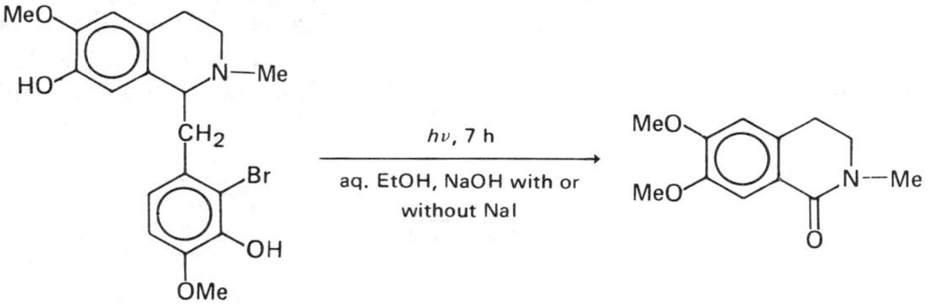

hν, 7 h; aq. EtOH, NaOH with or without NaI (379)

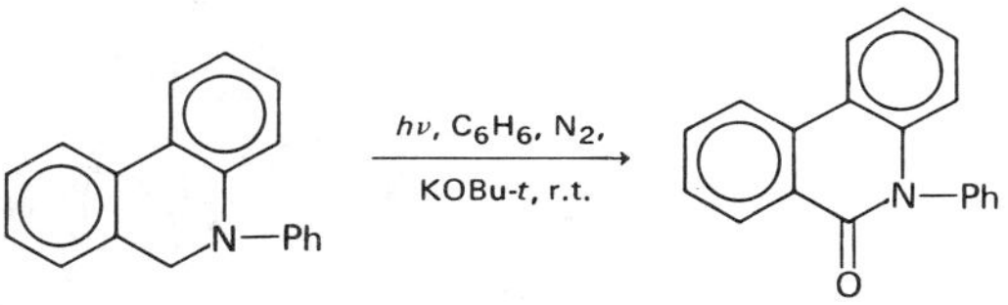

hν, C_6H_6, N_2, KOBu-*t*, r.t. (380)

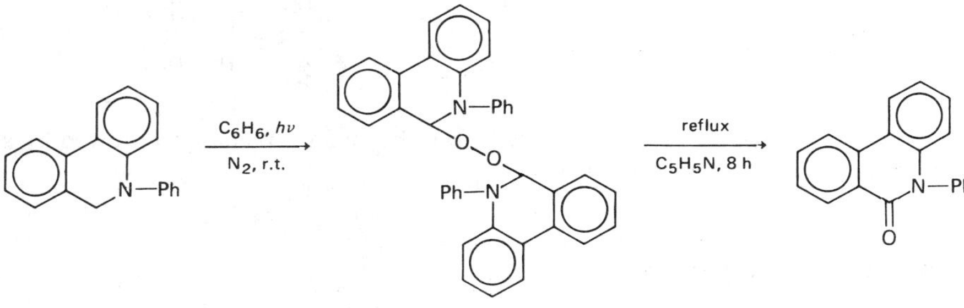

C_6H_6, hν; N_2, r.t. → 38% → reflux; C_5H_5N, 8 h (381)

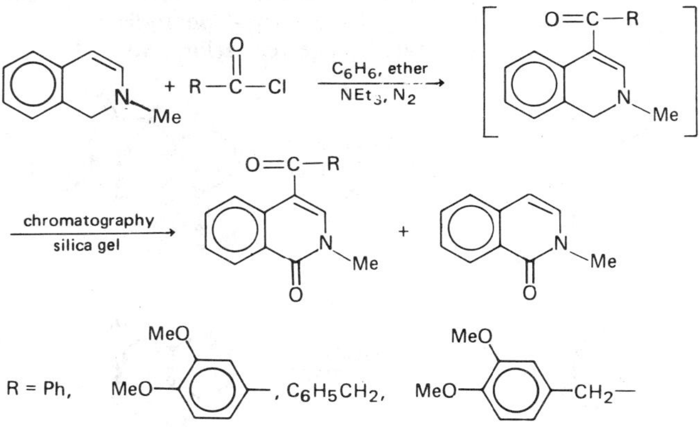

C_6H_6, ether; NEt_3, N_2 → chromatography silica gel (382)

R = Ph, 3,4-(MeO)$_2C_6H_3$—, $C_6H_5CH_2$, 3,4-(MeO)$_2C_6H_3CH_2$—

6. *Via autooxidation*

Attempted acylation of 2-methyl-1,2-dihydroisoquinoline using benzoyl,3,4-dimethoxybenzoyl, phenacetyl and 3,4-dimethoxyphencaetyl chlorides has been reported[702] to give acetylated isocarbostyrils in all cases (equation 382). These products arise when the initial reaction products are oxidized by exposure to air for several days followed by chromatography on silica gel[702]. Similar results are obtained[715] when 1,2-dihydro-4-methyl-3-phenylisoquinoline is exposed to air for several days followed by chromatography on alumina (equation 383). The

1. air
2. chromatography on alumina

61%

$-H_2O$

O_2

(383)

mechanism for these conversions appears[716] to be an autooxidation followed by a dehydration of the intermediate peroxide.

7. *Using miscellaneous reagents*

A variety of lactams have been prepared via oxidation using a variety of miscellaneous reagents. For example, potassium hexacyanoferrate has been used to oxidize *d*-lupanine to *d*-oxylupainine (equation 384)[704], *l*-sparteine to *l*-oxysparteine (equation 385)[704] and 2,4-dimethyl-3-phenylisoquinolinium iodide to 2,4-dimethyl-3-phenylisoquinoline-1(2*H*)-one (equation 386)[715].

$K_3Fe(CN)_6$
NaOH, H_2O, 40 min

(384)

$K_3Fe(CN)_6$
NaOH, H_2O

(385)

Me, Ph, N^+, Me, I^- —[$K_3Fe(CN)_6$; KOH, H_2O, EtOH]→ Me, Ph, N, Me, O (386)

88%

Wasserman and Tremper[717] have reported that treatment of 1-substituted azetidine-2-carboxylic acid with oxalyl chloride affords the iminium salt shown in equation (387), which upon treatment with *m*-chloroperbenzoic acid in pyridine produces a 70–80% yield of 1-substituted β-lactams. This reaction is reported to be more convenient than the alternative procedure of low-temperature dianion oxygenation reported elsewhere[683] in this review.

COOH, N, R —[$(COCl)_2$, 0°C; then warm 45°C for 10 min]→ N^+, R, Cl^-

(387)

—[m-$ClC_6H_4CO_3H$, C_5H_5N; cold, stir 40 min]→ C_5H_5N, H, O—O—C(=O)—C_6H_4Cl-*m*, N, R ⟶ O, N, R

70–80%

R = Me_3C, $C_6H_5CH_2$, *p*-$MeOC_6H_4CH_2CH_2$, cyclohexyl, $C_6H_5CH_2CH_2$

Treatment of cyclic amines such as pyrrolidine with a hydroperoxide in the presence of a metal ion catalyst, such as manganic acetylacetonate, cobalt naphthenate or dicyclopentadienyltitanium dichloride affords the corresponding lactam (equation 388)[718].

N, H —[Me_3C—CO_3H, −4°C; manganic acetylacetonate]→ N, H, O (388)

HO—$C(R^1)(R^2)$—$(CH_2)_n$—N(ring)X —[$Hg(OAc)_2$, EDTA, 1% HOAc, 2 h, 100°C]→ HO—$C(R^1)(R^2)$—$(CH_2)_n$—N(ring, C=O)X (389)

R^1	R^2	n	X	Yield (%)
H	Ph	2	O	–
H	Ph	2	$(CH_2)_2$	71.2
H	$PhOCH_2$	1	$(CH_2)_2$	86.2
Ph	Ph	2	CH_2	45.0

An interesting preparation of lactams, which appears formally to be an oxidation but which in reality is a dehydrogenation, has also been reported[719] using the Hg(II) salt of ethylenediaminetetraacetate (EDTA) (equation 389).

G. Miscellaneous Lactam Syntheses

The following methods do not qualify for inclusion in one of the foregoing sections, but appear to have sufficient generality to serve as useful, albeit somewhat specialized, synthetic procedures.

Condensations of 4-arylmethylene-2,3-pyrrolidinediones with β-aminocrotonate or with 4-amino-3-penten-2-one result in addition of the nucleophilic vinyl carbon of the enamine to the arylmethylene function, accompanied by cyclization of the amino groups of the addend with the 3-carbonyl group of the pyrrolidinedione. The resulting dihydropyrrolo[3,4,*b*]pyridin-7-ones can be oxidized by bromine to afford the pyridine-fused δ-lactams shown in equation (390)[720]. When *N*-phenacylpyridinium bromide was allowed to react with the pyrrolidinediones, the aromatic δ-lactams were formed directly. In the same study[720] it was found that when 4-(*o*-nitrobenzylidene) derivatives of 1-substituted 2,3-pyrrolidinediones were treated with tin(II) chloride or with sodium dithionate, reductive cyclization took place to afford 1,2-dihydropyrrolo[3,4,*b*]quinolin-3-ones (equation 391).

ArCH, N, O, O, R^1 —[NH_2, MeC=CHR^2]→ Ar H, R^2, Me, N, H, N—R^1, O —[Br_2]→ Ar, R^2, Me, N, N—R^1, O (390)

R^1 = c-C_6H_{11}, t-Bu, $C_6H_5CH_2$, $C_6H_5(CH_2)_2$, $MeO_2C(CH_2)_2$

R^2 = COOEt, MeCO

NO_2, CH, O, N, O, R^1 —[$SnCl_2$]→ N, N—R^1, O (391)

R^1 = c-C_6H_{11}, $C_6H_5(CH_2)_2$

A convenient synthesis of 5-hydroxy- and 5-methoxy-3-pyrrolin-2-ones has been carried out via singlet oxygen addition to an appropriate furan derivative followed by ammonolysis of the resulting pseudo ester (equation 392)[721].

R^2, R^1, O —[O_2^1, MeOH]→ R^2, R^1, O, O, OMe —[NH_3]→ R^2, R^1, N, O, OH, H (392)

Anils of cycloalkanones have been found[722] to react with oxalyl chloride to afford 1-phenyl-4,5-polymethylene-2,3-pyrrolidinediones (equation 393). When

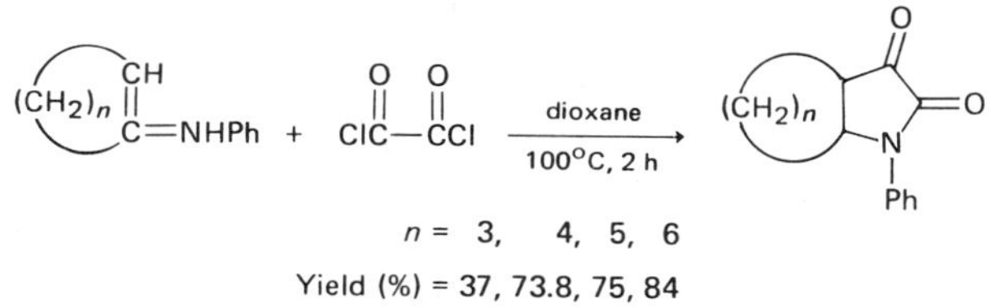

(393)

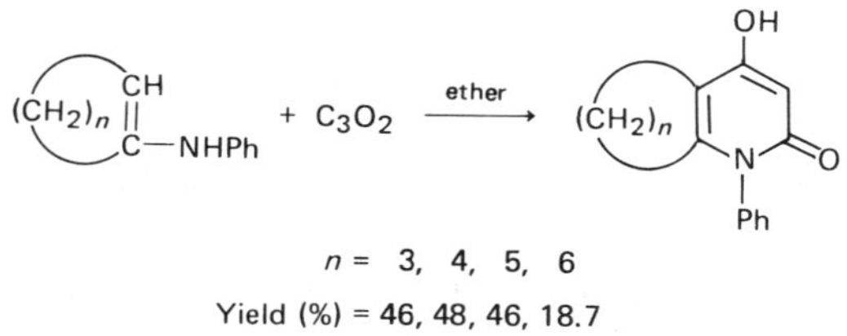

n = 3, 4, 5, 6

Yield (%) = 37, 73.8, 75, 84

(394)

n = 3, 4, 5, 6

Yield (%) = 46, 48, 46, 18.7

carbon suboxide is used instead of oxalyl chloride, the same anils afford 4-hydroxy-5,6-polymethylene-2-pyridones (equation 394)[723].

IV. ACKNOWLEDGMENTS

This undertaking was only possible with the dedicated help of our friend and typist, Mrs. Brenda Mills, who typed the entire manuscript and provided all of the structural drawings. Her steadfastness and pleasant nature contributed greatly to the completion of this work, and we are pleased to acknowledge her contributions. Miss Susan Stevens provided us with many hours of help in collecting and organizing the primary literature references used in this review. We are grateful to the Department of Chemistry for providing facilities and financial support during the writing of this chapter. We are also pleased to acknowledge the National Aeronautics and Space Administration (Grants NSG-1064 and NSG-1286) and the National Science Foundation (Grant CHE 74-20520) for support of our research programmes during the writing of this chapter.

V. REFERENCES

1. H. E. Zaugg, *Org. Reactions*, **8**, 305 (1954).
2. R. Filler, *Chem. Rev.*, **63**, 21 (1963).
3. Y. Etienne and N. Fischer, *The Chemistry of Heterocyclic Compounds* (Ed. A. Weissberger), Vol. 19, Interscience, New York, 1964, p. 729.
4. Y. S. Rao, *Chem. Rev.*, **64**, 353 (1964).
5. M. F. Ansell and M. H. Palmer, *Quart. Rev.*, **18**, 211 (1964).
6. N. P. Shusherina, N. D. Dmitrieva, E. A. Luk'yanets and R. Y. Levina, *Russ. Chem. Rev.*, **36**, 175 (1967).
7. B. Chemielarz, *Tluszcze Srodki Piorace Kosmet*, **12**, 21 (1968); *Chem. Abstr.*, **69**, 67715e (1968).
8. V. M. Dashunin, R. V. Maeva, G. A. Samatuga and V. N. Belov, *Mezhdunar. Kongr. Efirnym Maslam* (Ed. P. V. Naumenko) 4th ed., Vol. 1, Vses. Nauchno-Issled. Inst. Sint. Nat. Dishistykh Veshchestv, Moscow, 1971, p. 90; *Chem. Abstr.*, **78**, 135970x (1973).
9. J. L. Brash and D. J. Lyman, *High Polym.*, **26**, 147 (1972).
10. S. G. Cottis, *High Polym.*, **27**, 311 (1972).
11. P. A. Grieco, *Synthesis*, 67 (1975).

12. Y. S. Rao, *Chem. Rev.*, **76**, 625 (1976).
13. J. H. Gardner and C. A. Naylor, Jr, *Org. Syntheses*, Coll. Vol. II, 526 (1943).
14. C. G. Overberger and H. Kaye, *J. Amer. Chem. Soc.*, **89**, 5640 (1967).
15. J. V. Karabinos, *Org. Syntheses*, Coll. Vol. IV, 506 (1963).
16. W. S. Johnson, V. J. Bauer, J. L. Margrave, M. A. Frisch, L. H. Dreger and W. N. Hubbard, *J. Amer. Chem. Soc.*, **83** 606 (1961).
17. W. Cocker, L. O. Hopkins, T. B. H. McMurray and M. A. Nisbet, *J. Chem. Soc.*, 4721 (1961).
18. C. Collin-Asselineau, S. Bory and E. Lederer, *Bull. Soc. Chim. Fr.*, 1524 (1955).
19. J. A. Marshall, N. Cohen and A. R. Hochstetler, *J. Amer. Chem. Soc.*, **88**, **3408 (1966)**.
20. G. Lardelli, V. Lamberti, W. T. Weller and A. P. DeJonge, *Rec. Trav. Chim. Pays-Bas*, **86**, 481 (1967).
21. J. D. White, S. N. Lodwig, G. L. Trammell and M. P. Fleming, *Tetrahedron Letters*, 3263 (1974).
22. R. E. Ireland, D. A. Evans, D. Glover, G. M. Rubottom and H. Young, *J. Org. Chem.*, **34**, 3717 (1969).
23. G. Caron and J. Lessard, *Can. J. Chem.*, **51**, 981 (1973).
24. A. P. Krapcho and E. G. E. Jahngen, Jr, *J. Org. Chem.*, **39**, 1322 (1974).
25. W. Adam, J. Baeza and J.-C. Liu, *J. Amer. Chem. Soc.*, **94**, 2000 (1972).
26. A. P Krapcho and E. G. E. Jahngen, Jr, *J. Org. Chem.*, **39**, 1650 (1974).
27. M. V. S. Sultanbawa, *Tetrahedron Letters*, 4569 (1968).
28. R. B. Woodward, F. E. Bader, H. Bickel, A. J. Frey and R. W. Kierstead, *Tetrahedron*, **2**, 1 (1958).
29. J. Meinwald and E. Frauenglass, *J. Amer. Chem. Soc.*, **82**, 5235 (1960).
30. J. A. Marshall and N. Cohen, *Tetrahedron Letters*, 1997 (1964).
31. J. A. Marshall and N. Cohen, *J. Org. Chem.*, **30**, 3475 (1965).
32. V. R. Tadwalkar and A. S. Rao, *Indian J. Chem.*, **9**, 1416 (1971).
33. J. Finkelstein and A. Brossi, *J. Heterocyclic Chem.*, **4**, 315 (1967).
34. J. Finkelstein and A. Brossi, *Org. Syntheses*, **55**, 45 (1976).
35. M. G. Zalinyan, V. S. Arutyunyan and M. T. Dangyan, *Arm. Khim. Zh.*, **26**, 827 (1973); *Chem. Abstr.*, **80**, 82014w (1974).
36. A. S. Kyazimov, M. M. Movsumzade, A. L. Shabanov and A. A. Babaeva, *Azerb. Khim. Zh.*, **53**, (1974); *Chem. Abstr.*, **82**, 97725u (1975).
37. A. Hirshfeld, W. Taub and E. Glotter, *Tetrahedron*, **28**, 1275 (1972).
38. E. M. P. Fowler and H. B. Henbest, *J. Chem. Soc.*, 3642 (1950).
39. D. Kosterman, *Nature*, **166**, 787 (1950).
40. D. Kosterman, *Rec. Trav. Chim.*, **70**, 79 (1951).
41. J. Gobley and H. O'Rorke, *J. Pharm. Chim.*, 598 (1860).
42. M. Cuzent, *Compt. Rend.*, 205 (1861).
43. W. Borsche and W. Peitzsch, *Ber.*, **62**, 360 (1929).
44. W. Borsche and C. K. Bodenstein, *Ber.*, **62**, 2515 (1929) and later papers through 1933.
45. J. W. Patterson and J. E. McMurray, *Chem. Commun.*, 488 (1971).
46. J. B. Brown, H. B. Henbest and E. R. H. Jones, *J. Chem. Soc.*, 3634 (1950).
47. E. E. van Tamelen and S. R. Bach, *J. Amer. Chem. Soc.*, **80**, 3079 (1958).
48. O. A. Sarkisyan, A. N. Stepanyan, V. S. Arutyunyan, M. G. Zalinyan and M. T. Dangyan, *Zh. Org. Khim.*, **5**, 1648 (1969); *Chem. Abstr.*, **72**, 2982g (1970).
49. E. J. Corey and K. C. Nicolaou, *J. Amer. Chem. Soc.*, **96**, 5614 (1974).
50. R. C. Blume, *Tetrahedron Letters*, 1047 (1969).
51. P. F. Hudrlik, L. R. Rudnick and S. H. Korzeniowski, *J. Amer. Chem. Soc.*, **95**, 6848 (1973).
52. W. C. Bailey, Jr, A. K. Bose, R. M. Ikeda, R. H. Newan, H. V. Secor and C. Varsel, *J. Org. Chem.*, **33**, 2819 (1968).
53. A. M. Chalmers and A. J. Baker, *Tetrahedron Letters*, 4529 (1974).
54. G. R. Pettit, D. C. Fessler, K. D. Paull, P. Hofer and J. C. Knight, *J. Org. Chem.*, **35**, 1398 (1970).
55. G. S. King and E. S. Waight, *J. Chem. Soc., Perkin Trans. I*, 1499 (1974).

56. E. J. Clarke and R. P. Hildebrand, *J. Inst. Brew.*, **73**, 60 (1967); *Chem. Abstr.*, **67**, 32303a (1967).
57. G. I. Nikishin, M. G. Vinogradov and T. M. Fedorova, *Chem. Commun.*, 693 (1973).
58. Z. Horii, T. Yagami and M. Hanaoka, *Chem. Commun.*, 634 (1966).
59. R. Hodges and A. L. Porte, *Tetrahedron*, **20**, 1463 (1964).
60. T. Wada, *Chem. Pharm. Bull.* (*Tokyo*), **12**, 1117 (1964); and **13**, 43 (1965).
61. T. Sakan, S. Isol and S. B. Hyeon, *Tetrahedron Letters*, 1623 (1967).
62. J. Bricout, R. Viani, F. Muggler-Chawan, J. P. Marion, D. Reymond and R. H. Egli, *Helv. Chim. Acta*, **50**, 1517 (1967).
63. G. Agnes and G. P. Chiusoli, *Chim. Ind.* (*Milan*), **50**, 194 (1968); *Chem Abstr.*, **69**, 35352t (1968).
64. K. Tsukida, M. Ito and F. Ikeda, *Experientia*, **29**, 1338 (1973).
65. P. K. Christensen, *Acta Chem. Scand.*, **11**, 582 (1957).
66. N. A. Sorensen and K. Stavholt, *Acta Chem. Scand.*, **4**, 1080 (1950).
67. I. Bill, E. R. H. Jones and M. C. Whiting, *J. Chem. Soc.*, 1313 (1958).
68. J. Castaner and J. Pascual, *J. Chem. Soc.*, 3962 (1958).
69. J. Castaner and J. Pascual, *Anales Real. Soc. Espan. Fis. y Quim (Madrid)*, **53B**, 651 (1957); *Chem. Abstr.*, **54**, 3404b (1960).
70. C. Belil, J. Castella, J. Castells, R. Mestres, J. Pascual and F. Settatosa, *Anales Real. Soc. Espan. Fis. y Quim (Madrid)*, **57B**, 617 (1961); *Chem. Abstr.*, **57**, 12455e (1962).
71. C. Belil, J. Pascual and F. Serratosa, *Tetrahedron*, **20**, 2701 (1964).
72. G. I. Nikishin, M. G. Vinogradov and T. M. Fedorova, *Chem. Commun.*, 693 (1973).
73. J. Castells, R. Mestres and J. Pascual, *Anales Real. Soc. Espan. Fis. y Quim (Madrid)*, **60B**, 843 (1964); *Chem Abstr.*, **63**, 11536d (1965).
74. F. Serratosa, *Tetrahedron*, **16**, 185 (1961).
75. J. Bosch, J. Castells and J. Pascual, *Anales Real. Soc. Espan. Fis. y Quim* (*Madrid*), **57B**, 469 (1961); *Chem Abstr.*, **56**, 8628d (1962).
76. M. Alguero, J. Bosch, J. Castaner, J. Castella, J. Castells, R. Mestres, J. Pascual and F. Serratosa, *Tetrahedron*, **18**, 1381 (1962).
77. R. H. Wiley, T. H. Crawford and C. E. Staples, *J. Org. Chem.*, **27**, 1535 (1962).
78. R. H. Wiley and C. E. Staples, *J. Org. Chem.*, **28**, 3408 (1963).
79. Y. S. Rao, *Tetrahedron Letters*, 1457 (1975).
80. R. H. Wiley, C. H. Jarboe and F. N. Hayes, *J. Amer. Chem. Soc.*, **79**, 2602 (1957).
81. G. W. Kenner, M. A. Murray and C. M. B. Taylor, *Tetrahedron*, **1**, 259 (1957).
82. O. L. Chapman and W. R. Adams, *J. Amer. Chem. Soc.*, **89**, 4243 (1967).
83. G. O. Schenck, G. Koltzenburg and H. Grossman, *Angew. Chem.*, **69**, 177 (1957).
84. R. Dulou, M. Vilkas and M. Pfau, *Compt. Rend.*, **249**, 429 (1959).
85. M. Pfau, R. Dulou and M. Vilkas, *Compt. Rend.*, **251**, 2188 (1960); **254**, 1817 (1962).
86. G. Buchi and S. H. Feaisheller, *J. Org. Chem.*, **34**, 609 (1969).
87. S. Majeti, *J. Org. Chem.*, **37**, 2914 (1972).
88. M. Tokuda, Y. Kokoyama, T. Taguchi, A. Suzuki and M. Itoh, *J. Org. Chem.*, **37**, 1859 (1972).
89. M. Itoh, T. Taguchi, V. Van Chung, M. Tokuda and A. Suzuki, *J. Org. Chem.*, **37**, 2357 (1972).
90. J. Bougault, *Ann. Chim. Phys.*, **14**, 145 (1908); **15**, 296 (1908).
91. R. P. Linstead and C. J. May, *J. Chem. Soc.*, 2565 (1927).
92. A. W. Schrecker, G. Y. Greenburg and J. L. Hartwell, *J. Amer. Chem. Soc.*, **74**, 5669 (1952).
93. A. W. Schrecker and J. L. Hartwell, *J. Amer. Chem. Soc.*, **74**, 5676 (1952).
94. E. E. van Tamelen and M. Shamma, *J. Amer. Chem. Soc.*, **76**, 2315 (1954).
95. J. Klein, *J. Amer. Chem. Soc.*, **81**, 3611 (1959).
96. S. Beckmann and H. Geiger, *Ber.*, **92**, 2411 (1959).
97. J. A. Berson and A. Remanick, *J. Amer. Chem. Soc.*, **83**, 4947 (1961).
98. W. E. Barnett and J. C. McKenna, *Chem. Commun.*, 551 (1971).
99. W. E. Barnett and J. C. McKenna, *Tetrahedron Letters*, 2595 (1971).
100. W. E. Barnett and W. H. Sohn, *Tetrahedron Letters*, 1777 (1972).

101. M. M. Movsumzade, A. S. Kyazimov, A. L. Shabanov and Z. A. Safarova, *Dokl. Akad. Nauk SSSR*, **30**, 40 (1974); *Chem. Abstr.*, **82**, 111649f (1975).
102. A. A. Akhnazaryan, L. A. Khachatryan and M. T. Dangyan, *U.S.S.R. Patent,* 317,652; *Chem. Abstr.*, **76**, 112917e (1972).
103. S. Sarel, Y. Shalon and Y. Yanuka, *Tetrahedron Letters*, 957, 961 (1969).
104. S. Sarel, Y. Shalon and Y. Yanuka, *Chem. Commun.*, 80 (1970).
105. B. Berkov, L. Cuellar, R. Grezemkovsky, N. V. Avila and A. D. Cross, *Tetrahedron*, **24**, 2851 (1968).
106. S. Sarel, Y. Shalon and Y. Yanuka, *Chem. Commun.*, 81 (1970).
107. E. J. Corey and R. L. Danheiser, *Tetrahedron Letters*, 4477 (1973).
108. R. T. Arnold and K. L. Lindsay, *J. Amer. Chem. Soc.*, **75**, 1048 (1953).
109. K. W. Schulte and K. Baranowsky, *Pharm. Zentr.*, **98**, 403 (1959).
110. K. E. Schutze, J. Reisch and O. Heine, *Arch. Pharm.*, **294**, 234 (1961).
111. L. H. Klemm and K. W. Gopinath, *J. Heterocyclic Chem.*, **2**, 225 (1965).
112. L. H. Klemm, D. H. Lee, K. W. Gopinath and C. E. Klopfenstein, *J. Org. Chem.*, **31**, 2376 (1966).
113. L. H. Klemm and K. W. Gopinath, *Tetrahedron Letters*, 1243 (1963).
114. L. H. Klemm, K. W. Gopinath, D. H. Lee, F. W. Kelly, E. Trod and T. M. McGuire, *Tetrahedron*, **22**, 1797 (1966).
115. E. L. Ghisalferti, P. R. Jefferies and T. G. Payne, *Tetrahedron*, **30**, 3099 (1974).
116. F. Arndt, *Org. Syntheses*, Coll. Vol. III, 231 (1955).
117. N. R. Smith and R. H. Wiley, *Org. Syntheses*, Coll. Vol. IV, 549 (1963).
118. A. A. Avetisyan, Ts. A. Mangasaryan, G. S. Melikyan, M. T. Dangyan and S. G. Matsoyan, *Zh Org. Khim.*, 7, 962 (1971); *Chem. Abstr.*, **75**, 63047q (1971).
119. G. D. Zuidema, E. van Tamelen and G. Van Zyl, *Org. Syntheses,* Coll. Vol. IV, 10 (1963).
120. D. T. C. Yang and S. W. Pelletier, *Org. Prep. Proced. (Int)*, **7**, 221 (1975).
121. D. Seebach and H. Meyer, *Angew. Chem. (Int. Ed. Engl.)*, **13**, 77 (1974).
122. S. Sethna and R. Phadke, *Org. Reactions*, **7**, 1 (1953).
123. H. Wiley and N. R. Smith, *Org. Syntheses*, Coll. Vol. IV, 201 (1963).
124. H. von Pechmann, *Ann.*, **264**, 272 (1891).
125. E. H. Woodruff, *Org. Syntheses*, Coll. Vol. III, 581 (1955).
126. A. Russell and J. R. Frye, *Org. Syntheses*, Coll. Vol. III, 281 (1955).
127. E. B. Vliet, *Org. Syntheses*, Coll. Vol. I, 360 (1932).
128. U. Kraatz and F. Korte, *Ber.*, **106**, 62 (1973).
129. R. Adams and B. R. Baker, *J. Amer. Chem. Soc.*, **62**, 2405 (1940).
130. G. Powell and T. H. Bembry, *J. Amer. Chem. Soc.*, **62**, 2568 (1940).
131. M. Guyot and C. Mentzer, *Bull. Soc. Chim. Fr.*, 2558 (1965).
132. V. B. Piskov, *Zh. Obsh. Khim.*, **30**, 1390 (1960); *J. Gen. Chem. U.S.S.R.*, **30**, 1421 (1960).
133. P. Kurath and W. Cole, *J. Org. Chem.*, **26**, 1939 (1961).
134. P. Kurath and W. Cole, *J. Org. Chem.*, **26**, 4592 (1961).
135. P. Kurath, W. Cole, J. Tadanier, M. Freifelder, G. R. Stone and E. V. Schuber, *J. Org. Chem.*, **28**, 2189 (1963).
136. M. A. Bielefeld and P. Kurath, *J. Org. Chem.*, **34**, 237 (1969).
137. V. Valcavi and I. L. Sianesi, *Gazz. Chim. Ital.*, **93**, 803 (1963).
138. V. Valcavi, *Gazz. Chim. Ital.*, **93**, 794, 929 (1963).
139. M. Debono, R. M. Molloy and L. E. Patterson, *J. Org. Chem.*, **34**, 3032 (1969).
140. G. R. Pettit, B. Green and G. L. Dunn, *J. Org. Chem.*, **35**, 1367 (1970).
141. H. H. Inhoffen, H. Krösche, K. Radscheit, H. Dettmer and W. Rudolph, *Ann.*, **714**, 8 (1968).
142. H. H. Inhoffen, W. Kreiser and M. Nazir, *Ann.*, **755**, 1, 12 (1972).
143. K. Miyano, J. Ohfune, S. Azuma and T. Matsumoto, *Tetrahedron Letters*, 1545 (1974).
144. E. C. Horning, M. G. Horning and D. A. Dimmig, *Org. Synthese*, Coll. Vol. III, 165 (1955).
145. G. R. Pettit, J. C. Knight and C. L. Heard, *J. Org. Chem.*, **35**, 1393 (1970).
146. A. A. Avetisyan, G. S. Melikyan, M. T. Dangyan and S. G. Matsoyan, *Zh. Org. Khim.*, **8**, 274 (1972); *Chem. Abstr.*, **76**, 139902h (1972).

147. A. A. Avetisyan, G. E. Tatevosyan, Ts. A. Mangasaryan, S. G. Matsoyan and M. T. Dangyan, *Zh. Org. Khim.*, **6**, 962 (1970); *Chem. Abstr.*, **74**, 87430q (1971).
148. R. R. Russell and C. A. Vander Werf, *J. Amer. Chem. Soc.*, **69**, 11 (1947).
149. G. Van Zyl and E. E. van Tamelen, *J. Amer. Chem. Soc.*, **72**, 1357 (1950).
150. S. J. Cristol and R. F. Helmreich, *J. Amer. Chem. Soc.*, **74**, 4083 (1952).
151. E. E. van Tamelen and S. R. Bach, *J. Amer. Chem. Soc.*, **77**, 4683 (1955).
152. C. H. DePuy, F. W. Breitbeil and K. L. Eilers, *J. Org. Chem.*, **29**, 2810 (1964).
153. P. M. G. Bavin, D. P. Hansell and R. G. W. Spickett, *J. Chem. Soc.*, 4535 (1964).
154. L. K. Dalton and B. C. Elmes, *Australian J. Chem.*, **25**, 625 (1972).
155. L. K. Dalton, B. C. Elmes and B. V. Kolczynski, *Australian J. Chem.*, **25**, 633 (1972).
156. A. Takeda, S. Tsubor and Y. Oota, *J. Org. Chem.*, **38**, 4148 (1973).
157. A. Franke and G. Groeger, *Monatsh. Chem.*, **43**, 55 (1922).
158. N. Bensel, H. Marshall and P. Weyerstahl, *Ber.*, **108**, 2697 (1975).
159. J. R. Johnson, *Org. Reactions*, **1**, 210 (1942).
160. H. O. House, *Modern Synthetic Reactions*, 2nd ed., W. A. Benjamin, Menlo Park, California, 1972, Chap. 10.
161. H. E. Carter, *Org. Reactions*, **3**, 198 (1946).
162. E. Baltazzi, *Quart. Rev. (Lond.)*, **9**, 150 (1955).
163. M. Lounasmaa, *Acta Chem. Scand.*, **22**, 70 (1968); **25**, 1849 (1971); **27**, 708 (1973).
164. M. Lounasmaa, *Acta Chem. Scand.*, **26**, 2703 (1972).
165. R. Filler, E. J. Piasek and H. A. Leipold, *Org. Syntheses*, Coll. Vol. V, 80 (1973).
166. R. Weiss, *Org. Syntheses*, Coll. Vol. II. 61 (1943).
167. C. D. Gutsche, E. F. Jason, R. S. Coffey and H. E. Johnson, *J. Amer. Chem. Soc.*, **80**, 5756 (1958).
168. V. I. Bendall and S. S. Dharamski, *J. Chem. Soc., Perkin I*, 2732 (1972).
169. M. Protiva, V. Hnevsova-Seidlova, V. Jirkovsky, L. Novak and Z. J. Vejdelek, *Ceskoslov. Farm.*, **10**, 501 (1962); *Chem. Abstr.*, **57**, 7196f (1962).
170. W. S. Johnson and G. H. Daub, *Org. Reactions*, **6**, 1 (1951).
171. A. A. Avetisyan, K. G. Akopyan and M. T. Dangyan, *Khim. Geterotsikl. Soedin.*, 1604 (1973); *Chem. Abstr.*, **80**, 82006v (1974).
172. G. Saucy and R. Borer, *Helv. Chim. Acta*, **54**, 2121, 2517 (1971).
173. M. Rosenberger, T. P. Fraher and G. Saucy, *Helv. Chim. Acta*, **54**, 2857 (1971).
174. M. Rosenberger, D. Andrews, F. DiMaria, A. J. Duggan and G. Saucy, *Helv. Chim. Acta*, **55**, 249 (1972).
175. P. M. Hardy, A. C. Nicholls and H. N. Rudon, *J. Chem. Soc. (D)*, 565 (1969).
176. R. Sandberg, *Arkiv Kemi*, **16**, 255 (1960).
177. M. Mousseron, M. Mousseron, I. Neyrelles and Y. Beziat, *Bull. Soc. Chim. Fr.*, 1483 (1963).
178. W. R. Vaughan, S. C. Berstein and M. E. Lorber, *J. Org. Chem.*, **30**, 1790 (1965).
179. J. C. Dubois, J. P. Guette and H. B. Kagan, *Bull. Soc. Chim. Fr.*, 3008 (1966).
180. C. Gandolfi, G. Doria, M. Amendola and E. Dradi, *Tetrahedron Letters*, 3923 (1970).
181. E. P. Kohler and G. L. Heritage, *Am. Chem. J.*, **43**, 475 (1911); *Chem Abstr.*, **4**, 2270 (1911).
182. E. P. Kohler and H. Gilman, *J. Amer. Chem. Soc.*, **41**, 683 (1919).
183. M. T. Bertrand, G. Courtois and L. Miginioc, *Compt. Rend., Ser. C*, **280**, 999 (1975); *Chem. Abstr.*, **83**, 96358k (1975).
184. E. Ohler, K. Reininger and U. Schmidt, *Angew. Chem. (Int. Ed. Engl.)*, **9**, 457 (1970).
185. A. Löffler, R. Pratt, J. Pucknat, G. Gelbard and A. S. Dreiding, *Chimia*, **23**, 413 (1969).
186. W. H. Klohs, F. Keller and R. E. Williams, *J. Org. Chem.*, **24**, 1829 (1959).
187. J. W. Cornforth and R. H. Cornforth, 'Natural substances formed biologically from mevalonic acid,' *Biochemical Society Symposium No. 29*, (Ed. T. W. Goodwin), Academic Press, New York, 1970, p. 1
188. W. F. Gray, G. L. Deets and T. Cohen, *J. Org. Chem.*, **33**, 4352 (1968).
189. H. G. Floss, M. Tcheng-Lin, C. Chang, B. Naidov, G. E. Blair, C. I. Abou-Chaar and J. M. Cassady, *J. Amer. Chem. Soc.*, **96**, 1898 (1974).
190. L. Pichat, B. Blagoev and J. C. Hardouin, *Bull. Soc. Chim. Fr.*, 4489 (1968).
191. M. W. Rathke and A. Lindert, *J. Org. Chem.*, **35**, 3966 (1970).

192. R. A. Ellison and P. K. Bhatnagar, *Synthesis*, 719 (1974).
193. A. K. Sorensen and N. A. Klitgaard, *Acta Chem. Scand.*, **24**, 343 (1970).
194. D. B. Denney and L. C. Smith, *J. Org. Chem.*, **27**, 3404 (1962).
195. P. L. Creger, *J. Org. Chem.*, **37**, 1907 (1972).
196. T. Fujita, S. Watanabe and K. Suga. *Australian J. Chem.*, **27**, 2205 (1974).
197. A. I. Meyers, *Heterocycles in Organic Synthesis*, Wiley–Interscience, New York, 1974, Chap. 10.
198. A. I. Meyers and E. D. Mihelick, *Angew. Chem.* (*Int. Ed. Engl.*), **15**, 270 (1976).
199. A. I. Meyers, E. D. Mihelick and R. L. Nolen, *J. Org. Chem.*, **39**, 2783 (1974).
200. A. I. Meyers and E. D. Mihelick, *J. Org. Chem.*, **40**, 1186 (1975).
201. J. Thiele, R. Tischbein and E. Lossow, *Ann.*, **319**, 180 (1910).
202. W. F. von Oettingen, *J. Amer. Chem. Soc.*, **52**, 2024 (1930).
203. E. Erlenmeyer and E. Braun, *Ann.*, **333**, 254 (1904).
204. R. Walter and H. Zimmer, *J. Heterocyclic Chem.*, **1**, 205 (1964).
205. R. Walter, H. Zimmer and T. C. Purcell, *J. Org. Chem.*, **31**, 2854 (1966).
206. R. Walter, D. Theodoropoulas and T. C. Purcell, *J. Org. Chem.*, **32**, 1649 (1967).
207. H. Zimmer and J. Rothe, *J. Org. Chem.*, **24**, 28 (1959).
208. W. Reppe *et al.*, *Ann.*, **596**, 158 (1955).
209. P. G. Gassman and B. L. Fox, *J. Org. Chem.*, **31**, 982 (1966).
210. E. Piers, M. B. Geraghty and R. D. Smille, *Chem. Commun.*, 614 (1971).
211. G. H. Posner and G. L. Loomis, *Chem. Commun.*, 892 (1972).
212. J. L. Herrmann and R. H. Schlessinger, *Chem. Commun.*, 711 (1973).
213. A. E. Green, J. C. Muller and G. Ourisson, *Tetrahedron Letters*, 4147 (1971).
214. H. J. Reich and J. H. Renga, *Chem. Commun.*, 135 (1974).
215. T. Hirrami, I. Niki and T. Agawa, *J. Org. Chem.*, **39**, 3236 (1974).
216. R. L. Evans and H. E. Stavely, *U.S. Patent*, 3,248,392; *Chem. Abstr.*, **65**, 2322e (1966).
217. N. Minami and I. Kuwajima, *Tetrahedron Letters*, 1423 (1977).
218. M. Fujii and A. Sudo, *Japanese Patent*, 72–25,065; *Chem. Abstr.*, **77**, 100847q (1972).
219. C. Metzger, D. Borrmann and R. Wegler, *Ber.*, **100**, 1817 (1967).
220. D. Borrmann and R. Wegler, *Ber.*, **100**, 1575 (1967).
221. D. Borrmann and R. Wegler, *Ber.*, **99**, 1245 (1966).
222. D. Borrmann and R. Wegler, *Ber.*, **102**, 64 (1969).
223. A. S. Kende, *Tetrahedron Letters*, 2661 (1967).
224. J. Ciabattoni and H. W. Anderson, *Tetrahedron Letters*, 3377 (1967).
225. W. Ried and H. Mengler, *Ann.*, **678**, 113 (1964).
226. P. M. M. Van Haard, L. Thijs and B. Zwanenburg, *Tetrahedron Letters*, 803 (1975).
227. P. D. Bartlett and M. Stiles, *J. Amer. Chem. Soc.*, **77**, 2806 (1955).
228. P. S. Bailey, *Chem. Rev.*, **58**, 925 (1958).
229. P. S. Bailey and A. G. Lane, *J. Amer. Chem. Soc.*, **89**, 4473 (1967).
230. R. Wheland and P. D. Bartlett, *J. Amer. Chem. Soc.*, **92**, 6057 (1970).
231. G. D. Buckley and W. J. Levy, *J. Chem. Soc.*, 3016 (1951).
232. S. W. Chaikin and W. G. Brown, *J. Amer. Chem. Soc.*, **71**, 122 (1949).
233. W. G. Brown, *Org. Reactions*, **1**, 469 (1951).
234. N. G. Gaylord, *Reduction with Complex Metal Hydrides*, Interscience, New York, 1956, pp. 373–379.
235. R. P. Linstead and A. F. Milledge, *J. Chem. Soc.*, 478 (1936).
236. D. M. Bailey and R. E. Johnson, *J. Org. Chem.*, **35**, 3574 (1970).
237. J. J. Bloomfield and S. L. Lee, *J. Org. Chem.*, **32**, 3919 (1967).
238. F. Weygand, K. G. Kinkel and D. Tietjen, *Ber.*, **83**, 394 (1950).
239. V. Parrini, *Gazz. Chim. Ital.*, **87**, 1147 (1957); *Chem. Abstr.*, **52**, 10001d (1958).
240. R. Granger and H. Techer, *Compt. Rend.*, **250**, 142 (1960).
241. B. E. Cross, R. H. B. Galt and J. R. Hanson, *J. Chem. Soc.*, 5052 (1963).
242. W. R. Vaughan, C. T. Goetschel, M. H. Goodrow and C. L. Warren, *J. Amer. Chem. Soc.*, **85**, 2282 (1963).
243. D. G. Farmun and J. P. Snyder, *Tetrahedron Letters*, 3861 (1965).
244. B. E. Cross and J. C. Stewart, *Tetrahedron Letters*, 3589 (1968).
245. R. H. Schlessinger and I. S. Ponticello, *Chem. Commun.*, 1013 (1969).

246. S. S. G. Sircar, *J. Chem. Soc.*, 898 (1928).
247. S. Takano and K. Ogasawara, *Synthesis,* **42**, (1974).
248. B. E. Cross, R. H. B. Galt and J. R. Hanson, *J. Chem. Soc.*, 5052 (1963).
249. G. Snatzke and G. Zanati, *Ann.*, **684**, 62 (1965).
250. D. S. Noyce and D. B. Denney, *J. Amer. Chem. Soc.*, **72**, 5743 (1950).
251. C. S. Marvel and J. A. Fuller, *J. Amer. Chem. Soc.*, **74**, 1506 (1952).
252. C. Glaret, *Ann. Chim.* [12], 293 (1947).
253. H. Minato and I. Horibe, *Chem. Commun.*, 531 (1965).
254. J. A. Marshall, N. Cohen and F. R. Arenson, *J. Org. Chem.*, **30**, 762 (1965).
255. R. Pappo and C. J. Jung, *Tetrahedron Letters*, 365 (1962).
256. A. L. Nussbaum, F. E. Carlon, E. P. Oliveto, E. Townley, P. Kabasakalian and D. H. R. Barton, *Tetrahedron*, **18**, 373 (1962).
257. J. T. Edward and P. F. Morand, *Can. J. Chem.*, **38**, 1325 (1960).
258. C. C. Bolt, *Rec. Trav. Chim.*, **70**, 940 (1951).
259. T. Tsuda, K. Tanabe, I. Iwai and K. Funakoshi, *J. Amer. Chem. Soc.*, **79**, 5721 (1957).
260. K. J. Divakar, P. P. Sane and A. S. Rao, *Tetrahedron Letters,* 399 (1974).
261. K. Yamada, M. Kato, M. Iyoda and Y. Hirata, *Chem. Commun.,* 499 (1973).
262. E. Schwenk, D. Papa, H. Hankin and H. Ginsberg, *Org. Syntheses*, Coll. Vol. III, 742 (1955); D. Papa, E. Schwenk and H. Ginsberg, *J. Org. Chem.*, **16**, 253 (1951).
263. R. I. Longley, W. S. Emerson and T. C. Shafer, *J. Amer. Chem. Soc.*, **74**, 2012 (1952).
264. R. I. Longley and W. S. Emerson, *Org. Syntheses*, Coll. Vol. IV, 677 (1963).
265. L. P. Kyrides and F. B. Zienty, *J. Amer. Chem. Soc.*, **68**, 1385 (1946).
266. L. E. Schniepp and H. H. Geller, *J. Amer. Chem. Soc.*, **69**, 1545 (1947).
267. J. A. Berson and W. M. Jones, *J. Org. Chem.*, **21**, 1325 (1956).
268. E. S. Rothman, M. E. Wall and C. R. Eddy, *J. Amer. Chem. Soc.*, **76**, 527 (1954).
269. M. F. Murray, B. A. Johnson, R. L. Pederson and A. C. Ott, *J. Amer. Chem. Soc.*, **78**, 981 (1956).
270. P. E. Eaton, G. F. Cooper, R. C. Johnson and R. H. Mueller, *J. Org. Chem.*, **37**, 1947 (1972).
271. V. I. Sternberg and R. J. Perkins, *J. Org. Chem.*, **28**, 323 (1963).
272 M. Fetizon, M. Golfier and J.-M. Louis, *Tetrahedron*, **31**, 171 (1975).
273. S. Hauptmann and A. Blaskovits, *Z. Chem.*, **6**, 466 (1966); *Chem. Abstr.*, **66**, 55262e (1967).
274. P. Johnston, R. C. Sheppard, C. E. Stehr and S. Turner, *J. Chem. Soc. (C)*, 1847 (1966).
275. G. Defaye, M. Fetizon and M. C. Tromeur, *Compt. Rend. Ser. C.*, **265**, 1489 (1967).
276. R. C. Sheppard and S. Turner, *Chem. Commun.*, 682 (1968).
277. M. Fetizon, M. Golfier and J.-M. Louis, *Chem. Commun.*, 1118 (1969).
278. R. Hirschmann, N. G. Steinberg and R. Walker, *J. Amer. Chem. Soc.*, **84**, 1270 (1962).
279. D. H. R. Barton, J. M. Beaton, L. E. Geller and M. M. Pechet, *J. Amer. Chem. Soc.*, **83**, 4076 (1961).
280. C. H. Hassall, *Org. Reactions*, **9**, 73 (1957).
281. P. A. S. Smith in *Molecular Rearrangements*, Vol. 1. (Ed. P. de Mayo), Wiley–Interscience, New York, 1963, pp. 568–591.
282. K. Miyano, J. Ohfune, S. Azuma and T. Matsumoto, *Tetrahedron Letters*, 1545 (1974).
283. C. G. Overberger and H. Kaye, *J. Amer. Chem. Soc.*, **89**, 5640 (1967).
284. P. Soucy, T.-L. Ho and P. Deslongchamps, *Can. J. Chem.*, **50**, 2047 (1972).
285. G. Mehta and P. N. Pandey, *Synthesis*, 404 (1975).
286. V. R. Ghatak and B. Sanyal, *Chem. Commun.*, 876 (1974).
287. H. Levy and R. P. Jacobsen, *J. Biol. Chem.*, **171**, 171 (1947).
288. S. Mori and F. Mukawa, *Bull. Chem. Soc. Japan*, **27**, 479 (1954).
289. J. Meinwald, M. C. Seidel and B. C. Cadoff, *J. Amer. Chem. Soc.*, **80**, 6303 (1958).
290. R. R. Sauers, *J. Amer. Chem. Soc.*, **81**, 925 (1959).
291. R. R. Sauers and G. P. Ahearn, *J. Amer. Chem. Soc.*, **83**, 2759 (1961).
292. S. Hara, N. Matsumoto and M. Takeuchi, *Chem. Ind. (Lond.)*, 2086 (1962).
293. S. Hara, *Chem. Pharm. Bull. (Tokyo)*, **12**, 1531 (1964).
294. M. J. Bogdanowicz, T. Ambelang and B. M. Trost, *Tetrahedron Letters,* 923 (1973).
295. B. M. Trost and M. J. Bogdanowicz, *J. Amer. Chem. Soc.*, **95**, 5321 (1973).

296. J. A. Horton, M. A. Laura, S. M. Kalbag and R. C. Petterson, *J. Org. Chem.*, **34**, 3366 (1969).
297. D. C. Dittmer, R. A. Fouty and J. R. Potoski, *Chem. Ind.* (*Lond.*), 152 (1964).
298. K. Kirschke and H. Oberender, *German Patent*, 2,122,598; *Chem. Abstr.*, **76**, 126409c (1972).
299. E. J. Corey, Z. Arnold and J. Hutton, *Tetrahedron Letters*, 307 (1970).
300. P. A. Grieco, *J. Org. Chem.*, **37**, 2363 (1972).
301. N. M. Weinshenker and R. Stephensen, *J. Org. Chem.*, **37**, 3741 (1972).
302. E. J. Corey, N. M. Weinshenker, T. K. Schaaf and W. Huber, *J. Amer. Chem. Soc.*, **91**, 5675 (1969).
303. E. J. Corey and T. Ravindranathan, *Tetrahedron Letters*, 4753 (1961).
304. Y. Tsuda, T. Tanno, A. Ukai and K. Isobe, *Tetrahedron Letters*, 2009 (1971).
305. G. Buchi and I. M. Goldman, *J. Amer. Chem. Soc.*, **79**, 4741 (1957).
306. R. Hanna and G. Ourisson, *Bull. Soc. Chim. Fr.*, **10**, 3742 (1967).
307. K. Orito, R. H. Manske and R. Rodrigo, *J. Amer. Chem. Soc.*, **96**, 1944 (1974).
308. R. Sandmeier and C. Tamm, *Helv. Chim. Acta*, **56**, 2239 (1973).
309. R. D. Clark and C. H. Heathcock, *Tetrahedron Letters*, 1713, 2027 (1974).
310. M. Oyama and M. Ohno, *Tetrahedron Letters*, 5201 (1966).
311. F. Dubois, *Ann.*, **256**, 134 (1890).
312. R. P. Linstead and H. N. Rydon, *J. Chem. Soc.*, 580 (1933).
313. R. H. Cornforth, J. W. Cornforth and G. Popjak, *Tetrahedron*, **18**, 1351 (1962).
314. L. M. Berkowitz and P. N. Rylander, *J. Amer. Chem. Soc.*, **80**, 6682 (1958).
315. G. F. Reynolds, G. H. Rasmusson, L. Birladeanu and G. E. Arth, *Tetrahedron Letters*, 5057 (1970).
316. K. Heusler and A. Wettslein, *Helv. Chim. Acta*, **35**, 284 (1952).
317. C. S. Foote, M. T. Wuesthoff, S. Wexler, I. G. Burstain, R. Denny, G. O. Schenck and K.-H. Schute-Elte, *Tetrahedron*, **23**, 2583 (1967).
318. C. S. Foote, M. T. Wuesthoff and I. G. Burstain, *Tetrahedron*, **23**, 2601 (1967).
319. S. H. Schroeter, R. Appel, R. Brammer and G. O. Schenck, *Ann.*, **692**, 42 (1966).
320. E. Koch and G. O. Schenck, *Ber.*, **99**, 1984 (1966).
321. J. P. van der Merve and C. F. Garbess, *J. South African Inst.*, **17**, 149 (1964); *Chem. Abstr.*, **62**, 9088e (1965).
322. J. B. Bush, Jr and H. Finkbeiner, *J. Amer. Chem. Soc.*, **90**, 5903 (1968).
323. E. I. Heiba, R. M. Dessau and W. J. Koehl, Jr, *J. Amer. Chem. Soc.*, **90**, 5905 (1968).
324. A. Mee, *German Patent*, 1,927,233; *Chem. Abstr.*, **72**, 78456j (1970).
325. E. I. Heiba, R. M. Dessau and W. J. Koehl, Jr, *J. Amer. Chem. Soc.*, **90**, 5905 (1968).
326. E. I. Heiba, R. M. Dessau and W. J. Koehl, Jr, *J. Amer. Chem. Soc.*, **90**, 2706 (1968).
327. J. Grimaldi, M. Malacria and M. Bertrand, *Tetrahedron Letters*, 275 (1974).
328. J. Falbe, M. Huppes and F. Karte, *Ber.*, **97**, 863 (1964).
329. J. Falbe and F. Karte, *Angew. Chem.* (*Int. Ed Engl.*), **1**, 657 (1962).
330. E. R. H. Jones, T. Y. Shen. and M. C. Whiting, *J. Chem. Soc.*, 230 (1950).
331. J. Falbe, H.-J. Schulze-Steinen and F. Karte, *Ber.*, **98**, 886 (1965).
331. (a) A. Matsuda, *Bull. Chem. Soc. Japan*, **41**, 1876 (1968).
332. H. Koch and W. Haaf, *Angew. Chem.*, **70**, 311 (1958).
333. H. Koch and W. Haaf, *Angew. Chem.*, **72**, 628 (1960).
334. Y. Soma and H. Sano, *Japanese Patent*, 61,166 (1974); *Chem. Abstr.*, **81**, 120010f (1974).
335. M. Foa and L. Cassar, *Gazz. Chim. Ital.*, **103**, 805 (1973).
336. T. K. Das Gupta, D. Felix, U. M. Kempe and A. Eschenmoser, *Helv. Chim. Acta*, **55**, 2198 (1972).
337. M. Petrzilka, D. Felix and A. Eschenmoser, *Helv. Chim. Acta*, **56**, 2950 (1973).
338. K. Kondo and F. Mari, *Chem. Letters*, 741 (1974).
339. C. B. Chapleo, P. Hallett, B. Lythgoe and P. W. Wright, *Tetrahedron Letters*, 847 (1974).
340. R. R. Sauers and P. E. Sonnet, *Tetrahedron*, **20**, 1029 (1964).
341. E. W. Wornhoff and V. Dave, *Can. J. Chem.*, **44**, 621 (1966).
342. T. V. Mandelshtam, L. D. Kristol, L. A. Bogdanova and T. N. Ratnikova, *J. Org. Chem. U.S.S.R.*, **4**, 963 (1968).
343. J. Meinwald, A. Lewis and P. G. Gassman, *J. Amer. Chem. Soc.*, **84**, 977 (1962).

344. J. A. Marshall, F. N. Tuller and R. Ellison, *Synthetic Commun.*, **3**, 465 (1973).
345. J. Bus, H. Steinberg and Th. J. DeBoer, *Rec. Trav. Chem. Pays-Bas*, **91**, 657 (1972).
346. R. G. Carlson, J. H.-A. Huber and D. E. Henton, *Chem. Commun.*, 223 (1973).
347. R. K. Murray, Jr and D. L. Goff, *Chem. Commun.*, 881 (1973).
348. R. K. Murray, Jr, T. K. Morgan, Jr, J. A. S. Polley, C. A. Andruskiewicz, Jr and D. L. Goff, *J. Amer. Chem. Soc.*, **97**, 938 (1975).
349. R. C. Cookson, A. G. Edwards, J. Huder and M. Kingsland, *Chem. Commun.*, 98 (1965).
350. H. V. Hostettler, *Tetrahedron Letters*, 1941 (1965).
351. H. Nozaki, Z. Yamaguti and R. Noyari, *Tetrahedron Letters*, 37 (1965).
352. H. Nozaki, Z. Yamaguti, T. Okada, R. Noyari and M. Kawanisi, *Tetrahedron*, **23**, 3993 (1967).
353. D. H. R. Barton, A. L. J. Beckwith and A. Goosen, *J. Chem. Soc.*, 181 (1965).
354. B. Danieli, P. Manitto and G. Russo, *Chim. Ind. (Milan)*, **50**, 553 (1968); *Chem. Abstr.*, **69**, 43375t (1968).
355. W. Kirmse, H. Dietrich and H. W. Bücking, *Tetrahedron Letters*, 1833 (1967).
356. G. Lowe and J. Parker, *Chem. Commun.*, 577 (1971).
357. J. A. Kaufman and S. J. Weininger, *Chem. Commun.*, 593 (1969).
358. S. H. Schroeter, *Tetrahedron Letters*, 1591 (1969).
359. O. L. Chapman, P. W. Wojtkowski, W. Adam, O. Rodriquez and R. Rucktaschel, *J. Amer. Chem. Soc.*, **94**, 1365 (1972).
360. R. A. Dombro, *U.S. Patent*, 3,644,426; *Chem. Abstr.*, **76**, 139953a (1972).
361. H. Alper and W. G. Root, *Chem. Commun.*, 956 (1974).
362. Y. Mori and J. Tsuji, *Japanese Patent*, 73-68,544; *Chem. Abstr.*, **80**, 59724u (1974).
363. T. A. Pudova, F. K. Velichko, L. V. Vinogradova and R. Kh. Freidlina, *Izv. Akad. Nauk SSSR, Ser. Khim.*, 116 (1975); *Chem. Abstr.*, **82**, 111279k (1975).
364. A. Belanger and P. Brassard, *Chem. Commun.*, 863 (1972).
365. K. Jost and J. Rudinger, *Coll. Czech. Chem. Commun.*, **32**, 2485 (1967).
366. H. Sugano and M. Miyoshi, *Bull. Chem. Soc. Japan*, **46**, 669 (1973).
367. J. C. Sheehan and E. J. Corey, *Org. Reactions*, **9**, 388 (1957).
368. K. Dachs and E. Schwartz, *Angew. Chem.*, **1**, 430 (1962).
369. E. Testa, *Farmaco (Pavia) Ed. Sci.*, **17**, (1962); *Chem. Abstr.*, **57**, 9772c (1962).
370. R. Graf, G. Lohaus, K. Börner, E. Schmidt and H. Bestian, *Angew. Chem.*, **1**, 481 (1962).
371. E. Piovera, *Corriere Farm.*, **21**, 512 (1966); *Chem. Abstr.*, **66**, 84282s (1967).
372. L. L. Muller and J. Harmer, *1,2-Cycloaddition Reactions*, Wiley, New York, 1967, p. 173.
373. I. Lengyel and J. C. Sheehan, *Angew. Chem.*, **7**, 25 (1968).
374. A. L. J. Beckwith, *The Chemistry of Amides*, (Ed. J. Zabicky), Interscience, New York, 1970, Chap. 2, p. 73.
375. G. L'Abbé and A. Hassner, *Angew. Chem.*, **10**, 98 (1971).
376. F. Millich and K. V. Seshadri, *High Polymers*, (Ed. H. Mark, C. S. Marvel and H. W. Melville), Vol. 26 (Ed. K. C. Frisch), Wiley–Interscience, New York, 1972, Chap. 3, p. 179.
377. M. S. Manhas and A. K. Bose, *Chemistry of β-Lactams, Natural and Synthetic*, Part I, John Wiley and Sons, New York, 1971, Chap. 1.
378. E. G. E. Hawkins, *Angew. Chem.*, **12**, 783 (1973).
379. A. K. Mukerjee and R. C. Srivastava, *Synthesis*, 327 (1973).
380. W. M. Pearlman, *Org. Syntheses*, Coll. Vol. V, 670 (1973).
381. R. L. Frank, W. R. Schmitz and B. Zeidman, *Org. Syntheses*, Coll. Vol. III, 328 (1955).
382. M. Mityoshi, *Bull. Chem. Soc. Japan*, **46**, 212 (1973).
383. R. B. Moffett, *Org. Syntheses*, Coll. Vol. IV, 357 (1963).
384. P. Cefelin A. Frydrychova, J. Labsky, P. Schmidt and J. Sebenda, *Coll. Czech. Chem. Commun*, **32**, 2787 (1967).
385. P. Cefelin, J. Labsky and J. Sebenda, *Coll. Czech. Chem. Commun.*, **33**, 1111 (1968).
386. N. Anghelide, C. Draghici and D. Raileanu, *Tetrahedron*, **30**, 623 (1974).
387. R. F. Borch, C. V. Grudzinskas, P. A. Peterson and L. D. Weber, *J. Org. Chem.*, **37**, 1141 (1972).
388. R. H. Earle, Jr, D. T. Hurst and M. Viney, *J. Chem. Soc. (C)*, 2093 (1969).
389. R. Breckpot, *Bull. Soc. Chim. Belges.*, **32**, 412 (1923).
390. E. Testa, L. Fontanella and V. Aresi, *Ann.*, **673**, 60 (1964).

391. E. Testa, A. Bonati, G. Pagani and E. Gatti, *Ann.*, **647**, 92 (1961).
392. E. Testa, F. Fava and L. Fontanella, *Ann.*, **614**, 167 (1958).
393. A. Bonati, G. F. Christiani and E. Testa, *Ann.*, **647**, 83 (1961).
394. E. Testa and L. Fontanella, *Ann.*, **625**, 95 (1959).
395. E. Testa, L. Fontanella, G. F. Christiani and F. Fava, *Ann.*, **614**, 158 (1958).
396. E. Testa and L. Fontanella, *Ann.*, **661**, 187 (1963).
397. L. Fontanella and E. Testa, *Ann.*, **616**, 148 (1958).
398. E. Testa, L. Fontanella and G. F. Cristiani, *Ann.*, **626**, 121 (1959).
399. E. Testa, L. Fontanella and L. Mariani, *Ann.*, **660**, 135 (1962).
400. E. Testa, L. Fontanella, L. Mariani and G. F. Cristiani, *Ann.*, **639**, 157 (1961).
401. E. Testa, L. Fontanella and V. Aresi, *Ann.*, **656**, 114 (1962).
402. G. Cignarella, G. F. Cristiani and E. Testa, *Ann.*, **661**, 181 (1963).
403. J. L. Luche and H. B. Kagan, *Bull. Soc. Chim. Fr.*, 3500 (1969).
404. F. F. Blicke and W. A. Gould, *J. Org. Chem.*, **23**, 1102 (1958).
405. P. G. Gassman and T. J. van Bergen, *J. Amer. Chem. Soc.*, **95**, 2718 (1973).
406. P. G. Gassman, T. J. van Bergen and G. Gruetzmacher, *J. Amer. Chem. Soc.*, **95**, 6508 (1973).
407. P. G. Gassman and T. J. van Bergen, *J. Amer. Chem. Soc.*, **96**, 5508 (1974).
408. H. C. Beyerman and W. M. van den Brink, *Proc. Chem. Soc.*, 226 (1963).
409. H. T. Openshaw and N. Whittaker, *J. Chem. Soc. (C)*, 89 (1969).
410. A. R. Battersby, J. F. Beck and E. McDonald, *J. Chem. Soc., Perkin I*, 160 (1974).
411. K. P. Klein and H. K. Reimschuessel, *J. Polym. Sci., A-1*, **10**, 1987 (1972).
412. K. P. Klein and H. K. Reimschuessel, *J. Polym. Sci., A-1*, **9**, 2717 (1971).
413. H. K. Reimschuessel, K. P. Klein and G. J. Schmitt, *Macromolecules*, **2**, 567 (1969).
414. G. H. Cocalas and W. H. Hartung, *J. Amer. Chem. Soc.*, **79**, 5203 (1957).
415. G. H. Cocalas, S. Avakian and G. J. Martin, *J. Org. Chem.*, **26**, 1313 (1961).
416. K. L. Kirk and L. A. Cohen, *J. Org. Chem.*, **34**, 395 (1969).
417. K. L. Kirk and L. A. Cohen, *J. Amer. Chem. Soc.*, **94**, 8142 (1972).
418. H. E. Baumgarten, R. L. Zey and U. Krolls, *J. Amer. Chem. Soc.*, **83**, 4469 (1961).
419. H. E. Baumgarten, *J. Amer. Chem. Soc.*, **84**, 4975 (1962).
420. H. E. Baumgarten, J. F. Fuerholzer, R. D. Clark and R. D. Thompson, *J. Amer. Chem. Soc.*, **85**, 3303 (1963).
421. J. C. Sheehan and I. Lengyel, *J. Amer. Chem. Soc.*, **86**, 746, 1356 (1964).
422. I. Lengyel and J. C. Sheehan, *Angew. Chem. (Int. Ed. Engl.)*, **7**, 25 (1968).
423. E. R. Talaty, C. M. Utermoehlen and L. H. Stekoll, *Synthesis*, 543 (1971).
424. E. R. Talaty, J. P. Madden and L. H. Stekoll, *Angew. Chem. (Int. Ed. Engl.)*, **10**, 753 (1971).
425. M. S. Manhas and S. J. Jeng, *J. Org. Chem.*, **32**, 1246 (1967).
426. H. Wamhoff and F. Karte, *Ber.*, **100**, 2122 (1967).
427. E. Testa, B. J. R. Nicolaus, E. Bellasio and L. Mariani, *Ann.*, **673**, 71 (1964).
428. I. L. Knunyants and N. P. Gambaryan, *Izv. Akad. Nauk SSSR, Otdl. Khim. Nauk*, 1037 (1955).
429. I. L. Knunyants and N. P. Gambaryan, *Izv. Akad. Nauk SSSR, Otdl. Khim. Nauk*, 834 (1957).
430. I. L. Knunyants, E. E. Rytslin and N. P. Gambaryan, *Izv. Akad. Nauk SSSR, Otdl. Khim.* Nauk, 83 (1961).
431. A. K. Bose, B. Anjaneyulu, S. K. Bhattacharya and M. S. Manhas, *Tetrahedron*, **23**, 4769 (1967).
432. J. C. Sheehan and A. K. Bose, *J. Amer. Chem. Soc.*, **72**, 5158 (1950).
433. J. C. Sheehan and A. K. Bose, *J. Amer. Chem. Soc.*, **73**, 1761 (1951).
434. A. K. Bose, B. N. Ghosh-Mazumdar and B. G. Chatterjee, *J. Amer. Chem. Soc.*, **82**, 2382 (1960).
435. A. K. Bose and M. S. Manhas, *J. Org. Chem.*, **27**, 1244 (1962).
436. A. K. Bose, M. S. Manhas and B. N. Ghosh-Mazumdar, *J. Org. Chem.*, **27**, 1458 (1962).
437. B. G. Chatterjee and R. F. Abdulla, *Z. Naturforsch. (B)*, **26**, 395 (1971).
438. B. G. Chatterjee and N. L. Nyss, *Z. Naturforsch. (B)*, **26**, 395 (1971).
439. B. G. Chatterjee and V. V. Rao, *Tetrahedron*, **23**, 487 (1967).

440. B. G. Chatterjee, V. V. Rao, S. K.Roy and H. P. S. Chawla, *Tetrahedron*, **23**, 493 (1967).
441. B. G. Chatterjee, P. N. Moza and S. K. Roy, *J. Org. Chem.*, **28**, 1418 (1963).
442. B. G. Chatterjee, V. V. Rao and B. N. Ghosh-Mazumdar, *J. Org. Chem.*, **30**, 4101 (1965).
443. T. Okawara and K. Harada, *J. Org. Chem.*, **37**, 3286 (1972).
444. H. Bohme, S. Ebel and K. Hartke, *Ber.*, **98**, 1463 (1965).
445. A. K. Bose, M. S. Manhas and R. M. Ramer, *Tetrahedron*, **21**, 449 (1965).
446. M. W. Rutenberg and E. C. Horning, *Org. Syntheses*, Coll. Vol. IV, 620 (1963).
447. H. Plieninger and A. Muller, *Synthesis*, 586 (1970).
448. W. H. Puterbaugh and C. R. Hauser, *J. Org. Chem.*, **29**, 853 (1964).
449. R. L. Vaux, W. H. Puterbaugh and C. R. Hauser, *J. Org. Chem.*, **29**, 3514 (1964).
450. I. T. Barnish, C. -L. Mao, R. L. Gay and C. R. Hauser, *Chem. Commun.*, 564 (1968).
451. E. M. Levi, C. -L. Mao and C. R. Hauser, *Can. J. Chem.*, **47**, 3671 (1969).
452. C. -L. Mao, I. T. Barnish and C. R. Hauser, *J. Heterocyclic Chem.*, **6**, 83 (1969).
453. C. -L. Mao and C. R. Hauser, *J. Org. Chem.*, **35**, 3704 (1970).
454. P. A. Petyunin and G. P. Petyunin, *U.S.S.R. Patent*, 371221 (1973); *Chem. Abstr.*, **79**, 31833j (1973).
455. J. F. Bagli and H. Immer, *J. Org. Chem.*, **35**, 3499 (1970).
456. E. W. Colvin, J. Martin, W. Parker, R. A. Raphael, B. Shroot and M. Doyle, *J. Chem. Soc., Perkin I*, 860 (1972).
457. O. L. Chapman and W. R. Adams, *J. Amer. Chem. Soc.*, **89**, 4243 (1967).
458. O. L. Chapman and W. R. Adams, *J. Amer. Chem. Soc.*, **90**, 2333 (1968).
459. P. G. Cleveland and O. L. Chapman, *Chem. Commun.*, 1064 (1967).
460. M. Ogata and H. Matsumoto, *Chem. Phar. Bull.*, **20**, 2264 (1972).
461. Y. Ogata, K. Takagi and I. Ishino, *J. Org. Chem.*, **36**, 3975 (1971).
462. B. S. Thyagarajan, N. Kharasch, H. B. Lewis and W. Wolf, *Chem. Commun.*, 614 (1967).
436. D. H. Hey, G. H. Jones and M. J. Perkins, *J. Chem. Soc. (C)*, 120 (1971).
464. A. Mondon and K. Krohn, *Ber.*, **105**, 3726 (1972).
465. G. R. Lenz, *J. Org. Chem.*, **39**, 2839 (1974).
466. G. R. Lenz, *J. Org. Chem.*, **39**, 2846 (1974).
467. T. Hasegawa and H. Aoyama, *Chem. Commun.*, 743 (1974).
468. N. C. Yang, A. Shani and G. R. Lenz, *J. Amer. Chem. Soc.*, **88**, 5369 (1966).
469. Y. Kanaoka and K. Itoh, *Synthesis*, 36 (1972).
470. E. Winterfeldt and H. J. Altmann, *Angew. Chem. (Int. Ed. Engl.)*, **7**, 466 (1968).
471. I. Ninomiya and T. Naito, *Chem. Commun.*, 137 (1973).
472. S. M. Kupchan, J. L. Moniot, R. M. Kanojia and J. B. O'Brien, *J. Org. Chem.*, **36**, 2413 (1971).
473. I. Ninomiya, T. Naito and S. Higuchi, *Chem. Commun.*, 1662 (1970).
474. I. Ninomiya, T. Naito, T. Kiguchi and T. Mori, *J. Chem. Soc., Perkin I*, 1696 (1973).
475. I. Ninomiya, T. Naito and T. Kiguchi, *Tetrahedron Letters*, 4451 (1970).
476. I. Ninomiya, T. Naito and T. Mori, *Tetrahedron Letters*, 2259 (1969).
477. I Ninomiya, T. Naito and T Mori, *Tetrahedron Letters*, 3643 (1969).
478. O. Yonemitsu, T. Tokuyama, M. Chaykovsky and B. Witkop, *J. Amer. Chem. Soc.*, **90**, 776 (1968).
479. O. Yonemitsu, B. Witkop and I. L. Karle, *J. Amer. Chem. Soc.*, **89**, 1039 (1967).
480. O. Yonemitsu, Y. Okuno, Y. Kanaoka and B. Witkop, *J. Amer. Chem. Soc.*, **92**, 5686 (1970).
481. O. Yonemitsu, H. Nakai, Y. Kanaoka, I. L. Karle and B. Witkop, *J. Amer. Chem. Soc.*, **91**, 4591 (1969).
482. T. Kobayashi, T. F. Spande, H. Aoyagi and B. Witkop, *J. Med. Chem.*, **12**, 636 (1969).
483. O. Yonemitsu, P. Cerutti and B. Witkop, *J. Amer. Chem. Soc.*, **88**, 3941 (1966).
484. B. Akermark, N.-G. Johansson and B. Sjoberg, *Tetrahedron Letters*, 371 (1969).
485. K. R. Henery-Logan and C. G. Chen, *Tetrahedron Letters*, 1103 (1973).
486. R. R. Rando, *J. Amer. Chem. Soc.*, **92**, 6706 (1970).
487. R. H. Earle, Jr, D. T. Hurst and M. Viney, *J. Chem. Soc. (C)*, 2093 (1969).
488. G. Lowe and J. Parker, *Chem. Commun.*, 577 (1971).
489. E. J. Corey and A. M. Felix, *J. Amer. Chem. Soc.*, **87**, 2518 (1965).
490. D. M. Brunwin, G. Lowe and J. Parker, *Chem. Commun.*, 865 (1971).

491. D. M. Brunwin, G. Lowe and J. Parker, *J. Chem. Soc.* (*C*), 3756 (1971).
492. G. Lowe and M. V. J. Ramsay, *J. Chem. Soc., Perkin I*, 479 (1973).
493. M. Perelman and S. A. Mizsak, *J. Amer. Chem. Soc.*, **84**, 4988 (1962).
494. R. Graf, *Ber.*, **89**, 1071 (1956).
495. R. Graf, *Org. Syntheses*, Coll. Vol. V, 226 (1973).
496. R. Graf, *Angew. Chem.* (*Int. Ed. Engl.*), 7, 172 (1968).
497. H. Ulrich, *Chem. Rev.*, **65**, 369 (1965).
498. R. Graf, *Ann.*, **661**, 111 (1963).
499. E. J. Moriconi and W. C. Crawford, *J. Org. Chem.*, **33**, 370 (1968).
500. T. Durst and M. J. O'Sullivan, *J. Org. Chem.*, **35** 2043 (1970).
501. Farbwerke Hoechst A.-G., French Patent, 2016990 (1970); *Chem. Abstr.*, **75**, 63164a (1971).
502. R. Graf, *Org. Syntheses*, Coll. Vol. V, 673 (1973).
503. H. Bestian, H. Biener, K. Clauss and H. Heyn, *Ann.*, **718**, 94 (1968).
504. E. J. Moriconi and W. C. Meyer, *Tetrahedron Letters*, 3823 (1968).
505. E. J. Moriconi and P. H. Mazzochi, *J. Org. Chem.*, **31**, 1372 (1966).
506. E. J. Moriconi and J. F. Kelly, *J. Amer. Chem. Soc.*, **88**, 3657 (1966).
507. M. Perelman and S. A. Mizsak, *J. Amer. Chem. Soc.*, **84**, 4988 (1962).
508. G. Opitz and J. Koch, *Angew. Chem.*, **75**, 167 (1963).
509. W. P. Giering and M. Rosenblum, *J. Amer. Chem. Soc.*, **93**, 5299 (1971).
510. W. P. Giering, S. Raghu, M. Rosenblum, A. Cutler, D. Ehntholt and R. W. Fish, *J. Amer. Chem. Soc.*, **94**, 8251 (1972).
511. Naser-ud-din, J. Riegl and L. Skattebol, *Chem. Commun.*, 271 (1973).
512. G. Merault, P. Bourgesis and N. Duffaut, *Bull. Soc. Chim. Fr.*, 1949 (1974).
513. I. Yamamoto, S. Yanagi, A. Mamba and H. Gotoh, *J. Org. Chem.*, **39**, 3924 (1974).
514. I. Yamamoto, Y. Tabo, H. Gotoh, T. Minami, Y. Ohshiro and T. Agawa, *Tetrahedron Letters*, 2295 (1971).
515. H. Staudinger, *Die Ketene*, F. Enke Verlag, Stuttgart, 1912.
516. W. T. Brady, E. D. Dorsey and F. H. Parry, III, *J. Org. Chem.*, **34** 2846 (1969).
517. A. Gomes and M. M. Joullie, *Chem. Commun.*, 935 (1967).
518. J.-L. Luche and H. B. Kagan, *Bull. Soc. Chim. Fr.*, 2450 (1968).
519. J. C. Sheehan and J. J. Ryan, *J. Amer. Chem. Soc.*, **73**, 1204 (1951).
520. A. K. Bose, G. Spiegelman and M. S. Manhas, *Tetrahedron Letters*, 3167 (1971).
521. H. Staudinger, *Ber.*, **50**, 1035 (1917).
522. J. Berson and W. M. Jones, *J. Amer. Chem. Soc.*, 78, 1625 (1956).
523. R. Pfleger and A. Jäger, **90**, 2460 (1957).
524. R. Graf, *Ann.*, **661**, 111 (1963).
525. W. Kirmse and L. Horner, *Ber.*, **89**, 2759 (1956).
526. W. T. Brady and E. F. Hoff, Jr, *J. Amer. Chem. Soc.*, **90**, 6256 (1968).
527. H. Staudinger and H. W. Klever, *Ber.*, **40**, 1149 (1907).
528. A. D. Holley and R. W. Holley, *J. Amer. Chem. Soc.*, **73**, 3172 (1951).
529. H. Staudinger, H. W. Klever and P. Kober, *Ann.*, **374**, 1 (1910).
530. R. N. Pratt, G. A. Taylor and S. A. Proctor, *J. Chem. Soc.* (*C*), 1569 (1967).
531. S. A. Ballard, D. S. Melstrom and C. W. Smith, *The Chemistry of Penicillin* (Ed. H. T. Clarke, J. R. Johnson and R. Robinson), Princeton University Press, Princeton, New Jersey, 1949, pp. 977, 984, 991, 992.
532. J. Decazes, J. L. Luche and H. B. Kagan, *Tetrahedron Letters*, 3661, 3665 (1970).
533. H. Staudinger, *Ber.*, **40**, 1145 (1907).
534. H. Staudinger, *Ann.*, **356**, 51 (1907).
535. H. Staudinger and S. Jelagin, *Ber.*, **44** 365 (1911).
536. H. Staudinger, *Ber.*, **40**, 1147 (1907).
537. H. B. Kagan and J. L. Luche, *Tetrahedron Letters*, 3093 (1968).
538. R. Huisgen, B. A. Davis and M. Morikawa, *Angew. Chem. (Int. Ed. Engl.)*, **7**, 826 (1968).
539. M. D. Bachi and M. Rothfield, *J. Chem. Soc., Perkin I*, 2326 (1972).
540. M. D. Bachi and O. Goldberg, *J. Chem. Soc., Perkin I*, 2332 (1972).
541. E. Fahr, K. Döppert, K. Königsdorfer and F. Scheckenbach, *Tetrahedron*, **24**, 1011 (1968).

542. A. K. Bose and I. Kugajevsky, *Tetrahedron,* **23**, 957 (1967).
543. R. D. Kimbrough, Jr, *J. Org. Chem.*, **29**, 1242 (1964).
544. E. Fahr, K. H. Keil, F. Scheckenbach and A. Jung, *Angew. Chem.* (*Int. Ed. Engl.*), **3**, 646 (1964).
545. A. K. Bose, Y. H. Chiang and M. S. Manhas, *Tetrahedron Letters*, 4091 (1972).
546. A. K. Bose, C. S. Narayanan and M. S. Manhas, *Chem. Commun.*, 975 (1970).
547. M. S. Manhas, J. S. Chib, Y. H. Chiang and A. K. Bose, *Tetrahedron*, **25**, 4421 (1969).
548. A. K. Bose, B. Anjaneyulu, S. K. Bhattacharya and M. S. Manhas, *Tetrahedron*, **23**, 4769 (1967).
549. S. M. Deshpande and A. K. Mukerjee, *Indian J. Chem.*, **4**, 79 (1966).
550. E. Ziegler, Th. Wimmer and H. Mittelbach, *Monatsh. Chem.*, **99**, 2128 (1968).
551. D. A. Nelson, *Tetrahedron Letters*, 2543 (1971).
552. D. A. Nelson, *J. Org. Chem.*, **37**, 1447 (1972).
553. L. Paul, A Draeger and G Hilgetag, *Ber.*, **99**, 1957 (1966).
554. G. Hilgetag, L. Paul and A. Draeger, *Ber.*, **96**, 1697 (1963).
555. L. Paul and K. Zieloff, *Ber.*, **99**, 1431 (1966).
556. M. D. Bachi and O. Goldberg, *Chem. Commun.*, 319 (1972).
557. A. K. Bose, S. D. Sharma, J. C. Kapur and M. S. Manhas, *Synthesis*, 216 (1973).
558. J. N. Wells, and R. E. Lee, *J. Org. Chem.*, **34**, 1477 (1969).
559. R. Lattrell, *Angew. Chem.* (*Int. Ed. Engl.*), **12**, 925 (1973).
560. A. K. Bose, B. Dayal, H. P. S. Chawla and M. S. Manhas, *Tetrahedron Letters*, 2823 (1972).
561. J. L. Fahey, B. C. Lange, J. M. Van der Veen, G. R. Young and A. K. Bose, *J. Chem. Soc., Perkin I*, 1117 (1977).
562. A. K. Bose and J. L. Fahey, *J. Org. Chem.*, **39**, 115 (1974).
563. A. K. Bose, M. Tsai, S. D. Sharma and M. S. Manhas, *Tetrahedron Letters*, 3851 (1973).
564. A. K. Bose, M. Tsai, J. C. Kapur and M. S. Manhas, *Tetrahedron*, **29**, 2355 (1973).
565. A. K. Bose, J. L. Fahey and M. S. Manhas, *J. Heterocyclic Chem.*, **10**, 791 (1973).
566. A. K. Bose, M. S. Manhas, J. S. Chib, H. P. S. Chawla and B. Dayal, *J. Org. Chem.*, **39**, 2877 (1974).
567. R. W. Ratcliffe and B. G. Christensen, *Tetrahedron Letters*, 4649 (1973).
568. A. K. Bose, G. Spiegelman and M. S. Manhas, *J. Chem. Soc.* (*C*), 2468 (1971).
569. J. A. Erickson, *Ph.D. Thesis*, M.I.T., 1953, as reported in Reference 568.
570. A. K. Bose, G. Spiegelman and M. S. Manhas, *J. Chem. Soc.* (*C*), 188 (1971).
571. A. K. Bose, V. Sudarsanam. B. Anjaneyulu and M. S. Manhas, *Tetrahedron*, **25**, 1191 (1969).
572. A. K. Bose, G. Spiegelman and M. S. Manhas, *J. Amer. Chem. Soc.*, **90**, 4506 (1968).
573. A. K. Bose, G. Spiegelman and M. S. Manhas, *Chem. Commun.*, 321 (1968).
574. A. K. Bose and B. Anjaneyulu, *Chem. Int.* (*Lond.*), 903 (1966).
575. A. K. Bose, H. P. S. Chawla, B. Dayal and M. S. Manhas, *Tetrahedron Letters*, 2503 (1973).
576. A. K. Bose, J. C. Kapur, S. D. Sharma and M. S. Manhas, *Tetrahedron Letters*, 2319 (1973).
577. J. C. Sheehan and G. D. Laubach, *J. Amer. Chem. Soc.*, **73**, 4376 (1951).
578. J. C. Sheehan, E. L. Buhle, E. J. Corey, G. D. Laubach and J. J. Ryan, *J. Amer. Chem. Soc.*, **72**, 3828 (1950).
579. J. C. Sheehan and G. D. Laubach, *J. Amer. Chem. Soc.*, **73**, 4752 (1951).
580. J. C. Sheehan and E. J. Corey, *J. Amer. Chem. Soc.*, **73**, 4756 (1951).
581. J. C. Sheehan and J. J. Ryan, *J. Amer. Chem. Soc.*, **73**, 4367 (1951).
582. J. C. Sheehan, H. W. Hill, Jr and E. L. Buhle, *J. Amer. Chem. Soc.*, **73**, 4373 (1951).
583. S. M. Deshpande and A. K. Mukerjee, *J. Chem. Soc.* (*C*), 1241 (1966).
584. L. Paul, P. Polczynski and G. Hilgetag, *Ber.*, **100**, 2761 (1967).
585. J. C. Sheehan and M. Dadic, *J. Heterocyclic Chem.*, **5**, 779 (1968).
586. E. Ziegler and Th. Wimmer, *Ber.*, **99**, 130 (1966).
587. F. Duran and L. Ghosez, *Tetrahedron Letters*, 245 (1970).
588. A. K. Bose, B. Lal, B. Dayal and M. S. Manhas, *Tetrahedron Letters*, 2633 (1974).
589. R. Hull, *J. Chem. Soc.* (*C*), 1154 (1967).

590. R. W. Ratcliffe and B. G. Christensen, *Tetrahedron Letters*, 4653 (1973).
591. R. L. Bentley and H. Suschitzky, *J. Chem. Soc., Perkin I*, 1725 (1976).
592. E. Ziegler and G. Kleineberg, *Monatsh. Chem.*, **96**, 1296 (1965).
593. A. K. Bose, J. C. Kapur, B. Dayal and M. S. Manhas, *Tetrahedron Letters*, **39**, 3797 (1973).
594. A. K. Bose, J. C. Kapur, B. Dayal and M. S. Manhas, *J. Org. Chem.*, **39**, 312 (1974).
595. G. O. Schenck and N. Engelhord, *Angew. Chem.*, **68**, 71 (1956).
596. L. Horner, E. Spietschka and A. Gross, *Ann.*, **573**, 17 (1951).
597. A. H. Cook and D. G. Jones, *J. Chem. Soc.*, 184 (1941).
598. C. K. Ingold and S. D. Weaver, *J. Chem. Soc.*, 378 (1925).
599. L. Horner and E. Spietschka, *Ber.*, 89, 2765 (1956).
600. M. L. Scheinbaum. *J. Org. Chem.*, **29**, 2200 (1964).
601. F. F. Blicke and W. A. Gould, *J. Org. Chem.*, **23**, 1102 (1958).
602. H. B. Kagan, J.-J. Basselier and J.-L. Luche, *Tetrahedron Letters*, 941 (1964).
603. J.-L. Luche and H. B. Kagan, *Bull. Soc. Chim. Fr.*, 3500 (1969).
604. H. Gilman and M. Speeter, *J. Amer. Chem. Soc.*, **65**, 2255 (1943).
605. J.-L. Luche and H. B. Kagan, *Bull. Soc. Chim. Fr.*, 1680 (1969).
606. J.-L. Luche, H. B. Kagan, R. Parthasarathy, G. Tsoucaris, C. de Rango and C. Zeliver, *Tetrahedron*, **24**, 1275 (1968).
607. E. Ziegler and G. Kleineberg, *Monatash. Chem.*, **96, 1360 (1965).**
608. V. Gomez Aranda, J. Barluenga and V. Gotor, *Tetrahedron Letters*, 977 (1974).
609. M. De Poortere, J. Marchand-Bryaert and L. Ghosez, *Angew. Chem. (Int. Ed. Engl.)*, **13**, 267 (1974).
610. H. Staudinger and M. Miescher, *Helv. Chim. Acta*, **2**, 564 (1919).
611. T. W. J. Taylor, J. S. Owen and D. Whittaker, *J. Chem. Soc.*, 206 (1938).
612. C. H. Hassall and A. E. Lippman, *J. Chem. Soc.*, 1059 (1953).
613. M. Kinugasa and S. Hashimoto, *Chem. Commun.*, 466 (1972).
614. J. C. Sheehan and A. K. Bose, *J. Amer. Chem. Soc.*, **72**, 5158 (1950).
615. G. Lowe and D. D. Ridley, *Chem. Commun.*, 328 (1973).
616. G. Lowe and D. D. Ridley, *J. Chem. Soc., Perkin I*, 2024 (1973).
617. G. Stork and R. P. Szajewski, *J. Amer. Chem. Soc.*, **96**, 5787 (1974).
618. C. E. Hatch and P. Y. Johnson, *Tetrahedron Letters*, 2719 (1974).
619. M. F. Chasle and A. Foucaud, *Compt. Rend. (C)*, **270**, 1045 (1970).
620. L. G. Donaruma and W. Z. Heldt, *Org. Reactions*, **11**, 1 (1960).
621. P. A. S. Smith in *Molecular Rearrangements*, (Ed P. de Mayo), Part 1, Interscience, New York, 1963, pp. 483–507.
622. F. Uhlig and H. R. Snyder, *Adv. Org. Chem.*, **1**, 35–81 (1960).
623. Tayo Rayon Co., Ltd., *British Patent*, 1188217; *Chem. Abstr.*, **76**, 59482f (1972).
624. Y. Tamura, H. Fujiwara, K. Sumoto, M. Ikeda and Y. Kita, *Synthesis*, 215 (1973).
625. Inventa A.-G., *British Patent*, 1105805; *Chem. Abstr.*, **68**, 95349m (1968).
626. T. Sonoda, M. Kato and S. Wakamatsu, Japanese Patent, 74135985; *Chem. Abstr.*, **82**, 156964w (1975).
627. S. Yura and K. Horiguchi, Japanese Patent, 7503317; *Chem. Abstr.*, **82**, 156979e (1975).
628. N. V. Stamicarbon, *Dutch Patent*, 6607466; *Chem. Abstr.*, **69**, 18634u (1968).
629. C. G. Overberger and H. Jabloner, *J. Amer. Chem. Soc.*, **85**, 3431 (1963).
630. E. Wenkert and B. F. Barnett, *J. Amer. Chem. Soc.*, **82**, 4671 (1966).
631. F. J. Donat and A. L. Nelson, *J. Org. Chem.*, **22**, 1107 (1957).
632. O. D. Strizhakov, E. N. Zil'berman and S. V. Svetozarskii, *Zh. Obsh. Khim.*, **35**, 628 (1965); *Chem. Abstr.*, **63**, 5535a (1965).
633. E. C. Horning, V. L. Stromberg and H. A. Lloyd, *J. Amer. Chem. Soc.*, **74**, 5153 (1952).
634. R. S. Montgomery and G. Dougherty, *J. Org. Chem.*, **17**, 823 (1952).
635. Y. Tamura, Y. Kita, Y. Matsutaka and M. Terashima, *Chem. Ind. (Lond.)*, 1350 (1970).
636. R. H. Mazur, *J. Org. Chem.*, **26**, 1289 (1961).
637. Y. Tamura, Y. Kita and M. Terashima, *Chem. Pharm. Bull.*, 529 (1971).
638. T. H. Koch, M. A. Geigel and C.-C. Tsai, *J. Org. Chem.*, **38**, 1090 (1973).
639. A. Zabza, H. Kuczynski, Z. Chabudzinski and D. Sedzik-Hibner, *Bull. Acad. Pol. Sci., Ser. Sci. Chim.*, **20**, 841 (1972); *Chem. Abstr.*, **78**, 124728y (1973).

640. T. Duong, R. H. Prager, J. M. Tippett, A. D. Ward and D. I. B. Kerr, *Australian J. Chem.*, **29**, 2667 (1976).
641. P. Rosenmund, D. Sauer and W. Trommer, *Ber.*, **103**, 496 (1970).
642. R. H. Mazur, *J. Amer. Chem. Soc.*, **81**, 1454 (1959).
643. J. Takeuchi and F. Iwata, *Japanese Patent*, 7456972; *Chem. Abstr.*, **83**, 59749t (1975).
644. H. Wolff, *Org. Reactions*, **3**, 307 (1946).
645. P. A. S. Smith in *Molecular Rearrangements*, (Ed. P. de Mayo), Part 1, Interscience, New York, 1963, pp. 507–527.
646. H. Shechter and J. C. Kirk, *Amer. Chem. Soc.*, **73**, 3087 (1951).
647. R. T. Conley, *J. Org. Chem.*, **23**, 1330 (1958).
648. T. Duong. R. H. Prager, A. D. Ward and D. I. B. Kerr, *Australian J. Chem.*, **29**, 2651 (1976).
649. Y. Tamura and Y. Kita, *Chem. Pharm. Bull.*, **19**, 1735 (1971).
650. Y. Tamura, Y. Yoshimura and Y. Kita, *Chem. Pharm. Bull.*, **19**, 1068 (1971).
651. K. Folkers, D. Misiti and H. W. Moore, *Tetrahedron Letters*, 1071 (1965).
652. D. Misiti, H. W. Moore and K. Folkers, *Tetrahedron*, **22**, 1201 (1966).
653. R. W. Richards and R. M. Smith, *Tetrahedron Letters*, 2361 (1966).
654. C. R. Bedford, G. Jones and B. R. Webster, *Tetrahedron Letters*, 2367 (1966).
655. C. S. Barnes, D. H. R. Barton, J. S. Fawcett and B. R. Thomas, *J. Chem. Soc.*, 2339 (1952).
656. K. A. Mueller and W. Kirchhof, *German Patent*, 1242621; *Chem. Abstr.*, **67**, 100022k (1967).
657. H. H. Wasserman, R. E. Cochoy and M. S. Baird, *J. Amer. Chem. Soc.*, **91**, 2375 (1969).
658. H. H. Wasserman, H. W. Adickes and O. Espejo de Ochoa, *J. Amer. Chem. Soc.*, **93**, 5586 (1971).
659. H. H. Wasserman E. L. Glazer and M. J. Hearn, *Tetrahedron Letters*, 4855 (1973).
660. H. H. Wasserman and E. L. Glazer, *J. Org. Chem.*, **40**, 1505 (1975).
661. A. F. M. Iqbal, *Tetrahedron Letters*, 3381 (1971).
662. C. W. Capp, K. W. Denbigh, P. J. Durston and B. W. Harris, *U.S. Patent*, 3583982; *Chem. Abstr.*, **75**, 89054q (1971).
663. H. L. Ammon, P. H. Mazzocchi, W. J. Kopecky, Jr, H. J. Tamburin and P. H. Watts, Jr, *J. Amer. Chem. Soc.*, **95**, 1968 (1973).
664. P. H. Mazzocchi, T. Halchak and H. J. Tamburin, *J. Org. Chem.*, **41**, 2808 (1976).
665. E. Oliveros-Desherces, M. Riviere, J. Parello and A. Lattes, *Compt. Rend (C)*, **275**, 581 (1972).
666. E. Oliveros-Desherces, M. Riviere, J. Parello and A. Lattes, *Tetrahedron Letters*, 851 (1975).
667. H. Krimm. *Ber.* , **91**, 1057 (1958).
668. E. Schmitz, H. U. Heyne and S. Schramm, *German Patent*, 2055165; *Chem. Abstr.*, **75**, 152320r (1971).
669. S. T. Reid, J. N. Tucker and E. J. Wilcox, *J. Chem. Soc., Perkin I*, 1359 (1974).
670. G. Ribaldone and A. Nenz, *Chem. Ind. (Milan)*, **49**, 701 (1967); *Chem. Abstr.*, **67**, 116621r (1967).
671. G. Ribaldone, F. Smai and G. Borsotti, U.S. Patent, 3328394; *Chem. Abstr.*, **68**, 59137g (1968).
672. J. Deyrup and S. C. Clough, *J. Amer. Chem. Soc.*, **91**, 4590 (1969).
673. D. St. C. Black, F. W. Eastwood, R. Okraglik, A. J. Poynton, A. M. Wade and C. H. Welker, *Australian J. Chem.*, **25**, 1483 (1972).
674. P. G. Gassman and B. L. Fox, *J. Org. Chem.*, **31**, 982 (1966).
675. T. Durst and M. J. LeBelle, *Can. J. Chem.*, **50**, 3196 (1972).
676. T. Durst, R. Van Den Elzen and R. Legault, *Can. J. Chem.*, **52**, 3206 (1974).
677. B. M. Trost and R. A. Kunz, *J. Org. Chem.*, **39**, 2476 (1974).
678. B. M. Trost and R. A. Kunz, *J; Amer. Chem. Soc.*, **97**, 7152 (1975).
679. See E. W. H. Böhme, H. E. Applegate, J. B. Ewing, P. T. Funke, M. S. Puar and J. E. Dolfini, *J. Org. Chem.*, **38**, 230 (1973) and references cited therein.
680. E. H. W. Böhme, H. E. Applegate, B. Toplitz, J. E. Dolfini and J. Z. Gougoutas, *J. Amer. Chem. Soc.*, **93**, 4324 (1971).

681. R. A. Firestone, N. Schelechow, D. B. R. Johnston and B. G. Christensen, *Tetrahedron Letters*, 375 (1972).
682. G. V. Kaiser, C. W. Ashbrook and J. E. Baldwin, *J. Amer. Chem. Soc.*, **93**, 2342 (1971).
683. H. H. Wasserman and B. H. Lipshutz, *Tetrahedron Letters*, 4613 (1976).
684. H. Shinozaki and M. Tada, *Chem. Ind. (Lond.)* 177 (1975).
685. C. S. Marvel and W. W. Mayer, Jr, *J. Org. Chem.*, **22**, 1065 (1957).
686. W. Ziegenbein and W. Franke, *Ber.*, **90**, 2291 (1957).
687. R. E. Benson and T. L. Cairns, *J. Amer. Chem. Soc.*, **70**, 2115 (1948).
688. D. St. C. Black, F. W. Eastwood, R. Okraglik, A. J. Poynton, A. M. Wade and C. H. Walker, *Australian J. Chem.*, **25**, 1483 (1972).
689. D. G. Lee, 'Oxidation of Oxygen and Nitrogen-containing Functional Groups With Transition Metal Compounds', in *Oxidation*, (Ed. R. L. Augustine), M. Dekker, New York, Vol. 1, 1969, p. 53.
690. A. Pinner, *Ber.*, **25**, 2807 (1892).
691. A. Pinner, *Ber.*, **26**, 292 (1893).
692. A. Pinner, *Arch. Pharm.*, **28**, 378 (1893).
693. L. D. Quin and P. M. Quan, unpublished results, reported in Reference 695.
694. A. M. Duffield, H. Budzikiewicz and C. Djerassi, *J. Amer. Chem. Soc.*, **87**, 2926 (1965).
695. H. McKennis, Jr, E. R. Bowman, L. D. Quin and R. C. Denney, *J. Chem. Soc., Perkin I*, 2046 (1973).
696. A. Picot and X. Lusinchi, *Synthesis*, 109 (1975).
697. M. F. Bartlett, D. F. Dicke and W. I. Taylor, *J. Amer. Chem. Soc.*, **80**, 126 (1958).
698. G. Buchi, R. E. Manning and S. A. Monti, *J. Amer. Chem. Soc.*, **85**, 1893 (1963).
699. G. B. Guise, E. Ritchie and W. C. Taylor, *Australian. J. Chem.*, **18**, 1279 (1965).
700. A. Cave, C. Kan-Fan, P. Potier, J. Le Men and M. M. Janot, *Tetrahedron*, **23**, 4691 (1967).
701. T. R. Govindachari, B. R. Pai, S. Rajappa, N. Viswanathan, W. G. Kump, K. Nagarajan and H. Schmid, *Helv. Chim. Acta*, **45**, 1146 (1962).
702. M. Sainsbury, S. F. Dyke and A. R. Marshall, *Tetrahedron*, **22**, 2445 (1966).
703. G. R. Clemo and G. C. Leitch, *J. Chem. Soc.*, 1811 (1928).
704. O. E. Edwards, F. H. Clarke and B. Douglas, *Can. J. Chem.*, **32**, 235 (1954).
705. C. Djerassi, M. Cereghetti, H. Budzikiewicz, M. M. Janot, M. Plat and J. Le Men, *Helv. Chim. Acta*, **47**, 827 (1964).
706. F. Percheron, *Ann. Chim.*, **4**, 303 (1959).
707. L. M. Berkowitz and P. N. Rylander, *J. Amer. Chem. Soc.*, **80**, 6682 (1958).
708. J. C. Sheehan and R. W. Tulis, *J. Org. Chem.*, **39**, 2264 (1974).
709. N. Tangari and V. Tortorella, *Chem. Commun.*, 71 (1975).
710. K. Orito, R. H. Manske and R. Rodrigo, *J. Amer. Chem. Soc.*, **96**, 1944 (1974).
711. D. Herlem, Y. Hubert-Brierre, F. Khuong-Huu and R. Goutarel, *Tetrahedron*, **29**, 2195 (1973).
712. I. R. C. Bick, J. B. Brenner and P. Wiriyachitra, *Tetrahedron Letters*, 4795 (1971).
713. T. Kametani, H. Nemoto, T. Nakano, S. Shibuya and K. Fukumoto, *Chem. Ind. (Lond.)*, **28**, 788 (1971).
714. E. Höft, A. Rieche and H. Schultze, *Ann.*, **697**, 181 (1966).
715. J. R. Brooks and D. W. Harcourt, *J. Chem. Soc., Perkin Trans. I*, 2588 (1973).
716. A. G. Davies, *Organic Peroxides*, Butterworths, London, 1961, pp. 28–30.
717. H. H. Wasserman and A. W. Tremper, *Tetrahedron Letters*, 1449 (1977).
718. J. E. McKeon and D. J. Trecker, *U. S. Patent 3634346; Chem. Abstr.*, **76**, 72397b (1972).
719. H. Moehrle and R. Engelsing, *Arch. Phar.*, **303**, 1 (1970).
720. R. Madhav, *J. Chem. Soc., Perkin I*, 2108 (1974).
721. F. Farina, M. V. Martin and M. C. Paredes, *Synthesis*, 167 (1973).
722. E. Ziegler, F. Hradetzky and M. Eder, *Monatsh. Chem.*, **97**, 1391 (1966).
723. E. Ziegler, F. Hradetzky and M. Eder, *Monatsh. Chem.*, **97**, 1394 (1966).

CHAPTER **2**

Appendix to 'The synthesis of lactones and lactams'[†]

*I. INTRODUCTION 271
*II. SYNTHESIS OF LACTONES 271
*A. By Intramolecular Cyclization of Hydroxy Acids, Hydroxy Acid Derivatives and Related Compounds 271
*B. By Intramolecular Cyclization of Unsaturated Acids and Esters . . . 283
*1. Acid-catalysed cyclizations 283
*2. Photochemical and electrochemical cyclizations 287
*3. Halolactonization 288
*4. Intramolecular Diels–Alder reactions 292
5. Using miscellaneous reagents 293
*E. By Malonic Ester or Malonic Acid Condensation 294
*G. By Grignard and Reformatsky Reactions 294
*I. From α-Anions of Carboxylic Acids or Esters 298
*K. By Direct Functionalization of Preformed Lactones 299
*L. From Ketenes 303
*M. By Reduction of Anhydrides, Esters and Acids 303
*N. By Oxidation Reactions 314
*1. Oxidation of diols 314
*2. Oxidation of ketones 314
*3. Oxidation of ethers 314
*4. Oxidation of olefins 323
5. Oxidation of other functional groups 336
*O. By Carbonation and Carbonylation Reactions 341
*Q. By Rearrangement Reactions 346
*2. Carbonium ion rearrangements 346
*3. Photochemical rearrangements 348
4. Miscellaneous rearrangements 350
*R. Lactone Interconversions 351
*S. Miscellaneous Lactone Syntheses 368

[†]The material in this Appendix is divided in the same manner as in the original Chapter 19 in Supplement B. Corresponding section numbers in this Appendix are preceded by an asterisk. Note that some section numbers are omitted while some new ones (not preceded by an asterisk) have been added. Structures, equations, tables, schemes and references run continuously in the original chapter and this Appendix.

*10. From aldehydes and ketones 368
*11. From carboxylic acids 377
*12. From carboxylic acid esters 384
*13. From acid halides 390
*14. From miscellaneous reagents 394
*III. SYNTHESIS OF LACTAMS 397
1. Nomenclature 398
*A. By Ring-closure Reactions (Chemical) 398
*1. From amino acids and related compounds 398
*2. From halo, hydroxy, keto and other substituted amides 420
*B. By Ring-closure Reactions (Photochemical) 465
*1. Cyclization of α,β-unsaturated amides 465
*3. Cyclization of enamides 465
*4. Cyclization of *N*-chloroacetyl-β-arylamines 469
*6. Miscellaneous cyclizations 470
*C. By Cycloaddition Reactions 475
*1. Addition of isocyanates of olefins 475
*2. From imines 490
*a. Reaction of imines with ketenes, acid chlorides or mixed anhydrides 490
*b. Reformatsky reaction with imines 581
*c. Other imine cycloadditions 585
*3. From nitrones and nitroso compounds 626
*D. By Rearrangements 627
*1. Ring contractions 627
*a. Wolff rearrangement 627
*b. Miscellaneous ring contractions 630
*2. Ring expansions 639
*a. Beckmann rearrangement 639
*b. Schmidt rearrangement 644
*c. Miscellaneous ring expansions 646
*E. By Direct Functionalization of Preformed Lactams 662
1. Functionalization of lactam nitrogen 662
2. Functionalization of lactam ring other than at lactam nitrogen 696
3. Conversion of substituents directly attached to the lactam nitrogen 725
4. Conversion of substituents directly attached to the lactam ring other than at lactam nitrogen 802
a. Reactions at the C-2 and C-3 positions 802
b. Reactions at the C-4 and higher positions 891
*F. By Oxidation Reactions 952
*2. Using chromium oxides 952
*4. Using ruthenium oxides 952
*5. Via sensitized and unsensitized photooxidation 953
*7. Using miscellaneous reagents 957
*G. Miscellaneous Lactam Syntheses 970
*IV. ACKNOWLEDGMENTS 1009
*V. REFERENCES 1009

*I. INTRODUCTION

This Appendix on the synthesis of lactones and lactams covers the primary literature from 1975 through 1987. Recent references have been included for all approaches to the synthesis of these compounds, including references to the general preparative methods presented in the original chapter.

The Appendix organization follows identically that used in the original chapter, and the reader will find information on the same topic by successively reading the original section of the chapter and then referring to the same numbered section in the Appendix. In instances where little or no new information has been published regarding a specific synthetic approach discussed in the original chapter, that corresponding section has been eliminated in the Appendix. Conversely, new synthetic approaches not covered in the original chapter have been added to the appropriate sections of this Appendix under final subsections titled 'Miscellaneous...'.

Structures, equations, compound numbers, schemes and references are numbered in continuation of those in the original chapter.

*II. SYNTHESIS OF LACTONES

Recent review articles describing the synthesis of lactones report on a wide variety of synthetic approaches which were employed to prepare a wide variety of lactones.

Review articles have reported the synthesis of lactones and macrolides[724–726], α-peroxylactones[727], Prelog–Djerassi lactones[728] and the construction of the carbon skeletons for natural products by the use of small membered lactones[729]. Also reviewed was the synthesis and synthetic utility of halolactones[730], the stereoselective syntheses with β-hydroxycarboxylic acids and β-lactones[731], lactonization using diethyl azodicarboxylate and triphenylphosphine[732] and novel lactonization reactions[733].

On the subject of macrocyclic lactones, reviews have reported developments in syntheses of macrocyclic substances with musk odor[734], the synthesis of macrocyclic lactones describing approaches to complex macrolide antibiotics[735], the synthesis of macrolides[736], the synthesis of β-lactone homopolymers[737], preparation and polymerization of β-lactones[738], polymerization of pivalolactone and related lactones[739], the kinetics and mechanism of anionic polymerization of lactones[740] and the selective synthesis of new macrolides by the ring-opening polymerization of a bicyclic oxolactone, their structures and complexation with metal ions and organic molecules[741].

In the area of natural product and steroidal lactones, several reviews have recently appeared also. These reports discuss the natural product synthesis via π-allyltricarbonyliron lactone complexes[742], transformations of the pregnane side chain to γ-lactones[743], the synthesis of γ-lactones fused with the steroidal skeleton in the 1,2-position[744], the use of lactones as building blocks in alkaloid synthesis[745], steroid spirolactones and their biological activity[746] and the synthesis of biologically active compounds including Corey lactone analogues[747].

Finally, a review on the enantioselective synthesis of biologically active cyclopentanoids via enzyme catalysed asymmetric reactions has also been published[748].

*A. By Intramolecular Cyclization of Hydroxy Acids, Hydroxy Acid Derivatives and Related Compounds

Optically pure enantiomers of 4-alkyl-γ-lactones have been prepared[749] by acid catalysed cyclization of optically pure γ-hydroxycarboxylic acids (equation 395).

$$\text{RCH(OH)CH}_2\text{CH}_2\text{COOH} \xrightarrow{H^+} \text{(}\gamma\text{-lactone, R-substituted)} \tag{395}$$

optically pure — optically pure

R = Me; Et; n-C_5H_{11}; n-C_8H_{11}

$[\alpha]_D^{25} = \pm 31.4; +51.5; +45.8; +37.7$

By treatment[750] of *cis*- or *trans*-1-(2,4-dimethoxybenzyl)-3-(1-hydroxyethyl)-2-azetidinon-4-carboxylic acid with a catalytic amount of concentrated hydrochloric acid in tetrahydrofuran, stereoselectivity produces (equation 396) the fused ring *cis*-γ-lactone as the only lactone product.

conc. HCl, THF

cis R^1=H, R^2=CO_2H

or

trans R^1=CO_2H, R^2=H

62% from *cis*-azetidinone

14% from *trans*-azetidinone

Another example of the preparation[751] of fused ring γ-lactones by acid catalysed intramolecular cyclization is the formation of 3,3*a*,4,8*b*-tetrahydroindeno[1,2-*b*]furan-2-one from *trans*-2-hydroxyindan-1-yl acetic acid upon treatment with sulphuric acid (equation 397).

$$\xrightarrow[\text{HOAc, r.t., 24 h}]{H_2SO_4} \tag{397}$$

trans

Acid catalysed cyclization has also been used to prepare δ-lactones. Thus, treating a mixture of sodium salts of butyric, 2-ethylhexanoic and 2-ethyl-3-propyl-4-(hydroxymethyl)hexanoic acids with sulphuric acid produces[752] a mixture of acids and lactones from which 2,4-diethyl-3-(n-propyl) δ-lactone was selectively isolated (equation 398).

n-PrCOO$^-$Na$^+$

n-BuCH(Et)COO$^-$Na$^+$

HOCH$_2$CH(Et)CH(n-Pr)CH(Et)COO$^-$Na$^+$

$\xrightarrow{H_2SO_4}$ mixture of acids and lactones $\xrightarrow{\text{base}}$

(398)

Substituted and *cis*-fused δ-valerolactones have also been prepared[753] by acid catalysed cyclization of δ-hydroxycarboxylic acids, which were obtained from β-hydroxycarboxylic acids (equation 399).

OH, COOH + *i*-PrOH —($NaBH_4$, reflux 6 h)→ [CHO, CH_2COOH] —($NaBH_4$)→

CH_2OH, CH_2COOH —(HCl)→ O, O (399)

β-Hydroxyacid	δ-Lactone	Yield(%)
H, OH, COOH, H	H, O, O, H	70
H, OH, COOH, H	H, O, O, H	77
Me, R, OH, COOH	Me, R, O, O	
R = Me		60
R = t-BuCH_2		82
OH, Me, Me, COOH	Me, Me, O, O	81 (5:1 mixture of stereoisomers)

Acid treatment of hydroxyesters also produces lactones usually via hydrolysis of the ester moiety and intermediate formation of the corresponding hydroxyacid, which then cyclizes. This approach was reported[754] in the cyclization of the *t*-butyl and ethyl esters of δ-hydroxycarboxylic acids, where the *t*-butyl ester substrates were directly converted to the corresponding δ-lactones upon treatment with trifluoroacetic acid (equation 400). However, the ethyl esters required initial hydrolysis to δ-hydroxyacids using potassium

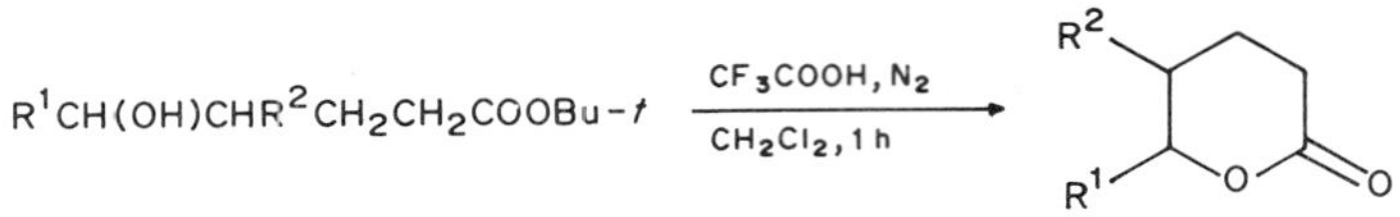

$$R^1CH(OH)CHR^2CH_2CH_2COOBu\text{-}t \xrightarrow[CH_2Cl_2,\ 1\ h]{CF_3COOH,\ N_2} \text{(δ-lactone, } R^1, R^2 \text{ substituted)} \qquad (400)$$

δ-Hydroxyesters	δ-Lactones	Yield (%)
n-$C_7H_{15}CH(OH)(CH_2)_3COOBu$-$t$	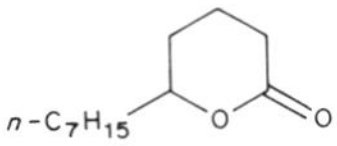 n-C_7H_{15} δ-lactone	100
$HOCH_2CH(CH_2Ph)(CH_2)_2COOBu$-$t$	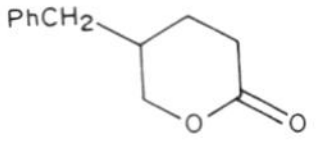 $PhCH_2$ δ-lactone	90
$Ph(CH_2)_2CH(OH)—CH(Me)(CH_2)_2COOBu$-$t$[a]	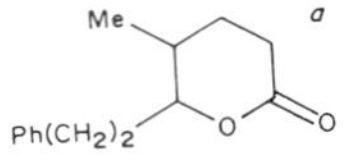 Me, $Ph(CH_2)_2$ δ-lactone[a]	72
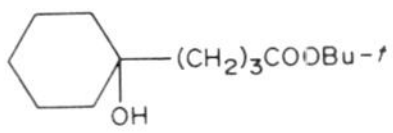 1-hydroxycyclohexyl-$(CH_2)_3COOBu$-t	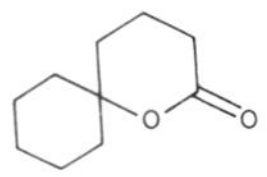 spiro δ-lactone	63

[a] A 1:1 mixture of diastereoisomers.

hydroxide before cyclization could be performed using acetic anhydride in pyridine (equation 401).

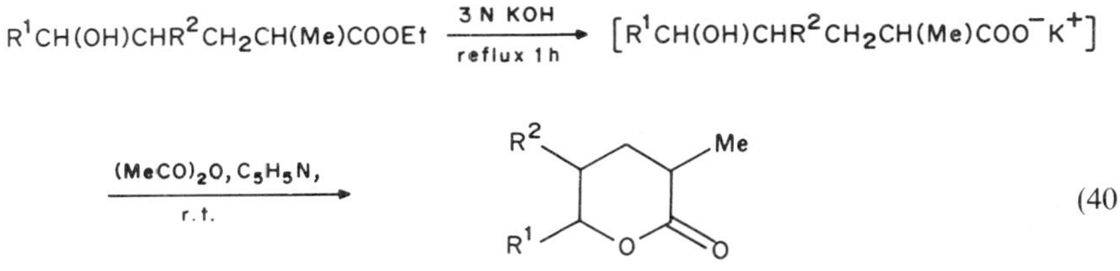

$$R^1CH(OH)CHR^2CH_2CH(Me)COOEt \xrightarrow[reflux\ 1\ h]{3\ N\ KOH} [R^1CH(OH)CHR^2CH_2CH(Me)COO^-K^+]$$

$$\xrightarrow[r.t.]{(MeCO)_2O,\ C_5H_5N,} \text{(δ-lactone with } R^1, R^2, Me) \qquad (401)$$

$R^1 = n\text{-}C_7H_{15}$; H
$R^2 = H$; $PhCH_2$
Yield (%) = 90[a] ; 73[a]

[a] A 1:1 mixture of diastereoisomers.

Acid treatment of hydroxyesters has also been reported[751] for the preparation of 3,3*a*,8,8*a*-tetrahydroindeno[2,1-*b*]furan-1-one from *trans*-2-carboxymethylindan-1-ol using sulphuric acid (equation 402), and 2-methylbut-2-enolide from methyl (2-chloro-4-

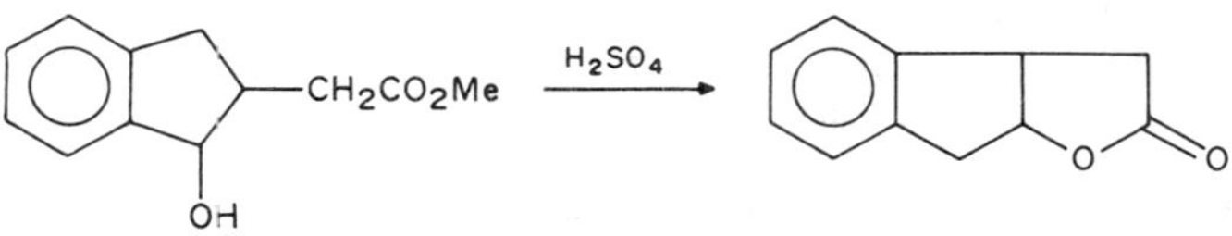

$$\text{1-hydroxyindan-2-yl-}CH_2CO_2Me \xrightarrow{H_2SO_4} \text{tetrahydroindenofuranone} \qquad (402)$$

$$MeCO_2CHClCH_2CMe(OH)CO_2Et \xrightarrow[\text{reflux, 4 h}]{\text{EtOH, HCl,}} \text{(3-methyl-2(5H)-furanone)} \qquad (403)$$

hydroxy-4-ethoxycarbonyl)pentanoate using hydrochloric acid (equation 403). Hydrochloric acid also catalyses the enantioselective synthesis[755] of the α-amino-γ-lactones derived from (2-methoxyethoxymethyl-protected) (*R*)-homoserine methyl esters, using this approach (equation 404).

$$R^1\text{—}C(COOMe)(NH_2)\text{—}CH(R^2)\text{—}CH(R^3)\text{—}CH_2OCH_2CH_2OMe \xrightarrow[\text{2. EtOH, reflux, 0.5 h}]{\text{1. 8}N\text{ HCl, reflux, 3 h}} \text{α-amino-γ-lactone} \cdot HCl \qquad (404)$$

R^1	R^2	R^3	Yield (%)	Stereochemical ratio
H	H	H	71	$(\alpha R):(\alpha S) = 95:5$
H	H	Me	47	$(\alpha R, \gamma RS):(\alpha S, \gamma RS) = 97.5:2.5$
H	Me	Me	44	$(\alpha R, \beta R, \gamma R):(\alpha R, \beta S, \gamma S) = 97.5:2.5$
Me	H	H	52	$(\alpha R):(\alpha S) = 97.5:2.5$

Acid treatment of *ortho* esters produces intermediate γ-hydroxyesters, which then cyclize upon further treatment with acid as was reported[756] with substituted 2,2-dialkoxytetrahydrofurans (equation 405). As reported above, the γ-lactone product could also be obtained[756] by initial treatment of the α-chloro *ortho* esters shown in equation (406) with potassium *t*-butoxide in tetrahydrofuran followed by treatment with sulfuric acid. The presence of the chloro group in the ring leads to the double bond found in the product.

$$\text{2,2-dimethoxy-tetrahydrofuran}(R^1,R^2,R^3,R^4) \xrightarrow[\text{high acid conc. and prolonged reaction time}]{H^+} R^1CH(OH)CHR^2CR^3R^4COOMe \longrightarrow \gamma\text{-lactone}(R^1,R^2,R^3,R^4) \qquad (405)$$

R^1, R^2, R^3 and R^4 unspecified

$$\text{4-Ph-3-Cl-2,2-diethoxytetrahydrofuran} \xrightarrow[\text{2. } H_2SO_4,\ 0\,^\circ C,\ 15\ \text{min}]{\text{1. } t\text{-BuOK, THF, 1 h}} \text{4-phenyl-2(5H)-furanone}\ (55\%) \qquad (406)$$

Finally, treatment of substituted 2,2-dialkoxytetrahydrofurans with commercially available aluminium *t*-butoxide affords the corresponding γ-lactone (equation 407),

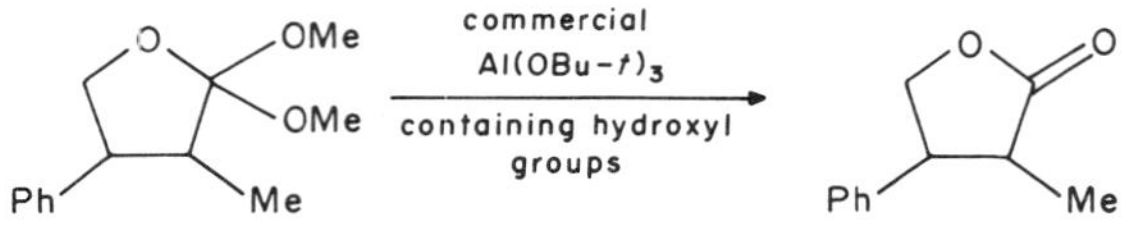

(407)

whereas treatment of the same starting material with freshly prepared aluminium *t*-butoxide affords[756] substituted 2-methoxy-4,5-dihydrofuran (equation 408).

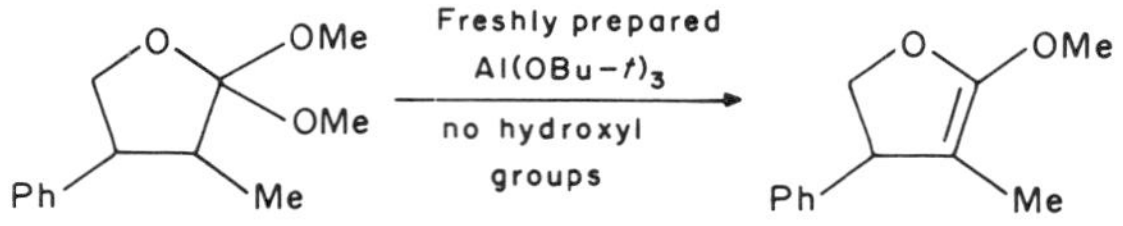

(408)

p-Toluenesulphonic acid and *p*-toluenesulphonyl chloride have both been used to catalyse cyclization of hydroxyacids and hydroxyesters to produce lactones. One example of their use is illustrated by the preparation[757] of a 12-membered enediyne lactone by *p*-toluenesulphonic acid catalysed lactonization of the corresponding ω-hydroxyacid precursor (equation 409).

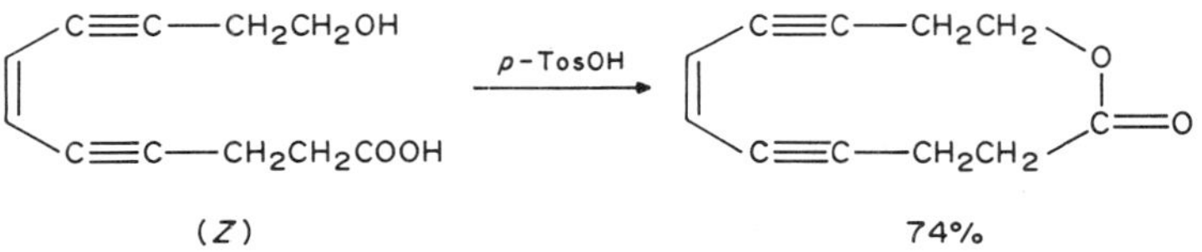

(409)

p-Toluenesulphonic acid has also been used to prepare[758] bicyclic spirolacones by cyclization of the corresponding intermediate bicyclic γ-hydroxyacids, which in turn were prepared via two methods as shown in equation (410).

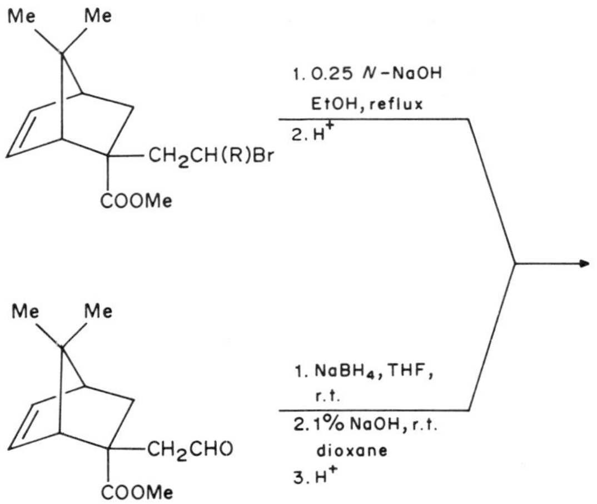

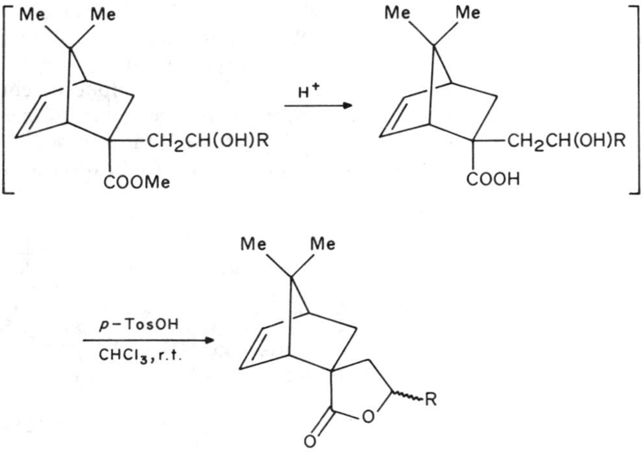

$\xrightarrow{H^+}$

$\xrightarrow[CHCl_3, r.t.]{p\text{-TosOH}}$ (410)

R=H; Me; Et; Ph

Yield (%)=63; 50–54

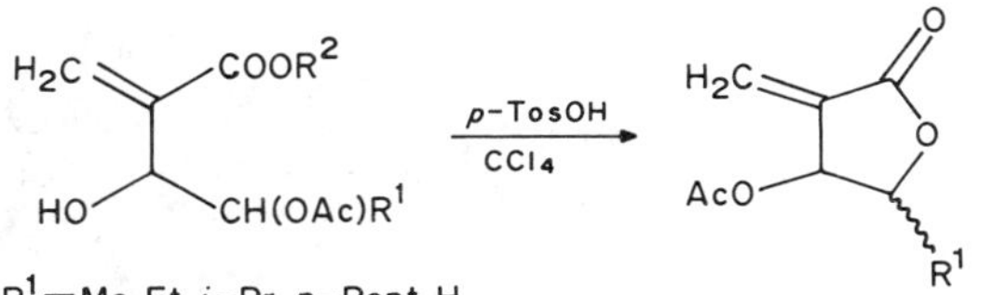

$\xrightarrow[CCl_4]{p\text{-TosOH}}$ (411)

R^1=Me, Et, *i*-Pr, *n*-Pent, H

R^2=Me, Et

yields range from 70–80%

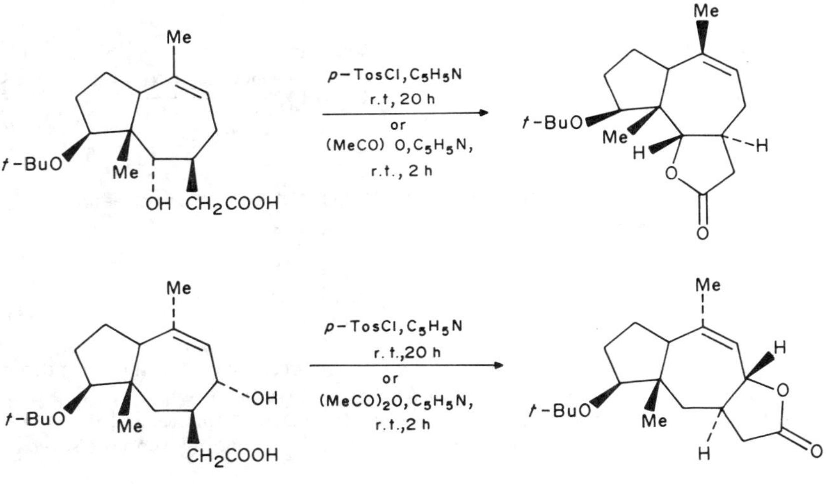

p-TosCl, C_5H_5N, r.t, 20 h or $(MeCO)_2O$, C_5H_5N, r.t., 2 h

(412)

p-TosCl, C_5H_5N, r.t., 20 h or $(MeCO)_2O$, C_5H_5N, r.t., 2 h

A similar reaction is observed[759] to occur when substituted allylic alcohols are treated with *p*-toluenesulphonic acid; however, this cyclization reaction to form the β-acetoxy-α-methylene-γ-lactone products is accompanied by acetyl migration (equation 411).

Two recent references illustrates the utilization of *p*-toluenesulphonyl chloride in the cyclization of hydroxyacids to produce lactones. In the first reference[760] *trans*-fused ring γ-butyrolactones are formed (equation 412), while in the second reference[761] monocyclic *R*-3-methyl-γ-butyrolactone was the product obtained from cyclization of the precursor shown in equation (413).

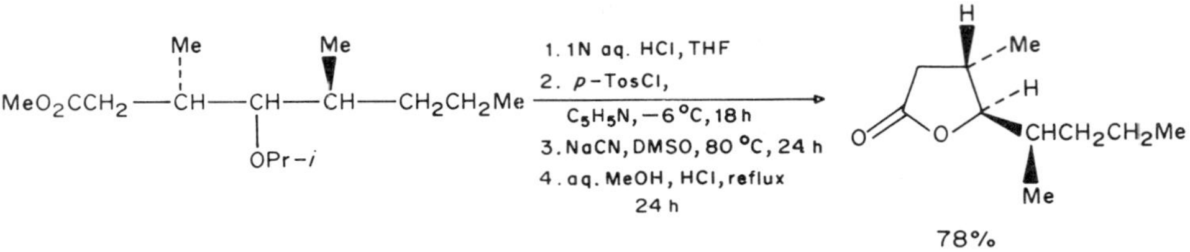

MeO_2CCH_2—CH(Me)—CH(OPr-*i*)—CH(Me)—CH_2CH_2Me → (1. 1N aq. HCl, THF; 2. *p*-TosCl, C_5H_5N, −6 °C, 18 h; 3. NaCN, DMSO, 80 °C, 24 h; 4. aq. MeOH, HCl, reflux 24 h) → lactone (H, Me, H, $CHCH_2CH_2Me$, Me) 78%

(413)

A reagent which is very similar in structure to *p*-toluenesulphonyl chloride and has also been used[762] to effect cyclization of γ-hydroxyacids to produce *cis*-fused ring γ-butyrolactones is mesitylenesulphonyl chloride (equation 414).

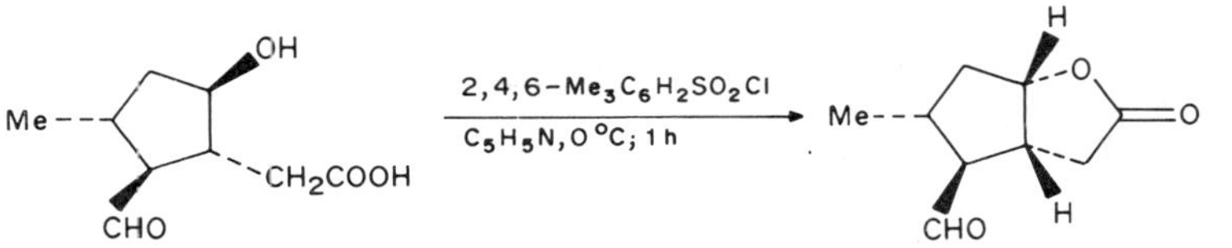

Me-cyclopentane (OH, CH_2COOH, CHO) → (2,4,6-$Me_3C_6H_2SO_2Cl$; C_5H_5N, 0 °C; 1 h) → Me-bicyclic lactone (H, O, =O, H, CHO) (414)

Alkaline hydrolysis of the acetate function of the α,β-unsaturated acids shown below produces[763] intermediate α-methylene γ- and δ-hydroxyacids which are converted directly into the corresponding α-methylene γ- and δ-lactones by treatment with *N,N'*-dicyclohexylcarbodiimide (DCCD) in pyridine (equation 415).

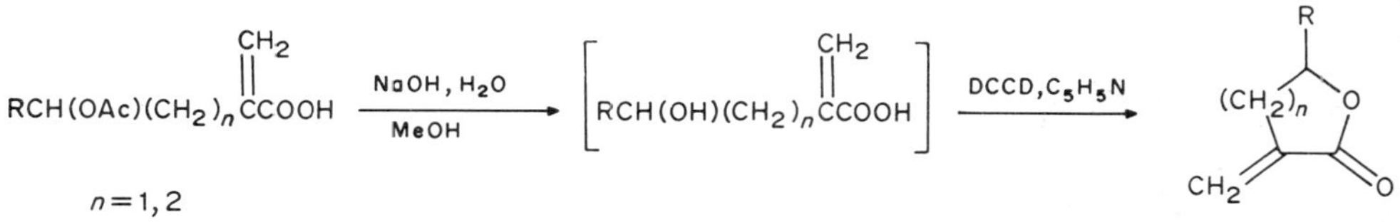

$RCH(OAc)(CH_2)_nC(=CH_2)COOH$ → (NaOH, H_2O, MeOH) → [$RCH(OH)(CH_2)_nC(=CH_2)COOH$] → (DCCD, C_5H_5N) → lactone (R, $(CH_2)_n$, O, CH_2=, =O)

$n = 1, 2$

R =	$Ph(CH_2)_2$	$CH_2{=}CH(CH_2)_8$	*n*-Hex	*n*-C_7H_{15}
n =	1	1	2	2
Yield (%) =	59	76	72	70

(415)

Cyanuric chloride has been reported[764] to be an effective reagent for the formation of macrocyclic lactones, since 13- to 19-member lactones have been prepared using this reagent (equation 416). When this reagent was used to cyclize aleuritic acid (9, 10, 16-trihydroxypalmitic acid) different products were observed[764] depending upon the solvent

$$RCH(OH)(CH_2)_nCOOH \xrightarrow[Et_3N,\ Me_2CO,\ r.t.]{\text{cyanuric chloride}} \text{(CH}_2)_n\text{-CHR-O-C=O (macrolactone)} \qquad (416)$$

R = *n*-Hex ; H ; H ; H
n = 10 ; 13 ; 14 ; 16
Yield (%) = 70 ; 68 ; 85 ; 33

used (equation 417). The proposed mechanisms[765] for these cyclizations with cyanuric chloride involves intermediate formation of the carboxylic acid chlorides.

$$HOCH_2(CH_2)_5CH(OH)CH(OH)(CH_2)_7COOH \;(\textit{threo}) \xrightarrow[Et_3N,\ Me_2CO]{\text{cyanuric chloride}} \text{acetonide macrolactone (86\%)} \qquad (417)$$

$$HOCH_2(CH_2)_5CH(OH)CH(OH)(CH_2)_7COOH \xrightarrow[Et_3N,\ CH_3CN]{\text{cyanuric chloride}} \text{HOCH-HOCH macrolactone}$$

Another nitrogen containing reagent which has been used to cyclize hydroxyacids to lactones is 2-chloro-1-methylpyridinium iodide. This reagent has been used very effectively to produce both macrocyclic lactones[766] (equation 418) and *trans*-fused lactones[767] from alkylidenehydroxyacids (equation 419) and other γ-hydroxyacids (equation 420).

$$HOCH_2(CH_2)_{13}COOH \xrightarrow[\text{MeCN, Et}_3\text{N, reflux 8 h}]{\text{2-chloro-1-methylpyridinium iodide}} (CH_2)_{14}\text{-O-C=O} \; (84\%) \qquad (418)$$

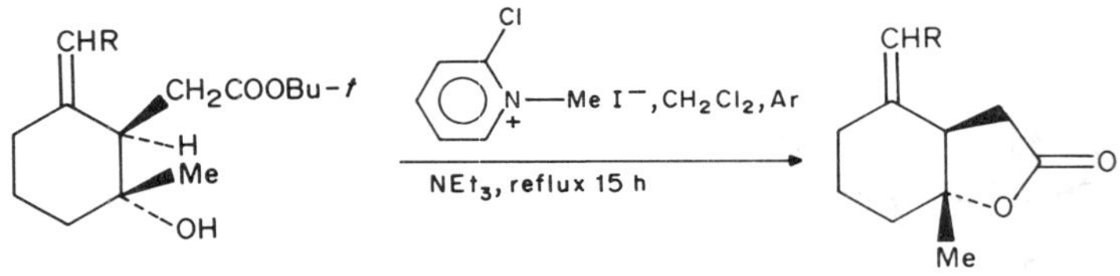

(419)

R = H ; CH_2SiMe_3 ; CH_2OH ; CH_2OCH_2SMe
Yield(%) = 97 ; 99 ; 98 ; 95

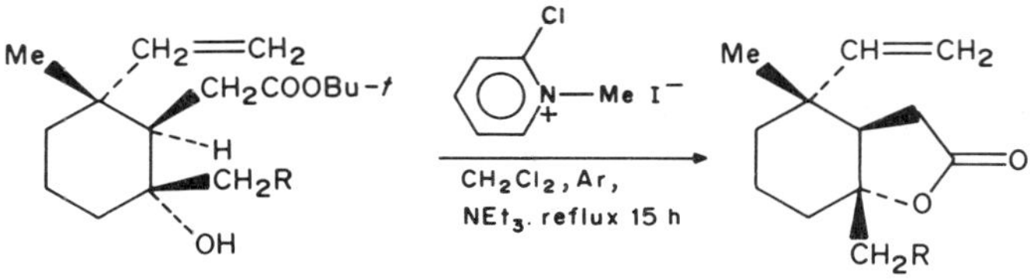

(420)

R = PhS ; PhS[a] ; H ; H[a]
Yield(%) = 95 ; 78 ; 96 ; 78

[a]Reagent used is *N,N'*-DCCD in pyridine.

Trans-fused lactones are also obtained from γ-hydroxynitriles[760] by a one-pot procedure which involves hydrolysis of the nitriles to acids followed by lactonization. The required γ-hydroxynitriles are produced by metal hydride reduction of acetate groups adjacent to exocyclic α,β-unsaturated nitriles (equation 421).

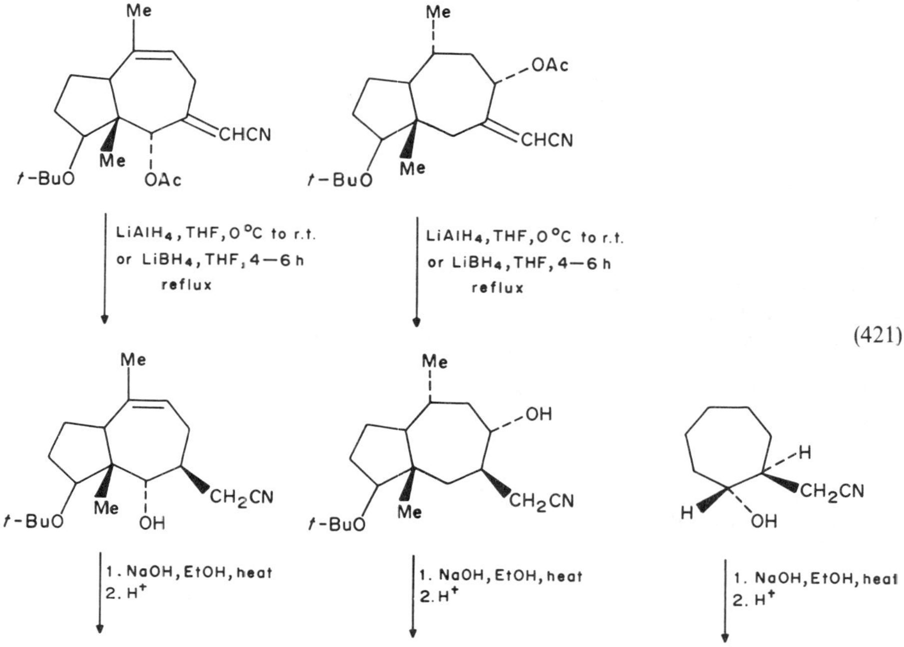

(421)

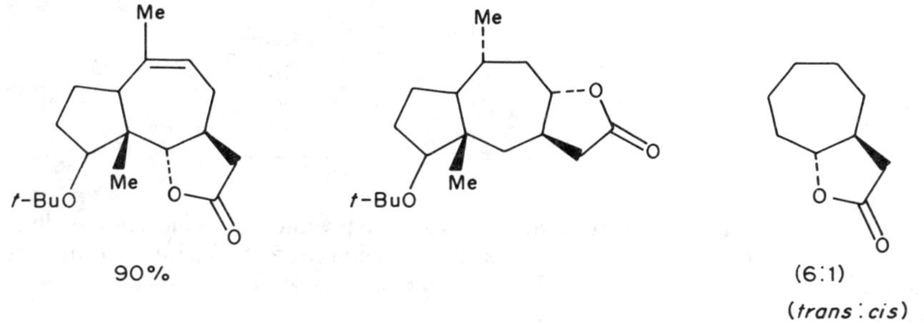

Me Me Me

t-BuO Me O O t-BuO Me O O O O

90% (6:1) (*trans* : *cis*) (421)

Me Me Me Me

H O O H PhCOOCH$_2$ O O H

PhCOOCH$_2$ (CH$_2$)$_3$COO(CH$_2$)$_3$OH H (CH$_2$)$_3$COO(CH$_2$)$_3$OH

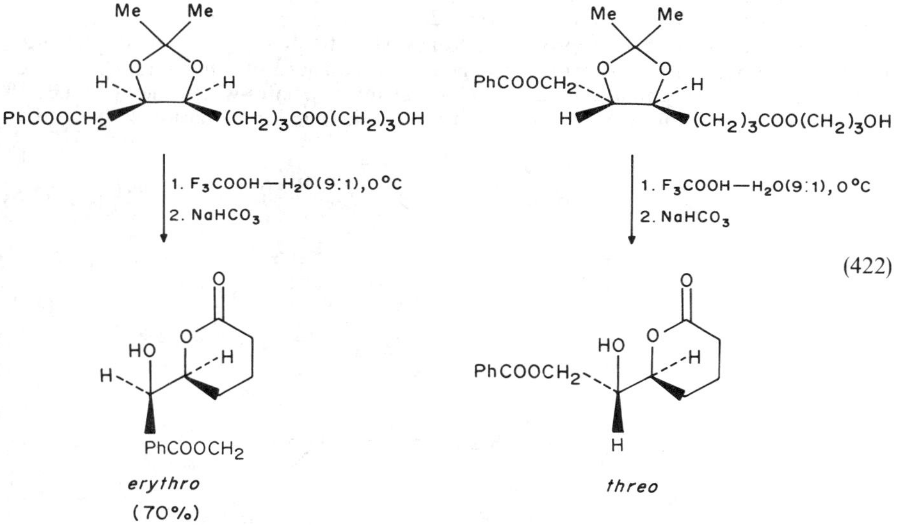

1. $F_3COOH-H_2O$(9:1), 0 °C
2. $NaHCO_3$

(422)

O O

HO O H HO O H

H PhCOOCH$_2$

PhCOOCH$_2$ H

erythro (70%) *threo*

(CH$_2$)$_n$OH

PhCONH—C—CONMe$_2$ $\xrightarrow[\text{toluene}]{\text{HCl,}}$

Ph

[Ph, $\overset{+}{H}N$—(CH$_2$)$_n$OH, Ph, O, O ⇌ PhCONH—C(Ph)(C≡O$^+$)—(CH$_2$)$_n$OH] Cl$^-$

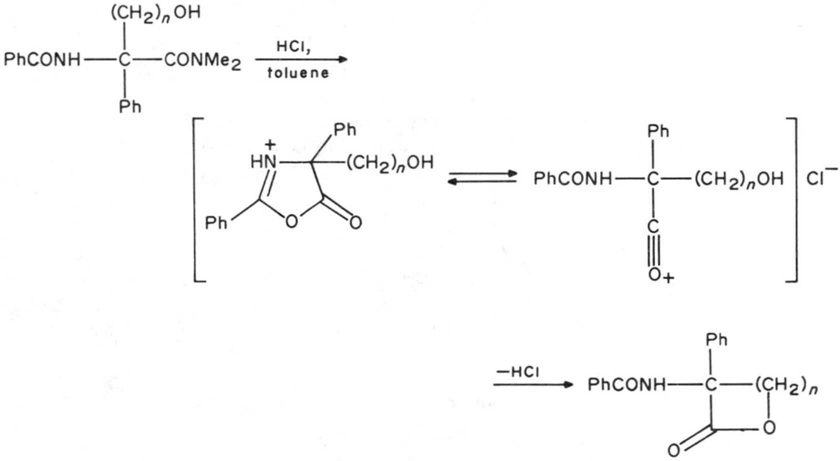

$\xrightarrow{-HCl}$ PhCONH—C(Ph)—(CH$_2$)$_n$ (423)

O O

n	Temp (°C)	Time (min)	Yield (%)
2	75–80	3	95
3	75–80	3	97
12	100	60	60

Diesters may also be used as starting materials for the production of lactones as illustrated by the reaction of the diester acetonides shown in equation (422) with a 9:1 mixture of trifluoroacetic acid–water, which affords[768] the corresponding *erythro*- or *threo*-hydroxyester lactones.

One class of hydroxyacid derivatives which produce lactones upon treatment with acid are *N*-acylaminoacid dimethylamides[769]. The lactones are produced[769] via 2-oxazolin-5-one intermediates as shown in equation (423).

ω-Halocarboxylic acids also produce lactones when treated[770] with an efficient base. When treated with the anion formed by the electroreduction of 2-pyrrolidone, ω-halocarboxylic acids form ammonium carboxylates as intermediates which then produce[770] macrolides by intramolecular reaction with the halide function (equation 424).

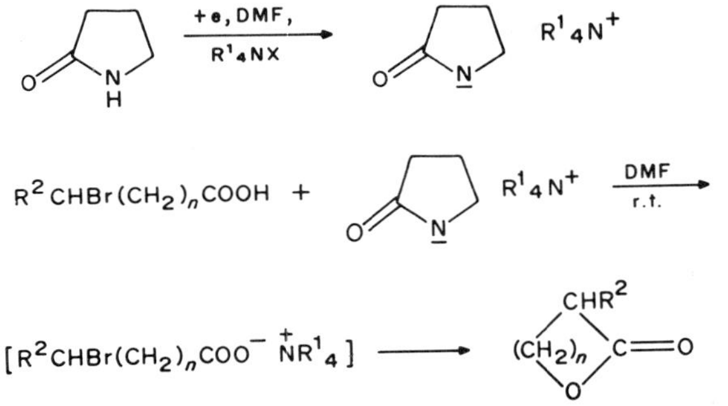

$$\text{2-pyrrolidone} \xrightarrow[R^1{}_4NX]{+e,\ DMF} \text{pyrrolidonide anion}\ R^1{}_4N^+$$

$$R^2CHBr(CH_2)_nCOOH + \text{pyrrolidonide anion}\ R^1{}_4N^+ \xrightarrow[r.t.]{DMF} \tag{424}$$

$$[R^2CHBr(CH_2)_nCOO^-\ \overset{+}{N}R^1{}_4] \longrightarrow (CH_2)_n\text{—}CHR^2\text{—}O\text{—}C{=}O \text{ (ring)}$$

R^1	R^2	n	Yield (%)
Et	H	11	48[a]
n-Bu	H	11	67[b]
n-C_8H_{17}	H	11	66
Et	H	15	77
n-Bu	n-Hex	10	77

[a] A 16% yield of the diolide shown below was also isolated.
[b] A 5% yield of the diolide shown below was also isolated.

$$\text{cyclo}[\text{—C(=O)—O—}(CH_2)_{11}\text{—C(=O)—O—}(CH_2)_{11}\text{—}]$$

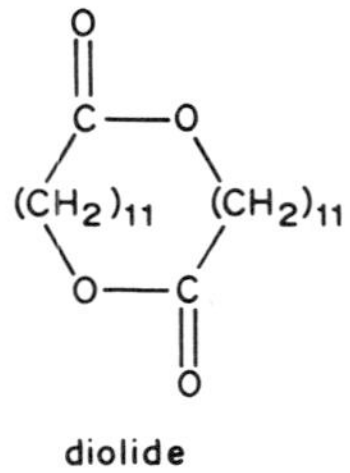

diolide

Treatment of the dihydroxycarboxylic acid precursor, dihydroxyvaleronitrile, with methanolic hydrogen chloride affords[771] 2,5-dideoxy-L-xylono-γ-lactone (equation 425), while hydrogenation of 5-hydroxymethyl-2-furancarboxylic acid using 5% rhodium on carbon affords[772] the corresponding *cis*-tetrahydro analogue, which upon treatment with the acid ion exchange resin Amberlyst 15 produces the bicyclic lactone 2-oxo-3,8-dioxabicyclo[3.2.1]octane (equation 426).

1. sat. HCl/MeOH reflux 2 h; 2. H_2O, r.t., overnight; 3. reflux 0.5 h (425)

5% Rh/C, H_2O, H_2(3 atm.), r.t. 5–6 h; 1. Amberlyst 15, N_2 benzene—dioxane stir 2.5 h; 2. Reflux 0.5 h

(426)

A mild and efficient reagent used[773] as a catalytically neutral esterification agent in the ring closure of ω-hydroxycarboxylic acids to macrolide-type lactones is di(n-butyl)tin oxide (equation 427).

$$HO(CH_2)_nCOOH + n\text{-}Bu_2SnO \xrightarrow[\text{reflux}]{\text{mesitylene}} (CH_2)_n \text{ lactone} \quad (427)$$

n = 7 ; 14 ; 15
Time(h) = 19 ; 23 ; 17
Yield(%) = 0 ; 63 ; 60

The mechanism proposed[773] for this conversion is presented in Scheme 6 and involves formation of a reactive stannylated intermediate (**58** or **60**) in low concentrations, followed by subsequent release of the tin reagent possibly by a template-driven extrusion process to afford the lactone.

B. By Intramolecular Cyclization of Unsaturated Acids and Esters

*1. Acid-catalysed cyclizations

Acid-catalysed cyclization to produce unsaturated lactones has been reported for a variety of substituted α,β-unsaturated carboxylic acids. The most common cyclizations have been reported for hydroxy substituted α,β-unsaturated acids as exemplified by the acid-catalysed lactonization of the optically pure-γ-alkyl-γ-hydroxy-α,β-unsaturated carboxylic acids shown in equation (428), which produce[774] the corresponding optically

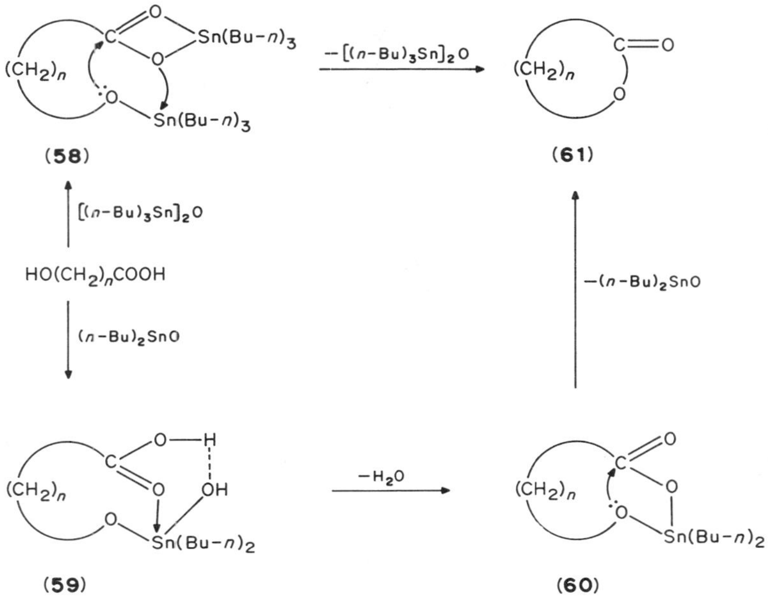

SCHEME 6

pure enantiomers of unsaturated 4-alkyl γ-lactones, and the acid-catalysed lactonization of both partially and completely hydrogenated γ-alkyl-γ-hydroxy-α,β-alkynoates which produce[775] both saturated and unsaturated 4-substituted γ-lactones as shown in equation (429).

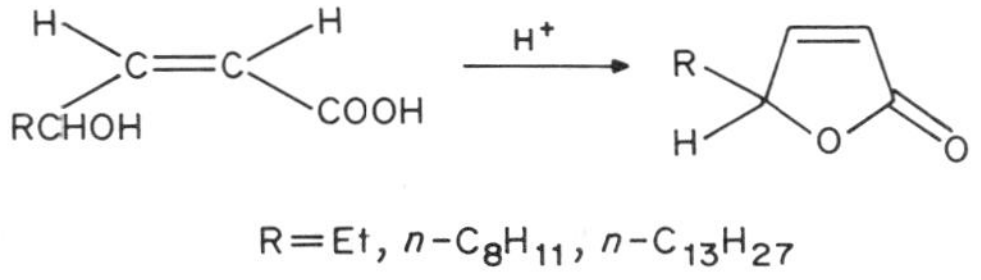

(428)

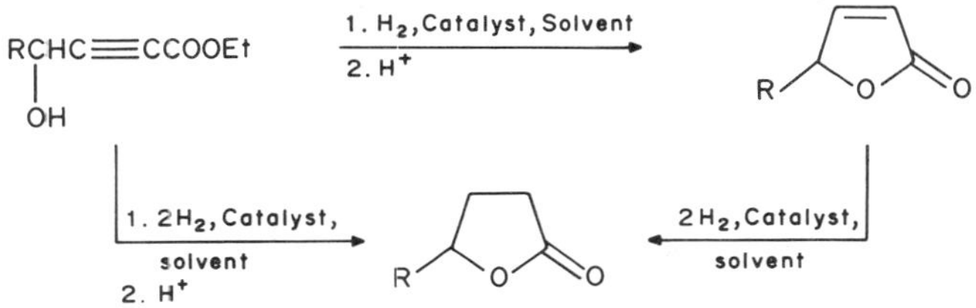

R=Et, n-C_8H_{11}, n-$C_{13}H_{27}$

(429)

R	Catalyst, Solvent	Product	%e.e.	Yield (%)
Et	Pd/C, MeOH	Et, O, O	87	90
(R) n-PentCH=CHCH$_2$ (Z)	Pd/BaSO$_4$, quinoline, MeOH	n-PentCH=CHCH$_2$, O, O	—	—
		↓ 'CuH$_2$'		
		n-PentCH=CHCH$_2$, O, O S(+)	88	—
(S) n-PentCH=CHCH$_2$ (Z)	Pd/BaSO$_4$, quinoline, MeOH	n-PentCH=CHCH$_2$, O, O	—	—
		↓ 'CuH$_2$'		
		n-PentCH=CHCH$_2$, O, O R(−)	79	—
n-C$_8$H$_{17}$CH=CH (Z)	'aged catalysts'	n-C$_8$H$_{17}$CH=CH, O, O	—	88
		↓ unspecified		
		n-C$_8$H$_{17}$CH=CH, O, O	—	—

Treatment of benzyl 4-(benzyloxy)-2-methylenebutanoate with a 25% solution of hydrogen bromide–acetic acid affords[776] 3-bromomethyl-2-furanone (equation 430).

$$\mathrm{CH_2{=}C(COOCH_2Ph)CH_2CH_2OCH_2Ph} \xrightarrow[\text{CH}_2\text{Cl}_2,\ 0\,^\circ\text{C, stir 4.5 h}]{\text{HBr—HOAc}} \text{3-(CH}_2\text{Br)-2-furanone (59.2\%)} \qquad (430)$$

Reaction[777] of the allylesters of bix(alkoxycarbonylamino)acetic acid with methanesulphonic acid at room temperature leads to a reactive intermediate which spontaneously cyclizes in an *exo–exo* trigonal fashion to give the corresponding mesylated butyrolactones (equation 431).

$$\mathrm{H_2C{=}CHCH_2OCOCH(NHCOOR)_2} \xrightarrow[\text{r.t.}]{\text{MeSO}_3\text{H}} \left[\mathrm{CH_2{=}CH{-}CH_2{-}O{-}C({=}O){-}HC{=}\overset{+}{N}HCOOR}\right] \longrightarrow$$

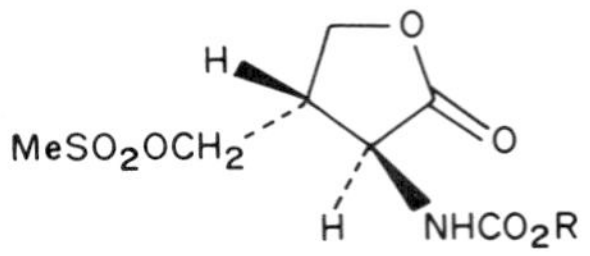

(431)

R=Me; *n*-Bu
Yield(%)=47; 64

Sulphuric acid treatment[778] of 1,4-pentadiene-1,2-dicarboxylates causes stereospecific cyclization to *Z*-α-methoxycarbonylmethylene γ-lactones in 80–85% yields, the required starting materials being produced[778] by an initial ene reaction of an olefinic component with dimethyl acetylenedicarboxylate (equations 432 and 433).

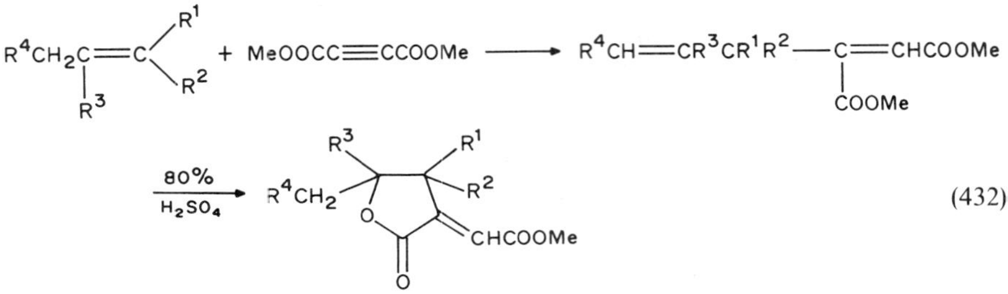

$R^4CH_2C(R^3)=CR^1R^2$ + MeOOCC≡CCOOMe ⟶ $R^4CH=CR^3CR^1R^2$—C(COOMe)=CHCOOMe

$\xrightarrow[H_2SO_4]{80\%}$ (432)

R^1	R^2	R^3	R^4
H	H	—$(CH_2)_4$—	
Me	H	Me	H[a]
Me	H	—$(CH_2)_4$—	
Me	Me	Me	H[b]

If anhydrous HCl in CH_2Cl_2 at 25 °C for 48 hours was used instead of 80% H_2SO_4:
[a]A 37% yield of E-lactone is obtained.
[b]A 60% yield of E-lactone is obtained.

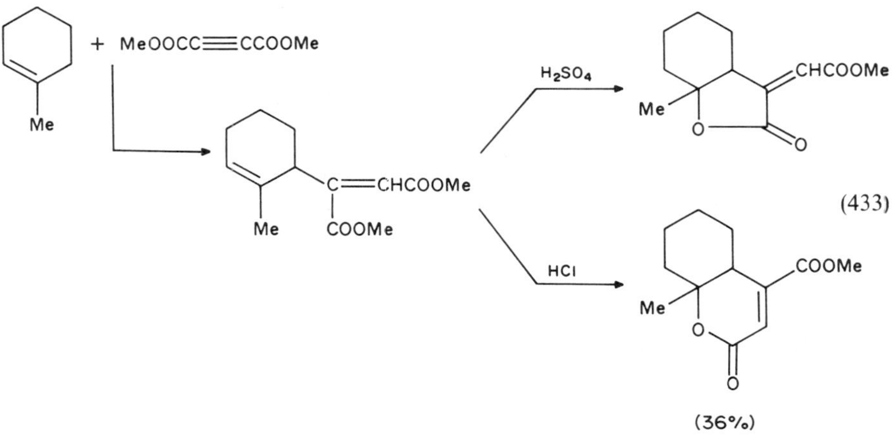

+ MeOOCC≡CCOOMe

H_2SO_4

HCl

(433)

(36%)

Acid treatment of methyl 4,5-epoxy-2-pentenoate as either the pure *cis*(*Z*) isomer or as a *trans*, *cis*(*E*, *Z*) isomer mixture also causes[779] cyclization and affords 4-hydroxymethyl $\Delta^{3,4}$-butenolide (equation 434).

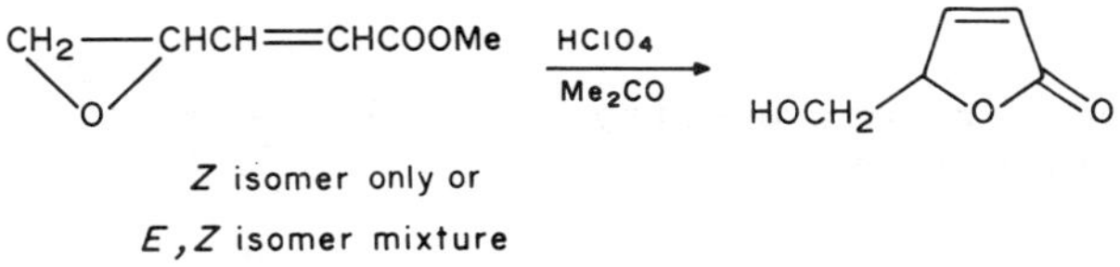

CH_2—CHCH=CHCOOMe (epoxide) $\xrightarrow[Me_2CO]{HClO_4}$ $HOCH_2$-butenolide (434)

Z isomer only or

E,*Z* isomer mixture

Reaction[780] of dimethyl *cis*-1,2,3,5-tetrahydro-1-oxo-4α (4*H*), 8α (8*H*)-naphthalenedicarboxylate with concentrated hydrochloric acid at reflux produces a 2:1 *trans*:*cis* mixture of a fused bicyclic lactone (equation 435).

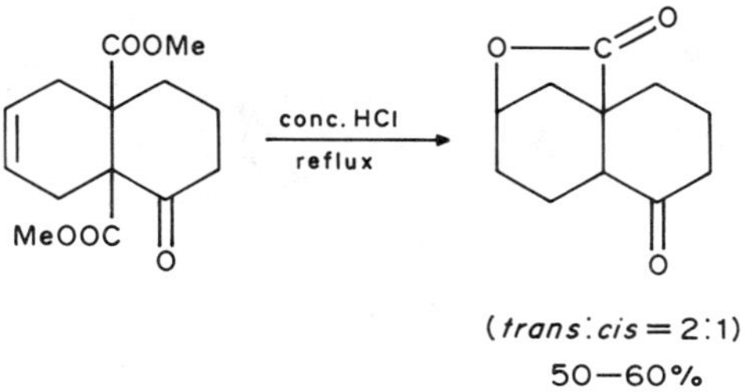

COOMe, MeOOC $\xrightarrow[reflux]{conc.\ HCl}$ (435)

(*trans*:*cis* = 2:1)

50–60%

Lithium borohydride reduction of the bicyclic lactone shown below produces[781] an intermediate dihydroxy diacid, which upon acid work-up affords a 60% yield of the spirolactone shown (equation 436).

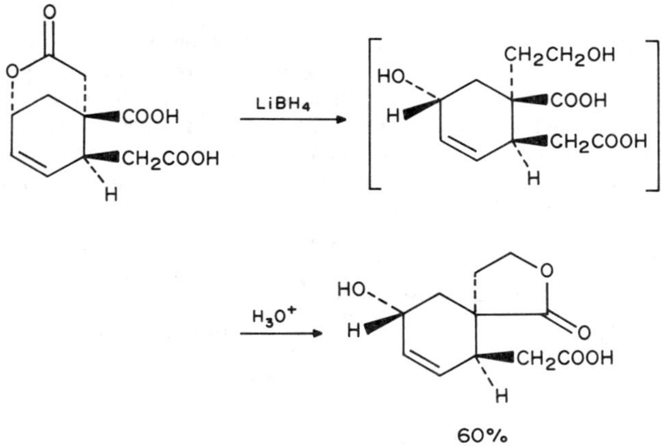

COOH, CH_2COOH, H $\xrightarrow{LiBH_4}$ [HO, H, CH_2CH_2OH, COOH, CH_2COOH, H]

$\xrightarrow{H_3O^+}$ HO, H, CH_2COOH, H (436)

60%

*2. *Photochemical and electrochemical cyclizations*

Exposure of 5-substituted-4-hydroxy-2-pentenoic acids to ultraviolet irradiation is reported [779] to produce 4-substituted $\Delta^{3,4}$-butenolides (equation 437).

RCH$_2$CH(OH)CH=CHCOOH (E) —UV, 400 W Hg, MeOH, H$^+$, r.t.→ 5-(RCH$_2$)-2(5H)-furanone (437)

R = HO, Br

Ultraviolet irradiation has also been used to produce[782] fused ring lactones from an (*E*)-furylacrylic acid (eqution 438).

2-furyl-CH=C(COOH)-C$_6$H$_4$-Me —UV, air, 48 h→ naphthofuran-COOH (Me) + epoxy-furopyranone (C$_6$H$_4$-Me) (438)

60% (1:1 mixture)

*3. *Halolactonization*

Examples of both bromo- and iodolactonizations have been reported and included among the reports of bromolactonization in the reaction[779] of 2,4-pentadienoic acid with bromine and sodium bicarbonate which produces 4-bromometyl-$\Delta^{3,4}$-butenolide (equation 439), the regio- and stereospecific cyclization[783] of 3-methyl-3-hydroxy-7-phenyl-4-heptenoic acid salt (equation 440), the cyclization[784] of 2-hydroxy-6-allylcyclohexanecarboxylic acid (equation 441), and a study[784] of the regioselectivity of the halolactonization of γ,δ-unsaturated acids. This latter study showed that bromolactonization of 4-hexenoic acids produced a greater percentage of δ-lactones than did iodolactonization, and that substituents located in the 3-position of the acids clearly favour the formation of δ-lactones while substituents located in the 6-position of the acids favour the formation of γ-lactones (equation 442).

CH$_2$=CHCH=CHCOOH —Br$_2$, H$_2$O, NaHCO$_3$→ 5-(BrCH$_2$)-2(5H)-furanone (439)

Me(HO)C(CH$_2$COO$^-$)CH=CHCH$_2$CH$_2$Ph —Br$_2$, MeOH, NaHCO$_3$, −78 °C→ δ-lactone (Me, HO, H, Br, H, CH$_2$CH$_2$Ph) (440)

Br_2, MeOH, $NaHCO_3$, −78 °C (441)

(78%)

$$\gamma,\delta\text{-unsaturated carboxylic acids} \xrightarrow[\text{conditions}]{\text{halolactonization}} \gamma\text{-lactones} + \delta\text{-lactones} \quad (442)$$

Acid	Halolactonization conditions	Ratio of γ:δ-lactones	Yield (%)
$McCH{=}CH(3H_2)_2COOH$ (*trans*)	I_2, KI, $NaHCO_3$	> 20:1	85
	I_2, ether, THF, $NaHCO_3$	24:1	71
	NBS, THF, HOAc	2.7:1	91
	Br_2, MeOH, $NaHCO_3$, −78 °C	1:1	46[a]
$Me(CH_2)_2CH{=}CH(CH_2)_2COOH$ (*trans*)	I_2, KI, $NaHCO_3$	> 15:1	91
	NBS, THF, HOAc	2.7:1	85
	Br_2, MeOH, $NaHCO_3$, −78 °C	2.7:1	50[a]
$MeCH{=}CHCHMeCH_2COOH$ (*trans*)	I_2, KI, $NaHCO_3$	6.5:1	88
	I_2, ether, THF, $NaHCO_3$	1.5:1	83[b]
	NBS, THF, HOAc	1:1.2	78[c]
	Br_2, MeOH, $NaHCO_3$, −78 °C	1:1.2	66
$MeCH{=}CHCMe_2CH_2COOH$ (*trans*)	I_2, KI, $NaHCO_3$	1.1:1	86
	I_2, ether, THF, $NaHCO_3$	1:1	81
	NBS, THF, HOAc	1:2.4	86
	Br_2, MeOH, $NaHCO_3$ −78 °C	1:2.5	93
$Me_2CHCH{=}CH(CH_2)_2COOH$ (*trans*)	I_2, KI, $NaHCO_3$	9:1	70
	NBS, THF, HOAc	9:1	76
	Br_2, MeOH, $NaHCO_3$, −78 °C	9:1	38
$Me(CH_2)_2CH{=}CHCH(OH)CH_2COOH$ (*trans*)	I_2, KI, $NaHCO_3$	5:1	70
	I_2, ether, THF, $NaHCO_3$	5:1	76[d]
	NBS, THF, HOAc	2.3:1	92[e]
	Br_2, MeOH, $NaHCO_3$, −78 °C	2:1	45
$Me(CH_2)_2CH{=}CHCMe(OH)CH_2COOH$ (*trans*)	I_2, KI, $NaHCO_3$	1:3.4	74
	I_2, ether, THF, $NaHCO_3$	1:3.5	85[f]
	NBS, THF, HOAc	< 1:9	93[g]
	Br_2, MeOH, $NaHCO_3$, −78 °C	< 1:9	89

[a]The methyl ester resulting from methanolysis of the γ-lactone was obtained.

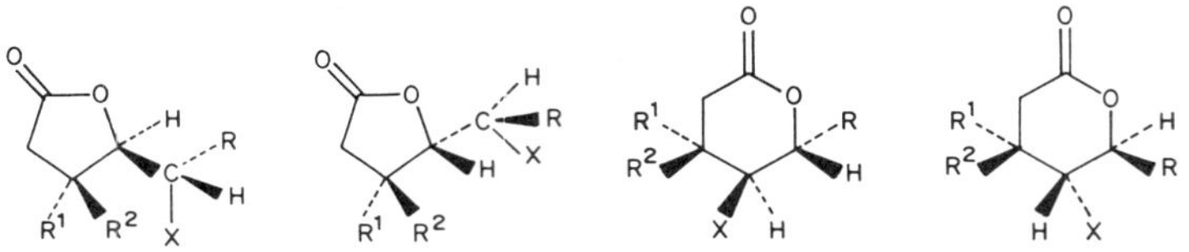

(62) (63) (64) (65)

[b]With R = Me, R^1 = H, R^2 = Me and X = I the percentage distribution of products **62**–**65** was: **62**, 36; **63**, 14; **64**, 4 and **65**, 29.
[c]With R = Me, R^1 = H, R^2 = Me and X = Br the percentage distribution of products **62**–**65** was: **62**, 22; **63**, 14; **64**, 5 and **65**, 37.
[d]With R = *n*-Pr, R^1 = H, R^2 = OH and X = I the percentage distribution of products **62**–**65** was: **62**, 60; **63**, 4; **64**, 8 and **65**, 4.
[e]With R = *n*-Pr, R^1 = H, R^2 = OH and X = Br the percentage distribution of products **62**–**65** was: **62**, 47; **63**, 0; **64**, 29 and **65**, 0.
[f]With R = *n*-Pr, R^1 = Me, R^2 = OH and X = I the percentage distribution of products **62**–**65** was: **62**, 19; **63**, 0; **64**, 66 and **65**, 0.
[g]With R = *n*-Pr, R^1 = Me, R^2 = OH and X = Br the percentage distribution of products **62**–**65** was: **62**, 9; **63**, 0; **64**, 84 and **65**, 0.

Iodolactonization has been used to prepare[785] γ-methylene butyrolactones (4-penten-4-olides) from their precursor 4-pentenoic acids and esters (equation 443), bicyclic lactones from 1-methyl-3-cyclohexenoic acid[784] (equation 444), from 1,2,3,4,4α,5,8,8α-octahydro-1-oxonaphthalene-4α-carboxylic acid[780] (equation 445), and from 4α,4,7,7α-tetrahydro-7α-carbomethoxymethyl-1-indanone[781] (equation 446). Spirolactones have been prepared from a monoacid[787] (equation 447) and a diester[785,786] (equation 448).

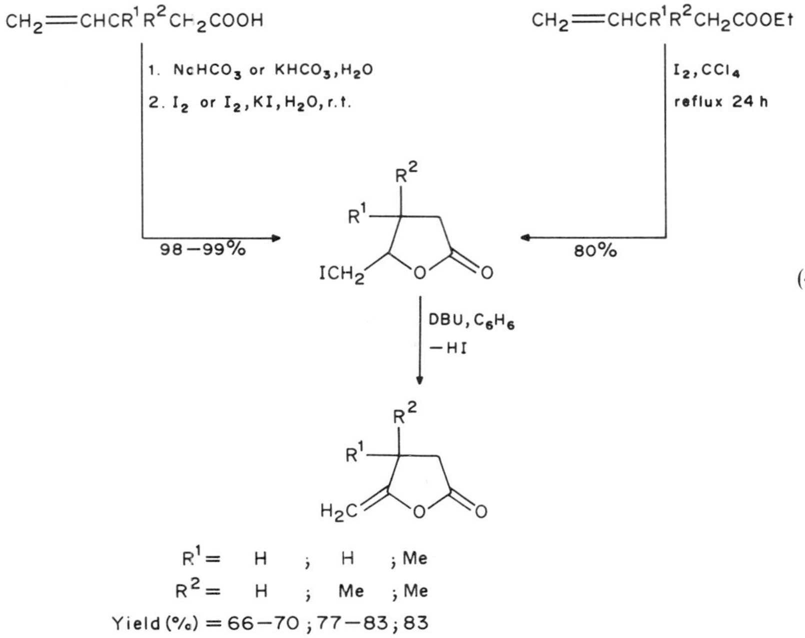

$CH_2{=}CHCR^1R^2CH_2COOH$ — 1. $NaHCO_3$ or $KHCO_3$, H_2O; 2. I_2 or I_2, KI, H_2O, r.t. — 98–99%

$CH_2{=}CHCR^1R^2CH_2COOEt$ — I_2, CCl_4, reflux 24 h — 80%

→ iodomethyl lactone (ICH_2) — DBU, C_6H_6, $-HI$ → $H_2C{=}$ methylene lactone (443)

R^1 =	H	H	Me
R^2 =	H	Me	Me
Yield (%) =	66–70	77–83	83

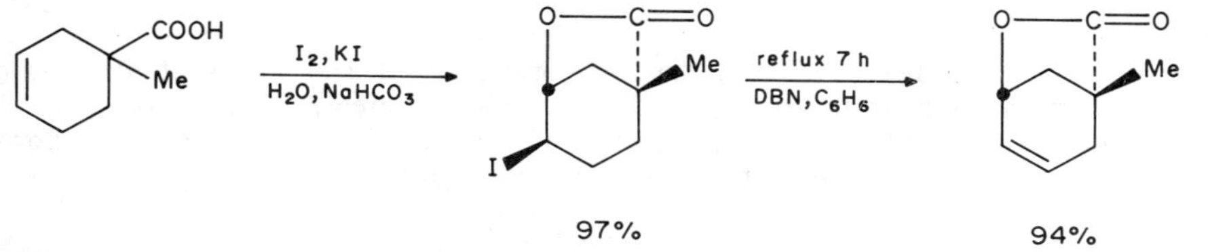

COOH, Me → I_2, KI / H_2O, $NaHCO_3$ → O—C=O, Me, I → reflux 7 h / DBN, C_6H_6 → O—C=O, Me

97% 94%

(444)

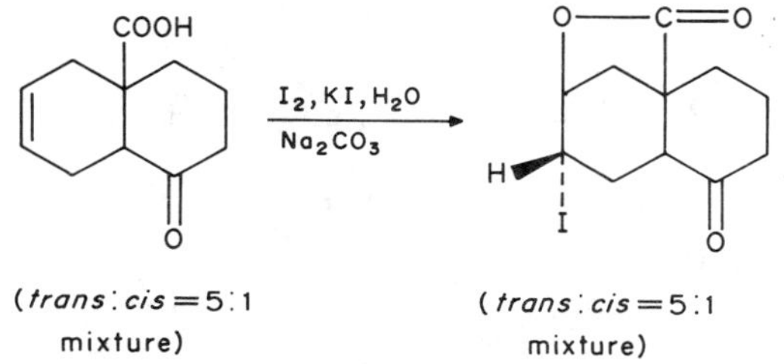

COOH, O → I_2, KI, H_2O / Na_2CO_3 → O—C=O, H, I, O (445)

(*trans* : *cis* = 5 : 1 mixture) (*trans* : *cis* = 5 : 1 mixture)

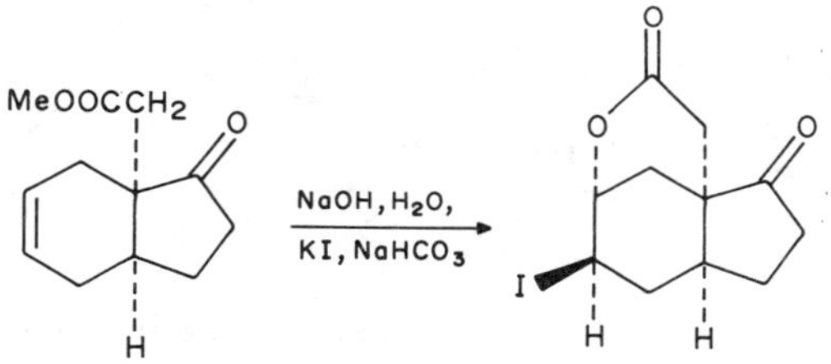

MeOOCCH$_2$, O, H → NaOH, H_2O, / KI, $NaHCO_3$ → O, O, O, I, H, H (446)

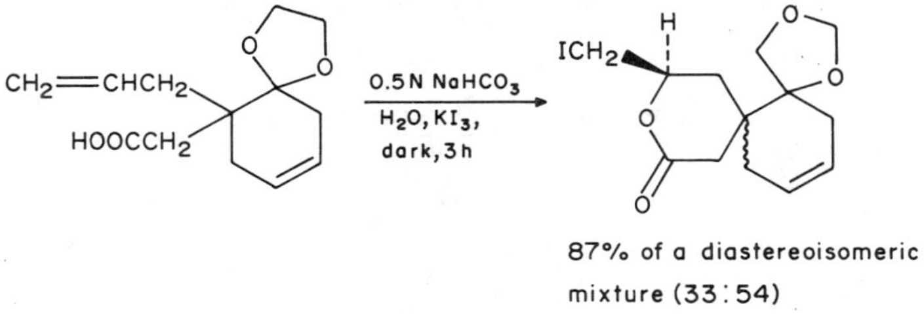

$CH_2{=}CHCH_2$, $HOOCCH_2$, O, O → 0.5 N $NaHCO_3$ / H_2O, KI_3, / dark, 3 h → ICH_2, H, O, O, O, O (447)

87% of a diastereoisomeric mixture (33 : 54)

$CH_2{=}CHCH_2C(COOEt)_2CH_2CH{=}CH_2$ → I_2, CCl_4 / 40 °C, 4 days →

ICH_2, O, O, O, O, CH_2I → DBN, C_6H_6, / 60 °C, 20 min → $CH_2{=}$, O, O, O, O, $=CH_2$

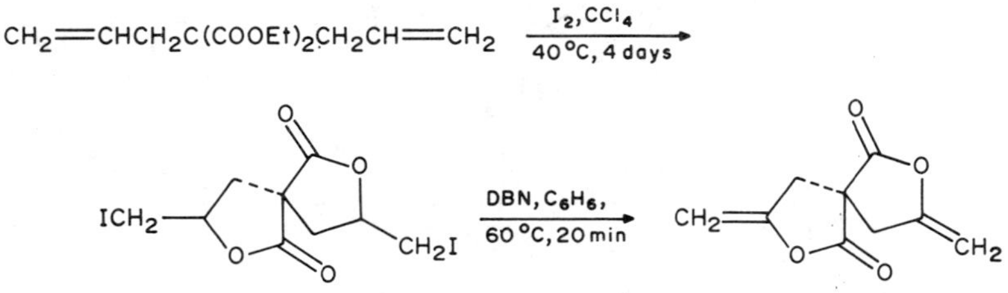

(448)

**4. Intramolecular Diels–Alder reactions*

Intramolecular Diels–Alder cyclization of acyclic triene esters derived from dichloromaleic anhydride are reported[788] to afford stereospecifically *trans*-bicyclic lactones (equation 449), while a similar reaction[788] of the corresponding free acid affords the *cis*-bicyclic lactone isomer stereospecifically (equation 450).

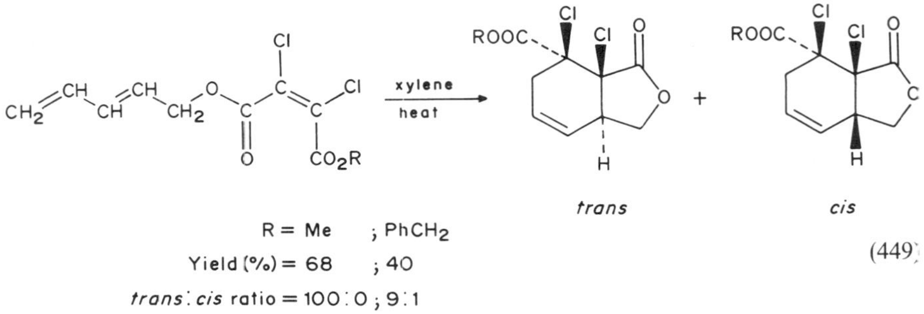

xylene, heat

trans *cis*

R = Me ; $PhCH_2$

Yield (%) = 68 ; 40

trans:*cis* ratio = 100:0 ; 9:1

(449)

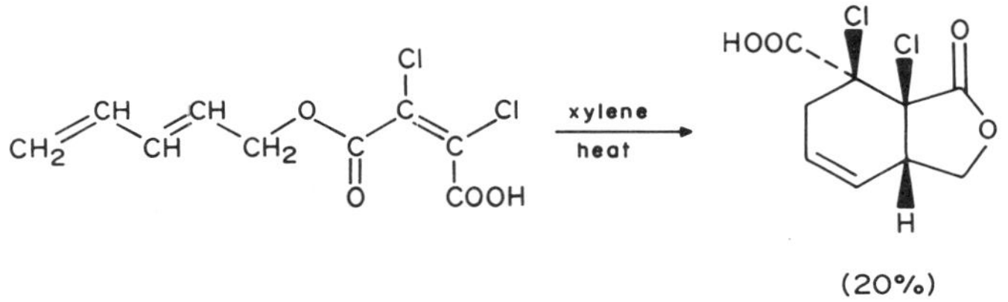

xylene, heat

(20%)

(450)

Another intramolecular Diels–Alder cyclization reaction reported[789], which shows much less stereoselectivity however, is the cycloaddition of 6-fulvenyl-2,2-dimethyl-3,5-hexadienoate to produce both a [6 + 4] and a [4 + 2] adduct (equation 451).

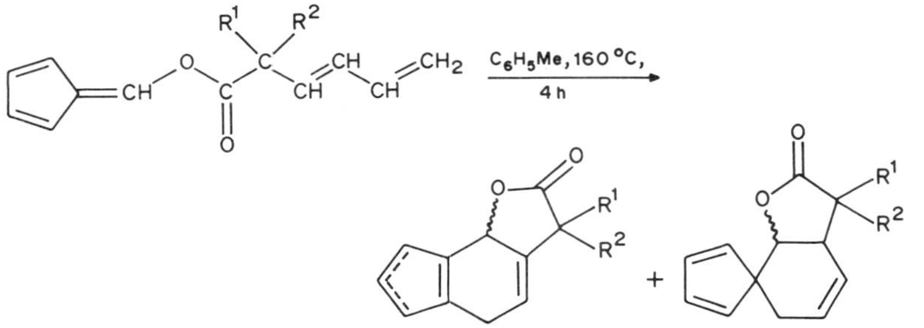

C_6H_5Me, 160 °C, 4 h

(66) **(67)**

(451)

R^1	R^2	Yield (%) **66**	**67**
H	H	52	14
Me	H	38	17
Me	Me	56	23

5. Using miscellaneous reagents

ω-Hydroxy unsaturated carboxylic acids have been converted to macrocyclic lactones using cyanuric chloride. Thus, treatement of *trans*-16-hydroxy-9-hexadecenoic acid with an acetonitrile solution of cyanuric chloride produces[764] a 70% yield of isoambrettolide (equation 452), while similar treatment of (*R*)-(+)-ricinelaidic acid with an acetone solution of cyanuric chloride produces[765] a 73% yield of (*R*)-(+)-ricinelaidic acid lactone (equation 453). Both products are formed without epimerization.

$HOCH_2(CH_2)_5CH{=}CH(CH_2)_7COOH$ —(cyanuric chloride, MeCN; Et_3N)→ isoambrettolide [H—C=C—H, $(CH_2)_6$, O, C=O, $(CH_2)_7$] 70% (452)

(*R*)-(+)-ricinelaidic acid [H, Hex-*n*, OH, OH, O] —(cyanuric chloride, Me_2CO, Et_3N, stir r.t., 6 h)→ lactone [H, Hex-*n*, O, O] 73% (453)

Reaction of β,γ-unsaturated carboxylic acids with phenylselenenyl chloride has been reported[790] to produce simple adducts which, upon treatment with silica gel, afford selenolactonization products, whose nature is highly sensitive to the substitution pattern of the substrate. Thus, acyclic β,γ-unsaturated carboxylic acids bearing a β-substituent, but no γ-alkyl groups, produce only decarboxylated products under a variety of basic conditions used for phenylselenolactonization, while in neutral solution these acids only produce 1:1 adducts. In contrast, the presence of no substituent, as in the case of vinylacetic acid (equation 454), β and γ substituents (equation 455), or γ-substituents alone (equation 456) all favour lactone formation in basic media.

$CH_2{=}CHCH_2COOH + PhSeCl \longrightarrow ClCH_2CH(SePh)CH_2COOH$ or

$PhSeCH_2CH(Cl)COOH \xrightarrow{\text{silica gel}}$ (PhSe-substituted γ-butyrolactone) 30% (454)

$$MeCH{=}CMeCH_2COOH + PhSeCl \longrightarrow MeCH(Cl)CMe(SePh)CH_2COOH \xrightarrow{\text{silica gel}}$$

$$MeCH{=}CMeCH_2COO^- Et_3\overset{+}{N}H + PhSeCl \longrightarrow \text{[3-Me-3-PhSe-4-Me-γ-butyrolactone] } 86\% \quad (455)$$

$$EtCH{=}CHCH_2COOH + PhSeCl \longrightarrow EtCH(Cl){-}CH(SePh)CH_2COOH \xrightarrow{\text{silica gel}}$$

$$EtCH{=}CHCH_2COO^- Et_3\overset{+}{N}H + PhSeCl \longrightarrow \text{[3-PhSe-4-Et-γ-butyrolactone] } 95\% \quad (456)$$

*E. By Malonic Ester or Malonic Acid Condensation

Reaction of malonic ester anion with (n-butoxymethyl)oxirane affords[779] a 53% yield of γ-(n-butoxymethyl)-α-carbomethoxy-α,β-butenolide (equation 457).

$$\text{(CH}_2\text{OBu-}n\text{)oxirane} + (MeOOC)_2\overset{-}{C}H\overset{+}{N}a \xrightarrow{MeOH} \text{[}n\text{-BuOCH}_2\text{-butenolide-COOMe] } 53\% \quad (457)$$

*G. By Grignard and Reformatsky Reactions

Grignard reagents are used to prepare lactones by employing their reaction with anhydrides. Thus, if two moles of methyl magnesium bromide in tetrahydrofuran are added to glutaric anhydride, a 43% yield[791] of α,δ-dimethylvalerolactone is obtained (equation 458).

$$\text{glutaric anhydride} + 2MeMgBr \xrightarrow[2.\ H^+]{1.\ THF} \text{[Me, Me-valerolactone] } 43\% \quad (458)$$

Similarly, reaction[792] of two moles of a primary Grignard reagent with 7-oxabicyclo[2.2.1]hept-5-ene-2,3-dicarboxylic anhydride affords the corresponding 4,4-dialkyl substituted bicyclo-γ-butanolides, which upon heating at 130 °C produce 4,4- disubstituted 2-butenolides by a retro-Diels–Alder reaction (equation 459).

+ 2RMgX → 1. N_2, ether, THF, or C_6H_6 2. H^+ → 130 °C →

R = *n*-Pr; *n*-Bu; *n*-Pent; *n*-Hex; Ph; $PhCH_2$

(459)

Reaction of the same substrate with a secondary Grignard reagent produces the corresponding 4-alkyl substituted bicyclo-γ-butanolides, which also undergo a retro-Diels–Alder reaction upon heating to 130 °C producing the 4-monosubstitued 2-butenolides (equation 460).

+ 2RMgX → 1. N_2, ether, THF, or C_6H_6 2. H^+ → 130 °C →

endo isomer preferred

R = *i*-Pr; Me(Et)CH

(460)

This approach has also been used[792] to produc 4-spirobicyclobutenolides by reaction of 7-oxabicyclo[2.2.1]hept-5-ene-2,3-dicarboxylic anhydride with α,ω-di(bromomagnesio)alkanes. These products also undergo retro-Diels–Alder reaction upon heating (equation 461).

+ $BrMg(CH_2)_nMgBr$ → 1. N_2, THF, stir overnight 2. H^+ → 130 °C →

$(CH_2)_n$ $(CH_2)_n$

$n = 4,5$

(461)

4-Spirobutenolides and bicyclobutenolides can also be regioselectively prepared in excellent yields by the one-step addition[793] of α,ω-di(bromomagnesio)alkanes to unsymmetrically substituted cyclic anhydrides. These regioselective syntheses occur with the nucleophilic addition oriented to the less hindered carbonyl group of the anhydride (equation 462), and the regioselectivity being observed to be greater in the preparation of the cyclohexane products rather than their cyclopentane analogues. With the heterocyclic analogue the yield was lower, but the regiospecificity was observed to be greater.

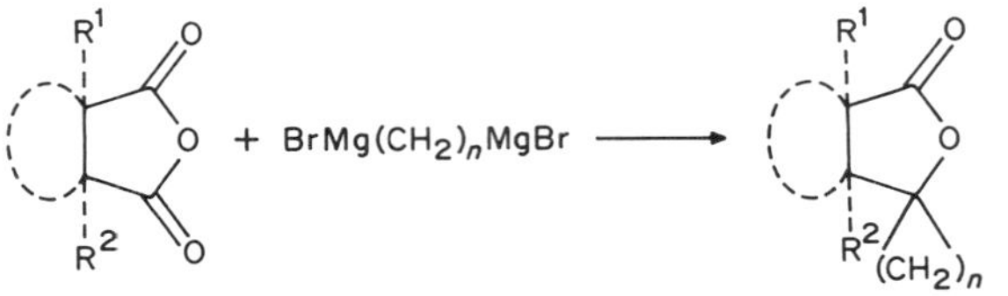

$$+ \ BrMg(CH_2)_nMgBr \longrightarrow \qquad (462)$$

Anhydride	n	Products (ratio)	Yield (%)
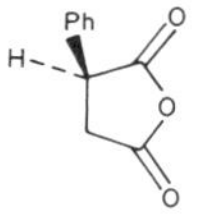	4	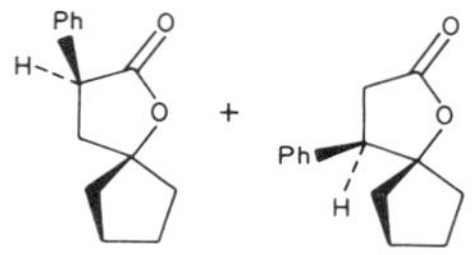 (3:1)	73
	5	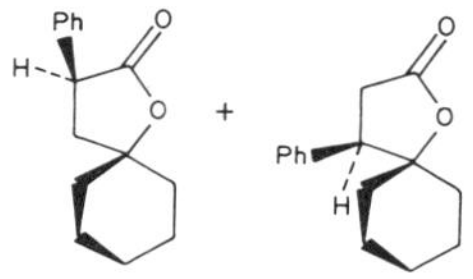 (3:1)	55
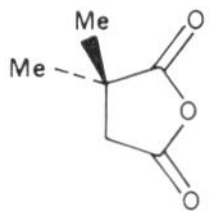	4	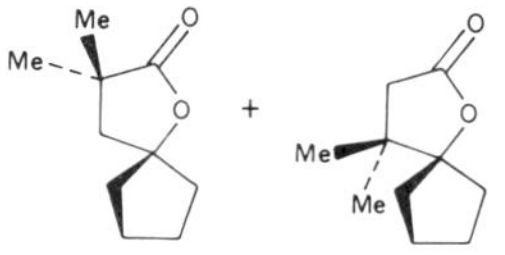 (3:2)	80
	5	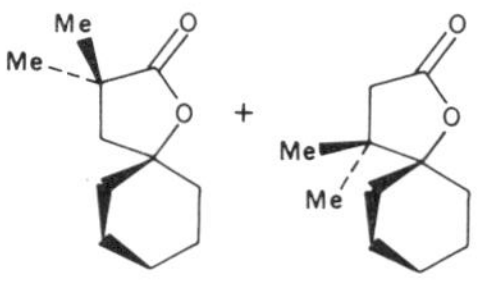 (4:1)	70
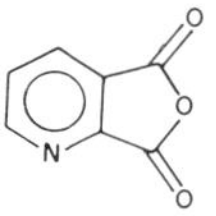	4	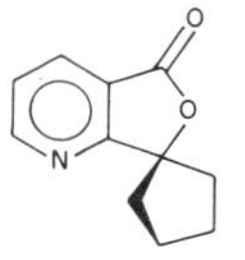	55

Anhydride	n	Products (ratio)	Yield (%)
	5	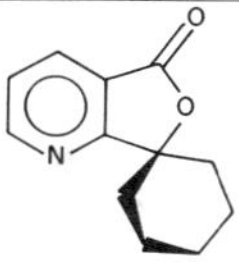	20
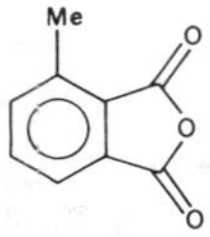	4	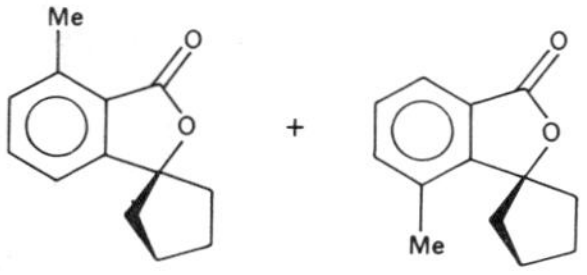 + (5:2)	75
	5	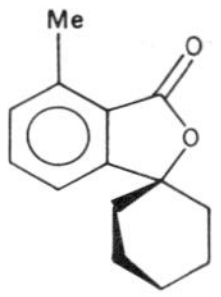	60
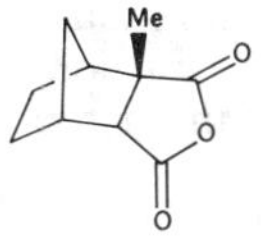	4	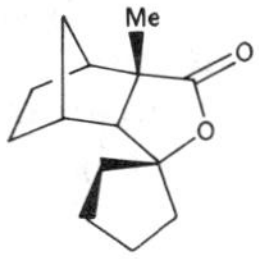	93
	5	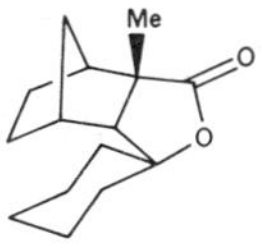	91

Reformatskii reaction of ethyl α-(bromomethyl)acrylate with acyl chlorides produces[794] α-methylene-γ-lactones via a double condensation followed by cyclization (equation 463).

$$BrCH_2C(=CH_2)COOEt + RCOCl \xrightarrow[-ZnX_2]{Zn, Et_2O} [RCOCH_2C(=CH_2)COOEt] \xrightarrow{[BrZnCH_2C(=CH_2)COOEt]}$$

$$CH_2{=}C(COOEt)CH_2\text{–}RC(OZnBr)\text{–}CH_2C(=CH_2)COOEt \longrightarrow \text{α-methylene-γ-lactone (R, } CH_2C(=CH_2)COOEt\text{)} \qquad (463)$$

R = Me; Ph; 3,4,5-$(MeO)_3C_6H_2$
Yield (%) = 62; 74; 67

*I. From α-Anions of Carboxylic Acids or Esters

Treatment[795] of ethyl 3,3,6-trimethyl-4-benzoyloxy-5-heptenoate with lithium di(isopropyl)amide produces an intermediate ester enolate which then attacks the carbonyl carbon of the benzoyl function. Rearrangement followed by cyclization affords the lactone shown in 80% yield (equation 464).

$$Me_2C{=}CHCH(O_2CPh)CMe_2CH_2COOEt \xrightarrow[2.\ r.t.,\ 5\ min]{1.\ LDA,\ -78\,^\circ C,\ THF} [Me_2C{=}CHCH(OC(=O)Ph)CMe_2\bar{C}HCOOEt]$$

$$\longrightarrow [Me_2C{=}CHCH(O^-)CMe_2CH(COPh)C(=O)OEt] \longrightarrow \text{lactone: } Me_2C{=}CH\text{, Me, Me, COPh} \quad 80\% \qquad (464)$$

Lithium di(isopropyl)amide has also been used to effect intramolecular alkylation of long-chain ω-halo β-keto esters, via an intermediate dianion, producing[796] macrocyclic β-keto lactones (equations 465, 466 and 467).

$$MeCOCH_2COO(CH_2)_nBr \xrightarrow[\text{2. r.t.}]{\text{1. LDA, THF, 0 °C}} \text{macrocyclic } \beta\text{-keto lactone: } (CH_2)_n\text{–}CH_2\text{–}C(=O)\text{–}CH_2\text{–}C(=O)\text{–}O \text{ ring} \tag{465}$$

$n = 5; 6; 7; 8; 9; 10; 11$

Yield (%) $= 0; 0; 0; 0; 43; 45; 49$

$$MeCOCH_2COOCHMe(CH_2)_8Br \xrightarrow[\text{2. r.t.}]{\text{1. LDA, THF, 0 °C}} \tag{466}$$

$$MeCOCH_2COO(CH_2)_{n-1}CH{=}CH(CH_2)_{n-1}Br \xrightarrow[\text{2. r.t.}]{\text{1. LDA, THF, 0 °C}} H_2C\text{–}C(=O)\text{–}CH_2\text{–}C(=O)\text{–}O\text{–}(CH_2)_{n-1}\text{–}CH{=}CH\text{–}(CH_2)_{n-1} \text{ ring} \tag{467}$$

$n = 10; 11$

Yield (%) $= 37; 37$

*K. By Direct Functionalization of Preformed Lactones

Structural elaboration of preformed lactones can be accomplished using a variety of reagents and reactions. Among the approaches reported is the alkylation of hydroxy lactones using alkyl lithium reagents which produce[797] the corresponding 4-alkyl Δ-2-butenolides via the mechanism shown (equation 468).

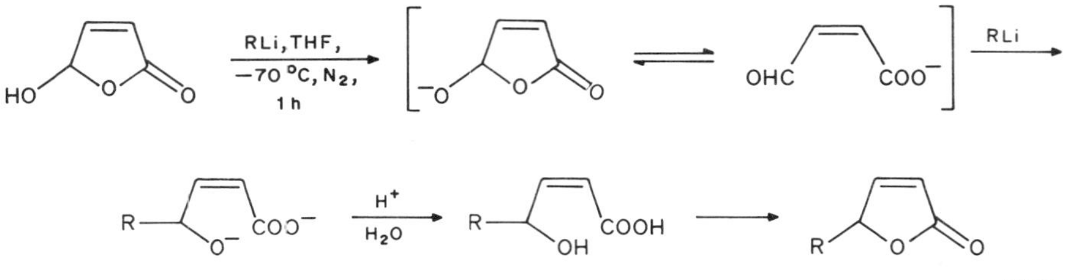

(468)

R = Me ; n-Bu ; s-Bu ; t-Bu ; Ph ; $CH_2{=}CH$; n-PrC≡C
Yield (%) = 71 ; 75 ; 72[a] ; 65 ; 45 ; 60 ; 75

[a] A 1:1 mixture of diastereomers.

Treatment of (3*S*)-3-benzyloxycarbonylamino-γ-butyrolactone with lithium diisopropylamide in tetrahedrofuran generates[798] a dianion, which upon treatment with various electrophiles produces the corresponding 2-alkylated products (equation 469). It can be seen from the results reported that the stereoselectivity in these alkylations is not very high, with the *trans* isomers favoured in about a 4 to 1 ratio.

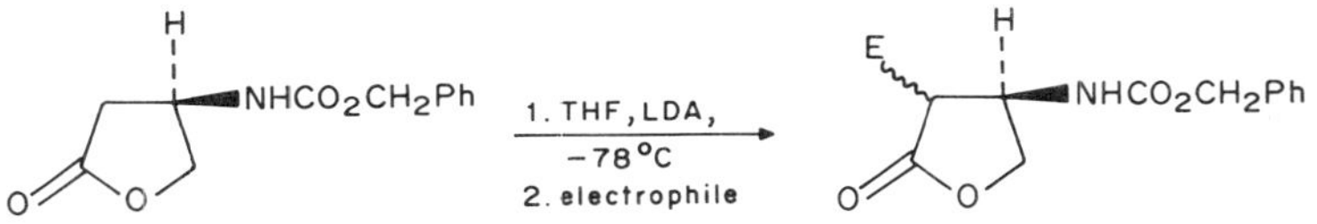

(469)

Electrophile	E	Temp (°C)	Yield (%)	Ratio *cis*:*trans*
MeI	Me	−78→40	70	1:4
EtI	Et[a]	−78→0	55	1:4
i-PrI	*i*-Pr[a]	−78→10	15	<5:>95
Me_2CO	$Me_2CH(OH)$	−78	84	<5:>95

[a] Alkylation performed in the presence of HMPA.

Conversion of the unsubstituted saturated lactone γ-butyrolactone into the α-methylene lactone tulipalin A has been reportedly effected[799] by reaction with methylene dimethylammonium iodide (Eschenmoser's salt). The mechanism of this reaction involves the formation of a Mannich intermediate (equation 470).

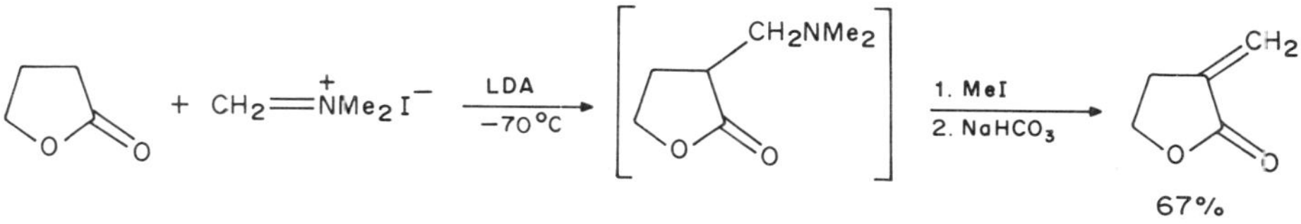

67%

(470)

A similar type of Hofmann degradation reaction has been reported[800] for the preparation of 2-(*cis*-2-hydroxycyclohexyl and heptyl)propenoic acid lactones (equation 471), from the corresponding α-anilinomethyl *cis*-fused γ-butyrolactone starting materials.

1. MeI, MeOH, reflux 5 h
2. 28% NH_4OH

$n = 4$, 88%
$n = 5$, 85%

MeI, MeOH
reflux 5 h

I^-

EtOH, EtONa
reflux 2 h

(471)

$n = 4$, 40%
$n = 5$, 45%

Although catalytic hydrogenation of unsaturated α-anilinomethyl-3,4-butenolides has been accomplished[800] using both the Adams and Raney nickel catalysts, different products result from the use of these two reagents. The point of difference is in the structure of the side chain, since reduction with Adams catalyst produces 2-(*cis*-2-hydroxycyclohexyl and heptyl)propanoic acid lactones (equation 472), while reduction with Raney nickel catalyst produces the corresponding α-anilinomethyl *cis*-fused γ-butyrolactones (equation 473).

Adams Cat.
EtOH, Pt, H_2,
70 °C, shake 6 h

(472)

$n = 4$, 94%
$n = 5$, 92%

Raney Ni, EtOH,
H_2, 70 °C, 6 h

(473)

$n = 4$, 95%
$n = 5$, 80%

Double bond formation within a saturated butanolide molecule has been accomplished[751] by heating the starting material with bromine in the presence of red-phosphorus (equation 474). The resulting 3,4-butenolide probably results from initial halogenation followed, under the conditions of the reaction, by dehydrohalogenation.

+ Br_2, P (red) → 1. 70 → 80 °C, 3 h; 2. H_2O, reflux, 4 h — 53% (474)

Two approaches to the halogenation of unsaturated butenolides have been reported in the recent literature[751]. One involves the conversion of a hydroxyl group to a halide by reaction with thionyl chloride (equation 475), while the second approach involves reaction of unsaturated alkyl substituted butenolides with thionyl chloride (equation 476) or *N*-bromosuccinimide (equation 477).

1. $SOCl_2$, glyme, Na_2CO_3, 40 °C, 3–24 h; 2. 80 °C, 1 h (475)

R^1 =	Me	Ph	Ph
R^2 =	H	H	Ph
Yield (%) =	84	97	—

$SOCl_2$, reflux 1 h (476)

NBS, CCl_4, N_2, benzoyl peroxide, reflux (477)

R^1 =	H	*t*-Bu	*t*-Bu
R^2 =	Me	H	H
R^3 =	H	H	*t*-Bu
reflux (h) =	—	3.5	12
Yield (%) =	—	—	85

A variety of fused ring furan-2-ones have reportedly[751] been formylated by reaction with ethyl formate (equation 478).

$$\text{(fused furan-2-one)} + HCO_2Et \xrightarrow[\text{stir, temp.}]{Et_2O,\ \text{base}} \text{(}\alpha\text{-CHOH furan-2-one)} \quad (478)$$

Fused ring	Base	Rx time (h)	Temp. (°C)
cyclopentene	NaOMe	24	−20
cyclopentene	NaOEt	—	—
indane	Na	18	r.t.

By condensation[751] of butenolides with the sodio derivatives of α-formyllactones and lactams, several analogues of the natural germination stimulant, strigol, for parasitic weeds of the genera *Striga* and *Orobanche*, have been prepared in yields ranging from 8 to 81% (equation 479 and Table 38).

$$\text{(}\alpha\text{-CHONa lactone/lactam)} + X\text{-butenolide} \xrightarrow{\text{conditions}} \text{(}\alpha\text{-CHO-butenolide lactone/lactam)} \quad (479)$$

X=halide or sulphonate

Y=O or N

*L. From Ketenes

Sulphoxide-directed lactonization of trisubstituted vinyl sulphoxides with dichloroketene has been reported[801] to proceed in a completely stereoselective manner to produce the corresponding γ-arylthio-γ-butyrolactones (equation 480).

Using a ketene synthesis which is extensively utilized for the preparation of lactams, and from which a lactam product was expected, benzylidene anthranilic acid was condensed[802] with phthalimidoacetyl chloride in the presence of triethylamine and produced the fused-ring lactone shown (equation 481).

*M. By Reduction of Anhydrides, Esters and Acids

Several reagents have been used to reduce anhydrides and produce lactones. Included among these reagents are lithium aluminium hydride and sodium borohydride which

TABLE 38. Synthetic analogues of strigol[751]

Reactants: Sodio salt	Reactants: Halide or sulphonate	Conditions	Product	Yield (%)
CHONa	Me, X		CHO, Me	
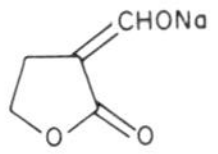	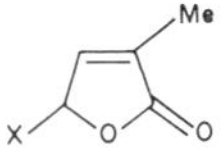 $X = OSO_2Me$	DME, r.t., 24h	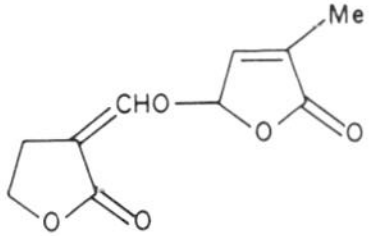	60
	X = Br	THF, HMPA, r.t.		—
	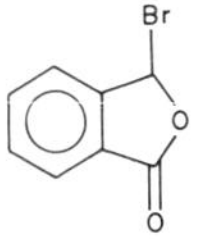 Br	DME, r.t.	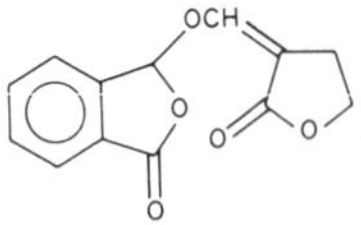 OCH	—
Me, CHONa	Br	DME	CHO, Me	75
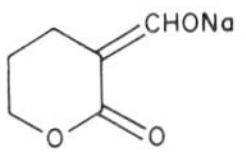 CHONa	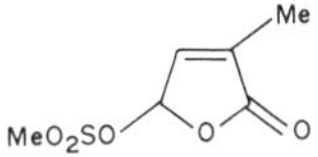 MeO_2SO, Me	DME, r.t., stir 4h	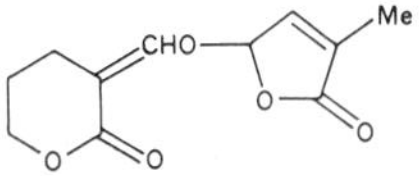 CHO, Me	71
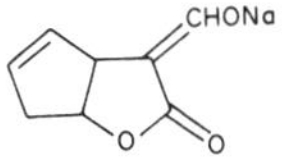 CHONa	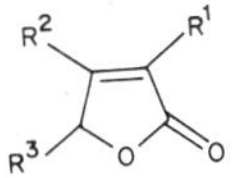 R^1, R^2, R^3; $R^1 = Me$, $R^2 = H$, $R^3 = Cl$	glyme, r.t., stir 24h	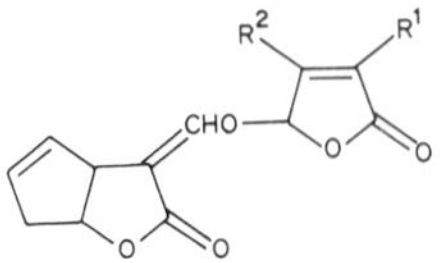 CHO, R^1, R^2	80

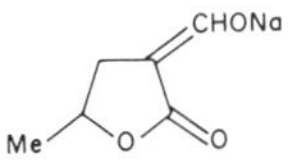

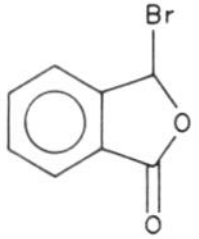

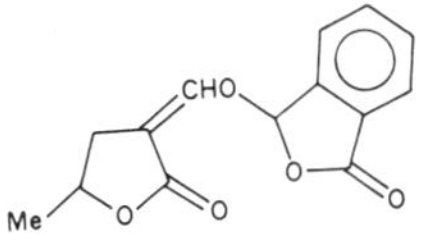

	$R^1 = H, R^2 = Me, R^3 = Br$	THF, 0 °C, N_2 stir overnight		—
	$R^1 = Ph, R^2 = H, R^3 = Cl$	THF, 70–75 °C, stir 24h		—
	$R^1 = R^2 = Ph, R^3 = Cl$	DME, reflux, 24h		64
	Br	DME, r.t., stir 18h	CHO	—
	Bu-*t*, Br, *t*-Bu	THF, reflux, 24h	Bu-*t*, *t*-Bu, CHO	81
CHONa	R^2, R^1, R^3			
	$R^1 = Me, R^2 = H, R^3 = Cl$	DME, 5 °C, stir 16h	R^2, R^1, CHO	—[a]
	$R^1 = H, R^2 = Me, R^3 = Br$	1. THF, O °C, 6h 2. r.t. 14h		49
	Br	DMF, r.t., 18h	CHO	—[a]
CHONa	R^1, R^2, Cl		R^1, R^2, CHO	
	$R^1 = H, R^2 = Me$	DME, r.t. stir 19h		—
	$R^1 = Me, R^2 = H$	THF, r.t., overnight		—
	$R^1 = H, R^2 = Ph$	1. DME, −50 °C 2. sitr r.t., overnight 3. reflux, 3h		—

(*continued*)

TABLE 38. (*continued*)

Reactants				
Sodio salt	Halide or sulphonate	Conditions	Product	Yield (%)
	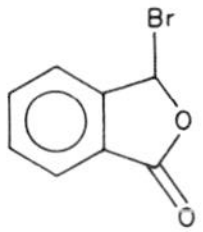	DME, r.t., stir 19h	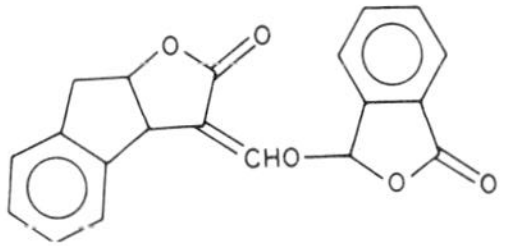	8[a]
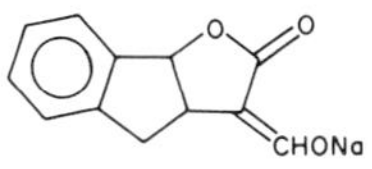	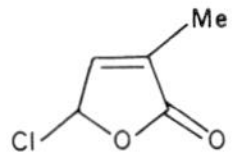	DME, r.t., stir 16h	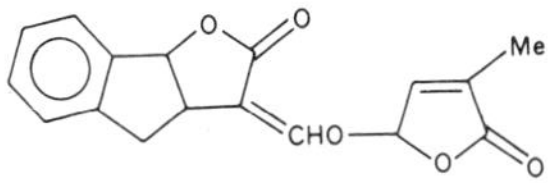	—[a]
	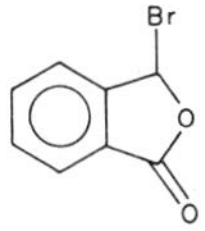	DME, r.t., stir	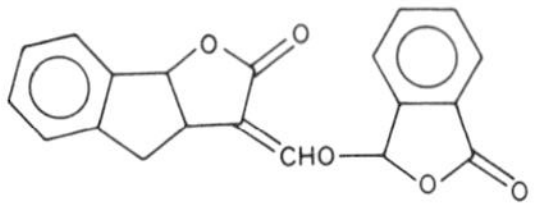	—
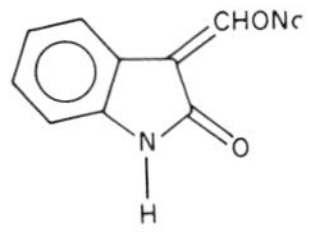	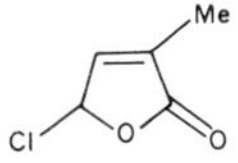	K_2CO_3, HMPA, N_2, r.t., stir 18h	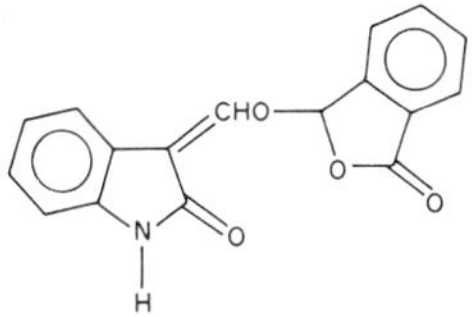	19
	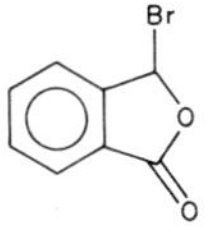	K_2CO_3, HMPA, N_2, r.t., stir	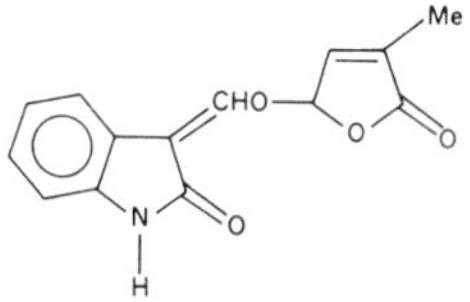	

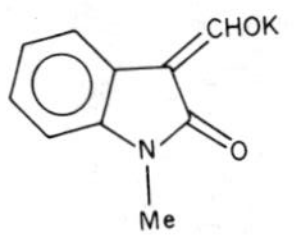	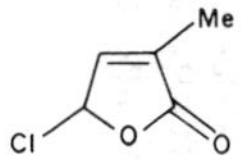	K_2CO_3, HMPA, N_2, r.t., stir	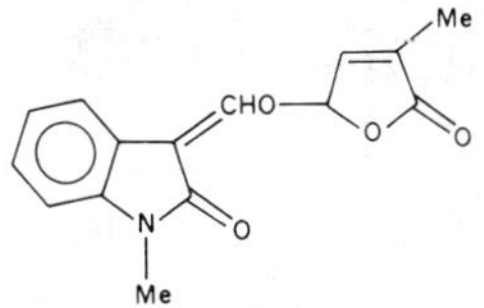	43
	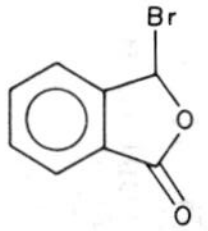	K_2CO_3, HMPA, N_2, r.t., stir	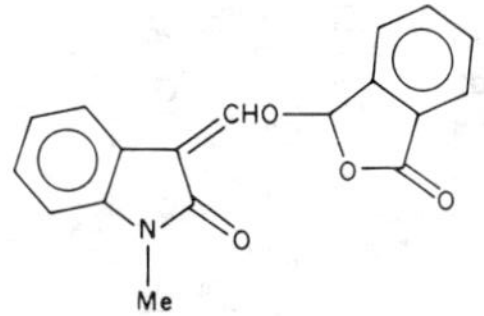	—
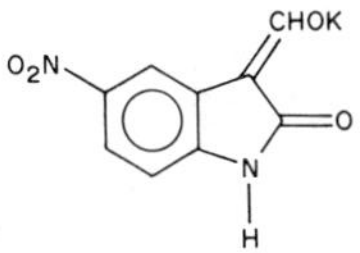	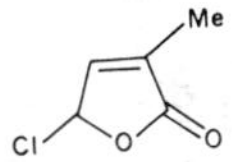	K_2CO_3, HMPA, N_2, r.t., stir	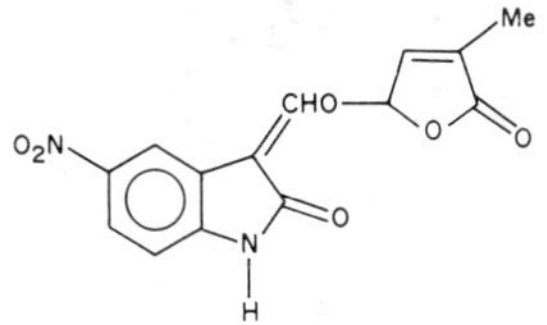	38
	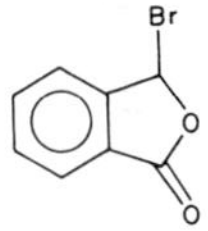	K_2CO_3, HMPA, N_2, r.t., stir	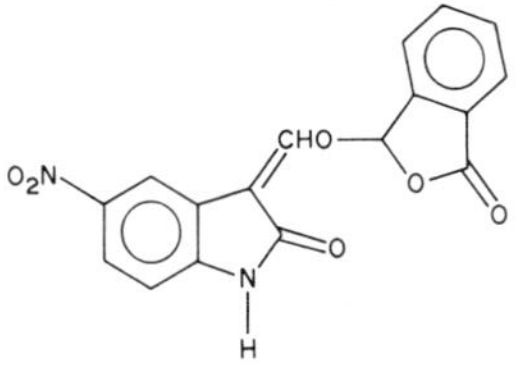	28
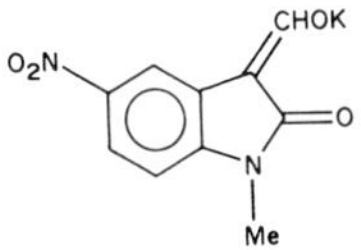	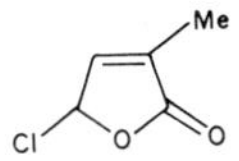	K_2CO_3, HMPA, N_2, r.t., stir	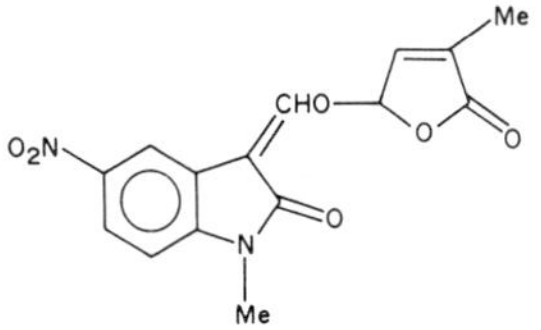	37

[a]Product obtained as a mixture of diastereoisomers.

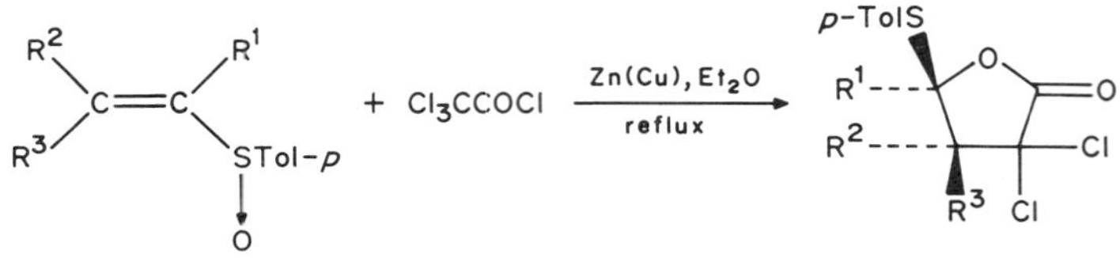

(480)

R^1	R^2	R^3	Yield (%)
3,4-$(MeO)_2C_6H_3$	Me	H	60
H	*n*-Bu	Me	70
n-Bu	Me	H(*R*)	75(3*R*, 4*R*)

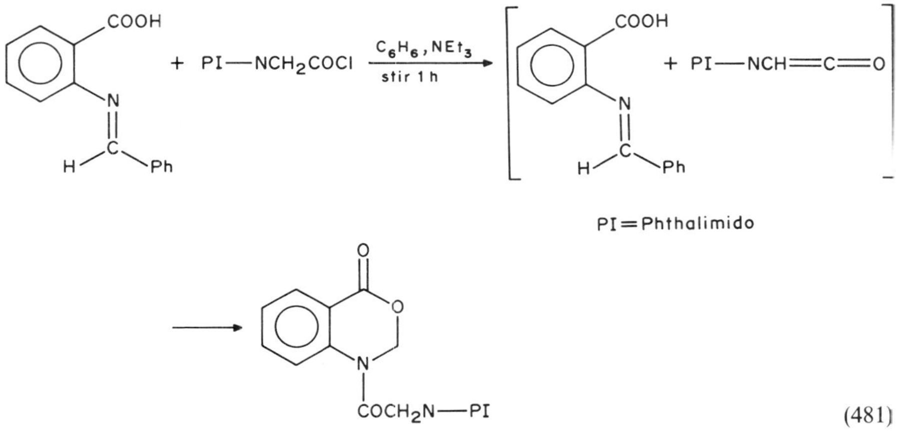

(481)

have been used[793] to regioselectively reduce unsymmetrically substituted cyclic anhydrides. These metal hydride reductions primarily occur at the carbonyl group adjacent to the more highly substituted carbon atom, except in the case of *exo*-2-methylbicyclo[2.2.2]oct-5-ene-*endo*-2,3-dicarboxylic anhydride and *exo*-2-methylbicyclo[2.2.1]hept-5-ene-*endo*-2,3-dicarboxylic anhydride (equation 482) which, by stereospecific addition, leads[793] to reduction of the carbonyl group adjacent to the least highly substituted carbon atom.

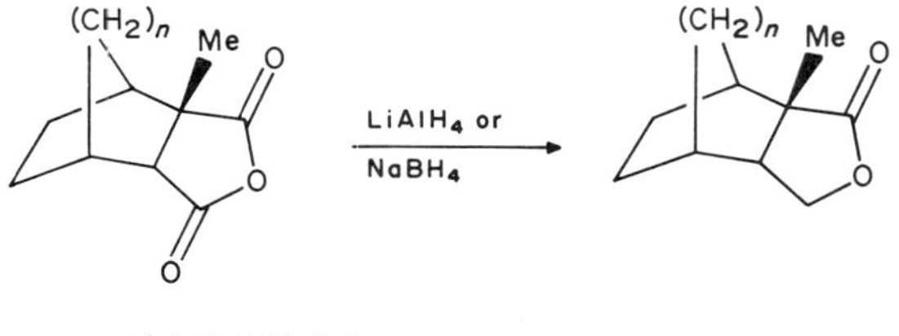

(482)

Two examples of the use of sodium borohydride to produce lactones from simpler anhydrides, in which the carbonyl groups adjacent to the more highly substituted carbon atom are reduced, are the preparation[798] of 3-methylbut-2-en-4-olide from citraconic anhydride (equation 483), and the preparation of (3*S*)-3-benzyloxycarboxylamino-γ-butyrolactone, by the regioselective reduction[798] of 3-(*N*-benzyloxycarbonylamino)-L-aspartic acid anhydride (equation 484).

Me

1. $NaBH_4$, THF, 0 °C, stir, 45 min
2. r.t., stir 45 min

Me

43% (483)

H $NHCO_2CH_2Ph$

1. THF, $NaBH_4$, 0 °C
2. stir r.t., 1 h
3. HCl, EtOH, reflux 1.5 h

H $NHCO_2CH_2Ph$

72% (484)

Bu-*t*

1. THF, $(t\text{-BuO})_3AlHLi$,
2. N_2, r.t.
3. reflux 1 h

Bu-*t*

40% (485)

\+ $RuH_2(PPh_3)_4$

C_6H_5Me, 50 °C, Argon or vacuum 50 °C

H, Ph_3P, PPh_3, Ru, O, PPh_3, O, C, R

dry HCl, ether or CO, ether

(486)

Anhydride	R in complex	Products
	CHO	
	$OHCCH_2CH_2$	
Me	$OHCCH_2CHMe$ + $OHCCHMeCH_2$ (2.8:1)	Me + Me (3:1)

Lithium tri-(*t*-butoxy)aluminium hydride has also been used[798] to produce a lactone (2-*t*-butylbut-2-en-4-olide) from *t*-butylmaleic anhydride, but in this case the carbonyl group adjacent to the least highly substituted carbon atom is again reduced (equation 485).

One class of reagents which have been used extensively in the recent literature to catalyse hydrogenation of anhydrides and produce lactones are ruthenium-containing reagents. Dihydridotetrakis(triphenylphosphine)ruthenium reacts with phthalic, succinic and methylsuccinic anhydrides to produce[803] an intermediate isolatable complex, which upon treatment with dry hydrogen chloride or carbon monoxide in ether quantitatively produces the corresponding lactones (equation 486).

Succinic[804,805] and phthalic[804] anhydrides can also be quantitatively converted to their corresponding lactones, γ-butyrolactone and phthalide, by catalytic hydrogenation using dichlorotris(triphenylphosphine)ruthenium (equation 487).

$Cl_2Ru(PPh_3)_3$

H_2, 100°C, Et_3N

(487)

$Cl_2Ru(PPh_3)_3$

H_2, 100°C, Et_3N

Regioselective hydrogenation of arylnaphthofurandiones has been accomplished[806] by using a variety of ruthenium complexes, but the best results were reported when tetrachlorotris[1,4-bis(diphenylphosphino)butane]diruthenium [$Ru_2Cl_4(dppb)_3$] was employed as the catalyst (equation 488).

R^1O R^1O OR^2 OR^2

C_6H_5Me, Ru Cat.

PPh_3, NEt_3, $MgSO_4$

stir 160°C, H_2

R^1O R^1O OR^2 OR^2

(68)

\+

R^1O R^1O OR^2 OR^2

(69)

(488)

R^1	R^2	Ruthenium catalyst	Time (h)	Yield (%) lactones	Ratio **68**:**69**
Me	Me	$RuCl_2(PPh_3)_3$	24	12	>99:1[a]
Me	Me	$RuH_2(PPh_3)_4$	24	28	93:7[a]
Me	Me	$RuH_2(PMe_2Ph)_4$	72	70	68:32[a]
Me	Me	$Ru_2Cl_4(dppb)_3$	24	88	>99:1[a]
—CH_2—	Me	$Ru_2Cl_4(dppb)_3$	24	89	>99:1[b]
—CH_2—	—CH_2—	$Ru_2Cl_4(dppb)_3$	24	82	>99:1[c]

[a] **68** = dehydrodimethylretrodendrin; **69** = dehydrodimethylconidendrin.
[b] **68** = chinensin; **69** = retrochinensin.
[c] **68** = taiwanin C; **69** = justicidin E.

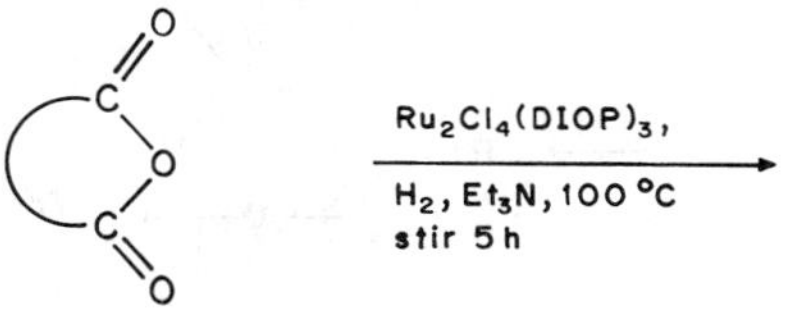

$Ru_2Cl_4(DIOP)_3$, H_2, Et_3N, 100 °C, stir 5 h (489)

Anhydride	Lactone	Yield (%)	% ee	Configuration
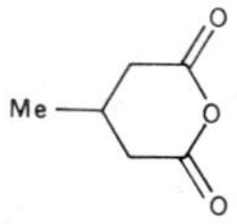 3-Me glutaric anhydride	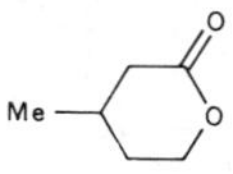 4-Me δ-lactone	32	16.4	*R*
		61[a]	19.4	*R*
		48[b]	14.0	*R*
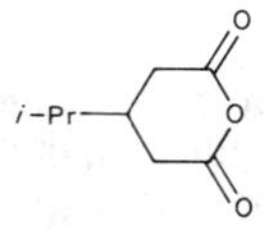 3-*i*-Pr glutaric anhydride	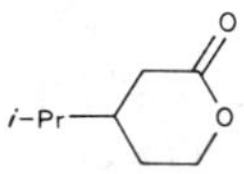 4-*i*-Pr δ-lactone	27	6.0	*R*
		79[a]	5.4	*R*
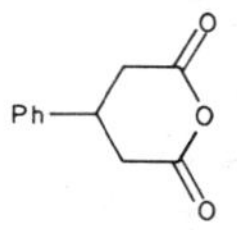 3-Ph glutaric anhydride	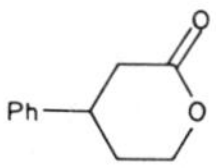 4-Ph δ-lactone	28	17.7	*R*
		56[a]	20.0	*R*
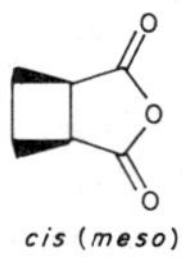 *cis* (*meso*)	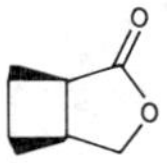 *cis*	35	12.8	1*R*, 2*S*
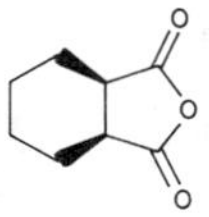 *cis*	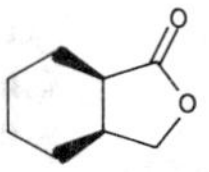 *cis*	42	—	—

[a] Free triphenylphosphine was added to the reaction mixture.
[b] Free (−)-2,3-*O*-isopropylidene-2,3-dihydroxy-1,4-bis(diphenylphosphino)-butane was added to the reaction mixture.

Asymmetric synthesis of chiral lactones has been reported[807] by the enantioselective reduction of a carbonyl group in cyclic anhydrides containing a prochiral carbon atom or containing two carbon centres of opposite chirality (*meso*) by using the chiral ruthenium(II) complex tetrachlorotris[(−)-2,3-*O*-isopropylidene-2,3-dihydroxy-1,4-bis(diphenylphosphino)butane]diruthenium [$Ru_2Cl_4(DIOP)_3$] as a catalyst (equation 489).

The mechanism proposed[807] for this asymmetric conversion involves ruthenium aldo-carboxylate complex formation caused by the cleavage of the C—O bond of the cyclic anhydrides, followed by the formation of an aldehyde group from a carbonyl group and a hydride ligand[803], and formation of the product. The absolute configuration of the product is determined by the selectivity of the C—O bond cleaved with the chiral ruthenium hydride species (equation 490).

M = Ru — DIOP

(490)

In the preparation of lactones by the reduction of esters, borohydride reagents have been used most often in recent reports. Using sodium borohydride, β-aldehydic α,β-ethylenic esters (equation 491) as well as β-aldehydic α,β-ethylenic acids (equation 492) have been converted[808] to their corresponding butenolides.

$R^1CH_2C(CHO){=}CHCOOR^2 \xrightarrow[MeOH]{NaBH_4}$ butenolide (R^1CH_2) (491)

R^1	R^2	Yield (%)
n-Bu	*t*-Bu	85
$Me_2C{=}CHCH_2$	Me	83
$Me_2C{=}CHCH_2$	*t*-Bu	81
$Me_2C{=}CH(CH_2)_2C(Me){=}CHCH_2$	*t*-Bu	87
2,6,6-trimethylcyclohexenyl-CH_2	*t*-Bu	80

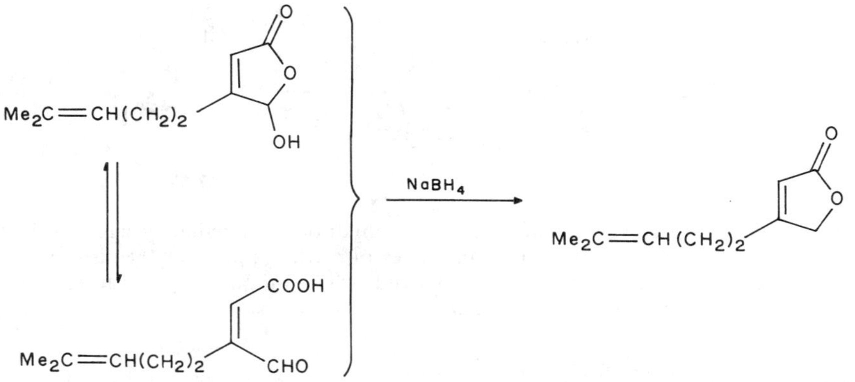

Me$_2$C=CH(CH$_2$)$_2$ ... OH ⇌ Me$_2$C=CH(CH$_2$)$_2$ ... COOH, CHO —NaBH$_4$→ Me$_2$C=CH(CH$_2$)$_2$ ...

(492)

Sodium borohydride reduction of the mixture of diastereomers shown in equation 493 results[809] in loss of the chrial auxiliary originally present and the formation of the corresponding lactones as an unequal mixture of enantiomers.

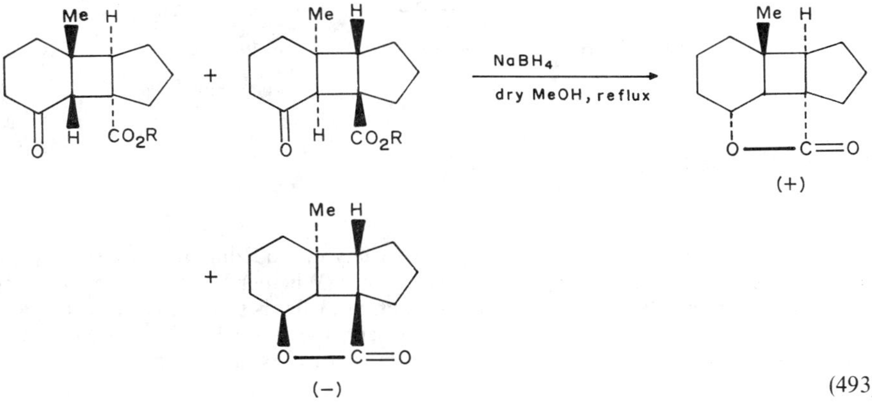

Me H ... H CO$_2$R + Me H ... H CO$_2$R —NaBH$_4$, dry MeOH, reflux→ Me H ... O—C=O (+) + Me H ... O—C=O (−)

(493)

R = (−) Me ... Me$_2$CPh ; (−) Me ... Me$_2$CH ; (−) Me Me Me

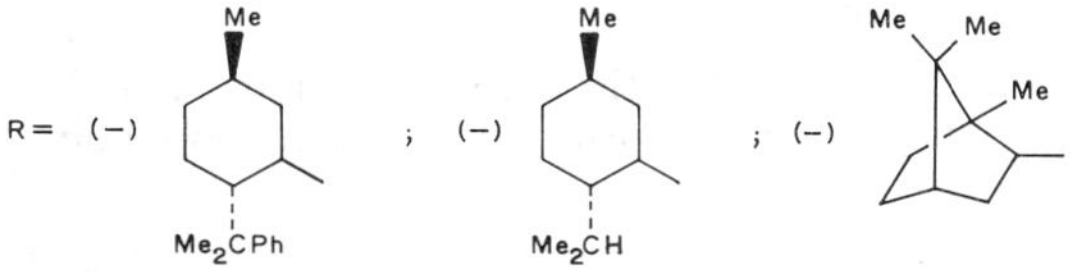

Reaction time (h) =	3 ;	1.5 ;	1.5
% ee[a]	57 ;	12 ;	18

[a]Value is % (+) minus % (−).

One step[810] in the conversion of methyl 2-methyl-3-phenyldimethylsilyl-5,5-dimethoxypentanoate into its corresponding δ-lactone involves the use of sodium borohydride as illustrated in equation 494.

$$(MeO)_2CHCH_2CH(SiMe_2Ph)CH(Me)COOMe \xrightarrow[\text{2. NaBH}_4\text{; 3. TosOH}]{\text{1. TosOH, Me}_2\text{CO}} \text{lactone (H, Me, H, SiMe}_2\text{Ph)}\ 47\% \quad (494)$$

One example[811] of the use of calcium borohydride as a reducing agent is in the conversion of the *R*(-(+)- and *S*-(−)-antipodes of methyl α-piperonylhemisuccinate to the crystalline, optically pure, (*R*)-(+)- and (*S*)-(−)-β-piperonyl-γ-butyrolactones, respectively, in high yields (equation 495).

$$\text{ArCH}_2\overset{2}{\text{C}}\text{H(COOMe)CH}_2\text{COOH} \xrightarrow{\text{Ca(BH}_4)_2} \text{ArCH}_2\text{-}\overset{3}{\text{C}}\text{H-butyrolactone}$$

(*R*)-(+)H-2α ⟶ (*R*)-(+)H-3α 82%

(*S*)-(−)H-2β ⟶ (*S*)-(−)H-3β 80%

(495)

*N. By Oxidation Reactions

*1. Oxidation of diols

Treatment of 1,4-, 1,5- and 1,6-diols with a variety of oxidizing agents results in the formation of their corresponding γ-, δ- or ε-lactones. Whether these reactions proceed via oxidation or dehydrogenation is a moot point, but what is sure is that these reagents do lead to the formation of lactones. Table 39 presents the diols which have been recently reported to be converted into lactones by treatment with oxidizing agents.

*2. Oxidation of ketones

Subjecting the substituted bicyclic ketone shown in equation 496 to a Baeyer–Villiger reaction produces[821] a quantitative mixture of the corresponding lactone regioisomers.

*3. Oxidation of ethers

Multiple examples of the oxidative conversion of ethers into lactones appear in the recent literature and may be divided into two categories. The first category includes examples of side-chain ether functions which, upon treatment with a variety of oxidizing agents, are converted into the ketone function of a lactone. Thus, oxidation of alkoxyfuran with Jones' reagent (CrO_3, dil. H_2SO_4/Me_2CO) converts[822] the alkoxy function into a ketone and produces furanone (equation 497).

Similar reactions[823] using Jones' reagent are employed to produce saturated (equation 498) and unsaturated (equation 499) β-oxy-γ-butyrolactones from cyclic acetal ethers, and the methylester of Prelog–Djerassi lactone[824] (equation 500).

TABLE 39. Preparation of lactones by oxidation of diols

Diols	Oxidizing agent	Product	Yield (%)	References
$HO(CH_2)_4OH$	CuO, 200 °C, liq. phase, absence of air, 15h		80	812
	$K_2Cr_2O_7$, DMF, r.t., 24h		70	813
$HO(CH_2)_5OH$	CuO, 230 °C, liq. phase, absence of air, 10h		41	812
$HO(CH_2)_6OH$	CuO, 230 °C, liq. phase, absence of air, 10h		5	812
	$K_2Cr_2O_7$, DMF, r.t., 6h		22 + 51	813
$HO(CH_2)_2CR^1R^2CH_2OH$				814
$R^1 = R^2 = Me$	$NiBr_2$, Bz_2O_2, MeCN, 60 °C		52 + 8	
	$Ph_3C^+BF_4^-$, MeCN, 60 °C		45 + 2	
$R^1 = Et$, $R^2 = Me$	$NiBr_2$, Bz_2O_2, MeCN, 60 °C		84 + 6	
	$Ph_3C^+BF_4^-$, MeCN, 60 °C		97 + 1	

(*continued*)

TABLE 39. (*continued*)

Diols	Oxidizing agent	Product	Yield (%)	References
$R^1 = R^2 = Ph$	$NiBr_2$, Bz_2O_2, MeCN, 60 °C		91 + 0	
	LiBr, Bz_2O_2, MeCN, 60 °C		45.5 + 45.5	
	$Me_4\overset{+}{N}Br^-$, Bz_2O_2, MeCN, 60 °C		48 + 50	
	Br_2, HMPA[a], CH_2Cl_2, 0 °C		6 + 42	
	$C_5H_5N{\cdot}HCl{\cdot}CrO_3$, CH_2Cl_2, 25 °C		69 + 30	
	$HCrO_4^-$, Me_2CO, H_2O, 25 °C		64 + 35	
	$Ph_3\overset{+}{C}BF_4^-$, MeCN, 60 °C		77 + 0	
$R^1R^2C(OH)CHR^3C(=CH_2)CH_2OH$[b]	MnO_2, CH_2Cl_2, r.t.	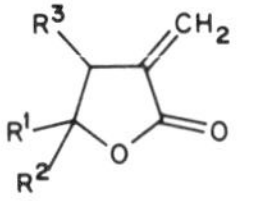 [structure: 5-R^1-5-R^2-4-R^3-3-methylene-dihydrofuran-2-one]		815, 816
$R^1 = Et$, $R^2 = R^3 = H$			65	815
$R^1 = n$-Hex, $R^2 = R^3 = H$			71	815
$R^1 = Et$, $R^2 = n$-Bu, $R^3 = H$			86	815
$R^1 = R^2 = H$, $R^3 = Ph$			77	815
$R^1 = R^3 = H$, $R^2 = n$-Hex			79	816
$R^1 = R^2 = Et$, $R^3 = H$			86	816
$R^1 = R^2 = n$-Bu, $R^3 = H$			91	816
$R^1R^2 = -(CH_2)_5-$, $R^3 = H$		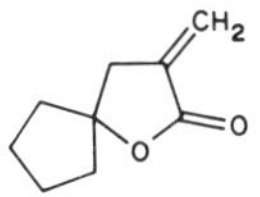 [structure: spiro α-methylene lactone]	85	816
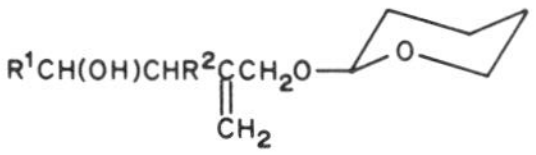 $R^1CH(OH)CHR^2C(=CH_2)CH_2O$-(tetrahydropyran-2-yl), $R^1 = Ph$, $R^2 = H$	1. p-TosOH, MeOH[c] 2. MnO_2, CH_2Cl_2	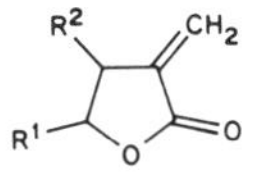 [structure: 5-R^1-4-R^2-3-methylene-dihydrofuran-2-one]		817

Substrate	Reagents/conditions	Product	Yield (%)	Ref.
$R^1 = Ph, R^2 = H$			49	
$R^1 = R^2 = Me$ (*cis*)		(*cis*)	56	
$R^1 = n$-Hex, $R^2 = H$			47	
$R^1R^2 = -(CH_2)_3-$		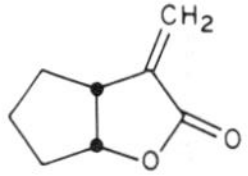 CH_2	40	
$R^1R^2 = -(CH_2)_4-$		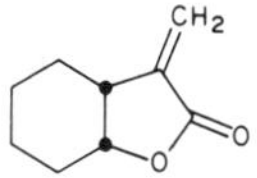 CH_2	27	
$R^1 = c$-Pent, $R^2 = H$		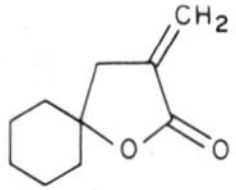 CH_2	50	
$R^1 = n$-$C_{13}H_{27}$, $R^2 = (t$-Bu$)Me_2SiOCH_2$ (*trans*)		*trans*	73	
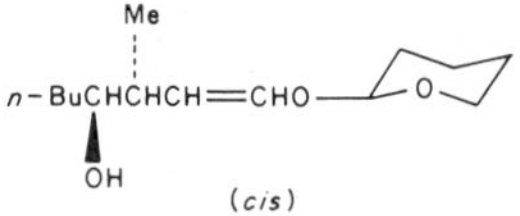 *n*-BuCHCHCH=CHO– (Me, OH) (*cis*)	1. 2N H_2SO_4, Me_2CO, 25 °C[d], 1 h 2. $KMnO_4$, 2N H_2SO_4, Me_2CO, 1 h, 10 °C 3. 25 °C	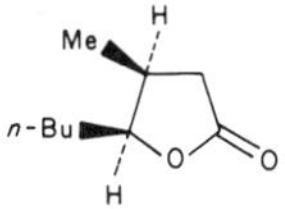 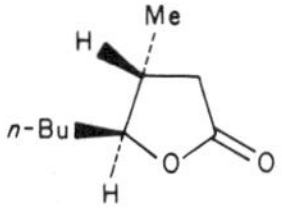 Me, H, *n*-Bu, H (*cis*) + H, Me, *n*-Bu, H (*trans*)	1 + 49	818
n-BuCHCHCH$_2$CH$_2$OH (Me, OH)	MnO_2, MeCN, 25 °C, 5h or $RuCl(PPh_3)_3$, C_6H_6, 25 °C, 5h		59 + 1	818
$HO(CH_2)_4OH$	$RuH_2(PPh_3)_4$, Me_2CO, C_6H_5Me, 180 °C, Ar, 3h	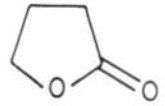	90	819
$HO(CH_2)_5OH$	$RuH_2(PPh_3)_4$, Me_2CO, C_6H_5Me, 180 °C, Ar, 2h	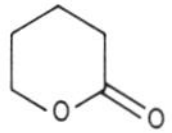	82	819

(*continued*)

TABLE 39. *(continued)*

Diols	Oxidizing agent	Product	Yield (%)	References
$HOCH_2CH{=}CHCH_2OH$ (*cis*)	$RuH_2(PPh_3)_4$, Me_2CO, C_6H_5Me, 180 °C, Ar, 3h	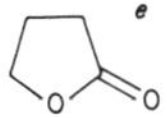 O, O [e]	88	819
$MeCH(OH)(CH_2)_3OH$	$RuH_2(PPh_3)_4$, Me_2CO, C_6H_5Me, 180 °C, Ar, 3h	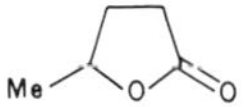 Me, O, O	92	819
$HO(CH_2)_4CH(OH)CH_2OH$	$RuH_2(PPh_3)_4$, Me_2CO, C_6H_5Me, 180 °C, Ar, 3h	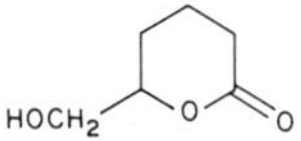 $HOCH_2$, O, O	71	819
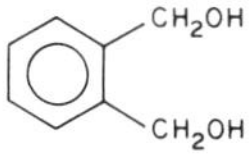 CH_2OH, CH_2OH	$RuH_2(PPh_3)_4$, C_6H_5Me, 180 °C, Ar, 12h	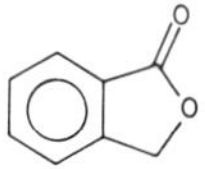 O, O	82	819
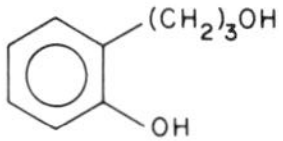 $(CH_2)_3OH$, OH	$RuH_2(PPh_3)_4$, Me_2CO, C_6H_5Me, 180 °C, Ar, 10h	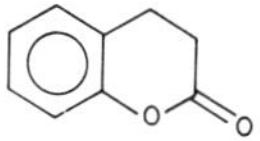 O, O	80	819
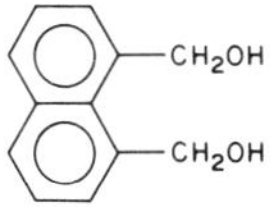 CH_2OH, CH_2OH	$RuH_2(PPh_3)_4$, Me_2CO, C_6H_5Me, 180 °C, Ar, 6h	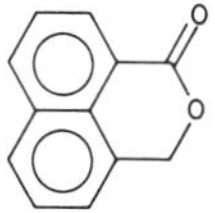 O, O	95	819
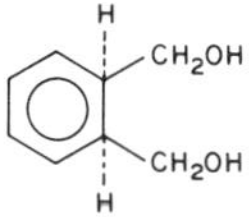 H, CH_2OH, CH_2OH, H	$RuH_2(PPh_3)_4$, C_6H_3Me, 180 °C, Ar, 15 h	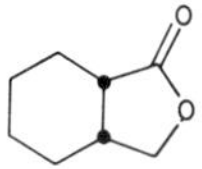 O, O	83	819

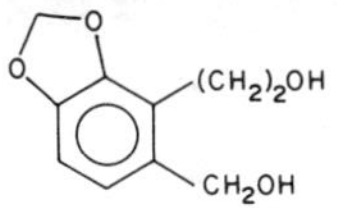	$RuH_2(PPh_3)_4$, Me_2CO, C_6H_5Me, 180 °C, Ar, 3h	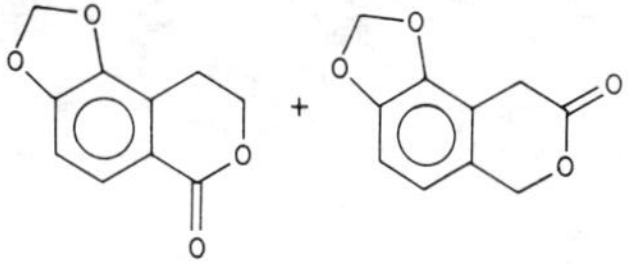 (2:1)	60	819
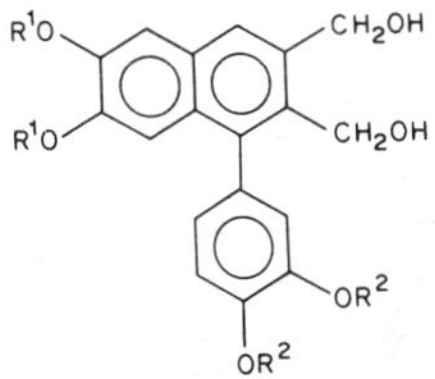	$RuH_2(PPh_3)_4$, C_6H_5Me, PhCH=CHCOMe, reflux, 10h	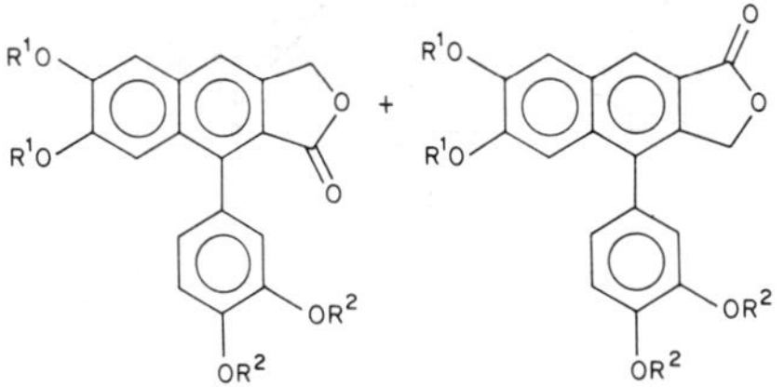		806
$R^1 = R^2 = Me$		(2:98)	68	
$R^1R^1 = -CH_2$-, $R^2 = Me$		(2:98)	94	
$R^1R^1 = R^2R^2 = -CH_2$-		(2:98)	84	
RCH(OH)CH=CHCH$_2$OH (*cis*)	Ag_2CO_3, C_6H_6, reflux			820
R = H	2h		78	
R = Et	3h		80	
R = $PhCH_2$	3h		56	
R = $PhCH_2OCMe_2(CH_2)_2$	7h		75	
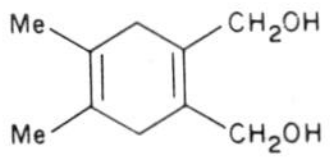	Ag_2CO_3, C_6H_6, reflux 8h	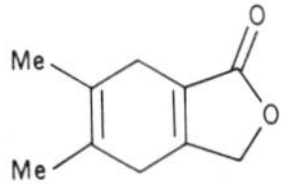	84	820

[a]Hexamethylphosphoric triamide.
[b]Probable mechanism: $R^1R^2C(OH)CHR^3C(=CH_2)CH_2OH \longrightarrow [R^1R^2C(OH)CHR^3C(=CH_2)CHO] \rightleftharpoons$ 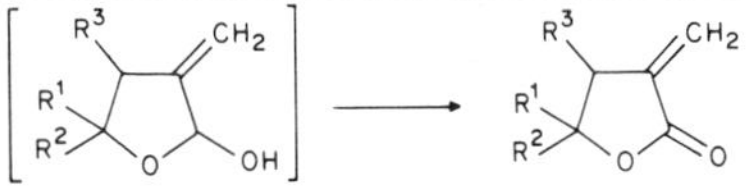
[c]First step was removal of tetrahydropyranyl group to produce $R^1CH(OH)CHR^2C(=CH_2)CH_2OH$.
[d]First step was removal of tetrahydropyranyl group to produce *n*-BuCH(OH)—CH(Me)CH=CHO.
[e]Hydrogenation accompanies lactonization.

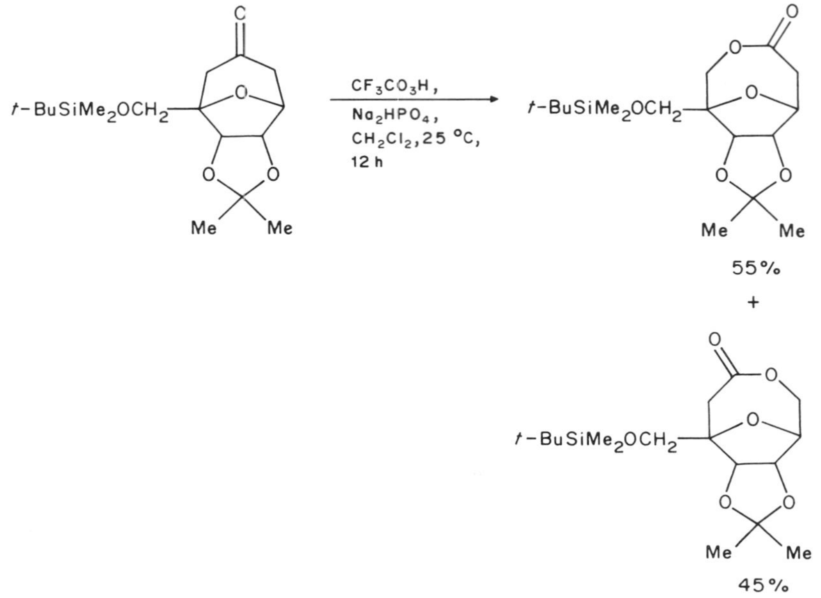

CF_3CO_3H, Na_2HPO_4, CH_2Cl_2, 25 °C, 12 h

t-BuSiMe$_2$OCH$_2$ … → t-BuSiMe$_2$OCH$_2$ … 55% + t-BuSiMe$_2$OCH$_2$ … 45% (496)

Me Me

Bu-n → Bu-n

n-BuO … Jones' reagent → … (497)

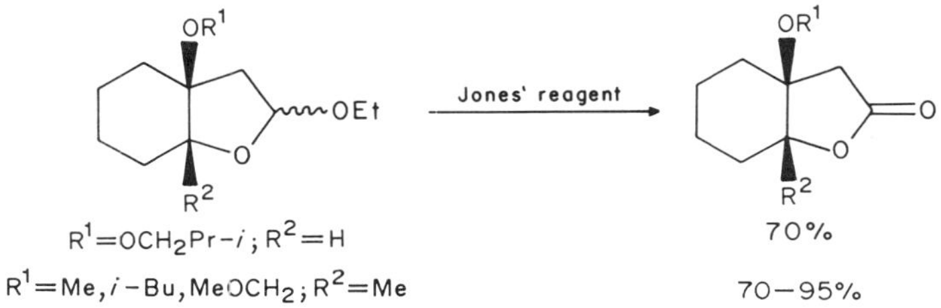

OR^1 … OEt — Jones' reagent → OR^1 … O (498)

R^1=OCH$_2$Pr-i; R^2=H — 70%

R^1=Me, i-Bu, MeOCH$_2$; R^2=Me — 70–95%

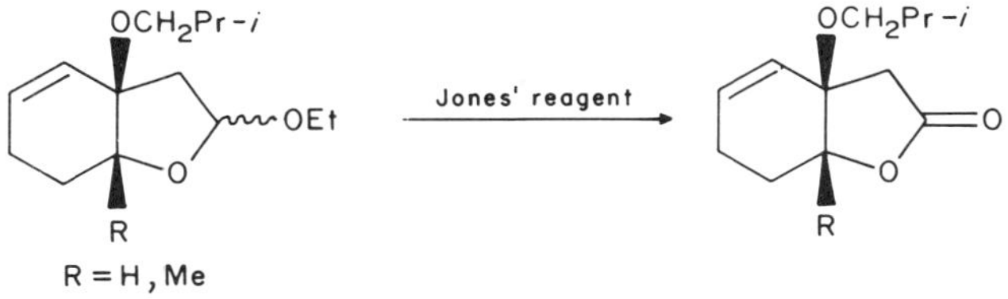

OCH$_2$Pr-i … OEt — Jones' reagent → OCH$_2$Pr-i … O (499)

R

R = H, Me

Jones' reagent, H_2CrO_4, H_2O, Me_2CO, r.t., 2 h; 76% (500)

By using a one-step procedure[825] γ-lactones may be regenerated from γ-lactol methyl ethers by the action of *m*-chloroperbenzoic acid and boron trifluoride etherate (equation 501). Although the overall reaction appears to be the oxidation of an ether

$BF_3 \cdot Et_2O$, 25 °C, *m*-$ClC_6H_4CO_3H$, CH_2Cl_2, 3 h (501)

Lactol	Product	Yield (%)
R = *n*-Bu		94
R = *n*-Hex		92
R = Ph		82
		86
		96
		83

function to a ketone, the detailed mechanism shows the reaction to be more complex, involving formation of an oxonium ion, which then reacts with the peracid to produce an intermediate perester, and finally fragments to afford the lactone product (equation 502). Application of this method to protected δ-lactols gave disappointingly low yields of δ-lactones[825].

$R'CO_3H$; + R'COOH (502)

This same approach has been used[826] with O-(4-hydroxy-1-alkenyl)carbamates, which upon treatment with methanesulphonic acid and mecuric acetate produce intermediate lactol methyl esters which are not isolated but are further treated with m-chloroperbenzoic acid and boron trifluoride etherate to produce the corresponding γ-lactones (equation 503). Thus, the overall reaction is a one-pot conversion of O-(4-hydroxy-1-alkenyl)carbamates to γ-lactones via the mechanism shown in equation 502. Using a modified Wacker oxidation (Pd_2Cl_4, CuCl, $MeSO_3H$, O_2) for the last step instead of the Grieco and coworkers procedure[825] proved to be less effective[826].

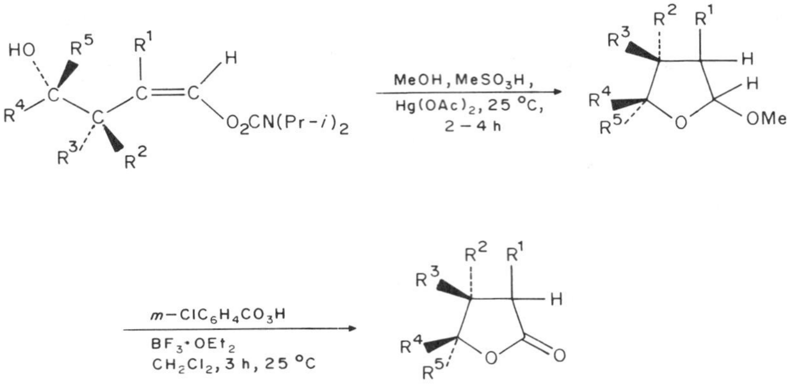

MeOH, $MeSO_3H$, $Hg(OAc)_2$, 25 °C, 2 – 4 h

m-$ClC_6H_4CO_3H$, $BF_3 \cdot OEt_2$, CH_2Cl_2, 3 h, 25 °C

(503)

Enol carbamate	Lactones	Yield (%)
$PhCMe(OH)CH_2CH{=}CHO_2CN(Pr\text{-}i)_2$	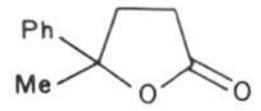	90
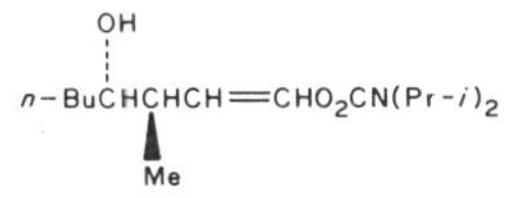 n-BuCH(OH)CH(Me)CH=CHO$_2$CN(Pr-i)$_2$	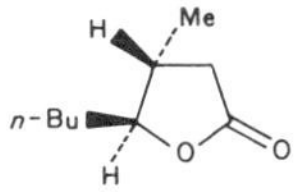	90
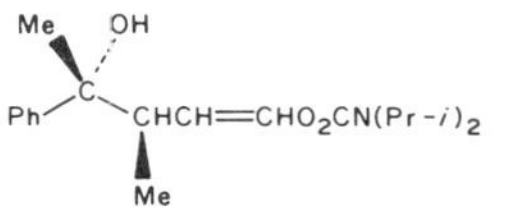	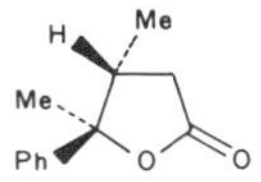	91
i-$PrCH(OH)CMe_2CH{=}CHO_2CN(Pr\text{-}i)_2$	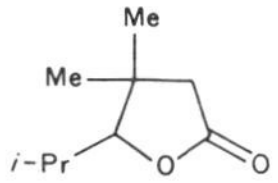	93
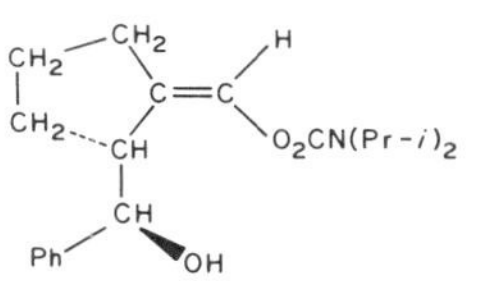	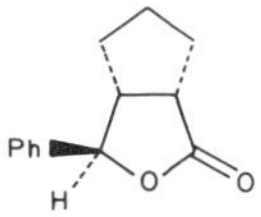	67[a]

[a] Mercuric acetate reaction was run for 3 h at 60 °C and the m-chloroperbenzoic acid reaction was run for 12 h at 25 °C.

The second category of oxidative conversion of ethers into lactones involves the ring expansion of cyclic ethers into cyclic lactones containing one additional carbon atom. This category is exemplified by the oxidative conversion[824] of furfuryl alcohols into hydro-3-pyranones (equation 504).

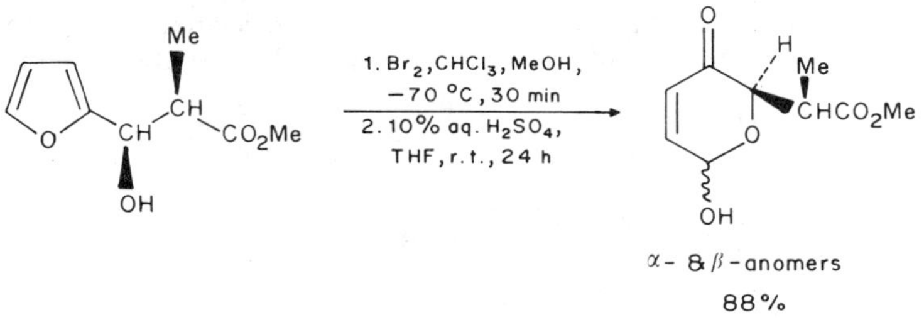
Me
CH
CH
CO2Me
OH
1. Br2, CHCl3, MeOH, −70 °C, 30 min
2. 10% aq. H2SO4, THF, r.t., 24 h
O
H
Me
CHCO2Me
O
OH
α- & β-anomers
88%
(504)

*4. *Oxidation of olefins*

Treatment of cyclic enol ethers with pyridinium chlorochromate converts[827] the enol ethers into lactones in excellent yields (equation 505). The mechanism proposed for this

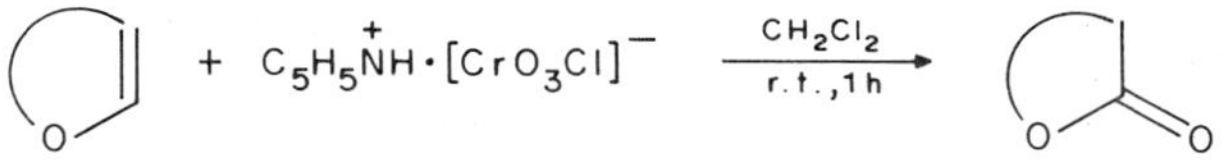
+ C5H5NH·[CrO3Cl]−
CH2Cl2
r.t., 1 h
O
O
O
(505)

Enol-ether	Lactone	Yield (%)
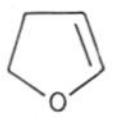	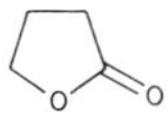	85
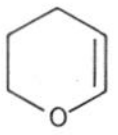	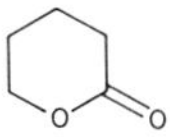	90

reaction involves initial electrophilic attack upon the olefin by the pyridinium chlorochromate to produce an unstable intermediate chromate complex, which then undergoes heterolytic cleavage of the Cr—O bond accompanied by a 1,2-hydride shift affording the lactone product (equation 506).

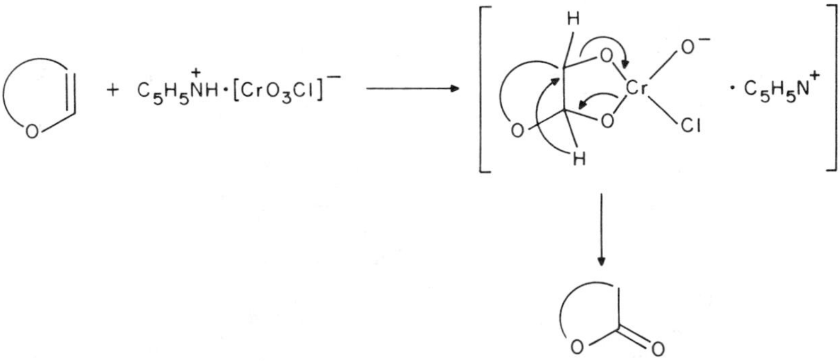
+ C5H5NH·[CrO3Cl]−
H
O
O−
Cr
Cl
O
O
H
· C5H5N+
O
O
(506)

Although pyridinium chlorochromate may also be used to prepare 2,3-unsaturated lactones from glycal and 2-acyloxyglycal esters[828], a more efficient and milder one-step procedure which has been used to effect this conversion is the boron trifluoride catalysed oxidation using *m*-chloroperbenzoic acid. Mechanistically, this oxidative elimination is initiated by boron trifluoride-induced removal of the allylic acyloxy function forming an allylcarboxonium ion, which then undergoes nucleophilic attack by the peroxyacid anion solely at C-1 and concludes by fragmentation of the peroxyester (equation 507).

m-ClC$_6$H$_4$CO$_3$H, BF$_3$(1:1), CH$_2$Cl$_2$, 0 °C to r.t. over 15–30 min

BF$_3$OAc$^-$

ArCO$_3^-$

Ar = m-ClC$_6$H$_4$

(507)

Ester	Lactone	Yield (%)
R^1 = CH$_2$OAc, R^2 = OAc		69
R^1 = CH$_2$OTs, R^2 = OAc		78
R^1 = CH$_2$OBz, R^2 = OBz		74
OAc, OAc, CH$_2$OAc	OAc, CH$_2$OAc	81
OAc, OAc, Me	OAc, Me	89
OR^2, R^{2}O, OR^2, R^1	R^{2}O, OR^2, R^1	—
R^1 = H, R^2 = Bz		78
R^1 = CH$_2$OAc, R^2 = Ac		67
R^1 = CH$_2$OBz, R^2 = Bz		91

Ester	Lactone	Yield (%)
OAc, OAc, CH_2OAc (glycal triacetate)	OAc, CH_2OAc	—
OAc, CH_2OAc, RO; R=MeO, EtO	OAc, CH_2OAc	—
OAc, RAcN, OAc, CH_2OAc; R=H, Ac	RAcN, CH_2OAc	83
OAc, AcO, OAc, AcO, CH_2OAc	OAc, AcO, OAc, CH_2OAc	85
OAc, AcO, O, CH_2OAc, AcO, OAc, OAc, CH_2OAc	OAc, AcO, O, CH_2OAc, AcO, OAc, OAc, CH_2OAc	87

Pyridinium chlorochromate is also less effective then other reagents when used to produce γ-lactones via oxidative cyclization of tertiary hydroxyolefins. This effect can be readily observed from the results[829] obtained for the product of *cis*-fused bicyclic γ-lactones from tertiary hydroxyolefins using pyridinium chlorochromate and $(BipyH_2)CrOCl_5$ (equation 508).

$$\xrightarrow[\text{or } (\mathrm{Bipy\,H_2})\mathrm{CrOCl_5}]{\mathrm{C_5H_5\overset{+}{N}H\cdot[CrO_3Cl]^-}} \tag{508}$$

Olefin	Reagent	Conditions	Yield (%)
(structure: Me, OH, R)	R = H		
	$C_5H_5\overset{+}{N}H\cdot[CrO_3Cl]^-$	CH_2Cl_2, 40 °C, 48 h	60
	(Bipy H_2)$CrOCl_5$	CH_2Cl_2, 40 °C, 8 h	80
	R = $CH_2CH{=}CH_2$		
	$C_5H_5\overset{+}{N}H\cdot[CrO_3Cl]^-$	CH_2Cl_2, 40 °C, 48 h	58
	(Bipy H_2)$CrOCl_5$	CH_2Cl_2, 40 °C, 8 h	75
(structure: Me, OH, Me, Me)	$C_5H_5\overset{+}{N}H\cdot[CrO_3Cl]^-$	CH_2Cl_2, 40 °C, 48 h	60
	(Bipy H_2)$CrOCl_5$	CH_2Cl_2, 40 °C, 8 h	75

The fact that pyridinium chlorochromate produces the results indicated above is reported[830] to be inconsistent with the findings obtained when tertiary γ- and δ-hydroxyalkenes are treated with pyridium chlorochromate. In this report[830] treatment of 1-(3-butenyl)-4,4-dimethylcyclohexanol with pyridinium chlorochromate, chromium(VI) oxide/pyridine or Jones' reagent under standard conditions, all had no effect upon the alcohol (equation 509), but reasonable yields of the corresponding γ- and δ-spirolactones are obtained when this and other tertiary γ- and δ-hydroxyalkenes are treated[830] with chromium trioxide in acetic acid–acetic anhydride (equation 510).

HO, $CH_2CH_2CH{=}CH_2$ (4,4-dimethylcyclohexanol; Me, Me) $\xrightarrow[\text{or } H_2CrO_4,\ H_2SO_4,\ Me_2CO,\ 0\ ^\circ C]{C_5H_5\overset{+}{N}H\cdot[CrO_3Cl]^- \text{ or } (C_5H_5N)_2CrO_3}$ N.R. (509)

HO, $(CH_2)_nCH{=}CH_2$ (C in ring) $\xrightarrow[Ac_2O,\ 10\ ^\circ C,\ 4\ h]{CrO_3,\ HOAc}$ O=C, O, $(CH_2)_n$, C (spirolactone) (510)

Alcohol	Lactone	Yield (%)
HO, $(CH_2)_2CH{=}CH_2$ (cyclohexanol; R, R)	(spirolactone; R, R)	
R = H		50
R = Me		57

Alcohol	Lactone	Yield (%)
HO, $(CH_2)_3CH{=}CH_2$ cyclohexane, R, R; R=H		30
R=Me		42
HO, $(CH_2)_2CH{=}CH_2$ cycloheptane		80
HO, $(CH_2)_3CH{=}CH_2$ cycloheptane		68

Oxidative cyclization of primary, secondary and tertiary γ- and δ-hydroxyolefins to produce γ- and δ-lactones containing one carbon less than the hydroxyolefin starting material has been accomplished[831] by using cetyltrimethylammonium permanganate (CTAP) (equation 511).

$$\text{OH, } CH_2{=}CH(CH_2)_n \xrightarrow[\substack{CH_2Cl_2 \text{ or } CHCl_3,\\ 25-50\,^\circ C}]{n\text{-}C_{16}H_{33}\overset{+}{N}Me_3\ MnO_4^-,} \text{lactone } (CH_2)_n \tag{511}$$

Alcohol	Reaction time(h)	Temp. (°C)	Solvent	Product	Yield (%)
	6	50	$CHCl_3$		76
	6.5	50	$CHCl_3$		71
Me, Me, OH	8	25	CH_2Cl_2	Me, Me	68
	3	50	$CHCl_3$		52

(continued)

Alcohol	Reaction time(h)	Temp. (°C)	Solvent	Product	Yield (%)
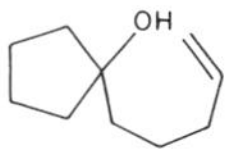	3	50	$CHCl_3$	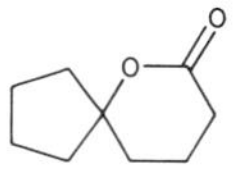	54
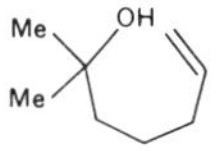	3	50	$CHCl_3$	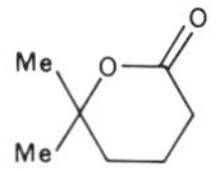	54
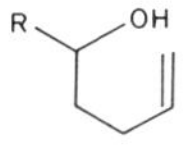 R = H	4.5	25	$CHCl_3$		65
R = Et	4	25	$CHCl_3$		69
R = *n*-Hex	5	25	$CHCl_3$		66
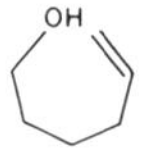	3	25	$CHCl_3$	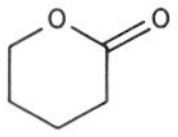	50
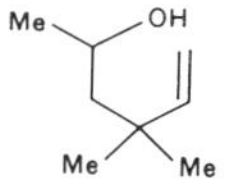	3.5	25	CH_2Cl_2	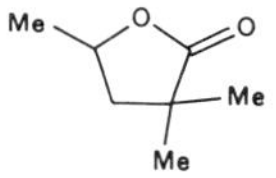	78
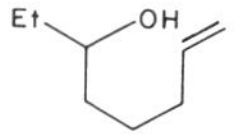	3	25	$CHCl_3$	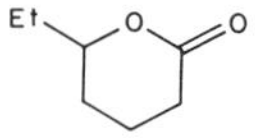	51
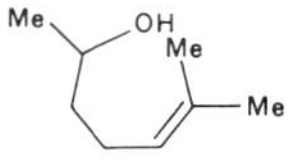	3.5	25	CH_2Cl_2		70
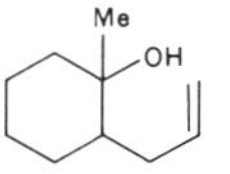	6	25	CH_2Cl_2	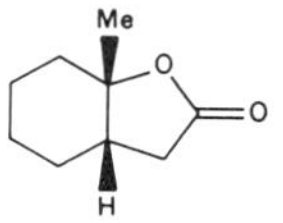	55
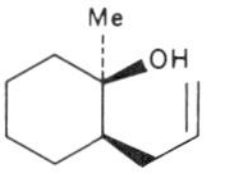	4	25	CH_2Cl_2	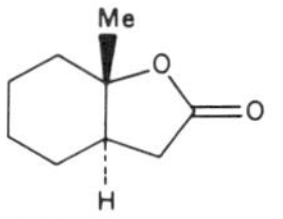	74

Two examples of the conversion of homoallylic alcohols to lactones which appear in the recent literature involve a two-step process, the second step of which in both cases involves the oxidation of an intermediate with pyridinium chlorochromate. In the first example[832] a hydroformylation–oxidation sequence in the presence of rhodium acetate dimer was used to efficiently convert the homoallylic alcohols into δ-lactones (equation 512), while in the second example[833] a hydroboration–oxidation sequence was used to convert homoallylic trimethylsilyl alcohols into γ-lactones (equation 513).

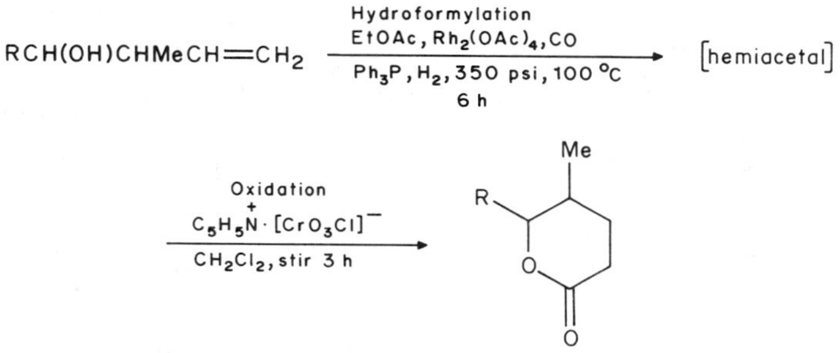

$$RCH(OH)CHMeCH{=}CH_2 \xrightarrow[\text{Ph}_3\text{P, H}_2\text{, 350 psi, 100 °C, 6 h}]{\text{Hydroformylation: EtOAc, Rh}_2\text{(OAc)}_4\text{, CO}} [\text{hemiacetal}] \xrightarrow[\text{CH}_2\text{Cl}_2\text{, stir 3 h}]{\text{Oxidation: C}_5\text{H}_5\overset{+}{\text{N}}\cdot[\text{CrO}_3\text{Cl}]^-} \text{lactone} \qquad (512)$$

Alcohol	Lactone	Yield (%)
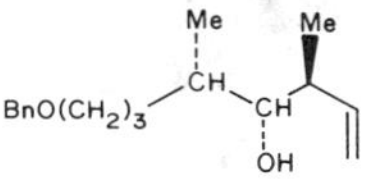 $BnO(CH_2)_3$	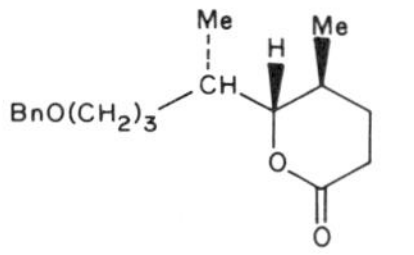 $BnO(CH_2)_3$	86
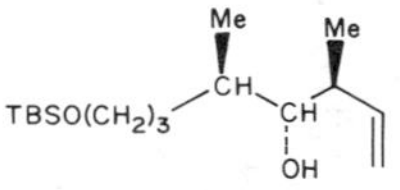 $TBSO(CH_2)_3$	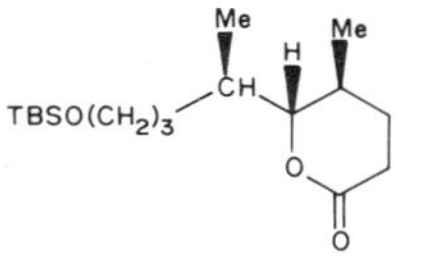 $TBSO(CH_2)_3$	80
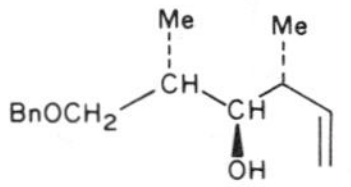 $BnOCH_2$	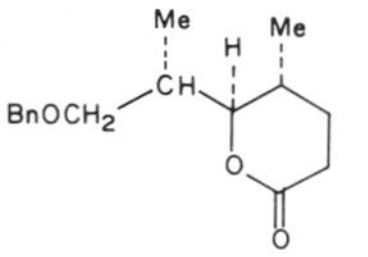 $BnOCH_2$	85
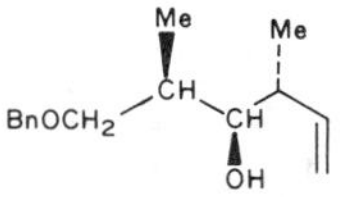 $BnOCH_2$	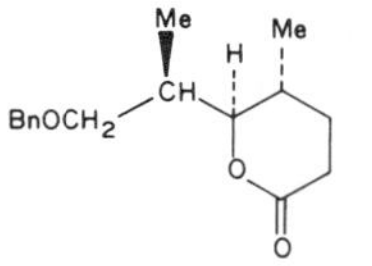 $BnOCH_2$	86
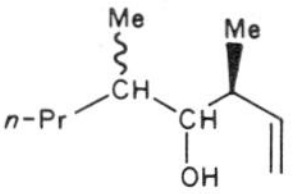 n-Pr	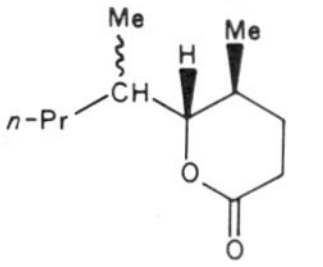 n-Pr	81

An interesting method[834] for the production of catharanthine lactone involves the hydroxylmercuration of the ene carboxylic acid of cartharanthine. Reduction of the initially formed mercuric acetate can be accomplished using either sodium amalgam or tetra(n-butyl)ammonium borohydride in carbon tetrachloride–methanol mixtures, with the products obtained from these reductions dependent upon the ratios of the solvents used (equation 514).

Both γ-lactones (equation 515)[835] and δ-lactones (equation 516)[810] have been produced from precursors by a multistep synthetic approach that includes ozonolysis,

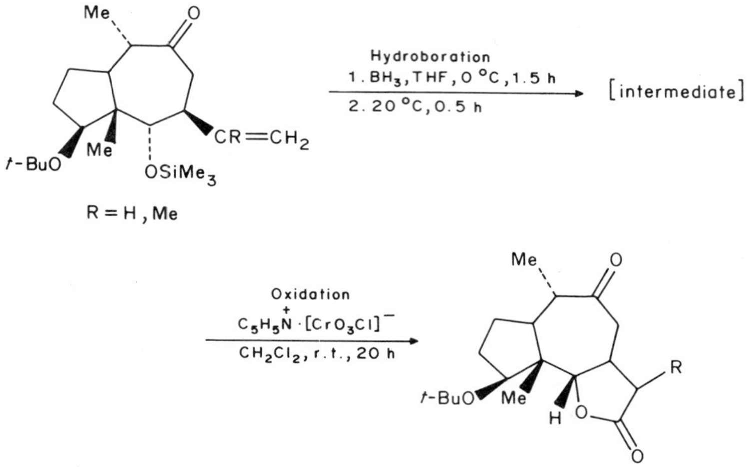
Me
O
t-BuO
Me
OSiMe3
CR=CH2
R = H, Me
Hydroboration
1. BH3, THF, 0 °C, 1.5 h
2. 20 °C, 0.5 h
[intermediate]
Oxidation
C5H5N⁺·[CrO3Cl]⁻
CH2Cl2, r.t., 20 h
Me
O
R
t-BuO
Me
H
O
O

(513)

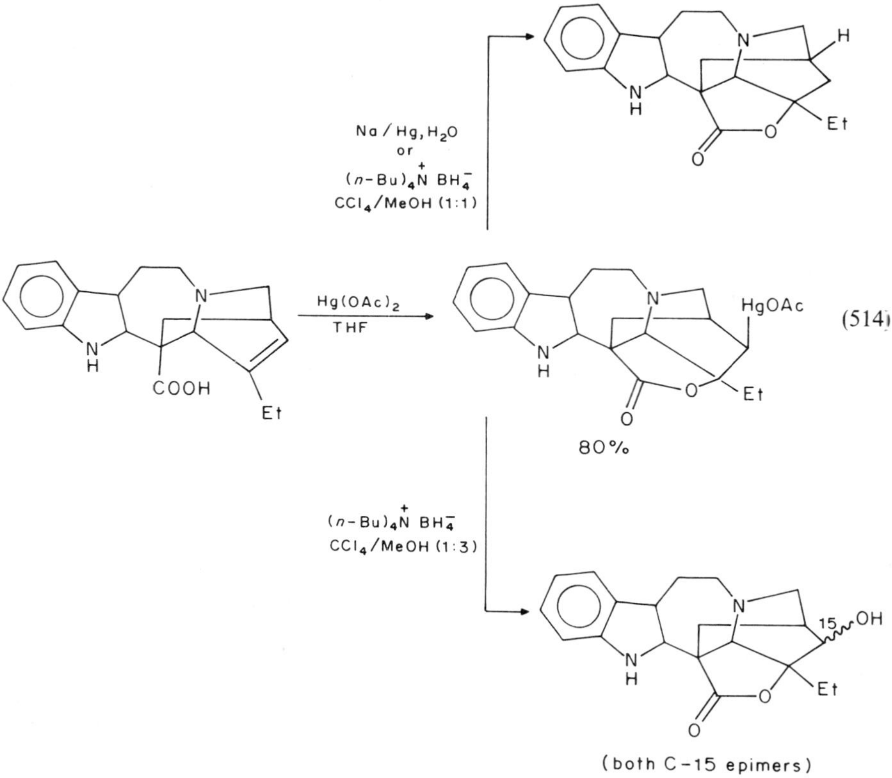
N
H
Et
H
O
O
Na / Hg, H2O
or
(n-Bu)4N⁺ BH4⁻
CCl4/MeOH (1:1)
N
H
COOH
Et
Hg(OAc)2
THF
HgOAc
80%
(n-Bu)4N⁺ BH4⁻
CCl4/MeOH (1:3)
15
OH
(both C-15 epimers)

(514)

1. O_3, CH_2Cl_2, −78 °C
2. Me_2S
3. Jones' oxidation

80%

(515)

$PhCH{=}CHCH(Me)CH_2CH(Me)CH(OAc)CH(Me)CH_2OAc$

1. O_3
2. Me_2S
3. Jones' reagent
4. K_2CO_3, MeOH
5. TosOH
6. PDC, DMF

28%

(516)

treatment with dimethyl sulphide and oxidation using Jones' reagent as all or part of the synthetic sequence.

Hydroxylation of γ-hydroxy α,β-unsaturated esters catalysed by osmium tetroxide leads to the highly stereoselective production[836] of the corresponding 3,4-dihydroxy-γ-lactones (equation 517). These results are explained[836] as a consequence of the conformation present in the transition state caused by the presence of the electron-withdrawing carbomethoxy function leading to interaction between the p orbitals of the double bond and an unshared electron pair on the γ-oxygen in the starting material.

Reaction of trimethylsilyl substitued unsaturated substrates with peroxides represents a recent and novel approach to the production of lactones. Because of the unsaturation present in the starting material, the intermediate formation of epoxides occurs regardless of the peroxide used. Thus, oxidation of 5-alkyl-2-trimethylsilyl furans with peracetic acid produces[837] good yields of Δ^3-butenolides (equation 518) via a mechanism which

E-isomer

OsO_4, H_2O, Me_2CO, r.t. *N*-methylmorpholine oxide

$EtCH(OH)C(Me){=}CHCOOMe$

Z-isomer

OsO_4, H_2O, Me_2CO, r.t. *N*-methylmorpholine oxide

(517)

1. NaOAc, $MeCO_3H$ (40%), CH_2Cl_2, 0 °C
2. Stir 3.5 h, 7 °C

(518)

R	Yield (%)
n-Hex	84
i-$PrCH_2CH_2$	78
$CH_2{=}CH(CH_2)_9$	65
$(EtO)_2CHCH_2CH_2$	73
n-$C_8H_{17}CH(OH)$	<30
n-$C_8H_{17}CH(OAc)$	trace
n-$C_8H_{17}CH[OSiMe_2(Bu$-$t)]$	64[a]
n-$C_8H_{17}CH(OH)CH_2$	36
n-$C_8H_{17}CH[OSiMe_2(Bu$-$t)]CH_2$	73[b]
n-BuO_2C	N.R.

[a]For 5 hours at room temperature.
[b]For 5.3 hours at 7 °C.

involves selective epoxidation of the starting material at the site bearing the electron-releasing trialkylsilyl group, followed by C—O bond fission of the epoxide with concomitant migration of the silyl group and hydrolysis under acidic conditions (equation 519). To establish the intermediacy of the silyl migrated structure the reaction was performed[837] using 5-(n-hexyl)-2-dimethyl(t-butyl)silylfuran and a 44% yield of the desired intermediate was obtained (equation 520). Attempts to perform the overall conversion using m-chloroperbenzoic acid as the oxidizing agent reduced the yield of lactone, while employing t-butyl hydroperoxide with titanium isopropoxide as the oxidizing agent was completely ineffective.

Me_3Si — $MeCO_3H$ → Me_3Si — → Me_3Si — $MeCO_2H$ → R

(519)

t-$BuMe_2Si$ — Hex-n — $MeCO_3H$, $NaHCO_3$ → t-$BuMe_2Si$ — Hex-n

44%

(520)

Similarly, hydrogen peroxide oxidation of β-allenic alcohols substituted by a trimethylsilyl group at the 1 or 3 position, respectively, has been reported[838] to give rise to γ- or δ-lactones via a mechanism which involves the initial intermediate formation of an epoxide in both cases (equation 521). In order to establish that a concerted mechanism is involved during the ring opening of the cyclopropane intermediate in the case of the γ-lactone formations (equation 522), the reaction was performed with a chiral β-allenic alcohol and the stereochemistry of the product formed noted (equation 523).

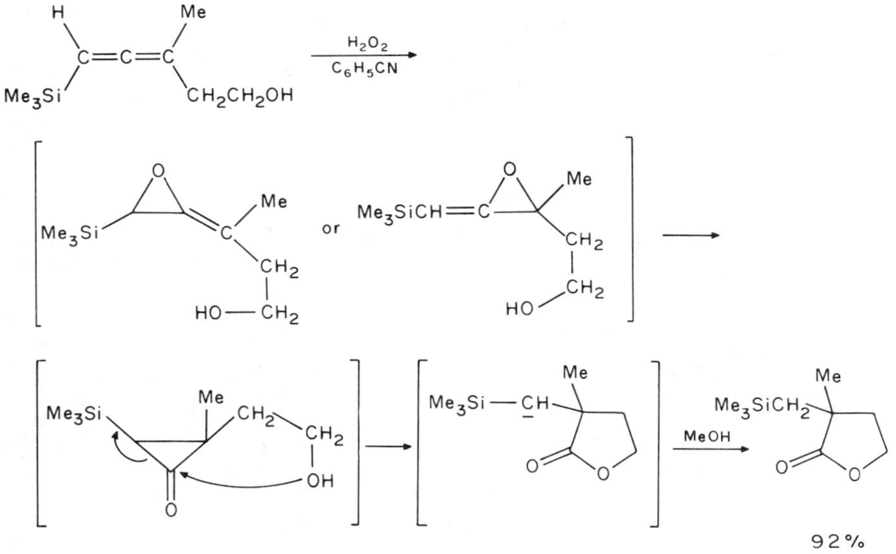

H_2O_2, C_6H_5CN; MeOH; 92%

(521)

H_2O_2, C_6H_5CN

$R^1=R^3=R^4=R^5=R^6=H, R^2=Me$
$R^1=Et, R^2=Me, R^3=R^4=R^5=R^6=H$
$R^1=R^5=R^6=H, R^2=R^3=R^4=Me$
$R^1=Et, R^2=R^5=R^6=Me, R^3=R^4=H$
all yields quantitative

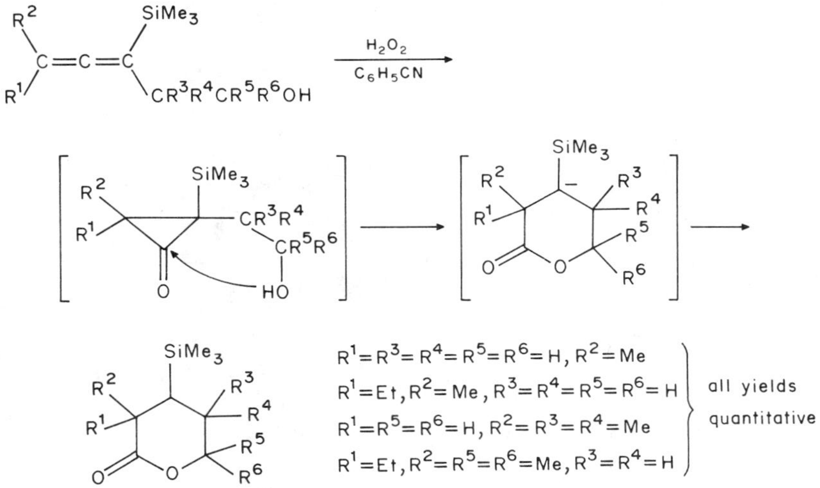

(522)

1. n-BuLi
2. $ClSiMe_3$
3. H_3O^+

$MeC(OEt)_3$, H^+

$LiAlH_4$

H_2O_2, C_6H_5CN

(523)

NBS, Me_2SO

(524)

NBS, Me_2SO

$R^1 = Ac, CPh_3$
$R^2 = Ac, 3\text{-}MeO\text{-}4\text{-}AcOC_6H_3CO$

Treatment of the natural iridoid glucoside Aucubin[762] and its substituted analogues[839] with *N*-bromosuccinimide in dimethyl sulphoxide produces the corresponding bromolactones as shown in equation 524.

Because of the oxidizing species used, the superoxide anion radical, the conversion[840] of 4,4-disubstituted-2-hydroxy-2,5-cyclohexadien-1-ones to the corresponding lactols represents a novel and interesting reaction (equation 525). The mechanism proposed[840]

KO_2, C_6H_6, r.t., 18-crown-6, 16 h (525)

R^1	R^2	Ratio of reactants KO_2:18-crown-6:substrate	Yield (%)
H	Me	2:1:1	50
OEt	Me	4:2:1	80
OMe	Me	4:2:1	80
Me	Me	2.5:1:1	50
OMe	$-(CH_2)_5-$[a]	4:2:1	90
H	Ph	10:5:1	N.R.

[a] $R^2R^2 = -(CH_2)_5-$

for this conversion involves the formation and reaction of the superoxide anion radical as a base. This is supported by the fact that the same lactol products are obtained using potassium hydroxide and *t*-butoxide, but the rates of the reaction decrease in the order

$O_2^{-\cdot}$; O_2 ; $-CO$; H^+ (526)

t-butoxide > superoxide > hydroxide. Thus, the basic superoxide initially deprotonates the enol hydrogen producing a substrate anion, which then combines with the oxygen generated producing the corresponding peroxy anion. Peroxy ring closure, followed by carbon monoxide elimination, protonation and lactol formation complete the sequence (equation 526).

5. Oxidation of other functional groups

In addition to the functional groups reported in the previous sections, recent reports reveal the conversion of acids, alcohols and aldehydes to lactones upon treatment with oxidizing agents.

$$R^1R^2CHCHR^3CH_2COOH \xrightarrow[CuCl_2]{Na_2S_2O_8} \text{(γ-lactone bearing } R^1, R^2, R^3\text{)} \quad 35-75\% \tag{527}$$

R^1	=	H	;	Me	;	Et	;	*n*-Pr	;	H	;	*n*-Bu
R^2	=	H	;	H	;	H	;	H	;	H	;	Et
R^3	=	H	;	H	;	H[a]	;	H[b]	;	Me	;	H

[a] With *n*-PentCOOH the 5-methyl δ-lactone is also obtained.
[b] With *n*-HexCOOH the 5-ethyl δ-lactone is also obtained.

$$\text{unsaturated carboxylic acid} \xrightarrow[\substack{AgNO_3,\ MeCN \\ H_2O,\ 80\,^\circ C}]{Na_2S_2O_8,\ CuX_2} \gamma\text{-lactones} \tag{528}$$

Acid	Product	Yield (%)
2-Me-benzoic acid (R substituted)	phthalide (R substituted)	
R = H		56
R = Me		33
$Me_2C{=}CHCOOH$	4-Me butenolide	36
$Et_2C{=}CHCOOH$	4-Et-5-Me butenolide	35
$(i\text{-}Pr)_2C{=}CHCOOH$	4-*i*-Pr-5,5-Me₂ butenolide	41
	+ 4-$Me_2C(OH)$-5,5-Me₂ butenolide	5

Sodium persulphate oxidation of carboxylic acids to produce lactones may be accomplished using copper(II) halide catalysts alone[841] (equation 527), or by using a silver(I) nitrate mediated copper(II) halide catalysed reaction[842] (equation 528). The mechanism proposed[842] for this conversion involves the formation of intermediate acyloxy radicals followed by a 1,5-hydrogen transfer (equation 529).

$S_2O_8^{-2}$, Ag^+

(529)

Photochemical oxidation of carboxylic acids has also been used[751] to produce 4-hydroxy substituted but-2-en-4-olides from correspondingly substituted furoic acids (equation 530).

O_2, eosin, $h\nu$ / EtOH, stir (530)

R^1 = Me ; Ph ; Ph

R^2 = H ; H ; Ph

Yield (%) = — ; 95 ; 89

An interesting oxidation of a carboxylic acid which leads to a fused-ring iodolactone has been reported[751] to occur when cyclopent-2-enylacetic acid is treated with hypoiodous acid (equation 531).

CH_2COOH —HOI→ (531)

With alcohols, three different oxidizing agents have been used to produce lactones. In the first example[843] catacondensed lactols are converted to their corresponding hypoiodites *in situ* upon reaction with an excess of mercury(II) oxide and iodine in benzene containing a small amount of pyridine. Irradiation of these hypoiodites readily leads to ring expansion to afford iodo substituted lactones by a regioselective scission of the catacondensed bond of the intermediate alkoxy radical. Removal of the iodine atom with tri-(n-butyl)tin hydride affords medium-sized lactones (equation 532).

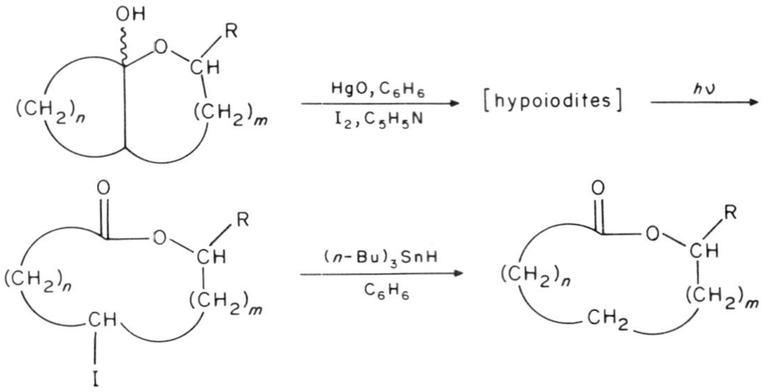

(532)

n	m	R	Iodolactone yield (%)
4	1	Ph	52
4	1	Me	53
4	2	H	79
4	3	Me[a]	76[b]

[a] A *cis–trans* mixture.
[b] An 82% yield of the lactone (+)-phoracantholide I was obtained after removal of the iodine.

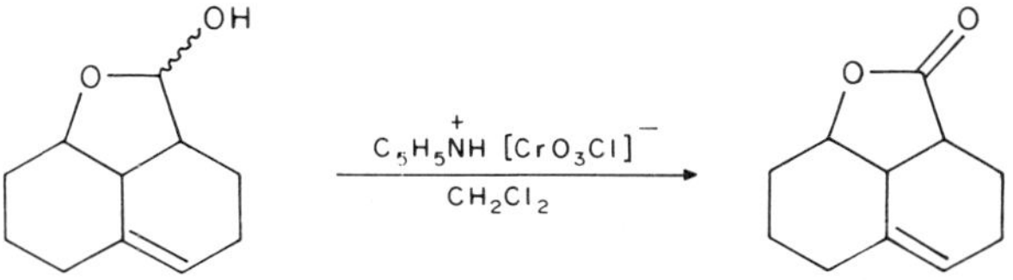

(533)

Lactol 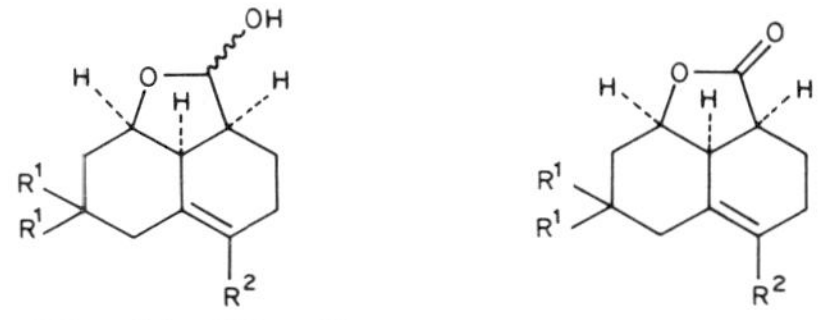	Product	Yield (%)
$R^1 = Me, R^2 = H$		98
$R^1 = R^2 = H$		89
$R^1 = Me, R^2 = CH{=}CH_2$		11
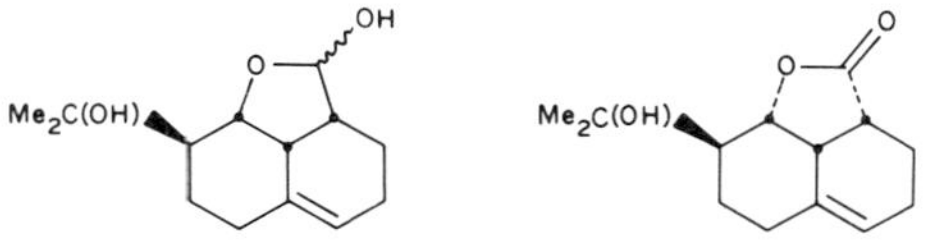 $Me_2C(OH)$	$Me_2C(OH)$	—

Reaction of tricyclic lactols with pyridinium chlorochromate produces[844] the corresponding tricyclic lactones (equation 533).

Finally, manganese dioxide oxidation of a mixture of alcohols resulting from the acid catalysed rearrangement of a bicyclic allylic alcohol produces[845] a good yield of the α-methylene-γ-butyrolactone of *trans*-1,3-dihydroxycyclohexane (equation 534).

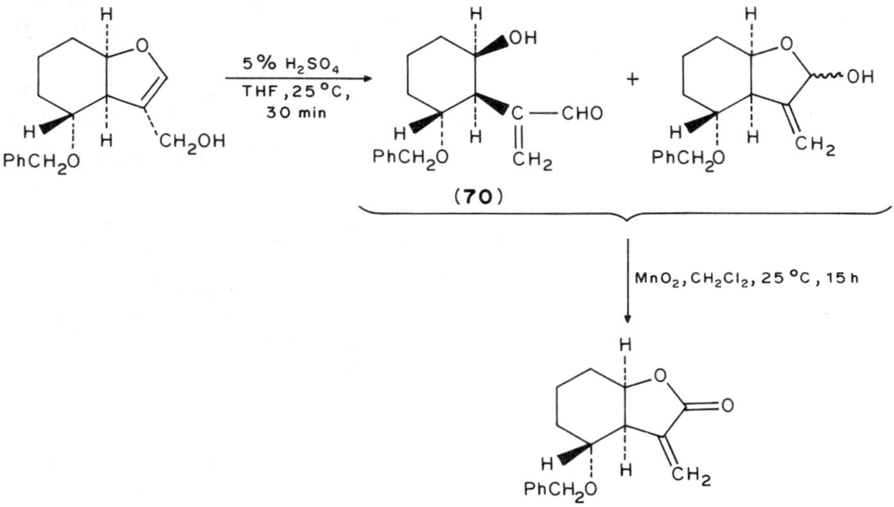

5% H_2SO_4, THF, 25 °C, 30 min → (70) + ; MnO_2, CH_2Cl_2, 25 °C, 15 h → 82%

(534)

A very similar reaction to the one discussed in equation 534 involves the oxidation of the *trans* analogue of the hydroxyaldehyde **70** using manganese dioxide and potassium cyanide in an 18-crown-6-tetrahydrofuran mixture and produces[815] the corresponding unsubstituted *trans* analogue of the α-methylene lactone prepared above. In this case, however, the reaction probably proceeds[815] via a hydroxy cyanoketone similar to the hydroxy cyanoketone intermediates involved in Corey's oxidation[846] of α,β-unsaturated aldehydes (equation 535).

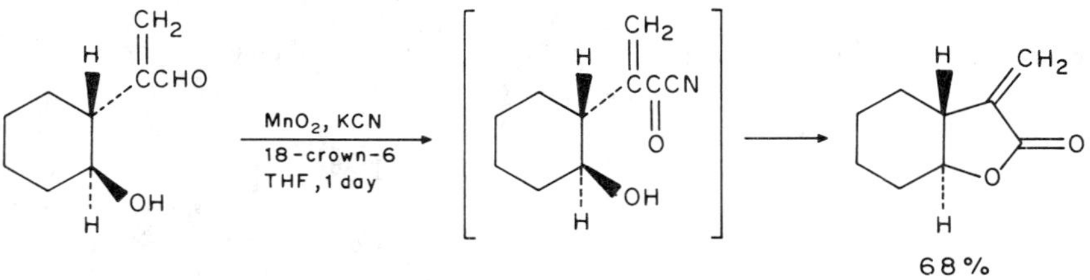

MnO_2, KCN, 18-crown-6, THF, 1 day → [ ] → 68%

(535)

Cyanohydrins of conjugated aldehydes, as their *O*-trimethylsilyl derivatives, have also been oxidized with pyridinium dichromate (PDC) in dimethylformamide (DMF) to produce[847] α,β-unsaturated γ-lactones (Δ²-butenolides) (equation 536). This approach is productive if (a) the β-carbon is disubstituted, and (b) the γ-carbon processes at least

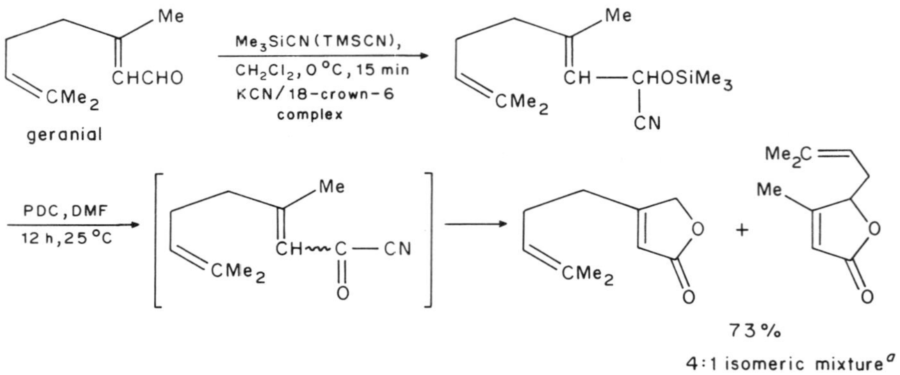

geranial

Me_3SiCN (TMSCN), CH_2Cl_2, 0 °C, 15 min
KCN/18-crown-6 complex

PDC, DMF
12 h, 25 °C

73%
4:1 isomeric mixture[a]

(536)

[a]Treatment of citral, a mixture of the geometric isomers geranial (citral A) and neral (citral B), as above gave the same mixture of products.

Aldehyde	Ratio (*E*:*Z*)	Product	Product Ratio	Yield (%)
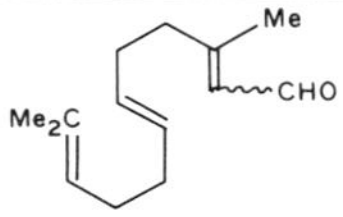	68:22	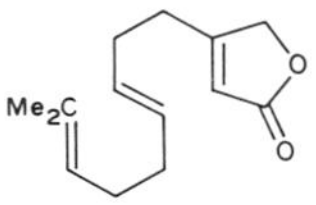	—	69
PhCMe=CHCHO	59:41	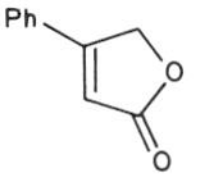	—	71
$PhCH_2CMe$=CHCHO	63:37	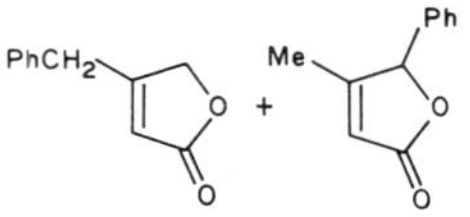	36:64	54
$PhCH_2CH_2CMe$=CHCHO	63:37	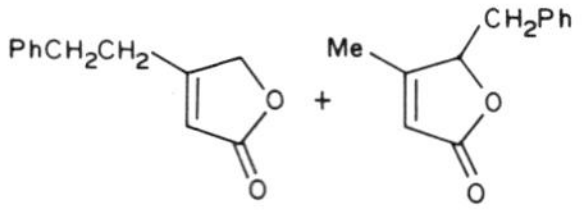	77:23	68
Me_2CHCMe=CHCHO	79:21	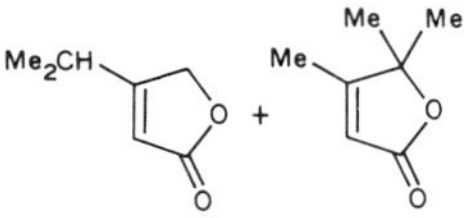	>95:5	75
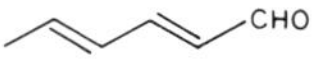	—	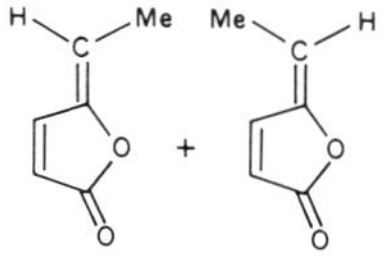	54:46	50
PhCH=CMeCHO	—	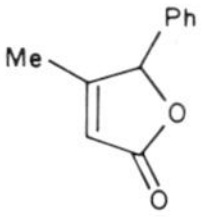	—	40

one hydrogen atom. The mechanism of this reaction[847] involves desilylation and oxidation of the chromate ester of the cyanohydrin.

Irradiation of an alcohol solution of furfural in the presence of oxygen produces[797] the corresponding 4-alkoxy-Δ^2-butenolide, which upon treatment with acid affords the Δ^2-lactol (equation 537).

$$\text{(furyl)–CHO} + O_2 \xrightarrow[h\nu]{ROH} \text{RO–(butenolide)} \xrightarrow{H_3O^+} \text{HO–(butenolide)} \qquad (537)$$

*O. By Carbonation and Carbonylation Reactions

Reaction of the Z-isomers of γ-hydroxyvinylstannanes with n-butyllithium followed by treatment with carbon dioxide and p-toluenesulphonic acid produces[848] α,β-unsaturated γ-lactones (equation 538).

$$(n\text{-Bu})_3Sn(R^1)C{=}CH{-}CH(OH)R^2 \xrightarrow[\text{2. } CO_2\text{; 3. } p\text{-TosOH, } C_6H_6\text{, reflux}]{\text{1. } n\text{-BuLi, THF}} \text{(3-}R^1\text{-5-}R^2\text{-butenolide)} \qquad (538)$$

R^1 =	n-Hex	n-Hex	$Ph(CH_2)_2$	$Ph(CH_2)_2$
R^2 =	n-Pent	$Ph(CH_2)_2$	n-Pent	$Ph(CH_2)_2$
Yield (%) =	97	66	54	64

Cobalt carbonyl compounds have been reported to catalyse the carbonylation of aryl halides[849] and alkynes[850]. With aryl halides the carbonylation reaction[849] is catalysed by an alkylcobalt carbonyl complex, either preformed or made *in situ*, and produces γ- or δ-benzolactones bearing substituents on both aromatic and aliphatic carbons (equation 539).

$$o\text{-Br}C_6H_4(CHR)_nOH + CO \xrightarrow[\text{EtOOCCH}_2\text{Co(CO)}_4\text{, base}]{\text{MeOH, 25 °C}} \text{benzolactone } [C_6H_4(CHR)_nOC(=O)] \qquad (539)$$

Aryl halide	Base	Product	Yield (%)
$o\text{-BrC}_6H_4CH_2OH$	K_2CO_3	phthalide	70
	K_2CO_3[a]		84
	NaOMe[b]		77
$o\text{-BrC}_6H_4CH(Me)OH$	NaOMe	3-Me-phthalide	65
	NaOH		60

(*continued*)

Aryl halide	Base	Product	Yield (%)
2-BrC$_6$H$_4$CH(Ph)OH	K_2CO_3	3-phenylphthalide	41
	NaOMe		42
MeO, OMe, CH_2OH, I substituted benzene	NaOMe	dimethoxyphthalide	77
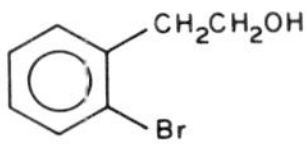 2-BrC$_6$H$_4$CH$_2$CH$_2$OH	K_2CO_3	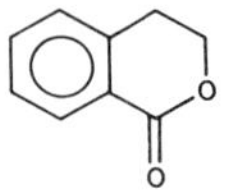 isochromanone	50

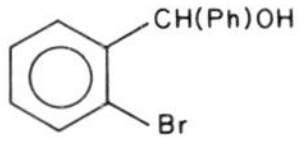

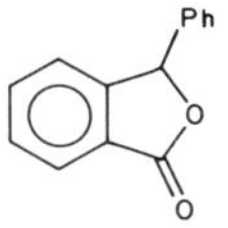

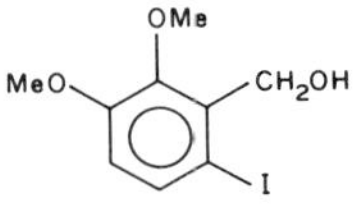

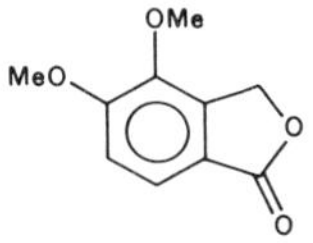

[a]Catalyst used was $FCH_2Co(CO)_4$.
[b]Catalyst used was $NaCo(CO)_4$ and $ClCH_2COOMe$ and the reaction run at 60 °C

Treatment of alkynes in a phase-transfer catalysed, cobalt carbonyl-catalysed reaction with carbon monoxide and methyl iodide produces a regiospecific synthesis[850] of hydroxybut-2-enolides (equation 540), via a mechanism involving the generation and reaction of the cobalt tetracarbonyl anion (equation 541).

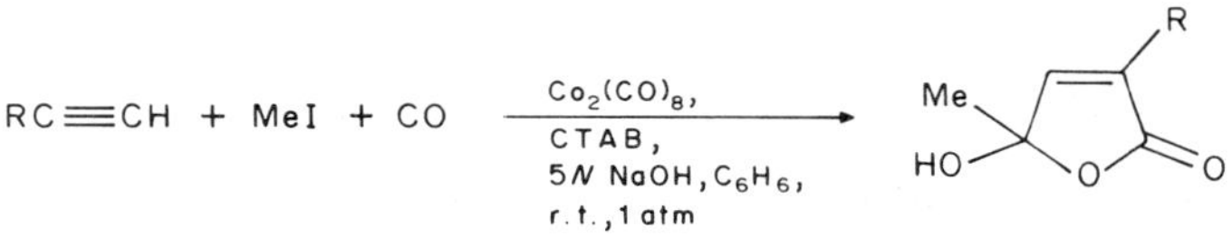

$$RC{\equiv}CH + MeI + CO \xrightarrow[\substack{CTAB,\\ 5N\ NaOH, C_6H_6,\\ r.t., 1\ atm}]{Co_2(CO)_8,} \text{(Me)(HO)-butenolide (R)} \qquad (540)$$

Alkyne	Product	Yield (%)
RC≡CH	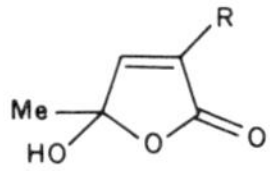	
R = Ph		44
R = *c*-Hex		18
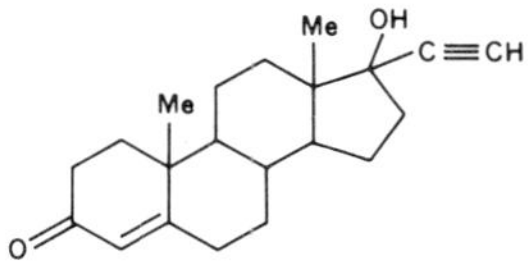 17α-ethynyl steroid (Me, OH, C≡CH)	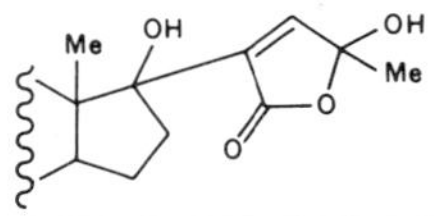	68

The reaction of 1,1-dibromo-2-(hydroxyalkyl)cyclopropanes with nickel tetracarbonyl in dimethylformamide produces[851] the corresponding bicyclic lactones via an intramolecular process (equation 542). If the same reaction is attempted[851] using 2,2-dichloro-3,3-dimethylcyclopropanemethanol, the gem-dichloride analogue of the

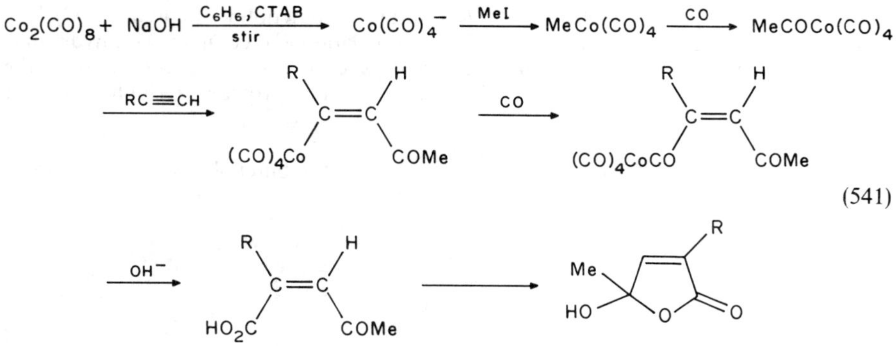

$$Co_2(CO)_8 + NaOH \xrightarrow[\text{stir}]{C_6H_6,\,CTAB} Co(CO)_4^- \xrightarrow{MeI} MeCo(CO)_4 \xrightarrow{CO} MeCOCo(CO)_4$$

$$\xrightarrow{RC\equiv CH} (CO)_4Co(R)C{=}C(H)COMe \xrightarrow{CO} (CO)_4CoCO(R)C{=}C(H)COMe \quad (541)$$

$$\xrightarrow{OH^-} HO_2C(R)C{=}C(H)COMe \longrightarrow \text{5-hydroxy-5-methyl-3-R-furan-2(5H)-one}$$

$$\text{2,2-dibromo-3,3-dimethylcyclopropylmethanol} + Ni(CO)_4 \xrightarrow[3\ h]{DMF,\ 75\ ^\circ C} \text{6,6-dimethyl-3-oxabicyclo[3.1.0]hexan-2-one} \ (73\%) \quad (542)$$

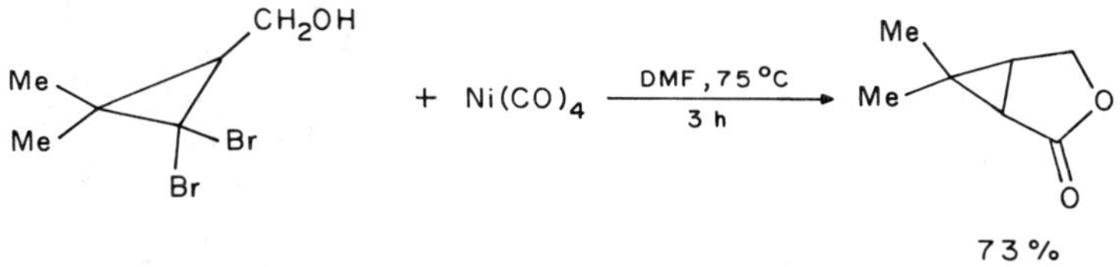

substrate shown above, at 75 °C for 5 hours or at 120 °C for 7 hours, no reaction is observed. Other gem-dibromides subjected to this reaction are illustrated below.

Gem-dibromide	$Ni(CO)_4$ (molor eq.)	Temp. (°C)	Time (h)	Product	Yield (%)
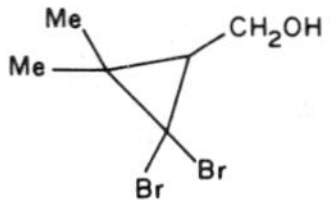 2,2-dibromo-3,3-dimethylcyclopropyl-CH_2OH	2.2	80	11	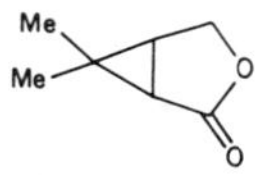 Me, Me bicyclic lactone	37
	7	75	3		73
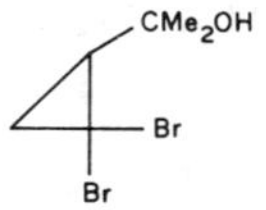 2,2-dibromocyclopropyl-CMe_2OH	7	75	3	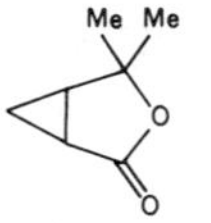 Me Me bicyclic lactone	70
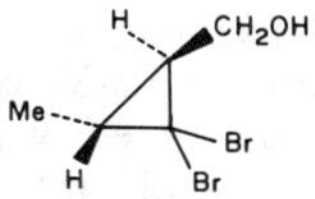 H, CH_2OH, Me, H, Br, Br (stereo) cyclopropane	1.2	80	11	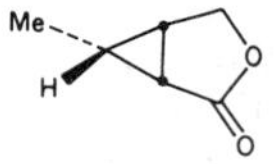 Me, H bicyclic lactone	43
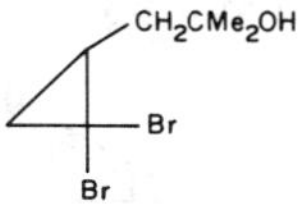 2,2-dibromocyclopropyl-CH_2CMe_2OH	7	75	3	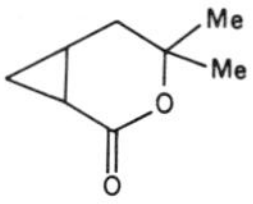 Me, Me bicyclic six-membered lactone	51

Palladium compounds also have been reported to catalyse the synthesis of $\Delta^{\alpha,\beta}$-butenolides[852] and *cis*-3-hydroxypyrrolidine-2-acetic acid lactone[853]. In the first report[852] carbon monoxide reacts with palladium(0) complexes of *Z*-iodoalkenols to produce the corresponding substituted α,β-unsaturated butenolides (equation 543). The palladium(0) is generated *in situ* by reduction of a palladium(II) complex with hydrazine. In addition, a base is needed to remove the acid generated in the reaction so that the palladium(0) complex will survive the reaction and provide a high turnover (equation 544).

$$\text{2-iodoalkenols} \xrightarrow[\text{2. CO, THF, } H_2NNH_2\text{, } 35\,^{\circ}C\text{, 2 days}]{\text{1. } PdCl_2(PPh_3)_2\text{, } K_2CO_3} \Delta^{\alpha,\beta}\text{-butenolides} \quad (543)$$

Iodoalkenol	Product	Yield (%)
$R^1CI{=}CHCH(OH)R^2$	[butenolide: R^1 at C-3, R^2 at C-5]	
$R^1 = R^2 = H$		76
$R^1 = Me$, $R^2 = H$		100
$R^1 = Ph$, $R^2 = Me$		68
$R^1 = H$, $R^2 = Me$		95
$R^1 = Me$, $R^2 = Ph$		47
$R^1 = R^2 = Me$		99
$EtCI{=}CHC(OH)EtMe$	[butenolide: Et at C-3, Me and Et at C-5]	85

$$PdCl_2(PPh_3)_2 \xrightarrow{N_2H_4} PdCO(PPh_3)_2 + R(OH)I \longrightarrow IPd(PPh_3)_2R(OH)$$

$$\xrightarrow{CO} IPd(PPh_3)COR(OH) \longrightarrow \text{[R–C(=O)–O ring]} + IPdH(PPh_3)_2 \quad (544)$$

In the second report[853], palladium is used to catalyse the stereoselective intramolecular aminocarbonylation of 3-hydroxypent-4-enylamide urethanes and tosylamides to produce *cis*-3-hydroxypent-4-enylamides urethanes and tosylamides to produce *cis*-3-hydroxypyrrolidine-2-acetic acid lactone (equation 545). This same reaction system was also used[853] to convert 4-penten-1,3-diols and 3-hydroxy-4-pentenoic acids into *cis*-3-hydroxytetrahydrofuran-2-acetic acid lactones and bis-lactones, respectively.

$$\text{HO–C}_3\text{(C}_4{=}\text{C}_5\text{)–C}_2\text{(R)–C}_1\text{–NH–X} \xrightarrow{\text{conditions}} \mathbf{(71)} + \mathbf{(72)} \quad (545)$$

R	X	Conditions	Yield (%) 71	72
H	CO_2Me	$PdCl_2$, $CuCl_2$, dry MeOH, r.t., CO, 1 day	35	24
H	SO_2Tol	$PdCl_2$, $CuCl_2$, dry MeOH, r.t., CO, 1 day	37	43
H	SO_2Tol	$PdCl_2$, $CuCl_2$, AcONa, HOAc, r.t., CO, 1 day	90	0
1-Ph	SO_2Tol[a]	$PdCl_2$, $CuCl_2$, AcONa, HOAc, r.t., CO, 2 days	74	0
2-Me	SO_2Tol[b]	$PdCl_2$, $CuCl_2$, AcONa, r.t., CO, 1 day	70	<5
2,2-Me_2	CO_2Me	$PdCl_2$, $CuCl_2$, dry MeOH, r.t., CO, 1 day	70	<5
3-Me	SO_2Tol	$PdCl_2$, $CuCl_2$, AcONa, HOAc, r.t., CO, 1 day	66	30
4-Me	SO_2Tol	$PdCl_2$, $CuCl_2$, AcONa, HOAc, r.t., CO, 2 days	No Reaction	
4-Me	CO_2Me	$PdCl_2$, $CuCl_2$, AcONa, HOAc, r.t., CO, 3 days	28	0

[a] 1-(*R*), 3-(*S*) isomer was used.
[b] A 1:1 diastereomeric mixture was used.

Finally, palladium acetate in the presence of triphenylphosphine and tri-(*n*-butyl)amine has been reported[776] to catalyse carbon monoxide insertion into vinyl halides bearing a secondary alcohol to produce 5-, 6- and 7-membered α-methylene lactones (equation 546).

$$\text{vinyl halide containing a —OH group} + Pd(OAc)_2 \xrightarrow[\text{HMPA, CO, 100 °C}]{Ph_3P, (n\text{-}Bu)_3N,} \alpha\text{-methylene lactones} \qquad (546)$$

Vinyl halide	Time (h)	Products	Yield (%)
$CH_2{=}C(Br)CH_2CH_2CH_2OH$	6	α-methylene-δ-valerolactone (H_2C=) +	27.6
		3-methyl-5,6-dihydro-2-pyranone (Me)	6.9
$CH_2{=}C(Br)CH_2CH_2CH(Me)OH$	3	α-methylene-δ-lactone (H_2C=, Me)	31

(continued)

Vinyl halides	Time (h)	Products	Yield (%)
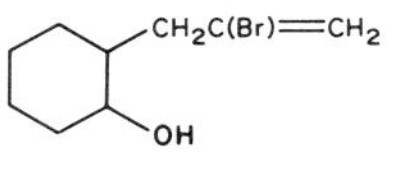 2-($CH_2C(Br){=}CH_2$)cyclohexanol	3	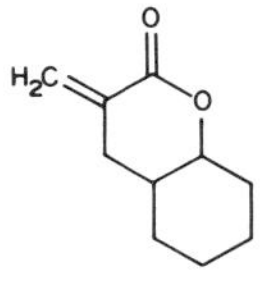 [structure] +	50 + 13
	6	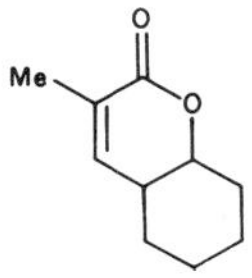 [structure]	41 + 10
$MeCl{=}CHCH_2CH_2CH_2OH$ +	6	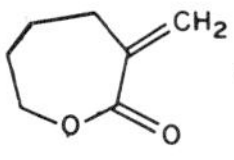 [structure] +	35.8 (2:1)
$H_2C{=}ClCH_2CH_2CH_2CH_2OH$		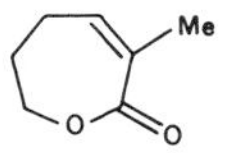 [structure]	

*Q. By Rearrangement Reactions

*2. Carbonium ion rearrangements

Carbonium ion rearrangement–lactonization of 3-hydroxy carboxylic acid esters produces[854] γ-lactones as illustrated in equation 547. Study of this reaction has established that the yields of lactone depend upon the structure of the R^1—R^5 substituents. High yields of lactone were obtained from 3-hydroxy ester substrates having secondary or tertiary groups in the 3-position and from esters containing $R^1 = t$-butyl.

$$R^1OOCCHR^2CR^3(OH)CHR^4R^5 \xrightarrow[r.t.]{H_2SO_4} \text{[lactone with } R^2, R^3, R^4, R^5\text{]} \qquad (547)$$

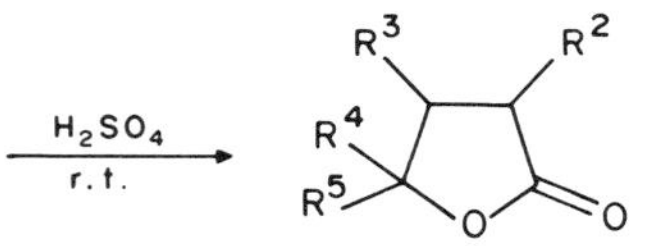

R^1 = Me, Et, *t*-Bu
R^2 = H, Et
R^3 = H, Me, *n*-Pr
R^4 = Me, Et, Ph
R^3-R^4 = alkylene
R^5 = H, Me, Ph

Treatment of tricyclo[5.2.2.0^{1,5}]undecane isomers with 85% sulphuric acid at room temperature causes a carbonium ion rearrangement–lactonization to produce[855] different products depending upon the isomer employed. Using methyl *endo*-5-methyltricyclo[5.2.2.0^{1,5}]undec-*endo*-4-hydroxy-*anti*-9-carboxylate affords a 95% yield

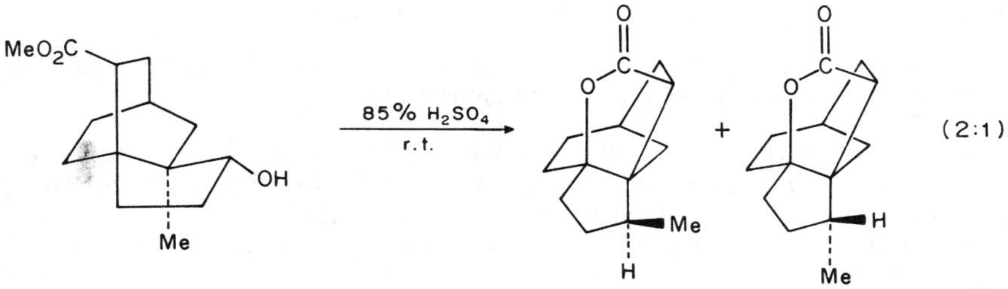

$$\text{(548)}$$

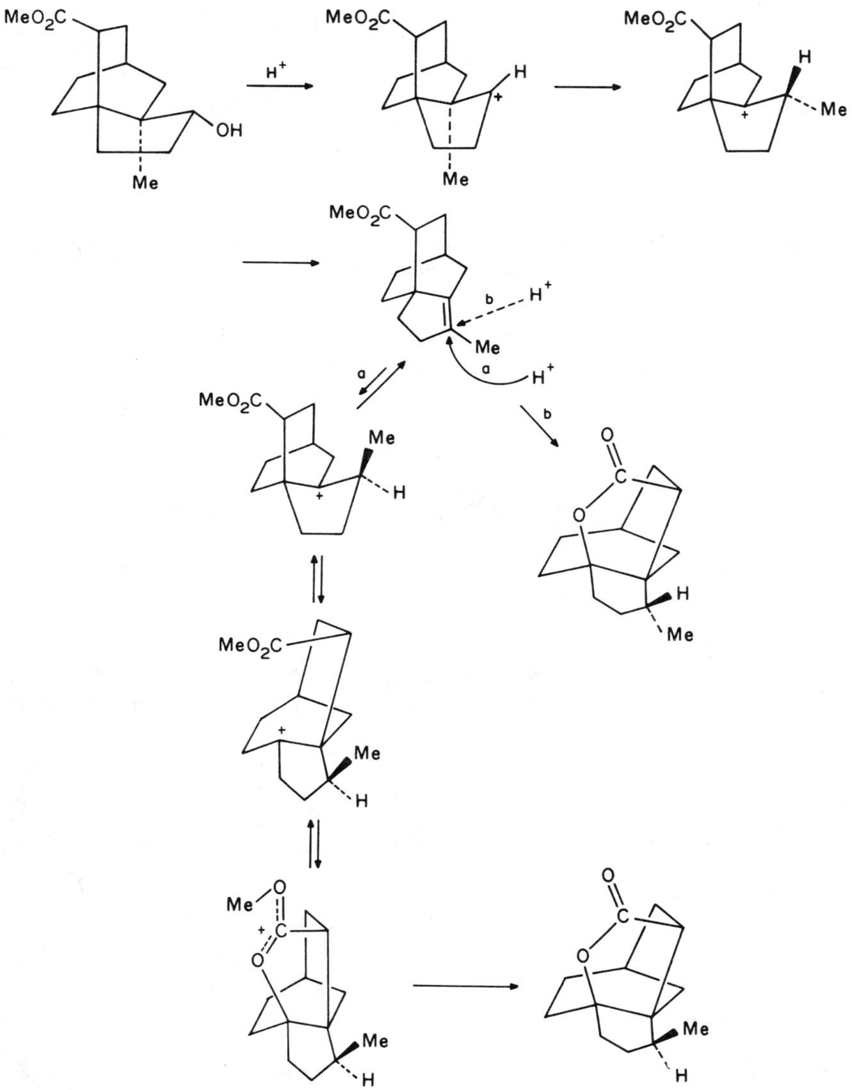

85% H_2SO_4, r.t. (2:1)

SCHEME 7

of a 2:1 mixture of 2β- and 2α-methyltricyclo[6.2.1.0^{1,5}]undecan-10-carboxy-5-hydroxylactone, respectively (equation 548). The mechanism proposed for this rearrangement is shown in Scheme 7.

Using the methyl *endo*-5-methyltricyclo[5.2.2.0^{1,5}]undec-*endo*-4-hydroxy-*syn*-9-carboxylate affords a 90% yield of a 1:1 mixture of methyl 4-methyltricyclo[5.2.2.0^{1,5}]-undec-4-en-9-*syn*-carboxylate and 5-methyltricyclo[5.2.2.0^{1,5}]undecan-9-carboxy-4-hydroxylactone (equation 549). The mechanism proposed for this rearrangement is similar to the one illustrated above for the *anti*-epimer.

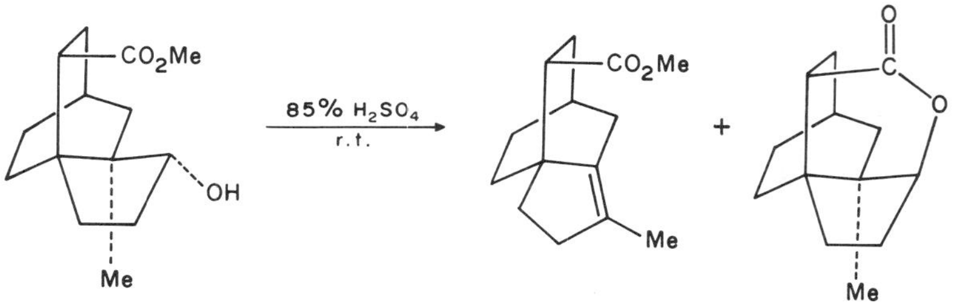

(549)

Acid catalysed carbonium ion rearrangement to produce lactones has also been reported[856] to occur with epoxy esters (equation 550).

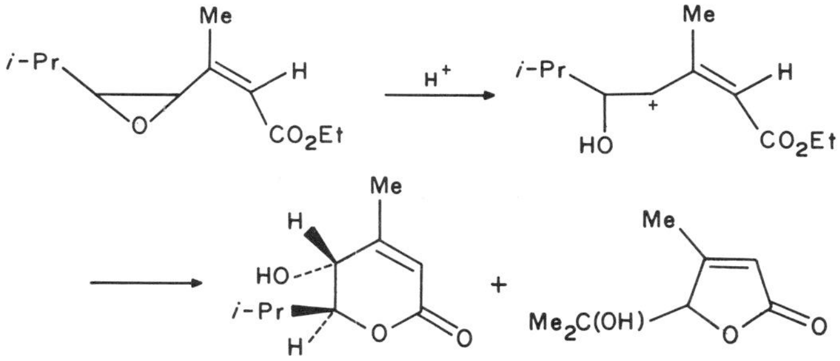

(550)

3. Photochemical rearrangements

Irradiation of 2,3-epoxy-1,4-cyclohexanediones in acetone produces[857,858] γ-alkylidene-γ-butyrolactones (equations 551 and 552), via a photo-induced rearrangement

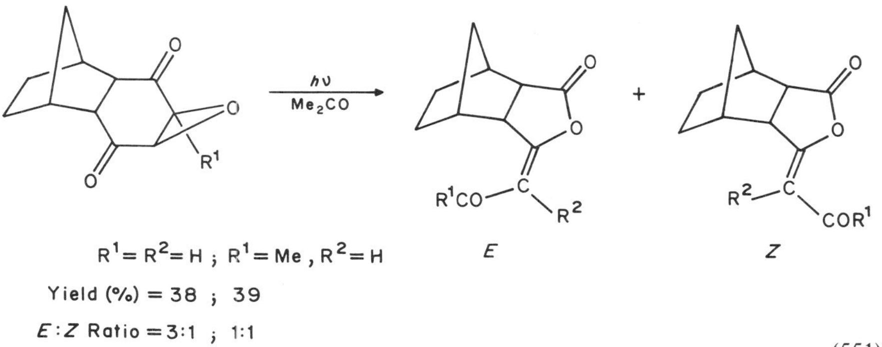

$R^1 = R^2 = H$; $R^1 = Me$, $R^2 = H$

Yield (%) = 38 ; 39

E:*Z* Ratio = 3:1 ; 1:1

(551)

mechanism probably involving[857] homolysis of the oxirane ring to produce a biradical, ring-opening of the cyclohexanedione ring, subsequent β-alkyl migration with an odd electron and coupling of the acyl radical and oxygen radical (C—O coupling) of the resultant biradical (equation 553).

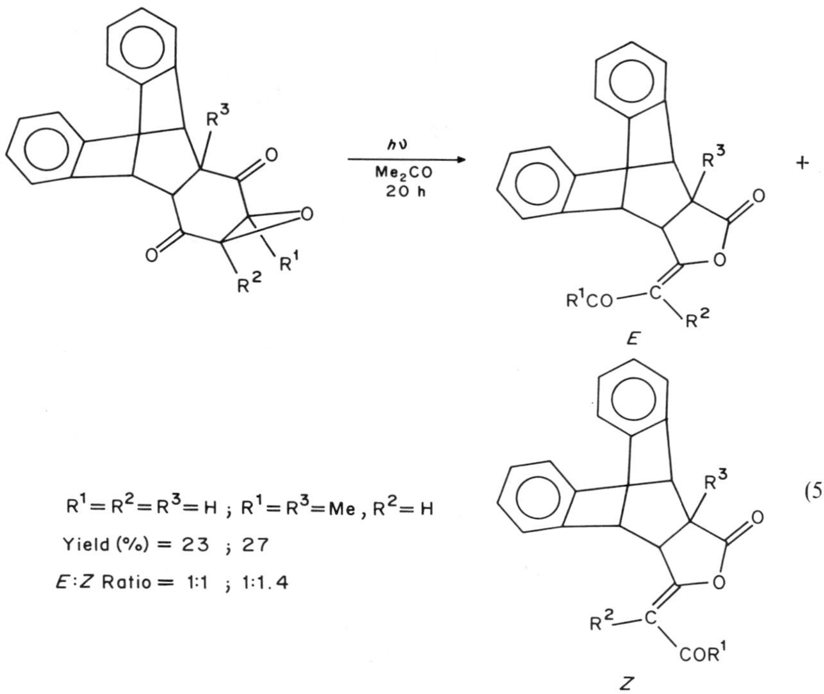

hν, Me_2CO, 20 h

$R^1 = R^2 = R^3 = H$; $R^1 = R^3 = Me$, $R^2 = H$

Yield (%) = 23 ; 27

E:*Z* Ratio = 1:1 ; 1:1.4

E + *Z* (552)

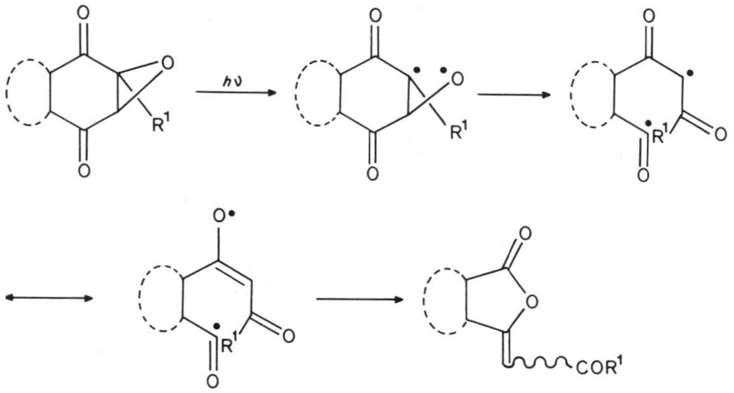

(553)

4. Miscellaneous rearrangements

The miscellaneous rearrangements recently reported all concern the formation of γ-butyrolactones from cyclopropanecarboxylic acids or esters.

In the first report[859], treatment of 2-phenyl-2-methoxy-3,3-dimethylcyclopropane-

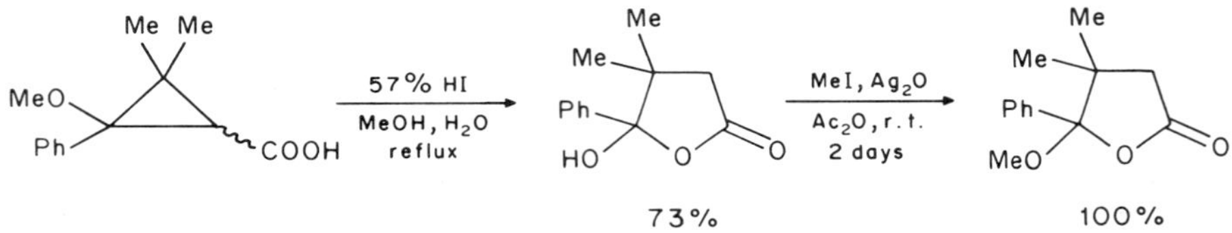
Me Me
MeO
Ph
COOH
57% HI
MeOH, H2O
reflux
Me
Me
Ph
HO
O
O
73%
MeI, Ag2O
Ac2O, r.t.
2 days
Me
Me
Ph
MeO
O
O
100%

(554)

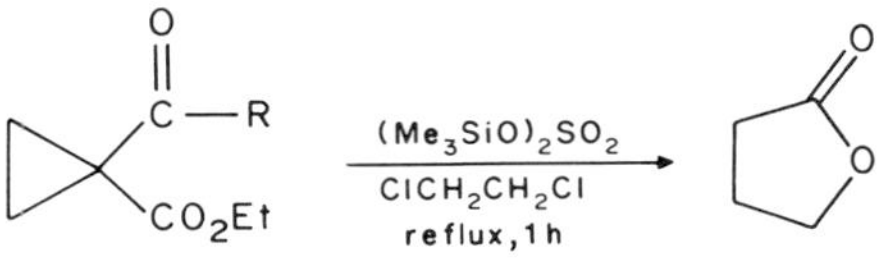
O
C — R
CO2Et
(Me3SiO)2SO2
ClCH2CH2Cl
reflux, 1 h
O
O

(555)

Substrate	Temp. (°C)	Product (ratio)	Yield (%)
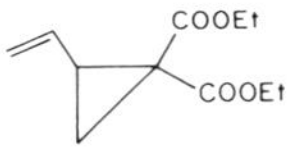	80	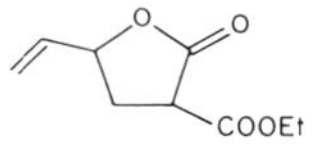	98[a]
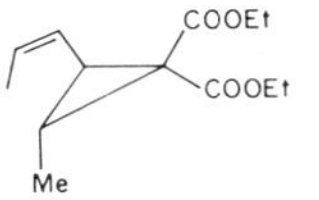	40	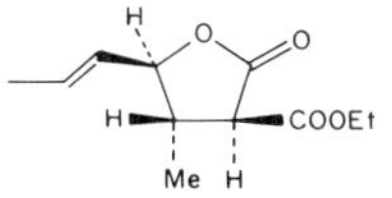	63
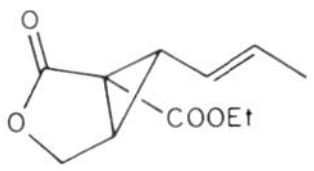	80	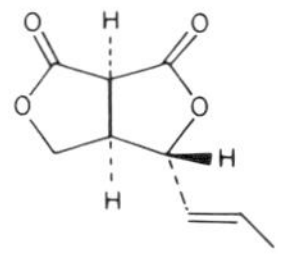	39
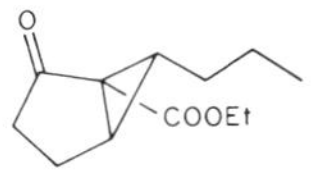	80	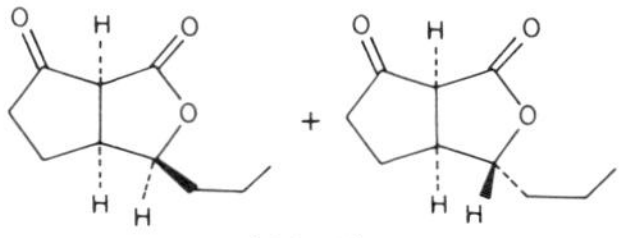 (82:18)	79
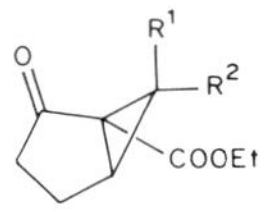		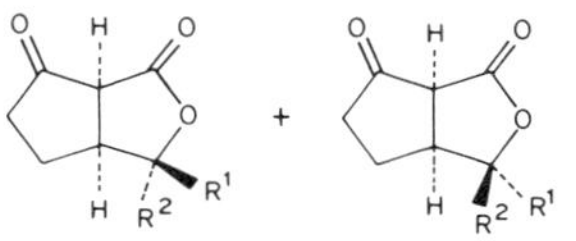	
$R^1 = Me, R^2 = H$	80	(23:77)	28
$R^1 = H, R^2 = Me$	80	(65:35)	71
$R^1 = Ph, R^2 = H$	80	0:100 or 100:0	28

[a]Other catalysts used and yields obtained for this conversion: H_2SO_4—61%; $Me_3SiOSO_2CF_3$—25%; $Me_3SiOClO_3$—39% in benzene and 30% in CH_2Cl_2; Me_3SiI—0%; $BF_3 \cdot OEt_2$—complex mixture of products: Et_2AlBF_4—0%; $TiCl_4$—no reaction.

carboxylic acid with 57% hydrogen iodide in a methanol–water mixture produces a 73% yield of 3,3-dimethyl-4-hydroxy-4-phenylbutyrolactone (equation 554). Further treatment of the hydroxy product with methyl iodide and silver oxide in acetic anhydride produces a quantitative conversion to the methoxy analogue.

In the second report[860], γ-lactonization of the ethyl esters of α-carbonyl substituted cyclopropanecarboxylic acids is catalysed by bis(trimethylsilyl)sulphate and produces good yields of lactones (equation 555). Using this approach requires that *gem*-dicarbonyl substituents be present on a cyclopropane carbon.

In the final report[861] α-methylene-γ-butyrolactones were prepared with high regio- and stereoselectivity from 1-(*N*,*N*-dimethylaminomethyl)cyclopropanecarboxylic esters by treatment with trimethylsilyl iodide followed by thermolysis (equation 556).

1. Me_3SiI, C_6H_6 or 1,2-DCE, 50–95 °C, several hours
2. distillative thermolysis

R^1 = Ph ; *n*-Hex ; $(CH_2)_4$; H
R^2 = H ; H ; (linked) ; H
Yield (%) = 62; 58[a] ; 64 ; 43[b] (556)

[a]Accompanied by 3-hydroxymethyl-2-methylenenonanoic lactone regio isomer of the product in 12% yield.
[b]The *N*,*N*-di(isopropyl)amine analogue was used as the starting material. The product formed is tulipalin.

*R. Lactone Interconversions

Several different types of lactone interconversions are discussed in this section beginning with the interconversion of substituents present on all size lactone rings.

Two kinds of lactone ring substituents have been converted into exomethylene functions on γ-butyrolactones and spiro bislactones. Treatment[862] of iodomethyl substituted γ-butyrolactones (equation 557) or bisiodomethyl spiro bislactones (equation 558) with diazabicycloundecene (DBU) in benzene causes an elimination of hydrogen

DBU, C_6H_6, 60–80 °C, 2h (557)

R^1 = H ; H ; Me
R^2 = H ; Me ; Me
Yield (%) = 66–70 ; 77–83 ; 83

ICH_2 ... CH_2I — DBU, C_6H_6, 60 °C, 20 min → H_2C ... CH_2 (558)

iodide with formation of the corresponding exomethylene function. Exomethylene functionality is also obtained[758] via a retro-Diels–Alder reaction of spiro norbornenyl γ-butyrolactones (equation 559).

Me Me ... R — C_6H_6, 140 °C, sealed tube, 30 min → H_2C ... R (559)

R = H, Me, Et, Ph

A similar lactone interconversion is represented by the introduction[821] of a dimethylaminomethylene function into the α-position of a bicyclo epoxylactone by reaction with t-butoxybis(dimethylamino)methane in dimethylformamide (equation 560).

$PhCOOCH_2$ ... Me Me — $(Me_2N)_2CHOBu$-t, DMF, 90 °C → $PhCOOCH_2$ ... $CHNMe_2$ ... Me Me, 91%

(560)

Depending upon the substrate structure and the reagents used, dechlorination and/or desulphurization reactions can lead to the production of either saturated or unsaturated lactone products. For example, dechlorination of α,α-dichloro-γ-thioaryl-γ-butyrolactones with aluminium amalgam in tetrahydrofuran produces[801] the corresponding saturated γ-butyrolactones with the γ-thioaryl function uneffected (equation 561), while dechlorination[788] of bicyclic lactones using zinc in acetic acid also proceeds without the introduction of any new unsaturation (equation 562). Dehydrochlorination of the same substrate using triethylamine in tetrahydrofuran produces[788] the corresponding aromatic bicyclic lactone with increased unsaturation (equation 563), while dehydrochlorination of the saturated substrate analog using the same reagents produces[788] a mixture of isomeric dienes (equation 564).

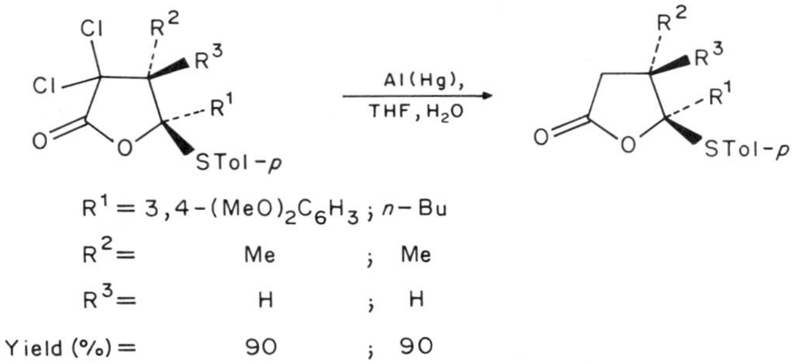
Cl
R2
R3
R1
O
STol-p
Al(Hg),
THF, H2O
R1 = 3,4-(MeO)2C6H3 ; n-Bu
R2 = Me ; Me
R3 = H ; H
Yield (%) = 90 ; 90

(561)

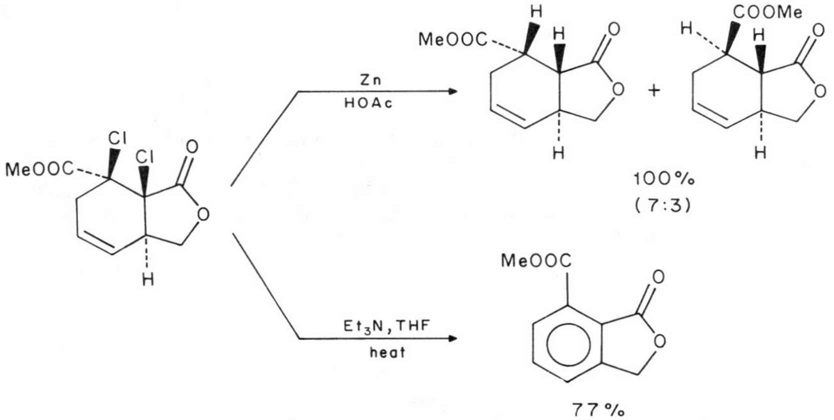
MeOOC
Cl
O
H
Zn
HOAc
COOMe
100%
(7:3)
Et3N, THF
heat
77%

(562)

(563)

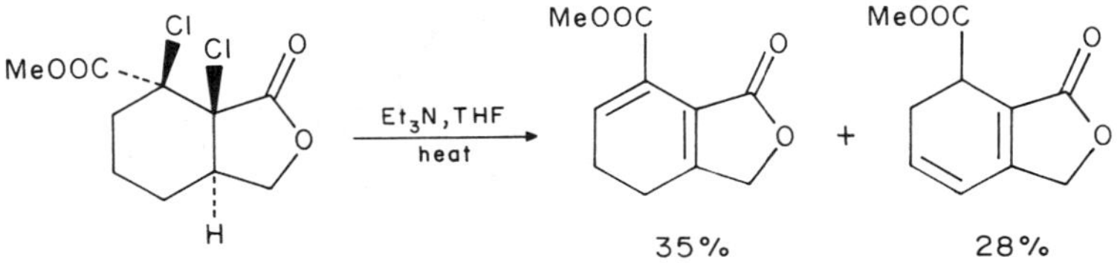
MeOOC
Cl
O
H
Et3N, THF
heat
35%
28%

(564)

Triethylamine in tetrahydrofuran has also been used[788] to effect dechlorodecarboxylation in the same bicyclic lactone series, with increased unsaturation observed in the products resulting from reaction of the completely unsaturated acid substrate (equation 565) or from the monoene acid substrate (equation 566). Reductive dechlorination was also accomplished[788] in this series (equation 567).

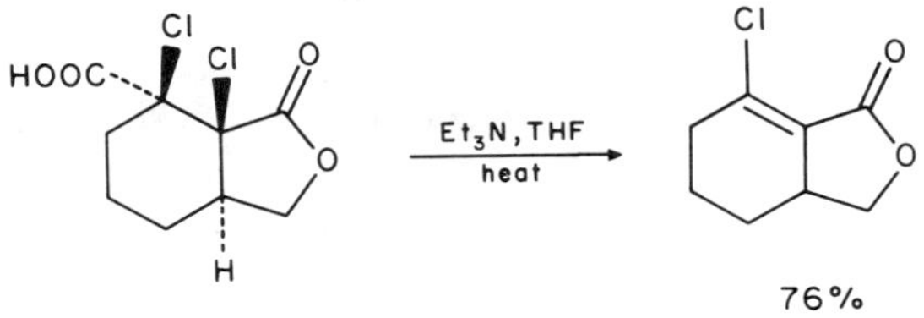
HOOC
Cl
O
H
Et3N, THF
heat
76%

(565)

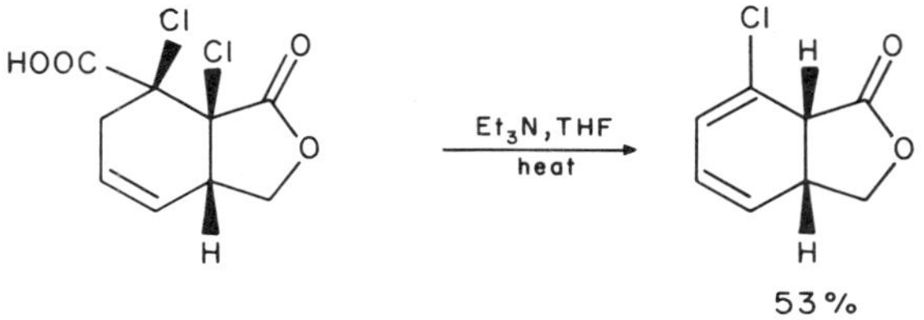

Et_3N, THF, heat

53%

(566)

H_2, Pd/C, EtOAc

98%

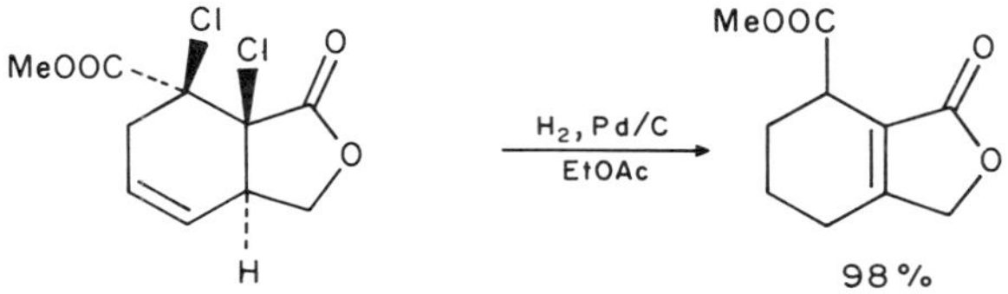

(567)

Desulphurization–dechlorination of α,α-dichloro-γ-thioaryl-γ-butyrolactones without the introduction of unsaturation has been reported[801] to occur using Raney nickel (equation 568) or tri(*n*-butyl)tin hydride (equation 569), while desulphurization only of γ-thioaryl-γ-butyrolactones using Raney nickel also proceeds[801] without the introduction of unsaturation but with a high degree of stereoselectivity (equation 570). Similar results are obtained[863] when α-sulphoxyphenyl-δ-lactones are treated with aluminium amalgam in tetrahydrofuran (equation 571).

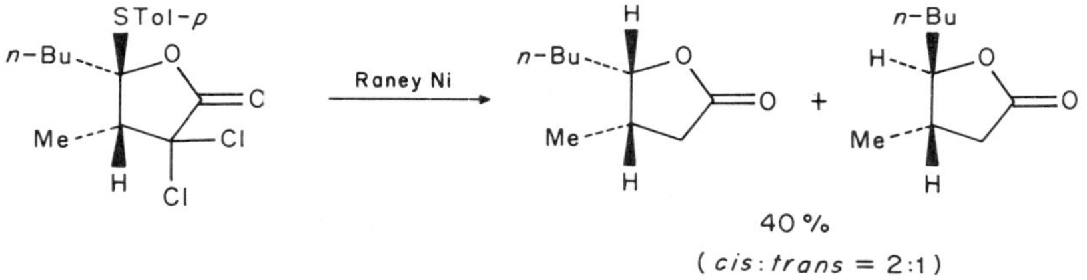

Raney Ni

40%
(*cis* : *trans* = 2:1)

(568)

$(n\text{-Bu})_3SnH$, AIBN, toluene

cis *trans*

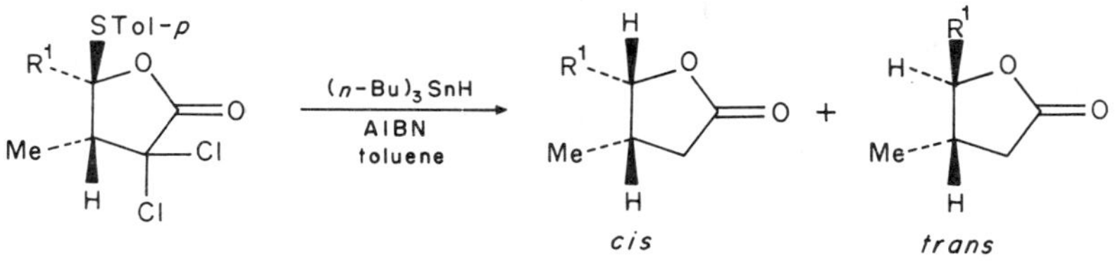

R^1 = 3,4-$(MeO)_2C_6H_3$; *n*-Bu ; *n*-Bu (3*R*, 4*R*)
Yield (%) = — ; — ; 70 (100% e.e.)
cis : *trans* ratio = 2·3:1 ; 1:1.8 ; (3*S*, 4*R*):(3*S*, 4*S*) = ~1.1:1

(569)

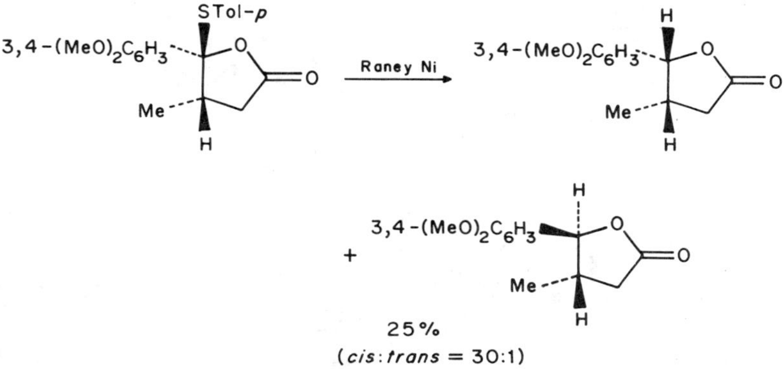

STol-p
3,4-(MeO)$_2$C$_6$H$_3$
O
O
Me
H
Raney Ni
H
3,4-(MeO)$_2$C$_6$H$_3$
O
O
Me
H

(570)

H
+ 3,4-(MeO)$_2$C$_6$H$_3$
O
O
Me
H
25%
(*cis*:*trans* = 30:1)

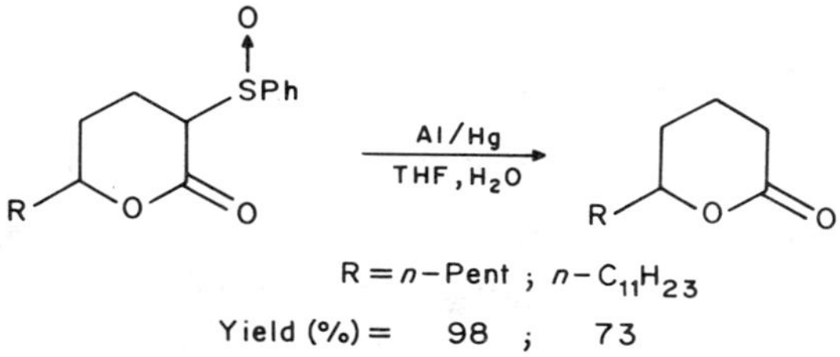

O
SPh
R
O
O
Al/Hg
THF, H$_2$O
R
O
O

(571)

R = *n*-Pent ; *n*-C$_{11}$H$_{23}$
Yield (%) = 98 ; 73

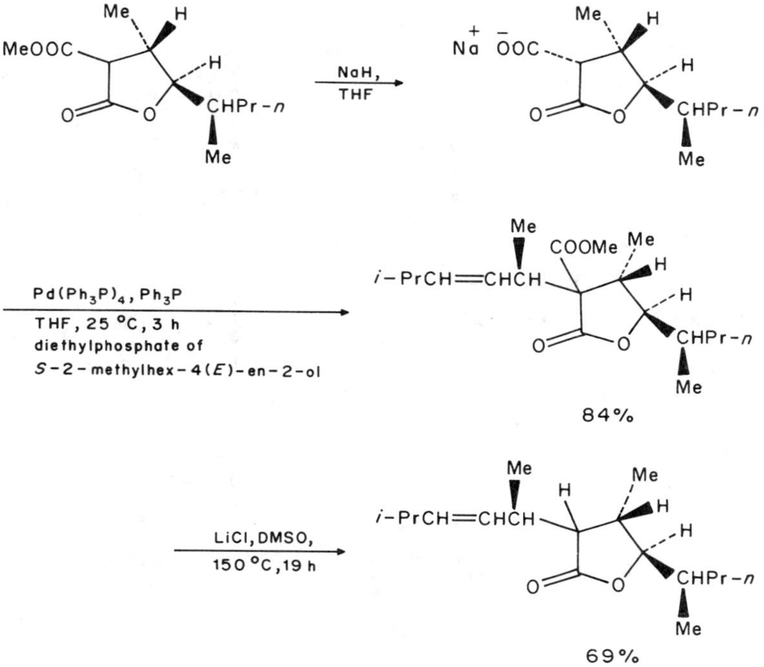

Me H
MeOOC
H
O
O
CHPr-*n*
Me
NaH, THF
Na$^+$ $^-$OOC
Me H
H
O
O
CHPr-*n*
Me

Pd(Ph$_3$P)$_4$, Ph$_3$P
THF, 25 °C, 3 h
diethylphosphate of
S-2-methylhex-4(*E*)-en-2-ol

Me
COOMe Me
i-PrCH=CHCH
H
H
O
O
CHPr-*n*
Me
84%

(572)

LiCl, DMSO,
150 °C, 19 h

Me
Me
H
i-PrCH=CHCH
H
H
O
O
CHPr-*n*
Me
69%

PhCH$_2$OCH$_2$CH(Me) ... LDA, MeI, HMPA, THF, −78 °C → 97% (1:1)

1. LDA, THF
2. H_3O^+
3. H_2, Pd/C

HOCH$_2$CH(Me) ...

RuCl$_3$, NaIO$_4$, MeCN, CCl$_4$

HOOCCH(Me) ...

(573)

MnO$_2$, CH$_2$Cl$_2$ → 84%

1. Me$_2$CuLi THF, 0 °C 2. aq. NH$_4$Cl → 51%

CH$_2$=CHMgBr CuI, THF, −78 °C → CH=CH$_2$ 47%

1. CuBr · SMe$_2$, CH$_2$=CHCH$_2$CH$_2$MgBr THF, Me$_2$S, −10 °C 2. aq. NH$_4$Cl → CH$_2$CH$_2$CH=CH$_2$ 74%

(574)

Alkylation of lactones is usually one step in an overall synthesis of complicated molecules which contain a lactone nucleus. Examples of extensive syntheses which include alkylation of a lactone as one step include: preparation[761] of precursors for the synthesis of (+)-invictolide (equation 572); the preparation[832] of the Prelog–Djerassi lactone (equation 573); elaboration[786] of *trans*-fused lactones (equation 574); silylation[781] of bicyclic keto lactones (equation 575); thiophenylation[779] of γ-(n-butoxymethyl)-α-chloro-γ-butyrolactone (equation 576) and γ-bromomethyl-α,β-butenolide (equation 577); phenylseleneylation[791] of δ,δ-dimethylvalerolactone (equation 578) and bromophthalide

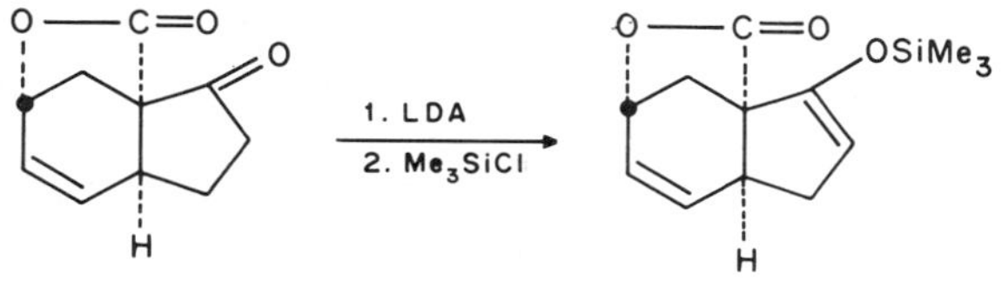

1. LDA
2. Me_3SiCl
OSiMe$_3$
(575)

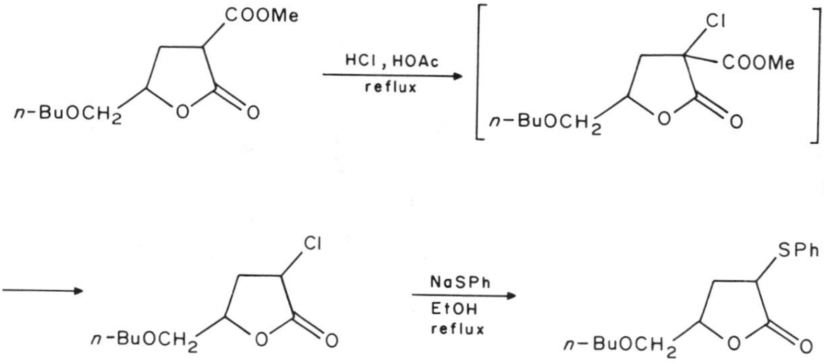

HCl, HOAc reflux
n-BuOCH$_2$ COOMe Cl
(576)

NaSPh EtOH reflux
SPh

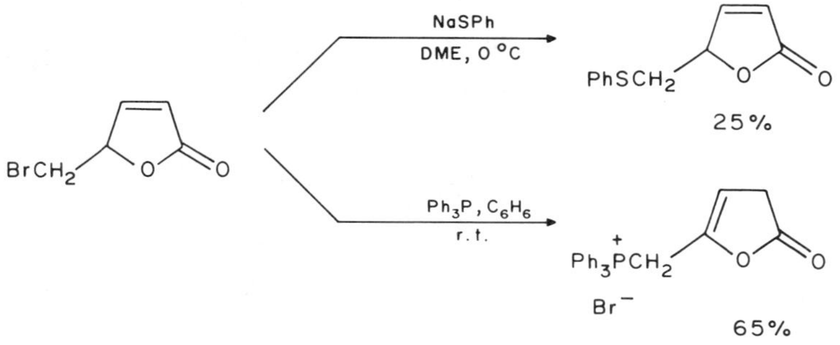

NaSPh DME, 0 °C
PhSCH$_2$ 25%
BrCH$_2$
Ph_3P, C_6H_6 r.t.
$Ph_3\overset{+}{P}CH_2$ Br^- 65%
(577)

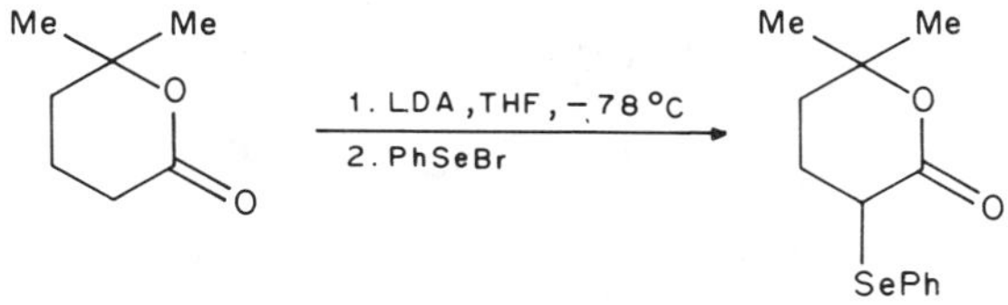

Me Me
1. LDA, THF, −78 °C
2. PhSeBr
SePh
(578)

(equation 579); and, reaction[864] of nucleophilic mono- and dianions with 4,6-dimethoxy-2-pyrone followed by demethylation to produce enol lactones of 6-substituted-4-methoxy-2-pyrones (equation 580). The mechanism proposed[864] for the pyrone

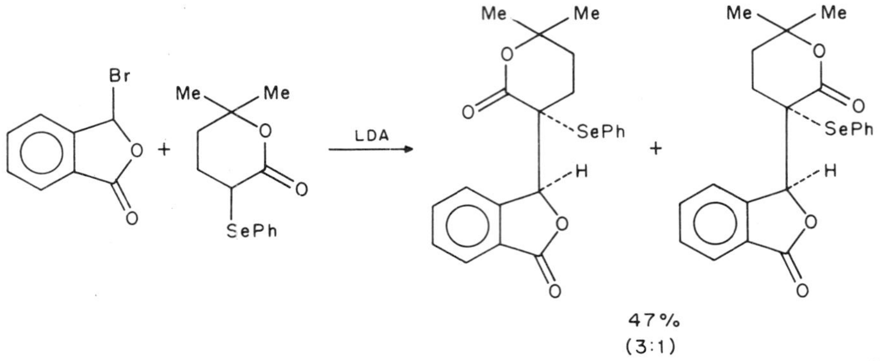

LDA

47%
(3:1)

(579)

THF, r.t.
4 h

Me_3SiI
$CHCl_3$

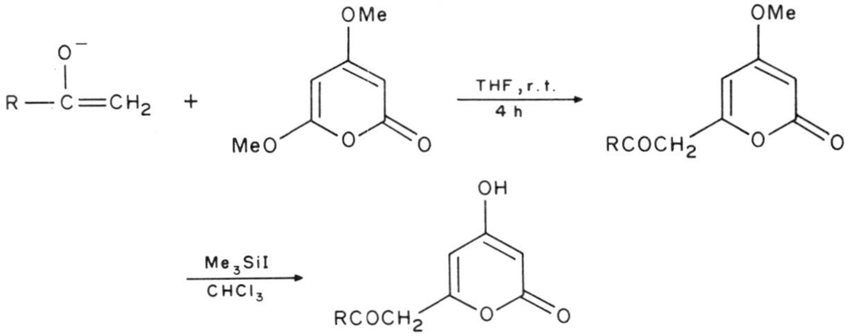

(580)

Nucleophile	% Yield of condensation product	Demethylation conditions	% Yield of Demethylated product
$Me-C(O^-)=CH_2$	(R = Me) 52	25 °C, 72 h	(R = Me) 54
$Ph-C(O^-)=CH_2$	(R = Ph) 67	50 °C, 40 h	(R = Ph) 52
$Me-C(O^-)=CH-C(O^-)=CH_2$	(R = $MeCOCH_2$) 82	25 °C, 48 h	(R = Me) 65
$Ph-C(O^-)=CH-C(O^-)=CH_2$	(R = $PhCOCH_2$) 74	25 °C, 26 h	(R = $PhCOCH_2$) 56

condensation involves enolate anion attack of the carbonyl group causing ring opening to give the enolate anion of an unsaturated δ-keto product which recyclizes by displacement of methoxide ion to give the condensation products (equation 581).

(581)

Other miscellaneous lactone substituent interconversions reported include: the conversion[779] of γ-hydroxymethyl-α,β-butenolide into γ-alkoxymethyl-α,β-butenolides (equation 582); conversion[821] of a t-butyldimethysilyl protecting group into the corresponding benzoate (equation 583); ring opening and silylation of a bicyclic epoxylactone[786] (equation 584); conversion of a spiro lactone into[787] a bicyclic lactone (equation 585);

$HOCH_2$-butenolide $\xrightarrow{\text{Conditions}}$ $ROCH_2$-butenolide (582)

R = Me ; Ph_3C ; CONHPh

Conditions = BF_3–ether ; Ph_3CCl, C_5H_5N ; PhNCO, C_6H_6, reflux

t-$BuMe_2SiOCH_2$ ... $\xrightarrow[\text{2. PhCOCl, } C_5H_5N\text{, 15 °C, 1 h}]{\text{1. } (n\text{-}Bu)_4\overset{+}{N}\ \overset{-}{F}\text{, THF, 20 °C, 30 min}}$ $PhCOOCH_2$ ...

(583)

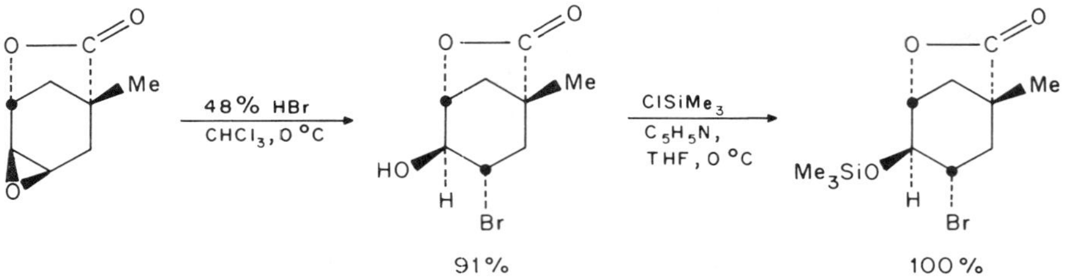

48% HBr / $CHCl_3$, 0 °C → 91%; $ClSiMe_3$ / C_5H_5N, THF, 0 °C → 100%

(584)

1. aq. NaOH, 65 °C, 4 h; 2. r.t., 24 h; 3. CO_2 → [$HOCH_2CH(OH)CH_2$ / $HOOCCH_2$]

aq. KI_3, dark, 6 h → $HOCH_2CH(OH)CH_2$ ...; aq. $NaIO_4$ / p-dioxane, r.t., 2.25 h → $OHCCH_2$ ... 82%

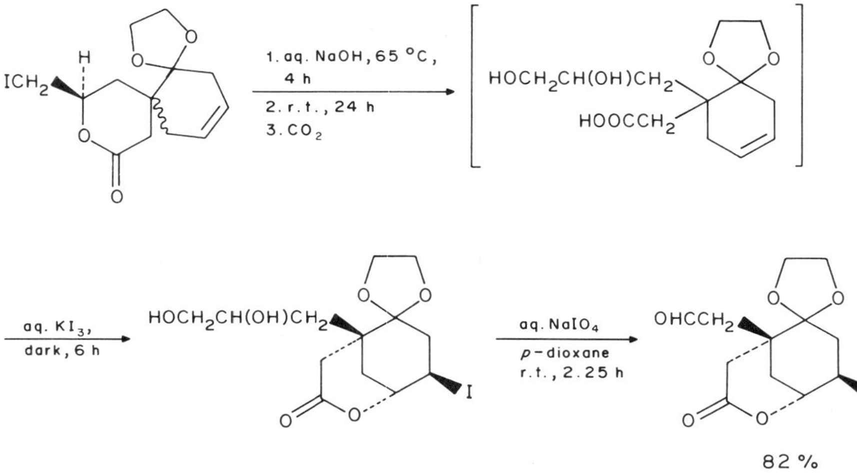

(585)

oxidation of substituent groups such as phenylthio on both γ-lactones[779,865] (equations 586 and 587) and δ-lactones[865] (equations 588 and 589), phenylseleno on δ-lactones[791] (equation 590), and O-silylated enolate of a bicyclic lactone[781] (equation 591); reduction of a β-keto function[796] (equation 592); and, dehydrogenation[780] using 2,3-dichloro-5,6-dicyanoquinone (DDQ, equation 593).

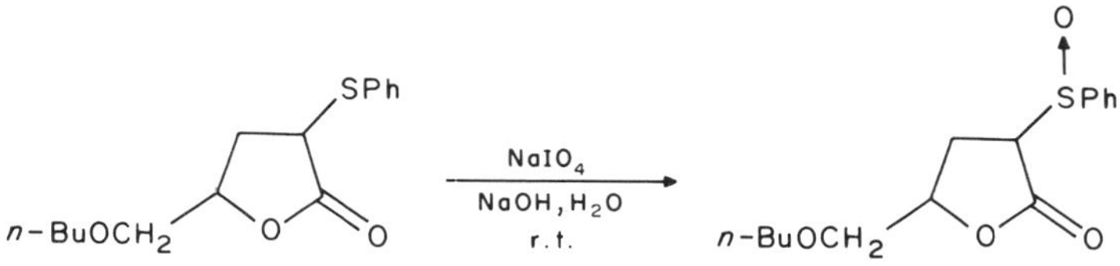

n-$BuOCH_2$ (SPh γ-lactone) —$NaIO_4$ / NaOH, H_2O, r.t.→ n-$BuOCH_2$ (S(O)Ph γ-lactone)

(586)[779]

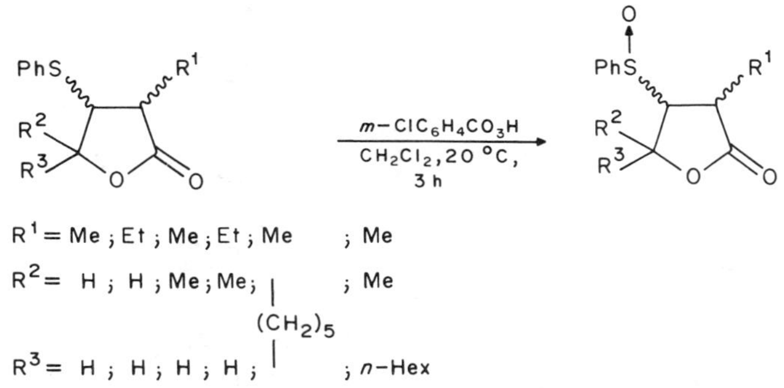

m-$ClC_6H_4CO_3H$, CH_2Cl_2, 20 °C, 3 h (587)[865]

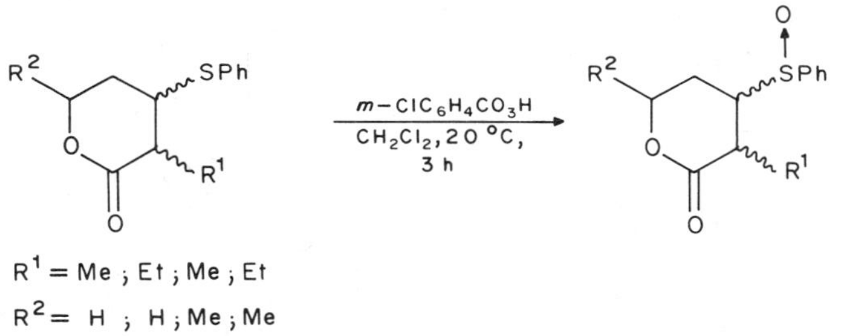

R^1 = Me ; Et ; Me ; Et ; Me ; Me
R^2 = H ; H ; Me ; Me ; $(CH_2)_5$; Me
R^3 = H ; H ; H ; H ; ; n-Hex

m-$ClC_6H_4CO_3H$, CH_2Cl_2, 20 °C, 3 h (588)

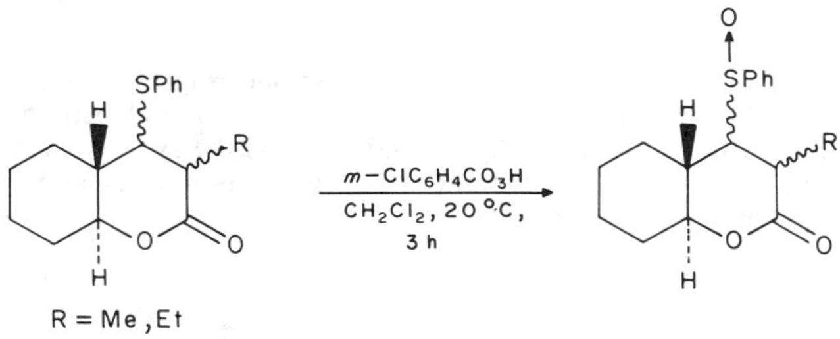

R^1 = Me ; Et ; Me ; Et
R^2 = H ; H ; Me ; Me

m-$ClC_6H_4CO_3H$, CH_2Cl_2, 20 °C, 3 h (589)

R = Me, Et

or

O_3, −78 °C

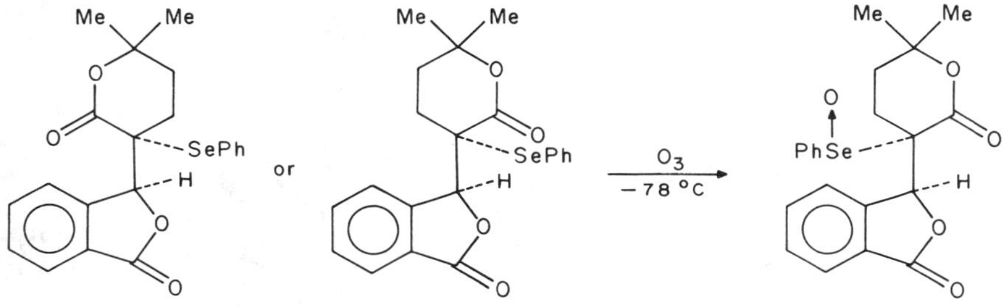

(590)

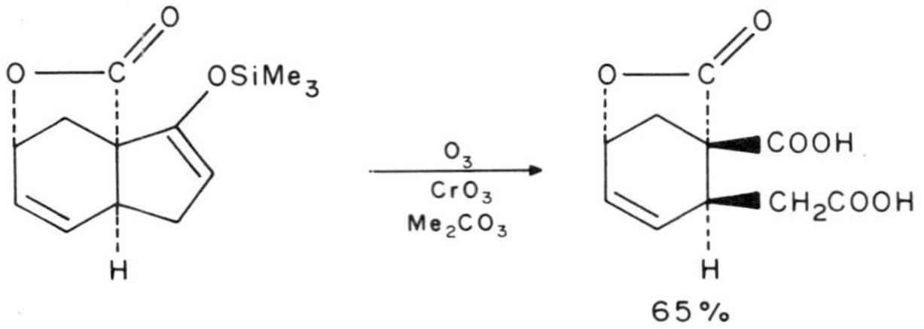

(591)

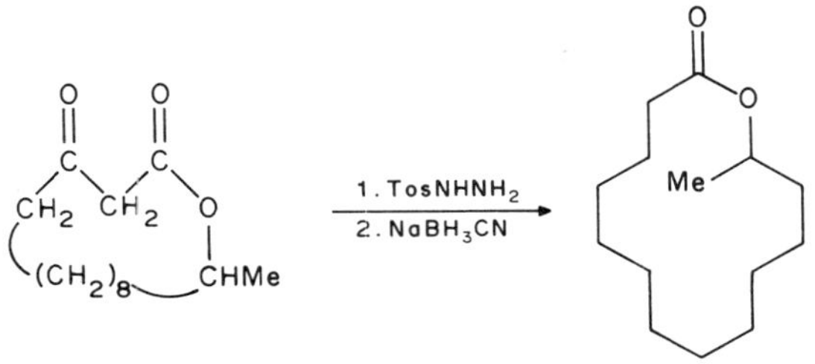

O_3, CrO_3, Me_2CO_3

65%

(592)

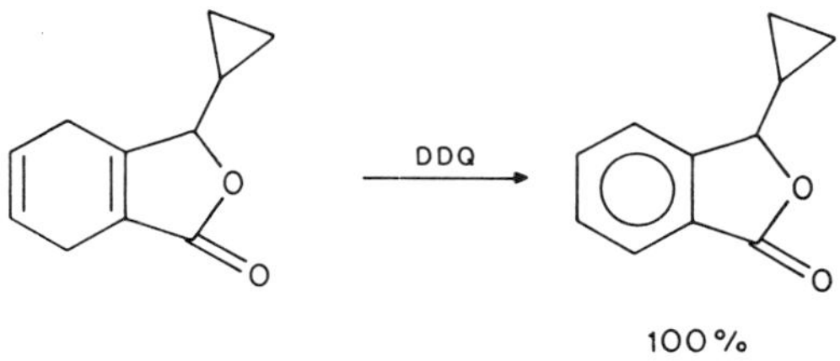

1. $TosNHNH_2$
2. $NaBH_3CN$

(593)

DDQ

100%

Formation of a double bond in a preformed lactone ring is another example of a lactone interconversion which has continued to receive considerable attention. Since the structure of the precursors vary greatly depending upon the synthetic route which was used to produce them, a variety of methods and reagents have been employed in the step which produces the double bond (Table 40).

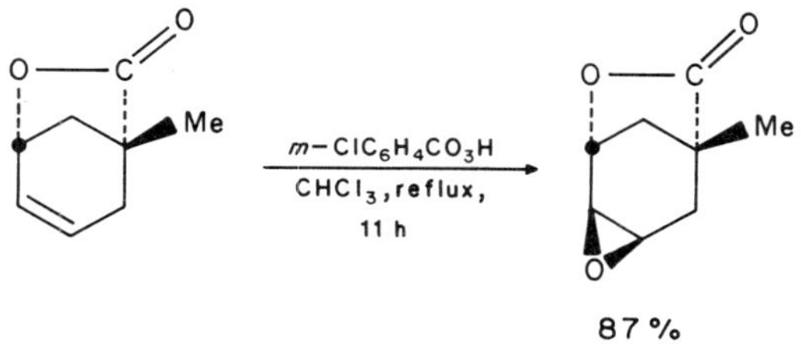

(594)

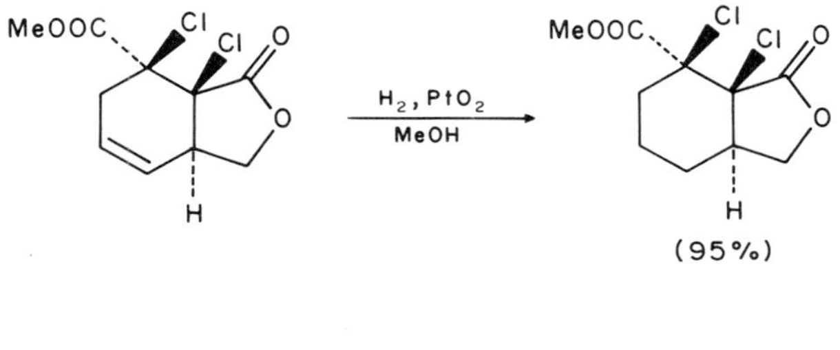

$m-ClC_6H_4CO_3H$, $CHCl_3$, reflux, 11 h

87%

(595)[788]

H_2, PtO_2, MeOH

(95%)

TABLE 40. Formation of a double bond in a preformed lactone ring

Substrate	Reagents and conditions	Product	Yield (%)	Reference
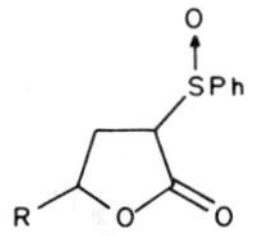 [structure: 3-(phenylsulfinyl)-5-R-tetrahydrofuran-2-one]		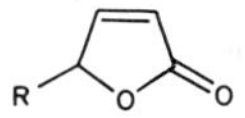 [structure: 5-R-furan-2(5H)-one]		
R = H	reduced pressure, 120 °C, 2 h		72	863
R = Et	reduced pressure, 120 °C, 2 h		68	863
R = *n*-Bu	reduced pressure, 120 °C, 2 h		61	863
R = *n*-Hex	reduced pressure, 120 °C, 2 h		76	863
R = *n*-C_7H_{15}	reduced pressure, 120 °C, 2 h		58	863
R = *n*-$C_{11}H_{23}$	reduced pressure, 120 °C, 2 h		80	863
R = *n*-$BuOCH_2$	toluene, reflux		—	779
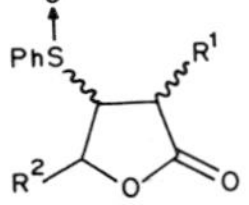 [structure: 4-(phenylsulfinyl)-3-R¹-5-R²-tetrahydrofuran-2-one]	DBU, CH_2Cl_2, 20 °C, 1 h	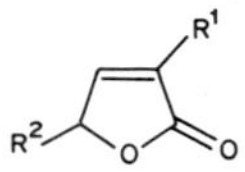 [structure: 3-R¹-5-R²-furan-2(5H)-one]		865
R^1 = Me; R^2 = H			92	
R^1 = Et; R^2 = H			90	
R^1 = R^2 = Me			91	
R^1 = Et; R^2 = Me			89	
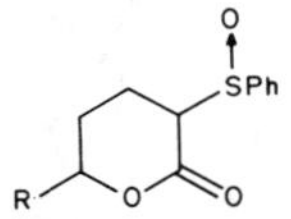 [structure: 3-(phenylsulfinyl)-6-R-tetrahydropyran-2-one]	reduced pressure, 120 °C, 2 h	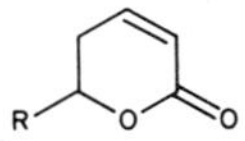 [structure: 6-R-5,6-dihydropyran-2-one]		863
R = H			85	
R = Ph			96	
R = *n*-Bu			86	
R = *n*-Pent			80	
R = *n*-Hex			85	
R = *n*-C_7H_{15}			80	
R = *n*-$C_{11}H_{23}$			87	
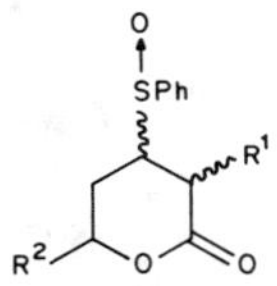 [structure: 4-(phenylsulfinyl)-3-R¹-6-R²-tetrahydropyran-2-one]	DBU, $CHCl_3$, 20 °C, 1 h	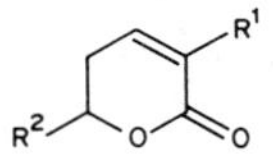 [structure: 3-R¹-6-R²-5,6-dihydropyran-2-one]		865
R^1 = Me; R^2 = H			85	
R^1 = Et; R^2 = H			90	

(continued)

TABLE 40. (*continued*)

Substrate	Reagents and conditions	Product	Yield (%)	Reference
$R^1 = R^2 = Me$			87	
$R^1 = Et$; $R^2 = Me$			92	
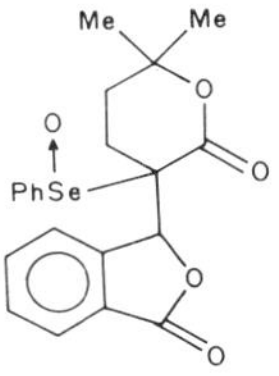	Et_3N, 25 °C	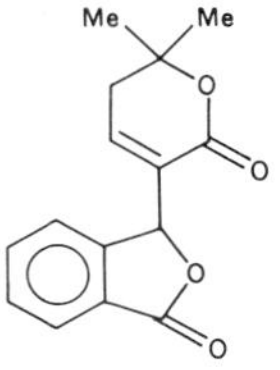	—	791
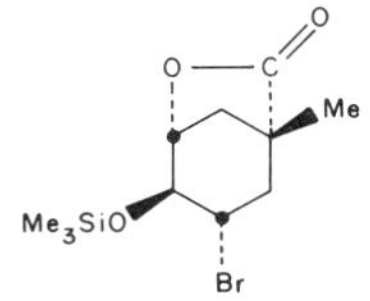	1. DBN, dioxane reflux 12 h 2. HCl, Me_2CO	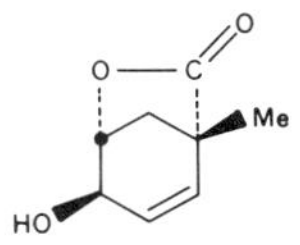	86	786
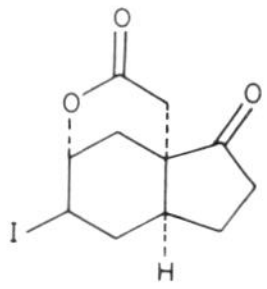	DBU	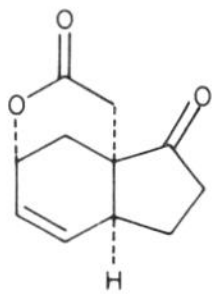	—	781
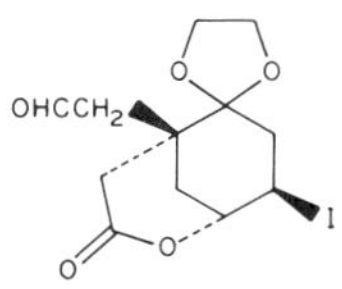	1. DBU, C_6H_6, 1 h 2. 65 °C, 3.5 h	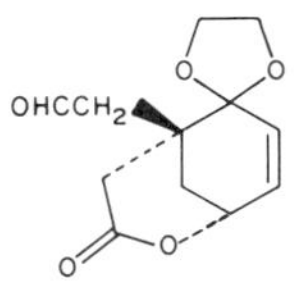	72	787

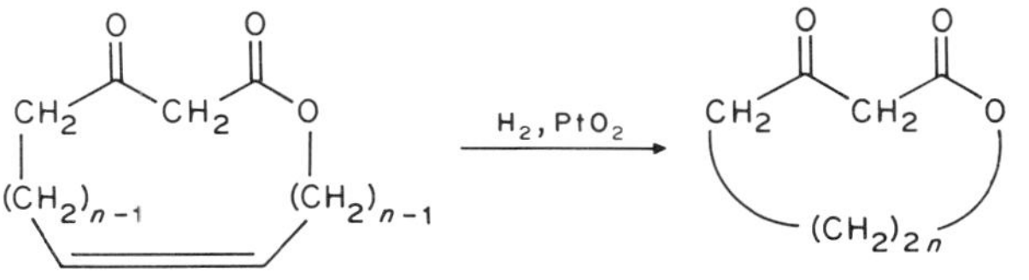

H_2, PtO_2 (596)[796]

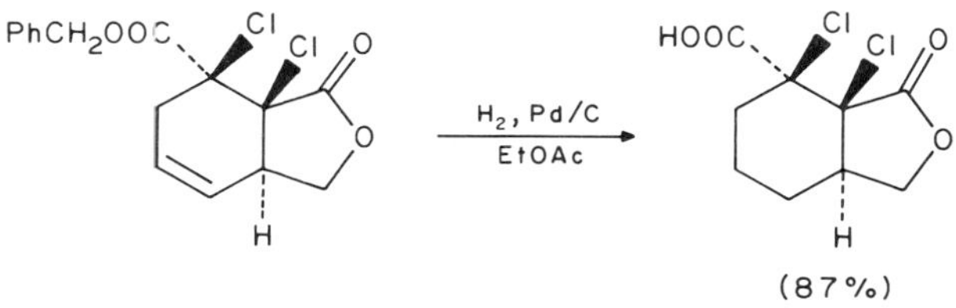

H_2, Pd/C, EtOAc (87%) (597)

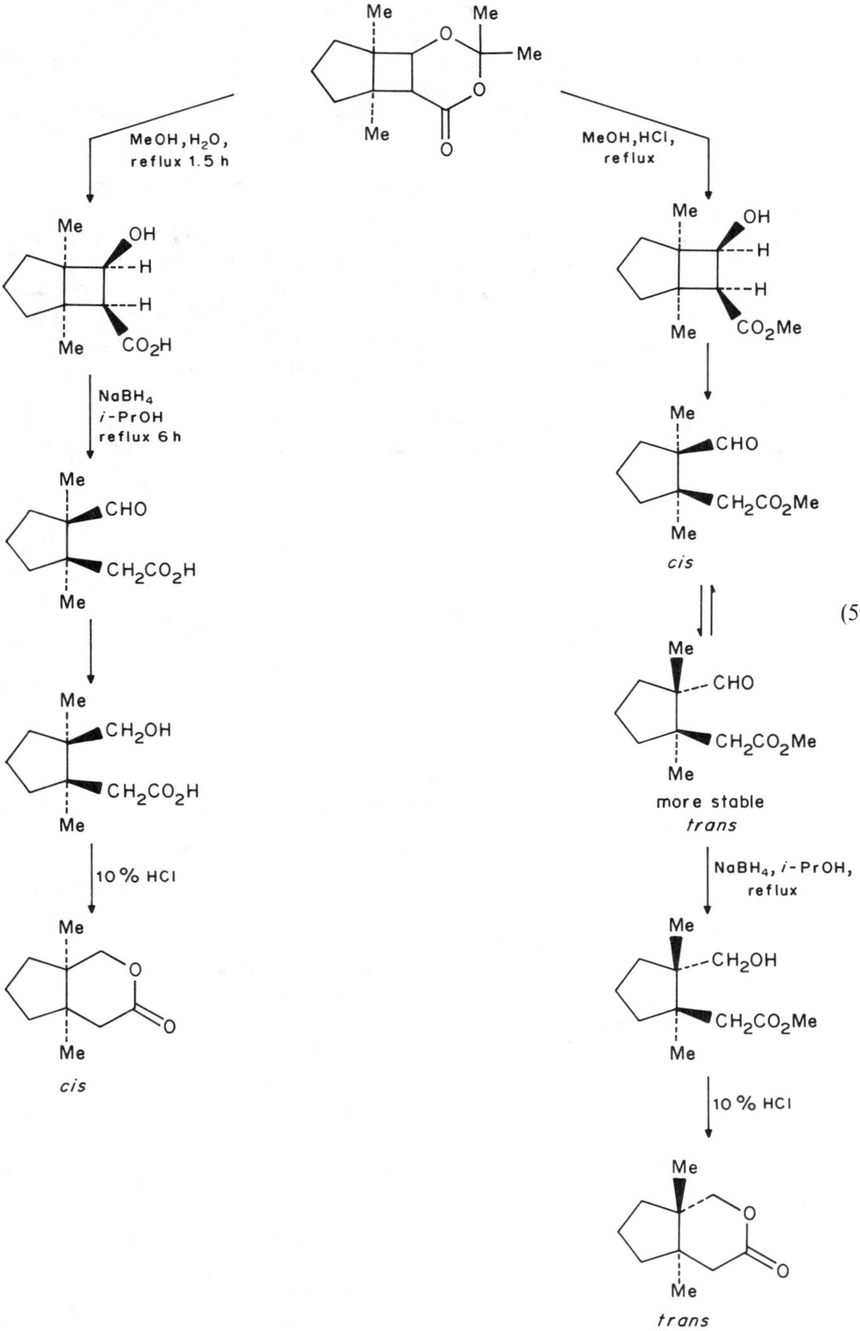

Me
Me
Me
Me
O
O
O
MeOH, H_2O,
reflux 1.5 h
MeOH, HCl,
reflux
Me
OH
H
H
Me
CO_2H
Me
OH
H
H
Me
CO_2Me
$NaBH_4$
i-PrOH
reflux 6 h
Me
CHO
CH_2CO_2H
Me
Me
CHO
CH_2CO_2Me
Me
cis
Me
CHO
CH_2CO_2Me
Me
more stable
trans
Me
CH_2OH
CH_2CO_2H
Me
$NaBH_4$, *i*-PrOH,
reflux
Me
CH_2OH
CH_2CO_2Me
Me
10% HCl
10% HCl
Me
O
O
Me
cis
Me
O
O
Me
trans

(598)

Recently, two kinds of reactions of lactone double bonds have been reported: epoxidation[786], which was accomplished using *m*-chloroperbenzoic acid (equation 594), and reduction[788] using hydrogen in the presence of platinum dioxide[788,796] (equations 595 and 596) or palladium on carbon[788] (equation 597).

Although not a lactone interconversion hydrolysis of the oxalactone shown in equation 598 can produce either *cis*- or *trans*-fused δ-lactones depending upon the reaction conditions[753].

Preparation of thionolactones from preformed lactones has been accomplished in two recent reports both of which utilize sodium hydrosulphide as the sulphur transfer agent in the reactions. In the first report *N,N*-dimethyliminolactonium fluoroborate salts, which were prepared from the starting lactones using the method of Deslongchamps and coworkers[866], were sulphydrolysed and acetylated to produce[867] the corresponding thionolactones (equation 599), while in the second report[868] *O*-alkylation of the starting lactone is accomplished using Meerwein's salt, followed by sulphydrolysis of the intermediate lactonium salt with anhydrous sodium hydrosulphide in acetonitrile at 0 °C (equation 600).

Finally, two reports have appeared, one of which achieved the conversion of a spirolactone into a fused δ-lactone[781] (equation 601), while in the second report[869]

1. Me_2NH or Me_2NAlMe
2. $SOCl_2$

Cl, CH_2, O, C—NMe_2

(599)

$Ag^+BF_4^-$

C=$\overset{+}{N}Me_2$ $^-BF_4$

1. NaSH, Me_2CO, −78 °C
2. MeCOCl, C_5H_5N, −78 °C

C=S

Lactonium salt	Thionolactones	Yield (%)
R–(5-ring)–O–C=$\overset{+}{N}Me_2$ $\bar{B}F_4$	R–(5-ring)–O–C=S	
R = H		78
R = Me		53
(6-ring) $\overset{+}{N}Me_2$ $\bar{B}F_4$	(6-ring) S	43
(7-ring) $\overset{+}{N}Me_2$ $\bar{B}F_4$	(7-ring) S	84

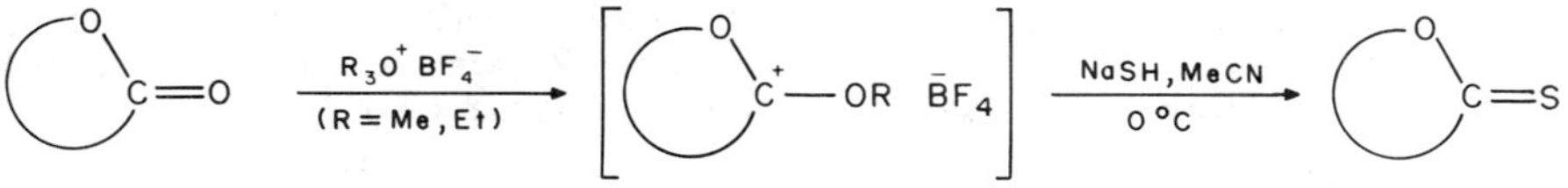

(600)

Lactone	R group in Meerwein salt	Time (h)	Thionolactone	Yield (%)
R′ = H	Me	2.5		90
R′ = Me	Et	1.5		78
R′ = H	Me	2		54
R′ = Et	Et	2.5		43
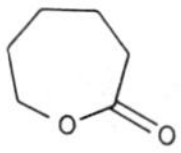	Et	2	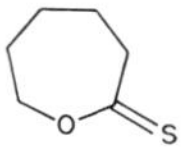	44

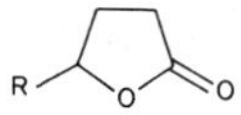

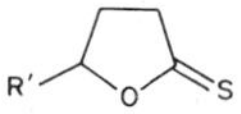

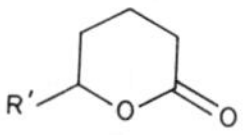

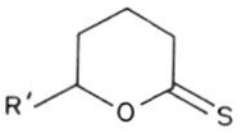

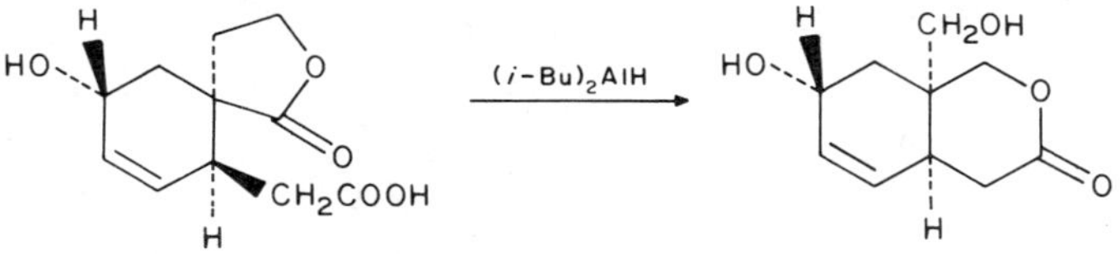

(601)

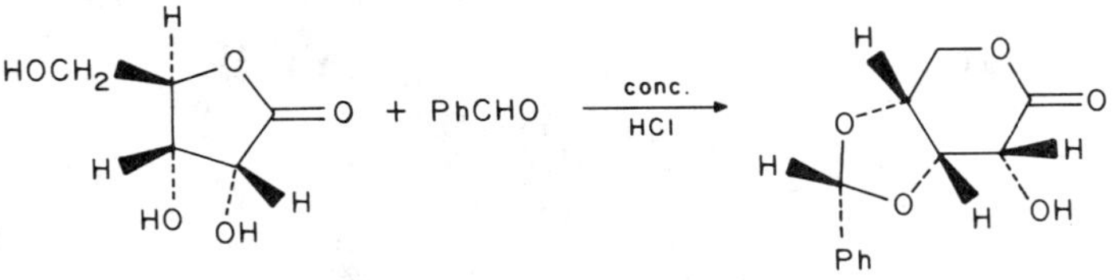

(602)

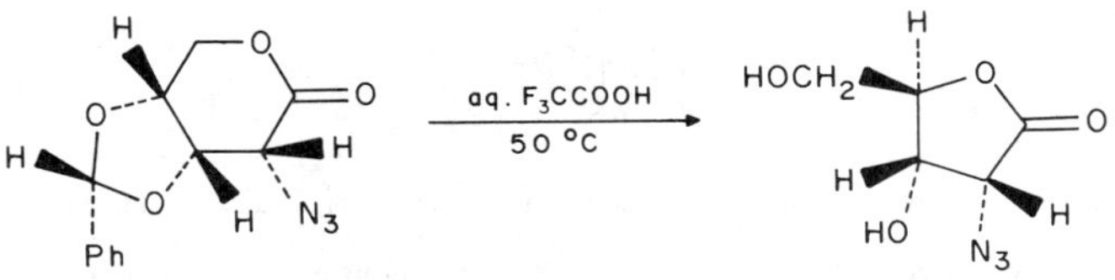

89%

(603)

94%

D-ribonolactone, a γ-lactone, was converted into Zinner's lactone, a δ-lactone (equation 602), and the azido substituted analogue of Zinner's lactone was converted into the azido substituted analogue of D-ribonolactone (equation 603).

*S. Miscellaneous Lactone Syntheses

*10. From aldehydes and ketones

Inter- and intramolecular reaction of aldehydes and ketones with mono- and dilithium salts of carboxylic acids produces lactones of varying types depending upon the structure of the substrates used.

Reaction[870] of lithium β-lithiopropionate with aldehydes or ketones offers a direct synthesis of γ-lactones of various structures in good to fair yields via a γ-hydroxy acid intermediate (equation 604).

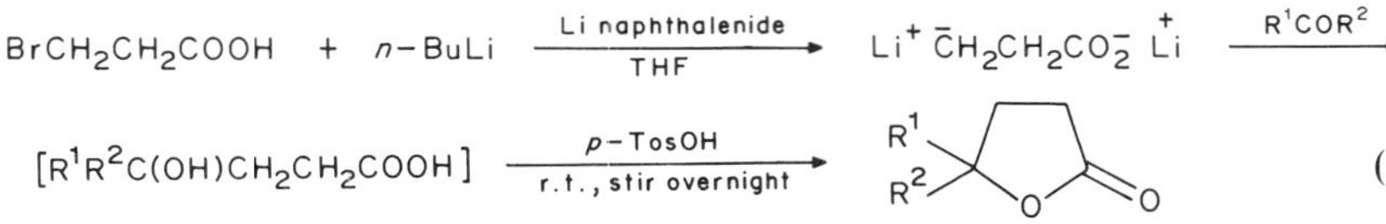

$$BrCH_2CH_2COOH + n\text{-}BuLi \xrightarrow[\text{THF}]{\text{Li naphthalenide}} Li^+ \bar{C}H_2CH_2CO_2^- \overset{+}{Li} \xrightarrow{R^1COR^2}$$

$$[R^1R^2C(OH)CH_2CH_2COOH] \xrightarrow[\text{r.t., stir overnight}]{p\text{-TosOH}} \text{(lactone: } R^1, R^2 \text{ on } \gamma\text{-carbon of } \gamma\text{-butyrolactone)} \quad (604)$$

Carbonyl compound	Lactone	Yield (%)
$Me(CH_2)_5COMe$	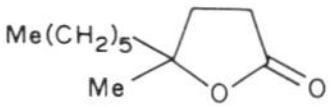 5-Me, 5-$Me(CH_2)_5$ γ-lactone	53
RCHO	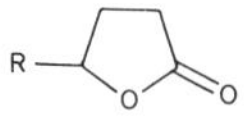 5-R γ-lactone	
R = *i*-Pr		57
R = Ph		56
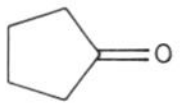 cyclopentanone	 spiro lactone	35
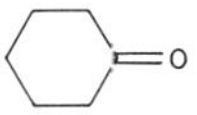 cyclohexanone	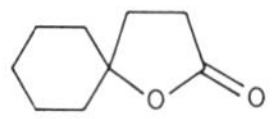 spiro lactone	51
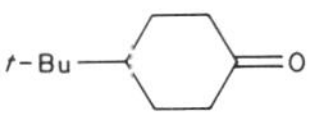 4-*t*-Bu-cyclohexanone	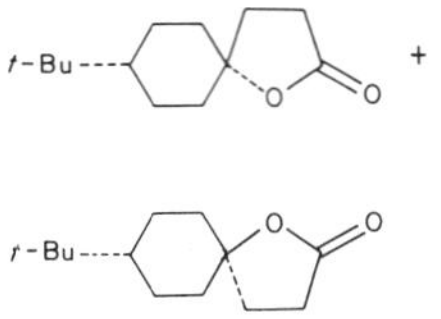 *t*-Bu spirolactones (syn + anti) (3.5 : 1)[a]	48
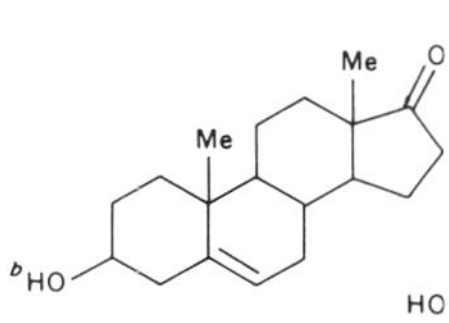 steroid 17-ketone (Me, Me, [b]HO)	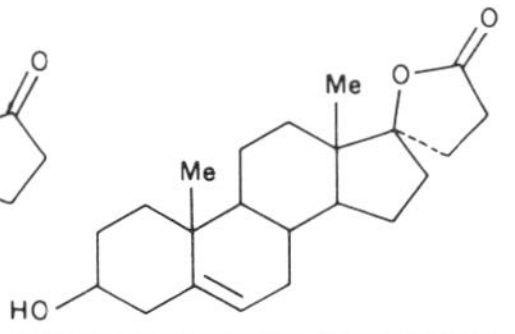 steroid 17-spirolactone (Me, Me, HO)	66

[a]A (3.5:1) ratio of *syn* to *anti* spirolactone was obtained from *equatorial* and *axial* addition, respectively, of the carbanionic reagent to the carbonyl group.

[b]The hydroxy group was protected as the 2-methoxypropanyl derivative prior to reaction.

A similar reaction[871] occurs when lithium β-lithioacrylates, prepared from Z-β-bromoacrylic acids, are allowed to react with carbonyl compounds (equation 605).

$$R^1CBr{=}CR^2CO_2H \xrightarrow[-78\,^\circ C,\,N_2,\ \text{stir } 1.5\text{ h}]{n\text{-BuLi, Et}_2\text{O,}} Li^+\,R^1\bar{C}{=}CR^2CO_2^-Li^+$$

$$Li^+\,R^1\bar{C}{=}CR^2CO_2^-Li^+ + R^3COR^4 \xrightarrow[2.\ \text{r.t., } H_2O]{1.\ \text{THF, stir, 2 h, } -78\,^\circ C}$$

$$[R^3R^4C(OH)CR^1{=}CR^2CO_2H] \longrightarrow \text{lactone } (R^1, R^2, R^3, R^4) \tag{605}$$

Carbonyl compound	R[1]	R[2]	Lactone	Yield (%)
cyclohexanone	H	H	spirolactone	48
	H	Me		71
	Me	H		57
PhCHO	H	H	Ph-lactone	49
	H	Me		70
	Me	H		52
t-Bu-cyclohexanone	H	H	t-Bu spirolactones + (58:42)[a,b]	48

[a]A (58:42) ratio of *syn* to *anti* spirolactone was obtained from *equatorial* and *axial* addition, respectively, of the carbanionic reagent to the carbonyl group.
[b]Treatment of this mixture with H_2/Pd on C affords the same mixture of saturated lactones.

Ortho lithiation of tertiary β-amino benzamides or *o*-toluamides followed by reaction with benzaldehyde or *N*,*N*-dimethylformamide and hydrolysis with aqueous acid produces[872] lactones in good yields (equations 606 and 607).

Medium- and large-ring lactones have reportedly[873] been prepared by the samarium diiodide induced cyclization of ω-(α-bromoacyloxy)aldehydes. Acetylation of the resulting lactones was accomplished using acetic anhydride and 4-dimethylaminopridine (DMAP) (equation 608).

$$\text{2-MeC}_6\text{H}_4\text{CONMe(CH}_2)_2\text{NMe}_2 \xrightarrow[\text{2. PhCHO, stir r.t., 30 min; 3. 6}N\text{ HCl; reflux 12 h}]{\text{1. LDA, }-78\,^\circ\text{C, THF, 1 h}} \text{3-phenyl-3,4-dihydroisocoumarin (53\%)} \qquad (606)$$

$$\text{PhCONEt(CH}_2)_2\text{NEt}_2 \xrightarrow[\text{2. PhCHO, stir r.t., 30 min; 3. 6}N\text{ HCl; reflux 12 h}]{\text{1. }s\text{-BuLi, TMEDA, THF, }-78\,^\circ\text{C, 1 h}} \text{3-phenylphthalide (50\%)}$$

$$\text{PhCONEt(CH}_2)_2\text{NEt}_2 \xrightarrow[\text{2. DMF; 3. 6}N\text{ HCl; reflux 48 h}]{\text{1. }s\text{-BuLi, TMEDA, THF, }-78\,^\circ\text{C, 1 h}} \text{3-hydroxyphthalide (62\%)} \qquad (607)$$

$$\text{OHC(CH}_2)_n\text{O}_2\text{CCHRBr} \xrightarrow[\text{2. (MeCO)}_2\text{O, DMAP}]{\text{1. SmI}_2\text{, THF, 0 °C, N}_2} \beta\text{-acetoxy lactone (ring O–CO–CHR–CH(OAc)–(CH}_2)_n\text{)} \qquad (608)$$

n	R	Yield[a] (%)	Isomeric ratio[b]
4	H	76	
4	Me	82	(38:62)
5	H	92	
5	Me	90	(31:69)
6	H	82	
6	Me	88 (75)[c]	(35:65)
7	H	86	
7	Me	80	(30:70)
8	H	85	
8	Me	82	(30:70)
9	H	80	
9	Me	91	(38:62)
10	H	84	

[a]Isolated yields after acetylation. Since medium-ring β-hydroxy lactones tend to be hydrolysed during work-up, the products were isolated as acetates.
[b]Configuration of the isomers was not determined.
[c]Isolated yield of β-hydroxy lactone without *O*-acetylation.

Macrocyclic lactones have also been prepared[874] by photochemical oxygenation of 4,5-diphenyl-2-hydroxyalkyloxazoles in the presence of Sensitox to produce intermediate dibenzoyltriamides, which upon treatment with *p*-toluenesulfonic acid cyclize to produce the desired macrocyclic lactones (equation 609).

$$\text{4,5-Ph}_2\text{-oxazole-2-}CH_2R \xrightarrow[\text{2. } ^1O_2, CH_2Cl_2, \text{ Sensitox, 30 min}]{\text{1. } H_3O^+} (PhCO)_2NCOCH_2R \xrightarrow[C_6H_6, \text{ reflux 24 h}]{p\text{-TosOH}} \text{macrocyclic lactone } (R^1, (CH_2)_n) \quad (609)$$

R	Lactone *n*	Lactone R^1	Yield (%)
$(CH_2)_{11}OH$	2	H	75
$(CH_2)_{12}OH$	3	H	76
$(CH_2)_{11}CH(OH)Me$	2	Me	64
$(CH_2)_5CH{=}CHCH_2CH(OH)Me$	(unsaturated macrolactone, Me)		55[a]

[a]A mixture of *E*- and *Z*-isomers.

Cathodic electroreductive hydrocoupling of unsaturated esters or acids with ketones or aldehydes in the presence of trimethylchlorosilane produces[875] a variety of γ-lactones in one step from readily available starting materials (equation 610). The reaction is performed in a cell with a lead cathode, a carbon rod anode and a ceramic diaphragm.

$$R^1R^2C{=}CR^3CO_2R^4 + R^5COR^6 \xrightarrow[\text{DMF, } Me_3SiCl, \ Et_4NOTos]{+e} \gamma\text{-lactone } (R^1, R^2, R^3, R^5, R^6) \quad (610)$$

α,β-Unsaturated ester or acid	Aldehyde or ketone	Product[a]	Yield (%)
$CH_2{=}CHCOOMe$	R^5CHO	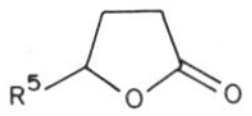	
	$R^5 = n\text{-Pr}$		73
	$R^5 = n\text{-Pent}$		86
	$R^5 = Me(CH_2)_6$		77
	$R^5 = EtCHMe$		51
$MeCH = CHCOOMe$	n-PrCHO	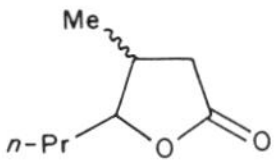	71
$CH_2 = CMeCOOMe$	n-PrCHO	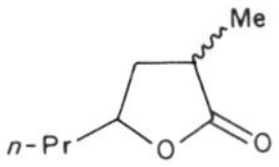	54
$CH_2{=}C(COOMe)CH_2COOMe$	n-PrCHO	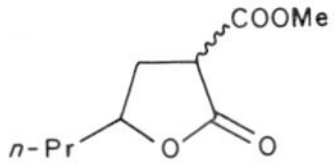	57
$MeOOCCH{=}CHCH_2COOMe$	$Me(CH_2)_7CHO$	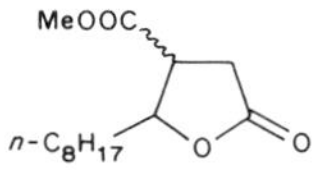	59
$CH_2{=}CHCOOMe$	R^5R^6CO	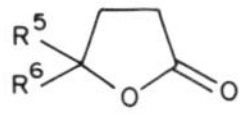	
	$R^5 = Me$, $R^6 = n\text{-Pent}$		76
	$R^5 = R^6 = Et$		67
	$R^5R^6 = {-}(CH_2)_5{-}$		78
$MeCH{=}CHCOOMe$	MeCOPent-n	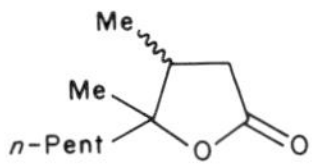	64
$CH_2{=}CMeCOOMe$	MeCOPent-n	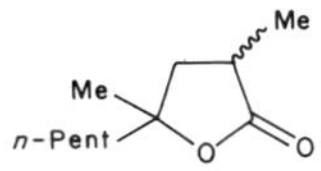	58
$CH_2{=}C(COOMe)CH_2COOMe$	MeCOPr-n	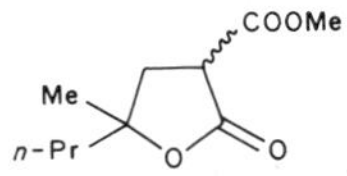	86
$CH_2{=}CHCOOH$	R^5R^6CO	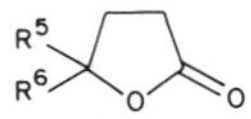	
	$R^5 = H$, $R^6 = n\text{-Pr}$		38
	$R^5 = Me$, $R^6 = n\text{-Pent}$		32

[a]Products are a mixture of stereoisomers, e.g. the products from methyl crotonate and butyraldehyde were a mixture of two stereoisomers in a 43:28 ratio though the exact assignment of configuration was not possible.

No coupled products were obtained using benzaldehyde, acetophenone or benzophenone as the carbonyl compound and using α,β-unsaturated acids instead of esters also produced lower yields of lactones. The mechanism proposed for this reaction involves initial formation of an anion radical and its reaction with either the trimethylchlorosilane activated carbonyl compound or a proton to form an anion which then reacts with the activated carbonyl compound (equation 611).

$CH_2{=}CHCOOMe$

+e

$\bar{C}H_2\dot{C}HCOOMe$

+e +H$^+$

$\bar{C}H_2CH_2COOMe$

$Cl^{\delta-}$ $^{\delta+}O$---SiMe$_3$ Me—CPent-n

Me n-Pent O SiMe$_3$ OMe =O

+e +H$^+$

Me n-Pent O SiMe$_3$ OMe =O

−Me$_3$SiOMe

Me n-Pent O =O

(611)

Using 2-acetoxyfuran as a substrate, three different methods have been used[876] to produce 4-substituted butenolides. The first method reported involves the substitution reaction of electrophiles such as aldehydes, acetals or carbonates with 2-acetoxyfuran in the presence of a Lewis acid (equation 612), the second method employed was a Friedel–Crafts-type acylation reaction (equation 613), while the third approach involved bromination followed by reaction of the intermediate 4-bromobutenolide with chrysanthemic acid (equation 614).

2-Acetoxyfuran + RCHO $\xrightarrow[\text{2. stand } -78\,^\circ\text{C, 2 h; 3. r.t.}]{\text{1. Lewis acid, } CH_2Cl_2,\ -78\,^\circ\text{C}}$ R(HO)CH-butenolide (alcohol) + R(AcO)CH-butenolide (acetate) (612)

Electrophile	Lewis acid (mmol)	Product	Yield (%)
n-PrCHO	$TiCl_4$ (4)	alcohol + acetate	70, 8
n-PrCHO	$TiCl_4$ (2)	alcohol + acetate	67, 6
n-PrCHO	$SnCl_4$ (4)	alcohol + acetate	18, 52
n-PrCHO	$BF_3 \cdot OEt_2$ (4)	alcohol + acetate	21, 55
n-C_7H_{15}CHO	$TiCl_4$ (4)	alchol	60
$CH_2(OMe)_2$	$TiCl_4$ (4)	$MeOCH_2$-butenolide	86
$CH(OMe)_3$	$TiCl_4$ (2)	$(MeO)_2CH$-butenolide	89

2-Acetoxyfuran + MeCOCl $\xrightarrow[\text{2. r.t. overnight}]{\text{1. } SnCl_4, CH_2Cl_2,\ \text{stir 35 h}, -78\,^\circ\text{C}}$ Me(AcO)C=butenolide, 50% (613)

2-Acetoxyfuran $\xrightarrow[-5 \text{ to } -10\,^\circ\text{C}]{Br_2, CCl_4}$ Br-butenolide

$\xrightarrow[\text{NaHCO}_3,\ \text{DMF}]{\text{(H)(Me}_2\text{)cyclopropane: CH=CMe}_2\text{, COOH}}$ $Me_2C{=}CH$–(2,2-dimethylcyclopropane)–CO_2–butenolide, 48% (614)

Acylations of an intramolecular type, via α-sulphinyl carbanions, have been used[863] for the preparation of α-phenylsulphinyl γ- and δ-lactones from sulphoxide carbonates (equation 615). The conversion of these products to α,β-unsaturated γ- and δ-lactones upon heating at 120 °C under reduced pressure is reported in Section *II.R.

$$MeO_2COCHR(CH_2)_2S(\rightarrow O)Ph \xrightarrow[\text{2. HOAc, stir, r.t., 1 h}]{\text{1. LDA, THF, 10 h, } -78\,^{\circ}\text{C to r.t.}} \text{5-R-3-(PhS(O))-}\gamma\text{-lactone}$$

(615)

$$MeO_2COCHR(CH_2)_3S(\rightarrow O)Ph \xrightarrow[\text{2. HOAc, stir, r.t., 1 h}]{\text{1. LDA, THF, 10 h, } -78\,^{\circ}\text{C to r.t.}} \text{6-R-3-(PhS(O))-}\delta\text{-lactone}$$

Carbonate	Product	Yield (%)
$RCH(OCO_2Me)(CH_2)_2S(\rightarrow O)Ph$	γ-lactone (R, SPh(O))	
R = H		53
R = Et		79
R = *n*-Bu		78
R = *n*-Hex		80
R = *n*-C_7H_{15}		72
R = *n*-$C_{11}H_{23}$		75
$RCH(OCO_2Me)(CH_2)_3S(\rightarrow O)Ph$	δ-lactone (R, SPh(O))	
R = H		53
R = Ph		68
R = *n*-Bu		85
R = *n*-Pent		55
R = *n*-Hex		87
R = *n*-C_7H_{15}		79
R = *n*-$C_{11}H_{23}$		83

Hydrolysis of the intermediate product formed[877] from reaction of unsaturated α-aminonitriles with ketones leads to the production of spirolactones (equation 616).

MeCH=CHCH(Z)CN —(LDA, THF, 0 °C)→ MeCH⁻CH⁻C(Z)CN + (cyclic ketone) C=O

Z = dimethylamino or piperidino group

→ Intermediate —(HCl or oxalic acid, 0 °C, 50% THF)→ spirolactone (616)

Intermediate	Product	Yield (%)
(cyclohexane) R = H		45[a]
		78[b]
R = Me		41[a]
		63[b]
(cycloheptane) R = H		53[a]
R = Me		16[a]
		50[b]
(decalin)		44[a]
		71[b]
(4-R-cyclohexane) R = *t*-Bu		43[a]
R = Ph		51[b]
(Ph, Me)		77[b]

[a] Method used was (α-aminonitrile + LDA) then ketone added.
[b] Method used was (α-aminonitrile + LDA) then (ketone + $ZnCl_2$) added.

11. From carboxylic acids

Carboxylic acids in combination with a wide variety of other reagents and catalysts have been used in the recent literature to produce lactones. Thus, reaction[878,879] of aliphatic acids containing an α-hydrogen atom with olefins in the presence of manganese, cerium or vanadium metal salts (manganese acetate dihydrate was found to be the best) produces a wide variety of substituted γ-butyrolactones, the structure of which depends upon the structure of the olefin and the aliphatic acid used (equation 617).

$$R^1R^2C{=}CR^3R^4 + R^5CH_2COOH \xrightarrow[KOAc]{Mn(OAc)_3\cdot 2H_2O} \text{lactone } (R^1, R^2, R^3, R^4, R^5) \quad (617)$$

Olefin	Acid	Lactone	Yield[a] (%)
$R^1R^2C{=}CH_2$	MeCOOH	(R^1, R^2)	
$R^1 = H, R^2 = n\text{-Hex}$			74
$R^1 = H, R^2 = n\text{-}C_8H_{17}$			66
$R^1 = H, R^2 = Ph$			60
$R^1 = H, R^2 = t\text{-Bu}$			48
$R^1 = Me, R^2 = Ph$			74
$R^1 = R^2 = Me$			30
$R^1CH{=}CHR^3$	MeCOOH	(R^3, R^1)	
$R^1 = R^3 = = n\text{-Pr}$ (*trans*)			44[b]
$R^1 = R^3 = Ph$ (*trans*)			16[c]
$R^1 = Ph, R^3 = Me$ (*trans*)			79[c]
$R^4R^5 =$ (cyclooctene)	MeCOOH		62
$R^1 = Ph, R^2 = COOMe$			45
$CH_2{=}CH(CH_2)_nCH{=}CH_2$	MeCOOH	$CH_2{=}CH(CH_2)_n$	
$n = 0$			30
$n = 2$			24
$n = 4$			26
$CH_2{=}CMeCH{=}CH_2$	MeCOOH	$CH_2{=}CHMe$—	13
		+	
		Me, $CH_2{=}CH$	37
$n\text{-PentC}{\equiv}CCH_2CH{=}CH_2$	MeCOOH	$n\text{-PentC}{\equiv}CCH_2$	50
$PhCH{=}CH_2$	R^5CH_2COOH	(R^5, Ph)	
	$R^5 = Me$		50
	$R^5 = NC$		41

(continued)

Olefin	Acid	Lactone	Yield[a] (%)
n-HexCH=CH$_2$	NCCH$_2$COOH	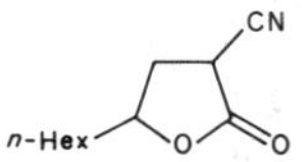 CN; n-Hex; O; O	60
PhCMe=CH$_2$	NCCH$_2$COOH	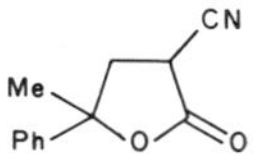 CN; Me; Ph; O; O	43
R^{1}CH=CHR^3	NCCH$_2$COOH	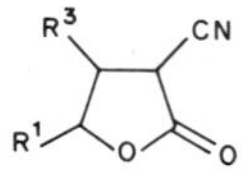 R^3; CN; R^1; O; O	
R^1 = R^3 = n-Pr			49
R^1 = Ph, R^3 = Me			51
CH$_2$=C(Me)CH=CH$_2$	NCCH$_2$COOH	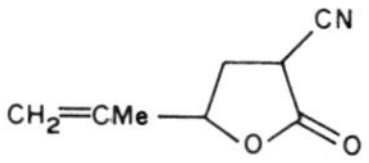 CN; CH$_2$=CMe; O; O	5
		+	
		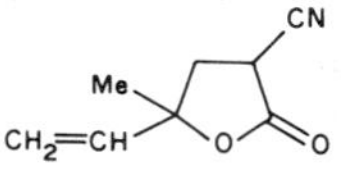 CN; Me; CH$_2$=CH; O; O	39
n-HexCH=CH$_2$	HOOC(CH$_2$)$_2$COOH	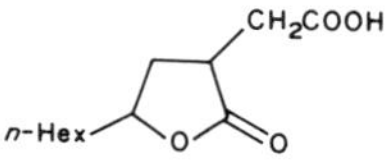 CH$_2$COOH; n-Hex; O; O	25

[a]Yield is based upon the manganese acetate dihydrate used.
[b]Two isomers in a 5:1 ratio were obtained.
[c]Only one isomer (presumably *trans*) was obtained.

The mechanism proposed for this reaction involves formation of an intermediate carboxymethyl radical which then reacts with the olefin (equation 618).

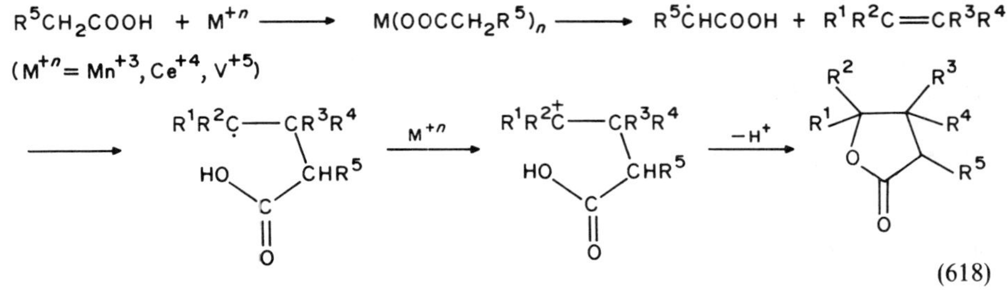

$$R^5CH_2COOH + M^{+n} \longrightarrow M(OOCCH_2R^5)_n \longrightarrow R^5\dot{C}HCOOH + R^1R^2C{=}CR^3R^4$$

$(M^{+n} = Mn^{+3}, Ce^{+4}, V^{+5})$

$$\longrightarrow R^1R^2\dot{C}\text{—}CR^3R^4(CHR^5COOH) \xrightarrow{M^{+n}} R^1R^2\overset{+}{C}\text{—}CR^3R^4(CHR^5COOH) \xrightarrow{-H^+} \text{lactone } (R^1, R^2, R^3, R^4, R^5) \tag{618}$$

When condensed with paraformaldehyde (P) or 1,3,5-trioxane (T) in boiling acetic acid containing catalytic amounts of aluminium chloride, cyclic 3,4-unsaturated carboxylic acids undergo an intramolecular cyclization to produce[880] fused 2,3-unsaturated δ-lactones (equation 619).

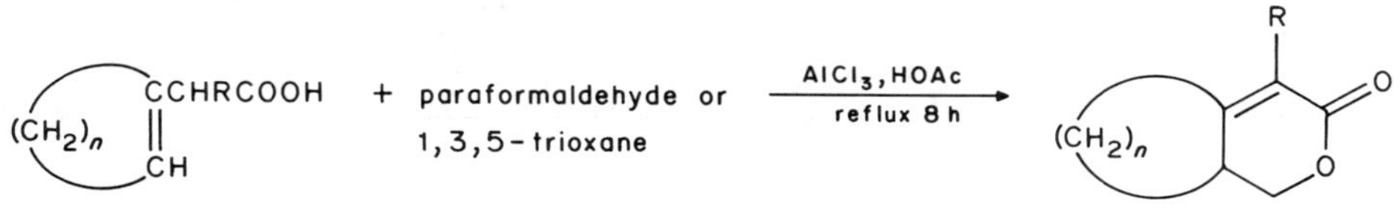

(619)

R	n	Reagent	Yield (%)
H	4	T	46
H	4	P	40
Me	4	T	76
Me	4	P	65
Et	4	T	65
Et	4	P	76
n-Pr	4	T	64
n-Pr	4	P	86
i-Pr	4	T	53
i-Pr	4	P	74
n-Bu	4	T	61
n-Bu	4	P	73
n-Hex	4	T	84
n-Hex	4	P	93
Me	10	T	96
Et	10	T	67

If lithium naphthalenide is allowed to react[881] with carboxylic acids which contain two hydrogens at the 2-position, in the presence of diethylamine, the dianion of the carboxylic acid is formed. Reaction of these dianions with epoxides produces the corresponding dianions of the 4-hydroxy carboxylic acids, which upon further reaction

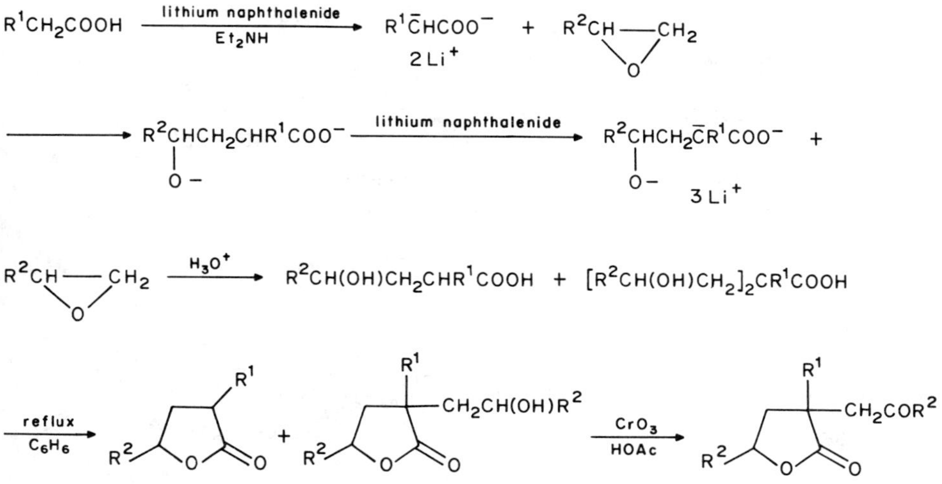

(73) **(74)** **(75)**

(620)

R^1	R^2	Yield (%) 73	74	75
Me	Me	37	16	—
Et	Me	63	21	26
n-Pr	Me	67	23	69
n-Bu	Me	66	24	68
n-Pent	Me	70	28	78
n-Hex	Me	53	32	87
i-Pr	Me	85	4	—
H	Et	37	31	59
Me	Et	49	34	78
Et	Et	62	30	77
n-Pr	Et	65	31	78
n-Bu	Et	63	26	81
n-Pent	Et	35	4	—
i-Pr	Et	89	2	—

with lithium naphthalenide form the corresponding trianion. Treatment of these trianions with additional epoxide followed by hydrolysis produces 4,4′-dihydroxycarboxylic acids, which readily cyclize to form γ-butyrolactones **74** containing hydroxy groups in their side chain (equation 620). Oxidation of these lactones using chromium trioxide in acetic acid produces γ-butyrolactones **75** containing a carbonyl group in their side chain.

β-Acyl β-ethoxycarbonyl-α-alkylpropionic acids have been used as substrates to produce[882] β-ethoxycarbonyl β,γ-unsaturated γ-lactones by dehydration using acetic anhydride (equation 621) or β-ethoxycarbonyl γ-lactones containing a β-ethylenic carbon chain branched on the γ-carbon of the lactone by reaction with allylzinc reagents (equation 622).

$R^1COCH(COOEt)CHR^2COOH$ —[$(MeCO)_2O, CH_2Cl_2$; H_3PO_4, reflux 4.5 h]→ EtOOC, R^2, R^1 substituted β,γ-unsaturated γ-lactone, 45 – 65% (621)

R^1 = Me ; Me ; Me ; *n*-Pr ; *n*-Pent ; *c*-Hex ; Ph ; *n*-Pent

R^2 = H ; Me ; Et ; H ; H ; H ; H ; Et

$R^1COCH(COOEt)CHR^2COOH + H_2C{=}CR^3CH_2ZnBr$ —[1. THF, −10 °C; 2. 20 °C, 4.5 h; 3. H_3O^+]→ γ-lactone bearing EtOOC, R^2, R^1 and $CH_2{=}CR^3CH_2$, 65 – 80% (*Z* and *E* isomers) (622)

R^1	R^2	R^3	Z/E ratio	Yield (%)
Me	Me	H	—	—
Me	Me	Me	—	—
Me	Ph	H	—	—
Me	Ph	Me	—	—
Me	H	H	30/70	80
Me	H	Me	40/80	75
Ph	H	H	50/50	65
Ph	H	Me	75/25	65

Treatment of ω-hydroxycarboxylic acids with N,N,N',N'-tetramethylchlorformamidinium chloride produces[883] macrolide lactones in yields ranging from 54 to 90% (equation 623).

$$HO(CH_2)_nCOOH + Cl(Me_2N)C{=}\overset{+}{N}Me_2\ Cl^- \xrightarrow[\text{MeCN, ether, r.t., stir}]{\text{Collidine}} (CH_2)_n\text{-lactone (O–C=O ring)} \tag{623}$$

n = 11 ; 12 ; 14
Reaction time (h) = 51 ; 47 ; 46
Yield (%) = 54 ; 68 ; 90

Condensation of carboxylic acids with o-hydroxyphenyl ketones followed by cyclization in the presence of phenyl dichlorophosphate produces[884] coumarins substituted in the 3- and/or 4-positions (equation 624).

$$o\text{-}HOC_6H_4COR^1 + R^2CH_2COOH \xrightarrow[\text{2. reflux}]{\text{1. PhOPOCl}_2\text{, NEt}_3\text{, r.t. stir, 30 min}} \text{4-}R^1\text{-3-}R^2\text{-coumarin} \tag{624}$$

R^1	R^2	Solvent	Time (h)	Yield (%)
H	Ph	CH_2Cl_2	1	95
Me	Ph	$ClCH_2CH_2Cl$	3.5	85
H	2-Thi	$ClCH_2CH_2Cl$	2	90
H	EtCOO	CH_2Cl_2	0.5	90
Me	EtCOO	$ClCH_2CH_2Cl$	2	70
H	PhO	1,2-$Cl_2C_6H_4$	2	30
H	PhO	1,2-$Cl_2C_6H_4$	2	60[a]
H	$PhCH_2$	1,2-$Cl_2C_6H_4$	2	50[a]
Me	3,4-$(MeO)_2C_6H_3$	$ClCH_2CH_2Cl$	4	85

[a]DBU was used instead of triethylamine.

Intramolecular sulphenyllactonization has been observed[885] when alkenoic acids are treated with dimethyl(methylthio)sulphonium fluoroborate and di(isopropyl)ethylamine in acetonitrile (equation 625).

$$H_2C{=}CH(CH_2)_nCOOH \xrightarrow[(i\text{-}Pr)_2NEt,\ MeCN,\ r.t.]{Me_2\overset{+}{S}SMe\ BF_4^-} \text{MeS-substituted lactone with ring }(CH_2)_n \quad (625)$$

Alkenoic acid	Time (h)	Product	Yield (%)
$CH_2{=}CH(CH_2)_2COOH$	72	$MeSCH_2$-γ-lactone	96
cyclopent-2-enyl-CH_2COOH	36	MeS bicyclic lactone	86
cyclohex-3-enyl-COOH	24	MeS bicyclic lactone	70
cyclohept-3-enyl-COOH	24	MeS bicyclic lactone	60

Reaction of carboxylic acids with trimethylsilyl chloride affords ketene bis(trimethylsilyl)-acetals (O-silylated enolate derivatives of carboxylic acids) which undergo zinc bromide catalysed alkylation with α-chloroalkyl phenyl sulphides to produce[865] phenylthio γ- and δ-lactones (equation 626).

$$R^1CH_2COOH + Me_3SiCl \longrightarrow R^1CH{=}C(OSiMe_3)_2 +$$

$$Me_3SiOCR^2R^3(CH_2)_nCH(Cl)SPh \xrightarrow[2.\ 2M\ HCl]{1.\ ZnBr_2, CH_2Cl_2,\ 20\ ^\circ C, 45\ min} \text{lactone with SPh, } R^1, (CH_2)_n, R^2, R^3 \quad (626)$$

R^1	R^2		R^3	Yield (%) $n=0$	Yield (%) $n=1$
Me	H		H	88	88
Et	H		H	84	93
Me	Me		H	81	90
Et	Me		H	88	90
Me		—$(CH_2)_5$—		89	—
Me	Me		n-Hex	75	—

The reaction may also be used to prepare[865] *trans*-fused ring and *trans*-fused ring spiro δ-lactones (equation 627), and spiro γ-lactones (equation 628). The conversion of all these phenylthio products to their α,β-unsaturated analogues has been discussed in Section *II.R.

$$+ \ RCH{=}C(OSiMe_3)_2 \xrightarrow[\text{20 °C, 45 min; 2. 2 M HCl}]{\text{1. } ZnBr_2, CH_2Cl_2}$$

R = Me, 92%
R = Et, 85%

(627)

$$+ \ C(OSiMe_3)_2 \xrightarrow[\text{20 °C, 45 min; 2. 2M HCl}]{\text{1. } ZnBr_2, CH_2Cl_2}$$

71%

$$C(OSiMe_3)_2 + n\text{-HexCMe(OSiMe}_3\text{)CHSPh(Cl)} \xrightarrow[\text{20 °C, 45 min; 2. 2M HCl}]{\text{1. } ZnBr_2, CH_2Cl_2,}$$

PhS, n-Hex, Me

85%

(628)

Reactive (procedure A) or unreactive (procedure B) unsaturated dicarboxylic acids have been bislactonized by reaction[886] with lead tetraacetate or by reaction (procedure C) of their tetra(n-butyl)ammonium diacid salts with lead tetraacetate (equation 629). Although the reaction of the disalt (procedure C) is slower than the bislactonization of the free diacid (procedure A), the yields obtained using procedure C are in general higher. The mechanism proposed for this conversion involves initial Pb(IV) induced plumbolactonization followed by either an S_N2 or an S_N1 displacement of lead, in the Pb(III) monolactone intermediate, by oxygen leading to the bislactone final product.

$$\begin{matrix}\text{diacid}\\ \text{or salt}\end{matrix} + Ph(OAc)_4 \xrightarrow[\text{Procedure A, B or C}]{\text{solvent, temp}} \text{bislactone} \qquad (629)$$

Procedure A: free diacid + $CHCl_3$ + $Ph(OAc)_4$ at 20–50 °C
Procedure B: free diacid + MeCN + excess $Pb(OAc)_4$ at 80 °C
Procedure C: $(n\text{-Bu})_4N^+$ salts of diacid + MeCN + excess $Ph(OAc)_4$ at 75–80 °C

Substrate	Procedure and conditions	Product	Yield (%)
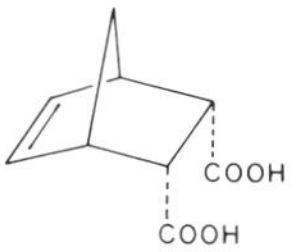	A, 30h, 23 °C C, 6h, 75 °C	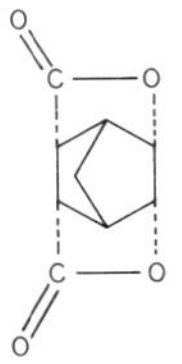	99+ 99+
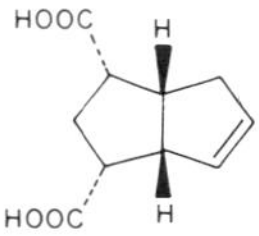	A, 48h, 50 °C C, 48h, 75 °C	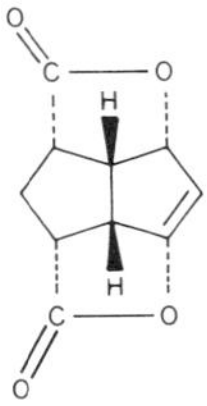	68 85
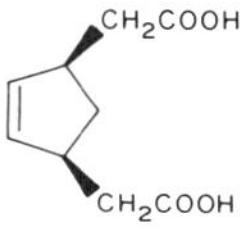	A, 72h, 23 °C C, 30h, 75 °C	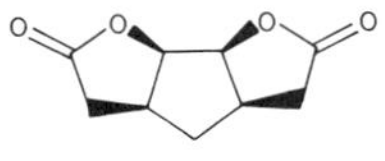	71 89
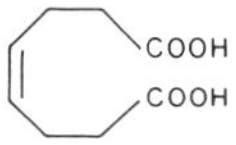	A, 60h, 23 °C[a] C, 26h, 75 °C[a]	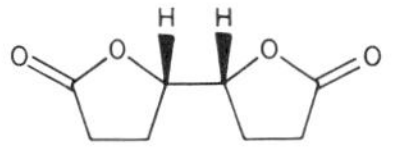	78 98
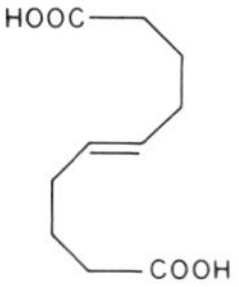	B, 50h, 80 °C[b] C, 48h, 80 °C[c]	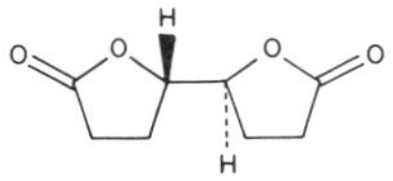	73 86

[a]This reaction afforded *meso*-bislactone stereospecifically.
[b]Product was a 1:1 mixture of *meso*- and *d,l*-bislactone.
[c]Product was a *d,l*-lactone with >20:1 stereoselectivity.

*12. From carboxylic acid esters

Base catalysed cyclization[887] of cyclopentanone-3-methylene methyl carbonate produces the corresponding *cis*-fused ring ketolactone shown in equation (630) used as a synthon in the preparation of dihydro-, dehydro- and jasmone.

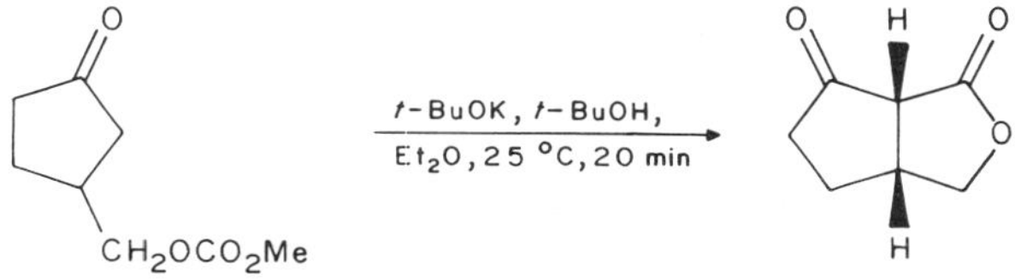

$$\xrightarrow[\text{Et}_2\text{O, 25 °C, 20 min}]{t\text{-BuOK, } t\text{-BuOH,}} \qquad (630)$$

CH_2OCO_2Me

N-Bromosuccinimide oxidation of dimethyl 2,6-dimethyl-4-(*o*-tolyl)-3,5-piperidine dicarboxylate converts one ester group to the vinylogous carbamate, which is then brominated on the C-2 methyl group to produce the corresponding bromomethyl α,β-unsaturated diester, and finally, under the conditions of the reaction, cyclizes to the unsaturated fused-ring piperidinolactone product[888] (equation 631).

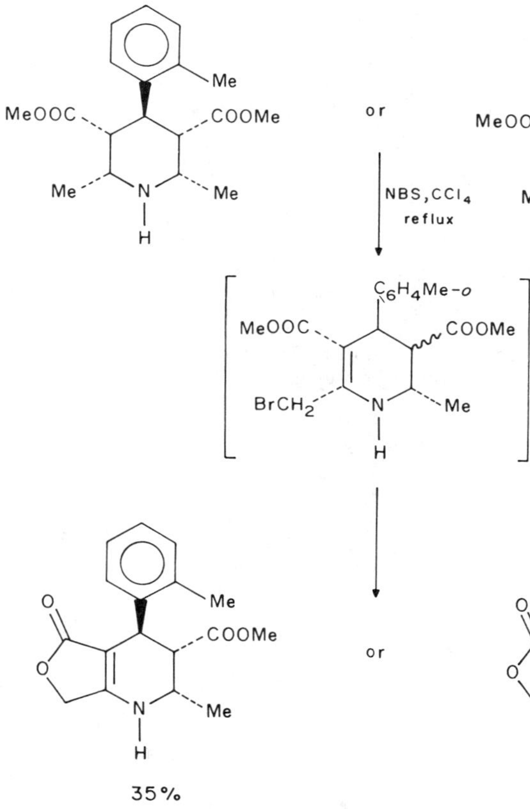

MeOOC, COOMe, Me, Me, Me, N–H or MeOOC, COOMe, Me, Me, Me, N–H

NBS, CCl_4, reflux

[C_6H_4Me-*o*, MeOOC, COOMe, $BrCH_2$, N–H, Me]

(631)

35% or 20%

Treatment of cyclopropyl esters with trimethylsilyl iodide causes ring opening to produce the corresponding β-iodocarboxylic acids, which upon treatment with base cyclizes to produce[889] the corresponding γ-butyrolactones (equation 632) in excellent overall yields. Hydrogen chloride may also be used[889] to produce lactones from cyclopropyl esters but in one step (equation 633), and this hydrogen chloride catalysed lactone formation allows the reversal of the regiochemistry of the lactone product obtained using trimethylsilyl iodide (equation 634).

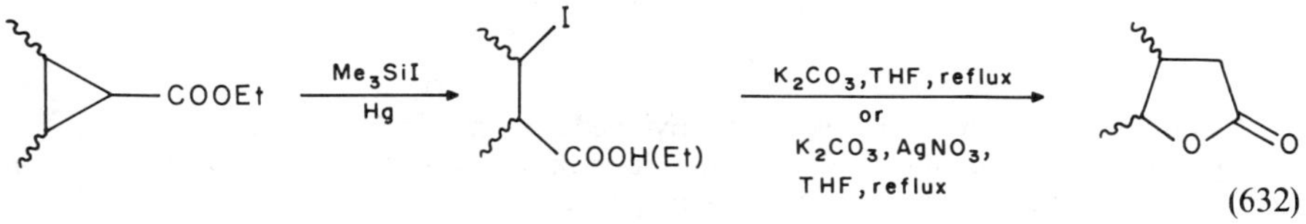

COOEt $\xrightarrow[\text{Hg}]{\text{Me}_3\text{SiI}}$ I, COOH(Et) $\xrightarrow{K_2CO_3, \text{THF, reflux or } K_2CO_3, AgNO_3, \text{THF, reflux}}$ O, O

(632)

Cyclopropyl ester	% Yield iodo compound	Lactone	% Yield lactone[a]
R = H	89		72[b]
R = n-Bu	100		89
	96		82
R = H	100		86
R = Me	74		89[b]

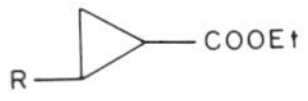

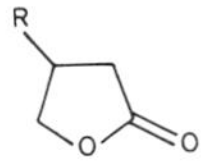

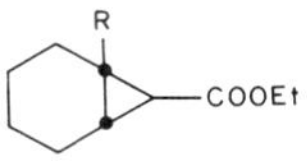

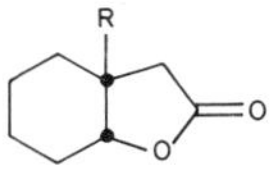

[a]Overall yield consisting of TMSI opening (using 2-6 eq. of TMSI for 2–4 days) of the cyclopropyl ester followed by ring closure using K_2CO_3 in refluxing THF for 1–4 days.
[b]$AgNO_3$ plus K_2CO_3 for 1–5 days to effect ring closure.

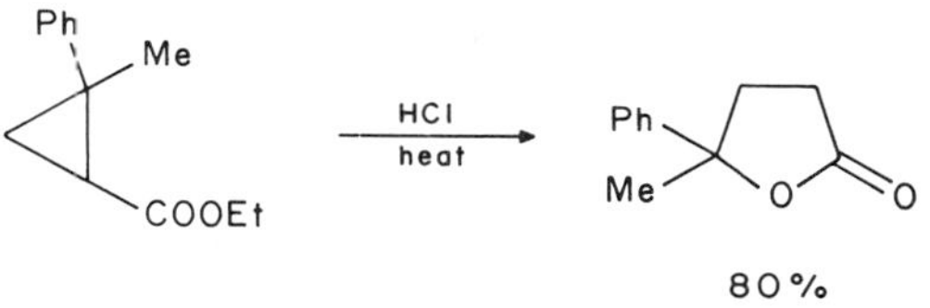

HCl, heat; 80% (633)

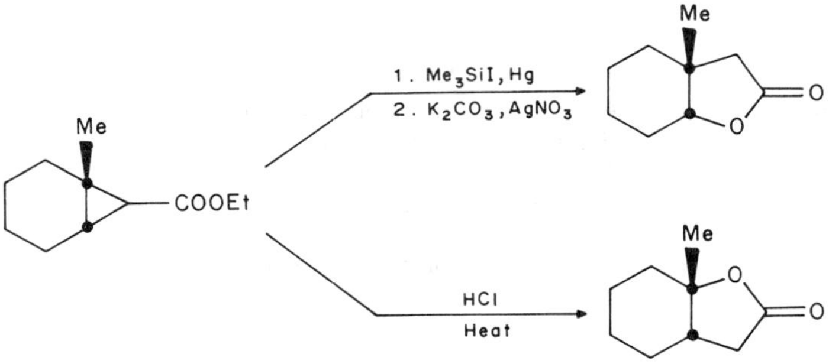

1. Me_3SiI, Hg; 2. K_2CO_3, $AgNO_3$; HCl, Heat (634)

Reaction of lithium enolates of methyl acetoacetate (equation 635) or methyl 3-oxohexadecanoate (equation 636) with a complex cationic enol ether–iron reagent produced[890] 3,4-disubstituted α-methylene γ-lactones.

$[Cp(CO)_2Fe{-}C(=CH_2)(OEt)COOEt]^+ BF_4^-$ (Fp) + $MeC(OLi){=}CHCOOMe$ —(1. THF, −78 °C, 0.5 h; 2. additional iron complex, 2 h)→ $[FpCH_2C(COOEt)(OEt){-}CH(COMe)COOMe]$ —(L selectride, ether, stir 2 h, −78 °C)→ lactone (EtO, $FpCH_2$, MeOOC, Me) 85% —(CH_2Cl_2, −78 °C, HBF_4, Et_2O, 0.5 h)→ [$H_2C{=}$(Fp) lactone, MeOOC, Me] —(Me_2CO, H_2O, r.t., 0.5 h)→ α-methylene lactone *trans* + *cis*

9:1(77%) (635)

$[Cp(CO)_2Fe{-}C(=CH_2)(OEt)COOEt]^+ BF_4^-$ (Fp) + $n{-}C_{13}H_{27}C(OLi){=}CHCOOMe$ —(toluene, −78 °C)→ $[FpCH_2C(COOEt)(OEt){-}CH(COC_{13}H_{27}{-}n)COOMe]$ 33%

—(LAH, Et_2O, −78 °C)→ lactone (EtO, $FpCH_2$, MeOOC, $C_{13}H_{27}{-}n$) *trans*: *cis* = 2:1, 47%

—(1. CH_2Cl_2, − 78 °C, HBF_4, Et_2O; 2. Me_2CO, H_2O)→ α-methylene lactone (MeOOC, $C_{13}H_{27}-n$)

trans only
(*d,l*-protolichesterinic ester)
95% (636)

A one-step conversion of *para*-substituted phenols into γ-lactones has been achieved[891] by zinc chloride catalysed alkylation with methyl 2-chloro-2-(butylthio- or phenylthio-)-propionate or acetate followed by desulphurization (equation 637).

$$\text{phenol} + \underset{\displaystyle\text{COOMe}}{\text{R}^1\text{SCR}^2\text{Cl}} \xrightarrow[\text{anh ZnCl}_2\text{, r.t., 1 h}]{\text{CH}_2\text{Cl}_2\text{, MeNO}_2\text{(1:1)}} \text{fused-ring } \gamma\text{-lactones} \qquad (637)$$

$R^1 = n$-Bu or Ph
$R^2 =$ H or Me

Phenol	Chloro-thio substrate	Product	Yield (%)
2-Me-3-(CH_2CH_2COOR)-phenol (HO, Me, CH_2CH_2COOR)	Me(Cl)C(n-BuS)(COOMe)	lactone with Me, CH_2CH_2COOR, Me, SBu-*n*	
R = H			44
R = Me			70
HO-C$_6$H$_4$-CH_2CH_2COOMe	*n*-BuSCH(Cl)COOMe	lactone with CH_2CH_2COOMe, SBu-*n*	58
HO, Me, CH_2CH_2COOMe	*n*-BuSCH(Cl)COOMe	lactone with Me, CH_2CH_2COOMe, SBu-*n* +	—
		lactone with *n*-BuS, Me, CH_2CH_2COOMe	—
HO-C$_6$H$_4$-Me	Me(Cl)C(PhS)(COOMe)	lactone with Me, PhS, Me	—
2-naphthol (OH)	Me(Cl)C(PhS)(COOMe)	naphthofuranone with PhS, Me	—

If the α-chlorosulphide functions are part of the substrate which also contains an ester group and an allylsilanyl moiety, an intramolecular condensation affording medium-sized lactones can be induced upon treatment[892] with a Lewis acid (equation 638).

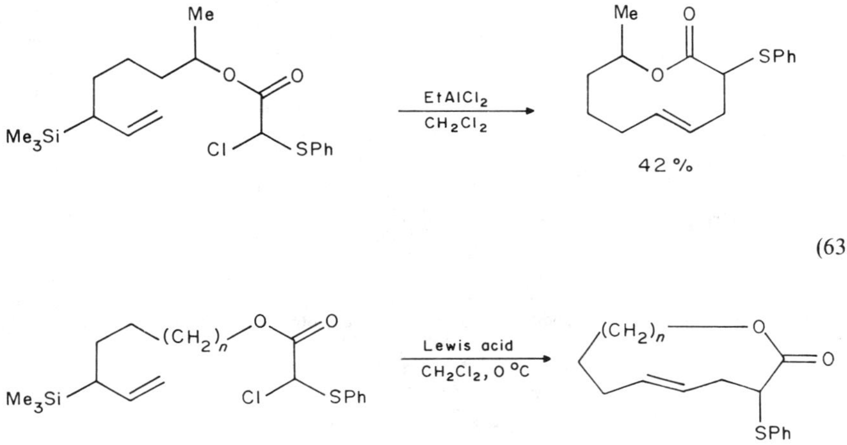

(638)

n	Lewis acid	Yield (%)
0	$SnCl_4$	18
0	$TiCl_4$	22
0	$ZnCl_2$	23
0	Et_2AlCl	33
0	$EtAlCl_2$	34
1	$EtAlCl_2$	48
2	$EtAlCl_2$	48
3	$EtAlCl_2$	55

A very similar intramolecular condensation has also been reported[893] using ω-haloalkyl 2-(phenylthiomethyl)benzoates upon treatment with the potassium salt of hexamethyldisilazane in tetrahydrofuran (equation 639).

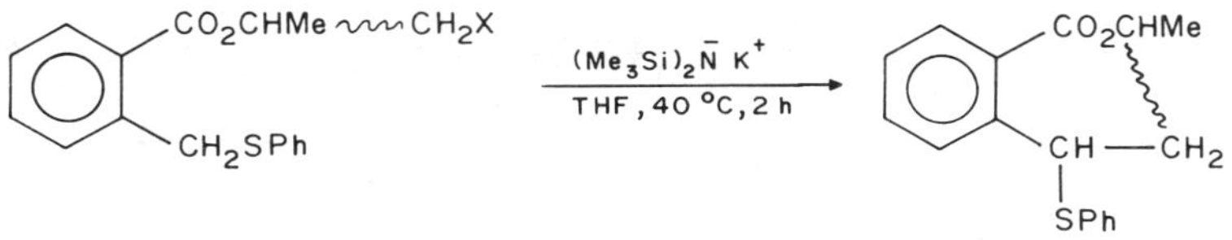

(639)

Substrate	Product	Yield (%)
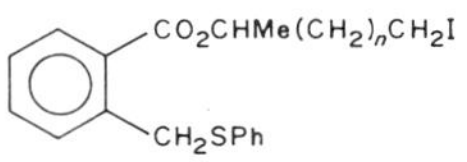 $C_6H_4(CO_2CHMe(CH_2)_nCH_2I)(CH_2SPh)$	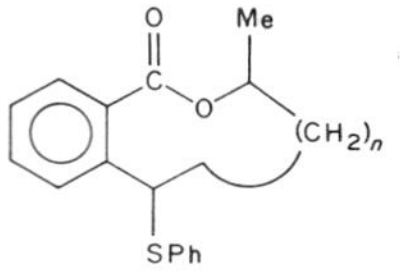	
$n = 7$		75[a]
$n = 8$		71
$n = 5$		N.R.
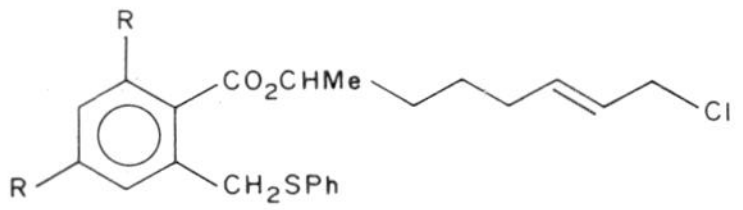	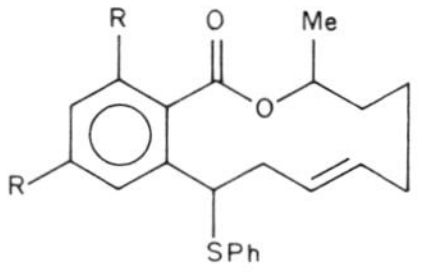	
R = H		41
R = MeO		40[b]

[a]Upon treatment of this product with $NaIO_4$, a quantitative yield is obtained of unsaturated lactone

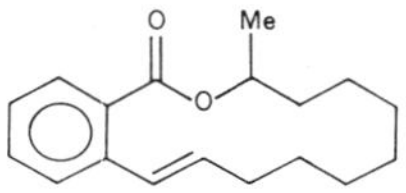

[b]Reaction of this product with Raney nickel in refluxing ethanol removes the phenylthio group and reduces the double bond to afford a 70% yield of the methoxy derivative of lasiodiplodin

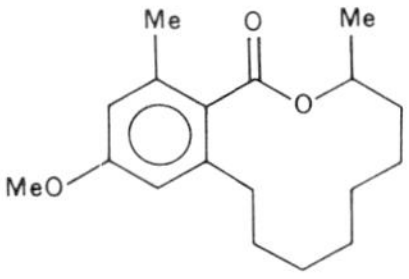

*13. *From acid halides*

By cyclocondensation of 2,2-dimethyl-2-sila-1,3-dithiacyclopentane with various diacid halides, macrocyclic tetrathiolactones can be prepared[894] albeit in low yields (equation 640).

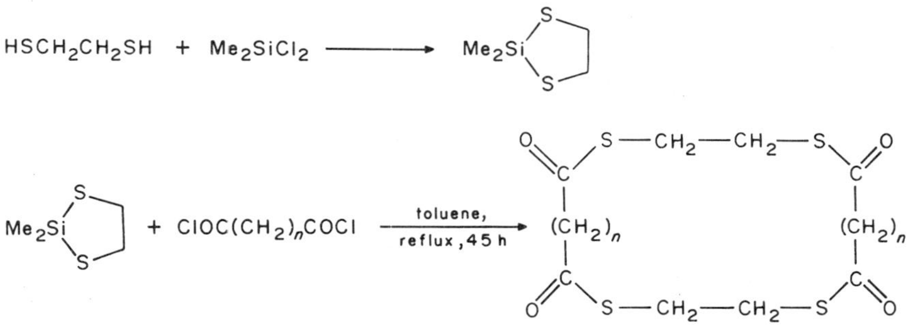

$$HSCH_2CH_2SH + Me_2SiCl_2 \longrightarrow Me_2Si(SCH_2CH_2S)$$

$$Me_2Si(SCH_2CH_2S) + ClOC(CH_2)_nCOCl \xrightarrow[\text{reflux, 45 h}]{\text{toluene,}} \text{cyclo}[-SCH_2CH_2S-CO(CH_2)_nCO-SCH_2CH_2S-CO(CH_2)_nCO-]$$

$n = 3;\ 5;\ 7;\ 8$

Yield (%) = 38; 9; 14; 18 (640)

Similar cyclocondensation of aromatic dicarboxylic acid chlorides with glycols or thioglycols affords[895] novel crown ether-lactones and -thiolactones but also in low yields (equation 641).

$$C_6H_4(COCl)_2 + HM(CH_2)_n[M(CH_2)_n]_mMH \xrightarrow[\text{heat}]{C_6H_6} C_6H_4\left(COM(CH_2)_n \ldots COM\left[(CH_2)_n\,M\right]_m\right) \quad (641)$$

$m = O, S$

Diacid chloride	Glycol or thioglycol	Product	Yield (%)
2-R-1,3-$C_6H_3(COCl)_2$	$HO(CH_2)_2[O(CH_2)_2]_3OH$	2-R-1,3-$C_6H_3[CO_2(CH_2)_2O(CH_2)_2]_2O$ (macrocycle)	
R = H			31
R = NO_2			26
1,4-$C_6H_4(COCl)_2$	$HO(CH_2)_2[O(CH_2)_2]_3OH$	1,4-$C_6H_4[CO_2(CH_2)_2O(CH_2)_2]_2O$ (macrocycle)	23
2,6-pyridine-$(COCl)_2$	$HO(CH_2)_2[O(CH_2)_2]_3OH$	2,6-pyridine-$[CO_2(CH_2)_2O(CH_2)_2]_2O$ (macrocycle)	70
2,6-pyridine-$(COCl)_2$	$HO(CH_2)_{10}OH$	bis(pyridine) macrocycle with two $CO_2(CH_2)_{10}O_2C$ bridges	48
2,6-pyridine-$(COCl)_2$	$HS(CH_2)_2[O(CH_2)_2]_3SH$	2,6-pyridine-$[COS(CH_2)_2O(CH_2)_2]_2O$ (macrocycle)	46
2,6-pyridine-$(COCl)_2$	$HS(CH_2)_2[S(CH_2)_2]_2SH$	2,6-pyridine-$[COS(CH_2)_2S]_2(CH_2)_2$ (macrocycle)	<10

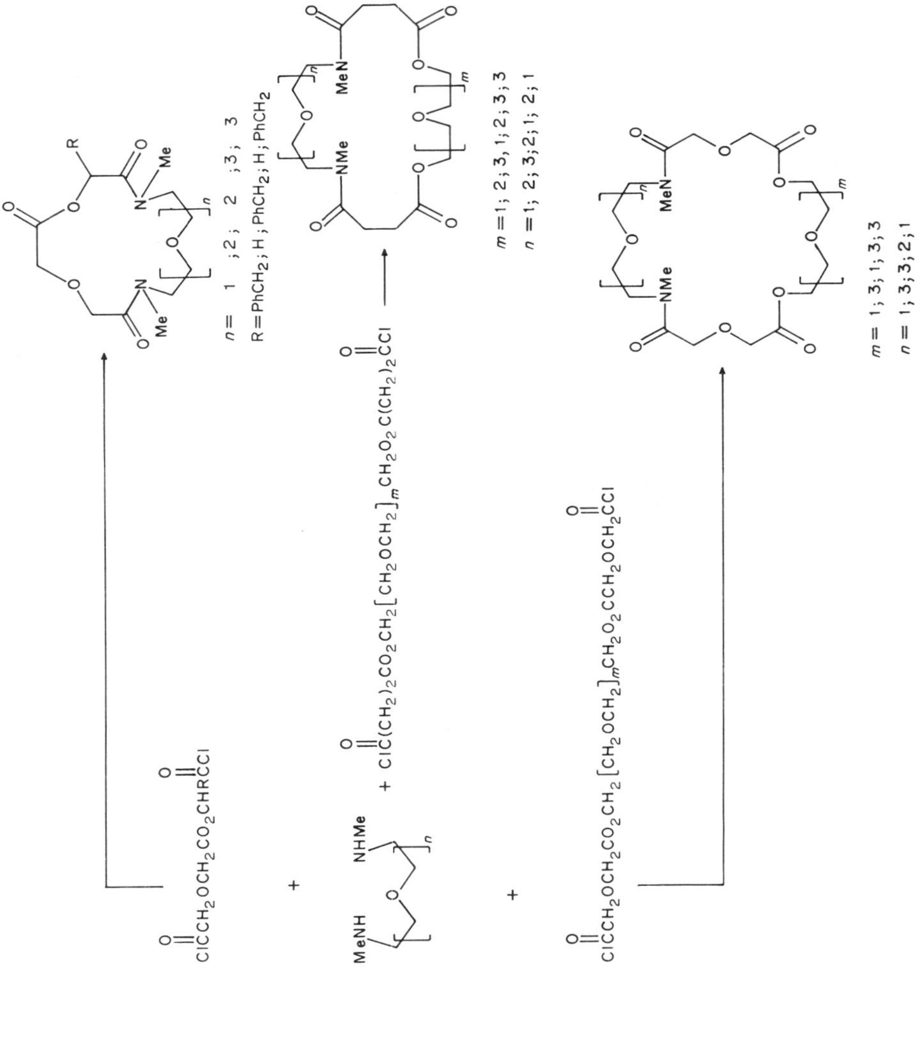

$ClCOCH_2OCH_2CO_2CHRCOCl$ + MeNH–$[CH_2CH_2OCH_2CH_2]_n$–NHMe → (cyclic diamide-diester)

$n = 1;\ 2;\ 2;\ 3;\ 3$

$R = PhCH_2;\ H;\ PhCH_2;\ H;\ PhCH_2$

MeNH–$[CH_2CH_2OCH_2CH_2]_n$–NHMe + $ClCO(CH_2)_2CO_2CH_2[CH_2OCH_2]_mCH_2O_2C(CH_2)_2COCl$ → (cyclic diamide-diester)

$m = 1;\ 2;\ 3,\ 1;\ 2;\ 3;\ 3$

$n = 1;\ 2;\ 3;\ 2;\ 1;\ 2;\ 1$

MeNH–$[CH_2CH_2OCH_2CH_2]_n$–NHMe + $ClCOCH_2OCH_2CO_2CH_2[CH_2OCH_2]_mCH_2O_2CCH_2OCH_2COCl$ → (cyclic diamide-diester)

$m = 1;\ 3;\ 1;\ 3;\ 3$

$n = 1;\ 3;\ 3;\ 2;\ 1$

(642)

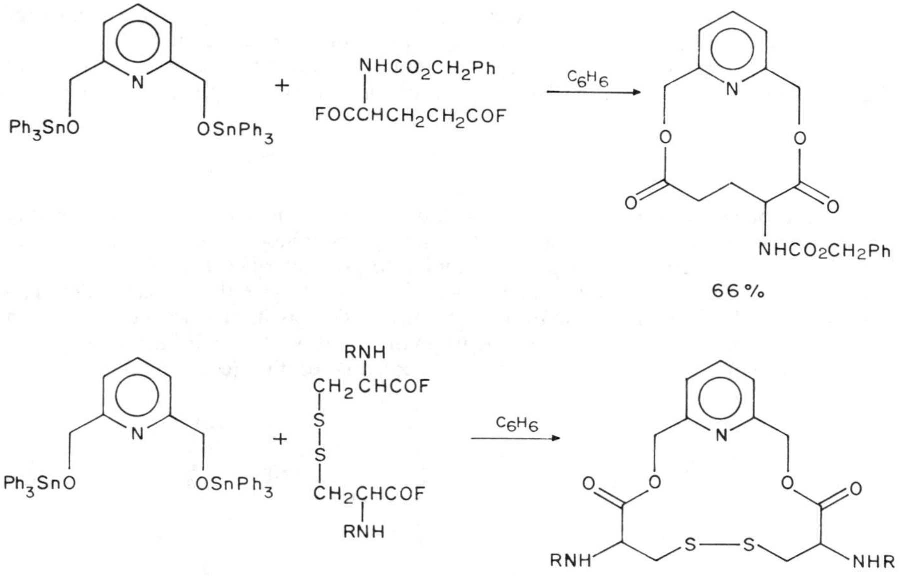

Ph_3SnO ... $OSnPh_3$ + $FOCCH(NHCO_2CH_2Ph)CH_2CH_2COF$ $\xrightarrow{C_6H_6}$ ... $NHCO_2CH_2Ph$

66%

Ph_3SnO ... $OSnPh_3$ + $(RNH)CH(COF)CH_2$–S–S–$CH_2CH(COF)(RNH)$ $\xrightarrow{C_6H_6}$ RNH ... S—S ... NHR

$R = CO_2CH_2Ph$; $CO_2CH_2C_6H_4NO_2$-*p*

Yield(%) = 57 ; 48

(643)

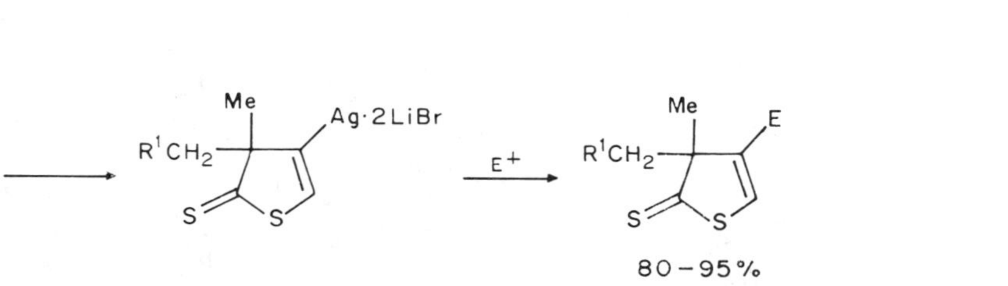

$H_2C{=}C(Me)C{\equiv}CH$ + $R^1Ag\cdot 2LiBr$ ⟶

$R^1CH_2(Me)C{=}C{=}C(H)Ag\cdot 2LiBr$ + CS_2 $\xrightarrow[\text{2. stir 1h, 25°C}]{\text{1. THF—HMPA, 1h, −50°C}}$ $[R^1CH_2(Me)C(CSSAg\cdot 2LiBr)—C{\equiv}CH]$

⟶ (Me, R^1CH_2, $Ag\cdot 2LiBr$, S, S) $\xrightarrow{E^+}$ (Me, R^1CH_2, E, S, S)

80–95%

R^1 = *i*-Pr; *n*-Bu; *t*-Bu; *t*-Bu

E = H ; H ; H ; D

(644)

Macrocyclic lactonolactams have been reportedly prepared[896] by the cyclization of the appropriate oxa diamines and dicarboxylic acid chlorides (equation 642).

Chiral macrocyclic dilactones containing L-glutamic acid or L-cystine moieties have also been synthesized[897] in good yields by a tin 'template-driven' process (equation 643).

*14. *From miscellaneous reagents*

1,4-Addition of alkylsilver(I) reagents to conjugated enynes produces the corresponding allenylsilver(I) products which react with carbon disulphide in a mixture of tetrahydrofuran and hexamethylphosphoric triamide to produce silver(I) 3-alkylnedithioate intermediates. These intermediates spontaneously cyclize to give the corresponding β,γ-unsaturated γ-dithiolactones (equation 644). This method is limited to the production of dithiolactones which bear two substituents on the γ-carbon. Allenyllithium compounds may also be employed[398] as starting materials for this reaction (equation 645).

i-PrCH$_2$(Me)C=C=C(H)Li → (1. CS$_2$; 2. AgBr·2LiBr; 3. H$_3$O$^+$) → i-PrCH$_2$(Me)-dithiolactone, 80% (645)

(646): NaH, DME, reflux 0.5–1 h; (CH$_2$)$_n$, (CH$_2$)$_3$OH, NO$_2$

n = 4 ; 5 ; 6 ; 10

Yield (%) = 92 ; 90 ; 95 ; 91

(647): Et$_3$N, MeCN, H$_2$O; (NH$_4$)$_2$[Ce(NO$_3$)$_6$]; reflux 2 days; (CH$_2$)$_n$, NO$_2$

n = 4 ; 5 ; 6 ; 10

Yield (%) = 78 ; 76 ; 81 ; 75

Reductive isomerization of 2-(3-hydroxypropyl)-2-nitrocycloalkanones in refluxing dimethoxyethane containing sodium hydride causes ring expansion and production[899] of macrocyclic nitro lactones (equation 646). Oxidation[899] of these lactones in refluxing acetonitrile containing triethylamine and ceric ammonium nitrate can be accomplished in two days to produce the corresponding keto lactones (equation 647).

Using a hetero-Diels–Alder reaction between aldehydes and 1,1-dimethoxy-3-trimethylsiloxy-1,3-butadiene produces[900] 6-substituted-2-methoxy-5,6-dihydro-γ-pyrones, which can be subsequently hydrolysed with dilute hychochloric acid in refluxing benzene to afford the corresponding substituted 3-oxo-δ-lactones in high yields (equation 648).

OMe OMe Me$_3$SiO CH$_2$ + RCHO —Eu(fod)$_3$, CH$_2$Cl$_2$, N$_2$, stir r.t., 48 h→ MeO OMe O Me$_3$SiO R

→ OMe O O R —dil HCl, C$_6$H$_6$, reflux→ O O O R (648)

R	Yield (%) Pyrone	Lactone
COOMe	73[a]	78
Ph	85	95
i-Pr	69	89
CH=CHMe (*trans*)	70	93
CMe=CHCl (*trans*)	87	90

[a]NaEu(fod)$_3$ catalyst used.

Most examples of the conversion of lactams into lactones which have appeared recently have had an epoxide ring as a reactive function. Thus, treatment[800,901] of lactams containing an epoxide substituent in the 3-position with methanesulphonic acid in benzene under reflux afforded 2-(1-anilinomethyl)but-2-enolides (equation 649).

R^{2}CH—CR^1 O R^3 O N Ph —MeSO$_3$H, C$_6$H$_6$, reflux 1.5 h→ R^1 CHR^{3}NHPh R^2 O O (649)

R^1	R^2	R^3	Product	Yield (%)	Reference
Me	H	H	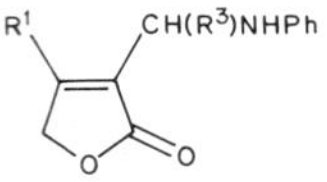 R[1], CH(R[3])NHPh	65	800
Me	H	*n*-Pr		53	901
Ph	H	H		70	800
Ph	H	Me		54	901
—$(CH_2)_3$—		H	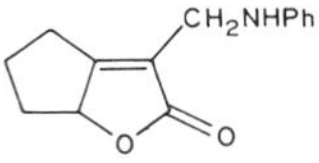 CH_2NHPh	25	800
—$(CH_2)_4$—		H	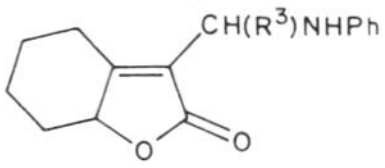 CH(R[3])NHPh	75	800
—$(CH_2)_4$—		Me		62	901
—$(CH_2)_4$—		*n*-Pr		55	901
—$(CH_2)_2CHMeCH_2$—		H	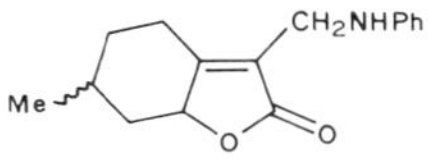 Me, CH_2NHPh	78	800
—$(CH_2)_5$—		H	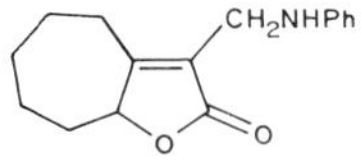 CH_2NHPh	75	800
Ph	Me	H	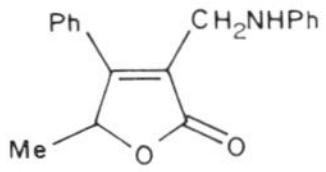 Ph, CH_2NHPh, Me	45	800

H_2N Ph N O + O —HOAc, 0 °C, stir 3 h→ H N Ph N H O O 60 % (650)

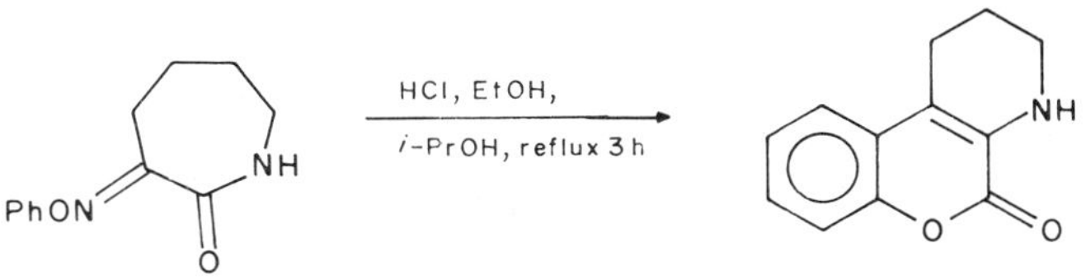

PhON NH O —HCl, EtOH, *i*-PrOH, reflux 3 h→ NH O O 69 % (651)

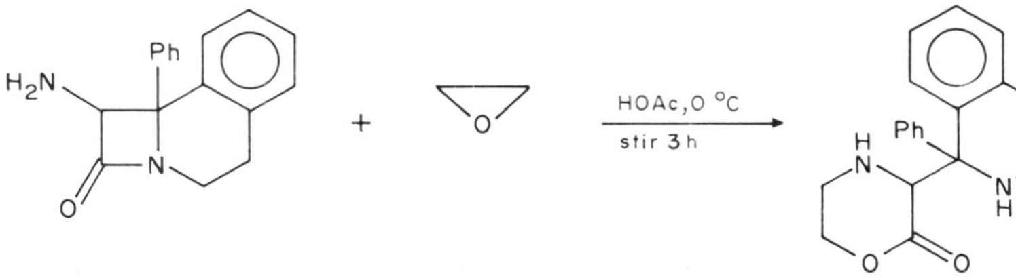

Another way in which an epoxide ring interacted with a lactam to produce a lactone is illustrated[902] in equation (650).

A rather interesting lactam-to-lactone conversion has been reported[903] when α-oxocaprolactam *O*-phenyloxime is treated with hydrochloric acid (equation 651).

*III. SYNTHESIS OF LACTAMS

Since 1976 a large number of both general and specific review articles concerning lactams have been published.

By far the largest number of review articles published dealing with lactams have been concerned with their synthesis. Topics reviewed include: strategy and design in synthesis[904], the synthesis of the β-lactam function[905–908], stereochemical study of lactams[909], the stereoselective synthesis of β-lactams[910], stereospecific construction of chiral β-lactams[911], saturated heterocyclic ring synthesis[912], the synthesis of α-methylene lactams[913], manipulation and transformation of penicillins to azetidinones[914,915], the synthesis of medium ring and macrocyclic lactams[725], synthesis of novel bicyclic β-lactam derivatives[916], synthesis and study of the structure of cyclopeptides and cyclic dilactams[917], prominent aspects of electroorganic synthesis in β-lactam chemistry[918], new reagents and methods for the synthesis of peptides, β-lactams and oligonucleotides[919], syntheses and uses of azetidinium salts[920], stereocontrolled annelation of imines to enantiomeric β-lactams[921], convenient stereoselective synthesis of isocephalosporins from threonine[922], β-lactam synthesis from 2-pyridone photoisomers[923] and the synthesis of novel fused β-lactams by intramolecular 1,3-dipolar cycloadditions[924].

One subset of the published reviews concerned with the synthesis of lactams is composed of those articles which deal with the preparation of antibiotics which contain β-lactam functions. In addition to the articles on the development of β-lactams as antibiotics[925–931], their chemistry[932–941] and history[942] as well as similar publications reviewing cephalosporins, penicillins and other β-lactams[943–947] which have appeared in print since 1976, several more specific review articles have been published. These articles discuss: β-lactamase and the new β-lactams[948], lactam analogues of cardioactive steroid lactones[949], aminoglycoside and the β-lactam antibiotics[950], the stereospecific[951] and enantioselective synthesis[952] of β-lactam antibiotics, the synthesis of 4-mercapto-azetidinones and their application to the preparation of β-lactam antibiotics[953,954] synthetic studies related to oral β-lactam antibiotics[955], some synthetic approaches to analogues of the β-lactam antibiotics[956], hydroxamate approach to the synthesis of β-lactam antibiotics[957], slightly water-soluble salts of β-lactam antibiotics[958], β-lactam antibiotics by fermentation and synthesis[959,960], general and stereocontrolled synthesis of carbapenem antibiotics[961], partial synthesis of nuclear analogues of cephalosporins[962], total synthesis of penicillins, cephalosporins and their nuclear analogues[963] and the total synthesis of penem FCE 22101[964].

Recent review articles which deal with the chemistry and reactions of lactams have included: an application of β-lactams to the syntheses of heterocyclic compounds[965], sulphur-containing heterocycles in new synthetic strategies for chiral β-lactam antibiotics[966], novel synthetic approaches to the biologically active heterocycles[967], expeditious synthesis of azetidinone-1-sulphonates and selected alternatively-activated analogues[968], chemistry of monobactams in comparison with cephalosporins[969], recent developments in the chemistry of β-lactam antibiotics[970–977], studies related to β-lactam compounds[978,979], structure, mechanism of action and therapeutic potential of monobactam antibiotics[980,981], the chemistry of δ-enaminolactams[982], chemical modifications of the β-lactam ring of penicillins and cephalosporins[983], advances in the chemistry of acetals of acid amides and lactams[984] and α-methoxylation of β-lactams by electrooxidation[985].

Only one review article[986] has appeared which deals with the biosynthesis of lactams and it discusses the industrial production, biosynthesis, bacterial enzymic inactivation, structure–activity relationships and allergic reactions to the β-lactam antibiotics.

Since β-lactam antibiotics comprise one of the more important classes of lactams, it is not surprising to find that most of the review articles published since 1976 which deal with naturally occurring lactams are concerned with the antibiotic β-lactams. The topics reviewed include: naturally occurring β-lactams[987], new naturally occurring β-lactam antibiotics and related compounds[988], β-lactam antibiotics, other sulphur-containing natural products and related compounds[989] and synthesis in the field of natural products of biological relevance[990].

Finally, several review articles dealing with the polymerization of lactams have appeared in press since 1976. These articles discuss: the thermodynamics and kinetics of lactam polymerization[991], the polymerizability of lactams[992], the mechanism of lactam polymerization[993], the capacity of lactams for polymerization in relation to their structure[994], recent progress in the polymerization of lactams[995] and promoters of the anionic polymerization of lactams[996].

1. Nomenclature

The nomenclature of lactams has been reviewed[997], and a new nomenclature has been proposed[998] which reportedly provides a convenient stereo-description of the diverse types of fused β-lactams which have appeared in the literature.

*A. By Ring-closure Reactions (Chemical)

*1. From amino acids and related compounds

Organic molecules containing a carboxylic acid or ester function and an appropriately positioned amino group have been treated with a variety of reagents to cause intramolecular cyclization resulting in lactam formation.

(76) $\xrightarrow[12-15\,h]{C_5H_5N}$ (78)

(77) $\xrightarrow[12-15\,h]{C_5H_5N}$ (79)

(652)

Bases are one variety of reagents which have been used to effect the above reaction, as examplified by the formation[999] of the bicyclic lactams **78** and **79** from the substitued thiazolidines **76** and **77** upon treatment with pyridine (equation 652).

Treatment of N^{α}-(*t*-butoxycarbonyl)-N^{δ}-(carboxymethyl)ornithine with dimethylformamide produces[1000] (*S*)-3-[(*t*-butoxycarbonyl)amino]-2-oxo-1-piperidineacetic acid (equation 653, $n = 1$). A similar reaction occurs with the homologous methyl ester using triethylamine (equation 653).

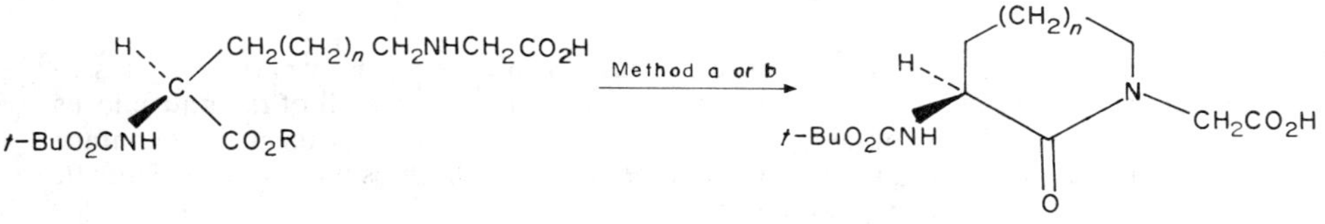

t-BuO$_2$CNH–C(H)(CO$_2$R)–CH$_2$(CH$_2$)$_n$CH$_2$NHCH$_2$CO$_2$H —Method a or b→ t-BuO$_2$CNH-substituted lactam, ring with (CH$_2$)$_n$, N–CH$_2$CO$_2$H, C=O

Method a: DMF, stir 55°C, 2 h;
b: MeCN, NEt_3, reflux 3 days

R = H ; Me
n = 1 ; 2
Yield (%) = 94; 25
Method = a ; b

(653)

Another reagent used to catalyse the cyclization of amino acids is methanesulphonyl chloride. When used as a condensing agent in the presence of a phase transfer catalyst and a base, this reagent effects cyclization of a variety of N-substituted amino acids to the corresponding N-substituted β-lactams (equation 654).

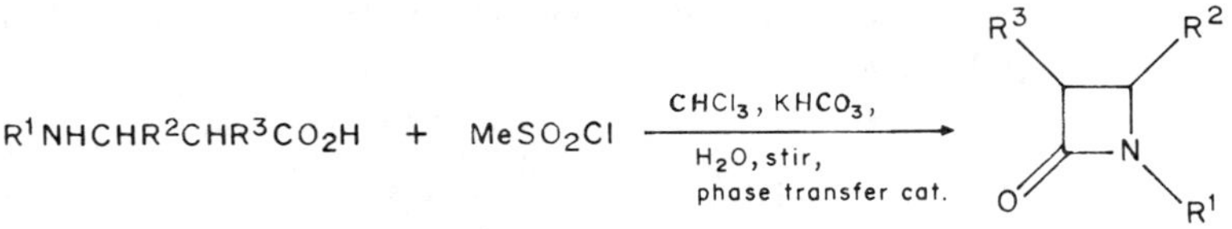

$$R^1NHCHR^2CHR^3CO_2H + MeSO_2Cl \xrightarrow[\text{H}_2\text{O, stir, phase transfer cat.}]{\text{CHCl}_3\text{, KHCO}_3\text{,}} \text{N-}R^1\text{ }\beta\text{-lactam (}R^2, R^3\text{)} \qquad (654)$$

R^1	R^2	R^3	Cat[a]	Configuration C-4	Reference
H	Me	*c*-Hex	A	—	1001
H	Me	$PhCH_2$	A	—	1001
H	Me	*n*-Hex	A	—	1001
H	*n*-Pr	$PhCH_2$	A	—	1001
H	Ph	$PhCH_2$	A	—	1001
H	MeO_2C	$PhCH_2$	A	—	1001
Me	H	$PhCH_2$	A	—	1001
PhCHMe	Me	H	B	*R* & *S*	1002
PhCHMe	*n*-Pr	H	B	*R*	1002
PhCHMe	*i*-Pr	H	B	*S*	1002
PhCHMe	Ph	H	B	*S*	1002
PhCHMe	$CH_2CH_2OCH_2OMe$	H	B	*R* & *S*	1002
PhCHMe	CH_2CO_2Me	H	B	*R*	1002

[a] A = $Et_4\overset{+}{N}\ HSO_4^-$; B = $(n\text{-Bu})_4\overset{+}{N}\ HSO_4^-$

Organometallic mediated cyclization of N-substituted amino acid esters has been employed to prepare a variety of β-lactams. The general reaction describing this approach is illustrated in equation 655, while the specific details are reported in Table 41.

$$R^1NHCHR^2CR^3R^4COOR^5 \xrightarrow[\text{reagent conditions}]{\text{organometallic}} \text{(azetidin-2-one: } R^3, R^4 \text{ at C-3; } R^2 \text{ at C-4; } R^1 \text{ on N)} \qquad (655)$$

The recent literature shows a marked increase in the use of phosphorus reagents to effect cyclization of amino acids to lactams. Interestingly, almost all of the phosphorus reagents employed either contain, or are used in conjunction with, reagents that contain, sulphur atoms. Thus, treatment of ω-amino acids with phosphorus pentasulphide produces[1010] thiolactams (equation 656).

$$H_2NCHR^2CHR^1CH_2COOH \xrightarrow[\text{reflux}]{P_2S_5} \text{(pyrrolidine-2-thione: } R^1 \text{ at C-4, } R^2 \text{ at C-5, NH)}$$

$R^1 = H,\ p\text{-}ClC_6H_4$

$R^2 = H,\ Me$ (656)

$$H_2N(CH_2)_4COOH \xrightarrow[\text{reflux}]{P_2S_5} \text{(piperidine-2-thione, NH)}$$

Treatment of 3-(*N*-benzylamino)propionic acid with phenyl[bis(2-oxo-3-thiazolidinyl)]-phosphine oxide in the presence of triethylamine affords[1011] a good yield of *N*-benzyl-2-azetidinone (equation 657).

$$PhCH_2NHCH_2CH_2COOH + \text{(2-oxo-3-thiazolidinyl)}_2P(O)Ph \xrightarrow[\text{reflux 1 h}]{NEt_3,\ MeCN} \text{N-}CH_2Ph\text{-azetidin-2-one}$$

73 %

(657)

Reaction of ω-aminocarboxylic acids with *o*-nitrophenyl thiocyanate and tri-*n*-butyl-phosphine in dimethylformamide causes intramolecular cyclization and formation of 5,6- or 7-membered lactams depending upon the structure of the amino acid employed (equation 658).[1012] The mechanism[1012] of this conversion involves reaction of the *o*-nitrophenyl thiocyanate with the phosphine to produce a thiaphosphonium cyanide, which then reacts with the amino acid to produce an acyl intermediate. This intermediate finally undergoes intramolecular cyclization, with elimination of tri-*n*-butylphosphine oxide, producing the lactam (equation 659).

TABLE 41. Organometallic mediated cyclization of amino acid esters

Starting Material	Organometallic Reagent	Conditions	Product	Yield (%)	Reference
Me_3SiNH–C(Ph)(H)–$CH_2COOSiMe_3$	EtMgBr	1. Et_2O, r.t., stir 3 h 2. stand overnight	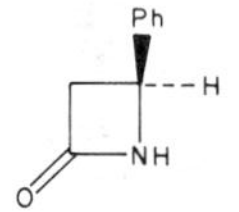	39	1003
$H_2NCHMeCMe_2COOMe$	EtMgBr	THF, N_2, 0°C, stir 3 h	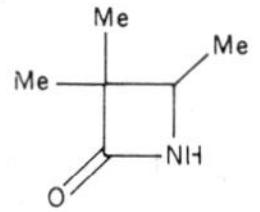	87	1004, 1005
$H_2NCHMeCH(n\text{-Hex})COOMe$	EtMgBr	THF, N_2, 0°C, stir 2 h	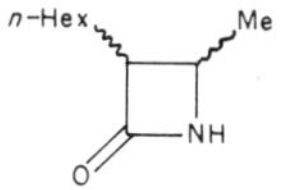	38[a]	1004, 1005
	PhMgBr	THF, N_2, 0°C, stir 1 h		51[a]	1004, 1005
$MeNHCH(R)CMe_2COOMe$	EtMgBr	THF, N_2, 0°C, stir 2 h	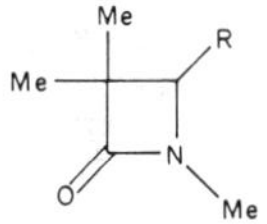		
R = H				87	1004, 1005
R = Me	EtMgBr	THF, N_2 0°C to r.t., stir 15 h		93	1004, 1005
R = Me	EtMgBr	Et_2O, N_2, 0° to 5°C, stir 4 h		92	1004, 1005
$i\text{-BuNHCH}(i\text{-Pr})CMe_2COOMe$	EtMgBr	THF, N_2, 0°C to r.t., stir 24 h	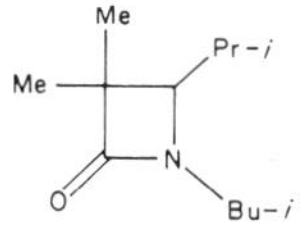	42	1004, 1005

(*continued*)

TABLE 41. (*continued*)

Starting Material	Organometallic Reagent	Conditions	Product	Yield (%)	Reference
$PhNHCH(R)CH_2COOEt$			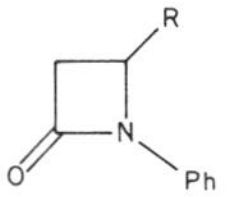 R, N, Ph, O		
R = Me	EtMgBr	Et_2O, THF, r.t., stir 12 h		61	901
R = *n*-Pr	EtMgBr	Et_2O, THF, r.t., stir 12 h		60	901
$PhCH_2NHCH(CH_2CH_2OCH_2Ph)CH(CH(OH)Me)COOMe$	RMgBr[b]	—[b]	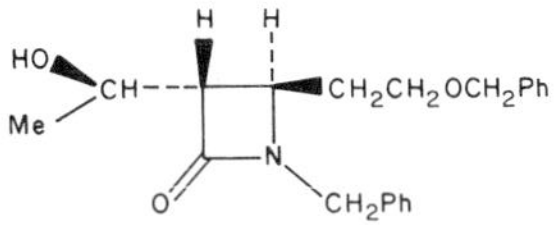 HO, Me, CH, H, H, $CH_2CH_2OCH_2Ph$, N, CH_2Ph, O	—	1006
$PhCH_2ONHCH_2CMe_2COOMe$	$(Me_3Si)_2NLi$	THF, −78°C, stir 1 h	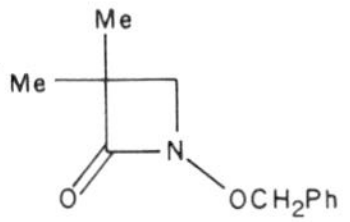 Me, Me, N, OCH_2Ph, O	76	1007
$PhCH_2ONHCHMeCH_2COOMe$	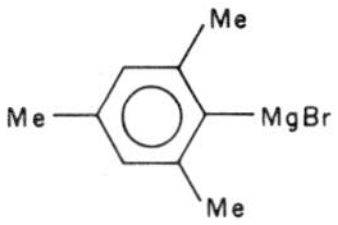 Me, Me, MgBr, Me (mesityl)	THF, 0°C, stir 1 h	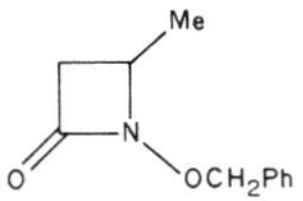 Me, N, OCH_2Ph, O	28	1007
$MeNHCHPhCH(CH(Me)OSiMe_3)COOMe$	EtMgBr	—[c]	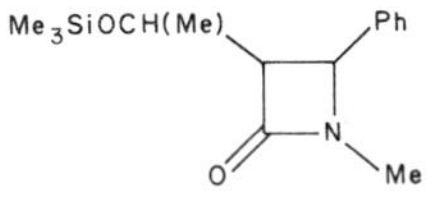 $Me_3SiOCH(Me)$, Ph, N, Me, O	—	1008
$(t\text{-Bu})Me_2SiOCH(CH_2OCOCH_3)CH_2CH(OH)CH(CH(NHPh)C_6H_4OMe\text{-}p)COOMe$	*t*-BuMgCl	THF, −10°C, r.t., stir overnight	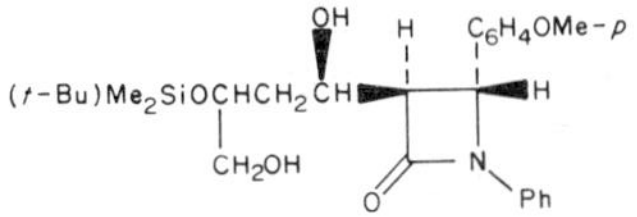 $(t\text{-Bu})Me_2SiOCH(CH_2OH)CH_2CH(OH)$, H, $C_6H_4OMe\text{-}p$, H, N, Ph, O	35	1009

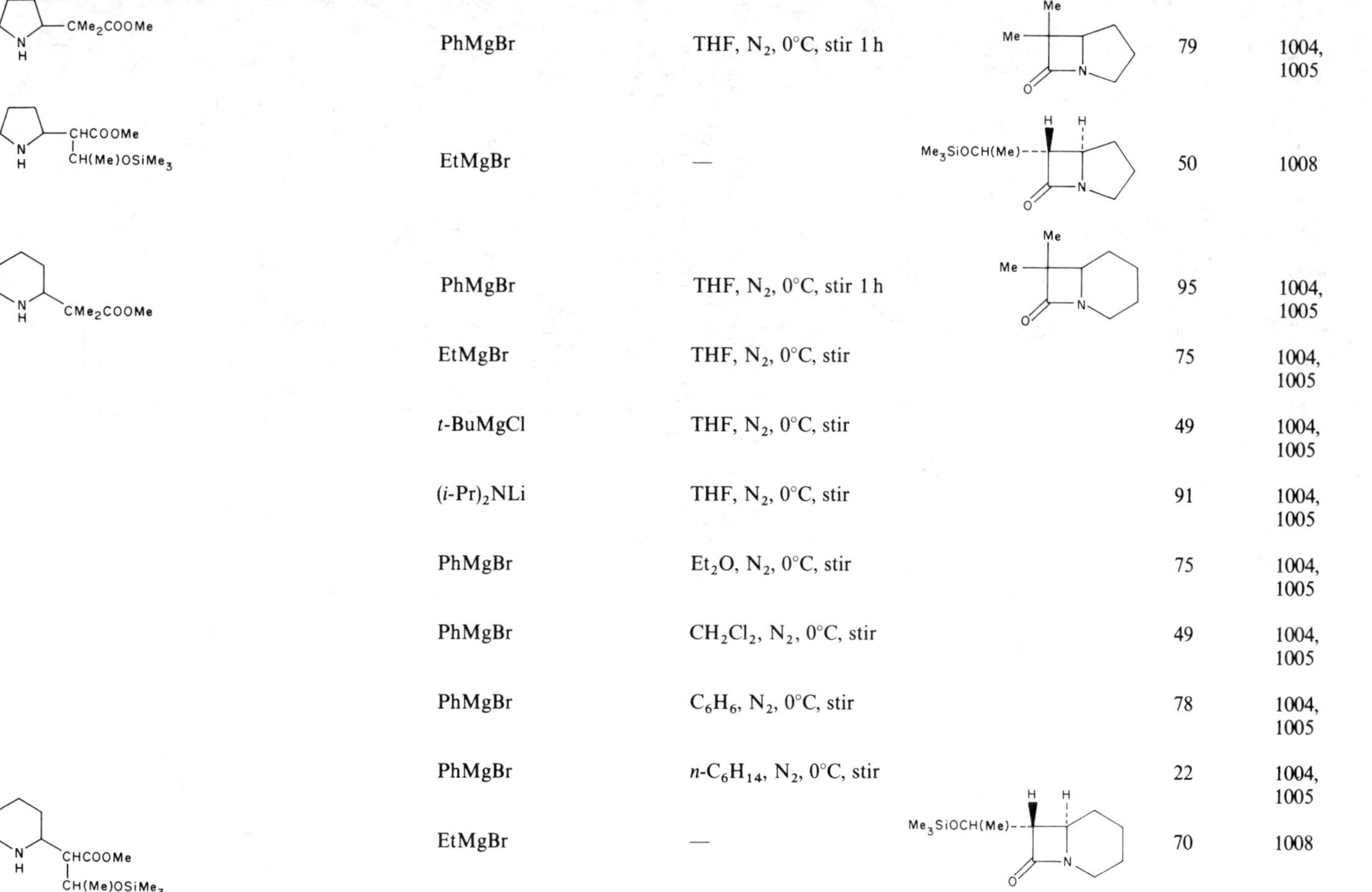

Substrate	Reagent	Conditions	Product	Yield (%)	Ref.
	PhMgBr	THF, N_2, 0°C, stir 1 h		79	1004, 1005
	EtMgBr	—		50	1008
	PhMgBr	THF, N_2, 0°C, stir 1 h		95	1004, 1005
	EtMgBr	THF, N_2, 0°C, stir		75	1004, 1005
	t-BuMgCl	THF, N_2, 0°C, stir		49	1004, 1005
	$(i\text{-Pr})_2NLi$	THF, N_2, 0°C, stir		91	1004, 1005
	PhMgBr	Et_2O, N_2, 0°C, stir		75	1004, 1005
	PhMgBr	CH_2Cl_2, N_2, 0°C, stir		49	1004, 1005
	PhMgBr	C_6H_6, N_2, 0°C, stir		78	1004, 1005
	PhMgBr	n-C_6H_{14}, N_2, 0°C, stir		22	1004, 1005
	EtMgBr	—		70	1008

[a]A mixture of two stereoisomers was obtained in a 2:1 ratio. [b]R group and conditions unspecified. [c]Conditions and product stereochemistry unspecified.

$$H_2N(CH_2)_nCOOH + \text{o-}NO_2C_6H_4SCN \xrightarrow[\text{DMF}]{(n\text{-}Bu)_3P,} (CH_2)_n\text{-lactam } (C{=}O, NH) \quad (658)$$

n	Time (h)	Yield (%)
3	19	88
4	22	73
5	23	97

$$\text{o-}NO_2C_6H_4SCN + (n\text{-}Bu)_3P \longrightarrow \text{o-}NO_2C_6H_4S\overset{+}{P}(Bu\text{-}n)_3 \; CN^-$$

$$\xrightarrow[-(\text{o-}NO_2C_6H_4S^- \text{ and } HCN)]{H_2N(CH_2)_nCOOH} H_2N(CH_2)_nC(=O)\text{—O—}\overset{+}{P}(Bu\text{-}n)_3 \xrightarrow{-H^+} (CH_2)_n\text{-lactam } (C{=}O, NH) + (n\text{-}Bu)_3PO \quad (659)$$

Of the phosphorus reagents employed recently to effect lactam formation by intramolecular cyclization of amino acids, triphenylphosphine and 2,2′-dipyridyl disulphide in acetonitrile have seen the most extensive use. Most of the reactions reported[798,1006,1013–1016] have used this mixture of reagents to prepare substituted β-lactams (equation 660 and Table 42), but larger ring lactams[1017] (equation 661), and bicyclic lactams[1018–1021] (equation 662 and Table 43) have also been prepared using this approach.

$$R^1NHCHR^2CHR^3COOH + \text{2-PyS—S-2-Py} \xrightarrow[\text{temp., time}]{Ph_3P, \text{ solvent,}} Ph_3PO + \text{pyridine-2(1H)-thione (NH, C=S)} + \beta\text{-lactam } (R^3, R^2, N\text{-}R^1, C{=}O) \quad (660)$$

$$H_2N(CH_2)_3N(Me)CH(Ph)CH_2COOH \xrightarrow[\text{MeCN, reflux (94\%)}]{Ph_3P, \text{ 2-PyS—S-2-Py}} \text{8-membered lactam (MeN, Ph, NH, C=O)}$$

$$H_2N(CH_2)_3N(Me)CH(Ph)CH_2COOH \xrightarrow[-15^\circ C\,(22\%)]{EtO_2CCl, \; NEt_3, \; DMF} \text{same lactam} \quad (661)$$

TABLE 42. Preparation of β-lactams from amino acids using triphenylphosphine and 2,2′-dipyridyl disulphide

R^1	R^2	R^3	Solvent	Temp. (°C)	Time (h)	Configuration of amino acid	Yield (%)	Reference
H	H	H	MeCN	55	24	—	39	1013
H	CH_2COOMe	H	MeCN	55	12	*S*	84	1013
H	CH_2COOMe	H	MeCN[a]	70	—[b]	—[b]	86	1014
H	CH_2COOMe	H	MeCN	reflux	12	*R*	82	1013
H	Ph	H	MeCN	55-reflux	4.5	*RS*	34–97[c]	1013
H	Ph	H	DMF	55	4.5	*RS*	9	1013
H	Ph	H	CH_2Cl_2	reflux	4.5	*RS*	trace	1013
H	Ph	H	$MeNO_2$	reflux	4.5	*RS*	26	1013
H	H	NH_2	MeCN	reflux	5.5	*RS*	56	1013
H	H	$(t\text{-Bu})Me_2SiOCHMe$	—[b]	—[b]	—[b]	*RS*	87	1015
$HOCH_2CH_2$	H	H	MeCN	reflux	4.5	—	68	1013
$PhCH_2$	H	H	MeCN	reflux	4.5	—	44–91[c]	1013
$PhCH_2$	Me	H	MeCN	reflux	4.5	*RS*	96	1013
H	$CH_2OSiMe_2(Bu\text{-}t)$	Et	MeCN	80	4.0	*RS*	51[d]	798
H	$CH_2OSiMe_2(Bu\text{-}t)$	$Me_2C(OH)$	MeCN	80	4.0	*RS*	17[e]	798
H	Et	$PhCH_2O_2CNH$	MeCN	reflux	18.0	—[b]	50[f]	1016
$PhCH_2$	$CH_2CH_2OCH_2Ph$	Me	MeCN	—[b]	—[b]	—[b]	89[g]	1006
$PhCH_2$	$CH_2CH_2OCH_2Ph$	*i*-Pr	MeCN	—[b]	—[b]	—[b]	88[h]	1006

[a] The reagents used were triphenylphosphine, 2-mercaptopyridine and manganese dioxide.
[b] Unspecified.
[c] Exact yield dependent upon the ratio of reactants used.
[d] A 4:1 *cis:trans* mixture of products was obtained.
[e] Only the *cis* product was obtained.
[f] Only the *trans* product was obtained.
[g] A 6:1 *trans:cis* mixture of products was obtained.
[h] A 9:1 *trans:cis* mixture of products was obtained.

TABLE 43. Preparation of bicyclic β-lactams from amino acids using triphenylphosphine and 2,2′-dipyridyl disulphide

Starting Material	Conditions	Product	Yields (%)	Reference
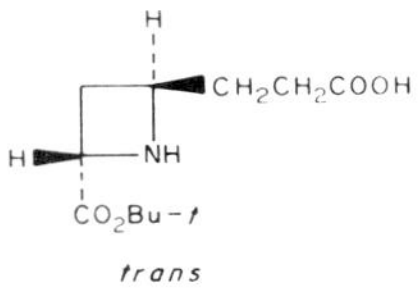 CH_2CH_2COOH, NH, CO_2Bu-*t*, H; *trans*	20 °C, stir 3 h	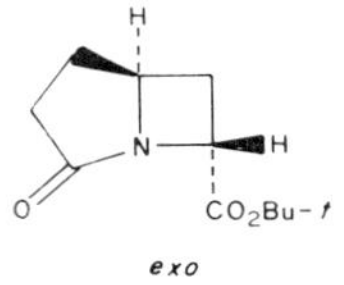 N, H, CO_2Bu-*t*, O; *exo*	58–91	1018, 1019
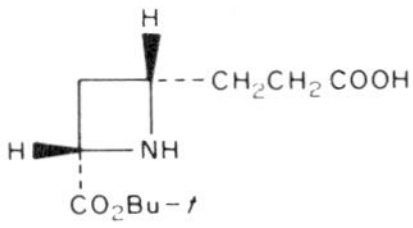 CH_2CH_2COOH, NH, CO_2Bu-*t*, H	20 °C, stir 3 h	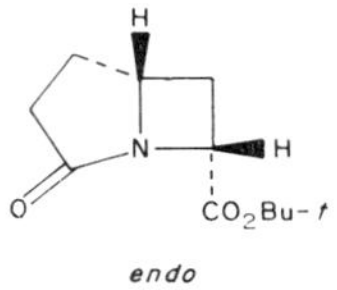 N, H, CO_2Bu-*t*, O; *endo*	55–75	1018, 1019
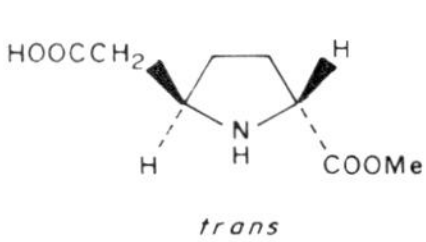 $HOOCCH_2$, N, H, COOMe; *trans*	reflux 8 h	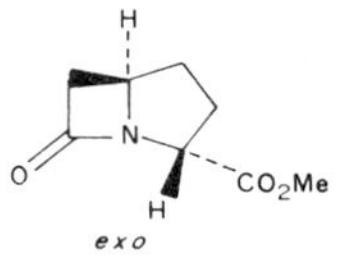 N, H, CO_2Me, O; *exo*	52	1020
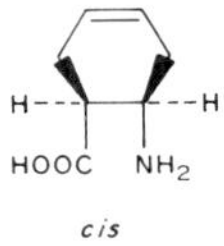 H, HOOC, NH_2; *cis*	reflux	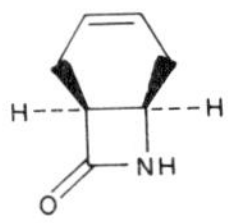 H, NH, O	—	1021

$(CH_2)_n$ / NH / $(CH_2)_mCOOH$ or $(CH_2)_n$ / NH_2 / $(CH_2)_mCOOH$ $\xrightarrow[\text{MeCN}]{Ph_3P,\ \text{2,2'-dipyridyl disulphide (py–S–S–py)}}$ $(CH_2)_n$ / N / $(CH_2)_m$ / O or $(CH_2)_n$ / NH / $(CH_2)_m$ / O (662)

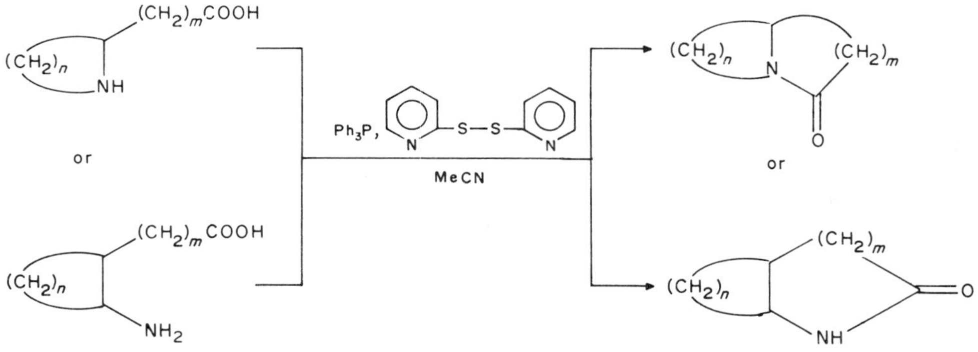

In a study of the stereoselectivity resulting from the cyclization of *threo*-β-amino acid derivatives, β-aminothiol esters were also converted[1006] to β-lactams using triphenylphosphine and 2,2′-dipyridyl disulphide in acetonitrile (equation 663).

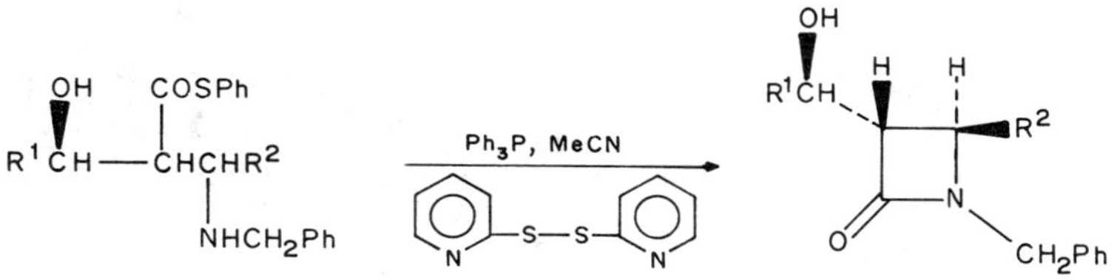

(663)

R^1	R^2	Yield (%)	Selectivity (%)
Me	$CH_2CH_2OCH_2Ph$	72	88[a]
Me	CH=CHPh (*trans*)	43	92[b]
Me	C≡CH	76	89[a]
Et	C≡CH	79	77[b]

[a]The stereochemistry of the major product was determined by conversion to a bicyclic lactam

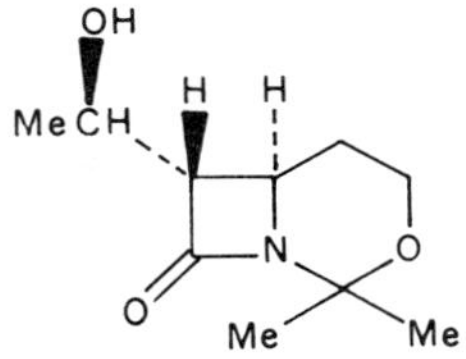

[b]The stereochemistry of the major product was determined by conversion to an ene lactam

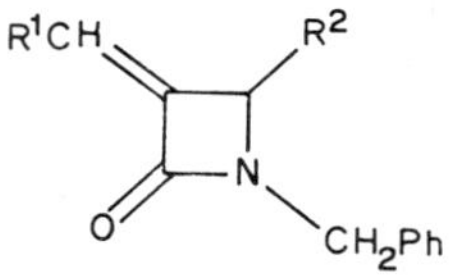

An example[1002] of a phosphorus reagent which does not contain a sulphur atom, but which has been used to produce β-lactams from β-amino acids, is benzotriazol-1-yloxytris(dimethylamino)phosphonium hexafluorophosphate (BOP). Although this reagent does not contain a sulphur atom, the starting materials used in conjunction with this reagent do all themselves contain a sulphur atom. Thus, treatment of a mixture of *cis* and *trans* isomers of 2-(5-acetyl-3,6-dihydro[2*H*]thiazine-1,3-yl-2)-*o*-carboxybenzamido-2-propionic acid with BOP produces[1022] a corresponding *cis,trans* mixture of 3-acetyl-7-methyl-7-phthalimidocephem (equation 664). This same conversion was also reported[1022] to occur using dicyclohexylcarbodiimide (DCCD) (equation 664).

One of the mildest condensing agents used to convert β-amino acids to β-lactams is 2-chloro-1-methylpyridinium iodide. Used in conjunction with triethylamine, this reagent has been employed[1023–1025] to produce N-substituted and unsubstituted β-lactams from the corresponding amino acids (equation 665 and Table 44).

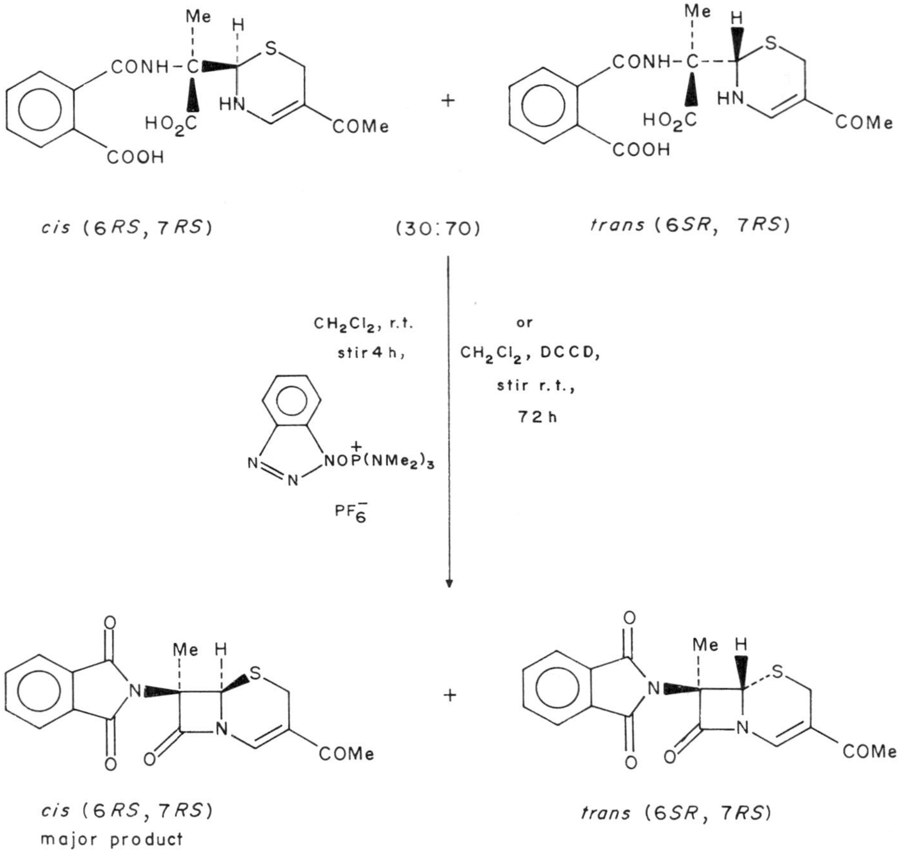

cis (6RS, 7RS) (30:70) trans (6SR, 7RS)

CH_2Cl_2, r.t. stir 4 h, $NOP^+(NMe_2)_3$ PF_6^- or CH_2Cl_2, DCCD, stir r.t., 72 h

cis (6RS, 7RS) major product trans (6SR, 7RS)

(664)

$R^1NHCHR^2CHR^3COOH$ + [2-chloro-1-methylpyridinium] I^- $\xrightarrow{Et_3N,\ solvent,\ conditions}$ [β-lactam with R^1, R^2, R^3]

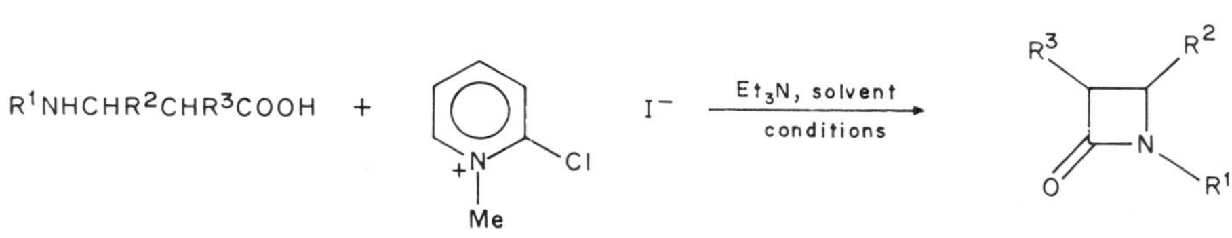

(665)

Tin mediated internal condensation of ω-aminocarboxylic acids has been reported[773,1026] to produce monocyclic lactams, containing from 5 to 13 members, in acceptable yields (equation 666). The tin reagent of choice for these reactions appears to be di(n-butyl)tin oxide, and it can be used to produce[1026] bicyclic as well monocyclic lactams (equation 667). The mechanism of these condensations is reported[773] to involve

TABLE 44. 2-Chloro-1-methylpyridinium iodide mediated cyclization of β-amino acids to β-lactams

R^1	R^2	R^3	Solvent	Conditions	Yield (%)	Reference
H	Me	H	MeCN	reflux 2.5 h	87	1023
H	Ph	H	MeCN	reflux 3 h	89	1023
H	$CH_2CHOSiMe_2(Bu\text{-}t)$ with $CH_2OSiMe_2(Bu\text{-}t)$ on CH	H	—	—	86	1024
$PhCH_2$	H	H	CH_2Cl_2	r.t. 1 h	60	1023
$PhCH_2$	H	H	CH_2Cl_2	r.t.[a]	86	1023
$PhCH_2$	H	Me	CH_2Cl_2	r.t. 2 h	83	1023
$PhCH_2$	H	Me	CH_2Cl_2	r.t.[a]	90	1023
$PhCH_2OCO$	H	MeCH(OH)	CH_2Cl_2	—	62	1025
$PhCH_2OCO$	H	$Ph(CH_2)_2CH(OH)$	CH_2Cl_2	—	80[b]	1025
$PhCH_2$	Me	H	CH_2Cl_2	r.t. 2 h	95	1023
$PhCH_2$	*n*-Pr	H	CH_2Cl_2	r.t. 2 h	94	1023
Ph	*p*-$MeOC_6H_4$	$(t\text{-Bu})Me_2SiO(CH_2)_2CH(OH)$	CH_2Cl_2	r.t.[c]	—	1009[c]
Ph	*p*-$MeOC_6H_4$	$(t\text{-Bu})Me_2SiOCH_2CHCH_2CH$ with $(t\text{-Bu})Me_2SiO$ on the first CH and OH (wedge) on the last CH	CH_2Cl_2	r.t., 2 h[c]	—	1009[c]

[a]Inverse addition. The β-amino acid was added to a suspension of the salt and triethylamine over 1 hour and further stirred for another 1 hour at room -temperature.
[b]The product was prepared stereospecifically with no epimerization observed.
[c]Only the *trans* product was obtained.

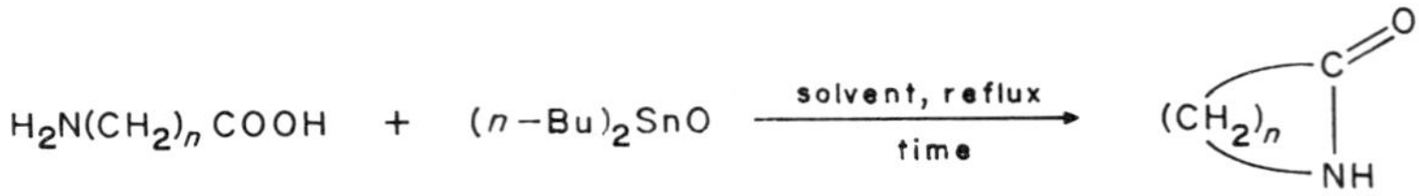

$$H_2N(CH_2)_nCOOH + (n\text{-}Bu)_2SnO \xrightarrow[\text{time}]{\text{solvent, reflux}} (CH_2)_n\text{-lactam (C=O, NH)} \qquad (666)$$

n	Solvent	Time (h)	Yield (%)	Reference
3	xylene	12	95	773, 1026
4	xylene	12	95	773, 1026
5	xylene	20, 12	95	773, 1026
6	mesitylene	6	8	1026
7	mesitylene	6	0	1026
10	mesitylene	24	22	773
11	mesitylene	24	25	773

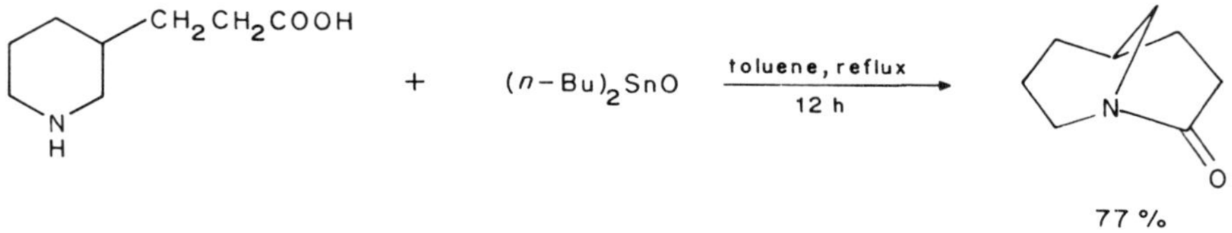

3-piperidyl-CH_2CH_2COOH + $(n\text{-}Bu)_2SnO$ $\xrightarrow[\text{12 h}]{\text{toluene, reflux}}$ bicyclic lactam, 77%

(667)

(80): $H_2N(CH_2)_nCOOH$ $\xrightarrow{(R_3Sn)_2O}$ (80) $\xrightarrow{-(R_3Sn)_2O}$ lactam

$H_2N(CH_2)_nCOOH$ $\xrightarrow{R_2SnO}$ intermediate $\xrightarrow{-H_2O}$ (81) $\xrightarrow{-R_2SnO}$ lactam

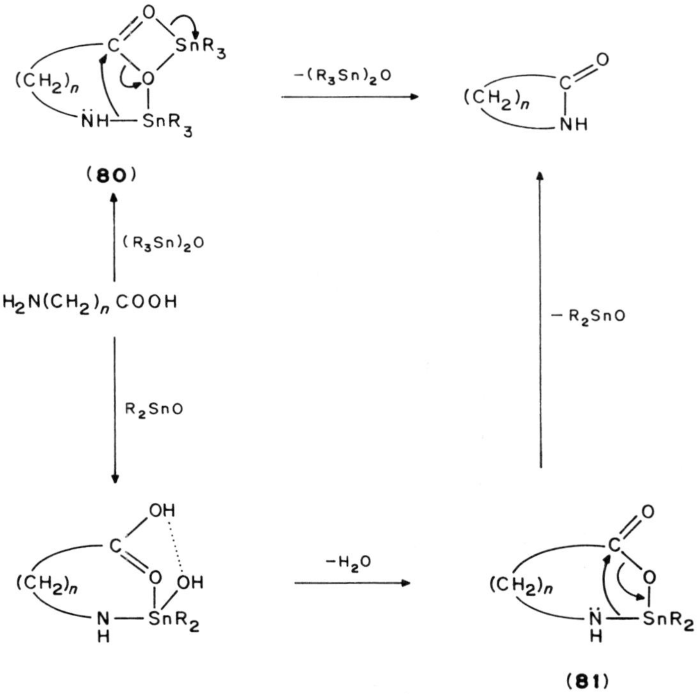

SCHEME 8

(Scheme 8) the reaction of the amino acid and the organotin oxide to produce a reactive stannylated **80** or **81**, with subsequent release of the tin reagent by a temperature-driven extrusion process. This same mechanism has been reported earlier to describe the formation of lactones from hydroxycarboxylic acids and di(n-butyl)tin oxide.

An interesting preparation of lactams involves the cyclization of ω-amino acids facilitated by γ-butyrolactone[1027] (equation 668). It appears that in the course of this reaction a depsipeptide is formed as an intermediate.

$$H_2N(CH_2)_nCOOH + \gamma\text{-butyrolactone} \xrightarrow[\text{reflux}]{\text{toluene}} [HO(CH_2)_3CONH(CH_2)_nCOOH] \longrightarrow (CH_2)_n\langle C{=}O, NH\rangle \quad n = 3,4,5 \tag{668}$$

ω-Amino acids and their tetra(n-butyl)ammonium salts have also been cyclized to their corresponding macrocylic lactams[1028] by using catecholborane in the presence of pyridine when amino acids were the starting materials (equation 669), and by using B-chlorocatecholborane in the presence of pyridine when amino acid ammonium salts were the starting materials (equation 670). Lactam dimer formation was not observed with the latter procedure.

Cyclodehydration of γ-, δ- and ε-amino acids to produce their corresponding lactams has been accomplished[1029] by the action of alumina or silica gel in boiling toluene (equation 671) with the aid of a Dean–Stark trap.

$$H_2N(CH_2)_nCOOH + \text{catecholborane (BH)} \xrightarrow{C_5H_5N,\ 80\,^\circ C} (CH_2)_n\langle C{=}O, NH\rangle + \text{cyclic dimer } [HN-(CH_2)_n-C{=}O \cdots NH-(CH_2)_n-C{=}O] \tag{669}$$

n	Monomer yield (%)	Dimer yield (%)
3	95	—
5	85	—
6	6	18
7	—	10
11	6	25
12	9	22
14	13	17

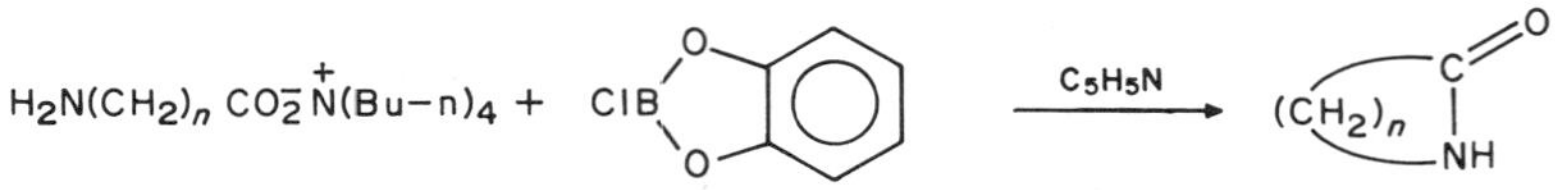

$$H_2N(CH_2)_nCO_2^-\overset{+}{N}(Bu\text{-}n)_4 + ClB(O_2C_6H_4) \xrightarrow{C_5H_5N} (CH_2)_n\langle C(=O)\text{-}NH \rangle \qquad (670)$$

n = 6; 12; 15
Yield(%) = 65; 15; 17

$$\omega\text{-amino acid} + \text{alumina or silica gel} \xrightarrow[\text{Dean–Stark trap}]{\text{tolune}} \text{lactam} \qquad (671)$$

Amino Acid	Time (h)	Support	Lactam product	Yield (%)
$H_2N(CH_2)_3COOH$	24	none	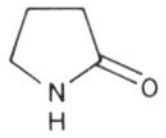	<1
	5	Al_2O_3 (neutral)		97
	5	Al_2O_3 (basic)		95
	5	SiO_2		97
	4	Florisil		70
(+)$HOOCCH(NH_2)(CH_2)_2COOH$	24	none		0
	72	Al_2O_3[a]	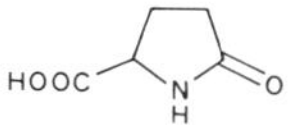	63[b]
(±)$H_2NCH_2CH(C_6H_4Cl\text{-}p)CH_2COOH$	24	none	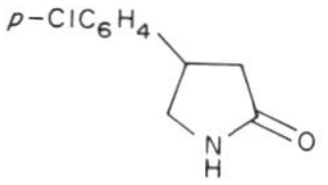	5
	5	Al_2O_3		88
(±)$H_2NCH_2CH(OH)CH_2COOH$	24	none		0
	24	Al_2O_3[a]	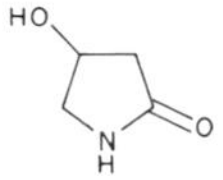	38
	24	Al_2O_3[c]		26
(±)$MeCH(NH_2)(CH_2)_2COOH$	24	none	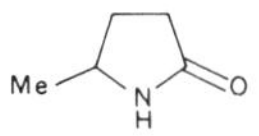	6
	5.5	Al_2O_3		93
$H_2N(CH_2)_4COOH$	24	none	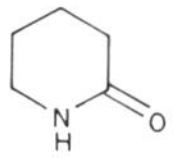	49
	1.5	Al_2O_3 (neutral)		76
	1.5	SiO_2		99
(±)$H_2N(CH_2)_3CH(NH_2\cdot HCl)COOH$[d]	24	none	—	0

Amino Acid	Time (h)	Support	Lactam product	Yield (%)
	3	Al_2O_3	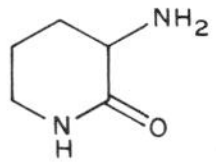 (3-amino-2-piperidinone)	79[e]
$H_2N(CH_2)_5COOH$	24	none	—	0
$H_2N(CH_2)_5COOH$	6	Al_2O_3 (basic)	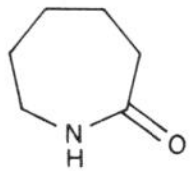 (caprolactam)	32
	6	Al_2O_3 (acid)		50
	6	Al_2O_3 (neutral)		53
	20–24	Al_2O_3 (neutral)		82
	6	SiO_2		82
	20	SiO_2		75
$(\pm)H_2N(CH_2)_4CH(NH_2\cdot HCl)COOH$[d]	24	none	—	0
	24	Al_2O_3	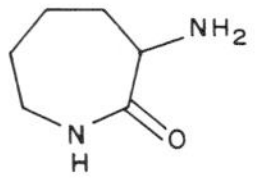 (3-amino-caprolactam)	70
	6	Al_2O_3 (basic)		25
	6	Al_2O_3 (neutral)		27[f]
	20	Al_2O_3		71[g]
	44	Al_2O_3		90[h]
	6	SiO_2		37

[a]A 3:1 toluene–pyridine mixture was used as solvent,
[b]Racemization (4%) occurred during cyclodehydration of the optically active glutamic acid.
[c]A water–Al_2O_3 (1:6) mixture was used.
[d]An equivalent amount of concentrated aqueous sodium hydroxide was used to free the amino acid and an extra quantity of solid support was used to absorb the added water and to facilitate stirring and disgregation of the cake initially formed.
[e]Racemization (75%) occurred during cyclodehydration of the optically active ornithine.
[f]Racemization (26%) occurred during cyclodehydration of the optically active lysine.
[g]Racemization (41%) occurred during cyclodehydration of the optically active lysine.
[h]Racemization (74%) occurred during cyclodehydration of the optically active lysine.

Treatment of amino acids or their hydrochloride salts with hexamethyldisilazane produces the corresponding trimethylsilyl amino esters, which upon reflux in xylene or acetonitrile, followed by dilution with methanol or ethanol and evaporation, causes *exo*-trigonal ring closure and formation[1030] of the corresponding γ-δ- or ε-lactams (equation 672). This one-pot procedure produces the lactams stereoselectively and in excellent yields.

In at least one reference[1031] cyanuric chloride in the presence of triethylamine has been used to convert N-substituted β-amino acids to the corresponding N-substituted β-lactams (equation 673).

$$H_2N(CH_2)_nCO_2H + (Me_3Si)_2NH \xrightarrow[\text{reflux; 2. dilute with MeOH or EtOH; 3. evaporate}]{\text{1. Method A or B, solvent}} (CH_2)_n\text{-lactam } (C{=}O, NH) \quad (672)$$

Method A: refluxing a mixture of the amino acid with hexamethyldisilazane in xylene.
Method B: refluxing a mixture of the hydrochloride salt of an amino acid with hexamethyldisilazane in acetonitrile.

Amino acid	Time (h)	Method	Product	Yields(%)
$H_2N(CH_2)_3COOH$	6	A	2-pyrrolidinone	87
S-(+)HOOCCH(NH_2)($CH_2)_2$COOH	8	A	pyroglutamic acid (HOOC, H)	93[a]
RS-$H_2NCH_2CH(OH)CH_2COOH$	4	A	4-hydroxy-2-pyrrolidinone (HO)	93[b]
R-(−)$H_2NCH_2CH(OH)CH_2COOH$	4	A	4-hydroxy-2-pyrrolidinone (HO, H)	89[b]
S-(−)$H_2NCH_2CH(OH)CH_2COOH$	4	A	4-hydroxy-2-pyrrolidinone (HO, H)	83[b]
RS-$H_2N(CH_2)_2CH(NH_2\cdot HCl)COOH$	48	B	3-amino-2-pyrrolidinone (NH_2)	95[c]
$H_2N(CH_2)_4COOH$	4	A	2-piperidinone	95
S-(+)$H_2N(CH_2)_3CH(NH_2\cdot HCl)COOH$	48	B	3-amino-2-piperidinone (H, NH_2)	91[d]
$H_2N(CH_2)_5COOH$	48	A	caprolactam	75
S-(+)$H_2N(CH_2)_4CH(NH_2)COOH$	48	A	α-amino-caprolactam (NH_2, H)	82[e]

[a]Isolated as the dicyclohexylamine salt.
[b]Product is the 4-trimethylsilyloxy derivative; desilylation achieved using MeCN, H_2O and HCl.
[c]The hydrochloride and the (*RS*)-3-(*p*-toluenesulphonamido)-2-pyrrolidinone were both prepared.
[d]The hydrochloride and the (*S*)-3-(*p*-toluenesulphonamido)-2-piperidinone were both prepared.
[e]The hydrochloride, (*S*)-α-(*p*-toluenesulphonamido)caprolactam and the salt of (*S*)-α-amino-ε-caprolactam with (*S*)-pyroglutamic acid were all prepared.

$$R^1NHCHR^2CH_2COOH + \text{(2,4,6-trichloro-1,3,5-triazine)} \xrightarrow[\text{r.t., 4 h}]{Et_3N,\ DMF,\ CH_2Cl_2} \text{(1-}R^1\text{-4-}R^2\text{-azetidin-2-one)}\quad 56\text{–}90\%$$

R^1 = $PhCH_2$, *n*-Bu, *n*-Pent, *n*-Hex, *c*-Hex
R^2 = Me, Ph

(673)

Reaction of a *β*-chlorovinyl sulphone with an *o*-methylaminothiocarboxylic acid produces an activated acid ester which, upon heating in toluene, affords[1032] a 15-membered lactam sulphide (equation 674).

$$o\text{-}(MeNH)C_6H_4S(CH_2)_{10}COOH + O_2N\text{-}C_6H_4\text{-}SO_2CH{=}CHCl \xrightarrow[\text{r.t., 30 min}]{DMF,\ Et_3N,} O_2N\text{-}C_6H_4\text{-}SO_2CH{=}CHO_2C(CH_2)_{10}S\text{-}C_6H_4\text{-}o\text{-}NHMe \xrightarrow[95\text{–}100\,^{\circ}C]{C_6H_5Me,\ N_2,} \text{(15-membered N-methyl lactam sulphide)}\quad 62.2\%$$

(674)

Ethylation of L-ornithine (2,5-diaminopentanoic acid) followed by cyclization has been reported to produce[1033] *S*-aminopiperidinone, but the details of the reaction are not given (equation 675).

$$H_2N(CH_2)_3CH(NH_2)COOH \xrightarrow[\text{2. cyclization}]{\text{1. ethylation}} \text{(3-aminopiperidin-2-one)} \qquad (675)$$

Excellent yields of saturated 5-, 6- and 7-membered lactams have been obtained[1034] by hydrogenation of cyanoalkanoates over a catalyst containing ruthenium and/or iron producing intermediate amino esters which then cyclize under the conditions of the reaction. An example of this procedure is shown in equation (676) for the preparation of pyrrolidone.

$$NC(CH_2)_2COOMe + H_2 \xrightarrow[\text{RuCl}_3\cdot H_2O,\ NaOH,\ 250\,^{\circ}C,\ 3\,h]{FeCl_3\cdot 6H_2O,}$$ 2-pyrrolidinone, 99% (676)

A similar reaction, but one which utilizes the cyano group as the source of the carbonyl function, is preparation[1035] of ε-caprolactam from 6-aminocapronitrile in the presence of Porasil A (equation 677).

$$H_2N(CH_2)_5CN \xrightarrow[150\,h,\ 300\,^{\circ}C]{\text{Porasil A},\ NH_3,}$$ ε-caprolactam, 87% (677)

Base catalysed cyclization of the *N*-benzyloxycarbonyl protected aldol products obtained from 3-(3-aminopropanoyl)thiazolidine-2-thione and a variety of aldehydes produces[1025] a novel mixed lactam–thiolactam (equation 678).

RCH(OH)CH(CH$_2$NHCOOCH$_2$Ph)CON(thiazolidine-2-thione) $\xrightarrow[K_2CO_3]{Et_3N,\ CsF\ \text{or}}$ 5-[RCH(OH)]-4-oxo-3-[(CH$_2$)$_2$SH]-1-(COOCH$_2$Ph)-tetrahydropyrimidine-2-thione (678)

R = Me, Ph, Ph(CH$_2$)$_2$, *n*-Pent

Recently, a variety of condensation reactions have been used to produce N-substituted lactams from N-unsubstituted amino acids. One example of this approach to the preparation[1036–1038] of β-lactams has been the four-site condensation of an amino acid, an aldehyde and an isocyanide (equation 679 and Table 45).

$$H_2NCHR^1CHR^2COOH - R^3CHO + R^4NC \xrightarrow[\text{r.t., stir, 10 h – 3 days}]{MeOH}$$ β-lactam (3-R^2, 4-R^1, N-CHR^{3}CONHR^4) (679)

Reaction of N-unsubstituted amino acid esters with cyclopropanone affords the corresponding *N*-(1-hydroxycyclopropan-1-yl) substituted amino acid ester, which upon reaction with *tert*-butyl hypochlorite produces an intermediate *N*-chloroamino ester. Further treatment of this intermediate with silver nitrate produces N-substituted β-lactams. These steps collectively are referred to as the *N*-chlorocarbinolamine method of β-lactam preparation and this method has been used [1039] in the recent literature (equation 680) to produce several β-lactams.

TABLE 45. Preparation of β-lactams *via* four-component condensations

R^1	R^2	R^3	R^4	Yield (%)	Reference
H	H	*i*-Pr	*o*-$N_3C_6H_4$	—	1036
H	H	*i*-Pr	*o*-$PhCH_2OC_6H_4$	—	1036
H	H	H	Ph_2CH	20	1037
H	H	*i*-Pr	Ph_2CH	54	1037
H	H	$HOCH_2$	Ph_2CH	48	1037
H	H	$ClCH_2$	Ph_2CH	30	1037
H	H	$(MeCO)_2CH$	Ph_2CH	48	1037
H	H	$PhCH_2OCO$	Ph_2CH	20	1037
H	H	MeCO	Ph_2CH	33	1037
H	H	3-O_2N-4-Cl-C_6H_3–C(H)=C(Me)–	Ph_2CH	29	1037
H	H	2-Fu	Ph_2CH	27	1037
H	H	Ph	Ph_2CH	49	1037
H	H	*p*-$MeOC_6H_4$	*n*-Bu	50	1037
H	H	*p*-$MeOC_6H_4$	*t*-Bu	51	1037
H	H	*p*-$MeOC_6H_4$	*c-Hex*	51	1037
H	H	*p*-$MeOC_6H_4$	Ph_2CH	50	1037
H	H	*p*-$MeOCH_2C_6H_4$	Ph_2CH	47	1037
H	H	*p*-$PhCH_2OC_6H_4$	Ph_2CH	36	1037
H	H	*p*-ClC_6H_4	Ph_2CH	35	1037
H	H	*p*-$O_2NC_6H_4$	Ph_2CH	31	1037
H	H	*m*-$O_2NC_6H_4$	Ph_2CH	31	1037
H	H	3,4,5-$(MeO)_3C_6H_2$	Ph_2CH	58	1037
H	HO	Ph	Ph_2CH	44	1037
p-$O_2NC_6H_4CH_2OCO$	H	$OHCH_2$	Ph_2CH	49[a]	1037
$HOCH_2$	N_3	H	*p*-$O_2NC_6H_4$	95[b]	1038
$HOCH_2$	N_3	$(EtO)_2CH$	*p*-$O_2NC_6H_4$	93[c]	1038

[a]Yield is made up of 32% of one isomer and 17% of another.
[b]Product obtained is exclusively *cis*.
[c]Product is a 1:1 diastereoisomeric mixture of the *cis* isomer.

Cyclopropanone + $H_2NCR^1R^2CO_2R^3$ —(Et_2O, N_2, −78 °C; CH_2Cl_2, 1 h)→ 1-hydroxy-1-($NHCR^1R^2CO_2R^3$)cyclopropane (OH, $NHCR^1R^2CO_2R^3$)

—(*t*-BuOCl, $NaHCO_3$; stir −10 °C, 40 min)→ [1-hydroxy-1-($N(Cl)CR^1R^2CO_2R^3$)cyclopropane (OH, $N(Cl)CR^1R^2CO_2R^3$)]

—(1. $AgNO_3$, MeCN; stir r.t., 1.5 h; 2. NH_4OH)→ azetidin-2-one (O, N–$CR^1R^2COOR^3$)

R^1 =	*p*-$MeOC_6H_4$;	*p*-$PhCH_2OC_6H_4$;	*p*-$PhCH_2OC_6H_4$
R^2 =	H ;	H ;	CO_2Me
R^3 =	Et ;	$PhCH_2$;	Me
Yield (%) =	52 ;	59 ;	38

(680)

Another condensation reaction which involves a cyclopropane derivative is the condensation of ε-(benzyloxycarbonyl)-L-lysine methyl ester hydrochloride with the electrophilic cyclopropane derivative shown to produce[1040] an α-carboxyl lactam as a mixture of diastereomers (equation 681). The product appears to be formed by initial attack of the lysine amino group at a cyclopropane methylene with opening of the 3-membered ring. The intermediate thus formed then cyclizes to one of the lactone carbonyls with expulsion of acetone.

HCl · H_2N—C(COOMe)(H)—$(CH_2)_4NHOCOCH_2Ph$ + (cyclopropane spiro-dioxanedione, Me, Me)

$\xrightarrow[65\,°C,\ 1\,h]{Et_3N,\ DMF,}$ (H, HOOC-substituted pyrrolidinone)N—C(COOMe)(H)—$(CH_2)_4NHOCOCH_2Ph$

(681)

Examples of base catalysed condensations to produce lactams have also been reported, and in one example[1041] phenylacetic acid is condensed with 1-amino-4-bromo-2-methylanthraquinone in the presence of pyridine and titanium tetrachloride (equation 682). This is one specific case of a more general condensation[1041] involving aromatic amines and substituted acetic acids (equation 683).

(1-amino-4-bromo-2-methylanthraquinone: O, NH_2, Me, O, Br) + $PhCH_2COOH$ $\xrightarrow{C_5H_5N,\ TiCl_4}$ (Ph, O, NH, Me, O, Br)

90 %

(682)

(R^{1}CO, NHR^4, R^2, R^3 aromatic amine) + R^5COOH $\xrightarrow[C_5H_5N]{Cat.}$ (O, R^5, N, R^4, R^1, R^2, R^3 quinolinone) (683)

R^1 = H, C_{1-10} alkyl

R^2, R^3 = alkylaryl, H, C_{1-10} alkyl, aryl, cycloalkyl

R^4 = H, C_{1-8} alkyl, aryl, alkylaryl, C_{6-10} aralkyl

Cat. = Ti(IV) or Si(IV)

The second example of a base catalysed condensation which produces lactams involves a one-pot condensation[1042] of β-haloacetyl chlorides with α-amino acids in the presence of aqueous sodium hydroxide (equation 684).

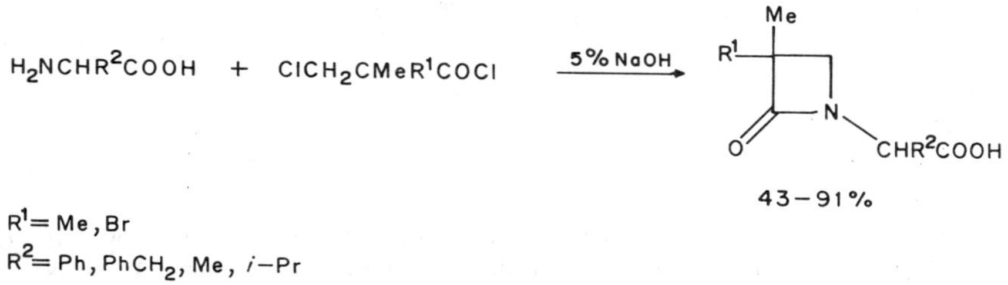
H2NCHR2COOH + ClCH2CMeR1COCl
5% NaOH
Me
R1
N
O
CHR2COOH
43–91%
R1 = Me, Br
R2 = Ph, PhCH2, Me, i-Pr

(684)

Condensation of methyl acrylate with L-cysteine hydrochloride or D-penicillamine affords[1043] the corresponding sulphur-containing 7-membered ring lactams (equation 685).

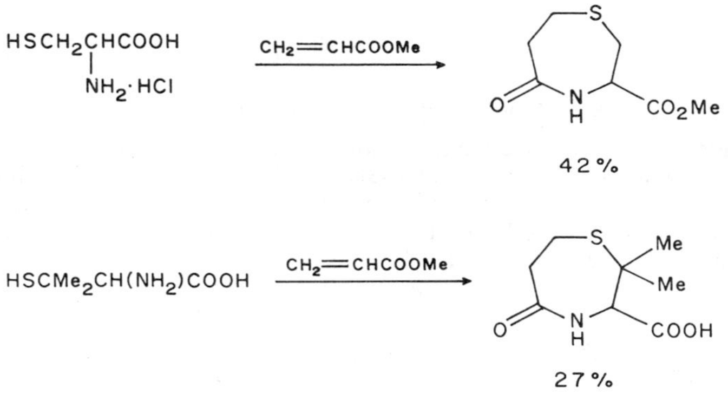
HSCH2CHCOOH
NH2·HCl
CH2=CHCOOMe
S
O
N
H
CO2Me
42%
HSCMe2CH(NH2)COOH
CH2=CHCOOMe
S
Me
Me
O
N
H
COOH
27%

(685)

Nucleophilic substitution of ethyl-4-bromo 3-(2-furyl)-2-butenoate with primary amines produces an intermediate amino acid ester which, under the conditions of the reaction, affords[1044] the corresponding furano-substituted γ-lactams (equation 686).

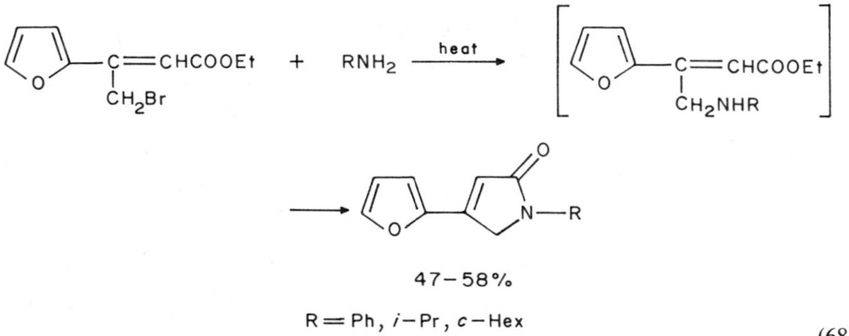
O
C=CHCOOEt
CH2Br
+ RNH2
heat
O
C=CHCOOEt
CH2NHR
O
O
N—R
47–58%
R = Ph, i-Pr, c-Hex

(686)

A very similar reaction has also been reported[1045] with the dimethyl ester of 2-bromoglutaric acid and benzylamine (equation 687).

$$MeOOC(CH_2)_2CHBrCOOMe + PhCH_2NH_2 \xrightarrow[20\ h]{MeOH,\ reflux} \text{N-benzyl-5-oxopyrrolidine-2-COOMe } (79\%) \qquad (687)$$

Finally, reaction of t-butyl carbamoyl-D,L-serine methyl ester with p-toluenesulphonyl chloride followed by condensation of the resulting product with hydrazine produces[1046] the corresponding t-butyl carbamoyl substituted pyrazolidinone (equation 688).

$$t\text{-}BuO_2CNHCH(CH_2OH)COOMe \xrightarrow[2.\ H_2NNH_2,\ CH_2Cl_2,\ r.t.,\ 16\ h]{1.\ p\text{-}TosCl} t\text{-}BuO_2CNH\text{-pyrazolidin-3-one } (60\%) \qquad (688)$$

*2. *From halo, hydroxy, keto and other substituted amides*

Halo amides have been converted to azetidinones by treatment with a variety of bases. The base most commonly used to effect these conversions[1039,1047–1051] is sodium hydride in a dimethylformamide–methylene chloride solvent mixture (equation 689 and Table 46). Using this approach, the best results, in most cases, are obtained at high dilutions using bromine as the halide[1047].

$$R^1NHCOCR^2R^3CH_2X \xrightarrow[temp.,\ time]{NaH,\ DMF/CH_2Cl_2\ (1:4),} \text{1-}R^1\text{-3-}R^2\text{-3-}R^3\text{-azetidin-2-one} \qquad (689)$$

$$ClCH_2C(Me)_2CONHCH(COOEt)_2 \xrightarrow[15\ h]{NaH,\ DMF-CH_2Cl_2,} \text{4,4-dimethyl-5-oxopyrrolidine-2,2-(COOEt)}_2 \ (49\%)$$

$$ClCH_2C(Me)_2CONHCH(COOEt)_2 \xrightarrow[2.\ r.t.,\ stir\ 12\ h]{1.\ LDA,\ THF,\ N_2,\ -78\,^{\circ}C} \text{3,3-dimethyl-1-}CH(COOEt)_2\text{-azetidin-2-one } (84\%) \qquad (690)$$

TABLE 46. Preparation of β-lactams from β-halopropionamides using sodium hydride[a]

R^1	R^2	R^3	X	Temp. (°C)	Time (h)	Yield (%)	Reference
c-Hex	H	H	Br	r.t.	2–5	41	1047
c-Hex	H	H	I	r.t.	2–5	56	1047
$PhCH_2CH_2$	H	H	Cl	r.t.	2–5	22	1047
$PhCH_2CH_2$	H	H	Br	r.t.	2–5	50	1047
p-$AnCH_2$	H	H	Cl	r.t.	2–5	26	1047
p-$AnCH_2$	H	H	Br	r.t.	2–5	60	1047
$PhCH_2O$	H	H	Cl	60	18	42	1048[b]
p-$PhCH_2OC_6H_4CHCO_2Et$	H	H	Br	r.t.	3	80	1039, 1047
p-$PhCH_2OC_6H_4CHCO_2Et$	H	H	Cl	r.t.	2–5	14	1047
p-$PhCH_2OC_6H_4CHCO_2CH_2Ph$	H	H	Br	r.t.	1	46	1039, 1047
p-$AnCHCO_2Me$	H	H	Br	r.t.	2–5	77	1039
p-$AnCHCO_2Et$	H	H	Br	r.t.	2–5	54	1039
$EtOOCCMe_2$	H	H	Cl	r.t.	2–5	86	1047
$EtOOCCMe_2$	H	H	Br	r.t.	2–5	66	1047
p-$PhCH_2OC_6H_4C(CO_2Me)_2$	H	H	Cl	r.t.	2–5	76	1039, 1047
p-$PhCH_2OC_6H_4C(CO_2Me)_2$	H	H	Br	r.t.	2–5	64	1047
$PhCH_2O$	Me	H	Br	70	18	98	1048[b]
$PhCH_2O$	$PhCH_2OCONH$	H	Cl	50	12	74–86	1049[c]
$PhCH_2O$	t-BuOCONH	H	Cl	50	12	75–88	1049[c]
$PhCH_2O$	Me	Me	Cl	r.t.	1	94	1049[b]
EtOOC(H)C=C(Me)–	Me	Me	Cl	r.t.	3	100	1050
$Me_2C{=}CCO_2Et$	Me	Me	Cl	r.t.	3	83	1050
2-Me-4-(COOBu-t)-phenyl (structure: Me, COOBu-t)	F	Br	Br	r.t.	2	78	1051
	F	F	Br	r.t.	2	72	1051
	Br	H	Br	r.t.	2	24	1051
	Br	Br	Br	r.t.	2	52	1051

[a] A 1:4 mixture of DMF to CH_2Cl_2 is used, unless otherwise stated.
[b] Reaction was performed using only DMF; no CH_2Cl_2 was used.
[c] A 1:1 mixture of DMF to CH_2Cl_2 was used.

One result which does not seem to fit the pattern established in equation (689) and Table 46, but which can be rationalized on the basis of the acidity of the proton involved, is observed[1050] when diethyl *N*-(β-chloropivaloyl)aminomalonate is treated with sodium hydride in the same dimethylformamide–methylene chloride reaction mixture (equation 690). Instead of obtaining the β-lactam product from this reaction, the product actually obtained is the corresponding pyrrolidinone in 49% yield with no β-lactam being detected. The β-lactam expected as the product from this reaction can indeed be obtained[1050] in 84% yield, however, by treatment of the aminomalonate with lithium diisopropylamide in tetrahydrofuran (equation 690).

Other examples of bases used to cyclize halo amides to lactams include triethylamine[1052] (equation 691) and lithium carbonate in dimethylformamide[1049] (equation 692).

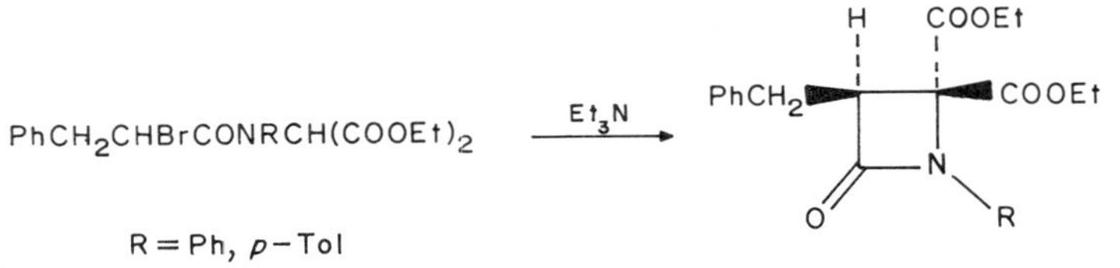

$$PhCH_2CHBrCONRCH(COOEt)_2 \xrightarrow{Et_3N} \text{[}\beta\text{-lactam: 3-}PhCH_2\text{, 4,4-}(COOEt)_2\text{, N-R]} \qquad (691)$$

R = Ph, *p*-Tol

$$ClCH_2CMe_2CONHO_2CBu\text{-}t \xrightarrow[20^\circ C,\ 24\,h]{Li_2CO_3,\ DMF} \text{[}\beta\text{-lactam: 3,3-}Me_2\text{, N-}O_2CBu\text{-}t\text{]} \qquad (692)$$

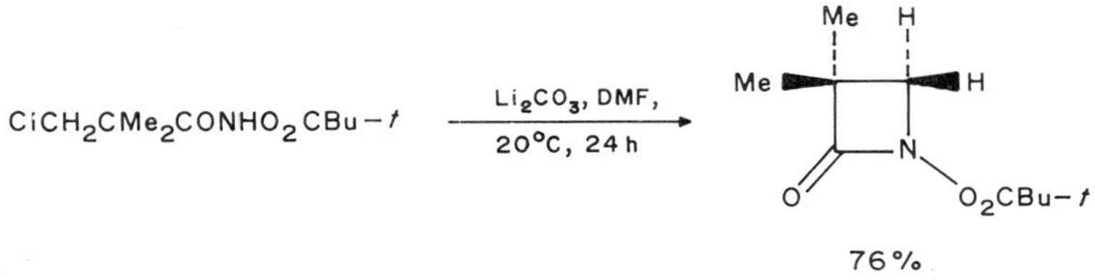

76%

An interesting set of results is obtained[750,1053] when 1,8-diazabicyclo[5.4.0]undec-7-ene (DBU) is used as the base to effect cyclization of halo amides to β-lactams. When (2*S*, 3*R*)-*N*-(2,4-dimethoxybenzyl)-*N*-[di(ethoxycarbonyl)methyl]-2-bromo-3-acetoxybutyramide is treated with DBU in benzene at 20°C for 14 hours, cyclization proceeds with inversion of configuration to produce[750] [3*S*-[3α(*S**)]-ethyl 1-(2,4-dimethoxybenzyl)-3-(1-acetoxyethyl)-2-azetidinone-4,4-dicarboxylate (equation 693). However, when (2*S*, 3*R*)-*N*-(2,4-dimethoxybenzyl)-*N*-(*t*-butoxycarbonylmethyl)-2-bromo-3-hydroxybutyramide is treated with DBU in tetrahydrofuran at 20°C for 4 hours, the cyclization which occurs does not produce a β-lactam, but instead produces[1053] (2*S*, 3*R*)-*N*-(2,4-dimethoxybenzyl)-*N*-(*t*-butoxycarbonylmethyl)-2,3-epoxybutyramide (equation 694). Further treatment of this product with lithium hexamethyldisilazide in tetrahydrofuran at −78°C affords[1053] the desired β-lactam in 22–28% yield, while similar treatment at 20°C produces[750,1053]

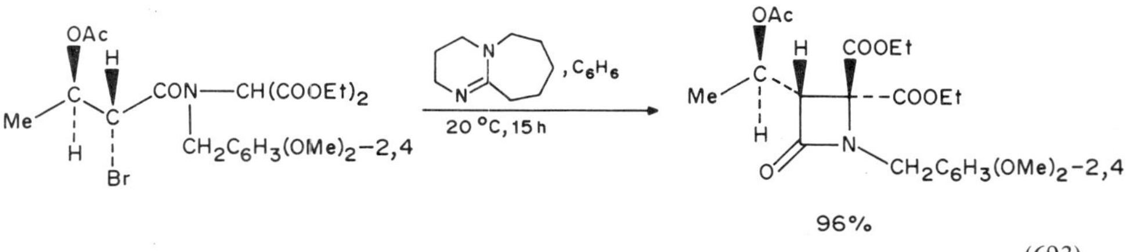

$$MeCH(OAc)CHBrCON(CH(COOEt)_2)CH_2C_6H_3(OMe)_2\text{-}2,4 \xrightarrow[20^\circ C,\ 15\,h]{DBU,\ C_6H_6} \text{[}\beta\text{-lactam: 3-}CH(OAc)Me\text{, 4,4-}(COOEt)_2\text{, N-}CH_2C_6H_3(OMe)_2\text{-}2,4\text{]} \qquad (693)$$

96%

the β-lactam in 61% yield (equation 694). The epoxy intermediate may be avoided entirely by treating the butyramide directly with two equivalents of lithium hexamethyldisilazide in tetrahydrofuran (equation 694). This approach was used[1053] to prepare several *trans*-alkyl N-substituted-3-1-(1-hydroxyethyl)-2-azetidinone-4-carboxylates as shown in equation (695).

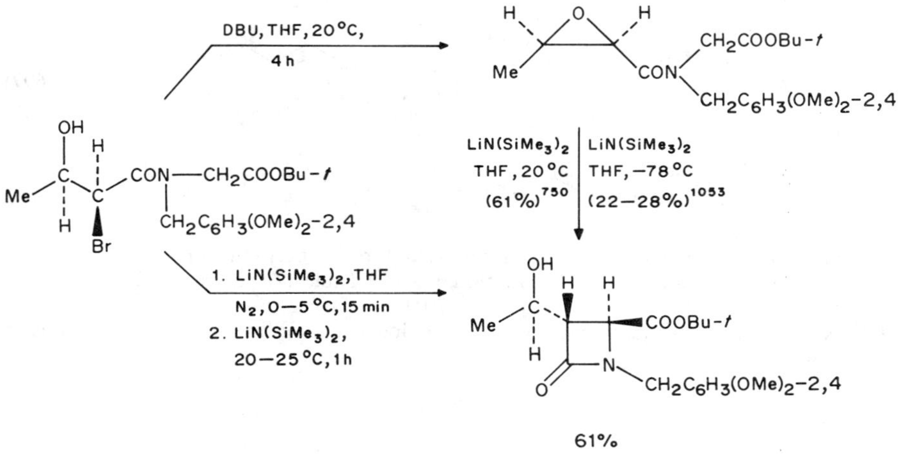

(694)

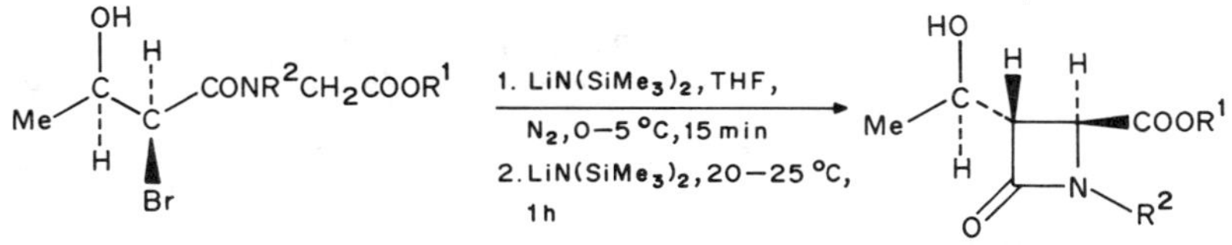

R^1 = $PhCH_2$; Me ; *t*-Bu ; *t*-Bu
R^2 = $CH_2C_6H_3(OMe)_2$-2,4 ; $CH_2C_6H_3(OMe)_2$-2,4 ; $PhCH_2$; *p*-An
Yield(%) = 40 ; 33 ; 52 ; 46

(695)

β-Lactams have also been prepared[1054] via a base promoted intramolecular nucleophilic substitution of the chloroamides shown in equation (696), but the authors failed to report the base used to effect this conversion.

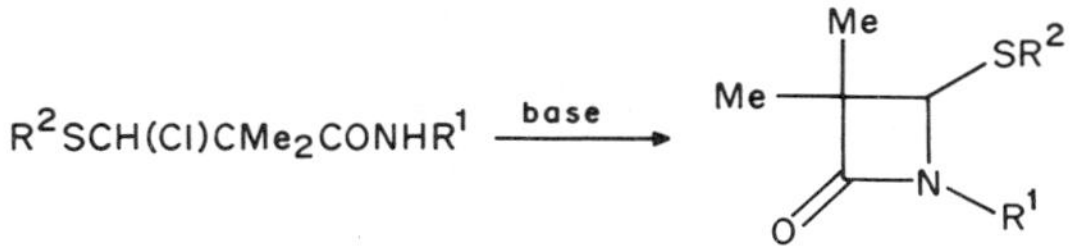

(696)

R^1 = $CH_2CH{=}CH_2$; $(CH_2)_2CH{=}CH_2$; $(CH_2)_3CH{=}CH_2$; *o*-$BuC_6H_4CH_2$;
R^2 = Ph ; Ph ; Ph ; Me ;
R^1 = *o*-$BrC_6H_4CH_2$; *o*-$BrC_6H_4CH_2$
R^2 = *t*-Bu ; Ph

Using sodium hydride in combination with sodium borohydride to treat several bis-[o-ω-bromocarboxamidophenyl)]disulphides affords[1055] medium-sized lactam sulphide products (equation 697).

$$\text{bis}[o\text{-}(Br(CH_2)_nCON(Me))C_6H_4]S\text{-}S \xrightarrow[NaBH_4]{NaH,} \text{benzothiazepinone (N-Me, C=O, S, } (CH_2)_n) \qquad (697)$$

Another approach which has been reported for the preparation of α-lactams[1056] (equation 698), β-lactams[1057–1059] (equation 699 and Table 47), piperazine-2,5-diones[1058] (equation 700), bis-β-lactams[1058] (equation 701) and larger ring lactams[1058] (equation 702) is the base catalysed cyclization of halocarboxamides under phase transfer conditions.

$$R^1R^2CHBrCONHR^3 \xrightarrow[\text{catalyst}]{\text{Phase transfer}} \text{aziridinone } (R^1, R^2, N\text{-}R^3) \qquad (698)$$

R^1	R^2	R^3	Phase transfer catalyst	Temp. (°C)	Time (h)	Yield (%)
t-Bu	H	t-Bu	a	20	12	80
t-Bu	H	t-Bu	b	20	125	19
t-Bu	H	t-Bu	c	20	135	17
t-Bu	H	t-Bu	d	20	110	⩽5
t-Bu	H	t-Bu	e	20	110	23
1-adamantyl	H	t-Bu	a	20	12	94
t-Bu	H	1-adamantyl	a	20	12	89
1-adamantyl	H	1-adamantyl	a	20	12	90
Me	Me	t-Bu	a	0	3.5	50

[a] KOH, C_6H_6 or C_6H_5Me, 18-crown-6 ether, stir.
[b] Aq. NaOH, CH_2Cl_2, $(n\text{-}Bu)_4\overset{+}{N}Br^-$.
[c] Aq. NaOH, CH_2Cl_2, $(n\text{-}Bu)_4\overset{+}{N}\ HSO_4^-$.
[d] Aq. NaOH, CH_2Cl_2, $(n\text{-}Bu)_4\overset{+}{N}I^-$.
[e] Aq. NaOH, CH_2Cl_2, $PhCH_2\overset{+}{N}Et_3Br^-$.

$$XCH_2CR^1R^2CONHR^3 \xrightarrow[\text{catalyst, stir, room temp.}]{\text{Phase transfer}} \text{azetidinone } (R^1, R^2, N\text{-}R^3) \qquad (699)$$

TABLE 47. Preparation of β-lactams from β-bromopropionamides using a phase transfer catalyst

R^1	R^2	R^3	X	Phase transfer catalyst	Solvent	Time (h)	Yield (%)	Reference
H	H	n-Pr	Br	*a*	CH_2Cl_2	0.5	67	1057
H	H	n-Pr	Br	*a*	THF	0.5	94	1057
H	H	c-Hex	Br	*a*	CH_2Cl_2	0.5	63	1057
H	H	c-Hex	Br	*a*	THF	0.5	74	1057
H	H	$PhCH_2CH_2$	Br	*a*	CH_2Cl_2	0.5	83	1057
H	H	$PhCH_2CH_2$	Br	*a*	THF	0.5	50	1057
H	H	$PhCH_2$	Br	*a*	CH_2Cl_2	0.5	86	1057
H	H	$PhCH_2$	Cl	*b*	CH_2Cl_2	100	trace	1058
H	H	p-$AnCH_2$	Br	*a*	CH_2Cl_2 + MeCN (19:1)	0.5	85	1057
H	H	PhCH(COOMe)	Br	*a*	CH_2Cl_2	0.5	83	1057
H	H	$MeOOCCH_2$	Br	*a*	CH_2Cl_2	0.5	50	1059
H	H	MeOOCCHMe	Br	*a*	CH_2Cl_2	0.5	52	1059
H	H	$MeOOCCH_2CHMe$	Br	*a*	CH_2Cl_2	0.5	60	1059
H	H	Ph	Cl	*b*	CH_2Cl_2	70	5	1058
H	H	Ph	Br	*a*	CH_2Cl_2	0.5	94	1057
H	H	p-An	Br	*a*	CH_2Cl_2	0.5	92	1057
H	H	p-ClC_6H_4	Br	*a*	CH_2Cl_2 + MeCN (19:1)	0.5	94	1057
H	H	p-$O_2NC_6H_4$	Br	*a*	CH_2Cl_2 + MeCN (19:1)	0.5	81	1057
H	H	α-naphthyl	Br	*a*	CH_2Cl_2 + MeCN (19:1)	0.5	91	1057
Me	Me	$PhCH_2$	Cl	*b*	C_6H_6	80	91	1058
Me	Br	$PhCH_2$	Br	*b*	C_6H_6	100	91	1058
Me	Br	$PhCH_2$	Br	*c*	C_6H_6	100	93	1058
Me	Br	$PhCH_2$	Br	none	C_6H_6	100	26	1058
Me	Me	Ph	Cl	*b*	C_6H_6	9	96	1058
Me	Me	Ph	Cl	*c*	C_6H_6	9	96	1058
Me	Br	Ph	Br	*b*	C_6H_6	4	95	1058

[a]Pulverized KOH + $(n\text{-Bu})_4\overset{+}{N}Br^-$.

[b]Durolite A-109 (Cl^- form, polystyrene quaternary type I).

[c]$PhCH_2\overset{+}{N}Et_3Cl^-$.

$$2\,R^1CHXCONHR^2 \xrightarrow[CH_2Cl_2,\ 50\%\ NaOH]{Cat.,\ r.t.,} \text{1,4-}R^2\text{-3,6-}R^1\text{-piperazine-2,5-dione} \quad (700)$$

R^1	X	R^2	Phase transfer catalyst	Time (h)	Yield (%)
H	Cl	$PhCH_2$	none	24	14
H	Cl	$PhCH_2$	*a*	24	88
H	Cl	$PhCH_2$	*b*	24	46
H	Cl	$PhCH_2$	*c*	24	51
H	Cl	Ph	*a*	18	64
Me	Br	$PhCH_2$	*a*	103	trace
Me	Br	$PhCH_2$	*c*	103	trace
Me	Br	Ph	*a*	24	64

[a]Durolite A-109 (Cl^- form, polystyrene quaternary type I).
[b]$(n\text{-Bu})_4\overset{+}{N}\,I^-$.
[c]$PhCH_2\overset{+}{N}Et_3\,Cl^-$.

$$(BrCH_2CMeBrCONH)_2Y \xrightarrow[50\%\ NaOH,\ CH_2Cl_2,\ r.t.]{Durolite\ A-109} \text{(3-bromo-3-methyl-2-oxoazetidin-1-yl)-Y-(3-bromo-3-methyl-2-oxoazetidin-1-yl)} \quad (701)$$

Y	Time (h)	Yield (%)
$(CH_2)_2$	3	97
1,2-C_6H_4	3	95
1,3-C_6H_4	1	98
1,4-C_6H_4	1	96
p-$C_6H_4SO_2C_6H_{4}$-p	1	89

$$BrCHR^1(CH_2)_nCHBrCONHR^2 \xrightarrow[50\%\ aq.\ NaOH,\ CH_2Cl_2]{Cat.,\ r.t.,} \text{3-Br-}N\text{-}R^2\text{-lactam ring with }(CH_2)_n\text{ and }R^1 \quad (702)$$

n	R^1	R^2	Phase transfer catalyst	Time (h)	Yield (%)
1	H	$PhCH_2$	none	27	65
1	H	$PhCH_2$	*a*	27	96
1	H	$PhCH_2$	*b*	27	91
1	H	Ph	*a*	2.5	94
1	Me	$PhCH_2$	*a*	45	53
1	Me	Ph	*a*	4.5	72
2	H	$PhCH_2$	*a*	34	89
2	H	Ph	*a*	5.5	92
3	H	$PhCH_2$	*a*	150	—
3	H	$PhCH_2$	*b*	150	—
3	H	Ph	*a*	95	63

[a]Durolite A-109 (Cl^- form, polystyrene quaternary type I).
[b]$PhCH_2\overset{+}{N}Et_3Cl^-$.

$$R^1O_2CCMe(CH_2Cl)CONHR^2 \xrightarrow[\text{stir r.t., 40 min}]{\text{KOH, THF,}} \text{oxirane(Me)(CONHR}^2) + R^1CO_2\text{-azetidinone(Me, N-R}^2) \quad (703)$$

R^1	R^2	Ratio of oxirane:azetedinone
Ph	*t*-Bu	4:1
Ph	*i*-Pr	1:9
PhCH=CH	*t*-Bu	4:1

Another approach to the synthesis of *β*-lactams under phase transfer conditions has been reported[1060,1061] which involves the cyclization of 2-acyloxy-3-chloropropan- and 2-acyloxy-3-chloro-2-chloromethylpropanamides by anionic activation with cesium fluoride. Whereas reaction of 2-acyloxy-3-chloropropanamides with potassium hydroxide in tetrahydrofuran produces[1060] a mixture of the corresponding oxiranes and azetidinones (equation 703), treatment[1060] of the same starting materials with cesium fluoride alone without solvent at 85–90 °C (equation 704), or with cesium fluoride in tetrahydrofuran in the presence of benzyltriethylammonium chloride as catalyst (equation 705), produces high yields of the azetidinones exclusively.

$$R^1CO_2CMe(CH_2Cl)CONHR^2 \xrightarrow[\text{6 h}]{\text{CsF, 85–90 °C}} R^1CO_2\text{-azetidinone(Me, N-R}^2)$$

R^1 = Ph ; Me
R^2 = *i*-Pr ; *t*-Bu
Yield(%) = 96 ; 95 (704)

$$R^3CO_2CR^1(CH_2R^2)CONHR^4 \xrightarrow[\text{reflux}]{\text{CsF, THF, stir, } PhCH_2\overset{+}{N}Et_3\,Cl^-} R^3CO_2\text{-(azetidin-2-one, 3-}R^1\text{, 4-}R^2\text{, N-}R^4) \quad (705)$$

R^1	R^{2a}	R^{3a}	R^4	Yield (%)
Me	H	Ph	*t*-Bu	95
Me	H	Ph	*i*-Pr	81
Me	H	Me	*t*-Bu	91
Me	H	*n*-Pr	*t*-Bu	85
Me	H	PhCH=CH	*t*-Bu	93
Me	Me	Ph	*t*-Bu	91[b]
Me	Me	Me	*t*-Bu	97[b]
Me	Me	*n*-Pr	*t*-Bu	97[b]
CH_2Cl	H	Ph	*t*-Bu	94
CH_2Cl	H	Me	*t*-Bu	94
Me	H	Me	*i*-Pr	78

[a]When $R^3 = H$ the mixture is refluxed for 1 h, when $R^2 = Me$ the mixture is refluxed for 12 h.
[b]A *cis, trans* mixture of products is obtained.

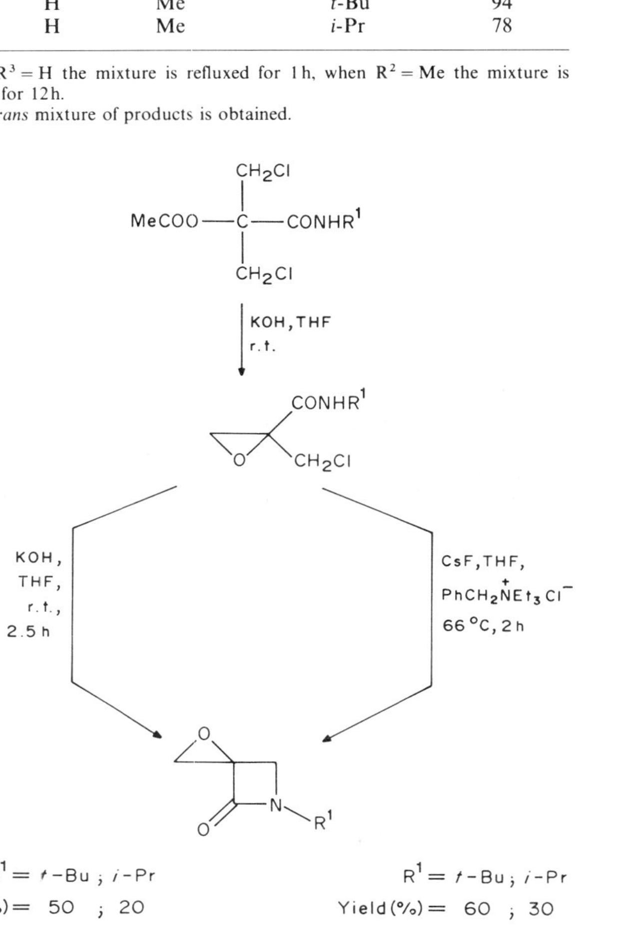

(706)

When 2-acyloxy-3-chloro-2-chloromethylpropanamides are used as the starting materials, slightly different results[1062] are obtained. Treatment with potassium hydroxide in tetrahydrofuran at room temperature converts the propanamides to the corresponding oxiranes which can be isolated, and then treated with either additional potassium hydroxide or cesium fluoride in tetrahydrofuran in the presence of benzyltriethylammonium chloride catalyst to afford epoxy substituted β-lactams (equation 706).

The two steps represented in equation (706) may be combined into a consecutive procedure[1061] by treating the 2-acyloxy-3-chloro-2-chloromethylpropanamides with potassium hydroxide, followed by treatment with cesium fluoride without isolation of any intermediate oxiranes (equation 707).

$$R^2COOC(CH_2Cl)_2CONHR^1 \xrightarrow[\text{r.t.}]{\text{KOH, THF,}} \left[\text{oxirane}(CONHR^1)(CH_2Cl)\right] \xrightarrow[\text{reflux}]{\text{CsF, THF},\ PhCH_2\overset{+}{N}Et_3Cl^-} \mathbf{(82)} \text{ and/or } \mathbf{(83)} \quad (707)$$

R^1	R^2	Time (h)	Product	Yield (%)
t-Bu	Me	6	**82**	80
t-Bu	Ph	1–5	**82**+	35
			83	30
i-Pr	Me	6	**82**	70
i-Pr	Ph	6	**82**+	50
			83	50
2,6-$Me_2C_6H_3$	Me	2	**82**	80

Finally, it is possible to prepare the β-lactam products directly from the 2-acyloxy-3-chloro-2-chloromethylpropanamides without intermediate formation or contamination with by-products, by treatment with cesium fluoride only (equation 708).

$$R^2COOC(CH_2Cl)_2CONHR^1 \xrightarrow[PhCH_2\overset{+}{N}Et_3Cl^-,\ 66\,^\circ C,\ 1\,h]{\text{CsF, THF,}} \text{3-(}R^2COO\text{)-3-(}ClCH_2\text{)-1-}R^1\text{-azetidin-2-one} \quad (708)$$

R^1	R^2	Yield (%)
t-Bu	Me	94
t-Bu	Ph	94
i-Pr	Me	96
i-Pr	Ph	96
2,6-$Me_2C_6H_3$	Me	82
CH_2COOEt	Me	68
CH_2COOEt	Ph	97

Cyclization of N-(alk-2-enyl)-α-chloro-α-(methylthio)acetamides can be accomplished[1062] by treatment with stannic chloride in methylene chloride at room temperature to produce the corresponding γ-lactams. It is interesting to note that when the N-(but-2-enyl) starting material is treated as indicated above, a 70% yield of 4-vinyl-1-methyl-3-(methylthio)pyrrolidin-2(1H)-one as a 71:29 mixture of two stereoisomers is obtained[1060] as the only product (equation 709), but when the N-(prop-2-enyl) starting material is used, two products, 1,4-dimethyl-3-methylthio-5H-pyrrol-2(1H)-one and 4-chloromethyl-1-methyl-3(methylthio)pyrrolidin-2-one, are obtained (equation 710). Also interesting is the fact that the N-(but-2-enyl) compound may be cyclized by chromatography on silica gel or simply by heating without solvent (equation 709), whereas similar treatment of the N-(prop-2-enyl) starting material gave only polymeric material[1062].

MeN(COCHClSMe)(CH$_2$CH=CHMe) *trans* → [$SnCl_4$, CH_2Cl_2, r.t. (70%); 130 °C (55%); silica gel (52%)] → 1-methyl-3-(SMe)-4-(CH=CH$_2$)pyrrolidin-2-one (709)

MeN(COCHClSMe)(CH$_2$CH=CH$_2$) → [$SnCl_4$, CH_2Cl_2, r.t.] → 1,4-dimethyl-3-(SMe)-pyrrol-2-one 18% + 1-methyl-3-(SMe)-4-(CH_2Cl)pyrrolidin-2-one 42% (710)

Using N-(but-3-enyl)-α-chloro-α-(methylthio)acetamide as the reactant afforded[1062] the corresponding δ-lactam products (equation 711).

MeN(COCH(Cl)SMe)(CH$_2$CH$_2$CH=CH$_2$) → [1. $SnCl_4$, CH_2Cl_2, 0 °C; 2. r.t., stir 2 h] → MeN lactam with SMe (unsaturated) 8% + MeN lactam with SMe, Cl 36%

(711)

N-Allyltrichloroacetamides have been reported[1063,1064] to undergo a novel copper or ruthenium-catalysed cyclization to produce β-butyrolactams[1063] or bicyclic lactams[1064] depending upon the specific structure of the acetamides used. With N-allyltrichloroacetamides where the allyl double bond is part of a linear system, reaction with a ruthenium catalyst according to procedure A or with a copper catalyst according to procedure B affords[1063] trichlorinated γ-butyrolactams (equation 712).

$$Cl_3CCONR^1CR^2_2CR^4{=}CR^3_2 \xrightarrow[\text{B}]{\text{Procedure A or}} \text{(trichlorinated γ-butyrolactam: ring bearing Cl, Cl; } R^4\text{, } CClR^3_2\text{; } R^2, R^2\text{; N–}R^1\text{; C=O)} \quad (712)$$

Procedure A: Benzene, heated in a sealed tube at 140 °C in the presence of $RuCl_2(PPh_3)_3$

B: Acetonitrile containing CuCl heated to 140 °C

R^1	R^2	R^3	R^4	Procedure	Temp. (°C)	Time (h)	Yield (%)
H	H	H	H	A	140	2	68
H	H	H	H	A	110	24	52
H	H	H	H	B	140	20	57
$CH_2CH{=}CH_2$	H	H	H	A	140	1	84
$CH_2CH{=}CH_2$	H	H	H	A	110	1	88
$CH_2CH{=}CH_2$	H	H	H	A	80	20	58
$CH_2CH{=}CH_2$	H	H	H	B	140	1	87
$CH_2CH{=}CH_2$	H	H	H	B	140	3	90
$CH_2CH{=}CH_2$	H	H	H	B	110	3	90
$CH_2CH{=}CH_2$	H	H	H	B	80	20	81
$PhCH_2$	H	H	H	A	140	3	68
H	Me	H	H	A	140	3	66
H	H	Me	H	A	140	3	82
H	H	H	Me	A	140	3	23[a]

[a]In this case a 17% yield of the pyrrolidone shown below was obtained.

(piperidone structure: ring bearing Cl, Cl; Me, Cl; C=O; N–H)

With N-allyltrichloroacetamides where the allylic double bond is part of a cyclobutenyl or a cyclopentenyl ring system and using the same two procedures discussed above, it produces[1064] bicyclic lactam products with an exclusively *cis* stereochemistry (equation 713). When the allylic double bond was part of a cyclohexenyl ring system a spiro-lactam

$$\text{(cycloalkenyl with } R^2\text{, } (CH_2)_n\text{, } Cl_3CCONR^1\text{)} \xrightarrow[\text{B}]{\text{Procedure A or}} \text{(bicyclic lactam: Cl, Cl; } R^2\text{; C(Cl)H; } (CH_2)_n\text{; N–}R^1\text{; C=O)} \quad (713)$$

n	R^1	R^2	Procedure	Temp. (°C)	Time (h)	Yield (%)
1	H	H	B	140	3	71
1	H	H	A	140	3	71
1	$PhCH_2$	H	B	110	1	89
1	$PhCH_2$	H	A	110	1	88
2	H	H	B	140	3	76
2	H	H	A	140	3	71
2	$PhCH_2$	H	B	110	1	91
2	$PhCH_2$	H	A	110	1	90
2	Me	Ph	B	120	3	45
2	Me	Ph	A	140	3	20
2	$PhCH_2COO$	Ph	B	110	1	78
2	$PhCH_2COO$	Ph	A	140	1	50
2	H	Ph	B		N.R.	
2	H	Ph	A		N.R.	
2	Me	3,4-$(MeO)_2C_6H_3$	B	120	2	47

was obtained[1064] (equation 714), whereas if the double bond was exocyclic to the cyclohexenyl ring a hexahydro-oxindole bearing an angular chloromethyl group was obtained[1064] (equation 715).

$Cl_3CCON(CH_2Ph)CH_2$–(cyclohexenyl) → [110 °C, 1 h; Procedure A (89%); Procedure B (81%)] → 3,3-dichloro-1-benzyl-spiro lactam bearing Cl on cyclohexane ring (714)

$Cl_3CCON(CH_2Ph)$–(2-methylenecyclohexyl) → [110 °C, 1 h; Procedure A (89%); Procedure B (81%)] → 3,3-dichloro-3a-(CH_2Cl)-1-benzyl-hexahydro-oxindole (715)

α-Methylene-β-lactams have also been prepared from halo amides using a base catalysed cyclization procedure with two different types of starting materials. Reaction of N-substituted 2-(bromomethyl)propenamides, prepared from N-substituted 2-[(tri-*n*-butylstannyl)methyl]propenamides by reaction with dioxane dibromide in tetrahydrofuran, with sodium hydride or potassium *t*-butoxide produces[1065] the corresponding N-substituted 3-methylene-2-azetidinones in good yields (equation 716). In at least one case a one-pot synthesis of the α-methylene-β-lactam was attempted[1065] using the stannyl starting material (equation 717).

$H_2C{=}C(CH_2Sn(Bu\text{-}n)_3)CONHR$ —(dioxane dibromide; THF, 0 °C, 1.5 h)→ $H_2C{=}C(CH_2Br)CONHR$ —(1. NaH or KOBu-*t*, THF, −78 °C; 2. r.t.)→ α-methylene-β-lactam ($H_2C{=}$, N–R) (716)

R	Base	Reaction conditions Temp. (°C)/Time (h)	Yield (%)
(*S*) PhCHMe	NaH	−78°/2 then 0°/1	87
(*R*) PhCHMe	*t*-BuOK	−78°/1 then 0°/1	89
(*S*) $PhCH_2CHCH_2OMe$	NaH	−78°/1 then r.t./2.5	86
(*R*) $PhCH_2CHCH_2OMe$	NaH	−78°/1 then r.t./2.5	80
(*S*) *i*-$PrCH_2CHCH_2OMe$	NaH	−78°/1 then r.t./18	60
(*R*) *i*-$PrCH_2CHCH_2OMe$	NaH	−78°/1 then r.t./18	63
(*S*) *i*-$PrCHCH_2OMe$	NaH	−78°/1 then r.t./18	82
(*R*) *i*-$PrCHCH_2OMe$	NaH	−78°/1 then r.t./18	89

$H_2C{=}C(CH_2Sn(Bu\text{-}n)_3)CONHCHMePh$ —(1. THF, 0 °C, dioxane dibromide, Ar, stir 1.5 h; 2. cool to −78 °C, KOBu-*t*, stir 1 h; 3. warm to 0 °C, stir 1 h)→ α-methylene-β-lactam ($H_2C{=}$, N–CHMePh) 78% (717)

The second type of starting material used to prepare α-methylene-β-lactams were N-substituted-β,β′-dibromoisobutyramides which were treated with sodium hydroxide under phase transfer conditions[1066,1067] (equation 718).

$(BrCH_2)_2CHCONHR$ —(40% NaOH, solvent, Phase transfer cat., stir r.t.)→ α-methylene-β-lactam ($H_2C{=}$, N–R) (718)

R	Solvent	Phase transfer catalyst[a]	Time (h)	Yield (%)	Reference
Et	CCl_4	A	18	18	1066
c-Hex	CCl_4	A	18	40	1066
i-Bu	CCl_4	A	18	56	1066
$PhCH_2$	CH_2Cl_2	B	41	23.6[b]	1067
Ph	CCl_4	A	18	86	1066
p-An	CCl_4	A	18	96	1066
p-$O_2NC_6H_4$	CCl_4	A	18	83	1066
p-NCC_6H_4	CCl_4	A	18	76	1066
2,6-$Me_2C_6H_3$	CCl_4	A	18	92	1066
2,6-$Cl_2C_6H_3$	CCl_4	A	18	78	1066
3,4-$Cl_2C_6H_3$	CCl_4	A	18	92	1066
2,4,6-$Br_3C_6H_2$	CCl_4	A	18	82	1066

[a] A = (*n*-Pent)$\overset{+}{N}Et_3Br^-$; B = $PhCH_2\overset{+}{N}Et_3Cl^-$.
[b] Also obtained was 35.2% of *N*-benzyl-3-bromomethyl-2-azetidinone.

At least one example of the use of a halo amide to produce a polycyclic lactam via an intramolecular Reformatsky reaction has been reported[1068] recently (equation 719).

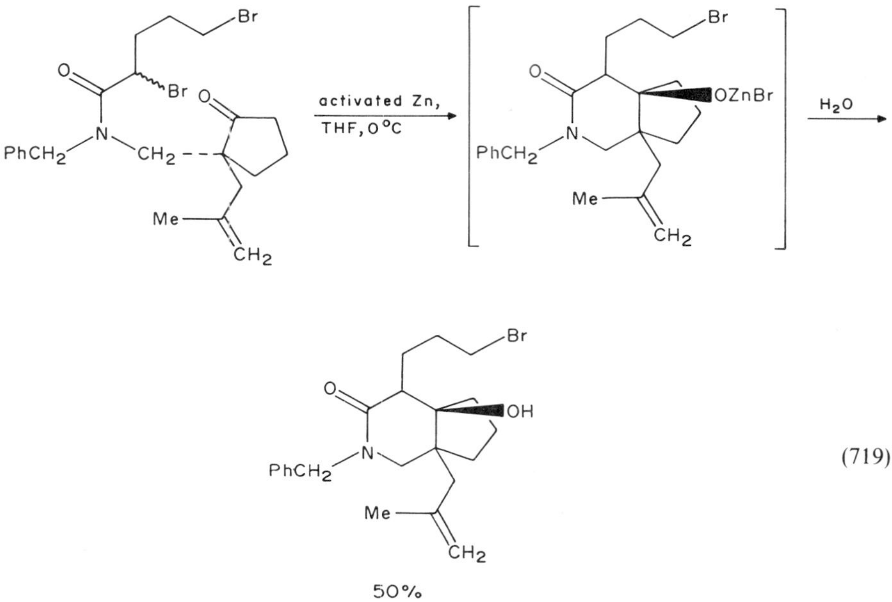

activated Zn, THF, 0 °C; H_2O

50%

(719)

Substituted haloacetamides have been used to produce[1069] substituted β-lactams by electrochemical cyclization and bond formation between carbons 3 and 4 of the azetidine ring. Depending upon the structure of the haloacetamide employed, two methods have been used to produce the corresponding β-lactams. In both methods the potential of the Calomel-type reference electrode was -0.029 V *vs* SCE, the cathode was a mercury pool under nitrogen and the carbolyte employed was dimethylformamide containing tetraethylammonium perchlorate. With acetamides having a halogen atom and a suitable leaving group, which may also be a halogen atom, attached to the α-carbon, the reaction was carried out by stepwise addition of the probase R—X in dimethylformamide to a

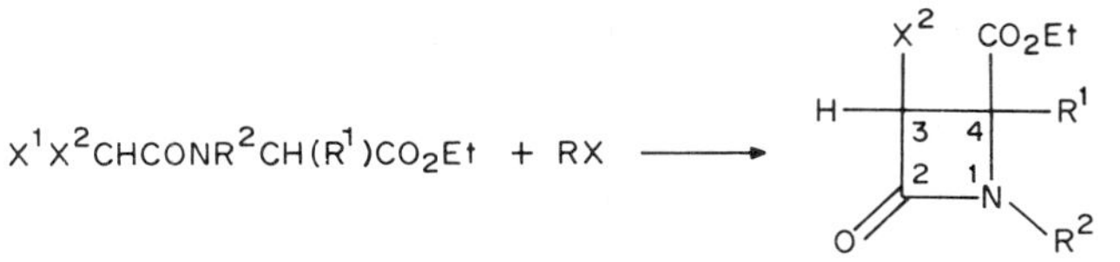

$$X^1X^2CHCONR^2CH(R^1)CO_2Et + RX \longrightarrow \qquad (720)$$

X^1	X^2	R^1	R^2	R—X	Yield (%)
Br	H	CO_2Et	CH_2CO_2Et	$BrCH(CO_2Et)_2$	90
Br	H	CO_2Et	*p*-An	$BrCH(CO_2Et)_2$	88
Cl	H	CO_2Et	*p*-An	$BrCH(CO_2Et)_2$	93
Cl	Cl	CO_2Et	*p*-An	$BrCH(CO_2Et)_2$	84
Cl	H	H	CH_2CO_2Me	$BrCMe_2CO_2Et$	67

solution of the haloacetamide in the carbolyte (equation 720). With acetamides containing a single halogen atom which acts as the leaving group attached to the α-carbon, the reaction was carried out by stepwise addition of the bromide in dimethylformamide to the carbolyte (equation 721).

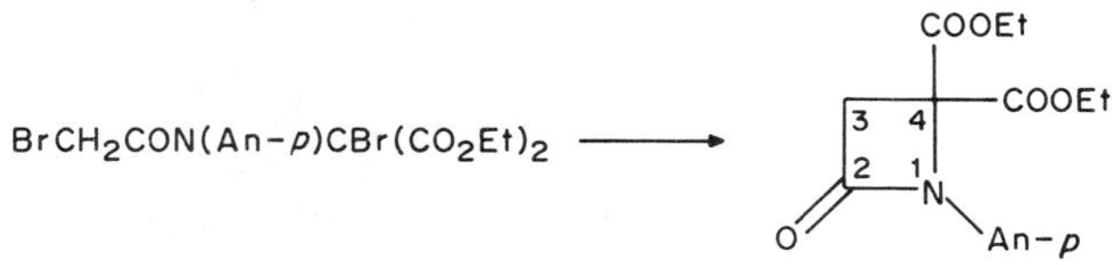

$$BrCH_2CON(An\text{-}p)CBr(CO_2Et)_2 \longrightarrow \text{1-(An-}p\text{)-4,4-bis(COOEt)azetidin-2-one} \quad 85\% \qquad (721)$$

Several of the methods used to produce lactams involve the *in situ* preparation of halo amides which, under the conditions of the reaction, directly cyclize to the lactam products. This approach is referred to in the literature as the halocyclization of amides and has been used to effect cyclization with a variety of starting materials, which include unsaturated monoamides[1070] (equation 722), unsaturated bisamides[1071] (equation 723) and acyl sulphonates[1072] (equations 724 and 725).

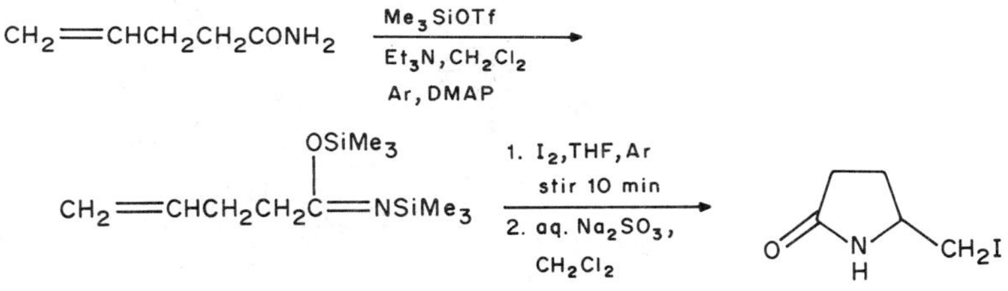

$$CH_2{=}CHCH_2CH_2CONH_2 \xrightarrow[\text{Ar, DMAP}]{Me_3SiOTf,\ Et_3N,\ CH_2Cl_2} CH_2{=}CHCH_2CH_2C(OSiMe_3){=}NSiMe_3 \xrightarrow[\text{2. aq. } Na_2SO_3,\ CH_2Cl_2]{\text{1. } I_2,\text{THF, Ar, stir 10 min}} \text{5-(}CH_2I\text{)pyrrolidin-2-one} \qquad (722)$$

64% (overall)

Other iodolactams prepared include:

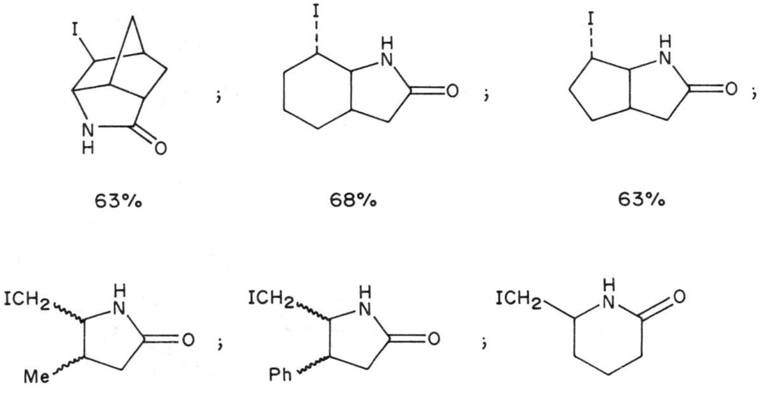

63%; 68%; 63%;

58% (*trans*:*cis* = 11:1); 52% (*trans*:*cis* = 22:1); 35%

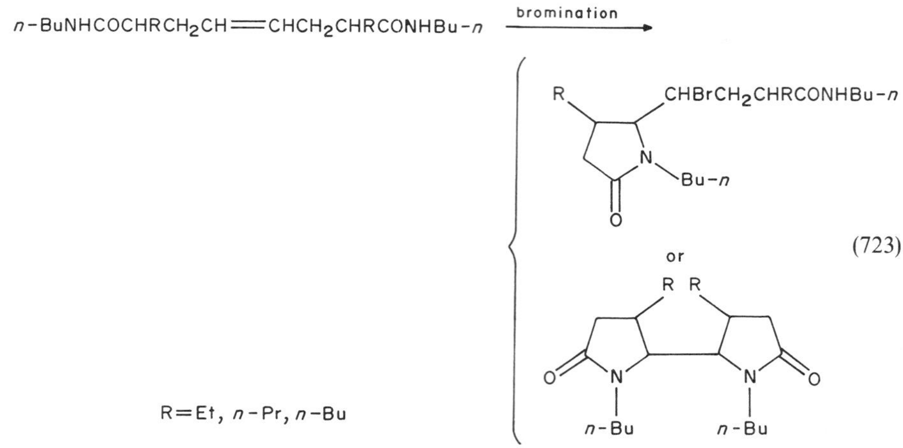

n-BuNHCOCHRCH$_2$CH=CHCH$_2$CHRCONHBu-n $\xrightarrow{\text{bromination}}$

R CHBrCH$_2$CHRCONHBu-n

N Bu-n

O

or (723)

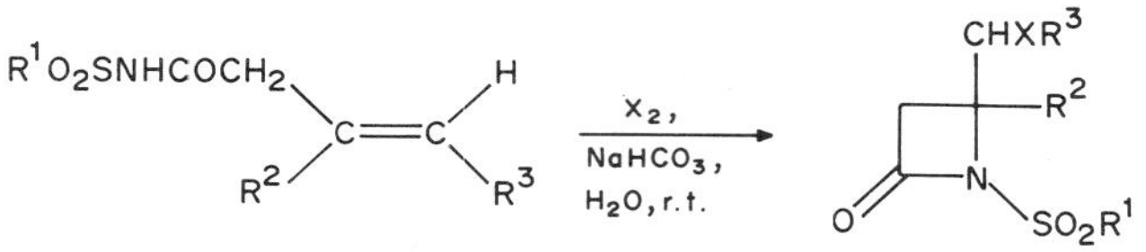

R R

O N N O

n-Bu n-Bu

R=Et, n-Pr, n-Bu

R^{1}O$_2$SNHCOCH$_2$ C=C H R^2 R^3 $\xrightarrow[\text{NaHCO}_3,\ \text{H}_2\text{O, r.t.}]{\text{X}_2,}$ CHXR^3 R^2 O N SO$_2$R^1 (724)

R^1	R^2	R^3	X	Yield (%)
p-An	H	H	Br	43
p-An	H	H	I	83
MeO	H	H	Br	12
MeO	H	H	I	12
EtO	H	H	I	77
Cl_3CCH_2O	H	H	Br	40
Cl_3CCH_2O	H	H	I	95
Cl_3CCH_2O	Me	H	I	95
Cl_3CCH_2O	H	Me	I	90[a]

[a]The product in this case was the iodolactam shown below.

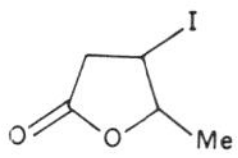

I

O O Me

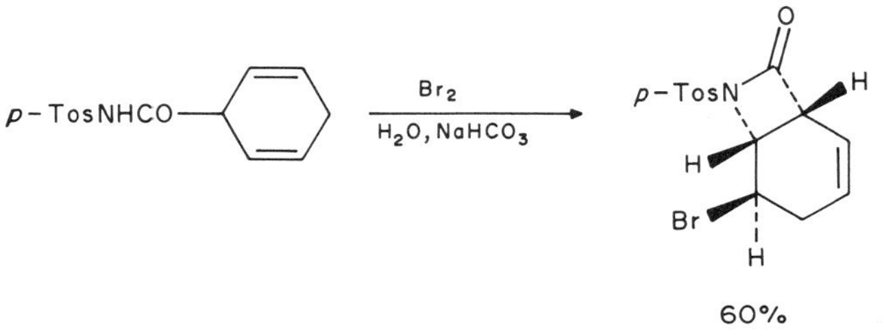

p-TosNHCO $\xrightarrow[\text{H}_2\text{O, NaHCO}_3]{\text{Br}_2}$ O p-TosN H H Br H (725)

60%

At least two condensation reactions which utilize halo amides to produce lactams have been reported in the recent literature[1073,1074]. In the first such reaction the halo amide is produced as a non-isolatable intermediate which immediately cyclizes to the lactam product[1073] (equation 726), while in the second report the α-bromo amide is present as one of the condensation reagents[1074] (equation 727). The mechanism for this conversion can be envisioned[1074] to occur via two possible pathways, initial cyclization to an α-lactam and subsequent reaction, or by direct conversion (equation 728).

$$Cl(CH_2)_{2+n}COCl + H_2NCN \longrightarrow [Cl(CH_2)_{2+n}CONHCN] \longrightarrow$$ (N-cyano lactam, ring with $(CH_2)_n$)

$n = 1$ (71%), 2 or 3

(726)

$$t\text{-BuCHBrCONHBu-}t + \overset{+}{Li}\,\overset{-}{C}\equiv CMe \longrightarrow$$ 1,2-di-t-Bu-5-Me-pyrrol-3-one

54%

(727)

t-BuCHBrCONHBu-t ⟶ t-BuCH–NBu-t (α-lactam, C=O)

Left pathway: $\overset{+}{Li}\,\overset{-}{C}\equiv CMe$ → t-Bu–CH(Br)–C(O⁻)(NHBu-t)–C≡CMe Li → t-BuCH(NHBu-t)COC≡CMe → ($\overset{+}{Li}\,\overset{-}{C}\equiv CMe$)

Right pathway: α-lactam → ($\overset{+}{Li}\,\overset{-}{C}\equiv CMe$) → t-BuCH(t-BuN⁻)–C(=O)–C≡C–Me Li⁺ → (H^+) → 1,2-di-t-Bu-5-Me-pyrrol-3-one

(728)

TABLE 48. Preparation of β-lactams from hydroxy amides[a]

Hydroxy amide	Conditions			β-Lactam product	Yield (%)	Reference
	Reactants	Temp. (°C)	Time			
PI—N, H, CH_2OH, $NHCH_2R$, O				PI—N, H, N, CH_2R, O		
R = CH_2Ph	DEAD, TPP	r.t.	5–10 min		26	1050
R = $CH_2C_6H_4OMe$-*p*	DEAD, TPP	r.t.	5–10 min		9	1050
R = CO_2Me	DEAD, TPP	r.t.	5–10 min		22	1050
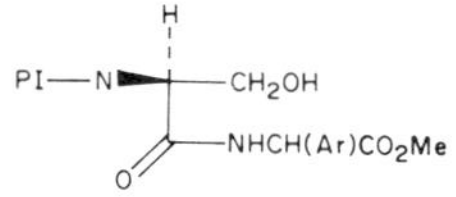 PI—N, H, CH_2OH, $NHCH(Ar)CO_2Me$, O				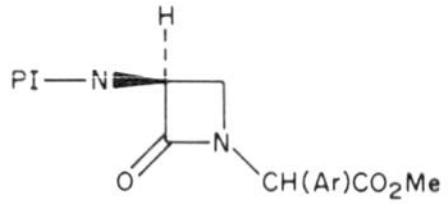 PI—N, H, N, $CH(Ar)CO_2Me$, O		
Ar = Ph	DEAD, TPP	r.t.	5–10 min		30–81[b]	1050
Ar = $C_6H_4OCH_2Ph$-*p*	DEAD, TPP	r.t.	5–10 min		87–93[b]	1050, 1075
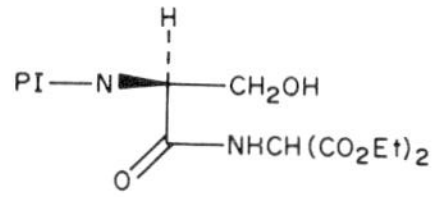 PI—N, H, CH_2OH, $NHCH(CO_2Et)_2$, O	DEAD, TPP	r.t.	5–10 min	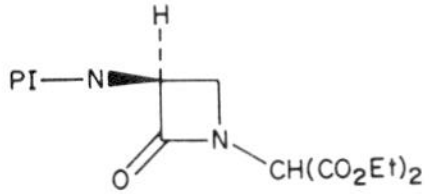 PI—N, H, N, $CH(CO_2Et)_2$, O	52	1050
	DEAD, TPP, DIAD	r.t.	1 h	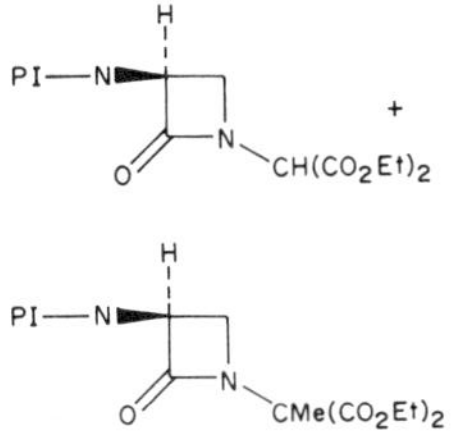 PI—N, H, N, $CH(CO_2Et)_2$, O + PI—N, H, N, $CMe(CO_2Et)_2$, O	58	1050

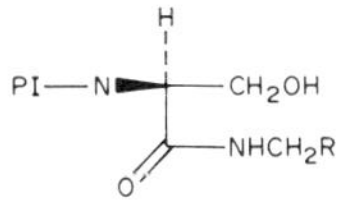

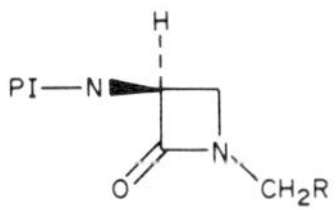

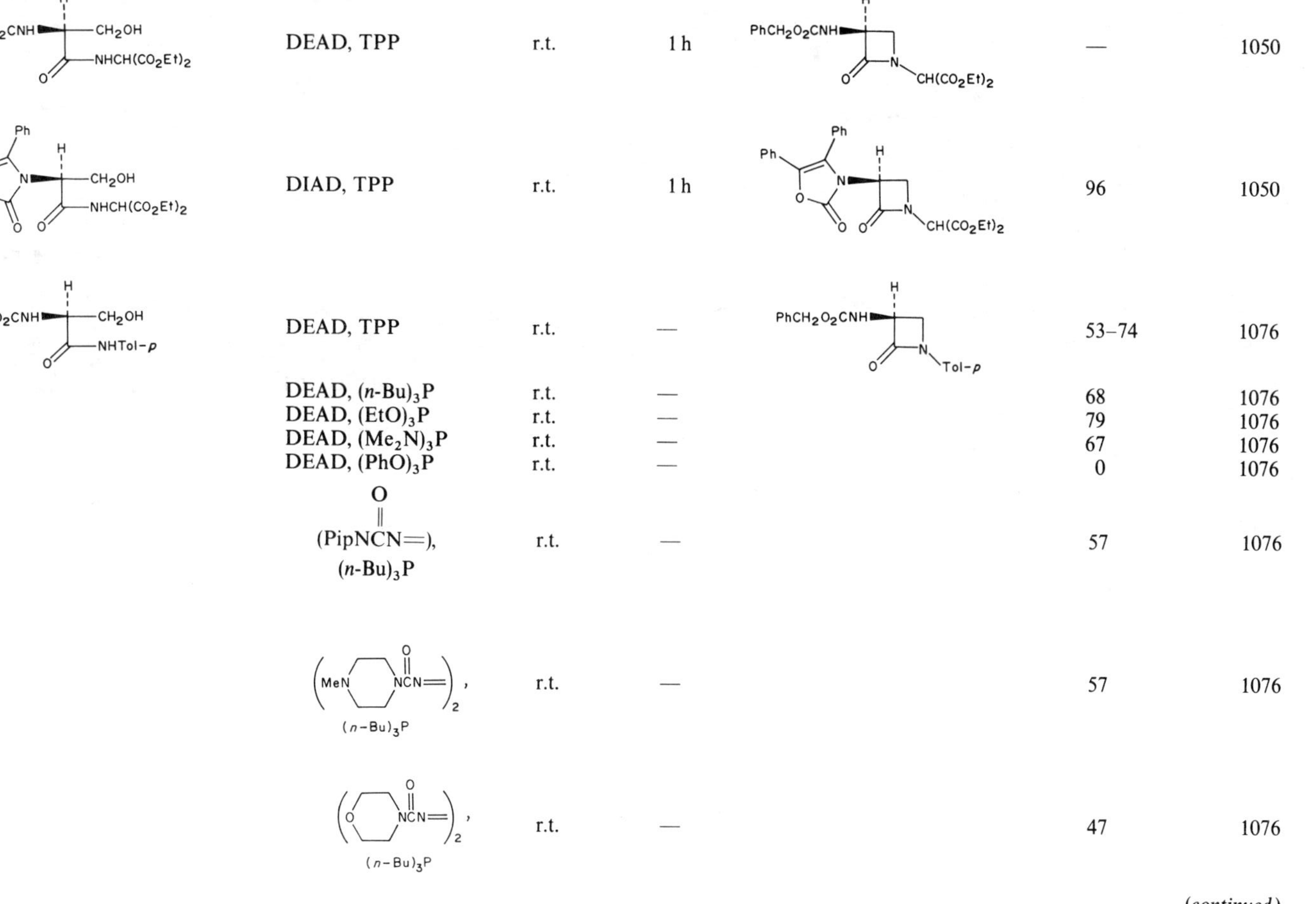

Substrate	Reagents	Temp.	Time	Product	Yield (%)	Ref.
$PhCH_2O_2CNH$–CH(CH$_2$OH)–C(=O)NHCH$(CO_2Et)_2$	DEAD, TPP	r.t.	1 h	azetidinone, N–CH$(CO_2Et)_2$	—	1050
(4,5-diphenyl-2-oxo-oxazolin-3-yl)–CH(CH$_2$OH)–C(=O)NHCH$(CO_2Et)_2$	DIAD, TPP	r.t.	1 h	azetidinone, N–CH$(CO_2Et)_2$	96	1050
$PhCH_2O_2CNH$–CH(CH$_2$OH)–C(=O)NHTol-*p*	DEAD, TPP	r.t.	—	azetidinone, N–Tol-*p*	53–74	1076
	DEAD, $(n\text{-}Bu)_3P$	r.t.	—		68	1076
	DEAD, $(EtO)_3P$	r.t.	—		79	1076
	DEAD, $(Me_2N)_3P$	r.t.	—		67	1076
	DEAD, $(PhO)_3P$	r.t.	—		0	1076
	(PipNC(=O)N=)$_2$, $(n\text{-}Bu)_3P$	r.t.	—		57	1076
	(MeN(CH$_2$CH$_2$)$_2$NC(=O)N=)$_2$, $(n\text{-}Bu)_3P$	r.t.	—		57	1076
	(O(CH$_2$CH$_2$)$_2$NC(=O)N=)$_2$, $(n\text{-}Bu)_3P$	r.t.	—		47	1076

(continued)

TABLE 48. (*continued*)

Hydroxy amide	Conditions: Reactants	Conditions: Temp. (°C)	Conditions: Time	β-Lactam product	Yield (%)	Reference
	DEAD, TPP	r.t.	—		42	1076
t-BuCO$_2$NH, H, CH$_2$OH, NHTol-p, O	DEAD, TPP	r.t.	—	t-BuCO$_2$NH, H, N, Tol-p, O	64	1076
t-BuCO$_2$NH, H, CH(OH)Ph, NHTol-p, O	DEAD, TPP	r.t.	—	t-BuCO$_2$NH, H, Ph, H, N, Tol-p, O	83	1076
PhCH$_2$O$_2$CNH, H, CH$_2$OH, NHN=CPh$_2$, O	DEAD, TPP	r.t.	—	PhCH$_2$O$_2$CNH, H, N, N=CPh$_2$, O	30	1076
PhCH$_2$O$_2$CNH, H, CH$_2$OH, NHPyr-2, O	DEAD, TPP	r.t.	3 h	PhCH$_2$O$_2$CNH, H, N, Pyr-2, O	44	1077
PhCH$_2$O$_2$CNH, H, CH$_2$OH, NH, N, S, O	DEAD, TPP	r.t.	3 h	PhCH$_2$O$_2$CNH, H, N, N, S, O	51	1077

$PhCH_2O_2CNH$, CH_2OH, NH–(pyrimidin-2-yl)	DEAD, TPP	r.t.	3 h	$PhCH_2O_2CNH$, N-(pyrimidin-2-yl) azetidinone	29	1077
$PhCH_2O_2CNH$, CH_2OH, NH–pyrimidine with $OCH_2C_6H_4NO_2$-*p*, $OCH_2C_6H_4NO_2$-*p*	DEAD, TPP	r.t.	3 h	$PhCH_2O_2CNH$, azetidinone N-pyrimidine with $OCH_2C_6H_4NO_2$-*p*, $OCH_2C_6H_4NO_2$-*p*	22	1077
$PhCH_2O_2CNH$, CH_2OH, NH–tetrazole (N–R)				$PhCH_2O_2CNH$, azetidinone N-tetrazole (N–R)		
R = H	DEAD, TPP	r.t.	3 h		24	1077
$R = CH_2CO_2CH_2C_6H_4NO_2$-*p*	DEAD, TPP	r.t.	3 h		68	1077
$PhCH_2O_2CNH$, CH_2OH, NH–imidazole (N–CPh_3)	DEAD, TPP	r.t.	3 h	$PhCH_2O_2CNH$, azetidinone N-imidazole (N–CPh_3)	50	1077
$PhCH_2O_2CNH$, CH_2OH, NH–triazole (NH, N–$CH_2CO_2CH_2C_6H_4NO_2$-*p*)	DEAD, TPP	r.t.	3 h	$PhCH_2O_2CNH$, azetidinone N-triazole (NH, N–$CH_2CO_2CH_2C_6H_4NO_2$-*p*)	72	1077
$CH(R)OH$, NHPh amide; R = Me	DEAD, TPP	r.t.	—	azetidinone, R, N–Ph	41	1076

(continued)

TABLE 48. (*continued*)

Hydroxy amide	Conditions: Reactants	Conditions: Temp. (°C)	Conditions: Time	β-Lactam product	Yield (%)	Reference
R = Ph	DEAD, TPP	r.t.	—		61	1076
PI—N, H, CH(R)OH, NHTol-*p*, O				PI—N, H, R, N, Tol-*p*, O		
R = H	DEAD, TPP	r.t.	—		78	1076
R = Me	DEAD, TPP	r.t.	—		41	1076
R = Ph	DEAD, TPP	r.t.	—		58	1076
Me, Me, CH_2OH, $NHCH(CO_2Et)_2$, O	DEAD, TPP	r.t.	1	Me, Me, N, $CH(CO_2Et)_2$, O	—	1050
$PhCH_2O_2CNH$, H, CH(Me)OH, NH, N—N, N, N, $CHPh_2$, O	DEAD, TPP	r.t.	3 h	$PhCH_2O_2CNH$, H, H, Me, N, N—N, N, N, $CHPh_2$, O	70	1077
RO_2CNH, H, CH_2OH, $NHOCH_2Ph$, O				RO_2CNH, H, N, OCH_2Ph, O		
R = *t*-Bu	DEAD, TPP	50	6 h		90	1049
		20	20 h		80	1049
$R = PhCH_2$	DEAD, TPP	50	6 h		82	1049
		20	20 h		54	1049

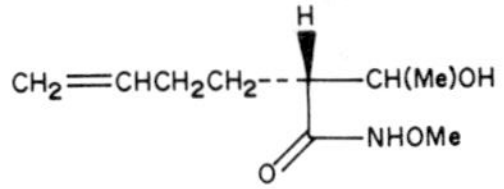 $CH_2{=}CHCH_2CH_2$, H, $CH(Me)OH$, $NHOMe$, O	DEAD, TPP	25	—	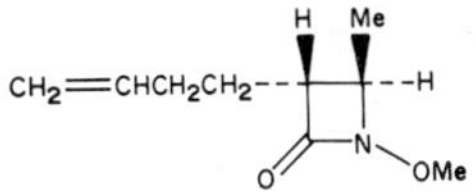 $CH_2{=}CHCH_2CH_2$, H, Me, H, N, OMe, O	75	1078
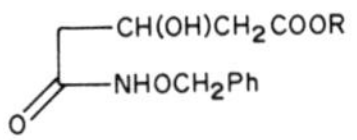 $CH(OH)CH_2COOR$, $NHOCH_2Ph$, O				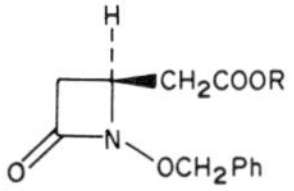 H, CH_2COOR, N, OCH_2Ph, O		
R = Me	DEAD, $(PhO)_3P$	r.t.	16 h		50	1079
R = $PhCH_2$	DEAD, $(PhO)_2PPh$	r.t.	—		35	1079
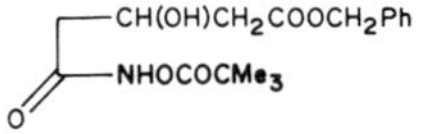 $CH(OH)CH_2COOCH_2Ph$, $NHOCOCMe_3$, O	DEAD, $(PhO)_3P$	r.t.	3–4 days	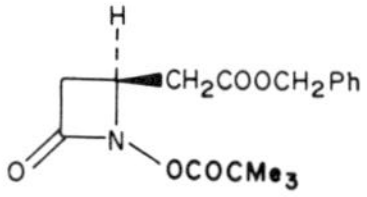 H, CH_2COOCH_2Ph, N, $OCOCMe_3$, O	70	1079
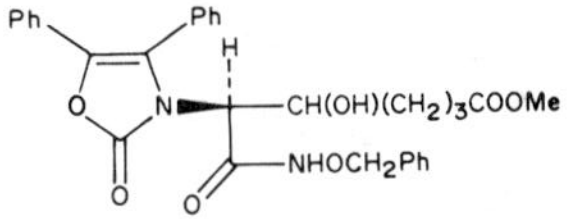 Ph, Ph, O, N, O, H, $CH(OH)(CH_2)_3COOMe$, $NHOCH_2Ph$, O	DEAD, TPP	r.t.	—	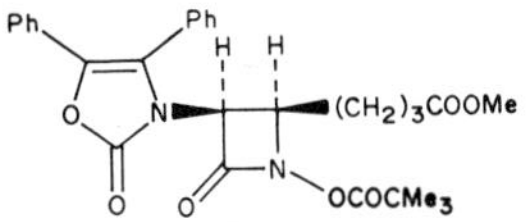 Ph, Ph, O, N, O, H, H, $(CH_2)_3COOMe$, N, $OCOCMe_3$, O	69	1080
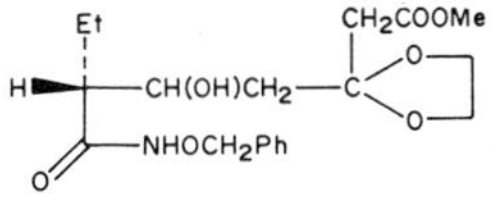 Et, H, $CH(OH)CH_2$, C, O, O, CH_2COOMe, $NHOCH_2Ph$, O	DEAD, TPP	r.t.	1.5 h	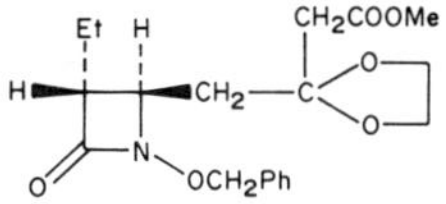 Et, H, H, CH_2, C, O, O, CH_2COOMe, N, OCH_2Ph, O	73	1080
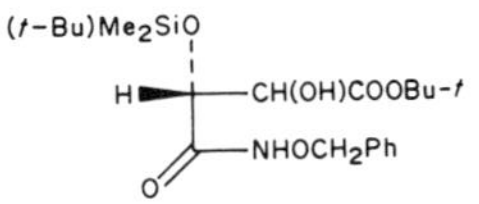 $(t\text{-}Bu)Me_2SiO$, H, $CH(OH)COOBu\text{-}t$, $NHOCH_2Ph$, O	$(MeO_2CN{=})_2$ TPP	r.t.	—	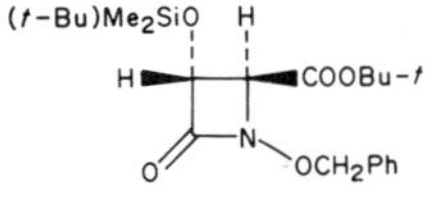 $(t\text{-}Bu)Me_2SiO$, H, H, $COOBu\text{-}t$, N, OCH_2Ph, O	40	1081
	DEAD, TPP	r.t.	—		22	1081
	DIAD, TPP	r.t.	—		7	1081

(*continued*)

TABLE 48. (*continued*)

Hydroxy amide	Conditions: Reactants	Conditions: Temp. (°C)	Conditions: Time	β-Lactam product	Yield (%)	Reference
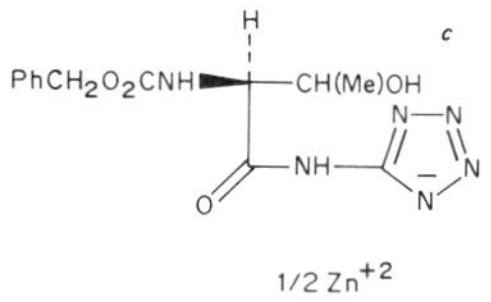 $PhCH_2O_2CNH$–CH(H)(CH(Me)OH)–C(=O)–NH–(tetrazolyl anion) · 1/2 Zn^{+2} [c]	DEAD, TPP	r.t.	12 h	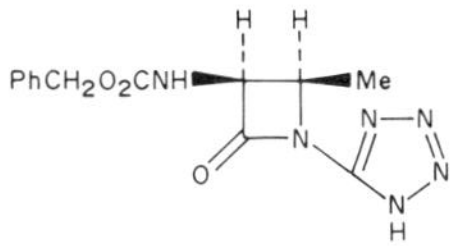 $PhCH_2O_2CNH$, H, H, Me β-lactam, N-(1H-tetrazol-5-yl)	63	1077

[a] PI—N = *N*-Phthalimido = (phthalimide, N—)

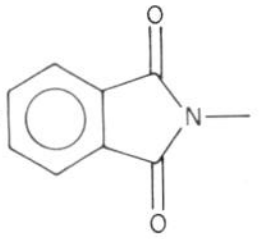

[b] Variation in yield due to amounts of reagent used.

[c] This starting material was prepared as shown below:

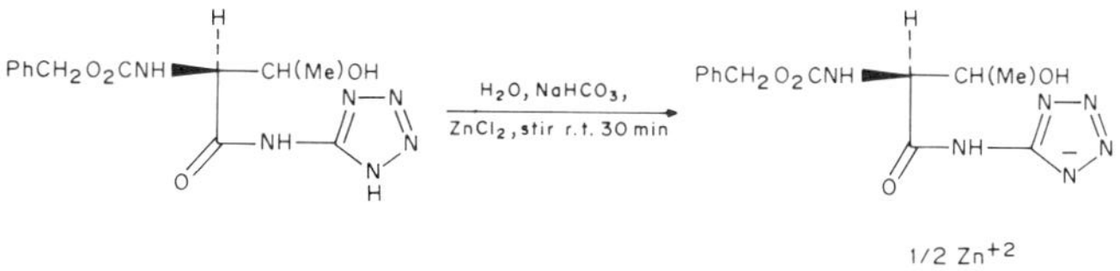

$PhCH_2O_2CNH$–CH(H)(CH(Me)OH)–C(=O)–NH–(1H-tetrazol-5-yl) $\xrightarrow[ZnCl_2,\ stir\ r.t.\ 30\ min]{H_2O,\ NaHCO_3,}$ $PhCH_2O_2CNH$–CH(H)(CH(Me)OH)–C(=O)–NH–(tetrazolyl anion) · 1/2 Zn^{+2}

In the recent literature when hydroxy amides are used as the starting materials, the most frequently prepared lactams are β-lactams, and the most common method of preparation has been reaction of the hydroxy amides with diethyl azodicarboxylate (DEAD) and triphenylphosphine (TPP), the so-called Mitsunobu conditions (equation 729 and Table 48).

$$HOCHR^1CR^2R^3CONHR^4 \xrightarrow[\text{THF, time, temp.}]{\text{DEAD, TPP}} \text{(2-azetidinone bearing } R^2, R^3 \text{ at C-3, } R^1 \text{ at C-4, } R^4 \text{ on N)} \quad (729)$$

In addition to the procedure reported in Table 48, (2*S*, 3*S*)-2-benzyloxycarbonylamino-3-hydroxy-*N*-(1*H*-tetrazol-5-yl)butyramide has been converted, albeit in lower yields, to the corresponding 2-azetidinone by treatment[1077] with trifluoroacetic acid (equation 730).

$$PhCH_2OCONH\text{–CH(CHOHMe)–CONH(tetrazolyl)} \xrightarrow[\text{r.t. stir 2h; 2. } Et_2O,\ H_2O,\ NaHCO_3]{\text{1. } C_6H_5OMe,\ CF_3COOH,} PhCH_2O_2CNH\text{–(azetidinone, Me, N-tetrazolyl)} \quad 20\% \quad (730)$$

A modification[1080–1082] of the procedure illustrated in equation 729 and Table 48 involves initial mesylation (Mes) or tosylation (Tos) of the hydroxy function, followed by treatment with a base (equation 731 and Table 49).

$$HCR^2(CHR^3OH)CONHR^1 \xrightarrow[p\text{-TosCl}]{\text{MesCl or}} HCR^2(CHR^3(\text{Mes or Tos-}p))CONHR^1 \xrightarrow[\text{conditions}]{K_2CO_3 \text{ or } KOH} \text{(2-azetidinone bearing } R^2, R^3, \text{ N-}R^1) \quad (731)$$

Two examples of lactam preparation have been reported[1083,1084] in which the hydroxy amides used as starting materials were produced *in situ*. In the first example[1083] methyl 5-oxohexanoate was hydrogenated in the presence of ammonia using a sodium-promoted catalyst (equation 732), while in the second example[1084] condensation of an amino diol

$$MeCO(CH_2)_3CO_2Me + H_2 \xrightarrow[\text{Pd}/\gamma\text{-}Al_2O_3 \text{ (Na promoted)}]{NH_3,\ 200-240\ ^\circ C} [MeCH(OH)(CH_2)_3CONH_2] \longrightarrow \text{6-methyl-2-piperidone} \quad 94\% \quad (732)$$

TABLE 49. Cyclization of mesylated or tosylated hydroxyamides to β-lactams

R^1	R^2	R^3	Mes or Tos	Base	Conditions	Stereochem. of Product	Yield (%)	Reference
MeO	Et	m-$MeOCH_2OC_6H_4CH_2$	Mes	K_2CO_3	Me_2CO, reflux	[a]	94	1082
$PhCH_2O$	Et	$MeOOCCH_2$—C—CH_2 (1,3-dioxolane ring: C bonded to two O)	Mes	K_2CO_3	MeOH, r.t., 7 h	*trans*	82	1080
$PhCH_2O$	(t-Bu)Me_2SiO	t-BuOOC	Mes	KOH or K_2CO_3	C_6H_6, DMSO, 50–60°C, 30 min	*trans*	50–60	1081
$PhCH_2O$	(t-Bu)Me_2SiO	t-BuOOC	Tos	KOH or K_2CO_3	C_6H_6, DMSO, 50–60°C, 48 h	*trans*	20	1081
MeO	(t-Bu)Me_2SiOCHMe	CH_2=$CHCH_2$	Mes	K_2CO_3	Me_2CO, reflux	*cis*	94	1061

[a] Unspecified.

with either 5-oxohexanoic acid (equation 733) or a racemic mixture of 2-phenyl 5-oxohexanoic acid (equation 734) leads to the formation of the corresponding di-alkylated bicyclic lactams indicated.

$$\text{PhCH(OH)CH(NH}_2\text{)CH}_2\text{OH} + \text{MeCO(CH}_2)_3\text{COOH} \xrightarrow[16\ \text{h}]{C_6H_6,\ \text{reflux},} [\text{PhCH(OH)CH(CH}_2\text{OH)NHCO(CH}_2)_3\text{COMe}] \longrightarrow \text{14\%} + \text{84\%} \tag{733}$$

$$\text{PhCH(OH)CH(NH}_2\text{)CH}_2\text{OH} + \text{MeCO(CH}_2)_2\text{CHPhCOOH} \xrightarrow[16\ \text{h}]{C_6H_6,\ \text{reflux}} [\text{PhCH(OH)CH(CH}_2\text{OH)NHCO(CH}_2)_2\text{CHPhCOMe}] \longrightarrow \text{20\%} + \text{80\%} \tag{734}$$

Similar results have been obtained[1085] when 2-(*p*-tolyl)-4-oxopentanoic acid was treated with the amino monool, S-valinol (equation 735).

When sulphur containing amides are used as starting materials for the preparation of lactams, a variety of methods have been employed to effect cyclization. One method employed involves α-acyl substituted α-thiocarbocations, readily formed from α-acyl sulphoxides under Pummerer reaction conditions, as initiators for cationic alkene cyclization to produce 4-, 5-, 6- and 7-membered lactams. The 4-membered ring lactams were prepared[1086] by treatment of 3-(phenylsulphinyl)propionamides with trimethylsilyl trifluoromethanesulphonate and triethylamine (equation 736).

$$\text{(CHMe}_2\text{)(H}_2\text{C)C(OH)–C(H)NH}_2 + \text{MeCOCH}_2\text{CH(}p\text{-Tol)COOH} \xrightarrow[16\ \text{h}]{C_6H_6,\ \text{reflux}}$$

91%

(as a 1:1 mixture of *endo*:*exo* aryl substituted diasteromers)

(735)

$$R^1NHCOCR^2R^3CH_2SOPh \xrightarrow[\text{Et}_3\text{N, CH}_2\text{Cl}_2,\ -20\ ^\circ\text{C, 30 min}]{\text{Me}_3\text{SiOTf}} \left[R^2R^3C(–CH{=}\overset{+}{S}Ph)–C(OSiMe_3){=}NR^1 \right] \longrightarrow$$

(736)

R^1	R^2	R^3	Yield (%)
H	H	H	41
$PhCH_2O$	H	H	14
H	H	Me[a]	41[b]
$PhCH_2O$	Me	Me	51

[a]A 1:1 mixture of diastereomers treated at 20°C, not −20°C.
[b]A 2.7:1 mixture of *cis*:*trans* isomers was obtained.

The 5- and 6-membered ring lactams were both obtained[1060,1087] from the same basic starting materials, *N*-alkenyl-α-(methylsulphinyl)acetamides using trifluoracetic anhydride, but dependent upon the substituent attached to the alkenyl carbons and the stability of the intermediate carbocation, 5-*exo*-trigonal cyclization occurred to produce γ-lactams (equation 737) or 6-*endo*-trigonal cyclization occurred to produce δ-lactams (equation 738).

$\xrightarrow[CH_2Cl_2,\ r.t.]{(CF_3CO)_2O}$ $^{-}OCOCF_3$

$(R^2 = Me)$

$(R^1 = Me)$

$-H^+$ $\bar{O}COCF_3$

(84)

(85)

(86)

(737)

$-H^+$

(87) +

(88)

(738)

R^1	R^2	R^3	Path (Equation No.)	Product	Yield (%)	Reference
H	Me	H	737	**86**	92[a]	1060, 1087
H	H	H	737	**84** (*trans*) + **85**	9 + 39	1060, 1087
H	Me	Me	737	**84** + **86**	—[b]	1060
Me	H	H	738	**87** + **88**	43 + 35	1060, 1087

[a]A 69:31 *trans*:*cis* mixture of stereoisomers was obtained.
[b]The **84**(*trans*):**84**(*cis*):**86** ratio obtained was 2:2:1.

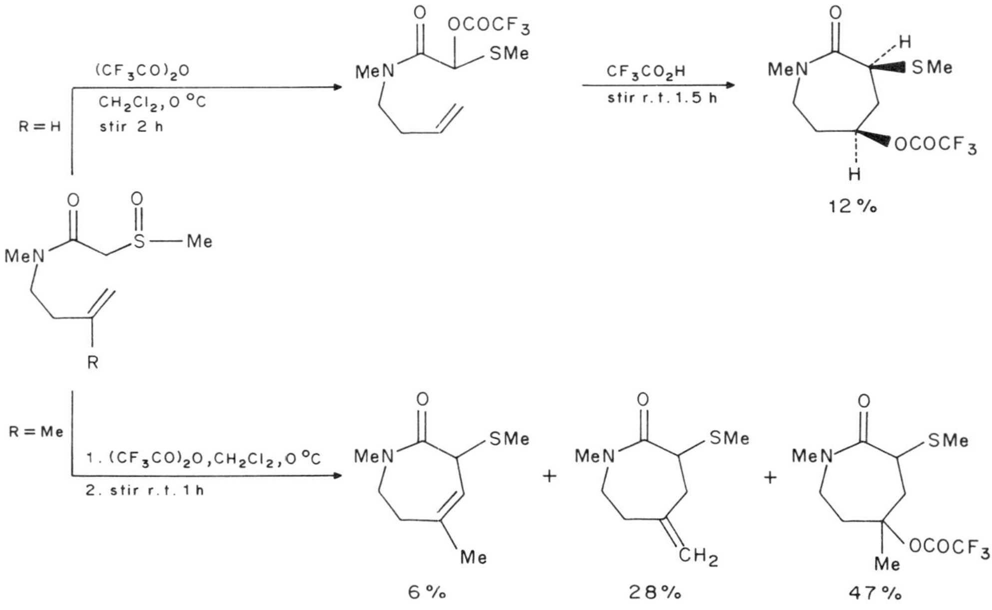

R = H: $(CF_3CO)_2O$, CH_2Cl_2, 0 °C, stir 2 h; then CF_3CO_2H, stir r.t. 1.5 h — 12%

R = Me: 1. $(CF_3CO)_2O$, CH_2Cl_2, 0 °C; 2. stir r.t. 1 h — 6% + 28% + 47%

(739)

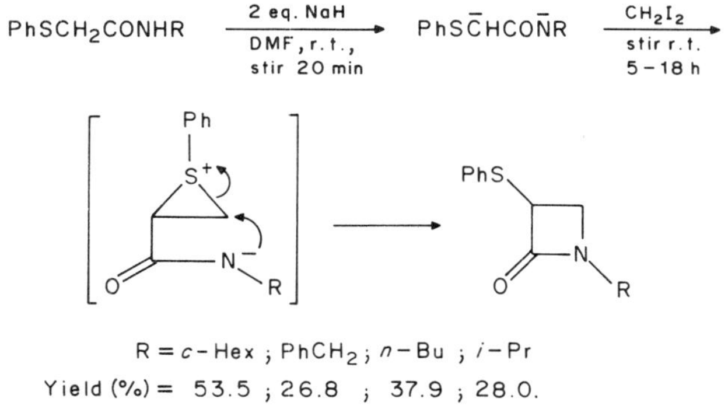

PhSCH$_2$CONHR $\xrightarrow[\text{DMF, r.t., stir 20 min}]{\text{2 eq. NaH}}$ PhS$\bar{C}$HCO$\bar{N}$R $\xrightarrow[\text{stir r.t. 5–18 h}]{CH_2I_2}$

(740)

R = *c*-Hex ; PhCH$_2$; *n*-Bu ; *i*-Pr
Yield (%) = 53.5 ; 26.8 ; 37.9 ; 28.0.

Seven-membered ring lactams were obtained[1060] from N-(3-butenyl)-N-methyl-α-(methylsulphinyl)acetamides upon treatment with trifluoroacetic anhydride (equation 739).

β-Lactams have also been prepared[1088] from sulphur containing amides by using the 1,3-dianion of α-(phenylthio)acetamide derivatives and methylene iodide. An episulphonium intermediate was proposed[1088] for this reaction (equation 740).

Treatment of 3-(phenylthio)propionamide derivatives with di-t-butyl peroxide and a copper catalyst, reagents which characteristically generate radical intermediates, has been reported[1043] to produce β-lactams albeit in very low yields (equation 741). Better yields were obtained[1043] when the 3-benzoyloxy derivatives were treated consecutively with hydrogen bromide in methylene chloride and potassium amide in liquid ammonia (equation 742). This route probably involves formation of an intermediate 3-bromo-compound which then undergoes intramolecular nucleophilic displacement of the bromide to produce the lactam.

Me; Me; CH_2SPh; NHMe; O

$(t\text{-BuO})_2$, $CuBr_2$, C_6H_6, reflux 6 h (4%)

$(t\text{-BuO})_2$, $CuBr_2$, naphthalene, 120–130 °C, 4 h (18%)

$(t\text{-BuO})_2$, $CuBr_2$, C_6H_6, $h\nu$, reflux 3 h (15%)

Me; Me; SPh; N; Me; O (741)

R^1; R^1; SPh; CHO_2CPh; NHMe; O

1. HBr, CH_2Cl_2, 0–5 °C, stand 2 h
2. KNH_2, CH_2Cl_2, –78 °C, 1 h

R^1; R^1; SPh; N; Me; O (742)

R^1 = H ; Me

Yield (%) = 59 ; 73

H; CH_2—$CH_2\overset{+}{S}Me_2$ I^-; t-BuO_2CNH; C; C; O; NH; C; CO_2Me; H; R

NaH, DMF–CH_2Cl_2 (1:1), N_2, 0 °C, stir 2.5 h

H; t-BuO_2CNH; O; N; C; CO_2H; H; R; (S)

R = H ; i-$PrCH_2$; $PhCH_2$; $CH_2(CH_2)_3NHCO_2CH_2Ph$

Yield (%) = 51 ; 40 ; 51 ; 58 (743)

Two final methods which have been used to produce lactams from sulphur containing amides involve the treatment of the methylsulphonium iodide salts of substituted dipeptide methyl esters with sodium hydride[1000] (equation 743), and the reaction of 5-(acetamidomethyl)-N^{α}-phthalimido-(*R*)-cysteinylglycine methyl ester with *p*-toluenesulphonic acid and paraformaldehyde[1000] (equation 744).

(744)

Keto substituted amides have also found use as starting materials for the preparation of lactams of varying structure. Thus, reaction of substituted pyruvamides with diethyl (diazomethyl)-phosphonate under basic conditions produces[1089] γ-lactams which result from the 1,5-carbon–hydrogen insertion of the intermediate alkylidenecarbene (equation 745).

(745)

R^1	R^2	R^3	Yield (%)	Ratio **89:90**
H	H	H	50^a	—
H	*n*-Pr	H	43^b	1.5:1
H	Ph	H	56^b	2.6:1
Me	Me	H	67^a	—
—$(CH_2)_3$—		H	67^a	—
—$(CH_2)_3$—		Me	63^b	1:1.2

^{a}Yield determined by PMR.
^{b}Isolated yield.

Refluxing cycloalkanone esters in an excess of a primary amine and a catalytic amount of the corresponding hydrochloride for 12 to 18 hours, produces[1090] the corresponding bicyclic lactam (equation 746).

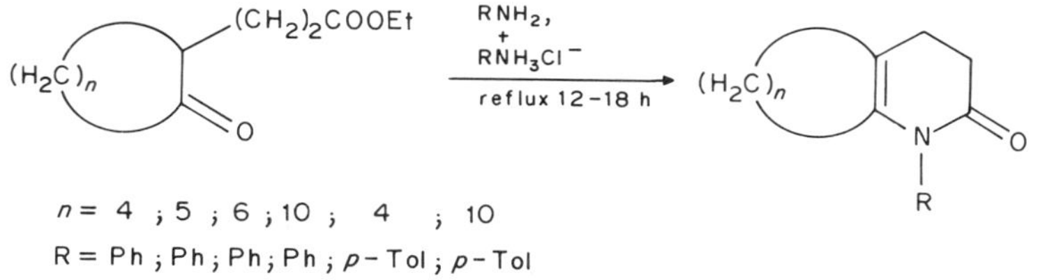

(746)

n = 4 ; 5 ; 6 ; 10 ; 4 ; 10
R = Ph ; Ph ; Ph ; Ph ; p-Tol ; p-Tol

N-Acyliminium cyclizations onto benzenoid rings or heterocycles, such as pyrrole, thiophene or indole, produce[1091] tetrahydroisoquinoline ring systems in fair to good yields. The precursors for these reactions were prepared by either of two routes: condensation of a mixed carbonic anhydride, ester or ene lactone with a 2-arylethylamine to give a keto amide or enamide; or condensation of succinic anhydride with a 2-arylethylamine to give an imide. Cyclization of the precursors was effected by exposure to acid, either polyphosphoric (PPA), hydrochloric or pyridinium polyhydrogen fluoride (PHF). Below are recorded several examples of this approach to the preparation of lactams.

Reaction of the mixed carbonic anhydride of levulinic acid with 2,2-diphenylethylamine produces a keto amide as a mixture of ring and chain tautomers, which upon treatment with polyphosphoric acid affords the corresponding fused ring lactam (equation 747). The overall yield obtained[1091] was 35%, of a mixture of substituted 6α- and 6β-pyrrolo[2,1-a]isoquinoline diastereomers in a 94:6 ratio. When angelica lactone was used instead of the mixed anhydride, the overall yield improved to 65% of the same 94:6 mixture of products.

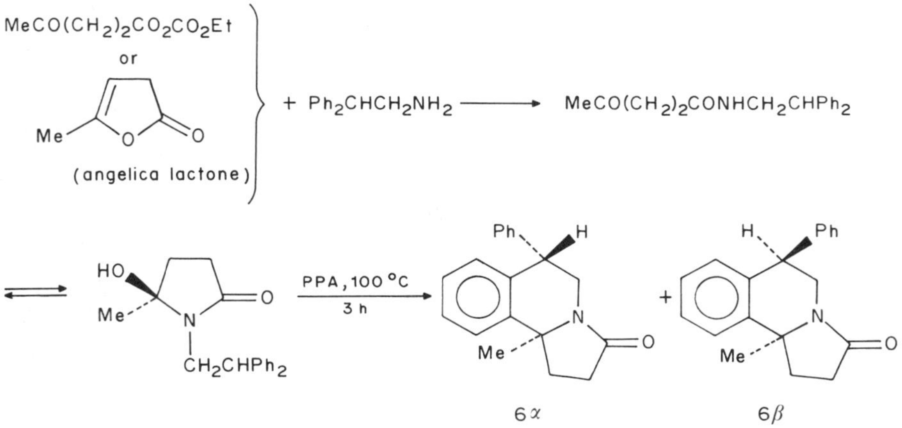

(747)

Allowing the mixed carbonic anhydride of 5-oxohexanoic acid to react with 2,2-diphenylethylamine under the same conditions as were used for the above reaction produced[1091] 7α- and 7β-benzo[a]quinolizidines in 74% as a 94:6 mixture of diastereomers (equation 748).

Reaction of α-angelica lactone with 2-(3,4-dimethoxyphenyl)-2-phenylethylamine in the presence of hydrogen chloride produces a 40% mixture of substituted 6α- and 6β-pyrrolo[2,1-a]isoquinoline diastereomers in a 90:10 ratio (equation 749). Similar results[1091], differing only in the yield of products and their ratios, were obtained with

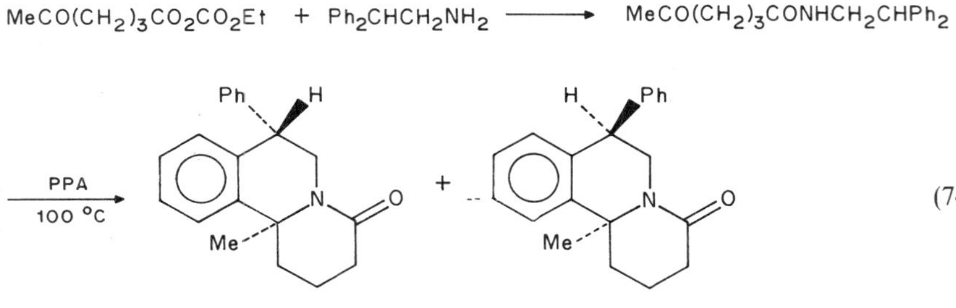

$MeCO(CH_2)_3CO_2CO_2Et$ + $Ph_2CHCH_2NH_2$ ⟶ $MeCO(CH_2)_3CONHCH_2CHPh_2$

$\xrightarrow[100\ ^\circ C]{PPA}$ 7α + 7β (748)

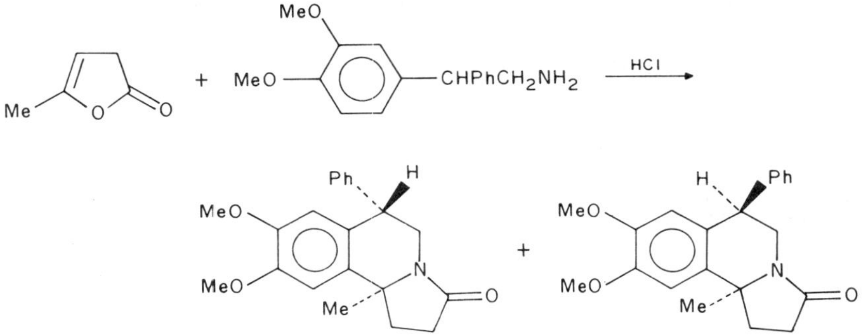

Me, O, Ph, H, N

MeO, MeO, $CHPhCH_2NH_2$, $\xrightarrow{HCl}$

6α + 6β (749)

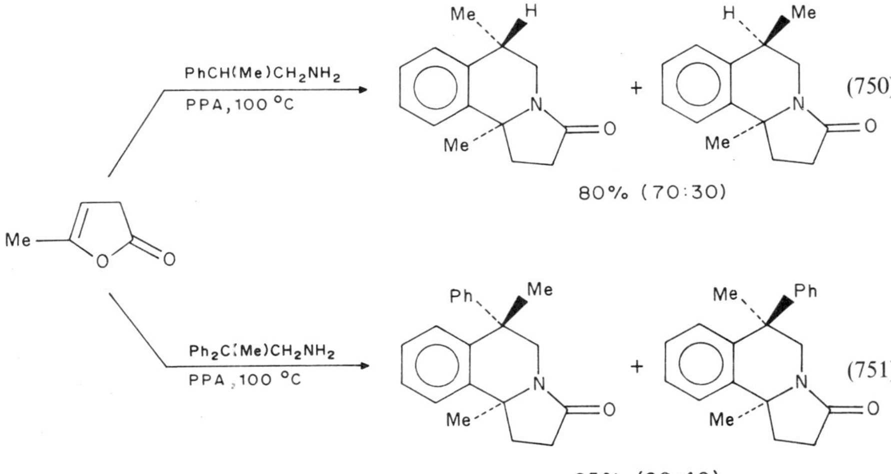

$\xrightarrow[PPA, 100\ ^\circ C]{PhCH(Me)CH_2NH_2}$ (750)

80% (70:30)

$\xrightarrow[PPA, 100\ ^\circ C]{Ph_2C(Me)CH_2NH_2}$ (751)

65% (60:40)

2-phenylpropylamine (equation 750) and 2,2-diphenylpropylamine (equation 751) using polyphosphoric acid.

When cyclohexanone-2-acetic acid ester was heated with 2,2-diphenylethylamine, a mixture of 3,3*a*- and 7,7*a*- enamides was obtained in a 3:1 ratio, which upon treatment with polyphosphoric acid afforded a 60% yield of 11α- and 11β-phenylerythrinan-8-one in a 95:5 mixture (equation 752). Similar reaction with 2-(3,4-dimethoxyphenyl)-2-phenylethylamine afforded a 98% yield of the corresponding 11α-isomer only (equation 753), while using cyclohexanone-3-propionic ester and 2,2-diphenylethylamine afforded mainly the 4*a*,8*a*-olefin isomer of the initial enamide, which upon treatment with polyphosphoric acid produced a 50% yield of a 95:5 mixture of 11α-, 11β-phenyl-β-homoerythrinane (equation 754).

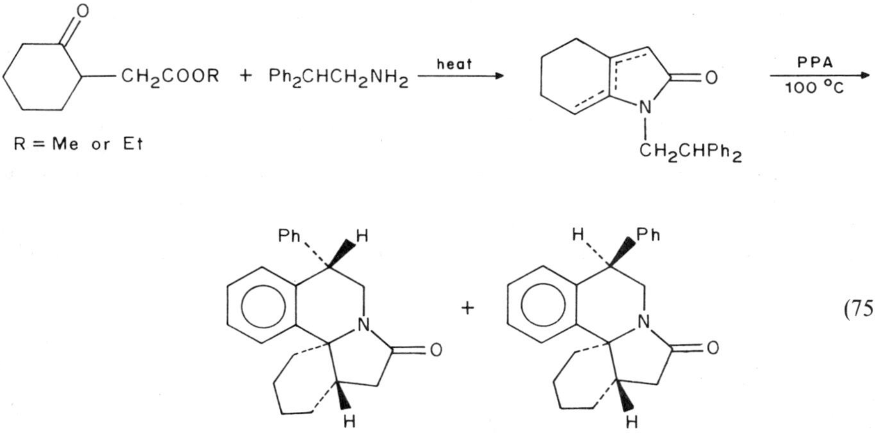

CH_2COOR + $Ph_2CHCH_2NH_2$ —heat→ [enamide, N–CH_2CHPh_2] —PPA, 100 °C→

R = Me or Et

11α + 11β (752)

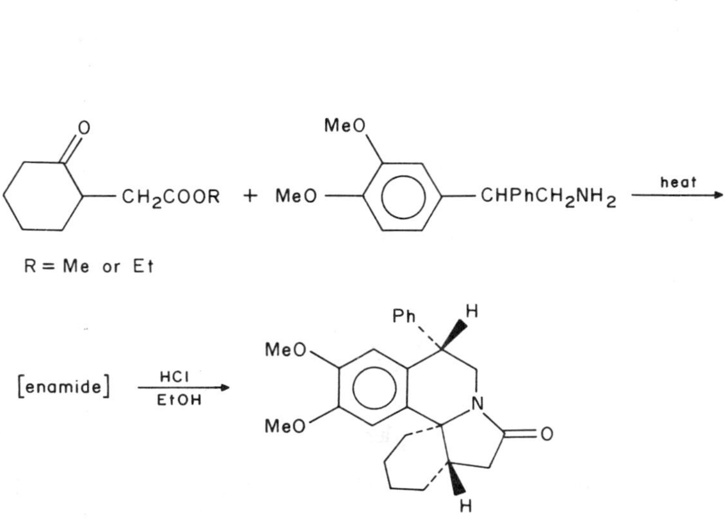

CH_2COOR + MeO, MeO–C$_6$H$_3$–$CHPhCH_2NH_2$ —heat→

R = Me or Et

[enamide] —HCl, EtOH→ 11α (753)

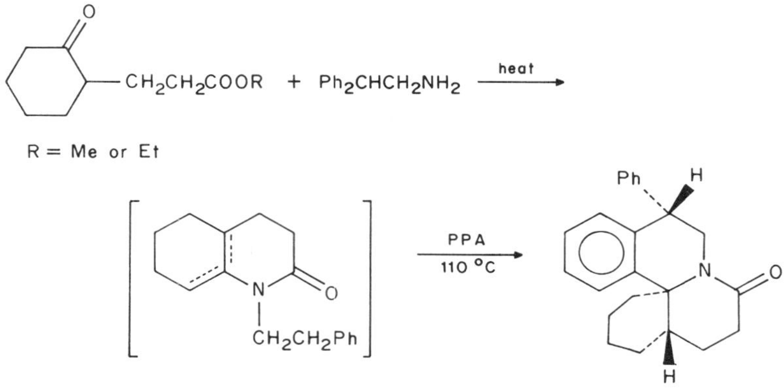

(754)

In the recent literature, amides which contain double bonds have been a rich source of starting materials for the preparation of lactams, and most of the procedures used are represented by but a single reference. For example, allylamides of bis(alkoxycarbonyl)-aminoacetic acids have been reported[777] to cyclize, via an intramolecular amidoalkylation of the resident olefin, in methanesulphonic acid at room temperature to produce the five-membered pyrrolidones rather than the corresponding six-membered piperidinones (equation 755).

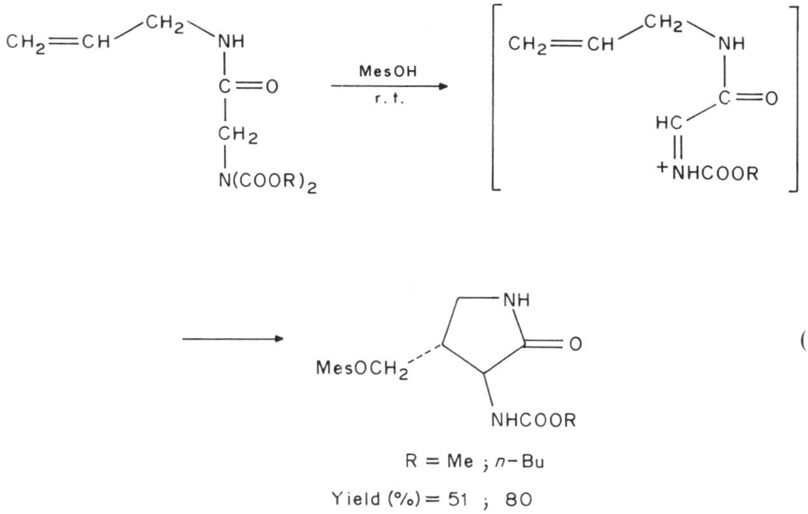

R = Me ; n-Bu

Yield (%) = 51 ; 80

(755)

The N-allylamide derivatives of α-ethylenic acids upon reflux with 1.2 equivalents of sodium hydride in xylene followed by aqueous hydrolysis, afford[1092] good yields of the corresponding ε-lactams **91**. When the reaction mixture was treated with methyl iodide instead of the hydrolysis step, the N-methyl derivatives of the ε-lactams **92** were obtained (equation 756).

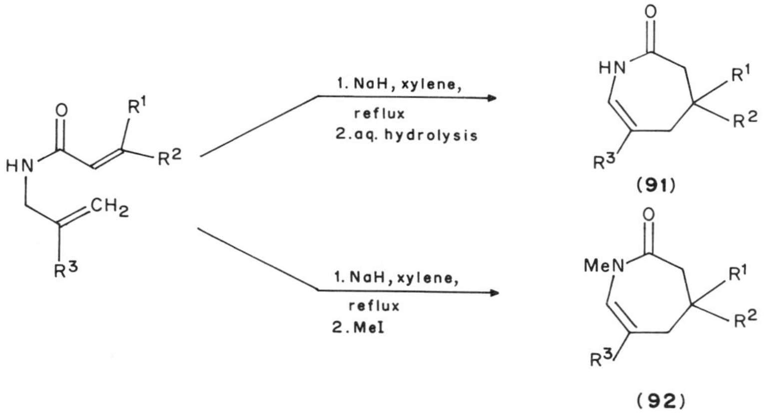

(756)

R^1	R^2	R^3	H_2O or MeI	Product	Yield (%)
Me	Me	H	H_2O	**91**	70
Me	Me	H	MeI	**92**	70
Me	Ph	H	MeI	**92**	80
Ph	Ph	H	MeI	**92**	73
Me	Me	Me	H_2O	**91**	78
Me	Me	Me	MeI	**92**	72
H	Me	H	H_2O	—	—
H	Me	H	MeI	—	—
H	Ph	H	H_2O	—	—
H	Ph	H	MeI	—	—

As the results indicate, there was no reaction if $R^1 = H$, also if the nitrogen was trisubstituted, or if excess sodium hydride was not used. The mechanism probably involves removal of the hydrogen attached to the nitrogen followed by further removal of the hydrogen from the allyl moiety. Similar results are obtained[1092] when the allylic double bond is part of a cyclic system (equation 757).

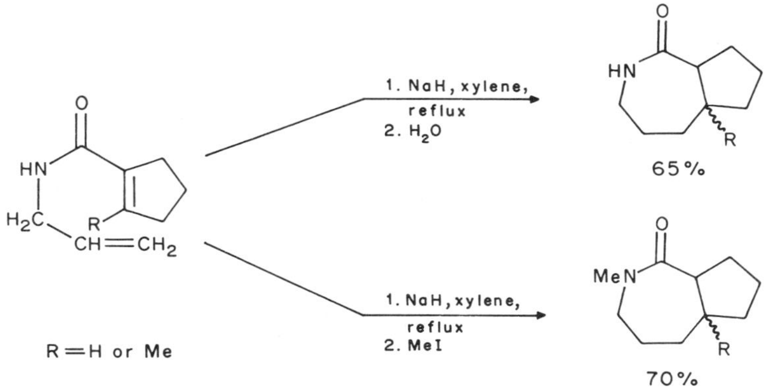

(757)

TABLE 50. Synthesis of γ-lactams by Diels–Alder addition of allyl- and propargylamides of 2,4-dienecarboxylic acids[1093]

Starting material	Conditions	Products	Addition mode	Yield (%)
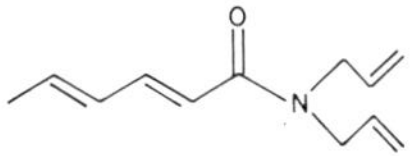	DMF, 156°C, 3 h	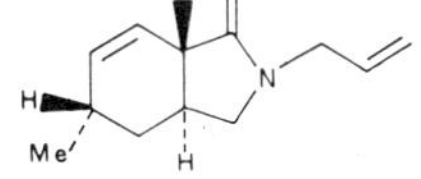	*exo*	32
		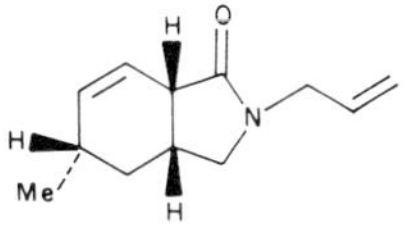 +	*endo*	44
		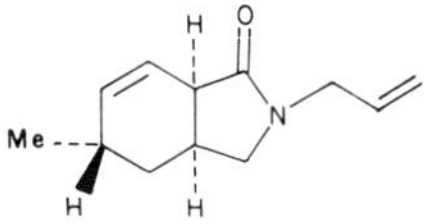 +	*endo*	4
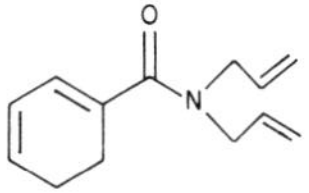	DMF, 156°C, 1.5 h	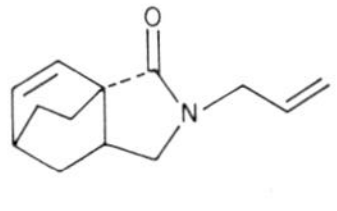	*endo*	56
		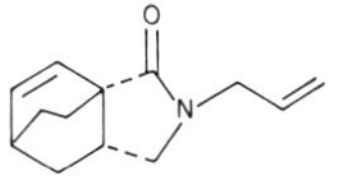 +	*exo*	24
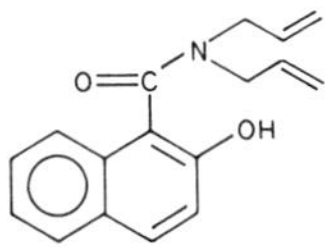	$C_6H_5NEt_2$, 218°C, 3.5 h	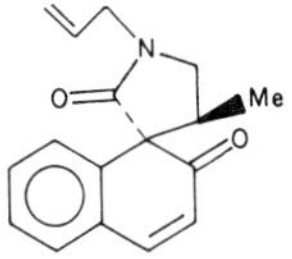 +	intramolecular ene reaction	48

		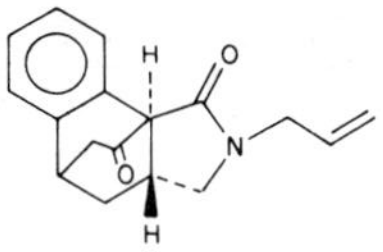	*exo*	22
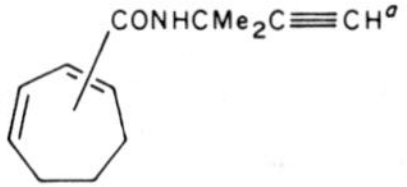 CONHCMe$_2$C≡CH[a]	$C_6H_5NEt_2$, 218°C, 3 h	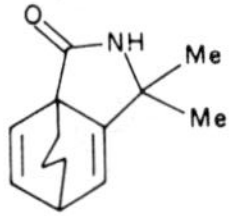	—	71
		+ 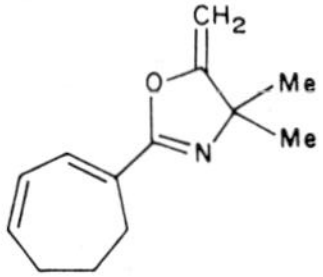	—	7
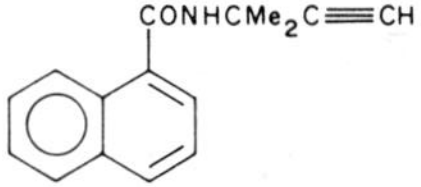 CONHCMe$_2$C≡CH	$C_6H_5NEt_2$, 218°C, 3.5 h	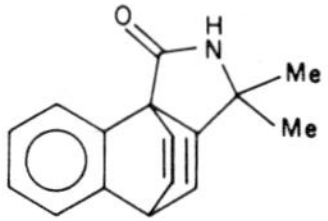	—	72
		+ 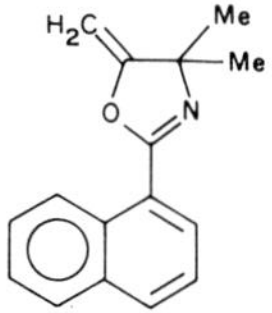	—	4

[a] Mixture of three isomers.

Allyl- and propargylamides of 2,4-dienecarboxylic acids have also been used as starting materials for the synthesis[1093] of γ-lactams via an intramolecular Diels–Alder addition (Table 50).

Synthesis of 5- or 6-membered lactam derivatives may be accomplished[1094] by intramolecular ene insertion of acylazocarboxylates having γ,δ- or δ,ε-unsaturation. The acylazocarboxylates required are generated by manganese dioxide oxidation of the parent hydrazines (equation 758 and 759).

NHNHCO₂Et → (Method A, B or C) → [N=N–COOEt, H] → N–NHCO₂Et (758)

Method A: MnO_2, CH_2Cl_2, 20 °C, stir gave 67% yield

B: n-BuLi, NBS, 0°C, THF gave 21% yield

C: $Pb(OAc)_4$, CH_2Cl_2, 0 °C, gave 38% yield

NHNHCO₂Et → (MnO_2, CH_2Cl_2, stir r.t. 2 h) → N–NHCO₂Et (759)

60–80%

Several examples of condensation reactions with alkenamides have also been reported to produce lactams. These examples include the formation of γ- or δ-lactams by the condensation[1095] of N-alkylalkenamides with benzeneselenenyl chloride via the intramolecular attack of the nitrogen atom of the amide group on the episelenonium ion intermediate formed during the reaction (equation 760). As can be seen from the results reported, the substituent on nitrogen and the substituent on the carbon atoms between the carbonyl group and the double bond play an important role in determining whether or not the reaction will proceed. The best indication of the sensitivity of this reaction to substituents and their location is obtained by the results recorded for the last two entries in the chart below equation 760.

Condensation of acrylamide with an enamine prepared from cyclohexanones or cyclopentanones in an anhydrous solvent produces[1096] the corresponding bicyclic 2-piperidones (equation 761). Similar results were also obtained[1096] using ethyl 2-(1-pyrrolidinyl)-2-cyclohexene-1-propanoate, ethyl 2-(1-pyrrolidinyl-2-cyclohexene-1-ethanoates or ethyl 2-(1-pyrrolidinyl)-2-cyclopentene-1-ethanoate. By condensing these materials with primary amines the corresponding N-substituted 2-piperidones and the N-substituted 2-pyrrolidones were obtained (equation 762).

Reaction of dimethylaminoformaldehyde diethyl acetal with enaminoamides produces[1097] the corresponding pyridoazepines and furopyridones (equation 763).

$$N\text{-alkylalkenamide} + \text{PhSeCl} \xrightarrow[\text{stir}]{\text{MeCN, r.t.}} \gamma\text{- or } \delta\text{-lactams} \quad (760)$$

Starting material	Time (h)	Product	Yield (%)
$CH_2{=}CRCH_2CHEtCONHBu$-*n*			
R = H	1	PhSeCH2, R, N—Bu-n, Et, O	87[a]
R = Me	1		94[a]
$PhCH{=}CHCH_2CHEtCONHBu$-*n*	24	PhSe, H, H, Ph, N, Bu-n, Et, H, O	73
(cyclopentenyl)—CHEtCONHBu-*n*	2	PhSe, H, H, Bu-n, N, O, H, Et	73[b]
3-($CH_2CH{=}CH_2$)-2-piperidinone (N-H, O)	10	O, N, CH2SePh, H	68
		+	
		O, N, H, CH2SePh	23
$CH_2{=}CHCH_2CHEtCONHPh$		No reaction	
$CH_2{=}CHCH_2CH_2CONHBu$-*n*		No reaction	

[a]Product is a 1:1 mixture of stereoisomers.
[b]Product is a 77:23 mixture of stereoisomers.

$$\text{(enamine: } R^1, (CH_2)_n, R^2, R^3, \text{N-pyrrolidinyl)} + CH_2{=}CHCONH_2 \xrightarrow[\text{reflux 3h}]{\text{anhy. solvent}} \text{(bicyclic lactam: } R^1, (CH_2)_n, R^2, R^3, \text{NH, O)} \quad (761)$$

R^1	R^2	R^3	*n*	Solvent	Yield (%)
H	H	H	1	dioxane	100[a]
Me	H	H	1	dioxane	90[a]
H	*n*-Pr	H	1	benzene	73
H	H	H	0	benzene	85[b]
Me	Me	Me	0	benzene	72

[a]A mixture of isomers of 3,4,4a,5,6,7-hexahydro- and 3,4,5,6,7,8-hexahydro-2(1*H*)-quinolone.
[b]Only $\Delta^{4a(7a)}$ hexahydro-2(1*H*)pyridone was formed.

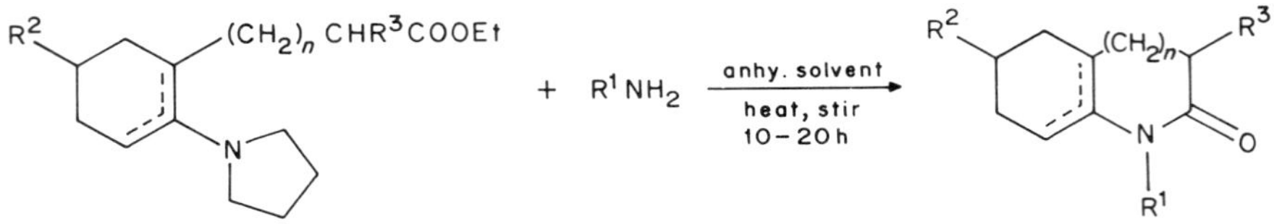

(762)

R^2	R^3	n	R^1	Solvent	Temp (°C)	Time (h)	Yield (%)
H	H	1	$PhCH_2$	DMF	150	15	16
H	Me	0	n-Pr	C_6H_6	80	10	52[a], 42[b]
H	Me	0	c-Hex	dioxane	100	10	52[c]
H	Me	0	$PhCH_2$	C_6H_6	80	20	34[a], 50[b]
Me	Me	0	$PhCH_2$	DMF	150	10	32[c]

[a]Yield of the $\Delta^{7(7a)}$-tetrahydro-2-indolone isomer.
[b]Yield of the $\Delta^{3a(7a)}$-tetrahydro-2-indolone isomer.
[c]Yield of an isomer mixture.

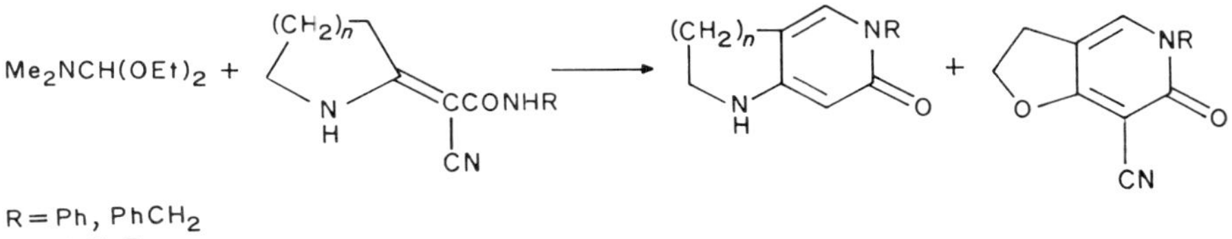

$R = Ph, PhCH_2$
$n = 1, 2, 3$

(763)

A variety of other substituted amides have also been used recently as starting materials for the preparation of lactams. For example, treatment of diazo compounds with a catalytic amount of rhodium(II) acetate in benzene reportedly[1098] produces aza-β-lactams in high yields (equation 764). The mechanism for this conversion appears to involve formation of a rhodium carbenoid intermediate, not a carbene, followed by nucleophilic attack by the lactam nitrogen on the rhodium carbenoid, rather than an insertion into the lactam nitrogen–hydrogen bond, to afford ring closure.

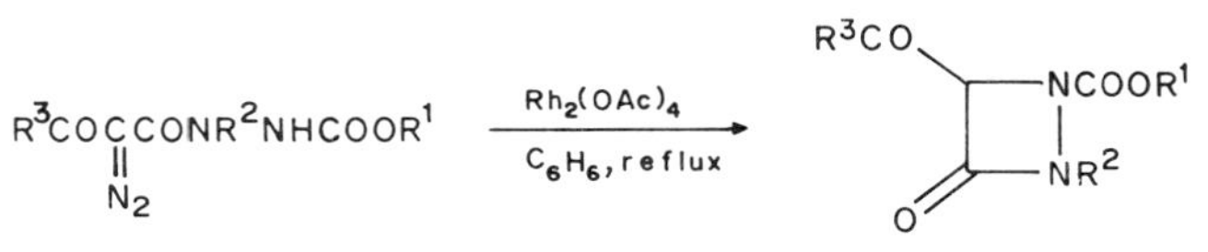

(764)

R^1	R^2	R^3	Yield (%)
t-Bu	$PhCH_2$	OEt	91
t-Bu	$PhCH_2$	Me	100
t-Bu	CH_2COOEt	OEt	95
t-Bu	CH_2COOEt	Me	75
$PhCH_2$	CH_2COOEt	OEt	93
$PhCH_2$	CH_2COOEt	Me	82

Reaction of the substituted urethanes shown in equation (765) with phosphorus oxychloride and a pinch of phosphorus pentoxide in dry xylene produces[1099] the corresponding substituted 1-oxo-1,2,3,4-tetrahydroisoquinolines.

$R^1 = MeO$; $R^1R^1 = CH_2$ (O—CH₂—O bridge)
Yield(%) = 45 ; 25

Reagents: $POCl_3$, P_2O_5, xylene, reflux, 3 h — starting material: $CH_2CH_2NHCO_2Et$ substituted benzene; product: NH lactam (765)

In 1980, a new synthesis of oxindoles (equation 766), isoquinolones (equation 767) and benzazepinones (equation 768) was reported[1100] which involved intramolecular amidoalkylation of aromatic amides of bis(methoxycarbonylamino)acetic acid.

R—C$_6$H$_4$—$NHCOCH(NHCO_2Me)_2$ $\xrightarrow[\text{r.t.}]{CF_3COOH,}$ oxindole with NHCOOMe (766)

R = H ; Me
Yield(%) = 67 ; 95

$PhCHRNHCOCH(NHCO_2Me)_2$ $\xrightarrow[\text{r.t.}]{MesOH}$ isoquinolone with $NHCO_2Me$ (767)

R = H ; Me ; COOMe
Yield(%) = 84 ; 83[a] ; 85[a]

[a]Product was a mixture of two isomers.

$CH_2CH_2NHCOCH(NHCO_2Me)_2$ (R, R substituted benzene) $\xrightarrow[\text{MesOH}]{CF_3COOH \text{ or}}$ benzazepinone with NH, O, $NHCO_2Me$

R = H[a] , OMe[b]
Yield(%) = 67 , 64

[a]MesOH used.
[b]CF_3COOH used.

(768)

Cyclization of phenylalkanehydroxamic acids using polyphosphoric acid (PPA) affords[1101] the corresponding benzolactams (equation 769).

$$Ph(CH_2)_n\,CONHOH \xrightarrow{PPA} \text{[Ph-substituted lactam ring, } (CH_2)_{n-1}\text{, NH, C=O]} \qquad (769)$$

$n = 2, 3$

Two interesting cyclocondensation reactions have also been used to produce dilactams and imidazolidinones. In the first cyclocondensation, 3-acetylhexane- and 4-acetylheptanedinitriles were refluxed with potassium carbonate in ethanol to produce[1102] hexahydro-1H-pyrrolo[2,3-b]pyridines (equation 770) and octahydro-1H-1,8-naphthyridines (equation 771), respectively, while in the second cyclocondensation, amino acid amides

$$3,4\text{-}R_2C_6H_3C(Ac)(CH_2CN)CH_2CH_2CN \xrightarrow[80\,^{\circ}C]{K_2CO_3,\ EtOH} \text{[bicyclic dilactam, } C_6H_3R_2\text{-}3,4\text{, Me]} \qquad (770)$$

$R = H, OMe$ 91–93%

$$3,4\text{-}R_2C_6H_3C(Ac)(CH_2CH_2CN)CH_2CH_2CN \xrightarrow[80\,^{\circ}C]{K_2CO_3,\ EtOH} \text{[bicyclic dilactam, } C_6H_3R_2\text{-}3,4\text{]} \qquad (771)$$

$R = H, OMe$ 34–36%

were refluxed with aldehydes or ketones in the presence of p-toluenesulphonic acid to produce[1103] the corresponding N-acetylated 4-imidazolidinones (equations 772, 773 and 774).

$$R^1CH(NHSO_2Me)CONHR^2 + R^3COR^4 + p\text{-TosOH} \xrightarrow[2.\ C_6H_5Me,\ \text{reflux } 1\text{–}24\text{ h}]{1.\ C_6H_6,\ \text{reflux } 3\text{ h}} \text{[imidazolidinone: } MeSO_2N,\ N\text{–}R^2,\ H,\ R^1,\ R^3,\ R^4,\ C{=}O\text{]}$$

(S)

R^1 = Me ; Ph ; i-Pr ; i-Bu ; $PhCH_2$; Me ; Me ; Me

R^2 = Me ; Me ; Me ; Me ; Me ; H ; H ; Me

R^3,R^4 = H[a] ; H[a] ; H[a] ; H[a] ; H[a] ; Me ; Et ; Me, H[b]

(772)

[a]Paraformaldehyde was used as the carbonyl reactant.

[b]Paraldehyde was used as the aldehyde reactant and both the *cis* and *trans* isomers of the product were obtained.

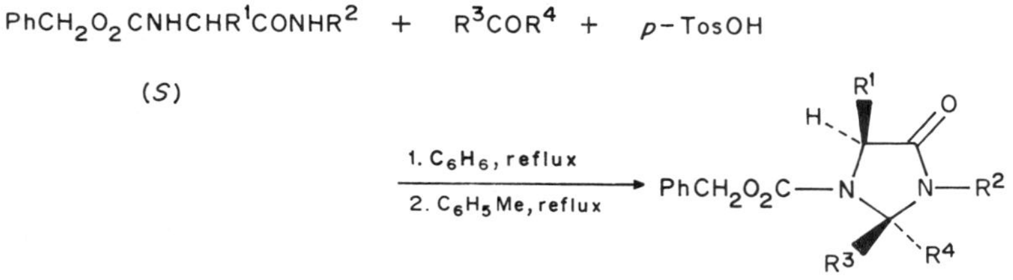

$PhCH_2O_2CNHCHR^1CONHR^2$ + R^3COR^4 + p-TosOH (S) → (1. C_6H_6, reflux; 2. C_6H_5Me, reflux) $PhCH_2O_2C$–N(ring with R^1, H, O, N–R^2, R^3, R^4) (773)

R^1 =	Me	$PhCH_2$	Me
R^2 =	H	Me	Me
R^3,R^4 =	Me	H[a]	H[a]

[a]Paraformaldehyde was used as the carbonyl reactant.

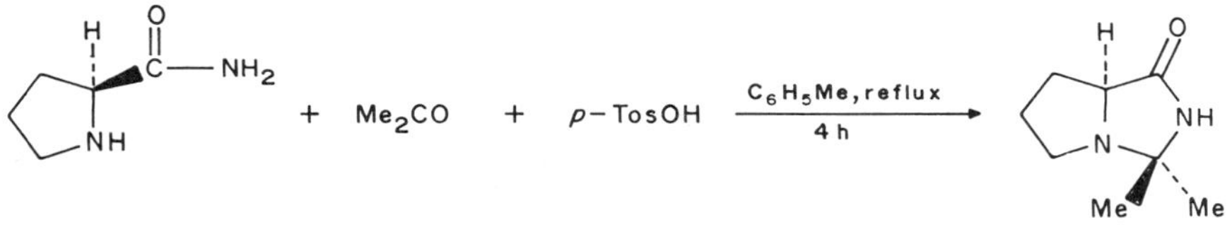

(774)

*B. By Ring-closure Reactions (Photochemical)

*1. Cyclization of α,β-unsaturated amides

Irradiation of *N*-(thiobenzoyl)methacrylamides in benzene affords[1104] thietan-fused β-lactams by a (2 + 2) photochemical cycloaddition (equation 775).

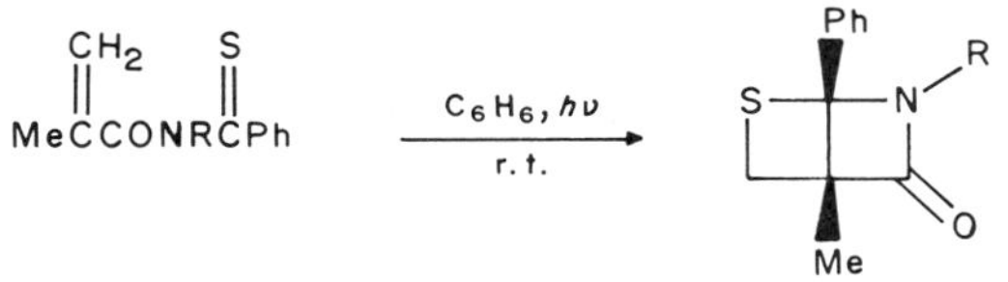

CH_2=C(Me)CONRC(=S)Ph → (C_6H_6, hν, r.t.) (775)

R =	Me	Et	i-Pr	$PhCH_2$	Ph
Yield(%) =	55	96	73	95	77

*3. Cyclization of enamides

Non-oxidative photocyclization of 2-aroyl-1-methylene-6,7-dimethoxy-1,2,3,4-tetrahydroisoquinolines, enamines of 6,7-dimethoxy-1-methyl-3,4-dihydroisoquinoline, produces[1105] a variety of the corresponding fused-ring dehydrolactams as illsutrated in equation (776) and the following chart. A similar reaction occurs[1105] when the 2-aroyl group is β-naphthylene (equation 777).

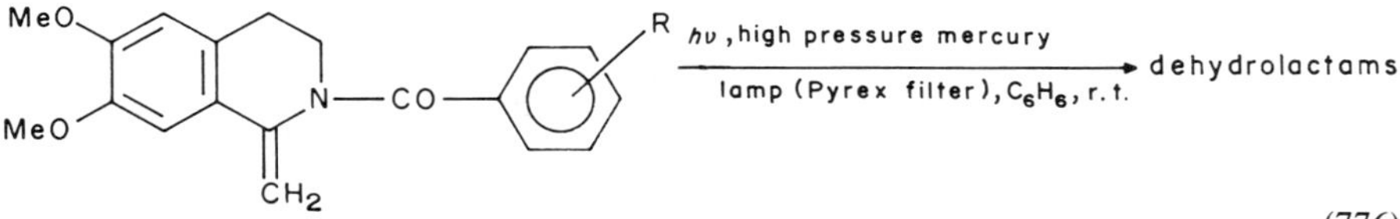

(MeO, MeO, N–CO–aryl-R, CH_2) → (hν, high pressure mercury lamp (Pyrex filter), C_6H_6, r.t.) dehydrolactams (776)

R	Temp. (°C)	Product(s)	Yield (%)
H	r.t.	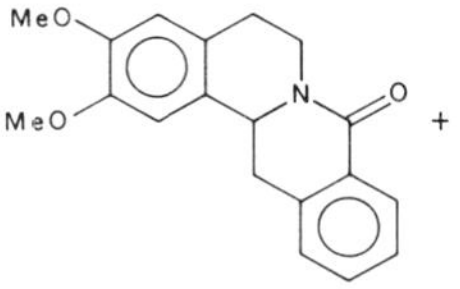 (93) +	25[a]
		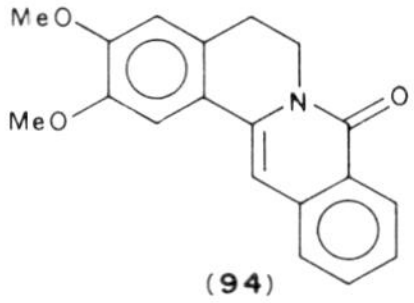 (94)	15[a]
o-MeO	r.t.	**94**	60
o-MeO	6	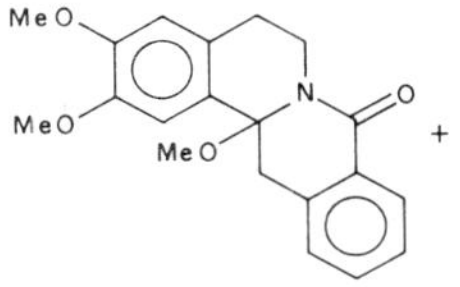 +	31
		94	30
m-MeO	r.t.	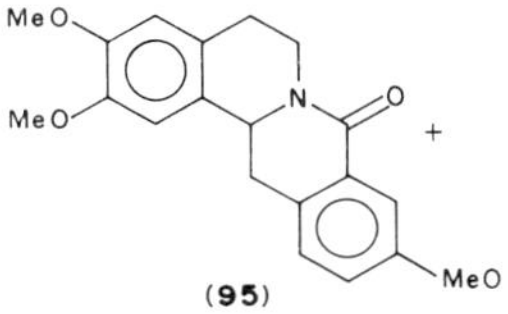 (95) +	13
		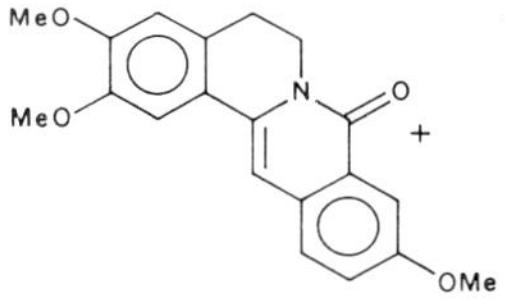 (96) + 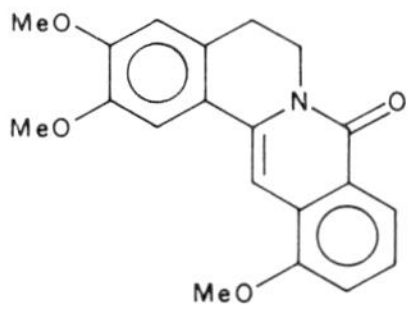 (97)	45
m-MeO	6	**95** +	15
		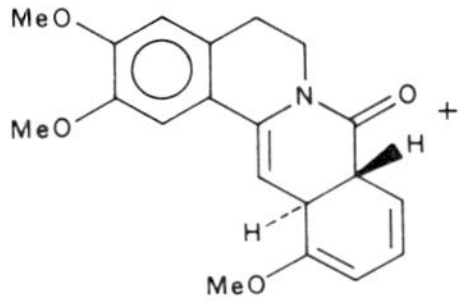 +	30

R	Temp. (°C)	Product(s)	Yield (%)
		{**96** + **97**}	20
p-MeO	r.t.	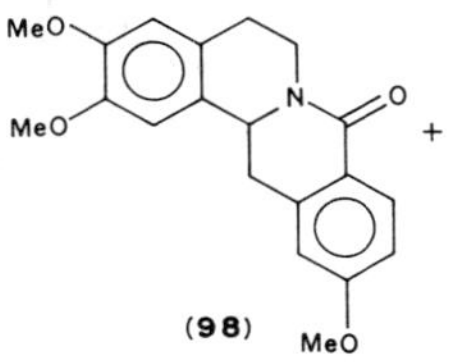 (**98**) +	27
		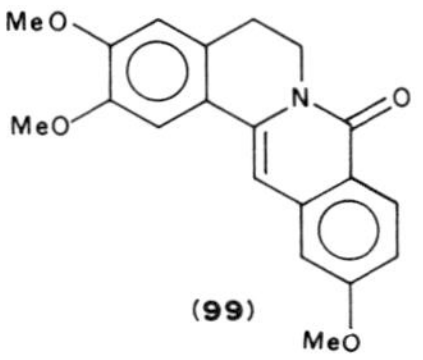 (**99**)	40
p-MeO	6	**98** +	29
		99 +	23
		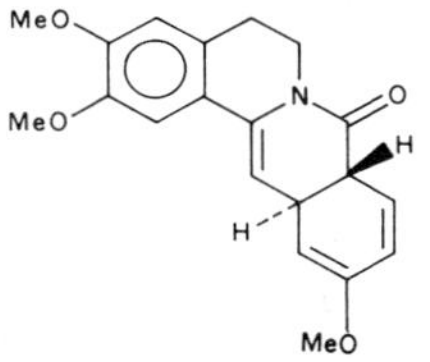	22
3,4-$(MeO)_2$	r.t.	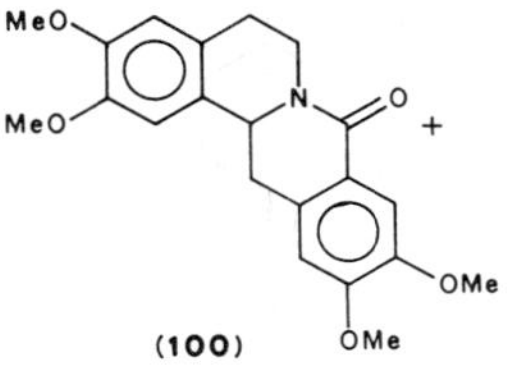 (**100**) +	5
		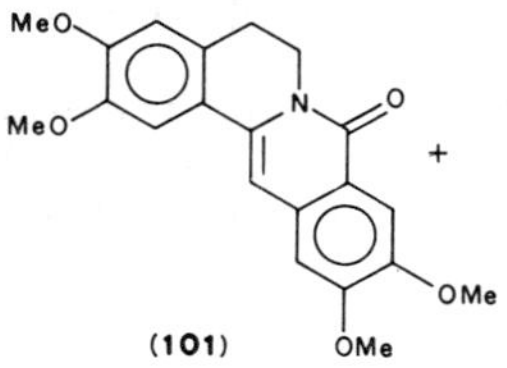 (**101**) +	40
		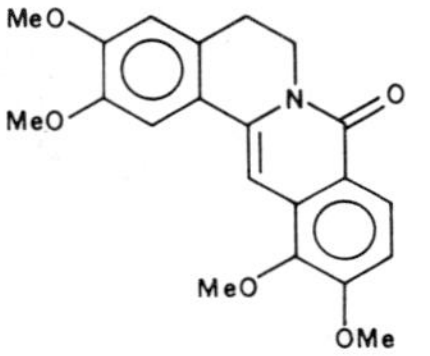	5
3,4-$(MeO)_2$	6	**100** +	25
		101 +	35

(*continued*)

R	Temp. (°C)	Product(s)	Yield (%)
		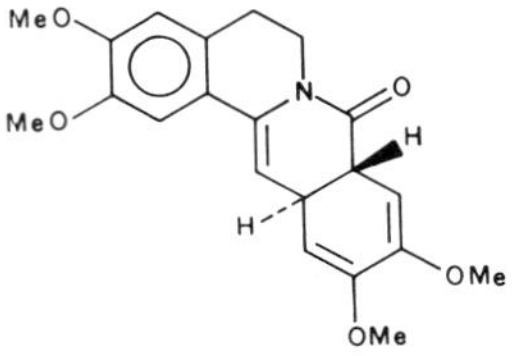	4

[a]The same products and yields were obtained when the reaction was performed at 6 °C.

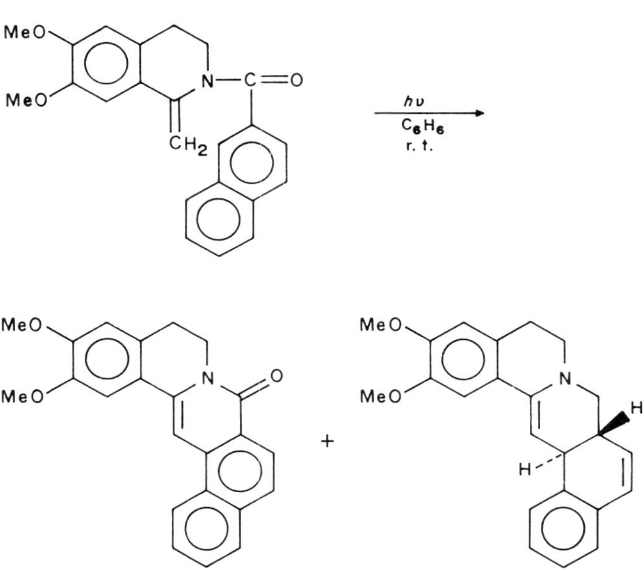

$h\nu$, C_6H_6, r. t.

(777)

The mechanism proposed for these conversions involves a [1,5]-sigmatropic proton shift (Scheme 9).

If the irradiation is performed[1105] in the presence of sodium borohydride in a benzene solution containing a small amount of methanol, the formation of the dihydrobenzene products is suppressed and only the more stable saturated lactam products are obtained (equation 778).

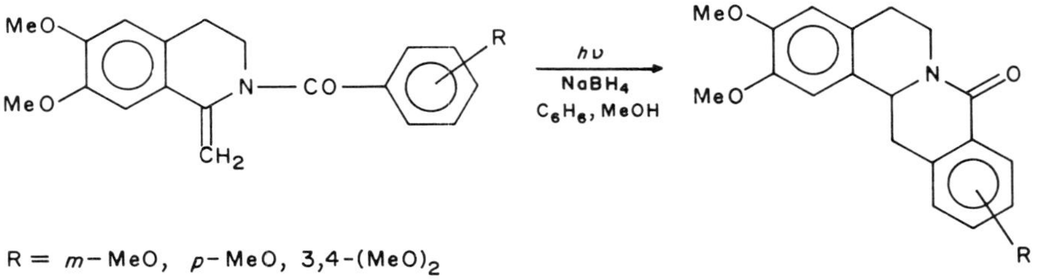

$h\nu$, $NaBH_4$, C_6H_6, MeOH

R = *m*-MeO, *p*-MeO, 3,4-$(MeO)_2$

(778)

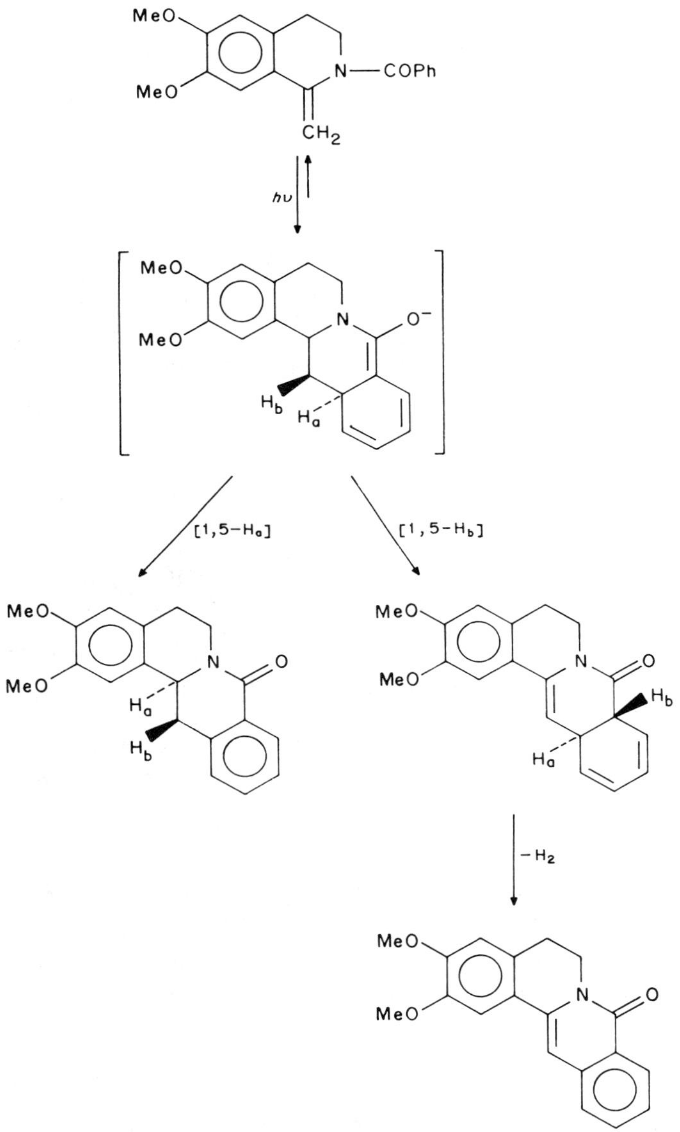

SCHEME 9

*4. *Cyclization of N-chloroacetyl-β-arylamines*

Photocyclization of *N*-chloracetyl derivatives of 4-(3,4-dimethoxyphenyl)butylamine and 5-(3,4-dimethoxyphenyl)pentylamine in 50% aqueous acetonitrile affords[1106] a variety of azepinoindole products (equation 779).

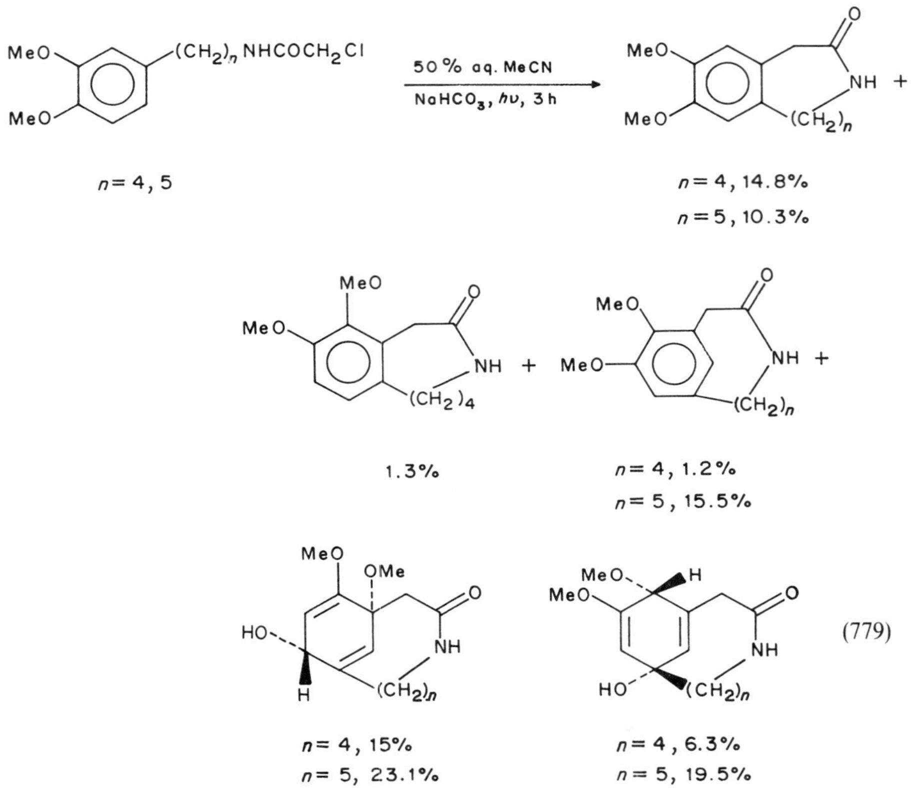

(779)

6. Miscellaneous cyclizations

A number of miscellaneous, photochemically induced cyclization reactions has been reported in the recent literature, the majority of which have been concerned with the preparation of β-lactams. Thus, photoirradiation of 2-N-acyl-(N-alkylamino)cyclohex-2-enones in ether or acetone under a nitrogen atmosphere produces[1107] the corresponding N-alkyl-1-azaspiro[3.5]nonane-2,5-diones via a 1,4-diradical intermediate (equation 780).

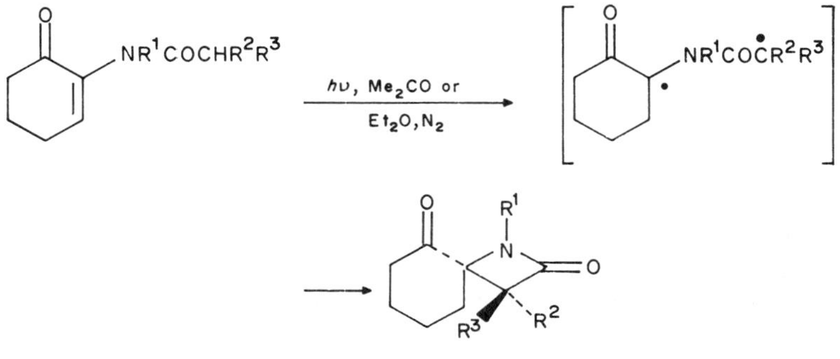

(780)

R^1	R^2		R^3	Yield (%)
Me	Me		Me	28
i-Pr	Me		Me	45
i-Pr		$—(CH_2)_5—$		48
i-Pr	MeS		H	45
i-Pr	Ph		H^a	45
$PhCH_2$	MeS		H	57
$PhCH_2$	Ph		H^a	65

[a]Stereochemistry uncertain.

Another photochemical approach[1108] to the preparation of β-lactams involves the irradiation of acyclic monothioimides in benzene under an argon atmosphere with a 1 kW high-pressure mercury lamp. However, since the β-lactams produced were too unstable to isolate directly, they were benzoylated with benzoyl chloride in the presence of triethylamine, and the resulting *S*-benzoyl β-lactams isolated (equation 781). The mechanism proposed[1108] for this reaction involves the intermediate formation of either a diradical or a zwitterion (equation 782).

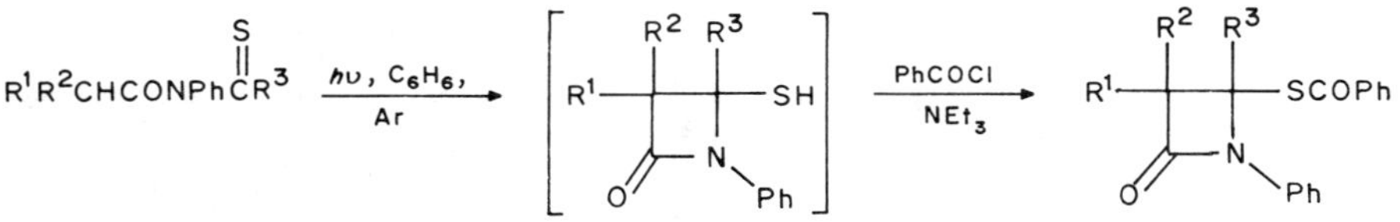

$R^1R^2CHCONPhC(=S)R^3$ —($h\nu$, C_6H_6, Ar)→ [β-lactam with R^1, R^2, R^3, SH, N–Ph] —(PhCOCl, NEt_3)→ β-lactam with R^1, R^2, R^3, SCOPh, N–Ph

(781)

R^1	R^2	R^3	Yield (%)
Me	Me	Ph	29
Me	Me	Me	16
H	Ph	Me	35^a
H	MeO	Me	76^b
H	MeO	Ph	90^c

[a]The *S*-benzoyl β-lactam product was obtained as a 1:1 mixture of stereoisomers. In addition a 13% yield of the *S*-phenylacetyl β-lactam shown below was also obtained.

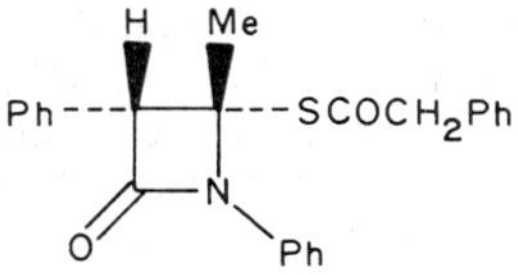

[β-lactam: H, Me; Ph, $SCOCH_2Ph$; N–Ph]

[b]The product obtained was the *cis* isomer only.
[c]The product obtained was a 63:27 *cis*:*trans* mixture.

The intermediacy of a zwitterion is also proposed[1109] in the mechanistic explanation of the formation of optically active β-lactams via solid state photolysis of inclusion complexes of *N*,*N*-dialkylpyruvamides with desoxycholic acid (equation 783). Although the enantiomeric excesses for the β-lactams obtained are not high, optically active products are obtained by this method which are not obtainable from the corresponding photolysis in solution[1109]. The 4:1 inclusion complexes of acid to amide required as the starting materials for these reactions were prepared[1109] by crystallizing the acid using the amides as the solvents.

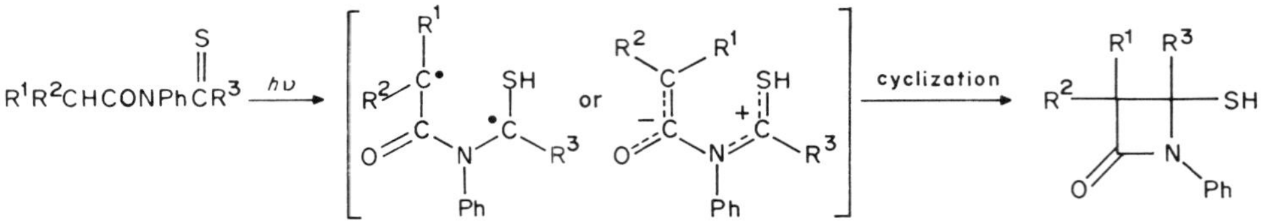

(782)

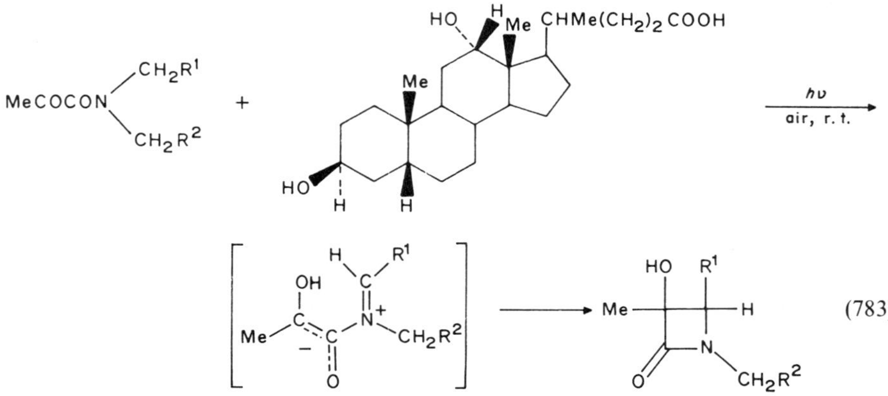

(783)

R^1	R^2	Enantiomeric excess (%)	Yield (%)
H	H	15[a]	42
Me	Me	9[a]	74[b]
—$(CH_2)_3$—		15	52

[a]3,5-Dinitrobenzoate derivative.
[b]Product obtained as a 1:1 mixture of stereoisomers.

Larger ring lactams have also been prepared by photochemically induced cyclization reactions as evidenced by the synthesis of γ- and δ-lactams. Irradiation of 2'-, 3'- or 4'-substituted biphenyl-2-carboxamides with *t*-butyl hypoiodite in *t*-butyl alcohol produces[1110] carboxamidyl radicals which cyclize intramolecularly to afford the γ- and δ-lactam products (equation 784). The proportion of each product obtained depends upon[1110] the steric and electronic effects of the substituents present.

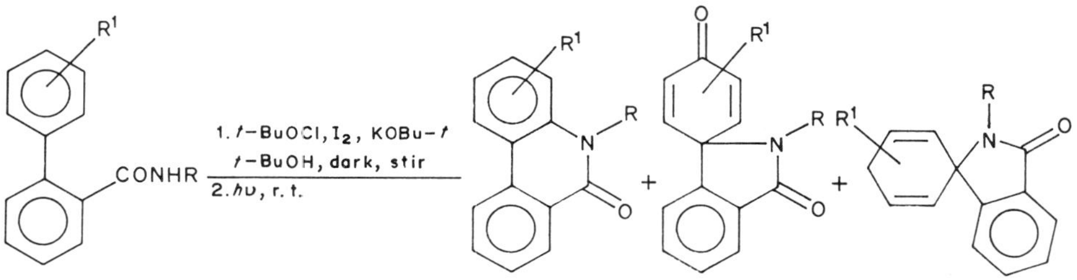

(784)

Starting material	Time (h)	Product	Yield (%)
OMe; CONHMe	2.5	OMe; N–Me; O + **(102)**	17
		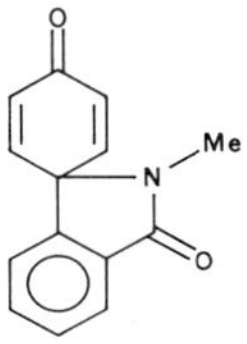 O; N–Me; O **(103)**	65
	1	**102** + **103** 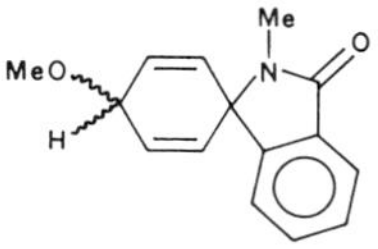 +	14 + 17
		Me; MeO; H; N; O	56
OMe; $CONH_2$	2.5	O; NH; O	68
R; CONHMe		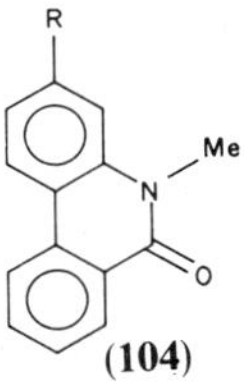 R; N–Me; O **(104)**	
R = I	4	+	17
		103	70
R = Cl	8	**104** +	17
		103	40
R = Me	3	**104**	15
R = CN	3	N.R.	—

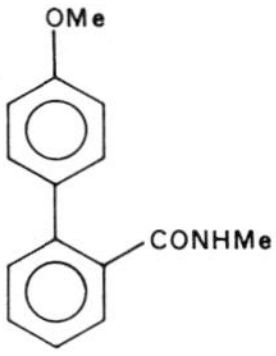

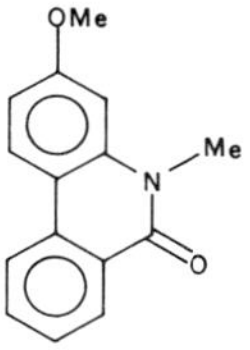

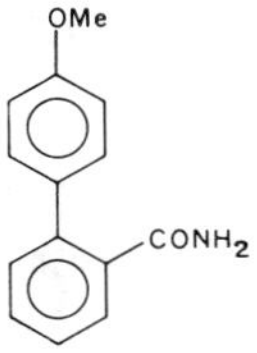

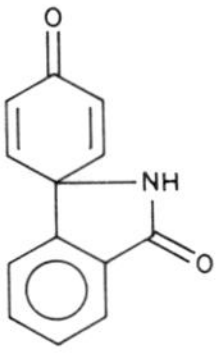

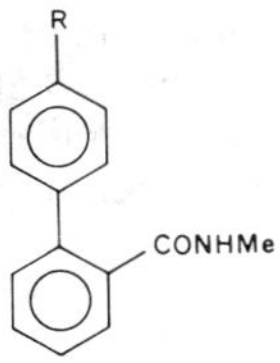

Starting material	Time (h)	Product	Yield (%)
MeO, CONHMe	2.5	MeO, N, Me, O +	23
		OMe, N, Me, O	17
MeO, CONHMe	2.5	O, MeO, N, Me, O	96
MeO, $CONH_2$	2.5[a]	O, MeO, NH, O	26[b,c,d]

[a]Condition used: *t*-BuOI, I_2, *t*-BuOH, r.t.
[b]Same conditions as in a but without I_2 gave a 22% yield of product.
[c]Condition used: *t*-BuOI, ICl, I_2, CCl_4, r.t. produced no reaction.
[d]Condition used: *t*-BuOI, ICl, I_2, *t*-BuOH, r.t. produced a 100% yield of product.

Macrocyclic lactams ranging in size from 26 to 28 members have been prepared[1111] in moderate yields by remote photocyclization of sulphur containing phthalimides (PI—NR) (equation 785). If analogues of the starting material shown in equation (785) are used, 31- and 38-membered macrocyclic lactams were prepared[1111], while similar reaction with an ester analogue produced[1111] 16- to 27-membered lactams depending upon the value of *m* and *n* in equation (786).

$$\mathrm{PI{-}N(CH_2)_{12}NHCO(CH_2)_{10}SMe} \xrightarrow[\text{30 min}]{\text{MeCN, } h\nu}$$

$$\text{26-membered monocyclic lactam (17\%)} + \text{28-membered monocyclic lactam (57\%)} \quad (785)$$

$$\mathrm{PI{-}N(CH_2)_mCO_2(CH_2)_nSMe} \xrightarrow{\text{MeCN, } h\nu} \text{16-, 18-, 22-, 25- and 27-membered lactams} \quad (786)$$

*C. By Cycloaddition Reactions

*1. Addition of isocyanates to olefins

The reaction of N-chlorosulphonyl isocyanates (CSI) with olefins to produce N-chlorosulphonyl substituted β-lactams, and their subsequent reduction to N-unsubstituted β-lactams (equation 787, $R^3 = SO_2Cl$), has again been used in the recent literature to produce substituted β-lactams. In addition to chlorosulphonyl isocyanate, a variety of other activated isocyanates have also been employed as reagents for the preparation of β-lactams via essentially the same two-step procedure used with chlorosulphonyl. isocyanate. Table 51 lists the various β-lactams prepared using these reagents.

$R^1CH{=}CHR^2$ + R^3NCO ⇌ [π-complex] → $R^{1(2)}$—CH—$\overset{+}{C}HR^{2(1)}$ / O=C—$\bar{N}$—R^3

→ [β-lactam: $R^{1(2)}$, $R^{2(1)}$, N—R^3, O] —reduction→ [β-lactam: $R^{1(2)}$, $R^{2(1)}$, NH, O] (787)

Using N-chlorosulphonyl isocyanate (CSI), an interesting series of reactions was observed[1120] with 4-methylenespirol[2.X]alkanes. With substituents such as methoxycarbonyl- or cyano- attached to the cyclopropyl portion of the spiroalkane, the reaction with CSI proceeded in a normal manner to produce the N-chlorosulphonyl spiro-β-lactam initially, which upon reduction with sodium sulphite produced the N-unsubstituted spiro-β-lactam (equation 788). However, when a similar reaction was attempted[1121] with unsubstituted 4-methylenespiro[2.X]alkane analogues no sign of initial β-lactam formation was observed and the initial products isolated were five- and six-membered ring N-chlorosulphonyl lactams, which upon reduction with thiophenol-pyridine in acetone afforded the corresponding N-unsubstituted lactams (equation 789). Three

[$(CH_2)_n$ spiro compound with CH_2, X] + CSI —0 °C→ [spiro-β-lactam, N—SO_2Cl, X] —Na_2SO_3→ [spiro-β-lactam, NH, X] (788)

$n = 2,3$
$X = CN, CO_2Me$

[$(CH_2)_n$ spiro compound with CH_2] + CSI —CH_2Cl_2, −70 °C→ [$(CH_2)_n$ lactam, N—SO_2Cl] —$C_6H_5SH-C_5H_5N$, Me_2CO, 0 °C, 1 h→ [$(CH_2)_n$ lactam, NH] (789)

TABLE 51. Preparation of β-lactams by addition of activated isocyanates of olefins

Olefin	Isocyanate (R^3 =)	Conditions	Initial product (%) yield	Reducing agent	Product	Yield (%)	Reference
(*t*-Bu)Me_2SiOCHMe | CH=CHSPh	SO_2Cl	1. Et_2O, 0 °C 2. r.t., 4 h	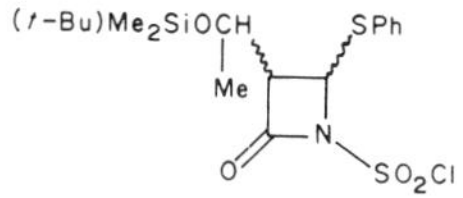	PhSH, C_5H_5N, Me_2CO	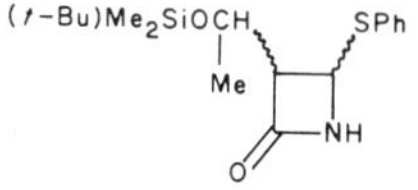	—	1112
CH_2=CH$(CH_2)_3$CH=CH_2	SO_2Cl	—	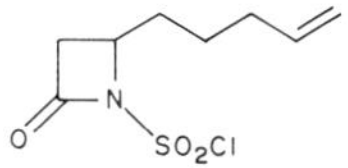	—	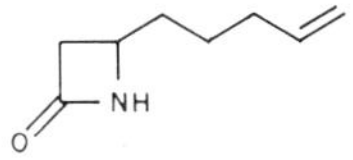	—	1113
PhCH=CH_2	SO_2Cl	—	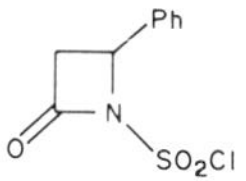	THF, Na_2SO_3, $NaHCO_3$	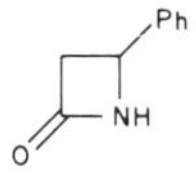	—	1003
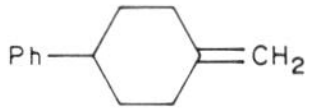	SO_2Cl	Et_2O, 0 °C	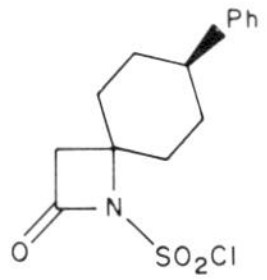 (**105**) 67%	THF, $NaSO_3$, $NaHCO_3$, H_2O, stir overnight	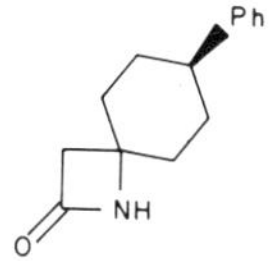 (**106**)	78	1114
			105	bis(ethylene-diamine) chromium(II) perchlorate, DMF, r.t., stir overnight	106	38	1114
			105	Zn/Cu, THF, stir 55 °C, 2 days	106	42	1114

Allene	R	Conditions	Intermediate	Conditions	Product	Yield (%)	Ref.
$Me_2C{=}C{=}C(SC_6H_4Cl\text{-}p)Me$	SO_2Cl	Et_2O, 0 °C	3-($Me_2C{=}$)-4-($SC_6H_4Cl\text{-}p$)-4-Me-1-SO_2Cl-azetidin-2-one	Na_2SO_3, Et_2O, K_2HPO_4, stir r.t., 1 h	3-($Me_2C{=}$)-4-($SC_6H_4Cl\text{-}p$)-4-Me-azetidin-2-one (NH)	55	1115
$Me_2C{=}C{=}C(SC_6H_4Cl\text{-}p)SiMe_3$	SO_2Cl	Et_2O, 0 °C	3-($Me_2C{=}$)-4-($SC_6H_4Cl\text{-}p$)-4-$SiMe_3$-1-SO_2Cl-azetidin-2-one	Na_2SO_3, Et_2O, K_2HPO_4, stir r.t., 1 h	3-($Me_2C{=}$)-4-($SC_6H_4Cl\text{-}p$)-4-$SiMe_3$-azetidin-2-one (NH)	87	1115
$Me_2C{=}C{=}C(SC_6H_4Cl\text{-}p)H$	SO_2Cl	Et_2O, 0 °C	3-($Me_2C{=}$)-4-($SC_6H_4Cl\text{-}p$)-4-H-1-SO_2Cl-azetidin-2-one	Na_2SO_3, Et_2O, K_2HPO_4, stir r.t., 1 h	3-($Me_2C{=}$)-4-($SC_6H_4Cl\text{-}p$)-4-H-azetidin-2-one (NH)	20	1115
$(Me)(Me_3Si)C{=}C{=}C(SC_6H_4Cl\text{-}p)SiMe_3$	SO_2Cl	Et_2O, 0 °C	3-[$(p\text{-}ClH_4C_6S)(Me_3Si)C{=}$]-4-Me-4-$SiMe_3$-1-$SO_2Cl$-azetidin-2-one	Na_2SO_3, Et_2O, K_2HPO_4, stir r.t., 1 h	3-[$(p\text{-}ClH_4C_6S)(Me_3Si)C{=}$]-4-Me-4-$SiMe_3$-azetidin-2-one (NH)	54	1115
$(Me)(H)C{=}C{=}C(SC_6H_4Cl\text{-}p)SiMe_3$	SO_2Cl	Et_2O, 0 °C	3-[$(Me)(H)C{=}$]-4-($SC_6H_4Cl\text{-}p$)-4-$SiMe_3$-1-SO_2Cl-azetidin-2-one + 3-[$(H)(Me)C{=}$]-4-($SC_6H_4Cl\text{-}p$)-4-$SiMe_3$-1-SO_2Cl-azetidin-2-one (1:2)	Na_2SO_3, Et_2O, K_2HPO_4, stir r.t., 1 h	3-[$(Me)(H)C{=}$]-4-($SC_6H_4Cl\text{-}p$)-4-$SiMe_3$-azetidin-2-one (NH) + 3-[$(H)(Me)C{=}$]-4-($SC_6H_4Cl\text{-}p$)-4-$SiMe_3$-azetidin-2-one (NH) (1:2)	10	1115

(continued)

TABLE 51. (*continued*)

Olefin	Isocyanate ($R^3 =$)	Conditions	Initial product (%) yield	Reducing agent	Product	Yield (%)	Reference
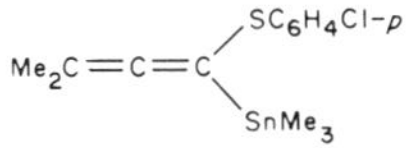	SO_2Cl	Et_2O, 0 °C	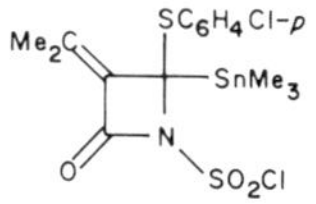	Na_2SO_3, Et_2O K_2HPO_4, stir r.t., 1 h	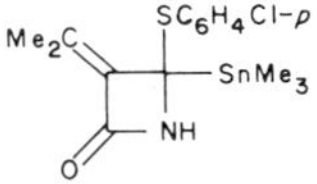	22	1115
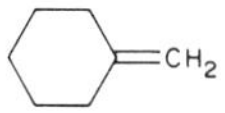	$Cl_3CCH_2SO_2$	$CHCl_3$, r.t.	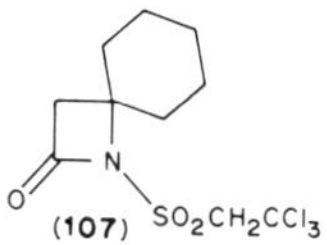 (107)	Zn dust aq. THF, NH_4Cl, stir 5 weeks	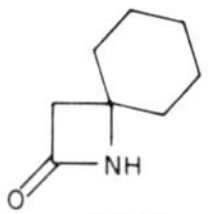 (108)	19–22	1114, 1116
			107	$Na_2S_2O_6$, $(n\text{-Bu})_4\overset{+}{N}\overset{-}{I}$, DMF, stir 2 weeks	108	24	1114
			107	NaS_2O_6, $(n\text{-Bu})_4\overset{+}{N}\overset{-}{I}$, 15-crown-5	**108**	72	1114
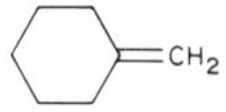	*p*-Tos	$CHCl_3$, r.t., 4 weeks	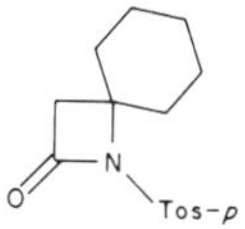	—	—	87	1114
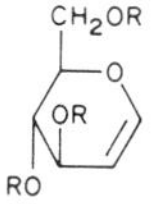	*p*-Tos	$CDCl_3$, r.t.	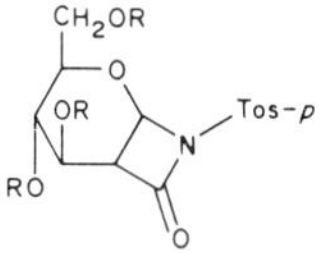	—	—	—	1117

$R = Me_3Si$		2 h 6 h 22 h 30 h	55 77 76 75	silica gel	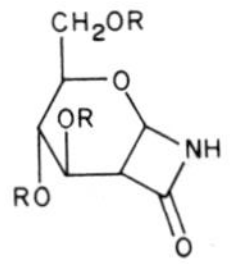 CH_2OR, OR, RO, NH, O (109)	30	
$R = (t\text{-}Bu)Me_2Si$		2 h 6 h 22 h 50 h	14 37 50 90	silica gel		50	
$R = Me$		2 h 6 h 22 h 50 h	17 41 75 71	—	—	—	
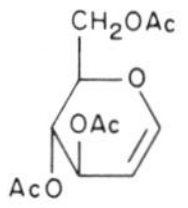 CH_2OAc, OAc, AcO, O	*p*-Tos	Et_2O, r.t. 10 kbar, 18 h	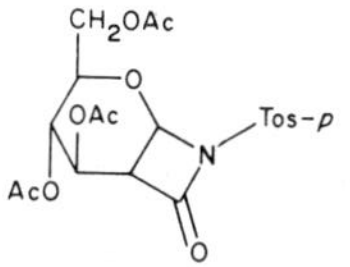 CH_2OAc, OAc, AcO, N, Tos-*p*, O	—	—	60–70	1118
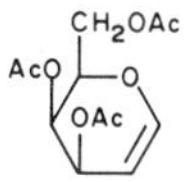 CH_2OAc, AcO, OAc, O	*p*-Tos	Et_2O, r.t. 10 kbar, 18 h	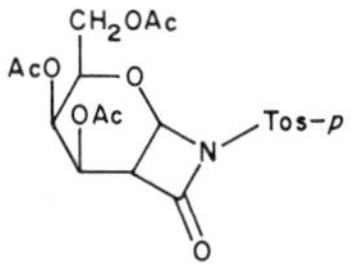 CH_2OAc, AcO, OAc, N, Tos-*p*, O	—	—	60–70	1118
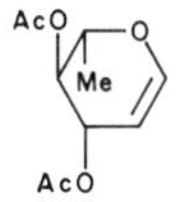 AcO, Me, AcO, O	*p*-Tos	Et_2O, r.t. 10 kbar, 18 h	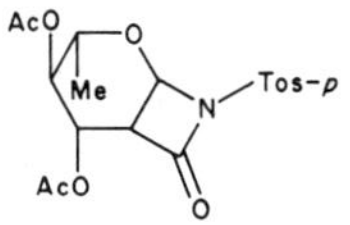 AcO, Me, AcO, N, Tos-*p*, O	—	—	60–70	1118

(continued)

TABLE 51. (*continued*)

Olefin	Isocyanate (R^3 =)	Conditions	Initial product (%) yield	Reducing agent	Product	Yield (%)	Reference
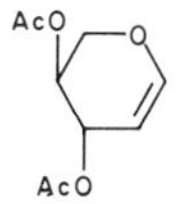 AcO, AcO	*p*-Tos	Et_2O, r.t. 10 kbar, 18 h	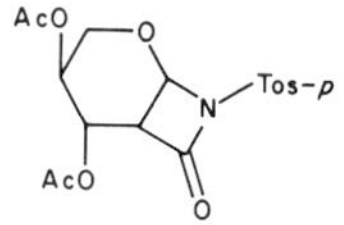 AcO, AcO, N–Tos-*p*	—	—	60–70	1118
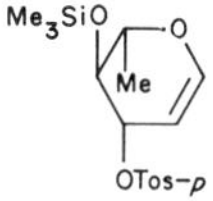 Me_3SiO, Me, OTos-*p*	*p*-Tos	$CDCl_3$, r.t.	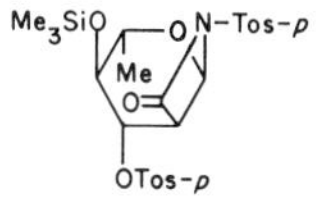 Me_3SiO, Me, N–Tos-*p*, OTos-*p*	—	—	—	1117
		2 h	59				
		6 h	73				
		22 h	75				
		50 h	72				
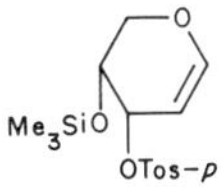 Me_3SiO, OTos-*p*	*p*-Tos	$CDCl_3$, r.t.	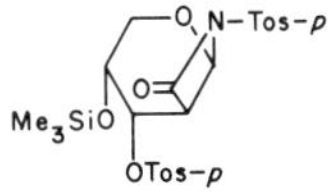 Me_3SiO, N–Tos-*p*, OTos-*p*	—	—	—	1117
		2 h	69				
		6 h	75				
		22 h	78				
		50 h	74				
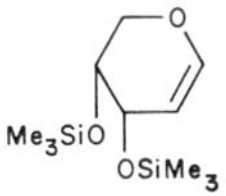 Me_3SiO, $OSiMe_3$	*p*-Tos	$CDCl_3$, r.t.	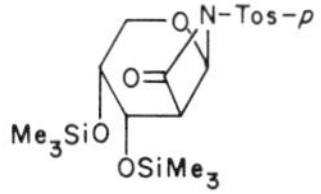 Me_3SiO, N–Tos-*p*, $OSiMe_3$	silica gel	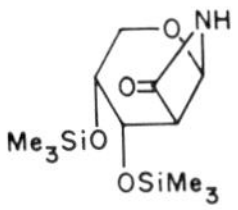 Me_3SiO, NH, $OSiMe_3$	40	1117
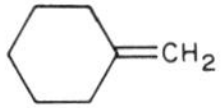 $=CH_2$	*o*-$O_2NC_6H_4SO_2$	r.t., 3 days	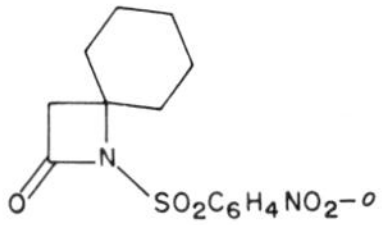 N, $SO_2C_6H_4NO_2$-*o* (98%)	10% Pd/C, EtO, H_2 12 h	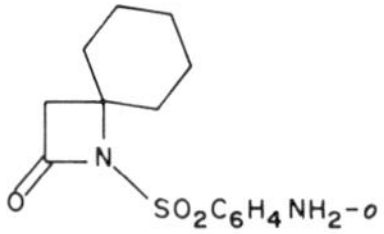 N, $SO_2C_6H_4NH_2$-*o*	89	1114

PhCHCH$_2$	Cl$_3$CCH$_2$OSO$_2$	Et$_2$O, reflux 2 days	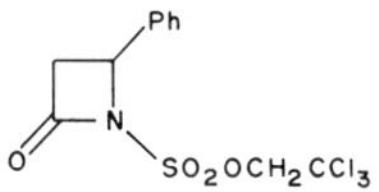 (**109**) (54–60%)	Zn/Cu, THF, stir overnight	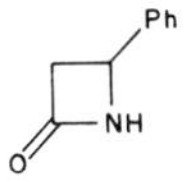 (**110**)	98	1114, 1116
			109	Cr(ClO$_4$)$_2$, DMF, H$_2$O, 2 min	**110**	96	1114
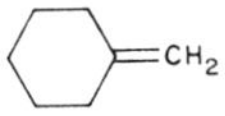	C$_3$CCH$_2$OSO$_2$	CCl$_4$, r.t.	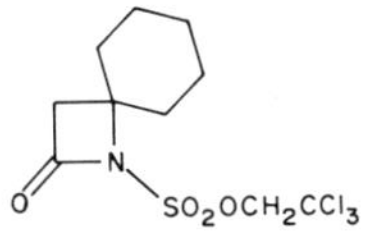	Zn dust, aq. THF, NH$_4$Cl, stir 72 h	**108**	29–30	1114, 1116
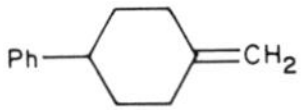	Cl$_3$CCH$_2$OSO$_2$	Et$_2$O, stand overnight	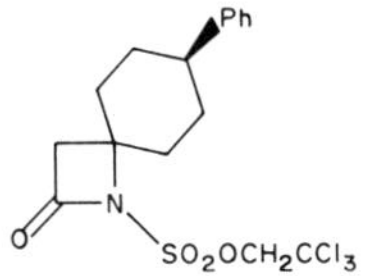 (67–86%)	Zn/Cu, THF	**106**	42	1114, 1116
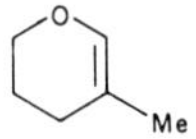	Cl$_3$CCH$_2$OSO$_2$	CHCl$_3$, r.t., stand 2 days	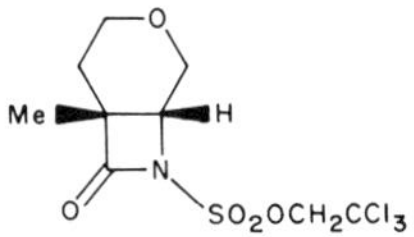		—	—	1114
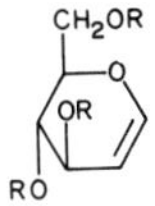	Cl$_3$CCH$_2$OSO$_2$	CDCl$_3$, r.t.	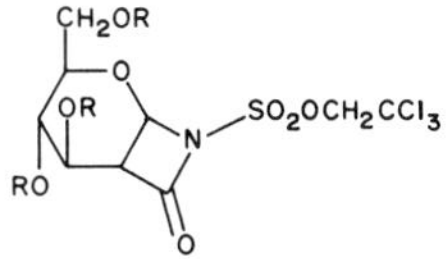	—	—	—	1117

(*continued*)

TABLE 51. *(continued)*

Olefin	Isocyanate ($R^3 =$)	Conditions	Initial product (%) yield	Reducing agent	Product	Yield (%)	Reference
$R = Me_3Si$		2 h 6 h	100 80	silica gel	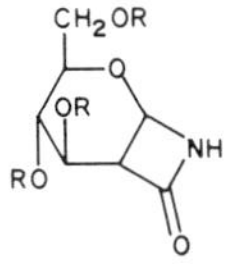 CH_2OR OR RO NH O (111)	30	
$R = (t\text{-Bu})Me_2Si$		2 h 6 h	100 100	silica gel	**111**	50	
$R = Me$		2 h	100	—	—	—	
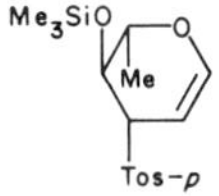 Me_3SiO O Me Tos-*p*	$Cl_3CH_2OSO_2$	$CDCl_3$, r.t., 2 h	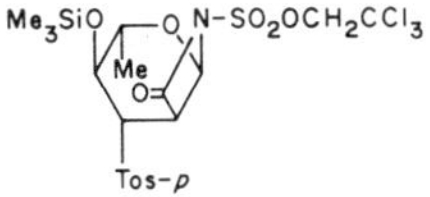 Me_3SiO O $N\text{-}SO_2OCH_2CCl_3$ Me O Tos-*p*	—	—	100	1117
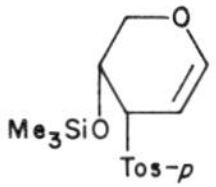 O Me_3SiO Tos-*p*	$Cl_3CH_2OSO_2$	$CDCl_3$, r.t.	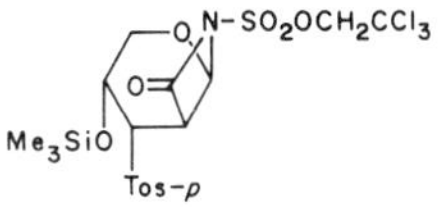 O $N\text{-}SO_2OCH_2CCl_3$ O Me_3SiO Tos-*p*	—	—	—	1117
		2 h 22 h	100 10				
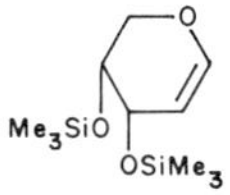 O Me_3SiO $OSiMe_3$	$Cl_3CH_2OSO_2$	$CDCl_3$, r.t.	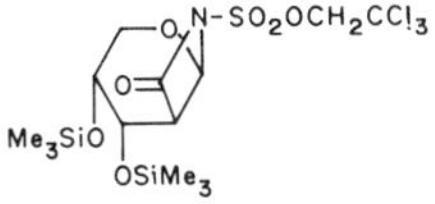 O $N\text{-}SO_2OCH_2CCl_3$ O Me_3SiO $OSiMe_3$	silica gel	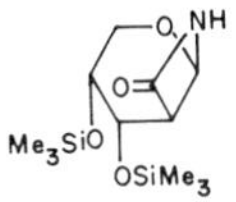 O NH O Me_3SiO $OSiMe_3$	40	1117

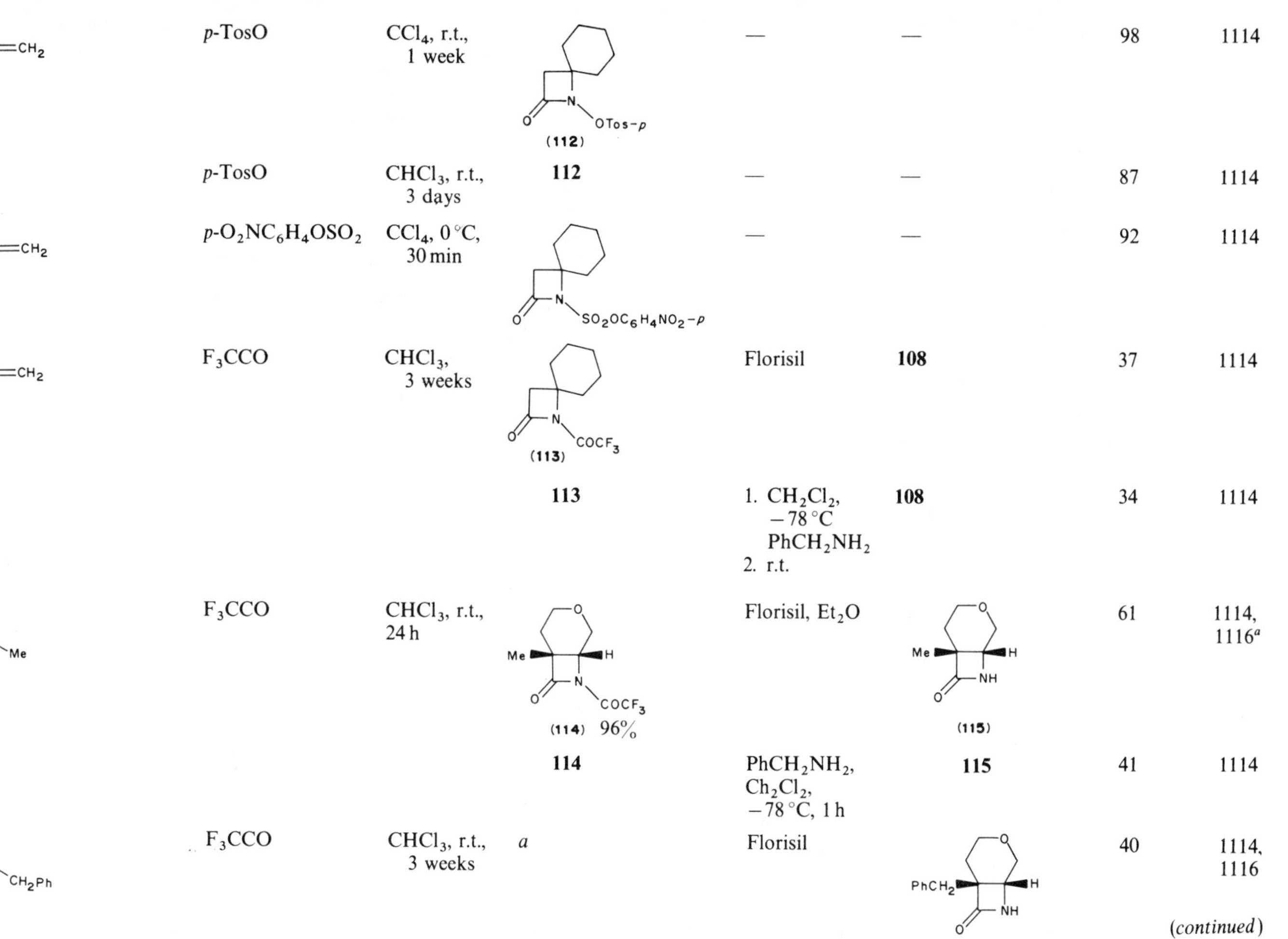

Methylenecyclohexane	p-TosO	CCl_4, r.t., 1 week	(112)	—	—	98	1114
	p-TosO	$CHCl_3$, r.t., 3 days	**112**	—	—	87	1114
Methylenecyclohexane	p-$O_2NC_6H_4OSO_2$	CCl_4, 0 °C, 30 min		—	—	92	1114
Methylenecyclohexane	F_3CCO	$CHCl_3$, 3 weeks	(113)	Florisil	**108**	37	1114
			113	1. CH_2Cl_2, −78 °C $PhCH_2NH_2$ 2. r.t.	**108**	34	1114
	F_3CCO	$CHCl_3$, r.t., 24 h	(114) 96%	Florisil, Et_2O	(115)	61	1114, 1116[a]
			114	$PhCH_2NH_2$, Ch_2Cl_2, −78 °C, 1 h	**115**	41	1114
	F_3CCO	$CHCl_3$, r.t., 3 weeks	*a*	Florisil		40	1114, 1116

(continued)

TABLE 51. (*continued*)

Olefin	Isocyanate ($R^3 =$)	Conditions	Initial product (%) yield	Reducing agent	Product	Yield (%)	Reference
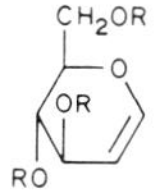	Cl_3CCO	$CDCl_3$,r.t.	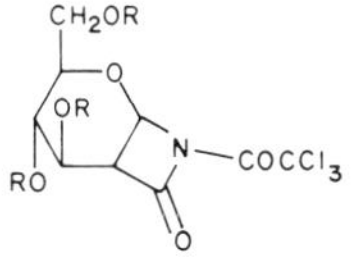	—	—	—	1117
R = Me_3Si		2 h 6 h 22 h 50 h	10[b] 25[c] 35[d] 35[d]	silica gel	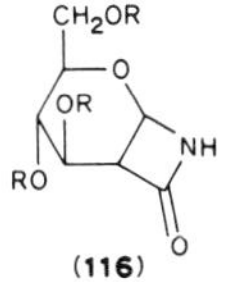 (116)	30	
R = (*t*-Bu)Me_2Si		2 h 6 h 22 h 50 h	2[e] 12[f] 28[g] 50[h]	silica gel	116	50	
R = Me		2 h 6 h 22 h 50 h	10[i] 25[i] 19[j] 19[k]		—		
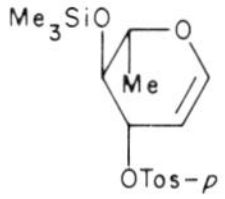	Cl_3CCO	$CDCl_3$, r.t.	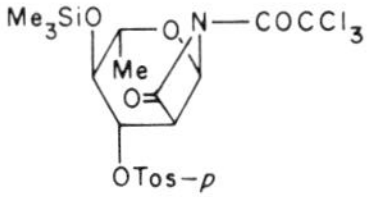	—	—	—	1117
		2 h 6 h 22 h 50 h	8[l] 23[m] 27[n] 10[o]	—	—	—	

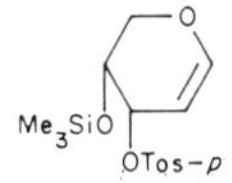 Me_3SiO, OTos-*p*	Cl_3CCO	$CDCl_3$, r.t.	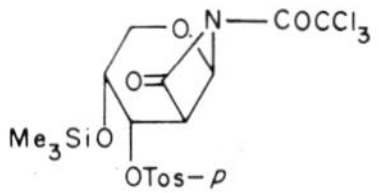 O, N—$COCCl_3$, O=, Me_3SiO, OTos-*p*	—	—	—	1117
		2 h 6 h 22 h 50 h	27[p] 42[q] 20[r] 7[s]	—	—	—	
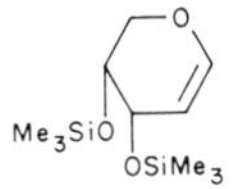 Me_3SiO, $OSiMe_3$	Cl_3CCO	$CDCl_3$. r.t.	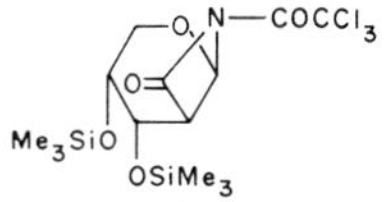 O, N—$COCCl_3$, O=, Me_3SiO, $OSiMe_3$	silica gel	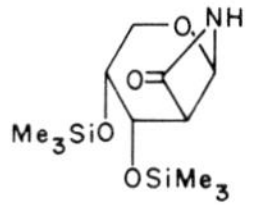 O, NH, O=, Me_3SiO, $OSiMe_3$	40	1117
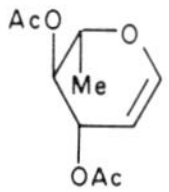 AcO, O, Me, OAc	Cl_3CCO	Et_2O 10 kbar, 18 h	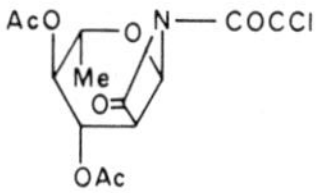 AcO, O, N—COCCl, Me, O=, OAc	—	—	12	1119
			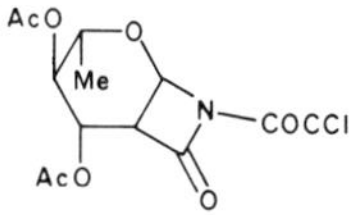 + AcO, O, Me, N—COCCl, AcO, O	—	—	9	
			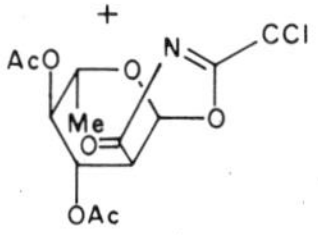 + AcO, O, N, CCl, O, Me, O=, OAc	—	—	43	

(*continued*)

TABLE 51. (*continued*)

Olefin	Isocyanate ($R^3 =$)	Conditions	Initial product (%) yield	Reducing agent	Product	Yield (%)	Reference
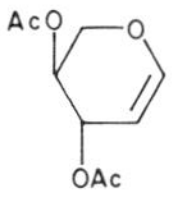 AcO, O, OAc	Cl_3CCO	Et_2O, 10 kbar, 18 h	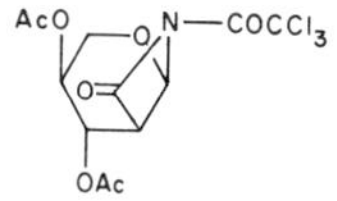 AcO, O, N—$COCCl_3$, O, OAc	—	—	6	1119
			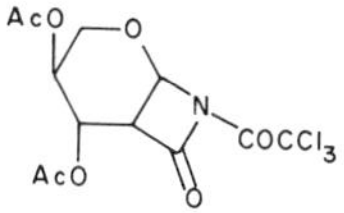 + AcO, O, N, $COCCl_3$, AcO, O	—	—	7.8	
			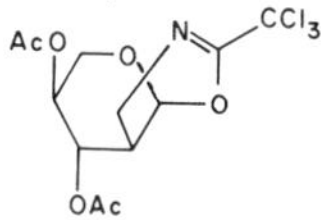 + AcO, O, N, CCl_3, O, OAc	—	—	7.8	

[a]Intermediate product was not isolated.
The product shown was also obtained in yields of:

CH_2OR, O, OR, O, CCl_3, RO, N, O

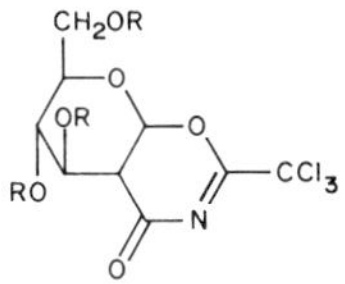

[b]15%. [c]40%. [d]65%. [e]3%. [f]18%. [g]26%. [h]18%.
[i]Unspecified mixture. [j]47%. [k]74%.

The product shown was also obtained in yields of:

Me_3SiO, O, N, CCl_3, Me, O, O, OTos-*p*

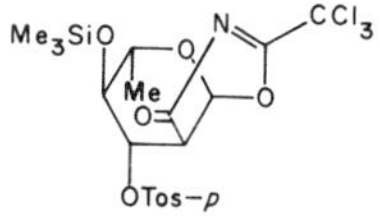

[l]18%. [m]51%. [n]70%. [o]85%.

The product shown was also obtained in yields of:

O, N, CCl_3, O, O, Me_3SiO, OTos-*p*

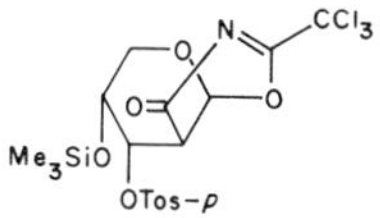

[p]31%. [q]50%. [r]76%. [s]88%.

mechanistic possibilities were proposed for this conversion. In route A, electrophilic attack of the CSI on the olefin generates fast equilibrating zwitterionic species which are stabilized through homoallylic charge delocalization, and a subsequent collapse into homo Diels–Alder adduct yields the product. Route B is a concerted, symmetry-allowed [2a + 4a]-type cycloaddition. Route C is a nearly concerted mechanism of the same symmetry-allowed [2a + 4a]-type (Scheme 10).

$(CH_2)_n$ CH_2 + O=C=N–SO_2Cl

Route A Route B Route C

$(CH_2)_n$ O $^-$N—SO_2Cl +

$(CH_2)_n$ O δ^- δ^+ N—SO_2Cl

$(CH_2)_n$ O N—SO_2Cl

SCHEME 10

Similar results were also obtained[1120] when an isopropenyl substitutent was present on the cyclopropyl ring and the [2 + 2]-cycloaddition reaction with the resulting *trans*-1-isopropenylspiro[2.X]alkane and CSI was performed. The products obtained[1120] from this reaction were seven- and nine-membered ring lactams (Scheme 11), resulting from electrophilic attack, preferentially at the methylene carbon rather than the isopropenyl carbon of the *cisoid–transoid* conformer of the spiroalkane starting material. The produced an intermediate spiro-β-lactam which was monitored by infrared spectroscopy, but could not be isolated because of the speed of its further addition to the spiro-cyclopropane and/or to the spiro-cyclopropylalkene to form the respective seven- and nine-membered ring lactams as the temperature was elevated.

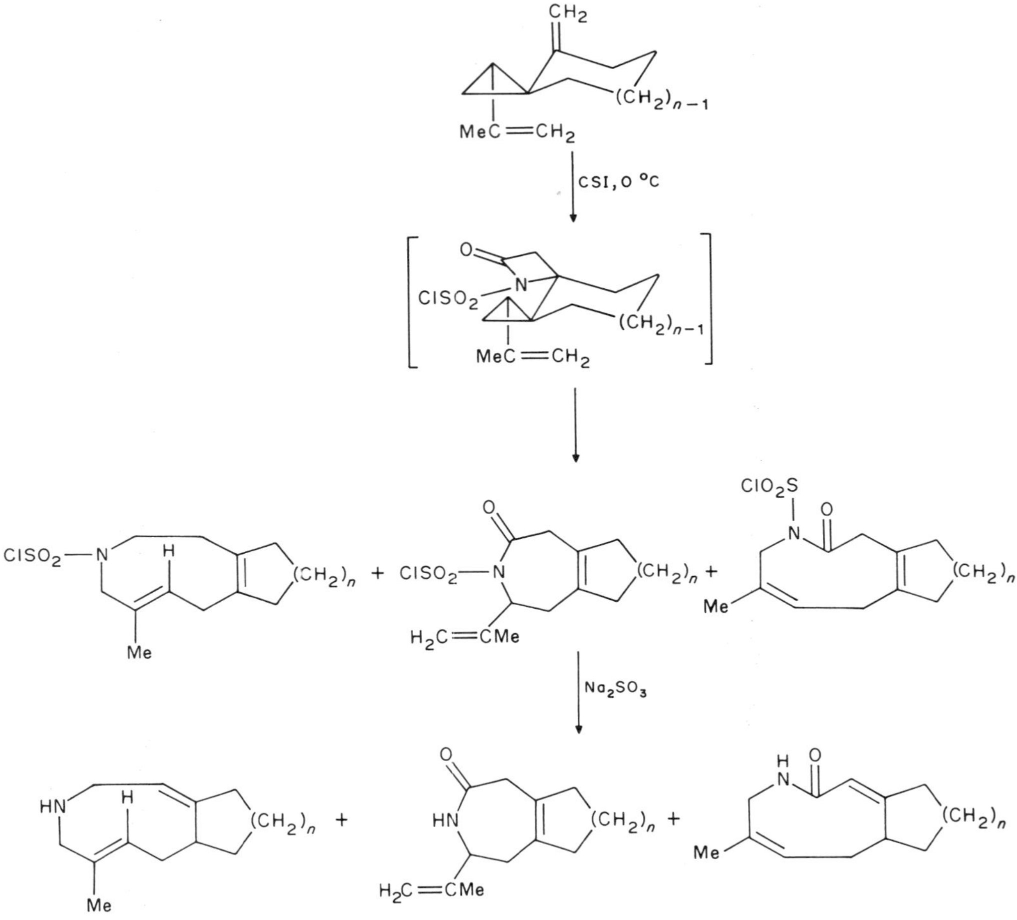

CSI, 0 °C

Na_2SO_3

with:

$n = 2$, a 68% yield of three isomeric 1:1 adducts is obtained in a 15:67:18 ratio,
$n = 3$, a 74% yield of three isomeric 1:1 adducts is obtained in a 11:74:15 ratio.

SCHEME 11

Recently, the use of transition metal complexes to form β-lactams was exemplified[1122] by a [1 + 1 + 2]-cycloaddition of iron carbene complexes with isocyanides followed by oxidation of the resulting adduct with potassium permanganate (equation 790).

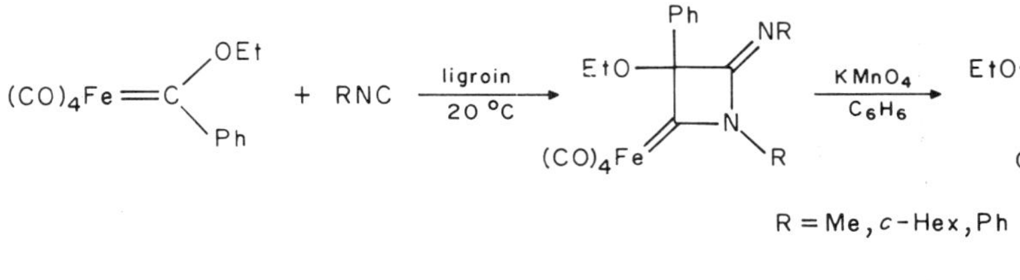

ligroin, 20 °C; $KMnO_4$, C_6H_6

R = Me, c-Hex, Ph (790)

Substituted β-lactams have also been synthesized by condensation[1123] of β-amino acids with various N-substituted isocyanates and formaldehyde (equation 791).

General:

$$HOOCCH_2CH(NH_2)CH_2COR^1 + R^2NCO + (37\%)\ CH_2O \xrightarrow{MeOH}$$ (β-lactam bearing CH_2COR^1 and N-CH_2CONHR^2)

R^1 = alkoxy, $PhCH_2O$, p-$O_2NC_6H_4CH_2O$, HO, PhS

R^2 = NHalkyl, $NHCH_2Ph$, alkoxy, $PhCH_2O$, p-$O_2NC_6H_4CH_2O$ (791)

Specific:

$$HOOCCH_2CH(NH_2)CH_2COOBu\text{-}t + MeNCO + CH_2O \xrightarrow{MeOH}$$ (β-lactam bearing CH_2COOBu-t and N-$CH_2CONHMe$)

66%

Intramolecular condensation of steroidyl isocyanates with an ester function has been reported[1124] to produce 2-aza-5α-cholestan-3-one, with the required isocyanate being prepared from the corresponding parent ester by reaction with diphenylphosphoryl azide (DPPA) to effect a Curtius rearrangment (equation 792).

Me, Me, Me, Me, H, Me, HO_2CCH_2, MeO_2CCH_2, H $\xrightarrow[\text{reflux 1 h, } C_6H_5Me]{DPPA, Et_3N,}$ $O{=}C{=}NCH_2$, Me, MeO_2CCH_2, H

$\xrightarrow[\text{reflux 21 h}]{DMF/H_2O}$ Me, Me, Me, Me, H, Me, HN, O, H (792)

Another example of intramolecular lactam formation is the Friedel–Crafts cyclization of 4-ω-phenylalkyl substituted-1H-pyrazol-5-isocyanates to produce the corresponding 7-, 8-, 9- and 10-membered lactams[1125] (equation 793).

Me, N, N, $N{=}C{=}O$, Me, $(CH_2)_nPh$ $\xrightarrow[\text{2. 215 °C, 1 h; 3. 145 °C, briefly}]{\text{1. } AlCl_3,\ o\text{-}Cl_2C_6H_4,\ 90\ °C,\ \text{stir 10–15 min}}$ Me, N, N, NHCO, Me, $(CH_2)_n$

n = 1 ; 2 ; 3 ; 4 ; 5

Yield (%) = 83 ; 85 ; 84 ; 47 ; 0 (793)

*2. From imines

**a. Reaction of imines with ketenes, acid chlorides or mixed anhydrides.* Recently, one review article[1126] on the chiral construction of β-lactams and other derivatives by the use of a chiral controlled Staudinger reaction and several other references on the preparation of β-lactams by the $[2+2]$ cycloaddition of imines and preformed or *in situ* generated ketenes (equation 794) have been published. The results of this approach, which truly involves ketene as a reactant, are reported in Table 52.

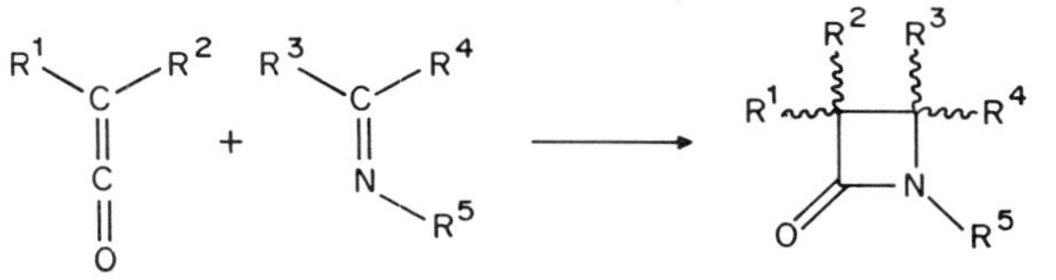

(794)

However, as perceived by recent reports, the most frequently used reaction for the preparation of β-lactams still remains the reaction of an imine with an acid halide in the presence of a tertiary amine. Also found in the recent literature is mounting supporting evidence[1126,1145–1147] that the mechanism for this reaction does not involve the inter-

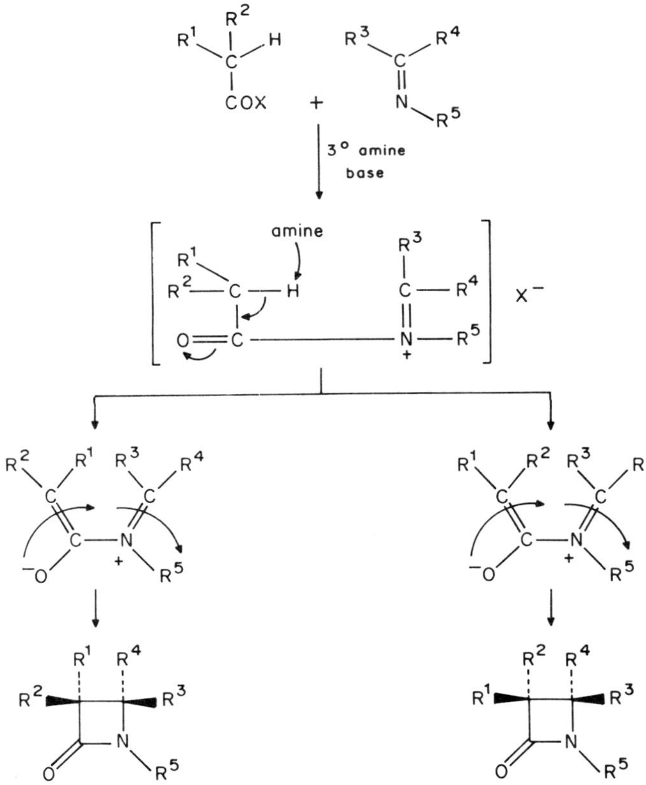

3° amine base

amine

X^-

SCHEME 12

TABLE 52. Production of lactams by reaction of ketenes with imines

Ketene	Imine	Conditions	Product (ratio)	Yield (%)	Reference
$CH_2{=}C{=}O$	Me; N–N; COR	r.t.	Me; N–N; COR; O		1127
	R = OEt	hexane		44	
	$R = NH_2$	C_6H_5Me		69	
$MeCH{=}C{=}O$	$R^1COCR^1 = NR^2$	C_6H_6, N_2, 30 min	Me; R^1; COR^1; N; R^2; O		1128
	$R^1 = R^2 = Ph$			45	
	$R^1 = Ph$, $R^2 = p$-An			50	
$MeCH{=}C{=}O$	Me; N–N; CO_2CH_2R	C_6H_5Me, r.t.	Me; H; Me; H; N–N; O; CO_2CH_2R		1129
	$R = CH_2SiMe_3$			94	
	$R = C_6H_4NO_2$-*p*			98	
$N_3CH{=}C{=}O$	$RCH{=}NCHCO_2Et$ with SEt on the CH	—	N_3; R; N–$CHCO_2Et$; SEt; O		1130
	R = H, Me, Ph			—	
$ClCH{=}C{=}O$	N=N; Me; Me; S; NPh	C_6H_6, reflux	Cl; O; Me; Me; S; N; Ph	54	1131

(continued)

TABLE 52. *(continued)*

Ketene	Imine	Conditions	Product (ratio)	Yield (%)	Reference
$ClCH{=}C{=}O$[a]	$R^1COCR^1=NR^2$	C_6H_6, N_2, 30 min	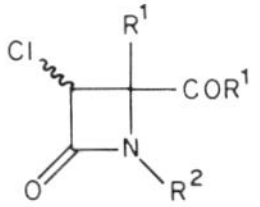 [structure]		1128
	$R^1 = R^2 = Ph$			40	
	$R^1 = Me$, $R^2 = p$-An			43	
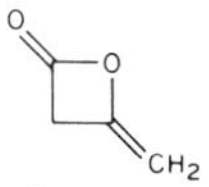 (diketene)	$R^1OCH_2CH(Me)CH{=}NR^2$	imidazole, THF, $-30°C$	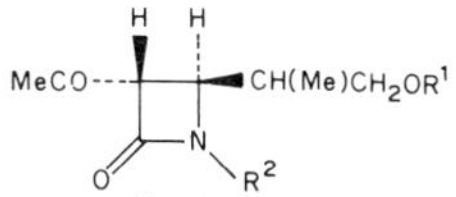 (117) + 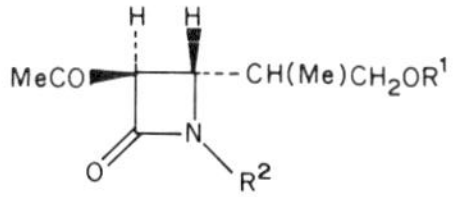 (118)		1132
	$R^1 = (t\text{-Bu})Me_2Si$, $R^2 = p$-An	64 h	**117**:**118** 0.25:1	33	
	$R^1 = (t\text{-Bu})Ph_2Si$, $R^2 = p$-An	15 h	0:1	10	
	$R^1 = MeO(CH_2)_2OCH_2$, $R^2 = p$-An	60 h	1.5:1	41	
	$R^1 = MeSCH_2$, $R^2 = p$-An	24 h	1.5:1	38	
	$R^1 = PhCH_2$, $R^2 = p$-An	60 h	1.6:1	58	
	$R^1 = Ph_3C$, $R^2 = (p\text{-An})_2CH$	34 h	0.25:1	12	
	$R^1 = MeOCH_2$, $R^2 = (p\text{-An})_2CH$	77 h	1.7:1	44	
	$R^1 = Me(EtO)CH$,	120 h	2.2:1	32	

	$R^1 = t$-Bu, $R^2 = (p\text{-An})_2CH$	60 h	3.3:1	26	
	$R^1 = PhCH_2$, $R^2 = (p\text{-An})_2CH$	60 h	2.5:1	47	
		DMF, 60 h	1.1:1	12	
		Hexane, THF (5:1), 60 h	3:1	36	
		$CHCl_3$, 39 h	3:1	25	
		Et_2O, 39 h	4.7:1	38	
		C_6H_5Me, 39 h	6.7:1	33	
		C_6H_5Me, 60 h, 4-Methylimidazole	11:1	52	
		C_6H_5Me, 90 h, 4-Methylimidazole	15:1	49[b]	
		C_6H_5Me, 96 h, Benzimidazole	—	~10	
		C_6H_5Me, 80 h, C_5H_5N	—	~10	
		C_6H_5Me, 80 h Et_3N	—	0	
PhCH=C=O[a]	$R^1COCR^1 = NR^2$	C_6H_6, N_2, 30 min	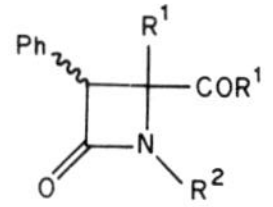		1128
	$R^1 = R^2 = Ph$			75	
	$R^1 = Ph$, $R^2 = p$-An			76	
	$R^1 = Ph$, $R^2 = p\text{-}ClC_6H_4$			65	
	$R^1 = Ph$, $R^2 = p$-Tol			74	
	$R^1 = Me$, $R^2 = Ph$			50	
	$R^1 = Me$, $R^2 = p$-An			60	
	$R^1 = Me$, $R^2 = p$-Tol			60	
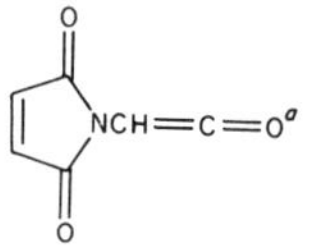	PhCOC(Ph)=NPh	C_6H_6, N_2, 30 min	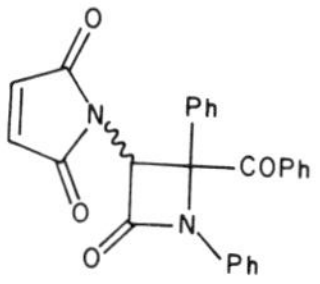	25	1128

(continued)

TABLE 52. (*continued*)

Ketene	Imine	Conditions	Product (ratio)	Yield (%)	Reference
$Me_2C{=}C{=}O$	$PhCH{=}NCH_2Ph$	—	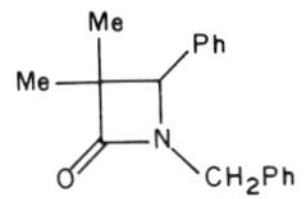 [structure]	—	1104
$Me_2C{=}C{=}O$	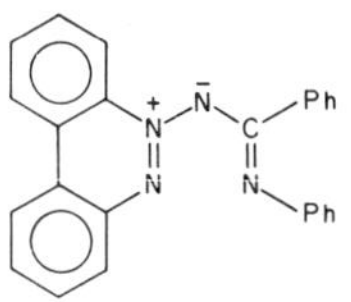 [structure]	C_6H_6, r.t.	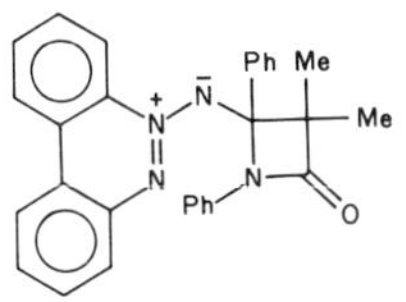 [structure]	—	1133
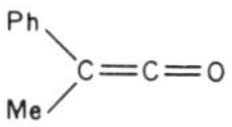 Ph(Me)C=C=O	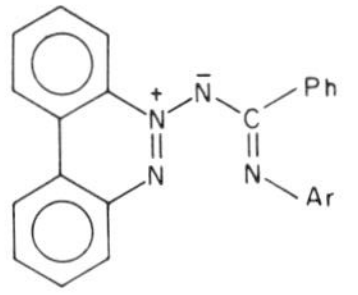 [structure] Ar = Ph, p-$O_2NC_6H_4$	C_6H_6, r.t.	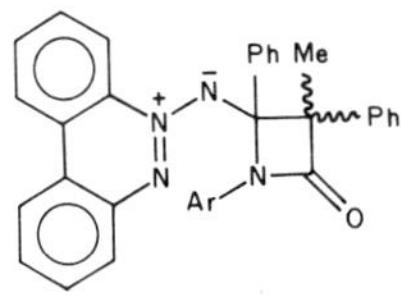 [structure]	—	1133
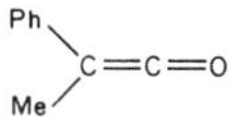 Ph(Me)C=C=O	$R^1N{=}C{=}NR^1$ $R^1 = (-)$-menthyl	Ar	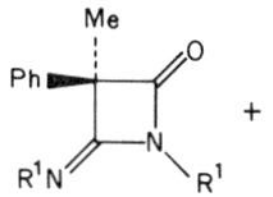 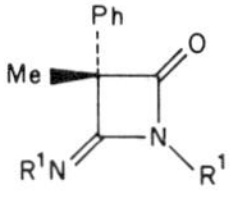 [structure] + [structure]	36.6 + 3.3	1134

(*continued*)

Ketene	Imine	Conditions	Product	Yield (%)	Ref.
EtOOC(R^2)C=C=O	$R^1N{=}C{=}NR^1$, $R^1 = (-)$-menthyl	Ar	EtOOC, R^2, R^1N, O, N, R^1 (β-lactam)		
$R^2 = t$-Bu				63.6	1134
$R^2 = Me_2CPh$				75	1134
MeOOC(MeS)C=C=O[c]	PhRC=NPh	$CDCl_3$, $h\nu$ or $Cl_2CHCHCl_2$, reflux	MeOOC, Ph, MeS, R, O, N, Ph (β-lactam)		1135
	R = H, SMe			—	
Ph(HO)C=C=O[d]	$R^1CH{=}NR^2$	C_6H_6, 80°C, $h\nu$, 3 h	HO, R^1, Ph, H, O, N, R^2 (β-lactam)		1136
	$R^1 = Ph$, $R^2 = PhCH_2$			73	
	$R^1 = p$-Tol, $R^2 = PhCH_2$			76	
	$R^1 = p$-An, $R^2 = PhCH_2$			50	
	$R^1 = p$-NCC_6H_4, $R^2 = PhCH_2$			77[e]	
	$R^1 = p$-ClC_6H_4, $R^2 = PhCH_2$			60	
	$R^1 = Ph$, $R^2 = p$-$TolCH_2$			46	
	$R^1 = Ph$, $R^2 = p$-$ClC_6H_4CH_2$			60	
	$R^1 = Ph$, $R^2 = Et$			56	
	$R^1 = Ph$, $R^2 = i$-Pr			27	
	$R^1 = Ph$, $R^2 = t$-Bu			45	
Ph(HO)C=C=O[d]	$R^1R^2C{=}NR^3$	C_6H_6, 80°C, $h\nu$, 3 h	HO, R^1, Ph, R^2, O, N, R^3 (β-lactam)		1136
	$R^1 = R^2 = Me$, $R^3 = i$-Pr			N.R.	
	$R^1 = R^2 = Ph$, $R^3 = H$			N.R.	

(continued)

TABLE 52. *(continued)*

Ketene	Imine	Conditions	Product (ratio)	Yield (%)	Reference
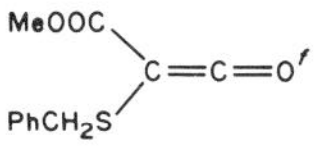 MeOOC(PhCH$_2$S)C=C=O[f]	PhCH=NPh	$CDCl_3$, *hv* or $Cl_2CHCHCl_2$, reflux	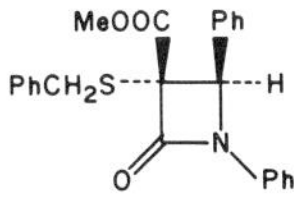	—	1135
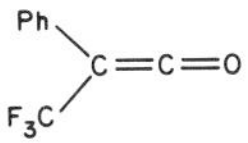 Ph(F$_3$C)C=C=O	$R^1N{=}C{=}NR^1$ $R^1 = (R)t$-BuCH(Me)	Ar	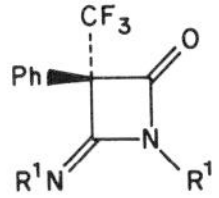 +	53 + 31	1134
$Ph_2C{=}C{=}O$	EtOCH=NPh	CH_2Cl_2, N_2 r.t. stir 12 h	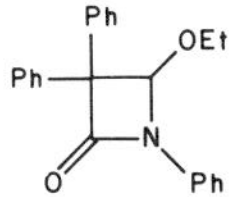	20	1137
$Ph_2C{=}C{=}O$	$R^1R^2C{=}NR^3$	Et_2O, N_2, r.t., 1 h	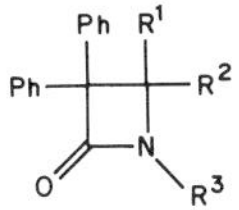		1138
	$R^1 = H$, $R^2 = PhCH{=}CH$, $R^3 = Ph$			74	
	$R^1 = H$, $R^2 = PhCH{=}CMe$, $R^3 = Ph$			72	
	$R^1 = Me$, $R^2 = PhCH{=}CH$, $R^3 = Ph$			63	

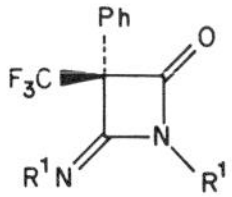

	$R^1 = H$, $R^2 = 2$-Furyl, $R^3 = Ph$			62	
	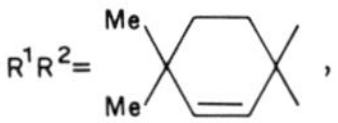 $R^1R^2 =$ (cyclic substituent, Me, Me), $R^3 = Ph$			76	
	$R^1 = H$, $R^2 = PhCH{=}CH$, $R^3 = i$-Pr			77	
	$R^1 = Me$, $R^2 = Me_2C{=}CH$, $R^3 = Ph$			70	
	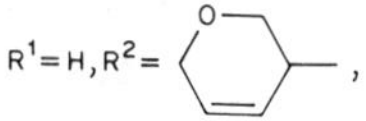 $R^1 = H$, $R^2 =$ (cyclic substituent, O), $R^3 = Ph$			84	
	$R^1 = H$, $R^2 = PhCH{=}CH$, $R^3 = c$-Hex			82	
	$R^1 = H$, $R^2 = Ph_2C{=}CH$, $R^3 = Ph$			84	
	$R^1 = H$, $R^2 = PhCH{=}CH$, $R^3 = t$-Bu			83	
$Ph_2C{=}C{=}O$[a]	$PhCOC(Ph){=}NR$	C_6H_6, N_2, 30 min	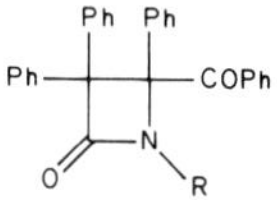		1128
	$R = PhCH_2$			85	
	$R = PhCHMe$			85	
$Ph_2C{=}C{=}O$	$R^1R^2C{=}NSiMe_3$	Molar ratio or reactants 1:1	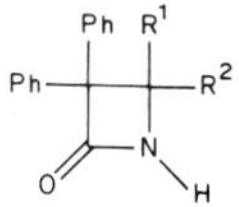		1139
	$R^1 = R^2 = H$, Ph			—	
	$R^1 = H$, $R^2 = Ph$, p-An			—	
	$R^1 = Ph$, p-An, $R^2 = H$			—	

(continued)

TABLE 52. (*continued*)

Ketene	Imine	Conditions	Product (ratio)	Yield (%)	Reference
$Ph_2C{=}C{=}O$	$R^1R^2C{=}NSiMe_3$ R^1 and R^2 as above	Molar ratio of reactants 2:1	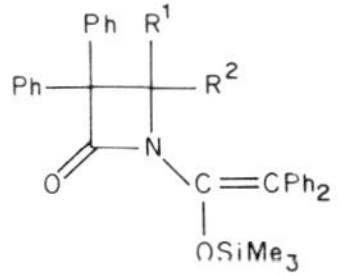	—	1139
$Ph_2C{=}C{=}O$	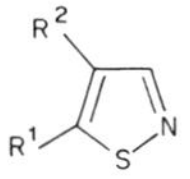	—	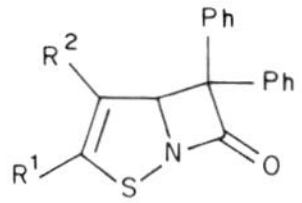		1140, 1141
	$R^1 = H$, $R^2 = Ph$, *p*-Tol			—	
	$R^1 = Ph$, $R^2 = H$			—	
$Ph_2C{=}C{=}O$	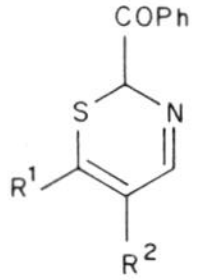	—	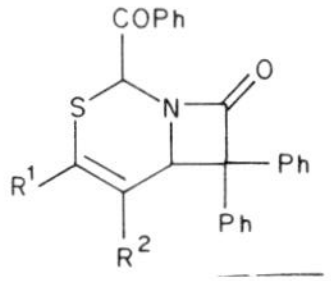		1140, 1141
	$R^1 = H$, $R^2 = Ph$, *p*-Tol			—	
	$R^1 = Ph$, $R^2 = H$			—	
$Ph_2C{=}C{=}O$	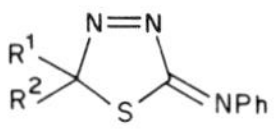	C_6H_6, reflux	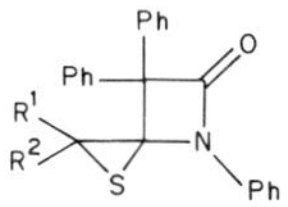		1131
	$R^1 = Me$, $R^2 = Me$			100	
	$R^1 = Me$, $R^2 = Et$			71	
	$R^1R^2 = -(CH_2)_5-$			100	

$Ph_2C{=}C{=}O$	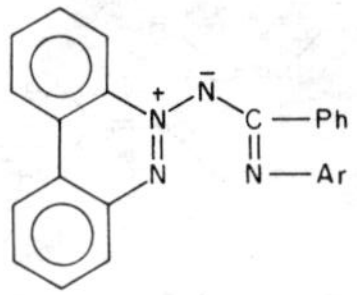 Ar = Ph, *p*-Tol	C_6H_6, r.t.	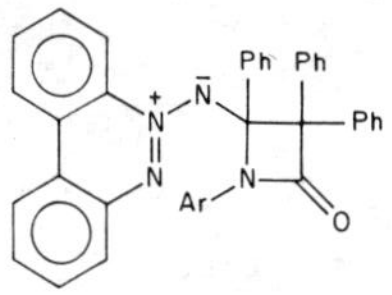	—	1133
$Ph_2C{=}C{=}O$	$PhN{=}CHCH{=}CHNMe_2$	—	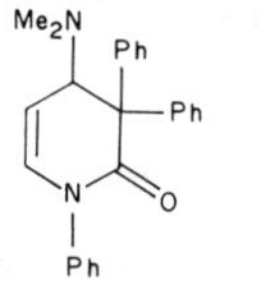	64	1138
$Cl_2C{=}C{=}O$[a]	PhCOC(Ph)=NAn-*p*	C_6H_6, N_2, 30 min.	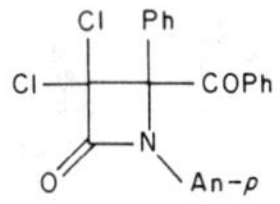	46	1128
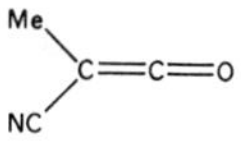 Me(NC)C=C=O	$R^1CH{=}NR^2$	C_6H_6, heat	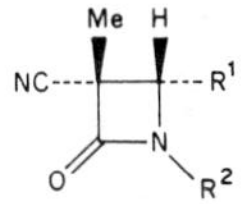		1142
	$R^1 = MeS$, $R^2 = c$-Hex			40	
	$R^1 = EtS$, $R^2 = c$-Hex			50	
	$R^1 = t$-BuS, $R^2 = c$-Hex			28	
	$R^1 = EtS$, $R^2 = Ph$			80	
	$R^1 = PhS$, $R^2 = Ph$			59	
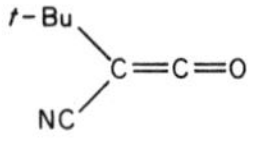 *t*-Bu(NC)C=C=O	$R^1CH{=}NR^2$	C_6H_6, heat	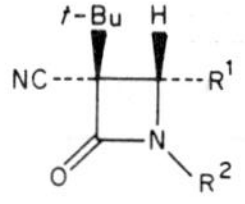		1142
	$R^1 = MeS$, $R^2 = c$-Hex			35	
	$R^1 = n$-BuSn, $R^2 = c$-Hex			29	

(*continued*)

TABLE 52. (*continued*)

Ketene	Imine	Conditions	Product (ratio)	Yield (%)	Reference
	$R^1 = t$-BuS, $R^2 = c$-Hex			34	
	$R^1 =$ EtS, $R^2 =$ Ph			20	
	$R^1 =$ PhS, $R^2 =$ Ph			31	
Cl(NC)C=C=O[g]	$R^1CH{=}NR^2$	C_6H_6, heat	β-lactam: NC, Cl; H, R^1; N–R^2; C=O		1142
	$R^1 =$ EtO, $R^2 = c$-Hex			94	
	$R^1 =$ EtS, $R^2 = c$-Hex			78	
	$R^1 =$ MeS, $R^2 = c$-Hex			85	
	$R^1 = i$-PrS, $R^2 = c$-Hex			50	
	$R^1 = n$-BuS, $R^2 = c$-Hex			82	
	$R^1 = t$-BuS, $R^2 = c$-Hex			67	
	$R^1 =$ PhS, $R^2 = c$-Hex			47	
	$R^1 =$ MeS, $R^2 =$ Ph			70	
	$R^1 =$ EtS, $R^2 =$ Ph			46	
	$R^1 = i$-PrS, $R^2 =$ Ph			56	
	$R^1 = n$-BuS, $R^2 =$ Ph			72	
	$R^1 = t$-BuS, $R^2 =$ Ph			56	
	$R^1 =$ PhS, $R^2 =$ Ph			36	
Cl(NC)C=C=O[g]	PhCH=NBu-*t*	C_6H_6, heat	β-lactam: NC, Cl; Ph, H; N–Bu-*t*; C=O	81	1143
Cl(NC)C=C=O	Ph(R^1)C=C(H)–C(H)=N–R^2	C_6H_6, heat			1144

$R^1 = H$, $R^2 = t$-Bu	Cl, NC, O, R^1, R^2, N, Ph (**119**) +	56 +
	NC, O, N, R^2, Ph (**120**)	19
$R^1 = H$, $R^2 = c$-Hex	**119** + **120**	71 + 21
$R^1 = H$, $R^2 = p$-An	Cl, H, NC, CH=CHPh, N, O, An-*p* +	17 +
	119 + **120**	42 + 22
$R^1 = Ph$. $R^2 = t$-Bu	NC, H, Cl, CH=CPh_2, N, O, Bu-*t*	86
$R^1 = Ph$, $R^2 = c$-Hex	Cl, H, NC, CH=CPh_2, N, O, Hex-*c* +	20 +
	NC, H, Cl, CH=CPh_2, N, O, Hex-*c* +	38 +
	119	36

(continued)

TABLE 52. (*continued*)

Ketene	Imine	Conditions	Product (ratio)	Yield (%)	Reference
	R^1 = Ph, R^2 = *p*-An		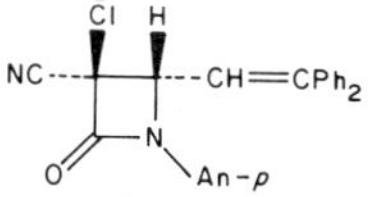	90	
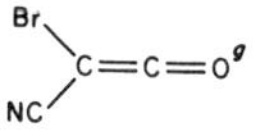 Br(NC)C=C=O[g]	RCH=NHex-*c*	C_6H_6, heat	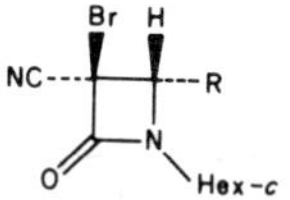		1142
	R = EtO			85	
	R = EtS			63	
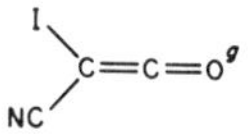 I(NC)C=C=O[g]	EtSCH=NHex-*c*	C_6H_6, heat	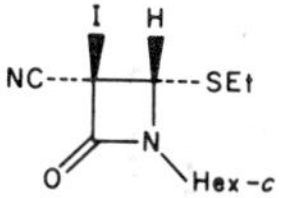	66	1142

[a]Two methods were used; Method A involved reaction of acid chloride with imine in the presence of triethylamine (see Table 53), and Method **B** was direct addition of ketene to the imine. Which method was employed with the different imine substitutents was not specified.
[b]Two equivalents of diketene used.
[c]Generated from MeO_2CCN_2COSMe via a sequence involving a carbene: C(COOMe)COSMe, a sulphonium ylide MeOOC—C̄—S̄Me and rearrangement to the ketene.

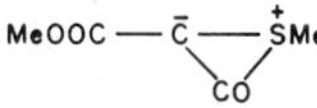

[d]Generated from PhCOCOOHex-*c*.
[e]A 6:4 mixture of two stereoisomers was obtained which could not be separated.
[f]Generated from $MeOOCCN_2COSCH_2Ph$ as discussed in c above.

[g]Generated by thermolysis of β-azido-α-halo-γ-methoxy-$\Delta^{\alpha,\beta}$-crotonolactones 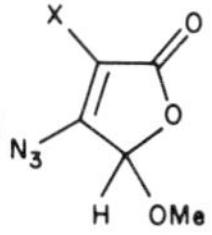 or 3,6-disubstituted-2,5-diazido-1,4-benzoquinones

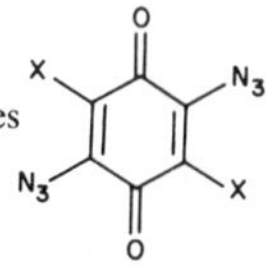

TABLE 53. Production of lactams by reaction of imines with acid halides

Acid halide	Imine	Conditions	Product	Yield (%)	Reference
$MeCH_2COCl$[a]	PhCO(Ph)C=NR	Et_3N, C_6H_5Me, reflux 24 h	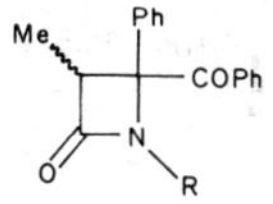		1128
	R = Ph			45	
	R = *p*-An			50	
RCH_2COCl	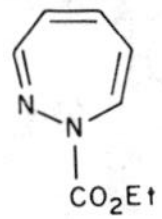	Et_3N, Et_2O or CH_2Cl_2	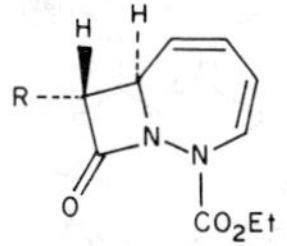		
R = Me				15	1127
R = Ph				56	1127
RCH_2COCl	PhCOCH=NAn-*p*	Et_3N, C_6H_6,	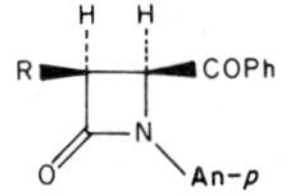		
R = Me		2–5 h stir r.t.		75	1148
R = Et				60	1148
R = *i*-Pr				65	1148
R = Ph				75	1148
$Ph_3COCH(Me)CH_2COCl$	2-FuCH=NAn-*p*	Et_3N	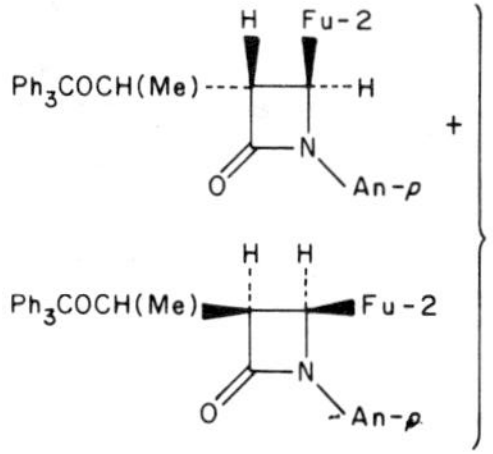	85	1149

(continued)

TABLE 53. (*continued*)

Acid halide	Imine	Conditions	Product	Yield (%)	Reference
$PhCH_2COCl^a$	$R^1COC(R^1){=}NR^2$	Et_3N, C_6H_5Me, reflux 24 h	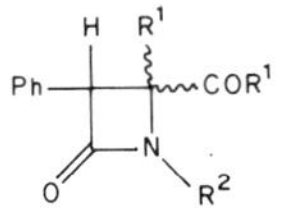		1128
	$R^1 = R^2 = Ph$			75	
	$R^1 = Ph$, $R^2 = An$-*p*			76	
	$R^1 = Ph$, $R^2 = C_6H_4Cl$-*p*			65	
	$R^1 = Ph$, $R^2 = Tol$-*p*			74	
	$R^1 = Me$, $R^2 = Ph$			50	
	$R^1 = Me$, $R^2 = An$-*p*			60	
	$R^1 = Me$, $R^2 = Tol$-*p*			60	
$PhCH_2COCl$	$PhCOCH{=}NAn$-*p*	Et_3N, C_6H_6, 2–5 h, stir r.t.	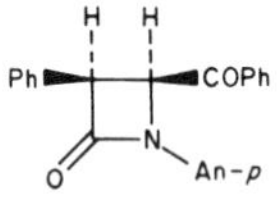	75	1148
$PhCH_2COCl$	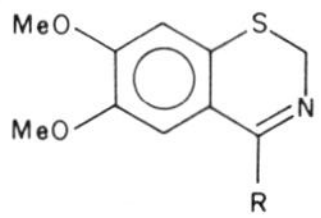	Et_3N, 5°C stand overnight	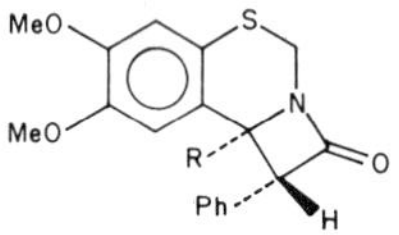		1150
	R = H	CH_2Cl_2		48	
	R = Ph	C_6H_6		—	
p-$HO_2CC_6H_4CH{=}NAn$-*p*	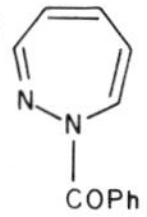	Et_3N, Et_2O or CH_2Cl_2	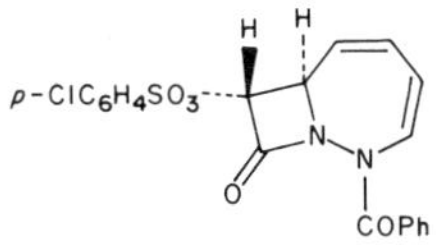	81	1127
$MeOCH_2COCl$	*p*-$HO_2CC_6H_4CH{=}NAn$-*p*	Et_3N, Me_3SiCl CH_2Cl_2	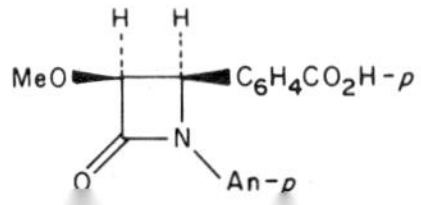	—	1151

$MeOCH_2COCl$	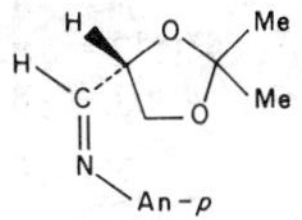	Et_3N	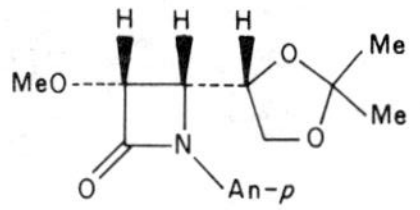	54	1152
$MeOCH_2COCl$	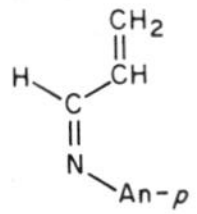	Et_3N, CH_2Cl_2	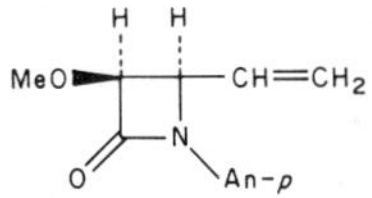	71	1153
$MeOCH_2COCl$	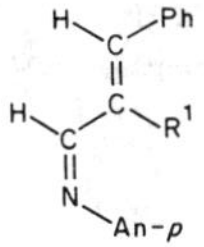	Et_3N	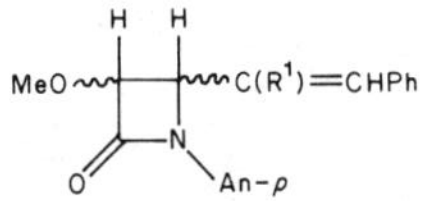		1144
	$R^1 = H$		*cis*:*trans* = 90:10	40[b]	
	$R^1 = Me$		*cis*:*trans* = 100:0	60[c], 40[b]	
$MeOCH_2COCl$	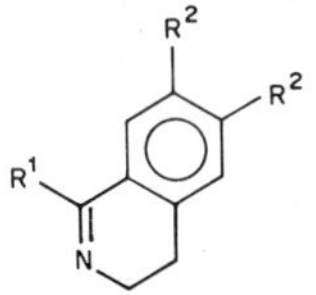	Et_3N, CH_2Cl_2, N_2 r.t., stir overnight	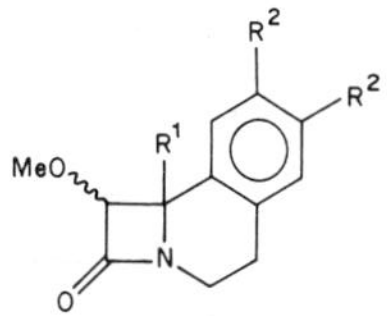		1154
	$R^1 = C_6H_4CN$-*p*, $R^2 = H$			60	
	$R^1 = C_6H_4NO_2$-*p*, $R^2 = H$			70	
	$R^1 = C_6H_4OH$-*p*, $R^2 = H$			50	
	$R^1 = C_6H_4Br$-*p*, $R^2 = OMe$			50	
$MeOCH_2COCl$	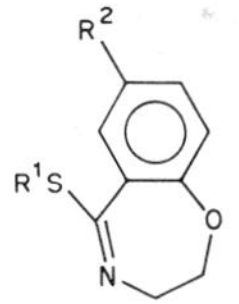	Et_3N, CH_2Cl_2, r.t. stir overnight	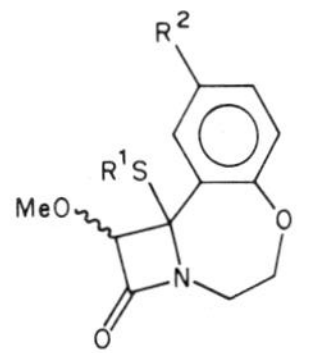		1155

(*continued*)

TABLE 53. (*continued*)

Acid halide	Imine	Conditions	Product	Yield (%)	Reference
	$R^1 = Me, R^2 = H$			53	
	$R^1 = i\text{-}Pr, R^2 = H$			89	
	$R^1 = Me, R^2 = Cl$			44	
	$R^1 = Et, R^2 = Cl$			43	
	$R^1 = i\text{-}Pr, R^2 = Cl$			51	
$MeOCH_2COCl$	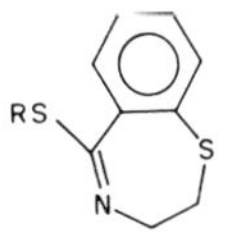 RS, S, N (structure)	Et_3N, CH_2Cl_2 r.t., stir overnight	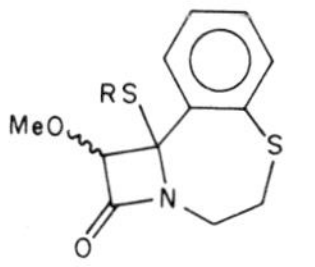 MeO, RS, S, N, O (structure)		1155
	$R = Me$			65	
	$R = i\text{-}Pr$			39	
$PhCH_2OCH_2COCl$	$R^1N{=}CR^2R^3$	Et_3N, CH_2Cl_2, r.t., stir 12 h	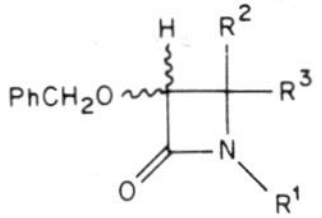 $PhCH_2O$, H, R^2, R^3, N, R^1, O (structure)		1156
	$R^1 = Tol\text{-}p$, $R^2 = An\text{-}p$, $R^3 = H$		*cis*	56	
	$R^1 = Tol\text{-}p$, $R^2 = 2\text{-}Fu$, $R^3 = H$		*cis*	70	
	$R^1 = Tol\text{-}p$, $R^2 = C_6H_4NO_2\text{-}p$, $R^3 = H$		*cis*	65	
	$R^1 = Tol\text{-}p$, $R^2 = C_6H_4CO_2H\text{-}p$, $R^3 = H$		*cis*	60	

	$R^1 = R^2 = Ph$, $R^3 = MeS$			70	
	$R^1 = R^2 = Ph$, $R^3 = H$		*cis* + *trans*	60	
	$R^1 = \alpha$-naphthyl, $R^2 = Ph$, $R^3 = H$		*trans*	70	
$PhCH_2OCH_2COCl$	*p*-TolN=CHAn-*p*	Et_3N, CH_2Cl_2, r.t., stir 12 h	[structure: PhCH₂O/An-*p* β-lactam, N-Tol-*p*]	56	902
$PhCH_2OCH_2COCl$	$R^1CH{=}NR^2$	Et_3N, CH_2Cl_2	[structure: PhCH₂O/R¹ β-lactam, N-R²]		1157
	$R^1 = 3,4\text{-}(MeO)_2C_6H_3$, $R^2 = Ph$, *i*-Pr, $(CH_2)_2COOEt$			—	
	$R^1 = C_6H_4F$-*p*, $C_6H_4OCOCH_2Ph$-*p*, $R^2 = Ph$			—	
	R^1 = [1-(CO₂CH₂Ph)indol-3-yl], R^2 = Ph			—	
$PhCH_2OCH_2COCl$	[2-phenyl-2-thiazoline]	Et_3N, CH_2Cl_2, r.t., stir 12 h	[structure: bicyclic β-lactam, PhCH₂O, Ph]	60	1156
$PhCH_2OCH_2COCl$	[5,5-dimethyl-4-CO₂Me-thiazoline]	Et_3N, CH_2Cl_2, r.t., stir 12 h	[structure: penam, PhCH₂O, Me, Me, CO₂Me]	30	1156

(*continued*)

TABLE 53. *(continued)*

Acid halide	Imine	Conditions	Product	Yield (%)	Reference
$PhCH_2OCH_2COCl$	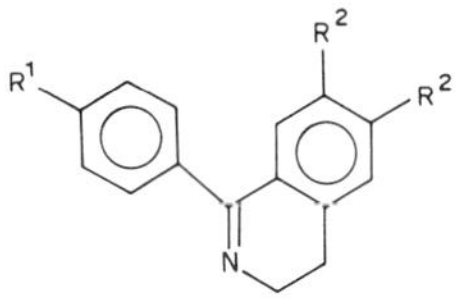 [structure: 1-(4-R¹-phenyl)-6,7-di-R²-3,4-dihydroisoquinoline]	Et_3N, CH_2Cl_2, r.t., stir 12 h	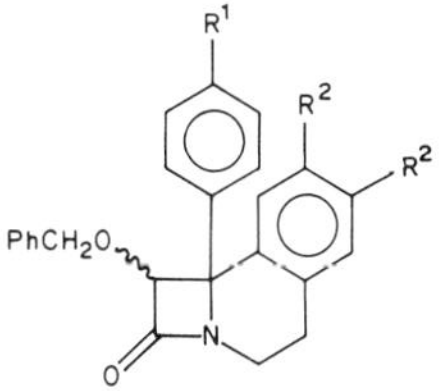 [structure: PhCH₂O-substituted azetidinone fused tetrahydroisoquinoline, R¹, R², R²]		902, 1156
	$R^1 = R^2 = H$			75	902
	$R^1 = Br$, $R^2 = OMe$			70	1156
	$R^1 = NH_2$, $R^2 = H$			68	1156
$PhCH_2OCH_2COCl$	$R^1CH{=}NCHR^2CO_2Me$	Et_3N, CH_2Cl_2	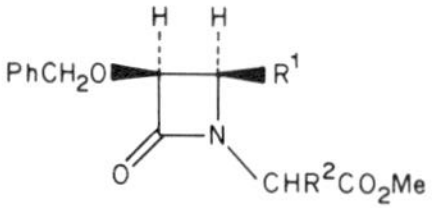 [structure: cis-3-PhCH₂O-4-R¹-1-(CHR^2CO_2Me)azetidin-2-one]		1158
	$R^1 = R^2 = Ph$			92	
	$R^1 = Ph$, $R^2 = i$-Pr			81	
	$R^1 = Ph$, $R^2 = Me$			93	
	$R^1 = Ph$, $R^2 = PhCH_2$			72	
	$R^1 = C_6H_4OCH_2Ph$-p, $R^2 = Ph$			80	
	$R^1 = 3,4$-$(PhCH_2O)_2C_6H_3$, $R^2 = Ph$			91	
$PhOCH_2COCl$	$R^1R^2C = NR^3$	Et_3N, CH_2Cl_2, r.t., stir overnight	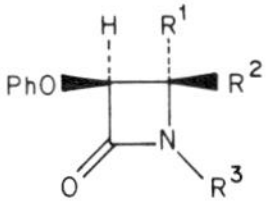 [structure: 3-PhO-4-R¹-4-R²-1-R³-azetidin-2-one]		902, 1143, 1151
	$R^1 = H$, $R^2 = C_6H_4NO_2$-p, $R^3 = PhCH_2$			58	1143

	$R^1 = H$, $R^2 = C_6H_4NO_2$-*p*, $R^3 = $ Tol-*p*			70	902
	$R^1 = H$, $R^2 = C_6H_4NO_2p$, $R^3 = Ph$			55	1143
	$R^1 = H$, $R^2 = C_6H_4NO_2$-*p*, $R^3 = \alpha$-naphthyl			50	1143
	$R^1 = R^2 = Ph$, $R^3 = PhCH_2$			70	1143
	$R^1 = H$, $R^2 = $ 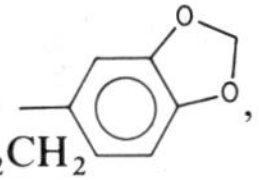 3,4-methylenedioxyphenyl, $R^3 = HOCH_2CH_2$	*d*		70	1143
	$R^1 = R^2 = Ph$, $R^3 = HOCH_2CH_2$	*d*		65	1143
	$R^1 = H$, $R^2 = C_6H_4CO_2H$-*p*, $R^3 = C_6H_4COMe$-*p*	*d*		—	1151
$PhOCH_2COCl$	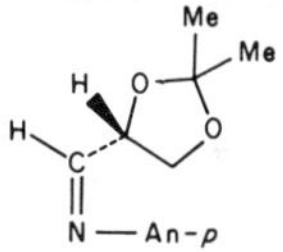 (Me, Me, H, H, O, O; $C{=}N$—An-*p*)	Et_3N, CH_2Cl_2	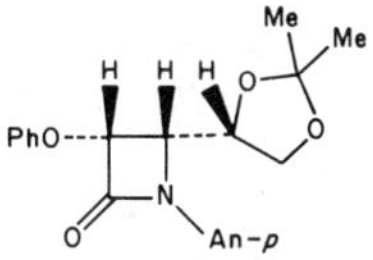 (PhO, H, H, H, O, N, An-*p*, Me, Me)	—	1152
$PhOCH_2COCl$	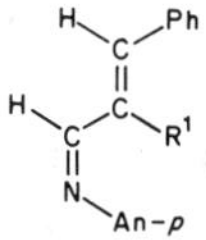 (H, Ph, H, R^1, $C{=}C$, $C{=}N$, An-*p*)	Et_3N[c]	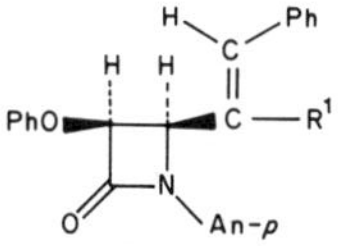 (PhO, H, H, H, Ph, $C{=}C$—R^1, O, N, An-*p*)		1144
	$R^1 = H$			45	
	$R^1 = Me$			70	
$PhOCH_2COCl$	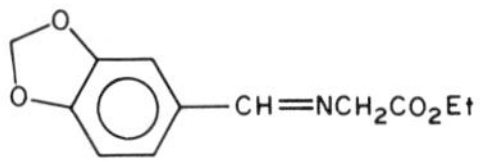 (3,4-methylenedioxyphenyl)—$CH{=}NCH_2CO_2Et$	Et_3N	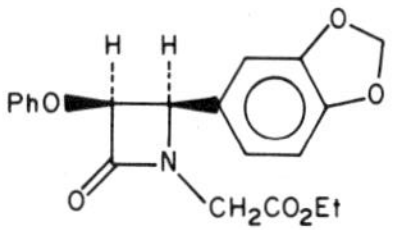 (PhO, H, H, O, N, CH_2CO_2Et)	—	1159

(*continued*)

TABLE 53. (*continued*)

Acid halide	Imine	Conditions	Product	Yield (%)	Reference
$PhOCH_2COCl$	$R^1OCH{=}NR^2$	Et_3N, CH_2Cl_2, N_2 r.t., stir 12 h	PhO, H, H, OR^1, O, N, R^2		
	$R^1 = Et$, $R^2 = Ph$			90	1137
	$R^1 = PhCH_2$, $R^2 = Ph$			98	1137
	R^1 = (2-methyl-1,3-dioxolan-2-yl)CH_2, $R^2 = Ph$			82	1137
	R^1 = (2-methyl-1,3-dioxolan-2-yl)CH_2, R^2 = An-*p*			92	1160
$PhOCH_2COCl$	$PhC(SMe){=}NCH_2CO_2Et$	Et_3N, C_6H_6	PhO, Ph, SMe, O, N, CH_2COOEt	60	1161
$PhOCH_2COCl$	$NCHR^2CO_2R^1$ ‖ HCNaph-2	N_2, $-5°C$, Et_3N^d	PhO, H, H, Naph-2, O, N, $CHR^2CO_2R^1$		1162
	$R^1 = Me$, $R^2 = H$, $PhCH_2$			—	
	$R^1 = R^2 = PhCH_2$			—	
$PhOCH_2COCl$	$NCHR^1COOMe$ ‖ HC (9-phenanthryl)	N_2, $-5°C$, Et_3N^d	PhO, H, H, (9-phenanthryl), O, N, CHR^1COOMe		1162

	$R^1 = H, PhCH_2$			—	
$PhOCH_2COCl$	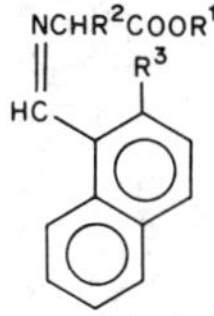	Et_3N, C_6H_6	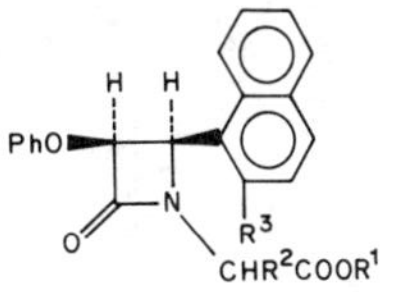		1162
	$R^1 = Me, R^2 = H$, $R^3 = H, MeO$			—	
	$R^1 = Me, R^2 = PhCH_2$, $R^3 = H, MeO$			—	
	$R^1 = R^2 = PhCH_2$, $R^3 = H$			—	
$PhOCH_2COCl$	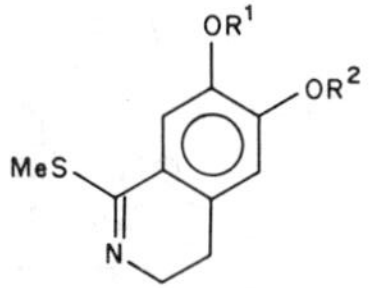	Et_3N, CH_2Cl_2, r.t., stir overnight	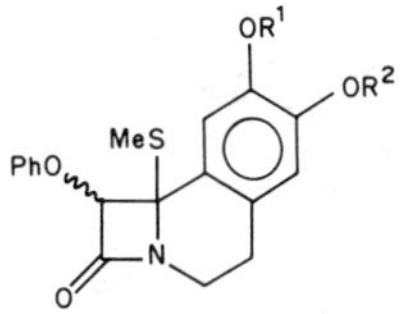		
	$R^1 = R^2 = MeO$			65	1099
	$R^1R^2 = CH_2$			60	1099
$PhOCH_2COCl$	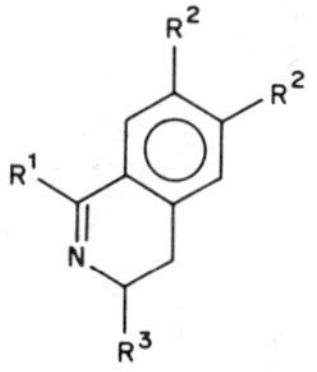	Et_3N, CH_2Cl_2, N_2, r.t., stir overnight	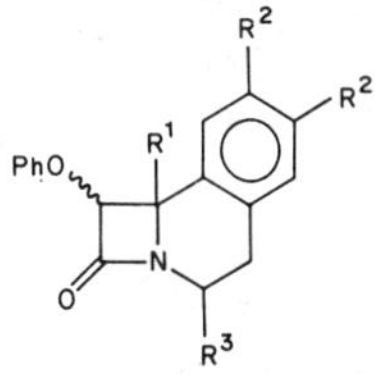		1154
	$R^1 = C_6H_4Br$-*p*, C_6H_4CN-*p*, $R^2 = MeO, R^3 = H$			75	
	$R^1 = Ph, R^2 = MeO$, $R^3 = CO_2Me$			60	

(continued)

TABLE 53. *(continued)*

Acid halide	Imine	Conditions	Product	Yield (%)	Reference
$PhOCH_2COCl$	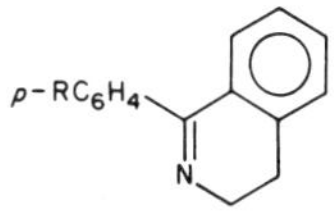 p-RC_6H_4 (structure)	1. Et_3N, CH_2Cl_2, $POCl_3$, 10–15°C 2. r.t., stir overnight	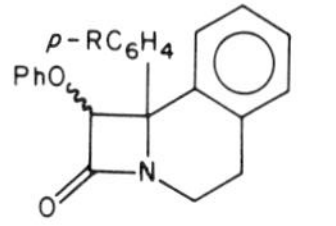 p-RC_6H_4, PhO (structure)		902, 1163
	R = Me			70	1163
	R = MeO			65	1163
	R = NO_2			65	1163
$PhOCH_2COCl$	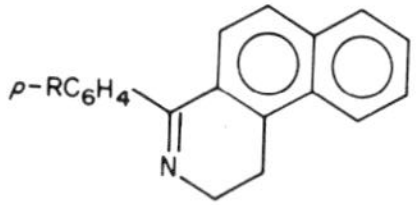 p-RC_6H_4 (structure)	1. Et_3N, CH_2Cl_2, 10–15°C, stir 2. r.t., stir overnight	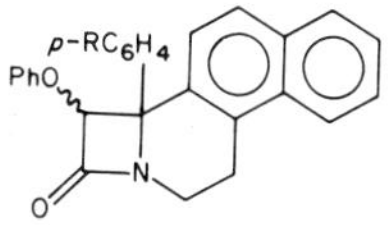 p-RC_6H_4, PhO (structure)		1163
	R = H			55	
	R = Me			60	
	R = MeO			55	
$PhOCH_2COCl$	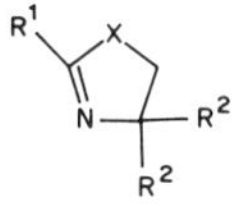 R^1, X, N, R^2, R^2 (structure)	Et_3N, CH_2Cl_2, stir overnight	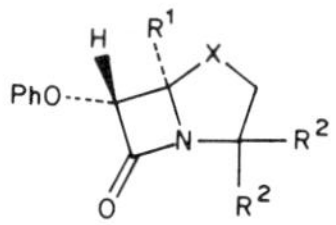 H, R^1, X, PhO, N, R^2, R^2, O (structure)		1164
	X = O, R^1 = Ph R^2 = H			32	
	X = O, R^1 = An-*p*, R^2 = H			31	
	X = O, R^1 = An-p, R^2 = Me			45	
	X = S, R^1 = Ph, R^2 = Me			44	
	X = S, R^1 = An-*p*, R^2 = H			42	
	X = S, R^1 = An-*p*, R^2 = Me			37	

$PhOCH_2COCl$	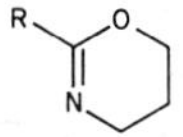	Et_3N, CH_2Cl_2, stir overnight	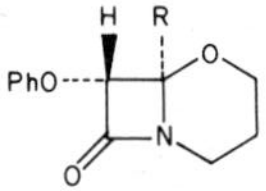		1164
	R = Ph			41	
	R = An-*p*			34	
$PhOCH_2COCl$	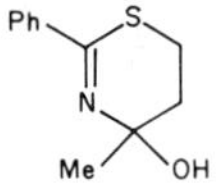	Et_3N, CH_2Cl_2, Me_3SiCl, stir overnight	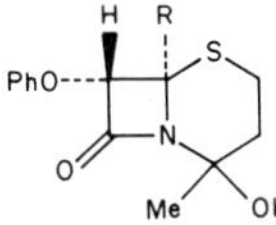	50	1164
$PhOCH_2COCl$	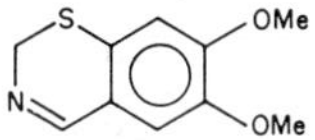	Et_3N, CH_2Cl_2, 5 °C, stand overnight	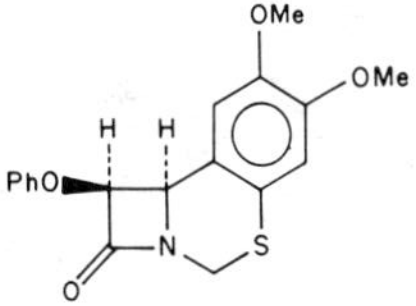	41	1150
$PhOCH_2COCl$	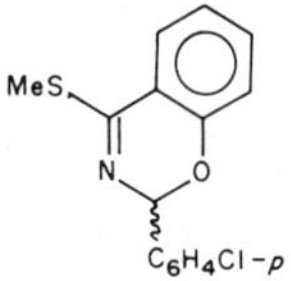	Et_3N, CH_2Cl_2, r.t., stir overnight	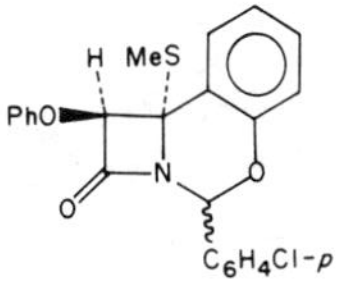	91	1165
$PhOCH_2COCl$	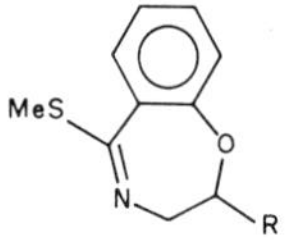	Et_3N, CH_2Cl_2, r.t. stir overnight	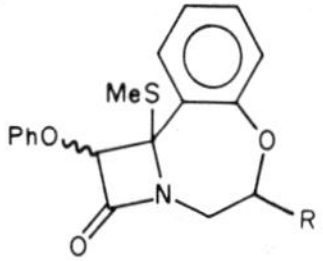		
	R = H			40	1155
	R = OMe			80	1155

(continued)

TABLE 53. (*continued*)

Acid halide	Imine	Conditions	Product	Yield (%)	Reference
$PhOCH_2COCl$	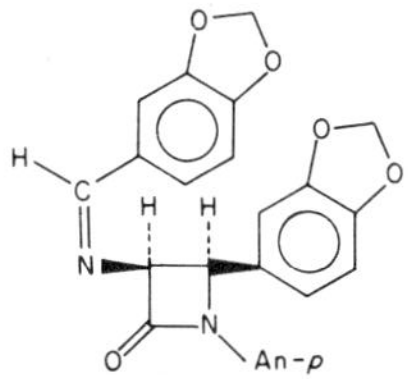	Et_3N, CH_2Cl_2	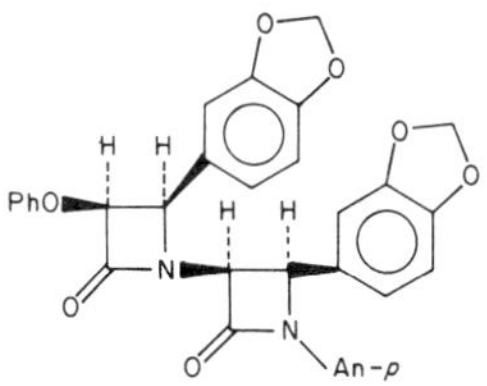	—	1159
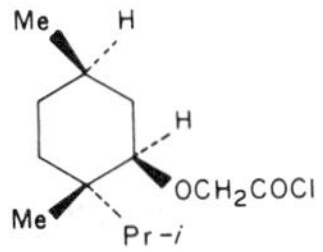	$PhCH{=}NR$	1. Et_3N, CH_2Cl_2, N_2, $-15\,^{\circ}C$ 2. r.t. stir overnight	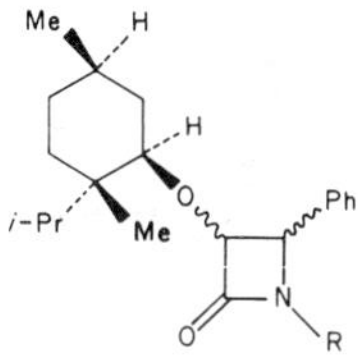		1166
	R = Ph			56	
	R = An-*p*			59	
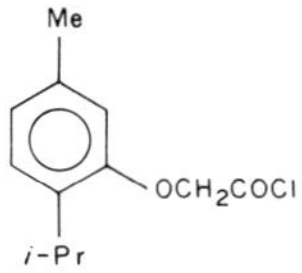	$PhCH{=}NR$	1. Et_3N, CH_2Cl_2, N_2, $-15\,^{\circ}C$ 2. r.t., stir overnight	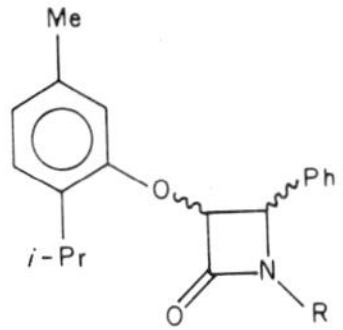		1166
	R = Ph			60	
	R = An-*p*			50	
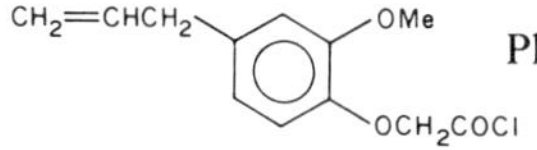	$PhCH{=}NR$	1. Et_3N, CH_2Cl_2, N_2, $-15\,^{\circ}C$ 2. r.t. stir overnight	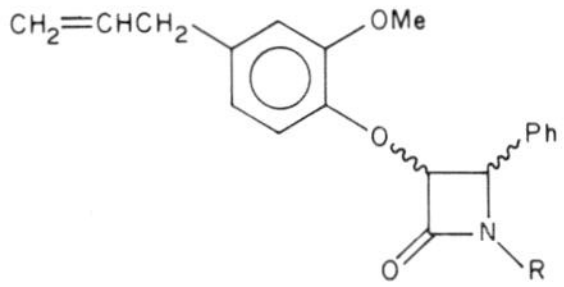		1166

	R = Ph			59	
	R = An-*p*			58	
N_3CH_2COCl	$R^1CH{=}NR^2$	1. Et_3N, CH_2Cl_2, N_2, −30 °C, stir 1 h 2. warm to 0 °C	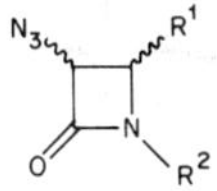 3-azido-1-R^2-4-R^1-azetidin-2-one		1167, 1168
	R^1 = Me, R^2 = PhMeCH		*trans*	~75	
	R^1 = Me, R^2 = Ph_2CH		*trans*	~75	
	R^1 = Me, R^2 = CH(An-*p*)$_2$		*cis*	20	
	R^1 = PhCH=CH, R^2 = $Me_3SiCH_2CH_2$, $CH_2{=}CHCH_2$, PhCH(Me), Ph_2CH, (*p*-An)$_2$CH		*cis*	—	
N_3CH_2COCl	$R^1CH{=}NR^2$	Et_3N, CH_2Cl_2	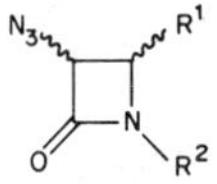 3-azido-1-R^2-4-R^1-azetidin-2-one		1157, 1167
	R^1 = C_6H_4F-*p*, $C_6H_4OCOCH_2Ph$-*p*, $C_6H_3(OMe)_2$-3,4, R^2 = Ph			—	
	R^1 = $C_6H_3(OMe)_2$-3,4, R^2 = $(CH_2)_2CO_2Et$, *i*-Pr, Ph			—	
	R^1 = 3,4-methylenedioxyphenyl,			—	
	R^2 = Ph				
	R^1 = 2-Fu, 2-Pyr, R^2 = Ph			—	

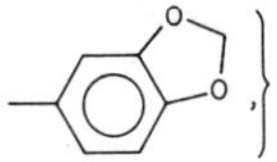

(continued)

TABLE 53. (*continued*)

Acid halide	Imine	Conditions	Product	Yield (%)	Reference
	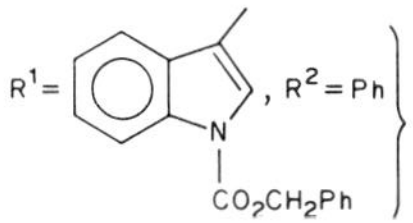 R^1 = [1-(benzyloxycarbonyl)indol-3-yl], R^2 = Ph			—	
	$R^1 = C_6H_3(OMe)$-3,4, R^2 = PhMeCH		(3*S*, 4*R*, 1′*R* + 3*R*, 4*S*, 1′*R*)	—	
N_3CH_2COCl	$PhCH{=}NCH_2CO_2Bu$-*t*	Et_3N, CH_2Cl_2	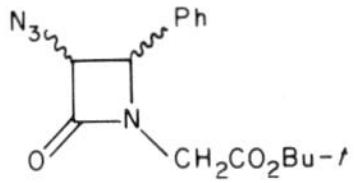 3-N_3-4-Ph-1-(CH_2CO_2Bu-*t*)-azetidin-2-one	56	1167, 1169
N_3CH_2COCl	*p*-$R^1C_6H_4CH{=}NCHR^2CO_2R^3$	Et_3N, CH_2Cl_2	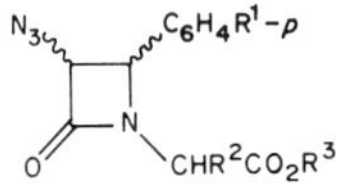 3-N_3-4-($C_6H_4R^1$-*p*)-1-($CHR^2CO_2R^3$)-azetidin-2-one		1169, 1170
	$R^1 = PhCH_2O$, R^2 = H, R^3 = Et			84	
	R^1 = H, $R^2 = CH_2Pr$-*i*, R^3 = *t*-Bu			—	
N_3CH_2COCl	*p*-$HO_2CC_6H_4CH{=}NAn$-*p*	Et_3N, CH_2Cl_2, Me_3SiCl^d	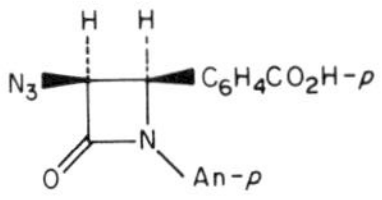 *cis*-3-N_3-4-($C_6H_4CO_2H$-*p*)-1-(An-*p*)-azetidin-2-one	—	1151, 1167
N_3CH_2COCl	$R^1CH{=}NCHR^2CO_2Me$	Et_3N, CH_2Cl_2	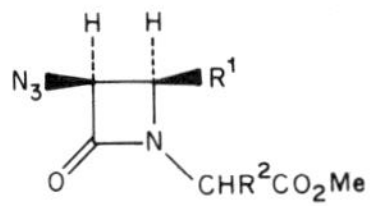 *cis*-3-N_3-4-R^1-1-(CHR^2CO_2Me)-azetidin-2-one		1158, 1167

	$R^1 = R^2 = Ph$			88	
	$R^1 = Ph$, $R^2 = i$-Pr			74	
	$R^1 = Ph$, $R^2 = PhCH_2$			88	
	$R^1 = C_6H_4OCH_2Ph$-*p*, $R^2 = Ph$			90	
	$R^1 = Ph$, $R^2 = Me$		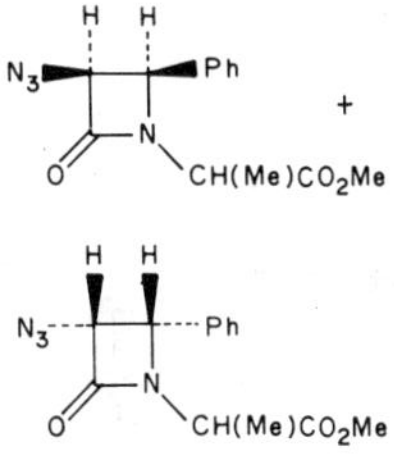 [structures] + [structures] (46:54)	75	
N_3CH_2COCl	PhCH=NBu-*t*	1. Et_3N, CH_2Cl_2, −78 °C 2. −78 °C to r.t., stand 14 h	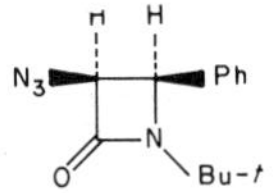 [structure]	—	1167, 1171
N_3CH_2COCl	PhCH=NR	1. Et_3N, CH_2Cl_2, −78 °C 2. −78 °C to r.t., stand 14 h	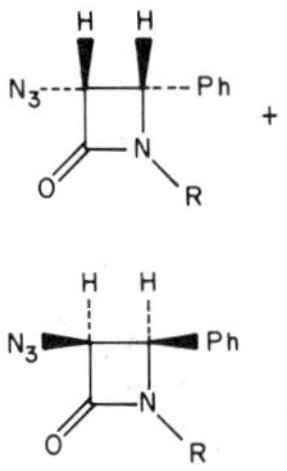 [structure] + [structure]		1167, 1171, 1172
	R = $MeCHCO_2Bu$-*t*		(41:39)	—	
	R = $CHCH_2Pr$-*i* (with CH_2OCH_2Ph)		(44:56)	87	

(continued)

TABLE 53. (*continued*)

Acid halide	Imine	Conditions	Product	Yield (%)	Reference
N_3CH_2COCl	*p*-$PhCH_2OC_6H_4CH$ $=NCH(Me)CO_2Bu$-*t*	Et_3N, CH_2Cl_2	N_3, H, H, $C_6H_4OCH_2Ph$-*p*, $CH(Me)CO_2Bu$-*t* + N_3, H, H, $C_6H_4OCH_2Ph$-*p*, $CH(Me)CO_2Bu$-*t* (55:45)	70	1167, 1172
N_3CH_2COCl	$MeSCH=NCMe_2CO_2CHPh_2$	1. Et_3N, C_6H_5Me, Ar, stir 12 h 2. Et_3N, C_6H_5Me, stir 16 h	H, SMe, N_3, H, $CMe_2CO_2CHPh_2$ +	83	1167, 1173
			N_3, SMe, H, H, $CMe_2CO_2CHPh_2$	6	
N_3CH_2COCl	$MeSCR^1=NR^2$	Et_3N, CH_2Cl_2, N_2, 30 °C, 2 h	SMe, N_3, R^1, R^2		1167, 1174
	$R^1 = H$, $R^2 = CH_2CO_2Me$		*trans*	70	
	$R^1 = MeS$, $R^2 = n$-Bu			92	
	$R^1 = MeS$, $R^2 = CH_2CO_2Me$			55	
	$R^1 = MeS$, $R^2 = CH_2CO_2Bu$-*t*			36	

	$R^1 = MeS$, $R^2 = CHPr\text{-}i$ (CO$_2$Me)			30	
	$R^1 = MeS$, $R^2 = C{=}CMe_2$ (CO$_2$CHPh$_2$)			61	
N_3CH_2COCl	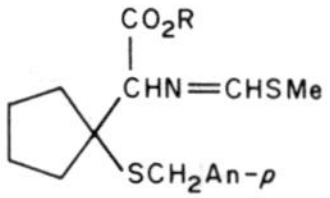 CO$_2$R, CHN=CHSMe, SCH$_2$An-*p*	Et_3N, C_6H_5Me, Ar, stir	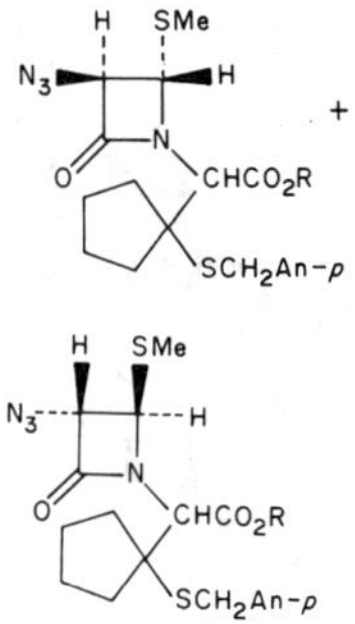 H, SMe, N_3, H, O, N, CHCO$_2$R, SCH$_2$An-*p* + H, SMe, N_3, H, O, N, CHCO$_2$R, SCH$_2$An-*p*		1167, 1175
	$R = CHPh_2$			30 + 25	
	$R = CH_2Ph$			28 + 11	
N_3CH_2COCl	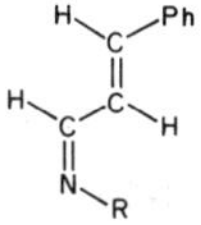 H, Ph, C, H, C, H, C, N, R	Et_3N, CH_2Cl_2	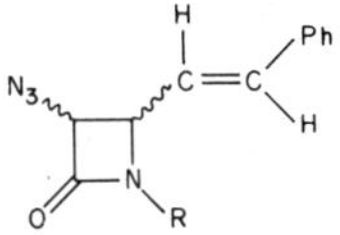 N_3, H, C=C, Ph, H, O, N, R		
	$R = C_6H_4COOH\text{-}o$	1. Me_3SiCl, $-5\,^\circ C$ stir 10 min. 2. add acid halide 3. r.t., stir 30 min.	*trans*		802, 1167

(continued)

TABLE 53. (*continued*)

Acid halide	Imine	Conditions	Product	Yield (%)	Reference
	$R = C_6H_4CH_2OSiMe_2(Bu\text{-}t)\text{-}o$	1. N_2, $-20\,^{\circ}C$ 2. r.t., stir 1 h	*cis*	74	802, 1167
	R = [structure: $COOSiMe_3$, OMe]	$-20\,^{\circ}C$, stir 30 min.	(*cis* + *trans*)[e]	—	1167, 1176
	R = [structure: $CH_2OSiMe_2(Bu\text{-}t)$, $OCOOCH_2Ph$]	$-20\,^{\circ}C$, N_2, stir 1 h	*cis*	84	1167, 1176
N_3CH_2COCl	$R^1R^2C{=}C(COOR^3)N{=}CH(SR^4)$	Et_3N, CH_2Cl_2, N_2, r.t.	[structure: azetidinone with N_3, H, SR^4, H; N-substituent $C(COOR^3){=}CR^1R^2$]		1167, 1177
	$(E)R^1 = R^2 = R^4 = Me$, $R^3 = Et$		*E*	93	
	$(E)R^1 = R^2 = R^4 = Me$, $R^3 = t\text{-Bu}$		*E*	87	
	$(E)R^1R^2 = -(CH_2)_4-$, $R^3 = R^4 = Me$		*E*	86	
	$(E/Z)R^1 = R^4 = Me$, $R^2 = Ph$, $R^3 = Et$		$E{:}Z = 4{:}6$	94	
	$(E/Z)R^1 = H$, $R^2 = R^3 = t\text{-Bu}$, $R^4 = Me$		$E{:}Z = >20$	85	

	mixture of $R^1 = R^2 = Me$, $R^3 = Et$, $R^4 = Ph_3C$ and $C_6H_4CHPh_2$-*p*				87 60% $R^4 = Ph_3C$ + 23% $R^4 = C_6H_4CHPh_2$-*p*	
	mixture of $R^1R^2 = -(CH_2)_4$-, $R^3 = Me$, $R^4 = Ph_3C$ and $C_6H_4CHPh_2$-*p*				88 58% $R^4 = Ph_3C$ + 22% $R^4 = C_6H_4CHPh_2$-*p*	
	(E/Z) mixture of $R^1 = Me$, $R^2 = Ph$, $R^3 = Et$, $R^4 = Ph_3C$, and $C_6H_4CHPh_2$-*p*			*E*:*Z* = 3:7	82 57% $R^4 = Ph_3C$ + 18% $R^4 = C_6H_4CHPh_2$-*p*	
	(E/Z) mixture of $R^1 = Me$, $R^2 = CH_2SPh$, $R^3 = Et$, $R^4 = Ph_3C$, and $C_6H_4CHPh_2$-*p*			*E*:*Z* = 3:1	81 57% $R^4 = Ph_3C$ + 20% $R^4 = C_6H_4CHPh_2$-*p*	
	(E/Z) mixture of $R^1 = H$, $R^2 = R^3 = t$-Bu, $R^4 = Ph_3C$, and $C_6H_4CHPh_2$-*p*			*E*:*Z* = > 20	87 45% $R^4 = Ph_3C$ + 25% $R^4 = C_6H_4CHPh_2$-*p*	
N_3CH_2COCl		Et_3N	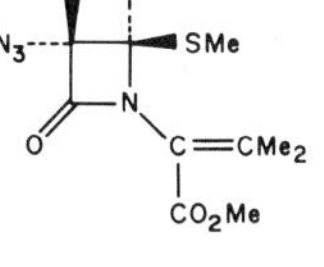		—	1167, 1178
N_3CH_2COCl[c]		1. Et_3N, CH_2Cl_2, N_2, $-5\,^{\circ}C$, stir 2. r.t., stir overnight	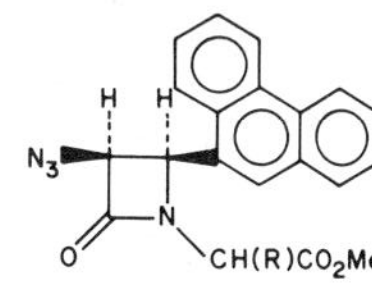			1162, 1167
	R = H, $PhCH_2$				—	

(continued)

TABLE 53. (*continued*)

Acid halide	Imine	Conditions	Product	Yield (%)	Reference
N_3CH_2COCl	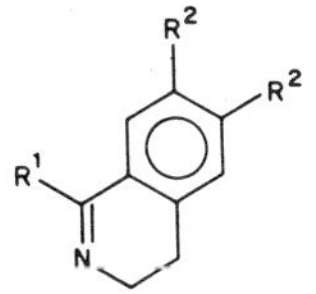	Et_3N, CH_2Cl_2, N_2, r.t., stir overnight	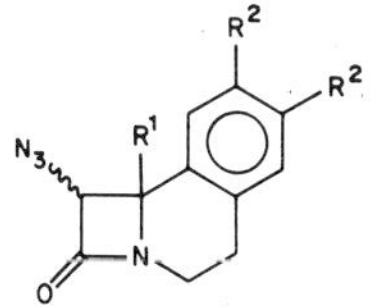		1154, 1167
	R^1 = Tol-*p*, R^2 = H			65	
	R^1 = An-*p*, R^2 = H			63	
	R^1 = R^2 = H			55	
	R^1 = C_6H_4Br-*p*, R^2 = OMe			60	
	R^1 = C_6H_4CN-*p*, R^2 = OMe			63	
	R^1 = An-*p*, R^2 = OMe			55	
	R^1 = Ph, R^2 = OMe			45	
N_3CH_2COCl	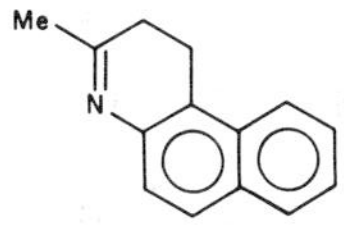	Et_3N	N.R.		1167, 1179
N_3CH_2COCl	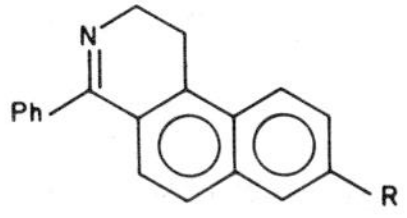	Et_3N	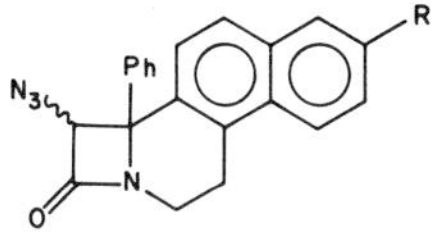		1167, 1179

Ketene precursor	Imine	Conditions	Product	Yield (%)	Ref.
	R = H, MeO			—	
N_3CH_2COCl	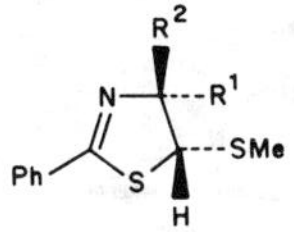	Et_3N, CH_2Cl_2	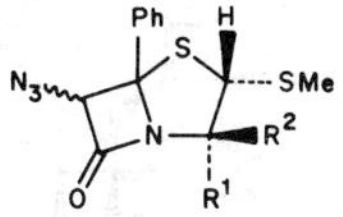		1145, 1167
	$R^1 = R^2 = Me$			50	
	$R^1 = i\text{-}Pr$, $R^2 = Me$		*cis*	35	
	$R^1 = Me$, $R^2 = i\text{-}Pr$		*trans*	84	
N_3CH_2COCl	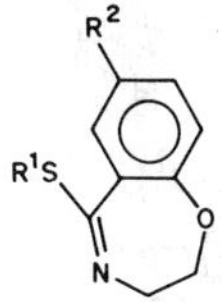	Et_3N, CH_2Cl_2, r.t.,	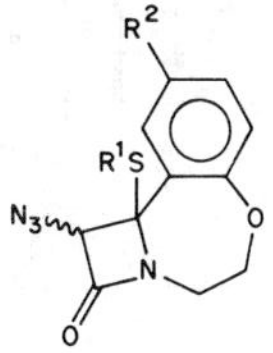		1155, 1167
	$R^1 = Me$, $R^2 = H$			65	
	$R^1 = Et$, $R^2 = H$			56	
	$R^1 = i\text{-}Pr$, $R^2 = H$			65	
	$R^1 = Me$, $R^2 = Cl$			63	
N_3CH_2COCl	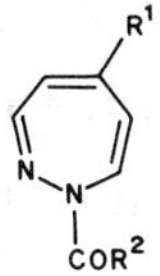	Et_3N, Et_2O or CH_2Cl_2	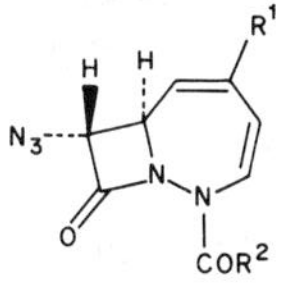		1127
	$R^1 = Me$, $R^2 = OEt$			86	
	$R^1 = H$, $R^2 = Ph$			55	
	$R^1 = Me$, $R^2 = Ph$			54	
N_3CH_2COCl	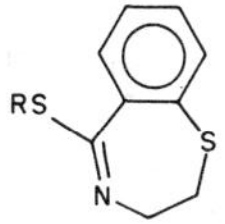	Et_3N, CH_2Cl_2, r.t. stir overnight	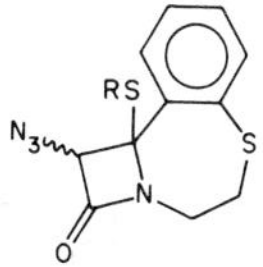		1155, 1167

(*continued*)

TABLE 53. (*continued*)

Acid halide	Imine	Conditions	Product	Yield (%)	Reference
	R = Me			50	
	R = *i*-Pr			30	
N_3CH_2COCl		Et_3N, CH_2Cl_2			
	R = CH_2CO_2Bu-*t*	—		58	1167, 1169
	R = $MeCHCO_2$Bu-*t*	1. − 78 °C, stir 2. r.t., 14 h		74	1167, 1171
N_3CH_2COCl		1. Et_3N, CH_2Cl_2, − 78 °C, stir 2. r.t., stir			1167 1171
	R = $MeCHCO_2$Bu-*t*			48	
	R = $CHCH_2$Pr-*i* (with CH_2OCH_2Ph)			46	
N_3CH_2COCl		1. Et_3N, CH_2Cl_2, − 78 °C, stir 2. r.t., stir		45	1167, 1171
N_3CH_2COCl		1. Et_3N, CH_2Cl_2, − 78 °C, stir 2. r.t., stir		60	1167, 1171

N_3CH_2COCl		1. Et_3N, CH_2Cl_2, $-78\,^{\circ}C$, stir 2. r.t., stir	+ (81:19)	65	1167 1171
N_3CH_2COCl		1. Et_3N, CH_2Cl_2, $-78\,^{\circ}C$, stir 2. r.t., stir	+ (32:68)	71	1167, 1171

(continued)

TABLE 53. (*continued*)

Acid halide	Imine	Conditions	Product	Yield (%)	Reference
$N_3CH(Me)COCl$	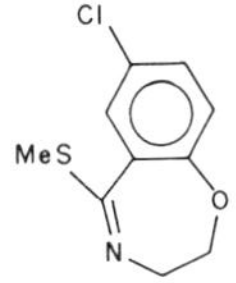	Et_3N, CH_2Cl_2, r.t. stir overnight	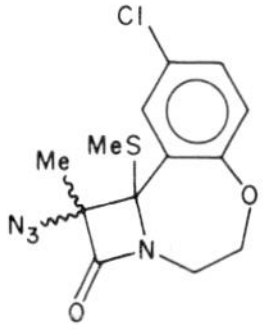	30	1155
MeO_2CCH_2COCl	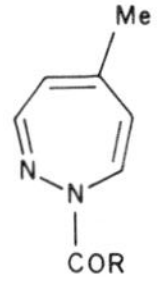	Et_3N, Et_2O or CH_2Cl_2	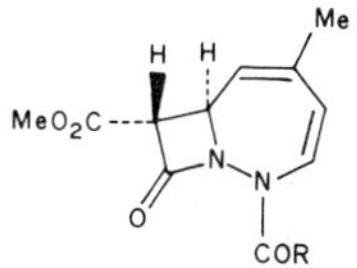		1127
	R = OEt			95	
	R = Ph			79	
RCH_2COCl	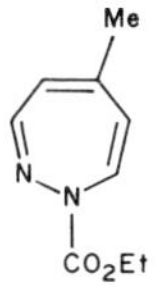	Et_3N, Et_2O or CH_2Cl_2	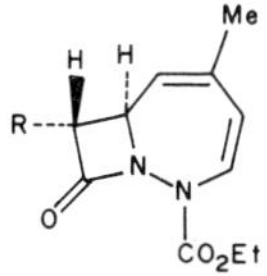		1127
R = p-$ClC_6H_4SO_3$				68	
R = $PhCH_2O_2CNH$				40	
$PhCH_2O_2$–$CNHCH_2COCl$	$R^1CH{=}NTol$-p	Et_3N, Et_2O, 0 °C, stir 3 h	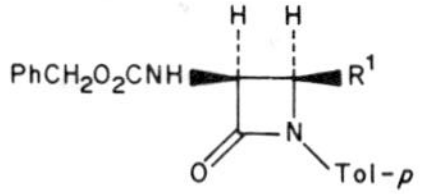		1180
	R^1 = p-An			15	
	R^1 = 2-Fur			22	
$PhCH_2O_2$–$CNHCH_2COCl$	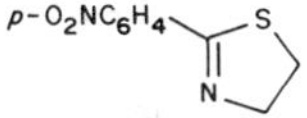	Et_3N, Et_2O, 0 °C, stir	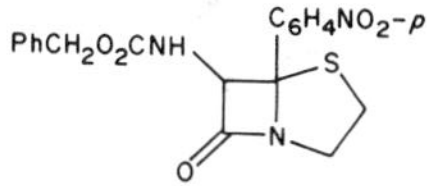	11	1180

$PhCH_2O_2$–$CNHCH_2COCl$	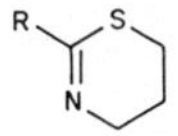 [structure: 2-R-5,6-dihydro-4H-1,3-thiazine]	Et_3N, Et_2O, r.t., stir 16 h	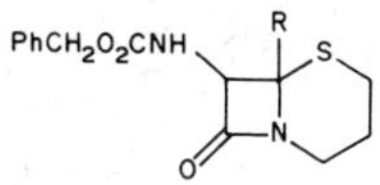 [structure: $PhCH_2O_2CNH$ bicyclic β-lactam, R]		1180
	R = Ph			60	
	$R = C_6H_4NO_2$-*p*			68	
$PhCH_2O_2$–$CNHCH_2COCl$	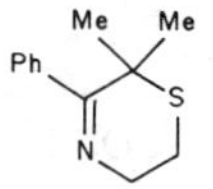 [structure: Ph, Me, Me thiazine]	Et_3N, Et_2O, r.t., stir	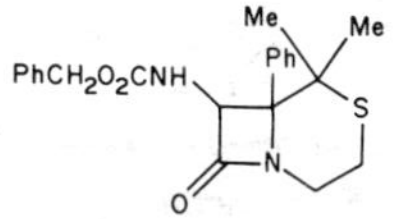 [structure: $PhCH_2O_2CNH$, Ph, Me, Me bicyclic β-lactam]		1180
p-$O_2NC_6H_4CH_2O_2C$–$NHCH_2COCl$	*p*-AnCH=NTol-*p*	Et_3N, Et_2O, 0 °C, stir	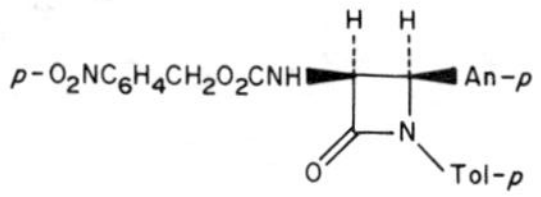 [structure: *p*-$O_2NC_6H_4CH_2O_2CNH$, An-*p*, Tol-*p* β-lactam]	87	1180
p-$O_2NC_6H_4CH_2O_2C$–$NHCH_2COCl$	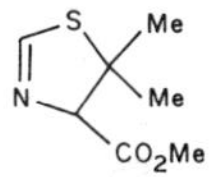 [structure: thiazoline, Me, Me, CO_2Me]	Et_3N, CH_2Cl_2, reflux 20h	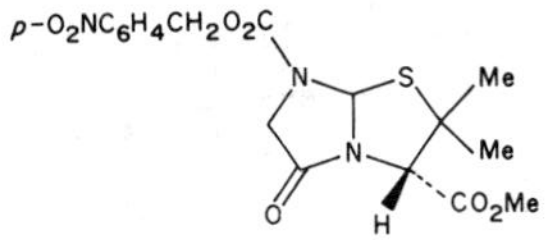 [structure: *p*-$O_2NC_6H_4CH_2O_2C$, Me, Me, CO_2Me]	80	1180
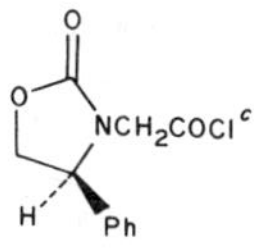 [structure: oxazolidinone NCH_2COCl[c], Ph]	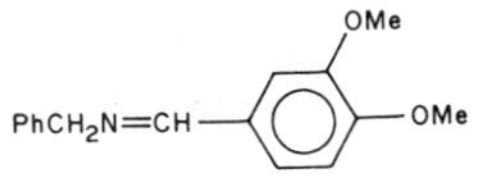 [structure: $PhCH_2N$=CH–dimethoxyphenyl, OMe, OMe]	1. Et_3N, CH_2Cl_2, −78 °C, stir 30 min. 2. r.t., stir overnight	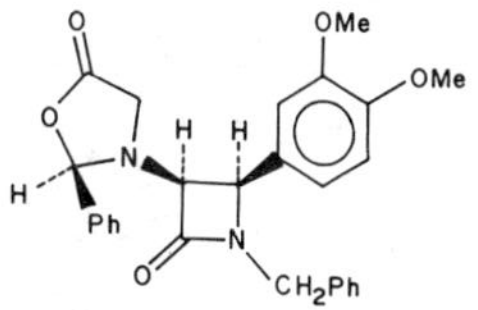 [structure: β-lactam, OMe, OMe, Ph, CH_2Ph]	90	1181
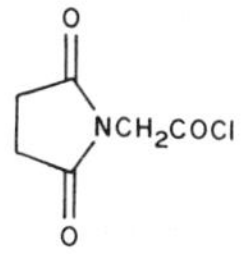 [structure: succinimide NCH_2COCl]	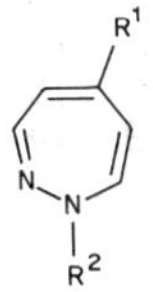 [structure: diazepine R^1, R^2]	Et_3N, CH_2Cl_2, stir 1 h	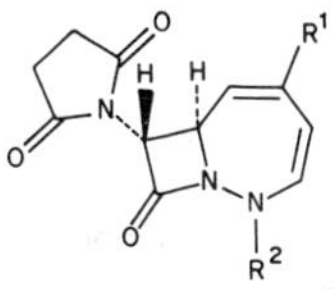 [structure: fused β-lactam R^1, R^2]		1182

(*continued*)

TABLE 53. (*continued*)

Acid halide	Imine	Conditions	Product	Yield (%)	Reference
	$R^1 = H$, $R^2 = CO_2Et$			62	
	$R^1 = Me$, $R^2 = CO_2Et$			78	
	$R^1 = H$, $R^2 = COPh$			61	
	$R^1 = Me$, $R^2 = COPh$			60	
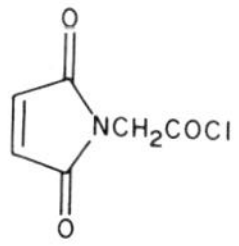 NCH_2COCl	PhCOC(Ph)=NPh	Et_3N, C_6H_5Me, reflux 24 h	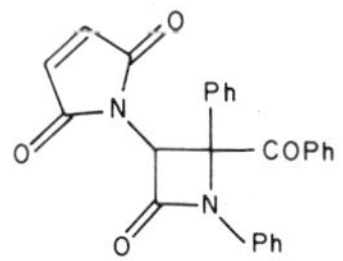 O, N, O, Ph, COPh, N, O, Ph	25[a]	1128
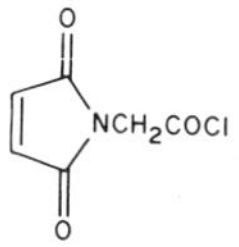 NCH_2COCl	PhCOCH=NR	Et_3N, C_6H_6, stir 2-5 h, 0 °C	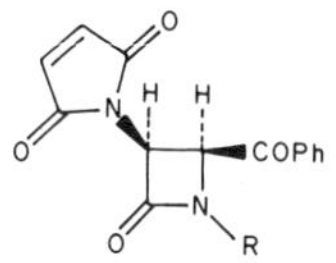 O, N, O, H, H, COPh, N, O, R		1148
	R = Ph			40	
	R = Tol-*p*			40	
	R = An-*p*			60	
$PI{-}NCH_2COCl$[f]	PhCOCH=NR	Et_3N, C_6H_6, stir 2-5 h, 0°C	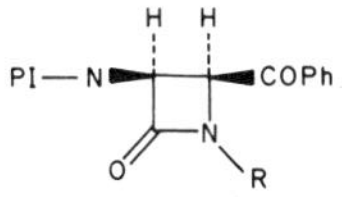 H, H, PI—N, COPh, N, O, R		1148
	R = Ph			30	
	R = Tol-*p*			55	
	R = An-*p*			75	
$PI{-}NCH_2COCl$	$R^1CH{=}NR^2$	1. Et_3N, CH_2Cl_2, Me_3SiCl, stir 2. acid halide, r.t., stir	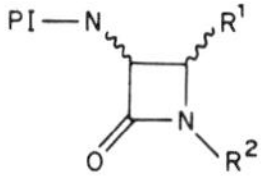 PI—N, R^1, N, O, R^2		

	$R^1 = Ph$, $R^2 = C_6H_4COOH$-*o*		*trans*	55	802
	$R^1 = C_6H_4COOH$-*p*, $R^2 = C_6H_4COMe$-*p*		*cis*	—	1151
PI—NCH_2COCl	R′CH=NR^2	Et_3N, CH_2Cl_2	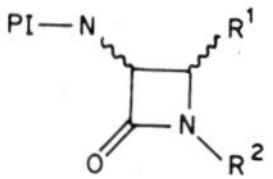 PI—N-substituted β-lactam (R^1, R^2)		
	$R^1 = C_6H_4OCH_2Ph$-*o*, $R^2 = PhCH_2$	−10 °C, stir overnight	*cis*	73	1165
	$R^1 = C_6H_4OCH_2Ph$-*o*, $R^2 = CH_2C_6H_3(OMe)_2$-2,5		*cis*	77	1183
	$R^1 = CHF_2$, $R^2 = MeCHPh$		*cis*	62	1184
	R^1 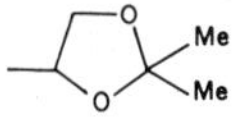 = 2,2-dimethyl-1,3-dioxolan-4-yl, $R^2 = CH_2C_6H_3(OMe)_2$-2,5		*cis*	—	1185
PI—NCH_2COCl[f]	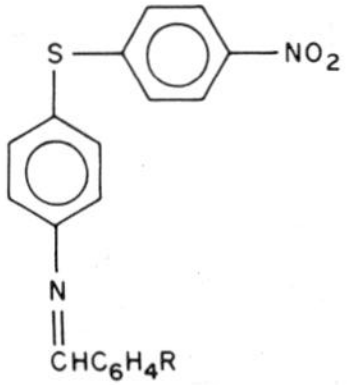 $O_2NC_6H_4SC_6H_4N{=}CHC_6H_4R$	1. Et_3N, dioxane[e], r.t., stir 5 h 2. stand 5 days	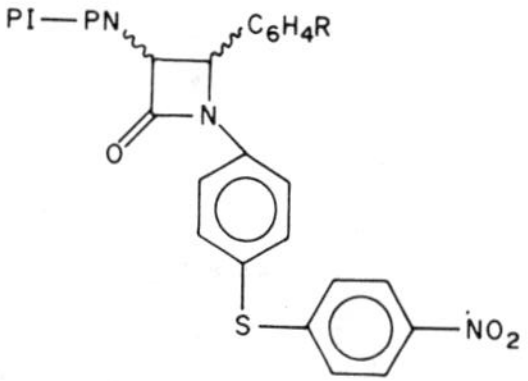 PI—PN-β-lactam (C_6H_4R, N-$C_6H_4SC_6H_4NO_2$)		1186
	R = H			45	
	R = *p*-NO_2			47	
	R = *p*-MeO			42	
	R = *p*-Cl			52	
	R = *p*-OH			55	
	R = *p*-Me_2N			57	

(*continued*)

TABLE 53. (*continued*)

Acid halide	Imine	Conditions	Product	Yield (%)	Reference
PI—NCH_2COCl[f]	$R^1OCH{=}NR^2$	Et_3N, CH_2Cl_2, N_2, r.t. stir	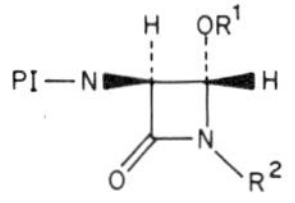		
	$R^1 = Me$, $R^2 = C_6H_4COOMe$-*o*	24 h		33	802
	$R^1 =$ 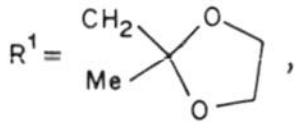 , $R^2 =$ An-*p*	3 h		30	1160
PI—NCH_2COBr[f]	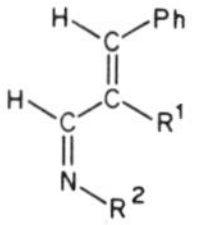	Et_3N	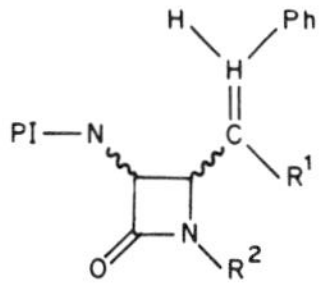		1144
	$R^1 = H$, $R^2 =$ An-*p*	*c*	*cis:trans* = 95:5	45	
	$R^1 = H$, $R^2 =$ An-*p*	*b*	*cis:trans* = 90:10	—	
	$R^1 = Me$, $R^2 =$ An-*p*	*b*	*cis:trans* = 100:0	64	
	$R^1 = H$, $R^2 = C_6H_4OSiMe_3$-*p*	*c*	*cis:trans* = 50:50	—	
	$R^1 = Me$, $R^2 = C_6H_4OSiMe_3$-*p*	*b*	*cis:trans* = 100:0	50[e]	

PI—NCH_2COCl^f	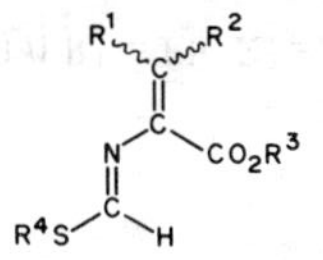 [structure: $R^1(R^2)C{=}C(CO_2R^3)N{=}C(H)SR^4$]	Et_3N, CH_2Cl_2, N_2, r.t.	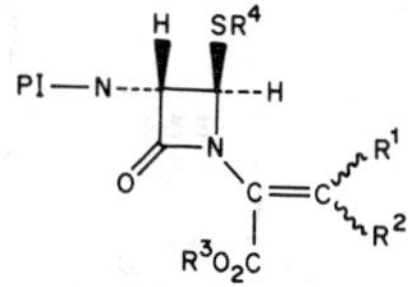 [structure: β-lactam PI—N, SR[4], N–C(CO_2R^3)=CR^1R^2]		1177
	(E/Z) mixture of $R^1 = R^4 = Me$, $R^2 = Ph$, $R^3 = Et$		$E/Z = 4{:}5$	48	
	$R^1 = R^2 = Me$, $R^3 = Et$, $R^4 = PhCH_2$			25	
	(E/Z) mixture of $R^1 = Me$, $R^2 = Ph$, $R^3 = Et$, $R^4 = Ph_3C$ and $C_6H_4CHPh_2$-*p*		$E/Z = 4{:}5$	64 (50% $R^4 = Ph_3C$)	
PI—NCH_2COCl^f	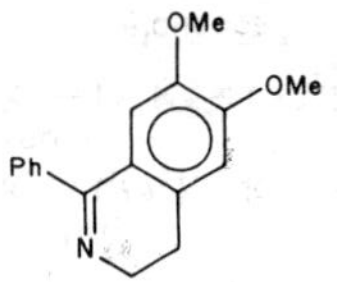 [structure: 1-Ph-6,7-(OMe)$_2$-3,4-dihydroisoquinoline]	Et_3N, C_6H_6, 0–5 °C	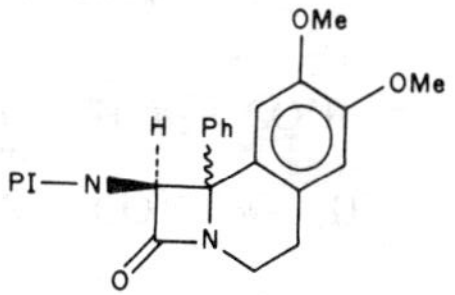 [structure: fused β-lactam, PI—N, H, Ph, OMe, OMe]	76	1187
PI—NCH_2COCl^f	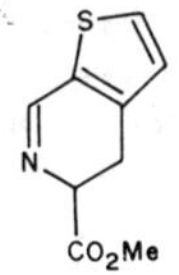 [structure: thieno-dihydropyridine with CO_2Me]	Et_3N, C_6H_6, 0–5 °C	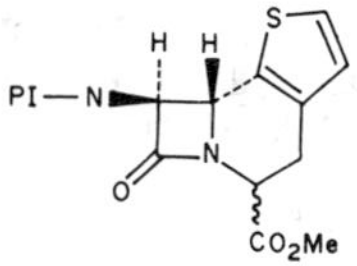 [structure: fused β-lactam, PI—N, H, H, CO_2Me]	24	1187
PI—NCH_2COCl^f	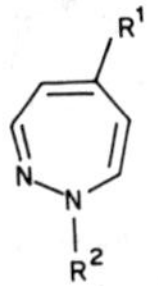 [structure: 1,2-diazepine with R^1, N–R^2]	Et_3N, CH_2Cl_2	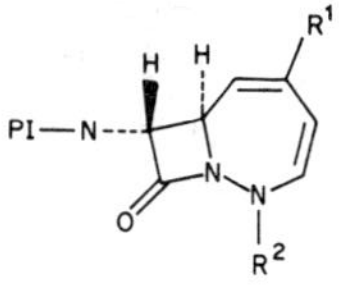 [structure: fused β-lactam diazepine, PI—N, H, H, R^1, N–R^2]		1182
	$R^1 = H$, $R^2 = CO_2Et$			61	
	$R^1 = Me$, $R^2 = CO_2Et$			91	

(continued)

TABLE 53. (*continued*)

Acid halide	Imine	Conditions	Product	Yield (%)	Reference
	$R^1 = H$, $R^2 = COPh$			62	
	$R^1 = Me$, $R^2 = COPh$			96	
PI—NCH_2COCl[f]	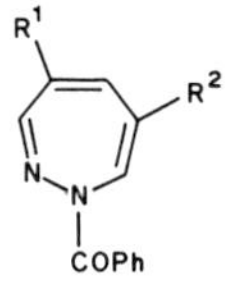 [structure: 1,2-diazepine with R^1, R^2, N–COPh]	Et_3N, CH_2Cl_2	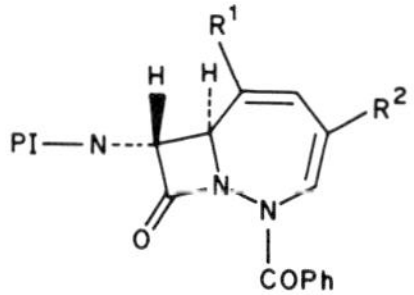 [structure: PI—N bicyclic β-lactam with R^1, R^2, N–COPh]		1182
	mixture of $R^1 = CO_2CH_2CH_2SiMe_3$, $R^2 = H$, and $R^1 = H$, $R^2 = CO_2CH_2CH_2SiMe_3$		$R^1 = CO_2CH_2CH_2SiMe_3$, $R^2 = H$, and $R^1 = CO_2H$, $R^2 = H$	37 + 12	
	mixture of $R^1 = CO_2Et$, $R^2 = H$, and $R^1 = H$, $R^2 = CO_2Et$		$R^1 = CO_2Et$, $R^2 = H$, and $R^1 = H$, $R^2 = CO_2Et$	20 + 6	
PI—NCH_2COCl—C_5H_5N[f] complex	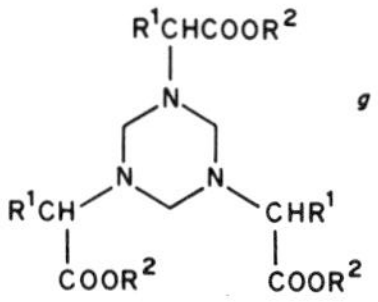 [structure: hexahydrotriazine with $R^1CHCOOR^2$ substituents][g]	1. $BF_3 \cdot OEt_2$, CH_2Cl_2 2. acid halide/pyridine complex	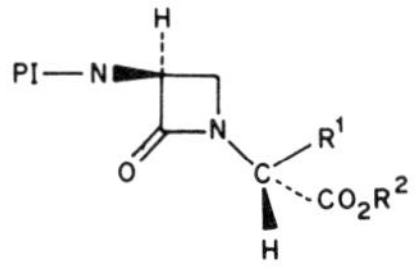 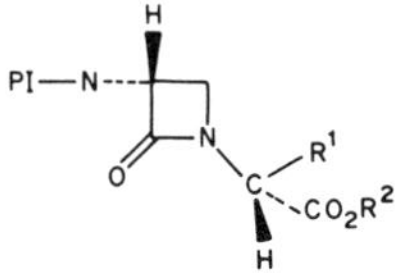 [structures: PI—N β-lactams (**121**) + (**122**) with $C(R^1)(CO_2R^2)H$]		1188
	(*d*) $R^1 = Ph$, $R^2 = Me$	0 °C, 2 h	(7:2)	80 (35% **121**)	

	(*d*) $R^1 = C_6H_4OCH_2Ph$-*p*, $R^2 = PhCH_2$	−78 to 0 °C, 3.5 h	(3:1)	87 (40% **121**)	
	(*dl*) R^1 = α-naphthyl, R^2 = Me	−78 to 0 °C, 2.5 h	(10:1)	51 (43% **121**)	
	(*dl*) R^1 = 2-Thi, R^2 = Me	−78 to 0 °C, 2.5 h	(7:2)	65 (32% **121**)	
	(*dl*) R^1 = 2-Fu, R^2 = Me	−78 to 0 °C, 2.5 h	(3:1)	39 (17% **121**)	
	R^1 = H, $R^2 = PhCH_2$	−78 to 0 °C, 2.5 h		(35% **121**)	
PI—NCH_2COCl[f]	2-($N{=}CHR^2$)-4-Ph-5-R^1-thiazole	Et_3N, C_6H_6, stir r.t. 1 h	PI—N-azetidinone (R^2), N-(4-Ph-5-R^1-thiazol-2-yl)		1189
	R^1 = H, $R^2 = C_6H_4OH$-*o*			22	
	R^1 = H, R^2 = An-*p*			64	
	R^1 = H, $R^2 = C_6H_4NO_2$-*o*			26	
	R^1 = H, $R^2 = C_6H_4NO_2$-*m*			25	
	R^1 = H, $R^2 = C_6H_4NO_2$-*p*			67	
	R^1 = H, $R^2 = C_6H_4OMe$-*p*			21	
	R^1 = H, R^2 = PhCH=CH			65	
	R^1 = Me, R^2 = An-*p*			11	
	R^1 = Me, $R^2 = C_6H_4NO_2$-*p*			14	
PI—NCH_2COCl[f]	2-(N=CHR)-4-(*p*-Tol)-thiazole	Et_3N, C_6H_6, stir r.t. 1 h	PI—N-azetidinone (R), N-(4-Tol-*p*-thiazol-2-yl)		1189

(*continued*)

TABLE 53. (*continued*)

Acid halide	Imine	Conditions	Product	Yield (%)	Reference
	R = C_6H_4OH-*o*			15	
	R = An-*p*			38	
	R = $C_6H_4NO_2$-*o*			29	
	R = $C_6H_4NO_2$-*m*			63	
	R = $C_6H_4NO_2$-*p*			59	
	R = PhCH=CH			54	
PI—NCHR[1]COCl[f]	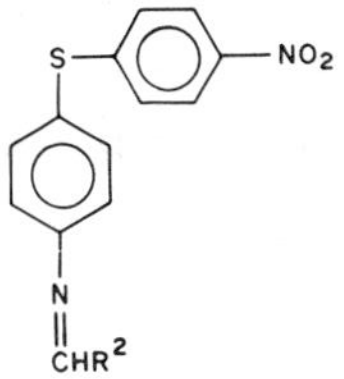 p-$O_2NC_6H_4S$-C_6H_4-$N{=}CHR^2$	1. Et_3N, dioxane, r.t. stir 5 h 2. stand 5 days	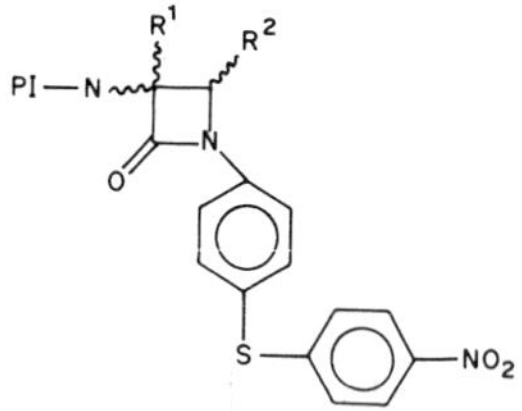 PI—N-β-lactam (R^1, R^2), N-C_6H_4-S-C_6H_4-NO_2		1186
R^1 = Me	R^2 = Ph			40	
R^1 = Me	R^2 = An-*p*			47	
R^1 = Me	R^2 = $C_6H_4NO_2$-*p*			43	
R^1 = *i*-Pr	R^2 = Ph			52	
R^1 = *i*-Pr	R^2 = An-*p*			57	
R^1 = *i*-Pr	R^2 = $C_6H_4NO_2$-*p*			60	
PI—NCHR[1]COCl[f]	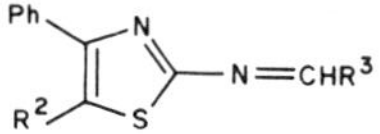 4-Ph-5-R^2-thiazol-2-yl-$N{=}CHR^3$	Et_3N, C_6H_6, stir r.t. 1 h	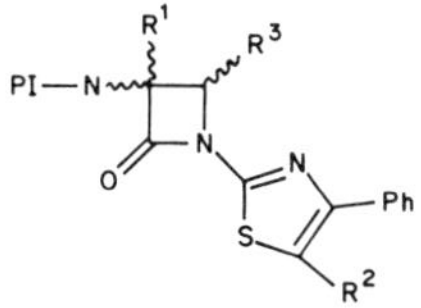 PI—N-β-lactam (R^1, R^3), N-(4-Ph-5-R^2-thiazol-2-yl)		1189
R^1 = Me	R^2 = H, R^3 = C_6H_4OH-*o*			45	
R^1 = Me	R^2 = H, R^3 = An-*p*			54	
R^1 = Me	R^2 = H, R^3 = $C_6H_4NO_2$-*o*			55	
R^1 = Me	R^2 = H, R^3 = $C_6H_4NO_2$-*m*			31	

$R^1 = Me$	$R^2 = H, R^3 = C_6H_4NO_2$-*p*			22	
$R^1 = Me$	$R^2 = H, R^3 = C_6H_4NMe_2$-*p*			40	
$R^1 = Me$	$R^2 = H, R^3 = PhCH{=}CH$			24	
$R^1 = Me$	$R^2 = Me, R^3 = $ An-*p*			25	
$R^1 = Me$	$R^2 = Me, R^3 = C_6H_4NO_2$-*p*			20	
$R^1 = i$-Pr	$R^2 = H, R^3 = C_6H_4OH$-*o*			33	
$R^1 = i$-Pr	$R^2 = H, R^3 = $ An-*p*			15	
$R^1 = i$-Pr	$R^2 = H, R^3 = C_6H_4NO_2$-*o*			45	
$R^1 = i$-Pr	$R^2 = H, R^3 = C_6H_4NO_2$-*m*			42	
$R^1 = i$-Pr	$R^2 = H, R^3 = C_6H_4NO_2$-*p*			14	
$R^1 = i$-Pr	$R^2 = H, R^3 = Me_2$NPyr-4			16	
$R^1 = i$-Pr	$R^2 = H, R^3 = PhCH{=}CH$			53	
PI—NCHR[1]COCl[f]	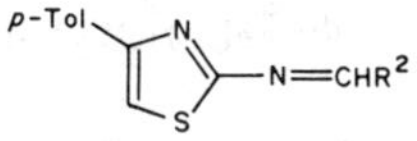 [structure: 2-(N=CHR[2])-4-(p-Tol)thiazole]	Et_3N, C_6H_6 stir r.t. 1 h	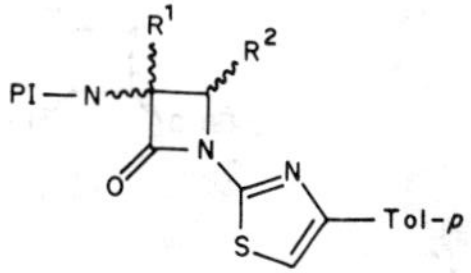 [structure: PI—N-azetidinone with R[1], R[2], N-(4-Tol-p-thiazol-2-yl)]		1189
$R^1 = Me$	$R^2 = C_6H_4OH$-*o*			60	
$R^1 = Me$	$R^2 = $ An-*p*			18	
$R^1 = Me$	$R^2 = C_6H_4NO_2$-*o*			54	
$R^1 = Me$	$R^2 = C_6H_4NO_2$-*m*			35	
$R^1 = Me$	$R^2 = C_6H_4NO_2$-*p*			57	
$R^1 = Me$	$R^2 = PhCH{=}CH$			24	
$R^1 = i$-Pr	$R^2 = C_6H_4OH$-*o*			14	
$R^1 = i$-Pr	$R^2 = $ An-*p*			28	
$R^1 = i$-Pr	$R^2 = C_6H_4NO_2$-*o*			31	
$R^1 = i$-Pr	$R^2 = C_6H_4NO_2$-*m*			20	
$R^1 = i$-Pr	$R^2 = C_6H_4NO_2$-*p*			15	
$R^1 = i$-Pr	$R^2 = PhCH{=}CH$			12	
$ClCH_2COCl$	$PhCH{=}NPh$	Et_3N, CH_2Cl_2	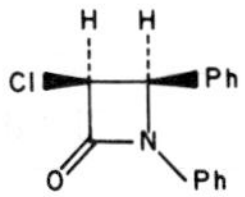 [structure: cis-3-Cl-4-Ph-1-Ph-azetidin-2-one, H, H]	—	1161

(*continued*)

TABLE 53. (*continued*)

Acid halide	Imine	Conditions	Product	Yield (%)	Reference
$ClCH_2COCl$	$EtO_2CCH_2N{=}C(Ph)SMe$	Et_3N, C_6H_6	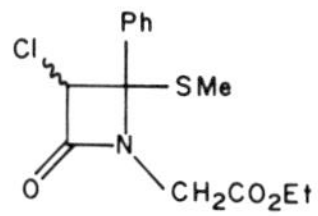	45	1161
$ClCH_2COCl$	$EtOCH{=}NPh$	Et_3N, CH_2Cl_2, N_2, r.t. stir 12 h	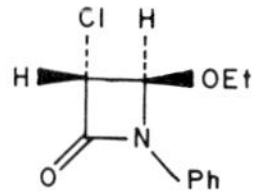	10	1137
$ClCH_2COCl$[a]	$R^1CO(R^1){=}NR^2$	Et_3N, C_6H_5Me reflux 24 h	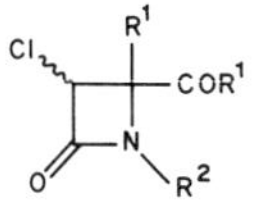		1128
	$R^1 = R^2 = Ph$			40	
	$R^1 = Me$, $R^2 = An\text{-}p$			43	
$ClCH_2COCl$	$PhCOCH{=}NAn\text{-}p$	Et_3N, C_6H_6, r.t. stir 2–5 h	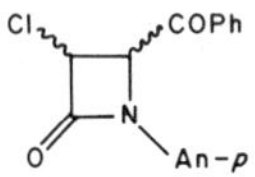	25	1148
$ClCH_2COCl$	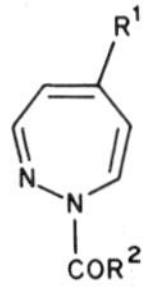	Et_3N, Et_2O, or CH_2Cl_2	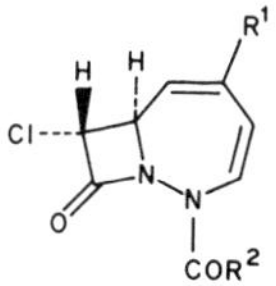		1127
	$R^1 = H$, $R^2 = OEt$			70	
	$R^1 = Me$, $R^2 = OEt$			61	
	$R^1 = CO_2Et$, $R^2 = OEt$			60	
	$R^1 = H$, $R^2 = Ph$			75	

$ClCH_2COCl$	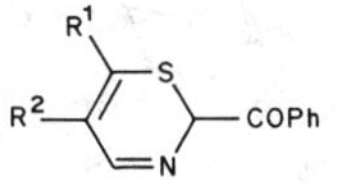	Et_3N	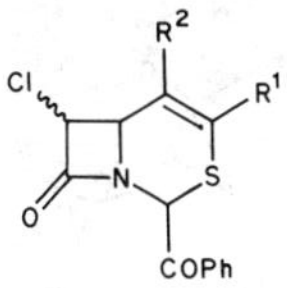		1140, 1141
	$R^1 = H$, $R^2 = Ph$, *p*-Tol			—	
	$R^1 = Ph$, $R^2 = H$			—	
$ClCH_2COCl$	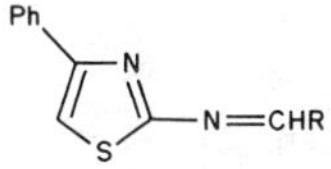	Et_3N, C_6H_6, r.t. stir 5 h	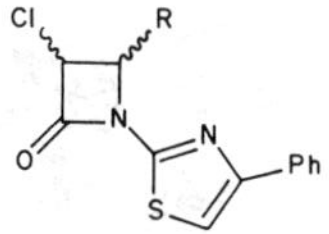		1189
	$R = C_6H_4OH$-*o*			25	
	$R =$ An-*p*			42	
	$R = C_6H_4NO_2$-*o*			38	
	$R = C_6H_4NO_2$-*m*			33	
	$R = Me_2NPyr$-4			37	
	$R = PhCH{=}CH$			73	
$ClCH_2COCl$	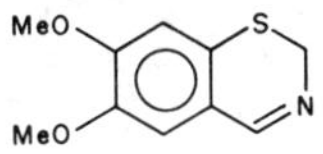	Et_3N, CH_2Cl_2, stand 5 °C overnight	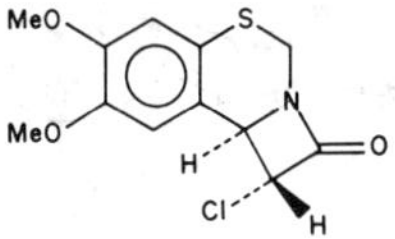	19	1150
$ClCH_2COCl$	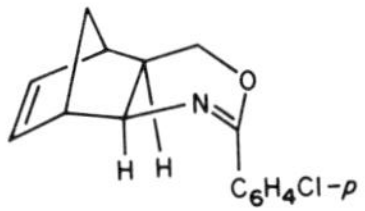	1. Et_3N, C_6H_6, stir 2. 50°C, 10 min	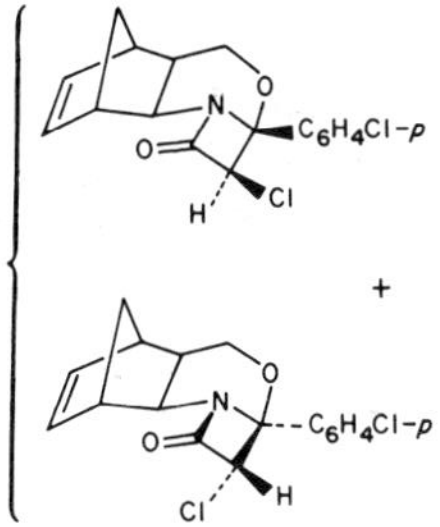 +	37^h	1190

(*continued*)

TABLE 53. (*continued*)

Acid halide	Imine	Conditions	Product	Yield (%)	Reference
$ClCH_2COCl$	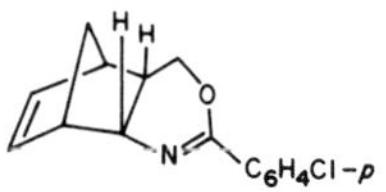 C_6H_4Cl-*p*	1. Et_3N, C_6H_6, stir 2. 50°C, 10 min	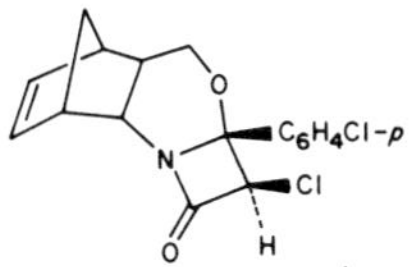 C_6H_4Cl-*p* +	18	1190
			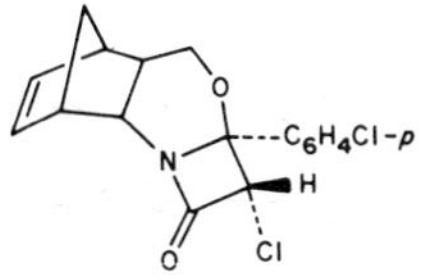 C_6H_4Cl-*p*	21	
$ClCH_2COCl$	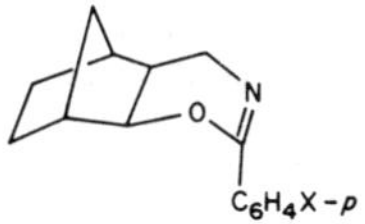 C_6H_4X-*p*	1. Et_3N, C_6H_6, stir 2. 50°C, 10 min	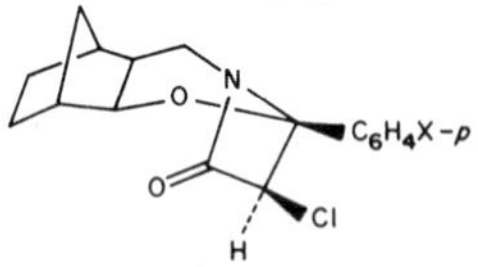 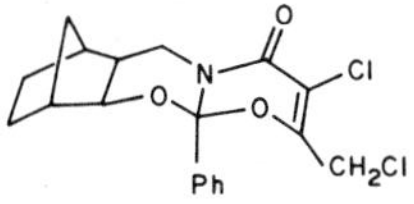 C_6H_4X-*p* (**123**) + Ph, CH_2Cl (**124**)		1190
	X = H			34(**123**) + 26(**124**)	
	X = Cl			39(**123**)	1190
$ClCH_2COCl$	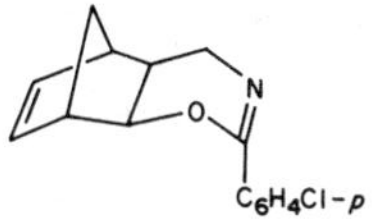 C_6H_4Cl-*p*	1. Et_3N, C_6H_6, stir 2. 50°C, 10 min	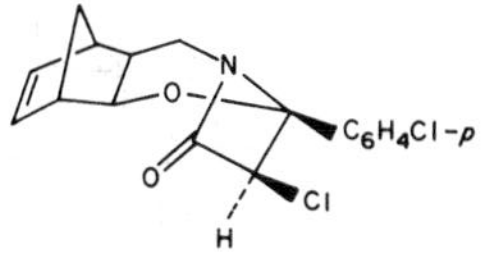 C_6H_4Cl-*p*	42	

$ClCH_2COCl$	(RCH=N–C$_6$H$_4$)$_2$SO$_2$	1. Et_3N, dioxane, r.t. stir 5h 2. r.t. stand 3 days	(3-Cl-4-R-2-oxoazetidin-1-yl-C$_6$H$_4$)$_2$SO$_2$		1191
	R = Ph			79	
	R = C_6H_4OH-*p*			55	
	R = C_6H_4OH-*m*			61	
	R = C_6H_4OH-*p*			69	
	R = 2-HO, 3-BrC_6H_3			71	
	R = 2-HO, 3,5-$Br_2C_6H_2$			85	
	R = 3-MeO, 4-HOC_6H_3			76	
	R = C_6H_4Cl-*o*			69	
	R = C_6H_4Cl-*p*			77	
	R = 2-Fu			79	
$ClCH_2COCl$	(RCH=N–C$_6$H$_3$Me)$_2$	1. Et_3N, dioxane r.t. stir 5h 2. r.t., stand 3 days	(3-Cl-4-R-2-oxoazetidin-1-yl-C$_6$H$_3$Me)$_2$		1191
	R = Ph			81	
	R = C_6H_4OH-*o*			65	
	R = C_6H_4OH-*m*			72	
	R = C_6H_4OH-*p*			69	
	R = 2-HO, 3-BrC_6H_3			70	
	R = 2-HO, 3,5-$Br_2C_6H_2$			85	
	R = 3-MeO, 4-HOC_6H_3			61	
	R = C_6H_4Cl-*o*			55	
	R = An-*p*			64	
	R = 2-Fu			71	
$ClCH_2COCl$	R^1–C$_6$H$_4$–N(azetidinone, R^2, R^3)–N=CHAn-*p*	1. Et_3N, dioxane r.t. stir 5h 2. r.t. stand 3 days	R^1–C$_6$H$_4$–(azetidinone, R^2, R^3)–N(azetidinone, An-*p*, Cl)		1186

(*continued*)

TABLE 53. *(continued)*

Acid halide	Imine	Conditions	Product	Yield (%)	Reference
	$R^1 = SC_6H_4NO_2$-*p*				
	$R^2 = H, R^3 = Ph$			41	
	$R^2 = H, R^3 = C_6H_4NO_2$-*p*			43	
	$R^2 = H, R^3 =$ An-*p*			45	
	$R^2 = H, R^3 = C_6H_4Cl$-*p*			42	
	$R^2 = H, R^3 = C_6H_4OH$-*p*			43	
	$R^2 = H, R^3 = Me_2NPyr$-4			40	
	$R^2 = Me, R^3 = Ph$			50	
	$R^2 = Me, R^3 =$ An-*p*			49	
	$R^2 = Me, R^3 = C_6H_4NO_2$-*p*			47	
	$R^2 =$ *i*-Pr, $R^3 = Ph$			53	
	$R^2 =$ *i*-Pr, $R^3 =$ An-*p*			58	
	$R^2 =$ *i*-Pr, $R^3 = C_6H_4NO_2$-*p*			60	
	$R^1 = SO_2C_6H_4NO_2$-*p*				
	$R^2 = H, R^3 = Ph$			26	
	$R^2 = H, R^3 = C_6H_4NO_2$-*p*			23	
	$R^2 = H, R^3 = C_6H_4Cl$-*p*			22	
	$R^2 = H, R^3 =$ An-*p*			22	
	$R^2 = H, R^3 = C_6H_4OH$-*o*			24	
	$R^2 = H, R^3 = Me_2NPyr$-4			36	
	$R^2 = Me, R^3 = Ph$			25	
	$R^2 = Me, R^3 =$ An-*p*			28	
	$R^2 = Me, R^3 = C_6H_4NO_2$-*p*			22	
	$R^2 =$ *i*-Pr, $R^3 = Ph$			21	
	$R^2 =$ *i*-Pr, $R^3 =$ An-*p*			23	
	$R^2 =$ *i*-Pr, $R^3 = C_6H_4NO_2$-*p*			22	
$BrCH_2COCl$	PhCH=NPh	Et_3N, CH_2Cl_2	H, Ph, Br, H, O, N, Ph	35	1161

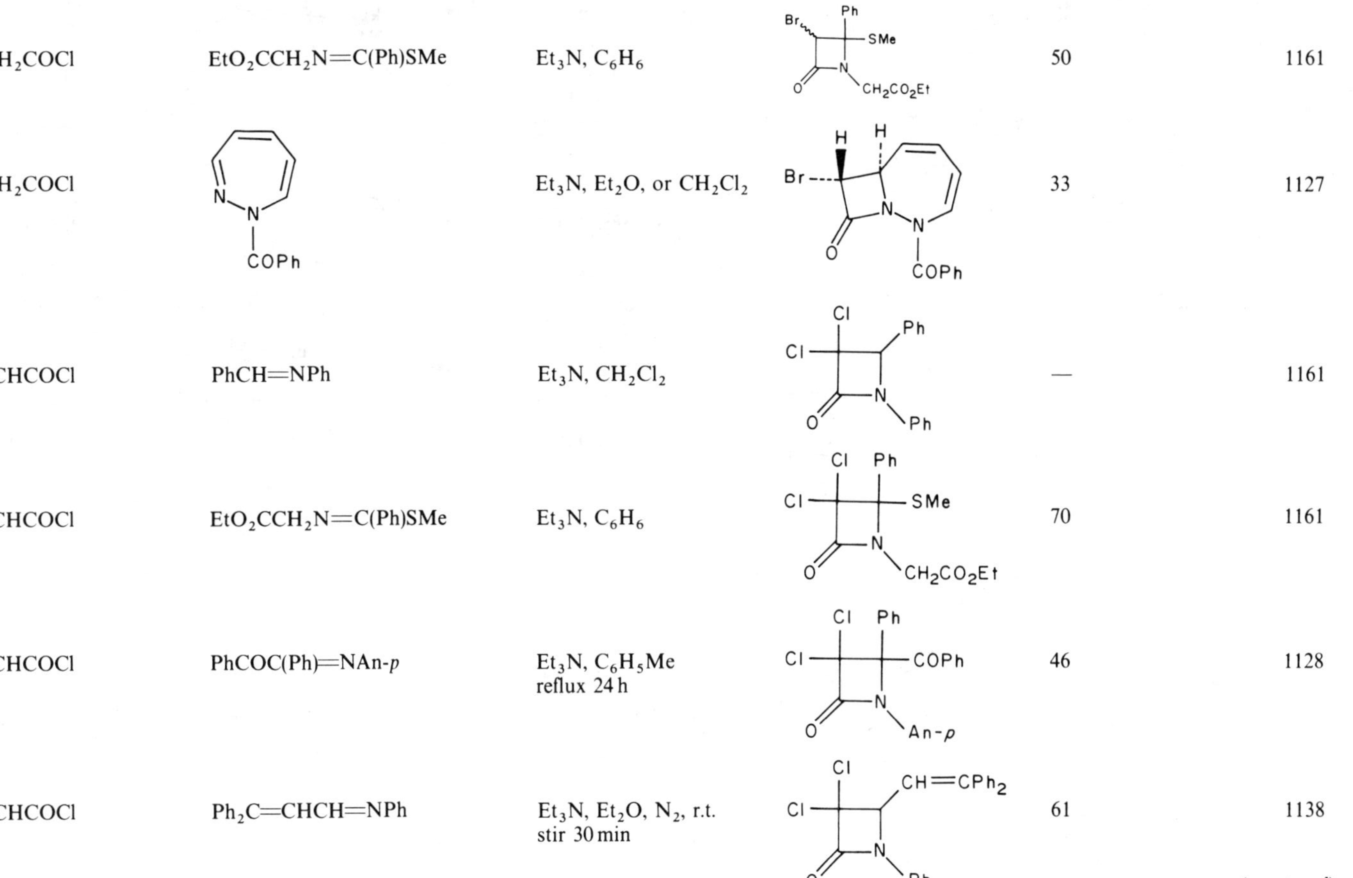

$BrCH_2COCl$	$EtO_2CCH_2N{=}C(Ph)SMe$	Et_3N, C_6H_6		50	1161
$BrCH_2COCl$		Et_3N, Et_2O, or CH_2Cl_2		33	1127
$Cl_2CHCOCl$	$PhCH{=}NPh$	Et_3N, CH_2Cl_2		—	1161
$Cl_2CHCOCl$	$EtO_2CCH_2N{=}C(Ph)SMe$	Et_3N, C_6H_6		70	1161
$Cl_2CHCOCl$	PhCOC(Ph)=NAn-*p*	Et_3N, C_6H_5Me reflux 24 h		46	1128
$Cl_2CHCOCl$	$Ph_2C{=}CHCH{=}NPh$	Et_3N, Et_2O, N_2, r.t. stir 30 min		61	1138

(continued)

TABLE 53. (*continued*)

Acid halide	Imine	Conditions	Product	Yield (%)	Reference
$Cl_2CHCOCl$	PhCH=CRCH=NPh	Et_3N, Et_2O, N_2, r.t., stir 30 min	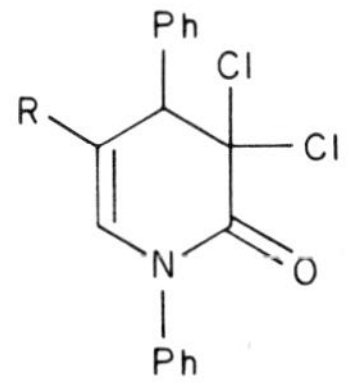		1138
	R = H			49	
	R = Me			66	
$Cl_2CHCOCl$	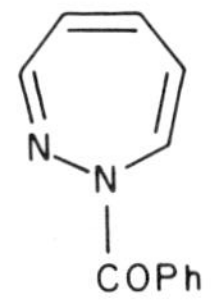	Et_3N, Et_2O	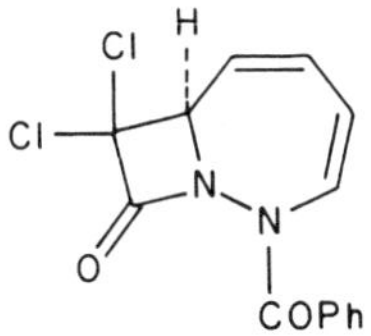	80	1127
$Cl_2CHCOCl$	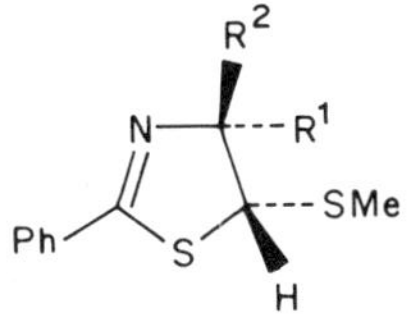	Et_3N, Hexane	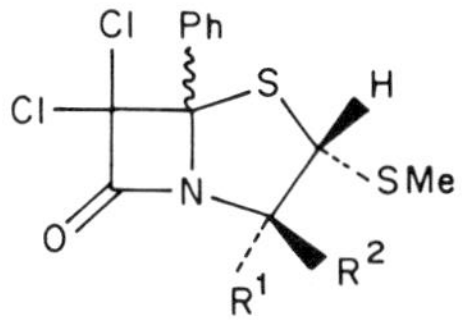		1145
	$R^1 = R^2 = Me$			10–37[i]	
	$R^1 = i\text{-}Pr$, $R^2 = Me$			94	
	$R^1 = Me$, $R^2 = i\text{-}Pr$			28	

$Cl_2CHCOCl$	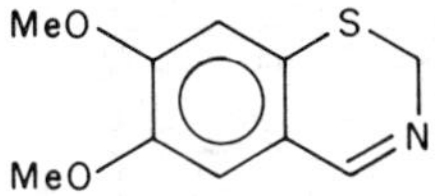	Et_3N, CH_2Cl_2, 50°C stand overnight	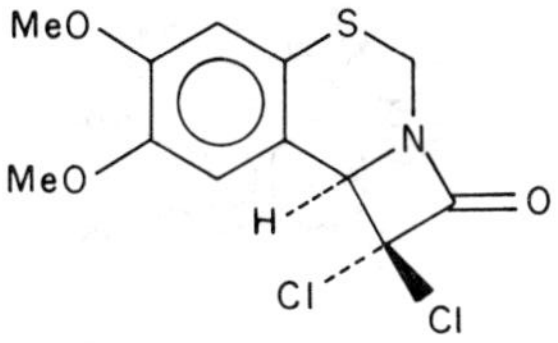	23	1150
$Cl_2CHCOCl$	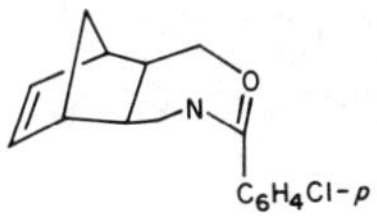	1. Et_3N, C_6H_6 stir r.t. 2. 50°C, 10 min	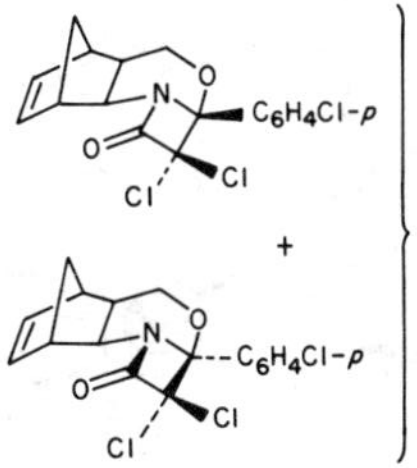 +	30[j]	1190
$Cl_2CHCOCl$	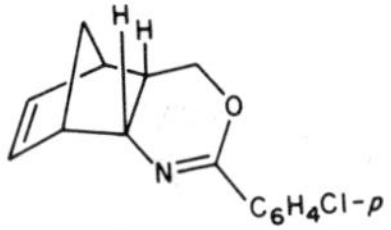	1. Et_3N, C_6H_6 stir r.t. 2. 50°C, 10 min	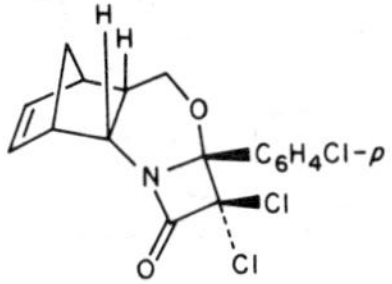	17	1190
			+ 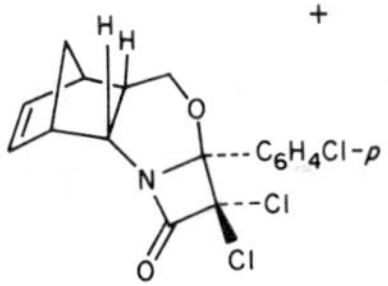	24	
$Cl_2CHCOCl$	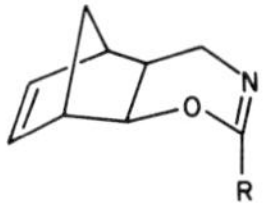	1. Et_3N, C_6H_6 stir r.t. 2. 50°C, 10 min	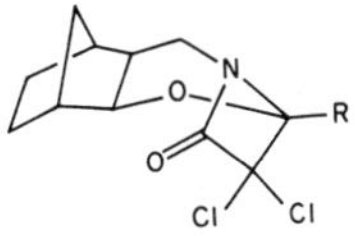		1190

(continued)

TABLE 53. (*continued*)

Acid halide	Imine	Conditions	Product	Yield (%)	Reference
	R = Ph			40	
	R = C_6H_4Cl-*p*			44	
$Cl_2CHCOCl$	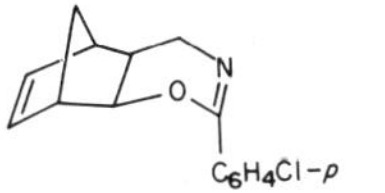 [structure: C_6H_4Cl-*p*]	1. Et_3N, C_6H_6 stir r.t. 2. 50°C, 10 min	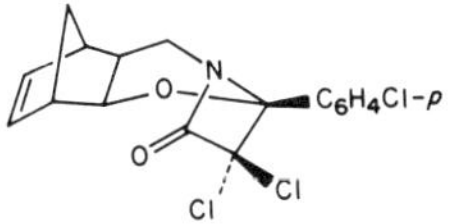 [structure: N, O, Cl, Cl, C_6H_4Cl-*p*]	37	1190
MeCH(Cl)COCl	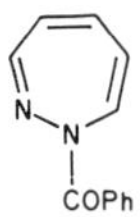 [structure: N–N, COPh]	Et_3N, Et_2O	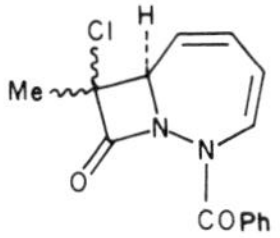 [structure: Me, Cl, H, N–N, COPh] (α:β choro = 65:35)	31	1127
PhCH(Cl)COCl	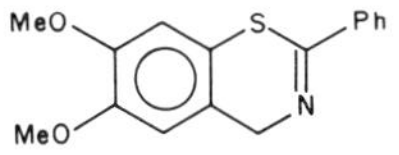 [structure: MeO, MeO, S, N, Ph]	Et_3N, C_6H_6, reflux 1 h	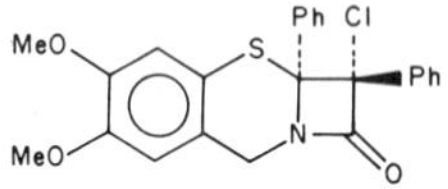 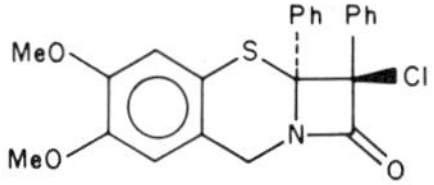 [structure: MeO, MeO, S, N, Ph, Cl, Ph, O] + [structure: MeO, MeO, S, N, Ph, Ph, Cl, O]	84 11	1150
PhCH(Cl)COCl	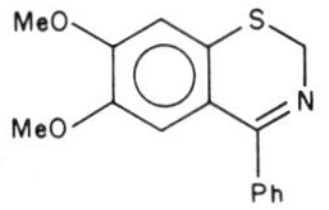 [structure: MeO, MeO, S, N, Ph]	Et_3N, C_6H_6, reflux 1 h	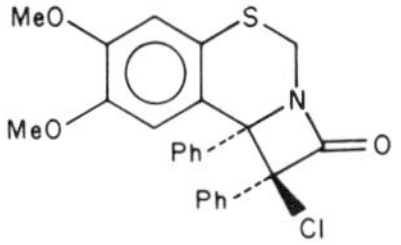 [structure: MeO, MeO, S, N, Ph, Ph, Cl, O]	89	1150

$Ph_2CHCOCl$	PhCOC(Ph)=NR	Et_3N, C_6H_4Me, reflux 24 h	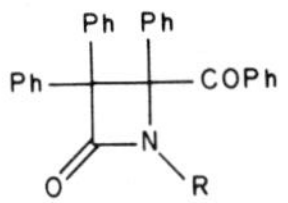 [β-lactam: Ph, Ph, Ph, COPh, N–R]		1128[a]
	R = $PhCH_2$			85	
	R = PhCHMe			85	
PhSeCH(Me)COCl	PhCH=NR	Et_3N, C_6H_6	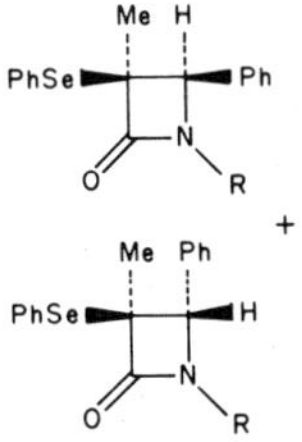 [β-lactam: PhSe, Me, H, Ph, N–R] + [β-lactam: PhSe, Me, Ph, H, N–R]		1192
	R = Me		(76:24)	50	
	R = *t*-Bu		(30:70)	52	
	R = Ph		(13:1)	56	
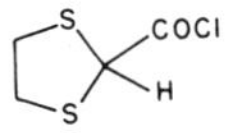 [2-chlorocarbonyl-1,3-dithiolane: S, S, COCl, H]	PhCH=NPh	Et_3N, CH_2Cl_2, −78°C	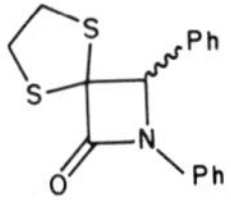 [spiro β-lactam: S, S, Ph, N–Ph]	66	1193
$MeCO_2CH_2COCl$	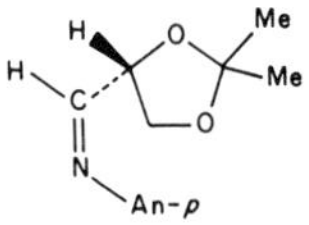 [imine: H, H, C=N–An-*p*, dioxolane O, O, Me, Me]	Et_3N, CH_2Cl_2	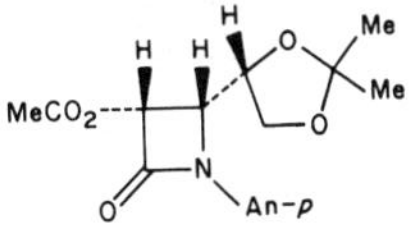 [β-lactam: $MeCO_2$, H, H, H, dioxolane O, O, Me, Me, N–An-*p*]	—	1152

(*continued*)

TABLE 53. (*continued*)

Acid halide	Imine	Conditions	Product	Yield (%)	Reference
$MeCO_2CH(Ph)COCl$	PhCH=NPh	Et_3N, CH_2Cl_2	[structure: β-lactam with $MeCO_2$, Ph, H, Ph, N–Ph, C=O]	—	1218

[a]Two methods were used: Method A involved reaction of acid chloride with imine in the presence of triethylamine and Method **B** was direct addition of ketene to the imine (see Table 52). Which method was employed with the different imine substituents was not specified.
[b]The triethylamine base was added to a solution of the acid halide and the imine.
[c]The acid halide was added to a solution of the imine and triethylamine.
[d]The imine was allowed to react with trimethylsilyl chloride and triethylamine in methylene chloride for 30 minutes, then the acid halide was added and the reaction mixture stirred overnight.
[e]Products are free acids.
[f]PI-N = Phthalimids.
[g]Hexahydro-s-triazine, the trimer of esters of substituted glycinates.
[h]The product is a 1:1 mixture of diastereomers.
[i]Yield varies depending upon the number of moles of acid chloride employed.
[j]The product is a 3:4 mixture of diastereomers.

TABLE 54. β-Lactam preparation by the reaction of imines with 'activated' carboxylic acids and derivatives

Carboxylic acid derivative	Imine	Conditions	Product	Yield (%)	Reference
t-BuO$_2$CNHCH$_2$CO$_2$CO$_2$Bu-i	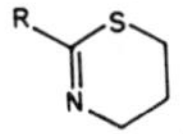 [2-R-5,6-dihydro-4H-1,3-thiazine]	1. Et_3N, THF, $-10\,°C$, stir 2h 2. r.t. stir 10h	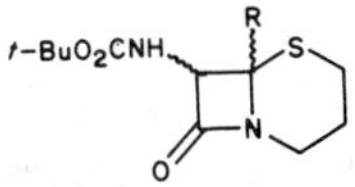 [t-BuO$_2$CNH-substituted bicyclic β-lactam, R at ring fusion]		1180
	R = Ph			64	
	R = $C_6H_4NO_2$-p			60	
$R^1CH_2CO_2COCF_3$	$R^2CH{=}NR^3$	1. Et_3N, CH_2Cl_2, reflux 1h 2. stir overnight	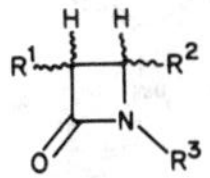 [β-lactam: R^1, R^2 on C-3, C-4; R^3 on N]	30–70	1194
$R^1 = N_3$	$R^2 = R^3 = Ph$		*cis/trans*		
$R^1 = PhO$	$R^2 = R^3 = Ph$		*cis/trans*		
$R^1 = N_3$	R^2 = An-p, $R^3 = C_6H_4CO_2H$-p		*trans*		
$R^1 = PhO$	R^2 = An-p, $R^3 = Ph_2CH$		*cis*		
$R^1 = N_3$	R^2 = An-p, $R^3 = C_6H_4CO_2SiMe_3$		—		
$N_3CH_2CO_2COCF_3$	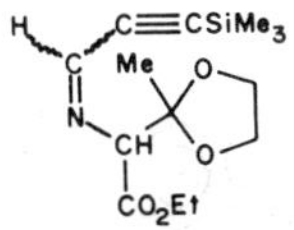 [H(C≡CSiMe$_3$)C=N–CH(CO$_2$Et)–C(Me)(OCH$_2$CH$_2$O)]	1. Et_3N, CH_2Cl_2, $0\,°C$ 2. stir 1h	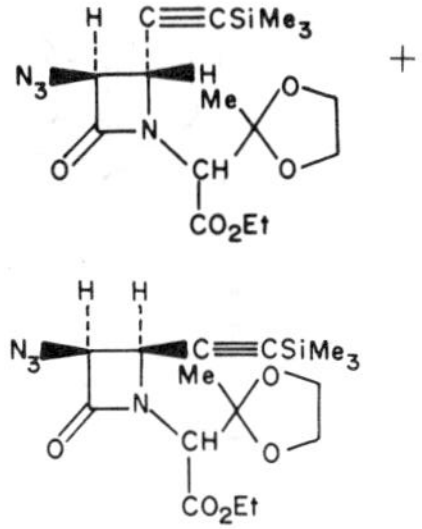 [3-N$_3$-4-(C≡CSiMe$_3$) β-lactam, N-CH(CO$_2$Et)C(Me)(OCH$_2$CH$_2$O), two stereoisomers] + [isomer] (4:1)	—	1195

(*continued*)

TABLE 54. (*continued*)

Carboxylic acid derivative	Imine	Conditions	Product	Yield (%)	Reference
$N_3CH_2CO_2COCF_3$	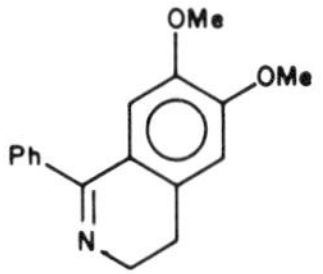	1. Et_3N, CH_2Cl_2, reflux 1 h 2. stir overnight	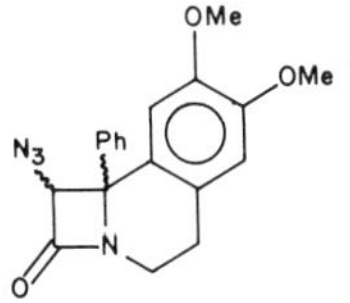	—	1194
$N_3CH_2CO_2COCF_3$	(*t*-Bu)$Ph_2SiOCH_2C{\equiv}C$ | *p*-($MeOCH_2O)C_6H_4N{=}CH$	Et_3N, CH_2Cl_2, Ar, 0 °C, 1.5 h	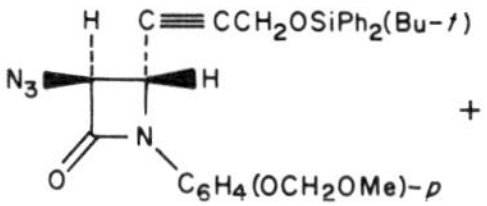 +	7	1196
			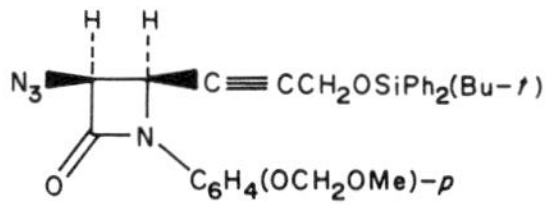	58	
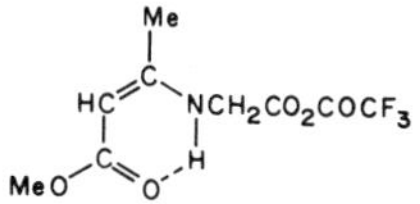	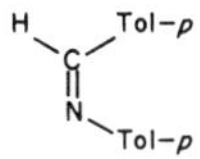	Et_2O, N_2	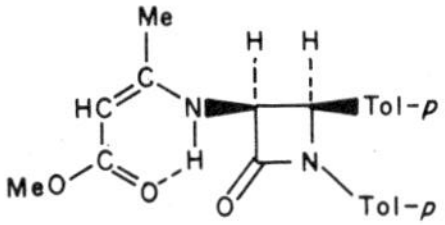	5	1178
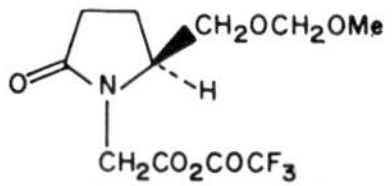	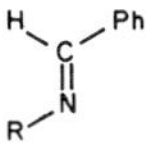	Et_3N, CH_2Cl_2, 20h	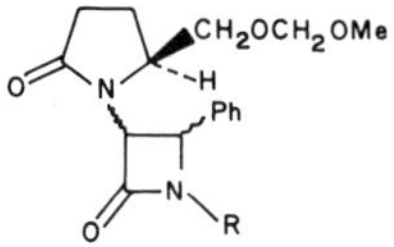		1197

	$R = Ph$	0 °C	*cis*:*trans* = 86:14[a]	71	
	$R = Ph$	−20 °C	*cis*:*trans* = 84:16[b]	62	
	$R = PhCH_2$	0 °C	*cis*:*trans* = 95:5[c]	55	
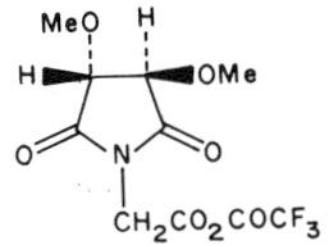 [succinimide, MeO, H, OMe, $N-CH_2CO_2COCF_3$]	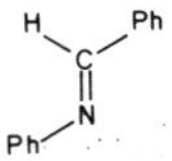 [H(Ph)C=N–Ph]	Et_3N, CH_2Cl_2, 20h	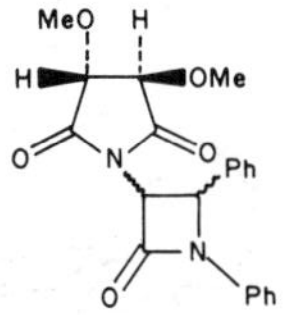 [succinimide–β-lactam product, Ph, N–Ph]		1197
		0 °C	100% *trans*[d]	47	
		−20 °C	100% *trans*[e]	40	
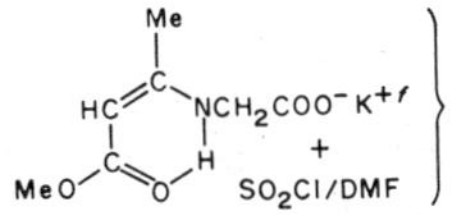 [MeO–C(=O)–CH=C(Me)–$NCH_2COO^-K^{+f}$ + SO_2Cl/DMF]	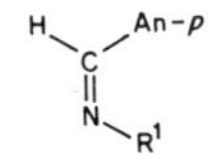 [H(An-*p*)C=N–R^1], R^1 = Ph or Tol-*p*	Et_3N, CH_2Cl_2, r.t. stir 20–24h	No reaction	—	1198
$R^1CH_2CO_2SO_2R^2$	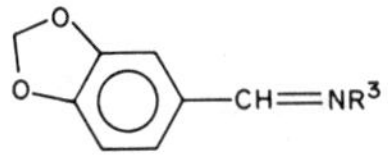 [3,4-methylenedioxyphenyl–CH=NR^3]	Et_3N, CH_2Cl_2, stir 50 °C. 30 min	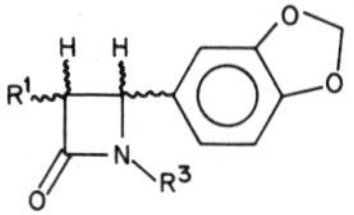 [R^1, H, H β-lactam, 3,4-methylenedioxyphenyl, N–R^3]		1199
$R^1 = PhO$, $R^2 = Me$	R^3 = An-*p*		*cis*	60	
$R^1 = PhO$, $R^2 = Ph$	R^3 = An-*p*		*cis*	60	
$R^1 = R^2 = Ph$	R^3 = An-*p*		*trans*	58	
$R^1 = R^2 = Ph$	R^3 = Tol-*p*		*trans*	50	
$R^1 = EtO$, $R^2 = Ph$	R^3 = An-*p*		*cis*	60	
$R^1 = PhO$, $R^2 = Ph$	$R^3 = PhCH_2$		*cis*	—	
$R^1 = PhO$, $R^2 = Ph$	$R^3 = C_6H_4NO_2$-*p*		*trans*	55	
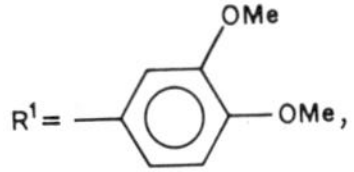 R^1 = [3,4-dimethoxyphenyl] (OMe, OMe), $R^2 = Ph$	R^3 = Tol-*p*		*trans*	50	

(continued)

TABLE 54. (*continued*)

Carboxylic acid derivative	Imine	Conditions	Product	Yield (%)	Reference
$R^1 = O$– (Cl, Cl),	R^3 = An-*p*		*cis*	60	
R^2 = Ph					
	$R^3 = C_6H_4NO_2$-*p*		*trans*	50	
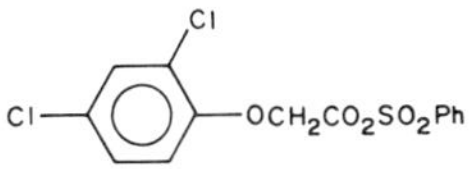 Cl, Cl, $OCH_2CO_2SO_2Ph$	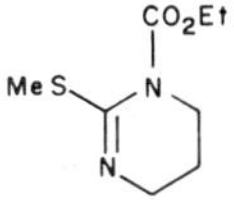 CO_2Et, MeS, N, N	Et_3N, CH_2Cl_2, stir	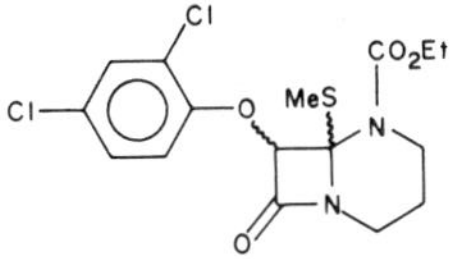 Cl, Cl, O, MeS, CO_2Et, N, N, O	—	1199
	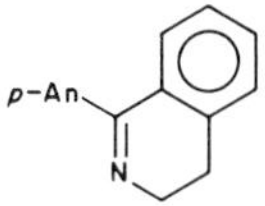 *p*-An, N	Et_3N, CH_2Cl_2, stir	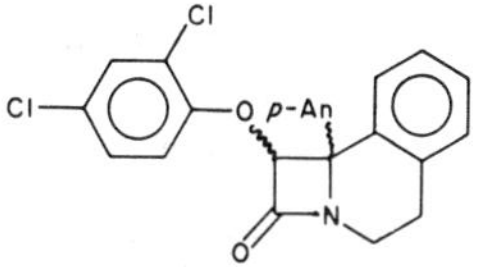 Cl, Cl, O, *p*-An, N, O	45	1199
$PhOCH_2CO_2SO_2Ph$	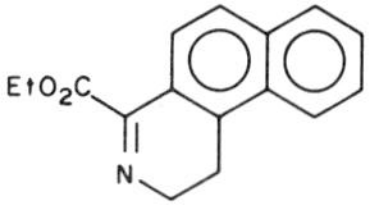 EtO_2C, N	Et_3N, CH_2Cl_2, stir	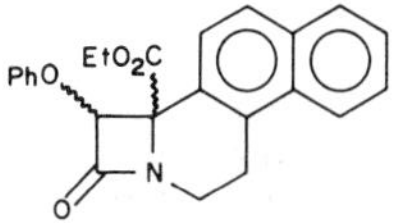 PhO, EtO_2C, N, O	65	1199

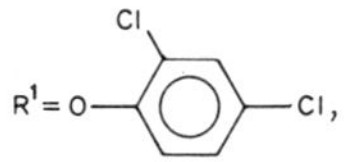

Reagent	Imine	Conditions	Product	Yield (%)	Ref.
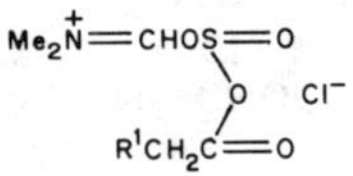 $Me_2\overset{+}{N}{=}CHOS({=}O)OC({=}O)CH_2R^1$ Cl^-	$R^2CH{=}NR^3$	Et_3N, CH_2Cl_2, r.t., stir 20–24 h	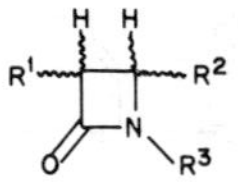 β-lactam: R^1, H / H, R^2 (wavy bonds), N–R^3, C=O		1200, 1201
$R^1 = MeO$[g]	$R^2 = R^3 = Ph$		*cis* = 9:1	60	
$R^1 = PhO$	$R^2 = R^3 = Ph$		*cis*:*trans* = 57:43	48	
$R^1 = PhO$	$R^2 = An\text{-}p$, $R^3 = EtO$		*trans*	60	
$R^1 = PhO$	$R^2 = An\text{-}p$, $R^3 = Tol\text{-}p$		*cis*:*trans* = 1:1	61	
$R^1 = Ph$	$R^2 = An\text{-}p$, $R^3 = Tol\text{-}p$		*cis*:*trans* = 65:35[h]	74	
$R^1 = Ph$	$R^2 = 2\text{-}Fu$, $R^3 = An\text{-}p$		*cis*	60	
$R^1 = p\text{-}An$	$R^2 = R^3 = Ph$		*trans*	35	
$R^1 = PI{-}N$	$R^2 = R^3 = Ph$		*trans*	50, 65	
$R^1 = PI{-}N$	$R^2 = An\text{-}p$, $R^3 = EtO$		*trans*	67	
$R^1 = PI{-}N$	$R^2 = An\text{-}p$, $R^3 = Tol\text{-}p$		*trans*	71, 74	
$R^1 = PI{-}N$	$R^2 = An\text{-}p$, $R^3 = Ph$		*trans*	70	
$R^1 = PI{-}N$	$R^2 = An\text{-}p$, $R^3 = \alpha\text{-naphthyl}$		*trans*	75	
$R^1 = PI{-}N$	$R^2 = An\text{-}p$, $R^3 = PhCH_2$		*trans*	60	
$R^1 = PI{-}N$	$R^2 = 2\text{-}Fu$, $R^3 = Ph$		*trans*	65	
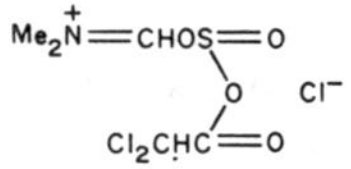 $Me_2\overset{+}{N}{=}CHOS({=}O)OC({=}O)CHCl_2$ Cl^-	$PhCH{=}NPh$	Et_3N, CH_2Cl_2, r.t., stir 20–24 h	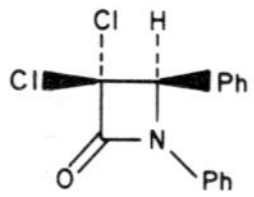 β-lactam: Cl, Cl / H, Ph, N–Ph, C=O	60	1201

(continued)

TABLE 54. (*continued*)

Carboxylic acid derivative	Imine	Conditions	Product	Yield (%)	Reference
$Me_2\overset{+}{N}{=}CHOS(=O)OC(=O)CH_2OPh$ Cl^-	$H(R^1)C{=}NCH_2CHR^2(OSiMe_3)$	1. Et_3N, CH_2Cl_2, 0 °C 2. r.t., stir 24h	β-lactam: PhO, H / H, R^1; N–$CH_2CH(OH)R^2$		1201
	$R^1 = Ph$, $R^2 = Me$		*cis*:*trans* = 75:25	67	
	R^1 = An-*p*, $R^2 = Me$		*cis*:*trans* = 1:1	60	
$Me_2\overset{+}{N}{=}CHOS(=O)OC(=O)CH_2R^1$ Cl^-	$H(R^2)C{=}NC_6H_4OSiMe_2R^3$-*p*	Et_3N, CH_2Cl_2, stir	β-lactam: R^1, H / H, R^2; N–$C_6H_4OSiMe_2R^3$-*p*		1201
$R^1 = PhO$	$R^2 = R^3 = Ph$	0–5 °C[i]	*cis*:*trans* = 95:5	77	
		[j]	*cis*:*trans* = 43:57	80	
		40 °C[i]	*cis*:*trans* = 95:5	77	
		[j]	*cis*:*trans* = 88:12	80	
$R^1 = PhO$	R^2 = An-*p*, R^3 = Tol-*p*	0–5 °C[i]	*cis*:*trans* = 90:10	80	
		[j]	*cis*:*trans* = 35:65	85	
		40 °C[i]	*cis*:*trans* = 85:15	87	
R^1 = PI—N	$R^2 = PhCH{=}CH$, $R^3 = Me$	40 °C[i,j]	*cis*:*trans* = 50:50	90	
R^1 = PI—N	$R^2 = PhCH{=}CH$, R^3 = *t*-Bu	40 °C[i]	*cis*:*trans* = 85:15	96	
$Me_2\overset{+}{N}{=}CHOCOCHOPh$ Cl^-	$H(R^1)C{=}NC_6H_4OSiMe_2R^2$-*p*	Et_3N, CH_2Cl_2, stir	β-lactam: PhO, H / H, R^1; N–$C_6H_4OSiMe_2R^2$-*p*		1201

	$R^1 = R^2 = Ph$	0–5 °C[i]	*cis*:*trans* = 97:3	99	
		[j]	*cis*:*trans* = 90:10	90	
	$R^1 = Ph$, $R^2 = t$-Bu	0–5 °C[i]	*cis*:*trans* = 100:0	96	
	$R^1 =$ An-*p*, $R^2 = t$-Bu	0–5 °C[i]	*cis*:*trans* = 100:0	96	
	$R^1 =$ An-*p*, $R^2 =$ Tol-*p*	0–5 °C[i]	*cis*:*trans* = 90:10	98	
		[j]	*cis*:*trans* = 70:30	96	
$R^1CH_2CO_2SO_2$An-*p*	$R^2CH{=}NR^3$	Et_3N, CH_2Cl_2, r.t. stir	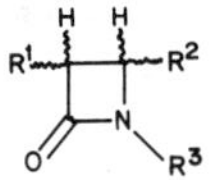		1202
$R^1 = p$-ClC_6H_4O	$R^2 =$ An-*p*, $R^3 =$ Tol-*p*			41	
$R^1 = p$-ClC_6H_4O	$R^2 = Ph$, $R^3 =$ Tol-*p*			41	
$R^1 =$ PI—N	$R^2 = R^3 = C_6H_4Cl$-*p*			48	
$R^1 =$ PI—N	$R^2 = Ph$, $R^3 = C_6H_4CO_2Et$-*p*			52	
$R^1CH_2CO_2SO_2$Tol-*p*	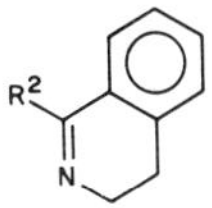	Et_3N, CH_2Cl_2, r.t. stir 48 h	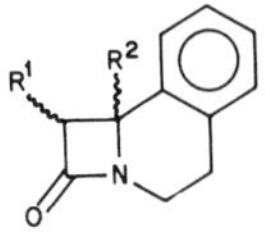		1201
$R^1 = p$-ClC_6H_4O	$R^2 = Ph$			49	
$R^1 = p$-ClC_6H_4O	$R^2 = C_6H_4NO_2$-*m*			55	
$R^1 =$ PI—N	$R^2 = C_6H_4NO_2$-*p*			53	

(*continued*)

TABLE 54. (*continued*)

Carboxylic acid derivative	Imine	Conditions	Product	Yield (%)	Reference
$R^1CH_2CO_2SO_2$Tol-*p*	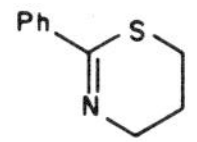	Et_3N, CH_2Cl_2, r.t. stir 48h	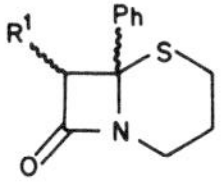		1201
R^1 = *p*-ClC_6H_4				39	
R^1 = *p*-ClC_6H_4O				46	
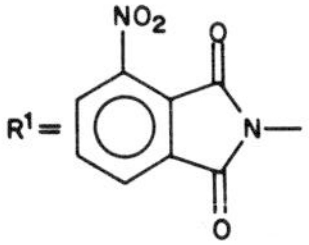 R^1 = (4-nitrophthalimido, N—)				64	
$R^1CH_2CO_2SO_2$Tol-*p*	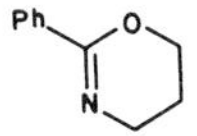	Et_3N, CH_2Cl_2, r.t. stir 48h	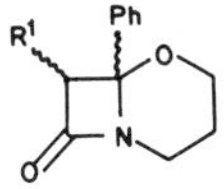		1201
R^1 = *p*-ClC_6H_4O				43	
R^1 = PI—N				44	
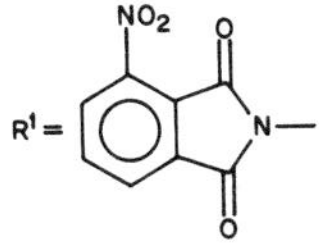 R^1 = (4-nitrophthalimido, N—)				51	
$PhCH_2O_2CNHCH_2CO_2$–$SO_2C_6H_4Cl$-*p*	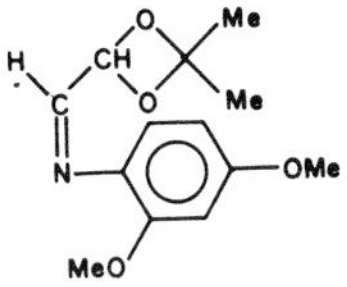	Et_3N	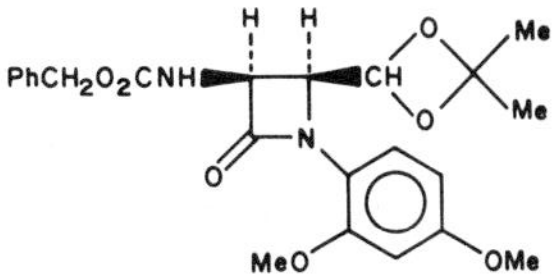 $PhCH_2O_2CNH$	62	1203

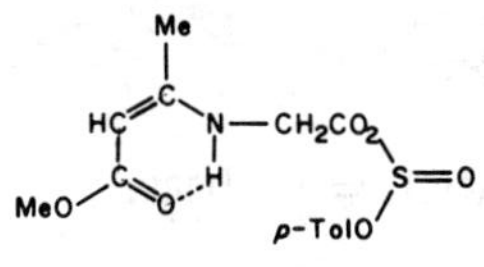 MeO–C(=O···H)–CH=C(Me)–N(H)–CH$_2$CO$_2$–S(=O)–O-Tol-p	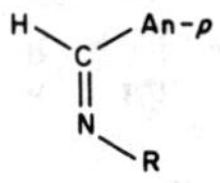 H–C(An-p)=N–R; R = Ph or Tol-p	1. Et_3N, CH_2Cl_2, 0 °C 2. r.t. stir	No reaction	—	1198, 1204
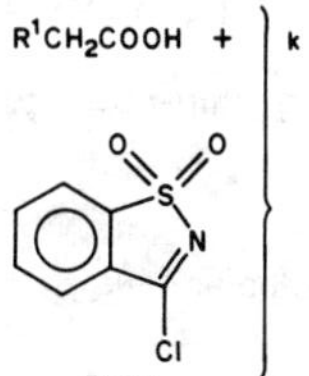 R^1CH_2COOH + (125) } k	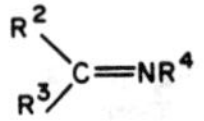 R^2R^3C=NR^4	Et_3N, CH_2Cl_2, reflux 10 h	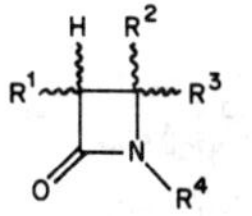 β-lactam: R^1, H / R^2, R^3 / N–R^4		1205
$R^1 = N_3$	$R^2 = R^4 = Ph$, $R^3 = MeS$		—	50	
$R^1 = MeO$	$R^2 = H$, $R^3 = R^4 = Ph$		*cis*	55	
$R^1 = PhCH_2O$	$R^2 = H$, $R^3 = R^4 = Ph$		*cis/trans*	60	
$R^1 = PhCH_2O$	$R^2 = H$, $R^3 = R^4 = An\text{-}p$		*cis*	65	
$R^1 = t\text{-}BuO$	$R^2 = H$, $R^3 = R^4 = Ph$		*cis*:*trans* = 4:1	60	
$R^1 = t\text{-}BuO$	$R^2 = H$, $R^3 = R^4 = An\text{-}p$		*cis*	60	
$R^1 = PhO$	$R^2 = H$, $R^3 = R^4 = Ph$		*cis*	60	
R^1 = PI—N	$R^2 = H$, $R^3 = R^4 = An\text{-}p$		*trans*	65	

(*continued*)

TABLE 54. (*continued*)

Carboxylic acid derivative	Imine	Conditions	Product	Yield (%)	Reference
$R^1 = PI—N$	$R^2 = H$ $R^3 =$ (3,4-methylenedioxyphenyl), $R^4 = C_6H_4Br$-*p*		*trans*	70	
$PhOCH_2COOH^k$ + **125**	Ph, S, N	Et_3N, CH_2Cl_2, reflux	PhO, Ph, S, N, O	70	1205
Me, HC, C, N—$CH_2CO_2^-K^+$ [k], H, MeO, C, O + **125**	H, C, An-*p*, N, R^1 R^1=Ph or Tol-*p*	1. Et_3N, CH_2Cl_2, 0 °C 2. r.t. stir	No reaction	—	1198
Me, HC, C, N—CH_2CO_2COR, H, MeO, C, O R = Me, Et, *i*-Bu, *t*-Bu	H, C, Tol-*p*, N, Tol-*p*	Et_3N, Et_2O, N_2	Me, HC, C, N, H, MeO, C, O, H, H, Tol-*p*, O, N, Tol-*p*	40–60	1178
$R^1CH_2CO_2CO_2Et^l$	$R^2CH{=}NR^3$	1. Et_3N, CH_2Cl_2, 0 °C, stir 2 h 2. r.t. stir 10–12 h	H, H, R^1, R^2, O, N, R^3		1178, 1206, 1207
$R^1{=}MeO_2CCH{=}C(Me)NH$	R^2 = Ph, R^3 = 3,4-$(MeO)_2C_6H_3CH_2$			65	1178, 1206
	R^2 = Ph, R^3 = 2,4-$(MeO)_2C_6H_3CH_2$			60	1178

	$R^2 = 2\text{-Fu}$, $R^3 = 3,4\text{-}(MeO)_2C_6H_3CH_2$			50	1178
	$R^2 = 2\text{-Fu}$, $R^3 = 2,4\text{-}(MeO)_2C_6H_3CH_2$			50	1178
	$R^2 = 2\text{-Fu}$, $R^3 = \text{An-}p$			50	1178
	$R^2{=}PhCH{=}CH$ (*trans*), $R^3 = 3,4\text{-}(MeO)_2C_6H_3CH_2$			46, 49	1178, 1207
	$R^2 = \text{An-}p$, $R^3 = \text{Tol-}p$			—	1178
$R^1 = PhOCH_2CONH$	$R^2 = Ph$, $R^3 = 3,4\text{-}(MeO)_2C_6H_3CH_2$			50	1178
	$R^2 = 2\text{-Fu}$, $R^3 = 3,4\text{-}(MeO)_2C_6H_3CH_2$			50	1178
$R^1 = PhCH_2CONH$	$R^2 = 2\text{-Fu}$, $R^3 = 3,4\text{-}(MeO)_2C_6H_3CH_2$			60	1178
$R^1CH_2CO_2CO_2Et$[l]	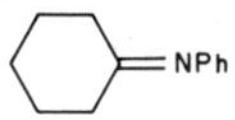	1. Et_3N, CH_2Cl_2, 0 °C, stir 2h 2. r.t. stir 10–12h	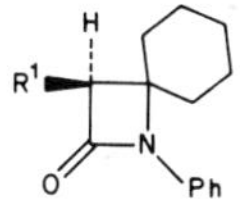		1178
$R^1{=}MeO_2CH{=}C(Me)NH$				60	
$R^1{=}PhOCH_2CONH$				65	
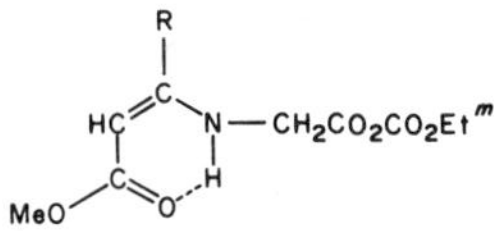 [m]	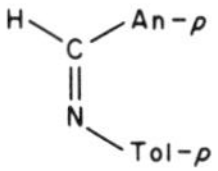	Et_3N, THF, N_2, −25 °C stir 2h	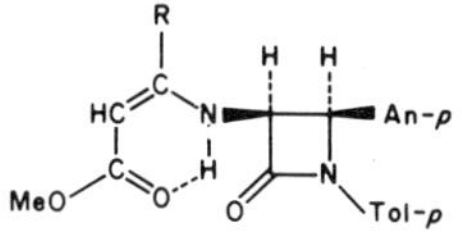		1208
$R^1 = PhOCH_2C(NH){=}CHCO_2Me$				20	

(continued)

TABLE 54. (*continued*)

Carboxylic acid derivative	Imine	Conditions	Product	Yield (%)	Reference
$R^1 = PhCH_2C(NH)=CHCO_2Me$				45	
$R^1 = MeC(NH)=CHCO_2Me$				61	
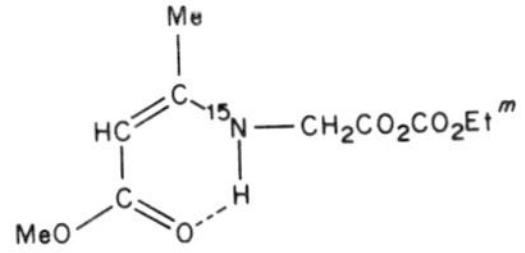 MeO–C(=O)–CH=C(Me)–^{15}NH–$CH_2CO_2CO_2Et$ [m]	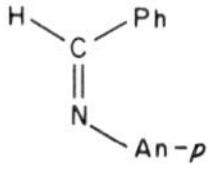 H–C(Ph)=N–An-*p*	1. Et_3N, CH_2Cl_2, 0 °C stir 2h 2. r.t. stir 10–12h	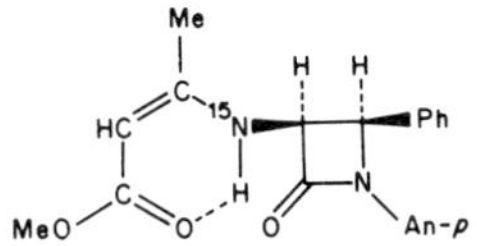 MeO–C(=O)–CH=C(Me)–^{15}NH–β-lactam (Ph, N–An-*p*)	50	1178
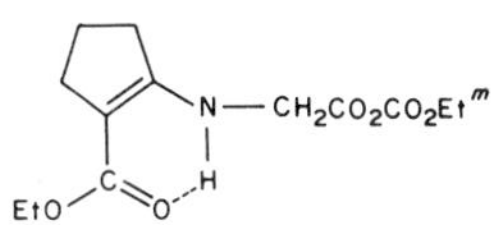 Cyclopentene–CO_2Et, N(H)–$CH_2CO_2CO_2Et$ [m]	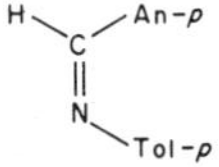 H–C(An-*p*)=N–Tol-*p*	1. Et_3N, CH_2Cl_2, 0 °C stir 2h 2. r.t. stir 10–12h	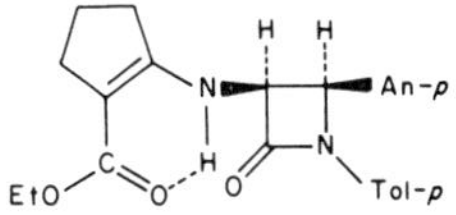 EtO–C(=O)-cyclopentenyl–NH–β-lactam (An-*p*, N–Tol-*p*)	—	1178
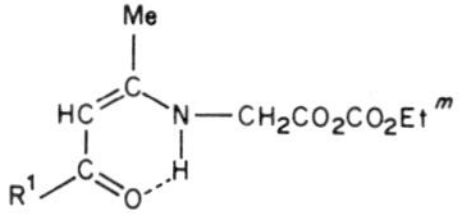 R^1–C(=O)–CH=C(Me)–NH–$CH_2CO_2CO_2Et$ [m]	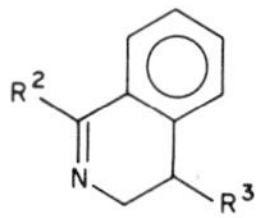 3,4-Dihydroisoquinoline (R^2, R^3)	1. Et_3N, CH_2Cl_2, 0 °C stir 2h 2. r.t. stir 10–12h	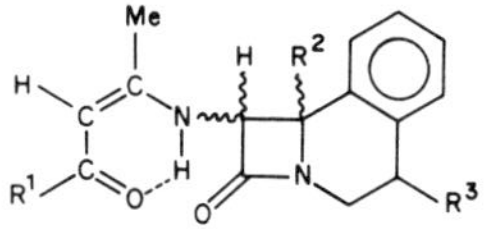 H–C(C(=O)R^1)=C(Me)–NH–β-lactam fused to tetrahydroisoquinoline (R^2, R^3)	—	1178
$R^1 = MeO$	$R^2 = Ph$, $R^3 = H$			40	
$R^1 = MeO$	$R^2 = MeS$, $R^3 = H$			40	
$R^1 = MeO$	$R^2 = MeS$, $R^3 = Me$			60	
$R^1 = MeO$	$R^2 = Ph$, $R^3 = H$			45	

$R^1 = t\text{-BuO}$	$R^2 = Ph$, $R^3 = H$			33	
$R^1 = MeO$	$R^2 = An\text{-}p$, $R^3 = H$		*trans*	80	
$R^1 = EtO$	$R^2 = An\text{-}p$, $R^3 = H$		*trans*	80	
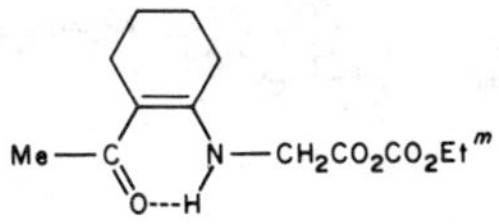 Me—C(=O···H)—N—$CH_2CO_2CO_2Et$ [m]	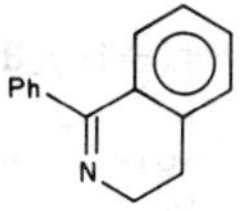	1. Et_3N, CH_2Cl_2, 0 °C stir 2 h 2. r.t. stir 10–12 h	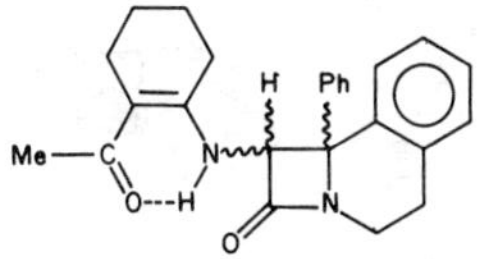	54	1178
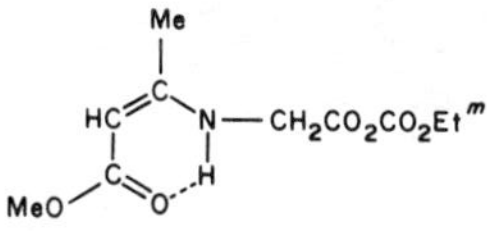 MeO—C(=O···H)—HC=C(Me)—N—$CH_2CO_2CO_2Et$ [m]	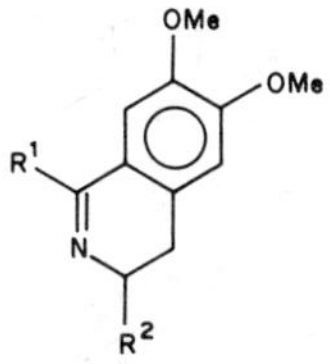	1. Et_3N, CH_2Cl_2, 0 °C stir 2 h 2. r.t. stir 10–12 h	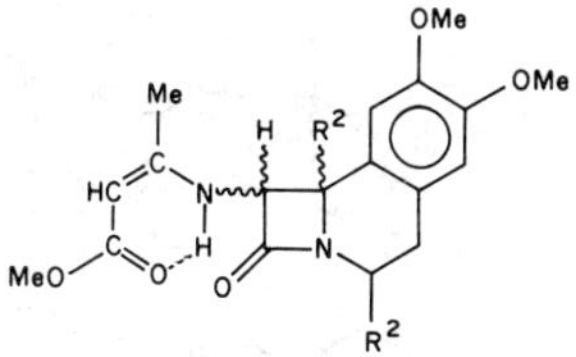		1178
	$R^1 = Ph$, $R^2 = CO_2Me$			80	
	$R^1 = C_6H_4Br\text{-}p$, $R^4 = H$			55	
	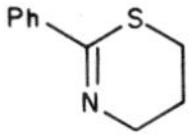	1. Et_3N, CH_2Cl_2, 0 °C stir 2 h 2. r.t. stir 10–12 h	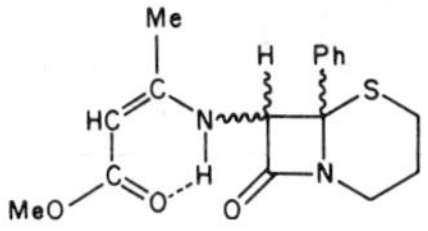	55	1178

(continued)

TABLE 54. (*continued*)

Carboxylic acid derivative	Imine	Conditions	Product	Yield (%)	Reference
	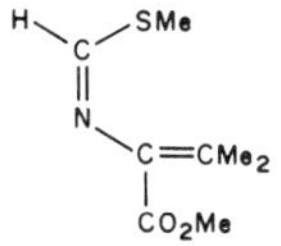 [structure]	1. Et_3N, CH_2Cl_2, 0 °C stir 2 h 2. r.t. stir 10–12 h	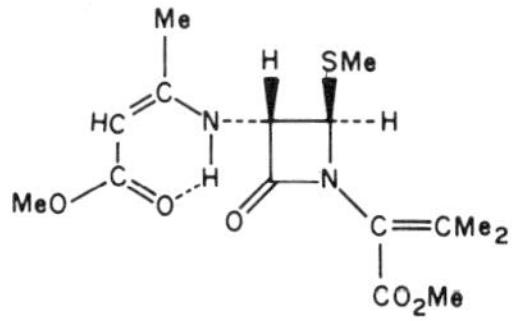 [structure]	—	1178
	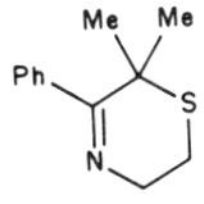 [structure]	1. Et_3N, CH_2Cl_2, 0 °C stir 2 h 2. r.t. stir 10–12 h	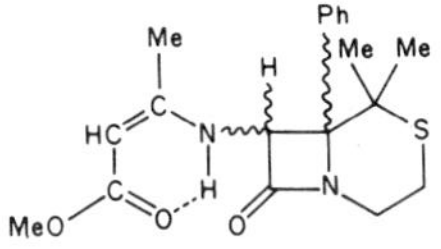 [structure]	—	1178
	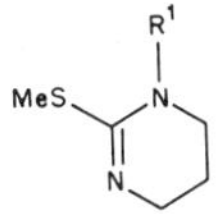 [structure]	1. Et_3N, CH_2Cl_2, 0 °C stir 2 h 2. r.t. stir 10–12 h	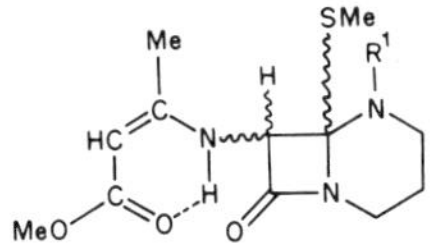 [structure]	—	1178
	R^1 = PI—NCH_2CO,			—	
	R^1 = *p*-Tos			—	
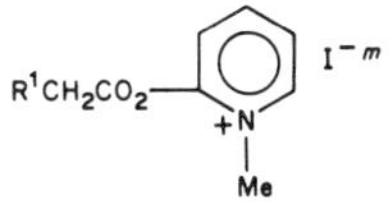 [structure]	$R^2CH = NTol$-*p*	Et_3N, CH_2Cl_2, reflux overnight	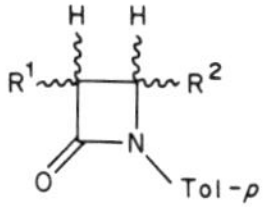 [structure]		1209
$R^1 = N_3$	R^2 = An-*p*		*cis*	35	
R^1 = PhO	R^2 = An-*p*		*cis*	45	
R^1 = PhS	R^2 = An-*p*		*trans*	34	
R^1 = PI—N	R^2 = An-*p*		*trans*	21	
$R^1 = PhCH_2O_2CNH$	R^2 = An-*p*		*cis*	35	
$R^1 = MeCO_2$	R^2 = An-*p*		*cis*	25	

$R^1 = MeO$	$R^2 =$ 2-Thi		*cis*	45	
$R^1 = Me_3CO$	$R^2 = 3,4\text{-}(MeO)_2C_6H_3$		—	50	
$R^1 = PhO$	$R^2 = 3,4\text{-}(MeO)_2C_6H_3$		*cis*	48	
$R^1 = PhO$	$R^2 =$ 2-Thi		*cis*	55	
$R^1 = PhCONH$	$R^2 =$ An-*p*		No reaction	—	
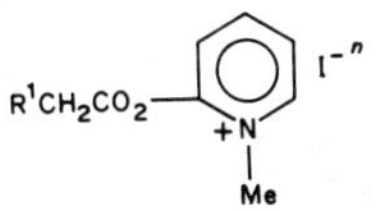 $R^1CH_2CO_2$, +N, Me, I^- [n]	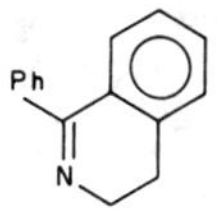 Ph, N	Et_3N, CH_2Cl_2, reflux[i]	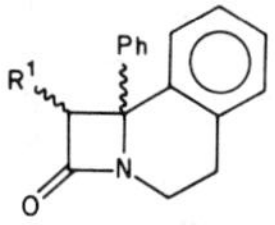 R^1, Ph, N, O		1209
$R^1 = PhO$			—	45	
$R^1 = PhCONH$			No reaction	—	
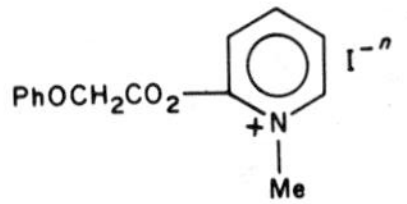 $PhOCH_2CO_2$, +N, Me, I^- [n]	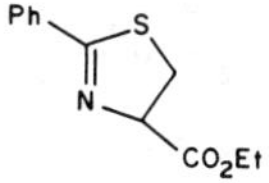 Ph, S, N, CO_2Et	Et_3N, CH_2Cl_2, reflux[i]	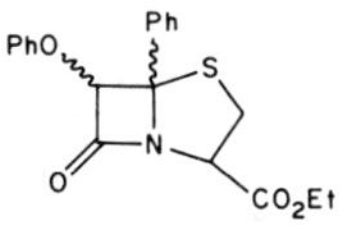 PhO, Ph, S, N, O, CO_2Et	15	1209
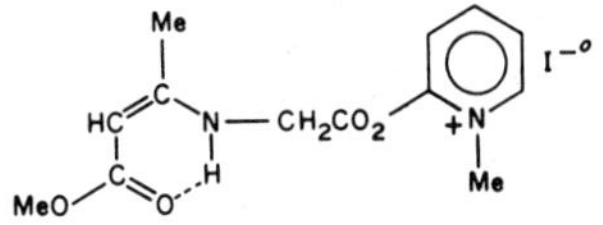 Me, HC, C, N, H, MeO, O, CH_2CO_2, +N, Me, I^- [o]	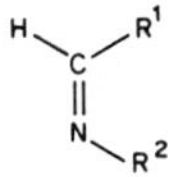 H, C, R^1, N, R^2	Et_3N, CH_2Cl_2, r.t., 20–24 h	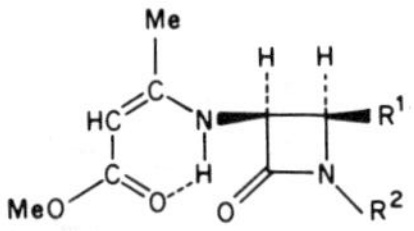 Me, H, H, HC, C, N, R^1, H, MeO, O, O, N, R^2		1198
	$R^1 = R^2 = Ph$			30	
	$R^1 =$ An-*p*, $R^2 = Ph$			35	
	$R^1 =$ An-*p*, $R^2 =$ Tol-*p*			50	
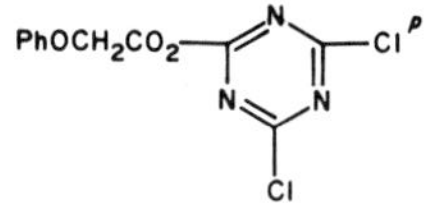 $PhOCH_2CO_2$, N, N, N, Cl [p], Cl	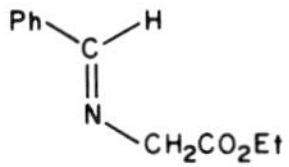 Ph, C, H, N, CH_2CO_2Et	1. Et_3N, CH_2Cl_2, ethyl glycinate N_2, 0 °C 2. r.t. stir overnight	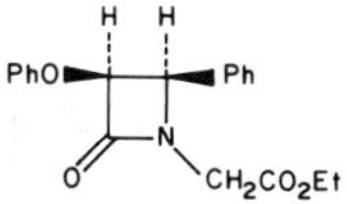 H, H, PhO, Ph, O, N, CH_2CO_2Et	58	1210

(*continued*)

TABLE 54. (*continued*)

Carboxylic acid derivative	Imine	Conditions	Product	Yield (%)	Reference
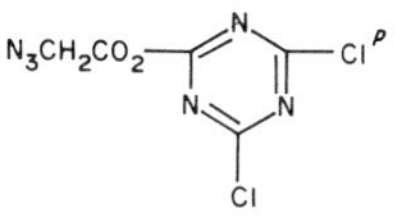 $N_3CH_2CO_2$, N, N, N, Cl^p, Cl	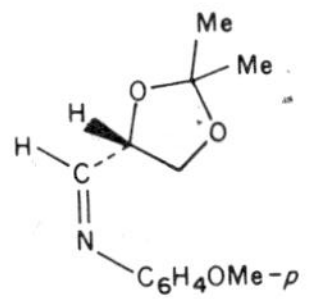 Me, Me, O, O, H, H, C, N, C_6H_4OMe-*p*	Et_3N, CH_2Cl_2	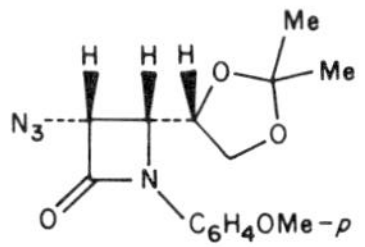 H, H, H, N_3, O, O, Me, Me, N, O, C_6H_4OMe-*p*	55	1152
	$(PhCH{=}N)_2CHPh$	1. Et_3N, CH_2Cl_2, 0 °C 2. r.t. stir 12 h	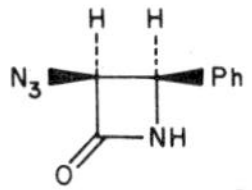 H, H, N_3, Ph, NH, O	46	1210
	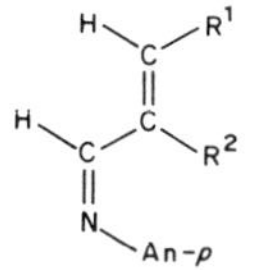 H, R^1, C, H, C, C, R^2, N, An-*p*	1. Et_3N, CH_2Cl_2, N_2, −20 °C 2. r.t. stir overnight	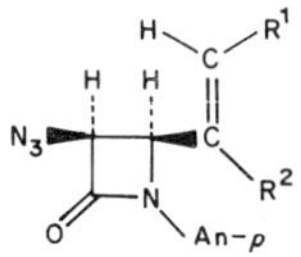 H, R^1, H, H, N_3, C, C, R^2, N, O, An-*p*		1153
	R^1 = Me, R^2 = H			68	
	R^1 = Ph, R^2 = H			73	
	R^1 = Ph, R^2 = Me			77	
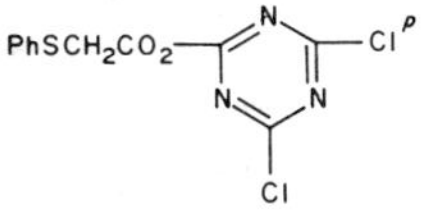 $PhSCH_2CO_2$, N, N, N, Cl^p, Cl	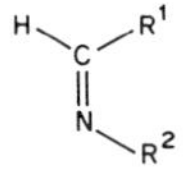 H, R^1, C, N, R^2	Et_3N, CH_2Cl_2	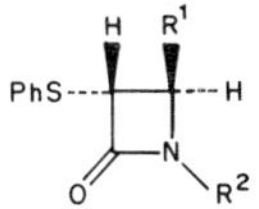 H, R^1, PhS, H, N, O, R^2		1211
	R^1 = Ph, R^2 = An-*p*			50–60	
	R^1 = R^2 = Ph			—	
	R^1 = Tol-*p*, C(Me)=CHPh, R^2 = An-*p*			—	
	R^1 = (3,4-methylenedioxyphenyl; O, O), R^2 = Ph			—	

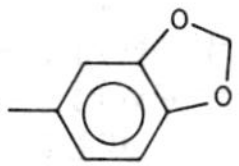

	$R^1 = $ Tol-*p*, $R^2 = $ An-*o*, C_6H_4Br-*p*			—	
$PhSCH_2CO_2$–(4,6-dichloro-1,3,5-triazin-2-yl) [p]	$HC(R^1)=C(R^2)–CH=N–$An-*p*	Et_3N, CH_2Cl_2, N_2, reflux 8–10 h	3-PhS-4-[$C(R^2)=CHR^1$]-1-(An-*p*)-azetidin-2-one		1153
	$R^1 = $ Me, $R^2 = $ H		*trans*	29	
	$R^1 = $ Ph, $R^2 = $ H		*trans*	32	
	$R^1 = $ Ph, $R^2 = $ Me		*cis*	37	
$MeO_2C–CH=C(Me)–NH–CH_2CO_2$–(4,6-dichloro-1,3,5-triazin-2-yl) [q]	$PhCH=CH–CH=N–CHCH(OH)Me(CO_2CH_2C_6H_4NO_2$-*p*)	1. Et_3N, CH_2Cl_2, N_2, $-20\,°C$ 2. r.t. stir overnight	3-[$MeO_2C–CH=C(Me)–NH$]-4-($CH=CHPh$)-1-[$CH(CO_2CH_2C_6H_4NO_2$-*p*)$CH(OH)Me$]-azetidin-2-one	40	1210
$MeO_2C–CH=C(Me)–NH–CH_2CO_2PCl_2$ [r]	$HC(C_6H_4OMe$-*p*$)=N–R^1$	Et_3N, CH_2Cl_2, r.t. 20–24 h	No Reaction	—	1198, 1204
	$R^1 = $ Ph or Tol-*p*				
$R^1CH_2CO_2\overset{+}{P}Ph_3Br^-$ [s,t]	$R^2CH{=}NR^3$	1. Et_3N, CH_2Cl_2, N_2, $-78\,°C$ stir 1 h 2. r.t. stir overnight	3-R^1-4-R^2-1-R^3-azetidin-2-one		1212, 1213
$R^1 = $ MeO	$R^2 = R^3 = $ Ph		*cis*	40	1213
$R^1 = $ MeO	$R^2 = $ An-*p*, $R^3 = $ Tol-*p*		*cis*	85	1212

(continued)

TABLE 54. (*continued*)

Carboxylic acid derivative	Imine	Conditions	Product	Yield (%)	Reference
$R^1 = PhCH_2O$	$R^2 = Ph$, $R^3 = Tol$-*p*		*cis*	85	1212
$R^1 = PhCH_2O$	$R^2 = Tol$-*p*, $R^3 = An$-*p*		*cis*	80	1212
$R^1 = PhO$	$R^2 = R^3 = Ph$		*trans*	50	1212
			cis	55	1213
$R^1 = PhO$	$R^2 = An$-*p*, $R^3 = CH_2CH_2OH$		*cis*	50	1213
$R^1 = PhO$	$R^2 = An$-*p*, $R^3 = CH_2CH(OH)Ph$		*cis*	55	1213
$R^1 = PhO$	$R^2 = An$-*p*, $R^3 = Tol$-*p*		*cis*	70	1212
$R^1 = An$-*p*	$R^2 = R^3 = Ph$		*trans*	46	1212
$R^1 = PhCH_2S$	$R^2 = An$-*p*, $R^3 = Tol$-*p*		*trans*	30	1212
$R^1 = PI—N$	$R^2 = R^3 = Ph$		*trans*	50	1213
$R^1 = PI—N$	$R^2 = An$-*p*, $R^3 = Ph$		*trans*	65	1213
$R^1 = PI—N$	$R^2 = Ph$, $R^3 = An$-*p*		*trans*	85	1212
$R^1 = PI—N$	$R^2 = An$-*p*, $R^3 = An$-*p*		*trans*	80	1212
$R^1 = PI—N$	$R^2 = C_6H_4NO_2$-*p*, $R^3 = Tol$-*p*		*trans*	70	1212
$R^1 = PhCH_2O_2CNH$	$R^2 = An$-*p*, $R^3 = Tol$-*p*		*cis*	50	1212
$MeOCH_2CO_2\overset{+}{P}Ph_3Br^{-s}$	PhC(SMe)=NPh	1. Et_3N, CH_2Cl_2, N_2, $-78\,^\circ C$, stir 1 h 2. r.t. stir overnight	MeO, H, SMe, Ph, N, Ph, O (β-lactam structure)	65	1212

$Cl_2CHCO_2\overset{+}{P}Ph_3Br^{-t}$	PhCH=NPh	Et_3N, CH_2Cl_2, r.t. stir 20–24 h	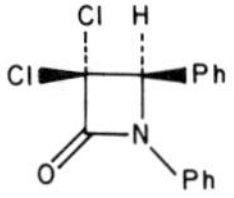	50	1213
$PhOCHCO_2\overset{+}{P}Ph_3Br^{-t}$	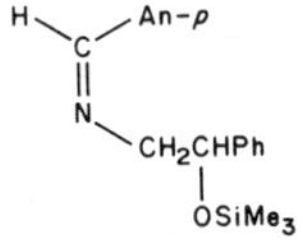	1. Et_3N, CH_2Cl_2, r.t. stir 20–24 h 2. H_2O	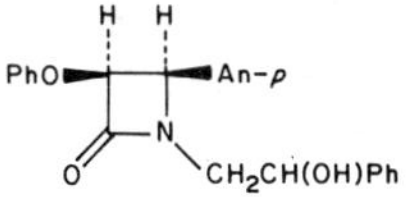	—	1213
$R_2CHCO_2\overset{+}{P}Ph_3Br^{-t}$	c-HexN=C=NHex-c	1. Et_3N, CH_2Cl_2, N_2, −78 °C stir 1 h 2. r.t. stir overnight	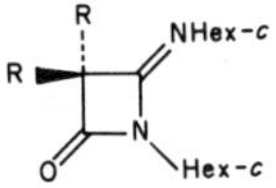		1213
R = Cl				45	
R = Ph				60	
$PhCH_2OCH_2CO_2\overset{+}{P}Ph_3Br^{-s}$	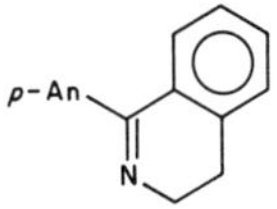	1. Et_3N, CH_2Cl_2, N_2, −78 °C stir 1 h 2. r.t. stir overnight	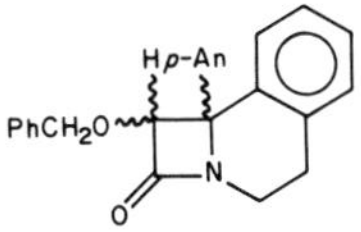	—	1212
$MeOCH_2CO_2\overset{+}{P}Ph_3Br^{-s}$	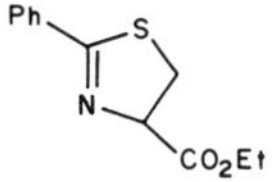	1. Et_3N, CH_2Cl_2, N_2, −78 °C stir 1 h 2. r.t. stir overnight	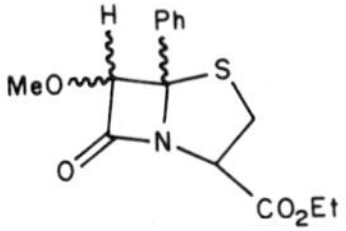	—	1212
$MeOCH_2CO_2\overset{+}{P}Ph_3Br^{-s}$	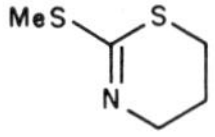	1. Et_3N, CH_2Cl_2, N_2, −78 °C stir 1 h 2. r.t. stir overnight	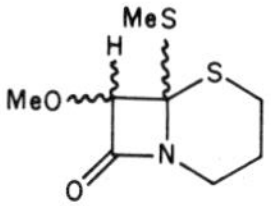	—	1212

(continued)

TABLE 54. (*continued*)

Carboxylic acid derivative	Imine	Conditions	Product	Yield (%)	Reference
$R^1CH_2CO_2\overset{+}{S}Me_2Br^{-}$[t]	PhCH=NPh	1. Et_3N, CH_2Cl_2, N_2, −78 °C stir 1 h 2. r.t. stir overnight	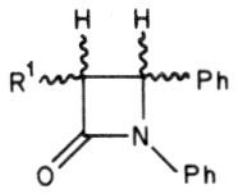 [β-lactam: R¹, H / H, Ph; N–Ph]	—	1213
R^1 = PhO			*cis*	25	
R^1 = PI—N			*trans*	20	
$Cl_2CHCO_2\overset{+}{S}Me_2Br^{-}$[u]	PhCH=NPh	1. Et_3N, CH_2Cl_2, N_2, −78 °C stir 1 h 2. r.t. stir overnight	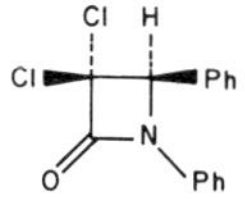 [β-lactam: Cl, Cl / H, Ph; N–Ph]	15	1213
$R^1CH_2CO_2POCl_2$[v]	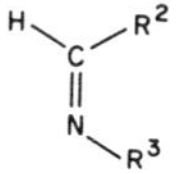 H(R²)C=N–R³	1. Et_3N, CH_2Cl_2, 10–15 °C 2. r.t. stir overnight	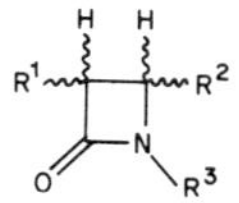 [β-lactam: R¹, H / H, R²; N–R³]	—	1214
R^1 = PhO	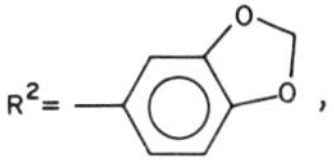 R^2 = 3,4-methylenedioxyphenyl, R^3 = An-*p*, Tol-*p*, $PhCH_2$		*cis*	—	
R^1 = PhO	$R^2 = R^3$ = Ph		*cis*/*trans*	—	
R^1 = PhO	R^2 = An-*p*, R^3 = Tol-*p* EtO_2CCHPh		*cis*	—	
R^1 = MeO	R^2 = An-*p*, R^3 = Ph_2CH, Tol-*p*		*cis*	—	

$PhCH_2O_2CNHCH_2CO_2POCl_2$ [v]	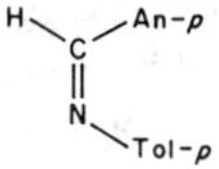 H(An-p)C=N–Tol-p	DMF, stir r.t. 1.5h	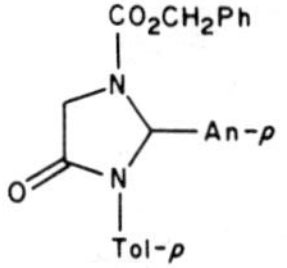 1-(CO_2CH_2Ph)-2-(An-p)-3-(Tol-p)-imidazolidin-4-one	59	1180
$Cl_2CHCO_2POCl_2$ [v]	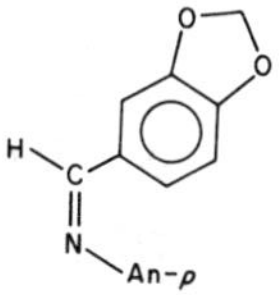 3,4-methylenedioxyphenyl–CH=N–An-p	1. Et_3N, CH_2Cl_2, 10–15 °C 2. r.t. stir overnight	No Reaction	—	1214
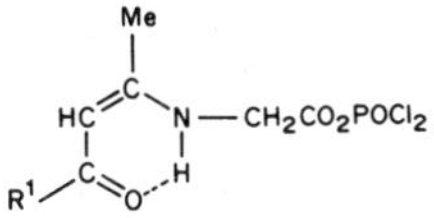 R^1–C(=O)–CH=C(Me)–N(H)–$CH_2CO_2POCl_2$	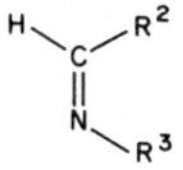 H(R^2)C=N–R^3	Et_3N, CH_2Cl_2, r.t., 20–24h	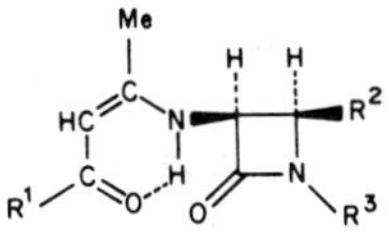 3-[R^1C(=O)CH=C(Me)NH]-4-R^2-1-R^3-azetidin-2-one (cis)		1143, 1159, 1198, 1215
$R^1 = Me$	$R^2 = An$-p, $R^3 = Tol$-p			35	1215
$R^1 = MeO$	$R^2 = R^3 = Ph$			16	1198
$R^1 = MeO$	$R^2 = An$-p, $R^3 = Ph$			22	1198
$R^1 = MeO$	$R^2 = An$-p, $R^3 = Tol$-p			45	1198
$R^1 = EtO$	$R^2 = R^3 = Ph$			—	1215
$R^1 = EtO$	$R^2 = An$-p, $R^3 = Tol$-p			—	1215
$R^1 = EtO$	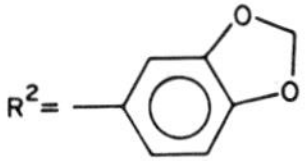 R^2 = 3,4-methylenedioxyphenyl, $R^3 = Tol$-p			—	1159, 1215
$R^1 = EtO$	$R^2 = C_6H_4NO_2$-p, $R^3 = PhCH_2$			63	1143

(continued)

TABLE 54. (*continued*)

Carboxylic acid derivative	Imine	Conditions	Product	Yield (%)	Reference
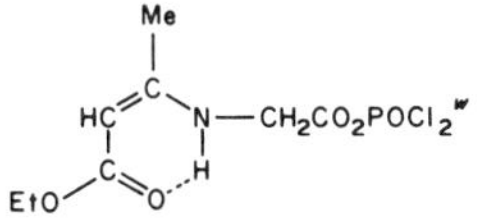 [structure: EtO–C(=O)–CH=C(Me)–NH–$CH_2CO_2POCl_2$][w]	$Ph_2C{=}NCH_2Ph$	Et_3N, CH_2Cl_2, reflux 1 h	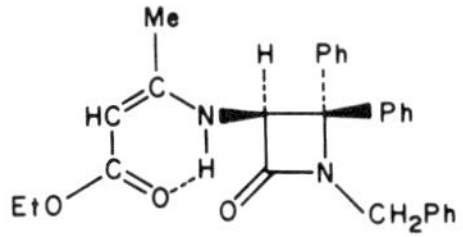 [structure]	63	1143
	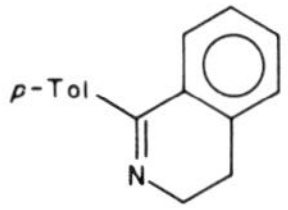 [structure: p-Tol dihydroisoquinoline]	Et_3N, CH_2Cl_2, reflux	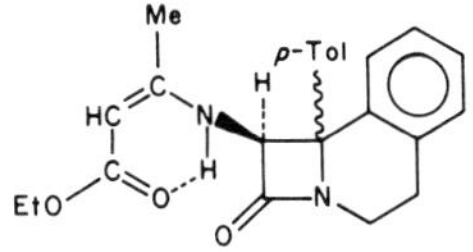 [structure]	45	1215
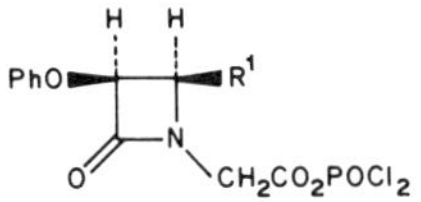 [structure: PhO, R[1] azetidinone N–$CH_2CO_2POCl_2$]	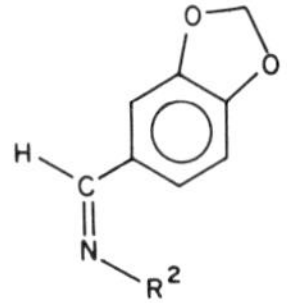 [structure: H–C(=N–R^2)–benzodioxolyl]	Et_3N	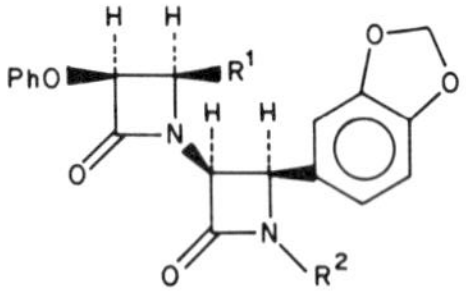 [structure]		1159
R^1 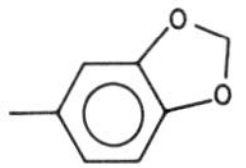 = [3,4-methylenedioxyphenyl]	$R^2 = $ An-*p*			55	
$R^1 = 3,4\text{-}(MeO)_2C_6H_3$	$R^2 = $ Tol-*p*			—	
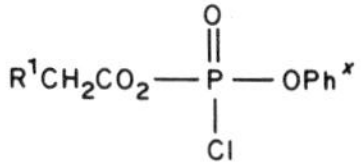 $R^1CH_2CO_2$–P(O)(Cl)–OPh[x]	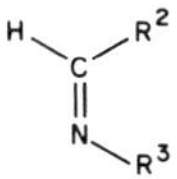 H–C(R^2)=N–R^3	Et_3N, CH_2Cl_2, r.t., stir 24h	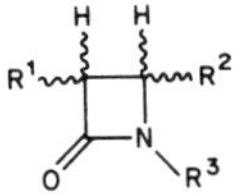 [structure: R^1, R^2 β-lactam, N–R^3]		1198, 1204
$R^1 = MeO$	$R^2 = R^3 = Ph$		*cis*	30	1198
$R^1 = PhO$	$R^2 = R^3 = Ph$		*cis*	75	1198
$R^1 = PhO$	$R^2 = $ An-*p*, $R^3 = Ph_2CH$		*cis*	84	1198

$R^1 = PhO$	$R^2 = An\text{-}p$, $R^3 = Tol\text{-}p$		*cis*	65	1198, 1204
$R^1 = PhCH_2O$	$R^2 = An\text{-}p$, $R^3 = Tol\text{-}p$		*cis*	70	1198
$R^1 = MeCO_2$	$R^2 = An\text{-}p$, $R^3 = Tol\text{-}p$		*cis*	50	1198
$R^1 = PI—N$	$R^2 = An\text{-}p$, $R^3 = Tol\text{-}p$		*trans*	70	1198, 1204
$PI—NCH_2CO_2—P(=O)(Cl)—OPh$ [x]	$H(R^2)C=N—R^3$	Et_3N, CH_2Cl_2	PI—N~(H)C–C(H)~R^2 (β-lactam: C=O, N–R^3)		1216
	$R^2 = R^3 = Ph$			66	
	$R^2 = An\text{-}p$, $R^3 = Ph$			60	
	$R^2 = An\text{-}p$, $R^3 = \alpha$-naphthyl			73	
	$R^2 = Me_2NPyr\text{-}4$, $R^3 = Ph$			50	
$\{PI—NCH_2CO_2CH=\overset{+}{N}Me_2$ Cl^- or $PhO—P(=O)(Cl)—O^-\}$ [y]	$H(R^2)C=N—R^3$	Et_3N, CH_2Cl_2, DMF	PI—N~(H)C–C(H)~R^2 (β-lactam: C=O, N–R^3)		1216
	$R^2 = R^3 = Ph$			60	
	$R^2 = An\text{-}p$, $R^3 = Ph$			75	
	$R^2 = An\text{-}p$, $R^3 = \alpha$-naphthyl			70	
	$R^2 = Me_2NPyr\text{-}4$, $R^3 = Ph$			75	

(*continued*)

TABLE 54. (*continued*)

Carboxylic acid derivative	Imine	Conditions	Product	Yield (%)	Reference
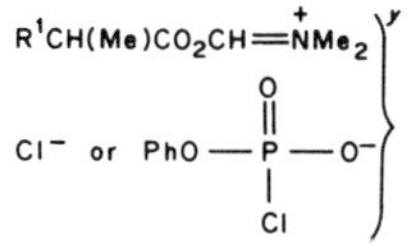 $R^1CH(Me)CO_2CH{=}\overset{+}{N}Me_2$ Cl^- or $PhO{-}P(=O)(Cl){-}O^-$ [y]	PhCH=NPh	Et_3N, CH_2Cl_2, DMF	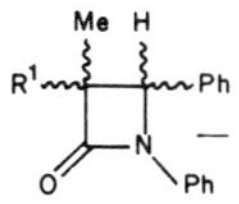 β-lactam: R^1, Me at C-3; H, Ph at C-4; N-Ph		1216
$R^1 = Br$				66	
$R^1 = PI{-}N$				51	
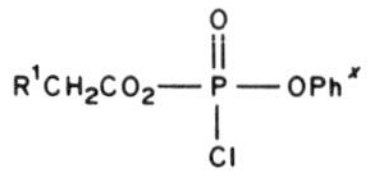 $R^1CH_2CO_2{-}P(=O)(Cl){-}OPh$ [x]	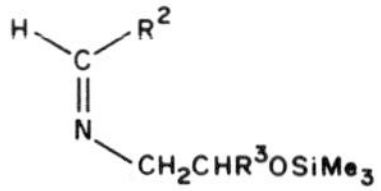 $H(R^2)C{=}N{-}CH_2CHR^3OSiMe_3$	1. Et_3N, CH_2Cl_2, r.t. stir 24h 2. H_2O, stir 15 min	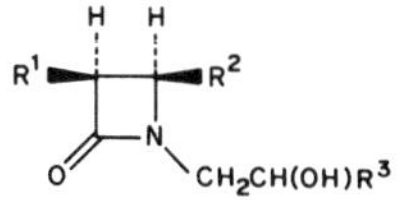 β-lactam: R^1, H at C-3; R^2, H at C-4; $N{-}CH_2CH(OH)R^3$		1198, 1201, 1204, 1217, 1218
$R^1 = Ph$	$R^2 = R^3 = Ph$			—	1217
$R^1 = Ph$	$R^2 = An$-*p*, $R^3 = Ph$			—	1217
$R^1 = PhO$	$R^2 = An$-*p*, $R^3 = H$			70	1198, 1204
$R^1 = PhO$	$R^2 = Ph$, $R^3 = Me$			75	1201
$R^1 = PhO$	$R^2 = An$-*p*, $R^3 = Me$			65	1201
$R^1 = PhO$	$R^2 = R^3 = Ph$			86	1198
$R^1 = PhO$	$R^2 = An$-*p*, $R^3 = Ph$			91	1198, 1204, 1218

$R^1 = PhO$	$R^2 = C_6H_4NO_2$-*p*, $R^3 = Ph$			80	1198
$R^1 = PI—N$	$R^2 =$ An-*p*, $R^3 = Ph$			60	1198, 1217
$R^1CH_2CO_2—P(=O)(Cl)—OPh^x$	$H(R^2N=)C—C(Me)=CHPh$	Et_3N, CH_2Cl_2, r.t. stir 20–24h	3-R^1-4-[C(Me)=CHPh]-1-R^2-azetidin-2-one (H, H)		1144
$R^1 = PhO$	$R^2 = MeO_2CCH_2$			—	
$R^1 = PI—N$	$R^2 = MeO_2CCH_2$			66	
$R^1 = PhO$	$R^2 = Me_3SiOCHPhCH_2$			—	
$R^1 = PI—N$	$R^2 = Me_3SiOCHPhCH_2$			—	
$Me_3SiOCH_2CO_2—P(=O)(Cl)—OPh^x$	$HC(R^1)=NR^2$	1. Et_3N, CH_2Cl_2, r.t. stir 20–24h 2. H_2O	3-HO-4-R^1-1-R^2-azetidin-2-one (H, H)		1218
	$R^1 = R^2 = Ph$		*cis*+ *trans*	35 10	
	$R^1 = Ph$, $R^2 =$ An-*p*		*cis*+ *trans*	41 8	
	$R^1 =$ An-*p*, $R^2 =$ Tol-*p*		*cis*+ *trans*	55 6.5	
	$R^1 = R^2 =$ An-*p*		*cis*	65	
	$R^1 =$ An-*p*, $R^2 = CH_2CHPhOSiMe_3$		*cis* ($R^2 = CH_2CH(OH)Ph$)	50	
$PhCH(OSiMe_3)CO_2—P(=O)(Cl)—OPh^x$	$HC($An-*p*$)=NR^1$	Et_3N, CH_2Cl_2, r.t. stir 20–24h	3-Me_3SiO-3-Ph-4-(An-*p*)-1-R^1-azetidin-2-one (H)		1218

(*continued*)

TABLE 54. (*continued*)

Carboxylic acid derivative	Imine	Conditions	Product	Yield (%)	Reference
	$R^1 = Ph$			82	
	$R^1 = An$-*p*			71	
	$R^1 = Ph_2CH$			55	
MeO–C(=O···H)–CH=C(Me)–N(H)–CH_2CO_2–P(=O)(Cl)–OPh[z]	H–C(R^1)=N–R^2	1. Et_3N, CH_2Cl_2, 0 °C 2. r.t. stir overnight	[β-lactam: N-(MeO–C(=O···H)–CH=C(Me)–NH)-3-substituted azetidin-2-one, 4-R^1, 1-R^2]		1144, 1198, 1204
	$R^1 = R^2 = Ph$			50	1198, 1204
	$R^1 = PhCH{=}CMe$ (*trans*), $R^2 = Ph$			55	1198, 1204
	$R^1 = Ph$, $R^2 = 3,4\text{-}(MeO)_2C_6H_3CH_2$			70	1198
	$R^1 = Ph$, $R^2 = 2,4\text{-}(MeO)_2C_6H_3CH_2$			50	1198
	$R^1 = An$-*p*, $R^2 = Ph$			55, 50	1198, 1204
	$R^1 = An$-*p*, $R^2 = Tol$-*p*			65, 62	1198, 1204
	$R^1 = PhCH{=}CMe$(*cis*), $R^2 = An$-*p*			50	1144
MeO–C(=O···H)–CH=C(Me)–N(H)–CH_2CO_2–P(=O)(Cl)–OPh[y]	H–C(R^1)=N–CH_2CHPh(OSiMe$_3$) $R^1 = Ph$ and An-*p*	1. Et_3N, CH_2Cl_2 2. H_2O	[β-lactam: N-(MeO–C(=O···H)–CH=C(Me)–NH)-3-substituted azetidin-2-one, 4-R^1, 1-$CH_2CH(OH)Ph$]	—	1217

$R^1CH_2CO_2P(O)(OR^2)_2$	$R^3CH{=}NR^4$	Et_3N, CH_2Cl_2, r.t. stir overnight	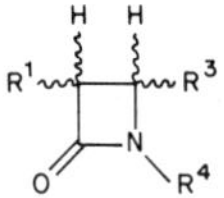 [β-lactam: H, H; R^1, R^3; N–R^4]		1219
$R^1 = Ph$, $R^2 = Et$[aa]	$R^3CH = NR^4$		*trans*	75	
$R^1 = R^2 = Ph$[ab]	$R^3 = R^4 = Ph$		*trans*	75	
$R^1 = Ph$, $(OR^2)_2$ 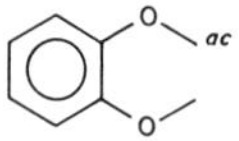 = [catechol O,O-diyl][ac]	$R^3 = R^4 = Ph$		*trans*	75	
$R^1 = MeO$, $R^2 = Ph$[ab]	$R^3 = R^4 = Ph$		*cis*:*trans* = 85:15	70	
$R^1 = N_3$, $R^2 = Et$[aa]	$R^3 = R^4 = Ph$		*cis*:*trans* = 9:1	65	
$R^1 = N_3$, $R^2 = Ph$[ab]	$R^3 = R^4 = Ph$		*cis*:*trans* = 9:1	65	
$R^1 = PhCH_2S$, $R^2 = Ph$[ab]	$R^3 = R^4 = Ph$		*trans*	80	
$R^1 = Br$, $R^2 = Et$[aa]	$R^3 = R^4 = Ph$		*trans*	50	
$R^1 = N_3$, $R^2 = Et$[aa]	$R^3 = C_6H_4NO_2$-*p*, R^4 = Tol-*p*		*cis*	60	
$R^1 = N_3$, $R^2 = Et$[aa]	R^3 = 2-Fu, R^4 = Tol-*p*		*cis*	50	
$R^1 = N_3$, $R^2 = Et$[aa]	$R^3 = C_6H_4NO_2$-*p*, $R^4 = Ph$		*cis*	60	
R^1 = PI—N, $R^2 = Et$[aa]	$R^3 = C_6H_4NO_2$-*p*, R^4 = Tol-*p*		*trans*	70	
$ROCH_2CO_2P(O)(OEt)_2$[aa]	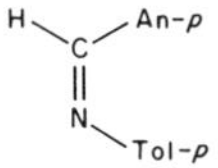 H(An-*p*)C=N–Tol-*p*	Et_3N, CH_2Cl_2, r.t. stir overnight	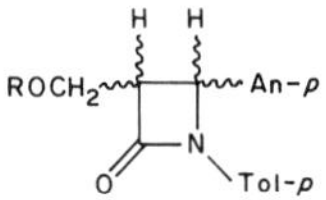 [β-lactam: H, H; $ROCH_2$, An-*p*; N–Tol-*p*]		902
$R = PhCH_2$, *t*-Bu				—	
t-$BuOCH_2CO_2P(O)(OEt)_2$[aa]	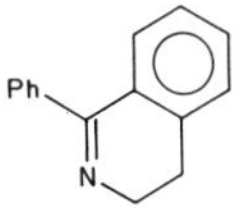 [1-Ph-3,4-dihydroisoquinoline]	Et_3N, CH_2Cl_2, r.t. stir overnight	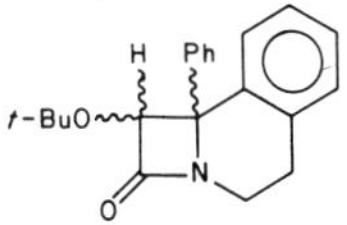 [fused β-lactam: H, Ph; *t*-BuO]	—	902

(*continued*)

TABLE 54. (*continued*)

Carboxylic acid derivative	Imine	Conditions	Product	Yield (%)	Reference
$R^1CH_2CO_2P(O)(OCH_2CCl_3)_2$[ad]	$R^2CH{=}NR^3$	Et_3N, CH_2Cl_2, r.t. stir 48h	[β-lactam: R¹, H, H, R², N–R³, C=O]		1220
R^1 = MeO	$R^2 = R^3$ = Ph		*cis*	43	
R^1 = PhO	$R^2 = R^3$ = Ph		*cis*	59	
R^1 = PhO	R^2 = Ph, $R^3 = PhCH_2$		*cis*	46	
R^1 = PhO	$R^2 = o\text{-}O_2NC_6H_4CH{=}CH$, R^3 = Ph		*cis*	52	
R^1 = PI—N	$R^2 = o\text{-}O_2NC_6H_4CH{=}CH$, $R^3 = PhCH_2$		*cis*	44	
R^1 = PI—N	$R^2 = R^3$ = Ph		*trans*	58	
$R^1CH_2CO_2P(O)(OPh)_2$[ab]	[1-R²-3,4-dihydroisoquinoline]	Et_3N, CH_2Cl_2, r.t. stir overnight	[fused β-lactam: R¹, H, R², N, O]	50–60	1219
$R^1 = N_3$	R^2 = Ph, *p*-Tol				
R^1 = PI—N	R^2 = An-*p*				
$R^1CH_2CO_2P(O)(OPh)_2$[ab]	[2-R²-5,6-dihydro-4H-1,3-thiazine]	Et_3N, CH_2Cl_2, r.t. stir overnight	[fused β-lactam: R¹, H, R², S, N, O]	50–60	1219
R^1 = MeO	R^2 = MeS				
R^1 = PhO	R^2 = MeS				
$R^1 = N_3$	R^2 = MeS				
$R^1 = N_3$	$R^2 = C_6H_4NO_2\text{-}p$				
R^1 = PI—N	$R^2 = C_6H_4NO_2\text{-}p$				

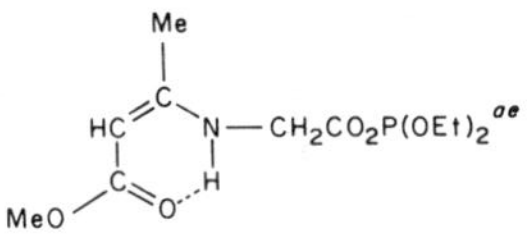 $MeOC(=O)CH=C(Me)NH-CH_2CO_2P(OEt)_2$ [ae]	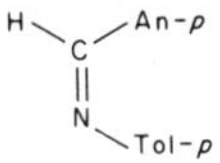 $H-C(An\text{-}p)=N-Tol\text{-}p$	Et_3N, CH_2Cl_2, r.t. stir	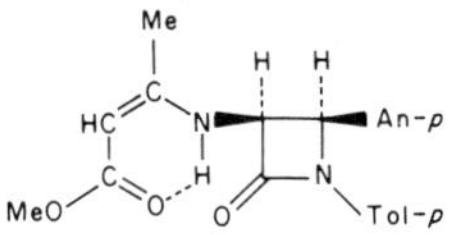 (β-lactam: N-(MeO₂C-CH=C(Me))-3-amino-4-An-p-1-Tol-p-azetidin-2-one)	25, 20	1198, 1204
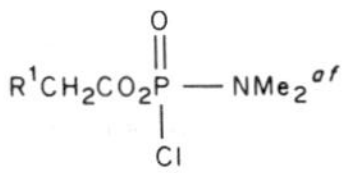 $R^1CH_2CO_2P(O)(Cl)NMe_2$ [af]	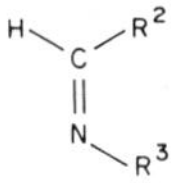 $H-C(R^2)=N-R^3$	Et_3N, CH_2Cl_2, r.t. stir 20–24h	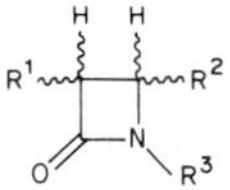 (β-lactam: 3-R^1-4-R^2-1-R^3-azetidin-2-one)		1221
$R^1 = MeO$	$R^2 = PhCH{=}CMe$, $R^3 = CH_2CO_2Me$		*cis*	55	
$R^1 = PhO$	$R^2 = 5\text{-Me-2-Fu}$, $R^3 = An\text{-}p$		*cis*	50	
$R^1 = PhO$	$R^2 = PhCH{=}CMe$, $R^3 = CH_2CO_2Me$		*cis*	60	
$R^1 = PhCH_2S$	$R^2 = R^3 = Ph$		*trans*	55	
$R^1 = PhS$	$R^2 = R^3 = Ph$		*trans*	12[ag]	
$R^1 = PhS$	$R^2 = Ph$, $R^3 = An\text{-}p$		*trans*	15	
$R^1 = Ph$	$R^2 = R^3 = Ph$		*trans*	15[ah]	
$R^1 = An\text{-}p$	$R^2 = R^3 = Ph$		*trans*	46[ai]	
$R^1 = 3,4\text{-}(MeO)_2C_6H_3$	$R^2 = R^3 = Ph$		*trans*	45	
$R^1 = MeCO_2$	$R^2 = Tol\text{-}p$, $R^3 = An\text{-}p$		*cis*	50	
$R^1 = PI{-}N$	$R^2 = PhCH{=}CH$, $R^3 = An\text{-}p$		*cis*	55	
$R^1 = PI{-}N$	$R^2 = PhCH{=}CMe$, $R^3 = CH_2CO_2Me$		*cis*	50	
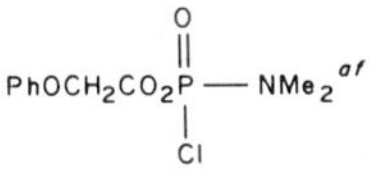 $PhOCH_2CO_2P(O)(Cl)NMe_2$ [af]	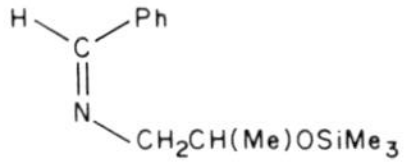 $H-C(Ph)=N-CH_2CH(Me)OSiMe_3$	1. Et_3N, CH_2Cl_2, r.t. stir 20–24h 2. H_2O	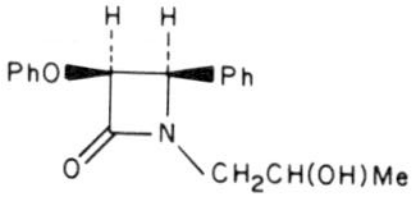 (β-lactam: 3-PhO-4-Ph-1-$CH_2CH(OH)Me$-azetidin-2-one)	40	1221

(continued)

TABLE 54. (*continued*)

Carboxylic acid derivative	Imine	Conditions	Product	Yield (%)	Reference
MeO–C(=O)–CH=C(Me)–N(H)–CH_2CO_2–P(=O)(Cl)–NMe_2 [aj]	H–C(R^1)=N–R^2	Et_3N, CH_2Cl_2, r.t. stir 20–24h	3-[MeO–C(=O)–CH=C(Me)–NH]-4-R^1-1-R^2-azetidin-2-one		1221
	R^1 = An-*p*, R^2 = Tol-*p*			40	
	R^1 = Ph, R^2 = 2,4-$(MeO)_2C_6H_3CH_2$			—	
	R^1 = MeCO, R^2 = CH_2CO_2Me			—	
MeO–C(=O)–CH=C(Me)–N(H)–CH_2CO_2–P(=O)(Cl)–NMe_2 [aj]	H–C(C(Me)=CHPh)=N–$CH_2CHPhOSiMe_3$	1. Et_3N, CH_2Cl_2, r.t. stir 20–24h 2. H_2O	3-[MeO–C(=O)–CH=C(Me)–NH]-4-C(Me)=CHPh-1-$CH_2CH(OH)Ph$-azetidin-2-one	—	1221
$(EtS)_2CHCO_2$P(=O)(Cl)–NMe_2 [af]	H–C(R^1)=N–R^2	1. Et_3N, CH_2Cl_2, 0 °C, 2. r.t. stir 24h	3,3-(EtS)$_2$-4-R^1-1-R^2-azetidin-2-one	—	1221
	R^1 = R^2 = Ph			90	
	R^1 = Ph, R^2 = An-*p*			80	
	R^1 = PhCH=CH, R^2 = An-*p*			85	
	R^1 = PhCH=CMe, R^2 = CH_2CO_2Me			65	

The structure drawings in this table are written below as linear formulas.

	R^1 = 3,4-(methylenedioxy)phenyl, $R^2 = Ph$			75	
$MeO{-}C(=O)CH{=}C(Me){-}NH{-}CH_2CO_2{-}P(O)(OPh){-}NHPh$ [ak]	$H{-}C(An\text{-}p){=}N{-}R$ $R = Ph$ or Tol-p	Et_3N, CH_2Cl_2, r.t. stir	No reaction	—	1198, 1204
$R^1CH_2CO_2P(O)(OPh){-}N(Me)Ph$ [al]	$H{-}C(R^2){=}N{-}R^3$	Et_3N, CH_2Cl_2, r.t. stir overnight	3-R^1-4-R^2-1-R^3-azetidin-2-one (H at C-3 and C-4, wavy bonds)		1222
$R^1 = PhCH_2O$	$R^2 = Ph$, $R^3 = C_6H_4Cl$-p			44	
$R^1 = PhO$	$R^2 =$ An-p, $R^3 = C_6H_4Cl$-p			60	
$R^1 = PhO$	$R^2 = C_6H_4OCH_2Ph$-p, $R^3 = C_6H_4Cl$-p			26	
$R^1 = PhO$	$R^2 = Ph$, $R^3 = C_6H_4Cl$-p			46	
$R^1 =$ PI-N	$R^2 = Ph$, $R^3 = C_6H_4Cl$-p			66	
$R^1 =$ PI-N	$R^2 = C_6H_4OMe$-p, $R^3 = C_6H_4Cl$-p			78	

(continued)

TABLE 54. (*continued*)

Carboxylic acid derivative	Imine	Conditions	Product	Yield (%)	Reference
$R^1 = PI—N$	$R^2 = Ph$, $R^3 = CH_2CO_2Et$			76	
$PI—NCH_2CO_2P(O)(OPh)—N(Me)Ph$ [al]	cyclohexylidene=NPh	Et_3N, CH_2Cl_2, r.t. stir overnight		41	1222
$R^1CH_2CO_2P(O)(OPh)—N(Me)Ph$ [al]	1-Ph-3,4-dihydroisoquinoline	Et_3N, CH_2Cl_2, r.t. stir overnight			1222
	$R^1 = PhO$			41	
	$R^1 = PI-N$			70	
$R^1CH_2CO_2P(O)(N\text{-oxazolidinone})_2$ [am]	$H(R^2)C{=}NR^3$	Et_3N, CH_2Cl_2, N_2, 10–20°C, 7–8 h			1223
$R^1 = PhCH_2O$	$R^2 = Ph$, $R^3 = C_6H_4Cl$-*p*		*cis*	50	
$R^1 = PhCH_2O$	$R^2 = C_6H_4OCH_2Ph$-*p*, $R^3 = C_6H_4Cl$-*p*		*cis*	50	
$R^1 = PhCH_2O$	$R^2 = C_6H_4Cl$-*p*, $R^3 =$ An-*p*		*cis*	45	
$R^1 = PhCH_2O$	$R^2 =$ An-*p*, $R^3 =$ Tol-*p*		*cis*	45	
$R^1 = PhO$	$R^2 = Ph$, $R^3 = C_6H_4Cl$-*p*		*cis*	55	
$R^1 = PhO$	$R^2 = R^3 = C_6H_4Cl$-*p*		*cis*	40	
$R^1 = PI—N$	$R^2 = Ph$, $R^3 = C_6H_4$-*p*		*trans*	65	

$R^1CH_2CO_2P(O)$–(N-oxazolidin-2-one)$_2$ [am]	Ph (1-phenyl-3,4-dihydroisoquinoline)	Et_3N, CH_2Cl_2, N_2, 10–20°C, 7–8 h	R^1, Ph, O, N (β-lactam fused to tetrahydroisoquinoline)		1223
$R^1 = PhCH_2O$				60	
$R^1 = PI\text{-}N$				80	
Me, HC, C, N—$CH_2CO_2P(OR^1)_2$, MeO, C, O, H	H, C, R^2, N, R^3	Et_3N	Me, HC, C, N, H, MeO, C, O, H, H, H, R^2, O, N, R^3		1178, 1198, 1204
$R^1 = Et$ [ae]	$R^2 = R^3 = Tol\text{-}p$	Et_2O, N_2		30	1178
$R^1 = Ph$ [an]	$R^2 = R^3 = Tol\text{-}p$	Et_2O, N_2		20	1178
$R^1 =$ (O, N—, O oxazolidin-2-one) [ao]	$R^2 = An\text{-}p$, $R^3 = Ph$	CH_2Cl_2, r.t. stir		30	1198
$R^1 =$ (O, N—, O oxazolidin-2-one) [ao]	$R^2 = An\text{-}p$, $R^3 = Tol\text{-}p$	CH_2Cl_2, r.t. stir		30, 10	1198, 1204

[a] The ratio of *cis*-diastereomers was 97:3 and the minor diastereomer of the *trans*-β-lactam was not detected. The product had 94% asymmetric induction.
[b] The ratio of *cis*-diastereomers was 98:2 and the minor diastereomer of the *trans*-β-lactam was not detected. The product had 96% asymmetric induction.
[c] The ratio of *cis*-diastereomers was 95:5 and the minor diastereomer of the *trans*-β-lactam was not detected. The product had 90% asymmetric induction.
[d] The ratio of *trans*-diastereomers was 84:16 and the product had 68% asymmetric induction.
[e] The ratio of *trans*-diastereomers was 87:13 and the product had 74% asymmetric induction.
[f] Dane salt activated using thionyl chloride in DMF.
[g] The reagent was prepared using $Me_2\overset{+}{N}{=}CHCl\ Cl^-$ and methoxyacetic acid.
[h] The triethylamine was added at −13°C.
[i] The 'activated' acid was added to a solution of the imine and triethylamine.
[j] The imine and triethylamine were added to a solution of the 'activated' carboxylic acid.
[k] Acid 'activated' with saccharyl chloride.
[l] Prepared from the substituted acetic acid and ethyl chloroformate.
[m] Prepared from the Dane salt and ethyl chloroformate.
[n] Prepared from the substituted acetic acid and 2-chloro-*N*-methylpyridinium iodide.

(*continued*)

TABLE 54. (*continued*)

[o] Prepared from the Dane salt and 2-chloro-*N*-methylpyridinium iodide.
[p] Prepared from the substituted acetic acid and cyanuric chloride.
[q] Prepared from the Dane salt and cyanuric chloride.
[r] Prepared from the Dane salt and phosphorus trichloride.
[s] Prepared by reacting the substituted acetic acid with the phosphonium salt $Ph_3\overset{+}{P}CBr_3\ Br^-$, prepared from triphenylphosphine and carbon tetrabromide.
[t] Prepared by reacting the substituted acetic acid with triphenyl phosphine dibromide.
[u] Prepared by reacting the substituted acetic acid with dimethyl sulphide dibromide.
[v] Prepared from the substituted acetic acid and phosphorus oxychloride.
[w] Prepared from the Dane salt and phosphous oxychloride.
[x] Prepared from the substituted acetic acid and phenyl dichlorophosphate.
[y] Prepared by the reaction of the phosphoric carboxylic mixed anhydride with dimethylformamide.
[z] Prepared from the Dane salt and phenyl dichlorophosphate.
[aa] Prepared from the substituted acetic acid and diethyl phosphorochloridate.
[ab] Prepared from the substituted acetic acid and diphenyl phosphorochloridate.
[ac] Prepared from the substituted acetic acid and *o*-phenylenephosphorochloridate.
[ad] Prepared from the substituted acetic acid and bis(2,2,2-trichloroethyl)phosphorochloridate.
[ae] Prepared from the Dane salt and diethyl phosphorochloridate.
[af] Prepared from the substituted acetic acid and *N*,*N*-dimethylphosphoramidic dichloride.
[ag] Yield increased to 15% by refluxing the reaction mixture.
[ah] Yield increased to 30% by refluxing the reaction mixture.
[ai] Yield increased to 50% by refluxing the reaction mixture.
[aj] Prepared from the Dane salt and *N*,*N*-dimethylphosphoramidic dichloride.
[ak] Prepared from the Dane salt and phenyl *N*-phenylphosphoramido chloridate.
[al] Prepared from the substituted acetic acid and *N*-methyl-*N*-phenylphosphoramido chloridate.
[am] Prepared from the substituted acetic acid and *N*,*N*-bis (2-oxo-3-oxazolidinyl)phosphorodiamidic chloride.
[an] Prepared from the Dane salt and diphenyl phosphorochloridate.
[ao] Prepared from the Dane salt and *N*,*N*-bis(2-oxo-3-oxazolidinyl)phosphorodiamidic chloride.

mediate formation of ketene, but instead involves substitution of the acid halide by the imine to produce an imine salt, which then undergoes proton abstraction by attack of the tertiary amine base on the C_3 proton (Scheme 12). The two zwitterions produced, then upon conrotatory ring closure, lead to the corresponding two streochemically distinct products possible for this reaction. In cases where both stereoisomers can be obtained, the preference for one over the other appears to be controlled by the steric interaction present between R^1 and R^3 or R^2 and R^3 in the zwitterions shown in Scheme 12. The result obtained using this approach to the synthesis of β-lactams are reported in Table 53.

Although the reactions of imines with anhydrides or 'activated' carboxylic acids or their salts are viable methods for the preparation of β-lactams, these approaches have found only limited use. The reagents used in these procedures are usually: mixed anhydrides made from two carboxylic acid components or a carboxylic and sulphonic acid component; a carboxylic acid 'activated' with cyanuric chloride or a chlorophosphate reagent; or Dane salts in the presence of a variety of 'activating agents' (equation 795). The results obtained using all of these approaches are summarized in Table 54.

$R^1CH_2CO_2COR^2$

$R^1CH_2CO_2SO_2R^2$

$R^1CH_2CO_2H$ + cyanuric chloride (2,4,6-trichloro-1,3,5-triazine)

$R^1CH_2CO_2H + R^2_2POCl$

$R^1CH_2CO_2^-K^+$ + $ClCO_2R$, $ClPOR^2_2$, or $(CF_3CO)_2O$

$$\xrightarrow[\text{conditions}]{R^3R^4C{=}NR^5}$$ β-lactam bearing R^1, R^3, R^4 and N–R^5 (795)

**b. Reformatsky reaction with imines.* Whereas the imine reactions discussed up to this point have been used to prepare β-lactams almost exclusively, the Reformatsky reaction with imines offers a method to prepare 4-, 5- or 6- membered lactams in usually good yields. The structures of the lactams produced from these reactions are solely dependent upon the structure of the imine used.

Reaction of ethyl α-bromoisobutyrate under Reformatsky conditions with carbodiimides produces[1224] the corresponding β-lactams in 51 to 81% yields (equation 796).

Ultrasound has been reported[1225] to promote the reaction between ethyl bromoacetate and imines under Reformatsky conditions to the extent that excellent yields of β-lactams were obtained in a few hours at room temperature (equation 797). Interestingly the substitution of α-bromopropionic or α-bromo-β-phenylpropionic esters for ethyl bromoacetate failed to produce any lactam product.

The Reformatsky reaction of ethyl (α-bromomethyl)acrylates with nitriles has been reported[794] to produce α-methylene-γ-lactams in good to excellent yields according to the mechanism shown in equation (798).

$R^1N{=}C{=}NR^2 + Me_2CBrCO_2Et \xrightarrow[\text{reflux}]{Zn,\ C_6H_5Me}$ (126) and/or (127) (796)

R^1	R^2	Yield (%) 126	Yield (%) 127
c-Hex	*c*-Hex	81	—
p-Tol	*p*-Tol	51	—
p-An	*p*-An	*a*	—
p-$O_2NC_6H_4$	*p*-$O_2NC_6H_4$	0	0
Ph	*n*-Bu	54	23
Ph	*i*-Pr	69	19
Ph	*t*-Bu	78[b]	—

[a]The main product was:

[b]Also formed:

$BrCH_2CO_2Et + R^1CH{=}NR^2 \xrightarrow[25\ ^\circ C,\ \text{ultrasound}]{Zn(I),\ \text{dioxane}}$ (797)

R^1	R^2	Time (h)	Yield (%)
Ph	Ph	5	70
Ph	*p*-An	5	82
p-Tol	*p*-An	4	95
		10	85
p-ClC_6H_4	*p*-An	6	77

$$BrCH_2C(=CH_2)CO_2Et + RC{\equiv}N \xrightarrow[\text{Reflux 2 h}]{\text{Zn, THF}} \left[R{-}C(=NZnBr)CH_2C(=CH_2)CO_2Et\right] \xrightarrow{BrZnCH_2C(=CH_2)CO_2Et}$$

$$\xrightarrow{-EtOZnBr} \qquad (798)$$

R	Yield (%)
Me	75.3
Et	96
i-Pr	99
n-Bu	96
$N{\equiv}CCH_2CH_2$	60
cyclopropyl	66
Ph	97
3,4,5-$(MeO)_3C_6H_2$	97.3

Fused ring γ-lactams are produced[1226] by the reaction of ethyl α-bromo-α- methylpropanoate with both cyanoesters (equation 799) and cyclic imides (equation 800).

$$\xrightarrow[\text{Zn, PhOMe, HgCl}_2]{Me_2CBrCOOEt} \quad 71\%$$

$$\xrightarrow[\text{Zn, PhOMe, HgCl}_2]{Me_2CBrCOOEt} \quad 83\% \qquad (799)$$

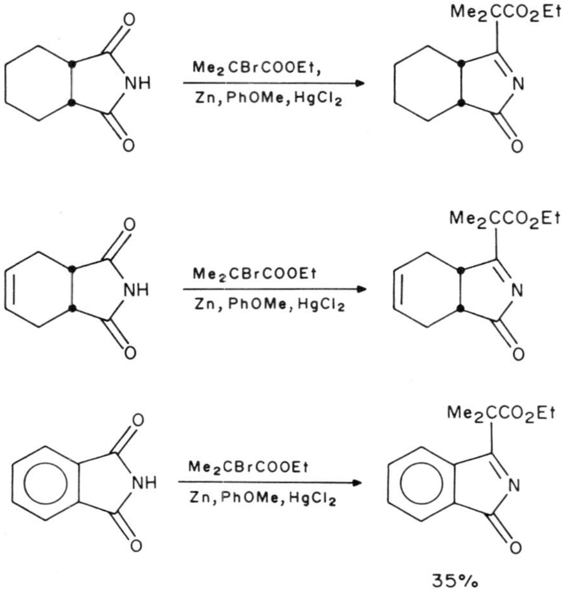
Me2CBrCOOEt,
Zn, PhOMe, HgCl2
Me2CCO2Et
NH
N
O
Me2CBrCOOEt
Zn, PhOMe, HgCl2
35%

(800)

δ-Lactams have been reportedly[1227] prepared by the reaction of ethyl α-bromoacetate and conjugated imines. This reaction involves formation of the zinc enolate anion which undergoes electrophilic attack by the α-carbon of the styrene moiety, followed by nucleophilic attack on the carbonyl carbon atom of the ester by the imino nitrogen, to give 1,4-cycloaddition, followed by elimination of ethoxide ion to form the 6-membered lactam (equation 801).

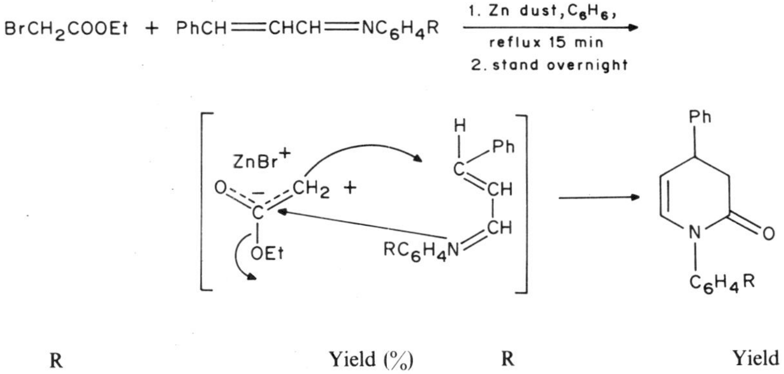
BrCH2COOEt + PhCH=CHCH=NC6H4R
1. Zn dust, C6H6, reflux 15 min
2. stand overnight
ZnBr+
CH2 +
OEt
Ph
CH
RC6H4N
C6H4R

(801)

R	Yield (%)	R	Yield (%)
H	62	*p*-Br	90
p-MeO	68	*p*-NO_2	16
p-EtO	74	*o*-OH	65
p-Me	34	*p*-SO_2NH_2	N.R.
p-Cl	70		

c. Other imine cycloadditions

Reaction of imines with a variety of chromium carbene complexes produces β-lactams. When [(methoxyalkyl or aryl) carbene] chromium complexes (**128**) are used in reaction with imines, irradiation by sunlight (or six 20-watt 'vitalite' fluorescent tubes) is required to produce[1228–1230] the β-lactams according to either of the two mechanistic pathways[1228,1229] shown in equation (802). The first pathway involves initial photocycloaddition of the imine to the chromium carbene to give the four-membered metallacycle, which then undergoes carbon monoxide insertion, followed by reductive elimination to produce the β-lactam product. The second pathway involves initial cycloaddition of the imine to the carbene carbon and an adjacent carbon monoxide moiety, to produce the acyl complex directly, which then undergoes reductive elimination producing the product. The overall reaction and its scope is shown in equation (803) and Table 55.

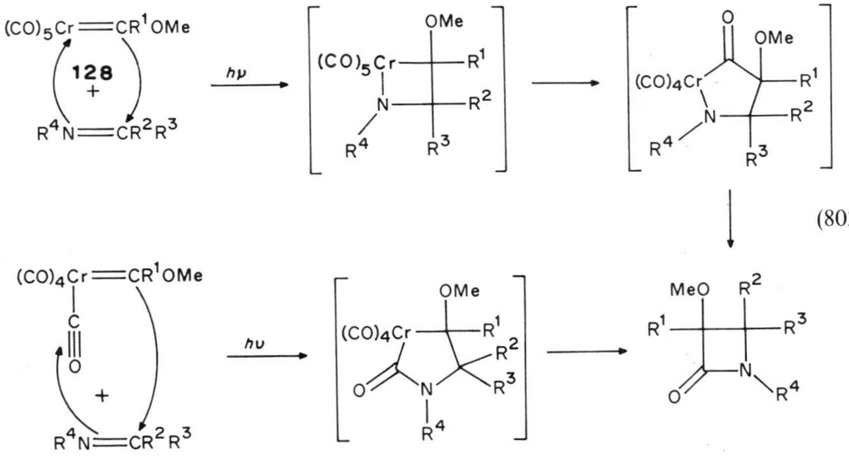

$(CO)_5Cr{=}CR^1OMe$ (**128**) + $R^4N{=}CR^2R^3$ $\xrightarrow{h\nu}$ [$(CO)_5Cr$–N metallacycle] ⟶ [$(CO)_4Cr$ acyl complex] ⟶ β-lactam

$(CO)_4Cr{=}CR^1OMe$ + $R^4N{=}CR^2R^3$ $\xrightarrow{h\nu}$ [$(CO)_4Cr$ acyl complex] ⟶ β-lactam (802)

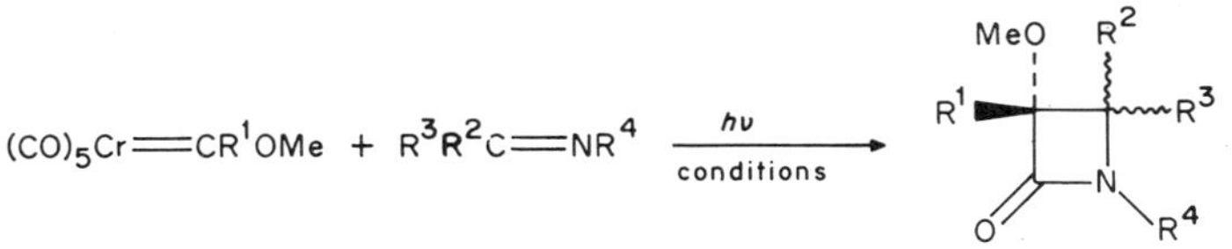

$$(CO)_5Cr{=}CR^1OMe + R^3R^2C{=}NR^4 \xrightarrow[\text{conditions}]{h\nu} \text{β-lactam} \quad (803)$$

If the tetramethylammonium salt of (1-hydroxyethylidene)pentacarbonylchromium is used as the carbene complex, irradiation is not required to produce[1231] the β-lactam products, but a second oxidation step is required to convert the [1-methyl-4-aryl-3(*E*)-(arylmethylene)azetidenylidene]pentacarbonylchromium initially formed[1231] to the final product (equation 804).

Molybdenum carbene complexes have also proven useful in the synthesis of lactams[1230], especially from oxazines (equation 805) and oxazolines (equation 806), which have been reported[1230] to be inert toward reaction with chromium carbene complexes.

Titanium tetrachloride has been used to catalyse both ketene bis (trimethylsilyl) acetals and ketene silyl acetals with a variety of Schiff bases to produce substituted β-lactams. Thus, reaction of ketene bis(trimethylsilyl) acetals with Schiff bases in the presence of titanium tetrachloride produces[1232] β-lactams in good yields, via a cross-aldol type condensation (equation 807).

TABLE 55. β-Lactams from reaction of chromium carbene complexes and imines.

Chromium carbene complex or R^1 in **128**	Imine	Conditions	Product	Yield (%)	Reference
Me	PhCH=NR R = Me	Hexane or Et_2O, Ar, 10–20°C, 1–3 h	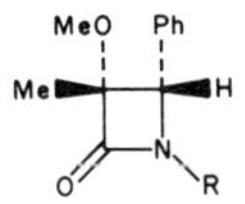	76	1228, 1229
	R = Ph			52	1228, 1229
	R = *p*-An	Et_2O, Ar, 2 h		66	1229
	R = $CH_2P(=O)(OEt)_2$	Et_2O, CH_2Cl_2		90	1230
	R = $CH(CO_2Me)P(=O)(OEt)_2$	Et_2O, CH_2Cl_2		80	1230
	R = $CH{=}CH_2$	Et_2O, Ar, 3 days		41	1230
Ph	PhCH=NMe	Hexane or Et_2O, Ar 10–20°C, 1–3 h	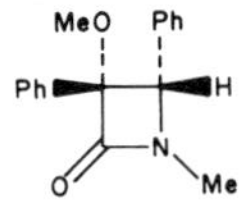	72	1228, 1229

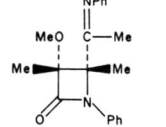

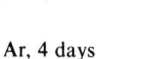

Me	PhCOC(Ph)=NR R = Ph	Et_2O, Ar, 4 days	(129) +	52 2:1	1230
			(130)		
	R = Me	Et_2O, 48 h		52 (1.5:1)	1230
Me		CH_2Cl_2, Ar, 3 h	+	40	1230
			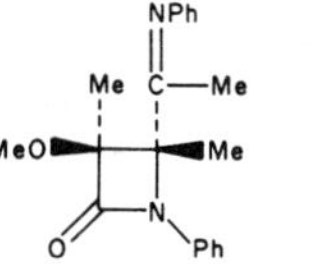 (0.25:1)		
	PhCH=NCH_2Ph	Et_2O, Ar, 4 days		53	1230 *(continued)*

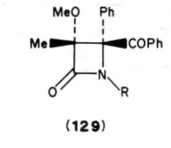

TABLE 55. (*continued*)

Chromium carbene complex or R^1 in **128**	Imine	Conditions	Product	Yield (%)	Reference
Me	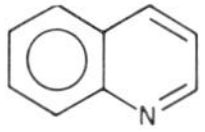	—	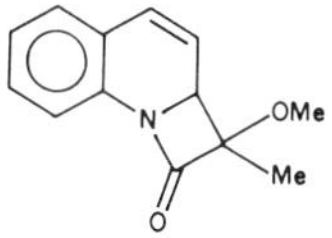	38	1229
Me	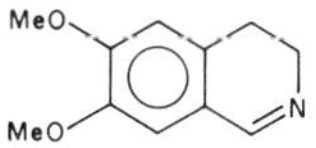	—	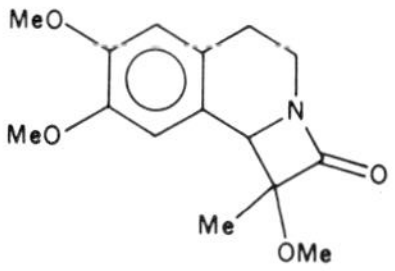	38, 45	1228, 1229
Me	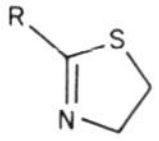	—	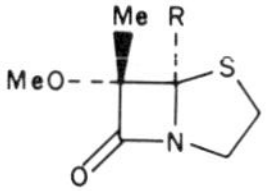		1228, 1229
	R=H			81	
	R=Ph			52	
Me	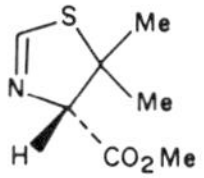	—	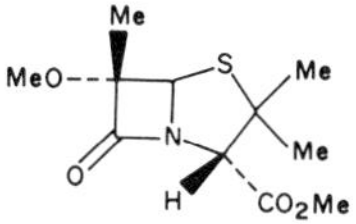	60	1229
Me	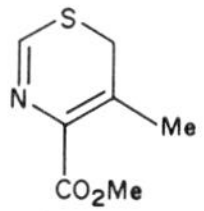	CH_2Cl_2	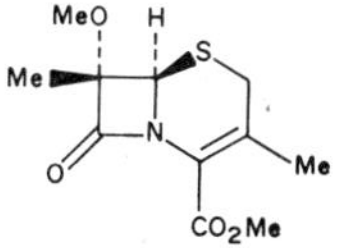	52	1230
Me	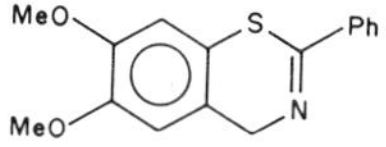	—	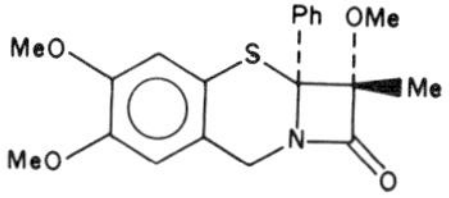	61	1229

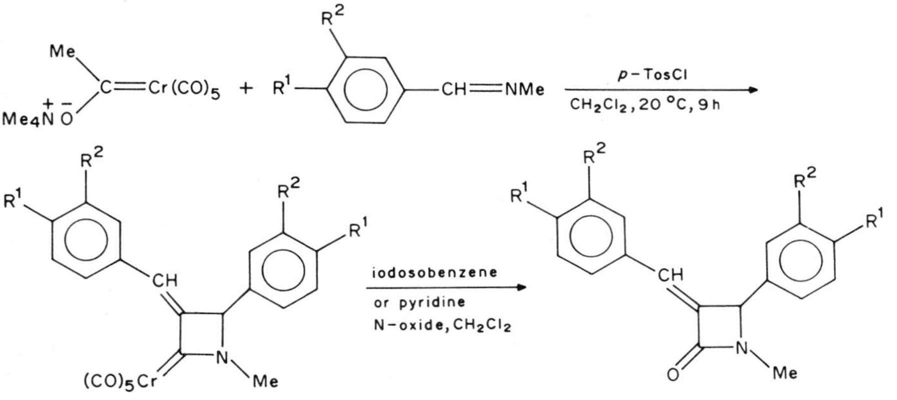

(804)

R^1	R^2	Oxidizing agent	Yield (%)
H	H	iodosobenzene	100
H	H	pyridine N-oxide	96
H	Me	pyridine N-oxide	87
Me	H	pyridine N-oxide	89

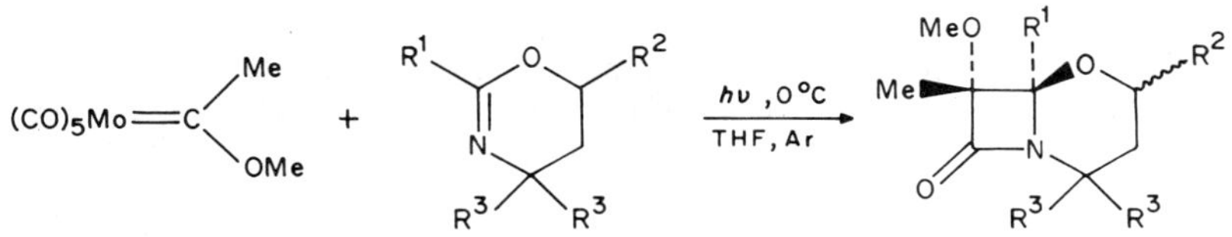

(805)

R^1	R^2	R^3	Reaction time (h)	Yield (%)
H	H	H	72	41
Et	Me	Me	48	42[a]
$PhCH_2$	Me	Me	72	46[a]

[a]The product consisted of two isomers.

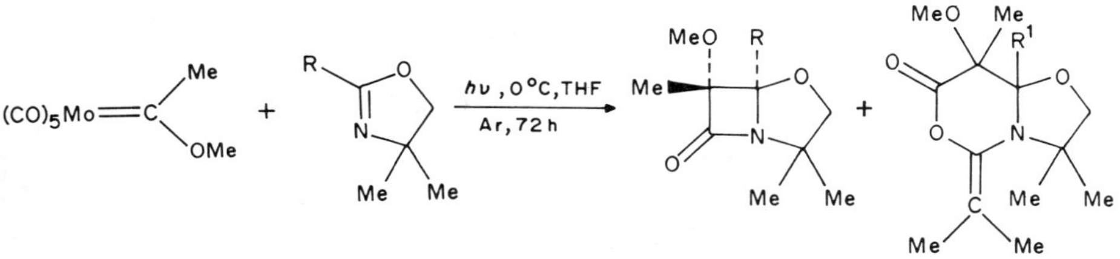

(131) **(132)**

(806)

R	Yield (%) 131	132
H	14	13
$PhCH_2$	13	13

$$R^1R^2C{=}C(OSiMe_3)_2 + R^3CH{=}NR^4 \xrightarrow[\text{2. stir r.t., 2 h}]{\text{1. } CH_2Cl_2, TiCl_4, \text{ r.t.}}$$

$$[Me_3SiO_2CCR^1R^2CHR^3NR^4TiCl_3] \longrightarrow \text{(azetidin-2-one: } R^1, R^2 \text{ at C-3; } R^3 \text{ at C-4; } N\text{–}R^4) \qquad (807)$$

R^1	R^2	R^3	R^4	Yield (%)
Me	Me	Ph	Ph	75
Me	Me	*i*-Pr	Ph	66
—$(CH_2)_5$—		Ph	Ph	60
Ph	H	Ph	Ph	69
Ph	H	*i*-Pr	Ph	65

Highly effective asymmetric induction by means of an intermediate 'titanium template' has been reported[1233,1234] to occur when dimethylketene silyl acetals react with (*S*)-alkylidene (1-arylethyl)amines[1233] (equation 808), and with the Schiff bases of chiral α-amino esters[1234] (equation 809).

$$R^1CH{=}N\overset{*}{C}H(Me)R^2 + Me_2C{=}C(OMe)(OSiMe_3) \xrightarrow[\text{2. r.t. stir 24 h}]{\text{1. }-78\,^\circ C\text{, stir 30 min, } TiCl_4, CH_2Cl_2\ (-Me_3SiCl)}$$

$$\left[R^1\text{—}\overset{*}{C}H(\text{—}C(Me_2)\text{—}COOMe)\text{—}N(TiCl_3)\text{—}\overset{*}{C}H(Me)R^2\right] \xrightarrow{-TiCl_3(OMe)}$$

$$(S,R) + (R,R) \qquad (808)$$

R^1	R^2	Ratio of (*S*, *R*):(*R*, *R*)	Asymmetric induction (%)	Yield (%)
Et	Ph	23:77	54	66
n-Pr	Ph	17:83	66	72
i-Pr	Ph	72:28	44	26
n-Bu	Ph	17:83	66	73
i-Bu	Ph	11:89	78	71
Et	α-naphthyl	23:77	56	69
n-Pr	α-naphthyl	19:81	62	70
i-Pr	α-naphthyl	24:76	52	10
n-Bu	α-naphthyl	19:81	62	69
i-Bu	α-naphthyl	15:85	70	72

$$R^1CH{=}N\overset{*}{C}HR^2COOMe + Me_2C{=}C(OMe)(OSiMe_3) \xrightarrow[2.\ r.t.]{1.\ TiCl_4, CH_2Cl_2, -78\,^\circ C} \left[R^1CH{=}N(\to TiCl_4)\overset{*}{C}H(R^2)C(OMe){=}O\cdots TiCl_4\right] \xrightarrow{Me_2C{=}C(OMe)(OSiMe_3)} (S,R) + (R,R) \qquad (809)$$

R^1	R^2	Ratio of (*S*, *R*):(*R*, *R*)	Asymmetric induction (%)	Yield (%)
Et	*i*-Pr	97:3	94	73
n-Pr	*i*-Pr	95:5	90	74
n-Bu	*i*-Pr	95:5	90	77
i-Bu	*i*-Pr	>99:<1	>98	81
Et	Me	73:27	46	28
Et	CH_2CO_2Me	57:43	14	53
Et	*i*-Bu	70:30	40	49
Et	$PhCH_2$	79:21	58	45

Similar results have also been reported[1235,1236] using substituted ketene silyl acetals and Schiff bases. However, in some of these reactions, an intermediate β-amino ester is initially formed which must be treated with lithium diisopropylamide in a second step to cause cyclization to the β-lactam. It appears that all reactions involving a Schiff base which have a phenyl substituent on the nitrogen atom require this second step (equation 810).

TABLE 56. Lactams from reaction of lithium enolates and imines

Ester							
R^1	R^2	R^3	Imine	Conditions	Product	Yield (%)	Reference
H	H	Et	$PhCH{=}NSiMe_3$	[a], 2 h	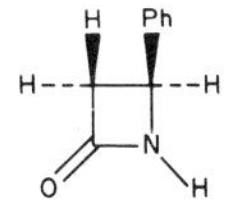 [structure]	14	1237
Me	H	Et	$PhCH{=}NSiMe_3$	[a], 3 h	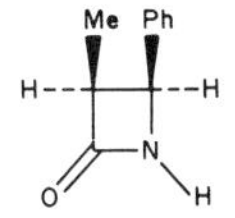 [structure] (**133**) +	41	1237
					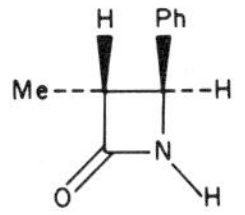 [structure] (**134**)	3	
Me	H	Et	$PhCH{=}NSiMe_3$	[b], 2 h	**133** +	19	1237
					134	25	
Me	H	Et	$PhCH{=}NSiMe_3$	[c], 2 h	**133** +	29	1237
					134	9	

Me	H	Et	PhCH=NPh	[d], 2 h	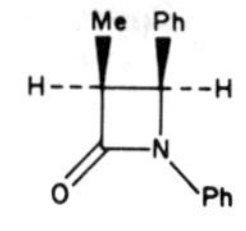 (135) +	45	1237
					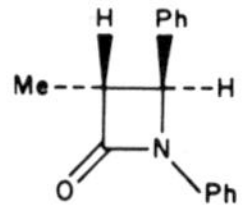 (136)	2	
Me	H	Et	PhCH=NPh	[b], 2 h	**135** +	12	1237
					136	19	
Me	H	Et	PhCH=NPh	[c], 2 h	**135** +	10	1237
					136	20	
Me	H	Et	PhCH=CHCH=NMe	1. THF, r.t. 20 h 2. H_2O	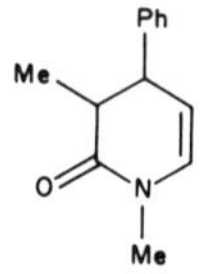 (137) +	58	1238[b]
					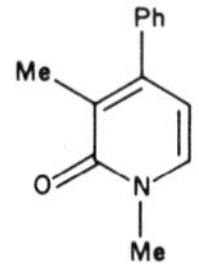 (138)	15	

(continued)

TABLE 56. (*continued*)

Ester							
R^1	R^2	R^3	Imine	Conditions	Product	Yield (%)	Reference
Me	H	Et	PhCH=CHCH=CHNMe	1. THF, r.t. 20 h 2. reflux 5 h	**137** +	41	1238[e]
					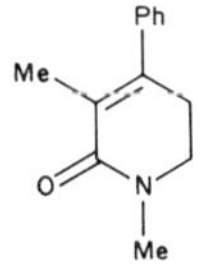 [structure: 3-Me-4-Ph-1-Me-dihydropyridin-2-one] +	11	
					138	8	
Et	H	Et	PhCH=NSiMe$_3$	[a], 2 h	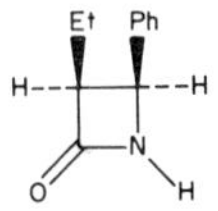 (139)	72	1237
Et	H	Et	PhCH=NSiMe$_3$	[b], 2h	**139** +	28	1237
					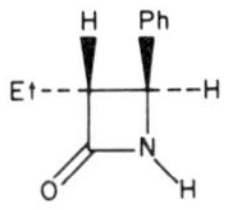 (140)	38	
Et	H	Et	PhCH=NSiMe$_3$	[c], 2 h	**139** +	40	1237
					140	16	

Et	H	Et	PhCH=NPh	[d], 2 h	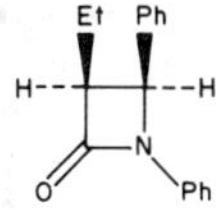 (141) +	69	1237
					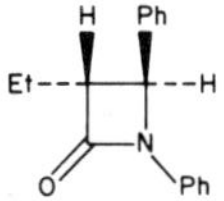 (142)	17	
Et	H	Et	PhCH=NPh	[b], 2 h	**141** +	5	1237
					142	42	
Et	H	Et	PhCH=NPh	[c], 2 h	**141** +	5	1237
					142	41	
MeCH(OH)	H	Et	PhSCH=CHCH=$NSiMe_3$	1. LDA, THF 2. HCl, H_2O	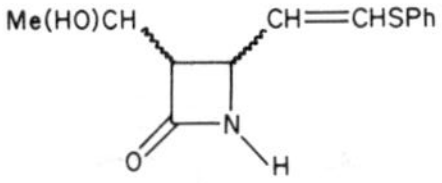 (mixture)	46	1237
MeCH(OH)	H	Et	$Me_3SiC≡CCH=NSiMe_3$	1. LDA, THF 2. HCl, H_2O	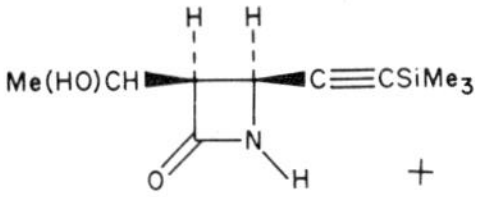 +	44	1237

TABLE 56. (*continued*)

Ester							
R^1	R^2	R^3	Imine	Conditions	Product	Yield (%)	Reference
					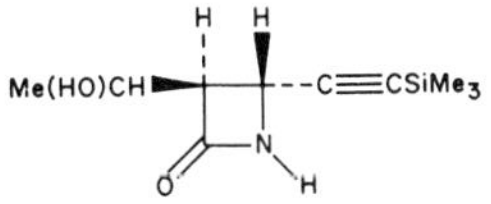 Me(HO)CH, H, H, C≡CSiMe₃ β-lactam +	5	
					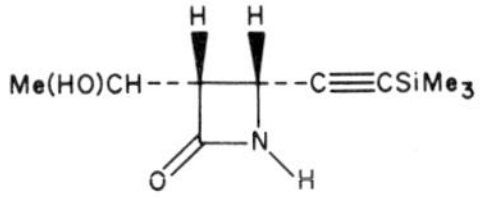 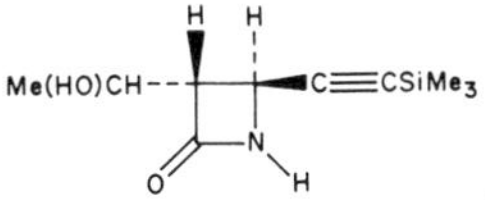 Me(HO)CH, H, H, C≡CSiMe₃ β-lactam + Me(HO)CH, H, H, C≡CSiMe₃ β-lactam	17	
i-Pr	H	Et	$PhCH{=}NSiMe_3$	[a], 1 h	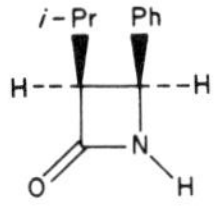 *i*-Pr, Ph β-lactam (143) +	80	1237
					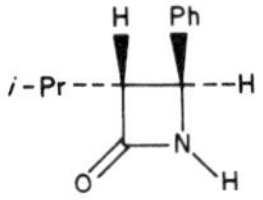 H, Ph, *i*-Pr β-lactam (144)	1	

i-Pr	H	Et	PhCH=NSiMe$_3$	[b], 2 h	**143** +	43	1237
					144	43	
i-Pr	H	Et	PhCH=NSiMe$_3$	[c], 2 h	**143** +	83	1237
					144	1	
i-Pr	H	Et	PhCH=NPh	Et$_3$Al, THF	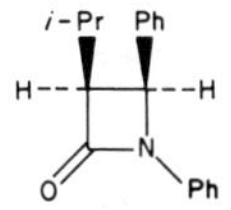 (**145**)	75	1239
i-Pr	H	Et	PhCH=NPh	[d], 2 h	**145** +	87	1237
					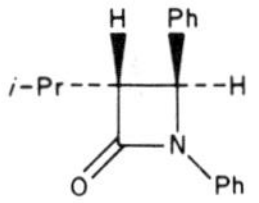 (**146**)	1	
i-Pr	H	Et	PhCH=NPh	[b], 2 h	**145** +	5	1237
					146	87	
i-Pr	H	Et	PhCH=NPh	[c], 2 h	**145** +	5	1237
					146	86	

(continued)

TABLE 56. (*continued*)

Ester							
R^1	R^2	R^3	Imine	Conditions	Product	Yield (%)	Reference
i-Pr	H	Et	PhCH=NAn-*p*	[d], 2 h	**(147)** +	82	1237
					(148)	2	
i-Pr	H	Et	PhCH=NAn-*p*	[b], 2 h at 0°C	**147** +	41	1237
					148	32	
i-Pr	H	Et	PhCH=NAn-*p*	[b], 2 h at r.t.	**147** +	4	1237
					148	80	
i-Pr	H	Et	PhCH=NAn-*p*	[c], 2 h	**147** +	5	1237
					148	79	

i-Pr	H	Et	$CH_2{=}CHCH{=}NSiMe_3$	[a], 1.5 h	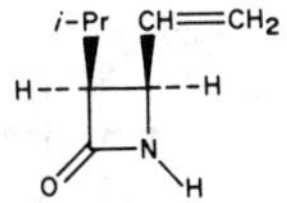	11	1237
i-Pr	H	Et	$Me_3SiC{\equiv}CCH{=}NSiMe_3$	[a], 2.5 h	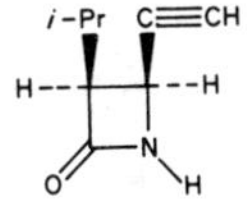 +	52	1237
					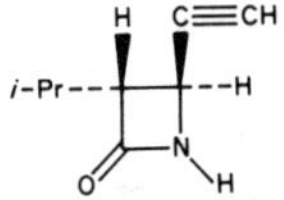	5	
i-Pr	H	Et	$PhSCH{=}CHCH{=}NSiMe_3$ (*trans*)	[a], 10 h	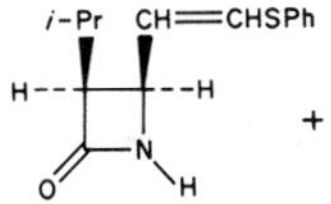 +	71	1237
					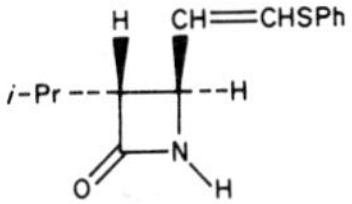	5	
i-Pr	H	Et	$2Fu\text{-}CH{=}NSiMe_3$	[a], 2.5 h	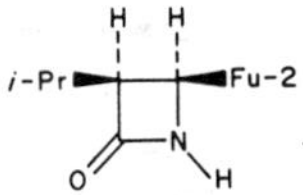 +	84	1237
					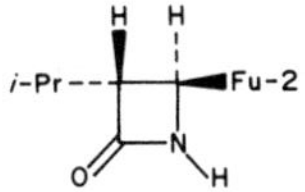	1	

(*continued*)

TABLE 56. *(continued)*

Ester							
R^1	R^2	R^3	Imine	Conditions	Product	Yield (%)	Reference
i-Pr	H	Et	*p*-AnC≡CCH=NSiMe$_3$	[a], 3 h	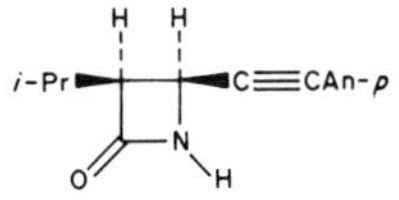 i-Pr, H, H, C≡CAn-p, O, N, H	81	1237
i-Pr	H	Et	Me$_3$SiC≡CCH=NAn-*p*	[a], 2 h	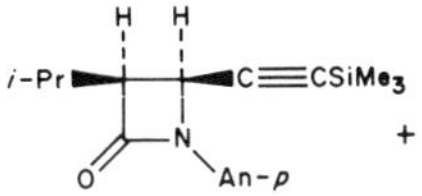 i-Pr, H, H, C≡CSiMe$_3$, O, N, An-p +	65	1237
					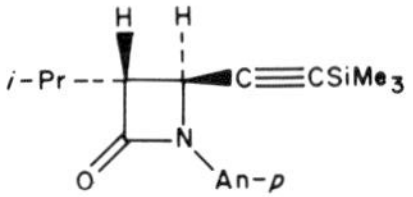 i-Pr, H, H, C≡CSiMe$_3$, O, N, An-p	6	
i-Pr	H	Et	MeCH=NSCPh$_3$	1. THF, −78°C to 25°C 2. 3 *N* HCl	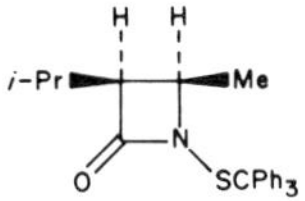 i-Pr, H, H, Me, O, N, SCPh$_3$ +	69	1241
					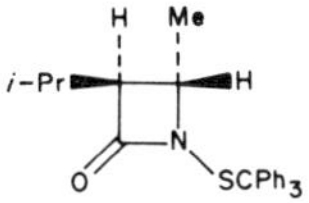 i-Pr, H, Me, H, O, N, SCPh$_3$	2	

i-Pr	H	Et	MeCH=NSPh	1. THF, −78°C to 25°C 2. 3 *N* HCl	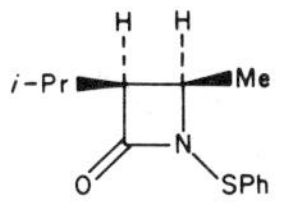 (i-Pr, H, H, Me, N–SPh β-lactam)	8	1241
i-Pr	H	Et	PhCH=NSPh	1. THF, −78°C to 25°C 2. 3 *N* HCl	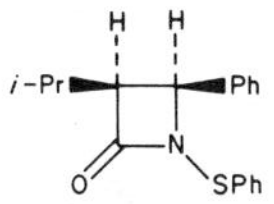 (i-Pr, H, H, Ph, N–SPh β-lactam)	37	1241
i-Pr	H	Et	$PhCH{=}NSCPh_3$	1. THF, −78°C to 25°C 2. 3 *N* HCl	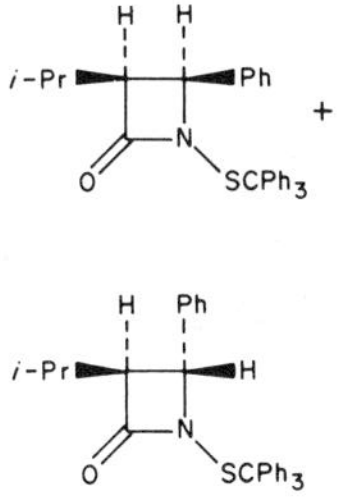 (i-Pr, H, H, Ph, N–$SCPh_3$ β-lactam) + (i-Pr, H, Ph, H, N–$SCPh_3$ β-lactam) (4.5:1)	70	1241
Ph	H	Et	PhCH=NPh	*f*	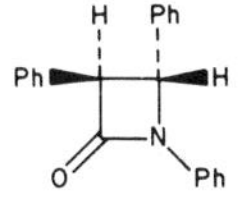 (Ph, H, Ph, H, N–Ph β-lactam)	35	1240
Ph	H	Et	PhCH=CHCH=NMe	1. THF, r.t., 65 h 2. H_2O	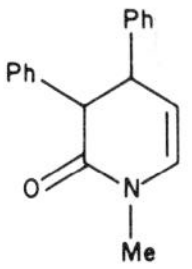 (149)	38	1238

(*continued*)

TABLE 56. (*continued*)

Ester							
R^1	R^2	R^3	Imine	Conditions	Product	Yield (%)	Reference
Ph	H	Et	PhCH=CHCH=NMe	1. THF, r.t., 20 h[e] 2. reflux 5 h 3. H_2O	**149** +	19	1238
					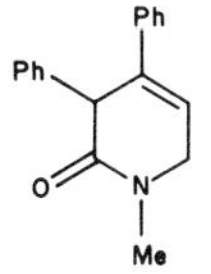 +	14	
					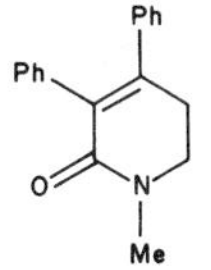 +	22	
					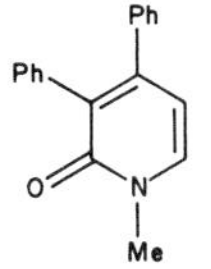	13	
Ph	H	Et	MeCH=CHCH=NPr-n	1. THF, r.t., 20 h 2. 45°C, 5 h 3. H_2O	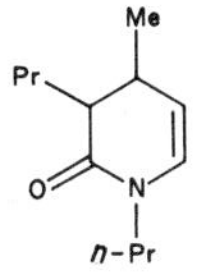	70	1238

PhS	H	Et	PhCH=$NSiMe_3$	1. THF, 3 h 2. HCl, H_2O	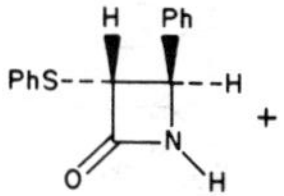 +	48	1237
					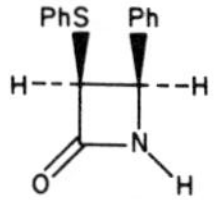	5	
$PhSCH_2CHMe$	H	Et	PhCH=$NSiMe_3$	1. THF 2. HCl, H_2O	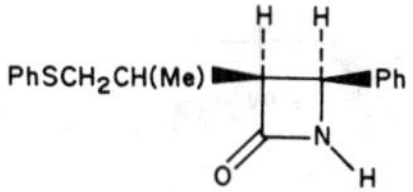	61	1237
Me_2NCH_2	H	Et	PhCH=$NSiMe_3$	1. THF 2. HCl, H_2O	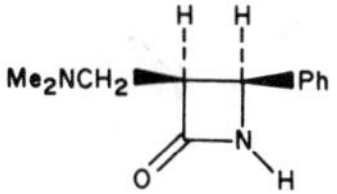	50	1237
PhCONH	H	Et	PhCH=NPh		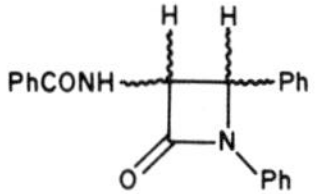	45	1240
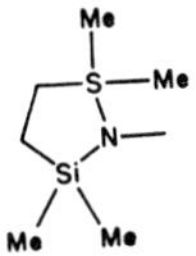	H	Et	MeCH=$NSCPh_3$	1. THF, −78° to 25°C, 4 h 2. 3 *N* HCl	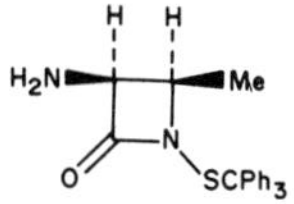 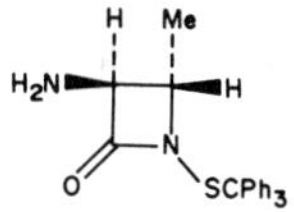 +	78	1241
					(5:1)		

(*continued*)

TABLE 56. (*continued*)

Ester							
R^1	R^2	R^3	Imine	Conditions	Product	Yield (%)	Reference
Me	Me	Et	p-$RC_6H_4CH{=}NPh$ R=H	[e]	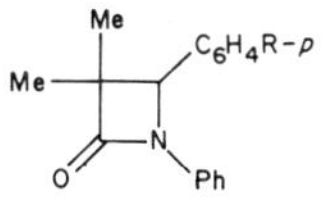	75	1240
			R = MeO	[e]		82	1240
			R = Cl	[e]		95	1240
			R = Me_2N	[e]		66	1240
Me	Me	Et	RCH=NTol-p R = Ph	[e]	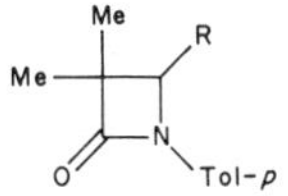	84	1240
			R = p-ClC_6H_4	[e]		80	1240
Me	Me	Et	RCH=NPh R = 2-Fu	[e]	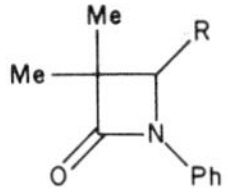	67	1240
			R = 2-Thi	[e]		89	1240
Me	Me	Et	RCH = $NSiMe_3$	1. THF 2. HCl, H_2O	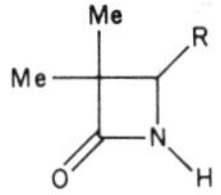		
			R = Ph			69–70	1237

			R = PhCH=CH (*trans*)	1. THF 2. HCl, H_2O		69–79	1237
			R = $Me_3SiC{\equiv}C$	1. THF 2. HCl, H_2O		69–70	1237
Me	Me	Et	PhCH=NSPh	1. THF, −78°C to 25°C 2. 3 *N* HCl	3,3-Me₂-4-Ph-1-SPh-azetidin-2-one	35	1241
Me	Me	Et	$PhCH{=}NSCPh_3$	1. THF, −78°C to 25°C 2. 3 *N* HCl	3,3-Me₂-4-Ph-1-$SCPh_3$-azetidin-2-one	87	1241
Me	Me	Et	PhCH=NCH=CHPh	1. THF, −78°C, 1 h 2. r.t., 20 h	3,3-Me₂-4-Ph-1-(CH=CHPh)-azetidin-2-one	83	1242
Me	Me	Et	EtCH=NN=CHEt	1. THF, −78°C, 1 h[h] 2. r.t., 20 h	3,3-Me₂-4-Et-1-(N=CHEt)-azetidin-2-one **(150)**	31	1242
Me	Me	Et	EtCH=NN=CHEt	1. THF, −78°C, 1 h 2. r.t., 1 h	**150** +	17	1242
					1,1′-bi(3,3-Me₂-4-Et-azetidin-2-one) **(151)**	36	

TABLE 56. (*continued*)

Ester							
R^1	R^2	R^3	Imine	Conditions	Product	Yield (%)	Reference
Me	Me	Et	EtCH=NN=CHEt	1. THF, −78°C, 1 h[i] 2. r.t., 1 h	**151**	76	1242
Me	Me	Et	PhCH=NN=CHPh	1. THF, −78°C, 1 h[h] 2. r.t., 20 h	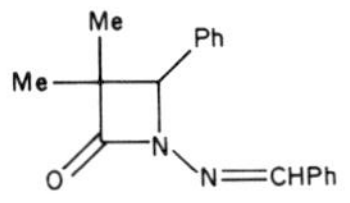	50	1242
Me	Me	Et	PhCH=NN=CHPh	1. THF, −78°C, 1 h[h] 2. r.t., 1 h	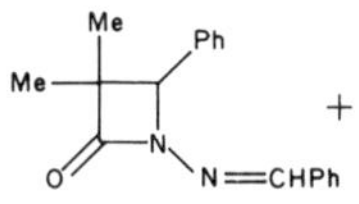 + (150)	59	1242
					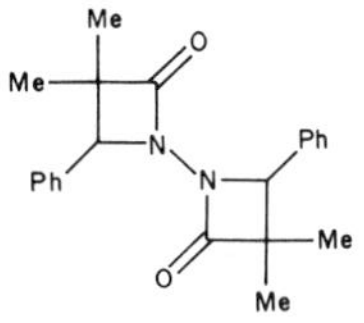 (153)	18	
Me	Me	Et	PhCH=NN=CHPh	1. THF, HMPA,[h] −78°C, 1 h 2. r.t., 1 h	**152**	85	1242

Me	Me	Et	PhCH=NN=CHPh	1. THF, −78°C, 1 h[h] 2. r.t., 1 h	**152** +	22	1242
					153	62	
Me	Me	Et	PhCH=CHCH=NR R = Me	1. THF, r.t., 20 h 2. H_2O	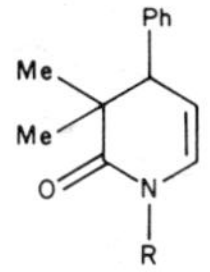 [structure: 3,3-dimethyl-4-phenyl-1-R-3,4-dihydropyridin-2-one]	63	1238
			R = *t*-Bu	1. THF, r.t., 20 h 2. H_2O		24	1238
			R = *t*-Bu	1. THF, diglyme, r.t., 20 h 2. H_2O		33	1238
			R = *t*-Bu	1. THF, HMPA, r.t., 20 h 2. H_2O		50	1238
Me	Me	Et	MeCH=CHCH=NPr-*n*	THF, r.t., 20 h 2. H_2O	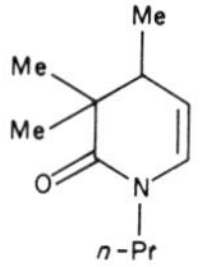 [structure: 3,3,4-trimethyl-1-*n*-Pr-3,4-dihydropyridin-2-one]	49	1238
Me	Me	Et	PhCH=C(Me)CH=NMe	1. THF, r.t., 20 h 2. H_2O	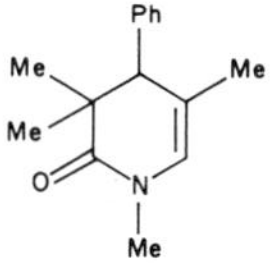 [structure: 1,3,3,5-tetramethyl-4-phenyl-3,4-dihydropyridin-2-one]	16	1238
				1. THF, HMPA, r.t., 20 h 2. H_2O		27	1238

(*continued*)

TABLE 56. (*continued*)

Ester							
R^1	R^2	R^3	Imine	Conditions	Product	Yield (%)	Reference
Me	Me	Et	$H_2C{=}CEtCH{=}NBu$-*t*	1. THF, HMPA, r.t., 20 h 2. H_2O	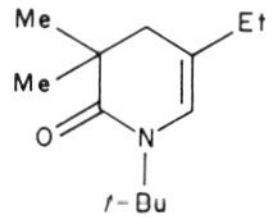	61	1238
				1. THF, r.t., 20 h 2. H_2O		29	1238
—$(CH_2)_5$—		Et	$PhCH{=}NCH{=}CHPh$	1. THF, −78 °C, 1 h 2. r.t., 20 h	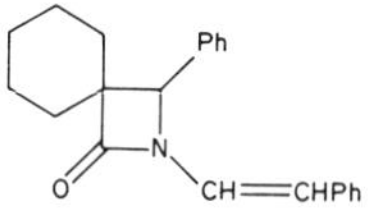	100	1242
—$(CH_2)_5$—		Et	$PhCH{=}CHCH{=}NMe$	1. THF, r.t., 20 h 2. H_2O	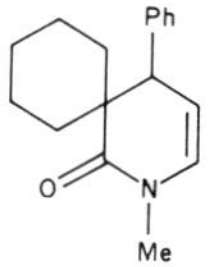	78	1238
—$(CH_2)_5$—		Et	$PhCH{=}NPh$	[e]	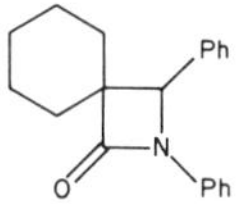	84	1240
Ph	Me	Me	$PhCH{=}NPh$	[e]	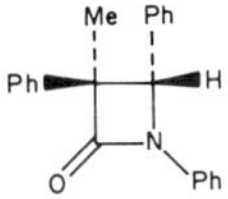	85[j]	1240

Ph	Me	Et	PhCH=NPh	[e]		90^{k}	1240
PhCONH	Me	Me	PhCH=NPh	[f]	(154)	75^{l}	1240
PhCONH	Me	Et	PhCH=NPh	[f]	**154**	91	1240
Ph	OH	Me	PhCH=NPh	[m]	(155)	74	1240
Ph	OH	Et	PhCH=NPh	[m]	**155**	70	1240
PhS	Me	Et	$PhCH=NSiMe_3$	1. THF 2. HCl, H_2O	+	18	1237
						41	
PhS	Et	Et	$PhCH=NSiMe_3$	1. THF 2. HCl, H_2O	+	2	1237
						6	

(continued)

TABLE 56. (*continued*)

Ester							
R^1	R^2	R^3	Imine	Conditions	Product	Yield (%)	Reference
PhS	*i*-Pr	Et	$PhCH{=}NSiMe_3$	1. THF 2. HCl, H_2O	No reaction	—	1237
	MeCOCOOEt or $Me_2C{=}CHCOOEt$		$PhCH{=}NSiMe_3$	1. THF 2. HCl, H_2O	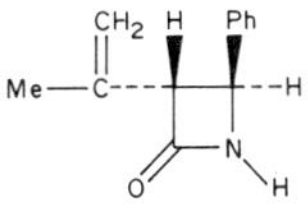	42	1237

[a]Hexamethyldisilazane in THF was added to *n*-BuLi at $-70°C$, stirred 10 min, then the ester in THF was added to the reaction mixture at a rate required to keep the temperature at $-60°C$. The mixture was stirred for 50 min, then the silyl imine was added and the resulting mixture stirred an additional 1 h at $-70°C$. The mixture was then allowed to warm to room temperature and stirred for the length of time indicated in the table.
[b]As in a above, but HMPA was added prior to the addition of the ester.
[c]As in a above, but HMPA was added after the preparation of the ester.
[d]As in a above, but using pure imine.
[e]The mechanism involves the initial preparation of 3,4-dihydropyridone which rearranges or dehydrogenates to product during the course of the reaction:

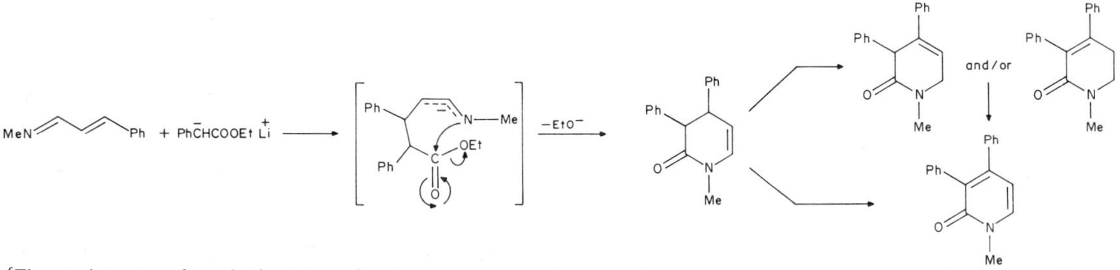

[f]The reaction was performed using 1.1 eq. of LDA per 2–5 mmoles of ester and 2–5 mmoles of imine; the imine was added to the reaction mixture at $-78°C$ and then the reaction mixture was stirred at room temperature for 3–8 h.
[g]The same as in *f* above, except that 2.2 eq. of was used.
[h]The mole ratio of 2,3-diazabutadiene: enolate used was 1:1.2.
[i]The mole ratio of 2,3-diazabutadiene: enolate used was 1:2.2.
[j]Observed 60% ee.
[k]An 8:1 mixture of diasteromers was obtained.
[l]Observed 4% ee.
[m]The same as in *g* above except that the reaction mixture was warmed to 0°C for 2 h and then recooled to $-78°C$ before addition of the imine.
[n]Observed 14% ee.

$$R^1CH{=}NR^2 + R^3R^4C{=}C(OMe)(OSiMe_3) \xrightarrow[\text{r.t. stir 1 h}]{TiCl_4,\ CH_2Cl_2} MeOOCCR^3R^4CHR^1NHR^2 \xrightarrow[\text{hexane—THF, 0 °C, stir 10 min}]{\text{THF, LDA,}} \beta\text{-lactam} \quad (810)$$

R^1	R^2	R^3	R^4	Yield (%) α-amino acid	Yield (%) β-lactam[a]
Ph	Ph	Me	Me	85	95
Ph	Ph	Me	H	83	90
Ph	Ph	—$(CH_2)_5$—		70	91
Ph	$PhCH_2$	Me	Me	92	77[b]
Ph	Me	Me	Me	—	72[c]
i-Pr	$PhCH_2$	Me	Me	—	43[c]
Et	Ph(Me)CH	Me	Me	—	54[c]

[a]Reaction also reported[1235] to occur when R^1 = 2-Fu, 3-Thi, 2-Pyr and R^3 = PhO, PhS but other R's and yields were not specified.

[b]The LDA treatment resulted in the formation of 3,3-dimethyl-4,5-diphenyl-2-pyrrolidone, the β-lactam was obtained upon treatment of the β-amino ester with potassium *tert*-butoxide in THF.

$$MeOOCCMe_2CH(Ph)NHCH_2Ph \xrightarrow{\text{LDA, THF}} \text{3,3-dimethyl-4,5-diphenyl-2-pyrrolidone}$$
$$MeOOCCMe_2CH(Ph)NHCH_2Ph \xrightarrow[\text{THF}]{t\text{-BuO}^-\text{K}^+,} \text{1-benzyl-3,3-dimethyl-4-phenyl-2-azetidinone}$$

[c]Under the reaction conditions cyclization occurred directly without the formation of the intermediate β-amino ester.

Another imine cycloaddition approach to the preparation of β-lactams which has been used effectively in the current literature involves the condensation of mono- and dianions of ester enolates with imines. The most general reaction in this category involves treatment of an enolizable ester of with lithium diisopropylamide in tetrahydrofuran to produce the enolate anion, which is then allowed to react with the imine (equation 811 and Table 56).

$$R^1R^2CHCOOR^3 \xrightarrow[-78\,^\circ C,\ \text{stir}]{\text{LDA, THF}} R^1R^2C{=}C(OR^3)O^-\ Li^+ \xrightarrow[\text{Conditions}]{R^4R^5C{=}NR^6} \beta\text{-lactam } (R^1, R^2, R^4, R^5 \text{ on ring C; } N\text{-}R^6) \quad (811)$$

Treatment of imines of α-aminoesters with sodium hydride in dimethyl sulphoxide produces the imine anions of the aminoesters. Alkylation of these anions with ω-halogenoesters produces substituted imines, which upon heating to 180 °C produce[1243] lactams of from 4 to 7 members depending upon the structure of the ω-halogenoester used in alkylation (equation 812).

PhCH=NCH(R)COOMe --(NaH, DMSO)--> [PhCH=N=C(R)=C(=O)OMe]⁻ Na⁺

--($X(CH_2)_nCOOMe$; X = unspecified)--> [PhCH=N–C(R)(COOMe)–$(CH_2)_n$–C(=O)–OMe]

--(180 °C, H_2O)--> lactam ($(CH_2)_n$, C(R)(COOMe), N–H, C=O) + MeOH + PhCHO (812)

R	n	Substituted imine yield (%)	Lactam yield (%)
Me	1	72	—
Me	2	80	70
Me	3	72	60
Me	4	58	47
H	2	60	—
H	3	45	22
H	4	43	—

Reaction of two equivalents of lithium diisopropylamide in tetrahydrofuran with α-hydroxyhippuric acid derivatives produces the dianions shown in equation (813). Allowing these dianions to react with Schiff bases prepared from an aromatic aldehyde and an arylamine produces[1244] single isomers of 3-methoxy-3-amido-2-azetidinones (equation 813).

R^1CONHCH(OMe)COOMe --(THF, LDA, −78 °C)--> [R^1CON̄–C(OMe)=C(O⁻)(OMe)]

--(1. R^2CH=NR^3, THF, −78 °C, stir 4 h; 2. r.t., stir 10 h)--> R^1CONH–C(OMe)–CH(R^2)–N(R^3)–C(=O) (azetidinone) (813)

R^1	R^2	R^3	Yield (%)
Ph	Ph	Ph	84
Ph	Ph	α-naphthyl	88
Ph	p-ClC_6H_4	Ph	91
$PhCH_2O$	Ph	α-naphthyl	84
$PhCH_2O$	p-ClC_6H_4	Ph	91

In an effort to induce asymmetry in the ester-imine condensation to produce β-lactams, the lithium enolates of the isoborneol 10-isopropylsulphonamide butyrates shown in equation (814) were allowed[1245] to react with Schiff bases. Treatment of the initial N-substituted β-lactam products with ceric ammonium nitrate afforded[1245] the corresponding N-unsubstituted β-lactams in 56–92% enantiomeric excess (equation 814).

Me Me OOCCH$_2$R^1 CH$_2$SO$_2$N(Pr-i)$_2$ —(1. LDA, THF; 2. R^{2}CH=NAn-p, −70 to 25 °C, 1.5 h)→ **(156)** —(CAN)→ **(157)**

(814)

R^1	R^2	Yield (%) **156**[a]	Yield (%) **157**[b]	e.e. (%)
Et	PhCH=CH (*trans*)	81 (10:1)	79	91
i-Pr	PhCH=CH (*trans*)	80 (21:1)	80	82
Et	Ph	88 (38:1)	85	92
i-Pr	Ph	70	86	56

[a]Combined yield of **156** and its *trans* diastereomers. The number in parentheses is the *cis:trans* product ratio.
[b]Yield after separation from the *trans* diastereomers.

Reaction of 3,4-diphenylisoxazole with *n*-butyl lithium in tetrahydrofuran at -78 °C produces[1246] an intense royal blue solution containing the corresponding anion which rearranges to the ynolate lithium-2-phenylethyn-1-olate. Treatment of this *in situ* generated ynolate with imines produces[1246] the corresponding β-lactams as shown in equation (815).

3,4-diphenylisoxazole + *n*-BuLi —(THF, N$_2$, −78 °C, stir 2 h)→ [anion] Li+

$$\xrightarrow{\ \ o\ \ } \left[PhC{\equiv}C\bar{O}\ \ Li^+\right] \xrightarrow[\substack{-78\,^\circ C,\ 2h \\ 2.\ -78\,^\circ C\ \text{to r.t.}}]{1.\ R^1CH{=}NR^2,\ THF,} \left[\ \text{Ph, }R^1\text{-azetine enolate, Li}^+\ \bar{O},\ N\text{–}R^2\ \right] \qquad (815)$$

$$\xrightarrow{R^1CH{=}NR^2} [\ \text{Li-chelated intermediate}\ ] \xrightarrow{H^+} \text{R}^2\text{NH(R}^1\text{)CH–azetidinone (Ph, R}^1\text{, N–R}^2\text{)}$$

R^1	R^2	Yield (%)
p-$O_2NC_6H_4$	*p*-$O_2NC_6H_4$	89
p-$O_2NC_6H_4$	*p*-$EtO_2CC_6H_4$	79
m-Tol	*p*-$O_2NC_6H_4$	58
Ph	*p*-$O_2NC_6H_4$	66

A convenient three-step one-pot process for the direct conversion of nitriles to 3,4-di-and trisubstituted azetidinones has been reported[1247] which involves, in the first step, the reduction of the nitrile with lithium triethoxyaluminium hydride (LTEA) to give an addition product. In the second step, addition of trimethylchlorosilane to the heterogeneous mixture of the addition product produces the corresponding intermediate trimethylsilylimine, which upon further treatment with ester enolates, in the third step, affords the azetidinones, albeit in low yields (equation 816).

$$R^1C{\equiv}N + LiAlH_4 + EtOAc \xrightarrow[0\,^\circ C,\ \text{stir 1 h}]{Ar,\ Et_2O,} \left[R^1C{\equiv}N + LiAl(OEt)_3H\right]$$

$$\longrightarrow R^1CH{=}NAl(OEt)_3Li \xrightarrow[\text{r.t., 2 h}]{Me_3SiCl} R^1CH{=}NSiMe_3$$

$$\xrightarrow[\substack{1.\ -78\ \text{to}\ -30\,^\circ C, \\ \text{stir overnight} \\ 2.\ NH_4Cl,\ H_2O,\ 0\,^\circ C}]{R^2R^3C{=}C(\bar{O}Li^+)(OEt)} \text{azetidin-2-one (}R^2, R^3, R^1, N\text{–H)} \qquad (816)$$

Nitrile	Enolate		*cis*:*trans*	
R^1	R^2	R^3	Ratio	Yield (%)
2-Fu	H	2,2,5,5-tetramethyl-1-aza-2,5-disilacyclopentan-1-yl ($-N(SiMe_2CH_2)_2$)	95:5	43[a]
2-Fu	H	Et	77:23	30[b]
2-Fu	H	Et	70:30	56[b]

Nitrile	Enolate			
R^1	R^2	R^3	*cis*:*trans* Ratio	Yield (%)
2-Fu	Me	Me	—	36
2-Thi	H	Et	50:50	50[b]
2-Thi	H	Et	44:56	29[b]
p-An	H	Et	66:34	30
Ph	Me	Me	—	12–57[b]
n-Pr	H	Et	50:50	12–40[b]
PhCH=CH (*trans*)	Me	Me	—	20
PhCH=CH (*trans*)	H	Et	86:14	25

[a]Isolated as 3-amino-4-(2′-furanoyl)azetidin-2-one.
[b]Different yields and *cis*:*trans* ratios result from different ratios of reactants.

A trimethylsilylimine intermediate is also involved in the production[1248] of 3-(α-hydroxyethyl)-4-(trimethylsilylethynyl)-2-azetidinone. Thus, addition of lithium bis (trimethylsilylamide) to trimethylsilylpropargylaldehyde produces the corresponding intermediate trimethylsilylimine, which upon reaction with ethyl acetoacetate oxime at -65 °C affords the β-lactam (equation 817).

$$Me_3SiC{\equiv}CCHO + Li^+\bar{N}(SiMe_3)_2 \longrightarrow Me_3SiC{\equiv}CCH{=}NSiMe_3 + MeC(=NOH)CH_2COOEt \xrightarrow{-65\,^\circ C} \text{3-(Me(HO)CH)-4-(C}{\equiv}\text{CSiMe}_3\text{)-azetidin-2-one (N–H)} \quad (3S,4S)\text{-}3R \qquad (817)$$

Cis and *trans* mixtures of β-lactams have been obtained[1249] by the reaction of copper acetylide with aldonitrones (equation 818).

$$PhC{\equiv}CCu + RCH{=}N(\rightarrow O)Ph \xrightarrow[\text{stir 2 h}]{C_5H_5N,\ r.t.,} \textit{cis}\text{-4-R-3-Ph-1-Ph-azetidin-2-one} + \textit{trans}\text{-4-R-3-Ph-1-Ph-azetidin-2-one} \qquad (818)$$

R	Yield (%) *Cis*	*Trans*
2-Fu	55	0
2-Thi	44	11
3-indolyl	40	15
PhCH=CH	46	10
$o\text{-}O_2NC_6H_4CH{=}CH$	40	0

Intramolecular cycloaddition of azoalkenes formed as transient intermediates by dehydrohalogenation of hydrazones derived from acid hydrazides and halocarbonyl compounds yields[1250] 4,4*a*,5,6-tetrahydropyridazin-7(3*H*)-ones (equation 819).

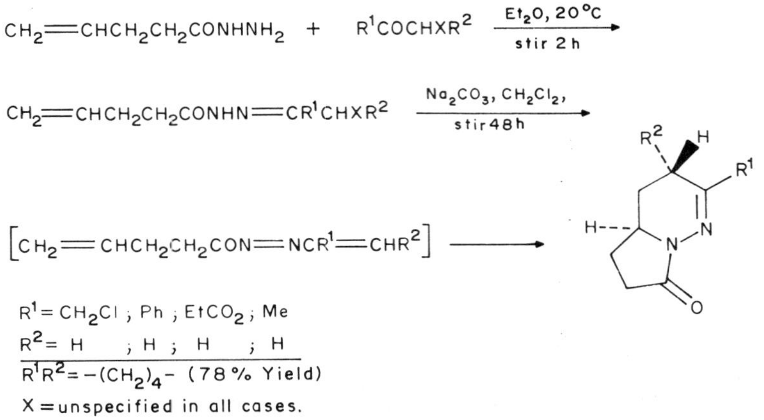

$$CH_2{=}CHCH_2CH_2CONHNH_2 + R^1COCHXR^2 \xrightarrow[\text{stir 2 h}]{Et_2O,\ 20\,°C} CH_2{=}CHCH_2CH_2CONHN{=}CR^1CHXR^2 \xrightarrow[\text{stir 48 h}]{Na_2CO_3,\ CH_2Cl_2,} [CH_2{=}CHCH_2CH_2CON{=}NCR^1{=}CHR^2] \longrightarrow \quad (819)$$

$R^1 = CH_2Cl$; Ph ; $EtCO_2$; Me
$R^2 =$ H ; H ; H ; H
$R^1R^2 = -(CH_2)_4-$ (78% Yield)
X = unspecified in all cases.

Similar reaction of phenylhydrazones with α-,β-,γ- and δ-haloacyl halides in the presence of a phase transfer catalyst produces[1251] *N*-phenylmethyleneamino-β,γ- and δ-lactams along with 1,4-(diphenylmethyleneamino)piperazine-2,5-diones as by-products (equation 820).

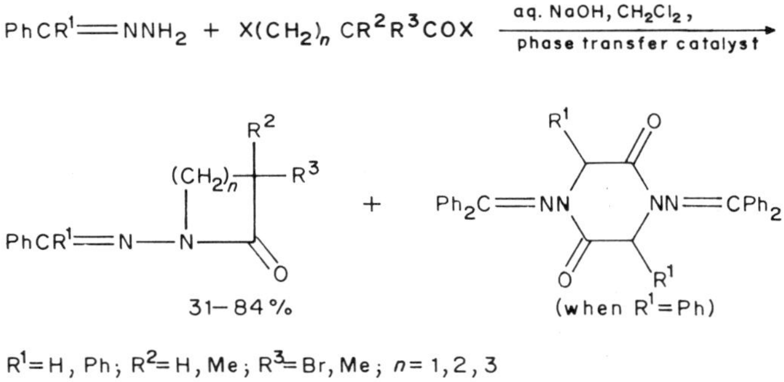

$$PhCR^1{=}NNH_2 + X(CH_2)_nCR^2R^3COX \xrightarrow[\text{phase transfer catalyst}]{aq.\ NaOH,\ CH_2Cl_2,} \quad (820)$$

31–84%

(when R^1 = Ph)

R^1 = H, Ph; R^2 = H, Me; R^3 = Br, Me; n = 1, 2, 3

An interesting preparation of fused-ring β-lactams has been reported[1252] which involves the condensation of a variety of carboxymethylthiazolidinium chlorides with 1-[3-(dimethylamino)propyl]-3-ethylcarbodiimide hydrochloride (equation 821–824). However, the corresponding fused-ring products formed range in yields from only 27 (equation 824) to 35% (equation 823).

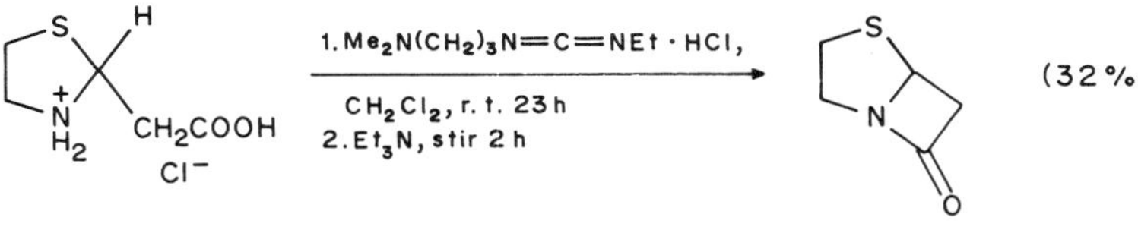

$$\xrightarrow[\text{2. }Et_3N,\ \text{stir 2 h}]{\text{1. }Me_2N(CH_2)_3N{=}C{=}NEt\cdot HCl,\ CH_2Cl_2,\ \text{r.t. 23 h}} \quad (32\%) \quad (821)$$

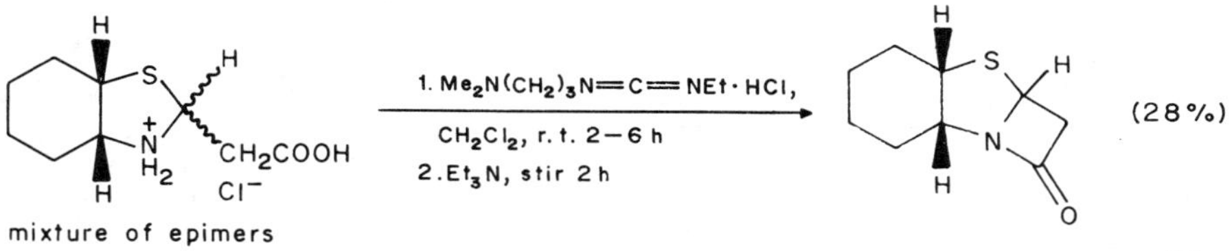

1. $Me_2N(CH_2)_3N{=}C{=}NEt \cdot HCl$, CH_2Cl_2, r.t. 2–6 h
2. Et_3N, stir 2 h

mixture of epimers

CH₂COOH → (28%)

(822)

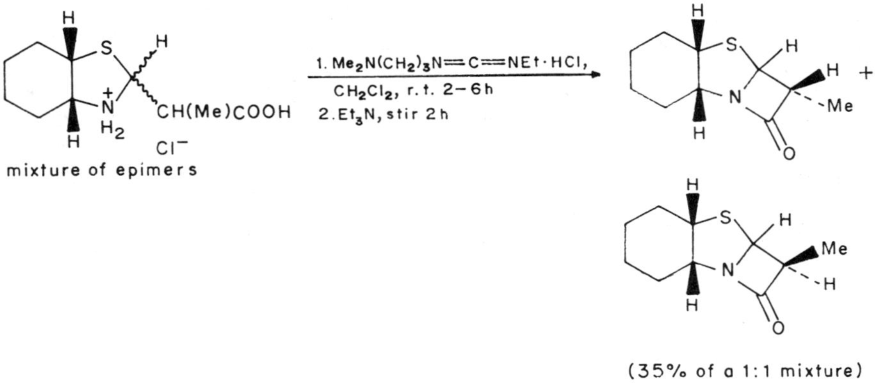

1. $Me_2N(CH_2)_3N{=}C{=}NEt \cdot HCl$, CH_2Cl_2, r.t. 2–6 h
2. Et_3N, stir 2 h

mixture of epimers

(35% of a 1:1 mixture)

(823)

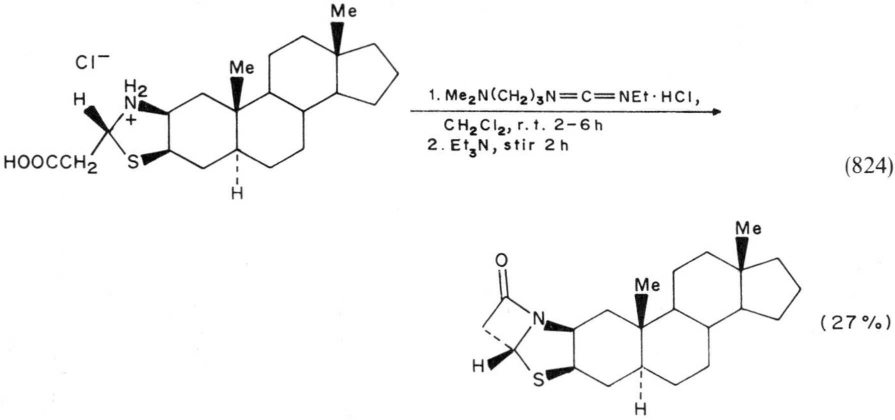

1. $Me_2N(CH_2)_3N{=}C{=}NEt \cdot HCl$, CH_2Cl_2, r.t. 2–6 h
2. Et_3N, stir 2 h

(824)

(27%)

In an effort to synthesize β-lactams asymmetrically, imines carrying a chiral substituent at nitrogen were allowed to react[1253,1254] with symmetrical or prochiral β-chloro iminium chlorides. The results indicated that a diastereoface-differentiating reaction had occurred to produce a mixture of diastereomeric or epimeric β-lactams. Also, when prochiral imines were allowed to react with chiral α-chloro iminium chlorides mixtures of diastereomeric β-lactams or their enantiomers with a clear selectivity were obtained[1253,1254] (equation 825 and Table 57).

TABLE 57. Epimeric β-lactams from imines and α-chloroiminium chlorides derived from amines **158**

Imine	Amide **158**	Product	Ratio				Yield (%)	Reference
			trans		*cis*			
			3(*S*), 4(*R*)	3(*R*), 4(*S*)	3(*R*), 4(*R*)	3(*S*), 4(*S*)		
PhCH=NCH(Me)Ph (*S*) − (+)	$MeCH_2CONMe_2$	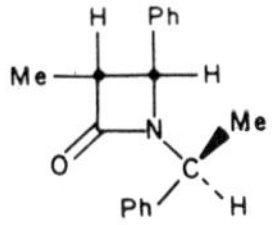 (**159**)	35:1	32:2	22:4	11:3	68	1253
PhCH=NCH(Me)Ph (*S*) − (+)	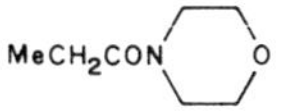 $MeCH_2CON$ (morpholine)	**159**	40:1	28:2	20:4	12:3	81	1253
PhCH=NCH(Me)Ph (*S*) − (+)	$EtCH_2CONMe_2$	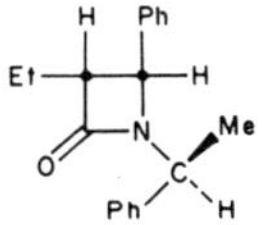	34:1	35:2	35:2	12:4	65	1253
PhCH=NCH(Me)Ph (*S*) − (+)	$PhCH_2CONMe_2$	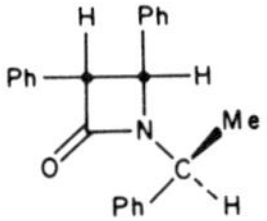	30:3	50:4	5:2	15:1	64	1253, 1254
PhCH=NCH(Me)Ph (*S*) − (+)	$PhCH(Me)CONMe_2$	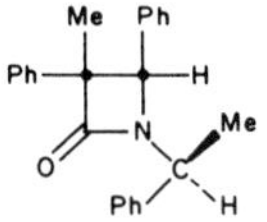	27:4	51:1	11:2	11:3	92	1253, 1254

PhCH=NCH(Me)Ph (S)−(+)	$Me_2CHCONMe_2$	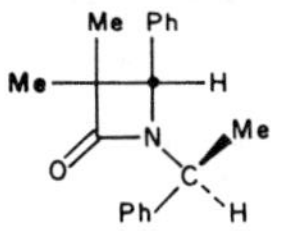 [structure]	4(*R*) = 33:1	4(*S*) = 67:2	81, 85	1253, 1254
MeCH=NCH(Me)Ph (S)−(+)	$MeCONMe_2$	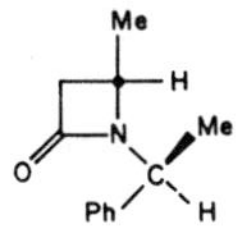 [structure]	4(*R*) = 47:1	4(*S*) = 53:2	90	1253
PhCH=NCH(Me)Ph (S)−(+)	$MeCONMe_2$	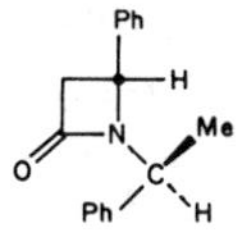 [structure]	4(*R*) = 27:1	4(*S*) = 73:2	87	1253, 1254
PhCH=NCH(Me)Ph (S)−(+)	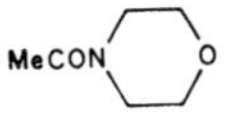 MeCON(morpholine)	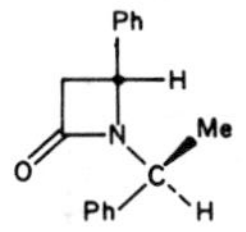 [structure]	4(*R*) = 40:1	4(*S*) = 60:2	58	1253
Ph(Me)C=NCH(Me)Ph (S)−(+)	*i*-$PrCONMe_2$	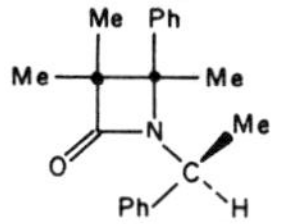 [structure]	4(*R*) = 49:1	4(*S*) = 51:2	63	1253
PhCH=NCH(Me)CH_2Ph (S)−(+)	*i*-$PrCONMe_2$	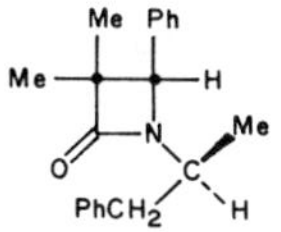 [structure]	4(*R*) = 67:1	4(*S*) = 33:2	53	1253
PhCH=NMe	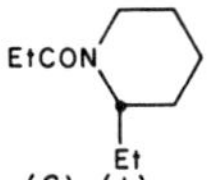 EtCON(2-Et-piperidine) (S)−(+)	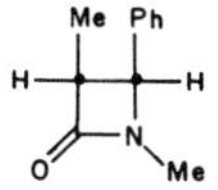 [structure]	*trans* = 68, *cis* = 32 with a 57:43 ratio of *trans* enantiomers		57	1253

(*continued*)

TABLE 57. (*continued*)

Imine	Amide **158**	Product	Ratio				Yield (%)	Reference
			trans		*cis*			
			3(*S*), 4(*R*)	3(*R*), 4(*S*)	3(*R*), 4(*R*)	3(*S*), 4(*S*)		
PhCH=NMe	$MeCH_2CON(Me)CH_2Ph$ (*S*)-(+)	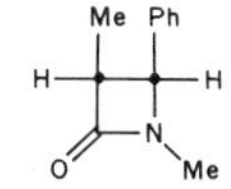	*trans* = 69, *cis* = 31 with a 64:36 ratio of *trans* enantiomers				69	1253
t-BuCH=NMe	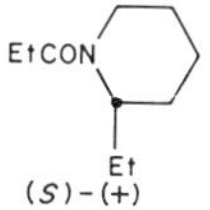 (*S*)-(+)	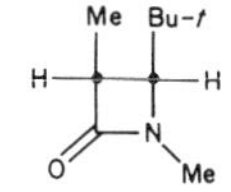	*trans* = 100 with a 88:12 ratio of *trans* enantiomers				54	1253
PhCH=NMe	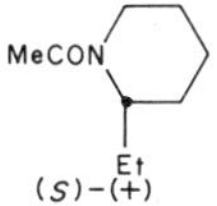 (*S*)-(+)	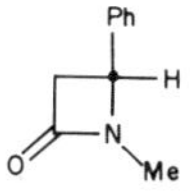 (**160**)	53:47 ratio of enantiomers				82	1253
PhCH=NMe	$MeCON(Me)CH(Me)CH_2Ph$ (*S*)-(+)	**160**	57:43 ratio of enantiomers				50	1253
PhCH=NMe	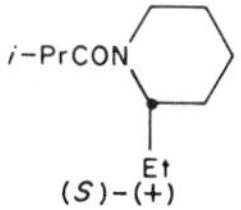 (*S*)-(+)	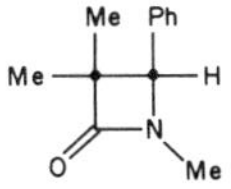 (**161**)	88:12 ratio of enantiomers				60	1253, 1254
PhCH=NMe	$Me_2CHCON(Me)CH(Me)CH_2Ph$ (*S*)-(+)	**161**	86:14 ratio of enantiomers				30	1253, 1254

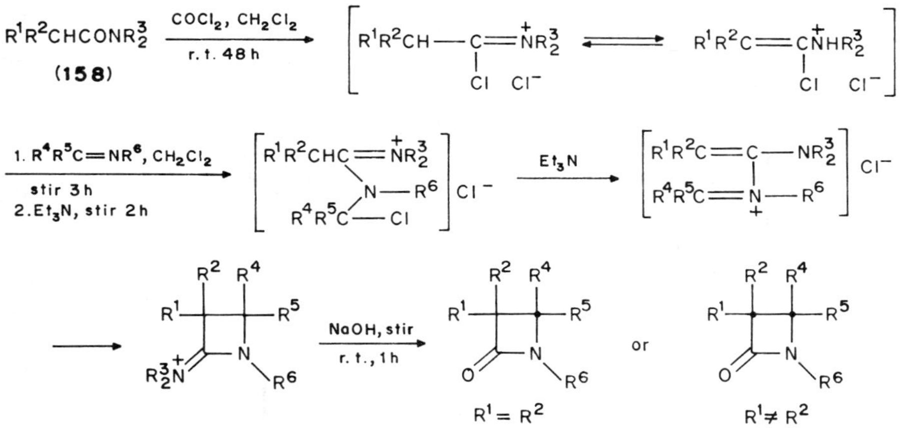

(825)

The bicyclic lactams 3,4,5,6,7,8-hexahydro-2-quinolinone and 2,3,5,6,7,8-hexahydro-4-quinolinone have been prepared[1255] by the reaction of cyclohexanone imine with β-propiolactone, acrylic, crotonic or methacrylic acids in refluxing chlorobenzene. The mechanism proposed for this reaction is shown in equation 826.

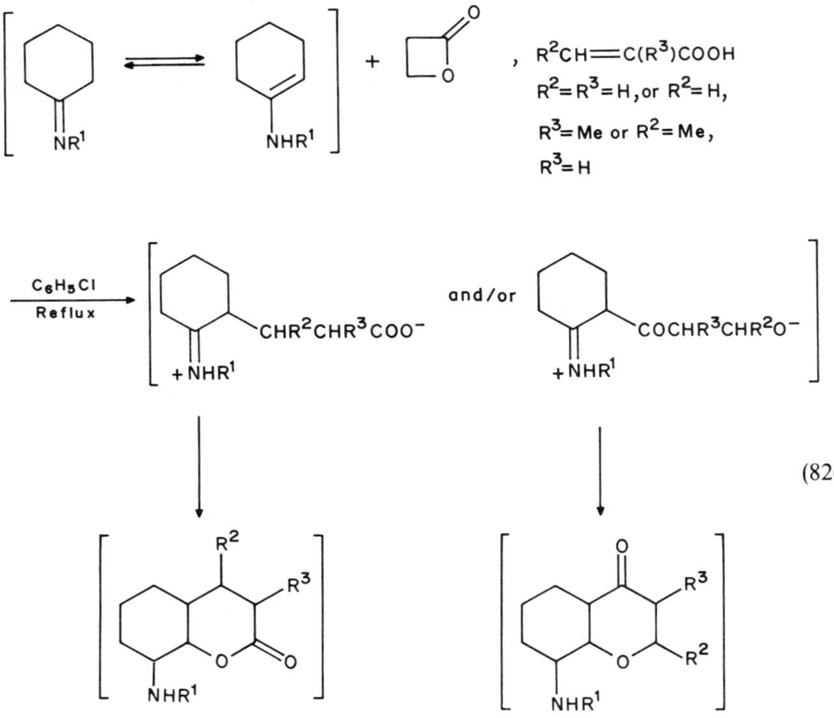

(826)

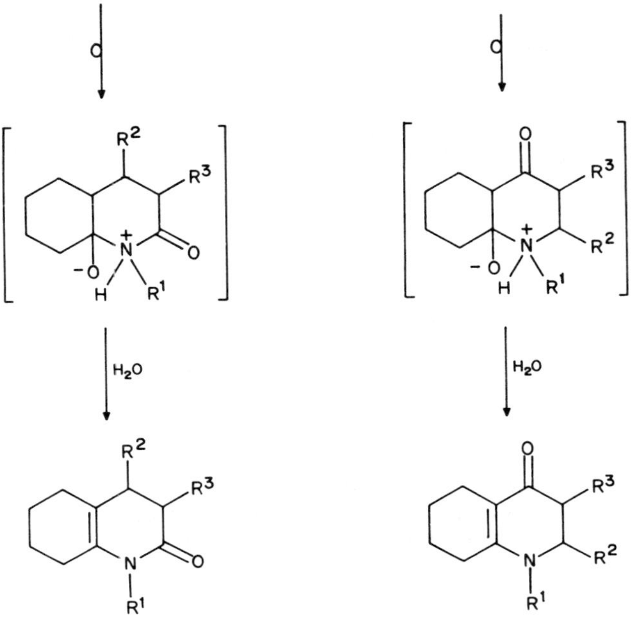

R^1	Lactone or acid	Reaction time (h)	Product	Yield (%)
n-Pr	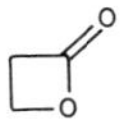 (β-propiolactone)	3	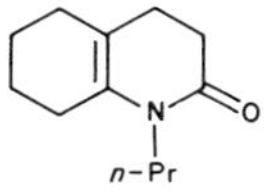 (**162**)	30
n-Pr	$CH_2{=}CHCOOH$	3	**162**	35
Ph	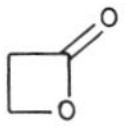 (β-propiolactone)	6	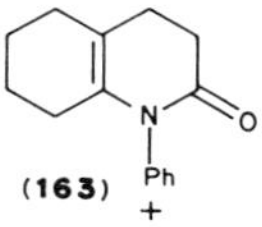 (**163**) +	12
			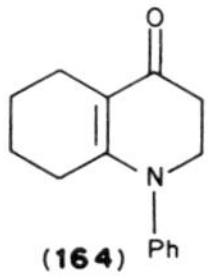 (**164**)	60
Ph	$CH_2{=}CHCOOH$	6	**163**+	65
			164	10

(*continued*)

R^1	Lactone or acid	Reaction time (h)	Product	Yield (%)
Ph	MeCH=CHCOOH (*trans*)	22		40
			+	30
Ph	CH_2=C(Me)COOH	20	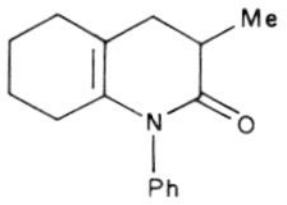	20
p-ClC_6H_4	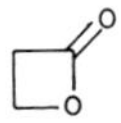	3	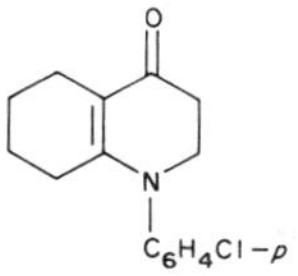	50
p-ClC_6H_4	CH_2=CHCOOH	3	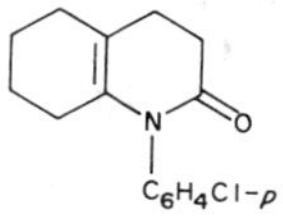	50

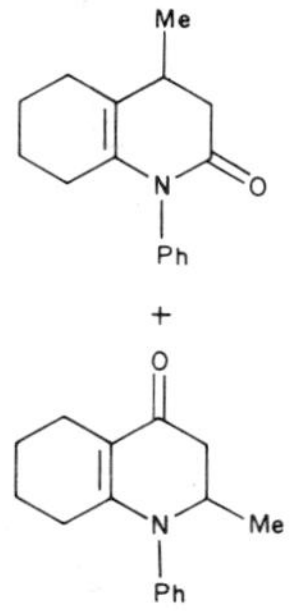

β-Lactams as well as tricyclic β-lactams have been prepared[1256] from 2-acyl-3-oxo-4,5-benzo-1,2-thiazoline 1,1-dioxide, with the structure of the final product governed by the structure of the second reactant. If imines are used as the second reactant, β-lactam products are obtained (equation 827), while with 1-substituted-3,4-dihydroisoquinolines, tricyclic β-lactams are produced (equation 828).

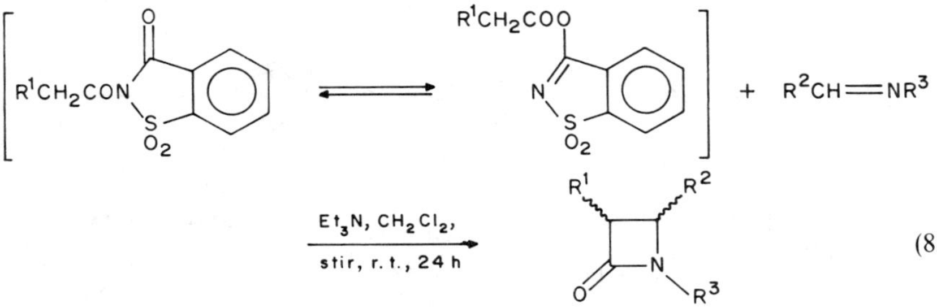

R^1CH_2CON ⇌ R^1CH_2COO + $R^2CH{=}NR^3$ $\xrightarrow[\text{stir, r.t., 24 h}]{Et_3N,\ CH_2Cl_2,}$ (827)

R^1	R^2	R^3	Yield (%)
PhO	PhCH=CH	Ph	65
p-ClC_6H_4O	PhCH=CH	Ph	67
p-ClC_6H_4O	Ph	c-Hex	77
PhO	o-$O_2NC_6H_4CH$=CH	p-ClC_6H_4	71
PI-N	p-An	p-ClC_6H_4	79
MeO	Ph	o-ClC_6H_4	66

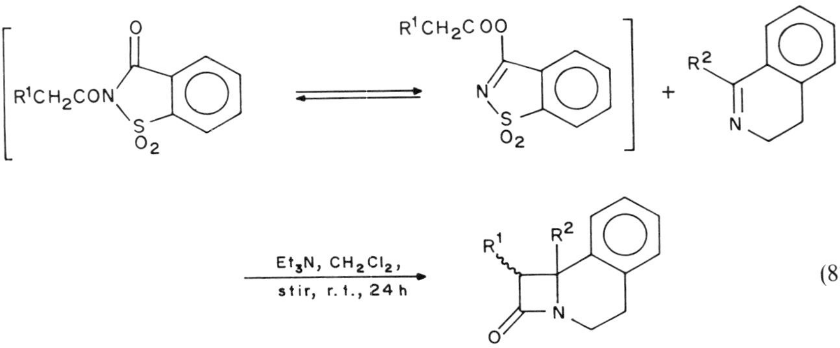

(828)

R^1	R^2	Yield (%)
PhO	m-$O_2NC_6H_4$	73
PhO	Ph	73
p-ClC_6H_4O	MeS	82
p-ClC_6H_4O	p-$O_2NC_6H_4$	76

3-p-Toluenesulphinyl-4-aryl-β-lactams were prepared[1257] in moderate yields and high stereoselectivity by condensation of aryl aldimines with either 2-p-toluenesulphinylacetic acid (equation 829) or with the p-nitrophenyl ester of 2-p-toluenesulphinylacetic acid (equation 830). In the former reaction carbonyl diimidazole is used to activate the acid before it reacts with the imine, while in the latter reaction imidazole was used as the base, because other catalysts which are more basic and less nucleophilic than imidazole (e.g. triethylamine, lithium diisopropylamide, sodium hydride or diazabicyclo[5.4.0]-undecene) gave low yields of the β-lactams, producing mainly the 2-p-toluenesulphinyl-acetic acid, probably via a ketene pathway.

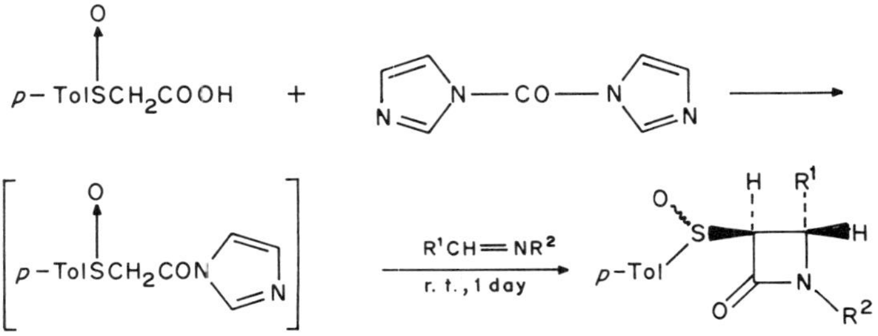

(829)

R^1	R^2	Diastereomeric ratio[a]	Yield (%)
Ph	Ph	84:16	44
Ph	*p*-An	80:20	38
Ph	*p*-ClC_6H_4	87:13	30
Ph	$PhCH_2$	44:56	45
p-ClC_6H_4	Ph	36:64	15
p-$MeCONHC_6H_4$	Ph	87:13	31

[a] Ratios of the diastereomers are for the two 3,4-*trans* isomers, the only ones formed from this reaction, but the specific structure of the ratio components were unassigned.

$$p\text{-TolS(O)CH}_2\text{COOC}_6\text{H}_4\text{NO}_2\text{-}p + \text{R}^1\text{CH{=}NR}^2 \xrightarrow[\text{r.t., 1 day}]{\text{imidazole, DMF,}}$$

(830)

R^1	R^2	Diastereomeric ratio[a]	Yield (%)
Ph	Ph	82:18	34
Ph	*p*-An	73:27	38
Ph	*p*-ClC_6H_4	87:13	30
Ph	$PhCH_2$	47:53	40
n-Hex	Ph	—	—

[a] Ratio of the diastereomers are for the two 3,4-*trans* isomers, the only ones formed from this reaction, but the specific structure of the ratio components were unassigned.

Highly substituted 3-amido-β-lactams have been synthesized[1258] by allowing imines to react with the mesoionic oxazoline 2,4-diphenyl-2-oxazolin-5-one. The mechanism proposed for this conversion involves the addition of the oxazolone in its carbanion form to the carbon–nitrogen double bond of the imine, followed by attack of the nucleophilic nitrogen (equation 831).

1. $R^1CH{=}NR^2$, C_6H_6, N_2, reflux 2 h
2. stand overnight, r.t.

PhCONH

(831)

R^1	R^2	Yield (%)
2-Thi	Me	70
2-Thi	Et	75
2-Fu	Me	60
2-Fu	Et	65
2-Fu	c-Hex	80
Ph	Me	78

An interesting synthesis of 4-(dimethylaminomethylene)-3-oxovalero- and -caprolactams has been reported[1259] which involves the condensation of piperidine- or hexamethyleneiminediones with dimethylaminoformaldehyde diethyl acetal to produce an intermediate, which is then allowed to react with an imine to produce the pyrido- or pyrimido-pyrimidines or azepines named above (equation 832).

$(CH_2)_n$ HN O O + $Me_2NCH(OEt)_2$ ⟶ $(CH_2)_n$ $CHNMe_2$ HN O O

$\xrightarrow{R-C(NH_2)=NH}$ $(CH_2)_n$ HN N N R O (832)

$n = 1$; R = NH_2, Me, Ph, SH, OH, p-$H_2NC_6H_4SO_2NH$;

Yield (%) = 28 − 86 (range)

$n = 2$; R = SH, NH_2, Me;

Yield (%) = 13 − 28 (range)

*3. *From nitrones and nitroso compounds*

Six-membered lactams have been prepared[1260] stereoselectively and regiospecifically by the intramolecular cyclization of nitrones joined to amides and olefins which were prepared *in situ* by reaction of the corresponding ketones with hydroxylamines (equations 833 and 834).

COMe N $COCHR^1$—C(H)=C(H)R^2 $\xrightarrow[\text{conditions}]{R^3NOH}$ R^3 N O H H R^2 Me H N O H R^1 (833)

R^1	R^2	R^3	Conditions	Yield (%)
$—(CH_2)_3—$		Me[a]	LDA, EtOH, reflux 34 h	51
$—(CH_2)_3—$		$PhCH_2$	C_6H_6, reflux 71 h	~60
$—(CH{=}CH)CH_2—$		Me[a]	LDA, EtOH, N_2, reflux 21 h	3[b]
$—(CH{=}CH)CH_2—$		$PhCH_2$	C_6H_6, reflux	~32
H	H	Me[a]	LDA, EtOH, N_2, reflux 25 h	40
				40
H	H	$PhCH_2$[c]	EtOH, reflux 4 days	21

[a]Used as HCl salt.
[b]Also isolated was:

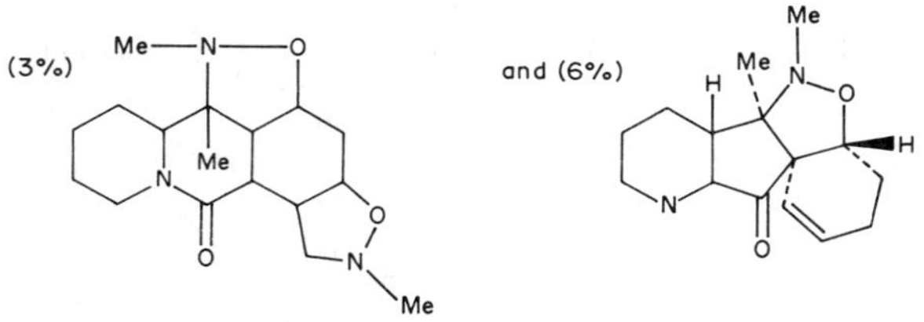

(3%) and (6%)

[c]Also isolated was:

MeNHOH · HCl, LDA, EtOH, reflux 19 h

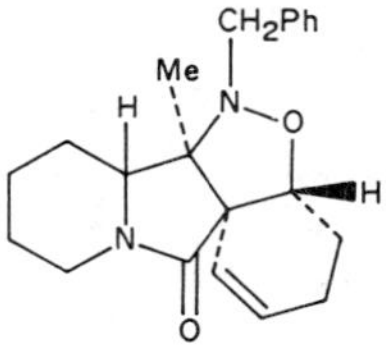

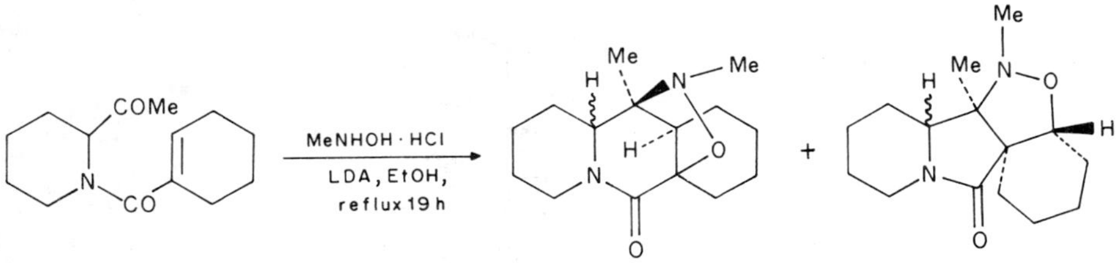

(5.3%) (64.7%)

(834)

*D. By Rearrangements

*1. Ring contractions

**a. Wolff rearrangement.* Regioselectivity has been observed[1261,1262] in the photochemical ring contraction of substituted 4-diazopyrazolidine-3,5-diones to produce mono- and bicyclic aza-β-lactams (equation 835). The relative migratory aptitude of the nitrogen groups in this Wolff rearrangement were observed[1262] to follow the order: $NPh > NCHPh_2 \sim NCH_2Ph \sim NMe > NCH_2COOR$.

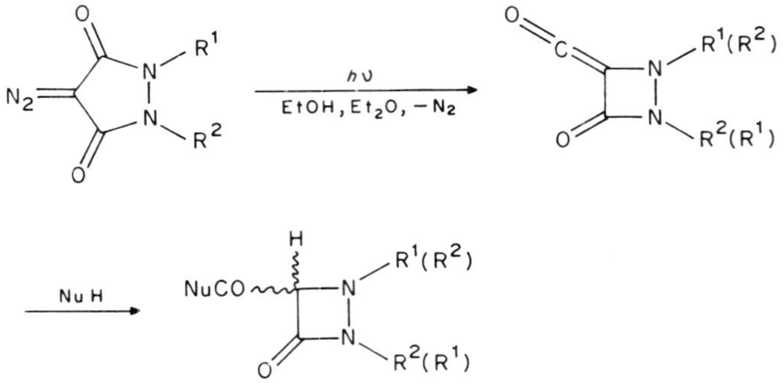

(835)

Diazodione	NuH	Product (ratio)	Yield (%)	Reference
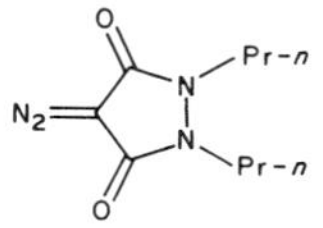	EtOH	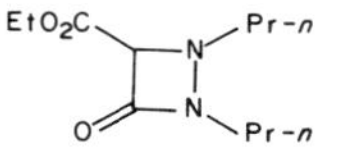	48	1261
	HOH	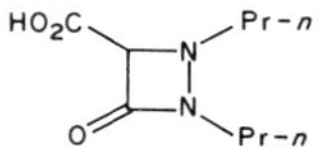	56	1261
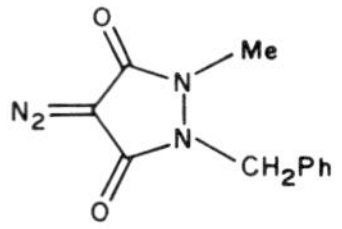	EtOH	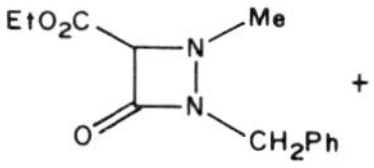 (165) 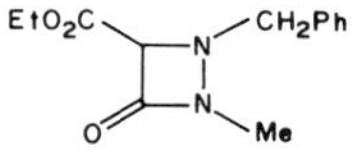 + (166) (1:1)	30	1262
	HOH	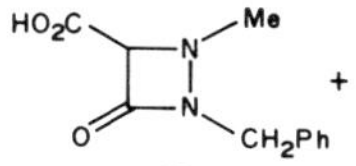 (167) 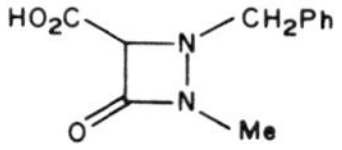 + (168) (1:1)	33	1262
	EtOH	**167** + **168**	36	1262
	HOH	**167** + **168**	33	1262
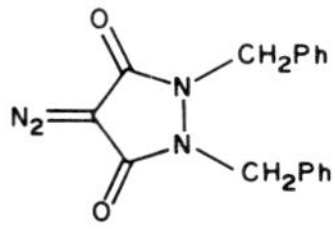	EtOH	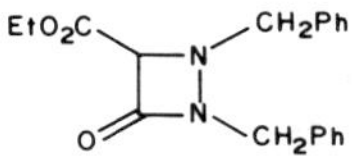	45	1261

(continued)

Diazodione	NuH	Product (ratio)	Yield (%)	Reference
	t-BuOH	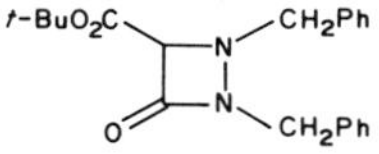 t-BuO_2C, N, CH_2Ph, N, CH_2Ph, O	30	1261
	HOH	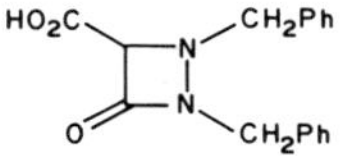 HO_2C, N, CH_2Ph, N, CH_2Ph, O	50	1261
	Et_2NH	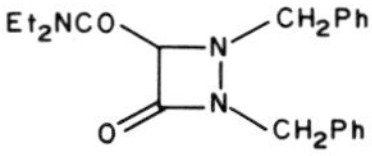 Et_2NCO, N, CH_2Ph, N, CH_2Ph, O	17	1261
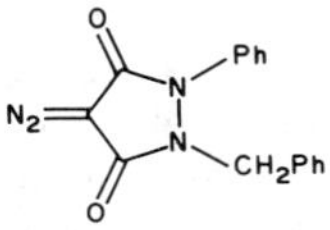 O, N_2, N, Ph, N, CH_2Ph, O	EtOH	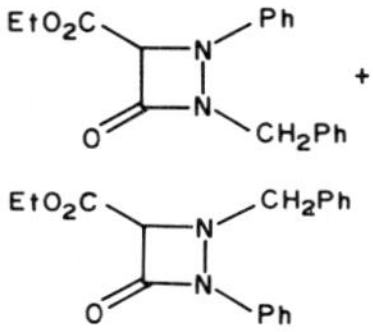 EtO_2C, N, Ph, N, CH_2Ph, O + EtO_2C, N, CH_2Ph, N, Ph, O (1.7:1)	72	1262
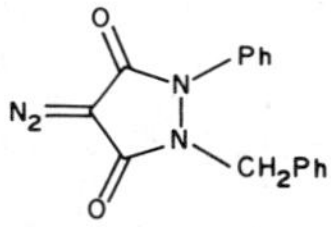 O, N_2, N, Ph, N, CH_2Ph, O	HOH	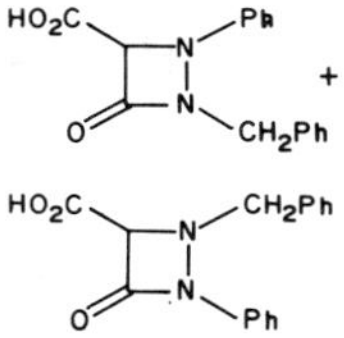 HO_2C, N, Ph, N, CH_2Ph, O + HO_2C, N, CH_2Ph, N, Ph, O (1.6:1)	55	1262
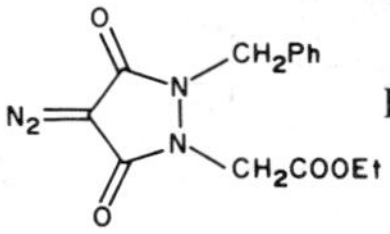 O, N_2, N, CH_2Ph, N, CH_2COOEt, O	HOH	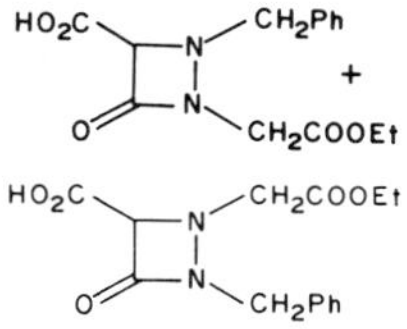 HO_2C, N, CH_2Ph, N, CH_2COOEt, O + HO_2C, N, CH_2COOEt, N, CH_2Ph, O (2.8:1)	47	1262
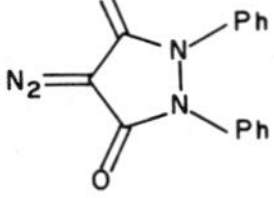 O, N_2, N, Ph, N, Ph, O	EtOH	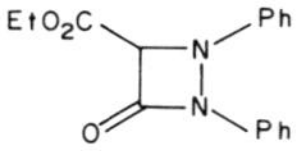 EtO_2C, N, Ph, N, Ph, O	0	1261
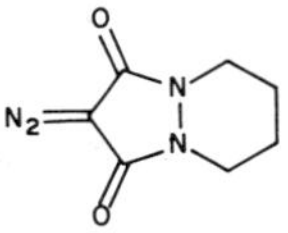 O, N_2, N, N, O	EtOH	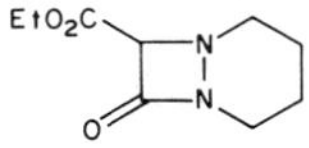 EtO_2C, N, N, O	4	1261
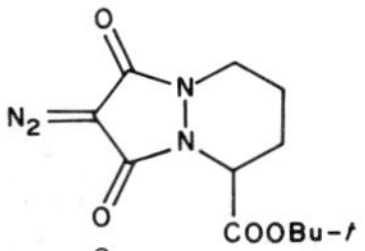 O, N_2, N, N, O, COOBu-t	EtOH	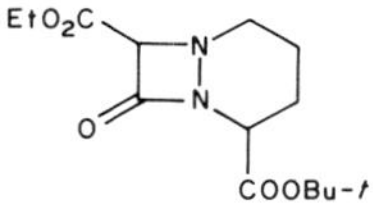 EtO_2C, N, N, O, COOBu-t	12	1262
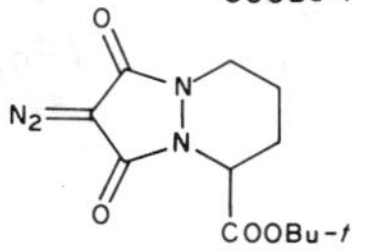 O, N_2, N, N, O, COOBu-t	HOH	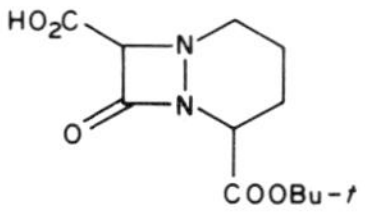 HO_2C, N, N, O, COOBu-t	14	1262

Similar photolysis, as a dilute THF solution in the presence of one equivalent of diisopropylamine at $-78\,^{\circ}$C, of the bicyclic diazopyrrolidinediones shown[1262] in equation 836 gave[1263] the corresponding bicyclic β-lactam as the *trans* isomer exclusively.

$$\xrightarrow[(i\text{-Pr})_2NH]{h\nu,\ THF,\ -78\,^{\circ}C} \quad (72\%) \tag{836}$$

Comparison of the photolysis of diazopyrrolidinediones and the pyrolysis of azidopyrrolidines as preparation methods for β-lactams have been made[1263] by studying a series of 5-alkynyl-4-diazo-5-methoxypyrrolidine-2,3-diones and 5-alkynyl-3-chloro-4-azido-5-methoxypyrrolidines, respectively (equation 837).

$$\xrightarrow[EtOH,\ 0\,^{\circ}C]{NaBH_4} \quad \xrightarrow[aq.\ HCl,\ CH_2Cl_2,\ 0\,^{\circ}C]{NaNO_2,} \quad \xrightarrow[C_6H_6,\ reflux]{} \quad \xrightarrow[MeOH,\ -78\,^{\circ}C,\ 1\,h]{h\nu\ (>3000\ \text{Å})} \tag{837}$$

Yield (%)	Ratio (Z:E)	R	Yield (%)	Ratio (Z:E)
87	3:1	Ph	74	5:1
60	5:1	$PhCH_2OCH_2$	57	6:1
55	1:1	$(CH_2)_2CH_2OTHP$	63	>10:1

**b. Miscellaneous ring contractions.* Thermolysis, photolysis, reaction with base, oxidation and reduction are all methods which have been used recently to produce lactams by ring contractions. Thus, thermolysis of 1,1-dioxo-4-thiazolidinones results in extrusion of sulphur dioxide and produces[1264] the corresponding substituted β-lactam products (equations 838, 839 and 840).

(838)

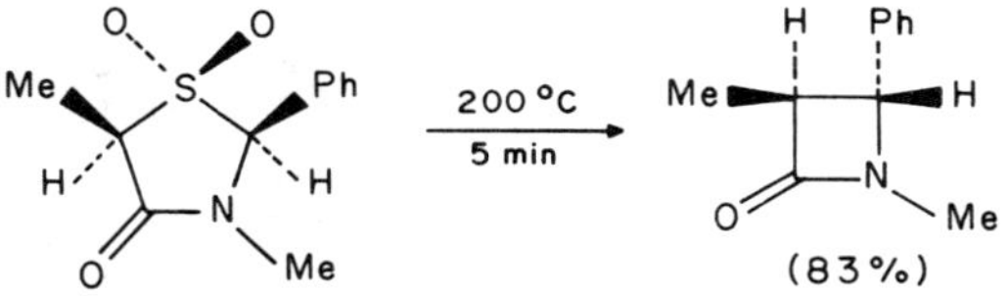

(839)

(840)

Heating *trans*-4-cyano-5-nitro-*trans*-3-phenyl-*N*-*tert*-butylisoxazolidine in methanol resulted[1141] in ring contraction and formation of 4-phenyl-3-cyano-*N*-tert-butyl-2-azetidinone (equation 841). The mechanism of this reaction reportedly involves removal of the acidic proton adjacent to the nitro group followed by nitrogen–oxygen bond cleavage. The initially produced acyl nitro intermediate then undergoes a subsequent cyclization under the reaction conditions used.

Thermolysis of N-substituted tetrahydro-1,2-oxazine-3,6-diones at 190 °C under an atmosphere of nitrogen affords[1265] the correspondingly N-substituted *β*-lactams as products (equation 842). A 1,4-diradical is proposed as an intermediate in this conversion.

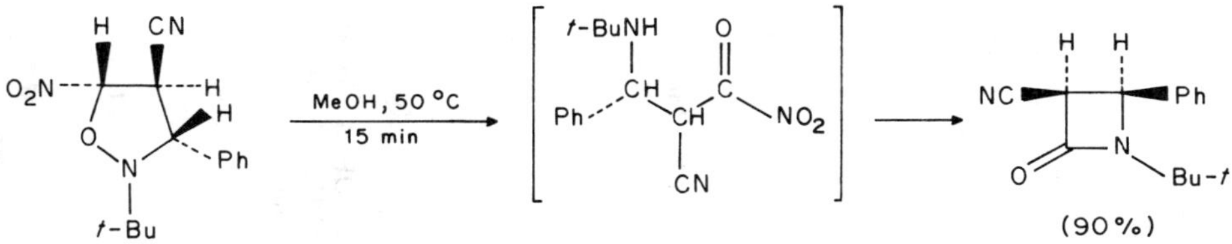

(841)

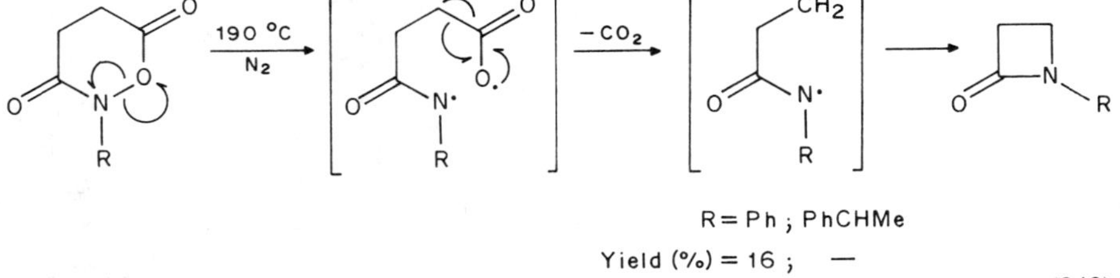

R = Ph ; PhCHMe

Yield (%) = 16 ; —

(842)

TABLE 58. Preparation of β-lactams by photolytic ring contraction

Starting material	Reaction conditions	Product	Yield (%)	Reference
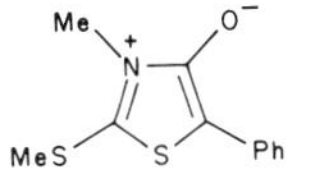	1. MeOH, $h\nu$, Ar, 50 °C, 5 h 2. THF, $NaBH_4$, H_2O, $NiCl_2 \cdot 6H_2O$, boric acid, EtOH, N_2, stir 15 min	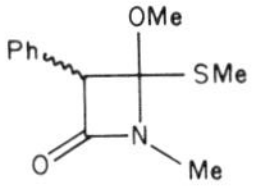 (73:27 mixture of diastereomers)	22	1266
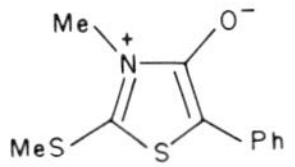	MeI, MeOH, Ar, $h\nu$, 6 h	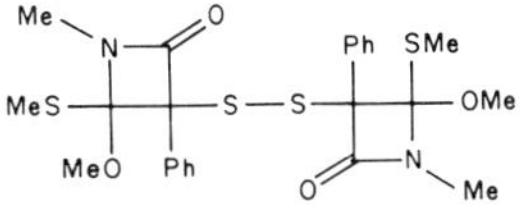	25	1266
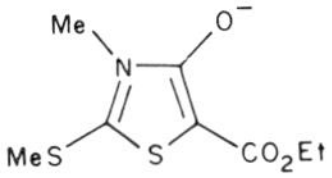	MeI or EtI, MeOH, $h\nu$, Ar, 15 h	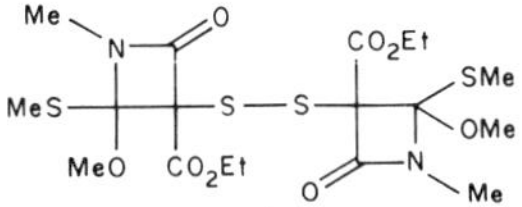 +	21	1266
		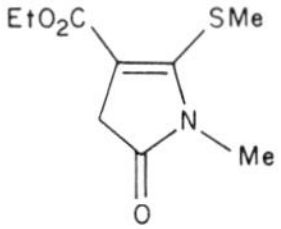	14	
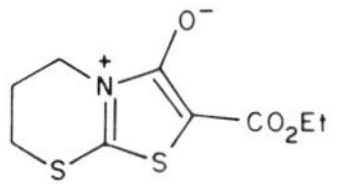	1. MeOH, Ar, $h\nu$, 50 °C, 34 h 2. C_6H_6 $(n\text{-}Bu)_3P$, reflux 5 min	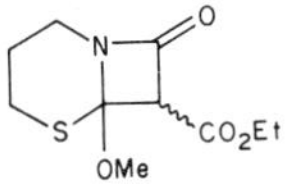 (82:18 mixture of diastereomers)	9	1266
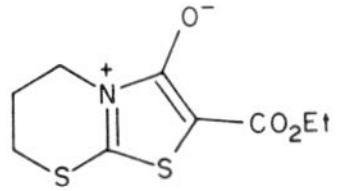	dioxane, $h\nu$, r.t., 36 h	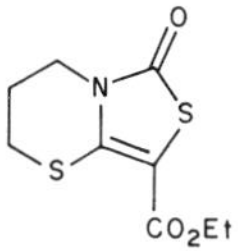	19	1266

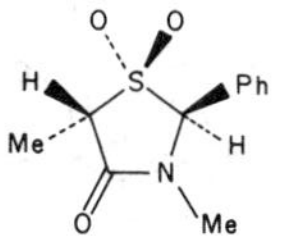	t-BuOH, MeCN, $h\nu$, 35 min	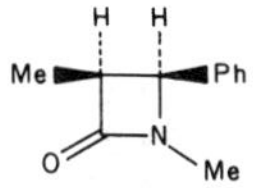 +	7	1264
		(169)		
		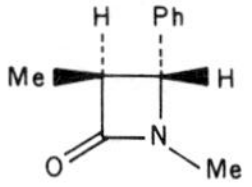	14	
		(170)		
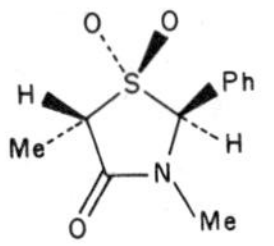	i-PrOH, $h\nu$, 2 h	**169** +	4	1264
		170	7.2	
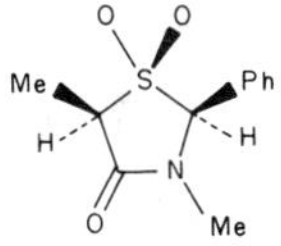	solvents, $h\nu$, 40 min	**169** + **170**		1264
	solvents = MeCN, t-BuOH (1:7)		31 + 8	
	solvents = i-PrOH, t-BuOH (1:1)		10 + 5	
	solvents = i-PrOH, t-BuOH (1:7)		14 + 7	
	solvent = i-PrOH		17 + 8	
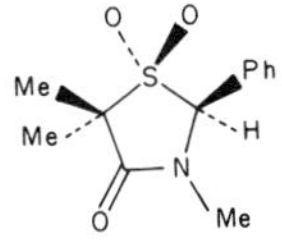	t-BuOH, MeCN, $h\nu$, 30 min	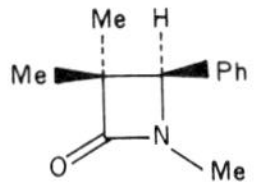	10	1264

(*continued*)

TABLE 58. *(continued)*

Starting material	Reaction conditions	Product	Yield (%)	Reference
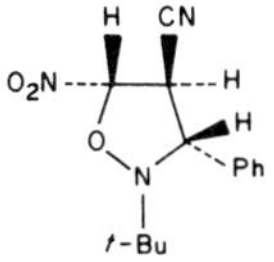	MeOH, $h\nu$, 2.5 H	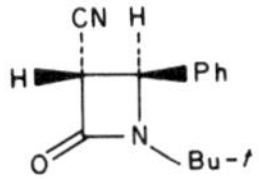	57	1141
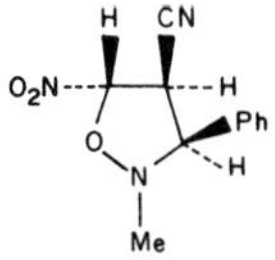	MeOH, $h\nu$ (3537 Å), Ar	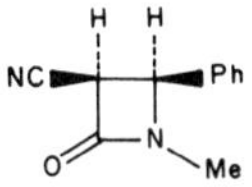	84[a]	1141
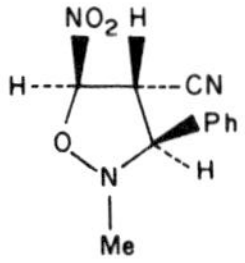	MeOH, $h\nu$ (3537 Å), 1.5 h	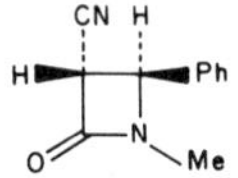	—	1141
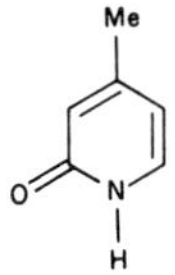	$h\nu$ (3100 Å)	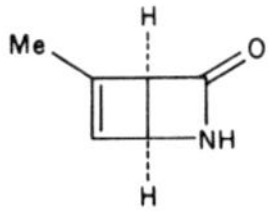	50–60	1267
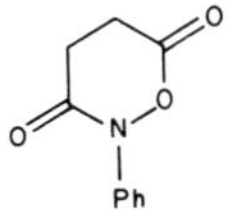	C_6H_6, $h\nu$, 40 min	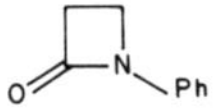	20	1265
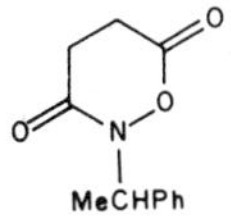	C_6H_6, $h\nu$, 220 h	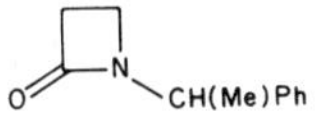	21	1265

Reactant	Conditions	Product	Yield (%)	Ref.
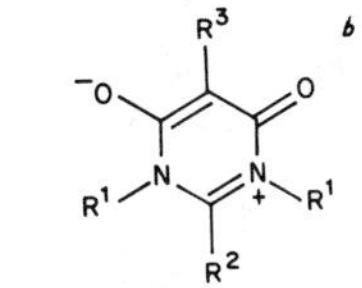 [structure] [b]	MeOH, $h\nu$, 20 °C	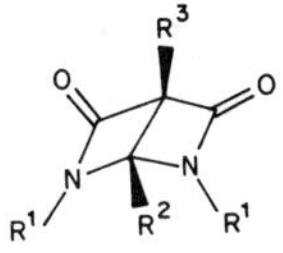 [structure]	—	1268
$R^1 = R^3 = Ph$; $R^2 = H$			96	
$R^1 = Ph$; $R^2 = Me$; $R^3 = H$			91	
$R^1 = Ph$; $R^2 = R^3 = Me$			83	
$R^1 = R^3 = Ph$; $R^2 = Me$			61	
$R^1 = Me$; $R^2 = EtS$; $R^3 = Ph$			92	
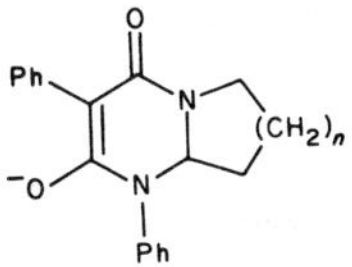 [structure]	MeOH, $h\nu$	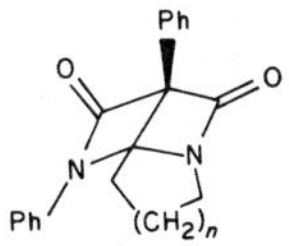 [structure]		1268
$n = 2$			64	
$n = 3$			92	
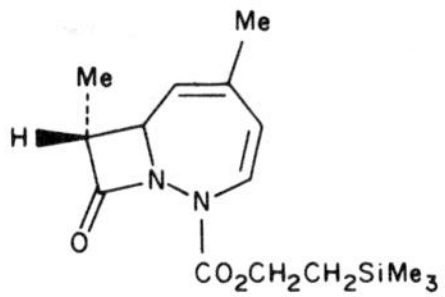 [structure] $CO_2CH_2CH_2SiMe_3$	CH_2Cl_2, $h\nu$, 5 h	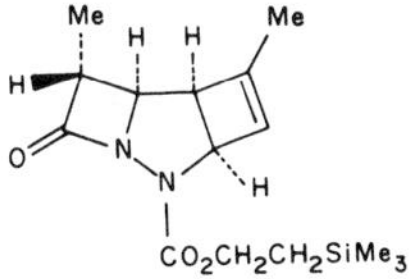 [structure] $CO_2CH_2CH_2SiMe_3$	73	1129
		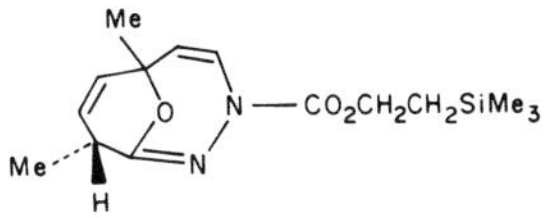 + [structure] $CO_2CH_2CH_2SiMe_3$	18	

(continued)

TABLE 58. (*continued*)

Starting material	Reaction conditions	Product	Yield (%)	Reference
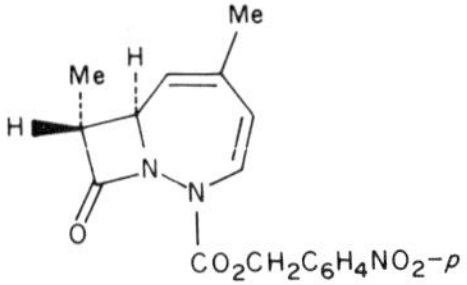 $CO_2CH_2C_6H_4NO_2$-*p*	CH_2Cl_2, *hv*	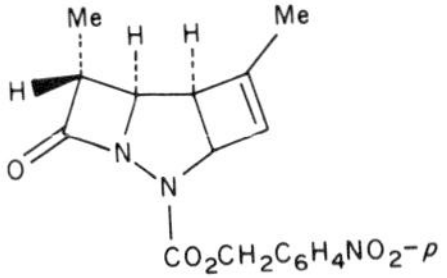 $CO_2CH_2C_6H_4NO_2$-*p*	40	1129

[a]Product isomerizes on work-up to the more stable *trans* isomer.

[b]Structure exists as an equilibrium mixture

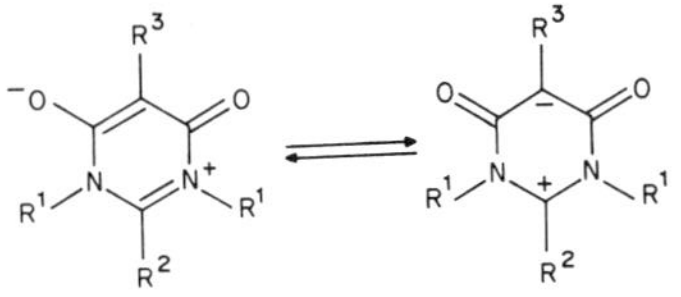

The variety of structures which have been used to produce lactams by photolytic ring-contraction reactions are reported in Table 58.

5-Nitro-substituted isoxazolidines, mono- and bicyclic-1-acylpyrazolidin-3-ones and Melillo's lactone all undergo ring contractions upon treatment with base to produce β-lactams. In the 5-nitroisoxazolidine compound series[1143] both potassium *tert*-butoxide (equation 843) and 1,5-diazabicyclo[4.3.0]-5-nonene (DBN, equations 844 and 845) have both been used as the base required to effect ring contractions.

In the monocyclic 1-acyl-5,5-dimethyl- (equation 846) and *cis*-bicyclo-1-acylpyrazolidin-3-one (equation 847) compound series[1269] a base in glyme is used in the first step to

DMSO, KOBu-*t*; r.t. stir, 18 h (11%) (843)

1. DBN, C_6H_6, stir, 25 °C, time (h); 2. H_2O, HCl (1:1)

R = Me ; *t*-Bu

Time (h) = 18 ; 12

Yield (%) = 50 ; —

(844)

DBN, C_6H_6, stir, 25 °C, 18 h (845)

R = Me ; *t*-Bu

Yield (%) = 69 ; 75

1. base, glyme; 2. HgO, 25 °C; 3. 2,4,6-$Me_3C_6H_2SO_2ONH_2$, CH_2Cl_2 (846)

R = Ph ; Cl_3CCH_2O

Yield (%) = 72 ; 39

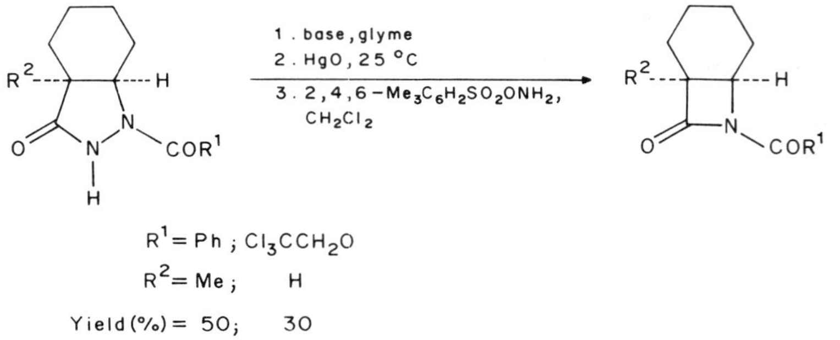

1. base, glyme
2. HgO, 25 °C
3. 2,4,6-$Me_3C_6H_2SO_2ONH_2$, CH_2Cl_2

R^1 = Ph ; Cl_3CCH_2O
R^2 = Me ; H
Yield (%) = 50; 30

(847)

produce the corresponding anion. Treatment of the anion with *o*-mesitylenesulphonylhydroxylamine causes amination of the anion, which upon subsequent oxidation with three equivalents of yellow mercuric oxide produces a *N*-nitrene intermediate. At this point two routes are possible to produce the β-lactam products, formation of a triazene intermediate or reaction by a dipolar mechanism, but which mechanism is operational is unknown at this time[1269] (equation 848).

Treatment of the optically active Melillo's lactone with dicyclohexylcarbodiimide produces[1270] the corresponding β-lactam (equation 849).

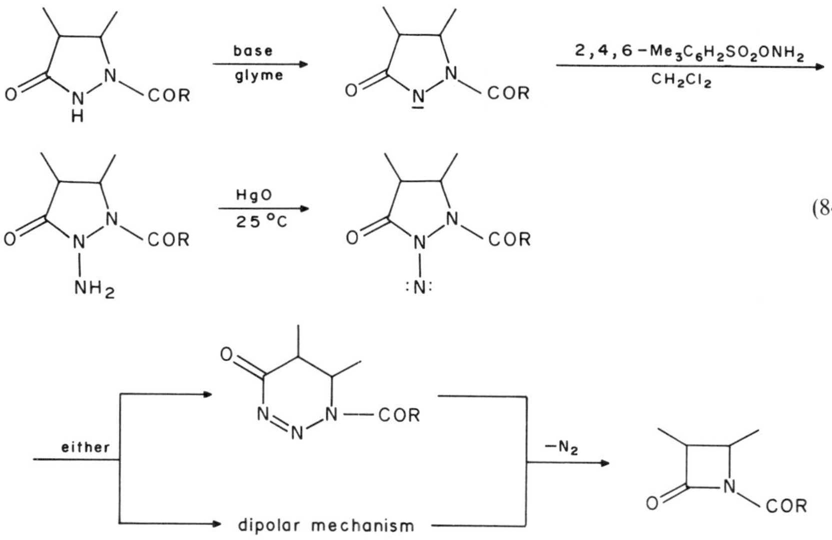

base, glyme; 2,4,6-$Me_3C_6H_2SO_2ONH_2$, CH_2Cl_2; HgO, 25 °C; either; dipolar mechanism; $-N_2$

(848)

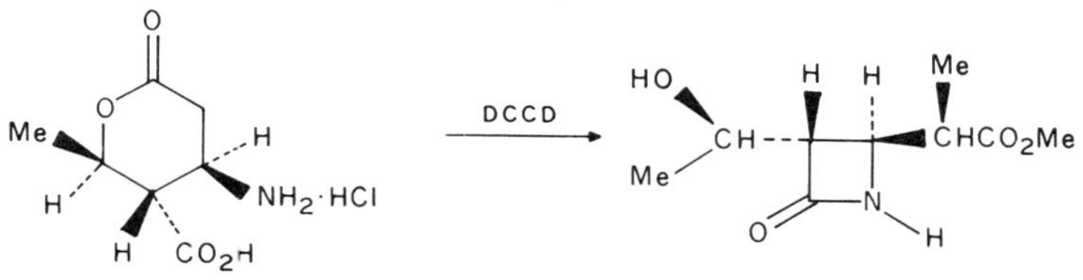

DCCD

(849)

Periodate oxidation of α-keto γ-lactams results[1271] in β-lactam formation by oxidative ring contraction. Application of this procedure to 7-substituted-8-hydroxy-9-oxo-1-azabicyclo[4.3.0]non-7-enes produces the corresponding bicyclic β-lactams (equation 850).

1. THF, $NaIO_4$, stir, 6–7 °C
2. $Et_3O^+ BF_4^-$, $NaHCO_3$, pH = 6–7

R = Br, Me

(850)

Reductive desulphurization of mesoionic thiazol-4-ones using Raney nickel affords[1272] stereospecific formation of *cis*-β-lactams through the intermediate formation of a dipolar ion (equation 851). Performing the same reaction in the presence of triphenylphosphine produced[1272] exclusively the *trans*-β-lactams through the intermediate formation of a structurally similar but configurationally different dipolar ion (equation 852).

Raney Ni, H_2, r.t., MeOH, THF or Me_2CO

(851)

R^1	R^2	Yield (%)
Ph	Ph	85
p-Tol	Ph	80
Ph	*o*-Tol	18
Ph	*p*-ClC_6H_4	78
Ph	$PhCH_2$	80

Raney Ni, H , r.t., Ph_3P

R = Ph ; *p*-ClC_6H_4

Yield (%) = 56 ; 52

(852)

*2. *Ring expansions*

**a. Beckmann rearrangement.* At least one review article dealing with the synthesis of heterocyclic systems from some hydroxylamine and hydrazine derivatives has been published[1273] recently, while a number of other articles published during this period discuss new apparatus and methods[1274,1275] and new catalysts which have been employed to effect a Beckmann rearrangement. The new catalysts employed range from a fluidized bed catalyst containing boric acid and aluminium oxide at 210–450°C[1276] to a boron-containing mixture prepared from boric acid and aluminium oxide[1277] to an acid chloride or oxychloride[1278] (equation 853).

NOH

1. C_6H_5Me, $SOCl_2$, 95 °C
2. reflux 5 min

NH_2 O

(95%)

(853)

Other sulphur containing catalysts which have been used to produce lactams from oximes via a Beckmann rearrangement include sulphuric acid[1279] (equation 854), sulphur trioxide in liquid sulphur dioxide[1280] (equation 855), diphenylsulphuryl anhydride[1281] (equation 856), and hydroxylamine-*O*-sulphuric acid[1282] (equation 857). In the triterpene series the oximes of methyl oleanonate (equation 858), methyl betulonate and lupenone (equation 859) have, upon treatment with *p*-toluenesulphonyl chloride or phosphorus oxychloride in pyridine, produced[1283] lactams via Beckmann rearrangement.

$(CH_2)_n$ NOH

$n = 1, 2$, and 3

$NCCH_2CH_2OH$
H_2SO_4, 125 °C, 30 min

$(CH_2)_{n+1}$ N H O

(854)

$NH_2 \cdot HCl$ NOH

SO_3, liq. SO_2

O N H NH_2

(67%)

(855)

NOH

C_6H_5Me, $(PhSO_2)_2O$,
reflux 95 °C
5 min
(also accomplished using $SOCl_2$)

NH O

(99%)

(856)

$$\text{ketone} + H_2NOSO_3H \longrightarrow [\,{>}{=}NOSO_2OH] \longrightarrow \text{lactam} \quad (857)$$

Ketone	Lactam
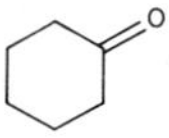	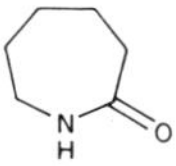 NH, O
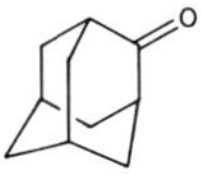 O	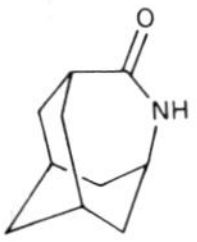 O, NH
 O	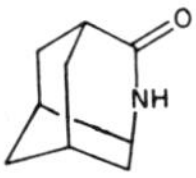 O, NH
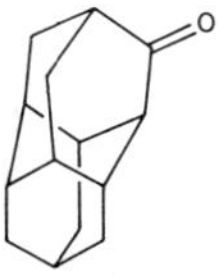 O	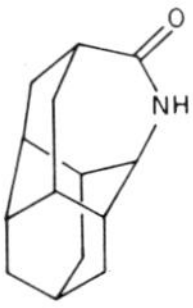 O, NH

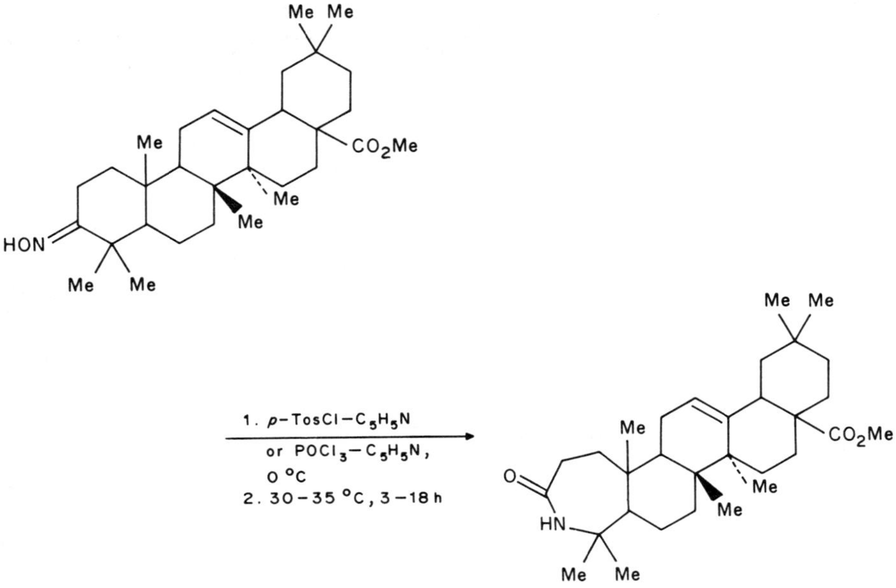

Me Me
Me
CO_2Me
Me
Me
HON
Me Me

1. p-TosCl$-C_5H_5N$ or $POCl_3-C_5H_5N$, 0 °C
2. 30–35 °C, 3–18 h

Me Me
Me
CO_2Me
O
Me
Me
HN
Me Me

(858)

1. p-TosCl–C_5H_5N or $POCl_3$–C_5H_5N, 0 °C
2. 30–35 °C, 3–18 h

R = CO_2Me (methyl betulonate), Me (lupenone)

(859)

In addition to the phosphorus oxychloride mentioned above, another phosphorus containing reagent which has been reported[1284] to catalyse Beckmann rearrangements is triphenylphosphine. Reactions of cycloalkanone oximes, 1-halo-1-nitroso- and 1-halo-1-nitrocycloalkanes with triphenylphosphine have produced lactams in high yields via a one-step conversion (equation 860). The mechanisms proposed for these conversions differ slightly depending upon the structure of the starting material. In the

$(CH_2)_n$ C=NOH: 1. C_6H_6, Ph_3P, X_2, r.t. 2. HCl, H_2O

$(CH_2)_n$ C(Cl)(NO): 1. C_6H_6, C_6H_5Me, Et_2O, or THF, Ph_3P, r.t. 2. HCl, H_2O

$(CH_2)_n$ C(Cl)(NO_2): 1. C_6H_6, Ph_3P, reflux 5 h 2. HCl, H_2O

→ $(CH_2)_n$ NH–C=O (860)

Starting material	Reagents	Product	Yield (%)
cyclopentanone oxime (=NOH)	Ph_3P, Cl_2	2-piperidone	76
cyclohexanone oxime (=NOH)	Ph_3P, Cl_2	caprolactam	86
	Ph_3P, Br_2		74
	Ph_3P, I_2		39

(continued)

Starting material	Reagents	Product	Yield (%)
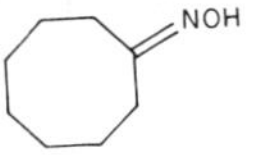 NOH	Ph_3P, Br_2	[a]	74
	Ph_3P, I_1	[a]	60
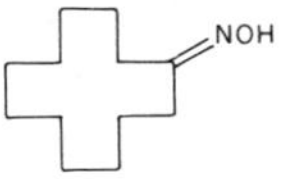 NOH	Ph_3P, Br_2	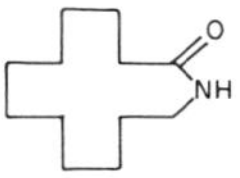 O, NH	81
	Ph_3P, I_2		48
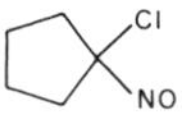 Cl, NO	Ph_3P	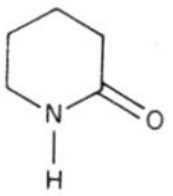 N, O, H	57
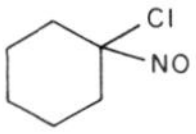 Cl, NO	Ph_3P	[a]	96
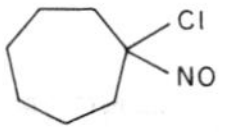 Cl, NO	Ph_3P	[a]	76
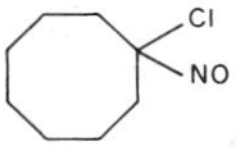 Cl, NO	Ph_3P	[a]	83
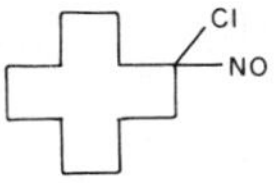 Cl, NO	Ph_3P	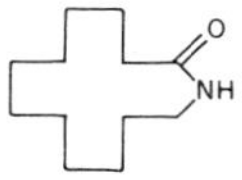 O, NH	78
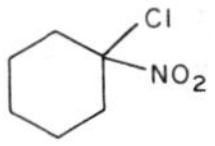 Cl, NO_2	Ph_3P	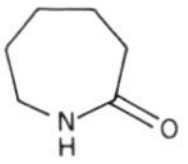 N, H, O	77
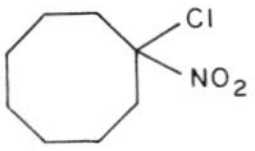 Cl, NO_2	Ph_3P	[a]	42
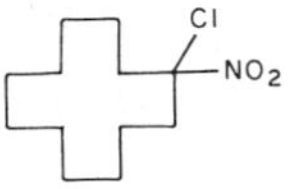 Cl, NO_2	Ph_3P	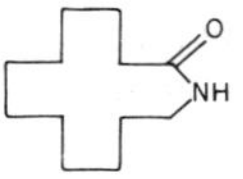 O, NH	32

[a] Perhydroazonin-2-one.

case of the 1-chloro-1-nitrosocycloalkanes the mechanisms proposed involve initial attack of phosphorus on oxygen to give a phosphonium salt, followed directly by thermal rearrangement to a chloro imine, which could not be isolated. Similar mechanistic steps are proposed for the conversions with cycloalkanone oximes (Scheme 13). In the case of 1-chloro-1-nitrocycloalkanes, the mechanism proposed involves initial formation of a Perkov-type intermediate, followed by transformation into a phosphonium salt, and then Beckmann rearrangement (Scheme 14).

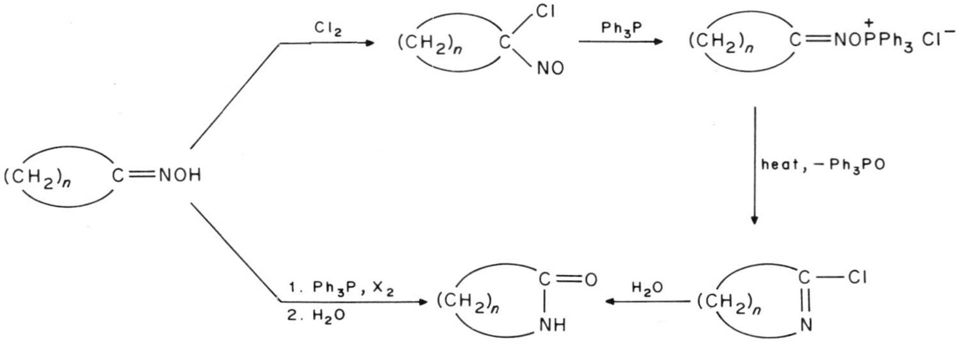

SCHEME 13

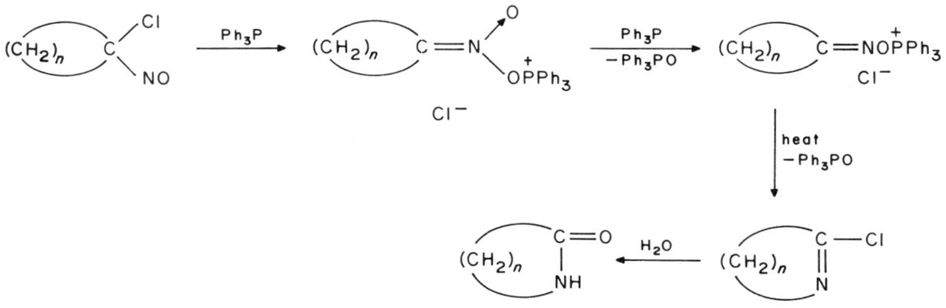

SCHEME 14

The photo-Beckmann rearrangement has been reported to occur with α-alkyl- (equation 861)[1285] 5-hydroxy- (equation 862)[1285,1286], β,γ-cyclopropyl- (equation 863)[1285] and β,γ-unsaturated (equation 864)[1285] steroidal ketone oximes to produce the corresponding lactams.

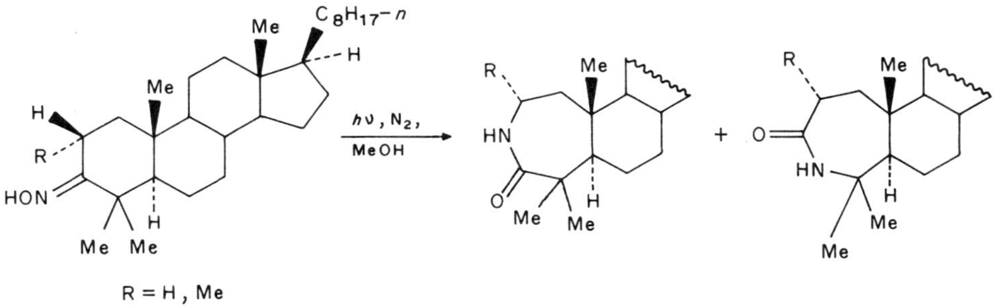

R = H, Me

(861)

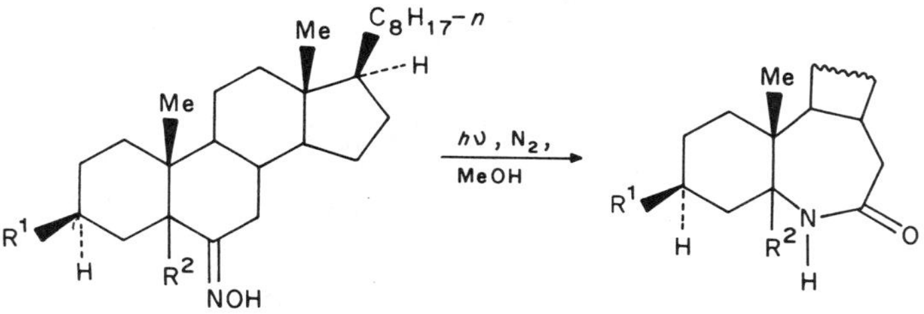

R^1 =	H	H	AcO	AcO
R^2 =	α-OH	β-OH	α-OH	β-OH
Rx. time (h) =	—	—	23	12.5
Yield (%) =	—	—	21 + 15	20 + 14

(862)

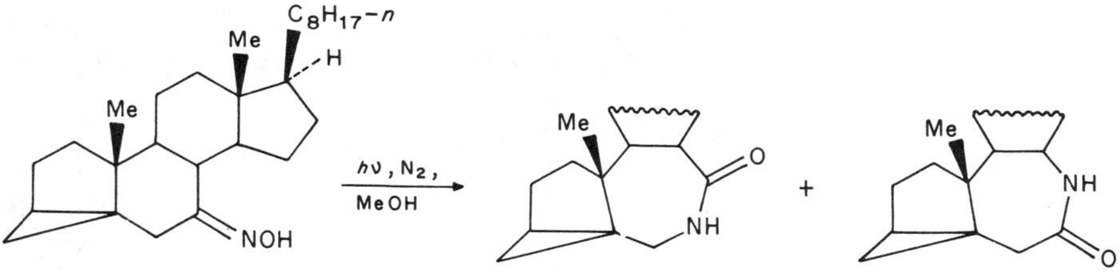

(863)

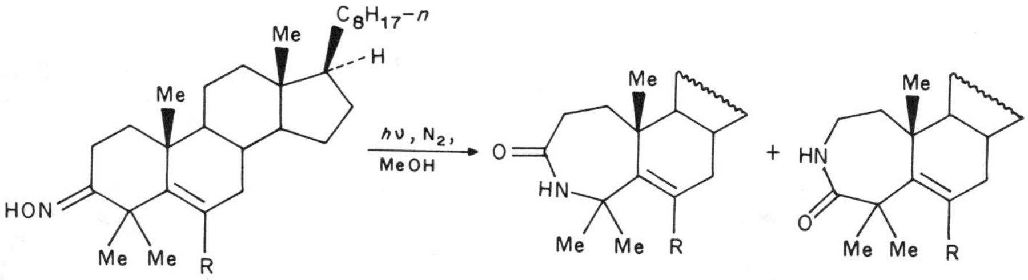

R = H, Me

(864)

******b.* *Schmidt rearrangement.* Reaction of thiochroman-4-one with hydrogen azide produces[1155] a 4:1 mixture o 2,3-dihydro-1,5-benzothiazepin-4(5*H*)-one and 3,4-dihydro-1,4-benzothiazepin-5(2*H*)-one via a Schmidt rearrangement (equation 865). The 2,3-dihydro-1,5-benzothiazepin-4(5*H*)-one was also prepared[1155] independently by a Beckmann rearrangement to confirm its structure (equation 866).

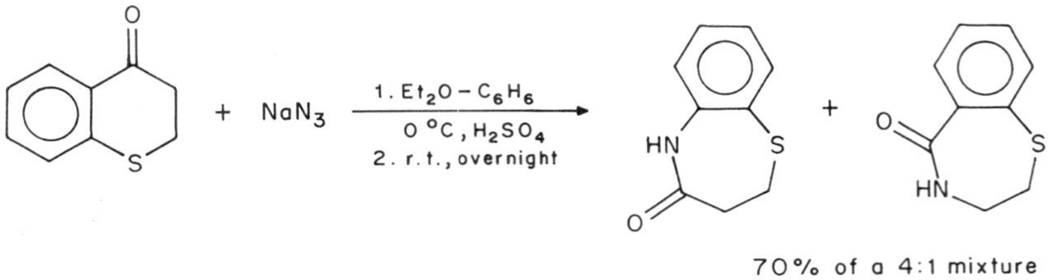
O
S
+ NaN3
1. Et2O – C6H6
0 °C, H2SO4
2. r.t., overnight
HN
S
O
+
O
S
HN
70% of a 4:1 mixture

(865)

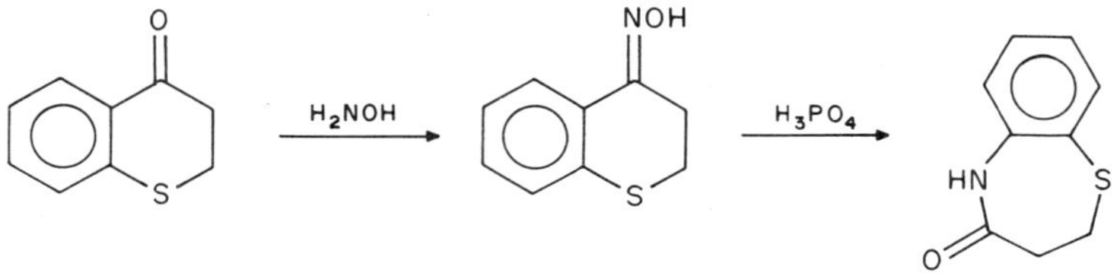
O
S
H2NOH
NOH
S
H3PO4
HN
S
O

(866)

Schmidt rearrangements have also been performed[1183] on the same series of triterpenes whose oximes were reported in Section *III.D.2.a. to undergo Beckmann rearrangements—methyl oleanonate (equation 867), methyl betulonate and lupenone (equation 868).

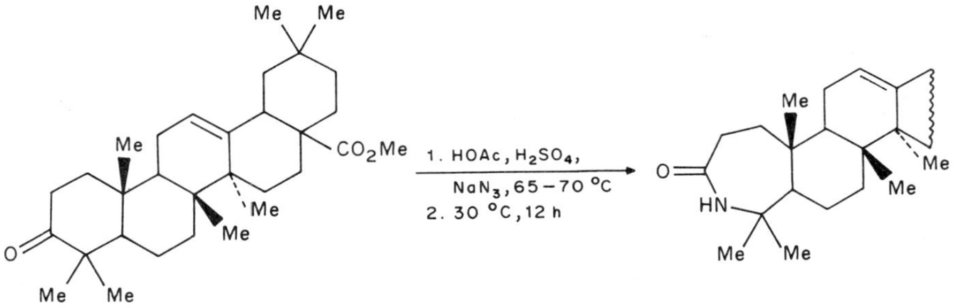
Me Me
Me
CO2Me
Me
Me
O
Me Me
1. HOAc, H2SO4,
NaN3, 65–70 °C
2. 30 °C, 12 h
Me
O
HN
Me
Me
Me Me

(867)

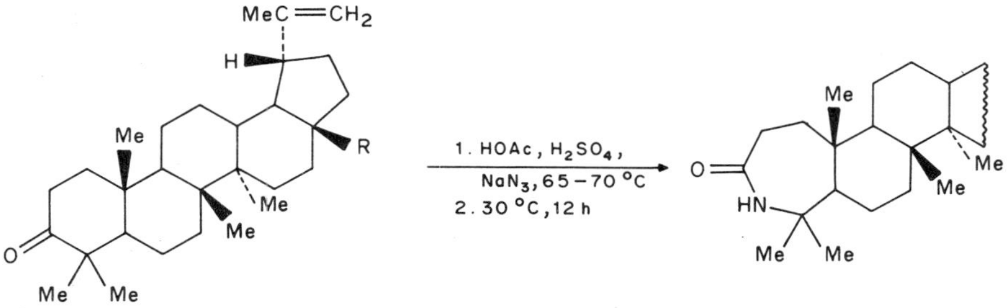
MeC=CH2
H
R
Me
Me
Me
O
Me Me
1. HOAc, H2SO4,
NaN3, 65–70 °C
2. 30 °C, 12 h
Me
O
HN
Me
Me
Me Me
R = CO2Me (methyl betulonate), Me (lupenone)

(868)

An interesting Schmidt-type rearrangement has been reported[1287] to occur with α-azido sulphides, prepared by reaction of the corresponding thioketal with iodine azide (equations 869 and 870). The lactam product obtained from this rearrangement results from exclusively migration of the more highly substituted carbon and is consistent with the migration to an electron-deficient species which is observed in typical Schmidt rearrangments.

n = 3 ; 4 ; 5 ; 6 ; 4.

R = H ; H ; H ; H ; Me.

Yield (%) = 88 ; 79 ; 64 ; 75 ; 74.

(869)

(66%)

(870)

*c. *Miscellaneous ring expansions.* The synthesis of polyamino lactams by ring enlargement using the 'zip' reaction has been reviewed[1288].

A variety of approaches have been used to prepare lactams by ring expansions other than the Beckmann and Schmidt rearrangements. In this section the ring expansion

Rh (I) complex + CO (30 atm.)

C_6H_6, 40 °C or

Co complex, C_6H_6, N_2, 65 °C

(871)

R^1	R^2	Complex	Yield (%)
t-Bu	*t*-Bu	$[Rh(CO)_2Cl]_2$	100
t-Bu	*t*-Bu	(1,5-hexadienyl $RhCl)_2$	70
t-Bu	*t*-Bu	(1,5-cyclooctadienyl $RhCl)_2$	75
t-Bu	*t*-Bu	$Co_2(CO)_8$	90
t-Bu	*t*-Bu	$Co_4(CO)_{12}$	84
t-Bu	1-adamantyl	$[Rh(CO)_2Cl]_2$	51
t-Bu	1-adamantyl	$Co_2(CO)_8$	51
1-adamantyl	*t*-Bu	$[Rh(CO)_2Cl]_2$	90
1-adamantyl	1-adamantyl	$[Rh(CO)_2Cl]_2$	80
1-adamantyl	1-adamantyl	$Co_2(CO)_8$	95
1-adamantyl	1-adamantyl	$Co_4(CO)_{12}$	100

reactions reported are discussed in order of the ring size, which means that the three- to four-membered ring expansions are discussed first, followed by the four- to five-membered ring expansions, etc.

Three methods have been used[1074,1289,1290] to expand three-membered lactams to larger ring lactams. In the first method, α-lactams are converted[1289] regiospecifically to azetidine-2,4-diones using rhodium(I) or cobalt complexes. However, the processes involved using these two reagents are significantly different. The rhodium reaction occurs using carbon monoxide and is catalytic while the cobalt reaction is inhibited by carbon monoxide and is not catalytic (equation 871). The mechanism proposed for the rhodium catalysed reaction involves insertion of the rhodium into the saturated carbon–nitrogen bond of the α-lactam to give a four-membered intermediate, which undergoes ligand migration to a five-membered intermediate, followed by addition of carbon monoxide and reductive elimination to afford the azetidine-2,4-diones (equation 872). The mechanism proposed for the stoichiometric cobalt carbonyl reaction involves the α-lactam functioning as a Lewis base and inducing disproportionation of the cobalt carbonyl to produce a new complex, of which the cationic portion rearranges to a metallacycle which undergoes ligand migration. Reaction of the rearranged complex with additional cobalt carbonyl and α-lactam produces the product, a new complex and carbon monoxide (equation 873)

+ $[Rh(CO)_2Cl]_2$ → [Cl—Rh(OC)(C=O) four-membered intermediate, R¹, R², N, O] → [Cl, OC, Rh, N, R², R¹, O, O five-membered intermediate] $\xrightarrow{CO}$ [OC, Cl—Rh, OC, N, R², R¹, O, O] → R¹ azetidine-2,4-dione N—R²

(872)

The second method used[1290] to enlarge three-membered lactams involves reaction of perfluoro-α-lactams with aldehydes or ketones to produce oxazolidinones (equation 874), or with nitriles to produce imidazolinones (equation 875).

Finally, treatment[1074] of N-substituted α-lactams with alkynyl lithium reagents produced 2-pyrrolin-4-ones (equation 876).

Beginning with β-lactams, several methods have been used[1133,1155,1261,1291–1293] to produce γ-lactams, while several additional methods have been used[1003,1017,1155,1294,1295] to produce larger ring lactams. Thus, reaction of 4-substituted azetidine 2-carboxylic acids with oxalyl chloride produces three products, an acid chloride (**171**), an iminium salt, identified as the azetidinone (**172**), and/or a ring expanded chloro-γ-lactam (**173**,

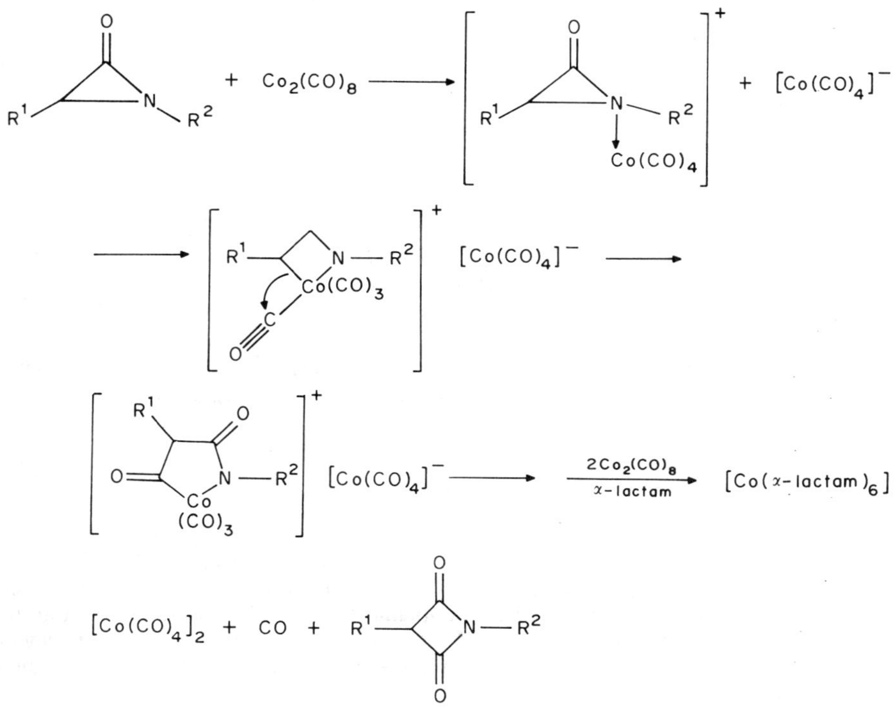

$+ Co_2(CO)_8 \longrightarrow [\ldots]^+ + [Co(CO)_4]^-$

$[\ldots]^+ [Co(CO)_4]^- \longrightarrow$

$[\ldots]^+ [Co(CO)_4]^- \longrightarrow \xrightarrow[\alpha\text{-lactam}]{2Co_2(CO)_8} [Co(\alpha\text{-lactam})_6]$

$[Co(CO)_4]_2 + CO +$

(873)

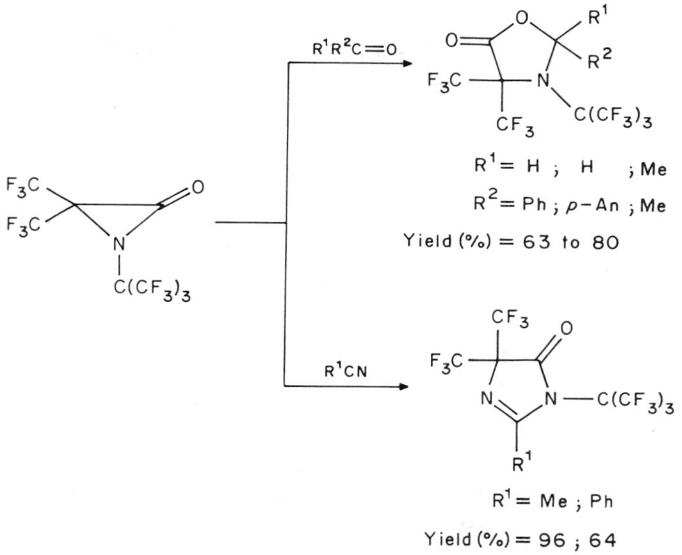

$R^1R^2C{=}O$

(874)

R^1 = H ; H ; Me

R^2 = Ph ; p-An ; Me

Yield (%) = 63 to 80

R^1CN

(875)

R^1 = Me ; Ph

Yield (%) = 96 ; 64

R^1 = t-Bu ; 1-adamantyl ; t-Bu ; 1-adamantyl.
R^2 = t-Bu ; 1-adamantyl ; 1-adamantyl; t-Bu.
Yield (%) = 54 ; — ; — ; —.

(876)

equation 877). In view of the stereochemistry observed in these reactions, the mechanism proposed[1291] involves the iminium salt formation through a transition state in which the lone pair on nitrogen is disposed antiperiplanar to the acid chloride carbonyl group permitting decarbonylation, as indicated in path a (Scheme 15); also, that rearrangement to the γ-lactam involves a fused-ring aziridinium salt which undergoes S_N2 ring opening by chloride ion as indicated in path b (Scheme 15).

(171) **(172)** **(173)**

(877)

R^1	R^2	R^3	Stereochemistry of starting acid	Yield (%) 171	172	173
Me	H	t-Bu	*trans*	—	—	69
H	Me	t-Bu	*cis*	78	—	—
Me	H	$PhCH_2$	*trans*	26	22	30
H	Me	$PhCH_2$	*cis*	57	22	—
Me	H	Me	*trans*	15	23	—
H	Me	Me	*cis*	54	30	—
MeO_2C	H	$PhCH_2$	*trans*	—	—	46
H	MeO_2C	$PhCH_2$	*cis*	—	—	68
MeO_2C	H	*c*-Hex	*trans*	—	—	58
H	MeO_2C	*c*-Hex	*cis*	—	—	59

SCHEME 15

Treatment of 4-benzoylazetidinone with alkyl halides in the presence of sodium hydride in dimethylformamide produces[1292] the ring expanded γ-lactams, alkoxypyrrolinones (equation 878).

R =	Me	Et	$PhCH_2$	$CH_2{=}CHCH_2$
X =	I	I	Cl	Br
Yield % =	7[a]	75	76	66[b]

(878)

[a]A mixture of both stereoisomers of 4-benzoyl-3-methyl-3,4-diphenyl-1-(*p*-methoxyphenyl) -2-oxazetidine was also obtained in 58% yield.
[b]One stereoisomer of 3-allyl-4-benzoyl-3,4-diphenyl-1-(*p*-methoxyphenyl)-2-oxoazetidine was also obtained.

Several interesting ring expansions of β-lactams to imidazolin-4-one have been reported[1133,1261,1293], the first involves reaction[1288] of *N,N'*-dibenzylaza-β-lactam with lithium diisopropylamide in tetrahydrofuran followed by treatment with methyl iodide (equation 879). The product presumably arises by rearrangement of the dipole stabilized anion formed by deprotonation of the *N*-2-benzyl group.

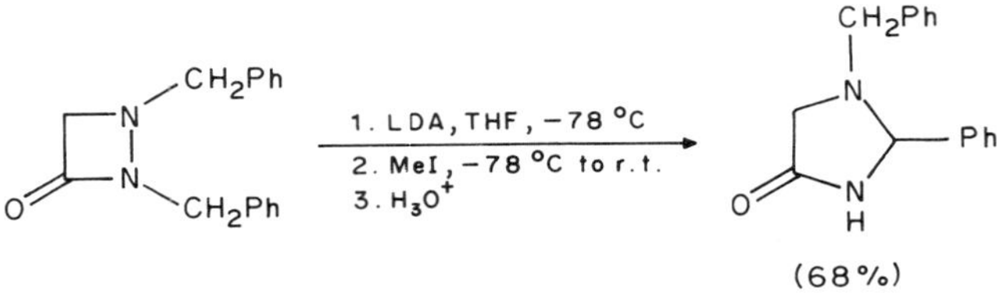

(879)

Refluxing the β-lactams, formed by reaction of imidoylazimines with ketones, produces[1133] benzocinnoline and good yields of 1,2,4,4-tetrasubstituted imidazolin-5-ones (equation 880).

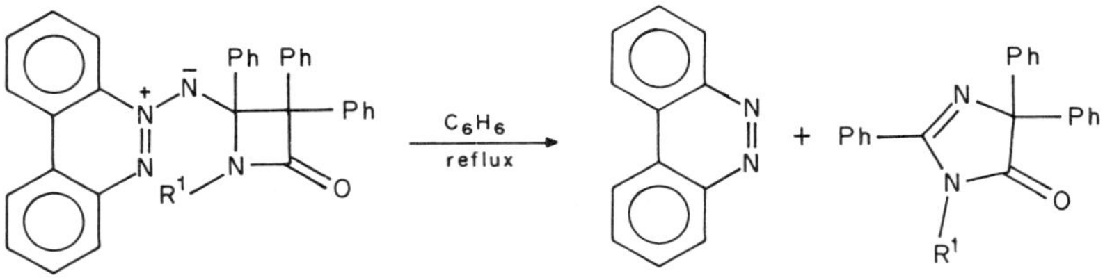

R^1 = Ph ; o-Tol

Yield (%) = 70 ; 75

(880)

Refluxing, this time in the presence of iodine, has also been used to produce[1293] 1,2,3-trisubstituted imidazolin-4-ones from amino substituted β-lactams (equation 881).

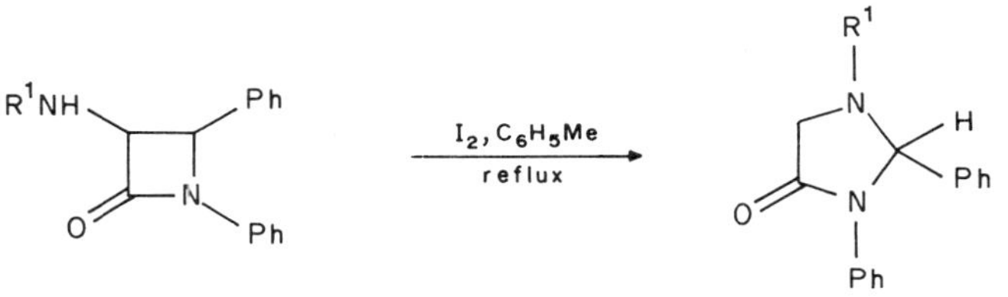

(881)

$R^1 = PhCH_2CO$; PhCO ; with p-Tos, no reaction was observed.

Transamidation of N-(haloalkyl)azetidin-2-ones by treatment with liquid ammonia or a primary amine in a sealed tube produces[1017,1294] the ring expanded seven-, eight- and nine-membered azalactams in good to excellent yields (equation 882, Table 59).

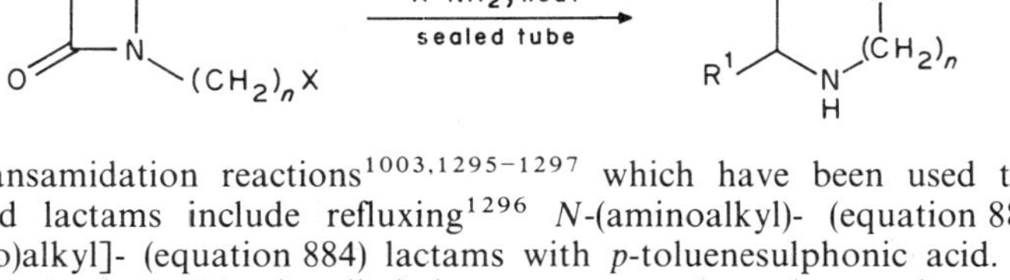

(882)

Other transamidation reactions[1003,1295–1297] which have been used to produce ring-enlarged lactams include refluxing[1296] N-(aminoalkyl)- (equation 883) or N-[(alkylamino)alkyl]- (equation 884) lactams with p-toluenesulphonic acid. The intermediates found when N-(aminoalkyl) lactams are used as the starting material are

TABLE 59. Transamidiation of *N*-(haloalkyl)azetidin-2-ones[1017,1293]

Azetidinone	Amine	Reaction conditions	Product	Yield (%)
4-Ph, N-$(CH_2)_2Br$	NH_3 (l)	45 °C, 14 days, sealed tube	Ph, O, NH, NH (7-membered ring)	70
4-Ph, N-$(CH_2)_3Cl$	NH_3 (l)	20 °C, 6 days, sealed tube	Ph, O, NH, NH (8-membered ring) **(174)**	90
4-Ph, N-$(CH_2)_3Br$	NH_3(l)	20 °C, 2 days, sealed tube	**174**	80
4-Ph, N-$(CH_2)_2CHMe$(Br)	NH_3 (l)	60 °C, 5 days, sealed tube	Ph, Me, O, NH, NH (8-membered ring), *trans* +	71
			4-Ph, N-$CH_2CH{=}CHMe$ azetidin-2-one	18
4-Ph, N-$(CH_2)_4Cl$	NH_3(l)	60 °C, 7 days, sealed tube	Ph, O, NH, NH (9-membered ring) **(175)**	67

(continued)

TABLE 59. *(continued)*

Azetidinone	Amine	Reaction conditions	Product	Yield (%)
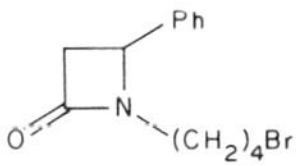 4-Ph, N-$(CH_2)_4Br$ azetidin-2-one	NH_3 (l)	55 °C, 8 days, sealed tube	**175**	65
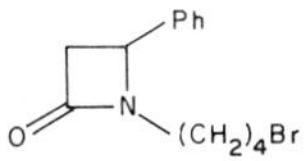 4-Ph, N-$(CH_2)_4Br$ azetidin-2-one	NH_3 (l)	20 °C, 5 days, sealed tube	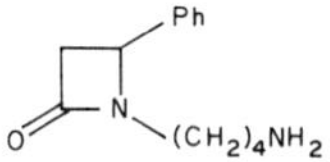 4-Ph, N-$(CH_2)_4NH_2$ azetidin-2-one	78
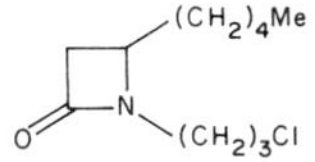 4-$(CH_2)_4Me$, N-$(CH_2)_3Cl$ azetidin-2-one	NH_3 (l)	50 °C, 3 days, sealed tube	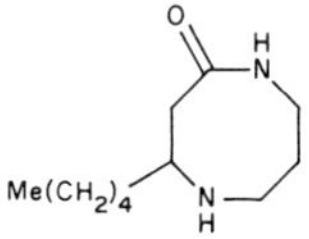 8-membered diazacyclic lactam (NH, NH) bearing $Me(CH_2)_4$	72
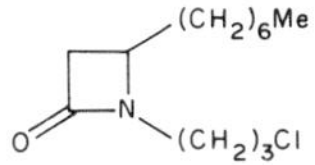 4-$(CH_2)_6Me$, N-$(CH_2)_3Cl$ azetidin-2-one	NH_3 (l)	50 °C, 3 days, sealed tube	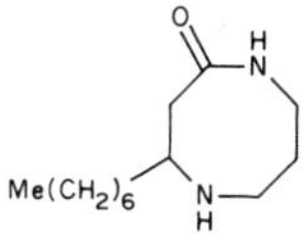 8-membered diazacyclic lactam (NH, NH) bearing $Me(CH_2)_6$	85
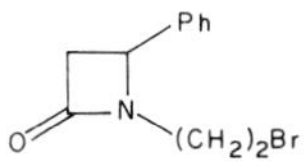 4-Ph, N-$(CH_2)_2Br$ azetidin-2-one	$EtNH_2$	45 °C, 7 days, sealed tube	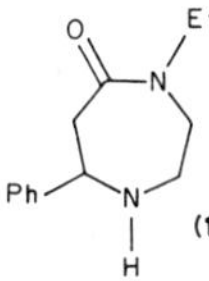 7-membered diazacyclic lactam (N-Et, NH) bearing Ph **(176)**	88
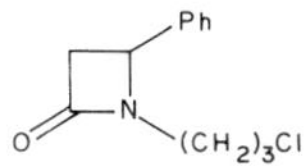 4-Ph, N-$(CH_2)_3Cl$ azetidin-2-one	$EtNH_2$	60 °C, 7 days, sealed tube	**176**	46

4-Ph-azetidin-2-one, N-$(CH_2)_3Cl$	$CH_2{=}CHCH_2NH_2$	85 °C, 7 days, sealed tube	Ph-substituted 1,5-diazocan-2-one, N-$CH_2CH{=}CH_2$, N–H	55
4-Ph-azetidin-2-one, N-$(CH_2)_5Br$	NH_3 (l)	65 °C, 7 days, sealed tube	4-Ph-azetidin-2-one, N-$(CH_2)_5NH_2$	60
4-Ph-azetidin-2-one, N-$(CH_2)_3Cl$	*i*-$PrNH_2$	45 °C, 4 days, sealed tube	4-Ph-azetidin-2-one, N-$(CH_2)_3NHPr$-*i*	85
4-Ph-azetidin-2-one, N-$(CH_2)_4Cl$	$CH_2{=}CHCH_2NH_2$	53 °C, 3 days, sealed tube	4-Ph-azetidin-2-one, N-$(CH_2)_4NHCH_2CH{=}CH_2$ → (toluene, reflux) Ph-substituted 1,6-diazonan-2-one, N-$CH_2CH{=}CH_2$, N–H	58

bicyclic amidines, while when N-[(alkylamino)alkyl] lactams are used, the intermediate involved is an amidinium salt. However, in both cases, partial hydrolysis using aqueous potassium hydroxide produces the starting lactam and a ring-enlarged lactam.

p-TosOH, xylene, reflux; H_2O, KOH; $CH_2CH_2NH_2$

(883)

Me; $(CH_2)_7$; C=O; $(CH_2)_3NHMe$; p-TosOH, xylene, reflux; H_2O, KOH; p-TosO⁻; HN; Me

(884)

Similarly, treatment of N-(aminoalkyl) lactams containing 8-, 9- and 13-membered rings with potassium 3-aminopropylamide in 1,3-propanediamine produced[1297] the corresponding 11-, 12-, 17- and 22-membered ring enlarged azalactams (equations 885 and 886).

$(CH_2)_n$; C=O; CH_2CH_3NHR + $\overset{+}{K}\ \bar{N}H(CH_2)_3NH_2$ $\xrightarrow[\text{time}]{H_2N(CH_2)_3NH_2}$ R = H, COMe

n = 6 ; 7

Time = 1 h ; 2 min

(885)

$(CH_2)_3NH(CH_2)_3NHR$ + $\overset{+}{K}\ \bar{N}H(CH_2)_3NH_2$ $\xrightarrow[\text{time}]{H_2N(CH_2)_3NH_2}$ $RNH(CH_2)_4N$; $(CH_2)_{11}$; N—R

R = H, COMe

(**177**)

+

Time =	5 min	30 min	7 h
Product =	**177**	**177** (R=H) + **178**	**178** (R=H)
Yield (%) =	–	41 + 12	31

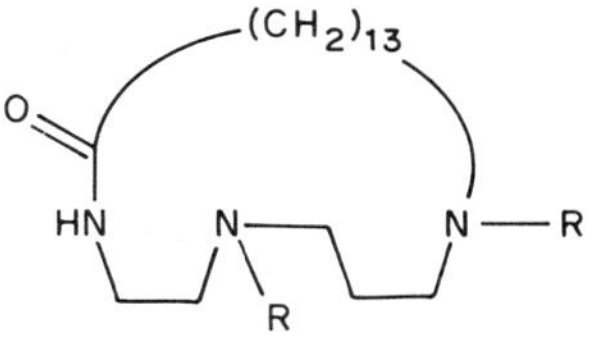

R = H, COMe

(168)

(886)

By refluxing 1,12-[(4*S*,4′*S*)bis(*N*-4-phenyl-2-oxoazetidino)]-4,9-diazadodecane in either quinoline or diphenyl ether, a transamidation ring expansion occurs to produce[1003] (−)-(2*S*, 2′*S*)-5,5′(tetramethylene)bis(2-phenyl-4-oxo-1,5-diazacyclooctane) (equation 887).

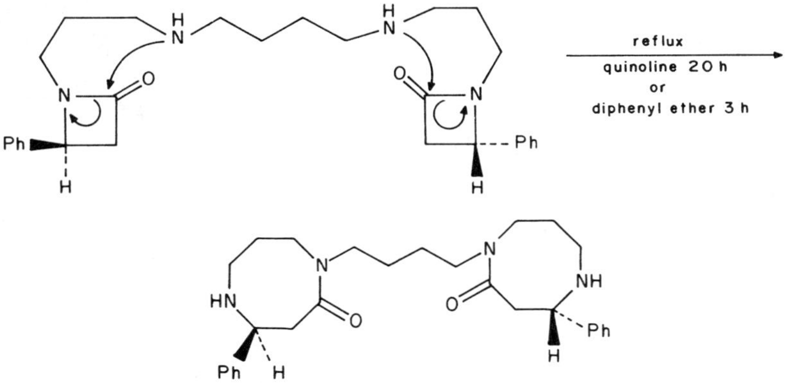

reflux, quinoline 20 h or diphenyl ether 3 h

with quinoline 25%;
with diphenyl ether 37%.

(887)

The final example of a transamidative ring expansion reaction is reported[1295] to occur when (2*S*, 3*S*)-3-[(2*R*)-2-amino-2-phenylacetamido]-2-methyl-4-oxo-1-azetidinesulphonic acid is treated with aqueous sodium hydroxide (pH = 7), at 35 °C for 5 days, to produce the correspondingly substituted 2,5-piperazinedione (equation 888).

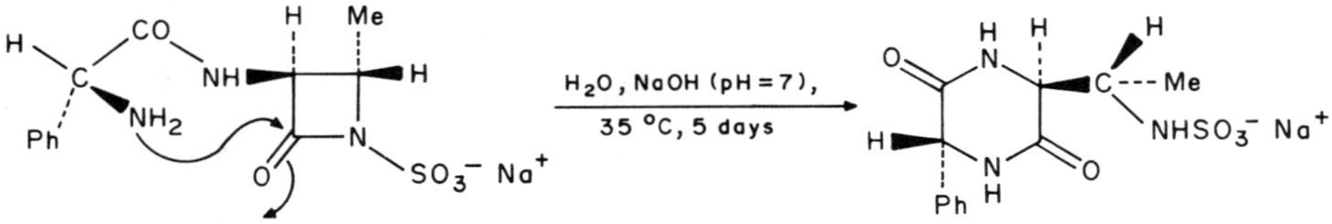

H_2O, NaOH (pH = 7), 35 °C, 5 days

(888)

Sodium peroxide oxidation of several substituted benzothiazepinones (equation 889) and benzoxazepinones (equation 890) produced[1155] the corresponding nine-membered benzothiazonindiones and benzoxazonindiones, while oxidation of enamine lactams using ozone or *m*-chloroperbenzoic acid (MCPA) produced[1090] ring-expanded medium- and macrocyclic dioxolactams in average yields (equation 891).

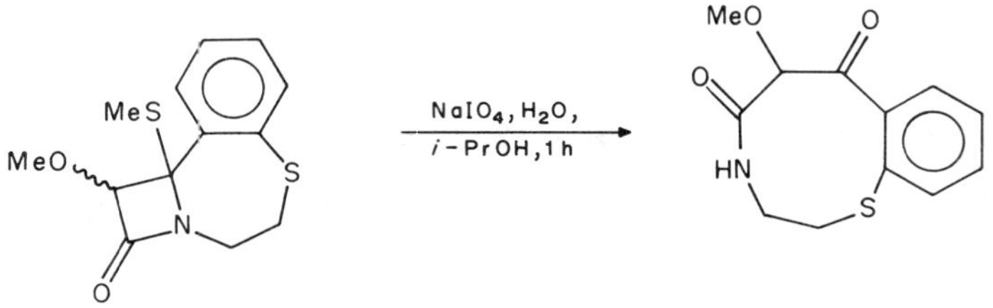

$\xrightarrow[i\text{-PrOH, 1 h}]{NaIO_4, H_2O,}$ (889)

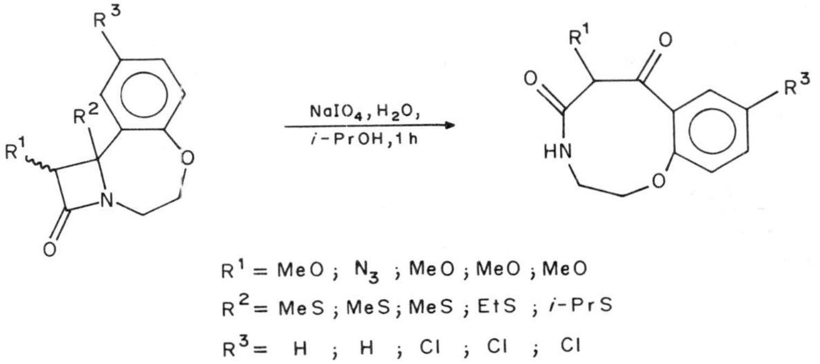

$\xrightarrow[i\text{-PrOH, 1 h}]{NaIO_4, H_2O,}$ (890)

R^1 = MeO ; N_3 ; MeO ; MeO ; MeO

R^2 = MeS ; MeS ; MeS ; EtS ; *i*-PrS

R^3 = H ; H ; Cl ; Cl ; Cl

1. O_3, CH_2Cl_2, 0 °C
2. Zn, HOAc, stir 2–4 h
or
m-$ClC_6H_4CO_3H, CH_2Cl_2$ heat

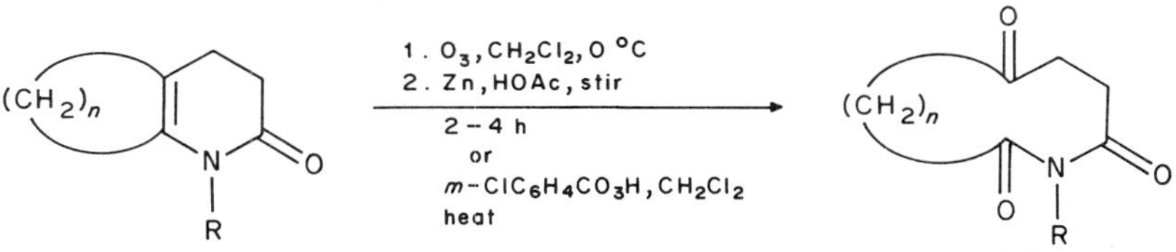

(891)

R	*n*	Oxidation method used	Yield (%)
Ph	4	ozone	60
Ph	4	MCPA	40
Ph	5	ozone	58
Ph	5	MCPA	45
Ph	6	ozone	62
Ph	6	MCPA	25
Ph	10	ozone	75
p-Tol	4	ozone	65
p-Tol	10	ozone	70

Reductive amination of a series of 3-(1-nitro-2-oxocycloalkyl)propanals produces[1298] the corresponding (aminopropyl)nitrocycloalkanones, which upon treatment with mild base affords from nine- to sixteen-membered macrocyclic lactams (equation 892).

$$\text{2-(CH}_2\text{CH}_2\text{CHO)-2-NO}_2\text{-cycloalkanone [(CH}_2)_n] \xrightarrow[\text{2. HCl, heat, time}]{\text{1. RNH}_2\text{, NaBH}_3\text{CN, THF}} \text{2-(CH}_2\text{CH}_2\text{CH}_2\overset{+}{\text{N}}\text{H}_2\text{R)-2-NO}_2\text{-cycloalkanone Cl}^- \xrightarrow[\text{MeOH, 20 °C}]{\text{NaHCO}_3\text{, H}_2\text{O}} \text{N-R nitro lactam [(CH}_2)_n] \quad (892)$$

R	n	Temp. (°C)	Time	Salt yield (%)	Lactam yield (%)
$PhCH_2$	1	90	20 min	—	50
H	3	—	—	—	2
$PhCH_2$	3	40	48 h	50	76[a]
n-Pr	3	—	—	48	0
H	7	—	—	—	41
$PhCH_2$	7	20	60 h	58	72
n-Pr	7	20	72 h	53	55
n-Pent	7	20	71 h	60	81
t-Bu	7	—	—	52	0
$HO(CH_2)_2CH_2$	7	20	50 h	60	69
$BOCNH(CH_2)_2CH_2$[b]	7	20	36 h	42	95

[a]Using $(n\text{-Bu})_4\overset{+}{N}\overset{-}{F}$ in THF.
[b]BOC = t-Butoxycarbonyl.

Other similar reactions reported[1296] are shown in equations 893–895.

$$\text{2-methylene-1-(CH}_2\text{CH}_2\text{CHO)-1-NO}_2\text{-cyclohexane} \xrightarrow[\text{THF; 2. HCl, heat}]{\text{1. NH}_3\text{, NaBH}_3\text{CN}} \text{nitro lactam (43\%)} \quad (893)$$

$$\text{cyclopentadecanone} \xrightarrow[\text{2. KH, BrCH}_2\text{Ph, 20 °C, 45 min}]{\text{1. NH}_2\text{OH, H}_2\text{SO}_4} \text{N-CH}_2\text{Ph lactam (51\%)} \quad (894)$$

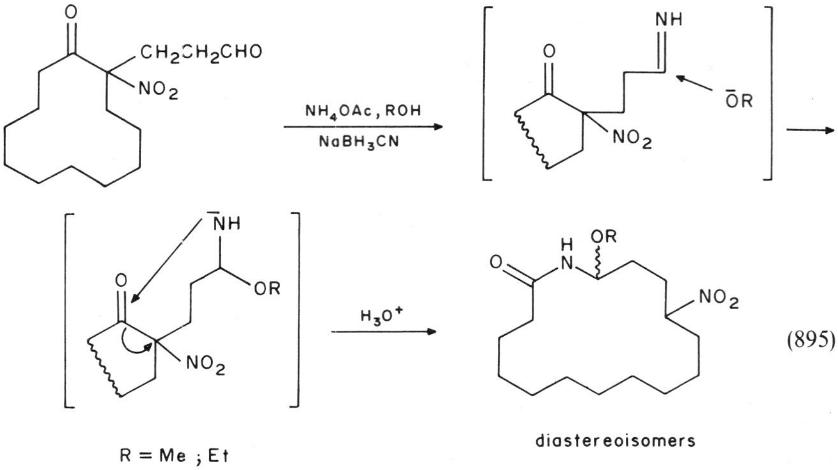

NH$_4$OAc, ROH
NaBH$_3$CN

H_3O^+

diastereoisomers

(895)

R = Me ; Et
Yield (%) = 46 ; 62

Medium-sized keto lactams containing from eight- to ten-membered rings were synthesized[1299] by a three-atom condensative ring expansion of the related fused tricyclic oxaziridines, which were prepared by peracetic acid oxidation of the precursor hydroxy imines (equations 896 and 897).

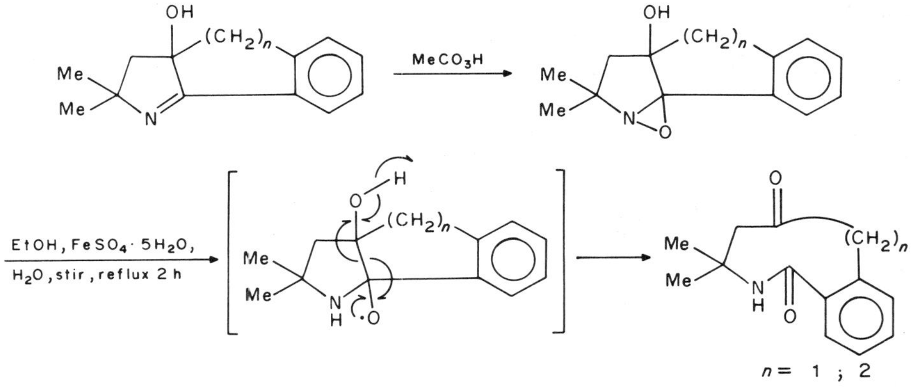

$MeCO_3H$

EtOH, FeSO$_4\cdot$5H$_2$O, H$_2$O, stir, reflux 2 h

n = 1 ; 2
Yield (%) = 29 ; 64

(896)

EtOH, FeSO$_4\cdot$5H$_2$O, H$_2$O, stir, reflux 2 h

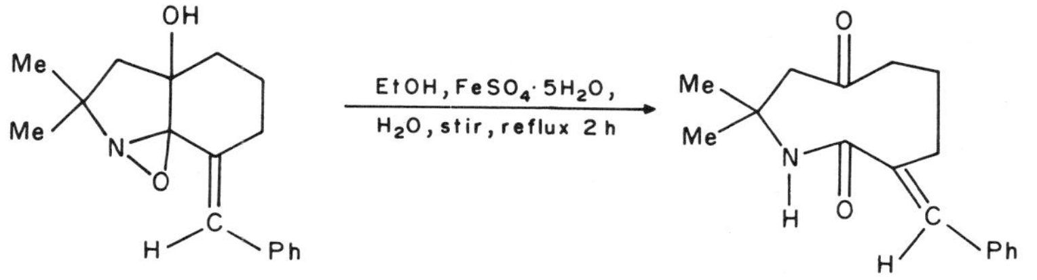

(897)

Attempting the same procedure[1299] with the corresponding hydroxy nitrones produced eight- to eleven-membered keto lactams, but the reaction required irradiation (equation 898).

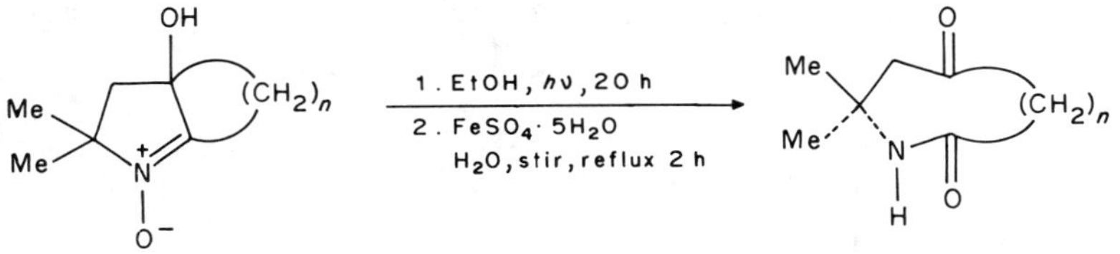

1. EtOH, $h\nu$, 20 h
2. $FeSO_4 \cdot 5H_2O$, H_2O, stir, reflux 2 h

n = 3 ; 4 ; 5 ; 6
Yield (%) = 30 ; 40 ; 42 ; 38

(898)

Irradiation has also been used in at least three other reports[1300–1302] to produce ring expanded products. Thus, irradiation of six-membered fused 4(3*H*)-pyrimidin-4-ones in the presence of methylamine affords[1300] eight- and nine-membered substitued lactams (equation 889), while irradiation of a 4-azaandrostenone produced[1301] a 5,10-secosteroid (equation 900).

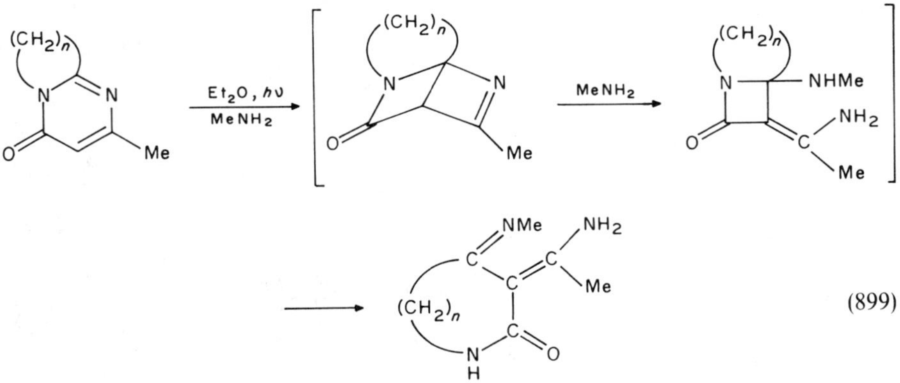

Et_2O, $h\nu$, $MeNH_2$; $MeNH_2$

n = 4 ; 5
Yield (%) = 30 ; 22.4

(899)

MeOH, $h\nu$, 36 h

(43 %)

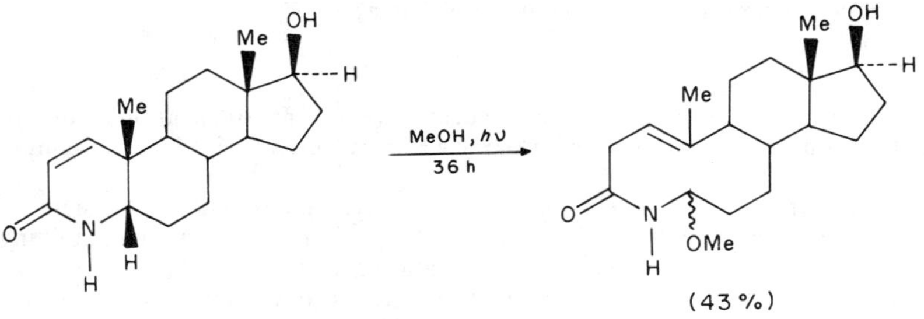

(900)

Photolysis of 3α-acetoxy-5α-androstan-17-one acetylhydrazone (**179**) and 3β-acetoxy-androst-5-en-17-one acetylhydrazone (**180**) in dioxane in the presence of oxygen produced[1302] 17-oxo-17α-aza-D-homosteroid and its 3α-isomer (equation 901). Under similar photolysis conditions 5α-cholestan-6-one acetylhydrazone afforded[1302] very low yields of 6-aza-D-homo-5α-cholestan-7-one and 7-aza-B-homo-5α-cholestan-6-one (equation 902).

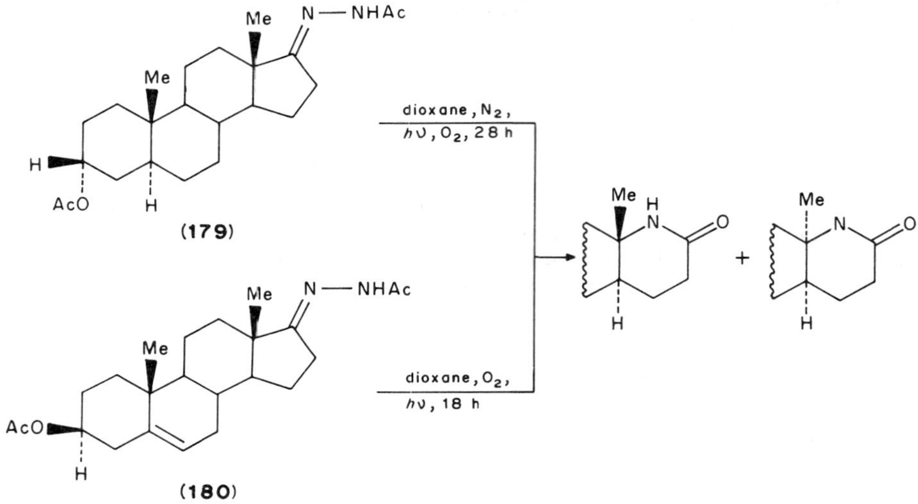

(179)

dioxane, N_2, $h\nu$, O_2, 28 h

(180)

dioxane, O_2, $h\nu$, 18 h

(901)

dioxane, O_2, $h\nu$, 14 h

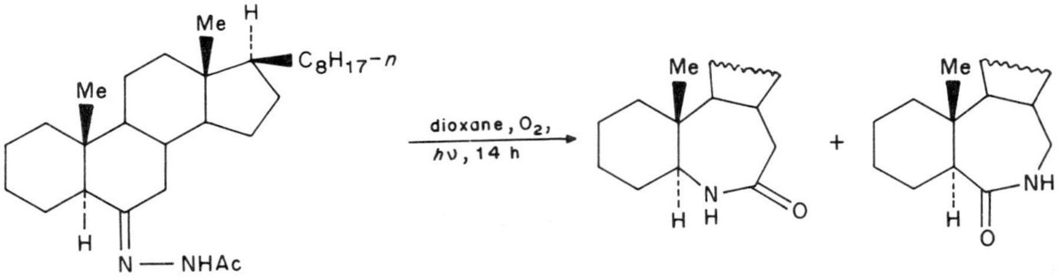

(902)

*E. By Direct Functionalization of Preformed Lactams

1. Functionalization of lactam nitrogen

The reactions discussed in this section all deal with the functionalization of an unsubstituted lactam nitrogen and begin with the reaction of lactams with halogen containing reagents.

Two general methods have been used to react alkyl halides and α-halo esters with a lactam nitrogen. Both methods involve initial reaction of a lactam containing an unsubstituted nitrogen with a base to form an intermediate anion, which is then allowed to condense with the halide containing reagent to produce the nitrogen-substituted lactam product. The difference between the two methods is the conditions used to effect reaction, since in one method only a base and solvent are used (equation 903, Table 60), whereas in the other method a quaternary amine salt is added to the reaction mixture to permit the reaction to be performed under phase transfer conditions (equation 904, Table 61).

1. base, solvent
2. RX
(903)

1. base, solvent
amine salt
2. RX
(904)

Many of the multistep synthetic procedures performed on lactams require that the hydrogen atom attached to the lactam nitrogen first be replaced before subsequent reactions are performed. The most common replacement procedure involves silylation of the lactam nitrogen using a variety of substituted silyl halides in the presence of a base (equation 905, Table 62).

1. base, solvent
2. $R^1R^2R^3SiX$
$SiR^1R^2R^3$
(905)

Unsubstituted lactam nitrogens have also been phosphorylated[1320] by treatment with *N*-butyl lithium followed by reaction of the resulting anion with diphenylchlorophosphate (equation 906).

$(CH_2)_2OTHP$
n-BuLi, N_2, −78 °C, THF, stir 10 min
$(CH_2)_2OTHP$ Li^+

$ClP(O)(OPh)_2$, −78 °C
stir 30 min
$(CH_2)_2OTHP$
$P(O)(OPh)_2$
(80%)
(906)

One of the more interesting methods used[1321] to *N*-alkylate lactams with alkyl halides involves electroreduction, performed in a divided cell equipped with a platinum cathode and anode and containing tetraethylammonium *p*-toluenesulphonate as a supporting electrolyte (equation 907).

TABLE 60. Lactam nitrogen substitution using base, solvent and halide

Lactam	Base	Solvent and conditions	Halide	Product	Yield (%)	Reference
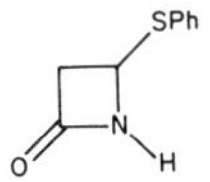 SPh, N, H, O	KNH_2	CH_2Cl_2, $-78°C$, 1 h	MeI	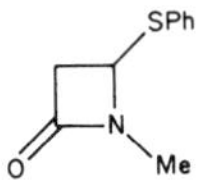 SPh, N, Me, O	63	1043
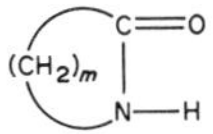 $(CH_2)_m$, C=O, N—H $m = 3,4,5$	—	—	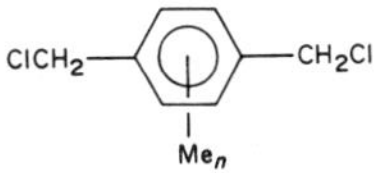 $ClCH_2$, CH_2Cl, Me_n $n = 0,1,2,3,4$	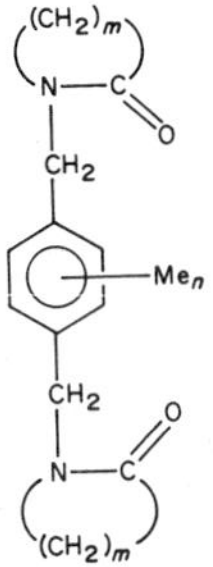 $(CH_2)_m$, N—C, O, CH_2, Me_n, CH_2, N—C, O, $(CH_2)_m$	—	1303
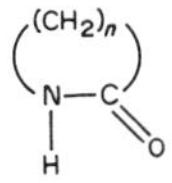 $(CH_2)_n$, N—C, H, O $n=3,7$	Na_2CO_3 or K_2CO_3	120–150°C	$ClCH_2SiMe_2R$ R = Et, n-Pr n-Bu, Ph	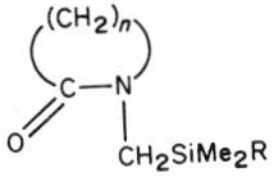 $(CH_2)_n$, C—N, O, CH_2SiMe_2R	—	1304
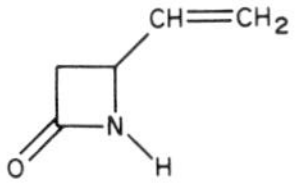 CH=CH_2, N, H, O	KOH	THF, DMF, 0°C, stir	$BrCH_2CO_2Bu$-t	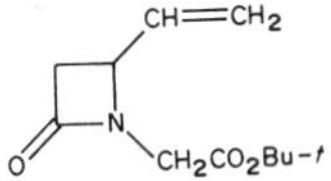 CH=CH_2, N, O, CH_2CO_2Bu-t	—	1195
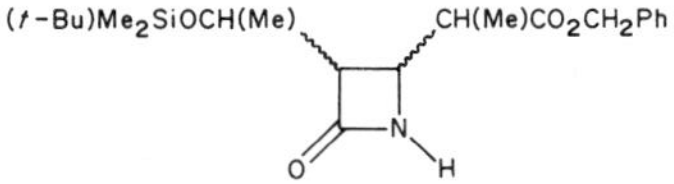 (t-Bu)Me_2SiOCH(Me), CH(Me)CO_2CH_2Ph, N, H, O		CH_2Cl_2	$BrCH_2CO_2Bu$-t	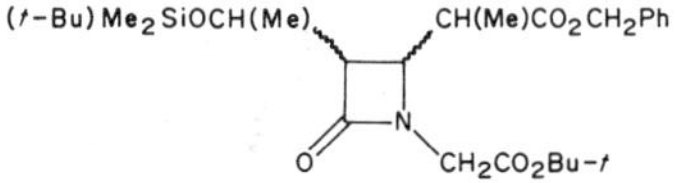 (t-Bu)Me_2SiOCH(Me), CH(Me)CO_2CH_2Ph, N, O, CH_2CO_2Bu-t	—	1305

Substrate	Base	Conditions	Reagent	Product	Yield (%)	Ref.
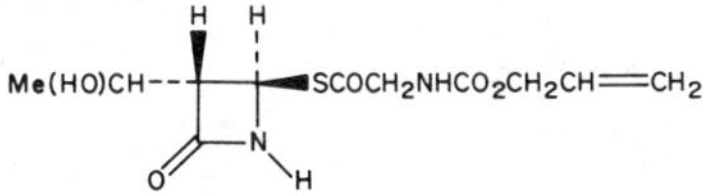 Me(HO)CH– azetidinone, N–H, $SCOCH_2NHCO_2CH_2CH{=}CH_2$	$EtN(Pr\text{-}i)_2$	—	$ClCOCO_2CH_2CH{=}CH_2$	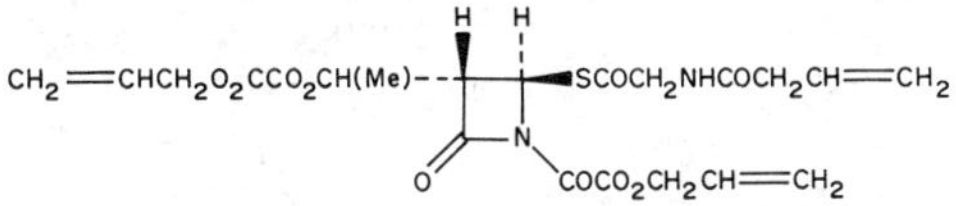 $CH_2{=}CHCH_2O_2CCO_2CH(Me)$– azetidinone, $SCOCH_2NHCOCH_2CH{=}CH_2$, N–$COCO_2CH_2CH{=}CH_2$	—	1306
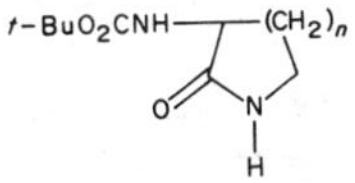 $t\text{-}BuO_2CNH$–lactam $(CH_2)_n$, N–H	NaH	THF, DMF, stir r.t., 3.5 h	$ICH_2CO_2CH_2Ph$	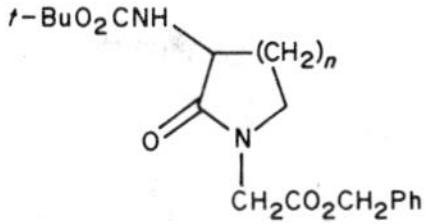 $t\text{-}BuO_2CNH$–lactam $(CH_2)_n$, N–$CH_2CO_2CH_2Ph$		1307
$n = 1$					46	
$n = 2$					56	
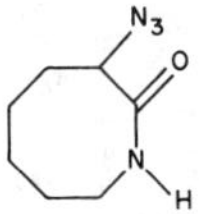 N_3–azocanone, N–H	NaH	THF, stir reflux 1.5 h	$ICH_2CO_2Bu\text{-}t$	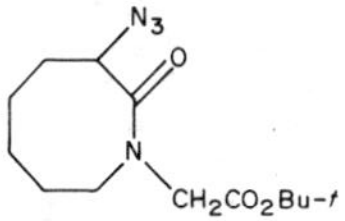 N_3–azocanone, N–$CH_2CO_2Bu\text{-}t$	76	1307
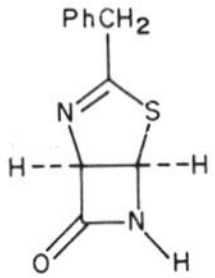 $PhCH_2$–thiazoline-fused azetidinone, N–H	Triton B	DMF, stir −30°C to 0°C	$BrCH_2CO_2R$	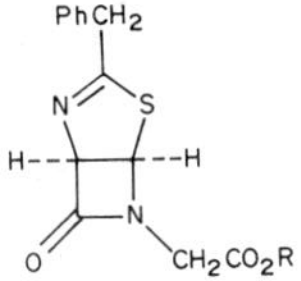 $PhCH_2$–thiazoline-fused azetidinone, N–CH_2CO_2R		1308
			R = Me		73	
			$R = PhCH_2$		47	
			$R = t\text{-}Bu$		75	
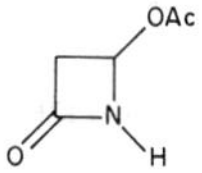 4-OAc-azetidin-2-one, N–H	NaH	THF, −50°C, 1 h	$BrCH(CO_2Et)_2$	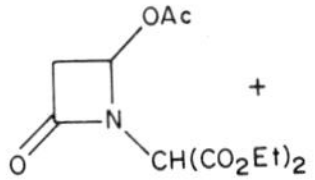 4-OAc-azetidinone, N–$CH(CO_2Et)_2$ +	22.5	1309
				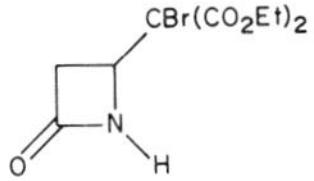 4-$CBr(CO_2Et)_2$-azetidinone, N–H	35.3	

(continued)

TABLE 60. (*continued*)

Lactam	Base	Solvent and conditions	Halide	Product	Yield (%)	Reference
Cl OAc H H N H O	NaH	THF, -50°C, 1 h	$BrCH(CO_2Et)_2$	Cl OAc H H N $CH(CO_2Et)_2$ O	32.5	1309
O O N H	DMAP	120°C	Cl_3CCO_2Et	O O N CO_2Et	—	1310
Me—N N—H Ph O	KOH	DMSO, 20°C	$Cl(CH_2)_4Br$	Me—N N—$(CH_2)_4Cl$ Ph O	—	1017
Me—N N—H Ph O	KOH	DMSO, 20°C	$Br(CH_2)_4Br$	Me Ph N N O $(CH_2)_4$ O N Ph N Me **(181)**	75	1017

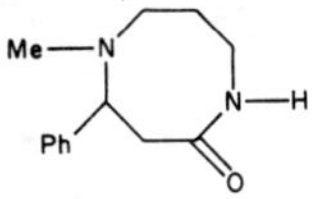 [structure]	KOH	DMSO, 20°C	$Br(CH_2)_4Br$	**181**	62[a]	1017
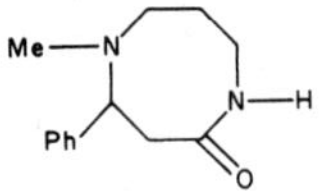 [structure]	KOH	DMSO, 20°C	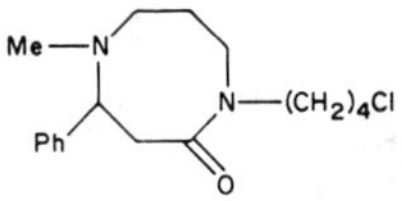 [structure]	**181**	—	1017

[a]A 1:1 mixture of (±)-homaline and *epi*-homaline was obtained.

TABLE 61. Lactam nitrogen substitution using phase transfer conditions

Lactam	Base	Amine salt	Solvent and conditions	Halide	Product	Yield (%)	Reference
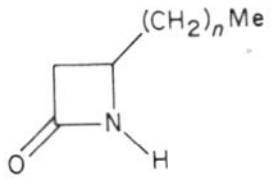 $(CH_2)_nMe$, N, H, O	KOH	$(n\text{-Bu})_4\overset{+}{N}\ H\overset{-}{S}O_4$	DMSO, 20°C	$Br(CH_2)_3Cl$	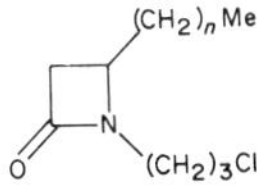 $(CH_2)_nMe$, N, $(CH_2)_3Cl$, O		1294
$n = 4$						52	
$n = 6$						60	
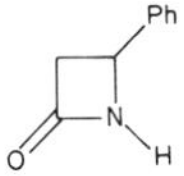 Ph, N, H, O	KOH	$(n\text{-Bu})_4\overset{+}{N}\ H\overset{-}{S}O_4$	THF, 20°C	$Br(CH_2)_nBr$	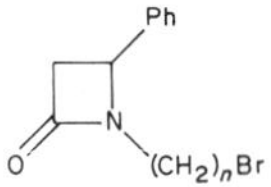 Ph, N, $(CH_2)_nBr$, O		1294
				$n = 2$		48	
				$n = 3$		68	
				$n = 5$		60	
	KOH	$(n\text{-Bu})_4\overset{+}{N}\ H\overset{-}{S}O_4$	THF, 20°C	$Br(CH_2)_3Cl$	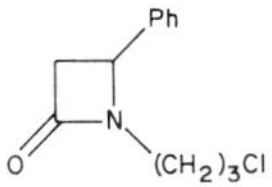 Ph, N, $(CH_2)_3Cl$, O	94	1017, 1294
	KOH	$(n\text{-Bu})_4\overset{+}{N}\ H\overset{-}{S}O_4$	DMSO, 20°C	$Br(CH_2)_4Cl$	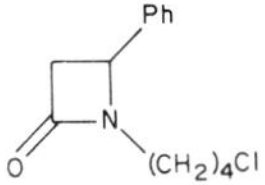 Ph, N, $(CH_2)_4Cl$, O	70	1294
	KOH	$(n\text{-Bu})_4\overset{+}{N}\ H\overset{-}{S}O_4$	THF, 20°C	$Br(CH_2)_nI$	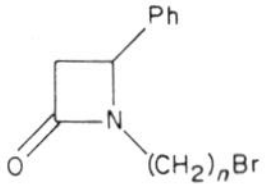 Ph, N, $(CH_2)_nBr$, O		1294
				$n = 3$		75	
				$n = 4$		81	

KOH	$(n\text{-Bu})_4\overset{+}{N}\ H\overset{-}{S}O_4$	THF, 20°C	$Br(CH_2)_2CH(Br)Me$	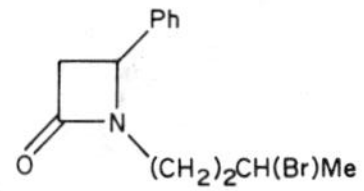 Ph, N, O, $(CH_2)_2CH(Br)Me$	75	1294
KOH	$(n\text{-Bu})_4\overset{+}{N}\ \overset{-}{Br}$	THF	$Me(CH_2)_5Br$	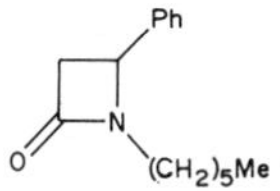 Ph, N, O, $(CH_2)_5Me$	83	1311
KOH	$(n\text{-Bu})_4\overset{+}{N}\ \overset{-}{Br}$	THF	$PhCH_2Cl$	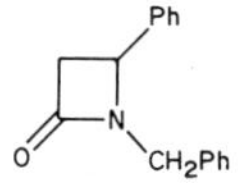 Ph, N, O, CH_2Ph	81	1311
KOH	$(n\text{-Bu})_4\overset{+}{N}\ \overset{-}{Br}$	THF	$Me(CH_2)_3CH(Br)Me$	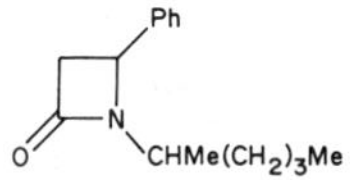 Ph, N, O, $CHMe(CH_2)_3Me$	49	1311
KOH	$(n\text{-Bu})_4\overset{+}{N}\ \overset{-}{Br}$	THF	$BrCH_2CO_2Et$	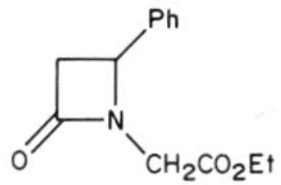 Ph, N, O, CH_2CO_2Et	78	1311
KOH	$(n\text{-Bu})_4\overset{+}{N}\ \overset{-}{Br}$	THF	$Br(CH_2)_6CO_2H$	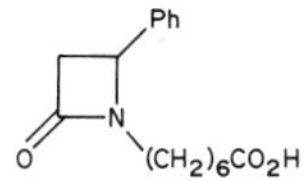 Ph, N, O, $(CH_2)_6CO_2H$	62	1311

(continued)

TABLE 61. (*continued*)

Lactam	Base	Amine salt	Solvent and conditions	Halide	Product	Yield (%)	Reference
	KOH	$(n\text{-Bu})_4\overset{+}{N}\ \overset{-}{Br}$	THF	$PhCH(Br)CO_2H$	Ph; O; N; $CH(Ph)CO_2H$	57	1311
$CH{=}CH_2$; O; N; H	KOH	$(n\text{-Bu})_4\overset{+}{N}\ \overset{-}{Br}$	THF	MeI	$CH{=}CH_2$; O; N; Me	86	1311
	KOH	$(n\text{-Bu})_4\overset{+}{N}\ \overset{-}{Br}$	THF	$R(CH_2)_nX$	$CH{=}CH_2$; O; N; $(CH_2)_nR$		1311
				R = Me, $n = 5$, X = Br		90	
				R = Ph, $n = 1$, X = Br		84	
				R = Ph, $n = 1$, X = Cl		83	
	KOH	$(n\text{-Bu})_4\overset{+}{N}\ \overset{-}{Br}$	THF	*i*-PrBr	$CH{=}CH_2$; O; N; Pr-*i*	45	1311

KOH	$(n\text{-}Bu)_4\overset{+}{N}\ \overset{-}{Br}$	THF	$Me(CH_2)_3CH(Br)Me$	$CH{=}CH_2$; O; N; $CHMe(CH_2)_3Me$	48	1311
KOH	$(n\text{-}Bu)_4\overset{+}{N}\ \overset{-}{Br}$	THF	p-$AnCH_2Cl$	$CH{=}CH_2$; O; N; $CH_2An\text{-}p$	78	1311
KOH	$(n\text{-}Bu)_4\overset{+}{N}\ \overset{-}{Br}$	THF	$I(CH_2)_4Cl$	$CH{=}CH_2$; O; N; $(CH_2)_4Cl$	79	1311
KOH	$(n\text{-}Bu)_4\overset{+}{N}\ \overset{-}{Br}$	THF	$Br(CH_2)_nCO_2Et$	$CH{=}CH_2$; O; N; $(CH_2)_nCO_2Et$		1311
			$n = 1$		77	
			$n = 6$		70	
KOH	$(n\text{-}Bu)_4\overset{+}{N}\ \overset{-}{Br}$	THF	$BrCH_2C{\equiv}C(CH_2)_3CO_2Me$	$CH{=}CH_2$; O; N; $CH_2C{\equiv}C(CH_2)_3CO_2Me$	71	1311
KOH	$(n\text{-}Bu)_4\overset{+}{N}\ \overset{-}{Br}$	THF	$BrCH(CO_2Et)_2$	$CH{=}CH_2$; O; N; $CH(CO_2Et)_2$	32	1311

(*continued*)

TABLE 61. *(continued)*

Lactam	Base	Amine salt	Solvent and conditions	Halide	Product	Yield (%)	Reference
	KOH	$(n\text{-Bu})_4\overset{+}{N}\,\overset{-}{Br}$	THF	$BrCH_2CH(OMe)_2$	4-($CH{=}CH_2$)-1-($CH_2CH(OMe)_2$)azetidin-2-one	30	1311
	KOH	$(n\text{-Bu})_4\overset{+}{N}\,\overset{-}{Br}$	THF	$Br(CH_2)_nCO_2H$	4-($CH{=}CH_2$)-1-($(CH_2)_nCO_2H$)azetidin-2-one		1311
				$n = 1$		65	
				$n = 6$		51	
	KOH	$(n\text{-Bu})_4\overset{+}{N}\,\overset{-}{Br}$	THF	$PhCH(Br)CO_2H$	4-($CH{=}CH_2$)-1-($CH(Ph)CO_2H$)azetidin-2-one	55	1311
	KOH	$(n\text{-Bu}_4\overset{+}{N}\,\overset{-}{Br}$	1. THF, N_2, stir 2. stir r.t.	$BrCH_2C(Br){=}CH_2$	4-($CH{=}CH_2$)-1-($CH_2C(Br){=}CH_2$)azetidin-2-one	73	1312
4-Me-4-($CH{=}CH_2$)azetidin-2-one (N–H)	KOH	$(n\text{-Bu})_4\overset{+}{N}\,\overset{-}{Br}$	—	$Br(CH_2)_nBr$ $n = 2,3$	4-Me-4-($CH{=}CH_2$)-1-($(CH_2)_nBr$)azetidin-2-one	—	1313

	KOH	$(n\text{-Bu})_4\overset{+}{N}\,\overset{-}{Br}$	—	$Br(CH_2)_nNO_2$ $n = 1,2$	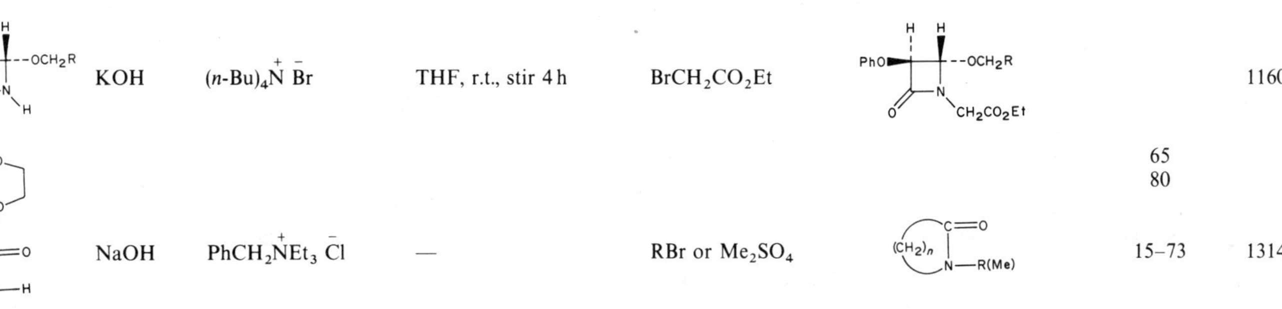	—	1313
	KOH	$(n\text{-Bu})_4\overset{+}{N}\,\overset{-}{HSO_4}$	1. THF, N_2 2. stir r.t., 5 h	$BrCH_2C(Br){=}CH_2$		67	1312
	KOH	$(n\text{-Bu})_4\overset{+}{N}\,\overset{-}{HSO_4}$	1. THF, N_2, stir 10°C 2. stir r.t., 5 h	$BrCH_2C(Br){=}CH_2$		61	1312
R = Ac R = (2-methyl-1,3-dioxolan-2-yl)	KOH	$(n\text{-Bu})_4\overset{+}{N}\,\overset{-}{Br}$	THF, r.t., stir 4 h	$BrCH_2CO_2Et$		65 80	1160
	NaOH	$PhCH_2\overset{+}{N}Et_3\,\overset{-}{Cl}$	—	RBr or Me_2SO_4 $R = n\text{-Bu}, PhCH_2$ $n\text{-}C_8H_{17}, n\text{-}C_{12}H_{25}$		15–73	1314

(continued)

TABLE 61. (*continued*)

Lactam	Base	Amine salt	Solvent and conditions	Halide	Product	Yield (%)	Reference
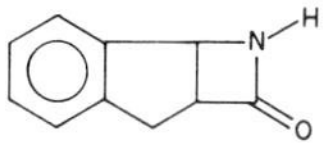 N–H	KOH	$(n\text{-Bu})_4\overset{+}{N}\ \overset{-}{Br}$	THF	$R(CH_2)_nX$	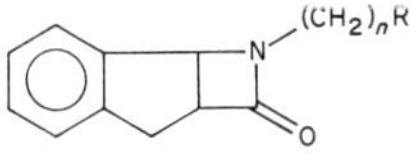 N–$(CH_2)_nR$		1311
				R = Me, $n = 5$, X = Br		70	
				R = Ph, $n = 1$, X = Cl		65	
	KOH	$(n\text{-Bu})_4\overset{+}{N}\ \overset{-}{Br}$	THF	$Me(CH_2)_3CH(Br)Me$	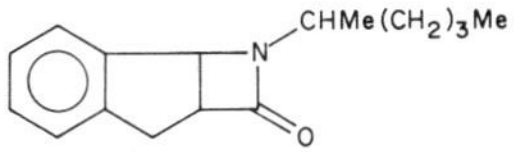 N–$CHMe(CH_2)_3Me$	33	1311
	KOH	$(n\text{-Bu})_4\overset{+}{N}\ \overset{-}{Br}$	THF	$BrCH_2CO_2Et$	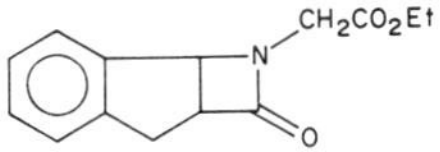 N–CH_2CO_2Et	67	1311
	KOH	$(n\text{-Bu})_4\overset{+}{N}\ \overset{-}{Br}$	THF	$Br(CH_2)_6CO_2H$	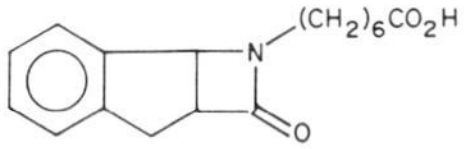 N–$(CH_2)_6CO_2H$	41	1311

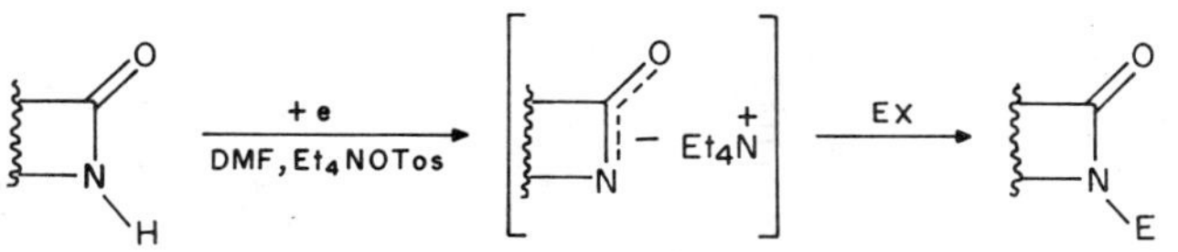

(907)

Lactam	EX[a]	Product	Yield (%)
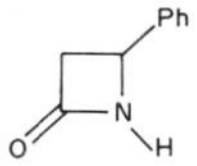	MeOMes	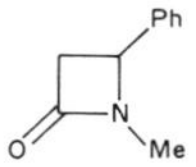	91
	s-BuOMes	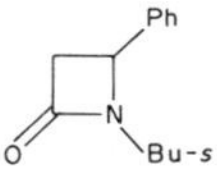	60
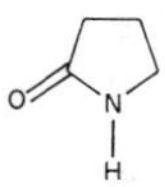	n-BuBr n-BuI	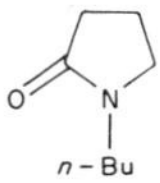	55 40
	s-BuCl s-BuOMes	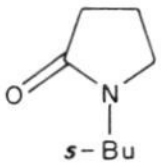	68 70
	$THPO(CH_2)_3Cl$	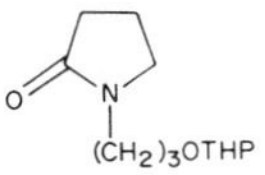	80
	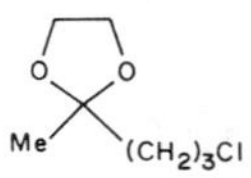	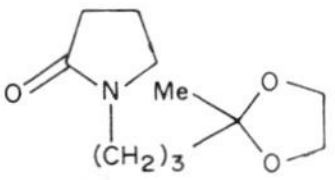	63
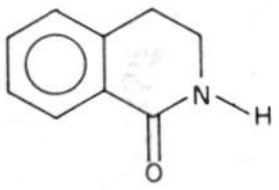	MeOMes	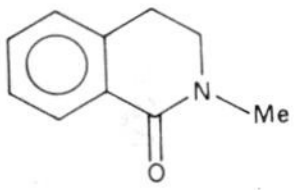	91
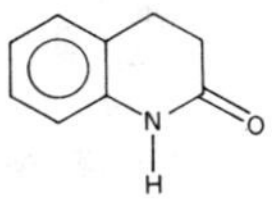	$PhCH_2Cl$	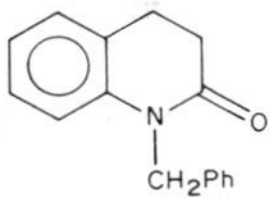	87

[a]Met = Methanesulphonyl

Lactams have also been N-alkylated by reaction[1322] with styrene in hexamethylphosphoramide (equation 908), and by the reaction[1003] of the sodium salt of (−)-4-phenyl-2-azetidinone with a ditosylate (equation 909).

TABLE 62. Silylation of lactam nitrogen

Lactam	Silylating agent	Base	Solvent and conditions	Product	Yield (%)	Reference
$CH{=}CH_2$; O; N; H	$(t\text{-Bu})Me_2SiCl$	Et_3N	DMF, 0 °C	$CH{=}CH_2$; O; N; $SiMe_2(Bu\text{-}t)$	—	1304
	$(t\text{-Bu})Ph_2SiCl$	Et_3N	—	$CH{=}CH_2$; O; N; $SiPh_2(Bu\text{-}t)$	—	1315
H H H; Ph; Et; C=C; H; O; N; H	$(t\text{-Bu})Me_2SiCl$	Et_3N	DMF	H H H; Ph; Et; C=C; H; O; N; $SiMe_2(Bu\text{-}t)$	92	1245
Me(HO)CH; $CH_2CH(OMe)_2$; O; N; H	$(t\text{-Bu})Me_2SiCl$	Et_3N	—	Me(HO)CH; $CH_2CH(OMe)_2$; O; N; $SiMe_2(Bu\text{-}t)$	—	1316
H H; $PhCH_2O_2CNH$; CO_2Me; O; N; H	$(t\text{-Bu})Me_2SiCl$	—	—	H H; $PhCH_2O_2CNH$; CO_2Me; O; N; $SiMe_2(Bu\text{-}t)$	—	1317

Substrate	Silylating agent	Base	Solvent	Product	Yield (%)	Ref.
OSiMe$_2$(Bu-t), MeCH, OSiMe$_3$, O, N, H (3R,5R)	(t-Bu)Me$_2$SiCl	Et$_3$N	DMF	OSiMe$_2$(Bu-t), MeCH, OSiMe$_3$, O, N, SiMe$_2$(Bu-t)	—	1318
Me(HO)CH, H, SPh, C=C, H, O, N, H (mixture)	(t-Bu)Me$_2$SiCl	Et$_3$N	DMF	(t-Bu)Me$_2$SiO, H, H, SPh, C=C, H, MeCH, H, O, N, SiMe$_2$(Bu-t) +	3	1237
				(t-Bu)Me$_2$SiO, H, H, H, SPh, MeCH, C=C, H, O, N, SiMe$_2$(Bu-t) +	18	
				(t-Bu)Me$_2$SiO, H, H, H, SPh, MeCH, C=C, H, O, N, SiMe$_2$(Bu-t)	66	
H, H, Me(HO)CH, C≡CH, O, N, H	(t-Bu)Me$_2$SiCl	Et$_3$N	DMF	(t-Bu)Me$_2$SiO, H, H, MeCH, C≡CH, O, N, SiMe$_2$(Bu-t)	94	1237
Me(HO)CH, C≡CSiMe$_3$, O, N, H (mixture)	1. (n-Bu)$_4\overset{+}{N}\overset{-}{F}$, THF 2. ($t$-Bu)Me$_2$SiCl	Et$_3$N	DMF	(t-Bu)Me$_2$SiO, H, H, MeCH, C≡CH, O, N, SiMe$_2$(Bu-t) +	0.8	1237

(*continued*)

TABLE 62. (*continued*)

Lactam	Silylating agent	Base	Solvent and conditions	Product	Yield (%)	Reference
				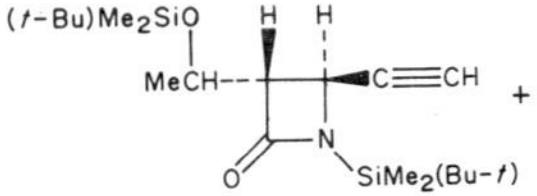 +	36	
				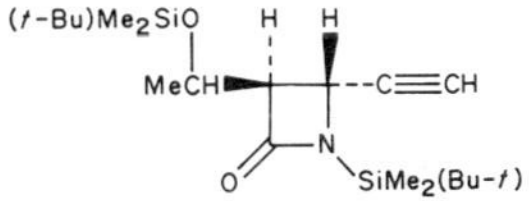	57	
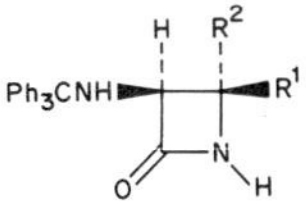 $R^1 = CH_2F$, $R^2 = H$ $R^1 = H$, $R^2 = CH_2F$ $R^1 = CO_2Me$, $R^2 = H$ $R^1 = CH_2OTHP$, $R^2 = H$ $R^1 = CHF_2$, $R^2 = H$	$(Me_3Si)_2$	—	THF, −40° 10 °C	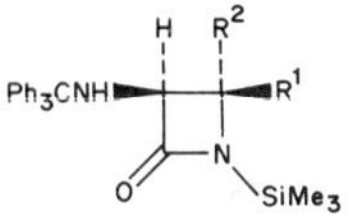	—	1184
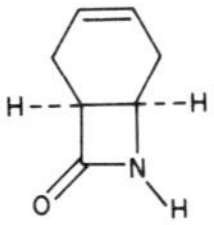	(*t*-Bu)Me_2SiCl	Et_3N	—	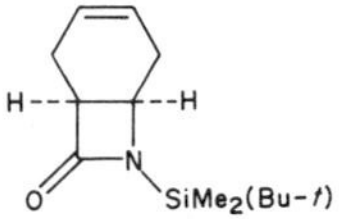	—	1021
	(*t*-Bu)Me_2SiCl	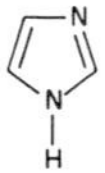	DMF		—	1319
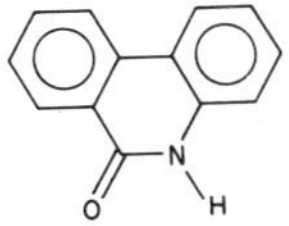	Me_3SiCl	Et_3N	dioxane, reflux 10 min	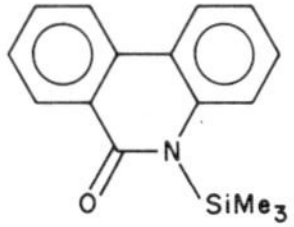	—	1320

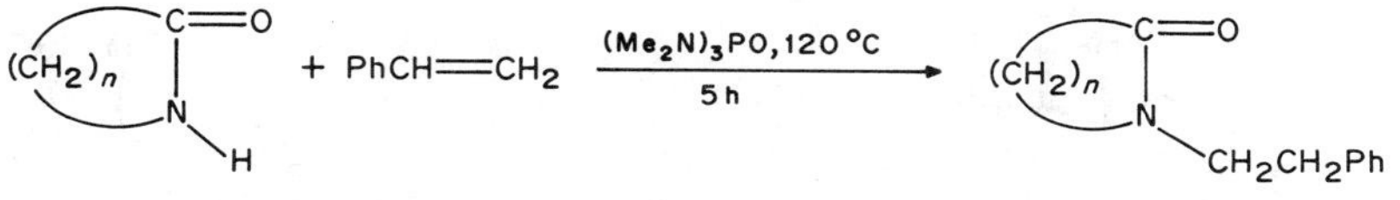

n = 3 ; 5
Yield (%) = 40–94; 43–51

(908)

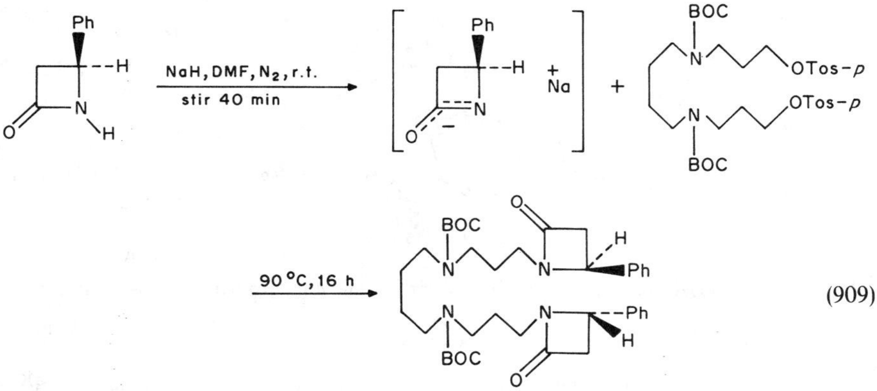

NaH, DMF, N_2, r.t., stir 40 min; 90 °C, 16 h

(63%)

BOC = t-Butoxycarbonyl

(909)

Carboxylic acids[1323] (equation 910), acid chlorides[1124,1324] (equations 911 and 912), acid anhydrides[1124,1155,1325] (equations 913, 914 and 915) and dimethylphosphor isocyanate[1326] (equation 916) have all been used to acylate an unsubstituted lactam nitrogen.

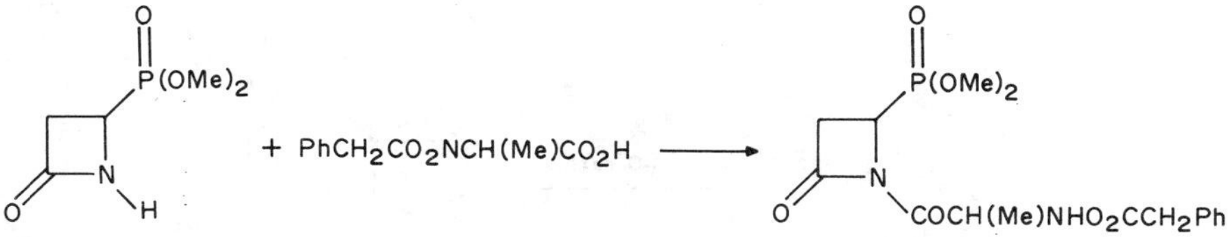

+ $PhCH_2CO_2NCH(Me)CO_2H$ → $COCH(Me)NHO_2CCH_2Ph$

(910)

+ R_2NCOCl (C_6H_5Me, NEt_3) → $CONR_2$

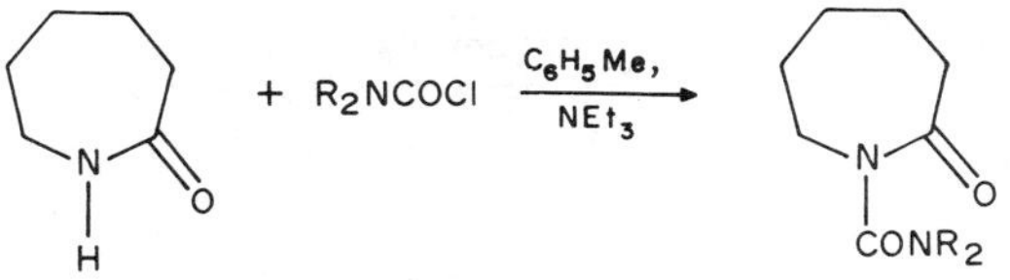

(911)[1324]

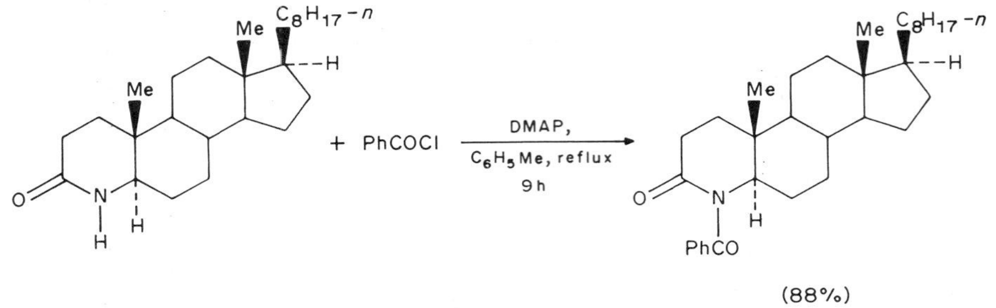

+ PhCOCl → DMAP, C_6H_5Me, reflux 9h

(88%)

(912)[1124]

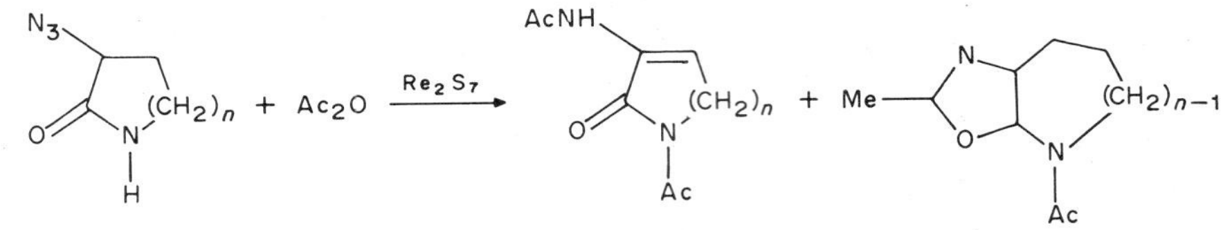

+ Ac_2O → Re_2S_7

$n = 1-4$, the oxazoloazocins are formed in addition to the dehydrolactams when $n = 3$ or 4 only

(913)[1325]

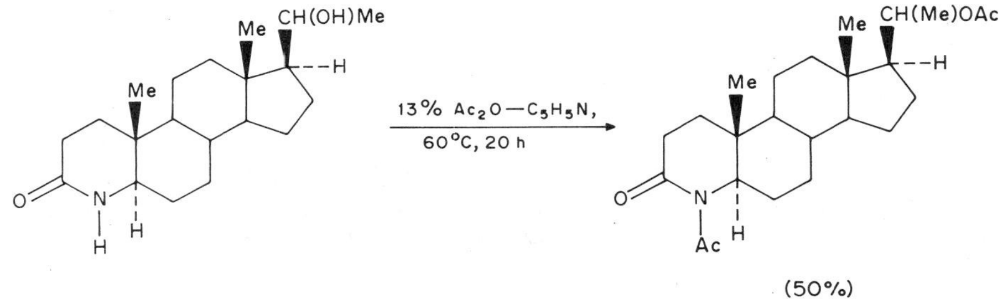

→ 13% $Ac_2O-C_5H_5N$, 60°C, 20 h

(50%)

(914)[1124]

→ Ac_2O, reflux 1.5 h

→ $(CF_3CO)_2O$

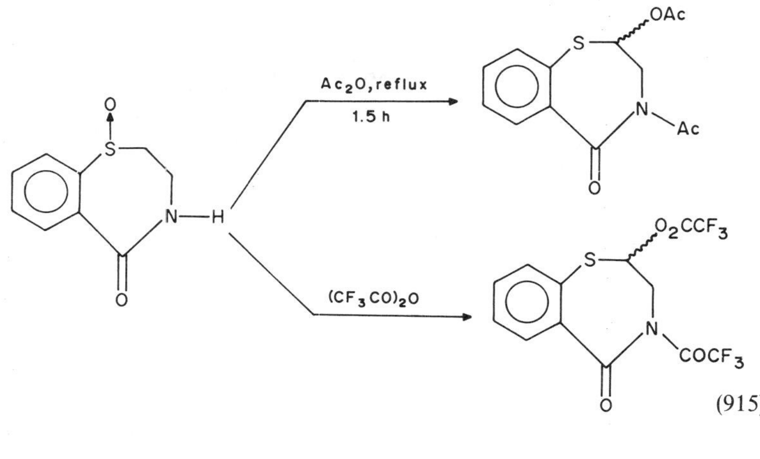

(915)[1155]

PhCH2CONH (azetidinone N–H) + $(MeO)_2P(O)NCO$ $\xrightarrow{Et_3N,\ MeCN,\ 45\,°C}$ PhCH2CONH (azetidinone N–CONHP(O)(OMe)$_2$)

(916)

When Grignard reagents are allowed to react with lactams containing unsubstituted ring nitrogens, an exchange reaction occurs producing the corresponding lactam Grignard reagent, which has been used in one case to react with succinic anhydride to produce[1327] an acyllactamcarboxylic acid (equation 917), and in another case to produce[1328] a bisazetidinone (equation 918).

+ EtMgBr → N–MgBr → N–$CO(CH_2)_2CO_2H$

(95%)

(917)

SO_2Ph azetidinone N–H + $BrMgCH_2CO_2Bu$-*t* → SO_2Ph azetidinone N–MgBr

→ (with SO_2Ph azetidinone N–H) bisazetidinone (SO_2Ph, N–H)

(51.7%)

(918)

Chloromethylation[1329] and chloroformylation[1330,1331] reactions have also been reported to produce N-haloalkyl and carbonyl substituted from N-unsubstituted lactams. For both reactions, the initial preparation of an *N*-silylated lactam was required. The chloromethylation reaction occurs in a single stage using a paraformaldehyde-trimethylchlorosilane system[1329] (equations 919 and 920), while the chloroformylation reaction requires two steps, the first being preparation of the *N*-trimethylsilyl substituted derivative, and the second being reaction with phosgene (equations 921 and 922).

$(CH_2O)_n$, Me_3SiCl, reflux (919)

$(CH_2O)_n$, Me_3SiCl, reflux (920)

1. Me_3SiCl, Et_3N, C_6H_5Me, 25 °C 2. stir 40 °C, 30 min → [N-SiMe₃ intermediate] → 1. cool to 0 °C 2. $COCl_2$, C_6H_5Me, 0 °C, stir 1 h → N-COCl (71.2%)

(921)[1330]

$R^1CH{=}N{-}N$ (imidazolidinone) NH $\xrightarrow{Me_3SiCl,\ Et_3N}$ [$R^1CH{=}N{-}N$ … $N{-}SiMe_3$] $\xrightarrow{COCl_2}$ $R^1CH{=}N{-}N$ … $N{-}COCl$ (922)[1331]

R^1 = 2-Fu, 2-Thi, 5-X, 2-Thi, 3-Pyr, PhCH=CH—, $C_6H_{5-n}R^2_n$

R^2_n = H, Y, Me, p-MeO, p-O_2N, p-CN, p-$MeSO_2$, 3,4-Cl_2

Phosgene, in conjunction with a 4:1 chloroform–pyridine mixture, has also been used[1332] to produce 1,1′-methylene di-2-pyrrolidone from 2-pyrrolidone. This preparation involves the reaction of the 2-pyrrolidone with the chloroform–pyridine mixture and phosgene to produce a cyclic iminochloride intermediate, which upon refluxing in anhydrous dimethyl sulphoxide affords[1332] a 55% yield of the *N,N′*-methylenedilactam product (equation 923). The multistep mechanism proposed to explain this reaction involves formation of an *N*-(methylthio)methylpyrrolidone (**184**) from a ylid (**183**) and a sulphoxonium intermediate (**182**) (Scheme 16). Alkylation of **184** by the iminochloride produces the sulphonium salt **185** which can undergo a carbon-to-nitrogen rearrangement to form **186**, or a cleavage to give a *N*-chloromethylpyrrolidone

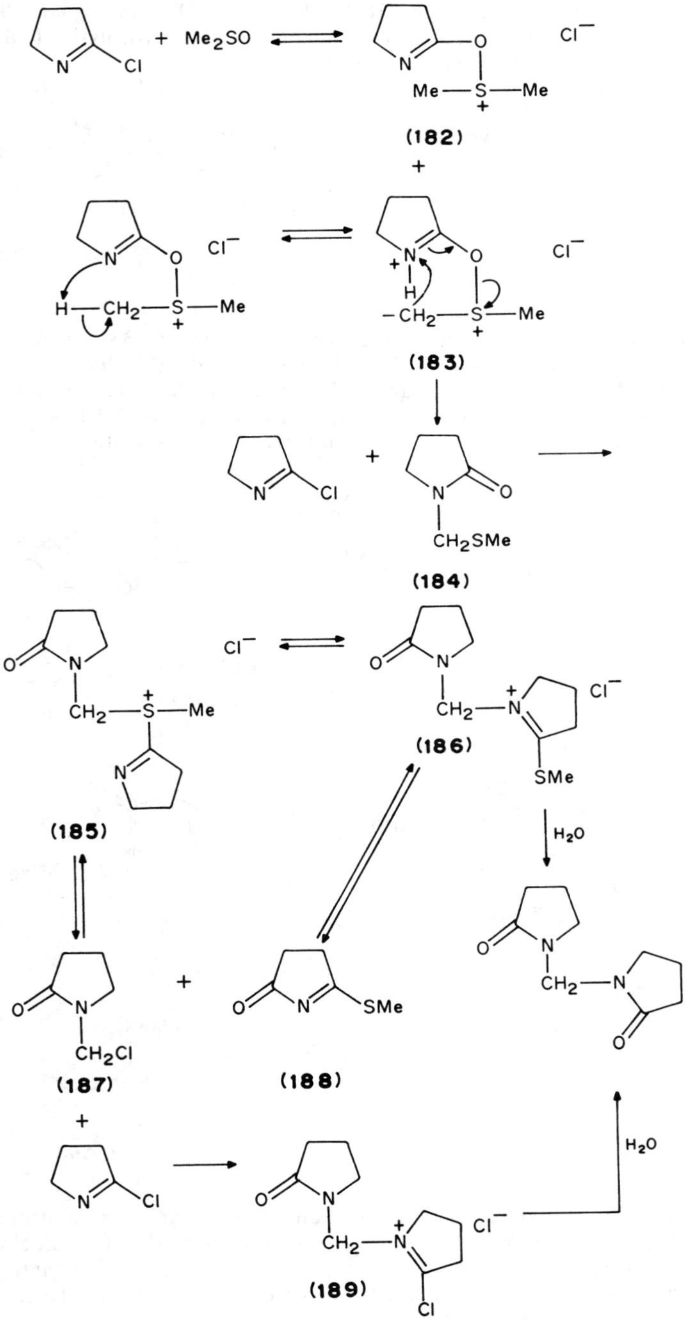

(182)

(183)

(184)

(185)

(186)

(187)

(188)

(189)

SCHEME 16

(**187**) and 2-methylthio-1-pyrroline (**188**). Intermediate **187** can alkylate **188** to form **186** or the iminochloride to form a immonium chloride **189**. Hydrolysis of **186** or **189** gives the 1,1′-methylenedi-2-pyrrolidone product.

1. $CHCl_3-C_5H_5N$(4:1) 0 °C, $COCl_2$ 2. stir overnight r.t.; DMSO, reflux 70 h; N—CH_2—N (55%)

(923)

Solutions of formaldehyde in the presence of secondary amines react with unsubstituted lactam nitrogens to produce[1297,1333] the corresponding *N*-dialkylaminomethyl substituted products (equations 924 and 925), while a similar solution of formaldehyde in the presence of sodium cyanoborohydride reacts with 4-phenyl-1,5-diazacyclooctan-2-one and reductively methylates the ring nitrogen in the 5-position[1017] (equation 926).

+ HCHO; NH, EtOH, reflux 2 h; CH_2—N (924)[1333]

R = $(CH_2)_2CH{=}CH_2$; $CH{=}CH_2$
Yield (%) = 92 ; 93

$(CH_2)_n$ + HCHO; $HNMe_2$, $MeCO_2H$; CH_2NMe_2 (925)[1297]

$n = 5,6,7$

HN, Ph, NH + HCHO; $NaBH_3CN$; MeN, Ph, NH (90%) (926)

4-(*t*-Butoxycarbonylamino)-1,2-pyrazolidin-3-one is another dinitrogen containing lactam which undergoes an N-substitution reaction with formaldehyde solutions. However, in this case the product obtained[1334] is a pyrazolidinium ylide (equation 927) which has been used to produce several bicyclic pyrazolidones. Their preparation is discussed at the end of this section.

t-BuO_2CNH–(pyrazolidinone) + HCHO $\xrightarrow{MeOH, H_2O}$ t-BuO_2CNH–(pyrazolidinone, $\overset{+}{N}{=}CH_2$, N^-)

(927)

A large variety of N-substituted lactams have been prepared by refluxing an N-unsubstituted lactam with a glyoxylate and in all cases a mixture of epimeric α-hydroxy esters is obtained (equation 928 and Table 63).

(β-lactam N–H) + OCHCOR $\xrightarrow{reflux}$ (β-lactam N–CH(OH)COR*)

mixture of epimers

(928)

At least two reports have appeared[1342,1343] which describe the N-sulphonation of lactams. In the first report[1342] 3-substituted 4-styryl β-lactams are N-sulphonated using a sulphur trioxide–pyridine complex (equation 929), while in the second report[1343] a mixture of sulphur trioxide and dimethylformamide was used to N-sulphonate (3*R*, 4*R*)- and (3*R*, 4*S*)-4-[(1-methyl-1*H*-tetrazol-5-yl)thio]-3-phenylacetamido-2-azetidinones (equations 930 and 931) as well as (3*S*, 4*S*)-4-(5-methyltetrazolyl)-3-phenylacetamido-2-azetidinone (equation 932).

1. $C_6H_5N, N_2, 80\,°C$
SO_3—C_5H_5N, stir., 2 h
2. $H_2O, (n\text{-}Bu_4)_4NHSO_4$

SO_3^- $(n\text{-}Bu)_4N^+$

(929)

R = N_3; $PhCH_2CONH$; Ph_3CNH–(thiazolyl)–C(=NOMe)CONH

Yield (%) = 80; 57 ; 51[a]

[a]This reaction required stirring for 10h; methanol and potassium 2-ethylhexanoate were used in the second step to produce the potassium salt.

$PhCH_2CONH$–(azetidinone)–S–(1-methyltetrazol-5-yl)

1. DMF, SO_3—DMF, 0 °C
2. r.t. stir 8 h
3. C_5H_5N, stir 5 min
4. Na^+ from Dowex 50 W

SO_3^- Na^+

34%

(930)

TABLE 63. Reaction of N-unsubstituted lactams with glyoxylates

Lactam	Glyoxylate	Reaction conditions	Product	Yield (%)	Reference
H, $CH{=}CH_2$, N–H, O	$OCHCO_2CH_2Ph$	C_6H_6, reflux	H, $CH{=}CH_2$, N–$CH(OH)CO_2CH_2Ph$, O	—	1196, 1335
$OCH_2CH{=}CHR$, N–H, O	$OCHCO_2Bu$-t	dry C_6H_5Me, reflux	$OCH_2CH{=}CHR$, N–$CH(OH)CO_2Bu$-t, O		1336
R = H		4 h		77	
R = Ph(*E*)		3 h		73	
$OCH_2CH{=}CH_2$, N–H, O	$(CH_2O)_n$	115 °C, Ar, stir 2 h	$OCH_2CH{=}CH_2$, N–CH_2OH, O	96	1336
$OCH_2CH{=}CH_2$, N–H, O	$(HO)_2C(CO_2Et)_2$	dry C_6H_5Me, reflux 3 h	$OCH_2CH{=}CH_2$, N–$C(OH)(CO_2Et)_2$, O	88	1336
$OCH_2C{\equiv}CR$, N–H, O	$OCHCO_2Bu$-t	dry C_6H_6, reflux	$OCH_2C{\equiv}CR$, N–$CH(OH)CO_2Bu$-t, O		1336

Substrate	Reagent	Conditions	Product	Yield (%)	Ref.
R = H		5 h		78	
R = Ph		16 h		86	
azetidinone N–H, $OCH_2CH{=}CHCO_2Me$ (E)	$OCHCO_2Bu$-*t*	dry C_6H_6, reflux 5 h	azetidinone, $OCH_2CH{=}CHCO_2Me$, N–$CH(OH)CO_2Bu$-*t*	79	1336
azetidinone N–H, $SCH{=}CHCO_2Et$; *E* or *Z* isomer	$OCHCO_2Bu$-*t*	dry C_6H_6, reflux 1.5 h	azetidinone, $SCH{=}CHCO_2Et$, N–$CH(OH)CO_2Bu$-*t*; two isomers (1:1)		1337
azetidinone N–H, S–C(CH_2CO_2Et)=HC–CO_2Et	$OCHCO_2R$ R = Me, $PhCH_2$, CH_2OAc	THF, Et_3N, r.t. or C_6H_6 reflux	azetidinone, S–C(CH_2CO_2Et)=HC–CO_2Et, N–$CH(OH)CO_2R$	70–80	1338
azetidinone N–H, S–C(CH_2CH_2OAc)=HC–CO_2Et	$OCHCO_2R$ R = Me, $PhCH_2$, CH_2OAc	THF, Et_3N, r.t. or C_6H_6 reflux	azetidinone, S–C(CH_2CH_2OAc)=HC–CO_2Et, N–$CH(OH)CO_2R$	—	1338
azetidinone N–H, H, H, $PhOCH_2CONH$, $OCH{=}CHMe$	$OCHCO_2CH_2Ph$	C_6H_6, Et_3N, dioxane reflux 1 h	azetidinone, H, H, $PhOCH_2CONH$, $OCH{=}CHMe$, N–$CH(OH)CO_2CH_2Ph$	—	1337

(continued)

TABLE 63. *(continued)*

Lactam	Glyoxylate	Reaction conditions	Product	Yield (%)	Reference
PhOCH2CONH, H, H, CH2CH=CH2, O, N, H	$OCHCO_2CH_2Ph$	1. C_6H_6, dioxane, reflux 2. Et_3N	PhOCH2CONH, H, H, CH2CH=CH2, O, N, CHCO2CH2Ph, OH	91	1196
PhOCH2CONH, H, H, CH2CH=CHCO2Me, O, N, H	$OCHCO_2CH_2Ph$	C_6H_6, reflux 14h	PhOCH2CONH, H, H, CH2CH=CHCO2Me, O, N, CHCO2CH2Ph, OH	82	1196
PhOCH2CONH, H, H, CH=CHCO2Me, O, N, H	$OCHCO_2CH_2Ph$	C_6H_6, reflux 14h	PhOCH2CONH, H, H, CH=CHCO2Me, O, N, CHCO2CH2Ph, OH (1:1 mixture of isomers)	88	1196
p-O2NC6H4CH2O2CO, H, C, Me, H, H, O2CMe, C, H, CH=CH2, O, N, H	$OCHCO_2CH_2C_6H_4NO_2$-*p*	—	p-O2NC6H4CH2O2CO, H, C, Me, H, H, O2CMe, C, H, CH=CH2, O, N, CHCO2CH2C6H4NO2-p, OH	—	1319
(t-Bu)Me2SiO, Me—CH, H, H, CH2I, O, N, H	$OCHCO_2CH_2C_6H_4NO_2$-*p*	C_6H_6, reflux 10h	(t-Bu)Me2SiO, Me—CH, H, H, CH2I, O, N, CH(OH)CO2CH2C6H4NO2-p	67	1339

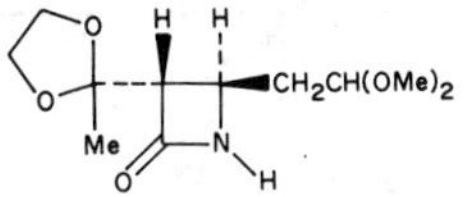	$OCHCO_2CH_2C_6H_4NO_2$-*p*	C_6H_6, reflux 1 h	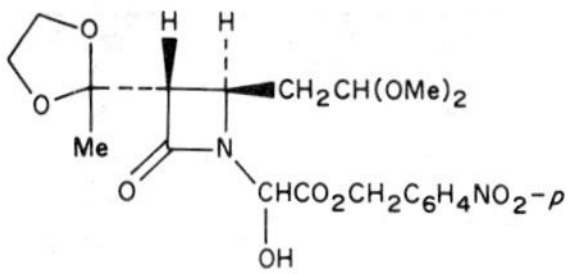	43	1340
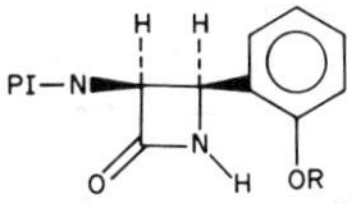	$OCHCO_2CH_2CCl_3$	C_6H_5Me, dioxane, Mol. Sieves, 90–100 °C	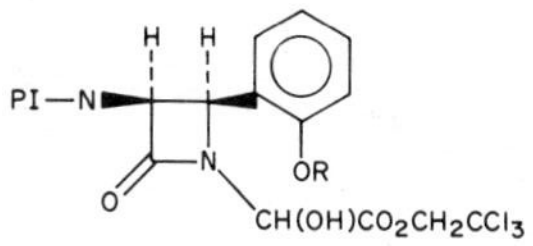		1183
R = H R = CH_2Ph			diasterisomeric mixture (2.7:1)	78.7 —	
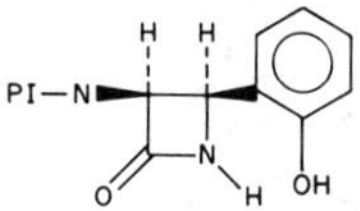	$OCHCO_2CH_2C_6H_4NO_2$-*p*	C_6H_6, dioxane, Mol. Sieves	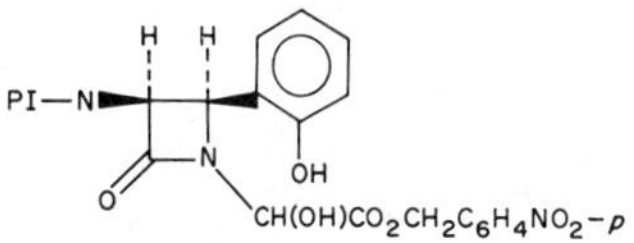	—	1183
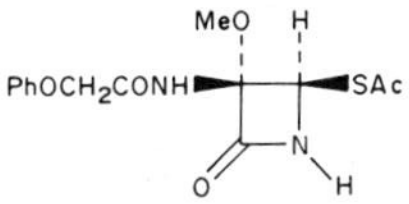	$OCHCO_2CH_2C_6H_4NO_2$-*p*	C_6H_6, Et_3N	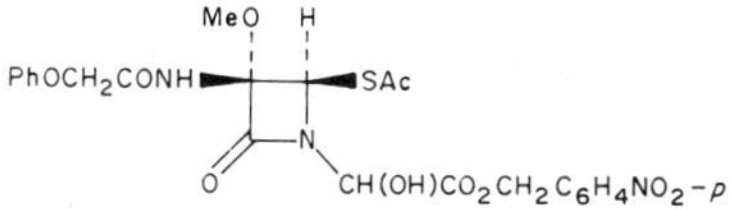	—	1341
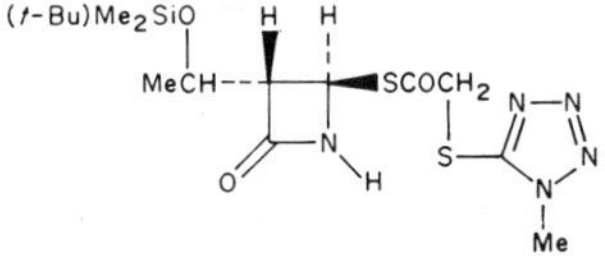	$OCHCO_2CH(OPh)Me$	—	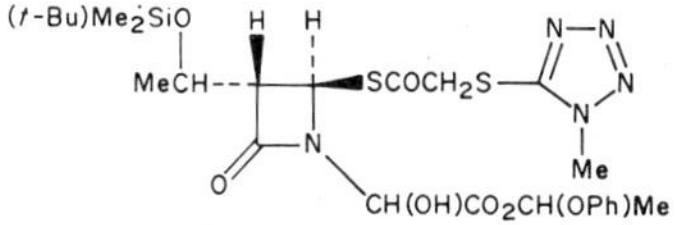	—	1342

(*continued*)

TABLE 63. (*continued*)

Lactam	Glyoxylate	Reaction conditions	Product	Yield (%)	Reference
(*t*-Bu)Me_2SiO, MeCH, H, H, SCOFu-2, O, N, H	$OCHCO_2CH_2CH{=}CH_2$	C_6H_6, reflux 25h	(*t*-Bu)Me_2SiO, MeCH, H, H, SCOFu-2, O, N, $CH(OH)CO_2CH_2CH{=}CH_2$	—	1344
(*t*-Bu)Me_2SiO, MeCH, H, H, COMe, O, N, H	$OCHCO_2CH_2CH{=}CH_2$	C_6H_6, reflux	(*t*-Bu)Me_2SiO, MeCH, H, H, COMe, O, N, $CH(OH)CO_2CH_2CH{=}CH_2$	—	1345
(*t*-Bu)Me_2SiO, MeCH, H, CH_2, C, COMe, O, N, H	$OCHCO_2CH_2C_6H_4NO_2$-*p*	C_6H_6, reflux	(*t*-Bu)Me_2SiO, MeCH, H, CH_2, C, COMe, O, N, $CH(OH)CO_2CH_2C_6H_4NO_2$-*p*	—	1345
O, N, H, SAc	$OCHCO_2CH_2C_6H_4NO_2$-*p*	C_6H_5Me, 110 °C, 1.5h	O, N, SAc, CH(OH), $CO_2CH_2C_6H_4NO_2$-*p*	—	1346
t-BuO_2CNH, SAc, N, H, O	$OCHCO_2CH_2C_6H_4NO_2$-*p*	C_6H_5Me, 110 °C, 1.5h	*t*-BuO_2CNH, SAc, N, CH(OH), $CO_2CH_2C_6H_4NO_2$-*p*, O	—	1346
PhO_2C, H, N, H, O, H	$OCHCO_2CH_2C_6H_4NO_2$-*p*	—	PhO_2C, H, N, $CH(OH)CO_2CH_2C_6H_4NO_2$-*p*, O, H	—	1319

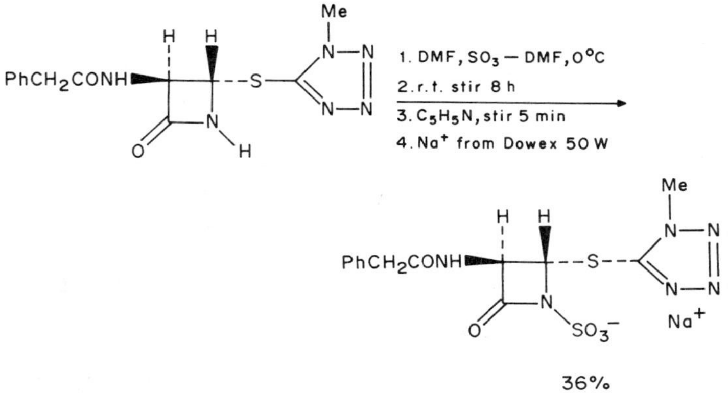

1. DMF, SO_3—DMF, 0 °C
2. r.t. stir 8 h
3. C_5H_5N, stir 5 min
4. Na^+ from Dowex 50 W

(931)

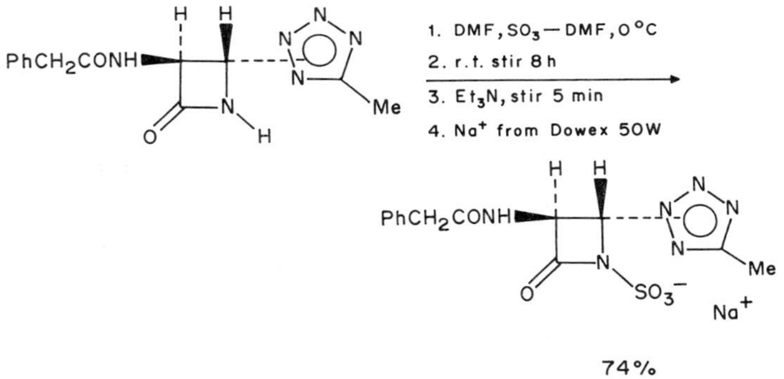

36%

1. DMF, SO_3—DMF, 0 °C
2. r.t. stir 8 h
3. Et_3N, stir 5 min
4. Na^+ from Dowex 50W

(932)

74%

Because of the reagents and conditions employed, many of the reactions used to produce N-substituted lactams also result in cyclization, thus affording bicyclic lactams as products. Two types of reactions have been reported which produce N-substitution and cyclization in one step. The first involves reaction of a substituted lactam where the substituent is located adjacent to the lactam nitrogen and contains a reactive site which reacts with the lactam nitrogen (equation 933), while the second involves reaction of a lactam with a bifunctional reagent which reacts simultaneously (or rapidly in two steps) with both the lactam nitrogen and an adjacent site on the lactam ring (equation 933). Table 64 shows the reactions reported to produce both N-substitution and cyclization.

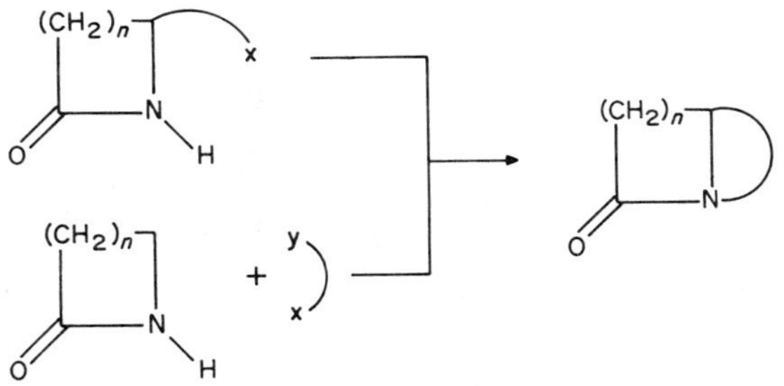

(933)

TABLE 64. N-substitution reactions accompanied by cyclization

Lactam	Reagent	Conditions	Product	Yield (%)	Reference
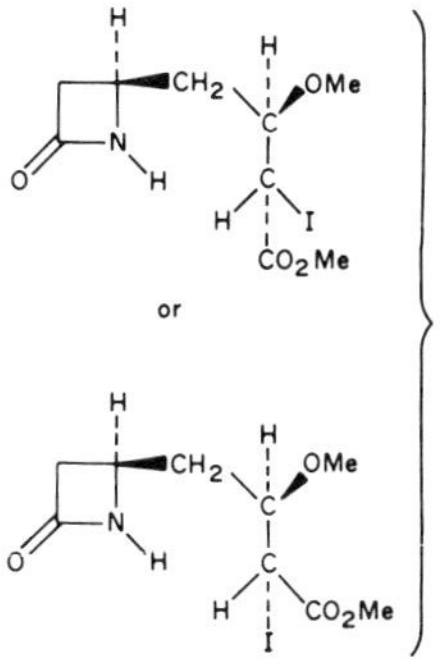 CH_2, OMe, I, CO_2Me or CH_2, OMe, CO_2Me, I	—	Triton B, MeCN, H_2O, 20 °C, 5 min	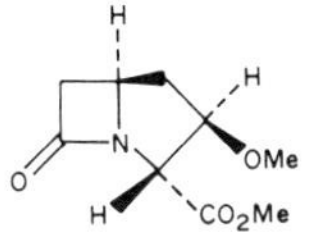 OMe, CO_2Me	80	1315
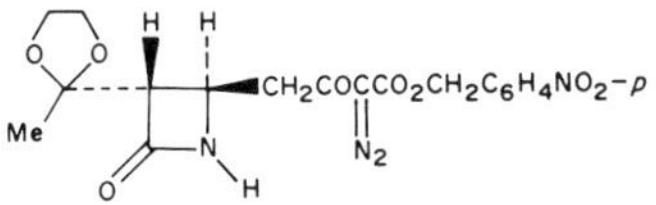 $CH_2COCCO_2CH_2C_6H_4NO_2$-*p* ($N_2$), Me	—	$Rh_2(OAc)_4 \cdot$THF	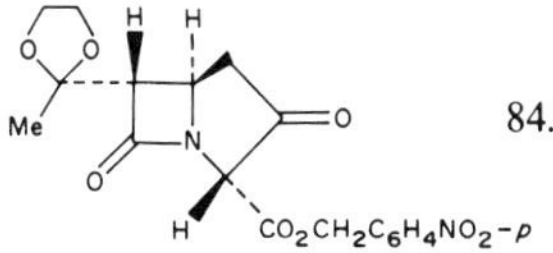 Me, $CO_2CH_2C_6H_4NO_2$-*p*	84.6	1347
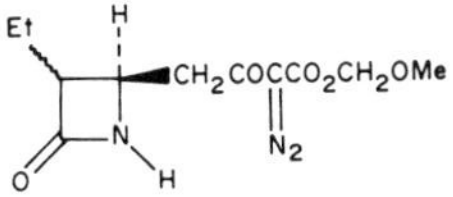 Et, $CH_2COCCO_2CH_2OMe$ (N_2)	—	$(n\text{-}C_7H_{15}CO_2)_4Rh_2$, $CHCl_3$, reflux 15 min	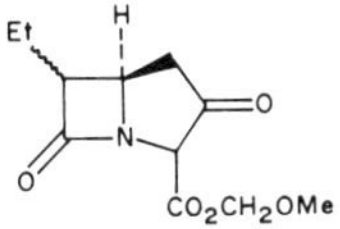 Et, CO_2CH_2OMe	—	1082
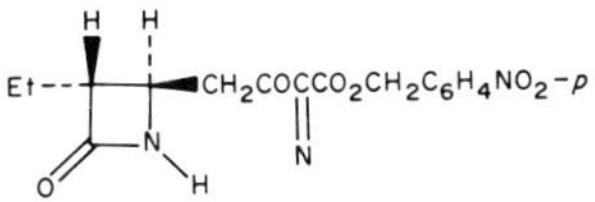 Et, $CH_2COCCO_2CH_2C_6H_4NO_2$-*p* (N)	—	$Rh_2(OAc)_4$, C_6H_6, heat	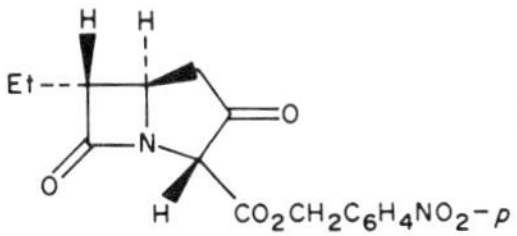 Et, $CO_2CH_2C_6H_4NO_2$-*p*	82	1245

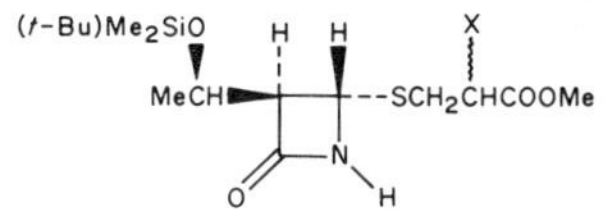 (mixture of diastereomers with X = Cl and OTs)	—	$2N$-NaOH, CH_2Cl_2, $(n\text{-}Bu)_4\overset{+}{N}Br^-$	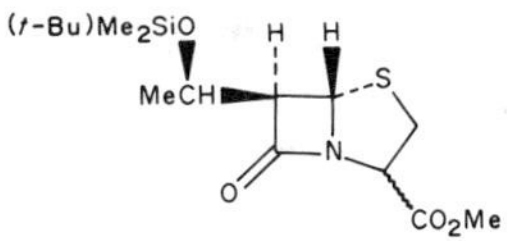	8–10	1348
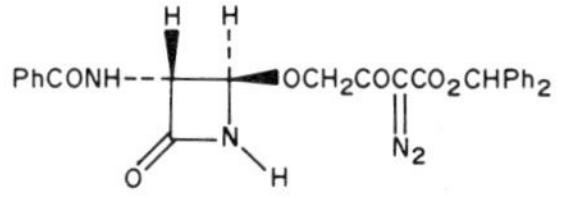	—	$Rh_2(OAc)_4$, C_6H_6, reflux	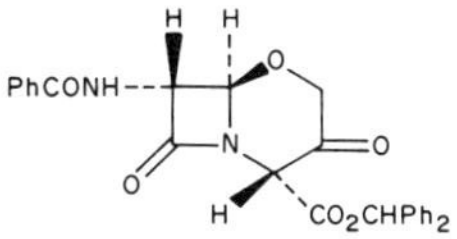	85	1349
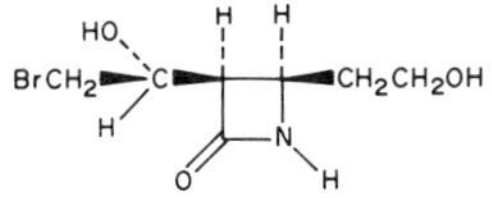	—	$Me_2C(OMe)_2$, CH_2Cl_2, $BF_3 \cdot OEt_2$	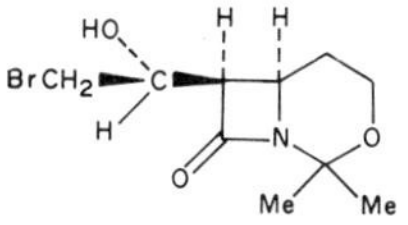	59	1021
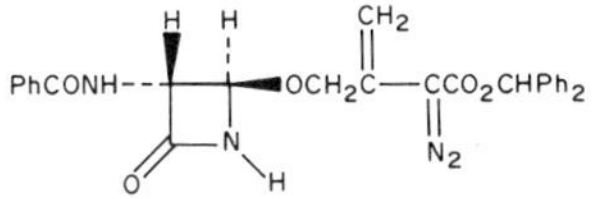	—	$Rh_2(OAc)_4$, C_6H_6, reflux	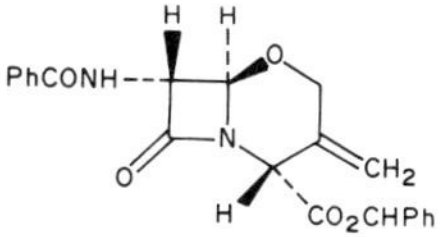	53	1349
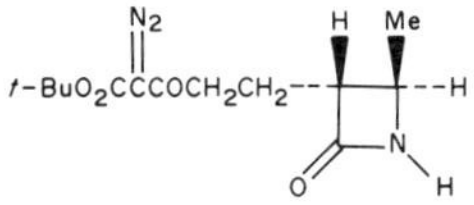	—	$Rh_2(OAc)_4$, C_6H_6, reflux 1 h	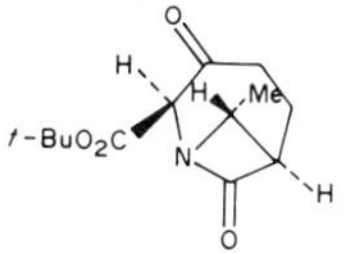	50	1078
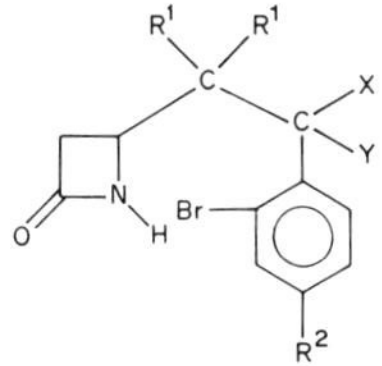	—	Cu, DMF, 100 °C	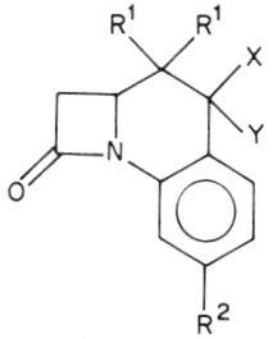		1350

(continued)

TABLE 64. (*continued*)

Lactam				Reagent	Conditions	Product	Yield (%)	Reference
R^1	R^2	X	Y					
CO_2Et	H	H	H				60	
CO_2Et	CO_2Et	H	H				45	
CO_2CH_2Ph	H	—O—					5	
H	H	—O—					13	
H	CO_2Me	—O—					10	
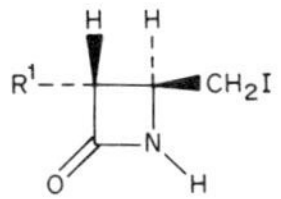 R^1, H, H, CH_2I, O, N, H				$(H)(MeO_2C)C{=}C(CO_2Me)(SPh)$	KH, THF 18-crown-6	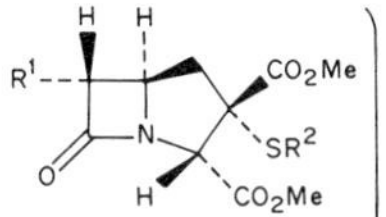 R^1, H, H, O, N, H, CO_2Me, SR^2, CO_2Me (**190**) R^2 = Ph 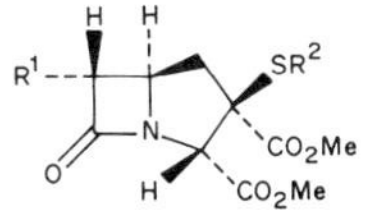 + R^1, H, H, O, N, H, SR^2, CO_2Me, CO_2Me (**191**) R^2 = Ph	—	1339
R^1 = H				−20 to −40 °C, stir 3 h			27	
R^1 = (*t*-Bu)Me_2SiOCHMe				-70 to −10 °C, stir 2.5 h			15	
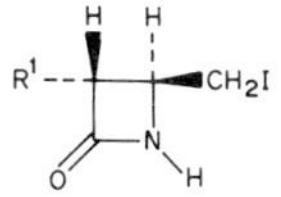 R^1, H, H, CH_2I, O, N, H				$(H)(MeO_2C)C{=}C(CO_2Me)(SMe)$	Ph_2CHK, THF 18-crown-6, −40 °C to −20 °C			1339
R^1 = H					stir 3 h	**190** + **191**, R^2 = Me	4	

$R^1 = (t\text{-Bu})Me_2SiOCHMe$		stir 2.75 h	**190** + **191**, $R^2 = Me$	18	
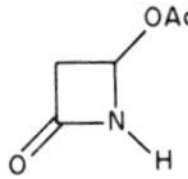	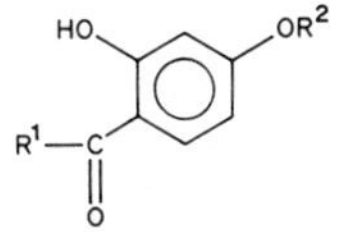	H_2O, dioxane, 0 °C, stir	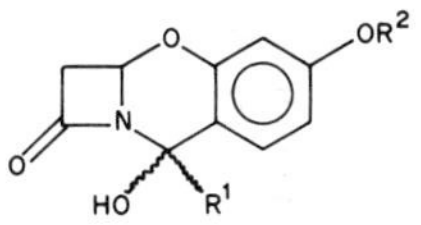		1351
	$R^1 = H$, $R^2 = CH_2Ph$	1 M NaOH, 4 h		33	
	$R^1 = EtO_2C$, $R^2 = CH_2Ph$	10% NaOH, 2 h		40	
	$R^1 = R^2 = CO_2CH_2Ph$	$CuCO_3$ or Na_2CO_3, 1 h		61	
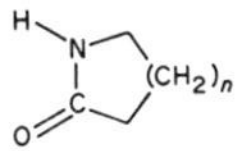 $n = 1,3$	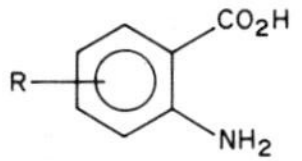 $R = H$, 3-Me, 4-NH_2 4-NO_2, 5-Br, 5-Cl, 5-I, 5-NH_2, 5-NO_2, 5-CO_2H	$POCl_3$, heat	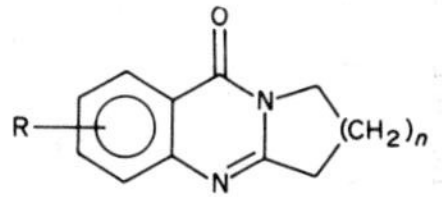	—	1352
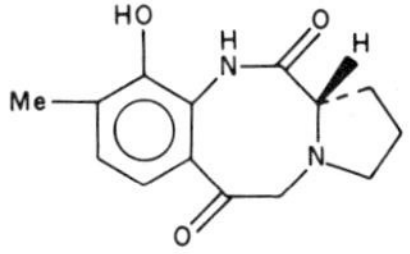	$PhCH(OMe)_2$	—	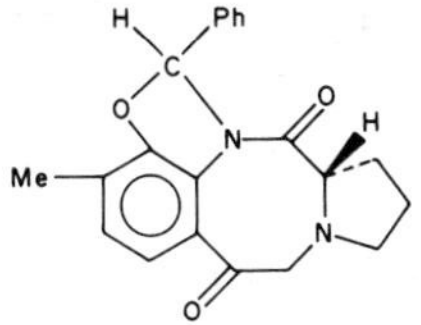	—	1353
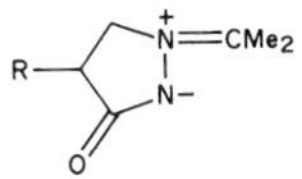	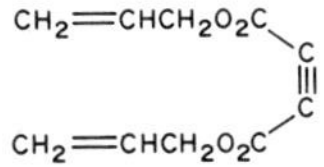		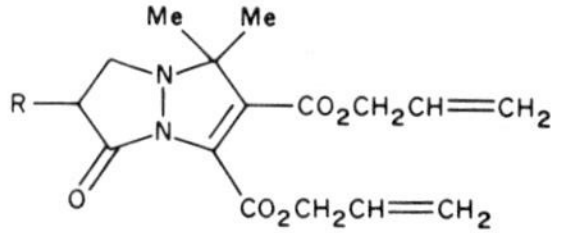	—	1046
$R = H$ $R = t\text{-BuO}_2CNH$		CH_2Cl_2, r.t., 72 h		67	1076
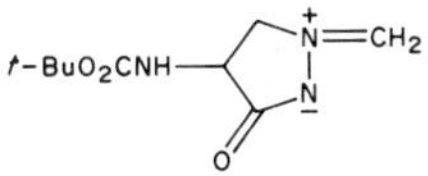	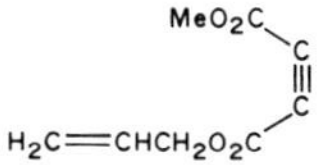	$ClCH_2CH_2Cl$, reflux	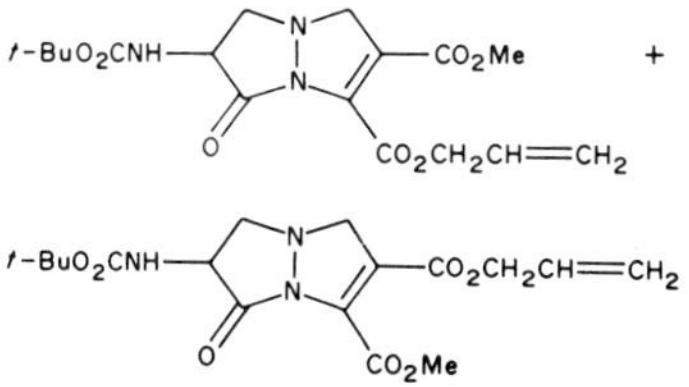 +	—	1334

2. Functionalization of lactam ring other than at lactam nitrogen

Functionalization of lactam ring sites other than at the lactam nitrogen has been accomplished using a variety of reactions many of which parallel the approaches discussed in the previous section. For example, reaction of N-substituted lactams with a base affords an anion at the site alpha to the ring carbonyl function and subsequent treatment of this enolate with halides, carbonyl compounds, olefins, *O*-(diphenylphosphinoyl)-hydroxylamine, *n*-propyl nitrate, phosphorus pentachloride, azides and other reagents produces a wide variety of 3-substituted lactams.

This section begins with a discussion of 3-substituted lactams formed from the reaction of a lactam anion and halides. 3-Silylation of 1,4-disubstitued *β*-lactams using trimethylsilyl chloride and a base produces[1354] initially *cis*-products, which isomerize to *trans*-products under the reaction conditions employed (equation 934).

base
Me_3SiCl
isomerization

R^1 = Ph; Me; *i*-Pr; Ph; *p*-Tol
R^2 = H; Ph; Ph; Ph; Ph

(934)

Asymmetric alkylations were reported[1181] to occur when (3*S*, 4*R*)-1-[(*S*)-1-(benzyloxy)-4-methylpent-2-yl]-3-(benzylideneamino)-4-phenylazetidin-2-one (equation 935) and (3*S*, 4*R*-1-benzyl-3-[2-oxo-4(*S*)-phenyloxazolidinyl]-4-(3,4-dimethoxyphenyl)azetidine-2-one (equation 936) were treated with lithium hexamethyldisilazide and the resulting enolates were quenched with alkyl halides.

1. THF, $(Me_3Si)_2Li$, −78 °C, stir 30 min
2. RBr, −78 °C
3. −78 °C to r.t. overnight

R = Me; $CH_2CH{=}CH_2$
Yield (%) = 94; 95

(935)

1. THF, $(Me_3Si)_2Li$, −78 °C, stir 1h
2. MeI, −78 °C
3. −78 °C to r.t. overnight

95%

(936)

TABLE 65. Alkylation and acylation of N-substituted lactams using organic halides

Lactam	Halide	Base	Reaction Conditions	Product	Yield (%)	Reference
PhS / O / N / Hex-*c*	RX	*n*-BuLi	THF, $-60\,^{\circ}C$	R / PhS / O / N / Hex-*c*	—	1089
	RX = MeI, EtBr, $CH_2{=}CHCH_2Br$					
H / SEt / NC / H / O / N / Hex-*c* (mixture of $Z{:}E = 2{:}1$)	RX	NaH	THF, $0\,^{\circ}C$	CN / SEt / R / H / O / N / Hex-*c*		1142
	RX = MeI				91	
	RX = $CH_2{=}CHCH_2Br$				59	
	RX = MeCOCl				29	
	RX = PhCOCl				43	
	RX = MeOCOCl				90	
H / OEt / NC / H / O / N / Ph (mixture of $Z{:}E = 2{:}1$)	MeI	NaH	THF, $0\,^{\circ}C$	CN / OEt / Me / H / O / N / Ph ($Z{:}E = 3{:}1$)	—	1355
H / HOOC / O / N / H / C / COOMe / $(CH_2)_4NHO_2CCH_2Ph$	$CH_2\overset{+}{N}Me_3$ / N / H / I^-	DBU	DMF, $80\,^{\circ}C$	COOH / CH_2 / N / H / O / N / H / C / COOMe / $(CH_2)_4NHO_2CCH_2Ph$ (1:1 mixture of diastereomers)	—	1040

(*continued*)

TABLE 65. (*Continued*)

Lactam	Halide	Base	Reaction Conditions	Product	Yield (%)	Reference
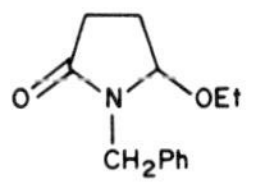 O N OEt CH_2Ph	$ICH_2CH_2CH=CCH_2SiMe_3$ (*cis*)	LDA	1. THF, $-78\,^{\circ}C$, 2 h 2. r.t.	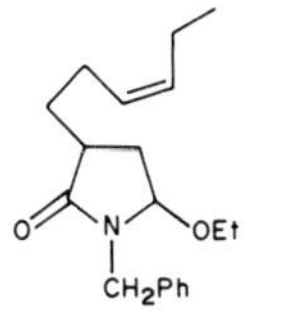 O N OEt CH_2Ph (60:40 mixture of stereoisomers)	85	1356
	$ICH_2CH_2C\equiv CCH_2SiMe_3$	LDA	1. THF, $-78\,^{\circ}C$, 2 h 2. r.t.	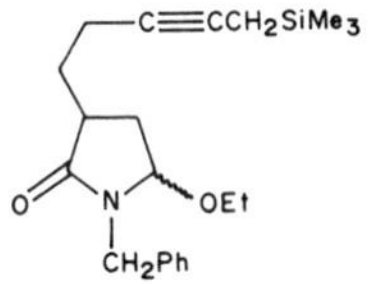 $C\equiv CCH_2SiMe_3$ O N OEt CH_2Ph (60:40 mixture of stereoisomers)	82	1356
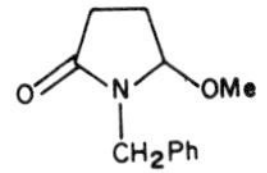 O N OMe CH_2Ph	$ICH_2(CH_2)_2CH=CHCH_2SiMe_3$ (*cis*)	LDA	1. THF, $-78\,^{\circ}C$, 2 h 2. r.t.	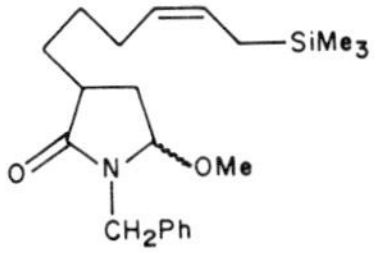 $SiMe_3$ O N OMe CH_2Ph (60:40 mixture of stereoisomers)	75	1356

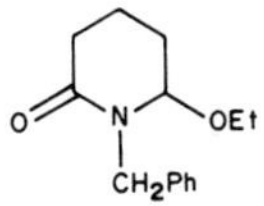 O=C–N(CH$_2$Ph)–CH(OEt) ring (piperidinone)	$ICH_2(CH_2)_nC{\equiv}CCH_2SiMe_3$	LDA	1. THF, −78 °C, 2 h 2. r.t.	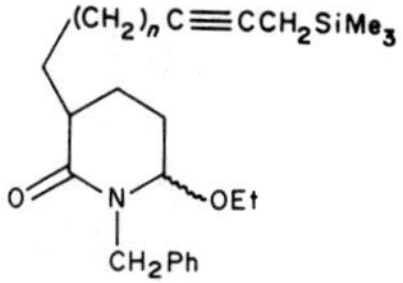 3-[$(CH_2)_nC{\equiv}CCH_2SiMe_3$]-substituted N-CH$_2$Ph, 6-OEt piperidinone (60:40 mixture of stereoisomers)		
	$n = 1$				75	1356
	$n = 2$				87	1336
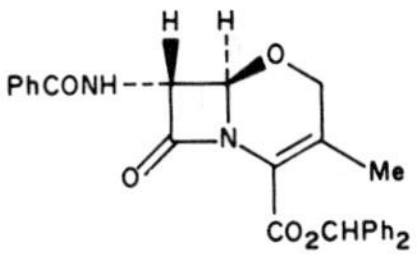 PhCONH, H, H, O, N, Me, CO_2CHPh_2 (bicyclic β-lactam)	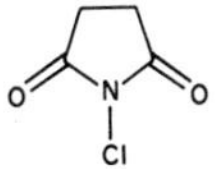 N-chlorosuccinimide (N–Cl)	DBU	—	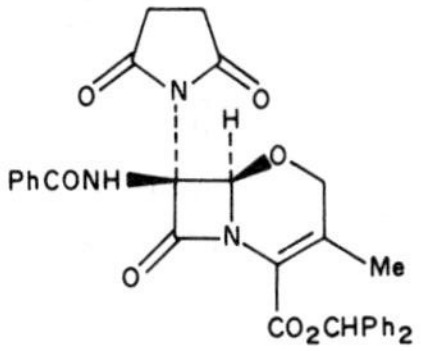 PhCONH, N-succinimidyl, H, O, N, Me, CO_2CHPh_2 (bicyclic β-lactam)	—	1357

Other examples of this approach to substitute lactams at the 3-position are reported in Table 65, where it should be noted that if the initial lactam substrate contains a substituent at the beta position, stereospecific alkylated or acylated products result.

An interesting alkylation of N-substituted γ-lactams occurs when they are treated[1322] with styrene in hexamethylphosphoramide (equation 937).

$$\text{N-R-pyrrolidin-2-one} + PhCH{=}CH_2 \xrightarrow[120\,^\circ C,\ 5\,h]{(Me_2N)_3PO} \text{3-}(PhCH_2CH_2)\text{-N-R-pyrrolidin-2-one} \tag{937}$$

$R = PhCH_2CH_2\,;\ CH_2{=}CH$

Yield (%) = 69 ; —

α,β-Unsaturated carbonyl compounds or nitriles have also been reported[1142] to react with β-lactam enolates in a Michael addition, to produce α-substituted lactams where the α-substituent is an alkyl group containing a γ-keto or γ-cyano group (equation 938).

$$\text{3-NC-4-SEt-1-(Hex-}c\text{)-azetidin-2-one } (Z:E = 2:1) \xrightarrow[2.\ CH_2{=}CHR]{1.\ NaH,\ THF,\ 0\,^\circ C} \text{3-CN-3-}(RCH_2CH_2)\text{-4-SEt-1-(Hex-}c\text{)-azetidin-2-one} \tag{938}$$

$R = CN;\ MeCO$

Yield (%) = 63; 92

$$\text{lactam (N-}R^1\text{)} \xrightarrow{\text{base conditions}} [\text{lactam enolate}]^- \longrightarrow \begin{cases} \xrightarrow{R^2CHO} \alpha\text{-}CH(OH)R^2 \text{ lactam} \\ \xrightarrow{R^2R^3C{=}O} \alpha\text{-}C(OH)R^2R^3 \text{ lactam} \\ \xrightarrow{R^2CO_2R^3} \alpha\text{-}COR^2 \text{ lactam} \\ \xrightarrow{R^2N{=}C{=}O} \alpha\text{-}CONHR^2 \text{ lactam} \end{cases} \tag{939}$$

TABLE 66. Reaction of lactam enolates with carbonyl compounds

Lactam	Base and RX. conditions	Carbonyl substrate	Product	Yield (%)	Reference
CN SEt, H, H, N–Hex-c, O (Z:E = 2:1)	NaH, THF, 0 °C	PhCHO	PhCH(OH)–, CN SEt, H, N–Hex-c, O	60	1142
CN SEt, H, H, N–Hex-c, O	NaH, THF, 0 °C	t-BuN=C=O	t-BuNHOC–, CN SEt, H, N–Hex-c, O	72	1142
H, –CH=CH$_2$, N–CH$_2$N(pyrrolidinyl), O	LDA, THF, −78 °C	MeCHO	RCH(OH)–, H H, CH=CH$_2$, N–H, O; (**192**) R = Me	58[a]	1333
H, –CH=CH$_2$, N–CH$_2$N(pyrrolidinyl), O	LDA, THF, −78 °C	MeCOMe	Me$_2$C(OH)–, H H, CH=CH$_2$, N–H, O	52[a]	1333
		PhCHO	(**192**) R = Ph (2:1 isomer mixture)	85[b]	1332
H, –CH$_2$CH$_2$CH=CH$_2$, N–CH$_2$N(pyrrolidinyl), O	LDA, THF, −78 °C	MeCHO	MeCR(OH)–, H H, CH$_2$CH$_2$CH=CH$_2$, N–H, O; (**193**) R = H	79[c]	1333 1333
		MeCOMe	(**193**) R = Me	74[a]	1333
H, –CH$_2$CH$_2$CH=CH$_2$, N–H, O	n-BuLi, THF, 0 °C	M~COMe	Me$_2$C(OH)–, H H, CH$_2$CH$_2$CH=CH$_2$, N–H, O (cis:trans = 1:1)	75	1333

(continued)

TABLE 66. (*continued*

Lactam	Base and RX. conditions	Carbonyl substrate	Product	Yield (%)	Reference
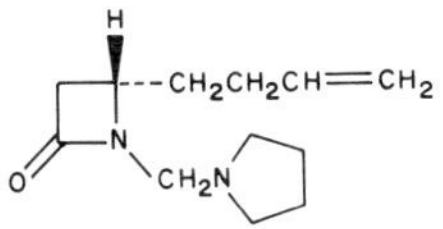 4-($CH_2CH_2CH{=}CH_2$)-1-(pyrrolidin-1-ylmethyl)azetidin-2-one	LDA, THF, −78 °C	$MeCO_2Et$	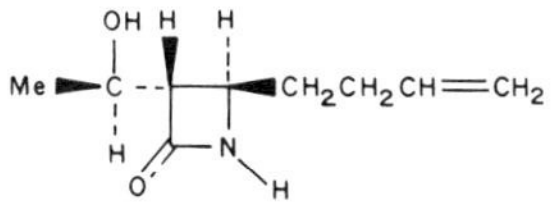 3-[C(OH)(Me)H]-4-($CH_2CH_2CH{=}CH_2$)azetidin-2-one (N–H)	68[c]	1333
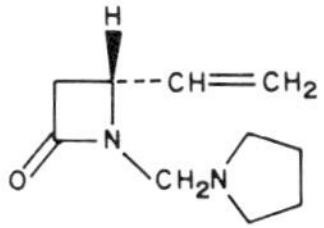 4-($CH{=}CH_2$)-1-(pyrrolidin-1-ylmethyl)azetidin-2-one	LDA, THF, −78 °C	$MeCO_2Et$	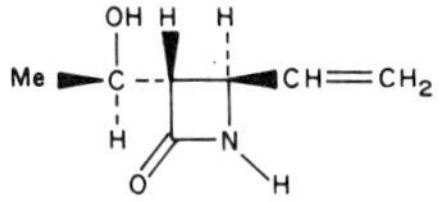 3-[C(OH)(Me)H]-4-($CH{=}CH_2$)azetidin-2-one (N–H)	60[d]	1333
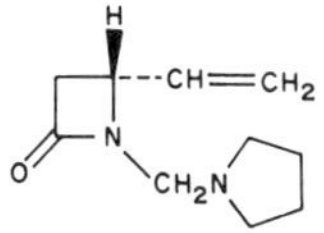 4-($CH{=}CH_2$)-1-(pyrrolidin-1-ylmethyl)azetidin-2-one	LDA, THF, −78 °C	PhCOPh	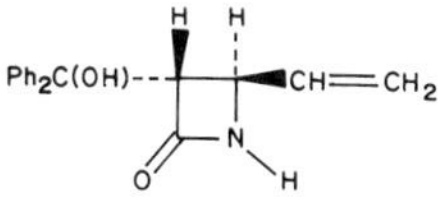 3-[$Ph_2C(OH)$]-4-($CH{=}CH_2$)azetidin-2-one (N–H)	65[f]	1333
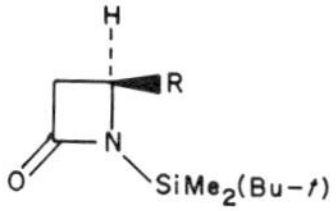 4-R-1-[$SiMe_2(Bu\text{-}t)$]azetidin-2-one	1. LDA, THF, −78 °C, 5 min 2. ketone, 10 min. 3. *t*-BuOK, *t*-BuOH, −78 °C to 0 °C, 10 min. 4. aq. NH_4Cl	$MeCOSiMe_2(Bu\text{-}t)$	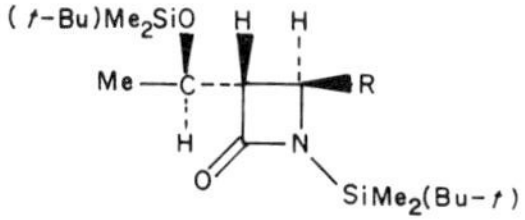 3-[C(Me)(H)(OSiMe₂(*t*-Bu))]-4-R-1-[$SiMe_2(Bu\text{-}t)$]azetidin-2-one		1358[g]
$R = CH{=}CH_2$			(*trans*-R:*cis*-R:*trans*-S = 85:8:7)	75	
$R = SCPh_3$			(*trans*-R)	79	
$R = CO_2H$	0 °C		(*trans*-R:*trans*-S = 87:13)	77[h]	

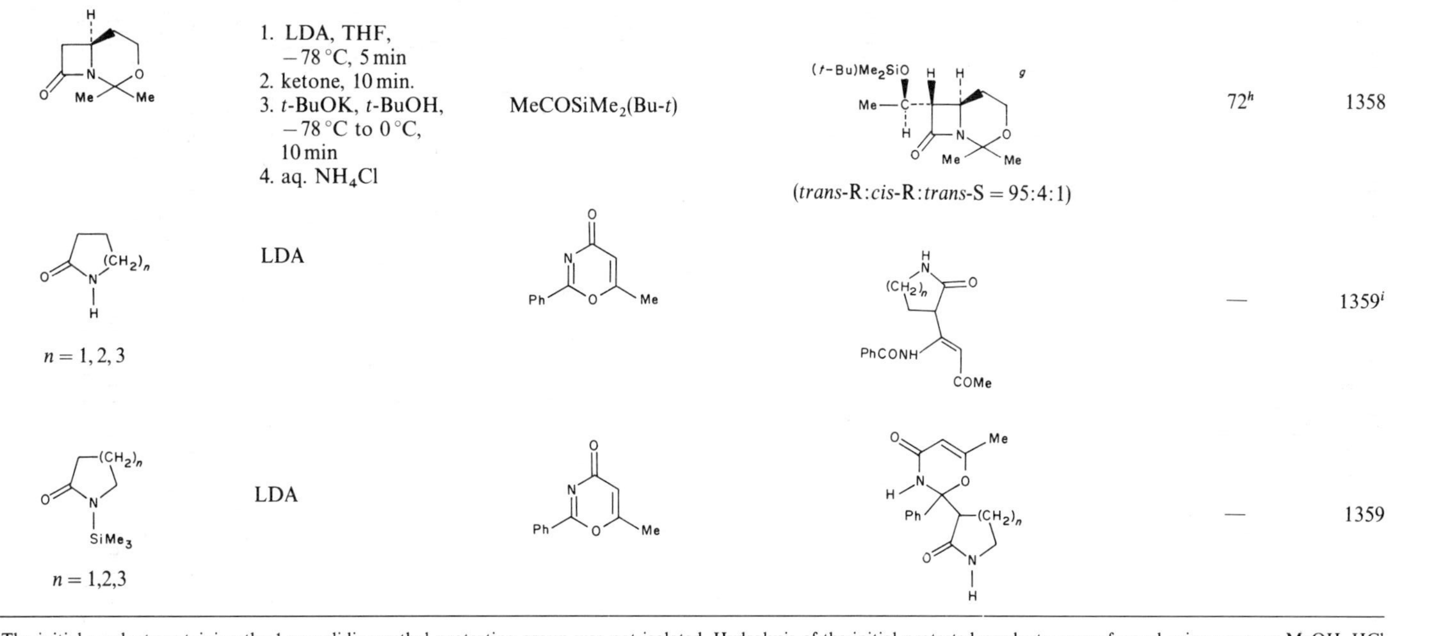

Substrate	Conditions	Reagent	Product	Yield (%)	Ref.
—	1. LDA, THF, −78 °C, 5 min 2. ketone, 10 min. 3. t-BuOK, t-BuOH, −78 °C to 0 °C, 10 min 4. aq. NH_4Cl	$MeCOSiMe_2(Bu\text{-}t)$	[g] (*trans*-R:*cis*-R:*trans*-S = 95:4:1)	72[h]	1358
$n = 1, 2, 3$	LDA	—	—	—	1359[i]
$n = 1,2,3$	LDA	—	—	—	1359

[a]The initial product containing the 1-pyrrolidinomethyl protecting group was not isolated. Hydrolysis of the initial protected product was performed using aqueous MeOH, HCl under reflux for 3 h and afforded the product shown in the yield reported.
[b]The initial protected product was isolated in 84% yield.
[c]As in a, except that the hydrolysis was performed using a 1:6 mixture of HF:HCl.
[d]The initial protected ketone product was isolated in 73% yield. After reduction with K-selectride followed by deprotection, the product was isolated as a 4:1 diastereomer mixture. Reduction was also accomplished using $NaBH_4$ affording 71% of a 3-hydroxyethyl derivative as a 1:2 diastereomer mixture, and by using aqueous MeOH, HCl giving the single diastereomer shown in 68% yield.
[e]The initial protected ketone product was isolated in 63% yield. After reduction with K-selectride followed by deprotection, the product was isolated as a 9:1 diastereomer mixture. Reduction was also accomplished using $NaBH_4$ affording a 72% yield of product, or by using aqueous MeOH, HCl which gave the single diastereomer shown in 60% yield.
[f]The initial protected product was isolated in 83% yield.
[g]The initial product formed is the *trans*-S silyl carbinol which underwent complete stereospecific rearrangement to the *trans*-R O-silyl ether upon treatment with KOBu-t in t-BuOH.
[h]The product isolated after fractional crystallization and/or chromatography was the *trans*-R product in the yield shown.
[i]When $n = 2$, a second product was also obtained.

Aldol condensations of lactam enolates with aldehydes, ketones, esters and isocyanates have been accomplished producing the corresponding α-hydroxylalkyl-, α-ketoalkyl- and α-alkylamido-substituted products (equation 939 and Table 66).

There has been at least one recent report[1018,1019] of the direct α-amination of a preformed lactam, involving the reaction of *t*-butyl 2-oxo-1-azabicyclo[3.2.0]heptane-7-*exo*-carboxylate with lithium hexamethyldisilazane. The formed intermediate monoanion upon quenching with *O*-(diphenylphosphinoy)hydroxylamine produces a 47% yield of *t*-butyl 3-*endo*-amino-2-oxo-1-azabicyclo[3.2.0]heptane-7-*exo*-carboxylate (equation 940).

$(Me_3Si)_2NLi$, THF, −70 °C, stir 1 h; Li+ ; CO_2Bu-*t*

$Ph_2PO_2NH_2$, −70 °C, stir 3 h; H_2N ; CO_2Bu-*t* (47%) (940)

A similar approach has been used[1360] to directly nitrate *N*-alkyl-2-pyrrolidones and involves preparation of the initial anion with lithium diisopropylamide, followed by quenching with *n*-propyl nitrate (equation 941).

LDA, THF, 0 °C, stir; Li^+; 1. *n*-$PrONO_2$, −25 °C, stir 30 min 2. H^+; NO_2

R = Me ; *c*-Hex ; *i*-Pr.

Yield (%) = 53 ; 61 ; 44 .

(941)

The *N*-alkyl-3-nitro-2-pyrrolidones formed as products from the above reaction serve as the starting materials for the preparation[1360] of *N*-alkyl-3-nitro-3-halo-2-pyrrolidones (equation 942).

Numerous other approaches to the halogenation of lactams have also been reported. The first of these reports involves the treatment[1333] of 1-pyrrolidinomethyl-4-vinylazetidin-2-one with lithium diisopropylamide followed by iodine (equation 943).

Reaction of 1-cyclohexyl-3-cyano-4-(ethylthio)azetidin-2-one with sodium hydride, followed by treatment with halogen, affords[1355] an isomeric mixture of the corresponding 3-halo-3-cyano derivative (equation 944).

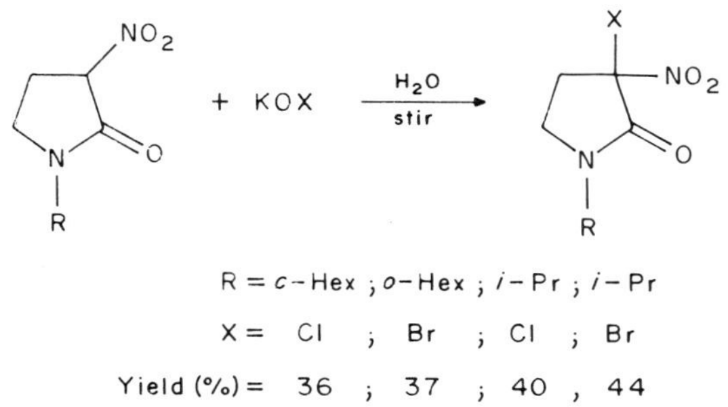

R = *c*-Hex ; *o*-Hex ; *i*-Pr ; *i*-Pr

X = Cl ; Br ; Cl ; Br

Yield (%) = 36 ; 37 ; 40 , 44

(942)

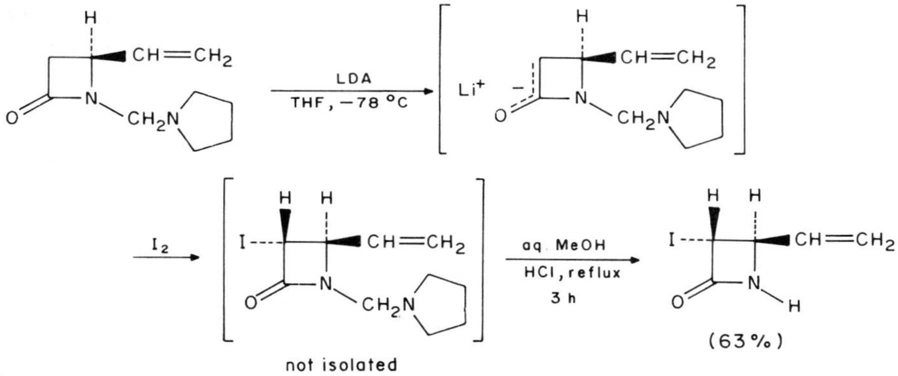

not isolated

(63%)

(943)

X = Cl ; Br ; I.

Z(%) = 40 ; 20 ; 8.

E (%) = 60 ; 80 ; 92.

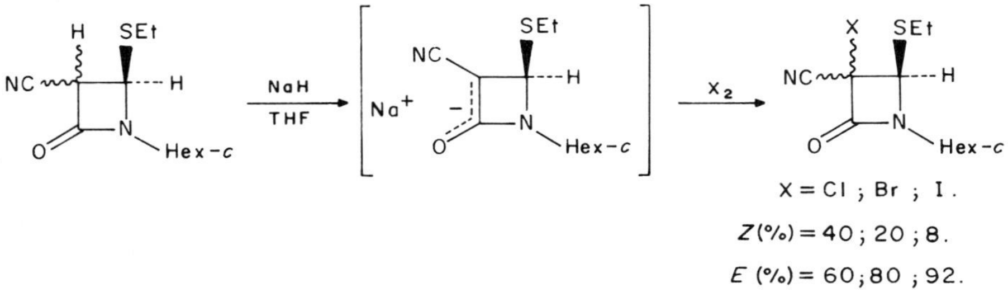

(944)

A similar reaction is observed[1141] when a 9:2 *cis*:*trans* mixture of 1-(*t*-butyl)-3-cyano-4-phenylazetidin-2-one is treated with lithium diisopropylamide, followed by treatment with *N*-chlorosuccinimide (equation 945).

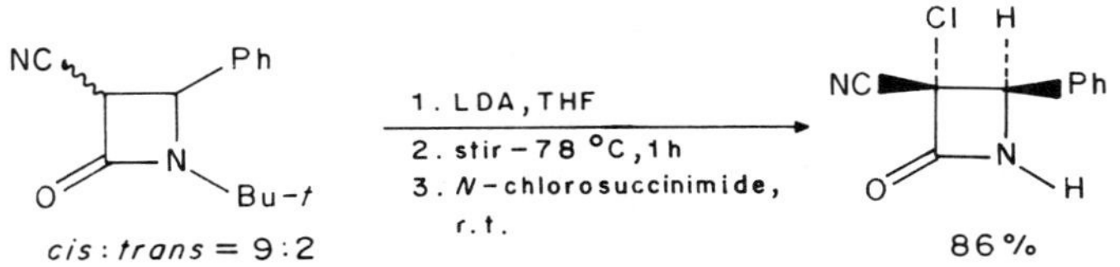

1. LDA, THF
2. stir −78 °C, 1 h
3. *N*-chlorosuccinimide, r.t.

cis : *trans* = 9:2

86%

(945)

Bromination of 7-, 8- and 9-membered lactams has been accomplished[1305] by reaction of the lactams with phosphorus pentabromide (equation 946).

$n = 3, 4, 5$ (946)

While sulphuryl chloride has been used[1211] to α-chlorinate α-phenylthio β-lactams (equation 947), thionyl chloride has been successfully employed[1155] to chlorinate the 7-membered benzothiazepinone sulphoxide shown in equation 948. Indeed, the presence of the sulphoxide was required to permit the chlorine substituent to be introduced on the aliphatic portion of the molecule, even though during the course of the reaction the sulphoxide is reduced to produce the α-chlorinated sulphide product shown.

(947)

R^1 = Ph ; Ph ; *p*-Tol ; 3,4-methylenedioxyphenyl ; C(Me)=CHPh ; *p*-Tol ; *p*-Tol

R^2 = *p*-An ; Ph ; *p*-An ; Ph ; *p*-An ; *o*-An ; *p*-BrC_6H_4

(948)

86%

The final approach for chlorination of lactams reported[1361] was a dichlorination–monodechlorination sequence of reactions, again using sulphuryl chloride, but this time in the presence of phosphorous pentachloride, performed on a series of *cis*- and *trans*-bicyclic lactams (equation 949 and 950).

One of the most common functions to be placed in the alpha position of a preformed lactam is an azide, and these reactions are all accomplished through an intermediate enolate formed by reaction of the N-substituted lactam with a base (equation 951 and Table 67).

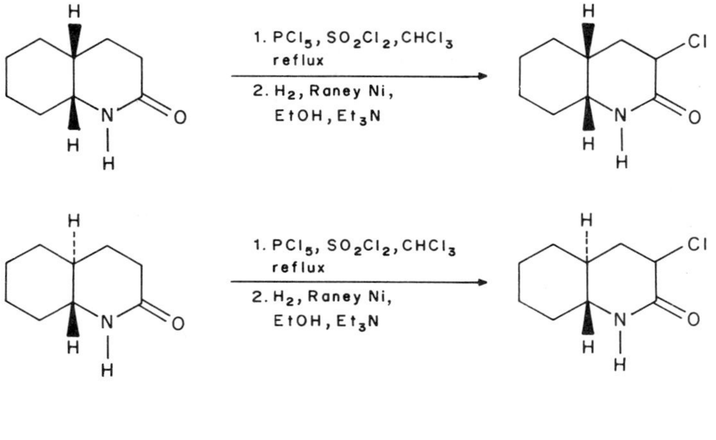

1. PCl_5, SO_2Cl_2, $CHCl_3$ reflux
2. H_2, Raney Ni, EtOH, Et_3N

1. PCl_5, SO_2Cl_2, $CHCl_3$ reflux
2. H_2, Raney Ni, EtOH, Et_3N

(949)

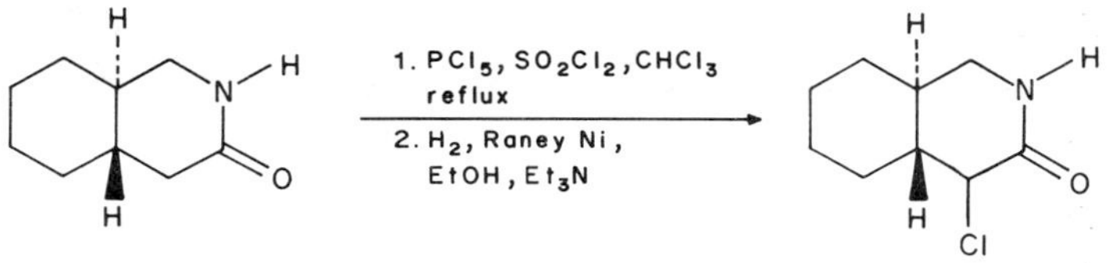

1. PCl_5, SO_2Cl_2, $CHCl_3$ reflux
2. H_2, Raney Ni, EtOH, Et_3N

(950)

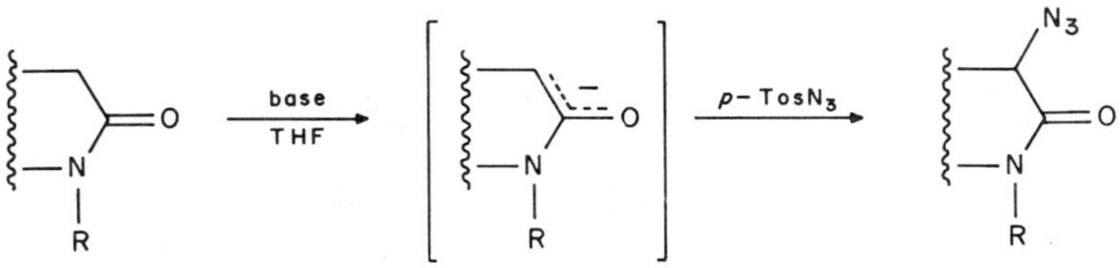

base / THF

p-$TosN_3$

(951)

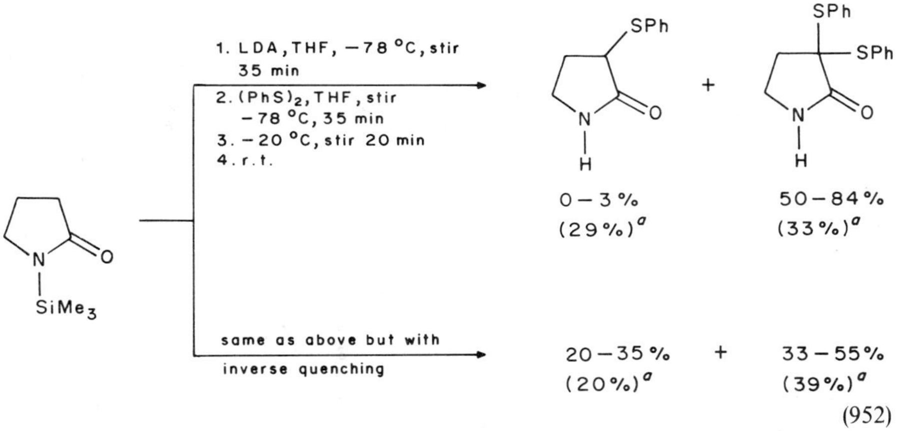

1. LDA, THF, −78 °C, stir 35 min
2. $(PhS)_2$, THF, stir −78 °C, 35 min
3. −20 °C, stir 20 min
4. r.t.

0−3% (29%)[a] + 50−84% (33%)[a]

same as above but with inverse quenching

20−35% (20%)[a] + 33−55% (39%)[a]

(952)

[a]Yields obtained when HMPA was added to the reaction mixture.

TABLE 67. Formation of α-azidolactams from lactams and *p*-tolylazide

Lactam	Azido reagent	Base and reaction conditions	Product	Yield (%)	Reference
Me; N–OCH_2Ph	*p*-$TosN_3$	—	N_3; Me; N–OCH_2Ph	—	1007
H; $CH_2CH{=}CH_2$; N–$SiMe_2(Bu\text{-}t)$	*p*-$TosN_3$	1. LDA, THF, −76 °C stir 2h 2. Me_3SiCl, THF, r.t. 17h	H H; N_3; $CH_2CH{=}CH_2$; N–$SiMe_2(Bu\text{-}t)$	42	1196, 1304
N–$CC_6H_4OCH_2Ph\text{-}p$; $(COOR)_2$	*p*-$TosN_3$	1. LDA, THF, N_2, −78 °C, 2 h 2. Me_3SiCl, THF, r.t., 1 h	N_3; N–$CC_6H_4OCH_2Ph\text{-}p$; $(COOR)_2$		1039
R = Me				76	
R = Et				48	
N; O; Me; Me	*p*-$TosN_3$	LDA, THF	N_3; N; O; Me; Me	69	1362

An interesting series of α-monophenylthio- and α,α-bis(phenylthio)-substituted lactams have been prepared[1363] by the base catalysed sulphenylation of 1-substituted-2-pyrrolidones. Thus, reaction of 1-trimethylsilyl-2-pyrrolidone with lithium diisopropylamide, followed by quenching the resulting anion with diphenyl disulphide produces 3-phenylthio- and 3,3-bis(phenylthio)-2-pyrrolidone in varying yields depending upon the ratio of reactants used (equation 952). Similar results[1363] were obtained using 1-methyl-2-pyrrolidone (equation 953) and 1-methyl-2-piperidone (equation 954).

1. LDA, THF, −78 °C, stir 35 min
2. $(PhS)_2$, THF, stir −78 °C, 35 min
3. −20 °C, stir 20 min
4. r.t.

SPh; SPh, SPh; N, O, Me

6−54% 2−83%

(953)

1. LDA, THF, −78 °C, stir 35 min
2. $(PhS)_2$, THF, stir −78 °C, 35 min −20 °C, stir 20 min
4. r.t.

SPh; SPh, SPh; N, O, Me

15−77% 0−62%

same as above but with inverse quenching

0% 76%

(954)

Sulphinylation of 1-trimethylsilyl- and 1-methyl-2-pyrrolidone was achieved[1363] using methyl benzenesulphinate (equations 955 and 956).

1. LDA, THF, −78 °C, stir 45 min
2. $PhSO_2Me$, THF, −78 °C, stir 1 h
3. r.t. stir

N, O, $SiMe_3$; SOPh, N, O, Me

67%

(955)

1. LDA, THF, −78 °C, stir 45 min
2. $PhSO_2Me$, THF, −78 °C, stir 1 h
3. r.t. stir

N, O, Me; SOPh, N, O, Me

96%

(956)

Four distinct methods have been reported to introduce an alkylidene function into the alpha position of a preformed lactam. The first method involves condensation[800,901] of the lactam with a ketone in the presence of lithium diisopropylamide (equation 957).

1. LDA, THF, −78 °C
2. Me_3SiCl, −78 °C, stir 10 min
3. R^2COR^3, THF
4. 40 °C, 20−30 min

(957)

R^1	R^2		R^3	Yield (%)	Reference
H	Me		Me	—	800
H	Me		Ph	—	800
H	Et		Ph	—	800
H		$-(CH_2)_4-$		—	800
H		$-(CH_2)_5-$		—	800
H		$-(CH_2)_2CHMe(CH_2)_2-$		—	800
H		$-(CH_2)_6-$		—	800
Me	Me		Ph	71	901
Me		$-(CH_2)_5-$		66	901
n-Pr	Me		Me	59	901
n-Pr		$-(CH_2)_5-$		60	901

The second method[1364,1365] involves a standard Vilsmeier formylation procedure applied to butyro- (equation 958)[1364,1365], valero- (equation 959)[1365] and caprolactams (equation 960)[1365].

DMF / $POCl_3$

$CHNMe_2$ + $CH(OH)OCH$

R = Me, *n*-Bu

(958)

DMF / $POCl_3$

CHCl + CHOCH

(959)

DMF / $POCl_3$

$CHNMe_2$ + CHO

(960)

Reaction of dichlorocarbene with lactams constitutes the third method[1366] which has also been applied to N-methylbutyro- (equation 961), valero- (equation 962) and caprolactams (equation 963). However, only in the reaction with butyro- and valerolactams could products be isolated and identified.

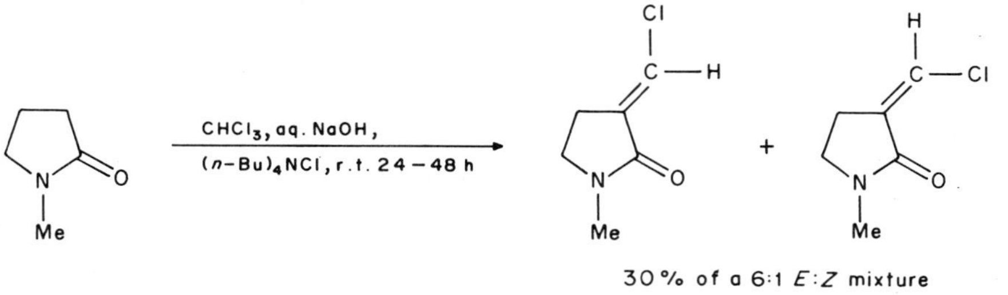

CHCl$_3$, aq. NaOH, (n-Bu)$_4$NCl, r.t. 24–48 h

30% of a 6:1 E:Z mixture

(961)

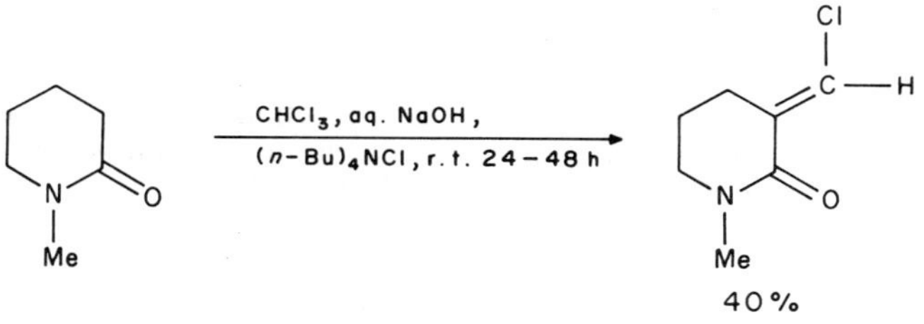

CHCl$_3$, aq. NaOH, (n-Bu)$_4$NCl, r.t. 24–48 h

40%

(962)

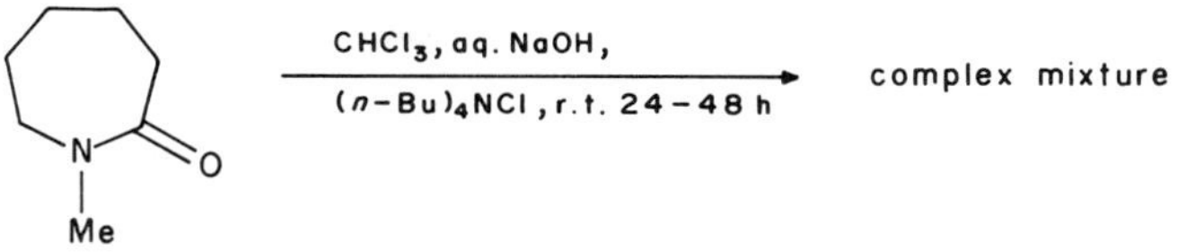

CHCl$_3$, aq. NaOH, (n-Bu)$_4$NCl, r.t. 24–48 h → complex mixture (963)

Finally, the last method employed[1367] utilize m-chloroperbenzoic acid oxidation of a thiophenyl substituent followed by thermal elimination of benzenesulphenic acid (equations 964 and 965).

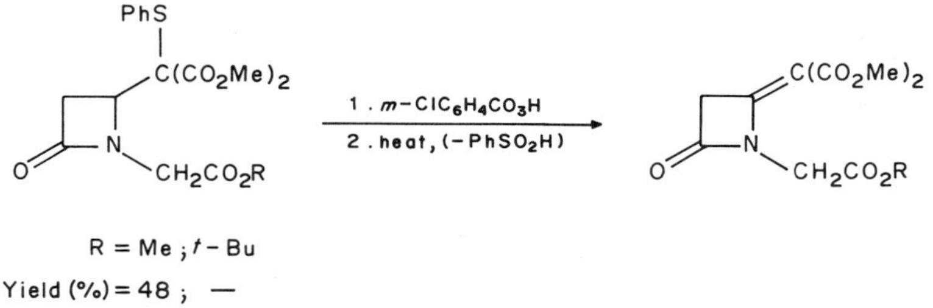

1. m-ClC$_6$H$_4$CO$_3$H
2. heat, (−PhSO$_2$H)

(964)

R = Me ; t-Bu
Yield (%) = 48 ; —

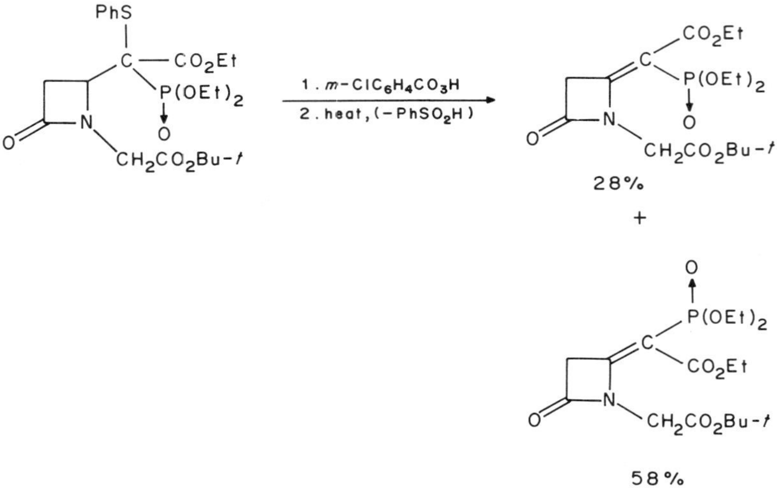

(965)

Michael addition[1066,1067] to the α-alkylidene function of a lactam results in the addition of a hydrogen to the alpha position of a preformed lactam, while the remaining reagent adds to the exocyclic carbon atom (equation 966).

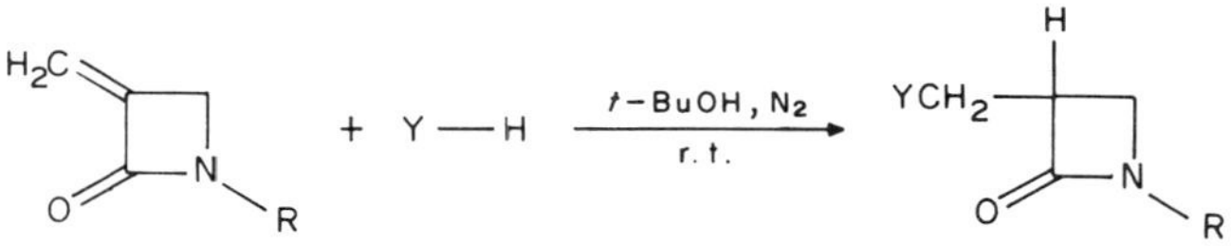

(966)

R	Y	Yield (%)	Reference
$PhCH_2$	$PhCH_2NH$	—	1067
$PhCH_2$	Pip	47	1067
Ph	Me_2N	—	1066
Ph	EtS	—	1066
Ph	H_2N	N.R.	1066
Ph	MeNH	N.R.	1066

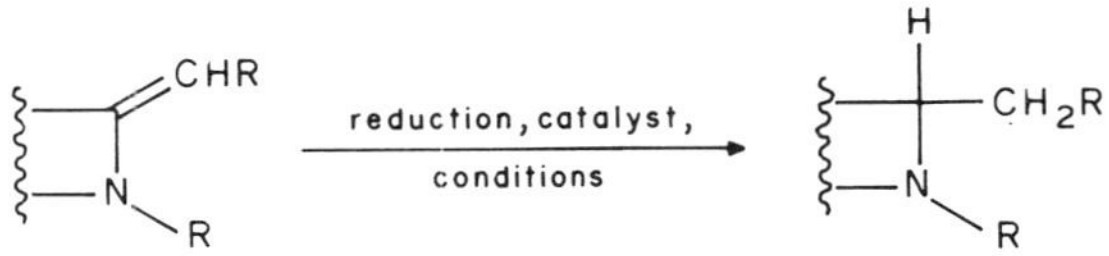

(967)

(968)

Typical reduction reactions of α-alkylidene or α-oxime functions have also been used to add a hydrogen to the alpha position of a preformed lactam (equations 967, 968 and Table 68).

A variety of methods have been used to methoxylate various sites on preformed lactam rings. α-Methoxylation of β-lactams has been accomplished by reaction of the β-lactam[1368] or penicillinate[1341] in tetrahydrofuran–methanol at −75 °C with t-butyl hypochlorite, followed by treatment with methanolic methoxide (equations 969 and 970).

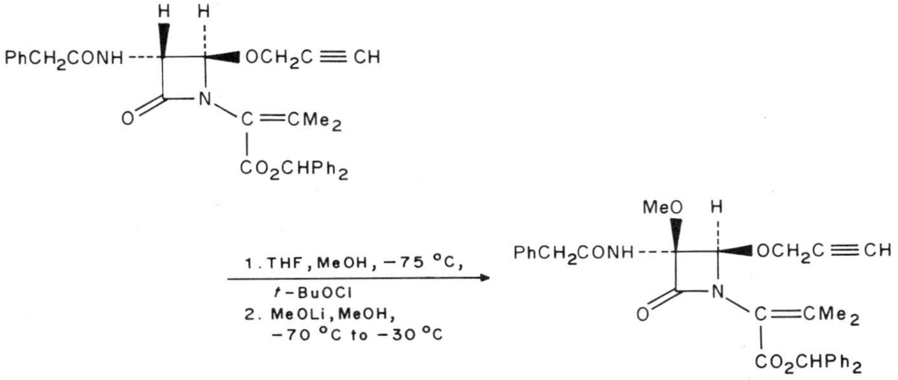

1. THF, MeOH, −75 °C, t-BuOCl
2. MeOLi, MeOH, −70 °C to −30 °C

(969)

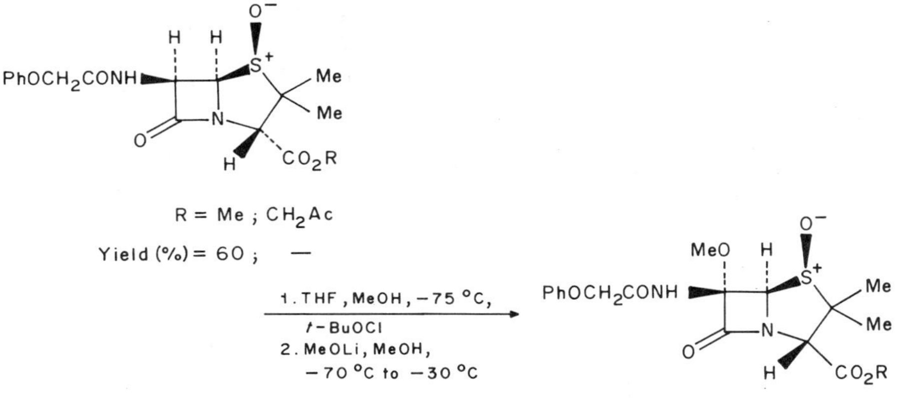

R = Me ; CH_2Ac

Yield (%) = 60 ; —

1. THF, MeOH, −75 °C, t-BuOCl
2. MeOLi, MeOH, −70 °C to −30 °C

(970)

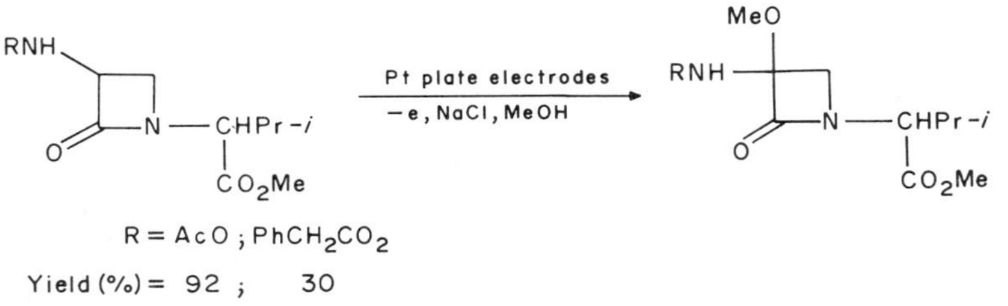

Pt plate electrodes
−e, NaCl, MeOH

(971)

R = AcO ; $PhCH_2CO_2$

Yield (%) = 92 ; 30

TABLE 68. Reduction of α-alkylidene and α-oxime functions of lactams

Substrate	Reducing agent	Conditions	Product	Yield (%)	Reference
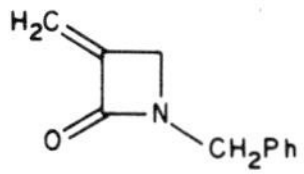 H2C, O, N, CH2Ph	$NaBH_4$	EtOH, stir r.t., stir 16h	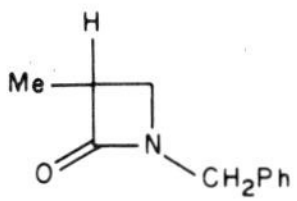 H, Me, O, N, CH2Ph	61.3	1067
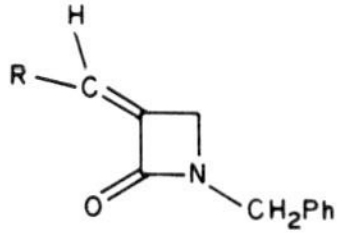 H, R, C, O, N, CH2Ph	H_2, PtO_2	EtOH, stir r.t.	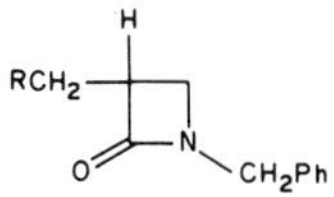 H, RCH2, O, N, CH2Ph		1067
R = H				86.3	
R = *n*-Bu				38	
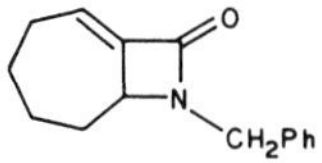 O, N, CH2Ph	H_2, PtO_2	EtOH, stir r.t.	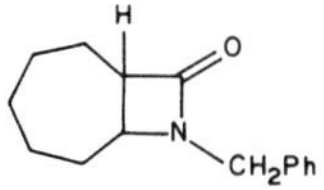 H, O, N, CH2Ph	92.2	1067
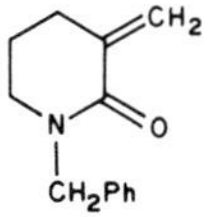 CH2, N, O, CH2Ph	H_2, PtO_2	EtOH, stir overnight	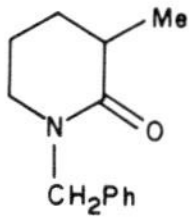 Me, N, O, CH2Ph	90	776
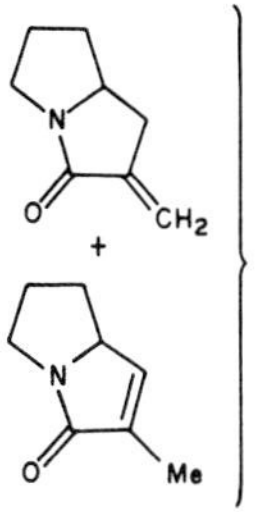 N, O, CH2 + N, O, Me	H_2, PtO_2	EtOH, stir r.t. 3h	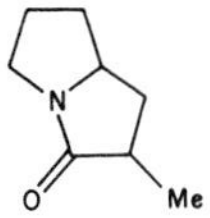 N, O, Me	100	776

Substrate	Reagent	Conditions	Product	Yield (%)	Ref.
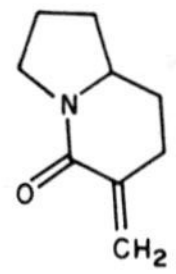 CH_2; O; N	H_2, PtO_2	EtOH, stir r.t. overnight	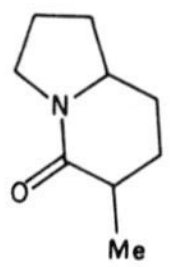 Me; O; N	88	776
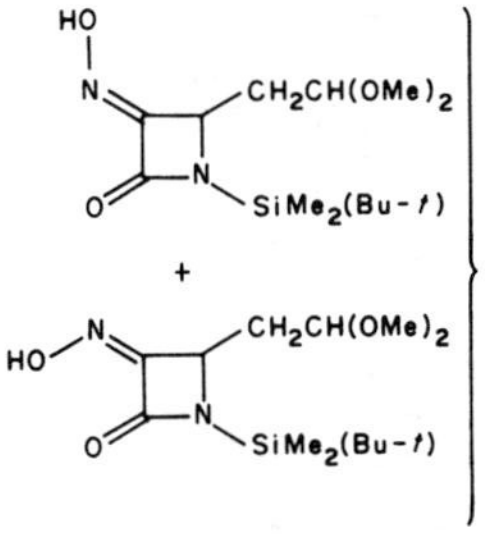 HO; N; $CH_2CH(OMe)_2$; O; N; $SiMe_2(Bu\text{-}t)$ + HO; N; $CH_2CH(OMe)_2$; O; N; $SiMe_2(Bu\text{-}t)$ (E:Z = 1:3.5)	H_2, 5% Rh on alumina	MeOH, 5 atm, 30 h	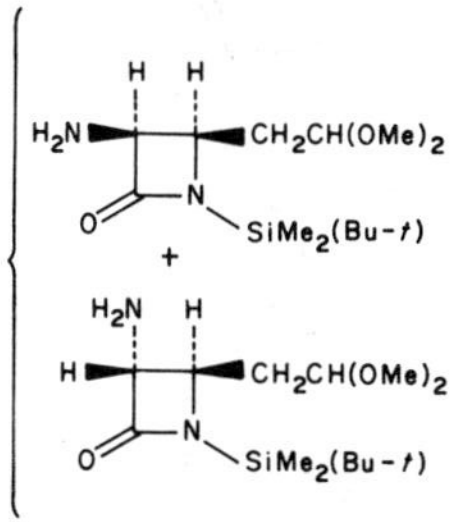 H; H; H_2N; $CH_2CH(OMe)_2$; O; N; $SiMe_2(Bu\text{-}t)$ + H_2N; H; H; $CH_2CH(OMe)_2$; O; N; $SiMe_2(Bu\text{-}t)$ (*cis*:*trans* = 5:1)	51	1316

Another method used[1369] to α-methoxylate β-lactams is indirect oxidation using an oxidizing reagent anodically generated *in situ* (equation 971). The mechanism proposed for this reaction is presented in Scheme 17.

SCHEME 17

In view of the reaction reported in equation 971, it is of interest to note that anodic α-alkoxylation of lactams which do not contain an α-amino function produce[1370,1371] products where the alkoxy function is located alpha to the lactam nitrogen and not alpha to the ring carbonyl function (equation 972). If the lactam nitrogen substituent contains a replaceable hydrogen, then methoxylation is also observed[1059] to occur at that site in addition to a site on the ring (equation 973 and 974).

(972)

$R^1 = H$, C_3—C_{10} alkyl with a secondary carbon in the α-position to the ring nitrogen
$n = 1$–10
$R^2 = C_1$ to C_4 alkyl

Salt = $Me_4\overset{+}{N}BF_4^-$, $(n\text{-}Pr)_4\overset{+}{N}PF_6^-$, $\overset{+}{K}BF_4^-$, $\overset{+}{Na}BF_4^-$

(194) (195) (973)

R^1	R^2	Ratio **195:195**	Yield (%)
H	CH_2COOMe	1:2.5	86
Me	COOMe	1:0.75	91
Me	CH_2COOMe	1:0.67	83

(196) (197)

(974)

R	Ratio **196**:**197**	Yield (%)
CH_2COOMe	1:13	66
CH_2OTHP	1:3.3	74
Ph	1:4.6	66

Reaction[1221] of 3-bis(ethylthio)-1-(4′-methoxyphenyl)-4-substituted β-lactams with iodine in refluxing methanol converts the ethylthio moieties into methoxy groups (equation 975).

(975)

R = Ph ; PhCH=CH

Yield (%) = 40 ; 44

Another type of reaction which occurs at the alpha position of a lactam ring is the formation of spiro-functions, and two approaches to their formation have been reported[1372,1373]. In the first approach[1372] N-substituted γ- and δ-lactams containing an ω-halogenated chain in the alpha position react with lithium diethylamide to produce spirolactams. However, reaction of similarly substituted *E* lactams with the same reagent furnishes a mixture of spiro- and ethylenic-products (equation 976).

(198) (199)

(976)

n	*m*	X	Yield **198** (%)	Yield **199**(%)
1	3	Cl	92	0
1	4	Cl	92	0
2	3	Cl	76	0
2	4	Cl	97	0
3	3	Cl	47	44
3	4	Cl	44	50
3	4	Br	0	85
3	4	I	0	83

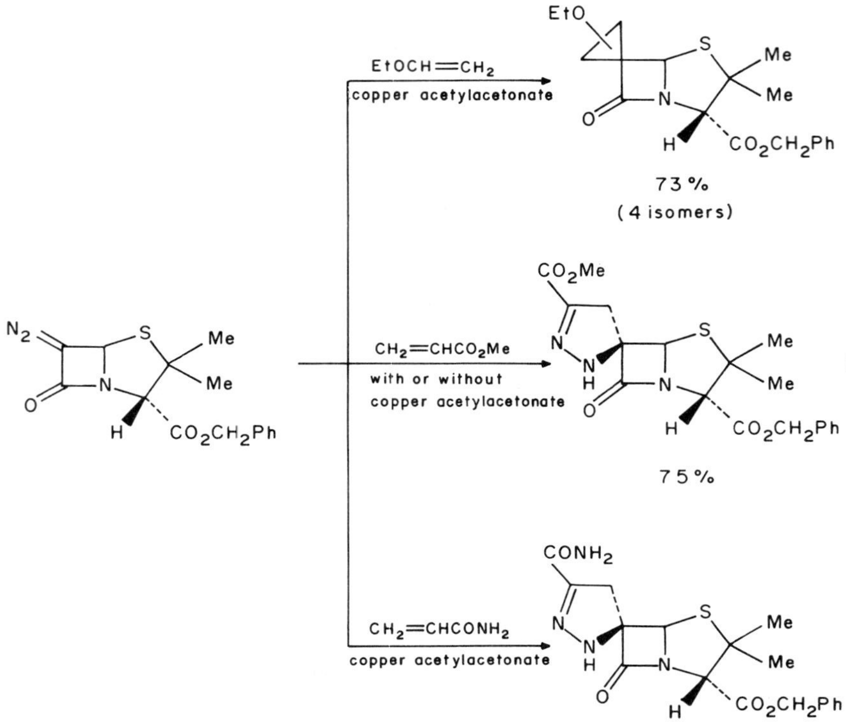

$EtOCH{=}CH_2$, copper acetylacetonate → 73% (4 isomers)

$CH_2{=}CHCO_2Me$, with or without copper acetylacetonate → 75%

$CH_2{=}CHCONH_2$, copper acetylacetonate

(977)

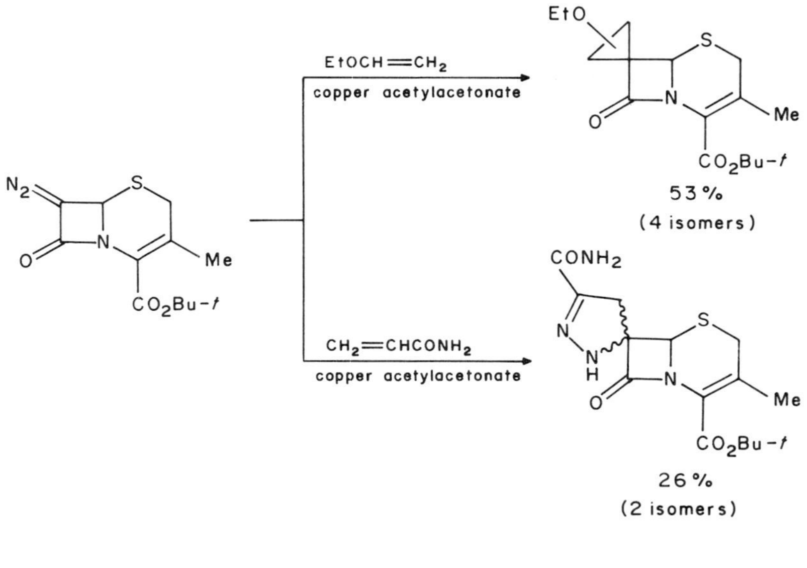

$EtOCH{=}CH_2$, copper acetylacetonate → 53% (4 isomers)

$CH_2{=}CHCONH_2$, copper acetylacetonate → 26% (2 isomers)

(978)

The second approach[1373] to the formation of α-spiro lactams involves the reaction of diazo-β-lactams with ethyl vinyl ether, methyl acrylate or acrylamide. With the vinyl ether, spirocyclopropyl derivatives are obtained, probably through a 1-pyrazoline intermediate, whereas with the acrylic acid spiropyrazolines are obtained, formed via a 1,3-dipolar cycloaddition (equations 977 and 978).

Although a variety of methods have been used to oxidize lactams, two distinct sets of products have been reported to be produced in each case. One set of products consists of hydroxy- or ketolactams with the substituent located alpha to the carbonyl carbon, while the second set of products contain the same substituents but has them located alpha to the ring nitrogen.

Introduction of an angular hydroxyl group into a bicyclic lactam was readily accomplished[1374] by treatment with lithium hexamethyldisilazane in tetrahydrofuran at −78 °C, followed by bubbling oxygen through the solution for one hour at 0 °C and then allowing the mixture to stand at room temperature for 2 hours in the presence of excess trimethyl phosphite (equation 979).

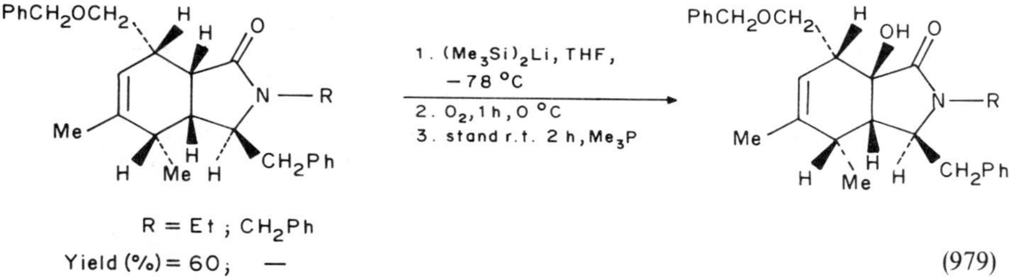

1. $(Me_3Si)_2Li$, THF, −78 °C
2. O_2, 1 h, 0 °C
3. stand r.t. 2 h, Me_3P

R = Et ; CH_2Ph
Yield (%) = 60 ; — (979)

Although not an oxidation reaction, treatment of 3-bis(ethylthio)-1,4-di-substituted β-lactams with *N*-bromosuccinimide in acetonitrile–water does hydrolyse the ethylthio functions to a keto function and produces the corresponding keto β-lactams[1221] (equation 980).

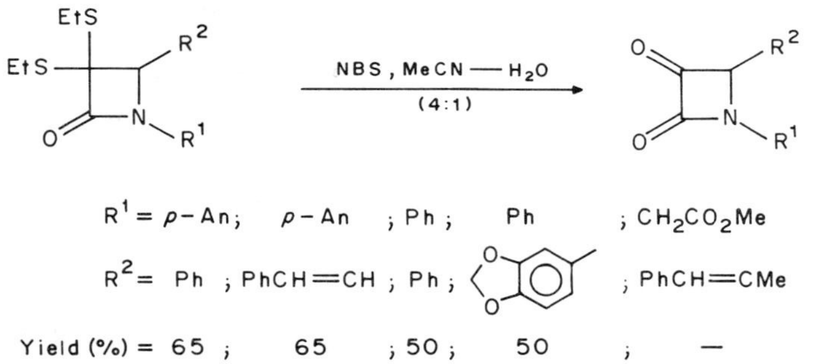

NBS, MeCN—H_2O (4:1) (980)

R^1 =	*p*-An ;	*p*-An ;	Ph ;	Ph ;	CH_2CO_2Me
R^2 =	Ph ;	PhCH=CH ;	Ph ;	3,4-methylenedioxyphenyl ;	PhCH=CMe
Yield (%) =	65 ;	65 ;	50 ;	50 ;	—

Controlled potential anode oxidation of *N*-alkyl lactams has been reported[1375] to occur regioselectively at the endocyclic carbon atom alpha to the nitrogen in both 5- and 6-membered lactam rings, and at the exocyclic alpha carbon atom in the 7-membered lactam rings to produce hydroxy-lactams, imides and dealkylation products (equation 981).

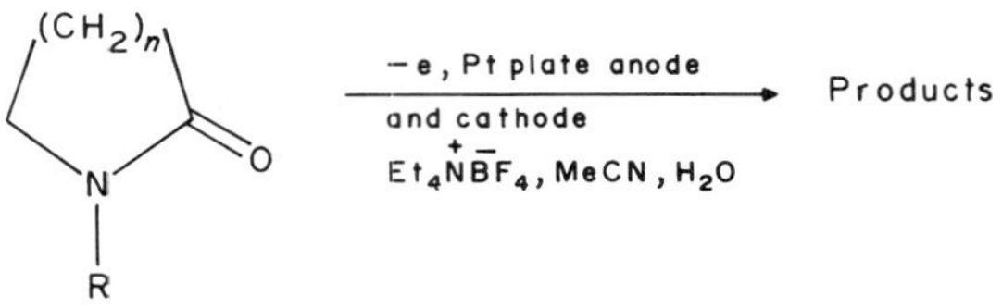

$$\text{(CH}_2)_n\text{-lactam (N-R)} \xrightarrow[\text{and cathode}, \ Et_4\overset{+}{N}\overset{-}{B}F_4, \ MeCN, \ H_2O]{-e, \ \text{Pt plate anode}} \text{Products} \qquad (981)$$

Lactam	Potential (V)	Product	Yield (%)
O, N, Me	2.2	O, N, Me, OH +	54.6
		O, N, Me, O	11.7
O, N, Et	2.0	O, N, Et, OH +	56.7
		O, N, Et, O	2.1
O, N, Me	2.2	O, N, Me, OH +	64.3
		O, N, Me, O	10.1
O, N, Et	2.0	O, N, Et, OH +	47.3
		O, N, Et, O	9.1
O, N, Me	2.4	O, N, H +	17.3
		O, N, CH_2OH +	35.0

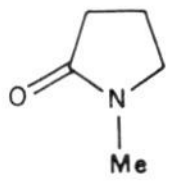

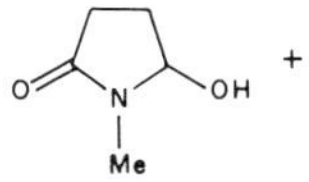

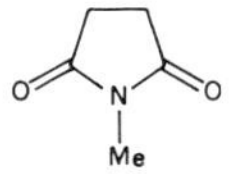

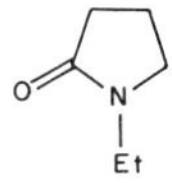

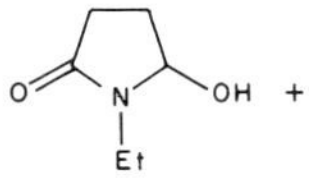

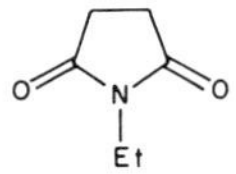

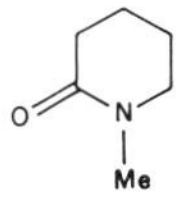

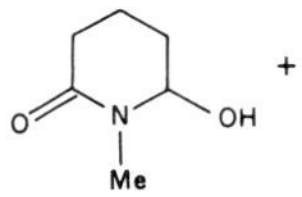

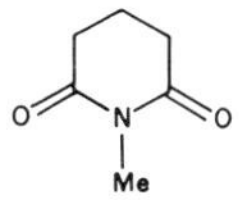

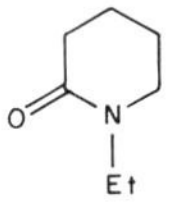

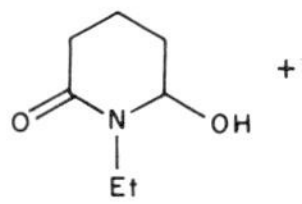

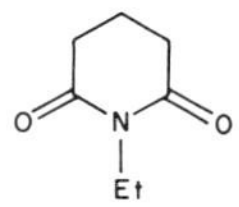

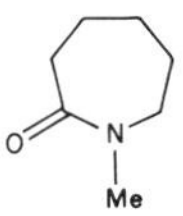

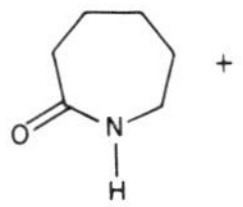

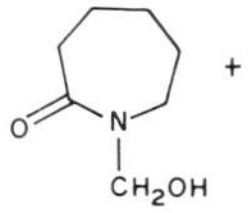

(continued)

Lactam	Potential (v)	Product	Yield (%)
		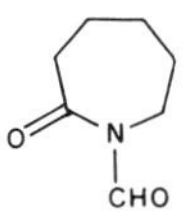 N-formylcaprolactam (N—CHO)	21.2
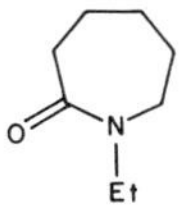 N-ethylcaprolactam (N—Et)	2.3	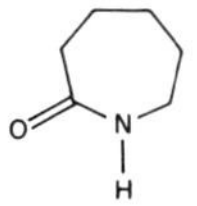 caprolactam (N—H) +	55.8
		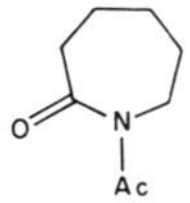 N-acetylcaprolactam (N—Ac)	3.1

Reaction of 3-aza-A-homo-4-cholesten-4-one with benzeneseleninic anhydride in benzene at reflux for 1.5 hours produced[1124] the corresponding α-ketolactam and the α-phenylseleno derivative in yields of 58% and 36%, respectively. Addition of triethylamine to the reaction mixture increased the yield of α-ketolactam to 80% (equation 982). Similar oxidation of the saturated azasteroid analogue with the same reagent afforded the analogous α-phenylseleno products in 65% and 24% yields, respectively (equation 983).

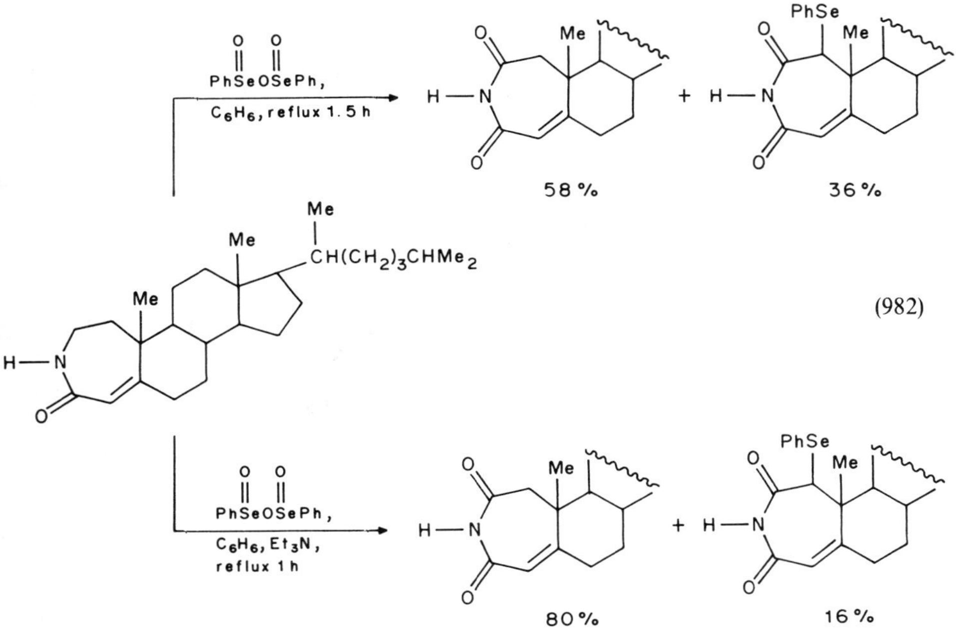

PhSe(O)OSe(O)Ph, C_6H_6, reflux 1.5 h → α-ketolactam (58%) + α-phenylseleno derivative (36%)

PhSe(O)OSe(O)Ph, C_6H_6, Et_3N, reflux 1 h → α-ketolactam (80%) + α-phenylseleno derivative (16%)

Me, Me, Me, $CH(CH_2)_3CHMe_2$, H—N, PhSe

(982)

PhSeOSePh, C_6H_6, reflux 2 h

65%

+

24%

(983)

One of the more interesting methods of oxidation which appears[1376] in the recent literature is microbial hydroxylation of 5-, 6- and 7-membered lactams by *Beauveria sulfurescens*. This oxidation occurs by two distinct processes, hydroxylation alpha to the ring nitrogen and hydroxylation of a non-activated carbon atom somewhere else in the lactam ring. The first process appears to occur when the second process is not favoured for geometric reasons (equation 984).

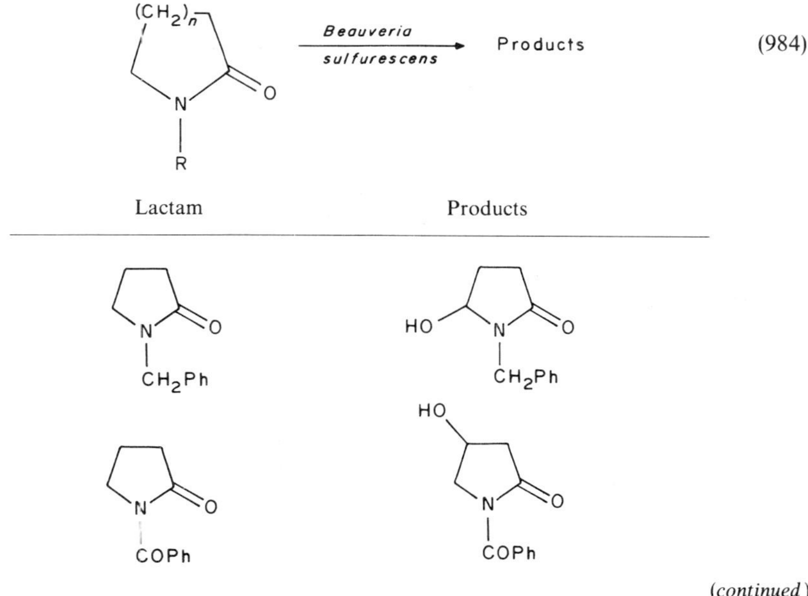

(continued)

Lactam	Products
N-benzyl-2-piperidinone (CH_2Ph)	4-hydroxy-N-benzyl-2-piperidinone + 6-hydroxy-N-benzyl-2-piperidinone
N-benzyl-caprolactam (CH_2Ph)	hydroxy-N-benzyl-caprolactam (HO) + hydroxy-N-benzyl-caprolactam (OH)

Photooxidation of lactams in the presence of oxygen has also been used[1377,1378] to derivatize 5-, 6- and 7-membered lactams and this approach also produces two kinds of products, imides and α-hydroperoxy- or α-hydroxylactams (equations 985–989). The products obtained depend upon the reaction conditions used and whether or not the substrate lactam bears a substituent alpha to the ring nitrogen atom. The mechanism for these transformations appears to be regioselective abstraction of a hydrogen atom alpha to the nitrogen atom by the benzophenone triplet to give radicals, which are then oxidized.

$$\text{N-R-pyrrolidin-2-one} \xrightarrow[\text{solvent, time}]{Ph_2CO,\ h\nu,\ O_2,} \text{N-R-succinimide} \tag{985}$$

R =	H	H	Me	Me
solvent =	C_6H_6	t-BuOH	C_6H_6	t-BuOH
time (h) =	30	80	60	60
Yield (%) =	20	60	20	60

$$\text{5-Me-N-R-pyrrolidin-2-one} \xrightarrow[\text{solvent, time}]{Ph_2CO,\ O_2,\ h\nu} \text{5-Me-5-HOO-N-R-pyrrolidin-2-one} \tag{986}$$

R =	H	Me	Me
solvent =	C_6H_6	C_6H_6	t-BuOH
time (h) =	42	37	32
Yield (%) =	60	6	26

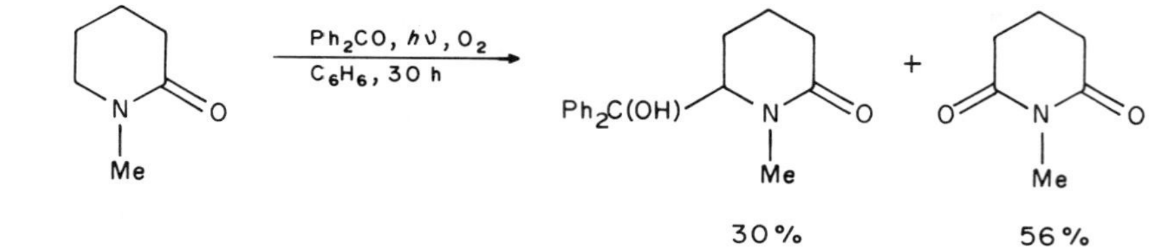

Ph$_2$CO, $h\nu$, O$_2$ / C$_6$H$_6$, 30 h

Ph$_2$C(OH) ... Me 30% + Me 56%

(987)

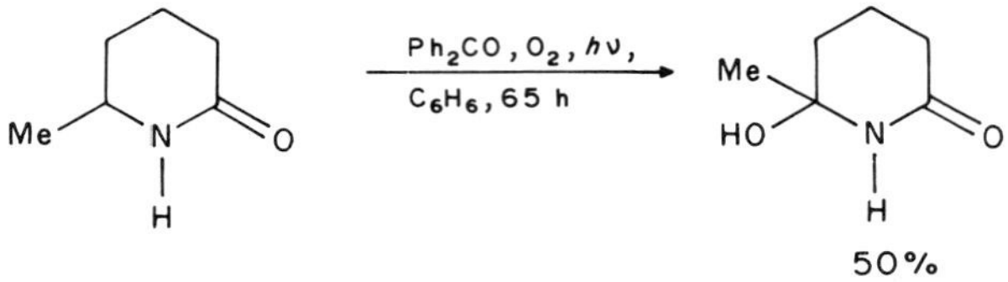

Ph$_2$CO, O$_2$, $h\nu$, / C$_6$H$_6$, 65 h

(988)

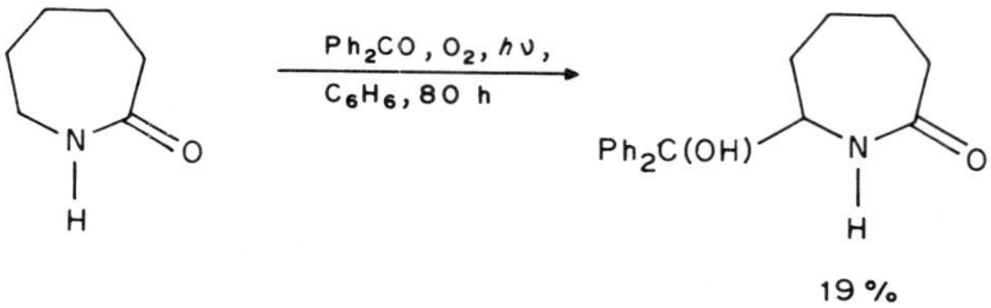

50%

Ph$_2$CO, O$_2$, $h\nu$, / C$_6$H$_6$, 80 h

(989)

Ph$_2$C(OH) 19%

β-Functionalization of a β-lactam has been achieved by ring cleavage accompanied by acyloxylation of a bicyclic lactam. Thus, refluxing methyl 6α-chloropenicillinate with two equivalents of mercuric acetate in acetic acid for 4 hours produces[1308] a separable mixture of 4α-acetoxy-3α-chloro-2-azetidinone and its 4β-isomer in the ratio of 3:2 and 26.3% yield (equation 990).

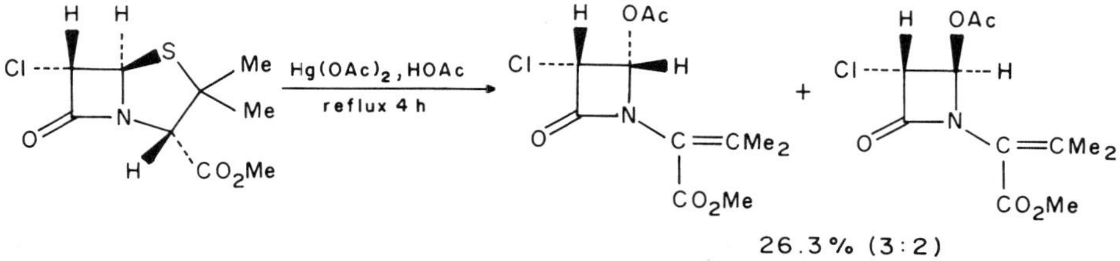

Hg(OAc)$_2$, HOAc / reflux 4 h

26.3% (3:2)

(990)

R = Me, CH$_2$Ph

n = 1, 2

53–84%

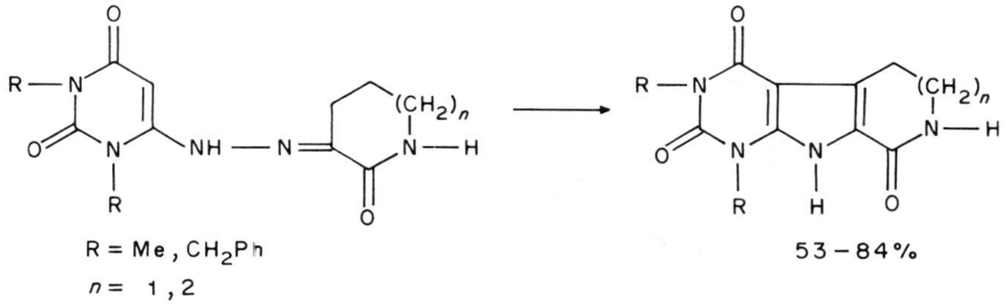

(991)

The converse of the reaction shown in equation 990, namely cyclization to form a bicyclic lactam, has also been used to functionalize lactams at both the beta- (equation 991)[1379] and gamma- sites (equation 992)[1380]. The mechanism of the latter reaction is explained by nucleophilic intramolecular conjugate addition of the hydroxy group to the pyrrole double bond bearing the ethoxycarbonyl function, to produce the 2-amino-4-carboxychroman-lactam derivatives, by annelation of a benzenoid ring.

HOAc, H_2SO_4, 20 °C, stand overnight (992)

R^1	R^2	R^3	R^4	Yield (%)
H	Ph	OH	H	76
Me	Ph	OH	H	60
$PhCH_2$	Ph	OH	H	80
H	*p*-Tol	OH	H	60
$PhCH_2$	*p*-Tol	OH	H	80
H	Ph	H	OH	38
Me	Ph	H	OH	55
$PhCH_2$	Ph	H	OH	75
H	Ph	Me	Me	63
Me	Ph	Me	Me	70
$PhCH_2$	Ph	Me	Me	57

The simplest alpha-substitution reaction of the lactam ring to be reported[1381] is the formation of labeled *N*-vinyllactams by treatment with deuterium or tritium oxide at pH 10.5–14 using an alkali metal hydroxide or quaternary ammonium hydroxide as a catalyst in an organic solvent or in the neat (equation 993).

D_2O or T_2O, pH = 10.5–14, alkali metal hydroxide or $R_4\overset{+}{N}\overset{-}{O}H$ (R = Me, Et), organic solvent or neat (993)

$n = 2-4$

3. Conversion of substituents directly attached to the lactam nitrogen

The most common reaction performed on preformed lactams which contain a substituent attached to the lactam nitrogen is removal of the substituent to form the N-unsubstituted product. As would be expected, the nature of the N-substituent, as well as the structure of any other substituents attached to the lactam, greatly influence the choice of reagents used to remove the group from the lactam nitrogen. The reagents used range from strong oxidizing agents, which have been used when the N-substituent contains a double bond, to quaternary ammonium salts, which have been used when the nitrogen contains a silyl substituent. The general reaction is represented in equation 994 with the details reported in Table 69.

$$\text{N–Y} \xrightarrow[\text{conditions}]{\text{reagent}} \text{N–H} \quad (994)$$

Another common conversion which is performed on substituents or functional groups directly attached to lactam nitrogens is halogenation, and three different approaches have been reported to accomplish this exocyclic halogenation. The first, and most common, approach is the conversion of a hydroxyl function to a chloride using thionyl chloride in 2,6-lutidine or pyridine (equation 995 and Table 70). In all cases the chlorinated products are obtained as mixtures of diastereomers which, in all but one case, were not isolated but were used directly in further reaction.

$$\text{N–}\overset{*}{C}H(OH)COR \xrightarrow[\text{2,6-lutidine or } C_5H_5N]{SOCl_2,} \text{N–}\overset{*}{C}H(Cl)COR \quad (995)$$

mixture of diastereomers

The second approach was the reaction of a lactam containing an activated carbon atom attached to the ring nitrogen with 4-(*N*,*N*-dimethylamino)pyridinium bromide perbromide ($DMAP \cdot HBr_3$)[1201,1213] (equation 996).

$$\text{PhO, R}^1\text{-azetidinone–N–}CH_2COR^2 \xrightarrow[MeCO_2H]{DMAP \cdot HBr_3} \text{PhO, R}^1\text{-azetidinone–N–}CH(Br)COR^2 \quad (996)$$

R^1 = Ph ; *p*-An ; *p*-An
R^2 = Ph ; Me ; Ph
Yield (%) = — ; — ; 86
Ref. = 1201 ; 1201 ; 1213

And finally, the last approach involves the addition of halogen to the exocyclic α-carbon of an α,β-unsaturated substituent attached to a lactam nitrogen, whether that substituent is part of a chain[1178] (equation 997) or a ring system[1398] (equations 998 and 999).

These exocyclic α-halogenated substrates have been used to effect the replacement of a variety of substituents at the alpha site by substitution for the attached halogen. Thus, reduction of the 3-chlorothiopenem produced in equation 998 above with either zinc–acetic acid in tetrahydrofuran or triphenylphosphine in aqueous methylene chloride affords[1398] the hydrogen substituted analogue in 55% yield (equation 1000).

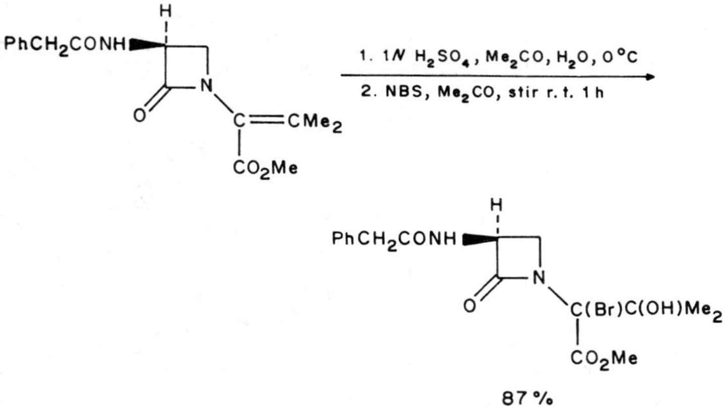
PhCH2CONH
C=CMe2
CO2Me
1. 1N H2SO4, Me2CO, H2O, 0 °C
2. NBS, Me2CO, stir r.t. 1 h
PhCH2CONH
C(Br)C(OH)Me2
CO2Me
87%

(997)

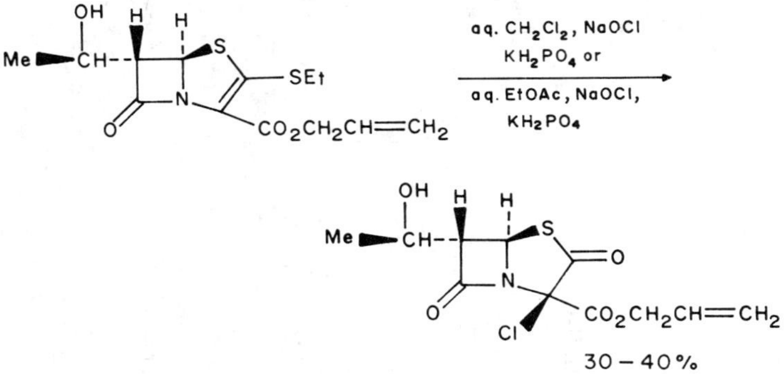
OH
Me
CH
SEt
CO2CH2CH=CH2
aq. CH2Cl2, NaOCl
KH2PO4 or
aq. EtOAc, NaOCl,
KH2PO4
CO2CH2CH=CH2
Cl
30 – 40%

(998)

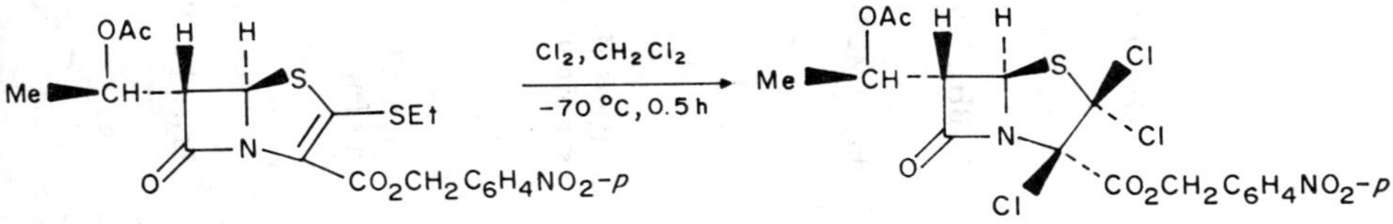
OAc
Me
CH
SEt
CO2CH2C6H4NO2-p
Cl2, CH2Cl2
-70 °C, 0.5 h
Cl
CO2CH2C6H4NO2-p

(999)

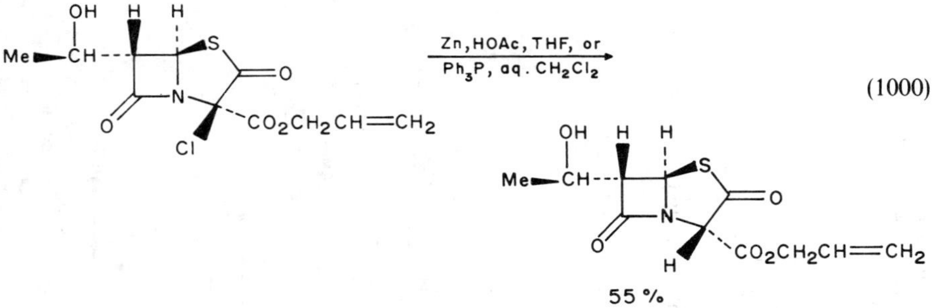
OH
Me
CH
CO2CH2CH=CH2
Cl
Zn, HOAc, THF, or
Ph3P, aq. CH2Cl2
CO2CH2CH=CH2
55%

(1000)

TABLE 69. Removal of substituent from lactam nitrogen

Substrate	Reagents and conditions	Product	Yield (%)	Reference
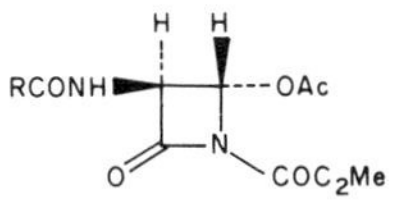 RCONH, H, H, OAc, N, O, COC_2Me	MeOH, H_2O	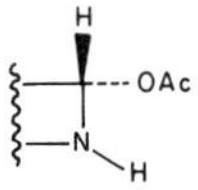 H, OAc, N, H	78–91	1382
R = $PhCH_2O$, *t*-BuO	$NaBH_4$, *i*-PrOH, H_2O		66–79	1382
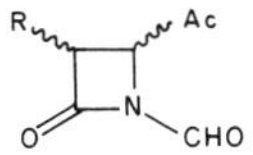 R, Ac, N, O, CHO	Et_3N, MeOH	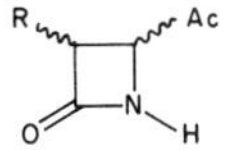 R, Ac, N, O, H		1146
R = PhO[a]			—	
R = PI—N			85	
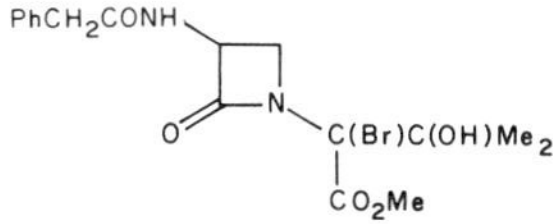 $PhCH_2CONH$, N, O, $C(Br)C(OH)Me_2$, CO_2Me	Et_3N, CH_2Cl_2, r.t., stir overnight	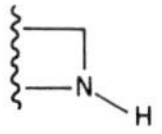 N, H	76	1178
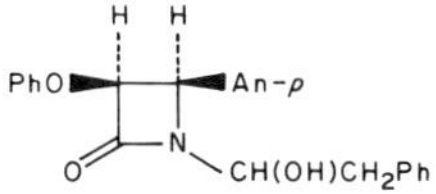 PhO, H, H, An-*p*, N, O, $CH(OH)CH_2Ph$	Et_3N, CH_2Cl_2, r.t., stir few h	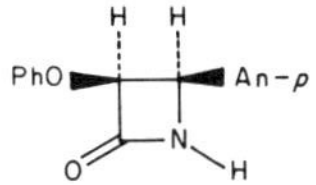 PhO, H, H, An-*p*, N, O, H	35	1213
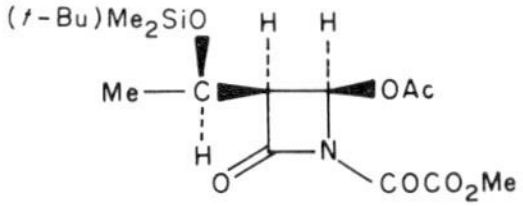 (*t*-Bu)Me_2SiO, Me—C, H, H, H, OAc, N, O, $COCO_2Me$	Et_3N, MeOH	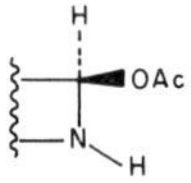 H, OAc, N, H	—	1348

Substrate	Conditions	Product	Yield (%)	Ref.
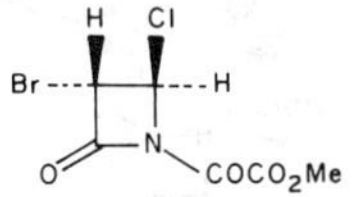	2,4-DNPH, THF, 20 °C, 18 h	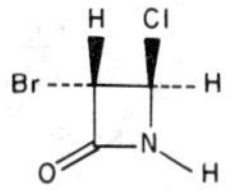	71	1383
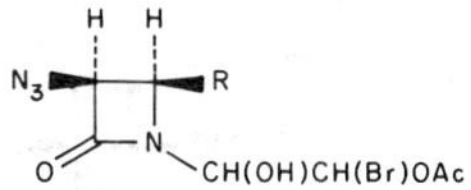	NH_2NH_2, stir r.t. 10 min	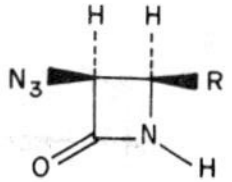		1384
R = PhCH=CH			66	
R = *p*-An			72	
R = 2,3,4-$(MeO)_3$-5-$(PhCH_2O)C_6H$			74	
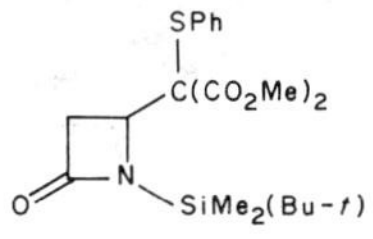	$(n\text{-Bu})_4N^+F^-$	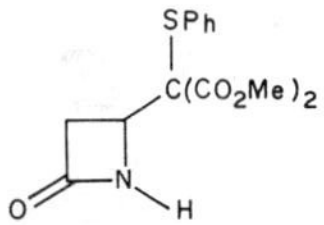	100	1367
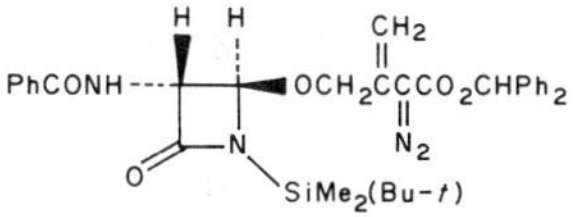	$Et_4N^+F^-$	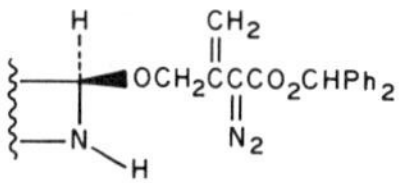	75	1349
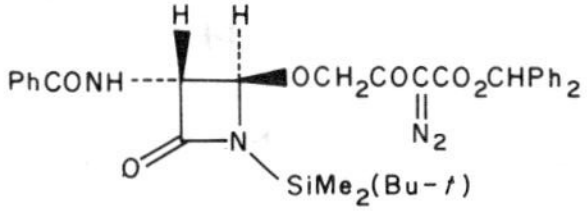	aq. HCl, EtOH	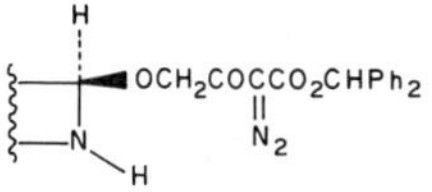	—	1349
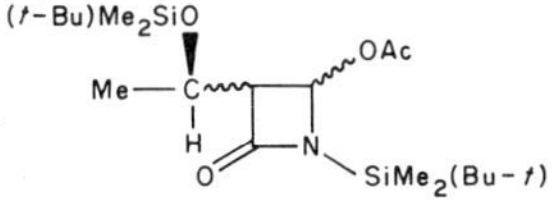	—	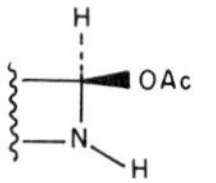 (3R, 4R)	—	1318

(continued)

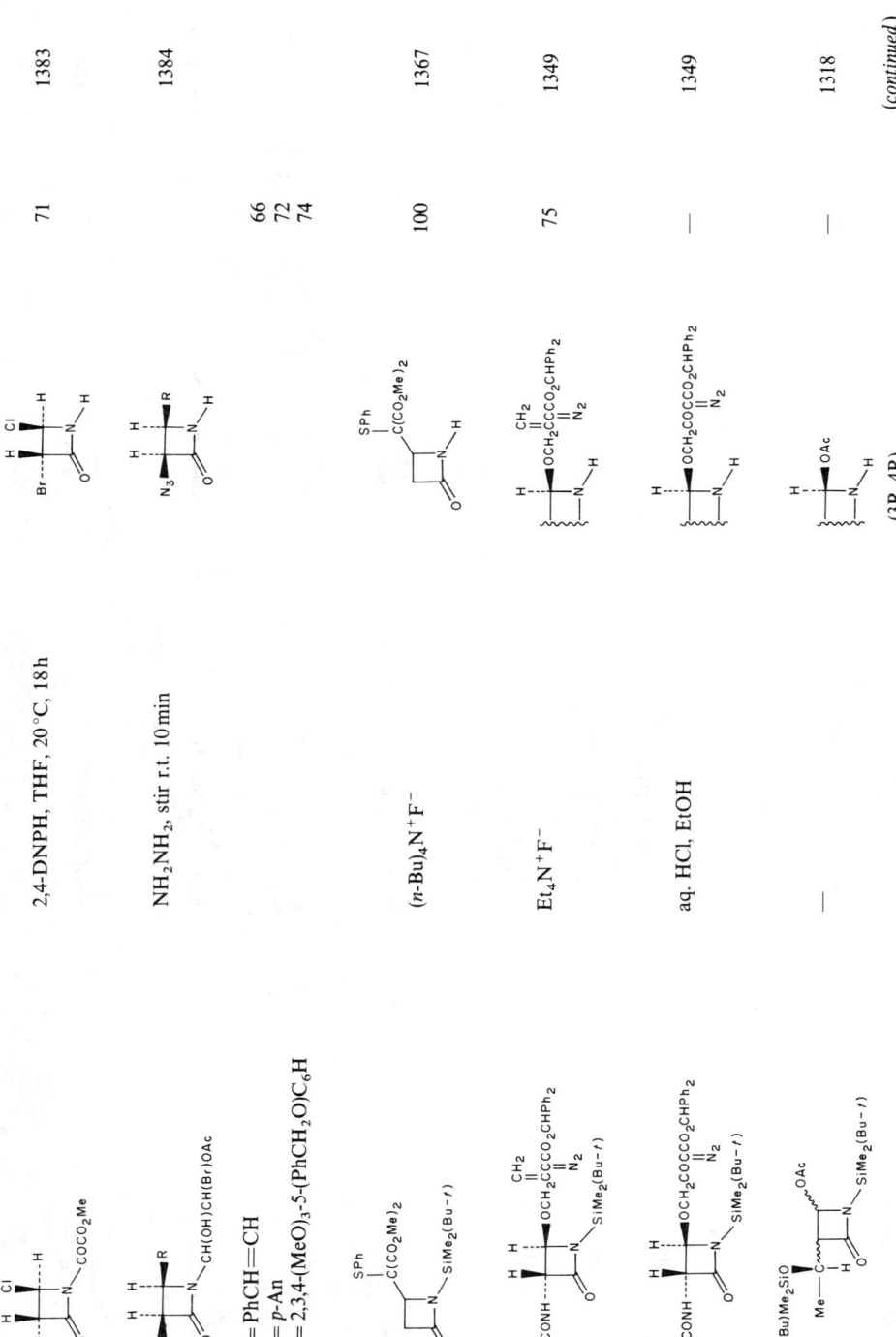

TABLE 69. (*continued*)

Substrate	Reagents and conditions	Product	Yield (%)	Reference
PhO, R^1 azetidinone, N–CH(Br)COR^2		PhO, R^1 azetidinone, N–H		1201
$R^1 = R^2 = Ph$	1. HOAc, stir r.t. 15 min 2. $NaHCO_3$, H_2O, Me_2CO, stir r.t. 48 h		42	
$R^1 = R^2 = Ph$	1. HOAc, stir r.t. 1 h 2. $NaHCO_3$, H_2O, Me_2CO, stir r.t. 2 h		40	
$R^1 = p$-An, $R^2 = Me$	1. HOAc, stir r.t. 1 h 2. $NaHCO_3$, H_2O, Me_2CO, stir r.t. 2 h		25	
R, H, Me, H pyrrolidinone, N–Tos-*p*	$Na^+\bar{C}_{10}H_8$, DME, 0 °C to r.t., 1 h	R, H, Me, H pyrrolidinone, N–H		1385
R = Ph			74	
R = *p*-An			63	
Br, H, H, Cl azetidinone, N–$COCO_2Me$	$Ph_3CS^-\overset{+}{Na}$, MeOH, −15 °C, 30 min	Br, H, H, $SCPh_3$ azetidinone, N–H	49	1383, 1386
PI–N, Ac azetidinone, N–CHO	NaOMe, MeOH	PI–N, Ac azetidinone, N–H	50	1146

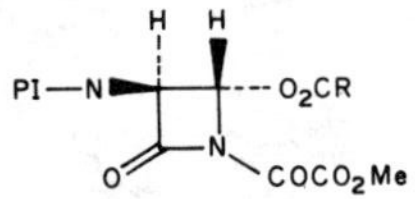 $R = Me$, $p\text{-}ClC_6H_4$	NaOMe, MeOH	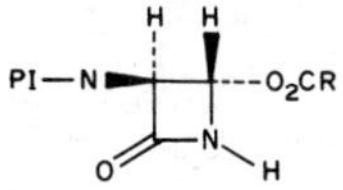	60–80	1387
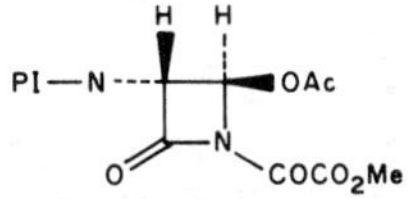	NaOMe, MeOH	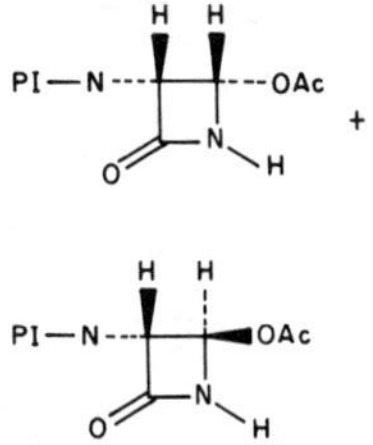 +	—	1387
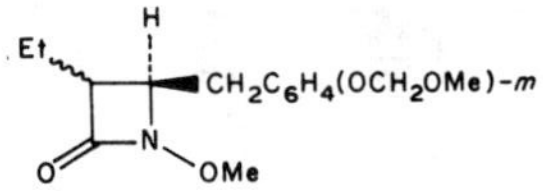	Li, NH_3:THF:t-BuOH (3:1:1), 78 °C, 1 h	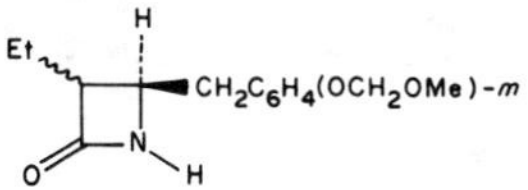	81	1082
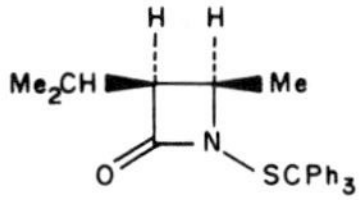	Li, liq. NH_3	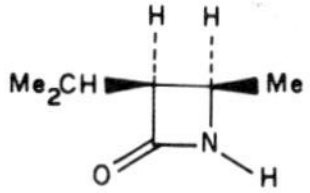	85	1241
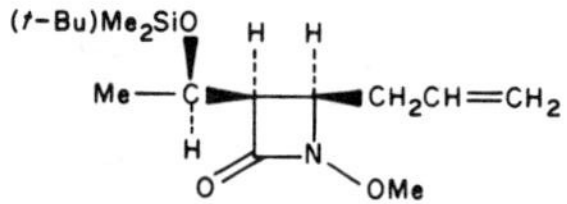	Li, liq NH_3, THF, -78 °C, 1 min	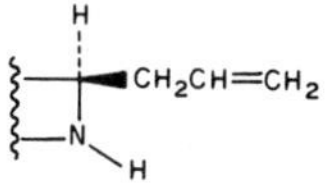	98	1015
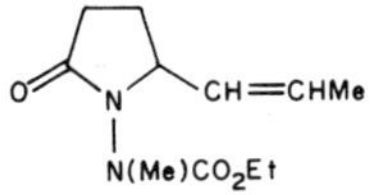 (*trans*)	Li, liq. NH_3, -33 °C, 0.25 h	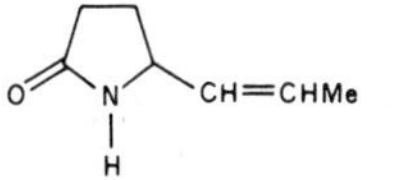 (*trans*)	90	1094

(continued)

TABLE 69. (*continued*)

Substrate	Reagents and conditions	Product	Yield (%)	Reference
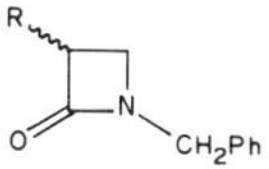 [structure: 3-R-1-benzyl-azetidin-2-one, N-CH_2Ph]	Na, liq. NH_3	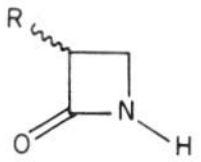 [structure: 3-R-azetidin-2-one, N–H]		1388
$R = CH_2{=}CMe$			88	
R 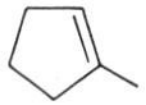 = [cyclopentenyl]			68	
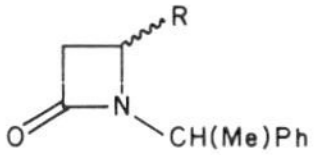 [structure: 4-R-1-CH(Me)Ph-azetidin-2-one]	Na, liq. NH_3	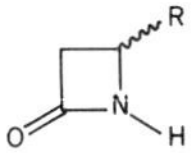 [structure: 4-R-azetidin-2-one, N–H]		1002
R / **Substituent configuration** / **Configuration at C4**				
Me / *R* / *R*			91	
Me / *S* / *S*			69	
n-Pr / *R* / *R*			50	
i-Pr / *R* / *S*			66	
Ph / *R* / *S*			32	
$CH_2CH_2OCH_2OCH_3$ / *R* / *R*			51	
$CH_2CH_2OCH_2OCH_3$ / *S* / *S*			39	
CH_2CO_2Me / *R* / *R*			—	
[structure: 3-(MeCH(OH))-4-($CH_2CH(OMe)_2$)-1-benzyl-azetidin-2-one, with H, H, OH]	Na, liq. NH_3, −78 °C	[structure: N–H azetidinone with H and $CH_2CH(OMe)_2$]	—	1389

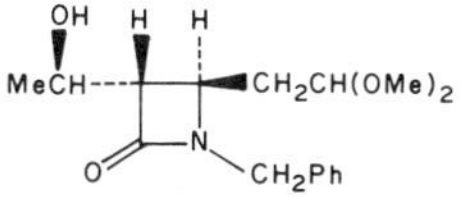

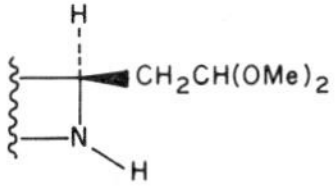

Substrate	Reagents and conditions	Product	Yield (%)	Ref.
MeC(=CH$_2$), H, H, CH$_2$CH(OMe)$_2$, O, N, CH$_2$Ph	Na, liq. NH_3, EtOH, $-78\,^{\circ}C$	H, CH$_2$CH(OMe)$_2$, N, H	100	1389
OH, MeCH, H, H, CH$_2$CH(OMe)$_2$, O, N, CH(Me)Ph	Na, liq. NH_3, EtOH, $-78\,^{\circ}C$, stir 10 min	H, CH$_2$CH(OMe)$_2$, N, H	83	1389
OH, MeCH, H, H, CH$_2$CH(OMe)$_2$, O, N, CH(Me)Ph	Na, liq. NH_3, EtOH, $-78\,^{\circ}C$, stir 10 min	H, CH$_2$CH(OMe)$_2$, N, H	100	1389
H, H, i-Pr, Me, O, N, SCPh$_3$	W-2 Raney Ni	H, H, i-Pr, Me, O, N, H	40	1241
Me, Me, Ph, O, N, SCPh$_3$	$(n\text{-Bu})_3P$, EtOH, THF, $115\,^{\circ}C$, 48h	Me, Me, Ph, O, N, H	75	1241
	Me_3SiI, CH_2Cl_2, $25\,^{\circ}C$, 7h		81	1241
H$_2$N, Me, O, N, OCH$_2$Ph	$TiCl_4$, heat	H$_2$N, Me, O, N, H	—	1007

(continued)

TABLE 69. (*continued*)

Substrate	Reagents and conditions	Product	Yield (%)	Reference
H; $CH_2CO_2CH_2Ph$; O; N; OH	20% aq. $TiCl_3$, N_2, THF/H_2O, NaOH, stir r.t. 1.5 h	H; $CH_2CO_2CH_2Ph$; O; N; H	50	1079
Tol-*p*; O; N; An-*p*	CAN, −5 °C, 2 h	Tol-*p*; O; N; H	60	1225
PhO; Ph; O; N; $C_6H_4OSiMe_2(Bu\text{-}t)\text{-}p$	1. CAN, KF, MeCN, H_2O, 70 °C 2. stir r.t. 1.25 h	PhO; Ph; O; N; H	54	1201
R; COPh; O; N; An-*p*	CAN, KF, MeCN, H_2O, r.t.	R; COPh; O; N; H		1148
R = Me			60	
R = Ph			68	
H H; PhO; CH_2; O O; C; Me; O; N; An-*p*	CAN	H H; PhO; CH_2; O O; C; Me; O; N; H	70	1160
H H; R; OCH_2COMe; O; N; An-*p*	CAN	H H; R; OCH_2COMe; O; N; H		1160
R = PhO			75	
R = PI—N			30	

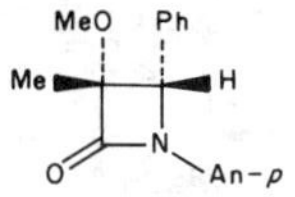	CAN	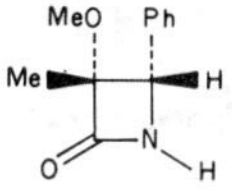	90	1229
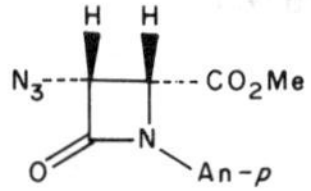	CAN, H_2O, MeCN, −5 °C, stir 2h	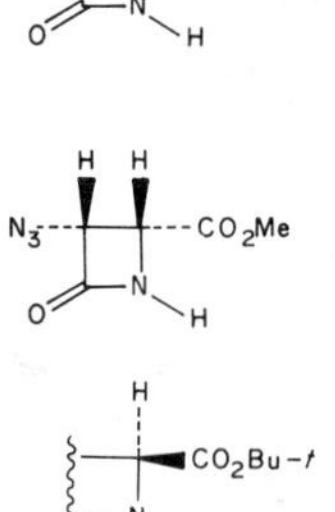	—	1152
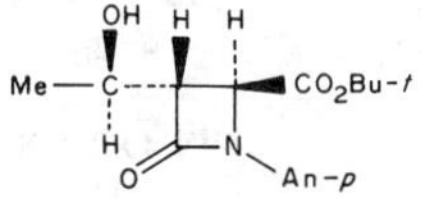	CAN, H_2O, Me_2CO, 5 °C, stir 30 min		53	1053
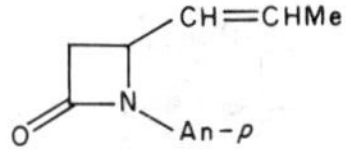	CAN, MeCN, H_2O, 0 °C, stir	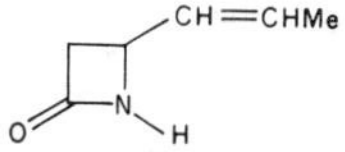	59	1153
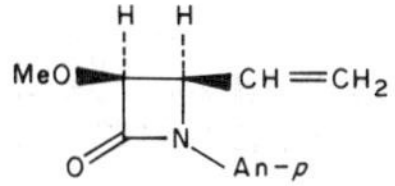	CAN, MeCN, H_2O, 0 °C, stir	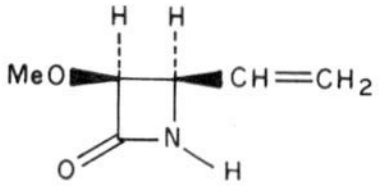	83	1153
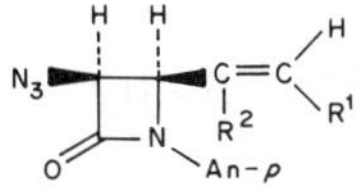	CAN, MeCN, H_2O, 0 °C, stir 1h	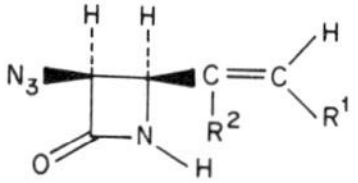		1153
$R^1 = Me$, $R^2 = H$			79	
$R^1 = Ph$, $R^2 = H$			72	
$R^1 = Ph$, $R^2 = Me$			89	

(continued)

TABLE 69. (*continued*)

Substrate	Reagents and conditions	Product	Yield (%)	Reference
$PhOCH_2CONH$, H, H, R, O, N, $C_6H_4(OCH_2OMe)$-*p*	1. CAN. THF, H_2O, 0 °C, 10 min 2. NaSH	H, R, N, H		
$R = CH_2CH(OMe)_2$			57	1196
				1196
$R = CH{=}CH_2$ $R = CH{=}CHCO_2Me$	1. CAN, THF, H_2O, 0 °C 2. NaSH 10 min		46 66[b]	
$PhOCH_2CONH$, H, H, $CH_2CH{=}CHR$, O, N, $C_6H_4(OCH_2OMe)$-*p*	1. CAN. THF, 0 °C, 2. NaSH			1196
R = H $R = CO_2Me$	1. H_2O 1. dioxane, stir 10 min		64 77[b]	
N_3, H, H, R, O, N, CH(An-*p*)$_2$	CAN, MeCN, − 5 °C, stir 45 min	N_3, H, H, R, O, N, H		1168
R = Me $R = CH_2{=}CHPh$			75 75	
Me—C, OH, H, H, H, CH(Me)CH_2OCH_2Ph, O, N, CH(An-*p*)$_2$	CAN, H_2O, MeCN, 0 °C	H, CH(Me)CH_2OCH_2Ph, N, H	91	1132

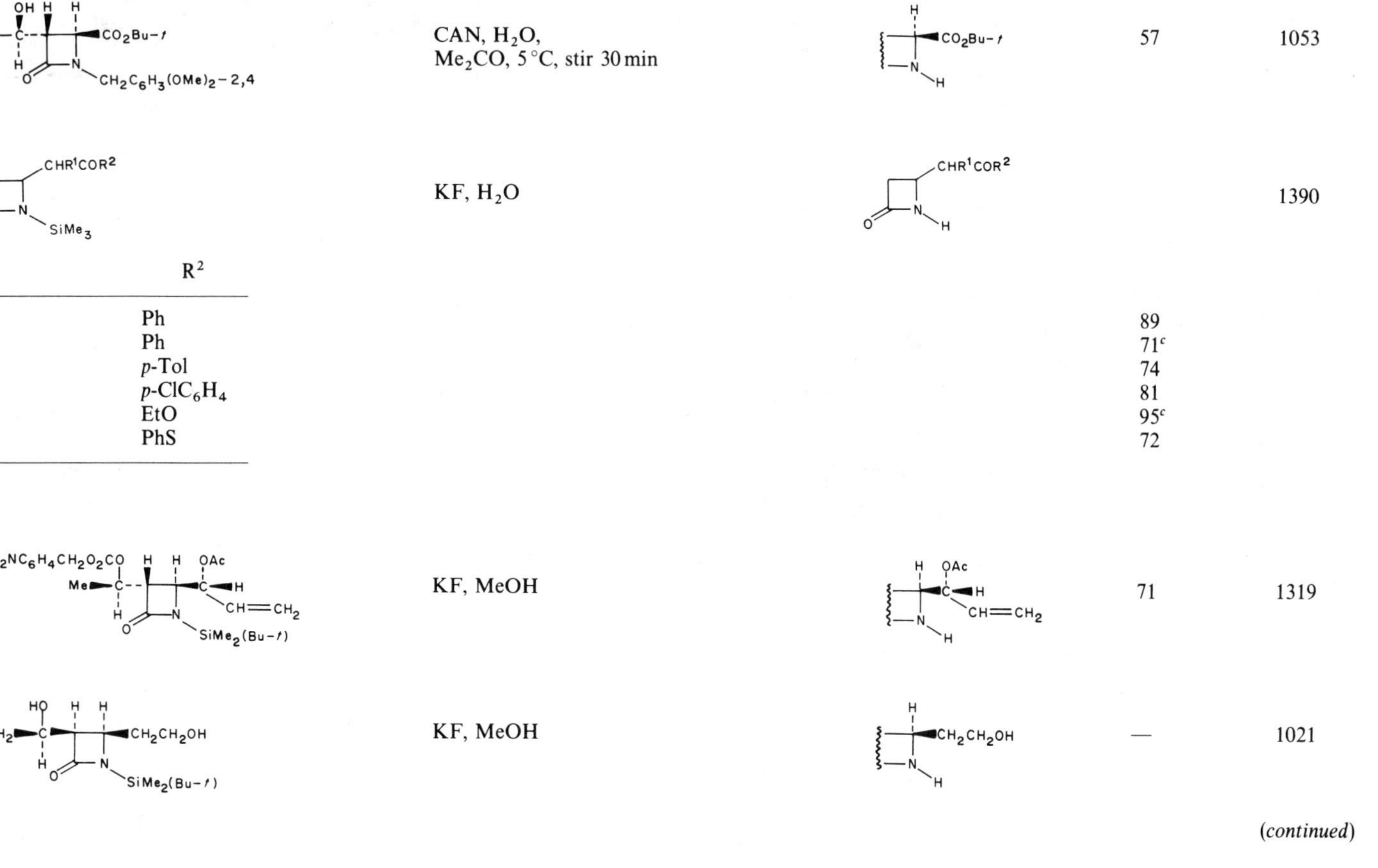

Substrate	Reagents	Product	Yield (%)	Ref.
	CAN, H_2O, Me_2CO, 5 °C, stir 30 min		57	1053
	KF, H_2O			1390
R^1 = H, R^2 = Ph			89	
R^1 = Me, R^2 = Ph			71[c]	
R^1 = H, R^2 = *p*-Tol			74	
R^1 = H, R^2 = *p*-ClC_6H_4			81	
R^1 = Me, R^2 = EtO			95[c]	
R^1 = H, R^2 = PhS			72	
	KF, MeOH		71	1319
	KF, MeOH		—	1021

(*continued*)

TABLE 69. *(continued)*

Substrate	Reagents and conditions	Product	Yield (%)	Reference
PhO_2C, $SiMe_2(Bu\text{-}t)$ + PhO_2C, $SiMe_2(Bu\text{-}t)$ + PhO_2C, $SiMe_2(Bu\text{-}t)$ + PhO_2C, $SiMe_2(Bu\text{-}t)$	KF, MeOH	N–H + N–H + N–H + N–H	—	1319
CH_2–C(H)(OMe)–C(I)(H)–CO_2Me, N–$SiPh_2(Bu\text{-}t)$ or CH_2–C(H)(OMe)–C(CO_2Me)(H)–I, N–$SiPh_2(Bu\text{-}t)$	HF, MeOH, 0 °C, stir 5 min	CH_2–C(H)(OMe)–C(I)(H)–CO_2Me, N–H or CH_2–C(H)(OMe)–C(CO_2Me)(H)–I, N–H	85	1315

Reactant	Conditions	Product	Yield (%)	Ref.
[structure: H, CO_2Me, N–CO_2Bu-*t*]	F_3CCOOH, CH_2Cl_2, stir r.t. 20 min	[structure: H, CO_2Me, N–H]	85	1391
[structure: PhS, Et, CO_2Bu-*n*, N–$CH_2C_6H_3(OMe)_2$-2,4]	F_3CCOOH, $BF_3\cdot Et_2O$, C_6H_5Me, 40 °C, stir 30 min	[structure: PhS, Et, CO_2Bu-*n*, N–H]	79	1392
[structure: H_2C=, N–CH(OMe)Ph]	*p*-TosOH, H_2O, Me_2CO	[structure: H_2C=, N–H]	50	1393
[structure: O, N–CH(OMe)Ar]	*p*-TosOH	[structure: O, N–H]		
Ar = Ph	Me_2CO, H_2O		57	1393
Ar = Ph	2-butanone		84	1393
Ar = Ph	H_2O, THF		—	1393
Ar = *p*-An	H_2O, THF, 8 h		100	1393
[structure: H, H, Et, CO_2H, N–$SiMe_2(Bu$-*t*)]	$Pb(OAc)_4$, DMF, $MeCO_2H$	[structure: H, Et, OAc, N–H] (*trans*:*cis* = 2:1)	89	1245

(continued)

TABLE 69. *(continued)*

Substrate	Reagents and conditions	Product	Yield (%)	Reference
RCH$_2$CONH, H, H, Ph, O, N, CH$_2$C$_6$H$_3$(OMe)$_2$-3,4	$K_2S_2O_8$, Na_2HPO_4, H_2O, MeCN, reflux 1.5h	H, Ph, N, H		1178, 1376
R = Ph			—	
R = PhO			25–35	
PI-N, H, H, CH$_2$F, O, N, C, H, Me, Ph	$(NH_4)_2S_2O_8$, MeCN, reflux 2h	PI-N, H, H, CH$_2$F, O, N, H	41	1184
PI-N, H, H, CHF$_2$, O, N, C, H, Me, Ph	$(NH_4)_2S_2O_8$, MeCN, reflux 2h	PI-N, H, H, CHF$_2$, O, N, H	50	1184
PI-N, H, H, C$_6$H$_4$(OCH$_2$Ph)-*o*, O, N, CH$_2$C$_6$H$_3$(OMe)$_2$-2,4	$K_2S_2O_8$, Na_2HPO_4, H_2O, MeCN, reflux 80 °C	PI-N, H, H, C$_6$H$_4$(OCH$_2$Ph)-*o*, O, N, H	74.3	1183
(*t*-Bu)Me$_2$SiO, Me, C, H, H, H, Ac, O, N, CH$_2$C$_6$H$_3$(OMe)$_2$-2,4	$K_2S_2O_8$, K_2HPO_4, H_2O, MeCN, Ar, 65 °C, 45 min	H, Ac, N, H	72	750

Substrate	Conditions	Product	Yield (%)	Ref.
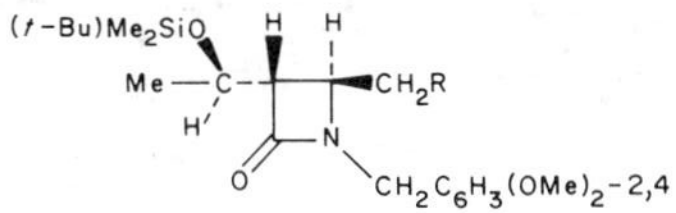 (t-Bu)Me_2SiO, Me—C—H; H, H; CH_2R; O, N—$CH_2C_6H_3(OMe)_2$-2,4	$K_2S_2O_8$, K_2HPO_4, H_2O, MeCN, Ar	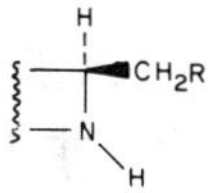 H; CH_2R; N—H		
R = OH	65 °C, 1.5 h		60	750
R = OH	75 °C, 1 h		60	1053
R = CN	75 °C, 1 h		83	1053
R = $COCH_2Cl$	65–70 °C, 70 min		55.6	1053
R = CO_2CH_2Ph	64 °C, 1 h		51	750
R = CO_2CH_2Ph	65–70 °C, 70 min		57	1053
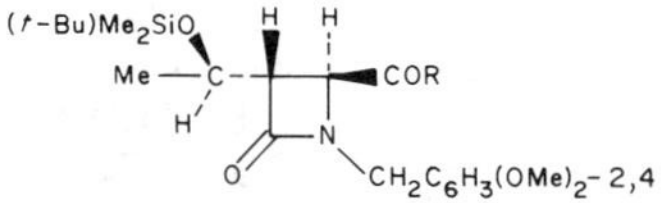 (t-Bu)Me_2SiO, Me—C—H; H, H; COR; O, N—$CH_2C_6H_3(OMe)_2$-2,4	$K_2S_2O_8$, K_2HPO_4, H_2O, MeCN, Ar, stir	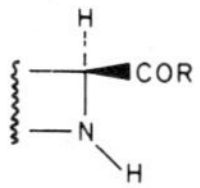 H; COR; N—H		
R = Me	65 °C, 45 min		82	750
R = OMe	75 °C, 1 h		69.3	1053
R = OBu-t	75 °C, 1 h		82	1053
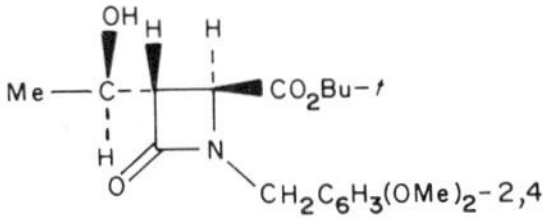 OH, Me—C—H; H, H; CO_2Bu-t; O, N—$CH_2C_6H_3(OMe)_2$-2,4	$K_2S_2O_8$, K_2HPO_4, H_2O, MeCN, Ar, stir	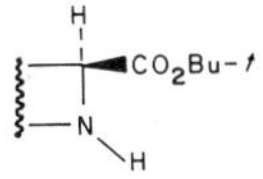 H; CO_2Bu-t; N—H	76.2	1053
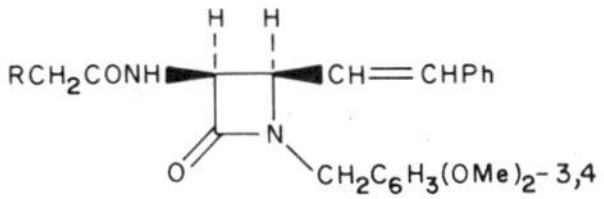 RCH_2CONH; H, H; CH=CHPh; O, N—$CH_2C_6H_3(OMe)_2$-3,4	$K_2S_2O_8$	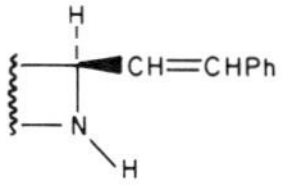 H; CH=CHPh; N—H		1377
R = PhO			21	
R = 2-Thi			18	

(continued)

TABLE 69. *(continued)*

Substrate	Reagents and conditions	Product	Yield (%)	Reference
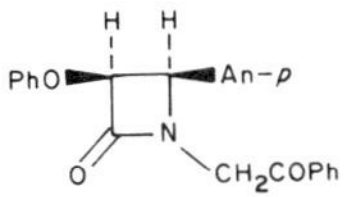 PhO, H, H, An-*p*, O, N, CH_2COPh	$KMnO_4$, Me_2CO, H_2O	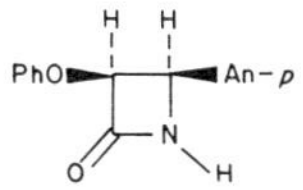 PhO, H, H, An-*p*, O, N, H	—	1218
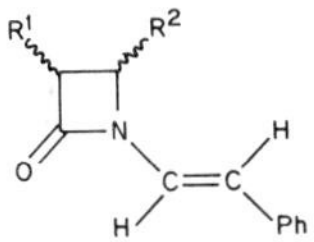 R^1, R^2, O, N, C=C, H, H, Ph	$KMnO_4$, Me_2CO, H_2O	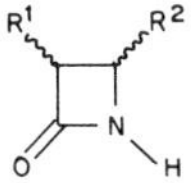 R^1, R^2, O, N, H		1217
R^1 — R^2				
PhO — Ph			77	
PhO — *p*-An			62	
PI—N — *p*-An		$(Z:E = 25:75)$	40	
$PhOCH_2CONH$ — Ph(*cis*)			80	
$PhOCH_2CONH$ — *p*-An(*cis*)			75	
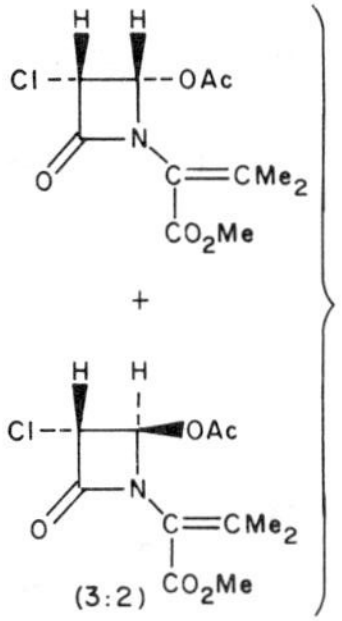 Cl, H, H, OAc, O, N, $C=CMe_2$, CO_2Me + Cl, H, H, OAc, O, N, $C=CMe_2$, CO_2Me (3:2)	$KMnO_4$, Me_2CO, H_2O, pH = 7, 0 °C, 30 min	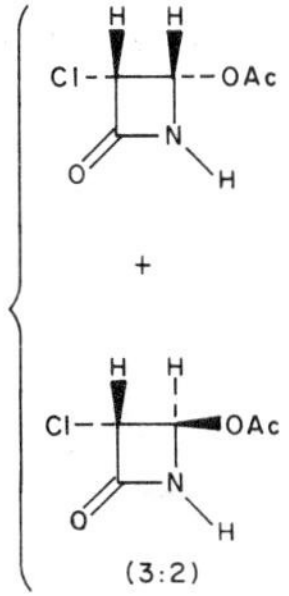 Cl, H, H, OAc, O, N, H + Cl, H, H, OAc, O, N, H (3:2)	50	1309

Substrate	Reagents	Product	Yield	Ref.
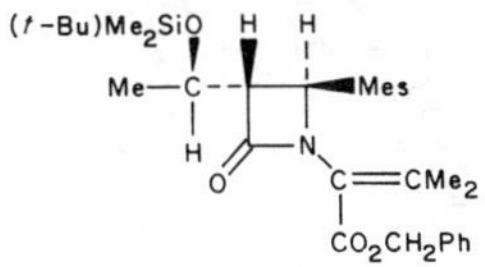	$KMnO_4$, Me_2CO, HOAc, 16h	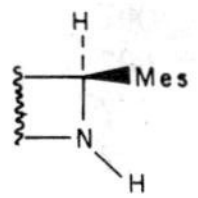	94	1339
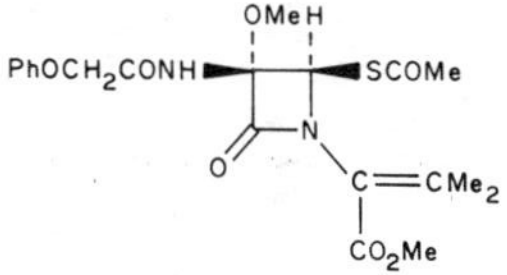	$KMnO_4$	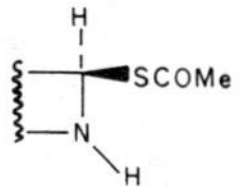	15	1341
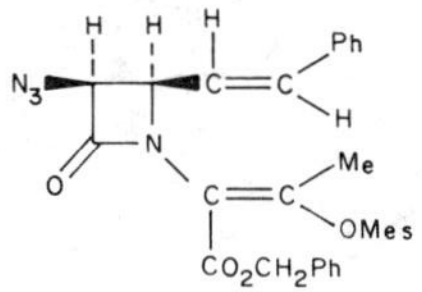	1. RuO_2, $NaIO_4$, Me_2CO, H_2O, r.t., 2h 2. CH_2N_2, Et_2O	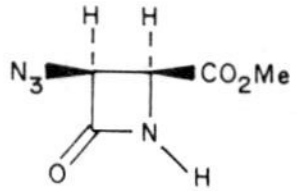	—	1152
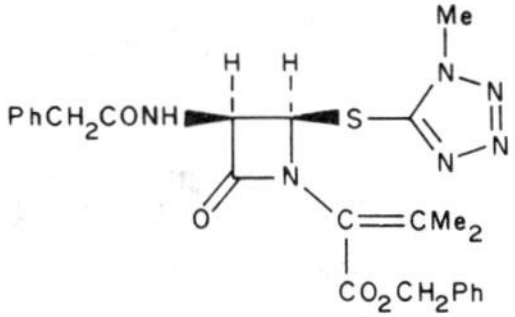	1. CH_2Cl_2, O_3, $-50\,^{\circ}C$ to $-60\,^{\circ}C$ 2. Me_2S, r.t., 1h 3. MeOH, NaOMe, stir 2h 4. HOAc	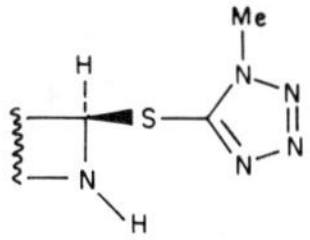	20.3	1343
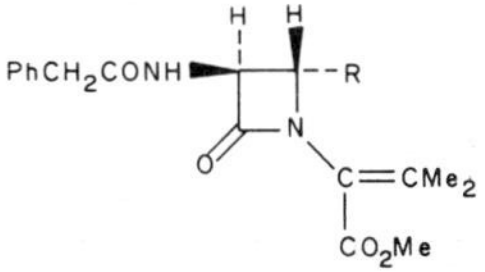	1. CH_2Cl_2, O_3, $-50\,^{\circ}C$ to $-60\,^{\circ}C$ 2. Me_2S, r.t., 1h 3. MeOH, NaOMe, stir 2h 4. HOAc	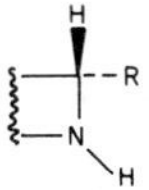		1343

(*continued*)

TABLE 69. (*continued*)

Substrate	Reagents and conditions	Product	Yield (%)	Reference
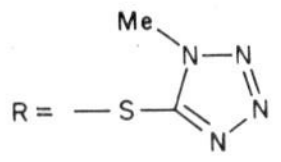 R = —S—(1-methyltetrazol-5-yl) [Me, N—N, N, N, S]			30	
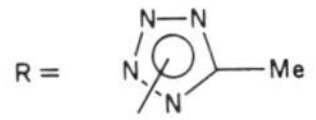 R = (5-methyltetrazol-2-yl) [N—N, N, N, Me]			46	
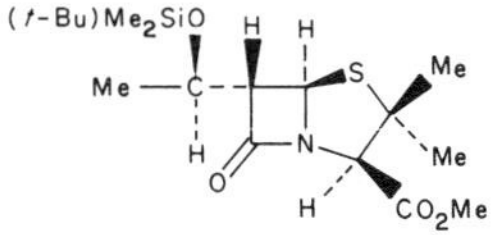 (*t*-Bu)Me_2SiO, Me—C, H, H, H, S, Me, Me, N, O, H, CO_2Me	1. $Hg(OAc)_2$, AcOH 2. $KMnO_4$, Me_2CO	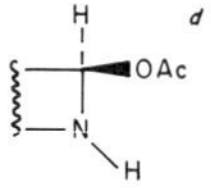 H, OAc, N, H [d]	—	1394
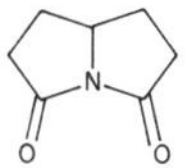 (bicyclic imide) N, O, O	$NaBH_4$, EtOH, 2*N* HCl, −5 °C	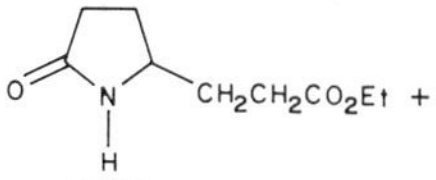 O, N, H, $CH_2CH_2CO_2Et$ + **(200)**	38	1395
		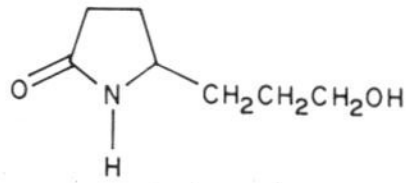 O, N, H, $CH_2CH_2CH_2OH$ **(201)**	48	
	$NaBH_4$, EtOH, r.t. stir 75 min	**200**	83	1395
	$Zn(BH_4)_2$, THF, Ar, stir 90 min	**201**	77	1395
	NaOEt, EtOH, r.t. stir 16 h	**200**	98	1395
	0.1 *N* HCl, EtOH, r.t. stir 13 h	**200**	96	1395

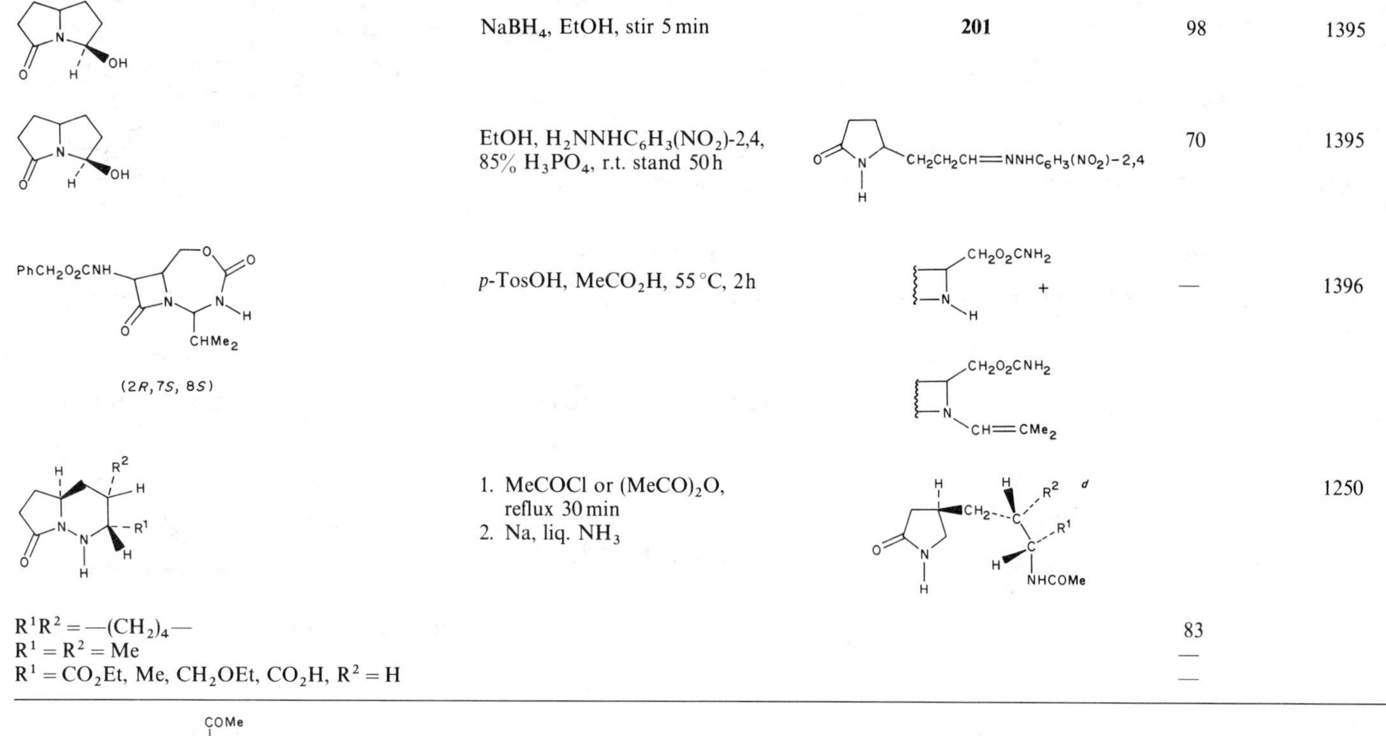

Substrate	Conditions	Product	Yield (%)	Ref.
(structure)	$NaBH_4$, EtOH, stir 5 min	**201**	98	1395
(structure)	EtOH, $H_2NNHC_6H_3(NO_2)$-2,4, 85% H_3PO_4, r.t. stand 50 h	(structure) $CH_2CH_2CH{=}NNHC_6H_3(NO_2)$-2,4	70	1395
$PhCH_2O_2CNH$ (structure) (2*R*,7*S*,8*S*)	*p*-TosOH, $MeCO_2H$, 55 °C, 2 h	(structure) $CH_2O_2CNH_2$ + (structure) $CH_2O_2CNH_2$, $CH{=}CMe_2$	—	1396
(structure)	1. MeCOCl or $(MeCO)_2O$, reflux 30 min 2. Na, liq. NH_3	(structure) NHCOMe [d]		1250
$R^1R^2 = —(CH_2)_4—$			83	
$R^1 = R^2 = Me$			—	
$R^1 = CO_2Et$, Me, CH_2OEt, CO_2H, $R^2 = H$			—	

[a]Also obtained

[b]Products isolated as a 2:1 mixture of geometric isomers.
[c]Product obtained as a mixture of diastereomers.
[d]Intermediates not identified or isolated.

TABLE 70. Lactam conversion of exocyclic α-alcohols to α-chlorides

Alcohol	Reagents and conditions	Product	Reference
H, R, O, N, $CH(OH)CO_2CH_2Ph$	$SOCl_2$, 2,6-$Me_2C_5H_3N$, THF,	H, R, O, N, $CH(Cl)CO_2CH_2Ph$	1196
R = CH=CH_2, CH=$CHCO_2Me$	−20 °C		
R = CH_2CH=CH_2[a], CH_2CH=$CHCO_2Me$[a]	0 °C		
$PhOCH_2CONH$, H, H, R, O, N, $CH(OH)CO_2CH_2Ph$	$SOCl_2$, 2,6-$Me_2C_5H_3N$, THF	H, R, N, $CH(Cl)CO_2CH_2Ph$	
R = CH=CH_2[a], CH=$CHCO_2Me$[a]	0 °C		1196
R = CH=CHMe	−15 °C		1337
R = CH_2CH=CH_2[a]	$SOCl_2$, 2,6-$Me_2C_5H_3N$, THF, 0 °C		1196
R = CH_2CH=$CHCO_2Me$[a]			1196
H, OCH_2CH=CH_2, O, N, CH_2OH	1. THF, Ar, 2,6-$Me_2C_5H_3N$, −15 °C 2. $SOCl_2$, −15 °C, stir 1 h	H, OCH_2CH=CH_2[a], O, N, CH_2Cl	1336
H, OCH_2R, O, N, $CH(OH)CO_2Bu$-*t*	1. THF, Ar, 2,6-$Me_2C_5H_3N$, −15 °C 2. $SOCl_2$, −15 °C, stir 1 h	H, OCH_2R, O, N, $CH(Cl)CO_2Bu$-*t* [a]	1336
R = CH=CH_2, (*E*)-CH=CHPh, (*E*)-CH=$CHCO_2Me$, C≡CH, C≡CPh			

Substrate	Conditions	Product	Ref.
4-(allyloxy)azetidin-2-one with N–C(OH)(CO_2Et)$_2$ [structure: $OCH_2CH{=}CH_2$, $C(CO_2Et)_2$, OH]	1. THF, Ar, 2,6-$Me_2C_5H_3N$, -15 °C 2. $SOCl_2$, −15 °C, stir 1 h	[structure: $OCH_2CH{=}CH_2$, $C(CO_2Et)_2$, Cl] [a]	1336
[structure: $SCH{=}CHCO_2Et$, $CH(OH)CO_2Bu$-*t*] (*cis* or *trans*)	$SOCl_2$, 2,6-$Me_2C_5H_3N$, THF, −15 °C	[structure: $SCH{=}CHCO_2Et$, $CH(Cl)CO_2Bu$-*t*] [a]	1337
[structure: S–C(=HC–CO_2Et)–CH_2CO_2Et, $CH(OH)CO_2R$] R = Me, $PhCH_2$, CH_2OAc	$SOCl_2$, C_5H_5N, THF, −50 °C to 0 °C	[structure: S–C(=HC–CO_2Et)–CH_2CO_2Et, $CH(Cl)CO_2R$] [b]	1338
[structure: S–C(=HC–CO_2Et)–CH_2CH_2OAc, $CH(OH)CO_2R$] R = Me, $PhCH_2$, CH_2OAc	$SOCl_2$, C_5H_5N, THF, −30 °C to 0 °C	[structure: S–C(=HC–CO_2Et)–CH_2CH_2OAc, $CH(Cl)CO_2R$] [a]	1338
[structure: (*t*-Bu)Me_2SiO, Me–C–H, R, $CH(OH)CO_2CH_2C_6H_4NO_2$-*p*] R = CH_2I	$SOCl_2$, 2,6-$Me_2C_5H_3N$, THF, −40 °C stir 20 min	[structure: R, $CH(Cl)CO_2CH_2C_6H_4NO_2$-*p*] [a]	1339
R = C(Ac)=CH_2	$SOCl_2$ [a]		1345

(*continued*)

TABLE 70. (*continued*)

Alcohol	Reagents and conditions	Product	Reference
R = (1-acetylcyclopropyl)	$SOCl_2$[a]		1345
$PhOCH_2CONH$, MeO, H, SAc, N–$CH(OH)CO_2R$ (azetidinone)	$SOCl_2$, 2,6-$Me_2C_5H_3N$, THF, −30 °C to 0 °C	H, SAc, N–$CH(Cl)CO_2R$ [a]	1341
R = Me, CH_2COMe, $CH_2C_6H_4NO_2$-*p*			
(*t*-Bu)Me_2SiO, Me–C–H, H, H, $SCOCH_2OSiPh_2(Bu$-*t*), N–$CH(OH)COCH_2Ac$	$SOCl_2$, C_5H_5N	H, $SCOCH_2OSiPh_2(Bu$-*t*), N–$CH(Cl)COCH_2Ac$ [a]	1397
(*t*-Bu)Me_2SiO, Me–C–H, H, H, SCOFu-2, N–$CH(OH)CO_2CH_2CH{=}CH_2$	$SOCl_2$, 2,6-$Me_2C_5H_3N$, THF, −10 °C	H, SCOFu-2, N–$CH(Cl)CO_2CH_2CH{=}CH_2$ [a]	1344
(*t*-Bu)Me_2SiO, Me–C–H, H, H, $SCOCH_2S$–(1-methyltetrazol-5-yl), N–$CH(OH)CH_2CH(OPh)Me$	$SOCl_2$	H, $SCOCH_2S$–(1-methyltetrazol-5-yl), N–$CH(Cl)CH_2CH(OPh)Me$ [a]	1342
p-$O_2NC_6H_4CH_2O_2CO$, Me–C–H, H, H, C(OAc)(H)–$CH{=}CH_2$, N–$CH(OH)CO_2CH_2C_6H_4NO_2$-*p*	$SOCl_2$	H, C(OAc)(H)–$CH{=}CH_2$, N–$CH(Cl)CO_2CH_2C_6H_4NO_2$-*p* [a]	1319

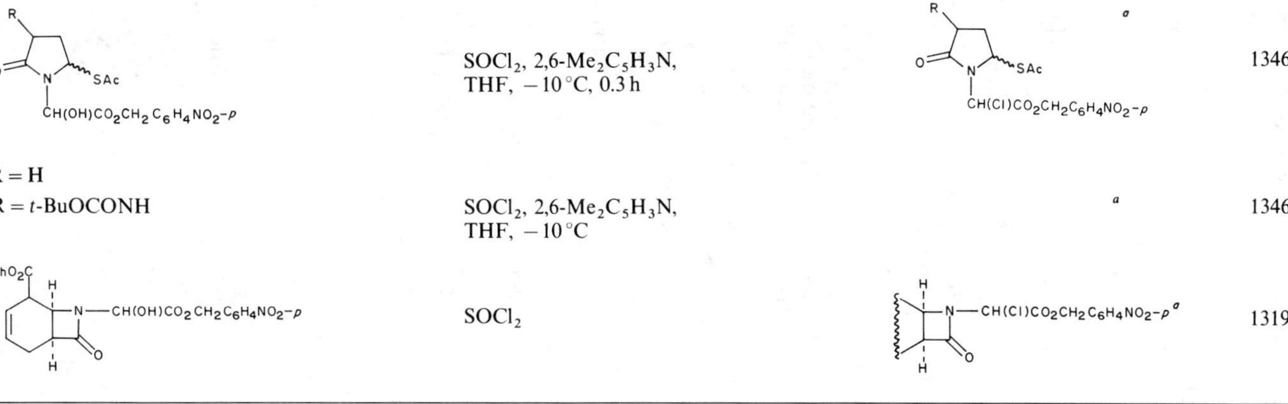

Substrate	Reagents and conditions	Product	Ref.
CH(OH)CO2CH2C6H4NO2-p (pyrrolidinone, R, SAc)	$SOCl_2$, 2,6-$Me_2C_5H_3N$, THF, −10 °C, 0.3 h	CH(Cl)CO2CH2C6H4NO2-p (pyrrolidinone, R, SAc) [a]	1346
R = H			
R = t-BuOCONH	$SOCl_2$, 2,6-$Me_2C_5H_3N$, THF, −10 °C	[a]	1346
PhO2C, H, H, N—CH(OH)CO2CH2C6H4NO2-p	$SOCl_2$	H, H, N—CH(Cl)CO2CH2C6H4NO2-p [a]	1319

[a] The product is not isolated.

[b] The products were isolated in 65–85% yields depending upon the substituent R.

Reaction of *N*-chloromethyl and *N*-siloxymethyl derivatives of lactams with *O*-silyl-substituted enols produces[1399] α-C-alkylated lactams in yields ranging from 65 to 92% (equation 1001).

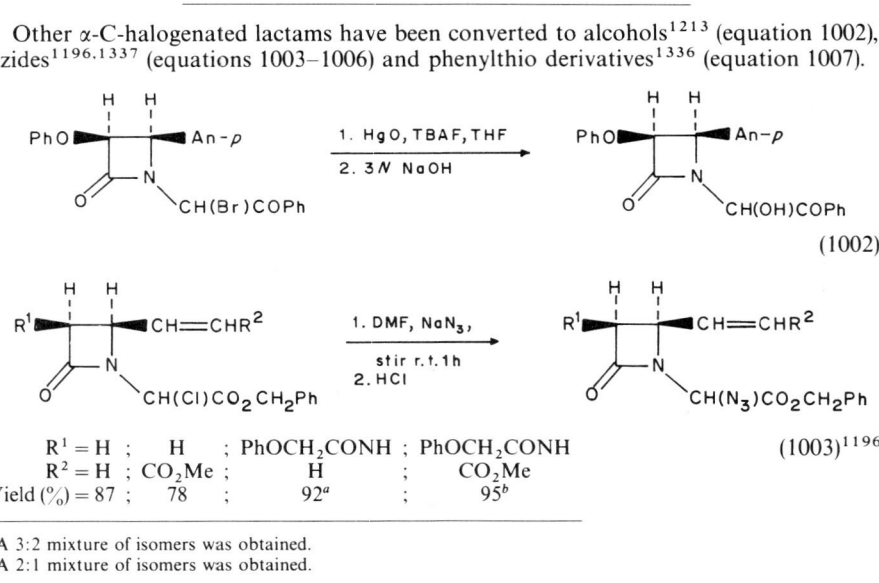

$$XCH_2N\text{(lactam, }(CH_2)_n\text{)} + R^1C(OSiMe_3){=}CHR^2 \xrightarrow{130-150^\circ C} R^1COCHR^2CH_2N\text{(lactam, }(CH_2)_n\text{)} \quad (1001)$$

X = Cl, Me_3SiO

n	R^1	R^2
2	*t*-Bu	H
2	H	*t*-Bu
2	1-Naph	H
3	1-Naph	H
4	H	*t*-Bu
4	1-Naph	H

Other α-C-halogenated lactams have been converted to alcohols[1213] (equation 1002), azides[1196,1337] (equations 1003–1006) and phenylthio derivatives[1336] (equation 1007).

$$\text{PhO, An-}p\text{ azetidinone–N–CH(Br)COPh} \xrightarrow[2.\ 3N\ NaOH]{1.\ HgO,\ TBAF,\ THF} \text{PhO, An-}p\text{ azetidinone–N–CH(OH)COPh} \quad (1002)$$

$$R^1\text{, }CH{=}CHR^2\text{ azetidinone–N–CH(Cl)}CO_2CH_2Ph \xrightarrow[\text{stir r.t. 1h; 2. HCl}]{1.\ DMF,\ NaN_3,} R^1\text{, }CH{=}CHR^2\text{ azetidinone–N–CH(}N_3\text{)}CO_2CH_2Ph \quad (1003)^{1196}$$

R^1 =	H	H	$PhOCH_2CONH$	$PhOCH_2CONH$
R^2 =	H	CO_2Me	H	CO_2Me
Yield (%) =	87	78	92[a]	95[b]

[a] A 3:2 mixture of isomers was obtained.
[b] A 2:1 mixture of isomers was obtained.

$$R^1\text{, }CH_2CH{=}CHR^2\text{ azetidinone–N–CH(Cl)}CO_2CH_2Ph \xrightarrow[\text{stir r.t.; 2. HCl}]{1.\ DMF,\ NaN_3,} R^1\text{, }CH_2CH{=}CHR^2\text{ azetidinone–N–CH(}N_3\text{)}CO_2CH_2Ph \quad (1004)^{1196}$$

R^1 =	H	H	$PhOCH_2CONH$	$PhOCH_2CONH$
R^2 =	H	CO_2Me	H	CO_2Me
Yield (%) =	80	—[a]	86	81

[a] Two azide epimers ($Z:E$ = 15:68) were obtained.

$$\text{(cis or trans) azetidinone: 4-}SCH{=}CHCO_2Et\text{, N-}CH(Cl)CO_2Bu\text{-}t \xrightarrow[\text{stir 10 min}]{DMF,\ NaN_3} \text{4-}SCH{=}CHCO_2Et\text{, N-}CH(N_3)CO_2Bu\text{-}t$$

[*cis* or *trans*, 2 isomers each (1:1)]

(1005)[1337]

$$\text{PhOCH}_2\text{CONH / OCH{=}CHMe azetidinone, N-}CH(Cl)CO_2CH_2Ph \xrightarrow[\text{stir 10 min}]{DMF,\ NaN_3} \text{PhOCH}_2\text{CONH / OCH{=}CHMe azetidinone, N-}CH(N_3)CO_2CH_2Ph$$

(1006)[1337]

$$\text{4-}OCH_2CH{=}CH_2\text{ azetidinone, N-}CH(Cl)CO_2Bu\text{-}t \xrightarrow[(n\text{-}Bu)_4N^+Br^-,\ 1N\ \text{aq. NaOH},\ 10\ ^\circ C,\ 10\ \text{min}]{\text{dry } C_6H_6,\ PhSH,} \text{4-}OCH_2CH{=}CH_2\text{ azetidinone, N-}CH(SPh)CO_2Bu\text{-}t$$

(1007)

One use for α-C-halogenated lactams has been in the preparation of exocyclic carbon–phosphorus double bonds which have been prepared by treating the halogenated lactams with triphenylphosphine and an amine base (equation 1008 and Table 71).

$$\text{N—CH(X)COR} \xrightarrow{Ph_3P,\ \text{amine}} \text{N—C(=PPh}_3\text{)COR} \qquad (1008)$$

Exocyclic α,β-double bond formation in lactams has also been accomplished[1146,1217] by dehydrohalogenation of β-C-halogenated lactams with 1,8-diazabicyclo[5.4.0]undec-7-ene (DBU) in refluxing benzene (equation 1009).

Upon treatment with lithium hexamethyldisilazide in tetrahydrofuran, esters of 2-oxo-4-vinylazetidin-1-ylacetic acid are converted[1304] into their enolates, which upon quenching with acetyl chloride provides the corresponding β-keto esters that exist primarily in their enol form (equation 1010).

TABLE 71. Formation of exocyclic lactam carbon to phosphorus double bonds

Halogenated lactam	Reagents and conditions	Product	Yield (%)	Reference
(t-Bu)Me$_2$SiO, H, Ac, Me—C, H, O, N, CH(Cl)CO$_2$CH$_2$CH=CH$_2$	Ph$_3$P, 2,6-Me$_2$C$_5$H$_3$N	Ac, N, C=PPh$_3$, CO$_2$CH$_2$CH=CH$_2$	63	1345
p-O$_2$NC$_6$H$_4$O$_2$CO, H, H, OAc, Me—C, C—CH=CH$_2$, H, H, O, N, CH(Cl)CO$_2$C$_6$H$_4$NO$_2$-p	Ph$_3$P	H, OAc, C—CH=CH$_2$, H, N, C=PPh$_3$, CO$_2$C$_6$H$_4$NO$_2$-p	—	1319
MeO, H, PhOCH$_2$CONH, SAc, O, N, CH(Cl)CO$_2$R	Ph$_3$P, silica gel, THF	H, SAc, N, C=PPh$_3$, CO$_2$R		1341
R = Me			60	
R = CH$_2$COMe			34	
R = CH$_2$C$_6$H$_4$NO$_2$-p			—	
(t-Bu)Me$_2$SiO, H, H, Me—C, SCOCH$_2$OSiPh$_2$(Bu-t), H, O, N, CH(Cl)CO$_2$CH$_2$COMe	Ph$_3$P, silica gel	H, SCOCH$_2$OSiPh$_2$(Bu-t), N, C=PPh$_3$, CO$_2$CH$_2$COMe	—	1397
(t-Bu)Me$_2$SiO, H, H, Me—C, SCOFu-2, H, O, N, CH(Cl)CO$_2$CH$_2$CH=CH$_2$	Ph$_3$P, 2,6-Me$_2$C$_5$H$_3$N, THF, 55–60 °C, 75 h	H, SCOFu-2, N, C=PPh$_3$, CO$_2$CH$_2$CH=CH$_2$	—	1344

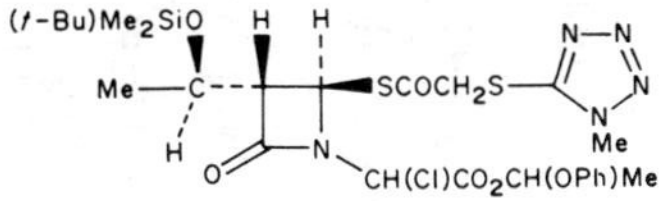 (*t*-Bu)Me_2SiO, Me, H, H, H, $SCOCH_2S$, N—N, N, N, Me, O, N, $CH(Cl)CO_2CH(OPh)Me$	Ph_3P	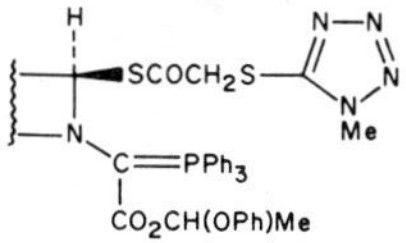 H, $SCOCH_2S$, N—N, N, N, Me, N, $C{=}PPh_3$, $CO_2CH(OPh)Me$	—	1342
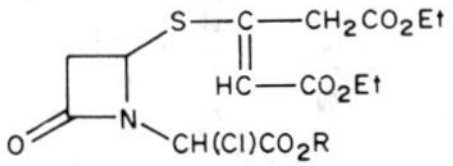 S, C, CH_2CO_2Et, HC, CO_2Et, O, N, $CH(Cl)CO_2R$ R = Me, $PhCH_2$, CH_2OAc	Ph_3P, C_5H_5N, THF, 25–50 °C	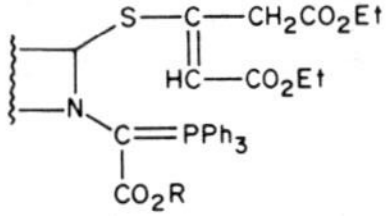 S, C, CH_2CO_2Et, HC, CO_2Et, N, $C{=}PPh_3$, CO_2R	—	1338
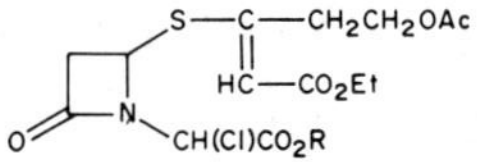 S, C, CH_2CH_2OAc, HC, CO_2Et, O, N, $CH(Cl)CO_2R$ R = Me, $PhCH_2$, CHOCOMe	Ph_3P, C_5H_5N, THF, 25–50 °C	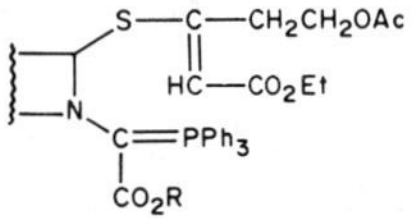 S, C, CH_2CH_2OAc, HC, CO_2Et, N, $C{=}PPh_3$, CO_2R	—	1338
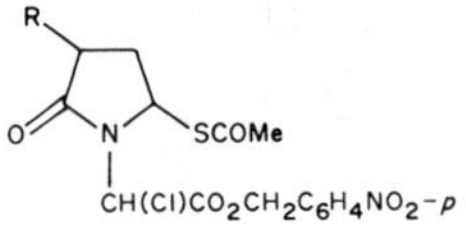 R, O, N, SCOMe, $CH(Cl)CO_2CH_2C_6H_4NO_2$-*p*	Ph_3P, dioxane, 2,6-$Me_2C_5H_3N$, 1 h	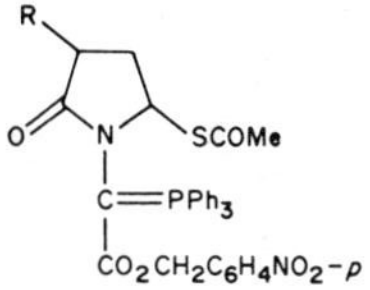 R, O, N, SCOMe, $C{=}PPh_3$, $CO_2CH_2C_6H_4NO_2$-*p*	—	1346
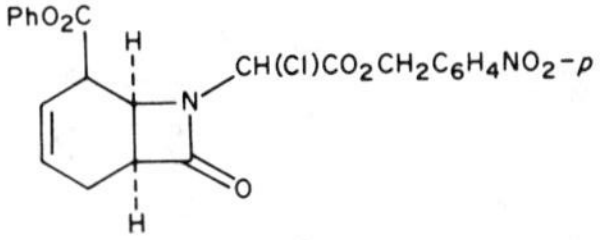 PhO_2C, H, H, N, $CH(Cl)CO_2CH_2C_6H_4NO_2$-*p*, O	Ph_3P	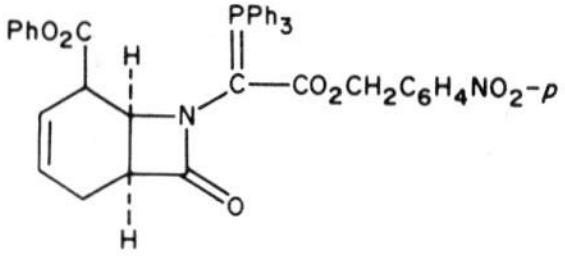 PhO_2C, H, H, N, PPh_3, C, $CO_2CH_2C_6H_4NO_2$-*p*, O	—	1319

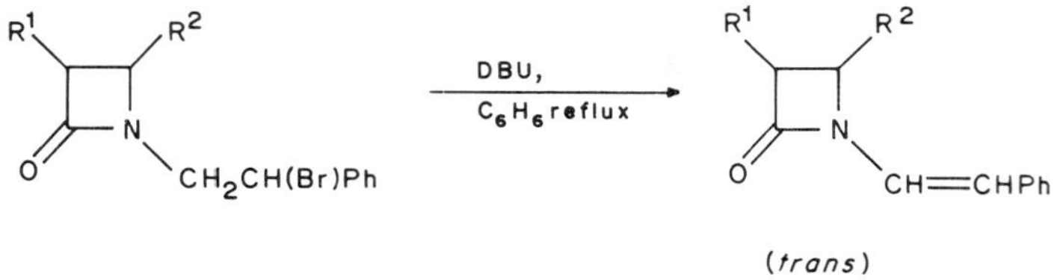

(1009)

R^1	R^2	Yield (%)	Reference
PhO	Ph	98	1385
PhO	*p*-An	90	1385
PI—N	*p*-An	94[a]	1385
$PhOCH_2CONH$	Ph	64[b]	1385
$PhOCH_2CONH$	*p*-An	53[b]	1385
PhO	CMe=CHPh	—	1319
PI—N	CMe=CHPh	—	1319

[a]Product was a $Z:E$ mixture of 25:75.
[b]Overall yield from alcohol, through bromide to olefin.

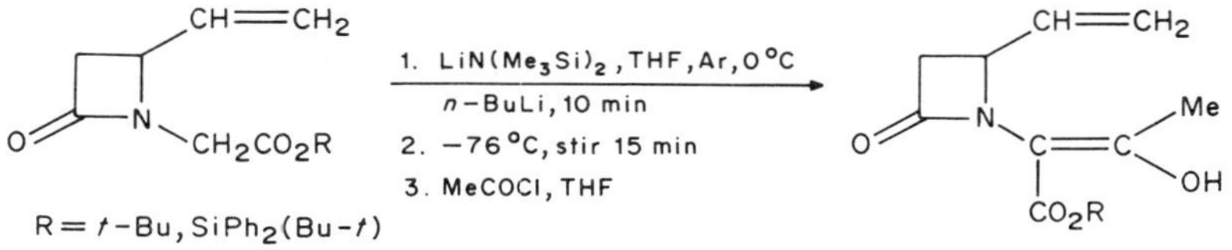

(1010)

Enols are also produced when analogous α-ethylenedioxy[1304] (equation 1011) or α-diethoxy[1039] acetals (equation 1012) are treated with 95% aqueous trifluoroacetic acid.

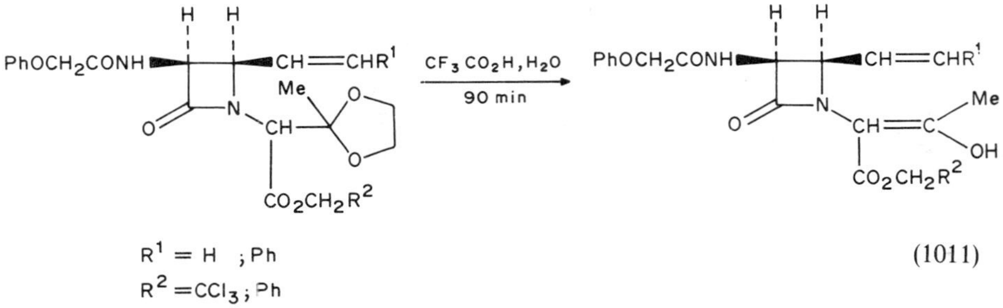

(1011)

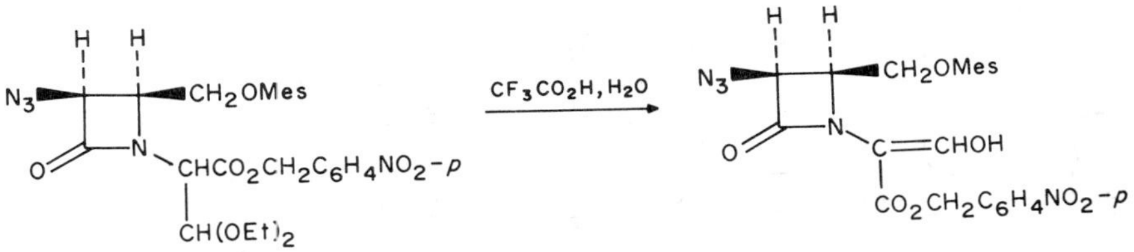

(1012)

Similar results were obtained[1152] when 1-(1-benzyloxycarbonyl-2-hydroxyprop-1-yl-3-azido-4-styrylazetidin 2-one was treated with Jones' reagent, followed by reaction of the intermediate ketone with methanesulphonyl chloride in 4-dimethylaminopyridine (equation 1013).

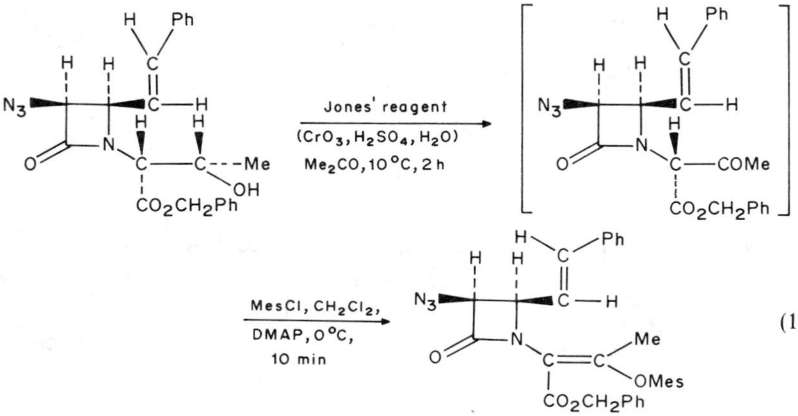

Jones' reagent (CrO_3, H_2SO_4, H_2O) Me_2CO, 10 °C, 2 h

MesCl, CH_2Cl_2, DMAP, 0 °C, 10 min

(1013)

Treatment of a mixture of structural isomers consisting of the methyl esters of 1-(1-methyl-2-carboxymethyl)-4-methoxyazetidin-2-one and 1-(1-methyl-1-methoxy-2-carboxymethyl)azetidin-2-one with *p*-toluenesulphonic acid in methanol produces[1060] a product mixture consisting of the recovered untreated 4-methoxy isomer and the exocyclic α,β-unsaturated product (equation 1014).

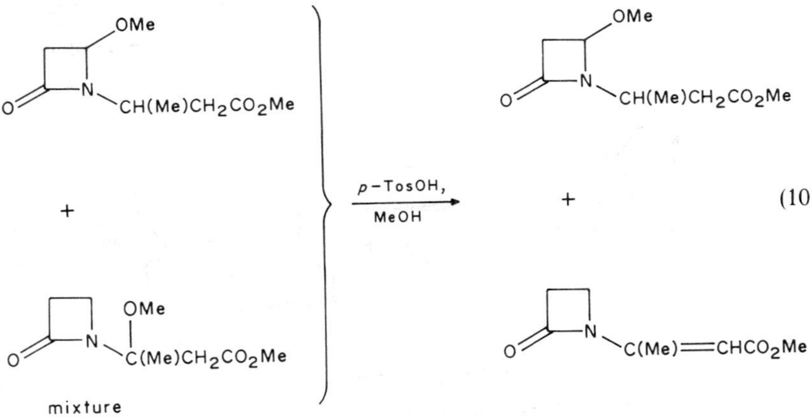

mixture

p-TosOH, MeOH

(1014)

An interesting approach to the preparation of N-side chain α,β-unsaturated azetidinones is illustrated by the reactions of α-*C*-2-methyloxiran-1-yl-substituted β-lactams. Subjection of these compounds to Mitsunobu reaction conditions (triphenylphosphine and diethyl azodicarboxylate in tetrahydrofuran at room temperature) produced[1400] two different α,β-unsaturated lactam products depending upon the ester function present in the starting substrate (equations 1015 and 1016).

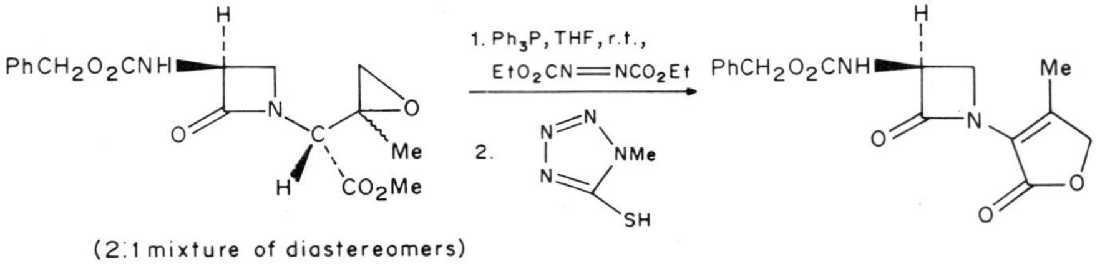

1. Ph_3P, THF, r.t., $EtO_2CN{=}NCO_2Et$

2. 1-methyl-5-mercaptotetrazole (NMe, SH)

(2:1 mixture of diastereomers)

(1015)

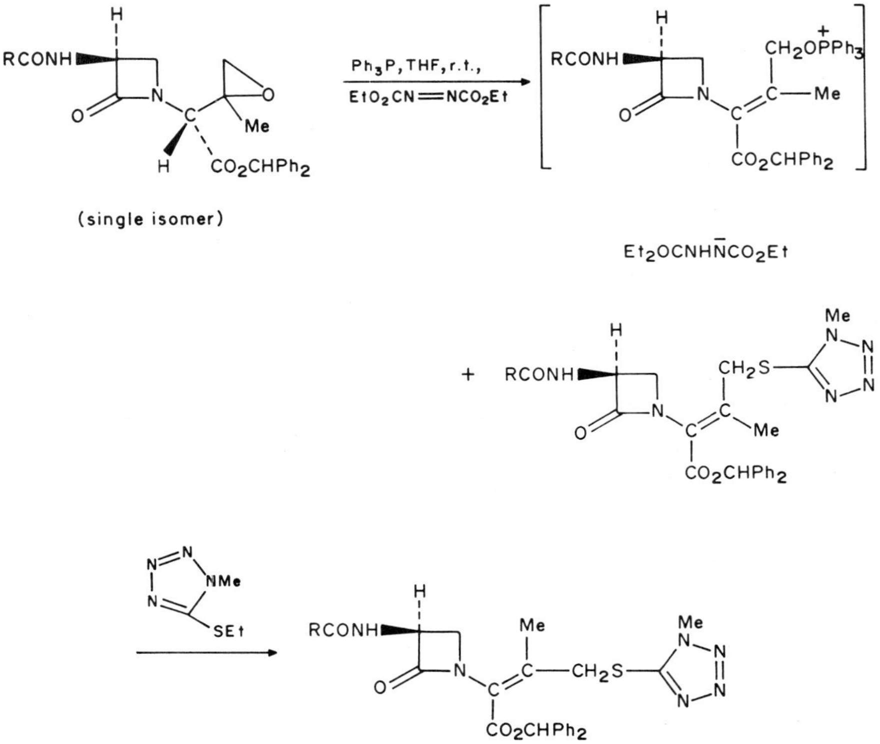

Ph_3P, THF, r.t., $EtO_2CN{=}NCO_2Et$

(single isomer)

$Et_2OCNH\bar{N}CO_2Et$

R = $PhCH_2$; $PhCH_2O$.

Yield (%) = 42 ; 48

Isomer mixture = 1:1 ; 1:1

(1016)

α,β-Double bond formation in bicyclic lactams has been reported[999,1082,1182,1398,1405] (equations 1017–1021).

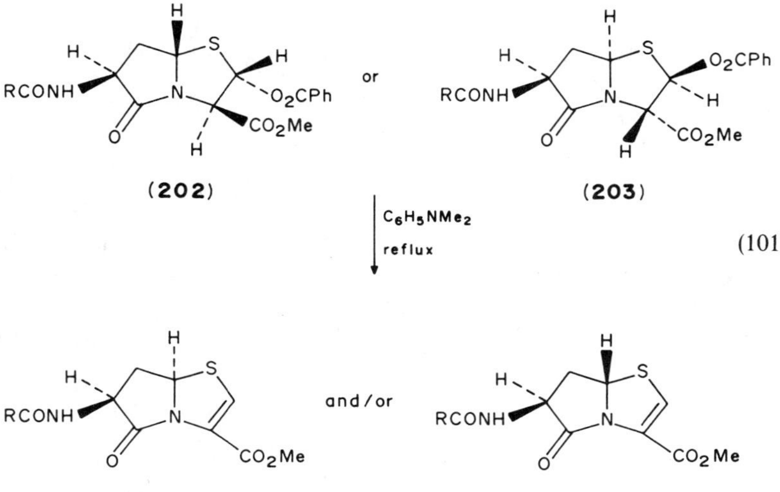

(202) or (203)

$C_6H_5NMe_2$, reflux

(1017)[999]

(204) and/or (205)

Substrate =	**202**	**203**	**203**
R =	$PhCH_2O$	$PhCH_2O$	$PhOCH_2$
Product(s) =	**204** + **205**	**205**	**205**
Yield (%) =	52[a]	—	—

[a]Product was a 4:1 isomer mixture of **204**:**205**

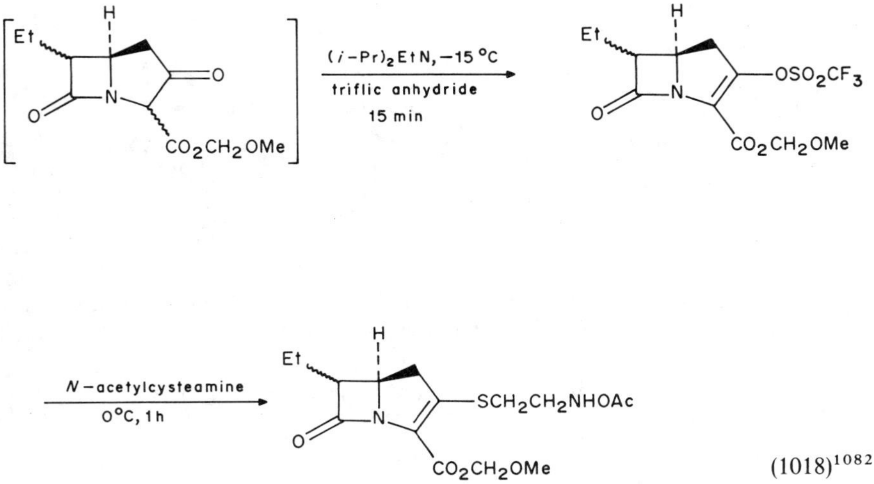

$(i\text{-Pr})_2EtN$, −15 °C, triflic anhydride, 15 min

N-acetylcysteamine, 0°C, 1 h

(1018)[1082]

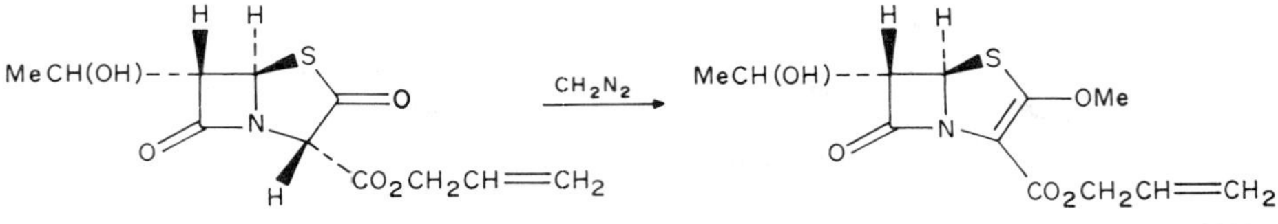

(1019)[1398]

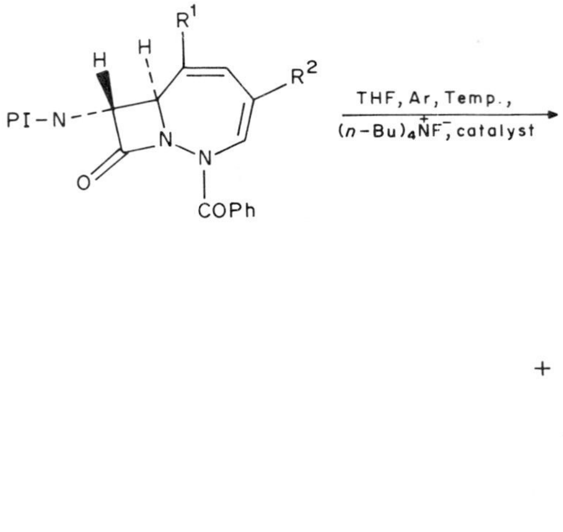

(206)

(207)

(1020)[1182]

R^1	R^2	Temp. (°C)	Catalyst	Product(s)	Yield (%)
$CO_2CH_2CH_2SiMe_3$	H	r.t.	none	**206**	52
$CO_2CH_2CH_2SiMe_3$	H	0	none	**206** + **207**	—
$CO_2CH_2CH_2SiMe_3$	H	−70	none	**206** + **207**	—
$CO_2CH_2CH_2SiMe_3$	H	r.t.	Et_3N	**206** + **207**	43 + 24
H	$CO_2CH_2CH_2SiMe_3$	r.t.	Et_3N	**206**	76

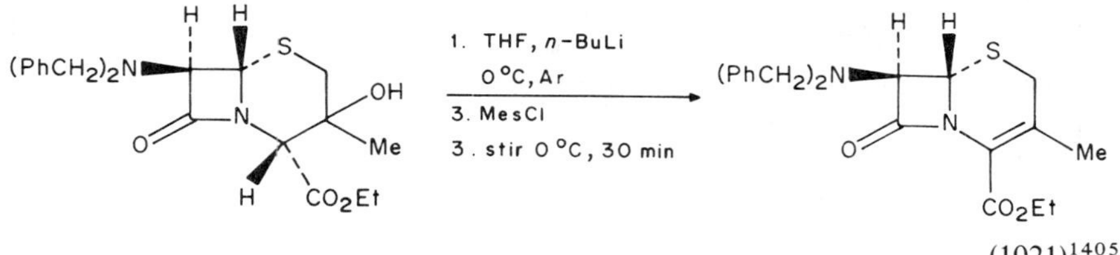

(1021)[1405]

Another method used to generate α,β-unsaturation in N-side chain azetidinones is by epimerization of β,γ-double bonds using triethylamine (equation 1022 and Table 72).

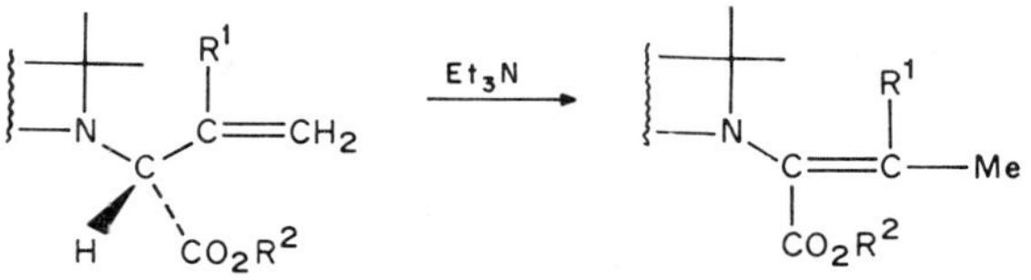

(1022)

TABLE 72.Epimerization of N-side chain β,γ- to α,β-double bonds

β,γ-Substrate	Product	Yield (%)	Reference
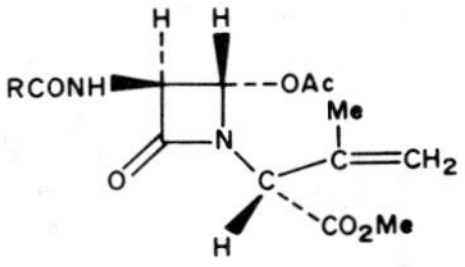 $R = t$-BuO, $PhCH_2O$	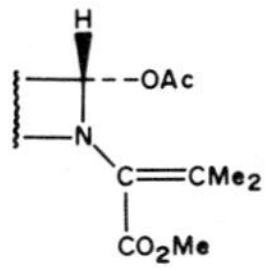	98–100	1382
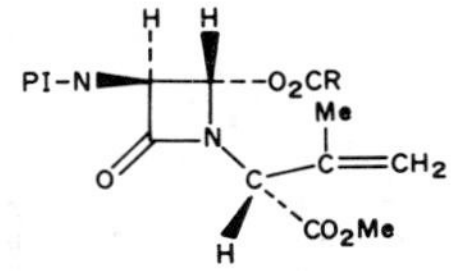 $R = Me$, p-ClC_6H_4	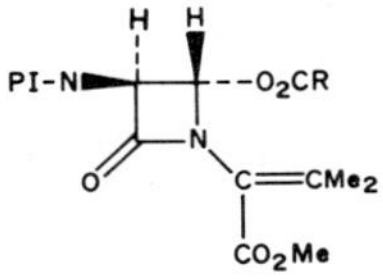	—	1387
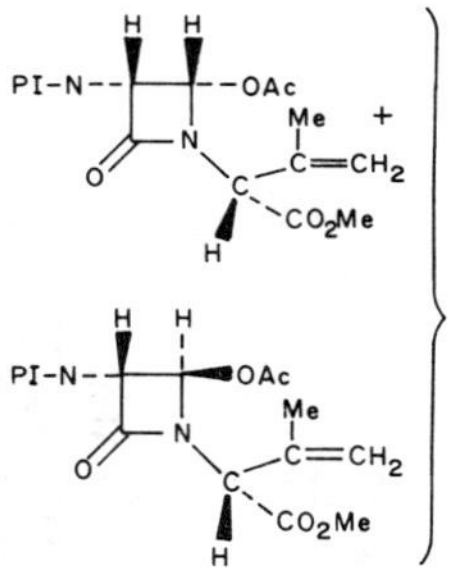	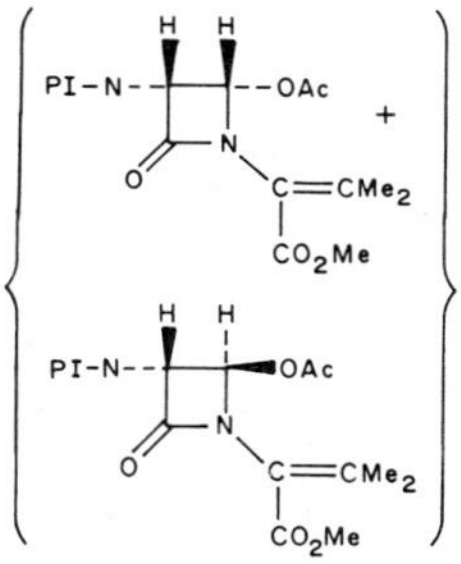	—	1387
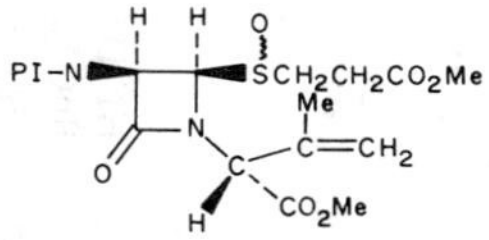 (2:1 isomer mixture)	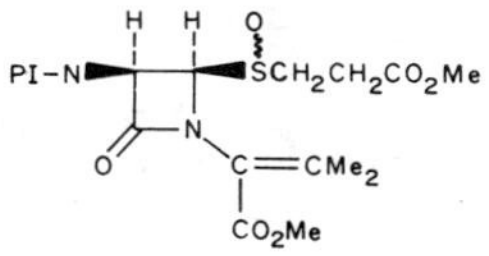 (2:1 isomer mixture)	—	1401
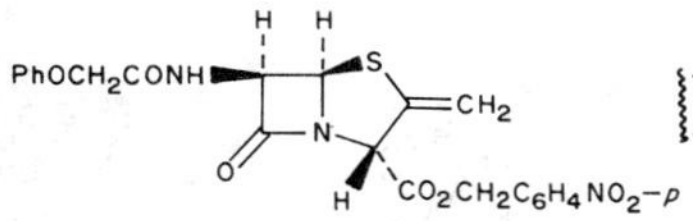	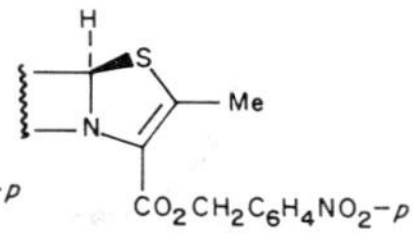	100	1402
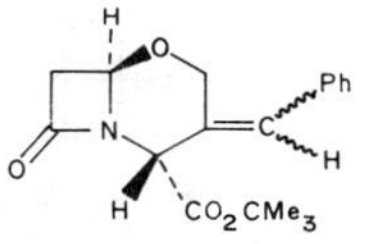 (mixture of Z and E isomers)	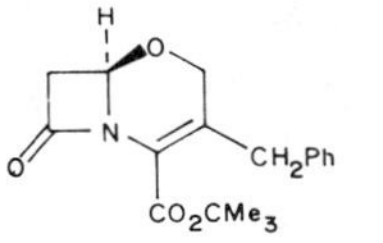	76	1336[a]

[a]The conditions used were: DMAP, CH_2Cl_2, Ar, r.t., 40 h.

TABLE 73. Ring opening of bicyclic lactams producing N-side chain α,β-unsaturation

Bicyclic substrate	Reagents and conditions	Product	Yield (%)	Reference
(t-Bu)Me₂SiO, Me, C, H, H, H, S, Me, Me, N, O, CO₂Me, H	$Hg(OAc)_2$	H, OAc, N, C=CMe₂, CO₂Me	—	1348
MeO₂CCONH, H, H, S, Me, Me, N, O, CO₂R, H R = $PhCH_2$, Ph_2CH	Cl_2, CH_2Cl_2, −20 °C	Cl, N, C=CMe₂, CO₂R	—	1403
PhCH₂CONH, H, H, S, Me, Me, N, O, CO₂CHPh₂, H	Cl_2, CH_2Cl_2, −20 °C	Cl, N, C=CMe₂, CO₂CHPh₂	—	1368
Br, H, H, S, Me, Me, N, O, CO₂Me, H	Cl_2, CH_2Cl_2, −78 °C to 0 °C, 1 h	Br, H, H, Cl, N, O, C=CMe₂, CO₂Me	86	1383
(t-Bu)Me₂SiO, Me, C, H, H, H, O, O, S, Me, Me, N, O, CO₂CH₂Ph, H	THF-DMF (2:1), MeI, t-BuOK, 30 min	H, SO₂Me, N, C=CMe₂, CO₂CH₂Ph	—	1339

Substrate	Reagents and conditions	Product	Yield (%)	Ref.
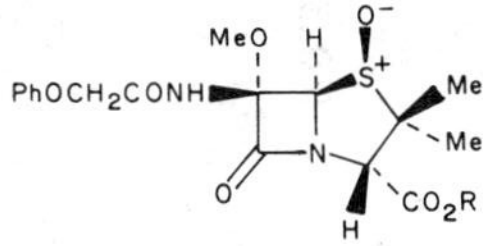 PhOCH$_2$CONH, MeO, H, O$^-$, S$^+$, Me, Me, N, O, H, CO$_2$R	1. C_6H_5Me–$(MeCO)_2O$ (5:1), Et_3P, heat, 4 h 2. Et_3N, EtOAc, heat	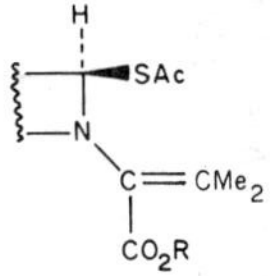 H, SAc, N, C=CMe$_2$, CO$_2$R		1341
R = Me			50	
R = CH_2Ac			—	
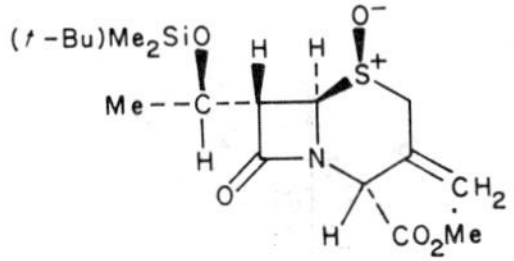 (t-Bu)Me$_2$SiO, Me–C–H, H, H, O$^-$, S$^+$, N, O, CH$_2$, H, CO$_2$Me	$(t\text{-Bu})Ph_2SiOCH_2COSH$, N_2, C_6H_5Me, reflux, 16 h	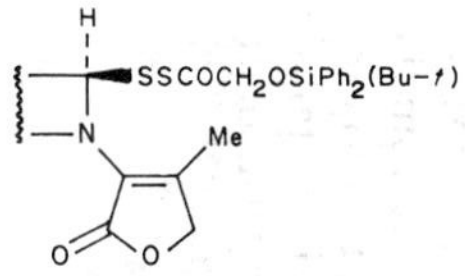 H, SSCOCH$_2$OSiPh$_2$(Bu-t), N, Me, O, O	—	1397
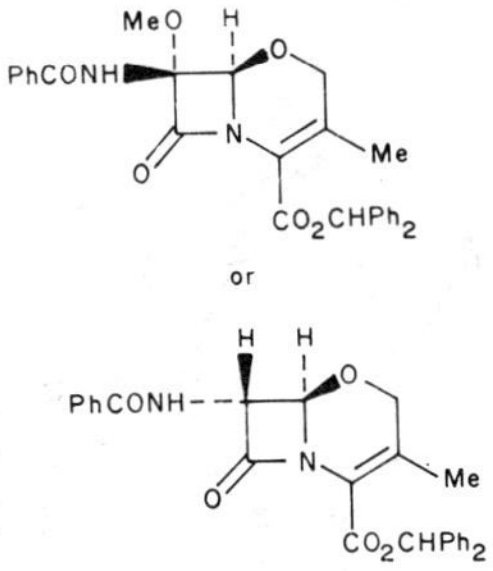 PhCONH, MeO, H, O, N, O, Me, CO$_2$CHPh$_2$ or PhCONH, H, H, O, N, O, Me, CO$_2$CHPh$_2$	NBS, AIBN, CCl_4, reflux	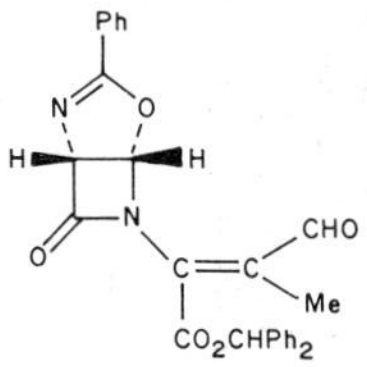 Ph, N, O, H, H, N, O, C=C, CHO, Me, CO$_2$CHPh$_2$	—	1357

Ring opening of bicyclic lactams under a variety of conditions also gives rise to the production of azetidinones containing α,β-unsaturation in the N-side chain (equation 1023 and Table 73).

reagent / conditions (1023)

The presence of an exocyclic double bond in lactams premits the preparation of a number of α-C-substituted compounds, which result from the variety of reactions which the double bond can undergo ranging from oxidation to cyclization. Beginning with the oxidation reactions, α,β-unsaturated N-side chain lactams have been converted into alcohols[1341] (equation 1024) by ozonolysis to intermediate ketones which were then reduced, aldehydes[1146] (equation 1025) or ketones[1348,1383,1387,1397] (equations 1026–1030) also by ozonolysis.

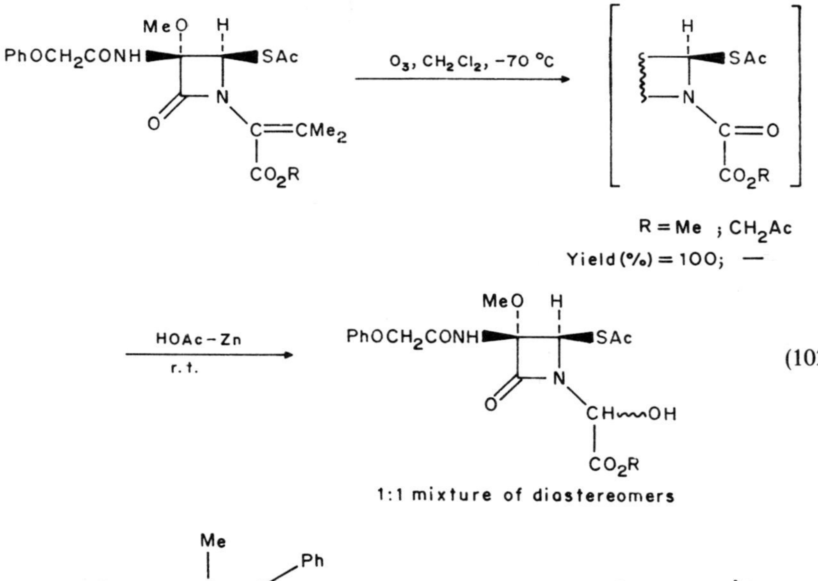

O_3, CH_2Cl_2, −70 °C

R = Me ; CH_2Ac
Yield (%) = 100; —

HOAc − Zn / r. t. (1024)

1:1 mixture of diastereomers

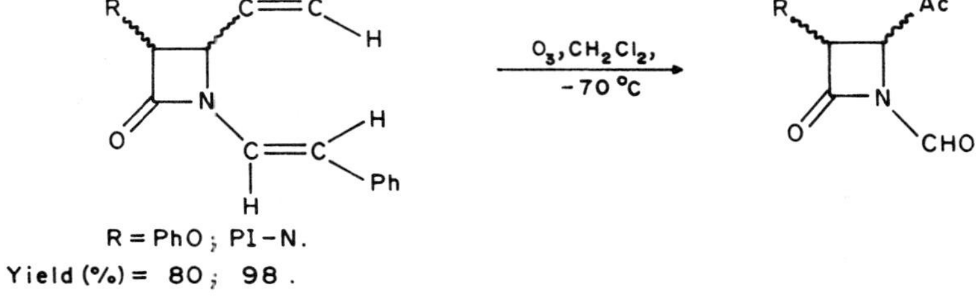

O_3, CH_2Cl_2, −70 °C (1025)

R = PhO ; PI−N.
Yield (%) = 80 ; 98 .

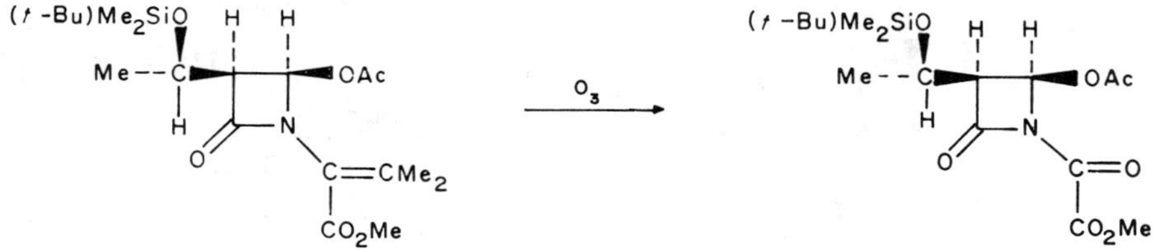

$(t\text{-Bu})Me_2SiO$ H H Me C OAc H O N $C{=}CMe_2$ CO_2Me —O_3→ $(t\text{-Bu})Me_2SiO$ H H Me C OAc H O N $C{=}O$ CO_2Me

(1026)[1348]

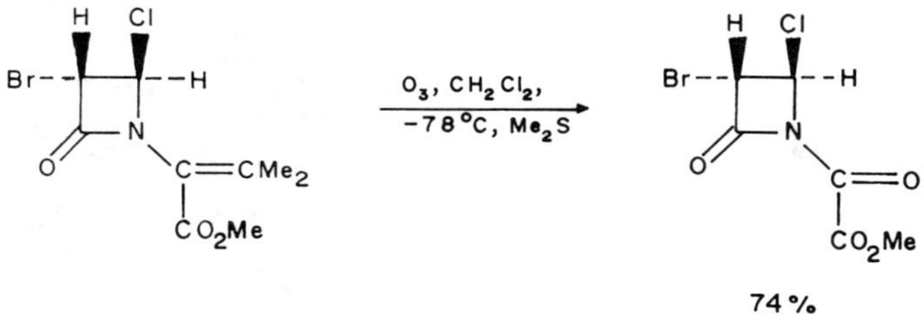

H Cl Br H O N $C{=}CMe_2$ CO_2Me —O_3, CH_2Cl_2, $-78\,^{\circ}C$, Me_2S→ H Cl Br H O N $C{=}O$ CO_2Me (1027)[1383]

74%

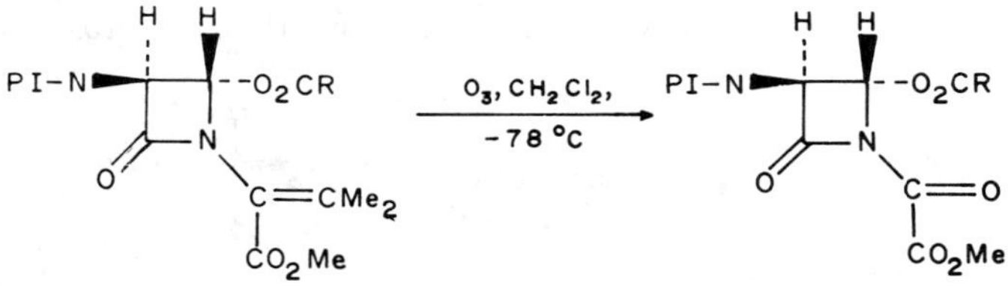

H H PI–N O_2CR O N $C{=}CMe_2$ CO_2Me —O_3, CH_2Cl_2, $-78\,^{\circ}C$→ H H PI–N O_2CR O N $C{=}O$ CO_2Me (1028)[1387]

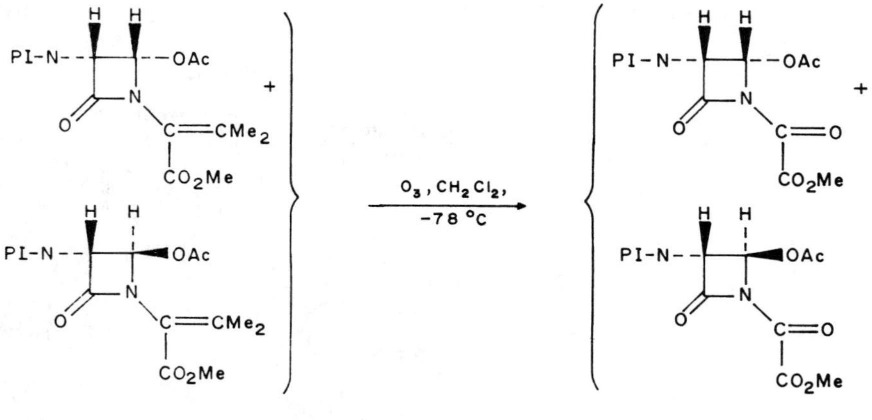

H H PI–N OAc O N $C{=}CMe_2$ CO_2Me + H H PI–N OAc O N $C{=}CMe_2$ CO_2Me —O_3, CH_2Cl_2, $-78\,^{\circ}C$→ H H PI–N OAc O N $C{=}O$ CO_2Me + H H PI–N OAc O N $C{=}O$ CO_2Me

(1029)[1387]

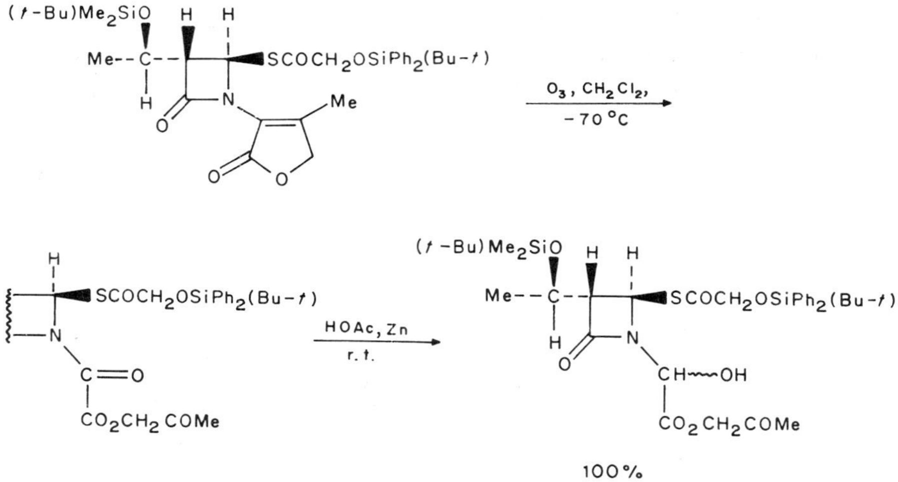

$$(1030)^{1397}$$

Treatment of N-side chain olefinic lactams with *N*-bromoacetamide (NBA) in aqueous acetronitrile affords[1384] the corresponding bromohydrins (equation 1031), while refluxing N-side chain vinyl azides produces[1304] azirines (equations 1032 and 1033).

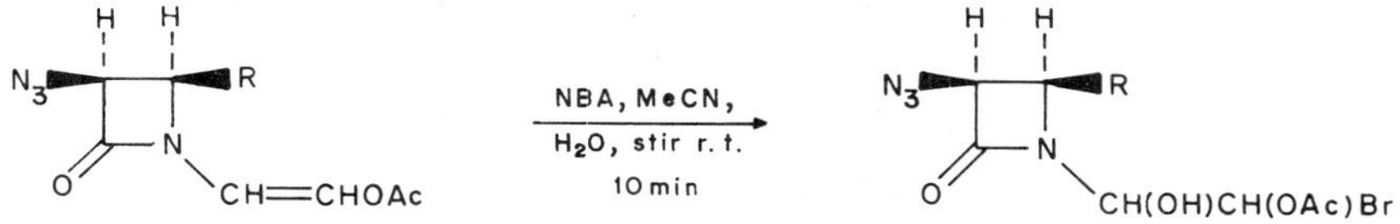

(mixture *cis*:*trans* = 2:1)

R = Ph, *p*-An, 2,3,4-$(MeO)_3$-5-$(PhCH_2O)C_6H$

(1031)

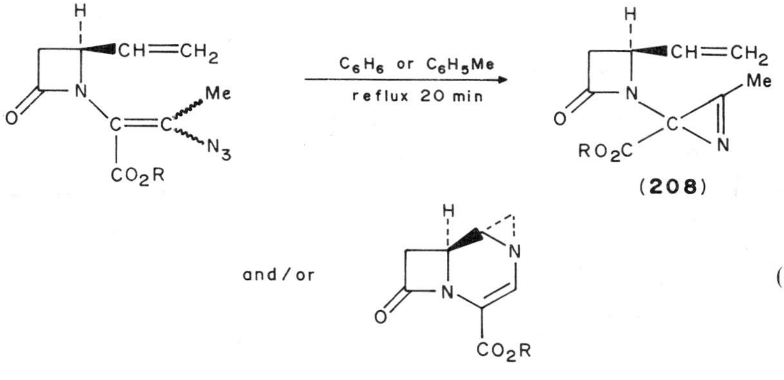

and/or

(208)

(209)

(1032)

R	Isomer	Solvent	Product(s)
t-Bu	*E*	C_6H_5Me	**208**
t-Bu	*Z*	C_6H_6	**209**
t-Bu	*E* + *Z* (2:1)	C_6H_6	**208** + **209**
$SiPh_2$(Bu-*t*)	*E* + *Z* (1:1)	C_6H_5Me	**208** + **209**
CH_2CCl_3	*E* + *Z* (9:1)	C_6H_6	**209**

C_6H_5Me, reflux

(2 isomers)

(1033)

In addition to the method illustrated in equations 1032 and 1033 to effect cyclization, fused-ring bicyclic lactams have also been prepared by several other cyclization methods most of which utilize a double bond present on the N-α-carbon atom. For example, lactams containing an α-carbon to phosphorus double bond have been used to prepare bicyclic lactams via an intramolecular Wittig reaction, where the new ring contains all carbons, sulphur or oxygen depending upon the atoms present in the carbonyl-containing side chain at the 4-position of the lactam ring (equation 1034 and Table 74).

reaction conditions

(1034)

X = $(CH_2)_n$, S, O

Similar cyclization reactions to produce bicyclic lactams have also reported to occur by reaction of lactams containing a saturated α-carbon side chain if a suitable reactive function is present on the α-carbon. Functions reported[1336] to be useful for this purpose

TABLE 74. Preparation of fused-ring lactams via intermolecular Wittig reactions

Lactam	Reaction conditions	Product	Yield (%)	Reference
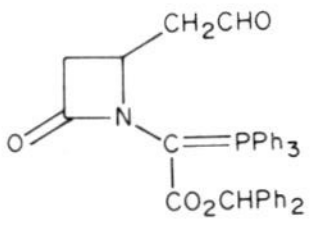 CH_2CHO, N, O, $C{=}PPh_3$, CO_2CHPh_2	aq. $NaHCO_3$, EtOAc	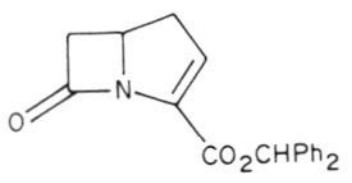 O, N, CO_2CHPh_2	51	1404
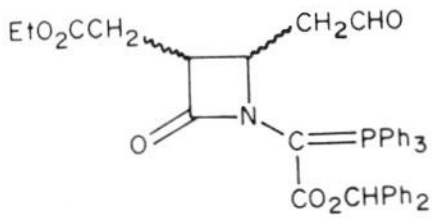 EtO_2CCH_2, CH_2CHO, O, N, $C{=}PPh_3$, CO_2CHPh_2 (6β, 5α or 6α, 5α)	aq. $NaHCO_3$, EtOAc	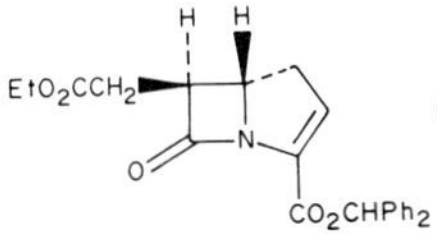 H, H, EtO_2CCH_2, O, N, CO_2CHPh_2 (from 6β, 5α) or	60	1404
		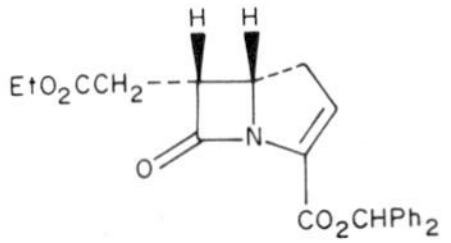 H, H, EtO_2CCH_2, O, N, CO_2CHPh_2 (from 6α, 5α)	64	1404
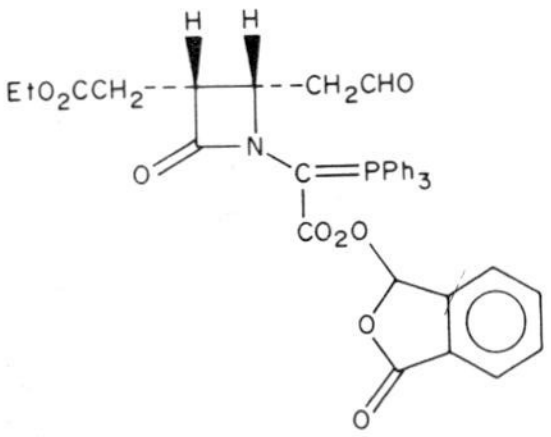 H, H, EtO_2CCH_2, CH_2CHO, O, N, $C{=}PPh_3$, CO_2O, O, O	aq. $NaHCO_3$, EtOAc	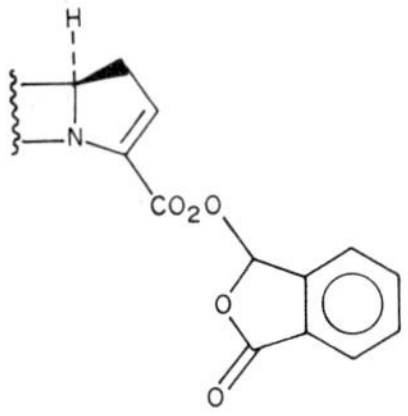 H, N, CO_2O, O, O	—	1404

Substrate	Conditions	Product	Yield (%)	Ref.
MeCH(OH), H, COMe, $C{=}PPh_3$, $CO_2CH_2CH{=}CH_2$	xylene, reflux 14 h	Me, $CO_2CH_2CH{=}CH_2$	—	1345
$(t\text{-Bu})Me_2SiO$, Me, H, Ac, $C{=}CH_2$, $C{=}PPh_3$, $CO_2CH_2C_6H_4NO_2\text{-}p$	C_6H_5Me, reflux 24 h	CH_2, Me, $CO_2CH_2C_6H_4NO_2\text{-}p$	79	1345
MeCH(OH), H, Ac, $C{=}CH_2$, $C{=}PPh_3$, $CO_2CH_2C_6H_4NO_2\text{-}p$	C_6H_5Me, reflux 24 h	No reaction	—	1345
$p\text{-}O_2NC_6H_4CH_2O_2CO$, Me, H, OAc, COS, N, $C{=}PPh_3$, $CO_2CH_2C_6H_4NO_2\text{-}p$	C_6H_5Me, reflux 2.5 h	H, OAc, S, N, $CO_2CH_2C_6H_4NO_2\text{-}p$	68	1319
$OHCCH_2$, H, O_2CPh, CHCHO, $C{=}PPh_3$, $CO_2CH_2C_6H_4NO_2\text{-}p$	$NaHCO_3$, heat	H, O_2CPh, $CO_2CH_2C_6H_4NO_2\text{-}p$	59	1319

(continued)

TABLE 74. (*continued*)

Lactam	Reaction conditions	Product	Yield (%)	Reference
$SCOCH_2CO_2Et$; N–C(=PPh_3)CO_2R	C_6H_5Me, reflux 2-3 h	=$CHCO_2Et$; CO_2R	75–82	1338
R = Me, $PhCH_2$, CH_2OAc		(*E*)		
$SCOCH_2CH_2OAc$; N–C(=PPh_3)CO_2R	C_6H_5Me, reflux	CH_2CH_2OAc; CO_2R	65	1338
R = Me, $PhCH_2$, CH_2OAc				
MeO, H, $PhOCH_2CONH$, SAc; N–C(=PPh_3)CO_2R	C_6H_5Me, heat	H, S, Me; CO_2R		1341
R = Me	80 °C, 4 h		52	
R = CH_2Ac	1. 83 °C, 20 h 2. 93 °C, 9 h		—	
R = $CH_2C_6H_4NO_2$-*p*	95 °C, 11 h		—	
(*t*-Bu)Me_2SiO, H, H, Me, H, $SCOCH_2OSiPh_2$(Bu-*t*); N–C(=PPh_3)CO_2CH_2COMe	C_6H_5Me, reflux 120 °C, 8 h	H, S, CH_2OSiPh_2(Bu-*t*); CO_2CH_2COMe	82	1397

Reactant	Conditions	Product	Yield (%)	Ref.
$CH_2{=}CHCH_2O_2CCO_2$, Me, H, H, $SCOCH_2NHCO_2CH_2CH{=}CH_2$, $C{=}PPh_3$, $CO_2CH_2CH{=}CH_2$	—	H, S, $CH_2NHCO_2CH_2CH{=}CH_2$, N, $CO_2CH_2CH{=}CH_2$	—	1306
$(t\text{-}Bu)Me_2SiO$, Me, H, H, SCOFu-2, $C{=}PPh_3$, $CO_2CH_2CH{=}CH_2$	C_6H_5Me, reflux 16 h	H, S, Fu-2, N, $CO_2CH_2CH{=}CH_2$	—	1344
$(t\text{-}Bu)Me_2SiO$, Me, H, H, $SCOCH_2S$–(N–N=N–N–Me tetrazolyl), $C{=}PPh_3$, $CO_2CH(OPh)Me$	—	H, S, CH_2S–(tetrazolyl, N–Me), N, $CO_2CH(OPh)Me$	—	1342
O, N, SAc, $C{=}PPh_3$, $CO_2CH_2C_6H_4NO_2\text{-}p$	C_6H_5Me, 80 °C, 17 h	S, Me, N, O, $CO_2CH_2C_6H_4NO_2\text{-}p$	—	1346
$t\text{-}BuO_2CONH$, O, N, SAc, $C{=}PPh_3$, $CO_2CH_2C_6H_4NO_2\text{-}p$	C_6H_5Me, 80 °C	H, H, S, Me, $t\text{-}BuO_2CONH$, N, O, $CO_2CH_2C_6H_4NO_2\text{-}p$ +	4	1346

(continued)

TABLE 74. (*continued*)

Lactam	Reaction conditions	Product	Yield (%)	Reference
		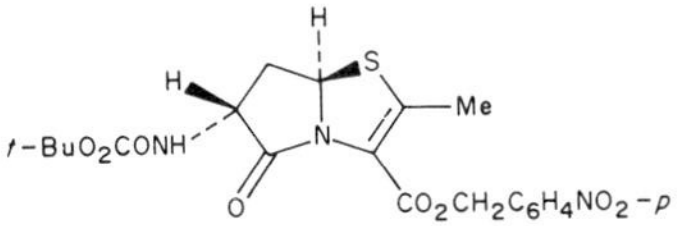 t-BuO_2CONH, H, S, Me, N, O, $CO_2CH_2C_6H_4NO_2$-p	15	
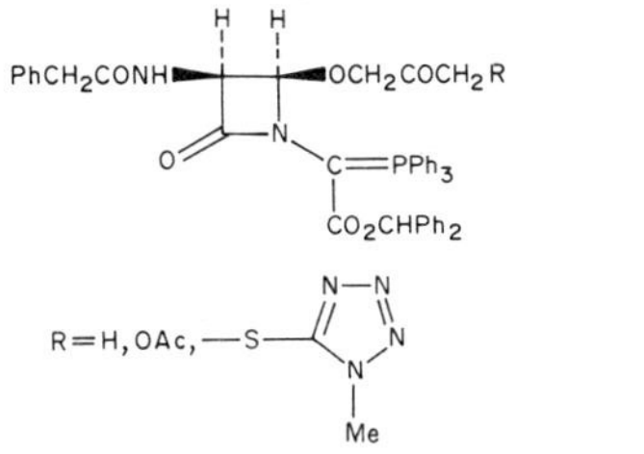 $PhCH_2CONH$, H, H, OCH_2COCH_2R, O, N, $C{=}PPh_3$, CO_2CHPh_2 $R{=}H, OAc, —S—$ (N—N, N, N, Me)	—	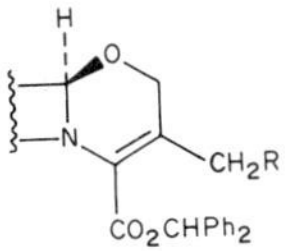 H, O, N, CH_2R, CO_2CHPh_2	—	1403
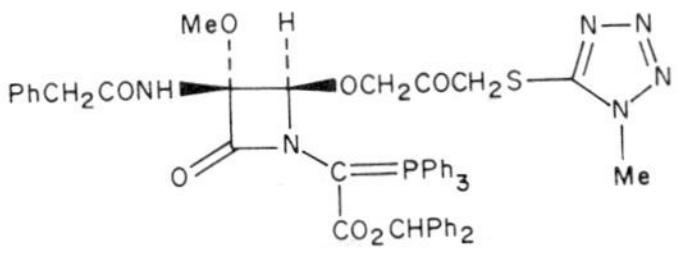 MeO, H, $PhCH_2CONH$, OCH_2COCH_2S— (N—N, N, N, Me), O, N, $C{=}PPh_3$, CO_2CHPh_2	dioxane, reflux	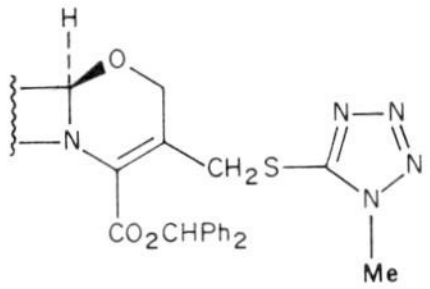 H, O, N, CH_2S— (N—N, N, N, Me), CO_2CHPh_2	—	1368

include halides, phenyl selenides and phenyl sulphides, all of which produce oxabicyclo β-lactams upon treatment with tri-(n-butyl)tin hydride and azobisisobutyronitrile (AIBN) (equation 1035 and Table 75).

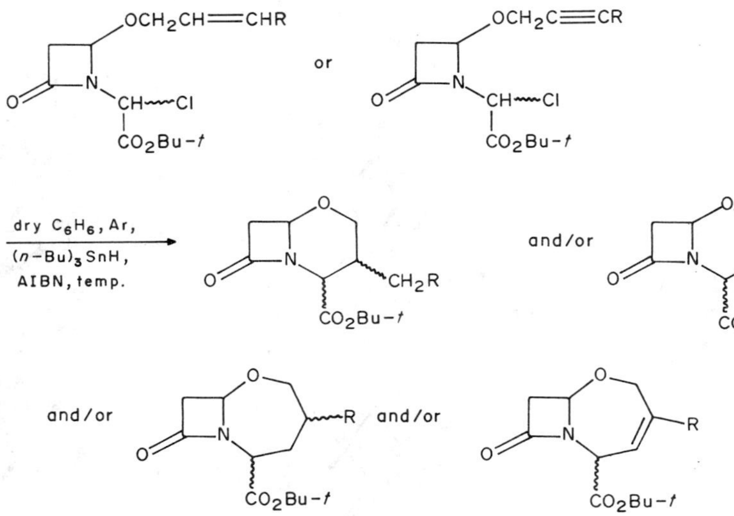

OCH2CH=CHR or OCH2C≡CR; CH~Cl, CO2Bu-t; dry C_6H_6, Ar, $(n\text{-Bu})_3SnH$, AIBN, temp.; CH2R and/or CHR and/or R and/or R; CO2Bu-t

(1035)

Another function found to be useful was the azido function, since thermolysis of β-lactams containing this function on the alpha carbon produced azabicyclo β-lactams[1196,1335] (equations 1036, 1037 and Table 76) and thia- and oxaazabicyclo β-lactams[1337] (equation 1038 and Table 76), all probably resulting from their corresponding 1,2,3-triazoline intermediates.

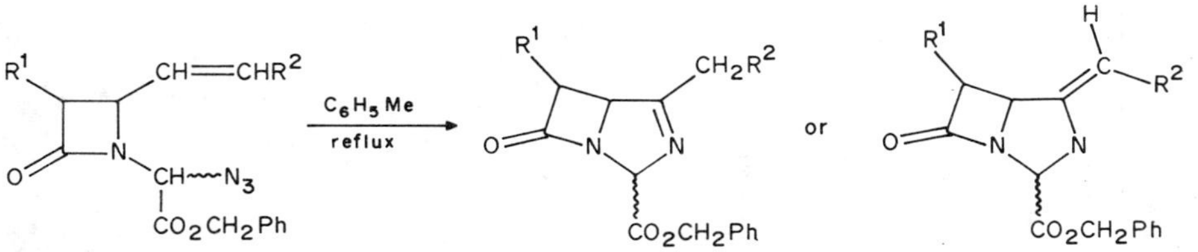

R^1, CH=CHR^2, CH~N3, CO2CH2Ph; C_6H_5Me, reflux; CH2R^2 or H, C, R^2; CO2CH2Ph

(1036)

R^1, CH2CH=CHR^2, CH~N3, CO2CH2Ph; C_6H_5Me, reflux; CH2R^2 or H, C, R^2, NH; CO2CH2Ph

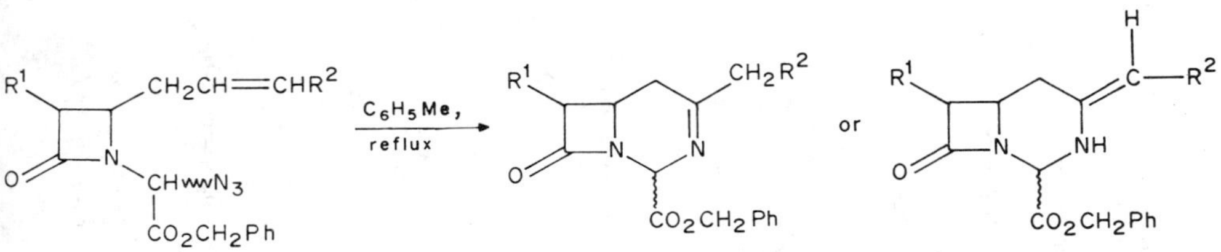

(1037)

TABLE 75. Preparation of oxabicyclo β-lactams[1336]

α-Chlorolactam	Temp. (°C)	Time (h)	Product	Yield (%)
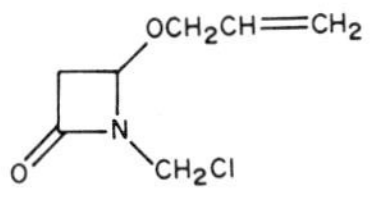 $OCH_2CH{=}CH_2$, O, N, CH_2Cl	80	44[a]	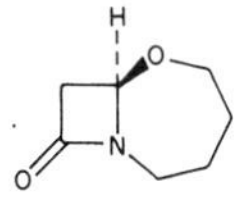 H, O, O, N	34–57
	80	90		56
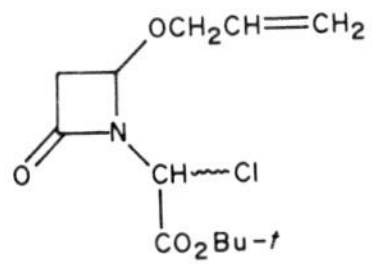 $OCH_2CH{=}CH_2$, O, N, CH—Cl, CO_2Bu-*t*	80	44	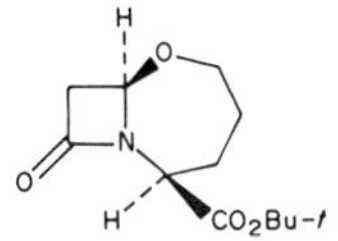 H, O, O, N, H, CO_2Bu-*t*	47
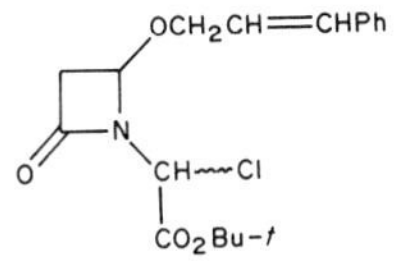 $OCH_2CH{=}CHPh$, O, N, CH—Cl, CO_2Bu-*t*	80	44	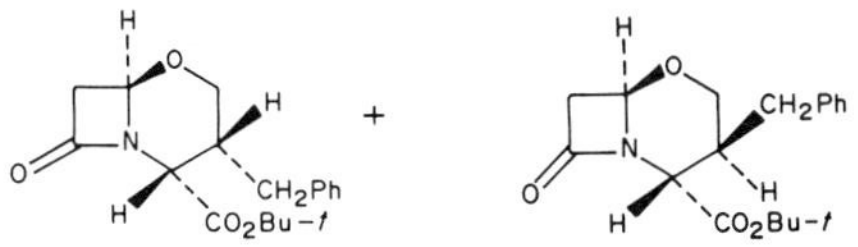 H, O, H, O, N, H, CH_2Ph, CO_2Bu-*t* + H, O, CH_2Ph, O, N, H, H, CO_2Bu-*t* (1:1 mixture)	68
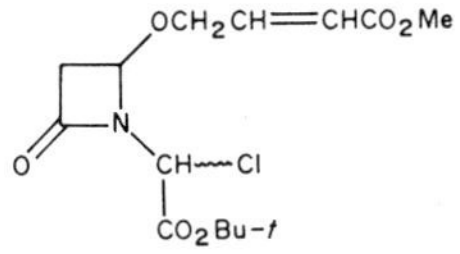 $OCH_2CH{=}CHCO_2Me$, O, N, CH—Cl, CO_2Bu-*t*	80	44	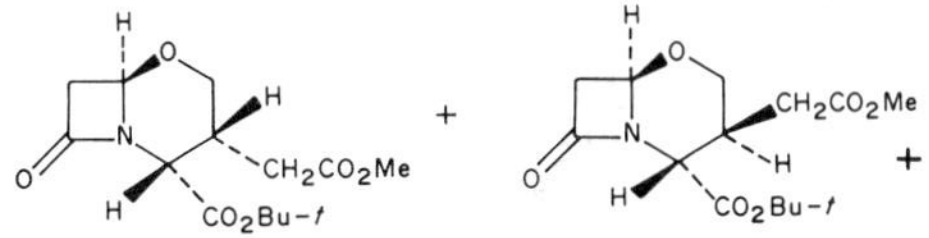 H, O, H, O, N, H, CH_2CO_2Me, CO_2Bu-*t* + H, O, CH_2CO_2Me, O, N, H, H, CO_2Bu-*t* + (3:3:1 mixture)	68

				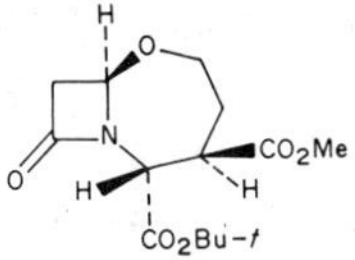	4
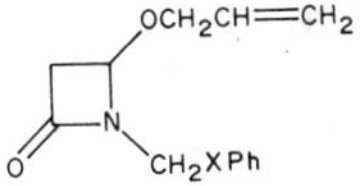	X = Se	80	44	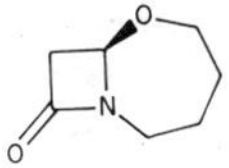	38
	X = S	15[b]	44		22
		80	44		41
		140	44		40
		80	120		56
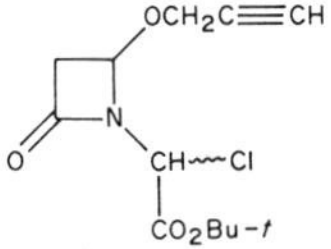		80	44	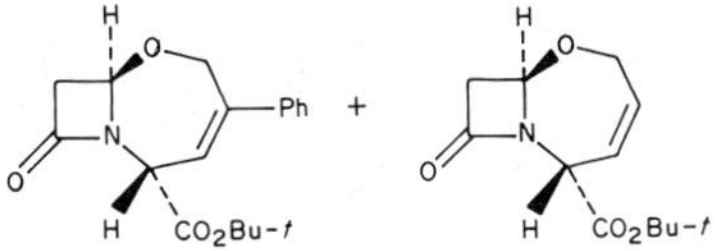 +	2.7 + 16
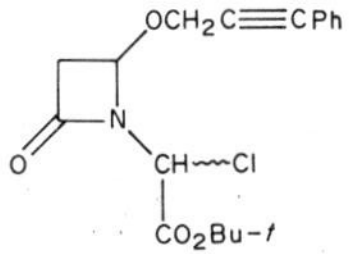		80	44	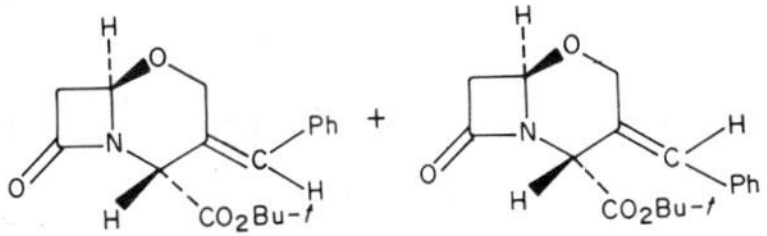 + (1.3:1 mixture)	64

[a]Concentration ratio of reactants was varied.
[b]Light was also employed.

TABLE 76. Preparation of aza-, thiaaza- and oxaazabicyclo β-lactams

α-Azidolactam	Reaction conditions	Products	Yield (%)	Reference
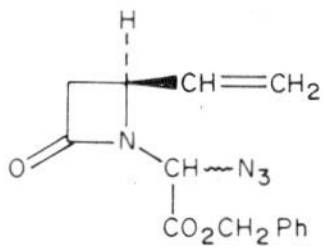	C_6H_5Me, Ar, reflux 31 h	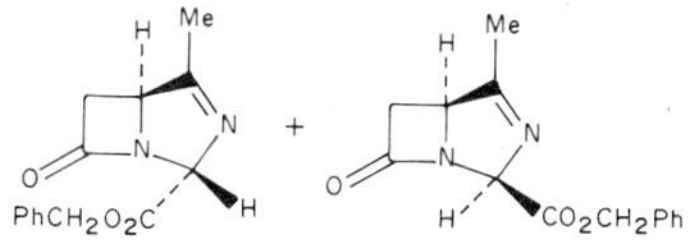 (2*RS*, 5*RS*) + (2*RS*, 5*SR*)	35 + 23	1196, 1335
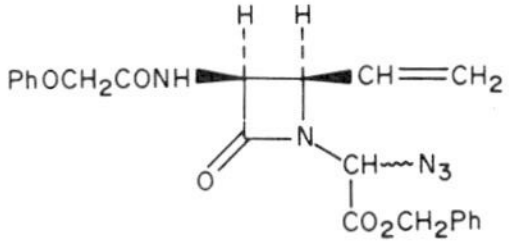	C_6H_5Me, Ar, reflux 34 h	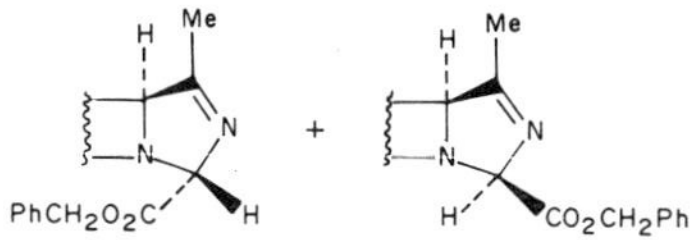 (2*RS*, 5*RS*, 6*RS*) + (2*RS*, 5*SR*, 6*SR*)	28 + 36	1196
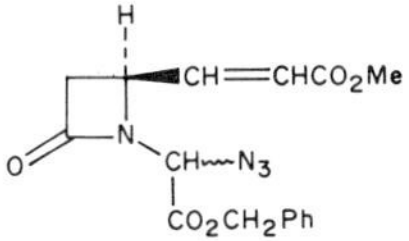	C_6H_5Me, Ar, 110 °C, 23–24 h	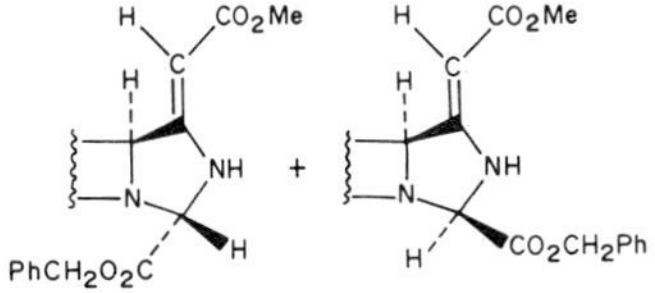 (2*RS*, 5*RS*) + (2*RS*, 5*SR*)	24 + 15	1196, 1335
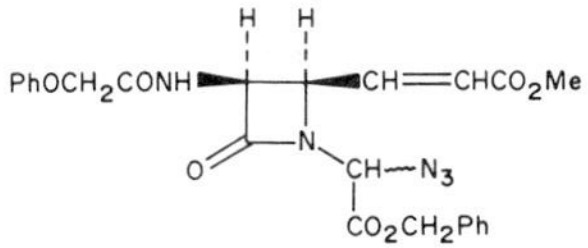	C_6H_5Me, Ar, reflux 25 h	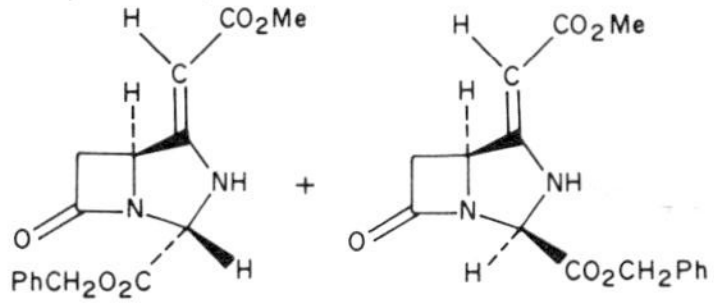 (2*RS*, 5*RS*, 6*RS*) + (2*RS*, 5*SR*, 6*SR*)	—	1196

Substrate	Conditions	Product(s)	Yield (%)	Ref.
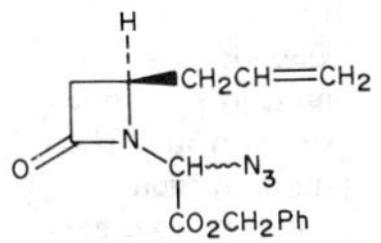 Azetidinone with $CH_2CH=CH_2$ and N–$CH(N_3)CO_2CH_2Ph$	C_6H_5Me, Ar, reflux 7 h	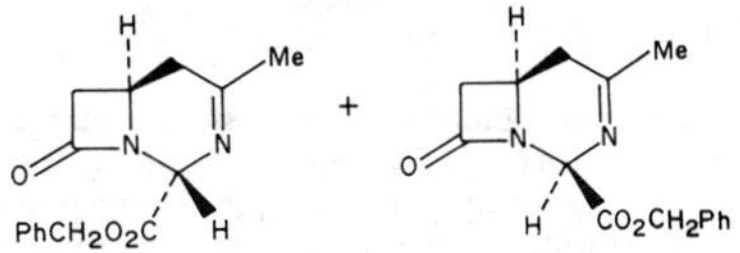 (2*RS*, 6*SR*) + (2*RS*, 6*RS*)	15 + 2[b]	1196, 1335
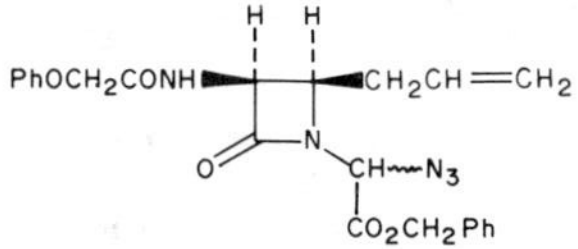 $PhOCH_2CONH$-azetidinone with $CH_2CH=CH_2$ and N–$CH(N_3)CO_2CH_2Ph$	C_6H_5Me, Ar, reflux 7 h	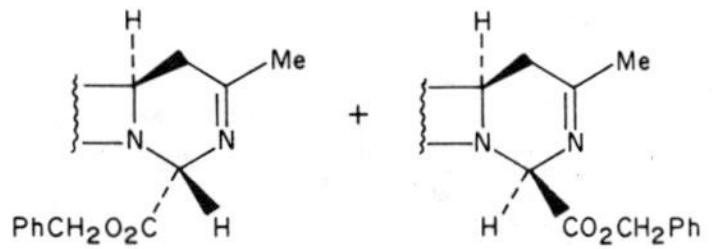 (2*RS*, 6*SR*, 7*RS*) + (2*RS*, 6*RS*, 7*SR*)	22 + 14	1196
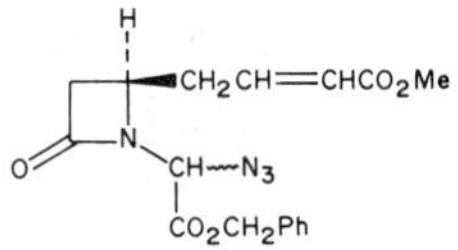 Azetidinone with $CH_2CH=CHCO_2Me$ and N–$CH(N_3)CO_2CH_2Ph$	C_6H_5Me, Ar, reflux 5.5 h	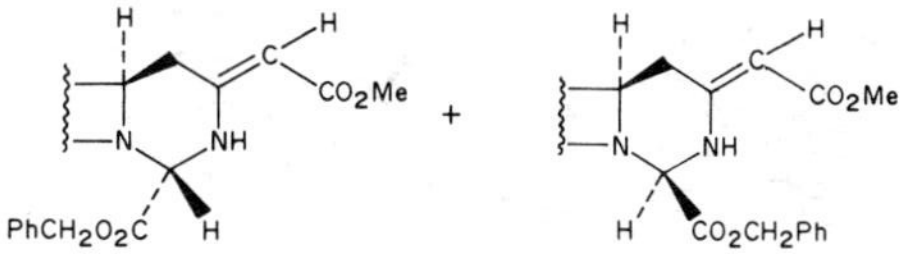 (2*RS*, 6*SR*) + (2*RS*, 6*RS*)	25 + 20[c]	1196, 1335
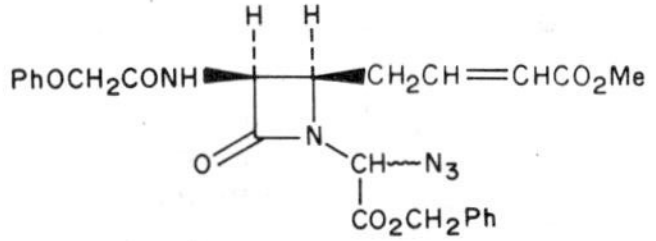 $PhOCH_2CONH$-azetidinone with $CH_2CH=CHCO_2Me$ and N–$CH(N_3)CO_2CH_2Ph$	C_6H_5Me, Ar, reflux 4 h	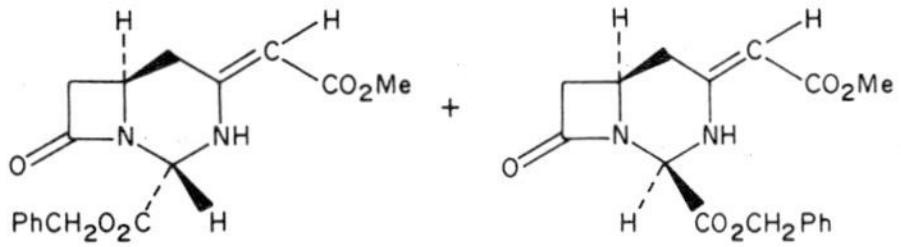 (2*RS*, 6*SR*, 7*RS*) + (2*RS*, 6*RS*, 7*SR*) (2:1 mixture)	72[d]	1196

(continued)

TABLE 76. (*continued*)

α-Azidolactam	Reaction conditions	Products	Yield (%)	Reference
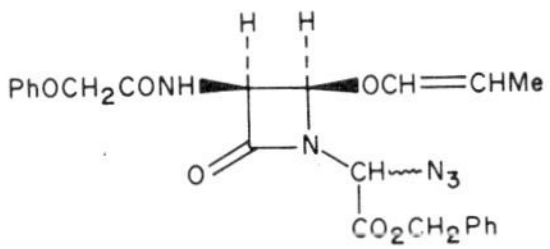 PhOCH$_2$CONH, OCH=CHMe, N, CH–N$_3$, CO$_2$CH$_2$Ph, O, H, H	1. C_6H_5Me, Ar, reflux 12 h 2. CH_2Cl_2, Ph_3P, 30 min	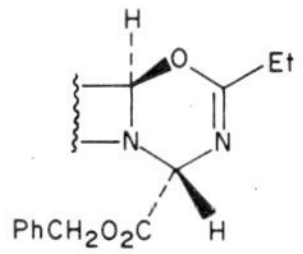 H, O, Et, N, PhCH$_2$O$_2$C, H (2*S*, 6*R*, 7*S*)	—	1337
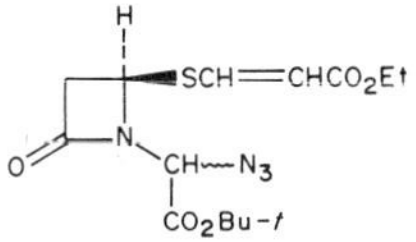 H, SCH=CHCO$_2$Et, N, CH–N$_3$, CO$_2$Bu-*t*, O *cis* or *trans*	xylene, Ar, reflux 2 h	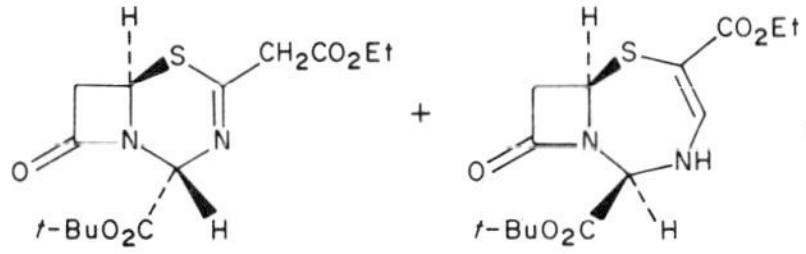 H, S, CH$_2$CO$_2$Et, N, *t*-BuO$_2$C, H, O (2*RS*, 6*SR*) + H, S, CO$_2$Et, N, NH, *t*-BuO$_2$C, H, O (2*RS*, 6*RS*) 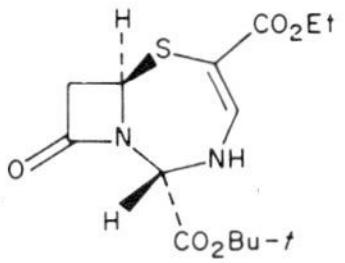 + H, S, CO$_2$Et, N, NH, H, CO$_2$Bu-*t*, O (2*RS*, 6*SR*)	—	1337

[a]Treatment of this product with DBU in CH_2Cl_2 at $-20\,°C$ for 1 h converts it to the (2*RS*, 5*RS*)-epimer shown in 50% yield.
[b]Treatment of a mixture of these two epimers as in a afforded a 30% yield of the (2*RS*, 6*SR*)-epimer exclusively.
[c]Treatment of a mixture of these two epimers as in a but at room temperature for 10 min produced a 77% yield of the (2*RS*, 6*SR*)-epimer exclusively.
[d]Total yield of both epimers which could not be separated. Treatment of this mixture as in a but at room temperature for 40 min produced a 90% yield of a 2:3 mixture of the same epimers.

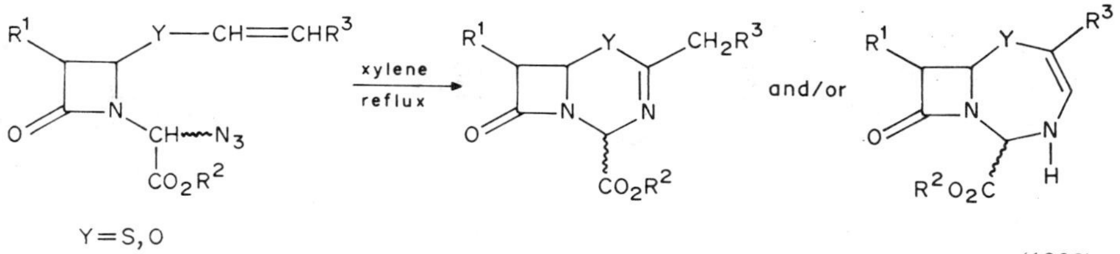

(1038)

Fused-ring bicyclic lactams where the β-lactam nitrogen atom occupies one of the fusion sites have been obtained when the active function is not attached to the alpha carbon on the nitrogen side chain but is located on a side chain carbon atom further removed from nitrogen, or is located in a side chain or on the lactam ring itself at the site adjacent to the nitrogen atom (equation 1039 and Table 77).

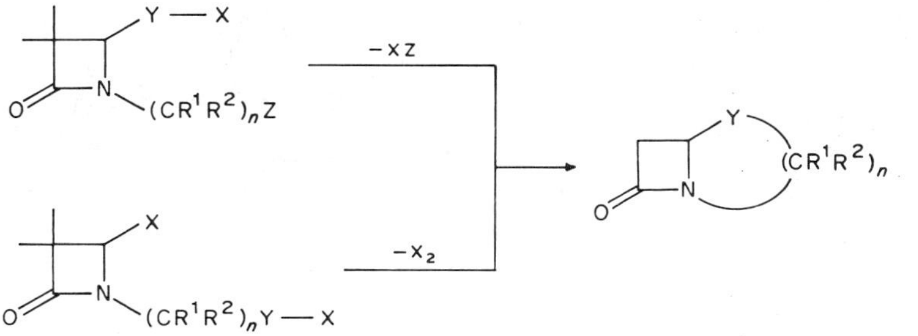

(1039)

Substitution of an alkyl of an alkoxycarbonyl group for an activated hydrogen attached to the alpha carbon of a lactam nitrogen side chain has been readily accomplished using a base. Thus, reaction of *N*-methoxycarbonylmethyl-4-methylthio-2-azetidinone in tetrahydrofuran with methyl chloroformate in the presence of two equivalents of lithium diisopropylamide (LDA) produced[1308] *N*-bis(methoxycarbonyl)methyl-4-methylthio-2-azetidinone (equation 1040).

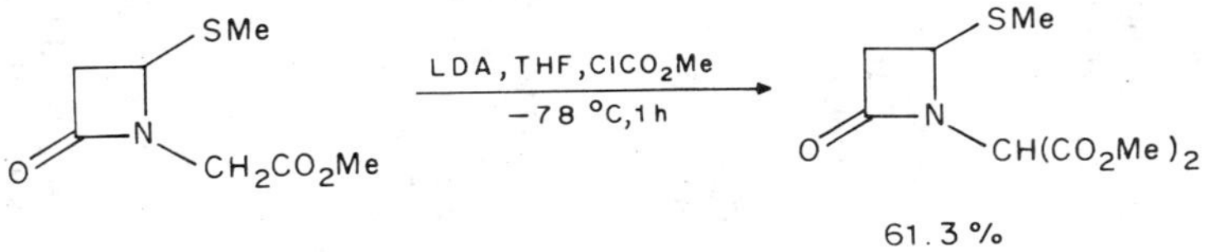

(1040)

A similar reaction is observed[1039] when ethyl α-[*p*-(benzyloxy)phenyl]-2-oxo-1-azetidineacetate is treated with one equivalent of lithium hexamethyldisilazide in tetrahydrofuran followed by one equivalent of ethyl chloroformate, producing diethyl α-[*p*-(benzyloxy)phenyl]-2-oxo-1-azetidinemalonate via selective carbethoxylation (equation 1041).

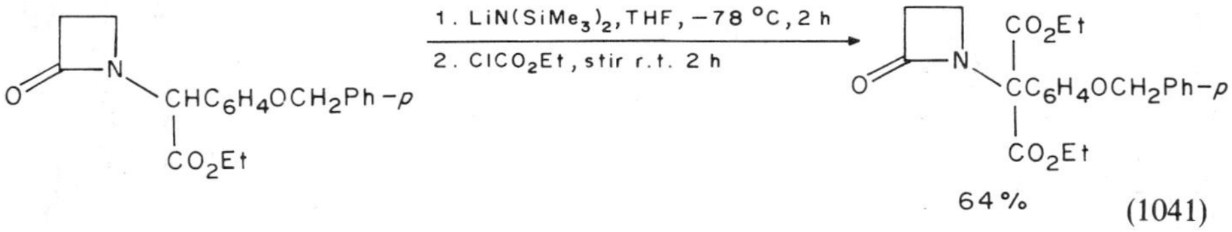

(1041)

TABLE 77. Preparation of fused ring lactams by miscellaneous methods

Substrate	Reaction conditions	Product	Yield (%)	Reference
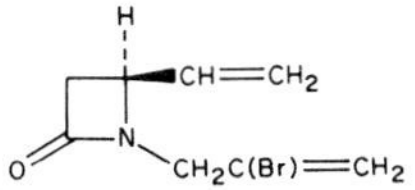 H; $CH{=}CH_2$; $CH_2C(Br){=}CH_2$	C_6H_6, $h\nu$, AIBN, $(n\text{-}Bu)_3SnH$, 6.5 h	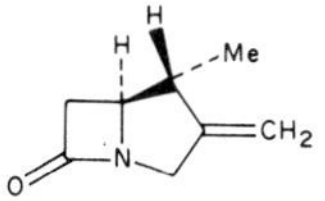 H; H; Me; CH_2	30–50	1311, 1406
	C_6H_5Me, AIBN, $(n\text{-}Bu)_3SnH$, reflux 3.5-4 days	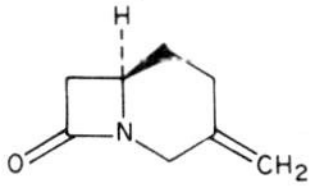 H; CH_2	58	1311, 1406
	C_6H_6, AIBN, $(n\text{-}Bu)_3SnH$, reflux 5.5 days		32	1311
	1. $Pd(OAc)_2$, Ph_3P, MeCN, N_2, r.t. 2. K_2CO_3, 80 °C, 5.5 h	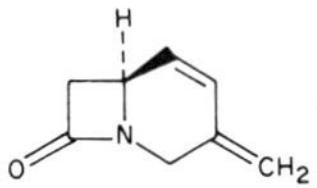 H; CH_2 +	35	1311, 1406
		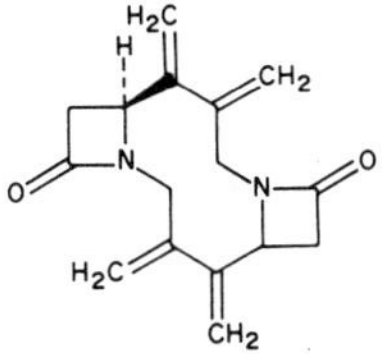 H_2C; H; CH_2; H_2C; CH_2	23	
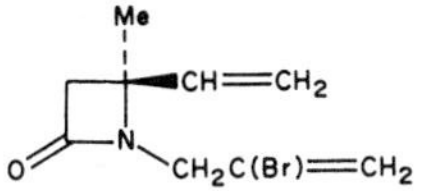 Me; $CH{=}CH_2$; $CH_2C(Br){=}CH_2$	C_6H_6, $h\nu$, AIBN, $(n\text{-}Bu)_3SnH$, 6.5 h	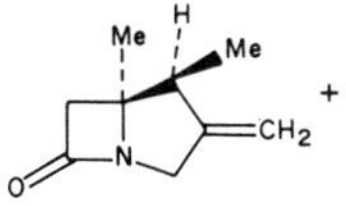 Me; H; Me; CH_2 + (210)	58	1311, 1406

Substrate	Conditions	Product	Yield (%)	Ref.
		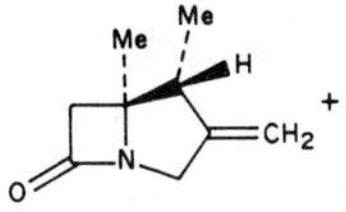 (211) +	10	
		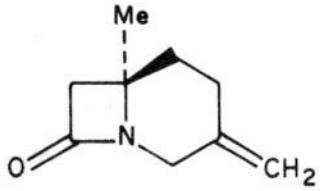 (212)	10	
	C_6H_5Me, AIBN, $(n\text{-}Bu)_3SnH$, stir reflux 2 days	**210** + **211** + **212**	59 + 3 + 30	1311, 1406
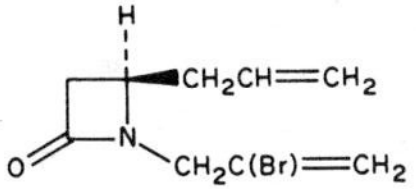	C_6H_6, $h\nu$, AIBN, $(n\text{-}Bu)_3SnH$, stir 3 h	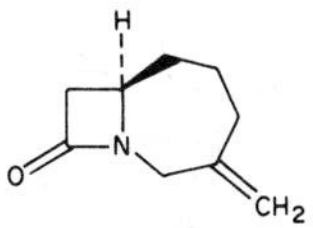	77	1311, 1406
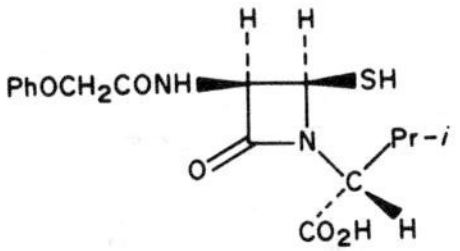	$FeSO_4$, ascorbic acid, EDTA, O_2 pH = 4.4, shake 37 °C, 2 h	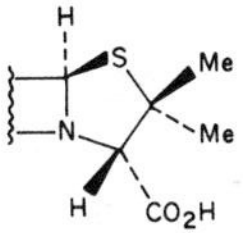 (213)	—	1407
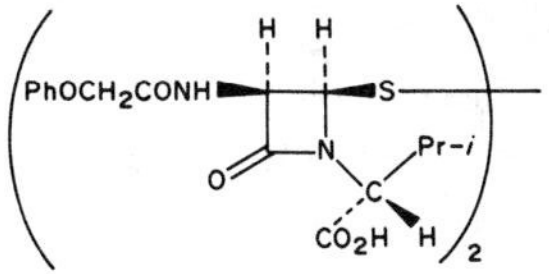	$FeSO_4$, ascorbic acid, EDTA, O_2, pH = 4.4, shake 37 °C, 2 h or as above but with H_2O_2 instead of O_2	**213**	—	1047

(*continued*)

TABLE 77. (*continued*)

Substrate	Reaction conditions	Product	Yield (%)	Reference
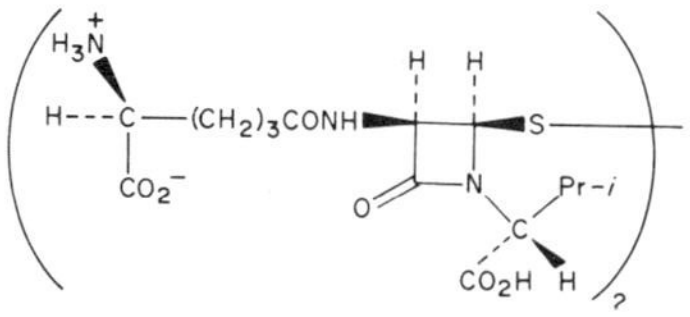 [structure]	$FeSO_4$, ascorbic acid, EDTA, O_2, pH = 4.4, shake	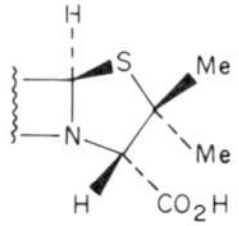 [structure]	—	1407
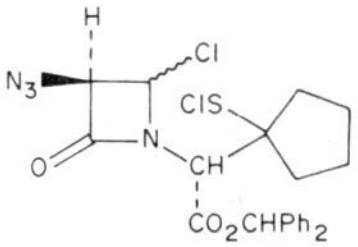 [structure]	1. dioxane, $SnCl_2$, r.t., 40 h 2. H_2S, 0 °C	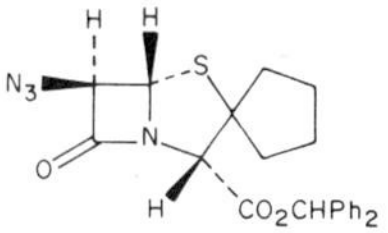 (214) +	23	1175
		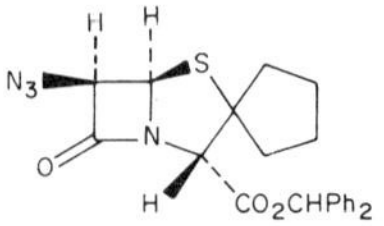 (215)	20	
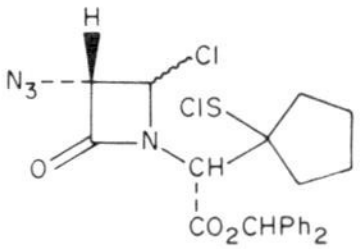 [structure]	dioxane, $SnCl_2$, r.t., 40 h	**215**	23	1175
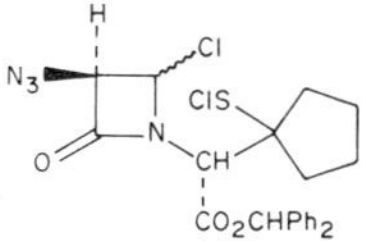 [structure]	1. dioxane, $SnCl_2$, r.t., 40 h 2. H_2S, 0 °C	**215**	28	1175

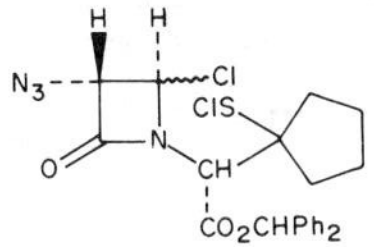	dioxane, $SnCl_2$, r.t. 40 h	**215**	28	1175
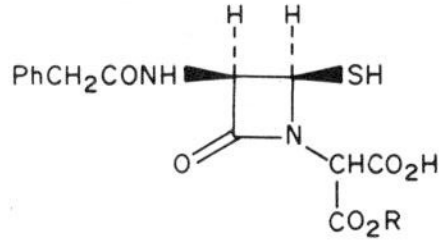	DICD, CH_2Cl_2, −60 °C to 20 °C	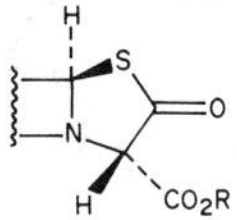		1307
R = Me			36	
R = $PhCH_2$			31	
R = *t*-Bu			65	
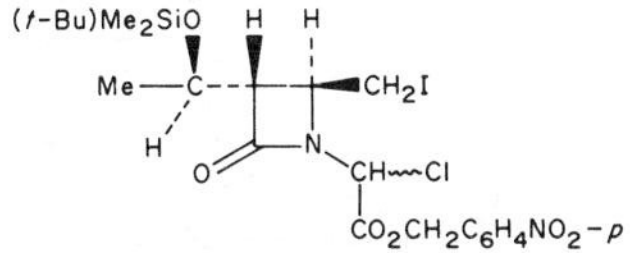	Et_3N, CH_2Cl_2, 0 °C, H_2S, 2 h	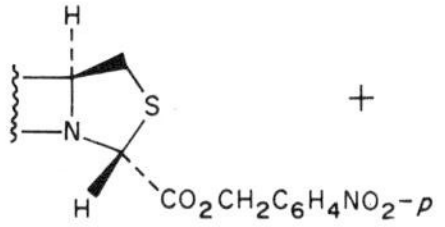 (2*R*, 5*S*, 6*S*) +	36	1339
		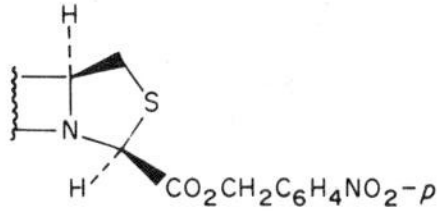 (2*S*, 5*S*, 6*S*)	60	
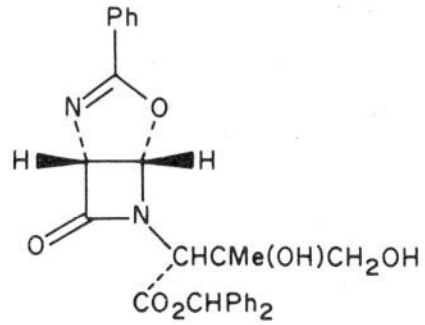	$BF_3 \cdot Et_2O$, Et_2O, CH_2Cl_2, 25 °C	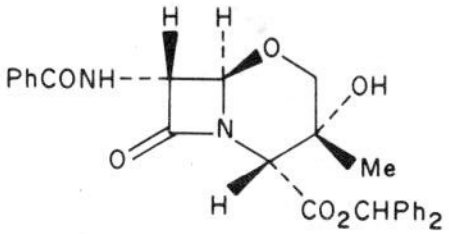 +	37	

(*continued*)

TABLE 77. *(continued)*

Substrate	Reaction conditions	Product	Yield (%)	Reference
		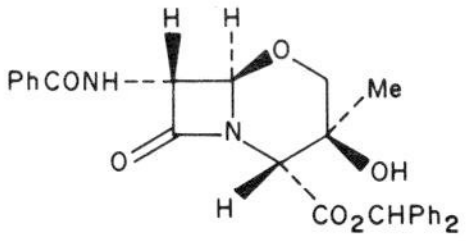	25	
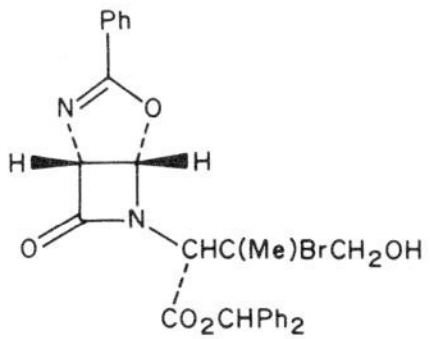	$BF_3 \cdot Et_2O$, EtOAc, CH_2Cl_2, 25 °C	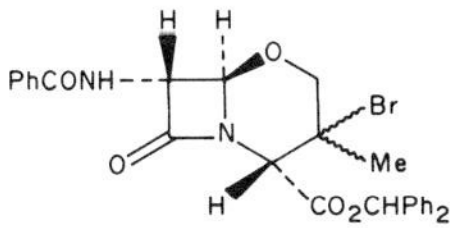	100	1357
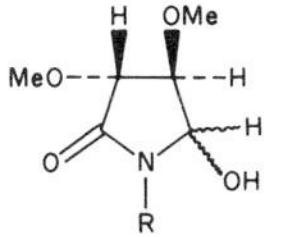	HCO_2H, r.t., 18 h			1408
$R = -(CH_2)_2C(R^2) = CHR^1$				
$R^1 = R^2 = H$		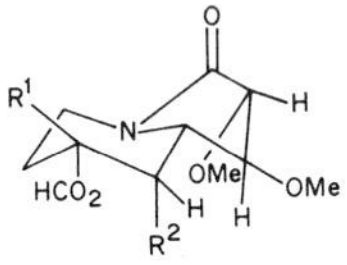 (216) + 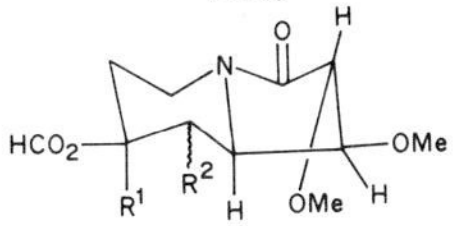 (217)	100 (2:1)	
$R^1 = Et$, $R^2 = H$		**216**	100	
$R^1 = H$, $R^2 = Me$		**216 + 217**	100 (1:5:1)	

Substrate	Conditions	Product	Yield (%)	Ref.
$R = -(CH_2)_2C\equiv CH$	HCO_2H, r.t., 118 h or HCO_2H, 43 °C, 120 h	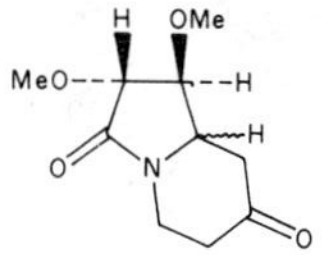	50	
$R = -(CH_2)_4C\equiv CH$	HCO_2H, 30 °C, 13 days	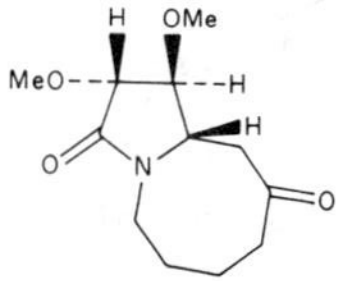	—	
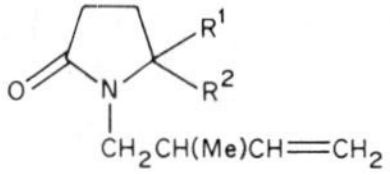	HCO_2H, r.t., 18 h	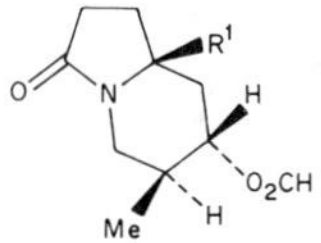		1409
$R^1 = H, R^2 = OEt$			78	
$R^1 = Me, R^2 = OH$			69	
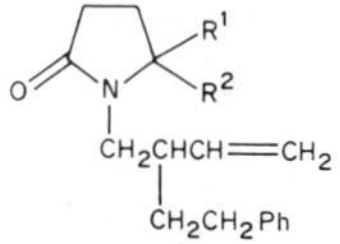	HCO_2H, r.t.	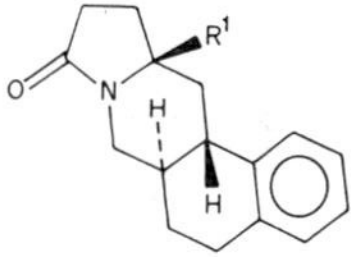		1409
$R^1 = H, R^2 = OEt$ $R^1 = Me, R^2 = OH$	18 h 120 h		91 100	
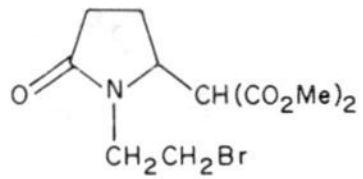	1. NaH, Et_2O 2. NaCN, DMF	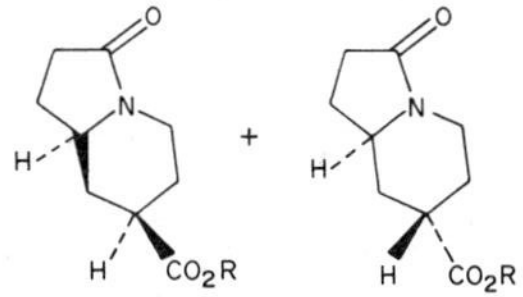 (1:1) (**218**) R = Me (**219**) R = Me	46	1410

(*continued*)

TABLE 77. (*continued*)

Substrate	Reaction conditions	Product	Yield (%)	Reference
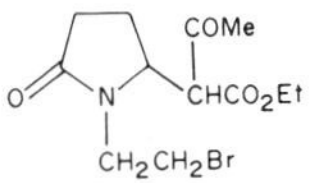 COMe; $CHCO_2Et$; N; O; CH_2CH_2Br	NaOEt, EtOH, r.t., 24 h	**218**, R = Et + **219**, R = Et (1:4)		1410
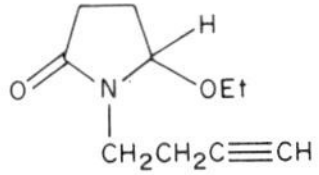 H; OEt; N; O; $CH_2CH_2C{\equiv}CH$	HCO_2H, r.t., 72 h	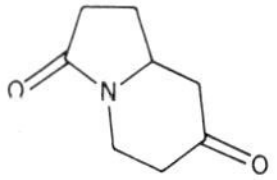 N; O; O	90	1411
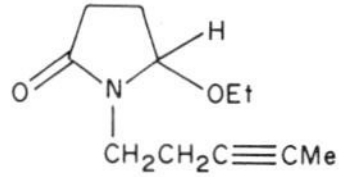 H; OEt; N; O; $CH_2CH_2C{\equiv}CMe$	HCO_2H, r.t., 72 h	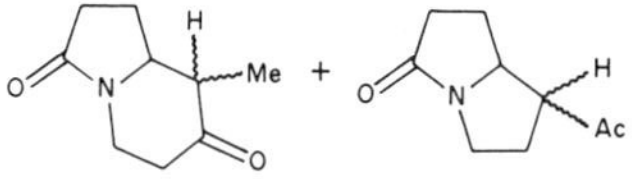 H; Me; N; O; O + H; Ac; N; O (9:1)	100	1411
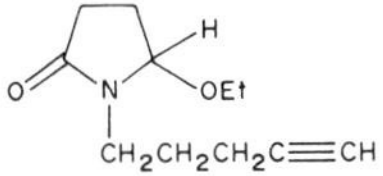 H; OEt; N; O; $CH_2CH_2CH_2C{\equiv}CH$	HCO_2H, r.t., 72 h	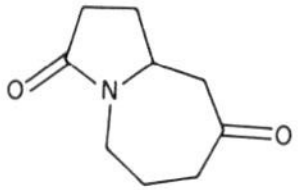 N; O; O	100	1411
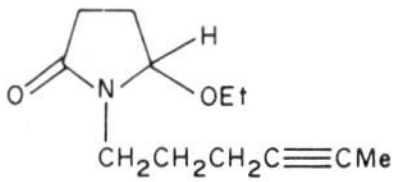 H; OEt; N; O; $CH_2CH_2CH_2C{\equiv}CMe$	HCO_2H, r.t., 72 h	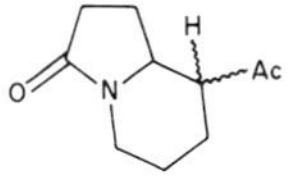 H; Ac; N; O (2 isomers 95:5)	90	1411
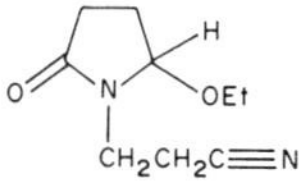 H; OEt; N; O; $CH_2CH_2C{\equiv}N$	HCO_2H, r.t., 75 h	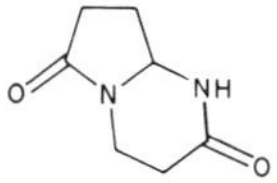 N; NH; O; O	50	1411
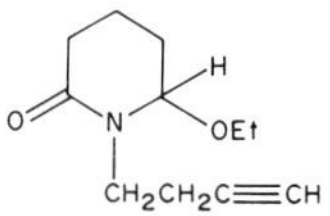 H; OEt; N; O; $CH_2CH_2C{\equiv}CH$	HCO_2H, r.t., 72 h	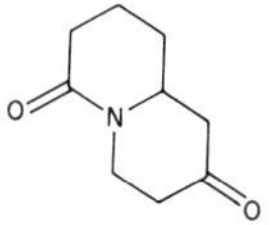 N; O; O +	90	1411

		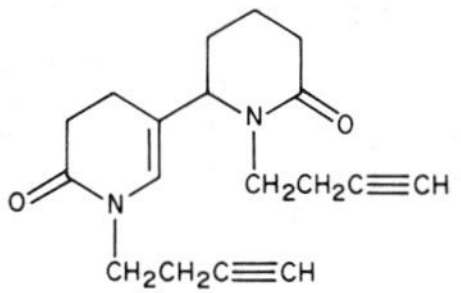	12	
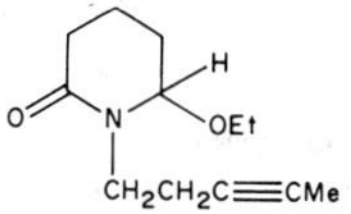	HCO_2H, r.t., 72 h	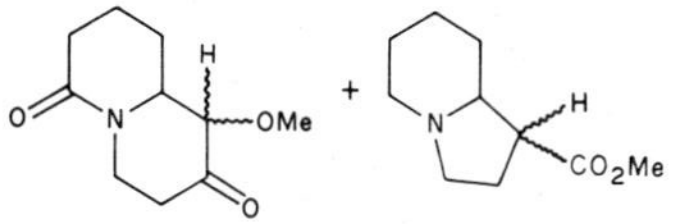 (16:84)	100	1411
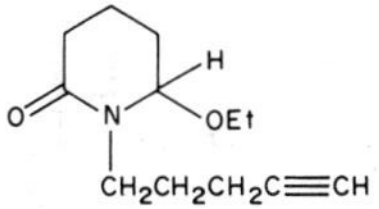	HCO_2H, r.t., 72 h	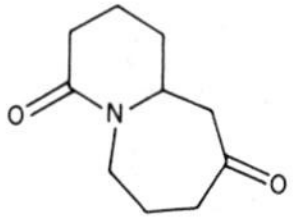	100	1411
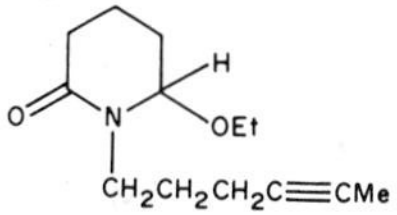	HCO_2H, r.t., 72 h	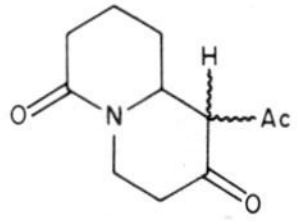 (2 isomers 85:15)	90	1411
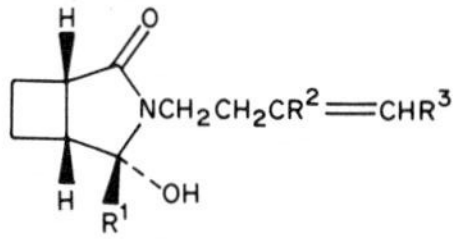	HCO_2H, r.t. stir	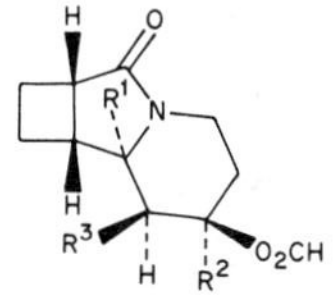		1412
$R^1 = R^2 = R^3 = H$	overnight	(C_7 epimers 90:10)	100	
$R^1 = Me$, $R^2 = R^3 = H$	overnight		100	
$R^1 = R^2 = Me$, $R^3 = H$	18 h		91	
$R^1 = R^2 = H$, $R^3 = Et$	24 h		94	

(*continued*)

TABLE 77. (*continued*)

Substrate	Reaction conditions	Product	Yield (%)	Reference
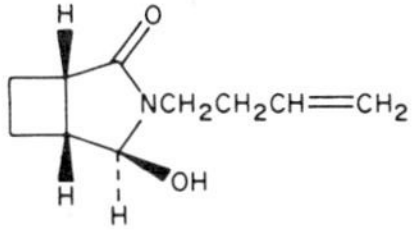 $NCH_2CH_2CH{=}CH_2$, OH	HCO_2H, r.t., stir overnight	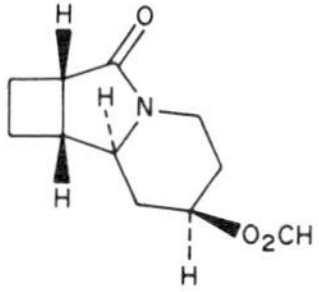 O_2CH (C$_7$ epimers 9:1)	100	1412
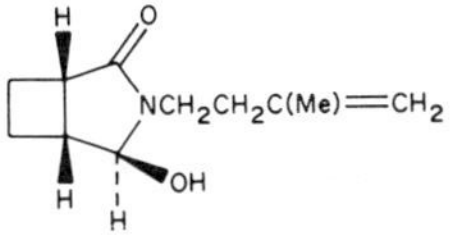 $NCH_2CH_2C(Me){=}CH_2$, OH	HCO_2H, r.t, stir 1 h	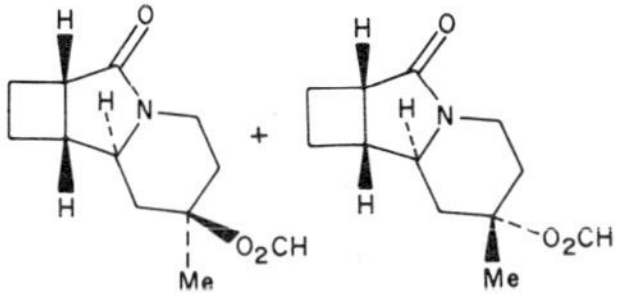 O_2CH, Me + O_2CH, Me (3:2)	100	1412
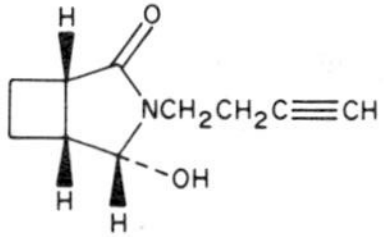 $NCH_2CH_2C{\equiv}CH$, OH	HCO_2H, r.t., stir 60 h	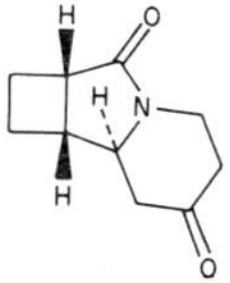	—	1412
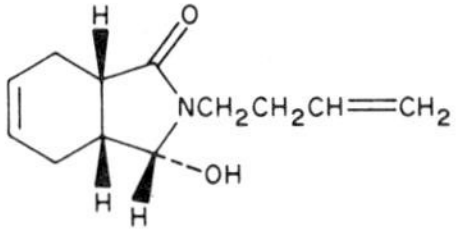 $NCH_2CH_2CH{=}CH_2$, OH	HCO_2H, r.t., stir overnight	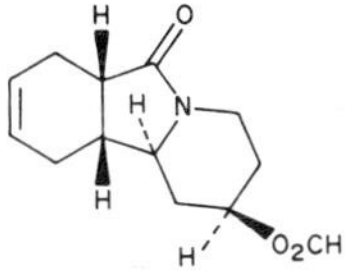 O_2CH	—	1412
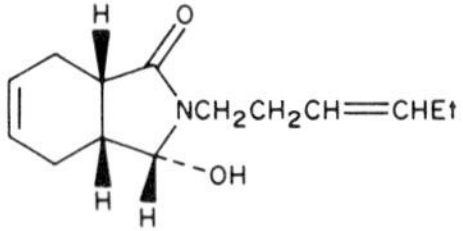 $NCH_2CH_2CH{=}CHEt$, OH	HCO_2H, r.t., stir 48 h	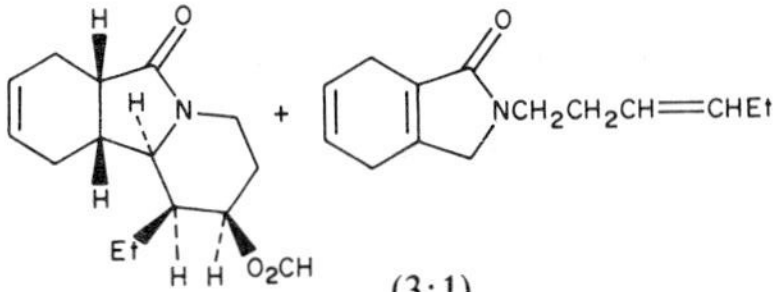 Et, O_2CH + $NCH_2CH_2CH{=}CHEt$ (3:1)	100	1412

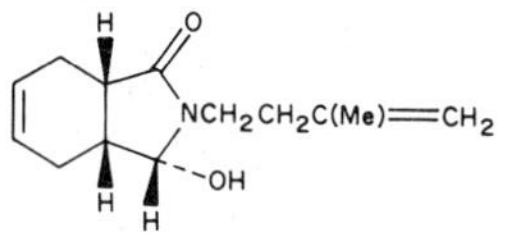 $NCH_2CH_2C(Me){=}CH_2$, OH	HCO_2H, r.t., stir 1.5 h	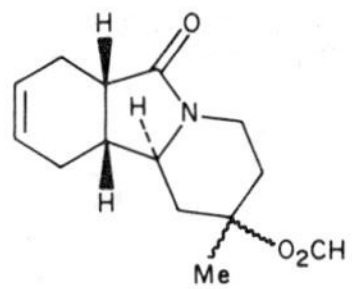 O_2CH, Me (C$_7$ epimer 3:2)	100	1412
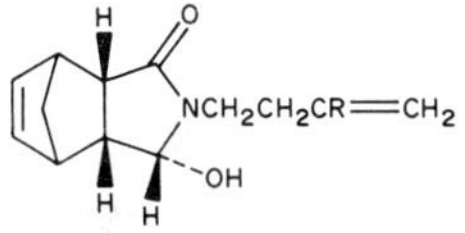 $NCH_2CH_2CR{=}CH_2$, OH	HCO_2H, r.t., stir	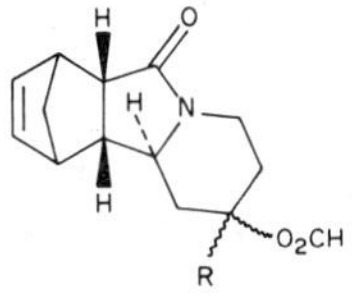 O_2CH, R		1412
R = H	overnight	(C$_7$ epimers 4:1)	—	
R = Me	12 h	(C$_7$ epimers 3:2)	100	
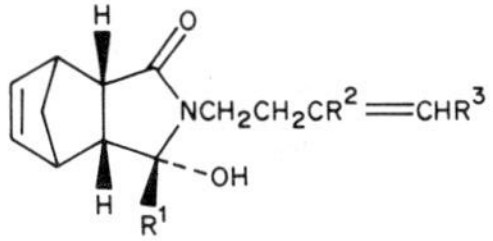 $NCH_2CH_2CR^2{=}CHR^3$, OH, R^1	HCO_2H, r.t., stir	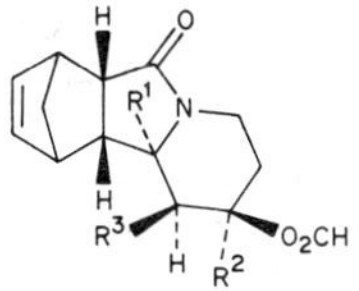 R^1, R^2, R^3, O_2CH		1412
$R^1 = Me$, $R^2 = R^3 = H$	65 h		85	
$R^1 = R^2 = Me$, $R^3 = H$	29 h		100	
$R^1 = R^2 = H$, $R^3 = Et$	18 h		100	
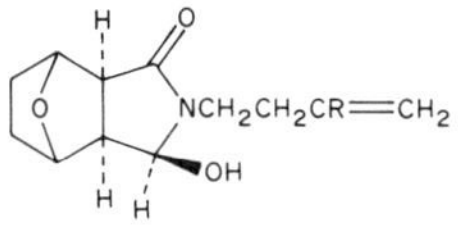 $NCH_2CH_2CR{=}CH_2$, OH	HCO_2H, r.t., stir	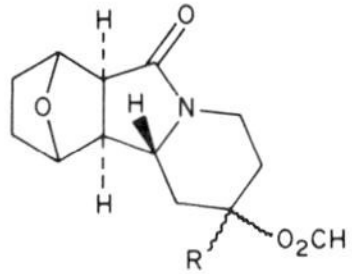 O_2CH, R		1412
R = H	overnight	(C$_7$ epimer 4:1)	100	
R = Me	1 h	(C$_7$ epimer 3:2)	100	

(continued)

TABLE 77. (*continued*)

Substrate	Reaction conditions	Product	Yield (%)	Reference
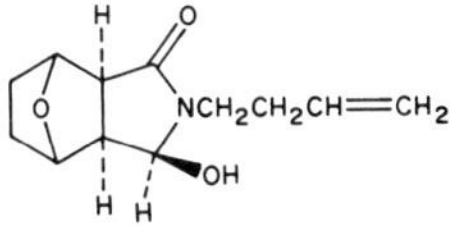 $NCH_2CH_2CH{=}CH_2$, OH	HCO_2H, r.t., stir overnight	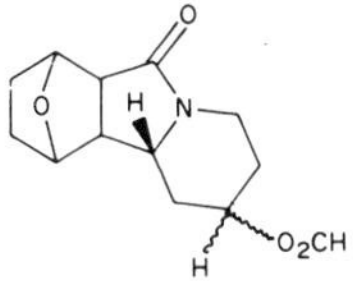 O_2CH (C$_7$ epimers 4:1)	100	1412
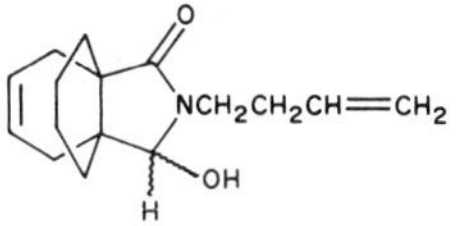 $NCH_2CH_2CH{=}CH_2$, OH	HCO_2H, r.t., stir 18 h	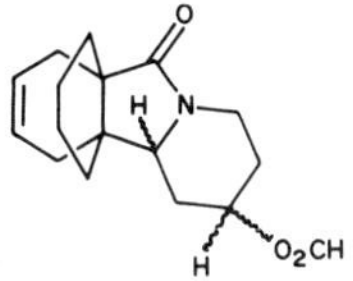 O_2CH	—	1412
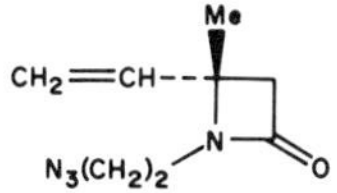 $CH_2{=}CH$, Me, $N_3(CH_2)_2$	C_6H_6, heat 14 h	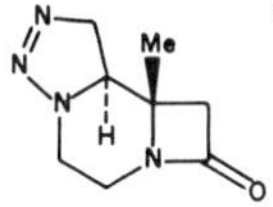 (*cis*) [a]	—	1312
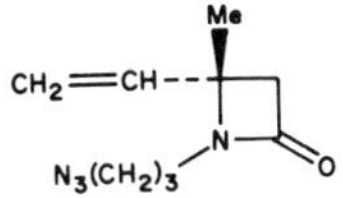 $CH_2{=}CH$, Me, $N_3(CH_2)_3$	C_6H_6, heat	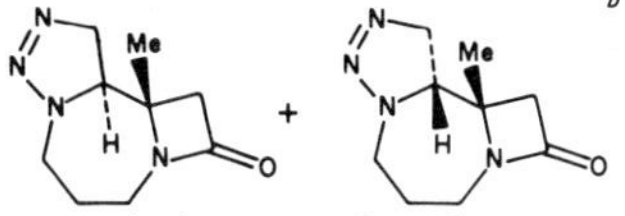 (*cis*) + (*trans*) (9:1) [b]	—	1312
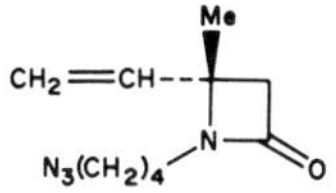 $CH_2{=}CH$, Me, $N_3(CH_2)_4$	C_6H_6, heat	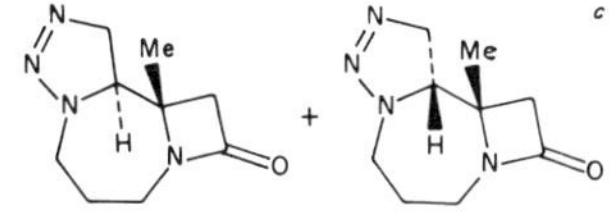 (*cis*) + (*trans*) (3:2) [c]	—	1312

Substrate	Conditions	Product(s)	Yield (%)	Ref.
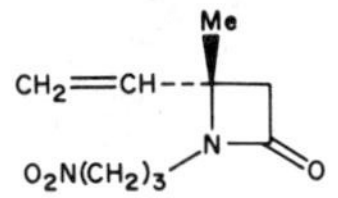 $CH_2{=}CH$, Me, $O_2N(CH_2)_3$	PhNCO, NEt_3	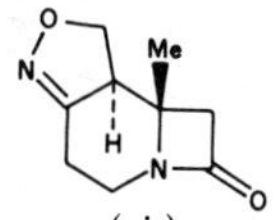 (*cis*)	—	1312
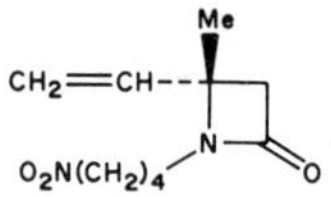 $CH_2{=}CH$, Me, $O_2N(CH_2)_4$	PhNCO, NEt_3	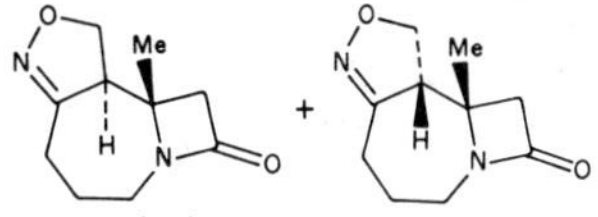 (*cis*) + (*trans*) (2:3)	—	1312
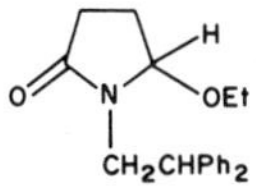 OEt, CH_2CHPh_2	Pyridinium polyhydrogen fluoride 25 °C, 30 min	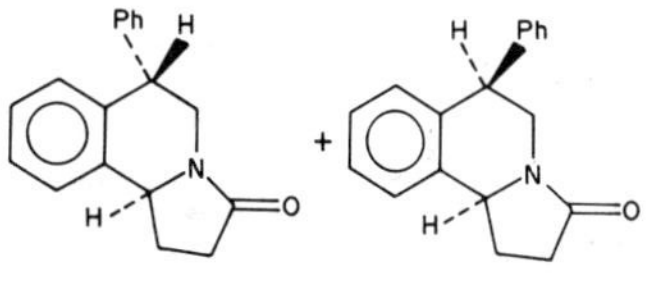 (93:7)	60	1091
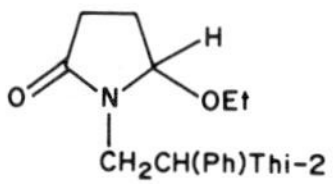 OEt, $CH_2CH(Ph)Thi$-2	HCl, EtOH	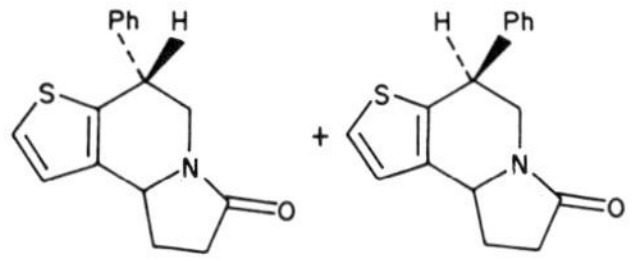 (96:4)	60	1091
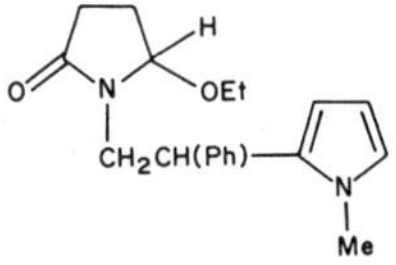 OEt, $CH_2CH(Ph)$, Me	$MeSO_3H$ (pH = 1), 20 °C	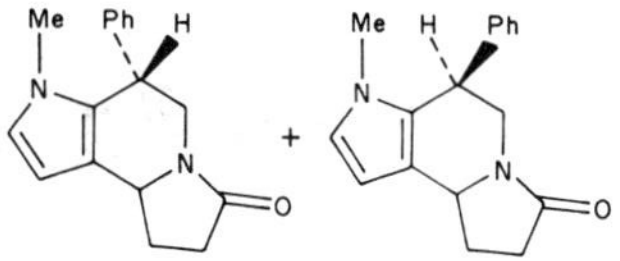 (74:26)	50	1091

(*continued*)

TABLE 77. (*continued*)

Substrate	Reaction conditions	Product	Yield (%)	Reference
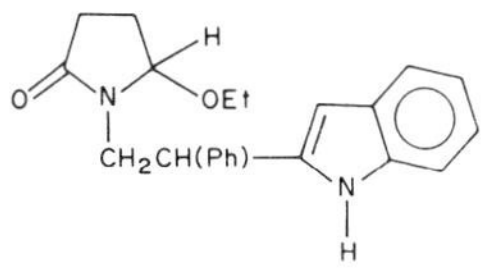 $CH_2CH(Ph)$	$MeSO_3H$ (pH = 1), 20 °C	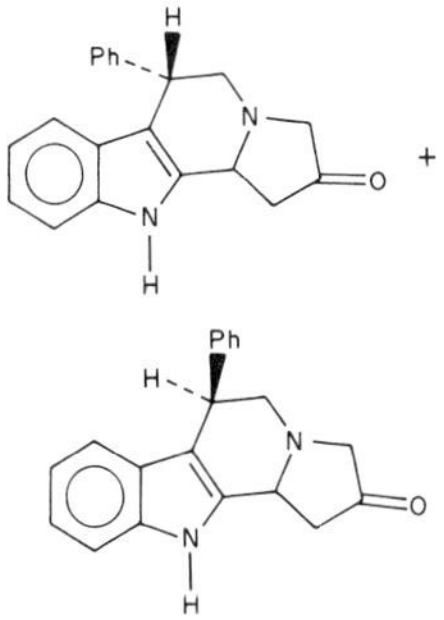 + (92:8)	70	1091
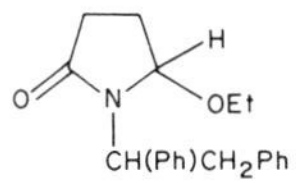 $CH(Ph)CH_2Ph$	polyphosphoric acid 100 °C	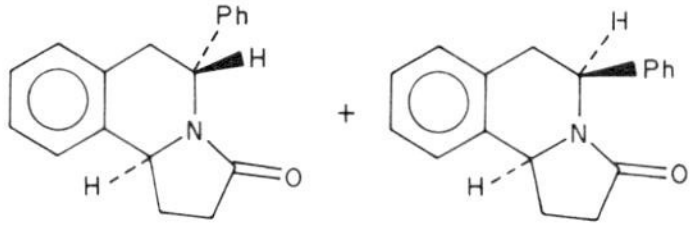 (major product) +	57	1091
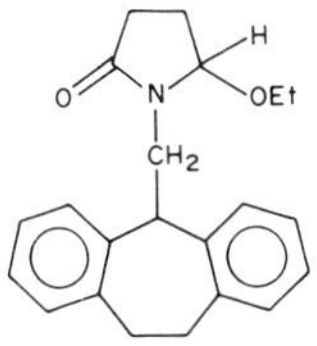 CH_2	polyphosphoric acid 100 °C	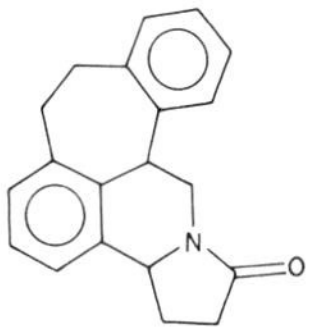 (α:β = 95:5)	45	1091
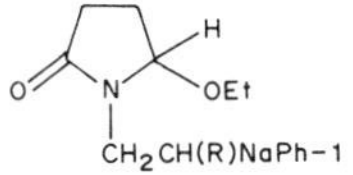 $CH_2CH(R)NaPh$-1, R = H, Me	*p*-TosOH	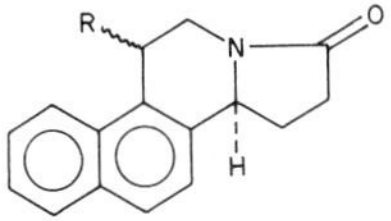	—	791

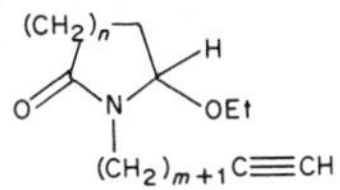 $(CH_2)_n$, H, OEt, N, O, $(CH_2)_{m+1}C\equiv CH$	HCO_2H, r.t. stir	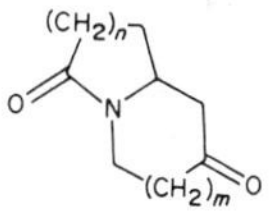 $(CH_2)_n$, N, O, $(CH_2)_m$		1413
$m = n = 1$	72 h		97	
$m = 1$, $n = 2$	92 h		89	
$m = 2$, $n = 1$	5 days		90	
$m = n = 2$	5 days		90	
$m = 3$, $n = 1$	14 days		80	
$m = 3$, $n = 2$	14 days		77	
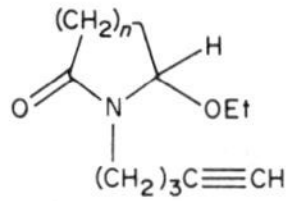 $(CH_2)_n$, H, OEt, N, O, $(CH_2)_3C\equiv CH$	HCO_2H, r.t. stir	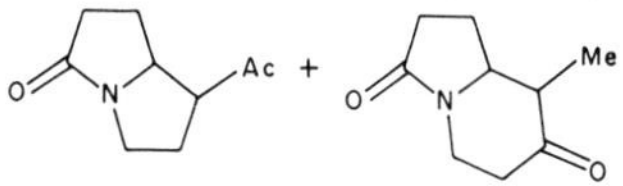 Ac + Me (O, N)		1413
$n = 1$	5 days	(1:9)	91	
$n = 2$	3 days	(1:5:8.5)	92	
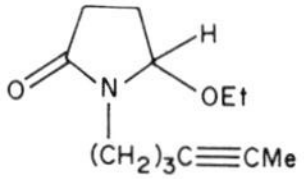 H, OEt, N, O, $(CH_2)_3C\equiv CMe$	HCO_2H, r.t. stir 24 h	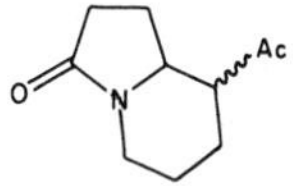 Ac (O, N) (C_5 epimer 9:1)[d]	92	1413
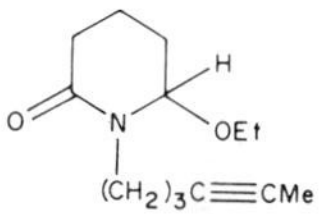 H, OEt, N, O, $(CH_2)_3C\equiv CMe$	HCO_2H, r.t. stir 72 h	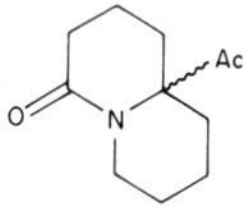 Ac (O, N) (C_7 epimer 9:1)[d]	88	1413
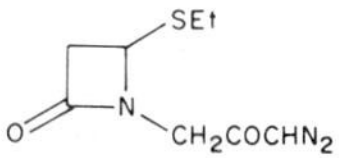 SEt, N, O, CH_2COCHN_2	C_6H_6, heat $Rh(OAc)_2$ or copper acetylacetonate	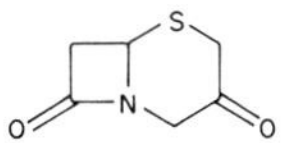 S, N, O, O	—	1414

(continued)

TABLE 77. (*continued*)

Substrate	Reaction conditions	Product	Yield (%)	Reference
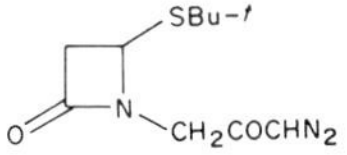 4-(SBu-*t*)-1-(CH_2COCHN_2)azetidin-2-one	C_6H_6, heat, $Rh(OAc)_2$ or copper acetylacetonate	N. R.	—	1414
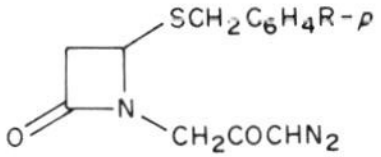 4-($SCH_2C_6H_4R$-*p*)-1-(CH_2COCHN_2)azetidin-2-one R = H, MeO	C_6H_5, heat, $Rh(OAc)_2$ or copper acetylacetonate	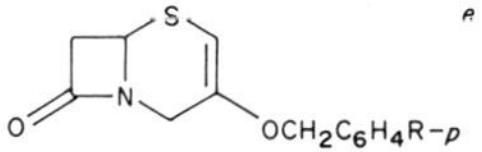 cepham with $OCH_2C_6H_4R$-*p* [a]	—	1414
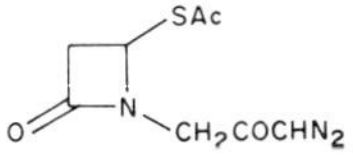 4-(SAc)-1-(CH_2COCHN_2)azetidin-2-one	C_6H_6, heat, $Rh(OAc)_2$ or copper acetylacetonate	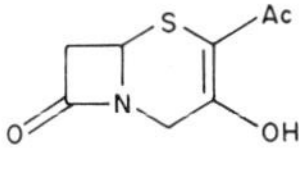 cepham with Ac, OH	20	1414
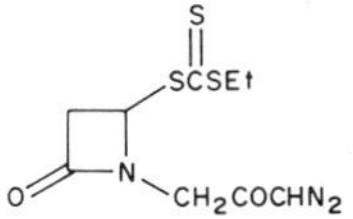 4-(SC(=S)SEt)-1-(CH_2COCHN_2)azetidin-2-one	C_6H_6, heat, $Rh(OAc)_2$ or copper acetylacetonate	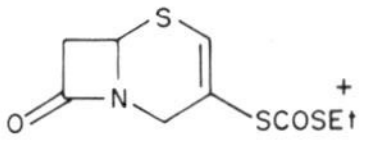 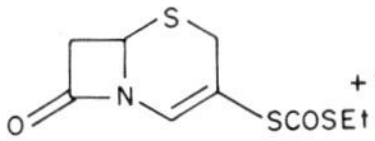 cephem with SCOSEt + isomeric cephem with SCOSEt	40	1414
		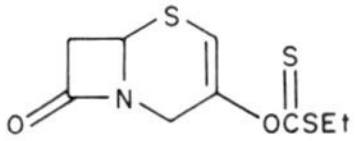 cephem with OC(=S)SEt	17	
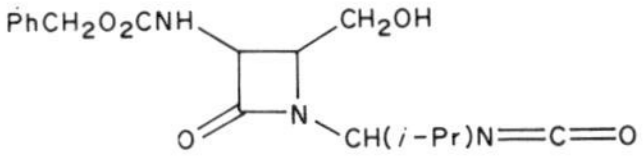 $PhCH_2O_2CNH$, CH_2OH, N-CH(*i*-Pr)N=C=O azetidin-2-one (1*S*, 3*S*, 4*S*)	MeCN, $H_2N(CH_2)_2NH_2$, r.t., 17 h	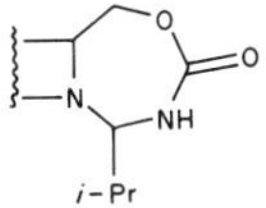 bicyclic cyclic carbamate with NH, *i*-Pr (2*R*, 7*S*, 8*S*)	—	1393

Substrate	Conditions	Product	Yield (%)	Ref.
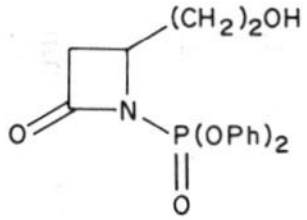 [structure: (CH2)2OH, N-P(OPh)2, O]	CsF, *t*-BuOH, stir, r.t., 2 h	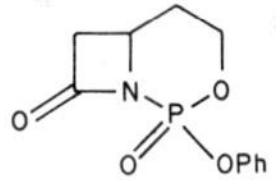 [structure: N, P, O, OPh]	20	1320
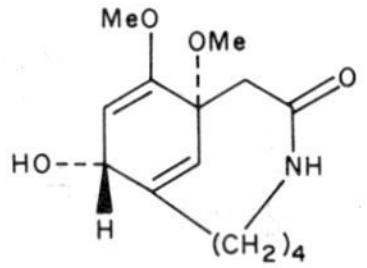 [structure: MeO, OMe, HO, H, NH, (CH2)4]	$BF_3 \cdot OEt_2$, THF, r.t., 20 h	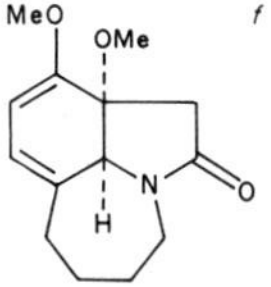 [structure: MeO, OMe, N, H] [f]	75	1105

[a] Further treatment of this product with silica gel afforded

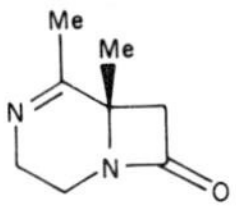

[b] Further treatment of this product mixture with silica gel produced a 7:3 mixture of +

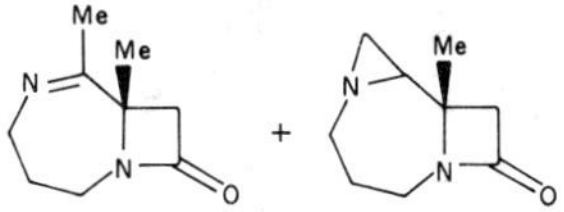

[c] Further treatment of the *cis* isomer with silica gel produced

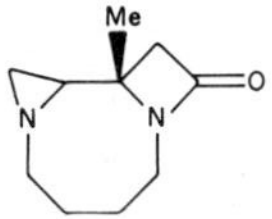

[d] Major epimer has acetyl group in the (pseudo) equatorial position.

[e] The mechanism for this cyclization involves formation of a carbene and its reaction with the sulphur atom to provide a ylide, which undergoes a[2,3]-sigmatropic rearrangement to produce a ketone intermediate which then undergoes a [3,3]-sigmatropic shift to produce the product.

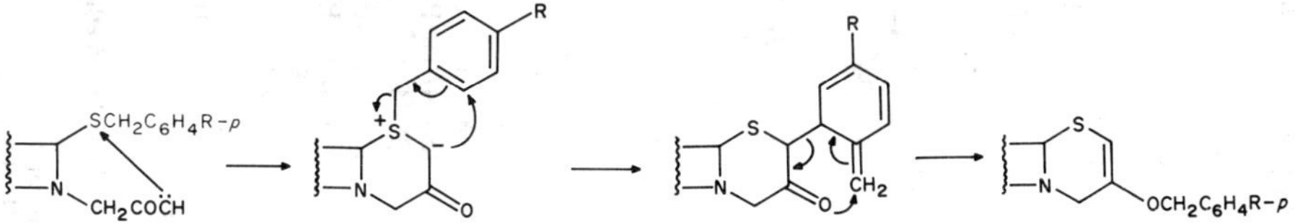

[scheme: $SCH_2C_6H_4R$-*p*, N, $CH_2CO\ddot{C}H$ → S+, R, N, O → S, R, N, O, CH_2 → S, N, $OCH_2C_6H_4R$-*p*]

[f] Reaction is an acid-catalysed transannular aromatization.

Saponification of lactam malonates, like the ones produced in equations 1040 and 1041, results[1039] in decarboxylation to the mono acid as a 1:1 mixture of diastereomers (equation 1042).

1. MeOH, NaOH, 0 °C, stir r.t. 1 h
2. HCl

71 %

(1042)

This type of substitution using alkyl groups is exemplified[1415] by the stereoselective alkylation of chiral β-lactam ester enolates via an intermediate chelate formed between the enolate with the β-lactam oxygen, followed by back-side attack of the electrophiles (equation 1043).

1. LDA, THF, 0–5 °C
2. −78 °C to −90 °C
3. R^2Br, 5 h

(1043)

R^1	β-lactam ester	R^2Br	Base[a]	Temp. (°C)	Product	Yield (%)	Stereo selectivity (% de)[b]
PhO	(3*S*, 4*R*)	$CH_2{=}CHCH_2Br$	LDA	−78	(3*S*, 4*R*)	95	>98 (*R*)
PhO	(3*S*, 4*R*)	$CH_2{=}CHCH_2Br$	LHDS	0–5	(3*S*, 4*R*)	94	95 (*R*)
PhO	(3*R*, 4*S*)	$CH_2{=}CHCH_2Br$	LDA	−78	(3*R*, 4*S*)	95	34 (*S*)
PhO	(3*S*, 4*R*)	$PhCH_2Br$	LDA	−78	(3*S*, 4*R*)	96	>98 (*R*)
PhO	(3*S*, 4*R*)	$PhCH_2Br$	LDA	0–5	(3*S*, 4*R*)	95	93 (*R*)
PhO	(3*S*, 4*R*)	$PhCH_2Br$	LDA	−10	(3*S*, 4*R*)	93	75 (*R*)
PhO	(3*S*, 4*R*)	$PhCH_2Br$	LDA	−90	(3*S*, 4*R*)	95	50 (*R*)
PhO	(3*R*, 4*S*)	EtBr	LDA	−78	(3*R*, 4*S*)	95	>98 (*R*)
PhO	(3*R*, 4*S*)	2,4-$(MeO)_2$· $C_6H_3CH_2Br$	LDA	0–5	(3*R*, 4*S*)	95	93 (*R*)
$PhCH_2$· O_2CNH	(3*R*, 4*S*)	$PhCH_2Br$	LDA[c]	−78–0	(3*R*, 4*S*)	—	93 (*S*)

[a]LHDS = lithium hexamethyldisilazane.
[b]*R* or *S* in parenthesis is the configuration of the newly formed quaternary center.
[c]Reaction conditions employed were: (1) LDA, −78 °C, THF, 3 min; (2) Me_3SiCl, −78 °C to 0 °C, 75 min; (3) LDA, THF, 0 °C, 1 min; (4) cooled to −78 °C; (5) $PhCH_2Br$, −78 °C, 2 h; (6) −78° to 0 °C 3 h, then 2 h at 0 °C.

Alkyl substitution has been reported[1094] when the alpha atom of the side chain is nitrogen instead of carbon (equation 1044).

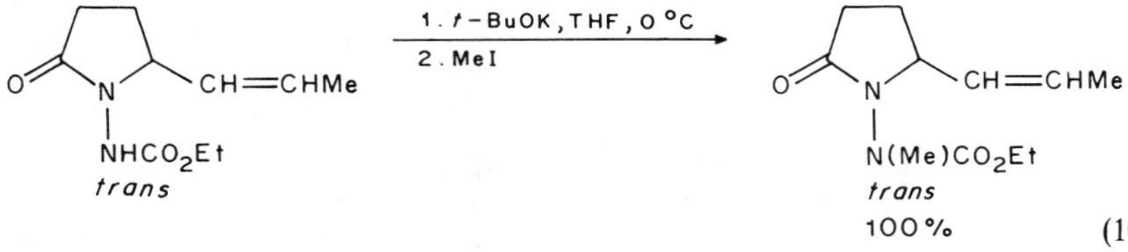

(1044)

Anodic oxidation of *N*-benzyl-3-methylene-β-lactams in methanol produces[1393] two methoxylated products, one where the methoxy group is substituted on the exocyclic carbon (**220**), and the other where the methoxy group is substituted on the endocyclic carbon (**221**), both carbons of which are alpha to nitrogen (equation 1045).

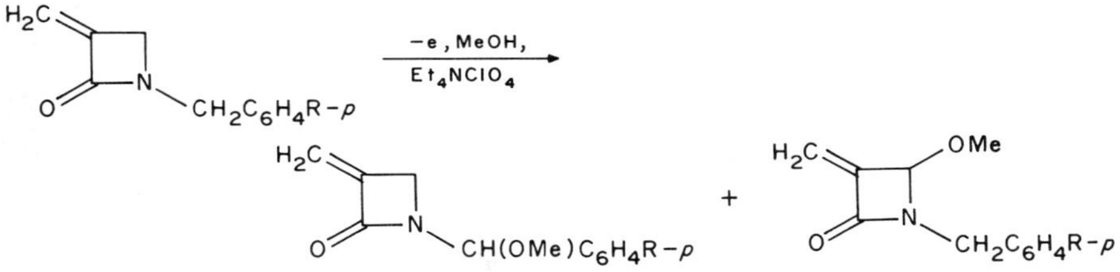

(1045)

R	Yield (%) 220	221	Recovered starting material
H	54	12	8
Me	21–39	—	39
MeO	54	—	17
COOMe	48	23	9

Similar results were obtained[1393] with the unsaturated fused-ring analogue shown in equation 1046.

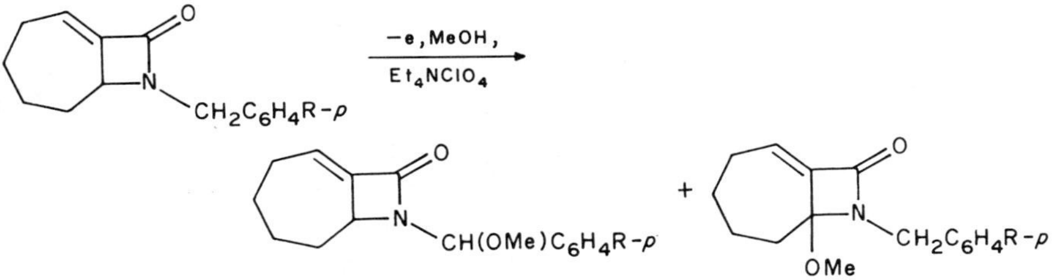

(1046)

R	Yield (%) 222	223
H	62	11
MeO	60	13

β-Lactams containing *O*-acyl, *O*-pivaloyl, or *O*-benzyl groups attached to the lactam nitrogen have been converted into the corresponding *N*-hydroxy β-lactams by different methods depending upon the structure of the group attached to oxygen. In the case of *N*-(pivaloyloxy) substituted 2-azetidinones aminolysis using benzylamine[1049] (equation 1047), or ammonium acetate in aqueous tetrahydrofuran[1097] (equation 1048) effected depivaloylation to the *N*-hydroxy analogue, while with perester[1416] (equation 1049) and benzyloxy[1048,1049,1080] (equation 1050) groups catalytic hydrogenolysis effected deacylation or debenzylation to produce the desired *N*-hydroxy analogues.

$PhCH_2NH_2$, THF (1047)

NH_4OAc, THF/H_2O, Me_2CO, stir r.t. 12 h; 60% (1048)

H_2, Pd/C; >90% (1049)

H_2, MeOH, 10% Pd/C (1050)

R^1	R^2	R^3	Yield (%)	Reference
H	H	H	86	1048
Me	H	H	80	1048
Me	Me	H	—	1049
t-BuOCONH	H	H	100	1049
H	Et	$-CH_2$ / MeO_2CCH_2 (1,3-dioxolane)	91	1080

At least one reaction of an N-hydroxy β-lactam has been reported[1417], and this involves the treatment of 4-carbophenethoxy N-hydroxy-2-azetidinone with diisopropyl carbodiimide to produce an 80% yield of the corresponding isourea (equation 1051).

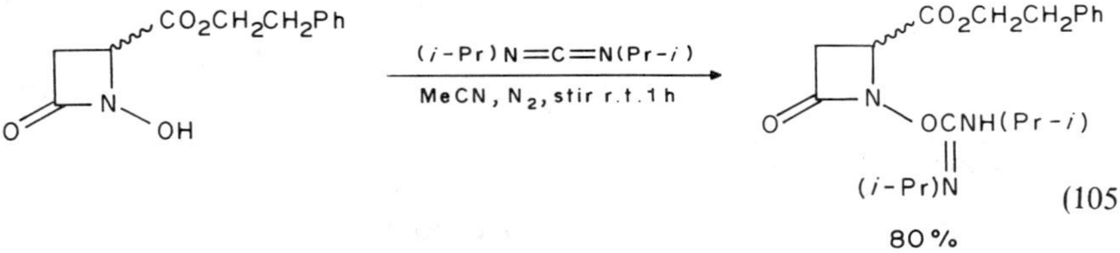

(1051)

N-Trimethylsilyl lactams have been used as starting materials for the preparation of N-(tetrazol-5-yl)[1184] (equation 1052) and N-(ω-isocyanatoacyl)- or N-(ω-isothiocyanatoacyl)-lactams[1320] (equations 1053–1055), by reacting them with 5-fluoro-1-benzyl-1H-tetrazole or isocyanato- or isothiocyanato-carboxylic acid chlorides, respectively.

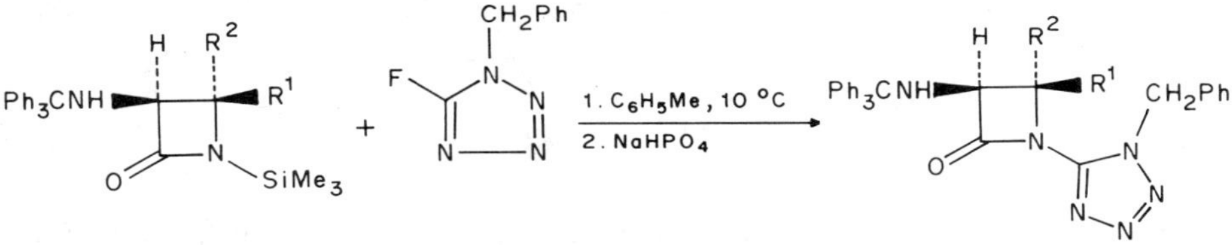

(1052)

R^1	R^2	Yield (%)	Product stereochemistry
CH_2F	H	70	(3S, 4S)
H	CH_2F	42	racemic
CO_2Me	H	50	racemic
CO_2OTHP	H	81	racemic
CHF_2	H	25	racemic

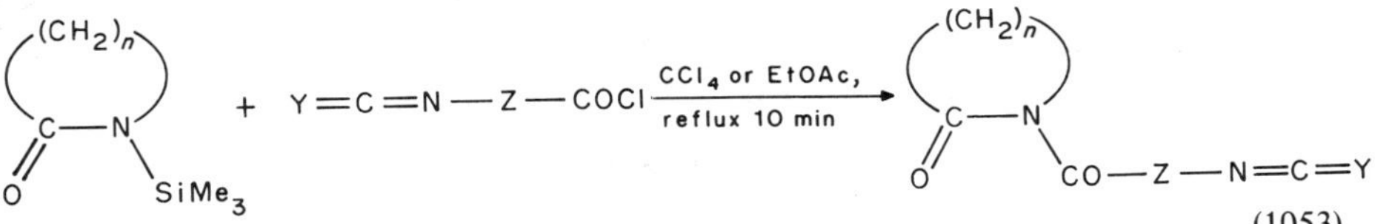

(1053)

n	Y	Z	Yield (%)
3	O	$—(CH_2)_2—$	93
3	S	$—(CH_2)_2—$	77
3	O	$—(CH_2)_3—$	72
3	S	p-C_6H_4	55
4	O	$—(CH_2)_2—$	76
5	O	$—(CH_2)_2—$	73
5	O	$—(CH_2)_3—$	77
5	S	p-C_6H_4	89

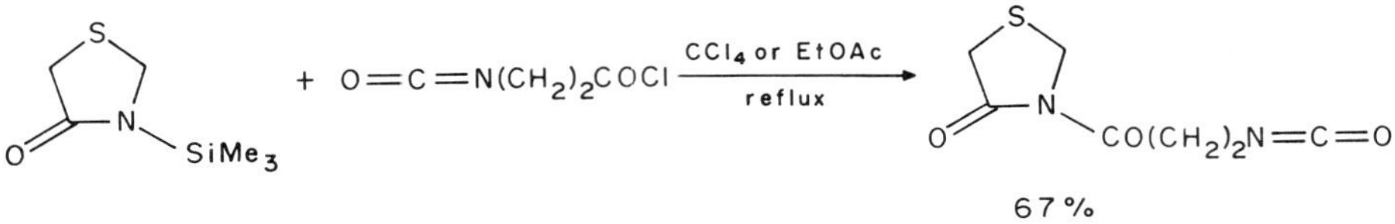

(1054)

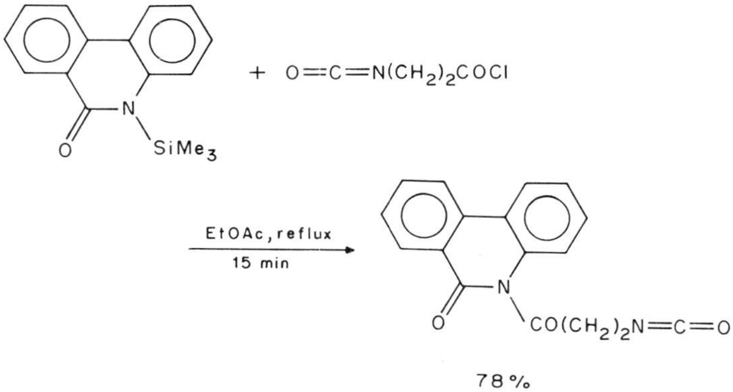

(1055)

Oxidation of alpha carbons adjacent to lactam nitrogens has been accomplished using ruthenium tetroxide[1418] (equation 1056) when the carbon is part of an alkyl group, and by pyridine chlorochromate[1351] when the carbon is part of an alcohol function (equation 1057). Using ruthenium tetroxide[1418], the oxidation of four- and eight-membered *N*-alkyllactams proceeds regioselectively to oxidize the exocyclic alpha carbon to produce *N*-acyllactams as illustrated in equation 1056. However, five- and six-membered lactams undergo endocyclic oxidation to yield cyclic imides, while seven-membered lactams yield a mixture of products arising from both exocyclic and endocyclic modes of oxidation.

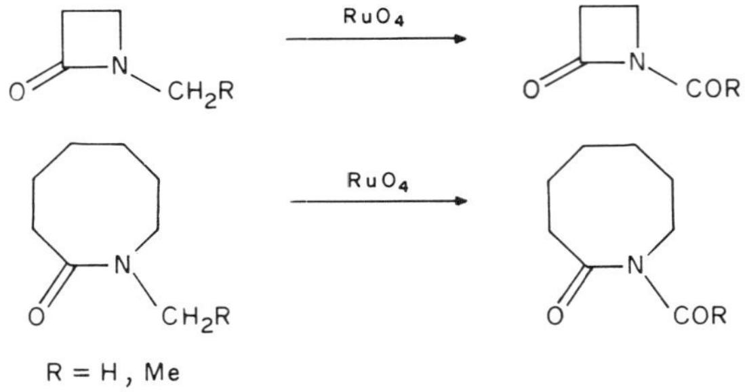

(1056)

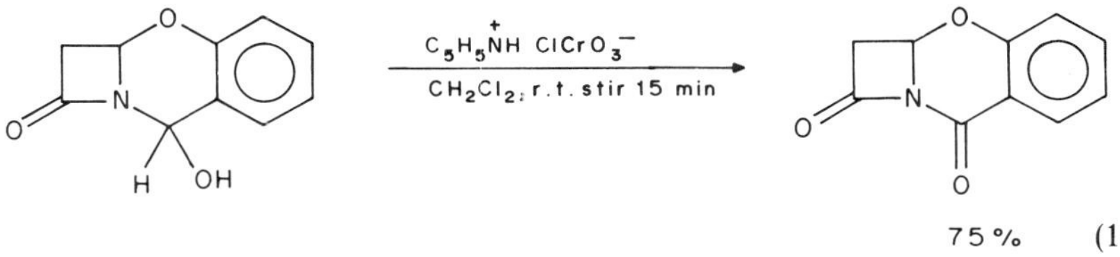

(1057)

Hydrogenolysis of an exocyclic alpha carbon which is part of an alcohol function has also been reported[1419] (equation 1058).

$$\text{N-(CH}_2\text{OH)-pyrrolidin-2-one} \xrightarrow[\text{autoclave, 150 °C, 100 kg/cm}^2\ \text{H}_2\text{, 4 h}]{\text{aq. H}_3\text{PO}_4\text{, Pd/C,}} \text{N-Me-pyrrolidin-2-one (87\%)} \qquad (1058)$$

Curtius rearrangement of an acyl azide attached to a lactam nitrogen alpha carbon produces[1396] the corresponding isocyanate (equation 1059).

$$\text{PhCH}_2\text{O}_2\text{CNH, CH}_2\text{OH azetidinone, N–CH(}i\text{-Pr)CON}_3\ (1S,3S,4S) \xrightarrow[\text{60 °C, 15 min}]{\text{ClCH}_2\text{CH}_2\text{Cl,}} \text{N–CH(}i\text{-Pr)N=C=O} \qquad (1059)$$

Warming 3,3-dimethyl-4-phenyl-*N*-tritylthioazetidin-2-one with two equivalents of anhydrous copper chloride in tetrahydrofuran-ethanol produces[1241] an interesting cleavage of only the sulphur–carbon bond, resulting in the formation of a 69% yield of the disulphide shown in equation 1060 as a mixture of diastereomers.

$$\text{Me, Me, Ph azetidinone N–SCPh}_3 \xrightarrow[\text{THF/EtOH, 75 °C, 5 h}]{\text{anhy. CuCl}_2\text{,}} (\text{azetidinone N–S})_2 \text{ (mixture of diastereomers) 69\%} \qquad (1060)$$

An interesting series of reactions has been reported[1129] to occur with [3α, 6α, 7α, 8α]-5,8-dimethyl-9-oxo-2-(substitued)ethoxycarbonyl-1,2-diazatricyclo[5.2.0.0^{3,6}]non-4-enes. Treatment of the 2-(2′-trimethylsilyl) derivative with tetra-(*n*-butyl)ammonium fluoride produces the N-2 deprotected product, which upon further successive treatment with a sulphur trioxide–dimethylformamide complex and tetra-(*n*-butyl)ammonium dihydrogenophosphate transforms the N-deprotected product into its N-sulphonated ammonium salt (equation 1061).

Furthermore, ozonation of these tricyclic compounds followed by photolysis of the secondary ozonides (**224**) obtained led[1129] to a mixture of unstable 2-azacarbapenems **225** and **226** (equation 1062).

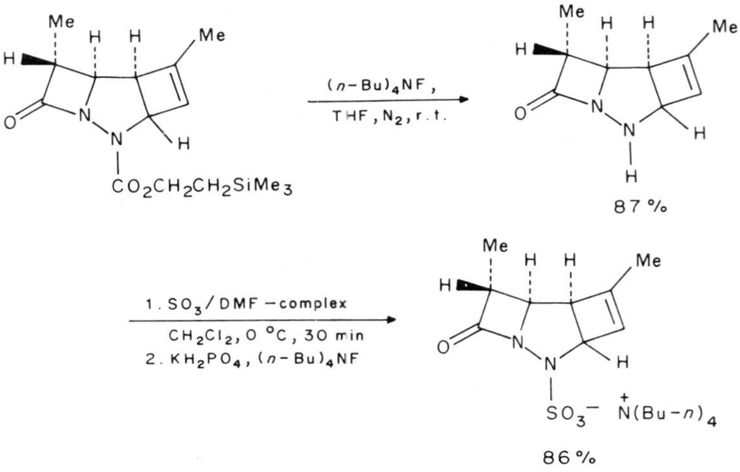
Me
H
H
Me
H
O
N
N
H
CO2CH2CH2SiMe3
(n-Bu)4NF,
THF, N2, r.t.
87 %
1. SO3/DMF-complex
CH2Cl2, 0 °C, 30 min
2. KH2PO4, (n-Bu)4NF
SO3⁻ N(Bu-n)4
86 %

(1061)

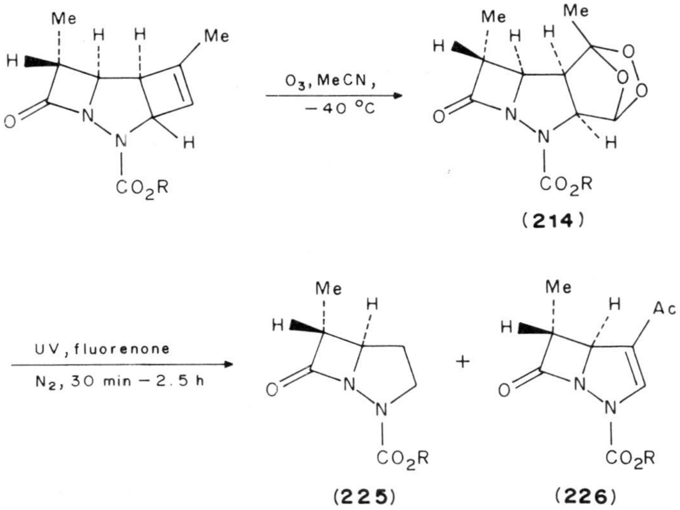
O3, MeCN,
−40 °C
CO2R
(214)
UV, fluorenone
N2, 30 min − 2.5 h
+
Ac
(225)
(226)

(1062)

R	Yield (%) 224	225	226
Et	60	32	12
$CH_2CH_2SiMe_3$	—[a]	34	13
$CH_2C_6H_4NO_2$-*p*	60	—[b]	—[b]

[a]Not isolated.
[b]Photolytic products could neither be isolated nor characterized.

Finally, interesting intramolecular rearrangements have been observed[1348] to occur with diazoketone substituents which are attached to lactam nitrogens. Thus treatment, either photochemically or with transition metals in hot benzene, of the diazoketones derived from penicillanic acid derivatives produced products derived from sulphur–ylid intermediates (equation 1063). A similar result was obtained[1348] when a C-6 hydroxyethyl substituted analogue was treated with copper acetylacetonate in hot benzene, but this result differed from the one obtained when the same starting material was treated photochemically, because in the latter case a lactone substituted β-lactam was formed via a Wolff rearrangement (equation 1064).

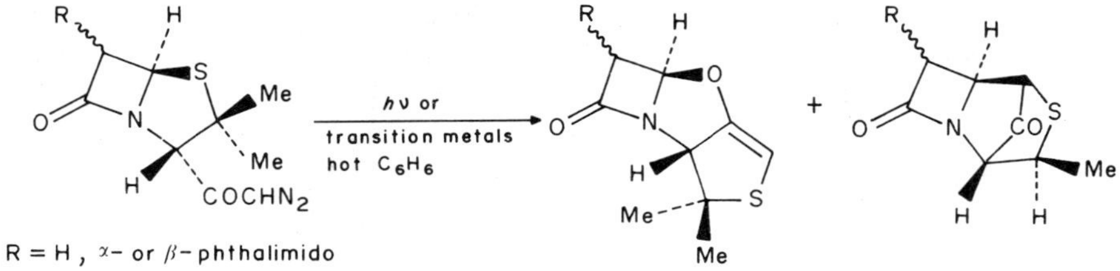

R = H, α- or β-phthalimido

hν or transition metals hot C_6H_6

(1063)

Cu acetylacetonate hot C_6H_6

31%

20%

$h\nu$, C_6H_6

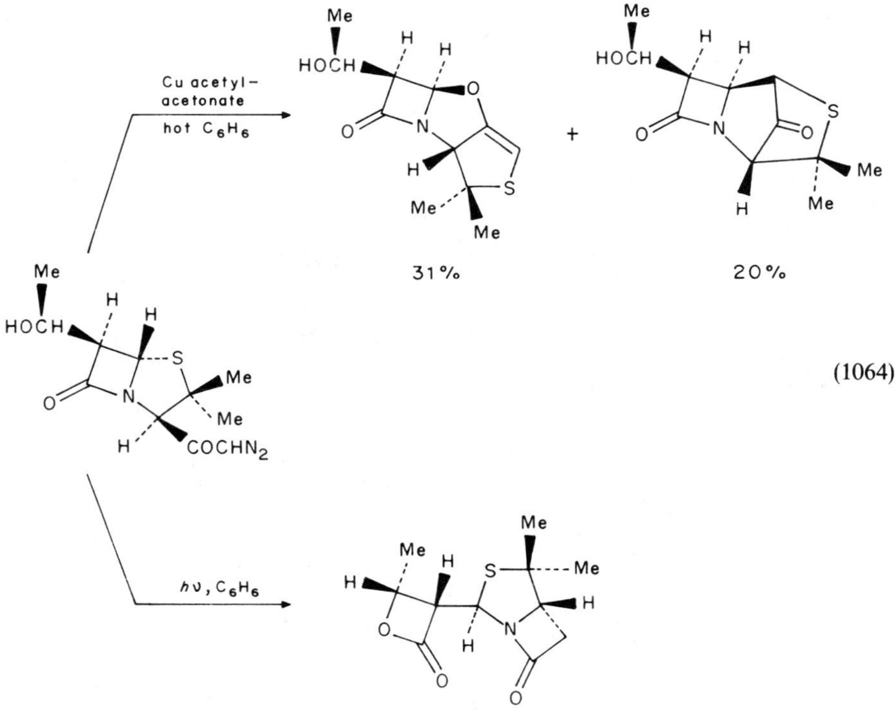

(1064)

This same type of Wolff-rearrangement was observed when the penicillin-derived diazoketone, which does not contain the *gem*-dimethyl groups, shown in equation 1065 is treated with rhodium acetate in hot benzene.

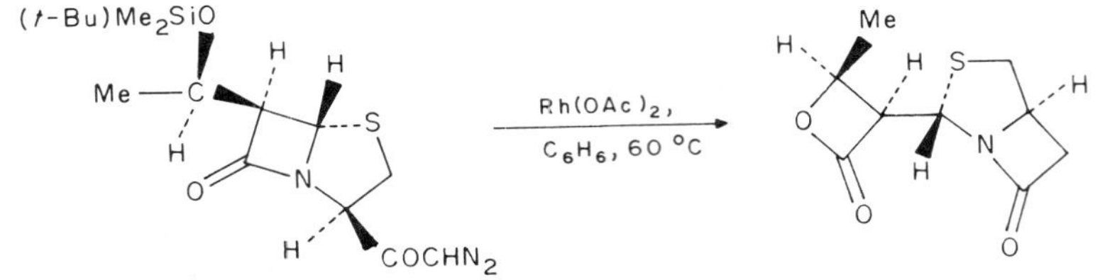

(1065)

It appears that the carbenes formed by transition metal-catalysed decomposition of penicillin derived diazoketones can follow different reaction pathways, depending upon the presence or absence of the *gem*-dimethyl groups. The mechanism proposed to explain the Wolff rearrangements[1420] is shown in equation 1066.

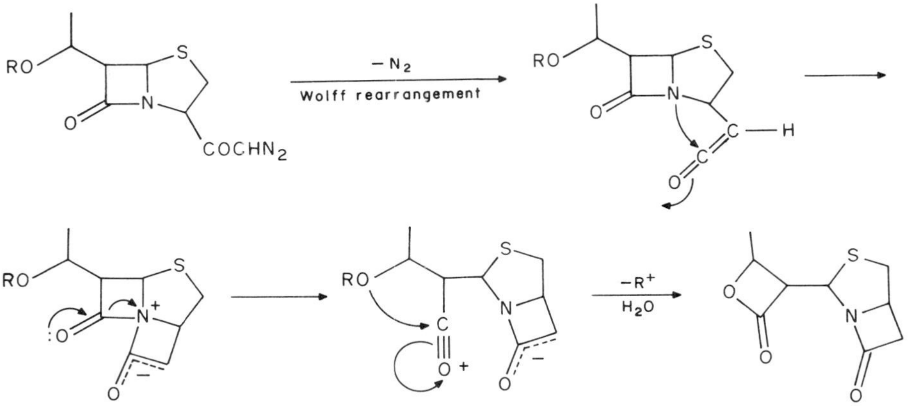

(1066)

4. Conversion of substituents directly attached to the lactam ring other than at lactam nitrogen

a. Reactions at the C-2 and C-3 positions The only reaction reported to occur at the lactam carbonyl site while still maintaining the integrity of the lactam ring is conversion to a thionated lactam. Two approaches have been used to convert a variety of lactams to thiolactams. The first approach involves reaction of the lactams with phosphorus pentasulphide, usually in combination with a base, while the second approach involves treatment of the lactam with Lawesson's reagent [2,4-bis(substituted)-1,3,2,4-dithiadiphosphetane-2,4-disulphide] (equation 1067). Table 78 reports the results obtained using both of these approaches.

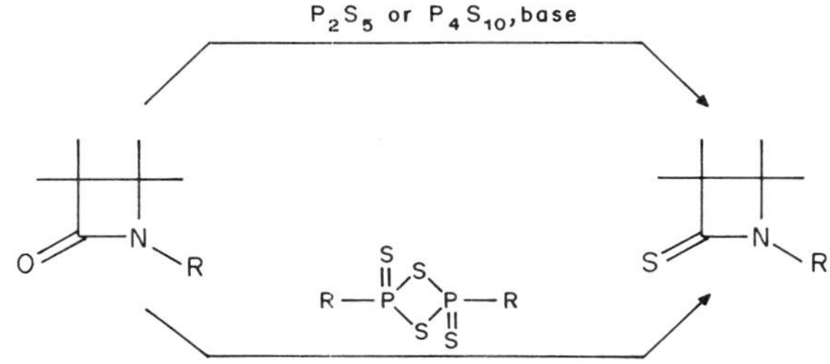

(1067)

One recent report[1424] compared the phosphorus sulphide and Lawesson's reagent methods for the conversion of lactams into thiolactams with a method involving O-alkylation of lactams using trialkyloxonium tetrafluoroborate, followed by treatment of the imidate tetrafluoroborate salt formed with either anhydrous sodium sulphide in acetone or hydrogen sulphide in pyridine (equation 1068).

$$\text{lactam (N-}R^1\text{)} + R^2_3O^+\,BF_4^- \xrightarrow[\text{r.t., 12 h}]{CH_2Cl_2,} \text{imidate } (N^+\text{-}R^1,\ C\text{-}OR)\ BF_4^-$$

$(R^2 = \text{Me or Et})$

$$\xrightarrow[\text{Me}_2\text{CO, 0 °C, 10 min; or } N_2,\ H_2S,\ C_5H_5N]{N_2,\ NaSH,} \text{thiolactam (N-}R^1\text{)} \qquad (1068)$$

Lactam	Thiolactam	Method[a], Yield (%) A	B	C	D	E
N-Me pyrrolidinone	N-Me pyrrolidinethione	100	96	73; 0[b]	35[c]; 50[d]	95
N-Me piperidinone	N-Me piperidinethione	98	—	63; 0[b]	23[d]	90
N-$(CH_2)_3F$ piperidinone	N-$(CH_2)_3F$ piperidinethione	55[e]	—	—	—	57; 9.5[f]
3-$(CH_2)_3F$-N-Me piperidinone	3-$(CH_2)_3F$-N-Me piperidinethione	68	—	—	—	65
N-Me caprolactam	N-Me thiocaprolactam	94	87	68	29[d]	95

[a]Methods: A, $Me_3O^+BF_4^-$ followed by treatment with anhydrous NaSH in acetone at 0 °C; B, $Et_3O^+BF_4^-$ followed by treatment with anhydrous NaSHi-Me_2CO at 0 °C; C, $Me_3O^+BF_4^-$ followed by treatment with H_2S in pyridine; D, P_4S_{10}; E, Lawesson's reagent [p-An(S=)P(μ-S)$_2$P(=S)An-p] in toluene at 110 °C for 3h.

[b]With sulphydrolysis conducted in acetone at 0 °C and in the absence of pyridine.

[c]Reflux in ether 5h.

[d]At 30 °C for 3h in acetonitrile.

[e]Based upon 76% and 88% yields, respectively, for the isolation of the crystalline imidate salts.

[f]In HMPA at 100 °C for 1.5h.

TABLE 78. Thionalation of lactams

Lactam	Sulphur reagent	Conditions	Product	Yield (%)	Reference
		C_6H_5Me, reflux		62	1421
	P_2S_5	C_6H_5Me, reflux		42	1421
		C_6H_5Me, reflux		47	1421
	P_2S_5	C_6H_5Me, reflux		21	1421
		DME			1422
	R = MeS	20–60 °C, 15 min		97	
	R = *p*-$PhOC_6H_4$	60 °C, 180 min		99	
	R = *p*-$PhSC_6H_4$	60 °C, 60 min		99	
	R = *p*-An	60 °C, 300 min		99	
					1422

Substrate	Reagent	Conditions	Product	Yield (%)	Ref.
	R = MeS	DME, 20–60 °C, 15 min		88	
	R = p-PhOC$_6$H$_4$	THF, 20 °C, 5 min		91	
	R = p-PhSC$_6$H$_4$	THF, 20 °C, 5 min		79	
	R = p-An	THF, 20 °C, 5 min		95	
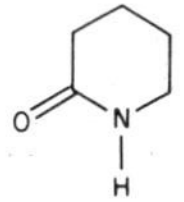	P_4S_{10}	n-BuLi, THF, hexane, < 10 °C to reflux, 16h	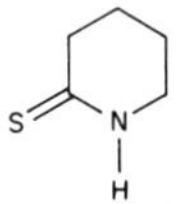	30	1423
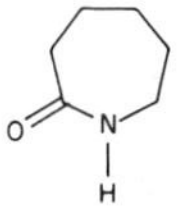	P_4S_{10}	n-BuLi, THF, hexane, < 10 °C to reflux, 16h	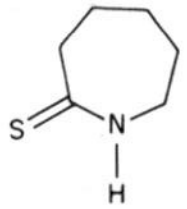	37	1423
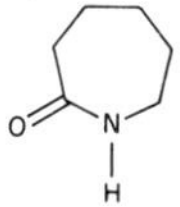	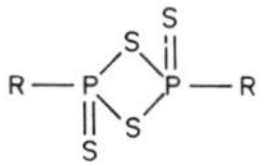		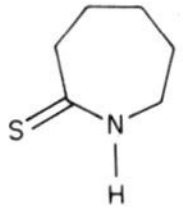		1422
	R = MeS	DMF, 20–60 °C, 15 min		85	
	R = p-PhOC$_6$H$_4$	THF, 20 °C, 5 min		98	
	R = p-PhSC$_6$H$_4$	THF, 20 °C, 5 min		79	
	R = p-An	THF, 20 °C, 5 min		98	
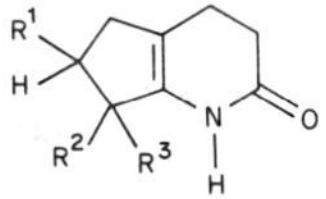	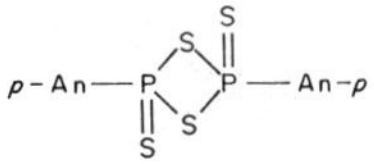	C_6H_6, stir r.t.	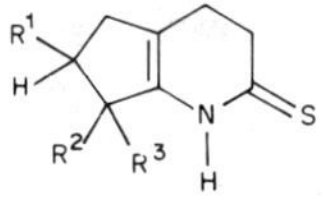		1096
$R^1 = R^2 = R^3 = H$				60	
$R^1 = R^2 = R^3 = Me$				69	

(continued)

TABLE 78. *(continued)*

Lactam	Sulphur reagent	Conditions	Product	Yield (%)	Reference
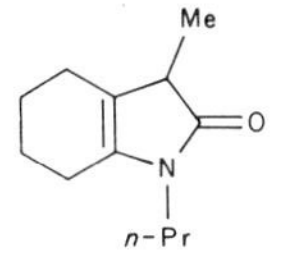 Me, O, N, n-Pr	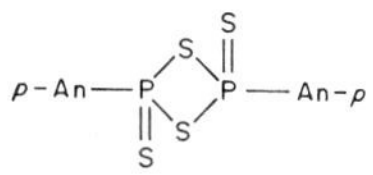 p-An—P(S)(μ-S)₂P(S)—An-p	C_6H_6, stir r.t.	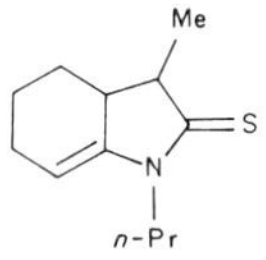 Me, S, N, n-Pr	20	1096
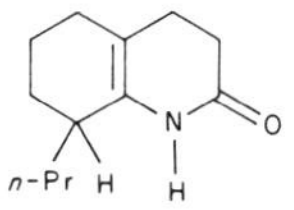 O, N, H, n-Pr, H	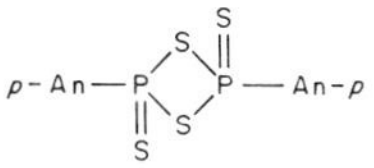 p-An—P(S)(μ-S)₂P(S)—An-p	C_6H_6, stir r.t.	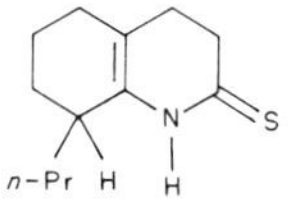 S, N, H, n-Pr, H	82	1096
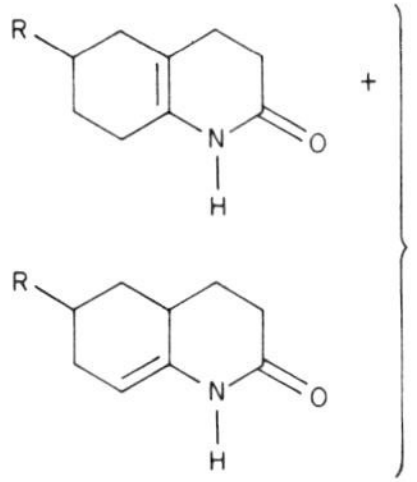 R, O, N, H + R, O, N, H	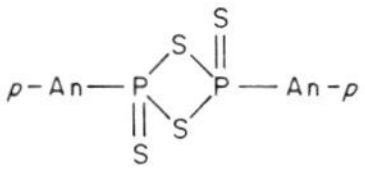 p-An—P(S)(μ-S)₂P(S)—An-p	C_6H_6, stir r.t.	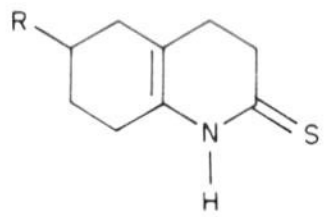 R, S, N, H		1096
R = H				65	
R = Me				50	
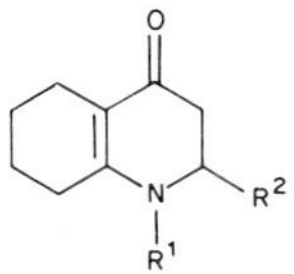 O, N, R^1, R^2	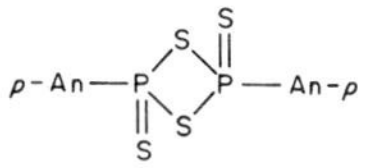 p-An—P(S)(μ-S)₂P(S)—An-p	C_6H_6, stir r.t. 15–60 min	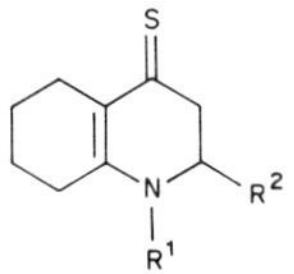 S, N, R^1, R^2		1225

Substrate	Reagent	Conditions	Product	Yield (%)	Ref.
$R^1 = Ph$, $R^2 = H$				95	
$R^1 = Ph$, $R^2 = Me$				88	
$R^1 = p\text{-}ClC_6H_4$, $R^2 = H$				70	
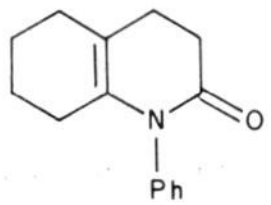 N-Ph lactam (O)	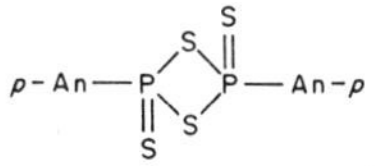 p-An–P(S)(μ-S)₂P(S)–An-p	C_6H_5Me, 80 °C, 1 h	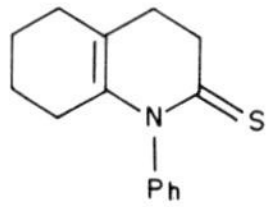 N-Ph thiolactam (S)	98	1255
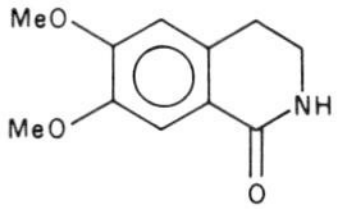 MeO, MeO; NH; O	P_2S_5	C_5H_5N, reflux, 2.5 h	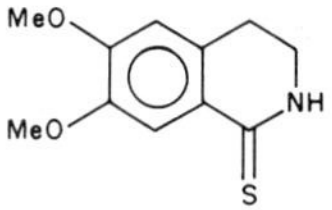 MeO, MeO; NH; S	54	1099
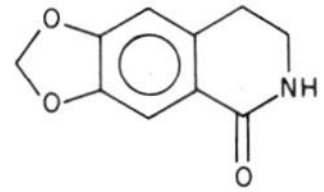 methylenedioxy; NH; O	P_2S_5	C_5H_5N, reflux, 2.5 h	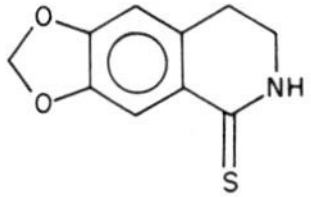 methylenedioxy; NH; S	50	1099
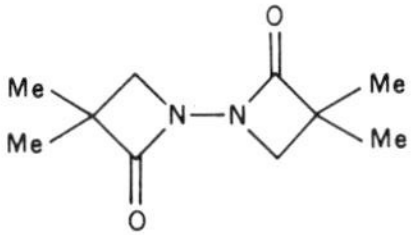 Me, Me; N–N; O, O; Me, Me	P_2S_5	C_6H_5Me, reflux	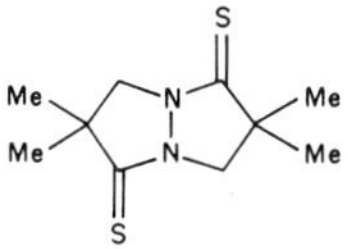 Me, Me; N–N; S, S; Me, Me	35	1421
	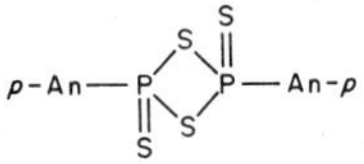 p-An–P(S)(μ-S)₂P(S)–An-p	C_6H_5Me, reflux	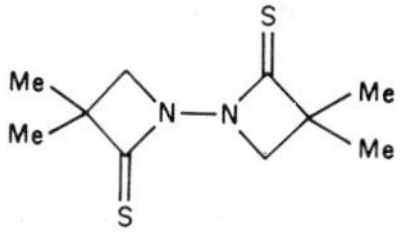 Me, Me; N–N; S, S; Me, Me	53	1421

(continued)

TABLE 78. (*continued*)

Lactam	Sulphur reagent	Conditions	Product	Yield (%)	Reference
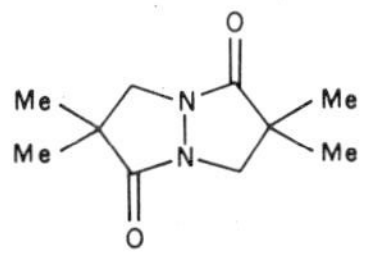 Me, Me, O, N, N, O, Me, Me		C_6H_5Me, reflux	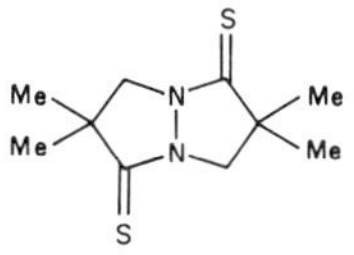 Me, Me, S, N, N, S, Me, Me		1421
	P_2S_5			37	
	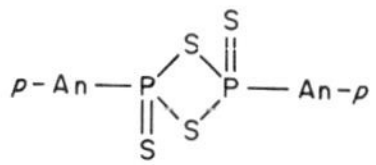 p-An—P(=S)(S)(S)P(=S)—An-p			61	
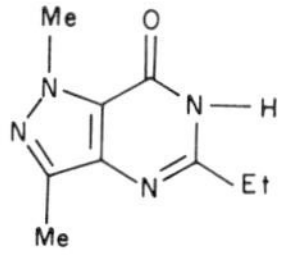 Me, O, N, N—H, N, N, Et, Me	P_4S_{10}	n-BuLi, THF, hexane, <10 °C to reflux, 16h	N.R.	—	1423
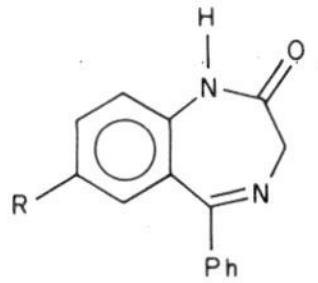 H, N, O, N, R, Ph	P_4S_{10}	n-BuLi, THF, hexane, <10 °C to reflux, 16h	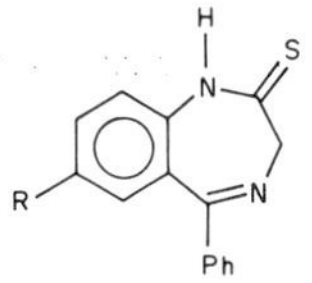 H, N, S, N, R, Ph	—	1423
R = H				65	
R = Cl				87	

Substrate	Reagent	Conditions	Product	Yield (%)	Ref.
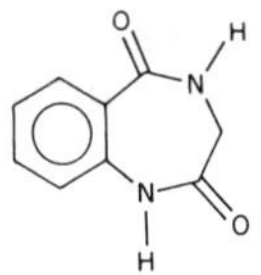	P_4S_{10}	*n*-BuLi, THF, hexane, <10 °C to reflux, 16h	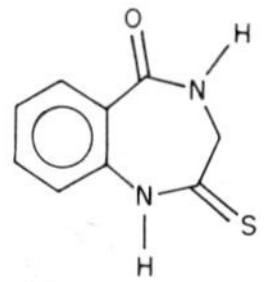	25	1423
	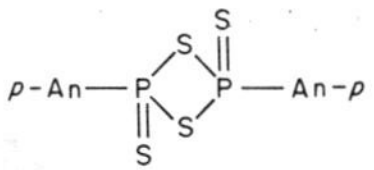 p-An—P(=S)(μ-S)₂P(=S)—An-p	HMPA		50	1423
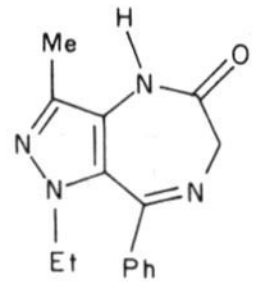	P_4S_{10}	*n*-BuLi, THF, hexane, <10 °C to reflux, 16h	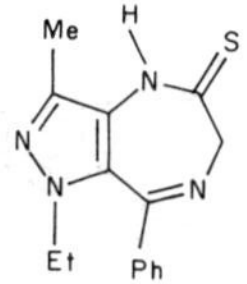	65	1423
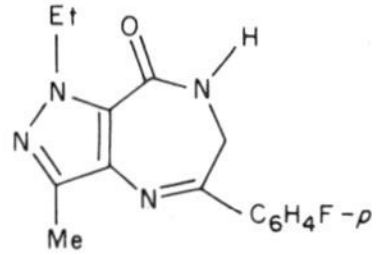	P_4S_{10}	*n*-BuLi, THF, hexane, <10 °C to reflux, 16h	N.R.	—	1423
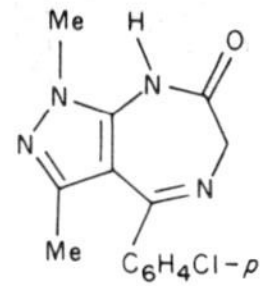	P_4S_{10}	*n*-BuLi, THF, hexane, <10 °C to reflux, 16h	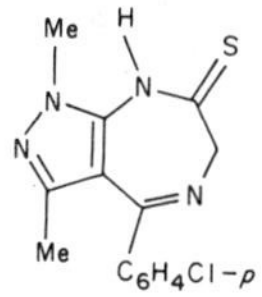	81	1423
	P_4S_{10}	MeLi, THF/ether		77	1423
	P_4S_{10}	PhLi. THF/ether/ cyclohexane		62	1423

(*continued*)

TABLE 78. (*continued*)

Lactam	Sulphur reagent	Conditions	Product	Yield (%)	Reference
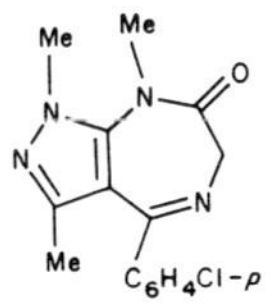 [structure]	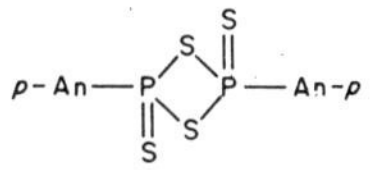 [structure]	C_5H_5N		90	1423
	P_4S_{10}	*n*-BuLi, THF, hexane, <10 °C to reflux, 16h	N.R.	—	1423
	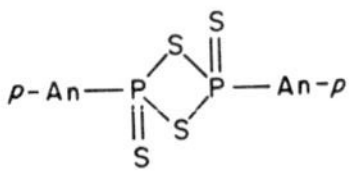 [structure]	HMPA	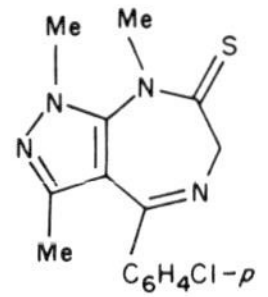 [structure]	10	1423
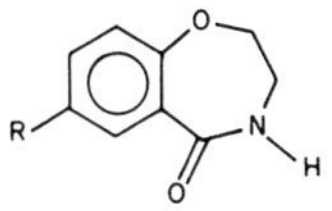 [structure]	P_2S_5	C_5H_5N, reflux 1–3h	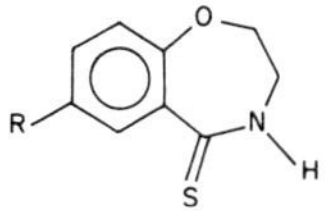 [structure]		1155
R = H				62	
R = Cl				—	
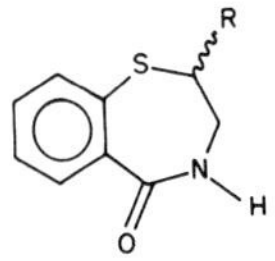 [structure]	P_2S_5	C_5H_5N, reflux 1–3 h	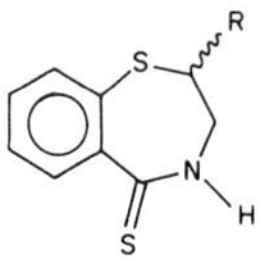 [structure]		1155
R = H				62–80	
R = MeO				75	

The authors claim[1424] that this O-alkylation method may, in principle, be applied to compounds containing esters, thiolactams, lactone, cyano, imino and epoxy functions. However, unlike the Lawesson's reagent method, the O-alkylation method is ineffective for the synthesis of secondary thiolactams and for the thionation of *N*,*N*-disubstituted amides.

The most common reaction reported for a substituent located on the lactam ring at a site other than on the ring nitrogen is acylation of an amine group located on the ring carbon alpha to the carbonyl function. Although this type of acylation may be performed in a number of ways, the most widely used method appears to be reaction of the amine substituent with an acid chloride in the presence of a mild base (equation 1069 and Table 79).

H_2N-azetidinone + RCOCl —(mild base)→ RCONH-azetidinone (1069)

Other methods used to perform this same type of acylation include treatment of the amine substituted lactams with an acid and *N*,*N*-dicyclohexylcarbodiimide (DCCD) (equation 1070 and Table 80), or with an acid and 1-ethoxycarbonyl-2-ethoxy-1,2-dihydroquinoline[1019] (equation 1071), reaction of an acid with a zwitterion intermediate[1168] (equation 1072), reaction with mixed anhydrides (equation 1073 and Table 81), reaction of the amine-substituted lactams with esters and a base (equation 1074 and Table 82), reaction with a diisocyanate[1425] (equation 1075), reaction of phosphinimino-2-azetidinone with an acid chloride[1173] (equations 1076 and 1077) and by enzymatic N-acylation using benzylpenicillin acylase copolymerized in a polyacrylamide matrix[1429] to effect the reaction of carboxylic acids and 3-amino-4α-methylmonobactamic acid (equation 1078).

H_2N-azetidinone + RCO_2H —(DCCD)→ RCONH-azetidinone (1070)

H_2N-bicyclic lactam (CO_2Bu-*t*) + $PhOCH_2CO_2H$ —(CH_2Cl_2, Na_2SO_4, 1-ethoxycarbonyl-2-ethoxy-1,2-dihydroquinoline (N-CO_2Et, OEt), stir 24 h)→ $PhOCH_2CONH$-bicyclic lactam (CO_2Bu-*t*) (1071)

85%

TABLE 79. Acid chloride acylation of amine group attached to lactams

Lactam	Acid chloride	Conditions	Product	Yield (%)	Reference
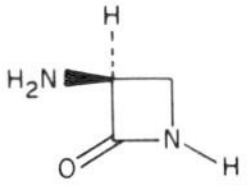 H, H_2N, N, H, O	$PhCH_2COCl$	Et_3N, DMF	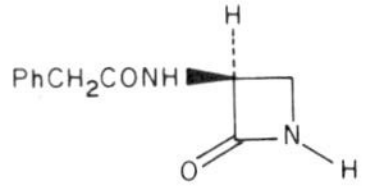 H, $PhCH_2CONH$, N, H, O	—	1326
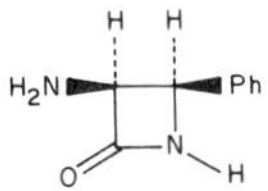 H, H, H_2N, Ph, N, H, O	$PhOCH_2COCl$	—	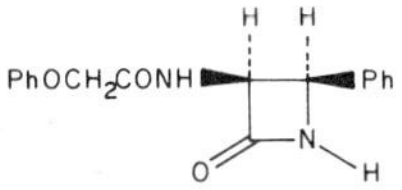 H, H, $PhOCH_2CONH$, Ph, N, H, O	—	1379
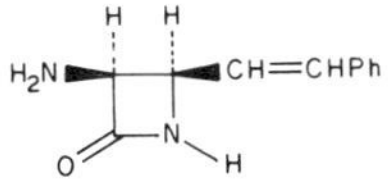 H, H, H_2N, CH=CHPh, N, H, O	$PhCH_2COCl$	C_5H_5N, CH_2Cl_2, N_2, 0 °C, stir 3 h	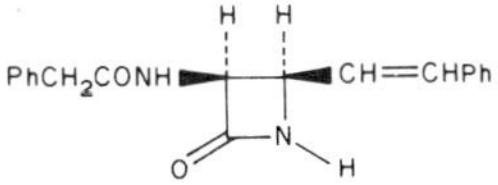 H, H, $PhCH_2CONH$, CH=CHPh, N, H, O	51	1168
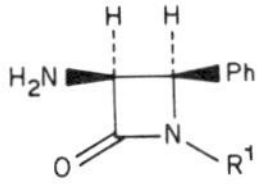 H, H, H_2N, Ph, N, R^1, O	R^2COCl	Et_3N, CH_2Cl_2	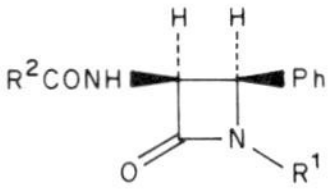 H, H, R^2CONH, Ph, N, R^1, O		
$R^1 = CH_2CH_2Cl$	$R^2 = Ph$			61	1204, 1372
$R^1 = CH_2CH_2Cl$	$R^2 = PhCH_2$			63	1204, 1372
$R^1 = H$	$R^2 = PhOCH_2$			30	1372
$R^1 = CH_2C_6H_3(OMe)_2$-2,4	$R^2 = PhCH_2$			61	1372
$R^1 = CH_2C_6H_3(OMe)_2$-2,4	$R^2 = PhOCH_2$			57	1372

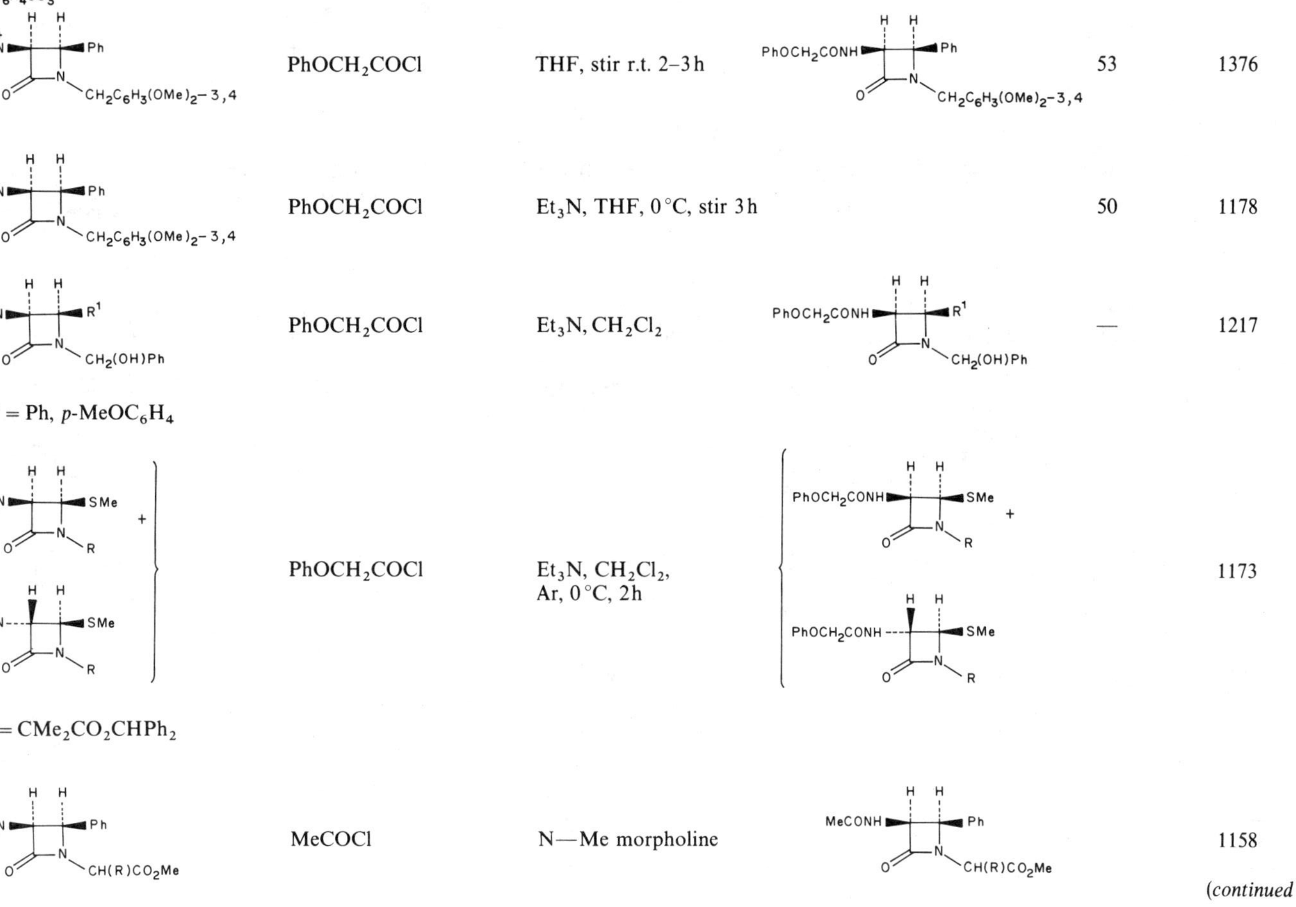

Substrate	Reagent	Conditions	Product	Yield (%)	Ref.
(p-$MeC_6H_4SO_3^-$ salt)	$PhOCH_2COCl$	THF, stir r.t. 2–3h		53	1376
	$PhOCH_2COCl$	Et_3N, THF, 0 °C, stir 3h		50	1178
R^1 = Ph, p-$MeOC_6H_4$	$PhOCH_2COCl$	Et_3N, CH_2Cl_2		—	1217
R = $CMe_2CO_2CHPh_2$	$PhOCH_2COCl$	Et_3N, CH_2Cl_2, Ar, 0 °C, 2h			1173
	MeCOCl	N—Me morpholine			1158

(continued)

TABLE 79. *(continued)*

Lactam	Acid chloride	Conditions	Product	Yield (%)	Reference
R = Me				82	
R = *i*-Pr				85	
R = $PhCH_2$				87	
R = Ph				80	
H_2N, H, H, Ph, O, N, $CH(Me)CO_2Me$	MeCOCl	N—Me morpholine	MeCONH, H, H, Ph, O, N, $CH(Me)CO_2Me$	—	1158
H_2N, H, H, Ph, O, N, $CH(R)CH_2CHMe_2$			N_3CH_2CONH, H, H, Ph, O, N, $CH(R)CH_2CHMe_2$		
R = CH_2OCH_2Ph	N_3CH_2COCl	N—Me morpholine, $CHCl_3$		85	1172
R = CO_2Bu-*t*	N_3CH_2COCl	N—Me morpholine		72	1170
H_2N, H, H, CH=CHPh, O, N, $CH_2C_6H_3(OMe)_2$-2,4	RCH_2COCl R = PhO	CH_2Cl_2	RCH_2CONH, H, H, CH=CHPh, O, N, $CH_2C_6H_3(OMe)_2$-2,4	50	1377
	R = z-Thi	CH_2Cl_2		56	1377

$R^1 = CH{=}CHPh$, $R^2 = Ph$ $R^1 = C{\equiv}CH$, $R^2 = CCl_3$	$PhOCH_2COCl$	Et_3N, CH_2Cl_2		—	1304
	$PhOCH_2COCl$	Et_3N, CH_2Cl_2, $-10\,^{\circ}C$			1196
$R^1 = CH{=}CHPh$				87	
$R^1 = CH_2CH(OMe)_2$				92	
$R^1 = CH_2CH_2NO_2$				59	
$R^1 = C{\equiv}CCH_2OSiPh_2(Bu\text{-}t)$				65	
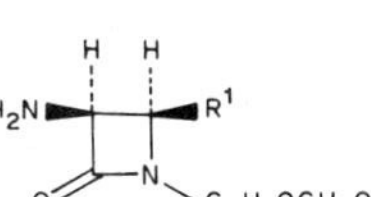 (227)	$PhOCH_2COCl$	1. Et_3N, CH_2Cl_2, $0\,^{\circ}C$ 2. r.t. stir 1 h	(229)	—	1316
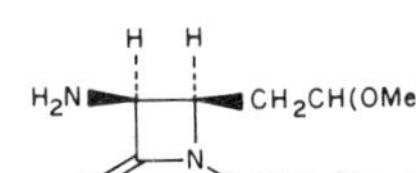 (228)	$PhOCH_2COCl$	1. Et_3N, CH_2Cl_3, $0\,^{\circ}C$ 2. r.t. stir 1 h	(230)	—	1316

(continued)

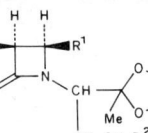

TABLE 79. (*continued*)

Lactam	Acid chloride	Conditions	Product	Yield (%)	Reference
227 + **228** (5:1)	$PhOCH_2COCl$	1. Et_3N, CH_2Cl_3, 0 °C 2. r.t. stir 1 h	**229** + **230** (1:1)	—	1316
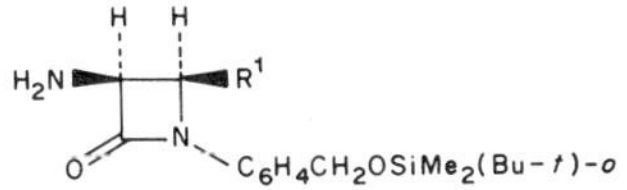 H H H_2N R^1 O N $C_6H_4CH_2OSiMe_2(Bu\text{-}t)\text{-}o$	$PhOCH_2COCl$	CH_2Cl_2	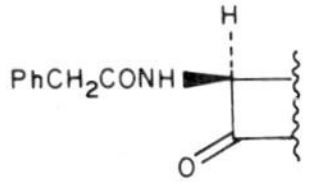 H $PhCH_2CONH$ O		802
$R^1 = CH_2OMe$		C_5H_5N, stir 20 min		65	
$R^1 = CH{=}CHPh$ (*trans*)		Et_3N, stir 30 min		60	
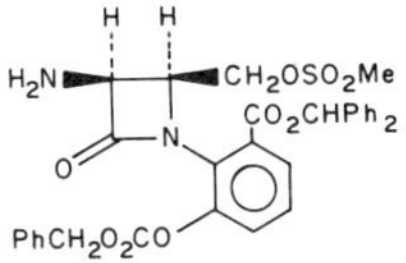 H H H_2N CH_2OSO_2Me CO_2CHPh_2 O N $PhCH_2O_2CO$	$PhCH_2COCl$	Et_3N, CH_2Cl_2, stir 0 °C, 1 h	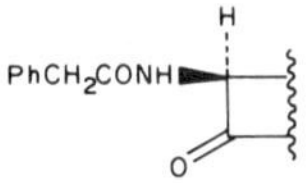 H $PhCH_2CONH$ O	64	1176
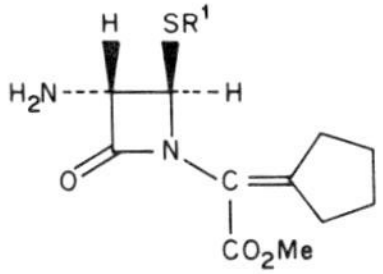 H SR^1 H_2N H O N C CO_2Me	R^2COCl	C_5H_5N, $CHCl_3$, 0 °C, 1 h	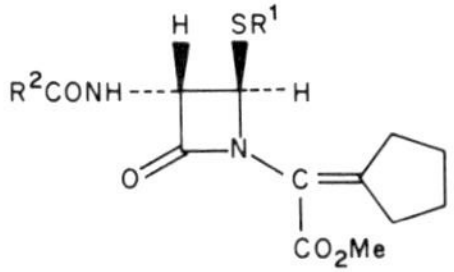 H SR^1 R^2CONH H O N C CO_2Me		1177
$R^1 = Me$	$R^2 = CH_2OPh$			83	
$R^1 = Ph_3C$	$R^2 = CH_2OPh$			68	
$R^1 = Me$	$R^2 = Ph$			73	

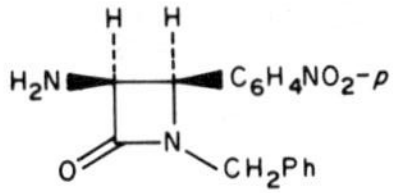 H H; H_2N; $C_6H_4NO_2$-*p*; O; N; CH_2Ph	$PhOCH_2COCl$	Et_3N, CH_2Cl_3, 5–10 °C, stir 4–5h	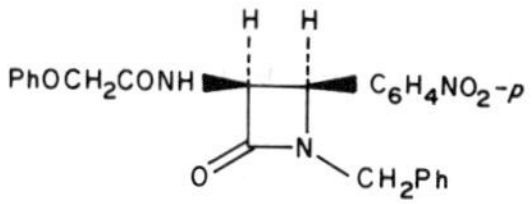 H H; $PhOCH_2CONH$; $C_6H_4NO_2$-*p*; O; N; CH_2Ph	50	1143
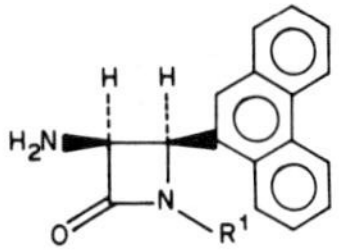 H H; H_2N; O; N; R^1	PhCOCl	Et_3N, CH_2Cl_2, stir 3h	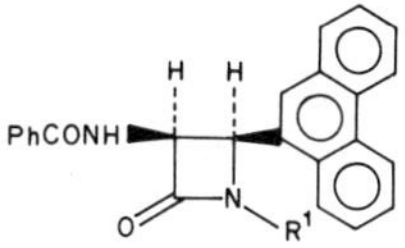 H H; PhCONH; O; N; R^1		1162
$R^1 = CH_2CO_2Me$				72	
$R^1 = CH(CH_2Ph)CO_2Me$				64	
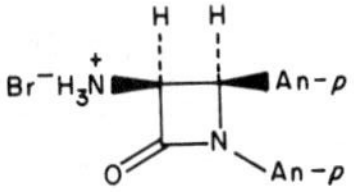 H H; $Br^-H_3\overset{+}{N}$; An-*p*; O; N; An-*p*	$PhOCH_2COCl$	Et_3N, CH_2Cl_2, stir r.t. 6h	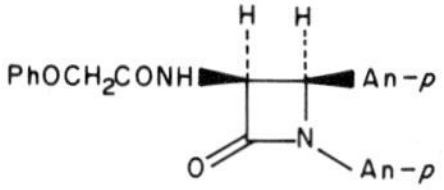 H H; $PhOCH_2CONH$; An-*p*; O; N; An-*p*	84	1180
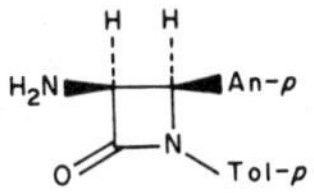 H H; H_2N; An-*p*; O; N; Tol-*p*	$PhOCH_2COCl$	—	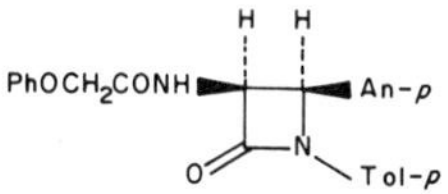 H H; $PhOCH_2CONH$; An-*p*; O; N; Tol-*p*	—	1214, 1215
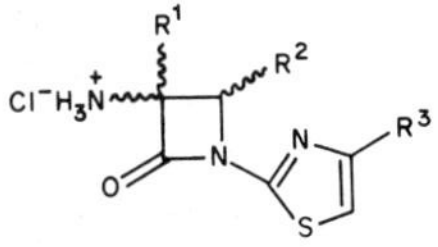 R^1; R^2; $Cl^-H_3\overset{+}{N}$; O; N; N; S; R^3	$PhCH_2COCl$	C_5H_5N, CH_2Cl_2	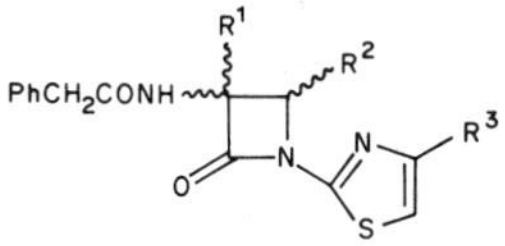 R^1; R^2; $PhCH_2CONH$; O; N; N; S; R^3		1189

(continued)

TABLE 79. (*continued*)

Lactam	Acid chloride	Conditions	Product	Yield (%)	Reference
R^1 / R^2 / R^3					
H / o-$O_2NC_6H_4$ / Ph				44	
H / p-An / Ph				48	
H / o-HOC_6H_4 / Ph				46	
H / m-$O_2NC_6H_4$ / p-MeC_6H_4				43	
Me / m-$O_2NC_6H_4$ / Ph				53	
i-Pr / PhCH=CH / Ph				38	
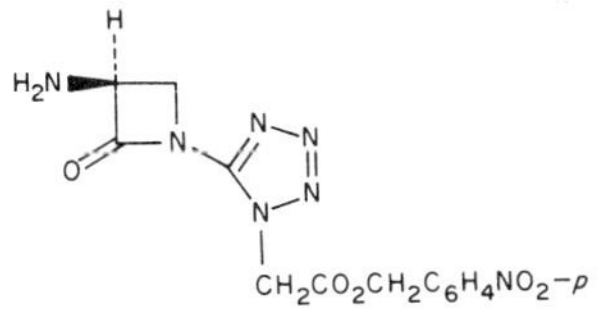 H, H_2N, N, N, N, N, $CH_2CO_2CH_2C_6H_4NO_2$-p, O	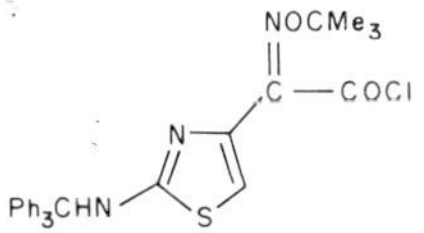 NOCMe$_3$, C—COCl, N, S, Ph_3CHN	1. Et_3N, DMF, −20 °C, stir 30 min 2. r.t. stir 1 h 3. THF, 50% aq. HCO_2H 40–50 °C, 1 h	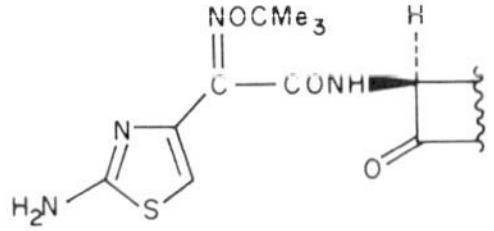 NOCMe$_3$, H, C—CONH, N, S, H_2N, O	41	1077
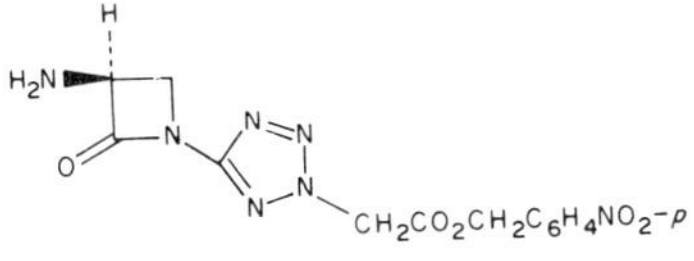 H, H_2N, N, N, N, N, $CH_2CO_2CH_2C_6H_4NO_2$-p, O	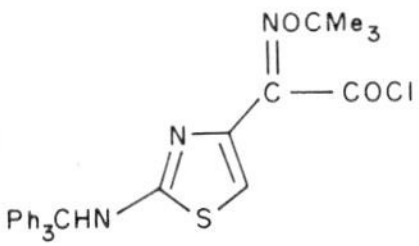 NOCMe$_3$, C—COCl, N, S, Ph_3CHN	1. Et_3N, DMF, −20 °C, stir 30 min 2. r.t. stir 1 h 3. THF, 50% aq. HCO_2H, 40–50 °C, 1 h	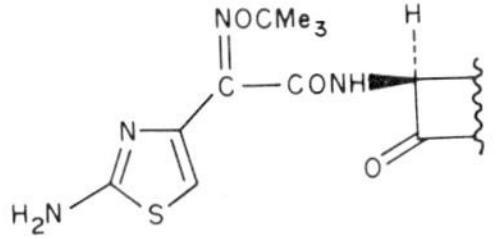 NOCMe$_3$, H, C—CONH, N, S, H_2N, O	47	1077
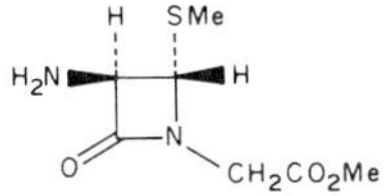 H, SMe, H_2N, H, N, CH_2CO_2Me, O	$PhCH_2COCl$	1. C_5H_5N, 0 °C, 15 min 2. stir r.t. 45 min	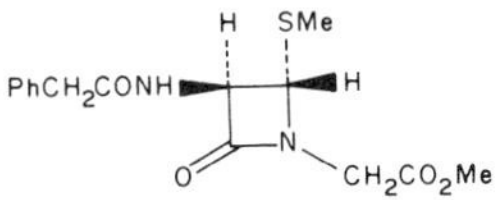 H, SMe, $PhCH_2CONH$, H, N, CH_2CO_2Me, O	—	1174

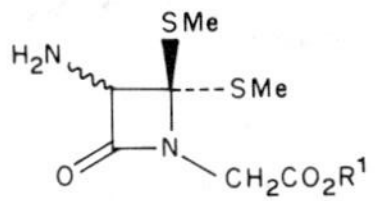	R^2CH_2COCl	1. C_5H_5N, 0 °C, 15 min 2. stir r.t. 45 min	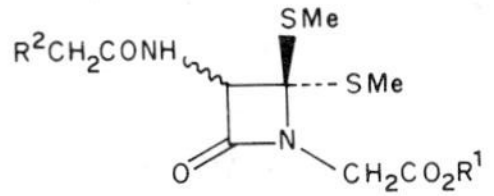		
$R^1 = t$-Bu	$R^2 = Ph$			59	1174
$R^1 = Me$	$R^2 = PhO$			35	1174
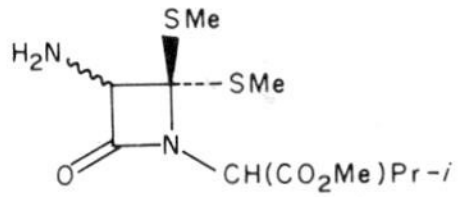	$PhCH_2COCl$	1. C_5H_5N, 0 °C, 15 min 2. stir r.t. 45 min	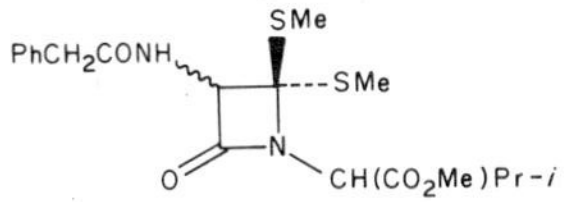	45	1174
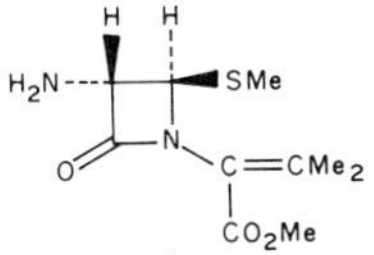	$PhCH_2COCl$	Et_3N, CH_2Cl_2, 0 °C	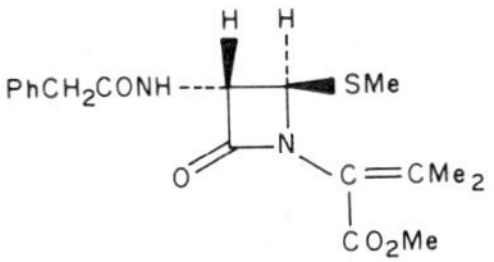	67	1178
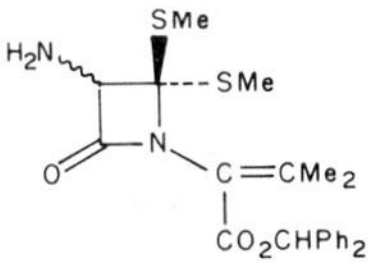	$PhOCH_2COCl$	C_5H_5N, 0 °C, 20 min	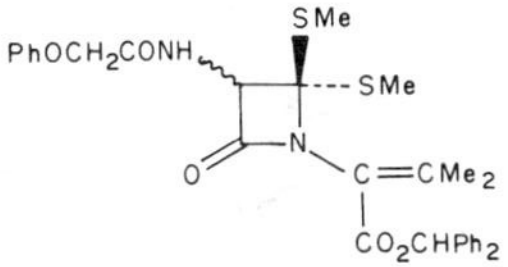	—	1174
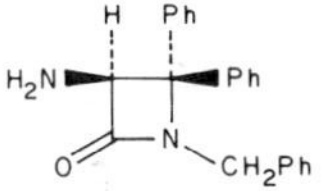	$PhOCH_2COCl$	1. Et_3N, CH_2Cl_2, 5–10 °C 2. r.t. stir 4–5h	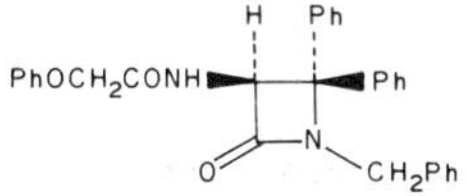	52	1143
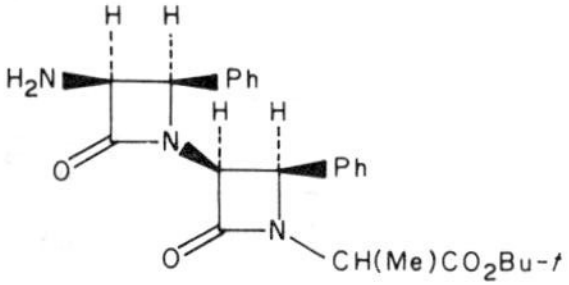	MeCOCl	N—Me morpholine, CH_2Cl_2, 0 °C, stir 1h	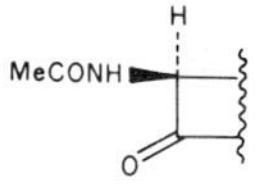	85	1171

(continued)

TABLE 79. (*continued*)

Lactam	Acid chloride	Conditions	Product	Yield (%)	Reference
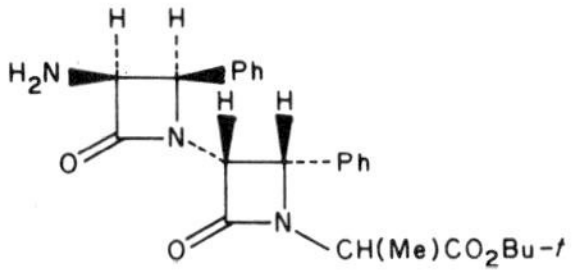 H_2N, H, H, Ph, H, H, Ph, $CH(Me)CO_2Bu$-*t*	MeCOCl	N—Me morpholine, CH_2Cl_2, 0 °C, stir 1 h	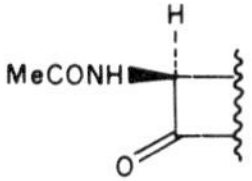 MeCONH, H	—	1171
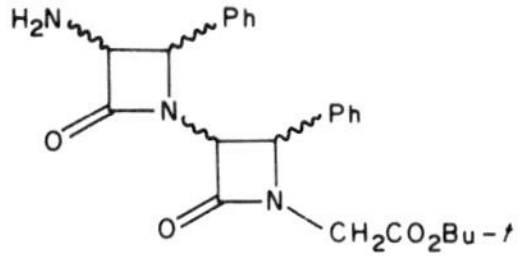 H_2N, Ph, Ph, CH_2CO_2Bu-*t*	MeCOCl	N—Me morpholine	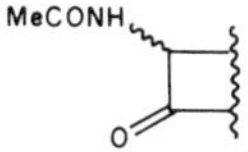 MeCONH	69	1169
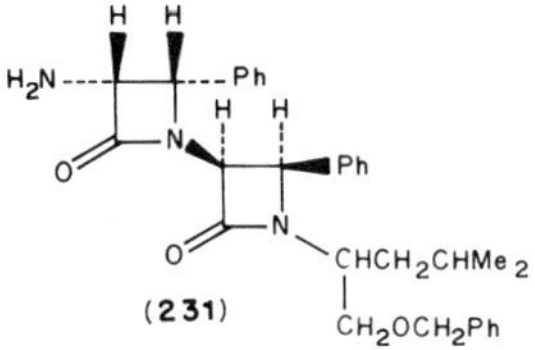 H_2N, H, H, Ph, H, H, Ph, $CHCH_2CHMe_2$, CH_2OCH_2Ph (**231**)	PhCOCl	N—Me morpholine	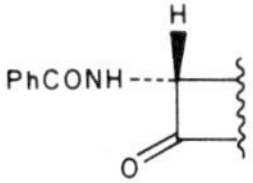 PhCONH, H	—	1171
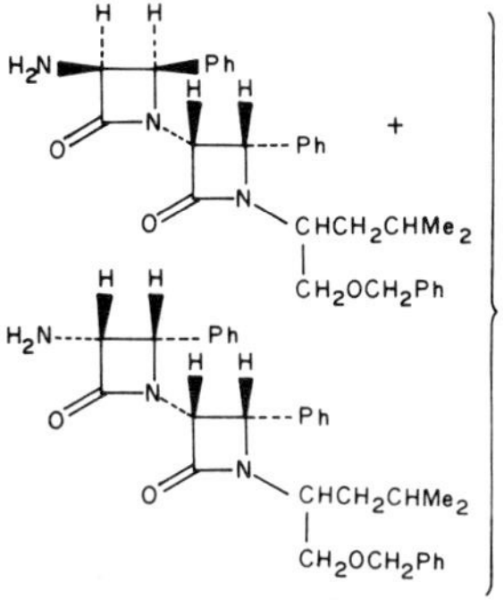 H_2N, H, H, Ph, H, H, Ph, $CHCH_2CHMe_2$, CH_2OCH_2Ph + H_2N, H, H, Ph, H, H, Ph, $CHCH_2CHMe_2$, CH_2OCH_2Ph	PhCOCl	N—Me morpholine	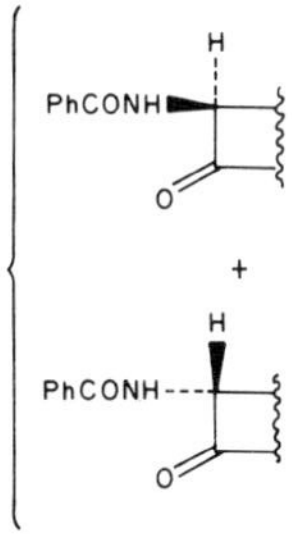 PhCONH, H + PhCONH, H	—	1171

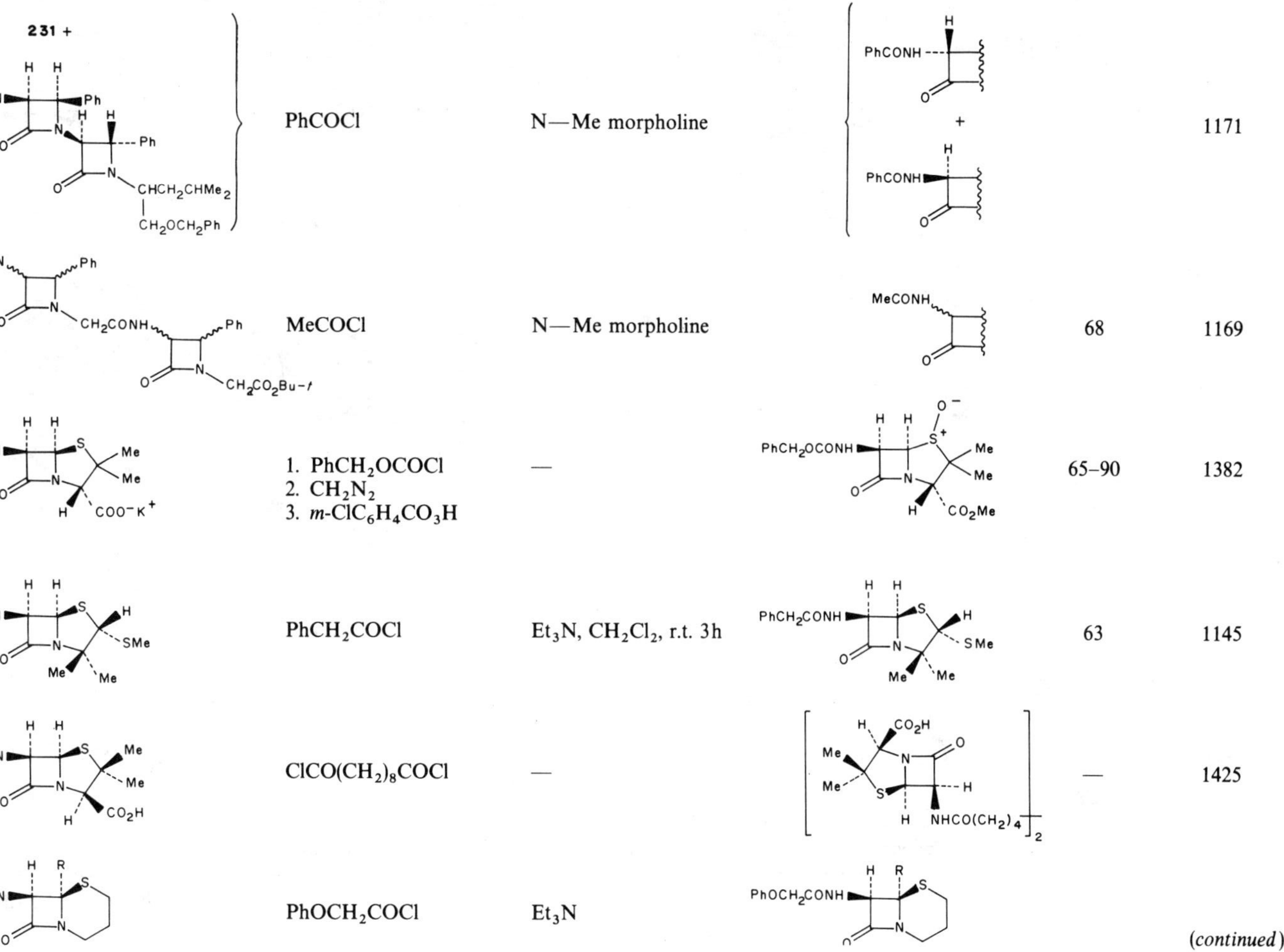

231 +	PhCOCl	N—Me morpholine			1171
	MeCOCl	N—Me morpholine		68	1169
	1. $PhCH_2OCOCl$ 2. CH_2N_2 3. *m*-$ClC_6H_4CO_3H$	—		65–90	1382
	$PhCH_2COCl$	Et_3N, CH_2Cl_2, r.t. 3h		63	1145
	$ClCO(CH_2)_8COCl$	—		—	1425
	$PhOCH_2COCl$	Et_3N			

(*continued*)

TABLE 79. *(continued)*

Lactam	Acid chloride	Conditions	Product	Yield (%)	Reference
R = Ph				85	1178, 1180
R = p-$O_2NC_6H_4$				—	1180
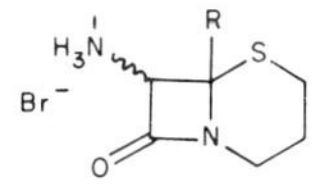	$PhOCH_2COCl$	Et_3N, CH_2Cl_2, r.t. stir 6h	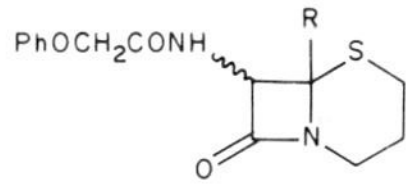		1180
R = Ph				68	
R = p-$O_2NC_6H_4$				65	
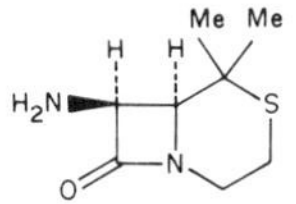	$PhOCH_2COCl$	Et_3N, CH_2Cl_2	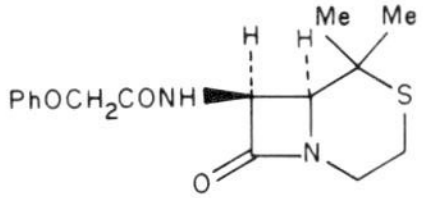	—	1178
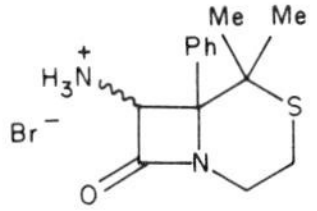	$PhOCH_2COCl$	Et_3N, CH_2Cl_2, r.t. stir	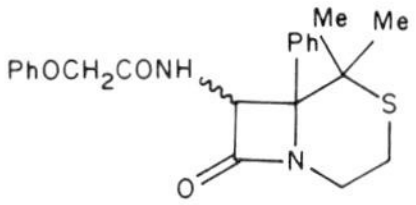	68	1180
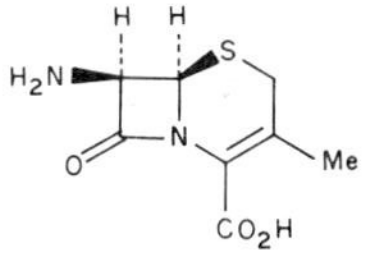	$ClCO(CH_2)_8COCl$	—	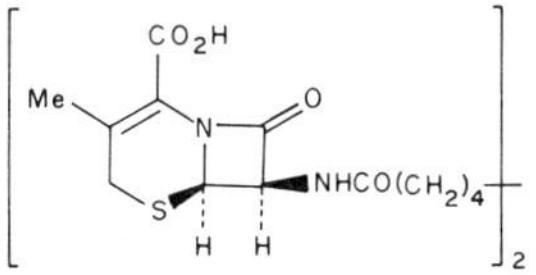	—	1425

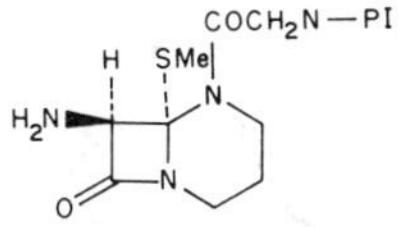	$PhOCH_2COCl$	Et_3N, CH_2Cl_2	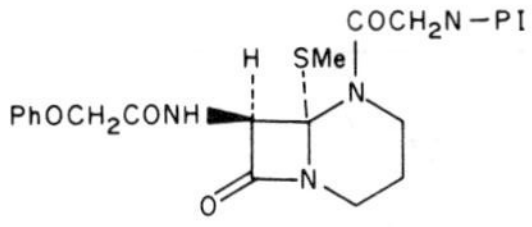	—	1178
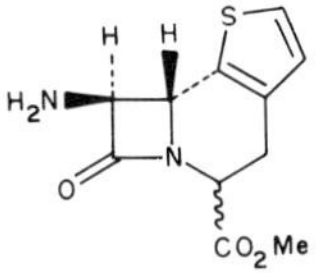	$PhCH_2COCl$	4% NaOH, CH_2Cl_2	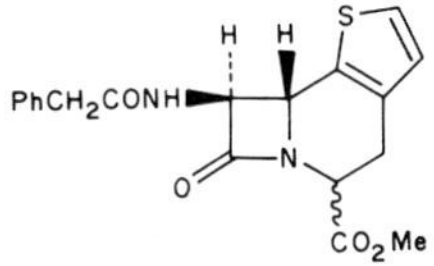	54	1187
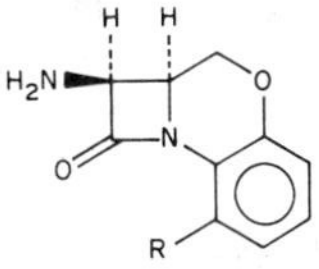	$PhCH_2COCl$	Et_3N, CH_2Cl_2, 0 °C, stir 1 h	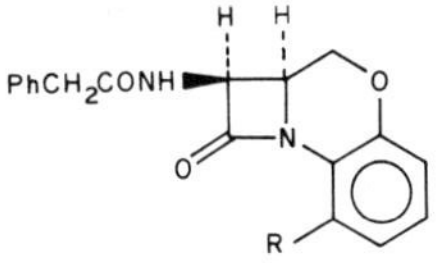		1176
$R = CO_2CHPh_2$				77	
$R = CH_2OSiMe_2(Bu\text{-}t)$				80	
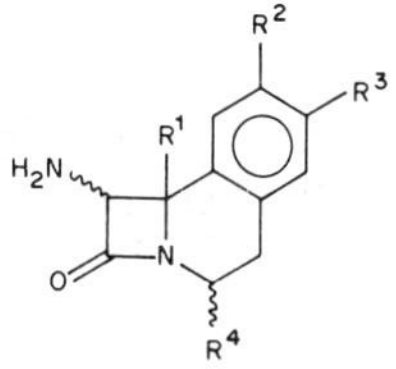	R^5COCl	Et_3N, CH_2Cl_2, stir overnight	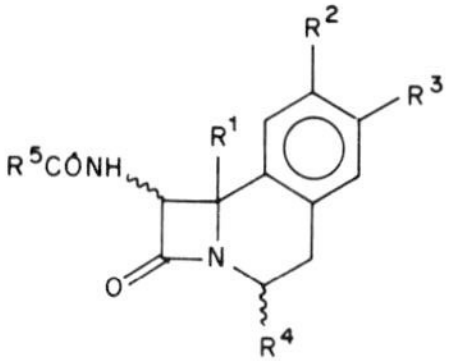		
$R^1 = p\text{-Tol}$, $R^2 = R^3 = R^4 = H$	$R^5 = PhCH_2$			75	1154
$R^1 = p\text{-An}$, $R^2 = R^3 = R^4 = H$	$R^5 = PhCH_2$			72	1154

(continued)

TABLE 79. *(continued)*

Lactam	Acid chloride	Conditions	Product	Yield (%)	Reference
$R^1 = R^2 = R^3 = R^4 = H$	$R^5 = Cl_2CH$			55	1154
$R^1 = p\text{-}BrC_6H_4$, $R^2 = R^3 = OMe$, $R^4 = H$	$R^5 = PhCH_2$			75	1154, 1178
$R^1 = p\text{-}An$ $R^2 = R^3 = OMe$, $R^4 = H$	$R^5 = PhCH_2$			78	1154
$R^1 = Ph$, $R^2 = R^3 = OMe$, $R^4 = CO_2Me$	$R^5 = PhCH_2$			70	1154, 1178
H_2N, N, N, R, O, $CO_2CH_2CH{=}CH_2$	$CH_2{=}CHCH_2O_2CNH$, S, N, $C{=}NOMe$, COCl	EtOAc, CH_2Cl_2	$CH_2{=}CHCH_2O_2CNH$, S, N, $C{=}NOMe$, CONH, O		1334
$R = CO_2Me$				42	
$R = CO_2CH_2CH{=}CH_2$				59	
H_2N, N, N, Me, Me, $CO_2CH_2CH{=}CH_2$, $CO_2CH_2CH{=}CH_2$, O	OHCHN, S, N, $C{=}NOMe$, COCl	$NaHCO_3$, Me_2CO, H_2O	OHCNH, S, N, $C{=}NOMe$, CONH, O	—	1046
H_2N, H, H, S, H, H, N, H, CO_2Me, O	$PhOCH_2COCl$	Et_3N, CH_2Cl_2	$PhOCH_2CONH$, H, O	—	999

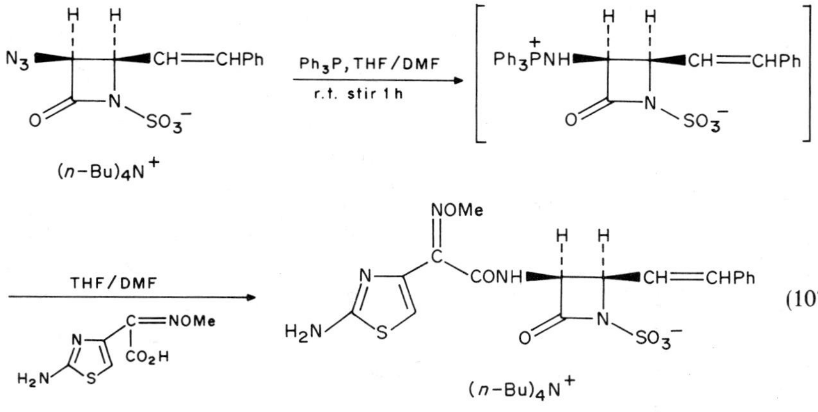

$\xrightarrow[\text{r.t. stir 1 h}]{Ph_3P,\ THF/DMF}$ $\xrightarrow[\text{THF/DMF}]{}$ (1072)

R^1CONH + $R^1CO_2COR^2$ $\xrightarrow{\text{base}}$ (1073)

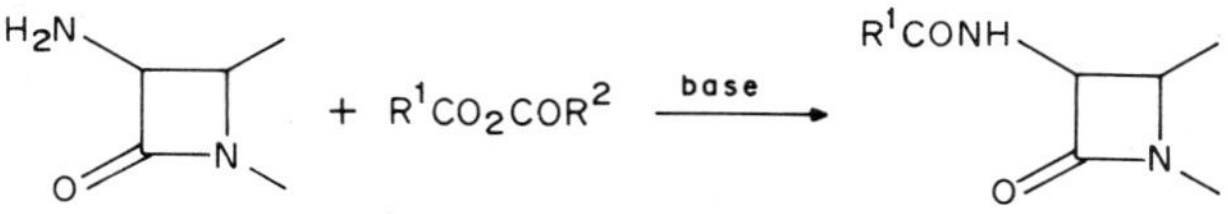

+ $R^1CO_2R^2$ $\xrightarrow{\text{base}}$ (1074)

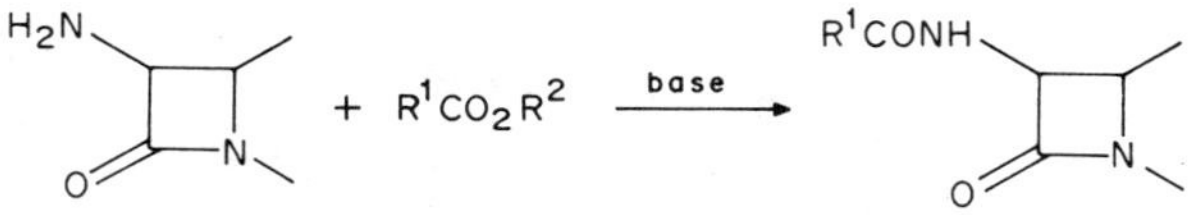

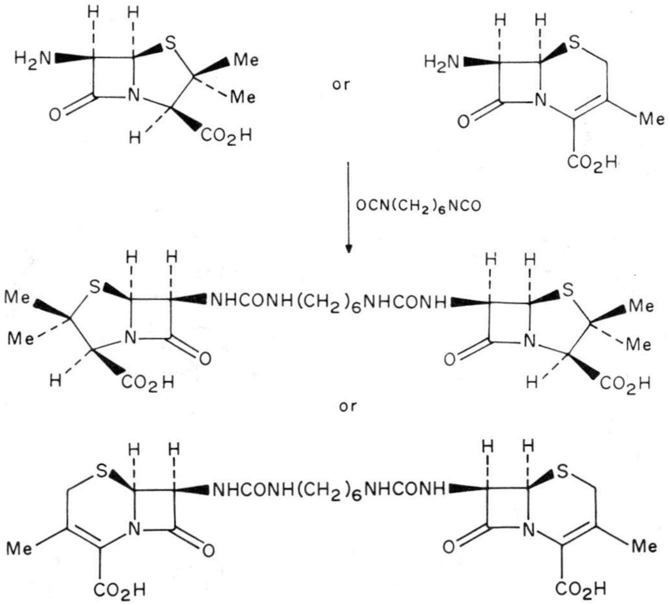

or

$\xrightarrow{OCN(CH_2)_6NCO}$

$NHCONH(CH_2)_6NHCONH$ (1075)

or

$NHCONH(CH_2)_6NHCONH$

TABLE 80. Acyclation of amino groups attached to lactams using a carboxylic acid and *N,N*-dicyclohexylcarbodiimide

Lactam	Carboxylic acid	Conditions	Product	Yield (%)	Reference
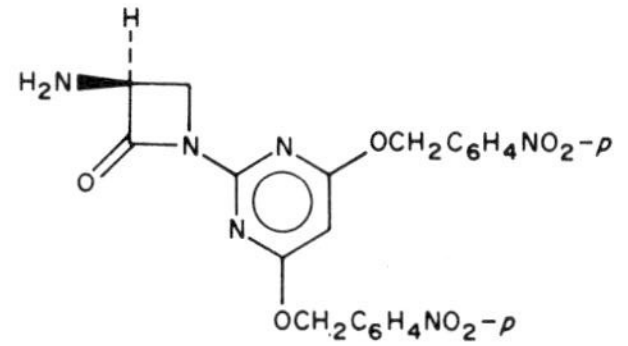 H_2N, H; $OCH_2C_6H_4NO_2$-*p*; $OCH_2C_6H_4NO_2$-*p*	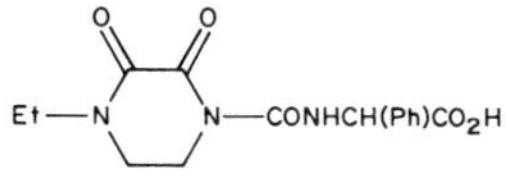 Et—N, N—CONHCH(Ph)CO_2H D-(−)	1. DCCD, DMF, 0 °C 2. stir r.t. 3h	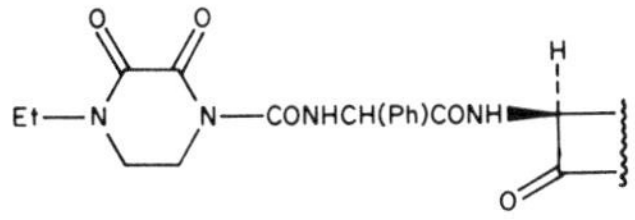 Et—N, N—CONHCH(Ph)CONH, H	21	1077
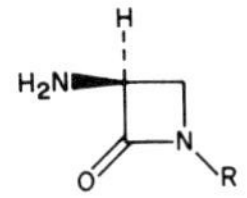 H_2N, H, N–R	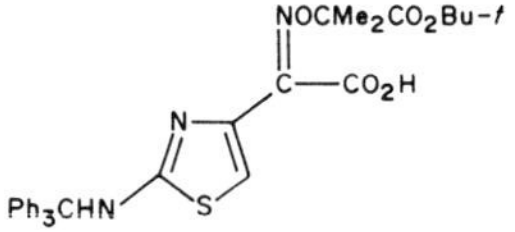 $NOCMe_2CO_2Bu$-*t*, C—CO_2H, Ph_3CHN	1. DCCD, DMF, 0 °C Mol. Sieves (4A) 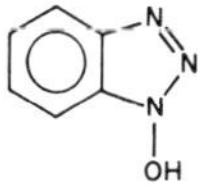 N, N, N–OH 2. stir r.t. 1h	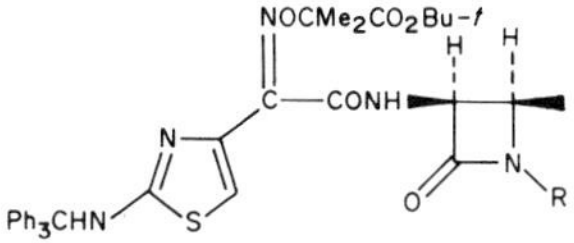 $NOCMe_2CO_2Bu$-*t*, C—CONH, H, H, N–R, Ph_3CHN	50–75	1077
R = 2-Pyr, 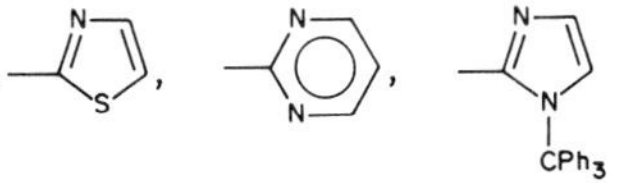 (thiazol-2-yl), (pyrimidin-2-yl), (1-CPh_3-imidazol-2-yl)					
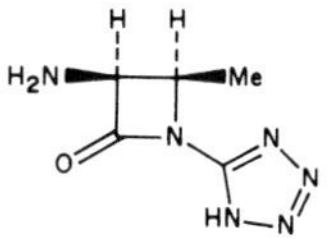 H_2N, H, H, Me, HN—N	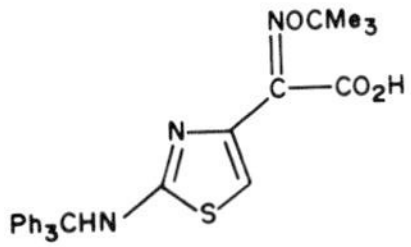 $NOCMe_3$, C—CO_2H, Ph_3CHN	1. DCCD, DMF, Et_3N, 0 °C, Mol. Sieves (4A) 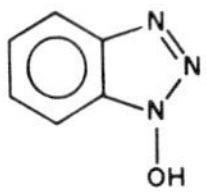 N, N, N–OH	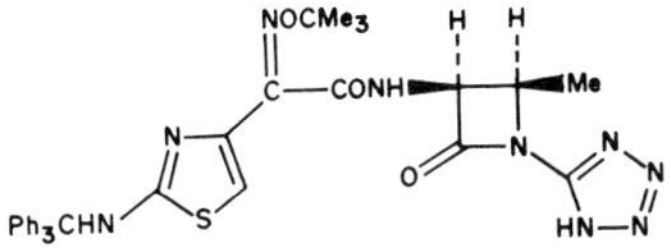 $NOCMe_3$, C—CONH, H, H, Me, HN—N, Ph_3CHN	49	1317

		2. stir r.t. 2h 3. H_2O, EtOAc, 2*N* HCl, (pH = 2)			
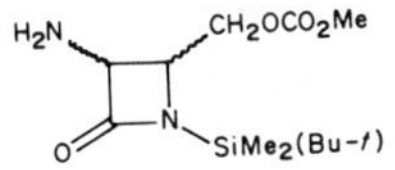	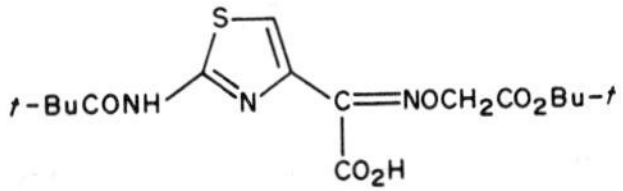	DCCD	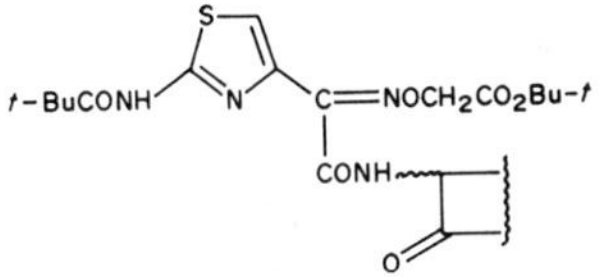	—	1317
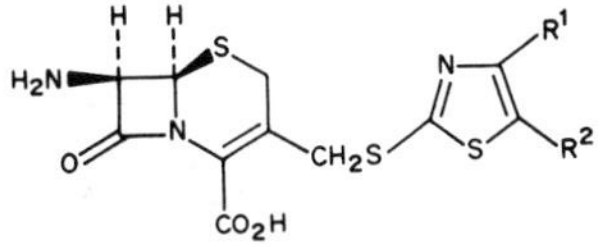	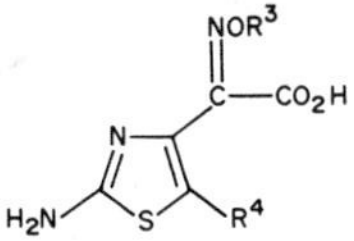	DCCD, H_2O, DMF, r.t. stir, 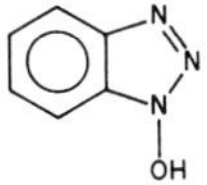	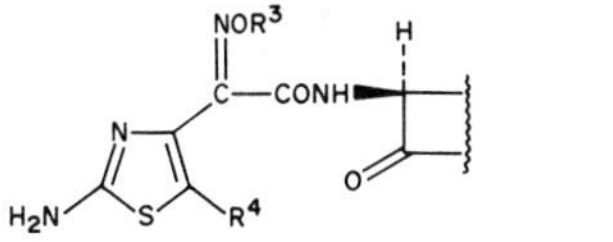	—	1426
$R^1 = Me$, $R^2 = CH_2CO_2Et$ $R^1 = Me$, $R^2 = CH_2CO_2Me$ $R^1 = Me$, $R^2 = CH_2CO_2Et$ $R^1 = R^2 = H$	$R^3 = Me$, $R^4 = H$ oxime = E $R^3 = R^4 = H$ oxime = Z $R^3 = R^4 = H$ oxime = Z $R^3 = Me$, $R^4 = Cl$				
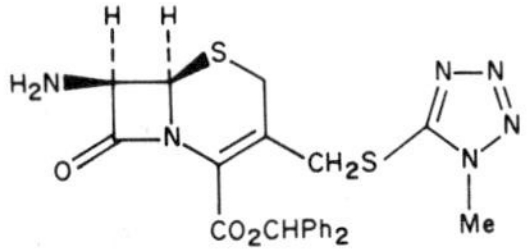	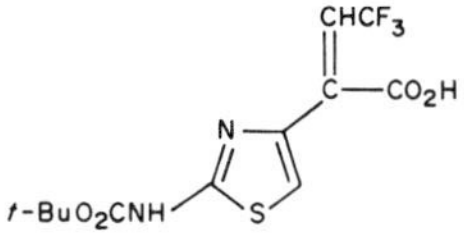	DCCD, CH_2Cl_2, r.t. 1.5h	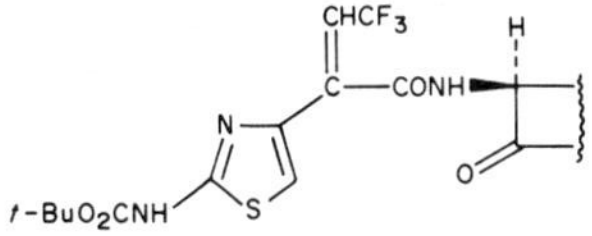	—	1427

(*continued*)

TABLE 80. (*continued*)

Lactam	Carboxylic acid	Conditions	Product	Yield (%)	Reference
H H S H_2N N O CO_2Me	2-ThiCH_2CO_2H	DCCD, CH_2Cl_2	H 2-ThiCH_2CONH O	61	1187
H H Me H_2N N N O CO_2Et	R^1CO_2H	DCCD. CH_2Cl_2, 0 °C, 2h	H R^1CONH O		1182
	R^1 = PhCH_2 R^1 = PhOCH_2 R^1 = *p*-$O_2NC_6H_4CH_2$ R^1 = 2-ThiCH_2			93 92 65 92	
H_2N Ph N O CH_2CO_2Bu-*t*	N_3 Ph N O CH_2CO_2H	DCCD	N_3 Ph N O CH_2CONH Ph N O CH_2CO_2Bu-*t*	42	1169
H H Cl^- $H_3\overset{+}{N}$ SHgCl N O CHCHMe_2 CO_2CH_2R	*p*-R$C_6H_4CH_2O_2C$NH *p*-R$C_6H_4CH_2O_2C$—C$(CH_2)_3CO_2H$ H	Et_3N, CH_2Cl_2. EEDQ[a], stir 24h	*p*-R$C_6H_4CH_2O_2C$NH *p*-R$C_6H_4CH_2O_2C$—C—C$(CH_2)_3$CONH H H O	—	1434

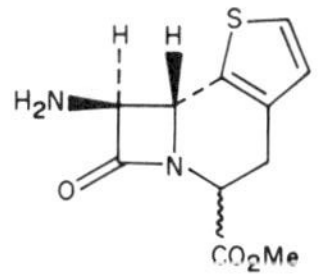

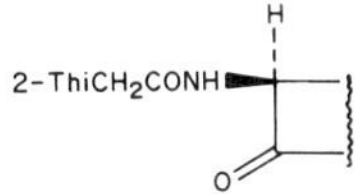

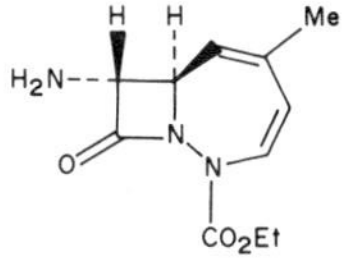

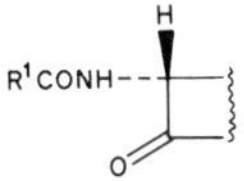

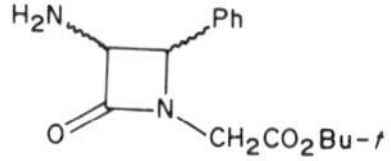

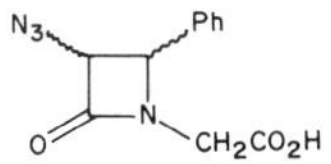

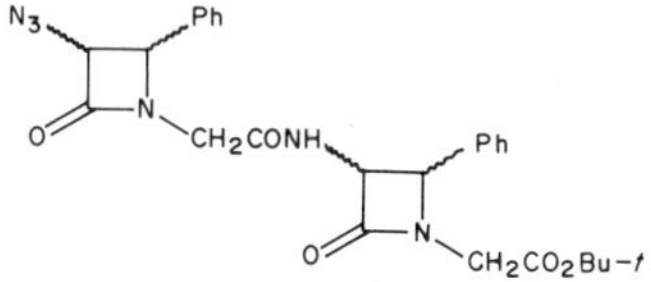

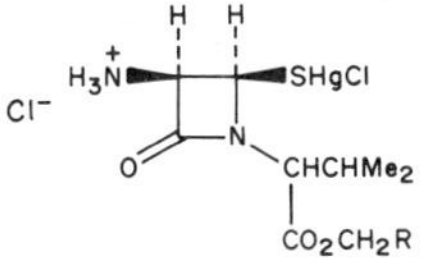

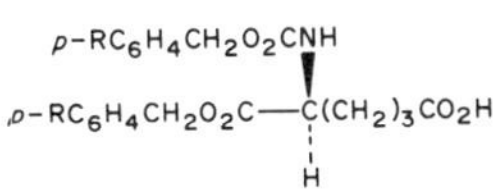

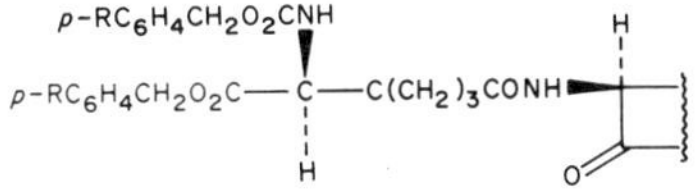

R = *p*-An, *p*-$O_2NC_6H_4$

[a]EEDQ = 2-ethoxy-1-ethoxycarbonyl-1,2-dihydroquinoline.

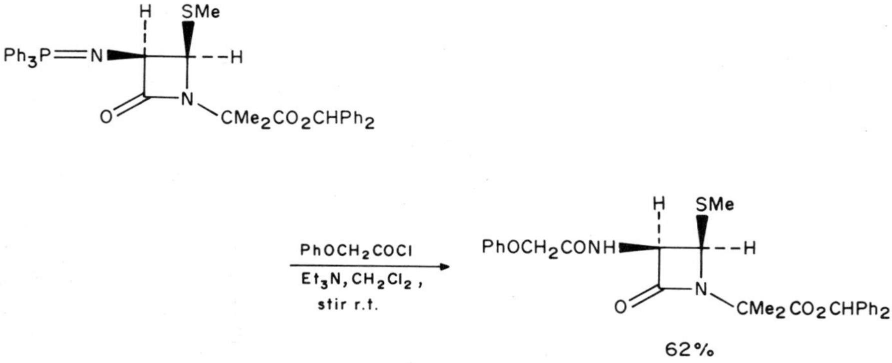

$Ph_3P{=}N$ ... $CMe_2CO_2CHPh_2$

$\xrightarrow[\text{Et}_3\text{N, CH}_2\text{Cl}_2\text{, stir r.t.}]{\text{PhOCH}_2\text{COCl}}$ $PhOCH_2CONH$ ... $CMe_2CO_2CHPh_2$

62%

(1076)

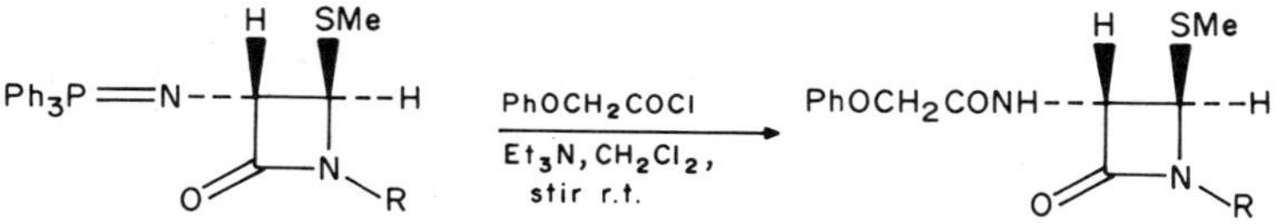

$Ph_3P{=}N$ ... R $\xrightarrow[\text{Et}_3\text{N, CH}_2\text{Cl}_2\text{, stir r.t.}]{\text{PhOCH}_2\text{COCl}}$ $PhOCH_2CONH$ ... R (1077)

R = CMe_2CO_2H ; $CMe_2CO_2CHPh_2$
Yield (%) = 44 ; 75

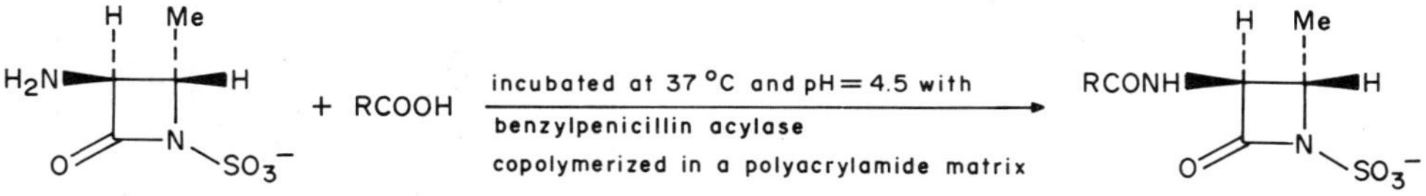

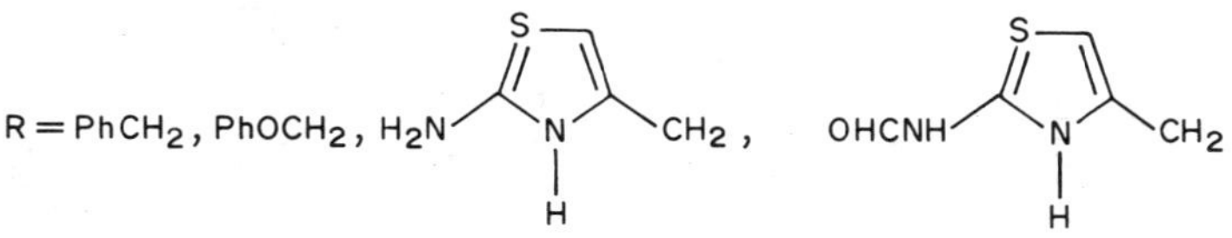

H_2N ... SO_3^- + RCOOH $\longrightarrow$ RCONH ... SO_3^-

incubated at 37 °C and pH = 4.5 with benzylpenicillin acylase copolymerized in a polyacrylamide matrix

R = $PhCH_2$, $PhOCH_2$, H_2N-(thiazolyl)-CH_2, OHCNH-(thiazolyl)-CH_2

No acylation occurred using the following acids:

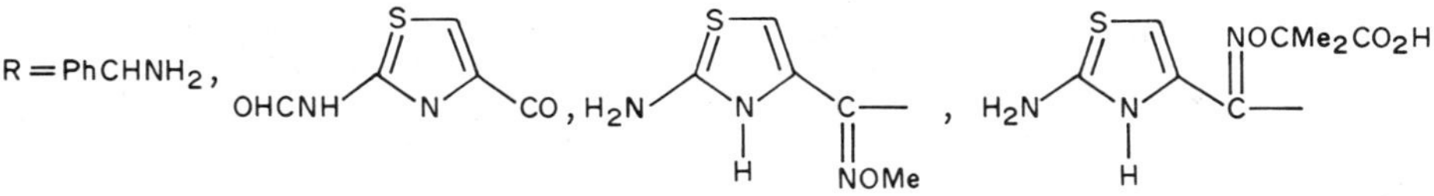

R = $PhCHNH_2$, OHCNH-(thiazolyl)-CO, H_2N-(thiazolyl)-C(=NOMe)—, H_2N-(thiazolyl)-C(=$NOCMe_2CO_2H$)—

(1078)

TABLE 81. Acylation of amino groups attached to lactams using a mixed anhydride

Lactam	Anhydride	Conditions	Product	Yield (%)	Reference
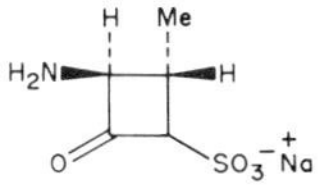	MeO_2CO_2C—CHPh, MeO_2CCH=CMeNH (NH bonded to CHPh)	1. Et_3N, DMF, −40 °C to −45 °C, stir 2h 2. r.t.	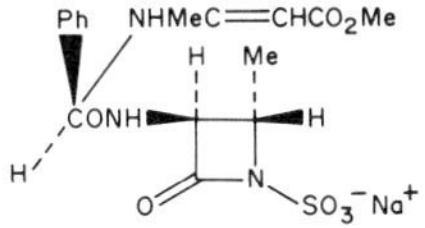	—	1295
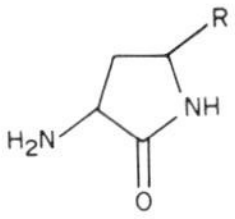	$(t\text{-}BuO_2C)_2O$	Et_3N, CH_2Cl_2, stir r.t. 3h	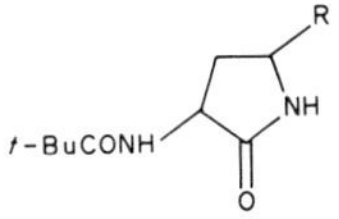		
R = H				85	1305
R = CO_2Et		CH_2Cl_2, 72h		—	1346
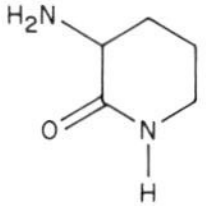	$(t\text{-}BuO_2C)_2O$	Et_3N, CH_2Cl_2, stir r.t.	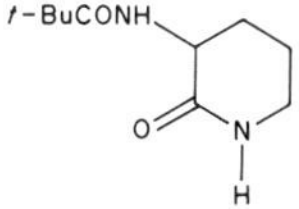	—	1305
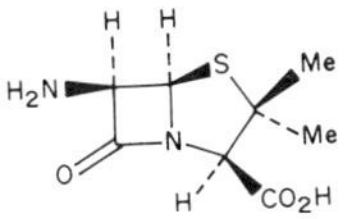	H_2C=CHCHCO$_2$CO$_2$Et, NHMeC=CHCO$_2$Me (NH bonded to CH)	Et_3N, Me_2CO	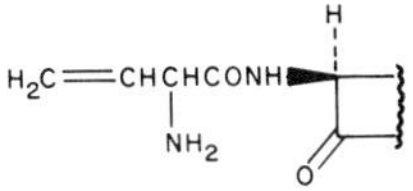	—	1428
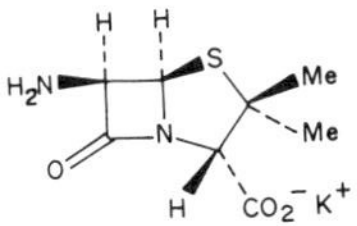	$(t\text{-}BuO_2C)_2O$	1. CH_2N_2 2. $m\text{-}ClC_6H_4CO_3H$	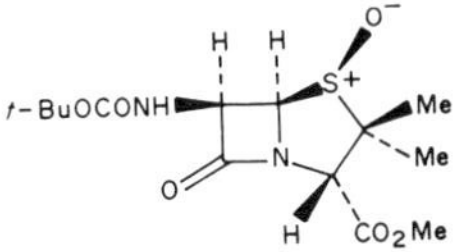	65–90	1382

TABLE 82. Acylation of amino groups attached to lactams using an ester

Lactam	Ester	Conditions	Product	Yield (%)	Reference
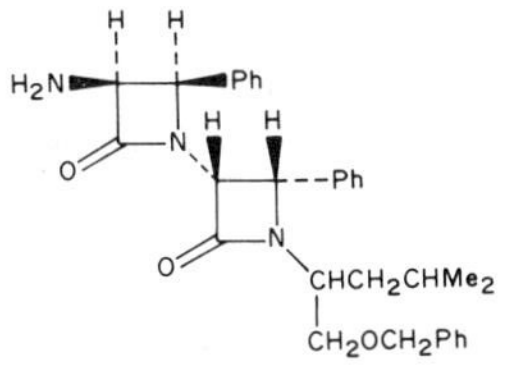	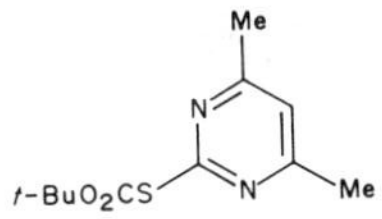	Et_3N, THF	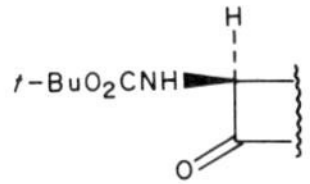	—	1171
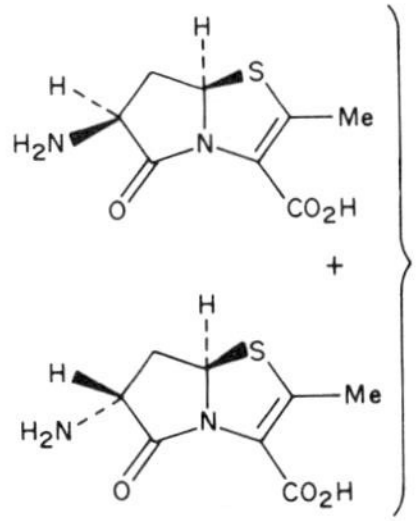	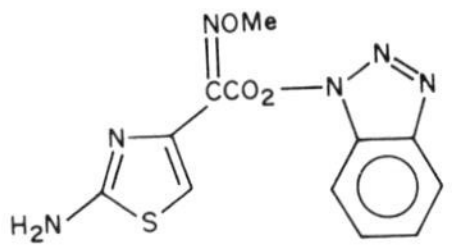	$NaHCO_3$, Me_2CO, H_2O	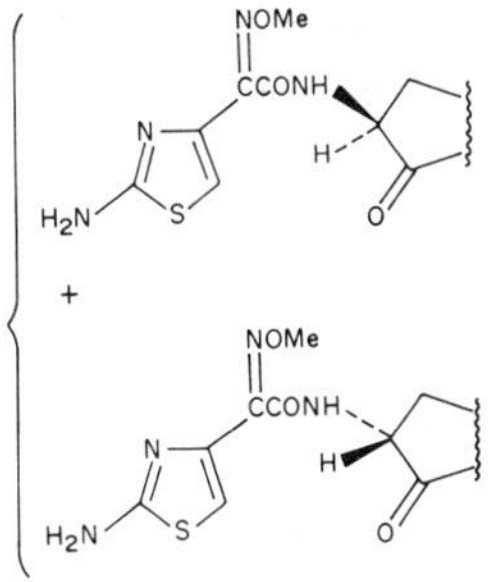	—	1346
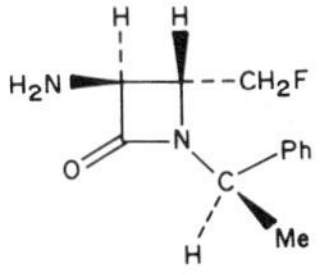	PI—NCO_2Et	THF, reflux 5h	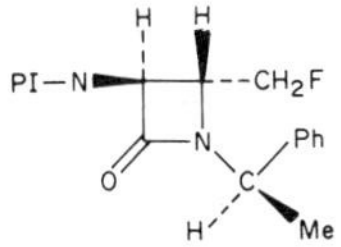	82	1184

TABLE 83. Formation of Schiff bases

Lactam	Aldehyde	Conditions	Product	Yield (%)	Reference
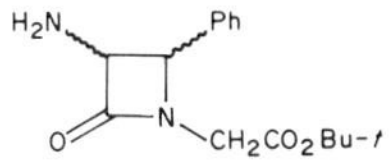	PhCHO	C_6H_6, Na_2SO_4	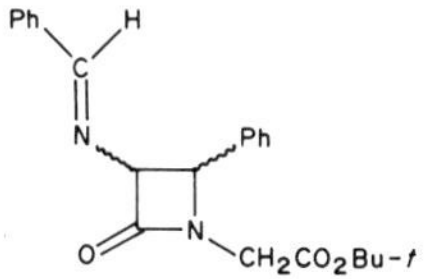	100	1169
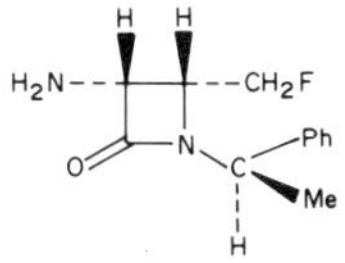	PhCHO	CH_2Cl_2, $MgSO_4$, 20 °C, 30 min	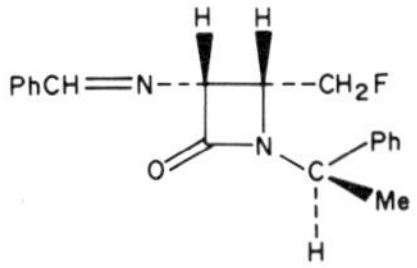	100	1184
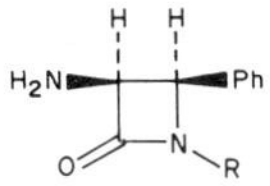	PhCHO	C_6H_6, Na_2SO_4, r.t. stir	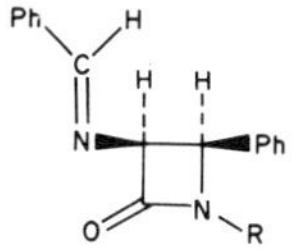		1171
R = t-Bu				—	
R = $MeCHCO_2Bu$-t		17h		100	
R = $—CHCH_2CHMe_2$ (with CH_2OCH_2Ph on CH)		17h		100	
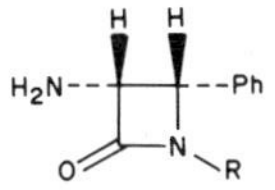	PhCHO	C_6H_6, Na_2SO_4, r.t. stir	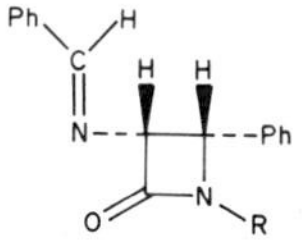		1171

Substrate	Reagent	Conditions	Product	Yield (%)	Ref.
R = $MeCHCO_2Bu$-*t*				100	
R = —$CHCH_2CHMe_2$ (with CH_2OCH_2Ph on the CH)		17h		100	
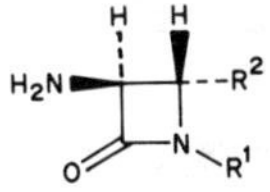 cis-3-amino-2-azetidinone (H_2N, H, H, R^2 at C-4, $N{-}R^1$, C=O)	 O_2N–C_6H_4–CHO (p)	C_6H_6, reflux	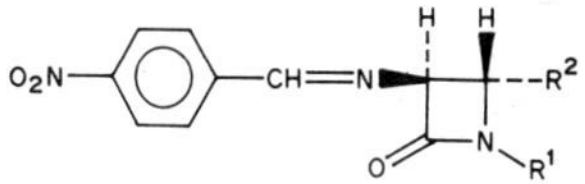 O_2N–C_6H_4–CH=N– (3-position of the same azetidinone: H, H, R^2, $N{-}R^1$, C=O)		802
R^1 = *o*-$MeOCH_2O_2CC_6H_4$, R^2 = Ph		1.5h		100	
R^1 = *o*-$MeO_2CC_6H_4$, R^2 = MeO		2h		—	
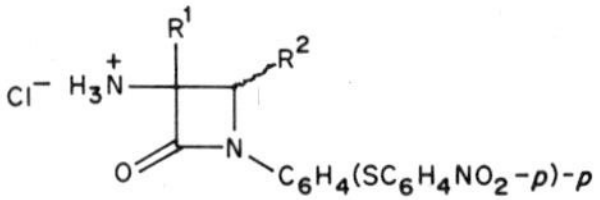 Cl^- $H_3\overset{+}{N}$– at C-3 (R^1) of an azetidin-2-one with R^2 at C-4 and N–$C_6H_4(SC_6H_4NO_2$-*p*)-*p*	*p*-AnCHO	EtOH or *n*-BuOH, C_5H_5N, reflux 2h	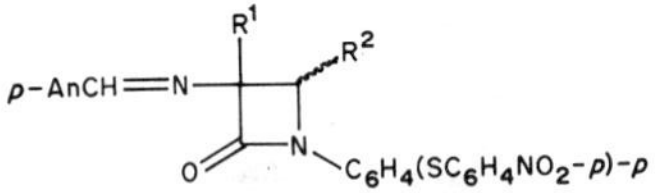 *p*-AnCH=N– at C-3 (R^1) of the same azetidin-2-one (R^2 at C-4, N–$C_6H_4(SC_6H_4NO_2$-*p*)-*p*)		1186
R^1 = H, R^2 = Ph				71	
R^1 = H, R^2 = *p*-$O_2NC_6H_4$				74	
R^1 = H, R^2 = *p*-An				73	
R^1 = H, R^2 = *p*-ClC_6H_4				75	
R^1 = H, R^2 = *o*-HOC_6H_4				78	
R^1 = H, R^2 = *p*-$Me_2NC_6H_4$				80	
R^1 = Me, R^2 = Ph				81	
R^1 = Me, R^2 = *p*-An				82	
R^1 = Me, R^2 = *p*-$O_2NC_6H_4$				80	
R^1 = *i*-Pr, R^2 = Ph				79	
R^1 = *i*-Pr, R^2 = *p*-An				78	
R^1 = *i*-Pr, R^2 = *p*-$O_2NC_6H_4$				75	

(continued)

TABLE 83. (*continued*)

Lactam	Aldehyde	Conditions	Product	Yield (%)	Reference
HN, N—NH_2, O	RCHO R = 2-Fu, 2-Thi 2-(5-X-Thi), 3-Pyr, PhCH=CH, Ph, *p*-An, *p*-$O_2NC_6H_4$, *p*-NCC_6H_4, *p*-MeTos, 3,4-$Cl_2C_6H_3$	—	HN, N—N=CHR, O	—	1331
H, H, Me, H_2N, O, N, N, CO_2Et	O_2N—C_6H_4—CHO	C_6H_6, r.t. 1h	O_2N—C_6H_4—CH=N, H, O	72	1182

Another common reaction which amine substituents located on a lactam ring carbon alpha to the carbonyl function undergo is conversion to a Schiff base by condensation with an aldehyde (equation 1079 and Table 83), or by reaction[1182] of 2-[([7α, 8α]-2-(ethoxycarbonyl or benzoyl)-5-(unsubstituted or methyl)-9-oxo-1,2-diazabicyclo[5.2.0]-3,5-dien-8-yl)aminocarbonyl]benzoic acid with dicyclohexylcarbodiimide (DCCD) (equation 1080).

H_2N + RCHO → RCH=N (1079)

DCCD, CH_2Cl_2, stir 1 h

R^1 =	H	Me	Me
R^2 =	CO_2Et	CO_2Et	COPh
Yield (%) =	66	90	85

(1080)

Similarly, reaction of 3-azido-2-azetidinones with triphenylphosphine produces[1168,1173] the corresponding 3-phosphino β-lactams (equations 1081[1168], 1082[1173] and 1083[1173]), while treatment of the *trans*-3-phosphinimino β-lactam resulting from equation 1083, with *p*-nitrobenzaldehyde, affords[1173] the Schiff base (equation 1084).

N_3 — CH=CHPh; THF, Ph_3P, N_2, stir 2 h → Ph_3P=N — CH=CHPh

(1081)[1168]

N_3 — SMe, N—$CMe_2CO_2CHPh_2$; Ph_3P, C_6H_6, stir 15 – 30 min → Ph_3P=N — SMe, N—$CMe_2CO_2CHPh_2$

(1082)[1173]

$$\text{3-azido-4-SMe-azetidin-2-one (N-R)} \xrightarrow[15-30\,\text{min}]{Ph_3P,\ C_6H_6,\ \text{stir}} Ph_3P{=}N\text{-azetidinone (N-R)} \quad (1083)^{1173}$$

$R = CMe_2CO_2CHPh_2$; CMe_2CO_2H

Yield (%) = 90 ; —

$$Ph_3P{=}N\text{-azetidinone (H, SMe; N-}CMe_2CO_2CHPh_2) + p\text{-}O_2N\text{-}C_6H_4\text{-}CHO \xrightarrow{2\,h} p\text{-}O_2NC_6H_4CH{=}N\text{-azetidinone (H, SMe; N-}CMe_2CO_2CHPh_2)\ \ 87\% \quad (1084)$$

One method used to produce the 3-amino substituted lactams required in the reactions described above is reduction of the 3-azido analogue (equation 1085 and Table 84), while a second method used involves deprotection of variously substituted 3-amino lactams (equation 1086 and Table 85).

$$N_3\text{-azetidin-2-one} \xrightarrow{\text{Reduction}} H_2N\text{-azetidin-2-one} \quad (1085)$$

$$RNH\text{-azetidin-2-one} \quad \text{or} \quad RCH{=}N\text{-azetidin-2-one} \xrightarrow{\text{Conditions}} H_2N\text{-azetidin-2-one} \quad (1086)$$

$R = R^1O_2C$, $R^1CH{=}CR^2N$, PI-N

TABLE 84. Preparation of 3-amino lactams by reduction of 3-azido lactams

Azido lactam	Reducing agent and conditions	Product	Yield (%)	Reference
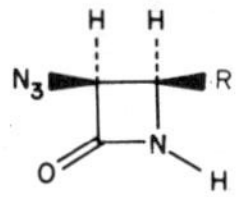		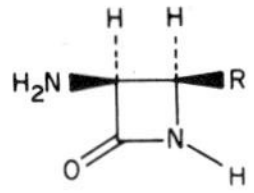		
R = Ph	Adams' catalyst, H_2, EtOH		—	1379
R = CH=CHPh	$NaBH_4$, EtOH, H_2, $Ni(OAc)_2$		30	1168
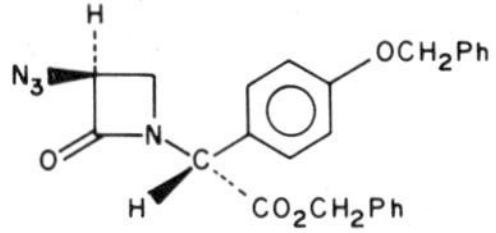	1. Zn, 90% aq. HOAc, stir 1h, r.t. 2. EtOAc, p-TosOH	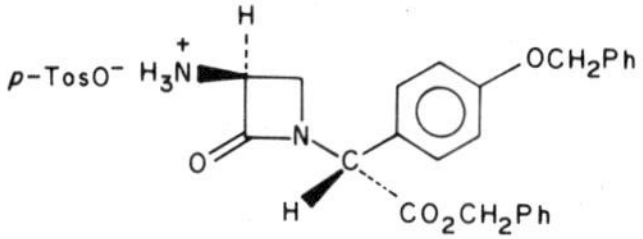	46	1039
	1. H_2S, CH_2Cl_2, Et_3N, 0 °C 2. EtOAc, p-TosOH		67	1039
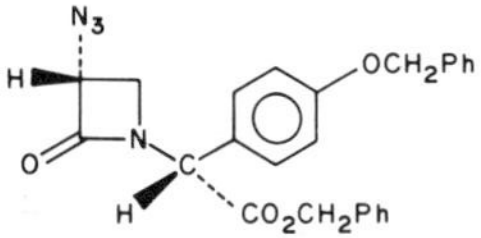	1. Zn, 90% aq. HOAc, stir 1h, r.t. 2. EtOAc, p-TosOH	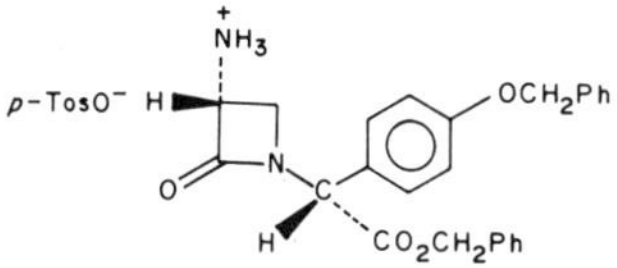	30	1039
	1. H_2S, CH_2Cl_2, Et_3N, 0 °C 2. EtOAc, p-TosOH		70	1039
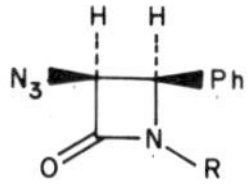		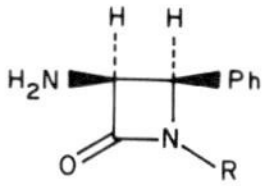		

(continued)

TABLE 84. *(continued)*

Azido lactam	Reducing agent and conditions	Product	Yield (%)	Reference
R = CH_2CO_2Bu-*t*	5% Pd/C, H_2, MeOH, r.t.		90	1169
R = $CH(Me)CO_2Me$	10% Pd/C, H_2, MeOH, r.t.		—	1158
R = $CH(Me)CO_2Bu$-*t*	5% Pd/C, H_2, MeOH, 0–5 °C, 6h		96	1171
R = CH(*i*-Pr)CO_2Me	10% Pd/C, H_2, MeOH, r.t.		—	1158
R = $PhCH_2CHCO_2Me$	10% Pd/C, H_2, MeOH, r.t.		—	1158
R = $CH(Ph)CO_2Me$	10% Pd/C, H_2, MeOH, r.t.		—	1158
R = $Me_2CHCH_2CHCO_2Bu$-*t*	5% Pd/C, H_2 (1 atm.), MeOH, r.t.		100	1170
R = *t*-Bu	5% Pd/C, H_2, MeOH, 0–5 °C		—	1171
R = $Me_2CHCH_2CHCH_2OCH_2Ph$	5% Pd/C, H_2, MeOH, 0–5 °C, 6h		100	1171
R = $Me_2CHCH_2CHCH_2OCH_2Ph$	5% Pd/C, H_2, MeOH, 25 °C		—	1172
N_3, Ph, H, H, O, N, R		H_2N, Ph, H, H, O, N, R		
R = $CH(Me)CO_2Me$	10% Pd/C, H_2, MeOH, r.t.		—	1158
R = $CH(Me)CO_2Bu$-*t*	5% Pd/C, H_2, MeOH, 0–5 °C, 6h		96	1171
R = $Me_2CHCH_2CHCH_2OCH_2Ph$	5% Pd/C, H_2, MeOH, 0–5 °C, 6h		100	1171
N_3, H, SMe, H, O, N, C=CMe_2, CO_2Me	Pd/C, H_2	H_2N, H, SMe, H, O, N, C=CMe_2, CO_2Me	—	1178
N_3, H, SR, H, O, N, C, MeO_2C	cat., H_2 or H_2S	H_2N, H, SR, H, O, N, C, MeO_2C	—	1177
R = Me, Ph_3C				

Substrate	Conditions	Product	Yield (%)	Ref.
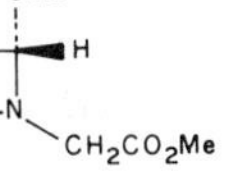	$(NH_4)_2S$, MeOH, NaCl, 0.5h	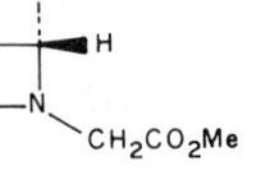	—	1174
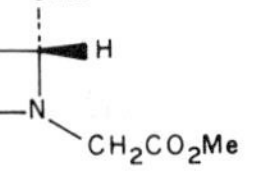	H_2S, CH_2Cl_2, Et_3N, 0 °C, stir r.t. 2h	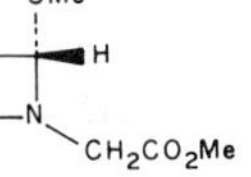		1162
R = H			91	
R = Ph			96	
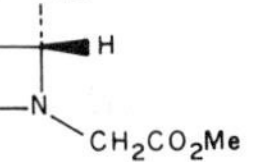	$(NH_4)_2S$, MeOH, NaCl, 0.5h	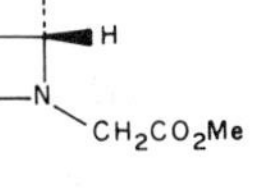	—	1174
R = CH_2CO_2Me, CH_2CO_2Bu-*t*, $Me_2CHCHCO_2Me$, $Me_2C{=}CCO_2CHPh_2$				
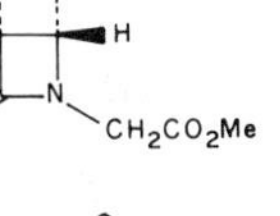	1. H_2S, CH_2Cl_2, Et_3N, 0 °C, 5 min 2. stand 1 h 0 °C 3. stand 45 min 10–15 °C	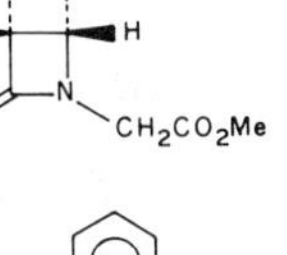 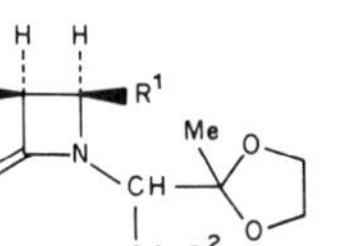	—	1304

R^1	R^2
CH=CHPh	CH_2Ph
C≡CH	CH_2CCl_3

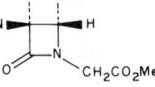

(continued)

TABLE 84. (*continued*)

Azido lactam	Reducing agent and conditions	Product	Yield (%)	Reference
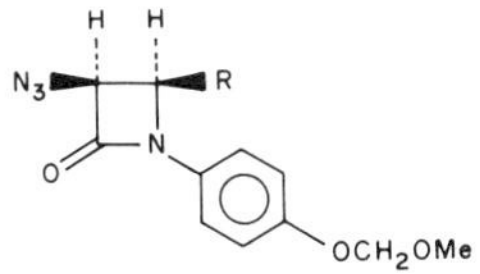 $R = CH{=}CHPh$, $CH_2CH(OMe)_2$, $CH_2CH_2NO_2$, $C{\equiv}CCH_2OSiPh_2(Bu\text{-}t)$	1. H_2S, CH_2Cl_2, Et_3N, 0 °C, 5 min 2. stand 0 °C 1 h	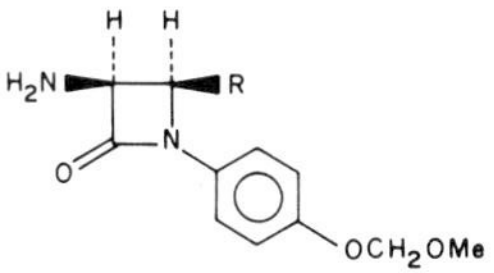	—	1196
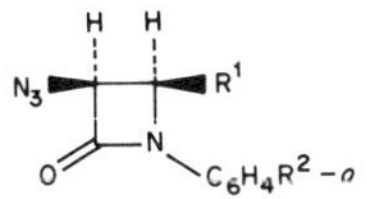 $R^1 = CH{=}CHPh$ (*trans*), CH_2OMe, $R^2 = CH_2OSiMe_2(Bu\text{-}t)$	H_2S, Et_3N, CH_2Cl_2, 1 h	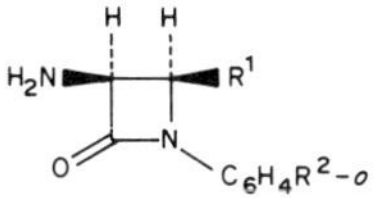	—	802
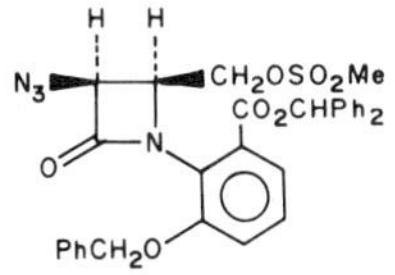	1. H_2S, CH_2Cl_2, 0 °C, 5 min 2. 20 °C, stir 1 h	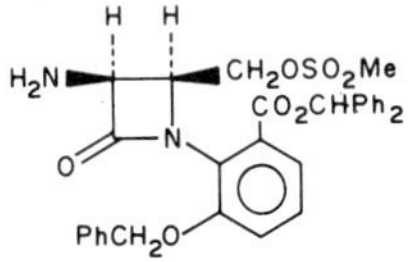	—	1176
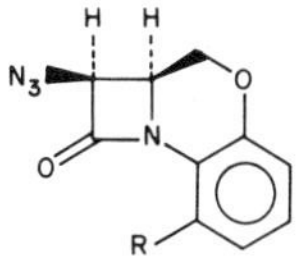 $R = CO_2CHPh_2$, $CH_2OSiMe_2(Bu\text{-}t)$	H_2S, Et_3N, CH_2Cl_2, 0 °C, 5 min	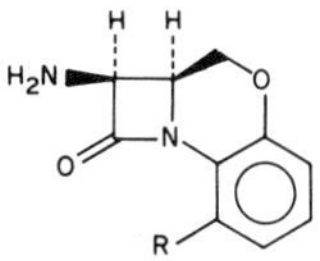	—	1176

Starting material	Reagents and conditions	Product	Yield (%)	Ref.
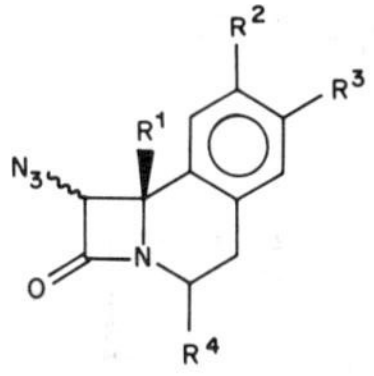 (azido tetrahydroisoquinoline-fused β-lactam)	Al(Hg), H_2O, THF, MeOH, stir r.t., 4h	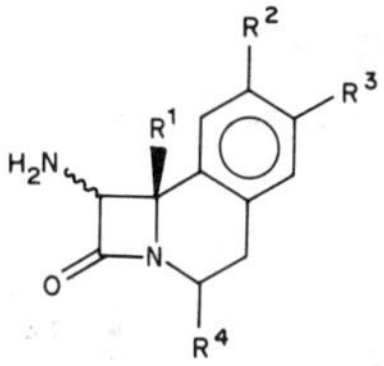 (amino tetrahydroisoquinoline-fused β-lactam)		1154

R^1	R^2	R^3	R^4	Yield (%)
H	H	H	H	55
p-Tol	H	H	H	60
p-An	H	H	H	60
p-An	MeO	MeO	H	55
p-NCC_6H_4	MeO	MeO	H	60
Ph	MeO	MeO	CO_2Me	75

Starting material	Reagents and conditions	Product	Yield (%)	Ref.
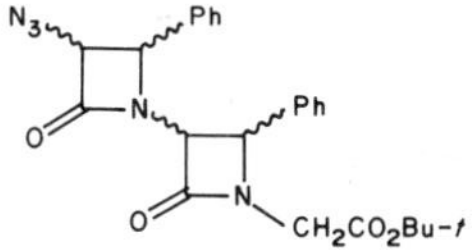 N_3 / Ph / Ph / CH_2CO_2Bu-*t*	10% Pd/C, H_2, EtOH, EtOAc, 50 °C	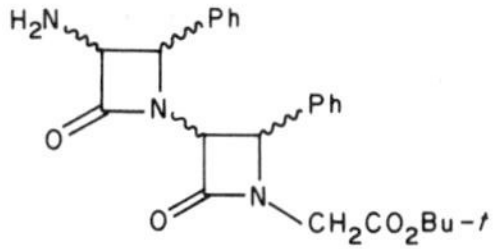 H_2N / Ph / Ph / CH_2CO_2Bu-*t*	99	1169
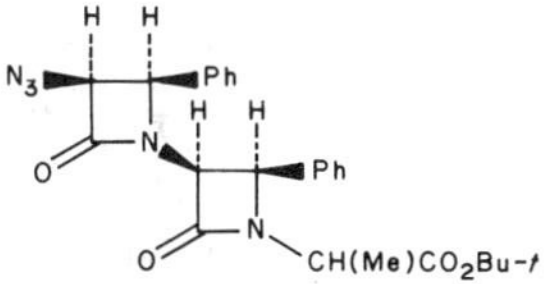 N_3 / Ph / Ph / $CH(Me)CO_2Bu$-*t*	5% Pd/C, H_2, EtOAc, EtOH, 0 °C, 18h	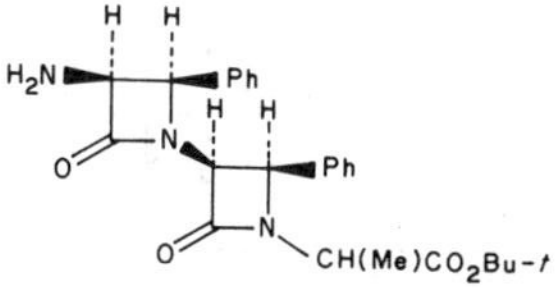 H_2N / Ph / Ph / $CH(Me)CO_2Bu$-*t*	—	1171
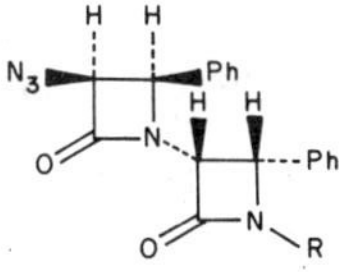 **(232)**		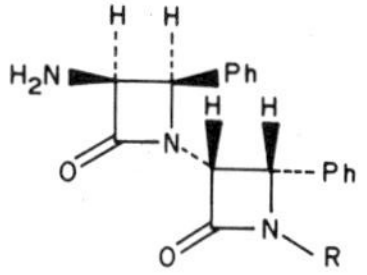 **(233)**	—	1171

(continued)

TABLE 84. (*continued*)

Azido lactam	Reducing agent and conditions	Product	Yield (%)	Reference
R = $CH(Me)CO_2Bu$-*t*	5% Pd/C, H_2, EtOAc, EtOH, 0 °C, 18h			
R = $Me_2CHCH_2CHCH_2OCH_2Ph$	5% Pd/C, H_2, MeOH, 0 °C, 18h	R = $Me_2CHCH_2CHCH_2OCH_2Ph$		
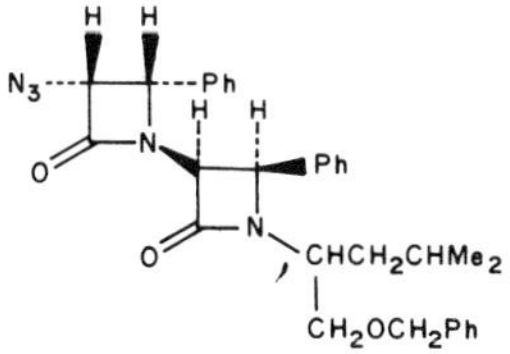 **(234)**	5% Pd/C, H_2, MeOH, 0 °C, 18h	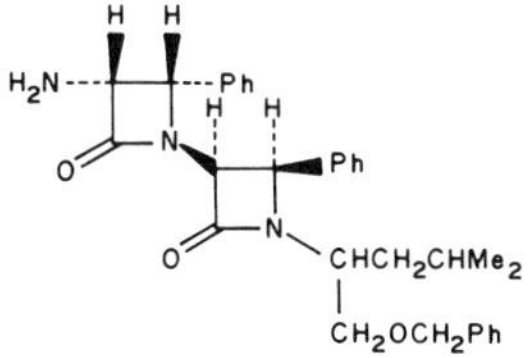 **(235)**	—	1171
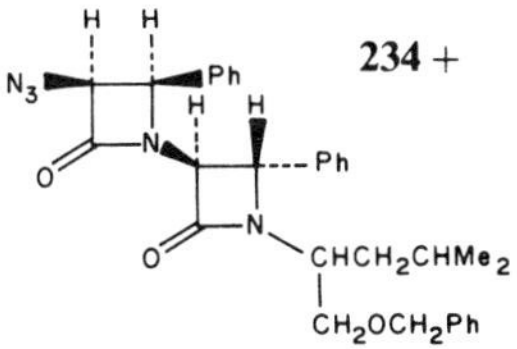 **234** + **(236)** (81:19)	5% Pd/C, H_2, MeOH, 0–5 °C	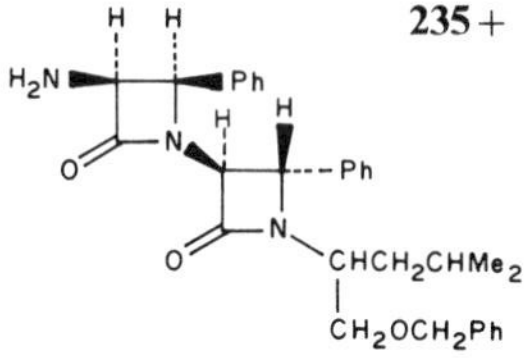 **235** + **(237)**	—	1171
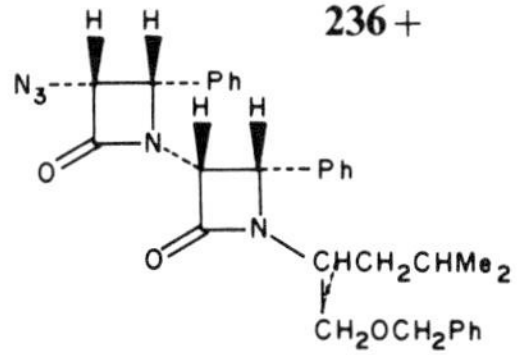 **236** + (32:68)	5% Pd/C, H_2, MeOH, 0–5 °C	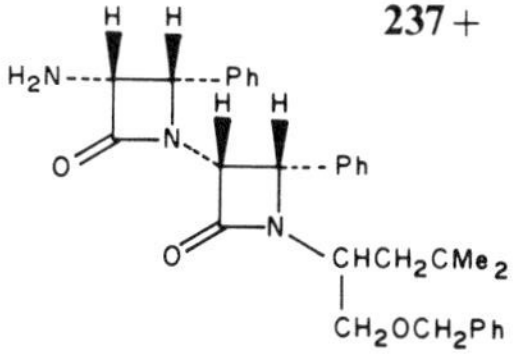 **237** +	—	1171

	5% Pd/C, H_2, EtOAc, EtOH, r.t.		98	1169
	PtO_2, C_6H_6, H_2, 14h			1145
R = Me			97	
R = *i*-Pr			94	
	H_2S, CH_2Cl_2, Et_3N		30	1182
	10% Pd/C, H_2		100	1305

TABLE 85. Deprotection of amino substituted lactams

Lactam	Conditions	Product	Yield (%)	Reference
Ph_3CNH, H, Me, O, N, $C{=}CMe_2$, CO_2CHPh_2	*p*-TosOH, H_2O, Me_2CO	H_2N, H, Me, O, N, $C{=}CMe_2$, CO_2CHPh_2	76[a]	1404
$PhCH_2O_2CNH$, H, O, N, H	10% Pd/C, H_2, EtOH	H_2N, H, O, N, H	—	1326
	Pd, H_2		100	1382
$PhCH_2O_2CNH$, H, H, CH_2O, O, O, N, H	5% Pd/C, H_2, EtOH, 20 °C, 15 min	H_2N, H, H, CH_2O, O, O, N, H	100	1184
$(PhCH_2)_2N$, H, H, Me, O, N, CH_2Ph + $(PhCH_2)_2N$, H, H, Me, O, N, CH_2Ph 6:4 mixture	10% Pd/C, H_2, EtOH, Ar, 3h	H_2N, H, H, Me, O, N, CH_2Ph + H_2N, H, H, Me, O, N, CH_2Ph 6:4 mixture	88	1405

Substrate	Conditions	Product	Yield (%)	Ref.
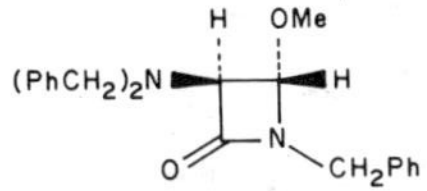 H, OMe, $(PhCH_2)_2N$, H, O, N, CH_2Ph	10% Pd/C, H_2, EtOH, 3h	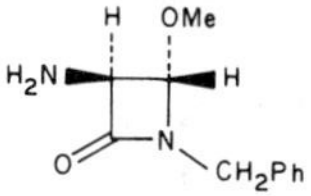 H, OMe, H_2N, H, O, N, CH_2Ph	75	1405
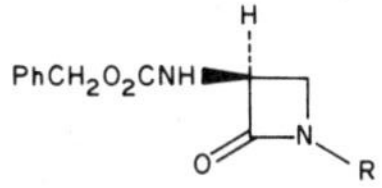 H, $PhCH_2O_2CNH$, O, N, R	10% Pd/C, H_2 (5 atm.), MeOH, r.t., 30 min	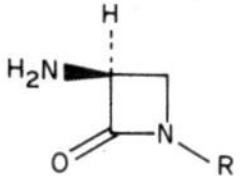 H, H_2N, O, N, R	30–90	1077
R = 2-Pyr, 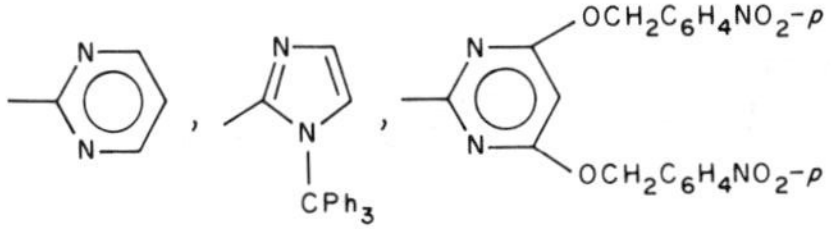 N, N, N, N, N, CPh_3, N, N, $OCH_2C_6H_4NO_2$-p, $OCH_2C_6H_4NO_2$-p				
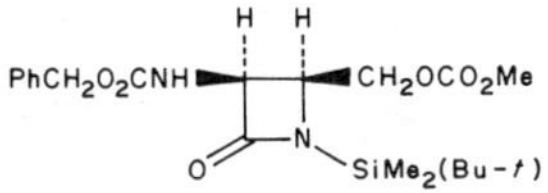 H, H, $PhCH_2O_2CNH$, CH_2OCO_2Me, O, N, $SiMe_2(Bu$-$t)$	cat., H_2	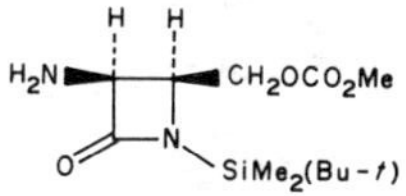 H, H, H_2N, CH_2OCO_2Me, O, N, $SiMe_2(Bu$-$t)$	—	1317
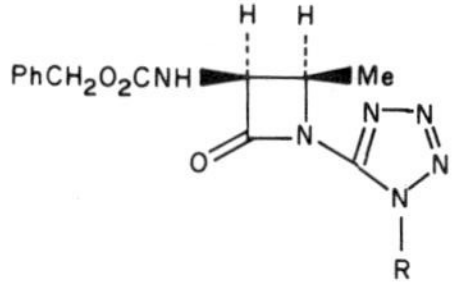 H, H, $PhCH_2O_2CNH$, Me, O, N, N–N, N, N, R	1. 5% Pd/C, H_2, MeOH, r.t., 1.5h 2. aq. $NaHCO_3$	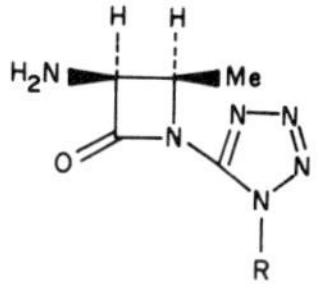 H, H, H_2N, Me, O, N, N–N, N, N, R		1077
R = H			83	
R = $CHPh_2$			30	
R = $CH_2CO_2CH_2C_6H_4NO_2$-p	10% Pd/C, H_2, MeOH, r.t., 30 min		77	
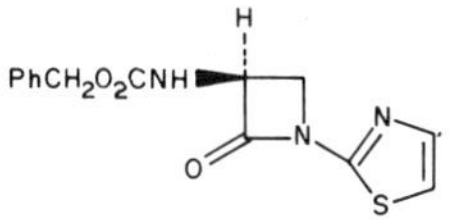 H, $PhCH_2O_2CNH$, O, N, N, S	$AlCl_3$, $MeNO_2$, CH_2Cl_2, C_6H_5OMe, −20 to −10 °C, stir 30 min	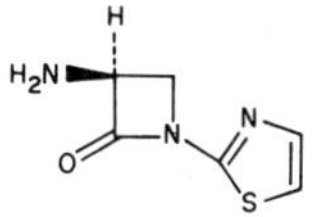 H, H_2N, O, N, N, S	29	1077

(continued)

TABLE 85. (*continued*)

Lactam	Conditions	Product	Yield (%)	Reference
$PhCH_2O_2CNH$ (azetidinone, H, H, An-*p*, N–An-*p*)	32% HBr, HOAc, r.t. stir	$H_3\overset{+}{N}$ Br^- (azetidinone, H, H, An-*p*, N–An-*p*)	86	1180
R^1O_2CNH (bicyclic, R^2, S, N)	32% HBr, HOAc, r.t., stir	$H_3\overset{+}{N}$ Br^- (bicyclic, R^2, S, N)		1180
$R^1 = PhCH_2$, $R^3 = Ph$			70	
$R^1 = PhCH_2$, $R^2 = p\text{-}O_2NC_6H_4$			—	
$R^1 = t\text{-Bu}$, $R^2 = Ph$, $p\text{-}O_2NC_6H_4$	32% HBr, HOAc, r.t., stir		—	
$PhCH_2O_2CNH$ (bicyclic, Ph, Me Me, S, N)	32% HBr, HOAc, r.t., stir	$H_3\overset{+}{N}$ Br^- (bicyclic, Ph, Me Me, S, N)	—	1180
$PhCH_2CONH$ (OMe, O, N, CH_2S–tetrazolyl N–Me, CO_2CHPh_2)	PCl_5, C_5H_5N, CH_2Cl_2, MeOH, Et_2NH, 3–10 °C	H_2N (OMe, O, N, CH_2S–tetrazolyl N–Me, CO_2CHPh_2)	54	1368
$(PhCH_2)_2N$ (H, H, O, N)	10% Pd/C, H_2, EtOH	H_2N (H, H, O, N)	94	1405

Starting material	Reagents	Product	Yield (%)	Ref.
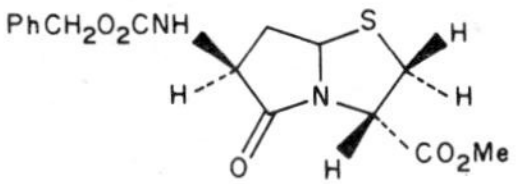 $PhCH_2O_2CNH$ … CO_2Me	45% HBr, HOAc	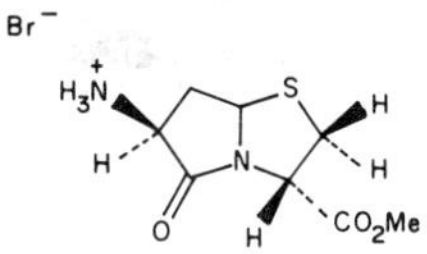 Br^- $H_3\overset{+}{N}$ … CO_2Me	—	999
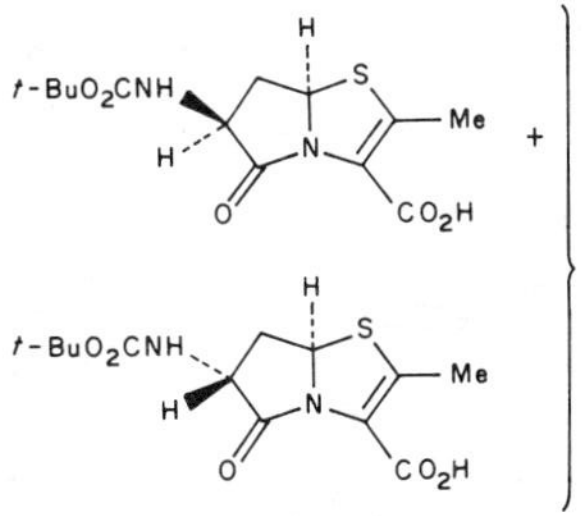 t-BuO_2CNH … Me, CO_2H + t-BuO_2CNH … Me, CO_2H	CF_3COOH	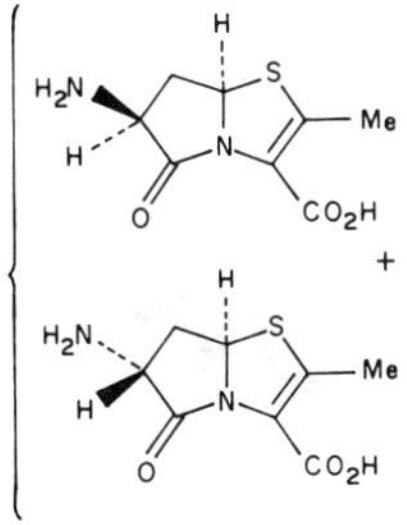 H_2N … Me, CO_2H + H_2N … Me, CO_2H	—	1346
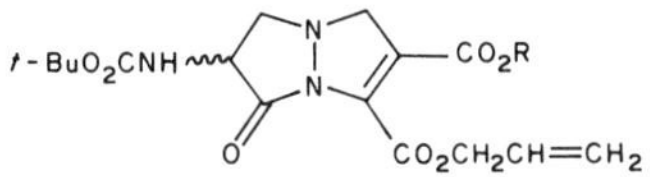 t-BuO_2CNH … CO_2R, $CO_2CH_2CH{=}CH_2$; R = Me, $CH_2CH{=}CH_2$	CF_3COOH	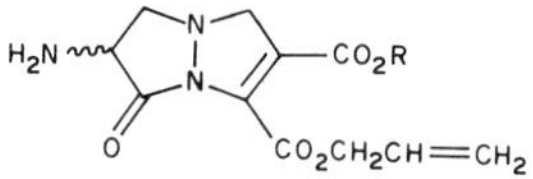 H_2N … CO_2R, $CO_2CH_2CH{=}CH_2$	—	1334
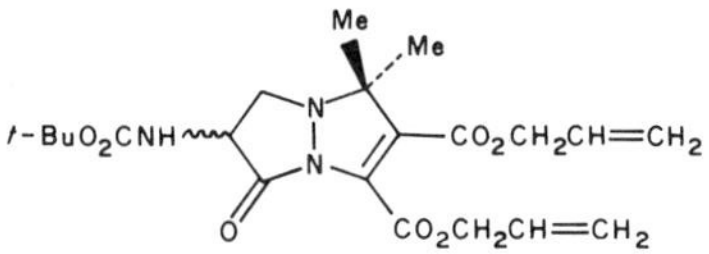 t-BuO_2CNH … Me, Me, $CO_2CH_2CH{=}CH_2$, $CO_2CH_2CH{=}CH_2$	CF_3COOH	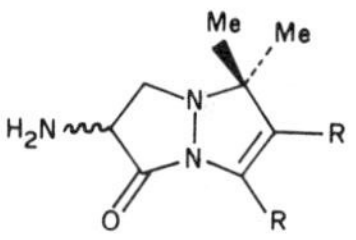 H_2N … Me, Me, R, R; $R = CO_2CH_2CH{=}CH_2$	—	1046
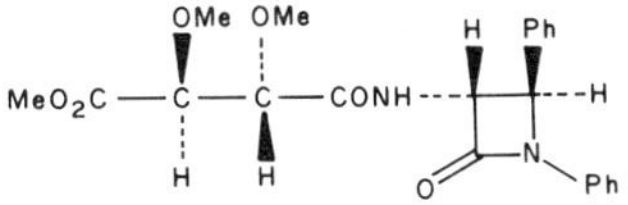 MeO_2C—C(OMe)(H)—C(OMe)(H)—CONH … H, Ph, H, N–Ph	PCl_5, C_5H_5N, CH_2Cl_2, MeOH, H^+	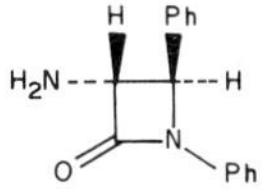 H_2N … H, Ph, H, N–Ph	40	1197

(continued)

TABLE 85. (*continued*)

Lactam	Conditions	Product	Yield (%)	Reference
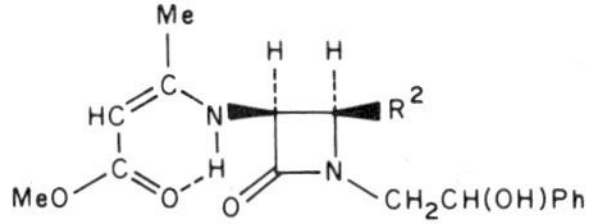 [structure: MeO–C(=O)–CH=C(Me)–NH–β-lactam, H H, R^2, N–$CH_2CH(OH)Ph$]	1. 6*N* HCl 2. NaOH	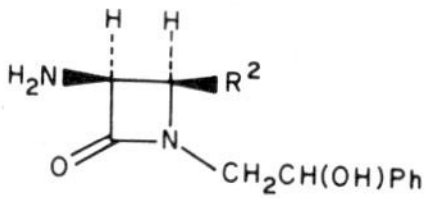 [structure: H_2N–β-lactam, H H, R^2, N–$CH_2CH(OH)Ph$]	—	1217
R = Ph, *p*-An				
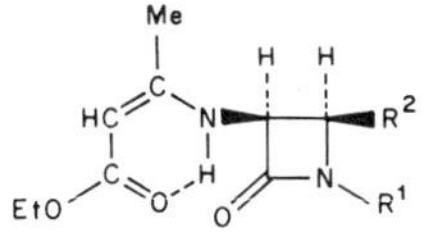 [structure: EtO–C(=O)–CH=C(Me)–NH–β-lactam, H H, R^2, N–R^1]	EtOH/HCl (2:1), r.t. stir	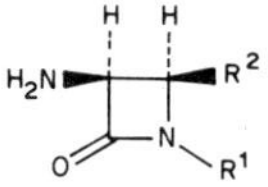 [structure: H_2N–β-lactam, H H, R^2, N–R^1]		
$R^1 = R^2 = Ph$			—	1214, 1215
$R^1 = p$-Tol, $R^2 = p$-An			—	1214, 1215
$R^1 = p$-Tol, $R^2 =$ 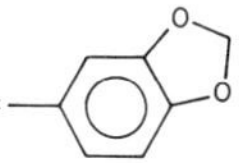 [3,4-methylenedioxyphenyl structure]			—	1214, 1215
$R^1 = p$-An, $R^2 =$ 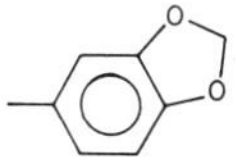 [3,4-methylenedioxyphenyl structure]			—	1159
$R^1 = PhCH_2$, $R^2 = p$-$O_2NC_6H_4$			50	1143
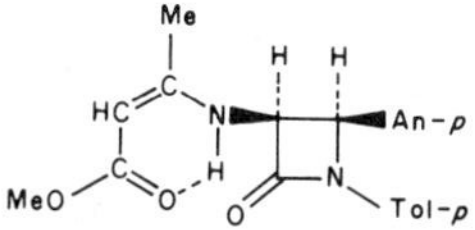 [structure: MeO–C(=O)–CH=C(Me)–NH–β-lactam, H H, An-*p*, N–Tol-*p*]	EtOH/HCl (2:1) r.t. stir	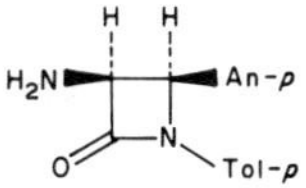 [structure: H_2N–β-lactam, H H, An-*p*, N–Tol-*p*]	—	1214, 1215

	HCl/H_2O, CH_2Cl_2, r.t. overnight		78	1372
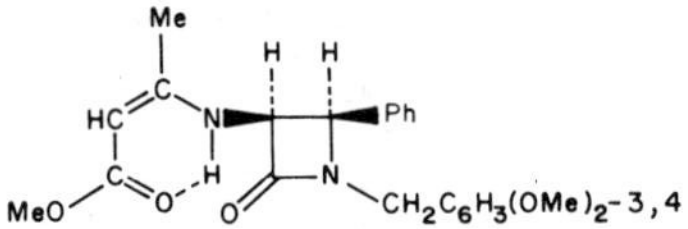	EtOH/HCl (2:1) r.t. 15 min	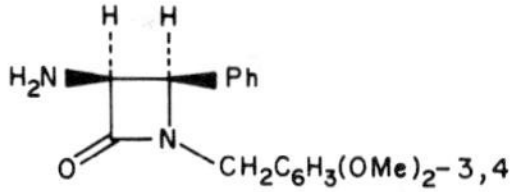	40	1372
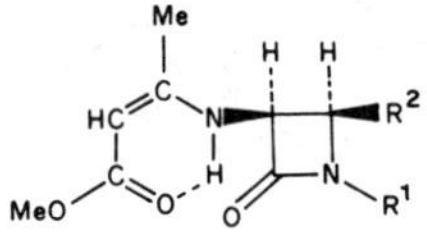	*p*-TosOH, Me_2CO, NaOH	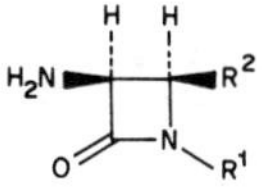		1204, 1372
R^1 = Ph, R^2 = PhCH=CH			57	
R^1 = *p*-Tol, R^2 = *p*-An			43	
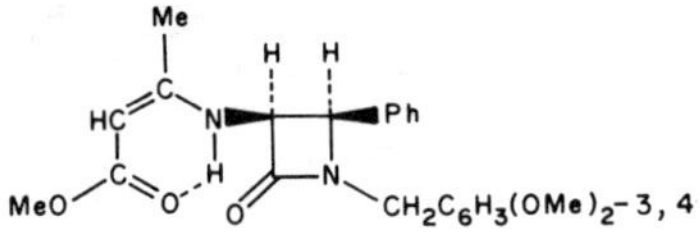	2*N* HCl, Me_2CO, stir r.t., 15 min	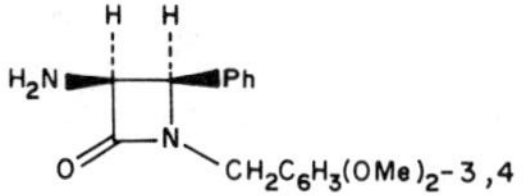	65	1178
	p-TosOH, Me_2CO, H_2O, r.t., stir overnight	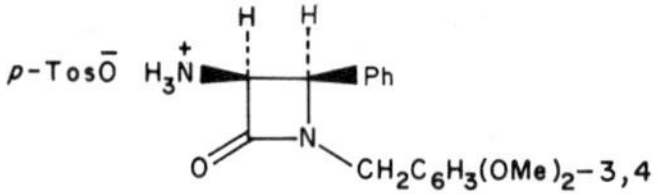	—	1176
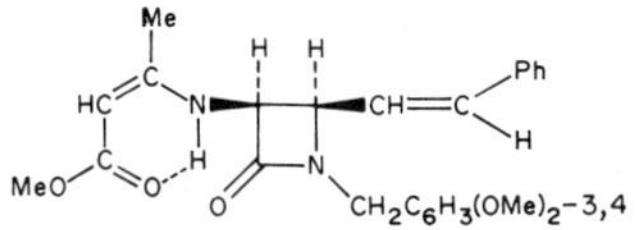	*p*-TosOH, H_2O, dioxane, r.t.	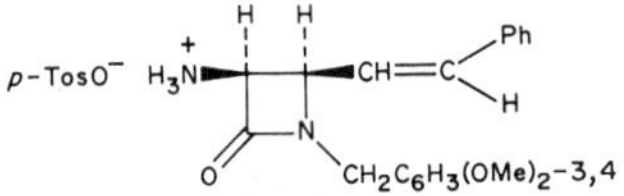		1377
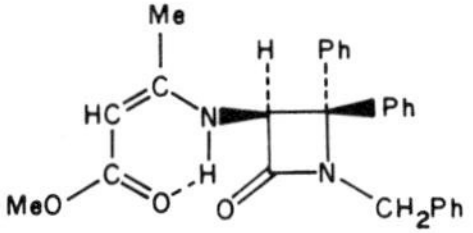	EtOH/HCl, r.t., stir 4 h	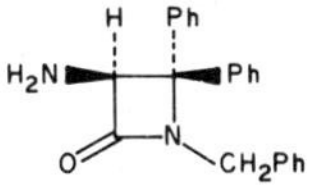	55	1143

(*continued*)

TABLE 85. *(continued)*

Lactam	Conditions	Product	Yield (%)	Reference
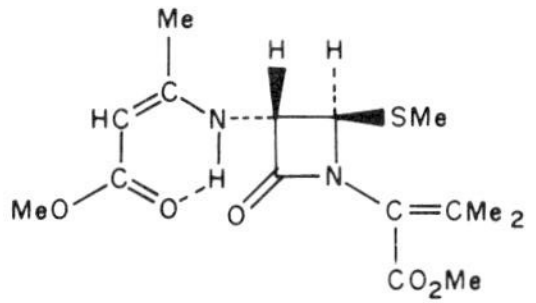	2*N* HCl, Me_2CO, r.t. stir	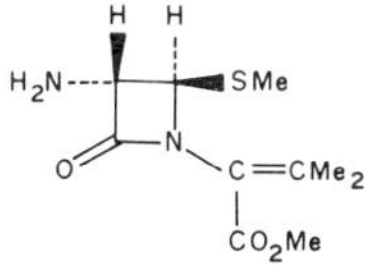	—	1178
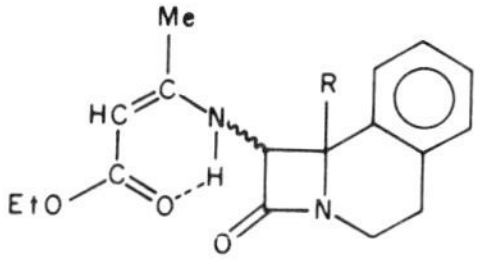 R = *p*-Tol	EtOH/HCl (2:1), r.t. stir	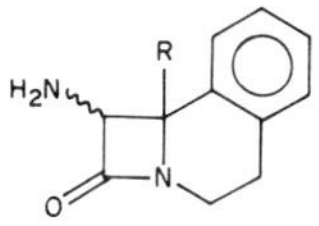	—	1214, 1215
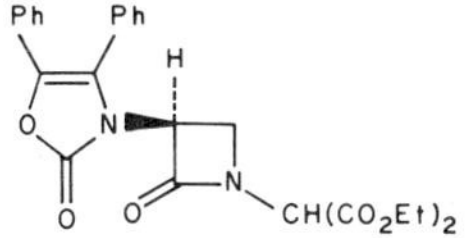	EtOH, 1.2*N* HCl, 10% Pd/C, H_2, 4.5h	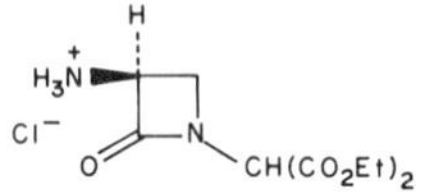	82	1050
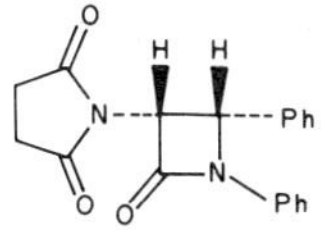	1. NaOMe, NaOH 2. PCl_5, C_5H_5N, CH_2Cl_2, MeOH, H^+	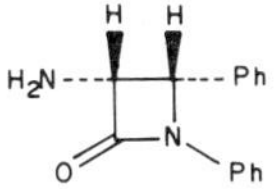	—	1197
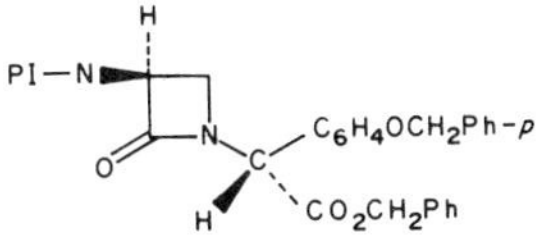	1. 10% Pd/C, H_2, EtOH, HOAc 2. $H_2N(CH_2)_3NMe_2$, MeOH	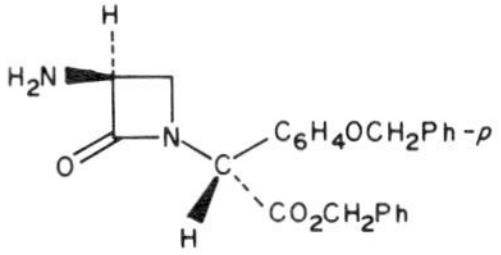	49	1188

Reactant	Conditions	Product	Yield (%)	Ref.
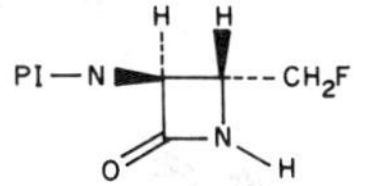	1. H_2NNH_2, dioxane, MeOH, 20 °C, 1 h 2. HCl	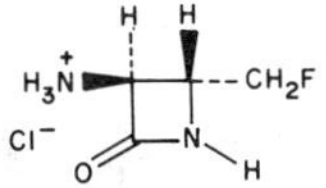	81	1184
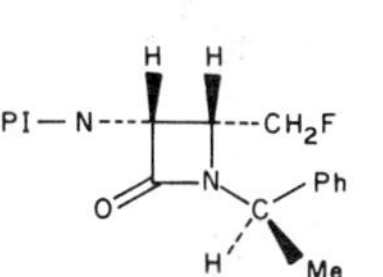	H_2NNH_2, dioxane, MeOH, 20 °C, 1 h	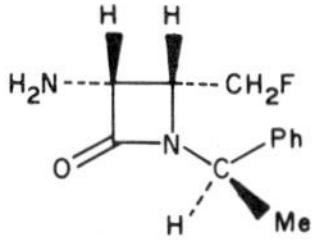	90	1184
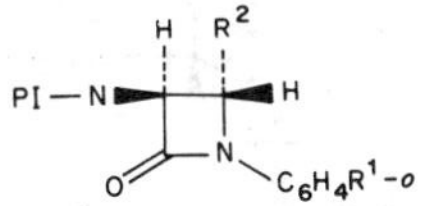	H_2NNH_2, CH_2Cl_2, r.t. stir	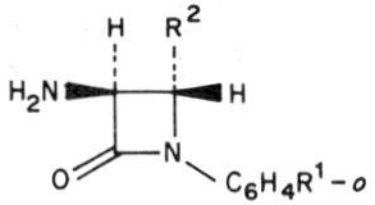		802
$R^1 = CO_2CH_2An$-*p*, R^2 = Ph	48 h		50	
$R^1 = CO_2Me$, R^2 = MeO	3 days		28	
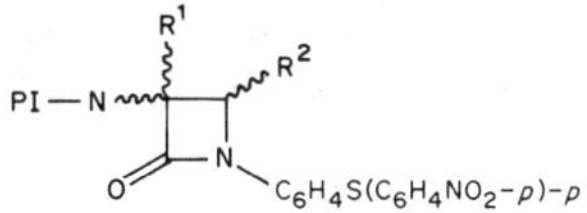	1. EtOH, H_2NNH_2, reflux 1 h 2. 2*N* HCl, 50 °C, 2 h 3. r.t. 30 min	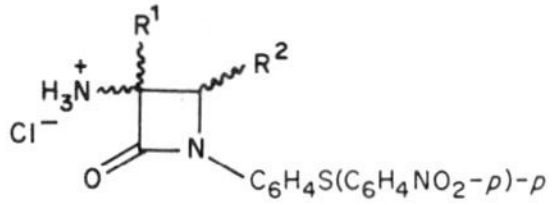		1186
R^1 = H, R^2 = Ph			39	
R^1 = H, R^2 = *p*-$O_2NC_6H_4$			41	
R^1 = H, R^2 = *p*-An			42	
R^1 = H, R^2 = *p*-ClC_6H_4			47	
R^1 = H, R^2 = *o*-HOC_6H_4			40	
R^1 = H, R^2 = *p*-$Me_2NC_6H_4$			42	
R^1 = Me, R^2 = Ph			47	
R^1 = Me, R^2 = *p*-An			49	
R^1 = Me, R^2 = *p*-$O_2NC_6H_4$			50	
R^1 = *i*-Pr, R^2 = Ph			58	

(continued)

TABLE 85. (*continued*)

Lactam	Conditions	Product	Yield (%)	Reference
$R^1 = i$-Pr, $R^2 = p$-An			60	
$R^1 = i$-Pr, $R^2 = p$-$O_2NC_6H_4$			59	
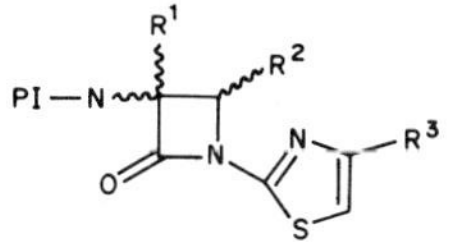 [structure: PI—N; R^1, R^2, N, R^3, S]	1. $H_2NNH_2 \cdot H_2O$, EtOH, reflux 2 h 2. HCl, stir 2h	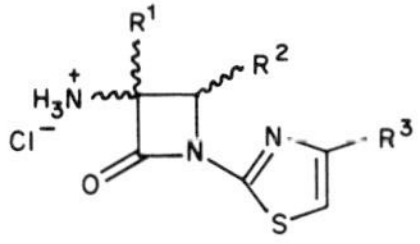 [structure: $H_3\overset{+}{N}$, Cl^-; R^1, R^2, N, R^3, S]		1189
R^1 / R^2 / R^3				
H / o-$O_2NC_6H_4$ / Ph			40	
Me / m-$O_2NC_6H_4$ / Ph			42	
i-Pr / PhCH=CH / Ph			46	
H / p-An / Ph			43	
H / o-HOC_6H_4 / Ph			45	
H / m-$O_2NC_6H_4$ / p-Tol			44	
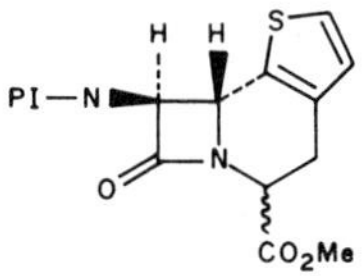 [structure: PI—N; H, H; S; CO_2Me]	$Me_2NCH_2CH_2CH_2NH_2$, MeOH, $CHCl_3$, r.t. 40h	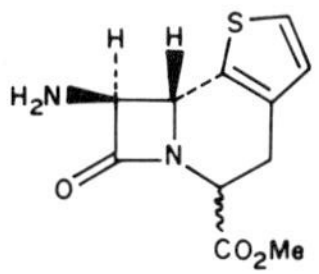 [structure: H_2N; H, H; S; CO_2Me]	87	1187
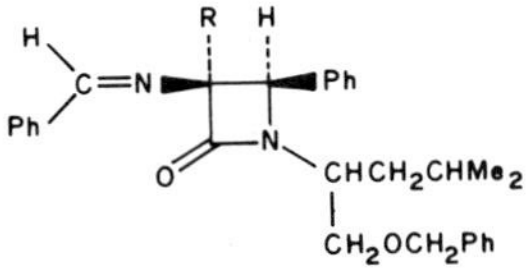 [structure: H, Ph, C=N; R, H, Ph; N—$CHCH_2CHMe_2$, CH_2OCH_2Ph]	HCl, MeOH, r.t. stir 3h	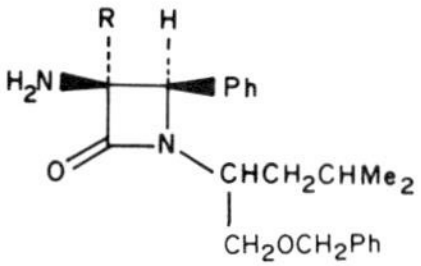 [structure: H_2N; R, H, Ph; N—$CHCH_2CHMe_2$, CH_2OCH_2Ph]		1181
R = Me			98	
$R = CH_2CH{=}CH_2$			98	

Starting material	Conditions	Product	Yield (%)	Ref.
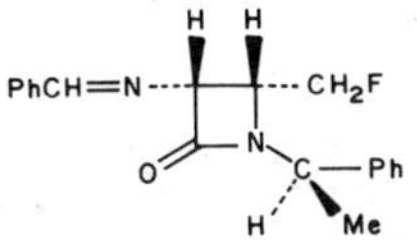	1. THF, $(Me_3Si)_2\overset{-}{N}\overset{+}{Na}$, $-50\,°C$, 15 min 2. HCl, $20\,°C$, 15 min 3. NH_4OH	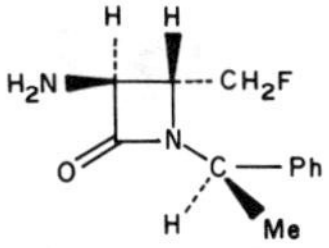	57	1184
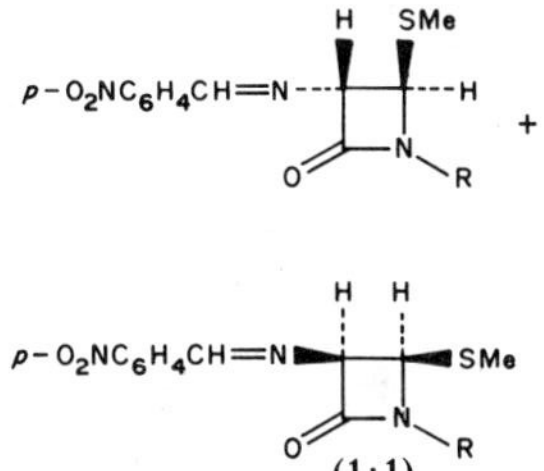 + (1:1) $R = CMe_2CO_2CHPh_2$	$2,4\text{-}(O_2N)_2C_6H_3NHNH_2$, *p*-TosOH, EtOH	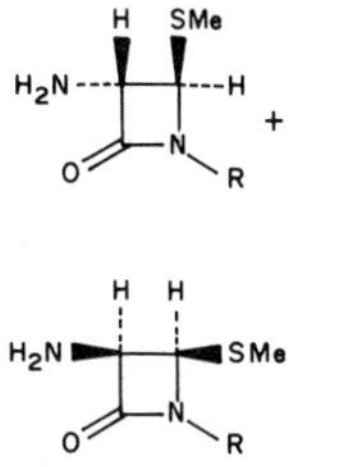 +	—	1173
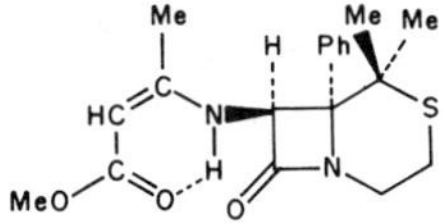	$2N$ HCl, Me_2CO_2, stir r.t.	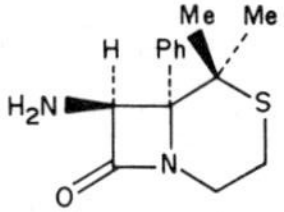	—	1178
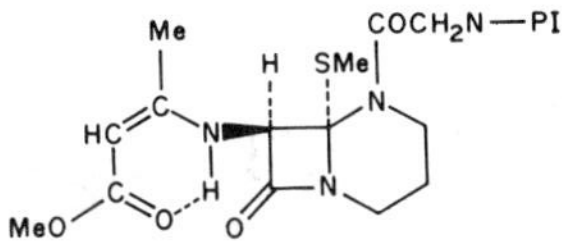	$2N$ HCl, Me_2CO, stir r.t.	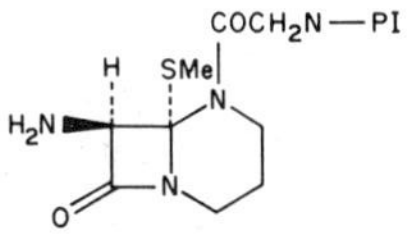	—	1178
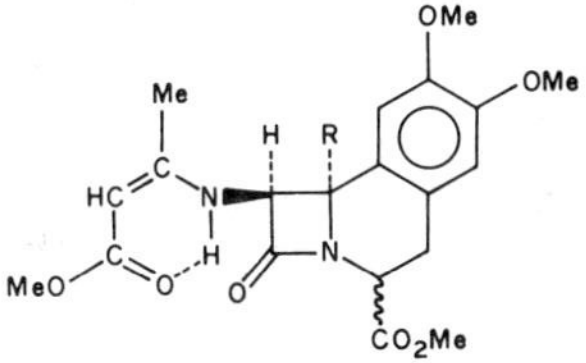 $R = Ph$, *p*-BrC_6H_4	$2N$ HCl, Me_2CO, stir r.t.	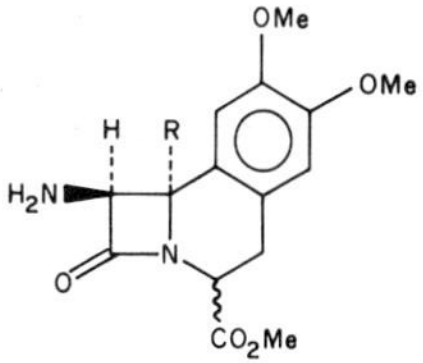	—	1178

(continued)

TABLE 85. *(continued)*

Lactam	Conditions	Product	Yield (%)	Reference
H H Me N O O CO_2Et	1. $MeNHNH_2$, THF, $-78\,^\circ C$, 1h 2. stand r.t.	H_2N H H Me O N N CO_2Et	84	1182
$Ph_3P{=}N$ H H CH=CHPh O N H	1. H_2O, stir overnight 2. *p*-TosOH, stir 3h	*p*-$TosO^-$ $H_3\overset{+}{N}$ H H CH=CHPh O N H	52	1168
$NHCO_2Bu$-*t* N O $CH_2CO_2CH_2Ph$	CF_3CO_2H, r.t., 2h	NH_2 N O CH_2CO_2H	—	1305
$NHCO_2Bu$-*t* N O CH_2CO_2R R = Ph, CH_2Ph	CF_3CO_2H, r.t., 2h	NH_2 N O CH_2CO_2H	—	1305

[a]The product is 31%—4β and 45%—4α.

In addition to the acylation reactions previously discussed which have been used to produce substituted α-amino lactams from α-amino lactams, alkylation reactions have also been employed to produce substituted α-amino lactams from α-amino lactams. Thus, reaction of the hydrochloride salts of 3-amino-4-substituted azetidin-2-ones with trityl chloride in the presence of triethylamine produces[1184] the corresponding 3-tritylamino substituted azetidin-2-ones (equation 1087).

Et_3N, CH_2Cl_2, 20 °C, 6 h

(1087)

R^1	R^2	Yield (%)
CH_2F	H	90
H	CH_2F	78
CO_2Me	H	89
CH_2OTHP	H	78
CHF_2	H	83

Reaction of (±)-1-[(benzyloxycarbonyl)methyl]-3-(*t*-butoxycarbonyl)amino-2-pyrrolidone or piperidone with ethyl 2-oxo-4-phenylbutyrate in the presence of sodium cyanoborohydride as reducing agent produces a reductive amination affording[1305] a mixture of diastereomeric diester products which were separated by using mediun-pressure chromatography over silica gel (equations 1088 and 1089). Similar results were obtained[1424] from the *t*-butyl esters of the seven-, eight- and nine-membered ring analogues using palladium on carbon and hydrogen (equation 1090).

1. CF_3CO_2H, r.t., 2 h
2. $PhCH_2CH_2COCO_2Et$, $NaBH_3CN$, NaOAc, EtOH

(1088)

1. CF_3CO_2H, r.t., 2 h
2. $PhCH_2CH_2COCO_2Et$, $NaBH_3CN$, NaOAc, EtOH

(1089)

Pd/C, H_2, HOAc
$PhCH_2CH_2COCO_2Et$

(1090)

n = 1–3

Mono- and disubstituted 3-amino lactams have been treated with a variety of reagents to effect structural changes in the substituent groups. Using this approach various Dane salts of 1-(*p*-tolyl)-4-(*p*-anisyl)azetidin-2-ones have been oxidized with ozone[1178] (equation 1091) or with ruthenium tetroxide[1208] (generated from ruthenium dioxide and sodium periodate in aqueous acetone; see equation 1092) to produce the corresponding acylated amidolactams.

O_3, CH_2Cl_2, −78 °C, 20 min → [ozonide]

Jones' reagent, Me_2CO, 0 °C, stir 15 min → MeCONH … An-*p* … Tol-*p* (1091)

RuO_4(RuO_2 + $NaIO_4$), Me_2CO, stir 2 h → RCONH … An-*p* … Tol-*p*

R = $PhOCH_2$; $PhCH_2$; Me
Yield (%) = 52 ; 40 ; 40 (1092)

Similar ozone oxidation of an analogous cyclic vinylamino-β-lactam produces[1178] a product containing an amide side chain with an α-keto ester as an additional functional group (equation 1093).

1. O_3, CH_2Cl_2, −78 °C; 2. Jones' reagent, Me_2CO, 0 °C, stir → $EtO_2CCO(CH_2)_3CONH$ … An-*p* … Tol-*p*, 50% (1093)

α-Succinimido[1182,1197] (equations 1094 and 1095) and α-phthalimido[1182] lactams (equation 1096) reportedly undergo side-chain ring opening to produce the corresponding acylated α-amino lactams as products when treated with base.

NaOH, MeOH

(1094)[1197]

1. phenolphthalein, dioxane
NaOH
2. HCl

$HO_2C(CH_2)_2CONH$

$R^1 =$	Me	H
$R^2 =$	CO_2Et	COPh
Yield (%)	70	90

(1095)[1182]

1. phenolphthalein, dioxane
NaOH
2. HCl

o-$HO_2C(C_6H_4)CONH$

$R^1 =$	H	Me	Me
$R^2 =$	CO_2Et	CO_2Et	COPh
Yield(%) =	82	90	95

(1096)[1182]

Treatment of *N*-benzyloxycarbonyl-*N*-hydroxymethyl substituted α-amino bicyclic γ-lactams with sodium carbonate in methanol removes the one-carbon substituent affording[1430] the monosubstituted *N*-benzyloxycarbonyl-α-amino products (equations 1097 and 1098).

Na_2CO_3, MeOH
stir r.t. 50 min

(1097)

At least one example of a modified Curtius rearrangement has been reported[1040] to occur with lactams and it involves the treatment of methyl 6-[(benzyloxycarbonyl)amino]-2(*S*)-[3-carboxy-3-(indol-3-ylmethyl)-2-oxo-1-pyrrolidinyl]hexanoate with diphenylphosphoryl azide (DPPA) in *t*-butyl alcohol producing a stereospecific conversion to methyl

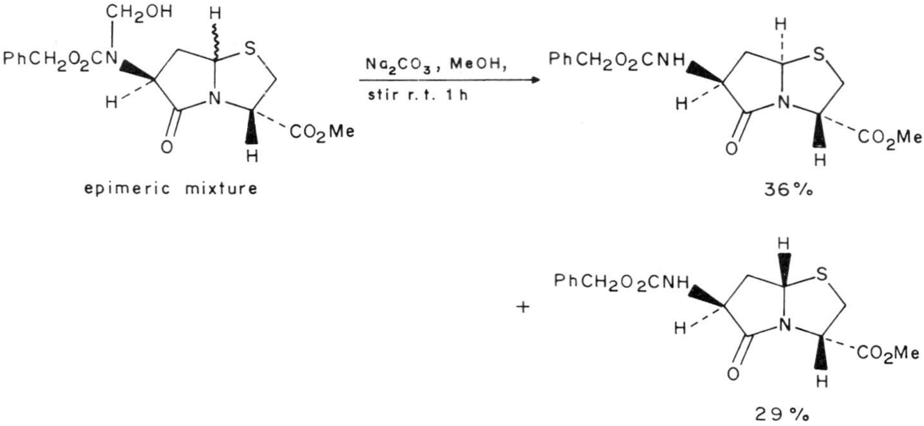

CH_2OH H

$PhCH_2O_2CN$ S

H N CO_2Me

O H

epimeric mixture

Na_2CO_3, MeOH, stir r.t. 1 h →

H

$PhCH_2O_2CNH$ S

H N CO_2Me

O H

36%

\+

H

$PhCH_2O_2CNH$ S

H N CO_2Me

O H

29%

(1098)

6-[(benzyloxycarbonyl)amino]-2(*S*)-[3(*S* or *R*)-((*t*-butyloxycarbonyl)amino)-3-(indol-3-ylmethyl)-2-oxo-1-pyrrolidinyl]hexanoate (equation 1099).

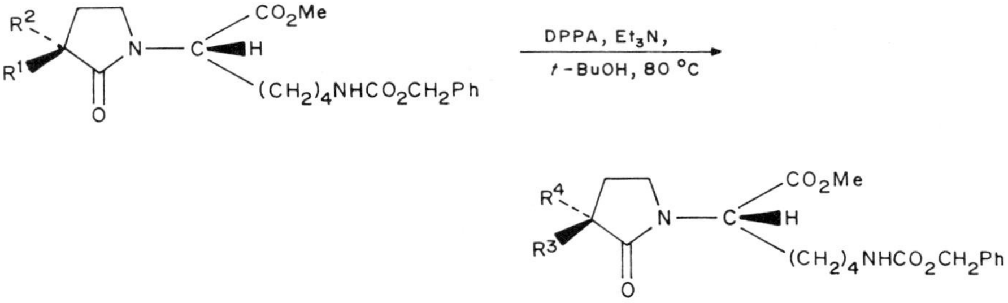

R^2 CO_2Me

N—C—H

R^1

$(CH_2)_4NHCO_2CH_2Ph$

O

DPPA, Et_3N, *t*-BuOH, 80 °C →

R^4 CO_2Me

N—C—H

R^3

$(CH_2)_4NHCO_2CH_2Ph$

O

R^1	R^2	R^3	R^4	Yield (%)
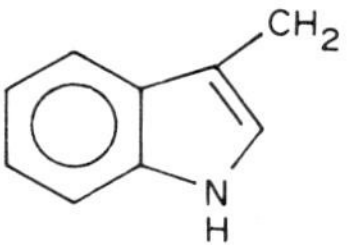 CH_2 N H	COOH	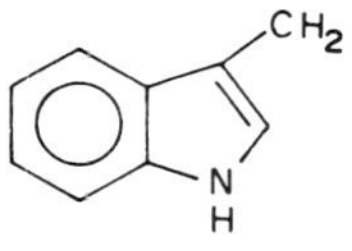 CH_2 N H	$NHCO_2Bu$-*t*	66
COOH	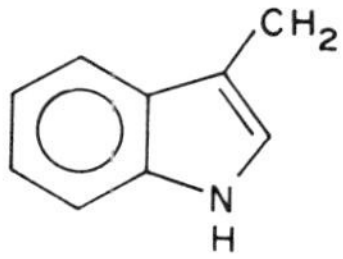 CH_2 N H	$NHCO_2Bu$-*t*	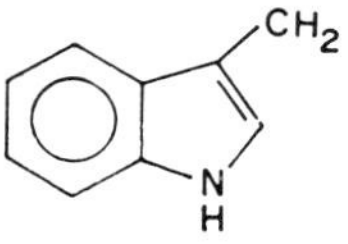 CH_2 N H	66

(1099)

Diazotization of methyl 6-aminopenicillinate followed by reaction of the diazonium salt with bromide ion affords[1394] methyl 6,6-dibromopenicillinate (equation 1100).

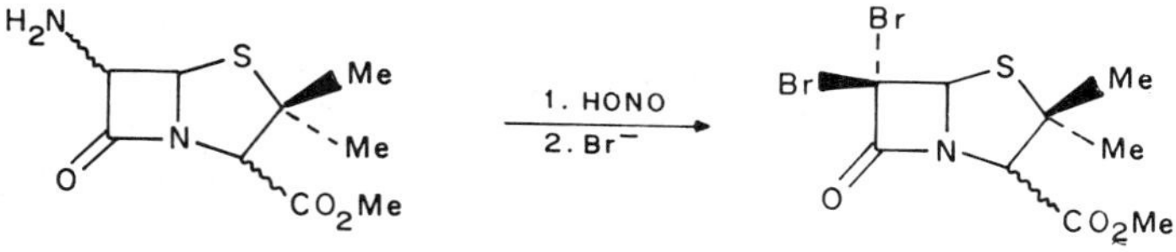

(1100)

Whether prepared by the method shown in equation 1100 or by some other synthetic route, α-mono- and α,α-dihalo lactams are useful starting materials for the preparation of variously substituted lactams. For example, *cis*-3-bromo-1,4-diphenylazetidin-2-one upon treatment with tetramethylguanidinium azide in dimethyl sulphoxide (DMSO) affords[1205] *trans*-3-azido-1,4-diphenylazetidin-2-one (equation 1101).

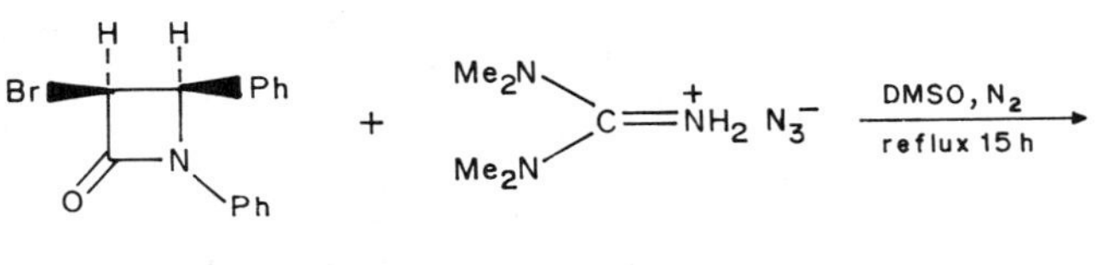

(1101)

Hydrolysis of α-chloro-α-phenylthio-β-lactams under very mild conditions affords[1211] the corresponding intermediate α-hydroxy α-phenylthio-β-lactam analogues which immediately eliminate thiophenol producing azetidine-2,3-diones as the final products (equation 1102).

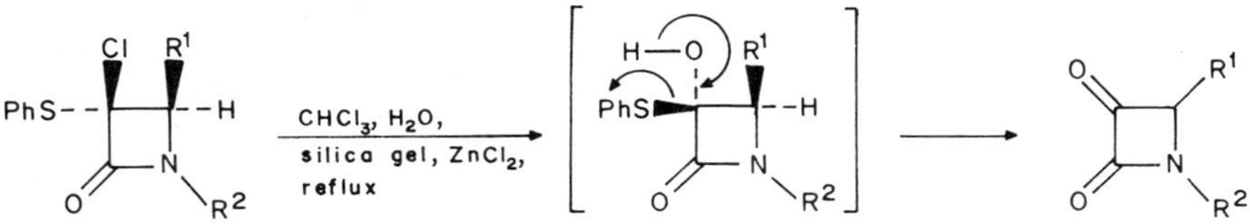

(1102)

R^1 =	Ph[a];	Ph;	*p*-Tol;	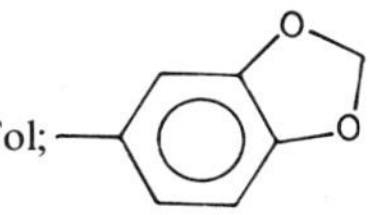 ;	—CMe=CHPh;	*p*-Tol;	*p*-Tol
R^2 =	*p*-An;	Ph;	*p*-An;	Ph ;	*p*-An ;	*o*-An;	*p*-BrC_6H_4

[a]Yield for this set of substituents was 89%.

Reaction of benzhydryl-*trans*-6-bromopenicillinate and an excess of substituted olefin with slow addition of tri-*n*-butyltin hydride (Method A) produces[1421] 30–35% reduction products and 43–67% of benzhydryl *trans*-6-alkyl substituted penicillinate via a chain reaction mediated by tributyl radicals (equation 1103).

Benzhydryl 6,6-dibromopenicillinate can be directly and stereoselectively transformed[1431] to benzhydryl 6α-(2′-cyanoethyl)penicillinate by a similar one-pot procedure which involves first refluxing a solution of the penicillinate with one equivalent of tributyltin hydride followed by addition of 15–20 equivalents of the substituted olefin and slow addition of more tributyltin hydride and azobisisobutyronitrile (AIBN) (Method B, equation 1104).

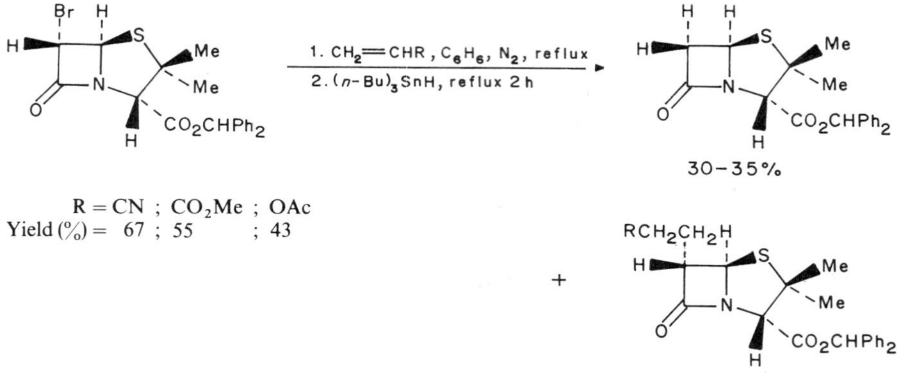

1. $CH_2=CHR$, C_6H_6, N_2, reflux
2. $(n\text{-}Bu)_3SnH$, reflux 2 h

30–35%

+

R = CN ; CO_2Me ; OAc
Yield (%) = 67 ; 55 ; 43

(1103)

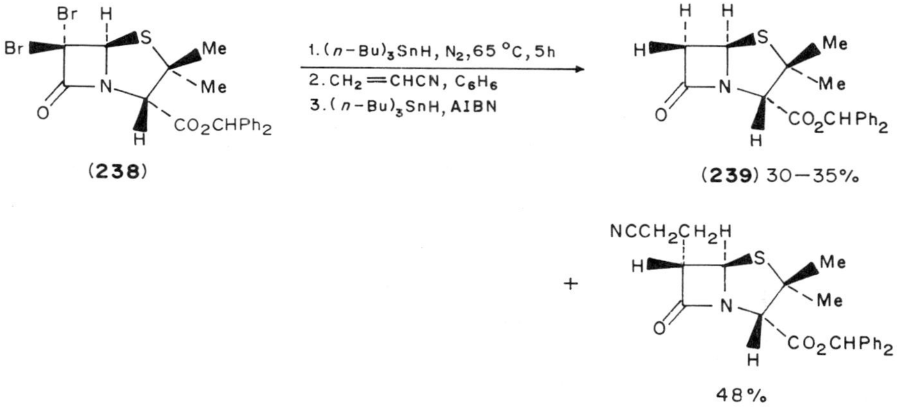

(**238**)

1. $(n\text{-}Bu)_3SnH$, N_2, 65 °C, 5h
2. $CH_2=CHCN$, C_6H_6
3. $(n\text{-}Bu)_3SnH$, AIBN

(**239**) 30–35%

+

48%

(1104)

238

1. C_6H_6, $CH_2=CHR$, N_2, heat
2. $(n\text{-}Bu)_3SnH$, C_6H_6, AIBN, N_2, reflux 3 h

239 + 40–55%

5–8%

+

35–47%

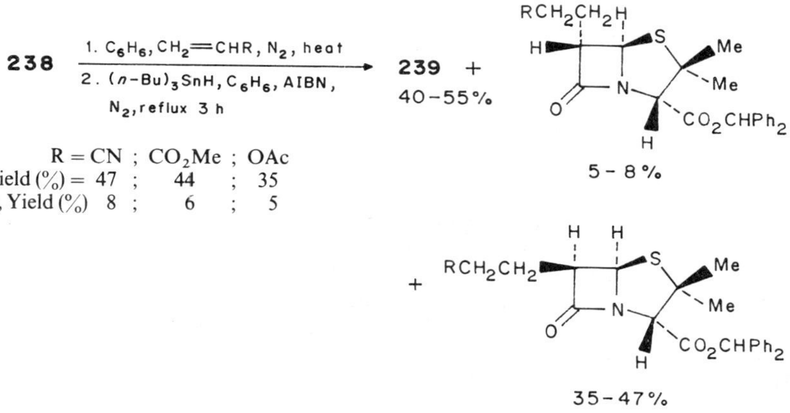

R = CN ; CO_2Me ; OAc
cis, Yield (%) = 47 ; 44 ; 35
trans, Yield (%) 8 ; 6 ; 5

(1105)

Alternatively, reaction of benzyhdryl 6,6-dibromopenicillinate in benzene with excess substituted olefin and one equivalent of tributyltin hydride and azobisisobutyronitrile followed by removal of the excess olefin and treatment of the residue with additional tributyltin hydride (Method C) affords[1431] 40–55% reduction product and 35–47% of the benzhydryl *cis*-6-alkyl substituted penicillinate as the major diastereomer (equation 1105).

Similarly, reaction of *N*-(*p*-methoxyphenyl)-3,3-dibromo-4-styrylazetidinone with methyl propenoate by Method A produces[1431] the *trans*-*N*-(*p*-methoxyphenyl)-3α-(2′-carbomethoxyethyl)-4-styrylazetidinone (**240**) exclusively, while reaction by Method C produces both the *cis* and *trans* products with the *cis* product as the major diastereomer (equation 1106).

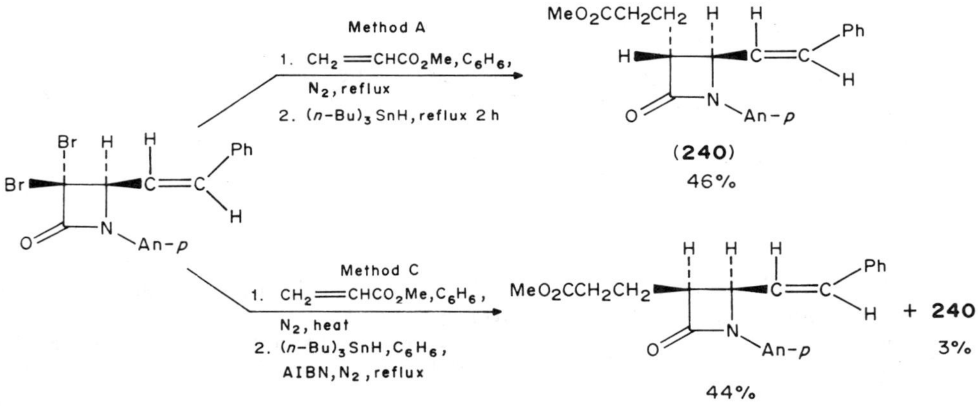

Method A: 1. $CH_2{=}CHCO_2Me, C_6H_6, N_2$, reflux; 2. $(n\text{-}Bu)_3SnH$, reflux 2 h → (**240**) 46%

Method C: 1. $CH_2{=}CHCO_2Me, C_6H_6, N_2$, heat; 2. $(n\text{-}Bu)_3SnH, C_6H_6$, AIBN, N_2, reflux → 44% + **240** 3%

(1106)

Reaction of either benzyhdryl 6α- or 6β-bromopenicillinate with allyltributyltin and azobisisobutyronitrile gave[1431] 95% of benzhydryl 6α-allylpenicillinate as the only detectable diastereomer (equation 1107).

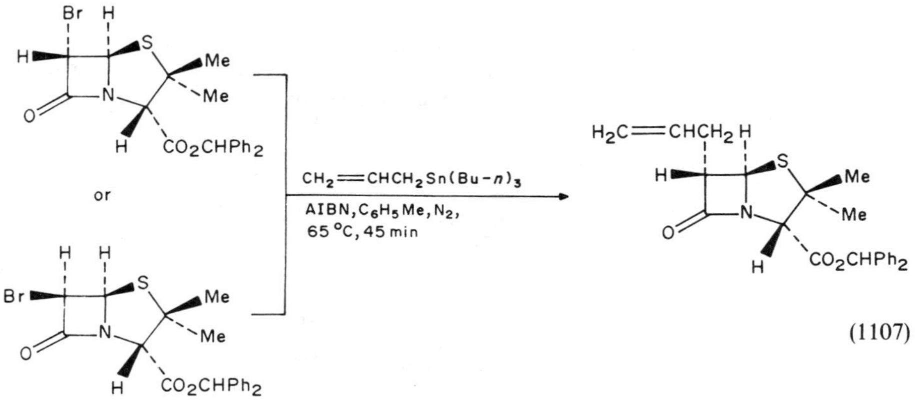

$CH_2{=}CHCH_2Sn(Bu\text{-}n)_3$, AIBN, C_6H_5Me, N_2, 65 °C, 45 min

(1107)

Using benzhydryl 6,6-dibromopenicillinate with the same tin reagent and reaction conditions produced[1431] 60% of benzhydryl 6β-bromo-6α-allylpenicillinate and 22% of benzhydryl 6,6-diallylpenicillinate, while reduction of this product mixture using tributyltin hydride transformed the 6β-bromo-6α-allyl-isomer into the 6β-allylpenicillinate product as the only diastereomer (equation 1108).

238 $\xrightarrow[\text{AIBN}]{CH_2{=}CHCH_2Sn(Bu\text{-}n)_3}$ [6β-bromo-6α-allyl benzhydryl penicillinate, CO_2CHPh_2] 60% +

[(**241**) benzhydryl 6,6-diallylpenicillinate] 22% $\xrightarrow[\text{AIBN}]{(n\text{-}Bu)_3SnH}$ [6β-allylpenicillinate, CO_2CHPh_2] + **241** (1108)

Another conversion of 6,6-dibromoazetidinones is exemplified by the metallation of methyl 6,6-dibrompenicillinate using methyl magnesium bromide, followed by condensation of the metallated intermediate with acetaldehyde to produce[1394] methyl- 6-bromo-6-(1-hydroxyethyl)penicillinate (equation 1109).

[methyl 6,6-dibromopenicillinate, CO_2Me] $\xrightarrow[\text{2. MeCHO}]{\text{1. MeMgBr}}$ [MeCH(OH)- 6-bromo penicillinate, CO_2Me] (1109)

By far the most common reaction performed with α-mono- and α,α-dihalolactams is reduction. Although a variety of reagents have been used to effect reduction of these compounds the overall effect in all cases is the replacement of the halogen(s) by hydrogen (equations 1110 and 1111, and Table 86).

$\xrightarrow[\text{conditions}]{\text{reduction reagent,}}$ (1110)

$\xrightarrow[\text{conditions}]{\text{reduction reagent,}}$ (1111)

X = halogen

TABLE 86. Reduction of α-mono- and α,α-dihalolactams

Halolactam	Reducing agent and conditions	Product	Yield (%)	Reference
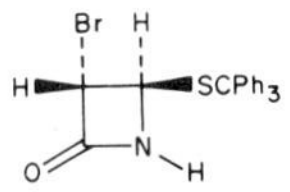 Br H H SCPh$_3$ O N H	Zn, HOAc, MeOH, −15 °C, 20 min	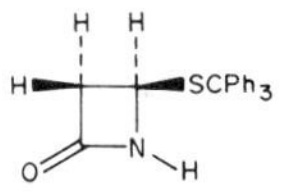 H H H SCPh$_3$ O N H	76	1383
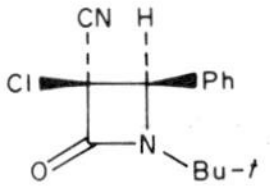 CN H Cl Ph O N Bu-*t*	1. Zn, HOAc, 0 °C, stir 2h 2. r.t.	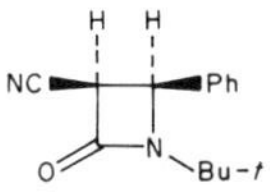 H H NC Ph O N Bu-*t* +	—	1141
		CN H H Ph O N Bu-*t*		
		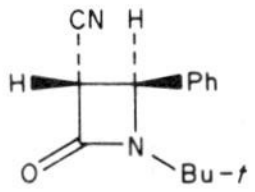 (9:2)		
CN SEt Cl H O N Hex-*c*	Zn, HOAc	CN SEt H H O N Hex-*c* +	66	1142
		H SEt NC H O N Hex-*c*	33	

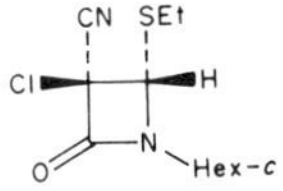

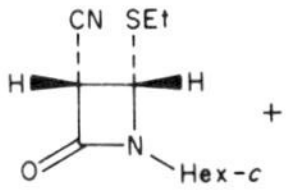

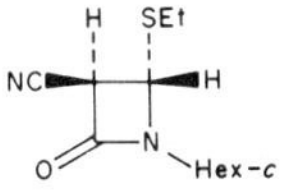

(*continued*)

TABLE 86. *(continued)*

Halolactam	Reducing agent and conditions	Product	Yield (%)	Reference
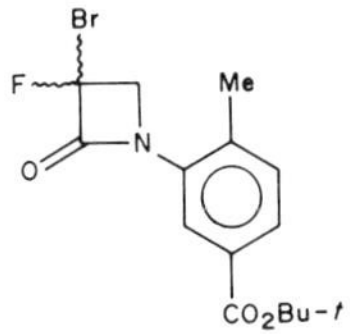	$(n\text{-Bu})_3SnH$, AIBH, C_6H_5Me, 60 °C, 90 min	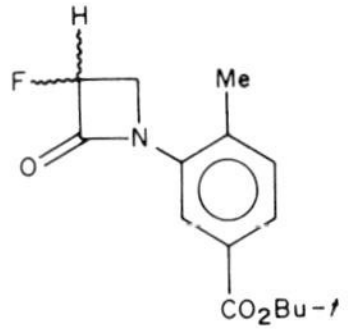	99	1051
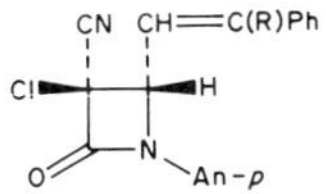 R = H, Ph	1. Zn, HOAc 2. NaH, THF	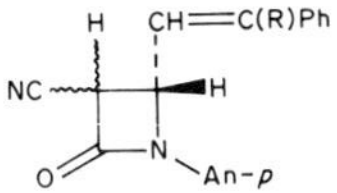	—	1144
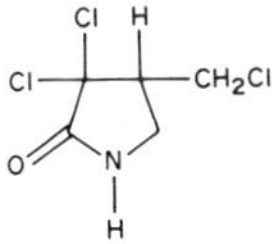	$(n\text{-Bu})_3SnH$, AIBN, C_6H_6, 80 °C, 0.5h	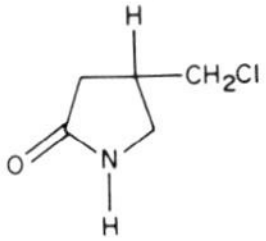	—	1063
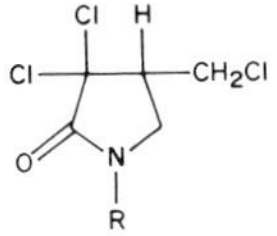 R = H, $CH_2CH{=}CH_2$	$(n\text{-Bu})_3SnH$, AIBN, 140 °C, 3h	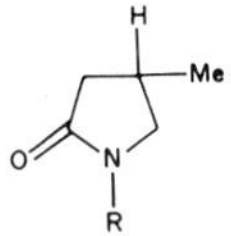	—	1063

Substrate	Conditions	Product	Yield (%)	Ref.
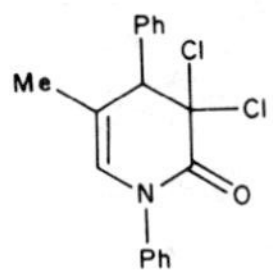	Pd black, C_6H_6, H_2, stir 8h	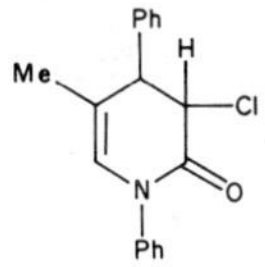	98	1138
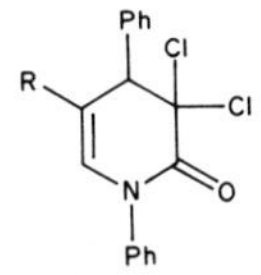	o-$Cl_2C_6H_4$, reflux, 3h	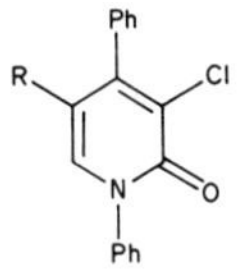		1138
R = H			98	
R = Me			100	
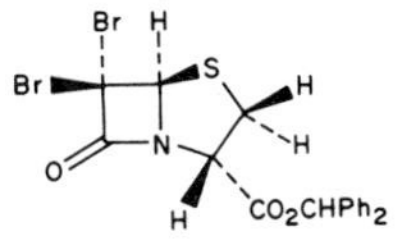	$(n\text{-Bu})_3SnH$, C_6H_6, 65 °C, N_2, 5h	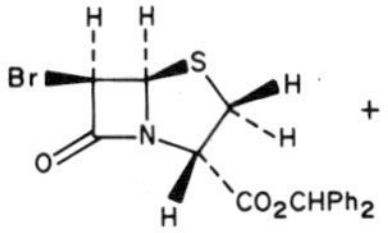 +	73	1431
		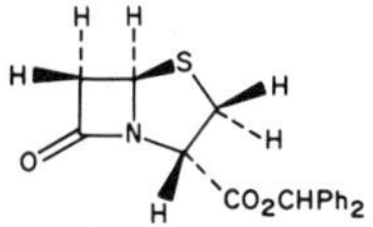	27	
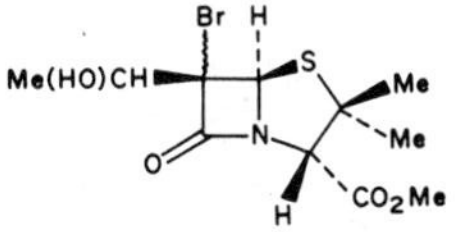	Zn, NH_4Cl, NH_4OH, Me_2CO	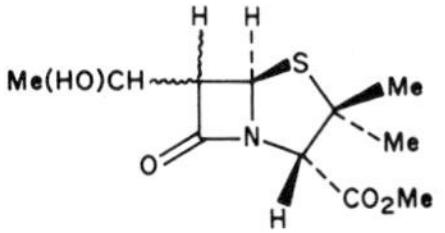	—	1394
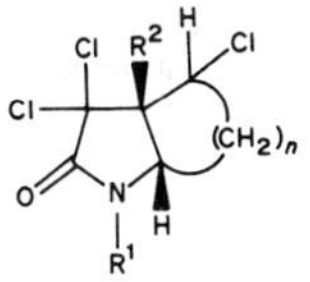	$(n\text{-Bu})_3SnH$, 140 °C, 1–5h	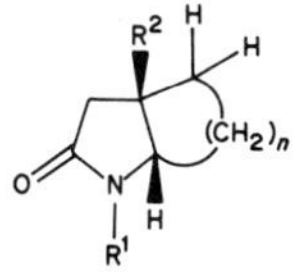		1064

(continued)

TABLE 86. *(continued)*

Halolactam			Reducing agent and conditions	Product	Yield (%)	Reference
n	R^1	R^2				
1	H	H			80	
1	$PhCH_2$	H			—	
2	H, $PhCH_2$	H			—	
2	H, Me	Ph			—	
2	$PhCH_2CO_2$	Ph			—	
2	Me	3,4-$(MeO)_2C_6H_3$			—	
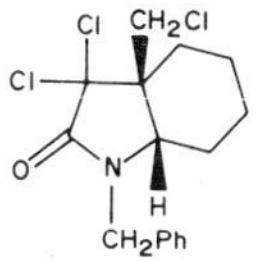 Cl, Cl, CH_2Cl, H, CH_2Ph			$(n\text{-Bu})_3SnH$, 140 °C	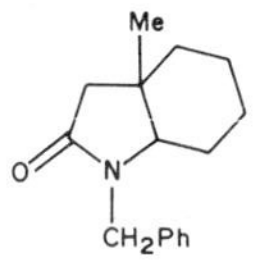 Me, O, N, CH_2Ph	—	1064
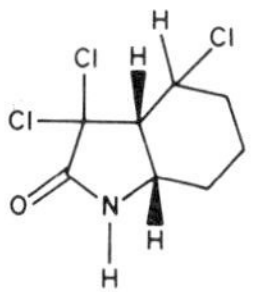 Cl, Cl, H, H, Cl, H, H			C_5H_5N, 120 °C, 3h	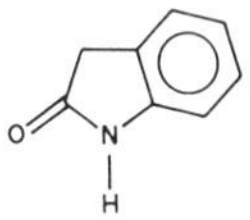 O, N, H	60	1064
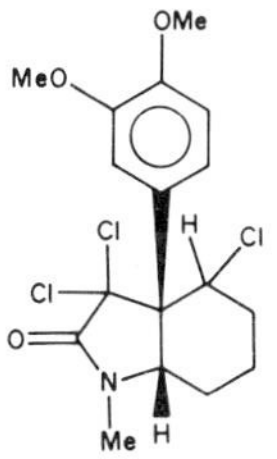 MeO, OMe, Cl, Cl, H, Cl, O, N, Me, H			$(n\text{-Bu})_3SnH$	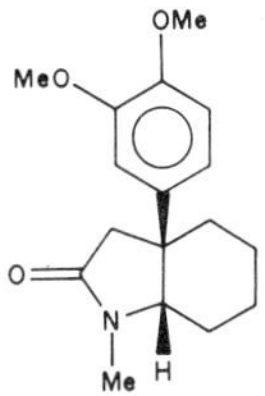 MeO, OMe, O, N, Me, H	—	1064

α-Mono- and α,α-dihalolactams have also been electrochemically reduce[1161] in aprotic solvents with or without added proton donors and/or electrophiles. Without added substrates, the carbanion arising from cleavage of the carbon–halogen bond undergoes protonation with a proton resulting mainly from the parent molecule. In addition to the protonation reaction, a competitive ring-opening reaction occurs which produces the corresponding dehalogenated β-lactams and α,β-unsaturated amides which, under the conditions of the reaction, are protonated and isolated as their saturated counterparts (equation 1112 and Table 87).

In the presence of a proton donor such as acetic acid, or an electrophile such as carbon dioxide, a protonation and coupling reaction, respectively, become predominant and the dehalogenated (equation 1113 and Table 87) or carboxylated (equation 1114 and Table 87) β-lactams are the main products.

TABLE 87. Electrochemical reduction of α-mono- and α,α-dihalolactams[1161]

Halolactam	Added substrate	E/V	Product	Yield (%)
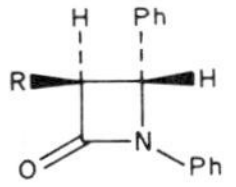 [structure: R, H, Ph, H; N-Ph β-lactam]	none		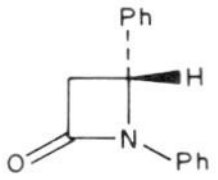 [structure: 3-unsubstituted 4-Ph, N-Ph β-lactam]	
R = Br		−1.5		64
		−1.8		45
		−2.0		35
R = Cl	none	−2.3		50
R = Br	$MeCO_2H$	−1.3		100
R = Cl	$MeCO_2H$	−2.2		99
R = Br	CO_2	−1.5	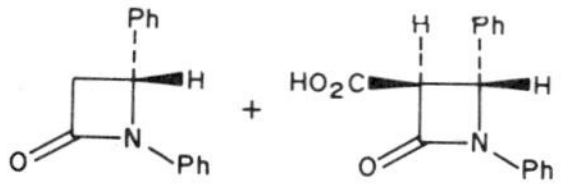 [structure] + [structure: HO_2C, H, Ph, H; N-Ph β-lactam]	8 + 90
R = Cl	CO_2	−2.0		5 + 95
R = Br	$BrCH_2CH_2CN$	−1.5	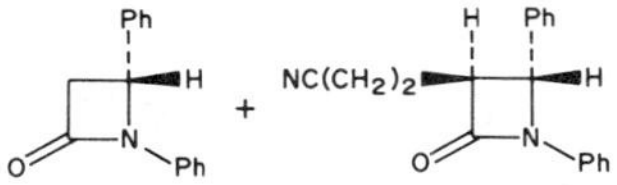 [structure] + [structure: $NC(CH_2)_2$, H, Ph, H; N-Ph β-lactam]	88 + 5
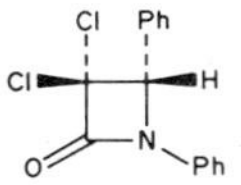 [structure: Cl, Cl, Ph, H; N-Ph β-lactam]	none		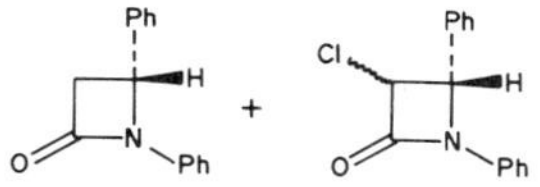 [structure] + [structure: Cl, Ph, H; N-Ph β-lactam]	
		−1.5		14 + 20
		−2.1		27 + 0
	$MeCO_2H$	−1.5		30 + 55
		−2.2		98 + 0
	CO_2	−1.5		3 + 7 +
			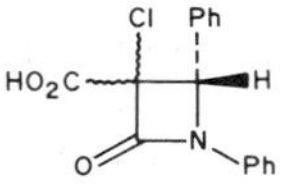 [structure: HO_2C, Cl, Ph, H; N-Ph β-lactam]	81

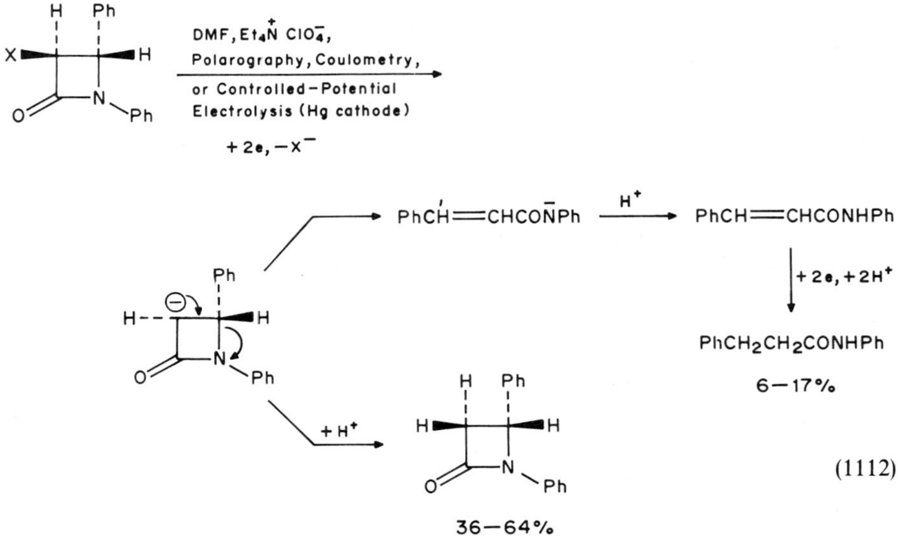

DMF, $Et_4\overset{+}{N}\,ClO_4^-$, Polarography, Coulometry, or Controlled-Potential Electrolysis (Hg cathode)

$+2e, -X^-$

$PhCH{=}CHCO\bar{N}Ph$ $\xrightarrow{H^+}$ $PhCH{=}CHCONHPh$ $\xrightarrow{+2e, +2H^+}$ $PhCH_2CH_2CONHPh$

6—17%

$+H^+$

36—64%

(1112)

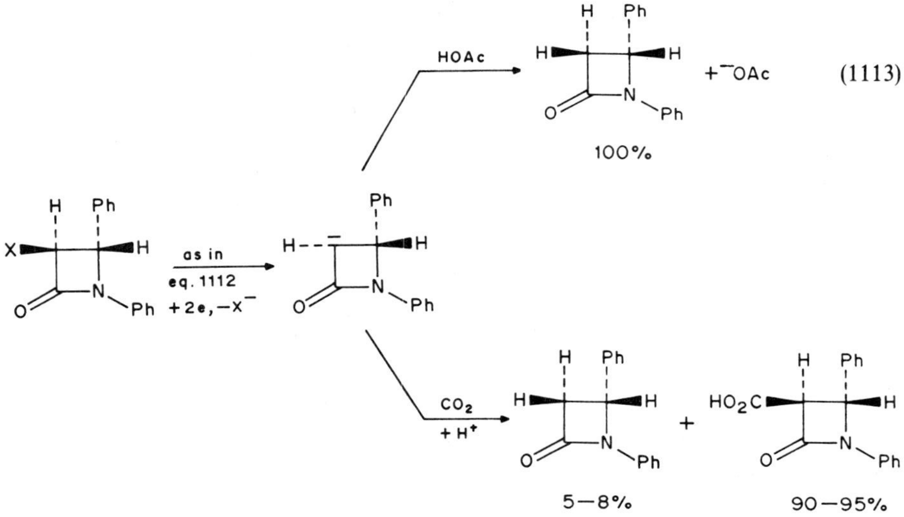

HOAc

$+ {}^-OAc$

100%

(1113)

as in eq. 1112 $+2e, -X^-$

CO_2 $+H^+$

5—8%

HO_2C

90—95%

(1114)

In the presence of 3-bromopropionitrile, a substrate which can behave both as an electrophile and as a proton donor, the protonation reaction is preferred, and the dehalogenated β-lactam predominates over the substitution product (equation 1115 and Table 87).

as in eq. 1112, $+2e, -X^-$

$Br(CH_2)_2CN$

5%

$+ Br^-$

(1115)

$+ BrCH_2\bar{C}HCN$

$-Br^-$

$CH_2{=}CHCN$

Other functional groups located at the 3-position of lactams which have been replaced by hydrogen include a carboxylic acid function attached to an aza-β-lactam which is quantitatively decarboxylated by heating in benzene[1261] (equation 1116), a methyl ester function in *trans*-methyl 1,4-diphenyl-3-methylthio-2-azetidinone-3-carboxylate which is quantitatively decarbomethoxylated by heating with lithium iodide in pyridine[1135] (equation 1117), a phenylthio function in *trans*-1-(4-methoxyphenyl)-3-phenylthio-4-(prop-1-enyl)azetidin-2-one[1153] (equation 1118) or in 1-cyclohexyl-3-(phenylthio)azetidin-2-one[1088] (equation 1119) which is desulphurized in both cases using Raney nickel, and a phenylseleno function from benzyl (2*S*, 5*R*)-6,6-bis(phenylseleno)-penicillinate[1339] which is partially deselenized using tributyltin hydride and azobisisobutyronitrile (equation 1120).

C_6H_6, reflux

100%

(1116)

LiI, C_5H_5N, reflux 6 h

100%

(50:50 mixture of stereoisomers)

(1117)

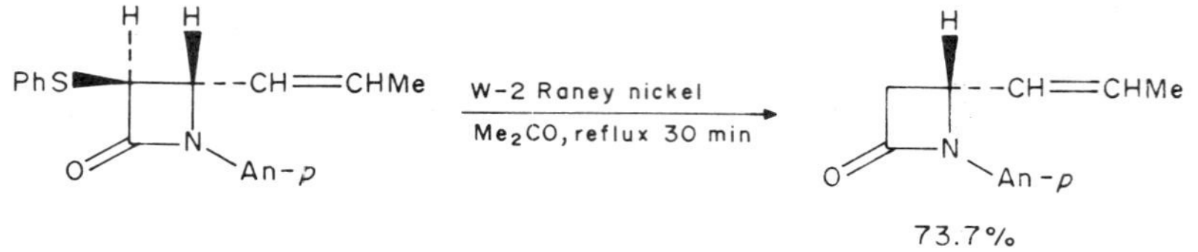
H
H
PhS
CH=CHMe
N
An-p
O
W-2 Raney nickel
Me2CO, reflux 30 min
H
CH=CHMe
N
An-p
O
73.7%

(1118)

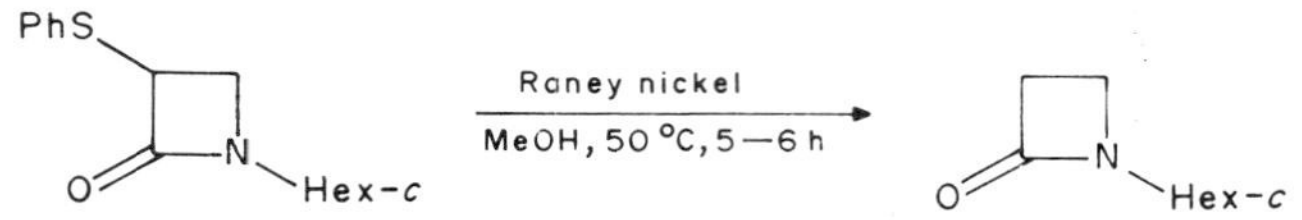
PhS
N
Hex-c
O
Raney nickel
MeOH, 50 °C, 5—6 h
N
Hex-c
O

(1119)

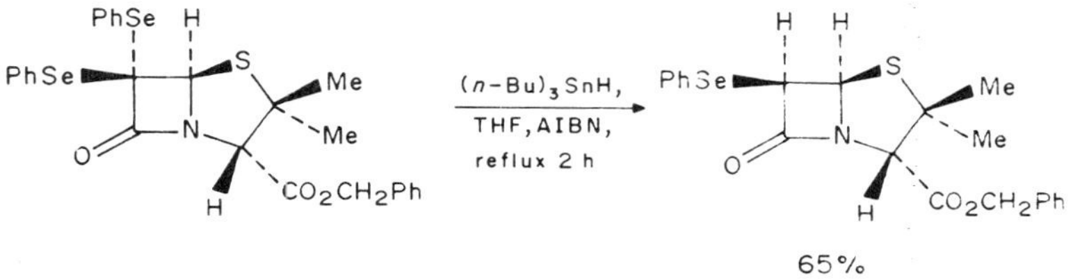
PhSe
H
PhSe
S
Me
Me
N
O
H
CO2CH2Ph
(n-Bu)3SnH,
THF, AIBN,
reflux 2 h
H
H
PhSe
S
Me
Me
N
O
H
CO2CH2Ph
65%

(1120)

Interconversion of oxygen containing functions located at the 3-position of lactams has also been reported, and mainly consists of two types of reactions, functionalization of a hydroxy substituent (equation 1121 and Table 88), or dealkylation of an alkoxy substituent to produce the hydroxy function (equation 1122 and Table 89).

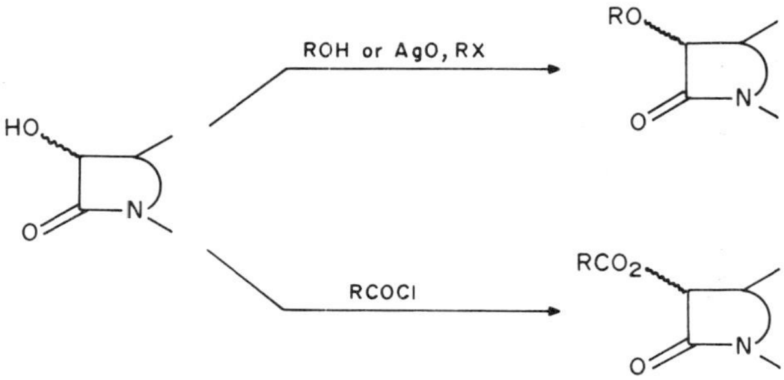
HO
N
O
ROH or AgO, RX
RO
N
O
RCOCl
RCO2
N
O

(1121)

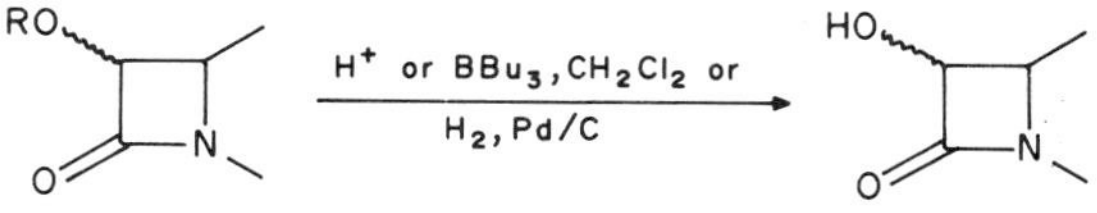
RO
N
O
H+ or BBu3, CH2Cl2 or
H2, Pd/C
HO
N
O

(1122)

α-Hydroxy-β-lactams have also been converted[1205] to α-bromo-β-lactams upon reaction with triphenylphosphine and carbon tetrabromide, and this conversion is accomplished by inversion of configuration at carbon-3 (equation 1123).

TABLE 88. Functionalization of α-hydroxy substituents on lactams

Hydroxylactam	Reagent	Conditions	Product	Yield (%)	Reference
HO-- (H, H) --An-p; O=, N-Tol-p	MeI	AgO, THF, reflux, 8h	MeO-- (H, H) --An-p; O=, N-Tol-p	80	1156
EtO_2C, OH, CH_2Ph, Me, N–R, O	MeO–, OMe (benzene)	AcOH, H_2SO_4, 20 °C, stand overnight	OMe, OMe, EtO_2C, CH_2Ph, Me, N–R, O		1380
R = H				48	
R = Me				41	
R = $PhCH_2$				58	
EtO_2C, OH, CH_2R^2, Me, N–R^1, O	HO–, R^3, R^4 (benzene)	AcOH, H_2SO_4, 20 °C, stand overnight	R^3, R^4, O, Me, N–R^1, CO_2Et, R^2CH_2, O		1380

(*continued*)

TABLE 88. (*continued*)

Hydroxylactam				Reagent	Conditions	Product	Yield (%)	Reference
R^1	R^2	R^3	R^4					
H	Ph	HO	H				76	
Me	Ph	HO	H				60	
$PhCH_2$	Ph	HO	H				80	
H	*p*-Tol	HO	H				60	
$PhCH_2$	*p*-Tol	HO	H				80	
H	Ph	H	H				38	
Me	Ph	H	HO				55	
$PhCH_2$	Ph	H	HO				75	
H	Ph	Me	Me				63	
Me	Ph	Me	Me				70	
$PhCH_2$	Ph	Me	Me				57	
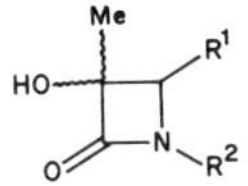				$3,5\text{-}(O_2N)_2C_6H_3COCl$		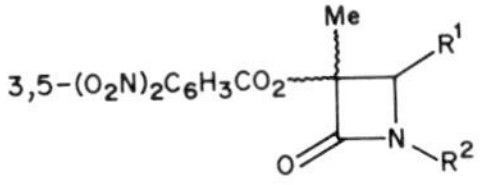		1109
$R^1 = H$, $R^2 = Me$					—		—	1109
$R^1 = Me$, $R^2 = Et$					—		—	
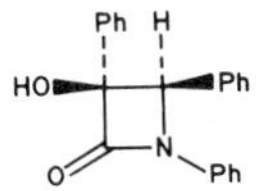				AcCl	Et_3N	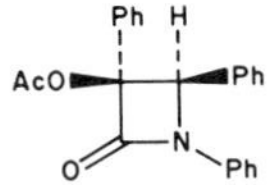	—	1218
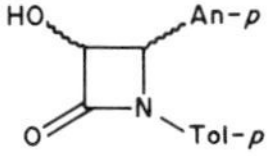				*p*-RC_6H_4COCl	Et_3N, CH_2Cl_2, r.t. stir overnight	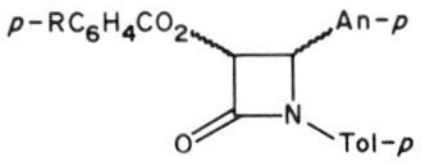		902

	$R = O_2N$			70	
	$R = PhOCH_2CONH$			80	
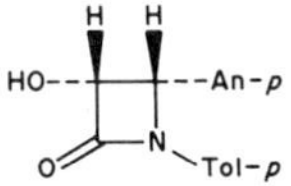	RCl	Et_3N, CH_2Cl_2, r.t. stir	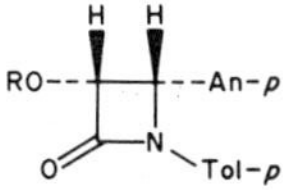		1156
	$R = PhCH_2CO$			75	
	$R = PhOCH_2CO$			80	
	$R = CF_3CH_2SO_2$			85	
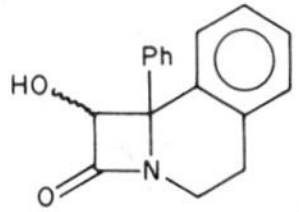	RCOCl	Et_3N, CH_2Cl_2, stir	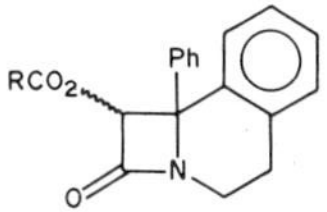		902
	$R = PhOCH_2$			80	
	$R = p\text{-}O_2NC_6H_4$			75	
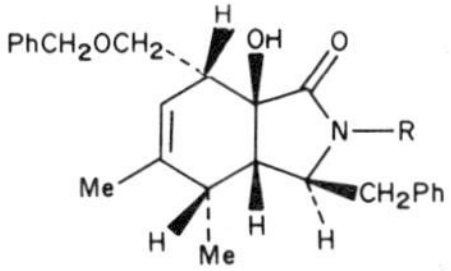	$(MeCO)_2O$	C_5H_5N, DMAP	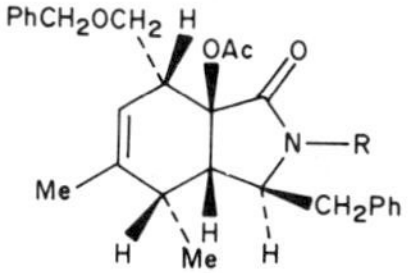	100	1374

$R = Et, CH_2Ph$

TABLE 89. Dealkylation of α-alkoxylactams

Alkoxylactam	Conditions	Product	Yield (%)	Reference
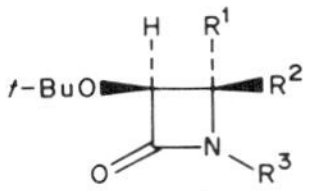	CF_3COOH, r.t., stir	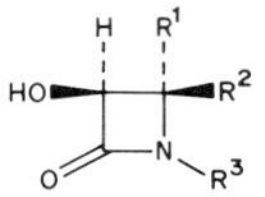		
$R^1 = H$, $R^2 = R^3 = Ph$			85	1205
$R^1 = R^3 = Ph$, $R^2 = H$			75	1205
$R^1 = H$, $R^2 = R^3 = p$-An			—	1205
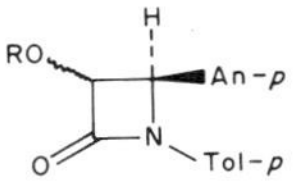		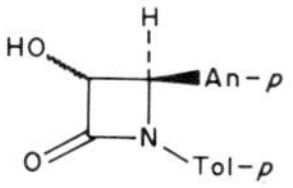		
R = *t*-BuO	CF_3COOH, 50 °C, 20 min		40	902
	BBr_3, CH_2Cl_2, 0 °C, stir 1 h		—	
R = $PhCH_2$	BBr_3, CH_2Cl_2, 0 °C, stir 1 h		80	902
	H_2(50 psi), Pd/C, THF, 12 h		90	
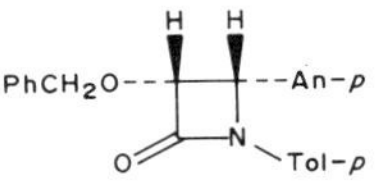	H_2, 10% Pd/C, THF, 12 h	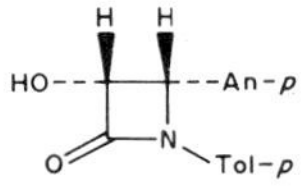	90	1156
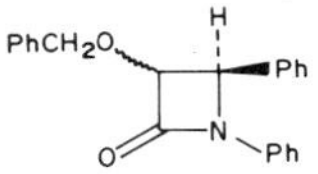 *cis–trans* mixture	H_2, 10% Pd/C, THF, 12 h	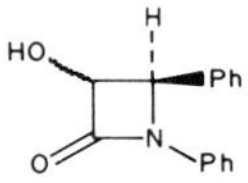 *cis–trans* mixture	80	1156
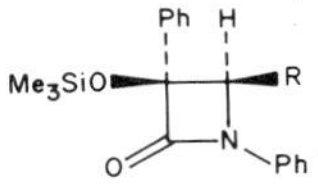		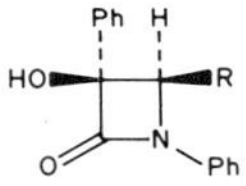		1218
R = An-*p*	Me_2CO, H_2O, 25 °C. stir 48 h or Me_2CO, H_2O, 70 °C, stir 4 h		N.R.	
R = An-*p*	HF, MeOH, CH_2Cl_2, 25 °C, stir 2 h		100	
	1*N* HCl, Me_2CO, 25 °C, 2 h		100	
R = Ph	HF, CH_2Cl_2		—	1218
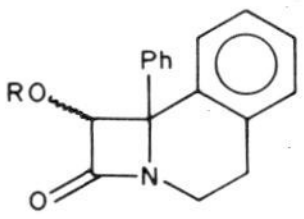 R = *t*-Bu, $PhCH_2$	BBr_3, CH_2Cl_2, 0 °C, stir	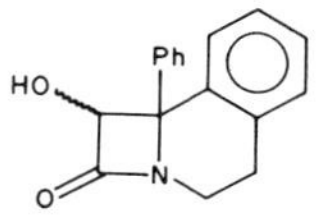	75	902

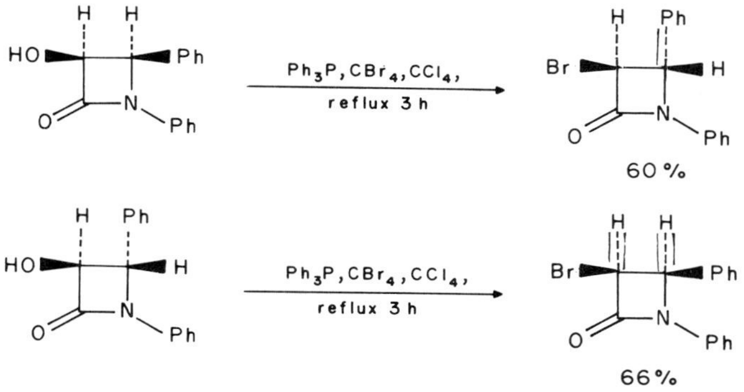

Ph_3P, CBr_4, CCl_4, reflux 3 h — 60%

Ph_3P, CBr_4, CCl_4, reflux 3 h — 66%

(1123)

Variously substituted γ- and δ-lactams have been converted into their α,β-unsaturated counterparts by reaction with a variety of reagents (equations 1124 and 1125 and Table 90).

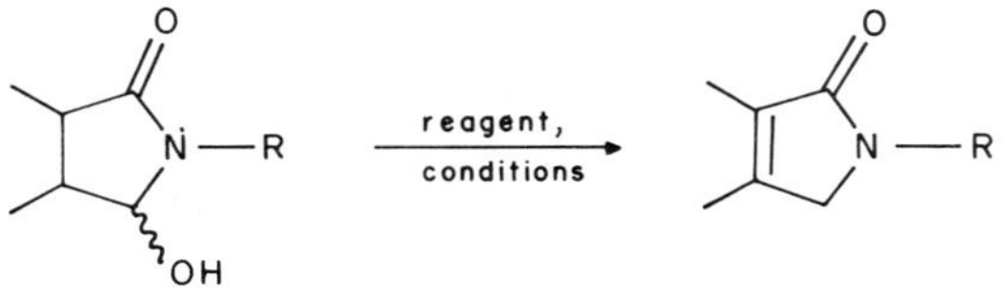

reagent, conditions (1124)

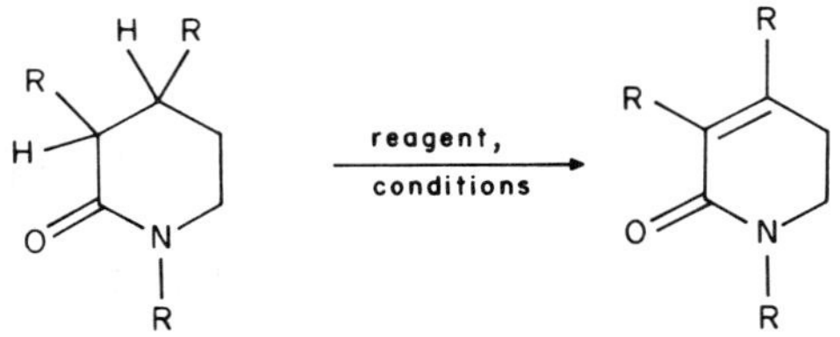

reagent, conditions (1125)

Numerous examples of the converse reaction, conversion of α,β-unsaturated lactams to saturated lactams, have also been reported (equation 1126 and Table 91).

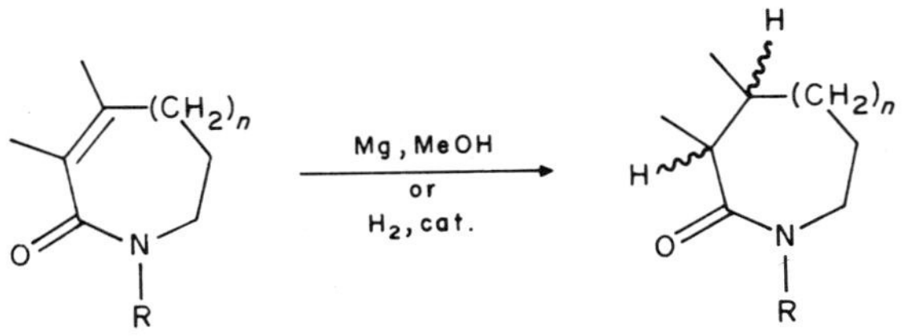

Mg, MeOH or H_2, cat. (1126)

One interesting method of converting[1385] α,β-unsaturated lactams to saturated lactams is by the conjugate addition of organocopper reagents to N-tosylated α,β-unsaturated lactams (equations 1127 and 1128).

TABLE 90. Formation of α,β-unsaturated lactams from saturated lactams

Lactam	Reagent and conditions	Product	Yield (%)	Reference
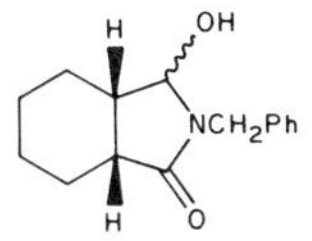	*p*-TosOH, C_6H_5Me, reflux 24h	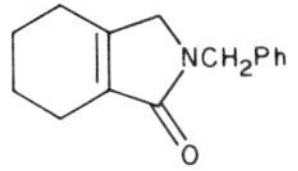	73	1432
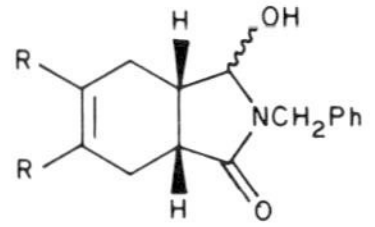	*p*-TosOH, C_6H_5Me, reflux	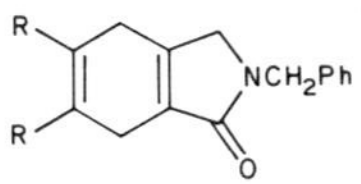		
R = H			55	1432
R = Me, Ph			50	1432
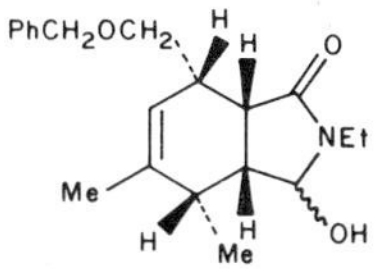	*p*-TosOH, MeOH, r.t.	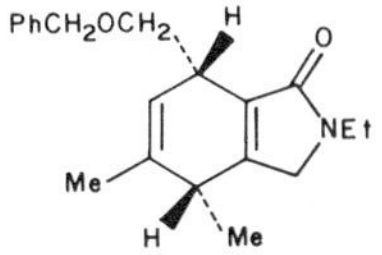	60	1374
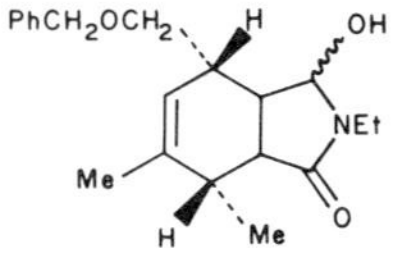	*p*-TosOH, MeOH, r.t.	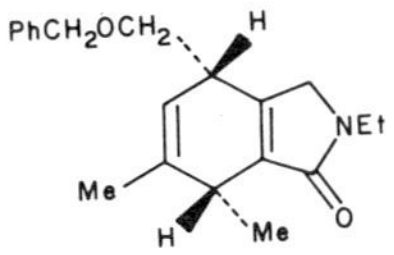	60	1374
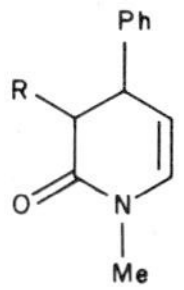	NaH, THF, reflux	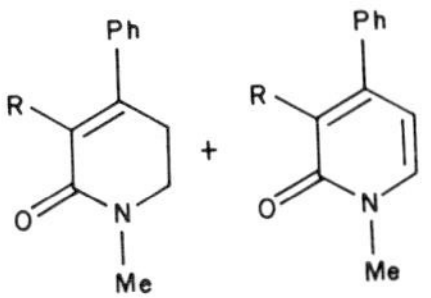		
R = Me			—	1238

Substrate	Conditions	Product	Yield (%)	Ref.
R = Ph	NaH, THF, reflux	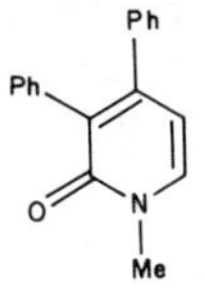 (242)	—	1238
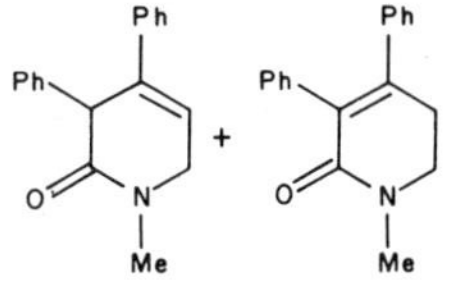 Ph Ph Me + Ph Ph Me	NaH, THF, reflux	242	—	1238
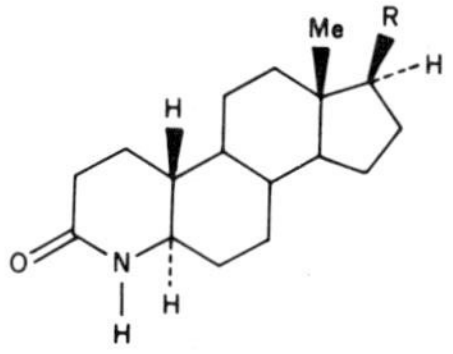	PhSe(O)OSe(O)Ph, diglyme, 120 °C	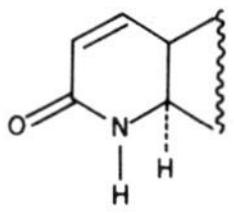		1124
$R = n\text{-}C_8H_{17}$	14h		88	
R = OAc	23h		64	
R = CH(Me)OAc	16h		—	
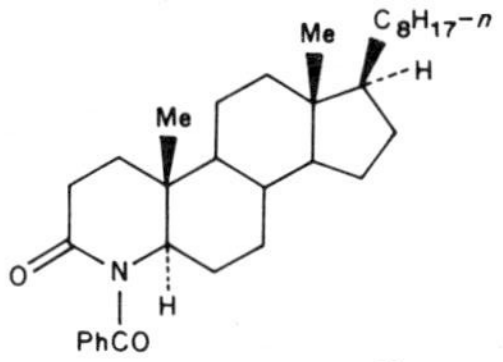 $C_8H_{17}\text{-}n$, PhCO	PhSe(O)OSe(O)Ph, diglyme, 120 °C, 3h	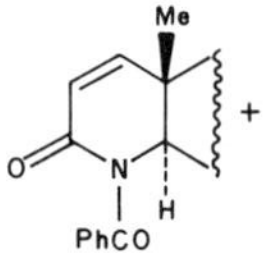 PhCO +	35	1124
		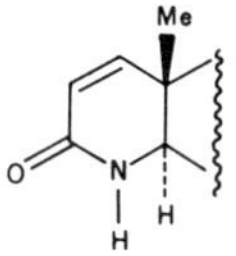	29	
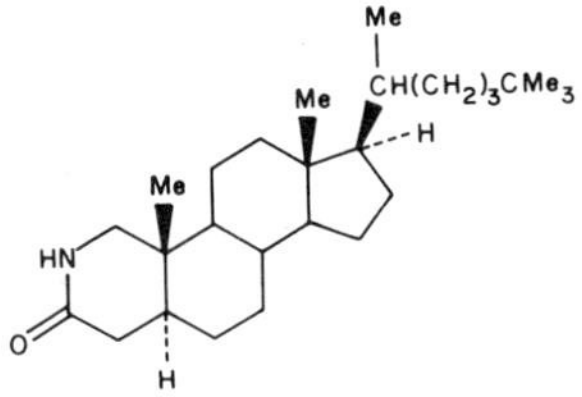 $CH(CH_2)_3CMe_3$, HN	PhSe(O)OSe(O)Ph, diglyme, 120 °C, 21h	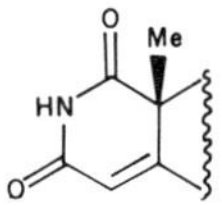	53	1124

TABLE 91. Reduction of α,β-unsaturated lactams

Lactam	Reagent and conditions	Product	Yield (%)	Reference
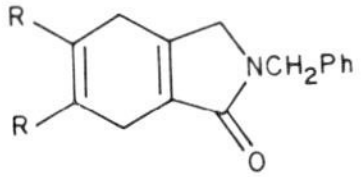 R, R; NCH_2Ph; O	Mg, MeOH, stir r.t.	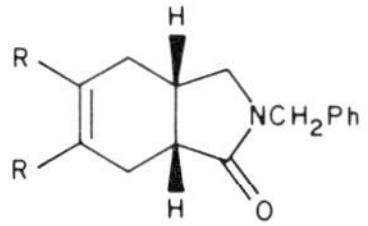 H; R, R; NCH_2Ph; H; O		1432, 1433
		cis:*trans* ratio		
R = H		2.8:1	15	
R = Me		4.0:1	61	
R = Ph		4.7:1	83	
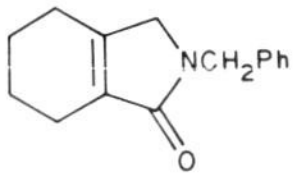 NCH_2Ph; O	Mg, MeOH, stir r.t.	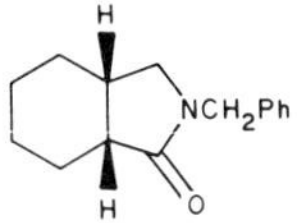 H; NCH_2Ph; H; O *cis* and *trans*	—	1432
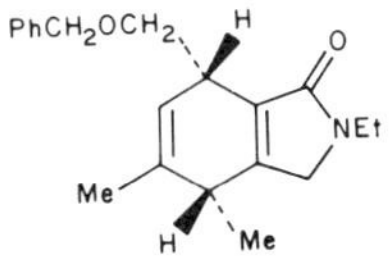 $PhCH_2OCH_2$; H; O; NEt; Me; H; Me	5% Pd/C, H_2, C_6H_5Me, reflux	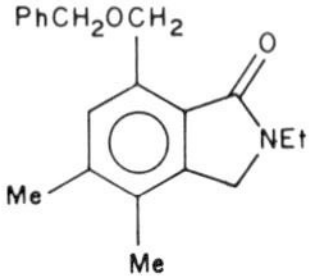 $PhCH_2OCH_2$; O; NEt; Me; Me	55	1374
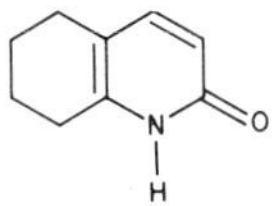 N; H; O	Mg, MeOH, stir r.t.	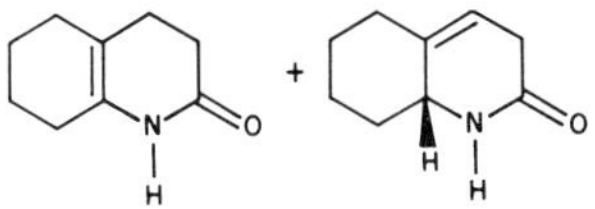 N; H; O + H; N; H; O	34 + 35	1432
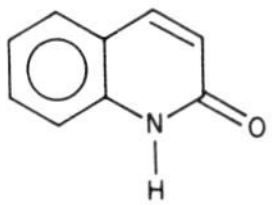 N; H; O	H_2(85 atm), W-2 Raney Ni, MeOH, autoclave, 140 °C, 5h	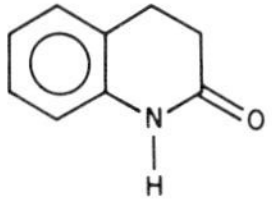 N; H; O	89	1432

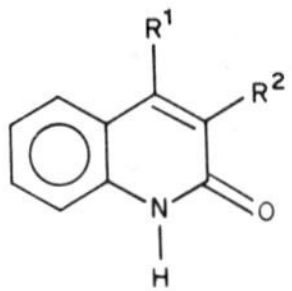 $R^1 = R^2 = H$ $R^1 = Me, R^2 = H$ $R^1 = R^2 = Me$	Mg, MeOH, r.t. stir	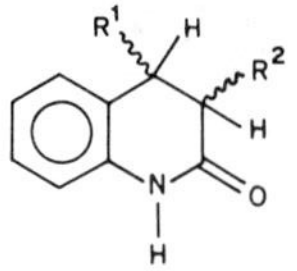 *cis* and *trans*	30 55 95	1432
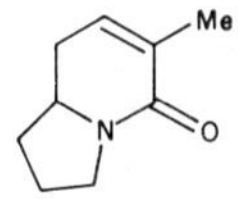	H_2, PtO_2, EtOH, r.t. stir overnight	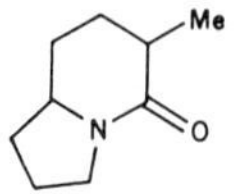	86	776
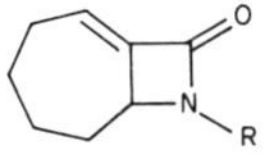 R = H, $Me_2\overset{\mid}{C}OH$	H_2, PtO_2, EtOH	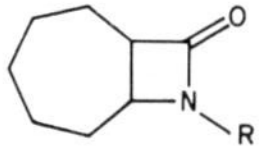	100	1393
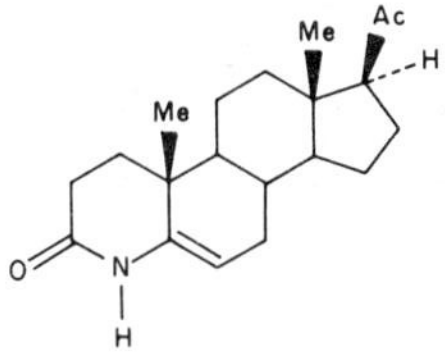	H_2, PtO_2, 10% HOAc–MeOH, 5 days	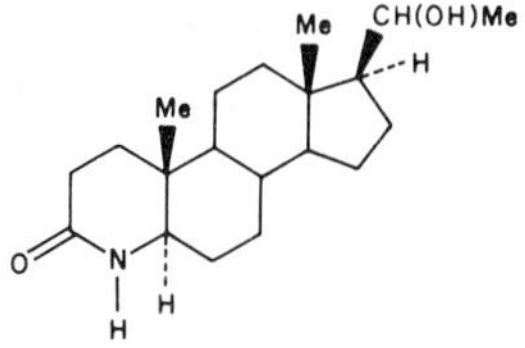	75	1124
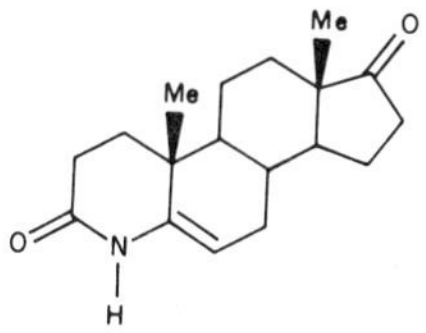	H_2, PtO_2, 10% HOAc–MeOH, 5 days	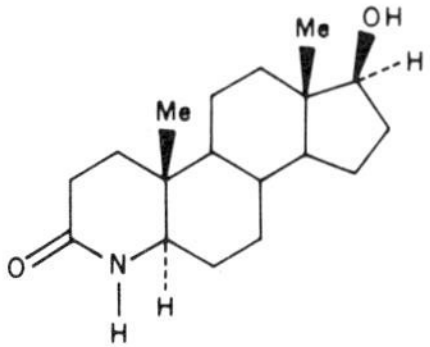	63	1124

+ R^2MgX/CuI —ether, −20 °C, stir 2.5 h→

R^1 = H ; Me ; Me

R^2 = Me; Ph ; p-An

Yield (%) = 73 ; 70 ; 74

(1127)

+ MeMgI/CuI —ether, −20 °C, stir 2.5 h→

(1128)

Numerous lactams which contain substituents at the alpha and beta sites have been reportedly isomerized upon treatment with base, heat or light. The general reaction for this conversion is illustrated in equation 1129 while the specific examples which have been reported are recorded in Table 92.

base, heat or light

(1129)

α-Mono-phenylthio or phenylseleno and α,α-bis-phenylthio or phenylseleno functions are useful substituents when attached to lactam rings, because they can be converted into a variety of other functional groups. A preparation of α-(phenylthio)azetidin-2-ones has already been discussed in Section III.A.2 and involved[1088] the reaction of the 1,3-dianion of α-(phenylthio)acetamide derivatives with methyl iodide. An episulphonium intermediate was proposed as the intermediate for that reaction (equation 740).

α-Mono-phenylthio and phenylseleno as well as α,α-bis-phenylthio and phenylseleno penicillinates have all been prepared[1339] from the corresponding diazopenicillinate (Scheme 18).

One example of the use of these substituents to effect functionalization at the alpha position is the acetylation shown in equation 1130[1088].

TABLE 92. Isomerization of α,β-substituted lactams

Lactam	Reagent and conditions	Product	Yield (%)	Reference
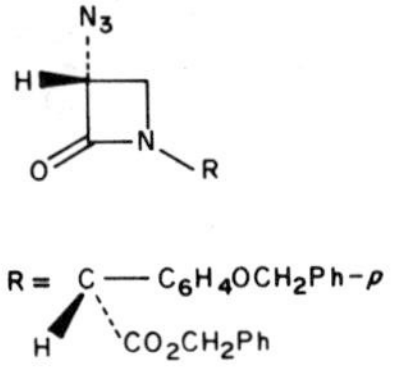 N_3, H, O, N, R; R = C—$C_6H_4OCH_2Ph$-*p*, H, CO_2CH_2Ph	KOBu-*t*,*t*-BuOH, THF, 0 °C, stir 2h	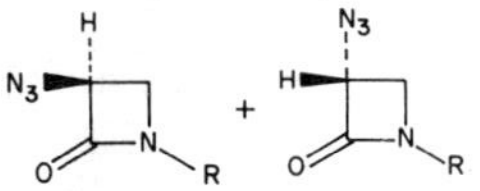 H, N_3, O, N, R + N_3, H, O, N, R (ratio 1:1)	100	1039
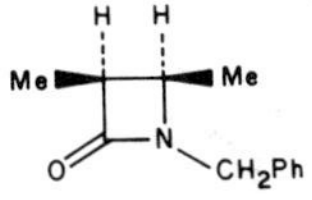 H, H, Me, Me, O, N, CH_2Ph **(243)**	KOBu-*t*,*t*-BuOH, 50 °C, 15h	**243** 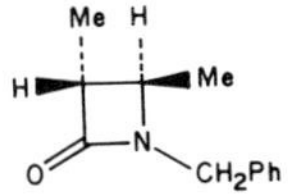 + Me, H, H, Me, O, N, CH_2Ph (ratio 1:2)	95	1386
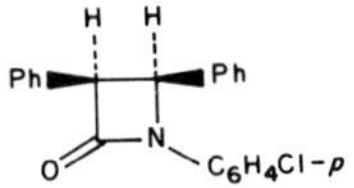 H, H, Ph, Ph, O, N, C_6H_4Cl-*p*	90–100 °C	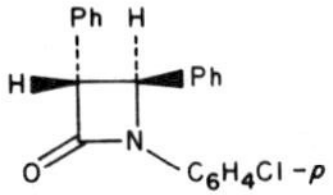 Ph, H, H, Ph, O, N, C_6H_4Cl-*p*	—	1272
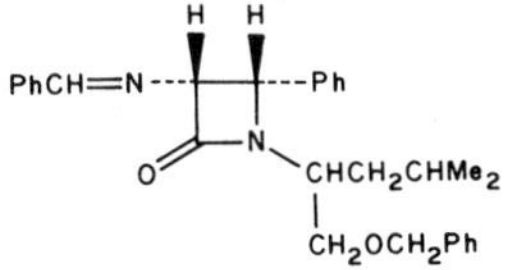 H, H, PhCH=N, Ph, O, N, $CHCH_2CHMe_2$, CH_2OCH_2Ph	1. THF, LDA, −95 °C, stir 4h 2. MeOH, −95 °C	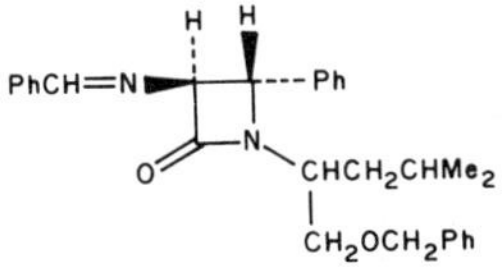 H, H, PhCH=N, Ph, O, N, $CHCH_2CHMe_2$, CH_2OCH_2Ph	90	1171
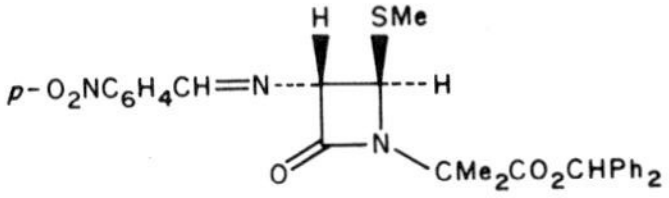 H, SMe, *p*-$O_2NC_6H_4CH$=N, H, O, N, $CMe_2CO_2CHPh_2$ **(244)**	1. THF, Ar, −68 °C, PhLi, 5 min 2. DMF, HOAc, H_2O, THF	**244** 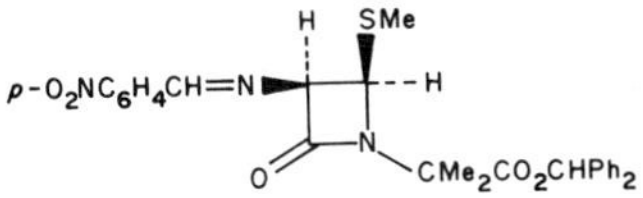 + H, SMe, *p*-$O_2NC_6H_4CH$=N, H, O, N, $CMe_2CO_2CHPh_2$ (ratio 1:1)	—	1173

(*continued*)

TABLE 92. *(continued)*

Lactam	Reagent and conditions	Product	Yield (%)	Reference
H H PhOCH$_2$CONH~ CH$_2$CH(OMe)$_2$ O N SiMe$_2$(Bu-*t*) (1:1-*cis,trans* mixture)	$(n\text{-Bu})_4$NF, THF, r.t., stir 5 min	H H PhOCH$_2$CONH CH$_2$CH(OMe)$_2$ O N SiMe$_2$(Bu-*t*)	81	1316
H H R COPh O N An-*p*		R H H COPh O N An-*p*	—	1148
R		*cis*:*trans* ratio		
Ph	*n*-BuLi, THF, −5 °C, 120 min	14:86		
Ph	*n*-BuLi, THF, 25 °C, 30 min	0:100		
Ph	NaOH, MeCN, H_2O, 25 °C, 15 min	63:37		
Ph	NaOH, MeCN, H_2O, 22 h	0:100		
Me	*n*-BuLi, THF, 25 °C, 30 min	100:0		
Me	NaOH, MeCN, H_2O, 25 °C, 15 min	68:32		
Me	NaOH, MeCN, H_2O, 25 °C, 22 h	42:58		
Me	NaOH, MeCN, H_2O, 25 °C, 72 h	28:72		
Me	NaOH, MeCN, H_2O, reflux, 1 h	40:60		
Et	*n*-BuLi, THF, 25 °C, 30 min	100:0		
Et	NaOH, MeCN, H_2O, 25 °C, 15 min	55:45		
Et	NaOH, MeCN, H_2O, 25 °C, 22 h	27:73		
Et	NaOH, MeCN, H_2O, 25 °C, 72 h	17:83		
i-Pr	*n*-BuLi, THF, 25 °C, 30 min	100:0		
i-Pr	NaOH, MeCN, H_2O, 25 °C, 15 min	42:58		
i-Pr	NaOH, MeCN, H_2O, 25 °C, 22 h	0:100		
H Ph NC H O N Bu-*t*		H Ph NC~ H O N Bu-*t*		1141

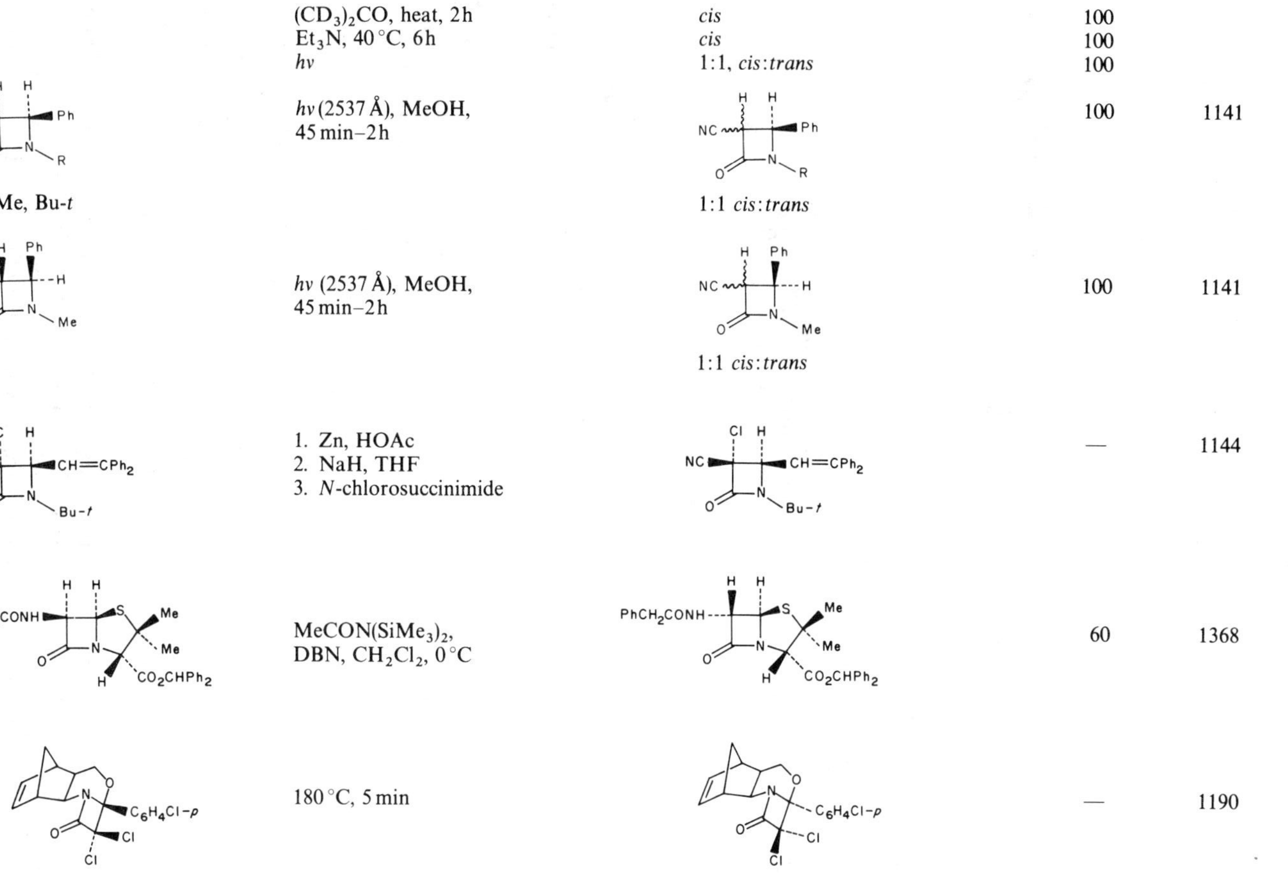

Substrate	Conditions	Product	Yield (%)	Ref.
	$(CD_3)_2CO$, heat, 2h	*cis*	100	
	Et_3N, 40 °C, 6h	*cis*	100	
	hν	1:1, *cis*:*trans*	100	
R = Me, Bu-*t*	*hν* (2537 Å), MeOH, 45 min–2h	1:1 *cis*:*trans*	100	1141
	hν (2537 Å), MeOH, 45 min–2h	1:1 *cis*:*trans*	100	1141
	1. Zn, HOAc 2. NaH, THF 3. *N*-chlorosuccinimide		—	1144
	$MeCON(SiMe_3)_2$, DBN, CH_2Cl_2, 0 °C		60	1368
	180 °C, 5 min		—	1190

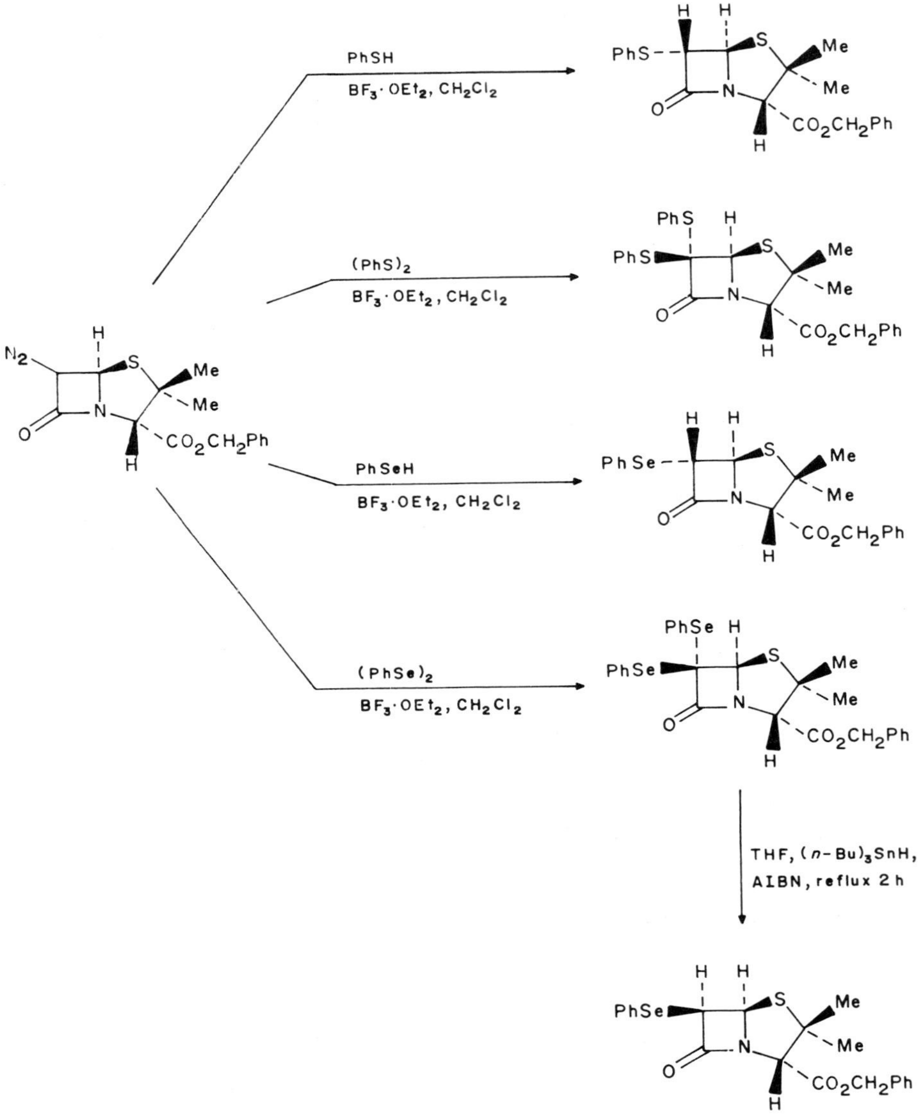

PhSH

$BF_3 \cdot OEt_2$, CH_2Cl_2

$(PhS)_2$

$BF_3 \cdot OEt_2$, CH_2Cl_2

PhSeH

$BF_3 \cdot OEt_2$, CH_2Cl_2

$(PhSe)_2$

$BF_3 \cdot OEt_2$, CH_2Cl_2

THF, $(n\text{-}Bu)_3SnH$, AIBN, reflux 2 h

SCHEME 18

Treatment of benzyl penicillinates containing these substituents with a base followed by treatment with acetaldehyde causes[1339] alpha-(*R*)-hydroxyethylation of the benzyl penicillinates (equation 1131), while treatment of the resulting benzyl (2*S*, 5*R*, 6*S*)-6-[(*R*)-1-hydroxyethyl]-6-(phenylseleno)penicillinate with tributyltin hydride and azobisisobutyronitrile reduces this product to benzyl (2*R*, 5*R*, 6*R*)-6-[(*R*)-1-hydroxyethyl]-penicillinate (equation 1132)[1339].

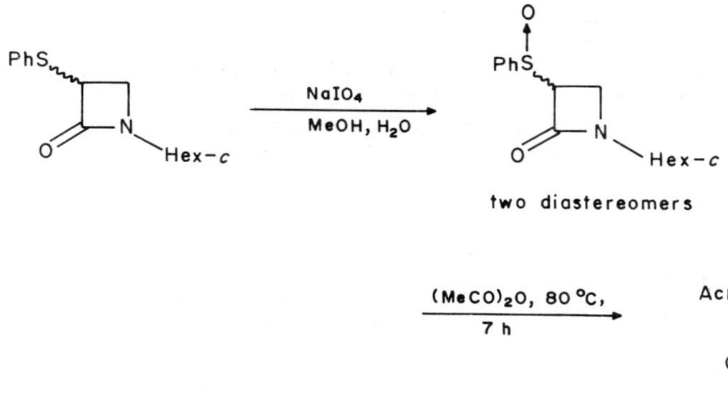

(1130)

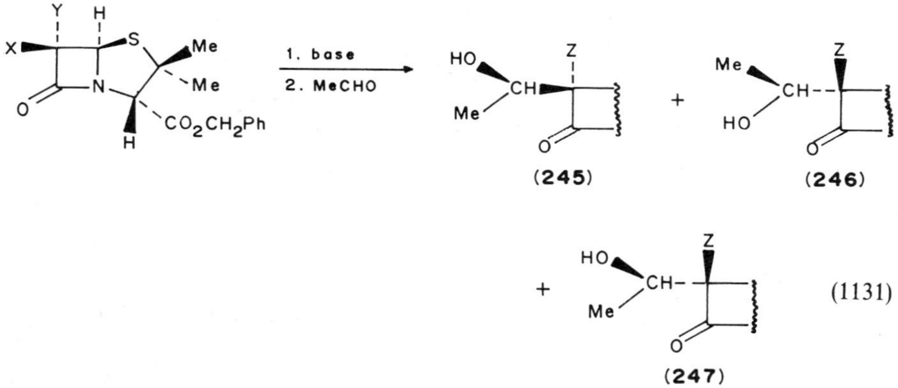

(245) (246) (247)

(1131)

X	Y	Z	Base	Yield (%)	Ratio **245:246:247**
PhS	PhS	—	MeMgBr	N.R.	—
PhS	PhS	—	*n*-BuLi	N.R.	—
PhSe	H	—	MeMgBr	N.R.	—
H	PhSe	PhSe	*n*-BuLi	48	1:2:1
H	PhSe	PhSe	MeMgBr	5	1:0:0
PhSe	PhSe	PhSe	*n*-BuLi	56	2:3:2
PhSe	PhSe	PhSe	MeMgBr	78	30:1:0
H	PhS	PhS	*n*-BuLi	38	0:1:0

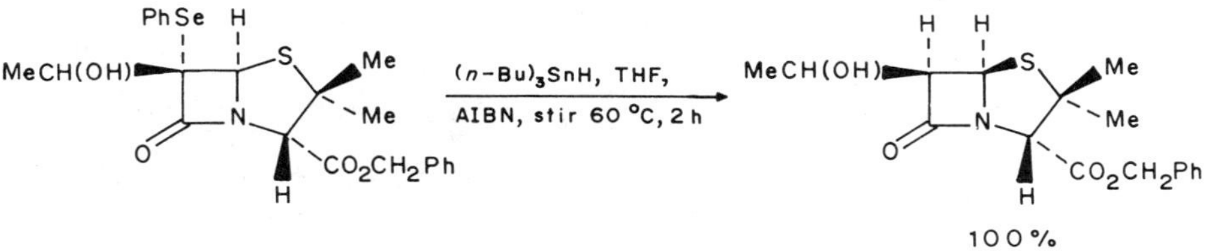

(1132)

α-Hydroxyethylated lactams have also been produced by simple hydrolysis of an acetoxy[750] (equation 1133) or dimethyl(*t*-butyl)silyl[1037,1339,1345] protecting function (equations 1134[1237], 1135[1339] and 1136[1345]), by hydrolysis of a 1,3-dioxolan function followed by reduction of the resulting[1340] ketones (equations 1137 and 1138) and by formation (accompanied by *cis* and *trans* conversion) and reduction[1389] (equations 1139 and 1140) or simple reduction[1132] (equation 1141) of 3-acetyl β-lactams.

AcO H H; Me—C—CO$_2$Et; N—CH$_2$C$_6$H$_3$(OMe)$_2$-2,4 —[2.1 eq. 1*N* NaOH, C$_5$H$_5$N, 20 °C, 15 h]→ HO H H; Me—C—CO$_2$H; N—CH$_2$C$_6$H$_3$(OMe)$_2$-2,4 + HO H CO$_2$H; Me—C—H; N—CH$_2$C$_6$H$_3$(OMe)$_2$-2,4

(1133)

(*t*-Bu)Me$_2$SiO H H; Me—C—CH=CH$_2$; N—SiMe$_2$(Bu-*t*) —[1 *N* HCl, H$_2$O, MeOH]→ HO H H; Me—C—CH=CH$_2$; N—SiMe$_2$(Bu-*t*)

(1134)[1237]

(*t*-Bu)Me$_2$SiO H H; Me—C; S; N—CO$_2$CH$_2$C$_6$H$_4$NO$_2$-*p*

mixture of 36% (2*R*) + 60% (2*S*)

—[THF, HOAc, (*n*-Bu)$_4$NF, 0 °C, stir overnight]→

HO H H; Me—C; S; N—CO$_2$CH$_2$C$_6$H$_4$NO$_2$-*p*

(1135)[1339]

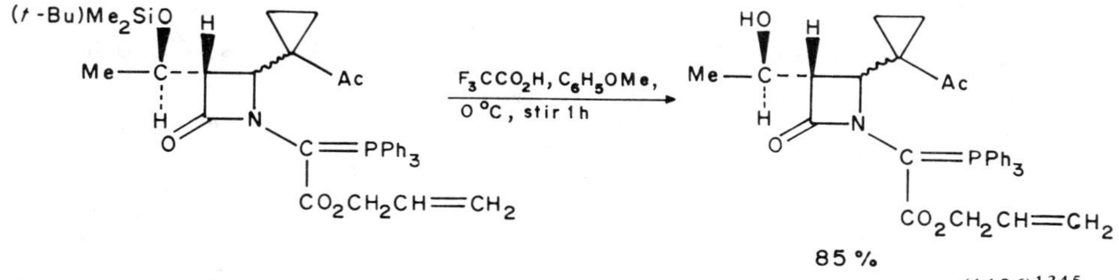

F_3CCO_2H, C_6H_5OMe, 0 °C, stir 1 h

85%

(1136)[1345]

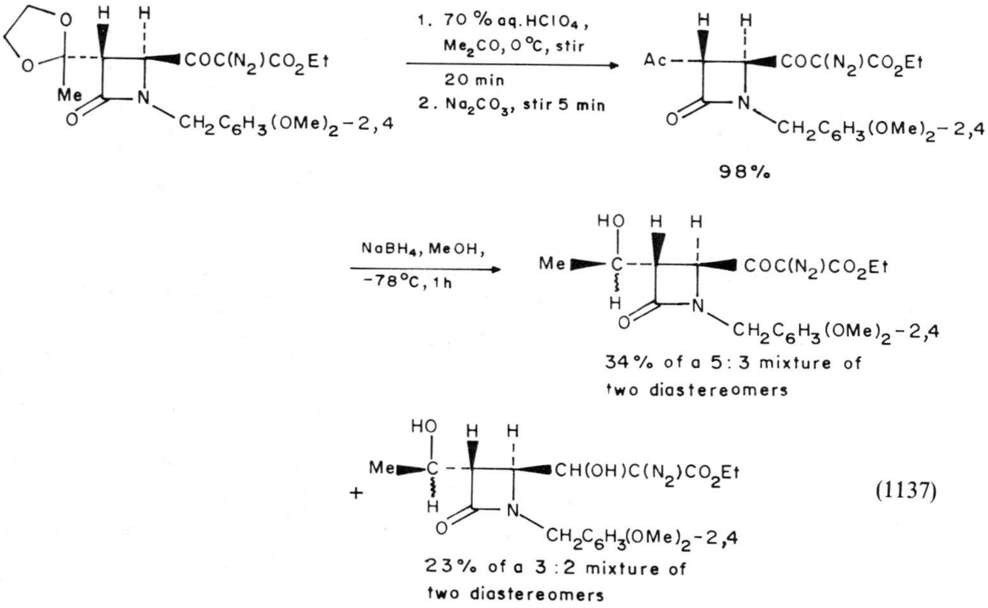

1. 70% aq. $HClO_4$, Me_2CO, 0 °C, stir 20 min
2. Na_2CO_3, stir 5 min

98%

$NaBH_4$, MeOH, −78 °C, 1 h

34% of a 5 : 3 mixture of two diastereomers

+

23% of a 3 : 2 mixture of two diastereomers

(1137)

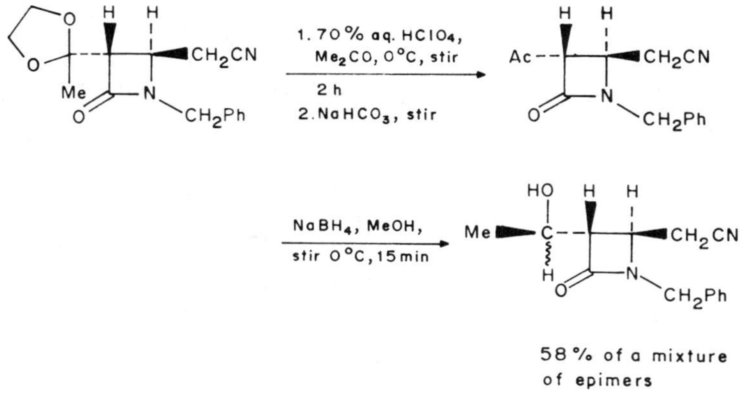

1. 70% aq. $HClO_4$, Me_2CO, 0 °C, stir 2 h
2. $NaHCO_3$, stir

$NaBH_4$, MeOH, stir 0 °C, 15 min

(1138)

58% of a mixture of epimers

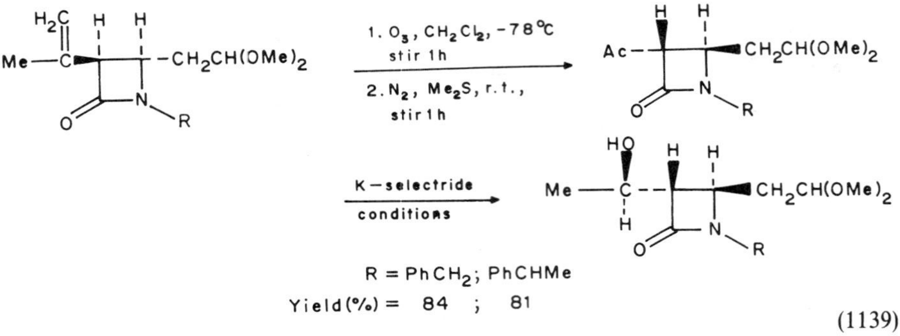

1. O_3, CH_2Cl_2, −78 °C, stir 1h
2. N_2, Me_2S, r.t., stir 1h

K-selectride / conditions

R = $PhCH_2$; PhCHMe
Yield (%) = 84 ; 81

(1139)

R	Conditions	Yield (%)
$PhCH_2$	THF, Et_2O, Ar, r.t., 24h	74
PhCHMe	1. KI, Et_2O, stir 0.5h 2. 0 °C, THF, stir 1h	71[a]

[a]A 9:1 mixture of C-1 epimers.

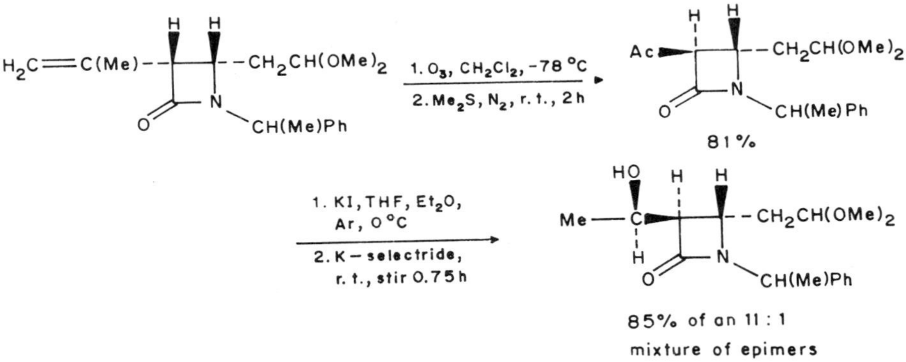

1. O_3, CH_2Cl_2, −78 °C
2. Me_2S, N_2, r.t., 2h

81%

1. KI, THF, Et_2O, Ar, 0 °C
2. K-selectride, r.t., stir 0.75h

85% of an 11 : 1 mixture of epimers

(1140)

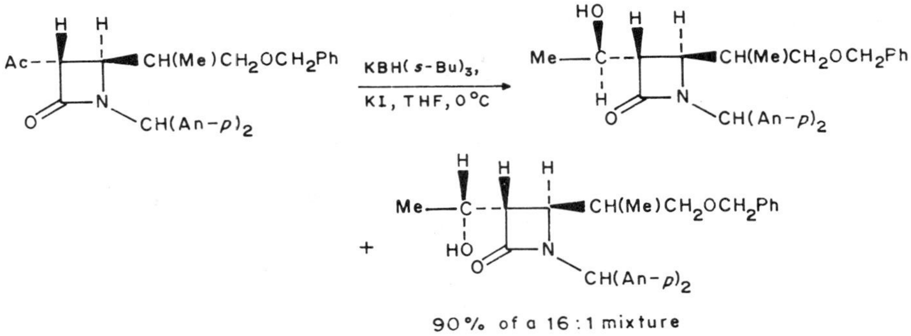

$KBH(s\text{-}Bu)_3$, KI, THF, 0 °C

90% of a 16 : 1 mixture

(1141)

Another example of the use of α-phenylseleno substituents to effect functionalization at the alpha position of lactams is illustrated by the formation[1192] of α-methylene β-lactams from *cis* or *trans* N-substituted 3-methyl-3-phenylseleno-4-phenylazetidin-2-ones (equation 1142).

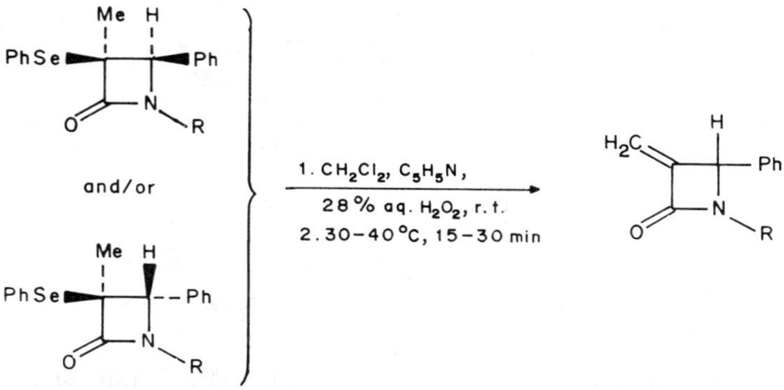

1. CH_2Cl_2, C_5H_5N, 28% aq. H_2O_2, r.t.
2. 30–40 °C, 15–30 min

(1142)

R =	Me	t-Bu	Ph
Starting material =	*cis* + *trans*	*cis* + *trans*	*cis* only
Yield (%) =	67	85	92

The reactions shown in Scheme 19 illustrate another sequence which has been used[1316] to produce an α-methylene β-lactam and other structures containing α-exocyclic double bonds.

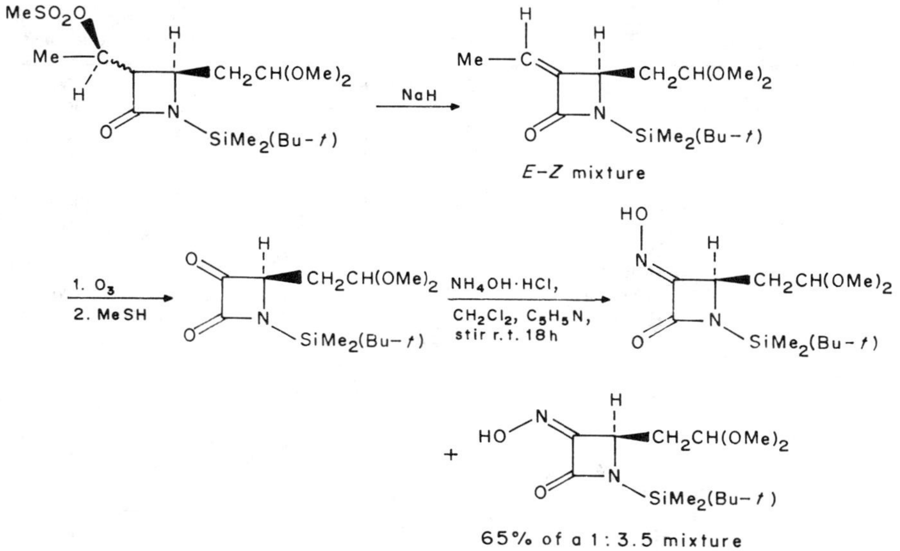

NaH

E–Z mixture

1. O_3
2. MeSH

$NH_4OH \cdot HCl$, CH_2Cl_2, C_5H_5N, stir r.t. 18 h

65% of a 1 : 3.5 mixture

SCHEME 19

Finally, *cis* N,4-diphenyl-3-methyl-3-(phenylseleno)azetidin-2-one has been converted[1192] to N,4-diphenyl-3,3-dimethylazetidin-2-one (equation 1143).

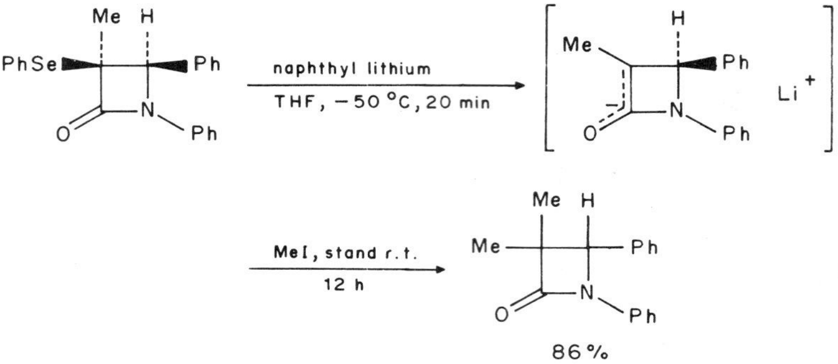

(1143)

An interesting reaction which occurs[800,901] when 3-alkylideneazetidin-2-ones are treated with lithium diisopropylamide is the isomerization of the exocyclic double bond (equation 1144). The resulting products being easily epoxidized[800,901] using *m*-chloroperbenzoic acid at room temperature (equation 1145).

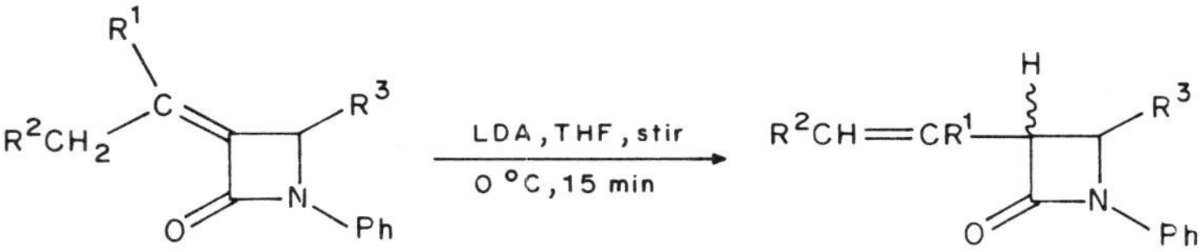

(1144)

R^1	R^2	R^3	Yield (%)	Reference
Me	H	H	80	800
Me	H	*n*-Pr	80	901
Ph	H	H	85	800
Ph	H	Me	80	901
Ph	Me	H	80[a]	800
$-(CH_2)_3-$		H	80	800
$-(CH_2)_4-$		H	80	800
$-(CH_2)_4-$		Me	85	901
$-(CH_2)_4-$		*n*-Pr	82	901
$-(CH_2)_2CH(Me)CH_2-$		H	80	800
$-(CH_2)_5-$		H	82	800

[a]Product was a 1:1 *E*:*Z* mixture.

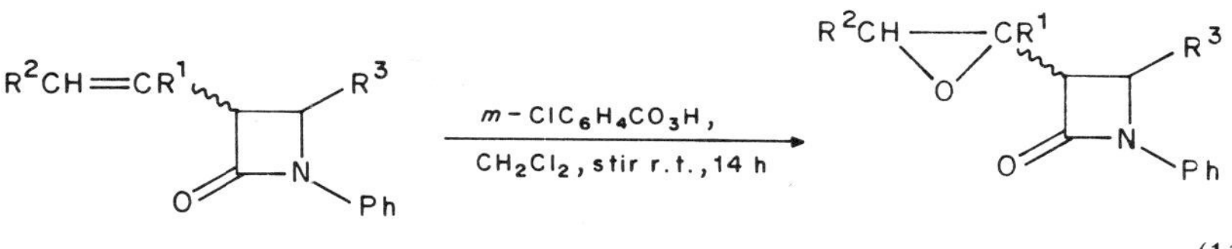

(1145)

R^1	R^2	R^3	Yield (%)	Reference
Me	H	H	90	800
Me	H	*n*-Pr	—	901
Ph	H	H	80	800
Ph	H	Me	—	901
Ph	Me	H	75	800
$—(CH_2)_3—$		H	95	800
$—(CH_2)_4—$		H	95	800
$—(CH_2)_4—$		Me	—	901
$—(CH_2)_4—$		*n*-Pr	—	901
$—(CH_2)_2CH(Me)CH_2—$		H	95	800
$—(CH_2)_5—$		H	93	800

Epoxide formation at the alpha site in β-lactams has also been used[1061] to produce α-spiro lactams. Thus, treatment of N-substituted 3-acyloxy-3-(chloromethyl)azetidin-2-ones with potassium hydroxide at room temperature affords 1-oxo-4-oxa-5-substituted azaspiro(2,3)hexanes (equation 1146).

$ClCH_2$, R^2CO_2 — azetidinone (N–R^1) $\xrightarrow{\text{KOH, THF, r.t.}}$ spiro-epoxide azetidinone (N–R^1) (1146)

R^1	R^2	Time (h)	Yield (%)
t-Bu	Me	2	90
t-Bu	Ph	18	60[a]
i-Pr	Me	2.5	95
i-Pr	Ph	2.5	5[b]

[a]40% of the starting material was recovered.
[b]95% of the starting material was recovered.

Functionalization of the alpha position of β-lactams has also been accomplished by the ring opening of α,β-fused ring containing lactams. An example of this approach is the reaction[1368] of the isopropylideneepioxazoline shown in equation (1147) with allyl or propargyl alcohols and a catalytic amount of trifluoromethanesulphonic acid. The alpha site may be further functionalized by stereoselective introduction of a methoxy group via the reaction shown.

Other examples of the α,β-fused ring lactam ring-opening approach to functionalization of the alpha position of lactams are shown in Table 93.

b. Reactions at the C-4 and higher positions. Hydroxy substituents attached to the C-4 or higher positions on lactams can be converted into a number of other functional groups depending upon the reagents used. By reaction of these hydroxyl substituents with an alcohol in the presence of an acid, alkoxy substituents are produced[1353,1356,1374] (equations 1148–1151).

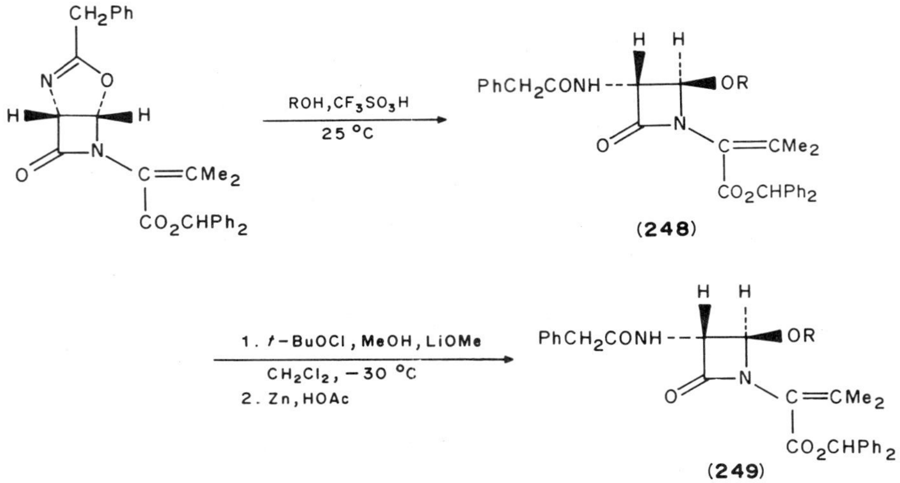

ROH, CF_3SO_3H, 25 °C

$PhCH_2CONH$ … OR … $C{=}CMe_2$ … CO_2CHPh_2

(**248**)

1. t-BuOCl, MeOH, LiOMe, CH_2Cl_2, −30 °C
2. Zn, HOAc

(**249**)

R = $CH_2{=}CHCH_2$; $HC{\equiv}CCH_2$

% Yield **248** = 80 ; —

% Yield **249** = 80 ; —

(1147)

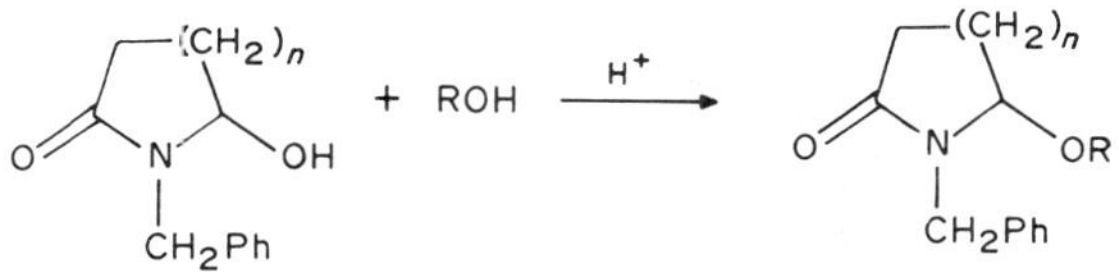

H^+

+ ROH

(1148)[1356]

n = 1 ; 1 ; 2 ; 2

R = Me ; Et ; Me ; Et

Yield (%) = — ; — ; 50–60 ; 80–90

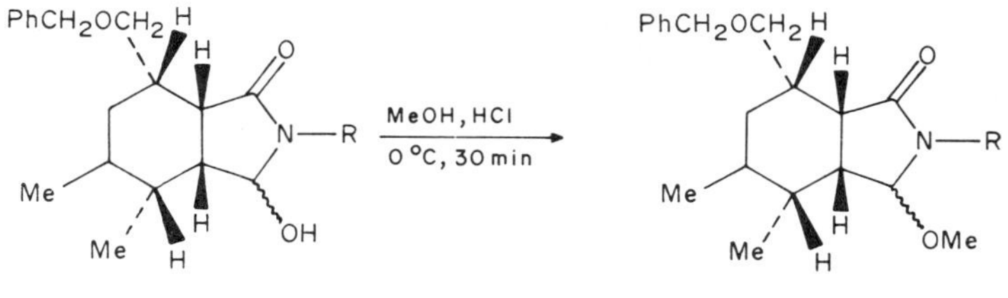

MeOH, HCl, 0 °C, 30 min

R = Et, $PhCH_2$

100%

(1149)[1374]

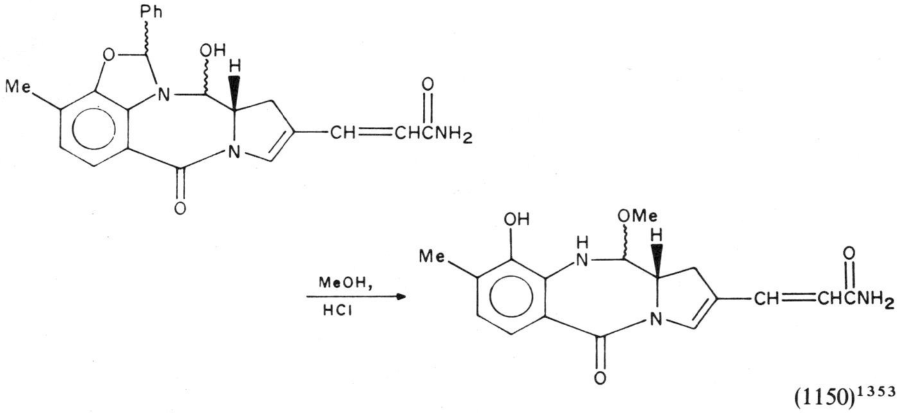

MeOH, HCl

(1150)[1353]

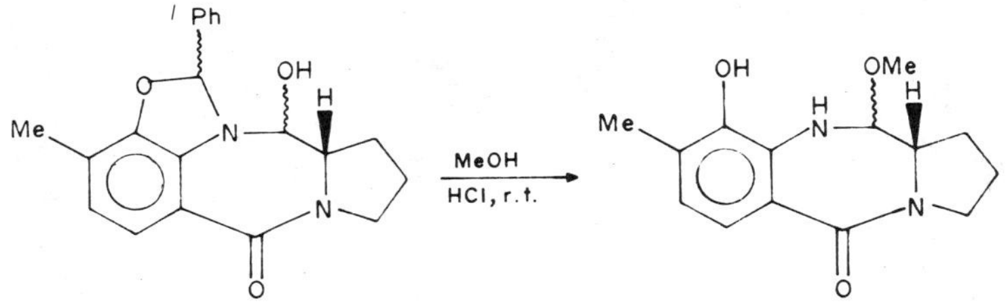

MeOH HCl, r.t.

(1151)[1353]

Active methylene compounds such as methyl ketones and nitromethane condense under the influence of base with ω-hydroxy lactams to produce[1436] the correspondingly substituted lactams (equation 1152). The ease of alkylation is reportedly[1436] determined by the position of a tautomeric equilibrium between the ω-hydroxy lactam and an open chain amide–aldehyde, which is dependent upon the type of N-substituent and upon the lactam ring size.

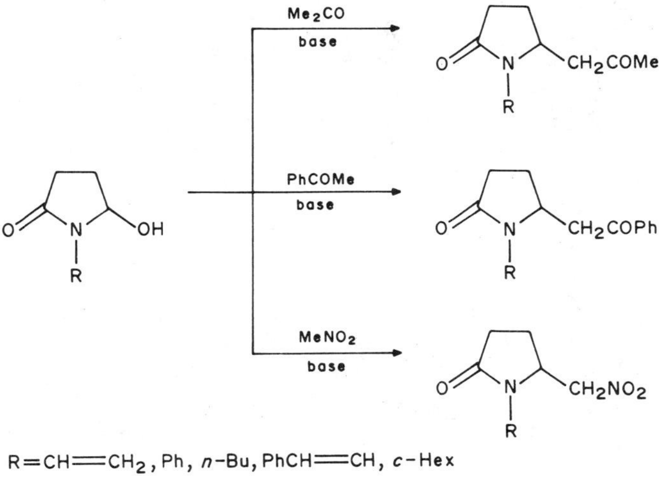

(1152)

R=CH=CH$_2$, Ph, n-Bu, PhCH=CH, c-Hex

TABLE 93. Functionalization of lactam α-position by ring opening

Fused-ring lactam	Conditions	Product	Yield (%)	Reference
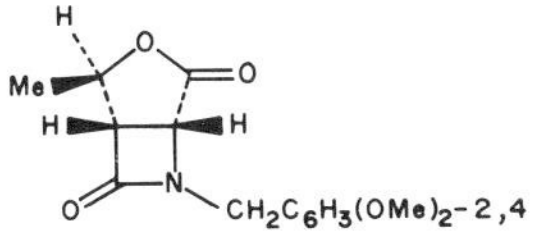 $CH_2C_6H_3(OMe)_2$-2,4	1. MeMgBr, THF, −78 °C, 30 min 2. (*t*-Bu)Me_2SiCl, DMF, 4-Me_2NPyr	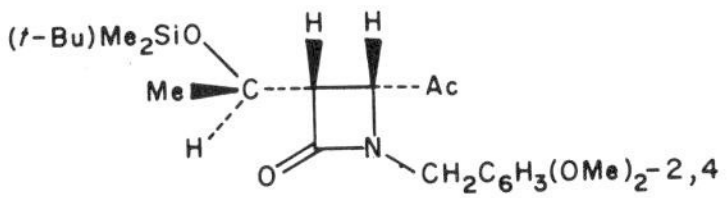 (*t*-Bu)Me_2SiO, Me, Ac, $CH_2C_6H_3(OMe)_2$-2,4	80	750
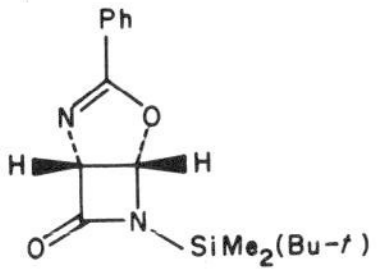 Ph, $SiMe_2(Bu\text{-}t)$	$HOCH_2C(=R^1)C(=R^2)CO_2CHPh_2$ EtOAc, r.t., $BF_3 \cdot Et_2O$	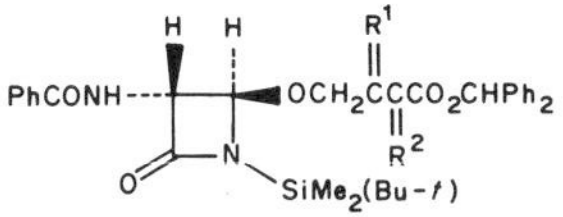 PhCONH, $OCH_2C(=R^1)C(=R^2)CO_2CHPh_2$, $SiMe_2(Bu\text{-}t)$		1349
	$R^1 = O, R^2 = N_2$		75	
	$R^1 = CH_2, R^2 = N_2$		34	
	$R^1 = CH_2, R^2 = NOH$		75	
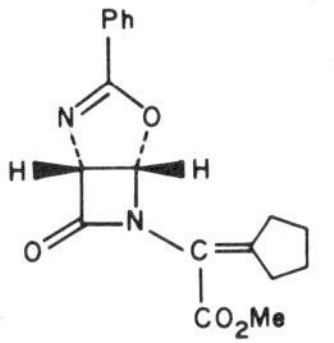 Ph, CO_2Me	MeOH, $SnCl_2$, 20 °C, 4h	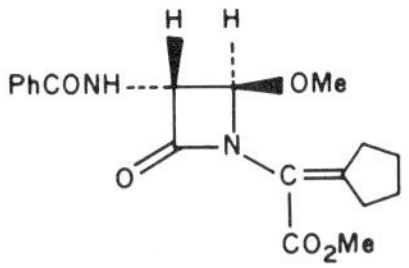 PhCONH, OMe, CO_2Me	88	1431
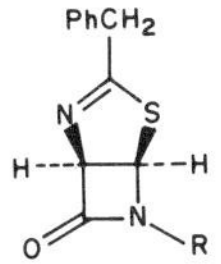 $PhCH_2$, R	1*N* HCl, MeOH, 20 °C, 30 min	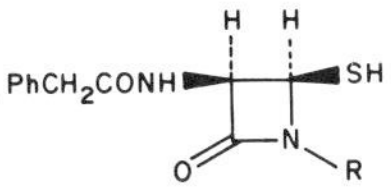 $PhCH_2CONH$, SH, R		1307
R = H			90	

$R = CHCO_2H$ (with CO_2R' on CH)	1 *N* HCl, MeOH EtOAc, 20 °C, 30 min		—	1307
$R' = Me$, $PhCH_2$, *t*-Bu				
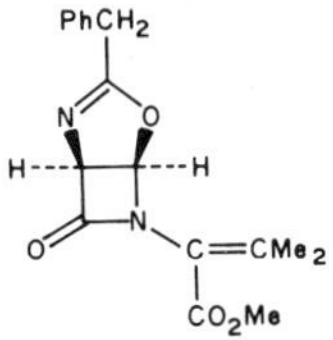 [structure: $PhCH_2$-oxazoline fused azetidinone, N–C(=CMe$_2$)CO_2Me]	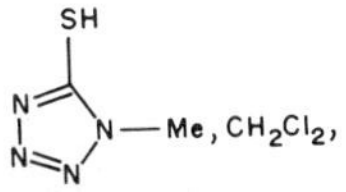 [1-methyl-5-mercaptotetrazole] SH, N—Me, CH_2Cl_2,	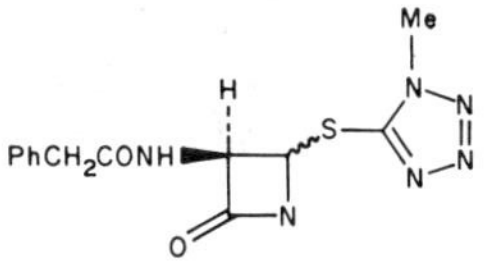 $PhCH_2CONH$–azetidinone–S–(1-methyltetrazol-5-yl) **(250)**		1343
		cis (3*R*, 4*R*) + *trans* (3*R*, 4*S*)		
	$BF_3 \cdot OEt_2$		25.2 + 68.4	
	$SnCl_4$		12.5 + 27.7	
	$TiCl_4$		16.2 + 24.3	
	$ZnCl_2$		18.1 + 42.5	
[structure: $PhCH_2$-oxazoline fused azetidinone (enantiomer), N–C(=CMe$_2$)CO_2Me]	[1-methyl-5-mercaptotetrazole] SH, N—Me, CH_2Cl_2, r.t. 1 h	**250** *cis* (3*S*, 4*S*) + *trans* (3*S*, 4*R*)		1343
	$BF_3 \cdot OEt_2$		20.4 + 64.0	
	$SnCl_4$		10.5 + 24.5	
	$TiCl_4$		16.3 + 24.3	
	$ZnCl_2$		15.2 + 35.7	

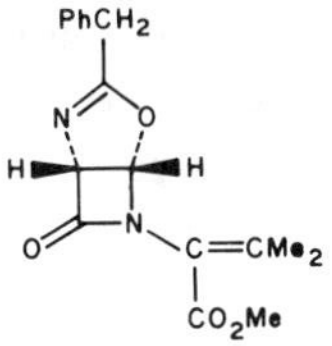

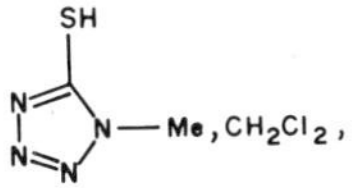

(continued)

TABLE 93. (*continued*)

Fused-ring lactam	Conditions	Product	Yield (%)	Reference
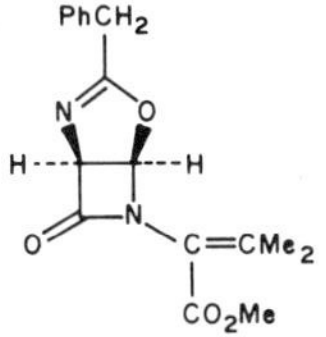 [structure: PhCH$_2$ oxazoline-fused β-lactam, N–C(=CMe$_2$)CO$_2$Me]	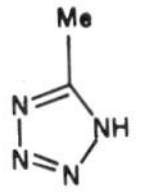 5-methyltetrazole [structure]	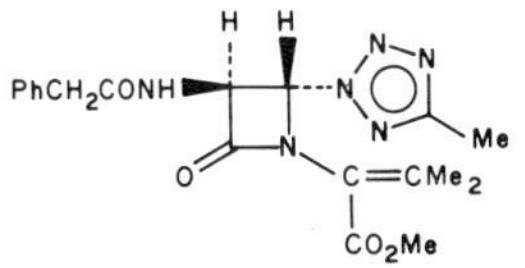 [structure: PhCH$_2$CONH β-lactam with 5-methyltetrazolyl, N–C(=CMe$_2$)CO$_2$Me] (3*S*, 4*R*)		1343
	$BF_3 \cdot OEt_2$, CH_2Cl_2 r.t., 1 h		12.9	
	$SiCl_4$, $CHCl_3$, reflux 1.5h		2.5	
	$TiCl_4$, $CHCl_3$, reflux 1.5h		1.7	
	$ZnCl_2$, CH_2Cl_2, r.t. 3h		6.8	
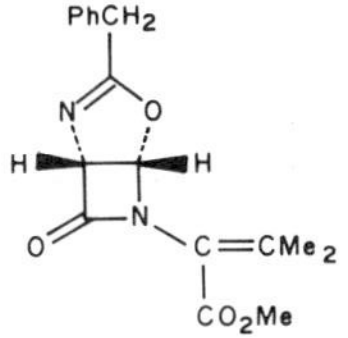 [structure: PhCH$_2$ oxazoline-fused β-lactam, N–C(=CMe$_2$)CO$_2$Me]	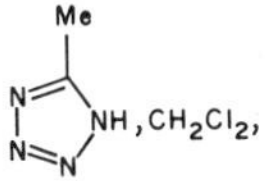 5-methyltetrazole [structure], CH_2Cl_2, r.t.	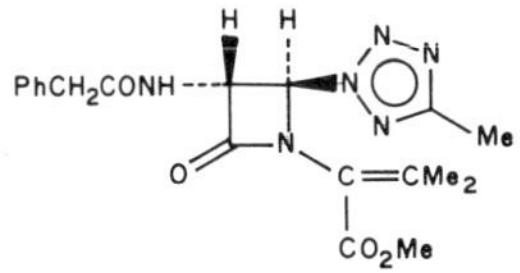 [structure: PhCH$_2$CONH β-lactam with 5-methyltetrazolyl, N–C(=CMe$_2$)CO$_2$Me] (3*R*, 4*R*)		1343
	$BF_3 \cdot OEt_2$, 0.5h		20.9	
	$SnCl_4$, overnight		1.8	
	$TiCl_4$, overnight		3.7	
	$ZnCl_2$, 3h		8.4	
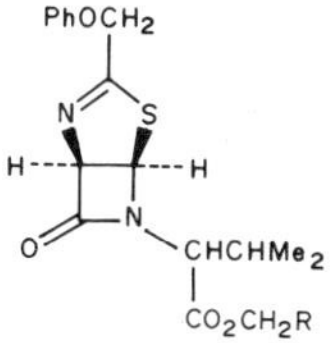 [structure: PhOCH$_2$ thiazoline-fused β-lactam, N–CHCHMe$_2$, CO$_2$CH$_2$R]		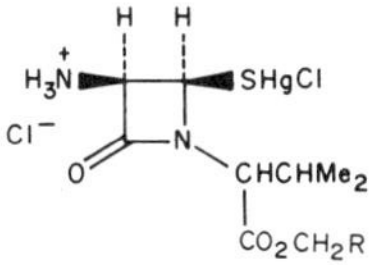 [structure: $H_3\overset{+}{N}$ Cl^- β-lactam with SHgCl, N–CHCHMe$_2$, CO$_2$CH$_2$R]		
R = $C_6H_4NO_2$-*p*	$HgCl_2$, $HOCH_2CMe_2CH_2OH$, CH_2Cl_2, stir 23h		95	1434
R = An-*p*	$HgCl_2$, $HOCH_2CMe_2CH_2OH$, CH_2Cl_2, stir 18h		87	1434

Substrate	Reagents and conditions	Product	Yield (%)	Ref.
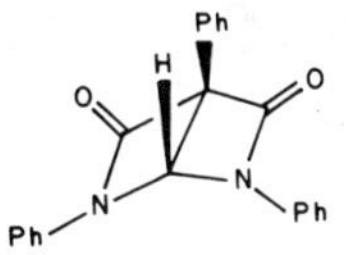	2,4,6-$Cl_3C_6H_2OH$, 235 °C	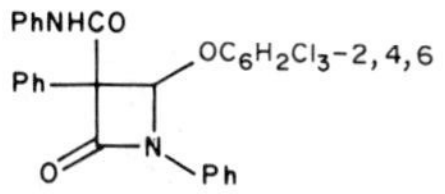	12	1268
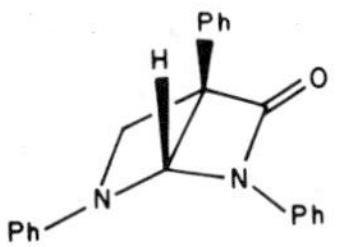	NaOMe, NaOH, reflux 15 min	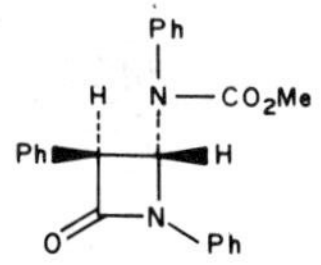	6	1268
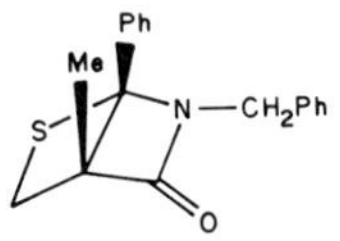	Raney nickel, H_2	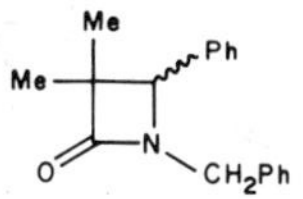	82	1104
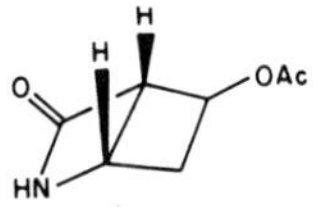	$NaBH_4$	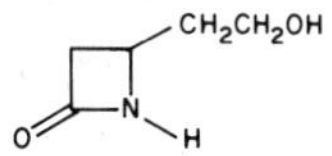	—	1435
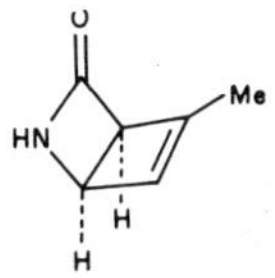	1. O_3, MeOH, −78 °C 2. $NaBH_4$	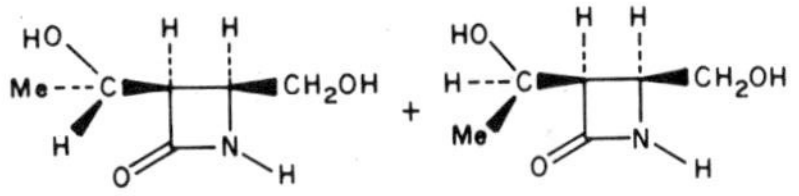 (1:1)	75	1267
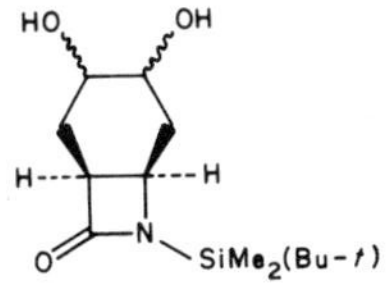	$HIO_4 \cdot 2H_2O$, THF	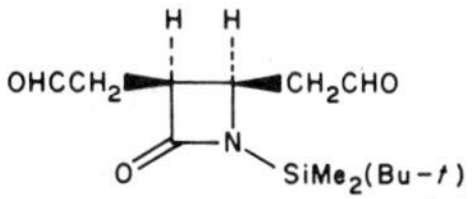	—	1021
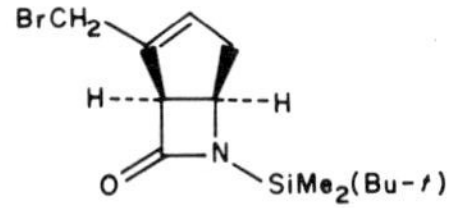	1. O_3, MeOH 2. NH_3, MeOH	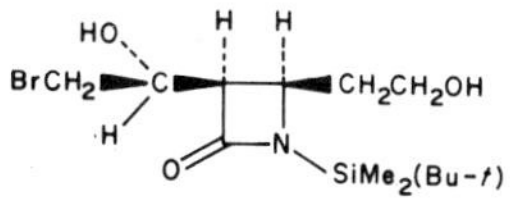	97	1021

Under the influence of acid, ω-hydroxy lactams have been found[1437] to condense with 1,3-dicarbonyl compounds and alpha-hydroxyalkylbenzenes via an intermolecular process which affords the corresponding ω-alkylated lactams (equation 1153).

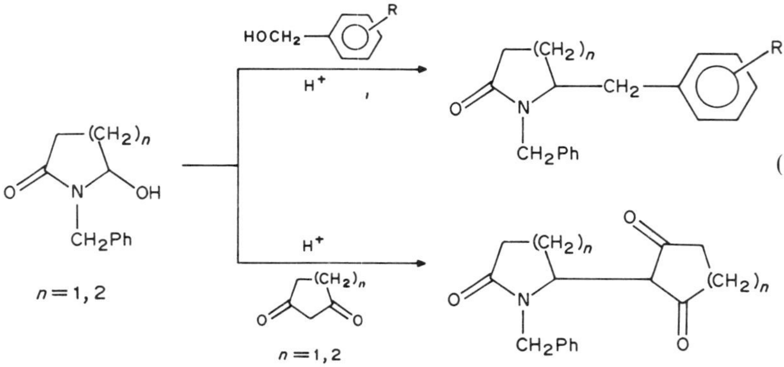

(1153)

Protected hydroxy groups located at the C-4 position of β-lactams also undergo conversion to other functional groups as illustrated by the reaction of (3*R*, 4*R*)-1-(*t*-butyldimethylsilyl)-3-[(*R*)-1-*t*-butyldimethylsilyloxyethyl)-4-(trimethylsilyloxy)azetidin-2-one with acetic anhydride in 4-(dimethylamino)pyridine to produce[1318] the acetoxy derivative (equation 1154).

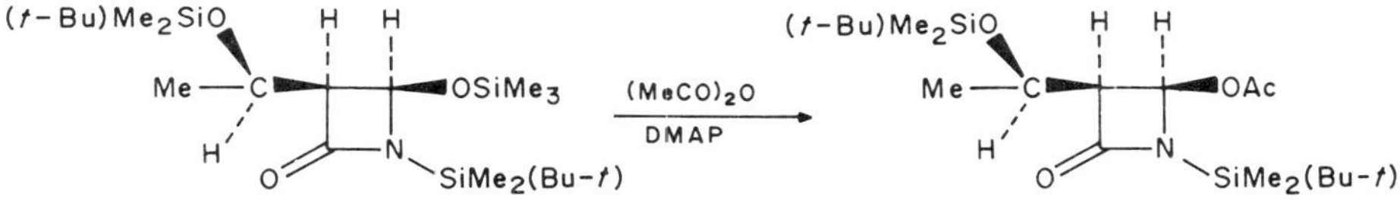

(1154)

Because of the extensive use made of the acetoxy substituent in interconversion reactions of lactams which is discussed next, it is appropriate at this point to report the other methods used to introduce this function into the lactam nucleus.

Baeyer–Villiger oxidation of 4-acetyl-substituted β-lactams, using *m*-chloroperbenzoic acid, converts[750,1146] these lactams into the corresponding 4-acetoxy-β-lactams (equations 1155–1157).

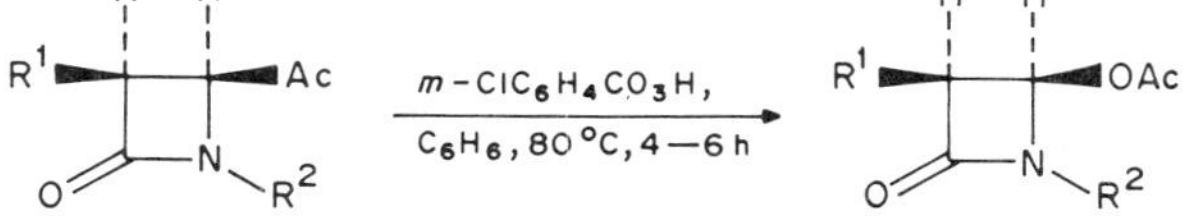

(1155)[1146]

R^1 = PhO; PI—N; PhO ; PI—N

R^2 = H ; H ; CH_2CO_2Me ; CH_2CO_2Me

(t-Bu)Me$_2$SiO, Me—C, H, H H, Ac, O, N, H $\xrightarrow[\text{CHCl}_3,\ 25\,^\circ\text{C},\ 18\ \text{h}]{m\text{-ClC}_6\text{H}_4\text{CO}_3\text{H},}$ (t-Bu)Me$_2$SiO, Me—C, H, H H, OAc, O, N, H

100%

(1156)[750]

(t-Bu)Me$_2$SiO, Me—C, H, H H, Ac, O, N, H $\xrightarrow[\text{CHCl}_3,\ 25\,^\circ\text{C}]{m\text{-ClC}_6\text{H}_4\text{CO}_3\text{H},}$ (t-Bu)Me$_2$SiO, Me—C, H, H H, OAc, O, N, H

84%

(1157)[750]

Treatment of C-4 acid substituted β-lactams with lead tetraacetate in dimethylformamide-acetic acid converts[1053,1245] the acid function into an acetoxy substituent (equations 1158 and 1159).

HO, Me—C, H, H H, CO_2H, O, N, $CH_2C_6H_3(OMe)_2$-2,4 $\xrightarrow[\text{HOAc, N}_2,\ 60\,^\circ\text{C}]{\text{Pb(OAc)}_4,\text{DMF},}$ HO, Me—C, H, H H, OAc, O, N, $CH_2C_6H_3(OMe)_2$-2,4

(1158)[1053]

H H, Et, CO_2H, O, N, $SiMe_2$(Bu-t) $\xrightarrow[\text{HOAc}]{\text{Pb(OAc)}_4,\text{DMF},}$ H OAc, Et, H, O, N, H + H H, Et, OAc, O, N, H

89% (2:1)

(1159)[1245]

Similar results are observed at the C-5 site of γ-lactams (equation 1160)[1346]

t-BuO$_2$CNH, CO_2H, N, H, O $\xrightarrow[67\,^\circ\text{C},\ 2\ \text{h}]{\text{Pb(OAc)}_4,\text{THF},}$ t-BuO$_2$CNH, OAc, N, H, O

(1160)

Ring-opening reactions of penicillin sulphoxide esters using trimethylphosphite and acetic acid affords[1382,1383] monocyclic C-4 acetoxy substituted β-lactams (equations 1161 and 1162).

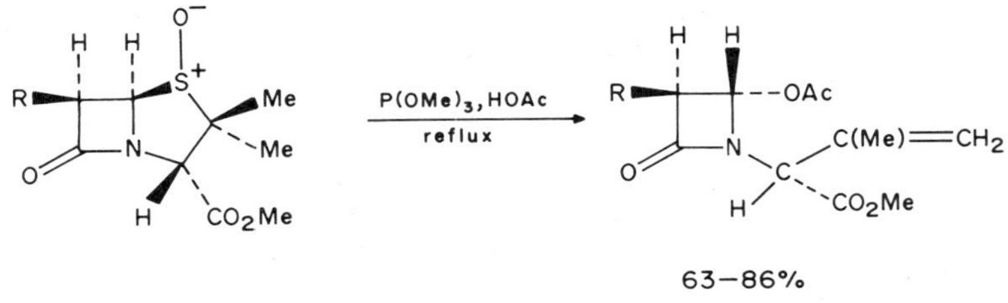

$P(OMe)_3$, HOAc, reflux

63–86%

R = t-BuO_2CNH ; $PhCH_2O$; $PhOCH_2CONH$; PI—N
Ref. = 1382 ; 1382 ; 1387 ; 1387

(1161)

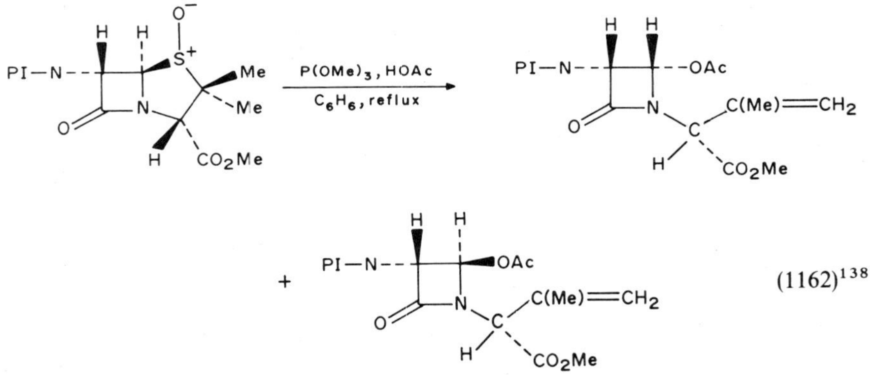

$P(OMe)_3$, HOAc, C_6H_6, reflux

(1162)[1387]

When the above reaction was performed[1387] using acids other than acetic acid, C-4 aryl ester substituted β-lactams were produced (equations 1163 and 1164).

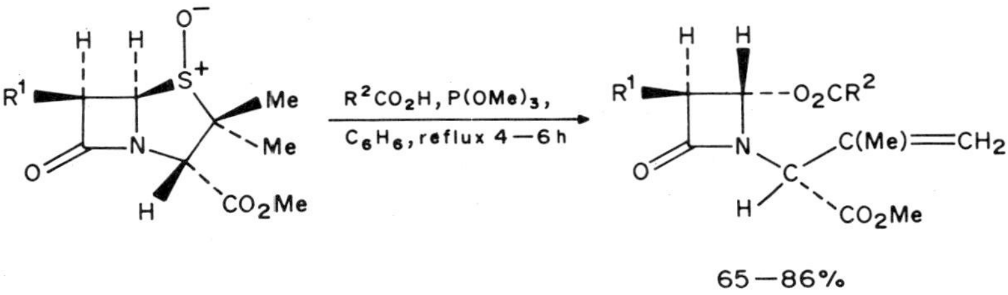

R^2CO_2H, $P(OMe)_3$, C_6H_6, reflux 4–6 h

65–86%

R^1 = $PhOCH_2CONH$; $PhOCH_2CONH$; PI—N ; PI—N
R^2 = p-ClC_6H_4 ; o-ClC_6H_4 ; Ph ; p-ClC_6H_4

(1163)

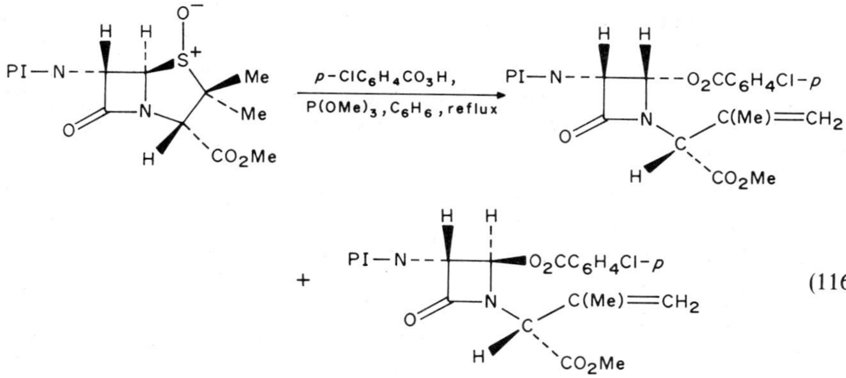

(1164)

Acetoxy functions located at the C-4 or C-5 positions of β- or γ-lactams have been extensively used as the starting function from which a variety of conversions have been achieved.

Simplest of the conversions reported[1382] is the replacement of the acetoxy function by hydrogen, which occurred in a straightforward manner using sodium borohydride when the lactam nitrogen was unsubstituted (equation 1165). In the case where the lactam nitrogen contained a carbomethoxycarbonyl function, oxamide hydrolysis accompanied reduction of the acetoxy function to also produce the unsubstituted nitrogen lactam (equation 1165).

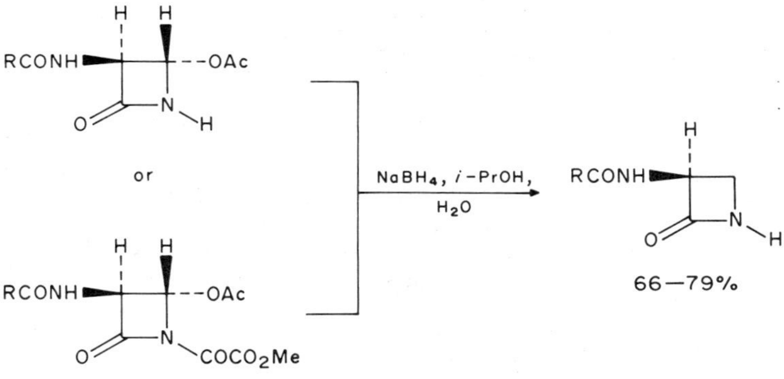

(1165)

To determine the scope of this reduction 4-acetoxyazetidin-2-one was reduced with potassium borohydride in water producing β-propiolactam in 69% yield (equation 1166)[1382].

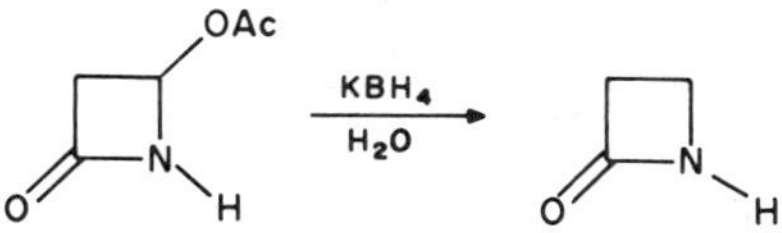

(1166)

The final question associated with the reduction processes reported above is the stereochemistry of the reduction, that is, does the nucleophilic hydride species enter into the β-lactam ring from the side of the acetoxy leaving group or does it attack from the opposite face? By using (3*S*, 4*S*)-4-acetoxy-3-(*t*-butoxycarbonyl)aminoazetidin-2-one and sodium borodeuteride it was found[1382] that the major product was the *trans* isomer which indicated an S_N1 mechanism with intermediate carbonium ion formation (equation 1167).

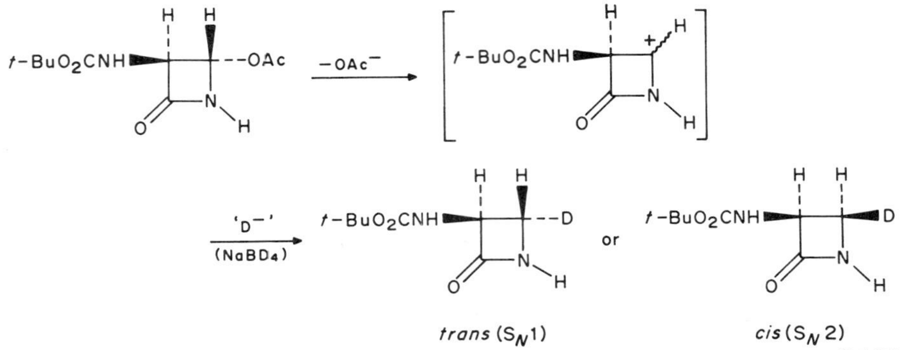

(1167)

Treatment of 4-acetoxyazetidin-2-ones with lithium organocuprates or Grignard reagents produces[1382] 4-alkyl or vinyl-azetidin-2-ones (equation 1168).

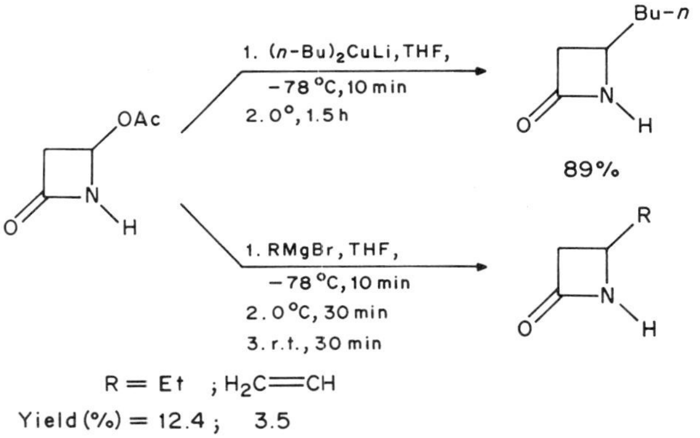

(1168)

Reactions of (3*S*, 4*R*)-3-[(*R*)-1-(*t*-butyldimethylsilyloxy)ethyl]-4-acetoxyazetidin-2-one with finely powdered potassium cyanide in tetrahydrofuran (THF) with added 18-crown-6 ether produces[1339] the corresponding 4-cyanoazetidin-2-one (equation 1169).

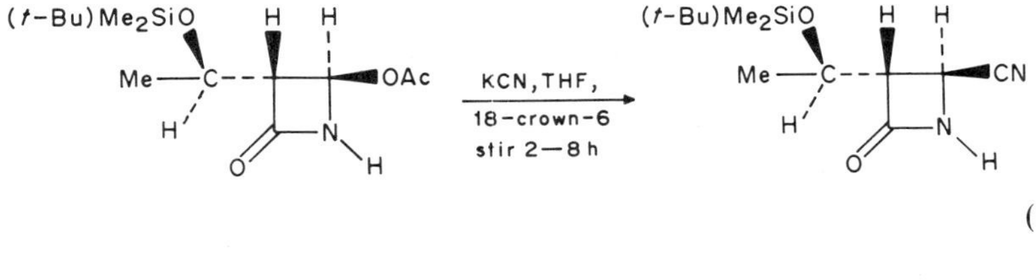

(1169)

Nucleophilic displacements of C-4 acetoxy groups have been accomplished using a variety of enolate reagents as represented in equation 1170 with the details reported in Table 94.

$$\text{4-OAc-azetidin-2-one (N-}R^1\text{)} + R^2CH{=}C(OR^4)R^3 \xrightarrow{\text{conditions}} \text{4-}CHR^2COR^3\text{-azetidin-2-one (N-}R^1\text{)} \quad (1170)$$

The substitution of C-4 acetoxy substituents by an alkenyloxy or alkynyloxy group to produce the corresponding 4-alkoxyazetidin-2-ones has been achieved[1336,1337,1387] via a zinc acetate catalysed reaction (equation 1171).

$$\text{3-}R^1\text{-4-}O_2CR^2\text{-azetidin-2-one (NH)} + HOR^3 \xrightarrow[\text{conditions}]{Zn(OAc)_2\cdot 2H_2O,} \text{3-}R^1\text{-4-}OR^3\text{-azetidin-2-one (NH)} \quad (1171)$$

R^1	R^2	R^3	Conditions	Yield (%)	Reference
H	Me	$CH_2CH{=}CH_2$	dry C_6H_6, reflux 20h	76	1336
H	Me	$CH_2CH{=}CHPh$ (*E*)	dry C_6H_6, reflux 23h	38	1336
H	Me	$CH_2CH{=}CHCO_2Me$ (*E*)	dry C_6H_6, reflux 21h	73	1336
H	Me	$CH_2C{\equiv}CH$	dry C_6H_6, reflux 20h	56	1336
H	Me	$CH_2C{\equiv}CPh$	dry C_6H_6, reflux 24h	70	1336
$PhOCH_2CONH$	Me	$CH_2CH{=}CH_2$	C_6H_5Me, 80 °C, stir 3h	—[a]	1337
PI—N	Me	CH_2CO_2Et	C_6H_6, reflux	—[b]	1387
PI—N	*p*-ClC_6H_4	CH_2CO_2Et	C_6H_6, reflux	—	1387
PI—N	Me	$CH_2CH(Br)CO_2Me$	C_6H_6, reflux	—[b,c]	1387
PI—N	*p*-ClC_6H_4	$CH_2CH(Br)CO_2Me$	C_6H_6, reflux	—[c]	1387

[a]The product obtained was a mixture of *cis* and *trans* isomers.

$$PhOCH_2CONH\text{-azetidinone-}OCH_2CH{=}CH_2\ (\textit{trans}) + PhOCH_2CONH\text{-azetidinone-}OCH_2CH{=}CH_2\ (\textit{cis})$$

[b]The same *trans* product was obtained from either the *cis* or *trans* starting material.
[c]The product obtained retained the bromine substituent in the side chain.

$$PI{-}N\text{-azetidinone-}OCH_2CH(Br)CO_2Me$$

TABLE 94. Displacement of C-4 acetoxy groups by enolates

Lactam	Reagent	Conditions	Product	Yield (%)	Reference
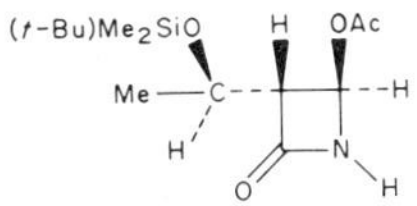	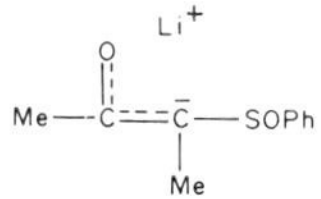	THF, $-20\,^{\circ}C$, 2h	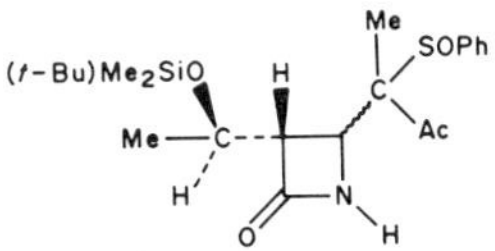	75^{a}	1345
	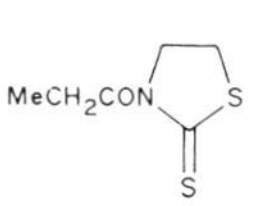	1. $Sn(O_3SCF_3)_2$, CH_2Cl_2, $C_5H_{10}NEt$, $-78\,^{\circ}C$ 2. r.t. stir 2h	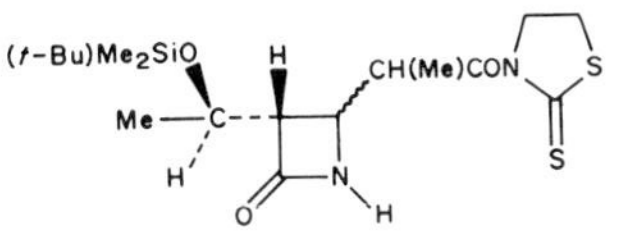 4:1 mixture of epimers	62	1437
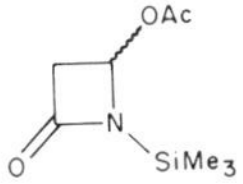	$R^1CH{=}C(OSiMe_3)R^2$	$F_3CSO_3SiMe_3$, CH_2Cl_2, -78 to $20\,^{\circ}C$	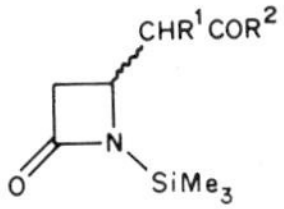		1390
		R^1 R^2			
		H Ph		89	
		Me Ph		71^{a}	
		H *p*-Tol		74	
		H *p*-ClC_6H_4		81	
		Me OEt		95^{a}	
		H PhS		72	

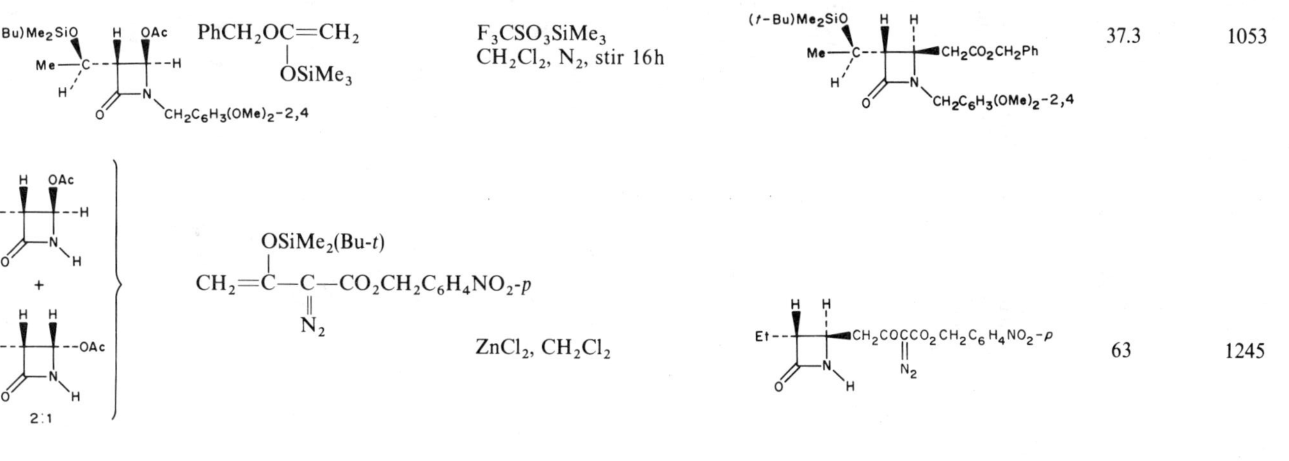

[a]Obtained as a mixture of diastereomers.

Treatment of ω-acyloxy containing β- and γ-lactams with substrates containing a mercapto function produce lactams with a sulphur attached side chain at the ω-site (equation 1172 and Table 95).

(CH$_2$)$_n$ ... O$_2$CR^2 + HSR^3 —conditions→ ... SR^3 (1172)

$n = 1, 2$

Performing the Arbusov reaction using 4-acetoxyazetidin-2-one or 4α-acetoxy-3β-phthalimidoazetidin-2-one with phosphites or phosphonites produced[1323] 2-oxoazetidin-4-yl phosphonates and phosphinates (equations 1173 and 1174).

OAc + $R^1P(OR^2)_2$ —heat→ P(O)(OR^2)R^1 (1173)

R^1	R^2	Time (h)	Time (°C)	Yield (%)
OMe	Me	1–2	110–120	high
OEt	Et	1–2	110–120	high
Me	Et	1	60	89[a]
OCH_2Ph	$PhCH_2$	7	110–120	46
OCH_2CCl_3	Cl_3CCH_2	—	110–120	N.R.
CH_2CCl_3	Cl_3CCH_2	—	120	42[a]

[a]Product was a mixture of diastereomers.

PI-N ... --OAc —P(OMe)$_3$, 110–120 °C→ PI-N ... P(O)(OMe)OMe

—MeP(OEt)$_2$→ PI-N ... P(O)(OEt)Me

cis, trans mixture (1174)

Other oxygen-containing substituents attached to lactams which undergo conversion reactions include: 5-methoxy-2-pyrrolidone, which upon solvolysis with neat thiolacetic

TABLE 95. Reaction of ω-acetoxylactams with mercapto containing substrates

Lactam	Mercapto substrate	Conditions	Product	Yield (%)	Reference
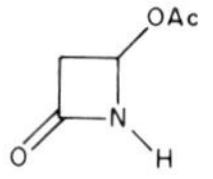 OAc, N, H, O	$\overset{+}{Na}\overset{-}{S}CH{=}CHCO_2Et$	Me_2CO, H_2O, 0 °C, 30 min	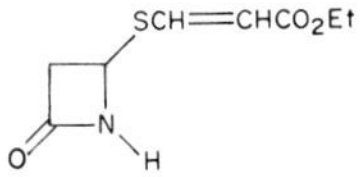 $SCH{=}CHCO_2Et$, N, H, O $E+Z$ isomers	—	1337
	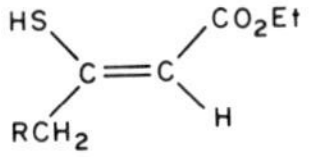 HS, CO_2Et, C=C, RCH_2, H E, $R = EtO_2C$	$NaHCO_3$, Me_2CO, H_2O, r.t.	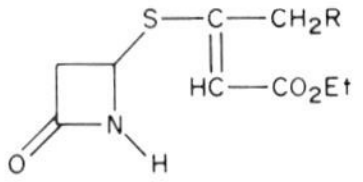 S—C—CH_2R, HC—CO_2Et, N, H, O	93	1338
	Z, $R = HOCH_2$			—	1338
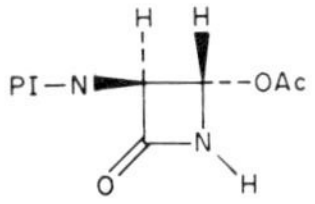 PI—N, H, H, OAc, N, H, O	$HSCH_2CO_2Me$	NaOH, Me_2CO, H_2O	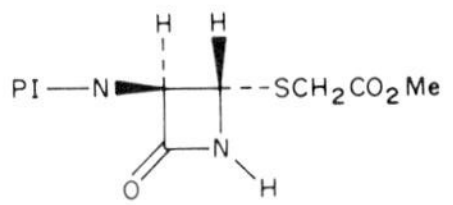 PI—N, H, H, SCH_2CO_2Me, N, H, O	—	1387
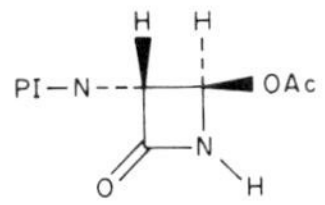 PI—N, H, H, OAc, N, H, O	$HSCH_2CO_2Me$	NaOH, Me_2CO, H_2O	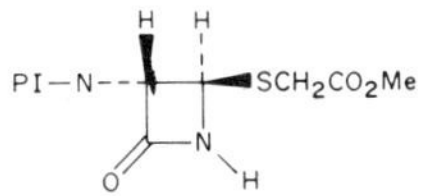 PI—N, H, H, SCH_2CO_2Me, N, H, O	—	1387
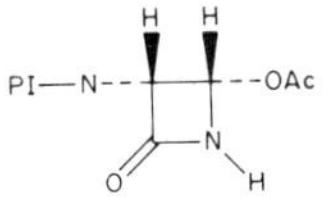 PI—N, H, H, OAc, N, H, O	$HSCH_2CO_2Me$	NaOH, Me_2CO, H_2O	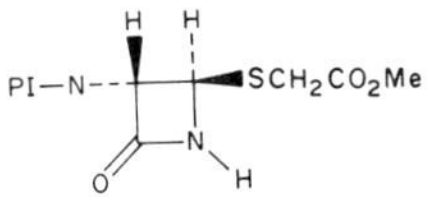 PI—N, H, H, SCH_2CO_2Me, N, H, O	—	1387
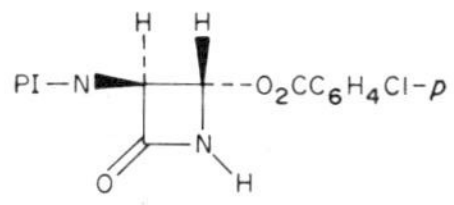 PI—N, H, H, $O_2CC_6H_4Cl$-p, N, H, O	$HSCH_2CO_2Me$	NaOH, Me_2CO, H_2O	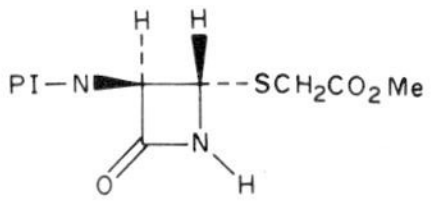 PI—N, H, H, SCH_2CO_2Me, N, H, O	—	1387

(continued)

TABLE 95. (*continued*)

Lactam	Mercapto substrate	Conditions	Product	Yield (%)	Reference
(t-Bu)Me$_2$SiO, Me—C, H; H, H; OAc; O, N, H	$Ph\bar{S}O_2\overset{+}{Na}$	dioxane/H_2O (1:1), 100 °C, 45 min	(t-Bu)Me$_2$SiO, Me—C, H; H, H; SO_2Ph; O, N, H	63	750
(t-Bu)Me$_2$SiO, Me—C, H; H, H; OAc; O, N, H	$Ph\bar{S}O_2\overset{+}{Na}$	dioxane/H_2O (1:1), 100 °C, 45 min	(t-Bu)Me$_2$SiO, Me—C, H; H, H; SO_2Ph; O, N, H	84	750
(t-Bu)Me$_2$SiO, Me—C, H; H, H; OAc; O, N, H	$HSCH_2CH(OH)CO_2Me$	—	(t-Bu)Me$_2$SiO, Me—C, H; H, H; $SCH_2CH(OH)CO_2Me$; O, N, H mixture of diastereomers	58	1348
HO, Me—C, H; H, H; O_2CPh; O, N, H	$HSCOCH_2NHCO_2CH_2CH{=}CH_2$	MeCN, H_2O	HO, Me—C, H; H, H; $SCOCH_2NHCO_2CH_2CH{=}CH_2$; O, N, H	—	1306
$PhCH_2CONH$; H, H; OAc; O, N—C(=CMe$_2$)CO$_2$Me	SH, N—Me (1-methyltetrazol-5-thiol)	CH_2Cl_2, r.t.	$PhCH_2CONH$; H; S—(1-methyltetrazol-5-yl); O, N—C(=CMe$_2$)CO$_2$Me *cis* (3*R*, 4*R*) + *trans* (3*R*, 4*S*)		1343
		$BF_3 \cdot OEt_2$, 3h		34.8 + 53.9	
		$SnCl_4$, 4h		8.2 + 12.5	

		$TiCl_4$, overnight $ZnCl_2$, overnight		4.6 + 10.4 5.2 + 7.8	
	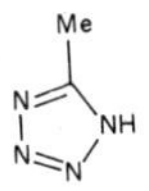	CH_2Cl_2, reflux 2h	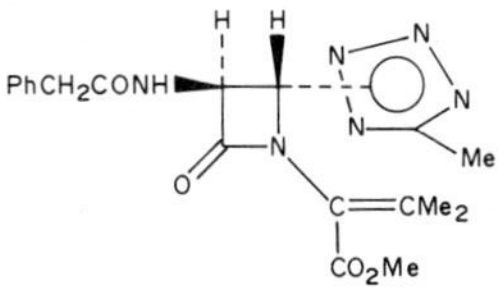 (3S, 4R)		1343
		$BF_3 \cdot OEt_2$ $SnCl_2$, $TiCl_4$ or $ZnCl_2$		1.1 0	
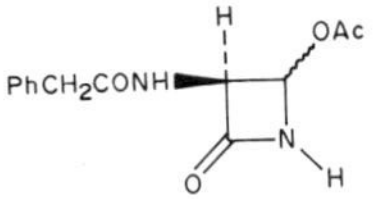	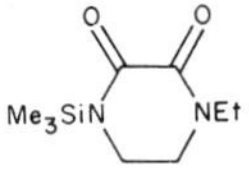	$CHCl_3$	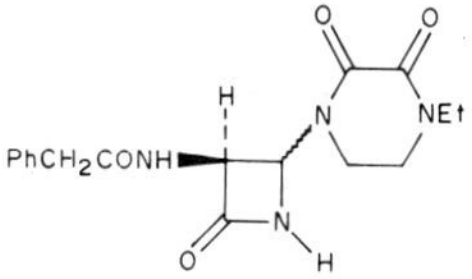 *cis* (3S, 4R) + *trans* (3S, 4S)		1343
		$BF_3 \cdot OEt_2$, r.t. overnight $BF_3 \cdot OEt_2$, reflux 1h $SnCl_4$, reflux 1h $TiCl_4$, r.t. 1h $ZnCl_2$, reflux 1h		1.8 + 18.3 2 + 18.1 13.7 + 30.5 5.8 + 5.8 9 + 26.4	
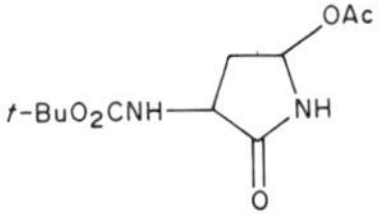	AcSH	reflux 20h	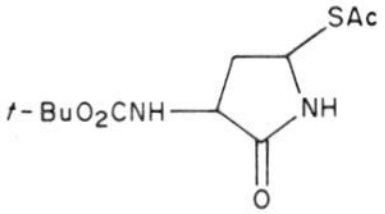	—	1346

acid produces[1346] the corresponding thioacetate (equation 1175); γ- and δ-lactams containing ω-ethoxy functions, which react with propargyl silane under the influence of boron trifluoride etherate to produce[1438] ω-allenyl lactams (equations 1176 and 1177); and with dimethyl malonate and ethyl acetoacetate in the presence of aluminium chloride to produce[1410] the corresponding amidoalkylation products (equations 1178 and 1179).

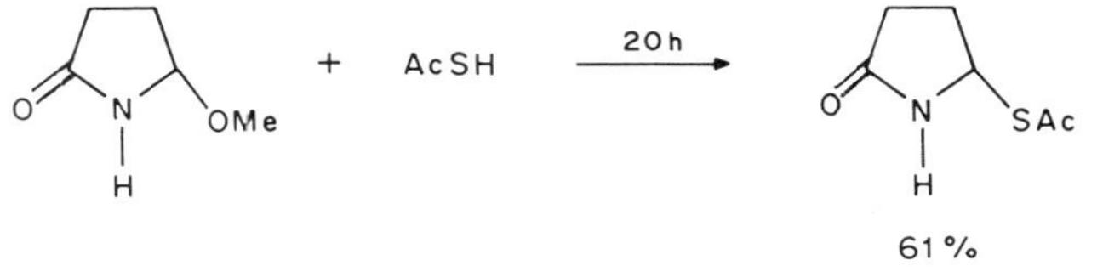

+ AcSH —20h→ 61% (1175)

$n = 1, 2$

$BF_3 \cdot OEt_2$, CH_2Cl_2, 0 °C to r.t.

$HC{\equiv}CCH_2SiMe_3$ → $CH{=}C{=}CH_2$ (1176)

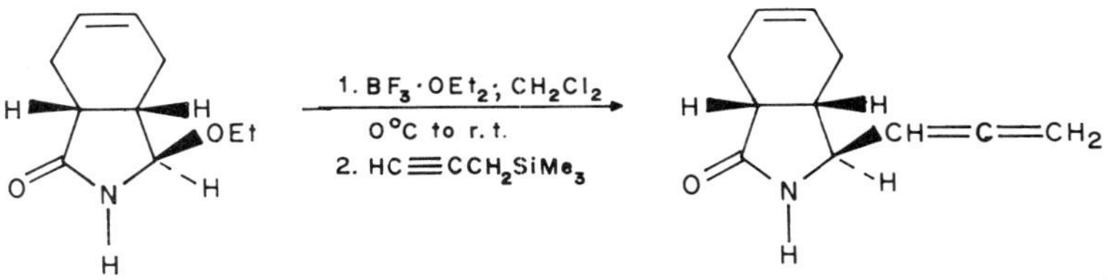

1. $BF_3 \cdot OEt_2$; CH_2Cl_2, 0 °C to r.t.
2. $HC{\equiv}CCH_2SiMe_3$

(1177)

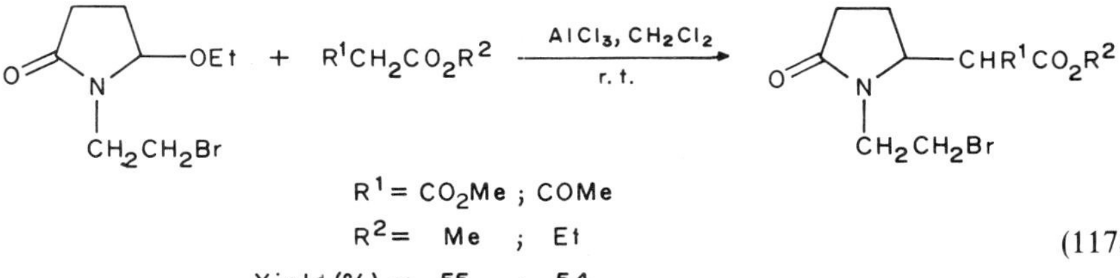

+ $R^1CH_2CO_2R^2$ —$AlCl_3$, CH_2Cl_2, r.t.→ $CHR^1CO_2R^2$, CH_2CH_2Br

R^1 = CO_2Me ; COMe
R^2 = Me ; Et
Yield (%) = 55 ; 54

(1178)

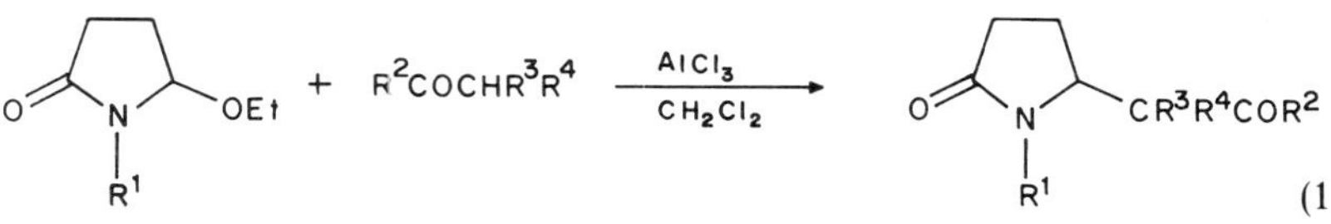

+ $R^2COCHR^3R^4$ —$AlCl_3$, CH_2Cl_2→ $CR^3R^4COR^2$

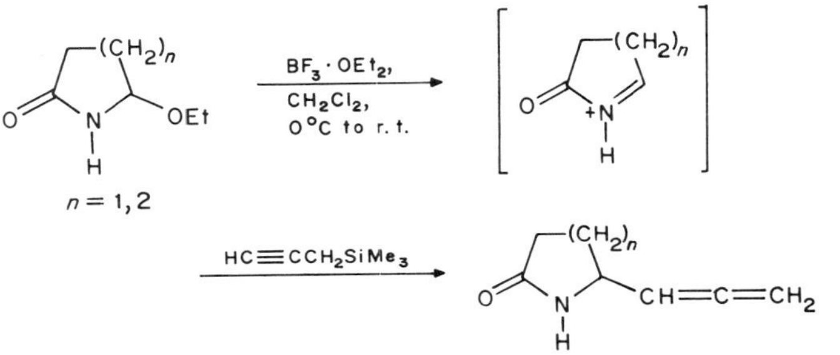

(1179)

R^1	R^2	R^3	R^4	Yield (%)
H	MeO	H	CO_2Me	60.5
H	EtO	CH_2CO_2Et	CO_2Et	55
H	EtO	$—CH_2CH_2O_2C—$		70.5
CH_2CH_2Br	MeO	H	NO_2	67
CH_2CH_2Br	EtO	CO_2Et	CO_2Et	51.6
$CH_2CH{=}CH_2$	MeO	H	CO_2Me	63
CH_2CH_2Br	Me	H	*p*-Tos	0
H	EtO	H	*p*-Tos	0
H	EtO	H	CO_2CH_2Ph	52.5

The generation of β, γ- and γ,δ-double bonds in lactam structures has been accomplished using a variety of methods. For example, reaction of 1-methylthiohydantoin with 1,3-dibromopropane and *N*-ethyldiisopropylamine produces[1266] 2-(3-bromopropylthio)-1-methylimidazolin-4-one (equation 1180), while dehydrogenation of 1-methyl-3,4-diphenyl-5,6-dihydro-2-pyridone with 2,3-dichloro-5,6-dicyano-1,4-benzoquinone (DDQ) in refluxing dioxane afforded[1238] 1-methyl-3,4-diphenyl-2-pyridone (equation 1181).

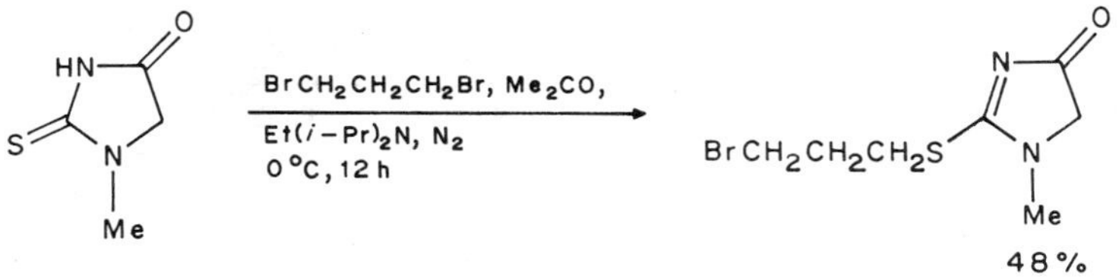

(1180)

(1181)

Finally, dehydrochlorination of 1-methyl-5,5,7-trichloro-1,3,4,5-tetrahydro-2*H*-1,4-benzodiazepin-2-one produced[1332] 5,7-dichloro-1-methyl-3*H*-1,4-benzodiazepin-2(1*H*)-one (equation 1182).

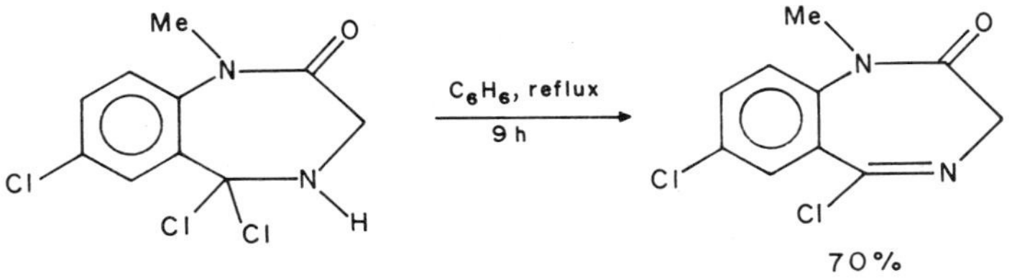

(1182)

At least one report[1439] in the recent literature discusses reduction of the γ,δ-double bond in a lactam thereby providing a stereospecific route to the preparation of *cis*- or *trans*-1,2,3,4,4*a*,5,6,10*b*-octahydrobenzo(*f*)quinolines from their hexahydro precursors (equation 1183).

H₂, Pd/C → cis, $R^1 = R^2 = R^3 = R^4 = H$

1. Et_3SiH, CH_2Cl_2 stir r.t. 10 min; 2. F_3CCOOH, 0 °C; 3. r.t. stir → trans

(1183)

R^1	R^2	R^3	R^4	Time (h)	Yield (%)
MeO	H	H	H	9	95
MeO	MeO	H	H	6.5	80
H	$PhCH_2O$	MeO	H	20	33
MeO	H	H	MeO	5	80

An interesting lactam interconversion involves[1298] the transformation of the nitro group of a 9-nonanelactam into a tautomeric mixture of the oxolactam and a hydroxy-1-azabicyclo[5.3.0]decan-2-one (equation 1184), while in a similar reaction[1298] performed using the nitro-substituted 15-pentadecanelactam analogue only the oxo derivative was formed (equation 1185).

LiOMe, $Na_2B_4O_7$, $KMnO_4$, H_2O, MeOH, 20 °C, 15 min — 68%

(1184)

LiOMe, $Na_2B_4O_7$, $KMnO_4$, H_2O, MeOH, 20 °C, 15 min — 2%

(1185)

Sulphur-containing groups located at the C-4 position of β-lactams have been used as the starting materials for the preparation of a variety of other C-4 substituted β-lactams. Reactions used to convert C-4 sulphur-containing side chains into new sulphur-containing C-4 substituents include: the reaction of (3*S*, 4*R*)-3[(*R*)-1-(*t*-butyldimethylsilyloxy)ethyl]-4-(methylsulphonyl)-2-azetidinone with sodium thiophenolate to produce[1339] (3*S*, 4*R*)-3[(*R*)-1-(*t*-butyldimethylsilyloxy)ethyl]-4-(phenylthio)-2-azetidinone (equation 1186); formation[1434] of *p*-nitrobenzyl (2*S*)-5-[(2*R*, 3*R*)-2-acetylthio-1-[(1*R*)-2-methyl-2-(4-nitrobenzyloxycarbonyl)propyl]-4-oxoazetidin-3-ylcarbamoyl]-2-(*p*-nitrobenzyloxycarbonylamino)-pentanoate by reaction of the corresponding 2-chloromercuriothio substituted starting material with acetyl chloride (equation 1187); generation[1434] of 1-[(1*R*)-carboxy-2-methylpropyl-(3*R*)-[(5*S*)-5-amino-5-carboxypentanamino]-(4*R*)-mercaptoazetidin-2-one by reaction of the 4-chloromercuriothio precursor with hydrogen sulphide at pH = 1.5 (equation 1188); preparation[1440] of (*trans*-3-azido-2-oxo-4-phenacylthio-1-azetidinyl) (cyclopentylidene)acetic acid methyl ester (equation 1189) and (*E*)- and (*Z*)-2-(*trans*-2-

EtOH, NaSPh; 5 °C, 2 h → 95%

(1186)

AcCl, C_5H_5N; stir 6 h

(1187)

H_2S (pH = 1.5)

(1188)

oxo-4-phenacylthio-3-phthalimidoazetidin-1-yl)-3-phenyl-2-butenoic acid ethyl ester (equation 1190) by the reaction of the corresponding C-4 tritylthio ethers with silver nitrate followed by treatment with phenacyl bromide; formation[1440] of bis-[4-[*trans*-3-azido-1-(*E*)-(1-*t*-butoxycarbonyl-3,3-dimethyl-1-butenyl)-2-oxoazetidin]disulphide by oxidative detritylation of (*trans*-3-azido-2-oxo-4-triphenylthioazetidin-1-yl)-4,4-dimethyl-2-pentenoic acid *t*-butyl ester (equation 1191); the reaction[1434] of iodine with *p*-nitrobenzyl (2*S*)-5-[(2*R*, 3*R*)-2-chloromercuriothio-1-[(1*R*)-2-methyl-1-(4-nitrobenzyloxycarbonyl)propyl]-4-oxoazetidin-3-ylcarbamoyl]-2-(*p*-nitrobenzyloxycarbonylamino)-pentanoate or 1-[(1*R*)-carboxy-2-methylpropyl]-(3*R*)-[(5*S*)-5-amino-5-carboxypentanamido]-(4*R*)-chloromercuriothioazetidin-2-one to produce the corresponding disulphide dimer of each (equation 1192); the reaction of iodine and sodium *p*-toluenesulphinate with (*E*)-2-(*trans*-3-azido-2-oxo-4-tritylthioazetidin-1-yl)-4,4-dimethyl-2-pentenoic acid

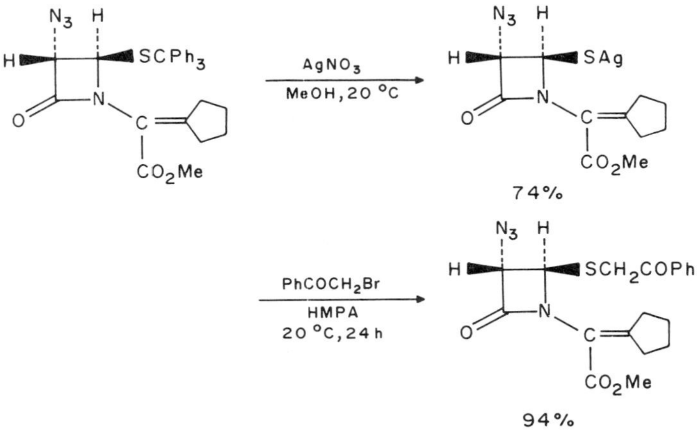

(1189)

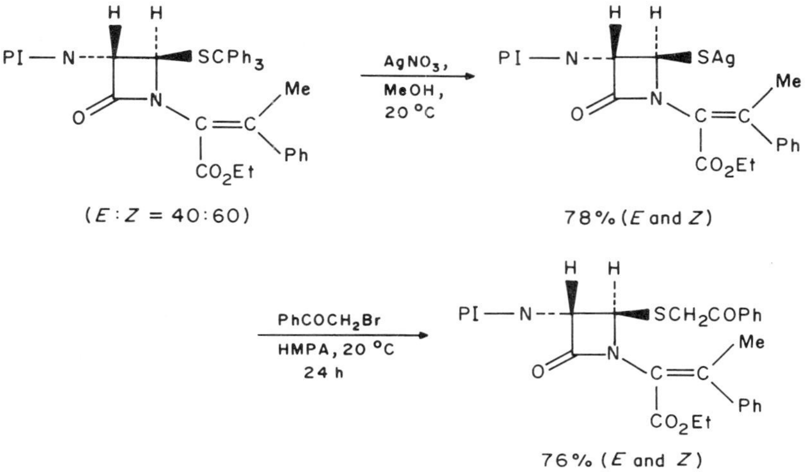

(1190)

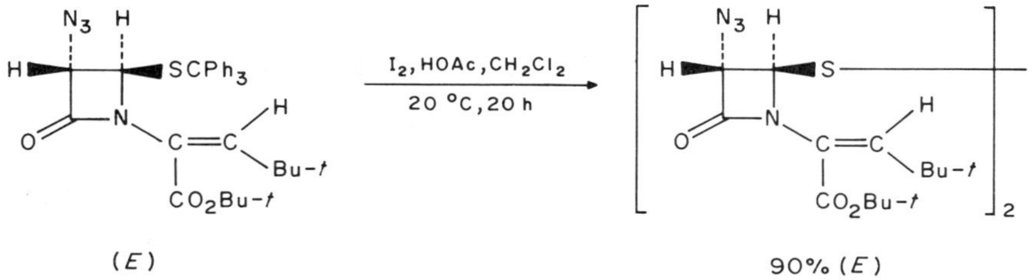

(1191)

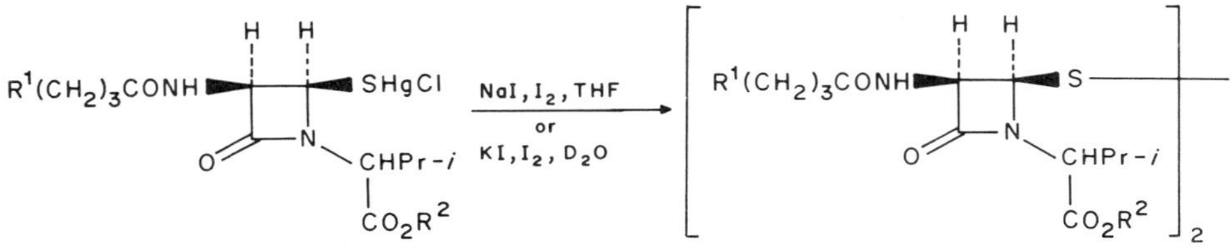

$R^1 = p\text{-}O_2NC_6H_4CH_2O_2C\overset{|}{C}HNHCO_2CH_2C_6H_4NO_2\text{-}p$; $\bar{O}_2C\overset{|}{C}H\overset{+}{N}H_3$

$R^2 = CH_2C_6H_4NO_2\text{-}p$; H

Yield (%) = 66 ; 80

(1192)

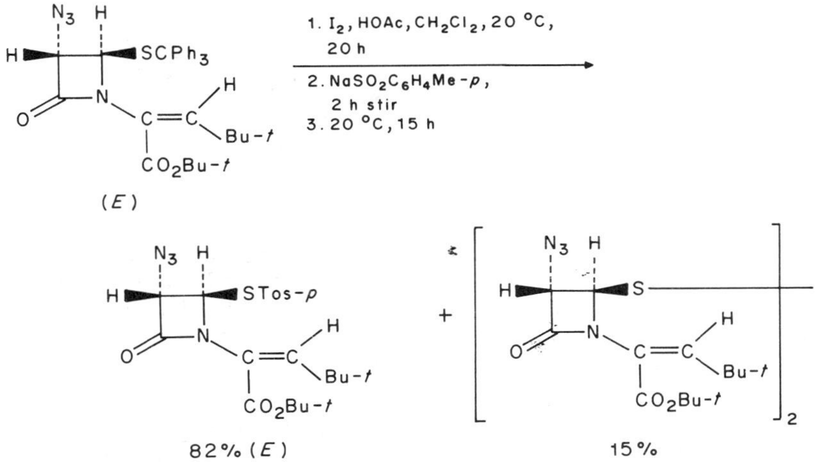

(1193)

t-butyl ester (equation 1193) or cyclopentylidene-(*trans*-2-oxo-3-phenoxyacetylamino-4-tritylthioazetidin-1-yr)acetic acid methyl ester (equation 1194) to produce[1440] the corresponding 4-(*p*-toluenesulphonylthio)-2-azetidinones; and the formation[1397] of the C-4 mercaptobenzothiazole substituted β-lactam shown in equation (1195) from the ring opening of the precursor 3-methylenecepham-1-oxide.

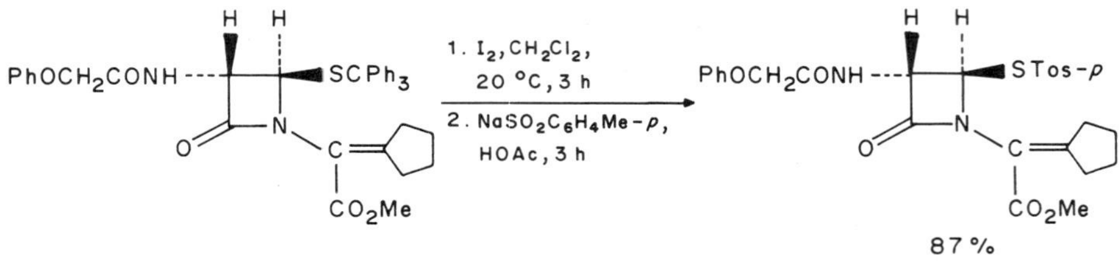

(1194)

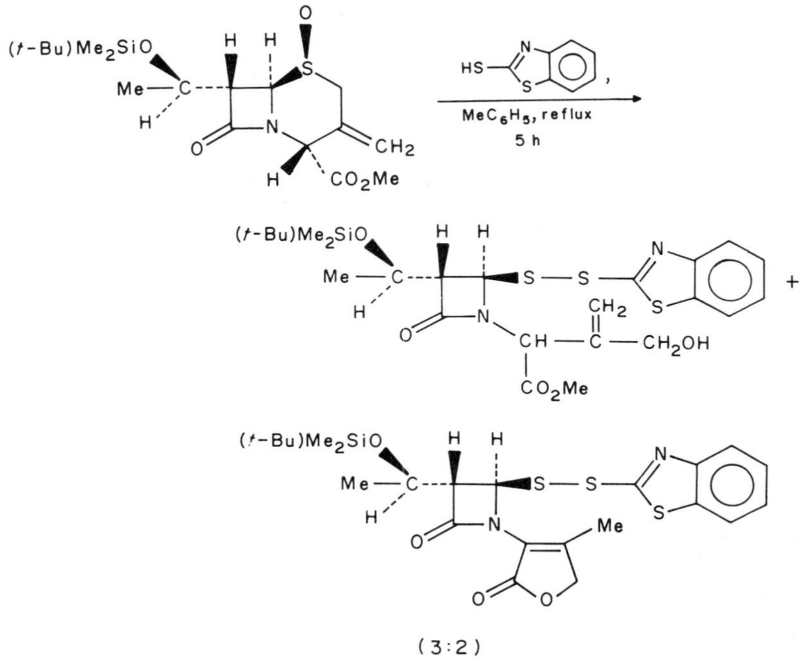

(1195)

(3:2)

Conversion of sulphur C-4 substituents to carbon substituents has also been accomplished and includes the preparation[1339] of (3*S*, 4*S*)-3[(*R*)-1-(*t*-butyldimethylsilyloxy)ethyl]-4-cyano-2-azetidinone from (3*S*, 4*R*)-3-[(*R*)-1-(*t*-butyldimethylsilyloxy)-ethyl]-4-(methylsulphonyl)-2-azetidinone by treatment with potassium cyanide and tetra-(*n*-butyl)ammonium bromide (equation 1196).

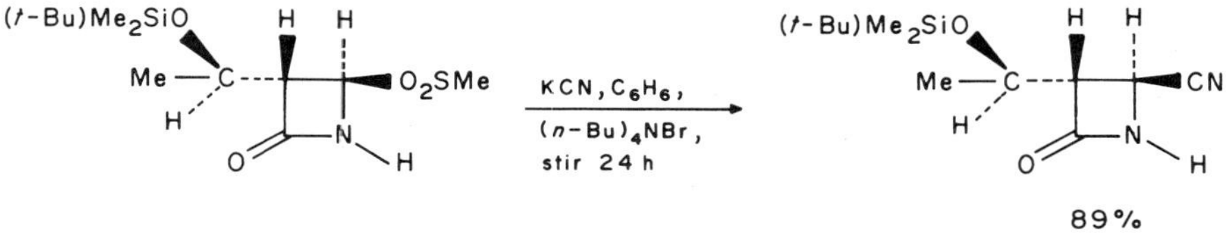

(1196)

4-Alkyl-, allyl-, vinyl- and ethynylazetidin-2-ones have been prepared[1328,1441] from 4-sulphonylazetidin-2-one using lithium organocuprates or Grignard reagents (equations 1197 and 1198).

1. R_2CuLi, THF, −78 °C, 10 min 2. 0 °C, 1.5 h

1. RMgBr, THF, −78 °C, 10 min 2. 0 °C, 30 min 3. r.t., 20 min

(1197)

Reagent	R in product	Yield (%)	Reference
$(n\text{-Bu})_2CuLi$	n-Bu	94	1328
EtMgBr	Et	72.2	1328
$H_2C{=}CHMgBr$	$H_2C{=}CH$	65.5	1328
$(H_2C{=}CHCH_2)_2CuLi$	$H_2C{=}CHCH_2$	100	1328
$H_2C{=}CHCH_2MgCl$	$H_2C{=}CHCH_2$	54.9	1328
$HC{\equiv}CCH_2MgBr$	$HC{\equiv}CCH_2$	—	1441
$EtOC{\equiv}CMgBr$	$EtOC{\equiv}C$	95.4	1328
$PhSC{\equiv}CMgBr$	$PhSC{\equiv}C$	68.9	1328

trans and *cis*

1. $PhC{\equiv}CMgBr$, THF, −30 °C, 30 min 2. r.t. 1 h

trans − 52.2%
cis − 22.3%

(1198)

Reaction of 4-arylthioazetidine-2-ones with dimethyl diazomalonate (equation 1199) or diazodiethylphosphonacetic acid ethyl ester (equation 1200) at 100 °C in the presence of rhodium(II) acetate produced[1367] the corresponding C-4 carbon extension products via a carbene insertion.

$+ N_2CH(CO_2Me)_2 \xrightarrow[100\,°C]{Rh(OAc)_2}$

R^1 = $PhCH_2$; Ph ; Ph ; Ph.
R^2 = $SiMe_2(Bu\text{-}t)$; $SiMe_2(Bu\text{-}t)$; CH_2CO_2Me ; $CH_2CO_2Bu\text{-}t$ (1199)
Yield (%) = — ; 60 ; — ; —

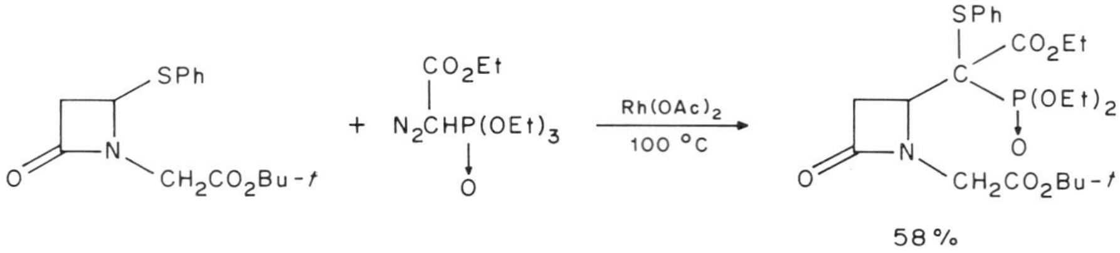

(1200)

Trityl thioether (equation 1201)[1440] and methylthio (equation 1202)[1155] C-4 substituted β-lactams have been converted into their methoxy analogues by reaction with silver nitrate[1440] or silver carbonate[1155] in the presence of methanol.

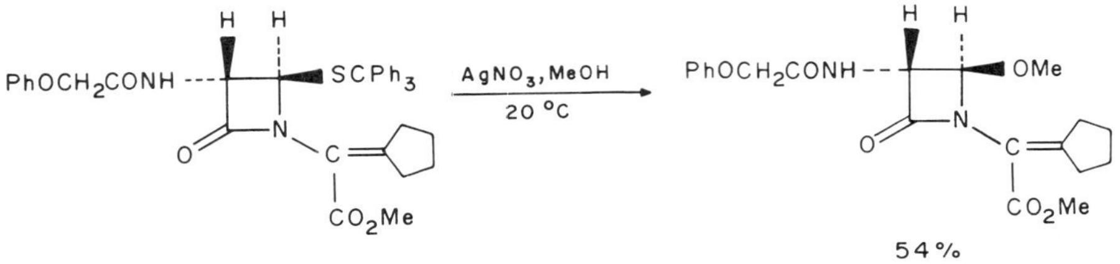

(1201)

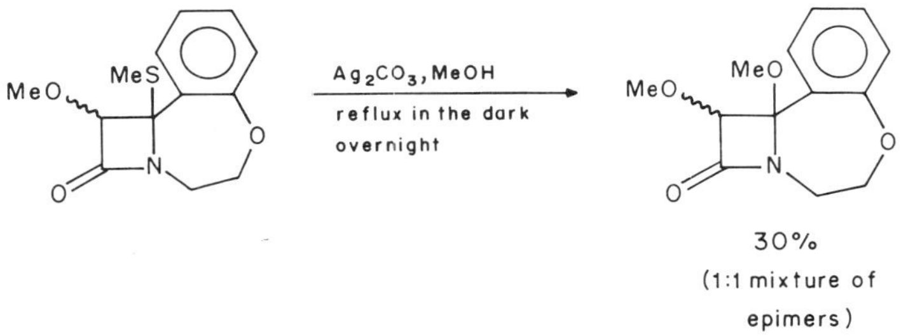

(1202)

Desulphurization of 4-arylthioazetidine-2-ones using hydride reagents[1115] or Raney nickel[1099,1165] occurs smoothly whether the cyclic moiety is an α-alkylidene β-lactam[1115] (equation 1203), a dimethoxy- or a methylenedioxybenzo[*a*]octem (equation 1204) or a 2-oxoazeto[1,2-*c*][1,3]benzoxazine[1339] (equation 1205). However, if the lactam nitrogen contains an allyl substituent, exclusive *endo* ring closure products are obtained[1054] via an intramolecular S_H2 process in addition to the direct reduction product (equation 1206), when the starting material is treated with tri-(*n*-butyl)stannane and a trace of azobisisobutyronitrile. A similar ring closure, but producing tricyclic products,

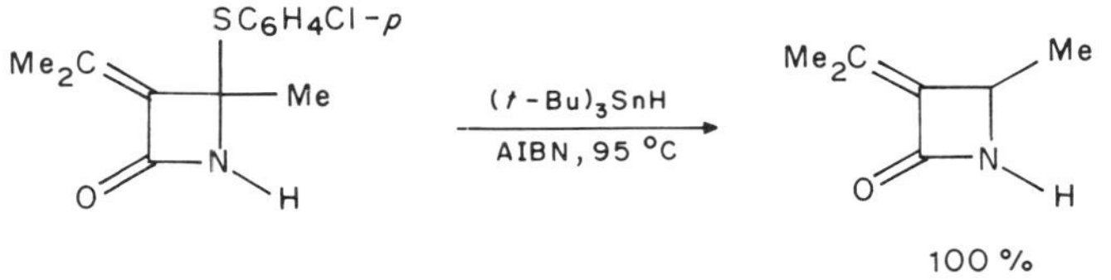

(1203)

was observed[1054] to occur when β-lactams containing *o*-bromophenylmethyl substituents attached to the lactam nitrogen were treated with tri-(*n*-butyl)stannane and azobisisobutyronitrile; however, in these cases both the monocyclic products and the tricyclic products retained the C-4 sulphur atom from the starting material (equation 1207).

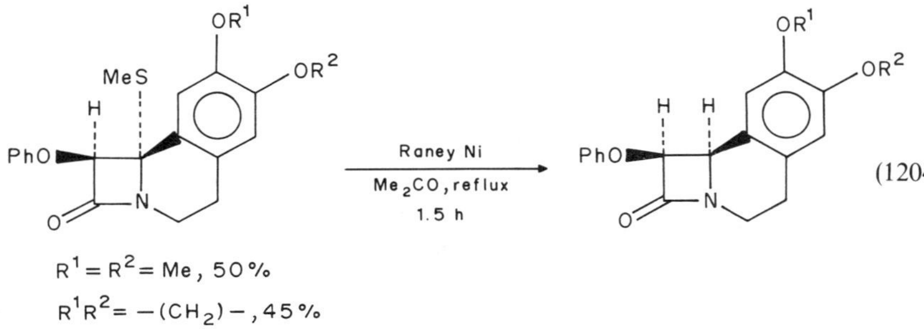

Raney Ni, Me_2CO, reflux, 1.5 h (1204)

$R^1 = R^2 = Me$, 50%

$R^1R^2 = -(CH_2)-$, 45%

Raney Ni, MeOH, C_6H_6, reflux, 15 min

C_6H_4Cl-*p*

C_6H_4Cl-*p* + Ph (1205)

26.6% 25.1%

$[(n\text{-}Bu)_3Sn]_2$, AIBN, C_6H_6, 80 °C

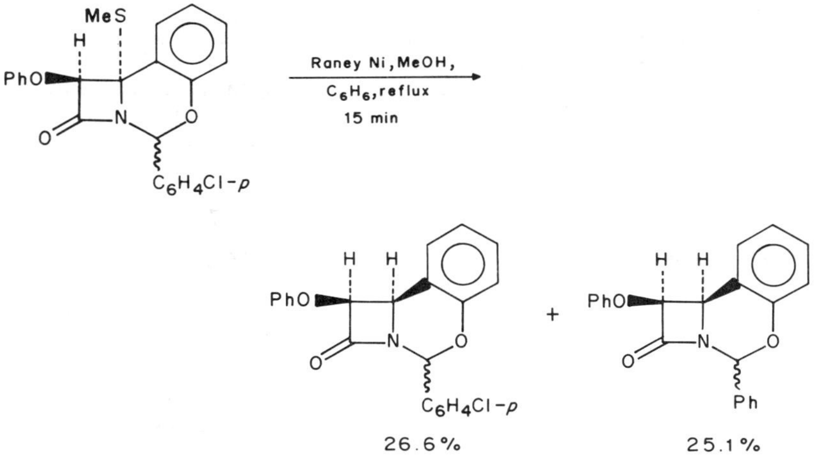

$(CH_2)_nCH{=}CH_2$

(**251**) (**252**)

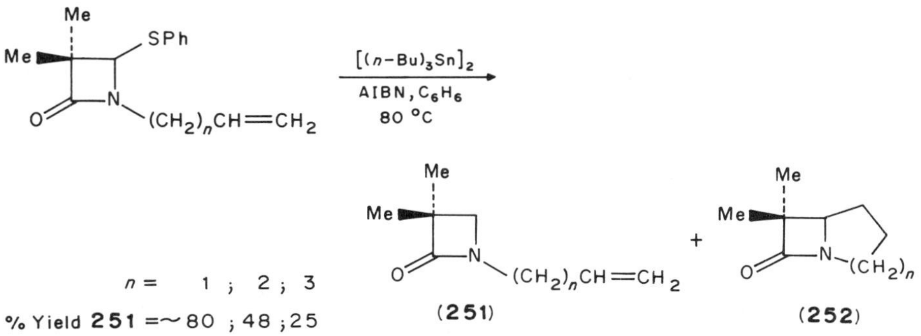

n =	1	2	3
% Yield **251** =	~80	48	25
% Yield **252** =	0	26	55

(1206)

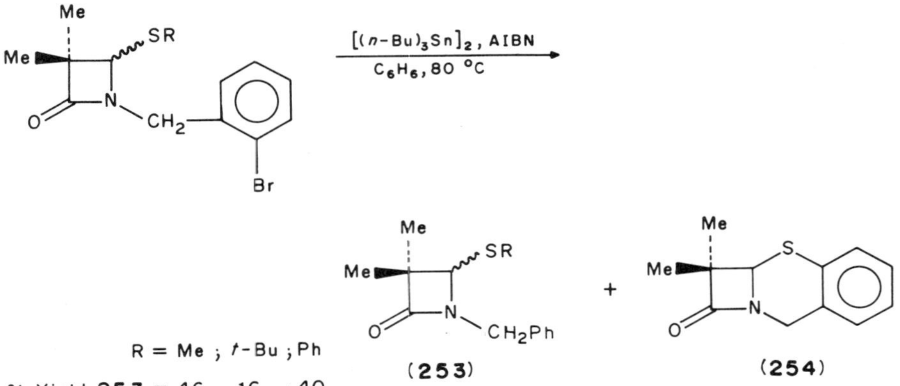

$[(n\text{-}Bu)_3Sn]_2$, AIBN

C_6H_6, 80 °C

R = Me ; t-Bu ; Ph

% Yield **253** = 46 ; 16 ; 40

% Yield **254** = 21 ; 42 ; 0

(**253**) (**254**)

(1207)

An interesting series of reactions has been reported[1401] using a 4-thioxo-2-azetidinone **255** prepared by thermolysis of 2-[*cis*-3-phthalimido-2-oxo-4-(2-carbomethoxyethylsulphoxyl)]-3-methyl-2-butenoic acid methyl ester (equation 1208). Ozonolysis produced[1401] the corresponding 2,4-azetidinedione (equation 1209), while 1,3-addition of diazoalkanes produced[1401] the thiadiazolines which spontaneously lost nitrogen affording the corresponding thiirans (equation 1210). Heating the thiirans in boiling benzene with triphenylphosphine affords[1401] the respective 4-alkylidene-2-azetidinones in excellent yields (equation 1211).

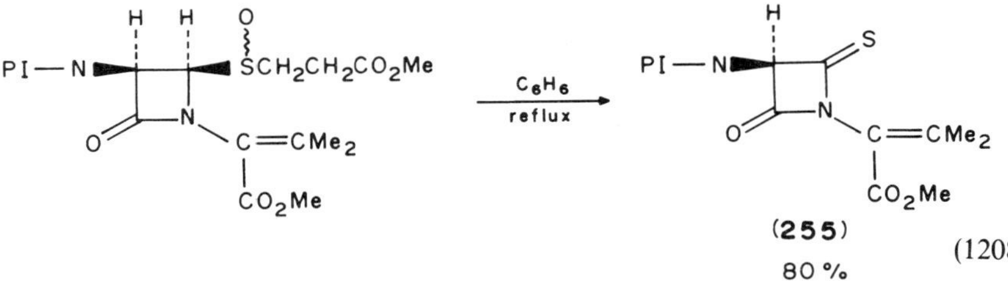

C_6H_6, reflux

(**255**) 80 %

(1208)

255

O_3, MeOH, 0 °C — 85 %

$m\text{-}ClC_6H_4CO_3H$, CH_2Cl_2, −50 °C — 25 %

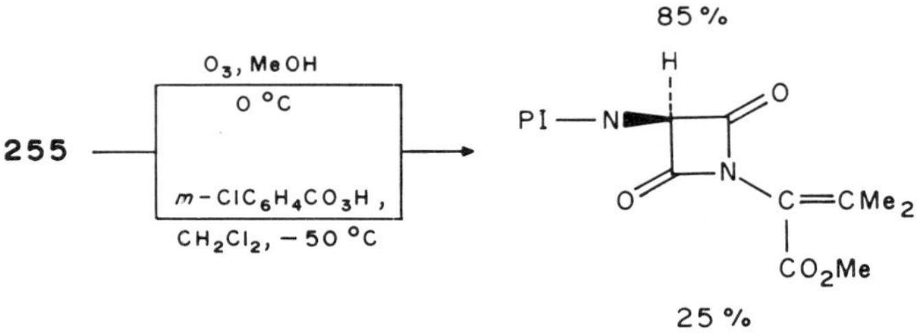

(1209)

Other methods used to effect this same type of conversion include refluxing the thiirans in dimethylformamide[1131] (equation 1212) and treatment[1131] with Raney nickel W-2 in boiling ethanol (equation 1213).

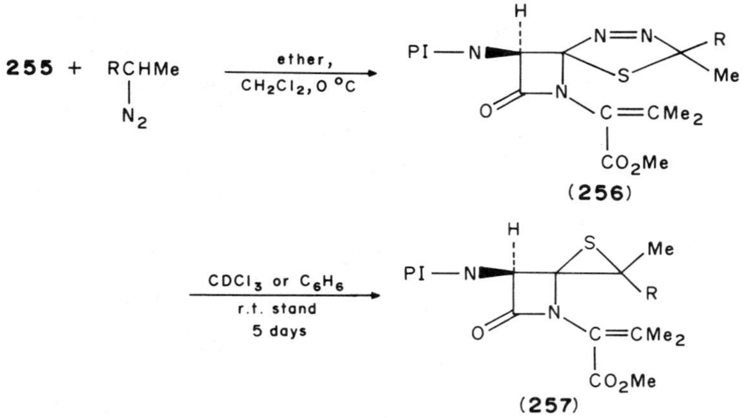

255 + RCHMe(N_2) —(ether, CH_2Cl_2, 0 °C)→ (**256**) —($CDCl_3$ or C_6H_6, r.t. stand, 5 days)→ (**257**) (1210)

R = H ; Me

% Yield **256** = – ; 75

% Yield **257** = 45 ; 100

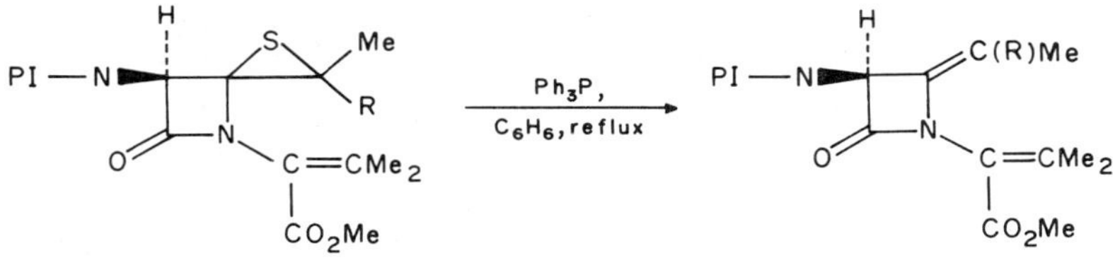

—(Ph_3P, C_6H_6, reflux)→ (1211)

R = H ; Me

Yield (%) = 95; 95

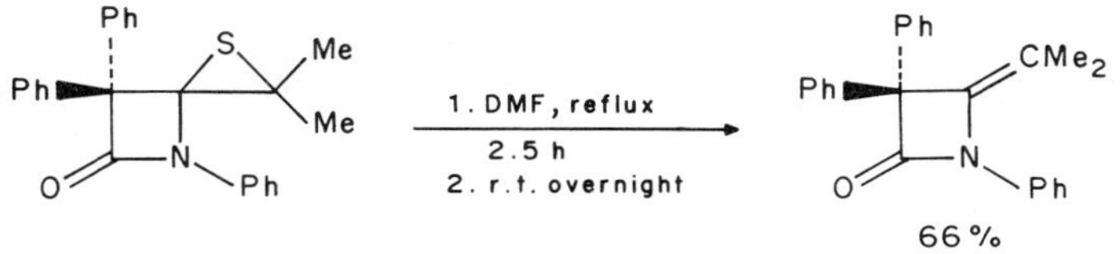

—(1. DMF, reflux, 2.5 h; 2. r.t. overnight)→ 66% (1212)

—(Raney Ni W-2, EtOH, reflux)→ (1213)

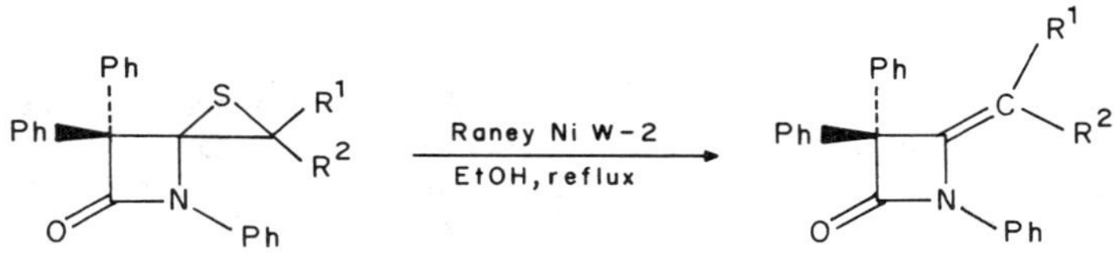

R^1 = Me ; Me ; $(CH_2)_5$

R^2 = Me ; Et ;

Yield (%) = 97 ; 90 ; 89

One of the more useful substituents into which the C-4 sulphur substituted β-lactams can be converted is the corresponding chloride. Thus, treatment of *N*-bis(methoxycarbonyl)methyl-4-methylthio-2-azetidinone[1308] (equation 1214) or various isomers of 3-azido-1-[diphenylmethoxycarbonyl-[(1-*p*-methoxybenzylthio)cyclopentyl]methyl]-4-(methylthio)azetidin-2-ones[1175] and 3-azido-1-[benzyloxycarbonyl-[(1-*p*-methoxybenzylthiocyclopentyl]methyl]-4-(methylthio)azetidin-2-ones[1175] (equation 1215) with chlorine produces the corresponding 4-chloroazetidinones.

SMe; $CH(CO_2Me)_2$ — Cl_2, CH_2Cl_2, 78 °C → Cl; $CH(CO_2Me)_2$ 100 % (1214)

Cl_2, CH_2Cl_2, CCl_4, Ar, stir 15 min (1215)

R = CH_2Ph, $CHPh_2$

Similar conversions to C-4 chloro β-lactams have been accomplished[1440] using sulphuryl chloride with 2-(*trans*-3-azido-4-methylthio-2-oxo-1-azetidinyl)-3-substituted-2-butenoic acid esters (equation 1216) and cyclopentylidene(*trans*-4-methylthio-2-oxo-3-phenoxyacetylamino-1-azetidinyl)acetic acid methyl ester (equation 1217). Attempting the same conversion[1440] with cyclopentyliden(*trans*-4-methylthio-2-oxo-3-phenacylamino-1-azetidinyl)acetic acid methyl ester, however, afforded only 9% of the corresponding *cis*-C-4 chloride substituted product, while the major product was the fused ring cyclopentyliden-*cis*-(7-oxo-3-phenyl-4-oxa-2,6-diazabicyclo[3.2.0]hept-2-en-6-yl)acetic acid methyl ester (equation 1218).

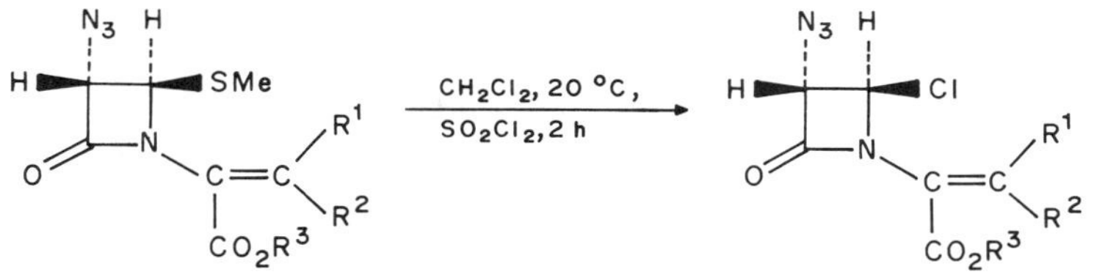

CH_2Cl_2, 20 °C, SO_2Cl_2, 2 h

(258)

+ (1216)

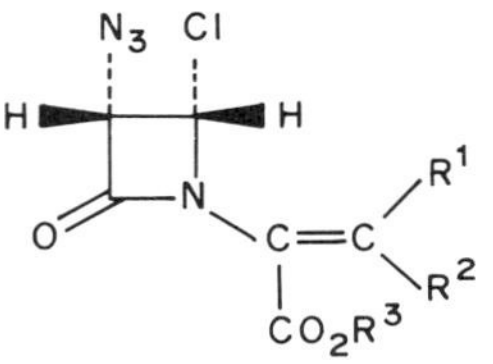

(259)

R^1		R^2	R^3	*trans*:*cis* Ratio	Yield (%)
Me		Me	*t*-Bu	10:1	90
	—$(CH_2)_4$—		Me	10:1	96

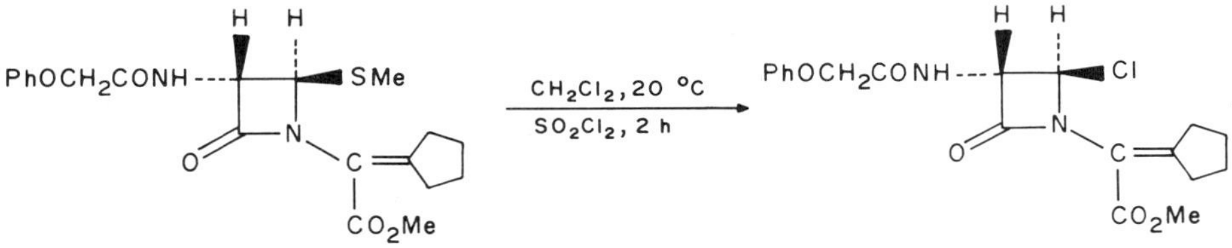

CH_2Cl_2, 20 °C, SO_2Cl_2, 2 h

(260)

+

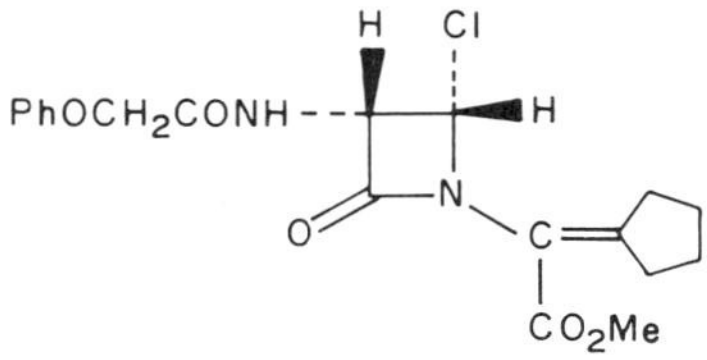

(261)

total yield = 90%

(1217)

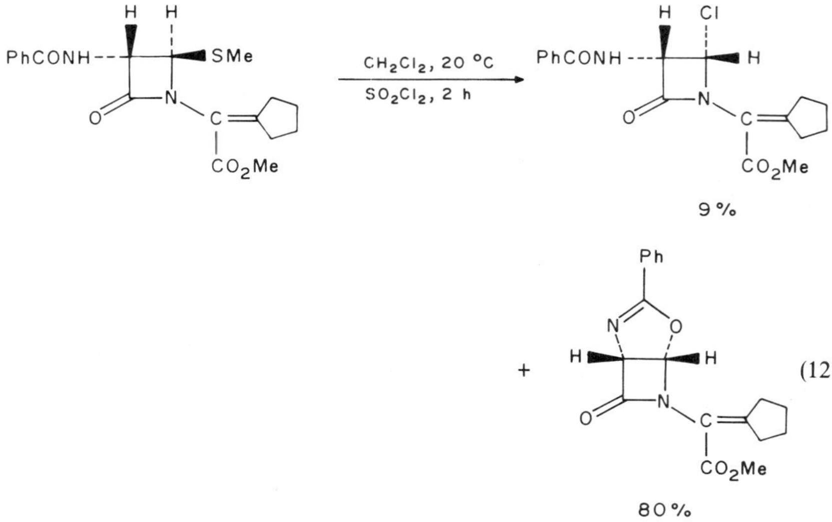

PhCONH---(H)(H)-SMe β-lactam, N-C(=cyclopentylidene)CO$_2$Me $\xrightarrow[SO_2Cl_2,\ 2\ h]{CH_2Cl_2,\ 20\ °C}$ PhCONH---(H)(Cl)(H) β-lactam, N-C(=cyclopentylidene)CO$_2$Me (9%)

+ Ph-oxazoline-fused β-lactam (H, H), N-C(=cyclopentylidene)CO$_2$Me (80%) (1218)

Similar results were obtained[1440] using thionyl chloride and the 4-methylsulphinyl analogues of the above compounds (equations 1219 and 1220).

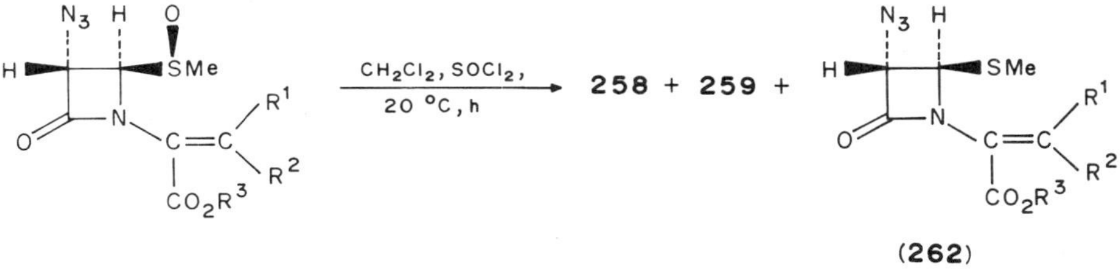

N_3, H, S(O)Me β-lactam, N-C(=CR^{1}R^2)CO$_2$R^3 $\xrightarrow[20\ °C,\ h]{CH_2Cl_2,\ SOCl_2,}$ **258** + **259** + N_3, H, SMe β-lactam, N-C(=CR^{1}R^2)CO$_2$R^3 **(262)** (1219)

R^1	R^2	R^3	Ratio 258 : 259 : 262	Yield (%)
Me	Me	Et	54 : 8 : 38	62
H	*t*-Bu	*t*-Bu	43 : 15 : 42	56

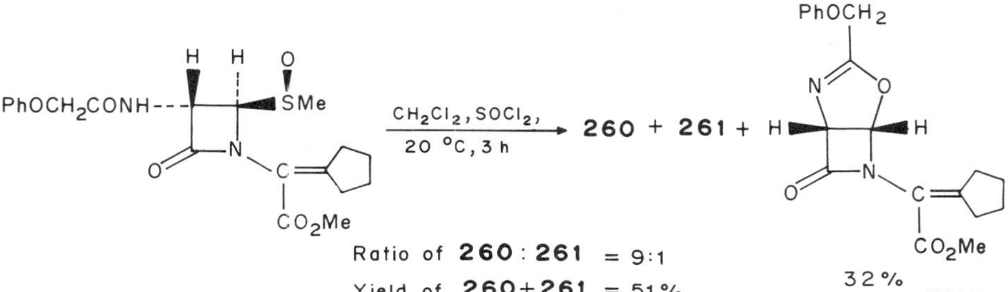

PhOCH$_2$CONH---(H)(H)-S(O)Me β-lactam, N-C(=cyclopentylidene)CO$_2$Me $\xrightarrow[20\ °C,\ 3\ h]{CH_2Cl_2,\ SOCl_2,}$ **260** + **261** + PhOCH$_2$-oxazoline-fused β-lactam (H, H), N-C(=cyclopentylidene)CO$_2$Me (32%) (1220)

Ratio of **260** : **261** = 9:1

Yield of **260**+**261** = 51%

The usefulness of the C-4 chloro substituent is exemplified by the following interconversions which utilize C-4 chloro-substituted lactams as starting materials.

By using organometallic reagents β-chloroazetidinones undergo a carbon–carbon coupling reaction to produce alkyl- and allyl-β-substituted lactams. (4*R*)-*n*-Butyl-3-(*S*)-bromo-2-azetidinone was produced[1383] in 57% yield by reaction of (3*S*)-bromo-(4*R*)-chloro-2-azetidinone with lithium di-(*n*-butyl)cuprate (equation 1221).

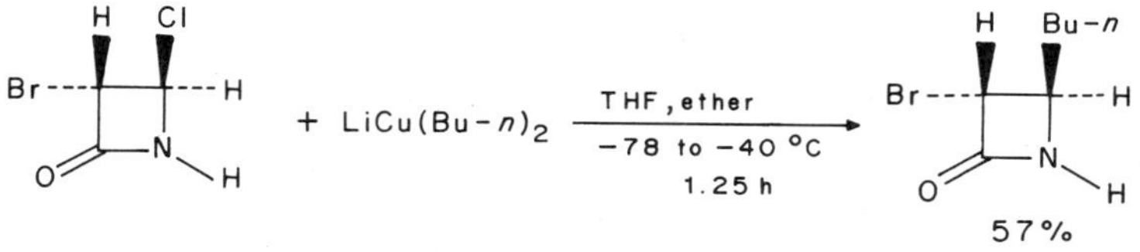
H
Cl
Br
H
O
N
H
+ LiCu(Bu-n)2
THF, ether
−78 to −40 °C
1.25 h
Bu-n
57%

(1221)

Lithium diallylcuprate[1383,1404], tetraallyltin[1383], 1-propen-3-yltributyltin[1442] and allylcoppers[1404] have all been used to prepare β-allylazetidin-2-ones as illustrated in the general equation 1222 with the details reported in Table 96.

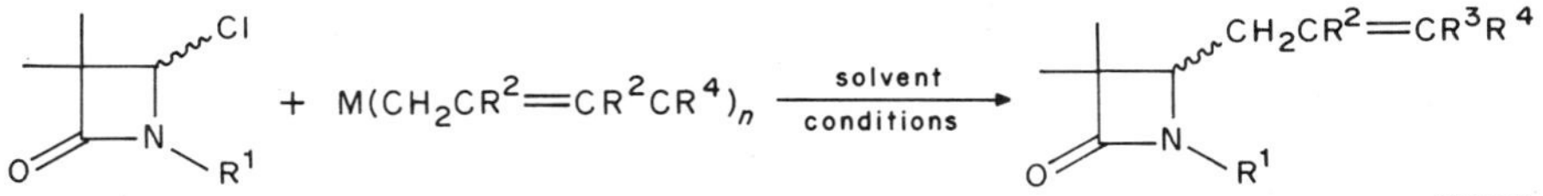
Cl
O
N
R1
+ M(CH2CR2=CR2CR4)n
solvent
conditions
CH2CR2=CR3R4

(1222)

Addition of (3*S*)-bromo-(4*R*)-chloro-2-azetidinone to a solution of sodium triphenylmethyl mercaptide in isopropanol–water produced[1383] the chiral *trans*-3-bromo-4-triphenylmethylthio-2-azetidinone in 25% yield (equation 1223).

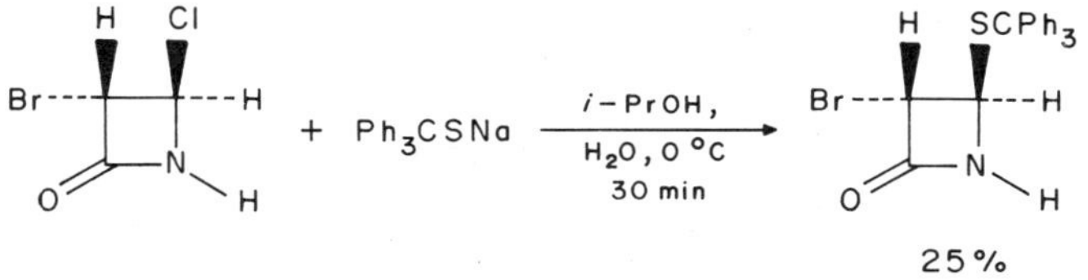
H
Cl
Br
H
O
N
H
+ Ph3CSNa
i-PrOH,
H2O, 0 °C
30 min
SCPh3
25%

(1223)

β- and δ-Chloro lactams have also been converted into β-methoxy lactams via methanolysis reactions catalysed by tin(II) chloride[1440] (equation 1224) or silver oxide[1155] (equation 1225).

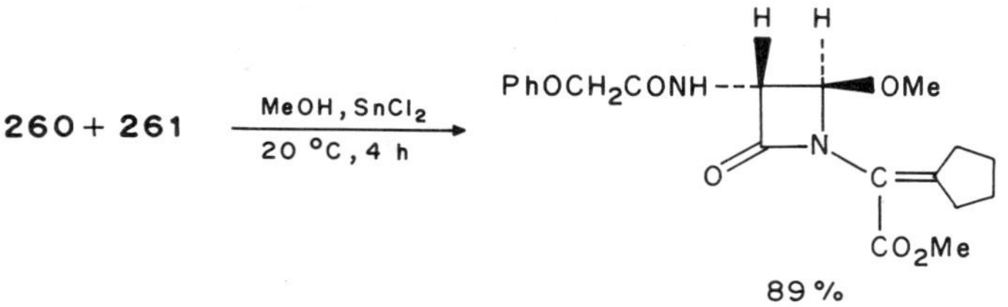
260 + 261
MeOH, SnCl2
20 °C, 4 h
H
H
PhOCH2CONH
OMe
O
N
C
CO2Me
89%

(1224)

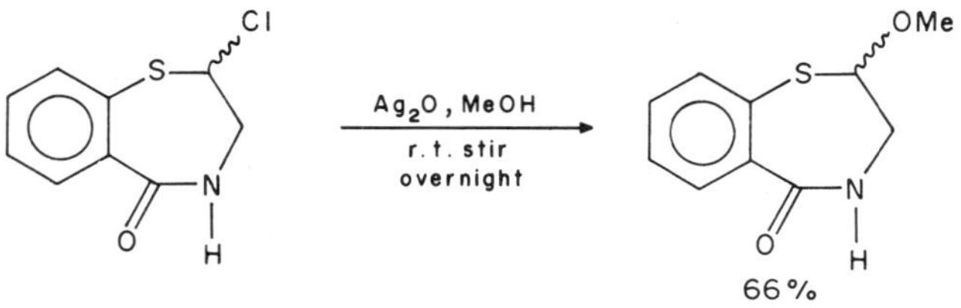
Cl
S
N
H
O
Ag2O, MeOH
r.t. stir
overnight
OMe
66%

(1225)

TABLE 96. Preparation of β-allylazetidin-2-ones from β-chloroazetidin-2-ones and organometallic reagents

β-Chlorolactam	Organometallic reagent	Conditions	Product	Yield (%)	Reference
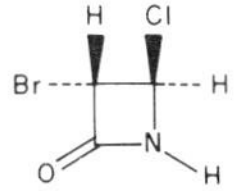	$LiCu(CH_2CH{=}CH_2)_2$	THF, ether, −78 °C, 20 min	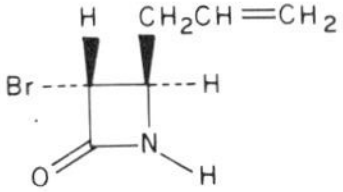	75	1383
	$Sn(CH_2CH{=}CH_2)_4$	CH_2Cl_2, −78 to 20 °C, 3.5 h	*trans*:*cis* = 95:5 *trans*:*cis* = 82:18	86	1383
	$Sn(CH_2CH{=}CH_2)_4$	THF, −78 to 20 °C, 3.5 h	*trans*:*cis* = 92.5:7.5	86	1383
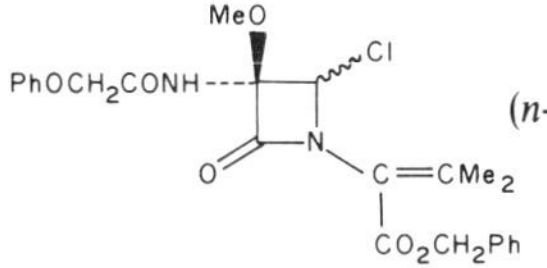 *cis* or *trans*	$(n\text{-}Bu)_3SnCH_2CH{=}CH_2$	C_6H_5Me, AIBN, *hv*, 2 h	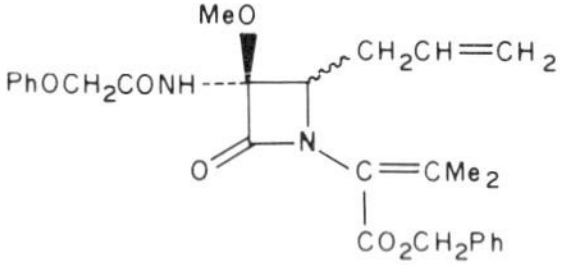 *cis* and *trans*	—	1442
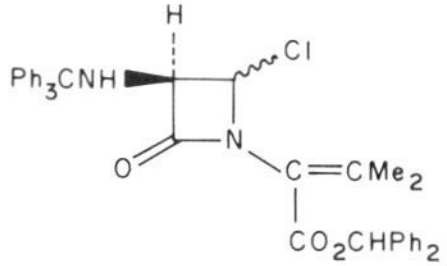 *cis* or *trans*	$LiCu(CH_2CR{=}CH_2)_2$ R = H, Me	THF	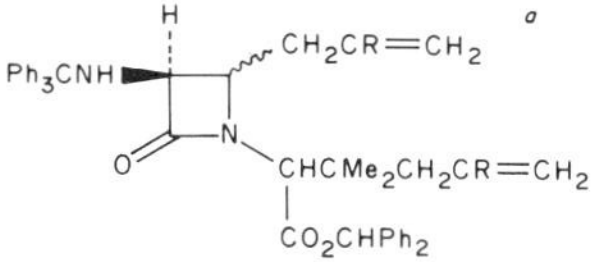 [a] *cis* and *trans*	—	1404
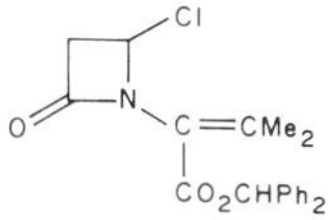	$CuCH_2CH{=}CHMe$	1. THF, −35 °C, 1–4 h 2. 0 °C, 1 h	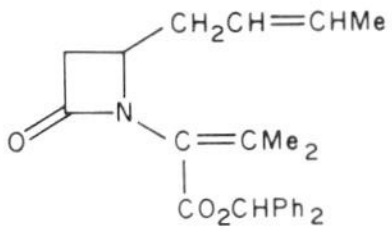	55	1404

Substrate	Reagent	Conditions	Product	Yield (%)	Ref.
Ph$_3$CNH (H, Cl) azetidinone, N–C(=CMe$_2$)CO$_2$CHPh$_2$	CuCH$_2$C(R^1)=C(R^3)R^2	1. THF, $-35\,^{\circ}$C, 1–4h 2. 0 °C, 1h	Ph$_3$CNH azetidinone, 4-CH$_2$C(R^1)=C(R^3)R^2, N–C(=CMe$_2$)CO$_2$CHPh$_2$		1404
	R^1 = R^2 = R^3 = H		*cis*:*trans* = 33:67	12.4	
	R^1 = R^3 = H, R^2 = Me		*cis*:*trans* = 45:55	66.1	
	R^1 = H, R^2 = R^3 = Me		*cis*:*trans* = 55:45	64	
	R^1 = Me, R^2 = R^3 = H		*cis*:*trans* = 33:67	63.6	
Ph$_3$CNH (H, Cl) azetidinone, N–C(=CMe$_2$)CO$_2$CHPh$_2$	CuCH$_2$C(Me)=CH$_2$	1. THF, $-35\,^{\circ}$C, 1–4h 2. 0 °C, 1h	Ph$_3$CNH azetidinone, 4-CH$_2$C(Me)=CH$_2$, N–C(=CMe$_2$)CO$_2$CHPh$_2$ *cis*:*trans* = 33:67	48.2	1404

[a]Conjugate addition products.

R = Ph, CH_2OCH_2Ph, $(CH_2)_3OTHP$

KN_3, MeCN, 18-crown-6, 10 days (left); dibenzo-18-crown-6, −12 °C, 4 days (right)

58–63% 40%

$NaBH_4$, EtOH, 0 °C (left); CH_2Cl_2, −12 °C (right)

80–90%

$NaNO_2$, aq. HCl, CH_2Cl_2, 0 °C

50–55% 62%

SCHEME 20

A novel sequence of reactions, which results in the ultimate formation of 4-diazo-2,3-diones, has been applied[1263] to both mono- and bicyclic-β-chloro lactams as shown in Scheme 20.

Finally, 1-methyl-5,7-dichloro-3*H*-1,4-benzodiazepin-2-(1*H*)-one upon dissolution in dimethyl sulphoxide at room temperature with stirring for 12 hours causes an exothermic reaction and results in the formation[1332] of 4,4′-methylenedi[7-chloro-1-methyl-3,4-dihydro-1*H*-1,4-benzodiazepine-2,5-dione] (equation 1226).

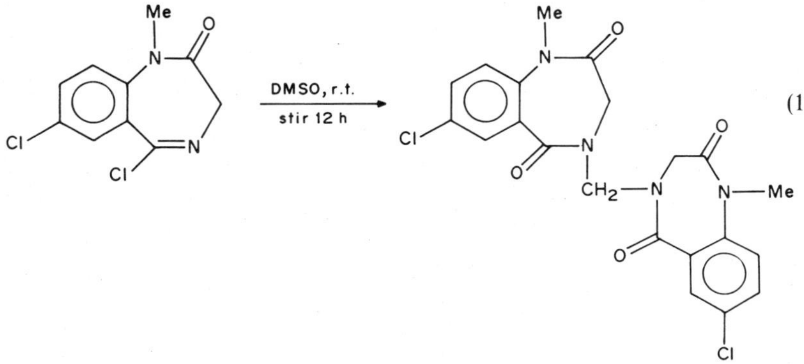

DMSO, r.t.
stir 12 h

(1226)

Another synthetically useful function located at the C-4 position of lactams consists of hydroxy containing methylene side chains. The simplest reaction these functions undergo is methanesulphonation or *p*-toluenesulphonation (equation 1227 and Table 97).

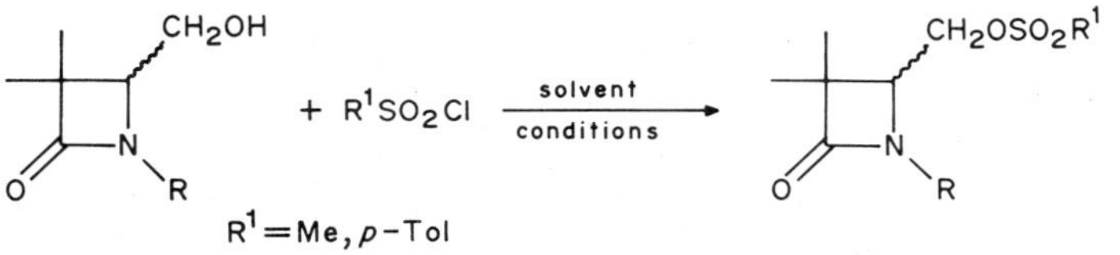

CH_2OH + R^1SO_2Cl —solvent, conditions→ $CH_2OSO_2R^1$

R^1 = Me, *p*-Tol

(1227)

Once formed, these methanesulphonates or *p*-toluenesulphonates can be converted to iodides[1053,1339] (equations 1228 and 1229) or used to produce[1176] ring closed products (equation 1230).

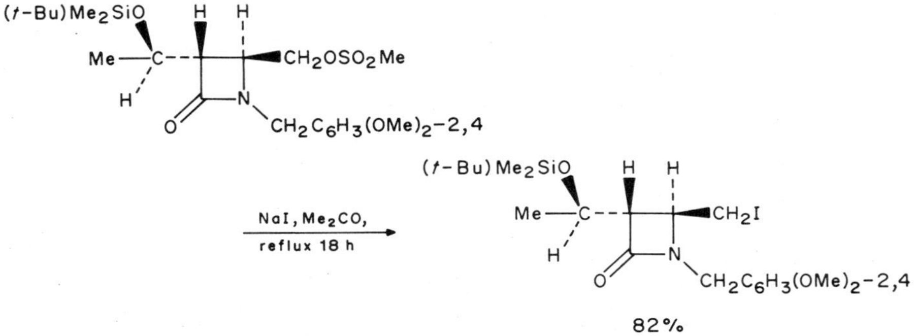

(*t*-Bu)Me_2SiO ... CH_2OSO_2Me ... $CH_2C_6H_3(OMe)_2$-2,4

NaI, Me_2CO,
reflux 18 h

(*t*-Bu)Me_2SiO ... CH_2I ... $CH_2C_6H_3(OMe)_2$-2,4

82%

(1228)

TABLE 97. Sulphonation of lactam β-hydroxyalkyl side chains

Lactam	Sulphonation reagent	Conditions	Product	Yield (%)	Reference
[structure: N_3, H, H, CH_2OH azetidinone; N–$CHCONHCH_2C_6H_4NO_2$-p, $CH(OEt)_2$]	$MeSO_2Cl$	1. Et_3N, THF, r.t. 10h 2. N_2O_4, $CHCl_3$, NaOAc, 0 °C, 1 h 3. CCl_4, reflux 3 h	[structure: N_3, H, H, CH_2OSO_2Me azetidinone; N–$CHCONHCH_2C_6H_4NO_2$-p, $CH(OEt)_2$]	57	1038
[structure: (t-Bu)Me_2SiO, Me, C, H, H, H, CH_2OH, N–R azetidinone]	$MeSO_2Cl$	1. Et_3N, CH_2Cl_2, 0–5 °C 2. r.t. stir 1 h	[structure: (t-Bu)Me_2SiO, Me, C, H, H, H, CH_2OH, R]		1053
R = H				83	
R = $CH_2C_6H_3(OMe)_2$-2,4				98	
[structure: N_3, H, H, CH_2OH, N–R azetidinone]	$MeSO_2Cl$	1. Et_3N, −78 °C 2. warm to 20 °C	[structure: N_3, H, H, CH_2OSO_2Me, N–R azetidinone]		1176

Substrate	Reagent	Conditions	Product	Yield (%)	Ref.
R = (aryl: CH$_2$OSiMe$_2$(Bu-*t*), OR)				100	
$R^1 = CO_2CH_2Ph$, $SiMe_2(Bu\text{-}t)$				95	
(*t*-Bu)Me$_2$SiO, Me—C, H, H, H, CH$_2$OH, O, N, H	*p*-TosCl	C_5H_5N, THF, CH_2Cl_2, stir overnight	(*t*-Bu)Me$_2$SiO, Me—C, H, H, H, CH$_2$OTos, O, N, H	68	1339
2-ThiCH$_2$CONH, H, H, CH$_2$OH, O, H	*p*-TosCl	C_5H_5N, CH_2Cl_2, 0 °C	2-ThiCH$_2$CONH, H, H, CH$_2$OTos, O, N, H	—	1377

$$\text{NaI, Me}_2\text{CO, }60\,^\circ\text{C, stir 7 h} \qquad (1229)$$

$$\text{1. MeCN, KF, 18-crown-6, stir r.t. 5 h; 2. KF, 18-crown-6, stir r.t. 4 h} \quad 92\% \qquad (1230)$$

Treatment of a C-4 hydroxymethylene function with diazomethane in the presence of boron trifluoride etherate produces[802] the corresponding methyl ether (equation 1231) while reaction of a beta 1-hydroxy-2-nitroethyl function with thionyl chloride and triethylamine converts[1196] it into a beta 2-nitrovinyl substituent (equation 1232).

$$\text{CH}_2\text{Cl}_2\text{, CH}_2\text{N}_2\text{, }0\,^\circ\text{C, BF}_3\cdot\text{Et}_2\text{O, ether, stir 1 h} \quad 74\% \qquad (1231)$$

$$\text{SOCl}_2\text{, NEt}_3\text{, THF, }-20\,^\circ\text{C} \quad 88\% \qquad (1232)$$

Other conversions of C-4 hydroxymethylene groups include acylation[1317] using methyl chloroformate (equation 1233) and oxidation with dimethyl sulphoxide (DMSO) in the presence of either dicyclohexylcarbodiimide (DCCD) which produced[1053] the corresponding aldehyde that was isolated directly in excellent yields (equation 1234), or trifluoroacetic acid anhydride and triethylamine which also produced[1340] the corresponding aldehyde that was not isolated, but that was treated with nitromethane to yield the nitro alcohol (**263**) as an epimeric mixture (equation 1235). O-Acylation of the nitro alcohol formed followed by sodium borohydride reduction produced[1340] the C-4 nitroethyl product shown in equation 1236.

$PhCH_2O_2CNH$ CH_2OH N $SiMe_2(Bu\text{-}t)$ —$ClCO_2Me$→ $PhCH_2O_2CNH$ CH_2OCO_2Me N $SiMe_2(Bu\text{-}t)$ (1233)

$(t\text{-}Bu)Me_2SiO$ H H Me—C CH_2OH H N $CH_2C_6H_3(OMe)_2\text{-}2,4$

—DMSO, DCCD, H_3PO_4; r.t. stir 18 h→ $(t\text{-}Bu)Me_2SiO$ H H Me—C CHO H N $CH_2C_6H_3(OMe)_2\text{-}2,4$

91.8%

(1234)

O O Me H H CH_2OH N $CH_2C_6H_3(OMe)_2\text{-}2,4$

—1. $(F_3CCO)_2O, CH_2Cl_2$, DMSO, −70 °C stir 0.5 h; 2. NEt_3, −70 °C to r.t.→ O O Me H H CHO N $CH_2C_6H_3(OMe)_2\text{-}2,4$

—$MeNO_2$, i-PrOH, NEt_3, r.t. 2 days→ O O Me H H OH $CHCH_2NO_2$ N $CH_2C_6H_3(OMe)_2\text{-}2,4$ (1235)

(**263**)
37%
(mixture of epimers)

263 $\xrightarrow[\text{DMAP}]{\text{Ac}_2\text{O, CH}_2\text{Cl}_2,}$ [azetidinone bearing dioxolanyl(Me) group, H, H, $CH(OAc)CH_2NO_2$, N–$CH_2C_6H_3(OMe)_2$-2,4]

$\xrightarrow[\text{0 °C, stir}]{\text{NaBH}_4\text{, MeOH,}}$ [azetidinone bearing dioxolanyl(Me) group, H, H, $CH_2CH_2NO_2$, N–$CH_2C_6H_3(OMe)_2$-2,4]

91%

(1236)

This same nitroethyl substituent may also be prepared[1196] by sodium borohydride reduction of (3*RS*, 4*SR*)-3-azido-1-(4-methoxymethoxyphenyl)-4-(2-nitrovinyl)azetidin-2-one (equation 1237).

[N_3, H, H, $CH{=}CHNO_2$ azetidinone, N–$C_6H_4OCH_2OMe$-*p*] $\xrightarrow[-20\ °\text{C}]{\text{NaBH}_4\text{, H}_2\text{O, THF,}}$ [N_3, H, H, $CH_2CH_2NO_2$ azetidinone, N–$C_6H_4OCH_2OMe$-*p*]

87%

(1237)

Many of the C-4 hydroxymethylene containing lactams discussed above were prepared by ozonolysis of C-4 carbon-to-carbon double functions. If the starting olefin is ozonized in the absence of other reagents, an aldehyde[1245] (equation 1238) or ketone[1146,1221] (equation 1239) is produced, depending upon whether or not a substituent is present on the alpha-carbon of the double bond. But if the initial ozonolysis reaction is followed by reaction with a reducing agent, the hydroxyl function is generated[802,1176] (equation 1240).

[Et, H, H, $CH{=}CH(Ph)$ azetidinone, N–$SiMe_2(Bu\text{-}t)$] $\xrightarrow[\text{2. Me}_2\text{S}]{\text{1. O}_3\text{, CH}_2\text{Cl}_2}$ [Et, H, H, CHO azetidinone, N–$SiMe_2(Bu\text{-}t)$] (1238)

84%

1. O_3, −70 °C
2. Me_2S

CH_2N_2, MeOH, r.t. 24 h

R = PhO ; MeO; PI—N
Yield(%)[1221] = 46 ; 40 ; N.R.

$KMnO_4$, NaH_2PO_4, Me_2CO, H_2O, r.t. or
1. O_3, CH_2Cl_2, −70 °C
2. Me_2S or C_5H_5N

R = PhO ; PI—N ; PI—N
Reagent used = $KMnO_4$; $KMnO_4$; O_3
Yield (%)[1146] = 80 ; 34 ; 55

(1239)

1. O_3, CH_2Cl_2, EtOH, −78 °C, time
2. $NaBH_4$, r. t.

(1240)

R	Olefin isomer	Time	Yield (%)	Reference
H	*trans*	1.5 h	76	802
$PhCH_2OCO_2$	*trans*	35 min	58	1176
(*t*-Bu)Me_2SiO	*cis*	45 min	53	1176

However, if the initial ozonolysis reaction is followed by treatment with nitromethane and triethylamine, a nitro alcohol function is obtained[1196] (equation 1241).

If the C-4 olefin oxidation reaction is performed using a stronger oxidizing agent such as permanganate[1377] (equation 1242), or if the initially formed C-4 aldehyde is oxidized further[1245] (equation 1243), a carboxylic acid (or ester, after treatment with diazomethane) is the product obtained.

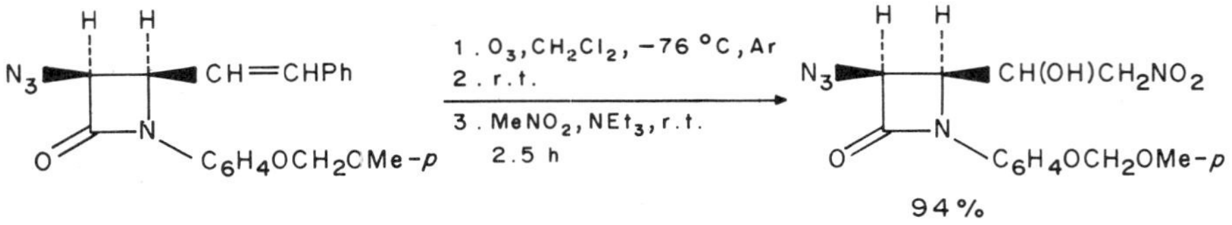

(1241)

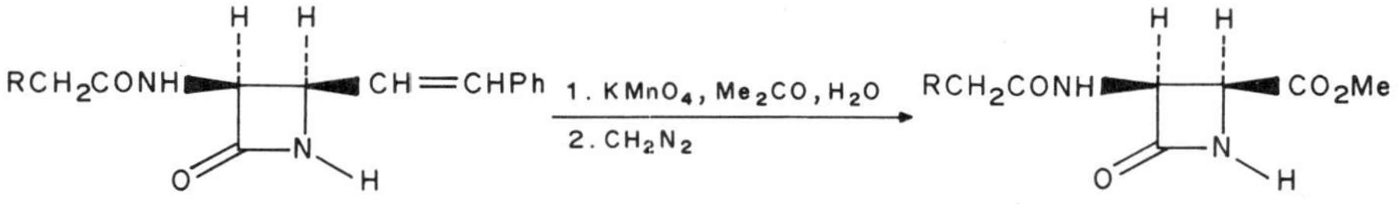

(1242)

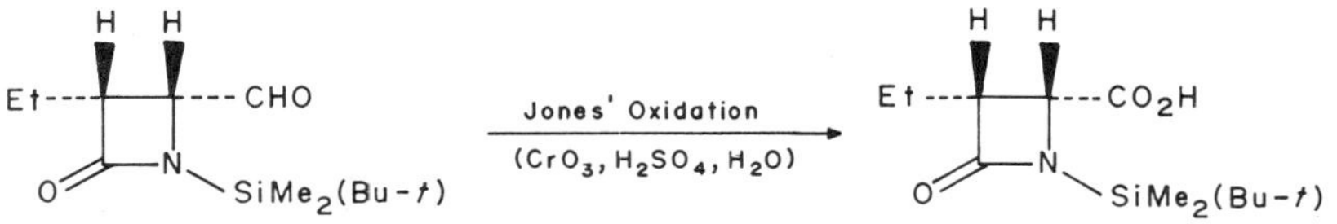

(1243)

Other preparations of C-4 carboxylic acids and esters include glycol cleavage[1152] (equation 1244), and hydrolysis[1339] of a cyano or amide function (equation 1245) followed by diazomethane esterification.

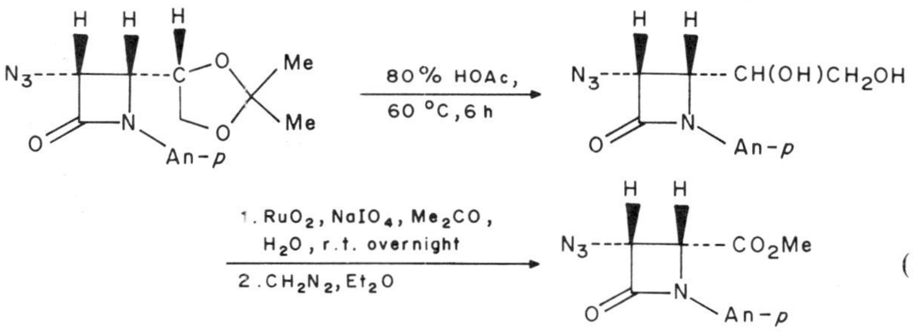

(1244)

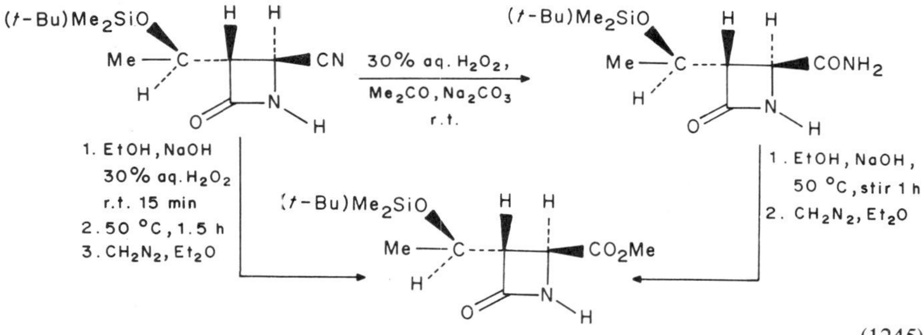

(1245)

TABLE 98. Generation of β-hydroxymethylene functions by reduction of carboxylic acids or esters

Lactam acid or ester	Reducing agent	Conditions	Product	Yield (%)	Reference
$(t\text{-Bu})Me_2SiO$, Me—C, H, H, H, CO_2R, O, N, H R = Me	$NaBH_4$	THF, H_2O, 0 °C stir 15 min	$(t\text{-Bu})Me_2SiO$, Me—C, H, H, H, CH_2OH, O, N, H	78	1339
R = t-Bu	$NaBH_4$	EtOH, 50 °C, stir		100	1053
	$NaBH_4$	EtOH, 50 °C, stir		86.6	1053
$(t\text{-Bu})Me_2SiO$, Me—C, H, H, H, CO_2R, O, N, $CH_2C_6H_3(OMe)_2$-2,4 R = H	$NaBH_4$	EtOH, H_2O (10:1) 70 °C, 10h	$(t\text{-Bu})Me_2SiO$, Me—C, H, H, H, CH_2OH, O, N, $CH_2C_6H_3(OMe)_2$-2,4	52	750
	$LiAlH_4$	THF, 0 °C, 45 min		50	
R = Me	$NaBH_4$	EtOH, 50 °C, stir 4h		100	1053
R = t-Bu	$NaBH_4$	EtOH, 70 °C, stir 18h		52.7	1053
	$LiAlH_4$	THF, 5 °C, stir 45 min		50	1053
2-ThiCH_2CONH, H, H, CO_2Me, O, N, H	$NaBH_4$	THF, H_2O, 0 °C	2-ThiCH_2CONH, H, H, CH_2OH, O, N, H	—	1377
$PhCH_2O_2CNH$, CO_2Me, O, N, $SiMe_2(Bu\text{-}t)$	$LiBH_4$	—	$PhCH_2O_2CNH$, CH_2OH, O, N, $SiMe_2(Bu\text{-}t)$	—	1317

Another method used to produce many of the C-4 hydroxymethyl containing lactams discussed above is reduction of a C-4 carboxylic acid or ester function (equation 1246 and Table 98).

$CO_2H(R)$ —reduction / reagent, conditions→ CH_2OH

(1246)

Reduction of [3*S*-[3α(*S**),4β]]-methyl 3-(1-*t*-butyldimethylsilyloxyethyl)-2-azetidinone-4-carboxylate with sodium bis(2-methoxyethoxy)aluminium hydride, however, afforded[1053] the corresponding 4-formyl product (equation 1247).

(*t*-Bu)Me_2SiO, Me, H, H, H, CO_2Me, O, N, H

$(MeOCH_2CH_2O)_2AlH_2Na$ / THF, −78 °C, N_2, stir 1 h →

(*t*-Bu)Me_2SiO, Me, H, H, H, CHO, O, N, H

78%

(1247)

In addition to these reductions, C-4 carboxylic acids also undergo other useful conversion reactions. Thus, β-lactams containing both a carboxylic acid and ester function at the C-4 position can be decarboxylated[750,1052] to produce the corresponding mono-ester (equations 1248 and 1249).

H, CO_2H, $PhCH_2$, CO_2Et, O, N, R

R = Ph, *p*-Tol

C_5H_5N, 125–130 °C →

H, H, $PhCH_2$, CO_2Et, O, N, R

(1248)[1052]

AcO, Me, C, H, H, CO_2H, CO_2Et, O, N, $CH_2C_6H_3(OMe)_2$-2,4

2,4,6-collidine / 160 °C, 45 min →

AcO, Me, C, H, H, H, CO_2Et, O, N, $CH_2C_6H_3(OMe)_2$-2,4

(1249)[750]

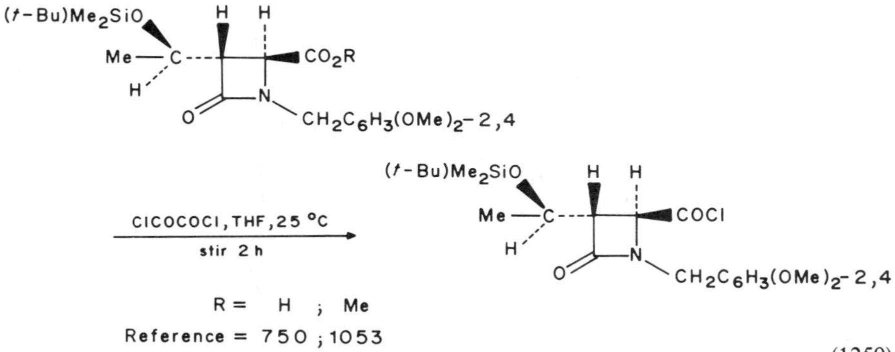

$$\xrightarrow[\text{stir 2 h}]{\text{ClCOCOCl, THF, 25 °C}}$$

R = H ; Me
Reference = 750 ; 1053

(1250)

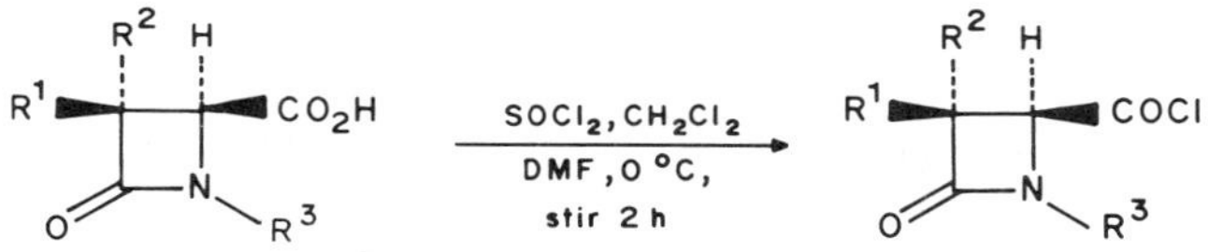

$$\xrightarrow[\text{DMF, 0 °C, stir 2 h}]{SOCl_2, CH_2Cl_2}$$

(1251)

R^1	R^2	R^3	Reference
H	H	$CH_2C_6H_3(OMe)_2$-2,4	1340[a]
$PhCH_2$	H	Ph	1052
$PhCH_2$	H	*p*-Tol	1052
H	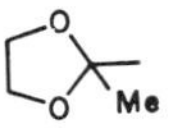 2-methyl-1,3-dioxolan-2-yl	$CH_2C_6H_3(OMe)_2$-2,4	1340[a]

[a]Product was not isolated but was subsequently used in another reaction.

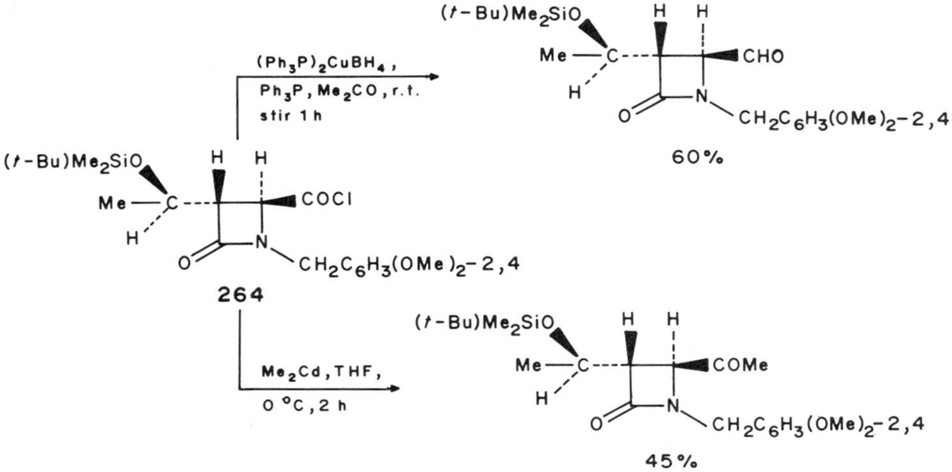

264

$$\xrightarrow{(Ph_3P)_2CuBH_4,\ Ph_3P,\ Me_2CO,\ \text{r.t., stir 1 h}}$$ CHO product, 60%

$$\xrightarrow{Me_2Cd,\ THF,\ 0\ °C,\ 2\ h}$$ COMe product, 45%

(1252)

Treatment of C-4 carboxylic acid or ester containing β-lactams with oxalyl chloride[750,1053] (equation 1250) or with thionyl chloride[1052,1340] (equation 1251) produces the corresponding acid chlorides, which were subsequently allowed to react with bis(triphenylphosphine)copper(I) tetrahydroborate to produce[1053] an aldehyde (equation 1252), dimethylcadmium to produce[750] a methyl ketone (equation 1252), aluminium chloride to produce[1052] intramolecular Friedel–Crafts ring closure products (equation 1253) or diazo substrates to produce[750,1053,1340] the corresponding diazoketones (equations 1254 and 1255).

$AlCl_3, CS_2$, 0 °C (1253)

R = Ph ; p-Tol

Yield (%) = 53 ; 55

264 → CH_2N_2, Et_2O, 0 – 5 °C, stir → (t-Bu)Me_2SiO, Me, H, $COCHN_2$, $CH_2C_6H_3(OMe)_2$-2,4

(265)

66 – 73%

(1254)[750,1053]

N_2CHCO_2Et, r.t., stir 70 h

R, COCl, $CH_2C_6H_3(OMe)_2$-2,4 → R, $COC(N_2)CO_2Et$, $CH_2C_6H_3(OMe)_2$-2,4

(266)

R = H ; Me

Yield (%) = 30 ; 45

(1255)[1340]

Treatment of the diazoketone products produced in equations 1254 and 1255 with light in the presence of a hydroxy-containing solvent affords[750,1053,1340] the Wolff rearrangement products (equations 1256 and 1257).

Hydrogenation or semi-hydrogenation of 4-ethynyl β-lactams affords 4-ethyl or 4-ethenyl products, respectively, depending upon the reagents and conditions used. Thus, catalytic hydrogenation of (3*R*, 4*S*)-3[(*R*)-1-(*t*-butyldimethylsilyloxy)ethyl]-4-ethynyl-1-(*t*-butyldimethylsilyl)-2-azetidinone over palladium on barium sulphate in pyridine produced[1237] the corresponding olefin exclusively in 91% yield (equation 1258).

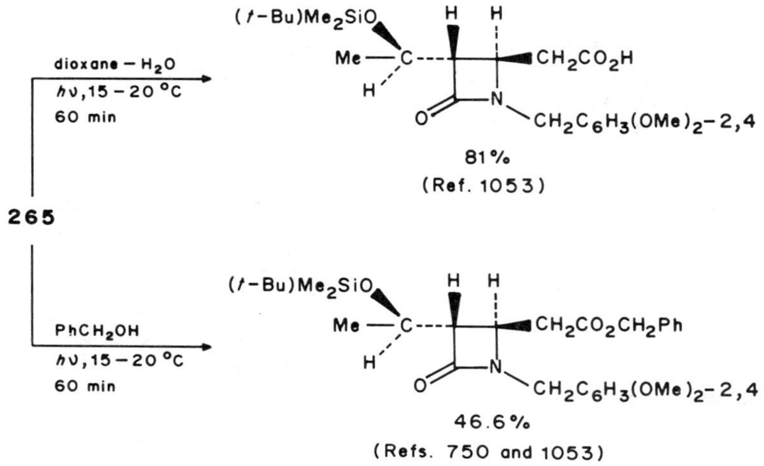

265 → (dioxane–H_2O, $h\nu$, 15–20 °C, 60 min) → (t-Bu)Me_2SiO, Me, H–C; H, H; CH_2CO_2H; O; N–$CH_2C_6H_3(OMe)_2$-2,4
81%
(Ref. 1053)

265 → ($PhCH_2OH$, $h\nu$, 15–20 °C, 60 min) → (t-Bu)Me_2SiO, Me, H–C; H, H; $CH_2CO_2CH_2Ph$; O; N–$CH_2C_6H_3(OMe)_2$-2,4
46.6%
(Refs. 750 and 1053)

(1256)

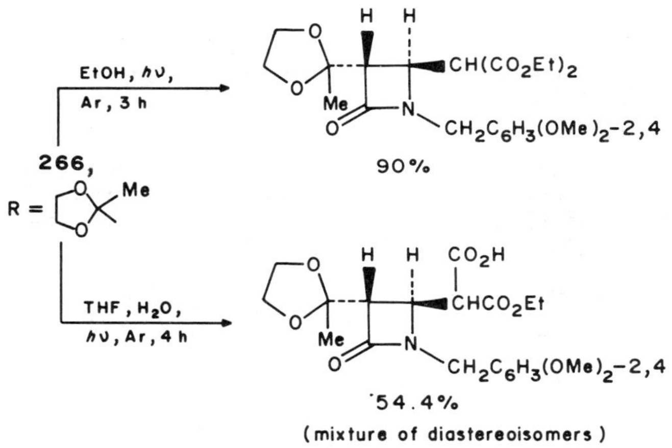

266, R = 2-methyl-1,3-dioxolan-2-yl (O, O, Me) → (EtOH, $h\nu$, Ar, 3 h) → O, O, Me; H, H; $CH(CO_2Et)_2$; O; N–$CH_2C_6H_3(OMe)_2$-2,4
90%

266 → (THF, H_2O, $h\nu$, Ar, 4 h) → O, O, Me; H, H; CO_2H, $CHCO_2Et$; O; N–$CH_2C_6H_3(OMe)_2$-2,4
54.4%
(mixture of diastereoisomers)

(1257)[1340]

(t-Bu)Me_2SiO, Me, H–C; H, H; C≡CH; O; N–$SiMe_2$(Bu-t) → (H_2, Pd–$BaSO_4$, C_5H_5N) → (t-Bu)Me_2SiO, Me, H–C; H, H; CH=CH_2; O; N–$SiMe_2$(Bu-t)
91%

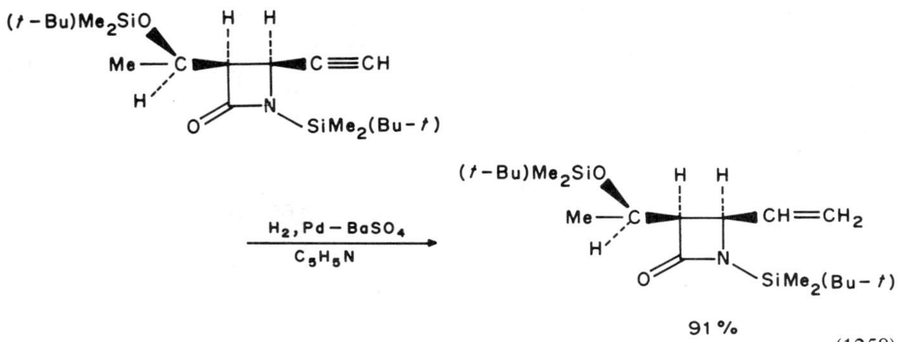

(1258)

Similar results are obtained[1196] when (3*RS*, 4*SR*)-1-(4-methoxymethoxyphenyl)-3-phenoxyacetamido-4-ethynylazetidin-2-one is hydrogenated using palladium on barium sulphate, but if the hydrogenation of performed[1196] in methanol using 10% palladium on calcium carbonate or in dioxane using 10% palladium on carbon, complete hydrogenation of the multiple bond occurs to produce the 4-ethylazetidinone (equation 1259).

H H
PhOCH$_2$CONH C≡CH
O N C$_6$H$_4$OCH$_2$OMe-*p*

H_2, 10% Pd–BaSO$_4$, dioxane, C_5H_5N, 1.5 h → PhOCH$_2$CONH (H H) CH=CH$_2$, N–C$_6$H$_4$OCH$_2$OMe-*p*

H_2, 10% Pd–CaCO$_3$, MeOH, 30 min → PhOCH$_2$CONH (H H) Et, N–C$_6$H$_4$OCH$_2$OMe-*p* 88%

H_2, 10% Pd–C, dioxane, 1 h

(1259)

However, a mixture of 4-vinyl- and 4-ethylazetidinones is obtained[1304] when (3*RS*, 4*SR*)-2-[2,2-ethylenedioxy-1-(2,2,2-trichloroethoxycarbonyl)propyl]-4-ethynyl-3-phenoxyacetamidoazetidin-2-one is hydrogenated using 10% palladium on barium sulphate (equation 1260).

PhOCH$_2$CONH (H H) C≡CH, N–CH(CO$_2$CH$_2$CCl$_3$)–C(Me)(OCH$_2$CH$_2$O)

H_2, 10% Pd–BaSO$_4$, dioxane, 1.5 h → PhOCH$_2$CONH (H H) CH=CH$_2$, N–CH(CO$_2$CH$_2$CCl$_3$)–C(Me)(OCH$_2$CH$_2$O) + PhOCH$_2$CONH (H H) Et, N–CH(CO$_2$CH$_2$CCl$_3$)–C(Me)(OCH$_2$CH$_2$O)

(1260)

Other interconversions reported to occur at the C-4 lactam site include the conversion of [3*S*-[3α(*S**),4β]]-1-(2,4-dimethoxybenzyl)-3-(1-*t*-butyldimethylsilyloxyethyl)-4-iodomethyl-2-azetidinone into its 4-cyanomethyl analogue upon reaction with potassium cyanide in dimethylformamide (equation 1261)[1053] and the dehalogenation[1072] of 4-halomethylazetidinones using tri-(*n*-butyl)tin hydride (equation 1262).

(t-Bu)Me$_2$SiO ... CH$_2$I ... CH$_2$C$_6$H$_3$(OMe)$_2$-2,4 —KCN, DMF, 50 °C stir 15 h→ (t-Bu)Me$_2$SiO ... CH$_2$CN ... CH$_2$C$_6$H$_3$(OMe)$_2$-2,4 93%

(1261)

CH$_2$X, R^2, SO$_2$R^1 —(n-Bu)$_3$SnH, THF, r.t.→ CH$_3$, R^2, SO$_2$R^1

(1262)

R^1	R^2	X	Yield (%)
p-Tol	H	Br	92
p-Tol	H	I	97
EtO	H	I	50[a]
Cl$_3$CCH$_2$O	H	I	100[b]
Cl$_3$CCH$_2$O	Me	I	99[b]

[a]Estimated yield only, product extremely unstable.
[b]During reaction, cleavage of one chlorine from the trichloroethoxy substituent occurred to give as the final product R^1 = OCH$_2$CHCl$_2$.

Two elimination reactions have been used to place a double bond at the carbon located at the C-4 position of β-lactams. In the first reaction[1443], the two active hydrogens of a C-4 methoxycarbonylmethyl substituent are sequentially replaced first by a methyl and then by a phenylseleno group. Oxidation of the resulting product with 30% hydrogen peroxide produces the enone product (equation 1263).

In the second reaction[1345], nucleophilic displacement of the C-4 acetoxy group of 3-(1-*t*-butyldimethylsilyloxyethyl)-4-acetoxy-2-azetidinone by the anion shown in equation (1264) produces the C-4 (α-phenylsulphoxy-α-acyl)ethyl substituent required. Refluxing this product in toluene causes elimination and production of the enone product. Further reaction of the enone product with methylene iodide under Simmons–Smith conditions produces a spirocyclopropane C-4 substituent (equation 1264).

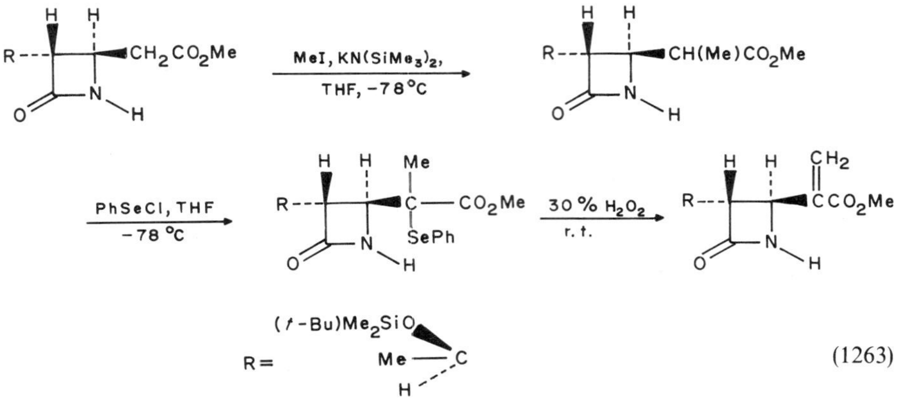

$$\text{R}= (t\text{-Bu})\text{Me}_2\text{SiO}(\text{Me})(\text{H})\text{C} \qquad (1263)$$

MeI, $KN(SiMe_3)_2$, THF, −78 °C; PhSeCl, THF, −78 °C; 30% H_2O_2, r.t.

(1263)

THF, −20 °C, stir; 75% (mixture of diastereoisomers); C_6H_5Me, reflux, 4 h; 80%; Zn−Cu, CH_2I_2, Et_2O; 60–70%

R = $(t\text{-Bu})Me_2SiO(Me)(H)C$

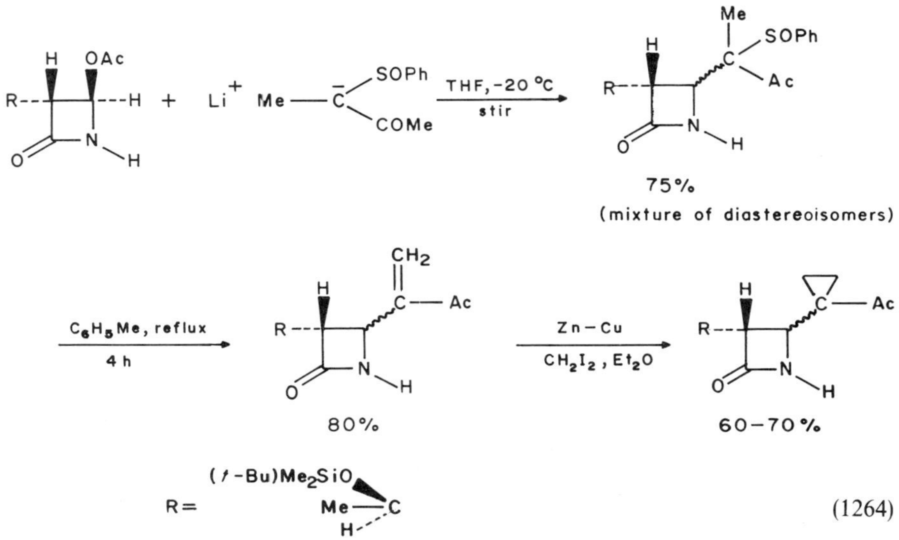

(1264)

4-Imidazolidinones containing a benzyloxycarbonyl protected amino group have been converted[1103] into their unprotected amino analogues by cleavage with hydrogen bromide in acetic acid (equation 1265).

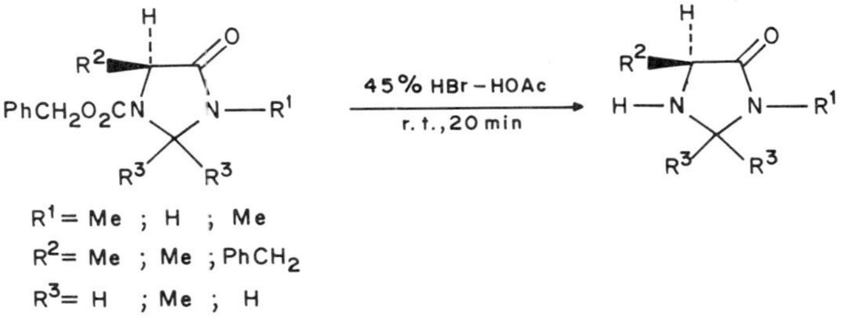

45% HBr − HOAc, r.t., 20 min

(1265)

R^1 = Me ; H ; Me
R^2 = Me ; Me ; $PhCH_2$
R^3 = H ; Me ; H

Interconversion of substituents at the C-4 position of β-lactams can be accomplished by ring opening and/or ring enlargement reactions of bicyclic lactams. Most of the bicyclic structures involved are members of the penam class of compounds and thus contain the β-lactam ring as one of the ring systems in the bicyclic structure. Almost all of the ring-opening reactions involve desulphurization of the five-membered ring of the penam system leaving a β-lactam with substituents located at the C-4 and/or the lactam nitrogen site. Examples[1164,1400] of the desulphurization ring-opening reactions are shown in equations 1266 and 1267.

Raney Ni, H_2 / Me_2CO, reflux 8 h (1266)[1164]

Raney Ni, H_2 / H_2O, heat (1267)[1400]

M = $\overset{+}{Na}$ or $\overset{+}{K}$

(267) (268)

R	Temp. (°C)	Time (min)	Yield (%) 267	Yield (%) 268
H	75	35	50[a]	50[a]
$PhCH_2CO$	175	15	40	60
$PhOCH_2CO$	175	15	36	64

[a]Separation was achieved after conversion to the *N*-carbobenzoxy derivatives.

A ring opening which does not involve desulphurization is the homolytic reductive debromination[1444] of 2β-bromomethyl penicillin and 3β-bromocepham using triphenylstannane under radical chain conditions. The products obtained are best explained by the mechanism shown in Scheme 21, while the conditions and reagents used and the specific products obtained are reported in Table 99.

The converse of the ring-opening reactions discussed above are ring-closure reactions which occur to form bicyclic ring systems by connecting positions C-3 and C-4 of a β-lactam ring. Depending upon the substituents located at the two positions involved,

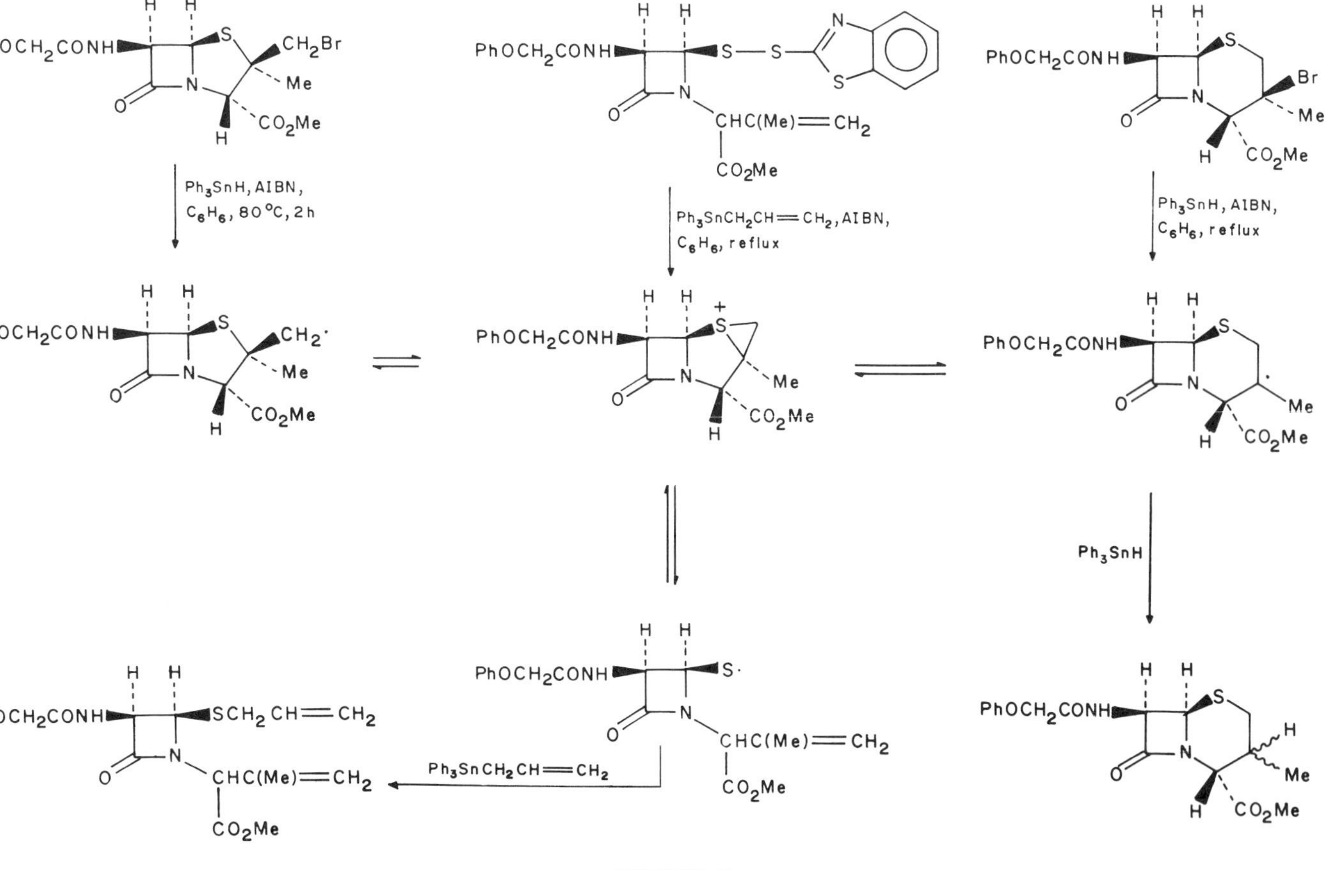
PhOCH2CONH
CH2Br
Me
CO2Me
Ph3SnH, AIBN, C6H6, 80°C, 2h
CH2·
S—S
CHC(Me)=CH2
Ph3SnCH2CH=CH2, AIBN, C6H6, reflux
Br
Ph3SnH, AIBN, C6H6, reflux
Ph3SnH
SCH2CH=CH2
Ph3SnCH2CH=CH2

SCHEME 21

the new ring system formed can contain none[1021] (equation 1268), one[750,1267] oxygen (equations 1269 and 1270) or two, nitrogen and sulphur[1434] (equation 1271) or nitrogen and oxygen[777,1368,1403] (equations 1272–1274), hetero atoms.

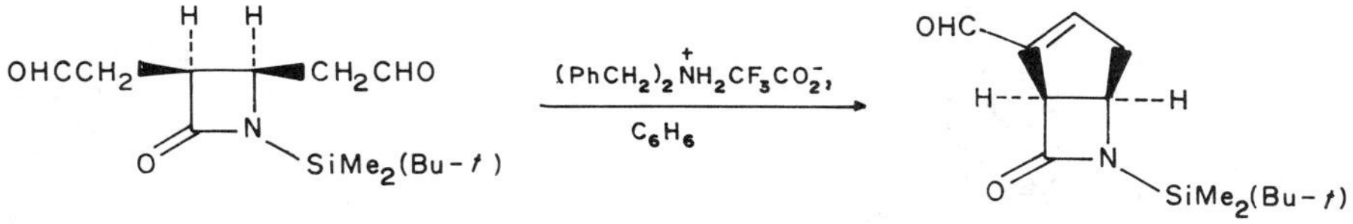

(1268)[1021]

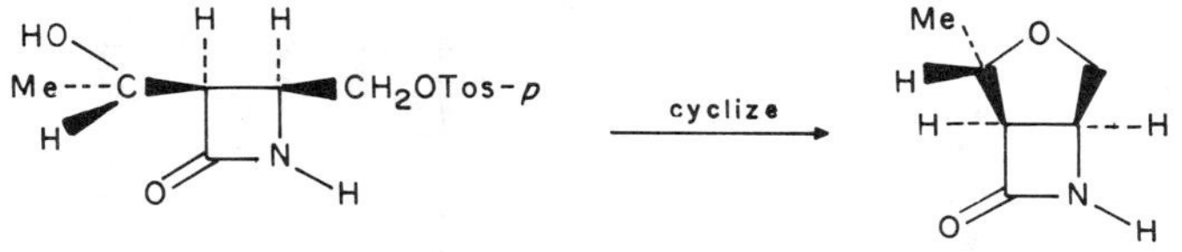

(1269)[1267]

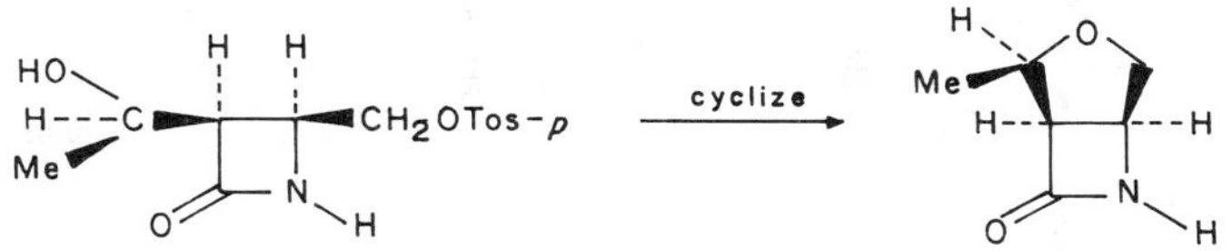

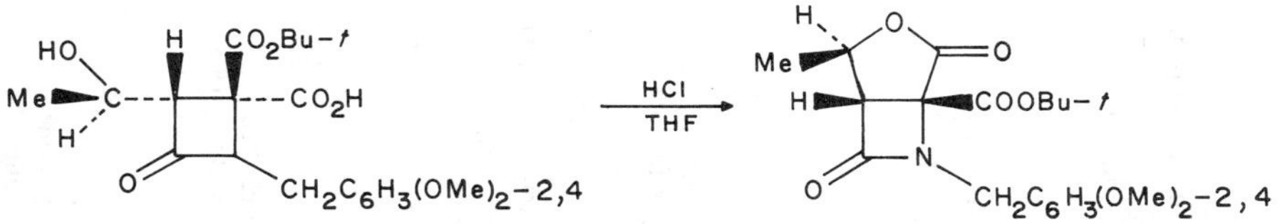

(1270)[750]

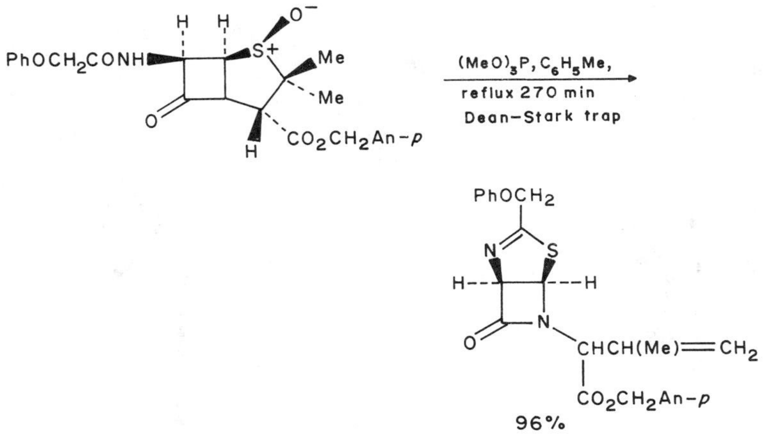

(1271)[1434]

TABLE 99. Results of reductive debromination of 2β-bromomethyl penicillin and 3β-bromocepham[1444]

Starting material	Conditions	Products	Yield (%)
PhOCH$_2$CONH (penicillin, CH_2Br, Me, CO_2Me)	Ph_3SnH, AIBN, C_6H_6, reflux 2h	PhOCH$_2$CONH (269) + PhOCH$_2$CONH (270)	40 + 35
	Ph_3SnH, benzoquinone, C_6H_6, reflux 2h	N.R.	
	$Ph_3SnCH_2CH{=}CH_2$ AIBN, C_6H_6, reflux 3h	PhOCH$_2$CONH, $SCH_2CH{=}CH_2$, $CHC(Me){=}CH_2$, CO_2Me (271)	94
PhOCH$_2$CONH (cepham, Br, Me, CO_2Me)	Ph_3SnH, AIBN, C_6H_6, reflux 2h	(269 + 270)	49 + 39
	Ph_3SnH, benzoquinone, C_6H_6, reflux 2h	N.R.	

Substrate	Conditions	Product	Yield (%)
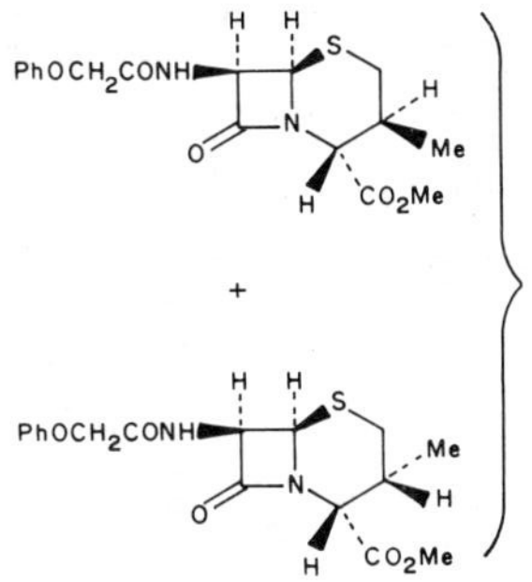 PhOCH$_2$CONH-azetidinone-thiazine (H, Me, CO$_2$Me) + PhOCH$_2$CONH-azetidinone-thiazine (Me, H, CO$_2$Me)	Ph$_3$SnCH$_2$CH=CH$_2$, AIBN, C$_6$H$_6$, reflux 3h	**271**	92
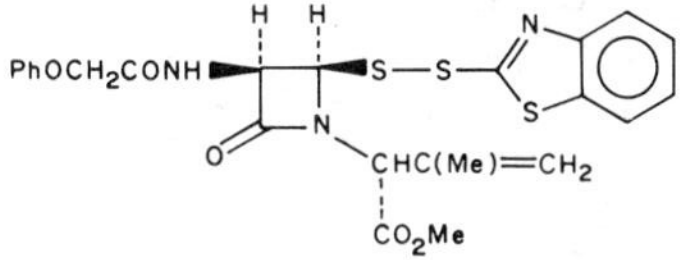 PhOCH$_2$CONH-azetidinone with S—S-benzothiazolyl, N-CHC(Me)=CH$_2$, CO$_2$Me	Ph$_3$SnH, AIBN, C$_6$H$_6$, reflux 2h	**269** + **270**	35 + 35
	Ph$_3$SnCH$_2$CH=CH$_2$, AIBN, C$_6$H$_6$, reflux 16h	**271**	85
	Ph$_3$SnCH$_2$CH=CH$_2$, benzoquinone or hydroquinone C$_6$H$_6$, reflux 16h	N.R.	

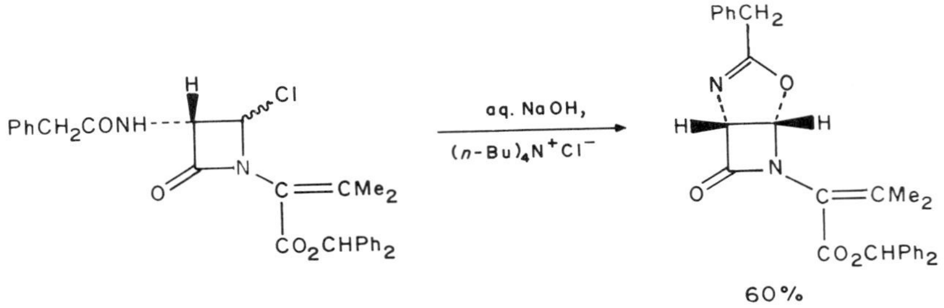

(1272)[1368]

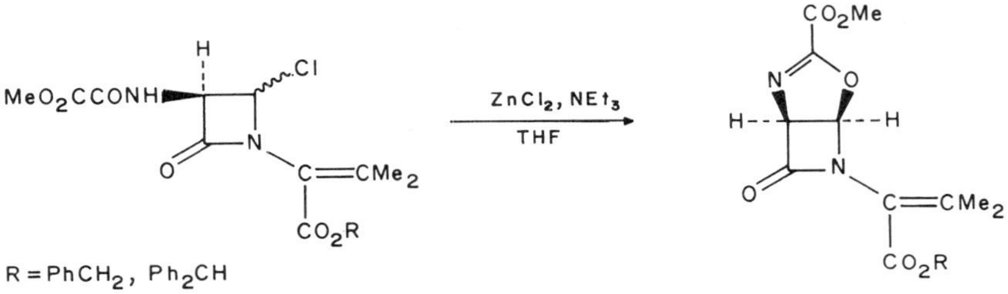

(1273)[1403]

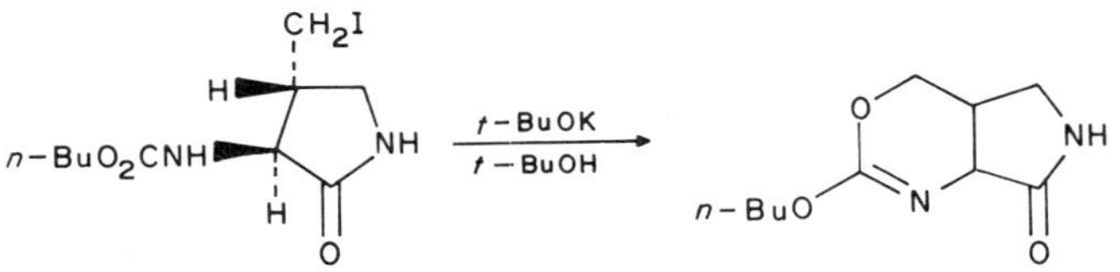

(1274)[777]

γ- and δ-Lactams containing an allyl or propargylsilane moiety at the C-3 position cyclize upon treatment with formic acid, trifluoroacetic acid or tin tetrachloride to give[1356] the azabicycles, 6-azabicyclo[3.2.1]octanes or 7-azabicyclo[4.2.1]nonanes by C-3 to C-5 ring closure, and 6-azabicyclo[4.2.2]decanes by C-3 to C-6 ring closure. All products obtained contain a vinyl or vinylidene substituent as shown in Table 100.

Even though many of the preparations of lactams are conducted using reagents that produce stereospecifically the *cis*-lactam isomers, because the *trans* isomers of most lactams are the more stable thermodynamically, it is relatively easy to isomerize most *cis*-lactam isomers into their *trans* forms, or at the very least produce a *cis,trans*-isomer mixture (equation 1275 and Table 101) where the substituent at the C-4 carbon changes direction.

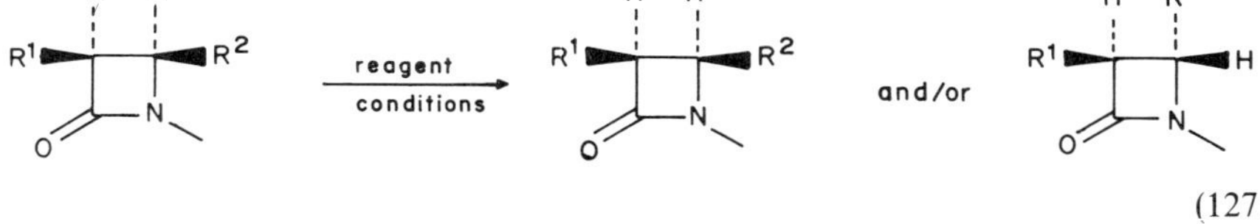

(1275)

TABLE 100. Ring closure of C-3 allyl or propargylsilane substituted γ- and δ-lactams[1356]

Lactam	Reagent and conditions	Product(s)	Yield (%)
$-C\equiv CCH_2SiMe_3$; O; N; OEt; CH_2Ph	CF_3COOH, 21 h	$PhCH_2$; N; CH_2; C; O	88
$SiMe_3$; O; N; OEt; CH_2Ph **(272)**	CF_3COOH, 21 h	$PhCH_2$; N; $CH=CH_2$; H; O **(273)** +	73
		$PhCH_2$; N; H; $CH=CH_2$; O **(274)** +	4
		$CH=CH_2$; O; N; CH_2Ph **(275)**	10
272	HCO_2H, 21 h	**273** + **274** + **275**	82 + 15 + < 2
CH_2SiMe_3; O; N; OEt; CH_2Ph	CF_3CO_2H, 105 h	$PhCH_2$; N; H; $CH=CH_2$; O +	67
		$PhCH_2$; N; $CH=CH_2$; H; O +	6
		$-CH=CH_2$; O; N; CH_2Ph	12

(continued)

TABLE 100. (*continued*)

Lactam	Reagent and conditions	Product(s)	Yield (%)
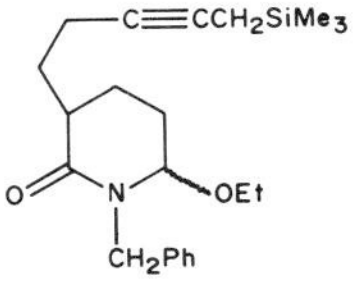	HCO_2H	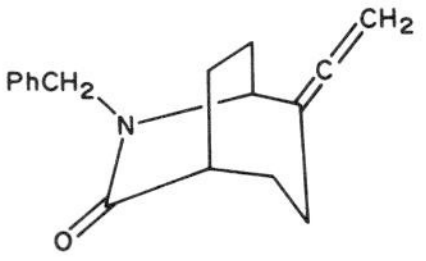	93
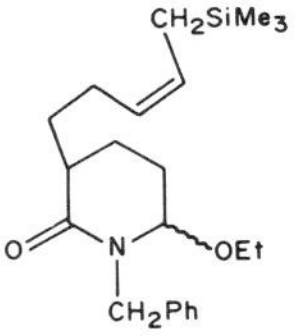	HCO_2H	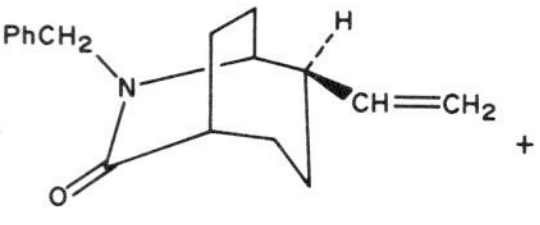 +	49
		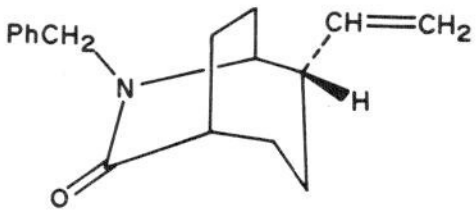	36
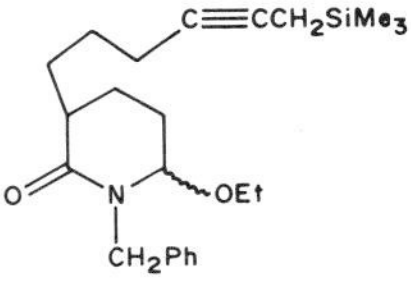	HCO_2H	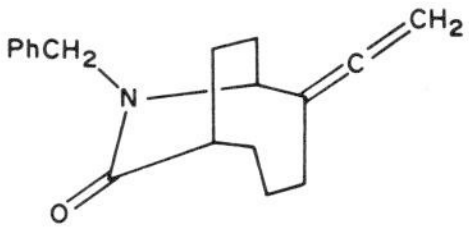	<10
	$SnCl_4$, CH_2Cl_2		81

*F. By Oxidation Reactions

*2. *Using chromium oxides*

Jones' oxidation of 3-acetamido-3-deoxy-1,2-*O*-isopropylidene-α-D-glucofuranose in water-saturated butanone produces[813] the tricyclic 3-acetamido-3-deoxy-1,2-*O*-isopropylidene-α-D-glucofuranurono-6,3-lactam (equation 1276). Further treatment of this product with manganese dioxide in acetone affords[813] 3-acetamido-3-deoxy-1,2-*O*-isopropylidene-α-D-xylo-5-hexulofuranurono-6,3-lactam, a product which can also be obtained directly from 3-acetamido-3-deoxy-1,2-*O*-isopropylidene-α-D-glucofuranose upon reaction with pyridinium dichromate in dimethylformamide (equation 1276).

*4. *Using ruthenium oxides*

Oxidation of *N*-acyl substituted 7- and 8-membered cyclic amines by ruthenium tetroxide affords[1445] 22–94% yields of the corresponding *N*-acyl substituted lactams (equation 1277). However, oxidation of *N*-benzyl cyclic amines of the same size and with the same reagent gives[1445] a mixture of *N*-benzyl and *N*-benzoyl substituted lactams (equation 1278).

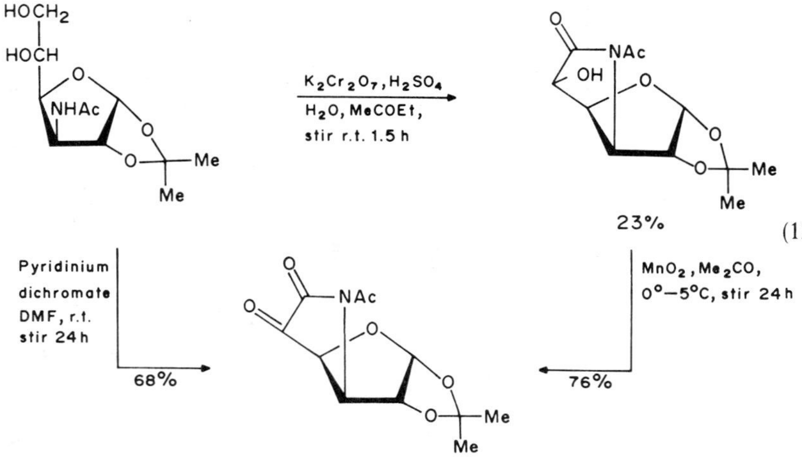

$$\xrightarrow[\text{stir r.t. 1.5 h}]{K_2Cr_2O_7, H_2SO_4 \atop H_2O, MeCOEt}$$ 23%

Pyridinium dichromate, DMF, r.t., stir 24 h, 68%; MnO_2, Me_2CO, 0°–5°C, stir 24 h, 76% (1276)

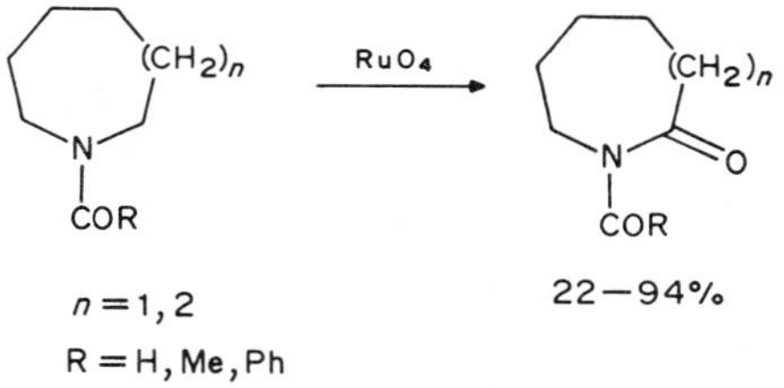

$\xrightarrow{RuO_4}$ 22–94% (1277)

$n = 1, 2$

$R = H, Me, Ph$

$\xrightarrow{RuO_4}$ (1278)

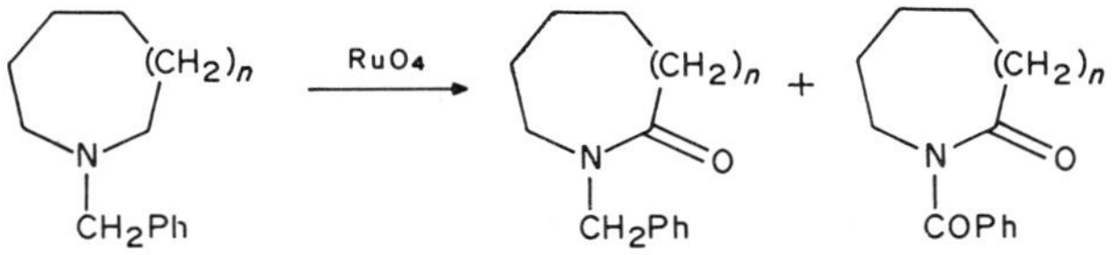

*5. *Via sensitized and unsensitized photooxidation*

Conversion of the esters of azetidinecarboxylic acids to their monoanions by reaction with lithium diisopropylamide in tetrahydrofuran followed by treatment of the resulting anion with *t*-butyldimethylchlorosilane, and then subjecting the protected enamino *O*-silylketene acetal to dye-sensitized photooxygenation, produces[1446,1447] β-lactams (equation 1279). This same procedure was found to be ineffective when applied to 5- and 6-membered ring analogues[1446].

Irradiation of an oxygen-saturated solution of *N*-formyl- or *N*-acylpyrrolidines (equation 1280) or piperidines (equation 1281) in benzene or *t*-butyl alcohol containing benzophenone produces[1378] the corresponding 5- or 6-membered N-substituted lactams. The mechanism of this conversion appears to involve the regioselective abstraction of the proton alpha to the nitrogen atom by the benzophenone triplet to give radicals which are then oxidized.

TABLE 101. Isomerization of lactams

Lactam	Reagent and conditions	Product(s)	Yield (%) (ratio)	Reference
PhO, Ac, N–CH_2CO_2Me β-lactam	DBU, r.t. 60 min	PhO, Ac, N–CH_2CO_2Me β-lactam + PhO, ---Ac, N–CH_2CO_2Me β-lactam	— (65:35)	1146
	DBU, r.t. 60 min		— (50:50)	1146
PI—N--, ---OAc, N–H β-lactam	Lewis acid, C_6H_6, reflux	PI—N--, OAc, N–H β-lactam	—	1387
PI—N, Ac, N–H β-lactam	DBU, CH_2Cl_2, r.t. 24 h	PI—N, Ac, N–H β-lactam + PI—N, ---Ac, N–H β-lactam	— (10:90)	1146
PI—N, Ac, N–CH_2CO_2Me β-lactam	DBU, r.t. 24 h	PI—N, Ac, N–CH_2CO_2Me β-lactam + PI—N, ---Ac, N–CH_2CO_2Me β-lactam	— (10:90)	1146

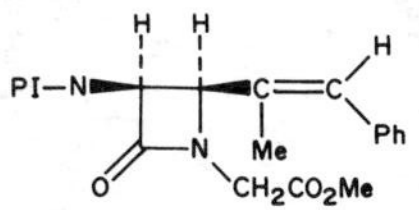 PI–N, Me, CH_2CO_2Me, Ph	DBU, r.t.	N.R.	—	1146
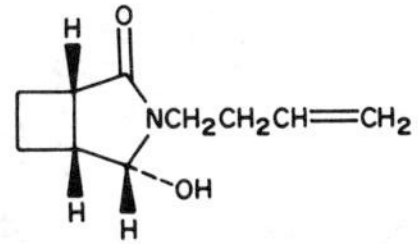 $NCH_2CH_2CH{=}CH_2$, OH	NaOEt, EtOH, r.t. stir 3 days	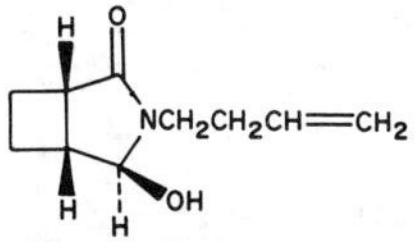 $NCH_2CH_2CH{=}CH_2$, OH	—	1412
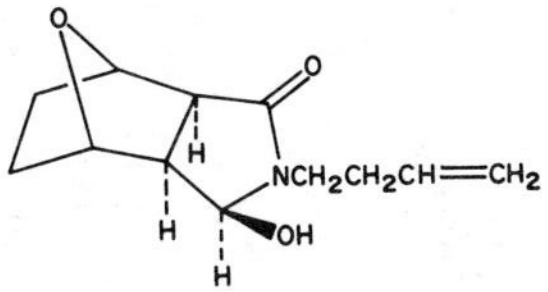 $NCH_2CH_2CH{=}CH_2$, OH	NaOEt, EtOH, r.t. stir 3 days	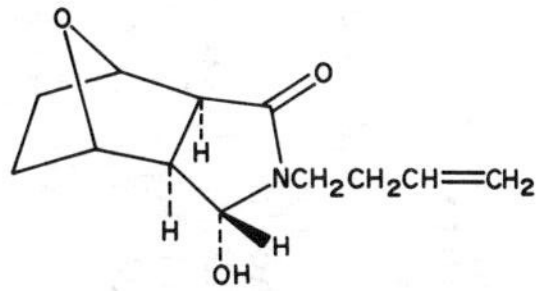 $NCH_2CH_2CH{=}CH_2$, OH	60	1412

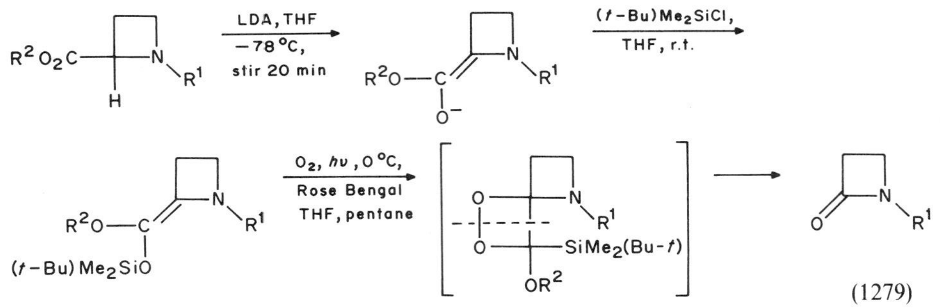

(1279)

R^1 =	2,4,6-$(MeO)_3C_6H_2CH_2$	$CH_2{=}CHCH_2$	p-$AnCH_2CH_2$	c-Hex	$(MeO)_2CHCH_2$	Ph_2CH
R^2 =	t-Bu	t-Bu	t-Bu	Me	Et	Et
Yield (%) =	56	48	66	46–50	56	55

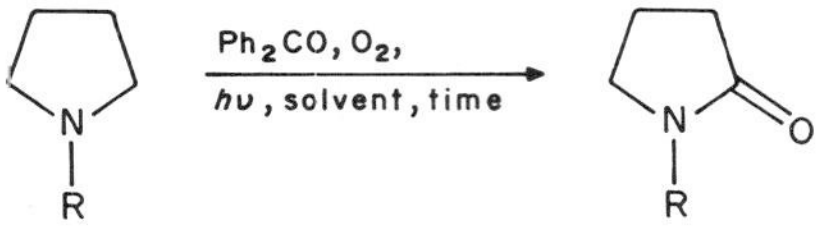

(1280)

R =	CHO	Ac	PhCO
Solvent =	C_6H_6	C_6H_6	t-BuOH
Time (h) =	45	18	25
Yield (%) =	18	28	60

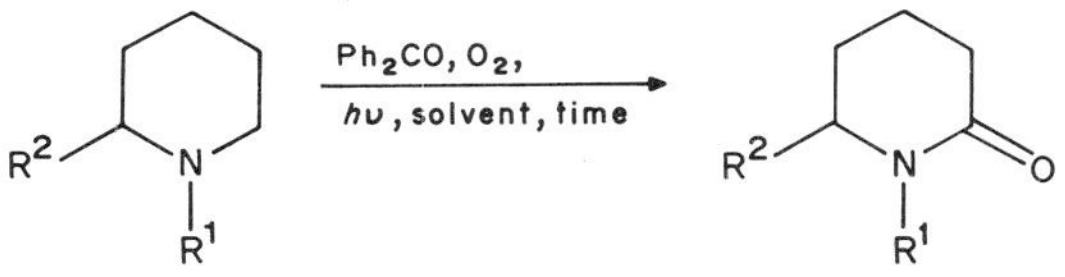

(1281)

R^1 =	CHO	CHO	Ac
R^2 =	H	Me	H
Solvent =	C_6H_6	C_6H_6	t-BuOH
Time (h) =	45	48	51
Yield (%) =	22	13	25

N-Acylpyrrolidines undergo photocatalytic oxidation upon irradiation in the presence of an aqueous suspension of titanium dioxide affording[1448] *N*-acyl-γ-lactams (equation 1282).

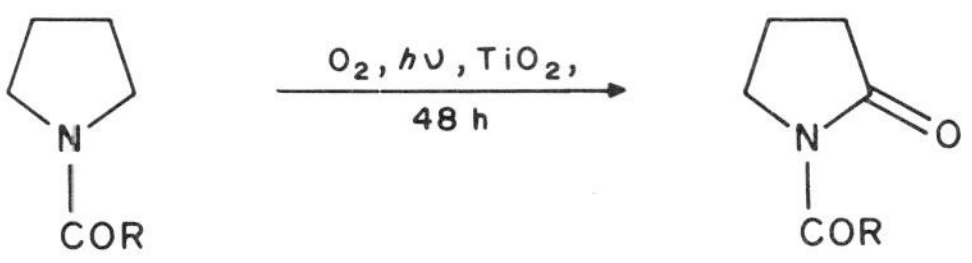

(1282)

R = Me, Et

α-Substituted β-lactams have been prepared[1405] by the unsensitized photoreaction of chromium(0) carbene complexes with imines, oxazines and thiazines (equation 1283 and Table 102).

$(CO)_5Cr{=}CHR$ + [cyclic imine] $\xrightarrow[h\nu]{MeCN,\ Ar,}$ [β-lactam] (1283)

*7. *Using miscellaneous reagents*

When *N*-phenylcarbamoylpyrrolidines, *N*-benzoylpyrrolidines and *N*-phenylcarbamoylpiperidines are oxidized with a mixture of iron(II) perchlorate, hydrogen peroxide and acetic acid in acetonitrile, the corresponding pyrrolidin- and piperidin-2-ones are obtained[1449] (equation 1284). The results from these reactions indicated that: (a) the methylene group at the α-position to the ring nitrogen rather than the methine group is preferentially attacked by the oxidant; (b) the derivatives of pyrrolidine are more reactive than those of piperidine; and (c) the presence of an α-substituent decreases the substrate reactivity.

$$\xrightarrow[Fe(ClO_4)_2\cdot 6H_2O,\ HOAc,\ N_2,\ stir\ 50\ min]{10\%\ H_2O_2,\ MeCN,} \qquad (1284)$$

n	R^1	R^2	Yield (%)
0	H	NHPh	75
0	H	Ph	40
0	CO_2Me	Ph	61
1	H	NHPh	56
1	Me	NHPh	52
1	Ph	NHPh	61

Similar reactions are observed[1449] when the same substrates are treated with molecular oxygen in the presence of other iron complexes (equation 1285).

$$\xrightarrow[O_2,\ Fe-complex,\ 20\ min]{aq.\ 90\%\ C_5H_5N,\ 20\ ^\circ C} \qquad (1285)$$

n	R^1	R^2	Iron complex[a]	Yield (%)
0	H	NHPh	A	34
0	H	Ph	A	37
0	CO_2Me	Ph	A	30
1	H	NHPh	A	34
1	H	NHPh	A[b]	29
1	H	NHPh	B	29
1	Me	NHPh	A	49
1	Me	NHPh	B	59
1	Ph	NHPh	A	42
1	Ph	NHPh	B	18

[a]Iron complex: A, $[Fe(II)Fe(III)_2O(OAc)_6(C_5H_5N)_3]$ generated from a mixture of iron powder, sodium sulphide and acetic acid in aqueous pyridine. B, $[Fe(salen)_2O]$ in the presence of sodium sulphide and acetic acid.
[b]The complex used was previously isolated.

TABLE 102. Photolytic reactions of chromium carbene complexes[1405]

Substrate	Chromium complex	Conditions	Product	Yield (%)
	$(CO)_5Cr{=}CHNMe_2$	MeCN, Ar, *hv*, 20h	+	30–40
				12–0
	$(CO)_5Cr{=}CHN(CH_2Ph)_2$	MeCN, Ar, *hv*, 7h	+	41
				9
	$(CO)_5Cr{=}C(OMe)Me$	MeCN, Ar, *hv*, 8h		70
	$(CO)_5Cr{=}C(OMe)Me$	MeCN, Ar, *hv*, 8h		63
	$(CO)_5Cr{=}CHN(CH_2Ph)_2$	MeCN, *hv*, 1 day		51

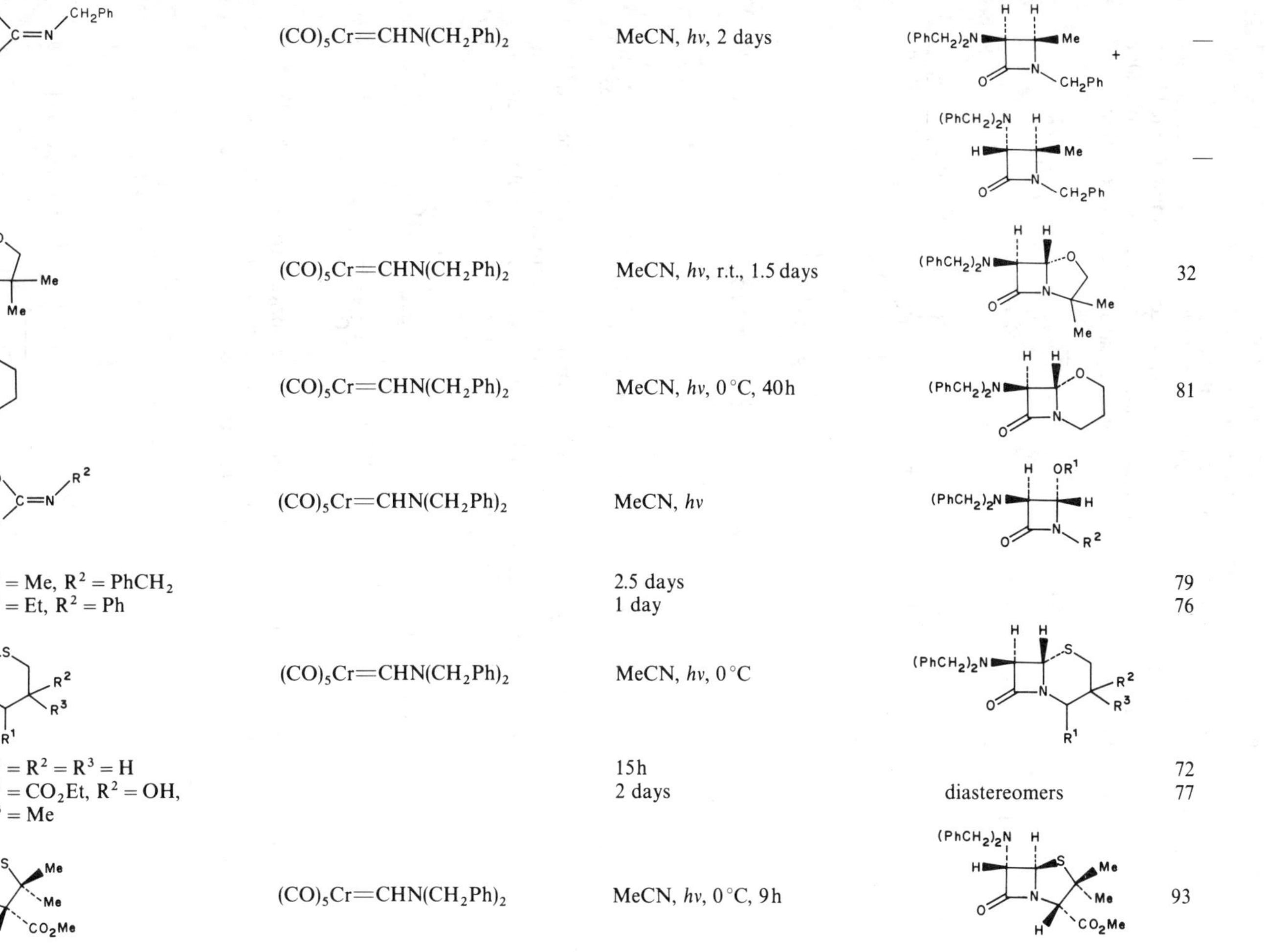

Substrate	Reagent	Conditions	Product	Yield (%)
	$(CO)_5Cr{=}CHN(CH_2Ph)_2$	MeCN, *hν*, 2 days	+	—
				—
	$(CO)_5Cr{=}CHN(CH_2Ph)_2$	MeCN, *hν*, r.t., 1.5 days		32
	$(CO)_5Cr{=}CHN(CH_2Ph)_2$	MeCN, *hν*, 0 °C, 40h		81
	$(CO)_5Cr{=}CHN(CH_2Ph)_2$	MeCN, *hν*		
$R^1 = Me$, $R^2 = PhCH_2$		2.5 days		79
$R^1 = Et$, $R^2 = Ph$		1 day		76
	$(CO)_5Cr{=}CHN(CH_2Ph)_2$	MeCN, *hν*, 0 °C		
$R^1 = R^2 = R^3 = H$		15h		72
$R^1 = CO_2Et$, $R^2 = OH$, $R^3 = Me$		2 days	diastereomers	77
	$(CO)_5Cr{=}CHN(CH_2Ph)_2$	MeCN, *hν*, 0 °C, 9h		93

The mechanism proposed[1449] for the iron complex catalysed oxidation discussed above involves initial one-electron transfer from the substrate to the active species generated in the reaction medium followed by oxidation.

Molecular oxygen has also been used to produce[1447,1450] β-lactams from azetidinecarboxylic acids by oxidative decarboxylation. In this approach the dianion formed from the reaction of the acid with lithium diisopropylamide is oxygenated with molecular oxygen followed by decomposition of the α-hydroperoxycarboxylic acid (equation 1286).

1. $Ba(OH)_2$, H_2O, reflux 1 h; 2. 0 °C, CO_2; LDA, THF, 0 °C, stir 1 h; O_2, ether, −78 °C; 1. THF, −78 °C, p-$MeC_6H_4SO_3H$; 2. 25 °C; + CO_2 + H_2O (1286)

R =	t-Bu	c-Hex	$CH_2CH(OMe)_2$	$Me(CH_2)_4$	c-C_8H_{15}	$Me_2N(CH_2)_3$	p-$AnCH_2CH_2$	$PhCH_2CH_2$
Yield (%)	60	47	50	61	52	45	47	55

Ozonolysis of the Wittig reaction products of penicillins[1451,1452], clavulanic acid[1451] and the 1,3-diaza analogue[1335] of these bicyclic ring systems converts them into their corresponding β-lactam ring containing derivatives. The general reaction illustrating this conversion is shown in equation 1287 while the specific details are reported in Table 103.

O_3, solvent, conditions (1287)

Treatment of azetidine carboxylic acids with oxalyl chloride produces[1039,1447,1453] an iminium salt which upon reaction with m-chloroperbenzoic acid in the presence of pyridine yields the corresponding β-lactam (equation 1288).

HOOC-azetidine + $(ClCO)_2$ — 1. 0 °C; 2. 45 °C, 10 min; 3. Et_2O, 70% $HClO_4$ → iminium ClO_4^-; m-$ClC_6H_4CO_3H$, CH_2Cl_2, C_5H_5N, 0 °C, stir 40 min → [m-$ClC_6H_4C(O)$—O—O— … C_5H_5N] → β-lactam

(1288)

TABLE 103. Ozonolysis of Wittig products of penicillins, clavulanic acid and 1,3-diaza analogues

Substrate	Solvent and conditions	Product	Yield (%)	Reference
$PhOCH_2CONH$, H, H, S, Me, Me, MeO_2CCH, N, H, CO_2CH_2Ph	1. EtOAc, $-70\,^{\circ}C$, O_3, 5 min 2. Ph_3P, r.t. 1 h	$PhOCH_2CONH$, H, H, S, Me, Me, O, N, H, CO_2CH_2Ph		
(*Z*)-isomer			95	1451, 1452
(E)-siomer			69	1451, 1452
$PhOCH_2CONH$, MeO, H, S, Me, Me, MeO_2C—C, H, N, H, CO_2CH_2Ph (*E*)	1. EtOAc, $-70\,^{\circ}C$, O_3 2. Ph_3P, r.t. 1 h	$PhOCH_2CONH$, MeO, H, S, Me, Me, O, N, H, CO_2CH_2Ph	33	1452
2-ThiCHCONH, $PhCH_2O_2C$, H, H, S, Me, Me, MeO_2CCH, N, H, CO_2CH_2Ph (*Z*+*E*)	1. EtOAc, $-70\,^{\circ}C$, O_3, Ar, 30 min 2. Ph_3P, r.t. stir 1 h	2-ThiCHCONH, $PhCH_2O_2C$, H, H, S, Me, Me, O, N, H, CO_2CH_2Ph	49	1452
H, O, CH_2OH, C, H, H—C, CO_2Me, N, H, CO_2CH_2Ph	O_3, EtOAc	H, O, O, O, N, H, CO_2CH_2Ph	—	1451
H, H, C—CO_2Me, O, N, NH, H, CO_2CH_2Ph	O_3, CH_2Cl_2, MeOH	H, O, O, N, NH, H, CO_2CH_2Ph	87	1335

R	Yield (%)	Reference
t-Bu	77	1447, 1453
c-Hex	80	1447, 1453
$PhCH_2$	80	1447, 1453
$PhCH_2CH_2$	71	1447, 1453
p-$AnCH_2CH_2$	77	1447, 1453
$EtO_2CC(Me)_2CHCO_2Et$	71	1447
$EtO_2CCHCHCHMe_2$ (with CO_2Et)	61	1447
$EtO_2CCHCH(An$-p$)CO_2Et$	61	1447
$EtO_2CCHCHC_6H_4OCH_2Ph$-p (with CO_2Et)	53	1039, 1447

m-Chloroperbenzoic acid has also been used to produce[1090] medium and macrocyclic N-acetylketolactams by oxidation of the precursor bicyclic N-acetylenamines (equation 1289).

$(H_2C)_n$ bicyclic N-COMe enamine $\xrightarrow[CH_2Cl_2,\ heat]{m\text{-}ClC_6H_4CO_3H}$ $(H_2C)_n$ N-COMe ketolactam (1289)

n = 4 ; 5 ; 6 ; 10

Yield (%) = 40 ; 32 ; 36 ; 42

Using bicyclic N-arylenamines and sodium periodate produces[1454] similar medium-sized cyclic N-arylketolactam products (equation 1290), while sodium periodate oxidation of 2-aryl-4,5,6,7-tetrahydroindoles and related compounds effects[1455] regiospecific cleavage of the central bond of the bicyclic pyrrole ring to afford 7- to 11-membered cyclic ketolactams (equation 1291).

$(H_2C)_n$ bicyclic N-R enamine $\xrightarrow[aq.\ MeOH,\ THF]{NaIO_4,}$ $(H_2C)_n$ N-R ketolactam (1290)

n = 1–3 ; 1

R = Ph ; p-Tol

53 – 66 %

X–C_6H_4–pyrrole(N-R)–$(CH_2)_n$ $\xrightarrow[H_2O,\ -5\ ^\circ C,\ stir\ 24\ h]{NaIO_4,\ MeOH}$ X–C_6H_4–ketolactam(N-R)–$(CH_2)_n$ (1291)

R	X	n	Yield (%)
H	H	0	67
H	H	2	70
Me	H	2	70
H	MeO	2	66
H	Me	2	75
H	Cl	2	72
H	Br	2	72
H	H	3	59
H	H	4	37

Reaction of chromium carbene complex[1231] or cationic iron vinylidenes[1456] with imines produces azetidine vinylidene chromium (equation 1292) or iron complexes (equations 1293–1295), which upon oxidation with iodosobenzene afford the corresponding 2-azetidinones.

$$(OC)_5Cr{=}C(CHR^1)O^- \ \overset{+}{N}Me_4 + R^2CH{=}NMe \ (R = aryl) \longrightarrow$$

$$\text{[azetidine: } (CO)_5Cr{=}C,\ {=}C{=}CHR^1,\ R^2,\ N\text{-Me]}\ (6-25\%) \xrightarrow[EtOH]{C_6H_5IO} \text{[3-(R}^1\text{CH=)-4-R}^2\text{-1-methyl-2-azetidinone]}\ (68-100\%) \qquad (1292)$$

$$[(R_3P)(Cp)(OC)\overset{+}{Fe}{=}C{=}CHPh]\ X^- + PhCH{=}NMe \xrightarrow[r.t.\ 24\ h]{CH_2Cl_2,}$$

R = MeO, Ph

X = BF_4, CF_3SO_3

$$\text{[azetidine: } (R_3P)(Cp)(OC)\overset{+}{Fe}{=}C,\ {=}C{=}CHPh,\ Ph,\ N\text{-Me]}\ X^- \xrightarrow[EtOH]{C_6H_5IO} \text{[3-(PhCH=)-4-Ph-1-methyl-2-azetidinone]}\ (68-100\%) \qquad (1293)$$

$$[(R_3P)(Cp)(OC)\overset{+}{Fe}{=}C{=}CMe_2]\ X^- + PhCH{=}NMe \xrightarrow[2.\ ClCH_2CH_2Cl,\ Amberlyst\ A\ 21,\ reflux\ 3\ h]{1.\ CH_2Cl_2,\ -78\ ^{\circ}C}$$

R = MeO, Ph

X = BF_4, CF_3SO_3

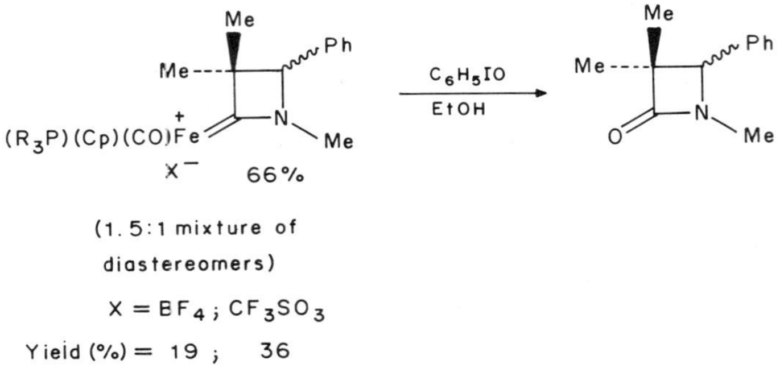

$$\text{(R}_3\text{P)(Cp)(CO)}\overset{+}{\text{Fe}}\text{=azetidinylidene (Me, Me, Ph, N–Me) X}^- \xrightarrow[\text{EtOH}]{C_6H_5IO} \text{3,3-dimethyl-4-phenyl-1-methylazetidin-2-one}$$ (1294)

66%

(1.5:1 mixture of diastereomers)

X = BF_4 ; CF_3SO_3

Yield (%) = 19 ; 36

$$[[(MeO)_3P](Cp)(OC)\overset{+}{Fe}{=}C{=}CMe_2]\ CF_3SO_3^- + \text{ethyl thiazoline-4-carboxylate} \xrightarrow[\text{reflux 3 h}]{ClCH_2CH_2Cl}$$

$$[(MeO)_3P](Cp)(OC)\overset{+}{Fe}\text{=bicyclic carbene (Me, Me, H, S, N, H, }CO_2Et\text{) } CF_3SO_3^- \xrightarrow[\text{EtOH}]{C_6H_5IO} \text{bicyclic }\beta\text{-lactam (Me, Me, H, S, N, H, }CO_2Et\text{)}$$

52%

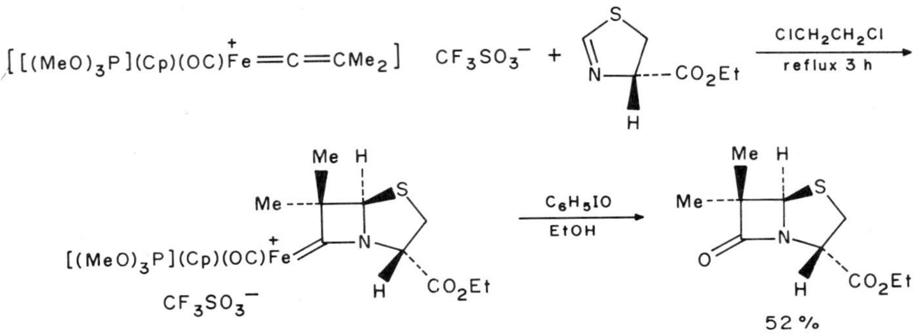

(1295)

Other iron complexes which, upon oxidation with ceric ammonium nitrate (CAN) or cupric chloride, lead to β-lactams are reported in Table 104.

By using acetates such as lead tetraacetate and mecuric acetate, *N*-hydroxyazetidines (equation 1296) and substituted piperidines (equation 1297) and pyridinium salts (equation 1298) can be converted into the corresponding β- or δ-lactams. Table 105 reports the details of the general reactions shown in equations (1296–1298).

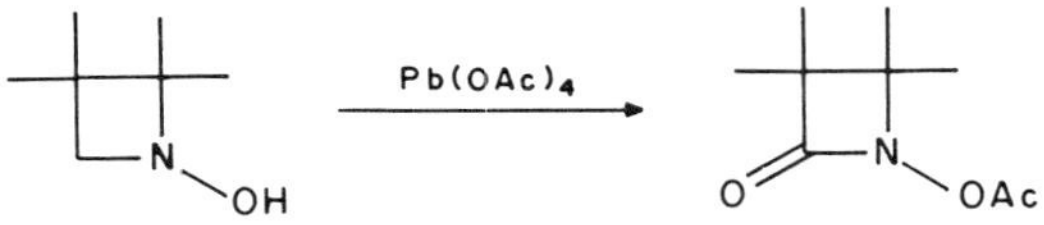

$$\text{N-hydroxyazetidine (N–OH)} \xrightarrow{Pb(OAc)_4} \text{N-acetoxyazetidin-2-one (N–OAc)}$$ (1296)

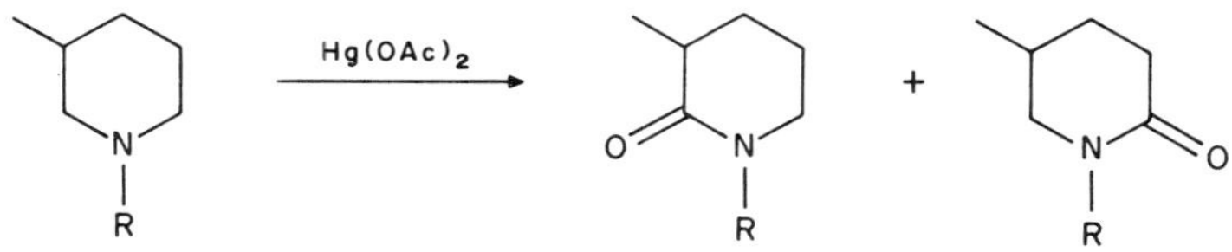

$$\text{3-methyl-N-R-piperidine} \xrightarrow{Hg(OAc)_2} \text{3-methyl-N-R-piperidin-2-one} + \text{5-methyl-N-R-piperidin-2-one}$$ (1297)

$$\text{3-methyl-N-R-pyridinium} \xrightarrow{Hg(OAc)_2} \text{3-methyl-N-R-pyridin-2-one} + \text{5-methyl-N-R-pyridin-2-one}$$ (1298)

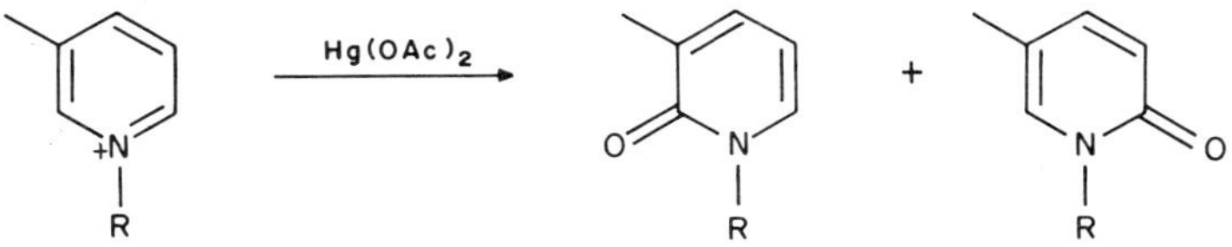

TABLE 104. Preparation of lactams by oxidation of iron complexes

Iron complex	Oxidizing agent and conditions	Product	Yield (%)	Reference
OC—Fe(Cp)(Ph$_3$P)—COCH$_2$C(H)(NHPh)(Ph)	CAN or $CuCl_2$	O, N—Ph, H, Ph	—	1457
(CO)$_3$Fe, Me, H_2C, O, NCH$_2$Ph, MeO, OMe	1. CAN, MeOH, −30 °C 2. −30 °C to r.t., 1 h	H_2C=C(Me), H, H, CH$_2$CH(OMe)$_2$, O, N, CH$_2$Ph +	64	1389
		(MeO)$_2$CHCH$_2$, N, CH$_2$Ph, Me, O	24	
O, C—Fe(CO)$_3$, Me, Me—CH(Ph)—N, CH$_2$, CH$_2$, CH(OMe)$_2$	1. CAN, MeOH, −30 °C 2. r.t., 2.5h	Me, O, N, CH—Ph, Me, H, CH$_2$, CH(OMe)$_2$	12	1389
(CO)$_3$Fe—C(=O), Me, Me, N—CH(Ph)(Me), H_2C, CH(OMe)$_2$	1. CAN, MeOH, −30 °C 2. r.t., 2.5h	H_2C=C(Me), H, H, CH$_2$CH(OMe)$_2$, O, N, CH—Ph, Me	88	1389

(*continued*)

TABLE 104. (*continued*)

Iron complex	Oxidizing agent and conditions	Product	Yield (%)	Reference
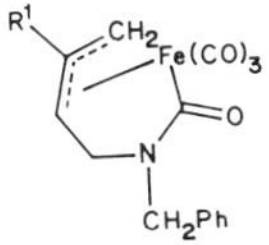 R^1, CH_2, $Fe(CO)_3$, O, N, CH_2Ph	CAN, EtOH, 1h	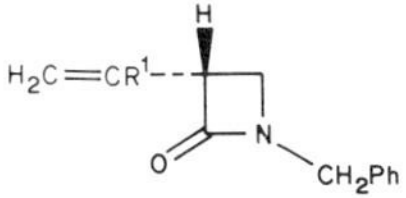 $H_2C{=}CR^1$, H, O, N, CH_2Ph		
$R^1 = H$	−30 °C to r.t.		72–75	1388, 1458
$R^1 = Me$	−30 °C to r.t.		88	1388
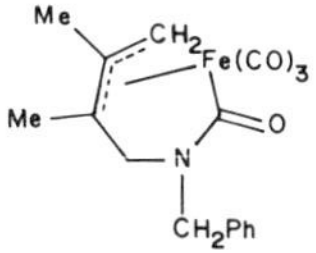 Me, Me, CH_2, $Fe(CO)_3$, O, N, CH_2Ph	CAN, EtOH, 1h 0 °C to r.t.	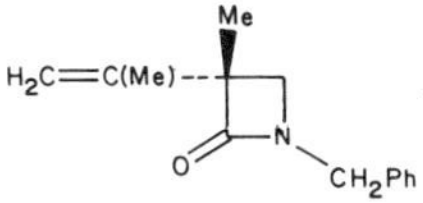 $H_2C{=}C(Me)$, Me, O, N, CH_2Ph +	34	1388, 1458
		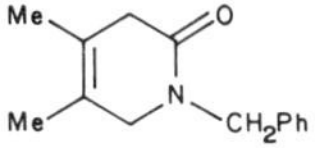 Me, Me, O, N, CH_2Ph	54–56	
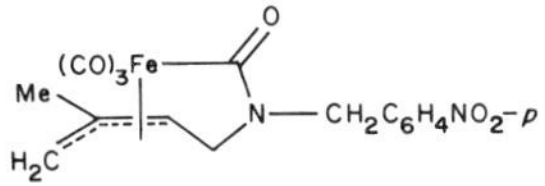 $(CO)_3Fe$, Me, H_2C, O, N—$CH_2C_6H_4NO_2$-*p*	CAN, EtOH	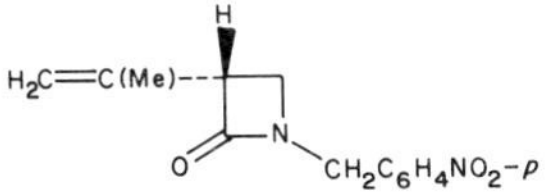 $H_2C{=}C(Me)$, H, O, N, $CH_2C_6H_4NO_2$-*p*	70	1388
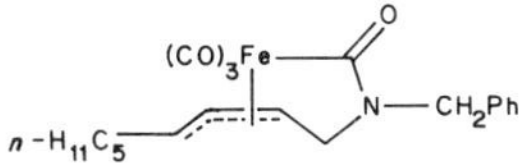 $(CO)_3Fe$, *n*-$H_{11}C_5$, O, N—CH_2Ph	CAN, EtOH, −5 °C, 1h	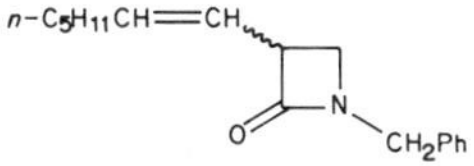 *n*-$C_5H_{11}CH{=}CH$, O, N, CH_2Ph 1:1 mixture	64	1388, 1458

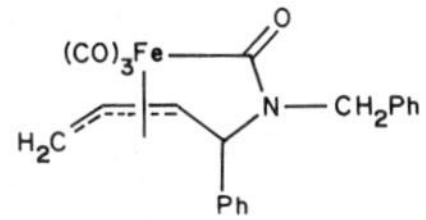	CAN, EtOH, −30 °C to r.t.	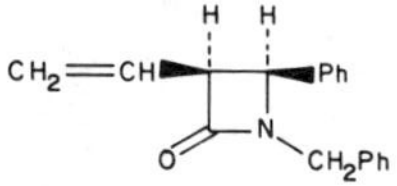	78	1388
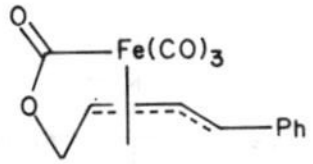	1. Et_2AlCl, $PhCH_2NH_2$, 0.5h 2. CAN, EtOH, −30 °C to r.t.	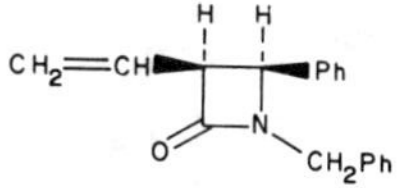	62	1388
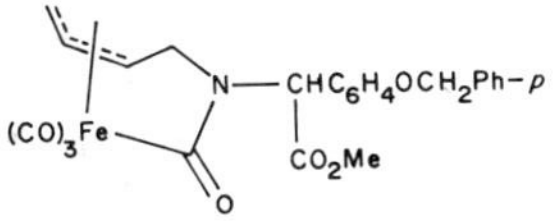	CAN, EtOH	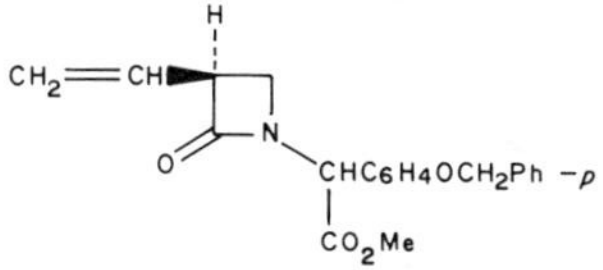	80	1458
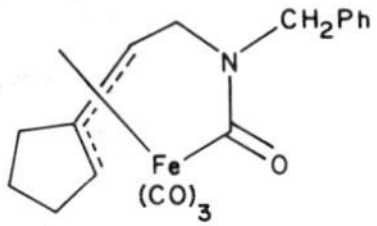	CAN, EtOH, r.t. 1h	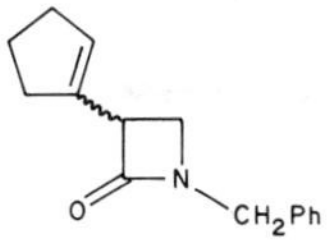	84–88	1388, 1458
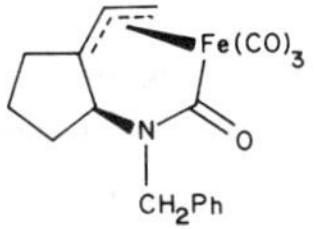	CAN, EtOH, −30 °C to r.t. 1h	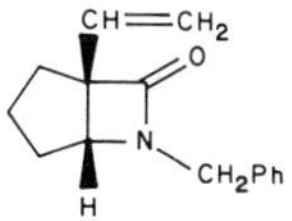	75	1388, 1458

TABLE 105. Acetate oxidation of azetidines, piperidines and pyridinium salts

Substrate	Reagent and conditions	Product	Yield (%)	Reference
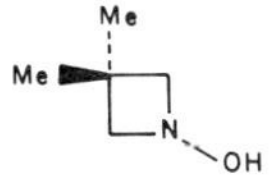	$Pb(OAc)_4$, C_6H_5Me, 0 °C, N_2, stir 30 min	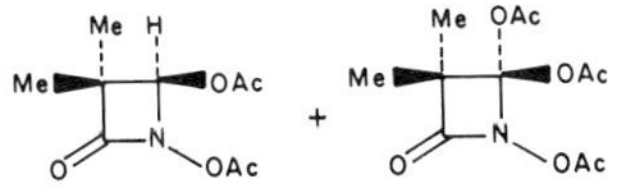 + (3:2)	59	1048
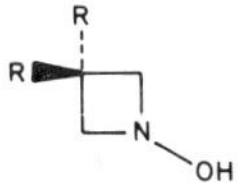	$Pb(OAc)_4$, C_6H_5Me, 0 °C, N_2, stir 30 min	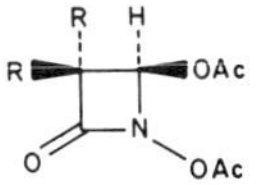		1048
R = Et			46	
R = *n*-Bu			42	
R = Ph			50	
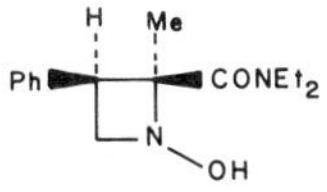	$Pb(OAc)_4$, C_6H_6, 6 °C,	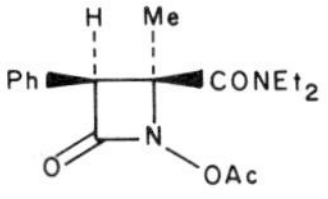	44	1457
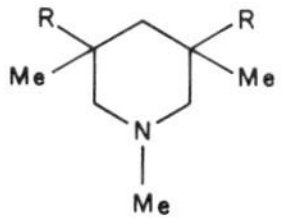 R = H, Me	$Hg(OAc)_2$ or $(NaO_2C)_2NCH_2CH_2N(CO_2)_2Hg$	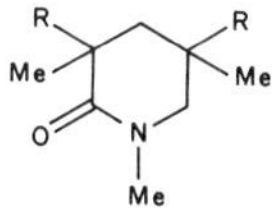	—	1459

Substrate	Conditions	Product(s)		Yield	Ref.
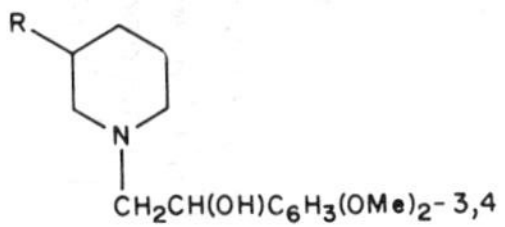 R (3-position), N–$CH_2CH(OH)C_6H_3(OMe)_2$-3,4	$Hg(OAc)_2$, EDTA	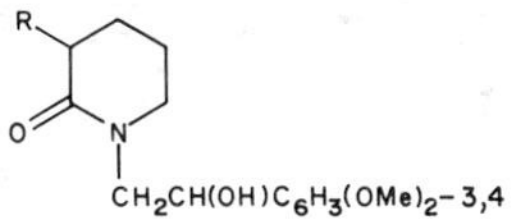 R (3-position), O, N–$CH_2CH(OH)C_6H_3(OMe)_2$-3,4	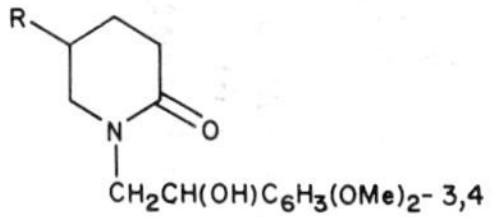 R, N, O, $CH_2CH(OH)C_6H_3(OMe)_2$-3,4	—	1460
R = Me, Et, *i*-Pr, *n*-Bu, Ac, Ph, $CONH_2$, CO_2Me, CH_2Ph					
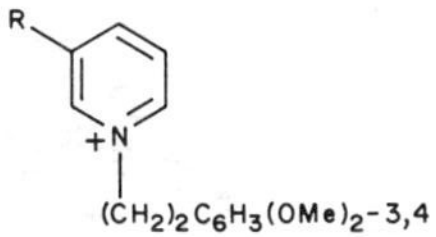 R, +N–$(CH_2)_2C_6H_3(OMe)_2$-3,4	alkaline ferricyanide, 32 °C	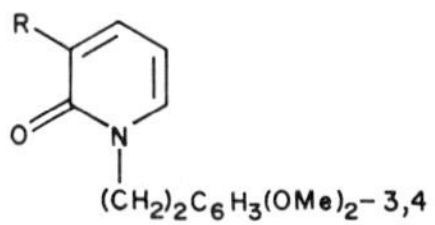 R, O, N–$(CH_2)_2C_6H_3(OMe)_2$-3,4	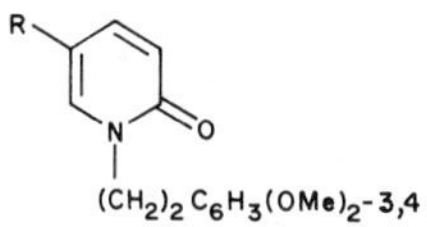 R, N, O, $(CH_2)_2C_6H_3(OMe)_2$-3,4	—	1460
R = Me, Et, *i*-Pr, *n*-Bu, Ac, Ph, $CONH_2$, CO_2Me, CH_2Ph					

When N-aryl 6- and 7-membered cyclic amines were treated[1461] with yellow mercuric oxide in ethylenediamine, the corresponding lactams were obtained (equation 1299). However, if the same reaction was attempted[1461] using N-(o-hydroxyphenyl)pyrrolidine only polymeric material was obtained.

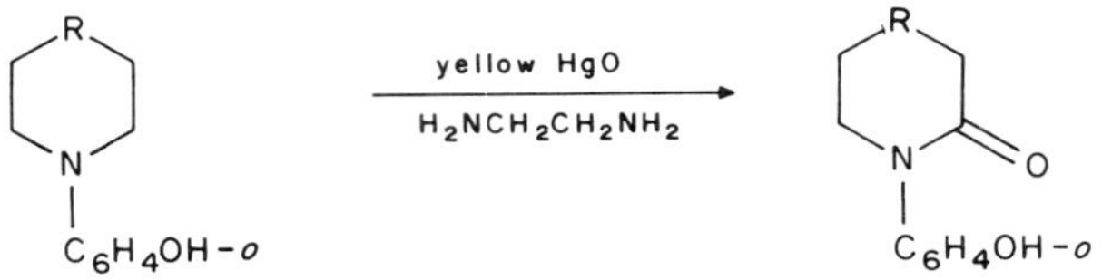

yellow HgO; $H_2NCH_2CH_2NH_2$ (1299)

$R = CH_2, CH_2CH_2, O$

N-Acylpyrrolidines can be converted[1462] into N-acyl-γ-lactams by electrochemical oxidation employing N-hydroxyphthalimide (NHPI) as a mediator or electron carrier in pyridine at 0.85 V versus SCE using a glassy-carbon anode contained in an H-type divided cell under oxygen (equation 1300).

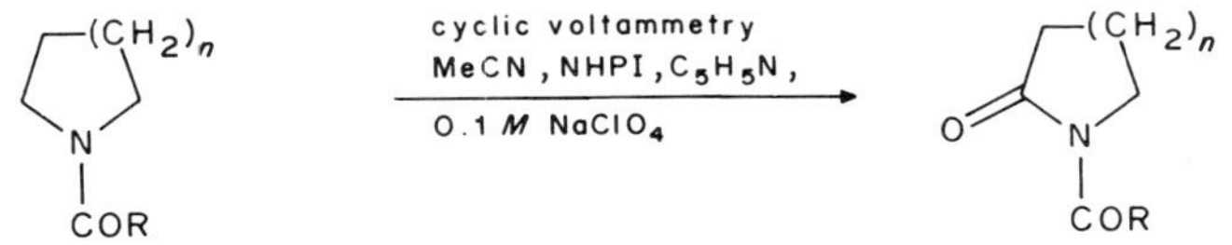

cyclic voltammetry; MeCN, NHPI, C_5H_5N, 0.1 M $NaClO_4$ (1300)

R =	Me	Ph	Ph
n =	1	1	2
Yield (%) =	80	79	9

The results indicate that the susceptibility toward oxidation of a 5-membered ring is much higher than that of a 6-membered ring.

*G. Miscellaneous Lactam Syntheses

One of the more common methods used to obtain lactams is the reduction of mono- and bicyclic imides and their analogues using a limited range of reducing agents. The general reaction involved is illustrated in equation 1301 while the details are reported in Table 106.

reducing agent; conditions (1301)

At least two examples of production of lactams by reductive cleavage have also been reported. In the first example[1346] 3,5-dicarboethoxy-2-pyrazoline is treated with Raney nickel to afford a mixture of aminopyrrolidones (equation 1302), while in the second example[1464] 1,1′-peroxydicyclohexylamine is treated with sodium in methanol to afford a mixture of caprolactam and cyclohexanone (equation 1303).

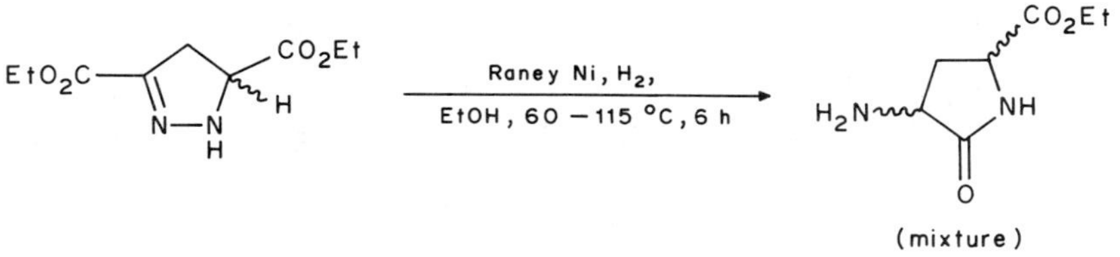

EtO2C, CO2Et, H; N—N, H; Raney Ni, H_2, EtOH, 60 – 115 °C, 6 h; H_2N, CO2Et, NH, O (mixture)

(1302)

NH, O—O; 1. MeOH, Na, reflux 1.5 h; 2. HCl, H_2O; N, H, O; + O

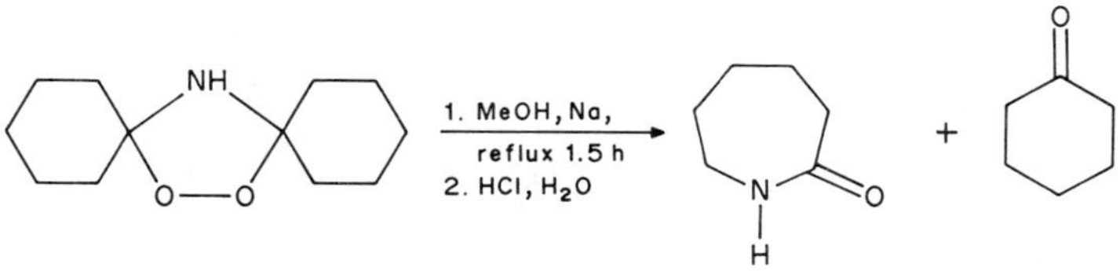

(1303)

Another method used to prepare lactams from imides and other dicarbonyl compounds involves reaction of these substrates with a variety of carbonyl reagents. Thus, reaction of 4-[(4α- and 4β-*trans*-2-butenyl)-2,3-dioxoazetidinyl]-5,5-diphenyl-2-methylpent-2-ene with the lithium salt of triethyl phosphonoacetate anion produces[1404] the corresponding α-alkylidene products (equations 1304 and 1305).

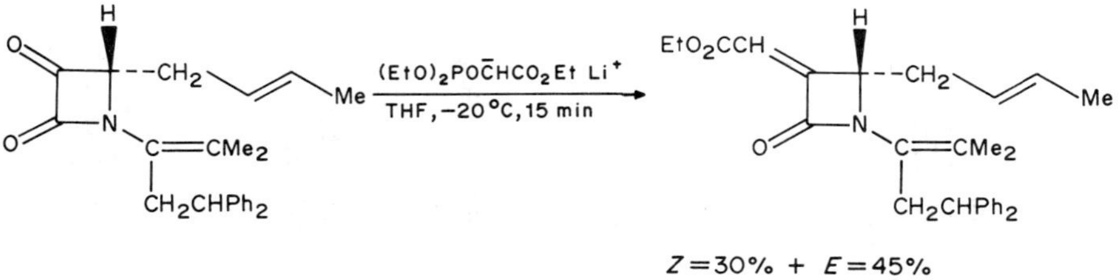

$(EtO)_2PO\bar{C}HCO_2Et\ Li^+$, THF, –20 °C, 15 min

$Z = 30\% + E = 45\%$

(1304)

$(EtO)_2PO\bar{C}HCO_2Et\ Li^+$, THF, –20 °C, 15 min

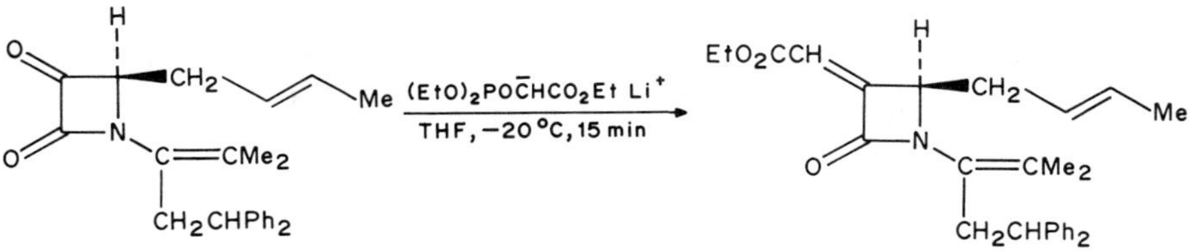

(1305)

Reaction of 3,4-dichloro-*N*-cyclohexylmaleimide with lithium acetylides produces[1263] the corresponding 5-alkynyl-5-hydroxy derivatives which were not isolated, but were subsequently treated with methyl iodide to produce the 5-alkynyl-5-methoxy products (equation 1306).

TABLE 106. Lactams by reduction of mono- and bicyclic imides

Imide	Reducing agent	Conditions	Product	Yield (%)	Reference
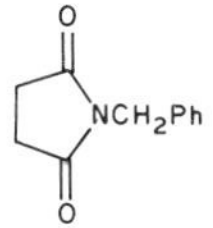 O=, O=, NCH_2Ph	$NaBH_4$	MeOH or EtOH	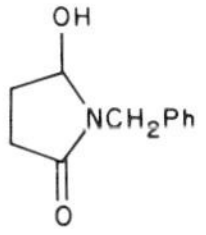 OH, O=, NCH_2Ph	—	1356
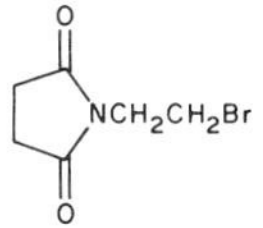 O=, O=, NCH_2CH_2Br	$NaBH_4$	EtOH, HCl	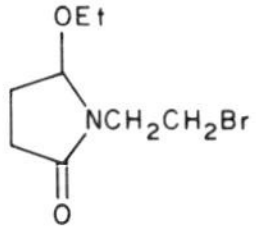 OEt, O=, NCH_2CH_2Br	70	1410
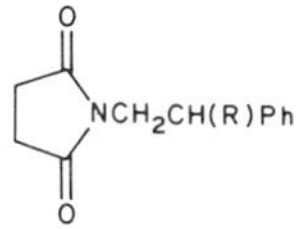 O=, O=, $NCH_2CH(R)Ph$	$NaBH_4$	EtOH, $MeSO_3H$, −5 to 0 °C	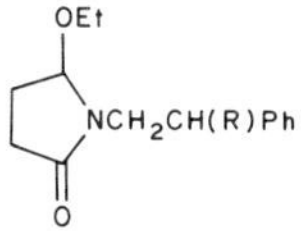 OEt, O=, $NCH_2CH(R)Ph$	—	1091
R = Ph	$NaBH_4$	EtOH, $MeSO_3H$ 20 °C		—	1091
R = 2-Thi				—	
R = 2(-*N*-methylpyrrolyl)				—	
R = 2-indolyl				—	
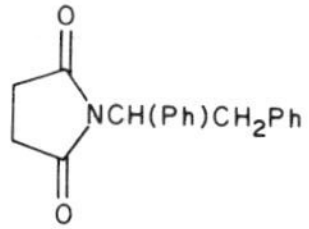 O=, O=, $NCH(Ph)CH_2Ph$	$NaBH_4$	EtOH, $MeSO_3H$	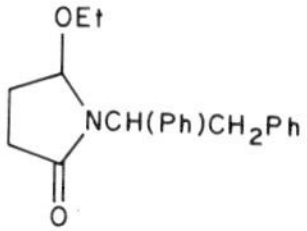 OEt, O=, $NCH(Ph)CH_2Ph$	—	1091

Substrate	Reagent	Conditions	Product	Yield (%)	Ref.
N-substituted succinimide: $NCH_2CH(R)C_{10}H_7$-α; R = H, Me	$NaBH_4$	EtOH, H^+ (pH = 3–4)	5-ethoxy-2-pyrrolidinone: $NCH_2CH(R)C_{10}H_7$-α	—	1179
N-substituted succinimide: NCH_2-(dibenzosuberyl)	$NaBH_4$	EtOH, $MeSO_3H$	5-ethoxy-2-pyrrolidinone: NCH_2-(dibenzosuberyl)	—	1091
N-substituted succinimide: $N(CH_2)_{n+1}C{\equiv}CR$	$NaBH_4$	EtOH, HCl, 0–5 °C, 4–5h	5-ethoxy-2-pyrrolidinone: $N(CH_2)_{n+1}C{\equiv}CR$		
$n = 0$, R = H				43	1413
$n = 1$, R = H				88	1411, 1413
$n = 1$, R = Me				60	1411, 1413
$n = 2$, R = H				87	1411, 1413
$n = 2$, R = Me				50	1411, 1413
$n = 3$, R = H				73	1413
3,4-dimethoxysuccinimide (MeO, H; MeO, H): NCH_2CH_2R; R = $CH{=}CH_2$; $CH{=}CHEt$; $C(Me){=}CH_2$; $C{\equiv}CH$; $CH_2CH_2C{\equiv}CH$	$NaBH_4$	H^+	5-hydroxy-3,4-dimethoxy-2-pyrrolidinone: NCH_2CH_2R (isomer ratio 20:1)	—	1408

(continued)

TABLE 106. (*continued*)

Imide	Reducing agent	Conditions	Product	Yield (%)	Reference
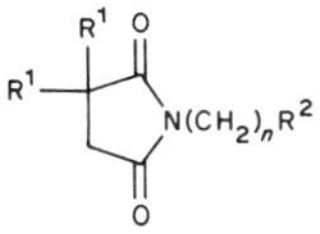 R¹,R¹-succinimide, N(CH$_2$)$_n$R^2	NaBH$_4$	H$^+$	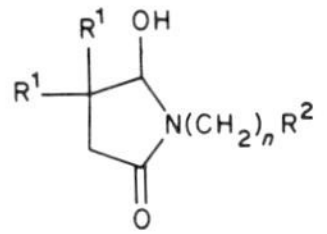 5-hydroxy-4,4-R¹-pyrrolidin-2-one, N(CH$_2$)$_n$R^2	—	1463

n	R^1	R^2
1	Me	3-cyclohexen-1-yl
2	H, Me	CH=C=CH$_2$
3	Me	CH=CH$_2$

Imide	Reducing agent	Conditions	Product	Yield (%)	Reference
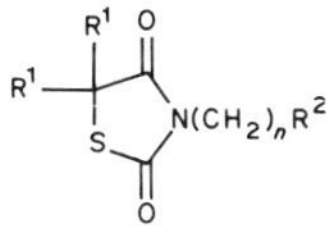 5,5-R¹-thiazolidine-2,4-dione, N(CH$_2$)$_n$R^2	NaBH$_4$	H$^+$	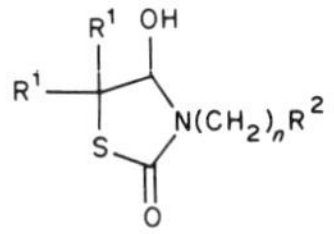 4-hydroxy-5,5-R¹-thiazolidin-2-one, N(CH$_2$)$_n$R^2	—	1463

n	R^1	R^2
1	H, Me	3-cyclohexen-1-yl
2	H, Me	CH=C=CH$_2$
2	H, Me	Ph
2	H, Me	$C_6H_3(OMe)_2$-3,4
2	H, Me	$C_6H_2(OMe)_3$-3,4,5
2	H, Me	CH=CH$_2$
2	H	C(Me)=CH$_2$
2	Me	CH=CHEt
2	Me	CH=CHCH$_2$CH$_2$Ph
2	Me	C≡CH
2	Me	C≡CCH$_2$CH$_2$Ph
3	Me	CH=CH$_2$
4	Me	C≡CH

Substrate	Reagent	Conditions	Product	Yield (%)	Ref.
5-R^1-5-Ph-3-R^2-thiazolidine-2,4-dione	$NaBH_4$	H^+	5-R^1-5-Ph-4-OH-3-R^2-thiazolidin-2-one	—	1463

R^1	R^2
H, Me	$CH_2CH_2CH{=}CHCH_2Me$ (*Z*)
CH_2Ph	Me
$CH_2C_6H_2(OMe)_3$-3,4,5	Me

Substrate	Reagent	Conditions	Product	Yield (%)	Ref.
N-CH_2Ph glutarimide	$NaBH_4$	—	N-CH_2Ph-6-OH-piperidin-2-one	—	1356
N-$(CH_2)_{n+1}C{\equiv}CR$ glutarimide	$NaBH_4$	EtOH, HCl, −15 to −10 °C 4–5h	N-$(CH_2)_{n+1}C{\equiv}CR$-6-OEt-piperidin-2-one		

n	R	Yield (%)	Ref.
0	H	85	1413
1	H	83	1411, 1413
1	Me	63	1411, 1413
2	H	68	1411, 1413
2	Me	46	1411, 1413
3	H	81	1413

(continued)

TABLE 106. (*continued*)

Imide	Reducing agent	Conditions	Product	Yield (%)	Reference
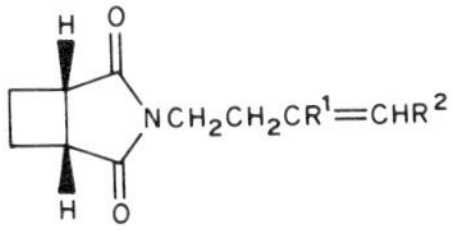 $NCH_2CH_2CR^1{=}CHR^2$	$NaBH_4$	THF, HCl, −20 °C stir 3h	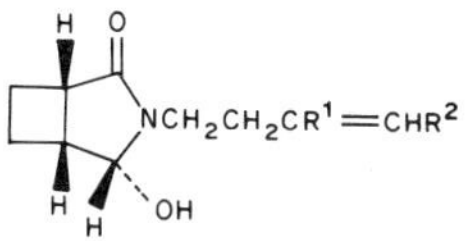 $NCH_2CH_2CR^1{=}CHR^2$ (OH)		1412
$R^1 = R^2 = H$				100	
$R^1 = Me, R^2 = H$				—	
$R^1 = H, R^2 = Et$				100	
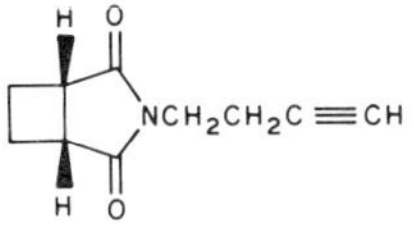 $NCH_2CH_2C{\equiv}CH$	$NaBH_4$	EtOH, HCl, −20 °C stir 5h	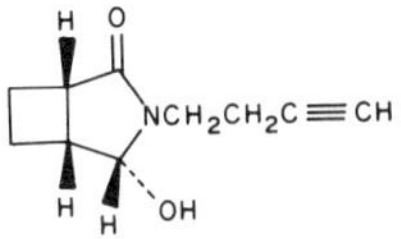 $NCH_2CH_2C{\equiv}CH$ (OH) (16:5 mixture of *endo* and *exo* isomers)	88	1412
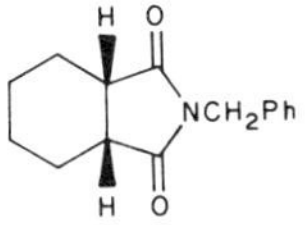 NCH_2Ph	$NaBH_4$	HCl, 0 °C, stir 4h	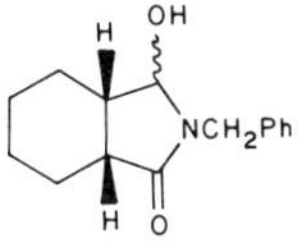 NCH_2Ph (OH) (2 isomeric *cis* products)	88	1432

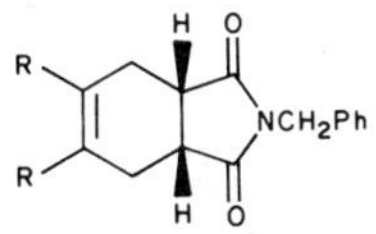 R = H, Me, Ph	$NaBH_4$	HCl, 0 °C, stir	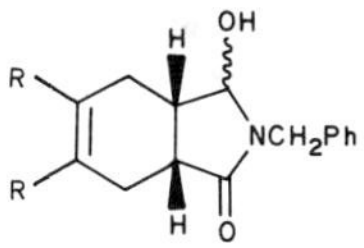	—	1432
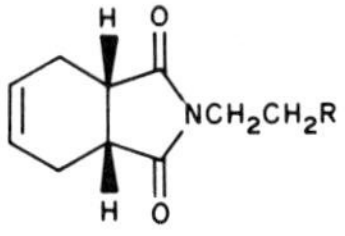 R = $CH{=}CH_2$; $C(Me){=}CH_2$; $CH{=}CHEt$	$NaBH_4$	EtOH, HCl, 0 °C, stir 5–6h	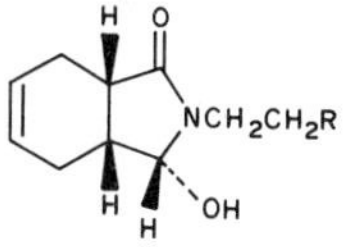	—	1412
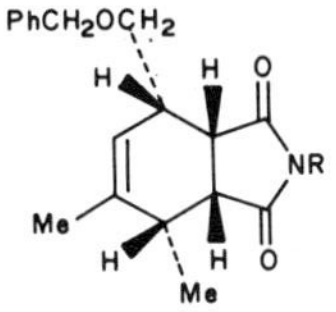 R = Et; CH_2Ph	$NaBH_4$	—	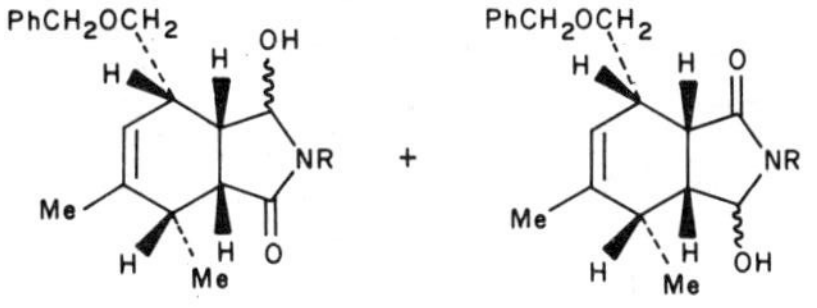 **(276)**	—	1374
	$(i\text{-}Bu)_2AlH$	C_6H_5Me, −78 °C	**276**	68	1374
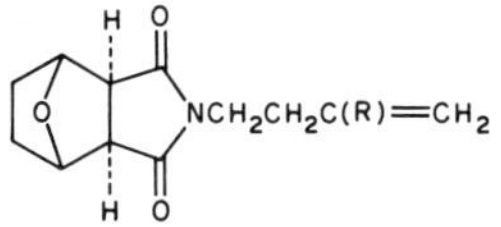	$NaBH_4$	EtOH, HCl, 0 °C, stir	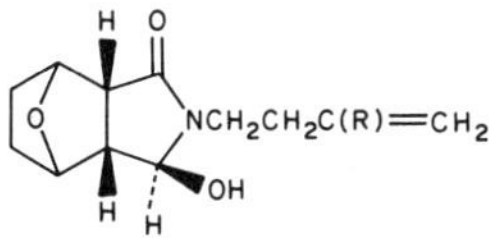		1412

(continued)

TABLE 106. (*continued*)

Imide	Reducing agent	Conditions	Product	Yield (%)	Reference
R = H		5h		83	
R = Me		THF, 4.5h		96	
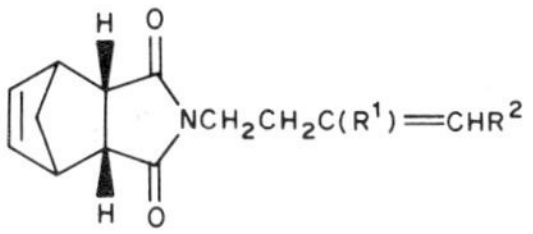 [structure] $NCH_2CH_2C(R^1)=CHR^2$	$NaBH_4$	EtOH, HCl, 0 °C stir	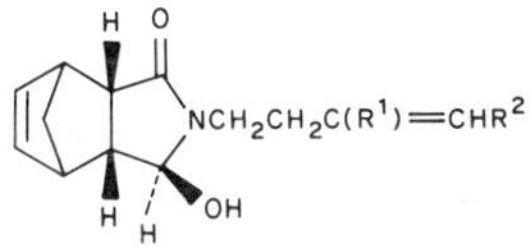 [structure] $NCH_2CH_2C(R^1)=CHR^2$, OH		1412
R^1 / R^2					
H / H		5h		100	
Me / H		5h		100	
H / Et		3.5h		100	
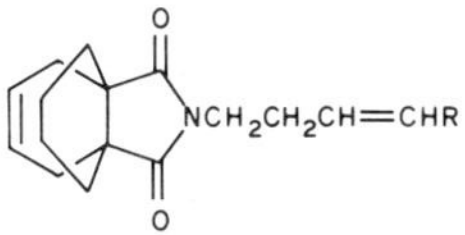 [structure] $NCH_2CH_2CH=CHR$	$NaBH_4$	EtOH, HCl, 0 °C stir	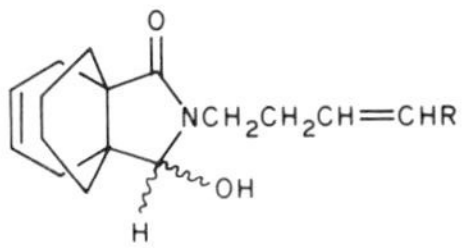 [structure] $NCH_2CH_2CH=CHR$, OH		1412
R = H		5h		50	
R = Et		12h		32	

Starting material	Reagent	Conditions	Product	Yield (%)	Ref.
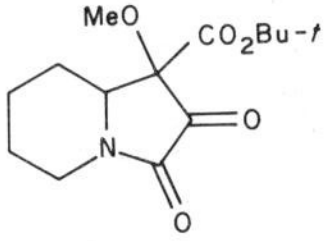	HOAc, HBr	r.t. stir 10 min	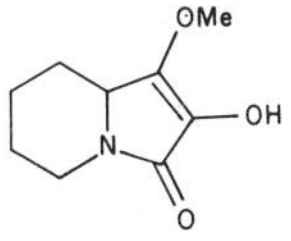	—	1271
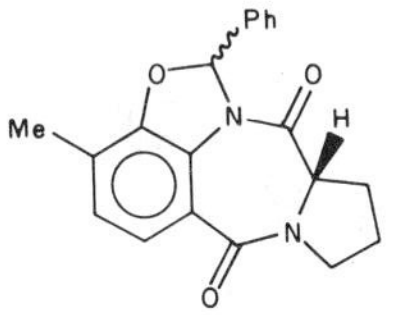	$NaBH_4$	MeOH, 4h	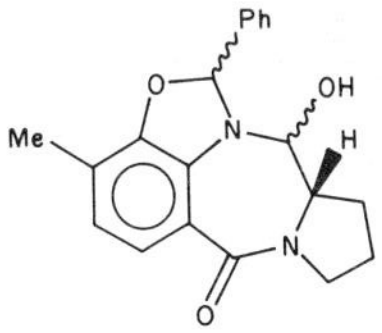	100	1353
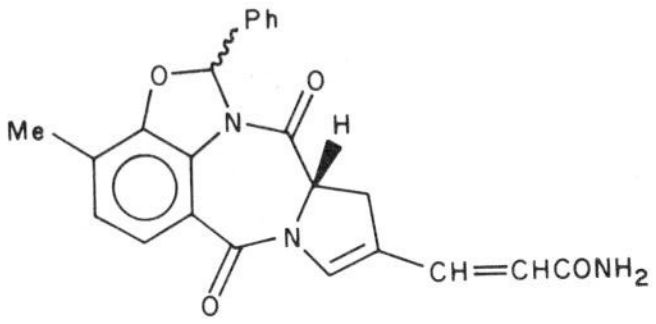	$NaBH_4$ or $LiAlH_4$	MeOH, 5 °C or THF, 60 °C	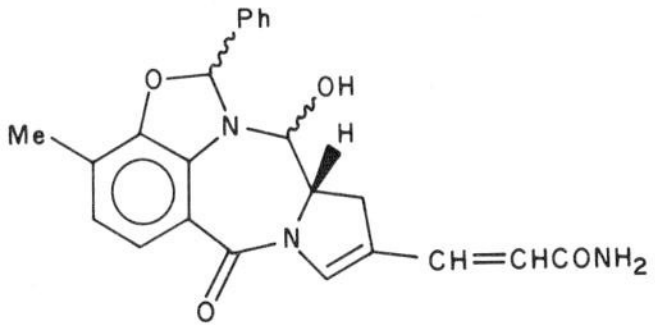	—	1353

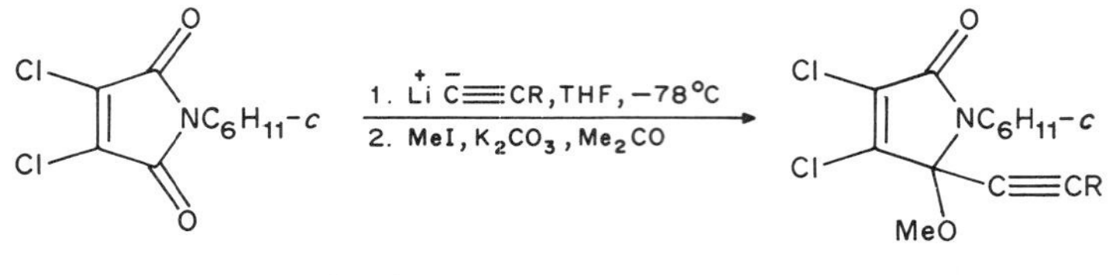

1. $Li^{+}\ ^{-}C{\equiv}CR$, THF, −78 °C; 2. MeI, K_2CO_3, Me_2CO (1306)

R = Ph, CH_2OCH_2Ph, $(CH_2)_2OTHP$ 80%

Treatment of bicyclic imides with methyl magnesium chloride produces[1412] the corresponding tertiary hydroxylactams (equations 1307 and 1308).

MeMgCl, THF; N_2, r.t. stir

R^1 =	H	Me	H
R^2 =	H	H	Et
Time (h) =	2	4.5	2.5
Yield (%) =	18	100	—[a]

(1307)

[a]Starting material and product are Z isomers.

MeMgCl, THF, N_2, r.t. stir

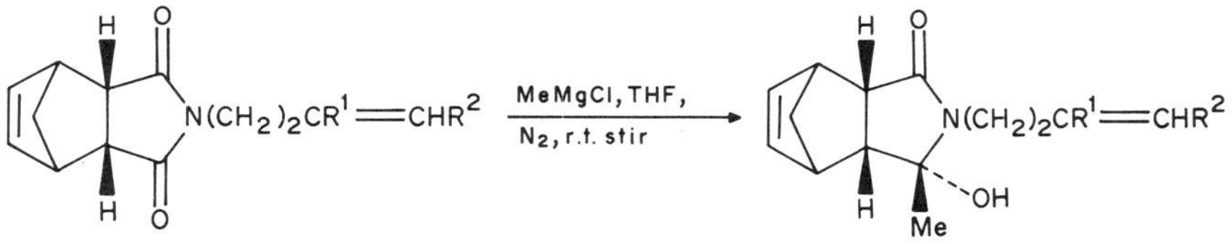

R^1 =	H	Me	H
R^2 =	H	H	Et
Time (h) =	4.5	5	2.5
Yield (%) =	—	87	100[a]

(1308)

[a]Starting material and product are Z isomers.

Condensation of a hydrazinouracil with an azepinedione produces[1379] a pyrimidinylhydrazone (equation 1309).

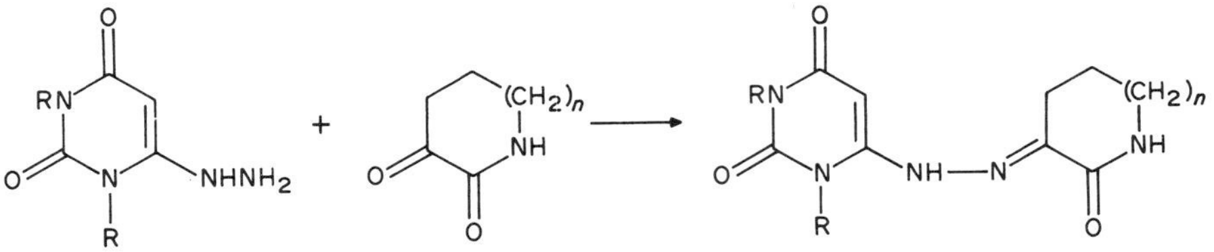

R = Me, $PhCH_2$
n = 1, 2

(1309)

Two examples of the preparation of seven-membered lactams from diones have been reported. In the first example α-oxocaprolactam is treated with phenoxyamine hydrochloride to produce[903] the α-*O*-phenyloxime, which can then be cyclized, quaternized and dehydrated to produce a tricyclic lactam product as shown in equation (1310). In the second example, 7-chloro-1-methyl-3,4-dihydro-1*H*-1,4-benzodiazepine-2,5-dione was treated with phosgene and afforded[1332] a 73% yield of 1-methyl-5,5,7-trichloro-1,3,4,5-tetrahydro-2*H*-1,4-benzodiazepin-2-one (equation 1311).

+ PhONH₂·HCl → PhON= ... NH, O (75%) — *i*-PrOH, HCl, EtOH, reflux 3 h → NH (69%) — MeI → $\overset{+}{N}$HMe I⁻ (33%) + OH, NH, O (26%) — CF_3CO_2H → NH, O (95%) (1310)

Me, N, O, Cl, N–H, O — 1. $CHCl_3$, 0°, $COCl_2$; 2. r.t. stand overnight → Me, N, O, Cl, N–H, Cl, Cl (73%) (1311)

Reaction of monothioimides with phosphorus ylides produces[1465] a mixture of ω-alkylidene lactams and ω-thiolactams (equation 1312).

An interesting approach to the preparation of optically pure β-lactams is the oxidative decomplexation of iron(cyclopentadienyl)carbonyl complexes[1386,1466,1467] (equation 1313).

$$\text{(1312)}$$

(Z) (Z)

$n = 1 ; 2 ; 3$
Lactam yield (%) = 66 ; 45 ; 51

1. $CH_2Cl_2, Br_2, -78\,°C$, stir 1 h
2. $Et_3N, -78\,°C$ to r.t.

NHCH₂Ph

(1313)

R^1	R^2	R^3	R^4	Yield (%)	Reference
CO	H	Me	Me	90	1466
Ph_3P	H	H	Me	65–68	1386, 1467
Ph_3P	Me	H	H	75	1467
Ph_3P	Me	H	Me	42–69	1386, 1467

Rhodium, palladium and cobalt catalysed carbonylation of a variety of substrates from aziridines to allylamines have been used to prepare 4-, 5-, 6- and 7-membered lactams and their fused ring analogues. The rhodium catalysts used include chlorodicarbonylrhodium(I) dimer which produces[1468,1469] regiospecific carbonylation of aziridines to β-lactams (equation 1314) in quantitative yields via a mechanism that involves[1468] oxidative addition of rhodium(I) to the more substituted carbon–nitrogen bond of the aziridine to give an intermediate rhodium(III) complex (**277**), which then undergoes ligand migration to produce **278**, followed by carbonylation to produce **279** (Scheme 22). The β-lactam product is then formed by reductive elimination, with or without the assistance of another molecule of aziridine.

+ CO(20 atm.) $\xrightarrow[C_6H_6,\ 90\,°C,\ \text{stir 2 days}]{[Rh(CO)_2Cl]_2}$ (1314)

100%

R^1	R^2	Reference
Ph	*t*-Bu	1468, 1469
Ph	1-adamantyl	1468
Ph	$SiMe_3$	1468
p-PhC_6H_4	*t*-Bu	1468
p-PhC_6H_4	1-adamantyl	1468
p-BrC_6H_4	*t*-Bu	1468

(277)

(278) (279)

SCHEME 22

A variety of rhodium catalysts has been used to effect ring closure of allylamines with carbonylation to produce[1470] γ-butyrolactam (equation 1315). Similar homogeneous rhodium catalysts have also been employed to produce[1470] γ-butyrolactam from allylic halides in the presence of carbon monoxide and ammonia (equation 1316) or *N*-alkyl substituted-2-pyrrolidones from allylic halides in the presence of carbon monoxide plus a primary alkylamine (equation 1317).

$$H_2C{=}CHCH_2NH_2 + CO\,(atm) \xrightarrow[C_6H_5Me,\ temp.,\ time]{Rh\ catalyst} \text{2-pyrrolidone} \tag{1315}$$

Rh catalyst	Temp. (°C)	Pressure CO (atm)	Time (h)	Yield (%)
$Rh(CO)(PPh_3)_2Cl$	150	136	2	67
$Rh(PPh_3)_3Cl$	150	136	2	68
$Rh_2Cl_2(C_2H_4)_4$	150	220	12	30
$Rh(C_5H_7O_2)_3$	120	136	12	28
$RhCl_3$	150	190	2	35

$$CH_2{=}CHCH_2Cl + NH_3 \xrightarrow[C_6H_5Me]{Rh(CO)(PPh_3)Cl} \text{2-pyrrolidone} \tag{1316}$$

$$CH_2{=}CHCH_2X + CO + MeNH_2 \xrightarrow{\text{catalyst}} \text{N-methyl-2-pyrrolidone} \quad (1317)$$

X = I or Cl

Catalyst = $Rh(acac)_3$ + KI or $RhCl(PPh_3)_3$

Palladium acetate in the presence of triphenylphosphine or tetrakis(triphenylphosphine)-palladium(0), both in the presence of carbon monoxide, have been used to carbonylate 2-halo-3-aminopropenes, 2-halo-4-aminobutenes and azirines producing the corresponding 4-, 5-, 6- or 7-membered lactams (equation 1318 and Table 107).

$$H_2C{=}CX(CH_2)_nNHR + CO(\text{atm.}) \xrightarrow[\text{time, conditions}]{\text{Pd catalyst}} \text{lactam } (H_2C{=}C\text{–}(CH_2)_n\text{–}NR\text{–}C{=}O \text{ ring}) \quad (1318)$$

The mechanism proposed[1471] for these conversions involves the formation of an intermediate assumed to be an acylpalladium complex generated from an enepalladium complex coordinated to carbon monoxide (equation 1319).

$$H_2C{=}CX(CH_2)_nNHR \xrightarrow{PdL_2} H_2C{=}C(PdXL_2)(CH_2)_nNHR \xrightarrow{CO(1\ \text{atm.})} \left[H_2C{=}C(C(O)PdL)(CH_2)_nN(H)R \right]^+ X^- \longrightarrow \text{product} \quad (1319)$$

L = Ph_3P, CO

When alkylcobalt carbonyl complexes, either preformed or produced 'in situ' in the presence of carbon monoxide and base, are allowed to react with N-substituted-*o*-bromobenzylamines, carbonylation ring closure occurs to produce[849] benzolactams (equation 1320). Although the starting amines may be primary or secondary, the best results were obtained using *N*-phenyl-*o*-bromobenzylamine.

$$o\text{-}BrC_6H_4CH_2NHR + CO\ (1\ \text{atm.}) \xrightarrow[\text{MeOH, 35 °C}]{EtO_2CCH_2Co(CO)_4,\ NaOMe} \text{N-R-isoindolin-1-one}$$

R =	H	Et	CH_2Ph	Ph	Ph[a]	Ph[b]
Yield (%) =	47	55	62	71	84	36

(1320)

[a]Using NaOEt as the base.
[b]Using $NCCH_2Co(CO)_4$ as the catalyst and EtOH as the solvent.

TABLE 107. Palladium catalysed carbonylation reactions producing lactams

Substrate	Pd catalyst	Pressure CO (atm.)	Conditions	Product	Yield (%)	Reference
$H_2C{=}C(Br)CH_2NHCH_2Ph$	$Pd(OAc)_2$–Ph_3P	1	HMPA, 100 °C, $(n\text{-}Bu)_3N$, stir 5h	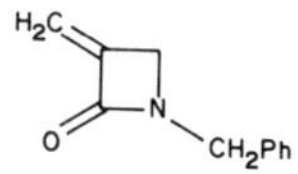 H_2C, O, N, CH_2Ph	67	1059, 1067, 1471
	$Pd(OAc)_2$–Ph_3P	1	HMPA, 100 °C, $(n\text{-}Bu)_3N$, stir 10h		53.5[a]	1067
	$Pd(OAc)_2$–Ph_3P	1	HMPA, 120 °C, $(n\text{-}Bu)_3N$, stir 4.5h		53.3[a]	1067
	$Pd(acac)_2$–Ph_3P	1	HMPA, 100 °C, $(n\text{-}Bu)_3N$, stir 7h		66.2	1067
$H_2C{=}C(Br)CH_2NH(CH_2)_2Ph$	$Pd(OAc)_2$–Ph_3P	1	HMPA, 100 °C, $(n\text{-}Bu)_3N$, stir 6h	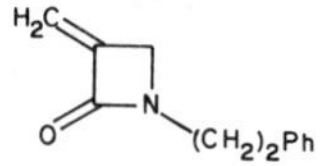 H_2C, O, N, $(CH_2)_2Ph$	61.9	1059, 1067, 1471
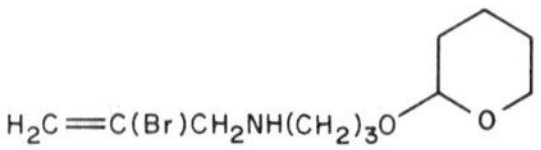 $H_2C{=}C(Br)CH_2NH(CH_2)_3O$, O	$Pd(OAc)_2$–Ph_3P	1	HMPA, 100 °C, $(n\text{-}Bu)_3N$, stir 4h	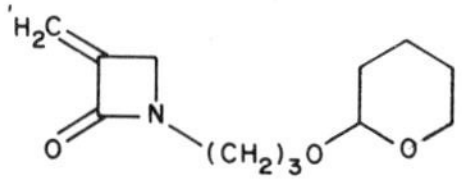 H_2C, O, N, $(CH_2)_3O$, O	62.9	1059, 1067, 1471
$H_2C{=}C(Br)CH_2NHCH(R)CO_2Me$	$Pd(OAc)_2$–Ph_3P	4		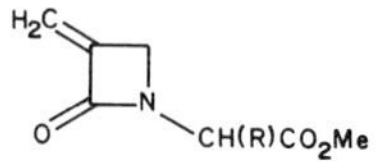 H_2C, O, N, $CH(R)CO_2Me$		

(*continued*)

TABLE 107. (*continued*)

Substrate	Pd catalyst	Pressure CO (atm.)	Conditions	Product	Yield (%)	Reference
R = H			HMPA, 100 °C, $(n\text{-}Bu)_3N$, stir 24h		60.1	1067
R = Me			HMPA, 100 °C, $(n\text{-}Bu)_3N$, stir 25.5h		40.6	1067
$H_2C{=}C(Br)CH_2NH(CH_2)_2CO_2Me$	$Pd(OAc)_2$–Ph_3P	1	HMPA, 100 °C, $(n\text{-}Bu)_3N$, stir 3.5h	H_2C; O; N; $(CH_2)_2CO_2Me$	37.6	1059, 1067, 1471
$H_2C{=}C(Br)CH_2NHCH(Me)\cdot CO_2CH_2Ph$	$Pd(OAc)_2$–Ph_3P	1	HMPA, 100 °C, $(n\text{-}Bu)_3N$, stir 5h	H_2C; O; N; $CH(Me)CO_2CH_2Ph$	20.4	1067
	$Pd(OAc)_2$–Ph_3P	4	HMPA, 100 °C, $(n\text{-}Bu)_3N$, stir 24h		44.7	1067
Ph, H, C=C, Br, CCH_2NHCH_2Ph (*Z*)	$Pd(acac)_2$–Ph_3P	1	HMPA, 100 °C, $(n\text{-}Bu)_3N$, stir 10h	H; Ph; C; O; N; CH_2Ph	79.9	1067, 1471

Substrate	Catalyst	Pressure (atm)	Conditions	Product	Yield (%)	Ref.
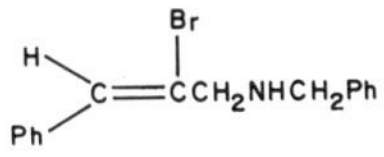 (H)(Ph)C=C(Br)CH$_2$NHCH$_2$Ph (E)	Pd(acac)$_2$–Ph$_3$P	1	HMPA, 100 °C, (n-Bu)$_3$N, stir 10h	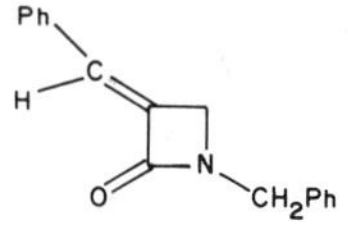 3-(PhCH=)-1-CH$_2$Ph-azetidin-2-one	89.5	1067, 1471
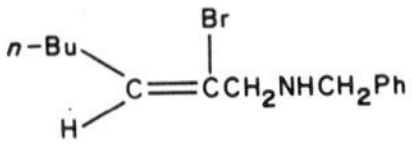 (n-Bu)(H)C=C(Br)CH$_2$NHCH$_2$Ph	Pd(acac)$_2$–Ph$_3$P	1	HMPA, 100 °C, (n-Bu)$_3$N, stir 8h	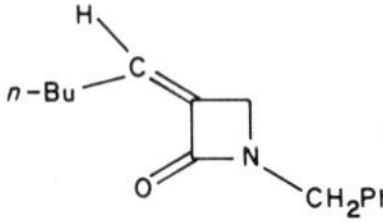 3-(n-BuCH=)-1-CH$_2$Ph-azetidin-2-one +	35.2	1067
				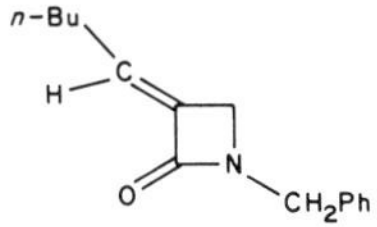 3-(n-BuCH=)-1-CH$_2$Ph-azetidin-2-one (isomer)	12.8	
H$_2$C=C(Br)CH$_2$NHCH$_2$C$_6$H$_4$R-p	Pd(OAc)$_2$–Ph$_3$P	1	HMPA, 100 °C, (n-Bu)$_3$N, stir	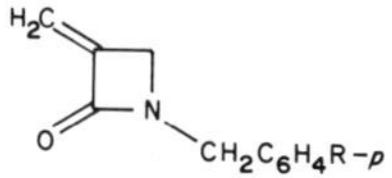 3-(H$_2$C=)-1-CH$_2$C$_6$H$_4$R-p-azetidin-2-one		1393
R = Me					57	
R = MeO					50	
R = CO$_2$Me					50	
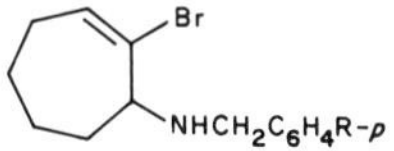 2-Br-cyclohept-2-enyl-NHCH$_2$C$_6$H$_4$R-p		1	HMPA, 100 °C, (n-Bu)$_3$N, stir	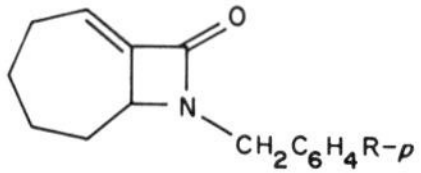 bicyclic β-lactam, N-CH$_2$C$_6$H$_4$R-p		

(continued)

TABLE 107. (*continued*)

Substrate	Pd catalyst	Pressure CO (atm.)	Conditions	Product	Yield (%)	Reference
R = H	$Pd(acac)_2$–Ph_3P		7h		85.7	1067
R = MeO	$Pd(OAc)_2$–Ph_3P		—		80	1393
$H_2C{=}C(Br)CH_2NHAn$-*p*	$Pd(OAc)_2$–Ph_3P	1	HMPA, 100 °C, $(n\text{-}Bu)_3N$, stir 5h	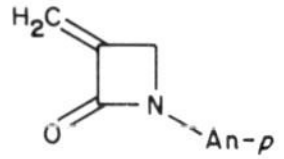	14.9	1067
$MeC(I){=}CHCH_2NHCH_2Ph$ + $H_2C{=}C(I)(CH_2)_2NHCH_2Ph$	$Pd(OAc)_2$–Ph_3P	1	HMPA, 100 °C, $(n\text{-}Bu)_3N$, stir	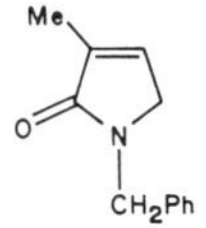 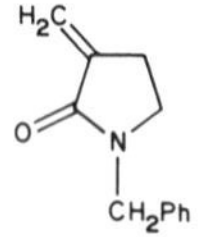 (1:10)	45.9	776
$H_2C{=}C(Br)(CH_2)_3NHCH_2Ph$	$Pd(OAc)_2$–Ph_3P	1	HMPA, 100 °C, $(n\text{-}Bu)_3N$, stir 5h	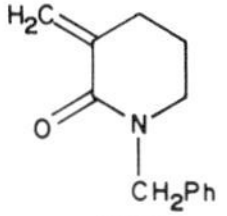 (**280**)	61	776
		—	HMPA, 120 °C, $(n\text{-}Bu)_3N$, stir 3h	**280** + 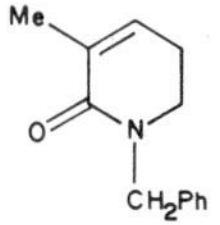	25 + 4	776

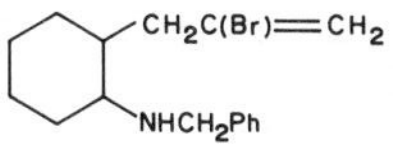 2-($CH_2C(Br)=CH_2$)-1-($NHCH_2Ph$)cyclohexane	$Pd(OAc)_2-Ph_3P$	1	HMPA, $(n\text{-}Bu)_3N$, 100 °C, 2.5h	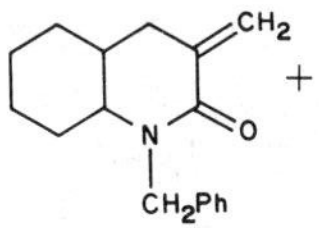 3-methylene-1-benzyl-octahydroquinolin-2-one (**281**) +	75	776
				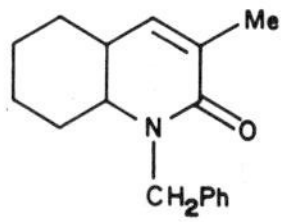 3-methyl-1-benzyl-hexahydroquinolin-2-one	7	
	$Pd(OAc)_2-Ph_3P$	1	HMPA, $(n\text{-}Bu)_3N$, 90 °C, 5h	**281** 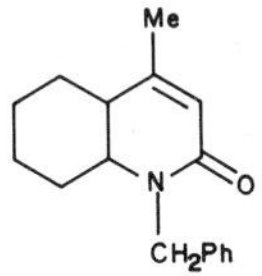 + 4-methyl-1-benzyl-hexahydroquinolin-2-one	80[b]	776
$Me(I)C=CH(CH_2)_3NHCH_2Ph$ + $H_2C=C(I)(CH_2)_4NHCH_2Ph$	$Pd(OAc)_2-Ph_3P$	1	HMPA, $(n\text{-}Bu)_3N$, 100 °C, stir	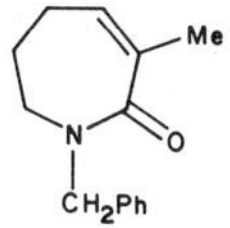 3-methyl-1-benzyl-azepin-2-one (unsaturated) +	23	776
				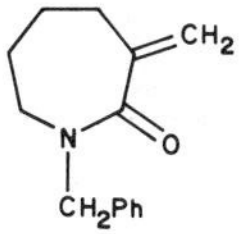 3-methylene-1-benzyl-azepan-2-one	28.4	

(continued)

TABLE 107. (*continued*)

Substrate	Pd catalyst	Pressure CO (atm.)	Conditions	Product	Yield (%)	Reference
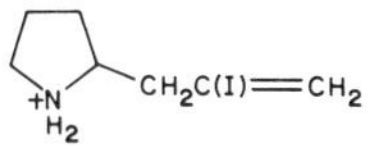 +N H_2 $CH_2C(I)=CH_2$ Br$^-$	$Pd(OAc)_2$–Ph_3P	1	HMPA, K_2CO_3 short reaction time	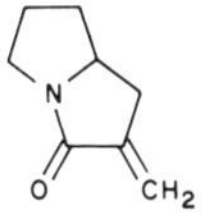 N O CH_2 (**282**)	36	776
			long reaction time	**282** 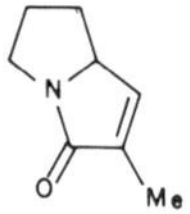 + N O Me	44[b]	776
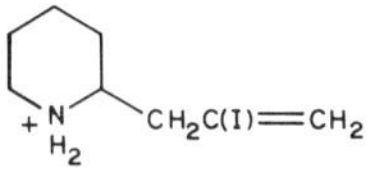 +N H_2 $CH_2C(I)=CH_2$ Br$^-$	$Pd(OAc)_2$–Ph_3P	1	HMPA, K_2CO_3	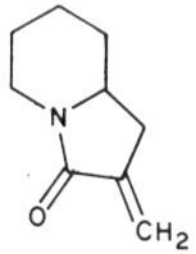 N O CH_2	58.7	776
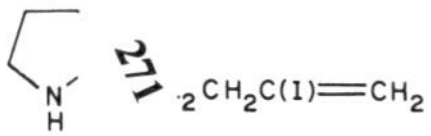 N H **271** ·2 $CH_2C(I)=CH_2$	$Pd(OAc)_2$–Ph_3P	1	HMPA, Et_3N	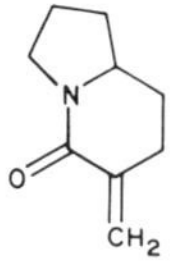 N O CH_2 +	26.4	776
				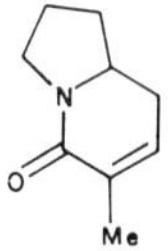 N O Me	26.8	

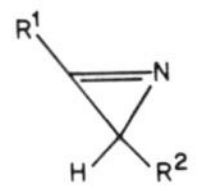

Substrate		Catalyst		Conditions	Product		Ref.
2H-azirine (R^1, H, R^2)		$Pd(PPh_3)_4$	1	40 °C	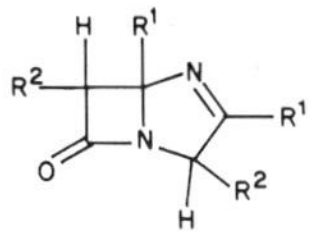 β-lactam product (R^1, R^2, H)	[c]	1472
R^1	R^2						
Ph	H			C_6H_6		63	
Ph	H			2,2′-bipyridyl, C_6H_6		29	
Ph	H			$MeCONMe_2$		16	
p-Tol	H			C_6H_6		50	
p-ClC_6H_4	H			C_6H_6		37	
p-BrC_6H_4	H			C_6H_6		55	
Ph	Me			C_6H_6		2.5	

[a]Lower concentration (mole %) of catalyst used.
[b]Yield of both products together.
[c]Substitution of $Ni(PPh_3)_4$ or $Pt(PPh_3)_4$ for $Pd(PPh_3)_4$ did not produce any β-lactam products even after a reaction time of 5 days.

Organometallic reagents have also been used to catalyse the formation[1262] of an aza-β-lactam by decomposition–cyclization of a diazo intermediate (equation 1321).

$$\mathrm{EtO_2CCH_2CON(Ph)N{=}CHPh} \xrightarrow{\mathrm{Pd/C,\,H_2}} [\mathrm{EtO_2CCH_2CON(Ph)NHCH_2Ph}]$$

$$\xrightarrow[\mathrm{CH_2Cl_2}]{\mathrm{TosN_3,\,Et_3N,}} \left[\mathrm{EtO_2C{-}C({=}N_2){-}CON(Ph)NHCH_2Ph}\right] \xrightarrow[\mathrm{C_6H_6}]{\mathrm{Rh_2(OAc)_4}} \text{1-benzyl-2-phenyl-3-ethoxycarbonyl-1,2-diazetidin-4-one}$$

(1321)

A large variety of other cyclization reactions have also been used to produce individual lactams, but because of the diversity of these methods it is nearly impossible to classify or categorize these reactions. For this reason the reactions are discussed essentially as individual approaches to the preparation of lactams by cyclization.

Ring closure of β-N-substituted amino thiol esters by refluxing in acetonitrile in the presence of triphenylphosphine and dipyridyl disulphide produces[1473] β-lactams as a mixture of isomers (equation 1322).

$$\mathrm{PhSCOCH(CH(OH)Me)CH(NHR)C{\equiv}CSiMe_3} \xrightarrow{\mathrm{KOH,\,THF}} \mathrm{HSCOCH(CH(OH)Me)CH(NHR)C{\equiv}CH}$$

$$\xrightarrow[\text{2,2'-dipyridyl disulphide, reflux}]{\mathrm{MeCN,\,Ph_3P,}} \text{3-(MeCH(OH))-4-(C}{\equiv}\text{CH)-N-R-azetidin-2-one (two isomers)}$$

R = CH_2Ph and other aralkyl groups.

(1322)

Oxidative cyclization of *O*-acyl vinylacetohydroxamates using bromine and base in aqueous acetonitrile provides a direct route[1416] to substituted 4-(bromomethyl)-*N*-acyloxy β-lactams (equation 1323).

$$\mathrm{HO_2CCH_2CH{=}CH_2} \xrightarrow[\text{3. }\mathrm{NH_2OH{\cdot}HCl,\,KOH,\,MeOH,\,0\,^\circ C}\text{; 4. stir r.t. 15–20 min}]{\text{1. }\mathrm{(COCl)_2,\,0\,^\circ C}\text{ stir 2 h; 2. r.t. stir 40–48 h}} [\mathrm{CH_2{=}CHCH_2CONHO^-\ K^+}]$$

$$\xrightarrow[\mathrm{C_5H_5N,\ THF\ r.t.\ stir\ 60{-}90\ min}]{\mathrm{RCOCl}} \mathrm{CH_2{=}CHCH_2CONHO_2CR} \xrightarrow[\mathrm{5{-}10\%\ aq.\ MeCN}]{\mathrm{K_2CO_3,\,Br_2,}} \text{4-(}\mathrm{CH_2Br}\text{)-1-(}\mathrm{O_2CR}\text{)-azetidin-2-one}$$

R =	Me	PhCH$_2$O	Me	*i*-Pr	*t*-Bu
Yield (%) =	73[a]	82–87	—	75–90	—

(1323)

[a]Using H_2O and CH_2Cl_2 instead of 5–10% aq. MeCN.

Treatment of the unsaturated α-cyano-β,γ-dialkyl-$\Delta^{\alpha,\beta}$-butenolides with primary amines in dioxane or xylene affords[1474] γ-lactams (equation 1324) as the major products.

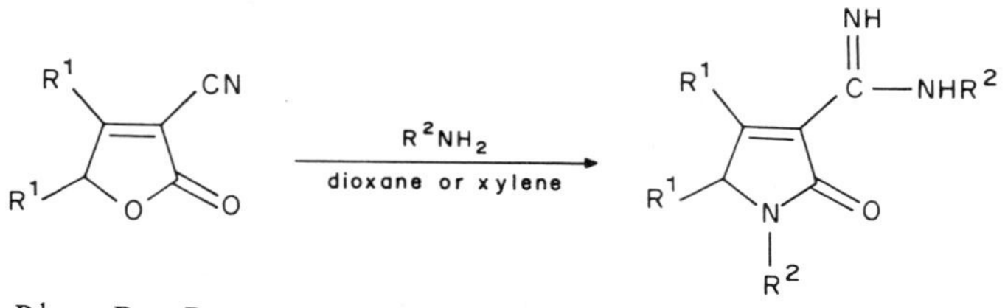

(1324)

$R^1 = n$-Pr, n-Bu.
$R^2 = PhCH_2$, n-Hex.

major product

β-Lactam ring closure of dihydrothiazines in the presence of 1-benzotriazolyloxytris-(dimethylamino)phosphonium hexafluorophosphate (**283**) affords[1475] cephems (equation 1325).

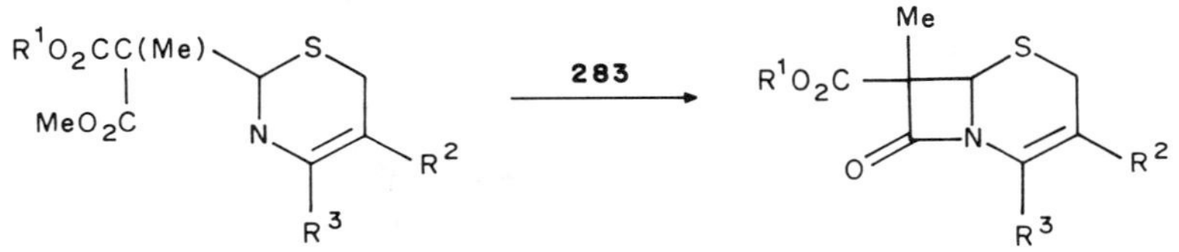

R^1 = H, t-Bu; R^2 = CHO, MeCO; R^3 = H, Me. (1325)

Catalytic hydrogenation resulting in cyclization and lactam formation has been observed to occur with hydroxysuccinimide esters[1476] (equations 1326 and 1327), the methyl ester of 5-azidomethyl-2-furanoic acid[1477] (equation 1328) and heterocyclic substituted nitro compounds[1478] (equations 1329–1332).

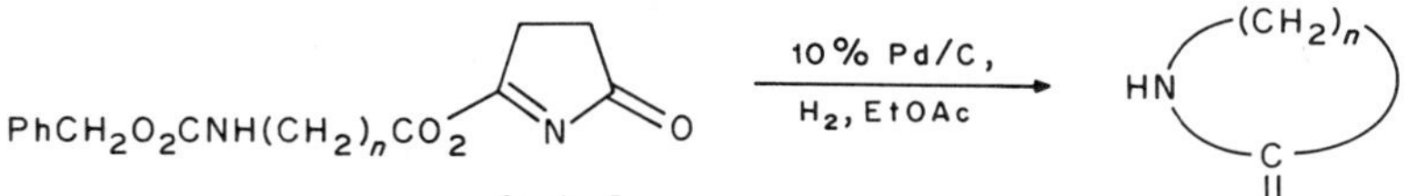

n = 3; 4; 5.
Yield (%) = 100; 95; 98. (1326)

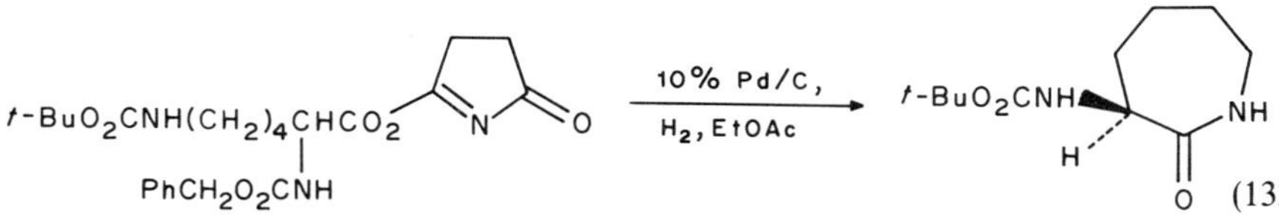

(1327)

90%

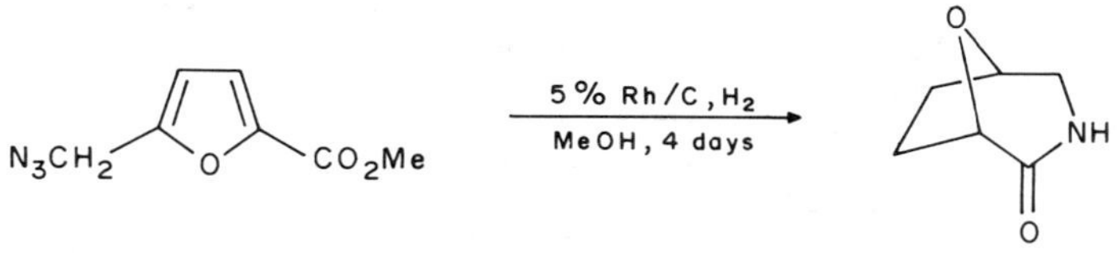

(1328)

70 %

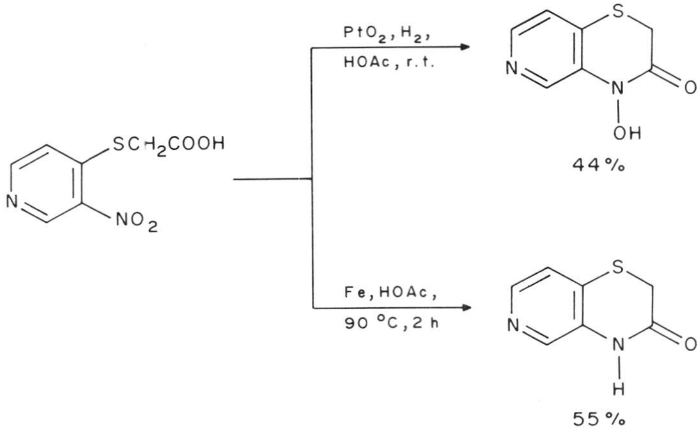

(1329)

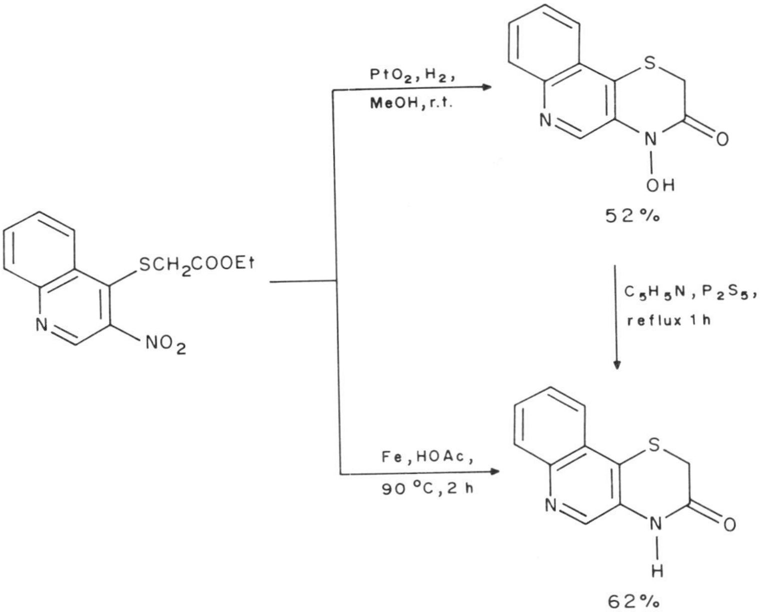

(1330)

Reactions leading to lactams, by cyclization which occurs after or during a condensation with another substrate, are also represented in the literature and are also essentially individual approaches to the preparation of lactams.

Intramolecular amidation of methyl 3-[2-(phenylsulphonyl)ethyl]quinoxaline-2-carboxylate 1,4-dioxide produces[1479] a series of novel 3,4-dihydropyrido[3,4-*b*]-quinoxalin-1(2*H*)-one 5,10-dioxides (equation 1333). The phenylsulphonylethyl side chain in the starting material may be viewed as a latent vinyl group which is unmasked by elimination of benzenesulphinic acid under the basic conditions of the reaction, leading to a Michael addition with the amine reagent, to produce an intermediate amino ester which then cyclizes to the lactam products.

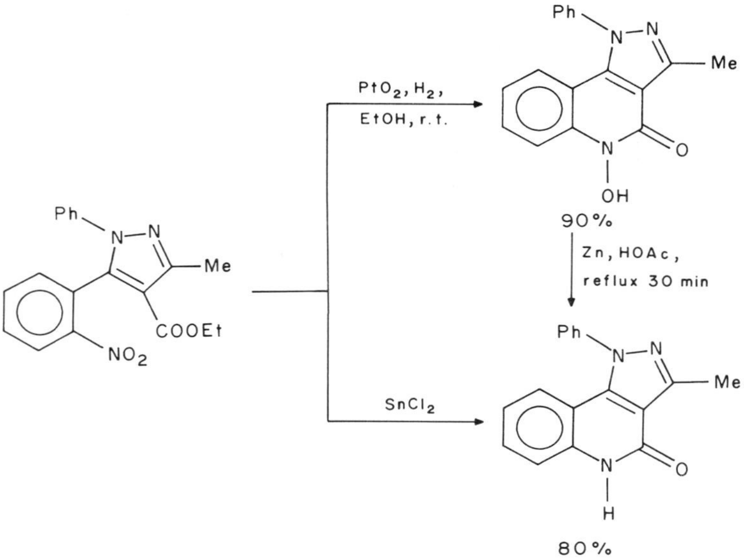

PtO$_2$, H$_2$, EtOH, r.t.

90%

Zn, HOAc, reflux 30 min

SnCl$_2$

80%

(1331)

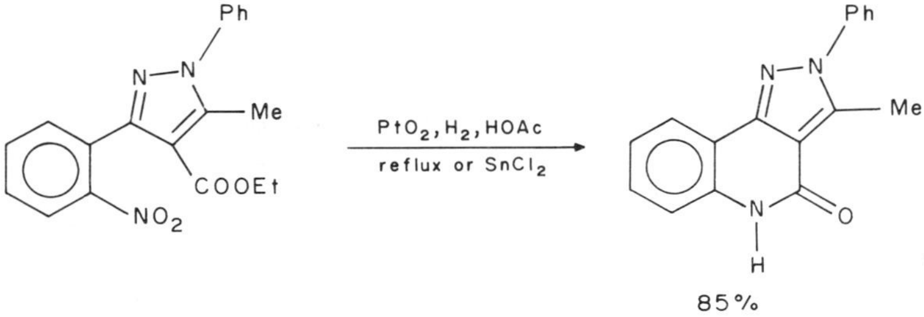

PtO$_2$, H$_2$, HOAc reflux or SnCl$_2$

85%

(1332)

1. RNH$_2$, MeCN
2. r.t. stir, 2–36 h

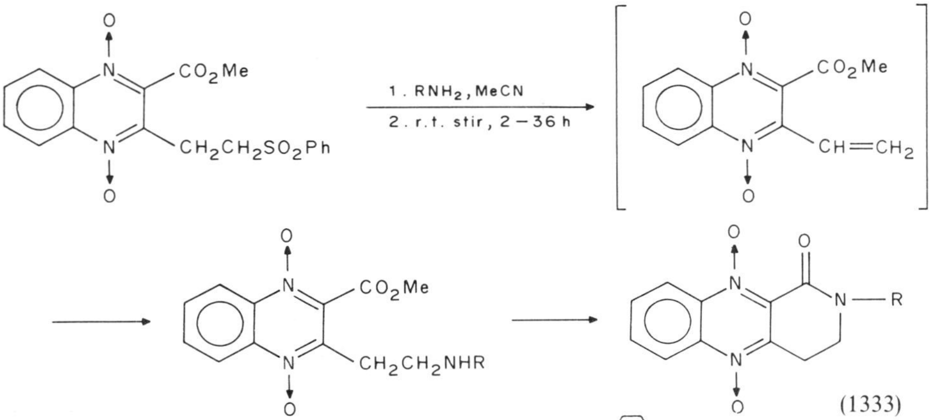

(1333)

R = H; Me; Et; CH$_2$CH$_2$OH; CH$_2$CH=CH$_2$; CH$_2$CH$_2$N(piperidino); c-Hex; CH$_2$CH$_2$OMe

Yield(%) = 64; 85; 51; 53; 50; 73; 59; 46

Aminolysis of bis(3-acylthiazolidine-2-thiones) with diamines affords[1480] macrolactams containing aromatic rings (equations 1334–1336).

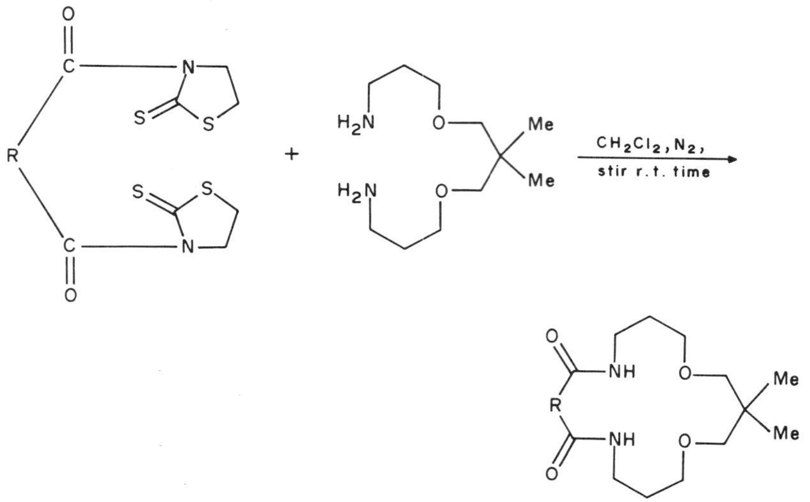

CH_2Cl_2, N_2, stir r.t. time

(1334)

R =	1,3-C_6H_4	3,5-C_5H_3N	3,4-Fu	2,2′-biphenyl
Time =	30 min	30 min	[a]	7 days
Yield (%) =	91	50	93	76

[a]Reaction mixture was worked-up immediately after the addition of the reactants was complete.

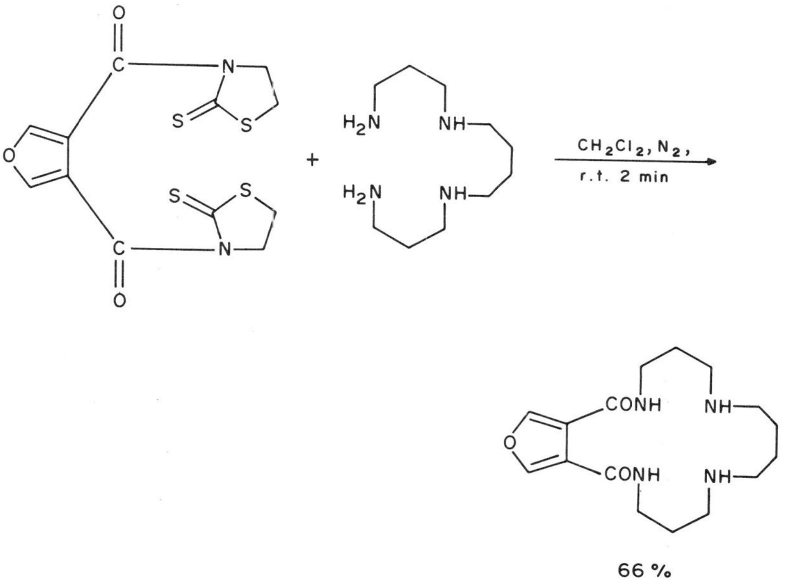

CH_2Cl_2, N_2, r.t. 2 min

66 %

(1335)

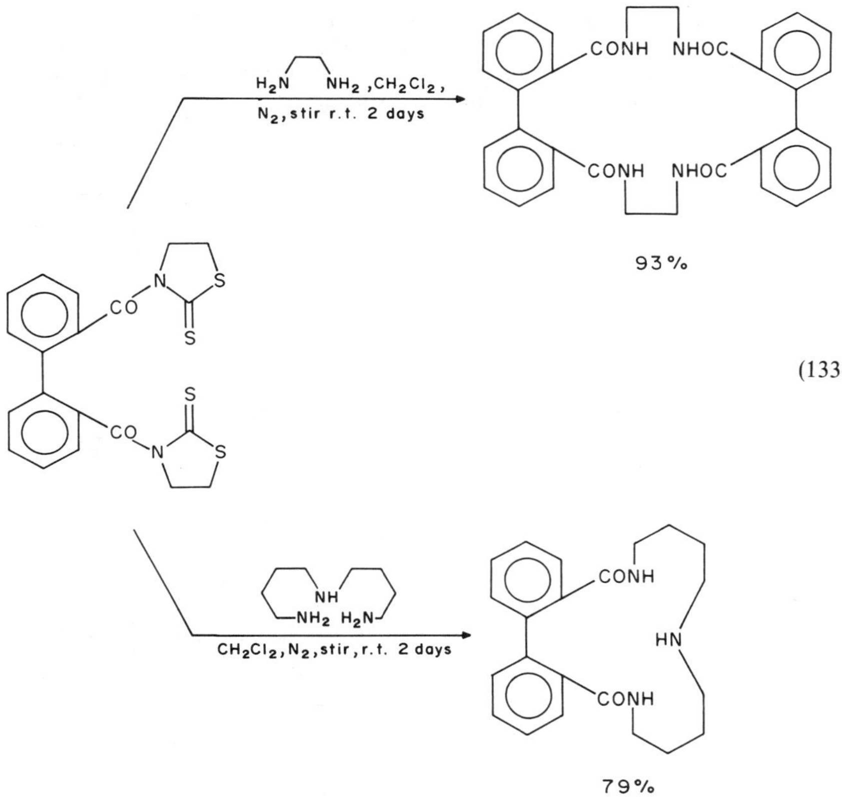

H_2N NH_2, CH_2Cl_2, N_2, stir r.t. 2 days

CONH NHOC

CONH NHOC

93%

CO N S S

CO N S S

(1336)

NH NH2 H2N

CH_2Cl_2, N_2, stir, r.t. 2 days

CONH HN CONH

79%

Condensation of α-haloacyl halides with 3-substituted thiosemicarbazides produces[1481] 4-substituted 4-aza-2-azetidinones (equation 1337), whereas condensation of α-chloroacyl chlorides with 1-substituted 2,3-dimethylthiosemicarbazides affords[1481] 2-imino-

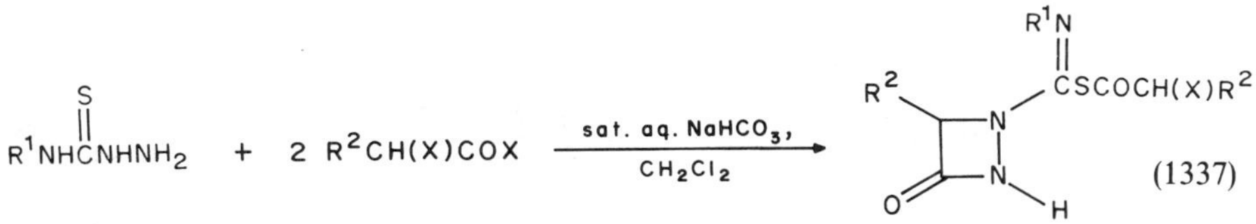

$$R^1NHC(=S)NHNH_2 + 2\ R^2CH(X)COX \xrightarrow[CH_2Cl_2]{\text{sat. aq. } NaHCO_3} \text{4-aza-2-azetidinone } (R^2\text{; N-}C(=NR^1)SCOCH(X)R^2\text{; N-H; C=O}) \qquad (1337)$$

R^1	R^2	X	Yield (%)
Me	H	Cl	44
n-Bu	H	Cl	68
c-Hex	H	Cl	72
$PhCH_2$	H	Cl	84
Ph	H	Cl	52
Me	Me	Br	64
$PhCH_2$	Me	Br	66

1,2,3,4-tetrahydro-3,4-dimethyl-1,3,4-thiadiazin-5-ones (equation 1338). However, if β-chloropivaloyl chloride is condensed with 1-methylthiourea a β-lactam is obtained[1482] (equation 1339), while condensation with 3-cyclohexylthiosemicarbazides produces[1481] β-lactams and 1,3,4-thiadiazepines (equation 1340).

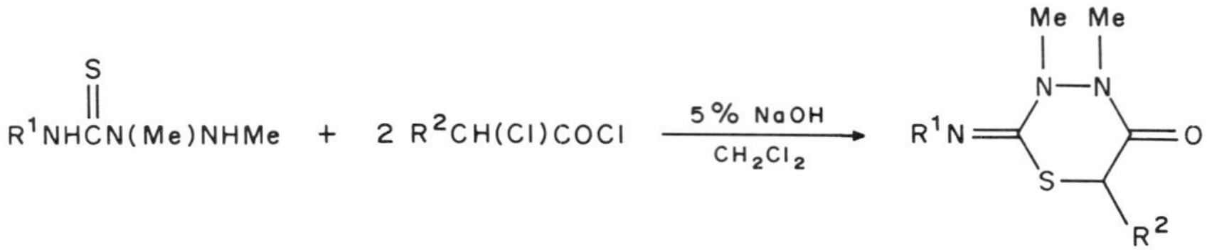

$R^1NHC(=S)N(Me)NHMe$ + 2 $R^2CH(Cl)COCl$ $\xrightarrow[CH_2Cl_2]{5\%\ NaOH}$

(1338)

R^1	R^2	Yield (%)
c-Hex	H	75
c-Hex	Et	52
c-Hex	Ph	55
Ph	H	81
Ph	Me	80
Ph	Ph	77
α-Naph	H	57

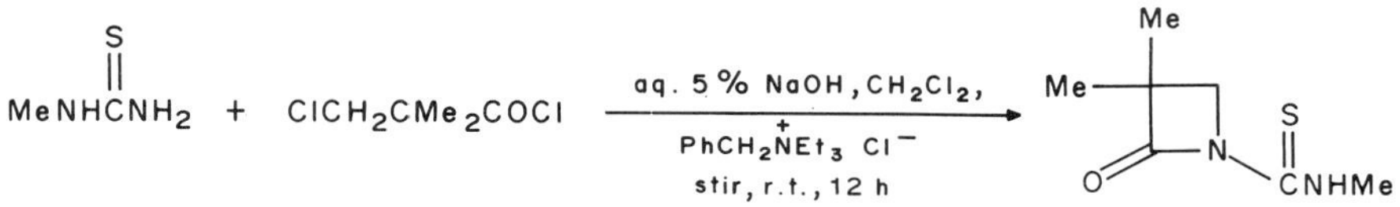

$MeNHC(=S)NH_2$ + $ClCH_2CMe_2COCl$ $\xrightarrow{aq.\ 5\%\ NaOH,\ CH_2Cl_2,\ PhCH_2\overset{+}{N}Et_3\ Cl^-,\ stir,\ r.t.,\ 12\ h}$ (β-lactam, N-C(=S)NHMe)

(1339)

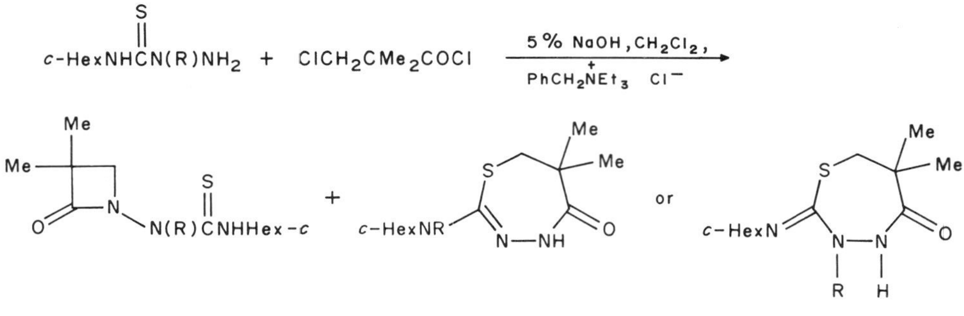

c-$HexNHC(=S)N(R)NH_2$ + $ClCH_2CMe_2COCl$ $\xrightarrow{5\%\ NaOH,\ CH_2Cl_2,\ PhCH_2\overset{+}{N}Et_3\ Cl^-}$

(284) + **(285)** or **(286)**

(1340)

R	Yield (%) 284	285	286
H	21	49	—
Me	56	—	20

Reaction of (4*S*)-3-(benzyloxycarbonyl)-4-(2-oxoethyl)-5-oxooxazolidine with L-cysteine methyl ester hydrochloride produce[1430] exclusively (2*R*, 5*S*, 7*S*)-1-aza-7-benzyloxycarbonyl(hydroxymethyl)amino-8-oxo-4-thiabicyclo[3.3.0]octane-2-carboxylic acid methyl ester as a single isomer (equation 1341), while reaction of the same oxazolidinone aldehyde with D-cysteine methyl ester hydrochloride produces the same product but as an epimeric mixture (equation 1341). Both reactions occur via a sequential double cyclization mechanism.

$\overset{+}{N}H_3\ Cl^-$ (L) $HSCH_2CHCO_2Me$ / C_5H_5N, r.t. stir 72 h → 21%

$\overset{+}{N}H_3\ Cl^-$ (D) $HSCH_2CHCO_2Me$ / C_5H_5N, r.t. stir 64 h → 19% (mixture of epimers)

PhCH2O2C, CH2CHO, PhCH2O2CN, HOCH2, CO2Me

(1341)

Aldehydes are also used as substrates along with lactam acetals for the preparation[1483] of 5-, 6- and 7-membered *erythro*- and *threo*-N-substituted 3-benzhydrol substituted lactams (equation 1342).

MeO, MeO, $(CH_2)_n$, N, R^1 + R^2CHO ⟶ $R^2(HO)CH$, $(CH_2)_n$, O, N, R^1 (1342)

R^1 = Me ; $PhCH_2$; *n*-Bu.
n = 1–3 ; 1 ; 1 .
R^2 = Ph; *o*-$O_2NC_6H_4$; *o*-$H_2NC_6H_4$; α-naphthyl; β-naphthyl; 2-pyrrolyl; *o*-HOC_6H_4; *p*-$HO_2CC_6H_4$; *o*-An; 3-Pyr; *p*-$O_2NC_6H_4$

Primary and secondary amines have also been used as substrates in condensation reactions to produce lactams. Thus, reaction of 6-bromobenzotropone with a selection of primary amines produces[1484] *N*-alkyl-6,9-dihydro-6,9-iminobenzocyclohepten-5-ones (**287**) and in some cases 3-acetonyl-2-alkylisoindolin-1-ones (equation 1343). The tricyclic products presumably result from 1,6-addition of the amines at position 9 of the bromobenzotropone, followed by displacement of the bromine by attack of the nitrogen,

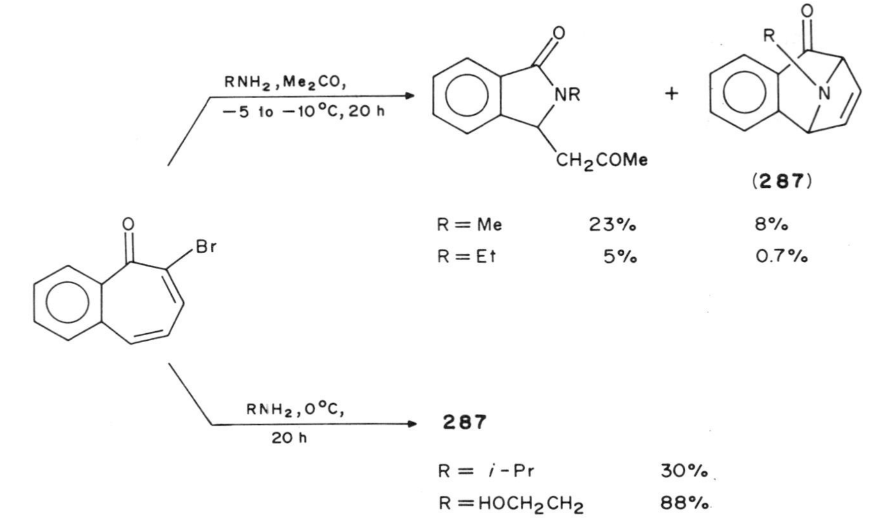
RNH2, Me2CO,
−5 to −10°C, 20 h
NR
CH2COMe
+
N
R
(287)
R = Me 23% 8%
R = Et 5% 0.7%
Br
RNH2, 0°C,
20 h
287
R = i-Pr 30%
R = HOCH2CH2 88%

(1343)

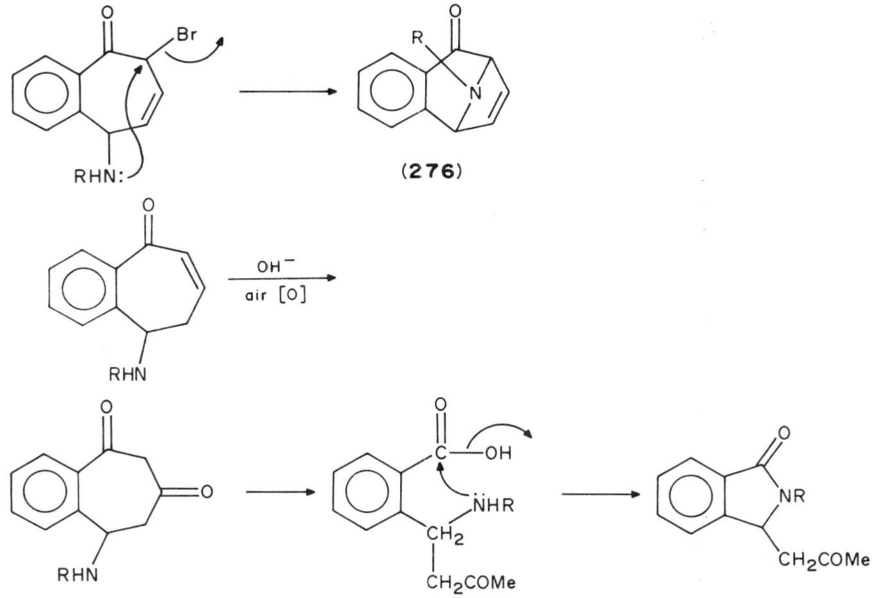
Br
RHN:
R
N
(276)
RHN
OH−
air [O]
RHN
C
OH
NHR
CH2
CH2COMe
NR
CH2COMe

(1344)

whereas the bicyclic keto lactam products probably arise from base hydrolysis of the 1,6-addition intermediate (equation 1344).

An example of the use of a secondary amine in a condensation reaction with both methyl valerolactim and methyl valerothiolactim, to produce[1485] a lactam, is shown in equation 1345.

(1345)

An example of an amide used in a condensation-like reaction to produce β-lactams is illustrated by the solid state photolysis of the inclusion complexes formed from *N,N*-dialkylpyruvamides and deoxycholic acid. The 4:1 inclusion complexes of acid to amide used in these photolysis reactions were prepared by crystallizing the acid using the amides as the solvents, and the photolysis was then carried out in the presence of air at room temperature. The β-lactams (**289**) obtained[1109] were optically active, albeit the enantiomeric excessess are not high, unlike the results obtained from the photolysis in solution, where the main products obtained were oxazolidinones (**290**) (equation 1346).

The mechanism proposed[1109] for these reactions involves formation of the zwitterion **291** as an intermediate which cyclizes to the β-lactam products **289** (equation 1347).

O-Silyl derivatives of several heterocyclic systems have also been used in condensation reactions to produce lactams. Thus, reaction of the *O*-trimethylsilyl derivatives of *N*-trifluoroacetylhexahydro-2*H*-azepin-2-one with 1-(chloromethyl)-2-pyrrolidone in the presence of trimethylsilyl trifluoromethanesulphonate produces[1486] the α-(1-methyl-2-pyrrolidone) substituted product (equation 1348).

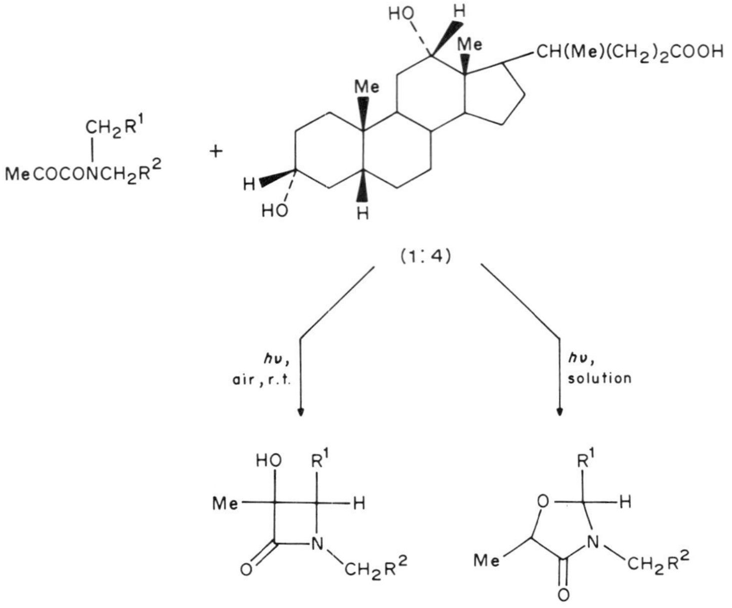

(1346)

(1:4)

hν, air, r.t. → (289)

hν, solution → (290)

R^1	R^2	% Yield **289**
H	H	42
Me	Me	74[a]
—$(CH_2)_3$—		

[a]Product is a 1:1 mixture of stereoisomers.

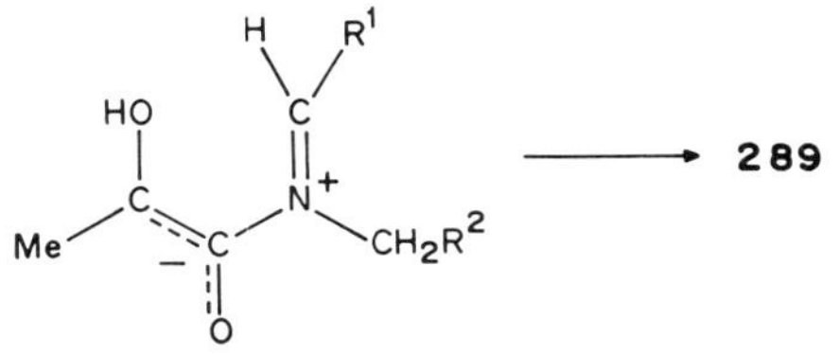

(291) → **289** (1347)

$CF_3SO_3SiMe_3$

66%

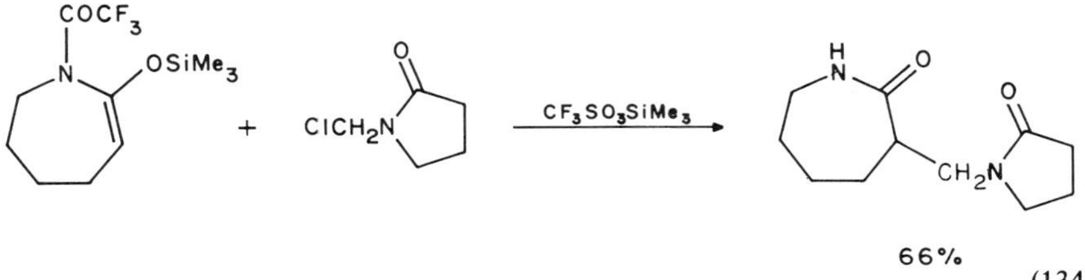

(1348)

Four-component condensation of 2-(t-butyl)dimethylsilyloxyphenylisocyanide, isobutyraldehyde and β-alanine in methanol produces[1487] the N-substituted β-lactam shown in equation (1349).

NC

OSiMe2(Bu-t) + i-PrCHO + $H_2NCH_2CH_2COOH$

1. MeOH, N_2, 50 °C 1 h

2. 20 °C, stir 3 days

i-Pr N HN O OSiMe2Bu-t

OSiMe2Bu-t O N CHCONH i-Pr

55%

(1349)

Although reaction of acyldioxanediones (acyl Meldrum's acids) and Schiff bases in the presence of chlorotrimethylsilane and triethylamine produces[1488] 2,3,6-trisubstituted 2,3-dihydro-1,3-oxazine-4-oxo-5-carboxylic acids, which also arise from isomerization of oxazinediones (equation 1350), reaction of the same acyldioxanediones or oxazinediones and Schiff bases in the presence of hydrochloric acid produces[1488] 1,4-disubstituted 3-acyl-β-lactams (equation 1351).

O COR1 O Me O O Me

(**292**)

R^1 = Me, Ph R^2 = n-Pr, i-Pr, $PhCH_2$, t-Bu, Ph

+ PhCH=NR2

1. Me_3SiCl, NEt_3

2. H_2O

75–90%

O (R4)R2 N COOH Ph O R1(R3)

O R4 N COR3 Ph O O

(**293**)

Me_3SiCl, NEt_3

(isomerization)

71–90%

R^3 = Me, Et, PhCH=CH, $PhCH_2$, c-Hex, Ph
R^4 = n-Pr, t-Bu, Ph

(1350)

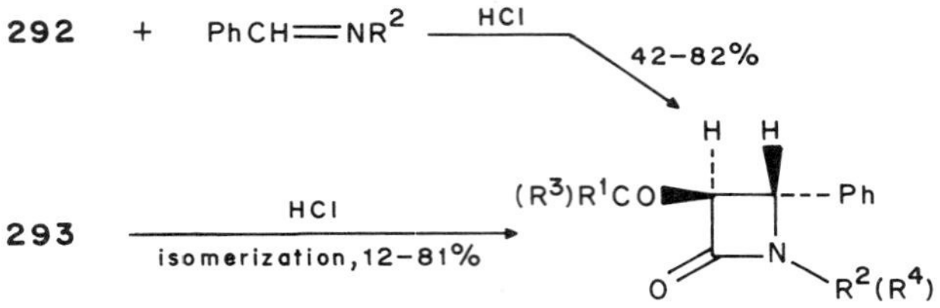

292 + PhCH=NR^2 $\xrightarrow{\text{HCl}}$ 42–82%

293 $\xrightarrow[\text{isomerization, 12–81\%}]{\text{HCl}}$ (R^3)R^{1}CO ... Ph ... N–R^2(R^4) (1351)

R^1—R^4 as in equation 1350

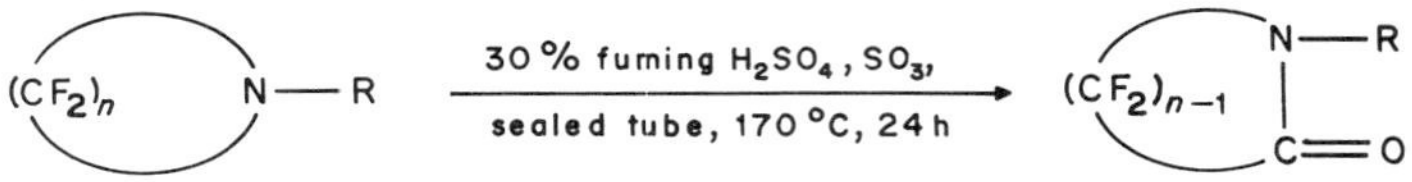

$(CF_2)_n$ N—R $\xrightarrow[\text{sealed tube, 170 °C, 24 h}]{\text{30\% fuming } H_2SO_4\text{, } SO_3\text{,}}$ $(CF_2)_{n-1}$ N—R, C=O (1352)

Substrate	Product	Yield (%)[a]
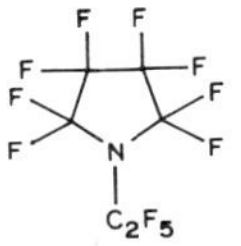 perfluoro-N-(C_2F_5)pyrrolidine	 lactam, N-C_2F_5	65.3
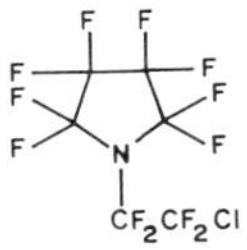 perfluoro-N-(CF_2CF_2Cl)pyrrolidine	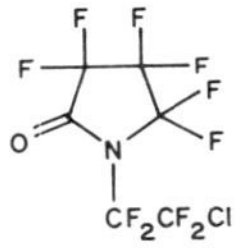 lactam, N-CF_2CF_2Cl	64.6
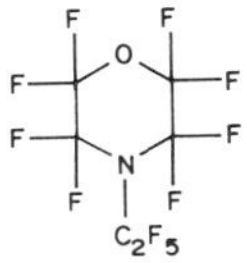 perfluoro-N-(C_2F_5)morpholine	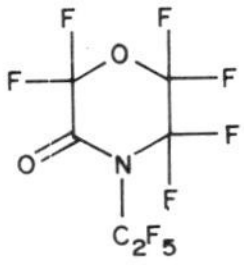 lactam, N-C_2F_5	51.2
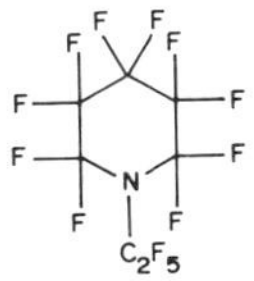 perfluoro-N-(C_2F_5)piperidine	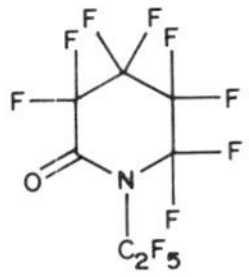 lactam, N-C_2F_5	61.7
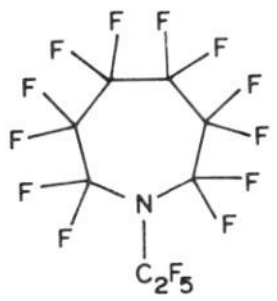 perfluoro-N-(C_2F_5)azepane	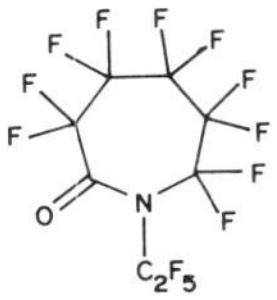 lactam, N-C_2F_5	49.2

[a]Yields based upon substrate consumed.

Included among the several other miscellaneous preparations of lactams which have been reported, is the hydrolysis[1489] of perfluoro(*N*-alkyl cyclic amines) using fuming sulphuric acid, which produces 5-, 6- and 7-membered lactams by hydrolysis of the fluorine atoms attached to the alpha carbon only (equation 1352).

Basic hydrolysis of (2-ethoxycarbonyl-5,8,8-tri- and 8,8-dimethyl-1,2-diazabicyclo-[5.2.0]nona-3,5-dien-9-yliden)dimethylammonium chloride produces[1127] the corresponding bicyclic lactams (equation 1353).

NaOH, CH_2Cl_2 (1353)

R = H ; Me

Yield(%) = 29; 25

Oxidative decarboxylation of cyclic amino acids produces β-lactams[1490], an example of which is shown in equation 1354.

1. LDA, THF, 0 °C
2. −78 °C
3. oxygen sat. ether at −78 °C containing *p*-TosOH, THF

60%

(1354)

Treatment of *N*-chlorocyclopropanolamines with silver nitrate in acetonitrile affords[1039] the corresponding β-lactams via a ring expansion reaction (equation 1355).

$AgNO_3$, MeCN, r. t., 1.5 h

R = $C(CO_2Et)_2C_6H_4OCH_2Ph$-*p* ; $CH(CO_2Et)An$-*p*; $CH(CO_2CH_2Ph)C_6H_4OCH_2Ph$-*p*

Yield (%) = 38 ; 52 ; 59

(1355)

Although the photoisomerization of 2- and 3-substituted phenazine *N*-oxides produces[1491] tricyclic lactams, the products and yields differ, depending upon the location of the substituent (equation 1356).

Two unsaturated peptide substrates upon incubation with the enzyme isopenicillin N synthase (IPNS) from *Cephalosporium acremonium* CO 728 each cyclize simultaneously to afford[1492] both desaturated and hydroxylated β-lactam products whose hydroxyl groups derive their oxygen from the cosubstrate dioxygen (equations 1357 and 1358).

(1356)

MeCN $h\nu$

MeCN $h\nu$

(294) (295) (296) (297)

Location of substitution	X	Yield (%) 294	295	296	297
2	Cl	15	15	14	—
2	MeO	36	14	11.5	—
2	Me	Total of **294** + **295** + **296** = 34.5			—
2	CN	24	13	21	—
3	Cl	—	7.5	6.5	10
3	MeO	—	2	1	2
3	Me	Total of **294** + **295** + **296** = 56			—

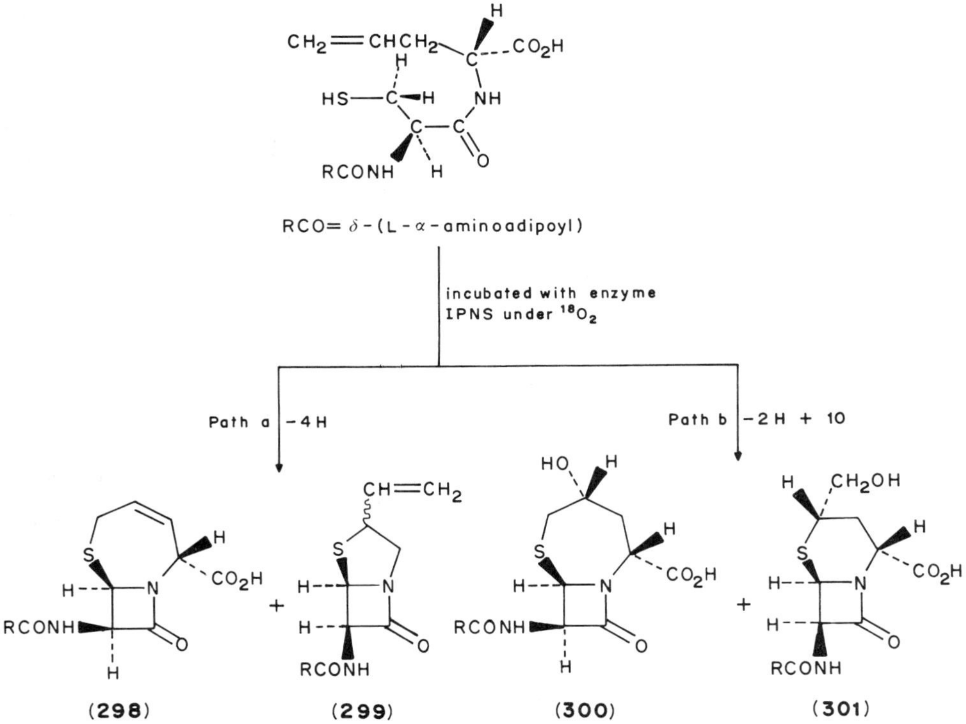

RCO= δ-(L-α-aminoadipoyl)

incubated with enzyme IPNS under $^{18}O_2$

Path a −4H

Path b −2H + 1O

(298) + **(299)** **(300)** + **(301)**

(1357)

For each substrate two pathways occur, Path a, a desaturative (−4H) route, and Path b, a hydroxylative (−2H + 1O) route, each path leading to different products. The products from Path b, **300**, **301**, and **303** show incorporation of ^{18}O and the details of the location and extent of this incorporation are presented in the original reference[1492], which also presents a proposed mechanism, involving iron-oxo species, to explain this phenomenon.

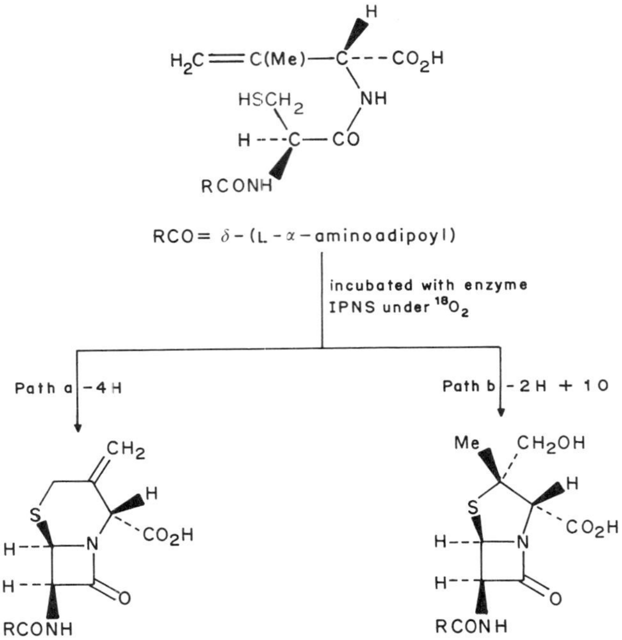

$H_2C{=}C(Me){-}CH(CO_2H){-}NH{-}CO{-}CH(CH_2SH)(NHCOR)$

RCO= δ-(L-α-aminoadipoyl)

incubated with enzyme IPNS under $^{18}O_2$

Path a −4 H → (302)

Path b −2 H + 1 O → (303)

(1358)

Novel phosphorus analogues of unsaturated β-lactams have been reportedly[1493] prepared by the carbonylation of a strained three-membered unsaturated ring containing phosphorus. Thus, carbonylation of phosphirene-chromium, -molybdenum and -tungsten pentacarbonyl complexes at 160 °C produce 2-oxo-1,2-dihydrophosphete complexes (equation 1359).

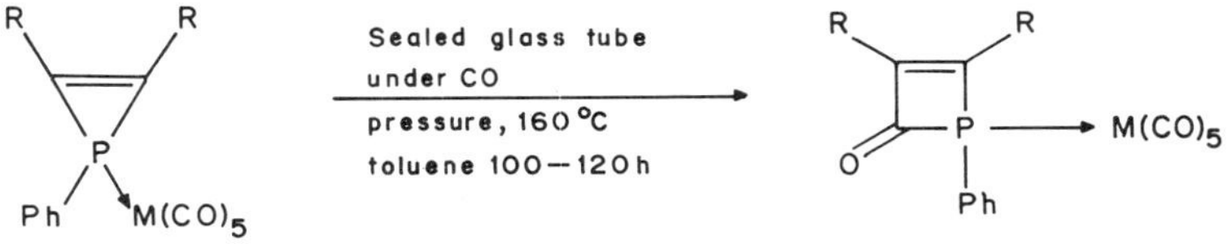

Sealed glass tube under CO pressure, 160 °C toluene 100–120 h

(1359)

R	M	Yield (%)
Ph	W	34
Et	W	43
Ph	Cr	8
Et	Cr	21
Ph	Mo	24

The mechanism proposed[1493] for these conversion involves the establishment of an equilibrium at high temperature between the phosphirene complex and 1-phospha-2-metallacyclobutenes, followed by carbonylation (equation 1360).

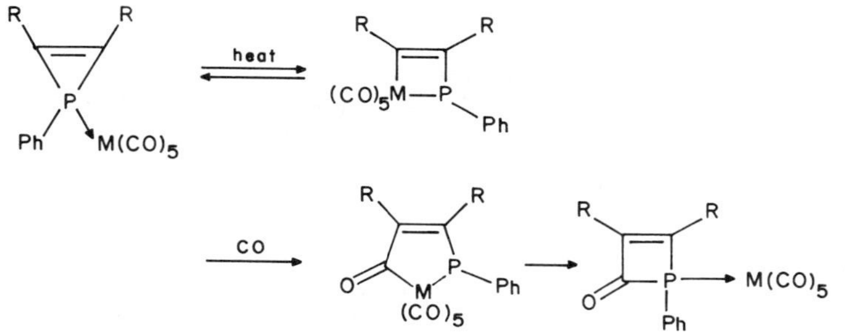

(1360)

The resulting 2-oxo-1,2-dihydrophosphete complexes may be decomplexed[1493] as shown in equation (1361).

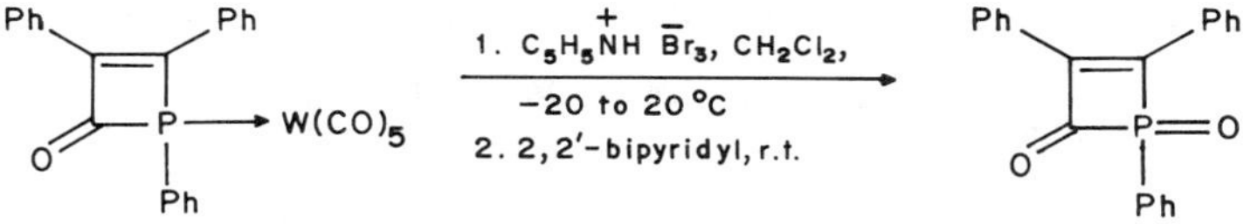

(1361)

*IV. ACKNOWLEDGMENTS

We are delighted to acknowledge our good colleague and superb typist, Mrs. Jeannie B. Turman, who gave many hours of overtime to skillful construction of this manuscript. We are grateful to the College of Arts and Sciences, the Department of Chemistry, the Office of the Senior Vice President and University Provost, and the Harvey W. Peters Research Center for Parkinson's Disease and Disorders of the Central Nervous System Foundation for financial support of this project and for providing us with a professional base of operations. The Peter's Foundation, the National Science Foundation, the Defense Advanced Research Program Administration, the National Institute of Neurological and Communicative Disorders and Stroke, and the National Aeronautics and Space Administration provided generous financial support of our research programs while this update was being written.

*V. REFERENCES

724. S. Glasgow, *Annu. Rep. Prog. Chem., Sect. B*, **72**, 199 (1975).
725. J. Raimahajan and H. Clemente de Arujo, *Cienc. Cult. (Sao Paulo)*, **32**, 893 (1980); *Chem. Abstr.*, **93**, 239274z (1980).
726. P. R. Jenkins, *Gen. Synth. Methods*, **6**, 98 (1983).
727. W. Adam, *Chem. Heterocycl. Compd.*, **42** (Part 3), 351 (1985).
728. S. Blechert, *Nachr. Chem. Tech. Lab.*, **28**, 218, 220 (1980); *Chem. Abstr.*, **93**, 26171b (1980).
729. T. Fujisawa and T. Sato, *Yuki Gossei Kagaku Kyokaishi*, **40**, 618 (1982); *Chem. Abstr.*, **97**, 198011y (1982).
730. M. D. Dowle and D. I. Davies, *Chem. Soc. Rev.*, **8**, 171 (1979).

731. J. Muzler, *Nachr. Chem. Tech. Lab.*, **29**, 614, 619 (1981); *Chem. Abstr.*, **95**, 149307z (1981).
732. O. Mitsunobu, *Kagaku Sosetsu*, **31**, 113 (1981); *Chem. Abstr.*, **95**, 149835v (1981).
733. T. Mukaiyama and K. Narasaka, *Kagaku Sosetsu*, **31**, 99 (1981); *Chem. Abstr.*, **95**, 149834u (1981).
734. A. Fajfer and J. Gora, *Pollena: Tluszcze, Srodki Piorace, Kosmet.*, **29**, 226 (1985); *Chem. Abstr.*, **105**, 60550a (1986).
735. T. G. Back, *Tetrahedron*, **33**, 3041 (1977).
736. K. C. Nicolaou, *Tetrahedron*, **33**, 683 (1977).
737. T. Aida and S. Inoue, *Yuki Gosei Kagaku Kyokaishi*, **43**, 300 (1985); *Chem. Abstr.*, **103**, 6729m (1985).
738. D. B. Johns and R. W. Lenz, *Polym. Prepr.*, **25**, 220 (1984).
739. W. H. Sharkey, *ACS Symp. Ser.*, **286**, 373 (1985); *Chem. Abstr.*, **104**, 88985z (1986).
740. S. Slomkowski and S. Penczek, *ACS Symp. Ser.*, **166**, 271 (1981); *Chem. Abstr.*, **96**, 20480s (1982).
741. H. Sumitomo and M. Okado, *Proc. IUPAC Macromol. Symp., 28th*, 216 (1982); *Chem. Abstr.*, **99**, 194827n (1983).
742. G. D. Annis, E. M. Hebblethwaite, S. T. Hodgson, A. M. Horton, D. M. Hollinshead, S. V. Ley, C. R. Self and R. Sivaramakrishnan, *Spec. Publ. R. Soc. Chem.*, (50), 148 (1984); *Chem. Abstr.*, **102**, 113092z (1985).
743. M. Kocor and W. Wojciechowska, *Symp. Pap. —IUPAC Int. Symp. Chem. Nat. Prod., 11th, Volume 2*, 157–159 (Eds. N. Marekov, I. Ognyanov and A. Orahovats), *Izd. BAN: Sofia; Chem. Abstr.*, **91** 193494z (1979).
744. M. Kocor, W. Kroszczynski and J. Pietrzak, *Symp. Pap. —IUPAC Int. Symp. Chem. Nat. Prod., 11th Volume 3*, 26–28 (Eds. N. Marekov. I. Ognyanov and A. Orahovats), *Izd. BAN: Sofia*; *Chem. Abstr.*, **91**, 193488a (1979).
745. G. D. Pandey and K. P. Tiwari, *Heterocycles*, **16**, 449 (1981).
746. A. V. Kamernitskii, I. G. Reshetova and K. Y. Chernyuk, *Khim.-Farm. Zh.*, **11**, 65 (1977); *Chem. Abstr.*, **88**, 152839k (1978).
747. C. Kaneko, M. Sato and N. Katagiri, *Yuki Gosei Kagaku Kyshaishi*, **44**, 1058 (1986); *Chem. Abstr.*, **107**, 154092w (1987).
748. H. J. Gais, *NATO Adv. Study Inst. Ser., Sec. C*, **178**, 97 (1986); *Chem. Abstr.*, **105**, 221885f (1986).
749. J. P. Vigneron and V. Bloy, *Tetrahedron Letters*, **21**, 1735 (1980).
750. M. Shiozaki, N. Ishida, T. Hiraoka and H. Yanagisawa, *Tetrahedron Letters*, **22**, 5205 (1981).
751. A. W. Johnson, G. Gowda, A. Hossanali, J. Knox, S. Monaco, Z. Razavi and G. Rosebery, *J.Chem. Soc., Perkin Trans. 1*, 1734 (1981).
752. V. V. Vanin, L. N. Krutskii, L. N. Sveshnikova and L. V. Krutskaya, *Zh. Prikl. Khim. (Leningrad)*, **57**, 628 (1984); *Chem. Abstr.*, **101**, 90718w (1984).
753. S. W. Baldwin and M. T. Crimmins, *Tetrahedron Letters*, 4197 (1978).
754. M. Yamaguchi, K. Shibato and I. Hirao, *Tetrahedron Letters*, **25**, 1159 (1984).
755. R. Gull and U. Schoellkopf, *Synthesis*, 1052 (1985).
756. J. W. Scheeren, F. J. M. Dahmen and C. G. Bakker, *Tetrahedron Letters*, 2925 (1979).
757. D. Guillerm and G. Linstrumelle, *Tetrahedron Letters*, **26**, 3811 (1985).
758. A. Ichihara, N. Nio and S. Sakamura, *Tetrahedron Letters*, **21**, 4467 (1980).
759. J.-P. Cobet and C. Benezra, *Tetrahedron Letters*, 4003 (1979).
760. P. T. L ansbury and J. P. Vacca, *Tetrahedron Letters*, **23**, 2623 (1982).
761. F. E. Ziegler, E. P. Stirchak and R. T. Wester, *Tetrahedron Letters*, **27**, 1229 (1986).
762. C. Bonini and R. Di Fabio, *Tetrahedron Letters*, **23**, 5199 (1982).
763. M. Ochiai, E. Fujita, M. Arimoto and H. Yamaguchi, *Tetrahedron Letters*, **24**, 777 (1983).
764. K. Venkataraman and D. R. Wagle, *Tetrahedron Letters*, **21**, 1893 (1980).
765. R. J. Lahoti and D. R. Wagle, *Indian J. Chem. Sect. B*, **20B**, 852 (1981).
766. T. Mukaiyama, M. Usui, K. Saigo and E. Shimada, *German Patent DE 2628941* (1977); *Chem. Abstr.*, **87**, 23086t (1977).
767. L. Strekowski, M. Visnick and M. A. Battiste, *Synthesis*, 493 (1983).
768. N. Cohen, B. L. Banner and R. J. Lopresti, *Tetahedron Letters*, **21**, 4163 (1980).
769. D. Obrecht and H. Heimgartner, *Tetrahedron Letters*, **25**, 1717 (1984).
770. T. Shono, O. Ishige, H. Uyama and S. Kashimura, *J. Org. Chem.*, **51**, 546 (1986).
771. R. M. Ortuno, D. Alonso and J. Font, *Tetrahedron Letters*, **27**, 1079 (1986).
772. J. A. Moore and J. E. Kelly, *J. Polym. Sci., Polym. Letters Ed.*, **13**, 333 (1975).

773. K. Steliou, A. Szczygielska-Nowasielska, A. Favre, M. A. Poupart and S. Hanessian, *J. Amer. Chem. Soc.*, **102**, 7578 (1980).
774. J. P. Vigneron and J. M. Blanchard, *Tetrahedron Letters*, **21**, 1739 (1980).
775. M. M. Midland and A. Tramontano, *Tetrahedron Letters*, **21**, 3549 (1980).
776. M. Mori, Y. Washioka, T. Urayama, K. Yoshiura, K. Chiba and Y. Ban, *J. Org. Chem.*, **48**, 4058 (1983).
777. D. Ben-Ishai, *J. Chem. Soc., Chem. Commun.*, 687 (1980).
778. A. W. McCulloch and A. G. McInnes, *Tetrahedron Letters*, 1963 (1979).
779. C. Estopa, J. Font, M. Moreno-Manas, F. Sanchez-Ferrando, S. Valle and L. Vilamajo, *Tetrahedron Letters*, **22**, 1467 (1981).
780. K. G. Bilyard and P. J. Garratt, *Tetrahedron Letters*, **22**, 1755 (1981).
781. T. V. Lee and J. Toczek, *Tetrahedron Letters*, **26**, 473 (1985).
782. G. Karminski-Zamola, L. Fiser-Jakic and K. Jakopcic, *Tetrahedron*, **38**, 1329 (1982).
783. A. Soto, A. Ogiso, H. Noguchi, S. Mitsui, I. Kaneko and Y. Shimada, *Chem. Pharm. Bull.*, **28**, 1509 (1980).
784. B. B. Snider and M. I. Johnston, *Tetrahedron Letters*, **26**, 5497 (1985).
785. V. Joeger and H. J. Guenther, *Tetrahedron Letters*, 2543 (1977).
786. G. Stork and E. W. Logusch, *Tetrahedron Letters*, 3361 (1979).
787. P. A. Zoretic, C. Bhakta and R. H. Khan, *Tetrahedron Letters*, **24**, 1125 (1983).
788. M. J. Batchelor and J. M. Mellor, *Tetrahedron Letters*, **26**, 5109 (1985).
789. Y. N. Gupta, R. T. Patterson, A. Z. Bimanand and K. N. Houk, *Tetrahedron Letters*, **27**, 295 (1986).
790. D. Goldsmith, D. Liotta, C. Lee and G. Zima, *Tetrahedron Letters*, 4801 (1979).
791. J. N. Marx and P. J. Dobrowolski, *Tetrahedron Letters*, **23**, 4457 (1982).
792. P. Canonne, M. Akssira and G. Lemay, *Tetrahedron Letters*, **22**, 2611 (1981).
793. P. Canonne, G. Lemay and D. Belanger, *Tetrahedron Letters*, **21**, 4167 (1980).
794. N. El Alami, C. Belaud and J. Villieras, *J. Organomet. Chem.*, **319**, 303 (1987).
795. J. Ficini, S. Falou and J. D'Angelo, *Tetrahedron Letters*, **24**, 375 (1983).
796. R. J. Sims, S. A. Tischler and L. Weiler, *Tetrahedron Letters*, **24**, 253 (1983).
797. F. W. Machado-Aracijo and J. Gore, *Tetrahedron Letters*, **22**, 1969 (1981).
798. Y. Takahashi, S. Hasegawa, T. Izawa, S. Kobayashi and M. Ohno, *Chem. Pharm. Bull.*, **34**, 3020 (1986).
799. J. L. Roberts, P. S. Borromeo and C. D. Poulter, *Tetrahedron Letters*, 1621 (1977).
800. S. Kano, T. Ebata, K. Fumaki and S. Shibuya, *J. Org. Chem.*, **44**, 3946 (1979).
801. J. P. Marino and R. Fernandez de la Prodilla, *Tetrahedron Letters*, **26**, 5381 (1985).
802. G. Just and R. Zamboni, *Can. J. Chem.*, **56**, 2720 (1978).
803. T. Ikariya, K. Osakada and S. Yoshikawa, *Tetrahedron Letters*, 3749 (1978).
804. J. E. Lyons, *U.S. Patent* US 3957827 (1976); *Chem. Abstr.*, **85**, 77672h (1976).
805. J. E. Lyons, *U.S. Patent* US 4485246 (1984); *Chem. Abstr.*, **102**, 131901s (1985).
806. Y. Ishii, T. Ikariya, M. Saburi and S. Yoshikawa, *Tetrahedron Letters*, **27**, 365 (1986).
807. K. Osakada, M. Obana, T. Ikariya, M. Saburi and S. Yoshikawa, *Tetrahedron Letters*, **22**, 4297 (1981).
808. M. Larcheveque, C. Legueut, A. Debal and J. Y. Lallemand, *Tetrahedron Letters*, **22**, 1595 (1981).
809. G. L. Lange and M. Lee, *Tetrahedron Letters*, **26**, 6163 (1985).
810. H. F. Chow and I. Fleming, *Tetrahedron Letters*, **26**, 397 (1985).
811. E. Brown and A. Daugan, *Tetrahedron Letters*, **26**, 3997 (1985).
812. B. Berthon, A. Forestiere, G. Leleu and B. Sillion, *Tetrahedron Letters*, **22**, 4073 (1981).
813. K. S. Kim and W. A. Szarek, *Carbohydr. Res.*, **104**, 328 (1982).
814. M. P. Doyle, R. L. Dow, V. Bagheri and W. J. Patrie, *Tetrahedron Letters*, **21**, 2795 (1980).
815. H. Nishiyama, H. Yokoyama, S. Narimatsu and K. Itoh, *Tetrahedron Letters*, **23**, 1267 (1982).
816. R. M. Carlson, *Tetrahedron Letters*, 111 (1978).
817. A. P. Kozikowski and A. K. Ghosh , *Tetrahedron Letters*, **24**, 2623 (1983).
818. E. Moret and M. Schlosser, *Tetrahedron Letters*, **25**, 4491 (1984).
819. S. Murahasji, K. Ito, T. Naota and Y. Maeda, *Tetrahedron Letters*, **22**, 5327 (1981).
820. T. K. Chakraborty and S. Chandrasekaran, *Tetrahedron Letters*, **25**, 2891 (1984).
821. T. Sato and R. Noyori, *Tetrahedron Letters*, **21**, 2535 (1980).

822. Y. Ueno, K. Chino, M. Watanabe, O. Moriya and M. Okawara, *J. Amer. Chem. Soc.*,**104**, 5564 (1982).
823. M. Ladlow and G. Pattendan, *Tetrahedron Letters*, **25**, 4317 (1984).
824. S. F. Martin and D. E. Guinn, *Tetrahedron Letters*, **25**, 5607 (1984).
825. P. A. Grieco, T. Oguri and Y. Yokayama, *Tetrahedron Letters*, 419 (1978).
826. D. Hoppe and A. Broenneke, *Tetrahedron Letters*, **24**, 1687 (1983).
827. G. Piancatelli, A. Scettri and M. D'Auria, *Tetrahedron Letters*, 3483 (1977).
828. P. Jarglis and F. W. Lichtenthaler, *Tetrahedron Letters*, **23**, 3781 (1982).
829. T. K. Chakraborty and S. Chandrasekaran, *Tetrahedron Letters*, **25**, 2895 (1984).
830. M. F. Schlecht and H. J. Kim, *Tetrahedron Letters*, **26**, 127 (1985).
831. R. Rathore, P. Vankar and S. Chandrasekaran, *Tetrahedron Letters*, **27**, 4079 (1986).
832. P. G. M. Wuts, M. L. Obrzut and P. A. Thompson, *Tetrahedron Letters*, **25**, 4051 (1984).
833. P. T. Lansbury and T. E. Nickson, *Tetrahedron Letters*, **23**, 2627 (1982).
834. P. Mangeney and Y. Langlois, *Tetrahedron Letters*, 3015 (1978).
835. M. Demuynck, A. A. Devreese, P. J. DeClercq and M. Vandewalle, *Tetrahedron Letters*, **23**, 2501 (1982).
836. G. Stork and M. Khan, *Tetrahedron Letters*, **24**, 3951 (1983).
837. I. Kuwajima and H. Urabe, *Tetrahedron Letters*, **22**, 5191 (1981).
838. M. Bertrand, J. P. Dulcere and G. Gerard, *Tetrahedron Letters*, **21**, 1945 (1980).
839. K. Wernges and H. Smuda, *Justus Liebigs Ann. Chem.*, 227 (1987).
840. A. A. Frimer, P. Gilinsky-Sharon and G. Aljadeff, *Tetrahedron Letters*, **23**, 1301 (1982).
841. G. I. Nikishin, E. I. Troyanskii and I. V. Svitanko, *Izv. Akad. Nauk SSSR, Ser. Khim.*, 1436 (1981); *Chem. Abstr.*, **95**, 114757u (1981).
842. M. P. Bertrand, A. Oumar-Mahamat and J. M. Surzur, *Tetrahedron Letters*, **26**, 1209 (1985).
843. A. Suginome and S. Yamada, *Tetrahedron Letters*, **26**, 3715 (1985).
844. K. Hayakawa, S. Ohsuki and K. Kanematsu, *Tetrahedron Letters*, **27**, 947 (1986).
845. M. E. Krafft, *Tetrahedron Letters*, **27**, 771 (1986).
846. E. J. Corey, N. W. Gilman and B. E. Ganem, *J. Amer. Chem. Soc.*, **90**, 5616 (1968).
847. E. J. Corey and G. Schmidt, *Tetrahedron Letters*, **21**, 731 (1980).
848. M. Ochiai, T. Ukita and E. Fujita, *Tetrahedron Letters*, **24**, 4025 (1983).
849. M. Foa, F. Francalanci, E. Bencini and A. Gardano, *J. Organomet. Chem.*, **285**, 293 (1985).
850. H. Alper, J. K. Currie and H. des Abbayes, *J. Chem. Soc., Chem. Commun.*, 311 (1978).
851. T. Hirao, Y. Harano, Y. Yamana, Y. Ohshiro and T. Agawa, *Tetrahedron Letters*, **24**, 1255 (1983).
852. A. Cowell and J. K. Stille, *Tetrahedron Letters*, 133 (1979).
853. Y. Tamaru, T. Kobayashi, S. Kawamura, H. Ochiai and Z. Yoshida, *Tetrahedron Letters*, **26**, 4479 (1985).
854. A. Dobrev, B. Dimitrova, T. Cholskova and K. Ivanov, *God. Sofii. Univ. Khim. Fak.*, 1975–1976, **70**, Pt. 2,5 (1979); *Chem. Abstr.*, **93**, 185696t (1980).
855. D. Spitzner, W. Weber and W. Kraus, *Tetrahedron Letters*, **23**, 2179 (1982).
856. B. Bardili, A. Marschall-Weyerstahl and P. Weyerstahl, *Justus Liebigs Ann. Chem.*, 275 (1985).
857. T. Kitamura, T. Imagawa and M. Kawanishi, *Tetrahedron Letters*, 3443 (1978).
858. T. Kitamura, Y. Kawakami, T. Imagawa and M. Kawanishi, *Tetrahedron Letters*, 4297 (1978).
859. H. Kunz and M. Lindig. *Chem. Ber.*, **116**, 220 (1983).
860. Y. Morizawa, T. Hiyama and H. Nozaki, *Tetrahedron Letters*, **22**, 2297 (1981).
861. T. Hiyama, H. Sarimoto, K. Nishio, M. Shinoda, H. Yamamoto and H. Nozaki, *Tetrahedron Letters*, 2043 (1979).
862. V. Jaeger and H. J. Guenther, *Tetrahedron Letters*, 2543 (1977).
863. M. Pohmakotr and P. Jarupan, *Tetrahedron Letters*, **26**, 2253 (1985).
864. J. A. Ray and T. M. Harris, *Tetrahedron Letters*, **23**, 1971 (1982).
865. A. H. Khan and I. Paterson, *Tetrahedron Letters*, **23**, 5083 (1982).
866. P. Deslongchamps, S. Dube, C. Lebreux, D. R. Patterson and R. J. Taillefer, *Can. J. Chem.*, **53**, 2791 (1975).
867. R. B. Nader and M. K. Kaloustian, *Tetrahedron Letters*, 1477 (1979).
868. M. K. Kaloustian and F. Khouri, *Tetrahedron Letters*, **22**, 413 (1981).
869. J. C. Oho, G. W. J. Fleet, J. M. Peach, K. Prout and P. W. Smith, *Tetrahedron Letters*, **27**, 3203 (1986).
870. D. Caine and A. S. Frobese, *Tetrahedron Letters*, 883 (1978).

871. D. Caine and A. S. Frobese, *Tetrahedron Letters*, 5167 (1978).
872. D. L. Comins and J. D. Brown, *J.Org. Chem.*, **51**, 3566 (1986).
873. T. Tabuchi, K. Kawamura, J. Inanoga and M. Yamaguchi, *Tetrahedron Letters*, **27**, 3889 (1986).
874. H. H. Wasserman, R. J. Gambale and M. J. Pulwer, *Tetrahedron Letters*, **22**, 1737 (1981).
875. T. Shono, H. Ohmizu, S. Kawakami and H. Sugiyama, *Tetrahedron Letters*, **21**, 5029 (1980).
876. T. Shono, Y. Matsumura and S. Yamane, *Tetrahedron Letters*, **22**, 3269 (1981).
877. R. M. Jacobson and J. W. Clader, *Tetrahedron Letters*, **21**, 1205 (1980).
878. E. I. Heiba, R. M. Dessau and P. G. Rodewald, *J. Amer. Chem. Soc.*, **96**, 7977 (1974).
879. E. I. Heiba, R. M. Dessau, A. L. Williams and P. G. Rodewald, *Org. Synth.*, **63**, 22 (1983).
880. T. Fujita, S. Watanabe, K. Suga, T. Kuramochi and F. Tsukogoshi, *J. Chem. Tech. Biotechnol*, **29**, 31 (1979); *Chem. Abstr.*, **92**, 58284m (1980).
881. T. Fujita, K. Suga, S. Watanabe, H. Nakayama and M. Hokyo, *Yukagaku*, **26**, 720 (1977); *Chem. Abstr.*, **88**, 104629v (1978).
882. F. Gaudemar-Bardone, M. Mladenova and R. Couffignal, *Tetrahedron Letters*, **25**, 1047 (1984).
883. T. Fujisawa, T. Mori, K. Fukumoto and T. Sato, *Chem. Letters*, 1891 (1982).
884. J. Gallastegui, J. M. Lago and C. Palomo, *J. Chem. Res.(S)*, 170 (1984).
885. G. J. O'Malley and M. P. Cava, *Tetrahedron Letters*, **26**, 6159 (1985).
886. E. J. Corey and A. W. Gross, *Tetrahedron Letters*, **21**, 1819 (1980).
887. D. J. Goldsmith and J. K. Thottathil, *Tetrahedron Letters*, **22**, 2447 (1981).
888. D. A. Claremon and S. D. Young, *Tetrahedron Letters*, **26**, 5417 (1985).
889. S. P. Brown, B. S. Bal and H. W. Pinnick, *Tetrahedron Letters*, **22**, 4891 (1981).
890. T. C. T. Chang and M. Rosenblum, *Tetrahedron Letters*, **24**, 695 (1983).
891. M. Kennedy, A. R. Maguire and M. A. McKervey, *Tetrahedron Letters*, **27**, 761 (1986).
892. M. Wada, J. Shigehisa and K. Akiba, *Tetrahedron Letters*, **26**, 5191 (1985).
893. T. Takahashi, K. Kasuga and J. Tsuji, *Tetrahedron Letters*, 4917 (1978).
894. A. Shanzer and E. Schwartz, *Tetrahedron Letters*, 5019 (1979).
895. K. Frensch and F. Voegtle, *Tetrahedron Letters*, 2573 (1977).
896. A. V. Bogatskii, N. G. Luk'yanenko, V. A. Shapkin, Y. A. Popkov, N. Y. Nazarova and Z. A. Chernotkach, *Zh. Org. Khim.*, **20**, 878 (1984); *Chem. Abstr.*, **101**, 130673m (1984).
897. C. Picard, L. Cazaux and P. Tisnes, *Tetrahedron Letters*, **25**, 3809 (1984).
898. H. Westmijze, K. Ruitenberg. J. Meijer and P. Vermeer, *Tetrahedron Letters*, **21**, 1771 (1980).
899. R. C. Cookson and P. S. Ray, *Tetrahedron Letters*, **23**, 3521 (1982).
900. S. Castellino and J. J. Sims, *Tetrahedron Letters*, **25**, 2307 (1984).
901. S. Kano, S. Shibuya and T. Ebata, *J. Chem. Soc., Perkin. Trans. 1*, 257 (1982).
902. M. S. Manhas, S. G. Amin and R. D. Glazer, *J. Heterocycl Chem.*, **16**, 283 (1979).
903. R. G. Glushkov, I. M. Zasosova, I. M. Ovcharova, N. P. Soloveva, O. S. Anisimova and Y. N. Sheinker, *Khim, Geterotsikl. Soedin.*, 1504 (1978); *Chem. Abstr.*, **90**, 87318q (1979).
904. S. Turner, *Gen. Synth. Methods*, **4**, 335 (1981).
905. G. A. Koppel, *Chem. Heterocycl. Cmpd.*, **42** (Small Ring Heterocycl., Part 2), 219 (1983).
906. N. S. Isaacs, *Chem. Soc. Rev.*, **5**, 181 (1976).
907. Y. Deng, *Kangshengsu*, **10**, 142 (1985); *Chem Abstr.*, **104**, 129664q (1986).
908. K. Hirai, *Yuki Gosei Kagaku Kyokaishi*, **38**, 97 (1980); *Chem. Abstr.*, **93**, 26170a (1980).
909. H. Ogura, *Lifechem* (*Tokyo*), **1**, 35 (1978); *Chem. Abstr.*, **94**, 102243q (1981).
910. A. K. Bose, B. Ram and M. S. Manhas, *Int. Congr. Ser. —Excerpta Med.*, **457** (Stereosel. Synth. Nat. Prod.), 181 (1979); *Chem. Abstr.*, **91**, 157625j (1979).
911. R. Labia and C. Morin, *J: Antibiot.*, **37**, 1103 (1984).
912. W. J. Ross, *Gen. Synth. Methods*, **4**, 279 (1981).
913. T. Shono and Y. Matsumura, *Yuki Gosei Kagaku Kyokaishi*, **39**, 358 (1981); *Chem. Abstr.*, **95**, 61328w (1981).
914. R. D. G. Cooper, in *Topics in Antibiotic Chemistry* (Ed. P. G. Sammes), Vol. 3, Part B, Ellis Harwood, Chichester, 1980, pp. 39–199.
915. F. J. Jung, R. Pilgrem J. P. Poyser and P. J. Sirt, in *Topics in Antibiotic Chemistry* (Ed. P. G. Sammes), Vol. 4, Ellis Harwood, Chichester, 1980, pp. 51–119.
916. R. J. Stoodley, *Spec. Publ. R. Soc. Chem.*, **52** (Recent Adv. Chem. β-Lactam Antibiot.), 183 (1985).
917. M. Kajtar, *Kem. Kozl.*, **58**, 379 (1982); *Chem. Abstr.*, **99**. 176235p (1983).

918. S. Torii. H. Tanaka, M. Sasaoka, N. Saitoh, T. Siroi and J. Nokami, *Bull. Soc. Chim. Belg.*, **91**, 951 (1982); *Chem. Abstr.*, **98**, 160472g (1983).
919. I. Ugi, H. Aigner, G. Glahol, R. Herges, G. Hering, R. Herrmann, G. Huebener, J. Goetz, P. Lemmen, et al., *Nat. Prod. Chem. Proc. Int. Symp. Pak. —U.S. Binatl. Workshop, 1st* (Ed. A. Ur. Rahman), Springer Berlin, 1984, pp. 457–484; *Chem. Abstr.*, **107**, 97063m (1987).
920. L. Ghosez, S. Bogdan, M. Caresiat, C. Frydrych, J. Marchand-Brynaert, M. Moya Portuguez and I. Huber, *Pure Appl. Chem.*, **59**, 393 (1987).
921. A. K. Bose, M. S. Manhas, J. M. Van der Veen, S. S. Bari, D. R. Wagle and V. R. Hegde, *Spec. Publ. R. Soc. Chem.*, **52** (Recent Adv. Chem β-Lactam Antibiot.), 387 (1985).
922. A. K. Bose, J. E. Vincent, I. F. Fernandez, K. Gala and M. S. Manhas, *Spec. Publ. R. Soc. Chem.*, **38** (Recent Adv. Chem. β-Lactam Antibiot.), 80 (1980).
923. C. J. Ashcroft, J. Brennan, W. J. Norris and C. A. Robson, *Spec. Publ. R. Soc. Chem.*, **52** (Recent Adv. Chem. β-Lactam Antibiot.), 151 (1985).
924. D. Davies and M. J. Pearson, *Spec. Publ. R. Soc. Chem.*, **38** (Recent Adv. Chem. β- Lactam Antibiot.), 88 (1980).
925. J. C. Pechere, *Presse Med.*, **13**, 1485 (1984); *Chem. Abstr.*, **101**, 78707j (1984).
926. E. Edelweiss, S. M. Martins, C. A. Aldabe and V. Severo, *Pesqui. Med.*, **12**, 269 (1977); *Chem. Abstr.*, **89**, 173186f (1978).
927. J. Marchand-Brynaert, *Chim. Nouv.*, **4**, 413, 420 (1986); *Chem. Abstr.*, **108**, 5702p (1988).
928. G. Lowe and S. Swain. *Spec. Publ. R. Soc. Chem.*, **52** (Recent Adv. Chem. β-Lactam Antibiot.), 209 (1985).
929. H. Sakai, *Yuki Gosei Kagaku Kyokaishi*, **39**, 243 (1981); *Chem. Abstr.*, **94**, 197453v (1981).
930. S. Shindai, *Gekkan Yakuji*, **20**, 311 (1978); *Chem. Abstr.*, **88**, 130585x (1978).
931. C. Ball, *FEMS Symp.*, **13**, (Overprod. Microb. Prod.), 515 (1982); *Chem. Abstr.*, **98**, 87499q (1983).
932. A. K. Mukerjee and A. K. Sing, *Tetrahedron*, **34**, 1731 (1978).
933. M. Neuman, *Drugs Exp. Clin. Res.*, **3**, 1 (1977); *Chem. Abstr.*, **89**, 16512s (1978).
934. G. Lowe, in *Comprehensive Organic Chemistry*, **5** (Ed. E. Haslam), Pergamon, Oxford, 1979, pp. 289–320.
935. S. Blechert, *Nachr. Chem. Tech. Lab.*, **27**, 127 (1979); *Chem. Abstr.*, **91**, 20351x (1979).
936. H. Tagawa, *Med. Pharm.*, **14**, 378 (1980).
937. C. M. Cimarusti and R. B. Sykes, *Chem. Ber.*, **19**, 302 (1983).
938. I. Nobuo, *Hoshasen Kagaku*, **28**, 53 (1985); *Chem. Abstr.*, **103**, 215082j (1985).
939. W.Nagata, *Gekkan Yakuji*, **26**, 55 (1984); *Chem. Abstr.*, **100**, 68045g (1984).
940. B. G. Christensen and R. W. Ratcliffe, *Annu. Rep. Med. Chem.*, **11**, 271 (1976).
941. T. Kametani, *Lect. Heterocycl. Chem.*, **8**, 1 (1985); *Chem. Abstr.*, **105**, 43120t (1986).
942. E. P. Abraham, *Beta Lactam Antibiot.* (Ed. S. Mitsukhashi), Japan Sci. Soc. Press, Tokyo, 1981, pp. 3–11; *Chem. Abstr.*, **95**, 186108f (1981).
943. A. K. Bose and M. S. Manhas, *Lect. Heterocycl. Chem.*, **3**, 45 (1976); *Chem. Abstr.*, **85**, 159926h (1976).
944. A. Zdunska, *Farm. Pol.*, **38**, 645 (1982); *Chem. Abstr.*, **98**, 215357f (1983).
945. D. Dzierzanowska, *Farm. Pol.*, **38**, 653 (1982); *Chem. Abstr.*, **98**, 215358q (1983).
946. Y. Nagao, *Kagaku (Kyoto)*, **42**, 190 (1987); *Chem. Abstr.*, **108**, 5705s (1988).
947. M. Alpegiani, C. Ballistini. A. Bedeschi, G. Franceschi, F. Giudici, E. Perrone, C. Scarafile and F. Zarini, *Chim. Ind. (Milan)*, **68**, 70 (1986); *Chem. Abstr.*, **106**, 175971b (1987).
948. M. H. Richmond, *Symp. Giovanni Lorenzini Found.*, **10**, (New Trends Anitibiot.: Res. Ther.), 113 (1981); *Chem. Abstr.*, **95**, 55473y (1981).
949. R. Megges, H. J. Portius and K. R. H. Repke, *Symp. Pap. —IUPAC Int. Symp. Chem. Nat. Prod.* (Eds. N. Morekov, I. Ognyanov and A. Orahovats), Vol. 2, Izd. BAN, Sofia, 1978, pp. 164–166; *Chem. Abstr.*, **91**, 193495a (1979).
950. H. Thrum, *Biol. Rundsch.*, **19**, 69 (1981); *Chem. Abstr.*, **95**, 55472x (1981).
951. C. Morin and R. Labia, *Actual. Chim.*, 31 (1984): *Chem. Abstr.*, **101**, 90621j (1984).
952. D. Hoppe, *Nachr. Chem., Tech. Lab.*, **30**, 24 (1982); *Chem. Abstr.*, **96**, 162370e (1982).
953. M. Narisada and W. Nagata, *Symp. Heterocycl.*, [*Pap.*] (Ed. T. Kametani), Sendai Inst. Heterocycl. Chem., Sendai, Japan, 1977, pp. 148–153.
954. M. Narisada and W. Nagata, *Heterocycles*, **6**, 1646 (1977).
955. M. Yoshioka, *Pure Appl. Chem.*, **59**, 1041 (1987).

956. R. J. Stoodely, *Int. Congr. Sci. —Excerpta Med.*, **457** (Stereosel. Synth. Nat. Prod.), 193 (1979); *Chem. Abstr.*, **91**, 157626k (1979).
957. M. J.Miller, *Acc. Chem. Res.*, **19**, 49 (1986).
958. T. Saesmaa and J. Halmekoski, *Acta Pharm. Fenn.*, **96**, 65 (1987); *Chem. Abstr.*, **108**, 101159n (1988).
959. A. Mangia and G. Pantini, *Tecnol. Chim.*, **6**, 54 (1986); *Chem. Abstr.*, **108**, 5704r (1988).
960. A. Mangia and G. Pantini, *Tecnol. Chim.*, **7**, 78 (1987); *Chem. Abstr.*, **107**, 197840f (1987).
961. K. Okano, M. Iimori, M. Kurihara, Y. Takahashi, T. Izawa, S. Kobayashi and M. Ohno, *Tennen Yuki Kagobutsu Toronkai Koen Yoshishu*, **26**, 461 (1983); *Chem. Abstr.*, **100**, 68044f (1984).
962. W. Nagata, M. Narisada and T. Yoshida, *Chem. Biol. β-Lactam Antibiot.*, **2** (Eds R. B. Morin and M. Gorman), 1982, pp, 1–98.
963. K. G. Holden, *Chem. Biol. β-Lactam Antibiot.*, **2** (Eds. R. B. Mortin and M. Gorman), 1982, pp. 99–164.
964. S. Hanessian, A. Bedeschi, C. Battistini and N. Mongelli, *Lect. Heterocycl. Chem.*, **8**, 43 (1985); *Chem. Abstr.*, **104**, 206962q (1986).
965. S. Kano, *Yuki Gosei Kagaku Kyokaishi*, **36**, 581 (1978); *Chem. Abstr.*, **89**, 163440v (1978).
966. Y. Nagao, *Stud. Org. Chem.* (*Amsterdam*), **28** (Perspect. Org. Chem. Sulfur), 52 (1987); *Chem. Abstr.*, **108**, 37417r (1988).
967. Y. Ban, T. Wakamatsu, M. Mori and T. Ohnuma, *Stud. Org. Chem.* (*Amsterdam*), **6** (New Synth. Methodol. Biol. Act. Subst.), 177 (1981); *Chem. Abstr.*, **96**, 68661d (1982).
968. C. M. Cimarusti, *Gazz. Chim. Ital.*, **116**, 169 (1986); *Chem. Abstr.*, **106**, 66944z (1987).
969. H. Breuer, *Fortschr, Antimikrob. Antineoplast. Chemother.*, **1**, (Cephalosporine 80th Year), 7 (1982); *Chem. Abstr.*, **101**, 6866n (1984).
970. P. Vellekoop, *Chem. Weekbl. Mag.*, (Aug.) 401 (1977); *Chem. Abstr.*, **90**, 6266w (1979).
971. G. I. Gregory (Ed.), *Spec. Publ. R. Soc. Chem.*, **38** (Recent Adv.·Chem. β-Lactam Antibiot.), 1980, pp. 1–378.
972. J. Brennan, *Amino Acids Pept.*, **17**, 171 (1986).
973. J. Brennan, and A. Sheppard, *Amino Acids Pept.*, **18**, 222 (1987).
974. P. G. Sammes, *Chem. Rev.*, **76**, 113 (1976).
975. R. B. Woodward, *Spec. Publ. R. Soc. Chem*, **28** (Recent Adv. Chem. β-Lactam Antibiot.), 167 (1977).
976. J. Elks (Ed.), *Spec. Publ. R. Soc. Chem.*, **28** (Recent Adv. Chem. β-Lactam Antibiot.), 1977, 313 pp.
977. T. Hiraoka and M. Arai, *Hakko to Kogyo*, **37**, 854 (1979); *Chem. Abstr.*, **92**, 6436t (1980).
978. S. Wolfe, *Curr. Trends Org. Synth., Proc. Int. Conf.*, 4th (Ed. H. Nozaki), Pergamon, Oxford, 1982, pp. 101–144.
979. H. Hassner (Ed.), *The Chemistry of Heterocyclic Compounds*, Vol. 42: *Small Ring Heterocycles*, Pt. 2: *Azetidines, β-Lactams, Diazetidines, and Diaziridines*, Wiley, New York, 1983, 656 pp.
980. H. Breuer, *Fortschr. Antimikrob. Antineoplast. Chemother.*, **4**, 1855 (1985); *Chem. Abstr.*, **104**, 161387n (1986).
981. M. Hejzlar, *Vnitr. Lek.*, **25**, 1154 (1979); *Chem. Abstr.*, **92**, 122484f (1980).
982. N. P. Shusherina, *Khim. Geterstsiki Soedin.*, 867 (1983); *Chem. Abstr.*, **99**, 87945r (1983).
983. J. C. Sheehan. *Spec. Publ. R. Soc. Chem.*, **28**, (Recent Adv. Chem. β-Lactam Antibiot.), 20 (1977).
984. V. G. Granik, A. M. Zhidkova and R. G. Glushkov, *Usp. Khim.*, **46**, 685 (1977); *Chem. Abstr.*, **87**, 21650y (1977).
985. T. Shono and Y. Matsumura, *Kagaku (Kyoto)*, **39**, 114 (1984); *Chem. Abstr.*, **100**, 175220g (1984).
986. D. Groeger, *Pharmazie*, **32**, 309 (1977); *Chem. Abstr.*, **87**, 106667j (1977).
987. R. Southgate and S. Elson, *Prog. Chem. Org. Nat. Prod.*, **47**, 1 (1985).
988. A. G. Brown, *J. Antimicrob. Chemother.*, **7**, 15 (1981).
989. J. G. Gleason and W. D. Kingsbury, *Org. Cmpd. Sulfur, Selenium, Tellurium*, **5**, 454 (1979).
990. F. Arcamone, *Bull. Soc. Chim. Belg.*, **91**, 1003 (1982); *Chem. Abstr.*, **98**, 160473h (1983).
991. A. K. Bonetskaya, *Polyamidy* 1975, Sb. Prednasek, 97–102. Dum Tech. CVTS: Pardubice, Czech.; *Chem. Abstr.*, **84**, 31512g (1976).
992. J. Sebenda, *Pure Appl. Chem.*, **48**, 329 (1976).

993. J. Sebenda, *Compr. Chem. Kinet*, **15**, (Eds. C. H. Bamford and C. F. H. Tipper), Elsevier Amsterdam, 1976, pp. 379–471.
994. V. V. Korshak, V. A. Kotelnikov, V. V. Kurashev and T. M. Frunze, *Usp. Khim*, **45**, 1673 (1976); *Chem. Abstr.*, **85**, 193145n (1976).
995. J. Sebenda, *Prog. Polym. Sci.*, **6**, 123 (1978).
996. T. M. Frunze, V. V. Kurashev, V. A. Kotelnikov and T. V. Volkova, *Usp. Khim.*, **48**, 1856 (1979); *Chem. Abstr.*, **92**, 6962m (1980).
997. H. Gebler, *PTA Prakt. Pharm.*, **8**, 190 (1979); *Chem. Abstr.*, **92**, 109918f (1980).
998. A. K. Bose, *J. Heterocycl. Chem.*, **13**, 93 (1976).
999. J. E. Baldwin, C. Lowe, C. J. Schonfield and E. Lee, *Tetrahedron Letters*, **27**, 3461 (1986).
1000. R. M. Freidinger, D. S. Perlow and D. F. Veber, *J. Org. Chem.*, **47**, 104 (1982).
1001. Nippon Chemiphar Co., Ltd., *Jpn. Kokai Tokkyo Koho* JP 57/159758 A2 (1982); *Chem. Abstr.*, **98**, 107076e (1983).
1002. S. W. Baldwin and J. Aube, *Tetrahedron Letters*, **28**, 179 (1987).
1003. H. H. Wasserman and G. D. Berger, *Tetrahedron*, **39**, 2459 (1983).
1004. T. Shono, K. Tsubata and N. Okinaga, *J. Org. Chem.*, **49**, 1056 (1984).
1005. T. Shono, T. Tsubata and N. Okinaga, *Kenkyu Hokoku-Asahi Garasu Kogyo Gijutsu Shoreikai*, **44**, 9 (1984); *Chem. Abstr.*, **103**, 53836d (1985).
1006. T. Iimori, Y. Ishida and M. Shibasaki, *Tetrahedron Letters*, **27**, 2153 (1986).
1007. K. Ikeda, K. Achiwa and M. Sekiya, *Tetrahedron Letters*, **24**, 4707 (1983).
1008. J. J. Tufariello, G. E. Lee, P. A. Senaratne and M. Al-Nuri, *Tetrahedron Letters*, 4359 (1979).
1009. I. Panfil, M. Chmielewski and C. Belzecki, *Heterocycles*, **24**, 1609 (1986).
1010. A. Blade-Font, *An. Quim., Ser. C.*, **78**, 266 (1982); *Chem. Abstr.*, **97**, 182141e (1982).
1011. Lederle (Japan) Ltd., *Jpn. Kokai Tokkyo Koho* JP 59/163367 (1984); *Chem. Abstr.*, **102**, 132469u (1985).
1012. P. A. Grieco, D. S. Clark and G. P. Withers, *J. Org. Chem.*, **44**, 2945 (1979).
1013. S. Kobayashi, T. Iimori, T. Izawa and M. Ohno, *J. Amer. Chem. Soc.*, **103**, 2406 (1981).
1014. Kanegafuchi Chemical Industry Co., Ltd., *Jpn. Kokai Tokkyo Koho* JP 58/144367 (1983); *Chem. Abstr.*, **100**, 68071n (1984).
1015. D. A. Evans and E. B. Sjogren, *Tetrahedron Letters*, **27**, 4961 (1986).
1016. S. Hanessian and S. P. Sahoo, *Can. J. Chem.*, **62**, 1400 (1984).
1017. L. Crombie, R. C. F. Jones, A. R. Mat-Zin and S. Osborne, *J. Chem. Soc., Chem. Commun.*, 960 (1983).
1018. J. E. Baldwin, A. M. Adlington, R. H. Jones, C. J. Schofield, C. Zarocostas and C. W. Greengrass, *J. Chem. Soc., Chem. Commun.*, 194 (1985).
1019. J. E. Baldwin, R. M. Adlington, R. H. Jones, C. J. Schofield, C. Zarocostas and C. W. Greengrass, *Tetrahedron*, **42**, 4879 (1986).
1020. T. Nagasaka, A. Tsukoda and F. Hamaguchi, *Heterocycles*, **24**, 2015 (1986).
1021. N. Tamura, Y. Kawano, Y. Matsushita, K. Yoshioka and M. Ochiai, *Tetrahedron Letters*, **27**, 3749 (1986).
1022. J. -C. Roze, J. P. Pradere, G. Duguay, A. Guevel, H. Quiniou and S. Poignant, *Can. J. Chem.*, **61**, 1169 (1983).
1023. H. Huang, N. Iwasawa and T. Mukaiyama, *Chem. Letters* 1465 (1984).
1024. M. Chmielewski and S. Maciejewski, *Carbohydr. Res.*, **157**, Cl (1986).
1025. N. Iwasawa, H. Huang and T. Mukaiyama, *Chem. Letters*, 1045 (1985).
1026. K. Steliou and M. A. Poupart, *J. Amer. Chem. Soc.*, **105**, 7130 (1983).
1027. A. Blade-Font, *Afinidad*, **37**, 445 (1980); *Chem. Abstr.*, **95**, 80695e (1981).
1028. D. B. Collum, S. C. Chen and B. Ganem, *J. Org. Chem.*, **43**, 4393 (1978).
1029. A. Blade-Font, *Tetrahedron Letters*, **21**, 2443 (1980).
1030. R. Pellegata, M. Pinza and G. Pifferi, *Synthesis*, 614 (1978).
1031. H. Huang and S. Wu, *Gaodeng Xuexiao Huaxue Xuebao*, **8**, 341 (1987); *Chem. Abstr.*, **108**, 94303z (1988).
1032. M. Shimagaki, H. Koshiji and T. Oishi, *Heterocycles*, **17** (Spec. Issue), 49 (1982).
1033. D. S. Karanewsky, *Ger. Often. DE 3411282 A1* (1984); *Chem. Abstr.*, **102**, 149523e (1985).
1034. F. A. Pesa and A. M. Graham, *Eur. Pat. Appl. EP 23751* (1981); *Chem. Abstr.*, **94**, 192178z (1981).
1035. F. Mares, T. R. Tang, J. E. Galle and R. M. Federici. *U.S. Patent US 4628085* (1986); *Chem. Abstr.*, **106**, 85220b (1987).

1036. J. Geller and I. Ugi, *Chem. Scr.*, **22**, 85 (1983); *Chem. Abstr.*, **99**, 70438s (1983).
1037. H. P. Iserning and W. Hofheinz, *Synthesis* 385 (1981).
1038. H. Nitta, M. Hatanaka and T. Ishimaru. *J. Chem. Soc., Chem. Commun.*, 51 (1987).
1039. H. H. Wasserman, D. J. Hlasta, A. W. Tremper and J. S. Wu, *J. Org. Chem.*, **46**, 2999 (1981).
1040. R. M. Freidinger. *J. Org. Chem.*, **50**, 3631 (1985).
1041. D. E. Keeley, *U. S. Patent US 4309543A* (1982); *Chem. Abstr.*, **97**, 38862y (1982).
1042. T. Okawara, T. Matsuda, Y. Noguchi and M. Furukawa, *Chem. Pharm. Bull.*, **30**, 1574 (1982).
1043. A. L. J. Beckwith and C. J. Easton, *Tetrahedron*, **39**, 3995 (1983).
1044. K. Spirkova, J. Kovac, I. Horsak and M. Dandarova, *Collect. Czech. Chem. Commun.*, **46** 1513 (1981); *Chem. Abstr.*, **95**, 132581m (1981).
1045. F. Effenberger and U. Burkard, *Justus Liebigs Ann. Chem.*, 334 (1986).
1046. L. N. Junghiem, S. K. Sigmund and N. D. Jones, *Tetrahedron Letters*, **28**, 289 (1987).
1047. H. H. Wassermann, D. J. Hlasta, A. W. Tremper and J. S. Wu, *Tetrahedron Letters*, 549 (1979).
1048. P. A. Van Elburg and D. N. Reinhoudt, *Heterocycles*, **26**, 437 (1987).
1049. P. G. Mattingly, J. F. Kerwin, Jr., and M. J. Miller, *J. Amer. Chem. Soc.*, **101**, 3983 (1979).
1050. M. J. Miller and P. G. Mattingly, *Tetrahedron*, **39**, 2563 (1983).
1051. R. Joyeau, H. Molines, R. Labia and M. Wakselman, *J. Med. Chem.*, **31**, 370 (1988).
1052. B. G. Chatterjee and D. P. Sahu, *Tetrahedron Letters*, 1129 (1977).
1053. M. Shiozaki, N. Ishida, T. Hiraoka, and H. Maruyama, *Tetrahedron*, **40**, 1795 (1984).
1054. A. L. J. Beckwith abd D. R. Boate, *Tetrahedron Letters*, **26**, 1761 (1985).
1055. Y. Ohtsuka and T. Oishi, *Chem. Pharm. Bull.*, **31**, 454 (1983).
1056. P. Scrimin, F. D'Angeli and A. C. Veronese, *Synthesis*, 586 (1982).
1057. H. Takahata, Y. Ohnishi and T. Yamazaki, *Heterocycles*, **14**, 467 (1980).
1058. T. Okawara, T. Matsuda and M. Furukawa, *Chem. Pharm. Bull.*, **30**, 1225 (1982).
1059. M. Okita, M. Mori, T. Wakamatsu and Y. Ban, *Heterocycles*, **23**, 247 (1985).
1060. S. Sebti and A. Foucaud, *Synthesis*, 546 (1983).
1061. S. Sebti and A. Foucaud, *Tetrahedron*, **40**, 3223 (1984).
1062. H. Ishibashi, M. Ikeda, H. Maeda, K. Ishiyama, M. Yoshida, S. Akai and Y. Tamura, *J. Chem. Soc., Perkin Trans. 1*, 1099 (1987).
1063. H. Nagashima, H. Wakamatsu and K. Itoh, *J. Chem. Soc., Chem. Commun.*, 652 (1984).
1064. H. Nagashima, K. Ara, H. Wakamatsu and K. Itoh, *J. Chem. Soc., Chem. Commun.*, 518 (1985).
1065. K. Tanaka, H. Yoda, K. Inoue and A. Kaji, *Synthesis*, 66 (1986).
1066. S. R. Fletcher and T. I. Kay, *J. Chem. Soc., Chem. Commun.*, 903 (1978).
1067. M. Mori, K. Chiba, M. Okita, I. Kay and Y. Ban, *Tetrahedron*, **41**, 375 (1985).
1068. R. B. Ruggeri and C. H. Heathcock, *J. Org. Chem.*, **52**, 5745 (1987).
1069. I. Corelli, A. Inesi, V. Carelli, M. A. Casadei, F. Liberatore and F. M. Moracci, *Synthesis*, 591 (1986).
1070. S. Knapp, K. E. Rodriques, A. T. Levorse and R. M. Ornaf, *Tetrahedron Letters*, **26**, 1803 (1985).
1071. A. A. Akhnazaryan, M. A. Manukyan and M. T. Dangyan, *Arm. Khim. Zh.*, **31**, 337 (1978); *Chem. Abstr.*, **89**, 108913b (1978).
1072. A. J. Biloski, R. D. Wood and B. Ganem, *J. Amer. Chem. Soc.*, **104**, 3233 (1982).
1073. D. Trachler and F. Lohse, *Eur. Pat. EP 49688* (1982); *Chem. Abstr.*, **97**, 72249p (1982).
1074. E. R. Talaty, A. R. Clague, M. O. Agho, M. N. Deshpandi, P. M. Courtney, D. H. Burger and E. F. Roberts, *J. Chem. Soc., Chem. Commun.*, 889 (1980).
1075. C. A. Townsend and L. T. Nguyen, *J. Amer. Chem. Soc.*, **103**, 4582 (1981).
1076. A. K. Bose, M. S. Manhas, D. P. Sahu and V. R. Hegde, *Can. J. Chem.*, **62**, 2498 (1984).
1077. C. Yoshida, K. Tanaka, J. Nakano, Y. Todo, T. Yamafuji, R. Hattori, Y. Fukuoka and I. Saikawa, *J. Antibiot.*, **39**, 76 (1986).
1078. R. M. Williams and B. H. Lee, *J. Amer. Chem. Soc.*, **108**, 6431 (1986).
1079. M. A. Morrison and M. J. Miller, *J. Org. Chem.*, **48**, 4421 (1983).
1080. M. Jung and M. J. Miller, *Tetrahedron Letters*, **26**, 977 (1985).
1081. T. Kolasa and M. J. Miller, *Tetrahedron Letters*, **28**, 1861 (1987).
1082. D. A. Evans and E. B. Sjogren, *Tetrahedron Letters*, **27**, 3119 (1986).
1083. R. E. Van der Stoel, M. A. R. Bosma, P. H. J. Janssen and G. M. C. Van der Moesdijk, *Eur. Pat. EP 80240* (1983); *Chem. Abstr.*, **99**, 158268x (1983).
1084. A. I. Meyers, B. A. Lefker, K. T. Wanner and A. R. Aitken, *J. Org. Chem.*, **51**, 1936 (1986).
1085. A. I. Meyers and B. A. Lefker, *J. Org. Chem.*, **51**, 1541 (1986).

1086. T. Kaneko, *J. Amer. Chem. Soc.*, **107**, 5490 (1985).
1087. Y. Tamura, H. Maeda, S. Akai, K. Ishiyama and H. Ishibashi, *Tetrahedron Letters*, **22**, 4301 (1981).
1088. K. Hirai and Y. Iwano, *Tetrahedron Letters*, 2031 (1979).
1089. J. C. Gilbert and B. K. Blackburn, *Tetrahedron Letters*, **25**, 4067 (1984).
1090. J. R. Mahajan, G. A. L. Ferreira, H. C. Araujo and B. J. Nunes, *Synthesis*, 112 (1976).
1091. B. E. Maryanoff, D. E. McComsey and B. A. Duhl-Emswiler, *J. Org. Chem.*, **48**, 5062 (1983).
1092. M. Bortolussi, R. Black and J. M. Conia, *Tetrahedron Letters*, 2289 (1977).
1093. G. Fratei, *Tetrahedron Letters*, 4517 (1976).
1094. E. Vedejas and G. P. Meier, *Tetrahedron Letters*, 4185 (1979).
1095. A. Toshimitsu, K. Terao and S. Uemura, *Tetrahedron Letters*, **25**, 5917 (1984).
1096. A. A. El-Barbary, S. Carlson and S. O. Lawesson, *Tetrahedron*, **38**, 405 (1982).
1097. L. V. Ershov and V. G. Granik, *Khim. Geterotsiki. Soedin.*, 929 (1985); *Chem. Abstr.*, **104**, 168389y (1986).
1098. C. J. Moody, C. J. Pearson and G. Lawton, *Tetrahedron Letters*, **26**, 3171 (1985).
1099. S. D. Sharma, U. Mehra and P. K. Gupta, *Tetrahedron*, **36**, 3427 (1980).
1100. D. Ben-Ishai, N. Peled and I. Sataty, *Tetrahedron Letters*, **21**, 569 (1980).
1101. J. Sliwinski, *Zesz. Nauk. Politech. Slask., Chem.*, **80**, 13 (1977); *Chem. Abstr.*, **89**, 109023y (1978).
1102. G. Metz, *Arch. Pharm. (Weinheim, Ger.)*, **320**, 285 (1987); *Chem. Abstr.*, **107**, 198122k (1987).
1103. T. Polonski, *Tetrahedron*, **41**, 611 (1985).
1104. M. Sakamoto, Y. Omote and H. Aoyama, *J. Org. Chem.*, **49**, 396 (1984).
1105. T. Naito, K. Katsumi, Y. Tada and I. Ninomiya, *Heterocycles*, **20**, 775 (1983).
1106. T. Hamada, M. Ohmori, and O. Yonemitsu, *Tetrahedron Letters*, 1519 (1977).
1107. M. Ikeda, T. Uchino, H. Ishibashi, Y. Tamura, and M. Kido, *J. Chem. Soc., Chem. Commun.*, 758 (1984).
1108. M. Sakamoto, H. Aoyama, and Y. Omote, *Tetrahedron Letters*, **26**, 4475 (1985).
1109. H. Aoyama, K. Miyozaki, M. Sakamoto and Y. Omote, *J. Chem. Soc., Chem. Commun.*, 333 (1983).
1110. S. A. Glover and A. Goosen, *J. Chem. Soc., Perkin Trans. 1*, 653 (1978).
1111. Y. Kanaoka, Y. Hatanoka, Y. Sato, M. Wada and H. Nakoi, *Kokagaku Toronkai Koen Yoshishu*, 266 (1979); *Chem. Abstr.*, **93**, 70455w (1980).
1112. M. Ishiguro, H. Iwata, T. Nakatsuka and M. Okitsu, *Jpn. Kokai Tokkyo Koho JP 61/207373 A2* (1986); *Chem. Abstr.*, **107**, 23171y (1987).
1113. A. G. M. Barrett, G. G. Graboski and M. A. Russell, *J. Org. Chem.*, **50**, 2603 (1985).
1114. A. G. M. Barrett, M. J. Betts and A. Fenwick, *J. Org. Chem.*, **50**, 169 (1985).
1115. J. D. Buynak, M. N. Rao, R. Y. Chandrasekaren, E. Haley, P. De Meester and S. C. Chu, *Tetrahedron Letters*, **26**, 5001 (1985).
1116. A. G. M. Barrett, A. Fenwick and M. J. Betts, *J. Chem. Soc., Chem. Commun.*, 299 (1983).
1117. M. Chmielewski and Z. Kaluza, *J. Org. Chem.*, **51**, 2395 (1986).
1118. M. Chmielewski, Z. Kaluza, C. Belzecki, P. Solanski and J. Jurczak, *Tetrahedron Letters*, **25**, 4797 (1984).
1119. M. Chmielewski, Z. Kaluza, C. Belzecki, and P. Solanski, *Heterocycles*, **24**, 285 (1986).
1120. M. Langbeheim and S. Sarel, *Tetrahedron Letters*, 2613 (1978).
1121. S. Sarel, A. Felzenstein and J. Yovell, *Tetrahedron Letters*, 451 (1976).
1122. A. Aumann and H. Heinen, *Chem. Ber.*, **120**, 1297 (1987).
1123. T. Ishimaru, K. Sagyo and K. Kenkyu, *Jpn. Kokai Tokkyo Koho JP 58/144368 A2* (1983); *Chem. Abstr.*, **100**, 68087x (1984).
1124. T. G. Back, *J. Org. Chem.*, **46**, 1442 (1981).
1125. D. E. Butler and S. M. Alexander, *J. Heterocycl. Chem.*, **19**, 1173 (1982).
1126. R. D. G. Cooper, B. W. Daugherty and D. B. Boyd, *Pure Appl. Chem.*, **59**, 485 (1987).
1127. R. Allmann, T. Debaerdemaeker, G. Kiehl, J. P. Luttringer, T. Tschamber, G. Wolff and J. Streith, *Justus Liebigs Ann. Chem.*, 1361 (1983).
1128. B. Alcaide, M. A. Leon-Santiago, R. Perez-Ossorio, J. Plumet, M. Sierra and M. C. DeLa Torre, *Synthesis*, 989 (1982).
1129. T. Tschamber, J. M. Henlin, D. Pipe and J. Streith, *Heterocycles*, **23**, 2589 (1985).
1130. S. Kobayashi, T. Izawa, Y. Nii, T. Iimori, K. Okano and M. Ohno, *Koen Yoshishu-Tennen Yuki Kagobutsu Toronkai*, 22nd, 470–476 (1979); *Chem. Abstr.*, **92**, 215167v (1980).

1131. I. Yamamoto, I. Abe, M. Nazawa, M. Kotani, J. Motoyoshiya, H. Gotoh and K. Matsuzaki, *J. Chem. Soc., Perkin Trans. 1*, 2297 (1983).
1132. T. Kawabata, Y. Kimura, Y. Ito, S. Terashima, A. Sasaki and M. Sunagawa, *Tetrahedron Letters*, **27**, 6421 (1986).
1133. J. J. Barr and R. C. Storr, *J. Chem. Soc., Chem. Commun.*, 788 (1975).
1134. C. Belzecki and Z. Krawczyk, *J. Chem. Soc., Chem. Commun.*, 302 (1977).
1135. V. Georgian, S. K. Boyer and B. Edwards, *Heterocycles*, **7**, 1003 (1977).
1136. H. Aoyama, M. Sakamoto, K. Yoshida and Y. Omote, *J. Heterocycl. Chem.*, **20**, 1099 (1983).
1137. M. Cardellini, F. Claudi and F. M. Mosacci, *Synthesis*, 1070 (1984).
1138. W. T. Brady and C. H. Shieh, *J. Org. Chem.*, **48**, 2499 (1983).
1139. L. Birkofer and J. Schramm, *Justus Liebigs Ann. Chem.*, 760 (1977).
1140. M. E. Hassan, *Bull. Soc. Chim. Belg.*, **94**, 149 (1985); *Chem. Abstr.*, **103**, 104750f (1985).
1141. M. E. Hassan, *Chem. Scr.*, **25**, 239 (1985); *Chem. Abstr.* **105**, 6468s (1986).
1142. D. M. Kunert, R. Chambers, F. Mercer, L. Hernandez, Jr. and H. W. Moore, *Tetrahedron Letters*, 929 (1978).
1143. A. Padwa, K. F. Koechler and A. Rodriguez, *J. Org. Chem.*, **49**, 282 (1984).
1144. H. W. Moore and G. Hughes, *Tetrahedron Letters*, **23**, 4003 (1982).
1145. S. D. Sharma, S. Kaur and U. Mehra, *Indian J. Chem., Sect. B*, **25B**, 141 (1986).
1146. J. M. Aizpurua, F. P. Cossio, B. Lecea and C. Palomo, *Tetrahedron Letters*, **27**, 4359 (1986).
1147. C. Jenny, A. Prewo, J. H. Bieri and H. Heimgartner, *Helv. Chim. Acta*, **69**, 1424 (1986).
1148. B. Alcaide, G. Dominguez, G. Escobar, V. Parreno and J. Plumet, *Heterocycles*, **24**, 1579 (1986).
1149. B. Ernst and D. Bellus, *Ger. Offen. DE 3620467 A1* (1987); *Chem. Abstr.*, **106**, 176045q (1987).
1150. P. Sohar, L. Fodor, J. Szabo and G. Bernath, *Tetrahedron*, **40**, 4387 (1984).
1151. A. K. Bose, *U. S. Patent US 3943123* (1976); *Chem. Abstr.*, **85**, 21077a (1976).
1152. A. K. Bose, V. R. Hedge, D. R. Wagle, S. S. Bari and M. S. Manhas, *J. Chem. Soc., Chem. Commun.*, 161 (1986).
1153. M. S. Manhas, V. R. Hegde, D. R. Wagle and A. K. Bose, *J. Chem. Soc., Perkin Trans. 1*, 2045 (1985).
1154. A. K. Bose, S. G. Amin, J. C. Kapur and M. S. Manhas, *J. Chem. soc., Perkin Trans. 1*, 2193 (1976).
1155. A. K. Bose, W. A. Hoffmann and M. S. Manhas, *J. Chem. Soc., Perkin Trans 1*, 2343 (1976).
1156. M. S. Manhas, S. G. Amin, H. P. S. Chawla and A. K. Bose, *J. Heterocycl. Chem.*, **15**, 601 (1978).
1157. I. Ojima, S. Suga and R. Abe, *Chem. Letters*, 853 (1980).
1158. I. Ojima, S. Suga and R. Abe, *Tetrahedron Letters*, **21**, 3907 (1980).
1159. S. D. Sharma, P. K. Gupta, J. Bindra and M. Sunita, *Tetrahedron Letters*, **21**, 3295 (1980).
1160. I. Antonini, M. Cardellini, F. Claudi and F. M. Moracci, *Synthesis*, 379 (1986).
1161. M. A. Casadei, F. M. Moracci and A. Inesi, *J. Chem. Soc., Perkin Trans. 2*, 419 (1986).
1162. M. Ghosh and B. G. Chatterjee, *J. Indian Chem. Soc.*, **62**, 457 (1985).
1163. S. D. Sharma, U. Mehra and P. K. Gupta, *Indian J. Chem., Sect. B*, **16B**, 461 (1978).
1164. S. D. Sharma, U. Mehra and S. Kaur, *Indian J. Chem., Sect. B*, **23B**, 857 (1984).
1165. T. Kametani, K. Kigasawa, M. Hiiragi, K. Wakisaka, H. Sugi and K. Tanigawa, *Heterocycles*, **12**, 735 (1979).
1166. R. L. Varma and C. S. Narayanan, *Indian J. Chem., Sect. B*, **24B**, 302 (1985).
1167. K. A. Prodan and F. M. Hershenson, *Chem. Eng. News*, **59**, 59 (1981).
1168. D. Dugat, G. Just and S. Sahoo, *Can. J. Chem.*, **65**, 88 (1987).
1169. N. Hatanaka and I. Ojima, *Chem. Letters*, 231 (1981).
1170. N. Hatanaka, R. Abe and I. Ojima, *Chem. Letters*. 1297 (1987).
1171. I. Ojima, K. Nakahashi, S. M. Brandstadter and N. Hatanaka, *J. Amer. Chem. Soc.*, **109**, 1798 (1987).
1172. N. Hatanaka, R. Abe and I. Ojima, *Chem. Letters*, 445 (1982).
1173. M. D. Bachi and J. Vaya, *J. Org. Chem.*, **44**, 4393 (1979).
1174. D. F. Sullivan, D. I. C. Scopes, A. K. Kluge and J. A. Edwards, *J. Org. Chem.*, **41**, 1112 (1976).
1175. M. D. Bachi, S. Sasson and J. Vaya, *J. Chem. Soc., Perkin Trans. 1*, 2228 (1980).
1176. G. Just and R. Zamboni, *Can. J. Chem.*, **56**, 2725 (1978).
1177. D. Hoope and M. Kloft, *Justus Liebigs Ann. Chem.*, 1512 (1980).
1178. A. K. Bose, M. S. Manhas, J. M. Van der Veen, S. G. Amin, I. F. Fernandez, K. Gala, R.

Gruska, J. C. Kapur, M. S. Khajavi, J. Kreder, L. Mukkavilli, B. Ram, M. Sugiura and J. F. Vincent, *Tetrahedron*, **37**, 2321 (1981).
1179. D. Nasipuri and G. Das, *Indian J. Chem., Sect. B*, **18B**, 205 (1979).
1180. A. K. Bose, M. S. Manhas, H. P. S. Chawla and B. Dayal, *J. Chem. Soc. Perkin Trans. 1*, 1880 (1975).
1181. I. Ojima, H. J. C. Chen and K. Nakahashi, *J. Amer. Chem. Soc.*, **110**, 278 (1988).
1182. J. Streith, T. Tschamber and G. Wolff, *Justus Liebigs Ann. Chem.*, 1374 (1983).
1183. T. Kametani, K. Kigasawa, M. Hiragi, K. Wakisaka, H. Sugi and K. Tanigawa, *Heterocycles*, **12**, 795 (1979).
1184. M. Klich and G. Teutsch, *Tetrahedron*, **42**, 2677 (1986).
1185. C. Hubschwerlen, *Eur. Pat. Appl. EP 120289 A2* (1984); *Chem. Abstr.*, **102**, 78641d (1985).
1186. M. A. Abbady, H. S. El-Kashef, H. A. Abd-Alla and M. M. Kandeel, *J. Chem. Technol. Biotechnol. Chem, Technol.*, **34A**, 62 (1984); *Chem. Abstr.*, **101**, 130464u (1984).
1187. T. Kametani, K. Fukumoto, K. Kigasawa, M. Hiiragi, K. Wakisaka, K. Tanigawa and H. Sugi, *Heterocycles*, **12**, 741 (1979).
1188. T. Kamiya, T. Oku, O. Nakaguchi, H. Takeno and M. Hashimoto, *Tetrahedron Letters*, 5119 (1978).
1189. A. M. Osman, K. M. Hassan, M. A. El-Maghraby, H. S. El-Kashef and A. M. Abdel-Mawgoud, *J. Prakt. Chem.*, **320**, 482 (1978).
1190. P. Sohar. G. Stajer, I. Pelczer, A. E. Szabo, J. Szunyog and G. Bernath, *Tetrahedron*, **41**, 1721 (1985).
1191. G. C. Kamdar, D. J. Bhatt and A. R. Porikh, *J. Indian Chem. Soc.*, **69**, 790 (1984).
1192. T. Agawa, M. Ishida and Y. Ohshiro, *Synthesis*, 933 (1980).
1193. W. Abramski, C. Belzecki and M. Chmielewski, *Bull. Pol. Acad. Sci. Chem.*, **33**, 451 (1985); *Chem. Abstr.*, **106**, 84433m (1987).
1194. A. K. Bose, J. C. Kapur, S. D. Sharma and M. S. Manhas, *Tetrahedron Letters*, 2319 (1973).
1195. M. J. Pearson and J. W. Tyler, *J. Chem. Soc., Perkin Trans. 1*, 1927 (1985).
1196. C. L. Branch and M. J. Pearson, *J. Chem. Soc., Perkin Trans. 1*, 1077 (1986).
1197. N. Ikota and A. Hanaki, *Heterocycles*, **22**, 2227 (1984).
1198. A. Arrieta, F. P. Cossio and C. Palomo, *Tetrahedron*, **41**, 1703 (1985).
1199. S. D. Sharma, U. Mehra and V. Kaur, *Indian J. Chem., Sect. B*, **25B**, 1061(1986).
1200. A. Arrieta, J. M. Aizpurua and C. Palomo, *Tetrahedron Letters*, **25**, 3365 (1984).
1201. A. Arrieta, B. Lecea and C. Palomo, *J. Chem. Soc., Perkin Trans. 1*, 845 (1987).
1202. M. Miyake, N. Tokutake and M. Kirisawa, *Synthesis*, 833 (1983).
1203. G. Schmid, *Eur. Pat. EP 138113 A1* (1985); *Chem. Abstr.*, **103**, 104792w (1985).
1204. J. M. Aizpurua, I. Ganboa, F. P. Cossio, A. Gonzalez, A. Arrieta and C. Palomo, *Tetrahedron Letters*, **25**, 3905 (1984).
1205. M. S. Manhas, H. P. S. Chawla, S. G. Amin and A. K. Bose, *Synthesis*, 407 (1977).
1206. A. K. Bose, B. Ram, S. G. Amin, L. Mukkavilli, J. E. Vincent and M. S. Manhas, *Synthesis*, 543 (1979).
1207. A. K. Bose. M. S. Manhas, S. G. Amin, J. C. Kapur, J. Kreder, L. Mukkavilli, B. Ram and J. E. Vincent, *Tetrahedron Letters*, 2771 (1979).
1208. A. K. Bose, M. S. Manhas, K. Gala, D. P. Sahu and V. Hegde, *J. Heterocycl. Chem.*, **17**, 1687 (1980).
1209. S. G. Amin, R. D. Glazer and M. S. Manhas, *Synthesis*, 210 (1979).
1210. M. S. Manhas, A. K. Bose and M. S. Khajavi, *Synthesis*, 209 (1981).
1211. M. S. Manhas, S. S. Bari, B. M. Bhawal and A. K. Bose, *Tetrahedron Letters*, **25**, 4733 (1984).
1212. M. S. Manhas, S. G. Amin, B. Ram and A. K. Bose, *Synthesis*, 689 (1976).
1213. F. P. Cassio, I. Ganboa and C. Polomo, *Tetrahedron Letters*, **26**, 3041 (1985).
1214. S. D. Sharma, G. Singh and P. K. Gupta, *Indian J. Chem.*, **16B**, 74 (1978).
1215. S. D. Sharma and P. K. Gupta, *Tetrahedron Letters*, 4587 (1978).
1216. A. Arrieta, J. M. Aizpurua and C. Palomo, *Synth. Commun.*, **12**, 967 (1982).
1217. F. P. Cossio and C. Palomo, *Tetrahedron Letters*, **26**, 4235 (1985).
1218. F. P. Cossio and C. Palomo, *Tetrahedron Letters*, **26**, 4239 (1985).
1219. M. S. Manhas, B. Lal, S. G. Amin and A. K. Bose, *Synth. Commun.*, **6**, 435 (1976).
1220. M. Miyake, M. Kirisawa and N. Tokutake, *Synthesis*, 1053 (1982).
1221. F. P. Cossio, I. Gamboa, J. M. Garcia, B. Lecea and C. Palomo, *Tetrahedron Letters*, **28**, 1945 (1987).

1222. D. R. Shridhar, B. Ram, V. L. Narayana, A. K. Awasthi and G. J. Reddy, *Synthesis*, 846 (1984).
1223. D. R. Shridhar, B. Ram, V. L. Narayana, *Synthesis*, 63 (1982).
1224. K. Piotrowska and D. Mostowicz, *J. Chem. Soc., Chem. Commun.*, 41 (1981).
1225. A. K. Bose, K. Gupta and M. S. Manhas, *J. Chem. Soc., Chem. Commun.*, 86 (1984).
1226. L. Arsenijevic, M. Bogavac, S. Pavlov and V. Arsenijevic, *Ark. Farm.*, **33**, 201 (1983); *Chem. Abst.*, **100**, 156476d (1984).
1227. K. Krishan, A. Singh, B. Singh and S. Kumar, *Synth. Commun.*, **14**, 219 (1984).
1228. M. A. McGuire and L. S. Hegedus, *J. Amer. Chem. Soc.*, **104**, 5538 (1982).
1229. L. S. Hegedus, *Pure Appl. Chem.*, **55**, 1745 (1983).
1230. L. S. Hegedus, *Tetrahedron*, **41**, 5833 (1985).
1231. A. G. M. Barrett, C.P. Brock and M. A. Sturgess, *Organometallics*, **4**, 1903 (1985).
1232. J. E. Dubois and G. Axiotis, *Tetrahedron Letters*, **25**, 2143 (1984).
1233. I. Ojima and S. Inaba, *Tetrahedron Letters*, **21**, 2077 (1980).
1234. I. Ojima and S. Inaba, *Tetrahedron Letters*, **21**, 2081 (1980).
1235. I. Ojima and S. Inaba, *Jpn. Kobai Tokkyo Yoho JP 53/108962* (1978); *Chem. Abstr.*, **90**, 87242k (1979).
1236. I. Ojima, S. Inaba and K. Yoshida, *Tetrahedron Letters*, 3643 (1977).
1237. D. C. Ha, D. J. Hart and T. K. Yang, *J. Amer. Chem. Soc.*, **106**, 4819 (1984).
1238. M. Komatsu, S. Yamamoto, Y. Ohshiro and T. Agawa, *Tetrahedron Letters*, **22**, 3769 (1981).
1239. N. Kokuni and S. Nakai, *Jpn. Kokai Tokkyo Koho JP 62/81368 A2* (1987); *Chem. Abstr.*, **108**, 94293w (1988).
1240. C. Gluchowski, L. Cooper, D. E. Bergbreiter and M. Newcomb, *J. Org. Chem.*, **45**, 3413 (1980).
1241. D. A. Burnett, D. J. Hart and J. Liu, *J. Org. Chem.*, **51**, 1929 (1986).
1242. M. Komatsu, S. Yamamoto, Y. Ohshiro and T. Agawa, *Heterocycles*, **23**, 677 (1985).
1243. A. Mkhairi and J. Hamelin, *Tetrahedron Letters*, **27**, 4435 (1986).
1244. A. K. Bose, M. S. Khajavi and M. S. Manhas, *Synthesis*, 407 (1982).
1245. D. J. Hart, C. S. Lu, W. H. Pirkle, M. H. Hyon and A. Tsipouras, *J. Amer. Chem. Soc.*, **108**, 6054 (1986).
1246. A. G. M. Barrett and P. Quayle, *J. Chem. Soc., Perkin Trans. 1*, 2193 (1982).
1247. P. Andreoli, G. Cainelli, M. Contento, D. Giacomini, G. Martelli and M. Panunzro, *Tetrahedron Letters*, **27**, 1695 (1986).
1248. T. Nakai and T. Chiba, *Eur. Pat. EP 171064 A1* (1986); *Chem. Abstr.*, **105**, 97251e (1986).
1249. D. K. Dutta, R. C. Boruah and J. S. Sandhu, *Indian J. Chem., Sect. B*, **25B**, 350 (1986).
1250. T. L. Gilchrist, D. Hughes and R. Wasson, *Tetrahedron Letters*, **28**, 1573 (1987).
1251. T. Okawara, R. Kato and M. Furukawa, *Chem. Pharm. Bull.*, **32**, 2426 (1986); *Chem. Abstr.*, **102**, 6409m (1985).
1252. R. C. Cambie, G. R. Clark, T. C. Jones, P. S. Rutledge, G. A. Strange and P. D. Woodgate, *Aust. J. Chem.*, **38**, 745 (1985).
1253. E. Rogalska and C. Belzecki, *J. Org. Chem.*, **49**, 1397 (1984).
1254. C. Belzecki and E. Rogalska, *J. Chem. Soc., Chem. Commun.*, 57 (1981).
1255. R. Shabana, J. B. Rasmussen, S. O. Olesan and S. D. Lawesson, *Tetrahedron*, **36**, 3047 (1980).
1256. M. Miyake, N. Tokutake and M. Kirisawa, *Synth. Commun.*, **14**, 353 (1984).
1257. G. Guanti, L. Banfi, E. Narisano and S. Thea, *J. Chem. Soc., Chem. Commun.*, 861 (1984).
1258. B. Sain and J. S. Sandhu, *Heterocycles*, **23**, 1611 (1985).
1259. R. G. Glushkov, O. Y. Belyaeva, V. G. Granik, M. K. Polievktov, A. B. Grigorev, V. E. Serokhvostova and T. F. Vlasova, *Khim. Geterotsikl. Soedin.*, 1640 (1976); *Chem. Abstr.*, **86**, 121287h (1977).
1260. R. Brambilla, R. Friary, A. Ganguly, M. S. Puar, B. R. Sunday, J. J. Wright, K. D. Onan and R. T. McPhail, *Tetrahedron*, **37**, 3615 (1981).
1261. G. Lawton, C. J. Moody and C. J. Pearson, *J. Chem. Soc., Chem. Commun.*, 754 (1984).
1262. C. J. Moody, C. J. Pearson and G. Lawton, *Tetrahedron Letters*, **26**, 3167 (1985).
1263. H. W. Moore and M. J. Arnold. *J. Org. Chem.*, **48**, 3365 (1983).
1264. M. R. Johnson, M. J. Fazio, D. L. Ward and L. R. Sousa, *J. Org. Chem.*, **48**, 494 (1983).
1265. J. Nally, N. H. R. Ordsmith and G. Procter, *Tetrahedron Letters*, **26**, 4107 (1985).
1266. D. H. R. Barton, E. Buschmann, J. Hauesler, C. W. Holzapfel, T. Sherodksy and D. A. Taylor, *J. Chem. Soc., Perkin Trans. 1*, 1107 (1977).
1267. J. Brennan, *J. Chem. Soc., Chem. Commun.*, 880 (1981).
1268. H. Gotthardt and K. H Schenk, *J. Chem. Soc., Chem. Commun.*, 687 (1986).

1269. P. Y. Johnson, N. R. Schmuff and C. E. Hatch, III, *Tetrahedron Letters*, 4089 (1975).
1270. M. Hatanaka and H. Nitta, *Tennen Yuki Kogobutsu Toronkai Koen Yoshishu*, 28th, 542 (1986); *Chem. Abstr.*, **108**, 21560w (1988).
1271. D. R. Bender, J. Brennan and H. Rapoport, *J. Org. Chem.*, **43**, 3354 (1978).
1272. T. Sheradsky and D. Zboida, *Tetrahedron Letters*, 2037 (1978).
1273. J. Suwinski, *Zesz. Nauk. Politech. Slask., Chem.*, **82**, 3 (1977); *Chem. Abstr.*, **90**, 72132a (1979).
1274. S. Matache, *Ital. Pat. Rom. RO 71846 B* (1977); *Chem. Abstr.*, **96**, 144917j (1982).
1275. S. Matache, *Ital. Pat. Rom. RO 71845 B* (1981); *Chem. Abstr.*, **97**, 8137y (1982).
1276. N. Petri, H. Bipp, K. Wintersberger, G. Wunsch and H. Fuchs, *Ger. Pat. DE 1951158* (1969); *Chem. Abstr.*, **89**, 146450f (1978).
1277. V. I. Gorshkov, A. S. Badrian, D. M. Dokholov and N. A. Dmitrieva, *U.S.S.R. Pat. SU 886966 A1* (1981); *Chem. Abstr.*, **96**, 110991p (1982).
1278. C. Poulain. R. Kern and D. Augustin, *Ger. Pat., DE 2534538* (1976); *Chem. Abstr.*, **84**, 165377a (1976).
1279. S. Eiga, K. Matsumoto and T. Furuta, *Jpn. Pat JP 50/20043* (1975); *Chem. Abstr.*, **84**, 58679d (1979).
1280. R. Fuhrmann, A. A. Tunick and S. Sifniades, *U.S. Pat. 3922265* (1975); *Chem. Abstr.*, **84**, 121193h (1976).
1281. H. Rademacher and H. -W. Voges, *Ger. Pat. DE 3538859 A1* (1987); *Chem. Abstr.*, **107**, 77652z (1987).
1282. E. F. Novoselov, S. D. Isaev, A. G. Yurchenko, L. Vodicka and J. Triska, *Zh. Org. Khim.*, **17**, 2558 (1981); *Chem. Abstr.*, **96**, 122607f (1982).
1283. T. Sundararamaiak, S. K. Ramraj, K. L. Rao and V. V. Bai, *J. Indian Chem. Soc.*, **53**, 664 (1976).
1284. I. Sakai, N. Kawabe and M. Ohno, *Bull. Chem. Soc. Jpn.*, **52**, 3381 (1979); *Chem. Abstr.*, **92**, 128701w (1980).
1285. H. Suginome, C.-M. Shea, A. Osada, Y. Takahashi, N. Miyata and T. Masamune, *Kokagaku Toronkai Koen Yoshishu*, 20–21, *Chem. Soc. Japan, Tokyo* (1979); *Chem. Abstr.*, **92**, 215617s (1980).
1286. H. Suginome and C. -M. Shea, *Synthesis*, 229 (1980).
1287. B. M. Trost, M. Vaultier and M. L. Santiago, *J. Amer. Chem. Soc.*, **102**, 7929 (1980).
1288. A. Guggisberg, B. Dabrowski, C. Heidelberger, U. Kramer, E. Stephanou and M. Hesse, *Symp. Pap. —IUPAC Int. Symp. Chem. Nat. Prod.*, **11**, Vol. 4, Issue Part I, p. 314 (1978); *Chem. Abstr.*, **92**, 59041s (1980).
1289. D. Roberts and H. Alper, *Organometallics*, **3**, 1767 (1984).
1290. D. P. Deltsova, N. P. Gambaryan and E. I. Mysov, *Izv. Akad. Nauk SSR, Ser. Khim.*, 2343 (1981); *Chem. Abstr.*, **96**, 52230b, (1982).
1291. H. H. Wasserman, W. T. Han, J. M. Schaus and J. W. Faller, *Tetrahedron Letters*, **25**, 3111 (1984).
1292. B. Alcaide, G. Dominguez, A. Martin-Domenech, J. Plumet, A. Monge and V. Perez-Garcia, *Heterocycles*, **26**, 1461 (1987).
1293. C. W. Bird, *Tetrahedron Letters*, 609 (1964).
1294. L. Crombie, R. C. F. Jones, S. Osborne and A. R. Mat-Zin, *J. Chem. Soc., Chem. Commun.*, 959 (1983).
1295. J. M. Indelicato, J. W. Fisher and C. E. Pasini, *J. Pharm. Sci.*, **75**, 304 (1986).
1296. C. Heidelberger, A. Guggisberg, E. Stephanou and M. Hesse, *Helv. Chim. Acta*, **64**, 399 (1981).
1297. E. Stephanou, A. Guggisberg and M. Hesse, *Helv. Chim. Acta*, **62**, 1932 (1979).
1298. R. Waelchli, S. Bienz and M. Hesse, *Helv. Chim. Acta*, **68**, 484 (1985).
1299. D. S. C. Black and L. M. Johnstone, *Aust. J. Chem.*, **37**, 599 (1984).
1300. Y. Hirai, T. Yamazaki, S. Hirokami and M. Nagota, *Tetrahedron Letters*, **21**, 3067 (1980).
1301. A. Canovas, J. Fonrodona, J. J. Bonet, M. C. Bianso and J. L. Brianso, *Helv. Chim. Acta*, **63**, 2380 (1980).
1302. H. Suginome and T. Uchido, *J. Chem. Soc., Perkin Trans. 1*, 1356 (1980).
1303. D. Dobrescu, M. Iovu, A. N. Cristea and M. A. E.-H. Omaima, *Romanian Pat. RO 82040 B* (1983); *Chem. Abstr.*, **103**, 87786c (1985).
1304. L. M. Khananshvili, D. S. Akhobadze, L. K. Dzhaniashvili and Z. S. Lomtatidze, *Khim.-Farm. Zh.*, **16**, 560 (1982); *Chem. Abstr.*, **97**, 110081n (1982).
1305. M. Sunagawa, A. Sasaki and K. Goda, *Eur. Pat. EP 188816 A1* (1986); *Chem. Abstr.*, **105**, 226191m (1986).

1306. E. Hungerbuehler and J. Kalvoda, *Eur. Pat. Appl. EP 215739 A1* (1987); *Chem. Abstr.*, **107**, 58743p (1987).
1307. E. D. Thorsett, E. E. Harris, S. D. Aster, E. R. Peterson. J. D. Snyder, J. P. Springer, J. Hirshfield, E. W. Tristram, A. A. Patchett, E. H. Ulm and T. C. Vassil, *J. Med. Chem.*, **29**, 251 (1986).
1308. J. Marchand-Brynaert, L. Ghosez and E. Cossement, *Tetrahedron Letters*, **21**, 3085 (1980).
1309. T. Kametani. S. Hirata, H. Nemoto, M. Ihara and K. Fukumoto, *Heterocycles*, **12**, 523 (1979).
1310. G. S. Poindexter, *U.S. Pat. U.S. 4416818 A* (1983); *Chem. Abstr.*, **100**, 68293m (1984).
1311. D. Reuschling. H. Pietsch and A. Linkies, *Tetrahedron Letters*, 615 (1978).
1312. J. Knight and P. J. Parsons, *J. Chem. Soc., Perkin Trans. 1*, 1237 (1987).
1313. K. S. K. Murthy and A. Hassner, *Tetrahedron Letters*, **28**, 97 (1987).
1314. J. Palecek and J. Kuthan, *Z.-Chem.*, **17**, 260 (1977); *Chem. Abstr.*, **87**, 151681y (1977).
1315. F. Dumas and J. d'Angelo, *Tetrahedron Letters*, **27**, 3725 (1986).
1316. T. Kametani, S. D. Chu, S. P. Huang and T. Honda, *Heterocycles*, **23**, 2693(1985).
1317. B. J. Magerlein, *PCT Int. Pat. NO. 86/6722 A1* (1986); *Chem. Abstr.*, **108**, 37489r (1988).
1318. T. Ohashi, K. Kazumori, I. Sada, A. Miyama and K. Watanabe, *Eur. Pat. EP 167154 A1* (1986); *Chem. Abstr.*, **105**, 97248j (1986).
1319. J. H. Bateson, A. M. Quinn and R. Southgate, *Tetrahedron Letters*, **28**, 1561 (1987).
1320. G. Just, D. Dugat and W. Y. Liu, *Can. J. Chem.*, **61**, 1730 (1983).
1321. T. Shono, S. Kashimura and H. Nagusa, *Chem. Letters*, 425 (1986).
1322. A. T. Malkhasyan, G. G. Sukiasyan, S. G. Matinyan and G. T. Martirosyan, *Arm. Khim. Zh.*, **29**, 458 (1976); *Chem. Abstr.*, **85**, 142966y (1976).
1323. M. M. Campbell and N. Carruthers, *J. Chem. Soc., Chem. Commun*, 730 (1980).
1324. V. M. Gavryushina, Y. A. Naum, A. A. Stepanova, V. P. Dremova and G. A. Pankratenko, *Nov. Khim. Sredstva Zashch. Rast.*, (Eds. Y. A. Kondratev and V. K. Promonenkov), Nütekhim, USSR, 1979, p. 34, *Chem. Abstr.*, **92**, 94220w (1980).
1325. T. Beisswenger and F. Effenberger, *Chem. Ber.*, **117**, 1513 (1984).
1326. F. Hoffmann-La Roche and Co. A.-G., *Jpn. Kokai Tokkyo Koho JP 60/56989 A2* (1985); *Chem. Abstr.*, **103**, 104794y (1985).
1327. Sumitomo Chemical Co. Ltd ., *Jpn. Kokai Tokkyo Koho JP 60/123467 A2* (1985); *Chem. Abstr.*, **104**, 51231z (1986).
1328. T. Kobayashi, N. Ishida and T. Hiraoka, *J. Chem. Soc., Chem. Commun.*, 736 (1980).
1329. A. G. Shipov, N. A. Orlova and Y. I. Baukov, *Zh. Obshch. Khim.*, **54**, 2645 (1984); *Chem. Abstr.*, **102**, 78704b (1985).
1330. M. Kovacevic, J. Herak and B. Gaspert, *Ger. Pat. DE 3120451 A1* (1982); *Chem. Abstr.*, **96**, 162556v (1982).
1331. H. B. Koenig, K. G. Metzger, M. Preiss and W. Schroeck, *Ger. Pat. DE 2512998* (1976); *Chem. Abstr.*, **86**, 72675x (1977).
1332. D. Habeck and W. J. Houlihan, *J. Heterocycl. Chem.*, **13**, 897 (1976).
1333. A. B. Hamlet and T. Durst, *Can. J. Chem.*, **61**, 411 (1983).
1334. L. N. Jungheim, S. K. Sigmund and J. W. Fisher, *Tetrahedron Letters*, **28**, 285 (1987).
1335. C. L. Branch and M. J. Pearson, *J. Chem. Soc., Chem. Commun.*, 946 (1981).
1336. M.D Bachi, F. Frolow and C. Hoornaert, *J. Org. Chem.*, **48**, 1841 (1983).
1337. C. L. Branch and M. J. Pearson, *J. Chem. Soc., Perkin Trans. 1*, 1097 (1986).
1338. P. Lombardi, G. Franceschi and F. Arcamone, *Tetrahedron Letters*, 3777 (1979).
1339. K. Fujimoto, Y. Iwano and K. Hirai, *Bull, Chem. Soc. Jpn.*, **59**, 1887 (1986).
1340. J. Fetter, K. Lempert, M. Kajtar-Peredy, G. Simig and G. Hornyak, *J. Chem. Soc., Perkin Trans. 1*, 1453 (1986).
1341. E. Perrone and R. J. Stoodley, *J. Chem. Soc., Chem. Commun.*, 933 (1982).
1342. M. Alpegiani, A. Bedeschi, E. Perrone and C. Gandolfi, *Belg. Pat. BE 897183 A1* (1983); *Chem. Abstr.*, **100**, 191651d (1984).
1343. C. Yoshida, T. Hori, K. Momonoi, K. Tanaka, S. Kishimoto and I. Saikawa, *Agric. Biol. Chem.*, **50**, 839 (1986); *Chem. Abstr.*, **106**, 18192f (1987).
1344. M. Ishiguro, H. Iwata and T. Nakatsuka, *Jpn. Kohai Tokkyo Koho JP 61/207387 A2* (1986); *Chem. Abstr.*, **106**, 101960a (1987).
1345. C. U. Kim, P. F. Misco and B. Y. Luh, *Heterocycles*, **26**, 1193 (1987).
1346. D. B. Boyd, T. K. Elzey, L. D. Hatfield, M. D. Kinnick and J. M. Morin, Jr., *Tetrahedron Letters*, **27**, 3453 (1986).

1347. K. Lempert, G. Doleschall, J. Fetter, G. Hornyak, J. Nyitrai, G. Simig, K. Zauer, K. Harsanyi, G. Fekete, *et. al.*, *Ger. Offen. DE 3248676 A1* (1983); *Chem. Abstr.*, **99**, 175465b (1983).
1348. H. Ludescher, C. P. Mak, G. Schulz and H. Fliri, *Heterocycles*, **26**, 885 (1987).
1349. S. Yamamoto, H. Itani, H. Takahashi, T. Tsuji and W. Nagata, *Tetrahedron Letters*, **25**, 4545 (1984).
1350. R. Joyeau, Y. Dugenet and M. Wakselman, *J. Chem. Soc.*, *Chem. Commun.*, 431 (1983).
1351. A. Arnoldi, L. Merlini and L. Scaglioni, *J. Hetercycl. Chem.*, **24**, 75 (1987).
1352. K. M. Shakhidoyatov, A. Irisbaev, L. M. Yun, E. Oripov and C. S. Kadyrov, *Khim. Geterotsikl. Soedin.*, 1564 (1976); *Chem. Abstr.*, **86**, 106517q (1977).
1353. D. E. Thurston, P. T. P. Kaumoya and L. H. Hurley, *Tetrahedron Letters*, **25**, 2649 (1984).
1354. H. J. Bergmann, R. Mayrhofer and H. H. Otto, *Arch. Pharm.* (*Weinheim, Ger.*), **319**, 203 (1986); *Chem. Abstr.*, **105**, 78774m (1986).
1355. R. Chambers, D. M. Kunert, L. Hernandez, Jr., F. Mercer and H. W. Moore, *Tetrahedron Letters*, 933 (1978).
1356. H. Hiemstra, W. J. Klaver and W. N. Speckamp, *J. Org. Chem.*, **49**, 1149 (1984).
1357. T. Aoki, M. Yoshioka, Y. Sendo and W. Nagata, *Tetrahedron Letters*, 4327 (1979).
1358. F. A. Bouffard and T. N. Solzmann, *Tetrahedron Letters*, **26**, 6285 (1985).
1359. Y. Yamamoto and Y. Morita, *Chem. Pharm. Bull.*, **32**, 2555 (1984).
1360. H. Feuer, C. S. Panda, L. Hou and H. S. Bevinakatti, *Synthesis*, 187 (1983).
1361. R. Henning and H. Urbach, *Tetrahedron Letters*, **24**, 5339 (1983).
1362. Kanegufuchi Chemical Industry Co. Ltd., *Jpn. Kokai Tokkyo Koho JP 58/131986 A2* (1983); *Chem. Abstr.*, **99**, 194716a (1983).
1363. P. A. Zoretic, P. Soja and N. D. Sinha, *J. Org. Chem.*, **43**, 1379 (1978).
1364. G. P. Tokmakov and I. I. Grandberg, *Khim. Dikarbonilnykh Soedin.*, *Tezisy Dokl. Veses.*, *Konf.*, 4th Meeting, 1975, p. 165; *Chem. Abstr.*, **87**, 53119y (1977).
1365. G. P. Tokmakov and I. I. Grandberg, *Izv. Timiryazevak, S-kk. Akad.*, 151 (1979); *Chem. Abstr.*, **92**, 94219c (1980).
1366. M. Marcos, J. L. Castro, L. Castedo and R. Riguera, *Tetrahedron*, **42**, 649 (1986).
1367. K. Prasad, P. Kneussel, G. Schulz and P. Stuetz, *Tetrahedron Letters*, **23**, 1247, (1982).
1368. S. Uyeo, I. Kikkawa, Y. Hamashima, H. Ona, Y. Nishitani, K. Okada, T. Kubata, K. Ishikura, Y. Ide, K. Nakano and W. Nagata, *J. Amer. Chem. Soc.*, **10**, 4403 (1979).
1369. T. Shono, Y. Matsummura and K. Inoue, *J. Org. Chem.*, **48**, 1388 (1983).
1370. Hoechst A.-G., *Belg. BE 849625* (1977); *Chem. Abstr.*, **88**, 136471d (1978).
1371. Hoechst A.-G., *Ger. Offen. DE 2557765* (1977); *Chem. Abstr.*, **87**, 134997k (1977).
1372. T. Cuvigny, P. Hullot, P. Mulot, M. Larcheveque and H. Normant, *Can. J. Chem.*, **57**, 1201 (1979).
1373. M. M. Campbell, R. G. Harcus and S. J. Ray, *Tetrahedron Letters*, 1441 (1979).
1374. M. Y. Kim and S. M. Weinreb, *Tetrahedron Letters*, 579 (1979).
1375. M. Okita, T. Wakamatsu and Y. Ban, *J. Chem. Soc.*, *Chem. Commun.*, 749 (1979).
1376. A. Archelas, R. Furstoss, D. Srairi and G. Maury, *Bull. Soc.*, *Chim. Fr.*, 234 (1986).
1377. J. C. Gramain, R. Remuson and Y. Troin, *J. Chem. Soc.*, *Chem. Commun.*, 194 (1976).
1378. J. C. Gramain, R. Remuson and Y. Troin, *Tetrahedron*, **35**, 759 (1979).
1379. R. G. Glushkov, I. M. Zasosova and I. M. Ovcharova, *Khim. Geterotsikl. Soedin.*, 1398 (1977); *Chem. Abstr.*, **88**, 50781t (1978).
1380. C. Deshayes and S. Gelin, *Synthesis*, 466 (1981).
1381. A. V. Kamaev, G. V. Ryazantsev, V. V. Sedov, Y. Y. Bairamov, V. P. Evdakov, V. I. Svergun, V. P. Panov, V. M. Fedoseev and P. M. Kochergin, *PCT Int. Appl. WD 801965* (1980); *Chem. Abstr.*, **94**, 65491w (1981).
1382. H. R. Pfaendler and H. Hoppe, *Heterocycles*, **23**, 265 (1985).
1383. A. Martel, J. P. Daris, C. Bachand, M. Menard, T. Durst and B. Belleau, *Can. J. Chem.*, **61**, 1899 (1983).
1384. T. Fukuyama, A. A. Laird and C. A. Schmidt, *Tetrahedron Letters*, **25**, 4709 (1984).
1385. N. Nagashima, N. Ozaki, M. Washiyama and K. Itoh, *Tetrahedron Letters*, **26**, 657(1985).
1386. S. G. Davies, I. M. Dordor-Hedgecock, K. H. Sutton, J. C. Walker, R. H. Jones and K. Prout, *Tetrahedron Letters*, **42**, 5123 (1986).
1387. A. Suarato, P. Lombardi, C. Galliani and G. Franceschi, *Tetrahedron Letters*, 4059 (1978).
1388. G. D. Annis, E. M. Hebblethwaite, S. T. Hodgson, D. M. Hollinshead and S. V. Levy, *J. Chem. Soc.*, *Perkin Trans. 1*, 2851 (1983).

1389. S. T. Hodgson, D. M. Hollinshead and S. V. Ley, *Tetrahedron*, **41**, 5871 (1985).
1390. A. G. M. Barrett and P. Quayle, *J. Chem. Soc., Chem. Commun.*, 1076 (1981).
1391. K. Tanaka, S. Yoshifuji and Y. Nitta, *Heterocycles*, **24**, 2539 (1986).
1392. M. Sunagawa, H. Matsumura, T. Inoue and T. Hirohashi, *Eur. Pat. Appl. EP 23097* (1981); *Chem. Abstr.*, **94**, 192112y (1981).
1393. M. Mori and Y. Ban, *Heterocycles*, **23**, 317 (1985).
1394. Y. M. Goo, Y. B. Lee, H. H. Kim, Y. Y. Lee and W. Y. Lee, *Bull. Korean Chem. Soc.*, **8**, 15 (1987); *Chem. Abstr.*, **108**, 21573c (1988).
1395. P. Buchs, A. Brossi and J. L. Flippen-Anderson, *J. Org. Chem.*, **47**, 719 (1982).
1396. K. Yoshioka and Y. Kawano, *Jpn. Kokai Tokkyo Koho JP 61/207386* (1986); *Chem. Abstr.*, **106**, 119551c (1987).
1397. M. Alpegiani, A. Bedeschi, E. Perrone and G. Franceschi, *Tetrahedron Letters*, **27**, 3453 (1986).
1398. V. M. Girijavallabhan, S. W. McCombie, P. Pinto, S. I. Lin and R. Versace, *J. Chem. Soc., Chem. Commun.*, 691 (1987).
1399. N. A. Anisimova, N. K. Orlova, A. G. Shipov, I. Y. Belavin and Y. I. Baukov, *Zh. Obshch. Khim.*, **54**, 1433 (1984); *Chem. Abstr.*, **101**, 230339a (1984).
1400. C. C. Wei, J. Borgese and M. Weigele, *Tetrahedron Letters*, **24**, 1875 (1983).
1401. M. D. Bachi, D. Goldberg and A. Gross, *Tetrahedron Letters*, 4167 (1978).
1402. J. E. Baldwin, A. K. Forrest, S. Ko and L. N. Sheppard, *J. Chem. Soc., Chem. Commun.*, 81 (1987).
1403. M. Yoshioka, I. Kikkawa, T. Tsuji, Y. Nishitani, S. Mori, K. Okada, M. Murakami, F. Matsubara, M. Yamaguchi and W. Nagata, *Tetrahedron Letters*, 4287 (1979).
1404. H. Onoue, M. Narisada, S. Uyeo, H. Matsumura, K. Okada, T. Yano and W. Nagata, *Tetrahedron Letters*, 3867 (1979).
1405. C. Borel, L. S. Hegedus, J. Krebes and Y. Satoh, *J. Amer. Chem. Soc.*, **109**, 1101 (1987).
1406. J. Knight, P. J. Parsons and R. Southgate, *J. Chem. Soc., Chem. Commun.*, 78 (1986).
1407. J. E. Baldwin, R. M. Adington and R. Bohlmann, *J. Chem. Soc., Chem. Commun.*, 357 (1985).
1408. B. P. Wijinberg and W. N. Speckamp, *Tetrahedron Letters*, **21**, 1987 (1980).
1409. P. M. M. Nossin and W. N. Speckamp, *Tetrahedron Letters*, **21**, 1991 (1980).
1410. G. A. Kraus and K. Neuenschwander, *Tetrahedron Letters*, **21**, 3841 (1980).
1411. T. Boer-Terpstra, J. Dijkink, H. E. Schoemaker and W. N. Speckamp, *Tetrahedron Letters*, 939 (1977).
1412. B. P. Wijinberg, W. N. Speckamp and A. R. C. Oostween, *Tetrahedron*, **38**, 209 (1982).
1413. H. E. Schoemaker, T. Boer-Terpstra, J. Dijkink and W. N. Speckamp, *Tetrahedron*, **36**, 143 (1980).
1414. H. Fliri, C. P. Mak, K. Prasad, G. Schulz and P. Stuetz, *Heterocycles*, **20**, 205 (1983).
1415. I. Ojima and X. Qiu, *J. Amer. Chem. Soc.*, **109**, 6357 (1987).
1416. G. Rajendra and M. J. Miller, *Tetrahedron Letters*, **26**, 5385 (1985).
1417. A. Biswas, C. Eigenbrot and M. J. Miller, *Tetrahedron*, **42**, 6421 (1986).
1418. S. Yoshifuji, Y. Arakawa and Y. Nitta, *Chem. Pharm. Bull.*, **35**, 357 (1987).
1419. M. Otake, I. Fukushima and K. Fujita, *Jpn. Kokai Tokkyo Koho JP 62/120360 A2* (1987); *Chem. Abstr.*, **107**, 198079b (1987).
1420. I. R. Trehan, K. Bala and J. B. Singh, *Indian J. Chem., Sect. B.*, **18B**, 295 (1979).
1421. C. Verkoyen and P. Rademacher, *Chem. Ber.*, **118**, 653 (1985).
1422. B. Yde, N. M. Yousif, U. Pedersen, I. Thomsen and S. O. Lawesson, *Tetrahedron*, **40**, 2047 (1984).
1423. O. P. Goel and U. Krolls, *Synthesis*, 162 (1987)
1424. J. J. Bodine and M. K. Kaloustian, *Synth. Commun.*, **12**, 787 (1982).
1425. J. I. Jin, M. S. Chang and S. H. Min, *Yakhak Hoechi*, **30**, 294 (1986); *Chem. Abstr.*, **107**, 236287c (1987).
1426. J. Blumback, W. Duerckheimer, E. Ehlers, K. Fleischmann, N. Klesel, M. Limbert, B. Mencke, J. Reden, K. H. Scheunermann, E. Schrinner, G. Selbert, M. Wieduwlet and M. Worm, *J. Antibiot.*, **40**, 29 (1987).
1427. D. Tunemoto, K. Nishide, T. Kohori and K. Kondo, *PCT Int. Appl. WO 86/5182 A1* (1986); *Chem. Abstr.*, **106**, 119550b (1987).
1428. G. Burton and M. J. Pearson, *Eur. Pat. EP 209237 A2* (1987); *Chem. Abstr.*, **107**, 39509a (1987).
1429. J. O'Sullivan and C. A. Aklonis, *J. Antibiot.*, **37**, 804 (1984).
1430. J. E. Baldwin and E. Lee, *Tetrahedron*, **42**, 6551 (1986).

1431. G. Sacripante and G. Just, *J. Org. Chem.*, **52**, 3659 (1987).
1432. R. Brettle and S. M. Shibib, *J. Chem. Soc., Perkin Trans. 1*, 2912 (1981).
1433. R. Brettle and S. M. Shibib, *Tetrahedron Letters*, **21**, 2915 (1980).
1434. J. E. Baldwin, E. P. Abraham, R. H. Adlington, M. J. Crimmin, L. D. Field, G. S. Jayatilake, R. L. White and J. J. Usher, *Tetrahedron*, **40**, 1907 (1984).
1435. C. Kaneko, T. Naito and R. Saito, *Jpn. Kokai Tokkyo JP 61/67 A2* (1986); *Chem. Abstr.*, **105**, 97249k (1986).
1436. W. N. Speckamp and J. J. J. DeBoer, *Recl. J. R. Neth. Chem. Soc.*, **102**, 410 (1983); *Chem. Abstr.*, **100**, 51413c (1984).
1437. A. Terajima, Y. Ito, Y. Kimura, K. Sakai and T. Hiyama, *Jpn. Kokai Tokkyo JP 62/158277* (1987); *Chem. Abstr.*, **108**, 55764f (1988).
1438. H. Hiemstra, H. P. Fortgens and W. N. Speckamp, *Tetrahedron Letters*, **25**, 3114 (1984).
1439. J. G. Cannon, Y. Chang, V. E. Amoo and K. A. Walker, *Synthesis*, 494 (1986).
1440. D. Hoope and M. Kloft, *Justus Liebigs Ann. Chem.*, 1527 (1980).
1441. A. Nishida, M. Shibasaki and S. Ikegami, *Chem. Pharm. Bull.*, **34**, 1423 (1986).
1442. L. C. Blaszczak, *Eur. Patt. Appl. EP 163452 A1* (1985); *Chem. Abstr.*, **105**, 114822m (1986).
1443. I. Shinkai, A. King and L. M. Fuentes, *Eur. Pat. Appl. EP 230792 A1* (1987); *Chem. Abstr.*, **107**, 217366m (1987).
1444. J. E. Baldwin, R. M. Adington, T. W. Kang, E. Lee, and C. J. Schofield, *J. Chem. Soc., Chem. Commun.*, 104 (1987).
1445. N. Tangari, M. Giovine, F. Morlacehi and C. Vetuschi, *Gazz. Chim. Ital.*, **115**, 325 (1985); *Chem. Abstr.*, **104**, 50776u (1986).
1446. H. H. Wasserman, B. H. Lipshutz and J. S. Wu, *Heterocycles*, **7**, 321 (1977).
1447. H. H. Wasserman, B. H. Lipshutz, A. W. Tremper and J. S. Wu, *J. Org. Chem.*, **46**, 2991 (1981).
1448. J. W. Pavlik and S. Tantayanon, *J. Amer. Chem. Soc.*, **103**, 6755 (1981).
1449. S. Murata, M.Miura and M. Nomura, *J. Chem. Soc., Perkin Trans. 1*, 1259 (1987).
1450. H. H. Wasserman and B. H. Lipshutz, *Tetrahedron Letters*, 4613 (1976).
1451. M. L. Gilpin, J. B. Harbridge, T. T. Howarth and T. J. King, *J. Chem. Soc., Chem. Commun.*, 929 (1981).
1452. M. L. Gilpin, J. B. Harbridge and T. T. Howarth *J. Chem. Soc., Perkin Trans. 1*, 1369 (1987).
1453. H. H. Wasserman and A. W. Tremper, *Tetrahedron Letters*, 1449 (1977).
1454. J. R. Mahajan, B. J. Nunes, H. C. Araujo and G. A. L. Ferreira *J. Chem. Res.(S)*, 284 (1979); *Chem. Abstr.*, **92**, 110499b (1980).
1455. Y. Otsuji, S. Nakanishi, N. Ohmura and K. Mizuno, *Synthesis*, 390 (1983).
1456. A. G. M. Barrett and M. A. Sturgess, *J. Org. Chem.*, **52**, 3940 (1987).
1457. M. L. M. Pennings and D. N. Reinhoudt, *Tetrahedron Letters*, **23**, 1003 (1982).
1458. G. D. Annis, E. M. Hebblethwaite and S. V. Ley, *J. Chem. Soc., Chem. Commun.*, 297 (1980).
1459. H. Molhrle and M. Claas, *Pharmazie*, **41**, 553 (1986); *Chem. Abstr.*, **108**, 37593v (1988).
1460. T. Fujii, M. Ohba and S. Yoshifuji, *Hukusokan Kogaku Toronkai Koen Yoshishu*, 8th, p. 134. Pharm. Inst., Tohoku Univ. Sendai, 1975; *Chem. Abstr.*, **84**, 164580z (1976).
1461. H. Moehrle and J. Gerloff, *Arch. Pharm. (Weinheim, Ger.)*, **311**, 672 (1978); *Chem. Abstr.*, **89**, 163365z (1978).
1462. M. Masui, S. Hara and S. Ozaki, *Chem. Pharm. Bull.*, **34**, 975 (1986); *Chem. Abstr.*, **106**, 66593j (1987).
1463. J. A. M. Hamersma and W. N. Speckamp, *Tetrahedron*, **38**, 3225 (1982).
1464. E. G. E. Hawkins, *U. S. Patent U. S. 3947406* (1976); *Chem. Abstr.*, **86**, 30276n (1977).
1465. A. Gossauer, R.-P. Hinze and H. Zilch, *Angew. Chem.*, **89**, 429 (1977).
1466. I. Ojima and H. B. Kwon, *Chem. Letters*, 1327 (1985).
1467. S. G. Davies, I. M. Dordor-Hedgecock, K. H. Sutton and J. C. Walker, *Tetrahedron Letters*, **27**, 3787 (1986).
1468. H. Alper, F. Urso and D. J. H. Smith, *J. Amer. Chem. Soc.*, **105**, 6737 (1983).
1469. H. Alper and D. J. H. Smith, *Brit. Pat. GB 2143819* (1985); *Chem. Abstr.*, **103**, 104835n (1985).
1470. J. F. Knifton, *J. Organomet. Chem.*, **188**, 223 (1980).
1471. M. Mori, K. Chiba, M. Okita and Y. Ban, *J. Chem. Soc., Chem. Commun.*, 698 (1979).
1472. H. Alper and C. P. Perera, *Organometallics*, **1**, 70 (1982).
1473. M. Shibasaki and T. Iimori, *Eur. Pat. EP 209886* (1987); *Chem. Abstr.*, **106**, 196123s (1987).
1474. A. A. Avetisyan, S. K. Karaglz, M. T. Dangyan and D. I. Gezalyan, *Arm. Khim. Zh.*, **32**, 389 (1979); *Chem. Abstr.*, **92**, 215023v (1980).

1475. F. Reliquet, A. Reliquet, F. Sharrard, J. C. Meslin and H. Quiniou, *Phosphorus Sulfur*, **24**, 279 (1985); *Chem. Abstr.*, **104**, 129676v (1986).
1476. H. Ogura and K. Takeda, *Heterocycles*, **15**, 467 (1981).
1477. J. A. Moore and E. M. Partain, III, J. *Polym. Sci., Polym. Letters. Ed.*, **20**, 521 (1982).
1478. D. Sicker, D. Reifegeiste, S. Hauptmann, H. Wilde and G. Mann, *Synthesis*, 331 (1985).
1479. E. A. Glazer and J. E. Presslitz, *J. Med. Chem.*, **25**, 868 (1982).
1480. Y. Nagao, T. Miyasaka, K. Seno and E. Fujita, *Heterocycles*, **15**, 1037 (1981).
1481. T. Okawara, R. Kato, T. Yamasaki, N. Yasuda and M. Furukawa, *Heterocycles*, **24**, 885 (1986).
1482. T. Okawara, K. Nakayama, T. Yamasaki and M. Furukawa, *J. Chem. Res. (S)*, 188 (1985).
1483. V. Virmani, J. Sing, P. C. Jain and N. Anand, *J. Chem. Soc., Pak.*, **1**, 109 (1979).
1484. J. Iriarte, C. Camargo and P. Crabbe, *J. Chem. Soc., Perkin Trans.* **1**, 2077 (1980).
1485. H. Takahata, M. Ishikura, K. Nagai, M. Nagata and T. Yamazaki, *Chem. Pharm. Bull.*, **29**, 366 (1981).
1486. E. P. Kramarova, A. G. Shipov, O. B. Artamkina and Y. I. Baukov, *Zh. Obshch. Khim.*, **54**, 1921 (1984); *Chem. Abstr.*, **102**, 62311k (1985).
1487. R. Obrecht, S. Toure and I. Ugi, *Heterocycles*, **21**, 271 (1984).
1488. Y. Yamamoto and Y. Watanabe, *Chem. Pharm. Bull.*, **35**, 1871 (1987).
1489. E. Hayashi, T. Abe and S. Nagase, *Chem. Letters*, 375 (1985).
1490. H. H. Wasserman and B. H. Lipshutz, *Ger. Pat. DE 2747494* (1978); *Chem. Abstr.*, **89**, 129384v (1978).
1491. A. Albini, A. Borinotti, G. F. Bettinetti and S. Pietra, *J. Chem. Soc., Perkin Trans. 2*, 238 (1977).
1492. J. E. Baldwin, R. M. Adlington, S. L. Flitsch, H. H. Ting and N. J. Turner, *J. Chem. Soc., Chem. Commun.*, 1305 (1986).
1493. A. Marinetti, J. Fischer and F. Mathey, *J. Amer. Chem. Soc.*, **107**, 5001 (1985).

Author index

This author index is designed to enable the reader to locate an author's name and work with the aid of the reference numbers appearing in the text. The page numbers are printed in normal type in ascending numerical order, followed by the reference numbers in parentheses. The numbers in *italics* refer to the pages on which the references are actually listed.

Abbady, M. A. 529, 534, 539, 833, 851 (1186), *1020*
Abbayes, H. des 341, 342 (850), *1012*
Abd-Alla, H. A. 529, 534, 539, 833, 851 (1186), *1020*
Abdel-Mawgoud, A. M. 533–535, 537, 817, 852 (1189), *1020*
Abdulla, R. F. 139 (437), *262*
Abe, I. 491, 498, 920 (1131), *1019*
Abe, R. 507 (1157), 508 (1158), 515 (1157), 516 (1158, 1170), 517, 518 (1172), 813 (1158), 814, 838 (1158, 1170, 1172), *1019*
Abe, T. 1005 (1489), *1027*
Abou-Chaar, C. I. 53 (189), *257*
Abraham, E. P. 397 (942), 828, 896, 913, 914, 947 (1434), *1014, 1026*
Abramski, W. 545 (1193), *1020*
Achiwa, K. 402, 708, 733 (1007), *1016*
Adam, W. 4, 8 (25), 130 (359), *254, 261*, 271 (727), *1009*
Adams, R. 36 (129), *256*
Adams, W. R. 18 (82), 144–146 (457, 458), *255, 263*
Adickes, H. W. 222, 223 (658), *267*
Adington, R. M. 779, 780 (1407), 945, 948 (1444), *1025, 1026*
Adlington, A. M. 404, 406, 704 (1018, 1019), 811 (1019), *1016*
Adlington, R. M. 828, 896, 913, 914, 947 (1434), 1005, 1007 (1492), *1026, 1027*
Agawa, T. 61, 66–68 (215), 168 (514), *258, 264*, 342 (851), 545 (1192), 593, 594, 601, 602 (1238), 605, 606 (1242), 607, 608 (1238, 1242), 876, 877 (1238), 889, 890 (1192), 911 (1238), *1012, 1020, 1021*
Agho, M. O. 437, 648 (1074), *1017*
Agnes, G. 17, 18 (63), *255*
Ahearn, G. P. 100, 101, 106, 125 (291), *259*
Aida, T. 271 (737), *1010*
Aigner, H. 397 (919), *1014*
Aitken, A. R. 445 (1084), *1017*
Aizpurua, J. M. 490 (1146), 551 (1200), 555, 568 (1204), 569, 570 (1204, 1216), 572, 575, 577, 579 (1204), 728, 730, 751, 762 (1146), 812, 849 (1204), 898, 934, 954, 955 (1146), *1019, 1020*
Akai, S. 429, 430 (1062), 448, 450 (1087), *1017, 1018*
Akermark, B. 144, 158 (484), *263*
Akhnazaryan, A. A. 28 (102), *256*, 435 (1071), *1017*
Akhobadze, D. S. 664, 676, 708, 751, 754, 764, 815, 839, 942 (1304), *1022*
Akiba, K. 389 (892), *1013*
Aklonis, C. A. 811 (1429), *1025*
Akopyan, K. G. 48 (171), *257*
Akssira, M. 295 (792), *1011*
Albini, A. 1005 (1491), *1027*
Alcaide, B. 491–493, 497, 499 (1128), 503 (1128, 1148), 504 (1128), 528, 536 (1128, 1148), 541, 545 (1128), 648, 651 (1292), 734, 882 (1148), *1018, 1019, 1022*
Aldabe, C. A. 397 (926), *1014*
Alexander, S. M. 489 (1125), *1018*
Alguero, M. 18, 20 (76), *255*
Aljadeff, G. 335 (840), *1012*
Allmann, R. 491, 503, 504, 523, 526, 536, 541, 542, 544, 1005 (1127), *1018*
Al-Nuri, M. 402, 403 (1008), *1016*
Alonso, D. 283 (771), *1010*

Alpegiani, M. 397 (947), 685, 689 (1342), 748 (1342, 1397), 752 (1397), 753 (1342), 761, 762, 764, 768 (1397), 769 (1342), 915 (1397), *1014, 1023, 1025*
Alper, H. 130 (361), *261*, 341, 342 (850), 648 (1289), 982, 983 (1468, 1469), 991 (1472), *1012, 1022, 1026*
Altmann, H. J. 144 (470), *263*
Ambelang, T. 102, 106 (294), *259*
Amendola, M. 51 (180), *257*
Amin, S. G. 397 (902), 505 (1154), 506 (1156), 507, 508 (902, 1156), 509 (902), 511 (1154), 512 (902), 521 (1178), 522 (1154), 548 (1178), 555 (1205), 556 (1178, 1205, 1206), 557 (1178, 1207), 558, 559 (1178), 560 (1178, 1209), 561 (1209), 563–565 (1212), 573 (902, 1219), 574 (1219), 579, 726, 728, 740, 813, 819, 822 (1178), 823, 824 (1154, 1178), 838 (1178), 841 (1154), 849, 850, 853, 856 (1178), 859, 870 (1205), 871 (1156), 872 (902), 873 (902, 1156), 874 (902, 1156, 1205), *1013, 1019, 1020*
Ammon, H. L. 230 (663), *267*
Amoo, V. E. 911 (1439), *1026*
Anand, N. 999 (1483), *1027*
Anderson, H. W. 69, 73 (224), *258*
Andreoli, P. 614 (1247), *1021*
Andrews, D. 48 (174), *257*
Andruskiewicz, C. A. Jr. 124 (348), *261*
Anghelide, N. 134 (386), *261*
Anisimova, N. A. 750 (1399), *1025*
Anisimova, O. S. 397, 981 (903), *1013*
Anjaneyulu, B. 142 (431), 170 (548, 571, 574), 183–185, 187, 190 (548), 192, 193 (574), 195 (571, 574), 196 (571), 197 (548), *262, 265*
Annis, G. D. 271 (742), 732 (1388), 966, 967 (1388, 1458), *1010, 1024, 1026*
Ansell, M. F. 3, 16 (5), *253*
Antonini, I. 510, 530, 673, 734 (1160), *1019*
Aoki, T. 699, 761, 782 (1357), *1024*
Aoyagi, H. 144, 157 (482), *263*
Aoyama, H. 144, 160, 162 (467), *263*, 465 (1104), 471 (1108, 1109), 494 (1104), 495 (1136), 872 (1109), 897 (1104), 1001 (1109), *1018, 1019*
Appel, R. 109, 110 (319), *260*
Applegate, H. E. 238 (679), 239 (680), *267*
Ara, K. 430–432, 865, 866 (1064), *1017*
Arai, M. 397 (977), *1015*
Arakawa, Y. 798 (1418), *1025*
Araujo, H. C. 452, 657, 962 (1090), *1018*
Arcamone, F. 398 (990), 687, 747, 753, 768, 907 (1338), *1015, 1023*
Archelas, A. 722, 740, 813 (1376), *1024*
Arenson, F. R. 83, 85, 91 (254), *259*
Aresi, V. 135 (401), 136 (390), 184 (401), *261, 262*
Arimoto, M. 278 (763), *1010*
Arndt, F. 34 (116), *256*
Arnold, M. J. 630, 929, 971 (1263), *1021*
Arnold, R. T. 30 (108), *256*
Arnold, Z. 104, 106 (299), *260*
Arnoldi, A. 695, 798 (1351), *1024*
Arrieta, A. 549 (1198), 551 (1200, 1201), 552–554 (1201), 555 (1198, 1204), 556, 561, 563, 567 (1198), 568 (1198, 1204), 569 (1198, 1204, 1216), 570 (1198, 1201, 1204, 1216), 571 (1198), 572, 575, 577, 579 (1198, 1204), 726, 730, 734 (1201), 812, 849 (1204), *1020*
Arsenijevic, L. 583 (1226), *1021*
Arsenijevic, V. 583 (1226), *1021*
Artamkina, O. B. 1001 (1486), *1027*
Arth, G. E. 109 (315), *260*
Arutyunyan, V. S. 10 (35), 13 (48), *254*
Ashbrook, C. W. 239 (682), *268*
Ashcroft, C. J. 397 (923), *1014*
Aster, S. D. 665, 781, 894, 895 (1307), *1023*
Aube, J. 399, 407, 732 (1002), *1016*
Augustin, D. 640 (1278), *1022*
Aumann, A. 488 (1122), *1018*
Avakian, S. 138 (415), *262*
Avetisyan, A. A. 34, 39 (118), 42 (146, 147), 48 (171), *256, 257*, 993 (1474), *1026*
Avila, N. V. 28 (105), *256*
Awasthi, A. K. 577, 578 (1222), *1021*
Axiotis, G. 585 (1232), *1021*
Azuma, S. 39 (143), 98, 106 (282), *256, 259*

Babaeva, A. A. 11 (36), *254*
Bach, S. R. 12 (47), 42 (151), 43 (47), *254, 257*
Bachand, C. 729, 730, 760, 762, 763, 863, 900, 925, 926 (1383), *1024*
Bachi, M. D. 170 (539, 540, 556), 180 (539, 540), 188 (540, 556), *264, 265*, 518 (1173), 519 (1175), 686, 687, 746, 747, 750 (1336), 759 (1336, 1401), 765, 772 (1336), 780, 781 (1175), 811, 813, 835, 853, 881 (1173), 903 (1336), 920 (1401), 922 (1175), *1019, 1023, 1025*
Back, T. G. 271 (735), 489, 679, 680, 721, 877, 879 (1124), *1010, 1018*
Bader, F. E. 9, 11, 13, 87, 125 (28), *254*
Badrian, A. S. 640 (1277), *1022*
Baeza, J. 4, 8 (25), *254*
Bagheri, V. 315 (814), *1011*
Bagli, J. F. 144 (455), *263*
Bai, V. V. 640 (1283), *1022*
Bailey, D. M. 76–78, 80–82 (236), *258*
Bailey, P. S. 76 (228, 229), *258*

Bailey, W. C. Jr. 13, 17 (52), *254*
Bairamov, Y. Y. 725 (1381), *1024*
Baird, M. S. 222 (657), *267*
Baker, A. J. 14, 16 (53), *254*
Baker, B. R. 36 (129), *256*
Bakker, C. G. 275, 276 (756), *1010*
Bal, B. S. 385 (889), *1013*
Bala, K. 802 (1420), *1025*
Baldwin, J. E. 239 (682), *268*, 399 (999), 404, 406, 704 (1018, 1019), 756, 757 (999), 759 (1402), 779, 780 (1407), 811 (1019), 824 (999), 828 (1434), 847 (999), 857 (1430), 896, 913, 914 (1434), 945 (1444), 947 (1434), 948 (1444), 999 (1430), 1005, 1007 (1492), *1016*, *1025–1027*
Baldwin, S. W. 273, 366 (753), 399, 407, 732 (1002), *1010*, *1016*
Ball, C. 397 (931), *1014*
Ballard, S. A. 170, 179, 180 (531), *264*
Ballistini, C. 397 (947), *1014*
Baltazzi, E. 46 (162), *257*
Ban, Y. 285, 345 (776), 397 (967), 424, 425 (1059), 433, 712 (1067), 714 (776, 1067), 715 (776), 716 (1059), 719 (1375), 739, 792, 795 (1393), 879 (776, 1393), 984 (1471), 985, 986 (1059, 1067, 1471), 987 (1067, 1393, 1471), 988 (776, 1067, 1393), 989, 990 (776), *1011*, *1015*, *1017*, *1024–1026*
Banfi, L. 624 (1257), *1021*
Banner, B. L. 282 (768), *1010*
Baranowsky, K. 30 (109), *256*
Bardili, B. 348 (856), *1012*
Bari, S. S. 397 (921), 505, 509, 545 (1152), 562 (1152, 1211), 706 (1211), 735, 743, 755 (1152), 859 (1211), 936 (1152), *1014*, *1019*, *1020*
Barluenga, J. 207 (608), *266*
Barnes, C. S. 222, 229 (655), *267*
Barnett, B. F. 214 (630), *266*
Barnett, W. E. 27 (98–100), 28 (100), *255*
Barnish, I. T. 143 (450, 452), *263*
Barr, J. J. 494, 499, 648, 651, 652 (1133), *1019*
Barrett, A. G. M. 476 (1113, 1114), 478 (1114, 1116), 480 (1114), 481, 483 (1114, 1116), 585 (1231), 613 (1246), 737, 904 (1390), 963 (1231, 1456), *1018*, *1021*, *1025*, *1026*
Bartlett, M. F. 242 (697), *268*
Bartlett, P. D. 76 (227, 230), *258*
Barton, D. H. R. 83 (256), 97 (256, 279), 127 (353), 222, 229 (655), *259*, *261*, *267*, 632, 911 (1266), *1021*
Basselier, J.-J. 171, 204, 205 (602), *266*
Batchelor, M. J. 292, 352, 353, 362, 366 (788), *1011*
Bateson, J. H. 678, 688, 690, 737, 738, 748, 749, 752–754, 767 (1319), *1023*
Battersby, A. R. 137 (410), *262*
Battiste, M. A. 279 (767), *1010*
Battistini, C. 397 (964), *1015*
Bauer, V. J. 4–6, 10, 11 (16), *254*
Baukov, Y. I. 681 (1329), 750 (1399), 1001 (1486), *1023*, *1025*, *1027*
Baumgarten, H. E. 138 (418–420), *262*
Bavin, P. M. G. 42 (153), *257*
Beaton, J. M. 97 (279), *259*
Beck, J. F. 137 (410), *262*
Beckmann, S. 27 (96), *255*
Beckwith, A. L. J. 127 (353), 133 (374), *261*, 419 (1043), 423 (1054), 451, 664 (1043), 918, 919 (1054), *1017*
Bedeschi, A. 397 (947, 964), 685, 689 (1342), 748 (1342, 1397), 752 (1397), 753 (1342), 761, 762, 764, 768 (1397), 769 (1342), 915 (1397), *1014*, *1015*, *1023*, *1025*
Bedford, C. R. 222 (654), *267*
Beisswenger, T. 679, 680 (1325), *1023*
Belanger, A. 131 (364), *261*
Belanger, D. 295, 308 (793), *1011*
Belaud, C. 298, 581 (794), *1011*
Belavin, I. Y. 750 (1399), *1025*
Belil, C. 18 (70, 71), *255*
Bellasio, E. 139 (427), *262*
Belleau, B. 729, 730, 760, 762, 763, 863, 900, 925, 926 (1383), *1024*
Bellus, D. 503 (1149), *1019*
Belov, V. N. 3 (8), *253*
Belyaeva, O. Y. 626 (1259), *1021*
Belzecki, C. 402, 409 (1009), 479, 480 (1118), 485, 486 (1119), 494–496 (1134), 545 (1193), 617–620 (1253, 1254), *1016*, *1018–1021*
Bembry, T. H. 36 (130), *256*
Bencini, E. 341, 984 (849), *1012*
Bendall, V. I. 47 (168), *257*
Bender, D. R. 639, 979 (1271), *1022*
Benezra, C. 278 (759), *1010*
Ben-Ishai, D. 285, 456 (777), 463 (1100), 947 (777), *1011*, *1018*
Bensel, N. 46 (158), *257*
Benson, R. E. 240 (687), *268*
Bentley, R. L. 170, 194, 199 (591), *266*
Bergbreiter, D. E. 601, 603, 604, 608, 609 (1240), *1021*
Bergen, T. J. van 135 (405–407), 137 (405, 407), *262*
Berger, G. D. 401, 476, 648, 652, 657, 675 (1003), *1016*
Bergmann, H. J. 696 (1354), *1024*
Berkov, B. 28 (105), *256*
Berkowitz, L. M. 109 (314), 245 (707), *260*, *268*

Bernath, G. 504, 513 (1150), 537 (1150, 1190), 538 (1190), 543, 544 (1150, 1190), 883 (1190), *1019, 1020*
Berson, J. 170 (522), *264*
Berson, J. A. 27 (97), 83, 89 (267), *255, 259*
Berstein, S. C. 49 (178), *257*
Berthon, B. 315 (812), *1011*
Bertrand, M. 111 (327), *260*, 333 (838), *1012*
Bertrand, M. P. 337 (842), *1012*
Bertrand, M. T. 51 (183), *257*
Bestian, H. 132 (370), 163 (370, 503), 164, 165 (503), *261, 264*
Bettinetti, G. F. 1005 (1491), *1027*
Betts, M. J. 476 (1114), 478 (1114, 1116), 480 (1114), 481, 483 (1114, 1116), *1018*
Bevinakatti, H. S. 704 (1360), *1024*
Beyerman, H. C. 137 (408), *262*
Beziat, Y. 49 (177), *257*
Bhakta, C. 290, 359, 364 (787), *1011*
Bhatnagar, P. K. 53 (192), *258*
Bhatt, D. J. 539 (1191), *1020*
Bhattacharya, S. K. 142 (431), 170, 183–185, 187, 190, 197 (548), *262, 265*
Bhawal, B. M. 562, 706, 859 (1211), *1020*
Bianso, M. C. 661 (1301), *1022*
Bick, I. R. C. 248 (712), *268*
Bickel, H. 9, 11, 13, 87, 125 (28), *254*
Bielefeld, M. A. 36 (136), *256*
Biener, H. 163–165 (503), *264*
Bienz, S. 658, 912 (1298), *1022*
Bieri, J. H. 490 (1147), *1019*
Bill, I. 18, 20 (67), *255*
Biloski, A. J. 435, 943 (1072), *1017*
Bilyard, K. G. 287, 290, 360 (780), *1011*
Bimanand, A. Z. 292 (789), *1011*
Bindra, J. 509, 514, 567, 568, 848 (1159), *1019*
Bipp, H. 640 (1276), *1022*
Bird, C. W. 648, 651–653 (1293), *1022*
Birkofer, L. 497, 498 (1139), *1019*
Birladeanu, L. 109 (315), *260*
Biswas, A. 797 (1417), *1025*
Black, D. S. C. 660, 661 (1299), *1022*
Black, D. St. C. 236, 237 (673), 240 (688), *267, 268*
Black, R. 456, 457 (1092), *1018*
Blackburn, B. K. 452, 697 (1089), *1018*
Blade-Font, A. 400 (1010), 411 (1027, 1029), *1016*
Blagoev, B. 53 (190), *257*
Blair, G. E. 53 (189), *257*
Blanchard, J. M. 283 (774), *1011*
Blaskovits, A. 83, 91, 92 (273), *259*
Blaszczak, L. C. 925, 926 (1442), *1026*
Blechert, S. 271 (728), 397 (935), *1009, 1014*
Blicke, F. F. 135 (404), 204, 205 (601), *262, 266*
Bloomfield, J. J. 76–80, 83 (237), *258*
Bloy, V. 271 (749), *1010*
Blumbach, J. 827 (1426), *1025*
Blume, R. C. 13 (50), *254*
Boate, D. R. 423, 918, 919 (1054), *1017*
Bodenstein, C. K. 11, 53 (44), *254*
Bodine, J. J. 803, 811, 855 (1424), *1025*
Boer-Terpstra, T. 784, 785 (1411), 791 (1413), 973, 975 (1411, 1413), *1025*
Bogatskii, A. V. 394 (896), *1013*
Bogavac, M. 583 (1226), *1021*
Bogdan, S. 397 (920), *1014*
Bogdanova, L. A. 123 (342), *260*
Bogdanowicz, M. J. 102 (294, 295), 103 (295), 106 (294, 295), *259*
Bohlmann, R. 779, 780 (1407), *1025*
Böhme, E. W. H. 238 (679), 239 (680), *267*
Bohme, H. 142 (444), *263*
Bolt, C. C. 83, 85 (258), *259*
Bonati, A. 136 (391), *262*
Bonet, J. J. 661 (1301), *1022*
Bonetskaya, A. K. 398 (991), *1015*
Bonini, C. 278, 335 (762), *1010*
Borch, R. F. 134 (387), *261*
Borel, C. 756, 758, 844–846, 956, 958 (1405), *1025*
Borer, R. 48 (172), *257*
Borgese, J. 755, 945 (1400), *1025*
Borinotti, A. 1005 (1491), *1027*
Börner, K. 132, 163 (370), *261*
Borrmann, D. 69 (219–222), 70 (219–221), 71 (221), 72 (221, 222), 73 (222), *258*
Borromeo, P. S. 300 (799), *1011*
Borsche, W. 11, 53 (43, 44), *254*
Borsotti, G. 234 (671), *267*
Bortolussi, M. 456, 457 (1092), *1018*
Boruah, R. C. 615 (1249), *1021*
Bory, S. 4 (18), *254*
Bosch, J. 18 (75, 76), 20 (76), *255*
Bose, A. K. 13, 17 (52), 133 (377), 138, 139 (432–436), 142 (431, 445), 163 (377), 169 (520), 170 (520, 542, 545–548, 557, 560–566, 568, 570–576, 588, 593, 594), 181 (542), 183 (520, 545, 547, 548), 184, 185 (545, 548), 187 (547, 548, 560, 563, 566), 188 (547, 557, 560, 564, 566), 189 (557), 190 (548), 192 (560, 565, 566, 568, 572, 574, 575), 193 (574), 194 (570, 573), 195 (560, 566, 571, 574, 575), 196 (571), 197 (548, 561, 562, 565, 576), 198 (588), 200 (593, 594), 201 (520), 210 (614), *254, 261–266*, 397 (910, 921, 922, 943), 398 (998), 439–442 (1076), 504 (1151), 505 (1152–1155), 506 (1155, 1156), 507 (1156), 508 (1151, 1156), 509 (1151, 1152), 511 (1154), 513 (1155), 516 (1151), 521 (1178), 522 (1154), 523 (1155), 526 (1155, 1180),

527 (1180), 529 (1151), 545 (1152), 547 (1180, 1194), 548 (1178, 1194), 555 (1205), 556 (1178, 1205, 1206), 557 (1178, 1207, 1208), 558, 559 (1178), 560 (1178, 1209), 561 (1209, 1210), 562 (1152, 1153, 1210, 1211), 563 (1153, 1210, 1212), 564, 565 (1212), 567 (1180), 573, 574 (1219), 579 (1178), 581 (1225), 612 (1244), 645, 648, 657, 679, 680 (1155), 695 (1076), 706 (1155, 1211), 726, 728 (1178), 734 (1225), 735 (1152, 1153), 740 (1178), 743, 755 (1152), 806 (1225), 810 (1155), 813 (1178), 817 (1180), 819 (1178), 822 (1178, 1180), 823, 824 (1154, 1178), 838 (1178), 841 (1154), 846 (1180), 849, 850, 853 (1178), 856 (1178, 1208), 859 (1205, 1211), 869 (1153), 870 (1205), 871, 873 (1156), 874 (1156, 1205), 918, 925 (1155), 936 (1152), *1013, 1014, 1016, 1017, 1019–1021*

Bosma, M. A. R. 445 (1083), *1017*
Bouffard, F. A. 702, 703 (1358), *1024*
Bougault, J. 19, 27 (90), *255*
Bourgesis, P. 167 (512), *264*
Bowman, E. R. 241, 242 (695), *268*
Boyd, D. B. 490 (1126), 690, 749, 753, 769, 830, 831, 847, 899, 909, 910, 970 (1346), *1018, 1023*
Boyer, S. K. 495, 496, 869 (1135), *1019*
Brady, W. T. 169 (516), 170 (516, 526), 172 (516), 173 (516, 526), 174, 177, 182 (516), *264*, 496, 499, 541, 542, 865 (1138), *1019*
Brambilla, R. 626 (1260), *1021*
Brammer, R. 109, 110 (319), *260*
Branch, C. L. 548 (1196), 686 (1196, 1335), 687 (1337), 688, 708, 736 (1196), 746 (1196, 1337), 747 (1337), 750 (1196, 1337), 751 (1337), 771 (1196, 1335, 1337), 774, 775 (1196, 1335), 776 (1337), 815, 840 (1196), 903, 907 (1337), 932, 934, 935, 942 (1196), 960, 961 (1335), *1020, 1023*
Brandstadter, S. M. 517, 524, 525, 819–821, 831, 832, 838, 841, 842, 881 (1171), *1019*
Brash, J. L. 3 (9), *253*
Brassard, P. 131 (364), *261*
Braun, E. 61 (203), *258*
Breckpot, R. 134 (389), *261*
Breitbeil, F. W. 42 (152), *257*
Brennan, J. 397 (923, 972, 973), 634 (1267), 639 (1271), 897, 947 (1267), 979 (1271), *1014, 1015, 1021, 1022*
Brenner, J. B. 248 (712), *268*
Brettle, R. 876 (1432), 878 (1432, 1433), 879, 976, 977 (1432), *1026*
Breuer, H. 397 (969, 980), *1015*
Brianso, J. L. 661 (1301), *1022*
Bricout, J. 17 (62), *255*
Brink, W. M. van den 137 (408), *262*
Brock, C. P. 585, 963 (1231), *1021*
Broenneke, A. 322 (826), *1012*
Brooks, J. R. 250 (715), *268*
Brossi, A. 10 (33, 34), *254*, 744, 745 (1395), *1025*
Brown, A. G. 398 (988), *1015*
Brown, E. 314 (811), *1011*
Brown, J. B. 12 (46), *254*
Brown, J. D. 369 (872), *1013*
Brown, S. P. 385 (889), *1013*
Brown, W. G. 76 (232, 233), *258*
Brunwin, D. M. 144 (490, 491), *263, 264*
Buchi, G. 18, 22 (86), 106 (305), 242, 245 (698), *255, 260, 268*
Buchs, P. 744, 745 (1395), *1025*
Bücking, H. W. 127, 128 (355), *261*
Buckley, G. D. 76 (231), *258*
Budzikiewicz, H. 241 (694), 244 (705), *268*
Buhle, E. L. 170, 193 (578, 582), *265*
Burger, D. H. 437, 648 (1074), *1017*
Burkard, U. 420 (1045), *1017*
Burnett, D. A. 600, 601, 603, 605, 731, 733, 799 (1241), *1021*
Burstain, I. G. 109, 110 (317, 318), *260*
Burton, G. 830 (1428), *1025*
Bus, J. 123 (345), *261*
Buschmann, E. 632, 911 (1266), *1021*
Bush, J. B. Jr. 110, 111 (322), *260*
Butler, D. E. 489 (1125), *1018*
Buynak, J. D. 477, 478, 918 (1115), *1018*

Cadoff, B. C. 100, 106 (289), *259*
Caine, D. 368 (870), 369 (871), *1012, 1013*
Cainelli, G. 614 (1247), *1021*
Cairns, T. L. 240 (687), *268*
Camargo, C. 999 (1484), *1027*
Cambie, R. C. 616 (1252), *1021*
Campbell, M. M. 679 (1323), 717, 719 (1373), 906 (1323), *1023, 1024*
Cannon, J. G. 911 (1439), *1026*
Canonne, P. 295 (792, 793), 308 (793), *1011*
Canovas, A. 661 (1301), *1022*
Capp, C. W. 223 (662), *267*
Cardellini, M. 496 (1137), 510 (1137, 1160), 530 (1160), 536 (1137), 673, 734 (1160), *1019*
Carelli, V. 434 (1069), *1017*
Carlon, F. E. 83, 97 (256), *259*
Carlson, R. G. 124 (346), *261*
Carlson, R. M. 316 (816), *1011*
Carlson, S. 460, 805, 806 (1096), *1018*
Caron, G. 8 (23), *254*
Carruthers, N. 679, 906 (1323), *1023*
Carter, H. E. 46 (161), *257*
Casadei, M. A. 434 (1069), 510, 535, 536, 540, 541, 867 (1161), *1017, 1019*
Cassady, J. M. 53 (189), *257*

Cassar, L. 118 (335), *260*
Cassio, F. P. 563–566, 726, 728, 750 (1213), *1020*
Castaner, J. 18 (68, 69, 76), 20 (68, 76), *255*
Castedo, L. 711 (1366), *1024*
Castella, J. 18 (70, 76), 20 (76), *255*
Castellino, S. 395 (900), *1013*
Castells, J. 18 (70, 73, 75, 76), 20 (76), *255*
Castro, J. L. 711 (1366), *1024*
Cava, M. P. 382 (885), *1013*
Cave, A. 242 (700), *268*
Cazaux, L. 394 (897), *1013*
Cefelin, P. 134 (384, 385), *261*
Cereghetti, M. 244 (705), *268*
Ceresiat, M. 397 (920), *1014*
Cerutti, P. 144, 157 (483), *263*
Chabudzinski, Z. 214, 219 (639), *266*
Chaikin, S. W. 76 (232), *258*
Chakraborty, T. K. 319 (820), 325 (829), *1011*, *1012*
Chalmers, A. M. 14, 16 (53), *254*
Chambers, R. 499, 500, 502 (1142), 697 (1142, 1355), 700, 701 (1142), 704 (1355), 863 (1142), *1019*, *1024*
Chandrasekaran, S. 325 (829), 327 (831), *1012*
Chandrasekaren, R. Y. 477, 478, 918 (1115), *1018*
Chandrasekeran, S. 319 (820), *1011*
Chang, C. 53 (189), *257*
Chang, M. S. 811, 821, 822 (1425), *1025*
Chang, T. C. T. 386 (890), *1013*
Chang, Y. 911 (1439), *1026*
Chapleo, C. B. 122 (339), *260*
Chapman, O. L. 18 (82), 130 (359), 144 (457–459), 145 (457, 458), 146 (457–459), 147 (459), *255*, *261*, *263*
Chasle, M. F. 211 (619), *266*
Chatterjee, B. G. 138 (434), 139 (434, 437–442), 140 (439), 141 (440), *262*, *263*, 422 (1052), 510, 511, 521, 817, 839 (1162), 938–940 (1052), *1017*, *1019*
Chawla, H. P. S. 139, 141 (440), 170 (560, 566, 575), 187, 188 (560, 566), 192, 195 (560, 566, 575), *263*, *265*, 506–508 (1156), 526, 527, 547 (1180), 555, 556 (1205), 567, 817, 822, 846 (1180), 859, 870 (1205), 871, 873 (1156), 874 (1156, 1205), *1019*, *1020*
Chaykovsky, M. 144, 155, 156 (478), *263*
Chemielarz, B. 3 (7), *253*
Chen, C. G. 144, 158 (485), *263*
Chen, H. J. C. 527, 696, 852 (1181), *1020*
Chen, S.-C. 411 (1028), *1016*
Chernotkach, Z. A. 394 (896), *1013*
Chernyuk, K. Y. 271 (746), *1010*
Chiang, Y. H. 170, 183 (545, 547), 184, 185 (545), 187, 188 (547), *265*
Chib, J. S. 170 (547, 566), 183 (547), 187, 188 (547, 566), 192, 195 (566), *265*
Chiba, K. 285, 345 (776), 433, 712 (1067), 714 (776, 1067), 715, 879 (776), 984 (1471), 985–987 (1067, 1471), 988 (776, 1067), 989, 990 (776), *1011*, *1017*, *1026*
Chiba, T. 615 (1248), *1021*
Chino, K. 314 (822), *1012*
Chiusoli, G. P. 17, 18 (63), *255*
Chmielewski, M. 402 (1009), 407 (1024), 409 (1009, 1024), 478 (1117), 479 (1118), 480 (1117, 1118), 481, 482, 484 (1117), 485 (1117, 1119), 486 (1119), 545 (1193), *1016*, *1018*, *1020*
Cholskova, T. 346 (854), *1012*
Chow, H. F. 313, 329 (810), *1011*
Christensen, B. G. 170 (567, 590), 183 (590), 195, 196 (567), 239 (681), *265*, *266*, *268*, 397 (940), *1014*
Christensen, P. K. 18, 20 (65), *255*
Christiani, G. F. 135 (402), 136 (395, 398, 400), *262*
Chu, S. C. 477, 478, 918 (1115), *1018*
Chu, S. D. 676, 715, 815, 816, 882, 889 (1316), *1023*
Ciabattoni, J. 69, 73 (224), *258*
Cignarella, G. 135 (402), *262*
Cimarusti, C. M. 397 (937, 968), *1014*, *1015*
Clader, J. W. 376 (877), *1013*
Clague, A. R. 437, 648 (1074), *1017*
Claremon, D. A. 385 (888), *1013*
Clark, D. S. 400 (1012), *1016*
Clark, G. R. 616 (1252), *1021*
Clark, R. D. 106 (309), 138 (420), *260*, *262*
Clarke, E. J. 16 (56), *255*
Clarke, F. H. 244, 250 (704), *268*
Class, M. 968 (1459), *1026*
Claudi, F. 496 (1137), 510 (1137, 1160), 530 (1160), 536 (1137), 673, 734 (1160), *1019*
Clauss, K. 163–165 (503), *264*
Clemente de Araujo, H. 271, 397 (725), *1009*
Clemo, G. R. 243 (703), *268*
Cleveland, P. G. 144, 146, 147 (459), *263*
Clough, S. C. 234 (672), *267*
Cocalas, G. H. 138 (414, 415), *262*
Cochoy, R. E. 222 (657), *267*
Cocker, W. 4, 6, 84, 86 (17), *254*
Coffey, R. S. 47 (167), *257*
Cohen, L. A. 138 (416, 417), *262*
Cohen, N. 4, 6 (19), 10, 11 (30, 31), 83 (254), 85 (19, 254), 86 (19), 91 (19, 30, 31, 254), *254*, *259*, 282 (768), *1010*
Cohen, T. 53 (188), *257*
Cole, W. 36 (133–135), *256*
Collin-Asselineau, C. 4 (18), *254*
Collum, D. B. 411 (1028), *1016*
Colvin, E. W. 144 (456), *263*
Comins, D. L. 369 (872), *1013*

Conia, J. M. 456, 457 (1092), *1018*
Conley, R. T. 222, 224, 225, 228 (647), *267*
Contento, M. 614 (1247), *1021*
Cook, A. H. 170, 202 (597), *266*
Cookson, R. C. 124, 125 (349), *261*, 395 (899), *1013*
Cooper, G. F. 83, 90 (270), *259*
Cooper, L. 601, 603, 604, 608, 609 (1240), *1021*
Cooper, R. D. G. 397 (914), 490 (1126), *1013*, *1018*
Corbet, J.-P. 278 (759), *1010*
Corelli, I. 434 (1069), *1017*
Corey, E. J. 13 (49), 30, 33 (107), 104 (299, 303), 106 (299, 302, 303), 132 (367), 144, 159 (489), 163 (367), 170 (578, 580), 184 (580), 193 (578, 580), *254*, *256*, *260*, *261*, *263*, *265*, 339 (846, 847), 341 (847), 382, 383 (886), *1012*, *1013*
Cornforth, J. W. 53 (187), 108 (313), *257*, *260*
Cornforth, R. H. 53 (187), 108 (313), *257*, *260*
Cossement, E. 665, 724, 777, 922 (1308), *1023*
Cossio, F. P. 490 (1146), 546 (1218), 549 (1198), 555 (1198, 1204), 556, 561, 563, 567 (1198), 568, 569 (1198, 1204), 570 (1198, 1204, 1217, 1218), 571 (1198, 1217, 1218), 572 (1198, 1204, 1217), 575 (1198, 1204, 1221), 576 (1221), 577, 579 (1198, 1204), 717, 719 (1221), 728, 730 (1146), 742 (1217, 1218), 751 (1146, 1217), 762 (1146), 812 (1204), 813, 848 (1217), 849 (1204), 872, 874 (1218), 898 (1146), 934 (1146, 1221), 954, 955 (1146), *1019*, *1020*
Cottis, S. G. 3 (10), *253*
Couffignal, R. 380 (882), *1013*
Courtney, P. M. 437, 648 (1074), *1017*
Courtois, G. 51 (183), *257*
Cowell, A. 343 (852), *1012*
Crabbe, P. 999 (1484), *1027*
Crawford, T. H. 18 (77), *255*
Crawford, W. C. 163, 165 (499), *264*
Creger, P. L. 54 (195), *258*
Crimmin, M. J. 828, 896, 913, 914, 947 (1434), *1026*
Crimmins, M. T. 273, 366 (753), *1010*
Cristea, A. N. 664 (1303), *1022*
Cristol, S. J. 42 (150), *257*
Crombie, L. 404 (1017), 648, 652 (1017, 1294), 653, 666, 667 (1017), 668 (1017, 1294), 669 (1294), 684 (1017), *1016*, *1022*
Cross, A. D. 28 (105), *256*
Cross, B. E. 76 (241, 244, 248), 78, 79 (241, 248), 81 (244), 102 (248), *258*, *259*
Cuellar, L. 28 (105), *256*
Currie, J. K. 341, 342 (850), *1012*
Cutler, A. 166 (510), *264*
Cuvigny, T. 717, 812, 849 (1372), *1024*
Cuzent, M. 11, 53 (42), *254*

Dabal, A. 312 (808), *1011*
Dabrowski, B. 647, 651 (1288), *1022*
Dachs, K. 132 (368), *261*
Dadic, M. 170, 196 (585), *265*
Dahmen, F. J. M. 275, 276 (756), *1010*
Dalton, L. K. 42 (154), 43 (155), 85 (154), 111 (154, 155), 117 (154), *257*
Dandarova, M. 419 (1044), *1017*
D'Angeli, F. 424 (1056), *1017*
d'Angelo, J. 298 (795), *1011*, 676, 692, 738 (1315), *1023*
Dangyan, M. T. 10 (35), 13 (48), 28 (102), 34, 39 (118), 42 (146, 147), 48 (171), *254*, *256*, *257*, 435 (1071), 993 (1474), (35, 48, 102, 118, 146, 147, 171), *1009*, *1017*, *1026*
Danheiser, R. L. 30, 33 (107), *256*
Danieli, B. 127 (354), *261*
Daris, J. P. 729, 730, 760, 762, 763, 863, 900, 925, 926 (1383), *1024*
Das, G. 522, 973 (1179), *1020*
Das Gupta, T. K. 118 (336), *260*
Dashunin, V. M. 3 (8), *253*
Daub, G. H. 48 (170), *257*
Daugan, A. 314 (811), *1011*
Daugherty, B. W. 490 (1126), *1018*
D'Auria, M. 323 (827), *1012*
Dave, V. 123 (341), *260*
Davies, A. G. 250 (716), *268*
Davies, D. 397 (924), *1014*
Davies, D. I. 271 (730), *1009*
Davies, S. G. 730, 881 (1386), 981 (1386, 1467), *1024*, *1026*
Davis, B. A. 170, 177 (538), *264*
Dayal, B. 170 (560, 566, 575, 588, 593, 594), 187, 188 (560, 566), 192, 195 (560, 566, 575), 198 (588), 200 (593, 594), *265*, *266*, 526, 527, 547, 567, 817, 822, 846 (1180), *1020*
Debaerdemaeker, T. 491, 503, 504, 523, 526, 536, 541, 542, 544, 1005 (1127), *1018*
DeBoer, J. J. J. 893 (1436), *1026*
DeBoer, Th. J. 123 (345), *261*
Debono, M. 38, 39 (139), *256*
Decazes, J. 170, 174, 176 (532), *264*
DeClercq, P. J. 329 (835), *1012*
Deets, G. L. 53 (188), *257*
Defaye, G. 83 (275), *259*
DeJonge, A. P. 4, 84 (20), *254*
DeLa Torre, M. C. 491–493, 497, 499, 503, 504, 528, 536, 541, 545 (1128), *1018*
Deltsova, D. P. 648 (1290), *1022*
De Meester, P. 477, 478, 918 (1115), *1018*
Demuynck, M. 329 (835), *1012*
Denbigh, K. W. 223 (662), *267*

Deng, Y. 397 (907), *1013*
Denney, D. B. 54 (194), 83, 85 (250), *258*, *259*
Denney, R. C. 241, 242 (695), *268*
Denny, R. 109, 110 (317), *260*
De Poortere, M. 207 (609), *266*
DePuy, C. H. 42 (152), *257*
Deshayes, C. 725, 871 (1380). *1024*
Deshpande, S. M. 170 (549, 583), 184 (549), 194, 195 (583), *265*
Deshpandi, M. N. 437, 648 (1074), *1017*
Deslongchamps, P. 99, 106, 108 (284), *259*, 366 (866), *1012*
Dessau, R. M. 110 (323, 325). 111 (323, 326), 112 (325), *260*, 377 (878, 879), *1013*
Dettmer, H. 39 (141), *256*
Devreese, A. A. 329 (835), *1012*
Deyrup, J. 234 (672), *267*
Dharamski, S. S. 47 (168), *257*
Dicke, D. F. 242 (697), *268*
Dietrich, H. 127, 128 (355), *261*
Di Fabio, R. 278, 335 (762), *1010*
Dijkink, J. 784, 785 (1411), 791 (1413), 973, 975 (1411, 1413), *1025*
DiMaria, F. 48 (174), *257*
Dimitrova, B. 346 (854), *1012*
Dimmig, D. A. 40 (144), *256*
Dittmer, D. C. 103, 106 (297), *260*
Divakar, K. J. 83 (260), *259*
Djerassi, C. 241 (694), 244 (705), *268*
Dmitrieva, N. A. 640 (1277), *1022*
Dmitrieva, N. D. 3, 28 (6), *253*
Dobrescu, D. 664 (1303), *1022*
Dobrev, A. 346 (854), *1012*
Dobrowolski, P. J. 294, 357, 360, 364, 790 (791), *1011*
Dokholov, D. M. 640 (1277), *1022*
Doleschall, G. 692 (1347), *1024*
Dolfini, J. E. 238 (679), 239 (680), *267*
Dombro, R. A. 130 (360), *261*
Dominguez, G. 503, 528, 536 (1148), 648, 651 (1292), 734, 882 (1148), *1019*, *1022*
Donaruma, L. G. 212 (620), *266*
Donat, F. J. 214, 216 (631), *266*
Döppert, K. 170, 180, 181 (541), *264*
Dordor-Hedgecock, I. M. 730, 881 (1386), 981 (1386, 1467), *1024*, *1026*
Doria, G. 51 (180), *257*
Dorsey, E. D. 169, 170, 172–174, 177, 182 (516), *264*
Dougherty, G. 214, 216, 217 (634), *266*
Douglas, B. 244, 250 (704), *268*
Dow, R. L. 315 (814), *1011*
Dowle, M. D. 271 (730), *1009*
Doyle, M. 144 (456), *263*
Doyle, M. P. 315 (814), *1011*
Dradi, E. 51 (180), *257*
Draeger, A. 170 (553, 554), 185, 188 (553), 189 (554), *265*
Draghici, C. 134 (386), *261*
Dreger, L. H. 4–6, 10, 11 (16), *254*
Dreiding, A. S. 51 (185), *257*
Dremova, V. P. 679 (1324), *1023*
Dube, S. 366 (866), *1012*
Dubois, F. 108 (311), *260*
Dubois, J. C. 49 (179), *257*
Dubois, J. E. 585 (1232), *1021*
Duerckheimer, W. 827 (1426), *1025*
Duffaut, N. 167 (512), *264*
Duffield, A. M. 241 (694), *268*
Dugat, D. 515 (1168), 663, 678 (1320), 736 (1168), 793, 797 (1320), 811, 812, 835, 837, 854 (1168), *1019*, *1023*
Dugenet, Y. 693 (1350), *1024*
Duggan, A. J. 48 (174), *257*
Duguay, G. 407 (1022), *1016*
Duhl-Emswiler, B. A. 453, 789, 790, 972, 973 (1091), *1018*
Dulcere, J. P. 333 (838), *1012*
Dulou, R. 18, 22 (84, 85), *255*
Dumas, F. 676, 692, 738 (1315), *1023*
Dunn, G. L. 39 (140), *256*
Duong, T. 214, 220 (640), 222 (640, 648), 224, 225 (640), 226, 227 (640, 648), 229 (640), 233 (648), 238, 240 (640, 648), *267*
Duran, F. 170, 188, 198 (587), *265*
Durst, T. 163–165 (500), 238 (675, 676), *264*, *267*, 684, 701, 702, 704 (1333), 729, 730, 760, 762, 763, 863, 900, 925, 926 (1383), *1023*, *1024*
Durston, P. J. 223 (662), *267*
Dutta, D. K. 615 (1249), *1021*
Dyke, S. F. 243, 250 (702), *268*
Dzhaniashvili, L. K. 664, 676, 708, 751, 754, 764, 815, 839, 942 (1304), *1022*
Dzierzanowska, D. 397 (945), *1014*

Earle, R. H. Jr. 134 (388), 144, 160 (487), *261*, *263*
Easton, C. J. 419, 451, 664 (1043), *1017*
Eastwood, F. W. 236, 237 (673), 240 (688), *267*, *268*
Eaton, P. E. 83, 90 (270), *259*
Ebata, T. 301 (800), 395, 396 (800, 901), 402 (901), 710, 890, 891 (800, 901), *1011*, *1013*
Ebel, S. 142 (444), *263*
Eddy, C. R. 83, 89, 98 (268), *259*
Edelweiss, E. 397 (926), *1014*
Eder, M. 252 (722), 253 (723), *268*
Edward, J. T. 83, 85 (257), *259*
Edwards, A. G. 124, 125 (349), *261*
Edwards, B. 495, 496, 869 (1135), *1019*
Edwards, J. A. 518, 818, 819, 839 (1174), *1019*

Edwards, O. E. 244, 250 (704), *268*
Effenberger, F. 420 (1045), 679, 680 (1325), *1017, 1023*
Egli, R. H. 17 (62), *255*
Ehlers, E. 827 (1426), *1025*
Ehntholt, D. 166 (510), *264*
Eiga, S. 640 (1279), *1022*
Eigenbrot, C. 797 (1417), *1025*
Eilers, K. L. 42 (152), *257*
El Alami, N. 298, 581 (794), *1011*
El-Barbary, A. A. 460, 805, 806 (1096), *1018*
El-Kashef, H. S. 529 (1186), 533 (1189), 534 (1186, 1189), 535, 537 (1189), 539 (1186), 817 (1189), 833, 851 (1186), 852 (1189), *1020*
Elks, J. 397 (976), *1015*
Ellison, R. 123 (344), *261*
Ellison, R. A. 53 (192), *258*
El-Maghraby, M. A. 533–535, 537, 817, 852 (1189), *1020*
Elmes, B. C. 42 (154), 43 (155), 85 (154), 111 (154, 155), 117 (154), *257*
Elson, S. 398 (987), *1015*
Elzey, T. K. 690, 749, 753, 769, 830, 831, 847, 899, 909, 910, 970 (1346), *1023*
Emerson, W. S. 83, 89 (263, 264), *259*
Engelhord, N. 170, 202 (595), *266*
Engelsing, R. 252 (719), *268*
Erickson, J. A. 170, 193 (569), *265*
Erlenmeyer, E. 61 (203), *258*
Ernst, B. 503 (1149), *1019*
Ershov, L. V. 460, 796 (1097), *1018*
Eschenmoser, A. 118 (336, 337), *260*
Escobar, G. 503, 528, 536, 734, 882 (1148), *1019*
Espejo de Ochoa, O. 222, 223 (658), *267*
Estopa, C. 287, 288, 294, 357, 359, 360, 363 (779), *1011*
Etienne, Y. 3 (3), *253*
Evans, D. A. 4, 7, 87 (22), *254*, 404 (1015), 445, 446, 692 (1082), 731 (1015, 1082), 756, 757 (1082), *1016, 1017*
Evans, R. L. 63, 68 (216), *258*
Evdakov, V. P. 725 (1381), *1024*
Ewing, J. B. 238 (679), *267*

Fahey, J. L. 170 (561, 562, 565), 192 (565), 197 (561, 562, 565), *265*
Fahr, E. 170 (541, 544), 180, 181 (541), *264, 265*
Fajfer, A. 271 (734), *1010*
Falbe, J. 111 (328, 329), 112 (331), 113 (328), 114, 115 (328, 329), 116, 117 (331), *260*
Faller, J. W. 648, 650 (1291), *1022*
Falou, S. 298 (795), *1011*
Farina, F. 252 (721), *268*
Farmun, D. G. 76, 79 (243), *258*
Fava, F. 136 (392, 395), *262*
Favre, A. 283, 408, 410 (773), *1011*
Fawcett, J. S. 222, 229 (655), *267*
Fazio, M. J. 630, 633 (1264), *1021*
Feaisheller, S. H. 18, 22 (86), *255*
Federici, R. M. 416 (1035), *1016*
Fedorova, T. M. 17 (57), 18 (72), *255*
Fedoseev, V. M. 725 (1381), *1024*
Fekete, G. 692 (1347), *1024*
Felix, A. M. 144, 159 (489), *263*
Felix, D. 118 (336, 337), *260*
Felzenstein, A. 475 (1121), *1018*
Fenwick, A. 476 (1114), 478 (1114, 1116), 480 (1114), 481, 483 (1114, 1116), *1018*
Fernandez, I. F. 397 (922), 521, 548, 556–560, 579, 726, 728, 740, 813, 819, 822–824, 838, 849, 850, 853, 856 (1178), *1014, 1019*
Fernandez de la Prodilla, R. 301, 303, 352, 354 (801), *1011*
Ferreira, G. A. L. 452, 657, 962 (1090), *1018*
Fessler, D. C. 16 (54), *254*
Fetizon, M. 83 (272, 275, 277), 90 (272), 93, 94 (272, 277), 95, 96 (272), *259*
Fetter, J. 689 (1340), 692 (1347), 886, 933, 939–941 (1340), *1023, 1024*
Feuer, H. 704 (1360), *1024*
Ficini, J. 298 (795), *1011*
Field, L. D. 828, 896, 913, 914, 947 (1434), *1026*
Filler, R. 3 (2), 47 (165), *253, 257*
Finkbeiner, H. 110, 111 (322), *260*
Finkelstein, J. 10 (33, 34), *254*
Firestone, R. A. 239 (681), *268*
Fischer, J. 1008, 1009 (1493), *1027*
Fischer, N. 3 (3), *253*
Fiser-Jakic, L. 288 (782), *1011*
Fish, R. W. 166 (510), *264*
Fisher, J. W. 648, 652, 657 (1295), 684, 695, 824 (1334), 830 (1295), 847 (1334), *1022, 1023*
Fleet, G. W. J. 367 (869), *1012*
Fleischmann, K. 827 (1426), *1025*
Fleming, I. 313, 329 (810), *1011*
Fleming, M. P. 4, 7 (21), *254*
Fletcher, S. R. 433, 712 (1066), *1017*
Flippen-Anderson, J. L. 744, 745 (1395), *1025*
Fliri, H. 693, 728, 760, 762, 763 (1348), 791, 792 (1414), 801, 908 (1348), *1024, 1025*
Flitsch, S. L. 1005, 1007 (1492), *1027*
Floss, H. G. 53 (189), *257*
Foa, M. 118 (335), *260*, 341, 984 (849), *1012*
Fodor, L. 504, 513, 537, 543, 544 (1150), *1019*
Folkers, K. 222 (651, 652), *267*
Fonrodona, J. 661 (1301), *1022*
Font, J. 283 (771), 287, 288, 294, 357, 359, 360, 363 (779), *1010, 1011*

Fontanella, L. 135 (401), 136 (390, 392, 394–400), 184 (401), *261, 262*
Foote, C. S. 109, 110 (317, 318), *260*
Forestiere, A. 315 (812), *1011*
Forrest, A. K. 759 (1402), *1025*
Fortgens, H. P. 910 (1438), *1026*
Foucaud, A. 211 (619), *266*, 427 (1060, 1061), 429, 446 (1061), 448, 450, 451, 755 (1060), 891 (1061), *1017*
Fouty, R. A. 103, 106 (297), *260*
Fowler, E. M. P. 11, 12 (38), *254*
Fox, B. L. 61 (209), 237 (674), *258, 267*
Fraher, T. P. 48 (173), *257*
Francalanci, F. 341, 984 (849), *1012*
Franceschi, G. 397 (947), 687 (1338), 731 (1387), 747 (1338), 748, 752 (1397), 753 (1338), 759 (1387), 761 (1397), 762 (1387, 1397), 763 (1387), 764 (1397), 768 (1338, 1397), 900, 903 (1387), 907 (1338, 1387), 915 (1397), 954 (1387), *1014, 1023–1025*
Frank, R. L. 133 (381), *261*
Franke, A. 45 (157), *257*
Franke, W. 240 (686), *268*
Fratei, G. 460 (1093), *1018*
Frauenglass, E. 10, 101 (29), *254*
Freidinger, R. M. 399 (1000), 418 (1040), 452 (1000), 697, 857 (1040), *1016, 1017*
Freidlina, R. Kh. 130 (363), *261*
Freifelder, M. 36 (135), *256*
Frensch, K. 391 (895), *1013*
Frey, A. J. 9, 11, 13, 87, 125 (28), *254*
Friary, R. 626 (1260), *1021*
Frimer, A. A. 335 (840), *1012*
Frisch, M. A. 4–6, 10, 11 (16), *254*
Frobese, A. S. 368 (870), 369 (871), *1012, 1013*
Frolow, F. 686, 687, 746, 747, 750, 759, 765, 772, 903 (1336), *1023*
Frunze, T. M. 398 (994, 996), *1016*
Frydrych, C. 397 (920), *1014*
Frydrychova, A. 134 (384), *261*
Frye, J. R. 36, 37 (126), *256*
Fuchs, H. 640 (1276), *1022*
Fuentes, L. M. 943 (1443), *1026*
Fuerholzer, J. F. 138 (420), *262*
Fuhrmann, R. 640 (1280), *1022*
Fujii, M. 69 (218), *258*
Fujii, T. 969 (1460), *1026*
Fujimoto, K. 688, 694, 743, 747, 760, 781, 869, 880, 884, 886, 902, 913, 916, 918, 929, 931, 936, 937 (1339), *1023*
Fujisawa, T. 271 (729), 381 (883), *1009, 1013*
Fujita, E. 278 (763), 341 (848), 996 (1480), *1010, 1012, 1027*
Fujita, K. 799 (1419), *1025*
Fujita, T. 54 (196), *258*, 378 (880), 379 (881), *1013*
Fujiwara, H. 214, 215, 217 (624), *266*
Fukumoto, K. 249 (713), *268*, 381 (883), 531 (1187), 665, 666, 742 (1309), 823, 828, 852 (1187), *1013, 1020, 1023*
Fukuoka, Y. 440–442, 444, 445, 818, 826, 845 (1077), *1017*
Fukushima, I. 799 (1419), *1025*
Fukuyama, T. 729, 764 (1384), *1024*
Fuller, J. A. 83 (251), *259*
Fumaki, K. 301, 395, 396, 710, 890, 891 (800), *1011*
Funakoshi, K. 83 (259), *259*
Funke, P. T. 238 (679), *267*
Furata, T. 640 (1279), *1022*
Furstoss, R. 722, 740, 813 (1376), *1024*
Furukawa, M. 419 (1042), 424, 425 (1058), 616 (1251), 997 (1481), 998 (1481, 1482), *1017, 1021, 1027*

Gais, H. J. 271 (748), *1010*
Gala, K. 397 (922), 521, 548, 556 (1178), 557 (1178, 1208), 558–560, 579, 726, 728, 740, 813, 819, 822–824, 838, 849, 850, 853 (1178), 856 (1178, 1208), *1014, 1019, 1020*
Gallastegui, J. 381 (884), *1013*
Galle, J. E. 416 (1035), *1016*
Galliani, C. 731, 759, 762, 763, 900, 903, 907, 954 (1387), *1024*
Galt, R. H. B. 76, 78, 79 (241, 248), 102 (248), *258, 259*
Gambale, R. J. 371 (874), *1013*
Gambaryan, N. P. 139 (428, 429), 140 (429, 430), *262*, 648 (1290), *1022*
Ganboa, I. 555 (1204), 563–566 (1213), 568–570, 572, 575, 577, 579 (1204), 575, 576, 717, 719, 934 (1221), 726, 728, 750 (1213), 812, 849 (1204), *1020*
Gandolfi, C. 51 (180), *257*, 685, 689, 748, 753, 769 (1342), *1023*
Ganem, B. 411 (1028), 435, 943 (1072), *1016, 1017*
Ganem, B. E. 339 (846), *1012*
Ganguly, A. 626 (1260), *1021*
Garbess, C. F. 109, 110 (321), *260*
Garcia, J. M. 575, 576, 717, 719, 934 (1221), *1020*
Gardano, A. 341, 984 (849), *1012*
Gardner, J. H. 3 (13), *254*
Garratt, P. J. 287, 290, 360 (780), *1011*
Gaspert, B. 681 (1330), *1023*
Gassman, P. G. 61 (209), 123 (343), 135 (405–407), 137 (405, 407), 237 (674), *258, 260, 262, 267*
Gatti, E. 136 (391), *262*
Gaudemar-Bardone, F. 380 (882), *1013*
Gavryushina, V. M. 679 (1324), *1023*

Gay, R. L. 143 (450), *263*
Gaylord, N. G. 76 (234), *258*
Gebler, H. 398 (997), *1016*
Geigel, M. A. 214, 216, 227 (638), *266*
Geiger, H. 27 (96), *255*
Gelbard, G. 51 (185), *257*
Gelin, S. 725, 871 (1380), *1024*
Geller, H. H. 83, 89 (266), *259*
Geller, J. 416, 417 (1036), *1017*
Geller, L. E. 97 (279), *259*
Georgian, V. 495, 496, 869 (1135), *1019*
Geraghty, M. B. 61 (210), *258*
Gerard, G. 333 (838), *1012*
Gerloff, J. 970 (1461), *1026*
Gezalyan, D. I. 993 (1474), *1026*
Ghatak, V. R. 99, 100, 106 (286), *259*
Ghisalferti, E. L. 30 (115), *256*
Ghosez, L. 170, 188, 198 (587), 207 (609), *265*, *266*, 397 (920), 665, 724, 777, 922 (1308), *1014*, *1023*
Ghosh, A. K. 317 (817), *1011*
Ghosh, M. 510, 511, 521, 817, 839 (1162), *1019*
Ghosh-Mazumdar, B. N. 138 (434, 436), 139 (434, 436, 442), *262*, *263*
Giacomini, D. 614 (1247), *1021*
Giering, W. P. 166 (509, 510), *264*
Gilbert, J. C. 452, 697 (1089), *1018*
Gilchrist, T. L. 616, 745 (1250), *1021*
Gilinsky-Sharon, P. 335 (840), *1012*
Gilman, H. 51 (182), 204 (604), *257*, *266*
Gilman, N. W. 339 (846), *1012*
Gilpin, M. L. 960, 961 (1451, 1452), *1026*
Ginsberg, H. 83, 86 (262), *259*
Giovine, M. 952 (1445), *1026*
Girijavallabhan, V. M. 726, 756, 758 (1398), *1025*
Giudici, F. 397 (947), *1014*
Glahol, G. 397 (919), *1014*
Glaret, C. 83, 89 (252), *259*
Glasgow, S. 271 (724), *1009*
Glazer, E. A. 994 (1479), *1027*
Glazer, E. L. 223 (659, 660), *267*
Glazer, R. D. 397, 507–509, 512 (902), 560, 561 (1209), 573, 872–874 (902), *1013*, *1020*
Gleason, J. G. 398 (989), *1015*
Glotter, E. 11 (37), *254*
Glover, D. 4, 7, 87 (22), *254*
Glover, S. A. 472 (1110), *1018*
Gluchowski, C. 601, 603, 604, 608, 609 (1240), *1021*
Glushkov, R. G. 397 (903), 398 (984), 626 (1259), 725, 812, 837, 980 (1379), 981 (903), *1013*, *1015*, *1021*, *1024*
Gobley, J. 11, 53 (41), *254*
Goda, K. 664, 706, 830, 843, 854, 855 (1305), *1022*
Goel, O. P. 805, 808–810 (1423), *1025*
Goetschel, C. T. 76, 79 (242), *258*
Goetz, J. 397 (919), *1014*
Goff, D. L. 124 (347, 348), *261*
Goldberg, D. 759, 920 (1401), *1025*
Goldberg, O. 170 (540, 556), 180 (540), 188 (540, 556), *264*, *265*
Goldman, I. M. 106 (305), *260*
Goldsmith, D. 293 (790), *1011*
Goldsmith, D. J. 384 (887), *1013*
Golfier, M. 83 (272, 277), 90 (272), 93, 94 (272, 277), 95, 96 (272), *259*
Gomes, A. 169, 170, 172 (517), *264*
Gomez Aranda, V. 207 (608), *266*
Gonzalez, A. 555, 568–570, 572, 575, 577, 579, 812, 849 (1204), *1020*
Goo, Y. M. 744, 858, 862, 865 (1394), *1025*
Goodrow, M. H. 76, 79 (242), *258*
Goosen, A. 127 (353), *261*, 472 (1110), *1018*
Gopinath, K. W. 31 (111, 112), 32 (113, 114), 33 (114), *256*
Gora, J. 271 (734), *1010*
Gore, J. 300, 341 (797), *1011*
Gorshkov, V. I. 640 (1277), *1022*
Gossauer, A. 981 (1465), *1026*
Gotoh, H. 168 (513, 514), *264*, 491, 498, 920 (1131), *1019*
Gotor, V. 207 (608), *266*
Gotthardt, H. 635, 897 (1268), *1021*
Gougoutas, J. Z. 239 (680), *267*
Gould, W. A. 135 (404), 204, 205 (601), *262*, *266*
Goutarel, R. 248 (711), *268*
Govindachari, T. R. 243 (701), *268*
Gowda, G. 272, 274, 302, 303, 337 (751), *1010*
Graboski, G. G. 476 (1113), *1018*
Graf, R. 132 (370), 162 (494, 495), 163 (370, 496, 498, 502), 164 (498), 170 (524), *261*, *264*
Graham, A. M. 415 (1034), *1016*
Gramain, J. C. 723 (1377, 1378), 741, 814, 849, 931, 935, 937 (1377), 953 (1378), *1024*
Grandberg, I. I. 710 (1364, 1365), *1024*
Granger, R. 76, 78 (240), *258*
Granik, V. G. 398 (984), 460 (1097), 626 (1259), 796 (1097), *1015*, *1018*, *1021*
Gray, W. F. 53 (188), *257*
Green, A. E. 61, 65 (213), *258*
Green, B. 39 (140), *256*
Greenburg, G. Y. 19 (92), *255*
Greengrass, C. W. 404, 406, 704 (1018, 1019), 811 (1019), *1016*
Gregory, G. I. 397 (971), *1015*
Grezemkovsky, R. 28 (105), *256*

Grieco, P. A. 3, 61 (11), 104. 106 (300), *253, 260*, 321 (825), 400 (1012), *1012, 1016*
Grigorev, A. B. 626 (1259), *1021*
Grimaldi, J. 111 (327), *260*
Groeger, D. 398 (986), *1015*
Groeger, G. 45 (157), *257*
Gross, A. 170, 202 (596), *266*, 759, 920 (1401), *1025*
Gross, A. W. 382, 383 (886), *1013*
Grossman, H. 18, 22 (83), *255*
Grudzinskas, C. V. 134 (387), *261*
Gruetzmacher, G. 135 (406), *262*
Gruska, R. 521, 548, 556–560, 579, 726, 728, 740, 813, 819, 822–824, 838, 849, 850, 853, 856 (1178), *1019*
Guanti, G. 624 (1257), *1021*
Guenther, H. J. 290 (785), 351 (862), *1011, 1012*
Guette, J. P. 49 (179), *257*
Guevel, A. 407 (1022), *1016*
Guggisberg, A. 647, 651 (1288), 652 (1296, 1297), 656 (1297), 659 (1296), 684 (1297), *1022*
Guillerm, D. 276 (757), *1010*
Guinn, D. E. 314, 323 (824), *1012*
Guise, G. B. 242, 244, 247 (699), *268*
Gull, R. 275 (755), *1010*
Gupta, K. 581, 734, 806 (1225), *1021*
Gupta, P. K. 463 (1099), 509 (1159), 511 (1099), 512 (1163), 514 (1159), 566 (1214), 567 (1159, 1214, 1215), 568 (1159, 1215), 807 (1099), 817 (1214, 1215), 848 (1159, 1214, 1215), 850 (1214, 1215), 918 (1099), *1018–1020*
Gupta, Y. N. 292 (789), *1011*
Gutsche, C. D. 47 (167), *257*
Guyot, M. 36 (131), *256*

Ha, D. C. 592–600, 603–605, 609, 610, 677, 886, 940 (1237), *1021*
Haaf, W. 112 (332, 333), *260*
Habeck, D. 682, 701, 911, 929, 981 (1332), *1023*
Halchak, T. 230 (664), *267*
Haley, E. 477, 478, 918 (1115), *1018*
Hallett, P. 122 (339), *260*
Halmekoski, J. 397 (958), *1015*
Hamada, T. 469 (1106), *1018*
Hamaguchi, F. 404, 406 (1020), *1016*
Hamashima, Y. 713, 760, 770, 846, 883, 891, 947 (1368), *1024*
Hamelin, J. 612 (1243), *1021*
Hamersma, J. A. M. 974, 975 (1463), *1026*
Hamlet, A. B. 684, 701, 702, 704 (1333), *1023*
Han, W. T. 648, 650 (1291), *1022*
Hanaki, A. 548, 549, 847, 850, 856, 857 (1197), *1020*
Hanaoka, M. 17 (58), *255*
Hanessian, S. 283 (773), 397 (964), 404, 405 (1016), 408, 410 (773), *1011, 1015, 1016*
Hankin, H. 83, 86 (262), *259*
Hanna, R. 106 (306), *260*
Hansell, D. P. 42 (153), *257*
Hanson, J. R. 76, 78, 79 (241, 248), 102 (248), *258, 259*
Hara, S. 101 (293), 106 (292, 293), *259*, 970 (1462), *1026*
Harada, K. 139, 141 (443), *263*
Harano, Y. 342 (851), *1012*
Harata, Y. 83, 85 (261), *259*
Harbridge, J. B. 960, 961 (1451, 1452), *1026*
Harcourt, D. W. 250 (715), *268*
Harcus, R. G. 717, 719 (1373), *1024*
Hardouin, J. C. 53 (190), *257*
Hardy, P. M. 48 (175), *257*
Harmer, J. 132, 163 (372), *261*
Harris, B. W. 223 (662), *267*
Harris, E. E. 665, 781, 894, 895 (1307), *1023*
Harris, T. M. 358 (864), *1012*
Harsanyi, K. 692 (1347), *1024*
Hart, D. J. 592–599 (1237), 600 (1237, 1241), 601 (1241), 603 (1237, 1241), 604 (1237), 605 (1237, 1241), 609, 610 (1237), 613, 676 (1245), 677 (1237), 692 (1245), 731, 733 (1241), 739 (1245), 799 (1241), 886 (1237), 899, 905, 934, 935 (1245), 940 (1237), *1021*
Hartke, K. 142 (444), *263*
Hartung, W. H. 138 (414), *262*
Hartwell, J. L. 19 (92, 93), *255*
Hasegawa, S. 300, 309, 310, 405 (798), *1011*
Hasegawa, T. 144, 160, 162 (467), *263*
Hashimoto, M. 532, 850 (1188), *1020*
Hashimoto, S. 209 (613), *266*
Hassall, C. H. 106 (280), 209 (612), *259, 266*
Hassan, K. M. 533–535, 537, 817, 852 (1189), *1020*
Hassan, M. E. 498, 537 (1140, 1141), 631, 634, 705, 863, 882, 883 (1141), *1019*
Hassner, A. 133 (375), *261*, 672, 673 (1313), *1023*
Hassner, H. 397 (979), *1015*
Hatanaka, M. 416, 417 (1038), 638 (1270), 930 (1038), *1017, 1022*
Hatanaka, N. 516 (1169, 1170), 517 (1171, 1172), 518 (1172), 524 (1169, 1171), 525 (1171), 814 (1170, 1172), 819 (1171), 820, 821 (1169, 1171), 828 (1169), 831 (1171), 832 (1169, 1171), 838 (1169–1172), 841 (1169, 1171), 842 (1171), 843 (1169), 881 (1171), *1019*
Hatanoka, Y. 474 (1111), *1018*
Hatch, C. E. 211 (618), *266*
Hatch, C. E. III 637, 638 (1269), *1022*

Hatfield, L. D. 690, 749, 753, 769, 830, 831, 847, 899, 909, 910, 970 (1346), *1023*
Hattori, R. 440–442, 444, 445, 818, 826, 845 (1077), *1017*
Hauesler, J. 632, 911 (1266), *1021*
Hauptmann, S. 83, 91, 92 (273), *259*, 993 (1478), *1027*
Hauser, C. R. 143 (448–452), 144 (453), *263*
Hawkins, E. G. E. 133 (378), *261*, 970 (1464), *1026*
Hayakawa, K. 339 (844), *1012*
Hayashi, E. 1005 (1489), *1027*
Hayes, F. N. 18, 21 (80), *255*
Heard, C. L. 40, 41 (145), *256*
Hearn, M. J. 223 (659), *267*
Heathcock, C. H. 106 (309), *260*, 434 (1068), *1017*
Hebblethwaite, E. M. 271 (742), 732 (1388), 966, 967 (1388, 1458), *1010, 1024, 1026*
Hegde, V. 557, 856 (1208), *1020*
Hegde, V. R. 397 (921), 439–442 (1076), 505 (1152, 1153), 509, 545 (1152), 562 (1152, 1153), 563 (1153), 695 (1076), 735 (1152, 1153), 743, 755 (1152), 869 (1153), 936 (1152), *1014, 1017, 1019*
Hegedus, L. S. 585, 586 (1228–1230), 587 (1230), 588 (1228–1230), 735 (1229), 756, 758, 844–846, 956, 958 (1405), *1021, 1025*
Heiba, E. I. 110 (323, 325), 111 (323, 326), 112 (325), *260*, 377 (878, 879), *1013*
Heidelberger, C. 647, 651 (1288), 652, 659 (1296), *1022*
Heimgartner, H. 282 (769), 490 (1147), *1010, 1019*
Heine, O. 30 (110), *256*
Heinen, H. 488 (1122), *1018*
Hejzlar, M. 397 (981), *1015*
Heldt, W. Z. 212 (620), *266*
Helmreich, R. F. 42 (150), *257*
Henbest, H. B. 11 (38), 12 (38, 46), *254*
Henery-Logan, K. R. 144, 158 (485), *263*
Henlin, J. M. 491, 635, 636, 799 (1129), *1018*
Henning, R. 706 (1361), *1024*
Henton, D. E. 124 (346), *261*
Herak, J. 681 (1330), *1023*
Herges, R. 397 (919), *1014*
Hering, G. 397 (919), *1014*
Heritage, G. L. 51 (181), *257*
Herlem, D. 248 (711), *268*
Hernandez, L. Jr. 499, 500, 502 (1142), 697 (1142, 1355), 700, 701 (1142), 704 (1355), 863 (1142), *1019, 1024*
Herrmann, J. L. 61, 64, 65 (212), *258*
Herrmann, R. 397 (919), *1014*
Hershenson, F. M. 515–525 (1167), *1019*
Hesse, M. 647, 651 (1288), 652 (1296, 1297), 656 (1297), 658 (1298), 659 (1296), 684 (1297), 912 (1298), *1022*
Heusler, K. 109 (316), *260*
Hey, D. H. 144, 145, 149, 150 (463), *263*
Heyn, H. 163–165 (503), *264*
Heyne, H. U. 232, 233 (668), *267*
Hiemstra, H. 698, 699, 891, 892 (1356), 910 (1438), 950, 951, 972, 975 (1356), *1024, 1026*
Higuchi, S. 144, 155 (473), *263*
Hiiragi, M. 513, 529 (1165), 531, 823, 828, 852 (1187), 918 (1165), *1019, 1020*
Hildebrand, R. P. 16 (56), *255*
Hilgetag, G. 170 (553, 554, 584), 185, 188 (553), 189 (554), 195 (584), *265*
Hill, H. W. Jr. 170, 193 (582), *265*
Hinze, R.-P. 981 (1465), *1026*
Hiragi, M. 529, 646, 689, 740 (1183), *1020*
Hirai, K. 397 (908), 451 (1088), 688, 694, 743, 747, 760, 781 (1339), 869, 880 (1088, 1339), 884, 886, 902, 913, 916, 918, 929, 931, 936, 937 (1339), *1013, 1018, 1023*
Hirai, Y. 661 (1300), *1022*
Hirao, I. 273 (754), *1010*
Hirao, T. 342 (851), *1012*
Hiraoka, T. 272 (750), 397 (977), 422 (750, 1053), 423 (1053), 681 (1328), 735, 737 (1053), 740 (750), 741 (750, 1053), 886, 894, 898 (750), 899 (750, 1053), 905 (1053), 908 (750), 917 (1328), 929, 930, 933 (1053), 937, 938, 940 (750, 1053), 943 (1053), 947 (750), *1010, 1015, 1017, 1023*
Hirata, S. 665, 666, 742 (1309), *1023*
Hirohashi, T. 739 (1392), *1025*
Hirokami, S. 661 (1300), *1022*
Hirrami, T. 61, 66–68 (215), *258*
Hirschmann, R. 83, 97 (278), *259*
Hirshfeld, A. 11 (37), *254*
Hirshfield, J. 665, 781, 894, 895 (1307), *1023*
Hiyama, T. 351 (860, 861), 898, 904 (1437), *1012, 1026*
Hlasta, D. J. 416 (1039), 420, 421 (1039, 1047), 708, 754, 777 (1039), 779 (1047), 794, 837, 881, 960, 962, 1005 (1039), *1017*
Hnevsova-Seidlova, V. 47 (169), *257*
Ho, T.-L. 99, 106, 108 (284), *259*
Hochstetler, A. R. 4, 6, 85, 86, 91 (19), *254*
Hodges, R. 17 (59), *255*
Hodgson, S. T. 271 (742), 732 (1388, 1389), 733, 886, 965 (1389), 966, 967 (1388), *1010, 1024, 1025*
Hofer, P. 16 (54), *254*
Hoff, E. F. Jr. 170, 173 (526), *264*
Hoffman, W. A. 505, 506, 513, 523, 526, 645, 648, 657, 679, 680, 706, 810, 918, 925 (1155), *1019*

Hofheinz, W. 416, 417, 886 (1037), *1017*
Höft, E. 249 (714), *268*
Hokyo, M. 379 (881), *1013*
Holden, K. G. 397 (963), *1015*
Holley, A. D. 170, 174 (528), *264*
Holley, R. W. 170, 174 (528), *264*
Hollinshead, D. M. 271 (742), 732 (1388, 1389), 733, 886, 965 (1389), 966, 967 (1388), *1010, 1024, 1025*
Holzapfel, C. W. 632, 911 (1266), *1021*
Honda, T. 676, 715, 815, 816, 882, 889 (1316), *1023*
Hoornaert, C. 686, 687, 746, 747, 750, 759, 765, 772, 903 (1336), *1023*
Hopkins, L. O. 4, 6, 84, 86 (17), *254*
Hoppe, D. 322 (826), 397 (952), 520, 531, 816, 838 (1177), 913–915, 918, 922, 924, 925 (1440), *1012, 1014, 1019, 1026*
Hoppe, H. 728, 759, 821, 830, 844, 900–902 (1382), *1024*
Hori, T. 685, 743, 895, 896, 908, 909 (1343), *1023*
Horibe, I. 83, 84 (253), *259*
Horiguchi, K. 214, 215 (627), *266*
Horii, Z. 17 (58), *255*
Horner, L. 170 (525, 596, 599), 173–176, 178, 179 (525), 202 (596, 599), *264, 266*
Horning, E. C. 40 (144), 142 (446), 214, 216, 220 (633), *256, 263, 266*
Horning, M. G. 40 (144), *256*
Hornyak, G. 689 (1340), 692 (1347), 886, 933, 939–941 (1340), *1023, 1024*
Horsak, I. 419 (1044), *1017*
Horton, A. M. 271 (742), *1010*
Horton, J. A. 103, 106 (296), *260*
Hossanali, A. 272, 274, 302, 303, 337 (751), *1010*
Hostettler, H. V. 124 (350), *261*
Hou, L. 704 (1360), *1024*
Houk, K. N. 292 (789), *1011*
Houlihan, W. J. 682, 701, 911, 929, 981 (1332), *1023*
House, H. O. 46 (160), *257*
Howarth, T. T. 960, 961 (1451, 1452), *1026*
Hradetzky, F. 252 (722), 253 (723), *268*
Huang, H. 407, 409 (1023, 1025), 413 (1031), 416 (1025), *1016*
Huang, S. P. 676, 715, 815, 816, 882, 889 (1316), *1023*
Hubbard, W. N. 4–6, 10, 11 (16), *254*
Huber, I. 397 (920), *1014*
Huber, J. H.-A. 124 (346), *261*
Huber, W. 106 (302), *260*
Hubert-Brierre, Y. 248 (711), *268*
Hubschwerlen, C. 529 (1185), *1020*
Huder, J. 124, 125 (349), *261*
Hudrlik, P. F. 13 (51), *254*
Huebener, G. 397 (919), *1014*
Hughes, D. 616, 745 (1250), *1021*
Hughes, G. 500, 505, 509, 530, 571, 572, 864, 883 (1144), *1019*
Huisgen, R. 170, 177 (538), *264*
Hull, R. 170, 198 (589), *265*
Hullot, P. 717, 812, 849 (1372), *1024*
Hungerbuehler, E. 665, 769, 908 (1306), *1023*
Huppes, M. 111, 113–115 (328), *260*
Hurley, L. H. 695, 891, 893, 979 (1353), *1024*
Hurst, D. T. 134 (388), 144, 160 (487), *261, 263*
Hutton, J. 104, 106 (299), *260*
Hyeon, S. B. 17 (61), *255*
Hyon, M. H. 613, 676, 692, 739, 899, 905, 934, 935 (1245), *1021*

Ichihara, A. 276, 352 (758), *1010*
Ide, Y. 713, 760, 770, 846, 883, 891, 947 (1368), *1024*
Ihara, M. 665, 666, 742 (1309), *1023*
Iimori, M. 397 (961), *1015*
Iimori, T. 402 (1006), 404 (1013), 405 (1006, 1013), 406 (1006), 491 (1130), 992 (1473), *1016, 1018, 1026*
Ikariya, T. 310 (803, 806), 312 (807), 319 (806), *1011*
Ikeda, F. 18 (64), *255*
Ikeda, K. 402, 708, 733 (1007), *1016*
Ikeda, M. 214, 215, 217 (624), *266*, 429, 430 (1062), 470 (1107), *1017, 1018*
Ikeda, R. M. 13, 17 (52), *254*
Ikegami, S. 917 (1441), *1026*
Ikota, N. 548, 549, 847, 850, 856, 857 (1197), *1020*
Imagawa, T. 348 (857, 858), 349 (857), *1012*
Immer, H. 144 (455), *263*
Inaba, S. 590 (1233, 1234), 591 (1235, 1236), 611 (1235), *1021*
Inanoga, J. 371 (873), *1013*
Indelicato, J. M. 648, 652, 657, 830 (1295), *1022*
Inesi, A. 434 (1069), 510, 535, 536, 540, 541, 867 (1161), *1017, 1019*
Ingold, C. K. 170 (598), *266*
Inhoffen, H. H. 39 (141, 142), *256*
Inoue, K. 432 (1065), 716 (1369), *1017, 1024*
Inoue, S. 271 (737), *1010*
Inoue, T. 739 (1392), *1025*
Iovu, M. 664 (1303), *1022*
Iqbal, A. F. M. 223 (661), *267*
Ireland, R. E. 4, 7, 87 (22), *254*
Iriarte, J. 999 (1484), *1027*
Irisbaev, A. 695 (1352), *1024*
Isaacs, N. S. 397 (906), *1013*
Isaev, S. D. 640 (1282), *1022*
Isenring, H. P. 416, 417, 886 (1037), *1017*

Ishibashi, H. 429, 430 (1062), 448, 450 (1087), 470 (1107), *1017, 1018*
Ishida, M. 545, 889, 890 (1192), *1020*
Ishida, N. 272 (750), 422 (750, 1053), 423 (1053), 681 (1328), 735, 737 (1053), 740 (750), 741 (750, 1053), 886, 894, 898 (750), 899 (750, 1053), 905 (1053), 908 (750), 917 (1328), 929, 930, 933 (1053), 937, 938, 940 (750, 1053), 943 (1053), 947 (750), *1010, 1017, 1023*
Ishida, Y. 402, 405, 406 (1006), *1016*
Ishige, O. 282 (770), *1010*
Ishiguro, M. 476 (1112), 690, 748, 752, 769 (1344), *1018, 1023*
Ishii, Y. 310, 319 (806), *1011*
Ishikura, K. 713, 760, 770, 846, 883, 891, 947 (1368), *1024*
Ishikura, M. 1000 (1485), *1027*
Ishimaru, T. 416, 417 (1038), 489 (1123), 930 (1038), *1017, 1018*
Ishino, I. 144, 145, 149 (461), *263*
Ishiyama, K. 429, 430 (1062), 448, 450 (1087), *1017, 1018*
Isobe, K. 106 (304), *260*
Isol, S. 17 (61), *255*
Itani, H. 693, 729, 894 (1349), *1024*
Ito, K. 317–319 (819), *1011*
Ito, M. 18 (64), *255*
Ito, Y. 492, 736, 886 (1132), 898, 904 (1437), *1019, 1026*
Itoh, K. 144 (469), *263*, 316, 339 (815), 430, 431 (1063, 1064), 432 (1064), 730, 754 (1385), 864 (1063), 865, 866 (1064), 875 (1385), *1011, 1017, 1024*
Itoh, M. 18 (88, 89), 23 (88), 24, 25 (88, 89), 26 (89), *255*
Ivanov, K. 346 (854), *1012*
Iwai, I. 83 (259), *259*
Iwano, Y. 451 (1088), 688, 694, 743, 747, 760, 781 (1339), 869, 880 (1088, 1339), 884, 886, 902, 913, 916, 918, 929, 931, 936, 937 (1339), *1018, 1023*
Iwasawa, N. 407, 409 (1023, 1025), 416 (1025), *1016*
Iwata, F. 222 (643), *267*
Iwata, H. 476 (1112), 690, 748, 752, 769 (1344), *1018, 1023*
Iyoda, M. 83, 85 (261), *259*
Izawa, T. 300, 309, 310 (798), 397 (961), 404 (1013), 405 (798, 1013), 491 (1130), *1011, 1015, 1016, 1018*

Jabloner, H. 214 (629), *266*
Jacobsen, R. P. 106 (287), *259*
Jacobson, R. M. 376 (877), *1013*
Jaeger, V. 351 (862), *1012*
Jäger, A. 170, 172–175, 178, 179 (523), *264*
Jahngen, E. G. E. Jr. 4 (26), 8 (24, 26), *254*
Jain, P. C. 999 (1483), *1027*
Jakopcic, K. 288 (782), *1011*
Janot, M. M. 242 (700), 244 (705), *268*
Janssen, P. H. J. 445 (1083), *1017*
Jarboe, C. H. 18, 21 (80), *255*
Jarglis, P. 324 (828), *1012*
Jarupan, P. 354, 363, 375 (863), *1012*
Jason, E. F. 47 (167), *257*
Jayatilake, G. S. 828, 896, 913, 914, 947 (1434), *1026*
Jefferies, P. R. 30 (115), *256*
Jelagin, S. 170, 178 (535), *264*
Jeng, S. J. 138, 139, 141, 142 (425), *262*
Jenkins, P. R. 271 (726), *1009*
Jenny, C. 490 (1147), *1019*
Jin, J. I. 811, 821, 822 (1425), *1025*
Jirkovsky, V. 47 (169), *257*
Joeger, V. 290 (785), *1011*
Johansson, N.-G. 144, 158 (484), *263*
Johns, D. B. 271 (738), *1010*
Johnson, A. W. 272, 274, 302, 303, 337 (751), *1010*
Johnson, B. A. 83 (269), *259*
Johnson, H. E. 47 (167), *257*
Johnson, J. R. 46 (159), *257*
Johnson, M. R. 630, 633 (1264), *1021*
Johnson, P. 83, 92, 93 (274), *259*
Johnson, P. Y. 211 (618), *266*, 637, 638 (1269), *1022*
Johnson, R. C. 83, 90 (270), *259*
Johnson, R. E. 76–78, 80–82 (236), *258*
Johnson, W. S. 4–6, 10, 11 (16), 48 (170), *254, 257*
Johnston, D. B. R. 239 (681), *268*
Johnston, M. I. 288, 290 (784), *1011*
Johnstone, L. M. 660, 661 (1299), *1022*
Jones, D. G. 170, 202 (597), *266*
Jones, E. R. H. 12 (46), 18, 20 (67), 111, 117 (330), *254, 255, 260*
Jones, G. 222 (654), *267*
Jones, G. H. 144, 145, 149, 150 (463), *263*
Jones, N. D. 420, 695, 824, 847 (1046), *1017*
Jones, R. C. F. 404 (1017), 648, 652 (1017, 1294), 653, 666, 667 (1017), 668 (1017, 1294), 669 (1294), 684 (1017), *1016, 1022*
Jones, R. H. 404, 406, 704 (1018, 1019), 730 (1386), 811 (1019), 881, 981 (1386), *1016, 1024*
Jones, T. C. 616 (1252), *1021*
Jones, W. M. 83, 89 (267), 170 (522), *259, 264*
Jost, K. 131 (365), *261*
Joullie, M. M. 169, 170, 172 (517), *264*
Joyeau, R. 420, 421 (1051), 693 (1350), 864 (1051), *1017, 1024*
Jung, A. 170 (544), *265*

Jung, C. J. 83, 85, 97 (255), *259*
Jung, F. J. 397 (915), *1013*
Jung, M. 443, 445, 446, 796 (1080), *1017*
Jungheim, L. N. 420 (1046), 684 (1334), 695, 824, 847 (1046, 1334), *1017, 1023*
Jurczak, J. 479, 480 (1118), *1018*
Just, G. 303 (802), 515 (1168), 519 (802), 520 (802, 1176), 529, 530 (802), 663, 678 (1320), 736 (1168), 793, 797 (1320), 811, 812 (1168), 816 (802, 1176), 823 (1176), 833 (802), 835, 837 (1168). 840 (802, 1176), 849 (1176), 851 (802), 854 (1168), 859, 861, 865, 894 (1431), 929, 930 (1176), 932 (802), 934 (802, 1176), *1011, 1019, 1023, 1026*

Kabasakalian, P. 83, 97 (256), *259*
Kadyrov, C. S. 695 (1352), *1024*
Kagan, H. B. 49 (179), 135 (403), 169 (518), 170 (518, 532, 537), 171 (602, 603, 605), 173 (518), 174 (518, 532), 176 (532), 178 (537), 203 (606), 204 (518, 602, 603), 205 (518, 602, 603, 606), *257, 262, 264, 266*
Kahn, M. 331 (836), *1012*
Kaiser, G. V. 239 (682), *268*
Kaji, A. 432 (1065), *1017*
Kajtar, M. 397 (917), *1013*
Kajtar-Peredy, M. 689, 886, 933, 939–941 (1340), *1023*
Kalbag, S. M. 103, 106 (296), *260*
Kaloustian, M. K. 366 (867, 868), 803, 811, 855 (1424), *1012, 1025*
Kaluza, Z. 478 (1117), 479 (1118), 480 (1117, 1118), 481, 482, 484 (1117), 485 (1117, 1119), 486 (1119), *1018*
Kalvoda, J. 665, 769, 908 (1306), *1023*
Kamaev, A. V. 725 (1381), *1024*
Kamdar, G. C. 539 (1191), *1020*
Kamernitskii, A. V. 271 (746), *1010*
Kametani, T. 249 (713), *268*, 397 (941), 513 (1165), 529 (1165, 1183), 531 (1187), 646 (1183), 665, 666 (1309), 676 (1316), 689 (1183), 715 (1316), 740 (1183), 742 (1309), 815, 816 (1316), 823, 828, 852 (1187), 882, 889 (1316), 918 (1165), *1014, 1019, 1020, 1023*
Kamiya, T. 532, 850 (1188), *1020*
Kanaoka, Y. 144 (469, 480, 481), 156 (480), 157 (481), *263*, 474 (1111), *1018*
Kandeel, M. M. 529, 534, 539, 833, 851 (1186), *1020*
Kaneko, C. 271 (747), 897 (1435), *1010, 1026*
Kaneko, I. 288 (783), *1011*
Kaneko, T. 447 (1086), *1018*
Kanematsu, K. 339 (844), *1012*
Kan-Fan, C. 242 (700), *268*
Kang, T. W. 945, 948 (1444), *1026*
Kano, S. 301 (800), 395, 396 (800, 901), 397 (965), 402 (901), 710, 890, 891 (800, 901), *1011, 1013, 1015*
Kanojia, R. M. 144 (472), *263*
Kapur, J. C. 170 (557, 564, 576, 593, 594), 188 (557, 564), 189 (557), 197 (576), 200 (593, 594), *265, 266*, 505, 511 (1154), 521 (1178), 522 (1154), 547 (1194), 548 (1178, 1194), 556 (1178), 557 (1178, 1207), 558–560, 579, 726, 728, 740, 813, 819, 822 (1178), 823, 824 (1154, 1178), 838 (1178), 841 (1154), 849, 850, 853, 856 (1178), *1019, 1020*
Karabinos, J. V. 3 (15), *254*
Karaglz, S. K. 993 (1474), *1026*
Karanewsky, D. S. 415 (1033), *1016*
Karle, I. L. 144 (479, 481), 157 (481), *263*
Karminski-Zamola, G. 288 (782), *1011*
Karte, F. 111 (328, 329), 112 (331), 113 (328), 114, 115 (328, 329), 116, 117 (331), 138, 142 (426), *260, 262*
Kashimura, S. 282 (770), 663 (1321), *1010, 1023*
Kasuga, K. 389 (893), *1013*
Katagiri, N. 271 (747), *1010*
Kato, M. 83, 85 (261), 214 (626), *259, 266*
Kato, R. 616 (1251), 997, 998 (1481), *1021, 1027*
Katsumi, K. 465, 468, 793 (1105), *1018*
Kaufman, J. A. 129 (357), *261*
Kaumoya, P. T. P. 695, 891, 893, 979 (1353), *1024*
Kaur, S. 490 (1145), 512, 513 (1164), 523, 542, 821, 843 (1145), 945 (1164), *1019*
Kaur, V. 549, 550 (1199), *1020*
Kawabata, T. 492, 736, 886 (1132), *1019*
Kawabe, N. 642 (1284), *1022*
Kawakami, S. 371 (875), *1013*
Kawakami, Y. 348 (858), *1012*
Kawamura, K. 371 (873), *1013*
Kawamura, S. 343, 344 (853), *1012*
Kawanishi, M. 348 (857, 858), 349 (857), *1012*
Kawanisi, M. 124, 125 (352), *261*
Kawano, Y. 404, 406, 678, 693, 737 (1021), 745, 799 (1396), 897, 947 (1021), *1016, 1025*
Kay, I. 433, 712, 714, 985–988 (1067), *1017*
Kay, T. I. 433, 712 (1066), *1017*
Kaye, H. 3 (14), 99, 106 (283), *254, 259*
Kazumori, K. 677, 729, 898 (1318), *1023*
Keeley, D. E. 418 (1041), *1017*
Keil, K. H. 170 (544), *265*
Keller, F. 53 (186), *257*
Kelly, F. W. 32, 33 (114), *256*
Kelly, J. E. 283 (772), *1010*
Kelly, J. F. 166, 188 (506), *264*

Kempe, U. M. 118 (336), *260*
Kende, A. S. 69 (223), *258*
Kenkyu, K. 489 (1123), *1018*
Kennedy, M. 388 (891), *1013*
Kenner, G. W. 18 (81), *255*
Kern, R. 640 (1278), *1022*
Kerr, D. I. B. 214, 220 (640), 222 (640, 648), 224, 225 (640), 226, 227 (640, 648), 229 (640), 233 (648), 238, 240 (640, 648), *267*
Kerwin, J. F. Jr. 420, 421, 442, 796 (1049), *1017*
Khachatryan, L. A. 28 (102), *256*
Khajavi, M. S. 521, 548, 556–560 (1178), 561–563 (1210), 579 (1178), 612 (1244), 726, 728, 740, 813, 819, 822–824, 838, 849, 850, 853, 856 (1178), *1019–1021*
Khan, A. H. 360, 363, 383 (865), *1012*
Khan, R. H. 290, 359, 364 (787), *1011*
Khananshvili, L. M. 664, 676, 708, 751, 754, 764, 815, 839, 942 (1304), *1022*
Kharasch, N. 144, 145, 149 (462), *263*
Khouri, F. 366 (868), *1012*
Khuong-Huu, F. 248 (711), *268*
Kido, M. 470 (1107), *1018*
Kiehl, G. 491, 503, 504, 523, 526, 536, 541, 542, 544, 1005 (1127), *1018*
Kierstead, R. W. 9, 11, 13, 87, 125 (28), *254*
Kigasawa, K. 513 (1165), 529 (1165, 1183), 531 (1187), 646, 689, 740 (1183), 823, 828, 852 (1187), 918 (1165), *1019, 1020*
Kiguchi, T. 144 (474, 475), 153 (475), 154 (474), *263*
Kikkawa, I. 713 (1368), 760, 770 (1368, 1403), 846, 883, 891 (1368), 947 (1368, 1403), *1024, 1025*
Kim, C. U. 690, 747, 748, 752, 767, 886, 904, 943 (1345), *1023*
Kim, H. H. 744, 858, 862, 865 (1394), *1025*
Kim, H. J. 326 (830), *1012*
Kim, K. S. 315, 952 (813), *1011*
Kim, M. Y. 719, 873, 876, 878, 891, 892, 977 (1374), *1024*
Kimbrough, R. D. Jr. 170, 179–182 (543), *265*
Kimura, Y. 492, 736, 886 (1132), 898, 904 (1437), *1019, 1026*
King, A. 943 (1443), *1026*
King, G. S. 16 (55), *254*
King, T. J. 960, 961 (1451), *1026*
Kingsbury, W. D. 398 (989), *1015*
Kingsland, M. 124, 125 (349), *261*
Kinkel, K. G. 76, 77 (238), *258*
Kinnick, M. D. 690, 749, 753, 769, 830, 831, 847, 899, 909, 910, 970 (1346), *1023*
Kinugasa, M. 209 (613), *266*
Kirchhof, W. 222, 229 (656), *267*
Kirisawa, M. 553 (1202), 574 (1220), 623 (1256), *1020, 1021*
Kirk, J. C. 222, 224, 225 (646), *267*
Kirk, K. L. 138 (416, 417), *262*
Kirmse, W. 127, 128 (355), 170, 173–176, 178, 179 (525), *261, 264*
Kirschke, K. 103, 106 (298), *260*
Kishimoto, S. 685, 743, 895, 896, 908, 909 (1343), *1023*
Kita, Y. 214 (624, 635, 637), 215 (624), 217 (624, 635, 637), 218 (635, 637), 219 (637), 222 (649, 650), 227 (649), 228 (650), *266, 267*
Kitamura, T. 348 (857, 858), 349 (857), *1012*
Klaver, W. J. 698, 699, 891, 892, 950, 951, 972, 975 (1356), *1024*
Klein, J. 19 (95), *255*
Klein, K. P. 137 (411–413), *262*
Kleineberg, G. 170, 200 (592), 203 (607), *266*
Klemm, L. H. 31 (111, 112), 32 (113, 114), 33 (114), *256*
Klesel, N. 827 (1426), *1025*
Klever, H. W. 170 (527, 529), *264*
Klich, M. 529, 678, 740, 797, 831, 832, 844, 851, 853, 855 (1184), *1020*
Klitgaard, N. A. 53, 54 (193), *258*
Kloft, M. 520, 531, 816, 838 (1177), 913–915, 918, 922, 924, 925 (1440), *1019, 1026*
Klohs, W. H. 53 (186), *257*
Klopfenstein, C. E. 31 (112), *256*
Kluge, A. F. 518, 818, 819, 839 (1174), *1019*
Knapp, S. 435 (1070), *1017*
Kneussel, P. 711, 729, 917 (1367), *1024*
Knifton, J. F. 983 (1470), *1026*
Knight, J. 672, 673 (1312), 778, 779 (1406), 788, 789 (1312), *1023, 1025*
Knight, J. C. 16 (54), 40, 41 (145), *254, 256*
Knox, J. 272, 274, 302, 303, 337 (751), *1010*
Knunyants, I. L. 139 (428, 429), 140 (429, 430), *262*
Ko, S. 759 (1402), *1025*
Kobayashi, S. 300, 309, 310 (798), 397 (961), 404 (1013), 405 (798, 1013), 491 (1130), *1011, 1015, 1016, 1018*
Kobayashi, T. 144, 157 (482), *263*, 343, 344 (853), 681, 917 (1328), *1012, 1023*
Kober, P. 170 (529), *264*
Koch, E. 109, 110 (320), *260*
Koch, H. 112 (332, 333), *260*
Koch, J. 166 (508), *264*
Koch, T. H. 214, 216, 227 (638), *266*
Kochergin, P. M. 725 (1381), *1024*
Kocor, M. 271 (743, 744), *1010*
Koechler, K. F. 500, 508, 509, 567, 568, 637, 817, 819, 848, 849 (1143), *1019*
Koehl, W. J. Jr. 110 (323, 325), 111 (323, 326), 112 (325), *260*
Koenig, H. B. 681, 834 (1331), *1023*
Kohler, E. P. 51 (181, 182), *257*

Kohori, T. 827 (1427), *1025*
Kokoyama, Y. 18, 23–25 (88), *255*
Kokuni, N. 597 (1239), *1021*
Kolasa, T. 443, 445, 446 (1081), *1017*
Kolczynski, B. V. 43, 111 (155), *257*
Koltzenburg, G. 18, 22 (83), *255*
Komatsu, M. 593, 594, 601, 602 (1238), 605, 606 (1242), 607, 608 (1238, 1242), 876, 877, 911 (1238), *1021*
Kondo, K. 120, 121 (338), *260*, 827 (1427), *1025*
Königsdorfer, K. 170, 180, 181 (541), *264*
Kopecky, W. J. Jr. 230 (663), *267*
Koppel, G. A. 397 (905), *1013*
Korshak, V. V. 398 (994), *1016*
Korte, F. 36 (128), *256*
Korzeniowski, S. H. 13 (51), *254*
Koshiji, H. 415 (1032), *1016*
Kosterman, D. 11 (39, 40), *254*
Kotani, M. 491, 498, 920 (1131), *1019*
Kotelnikov, V. A. 398 (994, 996), *1016*
Kovac, J. 419 (1044), *1017*
Kovacevic, M. 681 (1330), *1023*
Kozikowski, A. P. 317 (817), *1011*
Kraatz, U. 36 (128), *256*
Krafft, M. E. 339 (845), *1012*
Kramarova, E. P. 1001 (1486), *1027*
Kramer, U. 647, 651 (1288), *1022*
Krapcho, A. P. 4 (26), 8 (24, 26), *254*
Kraus, G. A. 783, 784, 910, 972 (1410), *1025*
Kraus, W. 346 (855), *1012*
Krawczyk, Z. 494–496 (1134), *1019*
Krebes, J. 756, 758, 844–846, 956, 958 (1405), *1025*
Kreder, J. 521, 548, 556 (1178), 557 (1178, 1207), 558–560, 579, 726, 728, 740, 813, 819, 822–824, 838, 849, 850, 853, 856 (1178), *1019, 1020*
Kreiser, W. 39 (142), *256*
Krimm, H. 232, 233 (667), *267*
Krishan, K. 584 (1227), *1021*
Kristol, L. D. 123 (342), *260*
Krohn, K. 144, 145, 150 (464), *263*
Krolls, U. 138 (418), *262*, 805, 808–810 (1423), *1025*
Krösche, H. 39 (141), *256*
Kroszczynski, W. 271 (744), *1010*
Krutskaya, L. V. 272 (752), *1010*
Krutskii, L. N. 272 (752), *1010*
Kuczynski, H. 214, 219 (639), *266*
Kugajevsky, I. 170, 181 (542), *265*
Kumar, S. 584 (1227), *1021*
Kump, W. G. 243 (701), *268*
Kunert, D. M. 499, 500, 502 (1142), 697 (1142, 1355), 700, 701 (1142), 704 (1355), 863 (1142), *1019, 1024*
Kunz, H. 350 (859), *1012*
Kunz, R. A. 238 (677, 678), *267*
Kupchan, S. M. 144 (472), *263*
Kuramochi, T. 378 (880), *1013*
Kurashev, V. V. 398 (994, 996), *1016*
Kurath, P. 36 (133–136), *256*
Kurihara, M. 397 (961), *1015*
Kuthan, J. 673 (1314), *1023*
Kuwajima, I. 63 (217), *258*, 331 (837), *1012*
Kwon, H. B. 981 (1466), *1026*
Kyazimov, A. S. 11 (36), 28 (101), *254, 256*
Kyrides, L. P. 83, 89 (265), *259*

L'Abbé, G. 133 (375), *261*
Labia, R. 397 (911, 951), 420, 421, 864 (1051), *1013, 1014, 1017*
Labsky, J. 134 (384, 385), *261*
Ladlow, M. 314 (823), *1012*
Lago, J. M. 381 (884), *1013*
Lahoti, R. J. 279, 293 (765), *1010*
Laird, A. A. 729, 764 (1384), *1024*
Lal, B. 170, 198 (588), *265*, 573, 574 (1219), *1020*
Lallemand, J. Y. 312 (808), *1011*
Lamberti, V. 4, 84 (20), *254*
Lane, A. G. 76 (229), *258*
Langbeheim, M. 475, 487 (1120), *1018*
Lange, B. C. 170, 197 (561), *265*
Lange, G. L. 313 (809), *1011*
Langlois, Y. 329 (834), *1012*
Lansbury, P. T. 278, 280 (760), 328 (833), *1010, 1012*
Larcheveque, M. 312 (808), 717, 812, 849 (1372), *1011, 1024*
Lardelli, G. 4, 84 (20), *254*
Lattes, A. 230 (665, 666), 232 (666), 233 (665, 666), *267*
Lattrell, R. 170, 188 (559), *265*
Laubach, G. D. 170, 193 (577–579), *265*
Laura, M. A. 103, 106 (296), *260*
Lawesson, S. O. 460 (1096), 621, 807 (1255), *1021*, 804 (1422), 805 (1096, 1422), 806 (1096), *1018, 1025*
Lawton, G. 462 (1098), 627–629 (1261, 1262), 630 (1262), 648, 651, 869 (1261), 992 (1262), *1018, 1021*
LeBelle, M. J. 238 (675), *267*
Lebreux, C. 366 (866), *1012*
Lecea, B. 490 (1146), 551–554, 570 (1201), 575, 576, 717, 719 (1221), 726 (1201), 728 (1146), 730 (1146, 1201), 734 (1201), 751, 762, 898 (1146), 934 (1146, 1221), 954, 955 (1146), *1019, 1020*
Lederer, E. 4 (18), *254*
Lee, B. H. 443, 693 (1078), *1017*
Lee, C. 293 (790), *1011*
Lee, D. G. 240 (689), *268*
Lee, D. H. 31 (112), 32, 33 (114), *256*

Lee, E. 399, 756, 757, 824, 847 (999), 857 (1430), 945, 948 (1444), 999 (1430), *1016, 1025, 1026*
Lee, G. E. 402, 403 (1008), *1016*
Lee, M. 313 (809), *1011*
Lee, R. E. 170, 189 (558), *265*
Lee, S. L. 76–80, 83 (237), *258*
Lee, T. V. 287, 290, 357, 364, 367 (781), *1011*
Lee, W. Y. 744, 858, 862, 865 (1394), *1025*
Lee, Y. B. 744, 858, 862, 865 (1394), *1025*
Lee, Y. Y. 744, 858, 862, 865 (1394), *1025*
Lefker, B. A. 445 (1084), 447 (1085), *1017*
Legault, R. 238 (676), *267*
Legueut, C. 312 (808), *1011*
Leipold, H. A. 47 (165), *257*
Leitch, G. C. 243 (703), *268*
Leleu, G. 315 (812), *1011*
Lemay, G. 295 (792, 793), 308 (793), *1011*
Le Men, J. 242 (700), 244 (705), *268*
Lemmen, P. 397 (919), *1014*
Lempert, K. 689 (1340), 692 (1347), 886, 933, 939–941 (1340), *1023, 1024*
Lengyel, I. 132 (373), 138 (421, 422), *261, 262*
Lenz, G. R. 144 (465, 466, 468), 145 (465, 466), 150 (465), 151 (465, 466), 153 (465), 161 (465, 468), 162 (468), *263*
Lenz, R. W. 271 (738), *1010*
Leon-Santiago, M. A. 491–493, 497, 499, 503, 504, 528, 536, 541, 545 (1128), *1018*
Lessard, J. 8 (23), *254*
Levi, E. M. 143 (451), *263*
Levina, R. Y. 3, 28 (6), *253*
Levorse, A. T. 435 (1070), *1017*
Levy, H. 106 (287), *259*
Levy, S. V. 732, 966, 967 (1388), *1024*
Levy, W. J. 76 (231), *258*
Lewis, A. 123 (343), *260*
Lewis, H. B. 144, 145, 149 (462), *263*
Ley, S. V. 271 (742), 732, 733, 886, 965 (1389), 966, 967 (1458), *1010, 1025, 1026*
Liberatore, F. 434 (1069), *1017*
Lichtenthaler, F. W. 324 (828), *1012*
Limbert, M. 827 (1426), *1025*
Lin, S. I. 726, 756, 758 (1398), *1025*
Lindert, A. 53 (191), *257*
Lindig, M. 350 (859), *1012*
Lindsay, K. L. 30 (108), *256*
Linkies, A. 669–672, 674, 778, 779 (1311), *1023*
Linstead, R. P. 19 (91), 76, 77 (235), 108 (312), *255, 258, 260*
Linstrumelle, G. 276 (757), *1010*
Liotta, D. 293 (790), *1011*
Lippman, A. E. 209 (612), *266*
Lipshutz, B. H. 238, 251 (683), *268*, 953 (1446, 1447), 960 (1447, 1450), 962 (1447), 1005 (1490), *1026, 1027*
Liu, J. 600, 601, 603, 605, 731, 733, 799 (1241), *1021*
Liu, J.-C. 4, 8 (25), *254*
Liu, W. Y. 663, 678, 793, 797 (1320), *1023*
Lloyd, H. A. 214, 216, 220 (633), *266*
Lodwig, S. N. 4, 7 (21), *254*
Löffler, A. 51 (185), *257*
Logusch, E. W. 290, 357, 359, 364, 366 (786), *1011*
Lohaus, G. 132, 163 (370), *261*
Lohse, F. 437 (1073), *1017*
Lombardi, P. 687 (1338), 731 (1387), 747, 753 (1338), 759, 762, 763 (1387), 768 (1338), 900, 903 (1387), 907 (1338, 1387), 954 (1387), *1023, 1024*
Lomtatidze, Z. S. 664, 676, 708, 751, 754, 764, 815, 839, 942 (1304), *1022*
Longley, R. I. 83, 89 (263, 264), *259*
Loomis, G. L. 61, 64, 65 (211), *258*
Lopresti, R. J. 282 (768), *1010*
Lorber, M. E. 49 (178), *257*
Lossow, E. 61 (201), *258*
Louis, J.-M. 83 (272, 277), 90 (272), 93, 94 (272, 277), 95, 96 (272), *259*
Lounasmaa, M. 46 (163, 164), *257*
Lowe, C. 399, 756, 757, 824, 847 (999), *1016*
Lowe, G. 128, 129 (356), 144 (488, 490–492), 159, 160 (488), 210, 211 (615, 616), *261, 263, 264, 266*, 397 (928, 934), *1014*
Lu, C. S. 613, 676, 692, 739, 899, 905, 934, 935 (1245), *1021*
Luche, J.-L. 169, 170 (518), 171 (602, 603, 605), 173, 174 (518), 203 (606), 204 (518, 602, 603), 205 (518, 602, 603, 606), *264, 266*
Luche, J. L. 135 (403), 170 (532, 537), 174, 176 (532), 178 (537), *262, 264*
Ludescher, H. 693, 728, 760, 762, 763, 801, 908 (1348), *1024*
Luh, B. Y. 690, 747, 748, 752, 767, 886, 904, 943 (1345), *1023*
Luk'yanenko, N. G. 394 (896), *1013*
Luk'yanets, E. A. 3, 28 (6), *253*
Lusinchi, X. 241 (696), *268*
Luttringer, J. P. 491, 503, 504, 523, 526, 536, 541, 542, 544, 1005 (1127), *1018*
Lyman, D. J. 3 (9), *253*
Lyons, J. E. 310 (804, 805), *1011*
Lythgoe, B. 122 (339), *260*

Machado-Aracijo, F. W. 300, 341 (797), *1011*
Maciejewski, S. 407, 409 (1024), *1016*
Madden, J. P. 138 (424), *262*
Madhav, R. 252 (720), *268*
Maeda, H. 429, 430 (1062), 448, 450 (1087), *1017, 1018*

Maeda, Y. 317–319 (819), *1011*
Maeva, R. V. 3 (8), *253*
Magerlein, B. J. 676, 826, 827, 845, 933, 937 (1317), *1023*
Maguire, A. R. 388 (891), *1013*
Mahajan, J. R. 452, 657, 962 (1090), *1018*
Majeti, S. 18, 22, 23 (87), *255*
Mak, C. P. 693, 728, 760, 762, 763 (1348), 791, 792 (1414), 801, 908 (1348), *1024, 1025*
Malacria, M. 111 (327), *260*
Malkhasyan, A. T. 675, 700 (1322), *1023*
Mamba, A. 168 (513), *264*
Mandelshtam, T. V. 123 (342), *260*
Mangasaryan, Ts. A. 34, 39 (118), 42 (147), *256, 257*
Mangeney, P. 329 (834), *1012*
Mangia, A. 397 (959, 960), *1015*
Manhas, M. S. 133 (377), 138, 139 (425, 435, 436), 141 (425), 142 (425, 431, 445), 163 (377), 169 (520), 170 (520, 545–548, 557, 560, 563–566, 568, 570–573, 575, 576, 588, 593, 594), 183 (520, 545, 547, 548), 184, 185 (545, 548), 187 (547, 548, 560, 563, 566), 188 (547, 557, 560, 564, 566), 189 (557), 190 (548), 192 (560, 565, 566, 568, 572, 575), 194 (570, 573), 195 (560, 566, 571, 575), 196 (571), 197 (548, 565, 576), 198 (588), 200 (593, 594), 201 (520), *261–266*, 397 (902, 910, 921, 922, 943), 439–442 (1076), 505 (1152–1155), 506 (1155, 1156), 507, 508 (902, 1156), 509 (902, 1152), 511 (1154), 512 (902), 513 (1155), 521 (1178), 522 (1154), 523 (1155), 526 (1155, 1180), 527 (1180), 545 (1152), 547 (1180, 1194), 548 (1178, 1194), 555 (1205), 556 (1178, 1205, 1206), 557 (1178, 1207, 1208), 558, 559 (1178), 560 (1178, 1209), 561 (1209, 1210), 562 (1152, 1153, 1210, 1211), 563 (1153, 1210, 1212), 564, 565 (1212), 567 (1180), 573 (902, 1219), 574 (1219), 579 (1178), 581 (1225), 612 (1244), 645, 648, 657, 679, 680 (1155), 695 (1076), 706 (1155, 1211), 726, 728 (1178), 734 (1225), 735 (1152, 1153), 740 (1178), 743, 755 (1152), 806 (1225), 810 (1155), 813 (1178), 817 (1180), 819 (1178), 822 (1178, 1180), 823, 824 (1154, 1178), 838 (1178), 841 (1154), 846 (1180), 849, 850, 853 (1178), 856 (1178, 1208), 859 (1205, 1211), 869 (1153), 870 (1205), 871 (1156), 872 (902), 873 (902, 1156), 874 (902, 1156, 1205), 918, 925 (1155), 936 (1152), *1013, 1014, 1017, 1019–1021*
Manitto, P. 127 (354), *261*
Mann, G. 993 (1478), *1027*
Manning, R. E. 242, 245 (698), *268*
Manske, R. H. 104, 106, 125 (307), 247 (710), *260, 268*
Manukyan, M. A. 435 (1071), *1017*
Mao, C.-L. 143 (450–452), 144 (453), *263*
Marchand-Brynaert, J. 207 (609), *266*, 397 (920, 927), 665, 724, 777, 922 (1308), *1014, 1023*
Marcos, M. 711 (1366), *1024*
Mares, F. 416 (1035), *1016*
Margrave, J. L. 4–6, 10, 11 (16), *254*
Mari, F. 120, 121 (338), *260*
Mariani, L. 136 (399, 400), 139 (427), *262*
Marinetti, A. 1008, 1009 (1493), *1027*
Marino, J. P. 301, 303, 352, 354 (801), *1011*
Marion, J. P. 17 (62), *255*
Marschall-Weyerstahl, A. 348 (856), *1012*
Marshall, A. R. 243, 250 (702), *268*
Marshall, H. 46 (158), *257*
Marshall, J. A. 4, 6 (19), 10, 11 (30, 31), 83 (254), 85 (19, 254), 86 (19), 91 (19, 30, 31, 254), 123 (344), *254, 259, 261*
Martel, A. 729, 730, 760, 762, 763, 863, 900, 925, 926 (1383), *1024*
Martelli, G. 614 (1247), *1021*
Martin, G. J. 138 (415), *262*
Martin, J. 144 (456), *263*
Martin, M. V. 252 (721), *268*
Martin, S. F. 314, 323 (824), *1012*
Martin-Domenech, A. 648, 651 (1292), *1022*
Martins, S. M. 397 (926), *1014*
Martirosyan, G. T. 675, 700 (1322), *1023*
Maruyama, H. 422, 423, 735, 737, 741, 899, 905, 929, 930, 933, 937, 938, 940, 943 (1053), *1017*
Marvel, C. S. 83 (251), 240 (685), *259, 268*
Marx, J. N. 294, 357, 360, 364, 790 (791), *1011*
Maryanoff, B. E. 453, 789, 790, 972, 973 (1091), *1018*
Masamune, T. 644 (1285), *1022*
Masui, M. 970 (1462), *1026*
Matache, S. 640 (1274, 1275), *1022*
Mathey, F. 1008, 1009 (1493), *1027*
Matinyan, S. G. 675, 700 (1322), *1023*
Matsoyan, S. G. 34, 39 (118), 42 (146, 147), *256, 257*
Matsubara, F. 760, 770, 947 (1403), *1025*
Matsuda, A. 112, 116, 117 (331a), *260*
Matsuda, T. 419 (1042), 424, 425 (1058), *1017*
Matsummura, Y. 716 (1369), *1024*
Matsumoto, H. 144, 147–149 (460), *263*
Matsumoto, K. 640 (1279), *1022*
Matsumoto, N. 106 (292), *259*
Matsumoto, T. 39 (143), 98, 106 (282), *256, 259*
Matsumura, H. 739 (1392), 766, 844, 925–927, 971 (1404), *1025*

Matsumura, Y. 373 (876), 397 (913), 398 (985), *1013, 1015*
Matsushita, Y. 404, 406, 678, 693, 737, 897, 947 (1021), *1016*
Matsutaka, Y. 214, 217, 218 (635), *266*
Matsuzaki, K. 491, 498, 920 (1131), *1019*
Mattingly, P. G. 420, 421 (1049, 1050), 422, 438, 439 (1050), 442 (1049, 1050), 796 (1049), 850 (1050), *1017*
Mat-Zin, A. R. 404 (1017), 648, 652 (1017, 1294), 653, 666, 667 (1017), 668 (1017, 1294), 669 (1294), 684 (1017), *1016, 1022*
Maury, G. 722, 740, 813 (1376), *1024*
May, C. J. 19 (91), *255*
Mayer, W. W. Jr. 240 (685), *268*
Mayrhofer, R. 696 (1354), *1024*
Mazur, R. H. 214 (636, 642), 216, 217 (636), 221, 222 (642), *266, 267*
Mazzocchi, P. H. 165 (505), *264*, 230 (663, 664), *267*
McCombie, S. W. 726, 756, 758 (1398), *1025*
McComsey, D. E. 453, 789, 790, 972, 973 (1091), *1018*
McCulloch, A. W. 286 (778), *1011*
McDonald, E. 137 (410), *262*
McGuire, M. A. 585, 586, 588 (1228), *1021*
McGuire, T. M. 32, 33 (114), *256*
McInnes, A. G. 286 (778), *1011*
McKenna, J. C. 27 (98, 99), *255*
McKennis, H. Jr. 241, 242 (695), *268*
McKeon, J. E. 251 (718), *268*
McKervey, M. A. 388 (891), *1013*
McMurray, J. E. 11, 12, 87, 88 (45), *254*
McMurray, T. B. H. 4, 6, 84, 86 (17), *254*
McPhail, R. T. 626 (1260), *1021*
Mee, A. 110 (324), *260*
Megges, R. 397 (949), *1014*
Mehra, U. 463 (1099), 490 (1145), 511 (1099), 512 (1163, 1164), 513 (1164), 523, 542 (1145), 549, 550 (1199), 807 (1099), 821, 843 (1145), 918 (1099), 945 (1164), *1018–1020*
Mehta, G. 99, 103, 104, 106 (285), *259*
Meier, G. P. 460, 731, 795 (1094), *1018*
Meijer, J. 394 (898), *1013*
Meinwald, J. 10 (29), 100 (289), 101 (29), 106 (289), 123 (343), *254, 259, 260*
Melikyan, G. S. 34, 39 (118), 42 (146), *256*
Mellor, J. M. 292, 352, 353, 362, 366 (788), *1011*
Melstrom, D. S. 170, 179, 180 (531), *264*
Menard, M. 729, 730, 760, 762, 763, 863, 900, 925, 926 (1383), *1024*
Mencke, B. 827 (1426), *1025*
Mengler, H. 69, 74 (225), *258*
Mentzer, C. 36 (131), *256*
Merault, G. 167 (512), *264*
Mercer, F. 499, 500, 502 (1142), 697 (1142, 1355), 700, 701 (1142), 704 (1355), 863 (1142), *1019, 1024*
Merlini, L. 695, 798 (1351), *1024*
Merve, J. P. van der 109, 110 (321), *260*
Meslin, J. C. 993 (1475), *1027*
Mestres, R. 18 (70, 73, 76), 20 (76), *255*
Metz, G. 464 (1102), *1018*
Metzger, C. 69, 70 (219), *258*
Metzger, K. G. 681, 834 (1331), *1023*
Meyer, H. 36 (121), *256*
Meyer, W. C. 165 (504), *264*
Meyers, A. I. 58 (197–199), 59 (199), 60 (200), *258*, 445 (1084), 447 (1085), *1017*
Midland, M. M. 284 (775), *1011*
Miescher, M. 207 (610), *266*
Miginioc, L. 51 (183), *257*
Mihelick, E. D. 58 (198, 199), 59 (199), 60 (200), *258*
Milledge, A. F. 76, 77 (235), *258*
Miller, M. J. 397 (957), 420, 421 (1049, 1050), 422, 438, 439 (1050), 442 (1049, 1050), 443 (1079–1081), 445, 446 (1080, 1081), 734 (1079), 796 (1049, 1080, 1416), 797 (1417), 850 (1050), 992 (1416), *1015, 1017, 1025*
Millich, F. 133 (376), *261*
Min, S. H. 811, 821, 822 (1425), *1025*
Minami, N. 63 (217), *258*
Minami, T. 168 (514), *264*
Minato, H. 83, 84 (253), *259*
Misco, P. F. 690, 747, 748, 752, 767, 886, 904, 943 (1345), *1023*
Misiti, D. 222 (651, 652), *267*
Mitsui, S. 288 (783), *1011*
Mitsunobu, O. 271 (732), *1010*
Mittelbach, H. 170, 185, 186, 201 (550), *265*
Mityoshi, M. 133 (382), *261*
Miura, M. 957, 960 (1449), *1026*
Miyake, M. 553 (1202), 574 (1220), 623 (1256), *1020, 1021*
Miyama, A. 677, 729, 898 (1318), *1023*
Miyano, K. 39 (143), 98, 106 (282), *256, 259*
Miyasaka, T. 996 (1480), *1027*
Miyata, N. 644 (1285), *1022*
Miyoshi, M. 131 (366), *261*
Miyozaki, K. 471, 872, 1001 (1109), *1018*
Mizsak, S. A. 162 (493), 166 (493, 507), *264*
Mizuno, K. 962 (1455), *1026*
Mkhairi, A. 612 (1243), *1021*
Mladenova, M. 380 (882), *1013*
Moehrle, H. 252 (719), *268*, 970 (1461), *1026*
Moffett, R. B. 133 (383), *261*
Molhrle, H. 968 (1459), *1026*
Molines, H. 420, 421, 864 (1051), *1017*

Molloy, R. M. 38, 39 (139), *256*
Momonoi, K. 685, 743, 895, 896, 908, 909 (1343), *1023*
Monaco, S. 272, 274, 302, 303, 337 (751), *1010*
Mondon, A. 144, 145, 150 (464), *263*
Monge, A. 648, 651 (1292), *1022*
Mongelli, N. 397 (964), *1015*
Moniot, J. L. 144 (472), *263*
Montgomery, R. S. 214, 216, 217 (634), *266*
Monti, S. A. 242, 245 (698), *268*
Moody, C. J. 462 (1098), 627–629 (1261, 1262), 630 (1262), 648, 651, 869 (1261), 992 (1262), *1018, 1021*
Moore, H. W. 222 (651, 652), *267*, 499 (1142), 500 (1142, 1144), 502 (1142), 505, 509, 530, 571, 572 (1144), 630 (1263), 697 (1142, 1355), 700, 701 (1142), 704 (1355), 863 (1142), 864, 883 (1144), 929, 971 (1263), *1019, 1021, 1024*
Moore, J. A. 283 (772), 993 (1477), *1010, 1027*
Moracci, F. M. 434 (1069), 510 (1160, 1161), 530 (1160), 535, 536, 540, 541 (1161), 673, 734 (1160), 867 (1161), *1017, 1019*
Moracehi, F. 952 (1445), *1026*
Morand, P. F. 83, 85 (257), *259*
Moreno-Manas, M. 287, 288, 294, 357, 359, 360, 363 (779), *1011*
Moret, E. 317 (818), *1011*
Morgan, T. K. Jr. 124 (348), *261*
Mori, M. 285, 345 (776), 397 (967), 424, 425 (1059), 433, 712 (1067), 714 (776, 1067), 715 (776), 716 (1059), 739, 792, 795 (1393), 879 (776, 1393), 984 (1471), 985, 986 (1059, 1067, 1471), 987 (1067, 1393, 1471), 988 (776, 1067, 1393), 989, 990 (776), *1011, 1015, 1017, 1025, 1026*
Mori, S. 106 (288), *259*, 760, 770, 947 (1403), *1025*
Mori, T. 144 (474, 476, 477), 154 (474, 477), 155 (476), *263*, 381 (883), *1013*
Mori, Y. 130 (362), *261*
Moriconi, E. J. 163 (499), 165 (499, 504, 505), 166, 188 (506), *264*
Morikawa, M. 170, 177 (538), *264*
Morin, C. 397 (911, 951), *1013, 1014*
Morin, J. M. Jr. 690, 749, 753, 769, 830, 831, 847, 899, 909, 910, 970 (1346), *1023*
Morita, Y. 703 (1359), *1024*
Moriya, O. 314 (822), *1012*
Morizawa, Y. 351 (860), *1012*
Morrison, M. A. 443, 734 (1079), *1017*
Mosacci, F. M. 496, 510, 536 (1137), *1019*
Mostowicz, D. 581 (1224), *1021*
Motoyoshiya, J. 491, 498, 920 (1131), *1019*
Mousseron, M. 49 (177), *257*
Movsumzade, M. M. 11 (36), 28 (101), *254, 256*
Moya Portuguez, M. 397 (920), *1014*
Moza, P. N. 139 (441), *263*
Mueller, K. A. 222, 229 (656), *267*
Mueller, R. H. 83, 90 (270), *259*
Muggler-Chawan, F. 17 (62), *255*
Mukaiyama, T. 271 (733), 279 (766), 407, 409 (1023, 1025), 416 (1025), *1010, 1016*
Mukawa, F. 106 (288), *259*
Mukerjee, A. K. 133, 163 (379), 170 (549, 583), 184 (549), 194, 195 (583), *261, 265*, 397 (932), *1014*
Mukkavilli, L. 521, 548 (1178), 556 (1178, 1206), 557 (1178, 1207), 558–560, 579, 726, 728, 740, 813, 819, 822–824, 838, 849, 850, 853, 856 (1178), *1019, 1020*
Muller, A. 142 (447), *263*
Muller, J. C. 61, 65 (213), *258*
Muller, L. L. 132, 163 (372), *261*
Mulot, P. 717, 812, 849 (1372), *1024*
Mulzer, J. 271 (731), *1010*
Murahashi, S. 317–319 (819), *1011*
Murakami, M. 760, 770, 947 (1403), *1025*
Murata, S. 957, 960 (1449), *1026*
Murray, M. A. 18 (81), *255*
Murray, M. F. 83 (269), *259*
Murray, R. K. Jr. 124 (347, 348), *261*
Murthy, K. S. K. 672, 673 (1313), *1023*
Mysov, E. I. 648 (1290), *1022*

Nader, R. B. 366 (867), *1012*
Nagai, K. 1000 (1485), *1027*
Nagao, Y. 397 (946, 966), 996 (1480), *1014, 1015, 1027*
Nagarajan, K. 243 (701), *268*
Nagasaka, T. 404, 406 (1020), *1016*
Nagase, S. 1005 (1489), *1027*
Nagashima, H. 430, 431 (1063, 1064), 432 (1064), 864 (1063), 865, 866 (1064), *1017*
Nagashima, N. 730, 754, 875 (1385), *1024*
Nagata, M. 1000 (1485), *1027*
Nagata, W. 397 (939, 953, 954, 962), 693 (1349), 699 (1357), 713 (1368), 729 (1349), 760 (1368, 1403), 761 (1357), 766 (1404), 770 (1368, 1403), 782 (1357), 844 (1404), 846, 883, 891 (1368), 894 (1349), 925–927 (1404), 947 (1368, 1403), 971 (1404), *1014, 1015, 1024, 1025*
Nagota, M. 661 (1300), *1022*
Nagusa, H. 663 (1321), *1023*
Naidov, B. 53 (189), *257*
Naito, T. 144 (471, 473–477), 152 (471), 153 (475), 154 (474, 477), 155 (473, 476), *263*, 465, 468, 793 (1105), 897 (1435), *1018, 1026*
Nakaguchi, O. 532, 850 (1188), *1020*

Nakahashi, K. 517, 524, 525 (1171), 527, 696 (1181), 819–821, 831, 832, 838, 841, 842 (1171), 852 (1181), 881 (1171), *1019, 1020*
Nakai, H. 144, 157 (481), *263*
Nakai, S. 597 (1239), *1021*
Nakai, T. 615 (1248), *1021*
Nakanishi, S. 962 (1455), *1026*
Nakano, J. 440–442, 444, 445, 818, 826, 845 (1077), *1017*
Nakano, K. 713, 760, 770, 846, 883, 891, 947 (1368), *1024*
Nakano, T. 249 (713), *268*
Nakatsuka, T. 476 (1112), 690, 748, 752, 769 (1344), *1018, 1023*
Nakayama, H. 379 (881), *1013*
Nakayama, K. 998 (1482), *1027*
Nakoi, H. 474 (1111), *1018*
Nally, J. 631, 634 (1265), *1021*
Naota, T. 317–319 (819), *1011*
Narasaka, K. 271 (733), *1010*
Narayana, V. L. 577 (1222), 578 (1222, 1223, 1223), 579 (1223, 1223), *1021*
Narayanan, C. S. 170 (546), *265*, 514 (1166), *1019*
Narimatsu, S. 316, 339 (815), *1011*
Narisada, M. 397 (953, 954, 962), 766, 844, 925–927, 971 (1404), *1014, 1015, 1025*
Narisano, E. 624 (1257), *1021*
Naser-ud-din 166 (511), *264*
Nasipuri, D. 522, 973 (1179), *1020*
Naum, Y. A. 679 (1324), *1023*
Naylor, C. A. Jr. 3 (13), *254*
Nazarova, N. Y. 394 (896), *1013*
Nazawa, M. 491, 498, 920 (1131), *1019*
Nazir, M. 39 (142), *256*
Nelson, A. L. 214, 216 (631), *266*
Nelson, D. A. 170, 183 (551, 552), 185 (552), 190 (551), 191 (551, 552), *265*
Nemoto, H. 249 (713), *268*, 665, 666, 742 (1309), *1023*
Nenz, A. 233, 234 (670), *267*
Neuenschwander, K. 783, 784, 910, 972 (1410), *1025*
Neuman, M. 397 (933), *1014*
Newan, R. H. 13, 17 (52), *254*
Newcomb, M. 601, 603, 604, 608, 609 (1240), *1021*
Neyrelles, I. 49 (177), *257*
Nguyen, L. T. 438 (1075), *1017*
Nicholls, A. C. 48 (175), *257*
Nickson, T. E. 328 (833), *1012*
Nicolaou, K. C. 13 (49), *254*, 271 (736), *1010*
Nicolaus, B. J. R. 139 (427), *262*
Nii, Y. 491 (1130), *1018*
Niki, I. 61, 66–68 (215), *258*
Nikishin, G. I. 17 (57), 18 (72), *255*, 337 (841), *1012*
Ninomiya, I. 144 (471, 473–477), 152 (471), 153 (475), 154 (474, 477), 155 (473, 476), *263*, 465, 468, 793 (1105), *1018*
Nio, N. 276, 352 (758), *1010*
Nisbet, M. A. 4, 6, 84, 86 (17), *254*
Nishida, A. 917 (1441), *1026*
Nishide, K. 827 (1427), *1025*
Nishio, K. 351 (861), *1012*
Nishitani, Y. 713 (1368), 760, 770 (1368, 1403), 846, 883, 891 (1368), 947 (1368, 1403), *1024, 1025*
Nishiyama, H. 316, 339 (815), *1011*
Nitta, H. 416, 417 (1038), 638 (1270), 930 (1038), *1017, 1022*
Nitta, Y. 739 (1391), 798 (1418), *1025*
Nobuo, I. 397 (938), *1014*
Noguchi, H. 288 (783), *1011*
Noguchi, Y. 419 (1042), *1017*
Nokami, J. 397 (918), *1014*
Nolen, R. L. 58, 59 (199), *258*
Nomura, M. 957, 960 (1449), *1026*
Normant, H. 717, 812, 849 (1372), *1024*
Norris, W. J. 397 (923), *1014*
Nossin, P. M. M. 783 (1409), *1025*
Novak, L. 47 (169), *257*
Novoselov, E. F. 640 (1282), *1022*
Noyari, R. 124 (351, 352), 125 (352), *261*
Noyce, D. S. 83, 85 (250), *259*
Noyori, R. 314, 352, 359 (821), *1011*
Nozaki, H. 124 (351, 352), 125 (352), *261*, 351 (860, 861), *1012*
Nunes, B. J. 452, 657, 962 (1090), *1018*
Nussbaum, A. L. 83, 97 (256), *259*
Nyitrai, J. 692 (1347), *1024*
Nyss, N. L. 139 (438), *262*

Obana, M. 312 (807), *1011*
Oberender, H. 103, 106 (298), *260*
Obrecht, D. 282 (769), *1010*
Obrecht, R. 1003 (1487), *1027*
O'Brien, J. B. 144 (472), *263*
Obrzut, M. L. 328, 357 (832), *1012*
Ochiai, H. 343, 344 (853), *1012*
Ochiai, M. 278 (763), 341 (848), 404, 406, 678, 693, 737, 897, 947 (1021), *1010, 1012, 1016*
Oettingen, W. F. von 61 (202), *258*
Ogasawara, K. 76, 82 (247), *259*
Ogata, M. 144, 147–149 (460), *263*
Ogata, Y. 144, 145, 149 (461), *263*
Ogiso, A. 288 (783), *1011*
Ogura, H. 397 (909), 993 (1476), *1013, 1027*
Oguri, T. 321 (825), *1012*
Ohashi, T. 677, 729, 898 (1318), *1023*
Ohba, M. 969 (1460), *1026*
Ohfune, J. 39 (143), 98, 106 (282), *256, 259*
Ohler, E. 51, 52 (184), *257*

Ohmizu, H. 371 (875), *1013*
Ohmori, M. 469 (1106), *1018*
Ohmura, N. 962 (1455), *1026*
Ohnishi, Y. 424, 425 (1057), *1017*
Ohno, M. 106, 108 (310), *260*, 300, 309, 310 (798), 397 (961), 404 (1013), 405 (798, 1013), 491 (1130), 642 (1284), *1011, 1015, 1016, 1018, 1022*
Ohnuma, T. 397 (967), *1015*
Oho, J. C. 367 (869), *1012*
Ohshiro, Y. 168 (514), *264*, 342 (851), 545 (1192), 593, 594, 601, 602 (1238), 605, 606 (1242), 607, 608 (1238, 1242), 876, 877 (1238), 889, 890 (1192), 911 (1238), *1012, 1020, 1021*
Ohsuki, S. 339 (844), *1012*
Ohtsuka, Y. 424 (1055), *1017*
Oishi, T. 415 (1032), 424 (1055), *1016, 1017*
Ojima, I. 507 (1157), 508 (1158), 515 (1157), 516 (1158, 1169, 1170), 517 (1171, 1172), 518 (1172), 524 (1169, 1171), 525 (1171), 527 (1181), 590 (1233, 1234), 591 (1235, 1236), 611 (1235), 696 (1181), 794 (1415), 813 (1158), 814 (1158, 1170, 1172), 819 (1171), 820, 821 (1169, 1171), 828 (1169), 831 (1171), 832 (1169, 1171), 838 (1158, 1169–1172), 841 (1169, 1171), 842 (1171), 843 (1169), 852 (1181), 881 (1171), 981 (1466), *1019–1021, 1025, 1026*
Okada, K. 713 (1368), 760 (1368, 1403), 766 (1404), 770 (1368, 1403), 844 (1404), 846, 883, 891 (1368), 925–927 (1404), 947 (1368, 1403), 971 (1404), *1024, 1025*
Okada, T. 124, 125 (352), *261*
Okado, M. 271 (741), *1010*
Okano, K. 397 (961), 491 (1130), *1015, 1018*
Okawara, M. 314 (822), *1012*
Okawara, T. 139, 141 (443), *263*, 419 (1042), 424, 425 (1058), 616 (1251), 997 (1481), 998 (1481, 1482), *1017, 1021, 1027*
Okinaga, N. 401, 403 (1004, 1005), *1016*
Okita, M. 424, 425 (1059), 433, 712, 714 (1067), 716 (1059), 719 (1375), 984 (1471), 985, 986 (1059, 1067, 1471), 987 (1067, 1471), 988 (1067), *1017, 1024, 1026*
Okitsu, M. 476 (1112), *1018*
Okraglik, R. 236, 237 (673), 240 (688), *267, 268*
Oku, T. 532, 850 (1188), *1020*
Okuno, Y. 144, 156 (480), *263*
Olesen, S. O. 621, 807 (1255), *1021*
Oliveros-Desherces, E. 230 (665, 666), 232 (666), 233 (665, 666), *267*
Oliveto, E. P. 83, 97 (256), *259*
Omaima, M. A. E.-H. 664 (1303), *1022*
O'Malley, G. J. 382 (885), *1013*
Omote, Y. 465 (1104), 471 (1108, 1109), 494 (1104), 495 (1136), 872 (1109), 897 (1104), 1001 (1109), *1018, 1019*
Ona, H. 713, 760, 770, 846, 883, 891, 947 (1368), *1024*
Onan, K. D. 626 (1260), *1021*
Onoue, H. 766, 844, 925–927, 971 (1404), *1025*
Oostween, A. R. C. 785–788, 955, 976–978, 980 (1412), *1025*
Oota, Y. 44, 45 (156), *257*
Openshaw, H. T. 137 (409), *262*
Opitz, G. 166 (508), *264*
Ordsmith, N. H. R. 631, 634 (1265), *1021*
Oripov, E. 695 (1352), *1024*
Orito, K. 104, 106, 125 (307), 247 (710), *260, 268*
Orlova, N. A. 681 (1329), *1023*
Orlova, N. K. 750 (1399), *1025*
Ornaf, R. M. 435 (1070), *1017*
O'Rorke, H. 11, 53 (41), *254*
Ortuno, R. M. 283 (771), *1010*
Osada, A. 644 (1285), *1022*
Osakada, K. 310 (803), *1011*
Osakoda, K. 312 (807), *1011*
Osborne, S. 404 (1017), 648, 652 (1017, 1294), 653, 666, 667 (1017), 668 (1017, 1294), 669 (1294), 684 (1017), *1016, 1022*
Osman, A. M. 533–535, 537, 817, 852 (1189), *1020*
O'Sullivan, J. 811 (1429), *1025*
O'Sullivan, M. J. 163–165 (500), *264*
Otake, M. 799 (1419), *1025*
Otsuji, Y. 962 (1455), *1026*
Ott, A. C. 83 (269), *259*
Otto, H. H. 696 (1354), *1024*
Oumar-Mahamat, A. 337 (842), *1012*
Ourisson, G. 61, 65 (213), 106 (306), *258, 260*
Ovcharova, I. M. 397 (903), 725, 812, 837, 980 (1379), 981 (903), *1013, 1024*
Overberger, C. G. 3 (14), 99, 106 (283), 214 (629), *254, 259, 266*
Owen, J. S. 208 (611), *266*
Oyama, M. 106, 108 (310), *260*
Ozaki, N. 730, 754, 875 (1385), *1024*
Ozaki, S. 970 (1462), *1026*

Pagani, G. 136 (391), *262*
Pai, B. R. 243 (701), *268*
Palecek, J. 673 (1314), *1023*
Palmer, M. H. 3, 16 (5), *253*
Palomo, C. 381 (884), 490 (1146), 546 (1218), 549 (1198), 551 (1200, 1201), 552–554 (1201), 555 (1198, 1204), 556, 561, 563, 567 (1198), 568 (1198, 1204), 569 (1198, 1204, 1216), 570 (1198, 1201, 1204,

1216–1218), 571 (1198, 1217, 1218), 572 (1198, 1204, 1217), 575 (1198, 1204, 1221), 576 (1221), 577, 579 (1198, 1204), 717, 719 (1221), 726 (1201), 728 (1146), 730 (1146, 1201), 734 (1201), 742 (1217, 1218), 751 (1146, 1217), 762 (1146), 812 (1204), 813, 848 (1217), 849 (1204), 872, 874 (1218), 898 (1146), 934 (1146, 1221), 954, 955 (1146), *1013, 1019, 1020*
Panda, C. S. 704 (1360), *1024*
Pandey, G. D. 271 (745), *1010*
Pandey, P. N. 99, 103, 104, 106 (285), *259*
Panfil, I. 402, 409 (1009), *1016*
Pankratenko, G. A. 679 (1324), *1023*
Panov, V. P. 725 (1381), *1024*
Pantini, G. 397 (959, 960), *1015*
Panunzro, M. 614 (1247), *1021*
Papa, D. 83, 86 (262), *259*
Pappo, R. 83, 85, 97 (255), *259*
Paredes, M. C. 252 (721), *268*
Parello, J. 230 (665, 666), 232 (666), 233 (665, 666), *267*
Parker, J. 128, 129 (356), 144 (488, 490, 491), 159, 160 (488), *261, 263, 264*
Parker, W. 144 (456), *263*
Parreno, V. 503, 528, 536, 734, 882 (1148), *1019*
Parrini, V. 76–78, 83 (239), *258*
Parry, F. H. III 169, 170, 172–174, 177, 182 (516), *264*
Parsons, P. J. 672, 673 (1312), 778, 779 (1406), 788, 789 (1312), *1023, 1025*
Partain, E. M. III 993 (1477), *1027*
Parthasarathy, R. 203, 205 (606), *266*
Pascual, J. 18 (68–71, 73, 75, 76), 20 (68, 76), *255*
Pasini, C. E. 648, 652, 657, 830 (1295), *1022*
Patchett, A. A. 665, 781, 894, 895 (1307), *1023*
Paterson, I. 360, 363, 383 (865), *1012*
Patrie, W. J. 315 (814), *1011*
Pattenden, G. 314 (823), *1012*
Patterson, D. R. 366 (866), *1012*
Patterson, J. W. 11, 12, 87, 88 (45), *254*
Patterson, L. E. 38, 39 (139), *256*
Patterson, R. T. 292 (789), *1011*
Paul, L. 170 (553–555, 584), 185, 188 (553), 189 (554), 191 (555), 195 (584), *265*
Paull, K. D. 16 (54), *254*
Pavlik, J. W. 956 (1448), *1026*
Pavlov, S. 583 (1226), *1021*
Pawda, A. 500, 508, 509, 567, 568, 637, 817, 819, 848, 849 (1143), *1019*
Payne, T. G. 30 (115), *256*
Peach, J. M. 367 (869), *1012*
Pearlman, W. M. 133 (380), *261*
Pearson, C. J. 462 (1098), 627–629 (1261, 1262), 630 (1262), 648, 651, 869 (1261), 992 (1262), *1018, 1021*
Pearson, M. J. 397 (924), 547 (1195), 548 (1196), 550, 664 (1195), 686 (1196, 1335), 687 (1337), 688, 708, 736 (1196), 746 (1196, 1337), 747 (1337), 750 (1196, 1337), 751 (1337), 771 (1196, 1335, 1337), 774, 775 (1196, 1335), 776 (1337), 815 (1196), 830 (1428), 840 (1196), 903, 907 (1337), 932, 934, 935, 942 (1196), 960, 961 (1335), *1014, 1020, 1023, 1025*
Pechere, J. C. 397 (925), *1014*
Pechet, M. M. 97 (279), *259*
Pechmann, H. von 36 (124), *256*
Pedersen, U. 804, 805 (1422), *1025*
Pederson, R. L. 83 (269), *259*
Peitzsch, W. 11, 53 (43), *254*
Pelczer, I. 537, 538, 543, 544, 883 (1190), *1020*
Peled, N. 463 (1100), *1018*
Pellegata, R. 413 (1030), *1016*
Pelletier, S. W. 34 (120), *256*
Penczek, S. 271 (740), *1010*
Pennings, M. L. M. 965, 968 (1457), *1026*
Percheron, F. 244 (706), *268*
Perelman, M. 162 (493), 166 (493, 507), *264*
Perera, C. P. 991 (1472), *1026*
Perez-Garcia, V. 648, 651 (1292), *1022*
Perez-Ossorio, R. 491–493, 497, 499, 503, 504, 528, 536, 541, 545 (1128), *1018*
Perkins, M. J. 144, 145, 149, 150 (463), *263*
Perkins, R. J. 83, 90 (271), *259*
Perlow, D. S. 399, 452 (1000), *1016*
Perrone, E. 397 (947), 685 (1342), 689 (1341, 1342), 713, 743 (1341), 748 (1341, 1342, 1397), 752 (1341, 1397), 753 (1342), 761, 762 (1341, 1397), 764 (1397), 768 (1341, 1397), 769 (1342), 915 (1397), *1014, 1023, 1025*
Pesa, F. A. 415 (1034), *1016*
Peterson, E. R. 665, 781, 894, 895 (1307), *1023*
Peterson, P. A. 134 (387), *261*
Petri, N. 640 (1276), *1022*
Petrzilka, M. 118 (337), *260*
Petterson, R. C. 103, 106 (296), *260*
Pettit, G. R. 16 (54), 39 (140), 40, 41 (145), *254, 256*
Petyunin, G. P. 144 (454), *263*
Petyunin, P. A. 144 (454), *263*
Pfaendler, H. R. 728, 759, 821, 830, 844, 900–902 (1382), *1024*
Pfau, M. 18, 22 (84, 85), *255*
Pfleger, R. 170, 172–175, 178, 179 (523), *264*
Phadke, R. 36 (122), *256*
Piancatelli, G. 323 (827), *1012*

Piasek, E. J. 47 (165), *257*
Picard, C. 394 (897), *1013*
Pichat, L. 53 (190), *257*
Picot, A. 241 (696), *268*
Piers, E. 61 (210), *258*
Pietra, S. 1005 (1491), *1027*
Pietrzak, J. 271 (744), *1010*
Pietsch, H. 669–672, 674, 778, 779 (1311), *1023*
Pifferi, G. 413 (1030), *1016*
Pilgrem, R. 397 (915), *1013*
Pinner, A. 240 (690–692), *268*
Pinnick, H. W. 385 (889), *1013*
Pinto, P. 726, 756, 758 (1398), *1025*
Pinza, M. 413 (1030), *1016*
Piotrowska, K. 581 (1224), *1021*
Piovera, E. 132, 184 (371), *261*
Pipe, D. 491, 635, 636, 799 (1129), *1018*
Pirkle, W. H. 613, 676, 692, 739, 899, 905, 934, 935 (1245), *1021*
Piskov, V. B. 36 (132), *256*
Plat, M. 244 (705), *268*
Plieninger, H. 142 (447), *263*
Plumet, J. 491–493, 497, 499 (1128), 503 (1128, 1148), 504 (1128), 528, 536 (1128, 1148), 541, 545 (1128), 648, 651 (1292), 734, 882 (1148), *1018, 1019, 1022*
Pohmakotr, M. 354, 363, 375 (863), *1012*
Poignant, S. 407 (1022), *1016*
Poindexter, G. S. 661 (1301), *1022*
Polczynski, P. 170, 195 (584), *265*
Polievktov, M. K. 626 (1259), *1021*
Polley, J. A. S. 124 (348), *261*
Polomo, C. 563–566, 726, 728, 750 (1213), *1020*
Polonski, T. 464, 944 (1103), *1018*
Ponticello, I. S. 76, 81 (245), *258*
Popjak, G. 108 (313), *260*
Popkov, Y. A. 394 (896), *1013*
Porikh, A. R. 539 (1191), *1020*
Porte, A. L. 17 (59), *255*
Portius, H. J. 397 (949), *1014*
Posner, G. H. 61, 64, 65 (211), *258*
Potier, P. 242 (700), *268*
Potoski, J. R. 103, 106 (297), *260*
Poulain, C. 640 (1278), *1022*
Poulter, C. D. 300 (799), *1011*
Poupart, M. A. 283 (773), 408, 410 (773, 1026), *1011, 1016*
Powell, G. 36 (130), *256*
Poynton, A. J. 236, 237 (673), 240 (688), *267, 268*
Poyser, J. P. 397 (915), *1013*
Pradere, J. P. 407 (1022), *1016*
Prager, R. H. 214, 220 (640), 222 (640, 648), 224, 225 (640), 226, 227 (640, 648), 229 (640), 233 (648), 238, 240 (640, 648), *267*
Prasad, K. 711, 729 (1367), 791, 792 (1414), 917 (1367), *1024, 1025*
Pratt, R. 51 (185), *257*
Pratt, R. N. 170 (530), *264*
Preiss, M. 681, 834 (1331), *1023*
Presslitz, J. E. 994 (1479), *1027*
Prewo, A. 490 (1147), *1019*
Procter, G. 631, 634 (1265), *1021*
Proctor, S. A. 170 (530), *264*
Prodan, K. A. 515–525 (1167), *1019*
Protiva, M. 47 (169), *257*
Prout, K. 367 (869), 730, 881, 981 (1386), *1012, 1024*
Puar, M. S. 238 (679), *267*, 626 (1260), *1021*
Pucknat, J. 51 (185), *257*
Pudova, T. A. 130 (363), *261*
Pulwer, M. J. 371 (874), *1013*
Purcell, T. C. 61 (205, 206), *258*
Puterbaugh, W. H. 143 (448, 449), *263*

Qiu, X. 794 (1415), *1025*
Quan, P. M. 241 (693), *268*
Quayle, P. 613 (1246), 737, 904 (1390), *1021, 1025*
Quin, L. D. 241 (693, 695), 242 (695), *268*
Quiniou, H. 407 (1022), 993 (1475), *1016, 1027*
Quinn, A. M. 678, 688, 690, 737, 738, 748, 749, 752–754, 767 (1319), *1023*

Rademacher, H. 640 (1281), *1022*
Rademacher, P. 804, 807, 808, 859 (1421), *1025*
Radscheit, K. 39 (141), *256*
Raghu, S. 166 (510), *264*
Raileanu, D. 134 (386), *261*
Raimahajan, J. 271, 397 (725), *1009*
Rajappa, S. 243 (701), *268*
Rajendra, G. 796, 992 (1416), *1025*
Ram, B. 397 (910), 521, 548 (1178), 556 (1178, 1206), 557 (1178, 1207), 558–560 (1178), 563–565 (1212), 577 (1222), 578 (1222, 1223), 579 (1178, 1223), 726, 728, 740, 813, 819, 822–824, 838, 849, 850, 853, 856 (1178), *1013, 1019–1021*
Ramer, R. M. 142 (445), *263*
Ramraj, S. K. 640 (1283), *1022*
Ramsay, M. V. J. 144 (492), *264*
Rando, R. R. 144, 159 (486), *263*
Rango, C. de 203, 205 (606), *266*
Rao, A. S. 10 (32), 83 (260), *254, 259*
Rao, K. L. 640 (1283), *1022*
Rao, M. N. 477, 478, 918 (1115), *1018*
Rao, V. V. 139 (439, 440, 442), 140 (439), 141 (440), *262, 263*
Rao, Y. S. 3 (4, 12), 18, 21 (79), *253–255*
Raphael, R. A. 144 (456), *263*

Rapoport, H. 639, 979 (1271), *1022*
Rasmussen, J. B. 621, 807 (1255), *1021*
Rasmusson, G. H. 109 (315), *260*
Ratcliffe, R. W. 170 (567, 590), 183 (590), 195, 196 (567), *265, 266*, 397 (940), *1014*
Rathke, M. W. 53 (191), *257*
Rathore, R. 327 (831), *1012*
Ratnikova, T. N. 123 (342), *260*
Ravindranathan, T. 104, 106 (303), *260*
Ray, J. A. 358 (864), *1012*
Ray, P. S. 395 (899), *1013*
Ray, S. J. 717, 719 (1373), *1024*
Razavi, Z. 272, 274, 302, 303, 337 (751), *1010*
Reddy, G. J. 577, 578 (1222), *1021*
Reden, J. 827 (1426), *1025*
Reich, H. J. 61 (214), *258*
Reid, S. T. 233 (669), *267*
Reifegeiste, D. 993 (1478), *1027*
Reimschuessel, H. K. 137 (411–413), *262*
Reinhoudt, D. N. 420, 421, 796 (1048), 965 (1457), 968 (1048, 1457), *1017, 1026*
Reininger, K. 51, 52 (184), *257*
Reisch, J. 30 (110), *256*
Reliquet, A. 993 (1475), *1027*
Reliquet, F. 993 (1475), *1027*
Remanick, A. 27 (97), *255*
Remuson, R. 723 (1377, 1378), 741, 814, 849, 931, 935, 937 (1377), 953 (1378), *1024*
Renga, J. H. 61 (214), *258*
Repke, K. R. H. 397 (949), *1014*
Reppe, W. 61 (208), *258*
Reshetova, I. G. 271 (746), *1010*
Reuschling, D. 669–672, 674, 778, 779 (1311), *1023*
Reymond, D. 17 (62), *255*
Reynolds, G. F. 109 (315), *260*
Ribaldone, G. 233 (670), 234 (670, 671), *267*
Richards, R. W. 222 (653), *267*
Richmond, M. H. 397 (948), *1014*
Ridley, D. D. 210, 211 (615, 616), *266*
Rieche, A. 249 (714), *268*
Ried, W. 69, 74 (225), *258*
Riegl, J. 166 (511), *264*
Riguera, R. 711 (1366), *1024*
Ritchie, E. 242, 244, 247 (699), *268*
Riviere, M. 230 (665, 666), 232 (666), 233 (665, 666), *267*
Roberts, D. 648 (1289), *1022*
Roberts, E. F. 437, 648 (1074), *1017*
Roberts, J. L. 300 (799), *1011*
Robson, C. A. 397 (923), *1014*
Rodewald, P. G. 377 (878, 879), *1013*
Rodrigo, R. 104, 106, 125 (307), 247 (710), *260, 268*
Rodriguez, A. 500, 508, 509, 567, 568, 637, 817, 819, 848, 849 (1143), *1019*
Rodriques, K. E. 435 (1070), *1017*
Rodriquez, O. 130 (359), *261*
Rogalska, E. 617–620 (1253, 1254), *1021*
Root, W. G. 130 (361), *261*
Rosebery, G. 272, 274, 302, 303, 337 (751), *1010*
Rosenberger, M. 48 (173, 174), *257*
Rosenblum, M. 166 (509, 510), *264*, 386 (890), *1013*
Rosenmund, P. 214, 220, 222, 225 (641), *267*
Ross, W. J. 397 (912), *1013*
Rothe, J. 61 (207), *258*
Rothfield, M. 170, 180 (539), *264*
Rothman, E. S. 83, 89, 98 (268), *259*
Roy, S. K. 139 (440, 441), 141 (440), *263*
Roze, J.-C. 407 (1022), *1016*
Rubottom, G. M. 4, 7, 87 (22), *254*
Rucktaschel, R. 130 (359), *261*
Rudinger, J. 131 (365), *261*
Rudnick, L. R. 13 (51), *254*
Rudolph, W. 39 (141), *256*
Rudon, H. N. 48 (175), *257*
Ruggeri, R. B. 434 (1068), *1017*
Ruitenberg, K. 394 (898), *1013*
Russell, A. 36, 37 (126), *256*
Russell, M. A. 476 (1113), *1018*
Russell, R. R. 42 (148), *257*
Russo, G. 127 (354), *261*
Rutenberg, M. W. 142 (446), *263*
Rutledge, P. S. 616 (1252), *1021*
Ryan, J. J. 169 (519), 170 (519, 578, 581), 171, 184 (519), 193 (578, 581), 201 (519), *264, 265*
Ryazantsev, G. B. 725 (1381), *1024*
Rydon, H. N. 108 (312), *260*
Rylander, P. N. 109 (314), 245 (707), *260, 268*
Rytslin, E. E. 140 (430), *262*

Saakamoto, M. 495 (1136), *1019*
Saburi, M. 310 (806), 312 (807), 319 (806), *1011*
Sacripante, G. 859, 861, 865, 894 (1431), *1026*
Sada, I. 677, 729, 898 (1318), *1023*
Saesmaa, T. 397 (958), *1015*
Safarova, Z. A. 28 (101), *256*
Sagyo, K. 489 (1123), *1018*
Sahoo, S. 515, 736, 811, 812, 835, 837, 854 (1168), *1019*
Sahoo, S. P. 404, 405 (1016), *1016*
Sahu, D. P. 422 (1052), 439–442 (1076), 557 (1208), 695 (1076), 856 (1208), 938–940 (1052), *1017, 1020*
Saigo, K. 279 (766), *1010*
Saikawa, I. 440–442, 444, 445 (1077), 685, 743 (1343), 818, 826, 845 (1077), 895, 896, 908, 909 (1343), *1017, 1023*
Sain, B. 625 (1258), *1021*
Sainsbury, M. 243, 250 (702), *268*

Saito, R. 897 (1435), *1026*
Saitoh, N. 397 (918), *1014*
Sakai, H. 397 (929), *1014*
Sakai, I. 642 (1284), *1022*
Sakai, K. 898, 904 (1437), *1026*
Sakamoto, M. 465 (1104), 471 (1108, 1109), 494 (1104), 872 (1109), 897 (1104), 1001 (1109), *1018*
Sakamura, S. 276, 352 (758), *1010*
Sakan, T. 17 (61), *255*
Samatuga, G. A. 3 (8), *253*
Sammes, P. G. 397 (974), *1015*
Sanchez-Ferrando, F. 287, 288, 294, 357, 359, 360, 363 (779), *1011*
Sandberg, R. 49 (176), *257*
Sandhu, J. S. 615 (1249), 625 (1258), *1021*
Sandmeier, R. 105, 106 (308), *260*
Sane, P. P. 83 (260), *259*
Sano, H. 112 (334), *260*
Santiago, M. L. 647 (1287), *1022*
Sanyal, B. 99, 100, 106 (286), *259*
Sarel, S. 28 (103, 104), 30 (104, 106), *256*, 475 (1120, 1121), 487 (1120), *1018*
Sarimoto, H. 351 (861), *1012*
Sarkisyan, O. A. 13 (48), *254*
Sasaki, A. 492 (1132), 664, 706 (1305), 736 (1132), 830, 843, 854, 855 (1305), 886 (1132), *1019, 1022*
Sasaoka, M. 397 (918), *1014*
Sasson, S. 519, 780, 781, 922 (1175), *1019*
Sataty, I. 463 (1100), *1018*
Sato, M. 271 (747), *1010*
Sato, T. 271 (729), 314, 352, 359 (821), 381 (883), *1009, 1011, 1013*
Sato, Y. 474 (1111), *1018*
Satoh, Y. 756, 758, 844–846, 956, 958 (1405), *1025*
Saucy, G. 48 (172–174), *257*
Sauer, D. 214, 220, 222, 225 (641), *267*
Sauers, R. R. 100 (290, 291), 101 (291), 106 (290, 291), 123 (340), 125 (290, 291), *259, 260*
Scaglioni, L. 695, 798 (1351), *1024*
Scarafile, C. 397 (947), *1014*
Scettri, A. 323 (827), *1012*
Schaaf, T. K. 106 (302), *260*
Schaus, J. M. 648, 650 (1291), *1022*
Scheckenbach, F. 170 (541, 544), 180, 181 (541), *264, 265*
Scheeren, J. W. 275, 276 (756), *1010*
Scheinbaum, M. L. 171 (600), *266*
Schelechow, N. 239 (681), *268*
Schenck, G. O. 18, 22 (83), 109, 110 (317, 319, 320), 170, 202 (595), *255, 260, 266*
Schenk, K. H. 635, 897 (1268), *1021*
Scheunemann, K. H. 827 (1426), *1025*
Schlecht, M. F. 326 (830), *1012*
Schlessinger, R. H. 61, 64, 65 (212), 76, 81 (245), *258*
Schlosser, M. 317 (818), *1011*
Schmid, G. 554 (1203), *1020*
Schmid, H. 243 (701), *268*
Schmidt, C. A. 729, 764 (1384), *1024*
Schmidt, E. 132, 163 (370), *261*
Schmidt, G. 339, 341 (847), *1012*
Schmidt, P. 134 (384), *261*
Schmidt, U. 51, 52 (184), *257*
Schmitt, G. J. 137 (413), *262*
Schmitz, E. 232, 233 (668), *267*
Schmitz, W. R. 133 (381), *261*
Schmuff, N. R. 637, 638 (1269), *1022*
Schniepp, L. E. 83, 89 (266), *259*
Schoellkopf, U. 275 (755), *1010*
Schoemaker, H. E. 784, 785 (1411), 791 (1413), 973, 975 (1411, 1413), *1025*
Schofield, C. J. 404, 406, 704 (1018, 1019), 811 (1019), 945, 948 (1444), *1016, 1026*
Schonfield, C. J. 399, 756, 757, 824, 847 (999), *1016*
Schramm, J. 497, 498 (1139), *1019*
Schramm, S. 232, 233 (668), *267*
Schrecker, A. W. 19 (92, 93), *255*
Schrinner, E. 827 (1426), *1025*
Schroeck, W. 681, 834 (1331), *1023*
Schroeter, S. H. 109, 110 (319), 129 (358), *260, 261*
Schuber, E. V. 36 (135), *256*
Schulte, K. W. 30 (109), *256*
Schultze, H. 249 (714), *268*
Schulz, G. 693 (1348), 711 (1367), 728 (1348), 729 (1367), 760, 762, 763 (1348), 791, 792 (1414), 801, 908 (1348), 917 (1367), *1024, 1025*
Schulze, K. E. 30 (110), *256*
Schulze-Steinen, H.-J. 112, 116, 117 (331), *260*
Schute-Elte, K.-H. 109, 110 (317), *260*
Schwartz, E. 132 (368), *261*, 390 (894), *1013*
Schwenk, E. 83, 86 (262), *259*
Scopes, D. I. C. 518, 818, 819, 839 (1174), *1019*
Scrimin, P. 424 (1056), *1017*
Sebenda, J. 134 (384, 385), *261*, 398 (992, 993, 995), *1015, 1016*
Sebti, S. 427 (1060, 1061), 429, 446 (1061), 448, 450, 451, 755 (1060), 891 (1061), *1017*
Secor, H. V. 13, 17 (52), *254*
Sedov, V. V. 725 (1381), *1024*
Sedriz-Hibner, D. 214, 219 (639), *266*
Seebach, D. 36 (121), *256*
Seidel, M. C. 100, 106 (289), *259*
Sekiya, M. 402, 708, 733 (1007), *1016*
Selbert, G. 827 (1426), *1025*

Self, C. R. 271 (742), *1010*
Senaratne, P. A. 402, 403 (1008), *1016*
Sendo, Y. 699, 761, 782 (1357), *1024*
Seno, K. 996 (1480), *1027*
Serokhvostova, V. E. 626 (1259), *1021*
Serratosa, F. 18 (71, 74, 76), 20 (74, 76), *255*
Seshadri, K. V. 133 (376), *261*
Sethna, S. 36 (122), *256*
Settatosa, F. 18 (70), *255*
Severo, V. 397 (926), *1014*
Shabana, R. 621, 807 (1255), *1021*
Shabanov, A. L. 11 (36), 28 (101), *254, 256*
Shafer, T. C. 83, 89 (263), *259*
Shakhidoyatov, K. M. 695 (1352), *1024*
Shalon, Y. 28 (103, 104), 30 (104, 106), *256*
Shamma, M. 19, 27 (94), *255*
Shani, A. 144, 161, 162 (468), *263*
Shanzer, A. 390 (894), *1013*
Shapkin, V. A. 394 (896), *1013*
Sharkey, W. H. 271 (739), *1010*
Sharma, S. D. 170 (557, 563, 576), 187 (563), 188, 189 (557), 197 (576), *265*, 463 (1099), 490 (1145), 509 (1159), 511 (1099), 512 (1163, 1164), 513 (1164), 514 (1159), 523, 542 (1145), 547, 548 (1194), 549, 550 (1199), 566 (1214), 567 (1159, 1214, 1215), 568 (1159, 1215), 807 (1099), 817 (1214, 1215), 821, 843 (1145), 848 (1159, 1214, 1215), 850 (1214, 1215), 918 (1099), 945 (1164), *1018–1020*
Sharrard, F. 993 (1475), *1027*
Shea, C.-M. 644 (1285, 1286), *1022*
Shechter, H. 222, 224, 225 (646), *267*
Sheehaan, J. C. 170 (577–582), 184 (580), 193 (577–582), *265*
Sheehan, J. C. 132 (367, 373), 138 (421, 422, 432, 433), 139 (432, 433), 163 (367), 169 (519), 170 (519, 585), 171, 184 (519), 196 (585), 201 (519), 210 (614), 245 (708), *261, 262, 264–266, 268*, 397 (983), *1015*
Sheinker, Y. N. 397, 981 (903), *1013*
Shen, T. Y. 111, 117 (330), *260*
Sheppard, A. 397 (973), *1015*
Sheppard, L. N. 759 (1402), *1025*
Sheppard, R. C. 83 (274, 276), 92 (274), 93 (274, 276), *259*
Sheradsky, T. 639, 881 (1272), *1022*
Sherodksy, T. 632, 911 (1266), *1021*
Shibasaki, M. 402, 405, 406 (1006), 917 (1441), 992 (1473), *1016, 1026*
Shibato, K. 273 (754), *1010*
Shibib, S. M. 876 (1432), 878 (1432, 1433), 879, 976, 977 (1432), *1026*
Shibuya, S. 249 (713), *268*, 301 (800), 395, 396 (800, 901), 402 (901), 710, 890, 891 (800, 901), *1011, 1013*
Shieh, C. H. 496, 499, 541, 542, 865 (1138), *1019*
Shigehisa, J. 389 (892), *1013*
Shimada, E. 279 (766), *1010*
Shimada, Y. 288 (783), *1011*
Shimagaki, M. 415 (1032), *1016*
Shindai, S. 397 (930), *1014*
Shinkai, I. 943 (1443), *1026*
Shinoda, M. 351 (861), *1012*
Shinozaki, H. 240 (684), *268*
Shiozaki, M. 272 (750), 422 (750, 1053), 423, 735, 737 (1053), 740 (750), 741 (750, 1053), 886, 894, 898 (750), 899 (750, 1053), 905 (1053), 908 (750), 929, 930, 933 (1053), 937, 938, 940 (750, 1053), 943 (1053), 947 (750), *1010, 1017*
Shipov, A. G. 681 (1329), 750 (1399), 1001 (1486), *1023, 1025, 1027*
Shono, T. 282 (770), 371 (875), 373 (876), 397 (913), 398 (985), 401, 403 (1004, 1005), 663 (1321), 716 (1369), *1010, 1013, 1015, 1016, 1023, 1024*
Shridhar, D. R. 577 (1222), 578 (1222, 1223), 579 (1223), *1021*
Shroot, B. 144 (456), *263*
Shusherina, N. P. 3, 28 (6), *253*, 397 (982), *1015*
Sianesi, I. L. 38 (137), *256*
Sicker, D. 993 (1478), *1027*
Sierra, M. 491–493, 497, 499, 503, 504, 528, 536, 541, 545 (1128), *1018*
Sifniades, S. 640 (1280), *1022*
Sigmund, S. K. 420 (1046), 684 (1334), 695, 824, 847 (1046, 1334), *1017, 1023*
Sillion, B. 315 (812), *1011*
Simig, G. 689 (1340), 692 (1347), 886, 933, 939–941 (1340), *1023, 1024*
Sims, J. J. 395 (900), *1013*
Sims, R. J. 299, 360, 366 (796), *1011*
Singh, A. 584 (1227), *1021*
Singh, A. K. 397 (932), *1014*
Singh, B. 584 (1227), *1021*
Singh, G. 566, 567, 817, 848, 850 (1214), *1020*
Singh, J. 999 (1483), *1027*
Singh, J. B. 802 (1420), *1025*
Sinha, N. D. 709 (1363), *1024*
Sircar, S. S. G. 76 (246), *259*
Siroi, T. 397 (918), *1014*
Sirt, P. J. 397 (915), *1013*
Sivaramakrishnan, R. 271 (742), *1010*
Sjoberg, B. 144, 158 (484), *263*
Sjogren, E. B. 404 (1015), 445, 446, 692 (1082), 731 (1015, 1082), 756, 757 (1082), *1016, 1017*
Skettebol, L. 166 (511), *264*
Sliwinski, J. 464 (1101), *1018*

Slomkowski, S. 271 (740), *1010*
Smai, F. 234 (671), *267*
Smille, R. D. 61 (210), *258*
Smith, C. W. 170, 179, 180 (531), *264*
Smith, D. J. H. 982, 983 (1468, 1469), *1026*
Smith, L. C. 54 (194), *258*
Smith, N. R. 34 (117), 36 (123), *256*
Smith, P. A. S. 106 (281), 212 (621), 222 (645), *259, 266, 267*
Smith, P. W. 367 (869), *1012*
Smith, R. M. 222 (653), *267*
Smuda, H. 335 (839), *1012*
Snatzke, G. 83 (249), *259*
Snider, B. B. 288, 290 (784), *1011*
Snyder, H. R. 212 (622), *266*
Snyder, J. D. 665, 781, 894, 895 (1307), *1023*
Snyder, J. P. 76, 79 (243), *258*
Sohar, P. 504, 513 (1150), 537 (1150, 1190), 538 (1190), 543, 544 (1150, 1190), 883 (1190), *1019, 1020*
Sohn, W. H. 27, 28 (100), *255*
Soja, P. 709 (1363), *1024*
Solanski, P. 479, 480 (1118), 485, 486 (1119), *1018*
Soloveva, N. P. 397, 981 (903), *1013*
Solzmann, T. N. 702, 703 (1358), *1024*
Soma, Y. 112 (334), *260*
Sonnet, P. E. 123 (340), *260*
Sonoda, T. 214 (626), *266*
Sorensen, A. K. 53, 54 (193), *258*
Sorensen, N. A. 18, 20 (66), *255*
Soto, A. 288 (783), *1011*
Soucy, P. 99, 106, 108 (284), *259*
Sousa, L. R. 630, 633 (1264), *1021*
Southgate, R. 398 (987), 678, 688, 690, 737, 738, 748, 749, 752–754, 767 (1319), 778, 779 (1406), *1015, 1023, 1025*
Spande, T. F. 144, 157 (482), *263*
Speckamp, W. N. 698, 699 (1356), 782 (1408), 783 (1409), 784 (1411), 785 (1411, 1412), 786–788 (1412), 791 (1413), 891, 892 (1356), 893 (1436), 910 (1438), 950, 951 (1356), 955 (1412), 972 (1356), 973 (1408, 1411, 1413), 974 (1463), 975 (1356, 1411, 1413, 1463), 976–978. 980 (1412), *1024–1026*
Speeter, M. 204 (604), *266*
Spickett, R. G. W. 42 (153), *257*
Spiegelman, G. 169 (520), 170 (520, 568, 570, 572, 573), 183 (520), 192 (568, 572), 194 (570, 573), 201 (520), *264, 265*
Spietschka, E. 170, 202 (596, 599), *266*
Spirkova, K. 419 (1044), *1017*
Spitzner, D. 346 (855), *1012*
Springer, J. P. 665, 781, 894, 895 (1307), *1023*
Srairi, D. 722, 740, 813 (1376), *1024*
Srivastava, R. C. 133, 163 (379), *261*
Stajer, G. 537, 538, 543, 544, 883 (1190), *1020*
Stamicarbon, N. V. 214, 215 (628), *266*
Staples, C. E. 18 (77, 78), *255*
Staudinger, H. 169 (515), 170 (515, 521, 527, 529, 533–536), 171 (515), 176 (521), 178 (533–536), 202 (515), 207 (610), *264, 266*
Stavely, H. E. 63, 68 (216), *258*
Stavholt, K. 18, 20 (66), *255*
Stehr, C. E. 83, 92, 93 (274), *259*
Steinberg, H. 123 (345), *261*
Steinberg, N. G. 83, 97 (278), *259*
Stekoll, L. H. 138 (423, 424), *262*
Steliou, K. 283 (773), 408, 410 (773, 1026), *1011, 1016*
Stepanova, A. A. 679 (1324), *1023*
Stepanyan, A. N. 13 (48), *254*
Stephanou, E. 647, 651 (1288), 652 (1296, 1297), 656 (1297), 659 (1296), 684 (1297), *1022*
Stephensen, R. 106 (301), *260*
Sternberg, V. I. 83, 90 (271), *259*
Stewart, J. C. 76, 81 (244), *258*
Stiles, M. 76 (227), *258*
Stille, J. K. 343 (852), *1012*
Stirchak, E. P. 278, 357 (761), *1010*
Stone, G. R. 36 (135), *256*
Stoodley, R. J. 397 (916, 956), 689, 713, 743, 748, 752, 761, 762, 768 (1341), *1013, 1015, 1023*
Stork, G. 211 (617), *266*, 290 (786), 331 (836), 357, 359, 364, 366 (786), *1011, 1012*
Storr, R. C. 494, 499, 648, 651, 652 (1133), *1019*
Strange, G. A. 616 (1252), *1021*
Streith, J. 491 (1127, 1129), 503, 504, 523, 526 (1127), 527, 531, 532 (1182), 536, 541, 542, 544 (1127), 635, 636 (1129), 756, 758 (1182), 799 (1129), 828, 834, 835, 843, 854, 856, 857 (1182), 1005 (1127), *1018, 1020*
Strekowski, L. 279 (767), *1010*
Strizhakov, O. D. 214, 216 (632), *266*
Stromberg, V. L. 214, 216, 220 (633), *266*
Stuetz, P. 711, 729 (1367), 791, 792 (1414), 917 (1367), *1024, 1025*
Sturgess, M. A. 585 (1231), 963 (1231, 1456), *1021, 1026*
Suarato, A. 731, 759, 762, 763, 900, 903, 907, 954 (1387), *1024*
Sudarsanam, V. 170, 195, 196 (571), *265*
Sudo, A. 69 (218), *258*
Suga, K. 54 (196), *258*, 378 (880), 379 (881), *1013*
Suga, S. 507 (1157), 508 (1158), 515 (1157), 516, 813, 814, 838 (1158), *1019*
Sugano, H. 131 (366), *261*

Sugi, H. 513 (1165), 529 (1165, 1183), 531 (1187), 646, 689, 740 (1183), 823, 828, 852 (1187), 918 (1165), *1019, 1020*
Suginome, A. 337 (843), *1012*
Suginome, H. 644 (1285, 1286), 661, 662 (1302), *1022*
Sugiura, M. 521, 548, 556–560, 579, 726, 728, 740, 813, 819, 822–824, 838, 849, 850, 853, 856 (1178), *1019*
Sugiyama, H. 371 (875), *1013*
Sukiasyan, G. G. 675, 700 (1322), *1023*
Sullivan, D. F. 518, 818, 819, 839 (1174), *1019*
Sultanbawa, M. V. S. 4 (27), *254*
Sumitomo, H. 271 (741), *1010*
Sumoto, K. 214, 215, 217 (624), *266*
Sunagawa, M. 492 (1132), 664, 706 (1305), 736 (1132), 739 (1392), 830, 843, 854, 855 (1305), 886 (1132), *1019, 1022, 1025*
Sundararamaiak, T. 640 (1283), *1022*
Sunday, B. R. 626 (1260), *1021*
Sunita, M. 509, 514, 567, 568, 848 (1159), *1019*
Surzur, J. M. 337 (842), *1012*
Suschitzky, H. 170, 194, 199 (591), *266*
Sutton, K. H. 730, 881 (1386), 981 (1386, 1467), *1024, 1026*
Suwinski, J. 640 (1273), *1022*
Suzuki, A. 18 (88, 89), 23 (88), 24, 25 (88, 89), 26 (89), *255*
Svergun, V. I. 725 (1381), *1024*
Sveshnikova, L. N. 272 (752), *1010*
Svetozarskii, S. V. 214, 216 (632), *266*
Svitanko, I. V. 337 (841), *1012*
Swain, S. 397 (928), *1014*
Sykes, R. B. 397 (937), *1014*
Szabo, A. E. 537, 538, 543, 544, 883 (1190), *1020*
Szabo, J. 504, 513, 537, 543, 544 (1150), *1019*
Szajewski, R. P. 211 (617), *266*
Szarek, W. A. 315, 952 (813), *1011*
Szczygielska-Nowasielska, A. 283, 408, 410 (773), *1011*
Szunyog, J. 537, 538, 543, 544, 883 (1190), *1020*

Tabo, Y. 168 (514), *264*
Tabuchi, T. 371 (873), *1013*
Tada, M. 240 (684), *268*
Tada, Y. 465, 468, 793 (1105), *1018*
Tadanier, J. 36 (135), *256*
Tadwalkar, V. R. 10 (32), *254*
Tagawa, H. 397 (936), *1014*
Taguchi, T. 18 (88, 89), 23 (88), 24, 25 (88, 89), 26 (89), *255*
Taillefer, R. J. 366 (866), *1012*
Takagi, K. 144, 145, 149 (461), *263*
Takahashi, H. 693, 729, 894 (1349), *1024*
Takahashi, T. 389 (893), *1013*
Takahashi, Y. 300, 309, 310 (798), 397 (961), 405 (798), 644 (1285), *1011, 1015, 1022*
Takahata, H. 424, 425 (1057), 1000 (1485), *1017, 1027*
Takano, S. 76, 82 (247), *259*
Takeda, A. 44, 45 (156), *257*
Takeda, K. 993 (1476), *1027*
Takeno, H. 532, 850 (1188), *1020*
Takeuchi, J. 222 (643), *267*
Takeuchi, M. 106 (292), *259*
Talaty, E. R. 138 (423, 424), *262*, 437, 648 (1074), *1017*
Tamaru, Y. 343, 344 (853), *1012*
Tamburin, H. J. 230 (663, 664), *267*
Tamelen, E. E. van 12 (47), 19, 27 (94), 42 (149, 151), 43 (47), *254, 255, 257*
Tamelen, E. van 34 (119), *256*
Tamm, C. 105, 106 (308), *260*
Tamura, N. 404, 406, 678, 693, 737, 897, 947 (1021), *1016*
Tamura, Y. 214 (624, 635, 637), 215 (624), 217 (624, 635, 637), 218 (635, 637), 219 (637), 222 (649, 650), 227 (649), 228 (650), *266, 267*, 429, 430 (1062), 448, 450 (1087), 470 (1107), *1017, 1018*
Tanabe, K. 83 (259), *259*
Tanaka, H. 397 (918), *1014*
Tanaka, K. 432 (1065), 440–442, 444, 445 (1077), 685 (1343), 739 (1391), 743 (1343), 818, 826, 845 (1077), 895, 896, 908, 909 (1343), *1017, 1023, 1025*
Tang, T. R. 416 (1035), *1016*
Tangari, N. 245, 247 (709), *268*, 952 (1445), *1026*
Tanigawa, K. 513 (1165), 529 (1165, 1183), 531 (1187), 646, 689, 740 (1183), 823, 828, 852 (1187), 918 (1165), *1019, 1020*
Tanno, T. 106 (304), *260*
Tantayanon, S. 956 (1448), *1026*
Tatevosyan, G. E. 42 (147), *257*
Taub, W. 11 (37), *254*
Taylor, C. M. B. 18 (81), *255*
Taylor, D. A. 632, 911 (1266), *1021*
Taylor, G. A. 170 (530), *264*
Taylor, T. W. J. 208 (611), *266*
Taylor, W. C. 242, 244, 247 (699), *268*
Taylor, W. I. 242 (697), *268*
Tcheng-Lin, M. 53 (189), *257*
Techer, H. 76, 78 (240), *258*
Terajima, A. 898, 904 (1437), *1026*
Terao, K. 460 (1095), *1018*
Terashima, M. 214, 217, 218 (635, 637), 219 (637), *266*
Terashima, S. 492, 736, 886 (1132), *1019*

Testa, E. 132 (369), 135 (401, 402), 136 (390–400), 139 (427), 184 (401), *261, 262*
Teutsch, G. 529, 678, 740, 797, 831, 832, 844, 851, 853, 855 (1184), *1020*
Thea, S. 624 (1257), *1021*
Theodoropoulas, D. 61 (206), *258*
Thiele, J. 61 (201), *258*
Thijs, L. 69 (226), *258*
Thomas, B. R. 222, 229 (655), *267*
Thompson, P. A. 328, 357 (832), *1012*
Thompson, R. D. 138 (420), *262*
Thomsen, I. 804, 805 (1422), *1025*
Thorsett, E. D. 665, 781, 894, 895 (1307), *1023*
Thottathil, J. K. 384 (887), *1013*
Thrum, H. 397 (950), *1014*
Thurston, D. E. 695, 891, 893, 979 (1353), *1024*
Thyagarajan, B. S. 144, 145, 149 (462), *263*
Tietjen, D. 76, 77 (238), *258*
Ting, H. H. 1005, 1007 (1492). *1027*
Tippett, J. M. 214, 220, 222, 224–227, 229, 238, 240 (640), *267*
Tischbein, R. 61 (201), *258*
Tischler, S. A. 299, 360, 366 (796), *1011*
Tisnes, P. 394 (897), *1013*
Tiwari, K. P. 271 (745), *1010*
Toczek, J. 287, 290, 357, 364, 367 (781), *1011*
Todo, Y. 440–442, 444, 445, 818, 826, 845 (1077), *1017*
Tokmakov, G. P. 710 (1364, 1365), *1024*
Tokuda, M. 18 (88, 89), 23 (88), 24, 25 (88, 89), 26 (89), *255*
Tokutake, N. 553 (1202), 574 (1220), 623 (1256), *1020, 1021*
Tokuyama, T. 144, 155, 156 (478), *263*
Toplitz, B. 239 (680), *267*
Torii, S. 397 (918), *1014*
Tortorella, V. 245, 247 (709), *268*
Toshimitsu, A. 460 (1095), *1018*
Toure, S. 1003 (1487), *1027*
Townley, E. 83, 97 (256), *259*
Townsend, C. A. 438 (1075), *1017*
Trachsler, D. 437 (1073), *1017*
Trammell, G. L. 4, 7 (21), *254*
Tramontano, A. 284 (775), *1011*
Trecker, D. J. 251 (718), *268*
Trehan, I. R. 802 (1420), *1025*
Tremper, A. W. 251 (717), *268*. 416 (1039), 420, 421 (1039, 1047), 708, 754, 777 (1039), 779 (1047), 794, 837, 881 (1039), 953 (1447), 960, 962 (1039, 1447, 1453), 1005 (1039), *1017, 1026*
Triska, J. 640 (1282), *1022*
Tristram, E. W. 665, 781, 894, 895 (1307), *1023*
Trod, E. 32, 33 (114), *256*
Troin, Y. 723 (1377, 1378), 741, 814, 849, 931, 935, 937 (1377), 953 (1378), *1024*
Tromeur, M. C. 83 (275), *259*
Trommer, W. 214, 220, 222, 225 (641), *267*
Trost, B. M. 102 (294, 295), 103 (295), 106 (294, 295), 238 (677, 678), *259, 267*, 647 (1287), *1022*
Troyanskii, E. I. 337 (841), *1012*
Tsai, C.-C. 214, 216, 227 (638), *266*
Tsai, M. 170 (563, 564), 187 (563), 188 (564), *265*
Tschamber, T. 491 (1127, 1129), 503, 504, 523, 526 (1127), 527, 531, 532 (1182), 536, 541, 542, 544 (1127), 635, 636 (1129), 756, 758 (1182), 799 (1129), 828, 834, 835, 843, 854, 856, 857 (1182), 1005 (1127), *1018, 1020*
Tsipouras, A. 613, 676, 692, 739, 899, 905, 934, 935 (1245), *1021*
Tsoucaris, G. 203, 205 (606), *266*
Tsubata, K. 401, 403 (1004, 1005), *1016*
Tsubor, S. 44, 45 (156), *257*
Tsuda, T. 83 (259), *259*
Tsuda, Y. 106 (304), *260*
Tsuji, J. 130 (362), *261*, 389 (893), *1013*
Tsuji, T. 693, 729 (1349), 760, 770 (1403), 894 (1349), 947 (1403), *1024, 1025*
Tsukida, K. 18 (64), *255*
Tsukoda, A. 404, 406 (1020), *1016*
Tsukogoshi, F. 378 (880), *1013*
Tucker, J. N. 233 (669), *267*
Tufariello, J. J. 402, 403 (1008), *1016*
Tulis, R. W. 245 (708), *268*
Tuller, F. N. 123 (344), *261*
Tunemoto, D. 827 (1427), *1025*
Tunick, A. A. 640 (1280), *1022*
Turner, N. J. 1005, 1007 (1492), *1027*
Turner, S. 83 (274, 276), 92 (274), 93 (274, 276), *259*, 397 (904), *1013*
Tyler, J. W. 547, 550, 664 (1195), *1020*

Uchido, T. 661, 662 (1302), *1022*
Uchino, T. 470 (1107), *1018*
Uemura, S. 460 (1095), *1018*
Ueno, Y. 314 (822), *1012*
Ugi, I. 397 (919), 416, 417 (1036), 1003 (1487), *1014, 1017, 1027*
Uhlig, F. 212 (622), *266*
Ukai, A. 106 (304), *260*
Ukita, T. 341 (848), *1012*
Ulm, E. H. 665, 781, 894, 895 (1307), *1023*
Ulrich, H. 163 (497), *264*
Urabe, H. 331 (837), *1012*
Urayama, T. 285, 345, 714, 715, 879, 988–990 (776), *1011*
Urbach, H. 706 (1361), *1024*
Urso, F. 982, 983 (1468), *1026*

Usher, J. J. 828, 896, 913, 914, 947 (1434), *1026*
Usui, M. 279 (766), *1010*
Utermoehlen, C. M. 138 (423), *262*
Uyama, H. 282 (770), *1010*
Uyeo, S. 713, 760 (1368), 766 (1404), 770 (1368), 844 (1404), 846, 883, 891 (1368), 925–927 (1404), 947 (1368), 971 (1404), *1024, 1025*

Vacca, J. P. 278, 280 (760), *1010*
Valcavi, V. 38 (137, 138), *256*
Valle, S. 287, 288, 294, 357, 359, 360, 363 (779), *1011*
Van Chung, V. 18, 24–26 (89), *255*
Van Den Elzen, R. 238 (676), *267*
Van der Moesdijk, G. M. C. 445 (1083), *1017*
Van der Stoel, R. E. 445 (1083), *1017*
Van der Veen, J. M. 170, 197 (561), *265*, 397 (921), 521, 548, 556–560, 579, 726, 728, 740, 813, 819, 822–824, 838, 849, 850, 853, 856 (1178), *1014, 1019*
Vander Werf, C. A. 42 (148), *257*
Vandewalle, M. 329 (835), *1012*
Van Elburg, P. A. 420, 421, 796, 968 (1048), *1017*
Van Haard, P. M. M. 69 (226), *258*
Vanin, V. V. 272 (752), *1010*
Vankar, P. 327 (831), *1012*
Van Zyl, G. 34 (119), 42 (149), *256, 257*
Varma, R. L. 514 (1166), *1019*
Varsel, C. 13, 17 (52), *254*
Vassil, T. C. 665, 781, 894, 895 (1307), *1023*
Vaughan, W. R. 49 (178), 76, 79 (242), *257, 258*
Vaultier, M. 647 (1287), *1022*
Vaux, R. L. 143 (449), *263*
Vaya, J. 518 (1173), 519, 780, 781 (1175), 811, 813, 835, 853, 881 (1173), 922 (1175), *1019*
Veber, D. F. 399, 452 (1000), *1016*
Vedejas, E. 460, 731, 795 (1094), *1018*
Vejdelek, Z. J. 47 (169), *257*
Velichko, F. K. 130 (363), *261*
Vellekoop, P. 397 (970), *1015*
Venkataraman, K. 278, 293 (764), *1010*
Verkoyen, C. 804, 807, 808, 859 (1421), *1025*
Vermeer, P. 394 (898), *1013*
Veronese, A. C. 424 (1056), *1017*
Versace, R. 726, 756, 758 (1398), *1025*
Vetuschi, C. 952 (1445), *1026*
Viani, R. 17 (62), *255*
Vigneron, J. P. 271 (749), 283 (774), *1010, 1011*
Vilamajo, L. 287, 288, 294, 357, 359, 360, 363 (779), *1011*
Vilkas, M. 18, 22 (84, 85), *255*
Villieras, J. 298, 581 (794), *1011*
Vincent, J. E. 397 (922), 556 (1206), 557 (1207), *1014, 1020*
Vincent, J. F. 521, 548, 556–560, 579, 726, 728, 740, 813, 819, 822–824, 838, 849, 850, 853, 856 (1178), *1019*
Viney, M. 134 (388), 144, 160 (487), *261, 263*
Vinogradov, M. G. 17 (57), 18 (72), *255*
Vinogradova, L. V. 130 (363), *261*
Virmani, V. 999 (1483), *1027*
Visnick, M. 279 (767), *1010*
Viswanathan, N. 243 (701), *268*
Vlasova, T. F. 626 (1259), *1021*
Vliet, E. B. 36, 37 (127), *256*
Vodicka, L. 640 (1282), *1022*
Voegtle, F. 391 (895), *1013*
Voges, H.-W. 640 (1281), *1022*
Volkova, T. V. 398 (996), *1016*

Wada, M. 389 (892), 474 (1111), *1013, 1018*
Wada, T. 17 (60), *255*
Wade, A. M. 236, 237 (673), 240 (688), *267, 268*
Waelchli, R. 658, 912 (1298), *1022*
Wagle, D. R. 278 (764), 279 (765), 293 (764, 765), 397 (921), 505 (1152, 1153), 509, 545 (1152), 562 (1152, 1153), 563 (1153), 735 (1152, 1153), 743, 755 (1152), 869 (1153), 936 (1152), *1010, 1014, 1019*
Waight, E. S. 16 (55), *254*
Wakamatsu, H. 430, 431 (1063, 1064), 432 (1064), 864 (1063), 865, 866 (1064), *1017*
Wakamatsu, S. 214 (626), *266*
Wakamatsu, T. 397 (967), 424, 425, 716 (1059), 719 (1375), 985, 986 (1059), *1015, 1017, 1024*
Wakisaka, K. 513 (1165), 529 (1165, 1183), 531 (1187), 646, 689, 740 (1183), 823, 828, 852 (1187), 918 (1165), *1019, 1020*
Wakselman, M. 420, 421 (1051), 693 (1350), 864 (1051), *1017, 1024*
Walker, C. H. 240 (688), *268*
Walker, J. C. 730, 881 (1386), 981 (1386, 1467), *1024, 1026*
Walker, K. A. 911 (1439), *1026*
Walker, R. 83, 97 (278), *259*
Wall, M. E. 83, 89, 98 (268), *259*
Walter, R. 61 (204–206), *258*
Wamhoff, H. 138, 142 (426), *262*
Wanner, K. T. 445 (1084), *1017*
Ward, A. D. 214, 220 (640), 222 (640, 648), 224, 225 (640), 226, 227 (640, 648), 229 (640), 233 (648), 238, 240 (640, 648), *267*
Ward, D. L. 630, 633 (1264), *1021*
Warren, C. L. 76, 79 (242), *258*
Washioka, Y. 285, 345, 714, 715, 879, 988–990 (776), *1011*

Washiyama, M. 730, 754, 875 (1385), *1024*
Wasserman, H. H. 222 (657, 658), 223 (658–660), 238 (683), 251 (683, 717), *267, 268*, 371 (874), 401 (1003), 416 (1039), 420, 421 (1039, 1047), 476 (1003), 648 (1003, 1291), 650 (1291), 652, 657, 675 (1003), 708, 754, 777 (1039), 779 (1047), 794, 837, 881 (1039), 953 (1446, 1447), 960 (1039, 1447, 1450, 1453), 962 (1039, 1447, 1453), 1005 (1039, 1490), *1013, 1016, 1017, 1022, 1026, 1027*
Wasson, R. 616, 745 (1250), *1021*
Watanabe, K. 677, 729, 898 (1318), *1023*
Watanabe, M. 314 (822), *1012*
Watanabe, S. 54 (196), *258*, 378 (880), 379 (881), *1013*
Watanabe, Y. 1003 (1488), *1027*
Watts, P. H. Jr. 230 (663), *267*
Weaver, S. D. 170 (598), *266*
Weber, L. D. 134 (387), *261*
Weber, W. 346 (855), *1012*
Webster, B. R. 222 (654), *267*
Wegler, R. 69 (219–222), 70 (219–221), 71 (221), 72 (221, 222), 73 (222), *258*
Wei, C. C. 755, 945 (1400), *1025*
Weigele, M. 755, 945 (1400), *1025*
Weiler, L. 299, 360, 366 (796), *1011*
Weininger, S. J. 129 (357), *261*
Weinreb, S. M. 719, 873, 876, 878, 891, 892, 977 (1374), *1024*
Weinshenker, N. M. 106 (301, 302), *260*
Weiss, R. 47 (166), *257*
Welker, C. H. 236, 237 (673), *267*
Weller, W. T. 4, 84 (20), *254*
Wells, J. N. 170, 189 (558), *265*
Wenkert, E. 214 (630), *266*
Wernges, K. 335 (839), *1012*
Wester, R. T. 278, 357 (761), *1010*
Westmijze, H. 394 (898), *1013*
Wettslein, A. 109 (316), *260*
Wexler, S. 109, 110 (317), *260*
Weyerstahl, P. 46 (158), *257*, 348 (856), *1012*
Weygand, F. 76, 77 (238), *258*
Wheland, R. 76 (230), *258*
White, J. D. 4, 7 (21), *254*
White, R. L. 828, 896, 913, 914, 947 (1434), *1026*
Whiting, M. C. 18, 20 (67), 111, 117 (330), *255, 260*
Whittaker, D. 208 (611), *266*
Whittaker, N. 137 (409), *262*
Wieduwlet, M. 827 (1426), *1025*
Wijinberg, B. P. 782 (1408), 785–788, 955 (1412), 973 (1408), 976–978, 980 (1412), *1025*
Wilcox, E. J. 233 (669), *267*
Wilde, H. 993 (1478), *1027*
Wiley, H. 36 (123), *256*
Wiley, R. H. 18 (77, 78, 80), 21 (80), 34 (117), *255, 256*
Williams, A. L. 377 (879), *1013*
Williams, R. E. 53 (186), *257*
Williams, R. M. 443, 693 (1078), *1017*
Wimmer, Th. 170 (550, 586), 185 (550), 186 (550, 586), 198 (586), 201 (550), *265*
Winterfeldt, E. 144 (470), *263*
Wintersberger, K. 640 (1276), *1022*
Wiriyachitra, P. 248 (712), *268*
Withers, G. P. 400 (1012), *1016*
Witkop, B. 144 (478–483), 155 (478), 156 (478, 480), 157 (481–483), *263*
Wojciechowska, W. 271 (743), *1010*
Wojtkowski, P. W. 130 (359), *261*
Wolf, W. 144, 145, 149 (462), *263*
Wolfe, S. 397 (978), *1015*
Wolff, G. 491, 503, 504, 523, 526 (1127), 527, 531, 532 (1182), 536, 541, 542, 544 (1127), 756, 758, 828, 834, 835, 843, 854, 856, 857 (1182), 1005 (1127), *1018, 1020*
Wolff, H. 222 (644), *267*
Wood, R. D. 435, 943 (1072), *1017*
Woodgate, P. D. 616 (1252), *1021*
Woodruff, E. H. 36, 37 (125), *256*
Woodward, R. B. 9, 11, 13, 87, 125 (28), *254*, 397 (975), *1015*
Worm, M. 827 (1426), *1025*
Wornhoff, E. W. 123 (341), *260*
Wright, J. J. 626 (1260), *1021*
Wright, P. W. 122 (339), *260*
Wu, J. S. 416 (1039), 420, 421 (1039, 1047), 708, 754, 777 (1039), 779 (1047), 794, 837, 881 (1039), 953 (1446, 1447), 960, 962 (1039, 1447), 1005 (1039), *1017, 1026*
Wu, S. 413 (1031), *1016*
Wuesthoff, M. T. 109, 110 (317, 318), *260*
Wunsch, G. 640 (1276), *1022*
Wuts, P. G. M. 328, 357 (832), *1012*

Yagami, T. 17 (58), *255*
Yamada, K. 83, 85 (261), *259*
Yamada, S. 337 (843), *1012*
Yamafuji, T. 440–442, 444, 445, 818, 826, 845 (1077), *1017*
Yamaguchi, H. 278 (763), *1010*
Yamaguchi, M. 273 (754), 371 (873), 760, 770, 947 (1403), *1010, 1013, 1025*
Yamaguti, Z. 124 (351, 352), 125 (352), *261*
Yamamoto, H. 351 (861), *1012*
Yamamoto, I. 168 (513, 514), *264*, 491, 498, 920 (1131), *1019*
Yamamoto, S. 593, 594, 601, 602 (1238), 605, 606 (1242), 607, 608 (1238, 1242), 693, 729 (1349), 876, 877 (1238), 894 (1349), 911 (1238), *1021, 1024*

Yamamoto, Y. 703 (1359), 1003 (1488), *1024*, *1027*
Yamana, Y. 342 (851), *1012*
Yamane, S. 373 (876), *1013*
Yamasaki, T. 997 (1481), 998 (1481, 1482), *1027*
Yamazaki, T. 424, 425 (1057), 661 (1300), 1000 (1485), *1017*, *1022*, *1027*
Yanagi, S. 168 (513), *264*
Yanagisawa, H. 272, 422, 740, 741, 886, 894, 898, 899, 908, 937, 938, 940, 947 (750), *1010*
Yang, D. T. C. 34 (120), *256*
Yang, N. C. 144, 161, 162 (468), *263*
Yang, T. K. 592–600, 603–605, 609, 610, 677, 886, 940 (1237), *1021*
Yano, T. 766, 844, 925–927, 971 (1404), *1025*
Yanuka, Y. 28 (103, 104), 30 (104, 106), *256*
Yasuda, N. 997, 998 (1481), *1027*
Yde, B. 804, 805 (1422), *1025*
Yoda, H. 432 (1065), *1017*
Yokayama, Y. 321 (825), *1012*
Yokoyama, H. 316, 339 (815), *1011*
Yonemitsu, O. 144 (478–481, 483), 155 (478), 156 (478, 480), 157 (481, 483), *263*, 469 (1106), *1018*
Yoshida, C. 440–442, 444, 445 (1077), 685, 743 (1343), 818, 826, 845 (1077), 895, 896, 908, 909 (1343), *1017*, *1023*
Yoshida, K. 495 (1136), 591 (1236), *1019*, *1021*
Yoshida, M. 429, 430 (1062), *1017*
Yoshida, T. 397 (962), *1015*
Yoshida, Z. 343, 344 (853), *1012*
Yoshifuji, S. 739 (1391), 798 (1418), 969 (1460), *1025*, *1026*
Yoshikawa, S. 310 (803, 806), 312 (807), 319 (806), *1011*
Yoshimura, Y. 222, 228 (650), *267*
Yoshioka, K. 404, 406, 678, 693, 737 (1021), 745, 799 (1396), 897, 947 (1021), *1016*, *1025*
Yoshioka, M. 397 (955), 699 (1357), 760 (1403), 761 (1357), 770 (1403), 782 (1357), 947 (1403), *1014*, *1024*, *1025*
Yoshiura, K. 285, 345, 714, 715, 879, 988–990 (776), *1011*
Young, G. R. 170, 197 (561), *265*
Young, H. 4, 7, 87 (22), *254*
Young, S. D. 385 (888), *1013*
Yousif, N. M. 804, 805 (1422), *1025*
Yovell, J. 475 (1121), *1018*
Yun, L. M. 695 (1352), *1024*
Yura, S. 214, 215 (627), *266*
Yurchenko, A. G. 640 (1282), *1022*

Zabza, A. 214, 219 (639), *266*
Zalinyan, M. G. 10 (35), 13 (48), *254*
Zamboni, R. 303, 519 (802), 520 (802, 1176), 529, 530 (802), 816 (802, 1176), 823 (1176), 833 (802), 840 (802, 1176), 849 (1176), 851 (802), 929, 930 (1176), 932 (802), 934 (802, 1176), *1011*, *1019*
Zanati, G. 83 (249), *259*
Zarini, F. 397 (947), *1014*
Zarocostas, C. 404, 406, 704 (1018, 1019), 811 (1019), *1016*
Zasosova, I. M. 397 (903), 725, 812, 837, 980 (1379), 981 (903), *1013*, *1024*
Zauer, K. 692 (1347), *1024*
Zaugg, H. E. 3 (1), *253*
Zboida, D. 639, 881 (1272), *1022*
Zdunska, A. 397 (944), *1014*
Zeidman, B. 133 (381), *261*
Zeliver, C. 203, 205 (606), *266*
Zey, R. L. 138 (418), *262*
Zhidkova, A. M. 398 (984), *1015*
Ziegenbein, W. 240 (686), *268*
Ziegler, E. 170 (550, 586, 592), 185 (550), 186 (550, 586), 198 (586), 200 (592), 201 (550), 203 (607), 252 (722), 253 (723), *265*, *266*, *268*
Ziegler, F. E. 278, 357 (761), *1010*
Zieloff, K. 170, 191 (555), *265*
Zienty, F. B. 83, 89 (265), *259*
Zil'berman, E. N. 214, 216 (632), *266*
Zilch, H. 981 (1465), *1026*
Zima, G. 293 (790), *1011*
Zimmer, H. 61 (204, 205, 207), *258*
Zoretic, P. A. 290, 359, 364 (787), 709 (1363), *1011*, *1024*
Zuidema, G. D. 34 (119), *256*
Zwanenburg, B. 69 (226), *258*

Index compiled by K. Raven

Subject index

Acetal ethers, oxidation of 314, 320
Acetamides — *see*
(Alkenyl)halo(methylthio)acetamides, *N*-Alkenyl-α-(methylsulphinyl)acetamides, *N*-Allyltrichloroacetamides
5-(Acetamidomethyl)cysteinylglycine esters, reactions of 452
Acetoacetates,
aldol condensation of 36
lithium enolates of, as lactone precursors 386, 387
Acetoglutarates, reactions with Grignard reagents 49
Acetoxyandrostanone acetylhydrazones, ring expansion of 662
Acetoxyfurans, reactions of 373, 374
4-Acetoxy-β-lactams,
reactions of 943, 944
with cyanides 902
with enolates 903–905
with mercapto compounds 906–909
with organometallics 902
with phosphorus compounds 906
reduction of 901, 902
synthesis of 898–900
5-Acetoxy-γ-lactams,
reactions of 906–909
synthesis of 899
Acetoxy-17-β-(6′α-pyronyl)androstanes, synthesis of 40–42
Acetylalkanedinitriles, cyclization of 464
Acetylenedicarboxylates, ene reactions of 286
Acetylenes, reactions with isocyanates 167, 168
Acetylenic acids, acid-catalysed cyclization of 18, 20, 21
β-Acetylenic alcohols, reactions with Reformatsky reagents 51
Acetylenic esters, Diels–Alder cyclization of 30–34
Acetylides, reactions with aldonitrones 615
N-Acetylketo lactams, synthesis of 962
3-Acetyl-β-lactams, reduction of 886, 888
4-Acetyl-β-lactams, oxidation of 898
4-Acetylthio-β-lactams, synthesis of 913
Acrylates — *see* α-(Haloalkyl)acrylates
Acrylic acids — *see* Furylacrylic acids
N-Acylaminoacid dimethylamides, cyclization of 281, 282
Acylazocarboxylates, ene insertion of 460
2-*N*-Acylcyclohex-2-enones, cyclization of 470, 471
Acyldioxanediones, reactions with Schiff bases 1003, 1004
β-Acyl-β-ethoxycarbonyl-α-alkylpropionic acids, as lactone precursors 380, 381
Acyl halides — *see also* Diacyl halides, Haloacyl halides
as acylating agents 811–824
N-Acyliminium cyclization 453
Acyllactam carboxylic acids, synthesis of 681
N-Acyllactams, synthesis of 798, 952, 953, 956, 970
3-Acyl-β-lactams, synthesis of 1003, 1004
2-Acyl-3-oxo-4, 5-benzo-1, 2-thiazoline dioxides, as lactam precursors 623, 624
(Acyloxy)halopropanamides, reactions of 427–429
N-Acylpiperidines, irradiation of 953, 956
Acylpyrazolidinones, ring contraction of 637, 638
N-Acylpyrrolidines,
irradiation of 953, 956
oxidation of 970
Adams catalyst, in hydrogenation of butenolides 301
Alcohols, unsaturated — *see* Unsaturated alcohols
Aldehydes — *see also* Benzaldehydes, ω-(α-Bromoacyloxy)aldehydes, Propanals
electroreductive hydrocoupling of 371–373
reactions of,
with acetoxyfurans 374
with amino acid amides 464, 465
with ketenes 63, 69–73
with lithium β-lithio esters 368, 369

Aldehydes (*cont.*)
reactions of, (*cont.*)
with Reformatsky reagents 51, 52
unsaturated — *see* Unsaturated aldehydes
Aldehydic acids, cyclization of 16
β-Aldehydic α,β-ethylenic acids/esters, reduction of 312, 313
Aldol condensation 36–40, 61
Aldonitrones, reactions with copper acetylide 615
Alkenamides, condensation of 460, 461
2-Alkene-1,4-diols, reactions with *ortho* esters 120–122
Alkenes — *see* Azoalkenes, Haloaminoalkenes, Olefins
Alkenoic acids — *see also* Dienoic acids, 4-Hexenoic acids, Hydroxyhexadecenoic acids, Pentenoic acids, Polyenoic acids
reactions with imines 621–623
sulphenyllactonization of 382
Alkenols — *see* Iodoalkenols
(Alkenyl)halo(methylthio)acetamides, reactions with stannic chloride 430
Alkenylmalonates, acid-catalysed cyclization of 17
N-Alkenyl-α-(methylsulphinyl)acetamides, as lactam precursors 448–451
Alkoxybutenolides, synthesis of 341
4-Alkoxycarbonylmethyl-β-lactams, as enone precursors 943, 944
1-Alkoxy-5,6-dihydro-γ-pyrones, hydrolysis of 395
Alkoxyfurans, oxidation of 314, 320
3-Alkoxy-β-lactams, dealkylation of 870, 874
4-Alkoxy-β-lactams, synthesis of 903, 918, 925
5-Alkoxy-γ-lactams, reactions of 906, 910
6-Alkoxy-δ-lactams, reactions of 910
γ-Alkoxymethyl-α,β-butenolides, synthesis of 294, 359
Alkoxymethyloxiranes, reactions with malonate anion 294
2-Alkoxyoxetanes, photolysis of 129
Alkoxypyrrolinones, synthesis of 651
α-Alkylamidolactams, synthesis of 704
N-[(Alkylamino)alkyl]lactams, transamidation of 652, 656
4-Alkyl-2-butenolides, synthesis of 300
Alkylcaprolactams, synthesis of 238, 240
γ-Alkylidene γ-butyrolactones, synthesis of 348, 349
α-Alkylidene lactams,
isomerization of exocyclic double bond in 890
Michael addition to 712
reduction of 713–715
synthesis of 710–712, 918, 963, 971
ω-Alkylidene lactams, synthesis of 981, 982
4-Alkylidene β-lactams, synthesis of 920, 921
N-Alkyllactams,
electrooxidation of 719–721
sulphenylation of 709
sulphinylation of 709
4-Alkyl-β-lactams, synthesis of 902, 917, 925, 940, 942
3-Alkyl-1-methyl-2-pyrrolidones, synthesis of 237
2-Alkyl-2-oxazolines, reactions of 58–61
Alkynes, carbonylation of 341–343
Alkynoates — *see* γ-Hydroxy-α,β-alkynoates
Alkynols, reactions with nitrous oxide 76
4-Alkynyl-β-lactams — *see also* 4-Ethynyl-β-lactams
hydrogenation of 940–942
synthesis of 630, 992
5-Alkynyl-5-methoxy-γ-lactams, synthesis of 971, 980
β-Allenic alcohols, oxidation of 333
ω-Allenyllactams, synthesis of 910
Allenyllithium compounds, as lactone precursors 394
Allylamides,
cyclization of 456, 457
Diels–Alder reaction of 458–460
Allylamines, ring closure of 983
Allyl bis(alkoxycarbonylamina)acetates, acid-catalysed cyclization of 285
Allyl halides, ring closure of 983
Allyl imidates, thermal rearrangement of 236, 237
N-Allyllactams, synthesis of 236, 237
4-Allyl-β-lactams, synthesis of 917, 925–927
π-Allyltricarbonylironlactone complexes 271
N-Allyltrichloroacetamides, cyclization of 430–432
Amides — *see also* Acetamides, Allylamides, Diazoamides, Hydroxypentenylamides, (Phenylthio)alkanamides, Propionamides, Pyruvamides, *N*-(Thiobenzoyl)methacrylamides, Toluamides
Barton reaction of 127
halocyclization of 435
sulphur-containing, as lactam precursors 447–452
α,β-unsaturated — *see* α,β-Unsaturated amides
Amidinium ions, as reaction intermediates 656
3-Amido-β-lactams, synthesis of 612, 625, 856, 857
4-Amido-β-lactams, hydrolysis of 936
Amines — *see* Allylamines, Benzylamines, Cyclic amines, *N*-Haloacyl-β-arylamines, Oxadiamines

Amino acid amides, reactions with carbonyls 464, 465
Amino acids,
cyclic — *see* Cyclic amino acids
cyclization of 133, 398–420
cyclodehydration of 411–413
α-Amino acids, reactions with β-haloacyl halides 419
β-Amino acids,
cyclization of 407–409, 413, 415
reactions with isocyanates 489
ω-Amino acids,
reactions of,
with γ-butyrolactone 411
with cathecolboranes 411, 412
with 2,2′-dipyridyl disulphide 404, 405
with *o*-nitrophenyl thiocyanate 400, 404
with phosphorus pentasulphide 400
with triphenylphosphine 404, 405
tin-mediated condensation of 408, 410, 411
N-(Aminoalkyl)lactams, transamidation of 652, 656, 657
Aminoanthraquinones, reactions of 418
β-Aminobenzamides, as lactone precursors 369, 370
2-Amino-4-carboxychroman lactams, synthesis of 725
Amino esters, cyclization of 133–137, 400–403
Aminoglycosides 397
3-Aminolactams,
acylation of,
using acyl halides 811–824
using benzylpenicillin acylase 811, 829
using carboxylic acids 811, 825–829
using carboxylic esters 811, 825, 831
using diisocyanates 811, 825
using mixed anhydrides 811, 825, 830
reactions of,
with aldehydes 832–835
with epoxides 396, 397
with ethyl 2-oxo-4-phenylbutyrate 855
ring expansion of 652
substituted, reactions of 855–858
synthesis of,
by deprotection of substituted 3-aminolactams 836, 844–854
from 3-azidolactams 836–843
4-Amino-β-lactams, synthesis of 166
α-Amino-γ-lactones, synthesis of 275
α-Aminonitriles, reactions with ketones 376
ω-Aminonitriles, as lactam precursors 416
6-Aminopenicillanic acids, diazotization and bromination of 858, 859
3-(2-Amino-2-phenylacetamido)-2-methyl-4-oxo-1-azetidinesulphonic acids, ring expansion of 657
Aminopiperidones, synthesis of 415
Aminopyrrolidones, synthesis of 970, 971
Aminothio acids, ring closure of 415
Amino thiol esters, ring closure of 406, 407, 992
β-Amino-*o*-toluamides, as lactone precursors 369, 370
Androstanolones, condensation with glyoxylic acid 38, 39
Androstanones — *see* Hydroxyandrostanones
Androstenones — *see* Azaandrostenones
Anhydrides,
as acylating agents 811, 825, 830
bicycloalkenedicarboxylic — *see* Bicycloalkenedicarboxylic anhydrides
carbonic — *see* Carbonic anhydrides
cyclic — *see* Cyclic anhydrides
reactions of 46, 47
with Grignard reagents 294–297
with imines 581
reduction of 76–83
Anilinoalkylbutenolides,
catalytic hydrogenation of 301
synthesis of 395
α-Anilinomethyl-γ-butyrolactones 301
Anthraquinones — *see* Aminoanthraquinones
Aralkylidinephthalides, synthesis of 47
Arbusov reaction 906
Aroylbenzaldehydes, reactions with isocyanates 167, 168
2-Aroyl-1-methylene-1,2,3,4-tetrahydroisoquinolines, cyclization of 465–469
Arylacetic acids, condensation with phthalic anhydride 47
N-(2-Arylethyl)lactams, synthesis of 240
Aryl halides, carbonylation of 341, 342
N-Arylketo lactams, synthesis of 962
Arylmethylphthalides, synthesis of 47
Arylnaphthofurandiones, reduction of 310
2-Aryl-4,5,6,7-tetrahydroindoles, oxidation of 962
γ-Arylthio-γ-butyrolactones, synthesis of 303, 308
Azaandrostenones, ring expansion of 661
Azabicyclo-β-lactams, synthesis of 771, 774–776
2-Azacarbapenems, formation of 799, 800
Azacholestanones, synthesis of 489
4-Aza-β-lactams,
ring expansion of 651, 652
synthesis of 462, 627–629, 992, 997
1-Azaspiro[3.5]nonane-2,5-diones, synthesis of 470, 471
Azepinediones, reactions of 980
Azepines — *see also* Pyridoazepines
synthesis of 626
Azepinoindoles, synthesis of 469, 470

Azepinones — *see* Benzazepinones
Azetidiminium salts, as lactam precursors 397
Azetidinecarboxylates, reactions of 953, 956
Azetidinecarboxylic acids, decarboxylation of 239, 251, 960
Azetidinediones, synthesis of 648, 649, 859, 920
Azetidines — *see* *N*-Hydroxyazetidines
Azetidinonecarboxylic acids — *see* Hydroxyazetidinonecarboxylic acids
Azetidinones — *see also* 1-(1-Benzyloxycarbonyl-2-hydroxypropyl)-3-styrylazetidinones, *N*-Bis(alkoxycarbonyl)methyl-4-methylthioazetidinones, Bisazetidinones, Diazetidinones, β-Lactams
  3-tritylamino-4-substituted 855
N-Azidoacyllactams, Curtius rearrangement of 799
3-Azidolactams,
  reactions with triphenylphosphine 835
  synthesis of 706–708, 859
Azidomethylfuranoates, hydrogenation of 993
Azidopyrrolidines, ring contraction of 630
α-Azido sulphides, ring expansion of 647
Aziridinecarboxylic acids, ring expansion of 235
Aziridines, carbonylation of 982, 983
Aziridinones — *see* α-Lactams
Azirines,
  carbonylation of 984
  synthesis of 764, 765
Azoalkenes, intramolecular cycloaddition of 616
Azocarboxylates — *see* Acylazocarboxylates, Azodicarboxylates
Azodicarboxylates, in lactonization 271

Baeyer–Villiger reaction 98–108, 314, 320, 898
Barton reaction 127
Beauveria sulfurescens 722, 723
Beckmann rearrangement 212–222, 639–645
Benzaldehydes — *see also* Aroylbenzaldehydes, Hydroxybenzaldehydes
  reactions of 40, 47
Benzamides — *see* Aminobenzamides
Benzanilides, cyclization of 145
Benzazepinones, synthesis of 453
Benzoates — *see* Hydroxymethylbenzoates
Benzodiazepinediones — *see* 4,4′-Methylenedi(3, 4-dihydro-1*H*-1,4-benzodiazepine-2, 5-diones)
Benzodiazepinones — *see also* 5,7-Dichloro-3*H*-1,4-benzodiazepinones
  reactions of 911
  synthesis of 981
Benzoic acids — *see* Phenylbenzoic acids
Benzolactams, synthesis of 464, 984
Benzolactones, synthesis of 341, 342
Benzoquinolines — *see* Octahydrobenzoquinolines
Benzo[*a*]quinolizidines, synthesis of 453
Benzoquinones — *see* Dimethoxy-*p*-benzoquinones
Benzothiazepinones, ring expansion of 657, 658
Benzothiazepinone sulphoxides, halogenation of 706
Benzothiazonindiones, synthesis of 657, 658
Benzotropones — *see* Halobenzotropones
Benzoxazepinones, ring expansion of 657, 658
Benzoxazonindiones, synthesis of 657, 658
N-Benzoyllactams, synthesis of 952, 953
S-Benzoyllactams, synthesis of 471
4-Benzoyloxy-5-heptenoates, reactions with LDA 298
Benzoylpropionic acids, reactions with benzaldehydes 47
Benzylamines — *see also* Halobenzylamines
  reactions with 2-bromoglutarates 420
Benzylaminopropionic acids, cyclization of 400
Benzylidene anthranilic acid, reactions of 303, 308
N-Benzyllactams, synthesis of 952, 953
N-Benzyl-3-methylene-β-lactams, anodic oxidation of 795
Benzyloxycarbonylamino-γ-butyrolactones, alkylation of 300
1-(1-Benzyloxycarbonyl-2-hydroxypropyl)-3-styrylazetidinones, reactions of 755
ε-(Benzyloxycarbonyl)-l-lysine methyl ester hydrochloride, reactions of 418
Benzyloxycarboxylamino-γ-butyrolactones, synthesis of 309
N-(Benzyloxy)-β-lactams, reactions of 796
4-(Benzyloxy)-2-methylene-butanoates, acid-catalysed cyclization of 285
α-[*p*-(Benzyloxy)phenyl]-2-oxo-1-azetidinemalonates, synthesis of 777
Berbamine, photooxidation of 248
Bicyclic *N*-acetylenamines, oxidation of 962
Bicyclic allylic alcohols, rearrangement of 339
Bicyclic *N*-arylenamines, oxidation of 962
Bicyclic diazopyrrolidinediones, ring contraction of 630
Bicyclic epoxylactones, reactions of 352, 359, 360
Bicyclic imides,
  reactions with organometallics 980
  reduction of 976–979
Bicyclic keto lactones, silation of 357

Bicyclic ketones, oxidation of 314, 320
Bicyclic lactams,
double-bond formation in 756–758
introduction of an angular hydroxyl group in 719
reactions of 857, 858
ring opening of 724, 891, 894–897, 945
producing N-side-chain α,β-unsaturation 760–762
synthesis of 397, 777–785, 791–793, 999, 1005
by intramolecular Wittig reactions 765–770
by N-substitutions/cyclizations 691–695
by ring closures 950–952
by ring contractions 630, 632, 635, 639
from allylhaloamides 430, 431
from amino acids 404, 406
from cyano esters 583
from cyclic imides 583
from cycloalkanone esters 452, 453
from enamines 460, 461
from hydroxyamides 447
from imines 498, 503, 504, 507, 513, 523, 526, 527, 531, 532, 536, 541, 542, 544, 547, 554, 556, 559–561, 565, 574, 616, 621–623
from 4-methylenespiro[2.X]alkanes 475, 487
from thiazolidines 398, 399
Bicyclic lactones — *see also* Bicyclic epoxylactones, Bicyclic keto lactones
halogenation of 706, 707
reduction of 287
synthesis of 290–292, 365–367, 388
from carboxymethylpyrrolidinols 11
from gem-dibromides 342, 343
from γ-hydroxy esters 10, 11
from spirolactones 359, 360
Bicyclic pyrazolidones, synthesis of 684
Bicycloalkenedicarboxylic anhydrides, reduction of 308
Bicyclo-γ-butanolides, synthesis of 295
Bicyclobutenolides, synthesis of 295
Biphenyl-2-carboxamides, cyclization of 472–474
Bis(3-acylthiazolidine-2-thiones), aminolysis of 996, 997
Bis(alkoxycarbonyl)aminoacetic acids, allylamides of, cyclization of 456
Bis(alkoxycarbonylamino)acetic acids, aromatic amides of, amidoalkylation of 463
N-Bis(alkoxycarbonyl)methyl-4-methylthioazetidinones, synthesis of 777
Bisazetidinones — *see also* Diazetidinones
synthesis of 681
3-Bis(ethylthio)-β-lactams, reactions of 717, 719
Bisiodomethyl spiro bislactones, reactions of 351, 352
Bislactams — *see also* Dilactams
synthesis of 424, 426
Bislactones — *see* Bisiodomethyl spiro bislactones
Bislactonization 383, 384
6,6-Bis(phenylseleno)penicillinates,
partial deselenization of 869, 870
synthesis of 880, 884
α,α-Bis(phenylthio)lactams, synthesis of 709
6,6-Bis(phenylthio)penicillinates, synthesis of 880, 884
ω-(α-Bromoacyloxy)aldehydes, cyclization of 371
α-Bromo-γ-butyrolactones, reactions of 61–63, 66–68
Bromocephams, reductive debromination of 945, 946, 948, 949
2-Bromoglutarates, reactions with benzylamine 420
Bromomethylpenicillins, reductive debromination of 945, 946, 948, 949
Bufadienolides, synthesis of 30, 39
Bufaenolides, synthesis of 28, 29
Butadienes — *see* 1,1-Dialkoxy-3-trimethylsiloxy-1,3-butadienes
Butanoates — *see* 4-(Benzyloxy)-2-methylene-butanoates
Butanolides — *see* Bicyclo-γ-butanolides
Butenolides — *see also* Alkoxybutenolides, γ-Alkoxymethyl-α,β-butenolides, 4-Alkyl-2-butenolides, Anilinoalkylbutenolides, Bicyclobutenolides, α-Cyano-α,β-butenolides, Halobutenolides, γ-Halomethyl-α,β-butenolides, Hydroxybutenolides, 4-Hydroxymethyl-2-butenolides, 4-Spirobutenolides
synthesis of 3, 34, 35
by Grignard reactions 295
by Perkin-like reactions 47
from acetoxyfurans 373, 374
from β-aldehydic α,β-ethylenic esters 131, 312
from anhydrides 309, 310
from butanolides 302, 363
from carbonyls 369
from furans 109
from hydroxy esters 274, 275
from hydroxypentenoic acids 287, 288
from γ-hydroxyvinylstannanes 341
from iodoalkenols 343, 344
from ketenes 69, 74, 75
from trimethylsilylfurans 331, 332
from unsaturated acids 336

Butenolides (*cont.*)
synthesis of (*cont.*)
from unsaturated aldehydes 339, 340
γ-(*N*-Butoxymethyl)-α-halo-γ-butyrolactones, thiophenylation of 357
Butyric acids, as lactone precursors 272
β-Butyrolactams, synthesis of,
from allylhaloamides 430
from amino esters 134, 135
γ-Butyrolactams,
reactions with dichlorocarbene 711
synthesis of,
from allylamines 983
from allylhaloamides 431
from allylic halides 983
Vilsmeier formylation of 710
Butyrolactone esters, synthesis of 130, 131
γ-Butyrolactones,
α-alkylation of 64, 65
mesylated 285, 286
β-methoxycarbonyl-γ-substituted 63
reactions with Eschenmoser's salt 300
synthesis of,
by aldol condensations 36
by malonate condensations 42–45
by Wittig-type reactions 54
from alkynols 76
from anhydrides 310
from carboxylic acid anions 58
from carboxylic acids 377–380
from cyclopropanecarboxylates 350, 351
from cyclopropanecarboxylic acids 122–124, 350, 351
from cyclopropyl esters 385, 386
from α-diazo esters 127–129
from ethers 109
from hemisuccinates 314
from hydroxy acids 277, 278
from hydroxy esters 12, 13
from oxazolines 58–61
from pentenoic acids 16

ε-Caprolactams — *see also* α-Oxo-caprolactams, Alkylcaprolactams
reactions with dichlorocarbene 711
synthesis of 223, 231–233, 416, 981
by Beckmann rearrangement 215, 222
by Schmidt rearrangement 224–228
from 6-aminocapronitrile 416
from amino esters 134
from azidomethylfuranoates 993
Vilsmeier formylation of 710
ε-Caprolactones, synthesis of 3
Carbalkoxylation 777
Carbamates — *see* *O*-(4-Hydroxy-1-alkenyl)carbamates
Carbapenem antibiotics, synthesis of 397
Carbene complexes, photoreactions of 585–589, 956–959
Carbodiimides, reactions of 581
Carboethoxycoumarins, synthesis of 40
Carboethoxymethylcyclohexanediones, as lactone precursors 14
3-Carbomethoxy-β-lactams, decarbomethoxylation of 869
Carbonic anhydrides, reactions with arylethylamines 453, 454
Carbonic esters, reactions with arylethylamines 453
Carbonium ion rearrangement 122–124
Carbonium ion rearrangement–lactonization 346–348
Carbonylation, metal-catalysed 982–991
Carbonyl compounds — *see* Aldehydes, Ketones, α,β-Unsaturated carbonyl compounds
Carbostyrils — *see also* Dihydrocarbostyrils
synthesis of 167
Carboxamides — *see* Biphenyl-2-carboxamides, α-Diazocarboxamides
Carboxamidyl radicals, cyclization of 472
α-Carboxylactams, synthesis of 418
3-Carboxy-β-lactams,
decarboxylation of 869
synthesis of 867, 868
Carboxylic acid anions, reactions with epoxides 54–58
Carboxylic acids — *see also* Dicarboxylic acids, ω-Halocarboxylic acids, Hydroxycarboxylic acids
activated, reactions with imines 547–581
as acylating agents 811, 825–829
oxidation of 336
reactions of,
with *o*-hydroxyphenyl ketones 381
with lithium naphthalenide 379, 380
with olefins 377, 378
with trimethylsilyl chloride 382, 383
unsaturated — *see* Alkenoic acids, 2, 4-Dienecarboxylic acids, Unsaturated carboxylic acids
Carboxylic esters — *see also* Acylazocarboxylates, Dicarboxylates, Hydroxycarboxylic esters
as acylating agents 811, 825, 831
cyclization of 384–390
reduction of 312–314
unsaturated — *see* Unsaturated carboxylic esters
Carboxymethylindanols, acid-catalysed cyclization of 274
Carboxymethyl radicals 378
Carboxymethylthiazolidinium salts, as lactam precursors 616, 617

Cardenolides, synthesis of 39
Catharanthine lactone, synthesis of 329
Cephalosporins, synthesis of 397
Cephalosporium acremonium 1005
Cephams — *see* Bromocephams
Cephems — *see also* Phthalimidocephems
synthesis of 993
Cetyltrimethylammonium permanganate, as oxidizing agent 327
Chinensin, synthesis of 311
Chiral lactones, synthesis of 312
4-Chloro-β-lactams,
reactions of 925
synthesis of 922, 923
4-Chloromercuriothio-β-lactams, reactions of 913, 915
5α-Cholestanone acetylhydrazones, ring expansion of 662
Cholestanones — *see* Azacholestanones
Chromanones — *see* Thiochromanones
Chromium carbene complexes, photoreactions of,
with imines 585–589, 956–959
with oxazines and thiazines 956, 957, 959
Chromium oxides, as oxidizing agents 952
Cinnamic acids, photocyclization of 18
Claisen rearrangement 120–122, 236, 237
Clavulanic acid, Wittig products of, ozonolysis of 960, 961
Cobalt carbonyls, as carbonylation catalysts 341, 342
Conanine lactam, synthesis of 242, 247, 248
Corey lactones 271
Cotinene, synthesis of 242
Cotinines, synthesis of 240, 241
Coumaranones — *see also* Hydroxybenzylidenecoumaranones
aldol condensation of 61
Coumarins — *see also* Carboethoxycoumarins
synthesis of 36, 37, 381
Cross-aldol condensation 585, 590
Crotonic acids, photocyclization of 18
Crown ether lactones, synthesis of 391
Curtius rearrangement 489, 799, 857, 858
Cyanoalkanoates, hydrogenation of 415, 416
α-Cyano-α,β-butenolides, reactions with amines 993
Cyano esters, as lactam precursors 583
3-Cyano-β-lactams, synthesis of 631
4-Cyano-β-lactams,
hydrolysis of 936
synthesis of 902, 916
4-Cyanomethyl-β-lactams, synthesis of 943
Cyclic amines,
oxidation of 241, 251, 952, 953, 970
reactions with valerolactims 1000, 1001
Cyclic amino acids, oxidative decarboxylation of 1005
Cyclic anhydrides, reduction of 303, 308–312
Cyclic enol ethers, oxidation of 323
Cyclic β,γ-epoxyketones, irradiation of 124
Cyclic ethers, oxidation of 314, 320–323
Cyclic imides,
reactions of 583
with acetylides 971, 980
with carbonyls 971
reduction of 970, 972–979
synthesis of 798
Cyclic ketones — *see also* Cyclic β,γ-epoxyketones
oxidation of 98–108
reactions with hydrazoic acid 222, 224–229
Cycloaddition reactions 162–210, 475–627
Cycloalkanecarboxylic acids — *see also* Cyclohexanecarboxylic acids, Cyclopropanecarboxylic acids
ring expansion of 233, 234
Cycloalkanone esters, reactions with amines 452, 453, 455
Cycloalkanones — *see* 2-Nitrocycloalkanones
Cycloheptenones — *see* Iminobenzocycloheptenones
Cyclohexadienones — *see* Hydroxycyclohexadienones
Cyclohexanecarboxylic acids — *see* 2-Hydroxy-6-allylcyclohexanecarboxylic acids
Cyclohexanediones — *see* Carboethoxymethylcyclohexanediones, 2, 3-Epoxy-1,4-cyclohexanediones
Cyclohexenoic acids, halolactonization of 290
Cyclohexenones — *see* 2-*N*-Acylcyclohex-2-enones
Cyclopentanone 3-methylcarboxylic acid, carbonate esters of, as keto lactone precursors 384
Cyclopentenylacetic acids, oxidation of 337
Cyclopeptides, synthesis of 397
Cyclopropanecarboxylates — *see also* Hydroxymethylcyclopropanecarboxylates
rearrangement of 350, 351
Cyclopropanecarboxylic acids, rearrangement of 122–124, 350, 351
Cyclopropanes — *see* 1,1-Dibromo-2-(hydroxyalkyl)cyclopropanes
Cyclopropyl esters, ring opening/cyclization of 385, 386
β,γ-Cyclopropyloximes, ring expansion of 644, 645
l-Cysteine hydrochloride, reactions with acrylates 419
Cysteinylglycine esters — *see* 5-(Acetamidomethyl)cysteinylglycine esters

Decarbomethoxylation 869
Decarboxylation 869, 938
oxidative 960, 1005
Dehydrodimethylconidendrin, synthesis of 311
Dehydrodimethylretrodendrin, synthesis of 311
Dehydrolactams, fused-ring, synthesis of 465–469
Dehydropregnelone acetates, reactions with Grignard or Reformatsky reagents 49, 51
Depsipeptides 411
Deselenization 869, 870
Desulphurization 869, 870
Diacyl halides, as lactone precursors 390–394
Dialkoxytetrahydrofurans, reactions of 275, 276
1,1-Dialkoxy-3-trimethylsiloxy-1,3-butadienes, reactions with aldehydes 395
Diaminopentanoic acids, ethylation/cyclization of 415
1,2-Diazatricyclo[5.2.0.0^{3,6}]non-4-enes, 2-substituted 799, 800
Diazetidinones — *see also* Bisazetidiones
synthesis of 170, 171, 202
Diazoamides, as lactam precursors 462
α-Diazocarboxamides, cyclization of 162
Diazo compounds, decomposition/cyclization of 992
4-Diazo-2,3-diones, synthesis of 929
α-Diazo esters, photolysis of 127–129
4-Diazoketo-β-lactams,
rearrangement of 940, 941
synthesis of 940
Diazoketones,
penicillin-derived, rearrangement of 801, 802
reactions with ketenes 69, 74, 75
3-Diazo-β-lactams, as α-spirolactam precursors 718, 719
Diazopyrazolidinediones, ring contraction of 627–630
Diazopyrrolidinediones, ring contraction of 210–212, 630
1,1-Dibromo-2-(hydroxyalkyl)cyclopropanes, carbonylation of 342, 343
3,5-Dicarboalkoxy-2-pyrazolines, reductive cleavage of 970, 971
Dicarboxylates — *see* Azodicarboxylates
Dicarboxylic acids,
reduction of 83
unsaturated — *see* Unsaturated dicarboxylic acids
5,7-Dichloro-3*H*-1,4-benzodiazepinones, reactions of 929
Diels–Alder reaction 30–34, 395, 458–460
Dienamino esters, cyclization of 134
2,4-Dienecarboxylic acids, allyl- and propargyl-amides of, Diels–Alder reaction of 458–460
Diene esters, Diels–Alder cyclization of 30, 34
Dienes, carbonylation of 112, 118
Dienoic acids — *see also* Pentadienoic acids
acid-catalysed cyclization of 17
α-Diethoxy acetals, reactions of 754
Dihaloketenes, reactions with vinyl sulphoxides 303, 308
6,6-Dihalopenicillinates, reactions of 859–862
α,α-Dihalo-γ-thioaryl-γ-butyrolactones,
dehalogenation of 352, 353
desulphurization–dehalogenation of 354
Dihydrobenzothiazepinones, synthesis of 464, 645
Dihydrocarbostyrils, synthesis of 144, 145
2,3-Dihydrofuranones, aldol condensation of 61
3,4-Dihydroisoquinolines, as lactam precursors 623, 624
2,3-Dihydro-1, 3-oxazine-4-oxo-5-carboxylic acids, synthesis of 1003
Dihydropyridones, synthesis of 207
3,4-Dihydropyrido[3,4-*b*]quinoxalin-1(2*H*)-one dioxides, synthesis of 994, 995
Dihydrospiro(androstenefuran)diones, synthesis of 54
Dihydrothiazines, ring closure of 993
Dihydroxy-γ-lactones, synthesis of 331
Dihydroxynitriles, as lactone precursors 283
Diisocyanates, as acylating agents 811, 825
β-Diketones,
irradiation of 124, 125
Wittig reaction of 53
Dilactams — *see also* Bislactams
synthesis of 397, 464, 657, 679, 681
from imines 514, 524, 525, 527, 528, 539, 568
Dimethoxy-*p*-benzoquinones, reactions with anhydrides 46
Dimethoxyphenylacetic acids, as lactone precursors 10
Dimethoxypyrones, reactions with nucleophile anions 358
Dimethylaminomethylene lactones, synthesis of 352
δ,δ-Dimethylvalerolactones, phenylselenylation of 357
Diols,
carbonylation of 112
oxidation of 83, 89–97, 314–319
Dioxanediones — *see* Acyldioxanediones
Dioxolactams, synthesis of 657, 658
1,2-Dioxolane-3,5-diones, photolysis of 130
Dioxothiazolidinones, ring contraction of 630, 631

Dipeptide methyl esters, methylsulphonium iodides of, reactions of 452
Disulphides, oxidation of 130
γ-Dithiolactones, β,γ-unsaturated — *see* β,γ-Unsaturated γ-dithiolactones

Enamides — *see also* Alkenamides
  cyclization of 145, 162, 465–469
Enamines — *see also* Bicyclic *N*-acetylenamines, Bicyclic *N*-arylenamines
  reactions of,
    with acrylamide 460, 461
    with isocyanates 166
Enaminoamides, reactions of 460, 462
δ-Enaminolactams, synthesis of 397
Ene lactones, reactions of 453–455
Enepalladium complexes 984
Enolates, reactions of,
  with 4-acetoxy-β-lactams 903–905
  with imines 592–611, 613
  with nitriles 614, 615
Enol ethers, cyclic — *see* Cyclic enol ethers
Enol lactams, synthesis of 754, 755
Enol lactones, synthesis of 13, 15, 16, 358, 359
Enynes, reactions of 393, 394
Enzyme-catalysed asymmetric reactions 271
Epichlorohydrin, reactions of 34
Episelenonium ions, as reaction intermediates 460, 461
Episulphonium ions, as reaction intermediates 451
Epoxides, condensation with malonates 42, 43
2,3-Epoxy-1,4-cyclohexanediones, photorearrangement of 348, 349
Epoxycyclooctenes, condensation with malonates 46
Epoxy esters, rearrangement of 348
3-Epoxy-β-lactams,
  as lactone precursors 395, 396
  synthesis of 428, 429
4,5-Epoxypentenoates, acid-catalysed cyclization of 287
Erythrinanones, synthesis of 455, 456
Ethers,
  cyclic — *see* Cyclic ethers
  oxidation of 109, 110
β-Ethoxycarbonyl-γ-lactones, synthesis of 380, 381
α-Ethylenedioxy acetals, reactions of 754
α-Ethylenic acids, allylamides of, cyclization of 456, 457
4-Ethynyl-β-lactams, synthesis of 917
Exocyclic double bonds, in N-substituted lactams,
  formation of 751–756
  reactions at 762–777

Fomannosin, synthesis of 39
4-Formyl-β-lactams, synthesis of 934, 938
α-Formyllactones,
  sodio derivatives of, reactions of 303–307
  synthesis of 303
N-Formylpiperidines, irradiation of 953, 956
N-Formylpyrrolidines, irradiation of 953, 956
Friedel–Crafts cyclization 489
Friedel–Crafts-type acylation 373, 374
Fufural, photooxidation of 341
6-Fulvenyl-2,2-dimethyl-3,5-hexadienoate, cycloaddition of 292
Furancarboxylic acids — *see* Hydroxymethylfurancarboxylic acids
Furandiones — *see* Arylnaphthofurandiones
Furanoates — *see* Azidomethylfuranoates
Furanones — *see also* 2,3-Dihydrofuranones, Tetrahydroindenefuranones
  formylation of 303
Furans — *see* Acetoxyfurans, Alkoxyfurans, Tetrahydrofurans, Trimethylsilylfurans
Furanurono-6,3-lactams, synthesis of 952
Furfuryl alcohols, oxidation of 323
Furoic acids, photooxidation of 337
Furopyridones, synthesis of 460, 462
Furylacrylic acids, UV irradiation of 288

Gibberellic acid, precursors of 30
Glutarates — *see* Acetoglutarates, 2-Bromoglutarates
Glutyraldehyde, reactions with Grignard reagents 48
Glycal esters, as lactone precursors 324
Glycol cleavage 936
Glycols, cyclocondensation with diacyl halides 391
Glycosides — *see* Aminoglycosides
Grignard reagents, reactions of 48–51, 294–297, 681
Gulonic acid, cyclization of 3
Gulonic-γ-lactones, synthesis of 3

N-Haloacyl-β-arylamines, cyclization of 162, 469, 470
Haloacyl halides, reactions of,
  with amino acids 419
  with phenylhydrazones 616
  with thiosemicarbazides 997, 998
4-Haloacyl-β-lactams,
  reactions of 939, 940
  synthesis of 939, 940
α-(Haloalkyl)acrylates, Reformatsky reaction of 298
Haloalkylfuranones, synthesis of 285

N-(Haloalkyl)-β-lactams, transamidation of 652–655
4-Haloalkyl-β-lactams — *see also* 4-Iodoalkyl-β-lactams
  dehalogenation of 943
ω-Haloalkyl 2-(phenylthiomethyl)benzoates, as lactone precursors 389, 390
Haloamides, cyclization of 138–143, 420–437
Haloaminoalkenes, carbonylation of 984
Halobenzotropones, reactions with amines 999, 1000
Halobenzylamines, carbonylation of 984
Halobutenolides,
  reactions of 373, 374
  synthesis of 302
ω-Halocarboxylic acids, as lactone precursors 282
Halocyclization, of amides 435
3-Halodihydropyrans, dehydrohalogenation of 131
Haloiminium halides, reactions with imines 617–620
ω-Halo-β-keto esters, reactions with LDA 299
3-Halolactams,
  hydrolysis of 859
  reduction of 862–866
    electrochemical 867–869
  substitution reactions of 859–862
  synthesis of 870, 875
Halolactones, synthesis of 19, 27–30, 334, 335, 337
Halolactonization 288–291
  of unsaturated acids and esters 19, 27–30
γ-Halomethyl-α,β-butenolides,
  synthesis of 288
  thiophenylation of 357
6-Halopenicillinates — *see also* 6, 6-Dihalopenicillinates
  reactions of 859–861
N-(*o*-Halophenylmethyl)-β-lactams, reactions of 919, 920
α-Halo-α-phenylthio-β-lactams, hydrolysis of 859
Halophthalides, phenylselenylation of 357, 358
(Halopivaloyl)aminomalonates, reactions of 420, 422
Halopivaloyl halides, as lactam precursors 998
Halopropionamides — *see also* (Acyloxy)halopropanamides
  reactions of,
    using a phase transfer catalyst 424, 426
    with sodium hydride 420, 421
Heptenoates — *see* 4-Benzoyloxy-5-heptenoates
Hexahydro-1*H*-pyrrolo[2,3-*b*]pyridines, synthesis of 464
Hexahydroquinolinones, synthesis of 621–623
Hexanoates — *see* Hydroxyhexanoates
Hexanoic acids — *see also* Hydroxyhexanoic acids
  as lactone precursors 272
4-Hexenoic acids, halolactonization of 288
Hofmann degradation 301
Homoallylic alcohols, oxidation of 328, 329
Homoallylic trimethylsilyl alcohols, hydroboration–oxidation of 328, 330
Homoserine lactone, synthesis of 131, 132
Homoserine methyl esters, acid-catalysed cyclization of 275
Hydrazinouracils, reactions of 980
Hydrazones — *see* Phenylhydrazones, Pyrimidinylhydrazones
α-Hydroperoxylactams, synthesis of 723
Hydro-3-pyranones, synthesis of 323
Hydroxamates — *see* Vinylacetohydroxamates
Hydroxamic acids — *see* Phenylalkanehydroxamic acids
O-(4-Hydroxy-1-alkenyl)carbamates, as lactone precursors 322
α-Hydroxyalkyllactams, synthesis of 700–704
4-Hydroxyalkyl-β-lactams,
  acylation of 933
  oxidation of 933
  reactions with diazomethane 932
  sulphonation of 929–931
  synthesis of 934, 935, 937, 938
Hydroxyalkyloxazoles, photooxygenation of 371
γ-Hydroxy-α,β-alkynoates, acid-catalysed cyclization of 284
2-Hydroxy-6-allylcyclohexanecarboxylic acids, cyclization of 288
Hydroxyamides,
  cyclization of 143, 144, 438–448
  mesylation or tosylation of 445, 446
Hydroxyandrostanones, aldol condensation of 38
n-Hydroxyazetidines, reactions with acetates 964, 968
Hydroxyazetidinonecarboxylic acids, acid-catalysed cyclization of 272
Hydroxybenzaldehydes, condensation of 40, 61
Hydroxybenzylidenecoumaranones, synthesis of 61
Hydroxybutenolides,
  reactions with alkyllithium 300
  synthesis of,
    from alkynes 342, 343
    from furoic acids 337
Hydroxycarboxylic acids, cyclization of 3–11, 271–279, 283
  using benzenesulphonyl chloride 4
  using borohydrides 273
  using cyanuric chloride 278, 279

using DCCD 10, 11
using halomethylpyridinium halides 279
using hydrochloric acid 272
using mesitylenesulphonyl chloride 278
using sodium acetate 9, 10
using sulphuric acid 272
using *p*-toluenesulphonic acid 4–7, 276, 277
using *p*-toluenesulphonyl chloride 278
ω-Hydroxycarboxylic acids,
as macrocyclic lactone precursors 13, 278, 279, 283, 381
unsaturated, cyclization of 293
Hydroxycarboxylic esters, cyclization of 3, 10–14, 273–275, 280
using boron trifluoride–etherate 11
using DCCD 11
using halomethylpyridinium halides 280
using hydrochloric acid 3, 275
using sulphuric acid 11, 274
using trifluoroactic acid 273, 274
Hydroxycycloalkylpropanoic acid lactones, synthesis of 301
Hydroxycycloalkylpropenoic acid lactones, synthesis of 301
Hydroxycyclohexadienones, oxidation of 335, 336
3-Hydroxyethyllactams, synthesis of 886–888
3-Hydroxy-4-heptenoic acid salts, cyclization of 288
Hydroxyhexadecenoic acids, cyclization of 293
Hydroxyhexanoates, as lactone precursors 10, 13
Hydroxyhexanoic acids, as lactone precursors 3, 272
α-Hydroxyhippuric acid dianions, reactions with Schiff bases 612
Hydroxyindanylacetic acids, acid-catalysed cyclization of 272
Hydroxyionolactone, synthesis of 17
α-Hydroxyketones, condensation with malonates 41
α-Hydroxylactam esters, synthesis of 685–690
α-Hydroxylactams,
functionalization of the hydroxy substituent in 870–873
synthesis of 723, 724
ω-Hydroxylactams,
reactions of 891–893, 898
synthesis of 719, 720
N-Hydroxy-β-lactams, reactions of 797
3-Hydroxy-β-lactams,
conversion to 3-halo-β-lactams 870, 875
synthesis of 1002
from 3-alkoxy-β-lactams 870, 874
5-Hydroxy-γ-lactams, synthesis of 980
Hydroxylactones, synthesis of 283, 370
Hydroxymethylbenzoates, as lactone precursors 3
4-Hydroxymethyl-2-butenolides, synthesis of 287
Hydroxymethylcyclopropanecarboxylates, as lactone precursors 13
Hydroxymethylfurancarboxylic acids, hydrogenation of 283
γ-Hydroxynitriles, as lactone precursors 280, 281
4-(1-Hydroxy-2-nitroethyl)-β-lactams, synthesis of 933, 936
Hydroxynitrones, ring expansion of 661
Hydroxyolefins, oxidative cyclization of 325–327
Hydroxyoximes, ring expansion of 644, 645
8-Hydroxy-9-oxo-1-azabicyclo[4.3.0]non-7-enes, ring contraction of 639
Hydroxypalmitic acids, cyclization of 278, 279
4-Hydroxy-2-pentenoic acids, UV irradiation of 287, 288
3-Hydroxy-4-pentenoic acids, carbonylation of 344
Hydroxypentenylamides, aminocarbonylation of 344, 345
o-Hydroxyphenyl ketones, reactions with carboxylic acids 381
Hydroxypropionic acids, condensation of 36, 38
Hydroxypyrrolidineacetic acid lactones, synthesis of 343–345
Hydroxysuccinimide esters, hydrogenation of 993
Hydroxytetrahydrofuran-2-acetic acid lactones, synthesis of 344
γ-Hydroxyvinylstannanes, reactions with alkyllithiums 341

Ibogaine lactam, synthesis of 242, 247
Ibogamine lactam, synthesis of 242
Iboluteine lactam, synthesis of 242
Iboquine lactam, synthesis of 242, 243
Imidazolidinones,
reactions of 944
synthesis of 464
Imidazolinones, synthesis of 648, 649, 651, 652, 911
Imides — *see also* Thioimides
as lactam precursors 453
cyclic — *see* Cyclic imides
synthesis of 719–721, 723
Imidoylazimines, reactions with ketones 652
Imines, reactions of 206, 207
with acetoacetate oximes 615
with activated carboxylic acids and derivatives 547–581
with acyl halides 490, 503–546, 581

Imines, reactions of (*cont.*)
with alkenoic acids 621–623
with anhydrides 581
with α-chloroiminium chlorides 617–620
with carbene complexes 585–589, 956–959
with ketenes 168–201, 490–502
with lactones 621–623
with lithium enolates 592–611, 613
with lithium ynolates 613, 614
with metal complexes 963, 964
with oxazolinones 625, 626
with Reformatsky reagents 171, 203–205, 581–584
with toluenesulphinylacetic acids 624, 625
Iminium halides — *see* Thioiminium halides
Iminobenzocycloheptenones, synthesis of 999, 1000
4-Imino-β-lactams, synthesis of 166, 167
2-Imino-1,2,3,4-tetrahydro-1,3,4-thiadiazin-5-ones, synthesis of 997, 998
Indanols — *see* Carboxymethylindanols
Indanones — *see* Tetrahydro-7α-carbomethoxymethyl-1-indanones
Indoles — *see* Azepinoindoles, Tetrahydroindoles
Invictolides, synthesis of 357
Iodoalkenols, carbonylation of 343, 344
Iodoalkyl-γ-butyrolactones, reactions of 351, 352
4-Iodoalkyl-β-lactams,
reactions of 943
synthesis of 929
Iron complexes,
oxidative decomplexation of 963–967, 981, 982
reactions with isocyanates 166
Isoambrettolide, synthesis of 293
Isobufadienolides, synthesis of 39, 40
Isocarbostyrils, synthesis of 243, 244, 249, 250
Isocardenolides, synthesis of 39
Isocephalosporins, synthesis of 397
Isochromanones, synthesis of 10
Isocyanates — *see also* Diisocyanates
reactions of,
with β-amino acids 489
with olefins 162–166, 475–489
N-(ω-Isocyanatoacyl)lactams, synthesis of 797, 798
N-Isocyanatoalkyllactams, synthesis of 799
Isoindolinones, synthesis of 999, 1000
Isopenicillin N synthase 1005
1-Isopropenylspiro[2.X]alkanes, reactions with isocyanates 487, 488
Isoquinolines — *see* 3,4-Dihydroisoquinolines, Pyrrolo[2,1-*a*]isoquinolines, Tetrahydroisoquinolines
Isoquinolinones, synthesis of 250
Isoquinolones, synthesis of 463
Isotetrandrine, photooxidation of 248
N-(ω-Isothiocyanatoacyl)lactams, synthesis of 797
Isoureas, synthesis of 797
Isoxazoles, reactions with *n*-butyllithium 613
Isoxazolidines — *see* Nitroisoxazolidines

Jasmones, synthesis of 384
Jones' reagent 755
as oxidizing agent 314, 320, 321, 326, 331
Justicidin, synthesis of 311

Kawains, synthesis of 12
Ketenes — *see also* Dihaloketenes
reactions of,
with azo compounds 170, 171, 202
with carbonyls 63, 69–73
with diazoketones 69, 74, 75
with imines 168–201, 490–502
with nitrones 207–209
with nitroso compounds 210
with ozone 76
Ketene silyl acetals, reactions of,
with alkylidene (1-arylethyl)amines 590
with α-chloroalkyl phenyl sulphides 382
with Schiff bases 585, 590, 591
Ketenimines, reactions with isocyanates 166
Keto acids,
conversion to enol lactones 13
reduction of 83–86
α-Keto alcohols, reactions of 34, 48
α-Ketoalkyllactams, synthesis of 704
Keto amides, cyclization of 144, 452, 453
Keto esters — *see also* ω-Halo-β-keto esters
reactions with amines 452, 453, 455
reduction of 83, 84, 86–88
Keto lactams,
oxidative ring contraction of 639
synthesis of 660, 661, 721, 962
Keto lactones,
bicyclic — *see* Bicyclic keto lactones
synthesis of 384, 394, 395
Ketones — *see also* Diazoketones, β-Diketones, α-Hydroxyketones
cyclic — *see* Cyclic ketones
electroreductive hydrocoupling of 371–373
o-hydroxyphenyl — *see* *o*-Hydroxyphenyl ketones
reactions of,
with amino acid amides 464, 465
with α-aminonitriles 376
with imidoylazimines 652
with ketenes 63, 69–73
with lithium β-lithio esters 368, 369
with Reformatsky reagents 51, 52
ring expansion of 641

α,β-unsaturated — *see* α,β-Unsaturated ketones
Kopsine lactam, synthesis of 242, 243

Lactam acetals 398
β-Lactam antibiotics,
allergic reactions to 398
synthesis of 397
β-Lactamase 397
Lactam carboxylic acid–esters, decarboxylation of 938
Lactam carboxylic acids — *see also* Acyllactam carboxylic acids
reduction of 937, 938
synthesis of 936
Lactam carboxylic esters,
reduction of 937–940
synthesis of 936
Lactam disulphides, synthesis of 799, 914, 915
Lactam enolates,
alkylation of 794
reactions of,
with carbonyls 700–704
with unsaturated nitriles 700
Lactam malonates,
saponification of 794
synthesis of 777
Lactam nitrogen,
acylation of 679–681
alkylation of 663, 675, 679
chloroformylation of 681, 682
chloromethylation of 681, 682
dialkylaminomethylation of 684
phosphorylation of 663
removal of substituent from 726, 728–745
silylation of 663, 676–678
substitution at,
accompanied by cyclization 691–695
using base, solvent and halide 662–667
using phase transfer conditions 662, 663, 668–674
sulphonation of 685, 691
β-Lactam ring, in penicillins and cephalosporins, chemical modification of 397
Lactams,
benzhydrol-substituted 999
bicyclic — *see* Bicyclic lactams
biosynthesis of 398
exocyclic α-halogenated,
reduction of 726, 727
replacement of halogen in 750–753
synthesis of 726, 727, 746–749
exocyclic β-halogenated,
dehydrohalogenation of 751, 754
isomerization of 950, 954, 955
macrocyclic — *see* Macrocyclic lactams
medium-ring — *see* Medium-ring lactams
nomenclature of 398
N-substituted,
exocyclic double-bond epimerization in 758, 759
exocyclic double-bond formation in 751–756
exocyclic halogenation of 726, 727, 746–749
hydrogenolysis of exocyclic α-carbons 799
oxidation of exocyclic α-carbons 798
reactions at exocyclic double bonds 762–777
synthesis of 685–690, 1003
polycyclic — *see* Polycyclic lactams
polymerization of 398
synthesis of 132, 133, 397, 398
by chemical ring closures 133–144, 398–465
by cycloadditions 162–210, 475–627
by direct functionalization of preformed lactams 237–240, 662–952
by metal-catalysed carbonylations 982–991
by oxidations 240–252, 952–970
by photochemical ring closures 144–162, 465–474
by rearrangements 210–237, 627–662
thionalation of 802–810
tricyclic — *see* Tricyclic lactams
unsaturated — *see* Unsaturated lactams
α-Lactams,
ring expansion of 648–650
synthesis of 132
from amino acids 133
from amino esters 134
from haloamides 138, 424
β-Lactams — *see also* Azetidinones, β-Butyrolactams
α-alkoxylation of 398, 713, 716
aryl ester substituted 900
α-functionalization of 891, 894–897
β-functionalization of 724
halogenation of 704–706
naturally occurring 398
optically active 471, 472
reactions of,
with azides 708
with glyoxylates 685–690
with Grignard reagents 681
with imines 397
ring expansion of 648–655
3-substituted, synthesis of 238, 956–959
3,4-substituted, ring closure of 945, 947, 950
α,β-substituted, isomerization of 880–883

β-Lactams (*cont.*)
  synthesis of 132, 133
    by four-component condensations 416, 417
    by oxidation of iron complexes 963–967
    by ring contractions 210–212, 627–639
    N-chlorocarbinolamine method for 416, 417
    electroorganic 397
    from amines 245
    from amino acids 133, 399, 404–409, 413, 415, 1005
    from amino esters 134–136, 400
    from aminomalonates 420, 422
    from β-aminothiol esters 406, 407
    from azetidinecarboxylates 953, 956
    from azetidinecarboxylic acids 239, 251, 960
    from aziridinecarboxylates 234, 235
    from aziridines 982, 983
    from β-chloropivaloyl chloride 998
    from cyclopropanolamines 1005
    from α-diazocarboxamides 162
    from glycolamides 144
    from haloamides 138–143, 420–437
    from hydroxyamides 438–445
    from *N*-hydroxyazetidines 964, 968
    from imines 169–207, 490–582, 585–620, 623–625
    from nitrones 209
    from olefins 162–166, 475–489
    from 2-piperidinylacetates 134
    from sulphur-containing amides 450, 451
    stereoselective 397
    using transition metal complexes 488
  thietan-fused 465
  thiophenyl-substituted, oxidation of 711, 712
γ-Lactams — *see also* γ-Butyrolactams
  conversion to unsaturated counterparts 875–877, 911
  furano-substituted 419
  microbial hydroxylation of 722
  photooxidation of 723
  reactions of,
    with azides 707
    with glyoxylates 690
  synthesis of,
    from alkenamides 460, 461
    from allyl- and propargyl-amides 458–460
    from amino acids 133
    from biphenyl-2-carboxamides 472–474
    from butenolides 993
    from cyclic imides 970, 972–979
    from haloamides 138, 141, 430
    from hydroxyamides 143
    from β-lactams 648, 650–652
    from nitro esters 137
    from pyrrolidines 953, 956, 957, 960
    from pyruvamides 452
    from sulphur-containing amides 448, 449
    from trimethylsilyl amino esters 413, 414
δ-Lactams — *see also* δ-Valerolactams
  conversion to unsaturated counterparts 875–877, 911
  microbial hydroxylation of 722, 723
  photooxidation of 723, 724
  sulphur-containing 452
  synthesis of,
    from alkenamides 460, 461
    from amino acids 133
    from amino esters 134, 413, 414
    from biphenyl-2-carboxamides 472–474
    from cyclic amines 970
    from cyclic imides 975
    from haloamides 138, 142, 430
    from hydroxyamides 143
    from imines 584
    from nitro esters 137
    from piperidines 953, 956, 957, 960, 964, 968, 969
    from pyridinium salts 964, 969
    from pyrrolidinediones 252
    from sulphur-containing amides 448, 449
ε-Lactams — *see also* ε-Caprolactams
  halogenation of 706
  microbial hydroxylation of 722, 723
  photooxidation of 723, 724
  reactions with Grignard reagents 681
  sulphur-containing 419
  synthesis of,
    from allylamides 456, 457
    from amino acids 133
    from trimethylsilyl amino esters 413, 414
Lactam sulphides, synthesis of 415
Lactam–thiolactams, synthesis of 416
Lactides, synthesis of 3
Lactims — *see* Valerolactims
γ-Lactol methyl ethers, oxidation of 321, 322
Lactols,
  oxidation of 337, 338
  synthesis of 335, 336
Lactone ring, formation of a double bond in 362–364
Lactones,
  α-alkylation of 61, 64, 65
  bicyclic — *see* Bicyclic lactones
  chiral — *see* Chiral lactones
  interconversion of 125, 351–368
  macrocyclic — *see* Macrocyclic lactones
  medium-ring — *see* Medium-ring lactones
  polymerization of 271
  reactions with imines 621–623

steroidal — *see* Steroidal lactones
synthesis of,
by acetate condensations 34–36
by aldol condensations 36–40
by carbonations 341–346
by carbonylations 111–118, 341–346
by cycloadditions 118–120
by direct functionalization of preformed lactones 61–63, 299–303
by Grignard and Reformatsky reactions 48–53, 294–298
by intramolecular cyclizations 3–34, 271–294
by malonic acid/ester condensations 40–46, 294
by oxidations 83, 89–111, 314–341
by Perkin and Stobbe reactions 46–48
by rearrangements 120–125, 346–351
by reduction of anhydrides, esters and acids 76–88, 303, 308–314
by Wittig-type reactions 53, 54
from carbonyls 368–376
from carboxylic acid anions 54–58, 298, 299
from carboxylic acids 377–384
from carboxylic esters 384–390
from diacyl halides 390–394
from ketenes 63, 69–76, 303
from lithio salts of 2-alkyl-2-oxazolines 58–61
trans-fused, elaboration of 357
tricyclic — *see* Tricyclic lactones
unsaturated — *see* Unsaturated lactones
α-Lactones, synthesis of 130
β-Lactones,
polymerization of 271
synthesis of 3, 271
by cyclizations 4, 9, 13, 18
by Perkin-like reactions 46, 47
from 2-alkoxyoxetanes 129
from allylallenes 111
from ketenes 63, 69–73
γ-Lactones — *see also* γ-Butyrolactones, γ-Valerolactones
synthesis of 329, 331
by condensations 34–36, 41, 42
by cyclizations 3, 17
by oxidations 314, 320–322, 325–327
from β-allenic alcohols 333
from carbonyls 368, 369, 371–373
from carboxylic acid anions 54–58
from dialkoxytetrahydrofurans 275, 276
from homoallylic trimethylsilyl alcohols 328, 330
from γ-hydroxy-α,β-alkynoates 284
from γ-hydroxycarboxylic acids 271, 272
from hydroxy esters 274, 275
from ketenes 69, 70
from olefins 110–112, 118–120
from unsaturated acids 18, 19, 22–26, 288–290, 336
trans-fused 46
α,β-unsaturated — *see* α,β-Unsaturated γ-lactones
δ-Lactones — *see also* δ-Valerolactones
cis/*trans*-fused 365, 366
synthesis of 130, 329, 331
by aldol condensations 36, 38
by cyclizations 10
by Grignard reactions 48, 49
by oxidations 314, 326, 327
from β-allenic alcohols 333
from homoallylic alcohols 328, 329
from hydroxycarboxylates 273, 274
from pentanoates 313, 314
from γ,δ-unsaturated carboxylic acids 288–290
unsaturated — *see* Unsaturated δ-lactones
ε-Lactones, synthesis of 3
by oxidation of diols 314
Laudanosine, photooxidation of 248
Lawesson's reagent 803–811
Lupanine lactam dimer, synthesis of 248
Lupanines — *see also* Oxylupanines
synthesis of 242
Lupenone, ring expansion of 646

Macrocyclic azalactams, synthesis of 656, 657
Macrocyclic dilactones, synthesis of 393, 394
Macrocyclic lactams — *see also* Macrocyclic azalactams
synthesis of 397
from ω-amino acids 411, 412
from *N*-(aminoalkyl)lactams 656, 657
from bis(3-acylthiazolidene-2-thiones) 996, 997
from 3-(1-nitro-2-oxocycloalkyl)propanals 658, 659
from sulphur-containing phthalimides 474
Macrocyclic lactones — *see also* Macrocyclic dilactones, Macrocyclic nitrolactones, Macrocyclic tetrathiolactones
synthesis of 3, 271
from catacondensed lactols 338
from hydroxy acids 278, 279, 283, 293, 381
from hydroxyalkyloxazoles 371
Macrocyclic lactonolactams, synthesis of 392, 394
Macrocyclic nitrolactones, synthesis of 394, 395
Macrocyclic tetrathiolactones, synthesis of 390
Macrolide antibiotics, synthesis of 271
Maleic anhydrides, reduction of 310

Malonates — *see also* Alkenylmalonates, α-[*p*-(Benzyloxyphenyl)]-2-oxo-1-azetidinemalonates, (Halopivalolyl)aminomalonates
condensation of 36, 40–46
Malonic acids, condensation of 40–46
Manganese dioxide, as oxidizing agent 339
Medium-ring lactams, synthesis of 397
Medium-ring lactones, synthesis of 389
Mellilo's lactone, ring contraction of 637, 638
Mercaptans, oxidation of 130
Mercapto compounds, reactions with acetoxylactams 906–909
4-Mercapto-β-lactams,
as precursors of β-lactam antibiotics 397
synthesis of 913
α-Methoxycarbonylmethylene γ-lactones, synthesis of 286
3-Methylenecepham-1-oxides, ring opening of 915, 916
4,4′-Methylenedi(3,4-dihydro-1*H*-1, 4-benzodiazepine-2,5-diones), synthesis of 929
Methylenedioxybenzo[*a*]octems, reactions of 918, 919
1,1′-Methylene di-2-pyrrolidone, synthesis of 682–684
α-Methylene lactams, synthesis of 397
α-Methylene β-lactams — *see also* *N*-Benzyl-3-methylene-β-lactams
synthesis of 889
from haloamides 432, 433
α-Methylene γ-lactams, synthesis of 581, 583
Methylene lactones — *see* Dimethylaminomethylene lactones
α-Methylene lactones, synthesis of 3, 278
from vinyl halides 345, 346
α-Methylene γ-lactones — *see also* α-Methoxycarbonylmethylene γ-lactones
β-acetoxy 278
synthesis of 300, 339, 352
by Reformatsky reactions 51, 52, 298
from allylic alcohols 277, 278
from cyclopropanecarboxylates 351
from 4-pentenoic acids/esters 290
4-Methylenespiro[2.X]alkanes, reactions with isocyanates 475, 487
4-Methylsulphinyl-β-lactams, reactions of 924
4-Methylsulphonyl-β-lactams, reactions of 913, 916, 917
4-Methylthio-β-lactams, reactions of 918, 922–924
Methyltricycloundecan(carboxy)hydroxylactones, synthesis of 348
d,l-Methysiticin, synthesis of 53
d,l-Mevalonolactone, synthesis of 53
Michael addition 700, 712, 994
Mitsunobu reaction 445, 755, 756
Molybdenum carbene complexes, photoreactions of 585, 589
Monobactam antibiotics 397

Naphthalenecarboxylates — *see* 1,2,3,5-Tetrahydro-1-oxonaphthalenecarboxylates
Naphthalenecarboxylic acids — *see* Octahydro-1-oxonaphthalene-4α-carboxylic acids
Naphthyridines — *see* Octahydro-1*H*-1,8-naphthyridines
Nicotine, bromination of 240, 241
Nitriles — *see also* Acetylalkanedinitriles, α-Aminonitriles, Dihydroxynitriles, γ-Hydroxynitriles
reactions of,
with (α-bromomethyl)acrylates 581, 583
with enolates 614, 615
2-Nitrocycloalkanones, reductive isomerization of 395
4-Nitroethyl-β-lactams, synthesis of 933, 934
3-Nitro-3-halo-2-pyrrolidones, synthesis of 704, 705
Nitro heterocycles, hydrogenation of 993–995
Nitroisoxazolidines, ring contraction of 637
Nitrolactams, reactions of 912
Nitrones — *see also* Aldonitrones, Hydroxynitrones
cyclization of 626, 627
reactions of,
with acetylides 209
with ketenes 207–209
3-(1-Nitro-2-oxocycloalkyl)propanals, reductive amination of 658, 659
3-Nitro-2-pyrrolidones — *see also* 3-Nitro-3-halo-2-pyrrolidones
synthesis of 704
Nitroso compounds, reactions with ketenes 210
4-(2-Nitrovinyl)-β-lactams,
reduction of 934
synthesis of 932
Norbornenyl acids, separation of *endo* and *exo* isomers of 27
21-Nor-5α-conanine-20-one, synthesis of 241

Octahydrobenzoquinolines, synthesis of 911, 912
Octahydro-1*H*-1,8-naphthyridines, synthesis of 464
Octahydro-1-oxonaphthalene-4α-carboxylic acids, halolactonization of 290
Olefinic acids, cyclization of 3
Olefins — *see also* Hydroxyolefins
cycloaddition to nitrones 118–120
oxidation of 110–112, 323–336
reactions of,
with carboxylic acids 377, 378

with diazonium salts 130
with isocyanates 162–166, 475–489
Oligonucleotides, synthesis of 307
Ortho esters, reactions with 2-alkene-1,4-diols 120–122
Oxaazabicyclo-β-lactams, synthesis of 771–777
Oxabicyclo-β-lactams, synthesis of 771–773
Oxadiamines, reactions with diacyl halides 294, 392
Oxalactones, hydrolysis of 365, 366
Oxazinediones — *see also* Tetrahydrooxazinediones
reactions of 1003, 1004
Oxazines — *see also* 2-Oxoazeto[1,2-*c*][1,3]benzoxazines
reactions with carbene complexes 585, 589, 956–959
Oxaziridines, tricyclic — *see* Tricyclic oxaziridines
Oxazoles — *see* Hydroxyalkyloxazoles
Oxazolidines — *see also* Oxooxazolidines
ring contraction of 631
Oxazolidinones, synthesis of 648, 649, 1001, 1002
Oxazolines — *see also* 2-Alkyl-2-oxazolines
reactions with carbene complexes 585, 589
Oxazolinones, reactions with imines 625, 626
Oxetanes — *see* 2-Alkoxyoxetanes
α-Oxime lactams, reduction of 713–715
Oximes — *see also* β,γ-Cyclopropyloximes, Hydroxyoximes
ring expansion of 212–222, 640–645
Oxindoles, synthesis of 463
from amino esters 135, 137
from haloamides 142, 143
Oxiranes — *see also* Alkoxymethyloxiranes, Spirooxiranes
synthesis of 427, 429
2-Oxo-1-azabicyclo[3.2.0]heptane-7-carboxylates, α-amination of 704
17-Oxo-17α-aza-D-homosteroids, synthesis of 662
2-Oxoazetidin-4-yl phosphinates, synthesis of 906
2-Oxoazetidin-4-yl phosphonates, synthesis of 906
2-Oxoazeto[1,2-*c*][1,3]benzoxazines, reactions of 918, 919
9-Oxobicyclo[6.3.0]undecenones, synthesis of 46
α-Oxocaprolactam *O*-phenyloxime, as lactone precursor 397
α-Oxocaprolactams, reactions of 981
3-Oxoconanine lactam, synthesis of 242
2-Oxo-1,2-dihydrophosphete complexes,
decomplexation of 1009
synthesis of 1008, 1009
Oxodioxabicyclooctanes, synthesis of 283
Oxohexadecanoates, lithium enolates of, as lactone precursors 386, 387
Oxolactones, synthesis of 36, 395
Oxonium ions, as reaction intermediates 321
Oxooxazolidines, reactions of 999
Oxopiperidineacetic acids, synthesis of 399
1-Oxo-1,2,3,4-tetrahydroisoquinolines, synthesis of 463
β-Oxy-γ-butyrolactones, synthesis of 314, 321
Oxylupanines, synthesis of 244, 250
Oxysparteines, synthesis of 250

Palladium compounds, in lactone synthesis 343–345
Palmitic acids — *see* Hydroxypalmitic acids
Penams, ring opening of 945
Penems — *see also* Azacarbapenems
synthesis of 397
Penicillanic acids — *see* 6-Aminopenicillanic acids
Penicillin acylases, as acylating agents 811, 829
D-Penicillinamine, reactions with acrylates 419
Penicillinates — *see also* 6-Halopenicillinates, 6-Phenylselenopenicillinates, 6-Phenylthiopenicillinates
α-methoxylation of 713
6-substituted, synthesis of 859–861, 880, 884
Penicillins — *see also* Bromomethylpenicillins
synthesis of 238, 239, 397
Wittig products of, ozonolysis of 960, 961
Penicillin sulphoxide esters, ring opening of 900
1,4-Pentadiene-1,2-dicarboxylates, acid-catalysed cyclization of 286
Pentadienoic acids, halolactonization of 288
Pentanoic acids — *see* Diaminopentanoic acids
4-Penten-1,3-diols, carbonylation of 344
Pentenoic acids,
acid-catalysed cyclization of 16
iodolactonization of 290
4-Penten-4-olides, synthesis of 290
Peptides — *see also* Cyclopeptides, Depsipeptides
synthesis of 397
Perfluoro(*N*-alkyl cyclic amines), hydrolysis of 1004, 1005
Perfluoro-α-lactams, ring expansion of 648, 649
Perkin reaction 46, 47
Perkov-type intermediates 643
α-Peroxylactones, synthesis of 271
Phenanthridones, synthesis of 145, 249
Phenazine *N*-oxides, photoisomerization of 1005, 1006

Phenols, reactions with 2-halo-2-alkylthio esters 388
4-Phenylacylthio-β-lactams, synthesis of 913, 914
Phenylalkanehydroxamic acids, cyclization of 464
4-ω-Phenylalkyl-1*H*-pyrazol-5-isocyanates, cyclization of 489
Phenylbenzoic acids, cyclization of 18
Phenylhydrazones, reactions with haloacyl halides 616
4-Phenyl-β-lactams, synthesis of 945
N-Phenylmethyleneaminolactams 616
α-Phenylselenolactams, synthesis of 721, 722
3-Phenylseleno-β-lactams,
as 3,3-dimethyl-β-lactam precursors 890
as α-methylene β-lactam precursors 889
6-Phenylselenopenicillinates — *see also* 6,6-Bis(phenylseleno)penicillinates
α-hydroxyethylation of 885
synthesis of 880, 884
3-Phenylsulphinyl-γ-lactams, synthesis of 709
α-Phenylsulphinyllactones, synthesis of 375
(Phenylsulphinyl)propionamides, as lactam precursors 447, 448
(Phenylthio)alkanamides, as lactam precursors 450, 451
α-Phenylthiolactams, synthesis of 707, 709
3-Phenylthio-β-lactams,
acetylation of 880, 885
desulphurization of 869, 870
halogenation of 706
synthesis of 880
4-Phenylthio-β-lactams, synthesis of 913
Phenylthiolactones, synthesis of 382
6-Phenylthiopenicillinates — *see also* 6,6-Bis(phenylthio)penicillinates
α-hydroxyethylation of 885
synthesis of 880, 884
3-Phosphinimino-β-lactams,
acylation of 811, 829
as Schiff base precursors 835, 836
synthesis of 835
Phosphirene–metal pentacarbonyl complexes, carbonylation of 1008, 1009
Phosphorus–carbon double bonds, exocyclic lactam, formation of 751–753
Phosphorus sulphide, as thionalation reagent 803–805, 807–810
Photochemical rearrangements 124, 125, 644, 645
Phthalic anhydrides, reduction of 310
Phthalides — *see also* Aralkylidenephthalides, Arylmethylphthalides, Halophthalides
synthesis of 3, 310
Phthalimides, sulphur-containing, cyclization of 474
Phthalimidines, synthesis of 168
Phthalimidocephems, synthesis of 407, 408
α-Phthalimidolactams, side-chain ring opening of 856, 857
Piperazinediones, synthesis of 424, 426, 657
Piperidine dicarboxylates, oxidation of 385
Piperidines — *see also* *N*-Acylpiperidines, *N*-Formylpiperidines
oxidation of 957, 960, 964, 968, 969
Piperidinolactones, synthesis of 385
2-Piperidinylacetates, cyclization of 134
Piperidones — *see also* Aminopiperidones
synthesis of 245–247, 460–462
β-Piperonyl-γ-butyrolactones, synthesis of 314
Pivalolactones, polymerization of 271
N-(Pivaloyloxy)-β-lactams, reactions of 796
Plumbolactonization 383
Polyacetylenes, as lactone precursors 3
Polycyclic lactams, synthesis of 434
Polyenoic acids, acid-catalysed cyclization of 18
Prelog–Djerassi lactone 314, 321
synthesis of 271, 357
Propanals — *see also* 3-(1-Nitro-2-oxocycloalkyl)propanals
condensation with malonates 44, 45
Propargylamides, of 2,4-dienecarboxylic acids, Diels–Alder reaction of 458–460
Propionamides — *see* Halopropionamides, (Phenylsulphinyl)propionamides
Propionic acids — *see* β-Acyl-β-ethoxycarbonyl-α-alkylpropionic acids, Benzoylpropionic acids, Benzylaminopropionic acids, Hydroxypropionic acids
Pyrans — *see* 3-Halodihydropyrans
Pyrazolidinediones — *see* Diazopyrazolidinediones
Pyrazolidinium ylides, synthesis of 684, 685
Pyrazolidinones — *see also* Acylpyrazolidinones
synthesis of 420
Pyrazolidones, bicyclic — *see* Bicyclic pyrazolidones
Pyrazolines — *see* 3,5-Dicarboalkoxy-2-pyrazolines
Pyridazinones — *see* Tetrahydropyridazinones
Pyridines — *see* Hexahydro-1*H*-pyrrolo[2,3-*b*]pyridines
Pyridinium chlorochromate, as oxidizing agent 323–326, 328, 338, 339
Pyridinium salts, reactions with acetates 964, 969
Pyridoazepines, synthesis of 460, 462
Pyridones — *see also* Dihydropyridones, Furopyridones
as lactam precursors 397

synthesis of 911
from anils 253
from dienamino esters 134
from imines 206
Pyrimidines, synthesis of 626
Pyrimidinones, ring expansion of 661
Pyrimidinylhydrazones, synthesis of 980
Pyrones — *see also* 1-Alkoxy-5,6-dihydro-γ-pyrones, Dimethoxypyrones
synthesis of 3, 40, 53, 54, 131
Pyrrolidinediones — *see also* Diazopyrrolidinediones
as δ-lactam precursors 252
synthesis of 252, 253
Pyrrolidines — *see also* *N*-Acylpyrrolidines, Azidopyrrolidines, *N*-Formylpyrrolidines
oxidation of 957, 960
Pyrrolidinones, synthesis of 138, 245–247, 422, 430
2-(1-Pyrrolidinyl)-2-cycloalkene-1-alkanoates, reactions with amines 460, 462
Pyrrolidones — *see also* 3-Alkyl-1-methylpyrrolidones, Aminopyrrolidones, 3-Nitro-2-pyrrolidones, 1-Trimethylsilyl-2-pyrrolidones
N-substituted 1001, 1002
sulphenylation of 709
sulphinylation of 709
synthesis of 456, 460, 462
from amino esters 133, 134
from cyanoalkanoates 415, 416
3-Pyrrolin-2-ones, synthesis of 252
2-Pyrrolin-4-ones, synthesis of 648, 650
Pyrrolo[2, 1-*a*]isoquinolines, synthesis of 453, 454
Pyrrolones, synthesis of 430
Pyruvamides,
as lactam precursors 452
inclusion complexes with deoxycholic acid 471, 472, 1001, 1002

Quinolinones — *see* Hexahydroquinolinones
Quinoxaline dioxides, amidation of 994, 995

Raney nickel, as hydrogenation catalyst for butenolides 301
Reformatsky reaction 48–53, 171, 203–205, 294, 298, 434, 581–584
Reserpine, synthesis of 9
Retrochinensin, synthesis of 311
Retro-Diels–Alder reaction 295
Ribonolactones 367, 368
Ricinelaidic acid lactones, synthesis of 293
Ruthenium complexes, as reduction catalysts 310, 311
Ruthenium oxides, as oxidizing agents 952

Schiff bases,
reactions of 612, 613
with acyldioxanediones 1003, 1004
with ketene silyl acetals 585, 590, 591
synthesis of 832–836
Schmidt rearrangement 222–229, 645–647
Secosteroids, synthesis of 661
Selenolactonization 293, 294
Semicarbazides — *see* Thiosemicarbazides
2-Sila-1,3-dithiacyclopentanes, cyclocondensation with diacyl halides 390
Sparteine lactam, synthesis of 248
Sparteines — *see* Oxysparteines
4-Spirobutenolides, synthesis of 295
Spirolactams, synthesis of 431, 432, 717–719, 891
from imines 545, 557, 578, 608
from 4-methylenespiro[2.X]alkanes 475
Spirolactones,
reactions of 359, 360, 367
synthesis of 54, 290, 291, 383
from α-aminonitriles 376
from bicyclic lactones 287
from carbonyls 369
from spiro ethers 109
Spiro norbornenyl γ-butyrolactones, retro-Diels–Alder reaction of 352
Spirooxiranes, ring expansion of 230–233
Stannanes — *see* γ-Hydroxyvinylstannanes
Staudinger reaction 490
Steroidal lactones, cardioactive, lactam analogues of 397
Steroidyl isocyanates, reactions with esters 489
Stobbe reaction 46, 48
Strigol, analogues of, synthesis of 303–307
Succinates, condensation of 48
Succinic anhydrides, reduction of 310
α-Succinimidolactams, side-chain ring opening of 856, 857
Sulphenyllactonization 382
4-Sulphonyl-β-lactams, reactions of 913, 916, 917
α-Sulphoxyphenyl-δ-lactones, desulphurization of 354, 355

Taiwanin, synthesis of 311
Terpenylic acid, synthesis of 45
Tetrahydro-7α-carbomethoxymethyl-1-indanones, halolactonization of 290
Tetrahydrofurans — *see* Dialkoxytetrahydrofurans
Tetrahydroindenofuranones, synthesis of 272, 274
Tetrahydroindoles — *see* 2-Aryl-4,5,6,7-tetrahydroindoles

Tetrahydroisoquinolines — *see also* 2-Aroyl-1-methylene-1,2,3,4-tetrahydroisoquinolines, 1-Oxo-1,2,3,4-tetrahydroisoquinolines
synthesis of 453
Tetrahydrooxazinediones, ring contraction of 631
1,2,3,5-Tetrahydro-1-oxonaphthalenedicarboxylates, acid-catalysed cyclization of 287
Tetrahydropyridazinones, synthesis of 616
(Tetramethylene)bis(2-phenyl-4-oxo-1, 5-diazacyclooctanes), synthesis of 657
N-(Tetrazol-5-yl)lactams, synthesis of 797
Thalifoline, synthesis of 249
Thiaazabicyclo-β-lactams, synthesis of 771–777
1,3,4-Thiadiazepines, synthesis of 998
Thiadiazinones — *see* 2-Imino-1,2,3,4-tetrahydro-1,3,4-thiadiazin-5-ones
Thiadiazolidinium salts — *see* Carboxymethylthiadiazolidinium salts
Thiadiazoline lactams 920, 921
4-Thia-β-lactams, reactions of 917–919
Thiazepinones — *see* Dihydrobenzothiazepinones
Thiazines — *see also* Dihydrothiazines
reactions with carbene complexes 956–959
Thiazolidines, as lactam precursors 398, 399
Thiazolidinones — *see* Dioxothiazolidinones
Thiazoline dioxides — *see* 2-Acyl-3-oxo-4,5-benzo-1, 2-thiazoline dioxides
Thiazolones, ring contraction of 639
Thiiran lactams 920, 921
γ-Thioaryl-γ-butyrolactones, desulphurization of 354, 355
N-(Thiobenzoyl)methacrylamides, cyclization of 465
Thiochromanones, reactions with azides 645, 646
Thioimides,
cyclization of 471
reactions with phosphorus ylides 981, 982
Thiolactams, synthesis of,
from ω-amino acids 400
from lactams 802–810
ω-Thiolactams, synthesis of 981, 982
Thionolactones, synthesis of 366, 367
Thiosemicarbazides, reactions of 997, 998
4-Thioxo-β-lactams, reactions of 920
Ticonines, synthesis of 240, 241
Toluamides — *see* β-Amino-*o*-toluamides
Toluenesulphinyl-β-lactams, synthesis of 624, 625
4-(*p*-Toluenesulphonylthio)-β-lactams, synthesis of 915, 916
Transition metal complexes, in β-lactam synthesis 488
Triazolines, as reaction intermediates 771
Tricyclic lactams, synthesis of 786–790, 793
by N-substitutions/cyclizations 693, 695
by photolytic ring contractions 635, 636
from 2-acyl-3-oxo-4,5-benzothiazoline dioxides 623, 624
from angelica lactone 453
from carboxymethylthiazolidinium chlorides 616, 617
from imines,
by reactions with activated carboxylic acids 548, 553, 558, 559, 561, 565, 573, 574, 579
by reactions with acyl halides 505, 506, 511–513, 522, 523, 526, 531, 537, 543, 544
from keto amides 144
from α-oxocaprolactam 981
from phenazine *N*-oxides 1005, 1006
Tricyclic lactones, synthesis of 338, 339
Tricyclic oxaziridines, ring expansion of 660
Tricyclo[5.2.2.$0^{1,5}$]undecanes, rearrangement of 346–348
Triene esters, Diels–Alder cyclization of 292
3-Trimethylsilylallyllactams, ring closure of 52, 950, 951
Trimethylsilyl amino esters, cyclization of 413, 414
Trimethylsilylfurans, oxidation of 331, 332
N-Trimethylsilyllactams, reactions of 797
3-Trimethylsilylpropargyllactams, ring closure of 950–952
1-Trimethylsilyl-2-pyrrolidones,
sulphenylation of 707, 709
sulphinylation of 709
Triterpenes, ring expansion of 646
4-Tritylthio-β-lactams,
reactions of 914–916, 918
synthesis of 925
Tulipalins, synthesis of 300, 351

Unsaturated alcohols, carbonylation of 111, 112, 116, 117
Unsaturated aldehydes, oxidation of 339, 340
α,β-Unsaturated amides, cyclization of 144, 145, 465
α,β-Unsaturated carbonyl compounds, reactions with lactam enolates 700
Unsaturated carboxylic acids — *see also* Olefinic acids
electroreductive hydrocoupling of 371–373
ω-hydroxy, cyclization of 293
α,β-Unsaturated carboxylic acids,
γ-alkyl-γ-hydroxy, cyclization of 283, 284
oxidation of 336, 337

β,γ-Unsaturated carboxylic acids,
selenolactonization of 293, 294
γ,δ-Unsaturated carboxylic acids,
halolactonization of 288–290
intramolecular cyclization of 378, 379
oxidation of 337
Unsaturated carboxylic esters,
carbonylation of 111, 113–115
electroreductive hydrocoupling of 371–373
α,β-Unsaturated carboxylic esters, γ-hydroxy,
hydroxylation of 331
Unsaturated dicarboxylic acids, bislactonization
of 383, 384
β,γ-Unsaturated γ-dithiolactones, synthesis of
393, 394
α,β-Unsaturated ketones, reactions with
Grignard reagents 49
Unsaturated lactams, synthesis of 144
α,β-Unsaturated lactams, conversion to
saturated counterparts 875, 878–880
Unsaturated lactones,
epoxidation of 362, 366
reduction of 362, 364, 366
synthesis of 362–364
α,β-Unsaturated γ-lactones, synthesis of 39,
40, 284, 285, 375
β,γ-Unsaturated γ-lactones, β-ethoxycarbonyl,
synthesis of 380
α,β-Unsaturated δ-lactones, synthesis of 17,
324, 325, 375
β,γ-Unsaturated δ-lactones, synthesis of 378,
379
β,γ-Unsaturated steroidal oximes, ring
expansion of 644, 645
Uracils — *see* Hydrazinouracils
Urethanes, as lactam precursors 463

δ-Valerolactams,
reactions with dichlorocarbene 711
synthesis of 626
Vilsmeier formylation of 710
Valerolactims — *see also* Valerothiolactims
as lactam precursors 1000, 1001
γ-Valerolactones, synthesis of 34, 35
δ-Valerolactones — *see also* δ,δ-
Dimethylvalerolactones
α-alkylation of 64, 65
synthesis of 273
by Grignard reactions 294
by Reformatsky reactions 51
by Stobbe condensations 48
Valerothiolactims, as lactam precursors 1000,
1001
Vilsmeier formylation 710
Vinylacetic acid, selenolactonization of 293
Vinylacetohydroxamates, cyclization of 992
Vinylamino-β-lactams, oxidation of 856
β-Vinyl-γ-butyrolactones, synthesis of
120–122
Vinyl halides, containing a hydroxyl group,
carbonylation of 345, 346
N-Vinyllactams, synthesis of 725
4-Vinyl-β-lactams, synthesis of 902, 917, 941,
942
Vinyl sulphoxides, reactions with dihaloketenes
303, 308
Voacangine lactam, synthesis of 242, 244, 245,
247
Von Pechmann reaction 36, 37

Wacker oxidation 322
Wittig reaction,
in lactam synthesis 765–770
in lactone synthesis 53, 54
Wolff rearrangement 210, 211, 627–630, 801,
802, 940, 941

α-Ylidene-γ-butyrolactones, synthesis of 61,
66–68
Ynolates, reactions with imines 613, 614

Zinner's lactone 367, 368
Zip reaction 647
Zwitterions, as reaction intermediates 471,
1001, 1002

Index compiled by P. Raven